ANATOMY & PHYSIOLOGY

The Unity of Form and Function

Ninth Edition

KENNETH S. SALADIN

Distinguished Professor of Biology, Emeritus
Georgia College

Digital Authors

CHRISTINA A. GAN

Highline College

HEATHER N. CUSHMAN

Tacoma Community College

McGraw Hill

ANATOMY & PHYSIOLOGY: THE UNITY OF FORM AND FUNCTION, NINTH EDITION

Published by McGraw-Hill Education, 2 Penn Plaza, New York, NY 10121. Copyright ©2021 by McGraw-Hill Education. All rights reserved. Printed in the United States of America. Previous editions ©2018, 2015, and 2012. No part of this publication may be reproduced or distributed in any form or by any means, or stored in a database or retrieval system, without the prior written consent of McGraw-Hill Education, including, but not limited to, in any network or other electronic storage or transmission, or broadcast for distance learning.

Some ancillaries, including electronic and print components, may not be available to customers outside the United States.

This book is printed on acid-free paper.

7 8 9 LWI 24 23

ISBN 978-1-260-25600-0 (bound edition)
MHID 1-260-25600-6 (bound edition)
ISBN 978-1-260-79162-4 (loose-leaf edition)
MHID 1-260-79162-9 (loose-leaf edition)

Senior Portfolio Manager: *Matthew Garcia*
Senior Product Developer: *Donna Nemmers*
Marketing Manager: *Valerie Kramer*
Senior Content Project Manager: *Vicki Krug*
Lead Content Project Manager: *Brent dela Cruz*
Senior Buyer: *Sandy Ludovissy*
Designer: *Egzon Shaqiri*
Lead Content Licensing Specialist: *Lori Hancock*
Cover Image: *Mike Kemp/Rubberball/Getty*
Compositor: *MPS Limited*

All credits appearing on page or at the end of the book are considered to be an extension of the copyright page.

Library of Congress Cataloging-in-Publication Data

Cataloging-in-Publication Data has been requested from the Library of Congress.

BRIEF CONTENTS

About the Authors   iv

PART ONE

ORGANIZATION OF THE BODY   1

1   Major Themes of Anatomy and Physiology   1

ATLAS A General Orientation to Human Anatomy   27

2   The Chemistry of Life   40

3   Cellular Form and Function   74

4   Genes and Cellular Function   109

5   The Human Tissues   137

PART TWO

SUPPORT AND MOVEMENT   174

6   The Integumentary System   174

7   Bone Tissue   198

8   The Skeletal System   223

9   Joints   267

10   The Muscular System   299

ATLAS B Regional and Surface Anatomy   362

11   Muscular Tissue   384

PART THREE

INTERNAL COORDINATION AND CONTROL   420

12   Nervous Tissue   420

13   The Spinal Cord, Spinal Nerves, and Somatic Reflexes   459

14   The Brain and Cranial Nerves   492

15   The Autonomic Nervous System and Visceral Reflexes   542

16   Sense Organs   563

17   The Endocrine System   612

PART FOUR

CIRCULATION AND DEFENSE   655

18   The Circulatory System: Blood   655

19   The Circulatory System: Heart   689

20   The Circulatory System: Blood Vessels and Circulation   724

21   The Lymphatic and Immune Systems   782

PART FIVE

INTAKE AND OUTPUT   825

22   The Respiratory System   825

23   The Urinary System   866

24   Fluid, Electrolyte, and Acid–Base Balance   901

25   The Digestive System   925

26   Nutrition and Metabolism   974

PART SIX

REPRODUCTION AND THE LIFE CYCLE   1007

27   The Male Reproductive System   1007

28   The Female Reproductive System   1037

29   Human Development and Aging   1075

APPENDIX A: Answer Keys   A-1

APPENDIX B: Symbols, Weights, and Measures   A-14

APPENDIX C: Periodic Table of the Elements   A-16

APPENDIX D: The Genetic Code and Amino Acids   A-17

APPENDIX E: Medical Word Roots and Affixes   A-19

Glossary   G-1

Index   I-1

© Tim Vacula

KENNETH S. SALADIN is Distinguished Professor of Biology, Emeritus, at Georgia College in Milledgeville, Georgia. He received his B.S. in zoology at Michigan State University and a Ph.D. in parasitology at Florida State University, with interests especially in the sensory ecology of freshwater invertebrates. He joined the Georgia College faculty in 1977. His courses included human anatomy and physiology, introduction to medical physiology, histology, general zoology, parasitology, animal behavior, biomedical etymology, study abroad in the Galápagos Islands, and premedical seminars, among others. Ken was recognized as "most significant undergraduate mentor" nine times over the years by outstanding students inducted into Phi Kappa Phi. He received the university's Excellence in Research and Publication Award for the first edition of this book, and was named Distinguished Professor in 2001. Ken is a member of the Human Anatomy and Physiology Society, American Association for Anatomy, American Physiological Society, Society for Integrative and Comparative Biology, Authors' Guild, and Textbook and Academic Authors Association. He served as a developmental reviewer and wrote supplements for several other McGraw-Hill anatomy and physiology textbooks for a number of years before becoming a textbook writer. Ken has used the earnings from his textbooks to support the Charles Darwin Research Station and fund ecosystem conservation and restoration in the Galápagos Islands, to remodel and equip the Georgia College anatomy laboratories, to fund the Honors Program, and to endow student scholarships, the William Harvey Chair in Biomedical Science, the Annual William Harvey Lecture in Medicine and Society, and the William P. Wall Museum of Natural History. Ken and his wife Diane have two grown children and a nature-loving grandson in North Carolina.

© Chris Gan/Yuen Lui Studios

CHRISTINA A. GAN, digital coauthor for Connect®, has been teaching anatomy and physiology, microbiology, and general biology at Highline College in Des Moines, Washington, since 2004. Before that, she taught at Rogue Community College in Medford, Oregon, for 6 years. She earned her M.A. in biology from Humboldt State University, researching the genetic variation of mitochondrial DNA in various salmonid species, and is a member of the Human Anatomy and Physiology Society. When she is not in the classroom or developing digital media, she is climbing, mountaineering, skiing, kayaking, sailing, cycling, and mountain biking throughout the Pacific Northwest.

© JC Penney Portraits/Lifetouch Portrait Studios, Inc.

HEATHER N. CUSHMAN, digital coauthor for Connect®, teaches anatomy and physiology at Tacoma Community College in Tacoma, Washington, and is a member of the Human Anatomy and Physiology Society. She received her Ph.D. in neuroscience from the University of Minnesota in 2002, and completed a postdoctoral fellowship at the Vollum Institute at Oregon Health & Science University in Portland, Oregon, where she studied sensory transduction and the cellular and molecular mechanisms of muscle pain. She currently resides in Tacoma, Washington, and enjoys climbing, camping, and hiking with her husband Ken and their daughter Annika.

CONTENTS

About the Authors iv

PART ONE

ORGANIZATION OF THE BODY

CHAPTER **1**

MAJOR THEMES OF ANATOMY AND PHYSIOLOGY 1

1.1 The Scope of Anatomy and Physiology 2
1.2 The Origins of Biomedical Science 3
1.3 Scientific Method 6
1.4 Human Origins and Adaptations 9
1.5 Human Structure 11
1.6 Human Function 13
1.7 The Language of Medicine 19
1.8 Review of Major Themes 21
STUDY GUIDE 24

ATLAS **A**

GENERAL ORIENTATION TO HUMAN ANATOMY 27

A.1 General Anatomical Terminology 28
A.2 Major Body Regions 29
A.3 Body Cavities and Membranes 31
A.4 Organ Systems 35
STUDY GUIDE 38

CHAPTER **2**

THE CHEMISTRY OF LIFE 40

2.1 Atoms, Ions, and Molecules 41
2.2 Water and Mixtures 48
2.3 Energy and Chemical Reactions 52
2.4 Organic Compounds 55
STUDY GUIDE 71

CHAPTER **3**

CELLULAR FORM AND FUNCTION 74

3.1 Concepts of Cellular Structure 75
3.2 The Cell Surface 78
3.3 Membrane Transport 87
3.4 The Cell Interior 96
STUDY GUIDE 106

CHAPTER **4**

GENES AND CELLULAR FUNCTION 109

4.1 DNA and RNA—The Nucleic Acids 110
4.2 Genes and Their Action 114
4.3 DNA Replication and the Cell Cycle 123
4.4 Chromosomes and Heredity 128
STUDY GUIDE 134

CHAPTER **5**

THE HUMAN TISSUES 137

5.1 The Study of Tissues 138
5.2 Epithelial Tissue 141
5.3 Connective Tissue 148
5.4 Nervous and Muscular Tissues—Excitable Tissues 157
5.5 Cellular Junctions, Glands, and Membranes 160
5.6 Tissue Growth, Development, Repair, and Degeneration 165
STUDY GUIDE 171

PART TWO

SUPPORT AND MOVEMENT

CHAPTER **6**

THE INTEGUMENTARY SYSTEM 174

6.1 Skin and Subcutaneous Tissue 175
6.2 Hair and Nails 183
6.3 Cutaneous Glands 187
6.4 Skin Disorders 190
CONNECTIVE ISSUES 195
STUDY GUIDE 196

CHAPTER **7**

BONE TISSUE 198

7.1 Tissues and Organs of the Skeletal System 199
7.2 Histology of Osseous Tissue 201
7.3 Bone Development 206
7.4 Physiology of Osseous Tissue 211
7.5 Bone Disorders 216
CONNECTIVE ISSUES 220
STUDY GUIDE 221

CHAPTER **8**

THE SKELETAL SYSTEM 223

8.1 Overview of the Skeleton 224
8.2 The Skull 227
8.3 The Vertebral Column and Thoracic Cage 239
8.4 The Pectoral Girdle and Upper Limb 249
8.5 The Pelvic Girdle and Lower Limb 254
STUDY GUIDE 264

CHAPTER **9**

JOINTS 267

9.1 Joints and Their Classification 268
9.2 Synovial Joints 272
9.3 Anatomy of Selected Diarthroses 285
STUDY GUIDE 296

CHAPTER **10**

THE MUSCULAR SYSTEM 299

10.1 Structural and Functional Organization of Muscles 300
10.2 Muscles of the Head and Neck 308
10.3 Muscles of the Trunk 318

10.4 Muscles Acting on the Shoulder and Upper Limb 329

10.5 Muscles Acting on the Hip and Lower Limb 344

STUDY GUIDE 359

ATLAS B

REGIONAL AND SURFACE ANATOMY 362

B.1 Regional Anatomy 363

B.2 The Importance of Surface Anatomy 363

B.3 Learning Strategy 363

CHAPTER 11

MUSCULAR TISSUE 384

11.1 Types and Characteristics of Muscular Tissue 385

11.2 Skeletal Muscle Cells 386

11.3 The Nerve–Muscle Relationship 391

11.4 Behavior of Skeletal Muscle Fibers 394

11.5 Behavior of Whole Muscles 401

11.6 Muscle Metabolism 404

11.7 Cardiac and Smooth Muscle 409

CONNECTIVE ISSUES 416

STUDY GUIDE 417

PART THREE

INTERNAL COORDINATION AND CONTROL

CHAPTER 12

NERVOUS TISSUE 420

12.1 Overview of the Nervous System 421

12.2 Properties of Neurons 422

12.3 Supportive Cells 426

12.4 Electrophysiology of Neurons 433

12.5 Synapses 440

12.6 Neural Integration 446

CONNECTIVE ISSUES 455

STUDY GUIDE 456

CHAPTER 13

THE SPINAL CORD, SPINAL NERVES, AND SOMATIC REFLEXES 459

13.1 The Spinal Cord 460

13.2 The Spinal Nerves 468

13.3 Somatic Reflexes 482

STUDY GUIDE 489

CHAPTER 14

THE BRAIN AND CRANIAL NERVES 492

14.1 Overview of the Brain 493

14.2 Meninges, Ventricles, Cerebrospinal Fluid, and Blood Supply 497

14.3 The Hindbrain and Midbrain 503

14.4 The Forebrain 509

14.5 Integrative Functions of the Brain 516

14.6 The Cranial Nerves 527

STUDY GUIDE 539

CHAPTER 15

THE AUTONOMIC NERVOUS SYSTEM AND VISCERAL REFLEXES 542

15.1 General Properties of the Autonomic Nervous System 543

15.2 Anatomy of the Autonomic Nervous System 546

15.3 Autonomic Effects on Target Organs 553

15.4 Central Control of Autonomic Function 557

STUDY GUIDE 560

CHAPTER 16

SENSE ORGANS 563

16.1 Properties and Types of Sensory Receptors 564

16.2 The General Senses 566

16.3 The Chemical Senses 572

16.4 Hearing and Equilibrium 577

16.5 Vision 591

STUDY GUIDE 609

CHAPTER 17

THE ENDOCRINE SYSTEM 612

17.1 Overview of the Endocrine System 613

17.2 The Hypothalamus and Pituitary Gland 616

17.3 Other Endocrine Glands 623

17.4 Hormones and Their Actions 633

17.5 Stress and Adaptation 642

17.6 Eicosanoids and Other Signaling Molecules 643

17.7 Endocrine Disorders 644

CONNECTIVE ISSUES 651

STUDY GUIDE 652

PART FOUR

CIRCULATION AND DEFENSE

CHAPTER 18

THE CIRCULATORY SYSTEM: BLOOD 655

18.1 Introduction 656

18.2 Erythrocytes 661

18.3 Blood Types 667

18.4 Leukocytes 672

18.5 Platelets and the Control of Bleeding 678

STUDY GUIDE 686

CHAPTER 19

THE CIRCULATORY SYSTEM: HEART 689

19.1 Overview of the Cardiovascular System 690

19.2 Gross Anatomy of the Heart 692

19.3 Cardiac Muscle and the Cardiac Conduction System 700

19.4 Electrical and Contractile Activity of the Heart 702

19.5 Blood Flow, Heart Sounds, and the Cardiac Cycle 707

19.6 Regulation of Cardiac Output 713

STUDY GUIDE 721

CHAPTER 20

THE CIRCULATORY SYSTEM: BLOOD VESSELS AND CIRCULATION 724

20.1 General Anatomy of the Blood Vessels 725

20.2 Blood Pressure, Resistance, and Flow 733

20.3 Capillaries and Fluid Exchange 740

20.4 Venous Return and Circulatory Shock 744

20.5 Special Circulatory Routes 746
20.6 Anatomy of the Pulmonary Circuit 747
20.7 Systemic Vessels of the Axial Region 748
20.8 Systemic Vessels of the Appendicular Region 766
CONNECTIVE ISSUES 777
STUDY GUIDE 778

CHAPTER 21

THE LYMPHATIC AND IMMUNE SYSTEMS 782

21.1 The Lymphatic System 783
21.2 Innate Immunity 795
21.3 Adaptive Immunity— General Aspects 803
21.4 Cellular Immunity 808
21.5 Humoral Immunity 811
21.6 Immune System Disorders 815
CONNECTIVE ISSUES 821
STUDY GUIDE 822

PART FIVE

INTAKE AND OUTPUT

CHAPTER 22

THE RESPIRATORY SYSTEM 825

22.1 Anatomy of the Respiratory System 826
22.2 Pulmonary Ventilation 837
22.3 Gas Exchange and Transport 848
22.4 Respiratory Disorders 858
CONNECTIVE ISSUES 862
STUDY GUIDE 863

CHAPTER 23

THE URINARY SYSTEM 866

23.1 Functions of the Urinary System 867
23.2 Anatomy of the Kidney 869
23.3 Urine Formation I: Glomerular Filtration 875
23.4 Urine Formation II: Tubular Reabsorption and Secretion 881
23.5 Urine Formation III: Water Conservation 885
23.6 Urine and Renal Function Tests 888
23.7 Urine Storage and Elimination 891

CONNECTIVE ISSUES 897
STUDY GUIDE 898

CHAPTER 24

FLUID, ELECTROLYTE, AND ACID–BASE BALANCE 901

24.1 Fluid Balance 902
24.2 Electrolyte Balance 908
24.3 Acid–Base Balance 914
STUDY GUIDE 922

CHAPTER 25

THE DIGESTIVE SYSTEM 925

25.1 General Anatomy and Digestive Processes 926
25.2 The Mouth Through Esophagus 930
25.3 The Stomach 937
25.4 The Liver, Gallbladder, and Pancreas 946
25.5 The Small Intestine 952
25.6 Chemical Digestion and Absorption 956
25.7 The Large Intestine 963
CONNECTIVE ISSUES 970
STUDY GUIDE 971

CHAPTER 26

NUTRITION AND METABOLISM 974

26.1 Nutrition 975
26.2 Carbohydrate Metabolism 986
26.3 Lipid and Protein Metabolism 993
26.4 Metabolic States and Metabolic Rate 997
26.5 Body Heat and Thermoregulation 999
STUDY GUIDE 1004

PART SIX

REPRODUCTION AND THE LIFE CYCLE

CHAPTER 27

THE MALE REPRODUCTIVE SYSTEM 1007

27.1 Sexual Reproduction and Development 1008

27.2 Male Reproductive Anatomy 1013
27.3 Puberty, Hormonal Control, and Climacteric 1021
27.4 Sperm and Semen 1023
27.5 Male Sexual Response 1028
STUDY GUIDE 1034

CHAPTER 28

THE FEMALE REPRODUCTIVE SYSTEM 1037

28.1 Female Reproductive Anatomy 1038
28.2 Puberty and Menopause 1048
28.3 Oogenesis and the Sexual Cycle 1050
28.4 Female Sexual Response 1058
28.5 Pregnancy and Childbirth 1059
28.6 Lactation 1066
CONNECTIVE ISSUES 1071
STUDY GUIDE 1072

CHAPTER 29

HUMAN DEVELOPMENT AND AGING 1075

29.1 Fertilization and the Preembryonic Stage 1076
29.2 The Embryonic and Fetal Stages 1082
29.3 The Neonate 1091
29.4 Aging and Senescence 1095
STUDY GUIDE 1106

APPENDIX A: ANSWER KEYS A-1
APPENDIX B: SYMBOLS, WEIGHTS, AND MEASURES A-14
APPENDIX C: PERIODIC TABLE OF THE ELEMENTS A-16
APPENDIX D: THE GENETIC CODE AND AMINO ACIDS A-17
APPENDIX E: MEDICAL WORD ROOTS AND AFFIXES A-19

GLOSSARY G-1

INDEX I-1

THE EVOLUTION OF A
STORYTELLER

Courtesy of Ken Saladin

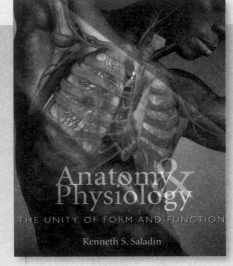

Ken in 1964

Ken Saladin's first step into authoring was a 318-page paper on the ecology of hydras written for his tenth-grade biology class. With his "first book," featuring 53 original India ink drawings and photomicrographs, a true story-teller was born.

When I first became a textbook writer, I found myself bringing the same enjoyment of writing and illustrating to this book that I first discovered when I was 15.

—Ken Saladin

Ken's "first book," *Hydra Ecology,* 1965
Courtesy of Ken Saladin

One of Ken's drawings from *Hydra Ecology*
Courtesy of Ken Saladin

Ken began working on his first book for McGraw-Hill in 1993, and in 1997 the first edition of *The Unity of Form and Function* was published. In 2020, the story continues with the ninth edition of Ken's best-selling A&P textbook.

The first edition (1997)

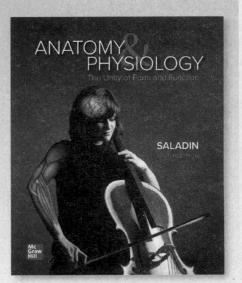

The story continues (2020)

Anatomy & Physiology: The Unity of Form and Function tells a story comprised of many layers, including core science, clinical applications, the history of medicine, and the evolution of the human body. Saladin combines this humanistic perspective on anatomy and physiology with vibrant photos and art to convey the beauty and excitement of the subject to beginning students.

To help students manage the tremendous amount of information in this introductory course, the narrative is broken into short segments, each framed by expected learning outcomes and self-testing review questions. This presentation strategy works as a whole to create a more efficient and effective way for students to learn A&P.

Writing Style and Level

Saladin's text is written using plain language for A&P students who may be taking this course early in their curricula. Careful attention has been given to word selection and paragraph structure to maintain the appropriate writing level for all students.

CHANGES TO THE NINTH EDITION

New Science

This edition draws on recent literature and scientific conferences attended by the author to update many topics, including but not limited to molecular, vascular, and brain imaging techniques; peroxisome and mitochondrial behavior; the DNA damage response; gene regulation; epigenetics; the tissue interstitium; regenerative medicine; osteoporosis; prosthetic joints; fibromyalgia; sleep physiology; trigeminal neuralgia; pain physiology; endocrine functions of osseous and adipose tissue; diabetes mellitus; cord blood transplants; thrombopoiesis; AIDS; prostate diseases; breast cancer; aging; life expectancy; and assisted reproductive technology.

New Deeper Insight sidebar essays have been added on cardiac tamponade; biopsy; stem-cell therapy; regenerative medicine; osteomalacia and rickets; vertebral disc herniation; rotator cuff injury; carpal tunnel syndrome; shinsplints; calcaneal tendon rupture; plantar fasciitis; brain connectomics and diffusion tensor imaging; lumbar puncture; stroke; blindness; alcoholic ascites; diverticulosis and diverticulitis; colorectal cancer; and cleft lip and palate.

While new science has been added, keeping up with such growth also means pruning back topics discredited by newer literature. For this edition, these include adult cerebral neurogenesis; endorphins and runner's high; human pheromones; pineal tumors and precocial puberty; prophylactic use of low-dose aspirin; myocardial regeneration; female ejaculation; and the free-radical DNA damage theory of senescence.

In consideration of user and reviewer suggestions to reduce detail in a few areas, this edition has more concise discussions of some topics: chromatin coiling; apoptosis; skin grafting; the hair cycle; calcium and phosphate homeostasis; and spinal cord tracts.

New Art and Photography

This edition features new drawings of epidermal histology, flat bone structure, lever mechanics, Parkinson disease, lumbar puncture, hand innervation, Bell palsy, the vagus nerve, olfactory pathways, erythropoiesis, cardiac innervation, regulation of cardiac output, air embolism, colonic histology, lipoprotein structure, cleft lip and palate, and senescent muscle atrophy.

New photos in this edition include digital subtraction angiography, molecular-scale cryo-EM imaging, diabetic gangrene, embryonic stem cells, albinism, jaundice, osteocyte SEM, rickets, muscle fiber histochemistry, diffusion tensor imaging of the brain connectome, shingles, cataracts, glaucoma, forelimb veins used for phlebotomy, kidney stones, gallstones, hepatic cirrhosis, MRI of obesity, and intracytoplasmic sperm injection.

Organizational Changes

For improved readability, narrative descriptions of some systems are moved from tables into chapter text; selected illustrations are moved outside of the tables; and tables are distilled to more concise summaries. These include the skeletal muscles (chapter 10), spinal nerve plexuses (chapter 13), cranial nerves (chapter 14), and blood vessels (chapter 20). A detailed list of changes by chapter follows.

Detailed List of Changes

Chapter 1, Major Themes of Anatomy and Physiology, now includes digital subtraction angiography among the common clinical imaging techniques.

Atlas A, General Orientation to Human Anatomy, has an added Deeper Insight A.1 on cardiac tamponade in relation to body cavities and membranes.

Chapter 2, The Chemistry of Life, has added the Nobel-winning new technique of cryo-electron microscopic imaging of biological structure at the atomic level.

Chapter 3, Cellular Form and Function, has enhanced discussions of limitations on cell size, the origin of peroxisomes, mitochondrial fusion and fission, and clinical mitochondrial transfer and three-parent babies.

Chapter 4, Genes and Cellular Function, updates protein processing by the Golgi complex, epigenetics, the DNA damage response, and the role of the nuclear lamina in gene silencing.

Chapter 5, The Human Tissues, has a new perspective on the tissue interstitium, updates on stem-cell therapy and regenerative medicine, and a new Deeper Insight on biopsy methods.

Chapter 6, The Integumentary System, has a new drawing of epidermal histology, new discussion of the evolutionary genetics of apocrine glands, an update on skin-grafting technology, and a simpler description of the hair growth cycle.

Chapter 7, Bone Tissue, gives a less detailed overview of calcium and phosphate homeostasis, adds a Deeper Insight on osteomalacia and rickets, and updates the pathology and treatment of osteoporosis.

Chapter 8, The Skeletal System, conforms the description of normal and pathological spinal curvatures to orthopedic terminology and has a new Deeper Insight on herniated discs.

Chapter 9, Joints, improves the discussion of joint biomechanics and updates the discussions of temporomandibular joint dysfunction and engineering of prosthetic joints.

Chapter 10, The Muscular System, pulls illustrations and narrative descriptions from the muscle tables, converts the narrative to easier-to-read normal text, and condenses the tables to more concise summaries. It updates inguinal hernias and adds new Deeper Insights on rotator cuff injury, shinsplints, calcaneal tendon rupture, and plantar fasciitis.

Chapter 11, Muscular Tissue, adds a photo of the histochemistry of fast glycolytic and slow oxidative muscle fiber types and updates the discussion of fibromyalgia.

Chapter 12, Nervous Tissue, includes updates on astrocyte functions, beta-endorphin and enkephalin, mutations affecting neurotransmitter reuptake and neurological disorders, and the implication of lipofuscin in some diseases. It introduces the frontier neuroscience of brain connectomics and the use of diffusion tensor imaging to visualize the connectome. There is now an illustration of the midbrain histological change and body posture characteristic of Parkinson disease.

Chapter 13, The Spinal Cord, Spinal Nerves, and Somatic Reflexes, adds a new Deeper Insight and illustration of lumbar puncture, reduces detail on spinal cord tracts, reformats the tables of spinal nerve plexuses, illustrates regional innervation of the hand by the major forearm nerves, and adds a photo of a shingles lesion.

Chapter 14, The Brain and Cranial Nerves, now adopts the concept of brainstem as excluding the diencephalon. It adds Deeper Insights on stroke and diffusion tensor imaging, and updates the Deeper Insight on trigeminal neuralgia and Bell palsy, adding an illustration of the latter. It updates sleep physiology and the functions of the midbrain colliculi and pretectal nuclei. It corrects a common misconception about the subdural space. The discussion and table of cranial nerves are reorganized.

Chapter 16, Sense Organs, has an updated discussion of pain physiology and includes phantom limb pain. It updates the genetics and functions of some taste sensations and flawed assumptions about human olfactory sensitivity. It deletes discredited or dubious views of endorphins and runner's high and human pheromones. It enhances the figure of olfactory projection pathways; adds the dorsal and ventral streams of visual processing pathways; adds photos of cataracts and glaucoma; adds macular degeneration and diabetic retinopathy to the Deeper Insight on blindness; and has better insights into the functions of the cornea, choroid, and vitreous body.

Chapter 17, The Endocrine System, updates the histology and cytology of the thyroid gland and pancreatic islets and the effects of melatonin; adds new information on hormones of osseous and adipose origin; updates the enteroendocrine system; and adds effects of lipocalin 2 on insulin action. It deletes the now-questionable idea about pineal tumors and precocial puberty. It updates the pathologies of Addison disease and myxedema, and the genetic, immunological, and treatment aspects of diabetes mellitus.

Chapter 18, The Circulatory System: Blood, now explains how blood is fractionated to obtain plasma and then serum, and the uses of blood serum. It has an enhanced explanation of the functional significance of the discoidal shape of erythrocytes, and includes cell proliferation in the illustration of erythropoiesis. It reports updated clinical research on the number of known blood groups and RBC antigens, cord blood transplants, other methods of bone marrow replacement, and pharmaceutical anticoagulants. It adds the surprising new discovery of abundant platelet production by megakaryocytes in the lungs and megakaryocyte migration between the lungs and bone marrow.

Chapter 19, The Circulatory System: Heart, is reorganized at section 19.1 to place figures closer to their references. Cardiac innervation is moved to section 19.6 on regulation of cardiac output, with a new illustration. The electrocardiogram is described with more detailed attention to interpretation of each wave, segment, and interval, with an added table. The section on cardiac arrhythmias includes a fuller explanation of atrial fibrillation.

Chapter 20, The Circulatory System: Blood Vessels and Circulation, has improved discussions of the vasa vasorum and metarterioles; describes the measurement of blood pressure in more depth; adds photos of edema, circulatory shock, and upper limb veins most often used for phlebotomy; and has a new drawing of air embolism. It discusses the difficulty of pancreatic surgery in light of the complex, delicate branches of the celiac trunk. The Deeper Insight on ascites is rewritten to relate it to alcoholism. The tables of blood vessels and routes of flow are now converted to normal, easier-to-read text.

Chapter 21, The Lymphatic and Immune Systems, updates bone marrow histology; the sources of macrophages; T cell diversity; asthma and AIDS mortality; and the obstacles to treating AIDS in pandemic countries. It adds the risk in splenectomy and the role of ATP and ADP as inflammatory chemoattractants.

Chapter 22, The Respiratory System, enhances descriptions of the nasal epithelium; the cricothyroid ligament in relation to emergency tracheotomy; the Deeper Insight on tracheotomy; cor pulmonale; and squamous cell carcinoma. It adds a mutational cause of Ondine's curse; discovery of pulmonary platelet production; and the potential of electronic cigarettes and legalization of recreational marijuana as emerging risk factors for lung cancer.

Chapter 23, The Urinary System, adds to the function of glomerular mesangial cells and has an improved Deeper Insight on kidney stones, with a new photo.

Chapter 24, Fluid, Electrolyte, and Acid–Base Balance, has further information on sodium and the effects of hypernatremia, and has added a new table summarizing the major electrolyte imbalances.

Chapter 25, The Digestive System, includes additions on the immune role of the omenta; dental proprioception; aspirin and peptic ulcer; the cell-signaling function of the intestinal mucous coat; anatomical variability of the colon and a new drawing of its histology; an updated Deeper Insight on gallstones, with a photo; a new Deeper Insight on diverticulosis and diverticulitis; a new Deeper Insight on colorectal cancer; and an improved description of intestinal lymphatic nodules.

Chapter 26, Nutrition and Metabolism, includes new MRI images of a morbidly obese individual compared to one of normal BMI; a new drawing of lipoprotein structure and chart of composition of the lipoprotein classes; new information on the effects of leptin on sympathetic nerve fibers and lipolysis; and a new photo of hepatic cirrhosis.

Chapter 27, The Male Reproductive System, has a new table and discussion of the composition of semen and function of the bulbourethral preejaculatory fluid, and updates on benign prostatic hyperplasia and prostate cancer. It adds discussion of zinc deficiency as a cause of infertility, hypothalamic maturation and GnRH in relation to the onset of puberty, and andropause in relation to declining androgen secretion.

Chapter 28, The Female Reproductive System, has improvements in hymen anatomy and the figure of ovarian structure; a new perspective on morning sickness as a possible factor mitigating birth defects; and updates on contraception and on breast cancer genes, risk factors, and mortality.

Chapter 29, Human Development and Aging, adds the role of the sperm centrosome in fertilization; chromosomal defects as a leading cause of first-trimester miscarriages; and the formation of monozygotic twins. It adds a new Deeper Insight and illustration of cleft lip and palate. It updates the telomere theory of senescence but deletes the now-doubtful theory of DNA damage by endogenous free radicals. It adds a new, MRI-based drawing of muscle atrophy in old age and a discussion of pineal gland senescence as a factor in the insomnia experienced by some older people. It updates statistics on human life expectancy and the major causes of death. The final Deeper Insight is retitled Assisted Reproductive Technology and has a new photo of intracytoplasmic sperm injection.

Appendix D, The Genetic Code and Amino Acids, now adds a table of the 20 amino acids and their symbols, and the structural formulae of the amino acids.

ACKNOWLEDGMENTS

Peer review is a critical part of the scientific process, and very important to ensure the content in this book continues to meet the needs of the instructors and students who use it. We are grateful for the people who agree to participate in this process and thank them for their time, talents, and feedback. The reviewers of this text (listed here) have contributed significant comments that help us refine and update the print and digital components of this program.

Christina Gan and Heather Cushman have updated the question bank and test bank to closely correlate with the intricate changes made in this ninth edition and have greatly increased the educational value of these books through their work to create self-assessment tools and align McGraw-Hill's Connect resources with the textbook. This has contributed significantly to student and instructor satisfaction with our overall package of learning media and to the students' success as they master A&P en route to their career aspirations.

I would also like to extend appreciation to members of the Life Sciences Book Team at McGraw-Hill Education who have worked with me on this project, including Matthew Garcia, Senior Portfolio Manager; Valerie Kramer, Marketing Manager; Donna Nemmers, Senior Product Developer; Vicki Krug, Senior Content Project Manager; Lori Hancock, Lead Content Licensing Specialist; Brent dela Cruz, Senior Content Project Manager; Egzon Shaqiri, Designer; and Jeanne Patterson, freelance copy editor. Their efforts have yielded another great edition of the text and its companion media suite of Connect products.

Timothy A. Ballard
University of North Carolina—Wilmington

Barry N. Bates
Atlanta Technical College

Christopher I. Brandon Jr.
Georgia Gwinnett College

Nickolas A. Butkevich
Schoolcraft College

John W. Campbell
Oklahoma City Community College

Jennifer Cochran Biederman
Winona State University

Mary B. Colon
Seminole State College of Florida

Abdeslem El Idrissi
College of Staten Island, City University of New York

Bagie George
Georgia Gwinnett College

Kyle P. Harris
Temple University

Karen L. Kandl
Western Carolina University

Stephen A. Kash
Oklahoma City Community College

Stephanie Matlock
Colorado Mesa University

Deborah T. Palatinus
Roane State Community College

Jeffrey Alan Pence
Excelsior College

Carla Perry
Community College of Philadelphia

Franz Sainvil
Broward College–Central Campus

Brian Stout
Northwest Vista College

Andrew Van Nguyen
The City University of New York–Queensborough Community College

Kimberly Vietti
Illinois Central College

Beth L. Williams
Wallace State Community College

Delon Washo-Krupps
Arizona State University

Samia Williams
Santa Fe College

INNOVATIVE CHAPTER SEQUENCING

Some chapters and topics are presented in a sequence that is more instructive than the conventional order.

Early Presentation of Heredity

Fundamental principles of heredity are presented in the last few pages of chapter 4 rather than at the back of the book to better integrate molecular and Mendelian genetics. This organization also prepares students to learn about such genetic traits and conditions as cystic fibrosis, color blindness, blood types, hemophilia, cancer genes, and sickle-cell disease by first teaching them about dominant and recessive alleles, genotype and phenotype, and sex linkage.

Urinary System Presented Close to Circulatory and Respiratory Systems

Most textbooks place this system near the end of the book because of its anatomical and developmental relationships with the reproductive system. However, its physiological ties to the circulatory and respiratory systems are much more important. Except for a necessary digression on lymphatics and immunity, the circulatory system is followed almost immediately with the respiratory and urinary systems, which regulate blood composition and whose functional mechanisms rely on recently covered principles of blood flow and capillary exchange.

Muscle Anatomy and Physiology Follow Skeleton and Joints

The functional morphology of the skeleton, joints, and muscles is treated in three consecutive chapters, 8 through 10, so when students learn muscle attachments, these come only two chapters after the names of the relevant bone features. When they learn muscle actions, it is in the first chapter after learning the terms for the joint movements. This order brings another advantage: The physiology of muscle and nerve cells is treated in two consecutive chapters (11 and 12), which are thus closely integrated in their treatment of synapses, neurotransmitters, and membrane electrophysiology.

BRIEF CONTENTS

About the Authors   iv

PART ONE
ORGANIZATION OF THE BODY   1

1   Major Themes of Anatomy and Physiology   1
    ATLAS A General Orientation to Human Anatomy   27
2   The Chemistry of Life   40
3   Cellular Form and Function   74
4   Genes and Cellular Function   109
5   The Human Tissues   137

PART TWO
SUPPORT AND MOVEMENT   174

6   The Integumentary System   174
7   Bone Tissue   198
8   The Skeletal System   223
9   Joints   267
10 The Muscular System   299
    ATLAS B Regional and Surface Anatomy   362
11 Muscular Tissue   384

PART THREE
INTERNAL COORDINATION AND CONTROL   420

12 Nervous Tissue   420
13 The Spinal Cord, Spinal Nerves, and Somatic Reflexes   459
14 The Brain and Cranial Nerves   492
15 The Autonomic Nervous System and Visceral Reflexes   542
16 Sense Organs   563
17 The Endocrine System   612

PART FOUR
CIRCULATION AND DEFENSE   655

18 The Circulatory System: Blood   655
19 The Circulatory System: Heart   689
20 The Circulatory System: Blood Vessels and Circulation   724
21 The Lymphatic and Immune Systems   782

PART FIVE
INTAKE AND OUTPUT   825

22 The Respiratory System   825
23 The Urinary System   866
24 Fluid, Electrolyte, and Acid–Base Balance   901
25 The Digestive System   925
26 Nutrition and Metabolism   974

PART SIX
REPRODUCTION AND THE LIFE CYCLE   1007

27 The Male Reproductive System   1007
28 The Female Reproductive System   1037
29 Human Development and Aging   1075

APPENDIX A: Answer Keys   A-1
APPENDIX B: Symbols, Weights, and Measures   A-14
APPENDIX C: Periodic Table of the Elements   A-16
APPENDIX D: The Genetic Code and Amino Acids   A-17
APPENDIX E: Medical Word Roots and Affixes   A-19

Glossary   G-1
Index   I-1

THE STORY OF
FORM AND FUNCTION

LEARNING TOOLS

Engaging Chapter Layouts

- Chapters are structured around the way students learn.
- Frequent subheadings and expected learning outcomes help students plan their study time and review strategies.

Deeper Insights highlight areas of interest and career relevance for students.

Chapter Outlines provide quick previews of the content.

CHAPTER 7

BONE TISSUE

A bone cell (osteocyte) surrounded by calcified bone matrix
Eye of Science/Science Source

CHAPTER OUTLINE

7.1 Tissues and Organs of the Skeletal System
  7.1a Functions of the Skeleton
  7.1b Bones and Osseous Tissue
  7.1c General Features of Bones

7.2 Histology of Osseous Tissue
  7.2a Bone Cells
  7.2b The Matrix
  7.2c Compact Bone
  7.2d Spongy Bone
  7.2e Bone Marrow

7.3 Bone Development
  7.3a Intramembranous Ossification
  7.3b Endochondral Ossification
  7.3c Bone Growth and Remodeling

7.4 Physiology of Osseous Tissue
  7.4a Mineral Deposition and Resorption
  7.4b Calcium Homeostasis
  7.4c Phosphate Homeostasis
  7.4d Other Factors Affecting Bone

7.5 Bone Disorders
  7.5a Fractures and Their Repair
  7.5b Other Bone Disorders

Connective Issues

Study Guide

DEEPER INSIGHTS

7.1 Medical History: Bone Contamination
7.2 Clinical Application: Achondroplastic Dwarfism
7.3 Clinical Application: Rickets and Osteomalacia
7.4 Clinical Application: Osteoporosis

Anatomy & Physiology Revealed 4.0

Module 5: Skeletal System

198

CHAPTER 3 Cellular Form and Function 75

BRUSHING UP

- The transport of matter through cell membranes follows the principles of flow down gradients (see section 1.6e).
- To adequately understand the structure of the cell surface, it is essential that you understand glycolipids and glycoproteins, as well as phospholipids and their amphipathic nature (see sections 2.4c and 2.4d).
- The proteins of cell membranes have a great variety of functions. To understand those depends on an acquaintance with the functions of proteins in general and how protein function depends on tertiary structure (see "Protein Structure" and "Protein Functions" in section 2.4e).

All organisms, from the simplest to the most complex, are composed of cells—whether the single cell of a bacterium or the trillions of cells that constitute the human body. These cells are responsible for all structural and functional properties of a living organism. A knowledge of cells is therefore indispensable to any true understanding of the workings of the human body, the mechanisms of disease, and the rationale of therapy. Thus, this chapter and the next one introduce the basic cell biology of the human body, and subsequent chapters expand upon this information as we examine the specialized cellular structure and function of specific organs.

3.1 Concepts of Cellular Structure

Expected Learning Outcomes

When you have completed this section, you should be able to

a. discuss the development and modern tenets of the cell theory;
b. describe cell shapes from their descriptive terms;
c. state the size range of human cells and discuss factors that limit their size;
d. discuss the way that developments in microscopy have changed our view of cell structure; and
e. outline the major components of a cell.

3.1a Development of the Cell Theory

Cytology,[1] the scientific study of cells, was born in 1663 when Robert Hooke observed the empty cell walls of cork and coined the word *cellulae* ("little cells") to describe them (see section 1.2). Soon he studied thin slices of fresh wood and saw living cells "filled with juices"—a fluid later named *cytoplasm*. Two centuries later, Theodor Schwann studied a wide range of animal tissues and concluded that all animals are made of cells.

[1] *cyto* = cell; *logy* = study of

Schwann and other biologists originally believed that cells came from nonliving body fluid that somehow congealed and acquired a membrane and nucleus. This idea of *spontaneous generation*—that living things arise from nonliving matter—was rooted in the scientific thought of the times. For centuries, it seemed to be simple common sense that decaying meat turned into maggots, stored grain into rodents, and mud into frogs. Schwann and his contemporaries merely extended this idea to cells. The idea of spontaneous generation wasn't discredited until some classic experiments by French microbiologist Louis Pasteur in 1859.

By the end of the nineteenth century, it was established beyond all reasonable doubt that cells arise only from other cells and every living organism is composed of cells and cell products. The cell came to be regarded, and still is, as the simplest structural and functional unit of life. There are no smaller subdivisions of a cell or organism that, in themselves, have all or most of the fundamental characteristics of life described in section 1.6a. Enzymes and organelles, for example, are not alive, although the life of a cell depends on their activity.

The development of biochemistry from the late nineteenth to the twentieth century made it further apparent that all physiological processes of the body are based on cellular activity and that the cells of all species exhibit remarkable biochemical unity. The various generalizations of these last two paragraphs now constitute the modern **cell theory.**

3.1b Cell Shapes and Sizes

We will shortly examine the structure of a generic cell, but the generalizations we draw shouldn't blind you to the diversity of cellular form and function in humans. There are about 200 kinds of cells in the human body, with a variety of shapes, sizes, and functions.

Descriptions of organ and tissue structure often refer to the shapes of cells by the following terms (**fig. 3.1**):

- **Squamous**[2] (SKWAY-mus)—a thin, flat, scaly shape, often with a bulge where the nucleus is, much like the shape of a fried egg "sunny side up." Squamous cells line the esophagus and form the surface layer (epidermis) of the skin.
- **Cuboidal**[3] (cue-BOY-dul)—squarish-looking in frontal sections and about equal in height and width; liver cells are a good example.
- **Columnar**—distinctly taller than wide, such as the inner lining cells of the stomach and intestines.
- **Polygonal**[4]—having irregularly angular shapes with four, five, or more sides.
- **Stellate**[5]—having multiple pointed processes projecting from the body of a cell, giving it a somewhat starlike shape. The cell bodies of many nerve cells are stellate.

[2] *squam* = scale; *ous* = characterized by
[3] *cub* = cube; *oidal* = like, resembling
[4] *poly* = many; *gon* = angles
[5] *stell* = star; *ate* = resembling, characterized by

Tiered Assessments Based on Key Learning Outcomes

- Chapters are divided into brief sections, enabling students to set specific goals for short study periods.
- Section-ending questions allow students to check their understanding before moving on.

Each chapter begins with **Brushing Up** to emphasize the interrelatedness of concepts, which is especially useful for adult students returning to the classroom, and serves as an aid for instructors when teaching chapters out of order.

Each major section begins with **Expected Learning Outcomes** to help focus the reader's attention on the larger concepts and make the course outcome-driven. This also assists instructors in structuring their courses around expected learning outcomes.

Questions in figure legends and **Apply What You Know** items prompt students to think more deeply about the implications and applications of what they have learned. This helps students practice higher order thinking skills throughout the chapter.

separation between the bones and length of the fibers give these joints more mobility than a suture or gomphosis has. An especially mobile syndesmosis exists between the shafts of the radius and ulna, which are joined by a broad fibrous *interosseous membrane.* This permits such movements as pronation and supination of the forearm. A less mobile syndesmosis is the one that binds the distal ends of the tibia and fibula together, side by side (see fig. 9.2c).

9.1c Cartilaginous Joints

A **cartilaginous joint** is also called an **amphiarthrosis**[7] (AM-fee-ar-THRO-sis). In these joints, two bones are linked by cartilage (**fig. 9.4**). The two types of cartilaginous joints are *synchondroses* and *symphyses.*

Synchondroses

A **synchondrosis**[8] (SIN-con-DRO-sis) is a joint in which the bones are bound by hyaline cartilage. An example is the temporary joint between the epiphysis and diaphysis of a long bone in a child, formed by the cartilage of the epiphysial plate. Another is the attachment of the first rib to the sternum by a hyaline costal cartilage

(fig. 9.4a). (The other costal cartilages are joined to the sternum by synovial joints.)

Symphyses

In a **symphysis**[9] (SIM-fih-sis), two bones are joined by fibrocartilage (fig. 9.4b, c). One example is the pubic symphysis, in which the right and left pubic bones are joined anteriorly by the cartilaginous interpubic disc. Another is the joint between the bodies of two vertebrae, united by an intervertebral disc. The surface of each vertebral body is covered with hyaline cartilage. Between the vertebrae, this cartilage becomes infiltrated with collagen bundles to form fibrocartilage. Each intervertebral disc permits only slight movement between adjacent vertebrae, but the collective effect of all 23 discs gives the spine considerable flexibility.

▶▶▶**APPLY WHAT YOU KNOW**
The intervertebral joints are symphyses only in the cervical through the lumbar region. How would you classify the intervertebral joints of the sacrum and coccyx in a middle-aged adult?

[7]*amphi* = on all sides; *arthr* = joined; *osis* = condition
[8]*syn* = together; *chondr* = cartilage; *osis* = condition

[9]*sym* = together; *physis* = growth

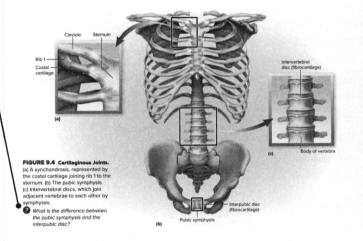

FIGURE 9.4 Cartilaginous Joints.
(a) A synchondrosis, represented by the costal cartilage joining rib 1 to the sternum. (b) The pubic symphysis. (c) Intervertebral discs, which join adjacent vertebrae to each other by symphyses.
❓ *What is the difference between the pubic symphysis and the interpubic disc?*

The end-of-chapter **Study Guide** offers several methods for assessment that are useful to both students and instructors.

Assess Your Learning Outcomes provides students a study outline for review, and addresses the needs of instructors whose colleges require outcome-oriented syllabi and assessment of student achievement of the expected learning outcomes.

End-of-chapter questions build on all levels of Bloom's Taxonomy in sections to
1. test simple recall and analytical thought;
2. build medical vocabulary; and
3. apply the basic knowledge to new clinical problems and other situations.

What's Wrong with These Statements? questions further address Bloom's Taxonomy by asking the student to explain *why* the false statements are untrue.

Testing Your Comprehension questions address Bloom's Taxonomy in going beyond recall to application of ideas.

▶ **Assess Your Learning Outcomes**

To test your knowledge, discuss the following topics with a study partner or in writing, ideally from memory.

9.1 Joints and Their Classification

1. The fundamental definition of *joint (articulation)* and why it cannot be defined as a point at which one bone moves relative to an adjacent bone

3. Three essential components of a lever
4. The meaning of *mechanical advantage* (MA); how the *MA* of a lever can be determined from measurements of its effort and resistance arms; and the respective advantages of levers in which the *MA* is greater than or less than 1.0
5. Comparison of first-, second-, and third-class levers, and anatomical examples of each

12. The same for flexion, extension, hyperextension, and lateral flexion of the spine, and right and left rotation of the trunk
13. The same for elevation, depression, protraction, retraction, and lateral and medial excursion of the mandible
14. The same for dorsiflexion, plantar flexion, inversion, eversion, pronation, and supination of the foot

▶ **Testing Your Recall**   *Answers in Appendix A*

1. Internal and external rotation of the humerus is made possible by a _____ joint.
   a. pivot
   b. condylar
   c. ball-and-socket
   d. saddle
   e. hinge

2. Which of the following is the least movable?
   a. a diarthrosis
   b. a synostosis
   c. a symphysis
   d. a synovial joint
   e. a condylar joint

3. Which of the following movements are unique to the foot?
   a. dorsiflexion and inversion
   b. elevation and depression
   c. circumduction and rotation
   d. abduction and adduction
   e. opposition and reposition

▶ **Building Your Medical Vocabulary**   *Answers in Appendix A*

State a meaning of each word element, and give a medical term from this chapter that uses it or a slight variation of it.

1. ab-
2. arthro-

3. -ate
4. cruci-
5. cruro-
6. -duc

7. kinesio-
8. men-
9. supin-
10. -trac

▶ **What's Wrong with These Statements?**   *Answers in Appendix A*

Briefly explain why each of the following statements is false, or reword it to make it true.

1. More people get rheumatoid arthritis than osteoarthritis.
2. A doctor who treats arthritis is called a kinesiologist.
3. Synovial joints are also known as synarthroses.

4. Menisci occur in the elbow and knee joints.
5. Reaching behind you to take something out of your hip pocket involves flexion of the shoulder.
6. The cruciate ligaments are in the feet.
7. The femur is held tightly in the acetabulum mainly by the round ligament.

8. The knuckles are amphiarthroses.
9. Synovial fluid is secreted by the bursae.
10. Like most ligaments, the periodontal ligaments attach one bone (the tooth) to another (the mandible or maxilla).

STUDY GUIDE

▶ **Testing Your Comprehension**

1. All second-class levers produce a mechanical advantage greater than 1.0 and all third-class levers produce a mechanical advantage less than 1.0. Explain why.

2. For each of the following joint movements, state what bone the axis of rotation passes through and which of the three anatomical planes contains the axis of rotation. You may find it helpful to produce some of these actions on an articulated laboratory

the first interphalangeal joint of the index finger. (Do not bend the fingers of a wired laboratory skeletal hand, because they can break off.)

3. In order of occurrence, list the joint actions (flexion, pronation, etc.) and the joints where they would occur as you (a) sit down at a table, (b) reach out and pick up an apple, (c) take a bite, and (d) chew it. Assume that you start in anatomical

arm. Imagine a person holding a weight in the hand and abducting the arm. On a laboratory skeleton, identify the fulcrum; measure the effort arm and resistance arm; determine the mechanical advantage of this movement; and determine which of the three lever types the upper limb acts as when performing this movement.

5. List the six types of synovial joints, and for each one, if possible, identify a joint in the

ARTWORK THAT INSPIRES LEARNING

The incredible art program in this textbook sets the standard in A&P. The stunning portfolio of art and photos was created with the aid of art focus groups and with feedback from hundreds of accuracy reviews.

Francis Leroy, Biocosmos/Science Source

Vivid Illustrations

Rich textures and shading and bold, bright colors bring structures to life.

Rebecca Gray/McGraw-Hill Education

Cadaver dissections are paired with carefully drawn illustrations to show intricate human detail.

Orientation Tools

Saladin art integrates tools to help students quickly orient themselves within a figure and make connections between ideas.

Palmaris longus tendon (cut)
Flexor carpi radialis tendon
Flexor pollicis longus tendon
Palmar carpal ligament (cut)
Radial artery
Median nerve
Trapezium

Flexor digitorum superficialis tendon
Flexor digitorum profundus tendon
Flexor carpi ulnaris tendon
Ulnar artery
Ulnar nerve
Flexor retinaculum covering carpal tunnel
Bursa
Superficial palmar arterial arch

(a) Anterior view

Anterior
Lateral — Medial
Posterior

Palmaris longus tendon
Thenar muscles
Flexor digitorum superficialis tendons
Trapezium
Flexor digitorum profundus tendons
Radial artery
Trapezoid
Scaphoid

Median nerve
Ulnar artery
Flexor retinaculum
Carpal tunnel
Hypothenar muscles
Ulnar bursa
Hamate
Capitate
Extensor tendons

(b) Cross section

Conducive to Learning

- Easy-to-understand process figures
- Tools for students to easily orient themselves

Process Figures

Saladin breaks complicated physiological processes into numbered steps for a manageable introduction to difficult concepts.

Aorta
Superior vena cava
Right pulmonary veins
Right atrium
Right AV valve
Right ventricle
Inferior vena cava

Left pulmonary artery
Pulmonary trunk
Left pulmonary veins
Left atrium
Aortic valve
Left AV valve
Left ventricle

1. Blood enters right atrium from superior and inferior venae cavae.

2. Blood in right atrium flows through right AV valve into right ventricle.

3. Contraction of right ventricle forces pulmonary valve open.

4. Blood flows through pulmonary valve into pulmonary trunk.

5. Blood is distributed by right and left pulmonary arteries to the lungs, where it unloads CO_2 and loads O_2.

6. Blood returns from lungs via pulmonary veins to left atrium.

7. Blood in left atrium flows through left AV valve into left ventricle.

8. Contraction of left ventricle (simultaneous with step 3) forces aortic valve open.

9. Blood flows through aortic valve into ascending aorta.

10. Blood in aorta is distributed to every organ in the body, where it unloads O_2 and loads CO_2.

11. Blood returns to right atrium via venae cavae.

Effective, efficient studying.

Connect helps you be more productive with your study time and get better grades using tools like SmartBook 2.0, which highlights key concepts and creates a personalized study plan. Connect sets you up for success, so you walk into class with confidence and walk out with better grades.

Study anytime, anywhere.

Download the free ReadAnywhere app and access your online eBook or SmartBook 2.0 assignments when it's convenient, even if you're offline. And since the app automatically syncs with your eBook and SmartBook 2.0 assignments in Connect, all of your work is available every time you open it. Find out more at **www.mheducation.com/readanywhere**

> *"I really liked this app—it made it easy to study when you don't have your text-book in front of you."*
>
> - Jordan Cunningham, Eastern Washington University

No surprises.

The Connect Calendar and Reports tools keep you on track with the work you need to get done and your assignment scores. Life gets busy; Connect tools help you keep learning through it all.

Calendar: owattaphotos/Getty Images

Learning for everyone.

McGraw-Hill works directly with Accessibility Services Departments and faculty to meet the learning needs of all students. Please contact your Accessibility Services office and ask them to email accessibility@mheducation.com, or visit **www.mheducation.com/about/accessibility** for more information.

McGraw-Hill Connect® empowers students to learn and succeed in the Anatomy and Physiology course with user-friendly digital solutions.

SMARTBOOK®

SmartBook 2.0 provides personalized learning to individual student needs, continually adapting to pinpoint knowledge gaps and focus learning on concepts requiring additional study. **The result? Students are highly engaged in the content and better prepared for lecture.**

LEARNSMART PREP®

LearnSmart Prep helps students thrive in college-level A&P by helping solidify knowledge in the key areas of cell biology, chemistry, study skills, and math. **The result? Students are better prepared for the A&P course.**

 Anatomy & Physiology Revealed® 4.0

Practice*ATLAS*

Practice Atlas for A&P is an interactive tool that pairs images of common anatomical models with stunning cadaver photography, allowing students to practice naming structures on both models and human bodies, anytime, anywhere. **The result? Students are better prepared, engaged, and move beyond basic memorization.**

Stop the Drop!

50% of the country's students are unable to pass the A&P course*

Anatomy & Physiology Revealed® (APR) 4.0 is an interactive cadaver dissection tool to enhance lecture and lab that students can use anytime, anywhere. **The result? Students are prepared for lab, engaged in the material, and utilize critical thinking.**

Ph.I.L.S.

Ph.I.L.S. 4.0 (Physiology Interactive Lab Simulations) software is the perfect way to reinforce key physiology concepts with powerful lab experiments. **The result? Students gain critical thinking skills and are better prepared for lab.**

Concept Overview Interactives are groundbreaking interactive animations that encourage students to explore key physiological processes and difficult concepts. **The result? Students are engaged and able to apply what they've learned while tackling difficult A&P concepts.**

*Statistic courtesy of *The New England Journal of Higher Education*

When I was a young boy, I became interested in what I then called "nature study" for two reasons. One was the sheer beauty of nature. I reveled in children's books with abundant, colorful drawings and photographs of animals, plants, minerals, and gems. It was this esthetic appreciation of nature that made me want to learn more about it and made me happily surprised to discover I could make a career of it. At a slightly later age, another thing that drew me still deeper into biology was to discover writers who had a way with words—who could captivate my imagination and curiosity with their elegant prose. Once I was old enough to hold part-time jobs, I began buying zoology and anatomy books that mesmerized me with their gracefulness of writing and fascinating art and photography. I wanted to write and draw like that myself, and I began teaching myself by learning from "the masters." I spent many late nights in my room peering into my microscope and jars of pond water, typing page after page of manuscript, and trying pen and ink as an art medium. My "first book" was a 318-page paper on some little pond animals called hydras, with 53 India ink illustrations that I wrote for my tenth-grade biology class when I was 16 (see page viii).

Fast-forward about 30 years, to when I became a textbook writer, and I found myself bringing that same enjoyment of writing and illustrating to the first edition of this book you are now holding. Why? Not only for its intrinsic creative satisfaction, but because I'm guessing that you're like I was—you can appreciate a book that does more than simply give you the information you need. You appreciate, I trust, a writer who makes it enjoyable for you through his scientific, storytelling prose and his concept of the way things should be illustrated to spark interest and facilitate understanding.

I know from my own students, however, that you need more than captivating illustrations and enjoyable reading. Let's face it— A&P is a complex subject and it may seem a formidable task to acquire even a basic knowledge of the human body. It was difficult even for me to learn (and the learning never ends). So in addition to simply writing this book, I've given a lot of thought to its pedagogy—the art of teaching. I've designed my chapters to make them easier for you to study and to give you abundant opportunity to check whether you've understood what you read—to test yourself (as I advise my own students) before the instructor tests you.

Each chapter is broken down into short, digestible bits with a set of Expected Learning Outcomes at the beginning of each section, and self-testing questions (Before You Go On) just a few pages later. Even if you have just 30 minutes to read during a lunch break or a bus ride, you can easily read or review one of these brief sections. There are also numerous self-testing questions in a Study Guide at the end of each chapter, in some of the figure legends, and the occasional Apply What You Know questions dispersed throughout each chapter. The questions cover a broad range of cognitive skills, from simple recall of a term to your ability to evaluate, analyze, and apply what you've learned to new clinical situations or other problems. In this era of digital publishing, however, learning aids go far beyond what I write into the book itself. SmartBook®, available on smartphones and tablets, includes all of the book's contents plus adaptive technology that can give you personalized instruction, target the unique gaps in your knowledge, and guide you in comprehension and retention of the subject matter.

I hope you enjoy your study of this book, but I know there are always ways to make it even better. Indeed, what quality you may find in this edition owes a great deal to feedback I've received from students all over the world. If you find any typos or other errors, if you have any suggestions for improvement, if I can clarify a concept for you, or even if you just want to comment on something you really like about the book, I hope you'll feel free to write to me. I correspond quite a lot with students and would enjoy hearing from you.

Ken Saladin
Georgia College
Milledgeville, GA 31061 (USA)
ksaladin2@windstream.net

CHAPTER

1

MAJOR THEMES OF ANATOMY AND PHYSIOLOGY

A colorized MRI scan of the human body
©Science Photo Library/Getty Images

CHAPTER OUTLINE

1.1 The Scope of Anatomy and Physiology

    1.1a   Anatomy—The Study of Form

    1.1b   Physiology—The Study of Function

1.2 The Origins of Biomedical Science

    1.2a   The Greek and Roman Legacy

    1.2b   The Birth of Modern Medicine

    1.2c   Living in a Revolution

1.3 Scientific Method

    1.3a   The Inductive Method

    1.3b   The Hypothetico–Deductive Method

    1.3c   Experimental Design

    1.3d   Peer Review

    1.3e   Facts, Laws, and Theories

1.4 Human Origins and Adaptations

    1.4a   Evolution, Selection, and Adaptation

    1.4b   Our Basic Primate Adaptations

    1.4c   Walking Upright

1.5 Human Structure

    1.5a   The Hierarchy of Complexity

    1.5b   Anatomical Variation

1.6 Human Function

    1.6a   Characteristics of Life

    1.6b   Physiological Variation

    1.6c   Negative Feedback and Homeostasis

    1.6d   Positive Feedback and Rapid Change

    1.6e   Gradients and Flow

1.7 The Language of Medicine

    1.7a   The History of Anatomical Terminology

    1.7b   Analyzing Medical Terms

    1.7c   Plurals, Adjectives, and Possessive Forms

    1.7d   Pronunciation

    1.7e   The Importance of Spelling

1.8 Review of Major Themes

Study Guide

DEEPER INSIGHTS

1.1 Evolutionary Medicine: Vestiges of Human Evolution

1.2 Clinical Application: Situs Inversus and Other Unusual Anatomy

1.3 Medical History: Men in the Oven

1.4 Medical History: Obscure Medical Word Origins

1.5 Clinical Application: Medical Imaging

Anatomy & Physiology
Revealed® 4.0

Module 1: Body Orientation

No branch of science hits as close to home as the science of our own bodies. We're grateful for the dependability of our hearts; we're awed by the capabilities of muscles and joints displayed by Olympic athletes; and we ponder with philosophers the ancient mysteries of mind and emotion. We want to know how our body works, and when it malfunctions, we want to know what's happening and what we can do about it. Even the most ancient writings of civilization include medical documents that attest to humanity's timeless drive to know itself. You are embarking on a subject that is as old as civilization, yet one that grows by thousands of scientific publications every week.

This book is an introduction to human structure and function, the biology of the human body. It is meant primarily to give you a foundation for advanced study in health care, exercise physiology, pathology, and other fields related to health and fitness. Beyond that purpose, however, it can also provide you with a deeply satisfying sense of self-understanding.

As rewarding and engrossing as this subject is, the human body is highly complex, and understanding it requires us to comprehend a great deal of detail. The details will be more manageable if we relate them to a few broad, unifying concepts. The aim of this chapter, therefore, is to introduce such concepts and put the rest of the book into perspective. We consider the historical development of anatomy and physiology, the thought processes that led to the knowledge in this book, the meaning of human life, some central concepts of physiology, and how to better understand medical terminology.

1.1   The Scope of Anatomy and Physiology

Expected Learning Outcomes

When you have completed this section, you should be able to

a. define *anatomy* and *physiology* and relate them to each other;

b. describe several ways of studying human anatomy; and

c. define a few subdisciplines of human physiology.

Anatomy is the study of structure, and **physiology** is the study of function. These approaches are complementary and never entirely separable. Together, they form the bedrock of the health sciences. When we study a structure, we want to know, What does it do? Physiology thus lends meaning to anatomy; conversely, anatomy is what makes physiology possible. This *unity of form and function* is an important point to bear in mind as you study the body. Many examples of it will be apparent throughout the book—some of them pointed out for you, and others you will notice for yourself.

1.1a Anatomy—The Study of Form

There are several ways to examine the structure of the human body. The simplest is **inspection**—simply looking at the body's appearance, as in performing a physical examination or making a clinical diagnosis from surface appearance. Physical examinations also involve touching and listening to the body. **Palpation**[1] means feeling a structure with the hands, such as palpating a swollen lymph node or taking a pulse. **Auscultation**[2] (AWS-cul-TAY-shun) is listening to the natural sounds made by the body, such as heart and lung sounds. In **percussion,** the examiner taps on the body, feels for abnormal resistance, and listens to the emitted sound for signs of abnormalities such as pockets of fluid, air, or scar tissue.

But a deeper understanding of the body depends on **dissection** (dis-SEC-shun)—carefully cutting and separating tissues to reveal their relationships. The very words *anatomy*[3] and *dissection*[4] both mean "cutting apart"; until the nineteenth century, dissection was called "anatomizing." In many schools of health science, one of the first steps in training students is dissection of the **cadaver,**[5] a dead human body. Many insights into human structure are obtained from **comparative anatomy**—the study of multiple species in order to examine similarities and differences and analyze evolutionary trends. Anatomy students often begin by dissecting other animals with which we share a common ancestry and many structural similarities. Many of the reasons for human structure become apparent only when we look at the structure of other animals.

Dissection, of course, is not the method of choice when studying a living person! It was once common to diagnose disorders through **exploratory surgery**—opening the body and taking a look inside to see what was wrong and what could be done about it. Any breach of the body cavities is risky, however, and most exploratory surgery has now been replaced by **medical imaging** techniques—methods of viewing the inside of the body without surgery, discussed at the end of this chapter (see Deeper Insight 1.5). The branch of medicine concerned with imaging is called **radiology.** Structure that can be seen with the naked eye—whether by surface observation, radiology, or dissection—is called **gross anatomy.**

Ultimately, the functions of the body result from its individual cells. To see those, we usually take tissue specimens, thinly slice and stain them, and observe them under the microscope. This approach is called **histology**[6] **(microscopic anatomy). Histopathology** is the microscopic examination of tissues for signs of disease. **Cytology**[7] is the study of the structure and function of individual cells. **Ultrastructure** refers to fine detail, down to the molecular level, revealed by the electron microscope.

1.1b Physiology—The Study of Function

Physiology[8] uses the methods of experimental science discussed later. It has many subdisciplines such as *neurophysiology* (physiology of the nervous system), *endocrinology* (physiology of

---

[1]*palp* = touch, feel; *ation* = process
[2]*auscult* = listen; *ation* = process
[3]*ana* = apart; *tom* = cut
[4]*dis* = apart; *sect* = cut
[5]from *cadere* = to fall down or die
[6]*histo* = tissue; *logy* = study of
[7]*cyto* = cell; *logy* = study of
[8]*physio* = nature; *logy* = study of

hormones), and *pathophysiology* (mechanisms of disease). Partly because of limitations on experimentation with humans, much of what we know about bodily function has been gained through **comparative physiology,** the study of how different species have solved problems of life such as water balance, respiration, and reproduction. Comparative physiology is also the basis for the development of new drugs and medical procedures. For example, a cardiac surgeon may learn animal surgery before practicing on humans, and a vaccine cannot be used on human subjects until it has been demonstrated through animal research that it confers significant benefits without unacceptable risks.

BEFORE YOU GO ON

Answer the following questions to test your understanding of the preceding section:

1. What is the difference between anatomy and physiology? How do these two sciences support each other?

2. Name the method that would be used for each of the following: listening to a patient for a heart murmur; studying the microscopic structure of the liver; microscopically examining liver tissue for signs of hepatitis; learning the blood vessels of a cadaver; and performing a breast self-examination.

1.2 The Origins of Biomedical Science

Expected Learning Outcomes

When you have completed this section, you should be able to

a. give examples of how modern biomedical science emerged from an era of superstition and authoritarianism; and

b. describe the contributions of some key people who helped to bring about this transformation.

Any science is more enjoyable if we consider not just the current state of knowledge, but how it compares to past understandings of the subject and how our knowledge was gained. Of all sciences, medicine has one of the most fascinating histories. Medical science has progressed far more in the last 50 years than in the 2,500 years before that, but the field didn't spring up overnight. It is built upon centuries of thought and controversy, triumph and defeat. We cannot fully appreciate its present state without understanding its past—people who had the curiosity to try new things, the vision to look at human form and function in new ways, and the courage to question authority.

1.2a The Greek and Roman Legacy

As early as 3,000 years ago, physicians in Mesopotamia and Egypt treated patients with herbal drugs, salts, physical therapy, and faith healing. The "father of medicine," however, is usually considered to be the Greek physician **Hippocrates** (c. 460–c. 375 BCE). He and his followers established a code of ethics for physicians, the Hippocratic Oath, which is still recited in modern form by graduating physicians at some medical schools. Hippocrates urged physicians to stop attributing disease to the activities of gods and demons and to seek their natural causes, which could afford the only rational basis for therapy.

Aristotle (384–322 BCE) was one of the first philosophers to write about anatomy and physiology. He believed that diseases and other natural events could have either supernatural causes, which he called *theologi,* or natural ones, which he called *physici* or *physiologi.* We derive such terms as *physician* and *physiology* from the latter. Until the nineteenth century, physicians were called "doctors of physic." In his anatomy book, *On the Parts of Animals,* Aristotle tried to identify unifying themes in nature. Among other points, he argued that complex structures are built from a smaller variety of simple components—a perspective that we will find useful later in this chapter.

▶▶▶ APPLY WHAT YOU KNOW

When you have completed this chapter, discuss the relevance of Aristotle's philosophy to our current thinking about human structure.

Claudius Galen (129–c. 200), physician to the Roman gladiators, wrote the most influential medical textbook of the ancient era—a book worshipped to excess by medical professors for centuries to follow. Cadaver dissection was banned in Galen's time because of some horrid excesses that preceded him, including public dissection of living slaves and prisoners. Aside from what he could learn by treating gladiators' wounds, Galen was therefore limited to dissecting pigs, monkeys, and other animals. Because he was not permitted to dissect cadavers, he had to guess at much of human anatomy and made some incorrect deductions from animal dissections. He described the human liver, for example, as having five fingerlike lobes, somewhat like a baseball glove, because that's what he had seen in baboons. But Galen saw science as a method of discovery, not a body of fact to be taken on faith. He warned that even his own books could be wrong and advised his followers to trust their own observations more than any book. Unfortunately, his advice was not heeded. For nearly 1,500 years, medical professors dogmatically taught what they read in Aristotle and Galen, seldom daring to question the authority of these "ancient masters."

1.2b The Birth of Modern Medicine

In the Middle Ages, the state of medical science varied greatly from one religious culture to another. Science was severely repressed in the Christian culture of Europe until about the sixteenth century, although some of the most famous medical schools of Europe were founded during this era. Their professors, however, taught medicine primarily as a dogmatic commentary on Galen and Aristotle, not as a field of original research. Medieval medical illustrations were crude representations of the body intended more to decorate a page than to depict the body realistically (**fig. 1.1a**). Some were astrological charts that showed which sign of the zodiac was thought to influence each organ of the body. From such pseudoscience came the word *influenza,* Italian for "influence."

Free inquiry was less inhibited in Jewish and Muslim culture during this time. Jewish physicians were the most esteemed practitioners of their art—and none more famous than *Moses ben Maimon* (1135–1204), known in Christendom as **Maimonides.** Born in Spain, he fled to Egypt at age 24 to escape antisemitic persecution. There he served the rest of his life as physician to the court of the sultan, Saladin. A highly admired rabbi, Maimonides wrote voluminously on Jewish law and theology, but also wrote 10 influential medical books and numerous treatises on specific diseases.

Among Muslims, probably the most highly regarded medical scholar was *Ibn Sina* (980–1037), known in the West as **Avicenna** or "the Galen of Islam." He studied Galen and Aristotle, combined their findings with original discoveries, and questioned authority when the evidence demanded it. Medicine in the Mideast soon became superior to European medicine. Avicenna's textbook, *The Canon of Medicine,* was the leading authority in European medical schools for over 500 years.

Chinese medicine had little influence on Western thought and practice until relatively recently; the medical arts evolved in China quite independently of European medicine. Later chapters of this book describe some of the insights of ancient China and India.

Modern Western medicine began around the sixteenth century in the innovative minds of such people as the anatomist Andreas Vesalius and the physiologist William Harvey.

(a)

(b)

FIGURE 1.1 The Evolution of Medical Art. Two illustrations of the skeletal system made about 500 years apart. (a) From an eleventh-century work attributed to Persian physician Avicenna. (b) From *De Humani Corporis Fabrica* by Andreas Vesalius, 1543.

a: Source: Wellcome Library, London/CC BY 4.0; **b:** Suzan Oschmann/Shutterstock

Andreas Vesalius (1514–64) taught anatomy in Italy. In his time, the Catholic Church relaxed its prohibition against cadaver dissection, in part to allow autopsies in cases of suspicious death. Furthermore, the Italian Renaissance created an environment more friendly to innovative scholarship. Dissection gradually found its way into the training of medical students throughout Europe. It was an unpleasant business, however, and most professors considered it beneath their dignity. In those days before refrigeration or embalming, the odor from the decaying cadaver was unbearable. Dissections were a race against decay. Bleary medical students had to fight the urge to vomit, lest they incur the wrath of an overbearing professor. Professors typically sat in an elevated chair, the *cathedra,* reading dryly in Latin from Galen or Aristotle while a lower-ranking *barber–surgeon* removed putrefying organs from the cadaver and held them up for the students to see. Barbering and surgery were considered to be "kindred arts of the knife"; today's barber poles date from this era, their red and white stripes symbolizing blood and bandages.

Vesalius broke with tradition by coming down from the cathedra and doing the dissections himself. He was quick to point out that much of the anatomy in Galen's books was wrong, and he was the first to publish accurate illustrations for teaching anatomy **(fig. 1.1b).** When others began to plagiarize them, Vesalius published the first atlas of anatomy, *De Humani Corporis Fabrica (On the Structure of the Human Body),* in 1543. This book began a rich tradition of medical illustration that has been handed down to us through such milestones as *Gray's Anatomy* (1856) and the vividly illustrated atlases and textbooks of today.

Anatomy preceded physiology and was a necessary foundation for it. What Vesalius was to anatomy, the Englishman **William Harvey** (1578–1657) was to physiology. Harvey is remembered especially for his studies of blood circulation and a little book he published in 1628, known by its abbreviated title *De Motu Cordis (On the Motion of the Heart).* He and **Michael Servetus** (1511–53) were the first Western scientists to realize that blood must circulate continuously around the body, from the heart to the other organs and back to the heart again. This flew in the face of Galen's belief that the liver converted food to blood, the heart pumped blood through the veins to all other organs, and those organs consumed it. Harvey's colleagues, wedded to the ideas of Galen, ridiculed Harvey for his theory, though we now know he was correct (see chapter 20 prologue). Despite persecution and setbacks, Harvey lived to a ripe old age, served as physician to the kings of England, and later did important work in embryology. Most importantly, Harvey's contributions represent the birth of experimental physiology—the method that generated most of the information in this book.

Modern medicine also owes an enormous debt to two inventors from this era, Robert Hooke and Antony van Leeuwenhoek, who extended the vision of biologists to the cellular level. **Robert Hooke** (1635–1703), an Englishman, designed scientific instruments of various kinds, including the compound microscope. This is a tube with a lens at each end—an *objective lens* near the specimen, which produces an initial magnified image, and an *ocular lens (eyepiece)* near the observer's eye, which magnifies the first image still further. Although crude compound microscopes had existed since 1595, Hooke improved the optics and invented several of the helpful features found in microscopes today—a stage to hold the specimen, an illuminator, and coarse and fine focus controls. His microscopes magnified only about 30 times, but with them, he was the first to see and name cells. In 1663, he observed thin shavings of cork and observed that they "consisted of a great many little boxes," which he called *cellulae* (little cells) after the cubicles of a monastery **(fig. 1.2).** He later observed living cells "filled with juices." Hooke became particularly interested in microscopic examination of such material as insects, plant tissues, and animal parts. He published the first comprehensive book of microscopy, *Micrographia,* in 1665.

Antony van Leeuwenhoek (an-TOE-nee vahn LAY-wenhook) (1632–1723), a Dutch textile merchant, invented a *simple* (single-lens) *microscope,* originally for the purpose of examining the weave of fabrics. His microscope was a beadlike lens mounted in a metal plate equipped with a movable specimen clip.

(a)   (b)

FIGURE 1.2 Hooke's Compound Microscope. (a) The compound microscope had a lens at each end of a tubular body. (b) Hooke's drawing of cork cells, showing the thick cell walls characteristic of plants.

a: Source: National Museum of Health and Medicine, Silver Spring, MD; **b:** Bettmann/Getty Images

Even though his microscopes were simpler than Hooke's, they achieved much greater useful magnification (up to 200×) owing to Leeuwenhoek's superior lens-making technique. Out of curiosity, he examined a drop of lake water and was astonished to find a variety of microorganisms—"little animalcules," he called them, "very prettily a-swimming." He went on to observe practically everything he could get his hands on, including blood cells, blood capillaries, sperm, muscular tissue, and bacteria from tooth scrapings. Leeuwenhoek began submitting his observations to the Royal Society of London in 1673. He was praised at first, and his observations were eagerly read by scientists, but enthusiasm for the microscope didn't last. By the end of the seventeenth century, it was treated as a mere toy for the upper classes, as amusing and meaningless as a kaleidoscope. Leeuwenhoek and Hooke had even become the brunt of satire. But probably no one in history had looked at nature in such a revolutionary way. By taking biology to the cellular level, the two men had laid an entirely new foundation for the modern medicine to follow centuries later.

The Hooke and Leeuwenhoek microscopes produced poor images with blurry edges (*spherical aberration*) and rainbow-like distortions (*chromatic aberration*). These problems had to be solved before the microscope could be widely used as a biological tool. In the nineteenth century, German inventors greatly improved the compound microscope, adding the condenser and developing superior optics. With improved microscopes, biologists began eagerly examining a wider variety of specimens. By 1839, botanist **Matthias Schleiden** (1804–81) and zoologist **Theodor Schwann** (1810–82) concluded that all organisms were composed of cells. Although it took another century for this idea to be generally accepted, it became the first tenet of the **cell theory,** added to by later biologists and summarized in section 3.1a. The cell theory was perhaps the most important breakthrough in biomedical history; all functions of the body are now interpreted as the effects of cellular activity.

Although the philosophical foundation for modern medicine was largely established by the time of Leeuwenhoek, Hooke, and Harvey, clinical practice was still in a dismal state. Few doctors attended medical school or received any formal education in basic science or human anatomy. Physicians tended to be ignorant, ineffective, and pompous. Their practice was heavily based on expelling imaginary toxins from the body by bleeding their patients or inducing vomiting, sweating, or diarrhea. They performed operations with filthy hands and instruments, spreading lethal infections from one patient to another and refusing, in their vanity, to believe that they themselves were the carriers of disease. Countless women died of infections acquired during childbirth from their obstetricians. Fractured limbs often became gangrenous and had to be amputated, and there was no anesthesia to lessen the pain. Disease was still widely attributed to demons and witches, and many people felt they would be interfering with God's will if they tried to treat it.

1.2c Living in a Revolution

This short history brings us only to the threshold of modern biomedical science; it stops short of such momentous discoveries as the germ theory of disease, the mechanisms of heredity,

and the structure of DNA. In the twentieth century, basic biology and biochemistry yielded a much deeper understanding of how the body works. Advances in medical imaging enhanced our diagnostic ability and life-support strategies. We witnessed monumental developments in chemotherapy, immunization, anesthesia, surgery, organ transplants, and human genetics. By the close of the twentieth century, we had discovered the chemical "base sequence" of every human gene and begun attempting gene therapy to treat children born with diseases recently considered incurable. As future historians look back on the turn of this century, they may exult about the Genetic Revolution in which you are now living.

Several discoveries of the nineteenth and twentieth centuries, and the men and women behind them, are covered in short historical sketches in later chapters. Yet, the stories told in this chapter are different in a significant way. The people discussed here were pioneers in establishing the scientific way of thinking. They helped to replace superstition with an appreciation of natural law. They bridged the chasm between mystery and medication. Without this intellectual revolution, those who followed could not have conceived of the right questions to ask, much less a method for answering them.

BEFORE YOU GO ON

Answer the following questions to test your understanding of the preceding section:

3. In what way did the followers of Galen disregard his advice? How does Galen's advice apply to you and this book?

4. Describe two ways in which Vesalius improved medical education and set standards that remain relevant today.

5. How is our concept of human form and function today affected by inventors from the seventeenth to the nineteenth centuries?

1.3 Scientific Method

Expected Learning Outcomes

When you have completed this section, you should be able to

a. describe the inductive and hypothetico–deductive methods of obtaining scientific knowledge;

b. describe some aspects of experimental design that help to ensure objective and reliable results; and

c. explain what is meant by *hypothesis, fact, law,* and *theory* in science.

Prior to the seventeenth century, science was done in a haphazard way by a small number of isolated individuals. The philosophers **Francis Bacon** (1561–1626) in England and **René Descartes** (1596–1650) in France envisioned science as a far greater, systematic enterprise with enormous possibilities for human health and welfare. They detested those who endlessly debated ancient philosophy without creating anything new. Bacon argued against biased thinking and for more objectivity in science. He outlined a systematic way of seeking similarities, differences, and trends in nature and drawing useful generalizations from observable facts. You will see echoes of Bacon's philosophy in the discussion of scientific method that follows.

Though the followers of Bacon and Descartes argued bitterly with one another, both men wanted science to become a public, cooperative enterprise, supported by governments and conducted by an international community of scholars rather than a few isolated amateurs. Inspired by their vision, the French and English governments established academies of science that still flourish today. Bacon and Descartes are credited with putting science on the path to modernity, not by discovering anything new in nature or inventing any techniques—for neither man was a scientist—but by inventing new habits of scientific thought.

When we say "scientific," we mean that such thinking is based on assumptions and methods that yield reliable, objective, testable information about nature. The assumptions of science are ideas that have proven fruitful in the past—for example, the idea that natural phenomena have natural causes and nature is therefore predictable and understandable. The methods of science are highly variable. **Scientific method** refers less to observational procedures than to certain habits of disciplined creativity, careful observation, logical thinking, and honest analysis of one's observations and conclusions. It is especially important in health science to understand these habits. This field is littered with more fads and frauds than any other. We are called upon constantly to judge which claims are trustworthy and which are bogus. To make such judgments depends on an appreciation of how scientists think, how they set standards for truth, and why their claims are more reliable than others (**fig. 1.3**).

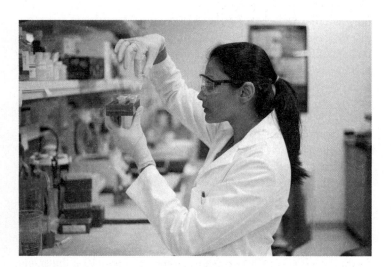

FIGURE 1.3 Biomedical Research. Research scientists employ habits of thought we call the scientific method to ensure the objectivity, reliability, and reproducibility of their results and conclusions.

AshTproductions/Shutterstock

1.3a The Inductive Method

The **inductive method,** first prescribed by Bacon, is a process of making numerous observations until one feels confident in drawing generalizations and predictions from them. What we know of anatomy is a product of the inductive method. We describe the normal structure of the body based on observations of many bodies.

This raises the issue of what is considered proof in science. We can never prove a claim beyond all possible refutation. We can, however, consider a statement as proven *beyond reasonable doubt* if it was arrived at by reliable methods of observation, tested and confirmed repeatedly, and not falsified by any credible observation. In science, all truth is tentative; there's no room for dogma. We must always be prepared to abandon yesterday's truth if tomorrow's facts disprove it.

1.3b The Hypothetico–Deductive Method

Most physiological knowledge was obtained by the **hypothetico–deductive method.** An investigator begins by asking a question and formulating a **hypothesis**—an educated speculation or possible answer to the question. A good hypothesis must be (1) consistent with what is already known and (2) capable of being tested and possibly falsified by evidence. **Falsifiability** means that if we claim something is scientifically true, we must be able to specify what evidence it would take to prove it wrong. If nothing could possibly prove it wrong, then it's not scientific.

▶▶▶**APPLY WHAT YOU KNOW**

The ancients thought that gods or invisible demons caused epilepsy. Today, epileptic seizures are attributed to bursts of abnormal electrical activity in nerve cells of the brain. Explain why one of these claims is falsifiable (and thus scientific), whereas the other claim is not.

The purpose of a hypothesis is to suggest a method for answering a question. From the hypothesis, a researcher makes a deduction, typically in the form of an "if–then" prediction: *If* my hypothesis on epilepsy is correct and I record the brain waves of patients during seizures, *then* I should observe abnormal bursts of activity. A properly conducted experiment yields observations that either support a hypothesis or require the scientist to modify or abandon it, formulate a better hypothesis, and test that one. Hypothesis testing operates in cycles of conjecture and disproof until one is found that is supported by the evidence.

1.3c Experimental Design

Doing an experiment properly involves several important considerations. What shall I measure and how can I measure it? What effects should I watch for and which ones should I ignore? How can I be sure my results are due to the variables that I manipulate and not due to something else? When working on human subjects, how can I prevent the subject's expectations or state of mind from influencing the results? How can I eliminate my own biases and be sure that even the most skeptical critics will have as much confidence in my conclusions as I do? Several elements of experimental design address these issues:

- **Sample size.** The number of subjects (animals or people) used in a study is the sample size. An adequate sample size controls for chance events and individual variations in response and thus enables us to place more confidence in the outcome. For example, would you rather trust your health to a drug that was tested on 5 people or one tested on 5,000? Why?

- **Controls.** Biomedical experiments require comparison between treated and untreated individuals so that we can judge whether the treatment has any effect. A **control group** consists of subjects that are as much like the **treatment group** as possible except with respect to the variable being tested. For example, there is evidence that garlic lowers blood cholesterol levels. In one study, volunteers with high cholesterol were each given 800 mg of garlic powder daily for 4 months and exhibited an average 12% reduction in cholesterol. Was this a significant reduction, and was it due to the garlic? It's impossible to say without comparison to a control group of similar people who received no treatment. In this study, the control group averaged only a 3% reduction in cholesterol, so garlic *seems* to have made a difference.

- **Psychosomatic effects.** Psychosomatic effects (effects of the subject's state of mind on his or her physiology) can have an undesirable effect on experimental results if we do not control for them. In drug research, it is therefore customary to give the control group a **placebo** (pla-SEE-bo)—a substance with no significant physiological effect on the body. If we were testing a drug, for example, we could give the treatment group the drug and the control group identical-looking sugar tablets. Neither group must know which tablets it is receiving. If the two groups showed significantly different effects, we could feel confident that it did not result from a knowledge of what they were taking.

- **Experimenter bias.** In the competitive, high-stakes world of medical research, experimenters may want certain results so much that their biases, even subconscious ones, can affect their interpretation of the data. One way to control for this is the **double-blind method.** In this procedure, neither the subject to whom a treatment is given nor the person giving it and recording the results knows whether that subject is receiving the experimental treatment or the placebo. A researcher may prepare identical-looking tablets, some with the drug and some with placebo; label them with code numbers; and distribute them to participating physicians. The physicians themselves do not know whether they are administering drug or placebo, so they cannot give the subjects even accidental hints of which substance they are taking. When the data are collected, the researcher can correlate them with the composition of the tablets and determine whether the drug had more effect than the placebo.

- **Statistical testing.** If you tossed a coin 100 times, you would expect it to come up about 50 heads and 50 tails. If it actually came up 48:52, you would probably attribute this to random error rather than bias in the coin. But what if it came up 40:60? At what point would you begin to suspect bias? This type of problem is faced routinely in research—how great a difference must there be between control and experimental groups before we feel confident that it was due to the treatment and not merely random variation? What if a treatment group exhibited a 12% reduction in cholesterol level and the placebo group a 10% reduction? Would this be enough to conclude that the treatment was effective? Scientists are well grounded in **statistical tests** that can be applied to the data—the chi-square test, the *t* test, and analysis of variance, for example. A typical outcome of a statistical test may be expressed, "We can be 99.5% sure that the difference between group A and group B was due to the experimental treatment and not to random variation." Science is grounded not in statements of absolute truth, but in statements of probability.

1.3d Peer Review

When a scientist applies for funds to support a research project or submits results for publication, the application or manuscript is submitted to **peer review**—a critical evaluation by other experts in that field. Even after a report is published, if the results are important or unconventional, other scientists may attempt to reproduce them to see if the author was correct. At every stage from planning to postpublication, scientists are therefore subject to intense scrutiny by their colleagues. Peer review is one mechanism for ensuring honesty, objectivity, and quality in science.

1.3e Facts, Laws, and Theories

The most important product of scientific research is understanding how nature works—whether it be the nature of a pond to an ecologist or the nature of a liver cell to a physiologist. We express our understanding as *facts, laws,* and *theories* of nature. It is important to appreciate the differences among these.

A scientific **fact** is information that can be independently verified by any trained person—for example, the fact that an iron deficiency leads to anemia. A **law of nature** is a generalization about the predictable ways in which matter and energy behave. It is the result of inductive reasoning based on repeated, confirmed observations. Some laws are expressed as concise verbal statements, such as the *law of complementary base pairing:* In the double helix of DNA, a chemical base called adenine always pairs with one called thymine, and a base called guanine always pairs with cytosine (see section 4.1a). Other laws are expressed as mathematical formulae, such as *Boyle's law,* used in respiratory physiology: Under specified conditions, the volume of a gas *(V)* is inversely proportional to its pressure *(P)*—that is,

$$V \propto 1/P.$$

A **theory** is an explanatory statement or set of statements derived from facts, laws, and confirmed hypotheses. Some theories have names, such as the *cell theory,* the *fluid-mosaic theory* of cell membranes, and the *sliding filament theory* of muscle contraction. Most, however, remain unnamed. The purpose of a theory is not only to concisely summarize what we already know but, moreover, to suggest directions for further study and to help predict what the findings should be if the theory is correct.

Law and *theory* mean something different in science than they do to most people. In common usage, a law is a rule created and enforced by people; we must obey it or risk a penalty. A law of nature, however, is a description; laws do not *govern* the universe—they *describe* it. Laypeople tend to use the word *theory* for what a scientist would call a hypothesis—for example, "I have a theory why my car won't start." The difference in meaning causes significant confusion when it leads people to think that a scientific theory (such as the theory of evolution) is merely a guess or conjecture, instead of recognizing it as a summary of conclusions drawn from a large body of observed facts. The concepts of gravity and electrons are theories, too, but this does not mean they are merely speculations.

▶▶▶**APPLY WHAT YOU KNOW**

Was the cell theory proposed by Schleiden and Schwann more a product of the hypothetico–deductive method or of the inductive method? Explain your answer.

BEFORE YOU GO ON

Answer the following questions to test your understanding of the preceding section:

6. Describe the general process involved in the inductive method.

7. Describe some sources of potential bias in biomedical research. What are some ways of minimizing such bias?

8. Is there more information in an individual scientific fact or in a theory? Explain.

1.4 Human Origins and Adaptations

Expected Learning Outcomes

When you have completed this section, you should be able to

a. explain why evolution is relevant to understanding human form and function;

b. define *evolution* and *natural selection;*

c. describe some human characteristics that can be attributed to the tree-dwelling habits of earlier primates; and

d. describe some human characteristics that evolved later in connection with upright walking.

If any two theories have the broadest implications for understanding the human body, they are probably the *cell theory* and the *theory of natural selection.* No understanding of human form and function is complete without an understanding of our evolutionary history, of how natural selection adapted the body to its ancestral habitat. As an explanation of how species originate and change through time, natural selection was the brainchild of **Charles Darwin** (1809–82)—certainly the most influential biologist who ever lived. His book, *On the Origin of Species by Means of Natural Selection* (1859), has been called "the book that shook the world." In presenting the first well-supported theory of how evolution works, it not only caused the restructuring of all of biology but also profoundly changed the prevailing view of our origin, nature, and place in the universe. In *The Descent of Man* (1871), Darwin directly addressed the issue of human evolution and emphasized features of anatomy and behavior that reveal our relationship to other animals. Here we will touch just briefly on how natural selection helps explain some of the distinctive characteristics seen in *Homo sapiens* today.

1.4a Evolution, Selection, and Adaptation

Evolution simply means change in the genetic composition of a population of organisms. Examples include the evolution of bacterial resistance to antibiotics, the appearance of new strains of the AIDS virus, and the emergence of new species of organisms.

Evolution works largely through the principle of **natural selection,** which states essentially this: Some individuals within a species have hereditary advantages over their competitors—for example, better camouflage, disease resistance, or ability to attract mates—that enable them to produce more offspring. They pass these advantages on to their offspring, and such characteristics therefore become more and more common in successive generations. This brings about the genetic change in a population that constitutes evolution.

Natural forces that promote the reproductive success of some individuals more than others are called **selection pressures.** They include such things as climate, predators, disease, competition, and food. **Adaptations** are features of anatomy, physiology, and behavior that evolve in response to these selection pressures and enable an organism to cope with the challenges of its environment.

Darwin could scarcely have predicted the overwhelming mass of genetic, molecular, fossil, and other evidence of human evolution that would accumulate in the twentieth century and further substantiate his theory. A technique called DNA hybridization, for example, reveals a difference of only 1.6% in DNA structure between humans and chimpanzees. Chimpanzees and gorillas differ by 2.3%. DNA structure thus suggests that a chimpanzee's closest living relative is not the gorilla—it is us, *Homo sapiens.*

Several aspects of our anatomy make little sense without an awareness that the human body has a history (see Deeper Insight 1.1). Our evolutionary relationship to other species is also important in choosing animals for biomedical research. If there were no issues of cost, availability, or ethics, we might test drugs on our close living relatives, the chimpanzees, before approving them for human use. Their genetics, anatomy, and physiology are most similar to ours, and their reactions to drugs therefore afford the best prediction of how the human body would react. On the other hand, if we had no kinship with any other species, the selection of a test species would be arbitrary; we might as well use frogs or snails. In reality, we compromise.

DEEPER INSIGHT 1.1

EVOLUTIONARY MEDICINE

Vestiges of Human Evolution

One of the classic lines of evidence for evolution, debated even before Darwin was born, is *vestigial organs.* These structures are the remnants of organs that apparently were better developed and more functional in the ancestors of a species. They now serve little or no purpose or, in some cases, have been converted to new functions.

Our bodies, for example, are covered with millions of hairs, each equipped with a useless little *arrector muscle.* In other mammals, these muscles fluff the hair and conserve heat. In humans, they merely produce goose bumps. Above each ear, we have three *auricularis muscles.* In other mammals, they move the ears to receive sounds better or to flick off flies and other pests, but most people cannot contract them at all. As Darwin said, it makes no sense that humans would have such structures were it not for the fact that we came from ancestors in which they were functional.

Rats and mice are used extensively for research because they are fellow mammals with a physiology similar to ours, but they present fewer of the aforementioned issues than chimpanzees or other mammals do. An animal species or strain selected for research on a particular problem is called a **model**—for example, a mouse model for leukemia.

1.4b Our Basic Primate Adaptations

We belong to an order of mammals called the Primates, which also includes the monkeys and apes. Some of our anatomical and physiological features can be traced to the earliest primates, which descended from certain squirrel-size, insect-eating, African mammals that took up life in the trees 55 to 60 million years ago. This **arboreal**[9] (treetop) habitat probably afforded greater safety from predators, less competition, and a rich food supply of leaves, fruit, insects, and lizards. But the forest canopy is a challenging world, with dim and dappled sunlight, swaying branches, shifting shadows, and prey darting about in the dense foliage. Any new feature that enabled arboreal animals to move about more easily in the treetops would have been strongly favored by natural selection. Thus, the shoulder became more mobile and enabled primates to reach out in any direction (even overhead, which few other mammals can do). The thumbs became fully **opposable**—they could cross the palm to touch the fingertips—and enabled primates to hold small objects and manipulate them more precisely than other mammals could. Opposable thumbs made the hands **prehensile**[10]—able to grasp objects by encircling them with the thumb and fingers **(fig. 1.4).** The thumb is so important that it receives highest priority in the repair of hand injuries. If the thumb can be saved, the hand can be reasonably functional; if it is lost, hand functions are severely diminished.

FIGURE 1.4 Human Adaptations Shared with Other Primates. Some major aspects of primate evolution are the opposable thumb, prehensile hand, forward-facing eyes, and stereoscopic vision. In humans, the hand became refined for increasingly sophisticated manipulation of objects.

Chimpanzee: Tim Davis/Science Source

The eyes of primates moved to a more forward-facing position, which allowed for **stereoscopic**[11] vision (depth perception). This adaptation provided better hand–eye coordination in catching and manipulating prey, with the added advantage of making it easier to judge distances accurately in leaping from tree to tree. Color vision, rare among mammals, is also a primate hallmark. Primates eat mainly fruit and leaves. The ability to distinguish subtle shades of orange and red enables them to distinguish ripe, sugary fruits from unripe ones. Distinguishing subtle shades of green helps them to differentiate between tender young leaves and tough, more toxic older foliage.

Various fruits ripen at different times and in widely separated places in the tropical forest. This requires a good memory of what will be available, when, and how to get there. Larger brains might have evolved in response to the challenge of efficient food finding and, in turn, laid the foundation for more sophisticated social organization.

---

[9]*arbor* = tree; *eal* = pertaining to
[10]*prehens* = to seize

[11]*stereo* = solid; *scop* = vision

None of this is meant to imply that humans evolved from monkeys—a common misconception about evolution that no biologist believes. Monkeys, apes, and humans do, however, share common ancestors. Our relationship is not like parent and child, but more like cousins who have the same grandparents. Observations of monkeys and apes provide insight into how primates adapt to the arboreal habitat and therefore how certain human adaptations probably originated.

1.4c Walking Upright

About 4 to 5 million years ago, parts of Africa became hotter and drier, and much of the forest was replaced by savanna (grassland). Some primates adapted to living on the savanna, but this was a dangerous place with more predators and less protection. Just as squirrels and monkeys stand briefly on their hind legs to look around for danger, so would these early ground dwellers. Being able to stand up not only helps an animal stay alert, but also frees the forelimbs for purposes other than walking. Chimpanzees sometimes walk upright to carry food, infants, or weapons (sticks and rocks), and it is reasonable to suppose that our early ancestors did so too.

These advantages are so great that they favored skeletal modifications that made **bipedalism**[12]—standing and walking on two legs—easier. Fossil evidence indicates that bipedalism was firmly established more than 4 million years ago. The anatomy of the human pelvis, femur, knee, great toe, foot arches, spinal column, skull, arms, and many muscles became adapted for bipedal locomotion (see Deeper Insight 8.5), as did many aspects of human family life and society. As the skeleton and muscles became adapted for bipedalism, brain volume increased dramatically, from 400 mL around 4 million years ago to an average of 1,350 mL today. It must have become increasingly difficult for a fully developed, large-brained infant to pass through the mother's pelvic outlet at birth. This may explain why humans are born in a relatively immature, helpless state compared with other mammals, before their nervous systems have matured and the bones of the skull have fused. The helplessness of human young and their extended dependence on parental care may help to explain why humans have such exceptionally strong family ties.

Most of the oldest bipedal primates are classified in the genus *Australopithecus* (aus-TRAL-oh-PITH-eh-cus). About 2.5 million years ago, hominids appeared with taller stature, greater brain volumes, simple stone tools, and probably articulate speech. These are the earliest members of the genus *Homo*. By at least 1.8 million years ago, *Homo erectus* migrated from Africa to parts of Asia. Anatomically modern *Homo sapiens*, our own species, originated in Africa about 200,000 years ago and is the sole surviving hominid species.

This brief account barely begins to explain how human anatomy, physiology, and behavior have been shaped by ancient selection pressures. Later chapters further demonstrate that the evolutionary perspective provides a meaningful understanding of why humans are the way we are. Evolution is the basis for comparative anatomy and physiology, which have been so fruitful for the understanding of human biology. If we weren't related to any other species, those sciences would be pointless.

The emerging science of **evolutionary medicine** analyzes how human disease and dysfunctions can be traced to differences between the artificial environment in which we now live, and the prehistoric environment to which *Homo sapiens* was biologically adapted. For example, we can relate sleep and mood disorders to artificial lighting and night-shift work, and the rise of asthma to our modern obsession with sanitation. Other examples in this book will relate evolution to obesity, diabetes, low-back pain, skin cancer, and other health issues.

BEFORE YOU GO ON

Answer the following questions to test your understanding of the preceding section:

9. Define *adaptation* and *selection pressure*. Why are these concepts important in understanding human anatomy and physiology?

10. Select any two human characteristics and explain how they might have originated in primate adaptations to an arboreal habitat.

11. Select two other human characteristics and explain how they might have resulted from later adaptation to a grassland habitat.

1.5 Human Structure

Expected Learning Outcomes

When you have completed this section, you should be able to

a. list the levels of human structure from the most complex to the simplest;

b. discuss the value of both reductionistic and holistic viewpoints to understanding human form and function; and

c. discuss the clinical significance of anatomical variation among humans.

Earlier in this chapter, we observed that human anatomy is studied by a variety of techniques—dissection, palpation, and so forth. In addition, anatomy is studied at several levels of detail, from the whole body down to the molecular level.

1.5a The Hierarchy of Complexity

Consider for the moment an analogy to human structure: The English language, like the human body, is very complex, yet an infinite variety of ideas can be conveyed with a limited number of words. All words in English are, in turn, composed of various combinations of just 26 letters. Between an essay and an alphabet are successively simpler levels of organization: paragraphs, sentences, words, and syllables. We can say that language exhibits a

---

[12]*bi* = two; *ped* = foot

Organism

Organ system   Organ   Tissue

Macromolecule

Organelle

Cell

Molecule   Atom

FIGURE 1.5 The Body's Structural Hierarchy.

hierarchy of complexity, with letters, syllables, words, and so forth being successive levels of the hierarchy. Humans have an analogous hierarchy of complexity, as follows **(fig. 1.5):**

The organism is composed of organ systems,
  organ systems are composed of organs,
    organs are composed of tissues,
      tissues are composed of cells,
        cells are composed partly of organelles,
          organelles are composed of molecules, and
            molecules are composed of atoms.

The **organism** is a single, complete individual.

An **organ system** is a group of organs with a unique collective function, such as circulation, respiration, or digestion. The human body has 11 organ systems, illustrated in atlas A immediately following this chapter: the integumentary, skeletal, muscular, nervous, endocrine, circulatory, lymphatic, respiratory, urinary, digestive, and reproductive systems. Usually, the organs of one system are physically interconnected, such as the kidneys, ureters, urinary bladder, and urethra, which compose the urinary system. Beginning with chapter 6, this book is organized around the organ systems.

An **organ** is a structure composed of two or more tissue types that work together to carry out a particular function. Organs have definite anatomical boundaries and are visibly distinguishable from adjacent structures. Most organs and higher levels of structure are within the domain of gross anatomy. However, there are organs within organs—the large organs visible to the naked eye often contain smaller organs visible only with the microscope. The skin, for example, is the body's largest organ. Included within it are thousands of smaller organs: Each hair, nail, gland, nerve, and blood vessel of the skin is an organ in itself. A single organ can belong to two organ systems. For example, the pancreas belongs to both the endocrine and digestive systems.

A **tissue** is a mass of similar cells and cell products that forms a discrete region of an organ and performs a specific function. The body is composed of only four primary classes of tissue: epithelial, connective, nervous, and muscular tissue. Histology, the study of tissues, is the subject of chapter 5.

Cells are the smallest units of an organism that carry out all the basic functions of life; nothing simpler than a cell is considered alive. A cell is enclosed in a *plasma membrane* composed of lipids and proteins. Most cells have one nucleus, an organelle that contains its DNA. *Cytology,* the study of cells and organelles, is the subject of chapters 3 and 4.

Organelles[13] are microscopic structures in a cell that carry out its individual functions. Examples include mitochondria, centrioles, and lysosomes.

Organelles and other cellular components are composed of **molecules.** The largest molecules, such as proteins, fats, and DNA, are called *macromolecules* (see chapter 2). A molecule is a particle composed of at least two **atoms,** the smallest particles with unique chemical identities.

The theory that a large, complex system such as the human body can be understood by studying its simpler components is called **reductionism.** First espoused by Aristotle, this has proved to be a highly productive approach; indeed, it is essential to scientific thinking. Yet the reductionistic view is not the only way of understanding human life. Just as it would be very difficult to predict the workings of an automobile transmission merely by looking at a pile of its disassembled gears and levers, one could never predict the human personality from a complete knowledge of the circuitry of the brain or the genetic sequence of DNA. **Holism**[14] is the complementary theory that there are "emergent properties" of the whole organism that cannot be predicted from the properties of its separate parts—human beings are more than the sum of their parts. To be most effective, a health-care provider treats not merely a disease or an organ system, but a whole person. A patient's perceptions, emotional responses to life, and confidence in the nurse, therapist, or physician profoundly affect the outcome of treatment. In fact, these psychological factors often play a greater role in a patient's recovery than the physical treatments administered.

---

[13]*elle* = little
[14]*holo* = whole, entire

1.5b Anatomical Variation

A quick look around any classroom is enough to show that no two humans are exactly alike; on close inspection, even identical twins exhibit differences. Yet anatomy atlases and textbooks can easily give the impression that everyone's internal anatomy is the same. This simply is not true. Books such as this one can teach you only the most common structure—the anatomy seen in about 70% or more of people. Someone who thinks that all human bodies are the same internally would make a very confused medical student or an incompetent surgeon.

Some people lack certain organs. For example, most of us have a *palmaris longus* muscle in the forearm and a *plantaris* muscle in the leg, but these are absent from others. Most of us have five lumbar vertebrae (bones of the lower spine), but some people have six and some have four. Most of us have one spleen and two kidneys, but some have two spleens or only one kidney. Most kidneys are supplied by a single *renal artery* and are drained by one *ureter,* but some have two renal arteries or ureters. **Figure 1.6** shows some common variations in human anatomy, and Deeper Insight 1.2 describes a particularly dramatic and clinically important variation.

▶▶▶APPLY WHAT YOU KNOW

People who are allergic to aspirin or penicillin often wear MedicAlert bracelets or necklaces that note this fact in case they need emergency medical treatment and are unable to communicate. Why would it be important for a person with situs inversus (see Deeper Insight 1.2) to have this noted on a MedicAlert bracelet?

DEEPER INSIGHT 1.2

CLINICAL APPLICATION

Situs Inversus and Other Unusual Anatomy

In most people, the spleen, pancreas, sigmoid colon, and most of the heart are on the left, while the appendix, gallbladder, and most of the liver are on the right. The normal arrangement of these and other internal organs is called *situs solitus* (SITE-us). About 1 in 8,000 people, however, is born with an abnormality called *situs inversus*—the organs of the thoracic and abdominal cavities are reversed between right and left. A selective right–left reversal of the heart is called *dextrocardia.* In *situs perversus,* a single organ occupies an atypical position—for example, a kidney located low in the pelvic cavity instead of high in the abdominal cavity.

Conditions such as dextrocardia in the absence of complete situs inversus can cause serious medical problems. Complete situs inversus, however, usually causes no functional problems because all of the viscera, though reversed, maintain their normal relationships to one another. Situs inversus is often discovered in the fetus by sonography, but many people remain unaware of their condition for decades until it is discovered by medical imaging, on physical examination, or in surgery. You can easily imagine the importance of such conditions in diagnosing appendicitis, performing gallbladder surgery, interpreting an X-ray, auscultating the heart valves, or recording an electrocardiogram.

BEFORE YOU GO ON

Answer the following questions to test your understanding of the preceding section:

12. In the hierarchy of human structure, what is the level between organ system and tissue? Between cell and molecule?

13. How are tissues relevant to the definition of an organ?

14. Why is reductionism a necessary but not sufficient point of view for fully understanding a patient's illness?

15. Why should medical students observe multiple cadavers and not be satisfied to dissect only one?

1.6 Human Function

Expected Learning Outcomes

When you have completed this section, you should be able to

a. state the characteristics that distinguish living organisms from nonliving objects;

b. explain the importance of physiological variation among persons;

c. define *homeostasis* and explain why this concept is central to physiology;

d. define *negative feedback,* give an example of it, and explain its importance to homeostasis;

e. define *positive feedback* and give examples of its beneficial and harmful effects; and

f. define *gradient,* describe the variety of gradients in human physiology, and identify some forms of matter and energy that flow down gradients.

1.6a Characteristics of Life

Why do we consider a growing child to be alive, but not a growing crystal? Is abortion the taking of a human life? If so, what about a contraceptive foam that kills only sperm? As a patient is dying, at what point does it become ethical to disconnect life-support equipment and remove organs for donation? If these organs are alive, as they must be to serve someone else, then why isn't the donor considered alive? Such questions have no easy answers, but they demand a concept of what life is—a concept that may differ with one's biological, medical, legal, or religious perspective.

From a biological viewpoint, life is not a single property. It is a collection of properties that help to distinguish living from nonliving things:

- **Organization.** Living things exhibit a far higher level of organization than the nonliving world around them. They expend a great deal of energy to maintain order, and a breakdown in this order is accompanied by disease and often death.

- **Cellular composition.** Living matter is always compartmentalized into one or more cells.

FIGURE 1.6 Variation in Anatomy of the Kidneys and the Major Arteries Near the Heart.

- **Metabolism.** Living things take in molecules from the environment and chemically change them into molecules that form their own structures, control their physiology, or provide them with energy. **Metabolism**[15] is the sum of all this internal chemical change. It inevitably produces chemical wastes, some of which are toxic if they accumulate. There is a constant turnover of molecules in the body. Although you sense a continuity of personality and experience from your childhood to the present, nearly every molecule of your body has been replaced within the past year.

- **Responsiveness** and **movement.** The ability to sense and react to **stimuli** (changes in the environment) is called *responsiveness* or *excitability*. It occurs at all levels from the single cell to the entire body, and it characterizes all living things from bacteria to you. Responsiveness is especially obvious in animals because of nerve and muscle cells that exhibit high sensitivity to environmental stimuli, rapid transmission of information, and quick reactions. Most living organisms are capable of self-propelled movement from place to place, and all organisms and cells are at least capable of moving substances internally, such as moving food along the digestive tract or moving molecules and organelles from place to place within a cell.

- **Homeostasis.** Although the environment around an organism changes, the organism maintains relatively stable internal conditions—for example, a stable temperature, blood pressure, and body weight. This ability to maintain internal stability, called *homeostasis,* is explored in section 1.6c.

- **Development.** Development is any change in form or function over the lifetime of the organism. In most organisms, it involves two major processes: (1) **differentiation,** the transformation of cells with no specialized function into cells that are committed to a particular task; and (2) **growth,** an increase in size. Some nonliving things grow, but not in the way your body does. If you let a saturated sugar solution evaporate, crystals will grow from it, but not through a change in the composition of the sugar. They merely add more sugar molecules from the solution to the crystal surface. The growth of the body, by contrast, occurs through chemical change (metabolism); for the most part, your body is not composed of the

---

[15]*metabol* = change; *ism* = process

molecules you ate but of molecules made by chemically altering your food.

- **Reproduction.** All living organisms can produce copies of themselves, thus passing their genes on to new, younger containers—their offspring.

- **Evolution.** All living species exhibit genetic change from generation to generation and therefore evolve. This occurs because *mutations* (changes in DNA structure) are inevitable and because environmental selection pressures favor the transmission of some genes more than others. Unlike the other characteristics of life, evolution is a characteristic seen only in the population as a whole. No single individual evolves over the course of its life.

Clinical and legal criteria of life differ from these biological criteria. A person who has shown no brain waves for 24 hours, and has no reflexes, respiration, or heartbeat other than what is provided by artificial life support, can be declared legally dead. At such time, however, most of the body is still biologically alive and its organs may be useful for transplant.

1.6b Physiological Variation

Earlier we considered the clinical importance of variations in human anatomy, but physiology is even more variable. Physiological variables differ with sex, age, weight, diet, degree of physical activity, genetics, and environment, among other things. Failure to consider such variation leads to medical mistakes such as overmedication of the elderly or medicating women on the basis of research done on young men. If a textbook states a typical human heart rate, blood pressure, red blood cell count, or body temperature, it is generally assumed, unless otherwise stated, that such values refer to a healthy 22-year-old weighing 58 kg (128 lb) for a female and 70 kg (154 lb) for a male, and a lifestyle of light physical activity and moderate caloric intake (2,000 and 2,800 kcal/day, respectively).

1.6c Negative Feedback and Homeostasis

The human body has a remarkable capacity for self-restoration. Hippocrates commented that it usually returns to a state of equilibrium by itself, and people recover from most illnesses even without the help of a physician. This tendency results from **homeostasis**[16] (HO-me-oh-STAY-sis), the body's ability to detect change, activate mechanisms that oppose it, and thereby maintain relatively stable internal conditions.

French physiologist **Claude Bernard** (1813–78) observed that the internal conditions of the body remain quite constant even when external conditions vary greatly. For example, whether it is freezing cold or swelteringly hot outdoors, the internal temperature of the body stays within a range of about 36° to 37°C (97°–99°F). American physiologist **Walter Cannon** (1871–1945) coined the

term *homeostasis* for this tendency to maintain internal stability. This has been one of the most enlightening theories in physiology. We now see physiology as largely a group of mechanisms for maintaining homeostasis, and the loss of homeostatic control as the cause of illness and death. Pathophysiology is essentially the study of unstable conditions that result when our homeostatic controls go awry.

Do not, however, overestimate the degree of internal stability. Internal conditions aren't absolutely constant but fluctuate within a limited range, such as the range of body temperatures noted earlier. The internal state of the body is best described as a **dynamic equilibrium** (balanced change), in which there is a certain **set point** or average value for a given variable (such as 37°C for body temperature) and conditions fluctuate slightly around this point.

The fundamental mechanism that keeps a variable close to its set point is **negative feedback**—a process in which the body senses a change and activates mechanisms that negate or reverse it. By maintaining stability, negative feedback is the key mechanism for maintaining health.

These principles can be understood by comparison to a home heating system **(fig. 1.7a).** Suppose it is a cold winter day and you have set your thermostat for 20°C (68°F)—the set point. If the room becomes too cold, a temperature-sensitive switch in the thermostat turns on the furnace. The temperature rises until it is slightly above the set point, and then the switch breaks the circuit and turns off the furnace. This is a negative feedback process that reverses the falling temperature and restores it to the set point. When the furnace turns off, the temperature slowly drops again until the switch is reactivated—thus, the furnace cycles on and off all day. The room temperature doesn't stay at exactly 20°C but fluctuates slightly—the system maintains a state of dynamic equilibrium in which the temperature averages 20°C and deviates only slightly from the set point. Because feedback mechanisms alter the original changes that triggered them (temperature, for example), they are often called **feedback loops.**

Body temperature is similarly regulated by a "thermostat"—a group of nerve cells in the base of the brain that monitor the temperature of the blood. If you become overheated, the thermostat triggers heat-losing mechanisms **(fig. 1.7b).** One of these is **vasodilation** (VAY-zo-dy-LAY-shun), the widening of blood vessels. When blood vessels of the skin dilate, warm blood flows closer to the body surface and loses heat to the surrounding air. If this isn't enough to return your temperature to normal, sweating occurs; the evaporation of water from the skin has a powerful cooling effect (see Deeper Insight 1.3). Conversely, if it is cold outside and your body temperature drops much below 37°C, these nerve cells activate heat-conserving mechanisms. The first to be activated is **vasoconstriction,** a narrowing of the blood vessels in the skin, which serves to retain warm blood deeper in your body and reduce heat loss. If this isn't enough, the brain activates shivering—muscle tremors that generate heat.

Let's consider one more example—a case of homeostatic control of blood pressure. When you first rise from bed in the morning, gravity causes some of your blood to drain away from

---

[16]*homeo* = the same; *stas* = to place, stand, stay

(a)

(b)

FIGURE 1.7 **Negative Feedback in Thermoregulation.**
(a) The negative feedback loop that maintains room temperature.
(b) Negative feedback usually keeps the human body temperature within about 0.5°C of a 37°C set point. Cutaneous vasoconstriction and shivering set in when the body temperature falls too low, and soon raise it. Cutaneous vasodilation and sweating set in when body temperature rises too high, and soon lower it.

? *How does vasodilation reduce the body temperature?*

your head and upper torso, resulting in falling blood pressure in this region—a local imbalance in your homeostasis **(fig. 1.8).** This is detected by sensory nerve endings called *baroreceptors* in large arteries near the heart. They transmit nerve signals to the brainstem, where we have a *cardiac center* that regulates the heart rate. The cardiac center responds by transmitting nerve signals to the heart, which speed it up. The faster heart rate quickly raises the blood pressure and restores normal homeostasis. In elderly people, this feedback loop is sometimes insufficiently responsive, and they may feel dizzy as they rise from a reclining

DEEPER INSIGHT 1.3
MEDICAL HISTORY

Men in the Oven

English physician Charles Blagden (1748–1820) staged a rather theatrical demonstration of homeostasis long before Cannon coined the word. In 1775, Blagden spent 45 minutes in a chamber heated to 127°C (260°F)—along with a dog, a beefsteak, and some research associates. Being dead and unable to maintain homeostasis, the steak was cooked. But being alive and capable of evaporative cooling, the dog panted, the men sweated, and all of them survived. History does not record whether the men ate the steak in celebration or shared it with the dog.

position and their cerebral blood pressure falls. This sometimes causes fainting.

This reflexive correction of blood pressure *(baroreflex)* illustrates three common, although not universal, components of a feedback loop: a receptor, an integrating center, and an effector. The **receptor** is a structure that senses a change in the body, such as the stretch receptors that monitor blood pressure. The **integrating (control) center,** such as the cardiac center of the

FIGURE 1.8 **Homeostatic Compensation for a Postural Change in Blood Pressure.**

brain, is a mechanism that processes this information, relates it to other available information (for example, comparing what the blood pressure is with what it should be), and makes a decision about what the appropriate response should be. The **effector** is the cell or organ that carries out the final corrective action. In the foregoing example, it is the heart. The response, such as the restoration of normal blood pressure, is then sensed by the receptor, and the feedback loop is complete.

1.6d Positive Feedback and Rapid Change

Positive feedback is a self-amplifying cycle in which a physiological change leads to even greater change in the same direction, rather than producing the corrective effects of negative feedback. Positive feedback is often a normal way of producing rapid change. When a woman is giving birth, for example, the head of the fetus pushes against her cervix (the neck of the uterus) and stimulates its nerve endings **(fig. 1.9)**. Nerve signals travel to the brain, which, in turn, stimulates the pituitary gland to secrete the hormone oxytocin. Oxytocin travels in the blood and stimulates the uterus to contract. This pushes the fetus downward, stimulating the cervix

③ Brain stimulates pituitary gland to secrete oxytocin

② Nerve impulses from cervix transmitted to brain

④ Oxytocin stimulates uterine contractions and pushes fetus toward cervix

① Head of fetus pushes against cervix

FIGURE 1.9 Positive Feedback in Childbirth.

❓ *Could childbirth as a whole be considered a negative feedback event? Discuss.*

still more and causing the positive feedback loop to be repeated. Labor contractions therefore become more and more intense until the fetus is expelled. Other cases of beneficial positive feedback are seen later in the book in, for example, blood clotting, protein digestion, and the generation of nerve signals.

Frequently, however, positive feedback is a harmful or even life-threatening process. This is because its self-amplifying nature can quickly change the internal state of the body to something far from its homeostatic set point. Consider a high fever, for example. A fever triggered by infection is beneficial up to a point, but if the body temperature rises much above 40°C (104°F), it may create a dangerous positive feedback loop. This high temperature raises the metabolic rate, which makes the body produce heat faster than it can get rid of it. Thus, temperature rises still further, increasing the metabolic rate and heat production still more. This "vicious circle" becomes fatal at approximately 45°C (113°F). Thus, positive feedback loops often create dangerously out-of-control situations that require emergency medical treatment.

1.6e Gradients and Flow

Another fundamental concept that will arise repeatedly in this book is that matter and energy tend to *flow down gradients*. This simple principle underlies processes as diverse as blood circulation, respiratory airflow, urine formation, nutrient absorption, body water distribution, temperature regulation, and the action of nerves and muscles.

A physiological **gradient** is a difference in chemical concentration, electrical charge, physical pressure, temperature, or other variable between one point and another. If matter or energy moves from the point where this variable has a higher value to the point with a lower value, we say it flows **down the gradient**—for example, from a warmer to a cooler point, or a place of high chemical concentration to one of lower concentration. Movement in the opposite direction is **up the gradient.**

Outside of biology, *gradient* can mean a hill or slope, and this affords us a useful analogy to biological processes **(fig. 1.10a)**. A wagon released at the top of a hill will roll down it ("flow") spontaneously, without need for anyone to exert energy to move it. Similarly, matter and energy in the body spontaneously flow down gradients, without the expenditure of metabolic energy. Movement up a gradient does require an energy expenditure, just as we would have to push or pull a wagon to move it uphill.

Consider some examples and analogies. If you open a water tap with a garden hose on it, you create a **pressure gradient;** water flows down the hose from the high-pressure point at the tap to the low-pressure point at the open end. Each heartbeat is like that, creating a gradient from high blood pressure near the heart to low pressure farther away; blood flows down this gradient away from the heart **(fig. 1.10b).** When we inhale, air flows down a pressure gradient from the surrounding atmosphere to pulmonary air passages where the pressure is lower. A pressure gradient also drives the process in which the kidneys filter water and waste products from the blood.

FIGURE 1.10 Flow Down Gradients. (a) A wagon rolling downhill (down a gradient) (left) is a useful analogy to spontaneous, gradient-driven physiological processes. Moving up a gradient (right) requires an energy input. (b) Blood flowing down a pressure gradient. (c) Dietary sugars flowing down a concentration gradient into an intestinal cell. (d) Sodium ions flowing down an electrical gradient into a cell. (e) Heat flowing down a thermal gradient to leave the body through the skin.

Chemicals flow down **concentration gradients.** When we digest starch, a high concentration of sugars accumulates in the small intestine. The cells lining the intestine contain only a low concentration of sugars, so sugars flow from the intestinal space into these cells, thus becoming absorbed into the body's tissues **(fig. 1.10c).** Water flows through cell membranes and epithelia by *osmosis,* from the side where it is more concentrated to the side where it is less so.

Charged particles flow down **electrical gradients.** Suppose there is a high concentration of sodium ions (Na^+) just outside a cell and much lower concentration inside, so the outer surface of the cell membrane has a relatively positive charge and the inner surface is relatively negative **(fig. 1.10d).** If we open channels in the membrane that will let sodium pass, sodium ions rush into the cell, flowing down their electrical gradient. Because each Na^+ carries a positive charge, this flow constitutes an electrical current through the membrane. We tap this current to make our nerves fire, our heart beat, and our muscles contract. In many cases, the flow of ions is governed by a combination of concentration and electrical charge differences between two points, and we say that ions flow down **electrochemical gradients.** These will be studied especially in connection with muscle and nerve action in chapters 11 and 12.

Heat flows down a **thermal gradient.** Suppose there is warm blood flowing through small arteries close to the skin surface, and the air temperature around the body is cooler **(fig. 1.10e).** Heat will flow from the blood to the surrounding air, down its thermal gradient, and be lost from the body. You will see in chapter 27 that heat flow is also important in preventing the testes from overheating, which would otherwise prevent sperm production.

Thus, you can see there are many applications in human physiology for this universal tendency of matter and energy to flow down gradients. This principle arises many times in the chapters to follow. We will revisit it next in chapter 3 when we consider how materials move into and out of cells through the cell membrane.

BEFORE YOU GO ON

Answer the following questions to test your understanding of the preceding section:

16. List four biological criteria of life and one clinical criterion. Explain how a person could be clinically dead but biologically alive.

17. What is meant by *dynamic equilibrium*? Why would it be wrong to say homeostasis prevents internal change?

18. Explain why stabilizing mechanisms are called *negative feedback*.

19. Explain why positive feedback is more likely than negative feedback to disturb homeostasis.

20. Active tissues generate carbon dioxide, which diffuses out of the tissue into the bloodstream, to be carried away. Is this diffusion into the blood a case of flow up a gradient, or down? Explain.

1.7 The Language of Medicine

Expected Learning Outcomes

When you have completed this section, you should be able to

a. explain why modern anatomical terminology is so heavily based on Greek and Latin;

b. recognize eponyms when you see them;

c. describe the efforts to achieve an internationally uniform anatomical terminology;

d. break medical terms down into their basic word elements;

e. state some reasons why the literal meaning of a word may not lend insight into its definition;

f. relate singular noun forms to their plural and adjective forms; and

g. discuss why precise spelling is important in anatomy and physiology.

One of the greatest challenges faced by students of anatomy and physiology is the vocabulary. In this book, you will encounter such Latin terms as *corpus callosum* (a brain structure), *ligamentum arteriosum* (a small fibrous band near the heart), and *extensor carpi radialis longus* (a forearm muscle). You may wonder why structures aren't named in "just plain English," and how you will ever remember such formidable names. This section will give you some answers to these questions and some useful tips on mastering anatomical terminology.

1.7a The History of Anatomical Terminology

The major features of human gross anatomy have standard international names prescribed by a book titled the *Terminologia Anatomica (TA)*. The *TA* was codified in 1998 by an international committee of anatomists and approved by professional associations of anatomists in more than 50 countries.

About 90% of today's medical terms are formed from just 1,200 Greek and Latin roots. Why those two languages? Scientific investigation began in ancient Greece and soon spread to Rome. The Greeks and Romans coined many of the words still used in human anatomy today: *duodenum, uterus, prostate, cerebellum, diaphragm, sacrum, amnion,* and others. In the Renaissance, the fast pace of discovery required a profusion of new terms to describe things. Anatomists in different countries began giving different names to the same structures. Adding to the confusion, they often named new structures and diseases in honor of their esteemed teachers and predecessors, giving us such nondescriptive terms as *fallopian tube* and *duct of Santorini.* Terms coined from the names of people, called **eponyms,**[17] afford little clue as to what a structure or condition is.

In hopes of resolving this growing confusion, anatomists began meeting as early as 1895 to devise a uniform international terminology. After several false starts, they agreed on a list of terms that rejected all eponyms and gave each structure a unique Latin name to be used worldwide. Even if you were to look at an anatomy atlas in Korean or Arabic, the illustrations may be labeled with the same Latin terms as in an English-language atlas. That list served for many decades until recently replaced by the *TA,* which prescribes both Latin names and accepted English equivalents. The terminology in this book conforms to the *TA* except where undue confusion would result from abandoning widely used, yet unofficial, terms.

1.7b Analyzing Medical Terms

The task of learning medical terminology seems overwhelming at first, but it is a simple skill to become more comfortable with the technical language of medicine. People who find scientific terms confusing and difficult to pronounce, spell, and remember often feel more confident once they realize the logic of how terms are composed. A term such as *hyponatremia* is less forbidding once we recognize that it is composed of three common word elements: *hypo-* (below normal), *natr-* (sodium), and *-emia* (blood condition). Thus, hyponatremia is a deficiency of sodium in the blood. Those word elements appear over and over in many other medical terms: *hypothermia, natriuretic, anemia,* and so on. Once you learn the meanings of *hypo-, natri-,* and *-emia,* you already have the tools to at least partially understand hundreds of other biomedical terms. In appendix E, you will find a lexicon of word elements commonly footnoted in this book.

Scientific terms are typically composed of one or more of the following elements:

- At least one *root (stem)* that bears the core meaning of the word. In *cardiology,* for example, the root is *cardi-* (heart). Many words have two or more roots. In *cardiomyopathy,* for example, the roots are *cardi-* (heart), *my-* (muscle), and *path-* (disease).

- *Combining vowels* that are often inserted to join roots and make the word easier to pronounce. In *cardiomyopathy,* each *o* is a combining vowel. Although *o* is the most common combining vowel, all vowels of the alphabet are used in this way, such as *a* in *ligament, e* in *vitreous, i* in *fusiform, u* in *ovulation,* and *y* in *tachycardia.* Some words, such as *intervertebral,* have no combining vowels. A combination of a root and combining vowel is called a *combining form;* for example, *chrom-* (color) + *o* (a combining vowel) make the combining form *chromo-,* as in *chromosome.*

- A *prefix* may be present to modify the core meaning of the word. For example, *gastric* (pertaining to the stomach or to the belly of a muscle) takes on a variety of new meanings when prefixes are added to it: *epigastric* (above the stomach), *hypogastric* (below the stomach), *endogastric* (within the stomach), and *digastric* (a muscle with two bellies).

- A *suffix* may be added to the end of a word to modify its core meaning. For example, *microscope, microscopy, microscopic,* and *microscopist* have different meanings because of their suffixes alone. Often two or more suffixes, or a

---

[17] *epo = epi* = upon, based upon; *nym* = name

root and suffix, occur together so often that they are treated jointly as a *compound suffix;* for example, *log* (study) + *y* (process) form the compound suffix *-logy* (the study of). Prefixes and suffixes are collectively called *affixes.*

To summarize these basic principles, consider the word *gastroenterology,* a branch of medicine dealing with the stomach and small intestine. It breaks down into *gastro/entero/logy:*

gastro	=	a combining form meaning "stomach"
entero	=	a combining form meaning "small intestine"
logy	=	a compound suffix meaning "the study of"

"Dissecting" words in this way and paying attention to the word-origin footnotes throughout this book will help you become more comfortable with the language of anatomy. Knowing how a word breaks down and knowing the meaning of its elements make it far easier to pronounce a word, spell it, and remember its definition.

There are a few unfortunate exceptions, however. The path from original meaning to current usage has often become obscured by history (see Deeper Insight 1.4). The foregoing approach also is no help with eponyms or **acronyms**[18]—words composed of the first letter, or first few letters, of a series of words. For example, a common medical imaging method is the PET scan, an acronym for *positron emission tomography.* Note that PET is a pronounceable word, hence a true acronym. Acronyms are not to be confused with simple abbreviations such as DNA and MRI, in which each letter must be pronounced separately.

1.7c Plurals, Adjectives, and Possessive Forms

A point of confusion for many beginning students is how to recognize the plural forms of medical terms. Few people would fail to recognize that *ovaries* is the plural of *ovary,* but the connection is harder to make in other cases: For example, the plural of *cortex* is *cortices* (COR-ti-sees), the plural of *corpus* is *corpora,* and the plural of *ganglion* is *ganglia.* **Table 1.1** will help you make the connection between common singular and plural noun terminals.

In some cases, what appears to the beginner to be two completely different words may be only the noun and adjective forms of the same word. For example, *brachium* denotes the arm, and *brachii* (as in the muscle name *biceps brachii*) means "of the arm." *Carpus* denotes the wrist, and *carpi,* a word used in several muscle names, means "of the wrist." Adjectives can also take different forms for the singular and plural and for different degrees of comparison. The *digits* are the fingers and toes. The word *digiti* in a muscle name means "of a single finger (or toe)," whereas *digitorum* is the plural, meaning "of multiple fingers (or toes)." Thus, the *extensor digiti minimi muscle* extends only the little finger, whereas the *extensor digitorum muscle* extends all fingers except the thumb.

DEEPER INSIGHT 1.4
MEDICAL HISTORY

Obscure Medical Word Origins

The literal translation of a word doesn't always provide great insight into its modern meaning. The history of language is full of twists and turns that are fascinating in their own right and say much about the history of human culture, but they can create confusion for students.

For example, the *amnion* is a transparent sac that forms around the developing fetus. The word is derived from *amnos,* from the Greek for "lamb." From this origin, *amnos* came to mean a bowl for catching the blood of sacrificial lambs, and from there the word found its way into biomedical usage for the membrane that emerges (quite bloody) as part of the afterbirth. The *acetabulum,* the socket of the hip joint, literally means "vinegar cup." Apparently the hip socket reminded an anatomist of the little cups used to serve vinegar as a condiment on dining tables in ancient Rome. The word *testicles* can be translated "little pots" or "little witnesses." The history of medical language has several amusing conjectures as to why this word was chosen to name the male gonads.

TABLE 1.1	Singular and Plural Forms of Some Noun Terminals	
Singular Ending	**Plural Ending**	**Examples**
-a	-ae	axilla, axillae
-en	-ina	lumen, lumina
-ex	-ices	cortex, cortices
-is	-es	diagnosis, diagnoses
-is	-ides	epididymis, epididymides
-ix	-ices	appendix, appendices
-ma	-mata	carcinoma, carcinomata
-on	-a	ganglion, ganglia
-um	-a	septum, septa
-us	-era	viscus, viscera
-us	-i	villus, villi
-us	-ora	corpus, corpora
-x	-ges	phalanx, phalanges
-y	-ies	ovary, ovaries
-yx	-yces	calyx, calyces

The English words *large, larger,* and *largest* are examples of the positive, comparative, and superlative degrees of comparison. In Latin, these are *magnus, major* (from *maior*), and *maximus.* We find these in the muscle names *adductor magnus* (a *large* muscle of the thigh), the *pectoralis major* (the *larger* of two pectoral muscles of the chest), and *gluteus maximus* (the *largest* of the three gluteal muscles of the buttock).

Some noun variations indicate the possessive, such as the *rectus abdominis,* a straight (*rectus*) muscle of the abdomen (*abdominis,* "of the abdomen"), and the *erector spinae,* a muscle that straightens (*erector*) the spinal column (*spinae,* "of the spine").

Anatomical terminology also frequently follows the Greek and Latin practice of placing the adjective after the noun. Thus, we have such names as the *stratum lucidum* for a clear *(lucidum)* layer *(stratum)* of the epidermis, the *foramen magnum* for a large *(magnum)* hole *(foramen)* in the skull, and the aforementioned *pectoralis major* muscle of the chest.

This is not to say that you must be conversant in Latin or Greek grammar to proceed with your study of anatomy. These few examples, however, may alert you to some patterns to watch for in the terminology you study and, ideally, will make your encounters with anatomical terminology less confusing.

1.7d Pronunciation

Pronunciation is another stumbling block for many beginning anatomy and physiology students. This book gives simple pro-NUN-see-AY-shun guides for many terms when they are first introduced. Read the syllables of these guides phonetically and accent the syllables in capital letters. You can also hear pronunciations of most of the anatomical terms within Anatomy & Physiology REVEALED®.

1.7e The Importance of Spelling

A final word of advice for your study of anatomy and physiology: Be accurate in your spelling and use of terms. It may seem trivial if you misspell *trapezius* as *trapezium,* but in doing so, you would be changing the name of a back muscle to the name of a wrist bone. Similarly, changing *occipitalis* to *occipital* or *zygomaticus* to *zygomatic* changes other muscle names to bone names. Changing *malleus* to *malleolus* changes the name of a middle-ear bone to the name of a bony protuberance of the ankle. And there is only a one-letter difference between *ileum* (the final portion of the small intestine) and *ilium* (part of the hip bone), and between *gustation* (the sense of taste) and *gestation* (pregnancy).

The health professions demand the utmost attention to detail and accuracy—people's lives may one day be in your hands. The habit of carefulness must extend to your use of language as well. Many patients have died simply because of tragic written and oral miscommunication in the hospital. Compared to this, it is hardly tragic if your instructor deducts a point or two for an error in spelling. It should be considered a lesson learned about the importance of accuracy.

BEFORE YOU GO ON

Answer the following questions to test your understanding of the preceding section:

21. Explain why modern anatomical terminology is so heavily based on Greek and Latin.

22. Distinguish between an eponym and an acronym, and explain why both of these present difficulties for interpreting anatomical terms.

23. Break each of the following words down into its roots, prefixes, and suffixes, and state their meanings, following the example of *gastroenterology* analyzed earlier: *pericardium, appendectomy, subcutaneous, phonocardiogram, otorhinolaryngology.* Consult the list of word elements in appendix E for help.

24. Write the singular form of each of the following words: *pleurae, gyri, ganglia, fissures.* Write the plural form of each of the following: *villus, tibia, encephalitis, cervix, stoma.*

1.8 Review of Major Themes

To close this chapter, let's distill a few major points from it. These themes can provide you with a sense of perspective that will make the rest of the book more meaningful and not just a collection of disconnected facts. These are some key unifying principles behind all study of human anatomy and physiology:

- **Unity of form and function.** *Form and function complement each other; physiology cannot be divorced from anatomy.* This unity holds true even down to the molecular level. Our very molecules, such as DNA and proteins, are structured in ways that enable them to carry out their functions. Slight changes in molecular structure can destroy their activity and threaten life.

- **Cell theory.** *All structure and function result from the activity of cells.* Every physiological concept in this book ultimately must be understood from the standpoint of how cells function. Even anatomy is a result of cellular function. If cells are damaged or destroyed, we see the results in disease symptoms of the whole person.

- **Evolution.** *The human body is a product of evolution.* Like every other living species, we have been molded by millions of years of natural selection to function in a changing environment. Many aspects of human anatomy and physiology reflect our ancestors' adaptations to their environment. Human form and function cannot be fully understood except in light of our evolutionary history.

- **Hierarchy of complexity.** *Human structure can be viewed as a series of levels of complexity.* Each level is composed of a smaller number of simpler subunits than the level above it. These subunits are arranged in different ways to form diverse structures of higher complexity. Understanding the simpler components is the key to understanding higher levels of structure.

- **Homeostasis.** *The purpose of most normal physiology is to maintain stable conditions within the body.* Human physiology is essentially a group of homeostatic mechanisms that produce stable internal conditions favorable to cellular function. Any serious departure from these conditions can be harmful or fatal to cells and thus to the whole body.

- **Gradients and flow.** Matter and energy tend to flow down gradients such as differences in chemical concentration, pressure, temperature, and electrical charge. This accounts for much of their movement in human physiology.

▶▶▶ **APPLY WHAT YOU KNOW**

Architect Louis Henri Sullivan coined the phrase, "Form ever follows function." What do you think he meant by this? Discuss how this idea could be applied to the human body and cite a specific example of human anatomy to support it.

DEEPER INSIGHT 1.5

CLINICAL APPLICATION

Medical Imaging

The development of techniques for looking into the body without having to do exploratory surgery has greatly accelerated progress in medicine. A few of these techniques are described here.

Radiography

Radiography, first performed in 1895, is the process of photographing internal structures with X-rays **(fig. 1.11a).** Until the 1960s, this was the only widely available imaging method; even today, it accounts for more than 50% of all clinical imaging. X-rays pass through the soft tissues of the body to a photographic film or detector on the other side, where they produce relatively dark images. They are absorbed, however, by dense matter such as bones,

teeth, tumors, and tuberculosis nodules, which leave the image lighter in these areas. The term *X-ray* also applies to an image *(radiograph)* made by this method. Radiography is commonly used in dentistry, mammography, diagnosis of fractures, and examination of the chest. Hollow organs can be visualized by filling them with a contrast medium that absorbs X-rays. Barium sulfate, for example, is given orally for examination of the esophagus, stomach, and small intestine or by enema for examination of the large intestine. Some disadvantages of radiography are that images of overlapping organs can be confusing and slight differences in tissue density are not easily detected. In addition, X-rays can cause mutations leading to cancer and birth defects. Radiography therefore cannot be used indiscriminately.

(a) X-ray (radiograph)

(b) Cerebral angiogram

(c) Computed tomographic (CT) scan

(d) Magnetic resonance image (MRI)

(e) Positron emission tomographic (PET) scan

FIGURE 1.11 Radiologic Images of the Head. (a) X-ray (radiograph) of the skull. (b) Digital subtraction angiogram (DSA) of the cerebral blood vessels. (c) CT scan at the level of the eyes. (d) MRI scan at the level of the eyes. The optic nerves appear in red and the muscles that move the eyes appear in green. (e) A PET scan of the brain of an unmedicated schizophrenic patient. Red areas indicate regions of high metabolic rate. In this patient, the visual center of the brain at the rear of the head (bottom of photo) was especially active during the scan.

❓ *What structures are seen better by MRI than by X-ray? What structures are seen better by X-ray than by PET?*

a: U.H.B. Trust/The Image Bank/Getty Images; b: pang_oasis/Shutterstock; c: Miriam Maslo/Science Source; d: UHB Trust/Getty Images; e: ISM/Sovereign/Medical Images

Blood vessels can be seen especially clearly with a radiographic method called *digital subtraction angiography (DSA)* **(fig. 1.11b).** This entails taking X-rays before and after injecting a contrast medium into a vessel. A computer then "erases" the first image from the second, leaving a clear, dark image of just the injected vessels without the overlying and surrounding tissues. This is useful for showing vascular blockages and anatomical malformations, abnormalities of cerebral blood flow, and narrowing (stenosis) of renal arteries, and as an aid in threading catheters into blood vessels. DSA is already being replaced in many clinics, however, by yet newer methods that are less invasive and avoid contrast medium and radiation exposure.

Computed Tomography

Computed tomography[19] (a *CT scan*) **(fig. 1.11c)** is a more sophisticated application of X-rays. The patient is moved through a ring-shaped machine that emits low-intensity X-rays on one side and receives them with a detector on the opposite side. A computer analyzes signals from the detector and produces an image of a "slice" of the body about as thin as a coin. The advantage of such thin planes of view is that there is little overlap of organs, so the image is much sharper than a conventional X-ray. It requires extensive knowledge of cross-sectional anatomy to interpret the images. CT scanning is useful for identifying tumors, aneurysms, cerebral hemorrhages, kidney stones, and other abnormalities.

Magnetic Resonance Imaging

Magnetic resonance imaging (MRI) **(fig. 1.11d)** is better than CT for visualizing some soft tissues. The patient lies in either a tube or an open-sided scanner surrounded by a powerful electromagnet. Hydrogen atoms in the patient's tissues alternately align themselves with this magnetic field and with a radio-frequency field turned on and off by the technologist. These changes in hydrogen alignment generate signals that are analyzed by computer to produce an anatomical image. MRI can "see" clearly through the skull and spine to produce images of the nervous tissue within, and it is better than CT for distinguishing between soft tissues such as the white and gray matter of the brain. MRI also avoids X-ray exposure and its risks.

MRI has disadvantages, however, such as the claustrophobic feeling some patients experience in the scanner, loud noises generated by the machine, and long exposure times that prevent sharp images being made of the constantly moving stomach and intestines. It requires a patient to lie still in the enclosed space for up to 45 minutes to scan one region of the body and may entail 90 minutes to scan multiple regions such as the abdominal and pelvic cavities. Some patients find they cannot tolerate this. Open-sided MRI machines are favored by some claustrophobic or obese patients, but have weaker magnetic fields, produce poorer images, and may miss important tissue abnormalities.

Functional MRI (fMRI) is a variation that visualizes moment-to-moment changes in tissue function. fMRI scans of the brain, for example, show shifting patterns of activity as the brain applies itself to a specific sensory, mental, or motor task. fMRI has lately replaced the PET scan as the most important method for visualizing brain function. The use of fMRI in brain imaging is further discussed in Deeper Insight 14.5.

Positron Emission Tomography

Positron emission tomography (the *PET scan*) **(fig. 1.11e)** is used to assess the metabolic state of a tissue and distinguish which tissues are most active at a given moment. The procedure begins with an injection of radioactively labeled glucose, which emits positrons (electron-like particles with a positive charge). When a positron and electron meet, they annihilate each other and give off a pair of gamma rays that can be detected by sensors and analyzed by computer. The computer displays a color image that shows which tissues were using the most glucose at the moment. PET scans are generally low-resolution, as in this photo, but nevertheless provide valuable diagnostic information. In cardiology, PET scans can show the extent of tissue death from a heart attack. Since it consumes little or no glucose, the damaged tissue appears dark. PET scans are also widely used to diagnose cancer and evaluate tumor status. The PET scan is an example of *nuclear medicine*—the use of radioactive isotopes to treat disease or to form diagnostic images of the body.

Sonography

Sonography[20] **(fig. 1.12)** is the second oldest and second most widely used method of imaging. A handheld device pressed against the skin produces high-frequency ultrasound waves and receives the signals that echo back from internal organs. Sonography isn't very useful for examining bones or lungs, but it is the method of choice in obstetrics, where the image *(sonogram)* can be used to locate the placenta and evaluate fetal age, position, and development. Sonography is also used to view tissues in motion, such as fetal movements, actions of the heart wall and valves, and blood ejection from the heart and flow through arteries and veins. Sonographic imaging of the beating heart is called *echocardiography.* Sonography avoids the harmful effects of X-rays, and the equipment is inexpensive and portable. Some disadvantages are that sonography can't penetrate bone and it usually doesn't produce a very sharp image.

(a)

(b)

FIGURE 1.12 Fetal Sonography. (a) Producing a sonogram. (b) Three-dimensional fetal sonogram at 32 weeks of gestation.

a: Kevin Brofsky/Getty Images; b: Ken Saladin

---

[19]*tomo* = section, cut, slice; *graphy* = recording process

[20]*sono* = sound; *graphy* = recording process

STUDY GUIDE

▶ Assess Your Learning Outcomes

To test your knowledge, discuss the following topics with a study partner or in writing, ideally from memory.

1.1 The Scope of Anatomy and Physiology

1. The meanings of *anatomy* and *physiology* and what it means to say these two sciences are complementary and inseparable
2. Methods of study in anatomy and clinical examination
3. Branches of anatomy that study the body at different levels of detail
4. How comparative physiology advances the understanding of human function

1.2 The Origins of Biomedical Science

1. Greek and Roman scholars who first gave medicine a scientific basis
2. Ways in which the work of Maimonides, Avicenna, Vesalius, and Harvey were groundbreaking in the context of their time and culture
3. Why medical science today owes such a great debt to Hooke, Leeuwenhoek, and other inventors
4. How Schleiden and Schwann revolutionized and unified the understanding of biological structure, ultimately including human anatomy and physiology

1.3 Scientific Method

1. How philosophers Bacon and Descartes revolutionized society's view of science, even though neither of them was a scientist
2. The essential qualities of the scientific method
3. The nature of the inductive and hypothetico–deductive methods, how they differ, and which areas of biomedical science most heavily employ each method
4. The qualities of a valid scientific hypothesis, the function of a hypothesis, and what is meant by *falsifiability* in science
5. How each of the following contributes to the reliability of a researcher's scientific conclusions and the trust that the public may place

in science: sample size, control groups, the double-blind method, statistical testing, and peer review
6. The distinctions between scientific facts, laws, and theories; the purpose of a theory; and how the scientific meanings of *law* and *theory* differ from the common lay meanings

1.4 Human Origins and Adaptations

1. The meanings of *evolution, natural selection, selection pressure,* and *adaptation,* with examples of each
2. The historical origin of the theory of natural selection and how this theory is relevant to a complete understanding of human anatomy and physiology
3. How the kinship among all species is relevant to the choice of model animals for biomedical research
4. Ecological conditions thought to have selected for such key characteristics of *Homo sapiens* as opposable thumbs, shoulder mobility, prehensile hands, stereoscopic vision, color vision, and bipedal locomotion
5. The meaning of *evolutionary medicine*

1.5 Human Structure

1. Levels of human structural complexity from organism to atom
2. Reductionism and holism; how they differ and why both ideas are relevant to the study of human anatomy and physiology and to the clinical care of patients
3. Examples of why the anatomy presented in textbooks is not necessarily true of every individual

1.6 Human Function

1. Eight essential qualities that distinguish living organisms from nonliving things
2. The meaning of *metabolism*
3. Clinical criteria for life and death, and why clinical and biological death are not exactly equivalent
4. The clinical importance of physiological variation between people, and the

assumptions that underlie typical values given in textbooks
5. The meaning of *homeostasis;* its importance for survival; and the historical origin of this concept
6. How negative feedback contributes to homeostasis; the meaning of *negative feedback loop;* how a receptor, integrating center, and effector are involved in many negative feedback loops; and at least one example of such a loop
7. How positive feedback differs from negative feedback; examples of beneficial and harmful cases of positive feedback
8. The concept of matter and energy flowing down gradients and how this applies to various areas of human physiology

1.7 The Language of Medicine

1. The origin and purpose of the *Terminologia Anatomica (TA)* and its relevance for anatomy students
2. How to break biomedical terms into familiar roots, prefixes, and suffixes, and why the habit of doing so aids in learning
3. Acronyms and eponyms, and why they cannot be understood by trying to analyze their roots
4. How to recognize when two or more words are singular and plural versions of one another; when one word is the possessive form of another; and when medical terms built on the same root represent different degrees of comparison (such as terms denoting *large, larger,* and *largest*)
5. Why accuracy in spelling and usage of medical terms can be a matter of life or death in a hospital or clinic, and how seemingly trivial spelling errors can radically alter meaning

1.8 Review of Major Themes

1. A description of six core themes of this book: unity of form and function, cell theory, evolution, hierarchy of complexity, homeostasis, and gradients and flow

STUDY GUIDE

▶ Testing Your Recall

Answers in Appendix A

1. Structure that can be observed with the naked eye is called
   a. gross anatomy.
   b. ultrastructure.
   c. microscopic anatomy.
   d. histology.
   e. cytology.

2. The word prefix *homeo-* means
   a. tissue.
   b. metabolism.
   c. change.
   d. human.
   e. same.

3. The simplest structures considered to be alive are
   a. organisms.
   b. organs.
   c. tissues.
   d. cells.
   e. organelles.

4. Which of the following people revolutionized the teaching of gross anatomy?
   a. Vesalius
   b. Aristotle
   c. Hippocrates
   d. Leeuwenhoek
   e. Cannon

5. Which of the following embodies the greatest amount of scientific information?
   a. a fact
   b. a law of nature
   c. a theory
   d. a deduction
   e. a hypothesis

6. An informed, uncertain, but testable conjecture is
   a. a natural law.
   b. a scientific theory.
   c. a hypothesis.
   d. a deduction.
   e. a scientific fact.

7. A self-amplifying chain of physiological events is called
   a. positive feedback.
   b. negative feedback.
   c. dynamic constancy.
   d. homeostasis.
   e. metabolism.

8. Which of the following is *not* a human organ system?
   a. integumentary
   b. muscular
   c. epithelial
   d. nervous
   e. endocrine

9. _____ means studying anatomy by touch.
   a. Gross anatomy
   b. Auscultation
   c. Osculation
   d. Palpation
   e. Percussion

10. The prefix *hetero-* means
    a. same.
    b. different.
    c. both.
    d. solid.
    e. below.

11. Cutting and separating tissues to reveal structural relationships is called _____.

12. A difference in chemical concentration between one point and another is called a concentration _____.

13. By the process of _____, a medical researcher predicts what the result of a certain experiment will be if his or her hypothesis is correct.

14. Physiological effects of a person's mental state are called _____ effects.

15. The tendency of the body to maintain stable internal conditions is called _____.

16. Blood pH averages 7.4 but fluctuates from 7.35 to 7.45. A pH of 7.4 can therefore be considered the _____ for this variable.

17. Self-corrective mechanisms in physiology are called _____ loops.

18. A/an _____ is the simplest body structure to be composed of two or more types of tissue.

19. Depth perception, or the ability to form three-dimensional images, is also called _____ vision.

20. Our hands are said to be _____ because they can encircle an object such as a branch or tool. The presence of an _____ thumb is important to this ability.

▶ Building Your Medical Vocabulary

Answers in Appendix A

State a meaning of each word element, and give a medical term from this chapter that uses it or a slight variation of it.

1. auscult-

2. dis-

3. homeo-

4. metabolo-

5. palp-

6. physio-

7. -sect

8. -stasis

9. stereo-

10. tomo-

STUDY GUIDE

▶ What's Wrong with These Statements?

Answers in Appendix A

Briefly explain why each of the following statements is false, or reword it to make it true.

1. The technique for taking a patient's pulse at the wrist is auscultation.

2. For a pregnant woman to have an MRI scan would expose her fetus to radiation that can potentially cause mutation and birth defects.

3. We usually depend on positive feedback to restore homeostatic balance and have a beneficial effect on the body.

4. There are far more cells than organelles in the body.

5. Matter doesn't generally move down a gradient in the body unless the body expends metabolic energy to move it.

6. Leeuwenhoek was a biologist who invented the simple microscope in order to examine organisms in lake water.

7. A scientific theory is just a speculation until someone finds the evidence to prove it.

8. In a typical clinical research study, volunteer patients are in the treatment group and the physicians and scientists who run the study constitute the control group.

9. Human evolution is basically a theory that humans came from monkeys.

10. Negative feedback usually has a negative (harmful) effect on the body.

▶ Testing Your Comprehension

1. Ellen is pregnant and tells Janet, one of her coworkers, that she is scheduled to get a fetal sonogram. Janet expresses alarm and warns Ellen about the danger of exposing a fetus to X-rays. Discuss why you think Janet's concern is warranted or unwarranted.

2. Which of the characteristics of living things are possessed by an automobile? What bearing does this have on our definition of life?

3. About 1 out of every 120 live-born infants has a structural defect in the heart such as a hole between two heart chambers. Such infants often suffer pulmonary congestion and heart failure, and about one-third of them die as a result. Which of the major themes in this chapter does this illustrate? Explain your answer.

4. How might human anatomy be different today if the forerunners of humans had never inhabited the forest canopy?

5. Suppose you have been doing heavy yard work on a hot day and sweating profusely. You become very thirsty, so you drink a tall glass of lemonade. Explain how your thirst relates to the concept of homeostasis. Which type of feedback—positive or negative—does this illustrate?

Colorized chest X-ray showing lung damage from tuberculosis
SPL/Science Source

A

GENERAL ORIENTATION TO HUMAN ANATOMY

ATLAS OUTLINE

A.1 General Anatomical Terminology

   A.1a Anatomical Position

   A.1b Anatomical Planes

   A.1c Directional Terms

A.2 Major Body Regions

   A.2a Axial Region

   A.2b Appendicular Region

A.3 Body Cavities and Membranes

   A.3a Cranial Cavity and Vertebral Canal

   A.3b Thoracic Cavity

   A.3c Abdominopelvic Cavity

   A.3d Potential Spaces

A.4 Organ Systems

Study Guide

DEEPER INSIGHTS

A.1 Clinical Application: Cardiac Tamponade

A.2 Clinical Application: Peritonitis

Anatomy & Physiology *Revealed* 4.0

Module 1: Body Orientation

A.1   General Anatomical Terminology

A.1a Anatomical Position

In describing the human body, anatomists assume that it is in **anatomical position (fig. A.1)**—that of a person standing upright with the feet flat on the floor, arms at the sides, and the palms and face directed forward. Without such a frame of reference, to say that a structure such as the sternum, thyroid gland, or aorta is "above the heart" would be vague, since it would depend on whether the subject was standing, lying face down *(prone),* or lying face up *(supine).* From the perspective of anatomical position, however, we can describe the thyroid as *superior* to the heart, the sternum as *anterior* to it, and the aorta as *posterior* to it. These descriptions remain valid regardless of the subject's position. Even if the body is lying down, such as a cadaver on the medical student's dissection table, to say the sternum is anterior to the heart invites the viewer to imagine the body is standing in anatomical position and not to call it "above the heart" simply because that is the way the body happens to be lying.

Unless stated otherwise, assume that all anatomical descriptions refer to anatomical position. Bear in mind that if a subject is facing you, the subject's left will be on your right and vice versa.

In most anatomical illustrations, for example, the left atrium of the heart appears toward the right side of the page, and although the appendix is located in the right lower quadrant of the abdomen, it appears on the left side of most illustrations.

A.1b Anatomical Planes

Many views of the body are based on real or imaginary "slices" called sections or planes. *Section* implies an actual cut or slice to reveal internal anatomy, whereas *plane* implies an imaginary flat surface passing through the body. The three primary anatomical planes are *sagittal, frontal,* and *transverse* (fig. A.1).

A **sagittal**[1] **plane** (SADJ-ih-tul) passes vertically through the body or an organ and divides it into right and left portions **(fig. A.2a).** The sagittal plane that divides the body or organ into equal halves is also called the **median (midsagittal) plane.** The head and pelvic organs are commonly illustrated on the median plane. Other sagittal planes parallel to this (off center) divide the body into unequal portions. Such planes are sometimes called *parasagittal*[2] planes.

A **frontal (coronal) plane** also extends vertically, but it is perpendicular to the sagittal plane and divides the body into anterior (front) and posterior (back) portions **(fig. A.2b).** A frontal section

---

[1]*sagitta* = arrow
[2]*para* = next to

(a) Sagittal section   **(b) Frontal section**

(c) Transverse section

FIGURE A.2 Body Sections Cut Along the Three Primary Anatomical Planes. (a) Sagittal section of the pelvic region. (b) Frontal section of the thoracic region. (c) Transverse section of the head at the level of the eyes.

FIGURE A.1 Anatomical Position and the Three Primary Anatomical Planes.

TABLE A.1	Directional Terms in Human Anatomy A&PR	
Term	**Meaning**	**Examples of Usage**
Ventral Dorsal	Toward the front* or belly Toward the back or spine	The aorta is ventral to the vertebral column. The vertebral column is dorsal to the aorta.
Anterior Posterior	Toward the ventral side* Toward the dorsal side*	The sternum is anterior to the heart. The esophagus is posterior to the trachea.
Cephalic Rostral Caudal	Toward the head or superior end Toward the forehead or nose Toward the tail or inferior end	The brain develops from the cephalic end of the neural tube. The forebrain is rostral to the brainstem. The spinal cord is caudal to the brain.
Superior Inferior	Above Below	The heart is superior to the diaphragm. The liver is inferior to the diaphragm.
Medial Lateral	Toward the median plane Away from the median plane	The heart is medial to the lungs. The eyes are lateral to the nose.
Proximal Distal	Closer to the point of attachment or origin Farther from the point of attachment or origin	The elbow is proximal to the wrist. The fingernails are at the distal ends of the fingers.
Ipsilateral Contralateral	On the same side of the body (right or left) On opposite sides of the body (right and left)	The liver is ipsilateral to the appendix. The spleen is contralateral to the liver.
Superficial Deep	Closer to the body surface Farther from the body surface	The skin is superficial to the muscles. The bones are deep to the muscles.

*In humans only; definition differs for other animals.

of the head, for example, would divide it into one portion bearing the face and another bearing the back of the head. Contents of the thoracic and abdominal cavities are most commonly shown as frontal sections.

A **transverse (horizontal) plane** passes across the body or an organ perpendicular to its long axis; it divides the body or organ into superior (upper) and inferior (lower) portions **(fig. A.2c).** CT scans are typically transverse sections (see fig. 1.11c).

A.1c Directional Terms

Words that describe the location of one structure relative to another are called the **directional terms** of anatomy. **Table A.1** summarizes those most frequently used. Most of these terms exist in pairs with opposite meanings: *anterior* versus *posterior, rostral* versus *caudal, superior* versus *inferior, medial* versus *lateral, proximal* versus *distal, ipsilateral* versus *contralateral,* and *superficial* versus *deep.* Intermediate directions are often indicated by combinations of these terms. For example, one's cheeks may be described as *inferolateral* to the eyes (below and to the side).

The terms *proximal* and *distal* are used especially in the anatomy of the limbs, with ***proximal*** used to denote something relatively close to the limb's point of attachment (the shoulder or hip) and ***distal*** to denote something farther away. These terms do have some applications to anatomy of the trunk, however—for example, in referring to certain aspects of the intestines and microscopic anatomy of the kidneys. But when describing the trunk and referring to a structure that lies above or below another, ***superior*** and ***inferior*** are the preferred terms. These terms are not usually used for the limbs. Although it may be technically correct, one would not generally say that the elbow is superior to the wrist, but proximal to it.

Because of the bipedal, upright stance of humans, some directional terms have different meanings for humans than they do for other animals. ***Anterior,*** for example, denotes the region of the body that leads the way in normal locomotion. For a four-legged animal such as a cat, this is the head end of the body; for a human, however, it is the front of the chest and abdomen. Thus, *anterior* has the same meaning as *ventral* for a human but not for a cat. ***Posterior*** denotes the region of the body that comes last in normal locomotion—the tail end of a cat but the dorsal side (back) of a human. In the anatomy of most other animals, *ventral* denotes the surface of the body closest to the ground and *dorsal* denotes the surface farthest away from the ground. These two words are too entrenched in human anatomy to completely ignore them, but we will minimize their use in this book to avoid confusion. You must keep such differences in mind, however, when dissecting other animals for comparison to human anatomy.

One vestige of the term *dorsal* is ***dorsum,*** used to denote the upper surface of the foot and the back of the hand. If you consider how a cat stands, the corresponding surfaces of its paws are uppermost, facing the same direction as the dorsal side of its trunk. Although these surfaces of the human hand and foot face entirely different directions in anatomical position, the term *dorsum* is still used.

A.2   Major Body Regions

Knowledge of the external anatomy and landmarks of the body is important in performing a physical examination and many other clinical procedures. For purposes of study, the body is divided into two major regions called the *axial* and *appendicular* regions. Smaller areas within the major regions are described in the following paragraphs and illustrated in **figure A.3.**

Cephalic
Facial
Cervical
Thoracic:
Sternal
Pectoral
Umbilical
Abdominal
Inguinal
Pubic
Femoral
Crural
Tarsal
Plantar

Acromial
Axillary
Brachial
Cubital
Antebrachial
Carpal
Palmar
Coxal
Patellar

(a) Anterior (ventral)

(b) Anterior (ventral)

Cranial
Nuchal
Interscapular
Scapular
Vertebral
Lumbar
Sacral
Gluteal
Perineal
Femoral
Popliteal
Crural
Tarsal
Calcaneal

(c) Posterior (dorsal)

(d) Posterior (dorsal)

FIGURE A.3 The Adult Female and Male Body Regions. (a) Female, anterior. (b) Male, anterior. (c) Female, posterior. (d) Male, posterior.

a–d: Joe DeGrandis/McGraw-Hill Education

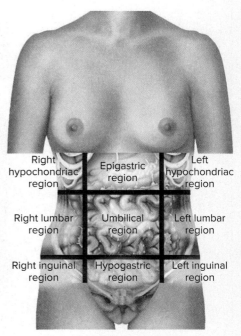

FIGURE A.4 The Four Quadrants and Nine Regions of the Abdomen. (a) External division into four quadrants. (b) External division into nine regions. **A&PR**

❓ *In what quadrant would the pain of appendicitis usually be felt?*

(a) Abdominopelvic quadrants

(b) Abdominopelvic regions

A.2a Axial Region

The **axial region** consists of the **head, neck (cervical[3] region)**, and **trunk.** The trunk is further divided into the **thoracic region** above the diaphragm and the **abdominal region** below it.

One way of referring to the locations of abdominal structures is to divide the region into quadrants. Two perpendicular lines intersecting at the umbilicus (navel) divide the abdomen into a **right upper quadrant (RUQ), right lower quadrant (RLQ), left upper quadrant (LUQ),** and **left lower quadrant (LLQ) (fig. A.4a).** The quadrant scheme is often used to describe the site of an abdominal pain or abnormality.

The abdomen also can be divided into nine regions defined by four lines that intersect like a tic-tac-toe grid **(fig. A.4b).** Each vertical line is called a *midclavicular line* because it passes through the midpoint of the clavicle (collarbone). The superior horizontal line is called the *subcostal[4] line* because it connects the inferior borders of the lowest costal cartilages (cartilage connecting the tenth rib on each side to the inferior end of the sternum). The inferior horizontal line is called the *intertubercular[5] line* because it passes from left to right between the tubercles *(anterior superior spines)* of the pelvis—two points of bone located about where the front pockets open on most pants. The three lateral regions of this grid, from upper to lower, are the **hypochondriac,[6] lumbar,** and **inguinal[7] (iliac) regions.** The three medial regions from upper to lower are the **epigastric,[8] umbilical,** and **hypogastric (pubic) regions.**

A.2b Appendicular Region

The **appendicular region** (AP-en-DIC-you-lur) of the body consists of the **upper** and **lower limbs** (also called *appendages* or *extremities*). The upper limb includes the **arm (brachial region)** (BRAY-kee-ul), **forearm (antebrachial[9] region)** (AN-teh-BRAY-kee-ul), **wrist (carpal region), hand,** and **fingers (digits).** The lower limb includes the **thigh (femoral region), leg (crural region)** (CROO-rul), **ankle (tarsal region), foot,** and **toes (digits).** In strict anatomical terms, *arm* refers only to that part of the upper limb between the shoulder and elbow. *Leg* refers only to that part of the lower limb between the knee and ankle.

A **segment** of a limb is a region between one joint and the next. The arm, for example, is the segment between the shoulder and elbow joints, and the forearm is the segment between the elbow and wrist joints. Flexing your fingers, you can easily see that your thumb has two segments (proximal and distal), whereas the other four digits have three segments (proximal, middle, and distal). The segment concept is especially useful in describing the locations of bones and muscles and the movements of the joints.

A.3 Body Cavities and Membranes

The body wall encloses multiple **body cavities (fig. A.5, table A.2),** each lined with a membrane and containing internal organs called **viscera** (VISS-er-uh) (singular, *viscus[10]*). Some of these membranes are two-layered, having one layer against the organ surface (such as the heart or lung) and one layer against a surrounding structure (forming, for example, the inner lining of the rib cage); there

---

[3]*cervic* = neck
[4]*sub* = below; *cost* = rib
[5]*inter* = between; *tubercul* = little swelling
[6]*hypo* = below; *chondr* = cartilage
[7]*inguin* = groin
[8]*epi* = above, over; *gastr* = stomach

[9]*ante* = fore, before; *brachi* = arm
[10]*viscus* = body organ

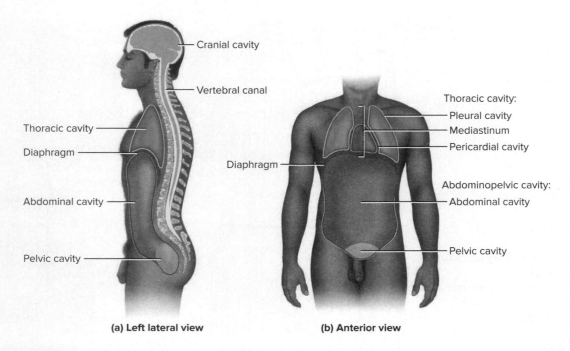

(a) Left lateral view   (b) Anterior view

FIGURE A.5 The Major Body Cavities. (a) Left lateral view. (b) Anterior view.

TABLE A.2	Body Cavities and Membranes **APR**	
Name of Cavity	**Associated Viscera**	**Membranous Lining**
Cranial cavity	Brain	Meninges
Vertebral canal	Spinal cord	Meninges
Thoracic cavity		
Pleural cavities (2)	Lungs	Pleurae
Pericardial cavity	Heart	Pericardium
Abdominopelvic cavity		
Abdominal cavity	Digestive organs, spleen, kidneys	Peritoneum
Pelvic cavity	Bladder, rectum, reproductive organs	Peritoneum

is only a thin film of liquid between them. In such cases, the inner layer, against the organ, is called the **visceral layer** (VISS-er-ul) of the membrane, and the more superficial or outer one, the **parietal**[11] **layer** (pa-RY-eh-tul).

A.3a Cranial Cavity and Vertebral Canal

The **cranial cavity** is enclosed by the cranium (braincase) and contains the brain. The **vertebral canal** is enclosed by the vertebral column (spine) and contains the spinal cord. The two are continuous with each other and are lined by three membrane layers called the **meninges** (meh-NIN-jeez). Among other functions, the meninges protect the delicate nervous tissue from the hard protective bone that encloses it.

A.3b Thoracic Cavity

The trunk of your body contains two major spaces, the thoracic cavity and abdominopelvic cavity, separated by a transverse muscular sheet, the **diaphragm.** Superior to the diaphragm, in your chest, is the **thoracic cavity,** and inferior to it, in your abdomen, is the **abdominopelvic cavity.** Both cavities are lined with thin **serous membranes,** which secrete a lubricating film of moisture similar to blood serum (hence their name).

The thoracic cavity is divided by a thick median wall called the **mediastinum**[12] (ME-dee-ah-STY-num) **(fig. A.5b).** This is the region between the lungs, extending from the base of the neck to the diaphragm. It is occupied by the heart, the major blood vessels connected to it, the esophagus, the trachea and bronchi, and a gland called the thymus.

The heart is enfolded in a two-layered membrane called the **pericardium.**[13] The inner layer of the pericardium forms the surface of the heart itself and is called the **visceral layer.** The outer layer is called the **parietal layer.** These layers are separated by a space called the **pericardial cavity (fig. A.6a),** which is lubricated by **pericardial fluid.** This space allows the heart freedom of movement during its contraction and relaxation, but can pose a life-threatening problem if it fills with serous fluid or blood (see Deeper Insight A.1).

[11]*pariet* = wall

[12]*mediastinum* = in the middle
[13]*peri* = around; *cardi* = heart

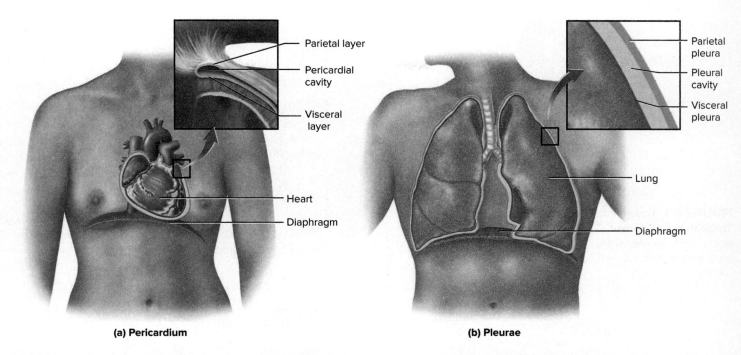

(a) Pericardium

(b) Pleurae

FIGURE A.6 Parietal and Visceral Layers of Double-Walled Membranes. (a) Relationship of the pericardium to the heart. (b) Relationship of the pleurae to the lungs.

DEEPER INSIGHT A.1

CLINICAL APPLICATION

Cardiac Tamponade

Being confined by the pericardium can cause a problem for the heart under some circumstances. If a heart wall weakened by disease should rupture, or if it suffers a penetrating injury such as a knife or gunshot wound, blood spurts from the heart into the pericardial cavity, filling the cavity more and more with each heartbeat. Diseased hearts also sometimes seep serous fluid into the pericardial sac. Either way, the effect is the same: The pericardial sac has little room to expand, so the accumulating fluid puts pressure on the heart, squeezing it and preventing it from refilling between beats. This condition is called *cardiac tamponade*. If the heart chambers cannot refill, then cardiac output declines and a person may die of catastrophic circulatory failure. A similar situation occurs if serous fluid or air accumulates in the pleural cavity, causing collapse of a lung.

The right and left sides of the thoracic cavity contain the lungs. Each lung is enfolded by a serous membrane called the **pleura**[14] (PLOOR-uh) **(fig. A.6b).** Like the pericardium, the pleura has visceral (inner) and parietal (outer) layers. The **visceral pleura** forms the external surface of the lung, and the **parietal pleura** lines the inside of the rib cage. The narrow space between them is called the **pleural cavity** (see fig. B.11 in atlas B, following chapter 10). It is lubricated by slippery **pleural fluid.**

Note that in both the pericardium and the pleura, the visceral layer of the membrane *covers* an organ surface and the parietal layer *lines* the inside of a body cavity. We will see this pattern elsewhere, including the abdominopelvic cavity.

A.3c Abdominopelvic Cavity

The abdominopelvic cavity consists of the **abdominal cavity** superiorly and the **pelvic cavity** inferiorly. The abdominal cavity contains most of the digestive organs as well as the spleen, kidneys, and ureters. It extends inferiorly to the level of a bony landmark called the *brim* of the pelvis (see figs. B.7 and 8.36). The pelvic cavity, below the brim, is continuous with the abdominal cavity (no wall separates them), but it is markedly narrower and tilts posteriorly (see fig. A.5a). It contains the rectum, urinary bladder, urethra, and reproductive organs.

The abdominopelvic cavity contains a two-layered serous membrane called the **peritoneum**[15] (PERR-ih-toe-NEE-um). Its outer layer, the **parietal peritoneum,** lines the cavity wall. Along the posterior midline, it turns inward and becomes another layer, the **visceral peritoneum,** suspending certain abdominal viscera from the body wall, covering their outer surfaces, and holding them in place. The **peritoneal cavity** is the space between the parietal and visceral layers. It is lubricated by **peritoneal fluid.**

Some organs of the abdominal cavity lie against the posterior body wall and are covered by peritoneum only on the side facing the peritoneal cavity. They are said to have a **retroperitoneal**[16] position **(fig. A.7).** These include the kidneys, ureters, adrenal glands, most of the pancreas, and abdominal portions of two major blood vessels—the aorta and inferior vena cava (see fig. B.6). Organs that are encircled by peritoneum and connected to the posterior body wall by peritoneal sheets are described as **intraperitoneal.**[17]

The visceral peritoneum is also called a **mesentery**[18] (MESS-en-tare-ee) at points where it forms a translucent, membranous

[14]*pleur* = rib, side

[15]*peri* = around; *tone* = stretched
[16]*retro* = behind
[17]*intra* = within
[18]*mes* = in the middle; *enter* = intestine

2nd lumbar vertebra
Kidney
Liver
Renal vein and artery
Inferior vena cava
Aorta
Intestine

Posterior

Back muscles
Spinal cord
Fat
Posterior mesentery
Parietal peritoneum
Visceral peritoneum (serosa)
Peritoneal cavity
Omentum or other anterior mesentery

Anterior

FIGURE A.7 Transverse Section Through the Abdomen. Shows the peritoneum, peritoneal cavity (with most viscera omitted), and some retroperitoneal organs.

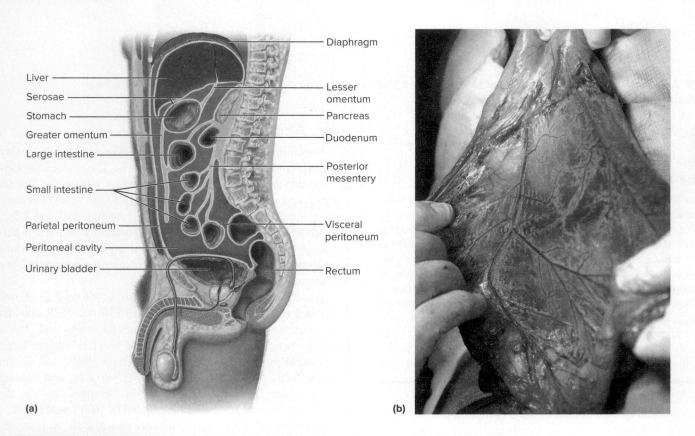

Liver
Serosae
Stomach
Greater omentum
Large intestine
Small intestine
Parietal peritoneum
Peritoneal cavity
Urinary bladder

Diaphragm
Lesser omentum
Pancreas
Duodenum
Posterior mesentery
Visceral peritoneum
Rectum

(a)

(b)

FIGURE A.8 Serous Membranes of the Abdominal Cavity. (a) Sagittal section, left lateral view. (b) Photo of the mesentery of the small intestine. Mesenteries contain blood vessels, lymphatic vessels, and nerves supplying the viscera. **APR**

? *Is the urinary bladder in the peritoneal cavity?*

b: MedicImage/Universal Images Group/Getty Images

curtain suspending and anchoring the viscera **(fig. A.8),** and a **serosa** (seer-OH-sa) at points where it enfolds and covers the outer surfaces of organs such as the stomach and small intestine. The intestines are suspended from the posterior (dorsal) abdominal wall by the **posterior mesentery.** The posterior mesentery of the large intestine is called the **mesocolon.** In some places, after wrapping around the intestines or other viscera, the mesentery continues toward the anterior body wall as the **anterior mesentery.** The most

significant example of this is a fatty membrane called the **greater omentum,**[19] which hangs like an apron from the inferolateral margin of the stomach and overlies the intestines (figs. A.8a and B.4). The greater omentum is unattached at its inferior border and can be lifted to reveal the intestines. A smaller **lesser omentum** extends from the superomedial margin of the stomach to the liver.

---

[19]*omentum* = covering

DEEPER INSIGHT A.2

CLINICAL APPLICATION

Peritonitis

Peritonitis is inflammation of the peritoneum. It is a critical, life-threatening condition necessitating prompt treatment. The most serious cause of peritonitis is a perforation in the digestive tract, such as a ruptured appendix or a gunshot wound. Digestive juices cause immediate chemical inflammation of the peritoneum, followed by microbial inflammation as intestinal bacteria invade the body cavity. Anything that perforates the abdominal wall can also lead to peritonitis, such as abdominal trauma or surgery. So, too, can free blood in the abdominal cavity, as from a ruptured aneurysm (a weak point in a blood vessel) or ectopic pregnancy (implantation of an embryo anywhere other than the uterus); blood itself is a chemical irritant to the peritoneum. Peritonitis tends to shift fluid from the circulation into the abdominal cavity. Death can follow within a few days from severe electrolyte imbalance, respiratory distress, kidney failure, and widespread blood clotting called *disseminated intravascular coagulation.*

A.3d Potential Spaces

Some of the spaces between body membranes are considered to be **potential spaces,** so named because under normal conditions, the membranes are pressed firmly together and there is no actual space between them. The membranes are not physically attached, however, and under unusual conditions, they may separate and create a space filled with fluid or other matter. Thus there is normally no actual space, but only a potential for membranes to separate and create one.

The pleural cavity is one example. Normally the parietal and visceral pleurae are pressed together without a gap between them, but under pathological conditions, air or serous fluid can accumulate between the membranes and open up a space. The internal cavity (**lumen**) of the uterus is another. In a nonpregnant uterus, the mucous membranes of opposite walls are pressed together so that there is no open space in the organ. In pregnancy, of course, a growing fetus occupies this space and pushes the mucous membranes apart.

A.4 Organ Systems

The human body has 11 **organ systems (fig. A.9)** and an immune system, which is better described as a population of cells that inhabit multiple organs rather than as an organ system. The organ systems are classified in the following list by their principal functions, but this is an unavoidably flawed classification. Some organs belong to two or more systems—for example, the male urethra is part of both the urinary and reproductive systems; the pharynx is part of the respiratory and digestive systems; and the mammary glands can be considered part of the integumentary and female reproductive systems. The organ systems are as follows:

Systems of protection, support, and movement

Integumentary system

Skeletal system

Muscular system

Systems of internal communication and control

Nervous system

Endocrine system

Systems of fluid transport

Circulatory system

Lymphatic system

Systems of intake and output

Respiratory system

Urinary system

Digestive system

Systems of reproduction

Male reproductive system

Female reproductive system

Some medical terms combine the names of two systems—for example, the *musculoskeletal system, cardiopulmonary system,* and *urogenital (genitourinary) system.* These terms serve to call attention to the close anatomical or physiological relationships between two systems, but these are not literally individual organ systems.

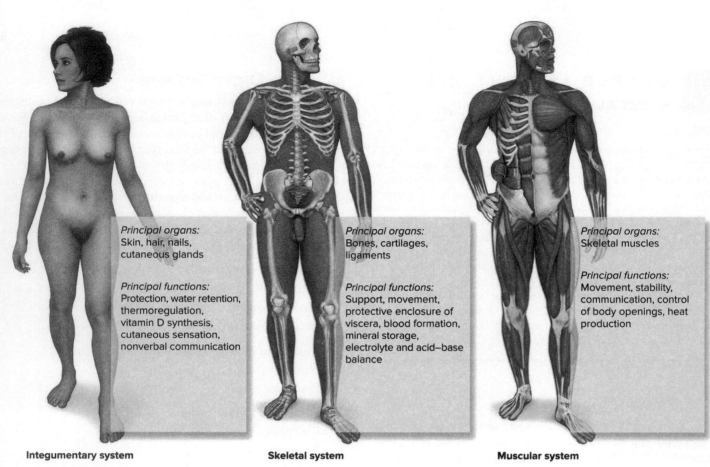

Integumentary system

Principal organs:
Skin, hair, nails,
cutaneous glands

Principal functions:
Protection, water retention,
thermoregulation,
vitamin D synthesis,
cutaneous sensation,
nonverbal communication

Skeletal system

Principal organs:
Bones, cartilages,
ligaments

Principal functions:
Support, movement,
protective enclosure of
viscera, blood formation,
mineral storage,
electrolyte and acid–base
balance

Muscular system

Principal organs:
Skeletal muscles

Principal functions:
Movement, stability,
communication, control
of body openings, heat
production

Lymphatic system

Principal organs:
Lymph nodes,
lymphatic vessels,
thymus, spleen, tonsils

Principal functions:
Recovery of excess
tissue fluid, detection of
pathogens, production
of immune cells, defense
against disease

Respiratory system

Principal organs:
Nose, pharynx, larynx,
trachea, bronchi, lungs

Principal functions:
Absorption of oxygen,
discharge of carbon
dioxide, acid–base
balance, speech

Urinary system

Principal organs:
Kidneys, ureters, urinary
bladder, urethra

Principal functions:
Elimination of wastes;
regulation of blood
volume and pressure;
stimulation of red blood
cell formation; control
of fluid, electrolyte,
and acid–base balance;
detoxification

FIGURE A.9 The Human Organ Systems.

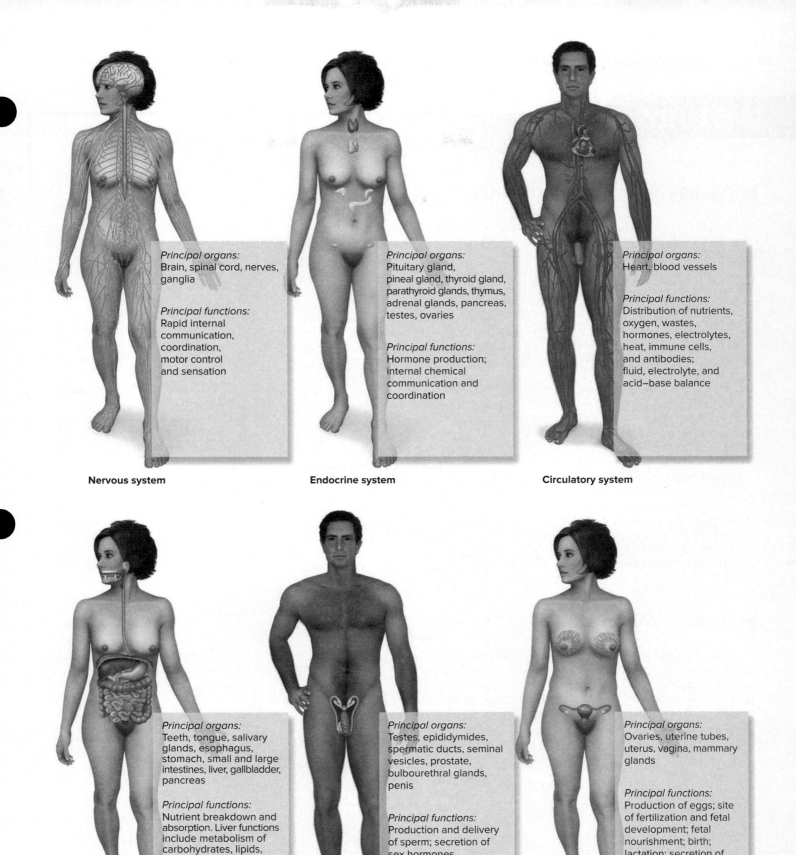

Nervous system

Principal organs:
Brain, spinal cord, nerves, ganglia

Principal functions:
Rapid internal communication, coordination, motor control and sensation

Endocrine system

Principal organs:
Pituitary gland, pineal gland, thyroid gland, parathyroid glands, thymus, adrenal glands, pancreas, testes, ovaries

Principal functions:
Hormone production; internal chemical communication and coordination

Circulatory system

Principal organs:
Heart, blood vessels

Principal functions:
Distribution of nutrients, oxygen, wastes, hormones, electrolytes, heat, immune cells, and antibodies; fluid, electrolyte, and acid–base balance

Digestive system

Principal organs:
Teeth, tongue, salivary glands, esophagus, stomach, small and large intestines, liver, gallbladder, pancreas

Principal functions:
Nutrient breakdown and absorption. Liver functions include metabolism of carbohydrates, lipids, proteins, vitamins, and minerals; synthesis of plasma proteins; disposal of drugs, toxins, and hormones; and cleansing of blood.

Male reproductive system

Principal organs:
Testes, epididymides, spermatic ducts, seminal vesicles, prostate, bulbourethral glands, penis

Principal functions:
Production and delivery of sperm; secretion of sex hormones

Female reproductive system

Principal organs:
Ovaries, uterine tubes, uterus, vagina, mammary glands

Principal functions:
Production of eggs; site of fertilization and fetal development; fetal nourishment; birth; lactation; secretion of sex hormones

FIGURE A.9 The Human Organ Systems (continued).

STUDY GUIDE

▶ Assess Your Learning Outcomes

To test your knowledge, discuss the following topics with a study partner or in writing, ideally from memory.

A.1 General Anatomical Terminology

1. Anatomical position and why it is important for anatomical description
2. Directions along which the body or an organ is divided by the sagittal, frontal, and transverse planes; how the median plane differs from other sagittal planes
3. Meanings of each of the following pairs or groups of terms, and the ability to describe the relative locations of two body parts using these terms: *ventral* and *dorsal; anterior* and *posterior; cephalic, rostral,* and *caudal; superior* and *inferior; medial* and *lateral; proximal* and *distal; superficial* and *deep*
4. Why the terms *ventral* and *dorsal* are ambiguous in human anatomy but less so in most other animals; what terms are used in their place in human anatomy; and reasons why they are occasionally appropriate or unavoidable in human anatomy

A.2 Major Body Regions

1. Distinctions between the axial and appendicular regions of the body
2. Subdivisions of the axial region and landmarks that divide and define them
3. The abdomen's four quadrants and nine regions; their defining landmarks; and why this scheme is clinically useful
4. The segments of the upper and lower limbs; how the anatomical meanings of *arm* and *leg* differ from the colloquial meanings

A.3 Body Cavities and Membranes

1. Locations and contents of the cranial cavity, vertebral canal, thoracic cavity, and abdominopelvic cavity; the membranes that line them; and the main viscera contained in each
2. Contents of the mediastinum and its relationship to the thoracic cavity as a whole
3. The pericardium, its two layers, the space and fluid between the layers, and its function
4. The pleurae, their two layers, the space and fluid between the layers, and their function

5. The two subdivisions of the abdominopelvic cavity and the skeletal landmark that divides them
6. The peritoneum; its functions; its two layers and their relationship to the abdominal viscera; and the peritoneal fluid
7. Mesenteries and serosae
8. Intraperitoneal versus retroperitoneal organs, examples of both, and how one would identify an organ as being intra- or retroperitoneal
9. Names and locations of the posterior and anterior mesenteries
10. The serosa of an abdominopelvic organ and how it relates to the peritoneum
11. Examples of potential spaces and why they are so named

A.4 Organ Systems

1. The 11 organ systems, the functions of each, and the principal organs of each system

▶ Testing Your Recall

Answers in Appendix A

1. Which of the following is *not* an essential part of anatomical position?
   a. feet together
   b. feet flat on the floor
   c. palms forward
   d. mouth closed
   e. arms down to the sides

2. A ring-shaped section of the small intestine would be a _____ section.
   a. sagittal
   b. coronal
   c. transverse
   d. frontal
   e. median

3. The tarsal region is _____ to the popliteal region.
   a. medial
   b. superficial
   c. superior
   d. dorsal
   e. distal

4. The greater omentum is _____ to the small intestine.
   a. posterior
   b. parietal
   c. deep
   d. superficial
   e. proximal

5. A _____ plane passes through the sternum, umbilicus, and mons pubis.
   a. central
   b. proximal
   c. midclavicular
   d. midsagittal
   e. intertubercular

6. The _____ region is immediately medial to the coxal region.
   a. inguinal
   b. hypochondriac
   c. umbilical
   d. popliteal
   e. cubital

7. Which of the following regions is not part of the upper limb?
   a. plantar
   b. carpal
   c. cubital
   d. brachial
   e. palmar

8. Which of these organs is within the peritoneal cavity?
   a. urinary bladder
   b. kidneys
   c. heart
   d. liver
   e. brain

9. In which area do you think pain from the gallbladder would be felt?
   a. umbilical region
   b. right upper quadrant
   c. hypogastric region
   d. left hypochondriac region
   e. left lower quadrant

STUDY GUIDE

10. Which organ system regulates blood volume, controls acid–base balance, and stimulates red blood cell production?
    a. digestive system
    b. lymphatic system
    c. nervous system
    d. urinary system
    e. circulatory system

11. The translucent membranes that suspend the intestines and hold them in place are called _____.

12. The superficial layer of the pleura is called the _____ pleura.

13. The right and left pleural cavities are separated by a thick wall called the _____.

14. The back of the neck is the _____ region.

15. The manual region is more commonly known as the _____ and the pedal region is more commonly known as the _____.

16. The cranial cavity is lined by membranes called the _____.

17. Organs that lie within the abdominal cavity but not within the peritoneal cavity are said to have a _____ position.

18. The sternal region is _____ to the pectoral region.

19. The pelvic cavity can be described as _____ to the abdominal cavity in position.

20. The anterior pit of the elbow is the _____ region, and the corresponding (but posterior) pit of the knee is the _____ region.

▶ Building Your Medical Vocabulary

Answers in Appendix A

State a meaning of each word element, and give a medical term from this atlas that uses it or a slight variation of it.

1. ante-

2. cervico-

3. epi-

4. hypo-

5. inguino-

6. intra-

7. parieto-

8. peri-

9. retro-

10. sagitto-

▶ What's Wrong with These Statements?

Answers in Appendix A

Briefly explain why each of the following statements is false, or reword it to make it true.

1. Both lungs could be shown in one sagittal section of the body.

2. A single frontal section of the head cannot include both eyes.

3. The knee is distal to the tarsal region.

4. The diaphragm is posterior to the lungs.

5. The esophagus is inferior to the stomach.

6. The liver is in the lumbar region.

7. The heart is in the space between the parietal and visceral pericardium, called the pericardial cavity.

8. The kidneys are in the peritoneal cavity of the abdomen.

9. The peritoneum lines the inside of the stomach and intestines.

10. The sigmoid colon is in the lower right quadrant of the abdomen.

▶ Testing Your Comprehension

1. Identify which anatomical plane—sagittal, frontal, or transverse—is the only one that could *not* show (a) both the brain and tongue, (b) both eyes, (c) both the hypogastric and gluteal regions, (d) both kidneys, (e) both the sternum and vertebral column, and (f) both the heart and uterus.

2. Laypeople often misunderstand anatomical terminology. What do you think people really mean when they say they have "planter's warts"?

3. Name one structure or anatomical feature that could be found in each of the following locations relative to the ribs: medial, lateral, superior, inferior, deep, superficial, posterior, and anterior. Try not to use the same example twice.

4. Based on the illustrations in this atlas, identify an internal organ that is (a) in the upper left quadrant and retroperitoneal, (b) in the lower right quadrant of the peritoneal cavity, (c) in the hypogastric region, (d) in the right hypochondriac region, and (e) in the pectoral region.

5. Why do you think people with imaginary illnesses came to be called hypochondriacs?

CHAPTER 2

THE CHEMISTRY OF LIFE

Coronavirus protein seen by cryo-electron microscopy (cryo-EM), using the electron microscope and a supercomputer to image supercooled proteins and reveal structure even to the level of individual atoms

NCI/Science Source

CHAPTER OUTLINE

2.1 Atoms, Ions, and Molecules

    2.1a The Chemical Elements

    2.1b Atomic Structure

    2.1c Isotopes and Radioactivity

    2.1d Ions, Electrolytes, and Free Radicals

    2.1e Molecules and Chemical Bonds

2.2 Water and Mixtures

    2.2a Water

    2.2b Solutions, Colloids, and Suspensions

    2.2c Acids, Bases, and pH

    2.2d Other Measures of Concentration

2.3 Energy and Chemical Reactions

    2.3a Energy and Work

    2.3b Classes of Chemical Reactions

    2.3c Reaction Rates

    2.3d Metabolism, Oxidation, and Reduction

2.4 Organic Compounds

    2.4a Carbon Compounds and Functional Groups

    2.4b Monomers and Polymers

    2.4c Carbohydrates

    2.4d Lipids

    2.4e Proteins

    2.4f Enzymes and Metabolism

    2.4g ATP, Other Nucleotides, and Nucleic Acids

Study Guide

DEEPER INSIGHTS

2.1 Medical History: Radiation and Madame Curie

2.2 Clinical Application: pH and Drug Action

2.3 Clinical Application: *Trans* Fats and Cardiovascular Health

2.4 Clinical Application: "Good" and "Bad" Cholesterol

2.5 Clinical Application: Blood Enzymes as Disease Markers

2.6 Clinical Application: Anabolic–Androgenic Steroids

Anatomy & Physiology *Revealed* 4.0

Module 2: Cells and Chemistry

Why is too much sodium or cholesterol harmful? Why does an iron deficiency cause anemia and an iodine deficiency cause a goiter? Why does a pH imbalance make some drugs less effective? Why do some pregnant women suffer convulsions after several days of vomiting? How can radiation cause cancer as well as cure it?

None of these questions can be answered, nor would the rest of this book be intelligible, without understanding the chemistry of life. A little knowledge of chemistry can help you choose a healthy diet, use medications more wisely, avoid worthless health fads and frauds, and explain treatments and procedures to your patients or clients. Thus, we begin our study of the human body with basic chemistry, the simplest level of the body's structural organization.

We will progress from general chemistry to **biochemistry**, study of the molecules that compose living organisms—especially molecules unique to living things, such as carbohydrates, fats, proteins, and nucleic acids. Most people have at least heard of these; it's common knowledge that we need proteins, fats, carbohydrates, vitamins, and minerals in our diet, and that we should avoid consuming too much saturated fat and cholesterol. But most people have only a vague concept of what these molecules are, much less how they function in the body. Such knowledge is very helpful in matters of personal fitness and patient education and is essential to the comprehension of the rest of this book.

2.1   Atoms, Ions, and Molecules

Expected Learning Outcomes

When you have completed this section, you should be able to

a. identify the elements of the body from their symbols;

b. distinguish between elements and compounds;

c. state the functions of minerals in the body;

d. explain the basis for radioactivity and the types and hazards of ionizing radiation;

e. distinguish between ions, electrolytes, and free radicals; and

f. define the types of chemical bonds.

2.1a The Chemical Elements

A chemical **element** is the simplest form of matter to have unique chemical properties. Water, for example, has unique properties, but it can be broken down into two elements, hydrogen and oxygen, that have unique chemical properties of their own. If we carry this process any further, however, we find that hydrogen and oxygen are made of protons, neutrons, and electrons—and none of these are unique. A proton of gold is identical to a proton of oxygen. Hydrogen and oxygen are the simplest chemically unique components of water and are thus elements.

Each element is identified by an *atomic number,* the number of protons in its nucleus. The atomic number of carbon is 6 and that of oxygen is 8, for example. The periodic table of the elements (see appendix C) arranges the elements in order by their atomic numbers. The elements are represented by one- or two-letter symbols, usually based on their English names: C for carbon, Mg for magnesium, Cl for chlorine, and so forth. A few symbols are based on Latin names, such as K for potassium *(kalium),* Na for sodium *(natrium),* and Fe for iron *(ferrum).*

There are 91 naturally occurring elements on earth, 24 of which play normal physiological roles in humans. **Table 2.1** groups these 24 according to their abundance in the body. Six of them account for 98.5% of the body's weight: oxygen, carbon, hydrogen, nitrogen, calcium, and phosphorus. The next 0.8% consists of another 6 elements: sulfur, potassium, sodium, chlorine, magnesium, and iron. The remaining 12 elements account for 0.7% of body weight, and no one of them accounts for more than 0.02%; thus, they are

TABLE 2.1	Elements of the Human Body		
Name and Symbol	**Percentage of Body Weight**	**Name and Symbol**	**Percentage of Body Weight**
Major Elements (Total 98.5%)			
Oxygen (O)	65.0	Nitrogen (N)	3.0
Carbon (C)	18.0	Calcium (Ca)	1.5
Hydrogen (H)	10.0	Phosphorus (P)	1.0
Lesser Elements (Total 0.8%)			
Sulfur (S)	0.25	Chlorine (Cl)	0.15
Potassium (K)	0.20	Magnesium (Mg)	0.05
Sodium (Na)	0.15	Iron (Fe)	0.006
Trace Elements (Total 0.7%) (Names and symbols only)			
Chromium (Cr)	Fluorine (F)	Molybdenum (Mo)	Tin (Sn)
Cobalt (Co)	Iodine (I)	Selenium (Se)	Vanadium (V)
Copper (Cu)	Manganese (Mn)	Silicon (Si)	Zinc (Zn)

known as **trace elements.** Despite their minute quantities, trace elements play vital roles in physiology. Other elements without natural physiological roles can contaminate the body and severely disrupt its functions, as in heavy-metal poisoning with lead or mercury.

Several of these elements are classified as **minerals**—inorganic elements extracted from the soil by plants and passed up the food chain to humans and other organisms. Minerals constitute about 4% of the human body by weight. Nearly three-quarters of this is Ca and P; the rest is mainly Cl, Mg, K, Na, and S. Minerals contribute significantly to body structure. The bones and teeth consist partly of crystals of calcium, phosphate, magnesium, fluoride, and sulfate ions. Many proteins include sulfur, and phosphorus is a major component of nucleic acids, ATP, and cell membranes. Minerals also enable enzymes and other organic molecules to function. Iodine is a component of thyroid hormone; iron is a component of hemoglobin; and some enzymes function only when manganese, zinc, copper, or other minerals are bound to them. The electrolytes needed for nerve and muscle function are mineral salts. The biological roles of minerals are discussed in more detail in chapters 24 and 26.

2.1b Atomic Structure

In the fifth century BCE, the Greek philosopher Democritus reasoned that we can cut matter such as a gold nugget into smaller and smaller pieces, but there must ultimately be particles so small that nothing could cut them. He called these imaginary particles **atoms**[1] ("indivisible"). Atoms were only a philosophical concept until 1803, when English chemist John Dalton began to develop an atomic theory based on experimental evidence. In 1913, Danish physicist Niels Bohr proposed a model of atomic structure similar to planets orbiting the sun **(fig. 2.1).** Although this *planetary model* is too simple to account for many of the properties of atoms, it remains useful for elementary purposes.

At the center of an atom is the *nucleus,* composed of protons and neutrons. **Protons (p$^+$)** have a single positive charge and **neutrons (n^0)** have no charge. Each proton or neutron weighs approximately 1 *atomic mass unit (amu).* The *atomic mass* of an element is approximately equal to its total number of protons and neutrons.

Around the nucleus are one or more clouds of **electrons (e$^-$),** tiny particles with a single negative charge and very low mass. It takes 1,836 electrons to equal the mass of one proton, so for most purposes we can disregard their mass. A person who weighs 64 kg (140 lb) contains less than 24 g (1 oz) of electrons. This hardly means that we can ignore electrons, however. They determine the chemical properties of an atom, thereby governing what molecules can exist and what chemical reactions can occur. The number of electrons equals the number of protons, so their charges cancel each other and an atom is electrically neutral.

Electrons swarm about the nucleus in concentric regions called *electron shells (energy levels).* The more energy an

Carbon (C)
6p$^+$, 6e$^-$, 6n^0
Atomic number = 6
Atomic mass = 12

Nitrogen (N)
7p$^+$, 7e$^-$, 7n^0
Atomic number = 7
Atomic mass = 14

Key

= Proton (p$^+$)

= Neutron (n^0)

= Electron (e$^-$)

Sodium (Na)
11p$^+$, 11e$^-$, 12n^0
Atomic number = 11
Atomic mass = 23

FIGURE 2.1 Bohr Planetary Models of Three Representative Elements. Note the filling of electron shells as atomic number increases.

electron has, the farther away from the nucleus its orbit lies. Each shell holds a limited number of electrons. The elements known to date have up to seven electron shells, but those ordinarily involved in human physiology do not exceed four. Electrons of the outermost shell, called **valence electrons,** determine the chemical bonding properties of an atom. Illustrations of atoms greatly understate the distances between their nuclei and electrons in order to fit the page. If you imagine the nucleus of an atom to be the size of a basketball, its nearest electron would be about 48 km (30 mi.) away. In other words, atoms (and your body) consist mostly of empty space!

2.1c Isotopes and Radioactivity

Dalton believed that every atom of an element was identical. We now know, however, that all elements have varieties called **isotopes,**[2] which differ from one another only in number of neutrons and therefore in atomic mass. Hydrogen atoms, for example, have only one proton. In the most common isotope, symbolized ^{1}H, that is all there is to the nucleus. Hydrogen has two other isotopes, however: *deuterium* (^{2}H) with one proton and one neutron, and *tritium* (^{3}H) with one proton and two neutrons **(fig. 2.3).** Over 99% of carbon atoms have an atomic mass of 12 (6p$^+$, 6n^0)

---

[1]*a* = not; *tom* = cut

[2]*iso* = same; *top* = place (same position in the periodic table)

DEEPER INSIGHT 2.1

MEDICAL HISTORY

Radiation and Madame Curie

In 1896, French scientist Henri Becquerel (1852–1908) discovered that uranium darkened photographic plates through several thick layers of paper. Marie Curie (1867–1934) and her husband Pierre Curie (1859–1906) discovered that polonium and radium did likewise. Marie Curie coined the term *radioactivity* for the emission of energy by these elements. Becquerel and the Curies shared a Nobel Prize in 1903 for this discovery.

Marie Curie **(fig. 2.2)** was not only the first woman in the world to receive a Nobel Prize but also the first woman in France to receive a Ph.D. She received a second Nobel Prize in 1911 for further work in radiation. Curie crusaded to train women for careers in science, and in World War I, she and her daughter, Irène Joliot-Curie (1897–1956), trained physicians in the use of X-ray machines. Marie pioneered radiation therapy for breast and uterine cancer.

In the wake of such discoveries, radium was regarded as a wonder drug. Unaware of its danger, people drank radium tonics and flocked to health spas to bathe in radium-enriched waters. Marie herself suffered extensive damage to her hands from handling radioactive minerals and died of radiation poisoning at age 67. The following year, Irène and her husband, Frédéric Joliot (1900–1958), were awarded a Nobel Prize for work in artificial radioactivity and synthetic radioisotopes. Apparently also a martyr to her science, Irène died of leukemia, possibly induced by radiation exposure.

FIGURE 2.2 Marie Curie (1867–1934). This portrait was made in 1911, when Curie received her second Nobel Prize.
Source: Library of Congress Prints and Photographs Division

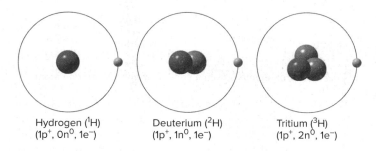

Hydrogen (^{1}H)
(1p$^+$, 0n^0, 1e$^-$)

Deuterium (^{2}H)
(1p$^+$, 1n^0, 1e$^-$)

Tritium (^{3}H)
(1p$^+$, 2n^0, 1e$^-$)

FIGURE 2.3 Isotopes of Hydrogen. The three isotopes differ only in the number of neutrons present.

and are called carbon-12 (^{12}C), but a small percentage of carbon atoms are ^{13}C, with seven neutrons, and ^{14}C, with eight. All isotopes of a given element behave the same chemically. Deuterium (^{2}H), for example, reacts with oxygen the same way ^{1}H does to produce water.

The *atomic weight (relative atomic mass)* of an element accounts for the fact that an element is a mixture of isotopes. If all carbon were ^{12}C, the atomic weight of carbon would be the same as its atomic mass, 12.000. But since a sample of carbon also contains small amounts of the heavier isotopes ^{13}C and ^{14}C, the atomic weight is slightly higher, 12.011.

Although different isotopes of an element exhibit identical chemical behavior, they differ in physical behavior. Many of

them are unstable and *decay* (break down) to more stable isotopes by giving off radiation. Unstable isotopes are therefore called **radioisotopes,** and the process of decay is called **radioactivity** (see Deeper Insight 2.1). Every element has at least one radioisotope. Oxygen, for example, has three stable isotopes and five radioisotopes. All of us contain radioisotopes such as ^{14}C and ^{40}K—that is, we are all mildly radioactive!

High-energy radiation, such as that emitted by radioisotopes, ejects electrons from atoms, converting atoms to ions; thus, it is called **ionizing radiation.** It destroys molecules and produces dangerous free radicals and ions in human tissues. In high doses, ionizing radiation is quickly fatal. In lower doses, it can be *mutagenic* (causing mutations in DNA) and *carcinogenic* (triggering cancer as a result of mutation).

Examples of ionizing radiation include ultraviolet rays, X-rays, and three kinds of radiation produced by nuclear decay: *alpha (α) particles, beta (β) particles,* and *gamma (γ) rays.* An alpha particle consists of two protons and two neutrons (equivalent to a helium nucleus), and a beta particle is a free electron. Alpha particles are too large to penetrate the skin, and beta particles penetrate only a few millimeters. They're relatively harmless when emitted by sources outside the body, but very dangerous when emitted by radioisotopes that have gotten into the body. Strontium-90 (^{90}Sr), for example, has been released by nuclear accidents and the atmospheric testing of nuclear weapons. It settles onto pastures and contaminates cow's milk. In the body, it behaves chemically like calcium, becoming incorporated into the bones, where it emits beta particles for years. Uranium and plutonium emit gamma rays. Because of their high penetrating power, these are very dangerous even when emitted by sources outside the body.

Each radioisotope has a characteristic **physical half-life,** the time required for 50% of its atoms to decay to a more stable isotope. One gram of ^{90}Sr, for example, would be half gone in 28 years. In 56 years, there would still be 0.25 g left, in 84 years 0.125 g, and so forth. Many radioisotopes are much longer-lived. The half-life of ^{40}K, for example, is 1.3 billion years. Nuclear power plants produce hundreds of radioisotopes that will be intensely radioactive for at least 10,000 years—longer than the life of any disposal container yet conceived.

The **biological half-life** of a radioisotope is the time required for half of it to disappear from the body. Some is lost by radioactive decay and even more by excretion from the body. Cesium-137, for example, has a physical half-life of 30 years but a biological half-life of only 17 days. Chemically, it behaves like potassium; it is quite mobile and rapidly excreted by the kidneys.

There are several ways to measure the intensity of ionizing radiation, the amount absorbed by the body, and its biological effects. To understand the units of measurement requires a grounding in physics beyond the scope of this book, but the standard international (SI) unit of radiation dosage is the *sievert*[3] *(Sv),* which takes into account the type and intensity of radiation and its biological effect. Doses of 5 Sv or more are usually fatal. The average person worldwide receives about 2.4 millisieverts (mSv) per year in *background radiation* from natural sources and another 0.6 mSv from artificial sources. The most significant natural source is *radon,* a gas produced by the decay of uranium in the earth; it can accumulate in buildings to unhealthy levels. Artificial sources of radiation exposure include medical X-rays, CT and PET scans, radiation therapy, and consumer products such as color televisions, smoke detectors, and luminous watch dials. A CT scan averages about 10 to 30 mSv for a full-body scan and 10 mSv for an abdominopelvic scan only; a typical mammogram is 0.4 to 0.6 mSv; a chest X-ray is about 0.1 mSv; and a hand or foot X-ray is about 0.001 mSv. Such voluntary exposure must be considered from the standpoint of its risk-to-benefit ratio. The benefits of a smoke detector or mammogram far outweigh the risk from the low levels of radiation involved. Radiation therapists and radiologists face a greater risk than their patients, however, and astronauts and airline flight crews receive more than average exposure. U.S. federal standards set a limit of 50 mSv/year as acceptable occupational exposure to ionizing radiation.

2.1d Ions, Electrolytes, and Free Radicals

Ions are charged particles with unequal numbers of protons and electrons. An ion can consist of a single atom with a positive or negative charge (such as Na^+ or Cl^-); a group of atoms such as

phosphate (PO_4^{3-}) and bicarbonate (HCO_3^-) ions; or a molecule as large as a protein with many charges on it.

Ions form because elements with one to three valence electrons tend to give them up, and those with four to seven electrons tend to gain more. If an atom of the first kind is exposed to an atom of the second, electrons may transfer from one to the other and turn both of them into ions. This process is called *ionization.* The particle that gains electrons acquires a negative charge and is called an **anion** (AN-eye-on). The one that loses electrons acquires a positive charge (because it then has a surplus of protons) and is called a **cation** (CAT-eye-on).

This happens, for example, when sodium and chlorine meet **(fig. 2.4).** One electron transfers from sodium to chlorine, producing a sodium ion with a unit positive charge (Na^+) and a chloride ion with a unit negative charge (Cl^-).

Some elements exist in two or more ionized forms. Iron, for example, has ferrous (Fe^{2+}) and ferric (Fe^{3+}) ions. Note that some ions have a single positive or negative charge, whereas others have

11 protons	
12 neutrons	Sodium
11 electrons	atom (Na)

17 protons	
18 neutrons	Chlorine
17 electrons	atom (Cl)

① Transfer of an electron from a sodium atom to a chlorine atom

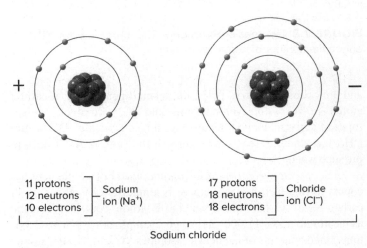

11 protons	
12 neutrons	Sodium
10 electrons	ion (Na⁺)

17 protons	
18 neutrons	Chloride
18 electrons	ion (Cl⁻)

Sodium chloride

② The charged sodium ion (Na^+) and chloride ion (Cl^-) that result

FIGURE 2.4 Ionization.

[3]Rolf Maximillian Sievert (1896–1966), Swedish radiologist

TABLE 2.2	Major Electrolytes and the Ions Released by their Dissociation	
Electrolyte		**Cations and Anions**
Calcium chloride ($CaCl_2$)	$\longrightarrow$	$Ca^{2+} + 2\ Cl^-$
Disodium phosphate (Na_2HPO_4)	$\longrightarrow$	$2\ Na^+ + HPO_4^{2-}$
Magnesium chloride ($MgCl_2$)	$\longrightarrow$	$Mg^{2+} + 2\ Cl^-$
Potassium chloride (KCl)	$\longrightarrow$	$K^+ + Cl^-$
Sodium bicarbonate ($NaHCO_3$)	$\longrightarrow$	$Na^+ + HCO_3^-$
Sodium chloride ($NaCl$)	$\longrightarrow$	$Na^+ + Cl^-$

charges of ±2 or ±3 because they gain or lose more than one electron. The charge on an ion is called its *valence*.

Ions with opposite charges are attracted to each other and tend to follow each other through the body. Thus, when Na^+ is excreted in the urine, Cl^- tends to follow it. The attraction of cations and anions to each other is especially important in the excitation of muscle and nerve cells, as we shall see in chapters 11 and 12.

Electrolytes are substances that ionize in water (acids, bases, or salts) and form solutions capable of conducting electricity **(table 2.2).** We can detect electrical activity of the muscles, heart, and brain with electrodes on the skin because electrolytes in the body fluids conduct electrical currents from these organs to the skin surface. Electrolytes are important for their chemical reactivity (as when calcium phosphate becomes incorporated into bone), osmotic effects (influence on water content and distribution in the body), and electrical effects (which are essential to nerve and muscle function). Electrolyte balance is one of the most important considerations in patient care (see section 24.2). Electrolyte imbalances have effects ranging from muscle cramps and brittle bones to coma and cardiac arrest.

Free radicals are unstable, highly reactive chemical particles with an odd number of electrons. For example, oxygen normally exists as a stable molecule composed of two oxygen atoms, O_2; but if an additional electron is added, it becomes a free radical called the *superoxide anion,* $O_2^-\bullet$. Free radicals are represented with a dot to symbolize the odd electron.

Free radicals are produced by some normal metabolic reactions of the body (such as the ATP-producing oxidation reactions in mitochondria, and a reaction that some white blood cells use to kill bacteria); by radiation (such as ultraviolet radiation and X-rays); and by chemicals (such as nitrites, used as preservatives in some wine, meat, and other foods). They are short-lived and combine quickly with molecules such as fats, proteins, and DNA, converting them into free radicals and triggering chain reactions that destroy still more molecules. Among the damages caused by free radicals are some forms of cancer and myocardial infarction, the death of heart tissue. One theory of aging is that it results in part from lifelong cellular damage by free radicals.

Because free radicals are so common and destructive, we have mechanisms for neutralizing them. An **antioxidant** is a chemical

that neutralizes free radicals. For example, the body produces an enzyme called *superoxide dismutase (SOD)* that converts superoxide into oxygen and hydrogen peroxide. Selenium, vitamin E (α-tocopherol), vitamin C (ascorbic acid), and carotenoids (such as β-carotene) are some antioxidants obtained from the diet. Dietary deficiencies of antioxidants have been associated with increased risk of heart attacks, sterility, muscular dystrophy, and other disorders.

2.1e Molecules and Chemical Bonds

Molecules are chemical particles composed of two or more atoms united by a chemical bond. The atoms may be identical, as in nitrogen (N_2), or different, as in glucose ($C_6H_{12}O_6$). Molecules composed of two or more elements are called **compounds.** Oxygen (O_2) and carbon dioxide (CO_2) are both molecules, because they consist of at least two atoms; but only CO_2 is a compound, because it has atoms of two different elements.

Molecules are represented by *molecular formulae* that identify their elements and show how many atoms of each are present. Molecules with identical molecular formulae but different arrangements of their atoms are called **isomers**[4] of each other. For example, both ethanol and ethyl ether have the molecular formula C_2H_6O, but they are certainly not interchangeable! To show the difference between them, we use *structural formulae* that show the location of each atom **(fig. 2.5).**

The **molecular weight (MW)** of a compound is the sum of the atomic weights of its atoms. Rounding to whole numbers, we can calculate the approximate MW of glucose ($C_6H_{12}O_6$), for example:

6 C atoms	×	12 amu each	=	72 amu
12 H atoms	×	1 amu each	=	12 amu
6 O atoms	×	16 amu each	=	96 amu
Molecular weight (MW)			=	180 amu

Molecular weight is needed to compute some measures of concentration discussed later (see section 2.2d and appendix B).

[4]*iso* = same; *mer* = part

	Structural formulae	Condensed structural formulae	Molecular formulae
Ethanol	H—C—C—OH	CH_3CH_2OH	C_2H_6O
Ethyl ether	H—C—O—C—H	CH_3OCH_3	C_2H_6O

FIGURE 2.5 Two Structural Isomers—Ethanol and Ethyl Ether. The molecular formulae are identical, but their structures, chemical properties, and physiological effects are different.

TABLE 2.3	Types of Chemical Bonds
Bond Type	**Definition and Remarks**
Ionic Bond	Relatively weak attraction between an anion and a cation. Easily disrupted in water, as when salt dissolves.
Covalent Bond	Sharing of one or more pairs of electrons between nuclei.
Single covalent	Sharing of one electron pair.
Double covalent	Sharing of two electron pairs. Often occurs between carbon atoms, between carbon and oxygen, and between carbon and nitrogen.
Nonpolar covalent	Covalent bond in which electrons are equally attracted to both nuclei. May be single or double. Strongest type of chemical bond.
Polar covalent	Covalent bond in which electrons are more attracted to one nucleus than to the other, resulting in slightly positive and negative regions in one molecule. May be single or double.
Hydrogen Bond	Weak attraction between polarized molecules or between polarized regions of the same molecule. Important in the three-dimensional folding and coiling of large molecules. Easily disrupted by temperature and pH changes.
Van der Waals Force	Weak, brief attraction due to random disturbances in the electron clouds of adjacent atoms. Weakest of all bonds individually, but can have strong effects collectively.

A molecule is held together, and molecules are attracted to one another, by forces called **chemical bonds.** The bonds of greatest physiological interest are *ionic bonds, covalent bonds, hydrogen bonds,* and *van der Waals forces* **(table 2.3).**

An **ionic bond** is the attraction of a cation to an anion. Sodium (Na^+) and chloride (Cl^-) ions, for example, are attracted to each other and form the compound sodium chloride ($NaCl$), common table salt. Ionic compounds can be composed of more than two ions, such as calcium chloride, $CaCl_2$. Ionic bonds are weak and easily dissociate (break up) in the presence of something more attractive, such as water. The ionic bonds of NaCl break down easily as salt dissolves in water, because both Na^+ and Cl^- are more attracted to water molecules than they are to each other.

▶▶▶APPLY WHAT YOU KNOW

Do you think ionic bonds are common in the human body? Explain your answer.

Covalent bonds form by the sharing of electrons. For example, two hydrogen atoms share valence electrons to form a hydrogen molecule, H_2 **(fig. 2.6a).** The two electrons, one donated by each atom, swarm around both nuclei in a dumbbell-shaped cloud. A *single covalent bond* is the sharing of a single pair of electrons. It is symbolized by a single line between atomic symbols, for example, H—H. A *double covalent bond* is the sharing of two pairs of electrons. In carbon dioxide, for example, a central carbon atom shares two electron pairs with each oxygen atom. Such bonds are symbolized by two lines—for example, O=C=O **(fig. 2.6b).**

When shared electrons spend approximately equal time around each nucleus, they form a *nonpolar covalent bond* **(fig. 2.7a),** the strongest of all chemical bonds. Carbon atoms bond to each other with nonpolar covalent bonds. If shared electrons

Hydrogen atom   Hydrogen atom   H—H
                                               Hydrogen molecule (H_2)

(a)

Oxygen atom   Carbon atom   Oxygen atom

O=C=O
Carbon dioxide molecule (CO_2)

(b)

FIGURE 2.6 Covalent Bonding. (a) Two hydrogen atoms share a single pair of electrons to form a hydrogen molecule. (b) A carbon dioxide molecule, in which a carbon atom shares two pairs of electrons with each oxygen atom, forming double covalent bonds.

spend significantly more time orbiting one nucleus than they do the other, they lend their negative charge to the region where they spend the most time, and they form a *polar covalent bond* **(fig. 2.7b).** When hydrogen bonds with oxygen, for example, the electrons are more attracted to the oxygen nucleus and orbit it more than they do the hydrogen. This makes the oxygen region

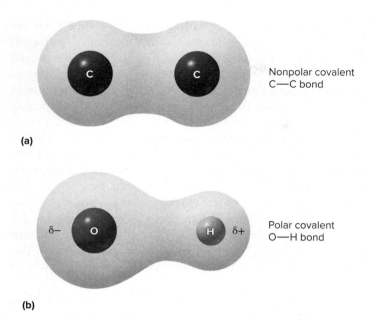

(a)

Nonpolar covalent
C—C bond

$\delta-$ O   H $\delta+$

(b)

Polar covalent
O—H bond

FIGURE 2.7 Nonpolar and Polar Covalent Bonds. (a) A nonpolar covalent bond between two carbon atoms, formed by electrons that spend an equal amount of time around each nucleus, as represented by the symmetric blue cloud. (b) A polar covalent bond, in which electrons orbit one nucleus significantly more than the other, as represented by the asymmetric cloud. This results in a slight negative charge ($\delta-$) in the region where the electrons spend most of their time, and a slight positive charge ($\delta+$) at the other pole.

Covalent bond

Hydrogen bond

Water molecule

FIGURE 2.8 Hydrogen Bonding of Water. The polar covalent bonds of water molecules enable each oxygen to form a hydrogen bond with a hydrogen of a neighboring molecule. Thus, the water molecules are weakly attracted to each other.

? *Why would this behavior raise the boiling point of water above that of a nonpolar liquid?*

of the molecule slightly negative and the hydrogen region slightly positive. The Greek delta (δ) is used to symbolize a charge less than that of one electron or proton. A slightly negative region of a molecule is represented $\delta-$ and a slightly positive region is represented $\delta+$.

A **hydrogen bond** is a weak attraction between a slightly positive hydrogen atom in one molecule and a slightly negative oxygen or nitrogen atom in another. Water molecules, for example, are weakly attracted to each other by hydrogen bonds **(fig. 2.8).** Hydrogen bonds also form between different regions of the same molecule, especially in very large molecules such as proteins and DNA. They cause such molecules to fold or coil into precise three-dimensional shapes. Hydrogen bonds are represented by dotted or broken lines between atoms: —C=O···H—N—. Hydrogen bonds are relatively weak, but they are enormously important to physiology.

Van der Waals[5] forces are important in protein folding, the binding of proteins to each other and to other molecules such as hormones, and the association of lipid molecules with each other in cell membranes. They are weak, brief attractions between neutral atoms. When electrons orbit a nucleus, they don't maintain a uniform distribution but show random fluctuations in density. If the electrons briefly crowd toward one side of an atom, they render that side slightly negative and the other side slightly positive for a

moment. If another atom is close enough to this one, the second atom responds with disturbances in its own electron cloud. Oppositely charged regions of the two atoms then attract each other for a very brief moment.

A single van der Waals force is only about 1% as strong as a covalent bond, but when two surfaces or large molecules meet, the van der Waals forces between large numbers of atoms can create a very strong attraction. This is how plastic wrap clings to food and dishes; flies and spiders walk across a ceiling; and even a heavy lizard or tree frog can walk up a windowpane.

BEFORE YOU GO ON

Answer the following questions to test your understanding of the preceding section:

1. Consider iron (Fe), hydrogen gas (H_2), and ammonia (NH_3). Which of them is or are atoms? Which of them is or are molecules? Which of them is or are compounds? Explain each answer.

2. Why is the biological half-life of a radioisotope shorter than its physical half-life?

3. Where do free radicals come from? What harm do they do? How is the body protected from free radicals?

[5]Johannes Diderik van der Waals (1837–1923), Dutch physicist

4. How does an ionic bond differ from a covalent bond?

5. What is a hydrogen bond? Why do hydrogen bonds depend on the existence of polar covalent bonds?

2.2   Water and Mixtures

Expected Learning Outcomes
When you have completed this section, you should be able to

a. define *mixture* and distinguish between mixtures and compounds;

b. describe the biologically important properties of water;

c. show how three kinds of mixtures differ from each other;

d. define *acid* and *base* and interpret the pH scale; and

e. discuss some ways in which the concentration of a solution can be expressed, and the kinds of information we can derive from the different units of measure.

Our body fluids are complex mixtures of chemicals. A **mixture** consists of substances that are physically blended but not chemically combined. Each substance retains its own chemical properties. To contrast a mixture with a compound, consider sodium chloride again. Sodium is a lightweight metal that bursts into flame if exposed to water, and chlorine is a yellow-green poisonous gas that was once used in chemical warfare. When these elements chemically react, they form common table salt. Clearly, the compound has properties much different from the properties of its elements. But if you were to put a little salt on your watermelon, the watermelon would taste salty and sweet because the sugar of the melon and the salt you added would merely form a mixture in which each compound retained its individual properties.

2.2a Water

Most mixtures in our bodies consist of chemicals dissolved or suspended in water. Water constitutes 50% to 75% of your body weight, depending on age, sex, fat content, and other factors. Its structure, simple as it is, has profound biological effects. Two aspects of its structure are particularly important: (1) Its atoms are joined by polar covalent bonds, and (2) the molecule is V-shaped, with a 105° bond angle **(fig. 2.9a).** This makes the molecule as a whole polar, because there is a slight negative charge (δ−) on the oxygen at the apex of the V and a slight positive charge (δ+) on each hydrogen. Like little magnets, water molecules are attracted to one another by hydrogen bonds (fig. 2.8). This gives water a set of properties that account for its ability to support life: *solvency, cohesion, adhesion, chemical reactivity,* and *thermal stability.*

Solvency is the ability to dissolve other chemicals. Water is called the *universal solvent* because it dissolves a broader range of substances than any other liquid. Substances that dissolve in water,

such as sugar, are said to be **hydrophilic**[6] (HY-dro-FILL-ic); the relatively few substances that do not, such as fats, are **hydrophobic**[7] (HY-dro-FOE-bic). Virtually all metabolic reactions depend on the solvency of water. Biological molecules must be dissolved in water to move freely, come together, and react. The solvency of water also makes it the body's primary means of transporting substances from place to place.

To be soluble in water, a molecule usually must be polarized or charged so that its charges can interact with those of water. When NaCl is dropped into water, for example, the ionic bonds between Na^+ and Cl^- are overpowered by the attraction of each ion to water molecules. Water molecules form a cluster, or *hydration sphere,* around each sodium ion with the $O^{\delta-}$ pole of each water molecule facing the sodium ion. They also form a hydration sphere around each chloride ion, with the $H^{\delta+}$ poles facing it. This isolates the sodium ions from the chloride ions and keeps them dissolved **(fig. 2.9b).**

Adhesion is the tendency of one substance to cling to another, whereas *cohesion* is the tendency of molecules of the same substance to cling to each other. Water adheres to the body's tissues and forms a lubricating film on membranes such as the pleura and pericardium. This reduces friction as the lungs and heart contract and expand and rub against these membranes. Water also is a very cohesive liquid because of its hydrogen bonds. This is why, when you spill water on the floor, it forms a puddle and evaporates slowly. By contrast, if you spill a nonpolar substance such as liquid nitrogen, it dances about and evaporates in seconds, like a drop of water in a hot dry skillet. This is because nitrogen molecules have no attraction for each other, so the little bit of heat provided by the floor is enough to disperse them into the air. The cohesion of water is especially evident at its surface, where it forms an elastic layer called the *surface film* held together by a force called *surface tension.* This force causes water to hang in drops from a leaky faucet and travel in rivulets down a window.

The *chemical reactivity* of water is its ability to participate in chemical reactions. Not only does water ionize many other chemicals such as acids and salts, but water itself ionizes into H^+ and OH^-. These ions can be incorporated into other molecules, or released from them, in the course of chemical reactions such as *hydrolysis* and *dehydration synthesis,* described later in this chapter.

The *thermal stability* of water helps to stabilize the internal temperature of the body. It results from the high *heat capacity* of water—the amount of heat required to raise the temperature of 1 g of a substance by 1°C. The base unit of heat is the **calorie**[8] **(cal)**—1 cal is the amount of heat that raises the temperature of 1 g of water 1°C. The same amount of heat would raise the temperature of a nonpolar substance such as nitrogen about four times as much. The difference stems from the presence or absence of hydrogen bonding. To increase in temperature, the molecules of a substance must move around more actively. The hydrogen bonds of water molecules inhibit their movement, so water can absorb

---
[6]*hydro* = water; *philic* = loving, attracted to
[7]*hydro* = water; *phobic* = fearing, avoiding
[8]*calor* = heat

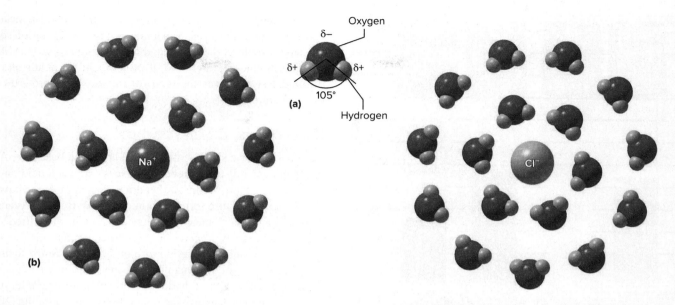

FIGURE 2.9 Water and Hydration Spheres. (a) A water molecule showing its bond angle and polarity. (b) Water molecules aggregate around a sodium ion with their negatively charged oxygen poles facing the Na^+ and aggregate around a chloride ion with their positively charged hydrogen poles facing the Cl^-.

a given amount of heat without changing temperature (molecular motion) as much.

The high heat capacity of water also makes it a very effective coolant. When it changes from a liquid to a vapor, water carries a large amount of heat with it. One milliliter of perspiration evaporating from the skin removes about 500 cal of heat from the body. This effect is very apparent when you are sweaty and stand in front of a fan.

▶▶▶ **APPLY WHAT YOU KNOW**

Why are heat and temperature not the same thing?

2.2b Solutions, Colloids, and Suspensions

Mixtures of other substances in water can be classified as *solutions, colloids,* and *suspensions.*

A **solution** consists of particles of matter called the **solute** mixed with a more abundant substance (usually water) called the **solvent.** The solute can be a gas, solid, or liquid—as in a solution of oxygen, sodium chloride, or alcohol in water, respectively. Solutions are defined by the following properties:

- The solute particles are under 1 nanometer (nm) in size. The solute and solvent therefore cannot be visually distinguished from each other, even with a microscope.

- Such small particles don't scatter light noticeably, so solutions are usually transparent (**fig. 2.10a**).

- The solute particles can pass through most selectively permeable membranes, such as dialysis tubing and cell membranes.

- The solute doesn't separate from the solvent when the solution is allowed to stand.

The most common **colloids**[9] in the body are mixtures of protein and water, such as the albumin in blood plasma. Many colloids can change from liquid to gel states—gelatin desserts, agar culture media, and the fluids within and between our cells, for example. Colloids are defined by the following physical properties:

- The colloidal particles range from 1 to 100 nm in size.

- Particles this large scatter light, so colloids are usually cloudy (**fig. 2.10b**).

- The particles are too large to pass through most selectively permeable membranes.

- The particles are still small enough, however, to remain permanently mixed with the solvent when the mixture stands.

The blood cells in our blood plasma exemplify a **suspension.** Suspensions are defined by the following properties:

- The suspended particles exceed 100 nm in size.

- Such large particles render suspensions cloudy or opaque.

- The particles are too large to penetrate selectively permeable membranes.

- The particles are too heavy to remain permanently suspended, so suspensions separate on standing. Blood cells, for example, form a suspension in the blood plasma and settle to the bottom of a tube when blood is allowed to stand without mixing (**fig. 2.10c, d**).

An **emulsion** is a suspension of one liquid in another, such as oil-and-vinegar salad dressing. The fat in breast milk is an emulsion, as are medications such as Kaopectate and milk of magnesia.

---

[9]*collo* = glue; *oid* = like, resembling

(a)   (b)   (c)   (d)

Solution   Colloid   Suspension

FIGURE 2.10 A Solution, a Colloid, and a Suspension. Top row: Photographs of a representative solution, colloid, and suspension. Bottom row: Symbolic representation of the particle sizes in each mixture. (a) In a copper sulfate solution, the solute particles are so small they remain permanently mixed and the solution is transparent. (b) In milk, the protein molecules are small enough to remain permanently mixed, but large enough to scatter light, so the mixture is opaque. (c) In blood, the red blood cells scatter light and make the mixture opaque. (d) Red blood cells are too large to remain evenly mixed, so they settle to the bottom as in this blood specimen that stood overnight.

Top: Ken Saladin

A single mixture can fit into more than one of these categories. Blood is a perfect example—it is a solution of sodium chloride, a colloid of protein, and a suspension of cells. Milk is a solution of calcium, a colloid of protein, and an emulsion of fat. **Table 2.4** summarizes the types of mixtures and provides additional examples.

2.2c Acids, Bases, and pH

Most people have some sense of what acids and bases are. Advertisements are full of references to excess stomach acid and pH-balanced shampoo. We know that drain cleaner (a strong base) and battery acid can cause serious chemical burns. But what exactly do "acidic" and "basic" mean, and how can they be quantified?

An **acid** is any *proton donor,* a molecule that releases a proton (H^+) in water. A **base** is a proton acceptor. Since hydroxide ions (OH^-) accept H^+, many bases are substances that release hydroxide ions—sodium hydroxide ($NaOH$), for example. A base doesn't have to be a hydroxide donor, however. Ammonia (NH_3) is also a base. It doesn't release hydroxide ions, but it readily accepts hydrogen ions to become the ammonium ion (NH_4^+).

Acidity is expressed in terms of **pH,** a measure derived from the molarity of H^+. Molarity (explained in the next section) is represented by square brackets, so H^+ molarity is symbolized $[H^+]$. pH is the negative logarithm of hydrogen ion molarity—that is, $pH = -\log [H^+]$. In pure water, 1 in 10 million molecules ionizes into hydrogen and hydroxide ions: $H_2O \rightleftharpoons H^+ + OH^-$. (The symbol $\rightleftharpoons$ denotes a reversible chemical reaction.) Pure water has a neutral pH because it contains equal amounts of H^+ and OH^-. Since 1 in 10 million molecules ionize, the molarity of H^+ and the pH of water are

$$[H^+] = 0.0000001 \text{ molar} = 10^{-7}M;$$
$$\log[H^+] = -7; \text{ and}$$
$$pH = -\log[H^+] = 7.$$

The pH scale **(fig. 2.11)** was invented in 1909 by Danish biochemist and brewer Sören Sörensen to measure the acidity of beer. The scale extends from 0.0 to 14.0. A solution with a pH of 7.0 is **neutral;** solutions with pH below 7 are **acidic;** and solutions with pH above 7 are **basic (alkaline).** The lower the pH value, the more hydrogen ions a solution has and the more acidic it is. Since the pH scale is logarithmic, a change of one whole number on the scale represents a 10-fold change in H^+ concentration. In other words, a solution with pH 4 is 10 times as acidic as one with pH 5 and 100 times as acidic as one with pH 6.

Slight disturbances of pH can seriously disrupt physiological functions and alter drug actions (see Deeper Insight 2.2), so it is important that the body carefully control its pH. Blood, for example, normally has a pH ranging from 7.35 to 7.45. Deviations from this range cause tremors, fainting, paralysis, or even death. Chemical solutions that resist changes in pH are called **buffers.** Buffers and acid–base balance are considered in detail in section 24.3.

▶▶▶ **APPLY WHAT YOU KNOW**

A pH of 7.20 is slightly alkaline, yet a blood pH of 7.20 is called acidosis. *Why do you think it is called this?*

TABLE 2.4	Types of Mixtures		
	Solution	Colloid	Suspension
Particle size	<1 nm	1–100 nm	>100 nm
Appearance	Clear	Often cloudy	Cloudy-opaque
Will particles settle out?	No	No	Yes
Will particles pass through a selectively permeable membrane?	Yes	No	No
Examples	Glucose in blood O_2 in water Saline solutions Sugar in coffee	Proteins in blood Intracellular fluid Milk protein Gelatin	Blood cells Cornstarch in water Fats in blood Kaopectate

2.2d Other Measures of Concentration

Solutions are often described in terms of their concentration—how much solute is present in a given volume of solution. Concentration is expressed in different ways for different purposes, briefly itemized here and more fully explained in appendix B.

- **Weight per volume.** This is the weight of solute (such as grams, g, or milligrams, mg) in a given volume of solution (such as liters, L, or deciliters, dL). For example, a typical serum cholesterol concentration is 200 mg/dL.

- **Percentage.** This is the weight of solute as a percentage of solution volume (weight per volume, w/v) or volume of a liquid as a percentage of total solution volume (volume per volume, v/v). For example, a common intravenous fluid is D5W, which means 5% w/v dextrose in distilled water. Ethanol is often used as a 70% v/v solution.

- **Molarity.** One mole of a chemical is the number of grams equal to its molecular weight, and molarity (M) is a measure of the number of moles of solute per liter of solution. This reflects not merely the weight of solute in the solution, but the number of molecules per volume. It is the number of molecules, not their total weight, that determines the physiological effect of a solution, so molarity is often the most meaningful measure of concentration. Body fluids are usually quantified in millimolar (mM) concentrations, since they are much less than 1 molar.

- **Milliequivalents per liter.** This unit of measure (expressed mEq/L) is used to express electrolyte concentrations; it takes into account not only the millimolar concentration of a solute but the electrical charge on its particles. This is important to processes such as nerve firing, the heartbeat, and muscle contractions, which are driven by electrical phenomena. The mEq/L concentration of electrolytes is critically important in giving intravenous fluids.

FIGURE 2.11 The pH Scale and Approximate pH of Some Familiar Household Substances. The pH is shown within the colored bar. H⁺ molarity increases 10-fold for every step down the scale.

DEEPER INSIGHT 2.2

CLINICAL APPLICATION

pH and Drug Action

The pH of our body fluids has a direct bearing on how we react to drugs. Depending on pH, drugs such as aspirin, phenobarbital, and penicillin can exist in charged (ionized) or uncharged forms. Whether a drug is charged or not can determine whether it will pass through cell membranes. When aspirin is in the acidic environment of the stomach, for example, it is uncharged and passes easily through the stomach lining into the bloodstream. Here it encounters a basic pH, whereupon it ionizes. In this state, it is unable to pass back through the membrane, so it accumulates in the blood. This effect, called *ion trapping (pH partitioning)*, can be controlled to help clear poisons from the body. The pH of the urine, for example, can be manipulated so that poisons become trapped there and are more rapidly excreted from the body.

BEFORE YOU GO ON

Answer the following questions to test your understanding of the preceding section:

6. What is the difference between a mixture and a compound?

7. What are hydrophilic and hydrophobic substances? Give an example of each.

8. Why would the cohesion and thermal stability of water be less if water did not have polar covalent bonds?

9. How do solutions, colloids, and suspensions differ from each other? Give an example of each in the human body.

10. If solution A had a H^+ concentration of 10^{-8} M, what would be its pH? If solution B had 1,000 times this H^+ concentration, what would be its pH? Would solution A be acidic or basic? What about solution B?

11. What information can we get from the molarity of a solution that we cannot know from its percentage concentration? What do we know from a concentration in mEq/L that we could not know from molarity alone?

2.3 Energy and Chemical Reactions

Expected Learning Outcomes

When you have completed this section, you should be able to

a. define *energy* and *work,* and describe some types of energy;

b. understand how chemical reactions are symbolized by chemical equations;

c. list and define the fundamental types of chemical reactions;

d. identify the factors that govern the speed and direction of a reaction;

e. define *metabolism* and its two subdivisions; and

f. define *oxidation* and *reduction,* and relate these to changes in the energy content of a molecule.

2.3a Energy and Work

Energy is the capacity to do work. To do **work** means to move something, whether it is a muscle or a molecule. Some examples of physiological work are breaking chemical bonds, building molecules, pumping blood, and contracting skeletal muscles. All of the body's activities are forms of work.

Energy is broadly classified as *potential* or *kinetic energy. Potential energy* is energy contained in an object because of its position or internal state but that is not doing work at the time. *Kinetic energy* is energy of motion, energy that is doing work. It is observed in musculoskeletal movements, the flow of ions into a cell, and vibration of the eardrum, for example. The water behind a dam has potential energy because of its position. Let the water flow through, and it exhibits kinetic energy that can be tapped for generating electricity. Like water behind a dam, ions concentrated on one side of a cell membrane have potential energy that can be released by opening gates in the membrane. As the ions flow through the gates, their kinetic energy can be tapped to create a nerve signal or make the heart beat.

Within the two broad categories of potential and kinetic energy, several forms of energy are relevant to human physiology. *Chemical energy* is potential energy stored in the bonds of molecules. Chemical reactions release this energy and make it available for physiological work. *Heat* is the kinetic energy of molecular motion. The temperature of a substance is a measure of rate of this motion, and adding heat to a substance increases molecular motion. *Electromagnetic energy* is the kinetic energy of moving "packets" of radiation called *photons*. The most familiar form of electromagnetic energy is light. *Electrical energy* has both potential and kinetic forms. It is potential energy when charged particles have accumulated at a point such as a battery terminal or on one side of a cell membrane; it becomes kinetic energy when these particles begin to move and create an electrical current—for example, when electrons move through your household wiring or sodium ions move through a cell membrane.

Free energy is the potential energy available in a system to do useful work. In human physiology, the most relevant free energy is the energy stored in the chemical bonds of organic molecules.

2.3b Classes of Chemical Reactions

A **chemical reaction** is a process in which a covalent or ionic bond is formed or broken. The course of a chemical reaction is symbolized by a **chemical equation** that typically shows the *reactants* on the left, the *products* on the right, and an arrow pointing from the reactants to the products. For example, consider this common occurrence: If you open a bottle of wine and let it stand for several days, it turns sour. Wine "turns to vinegar" because oxygen gets into the bottle and reacts with ethanol to produce acetic acid and water. Acetic acid gives the tart flavor to vinegar and spoiled wine. The equation for this reaction is

$$CH_3CH_2OH \quad + \quad O_2 \quad \longrightarrow \quad CH_3COOH \quad + \quad H_2O$$

Ethanol   Oxygen   Acetic acid   Water

(a) Decomposition reaction **(b) Synthesis reaction** **(c) Exchange reaction**

FIGURE 2.12 Decomposition, Synthesis, and Exchange Reactions. (a) In a decomposition reaction, large molecules are broken down into simpler ones. (b) In a synthesis reaction, smaller molecules are joined to form larger ones. (c) In an exchange reaction, two molecules exchange atoms.

❓ *To which of these categories does the digestion of food belong?*

Ethanol and oxygen are the reactants, and acetic acid and water are the products of this reaction. Not all reactions are shown with the arrow pointing from left to right. In complex biochemical equations, reaction chains are often written vertically or even in circles.

Chemical reactions can be classified as *decomposition, synthesis,* or *exchange reactions.* In **decomposition reactions,** a large molecule breaks down into two or more smaller ones **(fig. 2.12a);** symbolically, AB $\longrightarrow$ A + B. When you eat a potato, for example, digestive enzymes decompose its starch into thousands of glucose molecules, and most cells further decompose glucose to water and carbon dioxide. Starch, a very large molecule, ultimately yields about 36,000 molecules of H_2O and CO_2.

Synthesis reactions are just the opposite—two or more small molecules combine to form a larger one; symbolically, A + B $\longrightarrow$ AB **(fig. 2.12b).** When the body synthesizes proteins, for example, it combines 50 or more amino acids into one protein molecule.

In **exchange reactions,** two molecules exchange atoms or groups of atoms; AB + CD $\longrightarrow$ AC + BD **(fig. 2.12c).** For example, when stomach acid (HCl) enters the small intestine, the pancreas secretes sodium bicarbonate ($NaHCO_3$) to neutralize it. The reaction between the two is

$$NaHCO_3 + HCl \longrightarrow NaCl + H_2CO_3.$$

We could say the sodium atom has exchanged its bicarbonate group ($-HCO_3$) for a chlorine atom.

Reversible reactions can go in either direction under different circumstances. For example, carbon dioxide combines with water to produce carbonic acid, which in turn decomposes into bicarbonate ions and hydrogen ions:

$$\underset{\substack{\text{Carbon}\\\text{dioxide}}}{CO_2} + \underset{\text{Water}}{H_2O} \rightleftharpoons \underset{\substack{\text{Carbonic}\\\text{acid}}}{H_2CO_3} \rightleftharpoons \underset{\substack{\text{Bicarbonate}\\\text{ion}}}{HCO_3^-} + \underset{\substack{\text{Hydrogen}\\\text{ion}}}{H^+}$$

This reaction appears in this book more often than any other, especially as we discuss respiratory, urinary, and digestive physiology.

The direction in which a reversible reaction goes is determined by the relative abundance of substances on each side of the equation. If there is a surplus of CO_2, this reaction proceeds left to right and produces bicarbonate and hydrogen ions. If bicarbonate and hydrogen ions are present in excess, the reaction proceeds right to left and generates CO_2 and H_2O. Reversible reactions follow the **law of mass action:** They proceed from the reactants in greater quantity to the substances with the lesser quantity. This law will help to explain processes discussed in later chapters, such as why hemoglobin binds oxygen in the lungs yet releases it to muscle tissue.

In the absence of upsetting influences, reversible reactions exist in a state of **equilibrium,** in which the ratio of products to reactants is stable. The carbonic acid reaction, for example, normally maintains a 20:1 ratio of bicarbonate ions to carbonic acid molecules. This equilibrium can be upset, however, by a surplus of hydrogen ions, which drives the reaction to the left, or adding carbon dioxide and driving it to the right.

2.3c Reaction Rates

Chemical reactions are based on molecular motion and collisions. All molecules are in constant motion, and reactions occur when mutually reactive molecules collide with sufficient force and the right orientation. The rate of a reaction depends on the nature of the reactants and on the frequency and force of these collisions. Some factors that affect reaction rates are

- **Concentration.** Reaction rate increases when the reactants are more concentrated. This is because the molecules are more crowded and collide more frequently.

- **Temperature.** Reaction rate increases as the temperature rises. This is because heat causes molecules to move more rapidly and collide with greater force and frequency.

- **Catalysts** (CAT-uh-lists). These are substances that temporarily bind to reactants, hold them in a favorable position to react with each other, and may change the shapes of reactants in ways that make them more likely to react. By reducing the element of chance in molecular collisions, a catalyst speeds up a reaction. It then releases the products and is available to repeat the process with more reactants. The catalyst itself is not consumed or changed by the reaction. The most important biological catalysts are enzymes, discussed later in this chapter.

2.3d Metabolism, Oxidation, and Reduction

All the chemical reactions in the body are collectively called **metabolism.** Metabolism has two divisions: *catabolism* and *anabolism.* **Catabolism**[10] (ca-TAB-oh-lizm) consists of energy-releasing decomposition reactions. Such reactions break covalent bonds, produce smaller molecules from larger ones, and release energy that can be used for other physiological work. Energy-releasing reactions are called *exergonic*[11] reactions. If you hold a beaker of water in your hand and pour sulfuric acid into it, for example, the beaker will get so hot you may have to put it down. If you break down energy-storage molecules to run a race, you too will get hot. In both cases, the heat signifies that exergonic reactions are occurring.

Anabolism[12] (ah-NAB-oh-lizm) consists of energy-storing synthesis reactions, such as the production of protein or fat. Reactions that require an energy input, such as these, are called *endergonic*[13] reactions. Anabolism is driven by the energy that catabolism releases, so endergonic and exergonic processes, anabolism and catabolism, are inseparably linked.

Oxidation is any chemical reaction in which a molecule gives up electrons and releases energy. A molecule is *oxidized* by this process, and whatever molecule takes the electrons from it is an *oxidizing agent (electron acceptor).* The term *oxidation* stems from the fact that oxygen is often involved as the electron acceptor. Thus, we can sometimes recognize an oxidation reaction from the fact that oxygen has been added to a molecule. The rusting of iron, for example, is a slow oxidation process in which oxygen is added to iron to form iron oxide (Fe_2O_3). Many oxidation reactions, however, don't involve oxygen at all. For example, when yeast ferments glucose to ethanol, no oxygen is required; indeed, the ethanol *contains less oxygen* than the glucose originally did, but it is *more oxidized* than the glucose:

$$C_6H_{12}O_6 \longrightarrow 2\,CH_3CH_2OH + 2\,CO_2$$
$$\text{Glucose} \qquad\qquad \text{Ethanol} \qquad\qquad \text{Carbon dioxide}$$

Reduction is a chemical reaction in which a molecule gains electrons and energy. When a molecule accepts electrons, it is said to be *reduced;* a molecule that donates electrons to another is therefore called a *reducing agent (electron donor).* The oxidation of one molecule is always accompanied by the reduction of another, so these electron transfers are known as *oxidation–reduction (redox) reactions.*

It isn't necessary that *only* electrons be transferred in a redox reaction. Often, the electrons are transferred in the form of hydrogen atoms. The fact that a proton (the hydrogen nucleus) is also transferred is immaterial to whether we consider a reaction oxidation or reduction.

Table 2.5 summarizes these energy-transfer reactions. We can symbolize oxidation and reduction as follows, letting A and B

---

[10]*cato* = down, to break down
[11]*ex, exo* = out; *erg* = work
[12]*ana* = up, to build up
[13]*end, endo* = in; *erg* = work

TABLE 2.5	Energy-Transfer Reactions in the Human Body
Exergonic Reactions	Reactions in which there is a net release of energy. The products have less total free energy than the reactants did.
Oxidation	An exergonic reaction in which electrons are removed from a reactant. Electrons may be removed one or two at a time and may be removed in the form of hydrogen atoms (H or H_2). The product is then said to be oxidized.
Decomposition	A reaction such as digestion and cell respiration, in which larger molecules are broken down into smaller ones.
Catabolism	The sum of all decomposition reactions in the body.
Endergonic Reactions	Reactions in which there is a net input of energy. The products have more total free energy than the reactants did.
Reduction	An endergonic reaction in which electrons are donated to a reactant. The product is then said to be reduced.
Synthesis	A reaction such as protein and glycogen synthesis, in which two or more smaller molecules are combined into a larger one.
Anabolism	The sum of all synthesis reactions in the body.

symbolize arbitrary molecules and letting e⁻ represent one or more electrons that transfer from molecule A to molecule B:

$$Ae^- \quad + \quad B \quad \longrightarrow \quad A \quad + \quad Be^-$$

| High-energy reduced state | Low-energy oxidized state | Low-energy oxidized state | High-energy reduced state |

Ae⁻ is a reducing agent because it reduces B, and B is an oxidizing agent because it oxidizes Ae⁻.

BEFORE YOU GO ON

Answer the following questions to test your understanding of the preceding section:

12. Define *energy*. Distinguish potential energy from kinetic energy.

13. Define *metabolism, catabolism,* and *anabolism*.

14. What does *oxidation* mean? What does *reduction* mean? Which of them is endergonic and which is exergonic?

15. When sodium chloride forms, which element—sodium or chlorine—is oxidized? Which one is reduced?

2.4 Organic Compounds

Expected Learning Outcomes

When you have completed this section, you should be able to

a. explain why carbon is especially well suited to serve as the structural foundation of many biological molecules;

b. identify some common functional groups of organic molecules from their formulae;

c. discuss the relevance of polymers to biology and explain how they are formed and broken by dehydration synthesis and hydrolysis;

d. discuss the types and functions of carbohydrates;

e. discuss the types and functions of lipids;

f. discuss protein structure and function;

g. explain how enzymes function;

h. describe the structure, production, and function of ATP;

i. identify other nucleotide types and their functions; and

j. identify the principal types of nucleic acids.

2.4a Carbon Compounds and Functional Groups

Organic chemistry is the study of compounds of carbon. By 1900, biochemists had classified the large organic molecules of life into four primary categories: *carbohydrates, lipids, proteins,* and *nucleic acids*. We examine the first three in this chapter but describe the details of nucleic acids, which are concerned with genetics, in section 4.1.

Carbon is an especially versatile atom that serves as the basis of a wide variety of structures. It has four valence electrons, so

it bonds with other atoms that can provide it with four more to complete its valence shell. Carbon atoms readily bond with each other and can form long chains, branched molecules, and rings—an enormous variety of **carbon backbones** for organic molecules. Carbon also commonly forms covalent bonds with hydrogen, oxygen, nitrogen, and sulfur.

Carbon backbones carry a variety of **functional groups**—small clusters of atoms that determine many of the properties of an organic molecule. For example, organic acids bear a **carboxyl group** (car-BOC-sil), and ATP is named for its three **phosphate groups.** Other common functional groups include **hydroxyl, methyl,** and **amino groups (fig. 2.13).**

2.4b Monomers and Polymers

Since carbon can form long chains, some organic molecules are gigantic *macromolecules* with molecular weights that range

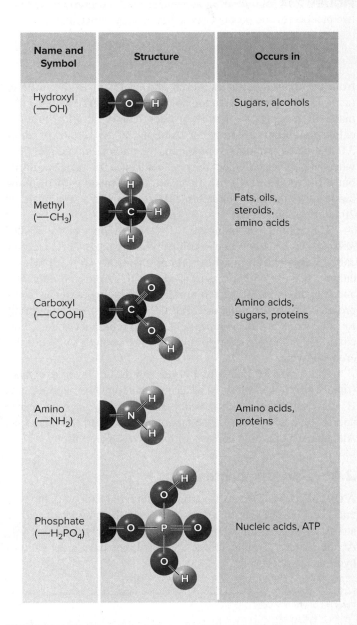

Name and Symbol	Structure	Occurs in
Hydroxyl (—OH)		Sugars, alcohols
Methyl (—CH₃)		Fats, oils, steroids, amino acids
Carboxyl (—COOH)		Amino acids, sugars, proteins
Amino (—NH₂)		Amino acids, proteins
Phosphate (—H₂PO₄)		Nucleic acids, ATP

FIGURE 2.13 Functional Groups of Organic Molecules.

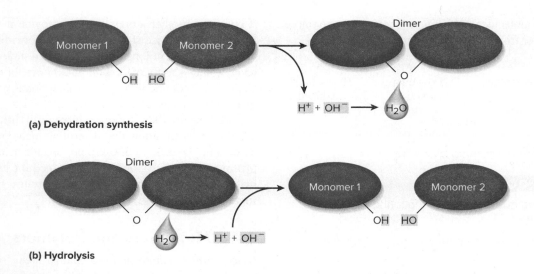

(a) Dehydration synthesis

(b) Hydrolysis

FIGURE 2.14 Dehydration Synthesis and Hydrolysis Reactions. (a) Dehydration synthesis, creating a new covalent bond between two monomers and producing water as a by-product. (b) Hydrolysis, consuming a water molecule to break a covalent bond.

from the thousands (as in starch and proteins) to the millions (as in DNA). Most macromolecules are **polymers**[14]—molecules made of a repetitive series of identical or similar subunits called **monomers.** Starch, for example, is a polymer of about 3,000 glucose monomers. In starch, the monomers are identical, whereas in other polymers they have a basic structural similarity but differ in detail. DNA, for example, is made of 4 different kinds of monomers (nucleotides), and proteins are made of 20 kinds (amino acids).

The joining of monomers to form a polymer is called *polymerization.* Living cells achieve this by means of a reaction called **dehydration synthesis (condensation) (fig. 2.14a).** An enzyme removes a hydroxyl group (—OH) from one monomer and a hydrogen (—H) from another, producing water as a by-product. The two monomers become joined by a covalent bond, forming a *dimer.* This is repeated for each monomer added to the chain, potentially leading to a chain long enough to be considered a polymer.

The opposite of dehydration synthesis is **hydrolysis**[15] (fig. 2.14b). In hydrolysis, a water molecule ionizes into OH⁻ and H⁺. An enzyme breaks the covalent bond linking one monomer to another, and adds OH⁻ to one monomer and H⁺ to the other one. All chemical digestion consists of hydrolysis reactions.

2.4c Carbohydrates

A **carbohydrate**[16] is a hydrophilic organic molecule with the general formula $(CH_2O)n$, where n represents the number of carbon atoms. In glucose, for example, $n = 6$ and the formula is $C_6H_{12}O_6$. As the generic formula shows, carbohydrates have a 2:1 ratio of hydrogen to oxygen.

▶▶▶**APPLY WHAT YOU KNOW**

Why is carbohydrate *an appropriate name for this class of compounds? Relate this name to the general formula of carbohydrates.*

The names of individual carbohydrates are often built on the word root *sacchar-* or the suffix *-ose,* both of which mean "sugar" or "sweet." The most familiar carbohydrates are sugars and starches.

The simplest carbohydrates are monomers called **monosaccharides**[17] (MON-oh-SAC-uh-rides), or simple sugars. The three of primary importance are **glucose, galactose,** and **fructose,** all with the molecular formula $C_6H_{12}O_6$; they are isomers of each other **(fig. 2.15).** We obtain these sugars mainly by the digestion of more complex carbohydrates. Glucose is the "blood sugar" that provides energy to most of our cells. Two other monosaccharides, ribose and deoxyribose, are important components of RNA and DNA, respectively.

Disaccharides are sugars composed of two monosaccharides. The three disaccharides of greatest importance are **sucrose** (made of glucose + fructose), **lactose** (glucose + galactose), and **maltose** (glucose + glucose) **(fig. 2.16).** Sucrose is produced by sugarcane and sugar beets and used as common table sugar. Lactose is milk sugar. Maltose is a product of starch digestion and is present in a few foods such as malt beverages and germinating grains.

Short chains of three or more monosaccharides are called **oligosaccharides,** and long chains (up to thousands of monosaccharides long) are called **polysaccharides.** There is no exact criterion for when a chain is long enough to be called a polysaccharide, but a chain of 10 or 20 monosaccharides would generally be considered an oligosaccharide, whereas a chain of 50 or more would generally be considered a polysaccharide. Polysaccharides

---

[14]*poly* = many; *mer* = part
[15]*hydro* = water; *lysis* = splitting apart
[16]*carbo* = carbon; *hydr* = water

[17]*mono* = one; *sacchar* = sugar

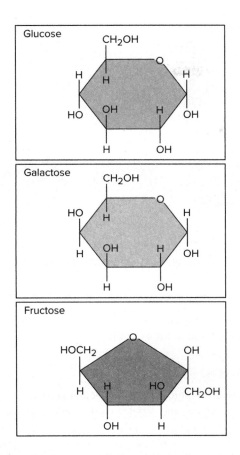

FIGURE 2.15 The Three Major Monosaccharides. Glucose, galactose, and fructose all have the molecular formula $C_6H_{12}O_6$, but with the atoms arranged differently as shown. Each angle in the rings represents a carbon atom except the one where oxygen is shown. This is a conventional way of representing carbon in the structural formulae of organic compounds.

FIGURE 2.16 The Three Major Disaccharides. Sucrose, lactose, and maltose all consist of two monosaccharides joined through an oxygen atom.

can be thousands of sugars long and may have molecular weights of 500,000 or more (compared with 180 for a single glucose). Three polysaccharides of interest to human physiology are glycogen, starch, and cellulose—all composed solely of glucose. Animals, including ourselves, make glycogen, whereas starch and cellulose are plant products.

Glycogen[18] is an energy-storage polysaccharide made by cells of the liver, muscles, brain, uterus, and vagina. It is a long, branched, glucose polymer **(fig. 2.17)**. The liver produces glycogen after a meal, when the blood glucose level is high, and then breaks it down between meals to maintain blood glucose levels when there is no food intake. Muscle stores glycogen for its own energy needs, and the uterus uses it in early pregnancy to nourish the embryo.

Starch is the corresponding energy-storage polysaccharide of plants. They store it when sunlight and nutrients are available and draw from it when photosynthesis is not possible (for example, at night and in winter, when a plant has shed its leaves). Starch is the only significant digestible polysaccharide in the human diet.

Cellulose is a structural polysaccharide that gives strength to the cell walls of plants. It is the principal component of wood, cotton, and paper. It is composed of a chain of a few thousand glucose monomers. Cellulose is the most abundant organic compound on earth and it is a common component of the diets of humans and other animals—yet we have no enzymes to digest it and thus derive no energy or nutrition from it. Nevertheless, it is important as dietary fiber ("bulk" or "roughage"). It swells with water in the digestive tract and helps move other materials through the intestine.

Carbohydrates are, above all, a source of energy that can be quickly mobilized. All digested carbohydrate is ultimately converted to glucose, and glucose is oxidized to make ATP, a high-energy compound discussed later. But carbohydrates have other functions as well **(table 2.6).** They are often **conjugated**[19] with (covalently bound to) proteins and lipids. Many lipid and protein molecules at the external surface of the cell membrane have chains of up to 12 sugars attached to them, thus forming **glycolipids** and **glycoproteins,** respectively. Among other functions, glycoproteins are a major component of mucus, which traps particles in the respiratory system, resists infection, and protects the digestive tract from its own acid and enzymes.

---

[18]*glyco* = sugar; *gen* = producing

[19]*con* = together; *jug* = join

(a)   **(b)**

FIGURE 2.17 Glycogen. This is the only polysaccharide found in human tissues. (a) Part of a glycogen molecule showing the chain of glucose monomers and branching pattern. (b) Detail of the molecule at a branch point.

Proteoglycans are macromolecules in which the carbohydrate component is dominant and a peptide or protein forms a smaller component. Proteoglycans form gels that hold cells and tissues together, form a gelatinous filler in the umbilical cord and eye, lubricate the joints of the skeletal system, and account for the tough rubbery texture of cartilage.

When discussing conjugated macromolecules, it is convenient to refer to each chemically different component as a **moiety**[20] (MOY-eh-tee). Proteoglycans have a protein moiety and a carbohydrate moiety, for example.

2.4d Lipids

A **lipid** is a hydrophobic organic molecule, usually composed only of carbon, hydrogen, and oxygen, with a high ratio of hydrogen to oxygen. A fat called *tristearin* (tri-STEE-uh-rin), for example, has the molecular formula $C_{57}H_{110}O_6$—more than 18 hydrogens for every oxygen. Lipids are less oxidized than carbohydrates, and thus have more calories per gram. Beyond these criteria, it is difficult to generalize about lipids; they are much more variable in structure than the other macromolecules we are considering. We consider here the five primary types of lipids in humans—*fatty acids, triglycerides, phospholipids, eicosanoids,* and *steroids* **(table 2.7).**

---

[20]*moiet* = half

TABLE 2.6	Carbohydrate Functions
Type	**Function**
Monosaccharides	
Glucose	Blood sugar—energy source for most cells
Galactose	Converted to glucose and metabolized
Fructose	Fruit sugar—converted to glucose and metabolized
Disaccharides	
Sucrose	Cane sugar—digested to glucose and fructose
Lactose	Milk sugar—digested to glucose and galactose; important in infant nutrition
Maltose	Malt sugar—product of starch digestion, further digested to glucose
Polysaccharides	
Cellulose	Structural polysaccharide of plants; dietary fiber
Starch	Energy storage in plant cells; energy source in human diet
Glycogen	Energy storage in animal cells (liver, muscle, brain, uterus, vagina)
Conjugated Carbohydrates	
Glycoprotein	Component of the cell surface coat and mucus, among other roles
Glycolipid	Component of the cell surface coat
Proteoglycan	Cell adhesion; lubrication; supportive filler of some tissues and organs

TABLE 2.7	Lipid Functions
Type	**Function**
Bile acids	Steroids that aid in fat digestion and nutrient absorption
Cholesterol	Component of cell membranes; precursor of other steroids
Eicosanoids	Chemical messengers between cells
Fat-soluble vitamins (A, D, E, and K)	Involved in a variety of functions including blood clotting, wound healing, vision, and calcium absorption
Fatty acids	Precursor of triglycerides; source of energy
Phospholipids	Major component of cell membranes; aid in fat digestion
Steroid hormones	Chemical messengers between cells
Triglycerides	Energy storage; thermal insulation; filling space; binding organs together; cushioning organs

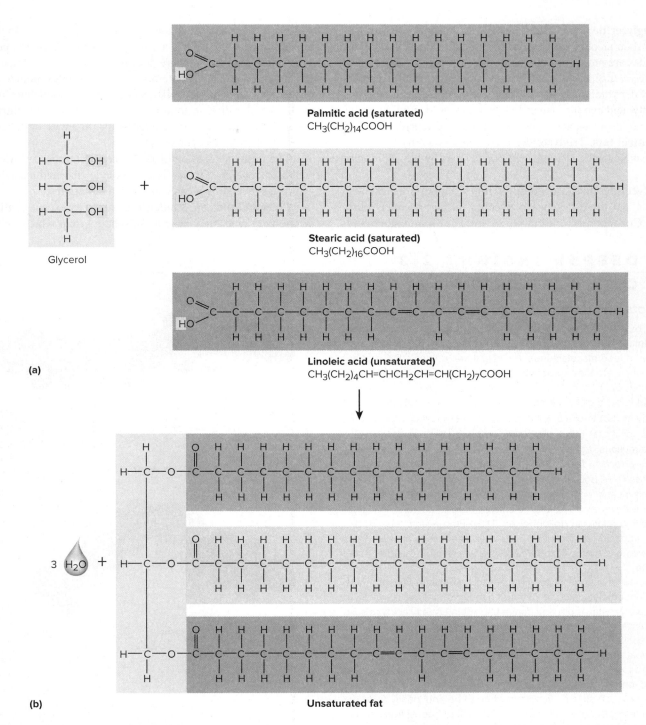

FIGURE 2.18 Triglyceride (Fat) Synthesis. The three fatty acids above the arrow (a) combine with glycerol (left) to produce the triglyceride (fat) below the arrow (b). Note the difference between saturated and unsaturated fatty acids and the production of 3 H_2O as a by-product of this dehydration synthesis reaction.

A **fatty acid** is a chain of usually 4 to 24 carbon atoms with a carboxyl group at one end and a methyl group at the other. Fatty acids and the fats made from them are classified as *saturated* or *unsaturated.* A **saturated fatty acid** such as palmitic acid has as much hydrogen as it can carry. No more could be added without exceeding four covalent bonds per carbon; thus, it is "saturated" with hydrogen. In **unsaturated fatty acids** such as linoleic acid, however,

some carbon atoms are joined by double covalent bonds (**fig. 2.18a**). Each of these could potentially share one pair of electrons with another hydrogen atom instead of the adjacent carbon, so hydrogen could be added to this molecule. **Polyunsaturated fatty acids** are those with multiple C=C bonds. Most fatty acids can be synthesized by the human body, but a few, called **essential fatty acids,** must be obtained from the diet because we cannot synthesize them.

A **triglyceride** (try-GLISS-ur-ide) is a molecule consisting of a three-carbon alcohol called **glycerol** linked to three fatty acids; triglycerides are more correctly, although less widely, also known as *triacylglycerols.* Each bond between a fatty acid and glycerol is formed by dehydration synthesis **(fig. 2.18b).** Once joined to glycerol, a fatty acid can no longer donate a proton to solution and is therefore no longer an acid. For this reason, triglycerides are also called **neutral fats.** Triglycerides are broken down by hydrolysis reactions, which split each of these bonds apart by the addition of water.

Triglycerides that are liquid at room temperature are also called *oils,* but the difference between a fat and an oil is fairly arbitrary. Coconut oil, for example, is solid at room temperature.

Animal fats are usually made of saturated fatty acids, so they are called *saturated fats.* They are solid at room or body temperature. Most plant triglycerides are *polyunsaturated fats,* which generally remain liquid at room temperature. Examples include peanut, olive, corn, and linseed oils. Saturated fats contribute more to cardiovascular disease than unsaturated fats, and for this reason it's healthier to cook with vegetable oils than with lard, bacon fat, or butter (but see Deeper Insight 2.3).

The primary function of fat is energy storage, but when concentrated in *adipose tissue,* it also provides thermal insulation and acts as a shock-absorbing cushion for vital organs.

Phospholipids are similar to neutral fats except that in place of one fatty acid, they have a phosphate group which, in turn, is

DEEPER INSIGHT 2.3

CLINICAL APPLICATION

Trans *Fats and Cardiovascular Health*

Trans fats have been a significant part of the American diet for more than a century. Yet in 2015, the U.S. Food and Drug Administration (FDA) ordered food manufacturers to discontinue their use. So what are *trans* fats, why were they so popular for so long, and why are they now almost entirely banished?

A *trans* fat is a triglyceride containing one or more *trans*-fatty acids. In such fatty acids, there is at least one unsaturated C═C double bond. On each side of that bond, the single covalent C─C bonds angle in opposite directions (*trans* means "across from") like a pair of bicycle pedals (see arrows in **fig. 2.19a**). This is in contrast to *cis*-fatty acids, in which the two C─C bonds adjacent to the C═C bond angle in the same direction (*cis* means "on the same side"). As you can see **(fig. 2.19b),** the *cis* configuration creates a kink in the chain, whereas the *trans* configuration results in a relatively straight chain. The kink prevents triglycerides with *cis* bonds from packing closely together; hence they are oils. *Trans* fats, however, are more densely packed and therefore solid at room temperature.

Small amounts of *trans* fat occur naturally in meat and dairy products, but in 1911, a food manufacturer invented a method for forcing hydrogen through liquid vegetable oils and converting liquid *cis*-fatty acids to solid *trans*-fatty acids. The resulting *partially hydrogenated oil (PHO),* sold as vegetable shortening, had several advantages and was heavily marketed for household and industrial use. It was easier to use in making such baked goods as pie crusts and biscuits; it gave products a desirable texture and longer shelf life; and to vegetarians, it was more acceptable than animal fat such as lard. *Trans* fats came to be abundantly used not only by home cooks but also in snack foods, baked goods, frosting, coffee creamer, margarine, and fast foods such as french fries and take-out fried chicken.

But *trans* fats resist enzymatic breakdown in the human body, remain in circulation longer, and have more tendency to deposit in the arteries than saturated and *cis* fats do. Therefore, they raise the risk of coronary heart disease (CHD). A study of more than 80,000 nurses who tracked their diets from 1980 to 1994 showed, among other things, that for every 2% increase in calories from *trans* fats as compared to carbohydrates, the women had a 93% elevated incidence of CHD. As the dangers of *trans* fats became widely known, their consumption in the United States declined greatly after 2003 and they were banned in some cities and states, until the FDA concluded in 2015 that they are "not safe for human consumption" and must be eliminated from processed and restaurant foods.

(a) A *trans*-fatty acid (elaidic acid)

(b) A *cis*-fatty acid (oleic acid)

FIGURE 2.19 *Trans-* and *Cis*-Fatty Acids. Each example is unsaturated at the C═C double bond in red. (a) In a *trans*-fatty acid, the bonds on opposite sides of the C═C bond angle in opposite directions (arrows). (b) In a *cis*-fatty acid, the bonds on opposite sides of the C═C bond angle in the same direction. The straight-chain *trans*-fatty acids pack more tightly together, thus remain solid (greasy) at room temperature. Oleic acid melts at 13.5°C (56.3°F), whereas elaidic acid melts at 46.5°C (115.7°F) and therefore would not liquify in the human body.

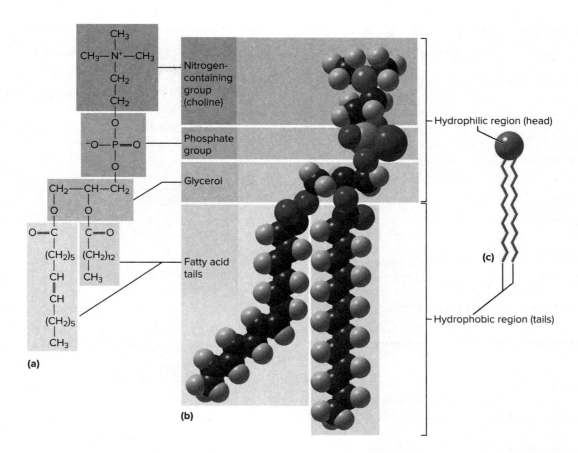

(a)

(b)

(c)

Nitrogen-containing group (choline)

Phosphate group

Glycerol

Fatty acid tails

Hydrophilic region (head)

Hydrophobic region (tails)

FIGURE 2.20 Lecithin, a Representative Phospholipid. (a) Structural formula. (b) A space-filling model that gives some idea of the actual shape of the molecule. (c) A simplified representation of the phospholipid molecule used in diagrams of cell membranes.

linked to other functional groups. Lecithin is a common phospholipid in which the phosphate is bonded to a nitrogenous group called *choline* (CO-leen) **(fig. 2.20).** Phospholipids have a dual nature. The two fatty acid "tails" of the molecule are hydrophobic, but the phosphate "head" is hydrophilic. Thus, phospholipids are said to be **amphipathic**[21] (AM-fih-PATH-ic). Together, the head and the two tails of a phospholipid give it a shape like a clothespin. The most important function of phospholipids is to serve as the structural foundation of cell membranes (see section 3.2a).

Eicosanoids[22] (eye-CO-sah-noyds) are 20-carbon compounds derived from a fatty acid called *arachidonic acid* (ah-RACK-ih-DON-ic). They function primarily as hormonelike chemical signals between cells. The most functionally diverse eicosanoids are **prostaglandins,** in which five of the carbon atoms are arranged in a ring **(fig. 2.21).** These were originally found in the secretions of bovine prostate glands—hence their name—but they are now known to be produced in almost all tissues. They play a variety of signaling roles in inflammation, blood clotting, hormone action, labor contractions, control of blood vessel diameter, and other processes.

A **steroid** is a lipid with 17 of its carbon atoms arranged in four rings **(fig. 2.22).** Cholesterol is the "parent" steroid from

COOH

OH   OH

FIGURE 2.21 Prostaglandin. This is a modified fatty acid with five of its carbon atoms arranged in a ring.

H₃C

CH₃

CH₃

CH₃

CH₃

HO

FIGURE 2.22 Cholesterol. All steroids have this basic four-ringed structure, with variations in the functional groups and locations of double bonds within the rings.

[21]*amphi* = both; *pathic* = feeling
[22]*eicosa* = 20

Some nonpolar amino acids

Some polar amino acids

(a)

which the other steroids are synthesized. The others include cortisol, progesterone, estrogens, testosterone, and bile acids. These differ from each other in the location of C=C bonds within the rings and in the functional groups attached to the rings.

We obtain dietary cholesterol only from foods of animal origin; plants make only trace amounts of no dietary importance. The average adult contains over 200 g (half a pound) of cholesterol. Cholesterol has a bad reputation as a factor in cardiovascular disease (see Deeper Insight 2.4), and it is true that hereditary and dietary factors can elevate blood cholesterol to dangerously high levels. Nevertheless, cholesterol is a natural product of the body and is necessary for human health. In addition to being the precursor of other steroids, it is an important component of cell membranes and is required for proper nervous system function. Only about 15% of our cholesterol comes from the diet; the other 85% is internally synthesized, primarily by the liver.

The principal lipids and their functions are summarized in table 2.7.

2.4e Proteins

The word *protein* is derived from the Greek word *proteios,* meaning "of first importance." Proteins are the most versatile molecules in the body, and many discussions in this book will draw on your understanding of protein structure and behavior.

Amino Acids and Peptides

A **protein** is a polymer of **amino acids.** An amino acid has a central carbon atom with an amino ($-NH_2$) and a carboxyl ($-COOH$) group bound to it **(fig. 2.23a).** The 20 amino acids used to make proteins are identical except for a third functional group called the *radical* (R group) attached to the central carbon. In the simplest amino acid, glycine, R is merely a hydrogen atom, whereas in the

FIGURE 2.23 Amino Acids and Peptides. (a) Four representative amino acids and their symbols. Note that they differ only in the R group, shaded in pink. For structures of all 20 amino acids, see appendix D. (b) The joining of two amino acids by a peptide bond, forming a dipeptide. Side groups R_1 and R_2 could be the groups indicated in pink in part (a), among other possibilities.

largest amino acids it includes rings of carbon. Some radicals are hydrophilic and some are hydrophobic. Being composed of many amino acids, proteins as a whole are therefore often amphipathic. The 20 amino acids involved in proteins are shown in appendix D along with their standard abbreviations.

A **peptide** is any molecule composed of two or more amino acids joined by **peptide bonds.** A peptide bond, formed by dehydration synthesis, joins the amino group of one amino acid to the carboxyl group of the next **(fig. 2.23b).** Peptides are named for the number of amino acids they have—for example, dipeptides have two and tripeptides have three. Chains of fewer than 10 or 15 amino acids are called **oligopeptides,**[23] and chains larger than that are called **polypeptides.** An example of an oligopeptide is the childbirth-inducing hormone oxytocin, composed of 9 amino acids. A representative polypeptide is adrenocorticotropic hormone (ACTH), which is 39 amino acids long. A protein is a polypeptide of 50 amino acids or more. A typical amino acid has a molecular weight of about 80 amu, and the molecular weights of the smallest proteins are around 4,000 to 8,000 amu. The average protein weighs in at about 30,000 amu, and some of them have molecular weights in the hundreds of thousands.

Protein Structure

Proteins have complex coiled and folded structures that are critically important to the roles they play. Even slight changes in their **conformation** (three-dimensional shape) can destroy protein function. Protein structure is now visible even to the level of individual atoms by the newly invented technique of cryo-electron microscopy (cryo-EM), pairing a transmission electron microscope with a supercomputer to image supercooled protein molecules (see this chapter's opening photo). This invention won the 2017 Nobel Prize in Chemistry for its inventors Joachim Frank, Richard Henderson, and Jacques Dubochet.

Protein molecules have three to four levels of complexity, from primary through quaternary structure **(fig. 2.24).**

Primary structure is the protein's sequence of amino acids, which is encoded in the genes **(fig. 2.25;** see also section 4.2b).

Secondary structure is a coiled or folded shape held together by hydrogen bonds between the slightly negative —C$=$O group of one peptide bond and the slightly positive —NH group of another one some distance away. The most common secondary structures are a springlike shape called the **alpha (α) helix** and a pleated, ribbonlike shape, the **beta (β) sheet (β-pleated sheet).** Many proteins have multiple α-helical and β-pleated regions joined by short segments with less orderly geometry. A single protein molecule may fold back on itself and have two or more β-pleated regions linked to one another by hydrogen bonds. Separate, parallel protein chains also may be hydrogen-bonded to each other through their β-pleated regions.

Tertiary[24] **structure** (TUR-she-air-ee) is formed by the further bending and folding of proteins into various globular and fibrous shapes. It results from hydrophobic radicals associating with each other and avoiding water while hydrophilic radicals are attracted to the surrounding water. Van der Waals forces play a significant role in stabilizing tertiary structure. *Globular proteins,* somewhat resembling a wadded ball of yarn, have a compact tertiary structure well suited for proteins embedded in cell membranes and proteins that must move around freely in the body fluids, such as enzymes and antibodies. *Fibrous proteins* such as

myosin, keratin, and collagen are slender filaments better suited for such roles as muscle contraction and providing strength to skin, hair, and tendons.

The amino acid cysteine (Cys), whose radical is —CH_2—SH (see fig. 2.23a), often stabilizes a protein's tertiary structure by forming covalent **disulfide bridges.** When two cysteines align with each other, each can release a hydrogen atom, leaving the sulfur atoms to form a disulfide (—S—S—) bridge. Disulfide bridges hold separate polypeptide chains together in such molecules as antibodies and insulin (fig. 2.25).

Quaternary[25] **structure** (QUA-tur-nare-ee) is the association of two or more polypeptide chains by noncovalent forces such as ionic bonds and hydrophilic–hydrophobic interactions. It occurs in only some proteins. Hemoglobin, for example, consists of four polypeptides: two identical alpha chains and two identical, slightly longer beta chains (fig. 2.24d).

One of the most important properties of proteins is their ability to change conformation, especially tertiary structure. This can be triggered by such influences as voltage changes on a cell membrane during the action of nerve cells, the binding of a hormone to a protein, or the dissociation of a molecule from a protein. Subtle, reversible changes in conformation are important to processes such as enzyme function, muscle contraction, and the opening and closing of pores in cell membranes. **Denaturation** is a more drastic conformational change in response to conditions such as extreme heat or pH. It is seen, for example, when you cook an egg and the egg white (albumen) turns from clear and runny to opaque and stiff. Denaturation makes a protein unable to perform its normal function. It is sometimes reversible, but usually it permanently destroys protein function.

Conjugated proteins have a non–amino acid moiety called a **prosthetic**[26] **group** covalently bound to them. Hemoglobin, for example, not only has the four polypeptide chains described earlier, but each chain also has a complex iron-containing ring called a *heme* moiety attached to it (fig. 2.24d). Hemoglobin cannot transport oxygen unless this group is present. In glycoproteins, as described earlier, the carbohydrate moiety is a prosthetic group.

Protein Functions

Proteins have more diverse functions than other macromolecules. These include

- **Structure.** *Keratin,* a tough structural protein, gives strength to the nails, hair, and skin surface. Deeper layers of the skin, as well as bones, cartilage, and teeth, contain an abundance of the durable protein *collagen.*

- **Communication.** Some hormones and other cell-to-cell signals are proteins, as are the receptors to which the signal molecules bind in the receiving cell. Any hormone or other molecule that reversibly binds to a protein is called a **ligand**[27] (LIG-and).

---

[23]*oligo* = a few
[24]*tert* = third

[25]*quater* = fourth
[26]*prosthe* = appendage, addition
[27]*lig* = to bind

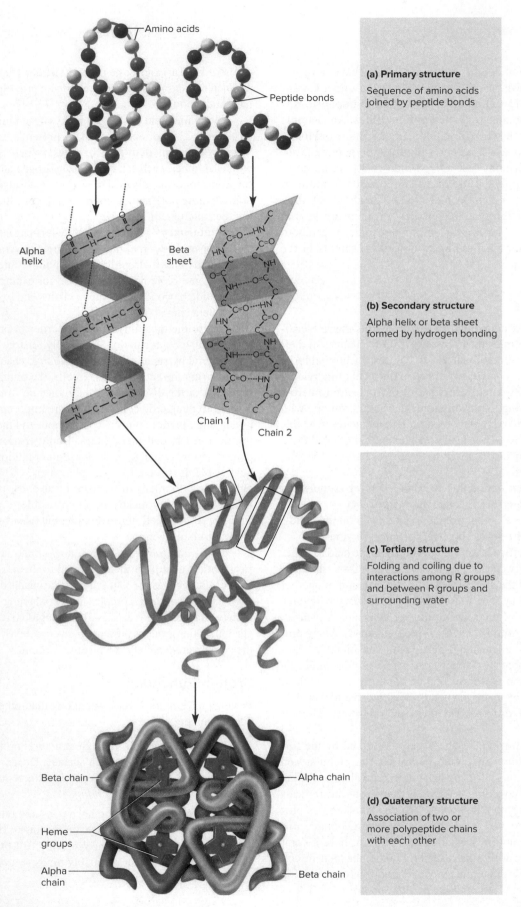

Amino acids

Peptide bonds

(a) Primary structure
Sequence of amino acids joined by peptide bonds

Alpha helix

Beta sheet

Chain 1

Chain 2

(b) Secondary structure
Alpha helix or beta sheet formed by hydrogen bonding

(c) Tertiary structure
Folding and coiling due to interactions among R groups and between R groups and surrounding water

Beta chain

Alpha chain

Heme groups

Alpha chain

Beta chain

(d) Quaternary structure
Association of two or more polypeptide chains with each other

FIGURE 2.24 Four Levels of Protein Structure. (a) Primary structure, the amino acid sequence. (b) Secondary structure, the alpha helix or beta sheet. (c) Tertiary structure, folding and coiling of the chain. (d) Quaternary structure, the association of two or more chains with each other. This example of quaternary structure is hemoglobin, composed of four polypeptide chains. The heme groups are iron-containing nonprotein moieties.

FIGURE 2.25 Primary Structure of Insulin. Insulin is composed of two polypeptide chains joined by three disulfide bridges (heavy bars). See appendix D to interpret the three-letter amino acid symbols.

- **Membrane transport.** Some proteins form channels in cell membranes that govern what passes through the membranes and when. Others act as carriers that briefly bind to solute particles and transport them to the other side of the membrane. Among their other roles, such proteins turn nerve and muscle activity on and off.

- **Catalysis.** Most metabolic pathways of the body are controlled by enzymes, which are globular proteins that function as catalysts.

- **Recognition and protection.** The role of glycoproteins in immune recognition was mentioned earlier. Antibodies and other proteins attack and neutralize organisms that invade the body. Clotting proteins protect the body against blood loss.

- **Movement.** Movement is fundamental to all life, from the intracellular transport of molecules to the galloping of a racehorse. Proteins, with their special ability to change shape repeatedly, are the basis for all such movement. Some proteins are called *molecular motors (motor proteins)* for this reason.

- **Cell adhesion.** Proteins bind cells to each other, which enables sperm to fertilize eggs, enables immune cells to bind to enemy cancer cells, and keeps tissues from falling apart.

2.4f Enzymes and Metabolism

Most **enzymes** are proteins that function as biological catalysts. (Some RNAs also play enzymatic roles.) They enable biochemical reactions to occur rapidly at normal body temperatures. Enzymes were initially given somewhat arbitrary names, some of which are still with us, such as *pepsin* and *trypsin*. The modern system of naming enzymes, however, is more uniform and informative. It identifies the substance the enzyme acts upon, called its **substrate;** sometimes refers to the enzyme's action; and adds the suffix *-ase.* Thus, *amylase* digests starch (*amyl-* = starch) and *carbonic anhydrase* removes water (*anhydr-*) from carbonic acid. Enzyme names may be further modified to distinguish different forms of the same enzyme found in different tissues (see Deeper Insight 2.5).

To appreciate the effect of an enzyme, think of what happens when paper burns. Paper is composed mainly of glucose (in the form of cellulose). The burning of glucose can be represented by the equation

$$C_6H_{12}O_6 + 6\ O_2 \longrightarrow 6\ CO_2 + 6\ H_2O.$$

Paper doesn't spontaneously burst into flame because few of its molecules have enough kinetic energy to react. Lighting the paper with a match, however, raises the kinetic energy enough to initiate combustion (rapid oxidation). The energy needed to get the reaction started, supplied by the match, is called the **activation energy (fig. 2.26a).**

In the body, we carry out the same reaction and oxidize glucose to water and carbon dioxide to extract its energy. We can't tolerate the heat of combustion in our bodies, however, so we must oxidize glucose in a more controlled way at a biologically feasible and safe temperature. Enzymes make this happen by lowering the activation energy—that is, by reducing the barrier to glucose oxidation **(fig. 2.26b)**—and by releasing the energy in small steps rather than a single burst of heat.

Enzyme Structure and Action

Figure 2.27 illustrates the action of an enzyme, using the example of *sucrase,* an enzyme that breaks sucrose down to glucose and fructose. The process occurs in three principal steps:

1. A substrate molecule (such as sucrose) approaches a pocket on the enzyme surface called the **active site.** Amino acid side groups in this region of the enzyme are arranged so as to bind functional groups on the substrate molecule. Many enzymes have two active sites, enabling them to bind two different substrates and bring them together in a way that makes them react more readily with each other.

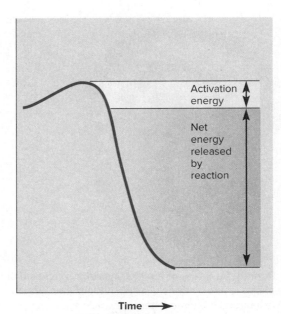

(a) Reaction occurring slowly without a catalyst

(b) Reaction occuring quickly with a catalyst

FIGURE 2.26 Effect of an Enzyme on Activation Energy. (a) Without catalysts, some chemical reactions proceed slowly because of the high activation energy needed to get molecules to react. (b) A catalyst facilitates molecular interaction, thus lowering the activation energy and making the reaction proceed more rapidly.

❓ *Does an enzyme release more energy from its substrate than an uncatalyzed reaction would release?*

FIGURE 2.27 The Three Steps of an Enzymatic Reaction.

② The substrate binds to the enzyme, forming an **enzyme–substrate complex.** The fit between a particular enzyme and its substrate is often compared to a lock and key. Just as only one key fits a particular lock, sucrose is the only substrate that fits the active site of sucrase. Sucrase cannot digest other disaccharides such as maltose or lactose. This selectivity is called **enzyme–substrate specificity.** Unlike a simple lock and key, however, the substrate slightly changes the shape of the enzyme to create a better fit between the two, as shown by the arrows in the figure.

③ Sucrase breaks the bond between the two sugars of sucrose, adding —H^+ and —OH^- groups from water. This hydrolyzes sucrose to two monosaccharides, glucose and fructose, which are then released by the enzyme as its **reaction products.** The enzyme remains unchanged and is ready to repeat the process if another sucrose is available.

Since an enzyme is not consumed by the reaction it catalyzes, one enzyme molecule can consume millions of substrate molecules, and at astonishing speed. A single molecule of carbonic anhydrase, for example, breaks carbonic acid (H_2CO_3) down to H_2O and CO_2 at a rate of 36 million molecules per minute.

Factors that change the shape of an enzyme—notably temperature and pH—tend to alter or destroy the ability of the enzyme to bind its substrate. They disrupt the hydrogen bonds and other weak forces that hold the enzyme in its proper conformation, essentially changing the shape of the "lock" so that the "key" no longer fits. Enzymes vary in optimum pH according to where in the body they normally function. Thus, salivary amylase, which digests starch in the mouth, functions best at pH 7 and is inactivated when it is exposed to stomach acid; pepsin, which works in the acidic environment of the stomach, functions best around pH 2; and trypsin, a digestive enzyme that works in the alkaline environment of the small intestine, has an optimum pH of 9.5. Our internal body temperature is nearly the same everywhere, however, and all human enzymes have a temperature optimum near 37°C at which they produce their fastest reaction rates.

▶▶▶ **APPLY WHAT YOU KNOW**

Why does enzyme function depend on homeostasis?

Cofactors

About two-thirds of human enzymes require a nonprotein partner called a **cofactor.** Inorganic cofactors include iron, copper, zinc, magnesium, and calcium ions. Some of these work by binding to the enzyme and inducing it to fold into a shape that activates its active site. **Coenzymes** are organic cofactors usually derived from niacin, riboflavin, and other water-soluble vitamins. They accept electrons from an enzyme in one metabolic pathway and transfer them to an enzyme in another. For example, cells partially oxidize glucose through a pathway called *glycolysis.* A coenzyme called NAD^+,[28] derived from niacin, shuttles electrons from this pathway to another one called *aerobic respiration,* which uses energy from the electrons to make ATP **(fig. 2.28).** If NAD^+ is unavailable, the glycolysis pathway shuts down.

Metabolic Pathways

A **metabolic pathway** is a chain of reactions with each step usually catalyzed by a different enzyme. A simple metabolic pathway can be symbolized

$$A \xrightarrow{\alpha} B \xrightarrow{\beta} C \xrightarrow{\gamma} D$$

---

[28]nicotinamide adenine dinucleotide

FIGURE 2.28 The Action of a Coenzyme. A coenzyme such as NAD^+ acts as a shuttle that picks up electrons from one metabolic pathway (in this case, glycolysis) and delivers them to another (in this case, aerobic respiration).

where A is the initial *reactant,* B and C are *intermediates,* and D is the *end product.* The Greek letters above the reaction arrows represent enzymes that catalyze each step of the reaction. A is the substrate for enzyme α, B is the substrate for enzyme β, and C for enzyme γ. Such a pathway can be turned on or off by altering the conformation of any of these enzymes, thereby activating or deactivating them. This can be done by such means as the binding or dissociation of a cofactor, or by an end product of the pathway binding to an enzyme at an earlier step (product D binding to enzyme α and shutting off the reaction chain at that step, for example). In these and other ways, cells are able to turn on metabolic pathways when their end products are needed and shut them down when the end products are not needed.

2.4g ATP, Other Nucleotides, and Nucleic Acids

Nucleotides are organic compounds with three principal components: a single or double carbon–nitrogen ring called a *nitrogenous base,* a monosaccharide, and one or more phosphate groups. One of the best-known nucleotides is ATP **(fig. 2.29a),** in which the nitrogenous base is a double ring called *adenine,* the sugar is *ribose,* and there are three phosphate groups.

Adenosine Triphosphate

Adenosine triphosphate (ATP) is the body's most important energy-transfer molecule. It briefly gains energy from exergonic reactions such as glucose oxidation and releases it within seconds for physiological work such as polymerization reactions, muscle contraction, and pumping ions through cell membranes. The second and third phosphate groups of ATP are attached to the rest of the molecule by high-energy covalent bonds traditionally indicated by a wavy line (~) in the structural formula. Since phosphate groups are negatively charged, they repel each other. It requires a high-energy bond to overcome that repellent force and hold them together—especially to add the third phosphate group to a chain that already has two negatively charged phosphates. Most energy transfers to and from ATP involve adding or removing that third phosphate.

Enzymes called **adenosine triphosphatases (ATPases)** are specialized to hydrolyze the third phosphate bond, producing **adenosine diphosphate (ADP)** and an inorganic phosphate group (P_i). This reaction releases 7.3 kilocalories (kcal) of energy for every mole (505 g) of ATP. Most of this energy escapes as heat, but we live on the portion of it that does useful work. We can summarize this as follows:

$$ATP + H_2O \xrightarrow{ATPase} ADP + P_i + Energy \begin{array}{l} \nearrow Heat \\ \searrow Work \end{array}$$

The free phosphate groups released by ATP hydrolysis are often added to enzymes or other molecules to activate them. This addition of P_i, called **phosphorylation,** is carried out by enzymes called **kinases.** The phosphorylation of an enzyme is sometimes the "switch" that turns a metabolic pathway on or off.

ATP is a short-lived molecule, usually consumed within 60 seconds of its formation. The entire amount in the body would

(a) **Adenosine triphosphate (ATP)**   (b) **Cyclic adenosine monophosphate (cAMP)**

FIGURE 2.29 Two Major Nucleotides. (a) Adenosine triphosphate (ATP). (b) Cyclic adenosine monophosphate (cAMP). The last two P~O bonds in ATP, indicated by wavy lines, are high-energy bonds.

support life for less than 1 minute if it weren't continually replenished. At a moderate rate of physical activity, a full day's supply of ATP would weigh twice as much as you do. Even if you never got out of bed, you would need about 45 kg (99 lb) of ATP to stay alive for a day. The reason cyanide is so lethal is that it halts ATP synthesis.

ATP synthesis is explained in detail in chapter 26, but you will find it necessary to become familiar with the general idea of it before you reach that chapter—especially in understanding muscle physiology (chapter 11). Much of the energy for ATP synthesis comes from glucose oxidation (**fig. 2.30**). The first stage in glucose oxidation (**fig. 2.31**) is the reaction pathway known as **glycolysis** (gly-COLL-ih-sis). This literally means "sugar splitting," and indeed its major effect is to split the six-carbon glucose molecule into two three-carbon molecules of *pyruvate*. A little ATP is produced in this stage (a net yield of 2 ATP per glucose), but most of the chemical energy of the glucose is still in the pyruvate.

What happens to pyruvate depends on how much oxygen is available relative to ATP demand. When the demand for ATP outpaces the oxygen supply, excess pyruvate is converted to lactate by a pathway called **anaerobic**[29] **fermentation** (AN-err-OH-bic). This pathway has two noteworthy disadvantages: First, it doesn't extract any more energy from pyruvate; second, the lactate it produces is toxic, so most cells can use anaerobic fermentation only as a temporary measure. The only advantage to this pathway is that it enables glycolysis to continue (for reasons explained in chapter 26) and thus enables a cell to continue producing a small amount of ATP.

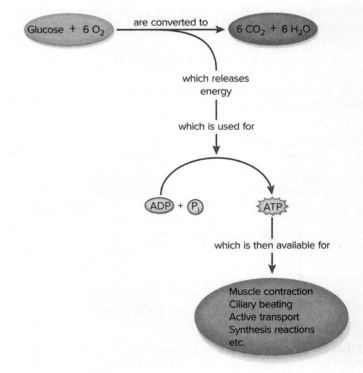

FIGURE 2.30 The Source and Uses of ATP.

---

[29]*an* = without; *aer* = air, oxygen; *obic* = pertaining to life

FIGURE 2.31 ATP Production. Glycolysis produces pyruvate and a net gain of two ATPs. Anaerobic fermentation converts pyruvate to lactate and permits glycolysis to continue producing ATP in the absence of oxygen. Aerobic respiration produces a much greater ATP yield but requires oxygen.

If enough oxygen is available, a more efficient pathway called **aerobic respiration** occurs. This breaks pyruvate down to carbon dioxide and water and generates up to 36 more molecules of ATP for each of the original glucose molecules. The reactions of aerobic respiration are carried out in the cell's *mitochondria*, described in chapter 3 (section 3.4b), so mitochondria are regarded as a cell's principal "ATP factories."

Other Nucleotides

Guanosine triphosphate (GTP) (GWAH-no-seen) is another nucleotide involved in energy transfers. In some reactions, it donates phosphate groups to other molecules. For example, it can donate its third phosphate group to ADP to regenerate ATP.

Cyclic adenosine monophosphate (cAMP) (fig. 2.29b) is a nucleotide formed by the removal of both the second and third phosphate groups from ATP. In many cases, when a hormone or other chemical signal ("first messenger") binds to a cell surface, it triggers an internal reaction that converts ATP to cAMP. The cAMP then acts as a "second messenger" to activate metabolic effects within the cell.

Nucleic Acids

Nucleic acids (new-CLAY-ic) are polymers of nucleotides. The largest of them, **deoxyribonucleic acid (DNA),** is typically 100 million to 1 billion nucleotides long. It constitutes our genes, gives instructions for synthesizing all of the body's proteins, and transfers hereditary information from cell to cell when cells divide and from generation to generation when organisms reproduce. Three forms of **ribonucleic acid (RNA),** which range from 70 to 10,000 nucleotides long, carry out those instructions and synthesize the proteins, assembling amino acids in the right order to produce each protein "described" by the DNA. The detailed structure of DNA and RNA and the mechanisms of protein synthesis and heredity are described in chapter 4.

 DEEPER INSIGHT 2.6

CLINICAL APPLICATION

Anabolic–Androgenic Steroids

It is routine news to hear of sports celebrities suspended or stripped of their honors for the use of anabolic steroids. Magazines of physical culture carry many tragic reports of the deaths of amateur athletes, or violent crimes committed by them, attributed to steroid abuse.

Anabolic steroids, as they are known on the street, are more properly called anabolic–androgenic[30] steroids. They are hormones derived from testosterone that stimulate muscle growth (the anabolic effect) and masculinize the body (the androgenic effect). Perhaps the earliest notion to put them to use arose in Nazi Germany, where testosterone was given to SS troops in an effort to make them more aggressive—but with no proven success. In the 1950s, however, when Soviet weight-lifting teams were routinely defeating American teams, it came to light that the Soviets were using testosterone as a performance enhancer. American team physician John Ziegler began experimenting with this back in the United

States. He disliked the androgenic side effects and approached the Ciba Pharmaceutical Company to develop a testosterone analog (a molecule of slightly altered structure) that would enhance the anabolic effect and weaken the androgenic effect. Ciba soon developed Dianabol. It produced spectacular effects in weight lifters and by the 1960s, several testosterone analogs were freely and legally available, designed to enhance anabolic potency, reduce androgenic effects, and prolong the half-life of the drug in the body. Some are taken orally and others by intramuscular (I.M.) injection.

In limited doses, these steroids have legitimate medical uses such as the treatment of anemia, breast cancer, osteoporosis, and some muscle diseases and to prevent the atrophy of muscles in immobilized patients. Some amateur and professional athletes, however, use them in doses 10 to 1,000 times stronger than therapeutic doses. While such doses bulk up the muscles **(fig. 2.32),** they have hidden, devastating effects on the body. They raise cholesterol levels, which promotes fatty degeneration of the arteries *(atherosclerosis).* This can lead to coronary artery disease, heart and kidney failure, and stroke. Deteriorating circulation

---

[30]*andro* = male; *genic* = producing

FIGURE 2.32 The Effect of Heavy Use of Anabolic–Androgenic Steroids.

Master1305/Shutterstock

produce growth of facial hair, enlargement of the clitoris, atrophy of the breasts and uterus, and irregularities of ovulation and menstruation. An enzyme called *aromatase* converts androgens to estrogens, and in men these often induce breast enlargement *(gynecomastia),* atrophy of the testes, impotence (inability to achieve or maintain an erection), low sperm count, and infertility.

Especially in men, steroid abuse can be linked to severe emotional disorders. Individuals vary in susceptibility, but the androgenic effects include heightened aggressiveness and unpredictable mood swings, so some abusers vacillate between depression and violence ("roid rage"), including physical abuse of family members and crimes as serious as homicide.

As the recreational use of anabolic–androgenic steroids became widespread, so did the tragic side effects of such heavy use—an outcome that Dr. Ziegler deeply regretted as the low point of his career. The U.S. Congress classified anabolic–androgenic steroids as a controlled substance in 1991. Their use in sports has been condemned by the American Medical Association and American College of Sports Medicine and banned by the International Olympic Committee, National Football League, Major League Baseball Players' Association, National Basketball Association, and National Collegiate Athletic Association.

Yet in spite of such warnings and bans, many continue to use steroids and related performance-enhancing drugs, which remain available through unscrupulous coaches, physicians, Internet sources, and foreign mail-order suppliers under a cloud of confusing trade names (Durabolin, Anadrol, Oxandrin, Dianabol, Winstrol, Primobolan, and others). By some estimates, as many as 80% of competitive weight lifters, 30% of college and professional athletes, and 20% of male high-school athletes now use anabolic–androgenic steroids. The National Institutes of Health find increasing use among high-school students and increasing denial that the steroids present a significant health hazard.

also sometimes results in gangrene, and many users have suffered amputation of the lower limbs as a result. As the liver attempts to dispose of the high concentration of steroids, liver cancer and other liver diseases may ensue. In addition, steroids suppress the immune system, so the user is more subject to infection and cancer. They cause a premature end to bone elongation, so people who use anabolic steroids in adolescence may never attain normal adult height.

Paradoxically, anabolic–androgenic steroids can have masculinizing effects on women and feminizing effects on men. In women, who are especially sensitive to the androgenic effect, the steroids commonly

---

BEFORE YOU GO ON

Answer the following questions to test your understanding of the preceding section:

16. Which reaction—dehydration synthesis or hydrolysis—converts a polymer to its monomers? Which one converts monomers to a polymer? Explain your answer.

17. What is the chemical name of blood sugar? What carbohydrate is polymerized to form starch and glycogen?

18. What is the main chemical similarity between carbohydrates and lipids? What are the main differences between them?

19. Explain the statement, All proteins are polypeptides but not all polypeptides are proteins.

20. Which is more likely to be changed by heating a protein, its primary structure or its tertiary structure? Explain.

21. Use the lock-and-key analogy to explain why excessively acidic body fluids (acidosis) could destroy enzyme function.

22. How does ATP change structure in the process of releasing energy?

23. What advantage and disadvantage does anaerobic fermentation have compared with aerobic respiration?

24. How is DNA related to nucleotides?

STUDY GUIDE

▶ Assess Your Learning Outcomes

To test your knowledge, discuss the following topics with a study partner or in writing, ideally from memory.

2.1 Atoms, Ions, and Molecules

1. The definition of *chemical element;* the six most abundant elements in the human body; and trace elements important in human physiology
2. The structure of an atom and the special functional relevance of its valence electrons
3. How isotopes of the same element differ from each other, and how radioisotopes differ from other isotopes
4. The clinical relevance of ionizing radiation; its three forms; the difference between the physical and biological half-life of a radioisotope; and the clinical relevance of that difference
5. The difference between an ion and an atom; how ions form; and the two types of ions and examples of each
6. How an electrolyte differs from an atom and from an ion; the most common ions that constitute electrolytes; and the functions and medical relevance of electrolytes
7. The definition of *free radical;* the medical relevance of free radicals; and how the body is partially protected against them
8. The definitions of *molecule* and *compound*
9. How isomers resemble and differ from each other
10. How a molecule's molecular weight is determined
11. The nature and distinguishing characteristics of ionic bonds, covalent bonds, hydrogen bonds, and van der Waals forces; how polar and nonpolar covalent bonds differ from each other; and how polar covalent bonds can give rise to hydrogen bonds

2.2 Water and Mixtures

1. How the biologically important properties of water arise from the polarity and bond angle of its molecules
2. The difference between a mixture and a compound
3. The differences between solutions, colloids, and suspensions, and examples of body fluids in each category
4. How pH is mathematically defined; the pH scale; and the meanings of *acid* and *base*

5. The action and physiological function of buffers
6. How various measures of chemical concentration differ from each other—weight per volume, percentage, molarity, and milliequivalents per liter—and how each unit of measure provides different kinds of information about the concentration of a solution

2.3 Energy and Chemical Reactions

1. The definition of *energy,* and the two basic forms of energy
2. The differences between decomposition, synthesis, and exchange reactions
3. What determines the direction of a reversible chemical reaction; the nature of a chemical equilibrium
4. Factors that determine the rate of a chemical reaction
5. The definition of *metabolism* and its two subdivisions
6. The difference between oxidation and reduction

2.4 Organic Compounds

1. The criterion for considering a compound to be organic
2. The difference between the carbon backbone and the functional group(s) of an organic molecule; the physiological relevance of functional groups
3. The structures of hydroxyl, methyl, carboxyl, amino, and phosphate functional groups
4. The difference between monomers and polymers; how dehydration synthesis and hydrolysis reactions convert one to the other; and the role of water in both types of reactions
5. The defining characteristics of carbohydrates and their principal roles in the body
6. The names of and basic structural differences between the three monosaccharides, three disaccharides, and three polysaccharides that are most abundant in the diet and most relevant in human physiology
7. The defining characteristics of lipids and their principal roles in the body
8. The major categories of lipids in human physiology, and the roles of each
9. How fatty acids and triglycerides are related through dehydration synthesis and hydrolysis reactions

10. The defining characteristics of amino acids and how the 20 amino acids involved in protein structure differ from each other
11. How amino acids are polymerized, and the structure of peptide bonds
12. Differences between a dipeptide, oligopeptide, polypeptide, and protein
13. The nature of the primary through quaternary levels of protein structure, and how an alpha helix differs from a beta sheet
14. Why protein function depends so strongly on the shape (conformation) of the molecule; how and why functionality is affected by denaturation; and the most common causes of protein denaturation
15. What defines a *conjugated* protein; the general term for the nonprotein component of such a molecule; and examples of conjugated proteins
16. The functions of proteins in human anatomy and physiology
17. How enzymes differ from other proteins, and the general role played by all enzymes
18. The general term for substances acted upon by enzymes; the relevance of active sites to enzyme action; why the active sites limit the range of substances on which an enzyme can act; and the name for this principle of selective enzyme action
19. The nature of cofactors and coenzymes, with examples
20. The term for a chain of linked enzymatic reactions; for its input and output; and for the intermediate steps between the input and output molecules
21. The basic structural components of adenosine triphosphate (ATP) and their organization in the molecule; the function of ATP and why life instantly ceases without it; and where in its molecular structure ATP carries the energy that is transferred to other chemicals
22. Differences between the aerobic and anaerobic mechanisms of producing ATP
23. How guanosine triphosphate (GTP) and cyclic adenosine monophosphate (cAMP) are related to ATP, and their functions
24. The basic chemical nature of the nucleic acids, DNA and RNA; their fundamental function; and how they are structurally related to ATP

STUDY GUIDE

▶ Testing Your Recall

Answers in Appendix A

1. A substance that _____ is considered to be a chemical compound.
   a. contains at least two different elements
   b. contains at least two atoms
   c. has a chemical bond
   d. has a stable valence shell
   e. has covalent bonds

2. An ionic bond is formed when
   a. two anions meet.
   b. two cations meet.
   c. an anion meets a cation.
   d. electrons are unequally shared between nuclei.
   e. electrons transfer completely from one atom to another.

3. The ionization of a sodium atom to produce Na^+ is an example of
   a. oxidation.
   b. reduction.
   c. catabolism.
   d. anabolism.
   e. decomposition.

4. The weakest and most temporary of the following chemical bonds are
   a. polar covalent bonds.
   b. nonpolar covalent bonds.
   c. hydrogen bonds.
   d. ionic bonds.
   e. double covalent bonds.

5. A substance capable of dissolving freely in water is
   a. hydrophilic.
   b. hydrophobic.
   c. hydrolyzed.
   d. hydrated.
   e. amphipathic.

6. A carboxyl group is symbolized
   a. —OH.
   b. —NH_2.
   c. —CH_3.
   d. —CH_2OH.
   e. —COOH.

7. The only polysaccharide synthesized in the human body is
   a. cellulose.
   b. glycogen.
   c. cholesterol.
   d. starch.
   e. prostaglandin.

8. The arrangement of a polypeptide into a fibrous or globular shape is called its
   a. primary structure.
   b. secondary structure.
   c. tertiary structure.
   d. quaternary structure.
   e. conjugated structure.

9. Which of the following functions is more characteristic of carbohydrates than of proteins?
   a. contraction
   b. energy storage
   c. catalyzing reactions
   d. immune defense
   e. intercellular communication

10. The feature that most distinguishes a lipid from a carbohydrate is that a lipid has
    a. more phosphate.
    b. more sulfur.
    c. a lower ratio of carbon to oxygen.
    d. a lower ratio of oxygen to hydrogen.
    e. a greater molecular weight.

11. When an atom gives up an electron and acquires a positive charge, it is called a/an _____.

12. Dietary antioxidants are important because they neutralize _____.

13. Any substance that increases the rate of a reaction without being consumed by it is a/an _____. In the human body, _____ serve this function.

14. All the synthesis reactions in the body form a division of metabolism called _____.

15. A chemical reaction that joins two organic molecules into a larger one and produces water as a by-product is called _____.

16. The suffix _____ denotes a sugar, whereas the suffix _____ denotes an enzyme.

17. The amphipathic lipids of cell membranes are called _____.

18. A chemical named _____ is derived from ATP and widely employed as a "second messenger" in cellular signaling.

19. When oxygen is too limited to meet a cell's ATP demand, a cell can employ a metabolic pathway called _____ to produce ATP.

20. A substance acted upon and changed by an enzyme is called the enzyme's _____.

▶ Building Your Medical Vocabulary

Answers in Appendix A

State a meaning of each word element, and give a medical term from this chapter that uses it or a slight variation of it.

1. a-
2. aero-
3. amphi-
4. caloro-
5. collo-
6. hydro-
7. -mer
8. mono-
9. oligo-
10. -philic

STUDY GUIDE

▶ What's Wrong with These Statements?

Answers in Appendix A

Briefly explain why each of the following statements is false, or reword it to make it true.

1. The monomers of a polysaccharide are called amino acids.

2. Most of the body's reserve energy is stored in the form of ATP.

3. Two molecules with the same atoms arranged in a different order are called isotopes.

4. The products of catabolism contain more chemical energy than the reactants did.

5. A polysaccharide is a chain of simple sugars joined by peptide bonds.

6. A saturated fat is defined as a fat to which no more carbon can be added.

7. Enzymes are often consumed by the reactions they catalyze.

8. The higher the temperature is, the faster an enzyme works.

9. Two percent sucrose and 2% sodium bicarbonate have the same number of molecules per liter of solution.

10. A solution of pH 8 has 10 times the hydrogen ion concentration of a solution with pH 7.

▶ Testing Your Comprehension

1. Suppose a pregnant woman with severe morning sickness has been vomiting steadily for several days. How will her loss of stomach acid affect the pH of her body fluids? Explain.

2. Suppose a person with a severe anxiety attack hyperventilates and exhales CO_2 faster than his body produces it. Consider the carbonic acid reaction in section 2.3b, and explain what effect this hyperventilation will have on his blood pH. (*Hint:* Remember the law of mass action.)

3. In one form of nuclear decay, a neutron breaks down into a proton and electron and emits a gamma ray. Is this an endergonic or exergonic reaction, or neither? Is it an anabolic or catabolic reaction, or neither? Explain both answers.

4. How would the body's metabolic rate be affected if there were no such thing as enzymes? Explain.

5. Some metabolic conditions such as diabetes mellitus cause disturbances in the acid–base balance of the body, which gives the body fluids an abnormally low pH. Explain how this could affect enzyme–substrate reactions and metabolic pathways in the body.

3

CELLULAR FORM AND FUNCTION

Dividing cancer cells from an adenocarcinoma. Adenocarcinoma is a tumor arising from glands in the mucous membrane of an organ such as the lung.

Eye of Science/Science Source

CHAPTER OUTLINE

3.1 Concepts of Cellular Structure
- 3.1a Development of the Cell Theory
- 3.1b Cell Shapes and Sizes
- 3.1c Basic Components of a Cell

3.2 The Cell Surface
- 3.2a The Plasma Membrane
- 3.2b The Glycocalyx
- 3.2c Extensions of the Cell Surface

3.3 Membrane Transport
- 3.3a Filtration
- 3.3b Simple Diffusion
- 3.3c Osmosis
- 3.3d Osmolarity and Tonicity
- 3.3e Carrier-Mediated Transport
- 3.3f Vesicular Transport

3.4 The Cell Interior
- 3.4a The Cytoskeleton
- 3.4b Organelles
- 3.4c Inclusions

Study Guide

DEEPER INSIGHTS

3.1 Clinical Application: Calcium Channel Blockers

3.2 Clinical Application: Cystic Fibrosis

3.3 Evolutionary Medicine: Mitochondria—Evolution and Clinical Significance

Anatomy & Physiology Revealed® 4.0

Module 2: Cells and Chemistry

All organisms, from the simplest to the most complex, are composed of cells—whether the single cell of a bacterium or the trillions of cells that constitute the human body. These cells are responsible for all structural and functional properties of a living organism. A knowledge of cells is therefore indispensable to any true understanding of the workings of the human body, the mechanisms of disease, and the rationale of therapy. Thus, this chapter and the next one introduce the basic cell biology of the human body, and subsequent chapters expand upon this information as we examine the specialized cellular structure and function of specific organs.

3.1 Concepts of Cellular Structure

Expected Learning Outcomes

When you have completed this section, you should be able to

a. discuss the development and modern tenets of the cell theory;

b. describe cell shapes from their descriptive terms;

c. state the size range of human cells and discuss factors that limit their size;

d. discuss the way that developments in microscopy have changed our view of cell structure; and

e. outline the major components of a cell.

3.1a Development of the Cell Theory

Cytology,[1] the scientific study of cells, was born in 1663 when Robert Hooke observed the empty cell walls of cork and coined the word *cellulae* ("little cells") to describe them (see section 1.2). Soon he studied thin slices of fresh wood and saw living cells "filled with juices"—a fluid later named *cytoplasm.* Two centuries later, Theodor Schwann studied a wide range of animal tissues and concluded that all animals are made of cells.

Schwann and other biologists originally believed that cells came from nonliving body fluid that somehow congealed and acquired a membrane and nucleus. This idea of *spontaneous generation*—that living things arise from nonliving matter—was rooted in the scientific thought of the times. For centuries, it seemed to be simple common sense that decaying meat turned into maggots, stored grain into rodents, and mud into frogs. Schwann and his contemporaries merely extended this idea to cells. The idea of spontaneous generation wasn't discredited until some classic experiments by French microbiologist Louis Pasteur in 1859.

By the end of the nineteenth century, it was established beyond all reasonable doubt that cells arise only from other cells and every living organism is composed of cells and cell products. The cell came to be regarded, and still is, as the simplest structural and functional unit of life. There are no smaller subdivisions of a cell or organism that, in themselves, have all or most of the fundamental characteristics of life described in section 1.6a. Enzymes and organelles, for example, are not alive, although the life of a cell depends on their activity.

The development of biochemistry from the late nineteenth to the twentieth century made it further apparent that all physiological processes of the body are based on cellular activity and that the cells of all species exhibit remarkable biochemical unity. The various generalizations of these last two paragraphs now constitute the modern **cell theory.**

3.1b Cell Shapes and Sizes

We will shortly examine the structure of a generic cell, but the generalizations we draw shouldn't blind you to the diversity of cellular form and function in humans. There are about 200 kinds of cells in the human body, with a variety of shapes, sizes, and functions.

Descriptions of organ and tissue structure often refer to the shapes of cells by the following terms (**fig. 3.1**):

- **Squamous**[2] (SKWAY-mus)—a thin, flat, scaly shape, often with a bulge where the nucleus is, much like the shape of a fried egg "sunny side up." Squamous cells line the esophagus and form the surface layer (epidermis) of the skin.

- **Cuboidal**[3] (cue-BOY-dul)—squarish-looking in frontal sections and about equal in height and width; liver cells are a good example.

- **Columnar**—distinctly taller than wide, such as the inner lining cells of the stomach and intestines.

- **Polygonal**[4]—having irregularly angular shapes with four, five, or more sides.

- **Stellate**[5]—having multiple pointed processes projecting from the body of a cell, giving it a somewhat starlike shape. The cell bodies of many nerve cells are stellate.

---

[1] *cyto* = cell; *logy* = study of

[2] *squam* = scale; *ous* = characterized by
[3] *cub* = cube; *oidal* = like, resembling
[4] *poly* = many; *gon* = angles
[5] *stell* = star; *ate* = resembling, characterized by

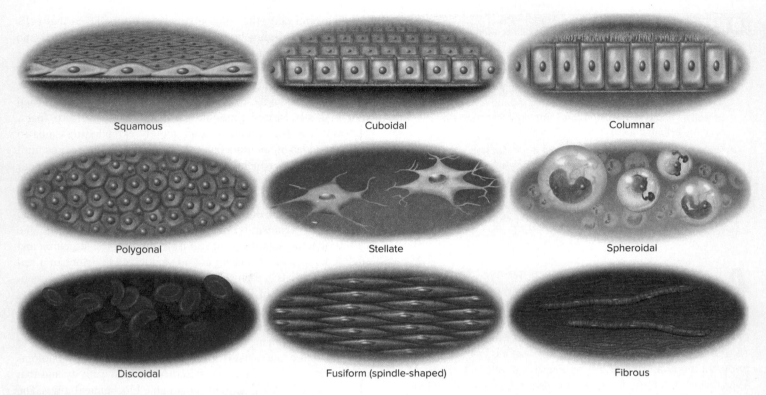

Squamous Cuboidal Columnar

Polygonal Stellate Spheroidal

Discoidal Fusiform (spindle-shaped) Fibrous

FIGURE 3.1 Common Cell Shapes. APR

- **Spheroidal** to **ovoid**—round to oval, as in egg cells and white blood cells.
- **Discoidal**—disc-shaped, as in red blood cells.
- **Fusiform**[6] (FEW-zih-form)—spindle-shaped; elongated, with a thick middle and tapered ends, as in smooth muscle cells.
- **Fibrous**—long, slender, and threadlike, as in skeletal muscle cells and the axons (nerve fibers) of nerve cells.

Some of these shapes refer to the way a cell looks in typical tissue sections, not to the complete three-dimensional shape of the cell. A cell that looks squamous, cuboidal, or columnar in a tissue section, for example, usually looks polygonal if viewed from its upper surface.

The most useful unit of measure for designating cell sizes is the **micrometer (μm),** formerly called the micron—one-millionth (10^{-6}) of a meter, one-thousandth (10^{-3}) of a millimeter. (See appendix B for units of measurement.) The smallest objects most people can see with the naked eye are about 100 μm, which is about one-quarter the size of the period at the end of a typical sentence of print. A few human cells fall within this range, such as the egg cell and some fat cells, but most human cells are about 10 to 15 μm wide. The longest human cells are nerve cells (sometimes over a meter long) and muscle cells

(up to 30 cm long), but both are usually too slender to be seen with the naked eye.

There are several factors that limit the size of cells. If a cell swelled to excessive size, it could rupture like an overfilled water balloon. In addition, cell size is limited by the relationship between its volume and surface area. The surface area of a cell is proportional to the square of its diameter, while volume is proportional to the cube of its diameter. Thus, for a given increase in diameter, volume increases much more than surface area. Picture a cuboidal cell 10 μm on each side **(fig. 3.2).** It would have a surface area of 600 μm^2 (10 μm × 10 μm × 6 sides) and a volume of 1,000 μm^3 (10 × 10 × 10 μm). Now, suppose it grew by another 10 μm on each side. Its new surface area would be 2,400 μm^2 (20 μm × 20 μm × 6) and its volume would be 8,000 μm^3 (20 × 20 × 20 μm). The 20 μm cell has eight times as much cytoplasm needing nourishment and waste removal, but only four times as much membrane surface through which wastes and nutrients can be exchanged. A cell that is too big cannot support itself.

Further, if a cell were too large, molecules couldn't diffuse from place to place fast enough to support its metabolism. The time required for diffusion is proportional to the square of distance, so if a cell diameter doubled, the travel time for molecules within the cell would increase fourfold. For example, if it took 10 seconds for a molecule to diffuse from the surface to the center of a cell with a 10 μm radius, then we increased the cell radius to 1 mm, it would take 278 hours to reach the center—far too slow to support the cell's life activities.

---

[6]*fusi* = spindle; *form* = shape

Large cell

Diameter	= 20 µm
Surface area	= 20 µm × 20 µm × 6 = 2,400 µm²
Volume	= 20 µm × 20 µm × 20 µm = 8,000 µm³

Small cell

Diameter	= 10 µm
Surface area	= 10 µm × 10 µm × 6 = 600 µm²
Volume	= 10 µm × 10 µm × 10 µm = 1,000 µm³

Effect of cell growth:

Diameter (D) increased by a factor of 2
Surface area increased by a factor of 4 (= D^2)
Volume increased by a factor of 8 (= D^3)

FIGURE 3.2 The Relationship Between Cell Surface Area and Volume. As a cell doubles in diameter, its volume increases eightfold, but its surface area increases only fourfold. A cell that is too large may have too little plasma membrane to serve the metabolic needs of the increased volume of cytoplasm.

Having organs composed of many small cells instead of fewer large ones has another advantage. The death of one or a few cells has less effect on the structure and function of the whole organ.

3.1c Basic Components of a Cell

In Schwann's time, little was known about cells except that they were enclosed in a membrane and contained a nucleus. The fluid between the nucleus and surface membrane, its **cytoplasm,**[7] was thought to be little more than a gelatinous mixture of chemicals and vaguely defined particles. The **transmission electron microscope (TEM),** invented in the mid-twentieth century, radically changed this concept. Using a beam of electrons in place of light, the TEM enabled biologists to see a cell's *ultrastructure,* a fine degree of detail extending even to the molecular level. The most important thing about a good microscope is not magnification but

[7]*cyto* = cell; *plasm* = formed, molded

resolution—the ability to reveal detail. Any image can be photographed and enlarged as much as we wish, but if enlargement fails to reveal any more useful detail, it is *empty magnification.* A big blurry image is not nearly as informative as one that is small and sharp. The TEM reveals far more detail than the light microscope (LM) **(fig. 3.3).** A later invention, the **scanning electron microscope (SEM),** produces dramatic three-dimensional images at high magnification and resolution (see fig. 3.10a), but can view only surface features.

A stunning application of SEM, often seen in this book, is the *vascular corrosion cast* technique for visualizing the blood vessels of an organ. The vessels are drained and flushed with saline, then carefully filled with a resin. After the resin solidifies, the actual tissue is dissolved with a corrosive agent such as potassium hydroxide. This leaves only a resin cast of the vessels, which is then photographed with the SEM. The resulting images are not only strikingly beautiful, but also give great insights into the blood supply to an organ from macro- to microscopic levels (see figs. 19.10c, 23.10a, and the opening page of chapter 17).

(a) Light microscope (LM)

(b) Transmission electron microscope (TEM)

FIGURE 3.3 Magnification Versus Resolution. Two white blood cells (neutrophils) shown at the same magnification. (a) Photographed with the light microscope (LM). (b) Photographed with the transmission electron microscope (TEM). Note the finer detail (resolution) obtained with the TEM.

a: Alvin Telser/McGraw-Hill Education; **b:** Biophoto Associates/Science Source

Table 3.1 gives the sizes of some cells and subcellular objects relative to the resolution of the naked eye, light microscope, and TEM. You can see why the very existence of cells was unsuspected until the light microscope was invented, and why little was known about their internal components until the TEM became available.

Figure 3.4 shows some major constituents of a typical cell. The cell is surrounded by a **plasma (cell) membrane** made of proteins and lipids. The composition and functions of this membrane can differ significantly from one region of a cell to another, especially between the basal, lateral, and apical (upper) surfaces of cells like the one pictured.

The cytoplasm is crowded with fibers, tubules, passages, and compartments. It contains the *cytoskeleton,* a supportive framework of protein filaments and tubules; an abundance of *organelles,* diverse structures that perform various metabolic tasks for the cell; and *inclusions,* which are foreign matter or stored cell products. A cell may have 10 billion protein molecules, including potent enzymes with the potential to destroy the cell if they're not contained and isolated from other cellular components. You can imagine the enormous problem of keeping track of all this material, directing molecules to the correct destinations, and maintaining order against nature's incessant trend toward disorder. Cells maintain order partly by compartmentalizing their contents in the organelles.

The cytoskeleton, organelles, and inclusions are embedded in a clear gel called the **cytosol**[8] or **intracellular fluid (ICF).** All body fluids not contained in the cells are collectively called the **extracellular fluid (ECF).** The ECF located amid the cells is also called **tissue (interstitial) fluid.** Some other extracellular fluids include blood plasma, lymph, and cerebrospinal fluid.

In summary, we regard cells as having the following major components:

Plasma membrane

Cytoplasm

    Cytoskeleton

    Organelles (including nucleus)

    Inclusions

    Cytosol

BEFORE YOU GO ON

Answer the following questions to test your understanding of the preceding section:

1. What are the basic principles of the cell theory?

2. What does it mean to say a cell is squamous, stellate, columnar, or fusiform?

3. Why can cells not grow to unlimited size?

4. What is the difference between cytoplasm and cytosol?

5. Define *intracellular fluid (ICF)* and *extracellular fluid (ECF).*

[8]*cyto* = cell; *sol* = dissolved matter

Object	Size
TABLE 3.1	**Sizes of Biological Structures in Relation to the Resolution of the Eye, Light Microscope, and Transmission Electron Microscope**
Visible to the Naked Eye (Resolution 70–100 μm)	
Human egg, diameter	100 μm
Visible with the Light Microscope (Resolution 200 nm)	
Most human cells, diameter	10–15 μm
Cilia, length	7–10 μm
Mitochondria, width × length	0.2 × 4 μm
Bacteria *(Escherichia coli),* length	1–3 μm
Microvilli, length	1–2 μm
Lysosomes, diameter	0.5 μm = 500 nm
Visible with the Transmission Electron Microscope (Resolution 0.5 nm)	
Nuclear pores, diameter	30–100 nm
Centriole, diameter × length	20 × 50 nm
Poliovirus, diameter	30 nm
Ribosomes, diameter	15 nm
Globular proteins, diameter	5–10 nm
Plasma membrane, thickness	7.5 nm
DNA molecule, diameter	2.0 nm
Plasma membrane channels, diameter	0.8 nm

3.2 The Cell Surface

Expected Learning Outcomes

When you have completed this section, you should be able to

a. describe the structure of the plasma membrane;

b. explain the functions of the lipid, protein, and carbohydrate components of the plasma membrane;

c. describe a second-messenger system and discuss its importance in human physiology;

d. explain the composition and functions of the glycocalyx that coats cell surfaces; and

e. describe the structure and functions of microvilli, cilia, flagella, and pseudopods.

Many physiologically important processes occur at the surface of a cell—immune responses, the binding of egg and sperm, cell-to-cell signaling by hormones, and the detection of tastes and smells, for example. A substantial part of this chapter is therefore

FIGURE 3.4 **Structure of a Representative Cell.**

concerned with the cell surface. Like explorers of a new continent, we will examine the interior only after we've investigated its coastline. In this section, we examine the structure of the plasma membrane, surface features such as cilia and microvilli, and methods of transport through the membrane.

3.2a The Plasma Membrane

The plasma membrane defines the boundaries of the cell, governs its interactions with other cells, and controls the passage of materials into and out of the cell. It appears to the electron microscope as a pair of dark parallel lines with a total thickness of about

7.5 nm **(fig. 3.5a).** The side that faces the cytoplasm is the *intracellular face* of the membrane, and the side that faces outward is the *extracellular face.* Similar membranes enclose most of a cell's organelles and control their uptake and release of chemicals.

Membrane Lipids

Figure 3.5b shows our current concept of the molecular structure of the plasma membrane—an oily film of lipids with proteins embedded in it. Typically about 98% of the membrane molecules are lipids, and about 75% of those are phospholipids. These amphipathic molecules arrange themselves into a bilayer, with their hydrophilic phosphate-containing heads facing the water on each side and their hydrophobic tails directed toward the center, avoiding the water. The phospholipids drift laterally from place to place, spin on their axes, and flex their tails. These movements keep the membrane fluid.

Plasma membrane of upper cell
Intercellular space
Plasma membrane of lower cell

Nuclear envelope
Nucleus

(a)

100 nm

▶▶▶ **APPLY WHAT YOU KNOW**

What would happen if the plasma membrane were made primarily of a hydrophilic substance such as carbohydrate? Which of the major themes at the end of chapter 1 does this point best exemplify?

Cholesterol molecules, found near the membrane surfaces amid the phospholipids, constitute about 20% of the membrane lipids. By interacting with the phospholipids and holding them still, cholesterol can stiffen the membrane (make it less fluid) in spots. Higher concentrations of cholesterol, however, can increase membrane fluidity by preventing phospholipids from packing closely together.

The remaining 5% of the membrane lipids are glycolipids—phospholipids with short carbohydrate chains on the extracellular face of the membrane. They contribute to the *glycocalyx,* a carbohydrate coating on the cell surface with multiple functions described shortly.

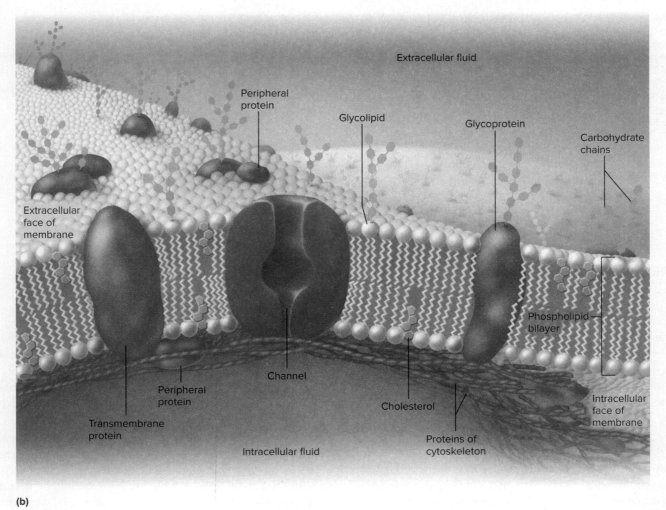

Extracellular fluid

Peripheral protein

Glycolipid

Glycoprotein

Carbohydrate chains

Extracellular face of membrane

Phospholipid bilayer

Intracellular face of membrane

Transmembrane protein

Peripheral protein

Channel

Cholesterol

Proteins of cytoskeleton

Intracellular fluid

(b)

FIGURE 3.5 The Plasma Membrane. (a) Plasma membranes of two adjacent cells (TEM). (b) Molecular structure of the plasma membrane.

a: Dr. Donald Fawcett/Science Source

Membrane Proteins

Although proteins are only about 2% of the molecules of the plasma membrane, they are larger than lipids and average about 50% of the membrane weight. There are two broad classes of membrane proteins: transmembrane and peripheral. **Transmembrane proteins** pass completely through the phospholipid bilayer. They have hydrophilic regions in contact with the water on both sides of the membrane, and hydrophobic regions that pass back and forth through the lipid **(fig. 3.6).** Most transmembrane proteins are glycoproteins, bound to oligosaccharides on the extracellular side of the membrane. Many of these proteins drift about freely in the phospholipid film, like ice cubes floating in a bowl of water. Others are anchored to the *cytoskeleton*—an intracellular system of tubules and filaments discussed later. **Peripheral proteins** do not protrude into the phospholipid layer but adhere to either the inner or outer face of the membrane. Those on the inner face are typically anchored to a transmembrane protein as well as to the cytoskeleton.

The functions of membrane proteins include:

- **Receptors (fig. 3.7a).** Many of the chemical signals by which cells communicate (epinephrine, for example) cannot enter the target cell but bind to surface proteins called receptors. Receptors are usually specific for one particular messenger, much like an enzyme that is specific for one substrate. Plasma membranes also have receptor proteins that bind chemicals and transport them into the cell, as discussed later in this chapter.

- **Second-messenger systems.** When a messenger binds to a surface receptor, it may trigger changes within the cell that produce a second messenger in the cytoplasm. This process involves both transmembrane proteins (the receptors) and peripheral proteins. Second-messenger systems are also discussed later in more detail.

- **Enzymes (fig. 3.7b).** Enzymes in the plasma membrane carry out the final stages of starch and protein digestion in the small intestine, help produce second messengers, and break down hormones and other signaling molecules whose job is done, thus stopping them from excessively stimulating a cell.

- **Channel proteins (fig. 3.7c).** Channels are passages that allow water and hydrophilic solutes to move through the membrane. A channel is a tunnel that passes through a complex of multiple proteins or between subunits of an individual protein. Some of them, called **leak channels,** are always open and allow materials to pass through continually. Others, called **gates (gated channels),** open and close under different circumstances and allow solutes through at some times, but not others **(fig. 3.7d).** These gates respond to three types of stimuli: **ligand-gated channels** respond to chemical messengers, **voltage-gated channels** to changes in electrical potential (voltage) across the plasma membrane, and **mechanically gated channels** to physical stress on a cell, such as stretch and pressure. By controlling the movement of electrolytes through the plasma membrane, gated channels play an important role in the timing of nerve signals and muscle contraction (see Deeper Insight 3.1). Some receptors double in function as gated channels. When a nerve stimulates a muscle, for example, a chemical from the nerve fiber binds to a receptor on the muscle fiber and the receptor opens to allow sodium and potassium ions to flow through and excite the muscle. Defects in channel proteins are responsible for a family of diseases called *channelopathies.*

FIGURE 3.6 Transmembrane Proteins. A transmembrane protein has hydrophobic regions embedded in the phospholipid bilayer and hydrophilic regions projecting into the intracellular and extracellular fluids. The protein may cross the membrane once (left) or multiple times (right). The intracellular regions are often anchored to the cytoskeleton by peripheral proteins.

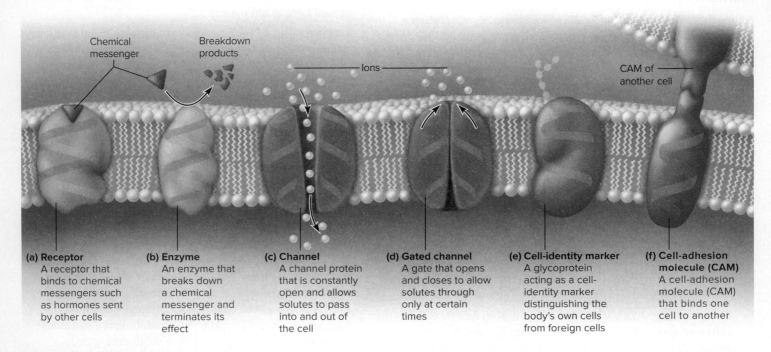

(a) Receptor
A receptor that binds to chemical messengers such as hormones sent by other cells

(b) Enzyme
An enzyme that breaks down a chemical messenger and terminates its effect

(c) Channel
A channel protein that is constantly open and allows solutes to pass into and out of the cell

(d) Gated channel
A gate that opens and closes to allow solutes through only at certain times

(e) Cell-identity marker
A glycoprotein acting as a cell-identity marker distinguishing the body's own cells from foreign cells

(f) Cell-adhesion molecule (CAM)
A cell-adhesion molecule (CAM) that binds one cell to another

FIGURE 3.7 Some Functions of Membrane Proteins.

- **Carriers** (see fig. 3.17). Carriers are transmembrane proteins that bind to glucose, electrolytes, and other solutes and transfer them to the other side of the membrane. Some carriers, called **pumps,** consume ATP in the process.

- **Cell-identity markers (fig. 3.7e).** Glycoproteins contribute to the glycocalyx, which acts like an "identification tag" that enables our bodies to tell which cells belong to it and which are foreign invaders.

DEEPER INSIGHT 3.1

CLINICAL APPLICATION

Calcium Channel Blockers

Calcium channel blockers are a class of drugs that show the therapeutic relevance of understanding gated membrane channels. The walls of the arteries contain smooth muscle that contracts or relaxes to change their diameter. These changes modify the blood flow and strongly influence blood pressure. Blood pressure rises when the arteries constrict and falls when they relax and dilate. Excessive, widespread vasoconstriction can cause hypertension (high blood pressure), and vasoconstriction in the coronary blood vessels of the heart can cause pain (angina) due to inadequate blood flow to the cardiac muscle. In order to contract, a smooth muscle cell must open calcium channels in its plasma membrane and allow calcium to enter from the extracellular fluid. Calcium channel blockers prevent these channels from opening and thereby relax the arteries, increase blood flow, relieve angina, and lower the blood pressure.

- **Cell-adhesion molecules (fig. 3.7f).** Cells adhere to one another and to extracellular material through membrane proteins called cell-adhesion molecules (CAMs). With few exceptions (such as blood cells and metastasizing cancer cells), cells don't grow or survive normally unless they're mechanically linked to the extracellular material. Special events such as sperm–egg binding and the binding of an immune cell to a cancer cell also require CAMs.

Second Messengers

Second messengers are of such importance that they require a closer look. You will find this information essential for your later understanding of hormone and neurotransmitter action. Let's consider how the hormone epinephrine stimulates a cell. Epinephrine, the "first messenger," cannot pass through the plasma membrane, so it binds to a surface receptor. The receptor is linked on the intracellular side to a peripheral **G protein (fig. 3.8).** G proteins are named for the ATP-like chemical, guanosine triphosphate (GTP), from which they get their energy. When activated by the receptor, a G protein relays the signal to another membrane protein, **adenylate cyclase** (ah-DEN-ih-late SY-clase). Adenylate cyclase removes two phosphate groups from ATP and converts it to **cyclic AMP (cAMP),** the second messenger (see fig. 2.29b). Cyclic AMP then activates cytoplasmic enzymes called **kinases** (KY-nace-es), which add phosphate groups to other cellular enzymes. This activates some enzymes and deactivates others, but either way, it triggers a great variety of physiological changes within the cell. Up to 60% of drugs work by altering the activity of G proteins.

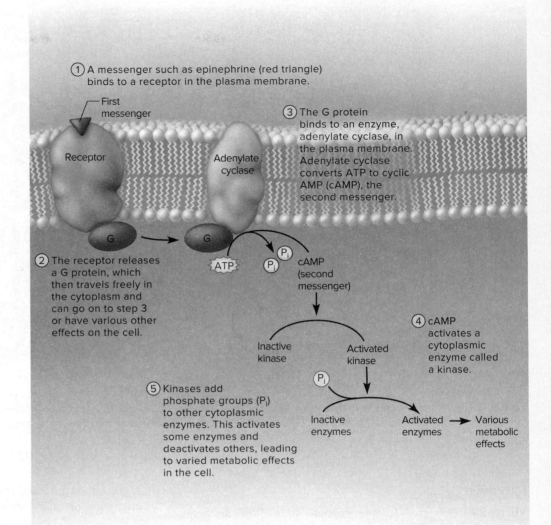

① A messenger such as epinephrine (red triangle) binds to a receptor in the plasma membrane.

First messenger

Receptor

Adenylate cyclase

③ The G protein binds to an enzyme, adenylate cyclase, in the plasma membrane. Adenylate cyclase converts ATP to cyclic AMP (cAMP), the second messenger.

② The receptor releases a G protein, which then travels freely in the cytoplasm and can go on to step 3 or have various other effects on the cell.

ATP

Pᵢ

Pᵢ

cAMP (second messenger)

④ cAMP activates a cytoplasmic enzyme called a kinase.

Inactive kinase

Activated kinase

⑤ Kinases add phosphate groups (Pᵢ) to other cytoplasmic enzymes. This activates some enzymes and deactivates others, leading to varied metabolic effects in the cell.

Pᵢ

Inactive enzymes

Activated enzymes → Various metabolic effects

FIGURE 3.8 A Second-Messenger System.

❓ *Is adenylate cyclase a transmembrane protein or a peripheral protein? What about the G protein?*

3.2b The Glycocalyx

External to the plasma membrane, all animal cells have a fuzzy coat called the **glycocalyx**[9] (GLY-co-CAY-licks) **(fig. 3.9),** composed of the carbohydrate moieties of membrane glycolipids and glycoproteins. It is chemically unique in everyone but identical twins, and acts like an identification tag that enables the body to distinguish its own healthy cells from transplanted tissues, invading organisms, and diseased cells. Human blood types and transfusion compatibility are determined by glycolipids. Functions of the glycocalyx are summarized in **table 3.2.**

3.2c Extensions of the Cell Surface

Many cells have surface extensions called *microvilli, cilia, flagella,* and *pseudopods.* These aid in absorption, movement, and sensory processes.

[9]*glyco* = sugar; *calyx* = cup, vessel

TABLE 3.2	Functions of the Glycocalyx
Protection	Cushions the plasma membrane and protects it from physical and chemical injury
Immunity to infection	Enables the immune system to recognize and selectively attack foreign organisms
Defense against cancer	Changes in the glycocalyx of cancerous cells enable the immune system to recognize and destroy them
Transplant compatibility	Forms the basis for compatibility of blood transfusions, tissue grafts, and organ transplants
Cell adhesion	Binds cells together so tissues do not fall apart
Fertilization	Enables sperm to recognize and bind to eggs
Embryonic development	Guides embryonic cells to their destinations in the body

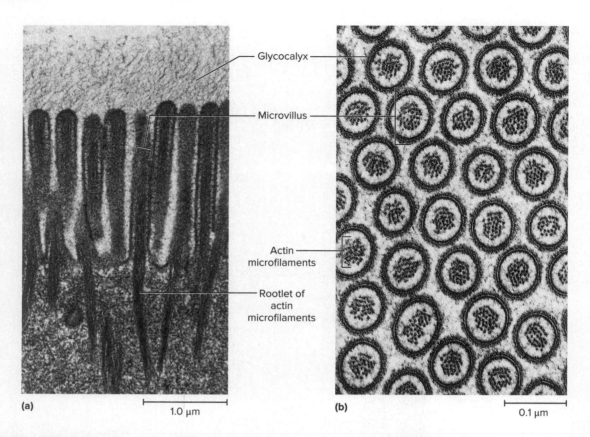

(a)

1.0 μm

(b)

0.1 μm

Glycocalyx

Microvillus

Actin microfilaments

Rootlet of actin microfilaments

FIGURE 3.9 Microvilli and the Glycocalyx (TEM). The microvilli are anchored by microfilaments of actin, which occupy the core of each microvillus and project into the cytoplasm. (a) Longitudinal section, perpendicular to the cell surface. (b) Cross section.

a: Don W. Fawcett/Science Source; **b:** Biophoto Associates/Science Source

Microvilli

Microvilli[10] (MY-cro-VIL-eye; singular, *microvillus*) are extensions of the plasma membrane that serve primarily to increase a cell's surface area (figs. 3.9 and 3.10a, c). They are best developed in cells specialized for absorption, such as the epithelial cells of the intestines and kidneys. They give such cells 15 to 40 times as much absorptive surface area as they would have if their apical surfaces were flat.

Individual microvilli cannot be distinguished very well with the light microscope because they are only 1 to 2 μm long. On some cells, they are very dense and appear as a fringe called the **brush border** at the apical cell surface. With the scanning electron microscope, they resemble a deep-pile carpet. With the transmission electron microscope, microvilli typically look like finger-shaped projections of the cell surface. They show little internal structure, but some have a bundle of stiff filaments of a protein called *actin*. Actin filaments attach to the inside of the plasma membrane at the tip of the microvillus, and at its base they extend a little way into the cell and anchor the microvillus to a protein mesh called the *terminal web*. When tugged by another protein in the cytoplasm, actin can shorten a microvillus to milk its absorbed contents downward into the cell.

In contrast to the long, shaggy microvilli of absorptive cells, those on many other cells are little more than tiny bumps on the surface. On cells of the taste buds and inner ear, they are well developed but serve sensory rather than absorptive functions.

Cilia and Flagella

Cilia (SIL-ee-uh; singular, *cilium*[11]) **(fig. 3.10)** are hairlike processes about 7 to 10 μm long. Nearly every human cell has a single, nonmotile *primary cilium* a few micrometers long. Its function in some cases is still a mystery, but many of them are sensory, serving as the cell's "antenna" for monitoring nearby conditions. In the inner ear, they play a role in the sense of balance; in the retina of the eye, they are highly elaborate and form the light-absorbing part of the receptor cells; and in the kidney, they're thought to monitor the flow of fluid as it is processed into urine. In some cases, they open calcium gates in the plasma membrane, activating an informative signal in the cell. Sensory cells in the nose have multiple nonmotile cilia that bind odor molecules. Defects in the development, structure, or function of cilia—especially these nonmotile primary cilia—are sometimes responsible for birth defects and hereditary diseases called

[10]*micro* = small; *villi* = hairs

[11]*cilium* = eyelash

FIGURE 3.10 Cilia. (a) Epithelium of the uterine (fallopian) tube (SEM). The short, mucus-secreting cells between the ciliated cells show bumpy microvilli on their surfaces. (b) Three-dimensional structure of a cilium. (c) Cross section of a few cilia and microvilli (TEM). (d) Cross-sectional structure of a cilium. Note the relative sizes of cilia and microvilli in parts (a) and (c).

a: Steve Gschmeissner/Science Photo Library/Getty Images; c: Don Fawcett/Science Source

ciliopathies. (See Testing Your Comprehension question 5 at the end of this chapter.)

Motile cilia are less widespread, but more numerous on the cells that do have them. They occur in the respiratory tract, uterine (fallopian) tubes, internal cavities (*ventricles*) of the brain, and short ducts (*efferent ductules*) associated with the testes. There may be 50 to 200 cilia on the surface of one cell. They beat in waves that sweep across the surface of an epithelium, always in the same direction **(fig. 3.11),** propelling such materials as mucus, an egg cell, or cerebrospinal fluid. Each cilium bends stiffly forward and produces a *power stroke* that pushes along the mucus or other matter. Shortly after a cilium begins its power stroke, the one just ahead of it begins, and the next and the next—collectively producing a wavelike motion. After a cilium completes its power stroke, it is pulled limply back by a *recovery stroke* that restores it to the upright position, ready to flex again.

Mucus
Saline layer
Epithelial cells

(a)

1　2　3　4　　　　5　6　　　7
Power stroke　　　　Recovery stroke

(b)

FIGURE 3.11 Ciliary Action. (a) Cilia of an epithelium moving mucus along a surface layer of saline. (b) Power and recovery strokes of an individual cilium. The cilium goes limp on the recovery stroke to return to its original position without touching the mucus above.

▶▶▶**APPLY WHAT YOU KNOW**

How would the movement of mucus in the respiratory tract be affected if cilia were equally stiff on both their power and recovery strokes?

Cilia couldn't beat freely if they were embedded in sticky mucus (see Deeper Insight 3.2). Instead, they beat within a saline (saltwater) layer at the cell surface. *Chloride pumps* in the apical plasma membrane produce this layer by pumping Cl⁻ into the extracellular fluid. Sodium ions follow by electrical attraction and water follows by osmosis. Mucus essentially floats on the surface of this layer and is pushed along by the tips of the cilia.

The structural basis for ciliary movement is a core called the **axoneme**[12] (ACK-so-neem), which consists of an array of thin

---

[12]*axo* = axis; *neme* = thread

DEEPER INSIGHT 3.2

CLINICAL APPLICATION

Cystic Fibrosis

The significance of chloride pumps becomes especially evident in *cystic fibrosis (CF),* a hereditary disease affecting primarily white children of European descent. CF is usually caused by a defect in which cells make chloride pumps but fail to install them in the plasma membrane. Consequently, there is an inadequate saline layer on the cell surface and the mucus is dehydrated and overly sticky. This thick mucus plugs the ducts of the pancreas and prevents it from secreting digestive enzymes into the small intestine, so digestion and nutrition are compromised. In the respiratory tract, the mucus clogs the cilia and prevents them from beating freely. The respiratory tract becomes congested with thick mucus, often leading to chronic infection and pulmonary collapse. The mean life expectancy of people with CF is about 30 years.

protein cylinders called *microtubules.* There are two central microtubules surrounded by a ring of nine microtubule pairs—an arrangement called the *9 + 2 structure.* In cross section, it is reminiscent of a Ferris wheel (fig. 3.10d). The central microtubules stop at the cell surface, but the peripheral microtubules continue a short distance into the cell as part of a **basal body** that anchors the cilium. In each pair of peripheral microtubules, one tubule has two little **dynein**[13] **arms** (DINE-een). Dynein, a motor protein, uses energy from ATP to crawl up the adjacent pair of microtubules. When microtubules on the front of the cilium crawl up the microtubules behind them, the cilium bends toward the front. The microtubules of a cilium also act like railroad tracks along which motor proteins carry materials up and down the cilium for use in its growth and maintenance. The primary cilia, which cannot move, lack the two central microtubules and dynein arms, but still have the nine peripheral pairs; they are said to have a *9 + 0 structure.*

The only functional **flagellum**[14] (fla-JEL-um) in humans is the whiplike tail of a sperm. It is much longer than a cilium and has an axoneme surrounded by a sheath of coarse fibers that stiffen the tail and give it more propulsive power. A flagellum does not beat with power and recovery strokes like those of a cilium, but in a more undulating, snakelike or corkscrew fashion. It is described in further detail as part of sperm structure in section 27.4c.

Pseudopods

Pseudopods[15] (SOO-do-pods) are cytoplasm-filled extensions of the cell varying in shape from fine, filamentous processes to blunt fingerlike ones **(fig. 3.12).** Unlike the other three kinds of surface extensions, they change continually. Some form anew as the cell surface bubbles outward and cytoplasm flows into a lengthening

---

[13]*dyn* = power, energy; *in* = protein
[14]*flagellum* = whip
[15]*pseudo* = false; *pod* = foot

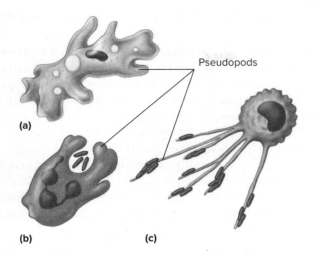

FIGURE 3.12 Pseudopods. (a) *Amoeba,* a freshwater organism that crawls and captures food by means of pseudopods. (b) A neutrophil (white blood cell) that similarly uses pseudopods for locomotion and capturing bacteria. (c) A macrophage extending filamentous pseudopods to snare and "reel in" bacteria (the red rods).

pseudopod, while others are retracted into the cell by disassembling protein filaments that supported them like a scaffold.

The freshwater organism *Amoeba* furnishes a familiar example of pseudopods, which it uses for locomotion and food capture. White blood cells called *neutrophils* crawl about like amebae by means of fingerlike pseudopods, and when they encounter a bacterium or other foreign particle, they reach out with their pseudopods to surround and engulf it. *Macrophages*—tissue cells derived from certain white blood cells—reach out with thin filamentous pseudopods to snare bacteria and cell debris and "reel them in" to be digested by the cell. Like little janitors, macrophages thereby keep our tissues cleaned up. Blood platelets reach out with thin pseudopods to adhere to each other and to the walls of damaged blood vessels, forming plugs that temporarily halt bleeding (see fig. 18.20).

BEFORE YOU GO ON

Answer the following questions to test your understanding of the preceding section:

6. How does the structure of a plasma membrane depend on the amphipathic nature of phospholipids?

7. Define *peripheral* versus transmembrane *proteins*.

8. Explain the differences between a receptor, pump, and cell-adhesion molecule.

9. How does a gate differ from other channel proteins? What three factors open and close membrane gates?

10. What related roles do cAMP, adenylate cyclase, and kinases play in cellular function?

11. Identify several reasons why the glycocalyx is important to human survival.

12. How do microvilli and cilia differ in structure and function?

<table>
<tr><td>**3.3**</td><td>**Membrane Transport**</td></tr>
</table>

Expected Learning Outcomes

When you have completed this section, you should be able to

a. explain what is meant by a *selectively permeable membrane;*

b. describe the various mechanisms for transporting material through cellular membranes; and

c. define *osmolarity* and *tonicity* and explain their importance.

One of the most important functions of cellular membranes is to control the passage of materials into and out of the organelles and the cell as a whole. The plasma membrane is both a barrier and gateway between the cytoplasm and ECF. It is **selectively permeable**—it allows some things through, such as nutrients and wastes, but usually prevents other things, such as proteins and phosphates, from entering or leaving the cell.

The methods of moving substances through the membrane can be classified in two overlapping ways: as passive or active mechanisms and as carrier-mediated or not. *Passive* mechanisms require no energy (ATP) expenditure by the cell. In most cases, the random molecular motion of the particles themselves provides the necessary energy. Passive mechanisms include filtration, diffusion, and osmosis. *Active* mechanisms, however, consume ATP. These include active transport and vesicular transport. *Carrier-mediated* mechanisms use a membrane protein to transport substances from one side of the membrane to the other, but some transport processes, such as osmosis, do not involve carriers.

3.3a Filtration

Filtration is a process in which a physical pressure forces fluid through a selectively permeable membrane. A coffee filter is an everyday example. The weight of the water drives water and dissolved matter through the filter, while the filter holds back larger particles (the coffee grounds). In physiology, the most important case of filtration is seen in the blood capillaries, where blood pressure forces fluid through gaps in the capillary wall **(fig. 3.13)**. This is how water, salts, nutrients, and other solutes are transferred from the bloodstream to the tissue fluid and how the kidneys filter wastes from the blood. Capillaries hold back larger particles such as blood cells and proteins. In most cases, water and solutes filter through narrow gaps between the capillary cells. In some capillaries, however, the cells have large *filtration pores* through them, like the holes in a slice of Swiss cheese, allowing for more rapid filtration of large solutes such as protein hormones.

3.3b Simple Diffusion

Simple diffusion is the net movement of particles from a place of high concentration to a place of lower concentration as a result of their constant, spontaneous motion. In other words, substances diffuse *down their concentration gradients* (see "Gradients and Flow," section 1.6e). Molecules move at astonishing speeds.

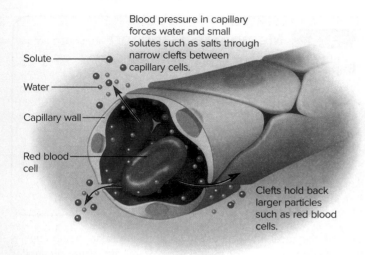

Blood pressure in capillary forces water and small solutes such as salts through narrow clefts between capillary cells.

Solute

Water

Capillary wall

Red blood cell

Clefts hold back larger particles such as red blood cells.

FIGURE 3.13 Filtration Through the Wall of a Blood Capillary. Water and small solutes pass through gaps between cells, while blood cells and other large particles are held back.

At body temperature, the average water molecule moves about 2,500 km/h (1,500 mi./h)! However, a molecule can travel only a short distance before colliding with another and careening off in a new direction, like colliding billiard balls. The rate of diffusion, therefore, is much slower than the speed of molecular motion.

Diffusion occurs readily in air or water and doesn't necessarily need a membrane—for example, when an odor spreads from its source to your nose. However, if there is a membrane in the path of the diffusing molecules, and if it is permeable to that substance, the molecules will pass from one side of the membrane to the other. This is how oxygen passes from the air we inhale into the bloodstream. Dialysis treatment for kidney patients is based on diffusion of solutes through artificial *dialysis membranes.*

Diffusion rates are important to cell survival because they determine how quickly a cell can acquire nutrients or rid itself of wastes. Some factors that affect the rate of diffusion through a membrane are as follows:

- **Temperature.** Diffusion is driven by the kinetic energy of the particles, and temperature is a measure of that kinetic energy. The warmer a substance is, the more rapidly its particles diffuse. This is why sugar diffuses more quickly through hot tea than through iced tea.

- **Molecular weight.** Heavy molecules such as proteins move more sluggishly and diffuse more slowly than light particles such as electrolytes and gases. Small molecules also pass through membrane pores more easily than large ones.

- **"Steepness" of the concentration gradient.** The steepness of a gradient refers to the concentration difference between two points. Particles diffuse more rapidly if there is a greater concentration difference.

- **Membrane surface area.** As noted earlier, the apical surface of cells specialized for absorption (for example, in the small intestine) is often extensively folded into microvilli.

This makes more membrane available for particles to diffuse through.

- **Membrane permeability.** Diffusion through a membrane depends on how permeable it is to the particles. For example, potassium ions diffuse more rapidly than sodium ions through a plasma membrane. Nonpolar, hydrophobic, lipid-soluble substances such as oxygen, nitric oxide, alcohol, and steroids diffuse through the phospholipid regions of a plasma membrane. Water and small charged, hydrophilic solutes such as electrolytes don't mix with lipids but diffuse primarily through channel proteins in the membrane. Cells can adjust their permeability to such a substance by adding channel proteins to the membrane, by taking them away, or by opening and closing membrane gates.

3.3c Osmosis

Osmosis[16] is the net flow of water from one side of a selectively permeable membrane to the other. It is crucial to the body's water distribution (fluid balance). Imbalances in osmosis underlie such problems as diarrhea, constipation, hypertension, and edema (tissue swelling); osmosis also is a vital consideration in intravenous (I.V.) fluid therapy.

Osmosis occurs through nonliving membranes, such as cellophane and dialysis membranes, and through the plasma membranes of cells. The usual direction of net movement is from the more watery side, with a lower concentration of dissolved matter, to the less watery side, with a greater concentration of solute. The reason for the accumulation of water on the high-solute side is that when water molecules encounter a solute particle, they tend to associate with it to form a *hydration sphere* (see fig. 2.9). Even though this is a loose, reversible attraction, it does make those water molecules less available to diffuse back across the membrane to the side from which they came. In essence, solute particles on one side of the membrane draw water away from the other side. Thus, water accumulates on the side with the most solute. All of this assumes that the solute molecules in question can't pass through the membrane, but stay on one side. The rate and direction of osmosis depend on the relative concentration of these nonpermeating solutes on the two sides of the membrane.

Significant amounts of water pass even through the hydrophobic, phospholipid regions of a plasma membrane, but water passes more easily through channel proteins called **aquaporins,** specialized for water. Cells can increase the rate of osmosis by installing more aquaporins in the membrane or decrease the rate by removing them. Certain cells of the kidney, for example, regulate the rate of urinary water loss by adding or removing aquaporins.

A cell can exchange a tremendous amount of water by osmosis. In red blood cells, for example, the amount of water passing through the plasma membrane every second is 100 times the volume of the cell.

[16]*osm* = push, thrust; *osis* = condition, process

Figure 3.14 is a conceptual model of osmosis. Imagine a chamber divided by a selectively permeable membrane. Side A contains distilled water and side B contains large particles of a *nonpermeating* solute—that is, a solute such as protein that cannot pass through the membrane pores because of its size or other properties. Water passes from side A to B (fig. 3.14a) and associates with the solute molecules on side B, hindering water movement back to side A.

Under such conditions, the water level in side A would fall and the level in side B would rise. It may seem as if this would continue indefinitely until side A dried up. This would not happen, however, because as water accumulated in side B, it would become heavier and exert more force, called **hydrostatic pressure,** on that side of the membrane. This would cause some filtration of water from B back to A. At some point, the rate of filtration would equal the rate of "forward" osmosis, water would pass through the membrane equally in both directions, and net osmosis would slow down and stop. At this point, an equilibrium (balance between opposing forces) would exist. The hydrostatic pressure required on side B to halt osmosis is called **osmotic pressure.** The more nonpermeating solute there is in B, the greater the osmotic pressure.

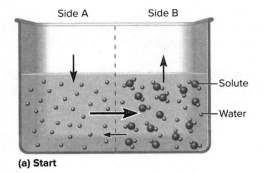

Side A   **Side B**

— Solute

— Water

(a) Start

— Osmotic pressure

— Hydrostatic pressure

(b) 30 minutes later

FIGURE 3.14 Osmosis. The dashed line represents a selectively permeable membrane dividing the chamber in half. (a) Initial net flow of water from side A to side B. (b) Equilibrium between osmotic and hydrostatic pressure, hence between osmosis and filtration, with no net flow. Water molecules aggregate around solute molecules and thus accumulate on side B.

▶▶▶**APPLY WHAT YOU KNOW**

If the solute concentration on side B was half what it was in the original experiment, would the fluid on that side reach a higher or lower level than before? Explain.

Reverse osmosis is a process in which a mechanical pressure applied to one side of the system can override osmotic pressure and drive water through a membrane against its concentration gradient. This principle is used to create highly purified water for laboratory use and to desalinate seawater, converting it to drinkable freshwater—handy for arid countries and ships at sea. The body's principal pump, the heart, drives water out of the smallest blood vessels (the capillaries) by reverse osmosis—a process called *capillary filtration.* The equilibrium between osmosis and filtration will be an important consideration when we study fluid exchange by the capillaries in section 20.3. Blood plasma also contains albumin. In the preceding discussion, side B is analogous to the high-protein bloodstream and side A to the low-protein tissue fluid surrounding the capillaries. Water leaves the capillaries by filtration, but this is approximately balanced by water reentering the capillaries by osmosis.

3.3d Osmolarity and Tonicity

The **osmolarity,** or osmotic concentration, of body fluids has such a great effect on cellular function that it is important to understand the units in which it is measured. Physiologists and clinicians usually express this in terms of **milliosmoles per liter (mOsm/L),** a unit of measure that expresses the quantity of nonpermeating particles per liter of solution. The basis of this unit of concentration is explained in appendix B. Blood plasma, tissue fluid, and intracellular fluids measure about 300 mOsm/L.

Tonicity is the ability of a solution to affect the fluid volume and pressure in a cell. If a solute cannot pass through a plasma membrane but remains more concentrated on one side of the membrane than on the other, it causes osmosis. A **hypotonic**[17] solution has a lower concentration of nonpermeating solutes than the intracellular fluid (ICF). Cells in a hypotonic solution absorb water, swell, and may burst *(lyse)* **(fig. 3.15a).** Distilled water is the extreme example; a sufficient quantity given to a person intravenously would lyse the blood cells, with dire consequences. A **hypertonic**[18] solution is one with a higher concentration of nonpermeating solutes than the ICF. It causes cells to lose water and shrivel *(crenate)* **(fig. 3.15c).** Such cells may die of torn membranes and cytoplasmic loss. In **isotonic**[19] solutions, the total concentration of nonpermeating solutes is the same as in the ICF—hence, isotonic solutions cause no change in cell volume or shape **(fig. 3.15b).**

It is essential for cells to be in a state of osmotic equilibrium with the fluid around them, and this requires that the ECF have the same concentration of nonpermeating solutes as the ICF. Intravenous fluids given to patients are usually isotonic solutions, but

---

[17]*hypo* = less; *ton* = tension
[18]*hyper* = more; *ton* = tension
[19]*iso* = equal; *ton* = tension

(a) Hypotonic   **(b) Isotonic**   **(c) Hypertonic**

FIGURE 3.15 Effects of Tonicity on Red Blood Cells (RBCs). (a) RBC swelling in a hypotonic medium such as distilled water. (b) Normal RBC size and shape in an isotonic medium such as 0.9% NaCl. (c) RBC shriveling in a hypertonic medium such as 2% NaCl.

a–c: David M. Phillips/Science Source

hypertonic or hypotonic fluids are given for special purposes. A 0.9% solution of NaCl, called *normal saline,* is isotonic to human blood cells.

It is important to note that osmolarity and tonicity are not the same. Urea, for example, is a small organic molecule that easily penetrates plasma membranes. If cells are placed in 300 mOsm/L urea, urea diffuses into them (down its concentration gradient), water follows by osmosis, and the cells swell and burst. Thus, 300 mOsm/L urea is not isotonic to the cells. Sodium chloride, by contrast, penetrates plasma membranes poorly. In 300 mOsm/L NaCl, there is little change in cell volume; this solution is isotonic to cells.

3.3e Carrier-Mediated Transport

The processes of membrane transport described up to this point don't necessarily require a cell membrane; they can occur as well through artificial membranes. Now, however, we come to processes for which a cell membrane is necessary, because they employ transport proteins, or carriers. Thus, the next three processes are classified as **carrier-mediated transport.** In these cases, a solute binds to a carrier in the plasma membrane, which then changes shape and releases the solute to the other side. Carriers can move substances into or out of a cell, and into or out of organelles within the cell. The process is very rapid; for example, one carrier can transport 1,000 glucose molecules per second across the membrane.

Carriers act like enzymes in some ways: The solute is a ligand that binds to a specific receptor site on the carrier, like a substrate binding to the active site of an enzyme. The carrier exhibits **specificity** for its ligand, just as an enzyme does for its substrate. A glucose carrier, for example, cannot transport fructose. Carriers also exhibit **saturation;** as the solute concentration rises, its rate of transport increases, but only up to a point. When every carrier is occupied, adding more solute can't

make the process go any faster. The carriers are saturated—no more are available to handle the increased demand, and transport levels off at a rate called the **transport maximum (T_m) (fig. 3.16).** You could think of carriers as analogous to buses. If all the buses on a given line are full ("saturated"), they can't carry any more passengers, regardless of how many people are waiting at the bus stop. As we'll see later in the book, the T_m explains why glucose appears in the urine of people with diabetes mellitus.

An important difference between a carrier and an enzyme is that carriers don't chemically change their ligands; they simply

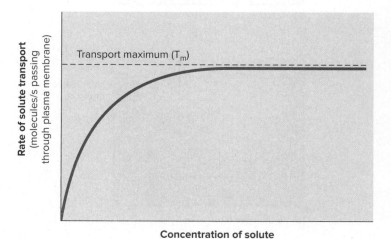

Concentration of solute

FIGURE 3.16 Carrier Saturation and Transport Maximum. Up to a point, increasing the solute concentration increases the rate of transport through a membrane. At the transport maximum (T_m), however, all carrier proteins are busy and cannot transport the solute any faster, even if more solute is added.

pick them up on one side of the membrane and release them, unchanged, on the other.

There are three kinds of carriers: uniports, symports, and antiports. A **uniport**[20] carries only one type of solute. For example, most cells pump out calcium by means of a uniport, maintaining a low intracellular concentration so calcium salts don't crystallize in the cytoplasm. Some carriers move two or more solutes through a membrane simultaneously in the same direction; this process is called **cotransport**[21] and the carrier protein that performs it is called a **symport**.[22] For example, absorptive cells of the small intestine and kidneys have a symport that takes up sodium and glucose simultaneously. Other carriers move two or more solutes in opposite directions; this process is called **countertransport** and the carrier protein is called an **antiport**.[23] For example, nearly all cells have an antiport called the *sodium–potassium pump* that continually removes Na$^+$ from the cell and brings in K$^+$.

There are three mechanisms of carrier-mediated transport: facilitated diffusion, primary active transport, and secondary active transport. **Facilitated**[24] **diffusion (fig. 3.17)** is the carrier-mediated transport of a solute through a membrane *down its concentration gradient*. It requires no expenditure of metabolic energy (ATP) by the cell. It transports solutes such as glucose

that cannot pass through the membrane unaided. The solute attaches to a binding site on the carrier, then the carrier changes conformation and releases the solute on the other side of the membrane.

Primary active transport is a process in which a carrier moves a substance through a cell membrane *up its concentration gradient* using energy provided by ATP. Just as rolling a ball up a ramp would require you to push it (an energy input), this mechanism requires energy to move material up its concentration gradient. ATP supplies this energy by transferring a phosphate group to the transport protein. The calcium pump mentioned previously uses this mechanism. Even though Ca^{2+} is already more concentrated in the ECF than within the cell, this carrier pumps still more of it out. Active transport also enables cells to absorb amino acids that are already more concentrated in the cytoplasm than in the ECF.

Secondary active transport also requires an energy input, but depends only indirectly on ATP. For example, certain kidney tubules have proteins called *sodium–glucose transporters (SGLTs)* that simultaneously bind sodium ions (Na$^+$) and glucose molecules and transport them into the tubule cells, saving glucose from being lost in the urine **(fig. 3.18)**. An SGLT itself doesn't use ATP. However, it depends on the fact that the cell actively maintains a low internal Na$^+$ concentration, so Na$^+$ will diffuse down its gradient into the cell. Glucose "hitches a ride" with the incoming Na$^+$. But what keeps the intracellular Na$^+$ concentration low is that the basal membrane of the cell has an ATP-driven sodium–potassium pump that constantly removes Na$^+$ from the cell. If not for this, the Na$^+$ and glucose inflow via the SGLT would soon cease. Therefore, the SGLT doesn't use ATP directly, but depends on ATP to drive the

---

[20]*uni* = one; *port* = carry

[21]*co* = together; *trans* = across; *port* = carry

[22]*sym* = together; *port* = carry

[23]*anti* = opposite; *port* = carry

[24]*facil* = easy

ECF

ICF

① A solute particle enters the channel of a membrane protein (carrier).

② The solute binds to a receptor site on the carrier and the carrier changes conformation.

③ The carrier releases the solute on the other side of the membrane.

FIGURE 3.17 Facilitated Diffusion. Note that the solute moves down its concentration gradient.

Lest you question the importance of the Na⁺–K⁺ pump, consider that half of the calories you use each day go to this purpose alone. The pump typically operates at about 10 cycles/s, but under certain conditions it can achieve 100 cycles/s. Various types of cells have from just a few hundred Na⁺–K⁺ pumps (red blood cells) to millions of them (nerve cells), so an average cell may exchange 30 million Na⁺ ions and 20 million K⁺ ions and consume 10 million ATPs per second. Beyond compensating for a leaky plasma membrane, the Na⁺–K⁺ pump has at least four functions:

1. **Secondary active transport.** It maintains a steep Na⁺ concentration gradient across the membrane. Like water behind a dam, this gradient is a source of potential energy that can be tapped to do other work. The secondary active transport described previously is an example of this.

2. **Regulation of cell volume.** Certain anions are confined to the cell and cannot penetrate the plasma membrane. These "fixed anions," such as proteins and phosphates, attract and retain cations. If there were nothing to correct for it, the retention of these ions would cause osmotic swelling and possibly lysis of the cell. Cellular swelling, however, elevates activity of the Na⁺–K⁺ pumps. Since each cycle of the pump removes one ion more than it brings in, the pumps are part of a negative feedback loop that reduces intracellular ion concentration, controls osmolarity, and prevents cellular swelling.

FIGURE 3.18 Secondary Active Transport. In this example, the sodium–glucose transporter (SGLT) at the apical cell surface carries out facilitated diffusion, but depends on active transport by the Na⁺–K⁺ pump at the base of the cell to keep it running.

Na⁺–K⁺ pump; it is therefore a secondary active transport protein. (*Secondary active transport* is an unfortunate name for this, as the SGLT is actually carrying out facilitated diffusion, but its dependence on a primary active transport pump has led to this name.)

The **sodium–potassium (Na⁺–K⁺) pump** itself **(fig. 3.19)** is a good example of primary active transport. It is also known as *Na⁺–K⁺ ATPase* because it is an enzyme that hydrolyzes ATP. The Na⁺–K⁺ pump binds three Na⁺ simultaneously on the cytoplasmic side of the membrane, releases these to the ECF, binds two K⁺ simultaneously from the ECF, and releases these into the cell. Each cycle of the pump consumes one ATP and exchanges three Na⁺ for two K⁺. This keeps the K⁺ concentration higher and the Na⁺ concentration lower within the cell than they are in the ECF. These ions continually leak through the membrane, and the Na⁺–K⁺ pump compensates like bailing out a leaky boat.

FIGURE 3.19 The Sodium–Potassium Pump (Na⁺–K⁺ ATPase).
? *Why would the Na⁺–K⁺ pump, but not osmosis, cease to function after a cell dies?*

3. **Maintenance of a membrane potential.** All living cells have an electrical charge difference called the *resting membrane potential* across the plasma membrane. Like the two poles of a battery, the inside of the membrane is negatively charged and the outside is positively charged. This difference stems from the unequal distribution of ions on the two sides of the membrane, maintained by the Na^+–K^+ pump. The membrane potential is essential to the excitability of nerve and muscle cells.

4. **Heat production.** When the weather turns chilly, we turn up not only the furnace in our home but also the "furnace" in our body. Thyroid hormone stimulates cells to produce more Na^+–K^+ pumps. As these pumps consume ATP, they release heat from it, compensating for the body heat we lose to the cold air around us.

▶▶▶ APPLY WHAT YOU KNOW

An important characteristic of proteins is their ability to change conformation in response to the binding or dissociation of a ligand (see "Protein Structure" in section 2.4e). Explain how this characteristic is essential to carrier-mediated transport.

In summary, carrier-mediated transport is any process in which solute particles move through a membrane by means of a transport protein. The protein is a uniport if it transports only one solute, a symport if it carries two types of solutes at once in the same direction, and an antiport if it carries two or more solutes in opposite directions. If the carrier doesn't depend on ATP at all and it moves solutes down their concentration gradient, the process is called facilitated diffusion. If the carrier itself consumes ATP and moves solutes up their concentration gradient, the process is called primary active transport. If the carrier doesn't directly use ATP, but depends on a concentration gradient produced by ATP-consuming Na^+–K^+ pumps elsewhere in the plasma membrane, the process is called secondary active transport.

3.3f Vesicular Transport

So far, we have considered processes that move one or a few ions or molecules at a time through the plasma membrane. **Vesicular transport** processes, by contrast, move large particles, droplets of fluid, or numerous molecules at once through the membrane, contained in bubblelike **vesicles** of membrane. Vesicular processes that bring matter into a cell are called **endocytosis**[25] (EN-doe-sy-TOE-sis) and those that release material from a cell are called **exocytosis**[26] (EC-so-sy-TOE-sis). These processes employ motor proteins whose movements are energized by ATP.

There are three forms of endocytosis: phagocytosis, pinocytosis, and receptor-mediated endocytosis. **Phagocytosis**[27] (FAG-oh-sy-TOE-sis), or "cell eating," is the process of engulfing particles such as bacteria, dust, and cellular debris—particles large enough to be seen with a microscope. For example, neutrophils (a class of white blood cells) protect the body from infection by phagocytizing and killing bacteria. A neutrophil spends most of its life crawling about in the connective tissues by means of its pseudopods. When a neutrophil encounters a bacterium, it surrounds it with pseudopods and traps it in a vesicle called a **phagosome**[28]—a vesicle in the cytoplasm surrounded by a unit membrane **(fig. 3.20).** A lysosome merges with the phagosome, converting it to a *phagolysosome,* and contributes enzymes that destroy the invader. Some other kinds of phagocytic cells are described in section 21.1b. In general, phagocytosis is a way of keeping the tissues free of debris and infectious microbes. Some cells called *macrophages* (literally, "big eaters") phagocytize the equivalent of 25% of their own volume per hour.

Pinocytosis[29] (PIN-oh-sy-TOE-sis), or "cell drinking," is the process of taking in droplets of ECF containing molecules of some use to the cell. While phagocytosis occurs in only a few specialized cells, pinocytosis occurs in all human cells. The process begins as the plasma membrane becomes dimpled, or caved in, at points. These pits soon separate from the surface membrane and form small membrane-bounded **pinocytotic vesicles** in the cytoplasm. The vesicles contain droplets of the ECF with whatever molecules happen to be there.

Receptor-mediated endocytosis (fig. 3.21) is a more selective form of either phagocytosis or pinocytosis. It enables a cell to take in specific molecules from the ECF with a minimum of unnecessary matter. Particles in the ECF bind to specific receptors on the plasma membrane. The receptors then cluster together and the membrane sinks in at this point, creating a pit coated with a peripheral membrane protein called *clathrin.*[30] The pit soon pinches off to form a *clathrin-coated vesicle* in the cytoplasm. Clathrin may serve as an "address label" on the coated vesicle that directs it to an appropriate destination in the cell, or it may inform other structures in the cell what to do with the vesicle.

One example of receptor-mediated endocytosis is the uptake of *low-density lipoproteins (LDLs)*—protein-coated droplets of cholesterol and other lipids in the blood (described in section 26.1g). The thin endothelial cells that line our blood vessels have LDL receptors on their surfaces and absorb LDLs in clathrin-coated vesicles. Inside the cell, the LDL is freed from the vesicle and metabolized, and the membrane with its receptors is recycled to the cell surface.

Endothelial cells also imbibe insulin by receptor-mediated endocytosis. Insulin is too large to pass through channels in the plasma membrane, yet it must somehow get out of the blood and

---

[25]*endo* = into; *cyt* = cell; *osis* = process
[26]*exo* = out of; *cyt* = cell; *osis* = process

[27]*phago* = eating; *cyt* = cell; *osis* = process
[28]*phago* = eaten; *some* = body
[29]*pino* = drinking; *cyt* = cell; *osis* = process
[30]*clathr* = lattice; *in* = protein

① A phagocytic cell encounters a particle of foreign matter.

Particle

Pseudopod

② The cell surrounds the particle with its pseudopods.

Nucleus

⑦ The indigestible residue is voided by exocytosis.

Residue

③ The particle is phagocytized and contained in a phagosome.

Phagosome

Lysosome

⑥ The phagolysosome fuses with the plasma membrane.

Vesicle fusing with membrane

④ The phagosome fuses with a lysosome and becomes a phagolysosome.

Phagolysosome

⑤ Enzymes from the lysosome digest the foreign matter.

FIGURE 3.20 Phagocytosis, Intracellular Digestion, and Exocytosis.

Extracellular molecules

Receptor

① Extracellular molecules bind to receptors on plasma membrane; receptors cluster together.

Coated pit

Clathrin

② Plasma membrane sinks inward, forms clathrin-coated pit.

Clathrin-coated vesicle

③ Pit separates from plasma membrane, forms clathrin-coated vesicle containing concentrated molecules from ECF.

FIGURE 3.21 Receptor-Mediated Endocytosis.

1–3: Courtesy of The Company of Biologists, Ltd.

reach the surrounding cells if it is to have any effect. Endothelial cells take up insulin by receptor-mediated endocytosis, transport the vesicles across the cell, and release the insulin on the other side, where tissue cells await it. Such transport of material across a cell (capture on one side and release on the other) is called **transcytosis**[31] **(fig. 3.22).** This process is especially active in muscle capillaries and transfers a significant amount of blood albumin into the tissue fluid.

Receptor-mediated endocytosis isn't always to our benefit; hepatitis, polio, and AIDS viruses trick our cells into engulfing them by receptor-mediated endocytosis, thus exploiting this mechanism to establish infection.

Exocytosis (fig. 3.23) is a process of discharging material from a cell. It occurs, for example, when endothelial cells release insulin to the tissue fluid, sperm cells release enzymes for penetrating an egg, mammary gland cells secrete milk sugar, and other gland cells release hormones. It bears a superficial resemblance to endocytosis in reverse. A secretory vesicle in the cell migrates to the surface and "docks" on peripheral proteins of the plasma membrane. These proteins pull the membrane inward and create a dimple that eventually fuses with the vesicle and allows it to release its contents.

The question might occur to you, If endocytosis continually takes away bits of plasma membrane to form intracellular vesicles, why doesn't the membrane grow smaller and smaller? Another purpose of exocytosis, however, is to replace plasma membrane that has been removed by endocytosis or that has become damaged or worn out. Plasma membrane is continually recycled from the cell surface into the cytoplasm and back to the surface.

Table 3.3 summarizes these mechanisms of transport.

[31]*trans* = across; *cyt* = cell; *osis* = process

FIGURE 3.22 Transcytosis. An endothelial cell of a capillary imbibes droplets of blood plasma at sites indicated by arrows along the left. This forms pinocytotic vesicles, which the cell transports to the other side. Here, it releases the contents by exocytosis at sites indicated by arrows along the right side of the cell.

Don Fawcett/Science Source

? *Why isn't transcytosis listed as a separate means of membrane transport, in addition to pinocytosis and the others?*

(a) ① A secretory vesicle approaches the plasma membrane and docks on it by means of linking proteins. The plasma membrane caves in at that point to meet the vesicle.

② The plasma membrane and vesicle unite to form a fusion pore through which the vesicle contents are released.

(b)

FIGURE 3.23 Exocytosis. (a) Stages of exocytosis. (b) Electron micrograph of exocytosis.

b: Courtesy Dr. Birgit Satir, Albert Einstein College of Medicine

TABLE 3.3	Methods of Membrane Transport
Transport Without Carriers	**Movement of Material Without the Aid of Carrier Proteins**
Filtration	Movement of water and solutes through a selectively permeable membrane as a result of hydrostatic pressure
Simple diffusion	Diffusion of particles through water or air or through a living or artificial membrane, down their concentration gradient, without the aid of membrane carriers
Osmosis	Net flow of water through a selectively permeable membrane, driven by either a difference in solute concentration or a mechanical force
Carrier-Mediated Transport	**Movement of Material Through a Cell Membrane by Carrier Proteins**
Facilitated diffusion	Transport of particles through a selectively permeable membrane, down their concentration gradient, by a carrier that does not directly consume ATP
Primary active transport	Transport of solute particles through a selectively permeable membrane, up their concentration gradient, by a carrier that consumes ATP
Secondary active transport	Transport of solute particles through a selectively permeable membrane, up their concentration gradient, by a carrier that doesn't use ATP itself but depends on concentration gradients produced by primary active transport elsewhere in the membrane
Cotransport	Simultaneous transport of two or more solutes in the same direction through a membrane by a carrier protein called a *symport,* using either facilitated diffusion or active transport
Countertransport	Transport of two or more solutes in opposite directions through a membrane by a carrier protein called an *antiport,* using either facilitated diffusion or active transport
Vesicular (Bulk) Transport	**Movement of Fluid and Particles Through a Plasma Membrane by Way of Membrane Vesicles; Consumes ATP**
Endocytosis	Vesicular transport of particles into a cell
Phagocytosis	Process of engulfing large particles by means of pseudopods; "cell eating"
Pinocytosis	Process of imbibing extracellular fluid in which the plasma membrane sinks in and pinches off small vesicles containing droplets of fluid; "cell drinking"
Receptor-mediated endocytosis	Phagocytosis or pinocytosis in which specific solute particles bind to receptors on the plasma membrane, and are then taken into the cell in clathrin-coated vesicles with a minimal amount of extraneous matter
Exocytosis	Process of eliminating material from a cell by means of a vesicle approaching the cell surface, fusing with the plasma membrane, and expelling its contents; used to release cell secretions, replace worn-out plasma membrane, and replace membrane that has been internalized by endocytosis

BEFORE YOU GO ON

Answer the following questions to test your understanding of the preceding section:

13. What is the importance of filtration to human physiology?

14. What does it mean to say a solute moves down its concentration gradient?

15. How does osmosis help to maintain blood volume?

16. Define *osmolarity* and *tonicity,* and explain the difference between them.

17. Define *hypotonic, isotonic,* and *hypertonic,* and explain why these concepts are important in clinical practice.

18. What do facilitated diffusion and active transport have in common? How are they different?

19. How does the Na⁺–K⁺ pump exchange sodium ions for potassium ions across the plasma membrane? What are some purposes served by this pump?

20. How does phagocytosis differ from pinocytosis?

21. Describe the process of exocytosis. What are some of its purposes?

3.4   The Cell Interior

Expected Learning Outcomes

When you have completed this section, you should be able to

a. describe the cytoskeleton and its functions;

b. list the main organelles of a cell, describe their structure, and explain their functions; and

c. give some examples of cell inclusions and explain how inclusions differ from organelles.

We now probe more deeply into the cell to study its internal structures. These are classified into three groups—*cytoskeleton, organelles,* and *inclusions*—all embedded in the clear, gelatinous cytosol.

3.4a The Cytoskeleton

The **cytoskeleton** is a network of protein filaments and cylinders that structurally support a cell, determine its shape, organize its contents, direct the movement of materials within the cell, and contribute to movements of the cell as a whole. It forms a dense supportive scaffold in the cytoplasm **(fig. 3.24).** It is connected to transmembrane proteins of the plasma membrane, and they in turn are connected to protein fibers external to the cell, creating a strong structural continuity from extracellular material to the cytoplasm. Cytoskeletal elements may even connect to chromosomes in the nucleus, enabling physical tension on a cell to move nuclear contents and mechanically stimulate genetic function.

The cytoskeleton is composed of *microfilaments, intermediate filaments,* and *microtubules.* If you think of intermediate filaments as being like the stiff rods of uncooked spaghetti, you could, by comparison, think of microfilaments as being like fine angel-hair pasta and microtubules as being like tubular penne pasta.

Microfilaments (thin filaments) are about 6 nm thick and are made of the protein *actin.* They are widespread throughout the cell but especially concentrated in a fibrous mat called the **terminal web (membrane skeleton)** on the cytoplasmic side of the plasma membrane. The phospholipids of the plasma membrane spread out over the terminal web like butter on a slice of bread. The web, like the bread, provides physical support, whereas the lipids, like butter, provide a permeability barrier. It is thought that without the support of the terminal web, the phospholipids would break up into little droplets and the plasma membrane would not hold together. As described earlier, actin microfilaments also form the supportive cores of the microvilli and play a role in cell movement. Through its role in cell motility, actin plays a crucial role in embryonic development, muscle contraction, immune function, wound healing, cancer metastasis, and other processes that involve cell migration.

Intermediate filaments (8–10 nm thick) are thicker and stiffer than microfilaments. They give the cell its shape, resist stress, and participate in junctions that attach cells to their neighbors. In epidermal cells, they are made of the tough protein *keratin* and occupy most of the cytoplasm. They are responsible for the strength of hair and fingernails.

Microtubules (25 nm in diameter) are cylinders made of 13 parallel strands called *protofilaments.* Each protofilament is a long chain of globular proteins called *tubulin* **(fig. 3.25).** Microtubules radiate from an area of the cell called the *centrosome.* They hold organelles in place, form bundles that maintain cell shape and rigidity, and act somewhat like monorail tracks. Motor proteins walk along these tracks carrying organelles and macromolecules to specific destinations in the cell. Microtubules form the axonemes of cilia and flagella and are responsible for their beating movements, and form the mitotic spindle that guides chromosome movement during cell division. Microtubules are not permanent structures. They come and go moment by moment as tubulin molecules assemble into a tubule and then suddenly break apart again to be used somewhere else in the cell. The microtubules in cilia, flagella, basal bodies, and centrioles, however, are more stable.

3.4b Organelles

Organelles are internal structures of a cell that carry out specialized metabolic tasks. Some are surrounded by membranes and are therefore referred to as *membranous organelles.* These are the nucleus, mitochondria, lysosomes, peroxisomes, endoplasmic reticulum, and Golgi complex. Organelles without membranes include ribosomes, proteasomes, centrosomes, centrioles, and basal bodies.

The Nucleus

The **nucleus (fig. 3.26)** is usually the largest organelle and the only one clearly visible with the light microscope. It contains the cell's chromosomes and is therefore the genetic control center of cellular activity. It is typically spheroidal to elliptical in shape and about 5 μm in diameter. Most cells have a single nucleus, but there are exceptions. Mature red blood cells have none; they are **anuclear.** A few cell types are **multinuclear,** having 2 to 50 nuclei. Examples include skeletal muscle cells, some liver cells, and certain bone-dissolving cells.

Figure 3.27 shows the details of nuclear structure. The nucleus is enclosed in a double membrane, the **nuclear envelope.** The envelope is perforated with **nuclear pores** formed by a ring of proteins called the *nuclear pore complex.* These proteins regulate molecular traffic through the envelope and act like a rivet to hold the two membrane layers together. Hundreds of molecules pass through the nuclear pores every minute. Coming into the nucleus are raw materials for DNA and RNA synthesis, enzymes that are made in the cytoplasm but function in the nucleus, and hormones and other chemical messengers that activate certain genes. Going the other way, RNA is made in the nucleus but leaves to perform its job in the cytoplasm.

Immediately inside the nuclear envelope is a narrow but densely fibrous zone called the **nuclear lamina,** composed of a web of intermediate filaments. It supports the nuclear envelope and pores, provides points of attachment and organization for the chromosomes inside the nucleus, and plays a role in regulating DNA replication and the cell life cycle. Abnormalities of its structure or function are associated with certain genetic diseases and premature cell death.

The material in the nucleus is called **nucleoplasm.** This includes **chromatin**[32] (CRO-muh-tin)—fine threadlike matter composed of DNA and protein—and one or more dark-staining

---

[32]*chromat* = colored; *in* = substance

FIGURE 3.24 The Cytoskeleton. (a) Components of the cytoskeleton. Few organelles are shown in order to emphasize the cytoskeleton. (b) Cells with their cytoskeletons labeled with fluorescent antibodies, photographed through a fluorescence microscope. The density of a typical cytoskeleton far exceeds even that shown in part (a). The pink bodies are the cell nuclei.

b: Dr. Torsten Wittmann/Science Source

FIGURE 3.25 **Microtubules.** (a) A microtubule is composed of 13 protofilaments. Each protofilament is a helical array of globular proteins called tubulin. (b) One of the nine microtubule pairs that form the axonemes of cilia and flagella, with the motor protein dynein attached. (c) One of the nine microtubule triplets that form a centriole.

FIGURE 3.27 **Structure of the Nucleus.** (a) Cutaway view showing the nuclear surface and contents of the nucleoplasm. (b) Detail of a nuclear pore complex.

❓ *Why do these nuclear pores have to be larger in diameter than the channels in the cell's plasma membrane? (See table 3.1.)*

masses called **nucleoli** (singular, *nucleolus*), where ribosomes are produced. The genetic function of the nucleus is described in section 4.2.

Endoplasmic Reticulum

Endoplasmic reticulum (ER) literally means "little network within the cytoplasm." It is a system of interconnected channels

(a) **Interior of nucleus**   2 μm

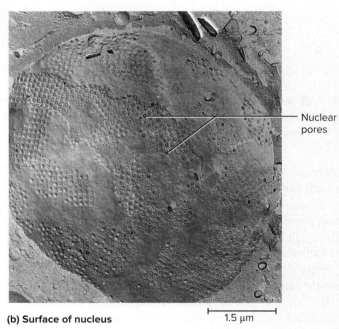

(b) **Surface of nucleus**   1.5 μm

FIGURE 3.26 **The Nucleus as Seen by Electron Microscopy.** These photomicrographs were made by different TEM methods to show the internal structure of the nucleus and surface of the nuclear envelope. (a) Interior of the nucleus. (b) Surface of the nucleus, showing clusters of nuclear pores.

a: Richard Chao; **b:** ©E.G. Pollock

Cisterns of rough ER

Nucleus

Ribosomes of rough ER

(a)

1 μm

Oil droplet (inclusion)

Smooth endoplasmic reticulum

(b)

1 μm

Rough endoplasmic reticulum

Ribosomes

Cisterns

(c)

Smooth endoplasmic reticulum

FIGURE 3.28 Endoplasmic Reticulum (ER). (a) Rough ER. (b) Smooth ER and an inclusion (oil droplet). (c) Structure of the endoplasmic reticulum, with rough and smooth regions.

a-b: Don Fawcett/Science Source

called **cisterns**[33] enclosed by a unit membrane **(fig. 3.28).** In areas called **rough endoplasmic reticulum,** the cisterns are parallel, flattened sacs covered with granules called *ribosomes.* Adjacent cisterns are connected by bridges to create one continuous internal space. The rough ER is continuous with the outer membrane of the nuclear envelope, and some authorities regard the nuclear envelope as simply a modified extension of it. In areas called **smooth endoplasmic reticulum,** the cisterns are more tubular, branch more extensively, and lack ribosomes. The cisterns of the smooth ER are continuous with those of the rough ER, so the two are different parts of the same network.

The ER synthesizes steroids and other lipids, detoxifies alcohol and other drugs, and manufactures nearly all membranes of the cell. Rough ER produces the phospholipids and proteins of the plasma membrane and synthesizes proteins that are either secreted from the cell or packaged in organelles such as lysosomes. Rough ER is most abundant in cells that synthesize large amounts of protein, such as antibody-producing cells and cells of the digestive glands. In such cells, the ER is often the largest organelle of all, although it isn't visible to the light microscope (LM) because its thin and closely spaced membranes are beyond the LM's limit of resolution.

Most cells have only a scanty smooth ER, but it is relatively abundant in cells that engage extensively in detoxification, such as liver and kidney cells. Long-term abuse of alcohol, barbiturates, and other drugs leads to tolerance partly because the smooth ER

---

[33]*cistern* = reservoir

proliferates and detoxifies the drugs more quickly. Smooth ER is also abundant in cells of the testes and ovaries that synthesize steroid hormones. Skeletal and cardiac muscle contain extensive networks of smooth ER that store calcium and release it to trigger muscle contraction.

Ribosomes

Ribosomes are small granules of protein and RNA found in the nucleoli, in the cytosol, in mitochondria, and on the outer surfaces of the rough ER and nuclear envelope. They "read" coded genetic messages (messenger RNA) and assemble amino acids into proteins specified by the code. The unattached ribosomes scattered throughout the cytoplasm make enzymes and other proteins for use within the cell. Free ribosomes within the nucleus and mitochondria make proteins for use in those organelles. Ribosomes attach to the rough ER when they make proteins destined to be packaged in lysosomes or to be secreted from the cell, such as digestive enzymes, antibodies, and some hormones.

Golgi Complex

The **Golgi**[34] **complex** (GOAL-jee) is a small system of cisterns that synthesize carbohydrates and put the finishing touches on protein and glycoprotein synthesis. The complex resembles a stack of pita bread. It consists of only a few cisterns, slightly separated from each other; each cistern is a flattened, often curved sac with swollen edges **(fig. 3.29).** The Golgi complex receives newly synthesized proteins from the rough ER. It sorts

them, cuts and splices some of them, and adds carbohydrate moieties to some. Finally, the most mature cistern with the finished cell product breaks up into membrane-bounded **Golgi vesicles,** which are abundant in the neighborhood of the Golgi complex. Some vesicles become *lysosomes,* the organelle discussed next; some migrate to the plasma membrane and fuse with it, contributing fresh protein and phospholipid to the membrane; and some become **secretory vesicles** that store a cell product, such as breast milk or digestive enzymes, for later release. The roles of the endoplasmic reticulum, ribosomes, and Golgi complex in protein synthesis and secretion are detailed in sections 4.2c and 4.2d.

Lysosomes

A **lysosome**[35] (LY-so-some) **(fig. 3.30a)** is a package of enzymes bounded by a membrane. Although often round or oval, lysosomes are extremely variable in shape. When viewed with the TEM, they often exhibit dark gray contents devoid of structure, but sometimes show crystals or parallel layers of protein. At least 50 lysosomal enzymes have been identified. They hydrolyze proteins, nucleic acids, complex carbohydrates, phospholipids, and other substrates. In the liver, lysosomes break down glycogen to release glucose into the bloodstream. White blood cells use lysosomes to digest phagocytized bacteria. Lysosomes also digest and dispose of surplus or nonvital organelles and other cell components in order to recycle their nutrients to more important cell needs; this process is called **autophagy**[36] (aw-TOFF-uh-jee). Lysosomes also aid in a process of "cell suicide." Some cells are meant to do a certain job and then destroy themselves. The uterus, for example, weighs about 900 g at full-term pregnancy and shrinks to 60 g within 5 or 6 weeks after birth. This shrinkage is due to **autolysis,**[37] the digestion of surplus cells by their own lysosomal enzymes.

Peroxisomes

Peroxisomes (fig. 3.30b) resemble lysosomes but contain different enzymes. They are produced by collaboration between the endoplasmic reticulum and mitochondria and by fission of preexisting peroxisomes. Their general function is to use molecular oxygen (O_2) to oxidize organic molecules. These reactions produce hydrogen peroxide (H_2O_2)—hence, the name of the organelle. H_2O_2 is then used to oxidize other molecules, and the excess is broken down to water and oxygen by an enzyme called *catalase.*

Peroxisomes occur in nearly all cells but are especially abundant in liver and kidney cells. They neutralize free radicals and detoxify alcohol, other drugs, and a variety of blood-borne toxins. Peroxisomes also decompose fatty acids into two-carbon fragments that the mitochondria use as an energy source for ATP synthesis.

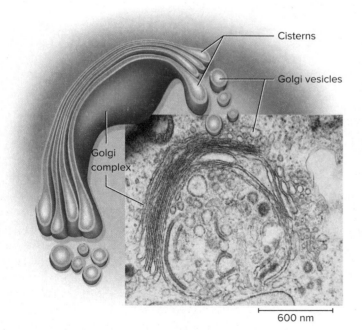

Cisterns

Golgi vesicles

Golgi complex

600 nm

FIGURE 3.29 The Golgi Complex.

David M. Phillips/Science Source

---

[34]Camillo Golgi (1843–1926), Italian histologist

[35]*lyso* = loosen, dissolve; *some* = body
[36]*auto* = self; *phagy* = eating
[37]*auto* = self; *lysis* = dissolving

(a) Lysosomes |← 1 µm →|

(b) Peroxisomes |← 0.3 µm →|

FIGURE 3.30 Lysosomes and Peroxisomes. (a) Lysosomes, produced from Golgi vesicles. (b) Peroxisomes, which look similar but are produced by mitochondria working with the endoplasmic reticulum, and by division of other peroxisomes.

a–b: Don Fawcett/Science Source

Proteasomes

Cells must tightly control the concentration of proteins in their cytoplasm. Therefore, they must not only synthesize new proteins, but also dispose of those that are no longer needed. Cells also need to rid themselves of damaged and nonfunctional proteins and foreign proteins introduced by such events as viral infection. Protein synthesis, we have seen, is the domain of the ribosomes; protein disposal is the function of another structurally simple organelle called a **proteasome.**

Proteasomes are hollow, cylindrical complexes of proteins located in both the cytoplasm and nucleus **(fig. 3.31).** A cell tags undesirable proteins for destruction and transports them to a proteasome. As the undesirable protein passes through the core of this organelle, the proteasome's enzymes unfold it and break it down into short peptides and free amino acids. These can be used to synthesize new proteins or be presented to the immune system for further degradation. Proteasomes degrade more than 80% of a cell's proteins.

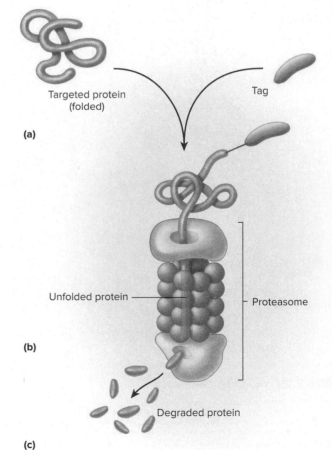

FIGURE 3.31 Protein Degradation by a Proteasome.
(a) The unwanted protein targeted for destruction is tagged and transported to a proteasome. (b) As the protein passes down the core of the proteasome, it is unfolded and cleaved into small peptides and free amino acids. (c) The degraded protein fragments are released from the other end of the proteasome.

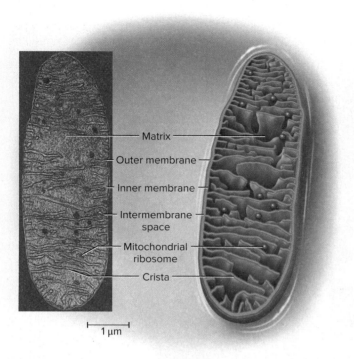

Matrix

Outer membrane

Inner membrane

Intermembrane space

Mitochondrial ribosome

Crista

1 μm

FIGURE 3.32 A Mitochondrion.

Keith R. Porter/Science Source

Mitochondria

Mitochondria[38] (MY-toe-CON-dree-uh) (singular, *mitochondrion*) are organelles specialized for synthesizing ATP. They have a variety of shapes—spheroidal, rod-shaped, kidney-shaped, or threadlike (**fig. 3.32**). They are quite mobile, squirming and changing shape continually. They sometimes undergo fusion (two mitochondria joining to become one) and fission (one mitochondrion dividing in two). Like the nucleus, a mitochondrion is surrounded by a double membrane. The inner membrane usually has folds called **cristae**[39] (CRIS-tee), which project like shelves across the organelle. The space between the cristae, called the **matrix,** contains ribosomes; enzymes used in ATP synthesis; and many small, circular DNA molecules called *mitochondrial DNA (mtDNA).* Mitochondria are the "powerhouses" of the cell. Energy is not *made* here, but it is extracted from organic compounds and transferred to ATP, primarily by enzymes located on the cristae. The role of mitochondria in ATP synthesis is explained in detail in section 26.2d, and some evolutionary and clinical aspects of mitochondria are discussed in Deeper Insight 3.3.

---

[38]*mito* = thread; *chondr* = grain
[39]*crista* = crest

DEEPER INSIGHT 3.3

EVOLUTIONARY MEDICINE

Mitochondria—Evolution and Clinical Significance

It is virtually certain that mitochondria evolved from bacteria that invaded another primitive cell, survived in its cytoplasm, and became permanent residents. The double membranes around the mitochondrion suggest that the original bacterium provided the inner membrane, and the host cell's phagosome provided the outer membrane when the bacterium was phagocytized.

Several comparisons show the apparent relationship of mitochondria to bacteria. Their ribosomes are more like bacterial ribosomes than those of eukaryotic (nucleated) cells. Mitochondrial DNA (mtDNA) is a small, circular molecule that resembles the circular DNA of bacteria, not the linear DNA of the cell nucleus. It replicates independently of nuclear DNA. Mitochondrial DNA codes for some of the enzymes employed in ATP synthesis. It consists of 16,569 *base pairs* (explained in section 4.1), comprising 37 genes, compared with over 3 billion base pairs and about 20,000 genes in nuclear DNA.

When a sperm fertilizes an egg, any mitochondria introduced by the sperm are destroyed and only those provided by the egg are passed on to the developing embryo. Therefore, mtDNA is inherited exclusively through the mother. While nuclear DNA is reshuffled in every generation by sexual reproduction, mtDNA remains unchanged except by random mutation. Because of the known pace of such mutations, biologists and anthropologists can use mtDNA as a "molecular clock" to trace evolutionary lineages in humans and other species. The amount of difference between the mtDNAs of related species affords a record of how much time has passed since they diverged from their last common ancestor. Anthropologists have gained evidence from mtDNA that of all the women who lived in Africa 200,000 years ago, only one has any descendants still living today.

This "mitochondrial Eve" is ancestor to us all. Mitochondrial DNA has also been used as evidence in criminal law and to identify the remains of soldiers killed in combat. It was used in 2001 to identify the remains of the famed bandit Jesse James, who was killed in 1882.

Mutations in mtDNA are responsible for various rare hereditary diseases and death in early childhood. Tissues and organs with the highest energy demands are the most vulnerable to mitochondrial dysfunctions—nervous tissue, the heart, the kidneys, and skeletal muscles, for example.

Mitochondrial myopathy is a degenerative muscle disease in which the muscle displays "ragged red fibers," cells with abnormal mitochondria that stain red with a particular histological stain. Another mtDNA disease is *Leber hereditary optic neuropathy (LHON)*, a form of blindness that usually appears in young adulthood as a result of damage to the optic nerve. *Kearns–Sayre syndrome (KSS)* involves paralysis of the eye muscles, degeneration of the retina, heart disease, hearing loss, diabetes, and kidney failure. Damage to mtDNA has also been implicated as a possible factor in Alzheimer disease, Huntington disease, and other degenerative diseases of old age.

Some women known to carry mtDNA mutations can now avoid passing such diseases to their children through in vitro fertilization (IVF) with mitochondrial replacement therapy (MRT). The techniques of MRT are complex and diverse, but the essence of it is to provide an egg of the patient with healthy mitochondria from another woman, an egg donor; fertilize the recipient egg with the father's sperm; and implant this in the patient's uterus. The successful result is a "three-parent baby" who has the father's nuclear DNA, the birth mother's nuclear DNA, and the mitochondria donor's mtDNA.

0.1 μm

(a) Cross section (TEM)

Microtubules

Protein link

Longitudinal view

Cross section

Microtubules

(b) Pair of centrioles

FIGURE 3.33 Centrioles. (a) Electron micrograph of a centriole as seen in cross section. (b) A pair of perpendicular centrioles.

a: Don W Fawcett/Getty Images

? *How does a centriole resemble the axoneme of a cilium? How does it differ?*

Centrioles

A **centriole** (SEN-tree-ole) is a short cylindrical assembly of microtubules, arranged in nine groups of three microtubules each **(fig. 3.33)**. Near the nucleus, most cells have a small, clear patch of cytoplasm called the **centrosome**[40] containing a pair of mutually perpendicular centrioles **(see fig. 3.24)**. Centrioles play a role in cell division described in section 4.3d. Each basal body of a flagellum or cilium is a single centriole oriented perpendicular to the plasma membrane. Basal bodies originate in a *centriolar organizing center* and migrate to the plasma membrane. Two microtubules of each triplet then elongate to form the nine pairs of peripheral microtubules of the axoneme. A cilium can grow to its full length in less than an hour.

3.4c Inclusions

Inclusions are of two kinds: accumulated cell products such as glycogen granules, pigments, and oil droplets (see fig. 3.28b); and

foreign bodies such as viruses, bacteria, and dust particles and other debris phagocytized by a cell. Inclusions are never enclosed in a membrane, and unlike the organelles and cytoskeleton, they are not essential to cell survival.

The major features of a cell are summarized in **table 3.4**.

> **BEFORE YOU GO ON**
>
> Answer the following questions to test your understanding of the preceding section:
>
> 22. Distinguish between organelles and inclusions. State two examples of each.
>
> 23. Briefly state how each of the following cell components can be recognized in electron micrographs: the nucleus, a mitochondrion, a lysosome, and a centriole. What is the primary function of each?
>
> 24. What three organelles are involved in protein synthesis?
>
> 25. In what ways do rough and smooth endoplasmic reticulum differ?
>
> 26. Define *centriole, microtubule, cytoskeleton,* and *axoneme.* How are these structures related to one another?

---

[40]*centro* = central; *some* = body

TABLE 3.4	Summary of Organelles and Other Cellular Structures	
Structure	**Appearance to TEM**	**Function**
Plasma membrane (fig. 3.5)	Two dark lines at cell surface, separated by a narrow light space	Prevents escape of cell contents; regulates exchange of materials between cytoplasm and extracellular fluid; involved in intercellular communication
Microvilli (fig. 3.9)	Short, densely spaced, hairlike processes or scattered bumps on cell surface; interior featureless or with bundle of microfilaments	Increase absorptive surface area; widespread sensory roles (hearing, equilibrium, taste)
Cilia (fig. 3.10)	Long hairlike projections of apical cell surface; axoneme with usually a 9 + 2 array of microtubules	Move substances along cell surface; widespread sensory roles (equilibrium, smell, vision)
Flagellum	Long, single, whiplike process with axoneme	Sperm motility
Microfilaments (figs. 3.9 and 3.24)	Thin protein filaments (6 nm diameter), often in parallel bundles or dense networks in cytoplasm	Support microvilli and plasma membrane; involved in muscle contraction and other cell motility, endocytosis, and cell division
Intermediate filaments (fig. 3.24)	Thicker protein filaments (8–10 nm diameter) extending throughout cytoplasm or concentrated at cell-to-cell junctions	Give shape and physical support to cell; anchor cells to each other and to extracellular material; compartmentalize cell contents
Microtubules (figs. 3.24 and 3.25)	Hollow protein cylinders (25 nm diameter) radiating from centrosome	Form axonemes of cilia and flagella, centrioles, basal bodies, and mitotic spindles; enable motility of cell parts; form trackways that direct organelles and macromolecules to their destinations within a cell
Nucleus (figs. 3.4, 3.26, and 3.27)	Largest organelle in most cells, surrounded by double membrane with nuclear pores	Genetic control center of cell; directs protein synthesis; shelters the DNA
Rough ER (fig. 3.28a)	Extensive sheets of parallel membranes with ribosomes on outer surface	Protein synthesis and manufacture of cellular membranes
Smooth ER (fig. 3.28b)	Branching network of tubules with smooth surface (no ribosomes); usually broken into numerous small segments in TEM photos	Lipid synthesis, detoxification, calcium storage
Ribosomes (fig. 3.28c)	Small dark granules free in cytosol, on surface of rough ER and nuclear envelope, and inside nucleus and mitochondria	Interpret the genetic code and synthesize polypeptides
Golgi complex (fig. 3.29)	Several closely spaced, parallel cisterns with thick edges, usually near nucleus, often with many Golgi vesicles nearby	Receives and modifies newly synthesized polypeptides; synthesizes carbohydrates; adds carbohydrates to glycoproteins; packages cell products into Golgi vesicles
Golgi vesicles (fig. 3.29)	Round to irregular sacs near Golgi complex, usually with light, featureless contents	Become secretory vesicles and carry cell products to apical surface for exocytosis, or become lysosomes
Lysosomes (fig. 3.30a)	Round to oval sacs with single enclosing membrane, often a dark featureless interior but sometimes with protein layers or crystals	Contain enzymes for intracellular digestion, autophagy, programmed cell death, and glucose mobilization
Peroxisomes (fig. 3.30b)	Similar to lysosomes; often lighter in color	Contain enzymes for detoxification of free radicals, alcohol, and other drugs; oxidize fatty acids
Proteasomes (fig. 3.31)	Small cytoplasmic granules composed of a cylindrical array of proteins	Degrade proteins that are undesirable or no longer needed by a cell
Mitochondria (fig. 3.32)	Round, rod-shaped, bean-shaped, or threadlike structures with double enclosing membrane and shelflike infoldings called cristae	ATP synthesis
Centrioles (fig. 3.33)	Short cylindrical bodies, each composed of a circle of nine triplets of microtubules	Form mitotic spindle during cell division; unpaired centrioles form basal bodies of cilia and flagella
Centrosome (fig. 3.24)	Clear area near nucleus containing a pair of centrioles	Organizing center for formation of microtubules of cytoskeleton and mitotic spindle
Basal body (fig. 3.10b)	Unpaired centriole at the base of a cilium or flagellum	Point of origin, growth, and anchorage of a cilium or flagellum; produces axoneme
Inclusions (fig. 3.28b)	Highly variable—fat droplets, glycogen granules, protein crystals, dust, bacteria, viruses; never enclosed in membranes	Storage products or other products of cellular metabolism, or foreign matter retained in cytoplasm

STUDY GUIDE

▶ Assess Your Learning Outcomes

To test your knowledge, discuss the following topics with a study partner or in writing, ideally from memory.

3.1 Concepts of Cellular Structure

1. The scope of cytology
2. Basic tenets of the cell theory
3. The nine common cell shapes
4. The size range of most human cells; some extremes outside this range; and some factors that limit cells from growing indefinitely large
5. The two kinds of electron microscopes; why they have enhanced the modern understanding of cells; and the distinction between magnification and resolution in microscopy
6. Basic structural components of a cell
7. The distinction between intracellular fluid (ICF) and extracellular fluid (ECF)

3.2 The Cell Surface

1. The molecules of the plasma membrane and how they are organized
2. The distinctive roles of phospholipids, glycolipids, cholesterol, integral proteins, peripheral proteins, and glycoproteins in membrane structure
3. Seven roles played by membrane proteins
4. Distinctions between leak channels and gated channels, and between ligand-gated, voltage-gated, and mechanically gated channels
5. The function of second-messenger systems associated with the plasma membrane; the specific roles of membrane receptors, G proteins, adenylate cyclase, cyclic adenosine monophosphate, and kinases in the cAMP second-messenger system
6. Composition and functions of the glycocalyx
7. Structure and functions of microvilli, and where they are found
8. Structure and functions of cilia, and where they are found

9. Structure and function of the only human flagellum
10. Structure and function of pseudopods

3.3 Membrane Transport

1. What it means to say that a plasma membrane is selectively permeable, and why this property is important for human survival
2. Filtration, where it occurs in the body, and why it depends on hydrostatic pressure
3. Simple diffusion, factors that determine its speed, and examples of its physiological and clinical relevance
4. Osmosis, examples of its physiological and clinical relevance, factors that determine its speed and direction, and the role of aquaporins
5. Reverse osmosis, where it occurs in the body, and the purpose it serves
6. In relation to osmosis, the meaning of *osmotic pressure, osmolarity, tonicity,* and *milliosmoles per liter (mOsm/L)*
7. Distinctions between hypotonic, hypertonic, and isotonic solutions; their effects on cells; and how this relates to intravenous fluid therapy
8. How carrier-mediated transport differs from other types of transport, and the relevance of specificity to this process
9. How carrier-mediated transport is limited by carrier saturation and the transport maximum (T_m)
10. Distinctions between a uniport, symport, and antiport; the meanings of *cotransport* and *countertransport*; and examples of where each is relevant in human physiology
11. Similarities and differences between facilitated diffusion and active transport
12. The distinction between primary and secondary active transport
13. The mechanism and roles of the sodium–potassium (Na^+–K^+) pump
14. How vesicular transport differs from other modes of membrane transport; the difference between endocytosis and exocytosis; different forms of endocytosis; and examples of the physiological relevance of each kind of vesicular transport
15. Of the preceding mechanisms of transport, which ones require a membrane; which ones require a plasma membrane and which ones can also occur through artificial membranes; which ones require ATP and cease if ATP is unavailable, as upon death

3.4 The Cell Interior

1. Distinctions between cytoplasm, cytosol, cytoskeleton, organelles, and inclusions; and the respective, general roles of each in the internal organization of a cell
2. Overall functions of the cytoskeleton and the differences between microfilaments, intermediate filaments, and microtubules
3. Which organelles are considered membranous and why, and which of these are enclosed in single or double membranes
4. Structure and function of the nucleus, especially the nuclear envelope and nuclear lamina
5. General structure of the endoplasmic reticulum (ER); the two types of ER and the structural and functional differences between them
6. The composition, location, and function of ribosomes
7. The structure and functions of the Golgi complex; the origin and destiny of Golgi vesicles
8. The structures and functions of lysosomes and peroxisomes, and the similarities and differences between them
9. Structure and function of proteasomes
10. Structure and function of mitochondria
11. Structures and functions of centrioles, the centrosome, and basal bodies; and how these relate to each other
12. How inclusions differ from organelles; the origins and types of inclusions

▶ Testing Your Recall

Answers in Appendix A

1. The clear, structureless gel in a cell is its
   a. nucleoplasm.
   b. protoplasm.
   c. cytoplasm.
   d. neoplasm.
   e. cytosol.

2. The Na^+–K^+ pump is
   a. a peripheral protein.
   b. a transmembrane protein.
   c. a G protein.
   d. a glycolipid.
   e. a phospholipid.

3. Which of the following processes could occur *only* in the plasma membrane of a living cell?
   a. facilitated diffusion
   b. simple diffusion
   c. filtration
   d. active transport
   e. osmosis

STUDY GUIDE

4. Cells specialized for absorption of matter from the ECF are likely to show an abundance of
   a. lysosomes.
   b. microvilli.
   c. mitochondria.
   d. secretory vesicles.
   e. ribosomes.

5. Aquaporins are transmembrane proteins that promote
   a. pinocytosis.
   b. carrier-mediated transport.
   c. active transport.
   d. facilitated diffusion.
   e. osmosis.

6. Membrane carriers resemble enzymes except for the fact that carriers
   a. are not proteins.
   b. do not have binding sites.
   c. are not selective for particular ligands.
   d. change conformation when they bind a ligand.
   e. do not chemically change their ligands.

7. The cotransport of glucose derives energy from
   a. a Na^+ concentration gradient.
   b. the glucose being transported.
   c. a Ca^{2+} gradient.

d. the membrane voltage.
   e. body heat.

8. The function of cAMP in a cell is
   a. to activate a G protein.
   b. to remove phosphate groups from ATP.
   c. to activate kinases.
   d. to bind to the first messenger.
   e. to add phosphate groups to enzymes.

9. Most cellular membranes are made by
   a. the nucleus.
   b. the cytoskeleton.
   c. enzymes in the peroxisomes.
   d. the endoplasmic reticulum.
   e. replication of existing membranes.

10. Matter can leave a cell by any of the following means *except*
    a. active transport.
    b. pinocytosis.
    c. an antiport.
    d. simple diffusion.
    e. exocytosis.

11. Most human cells are 10 to 15 _____ in diameter.

12. When a hormone cannot enter a cell, it activates the formation of a/an _____ inside the cell.

13. _____ channels in the plasma membrane open or close in response to changes in the electrical charge difference across the membrane.

14. The force exerted on a membrane by water is called _____.

15. A concentrated solution that causes a cell to shrink is _____ to the cell.

16. Fusion of a secretory vesicle with the plasma membrane and release of the vesicle's contents is a process called _____.

17. _____ and _____ are two granular organelles (enzyme complexes) that, respectively, synthesize and degrade proteins.

18. Liver cells can detoxify alcohol with two organelles, the _____ and _____.

19. An ion gate in the plasma membrane that opens or closes when a chemical binds to it is called a/an _____.

20. The space enclosed by the membranes of the Golgi complex and endoplasmic reticulum is called the _____.

▶ Building Your Medical Vocabulary

Answers in Appendix A

State a meaning of each word element, and give a medical term from this chapter that uses it or a slight variation of it.

1. anti-

2. chromato-

3. co-

4. cyto-

5. endo-

6. facil-

7. fusi-

8. -logy

9. -osis

10. phago-

▶ What's Wrong with These Statements?

Answers in Appendix A

Briefly explain why each of the following statements is false, or reword it to make it true.

1. If a cell were poisoned so it could not make ATP, osmosis through its membrane would cease.

2. Each cell of the human body has a single nucleus.

3. A cell's second messengers serve mainly to transport solutes through the membrane.

4. The Golgi complex makes all of a cell's lysosomes and peroxisomes.

5. Some membrane channels are peripheral proteins.

6. The plasma membrane consists primarily of protein molecules.

7. The brush border of a cell is composed of cilia.

8. Human cells placed in hypertonic saline will swell and burst.

9. The transport maximum (T_m) sets an upper limit on the rate of osmosis.

10. All of a cell's ribosomes are found on the nuclear envelope and the rough endoplasmic reticulum.

STUDY GUIDE

▶ Testing Your Comprehension

1. If someone bought a saltwater fish in a pet shop and put it in a freshwater aquarium at home, what would happen to the fish's cells? What would happen if someone put a freshwater fish in a saltwater aquarium? Explain.

2. A farmer's hand and forearm are badly crushed in a hay baler. When examined at the hospital, his blood potassium level is found to be abnormal. Would you expect it to be higher or lower than normal? Explain.

3. Many children worldwide suffer from a severe deficiency of dietary protein. As a result, they have very low levels of blood albumin. How do you think this affects the water content and volume of their blood? Explain.

4. It is often said, even in some textbooks, that mitochondria make energy for a cell. Why is this statement false?

5. Kartagener syndrome is a hereditary disease in which dynein arms are lacking from the axonemes of cilia and flagella. Predict the effect of Kartagener syndrome on a man's ability to father a child. Predict its effect on his respiratory health. Explain both answers.

Ribosomes (gray) reading the genetic instructions on a messenger RNA strand (red), generating proteins (green) with an amino acid sequence determined by the genetic code (TEM)

Dr. Elena Kiseleva/Science Source

CHAPTER 4

GENES AND CELLULAR FUNCTION

CHAPTER OUTLINE

4.1 DNA and RNA—The Nucleic Acids

   4.1a DNA Structure and Function

   4.1b Chromatin and Chromosomes

   4.1c RNA Structure and Function

4.2 Genes and Their Action

   4.2a What Is a Gene?

   4.2b The Genetic Code

   4.2c Protein Synthesis

   4.2d Protein Processing and Secretion

   4.2e Gene Regulation

   4.2f Synthesizing Compounds Other Than Proteins

4.3 DNA Replication and the Cell Cycle

   4.3a DNA Replication

   4.3b Errors and Mutations

   4.3c The Cell Cycle

   4.3d Mitosis

   4.3e Regulation of Cell Division

4.4 Chromosomes and Heredity

   4.4a The Karyotype

   4.4b Genes and Alleles

   4.4c Multiple Alleles, Codominance, and Incomplete Dominance

   4.4d Polygenic Inheritance and Pleiotropy

   4.4e Sex Linkage

   4.4f Dominant and Recessive Alleles at the Population Level

   4.4g Environmental Effects and Gene Expression

   4.4h Epigenetics

Study Guide

DEEPER INSIGHTS

4.1 Medical History: Discovery of the Double Helix

4.2 Clinical Application: Genomic Medicine

4.3 Clinical Application: Cancer

Anatomy & Physiology Revealed® 4.0

Module 2: Cells and Chemistry

Several chapters in this book discuss hereditary traits such as blood types and hair color and genetic disorders such as color blindness, cystic fibrosis, diabetes mellitus, and hemophilia. To understand such conditions, it is necessary to have some familiarity with DNA and genes. This chapter is intended to provide such preparation.

Heredity has been a matter of human interest dating even to biblical writings, but a scientific understanding of how traits pass from parents to offspring began with the Austrian monk Gregor Mendel (1822–84) and his famous experiments on garden peas. By the late nineteenth century, biologists had observed chromosomes and their behavior during cell division. In the early twentieth century, they rediscovered Mendel's work and realized the correlation between chromosome behavior and his laws of heredity. From that simple but insightful beginning, genetics has grown into a highly diverse science with several subdisciplines and is arguably the most dynamic of all natural sciences in these early decades of the twenty-first century.

Mendelian genetics deals with parent–offspring and larger family relationships to discern and predict patterns of inheritance within a family line. *Cytogenetics* uses the techniques of cytology and microscopy to study chromosomes and their relationship to hereditary traits. *Molecular genetics* uses the techniques of biochemistry to study the structure and function of DNA. *Genomic medicine* comprehensively studies the entire DNA endowment of an individual (the *genome*), how it influences health and disease, and how this knowledge can be used to prevent, treat, or cure diseases. We will examine these perspectives in this chapter.

4.1   DNA and RNA—The Nucleic Acids

Expected Learning Outcomes

When you have completed this section, you should be able to

a. describe the structure of DNA and relate this to its function;

b. explain how DNA and proteins are organized to form the chromosomes; and

c. describe the types of RNA, their structural and functional differences, and how they compare with DNA.

With improvements in the microscope, biologists of the late nineteenth century saw that cell division is immediately preceded by nuclear division, and during nuclear division, the chromosomes split neatly in two and distribute their halves to the two daughter cells. They came to suspect that the nucleus was the center of heredity and cellular control, and they began probing it for the biochemical secrets of heredity. Swiss biochemist Johann Friedrich Miescher (1844–95) studied the nuclei of white blood cells extracted from pus in hospital bandages, and later the nuclei of salmon sperm, since both cell types offered large nuclei with minimal amounts of contaminating cytoplasm. In 1869, he discovered an acidic, phosphorus-rich substance he named *nuclein.* He correctly believed this to be the cell's hereditary matter, although he was never able to convince other scientists of this. We now call this substance **deoxyribonucleic acid (DNA)** and know it to be the repository of our genes.

By 1900, biochemists knew the basic components of DNA—sugar, phosphate groups, and organic rings called nitrogenous bases—but they didn't have the technology to determine how these were put together. That understanding didn't come until 1953, in one of the twentieth century's most dramatic and important stories of scientific discovery (see Deeper Insight 4.1). The following description of DNA is largely the outcome of that work.

4.1a DNA Structure and Function

DNA is a long threadlike molecule with a uniform diameter of 2 nm, although its length varies greatly from the smallest to the largest chromosomes. Most human cells have 46 molecules of DNA totaling 2 m in length. This makes the average DNA molecule about 43 mm (almost 2 in.) long. To put this in perspective, imagine that an average DNA molecule was scaled up to the diameter of a utility pole (about 20 cm, or 8 in.). At this diameter, a pole proportionate to DNA would rise about 4,400 km (2,700 mi.) into space—far higher than the orbits of space shuttles (320–390 km) and the Hubble Space Telescope (600 km).

At the molecular level, DNA and other nucleic acids are polymers of **nucleotides.** A nucleotide consists of a sugar, a phosphate group, and a single- or double-ringed **nitrogenous base** **(fig. 4.1a).** Two of the bases in DNA—**cytosine (C)** and **thymine (T)**—have a single carbon–nitrogen ring and are classified as *pyrimidines* (py-RIM-ih-deens). The other two bases—**adenine (A)** and **guanine (G)**—have double rings and are classified as *purines* **(fig. 4.1b).**

The structure of DNA, commonly described as a *double helix,* resembles a spiral staircase **(fig. 4.2).** Each sidepiece is a backbone composed of phosphate groups alternating with the sugar *deoxyribose.* The steplike connections between the backbones are pairs of nitrogenous bases. The bases face the inside of the helix and hold the two backbones together with hydrogen bonds. Across from a purine on one backbone, there is a pyrimidine on the other. The pairing of each small, single-ringed pyrimidine with a large, double-ringed purine gives the DNA molecule its uniform 2 nm width.

A given purine cannot arbitrarily bind to just any pyrimidine. Adenine and thymine form two hydrogen bonds with each other, and guanine and cytosine form three, as shown in figure 4.2c. Therefore, where there is an A on one backbone, there is normally a T across from it, and every C is normally paired with a G. A—T and C—G are called the **base pairs.** The fact that one strand governs the base sequence of the other is called the **law of complementary base pairing.** It enables us to predict the base sequence of one strand if we know the sequence of the complementary strand.

(a)

(b)

FIGURE 4.1 Nucleotides and Nitrogenous Bases. (a) The structure of a nucleotide, one of the monomers of DNA and RNA. In RNA, the sugar is ribose. (b) The five nitrogenous bases found in DNA and RNA.

❓ *How would the uniform 2 nm diameter of DNA be affected if two purines or two pyrimidines could pair with each other?*

(a)

(b)

(c)

FIGURE 4.2 DNA Structure. (a) A molecular space-filling model of DNA giving some impression of its molecular geometry. (b) The "spiral staircase" structure. The two sugar–phosphate backbones twine around each other while complementary bases (colored bars) face each other on the inside of the double helix. (c) A small segment of DNA showing the composition of the backbone and complementary pairing of the nitrogenous bases.

DEEPER INSIGHT 4.1

MEDICAL HISTORY

Discovery of the Double Helix

Credit for determining the double-helical structure of DNA has gone mainly to James Watson and Francis Crick **(fig. 4.3).** The events surrounding their discovery form one of the most dramatic stories of modern science—the subject of many books and at least one movie. When Watson and Crick came to share a laboratory at Cambridge University in 1951, both had barely begun their careers. Watson, age 23, had just completed his Ph.D. in the United States, and Crick, 11 years older, was a doctoral candidate in England. Yet the two were about to become the most famous molecular biologists of the twentieth century, and the discovery that won them such acclaim came without a single laboratory experiment of their own.

Others were fervently at work on DNA, including Rosalind Franklin and Maurice Wilkins at King's College in London. Using a technique called X-ray diffraction, Franklin had determined that DNA had a repetitious helical structure with sugar and phosphate on the outside of the helix. Without her permission, Wilkins showed one of Franklin's best X-ray photographs to Watson. Watson said, "The instant I saw the picture my mouth fell open and my pulse began to race." It provided a flash of insight that allowed the Watson and Crick team to beat Franklin to the goal. Combining what they already knew with the molecular geometry revealed by Franklin's photo, they were quickly able to piece together a scale model from cardboard and sheet metal that fully accounted for the known geometry of DNA. They rushed a paper into print in 1953 describing the double helix, barely mentioning the importance of Franklin's 2 years of painstaking X-ray diffraction work in unlocking the mystery of life's most important molecule. Franklin published her findings in a separate paper back to back with theirs.

For this discovery and the ensuing decade of research on DNA that it opened up, Watson, Crick, and Wilkins shared the Nobel Prize for Physiology or Medicine in 1962. Nobel Prizes are awarded only to the living, and in the final irony of her career, Rosalind Franklin had died in 1958, at the age of 37, of a cancer possibly induced by the X-rays that were her window on DNA architecture.

(a)

(b)

(c)

FIGURE 4.3 Discoverers of the Double Helix. (a) Rosalind Franklin (1920–58), whose painstaking X-ray diffraction photographs revealed important information about the molecular geometry of DNA. (b) One of Franklin's X-ray photographs. (c) James Watson (1928–) (left) and Francis Crick (1916–2004) with their model of the double helix.

a: Courtesy of Cold Springs Harbor Laboratory; b: Courtesy of King's College, London; c: A. Barrington Brown/Science Source

▶▶▶ **APPLY WHAT YOU KNOW**

What would be the base sequence of the DNA strand across from ATTGACTCG? If a DNA molecule was known to be 20% adenine, predict its percentage of cytosine and explain your answer.

The essential function of DNA is to carry instructions, called *genes,* for the synthesis of proteins. At this point in the chapter, we will provisionally regard a gene as a segment of DNA that codes for a protein. Later, we shall have to confront the fact that this is an inadequate definition, and we'll examine the meaning of the word more deeply.

Humans are estimated to have about 22,300 genes. These constitute only about 2% of the DNA. The other 98% does not code for proteins, but apparently plays various roles in chromosome structure and regulation of gene activity.

4.1b Chromatin and Chromosomes

DNA doesn't exist as a naked double helix in the nucleus of a cell, but is complexed with proteins to form a fine filamentous material called **chromatin.** In most cells, the chromatin occurs as 46 long filaments called **chromosomes.** There is a stupendous amount of DNA in one nucleus—about 2 m (6 ft) of it in the first half of a cell's life cycle and twice as much when a cell has replicated its DNA in preparation for cell division. It is a prodigious feat to pack this much DNA into a nucleus only about 5 µm in diameter—and in such an orderly fashion that it doesn't become tangled, broken, and damaged beyond use.

In order to achieve such packing, DNA is extensively coiled and supercoiled. **Figure 4.4** shows the successive levels of this organization. The DNA first winds around spools of proteins called *histones* to form the little granules *(core particles)* visible in figure 4.4a. The chromatin then folds into successive zigzags, loops, and coils, getting thicker and shorter as it does so (fig. 4.4b). Ultimately the DNA, itself 2 nm in diameter, is consolidated into chromatin strands 150 times thicker and 1,000 times shorter than the naked DNA. Finally, each chromosome is packed into its own spheroidal region of the nucleus, called a *chromosome territory.* A chromosome territory is permeated with channels that allow regulatory chemicals to have access to the genes.

This is the state of the DNA in a nondividing cell. It's not a static structure, but changes from moment to moment according to the genetic activity of the cell as individual genes are turned on and off. Whole chromosomes migrate to new territories as a cell develops—for example, moving from the edge to the core of a nucleus as its genes are activated for a certain developmental task, or back to the nuclear lamina to silence some genes. This allows genes on different chromosomes to partner with each other in bringing about developmental changes in the cell.

When a cell is preparing to divide, it makes an exact copy of all its nuclear DNA by a process described later, increasing its allotment to about 4 m of DNA. Each chromosome then consists of two parallel filaments called **sister chromatids.** In the early stage of cell division *(prophase),* these chromatids coil some more until each one becomes, at its most compact, 10,000 times shorter but 350 times thicker than the DNA double helix. Only now are the chromosomes thick enough to be seen with a light microscope. This compaction not only allows the 4 m of DNA to

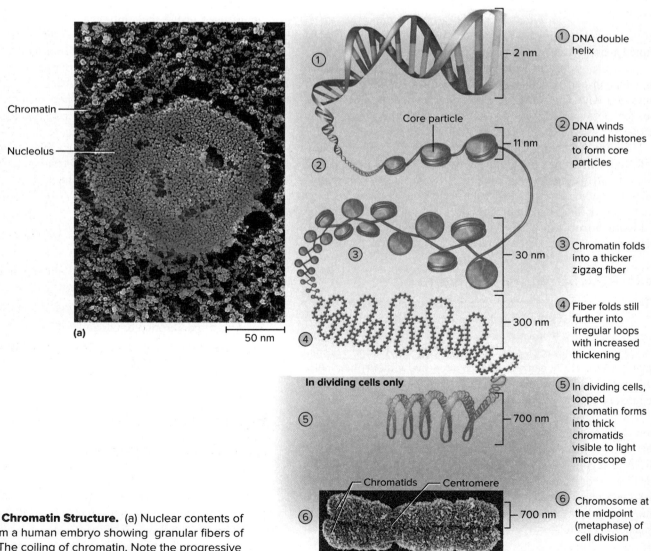

(a)   50 nm

① DNA double helix — 2 nm

Core particle

② DNA winds around histones to form core particles — 11 nm

③ Chromatin folds into a thicker zigzag fiber — 30 nm

④ Fiber folds still further into irregular loops with increased thickening — 300 nm

In dividing cells only

⑤ In dividing cells, looped chromatin forms into thick chromatids visible to light microscope — 700 nm

Chromatids Centromere

⑥ Chromosome at the midpoint (metaphase) of cell division — 700 nm

Chromatin
Nucleolus

FIGURE 4.4 Chromatin Structure. (a) Nuclear contents of a germ cell from a human embryo showing granular fibers of chromatin. (b) The coiling of chromatin. Note the progressive thickening of the fiber from top to bottom.

a: P. Motta & T. Naguro/SPL/Science Source; b: Biophoto Associates/Science Source   (b)

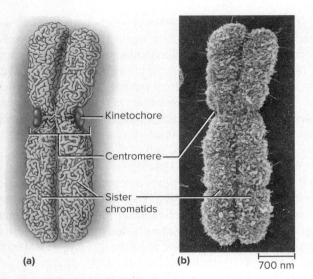

(a)

(b)

700 nm

FIGURE 4.5 Chromosome Structure at Metaphase.
(a) Drawing of a metaphase chromosome. (b) Scanning
electron micrograph.

b: Biophoto Associates/Science Source

TABLE 4.1	Comparison of DNA and RNA	
Feature	DNA	RNA
Sugar	Deoxyribose	Ribose
Types of nitrogenous bases	A, T, C, G	A, U, C, G
Number of nitrogenous bases	Averages 10^8 base pairs	70–10,000 bases, mostly unpaired
Number of nucleotide chains	Two (double helix)	One
Site of action	Functions in nucleus; cannot leave	Leaves nucleus; functions mainly in cytoplasm
Function	Codes for synthesis of RNA and protein	Carries out the instructions in DNA; assembles proteins

fit in the nucleus, but also enables the two sister chromatids to be pulled apart and carried to separate daughter cells.

Despite all this intricate packaging, the DNA of the average mammalian cell is damaged an astonishing 10,000 to 100,000 times per day! The consequences would be catastrophic were it not for DNA repair enzymes that detect and undo most of the damage.

Figure 4.5 shows the structure of a chromosome in early cell division, when it is compacted to its maximum extent. It consists of two genetically identical, rodlike sister chromatids joined together at a pinched spot called the **centromere.** On each side of the centromere, there is a protein plaque called a **kinetochore**[1] (kih-NEE-to-core), which plays a role in cell division.

4.1c RNA Structure and Function

The ribonucleic acids (RNAs) are smaller cousins of DNA. There are many forms of RNA with diverse functions in a cell, but we'll focus on the three that are directly involved in producing proteins: *messenger RNA (mRNA), ribosomal RNA (rRNA),* and *transfer RNA (tRNA).* DNA cannot produce proteins without their help. Other RNA types called *noncoding RNA (ncRNA)* play various regulatory and enzymatic roles beyond the scope of this book.

What do mRNA, rRNA, and tRNA have in common, and how do they differ from DNA? The most significant difference is that RNA is much smaller, ranging from about 70 to 90 bases in tRNA to slightly over 10,000 bases in the largest mRNA. DNA, by contrast, averages more than 100 million base pairs long **(table 4.1).** Also, whereas DNA is a double helix, RNA consists of only one nucleotide chain, not held together by complementary base pairs except in certain short regions where the molecule folds back on

[1]*kineto* = motion; *chore* = place

itself. The sugar in RNA is ribose instead of deoxyribose. RNA contains three of the same nitrogenous bases as DNA—adenine, cytosine, and guanine—but it has no thymine; a base called **uracil (U)** takes its place (see fig. 4.1b). Transfer RNA has more than 50 different nitrogenous bases, but for our purposes, we don't need to consider any except A, U, C, and G.

The essential function of the three principal RNAs is to interpret the code in DNA and use those instructions to synthesize proteins. RNA is a disposable molecule that works mainly in the cytoplasm, while DNA is irreplaceable and remains safely behind in the nucleus, "giving orders" from there. This process is described in the next section of the chapter.

BEFORE YOU GO ON

Answer the following questions to test your understanding of the preceding section:

1. What are the three components of a nucleotide? Which component varies from one nucleotide to another in DNA?

2. What governs the pattern of base pairing in DNA?

3. What is the difference between DNA and chromatin?

4. Summarize the structural and functional differences between DNA and RNA.

4.2 Genes and Their Action

Expected Learning Outcomes

When you have completed this section, you should be able to

a. give a working definition of the *gene* and explain why new discoveries in genetics have changed our concept of what a gene is;

b. explain what the human genome is and what relationship it has to the health sciences;

c. define *genetic code* and describe how DNA codes for protein structure;

d. describe the process of assembling amino acids to form a protein;

e. explain what happens to a protein after its amino acid sequence has been synthesized;

f. describe some ways that a gene can be turned on or off; and

g. explain how DNA indirectly regulates the synthesis of nonprotein molecules.

4.2a What Is a Gene?

As much as biologists talk about genes, the term is devilishly difficult to define. The classical concept of the gene, rooted in Mendel's studies of heredity in peas, was that it is an abstract "unit of heredity" by which a trait passes from parent to offspring. Following discovery of the double helix, molecular biologists worked out the *genetic code* contained in the four bases of DNA and considered a gene to be a segment of the DNA that carries the code for a particular protein. Now, however, we know that the human body has millions of different proteins but only 22,300 protein-coding genes; obviously there isn't a separate gene for every protein. In addition, we know now that several human genes produce only ncRNA molecules that never go on to direct the synthesis of a protein; RNA is their final product. Other recent discoveries have complicated our concept of the gene still more—genes overlapping each other, so some segments of DNA belong to two different genes; short genes embedded within longer ones; multiple related proteins encoded by a single gene; and other unexpected arrangements. As molecular biologists have learned more and more about DNA, the definition of the gene has become more and more frayed around the edges.

For the purposes of this introductory book, however, we can settle for an approximate meaning. We will define **gene** as an information-containing segment of DNA that codes for the production of a molecule of RNA, which in most cases goes on to play a role in the synthesis of one or more proteins. The amino acid sequence of a protein is determined by a nucleotide sequence in the DNA.

The 46 human chromosomes come in two sets of 23 each, one set from each parent. Some of these are gene-rich, such as chromosomes 17, 19, and 22, whereas others are gene-poor, such as 4, 8, 13, 18, 21, and the Y chromosome (see fig. 4.16). All the DNA, both coding and noncoding, in one 23-chromosome set is called the **genome.** The total genome consists of about 3.1 billion nucleotide pairs. Individual genes average about 3,000 nucleotides long, but range up to 2.4 million. Most genes are identical in every human being. All humans, worldwide, are at least 99.99% genetically identical, but even the 0.01% variation means that we can differ from one another in more than 3 million base pairs. Various combinations of these **single-nucleotide polymorphisms**[2] account for all human genetic variation.

Genomics is a relatively young science concerned with the comprehensive study of the genome and how its genes and noncoding DNA affect the structure and function of the organism. Among the other fruits of this research, we now know the chromosomal locations of more than 1,400 disease-producing mutations. This information has opened the door to a branch of medical diagnosis and therapy called **genomic medicine** (see Deeper Insight 4.2).

---

[2]*poly* = multiple; *morph* = form

DEEPER INSIGHT 4.2

CLINICAL APPLICATION

Genomic Medicine

Genomic medicine is the application of our knowledge of the genome to the prediction, diagnosis, and treatment of disease. It is relevant to disorders as diverse as cancer, Alzheimer disease, schizophrenia, obesity, and even a person's susceptibility to nonhereditary diseases such as AIDS and tuberculosis.

Genomic technology has advanced to the point that for less than $1,000, one can have one's entire genome scanned for markers of disease risk. Why would anyone want to? Because knowing one's genome could dramatically change clinical care. It may allow clinicians to forecast a person's risk of disease and to predict its course; mutations in a single gene can affect the severity of such diseases as hemophilia, muscular dystrophy, cancer, and cystic fibrosis. Genomics should also allow for earlier detection of diseases and for earlier, more effective clinical intervention. Drugs that are safe for most people can have serious side effects in others, owing to genetic variations in drug metabolism. Genomics has begun providing a basis for choosing the safest or most effective drugs and for adjusting dosages for different patients on the basis of their genetic makeup.

Knowing the sites of disease-producing mutations expands the potential for *gene-substitution therapy.* This is a procedure in which cells are removed from a patient with a genetic disorder, supplied with a normal gene in place of the defective one, and reintroduced to the body. The hope is that these genetically modified cells will proliferate and provide the patient with a gene product that he or she was lacking—perhaps insulin for a patient with diabetes or a blood-clotting factor for a patient with hemophilia. Researchers are currently exploring a new gene-editing technology called *CRISPR-Cas9* that enables one to locate a specific gene in the DNA and remove or replace it. Clinical applicability and safety were still under investigation and in early clinical trials as this textbook edition was written.

Genomics is introducing new problems in medical ethics and law. Should your genome be a private matter between you and your physician? Should an insurance company be entitled to know your genome before issuing health or life insurance to you so it can know your risk of contracting a catastrophic illness, adjust the cost of your coverage, or even deny coverage? Should a prospective employer have the right to know your genome before offering employment? These are areas in which biology, politics, and law converge to shape public policy.

4.2b The Genetic Code

It seems remarkable that the body can make millions of different proteins (called the **proteome**), all from the same 20 amino acids and all encoded by genes made of just 4 nucleotides (A, T, C, G). This is a striking illustration of how a great variety of complex structures can be made from a small variety of simpler components. The **genetic code** is a system that enables these 4 nucleotides to code for the amino acid sequences of all proteins.

It's not unusual for simple codes to represent complex information. Computers store and transmit complex information, including pictures and sounds, in a binary code with only the symbols 1 and 0. Thus, it should not be surprising that a mere 20 amino acids can be represented by a code of 4 nucleotides; all this requires is to combine these symbols in varied ways. It requires more than 2 nucleotides to code for each amino acid, because the A, U, C, and G of mRNA can combine in only 16 different pairs (AA, AU, AC, AG, UA, UU, and so on). The minimum code to symbolize 20 amino acids is 3 nucleotides per amino acid, and indeed, this is the case in DNA. A sequence of 3 DNA nucleotides that stands for 1 amino acid is called a **base triplet.** When messenger RNA is produced, it carries a coded message based on these DNA triplets. A 3-base sequence in mRNA is called a **codon.** The genetic code is expressed in terms of codons.

Table 4.2 shows a few illuminating examples of how the DNA triplets relate to the mRNA codons and how those, in turn, relate to the amino acids of a protein (see appendix D for the entire genetic code). You can see from the table that sometimes two or more codons represent the same amino acid. The reason for this is easy to explain mathematically. Four symbols (N) taken three at a time (x) can be combined in N^x different ways; that is, there are $4^3 = 64$ possible codons available to represent the 20 amino acids.

Only 61 of these, however, code for amino acids. The other 3 —UAG, UGA, and UAA—are called **stop codons;** they signal "end of message," like the period at the end of a sentence. A stop codon enables the cell's protein-synthesizing machinery to sense that it has reached the end of the instruction for a particular protein. The codon AUG plays two roles: It serves as a code for methionine and as a **start codon.** This dual function is explained shortly.

4.2c Protein Synthesis

We can now move on to an understanding of how DNA and RNA collaborate to produce proteins. Before studying the details, however, it will be helpful to consider the big picture. In brief, the genetic code in DNA specifies which proteins a cell can make. All the body's cells except the sex cells and some immune cells contain identical genes. However, different genes are activated in different cells; for example, genes for digestive enzymes are active in stomach cells but not in muscle cells. Any given cell uses only one-third to two-thirds of its genes; the rest remain dormant in that cell, but may be functional in other types of cells.

When a gene is activated, a **messenger RNA (mRNA)** is made—a mirror image of the gene, more or less. Most mRNA migrates from the nucleus to the cytoplasm, where it serves as a code for assembling amino acids in the right order to make a particular protein. In summary, you can think of the process of protein synthesis as DNA $\longrightarrow$ mRNA $\longrightarrow$ protein, with each arrow reading as "codes for the production of." The step from DNA to mRNA is called *transcription,* and the step from mRNA to protein is called *translation.* Transcription occurs in the nucleus, where the DNA is. Most translation occurs in the cytoplasm, but 10% to 15% of proteins may be synthesized in the nucleus, with both steps occurring there.

Transcription

DNA is too large to leave the nucleus and participate directly in cytoplasmic protein synthesis. It is necessary, therefore, to make a small mRNA copy that can migrate through a nuclear pore into the cytoplasm. Just as we might transcribe (copy) a document, **transcription** in genetics means the process of copying genetic instructions from DNA to RNA. An enzyme called **RNA polymerase** (po-LIM-ur-ase) binds to the DNA and assembles the RNA. It opens up the DNA helix about 17 base pairs at a time, reading the bases from one strand of the DNA and making a corresponding RNA. Where it finds a C on the DNA, it adds G to the RNA; where it finds an A, it adds U; and so forth. The enzyme then rewinds the DNA helix behind it. Another RNA polymerase may follow closely behind the first one; thus, a gene may be transcribed by several polymerase molecules at once, and numerous copies of the same RNA are made. At the end of the gene is a base sequence that serves as a terminator, which signals the polymerase to stop.

The RNA produced by transcription is an "immature" form called *pre-mRNA*. This molecule contains segments called *exons* that will be translated into a protein, and segments called *introns* that are removed before translation. Enzymes cut out the introns and splice the exons together into a functional mRNA molecule, which then usually leaves the nucleus. It may help you in remembering these if you think of *in*trons being removed while still *in* the nucleus and the *ex*ons being *ex*ported from the nucleus to

TABLE 4.2	Examples of the Genetic Code*		
Base Triplet of DNA	Codon of mRNA	Name of Amino Acid	Abbreviation for Amino Acid
CCT	GGA	Glycine	Gly
CCA	GGU	Glycine	Gly
CCC	GGG	Glycine	Gly
CGA	GCU	Alanine	Ala
CGT	GCA	Alanine	Ala
TGG	ACC	Threonine	Thr
TGC	ACG	Threonine	Thr
GTA	CAU	Histidine	His
TAC	AUG	Methionine	Met

*For the complete genetic code, see appendix D.

Gene (DNA)

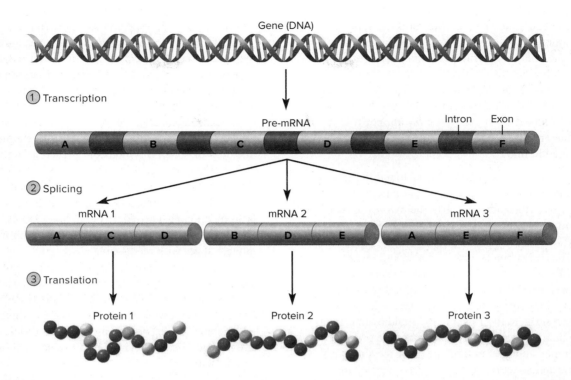

① Transcription

Pre-mRNA

Intron   Exon

| A | B | C | D | E | F |

② Splicing

mRNA 1

| A | C | D |

mRNA 2

| B | D | E |

mRNA 3

| A | E | F |

③ Translation

Protein 1

Protein 2

Protein 3

FIGURE 4.6 Alternative Splicing of mRNA. By splicing together different combinations of exons from a single pre-mRNA, a cell can generate multiple proteins from a single gene.

undergo translation in the cytoplasm. The introns are not entirely useless, but code for a variety of regulatory RNAs.

Through a mechanism called *alternative splicing,* one gene can code for more than one protein. Suppose a gene produced a pre-mRNA containing six exons separated by noncoding introns. As shown in **figure 4.6,** these exons can be spliced together in various combinations to yield codes for two or more proteins. This is a partial explanation of how the body can produce millions of different proteins with little more than 22,000 genes.

Translation

Just as we might translate a work from Spanish into English, genetic **translation** converts the language of nucleotides into the language of amino acids. This job is carried out by the following participants:

1. **Messenger RNA (mRNA),** which carries the genetic code from the nucleus to the cytoplasm. During its synthesis in the nucleus, mRNA acquires a protein cap that acts like a passport, permitting it to pass through a nuclear pore into the cytosol. The cap also acts as a recognition site that tells a ribosome where to begin translation.

2. **Transfer RNA (tRNA),** a relatively small RNA whose job is to bind a free amino acid in the cytosol and deliver it to the ribosome to be added to a growing protein chain. tRNA is a single-stranded molecule that turns back and coils on itself to form an angular L shape **(fig. 4.7).** One

Amino acid–
accepting end

A
C
C

U U A
Anticodon

FIGURE 4.7 Transfer RNA (tRNA).

loop of the molecule includes an **anticodon,** a series of three nucleotides complementary to a specific codon of mRNA. For the codon AUG, for example, the anticodon is UAC. The other end of the tRNA has an amino acid–accepting end that binds a specific amino acid corresponding to that codon. The first tRNA to bind to a ribosome at the start of translation is called the *initiator tRNA.* It always has the anticodon UAC and always carries the amino acid methionine.

3. **Ribosomes,** the little "reading machines" found in the cytosol and on the outside of the rough ER and nuclear envelope. Inactive ribosomes occur in the cytosol in two pieces—a **small subunit** and a **large subunit.** Each is composed of several enzymes and ribosomal RNA (rRNA) molecules. The two subunits come together only when translating mRNA. A ribosome has three pockets that serve as binding sites for tRNA. In the course of translation, a tRNA usually binds first to the *A site* on one side of the ribosome, then shifts to the *P site* in the middle, and finally to the *E site* on the other side. To remember the order of these sites, it may help you to think of *A* for the site that *accepts* a new amino acid; *P* for the site that carries the growing *protein;* and *E* for *exit.* (*A* and *P* actually stand for *aminoacyl* and *peptidyl* sites.)

Translation occurs in three steps called *initiation, elongation,* and *termination.* These are shown in **figure 4.8,** panels 1 through 3; panel 4 illustrates a further aspect of the production of proteins destined to be packaged into lysosomes or secretory vesicles.

① **Initiation.** mRNA passes through a nuclear pore into the cytosol and forms a loop. A small ribosomal subunit binds to a *leader sequence* of bases on the mRNA near the cap, then slides along the mRNA until it recognizes the start codon AUG. An initiator tRNA with the anticodon UAC pairs with the start codon and settles into the P (middle) site of the ribosome with its cargo of methionine (Met). The large subunit of the ribosome then joins the complex. The assembled ribosome now embraces the mRNA in a groove between the subunits and begins sliding along it, reading its bases.

② **Elongation.** The next tRNA arrives, carrying another amino acid; it binds to the A site of the ribosome and its anticodon pairs with the second codon of the mRNA—GCU, for example. A tRNA with the anticodon CGA would bind here, and according to the genetic code, it would carry alanine (Ala). An enzyme in the ribosome transfers the Met of the initiator tRNA to the Ala delivered by the second tRNA and creates a peptide bond between them, giving us a dipeptide, Met—Ala. The ribosome then slides down to read the next codon. This shifts the initiator tRNA (now carrying no amino acid) into the E site, where it leaves the ribosome. The second tRNA (now carrying Met—Ala) shifts into the P site. The now-vacant A site binds a third tRNA. Suppose the next codon is ACG. A tRNA with anticodon UGC would bind here, and would carry threonine

(Thr). (This is the state shown in the figure.) The ribosome transfers the Met—Ala to the Thr, creates another peptide bond, and we now have a tripeptide, Met—Ala—Thr. By repetition of this process, a larger and larger protein is produced. As the protein elongates, it folds into its three-dimensional shape.

Each time a tRNA leaves the E site, it goes off to pick up another amino acid from a pool of free amino acids in the cytosol. One ATP molecule is used in binding the amino acid to the tRNA; therefore, protein synthesis consumes one ATP for every amino acid added to the chain.

All new proteins, as we can see, begin with the amino acid methionine, carried by the initiator tRNA. This is often cleaved off in later processing, however, so not every finished protein starts with methionine.

Codon–anticodon pairing is less precise than just depicted; it tolerates some mismatches, especially at the third base of the codon. Therefore, UGC isn't necessarily the only anticodon that can pair with ACG. Due to this imprecision or "wobble" in the system, as few as 48 different tRNAs are needed to pair up with the 61 codons that represent the amino acids.

③ **Termination.** When the ribosome reaches a stop codon, its A site binds a protein called a *release factor* instead of a tRNA. The release factor causes the finished protein to break away from the ribosome. The ribosome then dissociates into its two subunits, but since these are now so close to the mRNA's leader sequence, they often reassemble on the same mRNA and repeat the process, making another copy of the same protein.

④ **Making proteins for packaging or export.** If a protein is to be packaged into a lysosome or secreted from the cell (such as a digestive enzyme), the ribosome docks on the rough endoplasmic reticulum and the new protein spools off into the cistern of the ER instead of into the cytosol. The ER modifies this protein and packages it into *transport vesicles* whose destiny we will examine later.

One ribosome can work very rapidly, adding about two to six amino acids per second. Most proteins take from 20 seconds to several minutes to make. But a ribosome doesn't work at the task alone. After one ribosome moves away from the leader sequence, another one often binds there and begins the process, following along behind the first—and then another and another, so that a single mRNA is commonly translated by 10 or 20 ribosomes at once. This cluster of ribosomes, all translating the same mRNA, is called a **polyribosome** (see the photo on the opening page of this chapter). The farther along the mRNA each ribosome is, the longer is the protein spooling from it. Not only is each mRNA translated by all these ribosomes at once, but a cell may have 300,000 identical mRNA molecules, each being simultaneously translated by some 20 ribosomes. With so many "factory workers" doing the same task, a cell may produce over 100,000 protein molecules per second—a remarkably productive factory! As much as 25% of the dry weight of liver cells, which are highly active in protein synthesis, consists of ribosomes.

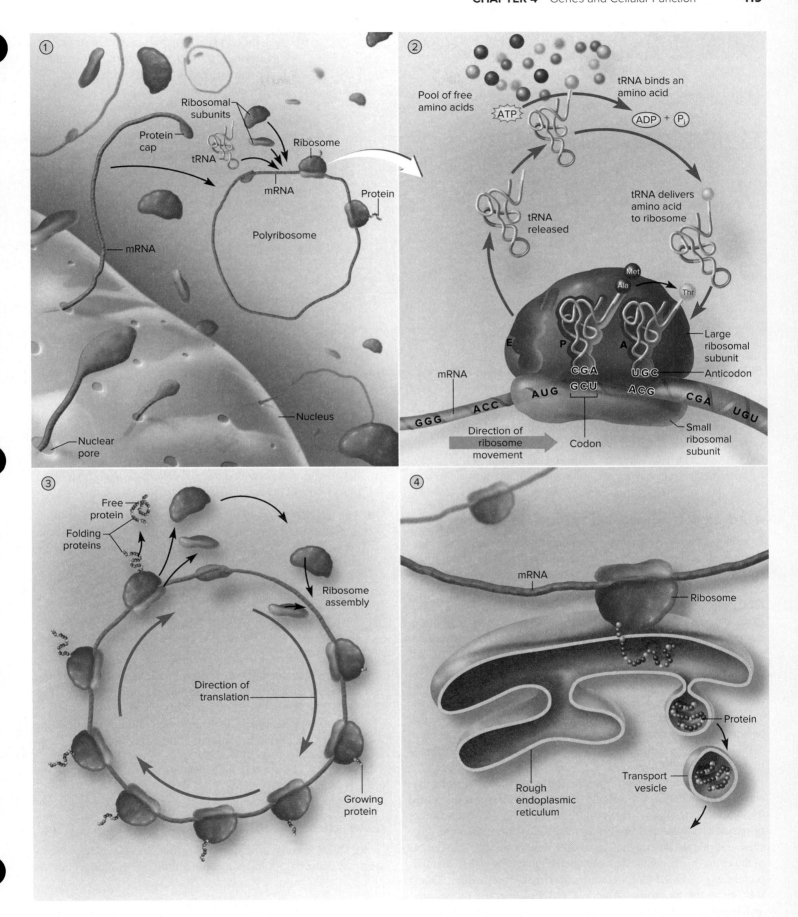

FIGURE 4.8 Translation of mRNA. See the text for explanation of the processes occurring at each step.

❓ *Why would translation not work if ribosomes could bind only one tRNA at a time?*

DNA double helix

Seven base triplets on the template strand of DNA

The corresponding codons of mRNA transcribed from the DNA triplets

The anticodons of tRNA that bind to the mRNA codons

The amino acids carried by those six tRNA molecules

The amino acids linked into a peptide chain

FIGURE 4.9 Relationship of a DNA Base Sequence to Peptide Structure.

Figure 4.9 summarizes transcription and translation and shows how a nucleotide sequence translates to a hypothetical peptide of 6 amino acids. An average protein is about 400 amino acids long; it would have to be represented, at a minimum, by a sequence of 1,203 nucleotides (3 for each amino acid, plus a stop codon).

4.2d Protein Processing and Secretion

Protein synthesis isn't finished when its amino acid sequence (primary structure) has been assembled. To be functional, it must coil or fold into precise secondary and tertiary structures; in some cases, it associates with other protein chains (quaternary structure) or binds with a nonprotein such as a vitamin or carbohydrate. As a new protein is assembled by a ribosome, it is often bound by an older protein called a **chaperone.** The chaperone guides the new protein in folding into the proper shape and helps to prevent improper associations between different proteins. As in the colloquial sense of the word, a chaperone is an older protein that escorts and regulates the behavior of the "youngsters." Some chaperones are also called *stress proteins* or *heat shock proteins* because they're produced in response to heat or other stress on a cell and help damaged proteins fold back into their correct functional shapes.

If a protein is going to be used in the cytosol (for example, the enzymes of glycolysis), it is likely to be made by free ribosomes in the cytosol. However, if it is going to be packaged into a lysosome or secreted from the cell (for example, insulin), the entire polyribosome migrates to the rough ER and docks on its surface. Assembly of the amino acid chain is then completed on the rough ER, and the protein is sent to the Golgi complex for final modification. Thus, we turn to the functions of the ER and Golgi in the processing and secretion of a protein. Compare the following description to **figure 4.10.**

(1) As a protein is assembled on the ER surface, it threads itself through a pore in the ER membrane and into the cistern. Enzymes in the cistern modify the new protein in a variety of ways—removing some amino acid segments, folding the protein and stabilizing it with disulfide bridges, adding carbohydrates, and so forth. Such changes are called **posttranslational modification.** Insulin, for example, is first synthesized as a protein 86 amino acids long. In posttranslational modification, the chain folds back on itself, 3 disulfide bridges are formed, and 35 amino acids are removed from the middle of the protein. The final insulin molecule is therefore made of 2 chains of 21 and 30 amino acids held together by disulfide bridges (see fig. 2.25).

(2) When the rough ER is finished with a protein, it pinches off bubblelike **transport vesicles** coated with a protein called *clathrin.* Clathrin apparently helps to select the proteins to be transported in the vesicles, and as a basketlike cage, it helps to mold the forming vesicles. Soon after the vesicles detach from the ER, they fuse into irregularly shaped clusters that carry their cargo to the Golgi complex.

(3) As they reach the complex, these clusters fuse and form a new Golgi cistern, called the *cis*[3] cistern because it is the one closest to the ER.

(4) This new cistern migrates through the complex toward the opposite *(trans)* face. It matures as it travels, producing new enzymes that modify the cargo in different ways. For example, it may add carbohydrate chains to the proteins, producing the *glycoproteins* mentioned in section 2.4c. Thyroid-stimulating hormone is one such glycoprotein made in this way.

(5) Finally, the *trans*[4] cistern—the one farthest away from the ER—breaks up into Golgi vesicles laden with the cell product.

(6) Some of the Golgi vesicles become lysosomes, while others become **secretory vesicles** that migrate to the plasma membrane and fuse with it, releasing the cell product by exocytosis. This is how a cell of a salivary gland, for example, secretes mucus and digestive enzymes, and how a cell of the pituitary gland releases thyroid-stimulating hormone.

Table 4.3 summarizes the diverse destinations and functions of newly synthesized proteins.

4.2e Gene Regulation

Genes don't simply produce their products at a steady, incessant pace, like a 24-hour nonstop manufacturing plant. They are

---

[3]*cis* = on the same side

[4]*trans* = across from

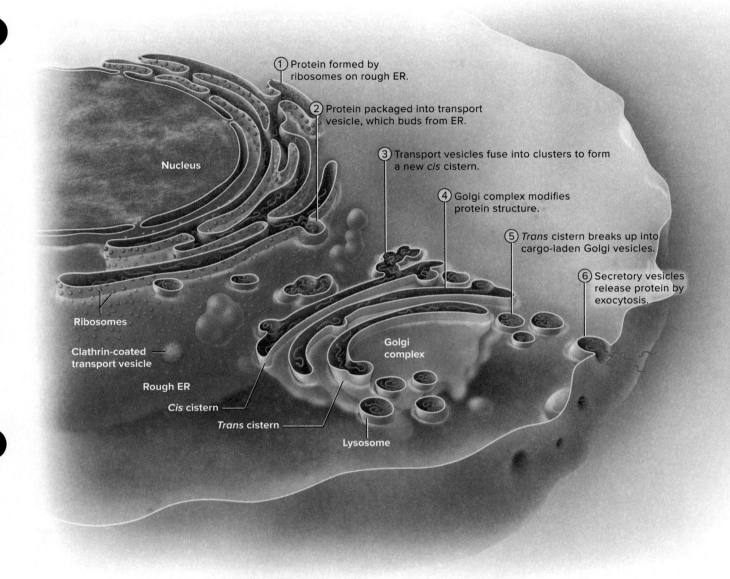

① Protein formed by ribosomes on rough ER.

② Protein packaged into transport vesicle, which buds from ER.

③ Transport vesicles fuse into clusters to form a new *cis* cistern.

④ Golgi complex modifies protein structure.

⑤ *Trans* cistern breaks up into cargo-laden Golgi vesicles.

⑥ Secretory vesicles release protein by exocytosis.

Nucleus

Ribosomes

Clathrin-coated transport vesicle

Rough ER

Cis cistern

Trans cistern

Golgi complex

Lysosome

FIGURE 4.10 Protein Processing and Secretion. See the text for further explanation of the numbered steps.

turned on and off from day to day, even hour to hour, as their products are needed or not, and many genes are permanently turned off in any given cell. The genes for hemoglobin and digestive enzymes, for example, are present but inactive in liver cells and skin cells.

There are several ways to turn genes on or off. We cannot consider all of them here, but an example can convey the general principle. Consider a woman who has just given birth to her first baby. In the ensuing days, the hormone *prolactin* stimulates cells of her mammary glands to synthesize the various components of breast milk, including the protein *casein*—something her body has never synthesized before. How is the gene for casein turned on at this point in her life? **Figure 4.11** shows the steps leading from prolactin stimulation to casein secretion.

TABLE 4.3	Some Destinations and Functions of Newly Synthesized Proteins
Destination or Function	**Proteins (Examples)**
Deposited as a structural protein within cells	Actin of cytoskeleton; keratin of epidermis
Used in the cytosol as a metabolic enzyme	ATPase; kinases
Returned to the nucleus for use in nuclear metabolism	Histones of chromatin; RNA polymerase
Packaged in lysosomes for autophagy, intracellular digestion, and other functions	Numerous lysosomal enzymes
Delivered to other organelles for intracellular use	Catalase of peroxisomes; mitochondrial enzymes
Delivered to plasma membrane to serve transport and other functions	Hormone receptors; sodium–potassium pumps
Secreted by exocytosis for extracellular functions	Digestive enzymes; casein of breast milk

① Prolactin binds to its receptor, a pair of proteins in the plasma membrane of the mammary gland cell.

② The receptor triggers the activation of a **regulatory protein (transcription activator)** in the cytoplasm.

③ The regulatory protein moves into the nucleus and binds to the DNA near the casein gene.

④ This binding enables RNA polymerase to bind to the gene and transcribe it, producing the mRNA for casein.

⑤ The casein mRNA moves into the cytoplasm and is translated by ribosomes on the rough endoplasmic reticulum.

⑥ The Golgi complex packages casein into secretory vesicles.

⑦ The secretory vesicles release the casein by exocytosis, and it becomes part of the milk.

At step 4, there are multiple ways that regulatory proteins can activate gene transcription. Some of them attract and position RNA polymerase so it can begin transcribing the gene.

Others modify the coiling of DNA in a nucleosome in a way that makes specific genes more accessible to RNA polymerase. To turn off a gene, a regulatory protein or ncRNA can coil the chromatin in a different way that makes the gene less accessible, thus preventing transcription. Moving the chromatin over to the nuclear lamina is another way of silencing some of its genes. There are several additional ways, beyond the scope of this book, for inducing or halting the production of a gene product, but the casein example shows how a certain gene may lie dormant in a person until, only a few times in one's life (and only if one bears children), it is activated by a stimulus such as a hormonal signal. Section 4.4h describes a further aspect of gene regulation called *epigenetics*.

4.2f Synthesizing Compounds Other Than Proteins

Cells, of course, make more than proteins—they also synthesize glycogen, fat, steroids, phospholipids, pigments, and many other compounds. There are no genes for these cell products, yet their synthesis is under indirect genetic control. How? They are produced by enzymatic reactions, and enzymes are proteins encoded by genes.

Consider the production of testosterone, for example **(fig. 4.12)**. This is a steroid; there is no gene for testosterone. But to make it, a cell of the testis takes in cholesterol and enzymatically converts it to testosterone. This can occur only if the genes for the enzymes are active. Yet a further implication of this is that genes may greatly affect such complex outcomes as behavior, since testosterone strongly influences such behaviors as aggression and sex drive. In short, DNA codes only for RNA and protein synthesis, yet it indirectly controls the synthesis of a much wider range of substances concerned with all aspects of anatomy, physiology, and behavior.

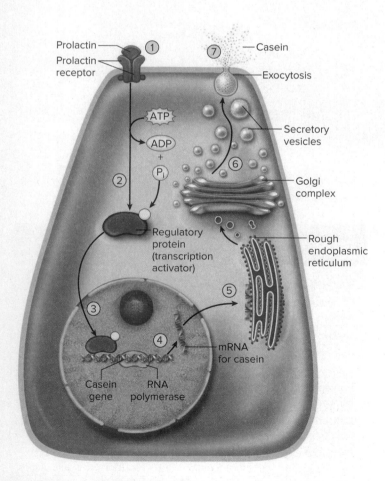

Prolactin
Prolactin receptor
Casein
Exocytosis
ATP
ADP + P𝗂
Secretory vesicles
Golgi complex
Regulatory protein (transcription activator)
Rough endoplasmic reticulum
mRNA for casein
Casein gene
RNA polymerase

FIGURE 4.11 A Mechanism of Gene Activation. The hormone prolactin triggers intracellular reactions that activate a regulatory protein and lead ultimately to the secretion of casein. See the text for further explanation of the numbered steps.

BEFORE YOU GO ON

Answer the following questions to test your understanding of the preceding section:

5. Define *gene, genetic code, codon,* and *anticodon.*

6. Describe the roles of RNA polymerase, ribosomes, and tRNA in producing a protein.

7. What is the difference between genetic transcription and translation?

8. Summarize the processing of a protein from the time a ribosome finishes its work to the time a protein is secreted from the cell. What roles do the endoplasmic reticulum and Golgi complex play in this?

9. Describe some ways a gene can be turned on or off at different times in a person's life.

10. Considering that genes can code only for RNA or proteins, how can the synthesis of nonprotein substances such as carbohydrates or steroids be under genetic control?

FIGURE 4.12 Indirect Control of Testosterone Synthesis by DNA. There is no gene for testosterone, but DNA regulates its production by coding for enzymes that convert cholesterol to testosterone.

4.3 DNA Replication and the Cell Cycle

Expected Learning Outcomes

When you have completed this section, you should be able to

a. describe how DNA is replicated;

b. discuss the consequences of replication errors;

c. describe the life history of a cell, including the events of mitosis; and

d. explain how the timing of cell division is regulated.

Before a cell divides, it must duplicate its DNA so it can give complete and identical copies of all of its genes to each daughter cell. Since DNA controls all cellular function, this replication process must be very exact. We now examine how it is accomplished and consider the consequences of mistakes.

4.3a DNA Replication

The law of complementary base pairing shows that we can predict the base sequence of one DNA strand if we know the sequence of the other. More importantly, it enables a cell to reproduce one strand based on information in the other. The fundamental steps of the replication process are as follows (**fig. 4.13**):

1. The double helix unwinds from the histones.

2. Like a zipper, an enzyme called **DNA helicase** opens up one short segment of the helix at a time, exposing its nitrogenous bases. The point where the DNA is opened up, like the two halves of a zipper separating, is called the **replication fork.**

3. Molecules of the enzyme **DNA polymerase** move along each strand, read the exposed bases, and like a matchmaker, arrange "marriages" with complementary free nucleotides. If the polymerase finds the sequence TCG, for example, it assembles AGC across from it. The two separated strands of DNA are copied by separate polymerase molecules, proceeding in opposite directions. On one strand (top of figure 4.13), the polymerase moves toward the replication fork and makes a long, continuous, new strand of DNA to complement the old one. On the other strand (bottom of the figure), another polymerase moves away from the replication fork and copies only a short segment of DNA at a time. The segments are then joined together by another enzyme called **DNA ligase.** Ultimately, from the old *parental DNA* molecule, two new *daughter DNA* molecules are made. Each daughter DNA

FIGURE 4.13 Semiconservative DNA Replication. (a) At the replication fork, DNA helicase unwinds the double helix and exposes the bases. DNA polymerases begin assembling new bases across from the existing ones. (b) On one strand, DNA polymerase moves toward the replication fork and makes one long, continuous new DNA strand. (c) On the other strand, DNA polymerases begin at the fork and move away from it, replicating the DNA in short segments with gaps between them. (d) DNA ligase closes the gaps to join the segments into a continuous double helix. (e) The ultimate result is two DNA double helices, each composed of one strand of the original DNA and one newly synthesized strand.

consists of one new helix synthesized from free nucleotides and one old helix conserved from the parental DNA. The process is therefore called **semiconservative replication.**

4. While DNA is synthesized in the nucleus, new histones are synthesized in the cytoplasm. Millions of histones are transported into the nucleus within a few minutes after DNA replication, and each new DNA helix wraps around them to make new nucleosomes.

Despite the complexity of this process, each DNA polymerase works at an impressive rate of about 100 base pairs per second. Even at this rate, however, it would take weeks for one polymerase molecule to replicate even one chromosome. But in reality, thousands of polymerase molecules work simultaneously on each DNA molecule, and all 46 chromosomes are replicated in a mere 6 to 8 hours.

4.3b Errors and Mutations

DNA polymerase is fast and accurate, but it makes mistakes. For example, it might read A and place a C across from it where it should have placed a T. If nothing were done to correct such errors, each generation of cells could have thousands of faulty proteins, coded for by DNA that had been miscopied. To prevent such catastrophic damage to the cell, there are multiple modes of correcting replication errors, collectively called the *DNA damage response (DDR)*. DNA polymerase itself double-checks the new base pair and tends to replace incorrect, biochemically unstable pairs with more stable, correct pairs—for example,

removing C and replacing it with T. As a result, only one mistake remains for every billion base pairs replicated—a very high degree of replication accuracy, if not completely flawless.

Changes in DNA structure, called **mutations,**[5] can result from replication errors or from environmental factors such as radiation, chemicals, and viruses. Uncorrected mutations can be passed on to the descendants of that cell, but many of them have no adverse effect. One reason is that a new base sequence sometimes codes for the same thing as the old one. For example, TGG and TGC both code for threonine (see table 4.2), so a mutation from G to C in the third place wouldn't necessarily change protein structure. Another reason is that a change in protein structure isn't always critical to its function. For example, the beta chain of hemoglobin is 146 amino acids long in both humans and horses, but 25 of these amino acids differ between the two species. Nevertheless, the hemoglobin is fully functional in both species. Furthermore, since 98% of the DNA doesn't code for any proteins, the great majority of mutations don't affect protein structure at all. Other mutations, however, may kill a cell, turn it cancerous, or cause genetic defects in future generations. When a mutation changes the sixth amino acid of β-hemoglobin from glutamic acid to valine, for example, the result is a crippling disorder called sickle-cell disease. Clearly some amino acid substitutions are more critical than others, and this affects the severity of a mutation. More than 30 diseases are known to result from defects in the DNA damage response, including some forms of anemia, cancer, immune deficiency, and brain defects.

---

[5]*muta* = change; *ation* = process

FIGURE 4.14 The Cell Cycle.

4.3c The Cell Cycle

Most cells periodically divide into two daughter cells, so a cell has a life cycle extending from one division to the next. This **cell cycle** is divided into four main phases: G_1, S, G_2, and M (**fig. 4.14**).

$\mathbf{G_1}$ is the **first gap phase,** an interval between cell division and DNA replication. During this time, a cell synthesizes proteins, grows, and carries out its preordained tasks for the body. Almost all of the discussion in this book relates to what cells do in the G_1 phase. Cells in G_1 also accumulate the materials needed to replicate their DNA in the next phase. In cultured cells called fibroblasts, which divide every 18 to 24 hours, G_1 lasts 8 to 10 hours.

$\mathbf{S}$ is the **synthesis phase,** in which a cell makes a duplicate copy of its centrioles and nuclear DNA. This is the point at which the cell carries out the semiconservative replication described earlier. The two identical sets of DNA molecules are then available to be divided up between daughter cells at the next cell division. This phase takes 6 to 8 hours in cultured fibroblasts.

$\mathbf{G_2}$, the **second gap phase,** is the interval (4–6 hours in fibroblasts) between DNA replication and cell division. In G_2, a cell exhibits further growth, makes more organelles, finishes replicating its centrioles, and synthesizes enzymes that control cell division. It also checks the fidelity of DNA replication and usually repairs any errors that are detected.

$\mathbf{M}$ is the **mitotic phase,** in which a cell replicates its nucleus and then pinches in two to form new daughter cells. In cultured fibroblasts, the M phase takes 1 to 2 hours. The details of this phase are considered in the next section. Phases G_1, S, and G_2 are collectively called **interphase**—the time between M phases.

The length of the cell cycle varies greatly from one cell type to another. Stomach and skin cells divide rapidly, whereas bone and cartilage cells divide slowly. Some cells leave the cell cycle for a "rest" and cease to divide for days, years, or the rest of one's life—mature neurons, skeletal muscle cells, and adipocytes, for example. Such cells are said to be in the $\mathbf{G_0}$ **(G-zero) phase.** The balance between cells that are actively cycling and those standing by in G_0 is an important factor in determining the number of cells in the body. An inability to stop cycling and enter G_0 is characteristic of cancer cells (see Deeper Insight 4.3).

▶▶▶ **APPLY WHAT YOU KNOW**

What is the maximum number of DNA molecules ever contained in a cell over the course of its life cycle? (Assume the cell has only one nucleus, and disregard mitochondrial DNA.)

4.3d Mitosis

Cells divide by two mechanisms called mitosis and meiosis. Meiosis, however, is restricted to one purpose, the production of eggs and sperm, and is therefore treated in section 27.4a on reproduction. **Mitosis** serves all the other functions of cell division:

- development of an individual, composed of some 50 trillion cells, from a one-celled fertilized egg;

- growth of all tissues and organs after birth;

- replacement of cells that die; and

- repair of damaged tissues.

Four phases of mitosis are recognizable: *prophase, metaphase, anaphase,* and *telophase* (**fig. 4.15**).

① **Prophase.**[6] At the outset of mitosis, the chromosomes shorten and thicken, eventually coiling into compact rods that are easier to distribute to daughter cells than the long, delicate chromatin of interphase. There are 46 chromosomes, each with two chromatids and one molecule of DNA per chromatid. The nuclear envelope disintegrates during prophase and releases the chromosomes into the cytosol. The centrioles begin to sprout elongated microtubules called **spindle fibers,** which push the centrioles apart as they grow. Eventually, a pair of centrioles comes to lie at each pole of the cell. Some spindle fibers grow toward the chromosomes and become attached to the kinetochore on each side of the centromere (see fig. 4.5). The spindle fibers then tug the chromosomes back and forth until they line up along the midline of the cell.

[6]*pro* = first

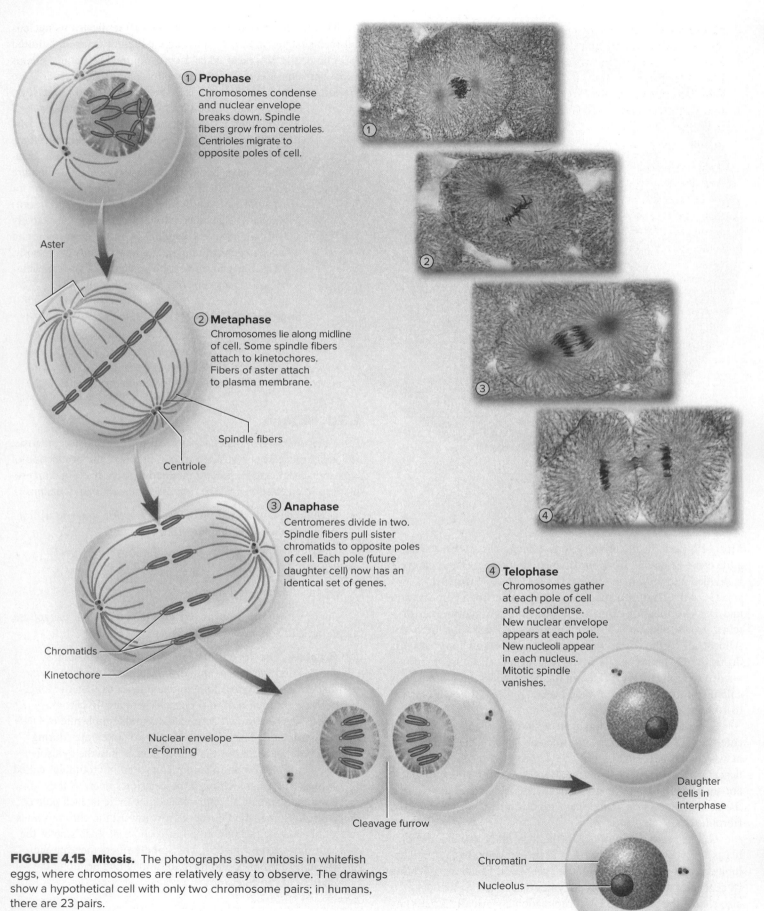

① Prophase
Chromosomes condense and nuclear envelope breaks down. Spindle fibers grow from centrioles. Centrioles migrate to opposite poles of cell.

Aster

② Metaphase
Chromosomes lie along midline of cell. Some spindle fibers attach to kinetochores. Fibers of aster attach to plasma membrane.

Spindle fibers

Centriole

③ Anaphase
Centromeres divide in two. Spindle fibers pull sister chromatids to opposite poles of cell. Each pole (future daughter cell) now has an identical set of genes.

④ Telophase
Chromosomes gather at each pole of cell and decondense. New nuclear envelope appears at each pole. New nucleoli appear in each nucleus. Mitotic spindle vanishes.

Chromatids

Kinetochore

Nuclear envelope re-forming

Cleavage furrow

Daughter cells in interphase

Chromatin

Nucleolus

FIGURE 4.15 Mitosis. The photographs show mitosis in whitefish eggs, where chromosomes are relatively easy to observe. The drawings show a hypothetical cell with only two chromosome pairs; in humans, there are 23 pairs.

1–4: ©Ed Reschke

(2) **Metaphase.**[7] The chromosomes are aligned on the cell equator, oscillating slightly and awaiting a signal that stimulates each of them to split in two at the centromere. The spindle fibers now form a lemon-shaped array called the **mitotic spindle.** Long microtubules reach out from each centriole to the chromosomes, and shorter microtubules form a star-like *aster,*[8] which anchors the assembly to the inside of the plasma membrane at each end of the cell.

(3) **Anaphase.**[9] This phase begins with activation of an enzyme that cleaves the two sister chromatids from each other at the centromere. Each chromatid is now regarded as a separate, single-stranded *daughter chromosome.* One daughter chromosome migrates to each pole of the cell, with its centromere leading the way and the arms trailing behind. Migration is achieved by means of motor proteins in the kinetochore crawling along the spindle fiber as the fiber itself is "chewed up" and disassembled at the chromosomal end. Since sister chromatids are genetically identical, and since each daughter cell receives one chromatid from each chromosome, the daughter cells of mitosis are genetically identical.

(4) **Telophase.**[10] The daughter chromosomes cluster on each side of the cell. The rough ER produces a new nuclear envelope around each cluster, and the chromosomes begin to uncoil and return to the thinly dispersed chromatin form. The mitotic spindle breaks up and vanishes. Each new nucleus forms nucleoli, indicating it has already begun making RNA and is preparing for protein synthesis.

Telophase is the end of nuclear division but overlaps with **cytokinesis**[11] (SY-toe-kih-NEE-sis), division of the cytoplasm into two cells. Early traces of cytokinesis are visible even at anaphase. It is achieved by the motor protein myosin pulling on microfilaments of actin in the terminal web of the cytoskeleton. This creates a crease called the *cleavage furrow* around the equator of the cell, and the cell eventually pinches in two. Interphase has now begun for these new cells. Be aware, however, that mitosis (nuclear division) can occur without cytokinesis (cellular division). This is why some cells acquire two or more nuclei or multiple identical sets of chromosomes.

4.3e Regulation of Cell Division

One of the most important questions in biology is what signals cells when to divide and when to stop. The activation and inhibition of cell division are subjects of intense research for obvious reasons such as management of cancer and tissue repair. Cells divide when (1) they grow large enough to have enough cytoplasm to distribute to their two daughter cells; (2) they have replicated their DNA, so they can give each daughter cell a duplicate set of genes; (3) they

receive an adequate supply of nutrients; (4) they are stimulated by **growth factors,** chemical signals secreted by blood platelets, kidney cells, and other sources; or (5) neighboring cells die, opening up space in a tissue to be occupied by new cells. Growth factors and their receptors are a central issue in understanding the uncontrolled growth of cancer (see Deeper Insight 4.3). Cells stop dividing when they snugly contact neighboring cells or when nutrients or growth factors are withdrawn. The cessation of cell division in response to contact with other cells is called **contact inhibition.** An absence of contact inhibition, leading to uncontrolled cell division, is one of the characteristics of cancer.

The cell cycle is regulated by a molecular timer and certain checkpoints at which the cell checks its own status before moving on to the next phase. Two key elements of the timer are families of proteins called **cyclins** and **cyclin-dependent kinases (Cdks).** Remember from section 2.4g that a kinase is an enzyme that phosphorylates (adds phosphate to) other proteins, thereby activating or suppressing their function. Cdks exist at fairly stable levels in cells, but without cyclins, they lie dormant and perform no function. Cyclin levels rise and fall through the cell cycle. In early interphase, cyclin genes are transcribed and cyclin levels rise. As they bind to Cdks, those enzymes phosphorylate proteins with various effects described shortly. At the end of mitosis, proteasomes degrade the cyclins, their levels fall, and the Cdks become dormant again until the next cycle.

Among other functions of cyclin–Cdk complexes, they control (1) the replication of DNA and centrioles in the S phase; (2) the condensation of chromosomes, breakdown of the nuclear envelope, formation of the mitotic spindle, and attachment of chromosomes to the spindle in prophase; and (3) splitting of the centromere and separation of the sister chromatids at anaphase.

At specific **checkpoints** during the cell cycle, a cyclin binds to a Cdk and activates a cascade of biochemical reactions that prepare a cell to move on to the next phase of the cycle. One of these, called the *Start* or G_1 *checkpoint,* either allows the cell to proceed toward the S phase or, if it doesn't, the cell goes into the noncycling G_0 phase. A G_2/M *checkpoint* late in the G_2 phase determines whether the cell is able to proceed to mitosis. A third checkpoint at the transition from metaphase to anaphase determines whether the cell can proceed to anaphase, leading to separation of its sister chromatids.

Malfunctions of this process can be disastrous to the cell and even to the life of the whole person. If it fails to allow a cell to progress through the cell cycle, the result can be a failure of tissue maintenance and repair. On the other hand, if the cell cycle progresses too readily, out of control, the result can be uncontrolled tissue growth (neoplasia) and cancer.

BEFORE YOU GO ON

Answer the following questions to test your understanding of the preceding section:

11. Describe the genetic roles of DNA helicase and DNA polymerase. Contrast the function of DNA polymerase with that of RNA polymerase.

12. Explain why DNA replication is called *semiconservative.*

---

[7]*meta* = next in a series
[8]*aster* = star
[9]*ana* = apart
[10]*telo* = end, final
[11]*cyto* = cell; *kinesis* = action, motion

13. Define *mutation*. Explain why some mutations are harmless and others can be lethal.

14. List the stages of the cell cycle and summarize what occurs in each one.

15. List the stages of mitosis and the main processes that occur in each one.

16. Concisely explain how cyclins, Cdks, and checkpoints control the cell cycle.

4.4   Chromosomes and Heredity

Expected Learning Outcomes

When you have completed this section, you should be able to

a. describe the paired arrangement of chromosomes in the human karyotype;

b. define *allele* and discuss how alleles affect the traits of an individual; and

c. discuss the interaction of heredity and environment in producing individual traits.

FIGURE 4.16 The Normal Human Karyotype. This is a micrograph of chromosomes stained to accentuate their banding patterns. The two chromosomes of each homologous pair exhibit similar size, shape, and banding. Each chromosome at this stage has two chromatids.
Omikron/Science Source

Heredity is the transmission of genetic characteristics from parent to offspring. In the following discussion, we will examine a few basic principles of normal heredity, thus establishing a basis for understanding hereditary traits in later chapters. Hereditary defects are described in section 29.3c along with nonhereditary birth defects.

4.4a The Karyotype

As we have seen, the agents of heredity are the genes, and the genes are carried on the chromosomes. When we lay the 46 chromosomes out in order by size and other physical features, we get a chart called a **karyotype**[12] **(fig. 4.16).** The chromosomes occur in 23 pairs; the two members of each pair are called **homologous**[13] **chromosomes** (ho-MOLL-uh-gus). One member of each pair is inherited from the individual's mother and one from the father. Except for the X and Y chromosomes, two homologous chromosomes look alike and carry the same genes, although they may have different varieties of those genes. Chromosomes X and Y are called the **sex chromosomes** because they determine the individual's sex; all the others are called **autosomes.** A female normally has a homologous pair of X chromosomes, whereas a male has one X and a much smaller Y.

▶▶▶ APPLY WHAT YOU KNOW

Why would a cell in metaphase be more useful than a cell in interphase for producing a karyotype?

Any cell with 23 pairs of chromosomes is said to be **diploid (2n).**[14] Sperm and egg cells, however, are **haploid (n),**[15] meaning they contain only 23 unpaired chromosomes. Sperm and eggs, and cells on their way to becoming sperm and eggs, are called **germ cells.** All other cells of the body are called **somatic cells.** In meiosis (see section 27.4a), the two homologous chromosomes of each pair become *segregated* into separate daughter cells leading to the haploid germ cells. At fertilization, one set of *paternal* (sperm) chromosomes unites with a set of *maternal* (egg) chromosomes, restoring the diploid number to the fertilized egg and the somatic cells that arise from it.

4.4b Genes and Alleles

The location of a particular gene on a chromosome is called its **locus.** Homologous chromosomes have the same gene at the same locus, although they may carry different forms of that gene, called **alleles**[16] (ah-LEELS), which produce alternative forms of a particular trait. Frequently, one allele is **dominant** and the other one **recessive.** If at least one chromosome carries the dominant allele, the corresponding trait is usually detectable in the individual. A dominant allele masks the effect of any recessive allele that may be present. Typically, but not always, dominant alleles code for a normal, functional protein and recessive alleles for a nonfunctional variant of the protein.

---

[12]*karyo* = nucleus
[13]*homo* = same; *log* = relation
[14]*diplo* = double
[15]*haplo* = half

[16]*allo* = different

A feature of the chin presents an example of dominant and recessive genetic effects. Some people have a *cleft chin,* with a deep dimple in the middle, whereas most people do not **(fig. 4.17a).** The allele for cleft chin is dominant; we will represent it with a capital *C.* Those who inherit this from one or both parents normally have a cleft chin. The allele for an uncleft chin is recessive, here represented with a lowercase *c.* (It is customary to represent a dominant allele with a capital letter and a recessive allele with its lowercase equivalent.) One must inherit the recessive allele from both parents to have an uncleft chin.

Individuals with two identical alleles, such as *CC* or *cc,* are said to be **homozygous**[17] (HO-mo-ZY-gus) for that trait. If the homologous chromosomes have different alleles for that gene *(Cc),* the individual is **heterozygous**[18] (HET-er-oh-ZY-gus). The paired alleles that an individual possesses for a particular trait constitute the **genotype** (JEE-no-type). An observable trait such as cleft or uncleft chin is called the **phenotype**[19] (FEE-no-type).

We say that an allele is **expressed** if it shows in the phenotype of an individual. Chin allele *c* is expressed only when it is present in a homozygous state *(cc);* allele *C* is expressed whether it is homozygous *(CC)* or heterozygous *(Cc).* The only way most recessive alleles can be expressed is for an individual to inherit them from both parents.

Recessive traits can "skip" one or more generations. A diagram called a *Punnett square* **(fig. 4.17b)** shows how two heterozygous parents with cleft chins can produce a child with an uncleft chin. Across the top are the two genetically possible types of eggs the mother could produce, and on the left side are the possible types of sperm from the father. The four cells of the square show the genotypes and phenotypes that would result from each possible combination of sperm and egg. You can see that three of the possible combinations would produce a child with a cleft chin (genotypes *CC* and *Cc*), but one combination *(cc)* would produce a child with an uncleft chin. Therefore, the uncleft chin trait skipped the parental generation in this case but could be expressed in their child.

This phenomenon becomes more significant when parents are heterozygous **carriers** of hereditary diseases such as cystic fibrosis—individuals who carry a recessive allele and may pass it on, but do not phenotypically express it in themselves. For some hereditary diseases, tests are available to detect carriers and allow couples to weigh their risk of having children with genetic disorders. *Genetic counselors* perform genetic testing or refer clients for tests, advise couples on the probability of transmitting genetic diseases, and assist people in coping with genetic disease.

▶▶▶ **APPLY WHAT YOU KNOW**

Would it be possible for a woman with an uncleft chin to have children with cleft chins? Use a Punnett square and one or more hypothetical genotypes for the father to demonstrate your point.

[17]*homo* = same; *zygo* = union, joined
[18]*hetero* = different; *zygo* = union, joined
[19]*pheno* = showing, evident

Cleft chin
CC, Cc

Uncleft chin
cc

(a)

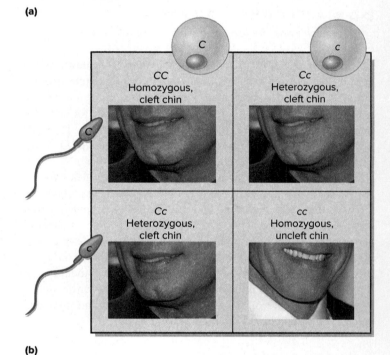

(b)

FIGURE 4.17 **Genetics of Cleft Chin.** (a) A cleft chin occurs if even one allele of the pair is dominant *(C).* The more common uncleft chin occurs only when both alleles are recessive *(c).* (b) A Punnett square shows why such a trait can "skip a generation." Both parents in this case have heterozygous genotypes *(Cc)* and cleft chins. An egg from the mother can carry either allele *C* or *c* (top), as can a sperm from the father (left). This gives their offspring a one-in-four chance of having an uncleft chin *(cc).*

a (left): Brad Barket/Getty Images; a (right): Kurt Krieger/Corbis Entertainment/Getty Images; b (top left, top right, bottom left): Brad Barket/Getty Images; b (bottom right): Kurt Krieger/Corbis Entertainment/Getty Images

4.4c Multiple Alleles, Codominance, and Incomplete Dominance

Some genes exist in more than two allelic forms—that is, there are **multiple alleles** within the collective genetic makeup, or **gene pool,** of the population as a whole. For example, there are over 100 alleles responsible for cystic fibrosis, and there are 3 alleles for

ABO blood types. Two of the ABO blood type alleles are dominant and symbolized with a capital *I* (for *immunoglobulin*) and a superscript: I^A and I^B. There is one recessive allele, symbolized with a lowercase *i*. Which two alleles one inherits determine the blood type, as follows:

Genotype	Phenotype
$I^A I^A$	Type A
$I^A i$	Type A
$I^B I^B$	Type B
$I^B i$	Type B
$I^A I^B$	Type AB
ii	Type O

▶▶▶ APPLY WHAT YOU KNOW

Why can't one person have all three of the ABO alleles?

Some alleles are equally dominant, or **codominant.** When both of them are present, both are phenotypically expressed. For example, a person who inherits allele I^A from one parent and I^B from the other has blood type AB. These alleles code for enzymes that produce the surface glycolipids of red blood cells. Type AB means that both A and B glycolipids are present, and type O means that neither of them is present.

Other alleles exhibit **incomplete dominance.** When two different alleles are present, the phenotype is intermediate between the traits that each allele would produce alone. Familial hypercholesterolemia, for example, is a disease in which people who inherit an abnormal allele from both parents have blood cholesterol levels up to six times higher than in the general population, whereas those who inherit it from only one parent have levels about two or three times normal. If untreated, those homozygous for the defective allele often die of heart attacks in childhood and untreated heterozygous individuals often die as young adults.

4.4d Polygenic Inheritance and Pleiotropy

Polygenic (multiple-gene) inheritance (fig. 4.18) is a phenomenon in which genes at two or more loci, or even on different chromosomes, contribute to a single phenotypic trait. Human eye and skin colors are normal polygenic traits, for example. They result from the combined expression of all the genes for each trait. Several diseases are also thought to stem from polygenic inheritance, including some forms of alcoholism, mental illness, cancer, and heart disease.

Pleiotropy[20] (ply-OT-roe-pee) is a phenomenon in which one gene produces multiple phenotypic effects. For example, about 1 in 200,000 people has a genetic disorder called *alkaptonuria,* caused by a mutation on chromosome 3. The mutation blocks the normal breakdown of the amino acid tyrosine, resulting in the accumulation of an intermediate breakdown product, homogentisic acid, in the body fluids and connective tissues. When

FIGURE 4.18 Polygenic Inheritance of Eye Color. A polygenic trait such as eye color is determined by the contributions of genes at multiple loci.

(top): Monica Lau/Getty Images; **(middle):** Anthony Saint James/Getty Images; **(bottom):** JupiterImages/Comstock/Getty Images

homogentisic acid oxidizes, it binds to collagen and turns the tissues gray to bluish black, and for unknown reasons, it causes degeneration of cartilages and other connective tissues. Among the multiple phenotypic effects of this disorder **(fig. 4.19)** are darkening of the skin; darkening and degeneration of the cartilages in

FIGURE 4.19 Pleiotropy in Alkaptonuria. Multiple phenotypic effects can result from a single gene mutation.

---

[20]*pleio* = more; *trop* = changes

places such as the ears, knees, and intervertebral discs; darkening of the urine when it stands long enough for the homogentisic acid to oxidize; discoloration of the teeth and the whites of the eyes; arthritis of the shoulders, hips, and knees; and damage to the heart valves, prostate gland, and other internal organs. Another well-known example of pleiotropy is sickle-cell disease, detailed in section 18.2.

4.4e Sex Linkage

Sex-linked traits are carried on the X or Y chromosome and therefore tend to be inherited by one sex more than the other. Men are more likely than women to have red–green color blindness or hemophilia, for example, because the allele for each is recessive and located on the X chromosome *(X-linked).* Women have two X chromosomes. If a woman inherits the recessive color-blindness allele *(c)* on one of her X chromosomes, there is still a good chance that her other X chromosome will carry a dominant allele *(C)* for normal color vision. Men, on the other hand, have only one X chromosome and normally express any allele found there **(fig. 4.20).** Ironically, even though color blindness is far more common among men than women, a man can inherit it only from his mother. Why? Because only his mother contributes an X chromosome to him. If he inherits *c* on his mother's X chromosome, he will be color-blind. He has no "second chance" to inherit a normal allele on a second X chromosome. A woman, however, gets an X chromosome from both parents. Even if one parent transmits the recessive allele to her, the chances are high that she will inherit a normal allele from her other parent. She would have to inherit it from both parents in order for her to have a trait such as red–green color blindness.

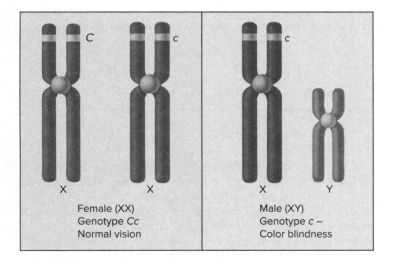

FIGURE 4.20 Sex-Linked Inheritance of Red–Green Color Blindness. Left: A female who inherits a recessive allele *(c)* for color blindness from one parent but a dominant allele *(C)* for normal vision from her other parent will have normal color vision. Right: A male who inherits *c* from his mother will exhibit red–green color blindness; the Y chromosome from his father has no corresponding gene to mask the effect of *c*.

The X chromosome carries about 260 genes, most of which have nothing to do with determining an individual's sex. There are so few functional genes on the Y chromosome—concerned mainly with development of the testes—that all proven sex-linked traits are associated with the X chromosome.

4.4f Dominant and Recessive Alleles at the Population Level

It is a common misconception that dominant alleles must be more common in the gene pool than recessive alleles. The truth is that dominance and recessiveness have little to do with how common an allele is. For example, type O is the most common ABO blood type in North America, but it is caused by the recessive allele *i*. Blood type AB, caused by the two dominant ABO alleles, is the rarest. *Polydactyly,*[21] the presence of extra fingers or toes, is a dominant trait, yet rare in the population. We also saw earlier that people with cleft chins are in a minority, even though the allele for that is dominant as well.

4.4g Environmental Effects and Gene Expression

It's one thing for any given cell to have all the usual genes in its nucleus, but it's a different matter whether those genes are turned on or off—whether the cell shows any detectable effects from them. For a gene to be turned on and have an effect on the individual is called **gene expression.** Even if an allele is dominant, it won't always be expressed in the phenotype. If a dominant allele is silenced and has no effect on some people who carry it, it is said to show **incomplete penetrance** in that population—that is, the genotype does not "penetrate to" and affect everyone who has it.

One reason the connection between genotype and phenotype is not inevitable is that environmental factors play an important role in the expression of all genes. At the very least, all gene expression depends on nutrition. Brown eyes, for example, require not only genes for the enzymes that synthesize the pigment melanin, but also the dietary raw material, phenylalanine, from which the melanin is made **(fig. 4.21).**

No gene can produce a phenotypic effect without nutritional and other environmental input, and no nutrients can produce a body or specific phenotype without genetic instructions that tell cells what to do with them. Just as you need both a recipe and ingredients to make a cake, it takes both heredity and environment to make a phenotype.

4.4h Epigenetics

It is often superficially believed that the only way gene expression can change is for the DNA base sequence to change—that is, through mutation. It is clear, however, that other mechanisms called **epigenetic**[22] **effects** can alter gene expression—reversibly

[21]*poly* = many; *dactyl* = fingers, toes
[22]*epi* = after, beyond, above the level of; *genet* = the gene or genome

FIGURE 4.21 The Roles of Environment and Heredity in Producing a Phenotype. Brown eye color requires phenylalanine from the diet (environment) and two genetically coded (hereditary) enzymes to convert phenylalanine to melanin, the eye pigment.

activating or silencing genes without changes in base sequence. Among these are variations in the higher-order packaging of chromatin, modification of the histones, gene silencing by noncoding RNA (ncRNA), and *DNA methylation*—adding methyl ($—CH_3$) groups to DNA bases, especially cytosine, with usually a gene-silencing effect.

Many epigenetic changes occur normally and affect such things as how stem cells end up as liver, nerve, or skin cells.

Different kinds of body cells are genetically identical, yet very different in structure and function because of epigenetic effects. Some epigenetic changes, however, may trigger such diseases and disorders as cancer, obesity, diabetes mellitus, or heart disease. Epigenetics is thought to be a possible culprit, for example, in the hereditary disorder *Prader–Willi syndrome,*[23] in which gene silencing leads to diverse effects including poor muscle tone, short stature, overeating, and childhood behavioral problems.

Particularly striking is the fact of **epigenetic inheritance;** we can pass on changes in gene expression to the next generation without changes in the structure of the genes themselves. Our habits and experiences today may affect gene expression in our children and perhaps even grandchildren. Anything from diet, smoking, or the use of recreational drugs to pollution, stress, epidemics, or famines can have epigenetic effects lasting for another generation or two. In light of epigenetics, we must now view heredity as entailing not only the genes passed from one generation to the next, but also whether other factors have abnormally activated or silenced those genes.

BEFORE YOU GO ON

Answer the following questions to test your understanding of the preceding section:

17. Why must the carrier of a genetic disease be heterozygous?

18. State at least three reasons why a person's phenotype can't always be determined from the genotype.

19. A man can inherit color blindness only from his mother, whereas a woman must inherit it from both her father and mother to show the trait. Explain this apparent paradox.

---

[23]Andrea Prader (1919–2001) and Heinrich Willi (1900–1971), Swiss pediatricians

 DEEPER INSIGHT 4.3

CLINICAL APPLICATION

Cancer

Anyone awaiting the results of a tumor biopsy hopes for the good news: benign! This means the tumor is slow-growing and contained in a fibrous capsule so it will not metastasize, and in most cases it is relatively easy to treat. The dreaded news is that it's malignant, meaning that it tends to grow rapidly and to *metastasize*—to give off cells that seed the growth of multiple tumors elsewhere, such as colon cancer metastasizing to the lungs and brain **(fig. 4.22).**

Oncology is a medical specialty that deals with both benign and malignant tumors, but only malignancies are called *cancer.* The word *cancer* literally means "crab." Hippocrates was the first to use the word this way, upon seeing a breast tumor with a tangle of blood vessels that reminded him of a crab's outstretched legs. Energy-hungry tumors often stimulate such ingrowth of blood vessels—a phenomenon called *tumor angiogenesis.*

Cancers are named for the tissue of origin: *carcinomas* originate in epithelial tissue; *lymphomas* in the lymph nodes; *melanomas* in pigment cells (melanocytes) of the epidermis; *leukemias* in blood-forming tissues such as bone marrow; and *sarcomas* in bone, other connective tissue, or muscle. About 90% of malignancies are carcinomas.

Only 5% to 10% of cancers are hereditary, but cancer is always a genetic disease. This is not as contradictory as it may seem. Most cases are due to mutations arising anew in the affected individual, not to genes inherited from a parent. Mutations can arise through errors in DNA replication or from exposure to carcinogens—radiation such as ultraviolet rays and X-rays; chemicals such as cigarette tar; and viruses such as human papillomavirus (HPV), hepatitis C, and type 2 herpes simplex.

FIGURE 4.22 Metastasis. The process by which malignant cells escape from a primary tumor, travel in the blood or lymph, and seed the growth of new (metastatic) tumors in other localities. Colon cancer, for example, can metastasize to the liver or brain by this method.

Oncologists are especially interested in two families of cancer genes called oncogenes and tumor suppressor genes. An *oncogene* is analogous to a stuck accelerator on a car—it causes cell division to accelerate out of control, sometimes by inducing the excessive secretion of growth factors that stimulate mitosis, or the production of excessive growth-factor receptors. An oncogene called *ras* underlies about one-quarter of human cancers, and *erbB2* is a common factor in breast and ovarian cancer. *Tumor suppressor (TS) genes* inhibit cancer by opposing oncogene action, coding for DNA-repair enzymes, and other means. Consequently, mutations that destroy their protective "braking" function can lead to cancer. Mutation of a TS gene called *p53*, for example, is involved in about 50% of cases of leukemia and colon, lung, breast, liver, brain, and esophageal tumors. Many human cancers are associated, however, not with mutation but with aberrant DNA methylation, which can, for example, silence one's TS genes and thus turn off their protective function.

FIGURE 4.23 Wilms Tumor. A malignant tumor of the kidney occurring especially in children.

Source: From the University of Alabama at Birmingham, Department of Pathology PEIR Digital Library © http://peir.net

Cancer seldom results from just one mutation. It usually requires 5 to 10 mutations at different gene loci. It takes time for so many mutations to accumulate, which is why cancer is more common in the elderly than in the young. In addition, as we age, we accrue more lifetime exposure to carcinogens, our DNA- and tissue-repair mechanisms become less efficient, and our immune system grows weaker and less able to detect and destroy malignant cells.

About one in every five of us in America will die of cancer. Cancer is almost always fatal if not treated. Malignant tumors replace functional tissue in vital organs **(fig. 4.23);** they steal nutrients from the rest of the body, sometimes causing a severe wasting away called *cachexia* (ka-KEX-ee-ah); they weaken one's immunity, opening the door to *opportunistic infections* that a healthier person could ward off; and they often invade blood vessels, lung tissue, or brain tissue, with such consequences as hemorrhage, pulmonary collapse, seizures, or coma. Mortality usually results not from the original (primary) tumor, but from metastasis.

Cancer is usually treated with surgery, chemotherapy, or radiation therapy. Two lively areas of cancer research today are the development of drugs to starve tumors by blocking tumor angiogenesis, and *cancer immunotherapy,* programming one's own immune cells (T cells) to selectively recognize and attack tumor cells.

STUDY GUIDE

▶ Assess Your Learning Outcomes

To test your knowledge, discuss the following topics with a study partner or in writing, ideally from memory.

4.1 DNA and RNA—The Nucleic Acids

1. The general name of the monomers that compose DNA and RNA; the three universal structural components of each monomer; the purines and pyrimidines involved in DNA and RNA structure; and how a purine differs from a pyrimidine

2. Structure of the double helix of DNA; what holds the two strands of the helix together; why DNA normally has only A—T and C—G base pairs (not A—C or G—T); and the law of complementary base pairing

3. The approximate number of genes in the human genome; what percentage of the DNA consists of genes; and to what extent, and in what ways, the remaining DNA may serve a purpose

4. How DNA and protein are combined to form chromatin; how the chromatin is coiled and supercoiled to fit into the compact cell nucleus and minimize damage; and the role of histones and nucleosomes in the organization of chromatin

5. The structure of a metaphase chromosome and how many chromosomes are in a typical cell

6. How RNA differs from DNA in structure and function, and the three types of RNA involved in protein synthesis

4.2 Genes and Their Action

1. The definition of *gene;* how genes relate to the amino acid sequence of proteins; why it cannot be said that every gene codes for protein structure; and why not every protein is encoded by its own unique gene

2. Definitions of *genome, genomics,* and *genomic medicine*

3. The organization of nucleotides into DNA triplets; how these relate to but differ from the codons of mRNA; the number of codons; and stop codons and their function

4. How the genetic code relates mRNA codons to protein structure

5. The process and outcome of genetic transcription; the enzyme that carries it out; the difference between pre-mRNA and mRNA; and the significance of introns and exons

6. How alternative splicing partially explains why the diversity of human proteins vastly exceeds the number of human genes

7. The process and outcome of genetic translation; the roles of mRNA, tRNA, and ribosomes in translation

8. Posttranslational modification of proteins, where it occurs, and how new proteins destined for extracellular use are packaged and released from a cell

9. How gene expression can be turned on or off according to the functions of different kinds of cells or physiological needs that change over time, such as a temporary need for breast milk

10. How the synthesis of nonproteins such as carbohydrates and steroids is regulated by the genes, even though genes code only for RNA or protein

4.3 DNA Replication and the Cell Cycle

1. Why every generation of cells must synthesize new DNA even though the chromosome number remains constant from generation to generation

2. Semiconservative replication, the enzymes that carry it out, and why it results in two DNA molecules that each contain one old and one new nucleotide polymer

3. What a mutation is and how a cell detects and corrects most mutations that arise during DNA replication; the varied consequences of uncorrected mutations

4. The four stages of the cell cycle, what occurs in each stage, and which stages occur during interphase

5. The functions of mitosis; its four stages; what occurs in each stage; and why the amount of DNA in a cell is halved by mitosis, yet the number of chromosomes remains the same

6. Cytokinesis and how it overlaps but differs from telophase

7. How mitosis can be either stimulated or inhibited according to the need for tissue maintenance or growth

8. How cyclins and cyclin-dependent kinases regulate the cell cycle

4.4 Chromosomes and Heredity

1. The definition of *heredity*

2. Organization of the karyotype; the number of homologous pairs of human chromosomes; why chromosomes occur in homologous pairs; and the differences between haploid and diploid cells and between germ cells and somatic cells

3. The differences between autosomes and sex chromosomes, and which sex chromosomes occur in males and females

4. Dominant and recessive alleles, the difference between homozygous and heterozygous individuals, and why a person can be homozygous for some alleles and heterozygous for others

5. The relationship of genotype to phenotype, the difference between the two, and why phenotype is not determined solely by genotype

6. Why a recessive trait can skip a generation, with examples; what is meant by a *carrier*

7. The differences between the genotype, genome, and gene pool; why a genotype can never be composed of more than two alleles of the same gene, but the gene pool can contain three or more alleles of one gene

8. How codominance and incomplete dominance differ from simple dominance of an allele

9. What polygenic inheritance and pleiotropy imply about the relationship between a certain phenotype and its associated genes; examples of both

10. Why some traits occur more in males than in females, the name of this phenomenon, and examples

11. Why some individuals may not exhibit the phenotype that one would predict from their genotype; the name for this incomplete expression of certain genotypes

12. Why it cannot be said that any human trait is exclusively the result of the genes or of the environment; an example that demonstrates this

13. Why it cannot be said that dominant alleles are the most common ones in a population and recessive alleles are less common; examples that support this point

14. The meaning of *gene expression* and the variety of factors that affect it

15. How heredity and health are affected not only by genetic changes (mutations) but also epigenetic ones

STUDY GUIDE

▶ Testing Your Recall

Answers in Appendix A

1. Production of more than one phenotypic trait by a single gene is called
   a. pleiotropy.
   b. genetic determinism.
   c. codominance.
   d. penetrance.
   e. genetic recombination.

2. When a ribosome reads a codon on mRNA, it must bind to the _____ of a corresponding tRNA.
   a. start codon
   b. stop codon
   c. intron
   d. exon
   e. anticodon

3. The normal functions of a liver cell—synthesizing proteins, detoxifying wastes, storing glycogen, and so forth—are done during its
   a. anaphase.
   b. telophase.
   c. G_1 phase.
   d. G_2 phase.
   e. synthesis phase.

4. Two genetically identical strands of a metaphase chromosome, joined at the centromere, are its
   a. kinetochores.
   b. centrioles.
   c. sister chromatids.
   d. homologous chromatids.
   e. nucleosomes.

5. Which of the following is *not* found in DNA?
   a. thymine
   b. phosphate
   c. cytosine
   d. deoxyribose
   e. uracil

6. Genetic transcription is performed by
   a. ribosomes.
   b. RNA polymerase.
   c. DNA polymerase.
   d. helicase.
   e. chaperones.

7. A chaperone comes into play in
   a. the folding of a new protein into its tertiary structure.
   b. keeping DNA organized within the nucleus.
   c. escorting sister chromatids to opposite daughter cells during mitosis.
   d. repairing DNA that has been damaged by mutagens.
   e. preventing malignant cells from metastasizing.

8. An allele that is not phenotypically expressed in the presence of an alternative allele of the same gene is said to be
   a. codominant.
   b. lacking penetrance.
   c. heterozygous.
   d. recessive.
   e. subordinate.

9. Semiconservative replication occurs during
   a. transcription.
   b. translation.
   c. posttranslational modification.
   d. the S phase of the cell cycle.
   e. mitosis.

10. Mutagens sometimes cause no harm to cells for all of the following reasons *except*
    a. some mutagens are natural, harmless products of the cell itself.
    b. most of the human DNA does not code for any proteins.
    c. the body's DNA repair mechanisms detect and correct genetic damage.

    d. change in a codon does not always change the amino acid encoded by it.
    e. some mutations change protein structure in ways that are not critical to normal function.

11. The cytoplasmic division at the end of mitosis is called _____.

12. At the checkpoints of the cell cycle, proteins called cyclins bind to _____ and activate reactions that enable the cell to proceed to the next stage of the cycle.

13. The pattern of nitrogenous bases that represents the 20 amino acids of a protein is called the _____.

14. Several ribosomes attached to one mRNA, which they are all transcribing, form a cluster called a/an _____.

15. The enzyme that produces pre-mRNA from the instructions in DNA is _____.

16. All the DNA in a haploid set of chromosomes is called a person's _____.

17. At prophase, a cell has _____ chromosomes, _____ chromatids, and _____ molecules of DNA.

18. The cytoplasmic granule of RNA and protein that reads the message in mRNA is a/an _____.

19. Cells are stimulated to divide by chemical signals called _____.

20. All chromosomes except the sex chromosomes are called _____.

▶ Building Your Medical Vocabulary

Answers in Appendix A

State a meaning of each word element, and give a medical term from this chapter that uses it or a slight variation of it.

1. allo-
2. dactylo-
3. diplo-
4. haplo-
5. hetero-
6. karyo-
7. meta-
8. morpho-
9. muta-
10. poly-

▶ What's Wrong with These Statements?

Answers in Appendix A

Briefly explain why each of the following statements is false, or reword it to make it true.

1. Proteins destined to be exported from a cell are made by ribosomes on the surface of the Golgi complex.

2. Steroids, carbohydrates, and phospholipids are encoded by different types of genes than are proteins.

3. Since uracil and thymine are very similar, a molecule of RNA would have about the same molecular weight as a stretch of DNA of the same length.

4. Each amino acid of a protein is represented by its own unique base pair in DNA.

5. Each gene codes for just one protein.

6. The law of complementary base pairing describes the way the bases in an mRNA codon pair up with the bases of a tRNA anticodon during translation.

7. All human DNA codes for the synthesis of proteins.

8. All mutations result in the production of defective proteins and are therefore harmful.

9. Males have only one sex chromosome whereas females have two.

10. A gene can be transcribed by only one RNA polymerase at a time.

▶ Testing Your Comprehension

1. Why would the supercoiled, condensed form of chromosomes seen in metaphase not be suitable for the G_1 phase of the cell cycle? Why would the finely dispersed chromatin of the G_1 phase not be suitable for mitosis?

2. Suppose a woman was heterozygous for blood type A and her husband had blood type AB. With the aid of a Punnett square, explain what blood type(s) their children could possibly have.

3. Given the information in this chapter, present an argument that evolution is not merely possible but inevitable. (*Hint:* Review the definition of *evolution* in chapter 1.)

4. What would be the minimum length (approximate number of bases) of an mRNA that coded for a protein 300 amino acids long?

5. Biology textbooks once taught the concept of "one gene, one protein"—that a gene is a segment of DNA that codes for just one protein, and every protein is represented by a separate, unique gene. Discuss the evidence that shows that this can no longer be regarded as true.

CHAPTER

5

THE HUMAN TISSUES

The lining of the small intestine, showing simple columnar epithelium with mucus-secreting goblet cells (magenta) (LM)

Victor P. Eroschenko

CHAPTER OUTLINE

5.1 The Study of Tissues
- 5.1a The Primary Tissue Classes
- 5.1b Embryonic Tissues
- 5.1c Interpreting Tissue Sections

5.2 Epithelial Tissue
- 5.2a Simple Epithelia
- 5.2b Stratified Epithelia

5.3 Connective Tissue
- 5.3a Overview
- 5.3b Fibrous Connective Tissue
- 5.3c Adipose Tissue
- 5.3d Cartilage
- 5.3e Bone
- 5.3f Blood

5.4 Nervous and Muscular Tissues—Excitable Tissues
- 5.4a Nervous Tissue
- 5.4b Muscular Tissue

5.5 Cellular Junctions, Glands, and Membranes
- 5.5a Cellular Junctions
- 5.5b Glands
- 5.5c Membranes

5.6 Tissue Growth, Development, Repair, and Degeneration
- 5.6a Tissue Growth
- 5.6b Tissue Development
- 5.6c Stem Cells
- 5.6d Tissue Repair
- 5.6e Tissue Degeneration and Death

Study Guide

DEEPER INSIGHTS

5.1 Clinical Application: Biopsy

5.2 Clinical Application: Marfan Syndrome—A Connective Tissue Disease

5.3 Clinical Application: Stem-Cell Therapy

5.4 Clinical Application: Regenerative Medicine

Anatomy & Physiology Revealed® 4.0

Module 3: Tissues

With its 50 trillion cells and thousands of organs, the human body may seem to be a structure of forbidding complexity. Fortunately for our health, longevity, and self-understanding, biologists of the past were not discouraged by this complexity, but discovered patterns that made it more understandable. One such pattern is the fact that these trillions of cells belong to only 200 types or so, and they are organized into tissues that fall into just four primary categories—*epithelial, connective, nervous,* and *muscular tissue*—although there are at least 23 subtypes of these four.

Organs derive their function not from their cells alone but from how the cells are organized into tissues. Cells are specialized for certain tasks: muscle contraction, defense, enzyme secretion, and so forth. No one cell type can carry out all of the body's vital functions. Cells therefore work together at certain tasks and form tissues that carry out a particular function, such as nerve signaling or nutrient digestion. An organ is a structure with discrete boundaries that is composed of two or more tissue types.

The study of tissues and how they are arranged into organs is called **histology,**[1] or **microscopic anatomy**—the subject of

[1]*histo* = tissue; *logy* = study of

this chapter. Here we study the four tissue classes; the variations within each class; how to recognize tissue types microscopically and relate their microscopic anatomy to their function; how tissues are arranged to form an organ; how tissues change as they grow, shrink, or change from one tissue type to another over the life of the individual; and modes of tissue degeneration and death. Histology bridges the gap between the cytology of the preceding chapters and the organ system approach of the chapters that follow.

5.1 The Study of Tissues

Expected Learning Outcomes

When you have completed this section, you should be able to

a. name the four primary classes into which all adult tissues are classified;

b. name the three embryonic germ layers and some adult tissues derived from each; and

c. visualize the three-dimensional shape of a structure from a two-dimensional tissue section.

5.1a The Primary Tissue Classes

A **tissue** is a group of similar cells and cell products that arise from the same region of the embryo and work together to perform a specific structural or physiological role in an organ. The four *primary tissues*—epithelial, connective, nervous, and muscular— are summarized in **table 5.1.** They differ in the types and functions of their cells, the characteristics of the **matrix (extracellular material)** that surrounds the cells, and the relative amount of space occupied by the cells and matrix. In muscle and epithelium, the cells are so close together that the matrix is scarcely visible, but in most connective tissues, the matrix occupies much more space than the cells do.

TABLE 5.1		The Four Primary Tissue Classes	
Type		**Definition**	**Representative Locations**
Epithelial		Tissue composed of layers of closely spaced cells that cover organ surfaces, form glands, and serve for protection, secretion, and absorption	Epidermis Inner lining of digestive tract Liver and other glands
Connective		Tissue with usually more matrix than cell volume, often specialized to support and protect organs and to bind other tissues and organs to each other	Tendons and ligaments Cartilage and bone Blood
Nervous		Tissue containing excitable cells specialized for rapid transmission of coded information to other cells	Brain Spinal cord Nerves
Muscular		Tissue composed of elongated, excitable muscle cells specialized for contraction	Skeletal muscles Heart (cardiac muscle) Walls of viscera (smooth muscle)

Photos (Epithelial): ©Ed Reschke/Getty Images; (Connective): Dennis Strete/McGraw-Hill Education; (Nervous, Muscular): Ed Reschke

The matrix is composed of fibrous proteins and, usually, a clear gel variously known as **ground substance, tissue fluid, extracellular fluid (ECF),** or **interstitial**[2] **fluid.** In cartilage and bone, it can be rubbery or stony in consistency. The ground substance contains water, gases, minerals, nutrients, wastes, hormones, and other chemicals. This is the medium from which all cells obtain their oxygen, nutrients, and other needs, and into which cells release metabolic wastes, hormones, and other products.

In summary, a tissue is composed of cells and matrix, and the matrix is composed of fibers and ground substance.

5.1b Embryonic Tissues

Human development begins with a single cell, the fertilized egg, which soon divides to produce scores of identical, smaller cells. The first tissues appear when these cells start to organize themselves into layers—first two, and soon three strata called the **primary germ layers,** which give rise to all of the body's mature tissues. The three primary germ layers are called *ectoderm, mesoderm,* and *endoderm.* The **ectoderm**[3] is an outer layer that gives rise to the epidermis and nervous system. The innermost

layer, the **endoderm,**[4] gives rise to the mucous membranes of the digestive and respiratory tracts and to the digestive glands, among other things. Between these two is the **mesoderm,**[5] a layer of more loosely organized cells. Mesoderm eventually turns to a gelatinous tissue called **mesenchyme,** composed of fine, wispy collagen (protein) fibers and branching *mesenchymal cells* embedded in a gelatinous ground substance. Mesenchyme gives rise to cardiac muscle, bone, and blood, among other tissues. (The development of the three primary tissues in the embryo is detailed at "Embryogenesis" in section 29.1f.) Most organs are composed of tissues derived from two or more primary germ layers. The rest of this chapter concerns the "mature" tissues that exist from infancy through adulthood.

5.1c Interpreting Tissue Sections

In your study of histology, you may be presented with various tissue preparations mounted on microscope slides. Most such preparations are thin slices called **histological sections.** The best anatomical insight depends on an ability to deduce the three-dimensional structure of an organ from these two-dimensional sections **(fig. 5.1).** This ability, in turn, depends on an awareness of how tissues are prepared for study.

---

[2]*inter* = between; *stit* = to stand
[3]*ecto* = outer; *derm* = skin

[4]*endo* = inner; *derm* = skin
[5]*meso* = middle; *derm* = skin

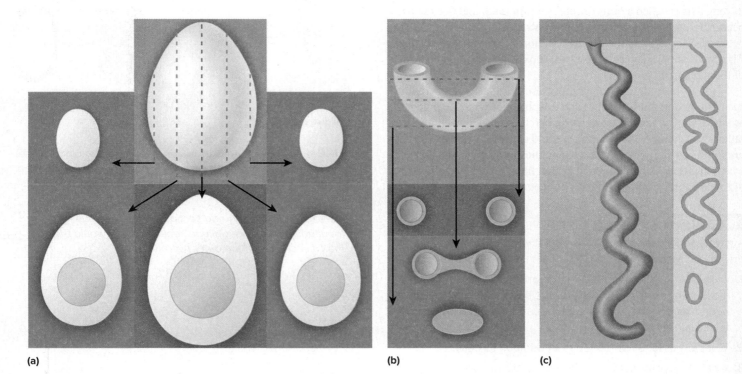

(a) (b) (c)

FIGURE 5.1 Three-Dimensional Interpretation of Two-Dimensional Images. (a) A boiled egg. Grazing sections (top left and right) would miss the yolk, just as a tissue section may miss a nucleus or other structure. (b) Elbow macaroni, which resembles many curved ducts and tubules. A section far from the bend would give the impression of two separate tubules; a section near the bend would show two interconnected lumina (cavities); and a section still farther down could miss the lumen completely. (c) A coiled gland in three dimensions and as it would look in a vertical tissue section of a tissue such as the lining of the uterus.

Histologists use a variety of techniques for preserving, sectioning (slicing), and staining tissues to show their structural details as clearly as possible. Tissue specimens are preserved in a **fixative**—a chemical such as formalin that prevents decay. After fixation, most tissues are cut into sections typically only one or two cells thick. Sectioning is necessary to allow the light of a microscope to pass through and so the image is not confused by too many layers of overlapping cells. The sections are then mounted on slides and artificially colored with histological **stains** to enhance detail. If they were not stained, most tissue sections would appear pale gray. With stains that bind to different components of a tissue, however, you may see pink cytoplasm; violet nuclei; and blue, green, or golden-brown protein fibers, depending on the stain used.

Sectioning a tissue reduces a three-dimensional structure to a series of two-dimensional slices. You must keep this in mind and try to translate the microscopic image into a mental image of the whole structure. Like the boiled egg and elbow macaroni in figure 5.1, an object may look quite different when it is cut at various levels, or *planes of section.* A coiled tube, such as a gland of the uterus (fig. 5.1c), is often broken up into multiple portions since it meanders in and out of the plane of section. An experienced viewer, however, recognizes that the separated pieces are parts of a single tube winding its way to the organ surface. Note that a grazing slice through a boiled egg might miss the yolk, just as a tissue section might miss the nucleus of a cell even though it was present.

Many anatomical structures are longer on one axis than another—the humerus and esophagus, for example. A tissue cut on its long axis is called a **longitudinal section (l.s.),** and one cut perpendicular to this is a **cross section (c.s.).** A section cut on a slant between a longitudinal and cross section is an **oblique section. Figure 5.2** shows how certain organs look when sectioned on each of these planes.

Not all histological preparations are sections. Liquid tissues such as blood and soft tissues such as spinal cord may be prepared as **smears,** in which the tissue is rubbed or spread across the slide rather than sliced. Some membranes and cobwebby tissues like the *areolar tissue* in figure 5.14 are sometimes mounted as **spreads,** in which the tissue is laid out on the slide, like placing a small square of tissue paper or a tuft of lint on a sheet of glass.

BEFORE YOU GO ON

Answer the following questions to test your understanding of the preceding section:

1. Classify each of the following into one of the four primary tissue classes: the skin surface, fat, the spinal cord, most heart tissue, bone, tendons, blood, and the inner lining of the stomach.

2. What are tissues composed of in addition to cells?

3. What embryonic germ layer gives rise to nervous tissue? To the liver? To muscle?

4. What is the term for a thin, stained slice of tissue mounted on a microscope slide?

(a) Longitudinal sections

(b) Cross sections

(c) Oblique sections

FIGURE 5.2 Three Planes of Section. A bone and blood vessel are used to relate two-dimensional sectioned appearance to three-dimensional structure. (a) Longitudinal sections. (b) Cross sections. (c) Oblique sections.

❓ *Would you classify the egg sections in the previous figure as longitudinal, cross, or oblique sections? How would the egg look if sectioned in the other two planes?*

5.2 Epithelial Tissue

Expected Learning Outcomes

When you have completed this section, you should be able to

a. describe the properties that distinguish epithelium from other tissue classes;

b. list and classify eight types of epithelium, distinguish them from each other, and state where each type can be found in the body;

c. explain how the structural differences between epithelia relate to their functional differences; and

d. visually recognize each epithelial type from specimens or photographs.

Epithelial[6] **tissue** consists of a sheet of closely adhering cells, one or more cells thick, with the upper surface usually exposed to the environment or to an internal space in the body. Epithelium covers the body surface, lines body cavities, forms the external and internal linings of many organs, and constitutes most gland tissue. The functions of epithelial tissue include

- **Protection.** Epithelia protect deeper tissues from invasion and injury. The epidermis of the skin, for example, is a barrier to infection, and the inner lining of the stomach protects its deeper tissues from stomach acid and enzymes.

- **Secretion.** Epithelia produce mucus, sweat, enzymes, hormones, and most of the body's other secretions; glands are composed largely of epithelial tissue.

- **Excretion.** Epithelia void wastes from the tissues, such as CO_2 across the pulmonary epithelium and bile from the epithelium of the liver.

- **Absorption.** Epithelia absorb chemicals from the adjacent medium; nutrients, for example, are absorbed through the epithelium of the small intestine.

- **Filtration.** All substances leaving the blood are selectively filtered through the epithelium that lines the blood vessels; all urinary waste is filtered through epithelia of the kidneys.

- **Sensation.** Epithelia are provided with nerve endings that sense stimulation ranging from a touch on the skin to irritation of the stomach.

The cells and extracellular material of an epithelium can be loosely compared to the bricks and mortar of a wall. The extracellular material ("mortar") is so thin, however, that it is barely visible with a light microscope, and the cells appear pressed very close together. Epithelia are *avascular*[7] (without blood vessels)—there is no room for them between the cells. Epithelia, however, usually lie on a vessel-rich layer of connective tissue, which furnishes them with nutrients and waste removal. Epithelial cells closest to the connective tissue typically exhibit a high rate of mitosis. This allows epithelia to repair themselves quickly—an ability of special importance in protective epithelia that are highly vulnerable to such injuries as skin abrasions and erosion by stomach acid.

Between an epithelium and the underlying connective tissue is a layer called the **basement membrane.** It contains collagen, glycoproteins, and other protein–carbohydrate complexes, and blends into other proteins of the connective tissue. The basement membrane serves to anchor an epithelium to the connective tissue; it controls the exchange of materials between the epithelium and the underlying tissues; and it binds growth factors from below that regulate epithelial development. The surface of an epithelial cell that faces the basement membrane is its **basal surface,** the one that faces away from it toward the body surface or the internal cavity (lumen) of an organ is the **apical surface,** and between these two, the "sidewall" of a cell is called the **lateral surface.**

Epithelia are classified into two broad categories—*simple* and *stratified*—with four types in each category. In a simple epithelium, every cell is anchored to the basement membrane, whereas in a stratified epithelium, some cells rest on top of other cells and do not contact the basement membrane **(fig. 5.3).**

Table 5.2 (containing **figs. 5.4** to **5.7**) summarizes the structural and functional differences between the four simple epithelia. In this and subsequent tables, each photograph is accompanied by a labeled drawing of the same specimen. The drawings clarify cell boundaries and other relevant features that may otherwise be difficult to see or identify in photographs or through the microscope. Each figure indicates the approximate magnification at which the original photograph was made. Each is enlarged much more than this when printed in the book, but selecting the closest magnification on a microscope should enable you to see a comparable level of detail (resolution).

5.2a Simple Epithelia

Generally, a **simple epithelium** has only one layer of cells, although this is a somewhat debatable point in the *pseudostratified columnar* type. Three types of simple epithelia are named for the shapes of their cells: **simple squamous**[8] (thin scaly cells), **simple cuboidal** (squarish or round cells), and **simple columnar** (tall narrow cells). In the fourth type, **pseudostratified columnar,** not all cells reach the surface; the shorter cells are covered by the taller ones. This epithelium looks stratified in most tissue sections, but careful examination, especially with the electron microscope, shows that every cell reaches the basement membrane—like trees in a forest, where some grow taller than others but all are anchored in the soil below.

Simple columnar and pseudostratified columnar epithelia often have wineglass-shaped **goblet cells** that produce protective

---

[6]*epi* = upon; *theli* = nipple, female
[7]*a* = without; *vas* = blood vessels

[8]*squam* = scale

FIGURE 5.3 Cell Shapes and Epithelial Types. (a) Frontal sections of three classes of epithelium. Pseudostratified columnar epithelium is a special type of simple epithelium that gives a false appearance of multiple cell layers. (b) Frontal and oblique views of three cell shapes—squamous, cuboidal, and columnar. Note that these traditional terms describe the frontal appearance, but all three are polygonal when viewed from above.

mucous coatings over the mucous membranes. These cells have an expanded apical end filled with secretory vesicles; their product becomes mucus when it is secreted and absorbs water. The basal part of the cell is a narrow stem, like that of a wineglass, that reaches to the basement membrane.

5.2b Stratified Epithelia

Stratified epithelia range from 2 to 20 or more layers of cells, with some cells resting directly on others and only the deepest layer attached to the basement membrane (**table 5.3,** containing **figs. 5.8** to **5.11**). Three of the stratified epithelia are named for the shapes of their surface cells: **stratified squamous, stratified cuboidal,** and **stratified columnar epithelia.** The deeper cells, however, may be of a different shape than the surface cells. The fourth type, **urothelium,** is named for the fact that it's unique to the urinary tract. It is sometimes called by an older name, *transitional epithelium,* that arose from a misunderstanding that it represented a transitional stage between stratified squamous and stratified columnar epithelium.

Stratified columnar epithelium is rare and of relatively minor importance—seen only in places where two other epithelial types meet, as in limited regions of the pharynx, larynx, anal canal, and male urethra. We will not consider this type any further.

The most widespread epithelium in the body is stratified squamous epithelium, which deserves further discussion. Its deepest layer of cells are cuboidal to columnar, and include mitotically active stem cells. Their daughter cells push toward the surface and become flatter (more scaly) as they migrate farther upward, until they finally die and flake off. Their loss is called **exfoliation (desquamation) (fig. 5.12);** the study of exfoliated

cells is called *exfoliate cytology.* You can easily study exfoliated cells by scraping your gums with a toothpick, smearing this material on a slide, and staining it with iodine for microscopic examination. A similar procedure is used in the *Pap smear,* an examination of exfoliated cells from the cervix for signs of uterine cancer (see fig. 28.5).

Stratified squamous epithelia are of two kinds—keratinized and nonkeratinized. A **keratinized (cornified)** epithelium, found in the epidermis, is covered with a layer of dead compressed cells. These cells, called **keratinocytes,** are packed with the durable protein *keratin* and coated with a water-repellent glycolipid. The skin surface is therefore relatively dry; it retards water loss from the body; and it resists penetration by disease organisms. (Keratin is also the protein of which animal horns are made, hence its name.[9]) The tongue, esophagus, vagina, and a few other internal membranes are covered with the **nonkeratinized** type, which lacks the surface layer of dead cells. This type provides a surface that is, again, abrasion-resistant, but also moist and slippery. These characteristics are well suited to resist stress produced by chewing and swallowing food and by sexual intercourse and childbirth.

Urothelium is another particularly interesting type of stratified epithelium. Why is it limited to the urinary tract? The answer relates to the fact that urine is usually acidic and hypertonic to the intracellular fluid. It would tend to draw water out of the cells by osmosis and kill them if there were nothing to protect them. The

(text continued after table 5.3)

---

[9]*kerat* = horn

TABLE 5.2	Simple Epithelia

Simple Squamous Epithelium	Simple Cuboidal Epithelium

(a)

(a)

Squamous epithelial cells Nuclei of smooth muscle

Basement membrane

(b)

Lumen of kidney tubule Cuboidal epithelial cells Basement membrane

(b)

FIGURE 5.4 Simple Squamous Epithelium. Serosa of the small intestine (×400). (a) Light micrograph. (b) Labeled drawing. **APR**

a: Dennis Strete/McGraw-Hill Education

FIGURE 5.5 Simple Cuboidal Epithelium. Kidney tubules (×400). (a) Light micrograph. (b) Labeled drawing. **APR**

a: Dennis Strete/McGraw-Hill Education

Microscopic appearance: Single layer of thin cells, shaped like fried eggs with bulge where nucleus is located; nucleus flattened in the plane of the cell, like an egg yolk; cytoplasm may be so thin it is hard to see in tissue sections; in surface view, cells have angular contours and nuclei appear round

Representative locations: Air sacs (alveoli) of lungs; glomerular capsules of kidneys; some kidney tubules; inner lining (endothelium) of heart and blood vessels; serous membranes of stomach, intestines, and some other viscera; surface mesothelium of pleura, pericardium, peritoneum, and mesenteries

Functions: Allows rapid diffusion or transport of substances through membrane; secretes lubricating serous fluid

Microscopic appearance: Single layer of square or round cells; in glands, cells often pyramidal and arranged like segments of an orange around a central space; spherical, centrally placed nuclei; with a brush border of microvilli in some kidney tubules; ciliated in bronchioles of lung

Representative locations: Liver, thyroid, mammary, salivary, and other glands; most kidney tubules; bronchioles

Functions: Absorption and secretion; production of protective mucous coat; movement of respiratory mucus

TABLE 5.2	Simple Epithelia *(continued)*

Simple Columnar Epithelium	Pseudostratified Columnar Epithelium
(a)	(a)
Brush border (microvilli) Connective tissue Basement membrane Nuclei Goblet cell Columnar cells (b)	Cilia Basement membrane Basal cells Goblet cell (b)

FIGURE 5.6 Simple Columnar Epithelium. Mucosa of the small intestine (×400). (a) Light micrograph. (b) Labeled drawing. **APR**

a: Ed Reschke/Getty Images

Microscopic appearance: Single layer of tall, narrow cells; oval or sausage-shaped nuclei, vertically oriented, usually in basal half of cell; apical portion of cell often shows secretory vesicles visible with TEM; often shows a brush border of microvilli; ciliated in some organs; may possess goblet cells
Representative locations: Inner lining of stomach, intestines, gallbladder, uterus, and uterine tubes; some kidney tubules
Functions: Absorption; secretion of mucus and other products; movement of egg and embryo in uterine tube

FIGURE 5.7 Ciliated Pseudostratified Columnar Epithelium. Mucosa of the trachea (×400). (a) Light micrograph. (b) Labeled drawing. **APR**

a: Dennis Strete/McGraw-Hill Education

Microscopic appearance: Looks multilayered; some cells do not reach free surface, but all cells reach basement membrane; nuclei at several levels in deeper half of epithelium; often with goblet cells; often ciliated
Representative locations: Respiratory tract from nasal cavity to bronchi; portions of male urethra
Functions: Secretes and propels mucus

TABLE 5.3	Stratified Epithelia

Stratified Squamous Epithelium—Keratinized	Stratified Squamous Epithelium—Nonkeratinized

(a)

(a)

Dead squamous cells   Living epithelial cells   Dense irregular connective tissue

Areolar tissue

(b)

Living epithelial cells   Connective tissue

(b)

FIGURE 5.8 Keratinized Stratified Squamous Epithelium.
Epidermis of the sole of the foot (×400). (a) Light micrograph.
(b) Labeled drawing. **APR**

a: ©Ed Reschke

FIGURE 5.9 Nonkeratinized Stratified Squamous Epithelium.
Mucosa of the vagina (×400). (a) Light micrograph. (b) Labeled
drawing. **APR**

a: ©Ed Reschke

Microscopic appearance: Multiple cell layers with cells becoming increasingly flat and scaly toward surface; surface covered with a layer of compact dead cells without nuclei; basal cells may be cuboidal to columnar	**Microscopic appearance:** Same as keratinized epithelium but without the surface layer of dead cells
Representative locations: Epidermis; palms and soles are especially heavily keratinized	**Representative locations:** Tongue, oral mucosa, esophagus, anal canal, vagina
Functions: Resists abrasion and penetration by pathogenic organisms; retards water loss through skin	**Functions:** Resists abrasion and penetration by pathogenic organisms

TABLE 5.3	Stratified Epithelia *(continued)*

Stratified Cuboidal Epithelium	Urothelium

(a)

(a)

Cuboidal cells   Epithelium   Connective tissue

(b)

Basement membrane   Connective tissue   Binucleate epithelial cell

(b)

FIGURE 5.10 Stratified Cuboidal Epithelium. Duct of a sweat gland (×400). (a) Light micrograph. (b) Labeled drawing. **APR**

a: Lester V. Bergman/Corbis NX/Getty Images

Microscopic appearance: Two or more layers of cells; surface cells square or round

Representative locations: Sweat gland ducts; egg-producing vesicles (follicles) of ovaries; sperm-producing ducts (seminiferous tubules) of testis

Functions: Contributes to sweat secretion; secretes ovarian hormones; produces sperm

FIGURE 5.11 Urothelium. Kidney (×400). (a) Light micrograph. (b) Labeled drawing. **APR**

a: Johnny R. Howze

Microscopic appearance: Somewhat resembles stratified squamous epithelium, but surface cells are rounded, not flattened, and often bulge at surface; typically five or six cells thick when relaxed and two or three cells thick when stretched; cells may be flatter and thinner when urothelium is stretched (as in a distended bladder); some cells have two nuclei

Representative locations: Urinary tract—part of kidney, ureter, bladder, part of urethra

Functions: Stretches to allow filling of urinary tract; protects underlying tissues from osmotic damage by urine

FIGURE 5.12 Exfoliation of Squamous Cells from the Mucosa of the Vagina.

David M. Phillips/Science Source

 Aside from the gums and vagina, name another epithelium in the body that would look like this to the scanning electron microscope.

domed surface cells of urothelium, however, have a unique protective property. They are called **umbrella cells.** On the upper surface of an umbrella cell, the outer phospholipid layer of the plasma membrane is thicker than usual and has dense patches called *lipid rafts* with embedded proteins called *uroplakins*. Uroplakins are impermeable to urine and protect the urothelium, including the cytoplasm of the umbrella cell itself. Lipid rafts are connected to each other by hinges of ordinary plasma membrane. When the bladder is empty and relaxed, these plaques fold at the hinges (like folding a laptop computer) and drop into the cell interior for storage, and the cell bulges upward as seen in figure 5.11. As the bladder fills with urine, the hinges open (like opening the computer), the plaques spread out over the surface to protect the cell, and the umbrella cells become thinner and flatter. Not surprisingly, this type of epithelium is best developed in the bladder, where it is subject to prolonged contact with stored urine.

BEFORE YOU GO ON

Answer the following questions to test your understanding of the preceding section:

5. Distinguish between simple and stratified epithelia, and explain why pseudostratified columnar epithelium belongs in the former category despite its superficial appearance.

6. Explain how to distinguish a stratified squamous epithelium from a urothelium.

7. What function do keratinized and nonkeratinized stratified squamous epithelia have in common? What is the structural difference between these two? How is this structural difference related to a functional difference between them?

8. How do the epithelia of the esophagus and stomach differ? How does this relate to their respective functions?

DEEPER INSIGHT 5.1

CLINICAL APPLICATION

Biopsy

Biopsy[10] means the removal and microscopic examination of a sample of living tissue. One purpose of a biopsy is diagnosis of diseases that can be identified only from the microscopic appearance of cells or tissues—for example, a malignancy such as colon cancer, an infection such as tuberculosis, or an inflammatory disease such as lupus. Diagnosis may include the *staging* of a cancer to express how advanced it is. (See Deeper Insight 28.1, Pap Smears and Cervical Cancer.) Another purpose is treatment planning. The appropriate drug for cancer chemotherapy, for example, may depend on the type of cancer revealed by the biopsy, and for cervical or breast cancer, it depends on the stage of the cancer. A third purpose is surgical guidance. A specimen from a patient undergoing lumpectomy for breast cancer, for example, may be examined for *clear margins* so the surgeon can know if all the malignant tissue has been removed or it is necessary to cut more widely.

Cutting out and examining an entire suspicious mass, such as a breast lump or lymph node, is called *excisional biopsy*. An *incisional* (core) biopsy is removal of just a portion of the mass, such as a suspected skin melanoma, for diagnostic examination. A *needle biopsy* may be done to obtain a sample of cells when it isn't necessary to maintain the exact structure of the tissue, such as a sample of bone marrow for the diagnosis of leukemia. Sucking out such a specimen with a needle is called *needle aspiration biopsy*. Some needle biopsies employ an instrument with jaws, inserted through a needle, to take a "bite" of the suspicious tissue or remove an entire mass such as a colon polyp. In some procedures called *CT guided needle biopsy*, exact placement of the needle is guided by watching it on a CT monitor.

In some cases, the biopsy specimen will be examined by an in-house pathologist while the patient is still under anesthesia and the surgeon awaits the lab report in order to know how to proceed. In outpatient cases, a specimen such as a lymph node may be sent to a medical diagnostics laboratory elsewhere, which will report the result back to the physician.

---

[10]*bio* = living; *opsy* = viewing

5.3 Connective Tissue

Expected Learning Outcomes

When you have completed this section, you should be able to

a. describe the properties that most connective tissues have in common;

b. discuss the types of cells found in connective tissue;

c. explain what the matrix of a connective tissue is and describe its components;

d. name and classify 10 types of connective tissue, describe their cellular components and matrix, and explain what distinguishes them from each other; and

e. visually recognize each connective tissue type from specimens or photographs.

5.3a Overview

Connective tissues are the most abundant, widely distributed, and histologically variable of the primary tissues. They include fibrous tissue, adipose tissue, cartilage, bone, and blood. Such diverse tissues may seem to have little in common, but as a rule, their cells occupy less space than the extracellular matrix. Usually their cells are not in direct contact with each other, but are separated by expanses of matrix. Connective tissues vary greatly in vascularity, from rich networks of blood vessels in the loose connective tissues to few or no blood vessels in cartilage.

The functions of connective tissue include

- **Binding of organs.** Tendons bind muscle to bone, ligaments bind one bone to another, fat holds the kidneys and eyes in place, and fibrous tissue binds the skin to underlying muscle.

- **Support.** Bones support the body; cartilage supports the ears, nose, larynx, and trachea; fibrous tissues form the framework of organs such as the spleen.

- **Physical protection.** The cranium, ribs, and sternum protect delicate organs such as the brain, lungs, and heart; fatty cushions protect the kidneys and eyes.

- **Immune protection.** Connective tissue cells attack foreign invaders, and connective tissue fiber forms a "battlefield" under the skin and mucous membranes where immune cells can be quickly mobilized against disease agents.

- **Movement.** Bones provide the lever system for body movement, cartilages are involved in movement of the vocal cords, and cartilages on bone surfaces ease joint movements.

- **Storage.** Fat is the body's major energy reserve; bone is a reservoir of calcium and phosphorus that can be drawn upon when needed.

- **Heat production.** Metabolism of brown fat generates heat in infants and children.

- **Transport.** Blood transports gases, nutrients, wastes, hormones, and blood cells.

The mesenchyme described earlier in this chapter is a form of embryonic connective tissue. The mature connective tissues fall into four broad categories: fibrous connective tissue, adipose tissue, supportive connective tissues (cartilage and bone), and fluid connective tissue (blood).

5.3b Fibrous Connective Tissue

Fibrous connective tissue is the most diverse type. Nearly all connective tissues contain fibers, but the tissues considered here are classified together because the fibers are so conspicuous. Fibers are, of course, just one component of the tissue, which also includes cells and ground substance. Before examining specific types of fibrous connective tissue, let's examine these components.

Components of Fibrous Connective Tissue

Cells The cells of fibrous connective tissue include the following types:

- **Fibroblasts.**[11] These are large, fusiform or stellate cells that often show slender, wispy branches. They produce the fibers and ground substance that form the matrix of the tissue.

- **Macrophages.**[12] These are large phagocytic cells that wander through the connective tissues, where they engulf and destroy bacteria, other foreign particles, and dead or dying cells of our own body. They also activate the immune system when they sense foreign matter called *antigens.* They arise from white blood cells called *monocytes* or from the same stem cells as monocytes.

- **Leukocytes,**[13] or **white blood cells (WBCs).** WBCs travel briefly in the bloodstream, then crawl out through the walls of small blood vessels and spend most of their time in the connective tissues. The two most common types are *neutrophils,* which wander about attacking bacteria, and *lymphocytes,* which react against bacteria, toxins, and other foreign agents. Lymphocytes often form dense patches in the mucous membranes.

- **Plasma cells.** Certain lymphocytes turn into plasma cells when they detect foreign agents. The plasma cells then synthesize disease-fighting proteins called *antibodies.* Plasma cells are rarely seen except in the wall of the intestines and in inflamed tissue.

- **Mast cells.** These cells, found especially alongside blood vessels, secrete a chemical called *heparin* that inhibits blood clotting, and one called *histamine* that increases blood flow by dilating blood vessels.

- **Adipocytes** (AD-ih-po-sites), or **fat cells.** These appear in small clusters in some fibrous connective tissues. When they dominate an area, the tissue is called *adipose tissue.*

---

[11]*fibro* = fiber; *blast* = producing
[12]*macro* = big; *phage* = eater
[13]*leuko* = white; *cyte* = cell

Extensor retinaculum

Tendons

FIGURE 5.13 Tendons and Ligament. The tendons of the hand and the ligamentous band (extensor retinaculum) of the wrist are composed of collagen, which has a white glistening appearance.

Rebecca Gray/McGraw-Hill Education

Fibers Three types of protein fibers are found in fibrous connective tissues:

- **Collagenous fibers** (col-LADJ-eh-nus). These fibers, made of collagen, are tough and flexible and resist stretching. Collagen is the body's most abundant protein, constituting about 25% of the total. It is the base of such animal products as gelatin, leather, and glue.[14] In fresh tissue, collagenous fibers have a glistening white appearance, as seen in tendons and some cuts of meat **(fig. 5.13)**; thus, they are often called *white fibers*. In tissue sections, collagen forms coarse, wavy bundles, often dyed pink, blue, or green by the most common histological stains. Tendons, ligaments, and the dermis of the skin are made mainly of collagen. Less visibly, collagen pervades the matrix of cartilage and bone.

- **Reticular**[15] **fibers.** These are thin collagen fibers coated with glycoprotein. They form a spongelike framework for

such organs as the spleen and lymph nodes and constitute part of the basement membranes underlying epithelia.

- **Elastic fibers.** These are thinner than collagenous fibers, and they branch and rejoin each other along their course. They are made of a protein called **elastin** coated with a glycoprotein *(fibrillin)*. The coiled structure of elastin allows it to stretch and recoil like a rubber band. Elastic fibers account for the ability of the skin, lungs, and arteries to spring back after they are stretched. (Elasticity is not the ability to stretch, but the tendency to recoil when tension is released.)

Ground Substance Amid the cells and fibers in some tissue sections, there appears to be a lot of empty space. In life, this space is occupied by the featureless **ground substance.** Ground substance usually has a gelatinous to rubbery consistency resulting from three classes of large molecules: glycosaminoglycans, proteoglycans, and adhesive glycoproteins. It absorbs compressive forces and, like the styrofoam packing in a shipping carton, protects the more delicate cells from mechanical injury.

A **glycosaminoglycan (GAG)** (GLY-co-seh-ME-no-GLY-can) is a long polysaccharide composed of unusual disaccharides called *amino sugars* and *uronic acid*. GAGs are negatively charged and thus tend to attract sodium and potassium ions, which in turn cause them to absorb and retain water. Thus, GAGs play an important role in regulating the water and electrolyte balance of tissues. The most common GAG is **chondroitin sulfate** (con-DRO-ih-tin). It is abundant in blood vessels and bones and gives cartilage its relative stiffness. Other GAGs that you will read of in this book are *heparin* (an anticoagulant) and *hyaluronic acid* (HY-uh-loo-RON-ic). The latter is a gigantic molecule up to 20 μm long, as large as most cells. It is a viscous, slippery substance that forms a lubricant in the joints and constitutes much of the jellylike *vitreous body* of the eyeball.

A **proteoglycan** is another gigantic molecule. It is shaped somewhat like a bottle brush, with a central core of protein and bristlelike outgrowths composed of GAGs. The entire proteoglycan may be attached to hyaluronic acid, thus forming an enormous molecular complex. Proteoglycans form thick colloids similar to those of gravy, gelatin, and glue. This gel slows the spread of pathogenic organisms through the tissues. Some proteoglycans are embedded in the plasma membranes of cells, attached to the cytoskeleton on the inside and to other extracellular molecules in the matrix. They create a strong structural bond between cells and extracellular macromolecules and help to hold tissues together.

Adhesive glycoproteins are protein–carbohydrate complexes that bind plasma membrane proteins to extracellular collagen and proteoglycans. They bind the components of a tissue together and mark paths that guide migrating embryonic cells to their destinations in a tissue.

Types of Fibrous Connective Tissue

Fibrous connective tissue is divided into two broad categories according to the relative abundance of fiber: *loose* and *dense connective tissue*. In loose connective tissue, much of the space is

---

[14]*colla* = glue; *gen* = producing
[15]*ret* = network; *icul* = little

occupied by ground substance, which dissolves out of the tissue during histological fixation and leaves empty space in prepared tissue sections. The loose connective tissues we will discuss are *areolar* and *reticular tissue* (**table 5.4**). In **dense connective tissue,** fiber occupies more space than the cells and ground substance, and appears closely packed in tissue sections. We will discuss two types: *dense regular* and *dense irregular connective tissue* (**table 5.5**).

Areolar[16] tissue (AIR-ee-OH-lur) exhibits loosely organized fibers, abundant blood vessels, and a lot of seemingly empty space. It possesses all six of the aforementioned cell types. Its fibers run in random directions and are mostly collagenous, but elastic and reticular fibers are also present. Areolar tissue is highly variable in appearance. In many serous membranes, it looks like **figure 5.14,** but in the skin and mucous membranes, it is more compact (see fig. 5.8) and sometimes difficult to distinguish from dense irregular connective tissue. Some advice on how to tell them apart is given after the discussion of dense irregular connective tissue.

Areolar tissue is found in tissue sections from almost every part of the body. It surrounds blood vessels and nerves and penetrates with them even into the small spaces of muscle, tendon, and other tissues. Nearly every epithelium rests on a layer of areolar tissue, whose blood vessels provide the epithelium with nutrition, waste removal, and a ready supply of infection-fighting leukocytes in times of need. Because of the abundance of open, fluid-filled space, leukocytes can move about freely in areolar tissue and can easily find and destroy pathogens.

Reticular tissue (**fig. 5.15**) is a mesh of reticular fibers and fibroblasts. It forms the framework (stroma) of such organs as the lymph nodes, spleen, thymus, and bone marrow. The space amid the fibers is filled with blood cells. If you imagine a sponge soaked with blood, the sponge fibers are analogous to the reticular tissue stroma.

Dense regular connective tissue (**fig. 5.16**) is named for two properties: (1) The collagen fibers are closely packed (dense) and leave relatively little open space, and (2) the fibers are parallel to each other (regular). It is found especially in tendons and ligaments. The parallel arrangement of fibers is an adaptation to the fact that musculoskeletal stresses pull tendons and ligaments in predictable directions. With minor exceptions such as blood vessels and sensory nerve fibers, the only cells in this tissue are fibroblasts, visible by their slender, violet-staining nuclei squeezed between bundles of collagen. This type of tissue has few blood vessels, so injured tendons and ligaments are slow to heal.

The vocal cords and some spinal ligaments are made of a dense regular connective tissue called **elastic tissue.** In addition to the densely packed collagen fibers, it exhibits branching elastic fibers and more fibroblasts. The fibroblasts have larger, more conspicuous nuclei than seen in most dense regular connective tissue. Elastic tissue also forms wavy sheets in large and medium arteries. When the heart pumps blood into the arteries, these sheets enable them to expand and relieve some of the pressure on smaller vessels downstream. When the heart relaxes, the arterial wall springs back and keeps the blood pressure from dropping too low between heartbeats. The importance of this elastic tissue becomes especially clear in diseases such as Marfan syndrome (see Deeper Insight 5.2) and arteriosclerosis, where the tissue is stiffened by lipid and calcium deposits (see Deeper Insight 19.4).

Dense irregular connective tissue (**fig. 5.17**) also has thick bundles of collagen and relatively little room for cells and ground substance, but the collagen bundles run in seemingly random directions. This arrangement enables the tissue to resist unpredictable stresses. This tissue constitutes most of the dermis, where it binds the skin to the underlying muscle and connective tissue. It forms a protective capsule around organs such as the kidneys, testes, and spleen and a tough fibrous sheath around the bones, nerves, and most cartilages.

It is sometimes difficult to judge whether a tissue is areolar or dense irregular. In the dermis, for example, these tissues occur side by side, and the transition from one to the other is not at all obvious (see fig. 5.8). A relatively large amount of clear space suggests areolar tissue, and thicker bundles of collagen and relatively little clear space suggest dense irregular connective tissue.

Fibrous connective tissue often forms a filler, or *interstitium,* around and between other structures such as blood vessels and gland ducts and within the walls of some organs. In some places, this interstitium is like a honeycomb of fluid-filled spaces supported by thick bundles of collagen. These are not seen in prepared slides, where the spaces are empty and the collagen is densely packed as in figure 5.17; they were recently discovered in surgical

DEEPER INSIGHT 5.2
CLINICAL APPLICATION

Marfan Syndrome—A Connective Tissue Disease

Marfan[17] *syndrome* is a hereditary defect in elastic fibers, usually resulting from a mutation in the gene for *fibrillin,* a glycoprotein that forms the structural scaffold for elastin. Clinical signs of Marfan syndrome include hyperextensible joints, hernias of the groin, and visual problems resulting from abnormally elongated eyes and deformed lenses. People with Marfan syndrome typically show unusually tall stature, long limbs, spidery fingers, abnormal spinal curvature, and a protruding "pigeon breast." More serious problems are weakened heart valves and arterial walls. The aorta, where blood pressure is highest, is sometimes enormously dilated close to the heart and may rupture. Marfan syndrome is present in about 1 out of 20,000 live births, and most victims die by their mid-30s. Abraham Lincoln's tall, gangly physique and spindly fingers led some authorities to suspect that he had Marfan syndrome, but the evidence is inconclusive. Some star athletes have died at a young age of Marfan syndrome, including Olympic volleyball champion Flo Hyman (1954–86), who died at the age of 31 of a ruptured aorta during a game in Japan.

---

[16]*areola* = little space

[17]Antoine Bernard-Jean Marfan (1858–1942), French physician

TABLE 5.4	Loose Connective Tissues
Areolar Tissue	**Reticular Tissue**

(a)

Ground substance   Elastic fibers   Collagenous fibers   Fibroblasts

(b)

Leukocytes   Reticular fibers

(b)

FIGURE 5.14 Areolar Tissue. Spread of the mesentery (×400). (a) Light micrograph. (b) Labeled drawing. **APR**

a: Dennis Strete/McGraw-Hill Education

Microscopic appearance: Loose arrangement of collagenous and elastic fibers; scattered cells of various types; abundant ground substance; numerous blood vessels
Representative locations: Underlying nearly all epithelia; surrounding blood vessels, nerves, esophagus, and trachea; fascia between muscles; mesenteries; visceral layers of pericardium and pleura
Functions: Loosely binds epithelia to deeper tissues; allows passage of nerves and blood vessels through other tissues; provides an arena for immune defense; blood vessels provide nutrients and waste removal for overlying epithelia

FIGURE 5.15 Reticular Tissue. Spleen (×400). (a) Light micrograph. (b) Labeled drawing. **APR**

a: McGraw-Hill Education/Al Telser, photographer

Microscopic appearance: Loose network of reticular fibers and cells, infiltrated with numerous leukocytes, especially lymphocytes
Representative locations: Lymph nodes, spleen, thymus, bone marrow
Function: Forms supportive stroma (framework) for lymphatic organs

TABLE 5.5	Dense Connective Tissues

Dense Regular Connective Tissue	Dense Irregular Connective Tissue

(a)

(a)

Collagen fibers   Ground substance   Fibroblast nuclei

Bundles of   Gland   Fibroblast   Ground
collagen fibers   ducts   nuclei   substance

(b)

(b)

FIGURE 5.16 Dense Regular Connective Tissue. Tendon
(×400). (a) Light micrograph. (b) Labeled drawing. **APR**

a: Dennis Strete/McGraw-Hill Education

Microscopic appearance: Densely packed, parallel, often wavy
collagen fibers; slender fibroblast nuclei compressed between
collagen bundles; scanty open space (ground substance); scarcity
of blood vessels

Representative locations: Tendons and ligaments

Functions: Ligaments tightly bind bones together and resist stress;
tendons attach muscle to bone and transfer muscular tension to
bones

FIGURE 5.17 Dense Irregular Connective Tissue. Dermis of the
skin (×400). (a) Light micrograph. (b) Labeled drawing. **APR**

a: McGraw-Hill Education/Dennis Strete, photographer

Microscopic appearance: Densely packed collagen fibers running in
random directions; scanty open space (ground substance); few visible
cells; scarcity of blood vessels

Representative locations: Deeper portion of dermis of skin; capsules
around viscera such as liver, kidney, spleen; fibrous sheaths around
cartilages and bones

Functions: Withstands stresses applied in unpredictable directions;
imparts durability to tissues

patients by a new microscopic imaging technique. They are found especially in areas subject to frequent compression, such as the dermis of the skin, the bile duct, and walls of the digestive tract and urinary bladder. They drain into the lymph nodes and may be important as a route of cancer metastasis.

5.3c Adipose Tissue

Adipose tissue, or **fat,** is tissue in which adipocytes are the dominant cell type (**table 5.6** and **fig. 5.18**). Adipocytes may also occur singly or in small clusters in areolar tissue. The space between adipocytes is occupied by areolar tissue, reticular tissue, and blood capillaries.

Fat is the body's primary energy reservoir. The quantity of stored triglyceride and the number of adipocytes are quite stable in a person, but this doesn't mean stored fat is stagnant. New triglycerides are constantly synthesized and stored as others are hydrolyzed and released into circulation. Thus, there is a constant turnover of stored triglyceride, with an equilibrium between synthesis and hydrolysis, energy storage and energy use.

There are two kinds of fat in humans—white (or yellow) fat and brown fat. **White fat** is the more abundant and is the most significant adipose tissue of the adult body. Its adipocytes are usually 70 to 120 μm in diameter, but may be five times as large in obese people. They have a single large, central globule of triglyceride. Their cytoplasm is restricted to a thin layer immediately beneath the plasma membrane, and the nucleus is pushed against the edge of the cell. Since the triglyceride is dissolved out by most histological fixatives, fat cells in most specimens look empty and somewhat collapsed, with a resemblance to chicken wire.

White fat provides thermal insulation, anchors and cushions such organs as the eyeballs and kidneys, and contributes to body contours such as the female breasts and hips. On average, women have more fat relative to body weight than men do. It helps to meet the caloric needs of pregnancy and nursing an infant, and having too little fat reduces female fertility.

Brown fat is found mainly in fetuses, infants, and children, but adults also have small deposits of brown fat; it accounts for up to 6% of an infant's weight, and is concentrated especially in fat pads in the shoulders, upper back, and around the kidneys. It stores lipid in the form of multiple globules rather than one large one. It gets its color from an unusual abundance of blood vessels and certain enzymes in its mitochondria. Brown fat is a heat-generating tissue. It has numerous mitochondria, but their oxidative pathway is not linked to ATP synthesis. Therefore, when these cells oxidize fats, they release all of the energy as heat. Hibernating animals accumulate brown fat in preparation for winter.

▶▶▶APPLY WHAT YOU KNOW

Why would infants and children have more need for brown fat than adults do? (Hint: Smaller bodies have a higher ratio of surface area to volume than larger bodies do.)

TABLE 5.6	Adipose Tissue

(a)

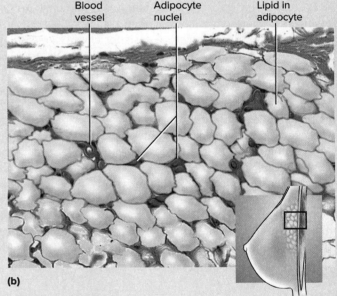

(b)

FIGURE 5.18 Adipose Tissue. Breast (×100). (a) Light micrograph. (b) Labeled drawing. **APR**

a: Dennis Strete/McGraw-Hill Education

Microscopic appearance: Dominated by adipocytes—large, empty-looking cells with thin margins; tissue sections often very pale because of scarcity of stained cytoplasm; adipocytes shrunken; nucleus pressed against plasma membrane; blood vessels present

Representative locations: Subcutaneous fat beneath skin; breast; heart surface; mesenteries; surrounding organs such as kidneys and eyes

Functions: Energy storage; thermal insulation; heat production by brown fat; protective cushion for some organs; filling space, shaping body

5.3d Cartilage

Cartilage is a relatively stiff connective tissue with a flexible rubbery matrix; you can feel its texture by folding and releasing the external ear or palpating the tip of your nose or your "Adam's apple" (the *thyroid cartilage* of the larynx). It is also easily seen in many grocery items—it is the milky-colored gristle at the ends of pork ribs and on chicken leg and breast bones, for example. Among other functions, cartilages shape and support the nose and ears and partially enclose the larynx (voice box), trachea (windpipe), and thoracic cavity.

Cartilage is produced by cells called **chondroblasts**[18] (CON-dro-blasts), which secrete the matrix and surround themselves with it until they become trapped in little cavities called **lacunae**[19] (la-CUE-nee). Once enclosed in lacunae, the cells are called **chondrocytes** (CON-dro-sites). Cartilage is devoid of blood capillaries, so nutrition and waste removal depend on solute diffusion through the stiff matrix. Because this is a slow process, chondrocytes have low rates of metabolism and cell division, and injured cartilage heals slowly.

The matrix is rich in glycosaminoglycans and contains collagen fibers that range from invisibly fine to conspicuously coarse. Differences in the fibers provide a basis for classifying cartilage into three types: *hyaline cartilage, elastic cartilage,* and *fibrocartilage* (shown in **table 5.7** and **figs. 5.19** through **5.21**).

Hyaline[20] **cartilage** (HY-uh-lin) is named for its clear, glassy appearance, which stems from the usually invisible fineness of its collagen fibers. **Elastic cartilage** is named for its conspicuous elastic fibers, and **fibrocartilage** for its coarse, readily visible bundles of collagen. Elastic cartilage and most hyaline cartilage are surrounded by a sheath of dense irregular connective tissue called the **perichondrium**[21] (PERR-ih-CON-dree-um). A reserve population of chondroblasts between the perichondrium and cartilage contributes to cartilage growth throughout life. Perichondrium is lacking from fibrocartilage and some hyaline cartilage, such as the cartilaginous caps at the ends of the long bones.

5.3e Bone

Bone, or **osseous tissue (table 5.8, fig. 5.22),** is a hard, calcified connective tissue that composes the skeleton. The term *bone* has two meanings in anatomy—an entire organ such as the femur and mandible, or just the osseous tissue. Bones are composed of not only **osseous tissue,** but also cartilage, bone marrow, dense irregular connective tissue, and other tissue types.

There are two forms of osseous tissue: (1) **Spongy bone** fills the heads of the long bones and forms the middle layer of flat bones such as the sternum and cranial bones. Although it is calcified and hard, its delicate slivers and plates give it a spongy appearance (see fig. 7.4a). (2) **Compact (dense) bone** is a denser calcified tissue with no spaces visible to the naked eye. It forms

the external surfaces of all bones, so spongy bone, when present, is always covered by a shell of compact bone.

Further differences between compact and spongy bone are described in sections 7.2c and 7.2d. Here, we examine only compact bone. Most specimens you study will probably be chips of dead, dried bone ground to microscopic thinness. In such preparations, the cells are absent but spaces reveal their former locations. Most compact bone is arranged in cylinders of tissue that surround **central (osteonic) canals,** which run longitudinally through the shafts of long bones such as the femur. Blood vessels and nerves travel through these canals. The bone matrix is deposited in **concentric lamellae**—onionlike layers around each canal. A central canal and its surrounding lamellae are called an **osteon.** Tiny lacunae between the lamellae are occupied by mature bone cells, or **osteocytes.**[22] Delicate channels called **canaliculi** radiate from each lacuna to its neighbors and allow the osteocytes to contact each other. The bone as a whole is covered with a tough fibrous **periosteum** (PERR-ee-OSS-tee-um) similar to the perichondrium of cartilage.

About one-third of the dry weight of bone is composed of collagen fibers and glycosaminoglycans, which enable a bone to bend slightly under stress. Two-thirds of its weight is minerals (mainly calcium and phosphate salts) that enable bones to withstand compression by the weight of the body.

5.3f Blood

Blood (table 5.9, fig. 5.23) is a fluid connective tissue that travels through tubular blood vessels. Its primary function is to transport cells and dissolved matter from place to place. It may seem odd that a tissue as fluid as blood and another as rock hard as bone are both considered connective tissues, but they have more in common than first meets the eye. Like other connective tissues, blood is composed of more ground substance than cells. Its ground substance is the **blood plasma** and its cellular components are collectively called the **formed elements.** Another factor placing blood in the connective tissue category is that it is produced by the connective tissues of the bone marrow and lymphatic organs. Unlike other connective tissues, blood doesn't exhibit fibers except when it clots.

The formed elements are of three kinds—erythrocytes, leukocytes, and platelets. **Erythrocytes**[23] (eh-RITH-ro-sites), or **red blood cells (RBCs),** are the most abundant. In stained blood films, they look like pink discs with thin, pale centers and no nuclei. Erythrocytes transport oxygen and carbon dioxide. **Leukocytes,** or **white blood cells (WBCs),** serve various roles in defense against infection and other diseases. They travel from one organ to another in the bloodstream and lymph but spend most of their lives in the connective tissues. Leukocytes are somewhat larger than erythrocytes and have conspicuous nuclei that usually appear violet in stained preparations. There are five kinds, distinguished partly by variations in nuclear shape: *neutrophils, eosinophils, basophils,*

---

[18]*chondro* = cartilage, gristle; *blast* = forming
[19]*lacuna* = lake, cavity
[20]*hyal* = glass
[21]*peri* = around; *chondri* = cartilage

[22]*osteo* = bone; *cyte* = cell
[23]*erythro* = red; *cyte* = cell

TABLE 5.7	Cartilage

Hyaline Cartilage	Elastic Cartilage	Fibrocartilage

FIGURE 5.19 Hyaline Cartilage.
Bronchus (×400). (a) Light micrograph.
(b) Labeled drawing. **APR**
a: ©Ed Reschke

Microscopic appearance: Clear, glassy matrix, often stained light blue or pink in tissue sections; fine, dispersed collagen fibers, not usually visible; chondrocytes enclosed in lacunae, often in small clusters of three or four cells *(cell nests);* usually covered by perichondrium
Representative locations: A thin *articular cartilage,* lacking perichondrium, over the ends of bones at movable joints; supportive rings and plates around trachea and bronchi; a boxlike enclosure around the larynx; much of the fetal skeleton; and a *costal cartilage* attaches the end of a rib to the breastbone
Functions: Eases joint movements; holds airway open during respiration; moves vocal cords during speech; a precursor of bone in the fetal skeleton and the growth zones of long bones of children

FIGURE 5.20 Elastic Cartilage.
External ear (×1,000). (a) Light micrograph.
(b) Labeled drawing. **APR**
a: ©Ed Reschke

Microscopic appearance: Elastic fibers form weblike mesh amid lacunae; always covered by perichondrium
Representative locations: External ear; epiglottis
Functions: Provides flexible, elastic support

FIGURE 5.21 Fibrocartilage.
Intervertebral disc (×400). (a) Light micrograph. (b) Labeled drawing. **APR**
a: Dr. Alvin Telser

Microscopic appearance: Parallel collagen fibers similar to those of tendon; rows of chondrocytes in lacunae between collagen fibers; never has a perichondrium
Representative locations: Pubic symphysis (anterior joint between two halves of pelvic girdle); intervertebral discs, which separate bones of vertebral column; menisci, or pads of shock-absorbing cartilage, in knee joint; at points where tendons insert on bones near articular hyaline cartilage
Functions: Resists compression and absorbs shock in some joints; often a transitional tissue between dense connective tissue and hyaline cartilage (for example, at some tendon–bone junctions)

TABLE 5.8	Bone

(a)

Lacunae   Canaliculi   Concentric lamellae of osteon   Central canal   Osteon

(b)

FIGURE 5.22 Compact Bone (×100). (a) Light micrograph. (b) Labeled drawing. **APR**

a: Dennis Strete/McGraw-Hill Education

Microscopic appearance (compact bone): Calcified matrix arranged in concentric lamellae around central canals; osteocytes in lacunae between adjacent lamellae; lacunae interconnected by delicate canaliculi

Representative location: Skeleton

Functions: Physical support of body; leverage for muscle action; protective enclosure of viscera; reservoir of calcium and phosphorus

TABLE 5.9	Blood

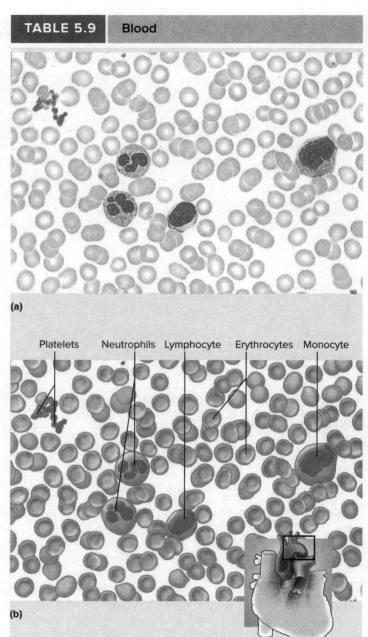

(a)

Platelets   Neutrophils   Lymphocyte   Erythrocytes   Monocyte

(b)

FIGURE 5.23 Blood Smear (×1,000). (a) Light micrograph. (b) Labeled drawing. **APR**

a: ©Ed Reschke

Microscopic appearance: Erythrocytes appear as pale pink discs with light centers and no nuclei; leukocytes are slightly larger, are much fewer, and have variously shaped nuclei, usually stained violet; platelets are cell fragments with no nuclei, about one-quarter the diameter of erythrocytes

Representative locations: Contained in heart and blood vessels

Functions: Transports gases, nutrients, wastes, chemical signals, and heat throughout body; provides defensive leukocytes; contains clotting agents to minimize bleeding; platelets secrete growth factors that promote tissue maintenance and repair *lymphocytes,* and *monocytes.* Their individual characteristics are considered in detail in table 18.6. **Platelets** are small cell fragments scattered amid the blood cells. They are involved in clotting and other mechanisms for minimizing blood loss, and in secreting growth factors that promote blood vessel growth and maintenance.

BEFORE YOU GO ON

Answer the following questions to test your understanding of the preceding section:

9. What features do most or all connective tissues have in common to set this class apart from nervous, muscular, and epithelial tissue?

10. List the cell and fiber types found in fibrous connective tissues and state their functional differences.

11. What substances account for the gelatinous consistency of connective tissue ground substance?

12. What is areolar tissue? How can it be distinguished from any other kind of connective tissue?

13. Discuss the difference between dense regular and dense irregular connective tissue as an example of the relationship between form and function.

14. Describe some similarities, differences, and functional relationships between hyaline cartilage and bone.

15. What are the three basic kinds of formed elements in blood, and what are their respective functions?

5.4 Nervous and Muscular Tissues—Excitable Tissues

Expected Learning Outcomes

When you have completed this section, you should be able to

a. explain what distinguishes excitable tissues from other tissues;

b. name the cell types that compose nervous tissue;

c. identify the major parts of a nerve cell;

d. visually recognize nervous tissue from specimens or photographs;

e. name the three kinds of muscular tissue and describe the differences between them; and

f. visually identify any type of muscular tissue from specimens or photographs.

Excitability is a characteristic of all living cells, but it is developed to its highest degree in nervous and muscular tissues, which are therefore described as **excitable tissues.** The basis for their excitation is an electrical charge difference (voltage) called the *membrane potential,* which occurs across the plasma membranes of all cells. Nervous and muscular tissues respond quickly to outside stimuli by means of changes in membrane potential. In nerve cells, these changes result

in the rapid transmission of signals to other cells. In muscle cells, they result in contraction, or shortening of the cell.

5.4a Nervous Tissue

Nervous tissue (table 5.10, fig. 5.24) is specialized for communication by means of electrical and chemical signals. It consists of **neurons** (NOOR-ons), or nerve cells, and a much greater number of **neuroglia** (noo-ROG-lee-uh), or **glial cells** (GLEE-ul), which protect and assist the neurons. Neurons detect stimuli, respond quickly, and transmit coded information rapidly to other cells. Each neuron has a prominent **neurosoma,** or cell body, that houses the nucleus and most other organelles. This is the cell's center of genetic control and protein synthesis. Neurosomas are usually round, ovoid, or stellate in shape. Extending from the neurosoma, there are usually multiple short, branched processes called **dendrites,**[24] which receive signals from other cells and conduct messages to the neurosoma; and a single, much longer **axon,** or **nerve fiber,** which sends outgoing signals to other cells. Some axons are more than a meter long and extend from the brainstem to the foot.

Glial cells constitute most of the volume of the nervous tissue. They are usually much smaller than neurons. There are six types of glial cells, described in section 12.3a, which provide a variety of supportive, protective, and "housekeeping" functions for the nervous system. Although they communicate with neurons and each other, they don't transmit long-distance signals.

Nervous tissue is found in the brain and spinal cord (central nervous system), nerves, and ganglia, which are knotlike swellings in nerves where the neurosomas outside the central nervous system are concentrated. Local variations in the structure of nervous tissue are described in chapters 12 to 16.

5.4b Muscular Tissue

Muscular tissue is specialized to contract when stimulated, and thus to exert a physical force on other tissues, organs, or fluids—for example, a skeletal muscle pulls on a bone, the heart contracts and expels blood, and the bladder contracts and expels urine. Not only do movements of the body and its limbs depend on muscle, but so do such processes as digestion, waste elimination, breathing, speech, and blood circulation. The muscles are also an important source of body heat.

There are three types of muscular tissue—*skeletal, cardiac, and smooth*—which differ in appearance, physiology, and function (as shown in **table 5.11** and **figs. 5.25** through **5.27**). **Skeletal muscle** consists of long threadlike cells called **muscle fibers.** Most skeletal muscle is attached to bones, but there are exceptions in the tongue, upper esophagus, some facial muscles, and some **sphincter**[25] (SFINK-tur) muscles (muscular rings or cuffs that open and close body passages). Each cell contains multiple nuclei adjacent to the plasma membrane. Skeletal muscle is described as striated and voluntary. The first term refers to alternating light and

---

[24]*dendr* = tree; *ite* = little

[25]*sphinc* = squeeze, bind tightly

TABLE 5.10	Nervous Tissue

(a)

Nuclei of glial cells   Axon   Neurosoma   Dendrites

(b)

FIGURE 5.24 Neuron and Glial Cells. Spinal cord smear (×400). (a) Light micrograph. (b) Labeled drawing. **A&PR**

a: ©Ed Reschke

Microscopic appearance: Most sections show a few large neurons, usually with rounded or stellate cell bodies (neurosomas) and fibrous processes (axon and dendrites) extending from the neurosomas; neurons are surrounded by a greater number of much smaller glial cells, which lack dendrites and axons.
Representative locations: Brain, spinal cord, nerves, ganglia
Function: Internal communication

dark bands, or **striations** (stry-AY-shuns), created by the overlapping pattern of cytoplasmic protein filaments that cause muscle contraction. The second term, *voluntary,* refers to the fact that we usually have conscious control over skeletal muscle.

Cardiac muscle is limited to the heart. It too is striated, but it differs from skeletal muscle in its other features. Its cells are much shorter, so they are commonly called **cardiomyocytes**[26] rather than fibers. They are branched or notched at the ends. They contain only one nucleus, which is located near the center and often surrounded by a light-staining region of glycogen. Cardiomyocytes are joined end to end by junctions called **intercalated discs**[27] (in-TUR-kuh-LAY-ted). Mechanical connections in these discs keep the cells from pulling apart when the heart contracts. Electrical junctions in the discs allow a wave of electrical excitation to travel rapidly from cell to cell so that all the cardiomyocytes of a heart chamber are stimulated and contract almost simultaneously. Intercalated discs appear as dark transverse lines separating each cell from the next. They may be only faintly visible, however, unless the tissue has been specially stained for them. Cardiac muscle is considered *involuntary* because it is not usually under conscious control; it contracts even if all nerve connections to it are severed.

Smooth muscle lacks striations and is involuntary. Smooth muscle cells are fusiform and relatively short. They have only one, centrally placed nucleus. Small amounts of smooth muscle are found in the iris of the eye and in the skin, but most of it, called **visceral muscle,** forms layers in the walls of the digestive, respiratory, and urinary tracts; blood vessels; the uterus; and other viscera. In locations such as the esophagus and small intestine, smooth muscle forms adjacent layers, with the cells of one layer encircling the organ and the cells of the other layer running longitudinally. When the circular smooth muscle contracts, it may propel contents such as food through the organ. By regulating the diameter of blood vessels, smooth muscle is very important in controlling blood pressure and flow. Both smooth and skeletal muscle form sphincters that control the emptying of the bladder and rectum.

▶▶▶**APPLY WHAT YOU KNOW**

How does the meaning of the word fiber *differ in the following uses:* muscle fiber, nerve fiber, *and* connective tissue fiber?

BEFORE YOU GO ON

Answer the following questions to test your understanding of the preceding section:

16. What do nervous and muscular tissue have in common? What is the primary function of each?

17. What kinds of cells compose nervous tissue, and how can they be distinguished from each other?

18. Name the three kinds of muscular tissue, describe how to distinguish them from each other in microscopic appearance, and state a location and function for each.

---

[26]*cardio* = heart; *myo* = muscle; *cyte* = cell
[27]*inter* = between; *calated* = inserted

TABLE 5.11	Muscular Tissue

Skeletal Muscle	Cardiac Muscle	Smooth Muscle
(a)	(a)	(a)
Nuclei Striations Muscle fiber (b)	Intercalated discs Striations Glycogen (b)	Nuclei Muscle cells (b)

FIGURE 5.25 Skeletal Muscle (×400). (a) Light micrograph. (b) Labeled drawing. **APR** a: ©Ed Reschke	**FIGURE 5.26 Cardiac Muscle (×400).** (a) Light micrograph. (b) Labeled drawing. **APR** a: ©Ed Reschke	**FIGURE 5.27 Smooth Muscle.** Intestinal wall (×1,000). (a) Light micrograph. (b) Labeled drawing. **APR** a: McGraw-Hill Education/Dennis Strete, photographer
Microscopic appearance: Long, threadlike, unbranched cells (fibers), relatively parallel in longitudinal tissue sections; striations; multiple nuclei per cell, near plasma membrane	**Microscopic appearance:** Short cells (cardiomyocytes) with notched or slightly branched ends; less parallel appearance in tissue sections; striations; intercalated discs; one nucleus per cell, centrally located and often surrounded by a light zone	**Microscopic appearance:** Short fusiform cells overlapping each other; nonstriated; one nucleus per cell, centrally located
Representative locations: Skeletal muscles, mostly attached to bones but also in the tongue, esophagus, and encircling the lips, eyelids, urethra, and anus	**Representative location:** Heart	**Representative locations:** Usually found as sheets of tissue in walls of blood vessels and viscera such as the digestive tract; also in iris and associated with hair follicles; involuntary sphincters of urethra and anus
Functions: Body movements, facial expression, posture, breathing, speech, swallowing, control of urination and defecation, and assistance in childbirth; under voluntary control	**Functions:** Pumping of blood; under involuntary control	**Functions:** Swallowing; contractions of stomach and intestines; expulsion of feces and urine; labor contractions; control of blood pressure and flow; control of respiratory airflow; control of pupillary diameter; erection of hairs; under involuntary control

5.5 Cellular Junctions, Glands, and Membranes

Expected Learning Outcomes

When you have completed this section, you should be able to

a. describe the junctions that hold cells and tissues together;

b. describe or define different types of glands;

c. describe the typical anatomy of a gland;

d. name and compare different modes of glandular secretion; and

e. describe the types and composition of the body's membranes.

5.5a Cellular Junctions

Most cells—except for blood cells, macrophages, and metastatic cancer cells—must be anchored to each other and to the matrix if they are to grow and divide normally. The connections between one cell and another are called **cellular junctions.** They enable the cells to resist stress, communicate with each other, and control the movement of substances through tissues. Without them, cardiac muscle cells would pull apart when they contracted, and every swallow of food would scrape away the lining of your esophagus. The main types of cellular junctions are shown in **figure 5.28.**

Tight Junctions

A **tight junction** completely encircles an epithelial cell near its apical surface and joins it tightly to the neighboring cells, somewhat like the plastic harness on a six-pack of beverage cans. At a tight junction, the plasma membranes of two adjacent cells come very close together and are linked by transmembrane cell-adhesion proteins. These zipperlike interlocking proteins seal off the intercellular space and make it difficult or impossible for substances to pass between cells.

In the stomach and intestines, tight junctions prevent digestive juices from seeping between epithelial cells and digesting the underlying connective tissue. They also help to prevent bacteria from invading the tissues, and they ensure that most nutrients pass *through* the epithelial cells and not *between* them. In addition, some membrane proteins function in the apical domain of the cell, and others in the lateral or basal domains; tight junctions limit how far drifting proteins can travel and keep them segregated in the appropriate domains of the membrane where they are needed to perform their tasks.

Desmosomes

A **desmosome**[28] (DEZ-mo-some) is a patch that holds cells together somewhat like the snap on a pair of jeans. They are not continuous and cannot prevent substances from passing around them and going between the cells, but serve to keep cells from pulling apart and

enable a tissue to resist mechanical stress. Desmosomes are common in the epidermis, the epithelium of the uterine cervix, other epithelia, and cardiac muscle. Hooklike J-shaped proteins arise from the cytoskeleton, approach the cell surface from within, and penetrate into a thick protein plaque on the inner face of the plasma membrane; then the short arm of the J turns back into the cell—thus anchoring the cytoskeleton to the membrane plaque. Proteins of the plaque are linked to transmembrane proteins that, in turn, are linked to transmembrane proteins of the next cell, forming a zone of strong cell adhesion. Each cell mirrors the other and contributes half of the desmosome. Such connections create a strong structural network that binds cells together throughout the tissue. The basal cells of an epithelium are similarly linked to the underlying basement membrane by half-desmosomes called **hemidesmosomes,** so an epithelium cannot easily peel away from the underlying tissue.

▶▶▶ APPLY WHAT YOU KNOW

Why would desmosomes not be suitable as the sole type of cell junction between epithelial cells of the stomach?

Gap Junctions

A **gap (communicating) junction** is formed by a *connexon,* which consists of six transmembrane proteins arranged in a ring, somewhat like the segments of an orange, surrounding a water-filled channel. Ions, glucose, amino acids, and other small solutes can pass directly from the cytoplasm of one cell into the next through the channel. In the embryo, nutrients pass from cell to cell through gap junctions until the circulatory system forms and takes over the role of nutrient distribution. In cardiac muscle and most smooth muscle, gap junctions allow electrical excitation to pass directly from cell to cell so that the cells contract in near unison. (Gap junctions are absent from skeletal muscle.) In the lens and cornea of the eye, which lack blood vessels, gap junctions allow nutrients and other material to pass from cell to cell.

5.5b Glands

A **gland** is a cell or organ that secretes substances for use elsewhere in the body or for elimination as waste. The gland product may be something synthesized by its cells (such as digestive enzymes) or something removed from the tissues and modified by the gland (such as urine and bile pigments). The product is called a **secretion** if it is useful to the body (such as an enzyme or hormone) and an **excretion** if it is a waste product (such as urine and bile). Glands are composed mostly of epithelial tissue, but usually have a supportive connective tissue framework and capsule.

Endocrine and Exocrine Glands

Glands are classified as endocrine or exocrine. Both types originate as invaginations of a surface epithelium **(fig. 5.29).** **Exocrine**[29] **glands** (EC-so-crin) usually maintain their contact with the surface by way of a **duct,** an epithelial tube that conveys

---

[28]*desmo* = band, bond, ligament; *som* = body

[29]*exo* = out; *crin* = to separate, secrete

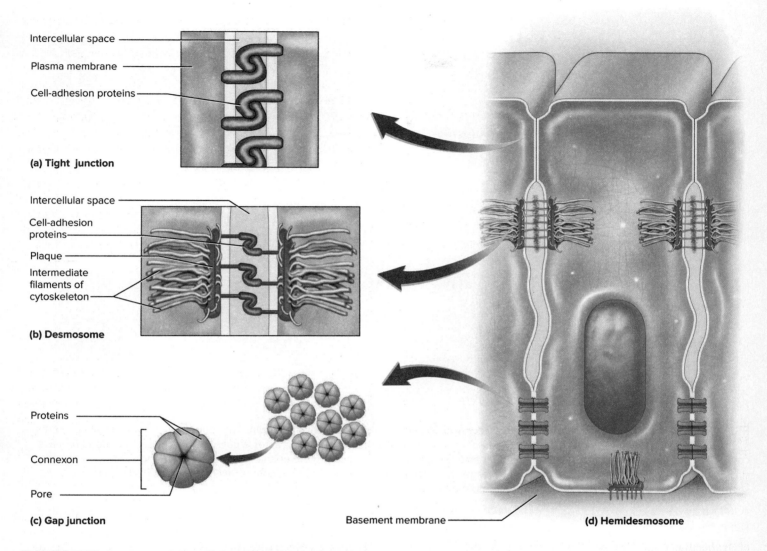

Intercellular space

Plasma membrane

Cell-adhesion proteins

(a) Tight junction

Intercellular space

Cell-adhesion proteins

Plaque

Intermediate filaments of cytoskeleton

(b) Desmosome

Proteins

Connexon

Pore

(c) Gap junction

Basement membrane

(d) Hemidesmosome

FIGURE 5.28 Structure of Four Kinds of Cellular Junctions. (a) Tight junction. (b) Desmosome. (c) Gap junction. (d) Hemidesmosome between cell and basement membrane.

? *Which of these junctions allows material to pass from one cell directly into the next?*

their secretion to the surface. The secretion may be released to the body surface, as in the case of sweat, mammary, and tear glands. More often, however, it is released into the cavity (lumen) of another organ such as the mouth or intestine; this is the case with salivary glands, the liver, and the pancreas.

Endocrine[30] **glands** lose contact with the surface and have no ducts. They do, however, have a high density of blood capillaries and secrete their products directly into the blood **(fig. 5.30a).** The secretions of endocrine glands, called *hormones,* function as chemical messengers to stimulate cells elsewhere in the body. Endocrine glands include the pituitary, thyroid, and adrenal glands.

The exocrine–endocrine distinction isn't always clear. The liver is an exocrine gland that secretes one of its products, bile, through a system of ducts, but secretes hormones, albumin, and other products directly into the blood. Several glands, such as the pancreas and kidney, have both exocrine and endocrine components. Nearly all of the viscera have at least some cells that secrete hormones, even though most of these organs are not usually thought of as glands (for example, the brain and heart).

Unicellular glands are secretory cells found in an epithelium that is predominantly nonsecretory. They can be endocrine or exocrine. For example, the respiratory tract, which is lined mainly by ciliated cells, also has a liberal scattering of exocrine goblet cells (see figs. 5.6 and 5.7). The stomach and small intestine have scattered endocrine cells, which secrete hormones that regulate digestion.

Endocrine glands are the subject of chapter 17 and are not further considered here.

Exocrine Gland Structure

Multicellular exocrine glands such as the pancreas and salivary glands are usually enclosed in a fibrous **capsule (fig. 5.30b).** The capsule often gives off extensions called **septa** (singular, *septum*),

[30]*endo* = in, into; *crin* = to separate, secrete

Epithelial cells

Connective tissue

(a) Exocrine gland

Duct

Blood capillary

(b) Endocrine gland

FIGURE 5.29 Development of Exocrine and Endocrine Glands. (a) An exocrine gland begins with epithelial cells proliferating into the connective tissue below. A form of cell death called *apoptosis* hollows out the core and creates a duct to the surface. The gland remains connected to the surface for life by way of this duct and releases its secretions onto the epithelial surface. (b) An endocrine gland begins similarly, but the cells connecting it to the surface degenerate while the secretory tissue becomes infiltrated with blood capillaries. The secretory cells secrete their products (hormones) into the blood.

or **trabeculae** (trah-BEC-you-lee), that divide the interior of the gland into compartments called **lobes,** which are visible to the naked eye. Finer connective tissue septa may further subdivide each lobe into microscopic **lobules.** Blood vessels, nerves, and the gland's own ducts generally travel through these septa. The connective tissue framework of the gland, called its **stroma,** supports and organizes the glandular tissue. The cells that perform the tasks of synthesis and secretion are collectively called the **parenchyma** (pa-REN-kih-muh). This is typically simple cuboidal or simple columnar epithelium.

Exocrine glands are classified according to their "architecture"—the branching of their ducts and the appearance and extent of their secretory portions **(fig. 5.31).** They are called **simple** if they have a single unbranched duct and **compound** if they have a branched duct. If the duct and secretory portion are of uniform diameter, the gland is called **tubular.** If the secretory cells form a dilated sac, the gland is called acinar and the sac is an **acinus**[31] (ASS-ih-nus), or **alveolus**[32] (AL-vee-OH-lus) **(fig. 5.30c).** A gland with secretory cells in both the tubular and acinar portions is called a **tubuloacinar (tubuloalveolar) gland.**

Types of Secretions

Glands are classified not only by their structure but also by the nature of their secretions. **Serous glands** (SEER-us) produce relatively thin, watery fluids such as perspiration, milk, tears, and digestive juices. **Mucous glands,** found in the oral and nasal cavities among other places, secrete a glycoprotein called *mucin* (MEW-sin). After it is secreted, mucin absorbs water and forms the sticky product *mucus.* Goblet cells are unicellular mucous glands. (Note that *mucus,* the secretion, is spelled differently from *mucous,* the adjective form of the word.) **Mixed glands,** such as the two pairs of salivary glands in the chin, contain both serous and mucous cells and produce a mixture of the two types of secretions.

Modes of Secretion

Exocrine glands are classified into *eccrine, apocrine,* and *holocrine* types according to how they release their secretions. **Eccrine**[33] **glands** (EC-rin), also called **merocrine**[34] **glands** (MERR-oh-crin), release their products by means of exocytosis. These include the tear glands, salivary glands, pancreas, and most others. Mammary glands

---

[31]*acinus* = berry
[32]*alveol* = cavity, pit

[33]*ec* = *ex* = out; *crin* = to separate, secrete
[34]*mero* = part; *crin* = to separate, secrete

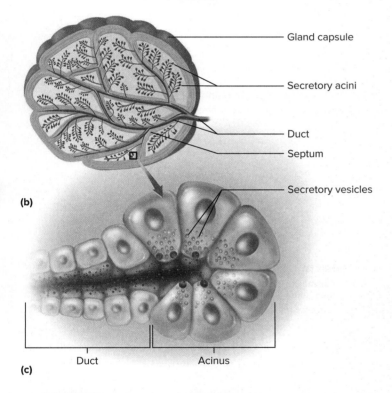

FIGURE 5.30 General Structure of Endocrine and Exocrine Glands. (a) Endocrine glands have no ducts but have a high density of blood capillaries and secrete their products (hormones) directly into the bloodstream. (b) Exocrine glands usually have a system of ducts, which often follow connective tissue septa, until their finest divisions end in saccular acini of secretory cells. (c) Detail of an acinus and the beginning of a duct.

? *What membrane transport process are the cells of the acinus in (c) carrying out? (Review section 3.3.)*

secrete milk sugar (lactose) and proteins (casein, lactalbumin) by this method **(fig. 5.32a),** but secrete the milk fat by another method called **apocrine**[35] secretion **(fig. 5.32b).** Lipids coalesce from the cytosol into a droplet that buds from the cell surface, covered by a layer of plasma membrane and a very thin film of cytoplasm. Sweat

glands of the axillary (armpit) region were once thought to use the apocrine method as well. Closer study showed this to be untrue; they are eccrine, but they're nevertheless different from other merocrine glands in function and histological appearance (see figure 6.10) and are still referred to as apocrine sweat glands.

In **holocrine**[36] **glands,** cells accumulate a product and then the entire cell disintegrates, *becoming* the secretion instead of *releasing* one **(fig. 5.32c).** Holocrine secretions tend to be relatively thick and oily, composed of cell fragments and the substances the cells had synthesized before disintegrating. Only a few glands use this method, such as the oil-producing glands of the scalp and other areas of skin, and certain glands of the eyelid.

5.5c Membranes

Atlas A (section A.3) describes the major body cavities and the membranes that line them and cover their viscera. We now consider some histological aspects of these membranes. Membranes may be composed of epithelial tissue only; connective tissue only; or epithelial, connective, and muscular tissue.

The largest membrane of the body is the **cutaneous membrane**—or more simply, the skin (detailed in chapter 6). It consists of a stratified squamous epithelium (epidermis) resting on a layer of connective tissue (dermis). Unlike the other membranes to be considered, it is relatively dry. It resists dehydration of the body and provides an inhospitable environment for the growth of infectious organisms.

The two principal kinds of internal membranes are mucous and serous membranes. A **mucous membrane (mucosa) (fig. 5.33a)** lines passages that open to the exterior environment: the digestive, respiratory, urinary, and reproductive tracts. A mucous membrane consists of two to three layers: (1) an epithelium; (2) an areolar connective tissue layer called the **lamina propria**[37] (LAM-ih-nuh PRO-pree-uh); and (3) often a layer of smooth muscle called the **muscularis mucosae** (MUSK-you-LAIR-iss mew-CO-see). Mucous membranes have absorptive, secretory, and protective functions. They are often covered with mucus secreted by goblet cells, multicellular mucous glands, or both. The mucus traps bacteria and foreign particles, which keeps them from invading the tissues and aids in their removal from the body. The epithelium of a mucous membrane may also include absorptive, ciliated, and other types of cells.

A **serous membrane (serosa)** is composed of a simple squamous epithelium resting on a thin layer of areolar connective tissue **(fig. 5.33b).** Serous membranes produce watery **serous fluid,** which arises from the blood and derives its name from the fact that it is similar to blood serum in composition. Serous membranes line the insides of some body cavities and form a smooth outer surface on some of the viscera, such as the digestive tract. The pleurae, pericardium, and peritoneum described in atlas A are serous membranes. Their epithelial component is called **mesothelium.**

The circulatory system is lined with a simple squamous epithelium called **endothelium,** derived from mesoderm. The endothelium rests on a thin layer of areolar tissue, which often rests in turn on an elastic sheet. Collectively, these tissues make

---

[35]*apo* = from, off, away; *crin* = to separate, secrete

[36]*holo* = whole, entire; *crin* = to separate, secrete
[37]*lamina* = layer; *propria* = of one's own

(a) **Simple coiled tubular gland**
   Example: Sweat gland

(b) **Compound acinar gland**
   Example: Mammary gland

(c) **Compound tubuloacinar**
   Example: Pancreas

Key
▱▱▱▱▱ Duct
▰▰▰▰ Secretory portion

FIGURE 5.31 Some Types of Exocrine Glands. (a) A simple coiled tubular gland such as a sweat gland. (b) A compound acinar gland such as the mammary gland. (c) A compound tubuloacinar gland such as the pancreas. These are only a few examples of the 10 or so types of gland architecture.

? *Predict and sketch the appearance of a simple acinar gland.*

Milk sugar and protein

Exocytosis

Secretory vesicle

Golgi complex

(a) Eccrine

Milk fat
Plasma membrane
Cytoplasm

(b) Apocrine

Secretion (sebum)

Gland capsule

Disintegrating cells

Mitosis in basal cells replaces cells that disintegrate and die

Intact basal cells

(c) Holocrine

FIGURE 5.32 The Three Modes of Exocrine Secretion. (a) Eccrine secretion in a cell of the mammary gland, secreting milk sugar (lactose) and proteins (casein, lactalbumin) by exocytosis. (b) Apocrine secretion of fat by a cell of the mammary gland. Fat droplets coalesce in the cytosol, then bud from the cell surface with a thin coat of cytoplasm and plasma membrane. (c) Holocrine secretion by a sebaceous (oil) gland of the scalp. In this method, entire gland cells break down and become the secretion (sebum).

? *Which of these three glands would require the highest rate of mitosis? Why?*

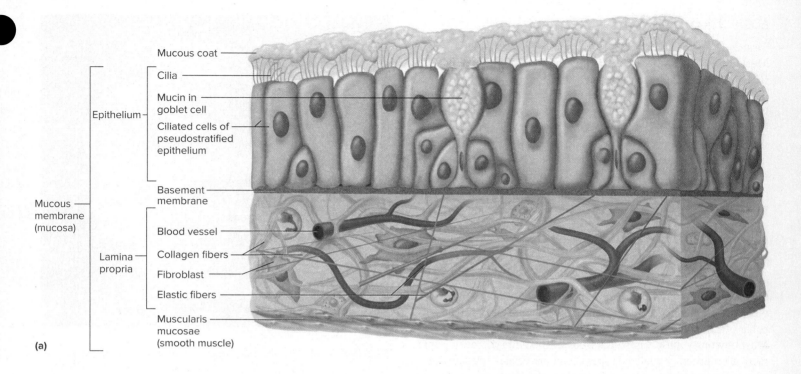

Epithelium
- Mucous coat
- Cilia
- Mucin in goblet cell
- Ciliated cells of pseudostratified epithelium
- Basement membrane

Mucous membrane (mucosa)

Lamina propria
- Blood vessel
- Collagen fibers
- Fibroblast
- Elastic fibers
- Muscularis mucosae (smooth muscle)

(a)

Serous membrane (serosa)
- Serous fluid
- Squamous cells
- Basement membrane
- Areolar tissue
- Smooth muscle

(b)

FIGURE 5.33 Histology of Mucous and Serous Membranes. (a) A mucous membrane such as the inner lining of the trachea. (b) A serous membrane such as the external surface of the small intestine.

up a membrane called the *tunica interna* of the blood vessels and *endocardium* of the heart.

The foregoing membranes are composed of two to three tissue types. By contrast, some membranes composed only of connective tissue include the *dura mater* around the brain, *synovial membranes* that enclose joints of the skeletal system, and the *periosteum* that covers each bone. Some membranes composed only of epithelium include the anterior surfaces of the lens and cornea of the eye. All of these are described in later chapters.

BEFORE YOU GO ON

Answer the following questions to test your understanding of the preceding section:

19. Compare the structure of tight junctions and gap junctions. Relate their structural differences to their functional differences.

20. Distinguish between a simple gland and a compound gland, and give an example of each. Distinguish between a tubular gland and an acinar gland, and give an example of each.

21. Contrast the eccrine, apocrine, and holocrine methods of secretion, and name a gland product produced by each method.

22. Describe the differences between a mucous and a serous membrane.

23. Name the layers of a mucous membrane, and state which of the four primary tissue classes composes each layer.

5.6 Tissue Growth, Development, Repair, and Degeneration

Expected Learning Outcomes

When you have completed this section, you should be able to

a. name and describe the modes of tissue growth;

b. define *adult* and *embryonic stem cells* and their varied degrees of developmental plasticity;

c. name and describe the ways that a tissue can change from one type to another;

d. name and describe the modes and causes of tissue shrinkage and death; and

e. name and describe the ways the body repairs damaged tissues.

5.6a Tissue Growth

Tissues grow because their cells increase in number or size. Most embryonic and childhood growth occurs by **hyperplasia**[38] (HY-pur-PLAY-zhuh)—tissue growth through cell multiplication. Skeletal muscles and adipose tissue grow, however, through **hypertrophy**[39] (hy-PUR-truh-fee)—the enlargement of preexisting cells. Even a very muscular or fat adult has essentially the same number of muscle fibers or adipocytes as he or she had in late adolescence, but the cells may be substantially larger. **Neoplasia**[40] (NEE-oh-PLAY-zhuh) is the development of a tumor (*neoplasm*)—whether benign or malignant—composed of abnormal, nonfunctional tissue.

5.6b Tissue Development

You have studied the form and function of more than two dozen discrete types of human tissue in this chapter. You should not leave this subject, however, with the impression that once these tissue types are established, they never change. Tissues are, in fact, capable of changing from one type to another within certain limits. Most obviously, unspecialized tissues of the embryo develop into more diverse and specialized types of mature tissue—mesenchyme to cartilage and bone, for example. This development of a more specialized form and function is called **differentiation.**

Some tissues can undergo **metaplasia,**[41] a change from one type of mature tissue to another. For example, the vagina of a young girl is lined with a simple cuboidal epithelium. At puberty, it changes to a stratified squamous epithelium, better adapted to the future demands of intercourse and childbirth. The long bones of a child are filled with blood-producing red bone marrow, but by adulthood, most of this changes to adipose tissue. In smokers, the pseudostratified columnar epithelium of the bronchi may transform into a stratified squamous epithelium.

►►►APPLY WHAT YOU KNOW

What functions of a ciliated pseudostratified columnar epithelium could not be served by a stratified squamous epithelium? In light of this, what might be some consequences of bronchial metaplasia in heavy smokers?

5.6c Stem Cells

The growth and differentiation of tissues depend upon a supply of reserve **stem cells.** These are undifferentiated cells that are not yet performing any specialized function, but have the potential to differentiate into one or more types of mature functional cells, such as liver, brain, cartilage, or skin cells. Such cells have various degrees of **developmental plasticity,** or diversity of mature cell types to which they can give rise.

There are two types of stem cells: embryonic and adult. **Embryonic stem cells** compose the early human embryo

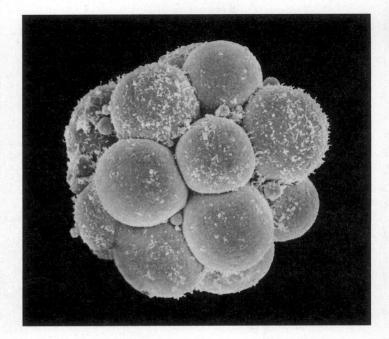

FIGURE 5.34 The Morula Stage of Human Development. This stage consists of about 16 to 32 totipotent stem cells with unlimited developmental plasticity.

Dr Yorgos Nikas/Science Photo Library/Getty Images

(fig. 5.34). In the early stages of development, these are called **totipotent stem cells,** because they have the potential to develop into any type of fully differentiated human cell—not only cells of the later embryonic, fetal, or adult body, but also cells of the temporary structures of pregnancy, such as the placenta and amniotic sac. Totipotency is unlimited developmental plasticity. About 4 days after fertilization, the developing embryo enters the *blastocyst* stage. The blastocyst is a hollow ball with an *outer cell mass* that helps form the placenta and other accessory organs of pregnancy, and an *inner cell mass (embryoblast)* that becomes the embryo itself (see fig. 29.4). Cells of the inner cell mass are called **pluripotent stem** cells; they can still develop into any cell type of the embryo, but not into the accessory organs of pregnancy. Thus their developmental plasticity is already somewhat limited.

Adult stem cells occur in small numbers in mature organs and tissues throughout a person's life. Typically an adult stem cell divides mitotically; one of its daughter cells remains a stem cell and the other one differentiates into a mature specialized cell. The latter cell may replace another that has grown old and died, contribute to the development of growing organs (as in a child), or help to repair damaged tissue. Some adult stem cells are **multipotent**—able to develop into two or more different cell lines, but not just any type of body cell. Certain multipotent bone marrow stem cells, for example, can give rise to red blood cells, five kinds of white blood cells, and platelet-producing cells. **Unipotent** stem cells have the most limited plasticity, as they can produce only one mature cell type. Examples include the cells that give rise to sperm, eggs, and keratinocytes (the majority cell type of the epidermis).

Both embryonic and adult stem cells have enormous potential for therapy, but advances in the field (see Deeper Insight 5.4) have

---

[38]*hyper* = excessive; *plas* = growth
[39]*hyper* = excessive; *trophy* = nourishment
[40]*neo* = new; *plas* = form, growth
[41]*meta* = change; *plas* = form, growth

DEEPER INSIGHT 5.3
CLINICAL APPLICATION

Stem-Cell Therapy

Cell biologists, clinicians, and hopeful patients look to stem cells as a possible treatment for diseases arising from the loss of functional tissue. Conceivably, with the right biochemical coaxing, adult stem cells (AS cells) may be induced to differentiate into various kinds of mature, differentiated cells to replace cardiac muscle damaged by heart attack, restore an injured spinal cord, cure neurodegenerative diseases such as parkinsonism, or cure diabetes by replacing lost insulin-secreting cells. AS cells have limited developmental potential, however, and may be unable to make many of the cell types needed to treat a broad range of degenerative diseases. In addition, they are present in very small numbers and are difficult to harvest from a tissue and culture in the quantities needed for therapy.

Embryonic stem cells (ES cells) could overcome these shortcomings. They are obtained from surplus embryos of about 16 to 32 cells (fig. 5.34) produced by *in vitro fertilization (IVF)* clinics, putting them to good use instead of discarding them. Unfortunately, though, stem-cell research has yet to produce many safe and proven therapies. Questions remain as to whether implanted stem cells may lodge in the wrong place in the body or grow into tumors instead of healthy functional tissue. Stem

cells have proven their value as a "workhorse" of laboratory research in mechanisms of disease, toxicity screening, and drug testing; so far, however, the only therapies proven safe and effective, and approved by the U.S. Food and Drug Administration, are treatments using bone marrow and umbililcal cord blood for certain leukemias and other blood- and immune-related diseases (see Deeper Insight 18.3). Typically, the lag from a medical discovery to clinical use is about 20 years, and stem-cell therapy seems to be no exception. We may be in for a long wait to see if stem-cell technology will yield wider applications.

Meanwhile, unfortunately, fradulent stem-cell "clinics" have sprung up around the world charging vulnerable people tens of thousands of dollars (in nonrefundable cash, of course) for worthless and even life-threatening treatments. Such reports have emerged as a patient blinded by supposed stem-cell injections into the eyes and another in whom injected stem cells developed into a spinal tumor. The FDA is attempting to crack down on illegal treatments in the United States, but many vulnerable patients continue to be victimized by illicit providers in Mexico, South America, India, East Asia, and elsewhere.

---

been slowed both by technical difficulties and controversy over the use of embryonic cells.

5.6d Tissue Repair

Damaged tissues can be repaired in two ways: *regeneration* or *fibrosis*. **Regeneration** is the replacement of dead or damaged cells by the same type of cells as before; it restores normal function to the organ. Most skin injuries (cuts, scrapes, and minor burns) heal by regeneration. The liver also regenerates remarkably well. **Fibrosis** is the replacement of damaged tissue with scar tissue, composed mainly of collagen produced by fibroblasts. Scar tissue helps to hold an organ together, but it doesn't restore normal function. Examples include the healing of severe cuts and burns, the healing of muscle injuries, and scarring of the lungs in tuberculosis.

Figure 5.35 illustrates the following stages in the healing of a cut in the skin, where both regeneration and fibrosis are involved:

① Severed blood vessels bleed into the cut. Mast cells and cells damaged by the cut release histamine, which dilates blood vessels, increases blood flow to the area, and makes blood capillaries more permeable. Blood plasma seeps into the wound, carrying antibodies and clotting proteins.

② A blood clot forms in the tissue, loosely knitting the edges of the cut together and inhibiting the spread of pathogens from the site of injury into healthy tissues. The surface of the blood clot dries and hardens in the air, forming a scab that temporarily seals the wound and blocks infection. Beneath the scab, macrophages begin to phagocytize and digest tissue debris.

③ New blood capillaries sprout from nearby vessels and grow into the wound. The deeper portions of the clot become infiltrated by capillaries and fibroblasts and transform into

a soft mass called **granulation tissue.** Macrophages remove the blood clot while fibroblasts deposit new collagen to replace it. This *fibroblastic (reconstructive) phase* of repair begins 3 to 4 days after the injury and lasts up to 2 weeks.

④ Surface epithelial cells around the wound multiply and migrate into the wounded area, beneath the scab. The scab loosens and eventually falls off, and the epithelium grows thicker. Thus, the epithelium *regenerates* while the underlying connective tissue undergoes *fibrosis,* or scarring. Capillaries withdraw from the area as fibrosis progresses. The scar tissue may or may not show through the epithelium, depending on the severity of the wound. The wound may exhibit a depressed area at first, but this is often filled in by continued fibrosis and remodeling from below, until the scar becomes unnoticeable. This *remodeling (maturation) phase* of tissue repair begins several weeks after injury and may last as long as 2 years.

5.6e Tissue Degeneration and Death

Atrophy[42] (AT-ro-fee) is the shrinkage of a tissue through a loss in cell size or number. It results from both normal aging (*senile atrophy*) and lack of use of an organ (*disuse atrophy*). Muscles that are not exercised exhibit disuse atrophy as their cells become smaller. This was a serious problem for the first astronauts who participated in prolonged microgravity space flights. Upon return to normal gravity, they were sometimes too weak from muscular atrophy to walk. Space stations and shuttles now include exercise equipment to maintain the crew's muscular condition. Disuse atrophy also occurs when a limb is immobilized in a cast or by

---

[42]*a* = without; *trophy* = nourishment

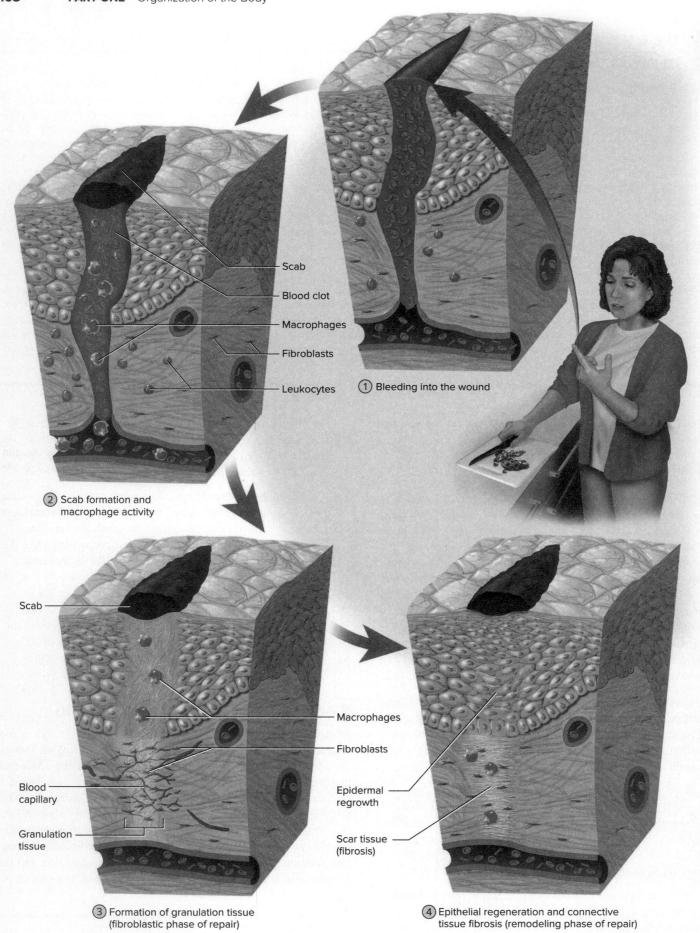

Scab

Blood clot

Macrophages

Fibroblasts

Leukocytes

① Bleeding into the wound

② Scab formation and
macrophage activity

Scab

Macrophages

Fibroblasts

Blood
capillary

Epidermal
regrowth

Granulation
tissue

Scar tissue
(fibrosis)

③ Formation of granulation tissue
(fibroblastic phase of repair)

④ Epithelial regeneration and connective
tissue fibrosis (remodeling phase of repair)

FIGURE 5.35 Stages in the Healing of a Skin Wound.

FIGURE 5.36 Dry Gangrene of the Foot Caused by Diabetes Mellitus.

Source: William Archibald/CDC

paralysis, or a person is confined to bed or a wheelchair by an illness or disability.

Necrosis[43] (neh-CRO-sis) is premature, pathological tissue death due to trauma, toxins, infection, and so forth; infarction and gangrene are two types of necrosis. **Infarction** is the sudden death of tissue, such as cardiac muscle *(myocardial infarction)* or brain tissue *(cerebral infarction),* that occurs when its blood supply is cut off. **Gangrene** is tissue necrosis resulting from infection or an obstructed blood supply. *Dry gangrene* often occurs in diabetics, especially in the feet. It is characterized by dry, shrunken skin with bluish-purple, brown, or black discoloration **(fig. 5.36).** A lack of sensation due to diabetic nerve damage can make a person oblivious to injury and infection, and poor blood circulation due to diabetic arterial damage results in slow healing and rapid spread of infection. This often necessitates the amputation of toes, feet, or legs. A **decubitus ulcer (bed sore** or **pressure sore)** is a form of dry gangrene that occurs when immobilized persons, such as those confined to a hospital bed or wheelchair, are unable to move, and continual pressure on the skin cuts off blood flow to an area. Pressure sores occur most often in places where a bone comes close to the body surface, such as the hips, sacral region, and ankles. Here,

the thin layer of skin and connective tissue is especially prone to compression between the bone and a bed or wheelchair.

Wet gangrene typically occurs in internal organs and involves neutrophil invasion, liquefaction of the tissue, pus, and a foul odor. It can result from appendicitis or an obstructed colon, for example. *Gas gangrene* is necrosis of a wound resulting from infection with certain bacteria of the genus *Clostridium,* usually introduced when a wound is contaminated with soil. The disorder is named for bubbles of gas (mainly hydrogen) that accumulate in the tissues. This is a deadly condition that requires immediate intervention, often including amputation.

Cells dying by necrosis usually swell, exhibit *blebbing* (bubbling) of their plasma membranes, and then rupture. The cell contents released into the tissues trigger an inflammatory response in which macrophages phagocytize the cellular debris.

Apoptosis[44] (AP-op-TOE-sis), or **programmed cell death,** is often the normal death of cells that have completed their function and best serve the body by dying and getting out of the way. Cells can be induced to undergo apoptosis in some pathological conditions, however. Cells undergoing apoptosis shrink and are quickly phagocytized by macrophages and other cells. The cell contents never escape, so there is no inflammatory response. Although billions of cells die every hour by apoptosis, they are engulfed so quickly that they are almost never seen except within macrophages.

Apparently, nearly every cell has a built-in "suicide program" that enables the body to dispose of it when necessary. In some cases, an extracellular suicide signal binds to a receptor protein in the plasma membrane, which then activates enzymes that degrade the cell's DNA and proteins. In other cases, cells seem to undergo apoptosis automatically if they stop receiving growth factors from other cells. For example, in embryonic development we produce about twice as many neurons as we need. Those that make connections with target cells survive, while the excess neurons die for lack of *nerve growth factor.* Apoptosis also dissolves the webbing between the fingers and toes during embryonic development; it frees the earlobe from the side of the head in people with the genotype for detached earlobes; and it causes shrinkage of the breasts after lactation ceases.

BEFORE YOU GO ON

Answer the following questions to test your understanding of the preceding section:

24. Distinguish between differentiation and metaplasia.

25. Tissues can grow through an increase in cell size or cell number. What are the respective terms for these two kinds of growth?

26. Distinguish between atrophy, necrosis, and apoptosis, and describe a circumstance under which each of these forms of tissue loss may occur.

27. Distinguish between regeneration and fibrosis. Which process restores normal cellular function? What good is the other process if it does not restore function?

---

[43]*necr* = death; *osis* = process

[44]*apo* = away; *ptosis* = falling

DEEPER INSIGHT 5.4

CLINICAL APPLICATION

Regenerative Medicine

As the population ages while science progresses in knowledge and technical skill, people look to the medical community with hopes of repairing or replacing body parts that are damaged, lost, defective from birth, or worn-out by age. Is it possible to repair brain or heart tissue; or a nose, ear, or finger; or even larger body parts lost to injury? The science of **regenerative medicine,** still in its infancy, seeks to do just that.

One of its approaches is **tissue engineering**—growing tissues and organs in the laboratory, ideally from the patient's own cells, and implanting them into the body. To produce something like a blood vessel or bronchus, tissue engineers start with an organ from a donor body and *decellularize* it—strip it of all living cells, leaving only its collagenous connective tissue framework, or scaffold. While preserving the shape and mechanical properties of the organ to be replaced, this cell-free scaffold eliminates the likelihood of immune rejection by the patient to receive it. It is then seeded with cells from the patient's body, such as chrondrocytes, fibroblasts, or keratinocytes. The product is maintained in an incubator called a *bioreactor,* which provides nutrients, oxygen, and growth factors while the cells repopulate the scaffold. Some treatments use synthetic polymer scaffolds and even artificial organs produced by 3D inkjet bioprinters. One of the most daunting problems in growing artificial organs is producing the microvascular blood supply needed to sustain an organ such as a liver.

Tissue engineering has succeeded in the laboratory in producing liver, bone, cartilage, ureter, tendon, intestine, and breast tissue, and even a beating rodent heart. "Bench to bedside" translation to the clinic and real patients is another matter. Early successes, however, include nasofacial reconstruction, engineered skin for covering burns and diabetic foot ulcers, repair of damaged knee cartilages, and even urinary bladders grown from patients' own cells and transplanted into their bodies. Preclinical trials are underway on producing bioengineered blood vessels for coronary bypass surgery, and hopes for future development include heart valves and patches of cardiac muscle.

STUDY GUIDE

▶ Assess Your Learning Outcomes

*To test your knowledge, discuss the following
topics with a study partner or in writing, ideally
from memory.*

5.1 The Study of Tissues

1. Two names for the branch of biology
   concerned with tissue structure
2. The four primary tissue classes that
   constitute the body
3. The roles of cells, matrix, fibers, and ground
   substance in tissue composition, and how
   these terms relate to each other
4. Primary germ layers of the embryo and their
   relationship to mature tissues
5. How and why tissues are prepared as stained
   histological sections; the three common
   planes of section; and some ways that tissues
   are prepared other than sectioning

5.2 Epithelial Tissue

1. Characteristics that distinguish epithelium
   from the other three primary tissue classes
2. Functions of the basement membrane and its
   relationship to an epithelium
3. Defining characteristics of a simple
   epithelium
4. Four types of simple epithelium and the
   appearance, functions, and representative
   locations of each
5. Defining characteristics of a stratified
   epithelium
6. Four types of stratified epithelium and the
   appearance, functions, and representative
   locations of each
7. Distinctions between keratinized and
   nonkeratinized stratified squamous
   epithelium, including differences in
   histology, locations, and functions
8. The special protective property and
   mechanism of urothelium
9. Epithelial exfoliation and its clinical
   relevance

5.3 Connective Tissue

1. Characteristics that distinguish connective
   tissue from the other three primary tissue
   classes
2. Functions of connective tissues

3. Cell types found in fibrous connective
   tissue, and the functions of each
4. Fiber types found in fibrous connective
   tissue, their composition, and the functions
   of each
5. The composition and variations in the
   ground substance of fibrous connective
   tissue
6. The appearance, functions, and locations of
   areolar, reticular, dense irregular, and dense
   regular connective tissue
7. The appearance, functions, and locations
   of adipose tissue, including the differences
   between white fat and brown fat
8. Defining characteristics of cartilage as
   a class; the three types of cartilage and
   how they differ in histology, function,
   and location; the relationship of the
   perichondrium to cartilage; and where
   perichondrium is absent
9. Defining characteristics of osseous tissue as
   a class; the distinction between spongy and
   compact bone; and the relationship of the
   periosteum to bone
10. The appearance of cross sections of compact
    bone
11. Why blood is classified as connective
    tissue; the term for its matrix; and the major
    categories of formed elements in blood

5.4 Nervous and Muscular Tissues—
Excitable Tissues

1. Why nervous and muscular tissues are called
   excitable tissues even though excitability is a
   property of all living cells
2. The two basic types of cells in nervous
   tissue and their functional differences
3. The general structure of neurons
4. Defining characteristics of muscular tissue
   as a class
5. Three types of muscle and how they differ in
   histology, function, and location

5.5 Cellular Junctions, Glands, and
Membranes

1. The general function of cellular junctions
2. Differences in the structure and function
   of tight junctions, desmosomes,
   hemidesmosomes, and gap junctions

3. The definition of a *gland* and the two basic
   functions of glands
4. The developmental, structural, and
   functional distinctions between exocrine
   and endocrine glands; examples of each;
   and why some glands cannot be strictly
   classified into one category or the other
5. Examples of unicellular glands in both the
   exocrine and endocrine categories
6. General histology of a typical exocrine gland
7. The scheme for classifying exocrine glands
   according to the anatomy of their duct
   systems and their distribution of secretory
   cells
8. Differences between serous, mucous, and
   mixed glands, and examples of each
9. Comparison of the mode of secretion of
   eccrine, apocrine, and holocrine glands
10. The variety of serous, mucous, and other
    membranes in the body, and names of some
    specialized membranes of the skin, blood
    vessels, and joints

5.6 Tissue Growth, Development, Repair,
and Degeneration

1. Differences between hyperplasia,
   hypertrophy, and neoplasia as normal and
   pathological modes of tissue growth
2. Differences between differentiation and
   metaplasia as modes of transformation from
   one tissue type to another
3. What stem cells are and how they relate
   to developmental plasticity; differences
   between embryonic and adult stem cells;
   and differences between totipotent,
   pluripotent, multipotent, and unipotent
   stem cells
4. Differences between regeneration and
   fibrosis as modes of tissue repair
5. Steps in the healing of a wound such as
   a cut in the skin
6. The general meaning of tissue *atrophy*
   and two forms or causes of atrophy
7. Differences between necrosis and
   apoptosis as modes of cell death and
   tissue shrinkage; some normal functions
   of apoptosis
8. Varieties of necrosis including infarction,
   dry gangrene, and gas gangrene

STUDY GUIDE

▶ Testing Your Recall

Answers in Appendix A

1. Urothelium is found in
   a. the urinary system.
   b. the respiratory system.
   c. the digestive system.
   d. the reproductive system.
   e. all of the above.

2. The external surface of the stomach is covered by
   a. a mucosa.
   b. a serosa.
   c. the parietal peritoneum.
   d. a lamina propria.
   e. a basement membrane.

3. Which of these is a primary germ layer?
   a. epidermis
   b. mucosa
   c. ectoderm
   d. endothelium
   e. epithelium

4. A seminiferous tubule of the testis is lined with _____ epithelium.
   a. simple cuboidal
   b. pseudostratified columnar ciliated
   c. stratified squamous
   d. transitional
   e. stratified cuboidal

5. _____ prevent fluids from seeping between epithelial cells.
   a. Glycosaminoglycans
   b. Hemidesmosomes
   c. Tight junctions
   d. Communicating junctions
   e. Basement membranes

6. A fixative serves to
   a. stop tissue decay.
   b. improve contrast.
   c. repair a damaged tissue.
   d. bind epithelial cells together.
   e. bind cardiac myocytes together.

7. The collagen of areolar tissue is produced by
   a. macrophages.
   b. fibroblasts.
   c. mast cells.
   d. leukocytes.
   e. chondrocytes.

8. Tendons are composed of _____ connective tissue.
   a. skeletal
   b. areolar
   c. dense irregular
   d. yellow elastic
   e. dense regular

9. The shape of the external ear is due to
   a. skeletal muscle.
   b. elastic cartilage.
   c. fibrocartilage.
   d. articular cartilage.
   e. hyaline cartilage.

10. The most abundant formed element(s) of blood is/are
    a. plasma.
    b. erythrocytes.
    c. platelets.
    d. leukocytes.
    e. proteins.

11. Any form of pathological tissue death is called _____.

12. The simple squamous epithelium that lines the peritoneal cavity is called _____.

13. Osteocytes and chondrocytes occupy little cavities called _____.

14. Muscle cells and axons are often called _____ because of their shape.

15. Tendons and ligaments are made mainly of the protein _____.

16. Of the three major categories of muscle, the only one that never has gap junctions is _____.

17. An epithelium rests on a layer called the _____ between its deepest cells and the underlying connective tissue.

18. Fibers and ground substance make up the _____ of a connective tissue.

19. A/An _____ adult stem cell can differentiate into two or more mature cell types.

20. Any epithelium in which every cell touches the basement membrane is called a/an _____ epithelium.

▶ Building Your Medical Vocabulary

Answers in Appendix A

State a meaning of each word element, and give a medical term from this chapter that uses it or a slight variation of it.

1. apo-

2. chondro-

3. ecto-

4. -gen

5. histo-

6. holo-

7. hyalo-

8. necro-

9. plas-

10. squamo-

STUDY GUIDE

▶ What's Wrong with These Statements?

Answers in Appendix A

Briefly explain why each of the following statements is false, or reword it to make it true.

1. The esophagus is protected from abrasion by a keratinized stratified squamous epithelium.

2. Only the basal cells of a pseudostratified columnar epithelium contact the basement membrane.

3. Skeletal muscle is defined by the fact that it is always attached to bones.

4. The secretions of a gland are produced by the cells of its stroma.

5. In all connective tissues, the matrix occupies more space than the cells do.

6. Adipocytes are limited to adipose tissue.

7. Tight junctions function primarily to prevent cells from pulling apart.

8. The development of mature tissue types from the immature tissues of a neonate (newborn) is called neoplasia.

9. Nerve and muscle cells are the body's only electrically excitable cells.

10. Cartilage is always covered by a fibrous perichondrium.

▶ Testing Your Comprehension

1. A woman in labor is often told to push. In doing so, is she consciously contracting her uterus to expel the baby? Justify your answer based on the muscular composition of the uterus.

2. A major tenet of the cell theory is that all bodily structure and function are based on cells. The structural properties of bone, cartilage, and tendons, however, are due more to their extracellular material than to their cells. Is this an exception to the cell theory? Why or why not?

3. When cartilage is compressed, water is squeezed out of it, and when pressure is taken off, water flows back into the matrix. This being the case, why do you think cartilage at weight-bearing joints such as the knees can degenerate from lack of exercise?

4. The epithelium of the respiratory tract is mostly of the pseudostratified columnar ciliated type, but in the alveoli—the tiny air sacs where oxygen and carbon dioxide are exchanged between the blood and inhaled air—the epithelium is simple squamous.

Explain the functional significance of this histological difference. That is, why don't the alveoli have the same kind of epithelium as the rest of the respiratory tract?

5. Which do you think would heal faster, cartilage or bone? Stratified squamous or simple columnar epithelium? Why?

CHAPTER

6

THE INTEGUMENTARY SYSTEM

A human hair emerging from its follicle (SEM)
SPL/Science Source

CHAPTER OUTLINE

6.1 Skin and Subcutaneous Tissue

  6.1a Functions of the Skin
  6.1b The Epidermis
  6.1c The Dermis
  6.1d The Hypodermis
  6.1e Skin Color
  6.1f Skin Markings

6.2 Hair and Nails

  6.2a Hair
  6.2b Nails

6.3 Cutaneous Glands

  6.3a Sweat Glands
  6.3b Sebaceous Glands
  6.3c Ceruminous Glands
  6.3d Mammary Glands

6.4 Skin Disorders

  6.4a Skin Cancer
  6.4b Burns

Connective Issues

Study Guide

DEEPER INSIGHTS

6.1 Clinical Application: Transdermal Absorption

6.2 Evolutionary Medicine: The Evolution of Skin Color

6.3 Clinical Application: UVA, UVB, and Sunscreens

6.4 Clinical Application: Skin Grafts and Artificial Skin

Anatomy & Physiology *Revealed*® 4.0

Module 4: Integumentary System

BRUSHING UP

- The epidermis of the skin is a keratinized stratified squamous epithelium, introduced in table 5.3.

- The dermis is composed of the areolar and dense irregular connective tissues described in tables 5.4 and 5.5. Its main protein is collagen, introduced in section 5.3b.

- The well-known durability of the epidermis depends in part on its desmosomes, depicted in figure 5.28b.

- Reviewing eccrine, apocrine, and holocrine gland types (see fig. 5.32) will help in understanding the types of sweat glands and oil glands of the skin in this chapter.

- Understanding some characteristics of the integumentary system, such as skin color and baldness, requires familiarity with dominant and recessive gene alleles (see section 4.4b).

We pay more daily attention to our skin, hair, and nails than to any other organ system. It is, after all, the most visible one, and its appearance strongly affects our social interactions. Few people venture out of the house without first looking in a mirror to see if their skin and hair are presentable. Social considerations aside, the integumentary system is important to one's self-image, and a positive self-image is important to the attitudes that promote overall good health. Care of the integumentary system is thus a particularly important part of the total plan of patient care.

Dermatology[1] is the branch of medicine specializing in the care and treatment of the skin. Even a general physical examination, however, should include inspection of the skin, hair, and nails. Their appearance can provide clues not only to their own health, but also to deeper disorders such as liver cancer, anemia, and heart failure. The skin also is the most vulnerable of our organs, exposed to radiation, trauma, infection, and injurious chemicals. Consequently, it needs and receives more medical attention than any other organ system.

6.1 Skin and Subcutaneous Tissue

Expected Learning Outcomes

When you have completed this section, you should be able to

a. list the functions of the skin and relate them to its structure;

b. describe the histological structure of the epidermis, dermis, and subcutaneous tissue;

c. describe the normal and pathological colors that the skin can have, and explain their causes; and

d. describe the common markings of the skin.

The **integumentary**[2] **system** consists of the skin, hair, nails, and their associated glands. The **skin (integument)** is the body's largest and heaviest organ. In adults, it covers an area of 1.5 to 2.0 m^2 and accounts for about 8% of the body weight. It consists of two layers: a stratified squamous epithelium called the *epidermis* and a deeper connective tissue layer called the *dermis* (**fig. 6.1).** Below the dermis is another connective tissue layer, the *hypodermis,* which is not part of the skin but is customarily studied in conjunction with it.

Most of the skin is 1 to 2 mm thick, but it ranges from less than 0.5 mm on the eyelids to 6 mm between the shoulder blades. The difference is due mainly to variation in thickness of the dermis, although skin is classified as thick or thin based on the relative thickness of the epidermis alone. **Thick skin** covers the palms, soles, and corresponding surfaces of the fingers and toes. Its epidermis alone is about 0.5 mm thick, due to a very thick surface layer of dead cells called the *stratum corneum* (see fig. 6.3). Thick skin has sweat glands but no hair follicles or sebaceous (oil) glands. The rest of the body is covered with **thin skin,** which has an epidermis about 0.1 mm thick, with a thin stratum corneum (see fig. 6.5). It possesses hair follicles, sebaceous glands, and sweat glands.

6.1a Functions of the Skin

The skin is much more than a container for the body. It has a variety of important functions that go well beyond appearance, as we shall see here.

1. **Resistance to trauma and infection.** The skin suffers the most physical injuries to the body, but it resists and recovers from trauma better than other organs do. The epidermal cells are packed with the tough protein **keratin** and linked by strong desmosomes that give this epithelium its durability. Few infectious organisms can penetrate the intact skin. Bacteria and fungi colonize the surface, but their numbers are kept in check by its relative dryness, its slight acidity (pH 4–6), and defensive antimicrobial peptides called *dermcidin* and *defensins.* The protective acidic film is called the *acid mantle.*

2. **Other barrier functions.** The skin is important as a barrier to water. It prevents the body from absorbing excess water when you are swimming or bathing, but even more importantly, it prevents the body from losing excess water. The epidermis is also a barrier to ultraviolet (UV) rays, blocking much of this cancer-causing radiation from reaching deeper tissue layers; and it is a barrier to many potentially harmful chemicals. It is, however, permeable to several drugs and poisons (see Deeper Insight 6.1).

3. **Vitamin D synthesis.** The skin carries out the first step in the synthesis of vitamin D, which is needed for bone development and maintenance. The liver and kidneys complete the process. This is further detailed in figure 7.16.

4. **Sensation.** The skin is our most extensive sense organ. It is equipped with a variety of nerve endings that react to heat,

---

[1]*dermat* = skin; *logy* = study of

[2]*integument* = covering

Epidermal ridge

Dermal papilla

Tactile corpuscle
(touch receptor)

Blood capillaries

Hair follicle

Sebaceous gland

Hair receptor

Apocrine sweat gland

Hair bulb

Sensory
nerve fibers

Arrector muscle

Lamellated (pacinian)
corpuscle (pressure receptor)

Hairs

Sweat pores

Epidermis

Dermis

Hypodermis
(subcutaneous fat)

Eccrine sweat
gland

Cutaneous blood
vessels

Motor nerve fibers

FIGURE 6.1 Structure of the Skin and Subcutaneous Tissue. The epidermis is peeled up at the upper left corner to show the dermal–epidermal boundary.

cold, touch, texture, pressure, vibration, and tissue injury (see section 16.2). These sensory receptors are especially abundant on the face, palms, fingers, soles, nipples, and genitals. There are relatively few on the back and in skin overlying joints such as the knees and elbows.

DEEPER INSIGHT 6.1

CLINICAL APPLICATION

Transdermal Absorption

The ability of the skin to absorb chemicals makes it possible to administer several medicines as ointments or lotions, or by means of adhesive patches that release the medicine steadily through a membrane. For example, inflammation can be treated with a hydrocortisone ointment, nitroglycerine patches are used to relieve heart pain, nicotine patches are used to help overcome tobacco addiction, and other medicated patches are used to control high blood pressure and motion sickness.

Unfortunately, the skin can also be a route for absorption of poisons. These include toxins from poison ivy and other plants; metals such as mercury, arsenic, and lead; and solvents such as acetone (nail polish remover) and paint thinner; and pesticides. Some of these can cause brain damage, liver failure, or kidney failure, which is good reason for using protective gloves when handling such substances.

5. **Thermoregulation.** The skin receives 10 times as much blood flow as it needs for its own maintenance, and is richly supplied with nerve endings called thermoreceptors, which monitor the body surface temperature. All of this relates to its great importance in regulating body temperature. In response to chilling, the body retains heat by constricting blood vessels of the dermis *(cutaneous vasoconstriction),* keeping warm blood deeper in the body. In response to overheating, it loses excess heat by dilating those vessels *(cutaneous vasodilation),* allowing more blood to flow close to the surface and lose heat through the skin. If this is insufficient to restore normal temperature, sweat glands secrete perspiration. The evaporation of sweat can have a powerful cooling effect. Thus, the skin plays roles in both warming and cooling the body.

6. **Nonverbal communication.** The skin is an important means of nonverbal communication. Humans, like most other primates, have much more expressive faces than other mammals **(fig. 6.2).** Complex skeletal muscles insert in the dermis and pull on the skin to create subtle and varied facial expressions. The general appearance of the skin, hair, and nails—whether the ravages of adolescent acne, the presence of a birthmark or scar, or just a "bad hair day"—is also important to social acceptance and to a person's self-image and emotional state.

(a)

(b)

FIGURE 6.2 Importance of the Skin in Nonverbal Expression. (a) Primates such as humans and this chimpanzee differ from other mammals in having very expressive faces due to facial muscles that insert on the collagen fibers of the dermis and move the skin. (b) Human facial expression, reflecting the actions of complex facial muscles surveyed in chapter 10.

a: DLILLC/Corbis/VCG/Getty Images; **b:** Joe DeGrandis/McGraw-Hill Education

6.1b The Epidermis

The **epidermis**[3] is a keratinized stratified squamous epithelium, as described in table 5.3. That is, its surface consists of dead cells packed with the tough protein keratin. Like other epithelia, the epidermis lacks blood vessels and depends on the diffusion of nutrients from the underlying connective tissue. It has sparse nerve endings for touch and pain, but most sensations of the skin are due to nerve endings in the dermis.

Cells of the Epidermis

The epidermis is composed of five types of cells **(fig. 6.3)**:

1. **Keratinocytes** (keh-RAT-ih-no-sites) are the great majority of epidermal cells. They are named for their role in synthesizing keratin. In ordinary histological specimens, nearly all visible epidermal cells are keratinocytes.

2. **Stem cells** are undifferentiated cells that divide and give rise to the keratinocytes. They are found only in the deepest layer of the epidermis, called the *stratum basale.*

3. **Melanocytes** also occur only in the stratum basale, amid the stem cells and deepest keratinocytes. They synthesize the brown to black pigment *melanin.* They have branching processes that spread among the keratinocytes and continually shed melanin-containing fragments *(melanosomes)* from their tips. The keratinocytes phagocytize these and gather the melanin granules on the "sunny side" of their nucleus. Like a parasol, the dark granules shield the DNA from ultraviolet rays.

4. **Tactile cells,** relatively few in number, are receptors for touch. They, too, are found in the basal layer of the epidermis and are associated with an underlying dermal nerve fiber. The tactile cell and its nerve fiber are collectively called a *tactile disc.*

5. **Dendritic**[4] **cells** are found in two layers of the epidermis called the *stratum spinosum* and *stratum granulosum* (described in the next section). They are immune cells that originate in the bone marrow but migrate to the epidermis and epithelia of the oral cavity, esophagus, and vagina. The epidermis has as many as 800 dendritic cells per square millimeter. They stand guard against toxins, microbes, and other pathogens that penetrate into the skin. When they detect such invaders, they alert the immune system so the body can defend itself (see section 21.1b).

Layers of the Epidermis

Cells of the epidermis are arranged in four to five zones, or strata (five in thick skin) (fig. 6.3). The following description progresses from deep to superficial, and from the youngest to the oldest keratinocytes.

1. The **stratum basale** (bah-SAY-lee) consists mainly of a single layer of cuboidal to low columnar stem cells and keratinocytes resting on the basement membrane. Scattered among these are the melanocytes, tactile cells, and stem cells. As stem cells divide, they give rise to keratinocytes that migrate toward the surface and replace lost epidermal cells.

2. The **stratum spinosum** (spy-NO-sum) consists of several layers of keratinocytes. In most places, this is the thickest stratum, but on the palms and soles it is usually exceeded by the stratum corneum. The deepest cells of the stratum spinosum continue dividing, but as they're pushed farther upward, they cease. Instead, they produce more and more keratin filaments, which cause the cells to flatten. Therefore, the higher up you look in the stratum spinosum, the flatter the cells appear. Dendritic cells are also found throughout the stratum spinosum, but are not usually identifiable in routinely stained tissue sections.

---

[3]*epi* = above, upon; *derm* = skin

[4]*dendr* = tree, branch

Stratum corneum

Stratum lucidum

Stratum granulosum

Stratum spinosum

Stratum basale

Dermis

(a)

Exfoliating dead cells

Sweat pore

Sweat duct

Dead keratinocytes

Living keratinocytes

Dendritic cell

Tight junctions

Tactile cell

Tactile nerve fiber

Stem cell in mitosis

Melanocyte

Melanized keratinocytes

Dermal blood vessel

Dermal papilla

(b)

FIGURE 6.3 Layers and Cell Types of the Epidermis. (a) Photo. (b) Drawing with added features of skin not visible in the photo. Gaps between the keratinocytes are an artifact created by cell shrinkage during fixation, but are typically seen in microscope slides of skin. The drawing accentuates the ladderlike tight junctions that bridge these gaps. **A&PR**

a: Biophoto Associates/Science Source

The stratum spinosum is named for an artificial appearance *(artifact)* created by the histological fixation of skin specimens. Keratinocytes are firmly attached to each other by numerous desmosomes, which partly account for the toughness of the epidermis. Histological fixatives shrink the keratinocytes so they pull away from each other, but they remain attached by the desmosomes—like two people holding hands while they step farther apart. The desmosomes thus create bridges from cell to cell, giving each cell a spiny appearance from which we derive the word *spinosum.* Keratinocytes are also bound to each other by tight junctions, which make an essential contribution to water retention by the skin.

3. The **stratum granulosum** consists of three to five layers of flat keratinocytes—more in thick skin than in thin skin. The keratinocytes of this layer contain coarse, dark-staining *keratohyalin granules* that give the layer its name.

4. The **stratum lucidum**[5] (LOO-sih-dum) is a thin zone superficial to the stratum granulosum, seen only in thick skin.

Here, the keratinocytes are densely packed with a clear protein named *eleidin* (ee-LEE-ih-din). The cells have no nuclei or other organelles. This zone has a pale, featureless appearance with indistinct cell boundaries.

5. The **stratum corneum** consists of up to 30 layers of dead, scaly, keratinized cells that form a durable surface layer. It is especially resistant to abrasion, penetration, and water loss.

The Life History of a Keratinocyte

Dead cells constantly flake off the skin surface. Because we constantly lose these epidermal cells, they must be continually replaced. Keratinocytes are produced deep in the epidermis by the mitosis of stem cells in the stratum basale. Some of the deepest keratinocytes in the stratum spinosum also continue dividing. Mitosis requires an abundant supply of oxygen and nutrients, which these deep cells acquire from the blood vessels in the nearby dermis. Once the epidermal cells migrate more than two or three cells away from the dermis, their mitosis ceases. Mitosis is seldom seen in prepared slides of the skin, because it occurs mainly at night and most tissue specimens are taken during the day.

[5]*stratum* = layer; *lucid* = light, clear

As new keratinocytes form, they push the older ones toward the surface. In 30 to 40 days, a keratinocyte makes its way to the surface and flakes off. This migration is slower in old age and faster in skin that has been injured or stressed. Injured epidermis regenerates more rapidly than any other tissue in the body. Mechanical stress from manual labor or tight shoes accelerates keratinocyte multiplication and results in *calluses* or *corns,* thick accumulations of dead keratinocytes on the hands or feet.

As keratinocytes are shoved upward by the dividing cells below, they flatten and produce more keratin filaments and lipid-filled **lamellar granules.** In the stratum granulosum, four important developments occur: (1) Keratohyalin granules release a protein called *filaggrin* that binds the cytoskeletal keratin filaments together into coarse, tough bundles. (2) The cells produce a tough layer of *envelope proteins* just beneath the plasma membrane, resulting in a nearly indestructible protein sac around the keratin bundles. (3) Lamellar granules release a lipid mixture that spreads out over the cell surface and waterproofs it. (4) Finally, as these barriers cut the keratinocytes off from the supply of nutrients from below, their organelles degenerate and the cells die, leaving just the tough waterproof sac enclosing coarse bundles of keratin. These processes, along with the tight junctions between keratinocytes, result in an **epidermal water barrier** that is crucial to the retention of body water.

The stratum corneum consists of compact layers of dead keratinocytes and keratinocyte fragments. The dead cells soon **exfoliate** (flake off) from the epidermal surface as specks called **dander.** Dander floats around as tiny white specks in the air, settles on household surfaces, and forms much of the house dust that accumulates there. *Dandruff* is composed of clumps of dander stuck together by sebum (oil).

A curious effect of the epidermal water barrier is the way our skin wrinkles when we linger in the bath or a lake. The keratin of the stratum corneum absorbs water and swells, but the deeper layers of the skin do not. The thickening of the stratum corneum forces it to wrinkle. This is especially conspicuous on the fingers and toes ("prune fingers") because they have such a thick stratum corneum and they lack the sebaceous glands that produce water-resistant oil elsewhere on the body. There may be more to the story than this, however, because the wrinkles do not form when the nerves to the fingers are severed, indicating some role for the nervous system. It has recently been hypothesized that this buckling of the skin may serve a function similar to the tread on a car tire, to channel water away and improve our grip when we press our fingertips to wet surfaces.

6.1c The Dermis

Beneath the epidermis is a connective tissue layer, the **dermis.** It ranges from 0.2 mm thick in the eyelids to about 4 mm thick in the palms and soles. It is composed mainly of collagen, but also contains elastic and reticular fibers, fibroblasts, and the other cells typical of fibrous connective tissue (described in section 5.3b). It is well supplied with blood vessels, cutaneous glands, and nerve endings. The hair follicles and nail roots are embedded in the dermis. In the face, skeletal muscles attach to dermal collagen fibers and produce such expressions as a smile, a wrinkle of the forehead, or the lifting of an eyebrow (see fig. 6.2).

The boundary between the epidermis and dermis is histologically conspicuous and usually wavy. The upward waves are fingerlike extensions of the dermis called **dermal papillae**[6] (see fig. 6.3), and the downward epidermal waves between the papillae are called **epidermal ridges.** The dermal and epidermal boundaries thus interlock like corrugated cardboard, an arrangement that resists slippage of the epidermis across the dermis. If you look closely at your hand and wrist, you will see delicate furrows that divide the skin into tiny rectangular to rhomboidal areas. The dermal papillae produce the raised areas between the furrows. On the fingertips, this wavy boundary forms the *friction ridges* that produce fingerprints. In highly sensitive areas such as the lips and genitals, exceptionally tall dermal papillae allow blood capillaries and nerve fibers to reach close to the surface. This imparts a redder color and more sensitivity to touch in such areas.

▶▶▶APPLY WHAT YOU KNOW

Dermal papillae are relatively high and numerous in palmar and plantar skin but low and few in number in the skin of the face and abdomen. What do you think is the functional significance of this difference?

There are two zones of dermis called the papillary and reticular layers **(fig. 6.4).** The **papillary layer** (PAP-ih-lerr-ee) is a thin zone of areolar tissue in and near the dermal papillae. This loosely organized tissue allows for mobility of leukocytes and other defenses against organisms introduced through breaks in the epidermis. This layer is especially rich in small blood vessels.

The **reticular**[7] **layer** of the dermis is deeper and much thicker. It consists of dense irregular connective tissue; it is much more fibrous than cellular, and thus tougher than the papillary layer. The boundary between the papillary and reticular layers is often vague. In the reticular layer, the collagen forms thicker bundles with less room for ground substance, and there are often small clusters of adipocytes. Stretching of the skin in obesity and pregnancy can tear the collagen fibers and produce *striae* (STRY-ee), or stretch marks. These occur especially in areas most stretched by weight gain: the thighs, buttocks, abdomen, and breasts.

There are extensive plexuses of blood vessels at the dermal–epidermal boundary, in mid-dermis, and between the dermis and hypodermis. When dermal blood vessels are damaged by such causes as burns and friction from tight shoes, serous fluid can seep out of the vessels and accumulate as a **blister,** separating the epidermis from the dermis until the fluid is either reabsorbed or expelled by rupture of the blister.

6.1d The Hypodermis

Beneath the skin is a layer called the **hypodermis**[8] or **subcutaneous tissue** (see fig. 6.1). The boundary between the dermis and hypodermis is indistinct, but the hypodermis generally has more areolar and adipose tissue. It pads the body and binds the skin to the underlying tissues. Drugs are introduced into the hypodermis

---

[6]*pap* = nipple; *illa* = little
[7]*reti* = network; *cul* = little
[8]*hypo* = below; *derm* = skin

(a)

(b) Papillary layer of dermis

(c) Reticular layer of dermis

FIGURE 6.4 The Dermis. (a) Light micrograph (LM) of axillary skin, with the epidermis stained pink and dermal collagen stained blue. (b) The papillary layer, composed of loosely organized, random-seeming bundles of collagen fibers; open fluid-filled space; and numerous small blood vessels that nourish both dermis and epidermis. (c) The reticular layer, composed of coarser, tougher bundles of collagen and less open space. Parts (b) and (c) are scanning electron micrographs (SEMs).

a: Dennis Strete/McGraw-Hill Education; b: David M. Phillips/ Science Source; c: Susumu Nishinaga/Science Source

by injection because the subcutaneous tissue is highly vascular and absorbs them quickly.

Subcutaneous fat is hypodermis composed predominantly of adipose tissue. It serves as an energy reservoir and thermal insulation, and it is compressible and protects deeper tissue by absorbing pressure and blows to the body. It is not uniformly distributed; for example, it is virtually absent from the scalp but relatively abundant in the abdomen, hips, thighs, and female breasts. The subcutaneous fat averages about 8% thicker in women than in men, and varies with age. Infants and elderly people have less subcutaneous fat than other people and are therefore more sensitive to cold.

Table 6.1 summarizes the layers of the skin and hypodermis, from superficial to deep.

6.1e Skin Color

The most significant factor in skin color is **melanin.** This is produced by the melanocytes but accumulates in the keratinocytes of the stratum basale and stratum spinosum (**fig. 6.5**). There are two forms of melanin: a brownish black **eumelanin**[9] and a reddish yellow sulfur-containing pigment, **pheomelanin.**[10] People of different skin colors have essentially the same number of melanocytes, but in dark skin, the melanocytes produce greater quantities of melanin, the melanin granules in the keratinocytes

[9] *eu* = true; *melan* = black; *in* = substance
[10] *pheo* = dusky; *melan* = black; *in* = substance

TABLE 6.1	Stratification of the Skin and Hypodermis
Layer	**Description**
Epidermis	Keratinized stratified squamous epithelium
Stratum corneum	Dead, keratinized cells of the skin surface
Stratum lucidum	Clear, featureless, narrow zone seen only in thick skin
Stratum granulosum	Two to five layers of cells with dark-staining keratohyalin granules; scanty in thin skin
Stratum spinosum	Many layers of keratinocytes, typically shrunken in fixed tissues but attached to each other by desmosomes, which give them a spiny look; progressively flattened the farther they are from the dermis. Dendritic cells are abundant here but are not distinguishable in routinely stained preparations.
Stratum basale	Single layer of cuboidal to columnar cells resting on basement membrane; site of most mitosis; consists of stem cells, keratinocytes, melanocytes, and tactile cells, but these are difficult to distinguish with routine stains. Melanin is conspicuous in keratinocytes of this layer in black to brown skin.
Dermis	Fibrous connective tissue, richly endowed with blood vessels and nerve endings. Sweat glands and hair follicles originate here and in hypodermis.
Papillary layer	Superficial one-fifth of dermis; composed of areolar tissue; often extends upward as dermal papillae
Reticular layer	Deeper four-fifths of dermis; dense irregular connective tissue
Hypodermis	Areolar or adipose tissue between skin and muscle

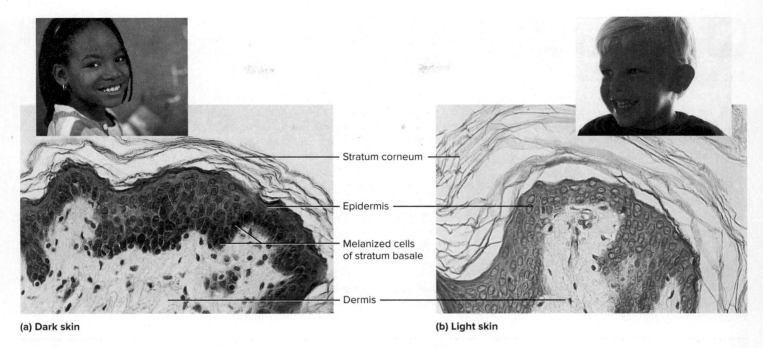

(a) Dark skin   **(b) Light skin**

FIGURE 6.5 Variations in Skin Pigmentation. (a) The stratum basale shows heavy deposits of melanin in dark skin. (b) Light skin shows little to no visible melanin.

❓ *Which of the five types of epidermal cells are the melanized cells in part (a)?*

a: Dennis Strete/McGraw-Hill Education; **(girl):** Tom & Dee Ann McCarthy/Corbis/Getty Images Plus; b: ©Dennis Strete/McGraw-Hill Education; **(boy):** Arthur Tilley/Stockbyte/Getty Images

are more spread out than tightly clumped, and the melanin breaks down more slowly. Thus, melanized cells may be seen throughout the epidermis, from stratum basale to stratum corneum. In light skin, the melanin is clumped near the keratinocyte nucleus, so it imparts less color to the cells. It also breaks down more rapidly, so little of it is seen beyond the stratum basale, if even there.

Skin color also varies with exposure to the UV rays of sunlight, which stimulate melanin synthesis and darken the skin. A suntan fades as melanin is degraded in older keratinocytes and as keratinocytes migrate to the surface and exfoliate. The amount of melanin also varies from place to place on the body. It is relatively concentrated in freckles and moles, on the dorsal surfaces of the hands and feet as compared with the palms and soles, on the nipple and surrounding area (areola) of the breast, around the anus, on the scrotum and penis, and on the lateral surfaces of the female genital folds (labia majora). The contrast between heavily melanized and lightly melanized regions of the skin is more pronounced in some people than others, but it exists to some extent in nearly everyone. Variation in ancestral exposure to UV is the primary reason for the geographic and ethnic variation in skin color today (see Deeper Insight 6.2).

Other factors in skin color are hemoglobin and carotene. **Hemoglobin,** the red pigment of blood, imparts reddish to pinkish hues as blood vessels show through the skin. Its color is lightened by the white of the dermal collagen. The skin is redder in places such as the lips, where blood capillaries come closer to the surface and the hemoglobin shows through more vividly.

Carotene[11] is a yellow pigment acquired from egg yolks and yellow and orange vegetables. Depending on the diet, carotene or related compounds can become concentrated to various degrees in the stratum corneum and subcutaneous fat, imparting a yellow color. This is often most conspicuous in skin of the heel and in calluses of the feet, because this is where the stratum corneum is thickest.

The skin may also exhibit abnormal colors of diagnostic value:

- **Cyanosis**[12] is blueness of the skin resulting from a deficiency of oxygen in the circulating blood. Oxygen deficiency turns the hemoglobin a reddish violet color, which is lightened to blue-violet as it shows through the white dermal collagen. Oxygen deficiency can result from conditions that prevent the blood from picking up a normal load of oxygen in the lungs, such as airway obstructions in drowning and choking, lung diseases such as emphysema, and respiratory arrest. Cyanosis also occurs in situations such as cold weather and cardiac arrest, when blood flows so slowly through the skin that the tissues consume its oxygen faster than freshly oxygenated blood arrives.

- **Erythema**[13] (ERR-ih-THEE-muh) is abnormal redness of the skin. It occurs in such situations as exercise, hot weather,

---

[11]*carot* = carrot
[12]*cyan* = blue; *osis* = condition
[13]*eryth* = red; *em* = blood

DEEPER INSIGHT 6.2

EVOLUTIONARY MEDICINE

The Evolution of Skin Color

One of the most conspicuous signs of human variation is skin color, which can range from the color of espresso or milk chocolate to café au lait or light peach. Such variation results from a combination of evolutionary selection pressures, especially differences in exposure to ultraviolet (UV) radiation. Environmental UV levels account for up to 77% of the variation in human skin color.

UV can have two adverse effects: It causes skin cancer and it breaks down folate, a B vitamin needed for normal cell division, fertility, and fetal development. It also has a desirable effect: It stimulates keratinocytes to synthesize vitamin D, which is needed for the absorption of dietary calcium and thus for healthy bone development. Too much UV raises the risk of cancer, infertility, and fetal deformities such as spina bifida; too little and one is at risk of bone deformities such as rickets. Consequently, populations native to the tropics and people descended from them tend to have well-melanized skin to screen out excessive UV. Populations native to far northern and southern latitudes, where the sunlight is weak, tend to have light skin to allow for

adequate UV penetration. Ancestral skin color is thus partly a compromise between vitamin D and folate requirements. Worldwide, women have skin averaging about 4% lighter than men do, perhaps because of their greater need for vitamin D and calcium to support pregnancy and lactation.

But for multiple reasons, there are exceptions to this trend. UV exposure is determined by more than latitude. It increases at higher elevations and in dry air, because the thinner, drier atmosphere filters out less UV. This helps to explain the darker skin of people indigenous to such localities as deserts, the Andes Mountains, and the high plateaus of Tibet and Ethiopia. Some other exceptions may be the result of human migrations from one latitude to another occurring too recently for their skin color to have adapted to the new level of UV exposure. Variation may also result from cultural differences in clothing and shelter, intermarriage among people of different geographic ancestries, and darwinian sexual selection—a preference in mate choice for partners of light or dark complexion, thereby perpetuating and accentuating ethnic differences in color.

---

sunburn, anger, and embarrassment. Erythema is caused by increased blood flow in dilated cutaneous blood vessels or by dermal pooling of red blood cells that have escaped from abnormally permeable capillaries, as in sunburn.

- **Pallor** is a pale or ashen color that occurs when there is so little blood flow through the skin that the white of the dermal collagen shows through. It can result from emotional stress, low blood pressure, circulatory shock, cold temperatures, or severe anemia.

- **Albinism**[14] **(fig. 6.6a)** is a genetic lack of melanin that usually results in milky white hair and skin, and blue-gray eyes. Melanin is synthesized from the amino acid tyrosine by the enzyme tyrosinase. People with albinism have inherited a recessive, nonfunctional tyrosinase allele from both parents.

- **Jaundice**[15] **(fig. 6.6b)** is yellowing of the skin and whites of the eyes resulting from high levels of bilirubin in the blood. Bilirubin is a hemoglobin breakdown product. When erythrocytes get old, they disintegrate and release their hemoglobin. The liver and spleen convert hemoglobin to bilirubin, which the liver excretes in the bile. Bilirubin can accumulate enough to discolor the skin, however, in such situations as a rapid rate of erythrocyte destruction; when diseases such as cancer, hepatitis, and cirrhosis compromise liver function; and in premature infants, where the liver is not well enough developed to dispose of bilirubin efficiently.

- A **hematoma**,[16] or bruise, is a mass of clotted blood showing through the skin. It is usually due to accidental trauma (blows to the skin), but it may indicate hemophilia, other metabolic or nutritional disorders, or physical abuse.

▶▶▶**APPLY WHAT YOU KNOW**

An infant brought to a clinic shows abnormally yellow skin. What sign could you look for to help decide whether this was due to jaundice or to a large amount of carotene from strained vegetables in the diet?

6.1f Skin Markings

The skin is marked by many lines, creases, ridges, and patches of accentuated pigmentation. **Friction ridges** are the markings on the fingertips that leave distinctive oily fingerprints on surfaces we touch. They are characteristic of most primates, though their function has long been obscure. When we stroke an uneven surface, friction ridges enhance fingertip sensitivity to texture by vibrating and stimulating sense organs called *lamellar corpuscles* deeper in the skin. They are also thought to improve one's grasp and aid in the manipulation of small and rough-surfaced objects. Friction ridges form during fetal development and remain essentially unchanged for life. Their patterns result from a combination of heredity and the surfaces that the fetus randomly touches with its fingertips before birth. Thus, everyone has a unique pattern of friction ridges; not even identical twins produce identical fingerprints.

Flexion lines (flexion creases) are the lines on the flexor surfaces of the digits, palms, wrists, elbows, and other places (see atlas B, fig. B.19). They mark sites where the skin folds during flexion of the joints. The skin is tightly bound to deeper connective tissues along these lines.

Freckles and moles are tan to black aggregations of melanized keratinocytes. **Freckles** are flat patches that vary with heredity and exposure to the sun. A **mole (nevus)** is an elevated patch of melanized skin, often with hair. Moles are harmless and sometimes even regarded as "beauty marks," but they should be watched for changes in color, diameter, or contour that may suggest malignancy (skin cancer) (see section 6.4a).

---

[14]*alb* = white; *ism* = state, condition
[15]*jaun* = yellow
[16]*hemat* = blood; *oma* = mass

(a)   **(b)**

FIGURE 6.6 Two Abnormal Skin Colors.
(a) Albinism. (b) Jaundice.

a: Jonathan Knowles/Getty Images; b: Dr P. Marazzi/Science Source

Birthmarks, or **hemangiomas,**[17] are patches of skin discolored by benign tumors of the blood capillaries. *Capillary hemangiomas* (strawberry birthmarks) usually develop about a month after birth. They become bright red to deep purple and develop small capillary-dense elevations that give them a strawberry-like appearance. About 90% of capillary hemangiomas disappear by the age of 5 or 6 years. *Cavernous hemangiomas* are flatter and duller in color. They are present at birth, enlarge up to 1 year of age, and then regress. About 90% disappear by the age of 9 years. A *port-wine stain* is flat and pinkish to dark purple in color. It can be quite large and remains for life.

BEFORE YOU GO ON

Answer the following questions to test your understanding of the preceding section:

1. What is the major histological difference between thick and thin skin? Where on the body is each type of skin found?

2. How does the skin help to adjust body temperature?

3. List the five cell types of the epidermis. Describe their locations and functions.

4. List the five layers of epidermis from deep to superficial. What are the distinctive features of each layer? Which layer is often absent?

5. What are the two layers of the dermis? What type of tissue composes each layer?

6. Name the pigments responsible for normal skin colors and explain how certain conditions can produce discolorations of the skin.

---

[17]*hem* = blood; *angi* = vessels; *oma* = mass, tumor

6.2 Hair and Nails

Expected Learning Outcomes

When you have completed this section, you should be able to

a. distinguish between three types of hair;

b. describe the histology of a hair and its follicle;

c. discuss some theories of the purposes served by various kinds of hair; and

d. describe the structure and function of nails.

The hair, nails, and cutaneous glands are the **accessory organs (appendages)** of the skin. Hair and nails are composed mostly of dead, keratinized cells. The stratum corneum of the skin is made of pliable **soft keratin,** but the hair and nails are composed mostly of **hard keratin.** This is more compact than soft keratin and is toughened by numerous cross-linkages between the keratin molecules.

6.2a Hair

A hair is also known as a **pilus** (PY-lus); in the plural, *pili* (PY-lye). It is a slender filament of keratinized cells that grows from an oblique tube in the skin called a **hair follicle (fig. 6.7).**

Distribution and Types

Hair occurs almost everywhere on the body except palms and soles; palmar, plantar, and lateral surfaces and distal segments of the fingers and toes; and the lips, nipples, and parts of the genitals. The limbs and trunk have about 55 to 70 hairs per square centimeter, and the face has about 10 times as many.

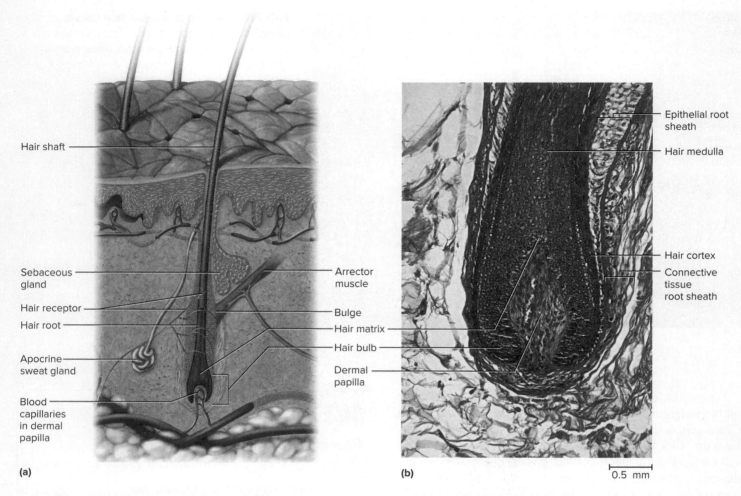

(a)

(b)

0.5 mm

FIGURE 6.7 Structure of a Hair and Its Follicle. (a) Anatomy of the follicle and associated structures. (b) Light micrograph of the base of a hair follicle. **APR**

b: Ed Reschke/Getty Images

There are about 30,000 hairs in a man's beard and about 100,000 hairs in the average person's scalp. The density of hair doesn't differ much from one person to another or even between the sexes; indeed, it is virtually the same in humans, chimpanzees, and gorillas. Differences in apparent hairiness are due mainly to differences in texture and pigmentation.

Not all hair is alike, even on one person. Over the course of our lives, we grow three kinds of hair: downy, vellus, and terminal hair. **Downy hair (lanugo[18])** is fine, unpigmented hair that appears on the fetus in the last 3 months of development. By the time of birth, most of it is replaced by similar fine, pale hair called **vellus hair.**[19] Vellus hair constitutes about two-thirds of the hair of women, one-tenth of the hair of men, and all the hair of children except for the eyebrows, eyelashes, and hair of the scalp. **Terminal hair** is longer, coarser, and usually more heavily pigmented. It forms the eyebrows and eyelashes and covers the scalp; after puberty, it also forms the axillary and pubic hair, all facial hair, and some of the hair on the trunk and limbs.

---

[18]*lan* = down, wool
[19]*vellus* = fleece

Structure of the Hair and Follicle

The portion of a hair above the skin is called the **shaft,** and all that beneath the surface is the **root.** The root penetrates deeply into the dermis or hypodermis and ends with a dilation called the **bulb.** The only living cells of a hair are in and near the bulb. The bulb grows around a bud of vascular connective tissue called the **dermal papilla,** which provides the hair with its sole source of nutrition. Immediately above the papilla is a region of mitotically active cells, the **hair matrix,** which is the hair's growth center. All cells higher up are dead.

In cross section, a hair reveals up to three layers. From the inside out, these are the medulla, cortex, and cuticle. The **medulla** is a core of loosely arranged cells and air spaces. It is most prominent in thick hairs such as those of the eyebrows, but narrower in hairs of medium thickness and absent from the thinnest hairs of the scalp and elsewhere. The **cortex** constitutes most of the bulk of a hair. It consists of several layers of elongated keratinized cells that appear cuboidal to flattened in cross sections. The **cuticle** is composed of multiple layers of thin, scaly cells that overlap each other like roof shingles with their free edges directed upward (see this chapter's opening photo). Cells lining the follicle are like shingles facing in the opposite direction. They interlock with the scales of the hair cuticle and resist pulling on the hair. When a hair is pulled out, this layer of follicle cells comes with it. The scaly cuticle also helps keep the hairs separated so they don't become matted together.

The follicle is a diagonal tube that contains the hair root. It has two principal layers: an **epithelial root sheath** and a **connective tissue root sheath.** The epithelial root sheath is an extension of the epidermis; it consists of stratified squamous epithelium and lies immediately adjacent to the hair root. Toward the deep end of the follicle, it widens to form a **bulge,** a source of stem cells for follicle growth. The connective tissue root sheath, which is derived from the dermis and composed of collagenous connective tissue, surrounds the epithelial sheath and is somewhat denser than the adjacent dermis.

Associated with the follicle are nerve and muscle fibers. Nerve fibers called **hair receptors** entwine each follicle and respond to hair movements. You can feel their effect by carefully moving a single hair with a pin or by lightly running your finger over the hairs of your forearm without touching the skin. Each hair has an **arrector muscle**—also known as a *pilomotor muscle* or *arrector pili*[20]—a bundle of smooth muscle cells extending from dermal collagen fibers to the connective tissue root sheath of the follicle

---

[20]*arrect* = erect; *pili* = of hair

(see figs. 6.1 and 6.7a). In response to cold, fear, touch, or other stimuli, the sympathetic nervous system stimulates the arrector to contract, making the hair stand on end and wrinkling the skin in such areas as the scrotum and areola. In other mammals, piloerection traps an insulating layer of warm air next to the skin or makes the animal appear larger and less vulnerable to a potential enemy. In humans, it pulls the follicles into a vertical position and causes "goose bumps," but serves no useful purpose.

Hair Texture and Color

The texture of hair is related to differences in cross-sectional shape **(fig. 6.8)**—straight hair is round, wavy hair is oval, and tightly curled hair is relatively flat. Hair color is due to pigment granules in the cells of the cortex. Brown and black hair are rich in eumelanin. Red hair has a slight amount of eumelanin but a high concentration of pheomelanin. Blond hair has an intermediate amount of pheomelanin but very little eumelanin. Gray and white hair result from a scarcity or absence of melanins in the cortex and the presence of air in the medulla.

Hair Growth and Loss

Each hair follicle goes through three developmental phases called the **hair cycle.** At any given time, about 90% of the scalp follicles

FIGURE 6.8 The Basis of Hair Color and Texture. Note the round cross-sectional shape of straight hair (a, b) compared to the flatter shape of wavy hair (c, d). Also note the proportions of pigments in the cortex in relation to hair color.

Which of the hair layers illustrated here corresponds to the scales seen in this chapter's opening photo?

a–d: Joe DeGrandis/McGraw-Hill Education

are in a growth stage called **anagen.**[21] Stem cells in the hair bulge multiply, the follicle deepens, and the hair grows in length during this time. Anagen lasts about 6 to 8 years for any given follicle. This is followed by **catagen,**[22] in which mitosis ceases, the follicle shrinks, and the hair dies and loses its anchorage. The hair becomes a **club hair,** named for a hard keratinized knot at its base. Club hairs are easily pulled out by brushing the hair, and the hard club can be felt at the end. Catagen lasts for about 2 to 3 weeks, then is followed by a **telogen**[23] stage, in which the follicle rests for 1 to 3 months. It then begins the process anew, starting over with anagen and the growth of a new hair, often alongside the old club hair. We lose 50 to 100 scalp hairs daily, but fortunately, the follicles are asynchronous, so we don't lose all our hairs at once.

Scalp hairs grow fastest from adolescence until the 40s, at a rate of about 1 mm per 3 days (10–18 cm/year). After that, an increasing percentage of follicles are in the catagen and telogen phases rather than the growing anagen phase. Follicles also shrink and begin producing wispy vellus hairs instead of thicker terminal hairs. Thinning of the hair, or baldness, is called **alopecia**[24] (AL-oh-PEE-she-uh). It occurs to some degree in both sexes and may be worsened by disease, poor nutrition, fever, emotional stress, radiation, or chemotherapy. In the great majority of cases, however, it is simply a matter of aging.

Pattern baldness is the condition in which hair is lost unevenly across the scalp rather than thinning uniformly. It results from a combination of genetic and hormonal influences. The relevant gene has two alleles: one for uniform hair growth and a baldness allele for patchy hair growth. The baldness allele is dominant in males and is expressed only in the presence of the high level of testosterone characteristic of men. In men who are either heterozygous or homozygous for the baldness allele, testosterone causes terminal hair to be replaced by vellus hair, beginning on top of the head and later the sides. In women, the baldness allele is recessive. Homozygous dominant and heterozygous women show normal hair distribution; only homozygous recessive women are at risk of pattern baldness. Even then, they exhibit the trait only if their testosterone levels are abnormally high for a woman (for example, because of a tumor of the adrenal gland, a woman's principal source of testosterone). Such characteristics in which an allele is dominant in one sex and recessive in the other are called *sex-influenced traits.*

Excessive or undesirable hairiness in areas that are not usually hairy, especially in women and children, is called **hirsutism.**[25] It tends to run in families and usually results from either masculinizing ovarian tumors or hypersecretion of testosterone by the adrenal cortex. It is often associated with menopause.

Contrary to popular misconceptions, hair and nails don't continue to grow after a person dies, cutting hair doesn't make it grow faster or thicker, and emotional stress cannot make the hair turn white overnight.

[21]*ana* = up; *gen* = build, produce
[22]*cata* = down
[23]*telo* = end
[24]*alopecia* = fox mange
[25]*hirsut* = shaggy

Functions of Hair

Compared with other mammals, the relative hairlessness of humans is so unusual that it raises the question, Why do we have any hair at all? What purpose does it serve? There are different answers for the different types of hair; furthermore, some of the answers would make little sense if we limited our frame of reference to industrialized societies, where barbers and hairdressers are engaged to alter the natural state of the hair. It is more useful to take a comparative approach to this question and consider the purposes hair serves in other species of mammals.

Most hair of the human trunk and limbs is probably best interpreted as vestigial, with little present function. Body hair undoubtedly served to keep our ancestors warm, but in modern humans it is too scanty. Stimulation of the hair receptors, however, alerts people to parasites crawling on the skin, such as fleas and ticks. Thus, we are less likely to become unknowingly infested with parasites.

The scalp is normally the only place where the hair is thick enough to retain heat. Heat loss from a bald scalp can be substantial and quite uncomfortable. The brain receives a rich supply of warm blood, and most of the scalp lacks an insulating fat layer. Heat is easily conducted through the bones of the skull and lost to the surrounding air. In addition, without hair there is nothing to break the wind and stop it from carrying away heat. Hair also protects the scalp from sunburn, since the scalp is otherwise most directly exposed to the sun's rays. These may be the reasons humans have retained thick hair on their heads while losing most of it from the rest of the body.

Tufts and patches of hair, sometimes with contrasting colors, are important among mammals for advertising species, age, sex, and individual identity. For the less groomed members of the human species, scalp and facial hair may play a similar role. The indefinitely growing hair of a man's scalp and beard, for example, could provide a striking contrast to a face that is otherwise almost hairless. It creates a badge of recognition instantly visible at a distance.

The beard and pubic and axillary hair signify sexual maturity and aid in the transmission of sexual scents. We will further reflect on this in a later discussion of apocrine sweat glands, whose distribution and function add significant evidence to support this theory.

Stout protective **guard hairs,** or **vibrissae** (vy-BRISS-ee), guard the nostrils and ear canals and prevent foreign particles from entering easily. The eyelashes and blink reflex shield the eye from dust and other windblown debris. The eyelashes of humans and other mammals also seem to be perfectly adapted for reducing airflow across the eye surface, thus reducing evaporation and dryness. In windy or rainy conditions, we can squint so that the eyelashes protect the eyes even further, but without completely obstructing our vision.

The eyebrows are often presumed to keep sweat or debris out of the eyes, but this seems negligible. More likely, they function mainly to enhance facial expression. Movements of the eyebrows are an important means of nonverbal communication in humans of all cultures, and we even have special *frontalis* muscles for this purpose. Eyebrow expressiveness is not unique to humans; even monkeys and apes use quick eyebrow flashes to greet each other, assert their dominance, and break up quarrels.

6.2b Nails

Fingernails and toenails are clear, hard derivatives of the stratum corneum. They are composed of thin, dead, scaly cells, densely packed together and filled with parallel fibers of hard keratin. Most mammals have claws, whereas flat nails are one of the distinguishing characteristics of humans and other primates. Flat nails serve as strong keratinized "tools" that can be used for grooming, picking apart food, and other manipulations. In addition, they allow for more fleshy and sensitive fingertips. Imagine touching a few grains of salt on a table. The nail, by providing a counterforce or resistance from the other side, enhances one's sensitivity to such tiny objects.

The hard part of the nail is the **nail plate (fig. 6.9),** which includes the **free edge** overhanging the tip of the finger or toe; the **nail body,** which is the visible attached part of the nail; and the **nail root,** which extends proximally under the overlying skin. The surrounding skin rises a bit above the nail as a **nail fold,** separated from the margin of the nail plate by a **nail groove.** The groove and the space under the free edge accumulate dirt and bacteria and require special attention when scrubbing for duty in an operating room or nursery.

The skin underlying the nail plate is the **nail bed;** its epidermis is called the **hyponychium**[26] (HIPE-o-NICK-ee-um). At the proximal end of the nail, the stratum basale thickens into a growth zone called the **nail matrix.** Mitosis in the matrix accounts for the growth of the nail—about 1 mm per week in the fingernails and slightly slower in the toenails. The thickness of the matrix obscures the underlying dermal blood vessels and is the reason why an opaque white crescent, the **lunule**[27] (LOON-yule), often appears at the proximal end of a nail. A narrow zone of dead skin, the **cuticle** or **eponychium**[28] (EP-o-NICK-ee-um), commonly overhangs this end of the nail.

The appearance of the fingertips and nails can be valuable in medical diagnosis. The fingertips become swollen or *clubbed* in response to long-term hypoxemia—a deficiency of oxygen in the blood stemming from conditions such as congenital heart defects and emphysema. Dietary deficiencies may be reflected in the appearance of the nails. An iron deficiency, for example, may cause them to become flat or concave (spoonlike) rather than convex. Contrary to popular belief, adding gelatin to the diet has no effect on the growth or hardness of the nails.

FIGURE 6.9 Anatomy of a Fingernail. A&PR

BEFORE YOU GO ON

Answer the following questions to test your understanding of the preceding section:

7. What is the difference between vellus hair and terminal hair?

8. State the functions of the hair papilla, hair receptors, and arrector muscle.

9. Describe what happens in the anagen, catagen, and telogen phases of the hair cycle.

10. State some reasonable theories for the different functions of hair of the eyebrows, eyelashes, scalp, nostrils, and axilla.

11. Define or describe the *nail plate, nail fold, eponychium, hyponychium,* and *nail matrix.*

6.3 Cutaneous Glands

Expected Learning Outcomes

When you have completed this section, you should be able to

a. name two types of sweat glands, and describe the structure and function of each;

b. describe the location, structure, and function of sebaceous and ceruminous glands; and

c. discuss the distinction between breasts and mammary glands, and explain their respective functions.

The skin has five kinds of glands: *eccrine sweat glands, apocrine sweat glands, sebaceous glands, ceruminous glands,* and *mammary glands.*

[26]*hypo* = below; *onych* = nail
[27]*lun* = moon; *ule* = little
[28]*ep* = above; *onych* = nail

6.3a Sweat Glands

Sweat glands are of two kinds: apocrine and eccrine (compare figure 5.32). **Apocrine sweat glands (fig. 6.10a)** occur in the groin, anal region, axilla, and areola, and in mature males, in the beard area. Axillary apocrine glands are scarce or absent from many East Asians, possibly because their ancestors, living in what is now Siberia, had little need to sweat in such a cold habitat. The absence of these glands is traceable to a mutation that arose at least 40,000 years ago. The ducts of apocrine glands lead into nearby hair follicles rather than directly to the skin surface. Apocrine glands, unfortunately, are misnamed because they don't use the apocrine mode of secretion; they use exocytosis, the same as eccrine sweat glands do. The secretory part of an apocrine gland, however, has a much larger lumen than that of an eccrine gland, so these glands continue to be called apocrine glands to distinguish them functionally and histologically from the eccrine type. Apocrine sweat is thicker and more milky than eccrine sweat because it has more fatty acids in it.

Apocrine sweat glands are scent glands that respond especially to stress and sexual stimulation. They are not active until puberty, and in women, they enlarge and shrink in phase with the menstrual cycle. These facts, as well as experimental evidence, suggest that they secrete *sex pheromones*—chemicals that exert subtle effects on the sexual behavior and physiology of other people. They apparently correspond to the scent glands that develop in other mammals as they approach sexual maturity. Fresh apocrine sweat doesn't have a disagreeable odor; indeed, it is considered attractive or arousing in some cultures, where it is as much a part of courtship as artificial perfume is to others. Clothing, however, traps stale sweat long enough for bacteria to degrade the secretion and release free fatty acids with a rancid odor. Disagreeable body odor is called *bromhidrosis*.[29] It occasionally indicates a metabolic disorder, but more often reflects inadequate hygiene.

Many mammals have apocrine scent glands associated with specialized tufts of hair. In humans, apocrine glands are found

---

[29]*brom* = stench; *hidros* = sweat

Lumen   Secretory cells

(a) Apocrine gland

Myoepithelial cells   Lumen

Secretory cells

(b) Eccrine gland

Gland   Hair follicle

(c) Sebaceous gland

FIGURE 6.10 Cutaneous Glands. (a) Apocrine sweat glands have a large lumen and a duct that conveys their aromatic secretion into a hair follicle. (b) Eccrine sweat glands have a relatively narrow lumen and a duct that opens by way of a pore on the skin surface. (c) Sebaceous glands have cells that break down in entirety to form an oily secretion that is released into the hair follicle.

a–c: Dennis Strete/McGraw-Hill Education

mainly in the regions covered by the pubic hair, axillary hair, and beard. This supports the interpretation that they are pheromone glands. The hair serves to retain the aromatic secretion and regulate its rate of evaporation from the skin. Thus, it seems no mere coincidence that women's faces lack both apocrine scent glands and a beard.

Eccrine (merocrine) sweat glands (EK-rin) (**fig. 6.10b**) are widely distributed over the entire body, but are especially abundant on the palms, soles, and forehead. Their primary function is to cool the body. Overheating activates sweating through the sympathetic nervous system. Then, because of the high heat capacity of water (see section 2.2), the evaporation of sweat carries away a substantial amount of body heat, cooling one down. Each is a simple tubular gland with a twisted coil in the dermis or hypodermis, and an undulating or coiled duct leading to a sweat pore on the skin surface. The duct is lined by a stratified cuboidal epithelium in the dermis and by keratinocytes in the epidermis.

In both apocrine and eccrine sweat glands, specialized **myoepithelial**[30] **cells** are found amid the secretory cells at the deep end of the gland. They have contractile properties similar to those of smooth muscle. The sympathetic nervous system stimulates them to contract, squeeze the base of the gland, and force perspiration up the duct—particularly under conditions of overheating, nervousness, or arousal.

Sweat production begins in the deep secretory portion of the gland. Protein-free fluid filters from the blood capillaries into the lumen of the gland. Most sodium chloride is reabsorbed from the filtrate as it passes up the duct, but some remains, along with potassium, urea, lactic acid, and ammonia. Some drugs are also excreted in the perspiration. Thus, the eccrine sweat glands excrete some of the same wastes as the kidneys. The contribution of an individual sweat gland to this is minuscule, but there are 3 or 4 million eccrine sweat glands in the adult skin, with a total mass about equal to one kidney.

Perspiration is about 99% water and has a pH ranging from 4 to 6, contributing the earlier-mentioned *acid mantle* that inhibits bacterial growth on the skin. Usually, perspiration evaporates about as fast as it is produced, so it isn't noticed; this is called **insensible perspiration.** But under such conditions as heat, exercise, and circulatory shock, more copious sweat is produced and noticeably wets the skin; this is called **diaphoresis**[31] (DY-uh-foe-REE-sis). Insensible perspiration typically amounts to about 500 mL/day, but in diaphoresis, a person can lose a liter or more of sweat per hour. In heavy sweating, fluid loss from the bloodstream can be so great as to cause circulatory shock. Sweating, however, isn't the only way the skin loses water. A significant amount also diffuses between the keratinocytes and evaporates from the skin surface—a process called *cutaneous transpiration* (see section 24.1b).

6.3b Sebaceous Glands

Sebaceous[32] **glands** (see-BAY-shus) produce an oily secretion called **sebum** (SEE-bum). They are flask-shaped, with short ducts that usually open into a hair follicle (**fig. 6.10c**), although some of them open directly onto the skin surface. These are holocrine glands with little visible lumen. Their secretion consists of broken-down cells that are replaced by mitosis at the base of the gland. Sebum keeps the skin and hair from becoming dry, brittle, and cracked. The sheen of well-brushed hair is due to sebum distributed by the hairbrush. Ironically, we go to great lengths to wash sebum from the skin, only to replace it with various skin creams and hand lotions made of little more than lanolin, which is sheep sebum.

6.3c Ceruminous Glands

Ceruminous glands (seh-ROO-mih-nus) are modified apocrine glands found only in the external ear canal. Their yellow, waxy secretion combines with sebum and dead epidermal cells to form earwax, or **cerumen.**[33] They are coiled, simple tubular glands with ducts leading to hair follicles or the skin surface of the ear canal. Cerumen keeps the eardrum pliable, waterproofs the canal, kills bacteria, and coats the guard hairs of the ear, making them sticky and more effective in blocking foreign particles from entering the canal.

6.3d Mammary Glands

The **mammary glands** and breasts *(mammae)* are often mistakenly regarded as one and the same. Breasts, however, are present in both sexes, and even in females they rarely contain more than small traces of mammary gland. The mammary glands, by contrast, are the milk-producing glands that develop within the female breast during pregnancy and lactation. They are modified apocrine sweat glands that produce a richer secretion and channel it through ducts to a nipple for more efficient conveyance to the offspring. The anatomy and physiology of the mammary gland are discussed in more detail in section 28.1c.

In most mammals, two rows of mammary glands form along lines called the *mammary ridges,* or *milk lines.* Primates have dispensed with all but two of these glands. A few people of both sexes, however, develop additional nipples or mammae along the milk line inferior to the primary mammae. In the Middle Ages and colonial America, this condition, called *polythelia,*[34] was used to incriminate women as supposed witches.

The glands of the skin are summarized in **table 6.2.**

> **BEFORE YOU GO ON**

Answer the following questions to test your understanding of the preceding section:

12. How do eccrine and apocrine sweat glands differ in structure and function?

13. What other type of gland is associated with hair follicles? How does its mode of secretion differ from that of sweat glands?

14. What is the difference between the breast and mammary gland?

---

[30]*myo* = muscle
[31]*dia* = through; *phoresis* = carrying
[32]*seb* = fat, tallow; *aceous* = possessing

[33]*cer* = wax
[34]*poly* = many; *theli* = nipples

TABLE 6.2	Cutaneous Glands
Gland Type	**Definition**
Sweat glands	Glands that produce perspiration
Eccrine glands	Sweat glands that function in evaporative cooling; widely distributed over the body surface; open by ducts onto the skin surface
Apocrine glands	Sweat glands that function as scent glands; found in the regions covered by the pubic, axillary, and male facial hair; open by ducts into hair follicles
Sebaceous glands	Oil glands associated with hair follicles
Ceruminous glands	Glands of the ear canal that contribute to the cerumen (earwax)
Mammary glands	Milk-producing glands located in the breasts

6.4 Skin Disorders

Expected Learning Outcomes

When you have completed this section, you should be able to

a. describe the three most common forms of skin cancer; and

b. describe the three classes of burns and the priorities in burn treatment.

Because it is the most exposed of all our organs, skin is not only the most vulnerable to injury and disease, but it is also the one place where we are most likely to notice anything out of the ordinary. Skin diseases become increasingly common in old age, and most people over age 70 have complaints about their integumentary system. Aging of the skin is discussed more fully in section 29.4a. The healing of cuts and other injuries to the skin occurs by the process described in section 5.6d. We focus here on two particularly common and serious disorders: skin cancer and burns. Other skin diseases are briefly summarized in **table 6.3.**

6.4a Skin Cancer

Skin cancer befalls about one out of five people in the United States at some time in their lives. Most cases are caused by UV radiation from the sun, which damages DNA and disables protective tumor suppressor genes in the epidermal cells. Consequently, most tumors occur on the head, neck, and hands, where exposure to the sun is greatest. It is most common in fair-skinned people and the elderly, who have had the longest lifetime UV exposure (see Deeper Insight 6.3). The ill-advised popularity of suntanning, however, has caused an alarming increase in skin cancer among younger people. Skin cancer is one of the most common cancers, but it is also one of the easiest to treat and has one of the highest survival rates when it is detected and treated early.

▶▶▶**APPLY WHAT YOU KNOW**

Skin cancer is relatively rare in people with dark skin. Other than possible differences in behavior, such as less intentional suntanning, why do you think this is so?

There are three types of skin cancer named for the epidermal cells in which they originate: *basal cell carcinoma, squamous cell carcinoma,* and *melanoma.* The three types are also distinguished from each other by the appearance of their **lesions**[35] (zones of tissue injury).

Basal cell carcinoma[36] **(fig. 6.11a)** is the most common type. It is the least deadly because it seldom metastasizes, but if neglected, it can severely disfigure the face. It arises from cells of the stratum basale and eventually invades the dermis. On the surface, the lesion first appears as a small, shiny bump. As the bump enlarges, it often develops a central depression and a beaded "pearly" edge.

Squamous cell carcinoma (fig. 6.11b) arises from keratinocytes of the stratum spinosum. Lesions usually appear on the scalp, ears, lower lip, or back of the hand. They have a raised, reddened, scaly appearance, later forming a concave ulcer with raised edges. The chance of recovery is good with early detection and surgical removal, but if it goes unnoticed or is neglected, this cancer tends to metastasize to the lymph nodes and can be lethal.

Melanoma (fig. 6.11c) is a skin cancer that arises from the melanocytes. It accounts for no more than 5% of skin cancers, but it is the sixth most frequently diagnosed cancer in the United States and is extremely aggressive. It can be treated surgically if caught early, but if it metastasizes—which it does quickly—it is unresponsive to chemotherapy. The prognosis for metastatic melanoma has generally been grim, with the average patient living only 6 months after diagnosis and only 5% to 14% surviving for 5 years. Hope is arising, however, from new targeted therapies for melanoma. One of these treatments uses the radioisotope rhenium-188 (^{188}Re) bonded to melanocyte-stimulating hormone (MSH). MSH selectively binds to melanocyte receptors, thus accumulating selectively in the very cells that compose metastatic melanoma. ^{188}Re emits alpha particles, a highly potent ionizing radiation that is lethal to the tumor cells. Yet alpha particles are too large and heavy to penetrate very far, and are thus harmless to healthy cells around the tumor. ^{188}Re, bound to other peptides, has also been used to treat liver cancer and is in clinical trials for pancreatic cancer.

The greatest risk factor for melanoma is a family history of the disease. It has a relatively high incidence in men, in redheads, and in people who experienced severe sunburns in childhood. About two-thirds of cases of melanoma in men result from an oncogene called *BRAF.* In women, *BRAF* does not appear to trigger melanoma, but it has been linked to some breast and ovarian cancers. *BRAF* mutations are commonly found in moles.

It is important to distinguish a mole from a melanoma. A mole usually has a uniform color and even contour, and it is no larger in diameter than the eraser of a new wooden pencil (about 6 mm). If it becomes malignant, however, it forms a large, flat, spreading

---

[35]*lesio* = injure

[36]*carcin* = cancer; *oma* = tumor

(a) Basal cell carcinoma

(b) Squamous cell carcinoma

(c) Melanoma

FIGURE 6.11 Skin Cancer. (a) Basal cell carcinoma. (b) Squamous cell carcinoma. (c) Melanoma.

❓ *Which of the ABCD rules can you identify in part (c)?*

a: jax10289/Shutterstock; **b:** DR P. Marazzi/Science Photo Library/Alamy; **c:** Source: National Cancer Institute (NCI)/U.S. Department of Health and Human Services (USHSS)

lesion with a scalloped border. The American Cancer Society suggests an "ABCD rule" for recognizing melanoma: *A* for asymmetry (irregular shape, unlike the rounded lesions of the other forms of skin cancer); *B* for border irregularity (the contour is not smooth but wavy or scalloped); *C* for color (often a mixture of brown, black, tan, and sometimes red and blue); and *D* for diameter (greater than 6 mm).

Depending on the type of cancer and its location (local lesion or metastatic spread), skin cancers are treated by surgical excision, radiation therapy, or destruction of the lesion by heat (electrodesiccation) or cold (cryosurgery).

6.4b Burns

Burns are usually caused by fires, kitchen spills, or excessively hot bath water, but they also can be caused by sunlight, ionizing radiation, strong acids and bases, or electrical shock. Burn deaths result primarily from fluid loss, infection, and the toxic effects of **eschar** (ESS-car)—the burned, dead tissue.

Burns are classified according to the depth of tissue involvement. **First-degree burns (fig. 6.12a)** involve only the epidermis and are marked by redness, slight edema, and pain. They heal in a few days and seldom leave scars. Most sunburns are first-degree burns.

Second-degree burns (fig. 6.12b) involve the epidermis and part of the dermis but leave at least some of the dermis intact. First- and second-degree burns are therefore also known as **partial-thickness burns.** A second-degree burn may be

red, tan, or white and is blistered and very painful. It may take from 2 weeks to several months to heal and may leave scars. The epidermis regenerates by division of epithelial cells in the hair follicles and sweat glands and around the edges of the lesion. Severe sunburns and many scalds are second-degree burns.

Third-degree burns (fig. 6.12c) are a leading cause of accidental death. They are also called **full-thickness burns** because the epidermis, all of the dermis, and often some deeper tissues (muscle and bone) are destroyed. (Some authorities call burns that extend to the bone *fourth-degree burns.*) Since no dermis remains, the skin can regenerate only from the edges of the wound. Third-degree burns often require skin grafts (see Deeper Insight 6.4). If a third-degree burn is left to itself to heal, contracture (abnormal connective tissue fibrosis) and severe disfigurement may result.

▶▶▶ APPLY WHAT YOU KNOW

A third-degree burn may be surrounded by painful areas of first- and second-degree burns, but the region of the third-degree burn is painless. Explain the reason for this lack of pain.

The two most urgent considerations in treating a burn patient are fluid replacement and infection control. A patient can lose several liters of water, electrolytes, and protein each day from the burned area. As fluid is lost from the tissues, more is transferred from the bloodstream to replace it, and the volume of circulating blood declines. A patient may lose up to 75% of the blood plasma

Partial-thickness burns

Full-thickness burns

(a) First degree

(b) Second degree

(c) Third degree

FIGURE 6.12 Burns. (a) First-degree burn, involving only the epidermis. (b) Second-degree burn, involving the epidermis and part of the dermis and often with blistering. (c) Third-degree burn, extending through the entire dermis and often involving even deeper tissue.

a: Dmitrii Kotin/Alamy Stock Photo; b: thawatchai_bandit/Shutterstock; c: Anukool Manoton/Shutterstock

within a few hours, potentially leading to circulatory shock and cardiac arrest—the principal cause of death in burn patients. Intravenous fluid must be given to make up for this loss. A severely burned patient may also require thousands of extra calories daily to compensate for protein loss and the demands of tissue repair. Supplemental nutrients are given intravenously or via gastric tube.

Infection is controlled by keeping the patient in an aseptic (germ-free) environment and administering antibiotics. The eschar is sterile for the first 24 hours, but then it quickly becomes infected and may have toxic effects on the digestive, respiratory, and other systems. Its removal, called **debridement**[37] (deh-BREED-ment), is essential to infection control.

BEFORE YOU GO ON

Answer the following questions to test your understanding of the preceding section:

15. What types of cells are involved in each type of skin cancer?

16. Which type of skin cancer is most dangerous? What are its early warning signs?

17. What are the differences between first-, second-, and third-degree burns?

18. What are the two most urgent priorities in treating a burn victim? How are these needs addressed?

---

[37]*de* = un; *bride* = bridle

TABLE 6.3	Some Disorders of the Integumentary System
Acne	Inflammation of the sebaceous glands, especially beginning at puberty; follicle becomes blocked with keratinocytes and sebum and develops into a whitehead *(comedo)* composed of these and bacteria; continued inflammation of follicle results in pus production and appearance of pimples, and oxidation of sebum turns a whitehead into a blackhead.
Dermatitis	Any inflammation of the skin, typically marked by itching and redness; often *contact dermatitis,* caused by exposure to toxic foliage such as poison ivy.
Eczema (ECK-zeh-mah)	Itchy, red, "weeping" skin lesions caused by an allergy, usually beginning before age 5; may progress to thickened, leathery, darkly pigmented patches of skin.
Psoriasis (so-RY-ah-sis)	Recurring, reddened plaques covered with silvery scale; sometimes disfiguring; possibly caused by an autoimmune response; runs in families.
Rosacea (ro-ZAY-she-ah)	A red rashlike area, often in the area of the nose and cheeks, marked by fine networks of dilated blood vessels; worsened by hot drinks, alcohol, and spicy food.
Seborrheic dermatitis (seb-oh-REE-ik)	Recurring patches of scaly white or yellowish inflammation often on the head, face, chest, and back; called *cradle cap* (a yellow, crusty scalp lesion) in infants. Cause unknown, but correlated with genetic and climatic factors.
Tinea	Any fungal infection of the skin; common in moist areas such as the axilla, groin, and foot *(athlete's foot).* Misnamed *ringworm* because of the circular, wormlike growth pattern sometimes exhibited.

You can find other integumentary system disorders described in the following places:

Genital warts in Deeper Insight 27.5; and in the present chapter, *pathological skin colors* and *birthmarks* in sections 6.1e and 6.1f; *baldness* and *hirsutism* in section 6.2a; *polythelia* in section 6.3d; and *skin cancer* and *burns* in section 6.4.

DEEPER INSIGHT 6.3

CLINICAL APPLICATION

UVA, UVB, and Sunscreens

Ultraviolet radiation is divided into two wavelength ranges, UVA and UVB. UVA, so-called "tanning rays," have the longer wavelength (320–400 nm) and lower energy of the two. UVA lies just below the visible spectrum of light, where 400 nm is on the borderline between black and the deepest violet we can see. UVB, so-called "burning rays," have a shorter wavelength (290–320 nm) and higher energy level. Tanning salons often advertise that they only use "safe" UVA rays, but public health authorities are skeptical. UVA can burn as well as tan, it is responsible for most of the undesirable photoaging effects on the skin (see section 29.4a), and it inhibits one's immune system. Both UVA and UVB are now thought to initiate skin cancer. As dermatologists like to say, there is no such thing as a healthy tan.

Many people buy sunscreen according to its sun protection factor (SPF), but may be misled by the concept. SPF is a laboratory measure of protection from UVB radiation. It is advisable to use a minimum SPF of 15 for meaningful protection. It might seem that an SPF of 30 would give twice as much protection as SPF 15, but the relationship between protection and SPF is not linear. An SPF 15 sunscreen protects the skin from 93% of UVB, but SPF 30 protects only slightly more—97%. Although manufacturers compete to produce high-SPF products, higher numbers create a false sense of much greater protection. SPF values above 30 are considered to be misleading by the Australian government.

Furthermore, effectiveness varies with the amount applied, frequency of reapplication, skin type, and sweating, and sunscreen is washed off by swimming. People tend to use only one-quarter to one-half as much as needed to provide the SPF rating on the label, and wait too long between reapplications. For effective protection, one should divide the SPF by 2 and reapply sunscreen that many minutes after the onset of sun exposure—that is, every 15 minutes for an SPF of 30. Some products claim "broad spectrum UVA/UVB protection," but there is no good evidence that they protect adequately against UVA.

It was once assumed that sunscreens protect against skin cancer, but more careful studies have cast doubt on this. Sunburn and skin cancer are caused by different mechanisms, and even if sunscreen protects against burning, it provides little protection against cancer. Ironically, as the sale of sunscreen has risen, so has the incidence of skin cancer—perhaps because people falsely assume that when they use sunscreen, they can safely stay in the sun longer.

DEEPER INSIGHT 6.4

CLINICAL APPLICATION

Skin Grafts and Artificial Skin

Third-degree burns leave no dermal tissue to regenerate what was lost; therefore, they generally require skin grafts. The ideal graft is an *autograft*[38]—taking epidermis and part of the dermis from an undamaged area of the patient's own body, such as the thigh or buttock, and grafting it to a burned area. This is called a *split-skin* graft because part of the dermis is left behind to proliferate and replace the epidermis that was removed—the same way a second-degree burn heals. The advantage of an autograft is that it is not rejected by the immune system. Relatively small areas of healthy skin can be made to cover much larger lesions by means of *meshed skin grafts* **(fig. 6.13).** The donor skin is cut into a mesh pattern and stretched over the lesion. Regenerating skin cells fill in the gaps in the mesh.

The next-best method, if the burns are too extensive for autografting, is to use skin from an identical twin to avoid immune rejection. Since identical twins are rare, however, the best one can usually hope for is another close relative as donor.

An *allograft*[39] is a graft from any other person, usually obtained from a skin bank. The immune system attempts to reject allografts, but they suffice as temporary coverings for the burned area. They can be replaced by autografts when the patient is well enough for healthy skin to be removed from an undamaged area of the body.

Pig skin is sometimes used on burn patients but presents the same problem of immune rejection. A graft of tissue from a different species is called a *xenograft*[40]. A xenograft is a short-term method of maintaining a patient until a better, long-term solution is possible. The immune reaction can be suppressed by drugs called *immunosuppressants,* but these lower one's resistance to infection, which is already compromised in a burn patient.

Some alternatives to skin grafts are also being used. Tiny keratinocyte patches cultured with growth stimulants have produced sheets of epidermal tissue as large as the entire body surface. These can replace large areas of burned tissue. Dermal fibroblasts also have been successfully cultured and used for autografts. A drawback to these approaches is that the culture process requires 3 or 4 weeks, which is too long a wait for some patients with severe burns.

Several bioengineering companies now produce artificial skin for both temporary and permanent coverings. A temporary silicone–nylon

FIGURE 6.13 A Meshed Skin Graft.

Barry Slaven/Science Source

synthetic membrane can provide covering and a moist environment for wound healing. Another method called *cultured epithelial autograft (CEA)* uses the patient's own keratinocytes from a small skin biopsy and cultures these in the laboratory for 2 to 3 weeks to produce sheets of tissue that can cover large burn areas. Another method cultures fibroblasts on a collagen gel to produce a dermis, then keratinocytes to produce an overlying epidermis. Still another technology available in some countries is a spray-on skin graft. A bit of epidermis the size of a postage stamp is taken from the patient and enzymatically broken up into separate cells. The cell suspension is then sprayed onto areas of second-degree burns or other skin lesions. Within a week, this can grow into an area of healthy skin the size of a textbook page.

---

[38]*auto* = self
[39]*allo* = different, other
[40]*xeno* = strange, alien, foreign

CONNECTIVE ISSUES

Effects of the INTEGUMENTARY SYSTEM on Other Organ Systems

ALL SYSTEMS

Skin covers the body and provides a barrier to pathogens and to excessive water loss; epidermal keratinocytes initiate synthesis of calcitriol, with effects on multiple other organ systems as noted.

SKELETAL SYSTEM

Bone growth and maintenance depend on calcium, which is absorbed from the diet under the influence of calcitriol.

MUSCULAR SYSTEM

Muscle contraction depends on calcium, which is absorbed from the diet under the influence of calcitriol.

NERVOUS SYSTEM

The transmission of nerve signals across synapses depends on calcium, which is absorbed from the diet under the influence of calcitriol.

ENDOCRINE SYSTEM

Hormone secretion depends on calcium as a trigger for exocytosis and, therefore, on calcitriol; the role of epidermal keratinocytes in synthesizing calcitriol is itself an endocrine function.

CIRCULATORY SYSTEM

The skin is a major blood reservoir; cutaneous vasoconstriction diverts blood to other organs; skin supports blood volume by retarding fluid loss; dermal vasoconstriction and vasodilation help to regulate blood temperature.

LYMPHATIC AND IMMUNE SYSTEMS

Dendritic cells of the skin alert the immune system when pathogens breach the epidermal barrier.

RESPIRATORY SYSTEM

Nasal guard hairs block some airborne debris from being inhaled; calcium is required for the secretion of respiratory mucus, which therefore depends on calcitriol.

URINARY SYSTEM

Skin complements the urinary system by excreting salts and some nitrogenous waste in the sweat; calcitriol promotes reabsorption of calcium by the kidneys.

DIGESTIVE SYSTEM

By their role in calcitriol synthesis, keratinocytes influence intestinal absorption of calcium; calcium is needed for the secretion of all digestive enzymes and mucus.

REPRODUCTIVE SYSTEM

Cutaneous nerve endings are important in sexual stimulation; mammary glands produce milk; apocrine sweat glands secrete pheromones that affect sexual behavior and physiology; skin stretches to accommodate abdominal growth in pregnancy.

STUDY GUIDE

▶ Assess Your Learning Outcomes

To test your knowledge, discuss the following topics with a study partner or in writing, ideally from memory.

6.1 The Skin and Subcutaneous Tissue

1. The name of the branch of medicine that deals with the integumentary system
2. The difference between the terms *integumentary system* and *integument;* organs that belong to the integumentary system
3. The two principal layers of the skin, and the name of the connective tissue layer that lies between the skin and the deeper muscle or other tissue
4. Normal thickness of the skin, and differences in histology and location between thick and thin skin
5. Functions of the skin
6. Five histological layers of thick skin and which of them is lacking from thin skin
7. Five kinds of epidermal cells, their respective functions, and the epidermal layers in which they occur
8. The life history of a keratinocyte from the time it is "born" by mitosis at the base of the epidermis to the time it exfoliates from the surface
9. Fiber and cell types of the dermis, other dermal structures, and the typical thickness of the dermis

10. Dermal papillae, epidermal ridges, their function, and their relationship to the surface appearance of skin
11. The two layers of the dermis, and how they differ histologically and functionally
12. Composition and functions of the hypodermis, and an alternative name for it when it is composed predominantly of adipose tissue
13. Factors that account for the variety of normal skin colors; abnormal skin colors and their causes
14. Friction ridges, flexion lines, freckles, moles, and hemangiomas

6.2 Hair and Nails

1. How the keratin of hair and nails differs from keratin of the epidermis
2. Three kinds of hair, including fetal and adult types
3. The three regions of a hair from base to tip, and the three layers of a hair from core to surface
4. Location of a hair's growth zone and of its source of nourishment
5. The two layers of a hair follicle; their composition; and the specialized nerve endings and smooth muscle associated with a follicle
6. The basis for differences between straight, wavy, and curly hair and for differences in hair color

7. Events of the anagen, catagen, and telogen stages of a hair's life; the typical life span and growth rate of scalp hairs
8. Alopecia, pattern baldness, and hirsutism
9. Functions and body distribution of the various kinds of hair
10. The anatomy of fingernails and toenails; location of their growth zone; and a typical rate of nail growth

6.3 Cutaneous Glands

1. Apocrine and eccrine sweat gland distribution, development, structure, and function
2. The same characteristics of sebaceous glands; the name of their product; and how their mode of secretion differs from that of the sweat glands
3. Ceruminous gland distribution, development, structure, and function
4. Mammary gland structure and development and how mammary glands relate to a type of sweat glands

6.4 Skin Disorders

1. Three forms of skin cancer and differences in their appearance, the cells in which they originate, their frequency of occurrence, and their severity
2. Three degrees of burns and how they are treated

▶ Testing Your Recall

Answers in Appendix A

1. Cells of the _____ are keratinized and dead.
   a. papillary layer
   b. stratum spinosum
   c. stratum basale
   d. stratum corneum
   e. stratum granulosum

2. The epidermal water barrier is formed at the point where epidermal cells
   a. pass from stratum basale to stratum spinosum.
   b. enter the telogen phase.
   c. pass from stratum spinosum to stratum granulosum.
   d. form the epidermal ridges.
   e. exfoliate.

3. Which of the following skin conditions or appearances would most likely result from liver failure?
   a. pallor
   b. erythema
   c. seborrheic dermatitis
   d. jaundice
   e. melanization

4. All of the following interfere with microbial invasion of the body *except*
   a. the acid mantle.
   b. melanization.
   c. dendritic cells.
   d. keratinization.
   e. sebum.

5. The hair on a 6-year-old's arms is
   a. vellus hair.
   b. terminal hair.
   c. alopecia.
   d. downy hair.
   e. rosacea.

6. Which of the following terms is *least* related to the rest?
   a. lunule
   b. nail plate
   c. hyponychium
   d. free edge
   e. cortex

STUDY GUIDE

7. Which of the following is a scent gland?
   a. eccrine gland
   b. sebaceous gland
   c. apocrine gland
   d. ceruminous gland
   e. eccrine gland

8. _____ are skin cells with a sensory role.
   a. Tactile cells
   b. Dendritic cells
   c. Stem cells
   d. Melanocytes
   e. Keratinocytes

9. Which of the following glands produce the acid mantle?
   a. eccrine sweat glands
   b. apocrine sweat glands
   c. mammary glands
   d. ceruminous glands
   e. sebaceous glands

10. Which of the following skin cells alert the immune system to pathogens?
    a. fibroblasts
    b. melanocytes
    c. keratinocytes
    d. dendritic cells
    e. tactile cells

11. _____ is sweating without noticeable wetness of the skin.

12. A muscle that causes a hair to stand on end is called a/an _____.

13. The process of removing burned skin from a patient is called _____.

14. Blueness of the skin due to low oxygen concentration in the blood is called _____.

15. Projections of the dermis toward the skin surface are called _____.

16. Cerumen is more commonly known as _____.

17. The holocrine glands that secrete into a hair follicle are called _____.

18. Hairs grow only during the _____ phase of the hair cycle.

19. A hair is nourished by blood vessels in a connective tissue projection called the _____.

20. A _____ burn destroys the entire dermis.

▶ Building Your Medical Vocabulary

Answers in Appendix A

State a meaning of each word element, and give a medical term from this chapter that uses it or a slight variation of it.

1. -in
2. albo-
3. dermato-
4. dia-
5. homo-
6. lesio-
7. melano-
8. -oma
9. onycho-
10. pilo-

▶ What's Wrong with These Statements?

Answers in Appendix A

Briefly explain why each of the following statements is false, or reword it to make it true.

1. Basal cell carcinoma is the rarest form of skin cancer and the least likely to metastasize.

2. Brown or black skin owes its color to a great abundance of melanocytes.

3. The dermis is composed mainly of keratin.

4. Vitamin D synthesis begins in certain cutaneous glands.

5. Epidermal cells multiply rapidly in the stratum granulosum to produce the thick, protective stratum corneum.

6. The hair cuticle is composed of dead cells; the living hair cells constitute the cortex.

7. The three layers of the skin are the epidermis, dermis, and hypodermis.

8. People of African descent have a much higher density of epidermal melanocytes than do people of northern European descent.

9. Pallor indicates a genetic lack of melanin.

10. Eccrine and apocrine sweat glands are present from birth.

▶ Testing Your Comprehension

1. Many organs of the body contain numerous smaller organs, perhaps even thousands. Describe an example of this in the integumentary system.

2. Certain aspects of human form and function are easier to understand when viewed from the perspective of comparative anatomy and evolution. Discuss examples of this in the integumentary system.

3. Explain how the complementarity of form and function is reflected in the fact that the dermis has two histological layers and not just one.

4. Cold weather does not normally interfere with oxygen uptake by the blood, but it can cause cyanosis anyway. Why?

5. Why is it important for the epidermis to be effective, but not *too* effective, in screening out UV radiation?

BONE TISSUE

A bone cell (osteocyte) surrounded by calcified bone matrix
Eye of Science/Science Source

CHAPTER OUTLINE

7.1 Tissues and Organs of the Skeletal System

  7.1a Functions of the Skeleton
  7.1b Bones and Osseous Tissue
  7.1c General Features of Bones

7.2 Histology of Osseous Tissue

  7.2a Bone Cells
  7.2b The Matrix
  7.2c Compact Bone
  7.2d Spongy Bone
  7.2e Bone Marrow

7.3 Bone Development

  7.3a Intramembranous Ossification
  7.3b Endochondral Ossification
  7.3c Bone Growth and Remodeling

7.4 Physiology of Osseous Tissue

  7.4a Mineral Deposition and Resorption
  7.4b Calcium Homeostasis
  7.4c Phosphate Homeostasis
  7.4d Other Factors Affecting Bone

7.5 Bone Disorders

  7.5a Fractures and Their Repair
  7.5b Other Bone Disorders

Connective Issues

Study Guide

DEEPER INSIGHTS

7.1 Medical History: Bone Contamination

7.2 Clinical Application: Achondroplastic Dwarfism

7.3 Clinical Application: Rickets and Osteomalacia

7.4 Clinical Application: Osteoporosis

Anatomy & Physiology
Revealed 4.0

Module 5: Skeletal System

In art and history, nothing has so often symbolized death as a skull or skeleton.[1] Bones and teeth are the most durable remains of a once-living body and the most vivid reminder of the impermanence of life.

The dry bones presented for laboratory study may wrongly suggest that the skeleton is a nonliving scaffold for the body, like the steel girders of a building. Seeing it in such a sanitized form makes it easy to forget that the living skeleton is made of dynamic tissues, full of cells—that it continually remodels itself and interacts physiologically with all of the other organ systems of the body. The skeleton is permeated with nerves and blood vessels, which attests to its sensitivity and metabolic activity.

Osteology,[2] the study of bone, is the subject of these next three chapters. In this chapter, we study bone as a tissue—its composition, its functions, how it develops and grows, how its metabolism is regulated, and some of its disorders. This will provide a basis for understanding the skeleton, joints, and muscles in the chapters that follow.

7.1 Tissues and Organs of the Skeletal System

Expected Learning Outcomes

When you have completed this section, you should be able to

a. name the tissues and organs that compose the skeletal system;

b. state several functions of the skeletal system;

c. distinguish between bone as a tissue and as an organ; and

d. describe the general features of a long bone and a flat bone.

The **skeletal system** is composed of bones, cartilages, and ligaments joined tightly to form a strong, flexible framework for the body. Cartilage, the forerunner of most bones in embryonic and childhood development, covers many joint surfaces in the mature skeleton. Ligaments hold bones together at the joints and are

discussed in chapter 9. Tendons are structurally similar to ligaments but attach muscle to bone; they are discussed with the muscular system in chapter 10. Here, we focus on the bones.

7.1a Functions of the Skeleton

The skeleton plays at least six roles:

1. **Support.** Bones of the limbs and vertebral column support the body; the mandible and maxilla support the teeth; and some viscera are supported by nearby bones.

2. **Protection.** Bones enclose and protect the brain, spinal cord, heart, lungs, pelvic viscera, and bone marrow.

3. **Movement.** Limb movements, breathing, and other movements are produced by the action of muscles on the bones.

4. **Electrolyte balance.** The skeleton stores calcium and phosphate ions and releases them into the tissue fluid and blood according to the body's physiological needs.

5. **Acid–base balance.** Bone tissue buffers the blood against excessive pH changes by absorbing or releasing alkaline phosphate and carbonate salts.

6. **Blood formation.** Red bone marrow is the major producer of blood cells, including cells of the immune system.

7.1b Bones and Osseous Tissue

Bone, or **osseous**[3] **tissue,** is a connective tissue in which the matrix is hardened by the deposition of calcium phosphate and other minerals. The hardening process is called **mineralization** or **calcification.** (Bone is not the hardest substance in the body; that distinction goes to tooth enamel.) Osseous tissue is only one of the tissues that make up a bone. Also present are blood, bone marrow, cartilage, adipose tissue, nervous tissue, and fibrous connective tissue. The word *bone* can denote an organ composed of all these tissues, or it can denote just the osseous tissue.

7.1c General Features of Bones

Bones have a wide variety of shapes correlated with their varied protective and locomotor functions. Most of the cranial bones are in the form of thin curved plates called **flat bones,** such as the paired parietal bones that form the dome of the top of the head. The sternum (breastbone), scapula (shoulder blade), ribs, and hip bones are also flat bones. The most important bones in movement are the **long bones** of the limbs—the humerus, radius, and ulna of the arm and forearm; the femur, tibia, and fibula of the thigh and leg; and the metacarpals, metatarsals, and phalanges of the hands and feet. Like crowbars, long bones serve as rigid levers that are acted upon by skeletal muscles to produce the major body movements. Various bones that don't fit the flat or long

---

[1]*skelet* = dried up
[2]*osteo* = bone; *logy* = study of

[3]*os, osse, oste* = bone

bone groups are sometimes called *short bones* (such as those of the wrist and ankle) or *irregular bones* (such as the vertebrae and some skull bones).

Figure 7.1 shows the general anatomy of a long bone. Much of it is composed of an outer shell of dense white osseous tissue called **compact (dense) bone** or **cortical bone.** The shell encloses a space called the **marrow cavity,** or **medullary cavity** (MED-you-lerr-ee), which contains bone marrow. At the ends of the bone, the central space is occupied by a more loosely organized form of osseous tissue called **spongy (cancellous) bone.** A narrow zone of spongy bone also occurs just inside the cortical bone of the shaft and in the middle of most flat, irregular, and short bones. The skeleton is about three-quarters compact bone and one-quarter spongy bone in dry weight. Spongy bone is always enclosed by a shell of more durable compact bone.

The principal features of a long bone are its shaft, called the **diaphysis**[4] (dy-AF-ih-sis), and an expanded head at each end called the **epiphysis**[5] (eh-PIF-ih-sis). The diaphysis provides leverage, and the epiphysis is enlarged to strengthen the joint and provide added surface area for the attachment of tendons and ligaments. Mature bones often exhibit an *epiphysial line* of slightly denser spongy bone between the epiphysis and diaphysis. This is a remnant of a childhood growth zone called an *epiphysial plate,* detailed later. The joint surface where one bone meets another is covered with a layer of hyaline cartilage called the **articular cartilage.** Together with a lubricating fluid secreted between the bones, this cartilage enables a joint to move far more easily than it would if one bone rubbed directly against the other. Blood vessels penetrate into the bone through minute holes called **nutrient foramina** (for-AM-ih-nuh); we will trace where they go when we consider the histology of bone.

Externally, a bone is covered with a sheath called the **periosteum.**[6] This has a tough, outer *fibrous layer* of collagen and an inner *osteogenic layer* of bone-forming cells described in the next section. Some collagen fibers of the outer layer are continuous with the tendons that bind muscle to bone, and some penetrate into the bone matrix as **perforating fibers.** The periosteum thus provides strong attachment and continuity from muscle to tendon to bone. The osteogenic layer is important to the growth of bone and healing of fractures. There is no periosteum over the articular cartilage.

A thin layer of reticular connective tissue called the **endosteum**[7] lines the internal marrow cavity, covers all the honeycombed surfaces of spongy bone, and lines a canal system found throughout the compact bone.

Flat bones have a sandwichlike construction with two layers of compact bone, called the *inner* and *outer tables,* enclosing a middle layer of spongy bone **(fig. 7.2).** The spongy layer in

FIGURE 7.1 Anatomy of a Long Bone. (a) The femur, with its soft tissues including bone marrow, articular cartilage, blood vessels, and periosteum. (b) A dried femur in longitudinal section.

❓ *What is the functional significance of a long bone being wider at the epiphyses than at the diaphysis?*

the cranium is called **diploe**[8] (DIP-lo-ee). A moderate blow to the skull can fracture the outer table of compact bone, but the diploe may absorb the impact and leave the inner table and brain unharmed. Both surfaces of a flat bone are covered with periosteum, and the marrow spaces amid the spongy bone are lined with endosteum.

---

[4]*dia* = across; *physis* = growth; originally named for a ridge on the shaft of the tibia
[5]*epi* = upon, above; *physis* = growth
[6]*peri* = around; *oste* = bone
[7]*endo* = within; *oste* = bone

[8]*diplo* = double

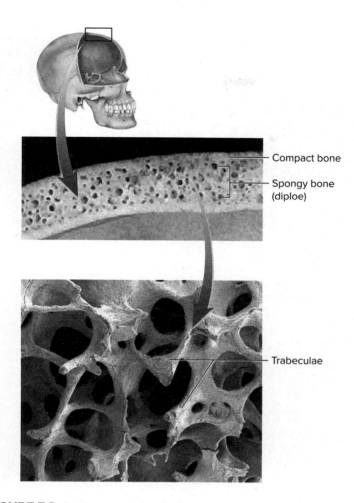

FIGURE 7.2 Anatomy of a Flat Bone.
Middle: Christine Eckel/McGraw-Hill Education; **Lower:** Steve Gschmeissner/Science Photo Library/Alamy Stock Photo

---

BEFORE YOU GO ON

Answer the following questions to test your understanding of the preceding section:

1. Name at least five tissues found in a bone.

2. List three or more functions of the skeletal system other than supporting the body and protecting some of the internal organs.

3. Describe the anatomical differences between compact and spongy bone, and their spatial relationship to each other in a long bone and a flat bone.

4. State the anatomical terms for the shaft, head, growth zone, and fibrous covering of a long bone.

7.2 | Histology of Osseous Tissue

Expected Learning Outcomes

When you have completed this section, you should be able to

a. list and describe the cells, fibers, and ground substance of bone tissue;

b. state the importance of each constituent of bone tissue;

c. compare the histology of the two types of bone tissue; and

d. distinguish between the two types of bone marrow.

7.2a Bone Cells

Like any other connective tissue, bone consists of cells, fibers, and ground substance. There are four principal types of bone cells (**fig. 7.3**):

1. **Osteogenic**[9] **cells** are stem cells that develop from embryonic mesenchyme and then give rise to most other bone cell types. They occur in the endosteum and inner layer of the periosteum. They multiply continually, and some go on to become the *osteoblasts* described next.

2. **Osteoblasts**[10] are bone-forming cells that synthesize the organic matter of the bone and then promote its mineralization. This bone-building activity is called **osteogenesis.** Osteoblasts form rows in the endosteum and inner layer of the periosteum and resemble a cuboidal epithelium on the bone surface (see fig. 7.8). They are nonmitotic, so the only source of new osteoblasts is the osteogenic cells. Stress and fractures stimulate accelerated mitosis of those cells and therefore a rapid rise in the number of osteoblasts, which then reinforce or rebuild the bone. Osteoblasts also have an endocrine function: They secrete the hormone *osteocalcin,* which stimulates insulin secretion by the pancreas, increases insulin sensitivity in adipocytes, and limits the growth of adipose tissue.

3. **Osteocytes** are former osteoblasts that have become trapped in the matrix they deposited (see this chapter's opening photo). They reside in tiny cavities called **lacunae,**[11] which are interconnected by slender channels called **canaliculi**[12] (CAN-uh-LIC-you-lye). Each osteocyte has delicate cytoplasmic processes that reach into the canaliculi to contact the processes from neighboring osteocytes. Some of them also contact osteoblasts on the bone surface. Neighboring osteocytes are connected by gap junctions where their processes meet, so they can pass nutrients and chemical signals to one another and pass their metabolic wastes to the nearest blood vessel for disposal.

   Osteocytes have multiple functions. Some resorb bone matrix and others deposit it, so they contribute to the homeostatic maintenance of bone density and blood concentrations of calcium and phosphate ions. Perhaps even more importantly, they are strain sensors. When a load is applied to a bone, it produces a flow in the extracellular fluid of the lacunae and canaliculi. This stimulates sensory cilia on the osteocytes and induces the cells to secrete signals that regulate bone remodeling—adjustments in bone shape and density to adapt to stress.

---

[9]*osteo* = bone; *genic* = producing
[10]*osteo* = bone; *blast* = form, produce
[11]*lac* = lake, hollow; *una* = little
[12]*canal* = canal, channel; *icul* = little

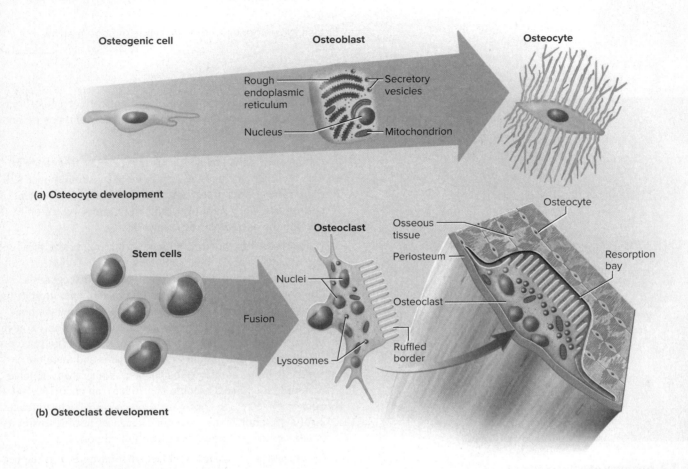

FIGURE 7.3 Bone Cells and Their Development. (a) Osteogenic cells give rise to osteoblasts, which deposit matrix around themselves and transform into osteocytes. (b) Bone marrow stem cells fuse to form osteoclasts.

4. **Osteoclasts**[13] are bone-dissolving cells on the bone surfaces. Their action is called **osteolysis,** the opposite of osteogenesis. They develop from the same bone marrow stem cells as blood cells. Thus, osteogenic cells, osteoblasts, and osteocytes all belong to one cell lineage, but osteoclasts have an independent origin (fig. 7.3). Several stem cells fuse to form each osteoclast, so osteoclasts are unusually large (up to 150 μm) and typically have 3 or 4 nuclei, but sometimes up to 50. The side of the osteoclast facing the bone surface has a *ruffled border* with many deep infoldings of the plasma membrane that increase surface area and the efficiency of bone resorption. Osteoclasts often reside in pits called *resorption bays* that they etch into the bone surface. Bone remodeling results from the combination of osteogenesis by osteoblasts and osteolysis by osteoclasts.

7.2b The Matrix

The matrix of osseous tissue averages, by dry weight, about one-third organic and two-thirds inorganic matter. The organic matter, synthesized by the osteoblasts, includes collagen and various protein–carbohydrate complexes such as glycosaminoglycans, proteoglycans, and glycoproteins (all described in section 2.4c). The inorganic matter is about 85% **hydroxyapatite,** a crystallized calcium phosphate salt $[Ca_{10}(PO_4)_6(OH)_2]$; 10% calcium carbonate $(CaCO_3)$; and lesser amounts of magnesium, sodium, potassium, fluoride, sulfate, carbonate, and hydroxide ions. Several foreign elements behave chemically like bone minerals and become incorporated into osseous tissue as contaminants, sometimes with deadly results (see Deeper Insight 7.1).

▶▶▶**APPLY WHAT YOU KNOW**

What two organelles do you think are especially prominent in osteoblasts? (Hint: *Consider the major substances that osteoblasts synthesize.*)

---

[13]*osteo* = bone; *clast* = destroy, break down

Bone is in a class of materials that engineers call a **composite,** a combination of two basic structural materials—in this case, a ceramic and a polymer. A composite can combine the optimal mechanical properties of each component. Consider a fiberglass fishing rod, for example, made of a ceramic (glass fibers) embedded in a polymer (resin). The resin alone would be too brittle and the fibers alone too flexible and limp to serve the purpose of a fishing rod, but together they produce a material of great strength and flexibility.

In bone, the polymer is collagen and the ceramic is hydroxyapatite and other minerals. The ceramic component enables a bone to support the weight of the body without sagging. When the bones are deficient in calcium salts, they are soft and bend easily. One way to demonstrate this is to soak a clean dry bone, such as a chicken bone, in vinegar for a few days. As the mild acid of the vinegar dissolves the minerals, the bone becomes flexible and rubbery. Such mineral deficiency and flexibility are the central problems in the childhood disease *rickets,* in which the soft bones of the lower limbs bend under the body's weight and become permanently deformed (see Deeper Insight 7.3).

The protein component gives bone a degree of flexibility. Without protein, a bone is excessively brittle, as in *osteogenesis imperfecta,* or *brittle bone disease* (see table 7.2). Without collagen, a jogger's bones would shatter under the impact of running. But normally, when a bone bends slightly toward one side, the tensile strength of the collagen fibers on the opposite side holds the bone together and prevents it from snapping like a stick of chalk. Collagen molecules have *sacrificial bonds* that break under stress, protecting a bone from fracture by dissipating some of the shock. The bonds re-form when the collagen is relieved of stress.

DEEPER INSIGHT 7.1
MEDICAL HISTORY

Bone Contamination

When Marie and Pierre Curie and Henri Becquerel received their 1903 Nobel Prize for the discovery of radioactivity (see Deeper Insight 2.1), radiation captured the public imagination. Not for several decades did anyone realize its dangers. For example, factories employed women to paint luminous numbers on watch and clock dials with radium paint. The women moistened their paint brushes with their tongues to keep them finely pointed and ingested radium in the process. The radium accumulated in their bones and caused many of them to develop a bone cancer called osteosarcoma. History remembers these women as the tragic "radium girls."

Even more horrific, in the wisdom of hindsight, was a deadly health fad in which people drank "tonics" made of radium-enriched water. One famous enthusiast was the millionaire playboy and championship golfer Eben Byers (1880–1932), who drank several bottles of radium tonic each day and praised its virtues as a wonder drug and aphrodisiac. Like the factory women, Byers contracted osteosarcoma. By the time of his death, holes had formed in his skull and doctors had removed his entire upper jaw and most of his mandible in an effort to halt the spreading cancer. Byers' bones and teeth were so radioactive they could expose photographic film in the dark. Brain damage left him unable to speak, but he remained mentally alert to the bitter end. His tragic decline and death shocked the world and helped put an end to the radium tonic fad.

Unlike fiberglass, bone varies from place to place in its ratio of minerals to collagen. The middle-ear bones, for example, are about 90% mineral, giving them the necessary stiffness to efficiently conduct sound vibrations to the inner ear. The limb bones, by contrast, have a higher percentage of collagen so they can bend slightly under the body's weight. Osseous tissue is thus adapted to different amounts of tension and compression exerted on different parts of the skeleton.

7.2c Compact Bone

The histological study of compact bone usually uses slices that have been dried, cut with a saw, and ground to translucent thinness. This procedure destroys the cells but reveals fine details of the matrix (**fig. 7.4**). Such sections show onionlike **concentric lamellae**—layers of matrix concentrically arranged around a **central (haversian[14]) canal** and connected with each other by canaliculi. A central canal and its lamellae constitute an **osteon (haversian system)**—the basic structural unit of compact bone. In longitudinal views and three-dimensional reconstructions, we can see that an osteon is actually a cylinder of tissue surrounding a central canal. Along their length, central canals are joined by transverse or diagonal passages called **perforating canals.** The central and perforating canals are lined with endosteum. Each osteon is separated from its neighbors by a thin *cement line,* which blocks microfractures of the bone from spreading and minimizes the chance of them causing a large-scale fracture.

Collagen fibers "corkscrew" down the matrix of a given lamella in a helical arrangement like the threads of a screw. The helices coil in one direction in one lamella and in the opposite direction in the next lamella (fig. 7.4b). This enhances the strength of bone on the same principle as plywood, made of thin layers of wood with the grain running in different directions from one layer to the next. In areas where the bone must resist tension (bending), the helix is loosely coiled like the threads on a wood screw and the fibers are more stretched out on the longitudinal axis of the bone. In weight-bearing areas where resistance to compression is more important, the helix is tightly coiled like the closely spaced threads on a bolt, and the fibers are more nearly transverse.

The skeleton receives about half a liter of blood per minute. Blood vessels, along with nerves, enter the bone tissue through nutrient foramina on the surface. These foramina open into the perforating canals that cross the matrix and feed into the central canals. The innermost osteocytes around each central canal receive nutrients from these blood vessels and pass them along through their gap junctions to neighboring osteocytes. They also receive wastes from their neighbors and convey them to the central canal for removal by the bloodstream. Thus, the cytoplasmic processes of the osteocytes maintain a two-way flow of nutrients and wastes between the central canal and the outermost cells of the osteon.

Not all of the matrix is organized into osteons. The inner and outer boundaries of dense bone are arranged in *circumferential lamellae* that run parallel to the bone surface. Between osteons, we can find irregular regions called *interstitial lamellae,* the remains of old osteons that broke down as the bone grew and remodeled itself.

[14]Clopton Havers (1650–1702), English anatomist

FIGURE 7.4 The Histology of Osseous Tissue. (a) Compact and spongy bone in a longitudinal section of the femur. (b) The three-dimensional structure of compact bone. Lamellae of one osteon are telescoped to show their alternating arrangement of collagen fibers. (c) Histology of decalcified spongy bone and red bone marrow. (d) Microscopic appearance of a cross section of an osteon of dried compact bone. The art inset relates osteocyte structure to the shapes of the lacunae and canaliculi of the bone. **APR**

a: B Christopher/Alamy Stock Photo; **c:** Biophoto Associates/Science Source; **d:** Ed Reschke/Getty Images

7.2d Spongy Bone

Spongy bone (fig. 7.4a–c) consists of a lattice of delicate slivers called **spicules**[15] (rods or spines) and **trabeculae**[16] (thin plates or beams). Although calcified and hard, it is named for its sponge-like appearance. It is covered with endosteum and permeated by spaces filled with bone marrow. The matrix is arranged in lamellae like those of compact bone, but there are few osteons. Central canals are not needed here because no osteocyte is very far from the marrow. Spongy bone is well designed to impart strength to a bone while adding a minimum of weight. Its trabeculae aren't randomly arranged as they may seem at a glance, but develop along the bone's lines of stress **(fig. 7.5)**. Spongy bone has much more surface area exposed to osteoclast action than compact bone does. Therefore, when osteoclasts resorb bone tissue, it comes largely from the spongy bone, as we see in osteoporosis (see Deeper Insight 7.4 at the end of this chapter).

7.2e Bone Marrow

Bone marrow is a general term for soft tissue that occupies the marrow cavity of a long bone, the spaces amid the trabeculae of spongy bone, and the larger central canals. There are two kinds of marrow—red and yellow. We can best appreciate their differences by considering how marrow changes over a person's lifetime.

In a child, the marrow cavity of nearly every bone is filled with **red bone marrow (myeloid tissue).** This is often described as *hematopoietic*[17] *tissue* (he-MAT-o-poy-ET-ic)—tissue that produces blood cells—but it is actually composed of multiple tissues in a delicate but intricate arrangement, and is properly considered an organ unto itself. Its structure is further described in section 21.1d.

In adults, most of the red marrow is replaced by fatty **yellow bone marrow,** like the fat at the center of a ham bone. Red marrow is then limited to the skull, vertebrae, ribs, sternum, part of the pelvic (hip) girdle, and the proximal heads of the humerus and femur **(fig. 7.6).** Yellow bone marrow no longer produces blood, although in the event of severe or chronic anemia, it can transform back into red marrow and resume its hematopoietic function.

BEFORE YOU GO ON

Answer the following questions to test your understanding of the preceding section:

5. Suppose you had unlabeled electron micrographs of the four kinds of bone cells and their neighboring tissues. Name the four cells and explain how you could visually distinguish each one from the other three.

6. Name three organic components of the bone matrix.

7. What are the mineral crystals of bone called, and what are they made of?

Greater trochanter

Head

Trabeculae of spongy bone

Compact bone

Lines of stress

Shaft (diaphysis)

FIGURE 7.5 Spongy Bone Structure in Relation to Mechanical Stress. In this frontal section of the femur, the trabeculae of spongy bone can be seen oriented along lines of mechanical stress applied by the weight of the body or the pull of a muscle.
B Christopher/Alamy Stock Photo

8. Sketch a cross section of an osteon and label its major parts.

9. What are the two kinds of bone marrow? What does *hematopoietic tissue* mean? Which type of bone marrow fits this description?

FIGURE 7.6 Adult Distribution of Red and Yellow Bone Marrow.

❓ *What would be the most accessible places to draw red bone marrow from an adult?*

---

[15]*spic* = dart, point; *ule* = little

[16]*trabe* = beam; *cul* = little

[17]*hemato* = blood; *poietic* = forming

7.3   Bone Development

Expected Learning Outcomes

When you have completed this section, you should be able to

a. describe two mechanisms of bone formation; and

b. explain how mature bone continues to grow and remodel itself.

The formation of bone is called **ossification** (OSS-ih-fih-CAY-shun) or **osteogenesis.** There are two methods of ossification—*intramembranous* and *endochondral*. Both begin with embryonic *mesenchyme* (MEZ-en-kime).

7.3a Intramembranous Ossification

Intramembranous[18] **ossification** (IN-tra-MEM-bra-nus) produces the flat bones of the skull, most of the clavicle (collarbone), and part of the mandible. Follow its stages in **figure 7.7** as you read the correspondingly numbered descriptions here.

(1) Mesenchyme first condenses into a soft sheet of tissue permeated with blood vessels—the *membrane* to which

intramembranous refers. Mesenchymal cells line up along the blood vessels, become osteoblasts, and secrete a soft collagenous *osteoid*[19] *tissue (prebone)* (**fig. 7.8**) in the direction away from the vessel. Osteoid tissue resembles bone but is not yet calcified.

(2) Calcium phosphate and other minerals crystallize on the collagen fibers of the osteoid tissue and harden the matrix. Continued osteoid deposition and mineralization squeeze the blood vessels and future bone marrow into narrower and narrower spaces. As osteoblasts become trapped in their own hardening matrix, they become osteocytes.

(3) While the foregoing processes are occurring, more mesenchyme adjacent to the developing bone condenses and forms a fibrous periosteum on each surface. The spongy bone becomes a honeycomb of slender calcified trabeculae.

(4) At the surfaces, osteoblasts beneath the periosteum deposit layers of bone, fill in the spaces between trabeculae, and create a zone of compact bone on each side as well as thicken the bone overall. This process gives rise to the sandwichlike structure typical of a flat cranial bone—a layer of spongy bone between two layers of compact bone.

---

[18]*intra* = within; *membran* = membrane

[19]*oste* = bone; *oid* = like, resembling

(1) Deposition of osteoid tissue into embryonic mesenchyme

Blood vessel
Osteoid tissue
Osteoblasts
Mesenchyme
Mesenchymal cell

(2) Calcification of osteoid tissue and entrapment of osteocytes

Osteoid tissue
Bone trabecula
Marrow spaces
Osteocyte
Osteoblast

(3) Honeycomb of spongy bone with developing periosteum

Mesenchyme
Periosteum
Spongy bone
Periosteum
Bone trabeculae

(4) Filling of space to form compact bone at surfaces, leaving spongy bone in middle

Periosteum
Compact bone
Spongy bone
Compact bone
Periosteum

FIGURE 7.7 Intramembranous Ossification. The figures are drawn to different scales, with the highest magnification and detail at the beginning and backing off for a broader overview at the end of the process.

❓ *With the aid of chapter 8, name at least two specific bones other than the clavicle that would form by this process.*

FIGURE 7.8 Intramembranous Ossification in the Fetal Cranium. Note the layers of osteoid tissue, osteoblasts, and fibrous periosteum on both sides of the bone.
Ken Saladin

Intramembranous ossification also plays an important role in the lifelong thickening, strengthening, and remodeling of the long bones discussed next. Throughout the skeleton, it is the method of depositing new tissue on the bone surface even past the age where the bones can no longer grow in length.

7.3b Endochondral Ossification

Endochondral[20] **ossification** (EN-doe-CON-drul) is a process in which a bone develops from a preexisting model composed of hyaline cartilage. It begins around the sixth week of fetal development and continues into a person's 20s. Most bones of the body develop in this way, including the vertebrae, ribs, sternum, scapula, pelvic girdle, and bones of the limbs.

Figure 7.9 shows the following steps in endochondral ossification. This figure uses a metacarpal bone from the palmar region of the hand as an example because of its relative simplicity, having only one *epiphysial plate* (growth center). Many other bones develop in more complex ways, having an epiphysial plate at both ends or multiple plates at each end, but the basic process is the same.

① Mesenchyme develops into a body of hyaline cartilage, covered with a fibrous perichondrium, in the location of a future bone. For a time, the perichondrium produces chondrocytes and the cartilage model grows in thickness.

② In a **primary ossification center** near the middle of this cartilage, chondrocytes begin to inflate and die, while the thin walls between them calcify. The perichondrium stops producing chondrocytes and begins producing osteoblasts. These deposit a thin collar of bone around the middle of the cartilage model,

[20]*endo* = within; *chondr* = cartilage

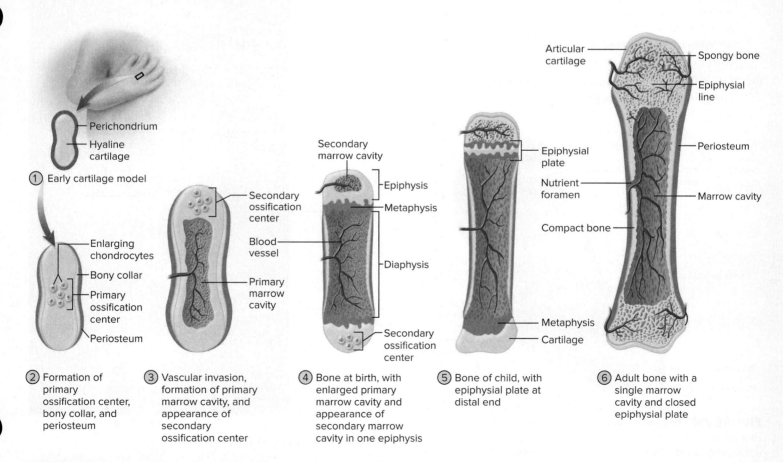

FIGURE 7.9 Stages of Endochondral Ossification. A metacarpal bone of the hand.

❓ *With the aid of chapter 8, name at least two specific bones that would have two epiphysial plates (proximal and distal) at stage 5.*

reinforcing it like a napkin ring. The former perichondrium is now considered to be a periosteum. As chondrocytes in the middle of the model die, their lacunae merge into a single cavity.

③ Blood vessels invade the primary ossification center, delivering blood-borne osteoclasts that digest the calcified tissue. This creates a hollowed-out center called the **primary marrow cavity.** Osteoblasts also arrive and deposit layers of bone lining the cavity, thickening the shaft. As the bony collar under the periosteum thickens and elongates, a wave of cartilage death progresses toward each end of the bone. Osteoclasts in the marrow cavity follow this wave, dissolving calcified cartilage remnants and enlarging the marrow cavity of the diaphysis. The region of transition from cartilage to bone at each end of the primary marrow cavity is called a **metaphysis** (meh-TAFF-ih-sis). Soon, chondrocyte enlargement and death and vascular invasion occur in the epiphysis of the model as well, creating a **secondary ossification center.** In the metacarpal bones, as illustrated in the figure, this occurs in only one epiphysis. In longer bones of the arms, forearms, legs, and thighs, it occurs at both ends.

④ The secondary ossification center hollows out by the same process as the diaphysis, generating a **secondary marrow cavity** in the epiphysis. This cavity expands outward from the center in all directions. At the time of birth, the bone typically looks like step 4 in the figure. In bones with two secondary ossification centers, one center lags behind the other, so at birth there is a secondary marrow cavity at one end while chondrocyte growth has just begun at the other. The joints of the limbs are still cartilaginous at birth, much as they are in the 12-week fetus in **figure 7.10.**

⑤ During infancy and childhood, the epiphyses fill with spongy bone. Cartilage is then limited to the articular

cartilage covering each joint surface, and to an **epiphysial plate** (EP-ih-FIZ-ee-ul), a thin wall of cartilage separating the primary and secondary marrow cavities at one or both ends of the bone. The plate persists through childhood and adolescence and serves as a growth zone for bone elongation.

⑥ By the late teens to early twenties, all remaining cartilage in the epiphysial plate is generally consumed and the gap between the epiphysis and diaphysis closes. The primary and secondary marrow cavities then unite into a single cavity.

7.3c Bone Growth and Remodeling

Ossification doesn't end at birth, but continues throughout life with the growth and remodeling of bones. Bones grow in two directions: length and width.

Bone Elongation

To understand growth in length, we must return to the epiphysial plates mentioned earlier (see fig. 7.9, step 5). From infancy through adolescence, an epiphysial plate is present at one or both ends of a long bone, at the junction between the diaphysis and epiphysis. On X-rays, it appears as a translucent line across the end of a bone, since it is not yet ossified (**fig. 7.11;** compare the X-ray of an adult hand in fig. 8.35). The epiphysial plate is a region of

FIGURE 7.10 The Fetal Skeleton at 12 Weeks. The red-stained regions are calcified at this age, whereas the elbow, wrist, knee, and ankle joints appear translucent because they are still cartilaginous.

❓ *Why are the joints of an infant weaker than those of an older child?*

Biophoto Associates/Science Source

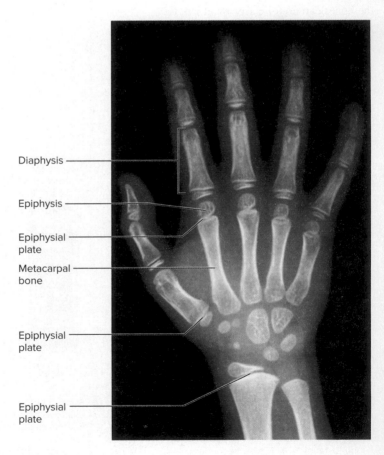

FIGURE 7.11 X-Ray of a Child's Hand. The cartilaginous epiphysial plates are evident at the ends of the long bones. Long bones of the hand and fingers develop only one epiphysial plate. Notice that the wrist is still largely cartilaginous (X-ray translucent).

Puwadol Jaturawutthichai/Shutterstock

transition from cartilage to bone, and functions as a growth zone where the bones elongate. Growth here is responsible for a person's increase in height.

The epiphysial plate consists of typical hyaline cartilage in the middle, with a transitional zone on each side where cartilage is being replaced by bone. The transitional zone, facing the marrow cavity, is called the **metaphysis** (meh-TAF-ih-sis). In figure 7.9, step 4, the cartilage is the blue region and each metaphysis is violet. **Figure 7.12** shows the histological structure of the metaphysis and the following steps in the replacement of cartilage by bone.

① **Zone of reserve cartilage.** This region, farthest from the marrow cavity, consists of typical hyaline cartilage with resting chondrocytes, not yet showing any sign of transformation into bone.

② **Zone of cell proliferation.** A little closer to the marrow cavity, chondrocytes multiply and arrange themselves into longitudinal columns of flattened lacunae.

③ **Zone of cell hypertrophy.** Next, the chondrocytes cease to multiply and begin to hypertrophy (enlarge), much like they

do in the primary ossification center of the fetus. The walls of matrix between lacunae become very thin.

④ **Zone of calcification.** Minerals are deposited in the matrix between the columns of lacunae and calcify the cartilage. These are not the permanent mineral deposits of bone, but only a temporary support for the cartilage that would otherwise soon be weakened by the breakdown of the enlarged lacunae.

⑤ **Zone of bone deposition.** Within each column, the walls between the lacunae break down and the chondrocytes die. This converts each column into a longitudinal channel (clear spaces in the figure), which is immediately invaded by blood vessels and marrow from the marrow cavity. Osteoblasts line up along the walls of these channels and begin depositing concentric lamellae of matrix, while osteoclasts dissolve the temporarily calcified cartilage.

The process of bone deposition in zone 5 creates a region of spongy bone at the end of the marrow cavity facing the metaphysis. This spongy bone remains for life, although with extensive lifelong

Multiplying chondrocytes

Enlarging chondrocytes

Breakdown of lacunae

Calcifying cartilage

Bone marrow

Osteoblasts

Osteocytes

Trabeculae of spongy bone

Zone ①
Zone ⑤

① **Zone of reserve cartilage**
Typical histology of resting hyaline cartilage

② **Zone of cell proliferation**
Chondrocytes multiplying and lining up in rows of small flattened lacunae

③ **Zone of cell hypertrophy**
Cessation of mitosis; enlargement of chondrocytes and thinning of lacuna walls

④ **Zone of calcification**
Temporary calcification of cartilage matrix between columns of lacunae

⑤ **Zone of bone deposition**
Breakdown of lacuna walls, leaving open channels; death of chondrocytes; bone deposition by osteoblasts, forming trabeculae of spongy bone

FIGURE 7.12 Zones of the Metaphysis. This micrograph shows the transition from cartilage to bone in the growth zone of a long bone.

❓ *Which two zones in this figure account for a child's growth in height?*

Victor Eroschenko

remodeling. But around the perimeter of the marrow cavity, continuing ossification converts this spongy bone to compact bone. Osteoblasts lining the aforementioned channels deposit layer after layer of bone matrix, so the channel grows narrower and narrower. These layers become the concentric lamellae of an osteon. Finally only a slender channel persists, the central canal of a new osteon. Osteoblasts trapped in the matrix become osteocytes.

▶▶▶APPLY WHAT YOU KNOW

In a given osteon, which lamellae are the oldest—those immediately adjacent to the central canal or those around the perimeter of the osteon? Explain your answer.

How does a child or adolescent grow in height? Chondrocyte multiplication in zone 2 and hypertrophy in zone 3 continually push the zone of reserve cartilage (1) toward the ends of the bone, so the bone elongates. In the lower limbs, this process causes a person to grow in height, while bones of the upper limbs grow proportionately.

Thus, bone elongation is really a result of cartilage growth. Cartilage growth from within, by the multiplication of chondrocytes and deposition of new matrix in the interior, is called **interstitial**[21] **growth.** The most common form of dwarfism results from a failure of cartilage growth in the long bones (see Deeper Insight 7.2).

In the late teens to early twenties, all the cartilage of the epiphysial plate is depleted. The primary and secondary marrow cavities now unite into one cavity. The junctional region where they meet is filled with spongy bone, and the site of the original epiphysial plate is marked with a line of slightly denser spongy bone called the **epiphysial line** (see figs. 7.1; 7.5; and 7.9, step 6). Often a delicate ridge on the bone surface marks the location of this line. When the epiphysial plate is depleted, we say that the epiphyses have "closed" because no gap between the epiphysis and diaphysis is visible on an X-ray. Once the epiphyses have all closed in the lower limbs, a person can grow no taller. The epiphysial plates close at different ages in different bones and in different regions of the same bone. The processes and rates of elongation and epiphysial closure are influenced especially by growth hormone and sex steroids (see section 7.4d). The state of closure in various bones of a subadult skeleton is often used in forensic science to estimate the individual's "bone age" at death.

Bone Widening and Thickening

Bones also continually grow in diameter and thickness. This involves a process called **appositional growth,**[22] the deposition of new tissue at the surface. Cartilages grow by both interstitial and appositional growth. In bone, however, osteocytes embedded in calcified matrix have little room to spare for the deposition of more matrix internally. Bone is therefore limited to appositional growth.

Appositional growth occurs by intramembranous ossification at the bone surface. Osteoblasts in the inner layer of periosteum deposit osteoid tissue on the bone surface, calcify it, and become trapped in it as osteocytes—much like the process in figure 7.8.

They lay down matrix in layers parallel to the surface, not in cylindrical osteons like those deeper in the bone. This process produces the surface layers of bone called *circumferential lamellae,* described earlier. As a bone increases in diameter, its marrow cavity also widens. This is achieved by osteoclasts of the endosteum dissolving tissue on the inner bone surface. Thus, flat bones develop by intramembranous ossificaton alone, whereas long bones develop by a combination of the intramembranous and endochondral methods.

Bone Remodeling

In addition to their growth, bones are continually remodeled throughout life by the absorption of old bone and deposition of new. This process replaces about 10% of the skeletal tissue per year. It releases minerals into the blood for uses elsewhere; reshapes bones in response to use and disuse; and repairs microfractures, preventing them from developing into catastrophic bone failure similar to metal fatigue.

Wolff's[23] **law of bone** states that the architecture of a bone is determined by the mechanical stresses placed upon it, and the bone thereby adapts to withstand them. Wolff's law is a fine example of the complementarity of form and function, showing that the form of a bone is shaped by its functional experience. It is admirably demonstrated by figure 7.5, in which we see that the trabeculae of spongy bone lie along the lines of stress placed on the femur. Wolff observed that these lines were similar to the ones engineers knew of in mechanical cranes. The effect of stress on bone development is quite evident in elite tennis players, in whom the cortical bone of the racket arm is up to 35% thicker than that of the other arm. Long bones of the limbs are thickest at midshaft, where they are subjected to the greatest stress.

Bone remodeling comes about through the collaborative action of osteoblasts and osteoclasts. If a bone is little used, osteoclasts remove matrix and get rid of unnecessary mass. If a bone is heavily used or stress is consistently applied to a particular region of a bone, osteoblasts deposit new osseous tissue and thicken it. Consequently, the comparatively smooth bones of an infant or toddler develop a variety of surface bumps, ridges, and spines (described in chapter 8) as the child begins to walk. The greater trochanter of the femur, for example (see figs. 7.5 and 8.39), is a massive outgrowth of bone stimulated by the pull of tendons from several powerful hip muscles employed in walking.

On average, bones have a greater density and mass in athletes and people engaged in heavy manual labor than they do in sedentary people. Anthropologists who study ancient skeletal remains use evidence of this sort to help distinguish between members of different social classes, such as distinguishing aristocrats from laborers. Even in studying modern skeletal remains, as in investigating a suspicious death, Wolff's law comes into play as the bones give evidence of a person's sex, race, height, weight, nutritional status, work or exercise habits, and medical history.

The orderly remodeling of bone depends on a precise balance between deposition and resorption, between osteoblasts and osteoclasts. If one process outpaces the other, or both processes occur

---

[21]*inter* = between; *stit* = to place, stand
[22]*ap* = *ad* = to, near; *posit* = to place

[23]Julius Wolff (1836–1902), German anatomist and surgeon

DEEPER INSIGHT 7.2

CLINICAL APPLICATION

Achondroplastic Dwarfism

Achondroplastic[24] *dwarfism* (a-con-dro-PLAS-tic) is a condition in which the long bones of the limbs stop growing in childhood, while the growth of other bones is unaffected. As a result, a person has a short stature but a normal-size head and trunk **(fig. 7.13).** As its name implies, achondroplastic dwarfism results from a failure of cartilage growth—specifically, failure of the chondrocytes in zones 2 and 3 of the metaphysis to multiply and enlarge. This is different from *pituitary dwarfism,* in which a deficiency of growth hormone stunts the growth of all of the bones, and a person has short stature but normal proportions throughout the skeletal system.

Achondroplastic dwarfism results from a spontaneous mutation that can arise any time DNA is replicated. Two people of normal height with no family history of dwarfism can therefore have a child with achondroplastic dwarfism. The mutant allele is dominant, so the offspring of a heterozygous achondroplastic dwarf have at least a 50% chance of exhibiting dwarfism, depending on the genotype of the other parent. Persons homozygous for the trait (those who inherit it from both parents) are usually stillborn or die soon after birth.

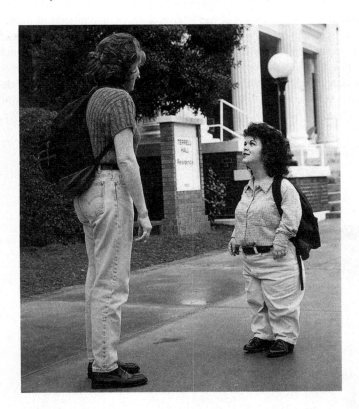

FIGURE 7.13 Achondroplastic Dwarfism. The student on the right, pictured with her roommate of normal height, is an achondroplastic dwarf with a height of about 122 cm (48 in.). Her parents were of normal height. Note the normal proportion of head to trunk but shortening of the limbs.

Joe DeGrandis/McGraw-Hill Education

---

[24]*a* = without; *chondr* = cartilage; *plast* = growth

too rapidly, various bone deformities, developmental abnormalities, and other disorders occur, such as *osteitis deformans* (Paget disease), *osteogenesis imperfecta* (brittle bone disease), and *osteoporosis* (see table 7.2 and Deeper Insight 7.4).

BEFORE YOU GO ON

Answer the following questions to test your understanding of the preceding section:

10. Describe the stages of intramembranous ossification. Name a bone that forms in this way.

11. Describe how a cartilage model transforms into a long bone in endochondral ossification.

12. Describe the five zones of a metaphysis and the major distinctions between them.

13. How does Wolff's law explain some of the structural differences between the bones of a young child and the bones of a young adult?

7.4   Physiology of Osseous Tissue

Expected Learning Outcomes

When you have completed this section, you should be able to

a. describe the processes by which minerals are added to and removed from bone tissue;

b. discuss the role of the bones in regulating blood calcium and phosphate levels; and

c. name the main hormones that regulate bone physiology, and describe their effects.

Even after a bone is fully formed, it remains a metabolically active organ with many roles to play. Not only is it involved in its own maintenance, growth, and remodeling, but it also exerts a profound influence on the rest of the body by exchanging minerals with the tissue fluid. Disturbances of calcium homeostasis in the skeleton can disrupt the functioning of other organ systems, especially the nervous and muscular systems. For reasons explained later, such disturbances can even cause death by suffocation.

7.4a   Mineral Deposition and Resorption

Mineral deposition (mineralization) is a crystallization process in which calcium, phosphate, and other ions are taken from the blood plasma and deposited in bone tissue, mainly as needlelike crystals of hydroxyapatite. Deposition begins in fetal ossification and continues throughout life.

Osteoblasts begin the process by laying down collagen fibers in a helical pattern along the length of the osteon. These fibers then become encrusted with minerals that harden the matrix. The first few hydroxyapatite crystals to form act as "seed crystals" that attract more calcium and phosphate from solution. The more hydroxyapatite that forms, the more it attracts additional minerals from the tissue fluid, until the matrix is thoroughly calcified.

▶▶▶APPLY WHAT YOU KNOW

What positive feedback process can you recognize in bone deposition?

Abnormal calcification of tissues, called **ectopic**[25] **ossification,** sometimes occurs in the lungs, brain, eyes, muscles, tendons, arteries, and other organs. One example of this is arteriosclerosis, or "hardening of the arteries," which results from calcification of the arterial walls. A calcified mass in an otherwise soft organ such as the lungs is called a **calculus.**[26]

Mineral resorption is the process of dissolving bone. It releases minerals into the blood and makes them available for other uses. Resorption is carried out by osteoclasts. They have surface receptors for calcium and respond to falling levels of calcium in the tissue fluid. Hydrogen pumps in the ruffled border of the osteoclast secrete hydrogen ions into the tissue fluid, and chloride ions follow by electrical attraction. The space between the osteoclast and the bone thus becomes filled with concentrated hydrochloric acid with a pH of about 4. The acid dissolves the bone minerals. The osteoclast also secretes an acid-tolerant enzyme (protease) that digests the collagen of the bone matrix.

When orthodontic appliances (braces) are used to reposition teeth, a tooth moves because osteoclasts dissolve bone ahead of the tooth (where the appliance creates greater pressure of the tooth against the bone) and osteoblasts deposit bone in the low-pressure zone behind it.

[25]*ec* = out of; *top* = place
[26]*calc* = stone; *ulus* = little

7.4b Calcium Homeostasis

The adult body contains about 1,100 g of calcium, with 99% of it in the bones. Calcium is needed for much more than bone structure. It also plays roles in communication among neurons and in muscle contraction, blood clotting, and exocytosis. Calcium is deposited in the skeleton when the supply is ample and withdrawn when needed for these other purposes. The skeleton exchanges about 18% of its calcium with the blood each year.

This exchange is tightly regulated by hormones to maintain a blood calcium concentration of 9.2 to 10.4 mg/dL. This is a narrow margin of safety. A calcium deficiency, called **hypocalcemia**[27] (HY-po-cal-SEE-me-uh), causes excessive excitability of the nervous system and can lead to muscle tremors, spasms, or tetany—inability of the muscles to relax. Tetany of the muscles of the larynx can cause death by suffocation. A calcium excess, called **hypercalcemia,**[28] makes nerve and muscle cells less excitable than normal. This can be manifested in nervous system depression, emotional disturbances, muscle weakness, sluggish reflexes, and sometimes cardiac arrest.

Hypercalcemia is rare, but hypocalcemia can result from a wide variety of causes including vitamin D deficiency, diarrhea, thyroid tumors, or underactive parathyroid glands. Pregnancy and lactation put women at risk of hypocalcemia because of the calcium demanded by ossification of the fetal skeleton and synthesis of milk.

You can see how critical blood calcium level is, but what causes it to deviate from the norm, and how does the body correct such imbalances? Calcium homeostasis depends on a balance between dietary intake, urinary and fecal losses, and exchanges with the osseous tissue **(fig. 7.14).** It is regulated by three hormones: *calcitriol, calcitonin,* and *parathyroid hormone.*

[27]*hypo* = below normal; *calc* = calcium; *emia* = blood condition
[28]*hyper* = above normal; *calc* = calcium; *emia* = blood condition

FIGURE 7.14 Hormonal Control of Calcium Balance. The central panel represents the blood reservoir of calcium and shows its normal (safe) range. Calcitriol and PTH regulate calcium exchanges between the blood and the small intestine and kidneys (left). Calcitonin, calcitriol, and PTH regulate calcium exchanges between blood and bone (right).

DEEPER INSIGHT 7.3

CLINICAL APPLICATION

Rickets and Osteomalacia

Rickets is a childhood disease in which the bones are soft and deformed as a result of vitamin D and calcium deficiency. It is characterized by fragile bones, bone pain, and frequent fractures; deformity of the chest, pelvis, and spine; and bowed legs **(fig. 7.15).** It is one of the most common childhood diseases of developing countries, but usually preventable by exposure to sunlight. Even 15 minutes per day is usually enough for adequate vitamin D synthesis. Vitamin D can also be obtained from fortified milk and dairy products, eggs, oily fish, and some mushrooms.

Rickets is common in areas wracked with warfare and famine. It also develops in infants who are breast-fed and kept from the sun without receiving supplemental vitamin D. In London during the early 1800s, it occurred in 80% to 90% of children who were kept indoors because of the heavy smog in the air and who were often exploited for long hours of factory work. The United States mandated the addition of vitamin D to milk in 1930 and outlawed child labor in 1937, greatly reducing rickets and other health problems. Yet rickets is increasing again as children spend more time indoors and even because of the increased use of sunscreen. The highest incidence today is in the Mideast, owing to covering of most or all of the body for religious or other cultural reasons.

Osteomalacia is a similar bone-softening, vitamin D deficiency disease of adults. It is especially common in nursing home residents and the homebound elderly. It often begins with lumbar aches and pains and

FIGURE 7.15 Rickets in a Boy of Eastern Kenya.
Jeff Rotman / Alamy Stock Photo

progresses to bone and joint pain, muscle weakness, difficulty walking and stair climbing, loss of height because of vertebral compression, a waddling gait due to spinal deformity, and pathological bone fractures.

Calcitriol

Calcitriol (CAL-sih-TRY-ol) is a form of vitamin D produced by the sequential action of the skin, liver, and kidneys **(fig. 7.16):**

1. Epidermal keratinocytes use ultraviolet radiation from sunlight to convert a steroid, 7-dehydrocholesterol, to previtamin D_3. Over another 3 days, the warmth of sunlight on the skin further converts this to vitamin D_3 *(cholecalciferol),* and a transport protein carries this to the bloodstream.

2. The liver adds a hydroxyl group, converting it to *calcidiol.*

3. The kidneys then add another hydroxyl group, converting calcidiol to calcitriol, the most active form of vitamin D.

Calcitriol behaves as a hormone—a blood-borne chemical messenger from one organ to another. It is called a vitamin only because it is added to the diet, mainly in fortified milk, as a safeguard for people who don't get enough sunlight to initiate adequate synthesis in the skin.

The principal function of calcitriol is to raise the blood calcium concentration. It does this in three ways (fig. 7.14), especially the first of these:

1. It increases calcium absorption by the small intestine, using mechanisms detailed in chapter 25 (see section 25.6f).

2. It increases calcium resorption from the skeleton. Calcitriol binds to osteoblasts, which then stimulate stem cells to differentiate into osteoclasts. The new osteoclasts liberate calcium and phosphate ions from bone.

3. It weakly promotes the reabsorption of calcium ions by the kidneys, so less calcium is lost in the urine.

Although calcitriol promotes bone resorption, it is also necessary for bone deposition. Without it, calcium and phosphate levels in the blood are too low for normal deposition. The result is a softness of the bones called **rickets** in children and **osteomalacia**[29] in adults (see Deeper Insight 7.3).

Calcitonin

Calcitonin is produced by *parafollicular (clear) cells* of the thyroid gland (see fig. 17.9). It is secreted when the blood calcium concentration rises too high, and it lowers the concentration by two principal mechanisms (figs. 7.14 and **7.17a):**

1. **Osteoclast inhibition.** Within 15 minutes after it is secreted, calcitonin reduces osteoclast activity by as much as 70%, so osteoclasts liberate less calcium from the skeleton.

2. **Osteoblast stimulation.** Within an hour, calcitonin increases the number and activity of osteoblasts, which deposit calcium into the skeleton.

Calcitonin plays an important role in children but has only a weak effect in most adults. The osteoclasts of children are highly active in skeletal remodeling and release 5 g or more of calcium into the blood each day. By inhibiting this, calcitonin

[29]*osteo* = bone; *malacia* = softening

FIGURE 7.16 Calcitriol (Vitamin D) Synthesis and Action. See text for explanation.

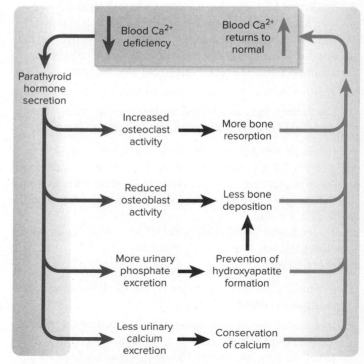

7-dehydrocholesterol

Ultraviolet light

Vitamin D₃ (cholecalciferol)

Calcidiol

Calcitriol

Bone resorption

Reduced excretion of Ca²⁺

Absorption of Ca²⁺ and phosphate

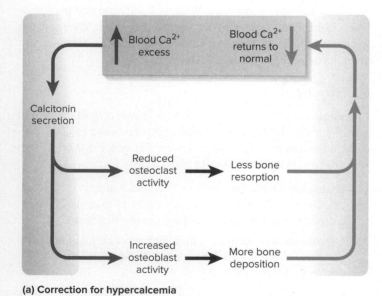

Blood Ca²⁺ excess

Blood Ca²⁺ returns to normal

Calcitonin secretion

Reduced osteoclast activity → Less bone resorption

Increased osteoblast activity → More bone deposition

(a) Correction for hypercalcemia

Blood Ca²⁺ deficiency

Blood Ca²⁺ returns to normal

Parathyroid hormone secretion

Increased osteoclast activity → More bone resorption

Reduced osteoblast activity → Less bone deposition

More urinary phosphate excretion → Prevention of hydroxyapatite formation

Less urinary calcium excretion → Conservation of calcium

(b) Correction for hypocalcemia

FIGURE 7.17 Negative Feedback Loops in Calcium Homeostasis. (a) The correction of hypercalcemia by calcitonin. (b) The correction of hypocalcemia by parathyroid hormone.

can significantly lower the blood calcium level in children. In adults, however, the osteoclasts release only about 0.8 g of calcium per day. Calcitonin cannot change adult blood calcium very much by suppressing this lesser contribution. Calcitonin deficiency isn't known to cause any adult disease. Calcitonin may, however, help to maintain bone density in pregnant and lactating women.

Parathyroid Hormone

Parathyroid hormone (PTH) is secreted by the parathyroid glands, which adhere to the posterior surface of the thyroid gland. These glands release PTH when blood calcium is low. A mere 1% drop in the calcium level doubles PTH secretion. PTH raises the calcium level by four mechanisms (figs. 7.14 and **7.17b**):

1. It binds to receptors on the osteoblasts, which in turn stimulate the osteoclast population and promote bone resorption.

2. It promotes calcium reabsorption by the kidneys, so less calcium is lost in the urine.

3. It promotes the final step of calcitriol synthesis in the kidneys, thus enhancing the calcium-raising effect of calcitriol.

4. It inhibits collagen synthesis by osteoblasts, thus inhibiting bone deposition.

7.4c Phosphate Homeostasis

The average adult has 500 to 800 g of phosphorus, of which 85% to 90% is in the bones. Phosphate is required not only for bone strength but also as a component of DNA, RNA, ATP, phospholipids, and other compounds. Plasma phosphorus concentration ranges from 3.5 to 4.0 mg/dL. Its level is not as tightly regulated as calcium. Nor, apparently, does it need to be; changes in plasma phosphate level are not associated with any immediate functional disorder. Calcitriol raises the phosphate level by promoting its absorption from the diet by the small intestine. This makes sense, because one effect of calcitriol is to promote bone deposition, and that requires both calcium and phosphate. Parathyroid hormone, on the other hand, lowers the blood phosphate level by promoting its urinary excretion.

▶▶▶APPLY WHAT YOU KNOW

While raising the blood calcium level, PTH lowers the phosphate level. Explain why this is important for achieving the purpose of PTH.

7.4d Other Factors Affecting Bone

At least 20 more hormones, growth factors, and vitamins affect osseous tissue in complex ways that are still not well understood **(table 7.1)**. Bone growth is especially rapid in

TABLE 7.1	Agents Affecting Calcium and Bone Metabolism
Name	**Effect**
Hormones	
Calcitonin	Promotes mineralization and lowers blood Ca^{2+} concentration in children, but usually has little effect in adults; may prevent bone loss in pregnant and lactating women
Calcitriol (vitamin D)	Promotes intestinal absorption of Ca^{2+} and phosphate; reduces urinary excretion of both; promotes both resorption and mineralization; stimulates osteoclast activity
Cortisol	Inhibits osteoclast activity, but if secreted in excess (Cushing disease), can cause osteoporosis by reducing bone deposition (inhibiting cell division and protein synthesis), inhibiting growth hormone secretion, and stimulating osteoclasts to resorb bone
Estrogen	Stimulates osteoblasts and adolescent growth; prevents osteoporosis
Growth hormone	Stimulates bone elongation and cartilage proliferation at epiphysial plate; increases urinary excretion of Ca^{2+} but also increases intestinal Ca^{2+} absorption, which compensates for the loss
Insulin	Stimulates bone formation; significant bone loss occurs in untreated diabetes mellitus
Parathyroid hormone	Indirectly activates osteoclasts, which resorb bone and raise blood Ca^{2+} concentration; inhibits urinary Ca^{2+} excretion; promotes calcitriol synthesis
Testosterone	Stimulates osteoblasts and promotes protein synthesis, thus promoting adolescent growth and epiphysial closure
Thyroid hormone	Essential to bone growth; enhances synthesis and effects of growth hormone, but excesses can cause hypercalcemia, increased Ca^{2+} excretion in urine, and osteoporosis
Growth Factors	At least 12 hormonelike substances produced in bone itself that stimulate neighboring bone cells, promote collagen synthesis, stimulate epiphysial growth, and produce many other effects
Vitamins	
Vitamin A	Promotes glycosaminoglycan (chondroitin sulfate) synthesis
Vitamin C (ascorbic acid)	Required for collagen synthesis, bone growth, and fracture repair
Vitamin D	Normally functions as a hormone (see calcitriol)

puberty and adolescence, when surges of growth hormone, estrogen, and testosterone promote ossification. These hormones stimulate rapid multiplication of osteogenic cells, matrix deposition by osteoblasts, and multiplication and hypertrophy of the chondrocytes in the metaphyses. Adolescent girls grow faster than boys and attain their full height earlier, not only because they begin puberty earlier but also because estrogen has a stronger effect than testosterone. Since males grow for a longer time, however, they usually grow taller. Sex steroids eventually deplete the cartilage of the epiphysial plates, bring about closure of the epiphyses, and put an end to one's growth in height. A deficiency or excess of these steroids can therefore cause abnormalities ranging from stunted growth to very tall stature. The use of anabolic steroids by adolescent athletes can cause premature closure and result in abnormally short adult stature (see Deeper Insight 2.6). The excessive consumption of cola (more than three 12-ounce servings per day) is associated with loss of bone density in women, but not in men. The effect is thought to be due to the phosphoric acid in cola, which binds intestinal calcium and interferes with its absorption. Other soft drinks don't contain phosphoric acid or affect bone density.

BEFORE YOU GO ON

Answer the following questions to test your understanding of the preceding section:

14. Describe the role of collagen and seed crystals in bone mineralization.

15. Why is it important to regulate blood calcium concentration within such a narrow range?

16. What effect does calcitonin have on blood calcium concentration, and how does it produce this effect? Answer the same questions for parathyroid hormone.

17. How is vitamin D synthesized, and what effect does it have on blood calcium concentration?

7.5   Bone Disorders

Expected Learning Outcomes

When you have completed this section, you should be able to

a. name and describe several bone diseases;

b. name and describe the types of fractures;

c. explain how a fracture is repaired; and

d. discuss some clinical treatments for fractures and other skeletal disorders.

Most people probably give little thought to their skeletal systems unless they break a bone. This section describes bone fractures, their healing, and their treatment, followed by a summary of other bone diseases. Bone disorders are among the concerns of

orthopedics,[30] a branch of medicine that originated as the treatment of skeletal deformities in children. It is now much more extensive and deals with the prevention and correction of injuries and disorders of the bones, joints, and muscles. It includes the design of artificial joints and limbs and the treatment of athletic injuries.

7.5a Fractures and Their Repair

There are multiple ways of classifying bone fractures. A **stress fracture** is a break caused by abnormal trauma to a bone, such as fractures incurred in falls, athletics, auto accidents, and military combat. A **pathological fracture** is a break in a bone weakened by some other disease, such as bone cancer or osteoporosis, usually caused by a stress that would not normally fracture a bone. Fractures are also classified according to the direction of the fracture line, whether the skin is broken, and whether a bone is merely cracked or broken into separate pieces. For example, a *nondisplaced* fracture is one in which the bone pieces remain in proper anatomical alignment, whereas a *displaced* fracture is one in which at least one piece is shifted out of alignment with the other **(fig. 7.18a, b).** A *comminuted* fracture is one in which a bone is broken into three or more pieces **(fig. 7.18c).** A *greenstick* fracture is one in which the bone is incompletely broken on one side but merely bent on the opposite side **(fig. 7.18d),** the way a green twig breaks only partially and not into separate pieces. Several other types of fractures are routinely taught in clinical and first aid courses.

The Healing of Fractures

An uncomplicated fracture heals in about 8 to 12 weeks, but complex fractures take longer and all fractures heal more slowly in older people. The healing process occurs in the following stages **(fig. 7.19):**

① **Formation of hematoma and granulation tissue.** A bone fracture severs blood vessels of the bone and periosteum, causing bleeding and the formation of a blood clot *(fracture hematoma).* Blood capillaries soon grow into the clot, while fibroblasts, macrophages, osteoclasts, and osteogenic cells invade the tissue from both the periosteal and medullary sides of the fracture. Osteogenic cells become very abundant within 48 hours of the injury. All of this capillary and cellular invasion converts the blood clot to a soft fibrous mass called **granulation tissue.**

② **Formation of a soft callus.**[31] Fibroblasts deposit collagen in the granulation tissue, while some osteogenic cells become chondroblasts and produce patches of fibrocartilage called the **soft callus.**

---

[30]*ortho* = straight; *ped* = child, foot
[31]*call* = hard, tough

(a) Nondisplaced

(b) Displaced

(c) Comminuted

(d) Greenstick

FIGURE 7.18 X-Rays of Representative Fracture Types.
(a) Nondisplaced fracture of the distal humerus in a 3-year-old.
(b) Displaced fracture of the tibia and fibula. (c) Comminuted
fracture of the tibia and fibula. (d) Greenstick fracture of the ulna.

a: Watney Collection/Medical Images; **b:** Howard Kingsnorth/The Image Bank/Getty
Images; **c:** Lester V. Bergman/Corbis NX/Getty Images; **d:** Biophoto Associates/
Science Source

③ **Conversion to hard callus.** Other osteogenic cells differentiate into osteoblasts, which produce a bony collar called the **hard callus** around the fracture. The hard callus is cemented to dead bone around the injury site and acts as a temporary splint to join the broken ends or bone fragments together. It takes about 4 to 6 weeks for a hard callus to form. During this period, it is important that a broken bone be immobilized by traction or a cast to prevent reinjury.

④ **Remodeling.** The hard callus persists for 3 to 4 months. Meanwhile, osteoclasts dissolve small fragments of broken bone, and osteoblasts deposit spongy bone to bridge the gap between the broken ends. This spongy bone gradually fills in to become compact bone, in a manner similar to intramembranous ossification. Usually the fracture leaves a slight thickening of the bone visible by X-ray; such thickenings may serve as forensic evidence of child abuse. In some cases, however, healing is so complete that no trace of the fracture can be found.

The Treatment of Fractures

Most fractures are set by **closed reduction,** a procedure in which the bone fragments are manipulated into their normal positions without surgery. **Open reduction** involves the surgical exposure of the bone and the use of plates, screws, or pins to realign the fragments **(fig. 7.20).** To stabilize the bone during healing, fractures are often set in casts. Traction is used to treat fractures of the femur in children. It aids in the alignment of the bone fragments by overriding the force of the strong thigh muscles. Traction is rarely used for elderly patients, however, because the risks from long-term confinement to bed outweigh the benefits. Hip fractures are usually pinned, and early ambulation (walking) is encouraged because it promotes blood circulation and healing. Fractures that take longer than 2 months to heal may be treated with electrical stimulation, which accelerates repair by suppressing the effects of parathyroid hormone.

① **Hematoma formation**
The hematoma is converted
to granulation tissue by invasion
of cells and blood capillaries.

② **Soft callus formation**
Deposition of collagen and
fibrocartilage converts granulation
tissue to a soft callus.

③ **Hard callus formation**
Osteoblasts deposit a temporary
bony collar around the fracture to
unite the broken pieces while
ossification occurs.

④ **Bone remodeling**
Small bone fragments are
removed by osteoclasts, while
osteoblasts deposit spongy
bone and then convert it to
compact bone.

FIGURE 7.19 The Healing of a Bone Fracture.

FIGURE 7.20 Open Reduction of an Ankle Fracture.
Southern Illinois University/Science Source

7.5b Other Bone Disorders

Several additional bone disorders are summarized in **table 7.2**. The most common bone disease, **osteoporosis,**[32] receives special consideration in Deeper Insight 7.4. The effects of aging on the skeletal system are described in section 29.4a.

BEFORE YOU GO ON

Answer the following questions to test your understanding of the preceding section:

18. Name and describe four types of bone fractures.

19. Why would osteomyelitis be more likely to occur in an open fracture than in a closed fracture?

20. What is a callus? How does it contribute to fracture repair?

TABLE 7.2	Bone Diseases
Osteitis deformans (Paget[33] disease)	Excessive proliferation of osteoclasts and resorption of excess bone, with osteoblasts attempting to compensate by depositing extra bone. This results in rapid, disorderly bone remodeling and weak, deformed bones. Osteitis deformans usually passes unnoticed, but in some cases it causes pain, disfiguration, and fractures. It is most common in males over the age of 50.
Osteomyelitis[34]	Inflammation of osseous tissue and bone marrow as a result of bacterial infection. This disease was often fatal before the discovery of antibiotics and is still very difficult to treat.
Osteogenesis imperfecta (brittle bone disease)	A defect in collagen deposition that renders bones exceptionally brittle, resulting in fractures present at birth or occurring with extraordinary frequency during childhood; also causing tooth deformity, and hearing loss due to deformity of middle-ear bones.
Osteosarcoma[35] (osteogenic sarcoma)	The most common and deadly form of bone cancer. It occurs most often in the tibia, femur, and humerus of males between the ages of 10 and 25. In 10% of cases, it metastasizes to the lungs or other organs; if untreated, death typically occurs within 1 year.

You can find other skeletal system disorders described in the following places:

Achondroplastic dwarfism in Deeper Insight 7.2; *ectopic ossification* in section 7.4; *rickets* and *osteomalacia* in Deeper Insight 7.3; *bone fractures* (in general) in section 7.5a; *osteoporosis* in Deeper Insight 7.4; *mastoiditis* in section 8.2a; *herniated discs* in Deeper Insight 8.4; *abnormal spinal curvatures* in Deeper Insight 8.3; *fractured clavicle* in section 8.4a; *fallen arches* in section 8.5b; *arthritis* in Deeper Insight 9.5; and cleft palate in Deeper Insight 29.4.

[32]*osteo* = bone; *por* = porous; *osis* = condition
[33]Sir James Paget (1814–99), English surgeon
[34]*osteo* = bone; *myel* = marrow; *itis* = inflammation
[35]*osteo* = bone; *sarc* = flesh; *oma* = tumor

DEEPER INSIGHT 7.4

CLINICAL APPLICATION

Osteoporosis

Osteoporosis is the most common of all bone diseases. It can be defined as a disorder in which bone density declines to the extent that the bones become brittle and subject to pathological fractures. Starting around age 40, bone resorption outpaces deposition, so we lose overall bone density. The loss comes especially from spongy bone, because it has the greatest surface area exposed to osteoclast action **(fig. 7.21a).** Proportionate amounts of organic matrix and minerals are lost, so the bone that remains is normal in composition. By old age, however, it may be insufficient in quantity to support the body's weight and withstand normal stresses.

People with osteoporosis are especially vulnerable to fractures of the hip, wrist, and spine. Their bones may break under stresses as slight as sitting down too quickly. Among the elderly, slowly healing hip fractures can impose prolonged immobility and lead to fatal complications such as pneumonia. Those who survive often face a long, costly recovery. Spinal deformity is also a common consequence of osteoporosis. As the bodies of the vertebrae lose spongy bone, they become compressed by body weight **(fig. 7.21b)** and the spine can develop an exaggerated thoracic curvature called hyperkyphosis **(fig. 7.21c).**

Postmenopausal women of European and Asian ancestry are at the greatest risk for osteoporosis. About 30% of U.S. women suffer a fracture due to osteoporosis at some point in their lives. Compared to men, they have less initial bone mass and begin to lose it at an earlier age. Since estrogen supports bone mass, the risk of osteoporosis rises sharply after menopause, when the ovaries cease to produce it. By age 70, the average white woman has lost 30% of her bone mass, and some as much as 50%. Osteoporosis is less common among women of African ancestry because of their greater initial bone density.

Other risk factors include family history; a light body build; dietary deficiencies of calcium, vitamin D, and protein; inadequate exercise; smoking; and overuse of alcohol. Osteoporosis is surprisingly common in young female runners, dancers, and gymnasts in spite of their vigorous

exercise. They sometimes have such a low percentage of body fat that they stop ovulating, and without developing follicles, their ovaries secrete low levels of estrogen. About 20% of osteoporosis sufferers are men. In early long-term space missions, astronauts developed osteoporosis because in a microgravity environment, their bones were subjected to too little of the stress that normally stimulates bone deposition. This and the prevention of muscle atrophy are reasons why exercise equipment is now standard on space stations.

Osteoporosis is now diagnosed by *bone mineral density (BMD)* tests using low-dose X-rays. This has enabled earlier diagnosis and more effective treatment. However, the severity of osteoporosis depends not on bone density alone, but also on the degree of connectivity between the spongy bone trabeculae, which is lost as trabeculae deteriorate. Neither BMD testing nor any other diagnostic method yet available can detect this.

Treatment is aimed at promoting bone deposition with drugs that either stimulate osteoblasts or slow the rate of bone resorption. Osteoblasts can be stimulated to build bone by pulsed treatment with a synthetic parathyroid hormone (PTH) or by selective estrogen–receptor modulators (SERMs), which mimic the effects of estrogen without producing undesirable estrogen side effects such as the risk of stroke or breast cancer. Resorption can be inhibited with a family of drugs called bisphosphanates. This is an intensive area of clinical research and new drugs are brought to market often.

As with so many other disorders, prevention is far preferable to treatment. Prevention is best begun in the young, bone-forming years of one's 20s and 30s, but continued even into old age. It includes weight-bearing exercise; ample dietary intake of calcium, vitamin D, and protein; and of course avoiding such risk factors as smoking and a sedentary lifestyle. For the elderly, such weight-bearing exercises as dancing, stair climbing, walking, and running can be pleasurable ways of minimizing the risk of osteoporosis.

(a)

(b)

(c)

FIGURE 7.21 Spinal Osteoporosis. (a) Spongy bone in the body of a vertebra in good health (left) and with osteoporosis (right). (b) X-ray of lumbar vertebrae severely damaged by osteoporosis. (c) Woman with severe hyperkyphosis due to compression of thoracic vertebrae.

a: Michael Klein/Photolibrary/Getty Images; b: Dr. P. Marazzi/Science Source; c: Phanie/Alamy

Effects of the **SKELETAL SYSTEM** on Other Organ Systems

INTEGUMENTARY SYSTEM
Bones lying close to the body surface support and shape the skin.

MUSCULAR SYSTEM
Bones are the attachment sites for most skeletal muscles and provide leverage for muscle action; calcium homeostasis, important for muscle contraction, is achieved partly through a balance between bone deposition and resorption.

NERVOUS SYSTEM
The cranium and vertebral column protect the brain and spinal cord; osseous tissue provides the calcium homeostasis needed for nerve function.

ENDOCRINE SYSTEM
Bones protect endocrine glands in the head, thorax, and pelvis; bones secrete the hormone osteocalcin, which promotes insulin action; hormone secretion depends on calcium homeostasis.

CIRCULATORY SYSTEM
Bone marrow forms blood cells and platelets; osseous tissue provides the calcium homeostasis needed for cardiac function and blood clotting.

LYMPHATIC AND IMMUNE SYSTEMS
White blood cells produced in the bone marrow carry out the body's immune functions.

RESPIRATORY SYSTEM
Ventilation of the lungs is achieved by musculoskeletal actions of the thoracic cage; the thoracic cage protects the delicate lungs from trauma; bones support and shape the nasal cavity.

URINARY SYSTEM
The thoracic cage partially protects the kidneys, and the pelvic girdle protects the lower urinary tract.

DIGESTIVE SYSTEM
Osseous tissue interacts with the digestive system in maintaining calcium homeostasis; the thoracic cage and pelvic girdle protect portions of the digestive tract; musculoskeletal movements are necessary for chewing.

REPRODUCTIVE SYSTEM
The pelvic girdle protects the internal reproductive organs; childbirth is adapted to the anatomy of the female pelvic girdle; ligaments anchor the penis, clitoris, uterus, and ovaries to the pelvic girdle.

STUDY GUIDE

▶ Assess Your Learning Outcomes

To test your knowledge, discuss the following topics with a study partner or in writing, ideally from memory.

7.1 Tissues and Organs of the Skeletal System

1. The branch of medicine and biology that deals with the skeleton and bone tissue
2. Organs and tissues that constitute the skeletal system
3. Functions of the skeletal system
4. Which primary tissue category includes bone, and how bone differs from other tissues in that category
5. The relationship of compact bone, spongy bone, and the marrow cavity in the anatomy of a long bone
6. Other anatomical features of a long bone including the diaphysis, epiphysis, epiphysial plate, articular cartilage, periosteum, and endosteum
7. Structure of a typical flat bone

7.2 Histology of Osseous Tissue

1. The four cell types in bone tissue; their functions, origins, and locations in the tissue
2. Organic and inorganic components of the bone matrix; their respective contributions to bone strength; and the significance of the helical arrangement of collagen fibers in bone
3. Osteon structure and the relationship of osteonic bone to interstitial and circumferential lamellae

4. The route by which nerves and blood vessels penetrate throughout a bone
5. Comparisons of the histology of spongy bone with that of compact bone; where spongy bone is found; and why bones are not composed solely of compact bone
6. Location and functions of the bone marrow; the composition and childhood versus adult distribution of the two types of marrow

7.3 Bone Development

1. Stages of intramembranous ossification; some bones that form in this way; and how far this process has progressed by birth
2. The same points concerning endochondral ossification
3. Histology, cell transformations, and tissue zones of the metaphysis; which zones and processes account for a child's or adolescent's growth in height
4. How stresses on bones remodel them throughout life; the difference between interstitial and appositional growth

7.4 Physiology of Osseous Tissue

1. The purpose and process of mineralization of osseous tissue, and the identity of the cells that carry it out
2. The purpose and process of bone resorption, and the identity of the cells and cell secretions that carry it out
3. Functions of calcium in the body; the normal range of blood calcium concentration;

and causes and consequences of hypocalcemia and hypercalcemia
4. The role of the skeleton as a calcium reservoir in regulating blood calcium levels
5. How calcitriol is synthesized and the mechanisms by which it supports or raises blood calcium level
6. The source of calcitonin and how it corrects hypercalcemia
7. The source of parathyroid hormone and multiple mechanisms by which it corrects hypocalcemia
8. Two forms of phosphate ions in the blood; the bodily functions of phosphate; and how calcitriol and parathyroid hormone affect blood phosphate levels
9. Effects of dietary vitamins A, C, and D on bone metabolism
10. Effects of cortisol, estrogen, testosterone, growth hormone, insulin, and thyroid hormone on bone metabolism

7.5 Bone Disorders

1. The difference between a stress fracture and a pathological fracture; stages in the healing of a fractured bone; and approaches to the clinical treatment of fractures
2. Causes of osteoporosis; its risk factors, pathological effects, diagnosis, treatment, and prevention

▶ Testing Your Recall

Answers in Appendix A

1. Which cells have a ruffled border and secrete hydrochloric acid?
   a. C cells
   b. osteocytes
   c. osteogenic cells
   d. osteoblasts
   e. osteoclasts

2. The marrow cavity of an adult bone may contain
   a. myeloid tissue.
   b. hyaline cartilage.
   c. periosteum.
   d. osteocytes.
   e. articular cartilages.

3. The spurt of growth in puberty results from cell proliferation and hypertrophy in
   a. the epiphysis.
   b. the epiphysial line.
   c. compact bone.
   d. the epiphysial plate.
   e. spongy bone.

4. Osteoclasts are most closely related, by common descent, to
   a. osteoprogenitor cells.
   b. osteogenic cells.
   c. blood cells.
   d. fibroblasts.
   e. osteoblasts.

5. The walls between cartilage lacunae break down in the zone of
   a. cell proliferation.
   b. calcification.
   c. reserve cartilage.
   d. bone deposition.
   e. cell hypertrophy.

6. Which of these is *not* an effect of PTH?
   a. rise in blood phosphate level
   b. reduction of calcium excretion
   c. increased intestinal calcium absorption
   d. increased number of osteoclasts
   e. increased calcitriol synthesis

STUDY GUIDE

7. A child jumps to the ground from the top of a playground "jungle gym." His leg bones do not shatter mainly because they contain
   a. an abundance of glycosaminoglycans.
   b. young, resilient osteocytes.
   c. an abundance of calcium phosphate.
   d. collagen fibers.
   e. hydroxyapatite crystals.

8. One long bone meets another at its
   a. diaphysis.
   b. epiphysial plate.
   c. periosteum.
   d. metaphysis.
   e. epiphysis.

9. Calcitriol is made from
   a. calcitonin.
   b. 7-dehydrocholesterol.

   c. hydroxyapatite.
   d. estrogen.
   e. PTH.

10. One sign of osteoporosis is
    a. osteosarcoma.
    b. osteomalacia.
    c. osteomyelitis.
    d. a spontaneous wrist fracture.
    e. hypocalcemia.

11. Calcium phosphate crystallizes in bone as a mineral called _____.

12. Osteocytes contact each other through channels called _____ in the bone matrix.

13. A bone increases in diameter only by _____ growth, the addition of new surface lamellae.

14. Seed crystals of hydroxyapatite form only when the levels of calcium and phosphate in the tissue fluid exceed the _____.

15. A calcium deficiency called _____ can cause death by suffocation.

16. _____ are cells that secrete collagen and stimulate calcium phosphate deposition.

17. The most active form of vitamin D, produced mainly by the kidneys, is _____.

18. The most common bone disease is _____.

19. The transitional region between epiphysial cartilage and the primary marrow cavity of a young bone is called the _____.

20. A pregnant, poorly nourished woman may suffer a softening of the bones called _____.

▶ Building Your Medical Vocabulary

Answers in Appendix A

State a meaning of each word element, and give a medical term from this chapter that uses it or a slight variation of it.

1. calc-
2. -clast
3. -malacia
4. myelo-
5. ortho-
6. osse-
7. osteo-
8. -physis
9. spic-
10. topo-

▶ What's Wrong with These Statements?

Answers in Appendix A

Briefly explain why each of the following statements is false, or reword it to make it true.

1. The flat cranial bones are composed of compact bone only, with no spongy bone.

2. In endochondral ossification, bone tissue is formed by the calcification of preexisting cartilage.

3. Fractures are the most common bone disorder.

4. The growth zone of the long bones of adolescents is the articular cartilage.

5. Osteoclasts develop from osteoblasts.

6. Osteoblasts are multipotent stem cells.

7. The protein of the bone matrix is called hydroxyapatite.

8. Osteocytes are nourished by blood capillaries in the canaliculi of the osteons.

9. Vitamin D promotes bone deposition, not resorption.

10. Parathyroid hormone stimulates bone deposition by osteoblasts.

▶ Testing Your Comprehension

1. Most osteocytes of an osteon are far removed from blood vessels, but still receive blood-borne nutrients. Explain how this is possible.

2. A 50-year-old business executive decides he has not been getting enough exercise for the last several years. He takes up hiking and finds that he really loves it. Within 2 years,

he is spending many of his weekends hiking with a heavy backpack and camping in the mountains. Explain what changes in his anatomy could be predicted from Wolff's law of bone.

3. How does the regulation of blood calcium concentration exemplify negative feedback and homeostasis?

4. Describe how the arrangement of trabeculae in spongy bone demonstrates the unity of form and function.

5. Identify two bone diseases you would expect to see if the epidermis were a completely effective barrier to UV radiation and a person took no dietary supplements to compensate for this. Explain your answer.

CHAPTER **8**

THE SKELETAL SYSTEM

X-ray of the cervical spine of a 20-year-old female
Science Photo Library - Zephyr/Getty Images

CHAPTER OUTLINE

8.1 Overview of the Skeleton

    8.1a Bones of the Skeletal System

    8.1b Anatomical Features of Bones

8.2 The Skull

    8.2a Cranial Bones

    8.2b Facial Bones

    8.2c Bones Associated with the Skull

    8.2d The Skull in Infancy and Childhood

8.3 The Vertebral Column and Thoracic Cage

    8.3a General Features of the Vertebral Column

    8.3b General Structure of a Vertebra

    8.3c Intervertebral Discs

    8.3d Regional Characteristics of Vertebrae

    8.3e The Thoracic Cage

8.4 The Pectoral Girdle and Upper Limb

    8.4a The Pectoral Girdle

    8.4b The Upper Limb

8.5 The Pelvic Girdle and Lower Limb

    8.5a The Pelvic Girdle

    8.5b The Lower Limb

Study Guide

DEEPER INSIGHTS

8.1 Clinical Application: Cleft Palate and Lip

8.2 Clinical Application: Cranial Assessment of the Newborn

8.3 Clinical Application: Abnormal Spinal Curvatures

8.4 Clinical Application: Herniated Disc

8.5 Evolutionary Medicine: Skeletal Adaptations for Bipedalism

Anatomy & Physiology Revealed 4.0

Module 5: Skeletal System

BRUSHING UP

- Anatomical descriptions of the skeletal system depend heavily on the directional terminology introduced in atlas A, table A.1.

- Understanding skeletal anatomy also depends on knowledge of the terms for body regions and cavities described in atlas A.

Knowledge of skeletal anatomy will be useful as you study later chapters. It provides a point of reference for studying the gross anatomy of other organ systems because many organs are named for their relationships to nearby bones. The subclavian artery and vein, for example, lie adjacent to the clavicle; the temporalis muscle is attached to the temporal bone; the ulnar nerve and radial artery travel beside the ulna and radius of the forearm; and the frontal, parietal, temporal, and occipital lobes of the brain are named for adjacent bones of the cranium. Understanding how the muscles produce body movements also depends on knowledge of skeletal anatomy. Additionally, the positions, shapes, and processes of bones can serve as landmarks for clinicians in determining where to give an injection or record a pulse, what to look for in an X-ray, and how to perform physical therapy and other clinical procedures.

8.1   Overview of the Skeleton

Expected Learning Outcomes

When you have completed this section, you should be able to

a. define the two subdivisions of the skeleton;

b. state the approximate number of bones in the adult body;

c. explain why this number varies with age and from one person to another; and

d. define several terms that denote surface features of bones.

The skeleton (**fig. 8.1**) is divided into two regions: the axial skeleton and appendicular skeleton. The **axial skeleton,** which forms the central supporting axis of the body, includes the skull, auditory ossicles (middle-ear bones), hyoid bone, vertebral column, and thoracic cage (ribs and sternum). The **appendicular skeleton** includes the bones of the upper limb and pectoral girdle and the bones of the lower limb and pelvic girdle.

8.1a Bones of the Skeletal System

It is often stated that there are 206 bones in the skeleton, but this is only a typical adult count, not an invariable number. At birth there are about 270, and even more bones form during childhood. With age, however, the number decreases as separate bones gradually fuse. For example, each side of a child's pelvic girdle has three bones—the *ilium, ischium,* and *pubis*—but in adults, these are fused into a single *hip bone* on each side. The

fusion of several bones, completed by late adolescence to the mid-20s, brings about the average adult number of 206. These bones are listed in **table 8.1.**

This number varies even among adults. One reason is the development of **sesamoid**[1] **bones**—bones that form within

[1]*sesam* = sesame seed; *oid* = resembling

TABLE 8.1	Bones of the Adult Skeletal System	
Axial Skeleton		
Skull (22 bones)		
Cranial bones		
Frontal bone (1)		Temporal bones (2)
Parietal bones (2)		Sphenoid bone (1)
Occipital bone (1)		Ethmoid bone (1)
Facial bones		
Maxillae (2)		Nasal bones (2)
Palatine bones (2)		Vomer (1)
Zygomatic bones (2)		Inferior nasal conchae (2)
Lacrimal bones (2)		Mandible (1)
Auditory ossicles (6 bones)		
Malleus (2)		Stapes (2)
Incus (2)		
Hyoid bone (1 bone)		
Vertebral column (26 bones)		
Cervical vertebrae (7)		Sacrum (1)
Thoracic vertebrae (12)		Coccyx (1)
Lumbar vertebrae (5)		
Thoracic cage (25 bones plus thoracic vertebrae)		
Ribs (24)		
Sternum (1)		
Appendicular Skeleton		
Pectoral girdle (4 bones)		
Scapulae (2)		Clavicles (2)
Upper limbs (60 bones)		
Humerus (2)		Carpal bones (16)
Radius (2)		Metacarpal bones (10)
Ulna (2)		Phalanges (28)
Pelvic girdle (2 bones)		
Hip bones (2)		
Lower limbs (60 bones)		
Femurs (2)		Tarsal bones (14)
Patellae (2)		Metatarsal bones (10)
Tibiae (2)		Phalanges (28)
Fibulae (2)		
Grand Total: 206 Bones		

Skull	Frontal bone
	Maxilla
	Mandible

Parietal bone
Occipital bone
Mandible

Pectoral girdle — Clavicle
Scapula

Clavicle
Scapula

Thoracic cage — Sternum
Ribs
Costal cartilages

Humerus

Vertebral column

Pelvis — Hip bone
Sacrum

Coccyx

Carpus

Metacarpal bones

Phalanges

Ulna
Radius

Femur

Patella

Fibula

Tibia

Tarsus

Metatarsal bones

Phalanges

(a) Anterior view

(b) Posterior view

FIGURE 8.1 The Adult Skeleton. (a) Anterior view. (b) Posterior view. The appendicular skeleton is colored green, and the rest is axial skeleton. **APR**

some tendons in response to strain. The patella (kneecap) is the largest of these; most of the others are small, rounded bones in such locations as the hands and feet (see fig. 8.35c).

Another reason for adult variation is that some people have extra bones in the skull called **sutural bones** (SOO-chur-ul) (see fig. 8.6).

8.1b Anatomical Features of Bones

Bones exhibit a variety of ridges, spines, bumps, depressions, canals, pores, slits, cavities, and articular surfaces. It is important to know the names of these *bone markings* because later descriptions of joints, muscle attachments, and the routes traveled by nerves and blood vessels are based on this terminology. Terms for the most common bone features are listed in **table 8.2,** and several are illustrated in **figure 8.2.**

TABLE 8.2	Anatomical Features (Markings) of Bones
Term	Description and Example
Articulations (Joint Surfaces)	
Condyle	A rounded knob that articulates with another bone (occipital condyles of the skull)
Facet	A smooth, flat, slightly concave or convex articular surface (articular facets of the vertebrae)
Head	The prominent expanded end of a bone, sometimes rounded (head of the femur)
Extensions and Projections	
Crest	A narrow ridge (iliac crest of the pelvis)
Epicondyle	An expanded region superior to a condyle (medial epicondyle of the femur)
Line	A slightly raised, elongated ridge (nuchal lines of the skull)
Process	Any bony prominence (mastoid process of the skull)
Protuberance	A bony outgrowth or protruding part (mental protuberance of the chin)
Spine	A sharp, slender, or narrow process (mental spines of the mandible)
Trochanter	Two massive processes unique to the femur
Tubercle	A small, rounded process (greater tubercle of the humerus)
Tuberosity	A rough elevated surface (tibial tuberosity)
Depressions	
Alveolus	A pit or socket (tooth socket)
Fossa	A shallow, broad, or elongated basin (mandibular fossa)
Fovea	A small pit (fovea capitis of the femur)
Sulcus	A groove for a tendon, nerve, or blood vessel (intertubercular sulcus of the humerus)
Passages and Cavities	
Canal	A tubular passage or tunnel in a bone (auditory canal of the skull)
Fissure	A slit through a bone (orbital fissures behind the eye)
Foramen	A hole through a bone, usually round (foramen magnum of the skull)
Meatus	A canal (external acoustic meatus of the ear)
Sinus	An air-filled space in a bone (frontal sinus of the forehead)

(a) Skull (lateral view)

(b) Scapula (posterior view)

(c) Femur (posterior view)

(d) Humerus (anterior view)

FIGURE 8.2 Anatomical Features of Bones. (a) Skull, lateral view. (b) Scapula, posterior view. (c) Femur, posterior view. (d) Humerus, anterior view. Most of these features also occur on many other bones of the body.

You will probably study both **articulated** skeletons (dried bones held together by wires and rods to show their spatial relationships to each other) and **disarticulated** bones (bones taken apart so their surface features can be studied in more detail). As you study this chapter, also use yourself as a model. You can easily palpate (feel) many of the bones and some of their details through the skin. Rotate your forearm, cross your legs, palpate your skull and wrist, and think about what is happening beneath the surface or what you can feel through the skin. You will gain the most from this chapter (and indeed, the entire book) if you are conscious of your own body in relation to what you are studying.

BEFORE YOU GO ON

Answer the following questions to test your understanding of the preceding section:

1. Name the major components of the axial skeleton. Name those of the appendicular skeleton.

2. Explain why an adult does not have as many bones as a child does. Explain why one adult may have more bones than another.

3. Briefly describe each of the following bone features: a condyle, crest, tubercle, fossa, sulcus, and foramen.

8.2 The Skull

Expected Learning Outcomes

When you have completed this section, you should be able to

a. distinguish between cranial and facial bones;

b. name the bones of the skull and their anatomical features;

c. identify the cavities in the skull and in some of its individual bones;

d. name the principal sutures that join the bones of the skull;

e. describe some bones that are closely associated with the skull; and

f. describe the development of the skull from infancy through childhood.

The skull is the most complex part of the skeleton. **Figures 8.3** through **8.6** present an overview of the skull's general anatomy. Although it may seem to consist only of the mandible (lower jaw) and "the rest," it is composed of 22 bones and sometimes more. Most of these are connected by immovable joints called **sutures** (SOO-chures), which are visible as seams on the surface (fig. 8.4). These are important landmarks in the descriptions that follow.

The skull contains several prominent cavities (**fig. 8.7**). The largest, with an adult volume of about 1,350 mL, is the **cranial cavity,** which encloses the brain. Other cavities include the **orbits** (eye sockets), **nasal cavity, oral (buccal) cavity, middle-** and **inner-ear cavities,** and **paranasal sinuses.** The sinuses are named for the bones in which they occur (**fig. 8.8**)—the **frontal, sphenoidal, ethmoidal,** and **maxillary sinuses.** They are connected with the nasal cavity, lined by mucous membranes, and filled with air. They lighten the anterior portion of the skull and act as chambers that add resonance to the voice. The latter effect can be sensed in the way your voice changes when you have a cold and mucus obstructs the travel of sound into the sinuses and back.

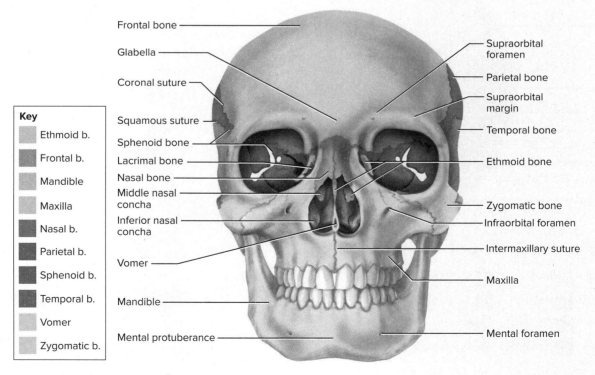

Key
- Ethmoid b.
- Frontal b.
- Mandible
- Maxilla
- Nasal b.
- Parietal b.
- Sphenoid b.
- Temporal b.
- Vomer
- Zygomatic b.

Frontal bone
Glabella
Coronal suture
Squamous suture
Sphenoid bone
Lacrimal bone
Nasal bone
Middle nasal concha
Inferior nasal concha
Vomer
Mandible
Mental protuberance

Supraorbital foramen
Parietal bone
Supraorbital margin
Temporal bone
Ethmoid bone
Zygomatic bone
Infraorbital foramen
Intermaxillary suture
Maxilla
Mental foramen

FIGURE 8.3 The Skull (Anterior View). A&PR

Coronal suture

Frontal bone

Temporal lines

Parietal bone

Lambdoid suture

Sphenoid bone

Occipital bone

Ethmoid bone

Nasal bone

Lacrimal bone

Squamous suture

Temporal bone

Zygomatic process

External acoustic meatus

Zygomaticofacial foramen

Infraorbital foramen

Zygomatic bone

Maxilla

Temporal process

Mastoid process

Styloid process

Mandibular condyle

Mandible

Mental foramen

(a) Right lateral view

Coronal suture

Frontal bone

Parietal bone

Squamous suture

Sphenoidal sinus

Frontal sinus

Crista galli

Cribriform plate of ethmoid bone

Occipital bone

Temporal bone

Sella turcica

Lambdoid suture

Internal acoustic meatus

Jugular foramen

Hypoglossal canal

Perpendicular plate of ethmoid bone

Nasal bone

Vomer

Palatine process of maxilla

Maxilla

Palatine bone

Mandible

Mental spines

Key

Ethmoid b.	Maxilla	Sphenoid b.
Frontal b.	Nasal b.	Temporal b.
Lacrimal b.	Occipital b.	Vomer
Mandible	Parietal b.	Zygomatic b.

(b) Median section

FIGURE 8.4   The Skull.   (a) Right lateral surface anatomy. (b) Internal anatomy, median section. **APR**

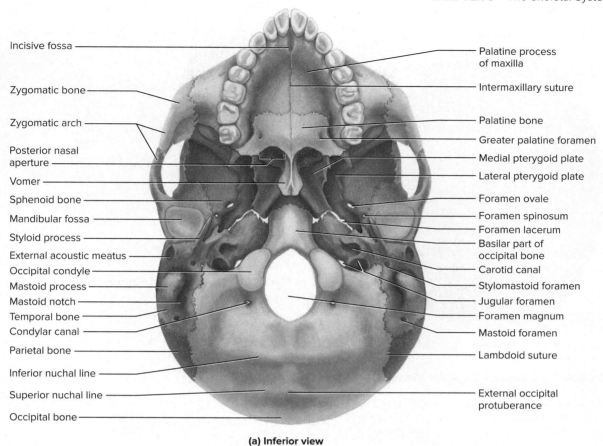

Incisive fossa

Zygomatic bone

Zygomatic arch

Posterior nasal aperture

Vomer

Sphenoid bone

Mandibular fossa

Styloid process

External acoustic meatus

Occipital condyle

Mastoid process

Mastoid notch

Temporal bone

Condylar canal

Parietal bone

Inferior nuchal line

Superior nuchal line

Occipital bone

Palatine process of maxilla

Intermaxillary suture

Palatine bone

Greater palatine foramen

Medial pterygoid plate

Lateral pterygoid plate

Foramen ovale

Foramen spinosum

Foramen lacerum

Basilar part of occipital bone

Carotid canal

Stylomastoid foramen

Jugular foramen

Foramen magnum

Mastoid foramen

Lambdoid suture

External occipital protuberance

(a) Inferior view

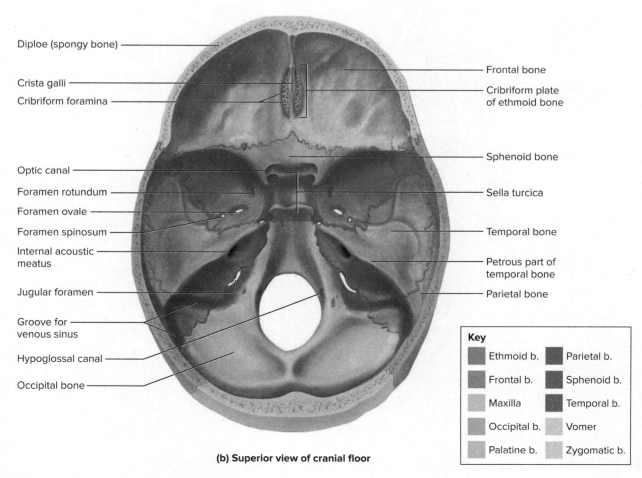

Diploe (spongy bone)

Crista galli

Cribriform foramina

Optic canal

Foramen rotundum

Foramen ovale

Foramen spinosum

Internal acoustic meatus

Jugular foramen

Groove for venous sinus

Hypoglossal canal

Occipital bone

Frontal bone

Cribriform plate of ethmoid bone

Sphenoid bone

Sella turcica

Temporal bone

Petrous part of temporal bone

Parietal bone

Key

Ethmoid b.

Frontal b.

Maxilla

Occipital b.

Palatine b.

Parietal b.

Sphenoid b.

Temporal b.

Vomer

Zygomatic b.

(b) Superior view of cranial floor

FIGURE 8.5 The Base of the Skull. (a) Inferior view. (b) Superior view of the cranial floor. **A&PR**

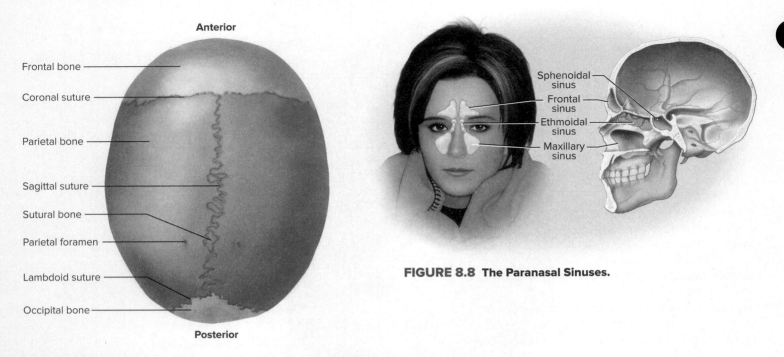

Anterior

Frontal bone

Coronal suture

Parietal bone

Sagittal suture

Sutural bone

Parietal foramen

Lambdoid suture

Occipital bone

Posterior

FIGURE 8.6 **The Calvaria (Skullcap) (Superior View).** **APR**

Sphenoidal sinus

Frontal sinus

Ethmoidal sinus

Maxillary sinus

FIGURE 8.8 **The Paranasal Sinuses.**

Frontal bone

Brain

Crista galli

Ethmoidal sinus

Temporalis muscle

Nasal cavity

Maxillary sinus

Nasal septum

Maxilla

Mandible

Frontal sinus

Muscles of eye movement

Optic nerve

Orbit with orbital fat

Nasal conchae

Zygomatic arch

Masseter muscle

Adipose tissue

Hard palate

Oral cavity

Tongue

FIGURE 8.7 **Frontal Section of the Head.** This shows the major cavities of the skull and their contents.

Bones of the skull have especially conspicuous **foramina**—singular, *foramen* (fo-RAY-men)—holes that allow passage for nerves and blood vessels. Some major foramina are summarized in **table 8.3.** Some of the details will mean more after you study cranial nerves and blood vessels in later chapters.

8.2a Cranial Bones

Cranial bones are those that enclose the brain; collectively, they compose the **cranium**[2] (braincase). The delicate brain tissue doesn't come directly into contact with the bones, but is separated from them by three membranes called the *meninges* (meh-NIN-jeez) (see section 14.2a). The thickest and toughest of these, the *dura mater*[3] (DUE-rah MAH-tur), is pressed against the inside of the cranium in most places and firmly attached to it at a few points.

The cranium is a rigid structure with an opening, the *foramen magnum* (literally "large hole"), where the spinal cord meets the brain. The cranium consists of two major parts—the calvaria and the base. The **calvaria**[4] (skullcap) isn't a single bone but simply the dome of the top of the skull; it is composed of parts of multiple bones that form the roof and walls (see fig. 8.6). In skulls prepared

for study, the calvaria is often sawed so that part of it can be lifted off for examination of the interior. This reveals the **base** (floor) of the cranial cavity (see fig. 8.5b), which exhibits three paired depressions called cranial fossae. These correspond to the contour of the inferior surface of the brain **(fig. 8.9).** The relatively shallow **anterior cranial fossa** is crescent-shaped and accommodates the frontal lobes of the brain. The **middle cranial fossa,** which drops abruptly deeper, is shaped like a pair of outstretched bird's wings and accommodates the temporal lobes. The **posterior cranial fossa** is deepest and houses a large posterior division of the brain called the cerebellum.

There are eight cranial bones:

1 frontal bone	1 occipital bone
2 parietal bones	1 sphenoid bone
2 temporal bones	1 ethmoid bone

The Frontal Bone

The **frontal bone** extends from the forehead back to a prominent *coronal suture,* which crosses the crown of the head from right to left and joins the frontal bone to the parietal bones (see figs. 8.3 and 8.4). The frontal bone forms the anterior wall and about one-third of the roof of the cranial cavity, and it turns inward to form nearly all of the anterior cranial fossa and the roof of the orbit.

---

[2]*crani* = helmet
[3]*dura* = tough, strong; *mater* = mother
[4]*calvar* = bald, skull

TABLE 8.3	Foramina of the Skull and the Nerves and Blood Vessels Transmitted Through Them
Bones and Their Foramina	**Structures Transmitted**
Frontal Bone	
Supraorbital foramen or notch	Supraorbital nerve, artery, and vein; ophthalmic nerve
Temporal Bone	
Carotid canal	Internal carotid artery
External acoustic meatus	Sound waves to eardrum
Jugular foramen	Internal jugular vein; glossopharyngeal, vagus, and accessory nerves
Occipital Bone	
Foramen magnum	Spinal cord; accessory nerve; vertebral arteries
Hypoglossal canal	Hypoglossal nerve to muscles of tongue
Sphenoid Bone	
Foramen ovale	Mandibular division of trigeminal nerve; accessory meningeal artery
Foramen rotundum	Maxillary division of trigeminal nerve
Optic canal	Optic nerve; ophthalmic artery
Superior orbital fissure	Oculomotor, trochlear, and abducens nerves; ophthalmic division of trigeminal nerve; ophthalmic veins
Maxilla	
Inferior orbital fissure	Infraorbital nerve; zygomatic nerve; infraorbital vessels
Infraorbital foramen	Infraorbital nerve and vessels
Mandible	
Mental foramen	Mental nerve and vessels
Mandibular foramen	Inferior alveolar nerves and vessels to the lower teeth

(a) **Superior view** (b) **Lateral view**

FIGURE 8.9 Cranial Fossae. (a) Superior view. (b) Lateral view showing how the three fossae conform to the contours of the base of the brain.

Deep to the eyebrows it has a ridge called the **supraorbital margin.** Each margin is perforated by a single **supraorbital foramen** (see figs. 8.3 and 8.14), which provides passage for a nerve, artery, and veins. In some people, the edge of this foramen breaks through the margin of the orbit and forms a *supraorbital notch.* A person may have a foramen on one supraorbital margin and a notch on the other. The smooth area of the frontal bone just above the root of the nose is called the **glabella.**[5] The frontal bone also contains the frontal sinus. You may not see this sinus on all skulls. On some, the calvaria is cut too high to show it, and some people simply don't have one. Along the cut edge of the calvaria, you can also see the diploe (DIP-lo-ee)—the layer of spongy bone in the middle of the cranial bones (see fig. 8.5b).

The Parietal Bones

The right and left **parietal bones** (pa-RY-eh-tul) form most of the cranial roof and part of its walls (see figs. 8.4 and 8.6). Each is bordered by four sutures that join it to the neighboring bones: (1) a **sagittal suture** between the parietal bones; (2) the **coronal**[6] **suture** at the anterior margin; (3) the **lambdoid**[7] **suture** (LAM-doyd) at the posterior margin; and (4) the **squamous suture** laterally. Small sutural bones are often seen along the sagittal and lambdoid sutures, like little islands of bone with the suture lines passing around them. Internally, the parietal and frontal bones have markings that look a bit like aerial photographs of river tributaries (see fig. 8.4b). These represent places where the bone molded itself around blood vessels of the meninges.

Externally, the parietal bones have few features. A **parietal foramen** sometimes occurs near the corner of the lambdoid and sagittal sutures (see fig. 8.6). It is an exit for a small vein from a blood sinus atop the brain. A pair of slight lateral thickenings, the superior and inferior **temporal lines,** form an arc across the parietal and frontal bones (see fig. 8.4a). They mark the attachment of the large, fan-shaped *temporalis muscle,* a chewing muscle that converges on the mandible.

The Temporal Bones

If you palpate your skull just above and anterior to the ear—that is, the temporal region—you can feel the **temporal bone,** which forms the lower wall and part of the floor of the cranial cavity **(fig. 8.10).** The temporal bone derives its name from the fact that people often develop their first gray hairs on the temples with the passage of time.[8] The relatively complex shape of the temporal bone is best understood by dividing it into four parts:

1. The **squamous**[9] **part** (which you just palpated) is relatively flat and vertical. It is encircled by the squamous suture. It bears two prominent features: (a) the **zygomatic process,** which extends anteriorly to form part of the *zygomatic arch,* described later; and (b) the **mandibular fossa,** a depression where the mandible articulates with the cranium.

2. The **tympanic**[10] **part** is a small ring of bone that borders the opening of the **external acoustic meatus** (me-AY-tus), or ear canal. It has a pointed spine on its inferior surface,

---

[5]*glab* = smooth
[6]*corona* = crown
[7]Shaped like the Greek letter lambda (λ)

[8]*tempor* = time
[9]*squam* = flat; *ous* = characterized by
[10]*tympan* = drum (eardrum); *ic* = pertaining to

FIGURE 8.10 The Right Temporal Bone. (a) The lateral surface, facing the scalp and external ear. (b) The medial surface, facing the brain. **APR**

❓ *List five bones that articulate with the temporal bone.*

the **styloid process,** named for its resemblance to the stylus used by ancient Greeks and Romans to write on wax tablets. The styloid process provides attachment for muscles of the tongue, pharynx, and hyoid bone.

3. The **mastoid**[11] **part** lies posterior to the tympanic part. It bears a heavy **mastoid process,** which you can palpate as a prominent lump behind the earlobe. It is filled with small air sinuses that communicate with the middle-ear cavity. These sinuses are subject to infection and inflammation *(mastoiditis),* which can erode the bone and spread to the brain. A groove called the **mastoid notch** lies medial to the mastoid process (see fig. 8.5a). It is the origin of the *digastric muscle,* which opens the mouth. The notch is perforated at its anterior end by the **stylomastoid foramen,** which is a passage for the facial nerve, and at its posterior end by the **mastoid foramen,** which passes a small artery and vein from the brain.

4. The **petrous**[12] **part** can be seen in the cranial floor, where it resembles a little mountain range separating the middle cranial fossa from the posterior fossa (fig. 8.10b). It houses the middle- and inner-ear cavities. The **internal acoustic meatus,** an opening on its posteromedial surface, allows passage of a nerve that carries signals for hearing and balance from the inner ear to the brain. On the inferior surface of the petrous part are two prominent foramina named for the major blood vessels that pass through them (see fig. 8.5a): (a) The **carotid canal** is a passage for the internal carotid artery, a major blood supply to the brain. This artery is so close to the inner ear that one can sometimes hear the pulsing of its blood when the ear is resting on a pillow or the heart is beating hard. (b) The **jugular foramen** is a large, irregular opening just medial to the styloid process, between the temporal and occipital bones. Blood from the brain drains through this foramen into the internal jugular vein of the neck. Three cranial nerves also pass through this foramen (see table 8.3).

The Occipital Bone

The **occipital bone** (oc-SIP-ih-tul) forms the rear of the skull *(occiput)* and much of its base (see fig. 8.5). Its most conspicuous feature, the **foramen magnum,** admits the spinal cord to the cranial cavity; the dura mater is attached to the rim of this foramen. An important consideration in head injuries is swelling of the brain. Since the cranium cannot expand, swelling puts pressure on the brain and results in even more tissue damage. Severe swelling can force the brainstem out through the foramen magnum, usually with fatal consequences.

The occipital bone continues anterior to this as a thick median plate, the **basilar part.** On either side of the foramen magnum is a smooth knob called the **occipital condyle** (CON-dile), where the skull rests on the vertebral column. Passing like a tunnel beneath each condyle is a **hypoglossal**[13] **canal,** named for the *hypoglossal nerve* that passes through it to innervate the muscles of the tongue. In some people, a **condylar canal** (CON-dih-lur) is found posterior to each occipital condyle. A small vein from a blood sinus of the brain passes through here.

Internally, the occipital bone displays impressions left by large venous sinuses that drain blood from the brain (see fig. 8.5b). One of these grooves travels along the midsagittal line. Just before reaching the foramen magnum, it branches into right and left grooves that wrap around the occipital bone like outstretched arms before terminating at the jugular foramina. The venous sinuses that occupy these grooves are described in table 20.4.

Other features of the occipital bone can be palpated on the back of your head. One is a prominent medial bump called the

---

[11]*mast* = breast; *oid* = shaped like

[12]*petr* = stone, rock; *ous* = like
[13]*hypo* = below; *gloss* = tongue

external occipital protuberance—the attachment for the **nuchal**[14] **ligament** (NEW-kul), which binds the skull to the vertebral column. A ridge, the **superior nuchal line,** can be traced horizontally from this protuberance toward the mastoid process (see fig. 8.5a). It defines the superior limit of the neck and provides attachment to the skull for several neck and back muscles. It forms the boundary where, in palpating the upper neck, you feel the transition from muscle to bone. By pulling down on the occipital bone, some of these muscles help to keep the head erect. The deeper **inferior nuchal line** provides attachment for some of the deep neck muscles. This inconspicuous ridge cannot be palpated on the living body but is visible on an isolated skull.

The Sphenoid Bone

The **sphenoid**[15] **bone** (SFEE-noyd) has a complex shape with a thick median **body** and outstretched **greater** and **lesser wings,** which give the bone as a whole a ragged mothlike shape. Most of it is best seen from the superior perspective **(fig. 8.11a).** In this view, the lesser wings form the posterior margin of the anterior cranial fossa and end at a sharp bony crest, where the sphenoid drops abruptly to the greater wings. The greater wings form about half of the middle cranial fossa (the temporal bone forming the rest) and are perforated by several foramina.

The greater wing also forms part of the lateral surface of the cranium just anterior to the temporal bone (see fig. 8.4a). The lesser wing forms the posterior wall of the orbit and contains the **optic canal,** which permits passage of the optic nerve and ophthalmic artery (see fig. 8.14). Superiorly, a pair of bony spines of the lesser wing called the **anterior clinoid processes** appears to guard the optic foramina. A gash in the posterior wall of the orbit, the **superior orbital fissure,** angles upward lateral to the optic canal. It serves as a passage for three nerves that supply the muscles of eye movement.

The body of the sphenoid bone **(fig. 8.11b)** contains a pair of sphenoidal sinuses and has a saddlelike surface feature named the **sella turcica**[16] (SEL-la TUR-sih-ca). The sella consists of a deep pit called the *hypophysial fossa,* which houses the pituitary gland (hypophysis); a raised anterior margin called the *tuberculum sellae* (too-BUR-cu-lum SEL-lee); and a posterior margin called the *dorsum sellae.* In life, the dura mater stretches over the sella turcica and attaches to the anterior clinoid processes. A stalk penetrates the dura to connect the pituitary gland to the base of the brain.

Lateral to the sella turcica, the sphenoid is perforated by several foramina (see fig. 8.5a). The **foramen rotundum** and **foramen ovale** (oh-VAY-lee) are passages for two branches of the trigeminal nerve. The **foramen spinosum,** about the diameter of

---

[14]*nucha* = back of the neck
[15]*sphen* = wedge; *oid* = resembling

[16]*sella* = saddle; *turcica* = Turkish

(a) Superior view

(b) Posterior view

FIGURE 8.11 The Sphenoid Bone. (a) Superior view. (b) Posterior view, as if looking from the back of the head into the opened skull.

APR

a pencil lead, provides passage for an artery of the meninges. An irregular gash called the **foramen lacerum**[17] (LASS-eh-rum) occurs at the junction of the sphenoid, temporal, and occipital bones. It is filled with cartilage in life and transmits no major vessels or nerves.

In an inferior view of the skull, the sphenoid can be seen just anterior to the basilar part of the occipital bone. The internal openings of the nasal cavity seen here are called the **posterior nasal apertures,** or **choanae**[18] (co-AH-nee). Lateral to each aperture, the sphenoid bone exhibits a pair of parallel plates—the **medial** and **lateral pterygoid**[19] **plates** (TERR-ih-goyd) (see fig. 8.5a). Each plate has a narrower inferior extension called the **pterygoid process.** These plates and processes provide attachment for some of the jaw muscles.

The Ethmoid Bone

The **ethmoid**[20] **bone** (ETH-moyd) is an anterior cranial bone located between the eyes (figs. 8.7 and **8.12**). It contributes to the medial wall of the orbit, the roof and walls of the nasal cavity, and the nasal septum. It is a very porous and delicate bone, with three major portions:

1. The vertical **perpendicular plate,** a thin median plate of bone that forms the superior two-thirds of the nasal septum (see fig. 8.4b). (The lower part is formed by the *vomer,*

---

[17]*lacerum* = torn, lacerated
[18]*choana* = funnel
[19]*pteryg* = wing; *oid* = resembling
[20]*ethmo* = sieve, strainer; *oid* = resembling

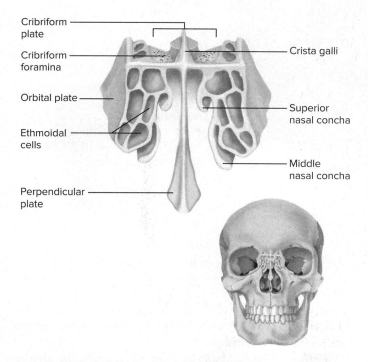

Cribriform plate
Cribriform foramina
Orbital plate
Ethmoidal cells
Perpendicular plate
Crista galli
Superior nasal concha
Middle nasal concha

FIGURE 8.12 The Ethmoid Bone (Anterior View). APR

❓ *List five bones that articulate with the ethmoid bone.*

discussed later.) The septum divides the nasal cavity into right and left air spaces called the **nasal fossae** (FOSS-ee). The septum is often curved, or deviated, toward one nasal fossa or the other.

2. A horizontal **cribriform**[21] **plate** (CRIB-rih-form), which forms the roof of the nasal cavity. This plate has a median blade called the **crista galli**[22] (GAL-eye), an attachment point for the dura mater. On each side of the crista is an elongated depressed area perforated with numerous holes, the **cribriform (olfactory) foramina.** A pair of *olfactory bulbs* of the brain, concerned with the sense of smell, rests in these depressions, and the foramina allow passage for olfactory nerves from the nasal cavity to the bulbs.

3. The **ethmoidal labyrinth,** a large mass on each side of the perpendicular plate. The labyrinth is named for the fact that internally, it has a maze of air spaces called the **ethmoidal cells.** Collectively, these constitute the *ethmoidal sinus* discussed earlier. The lateral surface of the labyrinth is a smooth, slightly concave **orbital plate** seen on the medial wall of the orbit (see fig. 8.14). The medial surface of the labyrinth gives rise to two curled, scroll-like plates of bone called the **superior** and **middle nasal conchae**[23] (CON-kee). These project into the nasal fossa from its lateral wall toward the septum (see figs. 8.7 and **8.13**). There is also a separate bone, the *inferior nasal concha,* discussed later. The three conchae occupy most of the nasal cavity, leaving little open space. By filling space and creating turbulence in the flow of inhaled air, they ensure that the air contacts the mucous membranes that cover these bones, which cleanse, humidify, and warm the inhaled air before it reaches the lungs. The superior concha and adjacent part of the nasal septum also bear the sensory cells of smell.

Usually, all that can be seen of the ethmoid is the perpendicular plate, by looking into the nasal cavity (see fig. 8.3); the orbital plate, by looking at the medial wall of the orbit **(fig. 8.14);** and the crista galli and cribriform plate, seen from within the cranial cavity (see fig. 8.5b).

8.2b Facial Bones

Facial bones do not enclose the brain but lie anterior to the cranial cavity. They support the orbital, nasal, and oral cavities, shape the face, and provide attachment for the muscles of facial expression and mastication. There are 14 facial bones:

2 maxillae	2 nasal bones
2 palatine bones	2 inferior nasal conchae
2 zygomatic bones	1 vomer
2 lacrimal bones	1 mandible

---

[21]*cribri* = sieve; *form* = in the shape of
[22]*crista* = crest; *galli* = of a rooster
[23]*concha* = conch (large marine snail)

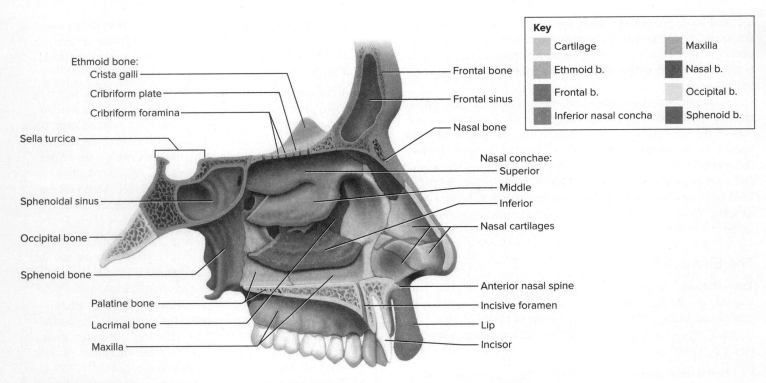

FIGURE 8.13 The Left Nasal Cavity, Sagittal Section with Nasal Septum Removed. A&PR

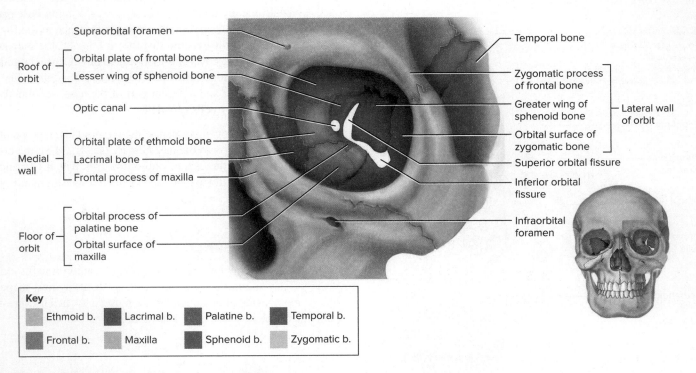

FIGURE 8.14 The Left Orbit (Anterior View). A&PR

The Maxillae

The **maxillae** (mac-SILL-ee) are the largest facial bones. They form the upper jaw and meet each other at a median *intermaxillary suture* (see figs. 8.3, 8.4a, and 8.5a). Small points of maxillary bone called **alveolar processes** grow into the spaces between the bases of the teeth. The root of each tooth is inserted into a deep socket, or **alveolus.** If a tooth is lost or extracted so that chewing no longer puts stress on the maxilla, the alveolar processes are resorbed and the alveolus fills in with new bone, leaving a smooth area on the maxilla.

Although they are preserved with the skull, the teeth are not bones. They are discussed in detail in section 25.2a.

▶▶▶ APPLY WHAT YOU KNOW

Suppose you were studying a skull with some teeth missing. How could you tell whether the teeth had been lost after the person's death or years before it?

Each maxilla extends from the teeth to the inferomedial wall of the orbit. Just below the orbit, it exhibits an **infraorbital foramen,** which provides passage for a blood vessel to the face and a nerve that receives sensations from the nasal region and cheek. This nerve emerges through the foramen rotundum into the cranial cavity. The maxilla forms part of the floor of the orbit, where it exhibits a gash called the **inferior orbital fissure** that angles downward and medially (fig. 8.14). The inferior and superior orbital fissures form a sideways V whose apex lies near the optic canal. The inferior orbital fissure is a passage for blood vessels and sensory nerves from the face.

The **palate** forms the roof of the mouth and floor of the nasal cavity. Its function is to separate the nasal cavity from the oral cavity, enabling us (and other mammals) to continue breathing while chewing. The high metabolic rate of humans requires rapid digestion of food, which in turn is aided by prolonged and thorough mastication into small, easily digested particles. This would be difficult if such prolonged mastication required an interruption of airflow.

The palate consists of a bony **hard palate** anteriorly and a fleshy **soft palate** posteriorly. Most of the hard palate is formed by horizontal extensions of the maxilla called **palatine processes** (PAL-uh-tine) (see fig. 8.5a). Just behind the incisors (front teeth) is a median pit, the **incisive fossa,** which is a passage for an artery to the palate and a nerve to the lower part of the nasal septum and the six front teeth of the maxilla. One or two pairs of *incisive foramina* open into this fossa (fig. 8.13) but are difficult to see. The palatine processes normally meet at the intermaxillary suture at about 12 weeks of gestation. Failure to join results in a *cleft palate* (see Deeper Insight 8.1).

DEEPER INSIGHT 8.1
CLINICAL APPLICATION

Cleft Palate and Lip

Failure of the fetal maxillae to join results in a cleft palate—a median fissure between the oral and nasal cavities, often accompanied by an upper lip cleft on one or both sides. A cleft palate makes it difficult for an infant to generate the suction needed for nursing, and may be accompanied by frequent ear infections and difficulties in hearing and speech. Cleft palate and lip can be surgically corrected with good cosmetic results, but may require follow-up speech therapy. Correction by 18 months of age can improve language acquisition and avoid the psychosocial problems often faced by school-age children with the condition.

The Palatine Bones

The **palatine bones** divide the oral and nasal cavities from each other posteriorly (fig. 8.13). Each has an L shape formed by a *horizontal plate* and a *perpendicular plate*. The horizontal plates form the posterior one-third of the bony palate. Each is marked by a large **greater palatine foramen,** a nerve passage to the palate. The perpendicular plate is a thin, delicate, irregularly shaped plate that forms part of the wall between the nasal cavity and the orbit (see figs. 8.5a and 8.13).

The Zygomatic Bones

The **zygomatic**[24] **bones,** colloquially called the cheekbones, form the angles of the cheeks at the inferolateral margins of the orbits and part of the lateral wall of each orbit; they extend about halfway to the ear (see figs. 8.4a and 8.5a). Each zygomatic bone has an inverted T shape and usually a small **zygomaticofacial foramen** (ZY-go-MAT-ih-co-FAY-shul) near the intersection of the stem and crossbar of the T. A nerve passes through here to supply the skin on the prominence of the cheek. The prominent zygomatic arch that flares from each side of the skull is formed mainly by the union of the zygomatic bone, temporal bone, and maxilla (see fig. 8.4a).

The Lacrimal Bones

The **lacrimal**[25] **bones** (LACK-rih-mul) form part of the medial wall of each orbit (fig. 8.14). They are the smallest bones of the skull, about the size of the little fingernail. A depression called the **lacrimal fossa** houses a membranous *lacrimal sac* in life. Tears from the eye collect in this sac and drain into the nasal cavity.

The Nasal Bones

Two small rectangular **nasal bones** form the bridge of the nose (see fig. 8.3) and support cartilages that shape its lower portion. If you palpate the bridge, you can easily feel where the nasal bones end and the cartilages begin. The nasal bones are often fractured by blows to the nose.

The Inferior Nasal Conchae

There are three conchae in the nasal cavity. The superior and middle conchae, as discussed earlier, are parts of the ethmoid bone. The **inferior nasal concha**—the largest of the three—is a separate bone (fig. 8.13).

The Vomer

The **vomer** forms the inferior half of the nasal septum (see figs. 8.3 and 8.4b). Its name literally means "plowshare," which refers to its resemblance to the blade of a plow. The superior half of the nasal septum is formed by the perpendicular plate of the ethmoid bone, as mentioned earlier. The vomer and perpendicular plate support a wall of *septal cartilage* that forms most of the anterior part of the septum.

---

[24]*zygo* = to join, unite
[25]*lacrim* = tear, to cry

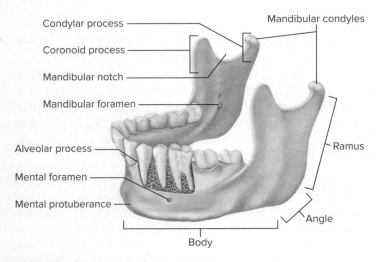

FIGURE 8.15 labels:
Condylar process
Coronoid process
Mandibular notch
Mandibular foramen
Alveolar process
Mental foramen
Mental protuberance
Mandibular condyles
Ramus
Angle
Body

FIGURE 8.15 The Mandible. APR

The Mandible

The **mandible (fig. 8.15)** is the strongest bone of the skull and the only one that can move significantly. It supports the lower teeth and provides attachment for muscles of mastication and facial expression. The horizontal portion, bearing the teeth, is called the **body;** the vertical to oblique posterior portion is the **ramus** (RAY-mus) (plural, *rami*); and these two portions meet at a corner called the **angle.** The mandible develops as separate right and left bones in the fetus, joined by a median cartilaginous joint called the **mental symphysis** (SIM-fih-sis) at the point of the chin. This joint ossifies in early childhood, uniting the two halves into a single bone. The point of the chin itself is called the **mental protuberance.** The inner (posterior) surface of the mandible in this region has a pair of small points, the **mental spines,** which serve for attachment of certain chin muscles (see fig. 8.4b).

On the anterolateral surface of the body, the **mental foramen** permits the passage of nerves and blood vessels of the chin. The inner surface of the body has a number of shallow depressions and ridges to accommodate muscles and salivary glands. The angle of the mandible has a rough lateral surface for insertion of the *masseter,* a muscle of mastication. Like the maxilla, the mandible has pointed alveolar processes between the teeth.

The ramus is somewhat Y-shaped. Its posterior branch, called the **condylar process** (CON-dih-lur), bears the **mandibular condyle**—an oval knob that articulates with the mandibular fossa of the temporal bone. The meeting of this condyle with the temporal bone forms a hinge, the **temporomandibular joint (TMJ).** The anterior branch of the ramus is a blade called the **coronoid process.** It is the point of insertion for the temporalis muscle, which pulls the mandible upward when you bite. The U-shaped arch between the two processes is the **mandibular notch.** Just below the notch, on the medial surface of the ramus, is the **mandibular foramen.** The nerve and blood vessels that supply the lower teeth enter this foramen and then travel through the bone of the mandibular body, giving off branches to each tooth along the way. Dentists commonly inject lidocaine near the mandibular foramen to deaden sensation from the lower teeth.

8.2c Bones Associated with the Skull

Seven bones are closely associated with the skull but not considered part of it. These are the three auditory ossicles in each middle-ear cavity and the hyoid bone beneath the chin. The **auditory ossicles**[26]—named the **malleus** (hammer), **incus** (anvil), and **stapes** (STAY-peez) (stirrup)—are discussed in connection with hearing in "Anatomy of the Ear" in section 16.4.

The **hyoid**[27] **bone** is a slender U-shaped bone between the chin and larynx **(fig. 8.16).** It is one of the few bones that doesn't articulate with any other. It is suspended from the styloid processes of the skull, somewhat like a hammock, by the small *stylohyoid muscles* and *stylohyoid ligaments.* The median **body** of the hyoid is flanked on either side by hornlike projections called the **greater** and **lesser horns (cornua).** The larynx (voice box) is suspended from the hyoid bone by a broad ligament (see fig. 22.4a), and the hyoid serves for attachment of several muscles that control the mandible, tongue, and larynx. Forensic pathologists look for a fractured hyoid as evidence of strangulation.

8.2d The Skull in Infancy and Childhood

The head of an infant couldn't fit through the mother's pelvic outlet at birth if not for the fact that the bones of its skull are not yet fused. The shifting of the cranial bones during birth may cause the infant's head to appear deformed, but it soon assumes a more normal shape. Spaces between the unfused cranial bones are called **fontanelles,**[28] after the fact that pulsation of the infant's blood can be felt there. The bones are joined at these points only by fibrous membranes, in which intramembranous ossification is completed later (see fig. 7.10). Four of these sites are especially prominent and regular in location: the **anterior, posterior, sphenoid (anterolateral),**

---

[26]*os* = bone; *icle* = little
[27]*hy* = the letter U; *oid* = resembling
[28]*fontan* = fountain; *elle* = little

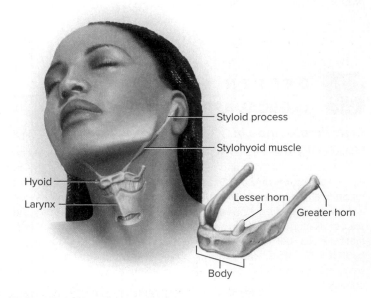

FIGURE 8.16 labels:
Styloid process
Stylohyoid muscle
Hyoid
Larynx
Lesser horn
Greater horn
Body

FIGURE 8.16 The Hyoid Bone.

DEEPER INSIGHT 8.2

CLINICAL APPLICATION

Cranial Assessment of the Newborn

Obstetric nurses routinely assess the fontanelles of newborns by palpation. In a difficult delivery, one cranial bone may override another along a suture line, which calls for close monitoring of the infant. Abnormally wide sutures may indicate hydrocephalus, the accumulation of excessive amounts of cerebrospinal fluid, which causes the cranium to swell. Bulging fontanelles suggest abnormally high intracranial pressure, while depressed fontanelles indicate dehydration.

and **mastoid (posterolateral) fontanelles (fig. 8.17).** Most fontanelles ossify by the time the infant is a year old, but the largest one—the anterior fontanelle—can still be palpated 18 to 24 months after birth.

The frontal bone and mandible are separate right and left bones at birth, but fuse medially in early childhood. The frontal

bones usually fuse by age 5 or 6, but in some children a *metopic*[29] *suture* persists between them. Traces of this suture are evident in some adult skulls.

The face of a newborn is flat and the cranium relatively large. To accommodate the growing brain, the skull grows more rapidly than the rest of the skeleton during childhood. It reaches about half its adult size by 9 months of age, three-quarters by age 2, and nearly final size by 8 or 9 years. The heads of babies and children are therefore much larger in proportion to the trunk than the heads of adults—an attribute thoroughly exploited by cartoonists and advertisers who draw big-headed characters to give them a more endearing or immature appearance. In humans and other animals, the large, rounded heads of the young are thought to promote survival by stimulating parental caregiving instincts.

BEFORE YOU GO ON

Answer the following questions to test your understanding of the preceding section:

4. Name the paranasal sinuses and state their locations. Name any four other cavities in the skull.

5. Explain the difference between a cranial bone and a facial bone. Give four examples of each.

6. Draw an oval representing a superior view of the calvaria. Draw lines representing the coronal, lambdoid, and sagittal sutures. Label the four bones separated by these sutures.

7. State which bone has each of these features: a squamous part, hypoglossal canal, greater horn, greater wing, condylar process, and cribriform plate.

8. Determine which of the following structures cannot normally be palpated on a living person: the mastoid process, crista galli, superior orbital fissure, palatine processes, zygomatic bone, mental protuberance, and stapes. You may find it useful to palpate some of these on your own skull as you try to answer.

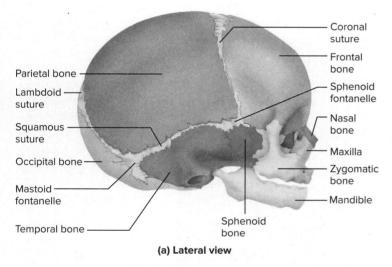

(a) **Lateral view**

Parietal bone
Lambdoid suture
Squamous suture
Occipital bone
Mastoid fontanelle
Temporal bone
Sphenoid bone
Coronal suture
Frontal bone
Sphenoid fontanelle
Nasal bone
Maxilla
Zygomatic bone
Mandible

(b) **Superior view**

Frontal bone
Anterior fontanelle
Sagittal suture
Parietal bone
Posterior fontanelle

FIGURE 8.17 The Fetal Skull Near the Time of Birth. (a) Lateral view. (b) Superior view of the calvaria.

8.3   The Vertebral Column and Thoracic Cage

Expected Learning Outcomes

When you have completed this section, you should be able to

a. describe the general features of the vertebral column and those of a typical vertebra;

b. describe the structure of the intervertebral discs and their relationship to the vertebrae;

c. describe the special features of vertebrae in different regions of the vertebral column, and discuss the functional significance of the regional differences; and

d. describe the anatomy of the sternum and ribs and how the ribs articulate with the thoracic vertebrae.

[29]*met* = beyond; *op* = the eyes

8.3a General Features of the Vertebral Column

The **vertebral column,** or **spine,** physically supports the skull and trunk, allows for their movement, protects the spinal cord, and absorbs stresses produced by walking, running, and lifting. It also provides attachment for the limbs, thoracic cage, and postural muscles. Although commonly called the backbone, it consists of not a single bone but a flexible chain of 33 **vertebrae** with **intervertebral discs** of fibrocartilage between most of them.

The adult vertebral column averages about 71 cm (28 in.) long, with the intervertebral discs accounting for about one-quarter of the length. Most people are about 1% shorter when they go to bed at night than when they first rise in the morning. This is because during the day, the weight of the body compresses the intervertebral discs and squeezes water out of them. When one is sleeping, with the weight off the spine, the discs reabsorb water and swell.

As shown in **figure 8.18,** the vertebrae are divided into five groups, usually numbering 7 *cervical vertebrae* (SUR-vih-cul)

in the neck, 12 *thoracic vertebrae* in the chest, 5 *lumbar vertebrae* in the lower back, 5 *sacral vertebrae* at the base of the spine, and 4 tiny *coccygeal vertebrae* (coc-SIDJ-ee-ul). To help remember the numbers of cervical, thoracic, and lumbar vertebrae—7, 12, and 5—think of a typical workday: Go to work at 7, have lunch at 12, and go home at 5. All mammals have 7 cervical vertebrae, even in the famously long necks of giraffes.

Variations in this arrangement occur in about 1 person in 20. For example, the last lumbar vertebra is sometimes incorporated into the sacrum, producing four lumbar and six sacral vertebrae. In other cases, the first sacral vertebra fails to fuse with the second, producing six lumbar and four sacral vertebrae. The coccyx usually has four but sometimes five vertebrae. The cervical and thoracic vertebrae are more constant in number.

Beyond the age of 3 years, the vertebral column has four bends that give it an undulating S shape **(fig. 8.19).** A bend that is concave in the anterior direction (curved outward toward the back) is called a **kyphosis.**[30] There are two of these, the *thoracic kyphosis* and *pelvic kyphosis.* A bend toward the front,

[30]*kypho* = crooked, bent; *osis* = condition

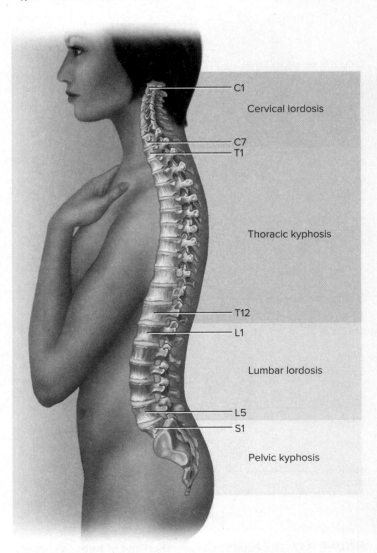

Atlas (C1)
Axis (C2)
Cervical vertebrae
C7
T1
Thoracic vertebrae
T12
L1
Lumbar vertebrae
L5
S1
Sacrum
S5
Coccyx   Coccyx

(a) Anterior view   **(b) Posterior view**

FIGURE 8.18 The Vertebral Column. (a) Anterior view. (b) Posterior view. **APR**

C1
Cervical lordosis
C7
T1
Thoracic kyphosis
T12
L1
Lumbar lordosis
L5
S1
Pelvic kyphosis

FIGURE 8.19 Curvatures of the Adult Vertebral Column.

concave posteriorly, is called a **lordosis,**[31] and there are also two of these—the *cervical lordosis* and *lumbar lordosis*. These are not present in the newborn, whose spine exhibits one continuous C-shaped curve, as it does in monkeys, apes, and most other four-legged animals **(fig. 8.20)**. As an infant begins to crawl and lift its head, the cervical region becomes curved toward the posterior side, enabling an infant on its belly to look up. As a toddler begins walking, another curve develops in the same direction in the lumbar region. The resulting **S** shape makes sustained bipedal walking possible (see Deeper Insight 8.5). The thoracic and pelvic kyphoses are called *primary curvatures* because they exist from birth. The cervical and lumbar lordoses are called *secondary curvatures* because they develop later, in the child's first few years of crawling and walking. Abnormal lateral and anteroposterior curvatures are among the most common disorders of the spine (see Deeper Insight 8.3).

8.3b General Structure of a Vertebra

A representative vertebra and intervertebral disc are shown in **figure 8.22.** The most obvious feature of a vertebra is the **body (centrum)**—a mass of spongy bone and red bone marrow covered with a thin shell of compact, cortical bone. This is the weight-bearing portion of the vertebra. Its rough superior and inferior surfaces provide firm attachment to the intervertebral discs.

FIGURE 8.20 Spinal Curvature of the Newborn Infant. At this age, the spine forms a single C-shaped curve.
Bob Coyle/McGraw-Hill Education

---

[31]*lordo* = backward; *osis* = condition

 DEEPER INSIGHT 8.3

CLINICAL APPLICATION

Abnormal Spinal Curvatures

Abnormal spinal curvatures **(fig. 8.21)** can result from disease, weakness or paralysis of the trunk muscles, poor posture, pregnancy, or congenital defects in vertebral anatomy. The most common deformity is an abnormal lateral curvature called *scoliosis*. It occurs most often in the thoracic region, particularly among adolescent girls. It sometimes results from a developmental abnormality in which the body and arch fail to develop on one side of a vertebra. If the person's skeletal growth is not yet complete, scoliosis can be corrected with a back brace.

An exaggerated thoracic curvature is called *hyperkyphosis*. It is usually a result of osteoporosis (see fig. 7.21), but it also occurs in people with osteomalacia or spinal tuberculosis and in adolescents who engage heavily in such spine-loading sports as wrestling and weight lifting. An exaggerated lumbar curvature is called *hyperlordosis*. It may have the same causes as hyperkyphosis, or it may result from added abdominal weight in pregnancy or obesity.

(a) Scoliosis **(b) Hyperkyphosis** **(c) Hyperlordosis**

Key
Normal
Pathological

FIGURE 8.21 Abnormal Spinal Curvatures. (a) Scoliosis. (b) Hyperkyphosis. (c) Hyperlordosis.

Posterior

(a) 2nd lumbar vertebra (L2)

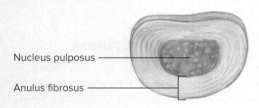

(b) Intervertebral disc

FIGURE 8.22 A Representative Vertebra and Intervertebral Disc (Superior Views). (a) A typical vertebra. (b) An intervertebral disc, oriented the same way as the vertebral body in part (a) for comparison.

▶▶▶**APPLY WHAT YOU KNOW**

The vertebral bodies and intervertebral discs get progressively larger as we look lower and lower on the vertebral column. What is the functional significance of this trend?

Posterior to the body of each vertebra is a triangular space called the **vertebral foramen.** The vertebral foramina collectively form the **vertebral canal,** a passage for the spinal cord. Each foramen is bordered by a bony **vertebral arch** composed of two parts on each side: a pillarlike **pedicle**[32] and platelike **lamina.**[33] Extending from the apex of the arch, a projection called the **spinous process** is directed posteriorly and downward. You can see and feel the spinous processes on a living person as a row of bumps along the spine. A **transverse process** extends laterally from the point where the pedicle and lamina meet. The spinous and transverse processes provide points of attachment for ligaments, ribs, and spinal muscles.

A pair of **superior articular processes** projects upward from one vertebra and meets a similar pair of inferior articular processes that projects downward from the vertebra above **(fig. 8.24a).** Each process has a flat articular surface (facet) facing that of the adjacent vertebra. These processes restrict twisting of the vertebral column, which could otherwise severely damage the spinal cord.

[32]*ped* = foot; *icle* = little
[33]*lamina* = plate

When two vertebrae are joined, they exhibit an opening on each side between their pedicles called the **intervertebral foramen.** These allow passage for spinal nerves that connect with the spinal cord at regular intervals. Each foramen is formed by an **inferior vertebral notch** in the pedicle of the upper vertebra and a **superior vertebral notch** in the pedicle of the lower one **(fig. 8.24b).**

8.3c Intervertebral Discs

An intervertebral disc is a cartilaginous pad located between the bodies of two adjacent vertebrae. It consists of an inner gelatinous **nucleus pulposus** surrounded by a ring of fibrocartilage, the **anulus fibrosus** (see fig. 8.22b). There are 23 discs—the first one between cervical vertebrae 2 and 3 (C2–C3) and the last one between the last lumbar vertebra and the sacrum (L5–S1). They help to bind adjacent vertebrae together, support the weight of the body, allow spinal mobility, and absorb shock. Discs can rupture painfully when subjected to excessive stress (see Deeper Insight 8.4).

DEEPER INSIGHT 8.4

CLINICAL APPLICATION

Herniated Disc

Stress on the vertebral column, as in lifting a heavy weight, compresses the intervertebral discs and makes them bulge laterally. Excessive stress can crack the anulus fibrosus, allowing the gelatinous nucleus pulposus to ooze out **(fig. 8.23).** This is called a *herniated disc* (colloquially, a "ruptured disc," or more inaccurately, "slipped disc"). Herniation usually occurs at the posterolateral corners of the discs because the anulus fibrosus is thinnest on the posterior side and corners are least supported by nearby spinal ligaments. About 95% of herniations are at the L4–L5 or L5–S1 level because these bear the most weight. Disc herniation rarely occurs in young people because of their well-hydrated, thick, resilient discs. Herniated discs are one of the most common causes of lower back pain, stemming from inflammation stimulated by the nucleus and pressure of the oozing nucleus on spinal nerve roots passing through this tight space. Over 80% of patients recover from the pain with only oral analgesics, requiring no surgery; up to 76% of cases resolve partially or completely within 1 year.

FIGURE 8.23 Herniated Disc.

(a) Posterior (dorsal) view

Transverse process
Body (centrum)
Intervertebral disc
Inferior articular process of L2
Superior articular process of L3
Lamina

L2
L3

(b) Left lateral view

Superior articular process of L1
Inferior vertebral notch of L1
Superior vertebral notch of L2
Intervertebral foramen
Spinous process
Intervertebral disc
Inferior articular process of L3

L1
L2
L3

FIGURE 8.24 Articulated Vertebrae. (a) Posterior view. (b) Left lateral view.

8.3d Regional Characteristics of Vertebrae

We are now prepared to consider how vertebrae differ from one region of the vertebral column to another and from the generalized anatomy just described. Knowing these variations will enable you to identify the region of the spine from which an isolated vertebra was taken. More importantly, these modifications in form reflect functional differences among the vertebrae.

The Cervical Vertebrae

The cervical vertebrae (C1–C7) are relatively small. Their function is to support the head and allow for its movements. The first two (C1 and C2) have unique structures for this purpose **(fig. 8.25).** Vertebra C1 is called the **atlas** because it supports the head in a manner reminiscent of Atlas, the giant of Greek mythology who was condemned by Zeus to carry the heavens

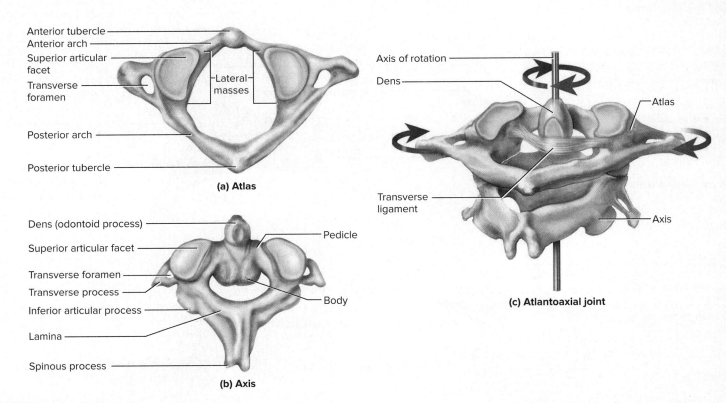

Anterior tubercle
Anterior arch
Superior articular facet
Transverse foramen
Posterior arch
Posterior tubercle

Lateral masses

(a) Atlas

Dens (odontoid process)
Superior articular facet
Transverse foramen
Transverse process
Inferior articular process
Lamina
Spinous process

Pedicle
Body

(b) Axis

Axis of rotation
Dens
Transverse ligament

Atlas
Axis

(c) Atlantoaxial joint

FIGURE 8.25 The Atlas and Axis, Cervical Vertebrae C1 and C2. (a) The atlas, superior view. (b) The axis, posterosuperior view. (c) Articulation of the atlas and axis and rotation of the atlas. This movement turns the head from side to side, as in gesturing "no." Note the transverse ligament holding the dens of the axis in place. **APR**

? *What serious consequence could result from a rupture of the transverse ligament?*

on his shoulders. It scarcely resembles the typical vertebra; it has no body, and is little more than a delicate ring surrounding a large vertebral foramen. On each side is a **lateral mass** with a deeply concave **superior articular facet** that articulates with the occipital condyle of the skull. A nodding motion of the skull, as in gesturing "yes," causes the occipital condyles to rock back and forth on these facets. The **inferior articular facets,** which are comparatively flat or only slightly concave, articulate with C2. The lateral masses are connected by an **anterior arch** and a **posterior arch,** which bear slight protuberances called the **anterior and posterior tubercle,** respectively.

Vertebra C2, the **axis,** allows rotation of the head as in gesturing "no." Its most distinctive feature is a prominent knob called the **dens** (pronounced "denz"), or **odontoid**[34] **process,** on

its anterosuperior side. No other vertebra has a dens. It begins to form as an independent ossification center during the first year of life and fuses with the axis by the age of 3 to 6 years. It projects into the vertebral foramen of the atlas, where it is nestled in a facet and held in place by a **transverse ligament** (fig. 8.25c). A heavy blow to the top of the head can cause a fatal injury in which the dens is driven through the foramen magnum into the brainstem.

The articulation between the atlas and the cranium is called the **atlanto–occipital joint;** the one between the atlas and axis is called the **atlantoaxial joint.**

The axis is the first vertebra to exhibit a spinous process. In vertebrae C2 through C6, the process is forked, or *bifid,*[35] at its tip **(fig. 8.26a).** This fork provides attachment for the *nuchal*

---

[34]*dens = odont* = tooth; *oid* = resembling

[35]*bifid* = branched, cleft into two parts

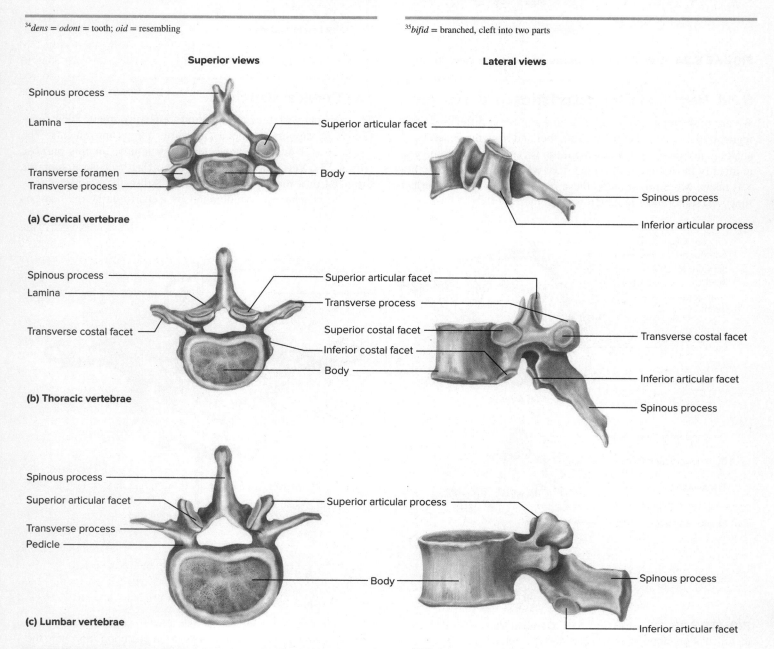

Superior views   **Lateral views**

(a) Cervical vertebrae

Spinous process — Lamina — Transverse foramen — Transverse process — Superior articular facet — Body — Spinous process — Inferior articular process

(b) Thoracic vertebrae

Spinous process — Lamina — Transverse costal facet — Superior articular facet — Transverse process — Superior costal facet — Inferior costal facet — Body — Transverse costal facet — Inferior articular facet — Spinous process

(c) Lumbar vertebrae

Spinous process — Superior articular facet — Transverse process — Pedicle — Superior articular process — Body — Spinous process — Inferior articular facet

FIGURE 8.26 Regional Differences in Vertebrae. (a) Cervical vertebrae. (b) Thoracic vertebrae. (c) Lumbar vertebrae. The left-hand figures are superior views and the right-hand figures are left lateral views. **APR**

ligament of the back of the neck. All seven cervical vertebrae have a prominent round **transverse foramen** in each transverse process. These foramina provide passage and protection for the *vertebral arteries,* which supply blood to the brain, and *vertebral veins,* which drain blood from various neck structures (but not from the brain). Transverse foramina occur in no other vertebrae and thus provide an easy means of recognizing a cervical vertebra.

Cervical vertebrae C3 through C6 are similar to the typical vertebra depicted in figure 8.22a, with the addition of the transverse foramina and bifid spinous processes. Vertebra C7 is a little different—its spinous process is not bifid, but is especially long and forms a prominent bump on the lower back of the neck. C7 is sometimes called the *vertebra prominens* because of this conspicuous spinous process. This feature is a convenient landmark for counting vertebrae by palpation. One can easily identify the largest bump on the neck as C7, then count up or down from there to identify others.

The Thoracic Vertebrae

There are 12 **thoracic vertebrae** (T1–T12), corresponding to the 12 pairs of ribs attached to them; no other vertebrae have ribs. One function of the thoracic vertebrae is to support the thoracic cage enclosing the heart and lungs. They lack the transverse foramina and bifid processes that distinguish the cervical vertebrae, but possess the following distinctive features of their own **(fig. 8.26b):**

- The spinous processes are relatively pointed and angle sharply downward.
- The body is somewhat heart-shaped, more massive than in the cervical vertebrae but less than in the lumbar vertebrae.
- The body has small, smooth, slightly concave spots called *costal facets* for attachment of the ribs.
- Vertebrae T1 through T10 have a shallow, cuplike **transverse costal**[36] **facet** at the end of each transverse process. These provide a second point of articulation for ribs 1 to 10. There are no transverse costal facets on T11 and T12 because ribs 11 and 12 attach only to the bodies of those two vertebrae.

Thoracic vertebrae vary among themselves mainly in the mode of articulation with the ribs (see fig. 8.30). In most cases, a rib inserts between two vertebrae, so each vertebra contributes one-half of the articular surface. A rib articulates with the **inferior costal facet** of the upper vertebra and the **superior costal facet** of the vertebra below that. This terminology may be a little confusing, but note that the superior and inferior facets are named for their position on the vertebral body, not for which part of the rib's articulation they provide. Vertebrae T1 and T10 through T12, however, have complete costal facets on the bodies for ribs 1 and 10 through 12, which articulate on the vertebral bodies instead of between vertebrae. Vertebrae T11 and T12, as noted, have no transverse costal facets. These variations will

be more functionally understandable after you have studied the anatomy of the ribs, so we will return then to the details of these articular surfaces.

Each thoracic vertebra has a pair of superior articular facets that face posteriorly and a pair of inferior articular facets that face anteriorly (except in vertebra T12). Thus, the superior facets of one vertebra articulate with the inferior facets of the next one above it. In vertebra T12, however, the inferior articular facets face somewhat laterally instead of anteriorly. This positions them to articulate with the medially facing superior articular facets of the first lumbar vertebra. T12 thus shows an anatomical transition between the thoracic and lumbar pattern, described next.

The Lumbar Vertebrae

There are five **lumbar vertebrae** (L1–L5). Their most distinctive features are a thick, stout body and a blunt, squarish spinous process for attachment of the strong lumbar muscles **(fig. 8.26c).** In addition, their articular processes are oriented differently than on other vertebrae. The superior processes face medially (like the palms of your hands about to clap), and the inferior processes face laterally, toward the superior processes of the next vertebra. This arrangement resists twisting of the lower spine. These differences are best observed on an articulated skeleton.

The Sacrum

The **sacrum** (SACK-rum, SAY-krum) is a bony plate that forms the posterior wall of the pelvic girdle **(fig. 8.27).** It was named *sacrum* for its prominence as the largest and most durable bone of the vertebral column.[37] There are five separate **sacral vertebrae** (S1–S5) in children, but they begin to fuse around age 16 and are fully fused by age 26.

The anterior surface of the sacrum is relatively smooth and concave and has four transverse lines that indicate where the five vertebrae have fused. This surface exhibits four pairs of large **anterior sacral foramina,** which allow for passage of nerves and arteries to the pelvic organs. The posterior surface is very rough. The spinous processes of the vertebrae fuse into a ridge called the **median sacral crest.** The transverse processes fuse into a less prominent **lateral sacral crest** on each side of the median crest. Again on the posterior side of the sacrum, there are four pairs of openings for spinal nerves, the **posterior sacral foramina.** The nerves that emerge here supply the gluteal region and lower limbs.

A **sacral canal** runs through the sacrum and ends in an inferior opening called the **sacral hiatus** (hy-AY-tus). This canal contains spinal nerve roots. On each side of the sacrum is an ear-shaped region called the **auricular**[38] **surface** (aw-RIC-you-lur). This articulates with a similarly shaped surface on the hip bone (see fig. 8.37b) and forms the strong, nearly immovable **sacroiliac**

---

[36]*costa* = rib; *al* = pertaining to

[37]*sacr* = great, prominent
[38]*auri* = ear; *cul* = little; *ar* = pertaining to

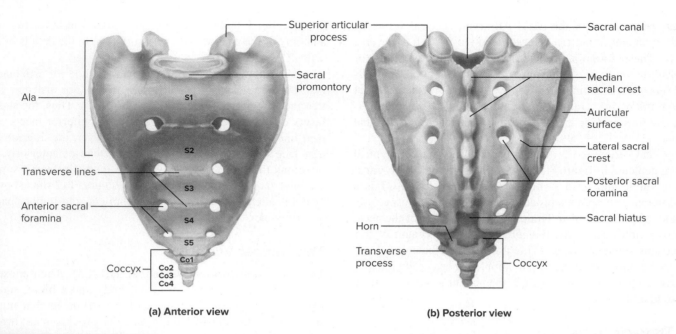

(a) Anterior view

(b) Posterior view

FIGURE 8.27 The Sacrum and Coccyx. (a) The anterior surface, which faces the viscera of the pelvic cavity. (b) The posterior surface. The processes of this surface can be palpated in the sacral region. **APR**

(SI) joint (SACK-ro-ILL-ee-ac). The body of vertebra S1 juts anteriorly to form a **sacral promontory,** which supports the body of vertebra L5. Lateral to the median sacral crest, S1 also has a pair of **superior articular processes** that articulate with vertebra L5. Lateral to these is a pair of large, rough, winglike extensions called the **alae**[39] (AIL-ee).

The Coccyx

Four (sometimes five) tiny **coccygeal vertebrae** (Co1 to Co4 or Co5) fuse by the age of 20 to 30 years to form the **coccyx**[40] (COC-six) (fig. 8.27), colloquially called the tailbone. Although it is indeed the vestige of an ancestral tail, it is not entirely useless; it provides attachment for the muscles of the pelvic floor. Vertebra Co1 has a pair of **horns (cornua)** that serve as attachment points for ligaments that bind the coccyx to the sacrum. The coccyx can be fractured by a difficult childbirth or a hard fall on the buttocks.

8.3e The Thoracic Cage

The **thoracic cage (fig. 8.28)** consists of the thoracic vertebrae, sternum, and ribs. It forms a roughly conical enclosure for the lungs and heart and provides attachment for the pectoral girdle and upper limb. It has a broad base and a somewhat narrower superior apex. Its inferior border is the arc of the lower ribs, called the **costal margin.** The cage protects not only the

thoracic organs but also the spleen, most of the liver, and to some extent the kidneys. Most important is its role in breathing; it is rhythmically expanded by the respiratory muscles to create a vacuum that draws air into the lungs, and then compressed to expel air.

The Sternum

The **sternum** (breastbone) is a bony plate anterior to the heart. It is subdivided into three regions: the manubrium, body, and xiphoid process. The **manubrium**[41] (ma-NOO-bree-um) is the broad superior portion, shaped like the knot of a necktie. It lies at the level of vertebrae T3 to T4. It has a median **suprasternal (jugular) notch,** which you can easily palpate between your clavicles (collarbones), and right and left **clavicular notches** where it articulates with the clavicles. The dagger-shaped **body,** or **gladiolus,**[42] is the longest part of the sternum, lying at the level of vertebrae T5 through T9. It joins the manubrium at the **sternal angle,** which can be palpated as a transverse ridge at the point where the sternum projects farthest forward. In some people, however, the angle is rounded or concave. The second rib attaches here, making the sternal angle a useful landmark for counting ribs by palpation. The manubrium and body have scalloped lateral margins where cartilages of the ribs are attached. At the inferior end (vertebral level T10 to T11) is a small, pointed **xiphoid**[43] **process** (ZIF-oyd) that provides attachment for some of the abdominal muscles. In cardiopulmonary resuscitation,

---

[39]*alae* = wings
[40]*coccyx* = cuckoo (named for resemblance to a cuckoo's beak)

[41]*manubrium* = handle
[42]*gladiolus* = sword
[43]*xipho* = sword; *oid* = resembling

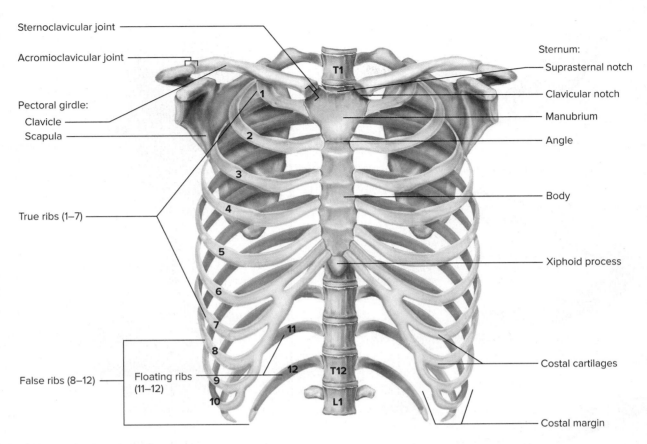

FIGURE 8.28 The Thoracic Cage and Pectoral Girdle (Anterior View). APR

improperly performed chest compressions can drive the xiphoid process into the liver and cause a fatal hemorrhage.

The Ribs

There are 12 pairs of **ribs,** with no difference in number between the sexes (despite popular religious belief). Each is attached at its posterior (proximal) end to the vertebral column, and most of them are also attached at the anterior (distal) end to the sternum. The anterior attachment is by way of a long strip of hyaline cartilage called the **costal cartilage** (COSS-tul).

As a rule, the ribs increase in length from 1 through 7 and become progressively smaller again through rib 12. They are increasingly oblique (slanted) in orientation from 1 through 9, then less so from 10 through 12. They also differ in their individual structure and attachments at different levels of the thoracic cage, so we will examine them in order as we descend the chest, taking note of their universal characteristics as well as their individual variations.

Rib 1 is peculiar. On an articulated skeleton, you must look for its vertebral attachment just below the base of the neck; much of this rib lies above the level of the clavicle (fig. 8.28). It is a short, flat, C-shaped plate of bone **(fig. 8.29a).** At the vertebral end, it exhibits a knobby **head** that articulates with the body of vertebra T1. On an isolated vertebra, you can find a smooth costal

facet for this attachment on the middle of the body. Immediately distal to the head, the rib narrows to a **neck** and then widens again to form a rough area called the **tubercle.** This is its point of attachment to the transverse costal facet of the same vertebra. Beyond the tubercle, the rib flattens and widens into a gently sloping bladelike **shaft.** The shaft ends distally in a squared-off, rough area. In the living individual, the costal cartilage begins here and spans the rest of the distance to the upper sternum. The superior surface of rib 1 has a pair of shallow grooves that serve as platforms for the subclavian artery and subclavian vein.

Ribs 2 through 7 present a more typical appearance **(fig. 8.29b).** At the proximal end, each exhibits a head, neck, and tubercle. The head is wedge-shaped and inserts between two vertebrae. Each margin of the wedge has a smooth surface called an *articular facet.* The **superior articular facet** joins the inferior costal facet of the vertebra above; the **inferior articular facet** joins the superior costal facet of vertebra below. The tubercle of the rib articulates with the transverse costal facet of each same-numbered vertebra. **Figure 8.30** details the rib–vertebra attachments typical of this region of the rib cage.

Beyond the tubercle, each rib makes a sharp curve around the side of the chest and then progresses anteriorly to approach the sternum (see fig. 8.28). The curve is called the **angle** of the rib and the rest of the bony blade distal to it is the **shaft.** The inferior margin of the shaft has a **costal groove** that marks the path of the

(a) Rib 1

(b) Ribs 2–10

(c) Ribs 11–12

FIGURE 8.29 Anatomy of the Ribs. (a) Rib 1 (superior view) is an atypical flat plate. (b) Typical features of ribs 2 to 10. (c) Appearance of the floating ribs, 11 and 12.

(a) Anterior view

(b) Superior view

FIGURE 8.30 Articulation of Rib 6 with Vertebrae T5 and T6. (a) Anterior view. Note the relationship of the articular facets of the rib with the costal facets of the two vertebrae. (b) Superior view. Note that the rib articulates with a vertebra at two points: the costal facet on the vertebral body and the transverse costal facet on the transverse process.

intercostal blood vessels and nerve. Each of these ribs, like rib 1, ends in a blunt, rough area where the costal cartilage begins. Each has its own costal cartilage connecting it to the sternum; because of this feature, ribs 1 through 7 are called **true ribs.**

Ribs 8 through 12 are called **false ribs** because they lack independent cartilaginous connections to the sternum. In 8 through 10, the costal cartilages sweep upward and end on the costal cartilage of rib 7 (see fig. 8.28). Rib 10 also differs from 2 through 9 in that it attaches to the body of a single vertebra (T10) rather than between vertebrae. Thus, vertebra T10 has a complete costal facet on its body for rib 10.

Ribs 11 and 12 are again unusual **(fig. 8.29c).** Posteriorly, they articulate with the bodies of vertebrae T11 and T12, but they do not have tubercles and do not attach to the transverse processes of the vertebrae. Those two vertebrae therefore have no transverse costal facets. At the distal end, these two relatively small, delicate ribs taper to a point and are capped by a small cartilaginous tip, but there is no cartilaginous connection to the sternum or to any of the higher costal cartilages. The ribs are merely embedded in lumbar muscle at this end. Consequently, 11 and 12 are also called **floating ribs.** In people of Japanese and some other ancestries, rib 10 is also usually

floating. **Table 8.4** summarizes these variations in rib anatomy and their vertebral and sternal attachments.

BEFORE YOU GO ON

Answer the following questions to test your understanding of the preceding section:

9. Discuss the contribution of the intervertebral discs to the length and flexibility of the spine.

10. Construct a three-column table headed C4, T4, and L4. In each column, list all anatomical features that would distinguish that vertebra from the other two.

11. Name the three parts of the sternum. How many ribs attach (directly or indirectly) to each part?

12. Describe how rib 5 articulates with the spine. How do ribs 1 and 12 differ from this and from each other in their modes of articulation?

13. Distinguish between true, false, and floating ribs. Which ribs fall into each category?

14. Name the three divisions of the sternum and list the sternal features that can be palpated on a living person.

TABLE 8.4	Articulations of the Ribs				
Rib	Type	Costal Cartilage	Articulating Vertebral Bodies	Articulating with a Transverse Costal Facet?	Rib Tubercle
1	True	Individual	T1	Yes	Present
2	True	Individual	T1 and T2	Yes	Present
3	True	Individual	T2 and T3	Yes	Present
4	True	Individual	T3 and T4	Yes	Present
5	True	Individual	T4 and T5	Yes	Present
6	True	Individual	T5 and T6	Yes	Present
7	True	Individual	T6 and T7	Yes	Present
8	False	Shared with rib 7	T7 and T8	Yes	Present
9	False	Shared with rib 7	T8 and T9	Yes	Present
10	False	Shared with rib 7	T10	Yes	Present
11	False, floating	None	T11	No	Absent
12	False, floating	None	T12	No	Absent

8.4 The Pectoral Girdle and Upper Limb

Expected Learning Outcome

When you have completed this section, you should be able to

a. identify and describe the features of the clavicle, scapula, humerus, radius, ulna, and bones of the wrist and hand.

8.4a The Pectoral Girdle

The **pectoral girdle** (shoulder girdle) supports the arm and links it to the axial skeleton. It consists of two bones on each side of the body: the *clavicle* (collarbone) and *scapula* (shoulder blade). The medial end of the clavicle articulates with the sternum at the **sternoclavicular joint,** and its lateral end articulates with the scapula at the **acromioclavicular joint** (see fig. 8.28). The scapula also articulates with the humerus at the **glenohumeral joint.** These are loose attachments that result in a shoulder far more flexible than that of most other mammals, but they also make the shoulder joint easy to dislocate.

▶▶▶APPLY WHAT YOU KNOW

How is the unusual flexibility of the human shoulder joint related to the habitat of our primate ancestors?

The Clavicle

The **clavicle**[44] **(fig. 8.31)** is slightly S-shaped, somewhat flattened from the upper to lower surface, and easily seen and palpated on

[44]*clav* = hammer, club, key; *icle* = little

the upper thorax (see atlas B, fig. B.1b). The superior surface is relatively smooth and rounded, whereas the inferior surface is flatter and marked by grooves and ridges for muscle attachment. The medial **sternal end** has a rounded, hammerlike head, and the lateral **acromial end** is markedly flattened. Near the acromial end is a rough tuberosity called the **conoid tubercle**—a ligament attachment that faces toward the rear and slightly downward.

The clavicle braces the shoulder, keeping the upper limb away from the midline of the body, and is an attachment for some of the muscles of head movement (sternocleidomastoid and trapezius). It also transfers force from the arm to the axial region of the body, as when doing pushups. It is thickened in people who do heavy manual labor, and in right-handed people, the right clavicle is usually stronger and shorter than the left because of the workload placed on it. Without the clavicles, the pectoralis major muscles would pull the shoulders forward and medially—which indeed happens when a clavicle is fractured. The clavicle is one of the

Sternal end — Acromial end

(a) Superior view

Conoid tubercle

Sternal end — Acromial end

(b) Inferior view

FIGURE 8.31 The Right Clavicle. (a) Superior view. (b) Inferior view. **APR**

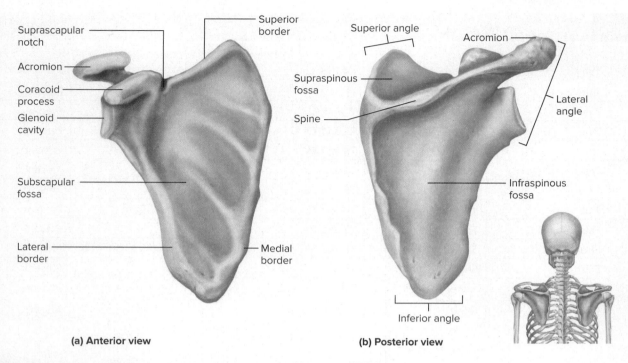

(a) Anterior view

(b) Posterior view

FIGURE 8.32 The Right Scapula. (a) Anterior view. (b) Posterior view. **A&PR**

most commonly fractured bones in the body because it is so close to the surface and because people often reach out with their arms to break a fall.

The Scapula

The **scapula**[45] **(fig. 8.32),** named for its resemblance to a spade or shovel, is a triangular plate that posteriorly overlies ribs 2 through 7. Its only direct attachment to the thorax is by muscles; it glides across the rib cage as the arm and shoulder move. The three sides of the triangle are called the **superior, medial (vertebral),** and **lateral (axillary) borders,** and its three angles are the **superior, inferior,** and **lateral angles.** A conspicuous **suprascapular notch** in the superior border provides passage for a nerve. The broad anterior surface of the scapula, called the **subscapular fossa,** is slightly concave and relatively featureless. The posterior surface has a transverse ridge called the **spine,** a deep indentation superior to the spine called the **supraspinous fossa,** and a broad surface inferior to it called the **infraspinous fossa.**[46]

The most complex region of the scapula is its lateral angle, which has three main features:

1. The **acromion**[47] (ah-CRO-me-on) is a platelike extension of the scapular spine that forms the apex of the shoulder. It articulates with the clavicle, which forms the sole bridge from the appendicular to the axial skeleton.

2. The **coracoid**[48] **process** (COR-uh-coyd) is named for a vague resemblance to a crow's beak, but is shaped more

like a bent finger; it provides attachment for tendons of the *biceps brachii* and other muscles of the arm.

3. The **glenoid**[49] **cavity** (GLEN-oyd) is a shallow socket that articulates with the head of the humerus, forming the gleno-humeral joint. The anatomy of this joint is detailed in section 9.3b and figure 9.24.

8.4b The Upper Limb

Each upper limb contains 30 bones distributed in the following three limb segments:

1. The arm proper (**brachial region** or **brachium**[50] [BRAY-kee-um]) extends from shoulder to elbow. It contains only one bone, the *humerus.*

2. The forearm (**antebrachial region** or **antebrachium**[51]) extends from elbow to wrist and contains two bones: the *radius* and *ulna.* In anatomical position, these bones are parallel and the radius is lateral to the ulna.

3. The hand consists of the **carpal**[52] **region,** with 8 small carpal bones arranged in two rows in the base of the hand; the **metacarpal region** in the palm, with 5 bones; and the fingers **(digits),** with 14 bones. The two hands together contain 54 of the 206 bones in the body—over one-quarter of the total.

Note that what we colloquially call the wrist—the narrow region where one may wear a bracelet or wristwatch—is not what

---

[45]*scap* = spade, shovel; *ula* = little
[46]*supra* = above; *infra* = below
[47]*acr* = extremity, point, apex; *omi* = shoulder
[48]*corac* = crow; *oid* = resembling

[49]*glen* = pit, socket; *oid* = resembling
[50]*brachi* = arm
[51]*ante* = before
[52]*carp* = wrist

anatomists call the wrist (carpal region): the thick, fleshy base of the hand proximal to the hollow of the palm.

The Humerus

The **humerus** has a hemispherical **head** that articulates with the glenoid cavity of the scapula (**fig. 8.33**). The smooth surface of the head (covered with articular cartilage in the living state) is bordered by a groove called the **anatomical neck.** Other prominent features of the proximal end are muscle attachments called the **greater** and **lesser tubercles** and an **intertubercular sulcus** between them that accommodates a tendon of the biceps muscle. The **surgical neck,** a common fracture site, is a narrowing of the bone just distal to the tubercles, at the transition from the head to the shaft. The shaft has a rough area called the **deltoid tuberosity** on its lateral surface. This is an insertion for the *deltoid muscle* of the shoulder.

The distal end of the humerus has two smooth condyles. The lateral one, called the **capitulum**[53] (ca-PIT-you-lum), is shaped like a wide tire and articulates with the radius. The medial one, called the **trochlea**[54] (TROCK-lee-uh), is pulleylike and articulates with

[53]*capit* = head; *ulum* = little
[54]*troch* = wheel, pulley

the ulna. Immediately proximal to these condyles, the humerus flares out to form two bony processes, the **lateral** and **medial epicondyles,** which are the bumps you can palpate at the widest point of your elbow. The medial epicondyle protects the *ulnar nerve,* which passes close to the surface across the back of the elbow. This epicondyle is popularly known as the "funny bone" because striking the elbow on the edge of a table stimulates the ulnar nerve and produces a sharp tingling sensation. Immediately proximal to the epicondyles, the margins of the humerus are called the **lateral** and **medial supracondylar ridges.** These are attachments for certain forearm muscles described in chapter 10.

The distal end of the humerus also shows three deep pits: two anterior and one posterior. The posterior pit, called the **olecranon fossa** (oh-LEC-ruh-non), accommodates a process of the ulna called the *olecranon* when the elbow is extended. On the anterior surface, a medial pit called the **coronoid fossa** accommodates the *coronoid process* of the ulna when the forearm is flexed. The lateral pit is the **radial fossa,** named for the nearby head of the radius.

The Radius

The **radius** has a distinctive discoidal **head** at its proximal end (**fig. 8.34**). When the forearm is rotated so the palm turns forward

(a) Anterior view   **(b) Posterior view**

FIGURE 8.33 The Right Humerus. (a) Anterior view. (b) Posterior view. **APR**

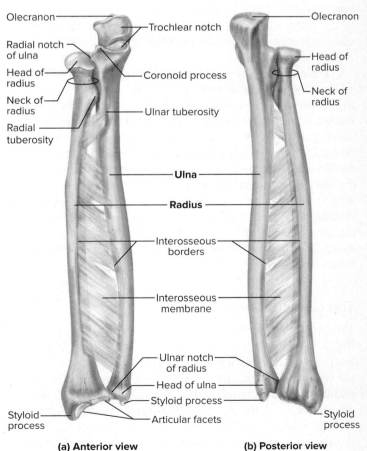

(a) Anterior view   **(b) Posterior view**

FIGURE 8.34 The Right Radius and Ulna. (a) Anterior view. (b) Posterior view. **APR**

and back, the circular superior surface of this disc spins on the capitulum of the humerus, and the edge of the disc spins on the radial notch of the ulna. Immediately distal to the head, the radius has a narrower **neck** and then widens to a rough prominence, the **radial tuberosity,** on its medial surface. The distal tendon of the biceps muscle terminates on this tuberosity.

The distal end of the radius has the following features, from lateral to medial:

1. a bony point, the **styloid process,** which can be palpated proximal to the thumb;

2. two shallow depressions (articular facets) that articulate with the scaphoid and lunate bones of the wrist; and

3. the **ulnar notch,** which articulates with the end of the ulna.

The Ulna

At the proximal end of the **ulna** (fig. 8.34) is a deep, C-shaped **trochlear notch** that wraps around the trochlea of the humerus. The posterior side of this notch is formed by a prominent **olecranon**—the bony point where you rest your elbow on a table. The anterior side is formed by a less prominent **coronoid process.** Laterally, the head of the ulna has a less conspicuous **radial notch,** which accommodates the edge of the head of the radius. At the distal end (**head**) of the ulna is a medial **styloid process.** The bony lumps you can palpate on each side of your wrist are the styloid processes of the radius and ulna. Notice that the "heads" of the radius and ulna are at opposite ends—the proximal end of the radius but distal end of the ulna.

The radius and ulna are attached along their shafts by a ligament called the **interosseous membrane (IM)** (IN-tur-OSS-ee-us), which is attached to an angular ridge called the **interosseous margin** on each bone. Most fibers of the IM are oriented obliquely, slanting upward from the ulna to the radius. If you lean forward on a table supporting your weight on your hands, about 80% of the force is borne by the radius. This tenses the IM, which pulls the ulna upward and transfers some of this force through the ulna to the humerus. The IM thereby enables two elbow joints (humeroradial and humeroulnar) to share the load; this reduces the wear and tear that one joint would otherwise have to bear alone. The IM also serves as an attachment for several forearm muscles.

The Carpal Bones

The **carpal bones** are arranged in two rows of four bones each (**fig. 8.35**). The short carpal bones allow movements of the hand from side to side and anterior to posterior. The carpal bones of the proximal row, starting at the lateral (thumb) side, are the **scaphoid, lunate, triquetrum** (tri-QUEE-trum), and **pisiform** (PY-sih-form)—Latin for "boat-," "moon-," "triangle-," and "pea-shaped," respectively. Unlike the other carpal bones, the pisiform is a sesamoid bone; it is not present at birth but develops around the age of 9 to 12 years within the tendon of the *flexor carpi ulnaris muscle.*

The bones of the distal row, again starting on the lateral side, are the **trapezium,**[55] **trapezoid, capitate,**[56] and **hamate.**[57] The hamate can be recognized by a prominent hook called the **hamulus** on the palmar side (fig. 8.35b). The hamulus is an attachment for the *flexor retinaculum,* a fibrous sheet in the wrist that covers the carpal tunnel (see fig. 10.31).

The Metacarpal Bones

The **metacarpal**[58] **bones** span the palmar region of the hand. Metacarpal I is located proximal to the base of the thumb and metacarpal V proximal to the base of the little finger. On a skeleton, the metacarpal bones look like extensions of the fingers, making the fingers seem much longer than they really are. The proximal end of a metacarpal bone is called the **base,** the shaft is called the **body,** and the distal end is called the **head.** The heads of the metacarpal bones form the knuckles when you clench your fist.

The Phalanges

The bones of the fingers are called **phalanges** (fah-LAN-jeez), in the singular, *phalanx* (FAY-lanks). There are two phalanges in the thumb and three in each of the other digits. Phalanges are identified by roman numerals preceded by *proximal, middle,* and *distal.* For example, proximal phalanx I is in the basal segment of the thumb (the first segment beyond the web between the thumb and palm); the left proximal phalanx IV is where people usually wear wedding rings; and distal phalanx V forms the tip of the little finger. The three parts of a phalanx are the same as in a metacarpal: base, body, and head. The anterior (palmar) surface of a phalanx is slightly concave from end to end and flattened from side to side; the posterior surface is rounder and slightly convex.

BEFORE YOU GO ON

Answer the following questions to test your understanding of the preceding section:

15. Describe how to distinguish the medial and lateral ends of the clavicle from each other, and how to distinguish its superior and inferior surfaces.

16. Name the three fossae of the scapula and describe the location of each.

17. What three bones meet at the elbow? Identify the fossae, articular surfaces, and processes of this joint and state to which bone each of these features belongs.

18. Name the four carpal bones of the proximal row from lateral to medial, then the four bones of the distal row in the same order.

19. Name the four long bones from the tip of the little finger to the base of the hand.

---

[55]*trapez* = table, grinding surface
[56]*capit* = head; *ate* = possessing
[57]*ham* = hook; *ate* = possessing
[58]*meta* = beyond; *carp* = wrist

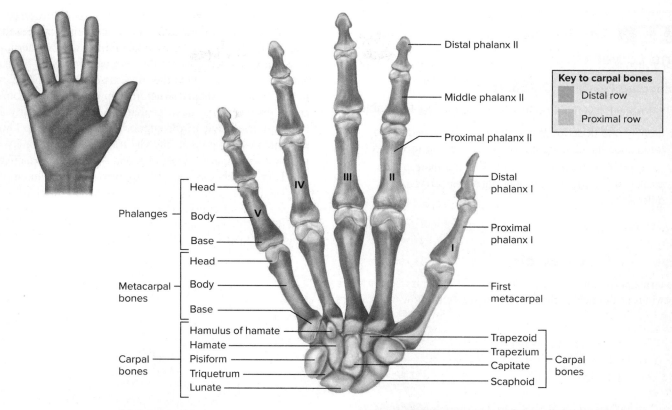

Distal phalanx II

Middle phalanx II

Key to carpal bones
Distal row
Proximal row

Proximal phalanx II

Distal phalanx I

Phalanges
- Head
- Body
- Base

Metacarpal bones
- Head
- Body
- Base

Proximal phalanx I

First metacarpal

Carpal bones
- Hamulus of hamate
- Hamate
- Pisiform
- Triquetrum
- Lunate

- Trapezoid
- Trapezium
- Capitate
- Scaphoid

Carpal bones

(a) Anterior view

Lateral

Hamulus

Articulation with triquetrum

Articulation with capitate

(b) Hamate bone

Sesamoid bone

(c) X-ray of adult hand

FIGURE 8.35 The Right Hand. (a) Carpal bones color-coded to distinguish the proximal and distal rows. (b) The right hamate bone, viewed from the palmar side to show its distinctive hook. This unique bone is a useful landmark for locating the others when studying the skeleton. (c) X-ray of an adult hand. Identify the unlabeled bones in the X-ray by comparing it with the drawing in part (a). A small sesamoid bone is visible at the base of the thumb. The pisiform bone in part (a) is also a sesamoid bone. **APR**

❓ *How does part (c) differ from the X-ray of a child's hand in figure 7.11?*

c: RNHRD NHS Trust/The Image Bank/Getty Images

8.5 The Pelvic Girdle and Lower Limb

Expected Learning Outcomes

When you have completed this section, you should be able to

a. identify and describe the features of the pelvic girdle, femur, patella, tibia, fibula, and bones of the foot; and

b. compare the anatomy of the male and female pelvic girdles and explain the functional significance of the differences.

8.5a The Pelvic Girdle

The terms *pelvis* and *pelvic girdle* are used in contradictory ways by various anatomical authorities. Here we will follow the practice of *Gray's Anatomy* and the *Terminologia Anatomica* in considering the **pelvic girdle** to consist of a complete ring composed of three bones (**fig. 8.36**)—two **hip (coxal) bones** and the sacrum (which, as you know, is also part of the vertebral column). The hip bones are also frequently called the *ossa coxae*[59] (OS-sa COC-see). The **pelvis**[60] is a bowl-shaped structure composed of these bones as well as the ligaments and muscles that line the pelvic cavity and form its floor. The pelvic girdle supports the trunk on the lower limbs and encloses and protects the viscera of the pelvic cavity—mainly the lower colon, urinary bladder, and internal reproductive organs.

Each hip bone is joined to the vertebral column at one point, the sacroiliac joint, where its ear-shaped **auricular surface** matches the auricular surface of the sacrum. The two hip bones articulate with each other on the anterior side of the pelvis, where they are joined by a pad of fibrocartilage called the **interpubic**

---

[59]*os* = bone; *coxae* = of the hip
[60]*pelv* = basin, bowl

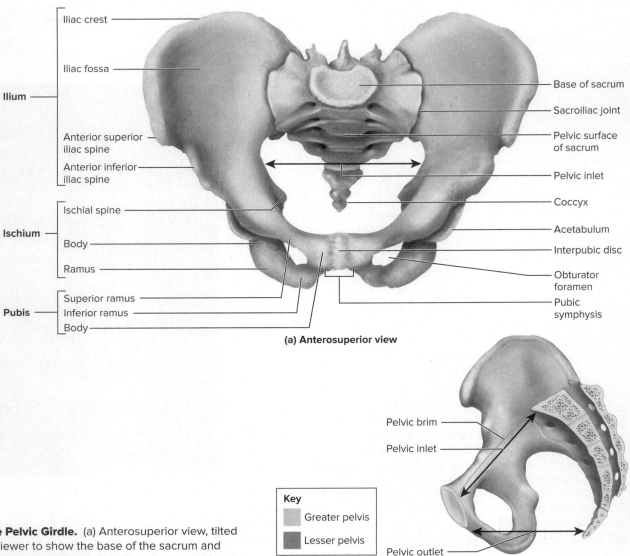

FIGURE 8.36 The Pelvic Girdle. (a) Anterosuperior view, tilted slightly toward the viewer to show the base of the sacrum and the pelvic inlet. (b) Median section, to show the greater and lesser pelvis and the pelvic inlet and outlet. **APR**

Labels in figure (a) Anterosuperior view:
- Iliac crest
- Iliac fossa
- Ilium
- Anterior superior iliac spine
- Anterior inferior iliac spine
- Ischial spine
- Ischium
- Body
- Ramus
- Pubis
- Superior ramus
- Inferior ramus
- Body
- Base of sacrum
- Sacroiliac joint
- Pelvic surface of sacrum
- Pelvic inlet
- Coccyx
- Acetabulum
- Interpubic disc
- Obturator foramen
- Pubic symphysis

Labels in figure (b) Median section:
- Pelvic brim
- Pelvic inlet
- Pelvic outlet

Key:
- Greater pelvis
- Lesser pelvis

disc. The disc and the adjacent region of each pubic bone constitute the **pubic symphysis,**[61] which can be palpated as a hard prominence immediately above the genitalia.

The pelvis is divided into the broad **greater (false) pelvis** between the flare of the hips and the narrower **lesser (true) pelvis** below. The two are separated by a round margin called the **pelvic brim.** The opening circumscribed by the brim is called the **pelvic inlet**—an entry into the lesser pelvis through which an infant's head passes during birth. The lower margin of the lesser pelvis is called the **pelvic outlet.**

The hip bones have three distinctive features that will serve as landmarks for further description. These are the **iliac**[62] **crest** (superior crest of the hip); **acetabulum**[63] (ASS-eh-TAB-you-lum) (the hip socket—named for its resemblance to vinegar cups used on ancient Roman dining tables); and **obturator**[64] **foramen** (a large round-to-triangular hole below the acetabulum, closed by a ligament called the *obturator membrane* in living persons).

The adult hip bone forms by the fusion of three childhood bones called the *ilium* (ILL-ee-um), *ischium* (ISS-kee-um), and *pubis* (PEW-biss), identified by color in **figure 8.37.** The largest of these is the **ilium,** which extends from the iliac crest to the center of the acetabulum. The iliac crest extends from an anterior point or angle called the **anterior superior iliac spine** to a sharp posterior angle called the **posterior superior iliac spine.** In a lean person, the anterior superior spines form visible anterior protrusions at a point where the front pockets usually open on a pair of pants, and the posterior superior spines are sometimes marked by dimples above the buttocks where connective tissue attached to the spines pulls inward on the skin (see atlas B, fig. B.15). The latter is a genetic trait like cheek dimples.

Below the superior spines are the **anterior** and **posterior inferior iliac spines.** Below the latter is a deep **greater sciatic notch** (sy-AT-ic), named for the thick sciatic nerve that passes through it and continues down the posterior side of the thigh.

The posterolateral surface of the ilium is relatively rough-textured because it serves for attachment of several muscles of the buttocks and thighs. The anteromedial surface, by contrast, is the smooth, slightly concave **iliac fossa,** covered in life by the broad *iliacus muscle.* Medially, the ilium exhibits an auricular surface that matches the one on the sacrum, so that the two bones form the sacroiliac (SI) joint.

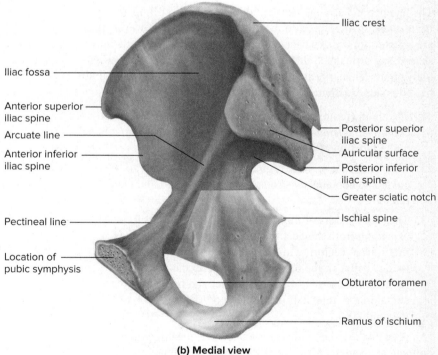

(a) Lateral view

(b) Medial view

FIGURE 8.37 The Right Hip Bone. (a) Lateral view. (b) Medial view. The three childhood bones that fuse to form the adult hip bone are identified by color according to the key. **A&PR**

The **ischium** is the inferoposterior portion of the hip bone. Its heavy **body** is marked with a prominent **ischial spine.** Inferior to the spine is a slight indentation, the **lesser sciatic notch,** and then the thick, rough-surfaced **ischial tuberosity,** which supports your body when you are sitting. The tuberosity can be palpated by

---

[61]*sym* = together; *physis* = growth
[62]*ili* = flank, loin; *ac* = pertaining to
[63]*acetabulum* = vinegar cup
[64]*obtur* = to close, stop up; *ator* = that which

sitting on your fingers. The **ramus** of the ischium joins the inferior ramus of the pubis anteriorly.

The **pubis (pubic bone)** is the most anterior portion of the hip bone. In anatomical position, it is nearly horizontal and serves as a platform for the urinary bladder. It has a **superior** and **inferior ramus** and a triangular **body.** The body of one pubis meets the body of the other at the pubic symphysis. The pubis and ischium encircle the obturator foramen. The pubis is often fractured when the pelvis is subjected to violent anteroposterior compression, as in seat-belt injuries.

The pelvis is the most *sexually dimorphic* part of the skeleton—that is, the one whose anatomy most differs between the sexes. In identifying the sex of skeletal remains, forensic scientists focus especially on the pelvis but on many other bones as well. The average male pelvis is more robust (heavier and thicker) than the female's owing to the forces exerted on the bones by stronger muscles. The female pelvis is adapted to the needs of pregnancy and childbirth. It is wider and shallower and has a larger pelvic inlet and outlet for passage of the infant's head. **Table 8.5** and **figure 8.38** summarize the most useful features of the pelvis in sex identification.

8.5b The Lower Limb

The number and arrangement of bones in the lower limb are similar to those of the upper limb. In the lower limb, however, they are adapted for weight bearing and locomotion and are therefore shaped and articulated differently. The femur and tibia are essentially pillars for supporting the weight of the body. Each lower limb has 30 bones distributed in the following three segments.

1. The thigh **(femoral region)** extends from hip to knee and contains the *femur.* The *patella* (kneecap) is a sesamoid bone at the junction of the femoral and crural regions.

2. The leg proper **(crural region)** extends from knee to ankle and contains two bones, the medial *tibia* and lateral *fibula.*

3. The foot consists of the **tarsal region,** with 7 tarsal bones extending from the heel to the midpoint of the foot arch; **metatarsal region,** with 5 bones extending from there to the "balls" of the feet just proximal to the toes; and toes **(digits),** with 14 bones. The feet and hands together contain more than half of all the body's bones (106 of the total 206).

As with the colloquial versus anatomical meaning of *wrist,* the colloquial meaning of *ankle* (the narrow point where one may wear an ankle bracelet) is different from the anatomical meaning: the posterior half of the foot containing the seven tarsal (ankle) bones. In anatomical terms, the wrist is part of the hand and the ankle is part (indeed, about half) of the foot.

The Femur

The **femur** (FEE-mur) is the longest and strongest bone of the body, measuring about one-quarter of one's height **(fig. 8.39).** It has a hemispherical head that articulates with the acetabulum of the pelvis, forming a quintessential *ball-and-socket joint.* A ligament extends from the acetabulum to a pit, the **fovea capitis**[65] (FOE-vee-uh CAP-ih-tiss), in the head of the femur. Distal to the head is a constricted **neck** and then two massive, rough processes called the **greater** and **lesser trochanters**[66] (tro-CAN-turs), which are insertions for the powerful muscles of the hip. The trochanters are connected on the posterior side by a thick oblique ridge of bone, the **intertrochanteric crest,** and on the anterior side by a more delicate **intertrochanteric line.**

The primary feature of the shaft is a posterior ridge called the **linea aspera**[67] (LIN-ee-uh ASS-peh-ruh) at its midpoint. At its upper end, the linea aspera forks into a medial **spiral (pectineal) line** and a lateral **gluteal tuberosity.** The gluteal tuberosity is a rough ridge (sometimes a depression) that serves for attachment of the powerful *gluteus maximus* muscle of the buttock. At its lower end, the linea aspera forks into **medial** and **lateral supracondylar lines,** which continue down to the respective epicondyles.

The **medial** and **lateral epicondyles** are the widest points of the femur, easily palpated at the knee. These and the supracondylar lines are attachments for certain thigh and leg muscles and knee ligaments. At the distal end of the femur are two smooth round surfaces of the knee joint, the **medial** and **lateral condyles,** separated by a groove called the **intercondylar fossa** (IN-tur-CON-dih-lur). During knee flexion and extension, the condyles rock on the superior surface of the tibia. On the anterior side of the femur, a smooth medial depression called the **patellar surface** articulates with the patella. On the posterior side is a flat or slightly depressed area called the **popliteal surface.**

The Patella

The **patella,**[68] or kneecap (fig. 8.39), is a roughly triangular sesamoid bone embedded in the tendon of the knee. It is cartilaginous at birth and ossifies at 3 to 6 years of age. It has a broad superior **base,** a pointed inferior **apex,** and a pair of shallow **articular facets** on its posterior surface where it articulates with the femur. The lateral facet is usually larger than the medial. The *quadriceps femoris tendon* extends from the anterior *quadriceps femoris muscle* of the thigh to the patella, and it continues as the *patellar ligament* from the patella to the tibia. This is a change in terminology more than a change in structure or function, as a tendon connects muscle to bone and a ligament connects bone to bone. Because of the way the quadriceps tendon loops over the patella, the patella acts like a pulley, modifying the direction of pull by the quadriceps muscle. This improves its efficiency in extending the knee, hence the efficiency of walking and running.

The Tibia

The leg has two bones: a thick strong tibia (TIB-ee-uh) on the medial side and a slender fibula (FIB-you-luh) on the lateral side

---

[65]*fovea* = pit; *capitis* = of the head
[66]*trochanter* = to run
[67]*linea* = line; *asper* = rough
[68]*pat* = pan; *ella* = little

TABLE 8.5	Comparison of the Male and Female Pelvic Girdles	
Feature	**Male**	**Female**
General appearance	More massive; rougher; heavier processes	Less massive; smoother; more delicate processes
Tilt	Upper end of pelvis relatively vertical	Upper end of pelvis tilted forward
Depth of greater pelvis	Deeper; ilium projects farther above sacroiliac joint	Shallower; ilium does not project as far above sacroiliac joint
Width of greater pelvis	Hips less flared; anterior superior spines closer together	Hips more flared; anterior superior spines farther apart
Pelvic inlet	Heart-shaped	Round or oval
Pelvic outlet	Smaller	Larger
Subpubic angle	Narrower, usually 90° or less	Wider, usually 100° or more
Pubic symphysis	Taller	Shorter
Body of pubis	More triangular	More rectangular
Greater sciatic notch	Narrower	Wider
Obturator foramen	Rounder	More oval to triangular
Acetabulum	Larger, faces more laterally	Smaller, faces slightly anteriorly
Sacrum	Narrower and deeper	Wider and shallower
Coccyx	Less movable; more vertical	More movable; tilted posteriorly

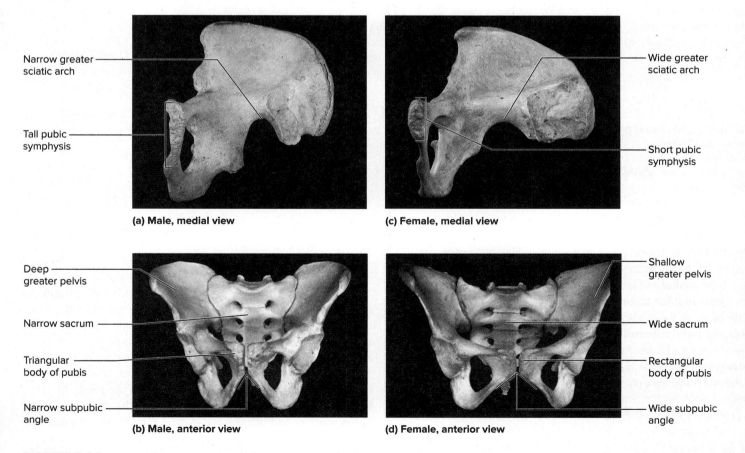

Narrow greater sciatic arch

Tall pubic symphysis

(a) Male, medial view

Wide greater sciatic arch

Short pubic symphysis

(c) Female, medial view

Deep greater pelvis

Narrow sacrum

Triangular body of pubis

Narrow subpubic angle

(b) Male, anterior view

Shallow greater pelvis

Wide sacrum

Rectangular body of pubis

Wide subpubic angle

(d) Female, anterior view

FIGURE 8.38 Comparison of the Male and Female Pelvic Girdles. (a) Male, medial view. (b) Male, anterior view. (c) Female, medial view. (d) Female, anterior view. Compare with table 8.5.

(top, both): David Hunt/specimens from the National Museum of Natural History, Smithsonian Institution; **(bottom, both):** VideoSurgery/Science Source

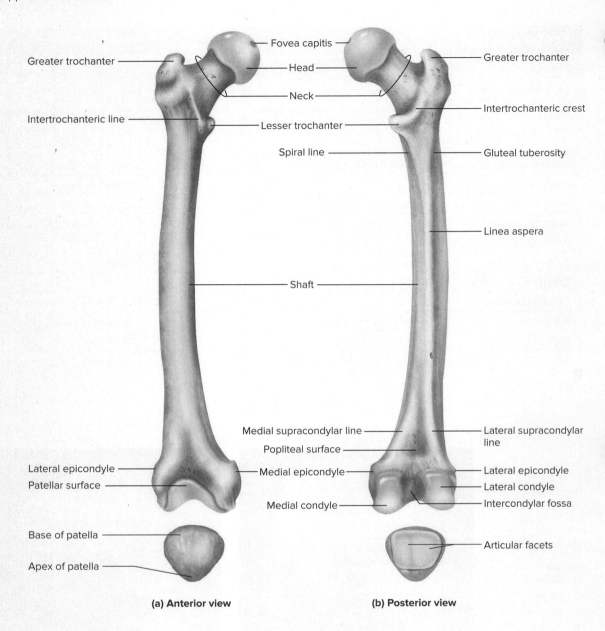

FIGURE 8.39 The Right Femur and Patella.
(a) Anterior view.
(b) Posterior view. **A&PR**

Greater trochanter
Intertrochanteric line
Fovea capitis
Head
Neck
Lesser trochanter
Spiral line
Greater trochanter
Intertrochanteric crest
Gluteal tuberosity
Linea aspera
Shaft
Lateral epicondyle
Patellar surface
Base of patella
Apex of patella
Medial supracondylar line
Popliteal surface
Medial epicondyle
Medial condyle
Lateral supracondylar line
Lateral epicondyle
Lateral condyle
Intercondylar fossa
Articular facets

(a) Anterior view   **(b) Posterior view**

(fig. 8.40). The **tibia** is the only weight-bearing bone of the crural region. Its broad superior head has two fairly flat articular surfaces, the **medial** and **lateral condyles,** separated by a ridge called the **intercondylar eminence.** The condyles of the tibia articulate with those of the femur. The rough anterior surface of the upper tibia, the **tibial tuberosity,** can be palpated just below the patella. This is an attachment for the powerful thigh muscles that extend (straighten) the knee. Distal to this, the shaft has a sharply angular **anterior border,** which can be palpated in the shin. At the ankle, just above the rim of a standard dress shoe, you can palpate a prominent bony knob on each side. These are the **medial** and **lateral malleoli**[69] (MAL-ee-OH-lie). The medial malleolus is part of the tibia, and the lateral malleolus is the part of the fibula.

The Fibula

The **fibula**[70] (fig. 8.40) is a slender lateral strut that helps to stabilize the ankle. It does not bear any of the body's weight; indeed, orthopedic surgeons sometimes remove part of the fibula and use it to replace damaged or missing bone elsewhere in the body. The fibula is somewhat thicker and broader at its proximal end, the **head,** than at the distal end. The point of the head is called the **apex.** The distal expansion is the lateral malleolus. Like the radius and ulna, the tibia and fibula are joined by an interosseous membrane along their shafts, and by shorter ligaments at the superior and inferior ends where the fibular head and apex contact the tibia.

[69]*malle* = hammer; *olus* = little

[70]*fibula* = little pin or clasp; from a garment pin in ancient Greece and Rome

FIGURE 8.40 **The Right Tibia and Fibula.** (a) Anterior view. (b) Posterior view. **APR**

The Ankle and Foot

The **tarsal bones** of the ankle are arranged in proximal and distal groups somewhat like the carpal bones of the wrist **(fig. 8.41).** Because of the load-bearing role of the ankle, however, their shapes and arrangement are conspicuously different from those of the carpal bones, and they are fully integrated into the structure of the foot. The largest tarsal bone is the **calcaneus**[71] (cal-CAY-nee-us), which forms the heel. Its posterior end is the point of attachment for the **calcaneal (Achilles) tendon** from the calf muscles. The second-largest tarsal bone, and the most superior, is the **talus.** It has three articular surfaces: an inferoposterior one that articulates with the calcaneus, a superior **trochlear surface** that articulates with the tibia, and an anterior surface that articulates with a short, wide tarsal bone called the **navicular.**[72] The talus, calcaneus, and navicular are considered the proximal row of tarsal bones.

The distal group forms a row of four bones. Proceeding from the medial to lateral, these are the **medial, intermediate,** and **lateral cuneiforms**[73] (cue-NEE-ih-forms) and the **cuboid.** The cuboid is the largest.

▶▶▶**APPLY WHAT YOU KNOW**

The upper and lower limbs each contain 30 bones, yet we have 8 carpal bones in the upper limb and only 7 tarsal bones in the lower limb. What makes up the difference in the lower limb?

The remaining bones of the foot are similar in arrangement and name to those of the hand. The proximal **metatarsal**[74] **bones** are similar to the metacarpals. They are **metatarsals I** through **V** from medial to lateral, metatarsal I being proximal to the great toe. Metatarsals I to III articulate with the first through third cuneiforms; metatarsals IV and V both articulate with the cuboid.

Bones of the toes, like those of the fingers, are called phalanges. The great toe contains only two bones, the proximal and distal phalanx I. The other toes each contain a proximal, middle, and distal phalanx, and are numbered II through V from medial to lateral. Thus middle phalanx V, for example, would be the middle bone of the smallest toe. The metatarsal and phalangeal bones each have a base, body, and head, like the bones of the hand. All of them, especially the phalanges, are slightly concave on the inferior (plantar) side.

Note that roman numeral I represents the *medial* group of bones in the foot but the *lateral* group in the hand. In both cases,

[71]*calc* = stone, chalk
[72]*navi* = boat; *cul* = little; *ar* = like
[73]*cunei* = wedge; *form* = in the shape of

[74]*meta* = beyond; *tars* = ankle

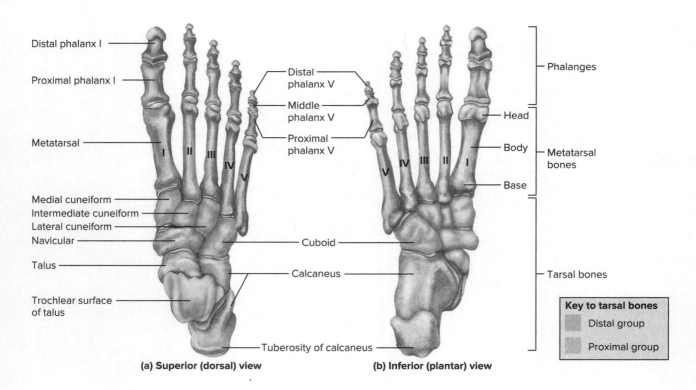

(a) Superior (dorsal) view

(b) Inferior (plantar) view

Key to tarsal bones
- Distal group
- Proximal group

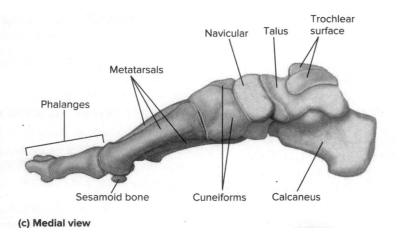

(c) Medial view

FIGURE 8.41 The Right Foot. (a) Superior (dorsal) view. (b) Inferior (plantar) view. (c) Medial view. **APR**

? *Contrast the tarsal bones with the carpal bones. Which ones are similar in location? Which ones are different?*

however, it refers to the largest digit of the limb. The reason for the difference between the hand and foot lies in a rotation of the limbs that occurs in the seventh week of embryonic development. Early in the seventh week, the limbs extend anteriorly from the body, the foot is a paddlelike *foot plate,* and the *hand plate* is also more or less paddlelike with the finger buds showing early separation **(fig. 8.42a).** The future thumb and great toe are both directed superiorly, and the future palms and soles face each other medially. But then each limb rotates about 90° in opposite directions. The upper limb rotates laterally. To visualize this, hold your hands straight out in front of you with the palms facing each other as if

(a) Seven weeks

(b) Eight weeks

FIGURE 8.42 Embryonic Limb Rotation. (a) Rotation of the hands and feet in opposite directions in week 7. (b) Resulting orientation of hands and feet establilshed by week 8.

Medial longitudinal arch
Transverse arch
Lateral longitudinal arch

Fibula Tibia Talus
Calcaneus
Navicular
Cuboid
Cuneiform
Metatarsal I
Proximal phalanx I

(a) Inferior (plantar) view   **(b) X-ray of right foot, lateral view**

FIGURE 8.43 Arches of the Foot. (a) Inferior (plantar) view. (b) X-ray of right foot, lateral view.
b: stockdevil/123RF

you were about to clap. Then rotate your forearms so the thumbs face away from each other (laterally) and the palms face upward. The lower limbs rotate in the opposite direction, medially, so that the soles face downward and the great toes become medial. So even though the thumb and great toe (digit I of the hand and foot) start out facing in the same direction, these opposite rotations result in their being on opposite sides of the hand and foot **(fig. 8.42b).** This rotation also explains why the elbow flexes posteriorly and the knee flexes anteriorly, and why (as you will see in chapter 10) the muscles that flex the elbow are on the anterior side of the arm, whereas those that flex the knee are on the posterior side of the thigh.

The foot normally doesn't rest flat on the ground, but has three springy arches that distribute the body's weight between the heel and the heads of the metatarsal bones and absorb the stress of walking **(fig. 8.43).** The **medial longitudinal arch,** which essentially extends from heel to great toe, is formed from the calcaneus, talus, navicular, cuneiforms, and metatarsals I to III. It is normally well above the ground, as evidenced by the shape of a wet footprint. The **lateral longitudinal arch** extends from heel to little toe and includes the calcaneus, cuboid, and metatarsals IV and V. The **transverse arch** includes the cuboid, cuneiforms, and proximal heads of the metatarsal bones. These arches are held together by short, strong ligaments. Excessive weight, repetitious stress, or

congenital weakness of these ligaments can stretch them, resulting in *pes planus* (commonly called flat feet or fallen arches). This condition makes a person less tolerant of prolonged standing and walking. A comparison of the flat-footed apes with humans underscores the significance of the human foot arches (see Deeper Insight 8.5).

BEFORE YOU GO ON

Answer the following questions to test your understanding of the preceding section:

20. Name the bones of the adult pelvic girdle. What three bones of a child fuse to form the hip bone of an adult?

21. Name any four structures of the pelvis that you can palpate and describe where to palpate them.

22. Describe several ways in which the male and female pelvic girdles differ.

23. What parts of the femur are involved in the hip joint? What parts are involved in the knee joint?

24. Name the prominent knobs on each side of your ankle. What bones contribute to these structures?

25. Name all the bones that articulate with the talus and describe the location of each.

DEEPER INSIGHT 8.5

EVOLUTIONARY MEDICINE

Skeletal Adaptations for Bipedalism

Some mammals can stand, hop, or walk briefly on their hind legs, but humans are the only mammals that are habitually bipedal. Footprints preserved in a layer of volcanic ash in Tanzania indicate that hominids walked upright as early as 3.6 million years ago. This bipedal locomotion is possible only because of several adaptations of the human feet, legs, spine, and skull. These features are so distinctive that paleoanthropologists (those who study human fossil remains) can tell with considerable certainty whether a fossil species was able to walk upright.

As important as the hand has been to human evolution, the foot may be an even more significant adaptation. Unlike other mammals, humans support their entire body weight on two feet. While apes are flat-footed, humans have strong, springy foot arches that absorb shock as the body jostles up and down during walking and running. The tarsal bones are tightly articulated with one another, and the calcaneus is strongly developed. The great toe is not opposable as it is in most Old World monkeys and apes, but it is highly developed so it provides the "toe-off" that pushes the body forward in the last phase of the stride **(fig. 8.44a).** For this reason, loss of the great toe has a more crippling effect than the loss of any other toe.

While the femurs of apes are nearly vertical, in humans they angle medially from hip to knee **(fig. 8.44b).** This places our knees closer together, beneath the body's center of gravity. We lock our knees when standing, allowing us to stand erect with little muscular effort. Apes cannot do this and cannot stand on two legs for very long without tiring—much as you would if you tried to maintain an erect posture with your knees slightly bent.

In apes and other quadrupedal (four-legged) mammals, the abdominal viscera are supported by the muscular abdominal wall. In humans, the viscera bear down on the floor of the pelvic cavity, and a bowl-shaped pelvis with an inturned floor is necessary to support their weight. This has resulted in a narrower pelvic outlet—a condition that creates pain and difficulty in giving birth to such large-brained infants. The pain of childbirth

(a) Foot   **(b) Knee**   **(c) Gluteal muscles**

Chimpanzee

Human

Chimpanzee   Human

Chimpanzee

Human

FIGURE 8.44 Skeletal Adaptations for Bipedalism. Human adaptations for bipedalism are best understood by comparison to our close living relative, the chimpanzee, which is not adapted for a comfortable or sustained erect stance. See the text for the relevance of each comparison. (a) Foot. (b) Knee. (c) Gluteal muscles. (d) Pelvis. (e) Vertebral column. (f) Skull.

(figure continues)

seems unique to humans and, one could say, is a price we must pay for having both a large brain and a bipedal stance.

The largest muscle of the buttock, the *gluteus maximus,* serves in apes primarily as an abductor of the thigh—that is, it moves the leg laterally. In humans, however, the ilium has expanded posteriorly, so the gluteus maximus originates behind the hip joint. This changes the function of the muscle—instead of abducting the thigh, it pulls the thigh back in the second half of a stride (pulling back on your right thigh, for example, when your left foot is off the ground and swinging forward). Two other buttock muscles, the *gluteus medius* and *gluteus minimus,* extend laterally in humans from the surface of the ilium to the greater trochanter of the femur **(fig. 8.44c).** In walking, when one foot is lifted from the ground, these muscles shift the body weight over the other foot so we don't fall over. The actions of all three gluteal muscles, and the corresponding evolutionary remodeling of the pelvis, account for the smooth, efficient stride of a human as compared with the awkward, shuffling gait of a chimpanzee or gorilla when walking upright. The posterior growth of the ilium **(fig. 8.44d)** is the reason the greater sciatic notch is so deeply concave.

The lumbar lordosis of the human spine allows for efficient bipedalism by shifting the body's center of gravity to the rear, above and slightly behind the hip joint **(fig. 8.44e).** Because of their C-shaped spines, chimpanzees cannot stand as easily. Their center of gravity is anterior to the hip joint when they stand; they must make a constant muscular effort to keep from falling forward, and they fatigue quickly. Humans, by contrast, require little effort to keep their balance. Our australopithecine ancestors probably could travel all day with relatively little fatigue.

The human head is balanced on the vertebral column with the gaze directed forward. The cervical curvature of the spine and remodeling of the skull have made this possible. The foramen magnum has moved to a more inferior and anterior location, and the face is much flatter than an ape's face **(fig. 8.44f),** so there is less weight anterior to the occipital condyles. Being balanced on the spine, the head doesn't require strong muscular attachments to hold it erect.

The forelimbs of apes are longer than the hindlimbs; indeed, some species such as the orangutan and gibbons hold their long forelimbs over their heads when they walk on their hind legs. By contrast, our forelimbs are shorter than our hindlimbs and far less muscular than the forelimbs of apes. No longer needed for locomotion, our forelimbs have become better adapted for carrying objects, holding things closer to the eyes, and manipulating them more precisely.

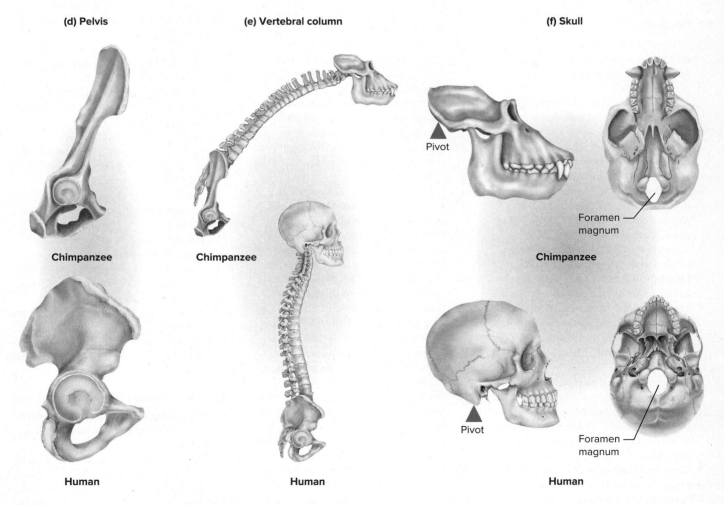

(d) Pelvis

Chimpanzee

Human

(e) Vertebral column

Chimpanzee

Human

(f) Skull

Pivot

Foramen magnum

Chimpanzee

Pivot

Foramen magnum

Human

FIGURE 8.44 Skeletal Adaptations for Bipedalism *(continued).*

STUDY GUIDE

▶ Assess Your Learning Outcomes

To test your knowledge, discuss the following topics with a study partner or in writing, ideally from memory.

8.1 Overview of the Skeleton

1. The difference between the axial and appendicular skeletons, and the bones in each category

2. The typical number of named bones in an adult; why this number differs in newborns and children; and why the number varies among adults

3. Sutural and sesamoid bones, and examples of the latter

4. Names of the various outgrowths, depressions, articular surfaces, cavities, and passages in bones

8.2 The Skull

1. The usual number of bones in the adult skull, and the collective name of the seams or joints that bind most of them together

2. Names and locations of the cavities that enclose the brain, nose, ears, and eyes, and of the paranasal sinuses

3. The collective function of the skull foramina; the location and function of the largest one, the foramen magnum

4. Major features of the cranium; the difference between its base and calvaria; the three cranial fossae and how they relate to brain anatomy

5. Names of the six different cranial bones; which ones are solitary and which are bilaterally paired; and what distinguishes a cranial bone from a facial bone

6. The location and extent of the frontal bone; the suture that binds it to the parietal bones; and the locations of its supraorbital margin and foramen, glabella, frontal sinus, and diploe

7. The location and extent of the parietal bones; the suture formed where they meet the occipital bone and the one that separates the parietal bones from each other; and the locations of the parietal foramen and temporal lines

8. The location and extent of the temporal bones; the suture that borders them; and the names and boundaries of their four major parts

9. The location and extent of the occipital bone, its basilar part, and the names and locations of its foramina, canals, and surface protrusions

10. The location and extent of the sphenoid bone; its wings, body, pterygoid plates, clinoid processes, and foramina; its relationships with the pituitary gland and nasal apertures

11. The location and extent of the ethmoid bone; its part in defining the nasal fossae; and the locations of its plates, foramina, air cells, and nasal conchae

12. Names of the eight different facial bones; which ones are solitary and which are bilaterally paired

13. The location and extent of the maxilla; its foramina, alveoli, and alveolar and palatine processes; and the suture that joins the right and left maxillae

14. The location and extent of the palatine bones; their foramina; and their part in partially defining the walls of the nasal cavity and orbit

15. Structure of the palate, including the hard and soft regions and the contributions of the palatine processes and palatine bones

16. The location and extent of the zygomatic bones, and the temporal process and main foramen of each

17. The three parts of the zygomatic arch

18. The locations and structures of the tiny lacrimal and nasal bones

19. The inferior nasal concha and why it is distinguished from the superior and middle conchae

20. The location and extent of the vomer; contributions of the vomer and ethmoid bone to the nasal septum

21. Structure of the mandible, including the body, ramus, and angle; its two main processes and the notch between them; its foramina, symphysis, protuberance, and spines

22. The bones, and their specific features, that form the temporomandibular joint

23. The locations and names of the auditory ossicles; location and features of the hyoid bone; and functions of these bones

24. Names and locations of the fontanelles of the neonatal skull; why they exist; and how a child's skull changes between birth and 9 years of age

8.3 The Vertebral Column and Thoracic Cage

1. The number of vertebrae and intervertebral discs in the vertebral column (spine)

2. Four curvatures of the adult spine; which ones are present at birth; and when and how the others develop

3. Features of a typical vertebra

4. The five classes of vertebrae and the number of vertebrae in each class; the system of numbering them; and why the number of vertebrae in a child differs from the number at 30 years of age and beyond

5. Features that identify an isolated vertebra as cervical, thoracic, or lumbar

6. How the anatomy of the first two vertebrae (C1–C2) relates to movements of the head

7. How the anatomy of the thoracic vertebrae (T1–T12) relates to the articulations of the ribs

8. The structure and function of intervertebral discs; which vertebrae have discs between them and which ones do not

9. Anatomical features of the sacrum, including its foramina, crests, canal and hiatus, auricular surface and sacroiliac joint, promontory, and alae

10. Features of the coccyx

11. Components and general shape of the thoracic cage

12. Three main regions of the sternum; its notches and sternal angle

13. The number of ribs; which ones are true, false, and floating ribs

14. All features seen in most of the ribs

15. Which ribs differ from that typical anatomy, and how

16. How the ribs articulate with the vertebrae, including variations from the top to bottom of the rib cage

8.4 The Pectoral Girdle and Upper Limb

1. Names and locations of the 4 bones of the pectoral girdle and 30 bones of each upper limb

2. Names of the joints at which the humerus articulates with the scapula, the scapula with the clavicle, and the clavicle with the axial skeleton

STUDY GUIDE

3. Features of the clavicle, including the sternal and acromial ends and conoid tubercle; function of the clavicles
4. Features of the scapula, including its borders and angles, fossae, suprascapular notch, acromion, coracoid process, and glenoid cavity
5. Names of the four regions of the upper limb and the bones contained in each
6. Features of the humerus, including the head, necks, tubercles, intertubercular sulcus, deltoid tuberosity, capitulum, trochlea, epicondyles, supracondylar ridges, and three fossae
7. Features of the radius, including the head, neck, radial tuberosity, styloid process, and ulnar notch
8. Features of the ulna, including the trochlear and radial notches, coronoid and styloid processes, and olecranon; and the relationship of the radius and ulna to the interosseous membrane
9. Names of the carpal bones, in order, from lateral to medial in the proximal row and lateral to medial in the distal row; the unusual structure of the hamate bone
10. The system of naming and numbering the 5 metacarpal bones of the palmar region and 5 sets of phalanges in the digits; why there

are 5 digits but only 14 phalanges; the base, body, and head of all 19 metacarpals and phalanges

8.5 The Pelvic Girdle and Lower Limb

1. Names and locations of the 3 bones of the pelvic girdle and 30 bones of each lower limb
2. Names of the joints at which the lower limb articulates with the pelvic girdle and the pelvic girdle articulates with the axial skeleton
3. The distinction between the pelvic girdle and pelvis
4. Three childhood bones that fuse to form each adult hip (coxal) bone, and the boundaries of each on the hip bone
5. Features of the hip (coxal) bones and pelvic girdle including the auricular surfaces; interpubic disc and pubic symphysis; greater and lesser pelves; pelvic brim, inlet, and outlet; iliac crest; acetabulum; obturator foramen; four ischial spines; two sciatic notches; iliac fossa; and parts of the ischium and pubis
6. Differences between the male and female pelvic girdles and the essential reason for these differences
7. Names of the four regions of the lower limb and the bones contained in each

8. Features of the femur, including the head, neck, fovea capitis, trochanters, intertrochanteric crest, gluteal tuberosity, condyles, intercondylar fossa, epicondyles, lines, and patellar and popliteal surfaces
9. Features of the patella, including the base, apex, and articular facets
10. Features of the tibia, including the lateral and medial condyles and intercondylar eminence; tibial tuberosity; anterior crest; and medial malleolus
11. Features of the fibula, including the head, apex, and lateral malleolus
12. Names of the tarsal bones from posterior to anterior, and from lateral to medial in the distal row; why they are more fully integrated into the foot than the carpal bones are into the hand
13. The system of naming and numbering the 5 metatarsal bones of the foot and 5 sets of phalanges in the digits; why there are 5 digits but only 14 phalanges; the base, body, and head of all 19 metatarsals and phalanges
14. Why the elbows and knees flex in opposite directions, and why the largest digit is lateral in the hand but medial in the foot
15. Names and locations of the three foot arches

▶ Testing Your Recall

Answers in Appendix A

1. Which of these is *not* a paranasal sinus?
   a. frontal
   b. temporal
   c. sphenoidal
   d. ethmoidal
   e. maxillary

2. Which of these is a facial bone?
   a. frontal
   b. ethmoid
   c. occipital
   d. temporal
   e. lacrimal

3. Which of these *cannot* be palpated on a living person?
   a. the crista galli
   b. the mastoid process
   c. the zygomatic arch
   d. the superior nuchal line
   e. the hyoid bone

4. All of the following are groups of vertebrae *except* for _____, which is a spinal kyphosis.
   a. thoracic   d. pelvic
   b. cervical   e. sacral
   c. lumbar

5. Thoracic vertebrae do *not* have
   a. transverse foramina.
   b. costal facets.
   c. spinous processes.
   d. transverse processes.
   e. pedicles.

6. The tubercle of a rib articulates with
   a. the sternal notch.
   b. the margin of the gladiolus.
   c. the costal facets of two vertebrae.
   d. the body of a vertebra.
   e. the transverse process of a vertebra.

7. The disc-shaped head of the radius articulates with the _____ of the humerus.
   a. radial tuberosity
   b. trochlea
   c. capitulum
   d. olecranon
   e. glenoid cavity

8. All of the following are carpal bones, *except* the _____, which is a tarsal bone.
   a. trapezium   d. triquetrum
   b. cuboid   e. pisiform
   c. trapezoid

9. The bone that supports your body weight when you are sitting down is
   a. the acetabulum.
   b. the pubis.
   c. the ilium.
   d. the coccyx.
   e. the ischium.

10. Which of these is the bone of the heel?
    a. cuboid   d. trochlear
    b. calcaneus   e. talus
    c. navicular

11. Gaps between the cranial bones of an infant are called _____.

12. The external auditory canal is a passage in the _____ bone.

13. Bones of the skull are joined along lines called _____.

14. The _____ bone has greater and lesser wings and protects the pituitary gland.

15. A herniated disc occurs when a ring called the _____ cracks.

16. The transverse ligament of the atlas holds the _____ of the axis in place.

17. The sacroiliac joint is formed where the _____ surface of the sacrum articulates with that of the ilium.

18. The _____ processes of the radius and ulna form bony protuberances on each side of the wrist.

19. Nerves and blood vessels pass through small holes in the skull called _____. The singular form of this word is _____.

20. The _____ arch of the foot extends from the heel to the great toe.

▶ Building Your Medical Vocabulary

Answers in Appendix A

State a meaning of each word element, and give a medical term from this chapter that uses it or a slight variation of it.

1. costo-

2. cranio-

3. dura

4. glosso-

5. -icle

6. masto-

7. pedo-

8. pterygo-

9. supra-

10. tarso-

▶ What's Wrong with These Statements?

Answers in Appendix A

Briefly explain why each of the following statements is false, or reword it to make it true.

1. The internal jugular vein passes out the foramen magnum of the skull and down the neck.

2. The hands have more phalanges than the feet.

3. As an adaptation to pregnancy, the female's pelvis is deeper than the male's.

4. A wristwatch or bracelet normally encircles the area of the carpal (wrist) bones.

5. On a living person, it would be possible to palpate the acromion and spine of the scapula but not the muscles of the infraspinous fossa.

6. If you rest your chin on your hands and your elbows on a table, the coronoid process of the ulna rests on the table.

7. The lumbar vertebrae do not articulate with any ribs and therefore do not have transverse processes.

8. The most frequently broken bone is the humerus.

9. In strict anatomical terminology, the words *arm* and *leg* both refer to regions with only one bone.

10. Sesamoid bones are found along the sutures between cranial bones in some people.

▶ Testing Your Comprehension

1. Most bones form joints with at least two others. For example, there are 12 joints between vertebra T6 and other, adjacent bones. Identify at least 10 of them.

2. By palpating the hind leg of a cat or dog or by examining a laboratory skeleton, you can see that cats and dogs stand on the heads of their metatarsal bones; the calcaneus does not touch the ground. How is this similar to the stance of a woman wearing high-heeled shoes? How is it different?

3. Between any two of the unfused vertebrae (cervical through lumbar), there is an intervertebral disc—except between C1 and C2. Give some reasons for the unique absence of a disc at that location.

4. In adolescents, trauma sometimes separates the head of the femur from the neck. Why do you think this is more common in adolescents than in adults?

5. Discuss all the ways you can in which the differences between vertebrae C5 and L3 exemplify this book's theme of the unity of form and function.

X-ray of knee replacement prosthesis, frontal and lateral views
Zephyr/Getty Images

CHAPTER OUTLINE

9.1 Joints and Their Classification

    9.1a Bony Joints

    9.1b Fibrous Joints

    9.1c Cartilaginous Joints

9.2 Synovial Joints

    9.2a General Anatomy

    9.2b Joints and Lever Systems

    9.2c Movements of Synovial Joints

9.3 Anatomy of Selected Diarthroses

    9.3a The Jaw Joint

    9.3b The Shoulder Joint

    9.3c The Elbow Joint

    9.3d The Hip Joint

    9.3e The Knee Joint

    9.3f The Ankle Joint

Study Guide

DEEPER INSIGHTS

9.1 Clinical Application: Exercise and Articular Cartilage

9.2 Clinical Application: TMJ Dysfunction

9.3 Clinical Application: Shoulder Dislocation

9.4 Clinical Application: Knee Injuries and Arthroscopic Surgery

9.5 Clinical Application: Arthritis and Artificial Joints

Anatomy & Physiology *Revealed* 4.0

Module 5: Skeletal System

Joints, or articulations, link the bones of the skeletal system into a functional whole—a system that supports the body, permits effective movement, and protects the softer organs. Joints such as the shoulder, elbow, and knee are remarkable specimens of biological design—self-lubricating, almost friction-less, and able to bear heavy loads and withstand compression while executing smooth and precise movements **(fig. 9.1).** Yet it is equally important that other joints be less movable or even immobile. Such joints are better able to support the body and protect delicate organs. The vertebral column, for example, is only moderately mobile, for it must allow for flexibility of the torso and yet protect the delicate spinal cord and support much of the body's weight. Bones of the cranium must protect the brain and sense organs, but need not allow for movement (except during birth); thus, they are locked together by immobile joints, the sutures studied in chapter 8.

In everyday life, we take the greatest notice of the most freely mobile joints of the limbs, and it is here that people feel most severely compromised by such disabling diseases as arthritis. Much of the work of physical therapists focuses on limb mobility. In this chapter, we will survey all types of joints, from the utterly immobile to the most mobile, but with an emphasis on the latter. This survey of joint anatomy and movements will provide a foundation for the study of muscle actions in chapter 10.

FIGURE 9.1 Joint Flexibility.
Rubberball/Erik Isakson/Getty Images

9.1   Joints and Their Classification

Expected Learning Outcomes

When you have completed this section, you should be able to

a. explain what joints are, how they are named, and what functions they serve;

b. name and describe the four major categories of joints;

c. describe the three types of fibrous joints and give an example of each;

d. distinguish between the three types of sutures;

e. describe the two types of cartilaginous joints and give an example of each; and

f. explain, with examples, why some joints change categories as a person ages.

Any point where two bones meet is called a **joint (articulation),** whether or not the bones are mobile at that interface. The science of joint structure, function, and dysfunction is called **arthrology.**[1] The study of musculoskeletal movement is **kinesiology**[2] (kih-NEE-see-OL-oh-jee).

The name of a joint is typically derived from the names of the bones involved. For example, the *atlanto–occipital joint* is where the atlas meets the occipital condyles; the *glenohumeral joint* is where the glenoid cavity of the scapula meets the humerus; and the *radioulnar joint* is where the radius meets the ulna.

Joints can be classified according to the manner in which the adjacent bones are bound to each other, with corresponding differences in how freely the bones can move. Authorities differ in their classification schemes, but one common view places the

---

[1]*arthro* = joint; *logy* = study of
[2]*kinesio* = movement; *logy* = study of

joints in four major categories: *bony, fibrous, cartilaginous,* and *synovial joints*. This section describes the first three of these and the subclasses of each. The remainder of the chapter will then be concerned primarily with synovial joints.

9.1a Bony Joints

A **bony joint,** or **synostosis**[3] (SIN-oss-TOE-sis), is an immobile joint formed when the gap between two bones ossifies and they become, in effect, a single bone. Bony joints can form by ossification of either fibrous or cartilaginous joints. An infant is born with right and left frontal and mandibular bones, for example, but these soon fuse seamlessly into a single frontal bone and mandible. Three childhood bones—the ilium, ischium, and pubis—fuse to form a single hip bone on each side of an adult. The epiphyses and

diaphyses of the long bones are joined by cartilaginous joints in childhood and adolescence, and these become bony joints in early adulthood. In old age, the first rib often fuses with the sternum and the two parietal bones sometimes fuse along the sagittal suture.

9.1b Fibrous Joints

A **fibrous joint** is also called a **synarthrosis**[4] (SIN-ar-THRO-sis). It is a point at which adjacent bones are bound by collagen fibers that emerge from one bone, cross the space between them, and penetrate into the other **(fig. 9.2).** There are three kinds of fibrous joints: *sutures, gomphoses,* and *syndesmoses*. In sutures and gomphoses, the fibers are very short and allow for little or no movement. In syndesmoses, the fibers are longer and the attached bones are more mobile.

---

[3]*syn* = together; *ost* = bone; *osis* = condition

[4]*syn* = together; *arthr* = joined; *osis* = condition

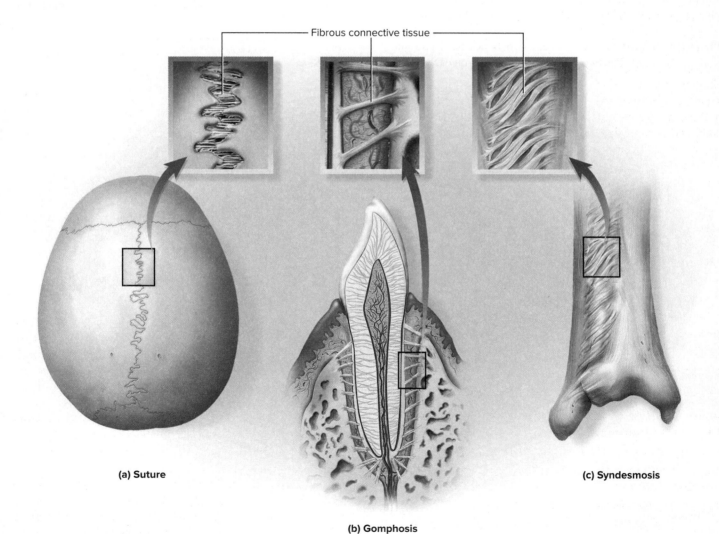

Fibrous connective tissue

(a) **Suture**

(b) **Gomphosis**

(c) **Syndesmosis**

FIGURE 9.2 Fibrous Joints. (a) A suture between the parietal bones. (b) A gomphosis between a tooth and the jaw. (c) A syndesmosis between the tibia and fibula.

❓ *If we narrowly define a joint as an attachment of one bone to another, one of these three would not qualify. Which one and why?*

Sutures

Sutures are immobile or only slightly mobile fibrous joints that closely bind the bones of the skull to each other; they occur nowhere else. In chapter 8, we didn't take much notice of the differences between one suture and another, but some differences may have caught your attention as you studied the diagrams in that chapter or examined laboratory specimens. Sutures can be classified as *serrate, lap,* and *plane sutures.* Readers with some knowledge of woodworking may recognize that the structures and functional properties of these sutures have something in common with basic types of carpentry joints **(fig. 9.3).**

Serrate sutures appear as wavy lines along which the adjoining bones firmly interlock with each other by their serrated margins, like pieces of a jigsaw puzzle. Serrate sutures are analogous to a dovetail wood joint. Examples include the coronal, sagittal, and lambdoid sutures that border the parietal bones.

Lap (squamous) sutures occur where two bones have overlapping beveled edges, like a miter joint in carpentry. On the surface, a lap suture appears as a relatively smooth (nonserrated) line. An example is the squamous suture where the temporal bone meets the sphenoid and parietal bones. The beveled edge of the temporal bone can be seen in figure 8.10b.

Plane (butt) sutures occur where two bones have straight nonoverlapping edges. The two bones merely border on each other, like two boards glued together in a butt joint. This type of joint is represented by the intermaxillary suture in the roof of the mouth (see fig. 8.5a).

Gomphoses

Even though the teeth are not bones, the attachment of a tooth to its socket is classified as a joint called a **gomphosis** (gom-FOE-sis). The term refers to its similarity to a nail hammered into wood.[5] The tooth is held firmly in place by a fibrous **periodontal ligament,** which consists of collagen fibers that extend from the bone matrix of the jaw into the dental tissue (see fig. 9.2b). The periodontal ligament allows the tooth to move or give a little under the stress of chewing. Along with associated nerve endings, this slight tooth movement allows us to sense how hard we're biting and to sense a particle of food stuck between the teeth.

Syndesmoses

A **syndesmosis**[6] (SIN-dez-MO-sis) is a fibrous joint at which two bones are bound by relatively long collagenous fibers. The

---

[5]*gomph* = nail, bolt; *osis* = condition
[6]*syn* = together; *desm* = band; *osis* = condition

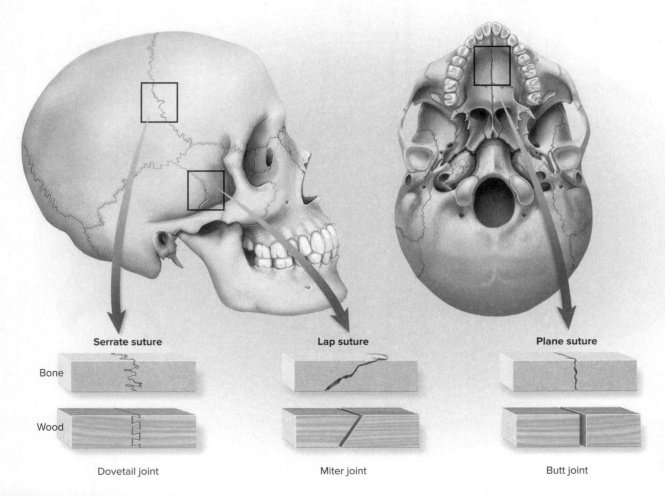

Serrate suture   Lap suture   Plane suture

Bone

Wood

Dovetail joint   Miter joint   Butt joint

FIGURE 9.3 Sutures. Serrate, lap, and plane sutures compared to some common wood joints.

separation between the bones and length of the fibers give these joints more mobility than a suture or gomphosis has. An especially mobile syndesmosis exists between the shafts of the radius and ulna, which are joined by a broad fibrous *interosseous membrane*. This permits such movements as pronation and supination of the forearm. A less mobile syndesmosis is the one that binds the distal ends of the tibia and fibula together, side by side (see fig. 9.2c).

9.1c Cartilaginous Joints

A **cartilaginous joint** is also called an **amphiarthrosis**[7] (AM-fee-ar-THRO-sis). In these joints, two bones are linked by cartilage **(fig. 9.4).** The two types of cartilaginous joints are *synchondroses* and *symphyses*.

Synchondroses

A **synchondrosis**[8] (SIN-con-DRO-sis) is a joint in which the bones are bound by hyaline cartilage. An example is the temporary joint between the epiphysis and diaphysis of a long bone in a child, formed by the cartilage of the epiphysial plate. Another is the attachment of the first rib to the sternum by a hyaline costal cartilage

(fig. 9.4a). (The other costal cartilages are joined to the sternum by synovial joints.)

Symphyses

In a **symphysis**[9] (SIM-fih-sis), two bones are joined by fibrocartilage (fig. 9.4b, c). One example is the pubic symphysis, in which the right and left pubic bones are joined anteriorly by the cartilaginous interpubic disc. Another is the joint between the bodies of two vertebrae, united by an intervertebral disc. The surface of each vertebral body is covered with hyaline cartilage. Between the vertebrae, this cartilage becomes infiltrated with collagen bundles to form fibrocartilage. Each intervertebral disc permits only slight movement between adjacent vertebrae, but the collective effect of all 23 discs gives the spine considerable flexibility.

▶▶▶**APPLY WHAT YOU KNOW**

The intervertebral joints are symphyses only in the cervical through the lumbar region. How would you classify the intervertebral joints of the sacrum and coccyx in a middle-aged adult?

---

[7]*amphi* = on all sides; *arthr* = joined; *osis* = condition
[8]*syn* = together; *chondr* = cartilage; *osis* = condition

[9]*sym* = together; *physis* = growth

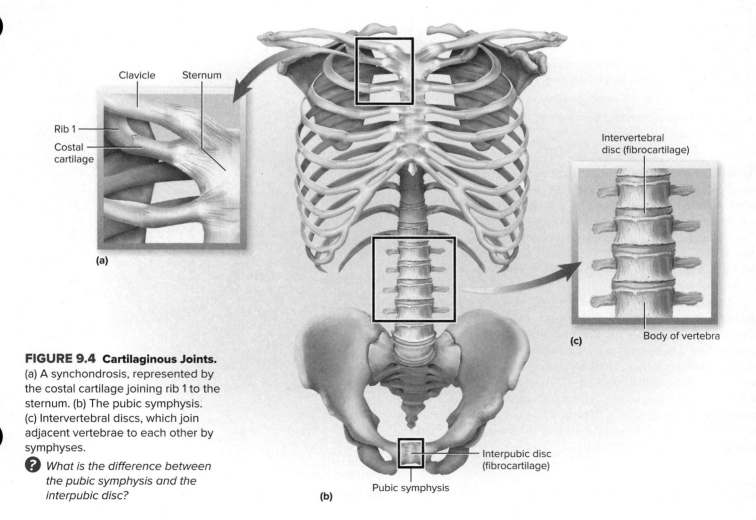

FIGURE 9.4 Cartilaginous Joints.
(a) A synchondrosis, represented by the costal cartilage joining rib 1 to the sternum. (b) The pubic symphysis. (c) Intervertebral discs, which join adjacent vertebrae to each other by symphyses.

❓ *What is the difference between the pubic symphysis and the interpubic disc?*

Clavicle   Sternum

Rib 1
Costal cartilage

(a)

Intervertebral disc (fibrocartilage)

(c)   Body of vertebra

Interpubic disc (fibrocartilage)

Pubic symphysis   (b)

Answer the following questions to test your understanding of the preceding section:

1. What is the difference between arthrology and kinesiology?

2. Distinguish between a synostosis, synarthrosis, and amphiarthrosis.

3. Define *suture, gomphosis,* and *syndesmosis,* and explain what these three joints have in common.

4. Name the three types of sutures and describe how they differ.

5. Name two synchondroses and two symphyses.

6. Give some examples of joints that become synostoses with age.

9.2 Synovial Joints

Expected Learning Outcomes

When you have completed this section, you should be able to

a. identify the anatomical components of a typical synovial joint;

b. classify any given joint action as a first-, second-, or third-class lever;

c. explain how mechanical advantage relates to the power and speed of joint movement;

d. discuss the factors that determine a joint's range of motion;

e. describe the primary axes of rotation that a bone can have and relate this to a joint's degrees of freedom;

f. name and describe six classes of synovial joints; and

g. use the correct standard terminology for various joint movements.

The most familiar type of joint is the **synovial joint** (sih-NO-vee-ul), also called a **diarthrosis**[10] (DY-ar-THRO-sis). Ask most people to point out any joint in the body, and they are likely to point to a synovial joint such as an elbow, knee, or knuckle. Many synovial joints, like these examples, are freely mobile. Others, such as the joints between the wrist and ankle bones and between the articular processes of the vertebrae, have more limited mobility.

Synovial joints are the most structurally complex type of joint and are the type most likely to develop uncomfortable and crippling dysfunctions. They are the most important joints for such professionals as physical and occupational therapists, athletic coaches, nurses, and fitness trainers to understand well. Their mobility makes the synovial joints especially important to the quality of life. Reflect, for example, on the performance extremes of a young athlete, the decline in flexibility that comes with age, and the crippling effect of rheumatoid arthritis. The rest of this chapter is concerned with synovial joints.

9.2a General Anatomy

In synovial joints, the facing surfaces of the two bones are covered with **articular cartilage,** a layer of hyaline cartilage up to 2 or 3 mm thick. These surfaces are separated by a narrow space, the **joint (articular) cavity,** containing a slippery lubricant called **synovial fluid (fig. 9.5).** This fluid, for which the joint is named, is rich in albumin and hyaluronic acid, which give it a viscous, slippery texture similar to raw egg white.[11] It nourishes the articular cartilages, removes their wastes, and makes movements at synovial joints almost friction-free. A connective tissue **joint (articular) capsule** encloses the cavity and retains the fluid. It has an outer **fibrous capsule** continuous with the periosteum of the adjoining bones, and an inner, cellular **synovial membrane.** The synovial membrane is composed mainly of fibroblast-like cells that secrete the fluid, and is populated by macrophages that remove debris from the joint cavity. Joint capsules and ligaments are well supplied with *lamellar corpuscles* (see section 16.2b) and other sensory nerve endings that enable the brain to monitor limb positions and joint movements.

In a few synovial joints, fibrocartilage grows inward from the joint capsule and forms a pad between the articulating bones. In the jaw (temporomandibular) joint, at both ends of the clavicle (sternoclavicular and acromioclavicular joints), and between the ulna and carpal bones, the pad crosses the entire joint capsule and

---

[11]*ovi* = egg

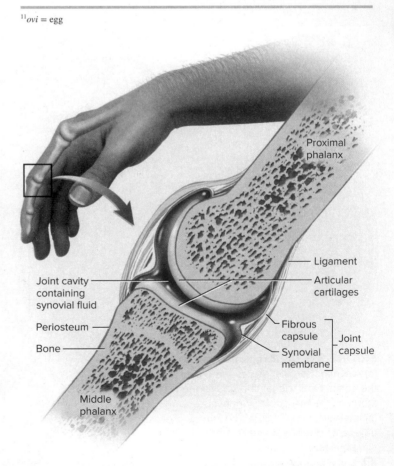

FIGURE 9.5 **Structure of a Simple Synovial Joint.**

 Why is a meniscus unnecessary in an interphalangeal joint?

---

[10]*dia* = separate, apart; *arthr* = joint; *osis* = condition

is called an **articular disc** (see fig. 9.23c). In the knee, two carti-lages extend inward from the left and right but don't entirely cross the joint (see fig. 9.28d). Each is called a **meniscus**[12] because of its crescent-moon shape. These cartilages absorb shock and pressure, guide the bones across each other, improve the fit between the bones, and stabilize the joint, reducing the chance of dislocation.

Accessory structures associated with a synovial joint include tendons, ligaments, and bursae. A **tendon** is a strip or sheet of tough collagenous connective tissue that attaches a muscle to a bone. Ten-dons are often the most important structures in stabilizing a joint. A **ligament** is a similar tissue that attaches one bone to another. Several ligaments are named and illustrated in our discussion of individual joints later in this chapter, and tendons are more fully considered in chapter 10 along with the gross anatomy of muscles.

A **bursa**[13] is a fibrous sac of synovial fluid located between adjacent muscles, where a tendon passes over a bone, or between bone and skin (see fig. 9.24). Bursae cushion muscles, help ten-dons slide more easily over the joints, and sometimes enhance the mechanical effect of a muscle by modifying the direction in which its tendon pulls. **Tendon (synovial) sheaths** are elongated cylindrical bursae wrapped around a tendon, seen especially in the hand and foot **(fig. 9.6).** They enable tendons to move back and forth more freely in such tight spaces as the wrist and ankle.

---

[12]*men* = moon, crescent; *isc* = little
[13]*bursa* = purse

DEEPER INSIGHT 9.1

CLINICAL APPLICATION

Exercise and Articular Cartilage

When synovial fluid is warmed by exercise, it becomes thinner (less viscous), like warm oil, and more easily absorbed by the articular carti-lage. The cartilage then swells and provides a more effective cushion against compression. For this reason, a warm-up period before vigor-ous exercise helps protect the articular cartilage from undue wear and tear.

Because cartilage is nonvascular, repetitive compression during exercise is important to its nutrition and waste removal. Each time a cartilage is compressed, fluid and metabolic wastes are squeezed out of it. When weight is taken off the joint, the cartilage absorbs synovial fluid like an expanding sponge, and the fluid carries oxygen and nutrients to the chondrocytes. Without exercise, articular cartilag-es deteriorate more rapidly from inadequate nutrition, oxygenation, and waste removal.

Weight-bearing exercise builds bone mass and strengthens the muscles that stabilize many of the joints, thus reducing the risk of joint dislocations. Excessive joint stress, however, can hasten the progression of osteoarthritis by damaging the articular cartilage (see Deeper Insight 9.5). Swimming and bicycling are good ways of exercising the joints with minimal damage.

FIGURE 9.6 Tendon Sheaths and Other Bursae in the Hand and Wrist.

9.2b Joints and Lever Systems

Many bones, especially the long bones, act as levers to enhance the speed or power of limb movements. A lever is any elongated, rigid object that rotates around a fixed point called the **fulcrum (fig. 9.7).** Rotation occurs when an effort applied to one point on the lever overcomes a resistance (load) at some other point. The portion of a lever from the fulcrum to the point of effort is called the **effort arm,** and the part from the fulcrum to the point of resistance is called the **resistance arm.** In skeletal anatomy, the fulcrum is a joint; the effort is applied by a muscle; and the resistance can be an object against which the body is working (as in weight lifting), the weight of the limb itself, or the tension in an opposing muscle.

Levers, Speed, and Force

The function of a lever is to produce a gain in the speed, distance, or force of a motion—either to exert more force against a resisting object than the force applied to the lever (for example, in moving a heavy object with a crowbar), or to move the resisting object farther or faster than the effort arm is moved (as in rowing a boat, where the blade of the oar moves much farther and faster than the handle). A single lever cannot confer both advantages. There is a trade-off between force on one hand and speed or distance on the other—as one increases, the other decreases.

The **mechanical advantage (MA)** of a lever is the ratio of its output force to its input force. If L_E is the length of the effort arm and L_R is the length of the resistance arm,

$$MA = L_E/L_R.$$

If MA is greater than 1.0, the lever produces more force, but less speed or distance, than the force exerted on it. If MA is less than 1.0, the lever produces more speed or distance, but less force, than the input **(fig. 9.8a, b).**

Consider, for example, the action of the brachialis muscle on the ulna when it flexes the elbow **(fig. 9.8c).** We will regard the ulna as the lever and the hand, along with whatever is in it, as the

FIGURE 9.7 The Basic Components of a Lever. This example is a first-class lever.

❓ *What would be the mechanical advantage of the lever shown here? Where would you put the fulcrum to increase the mechanical advantage without changing the lever class?*

FIGURE 9.8 Mechanical Advantage (MA). (a) $MA > 1$ in a first-class lever with $L_E > L_R$. (b) $MA < 1$ in a first-class lever with $L_E < L_R$. (c) $MA < 1$ in the forearm; when acted upon by the brachialis muscle, this is a third-class lever.

load or resistance being moved. The fulcrum of this lever is the humeroulnar joint. The brachialis tendon attaches to the ulna only slightly distal to the joint, so the effort arm is very short. The resistance arm extends from the joint to the load on the muscle (such as the weight in the hand), so L_E is much greater than L_R. Considering a typical ulna measuring 5 cm from the joint to the brachialis tendon and 33 cm from the joint to the load, the mechanical advantage would be $MA = 5/33 = 0.15$. As expected, with this $MA < 1$, the hand travels with more speed and distance than the point of brachialis attachment.

In chapter 10, you will often find that two or more muscles act on the same joint, seemingly producing the same effect. This may seem redundant, but it makes sense if the tendinous attachments of the muscles are at different points on a bone and produce different mechanical advantages. A sprinter taking off from the starting line, for example, uses "low-gear" (high-*MA*) muscles that don't generate much speed, but have the power to overcome the inertia of the body. The runner then "shifts into high gear" by using muscles with different attachments that confer a lower mechanical advantage but produce more speed at the feet. This is analogous to the way an automobile transmission uses one gear to get a car moving and other gears to cruise at higher speeds.

Types of Levers

There are three classes of levers that differ with respect to which component is in the middle—the fulcrum (F), effort (E), or resistance (R) **(fig. 9.9).**

1. A **first-class lever** is one with the fulcrum in the middle (EFR), such as a seesaw. An anatomical example is the atlanto–occipital joint of the neck, where the muscles of the back of the neck pull down on the occipital bone of the skull and oppose the tendency of the head to tip forward. Loss of muscle tone here can be embarrassing if you nod off in class. Rocking of the foot on the tibia as the toes are raised and lowered also exemplifies a first-class lever. (It is often misinterpreted as a second-class lever because of a superficial resemblance between standing on tiptoes and the wheelbarrow example that follows.)

2. A **second-class lever** has the resistance in the middle (FRE). Lifting the handles of a wheelbarrow, for example, causes it to pivot on the axle of the wheel at the opposite end and lift a load in the middle. If you sit in a chair and raise one thigh, like bouncing a small child on your knee, the femur pivots on the hip joint (the fulcrum), the quadriceps femoris muscle of the anterior thigh elevates the tibia like the wheelbarrow handles, and the resistance is the weight of the child or the thigh itself.

3. In a **third-class lever,** the effort is applied between the fulcrum and resistance (REF). For example, in paddling a canoe, the relatively stationary grip at the upper end of the paddle is the fulcrum, the effort is applied to the middle of the shaft, and the resistance is produced by the water against the blade. Most musculoskeletal levers are third class. The forearm acts as a third-class lever when you flex your elbow. The fulcrum is the joint between the ulna and humerus, the effort is applied partly by the biceps brachii muscle, and the resistance can be any weight in the hand or the weight of the forearm itself.

The classification of a lever changes as it performs different actions. We use the forearm as a third-class lever when we flex the elbow, as in weight lifting, but we use it as a first-class lever when we extend it, as in hammering nails. The mandible is a second-class lever when we open the mouth and a third-class lever when we close it to bite off a piece of food.

Range of Motion

One aspect of joint performance and physical assessment of a patient is a joint's flexibility, or **range of motion (ROM)**—the degrees through which a joint can move. The knee, for example, can flex through an arc of 130° to 140°, the metacarpophalangeal joint of the index finger about 90°, and the ankle about 74°. *ROM* obviously affects a person's functional independence and quality of life. It is also an important consideration in training for athletics or dance, in clinical diagnosis, and in monitoring the progress of rehabilitation. The *ROM* of a joint is normally determined by the following factors:

- **Structure of the articular surfaces of the bones.** In many cases, joint movement is limited by the shapes of the bone surfaces. For example, you cannot straighten your elbow beyond 180° or so because, as it straightens, the olecranon of the ulna swings into the olecranon fossa of the humerus and the fossa prevents it from moving any farther.

- **Strength and tautness of ligaments and joint capsules.** Some bone surfaces impose little if any limitation on joint movement. The articulations of the phalanges are an example; as one can see by examining a dry skeleton, an interphalangeal joint can bend through a broad arc. In the living body, however, these bones are joined by ligaments that limit their movement. As you flex one of your knuckles, ligaments on the anterior (palmar) side of the joint go slack, but ligaments on the posterior (dorsal) side tighten and prevent the joint from flexing beyond 90° or so. The knee is another case in point. In kicking a football, the knee rapidly extends to about 180°, but it can go no farther. Its motion is limited in part by a *cruciate ligament* and other knee ligaments described later. Gymnasts, dancers, and acrobats increase the *ROM* of their synovial joints by gradually stretching their ligaments during training. "Double-jointed" people have unusually large *ROM*s at some joints, not because the joint is actually double or fundamentally different from normal in its anatomy, but because the ligaments are unusually long or slack.

- **Action of the muscles and tendons.** Extension of the knee is also limited by the *hamstring muscles* on the posterior side of the thigh. In many other joints, too, pairs of muscles oppose each other and moderate the speed and range of joint motion. Even a resting muscle maintains a state of tension called *muscle tone,* which serves in many cases to stabilize a joint. One of the major factors preventing dislocation of the shoulder joint, for example, is tension in the *biceps brachii* muscle, whose tendons cross the joint, insert on the scapula, and hold the head of the humerus against the glenoid cavity. The nervous system continually monitors and adjusts joint angles and muscle tone to maintain joint stability and limit unwanted movements.

Axes of Rotation

In solid geometry, we recognize three mutually perpendicular axes, *x, y,* and *z.* In anatomy, these correspond to the transverse, frontal, and sagittal planes of the body. Just as we can describe any point

(a) First-class lever

(b) Second-class lever

(c) Third-class lever

FIGURE 9.9 The Three Classes of Levers. Left: The lever classes defined by the relative positions of the resistance (load), fulcrum, and effort. Center: Mechanical examples. Right: Anatomical examples. (a) Muscles of the back of the neck pull down on the occipital bone to oppose the tendency of the head to drop forward. The fulcrum is the occipital condyles. (b) The quadriceps muscle of the anterior thigh elevates the knee. The fulcrum is the hip joint. (c) In flexing the elbow, the biceps brachii muscle exerts an effort on the radius. Resistance is provided by the weight of the forearm or anything held in the hand. The fulcrum is the elbow joint.

in space by its *x, y,* and *z* coordinates, we can describe any joint movement by reference to the transverse, frontal, or sagittal anatomical planes.

A moving bone has a relatively stationary **axis of rotation** that passes through the bone in a direction perpendicular to the plane of movement. Think of a door for comparison; it moves horizontally as it opens and closes, and it rotates on hinges that are oriented on the vertical axis. Now consider the shoulder joint (**fig. 9.10**), where

the convex head of the humerus inserts into the concave glenoid cavity of the scapula. If you raise your arm to one side of your body, the head of the humerus rotates on an axis that passes from anterior to posterior; the arm rises in the frontal plane whereas its axis of rotation is in the sagittal plane. If you lift your arm to point at something straight in front of you, it moves through the sagittal plane whereas its axis of rotation is on the frontal plane, passing through the shoulder from lateral to medial. And if you swing your

(a) Abduction of arm

(c) Internal rotation of arm

(b) Flexion of arm

FIGURE 9.10 Axes of Joint Rotation. (a) Abduction of the arm in the frontal plane. (b) Flexion of the arm in the sagittal plane. (c) Internal rotation of the arm. All three axes are represented in movements of the multiaxial ball-and-socket joint of the shoulder.

arm in a horizontal arc, for example to grasp the opposite shoulder, the humeral head rotates in the transverse plane and its axis of rotation passes vertically through the joint.

Because the arm can move in all three anatomical planes, the shoulder joint is said to have three **degrees of freedom,** or to be a **multiaxial** joint. Other joints move through only one or two planes; they have one or two degrees of freedom and are called **monaxial** and **biaxial** joints, respectively. Degrees of freedom are a factor used in classifying the synovial joints.

Classes of Synovial Joints

There are six fundamental types of synovial joints, distinguished by the shapes of their articular surfaces and their degrees of freedom. We will begin by looking at these six types in simple terms, but then see that this is an imperfect classification for reasons discussed at the end. All six types can be found in the upper limb **(fig. 9.11).** They are listed here in descending order of mobility: one multiaxial type (ball-and-socket), three biaxial types (condylar, saddle, and plane), and two monaxial types (hinge and pivot).

1. **Ball-and-socket joints.** These are the shoulder and hip joints—the only multiaxial joints in the body. In both cases, one bone (the humerus or femur) has a smooth hemispherical head that fits into a cuplike socket on the other (the glenoid cavity of the scapula or the acetabulum of the hip bone).

2. **Condylar (ellipsoid) joints.** These joints exhibit an oval convex surface on one bone that fits into a complementary-shaped depression on the other. The radiocarpal joint of the wrist and metacarpophalangeal (MET-uh-CAR-po-fah-LAN-jee-ul)

joints at the bases of the fingers are examples. They are biaxial joints, capable of movement in two planes. To demonstrate this, hold your hand with the palm facing you. Make a fist, and these joints flex in the sagittal plane. Fan your fingers apart, and they move in the frontal plane.

3. **Saddle joints.** Here, both bones have a saddle-shaped surface—concave in one direction (like the front-to-rear curvature of a horse's saddle) and convex in the other (like the left-to-right curvature of a saddle). The clearest example of this is the trapeziometacarpal joint between the trapezium of the wrist and metacarpal I at the base of the thumb. Saddle joints are biaxial. The thumb, for example, moves in a frontal plane when you spread the fingers apart, and in a sagittal plane when you move it as if to grasp a tool such as a hammer. This range of motion gives us and other primates that invaluable anatomical hallmark, the opposable thumb. Another saddle joint is the sternoclavicular joint, where the clavicle articulates with the sternum. The clavicle moves vertically in the frontal plane at this joint when you lift a suitcase, and moves horizontally in the transverse plane when you reach forward to push open a door.

4. **Plane (gliding) joints.** Here the bone surfaces are flat or only slightly concave and convex. The adjacent bones slide over each other and have relatively limited movement. Plane joints are found between the carpal bones of the wrist, the tarsal bones of the ankle, and the articular processes of the vertebrae. Their movements, although slight, are complex. They are usually biaxial. For example, when the head is tilted forward and back, the articular facets of the vertebrae slide anteriorly and posteriorly; when the head is tilted from side to side, the facets slide laterally. Although any one joint moves only slightly, the combined action of the many joints in the wrist, ankle, and vertebral column allows for a significant amount of overall movement.

5. **Hinge joints.** These are essentially monaxial joints, moving freely in one plane with very little movement in any other, like a door hinge. Some examples are the elbow, knee, and interphalangeal (finger and toe) joints. In these cases, one bone has a convex (but not hemispherical) surface, such as the trochlea of the humerus and the condyles of the femur. This fits into a concave depression on the other bone, such as the trochlear notch of the ulna and the condyles of the tibia.

6. **Pivot joints.** These are monaxial joints in which a bone spins on its longitudinal axis like the axle of a bicycle wheel. There are two principal examples: the atlantoaxial joint between the first two vertebrae, and the radioulnar joint at the elbow. At the atlantoaxial joint, the dens of the axis projects into the vertebral foramen of the atlas and is held against the anterior arch of the atlas by the transverse ligament (see fig. 8.25c). As the head rotates left and right, the skull and atlas pivot around the dens. At the radioulnar joint, the anular ligament of the ulna wraps around the neck of the radius. During pronation and supination of the forearm, the disclike radial head pivots like a wheel turning on its axle. The edge of the wheel spins against the radial notch of the ulna like a car tire spinning in snow.

FIGURE 9.11 The Six Types of Synovial Joints. All six have representatives in the forelimb. Mechanical models show the types of motion possible at each joint.

Some joints cannot be easily classified into any one of these six categories. The jaw joint, for example, has some aspects of condylar, hinge, and plane joints. It clearly has an elongated condyle where it meets the temporal bone of the cranium, but it moves in a hingelike fashion when the mandible moves up and down in speaking, biting, and chewing; it glides slightly forward when the jaw juts forward to take a bite; and it glides from side to side to grind food between the molars. To observe the importance of the forward glide, try to open

your mouth while pushing the jaw posteriorly with the heel of your hand; it is difficult to open the mouth more than 1 or 2 cm when there is resistance to protraction of the mandible.

The knee is a classic hinge joint, but has an element of the pivot type; when we lock our knees to stand more effortlessly, the femur pivots slightly on the tibia. The humeroradial joint acts as a hinge joint when the elbow flexes and a pivot joint when the forearm pronates.

9.2c Movements of Synovial Joints

Kinesiology, physical therapy, and other medical and scientific fields have a specific vocabulary for the movements of synovial joints. The following terms form a basis for describing the muscle actions in chapter 10 and may also be indispensable to your advanced coursework or intended career. This section introduces the terms for joint movements, many of which are presented in pairs or groups with opposite or contrasting meanings. This section relies on familiarity with the three cardinal anatomical planes and the directional terms in atlas A, table A.1. All directional terms used here refer to a person in standard anatomical position. When one is standing in anatomical position, each joint is said to be in its **zero position.** Joint movements can be described as deviating from the zero position or returning to it.

Flexion and Extension

Flexion (fig. 9.12) is a movement that decreases a joint angle, usually in the sagittal plane. This is particularly common at hinge joints—for example, bending the elbow so that the arm and forearm go from a 180° angle to 90° or less. It occurs in other types of joints as well. For example, if you hold out your hands with the palms up, flexion of the wrist tips your palms toward you. The meaning of *flexion* is perhaps least obvious in the ball-and-socket joints of the shoulder and hip. At the shoulder, it means to raise your arm as if pointing at something directly in front of you or to continue in that arc and point toward the sky. At the hip, it means to raise the thigh, for example to place your foot on the next higher step when ascending a flight of stairs.

(a)

(c)

(b)

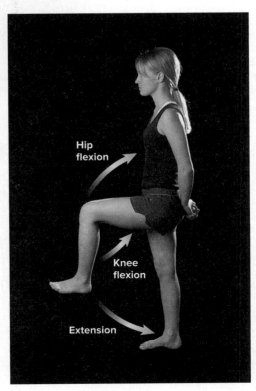

(d)

FIGURE 9.12 Flexion and Extension. (a) Flexion and extension of the elbow. (b) Flexion, extension, and hyperextension of the wrist. (c) Flexion and hyperextension of the shoulder. (d) Flexion and extension of the hip and knee.

Extension (fig. 9.12) is a movement that straightens a joint and generally returns a body part to the zero position—for example, straightening the elbow, wrist, or knee, or returning the arm or thigh back to zero position. In stair climbing, both the hip and knee extend when lifting the body to the next higher step.

Further extension of a joint beyond the zero position is called **hyperextension.**[14] For example, if you hold your hand in front of you with the palm down, then raise the back of your hand as if you were admiring a new ring, you hyperextend the wrist. Hyperextension of the upper or lower limb means to move the limb to a position behind the frontal plane of the trunk, as if reaching around with your arm to scratch your back. Each backswing of the lower limb when you walk hyperextends the hip.

Flexion and extension occur at nearly all diarthroses, but hyperextension is limited to only a few. At most diarthroses, ligaments or bone structures prevent hyperextension.

---

[14]*hyper* = excessive, beyond normal

Abduction and Adduction

Abduction[15] (ab-DUC-shun) **(fig. 9.13a)** is the movement of a body part in the frontal plane away from the midline of the body—for example, moving the feet apart to stand spread-legged, or raising an arm to one side of the body. **Adduction**[16] **(fig. 9.13b)** is movement in the frontal plane back toward the midline. Some joints can be **hyperadducted,** as when you stand with your ankles crossed, cross your fingers, or hyperadduct the shoulder to stand with your elbows straight and your hands clasped below your waist. You **hyperabduct** the arm if you raise it high enough to cross slightly over the front or back of your head.

Elevation and Depression

Elevation (fig. 9.14a) is a movement that raises a body part vertically in the frontal plane. **Depression (fig. 9.14b)** lowers a

---

[15]*ab* = away; *duc* = to lead or carry
[16]*ad* = toward; *duc* = to lead or carry

(a) Abduction **(b) Adduction**

FIGURE 9.13 Abduction and Adduction.
(a) Abduction of the limbs. (b) Adduction of the limbs, returning them to zero position.
a, b: Timothy L. Vacula/McGraw-Hill Education

(a) Elevation **(b) Depression**

FIGURE 9.14 Elevation and Depression.
(a) Elevation of the shoulders. (b) Depression of the shoulders.
a, b: Timothy L. Vacula/McGraw-Hill Education

(a) Protraction

(b) Retraction

FIGURE 9.15 Protraction and Retraction. (a) Protraction of the shoulder, as in pushing open a door. (b) Retraction of the shoulder.

a, b: Timothy L. Vacula/McGraw-Hill Education

body part in the same plane. For example, to lift a suitcase from the floor, you elevate your scapula; in setting it down again, you depress the scapula. These are also important jaw movements in biting.

Protraction and Retraction

Protraction[17] **(fig. 9.15a)** is the anterior movement of a body part in the transverse (horizontal) plane, and **retraction**[18] **(fig. 9.15b)** is posterior movement. Your shoulder protracts, for example, when you reach in front of you to push a door open. It retracts when you return it to the resting (zero) position or pull the shoulders back to stand at military attention. Such exercises as rowing a boat, bench presses, and push-ups involve repeated protraction and retraction of the shoulders.

Circumduction

In **circumduction**[19] **(fig. 9.16),** one end of an appendage remains fairly stationary while the other end makes a circular motion. If an artist standing at an easel reaches forward and draws a circle on a canvas, she circumducts the upper limb; the shoulder remains stationary while the hand moves in a circle. A baseball player winding up for the pitch circumducts the upper limb in a more extreme "windmill" fashion. One can also circumduct an individual finger, the hand, the thigh, the foot, the trunk, and the head.

FIGURE 9.16 Circumduction.

Timothy L. Vacula/McGraw-Hill Education

▶▶▶**APPLY WHAT YOU KNOW**

Choose any example of circumduction and explain why this motion is actually a sequence of flexion, abduction, extension, and adduction.

Rotation

Rotation (fig. 9.17) is a movement in which a bone spins on its longitudinal axis, like the axle of a bicycle wheel. For example, if you stand with bent elbow and move your forearm to grasp your opposite arm, your humerus spins in a motion called **medial (internal) rotation.** If you make the opposite action, so the forearm points away from your trunk, your humerus undergoes **lateral**

(a) Medial (internal) rotation   **(b) Lateral (external) rotation**

FIGURE 9.17 Medial (Internal) and Lateral (External) Rotation. (a) Medial (internal) rotation of the humerus and femur. (b) Lateral (external) rotation of both.

a, b: Timothy L. Vacula/McGraw-Hill Education

---

[17]*pro* = forward; *trac* = to pull or draw
[18]*re* = back; *trac* = to pull or draw
[19]*circum* = around; *duc* = to carry, lead

(external) rotation. Good examples of lateral and medial rotation of the humerus are its movements in the forehand and backhand strokes of tennis. The femur can also rotate. If you stand and turn your right foot so your toes point toward your left foot, then turn it so your toes point away from the left food, your femur undergoes medial and lateral rotation, respectively. Powerful left and right rotation at the waist is important in such actions as baseball pitching and golf. Other examples are given in the coming discussions of forearm and head movements.

Supination and Pronation

Supination and pronation are known primarily as forearm movements, but see also the later discussion of foot movements. **Supination**[20] (SOO-pih-NAY-shun) **(fig. 9.18a)** of the forearm is a movement that turns the palm to face anteriorly or upward; in anatomical position, the forearm is supinated and the radius is parallel to the ulna. **Pronation**[21] **(fig. 9.18b)** is the opposite movement, causing the palm to face posteriorly or downward, and the radius to cross the ulna like an X. During these movements, the concave end of the disc-shaped head of the radius spins on the capitulum of the humerus, and the edge of the disc spins in the radial notch of the ulna. The ulna remains relatively stationary.

As an aid to remembering these terms, think of it this way: You are *prone* to stand in the most comfortable position, which is with the forearm *pronated*. But if you were holding a bowl of *soup* in your palm, you would need to *supinate* the forearm to keep from spilling it.

Chapter 10 describes the muscles that perform these actions (see table 10.10). Of these, the *supinator* is the most powerful. Supination is the type of movement you would usually make with your right hand to turn a doorknob clockwise or to drive a screw into a piece of wood. The threads of screws and bolts are designed with the relative strength of the supinator in mind, so the greatest power can be applied when driving them with a screwdriver in the right hand.

We will now consider a few body regions that combine the foregoing motions, or that have unique movements and terminology.

Special Movements of the Head and Trunk

Flexion of the spine produces forward-bending movements, as in tilting the head forward or bending at the waist in a toe-touching exercise **(fig. 9.19a)**. *Extension* of the vertebral column straightens the trunk or the neck, as in standing up or returning the head to a forward-looking zero position. *Hyperextension* is employed in looking up toward the sky or bending over backward **(fig. 9.19b)**.

Lateral flexion is tilting the head or trunk to the right or left of the midline **(fig. 9.19c)**. Twisting at the waist or turning of the head is called **right rotation** or **left rotation** when the chest or the face turns to the right or left of the forward-facing

zero position **(fig. 9.19d, e)**. Powerful right and left rotation at the waist is important in baseball pitching, golf, discus throwing, and other sports.

(a) Supination

(b) Pronation

FIGURE 9.18 Supination and Pronation of the Forearm.
(a) Supination. (b) Pronation. Note the way these forearm rotations affect the relationship of the radius and ulna. Relative positions of muscles, nerves, and blood vessels are similarly affected.

a, b: Timothy L. Vacula/McGraw-Hill Education

---

[20]*supin* = to lay back
[21]*pron* = to bend forward

(a) Flexion

(b) Hyperextension

(c) Lateral flexion

(d) Right rotation

(e) Rotation

FIGURE 9.19 Movements of the Head and Trunk. (a) Anterior flexion of the spine. (b) Hyperextension of the spine. (c) Lateral flexion of the spine. (d) Right rotation at the waist. (e) Right and left rotation of the head.

❓ *In rotation of the head (e), what bone spins on its axis?*

a–e: Timothy L. Vacula/McGraw-Hill Education

Special Movements of the Mandible

Movements of the mandible are concerned especially with biting and chewing. Imagine taking a bite of raw carrot. Most people have some degree of overbite; at rest, the upper incisors (front teeth) overhang the lower ones. For effective biting, however, the chisel-like edges of the incisors must meet. In preparation to bite, we therefore *protract* the mandible to bring the lower incisors forward. After the bite is taken, we *retract* it **(fig. 9.20a, b).** To actually take the bite, we must *depress* the mandible to open the mouth, then *elevate* it so the incisors cut off the piece of food.

Next, to chew the food, we don't simply raise and lower the mandible as if hammering away at the food between the teeth; rather, we exercise a grinding action that shreds the food between the broad, bumpy surfaces of the premolars and molars. This entails a side-to-side movement of the mandible called **lateral excursion** (movement to the left or right of the zero position) and **medial excursion** (movement back to the median, zero position) **(fig. 9.20c, d).**

Special Movements of the Hand and Digits

The hand moves anteriorly and posteriorly by flexion and extension of the wrist. It can also move in the frontal plane. **Ulnar flexion** tilts the hand toward the little finger, and **radial flexion** tilts it toward the thumb **(fig. 9.21a, b).** We often use such motions when waving hello to someone with a side-to-side wave of the hand, or when washing windows, polishing furniture, or keyboarding.

Movements of the digits are more varied, especially those of the thumb. *Flexion* of the fingers is curling them; *extension* is straightening them. Most people cannot hyperextend their fingers. Spreading the fingers apart is *abduction* **(fig. 9.21c),** and bringing them together again so they touch along their surfaces is *adduction* (as in fig. 9.21a, b).

The thumb is different, however, because in embryonic development it rotates nearly 90° from the rest of the hand. If you hold your hand in a completely relaxed position (but not resting on a table), you will probably see that the plane that contains your thumb and index finger is about 90° to the plane that contains the index through little fingers. Much of the terminology of thumb movement therefore differs from that of the other four fingers. *Flexion* of the thumb is bending the joints so the tip of the thumb is directed toward the palm, and *extension* is straightening it. If you now place the palm of your hand on a tabletop with all five digits parallel and touching, the thumb is extended. Keeping your hand there, if you move your thumb away from the index finger so they

(a) Protraction   **(b) Retraction**

(c) Lateral excursion   **(d) Medial excursion**

FIGURE 9.20 Movements of the Mandible. (a) Protraction and (b) retraction. (c) Lateral and (d) medial excursion, as in chewing food.

a–d: Timothy L. Vacula/McGraw-Hill Education

form a 90° angle (but both are on the plane of the table), the thumb movement is called **radial abduction** (as in fig. 9.21c). Another movement, **palmar abduction,** moves the thumb away from the plane of the hand so it points anteriorly, as you would do if you were about to wrap your hand around a tool handle **(fig. 9.21d).** From either position—radial or palmar abduction—*adduction* of the thumb means to bring it back to zero position, touching the base of the index finger.

Two terms are unique to the thumb: **Opposition**[22] means to move the thumb to approach or touch the tip of any of the other four fingers **(fig. 9.21e). Reposition**[23] is the return to zero position.

Special Movements of the Foot

A few additional movement terms are unique to the foot. **Dorsiflexion** is a movement in which the toes are elevated, as you might do to trim your toenails **(fig. 9.22a).** In each step you take, the foot dorsiflexes as it comes forward. This prevents you from scraping your toes on the ground and results in the characteristic *heel strike* of human locomotion when the foot touches down in front of you. **Plantar flexion** is movement of the foot so the toes point downward, as in pressing the gas pedal of a car or standing on tiptoes. This motion also produces the *toe-off* in each step you take, as the heel of the foot behind you lifts off the ground. Plantar flexion can be a very powerful motion, epitomized by high jumpers and the jump shots of basketball players.

[22]*op* = against; *posit* = to place
[23]*re* = back; *posit* = to place

(a) Radial flexion   **(b) Ulnar flexion**   **(c) Abduction of fingers**

(d) Palmar abduction of thumb   **(e) Opposition of thumb**

FIGURE 9.21 Movements of the Hand and Digits.
(a) Radial flexion of the wrist. (b) Ulnar flexion of the wrist. (c) Abduction of the fingers. The thumb position in this figure is called *radial abduction.* Parts (a) and (b) show adduction of the fingers. (d) Palmar abduction of the thumb. (e) Opposition of the thumb; reposition is shown in parts (a) and (b).

a–e: Timothy L. Vacula/McGraw-Hill Education

(b) Inversion

(c) Eversion

(a) Flexion of ankle

FIGURE 9.22 Movements of the Foot. (a) Plantar flexion and dorsiflexion of the ankle. (b) Inversion of the feet. (c) Eversion of the feet.

a–c: Timothy L. Vacula/McGraw-Hill Education

Inversion[24] is a foot movement that tips the soles medially, somewhat facing each other, and **eversion**[25] is a movement that tips the soles laterally, away from each other **(fig. 9.22b, c).** These movements are important in walking on uneven surfaces such as a rocky trail. They are common in fast sports such as tennis and football, and sometimes cause ankle sprains. These terms also refer to congenital deformities of the feet, which are often corrected by orthopedic shoes or braces.

Pronation and *supination,* referring mainly to forearm movements, also apply to the feet but refer here to a more complex combination of movements. Pronation of the foot is a combination of dorsiflexion, eversion, and abduction—that is, the toes are elevated and turned away from the other foot and the sole is tilted away from the other foot. Supination of the foot is a combination of plantar flexion, inversion, and adduction—the toes are lowered and turned toward the other foot and the sole is tilted toward it. These may seem a little difficult to visualize and consciously perform, but they are common motions in walking, running, ballet, and crossing uneven surfaces.

You can perhaps understand why these terms apply to the feet if you place the palms of your hands on a table and pretend they are your soles. Tilt your hands so the inner edge (thumb side) of each is raised from the table. This is like raising the medial edge of your foot from the ground, and as you can see, it involves a slight supination of your forearms. Resting your hands palms down on a table, your forearms are already pronated; but if you raise the outer edges of your hands (the little finger side), like pronating the feet, you will see that it involves a continuation of the pronation movement of the forearm.

> **BEFORE YOU GO ON**
>
> Answer the following questions to test your understanding of the preceding section:
>
> 7. Describe the roles of articular cartilage and synovial fluid in joint mobility.

8. Give an anatomical example of each class of levers and explain why each example belongs in that class.

9. Give an example of each of the six classes of synovial joints and state how many axes of rotation each example has.

10. Suppose you reach overhead and screw a lightbulb into a ceiling fixture. Name each joint that would be involved and the joint actions that would occur.

11. Where are the effort, fulcrum, and resistance in the act of dorsiflexion? What class of lever does the foot act as during dorsiflexion? Would you expect it to have a mechanical advantage greater or less than 1.0? Why?

9.3 **Anatomy of Selected Diarthroses**

Expected Learning Outcomes

When you have completed this section, you should be able to

a. identify the major anatomical features of the jaw, shoulder, elbow, hip, knee, and ankle joints; and

b. explain how the anatomical differences between these joints are related to differences in function.

We now examine the gross anatomy of certain diarthroses. It is beyond the scope of this book to discuss all of them, but the ones selected here most often require medical attention and many of them have a strong bearing on athletic performance and everyday mobility.

9.3a The Jaw Joint

The **temporomandibular (jaw) joint (TMJ)** is the articulation of the condyle of the mandible with the mandibular fossa of the temporal bone **(fig. 9.23).** You can feel its action by pressing your fingertips against the jaw immediately anterior to the ear while opening and closing your mouth.

The synovial cavity of the TMJ is divided into superior and inferior chambers by an articular disc, which permits lateral and

---

[24]*in* = inward; *version* = turning
[25]*e* = outward; *version* = turning

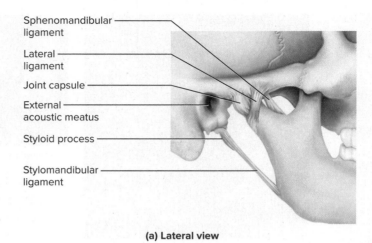

Sphenomandibular ligament

Lateral ligament

Joint capsule

External acoustic meatus

Styloid process

Stylomandibular ligament

(a) Lateral view

Sphenoidal sinus

Occipital bone

Sphenomandibular ligament

Styloid process

Stylomandibular ligament

(b) Medial view

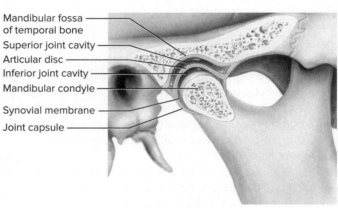

Mandibular fossa of temporal bone

Superior joint cavity

Articular disc

Inferior joint cavity

Mandibular condyle

Synovial membrane

Joint capsule

(c) Sagittal section

FIGURE 9.23 The Temporomandibular (Jaw) Joint (TMJ). (a) Lateral view. (b) Medial view of bisected skull. (c) Sagittal section of joint. **APR**

medial excursion of the mandible. Two ligaments support the joint. The **lateral ligament** prevents posterior displacement of the mandible. If the jaw receives a hard blow, this ligament normally prevents the condylar process from being driven upward and fracturing the base of the skull. The **sphenomandibular ligament** on

the medial side of the joint extends from the sphenoid bone to the ramus of the mandible. A *stylomandibular ligament* extends from the styloid process to the angle of the mandible but is not part of the TMJ proper.

A deep yawn or other strenuous depression of the mandible can dislocate the TMJ by making the condyle pop out of the fossa and slip forward. The joint is relocated by pressing down on the molars while pushing the jaw posteriorly.

DEEPER INSIGHT 9.2

CLINICAL APPLICATION

TMJ Dysfunction

Temporomandibular joint dysfunction (TMD) afflicts 20% to 30% of the adult population worldwide, and ranks second only to toothache among the most common causes of orofacial pain. It seems to be a cluster of disorders with symptoms that include clicking, popping, or grating noises; restricted jaw movements that may cause problems eating or speaking; and aching pain in the TMJ and associated chewing muscles that intensifies in chewing and yawning. The causes of TMD are so poorly understood that there is no agreement on the best treatment and there are, indeed, more than a dozen different names for it. Pain medication (analgesics) can be helpful. Since TMD is often associated with anxiety, depression, or stress, some patients and physicians report success with behavioral therapies as diverse as meditation, biofeedback, and yoga.

9.3b The Shoulder Joint

The **glenohumeral (humeroscapular) joint,** or shoulder joint, is where the hemispherical head of the humerus articulates with the glenoid cavity of the scapula **(fig. 9.24).** Together, the shoulder and elbow joints serve to position the hand for the performance of a task; without a hand, shoulder and elbow movements are almost useless. The relatively loose shoulder joint capsule and shallow glenoid cavity sacrifice joint stability for freedom of movement (see Deeper Insight 9.3). The cavity, however, has a ring of fibrocartilage called the **glenoid labrum**[26] around its margin, making it somewhat deeper than it looks on a dried skeleton.

---

[26]*glen* = socket; *oid* = resembling; *labrum* = lip

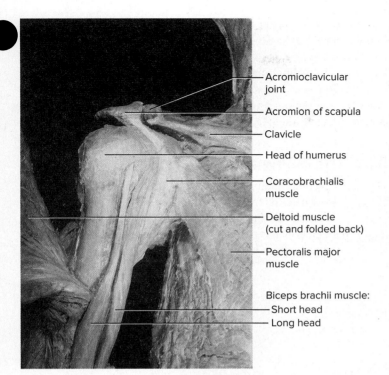

Acromioclavicular joint
Acromion of scapula
Clavicle
Head of humerus
Coracobrachialis muscle
Deltoid muscle (cut and folded back)
Pectoralis major muscle
Biceps brachii muscle:
Short head
Long head

(a) Anterior dissection

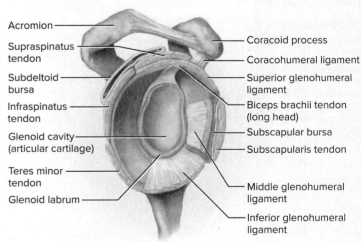

Acromion
Subacromial bursa
Supraspinatus tendon
Coracohumeral ligament
Subdeltoid bursa
Subscapularis tendon
Transverse humeral ligament
Tendon sheath
Biceps brachii tendon (long head)
Humerus

Acromioclavicular ligament
Clavicle
Coraco-clavicular ligament
Coraco-acromial ligament
Coracoid process
Subcoracoid bursa
Subscapular bursa
Glenohumeral ligaments

(b) Anterior view

Acromion
Subdeltoid bursa
Deltoid muscle
Humerus

Supraspinatus tendon
Capsular ligament
Glenoid labrum
Synovial membrane
Glenoid cavity of scapula
Glenoid labrum

(c) Frontal section

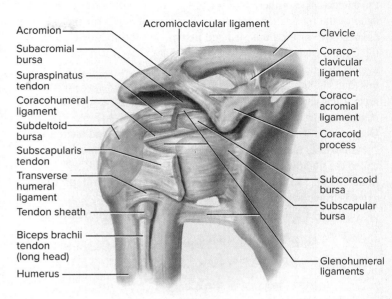

Acromion
Supraspinatus tendon
Subdeltoid bursa
Infraspinatus tendon
Glenoid cavity (articular cartilage)
Teres minor tendon
Glenoid labrum

Coracoid process
Coracohumeral ligament
Superior glenohumeral ligament
Biceps brachii tendon (long head)
Subscapular bursa
Subscapularis tendon
Middle glenohumeral ligament
Inferior glenohumeral ligament

(d) Lateral view, humerus removed

FIGURE 9.24 The Glenohumeral (Shoulder) Joint. (a) Anterior view of dissected cadaver shoulder. (b) Anterior view of shoulder ligaments and bursae. (c) Frontal section, showing joint cavity. (d) Lateral view of the shoulder socket (glenoid cavity of the scapula) with the humerus removed. **A&PR**

a: Rebecca Gray/McGraw-Hill Education

DEEPER INSIGHT 9.3

CLINICAL APPLICATION

Shoulder Dislocation

The anatomy and mobility of the shoulder joint make it especially susceptible to dislocation. Over 95% of cases are classified as *anterior dislocation* (displacement of the humeral head in the anterior direction). Such dislocations are usually caused when the arm is abducted and receives a blow from above—for example, by heavy objects falling from a shelf. A complex of nerves and blood vessels traverses the axillary region, and shoulder dislocation can easily damage the axillary nerve or

artery (see figs. 13.17 and 20.39). Left untreated, this can lead to muscle atrophy, weakness, or paralysis. Greek physician Hippocrates taught students to treat shoulder dislocation by placing a heel in the patient's axilla and pulling on the arm, but this can cause even worse nerve damage and is never done anymore by professionals. Because the shoulder is so easily dislocated, one also should never attempt to move an unconscious or immobilized person by pulling on his or her arm.

The shoulder is stabilized mainly by the biceps brachii muscle on the anterior side of the arm. One of its tendons arises from the *long head* of the muscle (see table 10.10), passes through the intertubercular groove of the humerus, and inserts on the superior margin of the glenoid cavity. It acts as a taut strap that presses the humeral head against the glenoid cavity. Four additional muscles help to stabilize this joint: the *supraspinatus, infraspinatus, teres minor,* and *subscapularis.* Their tendons form the **rotator cuff,** which is fused to the joint capsule on all sides except the inferior (see fig. 10.24). Table 10.9 and Deeper Insight 10.4 further describe the rotator cuff and its injuries.

Five principal ligaments also support this joint. Three of them, called the **glenohumeral ligaments,** are relatively weak and sometimes absent. The other two are the **coracohumeral ligament,** which extends from the coracoid process of the scapula to the greater tubercle of the humerus, and the **transverse humeral ligament,** which

extends from the greater to the lesser tubercle of the humerus and forms a tunnel housing the tendon from the long head of the biceps.

Four bursae occur at the shoulder. Their names describe their locations: the **subdeltoid, subacromial, subcoracoid,** and **subscapular bursae.** The *deltoid* is the large muscle that caps the shoulder, and the other bursae are named for parts of the scapula described in section 8.4a.

9.3c The Elbow Joint

The elbow is a hinge joint composed of two articulations: the **humeroulnar joint** where the trochlea of the humerus joins the trochlear notch of the ulna, and the **humeroradial joint** where the capitulum of the humerus meets the head of the radius **(fig. 9.25).** Both are enclosed in a single joint capsule. On the posterior side of the

FIGURE 9.25 The Elbow Joint. This region includes two joints that form the elbow hinge—the humeroulnar and humeroradial—and one joint, the radioulnar, not involved in the hinge. (a) Anterior elbow ligaments. (b) Sagittal section, showing joint cavity. (c) Medial elbow ligaments. (d) Lateral elbow ligaments. **A&PR**

elbow, there is a prominent **olecranon bursa** to ease the movement of tendons over the joint. Side-to-side motions of the elbow joint are restricted by a pair of ligaments: the **radial (lateral) collateral ligament** and **ulnar (medial) collateral ligament.**

Another joint occurs in the elbow region, the **proximal radioulnar joint,** but it is not involved in the hinge. At this joint, the edge of the disclike head of the radius fits into the radial notch of the ulna. It is held in place by the **anular ligament,** which encircles the radial head

and is attached at each end to the ulna. The radial head rotates like a wheel against the ulna as the forearm is pronated or supinated.

9.3d The Hip Joint

The **coxal (hip) joint** is the point where the head of the femur inserts into the acetabulum of the hip bone **(fig. 9.26).** Because the coxal joints bear much of the body's weight, they have deep sockets

(a) Anterior dissection

(b) Lateral view, femur retracted

(c) Anterior view

(d) Posterior view

FIGURE 9.26 The Coxal (Hip) Joint. (a) Anterior view of dissected cadaver hip. (b) Lateral view of hip joint with femur retracted from acetabulum. (c) Anterior hip ligaments. (d) Posterior hip ligaments.

a: Rebecca Gray/McGraw-Hill Education

and are much more stable than the shoulder joint. The depth of the socket is somewhat greater than you see on dried bones because of a horseshoe-shaped ring of fibrocartilage, the **acetabular labrum,** attached to its rim. Dislocations of the hip are therefore rare.

▶▶▶APPLY WHAT YOU KNOW

Where else in the body is there a structure similar to the acetabular labrum? What do those two locations have in common?

Ligaments that support the coxal joint include the **iliofemoral** and **pubofemoral ligaments** (ILL-ee-oh-FEM-o-rul, PYU-bo-FEM-o-rul) on the anterior side and the **ischiofemoral ligament** (ISS-kee-oh-FEM-o-rul) on the posterior side. The name of each ligament refers to the bones to which it attaches—the femur and the ilium, pubis, or ischium. When you stand up, these ligaments become twisted and pull the head of the femur tightly into the acetabulum. The head of the femur has a conspicuous pit called the **fovea capitis.** The **round ligament,** or **ligamentum teres**[27] (TERR-eez), arises here and attaches to the lower margin of the acetabulum. This is a relatively slack ligament, so it is doubtful that it plays a significant role in holding the femur in its socket. It does, however, contain an artery that supplies blood to the head of the femur. A **transverse acetabular ligament** bridges a gap in the inferior margin of the acetabular labrum.

9.3e The Knee Joint

The **tibiofemoral (knee) joint** is the largest and most complex diarthrosis of the body (**figs. 9.27** and **9.28**). It is primarily a hinge joint, but when the knee is flexed it is also capable of slight rotation and lateral gliding. The patella and patellar ligament also articulate with the femur to form a gliding **patellofemoral joint.**

The joint capsule encloses only the lateral and posterior aspects of the knee joint, not the anterior. The anterior aspect is covered by the patella, patellar ligament, and *lateral* and *medial patellar retinacula* (not illustrated). The retinacula are extensions of the tendon of the *quadriceps femoris muscle,* the large anterior muscle of the thigh. The knee is stabilized mainly by the quadriceps tendon in front and the tendon of the *semimembranosus muscle* on the rear of the thigh. Developing strength in these muscles therefore reduces the risk of knee injury.

The joint cavity contains two C-shaped cartilages called the **lateral** and **medial menisci** (singular, **meniscus**) joined by a **transverse ligament.** The menisci absorb the shock of the body weight jostling up and down on the knee and prevent the femur from rocking from side to side on the tibia.

The posterior **popliteal region** (pop-LIT-ee-ul) of the knee is supported by a complex array of *extracapsular ligaments* external to the joint capsule and two *intracapsular ligaments* within it. The extracapsular ligaments include two collateral ligaments that prevent the knee from rotating when the joint is extended—the **fibular (lateral) collateral ligament** and the **tibial (medial) collateral ligament**—and other ligaments not illustrated.

Lateral ◀━━│━━▶ Medial

Femur:
  Shaft
  Patellar surface
  Medial condyle
  Lateral condyle

Joint capsule
Joint cavity:
  Anterior cruciate ligament
  Medial meniscus
  Lateral meniscus

Tibia:
  Lateral condyle
  Medial condyle
  Tuberosity

Patellar ligament

Patella (posterior surface)

Articular facets

Quadriceps tendon (reflected)

FIGURE 9.27 The Right Knee, Anterior Dissection. The quadriceps tendon has been cut and folded (reflected) downward to expose the joint cavity and the posterior surface of the patella.
Rebecca Gray/McGraw-Hill Education

The two intracapsular ligaments lie deep within the joint. The synovial membrane folds around them, however, so that they are excluded from the fluid-filled synovial cavity. These ligaments cross each other in the form of an X; hence, they are called the **anterior cruciate**[28] **ligament (ACL)** (CROO-she-ate) and **posterior cruciate ligament (PCL).** These are named according to whether they attach to the anterior or posterior side of the tibia, not for their attachments to the femur. When the knee is extended, the ACL is pulled tight and prevents hyperextension. The PCL prevents the femur from sliding off the front of the tibia and prevents the tibia from being displaced backward. The ACL is one of the most common sites of knee injury (see Deeper Insight 9.4).

[27]*teres* = round

[28]*cruci* = cross; *ate* = characterized by

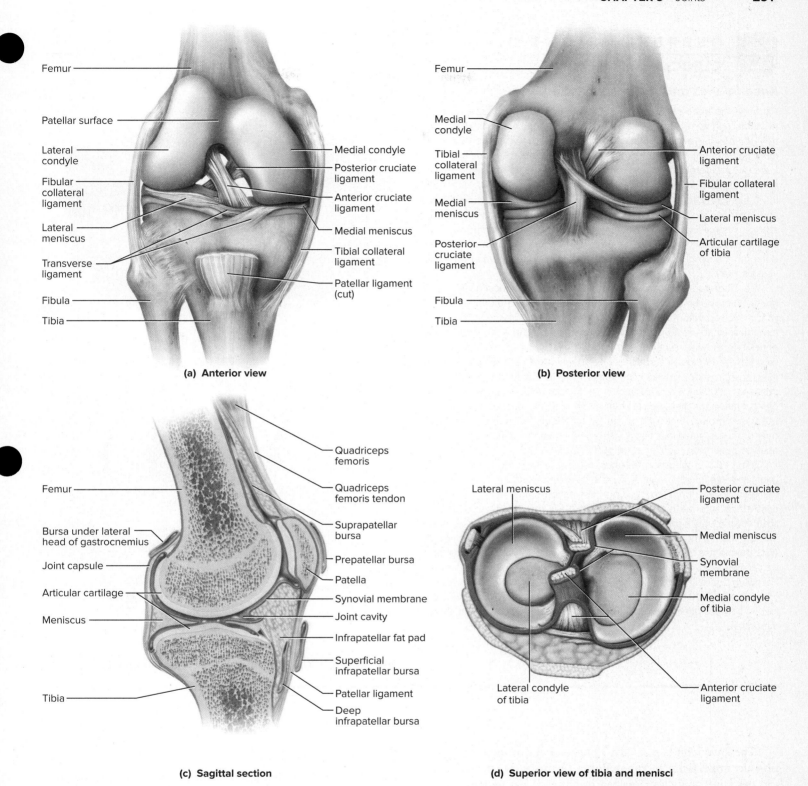

(a) Anterior view

Femur
Patellar surface
Lateral condyle
Fibular collateral ligament
Lateral meniscus
Transverse ligament
Fibula
Tibia
Medial condyle
Posterior cruciate ligament
Anterior cruciate ligament
Medial meniscus
Tibial collateral ligament
Patellar ligament (cut)

(b) Posterior view

Femur
Medial condyle
Tibial collateral ligament
Medial meniscus
Posterior cruciate ligament
Fibula
Tibia
Anterior cruciate ligament
Fibular collateral ligament
Lateral meniscus
Articular cartilage of tibia

(c) Sagittal section

Femur
Bursa under lateral head of gastrocnemius
Joint capsule
Articular cartilage
Meniscus
Tibia
Quadriceps femoris
Quadriceps femoris tendon
Suprapatellar bursa
Prepatellar bursa
Patella
Synovial membrane
Joint cavity
Infrapatellar fat pad
Superficial infrapatellar bursa
Patellar ligament
Deep infrapatellar bursa

(d) Superior view of tibia and menisci

Lateral meniscus
Lateral condyle of tibia
Posterior cruciate ligament
Medial meniscus
Synovial membrane
Medial condyle of tibia
Anterior cruciate ligament

FIGURE 9.28 The Right Tibiofemoral (Knee) Joint. (a) Anterior view of knee joint with patella removed. (b) Posterior view. (c) Sagittal section of knee showing joint cavity and bursae. (d) Superior view of tibia showing the menisci. **APR**

▶▶▶ **APPLY WHAT YOU KNOW**

What structure in the elbow joint serves the same function as the ACL of the knee?

An important aspect of human bipedalism is the ability to lock the knees and stand erect without tiring the extensor muscles of the leg. When the knee is extended to the fullest degree allowed by the ACL, the femur rotates medially on the tibia. This action locks the knee, and in this state, all the major knee ligaments are twisted and taut. To unlock the knee, the *popliteus* muscle rotates the femur laterally and untwists the ligaments.

DEEPER INSIGHT 9.4

CLINICAL APPLICATION

Knee Injuries and Arthroscopic Surgery

Although the knee can bear a lot of weight, it is highly vulnerable to rotational and horizontal stress, especially when the knee is flexed (as in skiing or running) and receives a blow from behind or from the side. The most common injuries are to a meniscus or the anterior cruciate ligament (ACL). Lateral blows to the knee or body, especially when the foot is firmly planted on the ground, sometimes cause a so-called "unhappy triad" of injuries: tears in the tibial collateral ligament, medial meniscus, and anterior cruciate ligament **(fig. 9.29).** These three structures are strongly interconnected, so an injurious force is easily transferred from one to another. Knee injuries heal slowly because ligaments and tendons have a scanty blood supply and cartilage usually has no blood vessels at all.

The diagnosis and surgical treatment of knee injuries have been greatly improved by *arthroscopy,* a procedure in which the interior of a joint is viewed with a pencil-thin instrument, the *arthroscope,* inserted through a small incision. The arthroscope has a light, a lens, and fiber optics that allow a viewer to see into the cavity and take photographs or video recordings. A surgeon can also withdraw samples of synovial fluid by arthroscopy or inject saline into the joint cavity to expand it and provide a clearer view. If surgery is required, additional small incisions can be made for the surgical instruments and the procedures can be observed through the arthroscope or on a monitor. Arthroscopic surgery produces much less tissue damage than conventional surgery and enables patients to recover more quickly.

Orthopedic surgeons now often replace a damaged ACL with a graft from the patellar ligament or a hamstring tendon. The surgeon "harvests" a strip from the middle of the patient's ligament (or tendon), drills a hole into the femur and tibia within the joint cavity, threads the ligament through the holes, and fastens it with biodegradable screws. The grafted ligament is more taut and "competent" than the damaged ACL. It becomes ingrown with blood vessels and serves as a substrate for the deposition of more collagen, which further strengthens it in time. Following arthroscopic ACL reconstruction, a patient typically must use crutches for 7 to 10 days and undergo physical therapy for 6 to 10 weeks, followed by self-directed exercise therapy. Healing is completed in about 9 months.

FIGURE 9.29 The "Unhappy Triad" of Knee Injuries.

The knee joint has about 13 bursae. Four of these are anterior: the **superficial infrapatellar, suprapatellar, prepatellar,** and **deep infrapatellar.** Located in the popliteal region are the *popliteal bursa* and *semimembranosus bursa* (not illustrated). At least seven more bursae are found on the lateral and medial sides of the knee joint. From figure 9.28c, your knowledge of the relevant word elements *(infra-, supra-, pre-),* and the terms *superficial* and *deep,* you should be able to work out the reasoning behind most of these names and develop a system for remembering the locations of these bursae.

9.3f The Ankle Joint

The **talocrural**[29] **(ankle) joint** includes two articulations—a medial joint between the tibia and talus and a lateral joint between the fibula and talus, both enclosed in one joint capsule **(fig. 9.30).** The malleoli of the tibia and fibula overhang the talus on each side like a cap and prevent most side-to-side motion. The ankle therefore has a more restricted range of motion than the wrist.

---

[29]*talo* = ankle; *crural* = pertaining to the leg

Fibula
Tibia
Anterior and posterior tibiofibular ligaments
Calcaneal tendon
Calcaneus

Lateral ligament:
Posterior talofibular ligament
Calcaneofibular ligament
Anterior talofibular ligament
Metatarsal V
Tendons of fibularis longus and brevis

(a) Lateral view

Fibula
Tibia
Anterior tibiofibular ligament
Talus
Dorsal talonavicular ligament
Calcaneocuboid ligament
Calcaneus
Cuboid
Long plantar ligament

(b) Lateral dissection

Medial ligament
Navicular
Metatarsal I
Tendons of tibialis anterior and posterior

Tibia
Calcaneal tendon
Calcaneus

(c) Medial view

Tibia
Fibula
Interosseous membrane
Medial malleolus
Calcaneus

Posterior tibiofibular ligament
Lateral malleolus
Posterior talofibular ligament
Calcaneofibular ligament

(d) Posterior view

FIGURE 9.30 The Talocrural (Ankle) Joint and Ligaments of the Right Foot. (a) Lateral view. (b) Lateral ligaments in dissected cadaver ankle. (c) Medial view. (d) Posterior view. **APR**

b: Christine Eckel/McGraw-Hill Education

The ligaments of the ankle include (1) **anterior** and **posterior tibiofibular ligaments,** which bind the tibia to the fibula; (2) a multipart **medial (deltoid[30]) ligament,** which binds the tibia to the foot on the medial side; and (3) a multipart **lateral (collateral) ligament,** which binds the fibula to the foot on the lateral side. The **calcaneal (Achilles) tendon** extends from the calf muscles to the calcaneus. It plantarflexes the foot and limits dorsiflexion. Plantar flexion is limited by extensor tendons on the anterior side of the ankle and by the anterior part of the joint capsule.

Sprains (torn ligaments and tendons) are common at the ankle, especially when the foot is suddenly inverted or everted to excess. They are painful and usually accompanied by immediate swelling. They are best treated by immobilizing the joint and reducing swelling with an ice pack, but in extreme cases may require a cast or surgery. Sprains and other joint disorders are briefly described in **table 9.1.**

BEFORE YOU GO ON

Answer the following questions to test your understanding of the preceding section:

12. What keeps the mandibular condyle from slipping out of its fossa in a posterior direction?

13. Explain how the biceps brachii tendon braces the shoulder joint.

14. Identify the three joints found at the elbow and name the movements in which each joint is involved.

15. What keeps the femur from slipping backward off the tibia?

16. What keeps the tibia from slipping sideways off the talus?

[30]*delt* = triangular, Greek letter delta (Δ); *oid* = resembling

TABLE 9.1	Some Common Joint Disorders
Arthritis	Broad term embracing more than 100 types of joint rheumatism.
Bursitis	Inflammation of a bursa, usually due to overuse of a joint.
Dislocation	Displacement of a bone from its normal position at a joint, usually accompanied by a sprain of the adjoining connective tissues. Most common at the fingers, thumb, shoulder, and knee.
Gout	A hereditary disease, most common in men, in which uric acid crystals accumulate in the joints and irritate the articular cartilage and synovial membrane. Causes gouty arthritis, with swelling, pain, tissue degeneration, and sometimes fusion of the joint. Most commonly affects the great toe.
Rheumatism	Broad term for any pain in the supportive and locomotory organs of the body, including bones, ligaments, tendons, and muscles.
Sprain	Torn ligament or tendon, sometimes with damage to a meniscus or other cartilage.
Strain	Painful overstretching of a tendon or muscle without serious tissue damage. Often results from inadequate warm-up before exercise.
Synovitis	Inflammation of a joint capsule, often as a complication of a sprain.
Tendinitis	A form of bursitis in which a tendon sheath is inflamed.

You can find other joint disorders in the following places:

Temporomandibular joint (TMJ) dysfunction in Deeper Insight 9.2; *jaw dislocation* in section 9.3a; *shoulder dislocation* in Deeper Insight 9.3; *knee injuries* in Deeper Insight 9.4; *arthritis* in Deeper Insight 9.5; *hip dislocation* in section 9.3d; and *rotator cuff injury* in Deeper Insight 10.4.

DEEPER INSIGHT 9.5

CLINICAL APPLICATION

Arthritis and Artificial Joints

Arthritis[31] is a broad term for pain and inflammation of a joint and embraces more than a hundred different diseases of largely obscure or unknown causes. In all of its forms, it is the most common crippling disease in the United States; nearly everyone past middle age develops arthritis to some degree. Physicians who treat arthritis and other joint disorders are called *rheumatologists*.

The most common form of arthritis is *osteoarthritis (OA)*, also called "wear-and-tear arthritis" because it is apparently a normal consequence of years of wear on the joints. As joints age, the articular cartilages soften and degenerate. As the cartilage becomes roughened by wear, joint movement may be accompanied by crunching or crackling sounds called *crepitus*. OA affects especially the fingers, intervertebral joints, hips, and knees. As the articular cartilage wears away, exposed bone tissue often develops spurs that grow into the joint cavity, restrict movement, and cause pain. OA rarely occurs before age 40, but it affects about 85% of people older than 70, especially those who are overweight. It usually doesn't cripple, but in severe cases it can immobilize the hip.

Rheumatoid arthritis (RA), which is far more severe than osteoarthritis, results from an autoimmune attack against the joint tissues. It begins when the body produces antibodies to fight an infection. Failing to recognize the body's own tissues, a misguided antibody known as *rheumatoid factor* also attacks the synovial membranes. Inflammatory cells accumulate in the synovial fluid and produce enzymes that degrade the articular cartilage. The synovial membrane thickens and adheres to the articular cartilage, fluid accumulates in the joint capsule, and the capsule is invaded by fibrous connective tissue. As articular cartilage degenerates, the joint begins to ossify, and sometimes the bones become solidly fused and immobilized, a condition called *ankylosis*[32] **(fig. 9.31).** The disease tends to develop symmetrically—if the right wrist or hip develops RA, so does the left.

Rheumatoid arthritis is named for the fact that symptoms tend to flare up and subside (go into remission) periodically.[33] It affects women far more often than men, and because RA typically begins as early as age 30 to 40, it can cause decades of pain and disability. There is no cure, but joint damage can be slowed with hydrocortisone or other steroids. Because long-term use of steroids weakens the bone, however, aspirin is the treatment of first choice to control the inflammation. Physical therapy is also used to preserve the joint's range of motion and the patient's functional ability.

Arthroplasty,[34] a treatment of last resort, is the replacement of a diseased joint with an artificial device called a *joint prosthesis*.[35] Joint prostheses were first developed to treat injuries in World War II and the Korean War. Total hip replacement (THR), first performed in 1963 by English orthopedic surgeon Sir John Charnley, is now the most common orthopedic procedure for the elderly. The first knee replacements were performed in the 1970s. Joint prostheses are now available for finger, shoulder, and elbow joints, as well as the hip and knee. Arthroplasty is performed on over 250,000 patients per year in the United States, primarily to relieve pain and restore function in elderly people with OA or RA.

---

[31]*arthr* = joint; *itis* = inflammation

[32]*ankyl* = bent, crooked; *osis* = condition
[33]*rheumat* = tending to change
[34]*arthro* = joint; *plasty* = surgical repair
[35]*prosthe* = something added

(a)

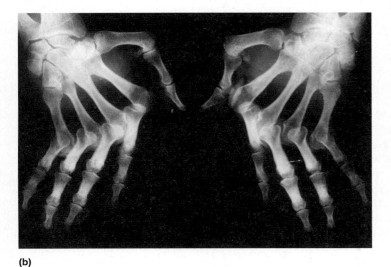

(b)

FIGURE 9.31 Rheumatoid Arthritis (RA). (a) A severe case with ankylosis of the joints. (b) X-ray of severe RA of the hands.

a: chaowalit407/iStock/Getty Images; b: Clinical Photography, Central Manchester University Hospitals NHS Foundation Trust, UK/Science Source

Arthroplasty presents ongoing challenges for biomedical engineering. An effective prosthesis must be strong, nontoxic, and corrosion-resistant. In addition, it must bond firmly to the patient's bones and enable a normal range of motion with a minimum of friction. The heads of long bones are usually replaced with prostheses made of very hard ceramics such as titanium carbide, cobalt–chromium, or other metal alloys. Joint sockets are made of polyethylene **(fig. 9.32).** Prostheses are bonded to the patient's bone with screws or bone cement.

Improvements in technology have resulted in long-lasting prostheses. Over 75% of artificial knees last 20 years, nearly 85% last 15 years, and over 90% last 10 years. The most common form of failure is detachment of the prosthesis from the bone. This problem has been reduced by using *porous-coated prostheses,* which become infiltrated by the patient's own bone and create a firmer bond. A prosthesis isn't as strong as a natural joint, however, and is not an option for many young, active patients.

Arthroplasty has been greatly improved by *computer-assisted design and manufacture (CAD/CAM).* A computer scans X-rays from the patient and presents several design possibilities for review. Once a design is selected, the computer generates a program to operate the machinery that produces the prosthesis. CAD/CAM has reduced the waiting period for a prosthesis from 12 weeks to about 2 weeks and has lowered the cost dramatically.

Prosthesis

(a)

Femur

Tibia

Fibula

(b)

FIGURE 9.32 Joint Prostheses. (a) Knee replacement surgery. (b) Knee prostheses bonded to natural bone of the femur and tibia. Compare to the X-ray on the opening page of this chapter.

a: Samrith Na Lumpoon/Shutterstock; b: Ron Mensching/Medical Images RM

STUDY GUIDE

▶ Assess Your Learning Outcomes

To test your knowledge, discuss the following topics with a study partner or in writing, ideally from memory.

9.1 Joints and Their Classification

1. The fundamental definition of *joint (articulation)* and why it cannot be defined as a point at which one bone moves relative to an adjacent bone
2. Relationships and differences between the sciences of arthrology, kinesiology, and biomechanics
3. The typical system for naming most joints after the bones they involve; examples of this
4. Basic criteria for classifying joints
5. Characteristics and examples of bony joints (synostoses)
6. Characteristics of fibrous joints (synarthroses) and each of their subclasses, with examples
7. Characteristics of cartilaginous joints (amphiarthroses) and each of their subclasses, with examples

9.2 Synovial Joints

1. The definition and anatomical features of a *synovial joint (diarthrosis),* examples of this type, and why this type is of greatest interest for kinesiology
2. General anatomy of tendons, ligaments, bursae, and tendon sheaths, and their contributions to joint function

3. Three essential components of a lever
4. The meaning of *mechanical advantage* (MA); how the *MA* of a lever can be determined from measurements of its effort and resistance arms; and the respective advantages of levers in which the *MA* is greater than or less than 1.0
5. Comparison of first-, second-, and third-class levers, and anatomical examples of each
6. Variables that determine a joint's range of motion *(ROM),* and the clinical relevance of *ROM*
7. Axes of rotation and degrees of freedom in joint movement, and how this relates to the classification of joints as monaxial, biaxial, or multiaxial
8. Six kinds of synovial joints; how each is classified as monaxial, biaxial, or multiaxial; imperfections in this classification; and examples of each type in the body
9. The concept of *zero position* and how it relates to the description of joint function
10. Examples of each of the following limb movements, including an ability to describe or demonstrate them: flexion, extension, hyperextension, abduction, adduction, hyperabduction, hyperadduction, circumduction, medial rotation, and lateral rotation
11. The same for supination, pronation, ulnar flexion, and radial flexion of the forearm and hand, and opposition, reposition, abduction, and adduction of the thumb

12. The same for flexion, extension, hyperextension, and lateral flexion of the spine, and right and left rotation of the trunk
13. The same for elevation, depression, protraction, retraction, and lateral and medial excursion of the mandible
14. The same for dorsiflexion, plantar flexion, inversion, eversion, pronation, and supination of the foot

9.3 Anatomy of Selected Diarthroses

1. Features of the jaw (temporomandibular) joint including the mandibular condyle, mandibular fossa, synovial cavity, articular disc, and principal ligaments
2. Features of the shoulder (glenohumeral) joint including the humeral head, glenoid cavity and labrum, five major ligaments and four bursae, and tendons of the biceps brachii and four rotator cuff muscles
3. Features of the elbow; the three joints that occur here; the olecranon bursa and four major ligaments
4. Features of the hip (coxal) joint including the femoral head, fovea capitis, acetabulum and labrum, and five principal ligaments
5. Features of the knee (tibiofemoral and patellofemoral joints), including the menisci, cruciate and other ligaments, and four major bursae around the patella
6. Features of the ankle (talocrural) joint, including the malleoli, calcaneal tendon, and major ligaments

▶ Testing Your Recall

Answers in Appendix A

1. Internal and external rotation of the humerus is made possible by a _____ joint.
   a. pivot
   b. condylar
   c. ball-and-socket
   d. saddle
   e. hinge

2. Which of the following is the least movable?
   a. a diarthrosis
   b. a synostosis
   c. a symphysis
   d. a synovial joint
   e. a condylar joint

3. Which of the following movements are unique to the foot?
   a. dorsiflexion and inversion
   b. elevation and depression
   c. circumduction and rotation
   d. abduction and adduction
   e. opposition and reposition

STUDY GUIDE

4. Which of the following joints cannot be circumducted?
   a. carpometacarpal
   b. metacarpophalangeal
   c. glenohumeral
   d. coxal
   e. interphalangeal

5. Which of the following terms denotes a general condition that includes the other four?
   a. gout
   b. arthritis
   c. rheumatism
   d. osteoarthritis
   e. rheumatoid arthritis

6. In the adult, the ischium and pubis are united by
   a. a synchondrosis.
   b. a diarthrosis.
   c. a synostosis.
   d. an amphiarthrosis.
   e. a symphysis.

7. In a second-class lever, the effort
   a. is applied to the end opposite the fulcrum.
   b. is applied to the fulcrum itself.
   c. is applied between the fulcrum and resistance.
   d. always produces an *MA* less than 1.0.
   e. is applied on one side of the fulcrum to move a resistance on the other side.

8. Which of the following joints has anterior and posterior cruciate ligaments?
   a. the shoulder
   b. the elbow
   c. the hip
   d. the knee
   e. the ankle

9. To bend backward at the waist involves _____ of the vertebral column.
   a. rotation
   b. hyperextension
   c. dorsiflexion
   d. abduction
   e. flexion

10. The rotator cuff includes the tendons of all of the following muscles *except*
    a. the subscapularis.
    b. the supraspinatus.
    c. the infraspinatus.
    d. the biceps brachii.
    e. the teres minor.

11. The lubricant of a diarthrosis is called _____.

12. A fluid-filled sac that eases the movement of a tendon over a bone is called a/an _____.

13. A _____ joint allows one bone to swivel on another.

14. _____ is the science of movement.

15. The joint between a tooth and the mandible is called a/an _____.

16. In a _____ suture, the articulating bones have interlocking wavy margins, somewhat like a dovetail joint in carpentry.

17. In kicking a football, what type of action does the knee joint exhibit?

18. The angle through which a joint can move is called its _____.

19. The menisci of the knee are functionally similar to the _____ of the temporomandibular joint.

20. At the ankle, both the tibia and fibula articulate with what tarsal bone?

▶ Building Your Medical Vocabulary

Answers in Appendix A

State a meaning of each word element, and give a medical term from this chapter that uses it or a slight variation of it.

1. ab-
2. arthro-
3. -ate
4. cruci-
5. cruro-
6. -duc
7. kinesio-
8. men-
9. supin-
10. -trac

▶ What's Wrong with These Statements?

Answers in Appendix A

Briefly explain why each of the following statements is false, or reword it to make it true.

1. More people get rheumatoid arthritis than osteoarthritis.

2. A doctor who treats arthritis is called a kinesiologist.

3. Synovial joints are also known as synarthroses.

4. Menisci occur in the elbow and knee joints.

5. Reaching behind you to take something out of your hip pocket involves flexion of the shoulder.

6. The cruciate ligaments are in the feet.

7. The femur is held tightly in the acetabulum mainly by the round ligament.

8. The knuckles are amphiarthroses.

9. Synovial fluid is secreted by the bursae.

10. Like most ligaments, the periodontal ligaments attach one bone (the tooth) to another (the mandible or maxilla).

STUDY GUIDE

▶ Testing Your Comprehension

1. All second-class levers produce a mechanical advantage greater than 1.0 and all third-class levers produce a mechanical advantage less than 1.0. Explain why.

2. For each of the following joint movements, state what bone the axis of rotation passes through and which of the three anatomical planes contains the axis of rotation. You may find it helpful to produce some of these actions on an articulated laboratory skeleton so you can more easily visualize the axis of rotation. (a) Plantar flexion; (b) flexion of the hip; (c) adduction of the thigh; (d) flexion of the knee; (e) flexion of the first interphalangeal joint of the index finger. (Do not bend the fingers of a wired laboratory skeletal hand, because they can break off.)

3. In order of occurrence, list the joint actions (flexion, pronation, etc.) and the joints where they would occur as you (a) sit down at a table, (b) reach out and pick up an apple, (c) take a bite, and (d) chew it. Assume that you start in anatomical position.

4. The deltoid muscle inserts on the deltoid tuberosity of the humerus and abducts the arm. Imagine a person holding a weight in the hand and abducting the arm. On a laboratory skeleton, identify the fulcrum; measure the effort arm and resistance arm; determine the mechanical advantage of this movement; and determine which of the three lever types the upper limb acts as when performing this movement.

5. List the six types of synovial joints, and for each one, if possible, identify a joint in the upper limb and a joint in the lower limb that fall into each category. Which of these six joints has/have no examples in the lower limb?

THE MUSCULAR SYSTEM

Colorized MRI scan showing muscles of the lumbar, pelvic, and upper femoral regions

Simon Fraser/Science Source

CHAPTER OUTLINE

10.1 Structural and Functional Organization of Muscles

10.1a The Functions of Muscles
10.1b Muscle Connective Tissues, Fascicles, and Compartments
10.1c Muscle Attachments
10.1d Functional Groups of Muscles
10.1e Innervation and Blood Supply
10.1f Muscle Names and Learning Strategy

10.2 Muscles of the Head and Neck

10.2a Muscles of Facial Expression
10.2b Muscles of Chewing and Swallowing
10.2c Muscles Acting on the Head and Neck

10.3 Muscles of the Trunk

10.3a Muscles of Respiration
10.3b Muscles of the Abdominal Wall
10.3c Muscles of the Back
10.3d Muscles of the Pelvic Floor

10.4 Muscles Acting on the Shoulder and Upper Limb

10.4a Muscles Acting on the Shoulder
10.4b Muscles Acting on the Arm
10.4c Muscles Acting on the Elbow and Forearm
10.4d Muscles Acting on the Wrist and Hand
10.4e Intrinsic Muscles of the Hand

10.5 Muscles Acting on the Hip and Lower Limb

10.5a Muscles Acting on the Hip and Thigh
10.5b Muscles Acting on the Knee and Leg
10.5c Muscles Acting on the Ankle and Foot
10.5d Intrinsic Muscles of the Foot

Study Guide

DEEPER INSIGHTS

10.1 Clinical Application: Compartment Syndrome

10.2 Clinical Application: Heavy Lifting and Back Injuries

10.3 Clinical Application: Hernias

10.4 Clinical Application: Rotator Cuff Injury

10.5 Clinical Application: Carpal Tunnel Syndrome

10.6 Clinical Application: Shinsplints

10.7 Clinical Application: Calcaneal Tendon Rupture

10.8 Clinical Application: Plantar Fasciitis

10.9 Clinical Application: Common Athletic Injuries

Anatomy & Physiology *Revealed* 4.0

Module 6: Muscular System

- Understanding of the skeletal muscles depends on a thorough knowledge of skeletal anatomy (chapter 8), including not just the bones but also their surface features (table 8.2), many of which are muscle attachments.

- This chapter describes the movements produced by muscles, called their *actions*, using the terminology of joint movements in section 9.2c.

Muscles constitute nearly half of the body's weight and occupy a place of central interest in several fields of health care and fitness. Physical and occupational therapists must be well acquainted with the muscular system to plan and carry out rehabilitation programs. Athletes and trainers, dancers and acrobats, and amateur fitness enthusiasts follow programs of resistance training to strengthen individual muscle groups through movement regimens based on knowledge of muscle, bone, and joint anatomy. Nurses employ their knowledge of the muscular system to give intramuscular injections correctly and to safely and effectively move patients who are physically incapacitated. Gerontological nurses are keenly aware of how deeply a person's muscular condition affects the quality of life in old age. The muscular system is highly important to biomedical disciplines even beyond the scope of the movement sciences. For example, it is the primary source of body heat in the moving individual, and loss of muscle mass can be a contributing factor in diabetes mellitus.

The muscular system is closely related to what we have covered in the preceding chapters. After this, we will examine the mechanisms of muscle contraction at the cellular and molecular levels in chapter 11, and chapter 12 will shed light on the relationship of the muscles to the nerves that control them.

10.1 Structural and Functional Organization of Muscles

Expected Learning Outcomes

When you have completed this section, you should be able to

a. describe the various functions of muscular tissue;

b. describe the connective tissue components of a muscle and their relationship to the internal organization of a muscle and compartmentalization of muscle groups;

c. relate muscle fascicles to the shapes and relative strengths of muscles;

d. name the types of muscle–bone attachments and explain the shortcoming of calling their attachments *origins* and *insertions*;

e. distinguish between intrinsic and extrinsic muscles;

f. describe the ways muscles work in groups to aid, oppose, and moderate each other's actions;

g. describe in general terms the nerve and blood supply to skeletal muscles; and

h. explain how the Latin names of muscles aid in visualizing and remembering them.

The word **muscle**[1] means "little mouse," and apparently was coined by one of the ancient Greek authorities who thought that skeletal muscles rippling under the skin resembled scurrying mice. As we saw in chapter 5, however, there are three kinds of muscle—skeletal, cardiac, and smooth. Despite their differences, all muscle is specialized for one fundamental purpose: to convert the chemical energy of ATP into the mechanical energy of motion. Muscle cells exert force on other tissues and organs, either to produce desirable movements or to prevent undesirable ones.

10.1a The Functions of Muscles

Collectively, the three types of muscle serve the following functions:

- **Movement.** Muscles enable us to move from place to place and to move individual body parts; they move body contents in the course of breathing, blood circulation, feeding and digestion, defecation, urination, and childbirth; and they serve various roles in communication—speech, writing, facial expressions, and other body language.

- **Stability.** Muscles maintain posture by preventing unwanted movements. Some are called *antigravity muscles* because, at least part of the time, they resist the pull of gravity and prevent us from falling or slumping over. Many muscles also stabilize the joints by maintaining tension on tendons and bones.

- **Control of body openings and passages.** Muscles encircling the mouth serve not only for speech but also for food intake and retention of food while chewing. In the eyelid and pupil, they regulate the admission of light to the eye. Internal muscular rings control the movement of food, bile, blood, and other materials within the body. Muscles encircling the urethra and anus control the elimination of waste. (Some of these muscles are called *sphincters,* but not all; this is clarified later.)

- **Heat production (thermogenesis).** The skeletal muscles produce 20% to 30% of the body's heat at rest and up to 85% during exercise. This body heat is vital to the functioning of enzymes and therefore to all metabolism.

- **Glycemic control.** This means the regulation of blood glucose concentration within its normal range. The skeletal muscles absorb, store, and use a large share of one's glucose and play a highly significant role in stabilizing its blood concentration. In old age, in obesity, and when muscles become deconditioned and weakened, people suffer an increased risk of type 2 diabetes mellitus because of the decline in this glucose-buffering function.

The rest of this chapter concerns only the skeletal muscles. The term **muscular system** refers only to these, not the other two types of muscle. The study of skeletal muscles is called **myology.**[2]

---

[1]*mus* = mouse; *cle* = little
[2]*myo* = muscle; *logy* = study of

There are about 600 muscles in the human muscular system, but fortunately you aren't expected to learn 600 names! Many have repetitious names, such as the right and left muscles of the same name and muscle series between ribs and vertebrae. This chapter describes fewer than one-third of the muscles.

10.1b Muscle Connective Tissues, Fascicles, and Compartments

A skeletal muscle consists of more than muscular tissue. It also includes connective tissue, nerves, and blood vessels. The connective tissue components, from the smallest to largest and from deep to superficial, are as follows (**fig. 10.1**):

- **Endomysium**[3] (EN-doe-MIZ-ee-um). This is a thin sleeve of loose connective tissue that surrounds each muscle fiber. It creates room for blood capillaries and nerve fibers to reach every muscle fiber, ensuring that no muscle cell is without stimulation and nourishment. The endomysium also provides the extracellular chemical environment for the muscle fiber and its associated nerve ending. Excitation of a muscle fiber is based on the exchange of calcium, sodium, and potassium ions between the endomysial tissue fluid and the nerve and muscle fibers.

[3]*endo* = within; *mys* = muscle

(a)

(b)

(c)

FIGURE 10.1 Connective Tissues of a Muscle. (a) The muscle–bone attachment and subdivisions of a skeletal muscle. (b) A cross section of the thigh showing the relationship of neighboring muscles to fasciae and bone. (c) Muscle fascicles in the tongue. Vertical fascicles passing between the superior and inferior surfaces of the tongue are seen alternating with cross-sectioned horizontal fascicles that pass from the rear to the tip of the tongue. A fibrous perimysium can be seen between the fascicles, and endomysium can be seen between the muscle fibers within each fascicle. (c.s. = cross section; l.s. = longitudinal section)

c: Victor Eroschenko

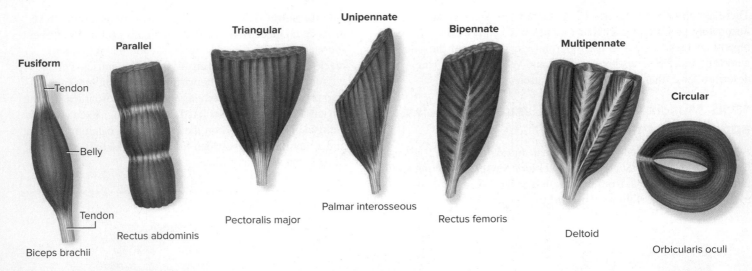

FIGURE 10.2 Classification of Muscles According to Fascicle Orientation. The fascicles are the grain visible in each illustration. Muscle types are named across the top and an example of each type across the bottom.

- **Perimysium.**[4] This is a thicker connective tissue sheath that wraps muscle fibers together in bundles called **fascicles**[5] (FASS-ih-culs). Fascicles are visible to the naked eye as parallel strands—the grain in a cut of meat; if you pull apart "fork-tender" roast beef, it separates along these fascicles. The perimysium carries the larger nerves and blood vessels as well as stretch receptors called muscle spindles.

- **Epimysium.**[6] This is a fibrous sheath that surrounds the entire muscle. On its outer surface, the epimysium grades into the fascia, and its inner surface issues projections between the fascicles to form the perimysium.

- **Fascia**[7] (FASH-ee-uh). This is a sheet of connective tissue that separates neighboring muscles or muscle groups from each other and from the subcutaneous tissue. Muscles are grouped in *compartments* separated from each other by fasciae.

The *fascicles* defined by the perimysium are oriented in a variety of ways that determine the strength of a muscle and the direction in which it pulls. According to fascicle orientation, muscles are classified as shown in **figure 10.2.**

- **Fusiform**[8] muscles are thick in the middle and tapered at each end. The *biceps brachii* of the arm and *gastrocnemius* of the calf are examples of this type. Muscle strength is proportional to the diameter of a muscle at its thickest point, and fusiform muscles are relatively strong.

- **Parallel muscles** have a fairly uniform width and parallel fascicles. Some of these are elongated straps, such as the *rectus abdominis* of the abdomen, *sartorius* of the thigh, and *zygomaticus major* of the face. Others are more squarish and are called *quadrilateral* (four-sided) muscles, such as the *masseter*

of the jaw. Parallel muscles can span long distances, such as from hip to knee, and they shorten more than other muscle types. However, having fewer muscle fibers than a fusiform muscle of the same mass, they produce less force.

- **Triangular (convergent) muscles** are fan-shaped—broad at one end and narrower at the other. Examples include the *pectoralis major* in the chest and the *temporalis* on the side of the head. Despite their small localized insertions on a bone, these muscles are relatively strong because they contain a large number of fibers in the wider part of the muscle.

- **Pennate**[9] muscles are feather-shaped. Their fascicles insert obliquely on a tendon that runs the length of the muscle, like the shaft of a feather. There are three types of pennate muscles: *unipennate,* in which all fascicles approach the tendon from one side (for example, the *palmar interosseous muscles* of the hand and *semimembranosus* of the thigh); *bipennate,* in which fascicles approach the tendon from both sides (for example, the *rectus femoris* of the thigh); and *multipennate,* shaped like a bunch of feathers with their quills converging on a single point (for example, the *deltoid* of the shoulder). These muscles generate more force than the preceding types because they fit more muscle fibers into a given length of muscle.

- **Circular muscles (sphincters)** form rings around certain body openings. When they contract, they constrict the opening and tend to prevent the passage of material through it. Examples include the *orbicularis oculi* of the eyelids and the *external urethral* and *anal sphincters.* Smooth muscle can also form sphincters—for example, the *pyloric valve* at the passage from the stomach to the small intestine and some sphincters of the urinary tract and anal canal.

The fasciae are not components of the muscles themselves, but package groups of functionally related muscles into **muscle compartments (fig. 10.3)** and stand between the muscles and the overlying

---

[4]*peri* = around; *mys* = muscle
[5]*fasc* = bundle; *icle* = little
[6]*epi* = upon, above; *mys* = muscle
[7]*fascia* = band
[8]*fusi* = spindle; *form* = shape

[9]*penna* = feather; *ate* = characterized by

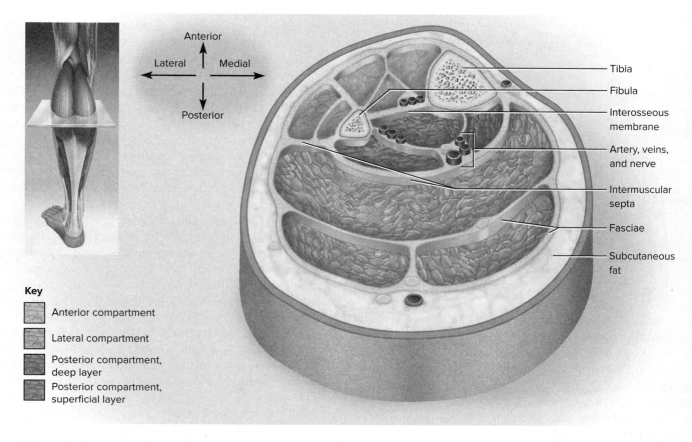

Key
- Anterior compartment
- Lateral compartment
- Posterior compartment, deep layer
- Posterior compartment, superficial layer

FIGURE 10.3 Muscle Compartments. A cross section of the left leg slightly above midcalf, oriented the same way as the reader's left leg.

hypodermis and skin. A compartment also contains the nerves and blood vessels that supply the muscle group. Such compartmentalization occurs in the thoracic and abdominal walls, pelvic floor, and limbs. Some of these fasciae are particularly thick and are called **intermuscular septa.** The tight binding of muscles by these fasciae contributes to a clinical problem described in Deeper Insight 10.1.

 DEEPER INSIGHT 10.1

CLINICAL APPLICATION

Compartment Syndrome

Muscle compartments are very snugly contained in their fasciae. If a blood vessel in a compartment is damaged by overuse or contusion (a bruising injury), blood and tissue fluid accumulate in the compartment. The inelastic fascia prevents the compartment from expanding to relieve the pressure. Mounting pressure on the muscles, nerves, and blood vessels triggers a sequence of degenerative events called *compartment syndrome*. Blood flow to the compartment is obstructed by pressure on its arteries. If *ischemia* (poor blood flow) persists for more than 2 to 4 hours, nerves begin to die, and after 6 hours, so does muscle tissue. Nerves can regenerate after the pressure is relieved, but muscle necrosis is irreversible. The breakdown of muscle releases myoglobin into the blood. *Myoglobinuria,* the presence of myoglobin in the urine, gives the urine a dark color and is one of the key signs of compartment syndrome and some other degenerative muscle disorders. Compartment syndrome is treated by immobilizing and resting the limb and, if necessary, making an incision *(fasciotomy)* to relieve the pressure.

10.1c Muscle Attachments

Connective tissue components of a muscle emerge from it as collagenous fibers that continue into its tendon or other bone attachment, then into the periosteum and matrix of the bone, creating very strong structural continuity from muscle to bone. Some muscles insert not on bones but on the fascia or tendon of another muscle or on collagen fibers of the dermis. The distal tendon of the biceps brachii, for example, inserts partly on the fascia of the forearm. Many muscles of the face insert in the skin of the lips, eyelids, and other areas, enabling them to produce expressions such as a smile.

Direct and Indirect Attachments

Muscles have two forms of attachment to bones—direct and indirect. In a **direct (fleshy) attachment,** such as in the *brachialis* and the lateral head of the *triceps brachii* in **figure 10.4,** there is so little separation between muscle and bone that to the naked eye, the red muscular tissue seems to emerge directly from the bone. At a microscopic level, however, the muscle fibers stop slightly short of the bone and the gap between muscle and bone is spanned by collagen fibers.

In an **indirect attachment,** the muscle ends visibly short of its bony destination, and the gap is bridged by a fibrous cord or band called a **tendon.** See, for example, the two ends of the biceps brachii in figure 10.4 and the photographs of tendons in figures 10.32b and 10.37. You can easily palpate tendons and feel their texture just above your heel (your *calcaneal* or *Achilles tendon*) and on the anterior side of your wrist (tendons of the *palmaris longus* and *flexor carpi radialis* muscles).

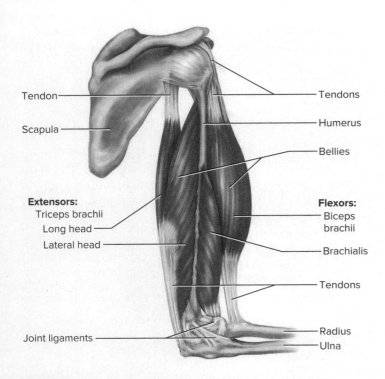

Extensors:
Triceps brachii
Long head
Lateral head

Tendon

Scapula

Joint ligaments

Tendons

Humerus

Bellies

Flexors:
Biceps brachii

Brachialis

Tendons

Radius
Ulna

FIGURE 10.4 Synergistic and Antagonistic Muscle Pairs. The biceps brachii and brachialis muscles are synergists in elbow flexion. The triceps brachii is an antagonist of those two muscles and is the prime mover in elbow extension.

? *Which of these muscles have direct attachments to the bones, and which have indirect attachments?*

In some cases, the tendon is a broad sheet called an **aponeurosis**[10] (AP-oh-new-RO-sis). This term originally referred to the tendon located beneath the scalp (hence the word root *neuro-*), but now it also refers to similar tendons associated with certain abdominal, lumbar, hand, and foot muscles. For example, the palmaris longus tendon passes through the wrist and then expands into a fanlike *palmar aponeurosis* beneath the skin of the palm (see fig. 10.29a).

In some places, groups of tendons from separate muscles pass under a band of connective tissue called a **retinaculum.**[11] One of these covers each surface of the wrist like a bracelet, for example. The tendons of several forearm muscles pass under them on their way to the hand (see figs. 5.13, 10.29a).

Moving and Stationary Attachments

Since most muscles span at least one joint and attach to a different bone at each end, their contraction moves one bone relative to the other. Traditionally, the attachment at the stationary end has been called the **origin** of the muscle and the attachment at the moving end has been called the **insertion.** *Gray's Anatomy* and other current authorities, however, have largely abandoned the *origin* and *insertion* terms because they are imperfect and sometimes misleading. There

are many cases in which the moving and nonmoving ends of the muscle are reversed when different actions are performed.

Consider the difference, for example, in the relative movements of the humerus and ulna when flexing the elbow to lift barbells as compared with flexing the elbow to perform chin-ups or scale a climbing wall. In weight lifting, the arm is relatively stationary and the forearm performs most of the motion, so the proximal end of the biceps would be considered the origin and its distal end the insertion. In chin-ups, by contrast, the forearm is more stationary and the arm moves more to lift the body, so the origin and insertion, if defined by relative motion, would be reversed.

Also consider the *quadriceps femoris* muscle on the anterior side of the thigh. It is a powerful extensor of the knee, connected at its proximal end mainly to the femur and at its distal end to the tibia, just below the knee. If you kick a soccer ball, the tibia moves more than the femur, so the tibia would be considered the insertion of the quadriceps and the femur would be considered its origin. But as you sit down in a chair, the femur moves more than the tibia, which remains relatively stationary. The quadriceps then acts as a brake so you don't sit down too abruptly and hard. By the foregoing definitions, the tibia would now be considered the origin of the quadriceps and the femur would be its insertion.

Some authorities therefore now speak of proximal and distal, medial and lateral, or superior and inferior muscle attachments, but no single system of naming applies throughout the body.

Even though *origin–insertion* terminology is becoming obsolete, it may still appear in some courses, board examinations, and other places. Therefore, for each muscle in tables 10.1 through 10.16, there are two bullet points (•) in the "Skeletal Attachments" column. The first bullet point denotes what has traditionally been called that muscle's origin, and the second one its insertion, enabling you to study the anatomy that way if your course so requires. These are aptly called skeletal attachments in the tables, because that applies to *nearly* all of them. Notwithstanding the table headers, though, be aware that a few of these muscles attach not to bone but to dermis, fasciae, or other soft connective tissues.

10.1d Functional Groups of Muscles

Consider the human hand. It requires subtle, finely controlled movement for such tasks as handwriting and keyboarding, but also powerful movements for such actions as climbing and tightly grasping tools. Muscles big enough to provide such power would be too bulky to fit within the hand itself, so they lie in the forearm and connect to the metacarpal bones and phalanges through long slender tendons that pass through the wrist. The more subtle movements are controlled by small muscles located entirely within the hand. Any muscle contained entirely within a region of interest is called an **intrinsic muscle.** A muscle that acts upon a designated organ or region such as the hand, but arises from another region such as the forearm, is called an **extrinsic muscle.** The intrinsic–extrinsic distinction applies not only to the hand but also to the muscles of the tongue, larynx, back, foot, and other regions. You will find several muscles classified this way in the tables in this chapter.

The effect produced by a muscle, whether it is to produce or prevent a movement, is called its **action.** Skeletal muscles seldom

---

[10]*apo* = upon, above; *neuro* = nervous system, brain
[11]*retinac* = retainer, bracelet; *cul* = little

act independently; instead, they function in groups whose combined actions produce the coordinated control of a joint. Muscles can be classified into four categories according to their actions, but it must be stressed that a particular muscle can act in a certain way during one joint action and in a different way during other actions of the same joint. Furthermore, the action of a given muscle depends on what other muscles are doing. For example, the *gastrocnemius* of the posterior calf usually flexes the knee, but if the *quadriceps* of the anterior thigh prevents knee flexion, the gastrocnemius flexes the ankle, causing plantar flexion. The following categories of muscle action are exemplified by muscles that act on the elbow (fig. 10.4):

1. The **prime mover (agonist)** is the muscle that produces most of the force during a particular joint action. In flexing the elbow, for example, the prime mover is the brachialis.

2. A **synergist**[12] (SIN-ur-jist) is a muscle that aids the prime mover. Two or more synergists acting on a joint can produce more power than a single larger muscle. The biceps brachii, for example, overlies the brachialis and works with it as a synergist to flex the elbow. The actions of a prime mover and its synergist aren't necessarily identical and redundant. If the prime mover worked alone at a joint, it could cause rotation or other undesirable movements of a bone. A synergist may stabilize a joint and restrict these movements, or modify the direction of a movement so that the action of the prime mover is more coordinated and specific.

3. An **antagonist**[13] is a muscle that opposes the prime mover. In some cases, it relaxes to give the prime mover almost complete control over an action. More often, however, the antagonist maintains some tension on a joint and thus limits the speed or range of the prime mover, preventing excessive movement, joint injury, or inappropriate actions. If you extend your arm to reach out and pick up a cup of tea, for example, your triceps brachii serves as the prime mover of elbow extension, and your brachialis acts as an antagonist to slow the extension and stop it at the appropriate point. If you extend your arm rapidly to throw a dart, however, the brachialis must be quite relaxed. The brachialis and triceps represent an **antagonistic pair** of muscles that act on opposite sides of a joint. We need antagonistic pairs at a joint because a muscle can only pull, not push—for example, a single muscle cannot flex *and* extend the elbow. Which member of the pair acts as the prime mover depends on the motion under consideration. In flexion of the elbow, the brachialis is the prime mover and the triceps is the antagonist; when the elbow is extended, their roles are reversed.

4. A **fixator** is a muscle that prevents a bone from moving. To *fix* a bone means to hold it steady, allowing another muscle attached to it to pull on something else. For example, consider again the flexion of the elbow by the biceps brachii. The biceps originates on the scapula, crosses both the shoulder and elbow joints, and inserts on the radius and forearm fascia. The scapula is loosely attached to the axial skeleton, so when the biceps contracts, it seems that it would pull the scapula

laterally. However, there are fixator muscles (the *rhomboids*) that attach the scapula to the vertebral column (see fig. 10.18). They contract at the same time as the biceps, holding the scapula firmly in place and ensuring that the force generated by the biceps moves the radius rather than the scapula.

10.1e Innervation and Blood Supply

A skeletal muscle cannot contract unless it is stimulated by a nerve fiber. If its nerve is severed, the muscle is paralyzed. The nerve that supplies a given muscle is called its **innervation.** Knowing the innervation to each muscle enables clinicians to diagnose nerve, spinal cord, and brainstem injuries from their effects on muscle function, and to set realistic goals for rehabilitation. The innervations described in this chapter will be more meaningful after you have studied the peripheral nervous system (chapters 13 and 14), but a brief orientation will be helpful here. The muscles are innervated by nerve branches that arise from these two groups:

- **Spinal nerves** arise from the spinal cord, emerge through the intervertebral foramina, and supply branches to muscles below the neck. Spinal nerves are identified by letters and numbers that refer to the adjacent vertebrae—for example, T6 for the sixth thoracic nerve and S2 for the second sacral nerve. Immediately after emerging from an intervertebral foramen, each spinal nerve branches into a *posterior* and *anterior ramus*. You will note references to nerve numbers and rami in many of the muscle tables. The term *plexus* in some of the tables refers to weblike networks of spinal nerves adjacent to the vertebral column. All of the spinal nerves named here are illustrated, and most are also discussed, in chapter 13 (see the four tables starting at table 13.3).

- **Cranial nerves** arise from the base of the brain, emerge through the skull foramina, and innervate muscles of the head and neck. Cranial nerves are identified by roman numerals (CN I to CN XII) and by names given in table 14.1, although not all 12 of them innervate skeletal muscles.

The muscular system has a very heavy demand for energy, obtained from ATP, and thus it receives a generous flow of blood delivering oxygen and organic nutrients. Even at rest, it receives about one-quarter of all the blood pumped by the heart, or about 1.24 L/min. During exercise, its share may rise as high as three-quarters of the cardiac output, or about 11.6 L/min. Blood capillaries branch extensively through the endomysium to reach every muscle fiber, sometimes so intimately associated with the muscle fibers that the fibers have surface indentations to accommodate them. The capillaries of skeletal muscle undulate or coil when the muscle is contracted, allowing them enough slack to stretch out straight, without breaking, when the muscle lengthens. Chapter 20 describes some special physiological properties of muscle circulation and names the major arteries that supply the skeletal muscle groups.

10.1f Muscle Names and Learning Strategy

Figures 10.5 and **10.6** show an overview of the major superficial muscles. Learning the names of these and other muscles may seem a forbidding task at first, especially when some of them have such

---

[12]*syn* = together; *erg* = work
[13]*ant* = against; *agonist* = actor, competitor

Superficial | Deep

Occipitofrontalis

Orbicularis oculi

Zygomaticus major

Masseter

Orbicularis oris

Sternocleidomastoid

Platysma

Trapezius

Pectoralis minor

Deltoid

Coracobrachialis

Pectoralis major

Serratus anterior

Brachialis

Biceps brachii

Rectus abdominis

Supinator

Flexor digitorum profundus

Brachioradialis

Flexor pollicis longus

Flexor carpi radialis

Transverse abdominal

External oblique

Internal oblique

Tensor fasciae latae

Pronator quadratus

Adductor longus

Adductors

Sartorius

Vastus lateralis

Rectus femoris

Vastus intermedius

Vastus lateralis

Gracilis

Vastus medialis

Fibularis longus

Gastrocnemius

Tibialis anterior

Soleus

Extensor digitorum longus

Extensor digitorum longus

FIGURE 10.5 The Muscular System, Anterior View. Major superficial muscles are shown on the anatomical right and some of the deeper muscles on the left. Muscles not labeled here are shown in more detail in later figures.

Deep | Superficial

Occipitofrontalis

Semispinalis capitis
Sternocleidomastoid
Splenius capitis
Levator scapulae
Supraspinatus
Rhomboid minor
Rhomboid major
Deltoid (cut)
Infraspinatus
Serratus anterior
Triceps brachii (cut)
Serratus posterior inferior
External oblique
Internal oblique
Erector spinae
Flexor carpi ulnaris
Extensor digitorum (cut)
Gluteus minimus
Lateral rotators
Adductor magnus
Iliotibial tract
Semimembranosus
Biceps femoris

Gastrocnemius (cut)
Soleus (cut)
Tibialis posterior
Flexor digitorum longus
Flexor hallucis longus
Fibularis longus
Calcaneal tendon

Trapezius

Infraspinatus
Teres minor
Teres major
Triceps brachii
Latissimus dorsi
Extensor carpi radialis longus and brevis
External oblique
Extensor digitorum
Gluteus medius
Extensor carpi ulnaris
Gluteus maximus

Gracilis
Semitendinosus
Iliotibial tract
Biceps femoris

Gastrocnemius

Soleus

FIGURE 10.6 The Muscular System, Posterior View. Major superficial muscles are shown on the right and some of the deeper muscles on the left. Muscles not labeled here are shown in more detail in later figures.

long Latin names as *depressor labii inferioris* and *flexor digiti minimi brevis*. Such names, however, typically describe some distinctive aspects of the structure, location, or action of a muscle, and become very helpful once we grow familiar with a few common Latin words. For example, the depressor labii inferioris is a muscle that lowers (depresses) the bottom (inferior) lip (labium), and the flexor digiti minimi brevis is a short (brevis) muscle that flexes the smallest (minimi) finger (digit). Muscle names are interpreted in footnotes throughout the chapter. Familiarity with these terms and attention to the footnotes will help you translate muscle names and remember the location, appearance, and action of the muscles.

In the remainder of this chapter, we consider about 160 muscles; most courses cover considerably fewer according to the choices of individual instructors. The following suggestions may help you develop a rational strategy for learning the muscular system:

- Examine models, cadavers, dissected animals, or a photographic atlas as you read about these muscles. Visual images are often easier to remember than words, and direct observation of a muscle may stick in your memory better than descriptive text or two-dimensional drawings.

- When studying a particular muscle, palpate it on yourself if possible. Contract the muscle to feel it bulge and sense its action. This makes muscle locations and actions less abstract. Atlas B, following this chapter, shows where you can see and palpate several of these muscles on the living body.

- Locate the attachments of muscles on an articulated skeleton. Some skeletons and models are painted and labeled to show these. This helps you visualize the locations of muscles and understand how they produce particular joint actions.

- Study the derivation of each muscle name; the name usually describes the muscle's location, appearance, attachments, or action.

- Say the names aloud to yourself or a study partner. It is harder to remember and spell terms you cannot pronounce, and silent pronunciation is not nearly as effective as speaking and hearing the names. Pronunciation guides based on the leading medical dictionaries are provided in the following muscle tables for all but the most obvious cases.

BEFORE YOU GO ON

Answer the following questions to test your understanding of the preceding section:

1. List some functions of the muscular system other than movement of the body.

2. Describe the relationship of endomysium, perimysium, and epimysium to each other. Which of these separates one fascicle from another? Which separates one muscle from another?

3. Distinguish between direct and indirect muscle attachments to bones.

4. Define *belly, action,* and *innervation.*

5. Describe the five basic muscle shapes (fascicle arrangements).

6. Distinguish between a synergist, antagonist, and fixator. Explain how each of these may affect the action of a prime mover.

10.2 Muscles of the Head and Neck

Expected Learning Outcomes

When you have completed this section, you should be able to

a. name and locate the muscles that produce facial expressions;

b. name and locate the muscles used for chewing and swallowing;

c. name and locate the neck muscles that move the head; and

d. identify the attachments, action, and innervation of these muscles.

The rest of this chapter consists largely of tables of the body's principal skeletal muscles. You will find a description and illustrations of each muscle group, with some generalizations and special-interest points about the group, then a table that gives the details:

- the name of each muscle in that group;

- the pronunciation of the name, unless it is self-evident or repeats words whose pronunciation has been provided in a recent entry;

- the actions of the muscle;

- the muscle's bony or soft-tissue attachments; and

- the nerve supply to that muscle.

We begin our survey of the muscular system with the head and neck (**fig. 10.7**). The first three sections divide these into muscles of facial expression, muscles of chewing and swallowing, and muscles that move the head as a whole.

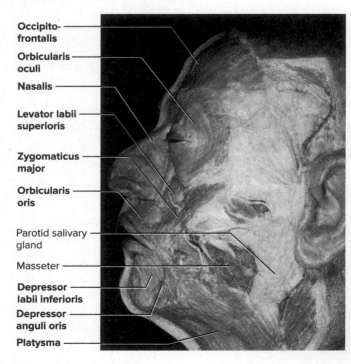

Occipito-frontalis
Orbicularis oculi
Nasalis
Levator labii superioris
Zygomaticus major
Orbicularis oris
Parotid salivary gland
Masseter
Depressor labii inferioris
Depressor anguli oris
Platysma

FIGURE 10.7 Some Facial Muscles of the Cadaver. Boldface labels indicate muscles employed in facial expression.

Rebecca Gray/McGraw-Hill Education

10.2a Muscles of Facial Expression

Humans have much more expressive faces than other mammals because of a complex array of muscles that insert in the dermis and subcutaneous tissues. These muscles tense the skin and produce such expressions as a pleasant smile, a flirtatious wink, a puzzled frown, or a threatening scowl (**fig. 10.8**). They add subtle shades of meaning to our spoken words. Facial muscles also contribute directly to speech, chewing, and other oral functions. All but one of them are innervated by the facial nerve (cranial nerve VII). This nerve is especially vulnerable to injury from lacerations and skull fractures, which can paralyze the muscles and cause parts of the face to sag. The only muscle not innervated by CN VII is the levator palpebrae superioris of the upper eyelid, innervated by the oculomotor nerve (CN III). **Table 10.1** groups these muscles into the following regions: the scalp; orbital and nasal regions (eye and nose); the oral region (encircling the mouth); and mental and buccal regions (chin and cheeks).

The Scalp

The **occipitofrontalis muscle** overlies the dome of the cranium. It is divided into a *frontal belly* in the forehead and *occipital belly* at the rear of the head (**fig. 10.9**), named for the underlying cranial bones. They are connected to each other by a broad aponeurosis, the **galea aponeurotica**[14] (GAY-lee-uh AP-oh-new-ROT-ih-cuh).

The Orbital and Nasal Regions

The **orbicularis oculi** is a sphincter of the eyelid that encircles and closes the eye. Deep to it, in the eyelid and orbit, is the **levator palpebrae superioris,** which opens the eye. Other muscles in this group move the eyelids and skin of the forehead and dilate the nostrils. Muscles within the orbit that move the eyeball itself are discussed in relation to vision in chapter 16.

---

[14]*galea* = helmet; *apo* = above; *neuro* = nervous system, brain

FIGURE 10.8 Expressions Produced by Several of the Facial Muscles. The ordinary actions of these muscles are usually more subtle than these demonstrations.

Joe DeGrandis/McGraw-Hill Education

❓ *Name an antagonist of each of these muscles: the depressor anguli oris, orbicularis oculi, and levator labii superioris.*

FIGURE 10.9 Muscles of Facial Expression.
(a) Anterior view. (b) Lateral view. Boldface
labels indicate muscles employed in facial
expression. **APR**

Superficial | Deep
← | →

Galea aponeurotica

**Frontal belly of
occipitofrontalis**

Corrugator supercilii

Orbicularis oculi

Nasalis

**Levator labii
superioris**

Levator anguli oris

Zygomaticus minor

Masseter

Zygomaticus major

Buccinator

Risorius

Modiolus

Orbicularis oris

**Depressor anguli
oris**

Mentalis (cut)

**Depressor labii
inferioris**

Platysma

(a) Anterior view

Galea aponeurotica

**Frontal belly of
occipitofrontalis**

Temporalis

Corrugator supercilii

Orbicularis oculi

**Occipital belly of
occipitofrontalis**

Nasalis

Levator labii superioris

Zygomatic arch

Zygomaticus minor

Zygomaticus major

Orbicularis oris

Masseter

Sternocleidomastoid

Levator scapulae

Modiolus

Inferior pharyngeal
constrictor

Risorius (cut)

Thyrohyoid

Mentalis

Sternothyroid

**Depressor labii
inferioris**

Omohyoid

Depressor anguli oris

Sternohyoid

(b) Lateral view

Buccinator

The Oral Region

The mouth is the most expressive part of the face, and lip movements are necessary for intelligible speech; thus, it's not surprising that the muscles here are especially diverse. The **orbicularis oris** is a complex of muscles in the lips that encircles the mouth; until recently it was misinterpreted as a sphincter, or circular muscle, but it is actually composed of four independent quadrants that interlace and give only an appearance of circularity. Other muscles in this region approach the lips from all directions and thus draw the lips or angles (corners) of the mouth upward, laterally, and downward. Some of these arise from a complex cord called the **modiolus**[15] just lateral to each angle of the lips (fig. 10.9b). Named for the hub of a cartwheel, the modiolus is a point of convergence of several muscles of the lower face. You can palpate it by inserting one finger just inside the corner of your lips and pinching the corner between the finger and thumb, feeling for a thick knot of tissue.

The Mental and Buccal Regions

Adjacent to the oral orifice are the *mental region* (chin) and *buccal region* (cheek). In addition to muscles already discussed that act on the lower lip, the mental region has a pair of small **mentalis muscles** extending from the upper margin of the mandible to the skin of the chin. In some people, they are especially thick and have a visible dimple between them called the mental cleft (see fig. 4.17). The **buccinator** is the muscle of the cheek. It has multiple functions in chewing, sucking, and blowing. If the cheeks are inflated with air, compression of the buccinators blows it out. Sucking is achieved by contracting the buccinators to draw the cheeks inward, then relaxing them. This action is especially important in nursing infants. To feel this action, hold your fingertips lightly against your cheeks as you make a kissing noise. You will notice relaxation of the buccinators at the moment air is sharply drawn in through the pursed lips. The **platysma** is a thin superficial muscle of the upper chest and lower face. It is relatively unimportant, but when men shave they tend to tense the platysma to make the concavity between the jaw and neck shallower and the skin tauter.

TABLE 10.1	Muscles of Facial Expression		
Name	**Action**	**Skeletal Attachments**	**Innervation**
The Scalp			
Occipitofrontalis, frontal belly (oc-SIP-ih-toe-frun-TAY-lis)	Elevates eyebrows in glancing upward and expressions of surprise or fright; draws scalp forward and wrinkles skin of forehead	• Galea aponeurotica • Subcutaneous tissue of eyebrows	Facial nerve
Occipitofrontalis, occipital belly	Retracts scalp; fixes galea aponeurotica so frontal belly can act on eyebrows	• Superior nuchal line and temporal bone • Galea aponeurotica	Facial nerve
The Orbital and Nasal Regions			
Orbicularis oculi[16] (or-BIC-you-LERR-is OC-you-lye)	Sphincter of the eyelids; closes eye in blinking, squinting, and sleep; aids in flow of tears across eye	• Lacrimal bone; adjacent regions of frontal bone and maxilla; medial angle of eyelids • Upper and lower eyelids; skin around margin of orbit	Facial nerve
Levator palpebrae superioris[17] (leh-VAY-tur pal-PEE-bree soo-PEER-ee-OR-is)	Elevates upper eyelid; opens eye	• Lesser wing of sphenoid in posterior wall of orbit • Upper eyelid	Oculomotor nerve
Corrugator supercilii[18] (COR-oo-GAY-tur SOO-per-SIL-ee-eye)	Draws eyebrows medially and downward in frowning and concentration; reduces glare of bright sunlight	• Medial end of supraorbital margin • Skin of eyebrow	Facial nerve
Nasalis[19] (nay-ZAY-lis)	Widens nostrils; narrows internal air passage between vestibule and nasal cavity	• Maxilla just lateral to nose • Bridge and alar cartilages of nose	Facial nerve
The Oral Region			
Orbicularis oris[20] (or-BIC-you-LERR-is OR-is)	Encircles mouth, closes lips, protrudes lips as in kissing; uniquely developed in humans for speech	• Modiolus of mouth • Submucosa and dermis of lips	Facial nerve

(continued)

---

[15]*modiolus* = hub
[16]*orb* = circle; *ocul* = eye
[17]*levator* = that which raises; *palpebr* = eyelid; *superior* = upper

[18]*corrug* = wrinkle; *supercilii* = of the eyebrow
[19]*nas* = of the nose
[20]*orb* = circle; *oris* = of the mouth

TABLE 10.1	Muscles of Facial Expression *(continued)*		

Name	Action	Skeletal Attachments	Innervation
The Oral Region *(continued)*			
Levator labii superioris[21] (leh-VAY-tur LAY-bee-eye soo-PEER-ee-OR-is)	Elevates and everts upper lip in sad, sneering, or serious expressions	• Zygomatic bone and maxilla near inferior margin of orbit • Muscles of upper lip	Facial nerve
Levator anguli oris[22] (leh-VAY-tur ANG-you-lye OR-is)	Elevates angle of mouth as in smiling	• Maxilla just below infraorbital foramen • Muscles at angle of mouth	Facial nerve
Zygomaticus[23] major (ZY-go-MAT-ih-cus)	Draws angle of mouth upward and laterally in laughing	• Zygomatic bone • Superolateral angle of mouth	Facial nerve
Zygomaticus minor	Elevates upper lip, exposes upper teeth in smiling or sneering	• Zygomatic bone • Muscles of upper lip	Facial nerve
Risorius[24] (rih-SOR-ee-us)	Draws angle of mouth laterally in expressions of laughing, horror, or disdain	• Zygomatic arch; fascia near ear • Modiolus	Facial nerve
Depressor anguli oris[25]	Draws angle of mouth laterally and downward in opening mouth or sad expressions	• Inferior margin of mandibular body • Modiolus	Facial nerve
Depressor labii inferioris[26]	Draws lower lip downward and laterally in chewing and expressions of melancholy or doubt	• Mandible near mental protuberance • Skin and mucosa of lower lip	Facial nerve
The Mental and Buccal Regions			
Mentalis (men-TAY-lis)	Elevates and protrudes lower lip in drinking, pouting, and expressions of doubt or disdain; elevates and wrinkles skin of chin	• Mandible near inferior incisors • Skin of chin at mental protuberance	Facial nerve
Buccinator[27] (BUC-sin-AY-tur)	Compresses cheek against teeth and gums; directs food between molars; retracts cheek from teeth when mouth is closing to prevent biting cheek; expels air and liquid from the mouth	• Alveolar processes on lateral surfaces of mandible and maxilla • Orbicularis oris; submucosa of cheek and lips	Facial nerve
Platysma[28] (plah-TIZ-muh)	Draws lower lip and angle of mouth downward in expressions of horror or surprise; may aid in opening mouth widely; tenses skin of the chin and neck	• Fascia of deltoid and pectoralis major • Mandible; skin and subcutaneous tissue of lower face	Facial nerve

10.2b Muscles of Chewing and Swallowing

The muscles in this section are concerned primarily with the manipulation of food, including tongue movements, chewing, and swallowing.

Muscles of the Tongue

The tongue is a very agile organ. It pushes food between the molars for chewing and later forces it into the pharynx for swallowing. It is also, of course, crucial to articulate speech. Both intrinsic and extrinsic muscles are responsible for its complex movements. The *intrinsic muscles* consist of a variable number of vertical fascicles that extend from the superior to the inferior side of the tongue, transverse fascicles that extend from left to right, and longitudinal fascicles than extend from front to rear (see figs. 10.1c, 25.5b).

Table 10.2 describes the *extrinsic muscles* that connect the tongue to other structures in the head (**fig. 10.10**).

Muscles of Chewing

Four pairs of muscles produce the biting and cutting movements of the mandible: the **temporalis, masseter,** and two pairs of **pterygoid muscles (fig. 10.11).** Their actions include depression of the mandible for opening the mouth to receive food, elevation for biting off a piece of food or crushing it between the teeth; protraction so the incisors meet in cutting off a piece of food; retraction to draw the lower incisors behind the upper ones and make the rear teeth (premolars and molars) meet; and lateral and medial excursion, the side-to-side movements that grind food between the rear teeth (see fig. 9.20).

---

[21]*levat* = to raise; *labi* = lip; *superior* = upper
[22]*angul* = angle, corner; *oris* = of the mouth
[23]*zygo* = join, unite (refers to zygomatic bone)
[24]*risor* = laughter

[25]*depress* = to lower; *angul* = angle, corner; *oris* = of the mouth
[26]*labi* = lip; *inferior* = lower
[27]*buccinator* = trumpeter
[28]*platy* = flat

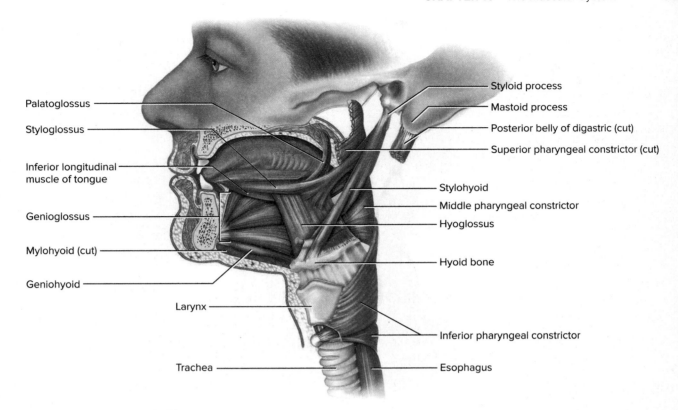

Palatoglossus

Styloglossus

Inferior longitudinal
muscle of tongue

Genioglossus

Mylohyoid (cut)

Geniohyoid

Larynx

Trachea

Styloid process

Mastoid process

Posterior belly of digastric (cut)

Superior pharyngeal constrictor (cut)

Stylohyoid

Middle pharyngeal constrictor

Hyoglossus

Hyoid bone

Inferior pharyngeal constrictor

Esophagus

FIGURE 10.10   Muscles of the Tongue and Pharynx.

Temporalis

Orbicularis oris

Buccinator

Masseter (cut)

(a) Lateral view

Lateral pterygoid plate

Medial pterygoid plate

Lateral pterygoid muscle

Medial pterygoid muscle

Interior of oral cavity

(b) Posterior view

FIGURE 10.11   Muscles of Chewing. Boldface labels indicate muscles that act on the mandible in its chewing movements. (a) Right lateral view. In order to expose the insertion of the temporalis muscle on the mandible, part of the zygomatic arch and masseter muscle are removed. (b) View of the pterygoid muscles looking into the oral cavity from behind the skull.

Suprahyoid Muscles

The hyoid bone, you will recall, is a slender U-shaped bone beneath the chin (see fig. 8.16). Eight pairs of *hyoid muscles* associated with this bone aid in chewing, swallowing, and speaking **(fig. 10.12).** Four of them, the *suprahyoid group,* are superior to the hyoid bone—the **digastric, geniohyoid, mylohyoid,** and

stylohyoid. The digastric is an unusual muscle named for its two bellies. Its *posterior belly* arises from the *mastoid notch* of the cranium and slopes downward and forward. The *anterior belly* arises from a trench called the *digastric fossa* on the inner surface of the mandibular body. It slopes downward and backward. The two bellies meet at a constriction, the *intermediate tendon.* This

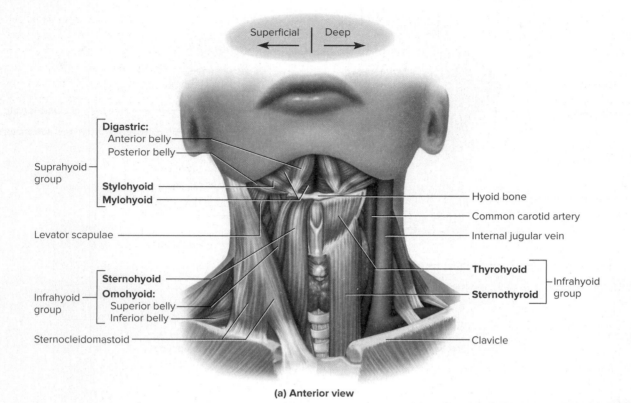

Superficial | Deep

Digastric:
Anterior belly
Posterior belly

Suprahyoid group

Stylohyoid
Mylohyoid

Levator scapulae

Hyoid bone

Common carotid artery

Internal jugular vein

Sternohyoid

Omohyoid:
Superior belly
Inferior belly

Infrahyoid group

Thyrohyoid

Sternothyroid

Infrahyoid group

Sternocleidomastoid

Clavicle

(a) Anterior view

Stylohyoid

Hyoglossus

Mylohyoid

Digastric
(anterior belly)

Hyoid bone

Thyrohyoid

Omohyoid
(superior belly)

Sternothyroid

Sternohyoid

Digastric (posterior belly)

Splenius capitis

Inferior pharyngeal constrictor

Sternocleidomastoid

Trapezius

Levator scapulae

Scalenes

Omohyoid (inferior belly)

(b) Lateral view

FIGURE 10.12 Muscles of the Neck. (a) Anterior view. (b) Lateral view. Boldface labels indicate muscles of the suprahyoid and infrahyoid groups. Another muscle of the suprahyoid group, the geniohyoid, lies deep to the mylohyoid and can be seen in figure 10.10. **APR**

tendon passes through a connective tissue loop, the *fascial sling*, attached to the hyoid bone. Therefore, when the two bellies of the digastric contract, they elevate the hyoid; but if the hyoid is fixed from below, the digastric aids in wide opening of the mouth. The lateral pterygoids are more important in mouth opening, however, with the digastrics coming into play only in extreme opening, as in yawning or taking a large bite from an apple.

Infrahyoid Muscles

The *infrahyoid muscles* are inferior to the hyoid bone. By fixing the hyoid from below, they enable the suprahyoid muscles to open the mouth. The **omohyoid** is unusual in that it arises from

the shoulder, passes under the sternocleidomastoid muscle, then ascends to the hyoid bone. Like the digastric, it has two bellies—superior and inferior. The **thyrohyoid,** named for the hyoid bone and the large shield-shaped thyroid cartilage of the larynx, helps prevent choking. It elevates the larynx during swallowing so its superior opening is sealed by a flap of tissue, the epiglottis. You can feel this effect by holding a fingertip on your "Adam's apple" (anterior prominence of the thyroid cartilage) and feeling it bob up as you swallow.

The infrahyoid muscles are also regarded as extrinsic muscles of the larynx, since they act on it from outside. The intrinsic muscles of the larynx are described in relation to breathing and speech in chapter 22. The *ansa cervicalis*[29] in table 10.2 is a loop

TABLE 10.2	Muscles of Chewing and Swallowing		
Name	Action	Skeletal Attachments	Innervation
Extrinsic Muscles of the Tongue			
Genioglossus[30] (JEE-nee-oh-GLOSS-us)	Unilateral action draws tongue to one side; bilateral action depresses midline of tongue or protrudes tongue	• Superior mental spine on posterior surface of mental protuberance • Inferior surface of tongue from root to apex	Hypoglossal nerve
Hyoglossus[31] (HI-oh-GLOSS-us)	Depresses tongue	• Body and greater horn of hyoid bone • Lateral and inferior surfaces of tongue	Hypoglossal nerve
Styloglossus[32] (STY-lo-GLOSS-us)	Draws tongue upward and posteriorly	• Styloid process of temporal bone and ligament from styloid process to mandible • Superolateral surface of tongue	Hypoglossal nerve
Palatoglossus[33] (PAL-a-toe-GLOSS-us)	Elevates root of tongue and closes oral cavity off from pharynx; forms palatoglossal arch at rear of oral cavity	• Soft palate • Lateral surface of tongue	Accessory and vagus nerves
Muscles of Chewing			
Temporalis[34] (TEM-po-RAY-liss)	Elevation, retraction, and lateral and medial excursion of the mandible	• Temporal lines and temporal fossa of cranium • Coronoid process and anterior border of mandibular ramus	Trigeminal nerve
Masseter[35] (ma-SEE-tur)	Elevation of the mandible, with smaller roles in protraction, retraction, and lateral and medial excursion	• Zygomatic arch • Lateral surface of mandibular ramus and angle	Trigeminal nerve
Medial pterygoid[36] (TERR-ih-goyd)	Elevation, protraction, and lateral and medial excursion of the mandible	• Medial surface of lateral pterygoid plate; palatine bone; lateral surface of maxilla near molar teeth • Medial surface of mandibular ramus and angle	Trigeminal nerve
Lateral pterygoid	Depression (in wide opening of the mouth), protraction, and lateral and medial excursion of the mandible	• Lateral surfaces of lateral pterygoid plate; greater wing of sphenoid • Neck of mandible (just below condyle); articular disc and capsule of temporomandibular joint	Trigeminal nerve

(continued)

[29]*ansa* = handle; *cervic* = neck; *alis* = of, belonging to
[30]*genio* = chin; *gloss* = tongue
[31]*hyo* = hyoid bone; *gloss* = tongue
[32]*stylo* = styloid process; *gloss* = tongue

[33]*palato* = palate; *gloss* = tongue
[34]*temporalis* = of the temporal region of the head
[35]*masset* = chew
[36]*pteryg* = wing; *oid* = resembling (refers to pterygoid plate of sphenoid bone)

TABLE 10.2	Muscles of Chewing and Swallowing (continued)		
Name	**Action**	**Skeletal Attachments**	**Innervation**
Suprahyoid Muscles			
Digastric[37]	Depresses mandible when hyoid is fixed; opens mouth widely, as when ingesting food or yawning; elevates hyoid when mandible is fixed	• Mastoid notch of temporal bone; digastric fossa of mandible • Hyoid bone via fascial sling	Posterior belly: facial nerve Anterior belly: trigeminal nerve
Geniohyoid[38] (JEE-nee-oh-HY-oyd)	Depresses mandible when hyoid is fixed; elevates and protracts hyoid when mandible is fixed	• Inferior mental spine of mandible • Hyoid bone	Spinal nerve C1 via hypoglossal nerve
Mylohyoid[39]	Spans mandible from side to side and forms floor of mouth; elevates floor of mouth in initial stage of swallowing	• Mylohyoid line near inferior margin • Hyoid bone	Trigeminal nerve
Stylohyoid	Elevates and retracts hyoid, elongating floor of mouth; roles in speech, chewing, and swallowing are not yet clearly understood	• Styloid process of temporal bone • Hyoid bone	Facial nerve
Infrahyoid Muscles			
Omohyoid[40]	Depresses hyoid after it has been elevated	• Superior border of scapula • Hyoid bone	Ansa cervicalis
Sternohyoid[41]	Depresses hyoid after it has been elevated	• Manubrium of sternum; medial end of clavicle • Hyoid bone	Ansa cervicalis
Thyrohyoid[42]	Depresses hyoid; with hyoid fixed, elevates larynx as in singing high notes	• Thyroid cartilage of larynx • Hyoid bone	Spinal nerve C1 via hypoglossal nerve
Sternothyroid	Depresses larynx after it has been elevated in swallowing and vocalization; aids in singing low notes	• Manubrium of sternum; costal cartilage • Thyroid cartilage of larynx	Ansa cervicalis
Pharyngeal Muscles			
Superior, middle, and inferior pharyngeal constrictors	During swallowing, contract in order from *superior* to *middle* to *inferior* constrictor to drive food into esophagus	• Medial pterygoid plate; mandible; hyoid; stylohyoid ligament; cartilages of larynx • Posteromedial seam of pharynx; basilar part of occipital bone	Glossopharyngeal and vagus nerves

of nerve on the side of the neck that innervates three of the infrahyoid muscles; it is formed by certain nerve fibers from cervical nerves 1 to 3 (see fig. 13.15).

Pharyngeal Muscles

Three pairs of **pharyngeal constrictors** encircle the pharynx on its posterior and lateral sides, forming a muscular funnel (see fig. 10.10). In swallowing, they push food from the pharynx into the esophagus.

10.2c Muscles Acting on the Head and Neck

Muscles that move the head arise from the vertebral column, thoracic cage, and pectoral girdle and end on the cranial bones **(table 10.3).** Their actions include flexion (tipping the head forward), extension (holding the head erect), hyperextension (as in looking upward), lateral flexion (tilting the head to one side), and rotation (turning the head to look left or right). Flexion, extension, and hyperextension involve simultaneous action of the right and left muscles of a pair; the other actions require the muscle

[37]*di* = two; *gastr* = bellies
[38]*genio* = chin
[39]*mylo* = mill, molar tooth

[40]*omo* = shoulder
[41]*sterno* = chest, sternum
[42]*thyro* = shield, thyroid cartilage

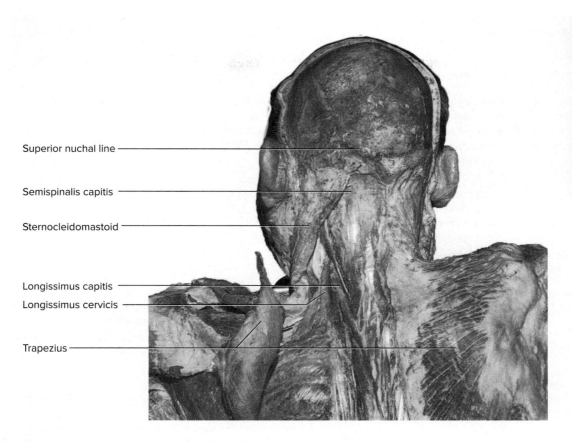

Superior nuchal line

Semispinalis capitis

Sternocleidomastoid

Longissimus capitis
Longissimus cervicis

Trapezius

FIGURE 10.13 Muscles of the Shoulder and Nuchal Regions of the Cadaver.
Rebecca Gray/Don Kincaid/McGraw-Hill Education

on one side to contract more strongly than its mate. Many head actions result from a combination of these movements—for example, looking up over the shoulder involves a combination of rotation and hyperextension.

Depending on its skeletal attachments, a muscle may cause a *contralateral* movement of the head (toward the opposite side, as when contraction of a muscle on the left turns the face toward the right) or an *ipsilateral* movement (toward the same side as the muscle, as when contraction of a muscle on the left tilts the head to the left).

Flexors of the Neck

The prime mover of neck flexion is the **sternocleidomastoid,** a thick muscular cord that extends from the upper chest (sternum and clavicle) to the mastoid process behind the ear (fig. 10.12). This is most easily seen when the head is rotated to one side and slightly extended. To visualize the action of a single sternocleidomastoid, place the index finger of your left hand on your left mastoid process and the index finger of your right hand on your suprasternal notch. Now contract your left sternocleidomastoid to bring your two fingertips as close together as possible. You will see that this tilts your face upward and to the right.

The three **scalenes**[43] on the side of the neck are named for their staircase-like arrangement. Their actions are similar so they are considered collectively in the table.

Extensors of the Neck

The extensors are located mainly in the nuchal region (back of the neck; **fig. 10.13**) and therefore tend to hold the head erect or draw it back. The **trapezius** is the largest and most superficial of these. It extends from the nuchal region over the shoulders and halfway down the back. It is named for the fact that the right and left trapezii together form a diamond or trapezoidal shape (see fig. 10.6). The *splenius* is a deeper, elongated muscle with **splenius capitis** and **splenius cervicis** regions in the head and neck, respectively. It is nicknamed the "bandage muscle" because of the way it wraps around still deeper neck muscles. One of those deeper muscles is the *semispinalis,* another elongated muscle with head, neck, and thoracic regions. Only the **semispinalis capitis** and **cervicis** are tabulated here; the *semispinalis thoracis* doesn't act on the neck, but is included in table 10.6.

▶▶▶APPLY WHAT YOU KNOW

Of the muscles you have studied so far, name three that you would consider intrinsic muscles of the head and three that you would classify as extrinsic. Explain your reason for each.

---

[43]*scal* = staircase

TABLE 10.3	Muscles Acting on the Head and Neck		
Name	**Action**	**Skeletal Attachments**	**Innervation**
Flexors of the Neck			
Sternocleidomastoid[44] (STIR-no-CLY-do-MAST-oyd)	Unilateral action tilts head slightly upward and toward the opposite side, as in looking over one's contralateral shoulder. The most common action is probably rotating the head to the left or right. Bilateral action draws the head straight forward and down, as when eating or reading. Aids in deep breathing when head is fixed.	• Manubrium of sternum; medial one-third of clavicle • Mastoid process; lateral half of superior nuchal line	Accessory nerve; spinal nerves C2–C4
Anterior, middle, and posterior scalenes (SCAY-leens)	Unilateral contraction causes ipsilateral flexion or contralateral rotation (tilts head toward same shoulder, or rotates face away), depending on action of other muscles. Bilateral contraction flexes neck. If spine is fixed, scalenes elevate ribs 1–2 and aid in breathing.	• Transverse processes of all cervical vertebrae (C1–C7) • Ribs 1–2	Anterior rami of spinal nerves C3–C8
Extensors of the Neck			
Trapezius[45] (tra-PEE-zee-us)	Extends and laterally flexes neck. See also roles in scapular movement in table 10.8.	• External occipital protuberance; medial one-third of superior nuchal line; nuchal ligament; spinous processes of vertebrae C7–T12 • Acromion and spine of scapula; lateral one-third of clavicle	Accessory nerve
Splenius capitis[46] and splenius cervicis[47] (SPLEE-nee-us CAP-ih-tiss, SIR-vih-sis)	Acting unilaterally, produce ipsilateral flexion and slight rotation of head; extend head when acting bilaterally	• Inferior half of nuchal ligament; spinous processes of vertebrae C7–T6 • Mastoid process and occipital bone just inferior to superior nuchal line; cervical vertebrae C1–C2 or C3	Posterior rami of middle cervical nerves
Semispinalis capitis and semispinalis cervicis (SEM-ee-spy-NAY-lis)	Extend and contralaterally rotate head	• Articular processes of vertebrae C4–C7; transverse processes of T1–T6 • Occipital bone between nuchal lines; spinous processes of vertebrae C2–C5	Posterior rami of cervical and thoracic nerves

BEFORE YOU GO ON

Answer the following questions to test your understanding of the preceding section:

7. Name two muscles that elevate the upper lip and two that depress the lower lip.

8. Name the four paired muscles of mastication and state where they insert on the mandible.

9. Distinguish between the functions of the suprahyoid and infrahyoid muscles.

10. List the muscles of neck extension and flexion.

10.3 Muscles of the Trunk

Expected Learning Outcomes

When you have completed this section, you should be able to

a. name and locate the muscles of respiration and explain how they affect airflow and abdominal pressure;

b. name and locate the muscles of the abdominal wall, back, and pelvic floor; and

c. identify the skeletal attachments, action, and innervation of these muscles.

In this section, we will examine muscles of the trunk of the body in four functional groups concerned with respiration, support of the abdominal wall, movement of the vertebral column, and support of the pelvic floor. In the illustrations, you will note some major muscles that are not discussed in the associated tables—for example, the pectoralis major and serratus anterior. Although they are *located* in the trunk, they *act upon* the limbs and limb girdles, and are further discussed in sections 10.4 and 10.5.

[44]*sterno* = chest, sternum; *cleido* = hammer, clavicle; *masto* = breastlike, mastoid process
[45]*trapez* = table, trapezoid
[46]*splenius* = bandage; *capitis* = of the head
[47]*cervicis* = of the neck

10.3a Muscles of Respiration

We breathe primarily by means of muscles that enclose the thoracic cavity—the diaphragm, external intercostal, internal intercostal, and innermost intercostal muscles **(fig. 10.14, table 10.4).**

The **diaphragm** is a muscular dome between the thoracic and abdominal cavities, bulging upward against the base of the lungs. It has openings for passage of the esophagus, major blood and lymphatic vessels, and nerves between the two cavities. Its fibers converge from the margins toward a fibrous **central tendon.** When the diaphragm contracts, it flattens slightly and enlarges the thoracic cavity, causing air intake (inspiration); when it relaxes, it rises and shrinks the thoracic cavity, expelling air (expiration).

Three layers of muscle lie between the ribs: the external, internal, and innermost intercostal muscles. The 11 pairs of **external intercostal muscles** constitute the most superficial layer. They extend from the rib tubercle posteriorly almost to the beginning of the costal cartilage anteriorly. Each one slopes downward and anteriorly from one rib to the next inferior one.

The 11 pairs of **internal intercostal muscles** lie deep to the external intercostals and extend from the margin of the sternum to the angles of the ribs. They are thickest in the region between the costal cartilages and grow thinner in the region where they overlap the internal intercostals. Their fibers slope downward and posteriorly from each rib to the one below, at nearly right angles to the external intercostals. Each is divided into an *intercartilaginous part* between the costal cartilages and an *interosseous part* between the bony part of the ribs. The two parts differ in their respiratory roles.

The **innermost intercostal muscles** vary in number, as they are sometimes absent from the upper thoracic cage. Their fibers run in the same direction as the internal intercostals, and they are presumed to serve the same function. The internal and innermost intercostals are separated by a fascia that allows passage for intercostal nerves and blood vessels (see fig. 13.14b).

The primary function of the intercostal muscles is to stiffen the thoracic cage during respiration so it doesn't cave inward when the diaphragm descends. However, they also contribute to enlargement and contraction of the thoracic cage and thus add to the air volume that ventilates the lungs.

Many other muscles of the chest and abdomen contribute significantly to breathing: the sternocleidomastoid and scalenes of the neck; pectoralis major and serratus anterior of the chest; latissimus dorsi of the lower back; internal and external obliques and transverse abdominal muscle; and even some of the anal muscles. The respiratory actions of all these muscles are described in section 22.2a.

▶▶▶**APPLY WHAT YOU KNOW**

A young thoracic surgery resident performing an operation for esophageal cancer accidentally severs the patient's left phrenic nerve. Predict the effect of this accident on the patient's respiration.

▶▶▶**APPLY WHAT YOU KNOW**

Carl and Carla are enjoying a platter of barbecued ribs. What muscles are they eating? What is the tough fibrous membrane between the meat and the bone?

External intercostals

Internal intercostals

(a) Lateral view of intercostal muscles

Xiphoid process of sternum

Inferior vena cava

Ribs

Esophagus

Central tendon of diaphragm

Aorta

Vertebral column

(b) Inferior view of diaphragm

FIGURE 10.14 Muscles of Respiration. (a) Lateral view of the intercostal muscles. (b) Inferior view of the diaphragm. **APR**

TABLE 10.4	Muscles of Respiration		
Name	Action	Skeletal Attachments	Innervation
Diaphragm[48] (DY-ah-fram)	Prime mover of inspiration (responsible for about two-thirds of air intake); contracts in preparation for sneezing, coughing, crying, laughing, and weight lifting; contraction compresses abdominal viscera and aids in childbirth and expulsion of urine and feces.	• Xiphoid process of sternum; ribs and costal cartilages 7–12; lumbar vertebrae • Central tendon of diaphragm	Phrenic nerves
External intercostals[49] (IN-tur-COSS-tulz)	When scalenes fix rib 1, external intercostals elevate and protract ribs 2–12, expanding the thoracic cavity and creating a partial vacuum causing inflow of air; exercise a braking action during expiration so that expiration is not overly abrupt.	• Inferior margins of ribs 1–11 • Superior margin of next lower rib	Intercostal nerves
Internal intercostals	In inspiration, the intercartilaginous part aids in elevating the ribs and expanding the thoracic cavity; in expiration, the interosseous part depresses and retracts the ribs, compressing the thoracic cavity and expelling air; the latter occurs only in forceful expiration, not in relaxed breathing.	• Superior margins and costal cartilages of ribs 2–12; margin of sternum • Inferior margin of next higher rib	Intercostal nerves
Innermost intercostals	Presumed to have the same action as the internal intercostals	• Superomedial surface of ribs 2–12; may be absent from upper ribs • Medial edge of costal groove of next higher rib	Intercostal nerves

10.3b Muscles of the Abdominal Wall

Unlike the thoracic cavity, the abdominal cavity has little skeletal support. It is enclosed, however, in layers of broad flat muscles whose fibers run in different directions, strengthening the abdominal wall on the same principle as the alternating layers of wood fibers in plywood **(table 10.5).**

Three layers of muscle enclose the lumbar region and extend about halfway across the anterior abdomen **(fig. 10.15).** The most superficial layer is the **external oblique muscle.** Its fibers pass downward and anteriorly. The next deeper layer is the **internal oblique muscle,** whose fibers pass upward and anteriorly, roughly perpendicular to those of the external oblique. The deepest layer is the **transverse abdominal muscle (transversus abdominis),** with horizontal fibers.

Anteriorly, a pair of vertical **rectus abdominis muscles** extends from sternum to pubis. These are divided into segments by three transverse *tendinous intersections,* giving them an appearance that body builders nickname the "six pack."

The tendons of the oblique and transverse muscles are *aponeuroses*—broad fibrous sheets that continue medially and inferiorly **(figs. 10.16, 10.17).** At the rectus abdominis, they diverge and pass around its anterior and posterior sides, enclosing the muscle in a vertical sleeve called the rectus sheath. They meet again at a median line called the **linea alba**[50] between the rectus muscles. Another line, the **linea semilunaris,**[51] marks the lateral boundary where the rectus sheath meets the aponeurosis. The aponeurosis of the external

FIGURE 10.15 **Cross Section of the Anterior Abdominal Wall.**

---

[48]*dia* = across; *phragm* = partition
[49]*inter* = between; *costa* = rib
[50]*linea* = line; *alba* = white
[51]*linea* = line; *semilunar* = half-moon

Pectoralis major

Tendinous intersections

Rectus sheath

Umbilicus

Linea semilunaris

Linea alba

Aponeurosis of external oblique

Latissimus dorsi

Serratus anterior

Rectus sheath (cut edges)

Transverse abdominal

Internal oblique (cut)

External oblique (cut)

Rectus abdominis

Inguinal ligament

(a) Superficial

Pectoralis minor

Serratus anterior

Rectus sheath

Internal oblique

Inguinal ligament

Subclavius

Pectoralis minor (cut)

Internal intercostals

External intercostals

Rectus abdominis (cut)

External oblique (cut)

Internal oblique (cut)

Posterior wall of rectus sheath (rectus abdominis removed)

Transverse abdominal (cut)

(b) Deep

FIGURE 10.16 Thoracic and Abdominal Muscles. (a) Superficial muscles. The left rectus sheath is cut away to expose the rectus abdominis muscle. (b) Deep muscles. On the anatomical right, the external oblique has been removed to expose the internal oblique and the pectoralis major has been removed to expose the pectoralis minor. On the anatomical left, the internal oblique has been cut to expose the transverse abdominal and the middle of the rectus abdominis has been cut out to expose the posterior rectus sheath. **APR**

? *Name at least three muscles that lie deep to the pectoralis major.*

FIGURE 10.17 Thoracic and Abdominal Muscles of the Cadaver. The rectus sheath has been removed on the anatomical left to expose the left rectus abdominis muscle. Inset shows area of dissection.

Christine Eckel/McGraw-Hill Education

TABLE 10.5	Muscles of the Abdominal Wall		
Name	**Action**	**Skeletal Attachments**	**Innervation**
External oblique	Supports abdominal viscera against pull of gravity; stabilizes vertebral column during heavy lifting; maintains posture; compresses abdominal organs, thus aiding in deep breathing, loud vocalizations such as singing and public speaking, and in expulsion of abdominopelvic contents during childbirth, urination, defecation, and vomiting; unilateral contraction causes contralateral rotation of the spine, as in twisting at the waist.	• Ribs 5–12 • Anterior half of iliac crest; symphysis and superior margin of pubis	Anterior rami of spinal nerves T7–T12
Internal oblique	Same as external oblique except that unilateral contraction causes ipsilateral rotation of waist	• Inguinal ligament; iliac crest; thoracolumbar fascia • Ribs 10–12; costal cartilages 7–10; pubis	Anterior rami of spinal nerves T7–L1
Transverse abdominal	Compresses abdominal contents, with same effects as external oblique, but does not contribute to movements of vertebral column	• Inguinal ligament; iliac crest; thoracolumbar fascia; costal cartilages 7–12 • Linea alba; pubis; aponeurosis of internal oblique	Anterior rami of spinal nerves T7–L1
Rectus[52] abdominis (REC-tus ab-DOM-ih-nis)	Flexes waist, as in bending forward or doing sit-ups; stabilizes pelvic region during walking; and compresses abdominal viscera	• Pubic symphysis and superior margin of pubis • Xiphoid process; costal cartilages 5–7	Anterior rami of spinal nerves T6–T12

[52]*rectus* = straight

oblique also forms a cordlike **inguinal ligament** at its inferior margin. This extends obliquely from the anterior superior spine of the ilium to the pubis. The linea alba, linea semilunaris, and inguinal ligament are externally visible on a person with good muscle definition (see atlas B, fig. B.8). Weak points in the abdominal wall can be sites of inguinal and umbilical hernias (see Deeper Insight 10.3).

▶▶▶**APPLY WHAT YOU KNOW**

Alice works out at a fitness center three times a week doing weight lifting and abdominal crunches. Martha prefers to sit on the sofa eating potato chips and watching

TV. Both become pregnant. Other things being equal, give one reason related to table 10.5 why Alice may have an easier time with her childbirth than Martha will.

10.3c Muscles of the Back

Muscles of the back (**table 10.6**) primarily extend, rotate, and laterally flex the vertebral column. The most prominent superficial back muscles are the *latissimus dorsi* and *trapezius* (**fig. 10.18**), but they are concerned with upper limb movements and covered in tables 10.8 and 10.9. Deep to these are the **serratus posterior superior** and **inferior** (figs. 10.18, **10.19**). They extend from the

Superficial | Deep

Left labels	Right labels
Sternocleidomastoid	Semispinalis capitis
Trapezius	Splenius capitis
	Levator scapulae
	Rhomboid minor
	Rhomboid major
	Supraspinatus
Deltoid	Infraspinatus
	Teres minor
	Teres major
Erector spinae	Serratus anterior
	Serratus posterior inferior
	External oblique
Latissimus dorsi	
External oblique	Internal oblique
Thoracolumbar fascia	
Gluteus medius	Gluteus minimus
Gluteus maximus	Lateral rotators

FIGURE 10.18 Neck, Back, and Gluteal Muscles. The most superficial muscles are shown on the left and the next deeper layer on the right. **APR**

Superior nuchal line

Longissimus capitis

Splenius capitis

Serratus posterior superior

Splenius cervicis

Erector spinae:
  Iliocostalis
  Longissimus
  Spinalis

Serratus posterior inferior

Internal oblique

External oblique (cut)

Semispinalis capitis

Semispinalis cervicis

Semispinalis thoracis

Multifidus

Quadratus lumborum

FIGURE 10.19 Muscles Acting on the Vertebral Column. These are deeper than the muscles in figure 10.18, and those on the right are deeper than those on the left.

vertebrae to the ribs. They aid in deep breathing and are further discussed in chapter 22.

Deep to these is a prominent muscle, the **erector spinae,** which runs vertically for the entire length of the back from the cranium to the sacrum **(fig. 10.20).** It is a thick muscle, easily palpated on each side of the vertebral column in the lumbar region. (Pork chops and T-bone steaks are cut from the erector spinae muscles of animals.) As it ascends, it divides in the upper lumbar region into three parallel columns. The most lateral of these is the **iliocostalis,** which from inferior to superior is divided into the *iliocostalis lumborum, iliocostalis thoracis,* and *iliocostalis cervicis* (lumbar, thoracic, and cervical regions).

The next column medially is the **longissimus,** divided from inferior to superior into the *longissimus thoracis, longissimus cervicis,* and *longissimus capitis* (thoracic, cervical, and cephalic regions). The most medial column is the **spinalis,** divided into *spinalis thoracis, spinalis cervicis,* and *spinalis capitis.* The functions of all three columns are sufficiently similar that we will treat them collectively as the erector spinae.

The major deep muscles are the **semispinalis thoracis** in the thoracic region and **quadratus lumborum** in the lumbar region (fig. 10.19). The erector spinae and quadratus lumborum are enclosed in a fibrous sheath called the **thoracolumbar fascia** (fig. 10.18), which is the origin of some of the abdominal and

Trapezius

Ribs

External intercostals

Erector spinae:

Spinalis thoracis

Iliocostalis thoracis

Longissimus thoracis

Latissimus dorsi

Iliocostalis lumborum

Thoracolumbar fascia

FIGURE 10.20 Deep Back Muscles of the Cadaver.

Rebecca Gray/Don Kincaid/McGraw-Hill Education

TABLE 10.6	Muscles of the Back		
Name	**Action**	**Skeletal Attachments**	**Innervation**
Erector spinae (eh-REC-tur SPY-nee)	Aids in sitting and standing erect; straightens back after one bends at waist, and is employed in arching the back; unilateral contraction flexes waist laterally; the longissimus capitis also produces ipsilateral rotation of the head.	• Nuchal ligament; ribs 3–12; thoracic and lumbar vertebrae; median and lateral sacral crests; thoracolumbar fascia • Mastoid process; cervical and thoracic vertebrae; all ribs	Posterior rami of cervical to lumbar spinal nerves
Semispinalis thoracis (SEM-ee-spy-NAY-liss tho-RA-sis)	Extension and contralateral rotation of vertebral column	• Vertebrae T6–T10 • Vertebrae C6–T4	Posterior rami of cervical and thoracic spinal nerves
Quadratus lumborum[53] (quad-RAY-tus lum-BORE-um)	Unilateral contraction causes ipsilateral flexion of lumbar spine; bilateral contraction extends lumbar spine. Aids respiration by fixing rib 12 and stabilizing attachments of diaphragm.	• Iliac crest; iliolumbar ligament • Rib 12; vertebrae L1–L4	Anterior rami of spinal nerves T12–L4
Multifidus[54] (mul-TIFF-ih-dus)	Stabilizes adjacent vertebrae, maintains posture, controls vertebral movement when erector spinae acts on spine	• Vertebrae C4–L5; posterior superior iliac spine; sacrum; aponeurosis of erector spinae • Laminae and spinous processes of vertebrae superior to origins	Posterior rami of cervical to lumbar spinal nerves

[53]*quadrat* = four-sided; *lumborum* = of the lumbar region   [54]*multi* = many; *fid* = branched, sectioned

DEEPER INSIGHT 10.2

CLINICAL APPLICATION

Heavy Lifting and Back Injuries

When you are fully bent over forward, as in touching your toes, the erector spinae is fully stretched. Because of the *length–tension relationship* explained in section 11.4e, muscles that are stretched to such extremes cannot contract very effectively. Standing up from such a position is therefore initiated by the hamstring muscles on the back of the thigh and the gluteus maximus of the buttocks. The erector spinae joins in the action when it is partially contracted.

Standing too suddenly or improperly lifting a heavy weight, however, can strain the erector spinae, cause painful muscle spasms, tear tendons and ligaments of the lower back, and rupture intervertebral discs. The lumbar muscles are adapted for maintaining posture, not for lifting. This is why it is important, in heavy lifting, to crouch and use the powerful extensor muscles of the thighs and buttocks to lift the load.

lumbar muscles. The **multifidus** is a collective name for a series of tiny muscles that connect adjacent vertebrae to each other from the cervical to lumbar region.

10.3d Muscles of the Pelvic Floor

The floor of the pelvic cavity (**fig. 10.21; table 10.7**) is a strong multilayered muscular sheet important for the support of the abdominal and pelvic viscera—especially in light of the bipedal stance of humans, since the abdominal wall no longer forms a supportive floor as it does in quadrupedal mammals. Weakness in the pelvic floor can result in urinary or fecal incontinence or the prolapse (dropping) of internal organs between the thighs. The muscles and skeletal landmarks here are also of special importance in obstetrics.

Viewed from within the pelvic cavity, its floor is formed mainly by an extensive muscle called the *levator ani*. Inferior to

Perineal body
Superficial transverse perineal
Coccyx

Ischiocavernosus
Bulbospongiosus
Levator ani
Gluteus maximus

(a) Superficial perineal space, inferior view, male

Clitoris
Vagina
Perineal body
Deep transverse perineal
External anal sphincter

Urogenital triangle
Urethra
Compressor urethrae
Anus
Anal triangle

(b) Deep perineal space, inferior view, female

Pubic symphysis
Levator ani:
  Pubococcygeus
  Iliococcygeus
  Ischiococcygeus
Piriformis
Nerve roots

Urethra
Vagina
Rectum
Fascia of obturator internus
Anococcygeal body
Sacrum

(c) Pelvic diaphragm, superior view, female

FIGURE 10.21 Muscles of the Pelvic Floor. (a) The superficial perineal space, male, viewed from below (the inferior aspect). (b) The deep perineal space, female, viewed from the same perspective. Other than the vaginal canal, the sexes are nearly identical at this level, including the urogenital and anal triangles. The root of the penis does not extend to this level. (c) The pelvic diaphragm, female, viewed from above (from within the pelvic cavity).

TABLE 10.7	Muscles of the Pelvic Floor		
Name	**Action**	**Skeletal Attachments**	**Innervation**
Superficial Perineal Space			
Ischiocavernosus[55] (ISS-kee-oh-CAV-er-NO-sus)	Maintains erection of the penis or clitoris by compressing deep structures of the organ and forcing blood forward into its body	• Ramus and tuberosity of ischium • Ensheaths internal structures of penis and clitoris	Pudendal nerve
Bulbospongiosus[56] (BUL-bo-SPUN-jee-OH-sus)	Expels residual urine from urethra after bladder has emptied. Aids in erection of penis or clitoris. In male, spasmodic contractions expel semen during ejaculation. In female, contractions constrict vaginal orifice and expel secretions of greater vestibular glands.	• Perineal body and median raphe • Male: ensheaths root of penis Female: pubic symphysis	Pudendal nerve
Deep Perineal Space			
Deep transverse perineal	Anchors perineal body, which supports other pelvic muscles; supports vaginal and urethral canals	• Ischiopubic rami • Perineal body	Pudendal nerve
Compressor urethrae (yu-REE-three)	Aids in urine retention; found in females only	• Ischiopubic rami • Right and left compressor urethrae meet each other inferior to external urethral sphincter	Pudendal nerve; spinal nerves S2–S4; pelvic splanchnic nerve
Anal Triangle			
External anal sphincter	Retains feces in rectum until voluntarily voided	• Coccyx, perineal body • Encircles anal canal and orifice	Pudendal nerve; spinal nerves S2–S4; pelvic splanchnic nerve
Pelvic Diaphragm			
Levator ani[57] (leh-VAY-tur AY-nye)	Compresses anal canal and reinforces external anal and urethral sphincters; supports uterus and other pelvic viscera; aids in falling away of the feces; vertical movements affect pressure differences between abdominal and thoracic cavities and thus aid in deep breathing.	• Inner surface of lesser pelvis from pubis through margin of obturator internus to spine of ischium • Coccyx via anococcygeal body; walls of urethra, vagina, and anal canal	Pudendal nerve; spinal nerves S2–S3

this is the **perineum** (PERR-ih-NEE-um), a diamond-shaped area between the thighs bordered by four bony landmarks: the pubic symphysis anteriorly, the coccyx posteriorly, and the ischial tuberosities laterally. The pelvic floor and perineum are penetrated by the anal canal, urethra, and vagina. The anterior half of the perineum is the **urogenital triangle** and the posterior half is the **anal triangle.**

The urogenital triangle is divided into two muscle compartments separated by a strong fibrous **perineal membrane.** The muscle compartment between this membrane and the skin is called *superficial perineal space,* and the compartment between the perineal membrane and levator ani is the *deep perineal space.* We will examine these structures beginning inferiorly, just beneath the skin, and progressing superiorly to the pelvic floor.

The Superficial Perineal Space

The **superficial perineal space** (fig. 10.21a) contains three pairs of muscles: the *ischiocavernosus, bulbospongiosus,* and *superficial transverse perineal muscle.* In females, this space also contains the clitoris; various glands and erectile tissues of the genitalia (see fig. 28.8); and adipose tissue, which extends into and fattens the mons pubis and labia majora. In males, it contains the root of the penis.

The **ischiocavernosus** muscles converge like a V from the ischial tuberosities toward the penis or clitoris. In males, the **bulbospongiosus (bulbocavernosus)** muscles form a sheath around the root of the penis, and in females they enclose the vagina like a pair of parentheses. *Cavernosus* in these names refers to the spongy, cavernous structure of tissues in the penis and clitoris.

The **superficial transverse perineal muscles** extend from the ischial tuberosities to a strong median fibromuscular anchorage, the **perineal body,** and a median seam called the **perineal raphe** (RAY-fee) that extends anteriorly from the perineal body. These muscles may help to anchor the perineal body, but they are weakly developed and not always present, therefore not included in table 10.7. The other two muscle pairs of this layer serve primarily sexual functions.

---

[55]*ischio* = ischium of hip bone; *cavernosus* = corpus cavernosum of the penis or clitoris
[56]*bulbo* = bulb of the penis; *spongiosus* = corpus spongiosum of the penis

[57]*levator* = that which elevates; *ani* = of the anus

DEEPER INSIGHT 10.3

CLINICAL APPLICATION

Hernias

A hernia is any condition in which the viscera protrude through a weak point in the muscular wall of the abdominopelvic cavity. The most common type to require treatment is an *inguinal hernia* **(fig. 10.22)**. In the male fetus, each testis descends from the pelvic cavity into the scrotum by way of a passage called the *inguinal canal* through the muscles of the groin. A long pouch of peritoneum descends with the testis, but usually disappears by birth. Hernias occur in infants and children when this pouch persists, allowing a loop of intestine to enter it and appear near or in the scrotum *(indirect hernia)*. Adult hernias may appear similar but are often due to a weakening of the inguinal canal *(direct hernia)*. This can result from repetitive stress such as lifting weights. When the diaphragm and abdominal muscles contract, pressure in the abdominal cavity can soar to 1,500 pounds per square inch—more than 100 times the normal pressure and quite sufficient to produce an inguinal hernia, or "rupture." Inguinal hernias in women occur through the pelvic floor and other nearby sites.

Two other sites of hernia are the diaphragm and navel. A *hiatal hernia* is a condition in which part of the stomach protrudes through the diaphragm into the thoracic cavity. This is most common in overweight people over 40. It may cause heartburn due to the regurgitation of stomach acid into the esophagus, but most cases go undetected. In an *umbilical hernia,* abdominal viscera protrude through the navel.

Aponeurosis of external oblique muscle

Inguinal canal

External inguinal ring

Herniated loop of small intestine

Upper scrotum

FIGURE 10.22 Inguinal Hernia. A loop of small intestine has protruded through the inguinal canal into a space beneath the skin.

The Deep Perineal Space

The **deep perineal space** (fig. 10.21b) contains a pair of **deep transverse perineal muscles** and, in females only, the **compressor urethrae muscles.** The deep transverse perineal muscles anchor the perineal body on the median plane; the perineal body, in turn, anchors other pelvic muscles. The female external urethral sphincter, long thought to be part of the deep perineal space, is now regarded as part of the urethra itself and not part of the pelvic floor musculature.

The Anal Triangle

The **anal triangle** contains the **external anal sphincter** and **anococcygeal ligament** (fig. 10.21b). The external anal sphincter is a tubular muscle surrounding the lower anal canal. The anococcygeal ligament is the median insertion of the levator ani muscles, and the ligament, in turn, inserts on the coccyx. It is therefore a major anchorage for the structures that compose the pelvic floor.

The Pelvic Diaphragm

The **pelvic diaphragm** (fig. 10.21c) is deep to the foregoing structures (uppermost in the pelvic floor) and is composed mainly

of the right and left **levator ani** muscles. (The piriformis, also illustrated, is primarily a lower limb muscle.) The levator ani spans most of the pelvic outlet and forms the floor of the lesser (true) pelvis. It is divided into three portions that are sometimes regarded as separate muscles—the *ischiococcygeus* (or *coccygeus*), *iliococcygeus,* and *pubococcygeus*. The left and right levator ani muscles converge on the anococcygeal ligament, through which they are indirectly anchored to the coccyx.

BEFORE YOU GO ON

Answer the following questions to test your understanding of the preceding section:

11. Which muscles are used more often, the external intercostals or internal intercostals? Explain.

12. Explain how pulmonary ventilation affects abdominal pressure and vice versa.

13. Name a major superficial muscle and two major deep muscles of the back.

14. Define *perineum, urogenital triangle,* and *anal triangle.*

15. Name one muscle in the superficial perineal space, one in the urogenital diaphragm, and one in the pelvic diaphragm. State the function of each.

10.4 Muscles Acting on the Shoulder and Upper Limb

Expected Learning Outcomes

When you have completed this section, you should be able to

a. name and locate the muscles that act on the pectoral girdle, shoulder, elbow, wrist, and hand;

b. relate the actions of these muscles to the joint movements described in chapter 9; and

c. describe the skeletal attachments, action, and innervation of these muscles.

The upper and lower limbs have numerous muscles that serve primarily for movement of the body and manipulation of objects. These muscles are organized into distinct compartments separated from each other by the interosseous membranes of the forearm and leg (see figs. 8.34, 8.40) and by intermuscular septa. In the ensuing tables, you will find muscles of the upper limb divided into anterior and posterior compartments, and those of the lower limb divided into anterior, posterior, medial, and lateral compartments. In most limb regions, the muscle groups are further subdivided by thinner fasciae into superficial and deep layers.

The upper limb is used for a broad range of both powerful and subtle actions, ranging from climbing, grasping, and throwing to writing, playing musical instruments, and manipulating small objects. It therefore has an especially complex array of muscles, but the muscles fall into logical groups that make their functional relationships and names easier to understand. The next five tables group these into muscles that act on the scapula, those that act on the humerus and shoulder joint, those that act on the forearm and elbow joint, extrinsic (forearm) muscles that act on the wrist and hand, and intrinsic (hand) muscles that act on the fingers.

10.4a Muscles Acting on the Shoulder

Muscles that act on the pectoral girdle **(table 10.8)** extend from the axial skeleton to the clavicle and scapula. The scapula is only loosely attached to the thoracic cage and is capable of considerable movement **(fig. 10.23)**—rotation (as in raising and lowering the apex of the shoulder), elevation and depression (as in shrugging and lowering the shoulders), and

Lateral rotation
Trapezius (superior part)
Serratus anterior

Medial rotation
Levator scapulae
Rhomboid major
Rhomboid minor

(a)

Elevation
Levator scapulae
Trapezius (superior part)
Rhomboid major
Rhomboid minor

Depression
Trapezius (inferior part)
Serratus anterior

(b)

Retraction
Rhomboid major
Rhomboid minor
Trapezius

Protraction
Pectoralis minor
Serratus anterior

(c)

FIGURE 10.23 Actions of Some Thoracic Muscles on the Scapula. (a) Lateral and medial rotation. (b) Elevation and depression. (c) Retraction and protraction. Note that an individual muscle can contribute to multiple actions, depending on which fibers contract and what synergists act with it.

TABLE 10.8	Muscles Acting on the Shoulder		
Name	Action	Skeletal Attachments	Innervation
Anterior Group			
Pectoralis minor (PECK-toe-RAY-liss)	With serratus anterior, draws scapula laterally and forward around chest wall; with other muscles, rotates scapula and depresses apex of shoulder, as in reaching down to pick up a suitcase	• Ribs 3–5 and overlying fascia • Coracoid process of scapula	Medial and lateral pectoral nerves
Serratus[58] anterior (serr-AY-tus)	With pectoralis minor, draws scapula laterally and forward around chest wall; protracts scapula, and is the prime mover in all forward-reaching and pushing actions; aids in rotating scapula to elevate apex of shoulder; fixes scapula during abduction of arm	• All or nearly all ribs • Medial border of scapula	Long thoracic nerve
Posterior Group			
Trapezius (tra-PEE-zee-us)	Stabilizes scapula and shoulder during arm movements; elevates and depresses apex of shoulder; acts with other muscles to rotate and retract scapula (see also roles in head and neck movements in table 10.3)	• External occipital protuberance; medial one-third of superior nuchal line; nuchal ligament; spinous processes of vertebrae C7–T12 • Acromion and spine of scapula; lateral one-third of clavicle	Accessory nerve
Levator scapulae (leh-VAY-tur SCAP-you-lee)	Elevates scapula if cervical vertebrae are fixed; flexes neck laterally if scapula is fixed; retracts scapula and braces shoulder; rotates scapula and depresses apex of shoulder	• Transverse processes of vertebrae C1–C4 • Superior angle to medial border of scapula	Spinal nerves C3–C4, and C5 via posterior scapular nerve
Rhomboid minor (ROM-boyd)	Retracts scapula and braces shoulder; fixes scapula during arm movements	• Spinous processes of vertebrae C7–T1; nuchal ligament • Medial border of scapula	Posterior scapular nerve
Rhomboid major	Same as rhomboid minor	• Spinous processes of vertebrae T2–T5 • Medial border of scapula	Posterior scapular nerve

protraction and retraction (pulling the shoulders forward and back). The clavicle braces the shoulder and moderates these movements.

Anterior Group

Muscles of the pectoral girdle fall into anterior and posterior groups (**figs. 10.24, 10.25**). The major muscles of the anterior group are the pectoralis minor and serratus anterior (see fig. 10.16b). The **pectoralis minor** arises by three heads from ribs 3 to 5 and converges on the coracoid process of the scapula. The **serratus anterior** is nicknamed the "boxer's muscle" because of its role in powerful thrusting movements of the arm such as a boxer's jab. It arises from separate heads on all or nearly all of the ribs, wraps laterally around the chest, passes across the back between the rib cage and scapula, and ends on the medial (vertebral) border of the scapula. Thus, when it contracts, the scapula glides laterally and slightly forward around the ribs.

[58]*serrate* = scalloped, zigzag

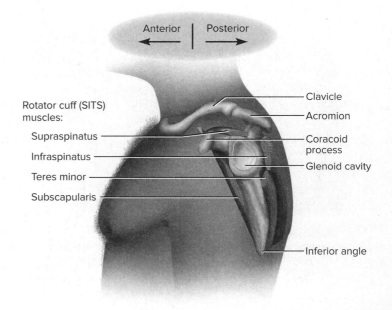

FIGURE 10.24 Rotator Cuff Muscles in Relation to the Scapula (Lateral View). For posterior and anterior views of these muscles, see figure 10.25, b and d.

Clavicle

Sternum

Pectoralis major

Deltoid

Coracobrachialis

Triceps brachii:
  Lateral head
  Long head
  Medial head
Biceps brachii
Brachialis
Brachioradialis

Supraspinatus

Spine of scapula

Greater tubercle of humerus

Infraspinatus

Humerus

Teres minor

Teres major

Triceps brachii:
  Lateral head
  Long head

Latissimus dorsi

(a) Anterior view

(b) Posterior view

Clavicle

Scapula

Biceps brachii:
  Long head
  Short head

Subscapularis

Coracobrachialis

Brachialis

(c) Anterior view

(d) Anterior view

FIGURE 10.25 Pectoral and Brachial Muscles. (a) Superficial muscles, anterior view. (b) Superficial muscles, posterior view. (c) The biceps brachii, the superficial flexor of the elbow. (d) The brachialis, the deep flexor of the elbow, and the coracobrachialis and subscapularis, which act on the humerus. **APR**

Posterior Group

The posterior muscles that act on the scapula include the large, superficial **trapezius,** already discussed (see table 10.3), and three deep muscles: the **levator scapulae, rhomboid minor,** and **rhomboid major.** The action of the trapezius depends on whether its superior, middle, or inferior fibers contract and whether it acts alone or with other muscles. The levator scapulae and superior fibers of the trapezius rotate the scapula in opposite directions if either of them acts alone. If both act together, their opposite rotational effects balance each other and they elevate the scapula and shoulder, as when you lift a suitcase from the floor. Depression of the scapula occurs mainly by gravitational pull, but the trapezius and serratus anterior can depress it more rapidly and forcefully, as in swimming, hammering, and rowing.

10.4b Muscles Acting on the Arm

Nine muscles cross the shoulder joint and insert on the humerus **(table 10.9).** Seven of them are considered *scapular muscles* because they arise from the scapula, and the other two are considered *axial muscles* because they arise primarily from the axial skeleton.

Rotator Cuff Muscles

Tendons from four of the scapular muscles form the **rotator cuff** (fig. 10.24), well known as a common site of musculoskeletal injury. These are nicknamed the "SITS muscles" for the first letters of their names—**supraspinatus, infraspinatus, teres minor,** and **subscapularis.** The first three of these lie on the posterior side of the scapula (fig. 10.25b). The supraspinatus and infraspinatus occupy the supraspinous and infraspinous fossae, above and below the scapular spine. The teres minor lies inferior to the infraspinatus. The subscapularis occupies the subscapular fossa on the anterior surface of the scapula, between the scapula and ribs (fig. 10.25d). The tendons of these muscles merge with the joint capsule of the shoulder as they cross it en route to the humerus. They insert on the proximal end of the humerus, forming a partial sleeve around it. The rotator cuff reinforces the joint capsule and holds the head of the humerus in the glenoid cavity. It is easily damaged by strenuous actions of the shoulder (see Deeper Insight 10.4).

Other Scapular Muscles

Of the remaining three scapular muscles, the most conspicuous is the **deltoid,** the thick triangular muscle that caps the shoulder (fig. 10.25a). Intramuscular drug injections are often given here. Its anterior, lateral, and posterior fibers act like three different

muscles. The **teres major** and **coracobrachialis** complete this group (fig. 10.25b, d).

Axial Muscles

The two axial muscles, both prominent, are the pectoralis major anteriorly and latissimus dorsi posteriorly (fig. 10.25a, b; **fig. 10.26**). The **pectoralis major** is the thick, fleshy muscle of the mammary region and the **latissimus dorsi** is a broad muscle of the back that extends from the waist to the axilla. These muscles bear the primary responsibility for attaching the arm to the trunk and are the prime movers of the shoulder joint.

10.4c Muscles Acting on the Elbow and Forearm

The elbow and forearm are capable of four motions—flexion, extension, pronation, and supination—carried out by muscles in both the brachium and antebrachium (arm and forearm) **(table 10.10).**

Muscles of the Brachial Region

The principal elbow flexors are in the anterior compartment of the arm—the **brachialis** and **biceps brachii** (fig. 10.25c, d). The biceps brachii appears as a large anterior bulge on the arm and commands considerable interest among body builders, but the brachialis underlying it generates about 50% more power and is thus the prime mover of elbow flexion. The biceps is not only a flexor but also a powerful forearm supinator. It is named for its two heads: a *short head* whose tendon arises from the coracoid process of the scapula, and a *long head* whose tendon arises from the superior margin of the glenoid cavity, loops over the shoulder, and braces the humerus against the glenoid cavity. The two heads converge close to the elbow on a single distal tendon that terminates on the radius and on the fascia of the medial side of the upper forearm. Note that *biceps* is the singular term; there is no such word as *bicep*.

The **triceps brachii** is a three-headed muscle on the posterior side of the humerus, and is the prime mover of elbow extension (fig. 10.25b).

Muscles of the Antebrachial Region

Most forearm muscles act on the wrist and hand, but two of them are synergists in elbow flexion and extension and three of them function in pronation and supination. The **brachioradialis** is the large fleshy mass of the lateral (radial) side of the forearm just distal to the elbow (fig. 10.25a; also see fig. 10.29a). It extends from the distal end of the humerus to the

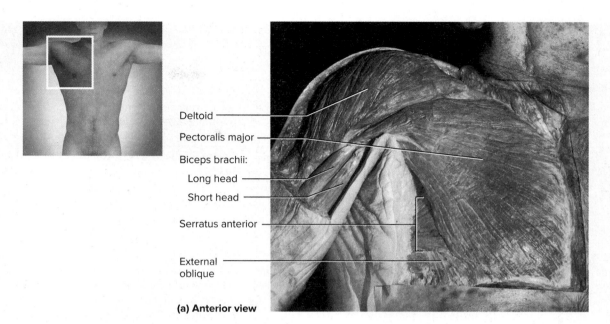

Deltoid

Pectoralis major

Biceps brachii:

Long head

Short head

Serratus anterior

External oblique

(a) Anterior view

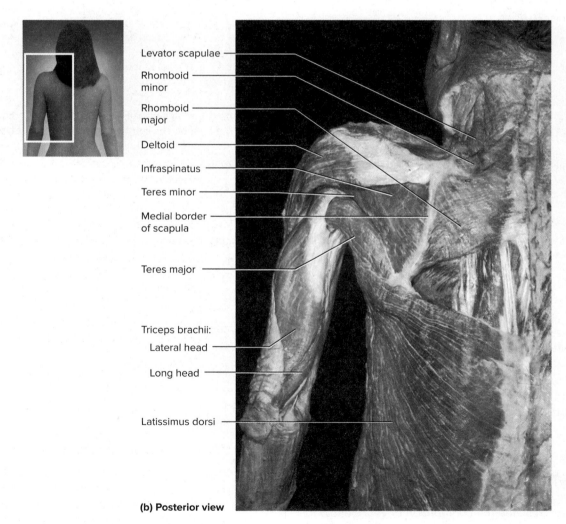

Levator scapulae

Rhomboid minor

Rhomboid major

Deltoid

Infraspinatus

Teres minor

Medial border of scapula

Teres major

Triceps brachii:

Lateral head

Long head

Latissimus dorsi

(b) Posterior view

FIGURE 10.26 **Pectoral, Brachial, and Upper Back Muscles of the Cadaver.** (a) Anterior view. (b) Posterior view.

a–b: Rebecca Gray/McGraw-Hill Education

TABLE 10.9	Muscles Acting on the Arm		
Name	Action	Skeletal Attachments	Innervation
Rotator Cuff Muscles			
Supraspinatus[59] (SOO-pra-spy-NAY-tus)	Aids deltoid in abduction of arm; resists downward slippage of humeral head when arm is relaxed or when carrying weight	• Supraspinous fossa of scapula • Greater tubercle of humerus	Suprascapular nerve
Infraspinatus[60] (IN-fra-spy-NAY-tus)	Modulates action of deltoid, preventing humeral head from sliding upward; rotates humerus laterally	• Infraspinous fossa of scapula • Greater tubercle of humerus	Suprascapular nerve
Teres minor (TERR-eez)	Modulates action of deltoid, preventing humeral head from sliding upward as arm is abducted; rotates humerus laterally	• Lateral border and adjacent posterior surface of scapula • Greater tubercle of humerus; posterior surface of joint capsule	Axillary nerve
Subscapularis[61] (SUB-SCAP-you-LERR-iss)	Modulates action of deltoid, preventing humeral head from sliding upward as arm is abducted; rotates humerus medially	• Subscapular fossa of scapula • Lesser tubercle of humerus; anterior surface of joint capsule	Upper and lower subscapular nerves
Other Scapular Muscles			
Deltoid	Anterior fibers flex and medially rotate arm; lateral fibers abduct arm; posterior fibers extend and laterally rotate arm; involved in arm swinging during such actions as walking or bowling, and in adjustment of hand height for various manual tasks	• Acromion and spine of scapula; clavicle • Deltoid tuberosity of humerus	Axillary nerve
Teres major (TERR-eez)	Extends and medially rotates humerus; contributes to arm swinging	• Inferior angle of scapula • Medial lip of intertubercular sulcus of humerus	Lower subscapular nerve
Coracobrachialis (COR-uh-co-BRAY-kee-AY-lis)	Flexes and medially rotates arm; resists deviation of arm from frontal plane during abduction	• Coracoid process • Medial aspect of humeral shaft	Musculocutaneous nerve
Axial Muscles			
Pectoralis major (PECK-toe-RAY-liss)	Flexes, adducts, and medially rotates humerus, as in climbing or hugging; aids in deep inspiration	• Medial half of clavicle; lateral margin of sternum; costal cartilages 1–7; aponeurosis of external oblique • Lateral lip of intertubercular sulcus of humerus	Medial and lateral pectoral nerves
Latissimus dorsi[62] (la-TISS-ih-mus DOR-sye)	Adducts and medially rotates humerus; extends the shoulder joint as in pulling on the oars of a rowboat; produces backward swing of arm in such actions as walking and bowling; with hands grasping overhead objects, pulls body forward and upward, as in climbing; aids in deep inspiration, sudden expiration such as sneezing and coughing, and prolonged forceful expiration as in singing or blowing a sustained note on a wind instrument	• Vertebrae T7–L5; lower three or four ribs; iliac crest; thoracolumbar fascia • Floor of intertubercular sulcus of humerus	Thoracodorsal nerve

▶▶▶**APPLY WHAT YOU KNOW**

Perform an action as if lifting a cup to your mouth to take a sip of tea. Describe the contribution of your deltoid to this action, using the terminology of joint movement in section 9.2c.

---

[59]*supra* = above; *spin* = spine of scapula
[60]*infra* = below, under; *spin* = spine of scapula

[61]*sub* = below, under
[62]*latissimus* = broadest; *dorsi* = of the back

DEEPER INSIGHT 10.4

CLINICAL APPLICATION

Rotator Cuff Injury

Rotator cuff injury is a tear in the tendon of any of the SITS (rotator cuff) muscles, most often the supraspinatus. Such injuries are caused by strenuous circumduction of the arm, shoulder dislocation, hard falls or blows to the shoulder, or repetitive use of the arm in a position above the horizontal. They are common among baseball pitchers and third basemen, bowlers, swimmers, and weight lifters, and in racquet sports. Recurrent inflammation of a SITS tendon can cause a tendon to degenerate and then to rupture in response to moderate stress. Injury causes pain and makes the shoulder joint unstable and subject to dislocation.

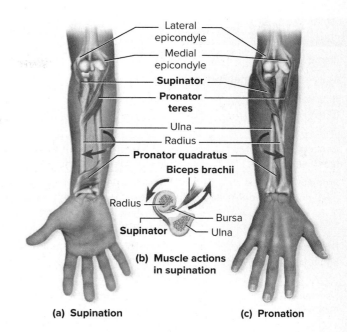

(a) Supination   **(c) Pronation**

FIGURE 10.27 Actions of the Rotator Muscles on the Forearm. (a) Supination. (b) Cross section just distal to the elbow, showing the synergistic action of the biceps brachii and supinator. (c) Pronation.

? *What do the names of the pronator teres and pronator quadratus muscles indicate about their shapes?*

distal end of the radius. With the latter attachment so far from the fulcrum of the elbow, it does not generate as much force as the brachialis and biceps; it is effective mainly when those muscles have already partially flexed the elbow. The **anconeus** is a weak synergist of elbow extension on the posterior side of the elbow (see fig. 10.30).

Pronation and supination are important forearm movements for such purposes as eating, manipulating and inspecting objects in the hands, and generating force in twisting movements of the hand (see fig. 9.18). The prime mover of pronation is the **pronator quadratus** near the wrist; the **pronator teres** near the elbow is a synergist. Supination is usually achieved by the **supinator** of the upper forearm, with the biceps brachii aiding when additional speed or power is required **(fig. 10.27).**

10.4d Muscles Acting on the Wrist and Hand

The hand is acted upon by extrinsic muscles in the forearm **(table 10.11)** and intrinsic muscles in the hand itself. The bellies of the extrinsic muscles, along with the brachioradialis,

TABLE 10.10	Muscles Acting on the Elbow and Forearm		
Name	**Action**	**Skeletal Attachments**	**Innervation**
Muscles of the Brachial Region			
Brachialis (BRAY-kee-AY-lis)	Prime mover of elbow flexion	• Anterior surface of distal half of humerus • Coronoid process and tuberosity of ulna	Musculocutaneous nerve; radial nerve
Biceps brachii (BY-seps BRAY-kee-eye)	Rapid or forceful supination of forearm; synergist in elbow flexion; slight shoulder flexion; tendon of long head stabilizes shoulder by holding humeral head against glenoid cavity	• Long head: superior margin of glenoid cavity Short head: coracoid process • Tuberosity of radius; fascia of forearm	Musculocutaneous nerve
Triceps brachii (TRI-seps BRAY-kee-eye)	Extends elbow; long head extends and adducts humerus	• Long head: inferior margin of glenoid cavity and joint capsule Lateral head: posterior surface of proximal end of humerus Medial head: posterior surface of entire humeral shaft • Olecranon; fascia of forearm	Radial nerve

(continued)

TABLE 10.10	Muscles Acting on the Elbow and Forearm (continued)		
Name	**Action**	**Skeletal Attachments**	**Innervation**
Muscles of the Antebrachial Region			
Brachioradialis (BRAY-kee-oh-RAY-dee-AY-lis)	Flexes elbow	• Lateral supracondylar ridge of humerus • Lateral surface of radius near styloid process	Radial nerve
Anconeus[63] (an-CO-nee-us)	Extends elbow; may help to control ulnar movement during pronation	• Lateral epicondyle of humerus • Olecranon and posterior surface of ulna	Radial nerve
Pronator quadratus (PRO-nay-tur quad-RAY-tus)	Prime mover of forearm pronation; also resists separation of radius and ulna when force is applied to forearm through wrist, as in doing push-ups	• Anterior surface of distal ulna • Anterior surface of distal radius	Median nerve
Pronator teres (PRO-nay-tur TERR-eez)	Assists pronator quadratus in pronation, but only in rapid or forceful action; weakly flexes elbow	• Humeral shaft near medial epicondyle; coronoid process of ulna • Lateral surface of radial shaft	Median nerve
Supinator (SOO-pih-NAY-tur)	Supinates forearm	• Lateral epicondyle of humerus; supinator crest and fossa of ulna just distal to radial notch; anular and radial collateral ligaments of elbow • Proximal one-third of radius	Radial nerve

form the fleshy roundness of the upper forearm; their tendons extend into the wrist and hand. Their actions are mainly flexion and extension of the wrist and digits, but also include radial and ulnar flexion, finger abduction and adduction, and thumb opposition. These muscles are numerous and complex, but their names often describe their location, appearance, and function.

Many of them act on the **metacarpophalangeal joints** between the metacarpal bones of the hand and the proximal phalanges of the fingers, and the **interphalangeal joints** between the proximal and middle or the middle and distal phalanges of the fingers (or between the proximal and distal phalanges in the thumb, which has no middle phalanx). The metacarpophalangeal joints form the knuckles at the bases of the fingers, and the interphalangeal joints form the second and third knuckles. Some tendons cross multiple joints before inserting on a middle or distal phalanx, and can flex or extend all the joints they cross.

Fasciae divide the forearm muscles into anterior and posterior compartments and each compartment into superficial and deep layers **(fig. 10.28).** The muscles in these four groups are described in the following sections.

The Anterior (Flexor) Compartment, Superficial Layer

Most muscles of the anterior compartment are wrist and finger flexors that arise from a common tendon on the humerus **(fig. 10.29).** The two prominent tendons you can palpate at the wrist belong to the **palmaris longus** on the medial side and the **flexor carpi radialis** on the lateral side (see fig. B.19a). The latter is an important landmark for finding the radial artery, where the pulse is usually taken. The palmaris longus is absent on one or both sides (most commonly the left) in about 14% of people. To see if you have one, flex your wrist and touch the tips of your thumb and little finger together. If present, the palmaris longus tendon will stand up prominently on the wrist.

Most tendons of these flexor muscles pass under a fibrous, braceletlike ligament called the **flexor retinaculum** on the anterior side of the wrist (see fig. 5.13); the palmaris longus tendon, unlike the rest, passes over it. The retinaculum prevents the tendons from standing up like taut bowstrings when the flexors contract. The **carpal tunnel** is a tight space between the flexor retinaculum and carpal bones. The flexor tendons passing through the tunnel are enclosed in tendon sheaths that enable them to slide back and forth quite easily, although continual repetitive motion of these tendons can cause the painful inflammation known as *carpal tunnel syndrome* (see Deeper Insight 10.5).

The Anterior (Flexor) Compartment, Deep Layer

Two anterior flexors constitute the deep layer—the **flexor digitorum profundus,** which flexes fingers II–V, and the **flexor pollicis longus,** which flexes only the thumb (fig. 10.29c). The latter is one of several muscles serving exclusively for thumb movements, attesting to the supreme importance of the thumb for hand function.

The Posterior (Extensor) Compartment, Superficial Layer

Superficial muscles of the posterior compartment are mostly wrist and finger extensors, and share a single tendon arising from the humerus **(fig. 10.30a).** The first of these, the **extensor digitorum,**

[63]*anconeus* = elbow

FIGURE 10.28 Serial Cross Sections Through the Upper Limb. Each section is taken at the correspondingly lettered level in the figure at the left and is pictured with the posterior muscle compartment facing the bottom of the page, as if viewing a person's right limb extended toward you with the palm up. (a) Superior arm at axilla. (b) Inferior arm. (c) Superior forearm.

❓ *Why are the extensor pollicis longus and extensor indicis not seen in part (c)?*

FIGURE 10.29 Flexors of the Wrist and Hand. Anterior views of the forearm. (a) Superficial flexors. (b) The intermediate flexor digitorum superficialis, deep to the muscles in part (a). (c) Deep flexors. Flexor muscles of each compartment are labeled in boldface. **APR**

FIGURE 10.30 Extensors of the Wrist and Hand. Posterior views of the forearm. Extensor muscles of each compartment are labeled in boldface. (a) Superficial extensors. (b) Deep extensors. **APR**

has four distal tendons that can easily be seen and palpated on the back of the hand when the fingers are strongly extended (see fig. B.19b). It serves digits II through V, and the other muscles in this group each serve a single digit. These superficial extensors are tabulated as follows and in the table from lateral to medial.

The Posterior (Extensor) Compartment, Deep Layer

The deep muscles that follow serve only the thumb and index finger **(fig. 10.30b).** By strongly abducting and extending the thumb into a hitchhiker's position, you may see a deep dorsolateral pit at the base of the thumb, with a taut tendon on each side of it (see fig. B.19b). This depression is called the *anatomical snuffbox* because it was once fashionable to place a pinch of snuff here and inhale it. It is bordered laterally by the tendons of the **abductor pollicis longus** and **extensor pollicis brevis,** and medially by the tendon of the **extensor pollicis longus.** Dedication of so many muscles to the thumb alone (to which *pollicis* refers) attests to the supreme importance of the thumb for so much of the hand's functionality; try to imagine how much less you could do with your hands without the thumbs and these muscles.

TABLE 10.11	Muscles Acting on the Wrist and Hand		
Name	**Action**	**Skeletal Attachments**	**Innervation**
The Anterior (Flexor) Compartment, Superficial Layer			
Flexor carpi radialis[64] (FLEX-ur CAR-pye RAY-dee-AY-lis)	Flexes wrist anteriorly; aids in radial flexion of wrist	• Medial epicondyle of humerus • Base of metacarpals II–III	Median nerve
Flexor carpi ulnaris[65] (ul-NAY-ris)	Flexes wrist anteriorly; aids in ulnar flexion of wrist	• Medial epicondyle of humerus; medial margin of olecranon; posterior surface of ulna • Metacarpal V; pisiform; hamate	Ulnar nerve
Flexor digitorum superficialis[66] (DIDJ-ih-TOE-rum SOO-per-FISH-ee-AY-lis)	Flexes wrist, metacarpophalangeal, and interphalangeal joints depending on action of other muscles	• Medial epicondyle of humerus; ulnar collateral ligament; coronoid process; superior half of radius • Middle phalanges II–V	Median nerve
Palmaris longus (pal-MERR-iss)	Anchors skin and fascia of palmar region; resists shearing forces when stress is applied to skin by such actions as climbing and tool use. Weakly developed and sometimes absent.	• Medial epicondyle of humerus • Flexor retinaculum, palmar aponeurosis	Median nerve
Anterior (Flexor) Compartment, Deep Layer			
Flexor digitorum profundus[67]	Flexes wrist, metacarpophalangeal, and interphalangeal joints; sole flexor of the distal interphalangeal joints	• Proximal three-quarters of ulna; coronoid process; interosseous membrane • Distal phalanges II–V	Median nerve; ulnar nerve
Flexor pollicis[68] longus (PAHL-ih-sis)	Flexes phalanges of thumb	• Radius; interosseous membrane • Distal phalanx I	Median nerve

(continued)

---

[64]*carpi* = of the wrist; *radialis* = of the radius
[65]*ulnaris* = of the ulna
[66]*digitorum* = of the digits; *superficialis* = shallow, near the surface

[67]*profundus* = deep
[68]*pollicis* = of the thumb

TABLE 10.11	Muscles Acting on the Wrist and Hand (continued)		
Name	**Action**	**Skeletal Attachments**	**Innervation**
Posterior (Extensor) Compartment, Superficial Layer			
Extensor carpi radialis longus	Extends wrist; aids in radial flexion of wrist	• Lateral supracondylar ridge of humerus • Base of metacarpal II	Radial nerve
Extensor carpi radialis brevis (BREV-iss)	Extends wrist; aids in radial flexion of wrist	• Lateral epicondyle of humerus • Base of metacarpal III	Radial nerve
Extensor digitorum	Extends wrist, metacarpophalangeal, and interphalangeal joints; tends to spread digits apart when extending metacarpophalangeal joints	• Lateral epicondyle of humerus • Dorsal surfaces of phalanges II–V	Radial nerve
Extensor digiti minimi[69] (DIDJ-ih-ty MIN-ih-my)	Extends wrist and all joints of little finger	• Lateral epicondyle of humerus • Proximal phalanx V	Radial nerve
Extensor carpi ulnaris	Extends and fixes wrist when fist is clenched or hand grips an object; aids in ulnar flexion of wrist	• Lateral epicondyle of humerus; posterior surface of ulnar shaft • Base of metacarpal V	Radial nerve
Posterior (Extensor) Compartment, Deep Layer			
Abductor pollicis longus	Abducts thumb in frontal (palmar) plane (radial abduction); extends thumb at carpometacarpal joint	• Posterior surfaces of radius and ulna; interosseous membrane • Trapezium; base of metacarpal I	Radial nerve
Extensor pollicis brevis	Extends metacarpal I and proximal phalanx of thumb	• Shaft of radius; interosseous membrane • Proximal phalanx I	Radial nerve
Extensor pollicis longus	Extends distal phalanx I; aids in extending proximal phalanx I and metacarpal I; adducts and laterally rotates thumb	• Posterior surface of ulna; interosseous membrane • Distal phalanx I	Radial nerve
Extensor indicis (IN-dih-sis)	Extends wrist and index finger	• Posterior surface of ulna; interosseous membrane • Middle and distal phalanges of index finger	Radial nerve

10.4e Intrinsic Muscles of the Hand

The intrinsic muscles of the hand (**table 10.12**) assist the flexors and extensors in the forearm and make finger movements more precise. They are divided into three groups: the *thenar group* at the base of the thumb, the *hypothenar group* at the base of the little finger, and the *midpalmar group* between these (**fig. 10.32**).

The Thenar Group

The **thenar group** of muscles forms the thick fleshy mass (*thenar eminence*) at the base of the thumb, and the **adductor pollicis** forms the web between the thumb and palm. All are concerned with thumb movements. The adductor pollicis has an oblique head that extends from the capitate bone of the wrist to the ulnar side of the base of the thumb, and a transverse head that extends from metacarpal III to the same termination as the oblique head.

The Hypothenar Group

The **hypothenar group** forms the fleshy mass (*hypothenar eminence*) at the base of the little finger. All of these muscles are concerned with movement of that digit.

The Midpalmar Group

The **midpalmar group** occupies the hollow of the palm. It has 11 small muscles divided into three groups—four **dorsal interosseous,** three **palmar interosseous,** and four **lumbrical muscles.**

[69] *digit* = finger; *minim* = smallest

DEEPER INSIGHT 10.5

CLINICAL APPLICATION

Carpal Tunnel Syndrome

Prolonged, repetitive motions of the wrist and fingers can cause tissues in the carpal tunnel to become inflamed, swollen, or fibrotic. Since the carpal tunnel cannot expand, swelling puts pressure on the median nerve of the wrist, which passes through the carpal tunnel with the flexor tendons **(fig. 10.31)**. This pressure causes tingling and muscular weakness in the palm and medial side of the hand and pain that may radiate to the arm and shoulder. This condition, called *carpal tunnel syndrome,* is common among keyboard operators, pianists, meat cutters, and others who spend long hours making repetitive wrist motions. Carpal tunnel syndrome is treated with aspirin and other anti-inflammatory drugs, immobilization of the wrist, and sometimes dividing the flexor retinaculum with a longitudinal incision to relieve pressure on the nerve.

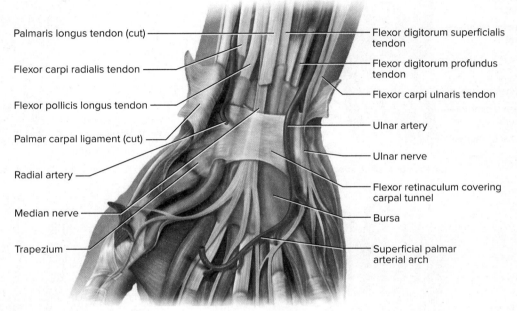

Palmaris longus tendon (cut)

Flexor carpi radialis tendon

Flexor pollicis longus tendon

Palmar carpal ligament (cut)

Radial artery

Median nerve

Trapezium

Flexor digitorum superficialis tendon

Flexor digitorum profundus tendon

Flexor carpi ulnaris tendon

Ulnar artery

Ulnar nerve

Flexor retinaculum covering carpal tunnel

Bursa

Superficial palmar arterial arch

(a) Anterior view

Anterior

Lateral   Medial

Posterior

Palmaris longus tendon

Thenar muscles

Flexor digitorum superficialis tendons

Trapezium

Flexor digitorum profundus tendons

Radial artery

Trapezoid

Scaphoid

Median nerve

Ulnar artery

Flexor retinaculum

Carpal tunnel

Hypothenar muscles

Ulnar bursa

Hamate

Capitate

Extensor tendons

(b) Cross section

FIGURE 10.31 The Carpal Tunnel. (a) Dissection of the wrist (anterior view) showing the tendons, nerve, and bursae that pass under the flexor retinaculum.   **A&PR** (b) Cross section of the wrist, viewed as if from the distal end of a person's right forearm extended toward you with the palm up. Note how the flexor tendons and median nerve are confined in the tight space between the carpal bones and flexor retinaculum. That tight packing and repetitive sliding of the flexor tendons through the tunnel contribute to carpal tunnel syndrome.

Tendon sheath

Tendon of flexor
digitorum profundus

Tendon of flexor
digitorum superficialis

Lumbricals

**Opponens
digiti minimi**

**Flexor digiti
minimi brevis**

**Abductor digiti
minimi**

Flexor retinaculum

Tendons of:
Flexor carpi ulnaris

Flexor digitorum
superficialis

Palmaris longus

**First dorsal
interosseous**

**Adductor
pollicis**

Tendon of flexor
pollicis longus

**Flexor pollicis
brevis**

**Abductor pollicis
brevis**

Opponens pollicis

Tendons of:
Abductor pollicis
longus
Flexor carpi
radialis
Flexor pollicis
longus

(a) Palmar aspect, superficial

Tendon of flexor
digitorum
superficialis

Lumbrical

Opponens
digiti minimi

**Flexor digiti
minimi brevis**

**Abductor digiti
minimi**

Pisiform bone

**Flexor digitorum
superficialis**

**Adductor
pollicis**

**Flexor pollicis
brevis**

**Abductor pollicis
brevis**

Tendon of extensor
pollicis brevis

Tendon of flexor
carpi radialis

(b) Palmar dissection, superficial

**Palmar
interosseous**

**Opponens digiti
minimi**

Flexor retinaculum
(cut)

Carpal tunnel

**Opponens
pollicis**

Tendons of:
Abductor pollicis
longus

Flexor carpi radialis

Flexor carpi
ulnaris

(c) Palmar aspect, deep

Adductor
pollicis

**Abductor
pollicis
brevis**

Common tendon
sheath of extensor
digitorum and
extensor indicis

Tendons of extensor
pollicis brevis and
abductor pollicis longus

Tendons of
extensor digitorum
(cut)

Dorsal interosseous

**Abductor digiti
minimi**

Extensor retinaculum

Tendons of:
Extensor digiti minimi
Extensor carpi ulnaris
Extensor pollicis longus
Extensor digitorum

(d) Dorsal aspect

FIGURE 10.32 Intrinsic Muscles of the Hand. (a) Superficial muscles of the palmar aspect. (b) Dissection of the palmar region
of the cadaver hand. (c) Deep muscles of the palmar aspect. (d) Muscles of the dorsal aspect. The boldface labels in parts (a),
(c), and (d) indicate the muscles that belong to the respective layers. **APR**

TABLE 10.12	Intrinsic Muscles of the Hand		
Name	**Action**	**Skeletal Attachments**	**Innervation**
Thenar Group			
Adductor pollicis	Draws thumb toward palm as in gripping a tool	• Capitate; bases of metacarpals II–III; anterior ligaments of wrist; tendon sheath of flexor carpi radialis • Medial surface of proximal phalanx I	Ulnar nerve
Abductor pollicis brevis	Abducts thumb in sagittal plane	• Mainly flexor retinaculum; also scaphoid, trapezium, and abductor pollicis longus tendon • Lateral surface of proximal phalanx I	Median nerve
Flexor pollicis brevis	Flexes metacarpophalangeal joint of thumb	• Trapezium; trapezoid; capitate; anterior ligaments of wrist; flexor retinaculum • Proximal phalanx I	Median nerve; ulnar nerve
Opponens pollicis (op-PO-nenz)	Flexes metacarpal I to oppose thumb to fingertips	• Trapezium; flexor retinaculum • Metacarpal I	Median nerve
Hypothenar Group			
Abductor digiti minimi	Abducts little finger, as in spreading fingers apart	• Pisiform; tendon of flexor carpi ulnaris • Medial surface of proximal phalanx V	Ulnar nerve
Flexor digiti minimi brevis	Flexes little finger at metacarpophalangeal joint	• Hamulus of hamate bone; flexor retinaculum • Medial surface of proximal phalanx V	Ulnar nerve
Opponens digiti minimi	Flexes metacarpal V at carpometacarpal joint when little finger is moved into opposition with tip of thumb; deepens palm of hand	• Hamulus of hamate bone; flexor retinaculum • Medial surface of metacarpal V	Ulnar nerve
Midpalmar Group			
Four dorsal interosseous[70] muscles (IN-tur-OSS-ee-us)	Abduct fingers; strongly flex metacarpophalangeal joints but extend interphalangeal joints, depending on action of other muscles; important in grip strength	• Each with two heads arising from facing surfaces of adjacent metacarpals • Proximal phalanges II–IV	Ulnar nerve
Three palmar interosseous muscles	Adduct fingers; other actions same as for dorsal interosseous muscles	• Metacarpals I, II, IV, V • Proximal phalanges II, IV, V	Ulnar nerve
Four lumbrical[71] muscles (LUM-brih-cul)	Extend interphalangeal joints; contribute to ability to pinch objects between fleshy pulp of thumb and finger, instead of these digits meeting by the edges of their nails	• Tendons of flexor digitorum profundus • Proximal phalanges II–V	Median nerve; ulnar nerve

BEFORE YOU GO ON

Answer the following questions to test your understanding of the preceding section:

16. Name a muscle that inserts on the scapula and plays a significant role in each of the following actions:

    a. pushing a stalled car,

    b. paddling a canoe,

    c. squaring the shoulders in military attention,

    d. lifting the shoulder to carry a heavy box on it, and

    e. lowering the shoulder to lift a suitcase.

17. Describe three contrasting actions of the deltoid muscle.

18. Name the four rotator cuff muscles and describe the scapular surfaces against which they lie.

19. Name the prime movers of elbow flexion and extension.

20. Identify three functions of the biceps brachii.

21. Name three extrinsic muscles and two intrinsic muscles that flex the phalanges.

---

[70]*inter* = between; *osse* = bones
[71]*lumbrical* = resembling an earthworm

10.5 Muscles Acting on the Hip and Lower Limb

Expected Learning Outcomes

When you have completed this section, you should be able to

a. name and locate the muscles that act on the hip, knee, ankle, and toe joints;

b. relate the actions of these muscles to the joint movements described in chapter 9; and

c. describe the skeletal attachments, action, and innervation of these muscles.

The largest muscles are found in the lower limb. Unlike those of the upper limb, they are adapted less for precision than for the strength needed to stand, maintain balance, walk, and run. Several of them cross and act upon two or more joints, such as the hip and knee. To avoid confusion in this discussion, remember that in the anatomical sense the word *leg* refers only to that part of the limb between the knee and ankle. The *foot* includes the tarsal region, metatarsal region, and toes. The next four tables group the muscles of the lower limb into those that act on the thigh and hip joint, those that act on the leg and knee joint, extrinsic (leg) muscles that act on the foot and ankle joint, and intrinsic (foot) muscles that act on the arches and toes.

10.5a Muscles Acting on the Hip and Thigh

The hip joint and thigh are acted on by muscles arising from both the pelvic girdle and the femur (**table 10.13**).

Anterior Muscles of the Hip

Most muscles that act on the femur originate on the hip bone. The two principal anterior muscles are the **iliacus,** which fills most of the broad iliac fossa of the pelvis, and the **psoas major,** a thick rounded muscle that arises mainly from the lumbar vertebrae (**fig. 10.33**). Collectively, they are called the **iliopsoas** and share a common tendon to the femur.

Lateral and Posterior Muscles of the Hip

On the lateral and posterior sides of the hip are the **tensor fasciae latae** and three gluteal muscles. The **fascia lata** is a fibrous sheath that encircles the thigh like a subcutaneous stocking and tightly binds its muscles. On the lateral surface, it combines with the tendons of the gluteus maximus and tensor fasciae latae to form the **iliotibial tract,** which extends from

FIGURE 10.33 Muscles That Act on the Hip and Femur (Anterior View). APR

the iliac crest to the lateral condyle of the tibia (see fig. 10.34; table 10.14). The tensor fasciae latae tautens the iliotibial tract and braces the knee, especially when the opposite foot is lifted.

The gluteal muscles are the **gluteus maximus, gluteus medius,** and **gluteus minimus (fig. 10.34).** The gluteus maximus is the largest of these and forms most of the lean mass of the buttock. It is an extensor of the hip joint that produces the backswing of the leg in walking and provides most of the lift when you climb stairs. It generates its maximum force when the thigh is flexed at a 45° angle to the trunk. This is the advantage in starting a foot race from a crouched position. The gluteus medius is deep and lateral to the gluteus maximus. Its name refers to its size, not its position. The gluteus minimus is the smallest and deepest of the three.

Iliac crest

Gluteus medius

Sacrum

Gluteus maximus

Coccyx

Gracilis

Iliotibial tract

Hamstring group:
  Biceps femoris
    Long head
    Short head
  Semitendinosus
  Semimembranosus

Popliteal fossa

Gluteus minimus

Lateral rotators:
  Piriformis
  Gemellus superior
  Obturator internus
  Obturator externus
  Gemellus inferior
  Quadratus femoris

Ischial tuberosity

Adductor magnus

Gracilis

Vastus lateralis

Gastrocnemius:
  Medial head
  Lateral head

FIGURE 10.34 Posterior Gluteal and Thigh Muscles. The left side shows the superficial muscles. On the right, the gluteus medius and maximus are removed to show the deeper gluteus minimus, lateral rotator group, and hamstring origins. The oblique white tendinous band across the semitendinosus is not always present. **A&PR**

❓ *Describe two everyday movements of the body that employ the power of the gluteus maximus.*

Medial (Adductor) Compartment of the Thigh

Fasciae divide the thigh into three compartments: the anterior (extensor) compartment, posterior (flexor) compartment, and medial (adductor) compartment (see fig. 10.41). Muscles of the anterior and posterior compartments function mainly as extensors and flexors of the knee, respectively, and are treated in table 10.14. The five muscles of the medial compartment act primarily as adductors of the thigh (see fig. 10.33), but some of them cross both the hip and knee joints and have additional actions as follows.

Lateral Rotators

Inferior to the gluteus minimus and deep to the other two gluteal muscles are six muscles called the **lateral rotators,** named for their action on the femur (fig. 10.34). Their action is most clearly visualized when you cross your legs to rest an ankle on your knee, causing your femur to rotate and the knee to point laterally. Thus, they oppose medial rotation by the gluteus medius and minimus. Most of them also abduct or adduct the femur. The abductors are important in walking because when one lifts a foot from the ground, they shift the body weight to the other leg and prevent falling.

TABLE 10.13	Muscles Acting on the Hip and Thigh		
Name	**Action**	**Skeletal Attachments**	**Innervation**
Anterior Muscles of the Hip			
Iliacus[72] (ih-LY-uh-cus)	Flexes thigh at hip when trunk is fixed; flexes trunk at hip when thigh is fixed, as in bending forward in a chair or sitting up in bed; balances trunk during sitting	• Iliac crest and fossa; superolateral region of sacrum; anterior sacroiliac and iliolumbar ligaments • Lesser trochanter and nearby shaft of femur	Femoral nerve
Psoas[73] major (SO-ass)	Same as iliacus	• Bodies and intervertebral discs of vertebrae T12–L5; transverse processes of lumbar vertebrae • Lesser trochanter and nearby shaft of femur	Anterior rami of lumbar spinal nerves
Lateral and Posterior Muscles of the Hip			
Tensor fasciae latae[74] (TEN-sur FASH-ee-ee LAY-tee)	Extends knee, laterally rotates tibia, aids in abduction and medial rotation of femur; during standing, steadies pelvis on femoral head and steadies femoral condyles on tibia	• Iliac crest; anterior superior spine; deep surface of fascia lata • Lateral condyle of tibia via iliotibial tract	Superior gluteal nerve
Gluteus maximus[75]	Extends thigh at hip as in stair climbing (rising to next step) or running and walking (backswing of limb); abducts thigh; elevates trunk after stooping; prevents trunk from pitching forward during walking and running; helps stabilize femur on tibia	• Posterior gluteal line of ilium, on posterior surface from iliac crest to posterior superior spine; coccyx; posterior surface of lower sacrum; aponeurosis of erector spinae • Gluteal tuberosity of femur; lateral condyle of tibia via iliotibial tract	Inferior gluteal nerve
Gluteus medius and gluteus minimus	Abduct and medially rotate thigh; during walking, shift weight of trunk toward limb with foot on the ground as other foot is lifted	• Most of lateral surface of ilium between crest and acetabulum • Greater trochanter of femur	Superior gluteal nerve
Medial (Adductor) Compartment of the Thigh			
Adductor brevis	Adducts thigh	• Body and inferior ramus of pubis • Linea aspera and spiral line of femur	Obturator nerve
Adductor longus	Adducts and medially rotates thigh; flexes thigh at hip	• Body and inferior ramus of pubis • Linea aspera of femur	Obturator nerve
Adductor magnus	Adducts and medially rotates thigh; extends thigh at hip	• Inferior ramus of pubis; ramus and tuberosity of ischium • Linea aspera, gluteal tuberosity, and medial supra-condylar line of femur	Obturator nerve; tibial nerve
Gracilis[76] (GRASS-ih-lis)	Flexes and medially rotates tibia at knee	• Body and inferior ramus of pubis; ramus of ischium • Medial surface of tibia just below condyle	Obturator nerve
Pectineus[77] (pec-TIN-ee-us)	Flexes and adducts thigh	• Superior ramus of pubis • Spiral line of femur	Femoral nerve
Lateral Rotators			
Gemellus[78] superior (jeh-MEL-us)	Laterally rotates extended thigh; abducts flexed thigh; sometimes absent	• Ischial spine • Greater trochanter of femur	Nerve to obturator internus
Gemellus inferior	Same actions as gemellus superior	• Ischial tuberosity • Greater trochanter of femur	Nerve to quadratus femoris
Obturator[79] externus (OB-too-RAY-tur)	Not well understood; thought to laterally rotate thigh in climbing	• External surface of obturator membrane; rami of pubis and ischium • Femur between head and greater trochanter	Obturator nerve

[72]*ili* = loin, flank
[73]*psoa* = loin
[74]*fasc* = band; *lat* = broad
[75]*glut* = buttock; *maxim* = largest

[76]*gracil* = slender
[77]*pectin* = comb
[78]*gemellus* = twin
[79]*obtur* = to close, stop up

TABLE 10.13	Muscles Acting on the Hip and Thigh (continued)		
Name	**Action**	**Skeletal Attachments**	**Innervation**
Lateral Rotators (continued)			
Obturator internus	Not well understood; thought to laterally rotate extended thigh and abduct flexed thigh	• Ramus of ischium; inferior ramus of pubis; antero-medial surface of lesser pelvis • Greater trochanter of femur	Nerve to obturator internus
Piriformis[80] (PIR-ih-FOR-mis)	Laterally rotates extended thigh; abducts flexed thigh	• Anterior surface of sacrum; gluteal surface of ilium; capsule of sacroiliac joint • Greater trochanter of femur	Spinal nerves L5–S2
Quadratus femoris[81] (quad-RAY-tus FEM-oh-ris)	Laterally rotates thigh	• Ischial tuberosity • Intertrochanteric crest of femur	Nerve to quadratus femoris

10.5b Muscles Acting on the Knee and Leg

The following muscles **(table 10.14)** form most of the mass of the thigh and produce their most obvious actions on the knee joint. Some of them, however, cross both the hip and knee joints and produce actions at both, moving the femur, tibia, and fibula.

Anterior (Extensor) Compartment of the Thigh

The anterior compartment of the thigh contains the large **quadriceps femoris,** the prime mover of knee extension and the most powerful muscle of the body **(figs. 10.35, 10.36).** As the name *quadriceps* implies, it has four heads: the **rectus femoris, vastus lateralis, vastus medialis,** and **vastus intermedius.** All four converge on a single **quadriceps (patellar) tendon,** which extends to the patella, then continues as the **patellar ligament** and inserts on the tibial tuberosity. (Remember that a tendon usually extends from muscle to bone, and a ligament from bone to bone.) The patellar ligament is struck with a rubber reflex hammer to test the knee-jerk reflex. The quadriceps extends the knee when you stand up, take a step, or kick a ball. One head, the rectus femoris, contributes to running by acting with the iliopsoas to flex the hip in each airborne phase of the leg's cycle of motion. The rectus femoris also flexes the hip in such actions as high kicks, stair climbing, or simply drawing the leg forward during a stride.

Crossing the quadriceps from the lateral side of the hip to the medial side of the knee is the narrow, straplike **sartorius,** the longest muscle of the body. It flexes the hip and knee joints and laterally rotates the thigh, as in crossing the legs. It is colloquially called the "tailor's muscle" after the cross-legged posture of a tailor supporting his work on the raised knee.

Posterior (Flexor) Compartment of the Thigh

The posterior compartment contains three muscles colloquially known as the **hamstring muscles;** from lateral to medial, they

FIGURE 10.35 Superficial Anterior Thigh Muscles of the Cadaver. Right thigh.

Rebecca Gray/McGraw-Hill Education

are the **biceps femoris, semitendinosus,** and **semimembranosus** (see fig. 10.34). The pit at the back of the knee, known anatomically as the **popliteal fossa,** is colloquially called the *ham.* The tendons of these muscles can be felt as prominent cords on both sides of the fossa—the biceps tendon on the lateral side and the

[80]*piri* = pear; *form* = shaped
[81]*quadrat* = four-sided; *femoris* = of the thigh or femur

Iliac crest

Iliopsoas:
  Iliacus
  Psoas major

L5

Anterior superior iliac spine

Tensor fasciae latae

Iliotibial tract

Medial compartment:
  Adductor magnus
  Pectineus
  Adductor brevis
  Adductor longus
  Gracilis

Anterior compartment:
  Sartorius

Quadriceps femoris:
  Vastus intermedius
  Rectus femoris
  Vastus lateralis
  Vastus medialis

Quadriceps femoris tendon

Patella

Patellar ligament

FIGURE 10.36 Anterior Muscles of the Thigh.
(a) Superficial muscles. (b) Rectus femoris and other muscles removed to expose the other three heads of the quadriceps femoris. **APR**

(a) Superficial

(b) Deep

semimembranosus and semitendinosus tendons on the medial side. When wolves attack large prey, they instinctively attempt to sever the hamstring tendons, because this renders the prey helpless. The hamstrings flex the knee, and aided by the gluteus maximus, they extend the hip during walking and running. The semitendinosus is named for its unusually long tendon. This muscle also is usually bisected by a transverse or oblique tendinous band.

The semimembranosus is named for the flat shape of its superior attachment.

Posterior Compartment of the Leg

Most muscles in the posterior compartment of the leg act on the ankle and foot and are reviewed in table 10.15, but the **popliteus** acts on the knee (see figs. 10.39, 10.40).

TABLE 10.14	Muscles Acting on the Knee and Leg		
Name	**Action**	**Skeletal Attachments**	**Innervation**
Anterior (Extensor) Compartment of the Thigh			
Quadriceps femoris (QUAD-rih-seps FEM-oh-ris)	Extends the knee, in addition to the actions of individual heads noted subsequently	• Varies; see individual heads • Patella; tibial tuberosity; lateral and medial condyles of tibia	Femoral nerve
Rectus femoris	Extends knee; flexes thigh at hip; flexes trunk on hip if thigh is fixed	• Ilium at anterior inferior spine and superior margin of acetabulum; capsule of hip joint • See quadriceps femoris	Femoral nerve
Vastus[82] lateralis	Extends knee; retains patella in groove on femur during knee movements	• Femur at greater trochanter and intertrochanteric line, gluteal tuberosity, and linea aspera • See quadriceps femoris	Femoral nerve
Vastus medialis	Same as vastus lateralis	• Femur at intertrochanteric line, spiral line, linea aspera, and medial supracondylar line • See quadriceps femoris	Femoral nerve
Vastus intermedius	Extends knee	• Anterior and lateral surfaces of femoral shaft • See quadriceps femoris	Femoral nerve
Sartorius[83]	Aids in knee and hip flexion, as in sitting or climbing; abducts and laterally rotates thigh	• On and near anterior superior spine of ilium • Medial surface of proximal end of tibia	Femoral nerve
Posterior (Flexor) Compartment of the Thigh			
Biceps femoris	Flexes knee; extends hip; elevates trunk from stooping posture; laterally rotates tibia on femur when knee is flexed; laterally rotates femur when hip is extended; counteracts forward bending at hips	• Long head: ischial tuberosity Short head: linea aspera and lateral supracondylar line of femur • Head of fibula	Tibial nerve; common fibular nerve
Semitendinosus[84] (SEM-ee-TEN-din-OH-sus)	Flexes knee; medially rotates tibia on femur when knee is flexed; medially rotates femur when hip is extended; counteracts forward bending at hips	• Ischial tuberosity • Medial surface of upper tibia	Tibial nerve
Semimembranosus[85] (SEM-ee-MEM-bran-OH-sus)	Same as semitendinosus	• Ischial tuberosity • Medial condyle and nearby margin of tibia; intercondylar line and lateral condyle of femur; ligament of popliteal region	Tibial nerve
Posterior Compartment of the Leg			
Popliteus[86] (pop-LIT-ee-us)	Rotates tibia medially on femur if femur is fixed (as in sitting down), or rotates femur laterally on tibia if tibia is fixed (as in standing up); unlocks knee to allow flexion; may prevent forward dislocation of femur during crouching	• Lateral condyle of femur; lateral meniscus and joint capsule • Posterior surface of upper tibia	Tibial nerve

[82]*vastus* = large, extensive
[83]*sartor* = tailor
[84]*semi* = half; *tendinosus* = tendinous

[85]*semi* = half; *membranosus* = membranous
[86]*poplit* = ham (pit) of the knee

10.5c Muscles Acting on the Ankle and Foot

The fleshy mass of the leg is formed by a group of crural muscles **(table 10.15),** which act on the foot **(fig. 10.37).** These muscles are tightly bound by fasciae that compress them and aid in the return of blood from the legs. The fasciae separate the crural muscles into anterior, lateral, and posterior compartments (see fig. 10.41b).

Anterior (Extensor) Compartment of the Leg

Muscles of the anterior compartment dorsiflex the ankle and prevent the toes from scuffing the ground during walking. From lateral to medial, these muscles are the **fibularis tertius, extensor digitorum longus** (extensor of toes II–V), **extensor hallucis longus** (extensor of the great toe), and **tibialis anterior (fig. 10.38).** Their tendons are held tightly against the ankle and kept from bowing by two **extensor retinacula** similar to the one at the wrist.

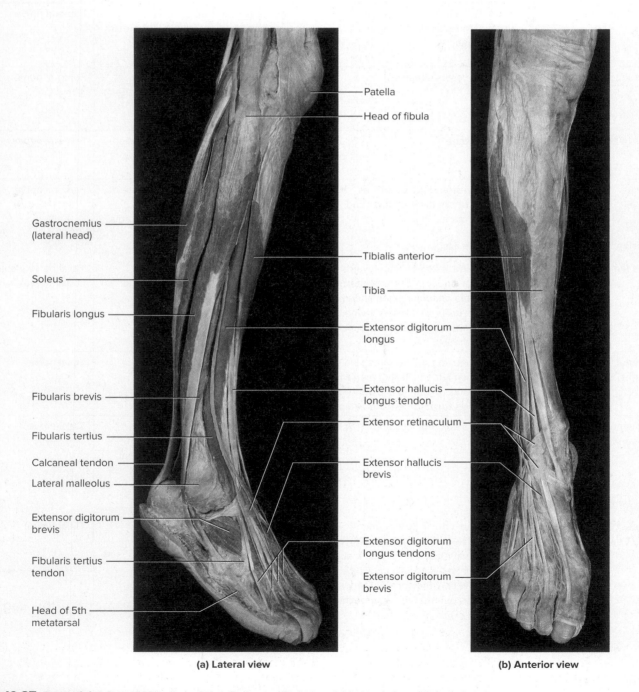

Patella
Head of fibula

Gastrocnemius (lateral head)

Soleus

Fibularis longus

Tibialis anterior

Tibia

Extensor digitorum longus

Fibularis brevis

Extensor hallucis longus tendon

Fibularis tertius

Extensor retinaculum

Calcaneal tendon

Lateral malleolus

Extensor hallucis brevis

Extensor digitorum brevis

Fibularis tertius tendon

Extensor digitorum longus tendons

Extensor digitorum brevis

Head of 5th metatarsal

(a) Lateral view

(b) Anterior view

FIGURE 10.37 Superficial Crural Muscles of the Cadaver. Right leg. (a) Lateral view. (b) Anterior view.

FIGURE 10.38 Muscles of the Leg, Anterior Compartment. Boldface labels indicate muscles belonging to the anterior compartment. (a) Superficial anterior view of the leg. Some muscles of the posterior and lateral compartments are also partially visible. (b)–(d) Individual muscles of the anterior compartment of the leg and dorsal aspect of the foot. **APR**

❓ *Palpate the hard anterior angle of your own tibia at midshaft, then continue medially until you feel muscle. What muscle is that?*

Inflammation of these tendons is one of the causes of *shinsplints* (see Deeper Insight 10.6).

Posterior (Flexor) Compartment of the Leg, Superficial Group

The posterior compartment has superficial and deep muscle groups **(fig. 10.39).** The three muscles of the superficial group are plantar flexors: the **gastrocnemius, soleus,** and **plantaris.** The first two of these, collectively known as the **triceps surae,**[87] insert on the calcaneus by way of the **calcaneal (Achilles) tendon.** This is the strongest tendon of the body but is nevertheless a common site of sports injuries resulting from sudden stress (see Deeper Insight 10.7). The *plantaris,* a weak synergist of the triceps surae, is a relatively unimportant muscle and is absent from many people; it is not tabulated here. Surgeons often use the plantaris tendon for tendon grafts needed in other parts of the body.

---

[87]*tri* = three; *ceps* = heads; *surae* = of the calf

 DEEPER INSIGHT 10.6

CLINICAL APPLICATION

Shinsplints

Shinsplints is a general term embracing several kinds of injury with pain in the crural region: tendinitis of the tibialis posterior muscle, inflammation of the tibial periosteum, and anterior compartment syndrome. Shinsplints can result from unaccustomed jogging, walking on snowshoes, or any vigorous activity of the legs after a period of relative inactivity.

Posterior (Flexor) Compartment of the Leg, Deep Group

There are four muscles in the deep group **(fig. 10.40).** The **flexor digitorum longus, flexor hallucis longus,** and **tibialis posterior** are plantar flexors. The fourth muscle, the *popliteus,* is described in table 10.14 because it acts on the knee rather than on the foot.

Plantaris

Heads of gastrocnemius (cut)

Popliteus

Fibularis longus

Gastrocnemius:

Medial head

Lateral head

Soleus

Tendon of plantaris

Tendon of gastrocnemius

Gastrocnemius (cut)

Fibularis longus

Fibularis brevis

Flexor digitorum longus

Flexor hallucis longus

Calcaneal tendon

Calcaneus

(a)   (b)

FIGURE 10.39 Superficial Muscles of the Leg, Posterior Compartment. (a) The gastrocnemius. (b) The soleus, deep to the gastrocnemius and sharing the calcaneal tendon with it. **APR**

 DEEPER INSIGHT 10.7

CLINICAL APPLICATION

Calcaneal Tendon Rupture

Calcaneal tendon rupture is a common injury experienced in amateur, strenuous sports such as basketball and tennis. When pushing off strongly from the ground as in a basketball jump shot, the tendon may rupture with a loud pop, described by some as sounding like a gunshot. The rupture causes pain just above the heel and inability to stand or walk on the affected leg. Two visible signs of a ruptured tendon are a large knot, the contracted triceps surae, in the upper calf; and abnormal dorsiflexion of the foot, caused by the action of the tibialis anterior suddenly unfettered by the opposing action of the triceps. Calcaneal tendon rupture is most common in overweight men, 30 to 40 years old, who are not in great physical condition but partake in an unaccustomed game such as pickup basketball or backyard football. A ruptured tendon can often heal on its own with rest and proper care, but may require surgical repair, especially in young physically active people not inclined to rest it for months.

Plantaris (cut)

Gastrocnemius (cut)

Popliteus

Soleus (cut)

Fibula

Tibialis posterior

Flexor digitorum longus

Fibularis longus

Flexor hallucis longus

Fibularis brevis

Calcaneal tendon (cut)

Calcaneus

(a)

Tibialis posterior

(b)

Flexor digitorum longus

(c)

Popliteus

Flexor hallucis longus

Plantar surface of the foot

(d)

FIGURE 10.40 Deep Muscles of the Leg, Posterior and Lateral Compartments. (a) Muscles deep to the soleus. (b)–(d) Exposure of some individual deep muscles with the foot plantar flexed (sole facing viewer). **APR**

TABLE 10.15	Muscles Acting on the Ankle and Foot		
Name	**Action**	**Skeletal Attachments**	**Innervation**
Anterior (Extensor) Compartment of the Leg			
Fibularis (peroneus[88]) tertius[89] (FIB-you-LERR-iss TUR-she-us)	Dorsiflexes and everts foot during walking; helps toes clear the ground during forward swing of leg	• Medial surface of lower one-third of fibula; interosseous membrane • Metatarsal V	Deep fibular (peroneal) nerve
Extensor digitorum longus (DIDJ-ih-TOE-rum)	Extends toes; dorsiflexes foot; tautens plantar aponeurosis	• Lateral condyle of tibia; shaft of fibula; interosseous membrane • Middle and distal phalanges II–V	Deep fibular (peroneal) nerve
Extensor hallucis longus (ha-LOO-sis)	Extends great toe; dorsiflexes foot	• Anterior surface of middle of fibula, interosseous membrane • Distal phalanx I	Deep fibular (peroneal) nerve
Tibialis[90] anterior (TIB-ee-AY-lis)	Dorsiflexes and inverts foot; resists backward tipping of body (as when standing on a moving boat deck); helps support medial longitudinal arch of foot	• Lateral condyle and lateral margin of proximal half of tibia; interosseous membrane • Medial cuneiform, metatarsal I	Deep fibular (peroneal) nerve
Posterior (Flexor) Compartment of the Leg, Superficial Group			
Gastrocnemius[91] (GAS-trock-NEE-me-us)	Plantar flexes foot, flexes knee; active in walking, running, and jumping	• Condyles, popliteal surface, and lateral supracondylar line of femur; capsule of knee joint • Calcaneus	Tibial nerve
Soleus[92] (SO-lee-us)	Plantar flexes foot; steadies leg on ankle during standing	• Posterior surface of head and proximal one-fourth of fibula; middle one-third of tibia; interosseous membrane • Calcaneus	Tibial nerve
Posterior (Flexor) Compartment of the Leg, Deep Group			
Flexor digitorum longus	Flexes phalanges of digits II–V as foot is raised from ground; stabilizes metatarsal heads and keeps distal pads of toes in contact with ground in toe-off and tiptoe movements	• Posterior surface of tibial shaft • Distal phalanges II–V	Tibial nerve
Flexor hallucis longus	Same actions as flexor digitorum longus, but for great toe (digit I)	• Distal two-thirds of fibula and interosseous membrane • Distal phalanx I	Tibial nerve
Tibialis posterior	Inverts foot; may assist in strong plantar flexion or control pronation of foot during walking	• Posterior surface of proximal half of tibia, fibula, and interosseous membrane • Navicular, medial cuneiform, metatarsals II–IV	Tibial nerve
Lateral (Fibular) Compartment of the Leg			
Fibularis (peroneus) brevis	Maintains concavity of sole during toe-off and tiptoeing; may evert foot and limit inversion and help steady leg on foot	• Lateral surface of distal two-thirds of fibula • Base of metatarsal V	Superficial fibular (peroneal) nerve
Fibularis (peroneus) longus	Maintains concavity of sole during toe-off and tiptoeing; everts and plantar flexes foot	• Head and lateral surface of proximal two-thirds of fibula • Medial cuneiform, metatarsal I	Superficial fibular (peroneal) nerve

[88]*perone* = pinlike (fibula)
[89]*fibularis* = of the fibula; *tert* = third
[90]*tibialis* = of the tibia

[91]*gastro* = belly; *cnem* = leg
[92]Named for its resemblance to a flatfish (sole)

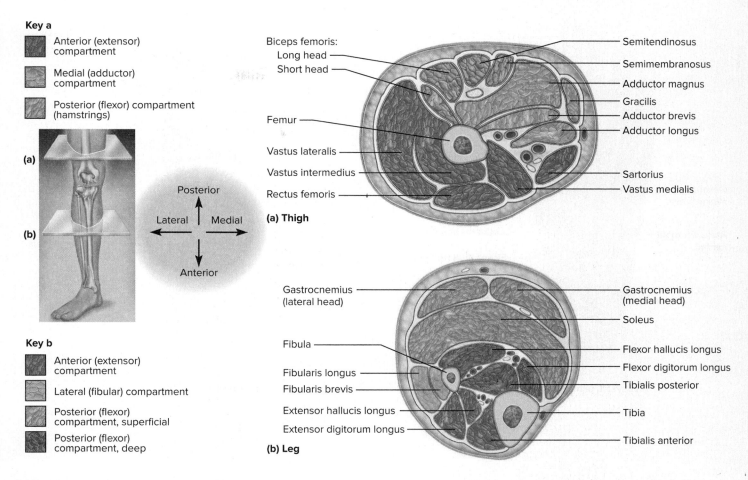

Key a

- Anterior (extensor) compartment
- Medial (adductor) compartment
- Posterior (flexor) compartment (hamstrings)

(a)

(b)

Posterior

Lateral — Medial

Anterior

Key b

- Anterior (extensor) compartment
- Lateral (fibular) compartment
- Posterior (flexor) compartment, superficial
- Posterior (flexor) compartment, deep

Biceps femoris:
Long head
Short head

Femur

Vastus lateralis

Vastus intermedius

Rectus femoris

(a) Thigh

Semitendinosus

Semimembranosus

Adductor magnus

Gracilis

Adductor brevis

Adductor longus

Sartorius

Vastus medialis

Gastrocnemius (lateral head)

Fibula

Fibularis longus

Fibularis brevis

Extensor hallucis longus

Extensor digitorum longus

(b) Leg

Gastrocnemius (medial head)

Soleus

Flexor hallucis longus

Flexor digitorum longus

Tibialis posterior

Tibia

Tibialis anterior

FIGURE 10.41 Serial Cross Sections Through the Lower Limb. Cross sections of the femur (a) and leg (b) are taken at the correspondingly lettered levels in the figure at the left.

Lateral (Fibular) Compartment of the Leg

The lateral compartment includes the **fibularis brevis** and **fibularis longus** (see figs. 10.37a, 10.38a; **fig. 10.41**). They plantar flex and evert the foot. Plantar flexion is important not only in standing on tiptoes but in providing lift and forward thrust each time you take a step.

▶▶▶ **APPLY WHAT YOU KNOW**

Suppose you tilt your foot upward to trim or paint your toenails. Identify as many muscles as you know that can contribute to this action.

▶▶▶ **APPLY WHAT YOU KNOW**

Suppose you are playing Frisbee and your Frisbee lands on the roof of your house. You have to climb a ladder to get it. Identify as many muscles as you can that would aid you in ascending from one rung of the ladder to the next.

10.5d Intrinsic Muscles of the Foot

The intrinsic muscles of the foot (**table 10.16**) help to support the arches and act on the toes in ways that aid locomotion. Several of them are similar in name and location to the intrinsic muscles of the hand.

Dorsal (Superior) Aspect of the Foot

Only one of the intrinsic muscles, the **extensor digitorum brevis,** is on the dorsal (superior) side of the foot. The medial slip of this muscle, serving the great toe, is sometimes called the *extensor hallucis brevis.*

Ventral Layer 1 (Most Superficial)

All remaining intrinsic muscles are on the ventral (inferior) aspect of the foot or between the metatarsal bones. They are grouped in four layers (**fig. 10.42).** Dissecting into the foot from the plantar surface, one first encounters a tough fibrous sheet, the **plantar aponeurosis,** between the skin and muscles. It diverges like a fan from the calcaneus to the bases of all the toes, and is a common site of painful inflammation (see Deeper Insight 10.8). Several ventral muscles arise from this aponeurosis. The ventral muscles include the stout **flexor digitorum brevis** on the midline of the foot, with four tendons that supply all digits except the great toe. It is flanked by the **abductor digiti minimi** laterally and the **abductor hallucis** medially.

Ventral Layer 2

The next deeper layer consists of the thick **quadratus plantae (flexor accessorius)** in the middle of the foot and the four **lumbrical muscles** located between the metatarsals.

Flexor digiti
minimi brevis

**Abductor digiti
minimi**

Abductor hallucis

**Flexor digitorum
brevis**

Plantar aponeurosis
(cut)

Calcaneus

(a) Layer 1, plantar view

Lumbricals

Flexor hallucis
longus tendon

Flexor digitorum
longus tendon

Abductor hallucis
(cut)

Quadratus plantae

Flexor digitorum
brevis (cut)

(b) Layer 2, plantar view

Adductor hallucis

Flexor hallucis brevis

Flexor hallucis
longus tendon (cut)

Abductor hallucis (cut)

Flexor digitorum
longus tendon (cut)

Flexor digiti
minimi brevis

Quadratus plantae
(cut)

(c) Layer 3, plantar view

**Plantar
interosseous**

**Dorsal
interosseous**

(d) Layer 4, plantar view

(e) Layer 4, dorsal view

FIGURE 10.42 Intrinsic Muscles of the Foot. (a)–(d) First through fourth layers, respectively, in ventral (plantar) views. (e) Fourth layer, dorsal view. The muscles belonging to each layer are shown in color and with boldface labels. **A&PR**

DEEPER INSIGHT 10.8

CLINICAL APPLICATION

Plantar Fasciitis

Plantar fasciitis (FASH-ee-EYE-tis) is inflammation of the plantar apo-neurosis, typically felt as a recurring pain in the heel and medial aspect of the foot. The pain is often most severe when one begins to walk after prolonged sitting or when rising from bed. It usually subsides in 5 to 10 minutes, but occurs again after the next period of rest. Dorsiflexion of the foot often intensifies the pain. Plantar fasciitis typically results from excessive or unaccustomed running and high-impact aerobics, especially in persons who don't wear properly supportive athletic footwear.

TABLE 10.16	Intrinsic Muscles of the Foot		
Name	**Action**	**Skeletal Attachments**	**Innervation**
Dorsal (Superior) Aspect of Foot			
Extensor digitorum brevis	Extends proximal phalanx I and all phalanges of digits II–IV	• Calcaneus; inferior extensor retinaculum of ankle • Proximal phalanx I, tendons of extensor digitorum longus to middle and distal phalanges II–IV	Deep fibular (peroneal) nerve
Ventral Layer 1 (Most Superficial)			
Flexor digitorum brevis	Flexes digits II–IV; supports arches of foot	• Calcaneus; plantar aponeurosis • Middle phalanges II–V	Medial plantar nerve
Abductor digiti minimi[93]	Abducts and flexes little toe; supports arches of foot	• Calcaneus; plantar aponeurosis • Proximal phalanx V	Lateral plantar nerve
Abductor hallucis	Abducts great toe; supports arches of foot	• Calcaneus; plantar aponeurosis; flexor retinaculum • Proximal phalanx I	Medial plantar nerve
Ventral Layer 2			
Quadratus plantae[94] (quad-RAY-tus PLAN-tee)	Same as flexor digitorum longus (table 10.15); flexion of digits II–V and associated locomotor functions	• Two heads on the medial and lateral sides of calcaneus • Distal phalanges II–V via flexor digitorum longus tendons	Lateral plantar nerve
Four lumbrical muscles (LUM-brih-cul)	Flex toes II–V	• Tendon of flexor digitorum longus • Proximal phalanges II–V	Lateral and medial plantar nerves
Ventral Layer 3			
Flexor digiti minimi brevis	Flexes little toe	• Metatarsal V, sheath of fibularis longus • Proximal phalanx V	Lateral plantar nerve
Flexor hallucis brevis	Flexes great toe	• Cuboid; lateral cuneiform; tibialis posterior tendon • Proximal phalanx I	Medial plantar nerve
Adductor hallucis	Adducts great toe	• Metatarsals II–IV; fibularis longus tendon; ligaments at bases of digits III–V • Proximal phalanx I	Lateral plantar nerve
Ventral Layer 4 (Deepest)			
Four dorsal interosseous muscles	Abduct toes II–IV	• Each with two heads arising from facing surfaces of two adjacent metatarsals • Proximal phalanges II–IV	Lateral plantar nerve
Three plantar interosseous muscles	Adduct toes III–V	• Medial aspect of metatarsals III–V • Proximal phalanges III–V	Lateral plantar nerve

---

[93]*digit* = toe; *minim* = smallest

[94]*quadrat* = four-sided; *plantae* = of the plantar region

Ventral Layer 3

The muscles of this layer serve only the great and little toes. They are the **flexor digiti minimi brevis, flexor hallucis brevis,** and **adductor hallucis.** The adductor hallucis has an oblique head that extends diagonally from the midplantar region to the base of the great toe, and a transverse head that passes across the bases of digits II–IV and meets the long head at the base of the great toe.

Ventral Layer 4 (Deepest)

This layer consists only of the small interosseous muscles located between the metatarsal bones—four dorsal and three plantar. Each **dorsal interosseous muscle** is bipennate and originates on two adjacent metatarsals. The **plantar interosseous muscles** are unipennate and originate on only one metatarsal each.

▶▶▶ APPLY WHAT YOU KNOW

Not everyone has the same muscles. From the information provided in this chapter, identify at least three muscles that are lacking in some people.

BEFORE YOU GO ON

Answer the following questions to test your understanding of the preceding section:

22. In the middle of a stride, you have one foot on the ground and you are about to swing the other leg forward. What muscles produce the movements of that leg?

23. Name the muscles that cross both the hip and knee joints and produce actions at both.

24. List the major actions of the muscles of the anterior, medial, and posterior compartments of the thigh.

25. Describe the roles of plantar flexion and dorsiflexion in walking. What muscles produce these actions?

DEEPER INSIGHT 10.9

CLINICAL APPLICATION

Common Athletic Injuries

Although the muscular system is subject to fewer diseases than most organ systems, it is particularly vulnerable to injuries resulting from sudden and intense stress placed on muscles and tendons. Each year, thousands of athletes from the high-school to professional level sustain some type of injury to their muscles, as do the increasing numbers of people who have taken up running and other forms of physical conditioning. Overzealous exertion without proper conditioning and warm-up is frequently the cause. This chapter has already described several examples: compartment syndrome, rotator cuff injury, shinsplints, calcaneal tendon rupture, and plantar fasciitis. A few more are described here.

Baseball finger—tears in the extensor tendons of the fingers resulting from the impact of a baseball with the extended fingertip.

Blocker's arm—abnormal calcification in the lateral margin of the forearm as a result of repeated impact with opposing players.

Charley horse—any painful tear, stiffness, and blood clotting in a muscle. A charley horse of the quadriceps femoris is often caused by football tackles.

Pitcher's arm—inflammation at the proximal attachment of the flexor carpi muscles resulting from hard wrist flexion in releasing a baseball.

Pulled groin—strain in the adductor muscles of the thigh; common in gymnasts and dancers who perform splits and high kicks.

Pulled hamstrings—strained hamstring muscles or a partial tear in their tendons, often with a hematoma (blood clot) in the fascia lata. This condition is frequently caused by repetitive kicking (as in football and soccer) or long, hard running.

Tennis elbow—inflammation at the proximal attachment of the extensor carpi muscles on the lateral epicondyle of the humerus. It occurs when these muscles are repeatedly tensed during backhand strokes and then strained by sudden impact with the tennis ball. Any activity that requires rotary movements of the forearm and a firm grip of the hand (for example, using a screwdriver) can cause the symptoms of tennis elbow.

Most athletic injuries can be prevented by proper conditioning. A person who suddenly takes up vigorous exercise may not have sufficient muscle and bone mass to withstand the stresses such exercise entails. These must be developed gradually. Stretching exercises keep ligaments and joint capsules supple and therefore reduce injuries. Warm-up exercises promote more efficient and less injurious musculoskeletal function in several ways. Most of all, moderation is important, as most injuries simply result from overuse of the muscles. "No pain, no gain" is a risky misconception.

Muscular injuries can be treated initially with "RICE": rest, ice, compression, and elevation. Rest prevents further injury and allows repair processes to occur; ice reduces swelling; compression with an elastic bandage helps to prevent fluid accumulation and swelling; and elevation of an injured limb promotes drainage of blood from the affected area and limits further swelling. If these measures are not enough, anti-inflammatory drugs may be employed, including corticosteroids as well as aspirin and other nonsteroidal agents. Serious injuries, such as compartment syndrome, require emergency attention by a physician.

STUDY GUIDE

▶ Assess Your Learning Outcomes

To test your knowledge, discuss the following topics with a study partner or in writing, ideally from memory.

10.1 Structural and Functional Organization of Muscles

1. Which muscles are included in the muscular system and which ones are not; the name of the science that specializes in the muscular system
2. Functions of the muscular system
3. The relationship of muscle structure to the endomysium, perimysium, and epimysium; what constitutes a fascicle of skeletal muscle and how it relates to these connective tissues; and the relationship of a fascia to a muscle
4. Classification of muscles according to the orientation of their fascicles
5. Muscle compartments, interosseous membranes, and intermuscular septa
6. Direct, indirect, moving, and stationary muscle attachments
7. The flaw in *origin–insertion* terminology of muscle attachments
8. The action of a muscle; how it relates to the classification of muscles as *prime movers, synergists, antagonists,* or *fixators;* why these terms are not fixed for a given muscle but differ from one joint movement to another, and examples to illustrate this point
9. Intrinsic versus extrinsic muscles, with examples
10. The nerve and blood supplies to muscles
11. Features to which the Latin names of muscles commonly refer, with examples

10.2 Muscles of the Head and Neck

Know the locations, actions, skeletal attachments, and innervations of the named muscles in each of the following groups, and be able to recognize them on laboratory specimens or models to the extent required in your course.

1. The occipitofrontalis muscle of the scalp, eyebrows, and forehead (table 10.1)
2. The orbicularis oculi, levator palpebrae superioris, and corrugator supercilii muscles, which move the eyelid and other tissues around the eye (table 10.1)
3. The nasalis muscle, which flares and compresses the nostrils (table 10.1)
4. The orbicularis oris, levator labii superioris, levator anguli oris, zygomaticus major and minor, risorius, depressor anguli oris, depressor labii inferioris, and mentalis muscles, which act on the lips (table 10.1)
5. The buccinator muscles of the cheeks (table 10.1)
6. The platysma, which acts upon the mandible and the skin of the neck (table 10.1)
7. The intrinsic muscles of the tongue in general, and specific extrinsic muscles: the genioglossus, hyoglossus, styloglossus, and palatoglossus muscles (table 10.2)
8. The temporalis, masseter, medial pterygoid, and lateral pterygoid muscles of biting and chewing (table 10.2)
9. The suprahyoid group: the digastric, geniohyoid, mylohyoid, and stylohyoid muscles (table 10.2)
10. The infrahyoid group: the omohyoid, sternohyoid, thyrohyoid, and sternothyroid muscles (table 10.2)
11. The superior, middle, and inferior pharyngeal constrictor muscles of the throat (table 10.2)
12. The sternocleidomastoid and three scalene muscles, which flex the neck, and the trapezius, splenius capitis, and semispinalis capitis muscles, which extend it (table 10.3)

10.3 Muscles of the Trunk

For the following muscles, know the same information as for section 10.2.

1. The diaphragm and the external intercostal, internal intercostal, and innermost intercostal muscles of respiration (table 10.4)
2. The external oblique, internal oblique, transverse abdominal, and rectus abdominis muscles of the anterior and lateral abdominal wall (table 10.5)
3. The superficial erector spinae muscle (and its subdivisions) and the deep semispinalis thoracis, quadratus lumborum, and multifidus muscles of the back (table 10.6)
4. The perineum, its two triangles, and their skeletal landmarks (table 10.7)
5. The ischiocavernosus and bulbospongiosus muscles of the superficial perineal space of the pelvic floor (table 10.7)
6. The deep transverse perineal muscle, and in females, the compressor urethrae of the deep perineal space of the pelvic floor, and the external anal sphincter of the anal triangle (table 10.7)
7. The levator ani muscle of the pelvic diaphragm, the deepest compartment of the pelvic floor (table 10.7)

10.4 Muscles Acting on the Shoulder and Upper Limb

For the following muscles, know the same information as for section 10.2.

1. The pectoralis minor, serratus anterior, trapezius, levator scapulae, rhomboid major, and rhomboid minor muscles of scapular movement (table 10.8)
2. Muscles that act on the humerus, including the pectoralis major, latissimus dorsi, deltoid, teres major, coracobrachialis, and four rotator cuff (SITS) muscles—the supraspinatus, infraspinatus, teres minor, and subscapularis (table 10.9)
3. The brachialis, biceps brachii, triceps brachii, brachioradialis, anconeus, pronator quadratus, pronator teres, and supinator muscles of forearm movement (table 10.10)
4. The relationship of the flexor retinaculum, extensor retinaculum, and carpal tunnel to the tendons of the forearm muscles
5. The palmaris longus, flexor carpi radialis, flexor carpi ulnaris, and flexor digitorum superficialis muscles of the superficial anterior compartment of the forearm, and the flexor digitorum profundus and flexor pollicis longus muscles of the deep anterior compartment (table 10.11)
6. The extensor carpi radialis longus, extensor carpi radialis brevis, extensor digitorum, extensor digiti minimi, and extensor carpi ulnaris muscles of the superficial posterior compartment (table 10.11)
7. The abductor pollicis longus, extensor pollicis brevis, extensor pollicis longus, and extensor indicis muscles of the deep posterior compartment (table 10.11)
8. The thenar group of intrinsic hand muscles: adductor pollicis, abductor pollicis brevis, flexor pollicis brevis, and opponens pollicis (table 10.12)
9. The hypothenar group of intrinsic hand muscles: abductor digiti minimi, flexor digiti minimi brevis, and opponens digiti minimi (table 10.12)
10. The midpalmar group of intrinsic hand muscles: four dorsal interosseous muscles, three palmar interosseous muscles, and four lumbrical muscles (table 10.12)

STUDY GUIDE

10.5 Muscles Acting on the Hip and Lower Limb

For the following muscles, know the same information as for section 10.2.

1. The iliopsoas muscle of the hip, and its two subdivisions, the iliacus and psoas major (table 10.13)
2. The tensor fasciae latae, gluteus maximus, gluteus medius, and gluteus minimus muscles of the hip and buttock, and the relationship of the first two to the fascia lata and iliotibial tract (table 10.13)
3. The lateral rotators: gemellus superior, gemellus inferior, obturator externus, obturator internus, piriformis, and quadratus femoris muscles (table 10.13)
4. The compartments of the thigh muscles: anterior (extensor), medial (adductor), and posterior (flexor) compartments
5. Muscles of the medial compartment of the thigh: adductor brevis, adductor longus, adductor magnus, gracilis, and pectineus (table 10.13)
6. Muscles of the anterior compartment of the thigh: sartorius and quadriceps femoris, and the four heads of the quadriceps (table 10.14)
7. The hamstring muscles of the posterior compartment of the thigh: biceps femoris, semitendinosus, and semimembranosus (table 10.14)
8. The compartments of the leg muscles: anterior, posterior, and lateral (table 10.15)
9. Muscles of the anterior compartment of the leg: fibularis tertius, extensor digitorum longus, extensor hallucis longus, and tibialis anterior muscles of the anterior compartment (table 10.15)
10. Muscles of the superficial posterior compartment of the leg: popliteus and triceps surae (gastrocnemius and soleus), and the relationship of the triceps surae to the calcaneal tendon and calcaneus (table 10.15)
11. Muscles of the deep posterior compartment of the leg: flexor digitorum longus, flexor hallucis longus, and tibialis posterior muscles of the deep posterior compartment
12. Muscles of the lateral compartment of the leg: fibularis brevis and fibularis longus (table 10.15)
13. The extensor digitorum brevis of the dorsal aspect of the foot (table 10.16)
14. The four muscle compartments (layers) of the ventral aspect of the foot, and the muscles in each: the flexor digitorum brevis, abductor digiti minimi, and abductor hallucis (layer 1); the quadratus plantae and four lumbrical muscles (layer 2); the flexor digiti minimi brevis, flexor hallucis brevis, and adductor hallucis (layer 3); and the four dorsal interosseous muscles and three plantar interosseous muscles (layer 4) (table 10.16)

▶ Testing Your Recall

Answers in Appendix A

1. Which of the following muscles does *not* contribute to the rotator cuff?
   a. teres minor
   b. teres major
   c. subscapularis
   d. infraspinatus
   e. supraspinatus

2. Make a fist, then straighten your fingers. The muscles that performed the latter action have tendons that pass through
   a. the flexor retinaculum.
   b. the extensor retinaculum.
   c. the carpal tunnel.
   d. the rotator cuff.
   e. the plantar region of the wrist.

3. Which of these is *not* a suprahyoid muscle?
   a. genioglossus
   b. geniohyoid
   c. stylohyoid
   d. mylohyoid
   e. digastric

4. Which of these muscles is an extensor of the neck?
   a. external oblique
   b. sternocleidomastoid
   c. splenius capitis
   d. iliocostalis
   e. latissimus dorsi

5. Which of these muscles of the pelvic floor is the deepest?
   a. superficial transverse perineal
   b. bulbospongiosus
   c. ischiocavernosus
   d. deep transverse perineal
   e. levator ani

6. Which of these actions is *not* performed by the trapezius?
   a. extension of the neck
   b. depression of the scapula
   c. elevation of the scapula
   d. rotation of the scapula
   e. adduction of the humerus

7. Both the hands and feet are acted upon by a muscle or muscles called
   a. the extensor digitorum.
   b. the abductor digiti minimi.
   c. the flexor digitorum profundus.
   d. the abductor hallucis.
   e. the flexor digitorum longus.

8. Which of the following muscles does *not* extend the hip joint?
   a. quadriceps femoris
   b. gluteus maximus
   c. biceps femoris
   d. semitendinosus
   e. semimembranosus

9. Both the gastrocnemius and _____ muscles insert on the heel by way of the calcaneal tendon.
   a. semimembranosus
   b. tibialis posterior
   c. tibialis anterior
   d. soleus
   e. plantaris

10. Which of the following muscles raises the upper lip?
    a. levator palpebrae superioris
    b. orbicularis oris
    c. zygomaticus minor
    d. masseter
    e. mentalis

11. What muscle is the prime mover in sucking through a soda straw or spitting out a mouthful of liquid?

12. A bundle of muscle fibers surrounded by perimysium is called a/an _____.

13. The _____ is the muscle that generates the most force in a given joint movement.

14. The three large muscles on the posterior side of the thigh are commonly known as the _____ muscles.

STUDY GUIDE

15. Connective tissue bands called _____ prevent flexor tendons of the forearm and leg from rising like bowstrings.

16. The anterior half of the perineum is a region called the _____ .

17. The abdominal aponeuroses converge on a median fibrous band on the abdomen called the _____ .

18. A muscle that works with another to produce the same or similar movement is called a/an _____ .

19. A muscle somewhat like a feather, with fibers obliquely approaching its tendon from both sides, is called a/an _____ muscle.

20. A circular muscle that closes a body opening is called a/an _____ .

▶ Building Your Medical Vocabulary

Answers in Appendix A

State a meaning of each word element, and give a medical term from this chapter that uses it or a slight variation of it.

1. capito-

2. ergo-

3. fasc-

4. labio-

5. lumbo-

6. mus-

7. mys-

8. omo-

9. penn-

10. tert-

▶ What's Wrong with These Statements?

Answers in Appendix A

Briefly explain why each of the following statements is false, or reword it to make it true.

1. Each skeletal muscle fiber is enclosed in a perimysium that separates it from adjacent fibers.

2. The orbicularis oris is a sphincter.

3. The biceps brachii is a bipennate muscle.

4. A synergist is a muscle whose action is opposite that of an agonist.

5. Each skeletal muscle is innervated by at least one spinal nerve.

6. One must contract the internal intercostal muscles in order to exhale.

7. In climbing stairs, the hamstrings provide much of the thrust that lifts the body to each higher step.

8. Severing the large trigeminal nerve would paralyze more facial muscles than severing the facial nerve.

9. The following intrinsic hand muscles are listed in order from strongest to weakest: abductor digiti minimi, adductor pollicis, dorsal interosseous.

10. The tibialis anterior and tibialis posterior are synergists.

▶ Testing Your Comprehension

1. Radical mastectomy, once a common treatment for breast cancer, involved removal of the pectoralis major along with the breast. What functional impairments would result from this? What synergists could a physical therapist train a patient to use to recover some lost function?

2. Removal of cancerous lymph nodes from the neck sometimes requires removal of the sternocleidomastoid on that side. How would this affect a patient's range of head movement?

3. Bicycles are designed so the rider leans forward at about a 45° angle rather than sitting upright. Aside from issues of wind resistance, explain the advantage of this in terms of musculoskeletal anatomy of the hip region.

4. Women who wear high heels most of the time may suffer painful "high heel syndrome" when they go barefoot or wear flat shoes. What muscle(s) and tendon(s) are involved? Explain.

5. A student moving out of a dormitory crouches, in correct fashion, to lift a heavy box of books. What prime movers are involved as he straightens his legs to lift the box?

REGIONAL AND SURFACE ANATOMY

How many muscles can you identify from their surface appearance?

Y Photo Studio/Shutterstock

ATLAS OUTLINE

B.1 Regional Anatomy

B.2 The Importance of Surface Anatomy

B.3 Learning Strategy

Figures B.1–B.2 The Head and Neck

Figures B.3–B.15 The Trunk

Figures B.16–B.19 The Upper Limb

Figures B.20–B.24 The Lower Limb

Figure B.25 Test of Muscle Recognition

Anatomy & Physiology *Revealed* 4.0

Module 6: Muscular System

B.1 Regional Anatomy

On the whole, this book takes a systems approach to anatomy, examining the structure and function of each organ system, one at a time, regardless of which body regions it may traverse. Physicians and surgeons, however, think and act in terms of regional anatomy. If a patient presents with pain in the lower right quadrant (see atlas A, fig. A.4a), the source may be the appendix, an ovary, or an inguinal muscle, among other possibilities. The question is to think not of an entire organ system (the esophagus is probably irrelevant to that quadrant), but of what organs are present in that region and what possibilities must be considered as the cause of the pain. This atlas presents several views of the body region by region so that you can see some of the spatial relationships that exist among the organ systems considered in their separate chapters.

B.2 The Importance of Surface Anatomy

In the study of human anatomy, it is easy to become so preoccupied with internal structure that we forget the importance of what we can see and feel externally. Yet external anatomy and appearance are major concerns in giving a physical examination and in many aspects of patient care. A knowledge of the body's surface landmarks is essential to one's competence in physical therapy, cardiopulmonary resuscitation, surgery, making X-rays and electrocardiograms, giving injections, drawing blood, listening to heart and respiratory sounds, measuring the pulse and blood pressure, and finding pressure points to stop arterial bleeding, among other procedures. A misguided attempt to perform some of these procedures while disregarding or misunderstanding external anatomy can be very harmful and even fatal to a patient.

Having just studied skeletal and muscular anatomy in the preceding chapters, this is an opportune time for you to study the body surface. Much of what we see there reflects the underlying structure of the superficial bones and muscles. A broad photographic overview of surface anatomy is given in atlas A (see fig. A.3), where it is necessary for providing a vocabulary for reference in subsequent chapters. This atlas shows this surface anatomy in closer detail so you can relate it to the musculoskeletal anatomy of the three preceding chapters.

B.3 Learning Strategy

To make the most profitable use of this atlas, refer back to earlier chapters as you study these illustrations. Relate drawings of the clavicle in figure 8.31 to the photograph in figure B.1, for example. Study the shape of the scapula in figure 8.32 and see how much of it you can trace on the photographs in figure B.9. See if you can relate the tendons visible on the hand (see fig. B.19) to the muscles of the forearm illustrated in figures 10.29 and 10.30, and the external markings of the pelvic girdle (see fig. B.15) to bone structure in section 8.5a.

For learning surface anatomy, there is a resource available to you that is far more valuable than any laboratory model or textbook illustration—your own body. For the best understanding of human structure, compare the art and photographs in this book with your body or with structures visible on a study partner. In addition to bones and muscles, you can palpate a number of superficial arteries, veins, tendons, ligaments, and cartilages, among other structures. By palpating regions such as the shoulder, elbow, or ankle, you can develop a mental image of the subsurface structures better than the image you can obtain by looking at two-dimensional textbook images. And the more you can study with other people, the more you will appreciate the variations in human structure and be able to apply your knowledge to your future patients or clients, who will not look quite like any textbook diagram or photograph you have ever seen. Through comparisons of art, photography, and the living body, you will get a much deeper understanding of the body than if you were to study this atlas in isolation from the earlier chapters.

At the end of this atlas, you can test your knowledge of externally visible muscle anatomy. The two photographs in figure B.25 have 30 numbered muscles and a list of 26 names, some of which are shown more than once in the photographs and some of which are not shown at all. Identify the muscles to your best ability without looking back at the previous illustrations, and then check your answers in appendix A.

(a) Lateral view

Occipital

Frontal

Temporal

Orbital

Nasal

Auricular

Oral

Mental

Buccal (cheek)

Cervical

Nuchal (posterior cervical)

(b) Anterior view

Frons (forehead)

Root of nose

Bridge of nose

Superciliary ridge

Lateral commissure

Superior palpebral sulcus

Medial commissure

Inferior palpebral sulcus

Dorsum nasi

Apex of nose

Auricle (pinna) of ear

Ala nasi

Philtrum

Mentolabial sulcus

Labia (lips)

Mentum (chin)

Sternoclavicular joints

Supraclavicular fossa

Clavicle

Suprasternal notch

Sternum

FIGURE B.1 **The Head and Neck.** (a) Anatomical regions of the head. (b) Features of the facial region and upper thorax.

a–b: Joe DeGrandis/McGraw-Hill Education

❓ *What muscle underlies the region of the philtrum? What muscle forms the slope of the shoulder?*

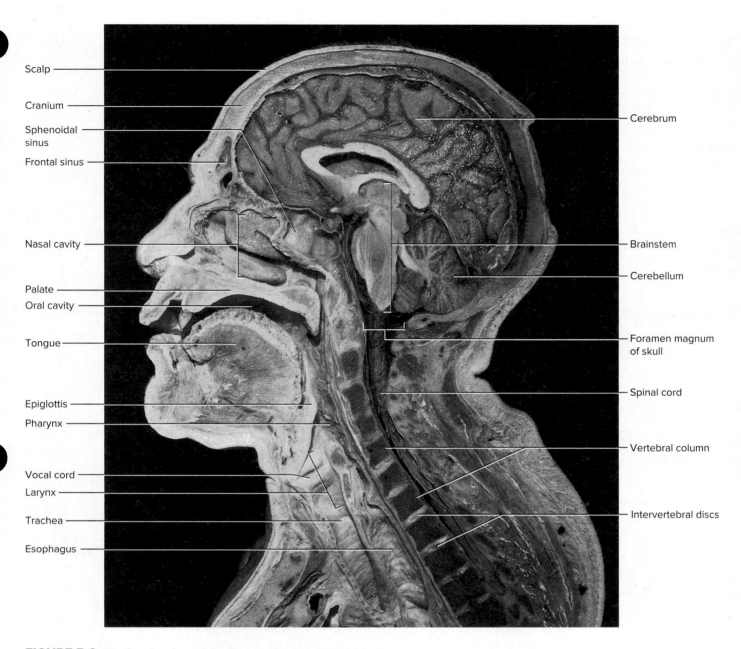

FIGURE B.2 Median Section of the Head. Shows contents of the cranial, nasal, and oral cavities.

Rebecca Gray/McGraw-Hill Education

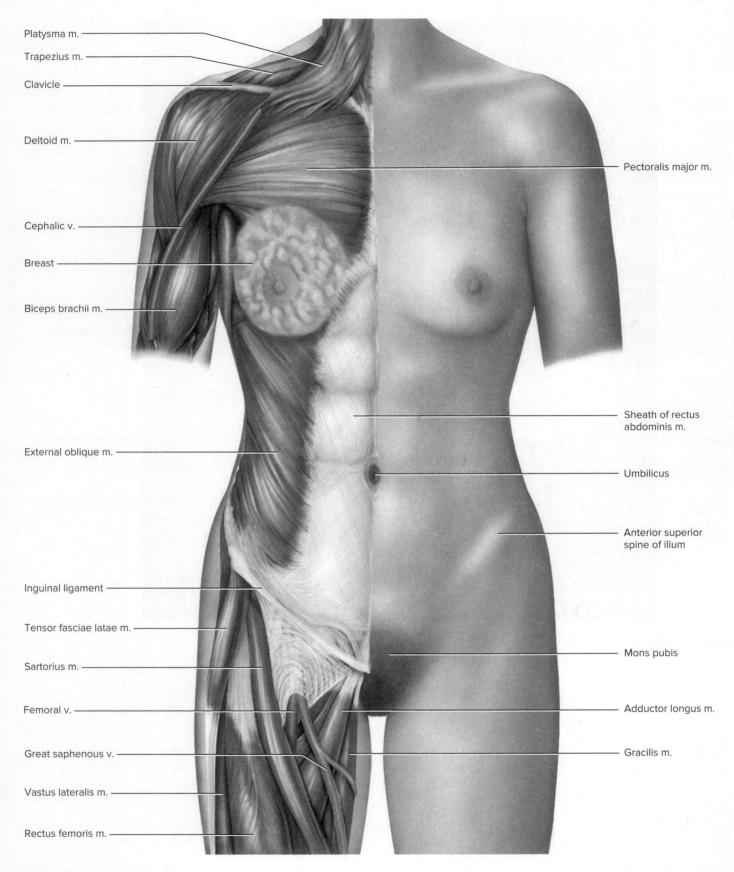

Platysma m.

Trapezius m.

Clavicle

Deltoid m.

Cephalic v.

Breast

Biceps brachii m.

External oblique m.

Inguinal ligament

Tensor fasciae latae m.

Sartorius m.

Femoral v.

Great saphenous v.

Vastus lateralis m.

Rectus femoris m.

Pectoralis major m.

Sheath of rectus abdominis m.

Umbilicus

Anterior superior spine of ilium

Mons pubis

Adductor longus m.

Gracilis m.

FIGURE B.3 Superficial Anatomy of the Trunk (Female). Surface anatomy is shown on the anatomical left, and structures immediately deep to the skin on the right.

Internal jugular v.

External jugular v.

Omohyoid m.

Clavicle

Internal intercostal mm.

External intercostal mm.

Costal cartilages

Liver

Gallbladder

External oblique m.

Internal oblique m.

Transverse abdominal m.

Greater omentum

Urinary bladder

Penis

Scrotum

Common carotid a.

Sternum

Sub-scapularis m.

Coraco-brachialis m.

Lung

Pericardium

Pleura

Diaphragm

Stomach

Large intestine

Femoral n.

Femoral a.

Femoral v.

FIGURE B.4 Anatomy at the Level of the Rib Cage and Greater Omentum (Male). The anterior body wall is removed, and the ribs, intercostal muscles, and pleura are removed from the anatomical left.

Thyroid cartilage of larynx

Thyroid gland

Brachial nerve plexus

Superior vena cava

Coraco-brachialis m.

Humerus

Lobes of lung

Small intestine

Cecum

Appendix

Tensor fasciae latae m.

Pectineus m.

Adductor longus m.

Gracilis m.

Adductor magnus m.

Rectus femoris m.

Brachio-cephalic v.

Subclavian v.

Subclavian a.

Aortic arch

Axillary v.

Axillary a.

Cephalic v.

Brachial v.

Brachial a.

Heart

Spleen

Stomach

Large intestine

Penis (cut)

Ductus deferens

Epididymis

Testis

Scrotum

FIGURE B.5 Anatomy at the Level of the Lungs and Intestines (Male). The sternum, ribs, and greater omentum are removed.

❓ *Name several viscera that are protected by the rib cage.*

Trachea

Superior vena cava

Bronchus

Lung (sectioned)

Esophagus

Thoracic aorta

Pleural cavity

Hepatic vv.

Spleen

Inferior vena cava

Adrenal gland

Splenic a.

Pancreas

Duodenum

Kidney

Superior mesenteric v.

Superior mesenteric a.

Abdominal aorta

Inferior mesenteric a.

Common iliac a.

Ureter

Ovary

Uterine tube

Tensor fasciae latae m. (cut)

Uterus

Sartorius m. (cut)

Urinary bladder

Rectus femoris m. (cut)

Pectineus m.

Adductor brevis m.

Gracilis m.

Vastus intermedius m.

Adductor longus m.

Adductor longus m. (cut)

Vastus lateralis m.

Vastus medialis m.

FIGURE B.6 Anatomy at the Level of the Retroperitoneal Viscera (Female). The heart is removed, the lungs are frontally sectioned, and the viscera of the peritoneal cavity and the peritoneum itself are removed.

Right common carotid a.

Right subclavian a.

Brachiocephalic trunk

External
intercostal m.

Ribs

Internal
intercostal m.

Quadratus
lumborum m.

Psoas major m.

Iliacus m.

Gluteus medius m.

Rectum

Vagina

Urethra

Femur

Gracilis m.

Adductor longus m.

Left common
carotid a.

Left subclavian a.

Thoracic aorta

Esophagus

Diaphragm

Abdominal aorta

Intervertebral disc

Lumbar vertebra

Iliac crest

Ilium

Sacrum

Anterior superior
spine of ilium

Brim of pelvis

Adductor magnus m.

Adductor brevis m.

FIGURE B.7 Anatomy at the Level of the Posterior Body Wall (Female). The lungs and retroperitoneal viscera are removed.

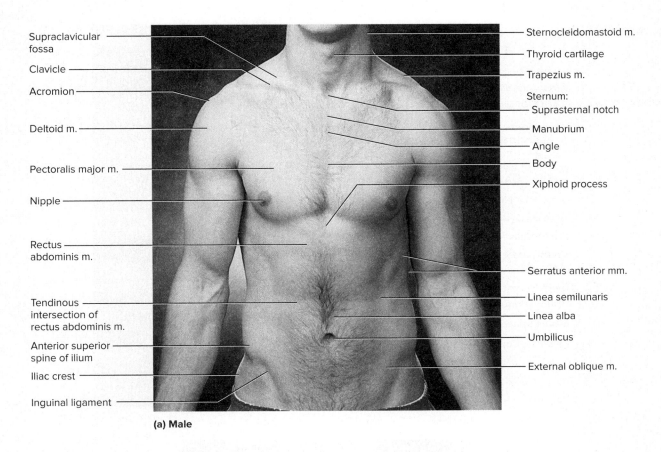

Supraclavicular fossa

Clavicle

Acromion

Deltoid m.

Pectoralis major m.

Nipple

Rectus abdominis m.

Tendinous intersection of rectus abdominis m.

Anterior superior spine of ilium

Iliac crest

Inguinal ligament

Sternocleidomastoid m.

Thyroid cartilage

Trapezius m.

Sternum:
Suprasternal notch

Manubrium

Angle

Body

Xiphoid process

Serratus anterior mm.

Linea semilunaris

Linea alba

Umbilicus

External oblique m.

(a) Male

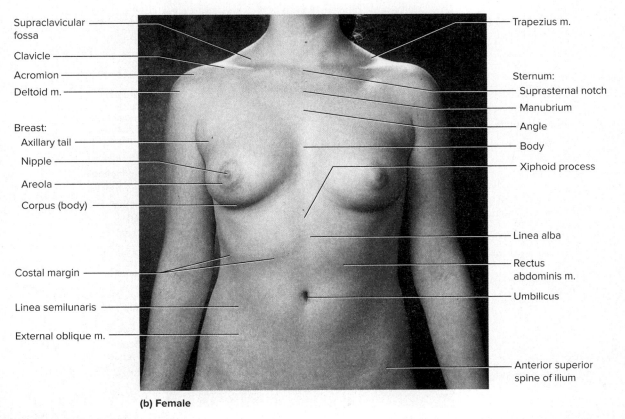

Supraclavicular fossa

Clavicle

Acromion

Deltoid m.

Breast:
Axillary tail

Nipple

Areola

Corpus (body)

Costal margin

Linea semilunaris

External oblique m.

Trapezius m.

Sternum:
Suprasternal notch

Manubrium

Angle

Body

Xiphoid process

Linea alba

Rectus abdominis m.

Umbilicus

Anterior superior spine of ilium

(b) Female

FIGURE B.8 The Thorax and Abdomen (Anterior View). (a) Male. (b) Female. All of the features labeled are common to both sexes, though some are labeled only on the photograph that shows them best.

a–b: Joe DeGrandis/McGraw-Hill Education

? *The V-shaped tendons on each side of the suprasternal notch in part (a) belong to what muscles?*

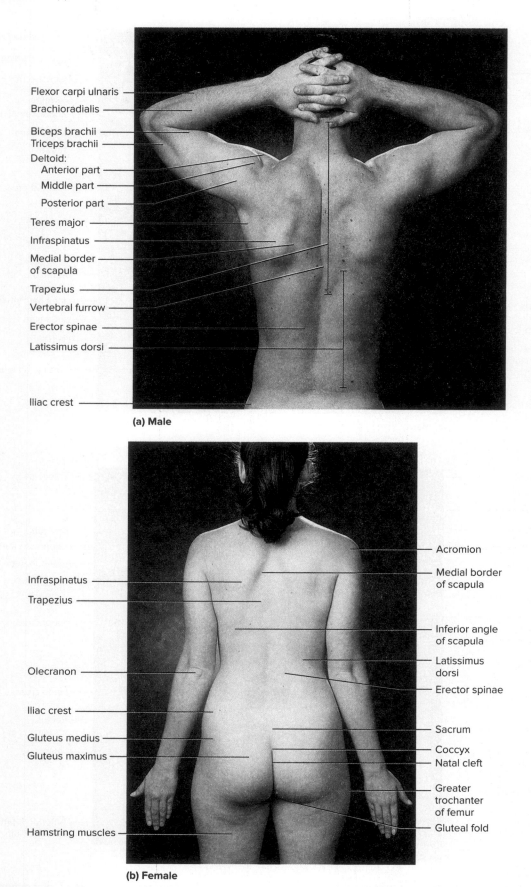

Flexor carpi ulnaris
Brachioradialis
Biceps brachii
Triceps brachii
Deltoid:
    Anterior part
    Middle part
    Posterior part
Teres major
Infraspinatus
Medial border
of scapula
Trapezius
Vertebral furrow
Erector spinae
Latissimus dorsi

Iliac crest

(a) Male

Infraspinatus
Trapezius

Olecranon

Iliac crest

Gluteus medius
Gluteus maximus

Hamstring muscles

Acromion
Medial border
of scapula

Inferior angle
of scapula

Latissimus
dorsi
Erector spinae

Sacrum
Coccyx
Natal cleft

Greater
trochanter
of femur
Gluteal fold

(b) Female

FIGURE B.9 The Back and Gluteal Region. (a) Male. (b) Female. All of the features labeled are common to both sexes, though some are labeled only on the photograph that shows them best.

a–b: Joe DeGrandis/McGraw-Hill Education

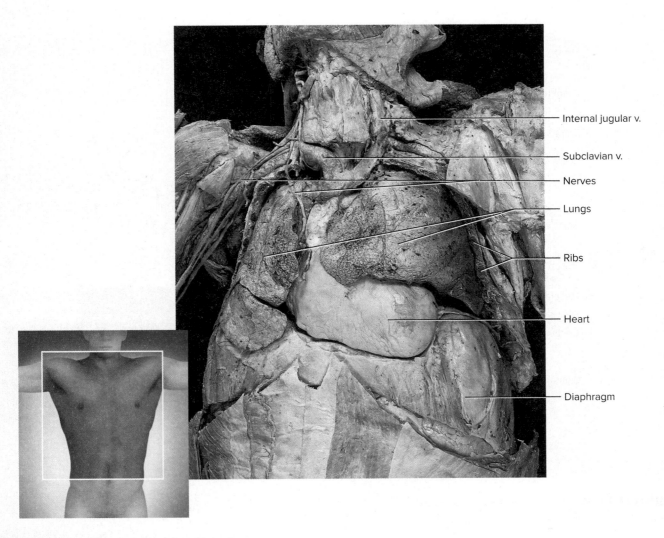

Internal jugular v.

Subclavian v.

Nerves

Lungs

Ribs

Heart

Diaphragm

FIGURE B.10 Frontal View of the Thoracic Cavity.

Rebecca Gray/McGraw-Hill Education

Anterior

Pectoralis major m.

Ventricles of heart

Pericardial cavity

Atria of heart

Left lung

Pleural cavity

Fat of breast

Sternum

Ribs

Right lung

Esophagus

Aorta

Vertebra

Spinal cord

Posterior

FIGURE B.11 Transverse Section of the Thorax. Section taken at the level shown by the inset and oriented the same as the reader's body.

Rebecca Gray/Don Kincaid/McGraw-Hill Education

❓ *In this section, which term best describes the position of the aorta relative to the heart: posterior, lateral, inferior, or proximal?*

FIGURE B.12 Frontal View of the Abdominal Cavity.

Rebecca Gray/Don Kincaid/McGraw-Hill Education

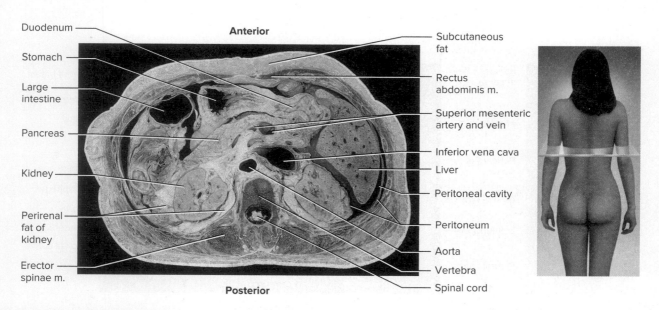

FIGURE B.13 Transverse Section of the Abdomen. Section taken at the level shown by the inset and oriented the same as the reader's body.

Rebecca Gray/Don Kincaid/McGraw-Hill Education

❓ *What tissue in this photograph is immediately superficial to the rectus abdominis muscle?*

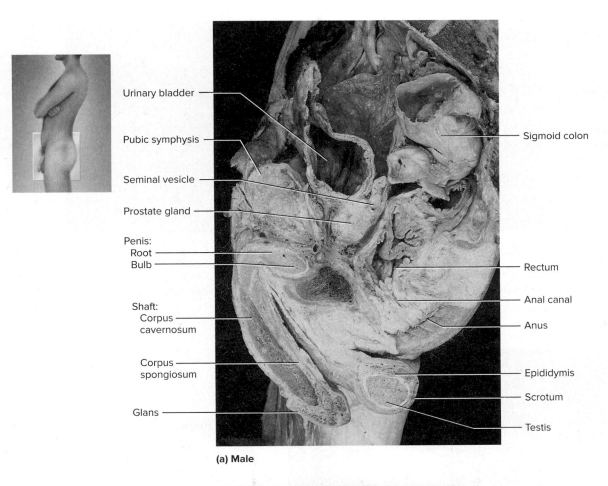

Urinary bladder

Pubic symphysis

Seminal vesicle

Prostate gland

Penis:
Root
Bulb

Shaft:
Corpus
cavernosum

Corpus
spongiosum

Glans

Sigmoid colon

Rectum

Anal canal

Anus

Epididymis

Scrotum

Testis

(a) Male

Mesentery

Small intestine

Uterus

Cervix

Urinary bladder

Pubic symphysis

Urethra

Vagina

Labium minus

Prepuce

Labium majus

Vertebra

Red bone marrow

Intervertebral disc

Sacrum

Sigmoid colon

Rectum

Anal canal

Anus

(b) Female

FIGURE B.14 Median Sections of the Pelvic Cavity (Left Lateral Views).
(a) Male. (b) Female.

a: Dennis Strete/McGraw-Hill Education;
b: Rebecca Gray/Don Kincaid/McGraw-Hill Education

(a) Anterior view

(b) Posterior view

FIGURE B.15 Pelvic Landmarks. (a) The anterior superior spines of the ilium are marked by anterolateral protuberances (arrows) at about the location where the front pockets usually open on a pair of pants. (b) The posterior superior spines are marked in some people by dimples in the sacral region (arrows).

a–b: Joe DeGrandis/McGraw-Hill Education

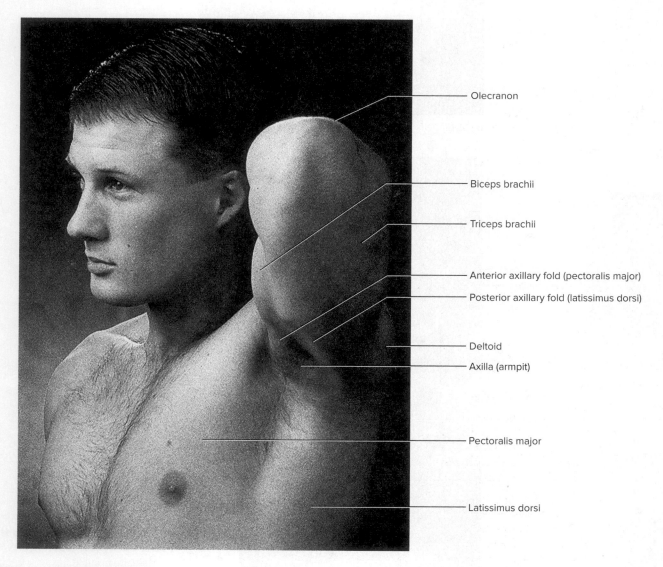

Olecranon

Biceps brachii

Triceps brachii

Anterior axillary fold (pectoralis major)

Posterior axillary fold (latissimus dorsi)

Deltoid

Axilla (armpit)

Pectoralis major

Latissimus dorsi

FIGURE B.16 The Axillary Region.

Joe DeGrandis/McGraw-Hill Education

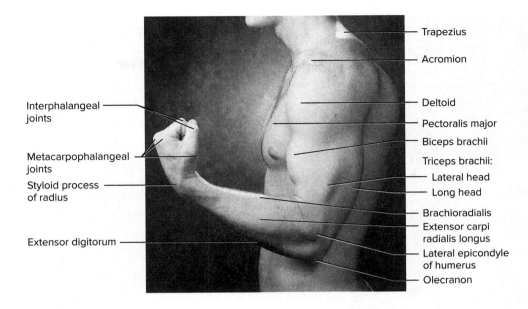

Trapezius
Acromion
Deltoid
Pectoralis major
Biceps brachii
Triceps brachii:
  Lateral head
  Long head
Brachioradialis
Extensor carpi radialis longus
Lateral epicondyle of humerus
Olecranon

Interphalangeal joints
Metacarpophalangeal joints
Styloid process of radius
Extensor digitorum

FIGURE B.17 The Upper Limb (Left Lateral View).

Joe DeGrandis/McGraw-Hill Education

Cubital fossa
Cephalic vein
Median cubital vein
Brachioradialis

Biceps brachii
Medial epicondyle of humerus

Flexor carpi radialis
Palmaris longus
Flexor carpi ulnaris

Styloid process of radius
Thenar eminence
Palmar surface of hand
Thumb

Styloid process of ulna
Hypothenar eminence
Flexion lines

Flexion lines
Volar surface of fingers

(a) Anterior view

Triceps brachii
Olecranon
Head of radius
Brachioradialis
Flexor carpi ulnaris
Extensor carpi ulnaris
Extensor digitorum
Tendons of extensor digitorum
Dorsum of hand

(b) Posterior view

FIGURE B.18 The Antebrachium (Forearm). (a) Anterior view. (b) Posterior view.

a–b: Joe DeGrandis/McGraw-Hill Education

❓ *Only two tendons of the extensor digitorum are labeled, but how many tendons does this muscle have in all?*

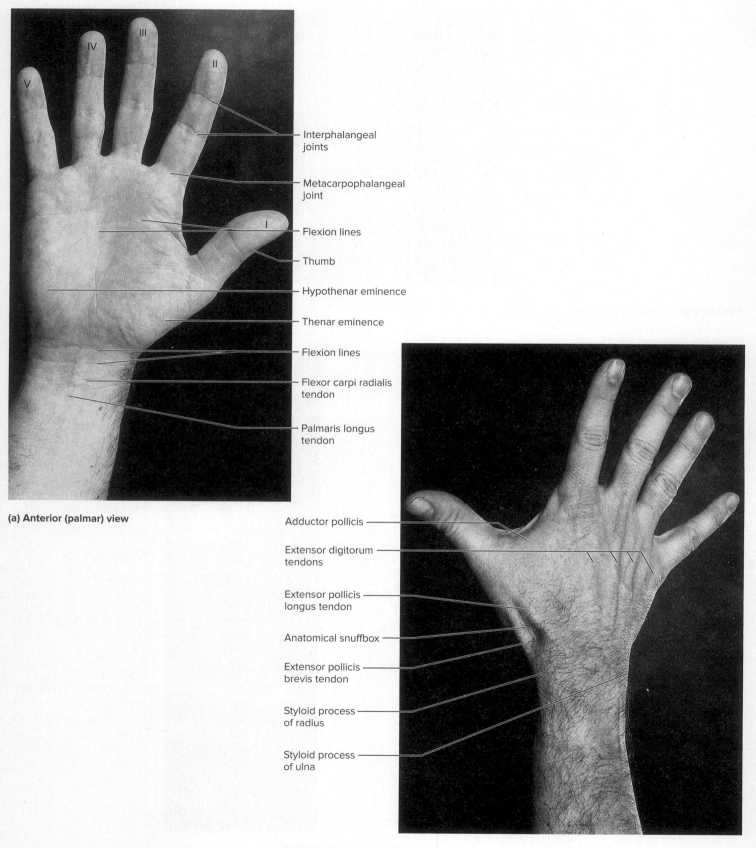

(a) Anterior (palmar) view

Interphalangeal joints

Metacarpophalangeal joint

Flexion lines

Thumb

Hypothenar eminence

Thenar eminence

Flexion lines

Flexor carpi radialis tendon

Palmaris longus tendon

Adductor pollicis

Extensor digitorum tendons

Extensor pollicis longus tendon

Anatomical snuffbox

Extensor pollicis brevis tendon

Styloid process of radius

Styloid process of ulna

(b) Posterior (dorsal) view

FIGURE B.19 The Wrist and Hand. (a) Anterior (palmar) view. (b) Posterior (dorsal) view.

a–b: Joe DeGrandis/McGraw-Hill Education

❓ *Mark the spot on one or both photographs where a saddle joint can be found.*

(a) Anterior view   (b) Posterior view

FIGURE B.20 The Thigh and Knee. (a) Anterior view. (b) Posterior view. Locations of posterior thigh muscles are indicated, but the boundaries of the individual muscles are rarely visible on a living person.

a–b: Joe DeGrandis/McGraw-Hill Education

❓ *Mark the spot on part (a) where the vastus intermedius would be found.*

(a) Lateral view

Vastus lateralis
Biceps femoris
Iliotibial band
Lateral epicondyle of femur
Head of fibula
Patellar ligament
Gastrocnemius, lateral head
Soleus
Fibularis longus
Tibialis anterior
Tendons of fibularis longus and brevis
Calcaneal tendon
Lateral malleolus of fibula
Calcaneus

Semimembranosus and tendon
Vastus medialis
Patella
Semitendinosus tendon
Medial epicondyle of femur
Medial condyle of tibia
Gastrocnemius, medial head
Soleus
Tibia
Medial malleolus of tibia
Tibialis anterior tendon
Medial longitudinal arch
Abductor hallucis
Head of metatarsal I

(b) Medial view

FIGURE B.21 The Leg and Foot. (a) Lateral view of left limb. (b) Medial view of right limb.

a–b: Joe DeGrandis/McGraw-Hill Education

❓ *The lateral malleolus is part of what bone?*

Medial | Lateral

Biceps femoris tendon
Semitendinosus tendon
Popliteal fossa

Gastrocnemius:
  Medial head
  Lateral head
Soleus
Fibularis longus

Tibialis anterior

Calcaneal tendon
Lateral malleolus
of fibula
Extensor digitorum brevis
Calcaneus

FIGURE B.22 **The Leg and Foot, Posterior View.**

Joe DeGrandis/McGraw-Hill Education

Great toe
Head of metatarsal I
Transverse arch
Head of metatarsal V
Abductor digiti minimi
Abductor hallucis
Medial longitudinal arch
Lateral longitudinal arch
Lateral malleolus
of fibula
Calcaneus

(a) Plantar view

Soleus
Tibia
Tibialis anterior
Medial malleolus
of tibia
Lateral malleolus
of fibula
Site for palpating dorsal pedal artery
Extensor hallucis longus tendon
Extensor digitorum longus tendons
Head of metatarsal I
Great toe

(b) Dorsal view

FIGURE B.23 **The Foot (Plantar and Dorsal Views).** (a) Plantar view. (b) Dorsal view. Compare the arches in part (a) to the skeletal anatomy in figure 8.43.

a–b: Joe DeGrandis/McGraw-Hill Education

Calcaneal tendon

Lateral malleolus
of fibula

Extensor digitorum brevis

Extensor digitorum longus
tendons

Lateral longitudinal arch

(a) Lateral view

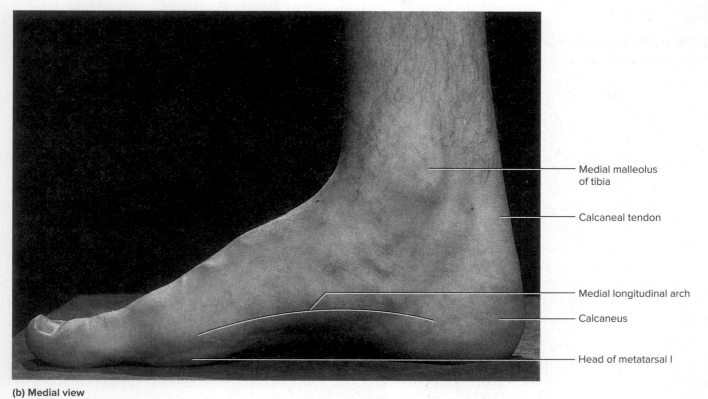

Medial malleolus
of tibia

Calcaneal tendon

Medial longitudinal arch

Calcaneus

Head of metatarsal I

(b) Medial view

FIGURE B.24 The Foot (Lateral and Medial Views). (a) Lateral view. (b) Medial view.

a–b: Joe DeGrandis/McGraw-Hill Education

❓ *Indicate the position of middle phalanx I on each photograph.*

(a) Anterior view

(b) Posterior view

FIGURE B.25 Test of Muscle Recognition. To test your knowledge of muscle anatomy, match the 30 labeled muscles on these photographs to the following alphabetical list of muscles. Answer as many as possible without referring back to the previous illustrations. Some of these names will be used more than once since the same muscle may be shown from different perspectives, and some of these names will not be used at all. The answers are in appendix A.

a–b: Joe DeGrandis/McGraw-Hill Education

Throughout these illustrations, the following abbreviations apply: a. = artery; m. = muscle; n. = nerve; v. = vein. Double letters such as mm. or vv. represent the plurals.

a. biceps brachii

b. brachioradialis

c. deltoid

d. erector spinae

e. external oblique

f. flexor carpi ulnaris

g. gastrocnemius

h. gracilis

i. hamstrings

j. infraspinatus

k. latissimus dorsi

l. pectineus

m. pectoralis major

n. rectus abdominis

o. rectus femoris

p. serratus anterior

q. soleus

r. splenius capitis

s. sternocleidomastoid

t. subscapularis

u. teres major

v. tibialis anterior

w. trapezius

x. triceps brachii

y. vastus lateralis

z. vastus medialis

MUSCULAR TISSUE

Neuromuscular junctions (SEM)
Dr. Donald Fawcett/Science Source

CHAPTER OUTLINE

11.1 Types and Characteristics of Muscular Tissue

11.1a Universal Characteristics of Muscle
11.1b Skeletal Muscle

11.2 Skeletal Muscle Cells

11.2a The Muscle Fiber
11.2b Myofilaments
11.2c Striations

11.3 The Nerve–Muscle Relationship

11.3a Motor Neurons and Motor Units
11.3b The Neuromuscular Junction
11.3c Electrically Excitable Cells

11.4 Behavior of Skeletal Muscle Fibers

11.4a Excitation
11.4b Excitation–Contraction Coupling
11.4c Contraction
11.4d Relaxation
11.4e The Length–Tension Relationship and Muscle Tone

11.5 Behavior of Whole Muscles

11.5a Threshold, Latent Period, and Twitch
11.5b Contraction Strength of Twitches
11.5c Isometric and Isotonic Contraction

11.6 Muscle Metabolism

11.6a ATP Sources
11.6b Fatigue and Endurance
11.6c Excess Postexercise Oxygen Consumption
11.6d Physiological Classes of Muscle Fibers
11.6e Muscular Strength and Conditioning

11.7 Cardiac and Smooth Muscle

11.7a Cardiac Muscle
11.7b Smooth Muscle

Connective Issues
Study Guide

DEEPER INSIGHTS

11.1 Clinical Application: Neuromuscular Toxins and Paralysis

11.2 Clinical Application: Rigor Mortis

11.3 Medical History: Galvani, Volta, and Animal Electricity

11.4 Clinical Application: Muscular Dystrophy and Myasthenia Gravis

Anatomy & Physiology Revealed 4.0

Module 6: Muscular System

Movement is a fundamental characteristic of all living organisms, from bacteria to humans. Even plants and other seemingly immobile organisms move cellular components from place to place. Across the entire spectrum of life, the molecular mechanisms of movement are very similar, involving motor proteins such as myosin and dynein. But in animals, movement has developed to the highest degree, with the evolution of **muscle cells** specialized for this function. A muscle cell is essentially a device for converting the chemical energy of ATP into the mechanical energy of movement.

The three types of muscular tissue—skeletal, cardiac, and smooth—are described and compared in table 5.11. Cardiac and smooth muscle are further described in this chapter, and cardiac muscle is discussed most extensively in chapter 19. Most of the present chapter, however, concerns skeletal muscle, the type that holds the body erect against the pull of gravity and produces its outwardly visible movements.

This chapter treats the structure, contraction, and metabolism of skeletal muscle at the molecular, cellular, and tissue levels of organization. Understanding muscle at these levels provides an indispensable basis for understanding such aspects of motor performance as quickness, strength, endurance, and fatigue. Such factors have obvious relevance to athletic performance, and they become very important when a lack of physical conditioning, old age, or injury interferes with a person's ability to carry out everyday tasks or meet the extra demands for speed or strength that we all occasionally encounter.

11.1 Types and Characteristics of Muscular Tissue

Expected Learning Outcomes

When you have completed this section, you should be able to

a. describe the physiological properties that all muscle types have in common;

b. list the defining characteristics of skeletal muscle; and

c. discuss the elastic functions of the connective tissue components of a muscle.

11.1a Universal Characteristics of Muscle

The functions of the muscular system were detailed in the preceding chapter: movement, stability, communication, control of body openings and passages, heat production, and glycemic control. To carry out those functions, all muscle cells have the following characteristics.

- **Excitability (responsiveness).** Excitability is a property of all living cells, but muscle and nerve cells have developed this property to the highest degree. When stimulated by chemical signals, stretch, and other stimuli, muscle cells respond with electrical changes across the plasma membrane.

- **Conductivity.** Stimulation of a muscle cell produces more than a local effect. Local electrical excitation sets off a wave of excitation that travels rapidly along the cell and initiates processes leading to contraction.

- **Contractility.** Muscle cells are unique in their ability to shorten substantially when stimulated. This enables them to pull on bones and other organs to create movement.

- **Extensibility.** In order to contract, a muscle cell must also be extensible—able to stretch again between contractions. Most cells rupture if they are stretched even a little, but skeletal muscle cells can stretch to as much as three times their contracted length.

- **Elasticity.** When a muscle cell is stretched and then released, it recoils to a shorter length. If it were not for this elastic recoil, resting muscles would be too slack or flabby. Owing to a *length–tension relationship* to be described in this chapter, they wouldn't contract very strongly when stimulated.

FIGURE 11.1 Skeletal Muscle Fibers. APR

Ed Reschke

❓ *What tissue characteristics evident in this photo distinguish this from cardiac and smooth muscle?*

11.1b Skeletal Muscle

Skeletal muscle may be defined as voluntary striated muscle that is usually attached to one or more bones. A skeletal muscle exhibits alternating light and dark transverse bands, or **striations (fig. 11.1),** that reflect an overlapping arrangement of their internal contractile proteins. Skeletal muscle is called **voluntary** because it is usually subject to conscious control. The other types of muscle are **involuntary** (not usually under conscious control), and they are never attached to bones.

A typical skeletal muscle cell is about 100 μm in diameter and 3 cm (30,000 μm) long; some are as thick as 500 μm and as long as 30 cm. Because of their extraordinary length, skeletal muscle cells are usually called **muscle fibers** or **myofibers.**

Recall that a skeletal muscle is composed not only of muscular tissue, but also of fibrous connective tissue: the *endomysium* that surrounds each muscle fiber, the *perimysium* that bundles muscle fibers together into fascicles, and the *epimysium* that encloses the entire muscle (see fig. 10.1). These connective tissues are continuous with the collagen fibers of tendons and those, in turn, with the collagen of the bone matrix. Thus, when a muscle fiber contracts, it pulls on these collagen fibers and typically moves a bone.

Collagen is neither excitable nor contractile, but it is somewhat extensible and elastic. When a muscle lengthens, for example during extension of a joint, its collagenous components resist excessive stretching and protect the muscle from injury. When a muscle relaxes, elastic recoil of the collagen may help to return the muscle to its resting length and keep it from becoming too flaccid. Some authorities contend that recoil of the tendons and other collagenous tissues contributes significantly to the power output and efficiency of a muscle. When you are running, for example, recoil of the calcaneal tendon may help to lift the heel and produce some of the thrust as your toes push off from the ground. (Such recoil contributes significantly to the long, efficient leaps of a kangaroo.) Others feel that the elasticity of these components is negligible in humans and that the recoil is produced entirely by certain intracellular proteins of the muscle fibers themselves.

BEFORE YOU GO ON

Answer the following questions to test your understanding of the preceding section:

1. Define *responsiveness, conductivity, contractility, extensibility,* and *elasticity.* State why each of these properties is necessary for muscle function.

2. How is skeletal muscle different from the other types of muscle?

3. Name and define the three layers of collagenous connective tissue in a skeletal muscle.

11.2 Skeletal Muscle Cells

Expected Learning Outcomes

When you have completed this section, you should be able to

a. describe the structural components of a muscle fiber;

b. relate the striations of a muscle fiber to the overlapping arrangement of its protein filaments; and

c. name the major proteins of a muscle fiber and state the function of each.

11.2a The Muscle Fiber

In order to understand muscle function, you must know how the organelles and macromolecules of a muscle fiber are arranged. Perhaps more than any other cell, a muscle fiber exemplifies the adage, Form follows function. It has a complex, tightly organized internal structure in which even the spatial arrangement of protein molecules is closely tied to its contractile function.

The plasma membrane of a muscle fiber is called the **sarcolemma,**[1] and its cytoplasm is called the **sarcoplasm.** The sarcoplasm is occupied mainly by long protein cords called **myofibrils** about 1 μm in diameter **(fig. 11.2).** It also contains an abundance of **glycogen,** a starchlike carbohydrate that provides energy for the cell during heightened levels of exercise, and the red oxygen-binding pigment **myoglobin,** which provides some of the oxygen needed for muscular activity.

Muscle fibers have multiple flattened or sausage-shaped nuclei pressed against the inside of the sarcolemma. This unusual multinuclear condition results from the embryonic development of a muscle fiber—several stem cells called **myoblasts**[2] fuse to produce each fiber, with each myoblast contributing one nucleus. Some myoblasts remain as unspecialized **satellite cells** between the muscle fiber and endomysium. These play an important role in the regeneration of damaged skeletal muscle.

---

[1]*sarco* = flesh, muscle; *lemma* = husk
[2]*myo* = muscle; *blast* = precursor

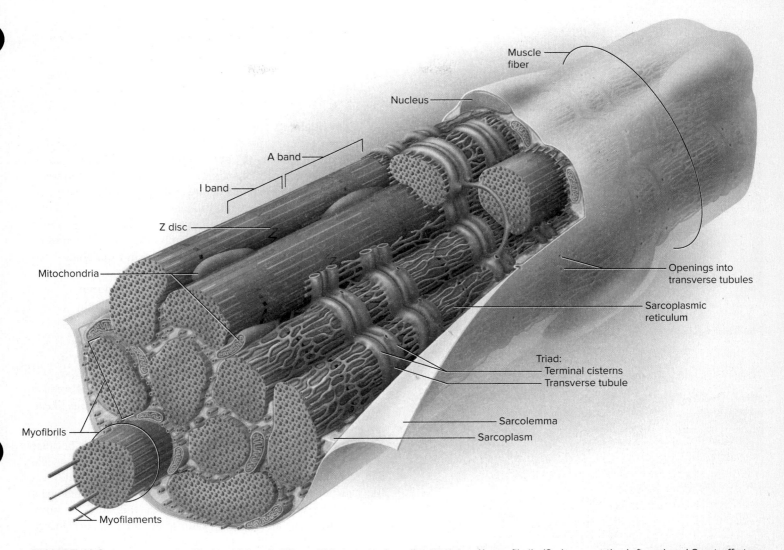

A band —

I band —

Z disc —

Mitochondria —

Myofibrils —

Myofilaments

Muscle fiber

Nucleus —

Openings into transverse tubules

Sarcoplasmic reticulum

Triad:
Terminal cisterns
Transverse tubule

Sarcolemma
Sarcoplasm

FIGURE 11.2 Structure of a Skeletal Muscle Fiber. This is a single cell containing 11 myofibrils (9 shown at the left end and 2 cut off at midfiber). A few myofilaments are shown projecting from the myofibril at the left. Most muscle fibers have from several dozen to a thousand or more myofibrils. Myofibril fine structure is shown in figure 11.3.

? *After reading a little farther in this chapter, explain why it is important for the transverse tubule to be so closely associated with the terminal cisterns.*

Most other organelles of the cell, such as mitochondria, are packed into the spaces between the myofibrils. The smooth endoplasmic reticulum, here called the **sarcoplasmic reticulum (SR),** forms a network around each myofibril (the blue web in fig. 11.2). It periodically exhibits dilated end sacs called **terminal cisterns,** which cross the muscle fiber from one side to the other. The sarcolemma has tubular infoldings called **transverse (T) tubules,** which penetrate through the cell and emerge on the other side. Each T tubule is closely associated with two terminal cisterns running alongside it, one on each side. The T tubule and the two cisterns associated with it constitute a *triad.*

Muscle has a high demand for ATP and therefore possesses an exceptionally large number of mitochondria wedged in between the myofibils. Muscle contraction also requires a lot of calcium ions (Ca^{2+}), as you will see, but this presents a problem. A high concentration of Ca^{2+} in the cytosol is lethal—it can react with phosphate ions to precipitate as calcium phosphate crystals, and can trigger cell death by apoptosis. Therefore, at rest, a muscle cell stores its Ca^{2+} in the sarcoplasmic reticulum, safely bound to

a protein called **calsequestrin.**[3] In a resting muscle fiber, Ca^{2+} is about 10,000 times as concentrated in the SR as it is in the cytosol. When the cell is stimulated, ion gates in the SR membrane open and Ca^{2+} floods into the cytosol to activate contraction. The T tubule signals the SR when to release these calcium bursts.

11.2b Myofilaments

Let's return to the myofibrils just mentioned—the long protein cords that fill most of the muscle cell—and look at their structure at a finer, molecular level. It is here that the key to muscle contraction lies. Each myofibril is a bundle of parallel proteins called **muscle filaments** or **myofilaments** (see the left end of fig. 11.2). There are three kinds of myofilaments:

1. **Thick filaments (fig. 11.3a, b)** are about 15 nm in diameter. Each is made of several hundred molecules of a motor

[3]*cal* = calcium; *sequestr* = to set apart or separate; *in* = protein

(a) **Myosin molecule**

Head

Tail

(b) **Thick filament**

Myosin head

(c) **Thin filament**

Tropomyosin   Troponin complex   G actin

(d) **Portion of a sarcomere showing the overlap of thick and thin filaments**

Thick filament

Thin filament

Bare zone

FIGURE 11.3 Molecular Structure of Thick and Thin Filaments. (a) A single myosin molecule composed of two intertwined proteins. (b) A thick filament composed of a bundle of myosin molecules. (c) A thin filament composed of actin, tropomyosin, and troponin. (d) A region of overlap between the thick and thin filaments.

protein called **myosin.** A myosin molecule is shaped like a golf club, with two chains intertwined to form a shaft-like *tail* and a double globular *head* projecting from it at an

angle. A thick filament consists of a bundle of 200 to 500 myosin molecules with their heads directed outward in a helical array around the bundle. The heads on one half of the thick filament angle to the left, and the heads on the other half angle to the right; in the middle is a *bare zone* with no heads.

2. **Thin filaments (fig. 11.3c, d),** 7 nm in diameter, are composed primarily of two intertwined strands of a protein called **fibrous (F) actin.** Each F actin is like a bead necklace—a string of subunits called **globular (G) actin.** Each G actin has an **active site** that can bind to the head of a myosin molecule. A thin filament also has 40 to 60 molecules of yet another protein, **tropomyosin.** When a muscle fiber is relaxed, each tropomyosin blocks the active sites of six or seven G actins and prevents myosin from binding to them. Each tropomyosin molecule, in turn, has a smaller, three-part calcium-binding protein called **troponin** bound to it.

3. **Elastic filaments** (see fig. 11.5b), 1 nm in diameter, are made of a huge springy protein called **titin.**[4] They run through the core of each thick filament and anchor it to structures called the *Z disc* at one end and *M line* at the other. Titin stabilizes the thick filament, centers it between the thin filaments, prevents overstretching, and recoils like a spring after a muscle is stretched.

Myosin and actin are called **contractile proteins** because they do the work of shortening the muscle fiber. Tropomyosin and troponin are called **regulatory proteins** because they act like a switch to determine when the fiber can contract and when it cannot. Several clues as to how they do this may be apparent from what has already been said—calcium ions are released into the sarcoplasm to activate contraction; calcium binds to troponin; troponin is also bound to tropomyosin; and tropomyosin blocks the active sites of actin, so that myosin cannot bind to it when the muscle is not stimulated. Perhaps you are already forming some idea of the contraction mechanism to be explained shortly.

At least seven other accessory proteins occur in the thick and thin filaments or are associated with them. Among other functions, they anchor the myofilaments, regulate their length, and keep them aligned with each other for optimal contractile effectiveness. The most clinically important of these is **dystrophin,** an enormous protein located between the sarcolemma and the outermost myofilaments. It links actin filaments to a peripheral protein on the inner face of the sarcolemma. Through a series of linking proteins **(fig. 11.4),** this leads ultimately to the fibrous endomysium surrounding the muscle fiber. Therefore, when the thin filaments move, dystrophin transfers the force to the basal lamina, endomysium, and ultimately to the tendon. Genetic defects in dystrophin are responsible for the disabling disease *muscular dystrophy* (see Deeper Insight 11.4).

---

[4]*tit* = giant; *in* = protein

Endomysium
Linking proteins
Basal lamina
Sarcolemma
Dystrophin
Thin filament
Thick filament
Elastic filament

FIGURE 11.4 Dystrophin. Dystrophin ultimately transfers the force of moving actin filaments to tissues leading to the tendon of a muscle.

11.2c Striations

Myosin and actin are not unique to muscle; they occur in nearly all cells, where they function in cellular motility, mitosis, and transport of intracellular materials. In skeletal and cardiac muscle they are especially abundant, however, and are organized in a precise array that accounts for the striations of these muscle types (**figs. 11.5,** 11.3d).

Striated muscle has dark **A bands** alternating with lighter **I bands.** (*A* stands for *anisotropic* and *I* for *isotropic,* which refer to the way these bands affect polarized light. To help remember which band is which, think "d**A**rk" and "l**I**ght.") Each A band consists of thick filaments lying side by side. Part of the A band, where thick and thin filaments overlap, is especially dark. In this region, each thick filament is surrounded by a hexagonal array of thin filaments. In the middle of the A band, there is a lighter region called the **H band,**[5] into which the thin filaments do not reach. In the middle of the H band, the thick filaments are linked to each other through a dark, transverse protein complex called the **M line.**[6]

Each light I band is bisected by a dark narrow **Z disc**[7] (**Z line**), which provides anchorage for the thin and elastic filaments. Each segment of a myofibril from one Z disc to the next is called a **sarcomere**[8] (SAR-co-meer), the functional contractile unit of the muscle

[5]*H = helle =* bright (German)
[6]*M = Mittel =* middle (German)
[7]*Z = Zwichenscheibe =* between disc (German)
[8]*sarco =* muscle; *mere =* part, segment

FIGURE 11.5 Muscle Striations and Their Molecular Basis. (a) Five myofibrils of a single muscle fiber, showing the striations in the relaxed state (TEM). (b) The overlapping pattern of thick and thin filaments that accounts for the striations seen in part (a). **APR**

a: Don W. Fawcett/Science Source

TABLE 11.1	The Structural Hierarchy of a Skeletal Muscle

Structural Level	Description
Muscle	A contractile organ, usually attached to bones by way of tendons. Composed of bundles (fascicles) of tightly packed, long, parallel cells (muscle fibers). Supplied with nerves and blood vessels and enclosed in a fibrous epimysium that separates it from neighboring muscles.
Fascicle	A bundle of muscle fibers within a muscle. Supplied by nerves and blood vessels and enclosed in a fibrous perimysium that separates it from neighboring fascicles.
Muscle Fiber	A single muscle cell. Slender, elongated, threadlike, enclosed in a specialized plasma membrane (sarcolemma). Contains densely packed bundles (myofibrils) of contractile protein myofilaments, multiple nuclei immediately beneath the sarcolemma, and an extensive network of specialized smooth endoplasmic reticulum (sarcoplasmic reticulum). Enclosed in a thin fibrous sleeve called endomysium.
Myofibril	A bundle of protein myofilaments within a muscle fiber; myofibrils collectively fill most of the cytoplasm. Each surrounded by sarcoplasmic reticulum and mitochondria. Has a banded (striated) appearance due to orderly overlap of protein myofilaments.
Sarcomere	A segment of myofibril from one Z disc to the next in the fiber's striation pattern. Hundreds of sarcomeres end to end compose a myofibril. The functional, contractile unit of the muscle fiber.
Myofilaments	Fibrous protein strands that carry out the contraction process. Two types: thick myofilaments composed mainly of myosin, and thin myofilaments composed mainly of actin. Thick and thin myofilaments slide over each other to shorten each sarcomere. Shortening of end-to-end sarcomeres shortens the entire muscle.

fiber. A muscle shortens because its individual sarcomeres shorten and pull the Z discs closer to each other, and dystrophin and the linking proteins pull on the extracellular proteins of the muscle. As the Z discs are pulled closer together, they pull on the sarcolemma to achieve overall shortening of the cell.

Table 11.1 reviews the organization of skeletal muscle at successive structural levels from the whole muscle to the myofilaments.

BEFORE YOU GO ON

Answer the following questions to test your understanding of the preceding section:

4. What special terms are given to the plasma membrane, cytoplasm, and smooth ER of a muscle cell?

5. What is the difference between a myofilament and a myofibril?

6. List five proteins of the myofilaments and describe their physical arrangement.

7. Sketch the overlapping pattern of myofilaments to show how they account for the A bands, I bands, H bands, and Z discs.

11.3 The Nerve–Muscle Relationship

Expected Learning Outcomes

When you have completed this section, you should be able to

a. explain what a motor unit is and how it relates to muscle contraction;

b. describe the structure of the junction where a nerve fiber meets a muscle fiber; and

c. explain why a cell has an electrical charge difference across its plasma membrane and, in general terms, how this relates to muscle contraction.

Skeletal muscle cannot contract unless it is stimulated by a nerve (or with electrodes). If its nerve connections are severed or poisoned, a muscle is paralyzed. If the connection is not restored, the paralyzed muscle wastes away in a shrinkage called *denervation atrophy.* Thus, muscle contraction cannot be understood without first understanding the relationship between nerve and muscle cells.

11.3a Motor Neurons and Motor Units

Skeletal muscles are served by nerve cells called *somatic motor neurons,* whose cell bodies are in the brainstem and spinal cord. Their axons, called **somatic motor fibers,** lead to the muscles. Each nerve fiber branches out to multiple muscle fibers, but each muscle fiber is supplied by only one motor neuron.

When a nerve signal approaches the end of an axon, it spreads out over all of its terminal branches and stimulates all muscle fibers supplied by them. Thus, these muscle fibers contract in unison; there is no way to stimulate some but not all of them. Since they behave as a single functional unit, one nerve fiber and all the

FIGURE 11.6 Motor Units.
(a) A slice of spinal cord depicting two motor neurons in its anterior horn. Some motor neurons (green) are small, relatively sensitive, and easily activated. Others (violet) are large, less sensitive, and activated only when greater muscular strength is needed. (b) In the muscle, the muscle fibers of small motor units are relatively small. Large motor units have larger and more numerous muscle fibers. Note that the muscle fibers of any given motor unit are distributed throughout the muscle and commingled with the fibers of other motor units (red), not clustered in one place.

muscle fibers innervated by it are called a **motor unit (fig. 11.6).** The muscle fibers of a motor unit are not clustered together but dispersed throughout a muscle. Therefore, when stimulated, they cause a weak contraction over a wide area—not just a local twitch in one small region. Effective muscle contraction usually requires the activation of many motor units at once.

On average, about 200 muscle fibers are innervated by each motor neuron, but motor units can be much smaller or larger than this to serve different purposes. Where fine control is needed, we have *small motor units.* These typically are supplied by small, relatively sensitive neurons. Where strength is more important than fine control, we have *large motor units.* These are innervated by less sensitive neurons with larger cell bodies, activated when strength is needed for a demanding task.

Consider a hand or eye muscle that requires fine motor control. Here, 1,000 muscle fibers may be innervated by 200 to 300 motor neurons with small motor units of only 3 to 5 muscle fibers each. Turning a few motor units on or off would produce small, subtle changes in muscle action. Then consider a powerful thigh or calf muscle such as the quadriceps femoris or gastrocnemius. Here, every 1,000 muscle fibers may be controlled by only 1 or 2 neurons with large motor units of 500 to 1,000 muscle fibers each. Turning a few motor units on or off would produce relatively large changes in muscle action, with large increments in strength but little fine control or subtlety.

In addition to adjustments in strength and control, another advantage of having multiple motor units in each muscle is that they work in shifts. Muscle fibers fatigue when subjected to continual stimulation. If all the fibers in one of your postural muscles fatigued at once, for example, you could collapse. To prevent this, other motor units take over while the fatigued ones recover, and the muscle as a whole can sustain long-term contraction.

11.3b The Neuromuscular Junction

The point where a nerve fiber meets any target cell is called a **synapse** (SIN-aps). When the target cell is a muscle fiber, the synapse is also called a **neuromuscular junction (NMJ)** or **motor end plate (fig. 11.7).** Each terminal branch of the nerve fiber within the NMJ forms a separate synapse with the muscle fiber. The sarcolemma of the NMJ is irregularly indented, a little like a handprint

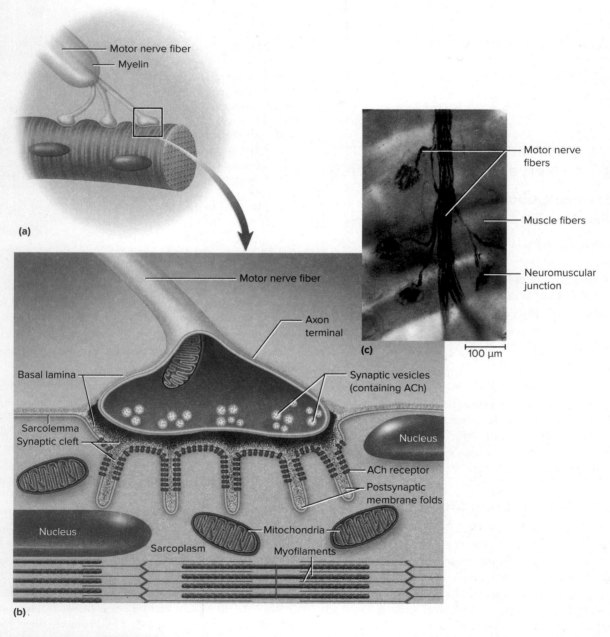

100 µm

(c)

FIGURE 11.7 Innervation of Skeletal Muscle. (a) Motor nerve ending on a muscle fiber. (b) Detail of the neuromuscular junction. (c) Light micrograph (LM) of neuromuscular junctions (compare the SEM photo on the opening page of this chapter). **APR**

c: Al Telser/McGraw-Hill Education

pressed into soft clay. If you imagine the nerve fiber to be like your forearm, branching out at the end like your fingers, the individual synapses would be like the dents where your fingertips press into the clay. Thus, one nerve fiber stimulates the muscle fiber at several nearby points within the NMJ.

At each synapse, the nerve fiber ends in a bulbous swelling called the **axon terminal.** The terminal doesn't directly touch the muscle fiber but is separated from it by a narrow space called the **synaptic cleft,** about 60 to 100 nm wide (scarcely wider than the thickness of one plasma membrane).

The axon terminal contains spheroidal organelles called **synaptic vesicles,** which are filled with a chemical, **acetylcholine (ACh)** (ASS-eh-till-CO-leen)—one of many *neurotransmitters* to be introduced in the next chapter. The electrical signal (nerve impulse) traveling down a nerve fiber cannot cross the synaptic cleft like a spark jumping between two electrodes—rather, it causes the synaptic vesicles to undergo exocytosis, releasing ACh into the cleft. ACh thus functions as a chemical messenger from the nerve cell to the muscle cell.

To respond to ACh, each synapse has about 50 million **ACh receptors**—proteins incorporated into the sarcolemma across from the axon terminals. To maximize the number of ACh receptors and thus its sensitivity to the neurotransmitter, the sarcolemma in this area has infoldings about 1 μm deep, called **postsynaptic membrane folds (junctional folds),** which increase the surface area of ACh-sensitive membrane. The muscle nuclei beneath the folds are specifically dedicated to the synthesis of ACh receptors and other proteins of the local sarcolemma. A deficiency of ACh receptors leads to muscle weakness in the disease *myasthenia gravis* (see Deeper Insight 11.4).

The entire NMJ is enclosed in a **basal lamina,** a mat of collagen and glycoprotein that separates the muscle fiber and nerve ending from the surrounding connective tissue. The basal lamina also passes through the synaptic cleft and virtually fills it. Both the sarcolemma and that part of the basal lamina in the cleft contain **acetylcholinesterase (AChE)** (ASS-eh-till-CO-lin-ESS-ter-ase). This is an enzyme that breaks down ACh after the ACh has stimulated the muscle cell; thus, it is important in turning off muscle contraction and allowing the muscle to relax (see Deeper Insight 11.1).

11.3c Electrically Excitable Cells

Muscle and nerve cells are regarded as *electrically excitable cells* because their plasma membranes exhibit voltage changes in response to stimulation. The study of the electrical activity of cells, called **electrophysiology,** is a key to understanding nerve activity, muscle contraction, the heartbeat and electrocardiogram, and other physiological phenomena. The details of electrophysiology are presented in section 12.4, but a few fundamental principles must be introduced here so you can understand muscle excitation.

The electrical activity of cells hinges on differences in the concentration of ions in the intracellular fluid (ICF) and extracellular fluid (ECF) adjacent to the plasma membrane. The ICF contains a greater concentration of negative anions than the ECF does—especially negatively charged proteins, nucleic acids, and phosphates, which are trapped in the cell and give its interior a net negative charge. That is, the membrane is **polarized,** like a little battery. Also contained in the ICF is a great excess of potassium ions (K^+), whereas the ECF contains a great excess of sodium ions (Na^+). The electrical events that initiate muscle contraction are driven by the movements of these two cations through the membrane when a muscle or nerve cell is excited.

DEEPER INSIGHT 11.1

CLINICAL APPLICATION

Neuromuscular Toxins and Paralysis

Toxins that interfere with synaptic function can paralyze the muscles. Organophosphate pesticides such as malathion are *cholinesterase inhibitors* that bind to AChE and prevent it from degrading ACh. Depending on the dose, this can prolong the action of ACh and produce *spastic paralysis,* a state in which the muscles contract and cannot relax; clinically, this is called a *cholinergic crisis.* It poses a danger of suffocation if it affects the laryngeal and respiratory muscles. Another example of spastic paralysis is *tetanus (lockjaw),* caused by the toxin of a soil bacterium, *Clostridium tetani.* In the spinal cord, a neurotransmitter called glycine normally stops motor neurons from producing unwanted muscle contractions. The tetanus toxin blocks glycine release and thus causes overstimulation and spastic paralysis of the muscles.

Flaccid paralysis is a state in which the muscles are limp and cannot contract. This too can cause respiratory arrest if it affects the thoracic muscles. Among the causes of flaccid paralysis are poisons such as curare (cue-RAH-ree), which competes with ACh for receptor sites but doesn't stimulate the muscle. Curare is extracted from certain plants and used by some South American natives to poison blowgun darts. It has been used to treat muscle spasms in some neurological disorders and to relax abdominal muscles for surgery, but other muscle relaxants have now replaced curare for most purposes.

Another cause of flaccid paralysis is *botulism,* a type of food poisoning caused by a neuromuscular toxin secreted by the bacterium *Clostridium botulinum.* Botulinum toxin blocks ACh release. Purified botulinum toxin is marketed as Botox Cosmetic. It is injected in small doses into specific facial muscles. Wrinkles gradually disappear as muscle paralysis sets in over the next few hours. The effect lasts about 4 months until the muscles retighten and the wrinkles return. Botox treatment has become the fastest-growing cosmetic medical procedure in the United States, as its usage has expanded to other conditions and many people go for cosmetic treatment every few months. It has had some undesirable consequences, however, as it is sometimes administered by unqualified practitioners. Even some qualified physicians use it for treatments not yet approved by the FDA, and some host festive "Botox parties" for treatment of patients in assembly-line fashion.

A difference in electrical charge from one point to another is called an *electrical potential,* or *voltage.* It typically measures 12 volts (V) for a car battery and 1.5 V for a flashlight battery, for example. On the sarcolemma of a muscle cell, the voltage is much smaller, about –90 millivolts (mV), but critically important to life. (The negative sign refers to the relatively negative charge on the intracellular side of the membrane.) This voltage is called the **resting membrane potential (RMP).** It is maintained by the sodium–potassium pump, as explained in section 12.4b.

When a nerve or muscle cell is stimulated, dramatic things happen electrically, as we shall soon see in the excitation of muscle. Ion channels in the plasma membrane open and Na^+ instantly flows into the cell, driven both by its concentration difference across the membrane and by its attraction to the negative charge of the cell interior—that is, it flows down an **electrochemical gradient.** These Na^+ cations override the negative charge just inside the membrane, so the inside of the membrane briefly becomes positive. This is called **depolarization** of the membrane. Immediately, Na^+ channels close and K^+ channels open. K^+ rushes out of the cell, partly because it is repelled by the positive sodium charge and partly because it is more concentrated in the ICF than in the ECF—that is, it flows down its own electrochemical gradient in the direction opposite from the sodium movement. The loss of K^+ ions from the cell turns the inside of the membrane negative again (**repolarization**). This quick up-and-down voltage shift, from the negative RMP to a positive value and then back to the RMP again, is called an **action potential.** The RMP is a stable voltage seen in a "waiting" cell, whereas the action potential is a quickly fluctuating voltage seen in an active, stimulated cell.

Action potentials have a way of perpetuating themselves—an action potential at one point on a plasma membrane causes another one to happen immediately in front of it, which triggers another one a little farther along, and so forth. A wave of action potentials spreading along a nerve fiber like this is called a *nerve impulse* or *nerve signal.* Such signals also travel along the sarcolemma of a muscle fiber. Next we will see how this leads to muscle contraction.

BEFORE YOU GO ON

Answer the following questions to test your understanding of the preceding section:

8. What differences would you expect to see between a motor unit where muscular strength is more important than fine control and another motor unit where fine control is more important?

9. State the functions of the axon terminal and its synaptic vesicles.

10. Distinguish between acetylcholine, an acetylcholine receptor, and acetylcholinesterase. State where each is found and describe the function it serves.

11. What accounts for the resting membrane potential seen in unstimulated nerve and muscle cells?

12. What is the difference between a resting membrane potential and an action potential?

11.4 Behavior of Skeletal Muscle Fibers

Expected Learning Outcomes

When you have completed this section, you should be able to

a. explain how a nerve fiber stimulates a skeletal muscle fiber;

b. explain how stimulation of a muscle fiber activates its contractile mechanism;

c. explain the mechanism of muscle contraction;

d. explain how a muscle fiber relaxes; and

e. explain why the force of a muscle contraction depends on the muscle's length prior to stimulation.

The process of muscle contraction and relaxation has four major phases: (1) excitation, (2) excitation–contraction coupling, (3) contraction, and (4) relaxation. Each phase occurs in several smaller steps, which we now examine in detail.

11.4a Excitation

Excitation is the process in which action potentials in the nerve fiber lead to action potentials in the muscle fiber. The steps in excitation are shown in **figure 11.8.** Follow these carefully as you read the following description to ensure understanding of the process.

1. A nerve signal arrives at the axon terminal and opens voltage-gated calcium channels. Calcium ions enter the terminal.

2. Calcium stimulates the synaptic vesicles to release acetylcholine (ACh) into the synaptic cleft. One action potential causes exocytosis of about 60 vesicles, and each vesicle releases about 10,000 molecules of ACh.

3. ACh diffuses across the synaptic cleft and binds to receptors on the sarcolemma.

4. These receptors are ligand-gated ion channels. Two ACh molecules must bind to each receptor to open the channel. When it opens, Na^+ flows quickly into the cell and K^+ flows out. The voltage on the sarcolemma quickly rises to a less negative value as Na^+ enters the cell, then falls back to the RMP as K^+ exits. This rapid up-and-down fluctuation in voltage at the motor end plate is called the **end-plate potential (EPP).**

5. Areas of sarcolemma next to the end plate have voltage-gated ion channels that open in response to the EPP. Some of these are specific for Na^+ and admit it to the cell, while others are specific for K^+ and allow it to leave. These ion movements create an *action potential.* The muscle fiber is now excited.

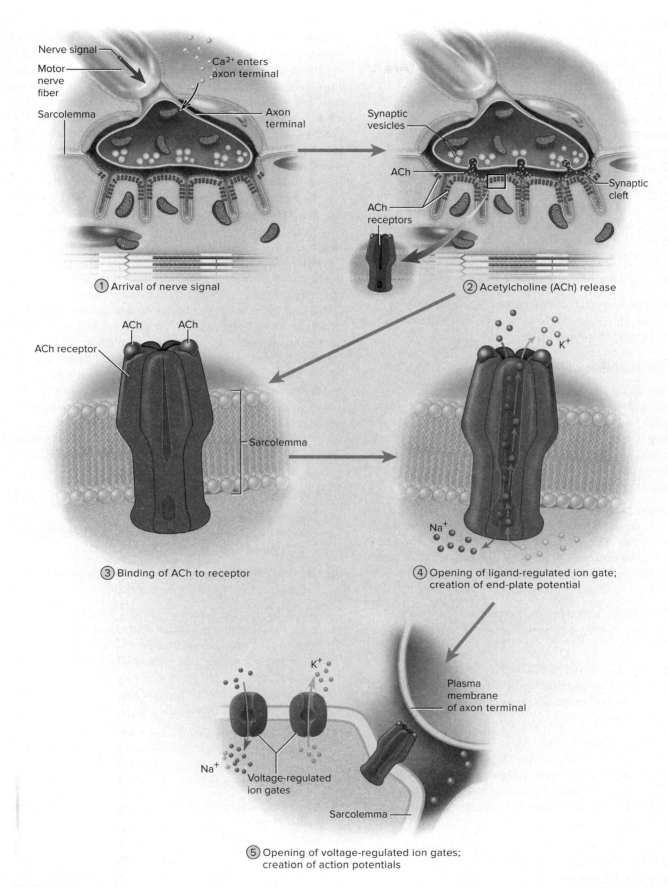

① Arrival of nerve signal

Nerve signal

Motor nerve fiber

Sarcolemma

Ca²⁺ enters axon terminal

Axon terminal

② Acetylcholine (ACh) release

Synaptic vesicles

ACh

ACh receptors

Synaptic cleft

③ Binding of ACh to receptor

ACh   ACh

ACh receptor

Sarcolemma

④ Opening of ligand-regulated ion gate; creation of end-plate potential

K⁺

Na⁺

⑤ Opening of voltage-regulated ion gates; creation of action potentials

K⁺

Na⁺

Voltage-regulated ion gates

Plasma membrane of axon terminal

Sarcolemma

FIGURE 11.8 Excitation of a Muscle Fiber. These events link action potentials in a nerve fiber to the generation of action potentials in the muscle fiber. See the corresponding numbered steps in the text for explanation.

11.4b Excitation–Contraction Coupling

Excitation–contraction coupling refers to events that link action potentials on the sarcolemma to activation of the myofilaments, thereby preparing them to contract. The steps in the coupling process are shown in **figure 11.9.**

⑥ A wave of action potentials spreads from the motor end plate in all directions, like ripples on a pond. When this wave of excitation reaches the T tubules, it continues down them into the cell interior.

⑦ Action potentials open voltage-gated ion channels in the T tubules. These are linked to calcium channels in the terminal cisterns of the sarcoplasmic reticulum (SR). Thus, channels in the SR open as well and calcium diffuses out of the SR, down its concentration gradient into the cytosol.

⑧ Calcium binds to the troponin of the thin filaments.

⑨ The troponin–tropomyosin complex changes shape and exposes the active sites on the actin. This makes them available for binding to myosin heads.

11.4c Contraction

Contraction is the step in which the muscle fiber develops tension and may shorten. The mechanism of contraction is called the **sliding filament theory.** It holds that the myofilaments don't become any shorter during contraction; rather, the thin filaments slide over the thick ones and pull the Z discs behind them, causing each sarcomere as a whole to shorten. The individual steps in this mechanism are shown in **figure 11.10.**

⑩ The myosin head must have an ATP molecule bound to it to initiate contraction. **Myosin ATPase,** an enzyme in the head, hydrolyzes this ATP into ADP and phosphate (P_i). The energy released by this process activates the head, which "cocks" into an extended, high-energy position. The head temporarily keeps the ADP and P_i bound to it.

⑪ The cocked myosin binds to an exposed active site on the thin filament, forming a **cross-bridge** between the myosin and actin.

⑫ Myosin releases the ADP and P_i and flexes into a bent, low-energy position, tugging the thin filament along with it. This is called the **power stroke.** The head remains bound to actin until it binds a new ATP.

⑬ The binding of a new ATP to myosin destabilizes the myosin–actin bond, breaking the cross-bridge. The myosin head now undergoes a **recovery stroke.** It hydrolyzes the new ATP, recocks (returns to step 10), and attaches to a new active site farther down the thin filament, ready for another power stroke.

It may seem as if releasing the thin filament at step 13 would simply allow it to slide back to its previous position, so nothing would have been accomplished. Think of the sliding filament mechanism, however, as being similar to the way you would pull in a boat anchor hand over hand. When the myosin head cocks, it is like your

hand reaching out to grasp the anchor rope. When it flexes back into the low-energy position, it is like your elbow flexing to pull on the rope and draw the anchor up a little bit. When you let go of the rope with one hand, you hold onto it with the other one, alternating hands until the anchor is pulled in. Similarly, when one myosin head releases actin in preparation for the recovery stroke, there are many other heads on the same thick filament holding onto the thin filament so it doesn't slide back. At any given moment during contraction, about half of the heads are bound to the thin filament and the other half are extending forward to grasp it farther down. That is, the myosin heads don't all stroke at once but contract sequentially.

Each head acts in a jerky manner, but hundreds of them working together produce a smooth, steady pull on the thin filament. This is similar to the locomotion of a millipede—a wormlike animal with a few hundred tiny legs. Watch an online video of a crawling millipede and you will see that each leg takes individual jerky steps, but all the legs working together produce a smooth gliding movement, like the glide of a thick muscle filament walking along a thin one. Note that even though the muscle fiber contracts, the *myofilaments don't become shorter* any more than a rope becomes shorter as you pull in an anchor. The thin filaments slide over the thick ones, as the name of the sliding filament theory implies.

A single cycle of power and recovery strokes by all myosin heads in a muscle fiber would shorten the fiber about 1%. A fiber, however, may shorten by as much as 40% of its resting length, so obviously the cycle of power and recovery must be repeated many times by each myosin head. Each head carries out about five strokes per second, and each stroke consumes one ATP.

11.4d Relaxation

When the nerve fiber stops stimulating it, a muscle fiber relaxes and returns to its resting length. This is achieved by the steps shown in **figure 11.11.**

⑭ Nerve signals stop arriving at the neuromuscular junction, so the axon terminal stops releasing ACh.

⑮ As ACh dissociates (separates) from its receptor, AChE breaks it down into fragments that cannot stimulate the muscle. The axon terminal reabsorbs these fragments for recycling. All of this happens continually while the muscle is stimulated, too, but when nerve signals stop, no more ACh is released to replace that which breaks down. Therefore, stimulation of the muscle fiber by ACh ceases.

⑯ From excitation through contraction, the SR simultaneously releases and reabsorbs Ca^{2+}; but when the nerve fiber stops firing and excitation ceases, Ca^{2+} release also ceases and only its reabsorption continues.

⑰ Owing to reabsorption by the SR, the level of free calcium in the cytosol falls dramatically. Now, when calcium dissociates from troponin, it is not replaced.

⑱ Tropomyosin moves back into the position where it blocks the active sites of the actin filament. Myosin can no longer bind to actin, and the muscle fiber ceases to produce or maintain tension.

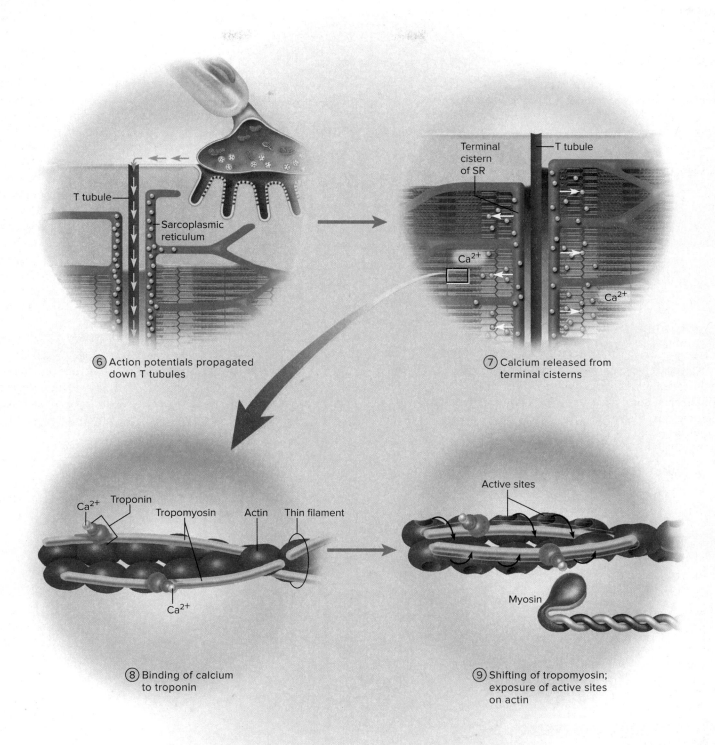

⑥ Action potentials propagated down T tubules

⑦ Calcium released from terminal cisterns

⑧ Binding of calcium to troponin

⑨ Shifting of tropomyosin; exposure of active sites on actin

FIGURE 11.9 Excitation–Contraction Coupling. These events link action potentials in the muscle fiber to the release and binding of calcium ions. See the corresponding numbered steps in the text for explanation. The numbers in this figure begin where figure 11.8 ended.

Troponin Tropomyosin

ADP
Pi

Myosin

⑩ Hydrolysis of ATP to ADP + Pi;
activation and cocking of myosin head
(recovery stroke)

Cross-bridge:
Actin
Myosin

ADP
Pi

⑪ Formation of myosin–actin cross-bridge

ATP

⑬ Binding of new ATP;
breaking of cross-bridge

ADP
Pi

⑫ Power stroke; sliding of thin
filament over thick filament

Z   Z

I   A   I

Appearance of contracting sarcomere

FIGURE 11.10 The Sliding Filament Mechanism of Contraction. This
is a cycle of repetitive events that cause a thin filament to slide over a
thick filament and generate tension in the muscle. See the corresponding
numbered steps in the text for explanation. The numbers in this figure begin
where figure 11.9 ended. Lower left: diagram of a contracting sarcomere.

14 Cessation of nervous stimulation and ACh release

15 ACh breakdown by acetylcholinesterase (AChE)

AChE

ACh

Terminal cistern of SR

Ca^{2+}

Ca^{2+}

Ca^{2+}

Ca^{2+}

ADP

P_i

16 Reabsorption of calcium ions by sarcoplasmic reticulum

17 Loss of calcium ions from troponin

Tropomyosin

ATP

18 Return of tropomyosin to position blocking active sites of actin

FIGURE 11.11 Relaxation of a Muscle Fiber. These events lead from the cessation of a nerve signal to the release of thin filaments by myosin. See the corresponding numbered steps in the text for explanation. The numbers in this figure begin where figure 11.10 ended.

Relaxation alone doesn't return a muscle to its resting length. That must be achieved by some force pulling the muscle and stretching it. For example, if the biceps flexes the elbow and then relaxes, it stretches back to its resting length only if the elbow is extended by contraction of the triceps or by the pull of gravity on the forearm.

▶▶▶**APPLY WHAT YOU KNOW**

One of the most important properties of proteins is their ability to change shape repeatedly (see section 2.4e). Identify at least two muscle proteins that must change shape in order for a muscle to contract and relax.

11.4e The Length–Tension Relationship and Muscle Tone

The tension generated by a muscle, and therefore the force of its contraction, depends on how stretched or contracted it was at the outset. This principle is called the **length–tension relationship.** The reasons for it can be seen in **figure 11.12.** If a fiber was already extremely contracted, its thick filaments would be rather close to the Z discs, as on the left side of the figure. The fiber couldn't contract much farther before the thick filaments would butt against the Z discs and stop. The contraction would be brief and weak. On the other hand, if a muscle fiber was extremely stretched, as on the right, there would be little overlap between the thick and thin filaments. The myosin heads would be unable to "get a grip" on the thin filaments, and again the contraction would be weak.

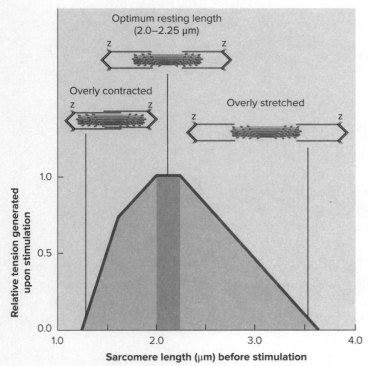

FIGURE 11.12 The Length–Tension Relationship.

DEEPER INSIGHT 11.2

CLINICAL APPLICATION

Rigor Mortis

Rigor mortis[9] is the hardening of the muscles and stiffening of the body that begins 3 to 4 hours after death. It occurs partly because the deteriorating sarcoplasmic reticulum releases calcium into the cytosol, and the deteriorating sarcolemma admits more calcium from the extracellular fluid. The calcium activates myosin–actin cross-bridging. Once bound to actin, myosin cannot release it without first binding an ATP molecule, and of course no ATP is available in a dead body. Thus, the thick and thin filaments remain rigidly cross-linked until the myofilaments begin to decay. Rigor mortis peaks about 12 hours after death and then diminishes over the next 48 to 60 hours.

Between these extremes, there is an optimum resting length at which a muscle responds with the greatest force. In this range (the flat top of the curve), the sarcomeres are 2.0 to 2.25 μm long. If the sarcomeres are less than 60% or more than 175% of their optimal length, they develop no tension at all in response to a stimulus.

The complete length–tension curve is derived from muscles isolated from an animal (often the frog gastrocnemius muscle) for laboratory stimulation. In reality, a muscle *in situ* (in its natural position in the living body) is never as extremely stretched or contracted as the far right and left sides of the figure depict. For one thing, the attachments of muscles to the bones and limitations on bone movement restrict muscle contraction to the midrange of the curve. For another, the central nervous system continually monitors and adjusts the length of the resting muscles, maintaining a state of partial contraction called **muscle tone.** This maintains optimum sarcomere length and makes the muscles ideally ready for action. The elastic filaments of the sarcomere also help to maintain enough myofilament overlap to ensure effective contraction when the muscle is called into action.

BEFORE YOU GO ON

Answer the following questions to test your understanding of the preceding section:

13. What change does ACh cause in an ACh receptor? How does this electrically affect the muscle fiber?

14. How do troponin and tropomyosin regulate the interaction between myosin and actin?

15. Describe the roles played by ATP in the power and recovery strokes of myosin.

16. What steps are necessary for a contracted muscle to return to its resting length?

[9]*rigor* = rigidity; *mortis* = of death

11.5 Behavior of Whole Muscles

Expected Learning Outcomes

When you have completed this section, you should be able to

a. describe the stages of a muscle twitch;

b. explain how successive muscle twitches can add up to produce stronger muscle contractions;

c. distinguish between isometric and isotonic contraction; and

d. distinguish between concentric and eccentric contraction.

Now you know how an individual muscle cell shortens. Our next objective is to move up to the organ grade of structure and consider how this relates to the action of the muscle as a whole.

11.5a Threshold, Latent Period, and Twitch

The timing and strength of a muscle's contraction can be shown in a chart called a **myogram (fig. 11.13).** A weak (subthreshold) electrical stimulus to a muscle produces no reaction. By gradually increasing the voltage and stimulating the muscle again, one can determine the **threshold,** or minimum voltage necessary to generate an action potential in the muscle fiber. At threshold or higher, a single stimulus causes a quick cycle of contraction and relaxation called a **twitch.**

There is a delay, or **latent period,** of about 2 milliseconds (ms) between the onset of the stimulus and onset of the twitch. This is the time required for excitation, excitation–contraction coupling, and tensing of the elastic components of the muscle. The force generated during this time is called *internal tension.* It isn't visible on the myogram because it causes no shortening of the muscle. On the left side, the myogram is therefore flat.

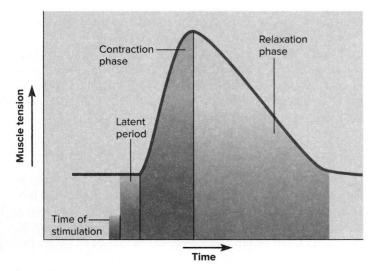

FIGURE 11.13 Idealized Myogram of a Muscle Twitch.

? *What role does ATP play during the relaxation phase?*

Once the elastic components are taut, the muscle begins to produce *external tension* and move a resisting object, or load, such as a bone or body limb. This is called the **contraction phase** of the twitch. By analogy, imagine lifting a weight suspended from a rubber band. At first, internal tension would only stretch the rubber band. Then, as the rubber band became taut, external tension would lift the weight.

The contraction phase is short-lived, because the sarcoplasmic reticulum quickly reabsorbs Ca^{2+} before the muscle develops maximal force. As the Ca^{2+} level in the cytoplasm falls, myosin releases the thin filaments and muscle tension declines. This is seen in the myogram as the **relaxation phase.** As shown by the asymmetry of the myogram, the muscle contracts more quickly than it relaxes. The entire twitch lasts from about 7 to 100 ms, so a muscle could theoretically complete about 10 to 140 twitches per second (if only the math mattered).

11.5b Contraction Strength of Twitches

We have seen that a subthreshold stimulus induces no muscle contraction at all, but at threshold intensity, a twitch is produced. Increasing the stimulus voltage still more, however, produces twitches no stronger than those at threshold. Superficially, the muscle fiber seems to be giving its maximum response once the stimulus intensity is at threshold or higher. However, even for a constant voltage, twitches vary in strength. This is so for a variety of reasons:

- Twitch strength depends on how stretched the muscle was just before it was stimulated, as we have just seen in the length–tension relationship.

- Twitches become weaker as a muscle fatigues, as discussed later in this chapter.

- Twitches vary with the temperature of the muscle; a warmed-up muscle contracts more strongly because enzymes such as the myosin heads work more quickly.

- Twitch strength varies with the muscle's state of hydration, which affects the spacing between thick and thin filaments and therefore the ability to form myosin–actin cross-bridges.
- Twitch strength varies with stimulus frequency; stimuli arriving close together produce stronger twitches than stimuli arriving at longer time intervals, as we will see shortly.

It should not be surprising that twitches vary in strength. Indeed, an individual twitch isn't strong enough to do any useful work. Muscles must contract with variable strength for different tasks, such as lifting a glass of champagne compared with lifting barbells at the gym.

Let us examine more closely the contrasting effects of stimulus *intensity* versus stimulus *frequency* on contraction strength. Suppose we apply a stimulating electrode to a motor nerve that supplies a muscle, such as a laboratory preparation of a frog sciatic nerve connected to its gastrocnemius muscle. Subthreshold stimulus voltages produce no response **(fig. 11.14)**. At threshold, we see a weak twitch (at *3* in the bottom row of the figure), and if we

Stimuli to nerve

Proportion of nerve fibers excited

Responses of muscle

FIGURE 11.14 The Relationship Between Stimulus Intensity (Voltage) and Muscle Tension. Top row: Nine stimuli of increasing strengths. The first two are subthreshold stimuli. Middle row: Cross section of a motor nerve with seven nerve fibers. The colored nerve fibers are the excited ones; note that none are excited by the subthreshold stimuli above. Bottom row: Graph of muscle tension. Subthreshold stimuli (1–2) produce no muscle contraction. When stimuli reach or exceed threshold (3–7), they excite more and more nerve fibers and motor units; thus, they produce stronger and stronger contractions. This is multiple motor unit summation (recruitment). Once all of the nerve fibers are stimulated (7–9), further increases in stimulus strength produce no further increase in muscle tension.

continue to raise the voltage, we see stronger twitches. The reason for this is that higher voltages excite more and more nerve fibers in the motor nerve (middle row of the figure), and thus stimulate more and more motor units to contract. The process of bringing more motor units into play is called **recruitment,** or **multiple motor unit (MMU) summation.** This is seen not just in artificial stimulation, but is part of the way the nervous system behaves naturally to produce varying muscle contractions. The neuromuscular system behaves according to the **size principle**—smaller, less powerful motor units with smaller, slower nerve fibers are activated first. This is sufficient for delicate tasks and refined movements, but if more power is needed, then larger motor units with larger, faster nerve fibers are subsequently activated.

But even when stimulus intensity (voltage) remains constant, twitch strength can vary with stimulus frequency. High-frequency stimulation produces stronger twitches than low-frequency stimulation. In **figure 11.15a,** we see that when a muscle is stimulated at low frequency, say 5 to 10 stimuli/s, it produces an identical twitch for each stimulus and fully recovers between twitches.

At higher stimulus frequencies, say 20 to 40 stimuli/s, each new stimulus arrives before the previous twitch is over. Each new twitch "rides piggyback" on the previous one and generates higher tension **(fig. 11.15b).** This phenomenon goes by two names: **temporal**[10] **summation,** because it results from two stimuli arriving close together in time, or **wave summation,** because it results from one wave of contraction added to another. Wave upon wave, each twitch reaches a higher level of tension than the one before, and the muscle relaxes only partially between stimuli. This effect produces a state of sustained fluttering contraction called **incomplete tetanus.**

In the laboratory, an isolated muscle can be stimulated at such high frequency that the twitches fuse into a single, nonfluctuating contraction called **complete (fused) tetanus (fig. 11.15c).** This doesn't happen in the body, however, because motor neurons don't fire that fast. Indeed, there is an inhibitory mechanism in the spinal cord that prevents them from doing so. Complete tetanus is injurious to muscle and associated soft tissues, so spinal inhibition protects the muscles by preventing complete tetanus. Muscle tetanus should not be confused with the disease of the same name caused by the tetanus toxin (see Deeper Insight 11.1).

Despite the fluttering contraction seen in incomplete tetanus, we know that a muscle taken as a whole can contract very smoothly. This is possible because motor units function asynchronously; when one motor unit relaxes, another contracts and takes over so that the muscle doesn't lose tension.

11.5c Isometric and Isotonic Contraction

In muscle physiology, "contraction" doesn't always mean the shortening of a muscle—it may mean only that the muscle produces internal tension while an external resistance causes it to stay the same length or even become longer. Thus, physiologists speak of different kinds of muscle contraction as *isometric* versus *isotonic* and *concentric* versus *eccentric* **(fig. 11.16).**

---

[10]*tempor* = time; *al* = pertaining to

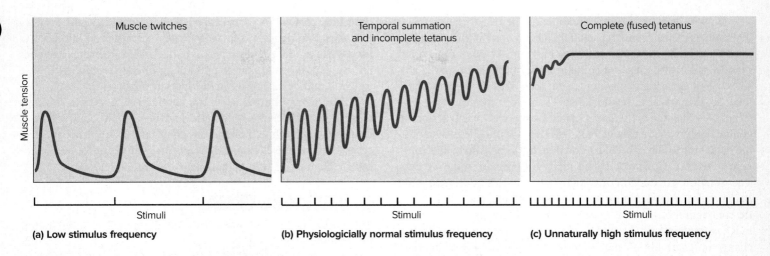

(a) Low stimulus frequency (b) Physiologicially normal stimulus frequency (c) Unnaturally high stimulus frequency

FIGURE 11.15 The Relationship Between Stimulus Frequency and Muscle Tension. (a) Twitch. At an unnaturally low stimulus frequency, as in laboratory preparations, the muscle relaxes completely between stimuli and shows twitches of uniform strength. (b) At a stimulus frequency within normal physiological range, the muscle doesn't have time to relax completely between twitches and the force of each twitch builds on the previous one, creating the state of incomplete tetanus. In this state, a muscle can attain three to four times as much tension, or force, as a single twitch produces. (c) At an unnaturally high stimulus frequency, attained only in laboratory preparations, the muscle can't relax at all between twitches, and twitches fuse into a state of complete tetanus.

(a) Isometric contraction (b) Isotonic concentric contraction (c) Isotonic eccentric contraction

FIGURE 11.16 Isometric and Isotonic Contraction. (a) Isometric contraction, in which a muscle develops tension but doesn't shorten. (b) Isotonic concentric contraction, in which the muscle shortens while maintaining a constant degree of tension. (c) Isotonic eccentric contraction, in which the muscle maintains tension while it lengthens, allowing a muscle to relax without going suddenly limp.

❓ *Name a muscle that undergoes eccentric contraction as you sit down in a chair.*

Suppose you lift a heavy dumbbell. When you first contract the muscles of your arm, you can feel the tension building in them even though the dumbbell isn't yet moving. At this point, your muscles are contracting at a cellular level, but their tension is absorbed by the elastic components and is resisted by the weight of the load; the muscle as a whole doesn't shorten or produce any external movement. This phase is called **isometric**[11] **contraction**—contraction without a change in length (fig. 11.16a). This occurs at the beginning of any muscle contraction, but is prolonged in lifting heavy weights. However, it isn't merely a prelude to movement. The isometric contraction of antagonistic muscles at a single joint is important in maintaining joint stability at rest, and the isometric contraction of postural muscles is what keeps us from sinking in a heap to the floor. **Isotonic**[12] **contraction**—contraction with a change in length but no change in tension—begins when internal tension builds to the point that it overcomes the resistance. The muscle now shortens, moves the load, and maintains essentially the same tension from then on (fig. 11.16b). Isometric and isotonic contraction are both phases of normal muscular action **(fig. 11.17).**

There are two forms of isotonic contraction: concentric and eccentric. In **concentric contraction,** a muscle shortens as it maintains tension—for example, when the biceps contracts and flexes the elbow. In **eccentric contraction,** a muscle lengthens as it maintains tension. If you set that dumbbell down again (fig. 11.16c), your biceps lengthens as you extend your elbow, but it maintains tension to act as a brake and keep you from simply dropping the weight. A weight lifter uses

[11]*iso* = same, uniform; *metr* = length; *ic* = pertaining to
[12]*iso* = same, uniform; *ton* = tension; *ic* = pertaining to

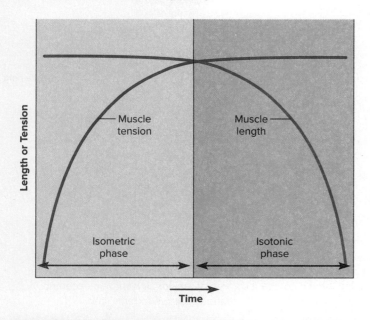

FIGURE 11.17 Isometric and Isotonic Phases of Contraction. At the beginning of a contraction (isometric phase), muscle tension rises but the length remains constant (the muscle does not shorten). When tension overcomes the resistance of the load, the tension levels off and the muscle begins to shorten and move the load (isotonic phase).

❓ *How would you extend this graph in order to show eccentric contraction?*

concentric contraction when lifting a dumbbell and eccentric contraction when lowering it. When weight lifters suffer muscle injuries, it is usually during the eccentric phase, because the sarcomeres and connective tissues of the muscle are pulling in one direction while the weight is pulling the muscle in the opposite direction.

In summary, during isometric contraction, a muscle develops tension without changing length, and in isotonic contraction, it changes length while maintaining constant tension. In concentric contraction, a muscle maintains tension as it shortens, and in eccentric contraction, it maintains tension while it lengthens.

BEFORE YOU GO ON

Answer the following questions to test your understanding of the preceding section:

17. State three or more reasons why muscle twitch strength can vary even when stimulus intensity remains constant.

18. Explain the role of tetanus in normal muscle action.

19. Describe an everyday activity not involving the arms in which your muscles would switch from isometric to isotonic contraction.

20. Describe an everyday activity not involving the arms that would involve concentric contraction and one that would involve eccentric contraction.

11.6 Muscle Metabolism

Expected Learning Outcomes

When you have completed this section, you should be able to

a. explain how skeletal muscle meets its energy demands during rest and exercise;

b. explain the basis of muscle fatigue and soreness;

c. discuss why extra oxygen is needed even after an exercise has ended;

d. distinguish between two physiological types of muscle fibers, and explain their functional roles;

e. discuss the factors that affect muscular strength; and

f. discuss the effects of resistance and endurance exercises on muscles.

11.6a ATP Sources

All muscle contraction depends on ATP; no other energy source can serve in its place. The supply of ATP depends, in turn, on the availability of oxygen and organic fuels such as glucose and fatty acids. To understand how muscle manages its ATP budget, you must be acquainted with the two main pathways of ATP synthesis: *anaerobic fermentation* and *aerobic respiration* (see fig. 2.31). Each of these has advantages and disadvantages. Anaerobic fermentation enables a cell to produce ATP without the need for oxygen, but the ATP yield is very limited and the process generates

FIGURE 11.18 Modes of ATP Synthesis During Exercise.

a toxic by-product, lactate (lactic acid), which must be removed from the muscle and disposed of by the liver. By contrast, aerobic respiration produces far more ATP and no lactate, but it requires a continual supply of oxygen. Although aerobic respiration is best known as a pathway for glucose oxidation, it is also used to extract energy from other organic compounds. In a resting muscle, most ATP is generated by the aerobic respiration of fatty acids.

During the course of exercise, different mechanisms of ATP synthesis are used depending on the exercise duration. We will view these from the standpoint of immediate, short-term, and long-term energy, but it must be stressed that muscle doesn't make sudden shifts from one mechanism to another like an automobile transmission shifting gears. Rather, these mechanisms blend and overlap as the exercise continues **(fig. 11.18)**.

Immediate Energy

In a short, intense exercise such as a 100 m dash, the myoglobin in a muscle fiber supplies oxygen for a limited amount of aerobic respiration at the outset, but this oxygen supply is quickly depleted. Until the respiratory and cardiovascular systems catch up with the heightened oxygen demand, the muscle meets most of its ATP needs by borrowing phosphate groups (P_i) from other molecules and transferring them to ADP. Two enzyme systems control these phosphate transfers **(fig. 11.19)**:

1. **Myokinase** (MY-oh-KY-nase) transfers P_i from one ADP to another, converting the latter to ATP that myosin can use.

2. **Creatine kinase** (CREE-uh-tin KY-nase) obtains P_i from a phosphate-storage molecule, **creatine phosphate (CP),** and donates it to ADP to make ATP. This is a fast-acting system that helps to maintain the ATP level while other ATP-generating mechanisms are being activated.

ATP and CP, collectively called the **phosphagen system,** provide nearly all the energy used for short bursts of intense activity. Muscle contains about 5 millimoles of ATP and 15 millimoles of CP per kilogram of tissue. Perhaps surprisingly, at the outset of an intense exercise, the amount of ATP in the muscle fibers changes

very little, but the amount of CP drops rapidly. The total supply of ATP + CP is enough to power about 1 minute of brisk walking or 6 seconds of sprinting or fast swimming. The phosphagen system is especially important in activities requiring short bursts of maximal effort, such as football, baseball, and weight lifting.

FIGURE 11.19 The Phosphagen System. (a) Myokinase, which obtains phosphates from ADP. (b) Creatine kinase, which obtains phosphates from creatine phosphate.

Short-Term Energy

As the phosphagen system is exhausted, the muscles transition to anaerobic fermentation to generate ATP by glycolysis. The point at which this occurs is called the **anaerobic threshold,** or sometimes the **lactate threshold** because one can begin to detect a rise in blood lactate levels at this time. (Athletes in training can detect this with a handheld device that measures lactate in a single drop of blood from a pinprick.) During the anaerobic phase, the muscles obtain glucose from the blood and their own stored glycogen and metabolize it to lactate. As we have seen earlier (see fig. 2.31), this pathway, the **glycogen–lactate system,** generates a net yield of 2 ATP for each glucose consumed. (*Anaerobic fermentation* is its final step, the conversion of pyruvate to lactate.) It can produce enough ATP for 30 to 40 seconds of maximum activity. Playing basketball or running completely around a baseball diamond, for example, depends heavily on this energy-transfer system.

Long-Term Energy

After 40 seconds or so, the respiratory and cardiovascular systems "catch up" and deliver oxygen to the muscles fast enough for aerobic respiration to once again meet most of the ATP demand. Aerobic respiration produces much more ATP than glycolysis does—typically another 30 ATP per glucose. Thus it is a very efficient means of meeting the ATP demands of prolonged exercise. One's rate of oxygen consumption rises for 3 to 4 minutes and then levels off at a *steady state* in which aerobic ATP production keeps pace with the demand. In exercises lasting more than 10 minutes, over 90% of the ATP is produced aerobically. For up to 30 minutes, the energy for this comes about equally from glucose and fatty acids; then, as glucose and glycogen are depleted, fatty acids become the more significant fuel.

11.6b Fatigue and Endurance

Muscle **fatigue** is the progressive weakness and loss of contractility that results from prolonged use of the muscles. For example, if you hold a heavy book at arm's length for a minute, you will feel your muscles growing weaker and soon you'll be unable to hold it up. Repeatedly squeezing a rubber ball, pushing a video game button, or trying to take lecture notes from a fast-talking professor produces fatigue in the hand muscles. In high-intensity, short-duration exercise, fatigue is thought to result from the following factors:

- **Potassium accumulation.** Each action potential releases K^+ from the sarcoplasm to the extracellular fluid. This lowers the membrane potential (hyperpolarizes the cell) and makes the muscle fiber less excitable. This is especially significant in the T tubules, where the low volume of ECF enables the K^+ concentration to rise to a high level and interfere with the release of calcium from the sarcoplasmic reticulum.

- **ADP/P_i accumulation.** The hydrolysis of ATP generates an ever-growing pool of ADP + P_i. ADP slows the cross-bridge cycling mechanism of contraction. Free phosphate (P_i) inhibits calcium release from the sarcoplasmic reticulum, calcium sensitivity of the contractile mechanism, and force production by the myofibrils. It is now thought to be a major contributor to muscle fatigue.

In exercise of long duration but low intensity, fatigue may result partially from the preceding causes, but is predominantly due to the following:

- **Fuel depletion.** Declining levels of muscle glycogen and blood glucose leave less fuel for ATP synthesis. Long-distance runners and cyclists call this "hitting the wall," and often endeavor to delay fatigue by means of high-carbohydrate diets before the race, loading the muscles with extra glycogen.

- **Electrolyte loss.** The loss of electrolytes through sweating can alter the ion balance of the extracellular fluid enough to reduce muscle excitability.

- **Central fatigue.** Exercising muscle generates ammonia, which is absorbed by the brain and inhibits motor neurons of the cerebrum. For this and other reasons not yet well understood, the central nervous system produces less signal output to the skeletal muscles. This is where psychological factors come into play, such as the will to complete a marathon.

Some former hypotheses on muscle fatigue have been discredited by more recent research. ATP depletion per se is no longer thought to cause fatigue; the ATP level in fatigued muscle is almost as great as in rested muscle. It was also long thought that the lactate from anaerobic fermentation contributed to fatigue by lowering the pH in the muscle fiber. Alterations in protein conformation by low pH could interfere with many processes required for contraction, such as Ca^{2+} binding by troponin, cross-bridge formation, and ATP hydrolysis by myosin, among others. Evidence has now shown, however, that lactate is removed to the liver about as fast as the muscles produce it, so it doesn't accumulate in the muscle tissue and probably has little or nothing to do with fatigue.

The ability to maintain high-intensity exercise for more than 4 or 5 minutes is determined in large part by one's **maximum oxygen uptake (Vo_2max)**—the point at which the rate of oxygen consumption reaches a plateau and increases no further with an added workload. Vo_2max is proportional to body size; it peaks around age 20; it is usually greater in males than in females; and it can be twice as great in a trained endurance athlete as in an untrained person. A typical sedentary adult uses a maximum of about 35 milliliters of oxygen per minute per kilogram of body weight. Such a person weighing 73 kg (160 lb) and exercising at maximum intensity could therefore "burn" oxygen at about 2.6 L/min., which sets a limit to his rate of ATP production. Elite endurance athletes can have a Vo_2max of about 70 mL/min./kg (5.2 L/min. for the same body weight).

11.6c Excess Postexercise Oxygen Consumption

You have probably noticed that you breathe heavily not only during strenuous exercise but also for several minutes afterward (fig. 11.18). This is to meet a metabolic demand called **excess postexercise oxygen consumption (EPOC),** also known by an older popularized term, **oxygen debt.** EPOC is the difference between the elevated rate of oxygen consumption at the end of an exercise and the normal rate at rest. It occurs in part because oxygen is needed to regenerate ATP aerobically, and that ATP goes in part to regenerate creatine phosphate. A small amount of oxygen serves to reoxygenate the muscle myoglobin, and the liver consumes oxygen in disposing of the lactate generated by exercise. In addition, exercise raises the body temperature and overall metabolic rate, which in itself consumes more oxygen.

ATP and CP are replenished in the early minutes of heaviest postexercise breathing, and oxygen consumption remains elevated for as much as an hour more as the liver uses it to oxidize lactate. EPOC can be as much as six times one's basal oxygen consumption, indicating that anaerobic mechanisms of ATP production during exercise allow six times as much physical exertion as would have been possible without those mechanisms.

11.6d Physiological Classes of Muscle Fibers

Not all muscle fibers are alike or adapted for the same tasks. For example, the weight-bearing and postural muscles of the back and lower limbs react slowly to stimulation and take up to 100 ms to reach peak tension. By contrast, muscles that control eye and hand movements react quickly to stimulation and reach peak tension in as little as 7.5 ms. The predominant fibers that compose these muscles are therefore called *slow-twitch* and *fast-twitch* fibers, respectively. We can find reasons for their differences in response time by looking at their cellular structure and biochemistry (**table 11.2**). The different fiber types can be identified in tissue sections with special histochemical staining (**fig. 11.20**).

Slow-twitch fibers are also called **slow oxidative (SO)** or **type I fibers.** They are well adapted for endurance and fatigue

FIGURE 11.20 **Skeletal Muscle Fiber Types.** (FG, fast glycolytic fibers; SO, slow oxidative fibers; FO, fast oxidative fibers)
G W Willis/Getty Images

resistance, so they are particularly important in muscles that support the body and maintain posture, such as the erector spinae and quadratus lumborum of the back. Their fatigue resistance stems from their *oxidative* mode of ATP production—that is, aerobic respiration. Oxidative metabolism, of course, requires a liberal supply of oxygen and the means to use it efficiently. Therefore, these fibers are surrounded by a dense network of blood capillaries, they are rich in mitochondria, and they have a high concentration of myoglobin, the red pigment that facilitates diffusion of oxygen from the blood into the muscle fiber. Slow-twitch fibers are also relatively thin, which minimizes the distance that oxygen must diffuse to even the deepest mitochondria. The slowness of these muscles is due to a sarcoplasmic reticulum that is relatively slow to release and reabsorb calcium, and a form of myosin ATPase that is relatively slow in its ATP hydrolysis and cross-bridge cycling. The high myoglobin concentration of these fibers gives them a relatively bright red color, so they are also called **red fibers.**

Fast-twitch fibers are also called **fast glycolytic (FG)** or **type II fibers.** They are well adapted for quick responses, so they are particularly important in the aforesaid eye and hand muscles, and in large muscles such as the gastrocnemius and biceps brachii, which we employ in such actions as jumping and elbow flexion. Their quickness stems from an especially extensive sarcoplasmic reticulum with fast release and reabsorption of calcium, and a form of myosin with very quick ATP hydrolysis and cross-bridge cycling. For energy, they depend primarily on glycolysis and anaerobic fermentation, which produces ATP more quickly (yet less efficiently) than aerobic respiration. To support this, they contain a high concentration of glycogen. They also have a high concentration of creatine phosphate, which you will

TABLE 11.2	Classification of Skeletal Muscle Fibers	
	Fiber Type	
Property	**Slow-Twitch (Slow Oxidative)**	**Fast-Twitch (Fast Glycolytic)**
Twitch duration	As long as 100 ms	As short as 7.5 ms
Motor unit size	Smaller	Larger
Motor neurons	Smaller, more excitable	Larger, less excitable
Motor unit strength	Weaker	Stronger
Relative diameter	Smaller	Larger
ATP synthesis	Aerobic	Anaerobic
Fatigue resistance	Good	Poor
ATP hydrolysis	Slow	Fast
Glycolysis	Moderate	Fast
Myoglobin content	Abundant	Low
Glycogen content	Low	Abundant
Mitochondria	Abundant and large	Fewer and smaller
Capillaries	Abundant	Fewer
Color	Red	White, pale
Representative muscles in which fiber type is predominant	Soleus Erector spinae Quadratus lumborum	Gastrocnemius Biceps brachii Muscles of eye movement

recall, aids in rapidly regenerating ATP. They have fewer mitochondria than slow-twitch fibers. Fast-twitch fibers are thicker than slow-twitch, because thick fibers are stronger and they have no need for especially rapid oxygen delivery to the deepest cytoplasm. Also, without need of such rapid oxygen uptake, they have less myoglobin. For this reason, fast-twitch fibers are relatively pale and are called **white fibers.** But the price paid for these fast-twitch mechanisms and anaerobic ATP production is that these fibers fatigue more easily, as you may know from writer's cramp or doing rapid biceps curls.

Some authorities recognize two subtypes of FG fibers called types IIA and IIB. Type IIB is the common type just described, whereas IIA (**intermediate** or **fast oxidative [FO] fibers**) combine fast-twitch responses with aerobic fatigue-resistant metabolism. Type IIA fibers, however, are known mainly from other species of mammals and are relatively rare in humans except in some endurance-trained athletes.

Nearly all muscles are composed of both SO and FG fibers, but usually one type or the other predominates according to the functions of that muscle—quick motility and reflexes, or sustained weight-bearing tension. People with different types and levels of physical activity differ in the proportion of one fiber type to another even in the same muscle, such as the *quadriceps femoris* of the anterior thigh (**table 11.3**). It is thought that people are born with a genetic predisposition for a certain ratio of fiber types. Those who go into competitive sports discover the sports at which they can excel and gravitate toward those for which heredity has best equipped them. One person might be a "born sprinter" and another a "born marathoner."

In humans, small motor units are composed of relatively small SO muscle fibers and supplied by small but easily excited motor neurons. These motor units are not as strong as large ones but produce more precise movements. Large motor units are composed of larger FG fibers; supplied by larger, less excitable neurons; and produce more power but less fine control. In the *size principle* discussed earlier (section 11.5b), the nervous system recruits small SO motor units first, then larger FG motor units only if more strength is needed for a particular task.

Muscles composed mainly of SO fibers are called *red muscles,* and those composed mainly of FG fibers are called *white muscles* because of the color difference stemming from their difference in myoglobin content. Anyone who eats chicken or turkey may be unwittingly

familiar with this distinction. The thighs are dark meat composed of SO fibers adapted to long periods of standing, and the breast is white meat composed of FG fibers adapted for short bursts of power when the birds take flight. Duck breast, however, is dark meat (red muscle) adapted for long-distance flight, because unlike chickens and turkeys, domestic ducks are descended from migratory ancestors.

We saw in chapter 10 that sometimes two or more muscles act across the same joint and seem to have the same function. We have already seen some reasons why such muscles are not as redundant as they seem. Another reason is that they may differ in the proportion of SO to FG fibers. For example, the gastrocnemius and soleus muscles of the calf both insert on the calcaneus, so they exert the same pull on the heel. The gastrocnemius, however, is a white, predominantly FG muscle adapted for quick, powerful movements such as jumping, whereas the soleus is a red, SO muscle that does most of the work in standing and in endurance exercises such as jogging and skiing.

11.6e Muscular Strength and Conditioning

We have far more muscular strength than we normally use. The gluteus maximus can generate 1,200 kg of tension, and all the muscles collectively can produce a total tension of 22,000 kg (nearly 25 tons). Indeed, the muscles can generate more tension than the bones and tendons can withstand—a fact that accounts for many injuries to the patellar and calcaneal tendons. Muscular strength depends on a variety of anatomical and physiological factors:

- **Muscle size.** The strength of a muscle depends primarily on its size; thicker muscles can form more myosin–actin cross-bridges and therefore generate more tension. A muscle can exert a tension of about 3 to 4 kg/cm^2 of cross-sectional area. This is why weight lifting increases both size and strength of a muscle.

- **Fascicle arrangement** (see fig. 10.2). Pennate muscles such as the rectus femoris are stronger than parallel muscles such as the sartorius, which in turn are stronger than circular muscles such as the orbicularis oculi.

- **Size of active motor units.** Large motor units produce stronger contractions than small ones.

- **Multiple motor unit summation.** When a stronger muscle contraction is desired, the nervous system activates more and larger motor units. Getting "psyched up" for athletic competition is partly a matter of multiple motor unit (MMU) summation.

- **Temporal summation.** Nerve impulses usually arrive at a muscle in a series of closely spaced action potentials. Because of the temporal summation described earlier, the greater the frequency of stimulation, the stronger the muscle contraction.

- **The length–tension relationship.** As noted earlier, a muscle resting at optimum length is prepared to contract more forcefully than a muscle that is excessively contracted or stretched. This is affected by one's posture, such as a runner's crouch, just before the onset of muscular effort.

TABLE 11.3	Proportion of Slow Oxidative (SO) and Fast Glycolytic (FG) Fibers in the Quadriceps Femoris Muscle of Male Athletes	
Sample Population	**SO**	**FG**
Marathon runners	82%	18%
Swimmers	74	26
Average males	45	55
Sprinters and jumpers	37	63

- **Fatigue.** Rested muscles contract more strongly than fatigued ones.

As fitness trainers and exercise enthusiasts know, there are two kinds of exercise with different effects on muscles—resistance and endurance exercise. **Resistance exercise,** such as weight lifting, is the contraction of muscles against a load that resists movement. A few minutes of resistance exercise at a time, a few times each week, is enough to stimulate muscle growth. Growth results primarily from cellular enlargement, not cellular division. The muscle fibers synthesize more myofilaments and the myofibrils grow thicker. Myofibrils split longitudinally when they reach a certain size, so a well-conditioned muscle has more myofibrils than a poorly conditioned one. Muscle fibers themselves are incapable of mitosis, but there is some evidence that as they enlarge, they too may split longitudinally. A small part of muscle growth may therefore result from an increase in the number of fibers, but most of it results from the enlargement of fibers that have existed since puberty.

Endurance (aerobic) exercise, such as jogging or swimming, improves the fatigue resistance of the muscles by enhancing the delivery and use of oxygen. Slow-twitch fibers, especially, produce more mitochondria and glycogen and acquire a greater density of blood capillaries as a result of conditioning. Endurance exercise also improves skeletal strength; increases the red blood cell count and the oxygen transport capacity of the blood; and enhances the function of the cardiovascular, respiratory, and nervous systems.

Endurance training doesn't significantly increase muscular strength, and resistance training doesn't improve endurance. Optimal performance and musculoskeletal health require **cross-training,** which incorporates elements of both types. If muscles aren't kept sufficiently active, they become *deconditioned*— weaker and more easily fatigued.

▶▶▶APPLY WHAT YOU KNOW

Is a weight lifter's muscle growth mainly the result of hypertrophy or hyperplasia?

BEFORE YOU GO ON

Answer the following questions to test your understanding of the preceding section:

21. From which two molecules can ADP borrow a phosphate group to become ATP? What is the enzyme that catalyzes each transfer?

22. In a long period of intense exercise, why does muscle generate ATP anaerobically at first and then switch to aerobic respiration?

23. List four causes of muscle fatigue.

24. List three causes of excess postexercise oxygen consumption.

25. What properties of fast glycolytic and slow oxidative fibers adapt them for different physiological purposes?

11.7 Cardiac and Smooth Muscle

Expected Learning Outcomes

When you have completed this section, you should be able to

a. describe the structural and physiological differences between cardiac muscle and skeletal muscle;

b. explain why these differences are important to cardiac function;

c. describe the structural and physiological differences between smooth muscle and skeletal muscle; and

d. relate the unique properties of smooth muscle to its locations and functions.

Cardiac and smooth muscle have special structural and physiological properties in common with each other, but different from those of skeletal muscle **(table 11.4).** These are related to their distinctive functions.

Any of the three types of muscle cells can be called **myocytes.** This term is preferable to *muscle fiber* for smooth and cardiac muscle because these two types of cells don't have the long fibrous shape of skeletal muscle cells. They are relatively short, and in further contrast to skeletal muscle fibers, they have only one or two nuclei. Cardiac muscle cells are also called **cardiomyocytes.**

Cardiac and smooth muscle are *involuntary* muscle tissues, not usually subject to our conscious control. They receive no innervation from somatic motor neurons, but cardiac muscle and some smooth muscle receive nerves from the sympathetic and parasympathetic divisions of the autonomic nervous system (see chapter 15).

11.7a Cardiac Muscle **APR**

Cardiac muscle is limited to the heart, where its function is to pump blood. Knowing that, we can predict the properties that it must have: (1) It must contract with a regular rhythm; (2) it must function in sleep and wakefulness, without fail or need of conscious attention; (3) it must be highly resistant to fatigue; (4) the cardiomyocytes of a given heart chamber must contract in unison so that the chamber can effectively expel blood; and (5) each contraction must last long enough to expel blood from the chamber. These functional necessities are the key to understanding how cardiac muscle differs structurally and physiologically from skeletal muscle (table 11.4).

Cardiac muscle is striated like skeletal muscle, but cardiomyocytes are shorter and thicker. They have one or two nuclei near the middle of the cell. Each cell is enclosed in an endomysium, but there is no perimysium or epimysium as in skeletal muscle. Cardiomyocytes branch slightly so each is joined end to end with several others. These intercellular connections, called **intercalated discs** (in-TUR-kuh-LAY-ted), appear as thick dark lines in stained tissue sections. An intercalated disc

TABLE 11.4	Comparison of Skeletal, Cardiac, and Smooth Muscle		
Feature	Skeletal Muscle	Cardiac Muscle	Smooth Muscle
Location	Associated with skeletal system	Heart	Walls of viscera and blood vessels, iris of eye, arrector muscle of hair follicles
Cell shape	Long threadlike fibers	Short, slightly branched cells	Short fusiform cells
Cell length	100 μm–30 cm	50–120 μm	30–200 μm
Cell width	10–500 μm	10–20 μm	5–10 μm
Striations	Present	Present	Absent
Nuclei	Multiple nuclei, adjacent to sarcolemma	Usually one nucleus, near middle of cell	One nucleus, near middle of cell
Connective tissues	Endomysium, perimysium, epimysium	Endomysium only	Endomysium only
Sarcoplasmic reticulum	Abundant	Present	Scanty
T tubules	Present, narrow	Present, wide	Absent
Gap junctions	Absent	Present in intercalated discs	Present in unitary smooth muscle
Autorhythmicity	Absent	Present	Present in unitary smooth muscle
Thin filament attachment	Z discs	Z discs	Dense bodies
Regulatory proteins	Tropomyosin, troponin	Tropomyosin, troponin	Calmodulin, myosin light-chain kinase
Ca^{2+} source	Sarcoplasmic reticulum	Sarcoplasmic reticulum and extracellular fluid	Mainly extracellular fluid
Ca^{2+} receptor	Troponin of thin filament	Troponin of thin filament	Calmodulin of thick filament
Innervation and control	Somatic motor fibers (voluntary)	Autonomic fibers (involuntary)	Autonomic fibers (involuntary)
Nervous stimulation required?	Yes	No	No
Effect of nervous stimulation	Excitatory only	Excitatory or inhibitory	Excitatory or inhibitory
Mode of tissue repair	Limited regeneration, mostly fibrosis	Limited regeneration, mostly fibrosis	Relatively good capacity for regeneration

has electrical *gap junctions* that allow each cardiomyocyte to directly stimulate its neighbors, and mechanical junctions that keep the cardiomyocytes from pulling apart when the heart contracts. The sarcoplasmic reticulum is less developed than in skeletal muscle, but the T tubules are larger and admit Ca^{2+} from the extracellular fluid. Damaged cardiac muscle is repaired by fibrosis. Cardiac muscle has no satellite cells, and even though mitosis has been detected in cardiomyocytes following heart attacks, it does not produce a significant amount of regenerated functional muscle.

Unlike skeletal muscle, cardiac muscle can contract without the need of nervous stimulation. The heart has a built-in **pacemaker** that rhythmically sets off a wave of electrical excitation, which travels through the muscle and triggers the contraction of the heart chambers. The heart is said to be **autorhythmic**[13]

because of this ability to contract rhythmically and independently. Stimulation by the autonomic nervous system, however, can increase or decrease the heart rate and contraction strength. Cardiac muscle does not exhibit quick twitches like skeletal muscle. Rather, it maintains tension for about 200 to 250 ms, giving the heart time to expel blood.

Cardiac muscle uses aerobic respiration almost exclusively. It is very rich in myoglobin and glycogen, and it has especially large mitochondria that fill about 25% of the cell, compared with smaller mitochondria occupying about 2% of a skeletal muscle fiber. Cardiac muscle is very adaptable with respect to the fuel used, but very vulnerable to interruptions in oxygen supply. Because it makes little use of anaerobic fermentation, cardiac muscle is highly resistant to fatigue.

11.7b Smooth Muscle A&PR

Smooth muscle is named for the fact that it has no striations, for a reason to be described shortly. Its myocytes are relatively small,

---
[13]*auto* = self

allowing for fine control of such tissues and organs as a single hair, the iris of the eye, and the tiniest arteries; yet, in the pregnant uterus, the myocytes become quite large and contribute to the powerful contractions of childbirth.

Smooth muscle isn't always innervated, but when it is, the nerve supply is autonomic, like that of the heart. Autonomic nerve fibers usually don't form precisely localized neuromuscular junctions with the myocytes. Rather, a nerve fiber has as many as 20,000 periodic swellings called **varicosities** along its length **(fig. 11.21).** Each varicosity contains synaptic vesicles from which it releases neurotransmitters—usually norepinephrine from the sympathetic fibers and acetylcholine from the parasympathetic fibers. The myocyte has no motor end plate, but instead has receptors for these neurotransmitters distributed over its surface. The varicosities simply release a flood of neurotransmitter into the tissue, and each myocyte may respond to more than one nerve fiber.

Whether innervated or not, smooth muscle responds to a wide variety of stimuli and often without any electrical excitation of the sarcolemma. It is much slower than skeletal and cardiac muscle to contract and relax, but it can remain contracted for a long time without fatigue and with minimal energy expenditure.

Smooth muscle doesn't usually form organs in itself, but forms layers in the walls of larger organs such as the stomach, intestines, uterus, and urinary bladder. In such cases, the muscle layer is quite variable in complexity. It can consist of as little as one cell in small arteries. The esophagus and intestines have an outer layer of longitudinal smooth muscle adjacent to a deeper, inner layer of circular muscle **(fig. 11.22).** When the longitudinal layer contracts, it shortens and dilates the organ; when the circular layer contracts, it constricts and lengthens the organ. In

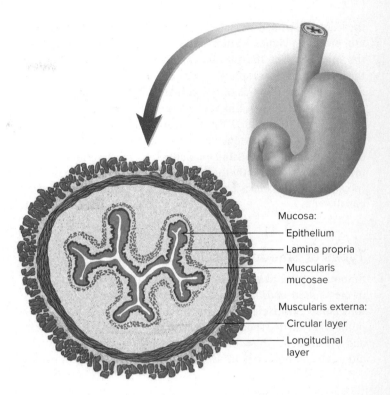

FIGURE 11.22 Layers of Visceral Muscle in a Cross Section of the Esophagus.

the stomach, urinary bladder, and uterus, smooth muscle forms three or more layers with bundles of myocytes running in multiple directions.

Smooth muscle can propel the contents of an organ, such as driving food through the digestive tract, voiding urine and feces, and expelling the infant in childbirth. By dilating or constricting the blood vessels and airway, it can modify the speed of air and blood flow, maintain blood pressure, and reroute blood from one pathway to another.

Unlike skeletal and cardiac muscle, smooth muscle is capable of not only hypertrophy (cellular growth) but also mitosis and hyperplasia (cell division). Thus, an organ such as the pregnant uterus grows by the addition of new myocytes as well as enlargement of existing ones. Injured smooth muscle regenerates well by mitosis.

Myocyte Structure

Smooth muscle myocytes have a fusiform shape, typically about 5 to 10 μm wide at the middle, tapering to a point at each end, and usually ranging from 30 to 200 μm long—but up to 500 μm long in the pregnant uterus. They are enclosed in endomysium but have no perimysium, fascicles, or epimysium. There is only one nucleus, located near the middle of the cell. The sarcoplasmic reticulum is scanty and there are no T tubules. Thick and thin filaments are present, but there are no striations, sarcomeres, or myofibrils because the myofilaments are not bundled and aligned with each other the way

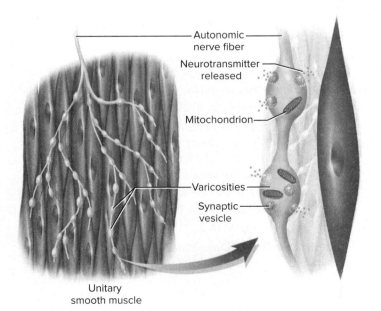

FIGURE 11.21 Varicosities of an Autonomic Nerve Fiber in Unitary Smooth Muscle.

they are in striated muscle. Z discs are absent. In their place are protein plaques called **dense bodies,** some adhering to the inner face of the plasma membrane and others dispersed throughout the sarcoplasm (see fig. 11.24). The membrane-associated dense bodies of one cell are often directly across from those of another, with linkages between them so that contractile force can be transmitted from cell to cell. Associated with the dense bodies is an extensive cytoskeletal network of intermediate filaments. Actin filaments attach to the intermediate filaments as well as directly to the dense bodies, so their movement (powered by myosin) is transferred to the sarcolemma and shortens the cell.

Types of Smooth Muscle

Smooth muscle tissue shows a range of types between two extremes called *multiunit* and *unitary* types **(fig. 11.23). Multiunit smooth muscle** occurs in some of the largest arteries and pulmonary air passages, arrector muscles of the hair follicles, and eye muscles that control the iris and lens. Its innervation, although autonomic, is to some degree similar to that of skeletal

Autonomic nerve fibers

Varicosities

Gap junctions

(a) Multiunit
smooth muscle

(b) Unitary
smooth muscle

FIGURE 11.23 Multiunit and Unitary Smooth Muscle.
(a) Multiunit smooth muscle, in which each muscle cell receives its own nerve supply and contracts independently. (b) Unitary smooth muscle, in which a nerve fiber passes through the tissue without synapsing with any specific muscle cell, and muscle cells are coupled by electrical gap junctions.

muscle—the terminal branches of a nerve fiber synapse with individual myocytes and form a motor unit. Each varicosity is associated with a particular myocyte, and each myocyte responds independently of all the others—hence the name *multiunit*. Multiunit smooth muscle does not, however, generate action potentials. It contracts in response to variable (graded) electrical changes in the sarcolemma or even in the absence of electrical excitation.

Unitary (single-unit) smooth muscle is more common. It occurs in most blood vessels and in the digestive, respiratory, urinary, and reproductive tracts—therefore, it is also called **visceral muscle.** This type forms the aforementioned layers in many of the hollow viscera. The names *unitary* and *single-unit* refer to the fact that the myocytes of this type of muscle are electrically coupled to each other by gap junctions. This allows them to directly stimulate each other, so numerous cells contract as a unit, almost as if they were a single cell. In this muscle type, the nerve varicosities are not associated with a specific myocyte, but stimulate several of them at once when they release neurotransmitter.

Excitation of Smooth Muscle

Whereas skeletal muscle contracts only in response to excitatory stimulation by a somatic motor fiber, smooth muscle can be stimulated in a multitude of ways. Some stimuli excite the myocyte and others inhibit it. Some produce action potentials in the sarcolemma, particularly in unitary smooth muscle, whereas others stimulate the myocyte by nonelectrical means. Some smooth muscle has no nerve supply at all. Modes of smooth muscle stimulation and some examples include

- **Autonomic nerve fibers and neurotransmitters.** For example, parasympathetic nerves secrete acetylcholine and stimulate gastrointestinal motility; sympathetic nerves secrete norepinephrine and dilate the bronchioles of the lungs.

- **Chemicals.** Smooth muscle reacts to hormones, carbon dioxide, oxygen, nitric oxide, low pH, and other chemical stimuli. The hormone oxytocin, for example, stimulates the labor contractions of the uterus, and histamine relaxes the smooth muscle of arteries.

- **Temperature.** Cold induces contraction of smooth muscle resulting in erection of the hairs and tautening of the skin in such regions as the areola and scrotum, whereas warmth relaxes smooth muscle in arteries of the skin.

- **Stretch.** The stomach and urinary bladder contract when stretched by food or urine.

- **Autorhythmicity.** Some unitary smooth muscle is autorhythmic, especially in the stomach and intestines. Some myocytes spontaneously depolarize at regular time intervals and set off waves of contraction throughout an entire layer of muscle. The rhythm is much slower than in cardiac muscle.

Regardless of how a myocyte is stimulated, however, the immediate trigger for contraction is the same as in skeletal and cardiac muscle—calcium ions. In some cases, the Ca^{2+} comes from the sarcoplasmic reticulum (SR), as it does in skeletal muscle. With a relatively sparse SR, however, smooth muscle usually gets most of its Ca^{2+} from the extracellular fluid by way of gated calcium channels in the sarcolemma. Compensating for the paucity of SR, the sarcolemma has numerous little pockets called **caveolae** (CAV-ee-OH-lee) where its calcium channels are concentrated. Calcium is 10,000 times as concentrated in the ECF as in the cytosol, so if these channels are opened, it diffuses quickly into the cell. Because smooth muscle cells are relatively small, the incoming Ca^{2+} quickly reaches all of the myofilaments.

What opens these gated channels? Some are mechanically gated and open in response to physical distortion such as stretch. This is the case in organs that periodically fill and empty, such as the stomach and urinary bladder. Some are voltage-gated and open in response to electrical depolarization of the sarcolemma. Still others are ligand-gated and open in response to chemicals originating outside the cell or from the cytosol. Acetylcholine and norepinephrine, for example, bind to surface receptors and activate the formation of second messengers within the smooth muscle cell. This leads to a series of intracellular events that open the surface calcium channels from within.

▶▶▶APPLY WHAT YOU KNOW

How is smooth muscle contraction affected by the drugs called calcium channel blockers? (See Deeper Insight 3.1.)

Contraction and Relaxation

Calcium ions are the immediate trigger for contraction, but unlike skeletal and cardiac muscle, smooth muscle has no troponin to bind it. Calcium binds instead to a similar protein called **calmodulin**[14] (cal-MOD-you-lin), associated with myosin. Calmodulin then activates an enzyme called **myosin light-chain kinase,** which adds a phosphate group to a small regulatory protein on the myosin head. This activates the myosin ATPase, enabling it to bind to actin and hydrolyze ATP. The myosin then produces repetitive power and recovery strokes like those of skeletal muscle.

As thick filaments pull on the thin ones, the thin filaments pull on the dense bodies and membrane plaques. Through the dense bodies and cytoskeleton, force is transferred to the plasma membrane and the entire cell shortens. When a smooth muscle cell contracts, it puckers and twists somewhat like wringing out a wet towel **(fig. 11.24).**

In skeletal muscle, there is typically a 2 ms latent period between stimulation and the onset of contraction. In smooth muscle, by contrast, the latent period is 50 to 100 ms long. Tension peaks about 500 ms (0.5 second) after the stimulus and then declines over

FIGURE 11.24 Smooth Muscle Contraction. (a) Relaxed cells. (b) Contracted cells.

a period of 1 to 2 seconds. The effect of all this is that compared with skeletal muscle, smooth muscle is very slow to contract and relax. It is slow to contract because its myosin ATPase is a slow enzyme. It is slow to relax because the pumps that remove Ca^{2+} from the cytosol are also slow. As the Ca^{2+} level falls, myosin releases its phosphate group and is no longer able to hydrolyze ATP and execute power strokes. However, it doesn't necessarily detach from actin immediately. It has a **latch-bridge mechanism** that enables it to remain attached to actin for a prolonged time without consuming more ATP.

Smooth muscle often exhibits tetanus and is very resistant to fatigue. It makes most of its ATP aerobically, but its ATP requirement is small and it has relatively few mitochondria. Skeletal muscle requires 10 to 300 times as much ATP as smooth muscle to maintain the same amount of tension. The fatigue resistance and latch-bridge mechanism of smooth muscle are important in enabling it to maintain a state of continual tonic contraction called **smooth muscle tone.** Muscle tone keeps the arteries in a state of partial constriction called *vasomotor tone.* A widespread loss of muscle tone in the arteries can cause a dangerous drop in blood pressure. Smooth muscle tone also keeps the intestines partially contracted. The intestines are much

---

[14]acronym for *calcium modulating protein*

TABLE 11.5	Some Disorders of the Muscular System
Contracture	Abnormal muscle shortening not caused by nervous stimulation. Can result from failure of the calcium pump to remove Ca^{2+} from the sarcoplasm or from contraction of scar tissue, as in burn patients.
Cramps	Painful muscle spasms caused by rapid firing of motor neurons; triggered by heavy exercise, cold, dehydration, electrolyte loss, low blood glucose, or lack of blood flow.
Crush syndrome	A shocklike state following the massive crushing of muscles; associated with high and potentially fatal fever, cardiac irregularities resulting from K^+ released from the muscle, and kidney failure resulting from blockage of the renal tubules with myoglobin released by the traumatized muscle. Myoglobinuria (myoglobin in the urine) is a common sign.
Delayed-onset muscle soreness	Pain, stiffness, and tenderness felt from several hours to a day after strenuous exercise. Associated with microtrauma to the muscles; with disrupted Z discs, myofibrils, and plasma membranes; and with elevated levels of myoglobin, creatine kinase, and lactate dehydrogenase in the blood.
Disuse atrophy	Reduction in the size of muscle fibers as a result of nerve damage or muscular inactivity, for example in limbs in a cast and in patients confined to a bed or wheelchair. Muscle strength can be lost at a rate of 3% per day of bed rest.
Fibromyalgia	Chronic pain of unknown cause, seeming to come from the muscles and bones but actually arising from abnormal processing of pain signals by the brain. Occurring twice as often in women as in men, especially between ages 30 and 50.
Myositis	Muscle inflammation and weakness resulting from infection or autoimmune disease.

You can find other muscle disorders described in the following places:

Compartment syndrome in Deeper Insight 10.1; *back injuries* in Deeper Insight 10.2; *hernia* in Deeper Insight 10.3; *common athletic injuries* in Deeper Insight 10.9; *paralysis* in Deeper Insights 11.1 and 13.5; and *muscular dystrophy* and *myasthenia gravis* in Deeper Insight 11.4.

longer in a cadaver than they are in a living person because of the loss of muscle tone at death.

Response to Stretch

Stretch alone sometimes causes smooth muscle to contract by opening mechanically gated calcium channels in the sarcolemma. Distension of the esophagus with food or the colon with feces, for example, evokes a wave of contraction called **peristalsis** (PERR-ih-STAL-sis) that propels the contents along the organ.

Smooth muscle exhibits a reaction called the **stress–relaxation** (or **receptive-relaxation**) **response.** When stretched, it briefly contracts and resists, but then relaxes. The significance of this response is apparent in the urinary bladder. If the stretched bladder contracted and didn't soon relax, it would expel urine almost as soon as it began to fill, thus failing to store the urine until an opportune time.

Remember that skeletal muscle cannot contract very forcefully if it is overstretched. Smooth muscle, however, is less limited by the length–tension relationship. It contracts forcefully even when greatly stretched, so hollow organs such as the stomach and bladder can fill and then expel their contents efficiently. Skeletal muscle must be within 30% of optimum length in order to contract strongly when stimulated. Smooth muscle, by contrast, can be anywhere from half to twice its resting length and still contract powerfully. There are three reasons for this: (1) There are no Z discs, so thick filaments cannot butt against them and stop the contraction; (2) since the thick and thin filaments are not arranged in orderly sarcomeres, stretching of the muscle doesn't cause a situation in which there is too little overlap for cross-bridges to form; and (3) the thick filaments of smooth muscle have myosin heads along their entire length (there is no bare zone), so cross-bridges can form anywhere, not just at the ends. Smooth muscle also exhibits **plasticity**—the ability to adjust its tension to the degree of stretch. Thus, a hollow organ such as the bladder can be greatly stretched yet not become flabby when it is empty.

The muscular system suffers fewer diseases than any other organ system, but several of its more common dysfunctions are listed in **table 11.5.** The effects of aging on the muscular system are described in section 29.4a.

BEFORE YOU GO ON

Answer the following questions to test your understanding of the preceding section:

26. Explain why intercalated discs are important to cardiac muscle function.

27. Explain why it is important for cardiac muscle to have longer-lasting contractions than skeletal muscle.

28. How do unitary and multiunit smooth muscle differ in innervation and contractile behavior?

29. How does smooth muscle differ from skeletal muscle with respect to its source of calcium and its calcium receptor?

30. Explain why the stress–relaxation response is an important factor in smooth muscle function.

DEEPER INSIGHT 11.4

CLINICAL APPLICATION

Muscular Dystrophy and Myasthenia Gravis

Muscular dystrophy[15] is a collective term for several hereditary diseases in which the muscles degenerate, weaken, and are gradually replaced by fat and fibrous scar tissue. The most common form of the disease is *Duchenne*[16] *muscular dystrophy (DMD),* a sex-linked recessive trait affecting about 1 out of every 3,500 live-born boys.

DMD is not evident at birth, but begins to exhibit its effects as a child shows difficulty keeping up with other children, falls frequently, and finds it hard to stand again. It is typically diagnosed between the ages of 2 and 10 years. It affects the muscles of the hips first; then the legs; and then progresses to the abdominal, spinal, and respiratory muscles as well as cardiac muscle. The muscles shorten as they atrophy, causing postural abnormalities such as scoliosis. Persons with DMD are usually wheelchair-dependent by the age of 10 or 12, and seldom live past the age of 20. For obscure reasons, they also frequently suffer a progressive decline in mental ability. Death usually results from respiratory insufficiency, pulmonary infection, or heart failure. DMD is incurable, but is treated with exercise to slow the atrophy of the muscles and with braces to reinforce the weakened hips and maintain posture.

The underlying cause of DMD is a mutation in the gene for the muscle protein dystrophin (see fig. 11.4)—a large gene highly vulnerable to mutation. Without dystrophin, there is no coupling between the thin myofilaments and the sarcolemma. The sarcomeres move independently of the sarcolemma, creating tears in the membrane. The torn membrane admits excess Ca^{2+} into the cell, which activates intracellular proteases (protein-digesting enzymes). These enzymes degrade the contractile proteins of the muscle, leading to weakness and cellular necrosis. Dying muscle fibers are replaced with scar tissue, which blocks blood circulation in the muscle and thereby contributes to still further necrosis. Muscle degeneration accelerates in a fatal spiral of positive feedback.

Genetic screening can identify heterozygous carriers of DMD, allowing for counseling of prospective parents on the risk of having a child with the disease. However, about one out of three cases arises by a new spontaneous mutation and therefore cannot be predicted by genetic testing.

Myasthenia gravis[17] *(MG)* (MY-ass-THEE-nee-uh GRAV-is) usually occurs in women between the ages of 20 and 40. It is an autoimmune disease in which antibodies attack the neuromuscular junctions and bind ACh receptors together in clusters. The muscle fiber then removes the clusters from the sarcolemma by endocytosis. As a result, the muscle fibers become less and less sensitive to ACh. The effects often appear first in the facial muscles and commonly include drooping eyelids (*ptosis,* **fig. 11.25**) and double vision (due to *strabismus,* inability to fixate on the same point with both eyes). The initial symptoms are often followed by difficulty in swallowing, weakness of the limbs, and poor physical endurance. Some people with MG die quickly as a result of respiratory failure, but others have normal life spans. One method of assessing the progress of the disease is to use *bungarotoxin,* a protein from cobra venom that binds to ACh receptors. The amount that binds is proportional to the number of receptors that are still functional. The muscle of an MG patient sometimes binds less than one-third as much bungarotoxin as normal muscle does.

Myasthenia gravis is often treated with cholinesterase inhibitors. These drugs retard the breakdown of ACh in the neuromuscular junction and enable it to stimulate the muscle longer. Immunosuppressive agents such as prednisone and azathioprine (Imuran) may be used to suppress the production of the antibodies that destroy ACh receptors. Since certain immune cells are stimulated by hormones from the thymus, removal of the thymus (*thymectomy*) helps to dampen the overactive immune response that causes myasthenia gravis. Also, a technique called *plasmapheresis* may be used to remove harmful antibodies from the blood plasma.

---

[15] *dys* = bad, abnormal; *trophy* = growth
[16] Guillaume B. A. Duchenne (1806–75), French physician

[17] *my* = muscle; *asthen* = weakness; *grav* = severe

| At start | At 60 seconds | At 90 seconds |

FIGURE 11.25 Test of Myasthenia Gravis. The subject is told to gaze upward. Within 60 to 90 seconds, there is obvious sagging (ptosis) of one eyelid due to inability to sustain stimulation of the orbicularis oculi muscle.

Effects of the **MUSCULAR SYSTEM** on Other Organ Systems

INTEGUMENTARY SYSTEM
Facial expressions result from the action of muscles on the skin.

SKELETAL SYSTEM
Muscles move and stabilize joints and produce stresses that affect ossification, bone remodeling, and the shapes of bones.

NERVOUS SYSTEM
Muscles give expression to thoughts, emotions, and motor commands that arise in the central nervous system.

ENDOCRINE SYSTEM
Skeletal muscles provide protective cover for some endocrine glands; muscle mass affects insulin sensitivity.

CIRCULATORY SYSTEM
Muscle contractions affect blood flow in many veins; exercise stimulates growth of new blood vessels.

LYMPHATIC AND IMMUNE SYSTEMS
Muscle contractions aid flow of lymph; exercise elevates levels of immune cells and antibodies in circulation; excessive exercise can inhibit immunity.

RESPIRATORY SYSTEM
Muscle contractions ventilate the lungs; muscles of the larynx and pharynx regulate airflow and speech; CO_2 generated by muscular activity stimulates respiration; abdominal muscles produce pressure bursts of coughing and sneezing and aid in deep breathing.

URINARY SYSTEM
A skeletal muscle sphincter retains urine in bladder until convenient for release; abdominal and pelvic muscles aid in compressing and emptying bladder; muscles of pelvic floor support bladder; bulbospongiosus muscle helps clear urine from male urethra.

DIGESTIVE SYSTEM
Muscles enable chewing and swallowing; muscles control voluntary aspect of defecation; abdominal muscles produce vomiting; abdominal and lumbar muscles protect digestive organs.

REPRODUCTIVE SYSTEM
Muscles are involved in sexual responses including erection and ejaculation; abdominal and pelvic muscles aid in childbirth.

STUDY GUIDE

▶ Assess Your Learning Outcomes

To test your knowledge, discuss the following topics with a study partner or in writing, ideally from memory.

11.1 Types and Characteristics of Muscular Tissue

1. Five physiological properties of all muscular tissue and their relevance to muscle function
2. Distinguishing characteristics of skeletal muscle
3. Dimensions of a typical skeletal muscle fiber and of the longest fibers
4. Connective tissues associated with a muscle fiber and their relationship to muscle–bone attachments

11.2 Skeletal Muscle Cells

1. The sarcolemma and sarcoplasm, and the roles of glycogen and myoglobin in the sarcoplasm
2. The role of myoblasts in the development of a muscle fiber, and how they give rise to the multinuclear condition of the muscle fiber and to the satellite cells external to the fiber
3. Structure and function of the sarcoplasmic reticulum and transverse tubules
4. Types of myofilaments that constitute a myofibril
5. Composition and molecular organization of a thick filament, and the structure of a myosin molecule
6. Composition of a thin filament; the organization of its actin, tropomyosin, and troponin; and the active sites of its actin monomers
7. Composition of elastic filaments and their relationship to the thick filaments and Z discs
8. The position and function of dystrophin in the muscle fiber
9. Names of the striations of skeletal and cardiac muscle and how they relate to the overlapping arrangement of thick and thin filaments
10. The definition of *sarcomere*

11.3 The Nerve–Muscle Relationship

1. Motor units; the meanings of *large* and *small motor units;* and the respective advantages of the two types
2. Structure of a neuromuscular junction and function of each of its components
3. The source, role, and fate of acetylcholine (ACh) in the neuromuscular junction
4. The role of acetylcholinesterase in neuromuscular function
5. How a nerve or muscle cell generates a resting membrane potential (RMP); the typical voltage of this potential in a skeletal muscle fiber
6. How an action potential differs from the RMP, and the effects of an action potential on a nerve or muscle cell

11.4 Behavior of Skeletal Muscle Fibers

1. Excitation of a muscle fiber; how a nerve signal leads to a traveling wave of electrical excitation in a muscle fiber
2. Excitation–contraction coupling; how electrical excitation of a muscle fiber leads to exposure of the active sites on the actin of a thin myofilament
3. The sliding filament mechanism of contraction; how exposure of the active sites leads to repetitive binding of myosin to actin and sliding of the thin filaments over the thick filaments
4. Muscle relaxation; how the cessation of the nerve signal leads to blockage of the active sites so myosin can no longer bind to them and maintain muscle tension
5. The roles of calcium, troponin, tropomyosin, and ATP in these processes
6. The length–tension relationship in muscle; why muscle would contract weakly if it was overcontracted or overstretched just prior to stimulation; and how this principle relates to the function of muscle tone

11.5 Behavior of Whole Muscles

1. Terms for the minimum stimulus intensity needed to make a muscle contract, and for the delay between stimulation and contraction
2. The phases of a muscle twitch
3. Reasons why muscle twitches vary in strength (tension)
4. How recruitment and tetanus are produced and how they affect muscle tension
5. Differences between isometric and isotonic contraction, and between the concentric and eccentric forms of isotonic contraction

11.6 Muscle Metabolism

1. Why a muscle cannot contract without ATP
2. Differences between aerobic respiration and anaerobic fermentation with respect to muscle function
3. The use of myoglobin and aerobic respiration to generate ATP at the outset of exercise
4. Two ways in which the phosphagen system generates ATP for continued exercise
5. How anaerobic fermentation generates ATP after the phosphagen system is depleted
6. Why a muscle is able to switch back to aerobic respiration to generate ATP after 40 seconds or so of exercise
7. Causes of muscle fatigue
8. Vo_2max, why it partially determines one's ability to maintain high-intensity exercise, and why it differs from one person to another
9. Why exercise is followed by a prolonged state of elevated oxygen consumption, and the name of that state
10. Differences between slow oxidative and fast glycolytic muscle fibers; the respective advantages of each; how they relate to the power and recruitment of motor units; and examples of muscles in which each type predominates
11. Factors that determine the strength of a muscle
12. Examples of resistance exercise and endurance exercise, and the effects of each on muscle performance

11.7 Cardiac and Smooth Muscle

1. Reasons why cardiac muscle must differ physiologically from skeletal muscle
2. Structural differences between cardiomyocytes and skeletal muscle fibers
3. The autorhythmicity of the heart and its ability to contract without nervous stimulation
4. The unusual fatigue resistance of cardiac muscle; structural and biochemical properties that account for it
5. Functional differences between smooth muscle and the two forms of striated muscle
6. How the innervation of smooth muscle differs from that of skeletal muscle
7. Variations in the complexity and anatomical organization of smooth muscle
8. Various functions of smooth muscle
9. Two modes of growth of smooth muscle tissue
10. The structure of smooth muscle cells and what takes the place of the absent Z discs and T tubules
11. Differences between multiunit and unitary smooth muscle, and the nerve–muscle relationship of each

STUDY GUIDE

12. Various modes of stimulation of smooth muscle
13. How excitation–contraction coupling in smooth muscle differs from that in skeletal muscle; the roles of calmodulin and myo-sin light-chain kinase in smooth muscle contraction
14. The nature and effect of the latch-bridge mechanism in smooth muscle
15. The role of smooth muscle in peristalsis
16. Benefits of the stress–relaxation response of smooth muscle, and of its absence of a length–tension relationship

▶ Testing Your Recall

Answers in Appendix A

1. To make a muscle contract more strongly, the nervous system can activate more motor units. This process is called
   a. recruitment.
   b. summation.
   c. incomplete tetanus.
   d. twitch.
   e. concentric contraction.

2. The functional unit of a muscle fiber is the _____, a segment from one Z disc to the next.
   a. myofibril
   b. I band
   c. sarcomere
   d. neuromuscular junction
   e. striation

3. Before a skeletal muscle fiber can contract, ATP must bind to
   a. a Z disc.
   b. the myosin head.
   c. tropomyosin.
   d. troponin.
   e. actin.

4. Before a skeletal muscle fiber can contract, Ca^{2+} must bind to
   a. calsequestrin.
   b. calmodulin.
   c. the myosin head.
   d. troponin.
   e. actin.

5. Which of the following muscle proteins is *not* intracellular?
   a. actin
   b. myosin
   c. collagen
   d. troponin
   e. dystrophin

6. Smooth muscle cells have _____, whereas skeletal muscle fibers do not.
   a. sarcoplasmic reticulum
   b. tropomyosin
   c. calmodulin
   d. Z discs
   e. myosin ATPase

7. ACh receptors are found mainly in
   a. synaptic vesicles.
   b. terminal cisterns.
   c. thick filaments.
   d. thin filaments.
   e. junctional folds.

8. Unitary smooth muscle cells can stimulate each other because they have
   a. a latch-bridge.
   b. diffuse junctions.
   c. gap junctions.
   d. tight junctions.
   e. cross-bridges.

9. A person with a high Vo_2max
   a. needs less oxygen than someone with a low Vo_2max.
   b. has stronger muscles than someone with a low Vo_2max.
   c. is less likely to show muscle tetanus than someone with a low Vo_2max.
   d. has fewer muscle mitochondria than someone with a low Vo_2max.
   e. experiences less muscle fatigue during exercise than someone with a low Vo_2max.

10. Slow oxidative fibers have all of the following *except*
    a. an abundance of myoglobin.
    b. an abundance of glycogen.
    c. high fatigue resistance.
    d. a red color.
    e. a high capacity to synthesize ATP aerobically.

11. The minimum stimulus intensity that will make a muscle contract is called _____.

12. Red muscles consist mainly of slow _____, or type I, muscle fibers.

13. Parts of the sarcoplasmic reticulum called _____ lie on each side of a T tubule.

14. Thick filaments consist mainly of the protein _____.

15. The neurotransmitter that stimulates skeletal muscle is _____.

16. Muscle contains an oxygen-binding pigment called _____.

17. The _____ of skeletal muscle play the same role as dense bodies in smooth muscle.

18. In autonomic nerve fibers that stimulate unitary smooth muscle, the neurotransmitter is contained in swellings called _____.

19. A state of continual partial muscle contraction is called _____.

20. _____ is an increase in muscle tension without a change in length.

▶ Building Your Medical Vocabulary

Answers in Appendix A

State a meaning of each word element, and give a medical term from this chapter that uses it or a slight variation of it.

1. astheno-
2. auto-
3. dys-
4. iso-
5. metri-
6. myo-
7. sarco-
8. temporo-
9. tono-
10. -trophy

STUDY GUIDE

▶ What's Wrong with These Statements?

Answers in Appendix A

Briefly explain why each of the following statements is false, or reword it to make it true.

1. Each motor neuron supplies one muscle fiber.

2. Somatic motor neurons excite skeletal muscle but inhibit cardiac muscle.

3. Fast glycolytic muscle fibers are relatively resistant to fatigue.

4. Thin myofilaments are found only in the I bands of striated muscle.

5. Thin myofilaments shorten when a muscle contracts.

6. Smooth muscle lacks striations because it does not have thick and thin myofilaments.

7. A muscle must contract to the point of complete tetanus if it is to move a load.

8. If no ATP were available to a muscle fiber, it could still be excited but it could not contract.

9. For the first 30 seconds of an intense exercise, muscle gets most of its energy from lactate.

10. Like cardiac muscle, all smooth muscle is autorhythmic.

▶ Testing Your Comprehension

1. Without ATP, relaxed muscle cannot contract and a contracted muscle cannot relax. Explain why.

2. Smooth muscle controls the curvature of the lens of the eye and the diameter of the pupil, but it would serve poorly for controlling eye movements as in tracking a flying bird or reading a page of print. Explain why.

3. Why would skeletal muscle be unsuitable for the wall of the urinary bladder? Explain how this illustrates the complementarity of form and function at a cellular and molecular level.

4. As skeletal muscle contracts, one or more bands of the sarcomere become narrower and disappear, and one or more of them remain the same width. Which bands will change—A, H, or I—and why?

5. Polio is a muscle-paralyzing disease caused by infection with poliovirus. This virus, however, never infects muscular tissue. (a) Explain how it could paralyze the muscles without invading them. (b) Explain why it does not paralyze cardiac muscle and stop the heart.

CHAPTER

12

NERVOUS TISSUE

A Purkinje cell, a neuron from the cerebellum of the brain
SPL/Science Source

CHAPTER OUTLINE

12.1 Overview of the Nervous System

12.2 Properties of Neurons
    12.2a Universal Properties
    12.2b Functional Classes
    12.2c Structure of a Neuron
    12.2d Axonal Transport

12.3 Supportive Cells
    12.3a Types of Neuroglia
    12.3b Myelin
    12.3c Unmyelinated Nerve Fibers
    12.3d Conduction Speed of Nerve Fibers
    12.3e Nerve Regeneration

12.4 Electrophysiology of Neurons
    12.4a Electrical Potentials and Currents
    12.4b The Resting Membrane Potential
    12.4c Local Potentials
    12.4d Action Potentials

12.4e The Refractory Period
12.4f Signal Conduction in Nerve Fibers

12.5 Synapses
    12.5a The Discovery of Neurotransmitters
    12.5b Structure of a Chemical Synapse
    12.5c Neurotransmitters and Related Messengers
    12.5d Synaptic Transmission
    12.5e Cessation of the Signal
    12.5f Neuromodulators

12.6 Neural Integration
    12.6a Postsynaptic Potentials
    12.6b Summation, Facilitation, and Inhibition
    12.6c Neural Coding
    12.6d Neural Pools and Circuits
    12.6e Memory and Synaptic Plasticity

Connective Issues
Study Guide

DEEPER INSIGHTS

12.1 Clinical Application: Glial Cells and Brain Tumors

12.2 Clinical Application: Diseases of the Myelin Sheath

12.3 Medical History: Nerve Growth Factor—From Home Laboratory to Nobel Prize

12.4 Clinical Application: Alzheimer and Parkinson Diseases

Anatomy & Physiology *Revealed* 4.0

Module 7: Nervous System

BRUSHING UP

- This chapter explains the electrophysiology of neurons, and must assume you are familiar with concentration and electrochemical gradients (see section 1.6e).

- This topic also assumes familiarity with "Membrane Proteins" in section 3.2a—especially receptors, enzymes, ligand- and voltage-gated ion channels, leak channels, and the sodium–potassium pump.

The nervous system is one of great complexity and mystery, and will absorb our attention for the next five chapters. It is the foundation of all our conscious experience, personality, and behavior. It profoundly intrigues biologists, physicians, psychologists, and even philosophers. Its scientific study, called **neurobiology,** is regarded by many as the ultimate challenge facing the behavioral and life sciences. We will begin at the simplest organizational level—the nerve cells *(neurons)* and cells called *neuroglia* that support their function in various ways. We will then progress to the organ level to examine the spinal cord (chapter 13), brain (chapter 14), autonomic nervous system (chapter 15), and sense organs (chapter 16).

12.1 Overview of the Nervous System

Expected Learning Outcomes

When you have completed this section, you should be able to

a. describe the overall function of the nervous system; and

b. describe its major anatomical and functional subdivisions.

If the body is to maintain homeostasis and function effectively, its trillions of cells must work together in a coordinated fashion. If each cell behaved without regard to what others were doing, the result would be physiological chaos and death. We have two organ systems dedicated to maintaining internal coordination—the **endocrine system** (see chapter 17), which communicates by means of chemical messengers (hormones) secreted into the blood, and the **nervous system (fig. 12.1),** which employs electrical and chemical means to send messages very quickly from cell to cell.

The nervous system carries out its coordinating task in three basic steps: (1) It receives information about changes in the body and external environment and transmits messages to the *central nervous system (CNS).* (2) The CNS processes this information and determines what response, if any, is appropriate to the circumstances. (3) The CNS issues commands primarily to muscle and gland cells to carry out such responses.

The nervous system has two major anatomical subdivisions **(fig. 12.2):**

- The **central nervous system (CNS)** consists of the brain and spinal cord, which are enclosed and protected by the cranium and vertebral column.

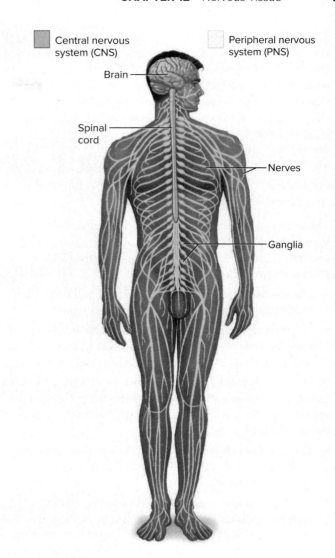

FIGURE 12.1 The Nervous System.

FIGURE 12.2 Subdivisions of the Nervous System.

- The **peripheral nervous system (PNS)** consists of all the rest; it is composed of nerves and ganglia. A **nerve** is a bundle of nerve fibers (axons) wrapped in fibrous connective tissue. Nerves emerge from the CNS through foramina of the skull and vertebral column and carry signals to and from other organs of the body. A **ganglion**[1] (plural, *ganglia*) is a knotlike swelling in a nerve where the cell bodies of peripheral neurons are concentrated.

The peripheral nervous system is functionally divided into *sensory* and *motor* divisions, and each of these is further divided into *somatic* and *visceral* subdivisions.

- The **sensory (afferent**[2]**) division** carries signals from various **receptors** (sense organs and simple sensory nerve endings) to the CNS. This pathway informs the CNS of stimuli within and around the body.
  - The **somatic**[3] **sensory division** carries signals from receptors in the skin, muscles, bones, and joints.
  - The **visceral sensory division** carries signals mainly from the viscera of the thoracic and abdominal cavities, such as the heart, lungs, stomach, and urinary bladder.
- The **motor (efferent**[4]**) division** carries signals from the CNS mainly to gland and muscle cells that carry out the body's responses. Cells and organs that respond to these signals are called **effectors.**
  - The **somatic motor division** carries signals to the skeletal muscles. This produces voluntary muscle contractions as well as automatic reflexes.
  - The **visceral motor division (autonomic**[5] **nervous system, ANS)** carries signals to glands, cardiac muscle, and smooth muscle. We usually have no voluntary control over these effectors, and the ANS operates at an unconscious level. The responses of the ANS and its effectors are *visceral reflexes*. The ANS has two further divisions:
    - The **sympathetic division** tends to arouse the body for action—for example, by accelerating the heartbeat and increasing respiratory airflow—but it inhibits digestion.
    - The **parasympathetic division** tends to have a calming effect—slowing the heartbeat, for example—but it stimulates digestion.

The foregoing terms may give the impression that we have several nervous systems—central, peripheral, sensory, motor, somatic, and visceral. These are just terms of convenience, however. There is only one nervous system, and these subsystems are interconnected parts of the whole.

---

[1]*gangli* = knot
[2]*af* = *ad* = toward; *fer* = to carry
[3]*somat* = body; *ic* = pertaining to
[4]*ef* = *ex* = out, away; *fer* = to carry
[5]*auto* = self; *nom* = law, governance

BEFORE YOU GO ON

Answer the following questions to test your understanding of the preceding section:

1. What is a receptor? Give two examples of effectors.
2. Distinguish between the central and peripheral nervous systems, and between visceral and somatic divisions of the sensory and motor systems.
3. What is another name for the visceral motor nervous system? What are its two subdivisions? What are their functions?

12.2 Properties of Neurons

Expected Learning Outcomes
When you have completed this section, you should be able to

a. describe three functional properties found in all neurons;
b. define the three most basic functional categories of neurons;
c. identify the parts of a neuron; and
d. explain how neurons transport materials between the cell body and tips of the axon.

12.2a Universal Properties

The communicative role of the nervous system is carried out by **nerve cells,** or **neurons.** These cells have three fundamental physiological properties that enable them to communicate with other cells:

1. **Excitability.** All cells are excitable—that is, they respond to environmental changes (**stimuli**). Neurons exhibit this property to the highest degree.
2. **Conductivity.** Neurons respond to stimuli by producing electrical signals that are quickly conducted to other cells at distant locations.
3. **Secretion.** When the signal reaches the end of a nerve fiber, the neuron secretes a *neurotransmitter* that crosses the gap and stimulates the next cell.

▶▶▶**APPLY WHAT YOU KNOW**

What basic physiological properties do a nerve cell and a muscle cell have in common? Name a physiological property of each that the other one lacks.

12.2b Functional Classes

There are three general classes of neurons (**fig. 12.3**) corresponding to the three major aspects of nervous system function listed earlier:

① **Sensory (afferent) neurons** are specialized to detect stimuli such as light, heat, pressure, and chemicals, and transmit information about them to the CNS. Such neurons begin in almost

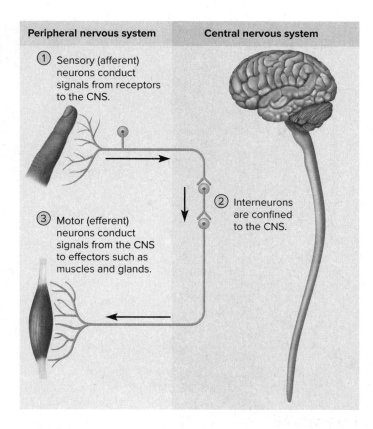

<table>
<tr><td>**Peripheral nervous system**</td><td>**Central nervous system**</td></tr>
</table>

① Sensory (afferent) neurons conduct signals from receptors to the CNS.

② Interneurons are confined to the CNS.

③ Motor (efferent) neurons conduct signals from the CNS to effectors such as muscles and glands.

FIGURE 12.3 The Three Functional Classes of Neurons. All neurons can be classified as sensory, motor, or interneurons depending on their location and the direction of signal conduction.

every organ of the body and end in the CNS; the word *afferent* refers to signal conduction *toward* the CNS. Some receptors, such as those for pain and smell, are themselves neurons. In other cases, such as taste and hearing, the receptor is a separate cell that communicates directly with a sensory neuron.

② **Interneurons** lie entirely within the CNS. They receive signals from many other neurons and carry out the integrative function of the nervous system—that is, they process, store, and retrieve information and "make decisions" that determine how the body responds to stimuli. About 90% of our neurons are interneurons. The word *interneuron* refers to the fact that they lie *between,* and interconnect, the incoming sensory pathways and the outgoing motor pathways of the CNS.

③ **Motor (efferent) neurons** send signals predominantly to muscle and gland cells, the effectors. They are called *motor* neurons because most of them lead to muscle cells, and *efferent* neurons to signify signal conduction *away from* the CNS.

12.2c Structure of a Neuron

There are several varieties of neurons, but a good starting point for discussion is a motor neuron of the spinal cord (**fig. 12.4**). The control center of the neuron is the **neurosoma,**[6] also called the

soma, cell body, or **perikaryon.**[7] It has a centrally located nucleus with a large nucleolus. The cytoplasm contains mitochondria, lysosomes, a Golgi complex, numerous inclusions, and an extensive rough endoplasmic reticulum and cytoskeleton. The cytoskeleton consists of a dense mesh of microtubules and **neurofibrils** (bundles of actin filaments), which compartmentalize the rough ER into dark-staining regions called **chromatophilic**[8] **substance** (fig. 12.4e). This is unique to neurons and a helpful clue to identifying them in tissue sections with mixed cell types. Mature neurons have no centrioles and cannot undergo any further mitosis after adolescence. Consequently, neurons that die are usually irreplaceable; surviving neurons cannot multiply to replace those lost. However, neurons are unusually long-lived cells, capable of functioning for over a hundred years.

The major inclusions in the neurosoma are glycogen granules, lipid droplets, melanin, and a golden brown pigment called *lipofuscin*[9] (LIP-oh-FEW-sin), produced when lysosomes degrade worn-out organelles and other products. Lipofuscin accumulates with age and pushes the nucleus to one side of the cell. Lipofuscin granules are also called "wear-and-tear granules" because they are most abundant in old neurons. They are also associated with certain degenerative diseases such as macular degeneration of the eye and amyotrophic lateral sclerosis (ALS).

The somas of most neurons give rise to a few thick processes that branch into a vast number of **dendrites**[10]—named for their striking resemblance to the bare branches of a tree in winter. Dendrites are the primary site for receiving signals from other neurons. Some neurons have only one dendrite and some have thousands. The more dendrites a neuron has, the more information it can receive and incorporate into its decision making. As tangled as the dendrites may seem, they provide exquisitely precise pathways for the reception and processing of neural information.

On one side of the neurosoma is a mound called the **axon hillock,** from which the **axon (nerve fiber)** originates. The axon is cylindrical and relatively unbranched for most of its length, although it may give rise to a few branches called *axon collaterals* near the soma, and most axons branch extensively at their distal end. An axon is specialized for rapid conduction of nerve signals to points remote from the soma. Its cytoplasm is called the **axoplasm** and its membrane the **axolemma.**[11] A neuron never has more than one axon, and some neurons have none.

Somas range from 5 to 135 μm in diameter, and axons from 1 to 20 μm in diameter and from a few millimeters to more than a meter long. (In the great blue whale, one nerve cell can be more than 30 m, or 100 ft, long.) The dimensions of a human neuron are more impressive when we scale them up to the size of familiar objects. If the soma of a spinal motor neuron were the size of a tennis ball, its dendrites would form a dense bushy mass that could

---

[6]*soma* = body

[7]*peri* = around, surrounding; *karyo* = nucleus
[8]*chromato* = color; *philic* = loving, attracting
[9]*lipo* = fat, lipid; *fusc* = dusky, brown
[10]*dendr* = tree, branch; *ite* = little
[11]*axo* = axis, axon; *lemma* = husk, peel, sheath

FIGURE 12.4 General Structure of a Neuron. (a) Structure of a multipolar neuron such as a spinal motor neuron. (b) Photograph of this neuron type. (c) Detail of the myelin sheath. (d) Neurofibrils of the neurosoma. (e) Chromatophilic substance, stained masses of rough ER separated by the bundles of neurofibrils shown in part (d). See section 12.3b for description of the myelin sheath.

b: Ed Reschke

fill a 30-seat classroom from floor to ceiling. Its axon would be up to a mile long but a little narrower than a garden hose. This is quite a point to ponder. The neuron must assemble molecules and organelles in its "tennis ball" soma and deliver them through its "mile-long garden hose" to the end of the axon. How it achieves this remarkable feat is explained shortly.

At the distal end, an axon usually has a **terminal arborization**[12]—an extensive complex of fine branches. Each branch ends in a bulbous **axon terminal (terminal button),** which forms a junction **(synapse**[13]**)** with the next cell. It contains **synaptic vesicles** full of neurotransmitter. In most autonomic neurons, however, the axon has numerous beads called **varicosities** along its length (see fig. 11.21). Each varicosity contains synaptic vesicles and secretes neurotransmitter.

Not all neurons fit the preceding description. Neurons are classified structurally according to the number of processes extending from the soma **(fig. 12.5):**

- **Multipolar neurons** are those, like the preceding, that have one axon and multiple dendrites. This is the most common type and includes most neurons of the brain and spinal cord.

- **Bipolar neurons** have one axon and one dendrite. Examples include olfactory cells of the nose, certain neurons of the retina, and sensory neurons of the ear.

- **Unipolar neurons** have only a single process leading away from the soma. They are represented by the neurons that carry signals to the spinal cord for such senses as touch and pain. They are also called *pseudounipolar* because they start out as bipolar neurons in the embryo, but their two processes fuse into one as the neuron matures. A short distance away from the soma, the process branches like a T into a *peripheral fiber* and a *central fiber.* The peripheral fiber begins with a sensory ending often far away from the soma—in the skin, for example. Its signals travel toward the soma, but bypass it and continue along the central fiber for a short remaining distance to the spinal cord. The dendrites are considered to be only the short receptive endings. The rest of the process, both peripheral and central, is the axon, defined by the presence of myelin and the ability to generate action potentials.

- **Anaxonic neurons** have multiple dendrites but no axon. They communicate locally through their dendrites and do not produce action potentials. Some anaxonic neurons are found in the brain, retina, and adrenal medulla. In the retina, they help in visual processes such as the perception of contrast.

12.2d Axonal Transport

All of the proteins needed by a neuron must be made in the soma, where the protein-synthesizing organelles such as the nucleus,

[12]*arbor* = tree
[13]*syn* = together; *aps* = to touch, join

(a) Multipolar neurons

(b) Bipolar neurons

(c) Unipolar neuron

(d) Anaxonic neuron

FIGURE 12.5 Variation in Neuron Structure. (a) Two multipolar neurons of the brain—a pyramidal cell *(left)* and a Purkinje cell. (b) Two bipolar neurons—a bipolar cell of the retina *(left)* and an olfactory neuron. (c) A unipolar neuron of the type involved in the senses of touch and pain. (d) An anaxonic neuron—an amacrine cell of the retina.

ribosomes, and rough endoplasmic reticulum are located. Yet many of these proteins are needed in the axon, for example to repair and maintain the axolemma, to serve as ion channels in the membrane, or to act in the axon terminal as enzymes and signaling molecules. Other substances are transported from the axon terminals back to the soma for disposal or recycling. The two-way passage of proteins, organelles, and other materials along an axon is called **axonal transport.** Movement away from the soma down the axon is called **anterograde**[14] **transport** and movement up the axon toward the soma is called **retrograde**[15] **transport.**

Materials travel along axonal microtubules that act like monorail tracks to guide them to their destination. But what is the "engine" that drives them along the tracks? Anterograde transport employs a motor protein called *kinesin*[16] and retrograde transport uses one called *dynein*[17] (the same protein responsible for the motility of cilia and flagella). These proteins carry materials "on their backs" while they reach out, like the myosin heads of muscle (see section 11.4c), to bind repeatedly to the microtubules and walk along them.

There are two types of axonal transport: fast and slow.

1. **Fast axonal transport** occurs at a rate of 200 to 400 mm/day and may be either anterograde or retrograde:

   • *Fast anterograde transport* moves mitochondria; synaptic vesicles; other organelles; components of the axolemma; calcium ions; enzymes such as acetylcholinesterase; and small molecules such as glucose, amino acids, and nucleotides toward the distal end of the axon.

   • *Fast retrograde transport* returns used synaptic vesicles and other materials to the soma and informs the soma of conditions at the axon terminals. Some pathogens exploit this process to invade the nervous system. They enter the distal tips of an axon and travel to the soma by retrograde transport. Examples include tetanus toxin and the herpes simplex, rabies, and polio viruses. In such infections, the delay between infection and the onset of symptoms corresponds to the time needed for the pathogens to reach the somas.

2. **Slow axonal transport** is an anterograde process that works in a stop-and-go fashion. If we compare fast axonal transport to an express train traveling nonstop to its destination, slow axonal transport is like a local train that stops at every station. When moving, it goes just as fast as the express train, but the frequent stops result in an overall progress of only 0.2 to 0.5 mm/day. It moves enzymes and cytoskeletal components down the axon, renews worn-out axoplasmic components in mature neurons, and supplies new axoplasm for developing or regenerating neurons. Damaged nerves regenerate at a speed governed by slow axonal transport.

▶▶▶**APPLY WHAT YOU KNOW**

The axon of a neuron has a dense cytoskeleton. Considering the functions of the cytoskeleton discussed in section 3.4a, give two reasons why this is so important to neuron structure and function.

BEFORE YOU GO ON

Answer the following questions to test your understanding of the preceding section:

4. Sketch a multipolar neuron and label its neurosoma, dendrites, axon, terminal arborization, axon terminals, and myelin sheath.

5. Explain the differences between a sensory neuron, motor neuron, and interneuron.

6. What is the functional difference between a dendrite and an axon?

7. How do proteins and other chemicals synthesized in the soma get to the axon terminals? By what process can a virus that invades a peripheral nerve fiber get to the soma of that neuron?

12.3 Supportive Cells

Expected Learning Outcomes

When you have completed this section, you should be able to

a. name the six types of cells that aid neurons, and state their respective functions;

b. describe the myelin sheath that is found around certain nerve fibers, and explain its importance;

c. describe the relationship of unmyelinated nerve fibers to their supportive cells; and

d. explain how damaged nerve fibers regenerate.

There are about a trillion (10^{12}) neurons in the nervous system—10 times as many neurons in your body as there are stars in our galaxy! Because they branch so extensively, they make up about 50% of the volume of the nervous tissue. Yet they are outnumbered at least 10 to 1 by cells called **neuroglia** (noo-ROG-lee-uh), or **glial cells** (GLEE-ul). Glial cells protect the neurons and help them function. The word *glia*, which means "glue," implies one of their roles—to bind neurons together and provide a supportive framework for the nervous tissue. In the fetus, they form a scaffold that guides young migrating neurons to their destinations. Wherever a mature neuron is not in synaptic contact with another cell, it is covered with glial cells. This prevents neurons from contacting each other except at points specialized for signal transmission, thus giving precision to their conduction pathways.

---

[14]*antero* = forward; *grad* = to walk, to step
[15]*retro* = back; *grad* = to walk, to step
[16]*kines* = motion; *in* = protein
[17]*dyne* = force; *in* = protein

12.3a Types of Neuroglia

There are six kinds of neuroglia, each with a unique function **(table 12.1)**. The first four types occur only in the central nervous system **(fig. 12.6)**:

1. **Oligodendrocytes**[18] (OL-ih-go-DEN-dro-sites) somewhat resemble an octopus; they have a bulbous body with as many as 15 arms. Each arm reaches out to a nerve fiber and spirals around it like electrical tape wrapped repeatedly around a wire. This wrapping, called the *myelin sheath,* insulates the nerve fiber from the extracellular fluid and speeds up signal conduction in the nerve fiber.

2. **Ependymal**[19] **cells** (ep-EN-dih-mul) resemble a cuboidal epithelium lining the internal cavities of the brain and spinal cord. Unlike true epithelial cells, however, they have no basement membrane and they exhibit rootlike processes that penetrate into the underlying tissue. Ependymal cells produce *cerebrospinal fluid (CSF),* a liquid that bathes the CNS and fills its internal cavities. They have patches of cilia on their apical surfaces that help to circulate the CSF. Ependymal cells and CSF are considered in more detail in section 14.2b.

3. **Microglia** are small macrophages that develop from white blood cells called monocytes. They wander through the CNS, putting out fingerlike extensions to constantly probe the tissue for cellular debris or other problems. They are thought to perform a complete checkup on the brain tissue

[18]*oligo* = few; *dendro* = branches; *cyte* = cell
[19]*ependyma* = upper garment

TABLE 12.1	Types of Glial Cells
Types	**Functions**
Neuroglia of CNS	
Oligodendrocytes	Form myelin in brain and spinal cord
Ependymal cells	Line cavities of brain and spinal cord; secrete and circulate cerebrospinal fluid
Microglia	Phagocytize and destroy microorganisms, foreign matter, and dead nervous tissue
Astrocytes	Cover brain surface and nonsynaptic regions of neurons; form supportive framework in CNS; induce formation of blood–brain barrier; nourish neurons; produce growth factors that stimulate neurons; promote the formation of synapses and neural circuitry; communicate electrically with neurons and may influence synaptic signaling; remove K^+ and some neurotransmitters from ECF of brain and spinal cord; help to regulate composition of ECF; form scar tissue to replace damaged nervous tissue
Neuroglia of PNS	
Schwann cells	Form neurilemma around all PNS nerve fibers and myelin around most of them; aid in regeneration of damaged nerve fibers
Satellite cells	Surround somas of neurons in the ganglia; provide electrical insulation and regulate chemical environment of neurons

FIGURE 12.6 Neuroglia of the Central Nervous System.

several times a day, phagocytizing dead tissue, microorganisms, and other foreign matter. They become concentrated in areas damaged by infection, trauma, or stroke. Pathologists look for clusters of microglia in brain tissue as a clue to sites of injury. Microglia also aid in synaptic remodeling, changing the connections between neurons.

4. **Astrocytes**[20] are the most abundant glial cells in the CNS and constitute over 90% of the tissue in some areas of the brain. They cover the entire brain surface and most nonsynaptic regions of the neurons in the gray matter. They are named for their many-branched, somewhat starlike shape. They have the most diverse functions of any glia:

   - They form a supportive framework for the nervous tissue.

   - They have extensions called *perivascular feet,* which contact the blood capillaries and stimulate them to form a tight, protective seal called the blood–brain barrier (see section 14.2c).

   - They monitor neuron activity, stimulate dilation and constriction of blood vessels, and thus regulate blood flow in the brain tissue to meet changing needs for oxygen and nutrients.

   - They convert blood glucose to lactate and supply this to the neurons for nourishment.

   - They secrete *nerve growth factors* that regulate nerve development (see Deeper Insight 12.3).

   - They promote synapse formation and fine-tune neural circuitry.

   - They communicate electrically with neurons and influence synaptic signaling between them.

   - They regulate the composition of the tissue fluid. When neurons transmit signals, they release neurotransmitters and potassium ions. Astrocytes absorb these and prevent them from reaching excessive levels in the tissue fluid.

   - When neurons are damaged, astrocytes form hardened scar tissue and fill space formerly occupied by the neurons. This process is called *astrocytosis* or *sclerosis.*

The other two types of glial cells occur only in the peripheral nervous system:

5. **Schwann**[21] **cells** (pronounced "shwon"), or **neurilemmocytes,** envelop nerve fibers of the PNS. In most cases, a Schwann cell winds repeatedly around a nerve fiber and produces a myelin sheath similar to the one produced by oligodendrocytes in the CNS. There are some important differences in myelin production between the CNS and PNS, which we consider shortly. Schwann cells also assist in the regeneration of damaged nerve fibers (see section 12.3e).

6. **Satellite cells** surround the somas in ganglia of the PNS. They provide insulation around the soma and regulate the chemical environment of the neurons.

---

[20]*astro* = star; *cyte* = cell
[21]Theodor Schwann (1810–82), German histologist

DEEPER INSIGHT 12.1

CLINICAL APPLICATION

Glial Cells and Brain Tumors

A tumor consists of a mass of rapidly dividing cells. Mature neurons, however, have little or no capacity for mitosis and seldom form tumors. Some brain tumors arise from the meninges (protective membranes of the CNS) or arise by metastasis from tumors elsewhere, such as malignant melanoma and colon cancer. Most adult brain tumors, however, are composed of glial cells, which are mitotically active throughout life. Such tumors are called *gliomas*[22] **(fig. 12.7).** Gliomas usually grow rapidly and are highly malignant. Because of the blood–brain barrier, brain tumors usually do not yield to chemotherapy and must be treated with radiation or surgery.

FIGURE 12.7 Brain Tumor. Frontal section of the brain showing a large glioma in the left cerebral hemisphere.
CNRI/Science Source

12.3b Myelin

The **myelin sheath** (MY-eh-lin) is a spiral layer of insulation around a nerve fiber, formed by oligodendrocytes in the CNS and Schwann cells in the PNS. Since it consists of the plasma membranes of glial cells, its composition is like that of plasma membranes in general. It is about 20% protein and 80% lipid, the latter including phospholipids, glycolipids, and cholesterol.

   Production of the myelin sheath is called **myelination.** It begins in the fourteenth week of fetal development, yet hardly any

---

[22]*glia* = glial cells; *oma* = tumor

myelin exists in the brain at the time of birth. Myelination proceeds rapidly in infancy and isn't completed until late adolescence. Since myelin has such a high lipid content, dietary fat is important to early nervous system development. It is best not to give children under 2 years old the sort of low-fat diets (skimmed milk, etc.) that may be beneficial to an adult.

In the PNS, a Schwann cell spirals repeatedly around a single nerve fiber, laying down up to 100 compact layers of its own membrane with almost no cytoplasm between the membranes **(fig. 12.8a).** These layers constitute the myelin sheath. The Schwann cell spirals outward as it wraps the nerve fiber, finally ending with a thick outermost coil called the **neurilemma**[23] (noor-ih-LEM-ah). Here, the bulging body of the Schwann cell contains its nucleus and most of its cytoplasm. External to the neurilemma is a basal lamina and then a thin sleeve of fibrous connective tissue called the *endoneurium.* To visualize this myelination process, imagine that you wrap an almost-empty tube of toothpaste tightly around a pencil. The pencil represents the axon, and the spiral layers of toothpaste tube represent the myelin. The toothpaste, like the cytoplasm of the cell, would be forced to one end of the tube and form a bulge on the external surface of the wrapping, like the body of the Schwann cell.

In the CNS, each oligodendrocyte reaches out to myelinate several nerve fibers in its immediate vicinity **(fig. 12.8b).** Since it is anchored to multiple nerve fibers, it can't migrate around any one of them like a Schwann cell does. It must push newer layers of myelin under the older ones, so myelination spirals inward toward the nerve fiber. Nerve fibers of the CNS have no neurilemma or endoneurium. The contrasting modes of myelination are called *centrifugal myelination* ("away from the center") in the PNS and *centripetal myelination* ("toward the center") in the CNS.

In both the PNS and CNS, a nerve fiber is much longer than the reach of a single glial cell, so it requires many Schwann cells or oligodendrocytes to cover one nerve fiber. Consequently, the myelin sheath is segmented. Each gap between segments is called a **myelin sheath gap** or **node of Ranvier**[24] (RON-vee-AY) (*node* for short); the myelin-covered segments from each node to the next are called **internodal segments** (see fig. 12.4a). These segments are about 0.2 to 1.0 mm long. The short section of nerve fiber between the axon hillock and the first glial cell is called the **initial segment.** Since the axon hillock and initial segment play an important role in initiating a nerve signal, they are collectively called the **trigger zone.**

12.3c Unmyelinated Nerve Fibers

Many nerve fibers in the CNS and PNS are unmyelinated. In the PNS, however, even the unmyelinated fibers are enveloped in Schwann cells. In this case, one Schwann cell harbors from 1 to 12 small nerve fibers in grooves in its surface **(fig. 12.9).** The Schwann cell's plasma membrane doesn't spiral repeatedly around the fiber as it does in a myelin sheath, but folds once around each fiber and may somewhat overlap itself along the edges. This wrapping is the neurilemma. Most nerve fibers travel through individual channels in

the Schwann cell, but small fibers are sometimes bundled together within a single channel, as on the right side of this figure. A basal lamina surrounds the entire Schwann cell along with its nerve fibers.

12.3d Conduction Speed of Nerve Fibers

The speed at which a nerve signal travels along a nerve fiber depends on two factors: the diameter of the fiber and the presence or absence of myelin. Signal conduction occurs along the surface of a fiber, not deep within its axoplasm. Large fibers have more surface area and conduct signals more rapidly than small fibers. Myelin further speeds signal conduction by a mechanism to be explained in section 12.4f. Nerve signals travel about 0.5 to 2.0 m/s in small unmyelinated fibers (2–4 μm in diameter); 3 to 15 m/s in myelinated fibers of the same size; and as fast as 120 m/s in large myelinated fibers (up to 20 μm in diameter). One may wonder why all of our nerve fibers aren't large, myelinated, and fast; but if this were so, our nervous system would be impossibly bulky or limited to far fewer fibers. Large nerve fibers require large somas and a large expenditure of energy to maintain them. The evolution of myelin allowed for the subsequent evolution of more complex and responsive nervous systems with smaller, more energy-efficient neurons. Slow unmyelinated fibers are quite sufficient for processes in which quick responses aren't

DEEPER INSIGHT 12.2
CLINICAL APPLICATION
Diseases of the Myelin Sheath

Multiple sclerosis and Tay–Sachs disease are degenerative disorders of the myelin sheath. In *multiple sclerosis*[25] *(MS),* the oligodendrocytes and myelin sheaths of the CNS deteriorate and are replaced by hardened scar tissue, especially between the ages of 20 and 40. Nerve conduction is disrupted, with effects that depend on what part of the CNS is involved—double vision, blindness, speech defects, neurosis, tremors, or numbness, for example. Patients experience variable cycles of milder and worse symptoms until they eventually become bedridden. The cause of MS remains uncertain; most hypotheses suggest that it is an autoimmune disorder triggered by a virus in genetically susceptible individuals. There is no cure. There is conflicting evidence of how much it shortens a person's life expectancy, if at all. A few die within 1 year of diagnosis, but many people live with MS for 25 or 30 years.

Tay–Sachs[26] disease is a hereditary disorder seen mainly in infants of Eastern European Jewish ancestry. It results from the abnormal accumulation of a glycolipid called GM$_2$ (ganglioside) in the myelin sheath. GM$_2$ is normally decomposed by a lysosomal enzyme, but this enzyme is lacking from people who are homozygous recessive for the Tay–Sachs allele. As GM$_2$ accumulates, it disrupts the conduction of nerve signals and the victim typically suffers blindness, loss of coordination, and dementia. Signs begin to appear before the child is a year old, and most victims die by the age of 3 or 4 years. Asymptomatic adult carriers can be identified by a blood test and advised by genetic counselors on the risk of their children having the disease.

---

[23]*neuri* = nerve; *lemma* = husk, peel, sheath
[24]L. A. Ranvier (1835–1922), French histologist and pathologist

[25]*scler* = hard, tough; *osis* = condition
[26]Warren Tay (1843–1927), English physician; Bernard Sachs (1858–1944), American neurologist

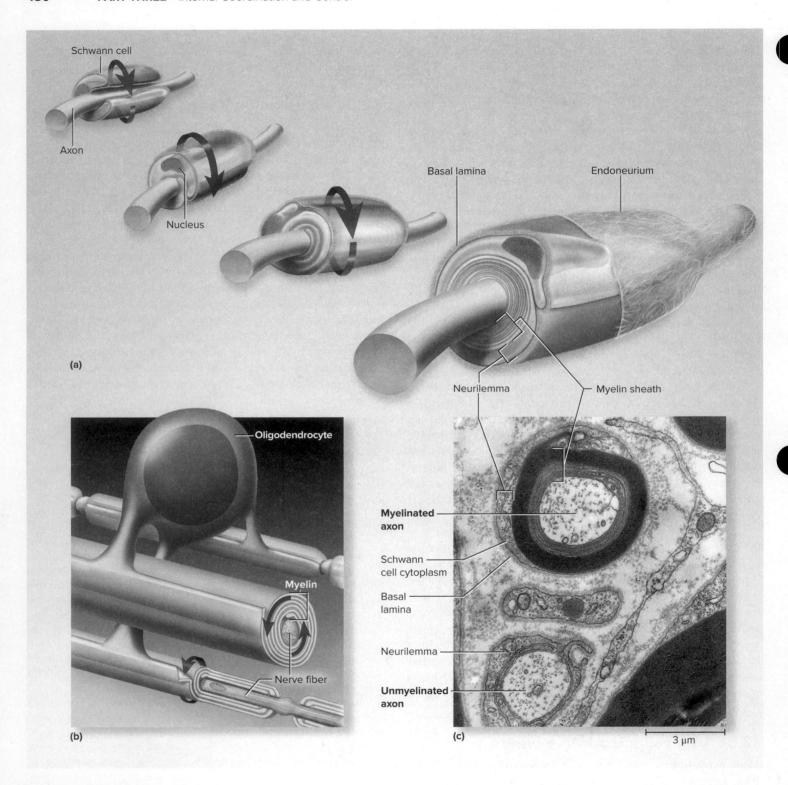

(a)

Schwann cell

Axon

Nucleus

Basal lamina

Endoneurium

Neurilemma

Myelin sheath

(b)

Oligodendrocyte

Myelin

Nerve fiber

(c)

Myelinated axon

Schwann cell cytoplasm

Basal lamina

Neurilemma

Unmyelinated axon

3 µm

FIGURE 12.8 Myelination. (a) A Schwann cell of the PNS, wrapping repeatedly around an axon to form the multilayered myelin sheath. The myelin spirals outward away from the axon as it is laid down. (b) An oligodendrocyte of the CNS wrapping around the axons of multiple neurons. Here, the myelin spirals inward toward the axon as it is laid down. (c) A myelinated axon (top) and unmyelinated axon (bottom) (TEM).

c: Dr. Dennis Emery, Iowa State University/McGraw-Hill Education

Unmyelinated nerve fibers

Schwann cell

Basal lamina

FIGURE 12.9 Unmyelinated Nerve Fibers.

❓ *What is the functional disadvantage of an unmyelinated nerve fiber? What is its anatomical advantage?*

particularly important, such as secreting stomach acid or dilating the pupil. Fast myelinated fibers are employed where speed is more important, as in motor commands to the skeletal muscles and sensory signals for vision and balance.

12.3e Nerve Regeneration

Nerve fibers of the PNS are vulnerable to cuts, crushing injuries, and other trauma. A damaged peripheral nerve fiber may regenerate, however, if its soma is intact and at least some neurilemma remains. **Figure 12.10** shows the process of regeneration, taking as its example a somatic motor neuron:

① In the normal nerve fiber, note the size of the soma and the size of the muscle fibers for comparison to later stages.

② When a nerve fiber is cut, the fiber distal to the injury can't survive because it is incapable of protein synthesis. Protein-synthesizing organelles are mostly in the soma. As the distal fiber degenerates, so do its Schwann cells, which depend on it for their maintenance. Macrophages clean up tissue debris at the point of injury and beyond.

③ The soma exhibits a number of abnormalities of its own, probably because it is cut off from the supply of nerve growth factors from the neuron's target cells (see

① **Normal nerve fiber**
Neuromuscular junction
Endoneurium
Myelin sheath
Muscle fiber

② **Injured fiber**
Local trauma
Macrophages
Degenerating terminal

③ **Degeneration of severed fiber**
Degenerating Schwann cells
Degenerating axon

④ **Early regeneration**
Growth processes
Schwann cells
Regeneration tube
Atrophy of muscle fibers

⑤ **Late regeneration**
Retraction of growth processes
Growth processes

⑥ **Regenerated fiber**
Regrowth of muscle fibers

FIGURE 12.10 Regeneration of a Damaged Nerve Fiber. See numbered steps in text for explanation. Nerve fibers of the PNS can regenerate if the neurosoma is intact.

DEEPER INSIGHT 12.3

MEDICAL HISTORY

Nerve Growth Factor—From Home Laboratory to Nobel Prize

It is remarkable what odds can be overcome by self-confident persistence. Neurobiologist Rita Levi-Montalcini **(fig. 12.11)** affords a striking example. Although born of a cultured and accomplished Italian Jewish family, she and her twin sister Paola were discouraged from considering a college education or career by their tradition-minded father. Determined to attend university anyway, they hired their own tutor to prepare them. Rita graduated summa cum laude in medicine and surgery in 1930 and embarked on advanced study in neurology. Paola became a renowned artist.

But then arose the sinister specter of anti-Semitism, as fascist dictator Mussolini barred Jews from professional careers. Rita despaired of pursuing medicine or research until a college friend reminded her of how much Cajal had achieved under very primitive conditions. That inspired her to set up a little laboratory in her bedroom, where she studied nervous system development in chick embryos. She had read of work by Viktor Hamburger in St. Louis, who believed that limb tissues secrete a chemical that attracts nerves to grow into them. Levi-Montalcini, however, believed that nerves grow into the limbs without such attractants, but die if they fail to receive a substance needed to sustain them.

Fleeing first from Allied bombing and then Hitler's invasion of Italy, Levi-Montalcini had to abandon her work as the family went underground until the end of the war. At war's end, Hamburger invited her to join him in America, where they found her hypothesis to be correct. She and Stanley Cohen isolated the nerve-sustaining substance and named it *nerve growth factor (NGF),* for which they shared a 1986 Nobel Prize. NGF is a protein secreted by muscle and glial cells. It prevents apoptosis in growing neurons and thus enables them to establish connections with their target cells. It was the first of many cell growth factors discovered, and launched what is today a vibrant field of research in the use of growth factors to stimulate tissue development and repair.

FIGURE 12.11 **Rita Levi-Montalcini (1909–2012).**
Olycom/SIPA/Newscom

Rita and Paola created the Levi-Montalcini Foundation to support the career development of young people and especially the scientific education of women in Africa. Rita also served as an honored member of the Italian Senate from 2001 until 2012, when she died at the age of 103.

Deeper Insight 12.3). The soma swells, the endoplasmic reticulum breaks up (so the chromatophilic substance disperses), and the nucleus moves off center. Not all damaged neurons survive; some die at this stage. But often, the axon stump sprouts multiple growth processes while the severed distal end shows continued degeneration of its axon and Schwann cells. Muscle fibers deprived of their nerve supply exhibit a shrinkage called *denervation atrophy.*

④ Near the injury, Schwann cells, the basal lamina, and the neurilemma form a **regeneration tube.** The Schwann cells produce cell-adhesion molecules and nerve growth factors that enable a neuron to regrow to its original destination. When one growth process finds its way into the tube, it grows rapidly (3–5 mm/day), and the other growth processes are retracted.

⑤ The regeneration tube guides the growing sprout back to the original target cells, reestablishing synaptic contact.

⑥ When contact is established, the soma shrinks and returns to its original appearance, and the reinnervated muscle fibers regrow.

Regeneration isn't perfect. Some nerve fibers connect to the wrong muscle fibers or never find a muscle fiber at all, and some damaged neurons simply die. Nerve injury is therefore often followed by some degree of functional deficit. Even when regeneration is achieved, the slow rate of axon regrowth means that some nerve function may take as long as 2 years to recover.

Schwann cells and endoneurium are required for nerve fiber regeneration. Both of these are lacking from the CNS, so damaged CNS nerve fibers cannot regenerate at all. However, since the CNS is encased in bone, it suffers less trauma than the peripheral nerves.

BEFORE YOU GO ON

Answer the following questions to test your understanding of the preceding section:

8. How is a glial cell different from a neuron? List the six types of glial cells and discuss their functions.

9. How is myelin produced? How does myelin production in the CNS differ from that in the PNS?

10. How can a severed peripheral nerve fiber find its way back to the cells it originally innervated?

12.4 Electrophysiology of Neurons

Expected Learning Outcomes
When you have completed this section, you should be able to

a. explain why a cell has an electrical charge difference (voltage) across its membrane;

b. explain how stimulation of a neuron causes a local electrical response in its membrane;

c. explain how local responses generate a nerve signal; and

d. explain how the nerve signal is conducted down an axon.

The nervous system has intrigued scientists and philosophers since ancient times. Galen, the preeminent physician of ancient Rome, thought that the brain pumped a vapor called *psychic pneuma* through hollow nerves and squirted it into the muscles to make them contract. The French philosopher René Descartes still argued for this theory in the seventeenth century. It finally fell out of favor in the eighteenth century, when Luigi Galvani discovered the role of electricity in muscle contraction (see Deeper Insight 11.3). Further progress had to await improvements in microscope technology and histological staining methods. Italian histologist Camillo Golgi (1843–1926) developed an important method for staining neurons with silver. This enabled Spanish histologist Santiago Ramón y Cajal (1852–1934), with tremendous skill and patience, to trace the course of nerve fibers over long distances through serial tissue sections. He demonstrated that the nervous pathway was not a continuous "wire" or tube, but a series of cells separated by the gaps we now call synapses. Golgi and Cajal, even though they intensely disliked each other, shared the 1906 Nobel Prize for Physiology or Medicine for these important discoveries.

Cajal's theory suggested another direction for research: How do neurons communicate? Two key issues in neurophysiology are (1) How does a neuron generate an electrical signal? and (2) How does it transmit a meaningful message to the next cell? These are the questions to which this section and the next are addressed.

12.4a Electrical Potentials and Currents

Neural communication, like muscle excitation, is based on electrophysiology—cellular mechanisms for producing electrical potentials and currents. An **electrical potential** is a difference in the concentration of charged particles between one point and another. It is a form of potential energy that, under the right circumstances, can produce a current. An electrical **current** is a flow of charged particles from one point to another. A new flashlight battery, for example, typically has a potential, or charge, of 1.5 volts (V). If a lightbulb and the two poles of the battery are connected by a wire, electrons flow through the wire from one pole to the other, creating a current that lights the bulb. As long as the battery has a potential (voltage), we say it is **polarized.**

Living cells are also polarized. The charge difference across the plasma membrane is called the **resting membrane potential (RMP).** It is much less than the potential of a flashlight battery—typically about –70 millivolts (mV) in an unstimulated, "resting"

neuron. The negative value means there are more negatively charged particles on the inside of the membrane than on the outside.

We don't have free electrons in the body as we do in an electrical circuit. Electrical currents in the body are created, instead, by the flow of ions such as Na^+ and K^+ through gated channels in the plasma membrane. Gated channels can be opened and closed by various stimuli, as we have seen earlier (see "Membrane Proteins" in section 3.2a and "Electrically Excitable Cells," section 11.3c). This enables cells to turn electrical currents on and off.

12.4b The Resting Membrane Potential

The reason a cell has a resting membrane potential is that electrolytes are unequally distributed between the extracellular fluid (ECF) on the outside of the plasma membrane and the intracellular fluid (ICF) on the inside. The RMP results from the combined effect of three factors: (1) the diffusion of ions down their concentration gradients through the membrane; (2) selective permeability of the membrane, allowing some ions to pass more easily than others; and (3) the electrical attraction of cations and anions to each other.

Potassium ions (K^+) have the greatest influence on the RMP because the plasma membrane is more permeable to K^+ than to any other ion. Imagine a hypothetical cell in which all the K^+ starts out in the ICF, with none in the ECF. Also in the ICF are a number of cytoplasmic anions that cannot escape from the cell because of their size or charge—phosphates, sulfates, small organic acids, proteins, ATP, and RNA. Potassium ions diffuse freely through leak channels in the plasma membrane, down their concentration gradient and out of the cell, leaving these cytoplasmic anions behind **(fig. 12.12).** As a result, the ICF grows more and more negatively charged. But as the ICF becomes more negative, it

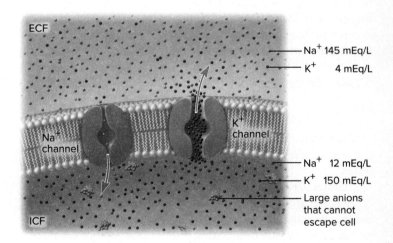

FIGURE 12.12 Ionic Basis of the Resting Membrane Potential. Note that sodium ions are much more concentrated in the extracellular fluid (ECF) than in the intracellular fluid (ICF), while potassium ions are more concentrated in the ICF. Large anions unable to penetrate the plasma membrane give the cytoplasm a negative charge relative to the ECF.

exerts a stronger attraction for the positive potassium ions and attracts some of them back into the cell. Eventually an *equilibrium* is reached in which K$^+$ is moving out of the cell (down its concentration gradient) and into the cell (by electrical attraction) at equal rates. There is no further *net* diffusion of K$^+$. At the point of equilibrium, K$^+$ is about 40 times as concentrated in the ICF as in the ECF.

If K$^+$ were the only ion affecting the RMP, it would give the membrane a potential of about –90 mV. However, sodium ions (Na$^+$) also enter the picture. Sodium is about 12 times as concentrated in the ECF as in the ICF. The resting plasma membrane is much less permeable to Na$^+$ than to K$^+$, but Na$^+$ does diffuse down its concentration gradient into the cell, attracted by the negative charge in the ICF. This sodium leak is only a trickle, but it is enough to cancel some of the negative charge and reduce the voltage across the membrane.

Sodium leaks into the cell and potassium leaks out, but the sodium–potassium (Na$^+$–K$^+$) pump continually compensates for this leakage. It pumps 3 Na$^+$ out of the cell for every 2 K$^+$ it brings in, consuming 1 ATP for each exchange cycle (see fig. 3.19). By removing more cations from the cell than it brings in, it contributes about –3 mV to the RMP. The resting membrane potential of –70 mV is the net effect of all these ion movements—K$^+$ diffusion out of the cell, Na$^+$ diffusion inward, and the Na$^+$–K$^+$ pump continually offsetting this ion leakage.

The Na$^+$–K$^+$ pump accounts for about 70% of the energy (ATP) requirement of the nervous system. Every signal generated by a neuron slightly upsets the distribution of Na$^+$ and K$^+$, so the pump must work continually to restore equilibrium. This is why nervous tissue has one of the highest rates of ATP consumption of any tissue in the body, and why it demands so much glucose and oxygen. Although a neuron is said to be resting when it is not producing signals, it is highly active maintaining its RMP and "waiting," as it were, for something to happen.

The uneven distribution of Na$^+$ and K$^+$ on the two sides of the plasma membrane pertains only to a very thin film of ions immediately adjacent to the membrane surfaces. The electrical events we are about to examine don't involve ions very far away from the membrane in either the ECF or the ICF.

12.4c Local Potentials

Stimulation of a neuron causes local disturbances in membrane potential. Typically (but with exceptions), the response begins at a dendrite, spreads through the soma, travels down the axon, and ends at the axon terminal. We consider the process in that order.

Various neurons can be stimulated by chemicals, light, heat, or mechanical forces. We'll take as our example a neuron being chemically stimulated (**fig. 12.13**). The chemical (ligand)—perhaps a pain signal from a damaged tissue or odor molecule in a breath of air—binds to receptors on the neuron. This opens ligand-gated sodium channels that allow Na$^+$ to flow into the cell. The Na$^+$ inflow cancels some of the internal negative charge, so the voltage across the membrane at that point drifts toward zero. Any such case in which the voltage shifts to a less negative value is called **depolarization.**

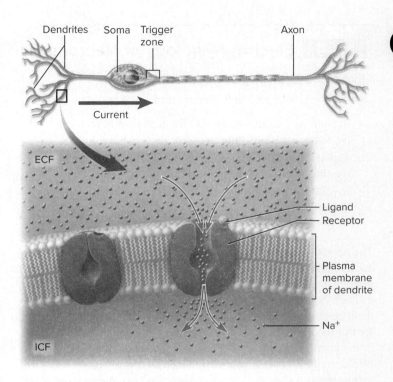

FIGURE 12.13 Excitation of a Neuron by a Chemical Stimulus.

The incoming Na$^+$ diffuses for short distances along the inside of the plasma membrane, creating a wave of excitation that spreads out from the point of stimulation, like ripples spreading across a pond when you drop a stone into it. This short-range change in voltage is called a **local potential.** There are four characteristics that distinguish local potentials from the action potentials we will study in the next section (**table 12.2**). You will appreciate these distinctions more fully after you have studied action potentials.

1. Local potentials are **graded,** meaning they vary in magnitude (voltage) according to the strength of the stimulus. An intense or prolonged stimulus opens more gated ion channels than a weaker stimulus, and they stay open longer. Thus, more Na$^+$ enters the cell and the voltage changes more than it does with a weaker stimulus.

2. Local potentials are **decremental,** meaning they get weaker as they spread from the point of origin. They decline in strength partly because the Na$^+$ leaks back out of the cell through channels along its path, and partly because as Na$^+$ spreads out under the plasma membrane and depolarizes it, K$^+$ flows out and reverses the effect of the Na$^+$ inflow. Therefore, the voltage shift caused by Na$^+$ diminishes rapidly with distance. This prevents local potentials from having long-distance effects.

3. Local potentials are **reversible,** meaning that if stimulation ceases, cation diffusion out of the cell quickly returns the membrane voltage to its resting potential.

4. Local potentials can be either **excitatory** or **inhibitory.** So far, we have considered only excitatory local potentials, which depolarize a cell and make a neuron more likely to

TABLE 12.2	Comparison of Local Potentials and Action Potentials
Local Potential	**Action Potential**
Produced by gated channels on the dendrites and soma	Produced by voltage-gated channels on the trigger zone and axon
May be a positive (depolarizing) or negative (hyperpolarizing) voltage change	Always begins with depolarization
Graded; proportional to stimulus strength	All or none; either does not occur at all or exhibits the same peak voltage regardless of stimulus strength
Reversible; returns to RMP if stimulation ceases before threshold is reached	Irreversible; goes to completion once it begins
Local; has effects for only a short distance from point of origin	Self-propagating; has effects a great distance from point of origin
Decremental; signal grows weaker with distance	Nondecremental; signal maintains same strength regardless of distance

produce an action potential. Acetylcholine usually has this effect. Other neurotransmitters, such as glycine, cause an opposite effect—they **hyperpolarize** a cell, or make the membrane more negative. This inhibits a neuron, making it less sensitive and less likely to produce an action potential. A balance between excitatory and inhibitory potentials is very important to information processing in the nervous system (see section 12.6b).

12.4d Action Potentials

An **action potential** is a more dramatic change produced by voltage-gated ion channels in the plasma membrane. Action potentials occur only where there is a high enough density of voltage-gated channels. Most of the soma has only 50 to 75 channels per square micrometer (μm^2), not dense enough to generate action potentials. The trigger zone, however, has 350 to 500 channels/μm^2. If an excitatory local potential spreads all the way to the trigger zone and is still strong enough when it arrives, it can open these channels and generate an action potential.

The action potential is a rapid up-and-down shift in voltage. **Figure 12.14a** shows an action potential numbered to correspond to the following description.

1. When the local current arrives at the axon hillock, it depolarizes the membrane at that point. This appears as a steadily rising local potential.

2. For anything more to happen, this local potential must rise to a critical voltage called the **threshold** (typically about –55 mV), the minimum needed to open voltage-gated channels.

(a)

(b)

FIGURE 12.14 An Action Potential. (a) Diagrammed with a distorted timescale to make details of the action potential visible. Numbers correspond to stages discussed in the text. (b) On a more accurate timescale, the local potential is so brief it is imperceptible, the action potential appears as a spike, and the hyperpolarization is very prolonged.

③ The neuron now "fires," or produces an action potential. At threshold, voltage-gated Na^+ channels open quickly, while gated K^+ channels open more slowly. The initial effect on membrane potential is therefore due to Na^+. Initially, only a few Na^+ channels open, but as Na^+ enters the cell, it further depolarizes the membrane. This stimulates still more voltage-gated Na^+ channels to open and admit even more Na^+, a positive feedback loop that causes the membrane voltage to rise even more rapidly.

④ As the rising potential passes 0 mV, Na^+ channels are *inactivated* and begin closing. By the time they all close and Na^+ inflow ceases, the voltage peaks at approximately +35 mV. (The peak is as low as 0 mV in some neurons and as high as 50 mV in others.) The membrane is now positive on the inside and negative on the outside—its polarity is reversed compared to the RMP.

⑤ By the time the voltage peaks, the slow K^+ channels are fully open. Potassium ions, repelled by the positive ICF, now exit the cell. Their outflow **repolarizes** the membrane—that is, it shifts the voltage back into the negative numbers. The action potential consists of the up-and-down voltage shifts that occur from the time the threshold is reached to the time the voltage returns to the RMP.

⑥ Potassium channels stay open longer than Na^+ channels, so slightly more K^+ leaves the cell than the amount of Na^+ that entered. Therefore, the membrane voltage drops to 1 or 2 mV more negative than the original RMP, producing a negative overshoot called *hyperpolarization.*

⑦ As you can see, Na^+ and K^+ switch places across the membrane during an action potential. During hyperpolarization, the membrane voltage gradually returns to the RMP because of Na^+ diffusion into the cell.

Figure 12.14a is deliberately distorted. To demonstrate the different phases of the local potential and action potential, the magnitudes of the local potential and hyperpolarization are exaggerated, the local potential is stretched out to make it seem longer, and the duration of hyperpolarization is shrunken so the graph will fit the page. When these events are plotted on a more realistic timescale, they look more like **figure 12.14b**. The local potential is so brief it is unnoticeable, and hyperpolarization is very long but only slightly more negative than the RMP. An action potential is often called a *spike;* it is easy to see why from this figure.

Figure 12.15 correlates these voltage changes with events in the plasma membrane. At the risk of being misleading, it is drawn as if most of the Na^+ and K^+ had traded places. In reality, only about one in a million ions crosses the membrane to produce an action potential, and an action potential involves only the thin layer of ions close to the membrane. If the illustration tried to represent these points accurately, the difference would be so slight you couldn't see it. Even after thousands of action potentials, the cytosol still has a higher concentration of K^+ and a lower concentration of Na^+ than the ECF does.

Earlier we saw that local potentials are graded, decremental, and reversible. We can now contrast this with action potentials.

- Action potentials follow an **all-or-none law.** If a stimulus depolarizes the neuron to threshold, the neuron fires at its maximum voltage (such as +35 mV); if threshold is not reached, the neuron doesn't fire at all. Above threshold, stronger stimuli don't produce stronger action potentials. Thus, action potentials are not graded (proportional to stimulus strength) like local potentials are.

- Action potentials are **nondecremental.** They don't get weaker with distance. The last action potential at the end of a nerve fiber is just as strong as the first one in the trigger zone, no matter how far away—even in a pain fiber that extends from your toes to your brainstem.

- Action potentials are **irreversible.** If a neuron reaches threshold, the action potential goes to completion; it can't be stopped once it begins.

In some respects, we can compare the firing of a neuron to the firing of a gun. As the trigger is squeezed, a gun either fires with maximum force or doesn't fire at all (analogous to the all-or-none law). You can't fire a fast bullet by squeezing the trigger hard or a slow bullet by squeezing it gently—once the trigger is pulled to its "threshold," the bullet always leaves the muzzle at the same velocity. And, like an action potential, the firing of a gun is irreversible once the threshold is reached; neither a bullet nor an action potential can be called back. Table 12.2 further contrasts a local potential with an action potential, including some characteristics of action potentials explained in the following sections.

12.4e The Refractory Period

During an action potential and for a few milliseconds after, it is difficult or impossible to stimulate that region of a neuron to fire again. This period of resistance to restimulation is called the **refractory period.** It is divided into two phases: an *absolute refractory period* in which no stimulus of any strength will trigger a new action potential, followed by a *relative refractory period* in which it is possible to trigger a new action potential, but only with an unusually strong stimulus **(fig. 12.16)**.

The absolute refractory period lasts from the start of the action potential until the membrane returns to the resting potential—that is, for as long as the Na^+ channels are open and subsequently inactivated. The relative refractory period lasts until hyperpolarization ends. During this period, K^+ channels are still open. A new stimulus tends to admit Na^+ and depolarize the membrane, but K^+ diffuses out through the open channels as Na^+ comes in, and thus opposes the effect of the stimulus. It requires an especially strong stimulus to override the K^+ outflow and depolarize the cell enough to set off a new action potential. By the end of hyperpolarization, K^+ channels are closed and the cell is as responsive as ever.

The refractory period refers only to a small patch of membrane where an action potential has already begun, not to the entire neuron. Other parts of the neuron can still be stimulated while a small area of it is refractory, and even this area quickly recovers once the nerve signal has passed.

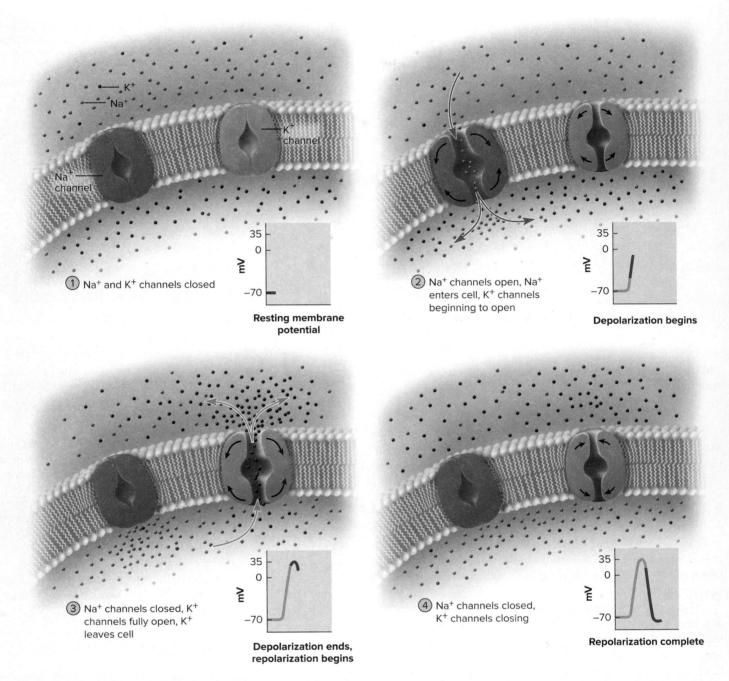

① Na⁺ and K⁺ channels closed

**Resting membrane
potential**

**② Na⁺ channels open, Na⁺
enters cell, K⁺ channels
beginning to open**

Depolarization begins

**③ Na⁺ channels closed, K⁺
channels fully open, K⁺
leaves cell**

**Depolarization ends,
repolarization begins**

**④ Na⁺ channels closed,
K⁺ channels closing**

Repolarization complete

FIGURE 12.15 Actions of the Sodium and Potassium Channels During an Action Potential. The red part of each graph shows the point in the action potential where the events of steps 1 through 4 occur. **APR**

12.4f Signal Conduction in Nerve Fibers

If a neuron is to communicate with another cell, a signal has to travel to the end of the axon. We now examine how this is achieved.

Unmyelinated Fibers and Continuous Conduction

Unmyelinated fibers present a relatively simple case of signal conduction, easy to understand based on what we have already covered **(fig. 12.17).** An unmyelinated fiber has voltage-gated channels along its entire length. When an action potential occurs

at the trigger zone, Na⁺ enters the axon and diffuses for a short distance just beneath the plasma membrane. The resulting depolarization excites voltage-gated channels immediately distal to the action potential. Sodium and potassium channels open and close just as they did at the trigger zone, and a new action potential is produced. By repetition, this excites the membrane immediately distal to that. This chain reaction continues until the traveling signal reaches the end of the axon. Because this produces an uninterrupted wave of electrical excitation all along the fiber, this mechanism is called **continuous conduction.**

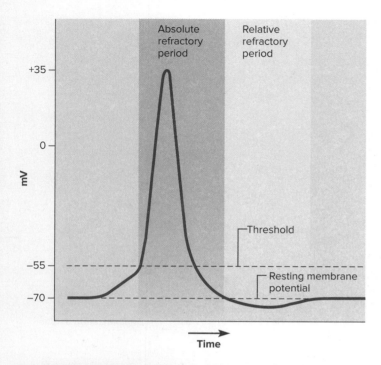

FIGURE 12.16 The Absolute and Relative Refractory Periods in Relation to the Action Potential.

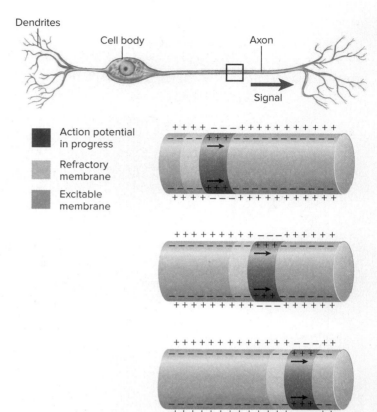

FIGURE 12.17 Continuous Conduction of a Nerve Signal in an Unmyelinated Fiber. Note that the membrane polarity is reversed in the region of the action potential (red). A region of membrane in its refractory period (yellow) trails the action potential and prevents the nerve signal from going backward toward the soma. The other membrane areas (green) are fully polarized and ready to respond. **APR**

Note that an action potential itself doesn't travel along an axon; rather, it stimulates the production of a new action potential in the membrane just ahead of it. Thus, we can distinguish an action potential from a nerve signal. The **nerve signal** is a traveling wave of excitation produced by self-propagating action potentials. It is like a line of falling dominoes. No one domino travels to the end of the line, but each domino pushes over the next one and there is a transmission of energy from the first domino to the last. Similarly, no one action potential travels to the end of an axon; a nerve signal is a chain reaction of action potentials, each triggering the next one ahead of it.

If one action potential stimulates the production of a new one next to it, you might think that the signal could also start traveling backward and return to the soma. This does not occur, however, because the membrane immediately behind the nerve signal is still in its refractory period and cannot be restimulated. Only the membrane ahead is sensitive to stimulation. The refractory period thus ensures that nerve signals are conducted in the proper direction, from the soma to the axon terminals.

A traveling nerve signal is an electrical current, but it isn't the same as a current traveling through a wire. A current in a wire travels millions of meters per second and is decremental—it gets weaker with distance. A nerve signal is much slower (not more than 2 m/s in unmyelinated fibers), but as already noted, it is nondecremental. To clarify this concept, we can compare the nerve signal to a burning fuse on a firecracker. When a fuse is lit, the heat ignites powder immediately in front of this point, and this repeats itself in a self-propagating fashion until the end of the fuse is reached. At the end, the fuse burns just as hotly as it did at the beginning. In a fuse, the combustible powder is the source of potential energy that keeps the process going in a nondecremental fashion. In an axon,

the potential energy comes from the ion gradient across the plasma membrane. Thus, the signal doesn't grow weaker with distance; it is self-propagating, like the burning of a fuse.

Myelinated Fibers and Saltatory Conduction

Myelinated fibers conduct signals in a very different manner called **saltatory**[27] **conduction**—meaning "leaping" or "jumping." These fibers cannot conduct a signal in continuous mode, like a burning fuse, because voltage-gated ion channels are too scarce in the myelin-covered internodal segments—fewer than $25/\mu m^2$ in these regions compared with 2,000 to $12,000/\mu m^2$ at the myelin sheath gaps (nodes of Ranvier). There would be little point in having ion channels in the internodal segments anyway—myelin insulates the fiber from the ECF in these segments, and Na^+ from the ECF couldn't flow into the cell even if more channels were present. Therefore, no action potentials can occur in the internodal segments, and the nerve signal requires some other way of traversing the distance from one node to the next.

[27]from *saltare* = to leap, to dance

1 Na⁺ inflow at node generates action potential.

2 Positive charge flows rapidly along axon and depolarizes membrane; signal grows weaker with distance.

3 Depolarization of membrane at next node opens Na⁺ channels, triggering new action potential.

(a)

Refractory membrane

Action potential in progress

Excitable membrane

(b)

FIGURE 12.18 Saltatory Conduction of a Nerve Signal in a Myelinated Fiber. (a) Sodium inflow at myelin sheath gaps, regenerating the flow of positive charges at each gap. (b) Action potentials can occur only at myelin sheath gaps, so the nerve signal appears as if it were jumping from gap to gap. The membrane (yellow) behind each action potential is temporarily in the refractive state, whereas the membrane (green) ahead of it is fully excitable and ready for action.

When Na⁺ enters the axon at a node, it diffuses for a short distance along the inner face of the membrane **(fig. 12.18a).** Each sodium ion has an electrical field around it. When one Na⁺ moves toward another, its field repels the other ion, which moves slightly and repels another, and so forth—like two magnets that repel each other if you try to push their north poles together. No one ion moves very far, but this energy transfer travels down the axon much faster and farther than any of the individual ions. The signal grows weaker with distance, however, partly because the axoplasm resists the movement of the ions and partly because Na⁺ leaks back out of the axon along the way. Therefore, with distance, there is a lower and

lower concentration of Na⁺ to relay the charge. Furthermore, with a surplus of positive charges on the inner face of the axolemma and a surplus of negative charges on the outer face, these cations and anions are attracted to each other through the membrane—like the opposite poles of two magnets attracting each other through a sheet of paper. This results in a "storage" (called *capacitance*) of unmoving or sluggishly moving charges on the membrane.

Myelin speeds up signal conduction in two ways. First of all, by wrapping tightly around the axon, it seals the nerve fiber and greatly increases its resistance to the leakage of Na⁺ out of the axon. Sodium ions therefore maintain a higher density on the inner face of the membrane and transfer energy from one to another more rapidly. Second, myelin creates a greater separation between the ICF and ECF. Cations and anions of the ICF and ECF are therefore less attracted to each other—like two magnets now separated by a thick sheet of plastic. Na^+ ions can therefore move more freely within the axon, transferring energy from one to another.

A signal propagated in the foregoing manner can't travel much farther than 1 mm before it becomes too weak to open any voltage-gated Na^+ channels. But fortunately, there is another node (gap) every millimeter or less along the axon, where the axolemma is exposed to ECF and there is an abundance of voltage-gated channels. When the internodal signal reaches this point, it is just strong enough to open these channels, admit more Na^+, and create a new action potential. This action potential has the same strength as the one at the previous node, so each node boosts the signal back to its original strength (+35 mV). However, the generation of action potentials is a relatively time-consuming process that slows down the nerve signal at the nodes.

Since action potentials occur only at the nodes, this mode of conduction creates a false impression that the nerve signal jumps from node to node (**fig. 12.18b**)—hence the expression *saltatory*.

You could think of saltatory conduction by analogy to a crowded subway car. The doors open (like the Na^+ gates at a node), 20 more people get on (like Na^+ flowing into the axon), and everyone has to push to the rear of the car to make room for them. No one passenger moves from the door to the rear, but the crowding and transfer of energy from person to person forces even those at the rear to move a little, like the sodium ions at the next node. Events at one node thus create excitation at the next node some distance away.

In summary, saltatory conduction is based on a process that is very fast in the internodal segments (transfer of energy from ion to ion), but decremental. In the nodes, conduction is slower but nondecremental. Since most of the axon is covered with myelin, conduction occurs mainly by the fast internodal process. This is why myelinated fibers conduct signals much faster (up to 120 m/s) than unmyelinated ones (up to 2 m/s).

▶▶▶APPLY WHAT YOU KNOW

You are about to have a dental procedure and the dentist "numbs you up" with an injection of lidocaine (Xylocaine). This is a local anesthetic that prevents voltage-gated Na^+ channels from opening. Explain why this mechanism would block the conduction of pain signals from your teeth to your brain.

BEFORE YOU GO ON

Answer the following questions to test your understanding of the preceding section:

11. What causes K^+ to diffuse out of a resting cell? What attracts it into the cell?

12. What happens to Na^+ when a neuron is stimulated on its dendrite? Why does the movement of Na^+ raise the voltage on the plasma membrane?

13. What does it mean to say a local potential is graded, decremental, and reversible?

14. How does the plasma membrane at the trigger zone differ from that on the soma? How does it resemble the membrane at a myelin sheath gap?

15. What makes an action potential rise to +35 mV? What makes it drop again after this peak?

16. List four ways in which an action potential is different from a local potential.

17. Explain why myelinated fibers conduct signals much faster than unmyelinated fibers.

12.5 Synapses

Expected Learning Outcomes
When you have completed this section, you should be able to

a. explain how messages are transmitted from one neuron to another;

b. give examples of neurotransmitters and neuromodulators and describe their actions; and

c. explain how stimulation of a postsynaptic cell is stopped.

All good things must come to an end; a nerve signal soon reaches the end of an axon and can go no farther. But in most cases, it triggers the release of a neurotransmitter that stimulates a new wave of electrical activity in the next cell across the synapse. The most thoroughly studied synapse is the neuromuscular junction described in chapter 11, but here we consider synapses between two neurons. Signals arrive at the synapse by way of the **presynaptic neuron,** which releases a neurotransmitter. The next neuron, which responds to it, is called the **postsynaptic neuron (fig. 12.19a).**

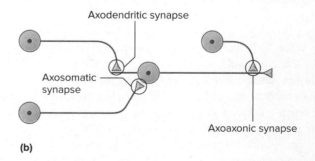

FIGURE 12.19 Synaptic Relationships Between Neurons.
(a) Pre- and postsynaptic neurons. (b) Types of synapses defined by the site of contact on the postsynaptic neuron.

Axon of presynaptic neuron

Axon terminal

Soma of postsynaptic neuron

FIGURE 12.20 Axon Terminals Synapsing with the Soma of a Neuron in a Marine Slug, *Aplysia* (SEM).

Omikron/Science Source/Getty Images

The presynaptic neuron may synapse with a dendrite, the soma, or the axon of a postsynaptic neuron, forming an *axodendritic, axosomatic,* or *axoaxonic synapse,* respectively **(fig. 12.19b)**. A neuron can have an enormous number of synapses **(fig. 12.20)**. For example, a spinal motor neuron is covered with about 10,000 axon terminals from other neurons—8,000 ending on its dendrites and another 2,000 on the soma. In a part of the brain called the cerebellum, one neuron can have as many as 100,000 synapses.

12.5a The Discovery of Neurotransmitters

In the early twentieth century, biologists assumed that synaptic communication was electrical—a logical hypothesis given that neurons seemed to touch each other and signals were transmitted so quickly from one to the next. Cajal's careful histological examinations, however, revealed a 20 to 40 nm gap between neurons—the **synaptic cleft**—casting doubt on the possibility of electrical transmission. Cajal was rudely criticized for such a "preposterous" idea, but he was eventually proved correct.

In 1921, German pharmacologist Otto Loewi conclusively demonstrated that neurons communicate by releasing chemicals. The *vagus nerves* supply the heart, among other organs, and slow it down. Loewi opened two frogs and flooded the hearts with saline to keep them moist. He stimulated the vagus nerve of one frog, and its heart rate dropped as expected. He then removed saline from that heart and squirted it onto the heart of the second frog. The solution alone reduced that frog's heart rate. Evidently it contained something released by the vagus nerve of the first frog. Loewi had discovered what we now call acetylcholine—the first known neurotransmitter.

▶▶▶APPLY WHAT YOU KNOW

As described, does the previous experiment conclusively prove that the second frog's heart slowed as a result of something released by the vagus nerves? If you were Loewi, what control experiment would you do to rule out alternative hypotheses?

Following Loewi's work, the idea of electrical communication between cells fell into disrepute. Now, however, we realize that some neurons, neuroglia, and cardiac and unitary smooth muscle do indeed have **electrical synapses,** where adjacent cells are joined by gap junctions and ions diffuse directly from one cell into the next. These junctions have the advantage of quick transmission because there is no delay for the release and binding of neurotransmitter. They are important in synchronizing the activity of local suites of neurons in certain regions of the brain. Their disadvantage, however, is that they cannot integrate information and make decisions. The ability to do that is a property of **chemical synapses,** in which neurons communicate by neurotransmitters. Chemical synapses are also the site of learning and memory, the target of many prescription drugs, and the site of action of drugs of addiction, among other things.

12.5b Structure of a Chemical Synapse

The axon terminal **(fig. 12.21)** contains synaptic vesicles, many of which are "docked" at release sites on the plasma membrane, ready to release neurotransmitter on demand. A reserve pool of synaptic vesicles is located a little farther away from the membrane, tethered to the cytoskeleton.

The postsynaptic neuron lacks these conspicuous specializations. At this end, the neuron has no synaptic vesicles and cannot release neurotransmitter. Its membrane does, however, have neurotransmitter receptors and ligand-gated ion channels.

12.5c Neurotransmitters and Related Messengers

More than 100 neurotransmitters have been identified since Loewi's time. Neurotransmitters can be defined as molecules that are synthesized by a neuron, released when a nerve signal reaches an axon terminal or varicosity of the nerve fiber, and have a specific effect on a receiving cell's physiology. Most of them are small organic molecules that are released by exocytosis and bind to specific receptors on the receiving cell, but there are exceptions. Some of the best-known neurotransmitters are listed in **table 12.3.** Parts of the brain referred to in this table will become familiar as you study chapter 14, and you may wish to refer back to this table then to enhance your understanding of brain function. Most neurotransmitters fall into the following categories **(fig. 12.22)**.

1. **Acetylcholine** is in a class by itself. It is formed from acetic acid (acetate) and choline.

2. **Amino acid** neurotransmitters include glycine, glutamate, aspartate, and γ-aminobutyric acid (GABA).

3. **Monoamines (biogenic amines)** are synthesized from amino acids by removal of the —COOH group. They retain the —NH$_2$ (amino group), hence their name. Some monoamine neurotransmitters are epinephrine, norepinephrine, dopamine, serotonin (5-hydroxytryptamine, or 5-HT), and histamine. The first three of these are in a subclass called **catecholamines** (CAT-eh-COAL-uh-meens).

4. **Purines** serving as neurotransmitters include adenosine and ATP (adenosine triphosphate).

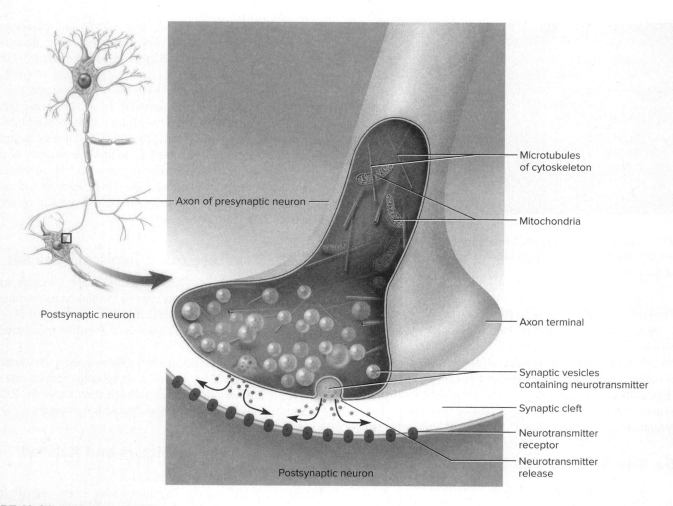

Microtubules of cytoskeleton

Axon of presynaptic neuron

Mitochondria

Postsynaptic neuron

Axon terminal

Synaptic vesicles containing neurotransmitter

Synaptic cleft

Neurotransmitter receptor

Neurotransmitter release

Postsynaptic neuron

FIGURE 12.21 Structure of a Chemical Synapse.

5. **Gases,** specifically nitric oxide (NO) and carbon monoxide (CO), are inorganic exceptions to the usual definition of neurotransmitters. They are synthesized as needed rather than stored in synaptic vesicles; they simply diffuse out of the axon terminal rather than being released by exocytosis; and they diffuse into the postsynaptic neuron rather than bind to a surface receptor.

6. **Neuropeptides** are chains of 2 to 40 amino acids. Some examples are cholecystokinin (CCK) and the endorphins. Neuropeptides are stored in *secretory granules (dense-core vesicles)* that are about 100 nm in diameter, twice as large as typical synaptic vesicles. Some neuropeptides also function as hormones or as *neuromodulators* (see section 12.5f). Some are produced not only by neurons but also by the digestive tract; thus, they are known as *gut–brain peptides.* Some of these cause cravings for specific nutrients such as fat, protein, or carbohydrates (see section 26.1b) and may be associated with certain eating disorders.

▶▶▶**APPLY WHAT YOU KNOW**

Unlike other neurotransmitters, neuropeptides can be synthesized only in the soma and must be transported to the axon terminals. Why is their synthesis limited to the soma?

We will see, especially in chapter 15, that a given neurotransmitter does not have the same effect everywhere in the body. There are multiple receptor types in the body for a particular neurotransmitter—over 14 receptor types for serotonin, for example—and it is the receptor that governs what effect a neurotransmitter has on its target cell. Most human and other mammalian neurons can secrete two or more neurotransmitters and can switch from one to another under different circumstances.

12.5d Synaptic Transmission

Some neurotransmitters are excitatory, some are inhibitory, and for some the effect depends on what kind of receptor the postsynaptic cell has. Some open ligand-gated ion channels and others act through second messengers. Bearing this diversity in mind, we will examine three kinds of synapses with different modes of action.

An Excitatory Cholinergic Synapse

A **cholinergic**[28] (CO-lin-UR-jic) synapse employs acetylcholine (ACh) as its neurotransmitter. ACh excites some postsynaptic cells (such as skeletal muscle) and inhibits others (such as

---

[28]*cholin* = acetylcholine; *erg* = work, action

TABLE 12.3	Neurotransmitters (Selected Examples)
Name	**Locations and Actions**
Acetylcholine (ACh)	Neuromuscular junctions, most synapses of autonomic nervous system, retina, and many parts of the brain; excites skeletal muscle, inhibits cardiac muscle, and has excitatory or inhibitory effects on smooth muscle and glands depending on location
Amino Acids	
Glutamate	Cerebral cortex and brainstem; accounts for about 75% of all excitatory synaptic transmission in the brain; involved in learning and memory
Aspartate	Spinal cord; effects similar to those of glutamate
Glycine	Inhibitory neurons of the brain, spinal cord, and retina; most common inhibitory neurotransmitter in the spinal cord
GABA	Thalamus, hypothalamus, cerebellum, occipital lobes of cerebrum, and retina; the most common inhibitory neurotransmitter in the brain
Monoamines	
Norepinephrine	Sympathetic nervous system, cerebral cortex, hypothalamus, brainstem, cerebellum, and spinal cord; involved in dreaming, waking, and mood; excites cardiac muscle; can excite or inhibit smooth muscle and glands depending on location
Epinephrine	Hypothalamus, thalamus, spinal cord, and adrenal medulla; effects similar to those of norepinephrine
Dopamine	Hypothalamus, limbic system, cerebral cortex, and retina; highly concentrated in substantia nigra of midbrain; involved in elevation of mood and control of skeletal muscles
Serotonin	Hypothalamus, limbic system, cerebellum, retina, and spinal cord; also secreted by blood platelets and intestinal cells; involved in sleepiness, alertness, thermoregulation, and mood
Histamine	Hypothalamus; also a potent vasodilator released by mast cells of connective tissue and basophils of the blood
Neuropeptides	
Substance P	Basal nuclei, midbrain, hypothalamus, cerebral cortex, small intestine, and pain-receptor neurons; mediates pain transmission
Enkephalins	Hypothalamus, limbic system, pituitary, pain pathways of spinal cord, and nerve endings of digestive tract; act as analgesics (pain relievers) by inhibiting substance P; inhibit intestinal motility; modulate immune responses
β-endorphin	Digestive tract, spinal cord, and many parts of the brain; also secreted as a hormone by the pituitary; suppresses pain; secretion rises sharply during labor and delivery and in response to other pain situations
Cholecystokinin	Cerebral cortex and small intestine; suppresses appetite

FIGURE 12.22 Classification of Some Neurotransmitters. The neuropeptides are chains of amino acids, each identified by its three-letter code. Appendix D explains the codes.

FIGURE 12.23 Transmission at a Cholinergic Synapse. Acetylcholine directly opens ion channels in the plasma membrane of the postsynaptic neuron. Numbered steps 1 through 5 correspond to the description in the adjacent text. Steps 6 through 8 pertain to cessation of the signal (section 12.5e). **A&PR**

cardiac muscle), but this discussion will describe an excitatory action. The steps in transmission at such a synapse are as follows **(fig. 12.23)**:

1. The arrival of a nerve signal at the axon terminal opens voltage-gated calcium channels.

2. Ca^{2+} enters the terminal and triggers exocytosis of the synaptic vesicles, releasing ACh.

3. Empty vesicles drop back into the cytoplasm to be refilled with ACh, while synaptic vesicles in the reserve pool move to the active sites and release their ACh—a bit like a line of Revolutionary War soldiers firing their muskets and falling back to reload as another line moves to the fore.

4. Meanwhile, ACh diffuses across the synaptic cleft and binds to ligand-gated channels on the postsynaptic neuron. These channels open, allowing Na^+ to enter the cell and K^+ to leave. Although illustrated separately, Na^+ and K^+ pass in opposite directions through the same gates.

5. As Na^+ enters, it spreads out along the inside of the plasma membrane and depolarizes it, producing a local voltage shift called the **postsynaptic potential.** Like other local potentials, if this is strong and persistent enough (that is, if enough current makes it to the axon hillock), it opens voltage-gated ion channels in the trigger zone and causes the postsynaptic neuron to fire.

Steps 6 through 8 in figure 12.23 concern the mechanism for halting transmission and are explained shortly (see section 12.5e).

An Inhibitory GABA-ergic Synapse

Synapses that employ γ-aminobutyric acid (GABA) as their neurotransmitter are called **GABA-ergic synapses.** Amino acid neurotransmitters such as GABA work by the same mechanism as ACh; they bind to ion channels and cause immediate changes in membrane potential. The release of GABA and binding to its receptor are similar to the preceding case. The GABA receptor, however, is a chloride channel. When it opens, Cl^- enters the cell and makes the inside even more negative than the resting membrane potential. The neuron is therefore inhibited, or less likely to fire.

An Excitatory Adrenergic Synapse

An **adrenergic synapse** employs the neurotransmitter norepinephrine (NE), also called noradrenaline. NE, other monoamines, and neuropeptides act through second-messenger systems such as cyclic AMP (cAMP). The receptor is not an ion channel but a transmembrane protein associated with a G protein on the inner face of the membrane. **Figure 12.24** shows some ways in which an adrenergic synapse can function, numbered to correspond to the following:

1. The unstimulated NE receptor is bound to a G protein.

2. Binding of NE to the receptor causes the G protein to dissociate from it.

3. The G protein binds to adenylate cyclase and activates this enzyme, which converts ATP to cAMP.

4. Cyclic AMP can induce several alternative effects in the cell.

5. One effect is to produce an internal chemical that binds to a ligand-gated ion channel from the inside, opening the channel and depolarizing the cell.

6. Another is to activate preexisting cytoplasmic enzymes, which can lead to diverse metabolic changes (for example, inducing a liver cell to break down glycogen and release glucose into the blood).

7. Yet another is for cAMP to induce genetic transcription, so that the cell produces new enzymes leading to diverse metabolic effects.

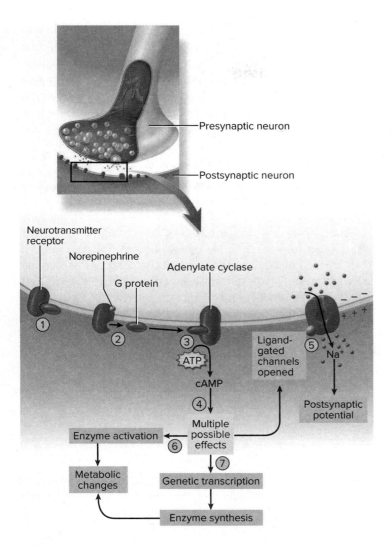

Presynaptic neuron

Postsynaptic neuron

Neurotransmitter receptor

Norepinephrine

G protein

Adenylate cyclase

① ② ③ ATP cAMP ④ Multiple possible effects ⑥ ⑦

Ligand-gated channels opened ⑤ Na⁺

Postsynaptic potential

Enzyme activation

Metabolic changes

Genetic transcription

Enzyme synthesis

FIGURE 12.24 Transmission at an Adrenergic Synapse. Numbered steps correspond to the description in nearby text.

Although slower to respond than cholinergic and GABA-ergic synapses, adrenergic synapses do have an advantage—**signal amplification.** A single NE molecule binding to a receptor can induce the formation of many cAMPs, each of those can activate many enzyme molecules or induce the transcription of a gene to generate numerous mRNA molecules, and each of those can result in the production of a vast number of enzyme molecules and metabolic products such as glucose molecules.

As complex as synaptic events may seem, they typically require only 0.5 ms or so—an interval called **synaptic delay.** This is the time from the arrival of a signal at the axon terminal of a presynaptic cell to the beginning of an action potential in the postsynaptic cell.

12.5e Cessation of the Signal

It is important not only to stimulate a postsynaptic cell but also to turn off the stimulus in due time. Otherwise the postsynaptic neuron could continue firing indefinitely, causing a breakdown in physiological coordination. But a neurotransmitter molecule binds to its receptor for only 1 ms or so, then dissociates from it. If the presynaptic

cell continues to release neurotransmitter, one molecule is quickly replaced by another and the postsynaptic cell is restimulated. This immediately suggests a way of stopping synaptic transmission—stop adding new neurotransmitter and get rid of that which is already there. The first step is achieved simply by the cessation of signals in the presynaptic nerve fiber. The second can be achieved in the following ways, pictured as steps 6 to 8 in figure 12.23:

⑥ **Neurotransmitter degradation.** An enzyme in the synaptic cleft breaks the neurotransmitter down into fragments that have no stimulatory effect on the postsynaptic cell. Depicted here is acetylcholinesterase (AChE) breaking ACh down into choline and acetate.

⑦ **Reuptake.** A neurotransmitter or its breakdown products are reabsorbed by transport proteins in the axon terminal, removing them from the synapse and ending their stimulatory effect. Choline from ACh is recycled to make new ACh. Amino acid and monoamine neurotransmitters are similarly reabsorbed, then broken down within the axon terminal by an enzyme called **monoamine oxidase (MAO).** Some antidepressant drugs work by inhibiting MAO (see Deeper Insight 15.2). Some cases of autism, attention-deficit/hyperactivity disorder (ADHD), parkinsonism, and depression stem from mutations that render some of these transport proteins nonfunctional.

⑧ **Diffusion.** Neurotransmitters or their breakdown products simply diffuse away from the synapse into the nearby extracellular fluid. As shown in figure 12.23, this is what happens to the acetate component of ACh, but in other cases, the neurotransmitter escapes the synapse intact. In the CNS, astrocytes absorb stray neurotransmitters and return them to the presynaptic neurons. In ganglia of the PNS, it seems that the satellite cells that surround neurons perform this role; this is an area of current scientific investigation.

12.5f Neuromodulators

Neurons sometimes secrete chemical signals that have long-term effects on entire groups of neurons instead of brief, quick effects at an individual synapse. Some call these **neuromodulators** to distinguish them from neurotransmitters; others use the term *neurotransmitter* broadly to include these. Neuromodulators adjust, or *modulate,* the activity of neuron groups in various ways: increasing the release of neurotransmitters by presynaptic neurons; adjusting the sensitivity of postsynaptic neurons to neurotransmitters; or altering the rate of neurotransmitter reuptake or breakdown to prolong their effects.

The simplest neuromodulator is the gas **nitric oxide (NO).** NO diffuses readily into a postsynaptic cell and activates second-messenger pathways with such effects as relaxing smooth muscle. This has the effect of dilating small arteries and increasing blood flow to a tissue; this is the basis for the action of drugs for erectile dysfunction (see Deeper Insight 27.4). The neuropeptides are neuromodulators; among these are the **enkephalins** and **endorphins,** which inhibit spinal neurons from transmitting pain signals to the brain (see section 16.2d). Other neuromodulators include hormones and some neurotransmitters such as dopamine, serotonin, and histamine. The last point may seem confusing, but the terms *neurotransmitter, hormone,* and *neuromodulator* define not so much the chemical itself, but the role it plays in a given context. One chemical can play two or more of these roles in different places and circumstances.

Answer the following questions to test your understanding of the preceding section:

18. Concisely describe five steps that occur between the arrival of an action potential at the axon terminal and the beginning of a new action potential in the postsynaptic neuron.

19. Contrast the actions of acetylcholine, GABA, and norepinephrine at their respective synapses.

20. Describe three mechanisms that stop synaptic transmission.

21. What is the function of neuromodulators? Compare and contrast neuromodulators and neurotransmitters.

12.6 Neural Integration

Expected Learning Outcomes

When you have completed this section, you should be able to

a. explain how a neuron "decides" whether or not to generate action potentials;

b. explain how the nervous system translates complex information into a simple code;

c. explain how neurons work together in groups to process information and produce effective output; and

d. describe how memory works at the cellular and molecular levels.

Synaptic delay slows the transmission of nerve signals; the more synapses there are in a neural pathway, the longer it takes information to get from its origin to its destination. You may wonder, therefore, why we have synapses—why a nervous pathway is not, indeed, a continuous "wire" as biologists believed before Cajal. The presence of synapses is not due to limitations on axon length—after all, one nerve fiber can reach from your toes to your brainstem. (In the neck of the giraffe, the *recurrent laryngeal nerve* has nerve fibers 4.6 m, or 15 ft, long. Nerve fibers from the hind foot to the brainstem are even longer by far.) We also have seen that cells communicate much more quickly through gap junctions than through chemical synapses. So why have chemical synapses at all?

What we value most about our nervous system is its ability to process information, store it, and make decisions—and chemical synapses are the decision-making devices of the system. The more synapses a neuron has, the greater its information-processing capability. At this moment, you are using certain *pyramidal cells* of the cerebral cortex (see fig. 12.5a) to read and comprehend this passage. Each pyramidal cell has about 40,000 synaptic contacts with other neurons. The cerebral cortex alone (the main

information-processing tissue of your brain) is estimated to have 100 trillion (10^{14}) synapses. To get some impression of this number, imagine trying to count them. Even if you could count two synapses per second, day and night without stopping, and you were immortal, it would take you 1.6 million years. The ability of your neurons to process information, store and recall it, and make decisions is called **neural integration.**

12.6a Postsynaptic Potentials

Neural integration is based on the postsynaptic potentials produced by neurotransmitters. Remember that a typical neuron has a resting membrane potential (RMP) of about –70 mV and a threshold of about –55 mV. A neuron has to be depolarized to this threshold in order to produce action potentials. Any voltage change in that direction makes a neuron more likely to fire and is therefore called an **excitatory postsynaptic potential (EPSP) (fig. 12.25a).** EPSPs usually result from Na$^+$ flowing into the cell and neutralizing some of the negative charge on the inside of the membrane.

(a)

(b)

FIGURE 12.25 Postsynaptic Potentials. (a) An excitatory postsynaptic potential (EPSP), which shifts the membrane voltage closer to threshold and makes the cell more likely to fire. (b) An inhibitory postsynaptic potential (IPSP), which shifts the membrane voltage farther away from threshold and makes the cell less likely to fire. The sizes of these postsynaptic potentials are greatly exaggerated here for clarity; compare figure 12.27.

? *Why is a single EPSP insufficient to make a neuron fire?*

In other cases, a neurotransmitter hyperpolarizes the postsynaptic cell and makes it more negative than the RMP. Since this makes the postsynaptic cell less likely to fire, it is called an **inhibitory postsynaptic potential (IPSP) (fig. 12.25b).** Some IPSPs are produced by a neurotransmitter opening ligand-gated chloride channels, causing Cl^- to flow into the cell and make the cytosol more negative. A less common way is to open selective K^+ channels, increasing K^+ diffusion out of the cell.

We must recognize that because of ion leakage through their membranes, all neurons fire at a certain background rate even when they are not being stimulated. EPSPs and IPSPs don't determine whether or not a neuron fires, but only change the rate of firing by stimulating or inhibiting the production of more action potentials.

Glutamate and aspartate are excitatory brain neurotransmitters that produce EPSPs. Glycine and GABA produce IPSPs and are therefore inhibitory. Acetylcholine (ACh) and norepinephrine are excitatory to some cells and inhibitory to others, depending on the type of receptors present. For example, ACh excites skeletal muscle but inhibits cardiac muscle because of different types of ACh receptors.

12.6b Summation, Facilitation, and Inhibition

One neuron may receive input from thousands of other neurons. Some incoming nerve fibers may produce EPSPs while others produce IPSPs. The neuron's response depends on whether the *net* input is excitatory or inhibitory. If EPSPs override the IPSPs, threshold may be reached and set off an action potential; if IPSPs prevail, they inhibit the neuron from firing. **Summation** is the process of adding up postsynaptic potentials and responding to their net effect. It occurs in the trigger zone.

Suppose, for example, you're working in the kitchen and accidentally touch a hot pot. EPSPs in your motor neurons might cause you to jerk your hand back quickly and avoid being burned. Yet a moment later, you might nonchalantly sip a cup of hot tea. Since you are expecting it to be hot, you don't jerk it away from your lips. You have learned that it won't injure you, so at some level of the nervous system, IPSPs prevail and inhibit the motor response.

It is fundamentally a balance between EPSPs and IPSPs that enables the nervous system to make decisions. A postsynaptic neuron is like a little cellular democracy acting on the "majority vote" of hundreds or thousands of presynaptic cells. In the tea example, some presynaptic neurons send messages that signify "Hot! Danger!" in the form of EPSPs that may activate a hand-withdrawal reflex, while at the same time, others produce IPSPs that signify "Safe" and suppress the reflex. Whether you jerk your hand away depends on whether the EPSPs override the IPSPs or vice versa.

Weak stimulation of a postsynaptic neuron may generate an EPSP, but it fades before reaching threshold. A typical EPSP is a voltage change of only 0.5 mV and lasts only 15 to 20 ms. If a neuron has an RMP of –70 mV and a threshold of –55 mV, it needs at least 30 EPSPs to reach threshold and fire. There are two ways in which EPSPs can add up to do this, and both may occur simultaneously.

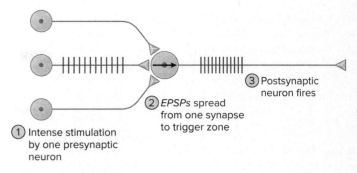

(a) **Temporal summation**

1. Intense stimulation by one presynaptic neuron
2. *EPSPs* spread from one synapse to trigger zone
3. Postsynaptic neuron fires

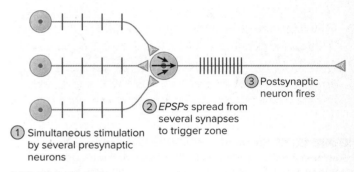

(b) **Spatial summation**

1. Simultaneous stimulation by several presynaptic neurons
2. *EPSPs* spread from several synapses to trigger zone
3. Postsynaptic neuron fires

FIGURE 12.26 Temporal and Spatial Summation. Vertical lines on the nerve fibers indicate relative firing frequency. Arrows within the postsynaptic neuron indicate the path of local graded potentials. (a) Temporal summation resulting from rapid firing of a single presynaptic neuron. (b) Spatial summation as an additive effect of multiple presynaptic neurons firing at moderate rates.

1. **Temporal summation (fig. 12.26a).** This occurs when a single synapse generates EPSPs so quickly that each is generated before the previous one fades. This allows the EPSPs to add up over time to a threshold voltage that triggers an action potential **(fig. 12.27).** Temporal summation can occur if even one presynaptic neuron stimulates the postsynaptic neuron at a fast enough rate.

2. **Spatial summation (fig. 12.26b).** This occurs when EPSPs from several synapses add up to threshold at the axon hillock. Any one synapse may generate only a weak signal, but several synapses acting together can bring the hillock to threshold. The presynaptic neurons collaborate to induce the postsynaptic neuron to fire.

Neurons routinely work in groups to modify each other's actions. **Presynaptic facilitation** is a process in which one neuron enhances the effect of another. **Figure 12.28** depicts such an interaction between three neurons we will call neuron F for a facilitating neuron, S for a stimulating neuron, and R for a responding neuron. Neuron F forms an axoaxonic synapse with the axon terminal of neuron S. When neuron F is quiescent, neuron S shows only a low level of activity, releasing some neurotransmitter but not necessarily enough to excite neuron R to threshold. But if circumstances call for enhancing information transmission across synapse S-R, neuron F can release a neurotransmitter

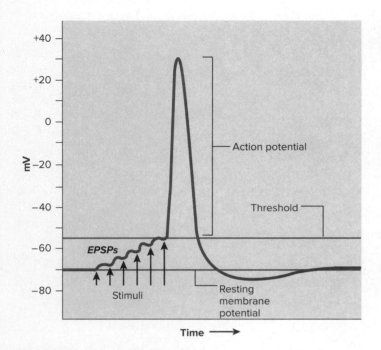

FIGURE 12.27 Summation of EPSPs. Each stimulus (arrow) produces one EPSP. If enough EPSPs arrive at the trigger zone faster than they fade, they can build on each other to bring the neuron to threshold and trigger an action potential.

such as serotonin. Serotonin makes voltage-gated calcium channels in the axon terminal of S remain open longer. Increased calcium inflow increases neurotransmitter release by S, which now excites neuron R. The facilitating neuron F thus accentuates synaptic transmission.

The spatial summation seen in figure 12.26b is another way of facilitating the firing of a postsynaptic neuron. One neuron acting alone may be unable to induce a postsynaptic neuron to fire, but when they collaborate, their combined "effort" does induce firing in the postsynaptic cell.

Presynaptic inhibition is the opposite of facilitation, a mechanism in which one presynaptic neuron suppresses another one. This mechanism is used to reduce or halt unwanted synaptic transmission. In **figure 12.29,** we see another three neurons in which the facilitating neuron is replaced by an inhibitory neuron, I. When neuron I is silent, neuron S releases its neurotransmitter and triggers a response in R. But when there is a need to block transmission across this pathway, neuron I releases the inhibitory neurotransmitter GABA. GABA prevents the voltage-gated calcium channels of neuron S from opening. Consequently, neuron S releases less neurotransmitter or none, and fails to stimulate neuron R.

12.6c Neural Coding

The nervous system must interpret and pass along both quantitative and qualitative information about its environment—whether a light is dim or bright, red or green; whether a taste is mild or intense, salty or sour; whether a sound is loud or soft, high-pitched or low. Considering the complexity of information to be communicated about conditions in and around the body, it seems a marvel that it can be done in the form of something as simple as action potentials—particularly since all the action potentials of a given neuron are identical. Yet when we considered the genetic code in chapter 4, we saw that complex messages can indeed be expressed in simple codes. The way in which the nervous system converts information to a meaningful pattern of action potentials is called **neural coding** (or *sensory coding* when it occurs in the sense organs).

The most important mechanism for transmitting qualitative information is the **labeled line code.** This code is based on the

(a)

(b)

FIGURE 12.28 Presynaptic Facilitation. F, facilitating neuron; S, stimulating neuron; R, responding postsynaptic neuron; +, excitation (EPSP). (a) Inactivity of neuron R owing to absence of stimulation by neuron F. (b) Activity in neuron R owing to stimulation (facilitation) of the S-R synapse by neuron F.

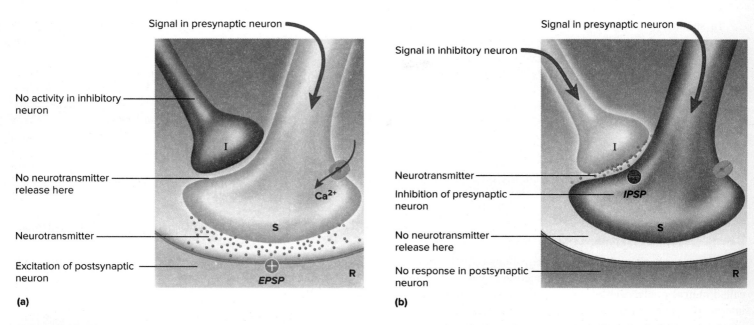

Signal in presynaptic neuron

No activity in inhibitory neuron

No neurotransmitter release here

Neurotransmitter

Excitation of postsynaptic neuron

I

S

Ca²⁺

EPSP

R

(a)

Signal in presynaptic neuron

Signal in inhibitory neuron

Neurotransmitter

Inhibition of presynaptic neuron

No neurotransmitter release here

No response in postsynaptic neuron

I

S

IPSP

R

(b)

FIGURE 12.29 Presynaptic Inhibition. S, stimulating neuron; I, inhibitory neuron; R, responding postsynaptic neuron; +, excitation (EPSP); −, inhibition (IPSP). (a) Activity in the S-R synapse in the absence of inhibition by neuron I. (b) Activity of neuron I inhibiting synaptic transmission between S and R.

fact that each nerve fiber to the brain leads from a receptor that specifically recognizes a particular stimulus type. Nerve fibers in the optic nerve, for example, carry signals only from light receptors in the eye; these fibers never carry information about taste or sound. The brain therefore interprets any signals in those fibers in terms of light—even if the signals result from artificial stimulation of the nerve. This effect is seen when you rub your eyelids and see flashes of light as the pressure on the eyeball mechanically stimulates optic nerve fibers. Electrical stimulation of the auditory nerve can enable deaf people to hear sounds of different frequencies, even when receptors in the inner ear are nonfunctional. Thus, each nerve fiber to the brain is a line of communication "labeled," or recognized by the brain, as representing a particular stimulus quality—the color of a light, the pitch of a sound, or the salty or sour quality of a taste, for example.

Quantitative information—information about the intensity of a stimulus—is encoded in two ways. One depends on the fact that different neurons have different thresholds of excitation. A weak stimulus excites sensitive neurons with the lowest thresholds, while a strong stimulus excites less sensitive high-threshold neurons. Bringing additional neurons into play as the stimulus becomes stronger is called **recruitment.** It enables the nervous system to judge stimulus strength by which neurons, and how many of them, are firing.

Another way of encoding stimulus strength depends on the fact that the more strongly a neuron is stimulated, the more frequently it fires. A weak stimulus may cause a neuron to generate 6 action potentials per second, and a strong stimulus, 600 per second. Thus, the central nervous system can judge stimulus strength from the firing frequency of afferent neurons **(fig. 12.30).**

There is a limit to how often a neuron can fire, set by its absolute refractory period. Think of an electronic camera flash by analogy. If you take a photograph and your flash unit takes 10 seconds to recharge, then you can't take more than six photos per minute.

Similarly, if a nerve fiber takes 1 ms to repolarize after it has fired, then it can't fire more than 1,000 times per second. Refractory periods may be as short as 0.5 ms, which sets a theoretical limit to firing frequency of 2,000 action potentials per second. The highest frequencies actually observed, however, are 500 to 1,000 per second.

Action potentials

2 g

5 g

10 g

20 g

Time

FIGURE 12.30 An Example of Neural Coding. This figure is based on recordings made from a sensory fiber of the frog sciatic nerve as the gastrocnemius muscle was stretched by suspending weights from it. As the stimulus strength (weight) and stretch increase, the firing frequency of the neuron increases. Firing frequency is a coded message that informs the CNS of stimulus intensity.

❓ *In what other way is the CNS informed of stimulus intensity?*

In summary, mild stimuli excite sensitive, low-threshold nerve fibers. As the stimulus intensity rises, these fibers fire at a higher and higher frequency, up to a certain maximum. If the stimulus intensity exceeds the capacity of these low-threshold fibers, it may recruit less sensitive, high-threshold fibers to begin firing. Still further increases in intensity cause these high-threshold fibers to fire at a higher and higher frequency.

▶▶▶**APPLY WHAT YOU KNOW**

How is neural recruitment related to the process of multiple motor unit summation (see section 11.5b)?

12.6d Neural Pools and Circuits

So far, we have dealt with interactions involving only two or three neurons at a time. Actually, neurons function in larger ensembles called **neural pools,** each of which may consist of thousands of interneurons concerned with a particular body function—one to control the rhythm of your breathing, one to move your limbs rhythmically as you walk, one to regulate your sense of hunger, and another to interpret smells, for example. At this point, we explore a few ways in which neural pools collectively process information.

Discharge and Facilitated Zones

Information arrives at a neural pool through one or more input neurons, which branch repeatedly and synapse with numerous interneurons in the pool. Some input neurons form multiple synapses with a single postsynaptic cell. They can produce EPSPs at all points of contact with that cell and, through spatial summation, make it fire more easily than if they synapsed with it at only one point. Within the **discharge zone** of an input neuron, that neuron acting alone can make the postsynaptic cells fire **(fig. 12.31).** But in a broader **facilitated zone,** it synapses with still other neurons in the pool, with fewer synapses on each of them. It can stimulate those neurons to fire only with the assistance of other input neurons; that is, it facilitates the others. It "has a vote" on what the postsynaptic cells in the facilitated zone will do, but it cannot determine the outcome by itself. Such arrangements, repeated thousands of times throughout the central nervous system, give neural pools great flexibility in integrating input from several sources and "deciding" on an appropriate output.

Types of Neural Circuits

The functioning of a radio can be understood from a circuit diagram showing its components and their connections. Similarly, the functions of a neural pool are partly determined by its **neural circuit**—the pathways among its neurons. Just as a wide variety of electronic devices are constructed from a relatively

FIGURE 12.31 Facilitated and Discharge Zones in a Neural Pool.

limited number of circuit types, a wide variety of neural functions result from the operation of four principal kinds of neural circuits **(fig. 12.32):**

1. In a **diverging circuit,** an individual neuron sends signals to multiple downstream neurons, or one neural pool may send output to multiple downstream neural pools. Each of those neurons or neural pools may communicate with several more, so input from just one pathway may produce output through hundreds of others. Such a circuit allows signals from one motor neuron of the brain, for example, to ultimately stimulate thousands of muscle fibers.

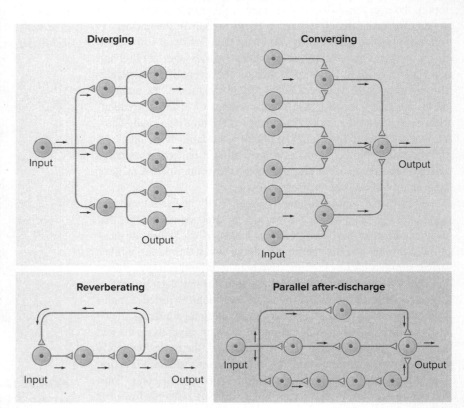

FIGURE 12.32 Four Types of Neural Circuits. Arrows indicate the direction of the nerve signal.

❓ *Which of these four circuits is likely to fire the longest after a stimulus ceases? Why?*

2. A **converging circuit** is the opposite of a diverging circuit—input from many nerve fibers or neural pools is funneled to fewer and fewer intermediate or output pathways. For example, you have a brainstem respiratory center that receives converging information from other parts of your brain, blood chemistry sensors in your arteries, and stretch receptors in your lungs. The respiratory center can then produce an output that takes all of these factors into account and sets an appropriate pattern of breathing.

3. In a **reverberating circuit,** neurons stimulate each other in a linear sequence from input to output neurons, but some of the neurons late in the path send axon collaterals back to neurons earlier in the path and restimulate them. As an exceedingly simplified model of such a circuit, consider a path such as A $\longrightarrow$ B $\longrightarrow$ C $\longrightarrow$ D, in which neuron C sends an axon collateral back to A. As a result, every time C fires it not only stimulates output neuron D, but also restimulates A and starts the process over. Such a circuit produces a prolonged or repetitive effect that lasts until one or more neurons in the circuit fail to fire, or an inhibitory signal from another source stops one of them from firing. A reverberating circuit sends repetitious signals to your diaphragm and intercostal muscles, for example, to make you inhale. Sustained output from the circuit ensures that the respiratory muscles contract for the 2 seconds or so that it normally takes to fill the lungs. When the circuit stops firing, you exhale; the next time it fires, you inhale again. Reverberating circuits may also be involved in short-term memory, as discussed in the next section, and they may play a role in the uncontrolled "storms" of neural activity that occur in epilepsy.

4. In a **parallel after-discharge** circuit, an input neuron diverges to stimulate several chains of neurons. Each chain has a different number of synapses, but eventually they all reconverge on one or a few output neurons. Since the chains differ in total synaptic delay, their signals arrive at the output neurons at different times, and the output neurons may go on firing for some time after input has ceased. Unlike a reverberating circuit, this type has no feedback loop. Once all the neurons in the circuit have fired, the output ceases. Continued firing after the stimulus stops is called *after-discharge*. It explains why you can stare at a lamp, then close your eyes and continue to see an image of it for a while. Such a circuit is also important in withdrawal reflexes, in which a brief pain produces a longer-lasting output to the limb muscles and causes you to draw back your hand or foot from danger.

Serial and Parallel Processing

In relation to these circuit types, the nervous system handles information in two modes called serial and parallel processing. In **serial processing,** neurons and neural pools relay information along a pathway in a relatively simple linear fashion and can process only one flow of information at a time. For example, you can read your e-mail or listen to a class lecture, and the language recognition centers of your brain can process one linguistic input or the other; however, you can't do both simultaneously. To understand the e-mail, you have to stop paying attention to the lecture; to understand the lecture, you have to stop checking your e-mail. You may jump back and forth between one and the other, only half understanding each, but you can't simultaneously process the written message in an e-mail and the spoken language of the lecture. Each must be processed separately and serially.

In **parallel processing,** information is transmitted along diverging circuits through different pathways that act on it simultaneously, to different purposes. For example, when you're driving your car, your visual system (eye and brain) must simultaneously process information about color, shape, depth of field, and motion in the scene before your eyes. This requires complex parallel processing circuits from the retinas of your eyes through the visual centers at the rear of your brain. At the same time, you must process traffic sounds and signals from your body's own joint and motion sensors to know, for example, how hard you are pressing the gas or brake pedal. Efficient simultaneous processing of such information is crucial to your own ability to drive safely.

Serial and parallel processing certainly occur as well in innumerable unconscious processes much simpler than the foregoing examples. In chapter 13, you should be able to see how they apply to certain spinal reflexes.

12.6e Memory and Synaptic Plasticity

You may have wondered as you studied this chapter, How am I going to remember all of this? It seems fitting that we end this chapter with the subject of how memory works, for you now have the information necessary to understand its cellular and chemical basis.

The things we learn and remember aren't stored in individual "memory cells" in the brain. You don't have a neuron assigned to remember your phone number and another to remember your grandmother's face, for example. Instead, the physical basis of memory is a *pathway* through the brain called a **memory trace (engram[29]),** in which new synapses have formed or existing synapses have been modified to make transmission easier. In other words, synapses aren't fixed for life; in response to experience, they can be added, taken away, or modified to make transmission easier or harder. Indeed, synapses can be created or deleted in as little as 1 or 2 hours. The ability of synapses to change is called **synaptic plasticity.**

Think about when you learned as a child to tie your shoes. The procedure was very slow, confusing, and laborious at first, but eventually it became so easy you could do it with little thought—like a motor program playing out in your brain without requiring your conscious attention. It became easier to do because the synapses in a certain pathway were modified to allow signals to travel more easily across them than across "untrained" synapses thus creating your *motor memory* for the task. The process of making transmission easier is called **synaptic potentiation** (one form of synaptic plasticity).

Neuroscientists still argue about how to classify the various forms of memory, but three kinds often recognized are *immediate*

---

[29]*en* = inner; *gram* = mark, trace, record

memory, short-term memory, and *long-term memory.* We also know of different modes of synaptic potentiation that last from just a few seconds to a lifetime, and we can correlate these at least tentatively with different forms of memory.

Immediate Memory

Immediate memory is the ability to hold something in mind for just a few seconds. By remembering what just happened, we get a feeling for the flow of events and a sense of the present. Immediate memory is indispensable to the ability to read; you must remember the earliest words of a sentence until you get to its end in order to extract any meaning from it. You couldn't make any sense of what you read if you forgot each word as soon as you moved on to the next one. Immediate memory may be based on reverberating circuits. Our impression of what just happened can thus echo in our minds for a few seconds as we experience the present moment and anticipate the next one.

Short-Term Memory

Short-term memory (STM) lasts from a few seconds to a few hours. Information stored in STM may be quickly forgotten if you stop mentally reciting it, you are distracted, or you have to remember something new. **Working memory** is a form of STM that allows you to hold an idea in mind long enough to carry out an action such as calling a telephone number you just looked up, working out the steps of a mathematics problem, or searching for a lost set of keys while remembering where you've already looked. It is limited to a few bits of information such as the digits of a telephone number. Evidence suggests that working memory resides in a circuit of facilitated synapses that remain quiescent (consuming little energy) most of the time, but are reactivated by new stimulation.

Making it easier to transmit signals across a synapse is called **synaptic facilitation** (different from the facilitation of one neuron by another that we studied earlier in the chapter). Synaptic facilitation can be produced by *tetanic stimulation,* the rapid arrival of repetitive signals at a synapse. Each signal causes a certain amount of Ca^{2+} to enter the axon terminal. If signals arrive rapidly, the neuron cannot pump out all the Ca^{2+} admitted by one action potential before the next action potential occurs. More and more Ca^{2+} accumulates in the terminal. Since Ca^{2+} is what triggers the release of neurotransmitter, each new signal releases more neurotransmitter than the one before. With more neurotransmitter, the EPSPs in the postsynaptic cell become stronger and stronger, and that cell is more likely to fire.

Memories lasting for a few hours, such as remembering what someone said to you earlier in the day or remembering an upcoming appointment, may involve **posttetanic potentiation.** In this process, the Ca^{2+} level in the axon terminal stays elevated for so long that another signal, coming well after the tetanic stimulation has ceased, releases an exceptionally large burst of neurotransmitter. That is, if a synapse has been heavily used in the recent past, a new stimulus can excite the postsynaptic cell more easily. Thus, your memory may need only a slight jog to recall something from several hours earlier.

Long-Term Memory

Long-term memory (LTM) lasts up to a lifetime and is less limited than STM in the amount of information it can store. LTM allows you to memorize the lines of a play, the words of a favorite song, or (one hopes!) textbook information for an exam. On a still longer timescale, it enables you to remember your name, the route to your home, and your childhood experiences.

There are two forms of long-term memory: explicit and implicit. **Explicit** or **declarative memory** is the retention of events and facts that you can put into words—numbers, names, dates, and so forth. You must think to remember these things. **Implicit memory** is the memory of things that come reflexively or unconsciously, including *emotional memories* (such as the fear of being stung if a wasp lands on you) and *procedural memory,* the retention of motor skills—how to tie your shoes, play a musical instrument, or type on a keyboard. These forms of memory involve different regions of the brain but are probably similar at the cellular level.

Some LTM involves the physical remodeling of synapses or the formation of new ones through the growth and branching of axon terminals and dendrites. In the pyramidal cells of the brain, the dendrites are studded with knoblike *dendritic spines* that increase the area of synaptic contact (see fig. 12.5a). Studies on fish and other experimental animals have shown that social and sensory deprivation causes these spines to decline in number, while a richly stimulatory environment causes them to proliferate—an intriguing clue to the importance of a stimulating environment to infant and child development. In some cases of LTM, a new synapse grows beside the original one, giving the presynaptic cell twice as much input into the postsynaptic cell.

LTM can also be grounded in molecular changes called **long-term potentiation (LTP).** This involves *NMDA*[30] *receptors,* which are glutamate-binding receptors found on the dendritic spines of pyramidal cells. NMDA receptors are usually blocked by magnesium ions (Mg^{2+}), but when they bind glutamate *and* are simultaneously subjected to high-frequency stimulation, they expel the Mg^{2+} and open to admit Ca^{2+} into the dendrite. When Ca^{2+} enters, it acts as a second messenger with multiple effects. A high Ca^{2+} level activates enzymes called *protein kinases,* which phosphorylate (add phosphate to) proteins employed in building and strengthening synapses. The neuron also produces even more NMDA receptors, making it more sensitive to glutamate, and it may send signals such as nitric oxide (NO) back to the presynaptic cell to enhance its release of neurotransmitter.

You can see that in all of these ways, long-term potentiation can increase transmission across "experienced" synapses. Remodeling a synapse or installing more neurotransmitter receptors has longer-lasting effects than facilitation or posttetanic potentiation.

---

[30]*N-methyl-D-aspartate,* a chemical similar to glutamate

How We Forget

Forgetting is arguably as important as remembering; we would likely be driven mad if we couldn't forget the myriad trivial things we encounter every day. People with a pathological inability to forget trivial information have great difficulty in reading comprehension and other functions that require us to distinguish what is important from what is not.

Immediate and short-term memories vanish simply as neural circuits cease to fire. Long-term memories can be erased by a process of **long-term depression (LTD).** Low-frequency stimulation of a synapse results in low levels of intracellular Ca^{2+}. This activates *protein phosphatases,* which dephosphorylate synaptic proteins such as actin microfilaments that support dendritic spines. These proteins are then degraded by *proteasomes* (see section 3.4b, fig. 3.31), which tear down dendritic spines and remove little-used synapses from the neural circuits.

The anatomical sites of memory in the brain are discussed in section 14.5d. Regardless of the sites, however, the cellular mechanisms are as described here.

DEEPER INSIGHT 12.4

CLINICAL APPLICATION

Alzheimer and Parkinson Diseases

Alzheimer and Parkinson diseases are the two most common degenerative disorders of the brain. Both are associated with neurotransmitter deficiencies.

Alzheimer[31] *disease (AD)* may begin before the age of 50 with signs so slight and ambiguous that early diagnosis is difficult. One of its first signs is memory loss, especially for recent events. A person with AD may ask the same questions repeatedly, show a reduced attention span, and become disoriented and lost in previously familiar places. Family members often feel helpless and confused as they watch their loved one's personality gradually deteriorate beyond recognition. The AD patient may become moody, confused, paranoid, combative, or hallucinatory. The patient may eventually lose even the ability to read, write, talk, walk, and eat. Death typically ensues from pneumonia or other complications of confinement and immobility.

AD affects about 11% of the U.S. population over the age of 65; the incidence rises to 47% by age 85. It accounts for nearly half of all nursing home admissions and is a leading cause of death among the elderly. AD claims about 100,000 lives per year in the United States.

Diagnosis of AD can be confirmed by autopsy. There is atrophy of some of the gyri (folds) of the cerebral cortex and the hippocampus, an important center of memory. Nerve cells exhibit *neurofibrillary tangles*—dense masses of broken and twisted cytoskeleton **(fig. 12.33).** Alois Alzheimer first observed these in 1907 in the brain of a patient who had died of senile dementia. The more severe the signs of disease, the more neurofibrillary tangles are seen at autopsy. In the intercellular spaces, there are *senile plaques* consisting of aggregations of cells, altered nerve fibers, and a core of *β-amyloid protein*—the breakdown product of a glycoprotein of plasma membranes. Amyloid protein is rarely seen in elderly people without AD. Oxidative stress (free radical injury) is now widely believed to underlie the formation of amyloid protein and neurofibrillary tangles, and all the other aspects of AD pathology. This has led to interest in whether dietary antioxidants or antioxidant therapy can reduce the incidence or progression of AD.

Shrunken gyri

Wide sulci

(a)

Neurons with neurofibrillary tangles

Senile plaque

(b)

FIGURE 12.33 Alzheimer Disease. (a) Brain of a person who died of AD. Note the shrunken folds of cerebral tissue (gyri) and wide gaps (sulci) between them. (b) Cerebral tissue from a person with AD. Neurofibrillary tangles are present within the neurons, and a senile plaque is evident in the extracellular matrix.

a: ©Science Source; b: Simon Fraser/Science Source

[31]Alois Alzheimer (1864–1915), German neurologist

Intense biomedical research efforts are currently geared toward identifying the causes of AD and developing treatment strategies. Three genes on chromosomes 1, 14, and 21 have been implicated in various forms of early- and late-onset AD. Interestingly, persons with Down syndrome (trisomy-21), who have three copies of chromosome 21 instead of the usual two, tend to show early-onset Alzheimer disease. Nongenetic (environmental) factors also seem to be involved.

As for treatment, considerable attention now focuses on trying to halt β-amyloid formation or stimulate the immune system to clear β-amyloid from the brain tissue, but clinical trials in both of these approaches have been suspended until certain serious side effects can be resolved. AD patients show deficiencies of acetylcholine (ACh) and nerve growth factor (NGF). Some patients show improvement when treated with NGF or cholinesterase inhibitors, but results so far have been modest.

Parkinson[32] *disease (PD),* also called *paralysis agitans* or *parkinsonism,* is a progressive loss of motor function beginning in a person's 50s or 60s. It is due to degeneration of dopamine-releasing neurons in a portion of the brainstem called the *substantia nigra* **(fig. 12.34).** A gene has recently been identified for a hereditary form of PD, but most cases are nonhereditary and of little-known cause; some authorities suspect environmental neurotoxins.

Dopamine (DA) is an inhibitory neurotransmitter that normally prevents excessive activity in motor centers of the brain called the *basal nuclei.* Degeneration of dopamine-releasing neurons leads to an excessive ratio of ACh to DA, causing hyperactivity of the basal nuclei. As a result, a person with PD suffers involuntary muscle contractions. These take such forms as shaking of the hands (tremor) and compulsive "pill-rolling" motions of the thumb and fingers. In addition, the facial muscles may become rigid and produce a staring, expressionless face with a slightly open mouth. The patient's range of motion diminishes. He or she takes smaller steps and develops a slow, shuffling gait with a forward-bent posture and a tendency to fall forward. Speech becomes slurred and handwriting becomes cramped and eventually illegible. Tasks such as buttoning clothes and preparing food become increasingly laborious.

Patients cannot be expected to recover from PD, but its effects can be alleviated with drugs and physical therapy. Treatment with dopamine is ineffective because it can't cross the blood–brain barrier, but its precursor, levodopa (L-dopa), does cross the barrier and has been used to treat PD since the 1960s. L-dopa affords some relief, but it doesn't slow progression of the disease and it has undesirable side effects on the liver and heart. It is effective for only 5 to 10 years of treatment. A newer drug, deprenyl, is a monoamine oxidase (MAO) inhibitor that retards neural degeneration and slows the development of PD.

A surgical technique called *pallidotomy* has been used since the 1940s to quell severe tremors. It involves the destruction of a small portion of cerebral tissue in an area called the *globus pallidus.* Pallidotomy fell out of favor in the late 1960s when L-dopa came into common use. By the early 1990s, however, the limitations of L-dopa had become apparent, while MRI- and CT-guided methods had improved surgical precision and reduced the risks of brain surgery. Pallidotomy has thus made a comeback. Other surgical treatments for parkinsonism target brain areas called the *subthalamic nucleus* and the *ventral intermediate nucleus* of the thalamus, and involve either the destruction of tiny areas of tissue or the implantation of a stimulating electrode. Such procedures are generally used only in severe cases that are unresponsive to medication.

(a)

(b)

(c)

FIGURE 12.34 Parkinson Disease (PD).
(a) Cross section of midbrain of a healthy person showing normal, dark band of substantia nigra at arrow. (b) Midbrain section of a PD patient showing absence of substantia nigra. (c) Dyskinesia, a difficulty in walking and forward-tilted stance characteristic of persons with PD.

a–b: ISM/Pr J.J. Hauw/Medical Images

[32]James Parkinson (1755–1824), British physician

Effects of the **NERVOUS SYSTEM** on Other Organ Systems

INTEGUMENTARY SYSTEM
Cutaneous nerves regulate piloerection, sweating, cutaneous vasoconstriction and vasodilation, and heat loss through the body surface, and provide for cutaneous sensations such as touch, itch, tickle, pressure, heat, and cold.

SKELETAL SYSTEM
Nervous stimulation maintains the muscle tension that stimulates bone growth and remodeling; nerves in the bones respond to strains and fractures.

MUSCULAR SYSTEM
Skeletal muscles cannot contract without nervous stimulation; the nervous system controls all body movements and muscle tone.

ENDOCRINE SYSTEM
The hypothalamus controls the pituitary gland; the sympathetic nervous system controls the adrenal medulla; neuroendocrine cells are neurons that secrete hormones such as oxytocin; sensory and other nervous input influences the secretion of numerous other hormones.

CIRCULATORY SYSTEM
The nervous system regulates the rate and force of the heartbeat, regulates blood vessel diameters, monitors and controls blood pressure and blood gas concentrations, routes blood to organs where needed, and influences blood clotting.

LYMPHATIC AND IMMUNE SYSTEMS
Nerves to lymphatic organs influence the development and activity of immune cells; emotional states influence susceptibility to infection and other failures of immunity.

RESPIRATORY SYSTEM
The brainstem regulates the rhythm of breathing, monitors blood pH and blood gases, and adjusts the respiratory rate and depth to control these within normal ranges.

URINARY SYSTEM
Sympathetic nerves modify the rate of urine production by the kidneys; nervous stimulation of urinary sphincters aids in urine retention in the bladder, and nervous reflexes control its emptying.

DIGESTIVE SYSTEM
The nervous system regulates appetite, feeding behavior, digestive secretion and motility, and defecation.

REPRODUCTIVE SYSTEM
The nervous system regulates sex drive, arousal, and orgasm; the brain regulates the secretion of pituitary hormones that control spermatogenesis in males and the ovarian cycle in females; the nervous system controls various aspects of pregnancy and childbirth; the brain produces oxytocin, which is involved in labor contractions and lactation.

STUDY GUIDE

▶ Assess Your Learning Outcomes

To test your knowledge, discuss the following topics with a study partner or in writing, ideally from memory.

12.1 Overview of the Nervous System

1. What the nervous and endocrine systems have in common
2. Three fundamental functions of the nervous system; the roles of receptors and effectors in carrying out these functions
3. Differences between the central nervous system (CNS) and peripheral nervous system (PNS); between the sensory and motor divisions of the PNS; and between the somatic and visceral subdivisions of both the sensory and motor divisions
4. The autonomic nervous system and its two divisions

12.2 Properties of Neurons

1. Three fundamental physiological properties of neurons
2. Differences between sensory (afferent) neurons, interneurons (association neurons), and motor (efferent) neurons
3. The parts of a generalized multipolar neuron, and their functions
4. Differences between multipolar, bipolar, unipolar, and anaxonic neurons; an example of each
5. Ways in which neurons transport substances between the neurosoma and the distal ends of the axon

12.3 Supportive Cells

1. Six kinds of neuroglia; the structure and functions of each; and which kinds are found in the CNS and which ones in the PNS
2. Structure of the myelin sheath, and how CNS and PNS glial cells produce it
3. How fiber diameter and the presence or absence of myelin affect the conduction speed of a nerve fiber
4. The regeneration of a damaged nerve fiber; the roles of Schwann cells, the basal lamina, and neurilemma in regeneration; and why CNS neurons cannot regenerate

12.4 Electrophysiology of Neurons

1. The meanings of *electrical potential* and *resting membrane potential* (RMP); the typical voltage of an RMP

2. What an electrical current is, and how sodium ions and gated membrane channels generate a current
3. How stimulation of a neuron generates a local potential; the physiological properties of a local potential
4. Special properties of the trigger zone and unmyelinated regions of a nerve fiber that enable these regions to generate action potentials
5. The mechanism of an action potential; how it relates to ion flows and the action of membrane channels; and what is meant by *depolarization* and *repolarization* of the plasma membrane during local and action potentials
6. The all-or-none law and how it applies to an action potential; other properties of action potentials in contrast to local potentials
7. The basis and significance of the refractory period that follows an action potential
8. How one action potential triggers another; how the continuous conduction seen in unmyelinated nerve fibers result from a chain reaction of action potentials; and what normally prevents the signal from traveling backward to the neurosoma
9. Saltatory conduction in a myelinated nerve fiber; differences in conduction mechanisms of the myelin sheath gaps and internodal segments; and why signals travel faster in myelinated fibers than in unmyelinated fibers of comparable size

12.5 Synapses

1. The structure and locations of synapses
2. The role of neurotransmitters in synaptic transmission
3. Categories of neurotransmitters and common examples of each
4. Why the same neurotransmitter can have different effects on different cells
5. Excitatory synapses; how acetylcholine and norepinephrine excite a postsynaptic neuron
6. Inhibitory synapses; how γ-aminobutyric acid (GABA) inhibits a postsynaptic neuron
7. How second-messenger systems function at synapses
8. Three ways in which synaptic transmission is ended
9. Neuromodulators, their chemical nature, and how they affect synaptic transmission

12.6 Neural Integration

1. Why synapses slow down nervous communication; the overriding benefit of synapses
2. The meaning of *excitatory* and *inhibitory postsynaptic potentials* (EPSPs and IPSPs)
3. Why the production of an EPSP or IPSP may depend on both the neurotransmitter released by the presynaptic neuron and the type of receptor on the postsynaptic neuron
4. How a postsynaptic neuron's decision to fire depends on the ratio of EPSPs to IPSPs
5. Temporal and spatial summation, where they occur, and how they determine whether a neuron fires
6. Mechanisms of presynaptic facilitation and inhibition, and how communication between two neurons can be strengthened or weakened by a third neuron employing one of these mechanisms
7. Mechanisms of neural coding; how a neuron communicates qualitative and quantitative information
8. Why the refractory period sets a limit to how frequently a neuron can fire
9. The meanings of *neural pool* and *neural circuit*
10. The difference between a neuron's discharge zone and facilitated zone, and how this relates to neurons working in groups
11. Diverging, converging, reverberating, and parallel after-discharge circuits of neurons; examples of their relevance to familiar body functions
12. The difference between serial and parallel processing of information and how they influence everyday body functions
13. The cellular basis of memory; what memory consists of in terms of neural pathways, and how it relates to synaptic plasticity and potentiation
14. Types of things remembered in immediate memory, short-term memory (STM), and long-term memory (LTM), and in the explicit and implicit forms of LTM
15. Neural mechanisms thought to be involved in these different forms of memory and in forgetting

STUDY GUIDE

▶ Testing Your Recall

Answers in Appendix A

1. The integrative functions of the nervous system are performed mainly by
   a. afferent neurons.
   b. efferent neurons.
   c. neuroglia.
   d. sensory neurons.
   e. interneurons.

2. The highest density of voltage-gated ion channels is found in the _____ of a neuron.
   a. dendrites
   b. neurosoma
   c. myelin sheath gaps
   d. internodal segments
   e. axon terminals

3. The neurosoma of a mature neuron lacks
   a. a nucleus.
   b. endoplasmic reticulum.
   c. lipofuscin.
   d. centrioles.
   e. ribosomes.

4. The glial cells that fight infections in the CNS are
   a. microglia.
   b. satellite cells.
   c. ependymal cells.
   d. oligodendrocytes.
   e. astrocytes.

5. Posttetanic potentiation of a synapse increases the amount of _____ in the axon terminal.
   a. neurotransmitter
   b. neurotransmitter receptors
   c. calcium
   d. sodium
   e. NMDA

6. An IPSP is _____ of the postsynaptic neuron.
   a. a refractory period
   b. an action potential
   c. a depolarization
   d. a repolarization
   e. a hyperpolarization

7. Saltatory conduction occurs only
   a. at chemical synapses.
   b. in the initial segment of an axon.
   c. in both the initial segment and axon hillock.
   d. in myelinated nerve fibers.
   e. in unmyelinated nerve fibers.

8. Some neurotransmitters can have either excitatory or inhibitory effects depending on the type of
   a. receptors on the postsynaptic cell.
   b. synaptic vesicles in the axon.
   c. synaptic potentiation that occurs.
   d. postsynaptic potentials on the axon terminal.
   e. neuromodulator involved.

9. Differences in the volume of a sound are likely to be encoded by differences in _____ in nerve fibers from the inner ear.
   a. neurotransmitters
   b. signal conduction velocity
   c. types of postsynaptic potentials
   d. firing frequency
   e. voltage of the action potentials

10. Motor effects that depend on repetitive output from a neural pool are most likely to use
    a. parallel after-discharge circuits.
    b. reverberating circuits.
    c. facilitated circuits.
    d. diverging circuits.
    e. converging circuits.

11. Neurons that convey information to the CNS are called sensory, or _____, neurons.

12. To perform their role, neurons must have the properties of excitability, secretion, and _____.

13. The _____ is a period of time in which a neuron is producing an action potential and cannot respond to another stimulus of any strength.

14. Neurons receive incoming signals by way of specialized extensions of the cell called _____.

15. In the CNS, myelin is produced by glial cells called _____.

16. A myelinated nerve fiber can produce action potentials only in specialized regions called _____.

17. The trigger zone of a neuron consists of its _____ and _____.

18. The neurotransmitter secreted at an adrenergic synapse is _____.

19. A presynaptic nerve fiber cannot cause other neurons in its _____ to fire, but it can make them more sensitive to stimulation from other presynaptic fibers.

20. _____ are substances released along with a neurotransmitter that modify the neurotransmitter's effect.

▶ Building Your Medical Vocabulary

Answers in Appendix A

State a meaning of each word element, and give a medical term from this chapter that uses it or a slight variation of it.

1. antero-

2. -aps

3. astro-

4. dendro-

5. -fer

6. gangli-

7. -grad

8. neuro-

9. sclero-

10. somato-

STUDY GUIDE

▶ What's Wrong with These Statements?

Answers in Appendix A

Briefly explain why each of the following statements is false, or reword it to make it true.

1. A neuron may have anywhere from one to several axons.

2. Astrocytes perform the same function in the brain as Schwann cells do in the peripheral nerves.

3. A resting neuron has a higher concentration of Na^+ in its cytoplasm than in the extracellular fluid surrounding it.

4. During an action potential, most of the Na^+ and K^+ exchange places across the plasma membrane.

5. Excitatory postsynaptic potentials lower the threshold of a neuron and thus make it easier to stimulate.

6. In theory, there is no upper limit to how often a neuron can fire if it is stimulated strongly enough.

7. A given neurotransmitter has the same effect no matter where in the body it is secreted.

8. Myelinated nerve fibers conduct signals more rapidly than unmyelinated ones because they have gaps in the myelin sheath.

9. Learning occurs by increasing the number of neurons in the brain tissue.

10. Dead neurons in the brain are quickly replaced by mitosis of surviving neurons.

▶ Testing Your Comprehension

1. Schizophrenia is sometimes treated with drugs such as chlorpromazine that inhibit dopamine receptors. A side effect is that patients begin to develop muscle tremors, speech impairment, and other disorders similar to Parkinson disease. Explain.

2. Hyperkalemia is an excess of potassium in the extracellular fluid. What effect would this have on the resting membrane potentials of the nervous system and on neural excitability?

3. Suppose a poison were to slow down the Na^+–K^+ pumps of nerve cells. How would this affect the resting membrane potentials of neurons? Would it make neurons more excitable than normal, or make them more difficult to stimulate? Explain.

4. The unity of form and function is an important concept in understanding synapses. Give two structural reasons why nerve signals cannot travel backward across a chemical synapse. What could be the consequences if signals did travel freely in both directions?

5. Positive feedback is usually harmful to the body, or even life-threatening, but there are a few cases where it is beneficial and necessary. Identify such a case in neuron electrophysiology. If necessary, review the defining characteristics of positive feedback in section 1.6d.

Cross section of the sciatic nerve, showing myelinated nerve fibers (large pale circles) and small unmyelinated fibers scattered among them

Science Photo Library/Alamy Stock Photo

THE SPINAL CORD, SPINAL NERVES, AND SOMATIC REFLEXES

CHAPTER OUTLINE

13.1 The Spinal Cord
- 13.1a Functions
- 13.1b Surface Anatomy
- 13.1c Meninges of the Spinal Cord
- 13.1d Cross-Sectional Anatomy
- 13.1e Spinal Tracts

13.2 The Spinal Nerves
- 13.2a General Anatomy of Nerves and Ganglia
- 13.2b Spinal Nerves
- 13.2c Nerve Plexuses
- 13.2d Cutaneous Innervation and Dermatomes

13.3 Somatic Reflexes
- 13.3a The Nature of Reflexes
- 13.3b The Muscle Spindle
- 13.3c The Stretch Reflex
- 13.3d The Flexor (Withdrawal) Reflex
- 13.3e The Crossed Extension Reflex
- 13.3f The Tendon Reflex

Study Guide

DEEPER INSIGHTS

- **13.1** Clinical Application: Spinal Tap
- **13.2** Clinical Application: Poliomyelitis and Amyotrophic Lateral Sclerosis
- **13.3** Clinical Application: Shingles
- **13.4** Clinical Application: Nerve Injuries
- **13.5** Clinical Application: Spinal Cord Trauma

Anatomy & Physiology *Revealed* 4.0

Module 7: Nervous System

Every year in the United States, thousands of people become paralyzed by spinal cord injuries, with devastating effects on their quality of life. The treatment of such injuries is one of the most lively areas of medical research today. Therapists in this specialty must know spinal cord anatomy and function to understand their patients' functional deficits and prospects for improvement and to plan a regimen of treatment. Such knowledge is necessary, as well, for understanding paralysis resulting from strokes and other brain injuries. The spinal cord is the "information highway" that connects the brain with the lower body; it contains the neural routes that explain why a lesion to a specific part of the brain results in a functional loss in a specific locality in the lower body.

In this chapter, we will study not only the spinal cord but also the spinal nerves that arise from it with ladderlike regularity at intervals along its length. Thus, we will examine components of both the central and peripheral nervous systems, but these are so closely related, structurally and functionally, that it is appropriate to consider them together. Similarly, the brain and cranial nerves will be considered together in the following chapter. Chapters 13 and 14 therefore elevate our study of the nervous system from the cellular level (chapter 12) to the organ and system levels.

13.1   The Spinal Cord

Expected Learning Outcomes

When you have completed this section, you should be able to

a. state the three principal functions of the spinal cord;

b. describe its gross and microscopic structure; and

c. trace the pathways followed by nerve signals traveling up and down the spinal cord.

13.1a Functions

The spinal cord serves four principal functions:

1. **Conduction.** It contains bundles of nerve fibers that conduct information up and down the cord, connecting different levels of the trunk with each other and with the brain. This enables sensory information to reach the brain, motor commands to reach the effectors, and input received at one level of the cord to affect output from another level.

2. **Neural integration.** Pools of spinal neurons receive input from multiple sources, integrate the information, and execute an appropriate output. For example, the spinal cord can integrate the stretch sensation from a full bladder with cerebral input concerning the appropriate time and place to urinate and execute control of the bladder accordingly.

3. **Locomotion.** Walking involves repetitive, coordinated contractions of several muscle groups in the limbs. Motor neurons in the brain initiate walking and determine its speed, distance, and direction, but the simple repetitive muscle contractions that put one foot in front of another, over and over, are coordinated by groups of neurons called **central pattern generators** in the cord. These neural circuits produce the sequence of outputs to the extensor and flexor muscles that cause alternating movements of the lower limbs.

4. **Reflexes.** Spinal reflexes play vital roles in posture, motor coordination, and protective responses to pain or injury.

13.1b Surface Anatomy

The **spinal cord (fig. 13.1)** arises from the brainstem at the foramen magnum of the skull. It passes through the vertebral canal as far as the inferior margin of the first lumbar vertebra (L1) or slightly beyond. In adults, it averages about 45 cm long and 1.8 cm thick (about as thick as one's little finger). Early in fetal development, the cord extends for the full length of the vertebral column. However, the vertebral column grows faster than the spinal cord, so the cord extends only to L3 by the time of birth and to L1 in an adult. Thus, it occupies only the upper two-thirds of the vertebral canal; the lower one-third is described shortly.

The cord gives rise to 31 pairs of *spinal nerves.* Although the spinal cord is not visibly segmented, the part supplied by each pair of nerves is called a *segment.* The cord exhibits longitudinal grooves on its anterior and posterior sides—the *anterior median fissure* and *posterior median sulcus,* respectively **(fig. 13.2).**

The spinal cord is divided into **cervical, thoracic, lumbar,** and **sacral regions.** It may seem odd that it has a sacral region when the cord itself ends well above the sacrum. These regions, however, are named for the level of the vertebral column from which the spinal nerves emerge, not for the vertebrae that contain the cord itself.

In two areas, the cord is a little thicker than elsewhere. In the inferior cervical region, a **cervical enlargement** gives rise to nerves of the upper limbs. In the lumbosacral region, there is a similar **lumbosacral enlargement** that issues nerves to the pelvic

(a)

(b)

FIGURE 13.1 The Spinal Cord, Posterior View.
(a) Overview of spinal cord structure. (b) Detail of the spinal cord and associated structures. **APR**

region and lower limbs. Inferior to the lumbosacral enlargement, the cord tapers to a point called the **medullary cone.** Arising from the lumbosacral enlargement and medullary cone is a bundle of nerve roots that occupy the vertebral canal from L2 to S5. This bundle, named the **cauda equina**[1] (CAW-duh ee-KWY-nah) for its resemblance to a horse's tail, innervates the pelvic organs and lower limbs.

▶▶▶APPLY WHAT YOU KNOW

Spinal cord injuries commonly result from fractures of vertebrae C5 to C6, but never from fractures of L3 to L5. Explain both observations.

13.1c Meninges of the Spinal Cord

The spinal cord and brain are enclosed in three fibrous membranes called **meninges**[2] (meh-NIN-jeez)—singular, *meninx* (MEN-inks) (fig. 13.2). These membranes separate the soft tissue of the central nervous system from the bones of the vertebrae and skull. From superficial to deep, they are the dura mater, arachnoid mater, and pia mater.

The **dura mater**[3] (DOO-ruh MAH-tur) forms a loose-fitting sleeve called the **dural sheath** around the spinal cord. It is a tough membrane about as thick as a rubber kitchen glove, composed of multiple layers of dense irregular connective tissue. The space between the sheath and vertebral bones, called the **epidural space,** is occupied by

---

[1]*cauda* = tail; *equin* = horse

[2]*menin* = membrane

[3]*dura* = tough; *mater* = mother, womb

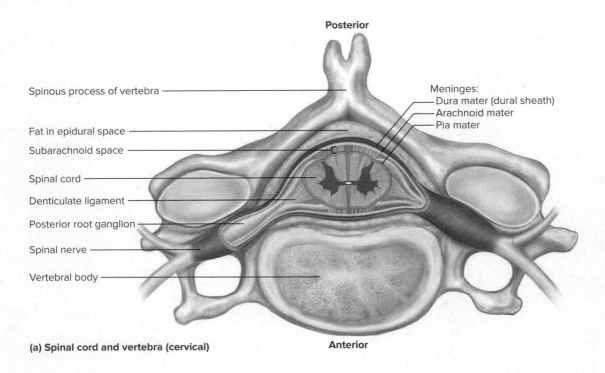

(a) Spinal cord and vertebra (cervical)

Posterior

Spinous process of vertebra

Meninges:
Dura mater (dural sheath)
Arachnoid mater
Pia mater

Fat in epidural space

Subarachnoid space

Spinal cord

Denticulate ligament

Posterior root ganglion

Spinal nerve

Vertebral body

Anterior

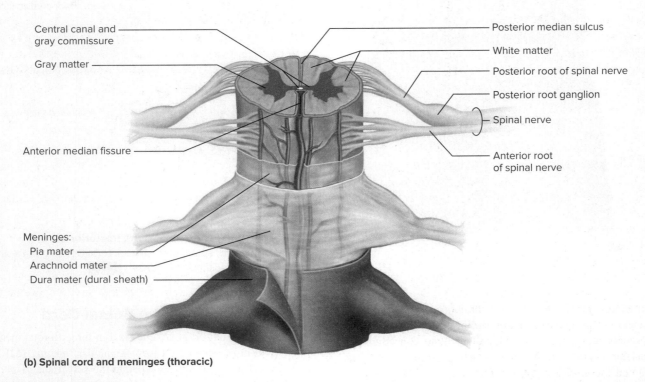

(b) Spinal cord and meninges (thoracic)

Central canal and gray commissure

Gray matter

Anterior median fissure

Posterior median sulcus

White matter

Posterior root of spinal nerve

Posterior root ganglion

Spinal nerve

Anterior root of spinal nerve

Meninges:
Pia mater
Arachnoid mater
Dura mater (dural sheath)

FIGURE 13.2 Cross-Sectional Anatomy of the Spinal Cord. (a) Relationship to the vertebra, meninges, and spinal nerves. (b) Detail of the spinal cord, meninges, and spinal nerve roots. See figure 13.3 for details of the gray and white matter.

blood vessels, adipose tissue, and loose connective tissue. Anesthetics can be introduced to this space to block pain signals during childbirth or surgery; this procedure is called *epidural anesthesia.*

The **arachnoid**[4] **mater** (ah-RACK-noyd) consists of the *arachnoid membrane*—five or six layers of squamous to cuboidal

cells adhering to the inside of the dura—and a looser array of cells and collagenous and elastic fibers spanning the gap between the arachnoid membrane and the pia mater. This gap, the **subarachnoid space,** is filled with **cerebrospinal fluid (CSF)** (see section 14.2b). Inferior to the medullary cone, the subarachnoid space is called the **lumbar cistern** and is occupied by the cauda equina and CSF. When a sample of CSF is needed for clinical purposes, it is

---

[4]*arachn* = spider, spider web; *oid* = resembling

FIGURE 13.3 Cross Section of the Spinal Cord (Lumbar Region).

Jose Luis Calvo/Shutterstock

taken from the lumbar cistern by a procedure called a **spinal tap** or **lumbar puncture** (see Deeper Insight 13.1).

The **pia**[5] **mater** (PEE-uh) is a delicate, transparent membrane composed of one or two layers of squamous to cuboidal cells and delicate collagenous and elastic fibers. It closely follows the contours of the spinal cord. It continues beyond the medullary cone as a fibrous strand, the **terminal filum,** within the lumbar cistern. At the level of vertebra S2, it exits the lower end of the cistern and fuses with the dura mater, and the two form a **coccygeal ligament** that anchors the cord and meninges to vertebra Co1. At regular intervals along the cord, extensions of the pia called **denticulate**[6] **ligaments** extend through the arachnoid to the dura, anchoring the cord and limiting side-to-side movements.

13.1d Cross-Sectional Anatomy

Figure 13.2a shows the relationship of the spinal cord to a vertebra and spinal nerve, and **figure 13.3** shows the cord itself in more detail. The spinal cord, like the brain, consists of two kinds of nervous tissue called gray and white matter. **Gray matter** has a relatively dull color because it contains little myelin. It contains the somas, dendrites, and proximal parts of the axons of neurons. It is the site of synaptic contact between neurons, and therefore the site of all neural integration in the spinal cord. **White matter,** by contrast, has a bright, pearly white appearance due to an abundance of myelin. It is composed of bundles of axons, called **tracts,** that carry signals from one level of the CNS to another. Both gray and white matter also have an abundance of glial cells. Nervous tissue is often histologically stained with silver compounds, which give the gray matter a brown or golden color and white matter a lighter tan to amber color on slides that you may study.

Gray Matter

The spinal cord has a central core of gray matter that looks somewhat butterfly- or H-shaped in cross sections. The core consists mainly of two **posterior (dorsal) horns,** which extend toward the posterolateral

surfaces of the cord, and two thicker **anterior (ventral) horns,** which extend toward the anterolateral surfaces. The right and left sides of the gray matter are connected by a median bridge called the **gray commissure.** In the middle of the commissure is the **central canal,** which is collapsed in most areas of the adult spinal cord, but in some places (and in young children) remains open, lined with ependymal cells, and filled with CSF.

The posterior horn receives sensory nerve fibers from the spinal nerves, which usually synapse with networks of interneurons in the horn. The anterior horn contains the large neurosomas of motor neurons whose axons lead out to the skeletal muscles. The interneurons and motor neurons are especially abundant in the cervical and lumbosacral enlargements and are quite conspicuous in histological sections from these levels. The high density of neurons in these regions is related to motor control and sensation in the upper and lower limbs.

An additional **lateral horn** is visible on each side of the gray matter from segments T2 through L1 of the cord. It contains neurons of the sympathetic nervous system, which send their axons out of the cord by way of the anterior root along with the somatic efferent fibers.

White Matter

The white matter of the spinal cord surrounds the gray matter. It consists of bundles of axons that course up and down the cord and provide avenues of communication between different levels of the CNS. These bundles are arranged in three pairs called **funiculi**[7] (few-NIC-you-lie)—a **posterior (dorsal), lateral,** and **anterior (ventral) funiculus** on each side. Each column consists of subdivisions called **tracts** or **fasciculi**[8] (fah-SIC-you-lye).

13.1e Spinal Tracts

Knowledge of the locations and functions of the spinal tracts is essential in diagnosing and managing spinal cord injuries. **Ascending tracts** carry sensory information up the

---

[5]*pia* = through mistranslation, now construed as tender, thin, or soft
[6]*denti* = tooth; *cul* = little; *ate* = resembling

[7]*funicul* = little rope, cord
[8]*fascicul* = little bundle

DEEPER INSIGHT 13.1

CLINICAL APPLICATION

Spinal Tap

Several neurological diseases are diagnosed in part by examining cerebrospinal fluid (CSF) for bacteria, blood, white blood cells, or abnormalities of chemical composition. CSF is obtained by a procedure called a *spinal tap,* or *lumbar puncture.* The patient leans forward or lies on one side with the spine flexed, thus spreading the vertebral laminae and spinous processes apart **(fig. 13.4)**. The skin over the lumbar vertebrae is anesthetized, and a needle is inserted between the spinous processes of L3 and L4 (sometimes L4 and L5). This is the safest place to obtain CSF because the spinal cord doesn't extend this far and isn't exposed to injury by the needle. At a depth of 4 to 6 cm, the needle punctures the dura mater and enters the lumbar cistern. CSF normally drips out at a rate of about 1 drop per second. A lumbar puncture is not performed if a patient has signs of high intracranial pressure, because the sudden release of pressure (causing CSF to jet from the puncture) can cause fatal herniation of the brainstem and cerebellum into the vertebral canal.

FIGURE 13.4 Spinal Tap (Lumbar Puncture).

cord, and **descending tracts** conduct motor impulses down **(fig. 13.5)**. All nerve fibers in a given tract have a similar origin, destination, and function. Many of these fibers have their origin or destination in a region called the *brainstem.* Described more fully in section 14.3 (see fig. 14.1), this is a vertical stalk that supports the large *cerebellum* at the rear of the head and, even larger, two *cerebral hemispheres* that dominate the brain. In the following discussion, you will find references to brainstem and other regions where spinal tracts begin and end. Spinal cord anatomy will grow in meaning as you study the brain.

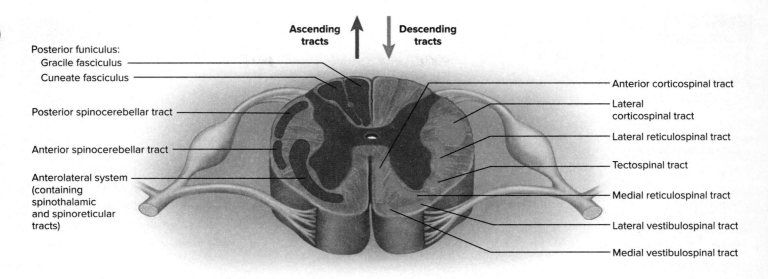

Ascending tracts

Descending tracts

Posterior funiculus:
Gracile fasciculus
Cuneate fasciculus

Posterior spinocerebellar tract

Anterior spinocerebellar tract

Anterolateral system
(containing
spinothalamic
and spinoreticular
tracts)

Anterior corticospinal tract

Lateral
corticospinal tract

Lateral reticulospinal tract

Tectospinal tract

Medial reticulospinal tract

Lateral vestibulospinal tract

Medial vestibulospinal tract

FIGURE 13.5 Tracts of the Spinal Cord. All of the illustrated tracts occur on both sides of the cord, but only the ascending sensory tracts are shown on the left (red), and only the descending motor tracts on the right (green).

❓ *If you were told that this cross section is either at level T4 or T10, how could you determine which is correct?*

Several of these tracts undergo **decussation**[9] (DEE-cuh-SAY-shun) as they pass up or down the brainstem and spinal cord—meaning that they cross over from the left side of the body to the right, or vice versa. As a result, the left side of the brain receives sensory information from the right side of the body and sends motor commands to that side, while the right side of the brain senses and controls the left side of the body. Therefore, a stroke that damages motor centers of the right side of the brain can cause paralysis of the left limbs and vice versa.

When the origin and destination of a tract are on opposite sides of the body, we say they are **contralateral**[10] to each other. When a tract doesn't decussate, its origin and destination are on the same side of the body and we say they are **ipsilateral**.[11]

The major spinal tracts are summarized in **table 13.1** and the following text. Bear in mind that each tract is repeated on the right and left sides of the spinal cord.

Ascending Tracts

Ascending tracts carry sensory signals up the spinal cord. Sensory signals typically travel across three neurons from their origin in the receptors to their destination in the brain: a **first-order neuron** that detects a stimulus and transmits a signal to the spinal cord or brainstem; a **second-order neuron** that continues as far as a "gateway" called the *thalamus* at the upper end of the brainstem; and a **third-order neuron** that carries the signal the rest of the way to the cerebral cortex. The axons of these neurons are called the *first-* through *third-order nerve fibers* (**fig. 13.6**).

The major ascending tracts are as follows. The names of most of them consist of the prefix *spino-* followed by a root denoting the destination of its fibers in the brain, although this naming system does not apply to the first two.

- The posterior funiculus consists of the **cuneate**[12] **fasciculus** (CUE-nee-ate) and **gracile**[13] **fasciculus** (GRAS-el) from level T6 to the brain, and only the gracile fasciculus from T7 down (fig. 13.6a). The gracile fasciculus carries sensory signals from the lower trunk and lower limbs, and the cuneate fasciculus from the chest and upper limbs. These signals include the senses of vibration, visceral pain, deep touch, and especially proprioception from the lower limbs and lower trunk. **Proprioception**[14] is one's nonvisual sense of the position and movements of the body and its parts, coming from sensory nerve endings in the muscles, tendons, and joints. The brain uses this information to adjust muscle tone and actions to maintain equilibrium (balance), posture, and coordination. The first-order nerve fibers of these fasciculi end in the medulla oblongata of the brainstem, but the second-order fibers decussate there to the other side of the brain. Thus, the right brain receives signals from the left side of the body and vice versa.

- The **spinothalamic tract** (SPY-no-tha-LAM-ic) (fig. 13.6b) and some smaller tracts form the *anterolateral system,* which passes up the anterior and lateral funiculi of the spinal cord. The spinothalamic tract carries signals for pain, temperature, pressure, tickle, itch, and light touch. Light touch is the

---

[9]*decuss* = to cross, form an X
[10]*contra* = opposite; *later* = side
[11]*ipsi* = the same; *later* = side

[12]*cune* = wedge
[13]*gracil* = thin, slender
[14]*proprio* = one's own; *ception* = sensation

TABLE 13.1	Major Spinal Tracts		
Tract	Funiculus	Decussation	Functions
Ascending (Sensory) Tracts			
Gracile fasciculus	Posterior	In medulla	Sensations of limb and trunk position and movement, deep touch, visceral pain, and vibration, below level T6
Cuneate fasciculus	Posterior	In medulla	Same as gracile fasciculus, from level T6 up
Spinothalamic	Lateral and anterior	In spinal cord	Sensations of light touch, tickle, itch, temperature, pain, and pressure
Spinoreticular	Lateral and anterior	In spinal cord (some fibers)	Sensation of pain from tissue injury
Posterior spinocerebellar	Lateral	None	Feedback from muscles (proprioception)
Anterior spinocerebellar	Lateral	In spinal cord	Same as posterior spinocerebellar
Descending (Motor) Tracts			
Lateral corticospinal	Lateral	In medulla	Fine control of limbs
Anterior corticospinal	Anterior	In spinal cord	Fine control of limbs
Tectospinal	Anterior	In midbrain	Reflexive head turning in response to visual and auditory stimuli
Lateral reticulospinal	Lateral	None	Balance and posture; regulation of awareness of pain
Medial reticulospinal	Anterior	None	Same as lateral reticulospinal
Lateral vestibulospinal	Anterior	None	Balance and posture
Medial vestibulospinal	Anterior	In medulla (some fibers)	Control of head position

sensation produced, for example, by stroking hairless skin with a feather or cotton wisp, without indenting the skin. In this pathway, first-order neurons end in the posterior horn of the spinal cord near the point of entry. Here they synapse with second-order neurons, which decussate and form the contralateral ascending spinothalamic tract. These fibers lead all the way to the thalamus. Therefore, here again, each hemisphere of the brain receives sensory input from the opposite side of the body.

- The **spinoreticular tract** carries pain signals up the anterolateral system. First-order pain fibers enter the posterior horn and immediately synapse with second-order neurons. These decussate to the opposite anterolateral system, ascend the cord, and end in a loosely organized core of gray matter called the *reticular*[15] *formation* in the medulla oblongata and pons. Pain pathways are further discussed in chapter 16.

- The **posterior** and **anterior spinocerebellar tracts** (SPY-no-SERR-eh-BEL-ur) travel through the lateral funiculus and carry proprioceptive signals from the limbs and trunk to the cerebellum at the rear of the brain. Their first-order neurons originate in muscles and tendons and end in the posterior horn of the spinal cord. Second-order neurons send their fibers up the spinocerebellar tracts and end in the cerebellum. Both tracts provide the cerebellum with feedback needed to coordinate muscle action.

Descending Tracts

Descending tracts carry motor signals down the brainstem and spinal cord. A descending motor pathway typically involves two neurons called the upper and lower motor neurons. The **upper motor neuron** begins with a neurosoma in the cerebral cortex or brainstem and has an axon that terminates on a **lower motor neuron** in the brainstem or spinal cord. The axon of the lower motor neuron then leads the rest of the way to the muscle or other target organ. The names of most descending tracts consist of a word root denoting the point of origin in the brain, followed by the suffix -*spinal*. The major descending tracts are described here.

- The **lateral** and **anterior corticospinal tracts** (COR-tih-co-SPY-nul) **(fig. 13.7)** carry motor signals from the cerebral cortex for precise, finely coordinated limb movements. The fibers of this system form ridges called *pyramids* on the anterior surface of the medulla oblongata, so these tracts were once called *pyramidal tracts*. Their fibers decussate either in the lower medulla or lower in the spinal cord, so each cerebral hemisphere controls muscles on the contralateral side of the body. The anterior tract gets smaller as it descends and gives off nerve fibers, and usually disappears by the midthoracic level.

- The **tectospinal**[16] **tract** (TEC-toe-SPY-nul) begins in the roof of the midbrain and descends the contralateral spinal

---

[15]*reticul* = little network

[16]*tectum* = roof (of the midbrain)

FIGURE 13.6 Some Ascending Pathways of the CNS. The spinal cord, medulla, and midbrain are shown in cross section and the cerebrum and thalamus (top) in frontal section. Nerve signals enter the spinal cord at the bottom of the figure and carry somatosensory information up to the cerebral cortex. (a) The cuneate fasciculus. (b) The spinothalamic tract.

cord only as far as the neck. It is involved in reflex turning of the head, especially in response to sights and sounds.

- The **lateral** and **medial reticulospinal tracts** (reh-TIC-you-lo-SPY-nul) originate in the reticular formation of the brainstem. They control muscles of the upper and lower

limbs, especially to maintain posture and balance. They also contain *descending analgesic pathways* that regulate the transmission of pain signals to the brain.

- The **lateral** and **medial vestibulospinal tracts** (vess-TIB-you-lo-SPY-nul) begin in the brainstem *vestibular nuclei,*

Motor cortex
(precentral gyrus)

Internal
capsule

Cerebrum

Midbrain

Cerebral
peduncle

Upper motor
neurons

Medulla

Medullary
pyramid

Decussation
in medulla

**Lateral
corticospinal
tract**

Spinal cord

**Anterior
corticospinal
tract**

Decussation in
spinal cord

Spinal cord

Lower motor
neurons

To skeletal muscles

FIGURE 13.7 Two Descending Pathways of the CNS. The lateral and anterior corticospinal tracts, which carry signals for voluntary muscle contraction. Nerve signals originate in the cerebral cortex at the top of the figure and carry motor commands down the spinal cord.

which receive signals for balance from the inner ear. The lateral vestibulospinal tract passes down the anterior funiculus and facilitates neurons that control extensor muscles of the limbs, thus inducing the limbs to stiffen and straighten. This is an important reflex in responding to body tilt and keeping one's balance. The medial vestibulospinal tract descends through the anterior funiculus on both sides of the cord and ends in the neck. It plays a role in the control of head position.

▶▶▶ APPLY WHAT YOU KNOW

You are blindfolded and either a tennis ball or an iron ball is placed in your right hand. What spinal tract(s) would carry the signals that enable you to discriminate between these two objects?

BEFORE YOU GO ON

Answer the following questions to test your understanding of the preceding section:

1. Name the four major regions and two enlargements of the spinal cord.

2. Describe the distal (inferior) end of the spinal cord and the contents of the vertebral canal from level L2 to S5.

3. Sketch a cross section of the spinal cord showing the anterior and posterior horns. Where are the gray and white matter? Where are the funiculi and tracts?

4. Give an anatomical explanation of why a stroke in the right cerebral hemisphere can paralyze the limbs on the left side of the body.

5. Identify each of the following spinal tracts—the gracile fasciculus and the lateral corticospinal, lateral reticulospinal, and spinothalamic tracts—with respect to whether it is ascending or descending; its origin and destination; and what sensory or motor purposes it serves.

13.2 The Spinal Nerves

Expected Learning Outcomes

When you have completed this section, you should be able to

a. describe the anatomy of nerves and ganglia in general;

b. describe the attachments of a spinal nerve to the spinal cord;

c. trace the branches of a spinal nerve distal to its attachments;

d. name the five plexuses of spinal nerves and describe their general anatomy;

e. name some major nerves that arise from each plexus and identify what they innervate; and

f. explain the relationship of dermatomes to the spinal nerves.

DEEPER INSIGHT 13.2

CLINICAL APPLICATION

Poliomyelitis and Amyotrophic Lateral Sclerosis

Poliomyelitis[17] and amyotrophic lateral sclerosis[18] (ALS) are two diseases that involve destruction of motor neurons. In both diseases, the skeletal muscles atrophy from lack of innervation.

Poliomyelitis is caused by the poliovirus, which destroys motor neurons in the brainstem and anterior horn of the spinal cord. Signs of polio include muscle pain, weakness, and loss of some reflexes, followed by paralysis, muscular atrophy, and sometimes respiratory arrest. The virus spreads by fecal contamination of water. Historically, polio afflicted many children who contracted the virus from swimming in contaminated pools. For a time, the polio vaccine nearly eliminated new cases, but the disease has lately begun to reemerge among children in some parts of the world because of antivaccination politics.

ALS is also known as Lou Gehrig[19] disease after the baseball player who had to retire from the sport because of it. It is marked not only by the degeneration of motor neurons and atrophy of the muscles, but also sclerosis (scarring) of the lateral regions of the spinal cord—hence its name. Most cases occur when astrocytes fail to reabsorb the neurotransmitter glutamate from the tissue fluid, allowing it to accumulate to a neurotoxic level. The early signs of ALS include muscular weakness and difficulty in speaking, swallowing, and using the hands. Sensory and intellectual functions remain unaffected, as evidenced by the accomplishments of astrophysicist and best-selling author Stephen Hawking **(fig. 13.8),** who was stricken with ALS in college. Despite near-total paralysis, he had a slowly progressing form of the disease, remained intellectually undiminished, and communicated with the aid of a speech synthesizer and computer. Tragically, many people are quick to assume that those who have lost most of their ability to communicate their ideas and feelings have few ideas and feelings to communicate. To a victim, this may be more unbearable than the loss of motor function itself.

FIGURE 13.8 Stephen Hawking (1942–2018). "When I was first diagnosed with ALS, I was given two years to live. Now 45 years later, I am doing pretty well" (CNN interview, 2010).

Geoff Robinson Photography/REX/Shutterstock

13.2a General Anatomy of Nerves and Ganglia

The spinal cord communicates with the rest of the body by way of the spinal nerves. Before we discuss those specific nerves, however, it is necessary to be familiar with the structure of nerves and ganglia in general.

A **nerve** is a cordlike organ composed of numerous nerve fibers (axons) bound together by connective tissue **(fig. 13.9).** If we compare a *nerve fiber* to a wire carrying an electrical current in one direction, a *nerve* would be comparable to an electrical cable composed of hundreds of wires carrying currents in opposite directions. A nerve contains anywhere from a few nerve fibers to (in the optic nerve) a million. Nerves usually have a pearly white color and resemble frayed string as they divide into smaller and smaller branches. As we move away from the spinal nerves proper, the smaller branches are called **peripheral nerves,** and their disorders are collectively called *peripheral neuropathy.*

Nerve fibers of the peripheral nervous system are ensheathed in Schwann cells, which form a neurilemma and often a myelin sheath around the axon (see section 12.3b). External to the neurilemma, each fiber is surrounded by a basal lamina and then a thin sleeve of loose connective tissue called the **endoneurium.** In most nerves, the fibers are gathered in bundles called **fascicles,** each wrapped in a sheath called the **perineurium.** The perineurium is composed of up to 20 layers of overlapping, squamous, epithelium-like cells. Several fascicles are then bundled together and wrapped in an outer **epineurium** to compose the nerve as a whole. The epineurium consists of dense irregular connective tissue and protects the nerve from stretching and injury. Nerves have a high metabolic rate and need a plentiful blood supply, which is furnished by blood vessels that penetrate these connective tissue coverings.

▶▶▶**APPLY WHAT YOU KNOW**

How does the structure of a nerve compare to that of a skeletal muscle? Which of the descriptive terms for nerves have similar counterparts in muscle histology?

---

[17]*polio* = gray matter; *myel* = spinal cord; *itis* = inflammation
[18]*a* = without; *myo* = muscle; *troph* = nourishment; *sclerosis* = hardening
[19]Lou Gehrig (1903–41), New York Yankees baseball player

FIGURE 13.9 Anatomy of a Nerve. (a) A spinal nerve and its association with the spinal cord. (b) Cross section of a nerve. Individual nerve fibers show as tiny red dots, each surrounded by a light ring of myelin.

b: PASIEKA/Science Photo Library/Getty Images

As we saw in section 12.1, peripheral nerve fibers are of two kinds: sensory (afferent) fibers carrying signals from sensory receptors to the CNS, and motor (efferent) fibers carrying signals from the CNS to muscles and glands. Both types can be classified as *somatic* or *visceral* and as *general* or *special* depending on the organs they innervate (**table 13.2).**

Purely **sensory nerves,** composed only of afferent fibers, are rare; they include nerves for smell and vision. **Motor nerves** carry only efferent fibers. Most nerves, however, are **mixed nerves,** which consist of both afferent and efferent fibers and therefore conduct signals in two directions. However, any one fiber in the nerve conducts signals in one direction only. Many nerves commonly described as motor are actually mixed because they carry sensory signals of proprioception from the muscle back to the CNS.

If a nerve resembles a thread, a **ganglion**[20] resembles a knot in the thread. A ganglion is a cluster of neurosomas outside the CNS. It is enveloped in an epineurium continuous with that of the nerve. Among the neurosomas are bundles of nerve fibers leading into and out of the ganglion. **Figure 13.10** shows a type of ganglion associated with the spinal nerves.

[20]*gangli* = knot

TABLE 13.2	The Classification of Nerve Fibers
Class	**Description**
Afferent fibers	Carry sensory signals from receptors to the CNS
Efferent fibers	Carry motor signals from the CNS to effectors
Somatic fibers	Innervate skin, skeletal muscles, bones, and joints
Visceral fibers	Innervate blood vessels, glands, and viscera
General fibers	Innervate widespread organs such as muscles, skin, glands, viscera, and blood vessels
Special fibers	Innervate more localized organs in the head, including the eyes, ears, olfactory and taste receptors, and muscles of chewing, swallowing, and facial expression

13.2b Spinal Nerves

There are 31 pairs of **spinal nerves:** 8 cervical (C1–C8), 12 thoracic (T1–T12), 5 lumbar (L1–L5), 5 sacral (S1–S5), and 1 coccygeal (Co1) **(fig. 13.11).** The first cervical nerve emerges between the skull and atlas, and the others emerge through intervertebral foramina, including the anterior and posterior foramina of the sacrum and the sacral hiatus. Thus, spinal nerves C1 through C7 emerge superior to the correspondingly numbered vertebrae (nerve C5 above vertebra C5, for example); nerve C8 emerges inferior to vertebra C7; and below this, all the remaining nerves emerge inferior to the correspondingly numbered vertebrae (nerve L3 inferior to vertebra L3, for example).

Proximal Branches

Each spinal nerve arises from two points of attachment to the spinal cord. In each segment of the cord, six to eight nerve **rootlets** emerge from the anterior surface and converge to form the **anterior (ventral) root** of the spinal nerve. Another six to eight rootlets emerge from the posterior surface and converge to form the **posterior (dorsal) root (figs. 13.12, 13.13).** A short distance away from the spinal cord, the posterior root swells into a **posterior (dorsal) root ganglion,** which contains the neurosomas of sensory neurons (fig. 13.10). There is no corresponding ganglion on the anterior root because the neurosomas are in the anterior horns of the spinal cord.

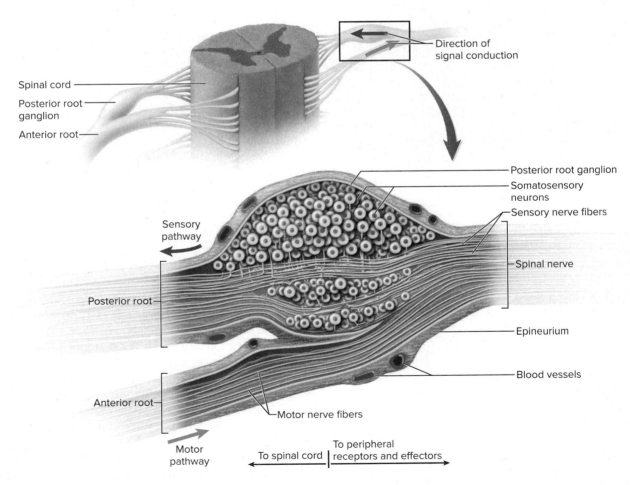

FIGURE 13.10 Anatomy of a Ganglion (Longitudinal Section). The posterior root ganglion contains the neurosomas of unipolar sensory neurons conducting signals from peripheral sense organs toward the spinal cord. Below this is the anterior root of the spinal nerve, which conducts motor signals away from the spinal cord, toward peripheral effectors. (The anterior root is not part of the ganglion.)

❓ *Where are the neurosomas of the neurons that give rise to the motor nerve fibers seen here?*

Vertebra C1 (atlas)

Cervical plexus (C1–C5)

Cervical nerves (8 pairs)
Cervical enlargement

Brachial plexus (C5–T1)

Vertebra T1

Intercostal (thoracic) nerves (T1–T12)

Thoracic nerves (12 pairs)

Lumbosacral enlargement

Vertebra L1

Medullary cone

Lumbar plexus (L1–L4)

Lumbar nerves (5 pairs)

Cauda equina

Sacral plexus (L4–S4)

Sacral nerves (5 pairs)

Coccygeal plexus (S4–Co1)

Coccygeal nerves (1 pair)
Sciatic nerve

FIGURE 13.11 The Spinal Nerve Roots and Plexuses, Posterior View.

Slightly distal to the posterior root ganglion, the anterior and posterior roots merge, leave the dural sheath, and form the spinal nerve proper (fig. 13.12). The nerve then exits the vertebral canal through the intervertebral foramen. The spinal nerve is a mixed nerve, carrying sensory signals to the spinal cord by way of the posterior root and ganglion, and motor signals out to more distant parts of the body by way of the anterior root.

The anterior and posterior roots are shortest in the cervical region and become longer inferiorly. The roots that arise from segments L2 to Co1 of the cord form the cauda equina. Some viruses invade the CNS by way of the spinal nerve roots (see Deeper Insight 13.3).

Distal Branches

Distal to the vertebrae, the branches of a spinal nerve are more complex **(fig. 13.14).** Immediately after emerging from the intervertebral foramen, the nerve divides into an **anterior ramus**[21]

[21]*ramus* = branch

(RAY-mus), **posterior ramus,** and a small **meningeal branch.** Thus, each spinal nerve branches on both ends—into anterior and posterior *roots* approaching the spinal cord, and anterior and posterior *rami* leading away from the vertebral column.

The meningeal branch (see fig. 13.12) reenters the vertebral canal and innervates the meninges, vertebrae, and spinal ligaments with sensory and motor fibers. The posterior ramus innervates the muscles and joints in that region of the spine and the skin of the back. The larger anterior ramus innervates the anterior and lateral skin and muscles of the trunk, and gives rise to nerves of the limbs.

The anterior ramus differs from one region of the trunk to another. In the thoracic region, it forms an **intercostal nerve,** which travels along the inferior margin of a rib and innervates the skin and intercostal muscles (thus contributing to breathing). Sensory fibers of the intercostal nerve branches to the skin are the most common routes of viral migration in the painful disease known as shingles (see Deeper Insight 13.3). Motor fibers of the intercostal nerves innervate the internal oblique, external oblique, and transverse abdominal muscles. All other anterior rami form the *nerve plexuses,* described next.

FIGURE 13.12 Branches of a Spinal Nerve in Relation to the Spinal Cord and Vertebra (Cross Section). APR

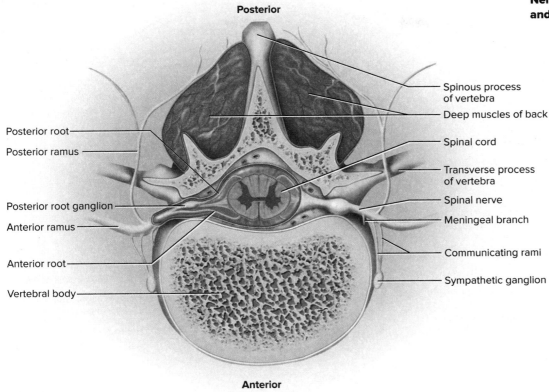

Posterior

Spinous process of vertebra

Deep muscles of back

Posterior root

Posterior ramus

Spinal cord

Transverse process of vertebra

Posterior root ganglion

Spinal nerve

Anterior ramus

Meningeal branch

Communicating rami

Anterior root

Sympathetic ganglion

Vertebral body

Anterior

FIGURE 13.13 The Point of Entry of Two Spinal Nerves into the Spinal Cord. Posterior (dorsal) view with vertebrae cut away. Note that each posterior root divides into several rootlets that enter the spinal cord. A segment of the spinal cord is the portion receiving all the rootlets of one spinal nerve.

Courtesy of Dr. Robert A. Chase, M.D.

? *In the labeled rootlets of spinal nerve C5, are the nerve fibers afferent or efferent? How do you know?*

Posterior median sulcus

Gracile fasciculus

Cuneate fasciculus

Lateral funiculus

Segment C5

Cross section

Arachnoid mater

Dura mater

Neural arch of vertebra C3 (cut)

Spinal nerve C4

Vertebral artery

Spinal nerve C5:

Rootlets

Posterior root

Posterior root ganglion

Anterior root

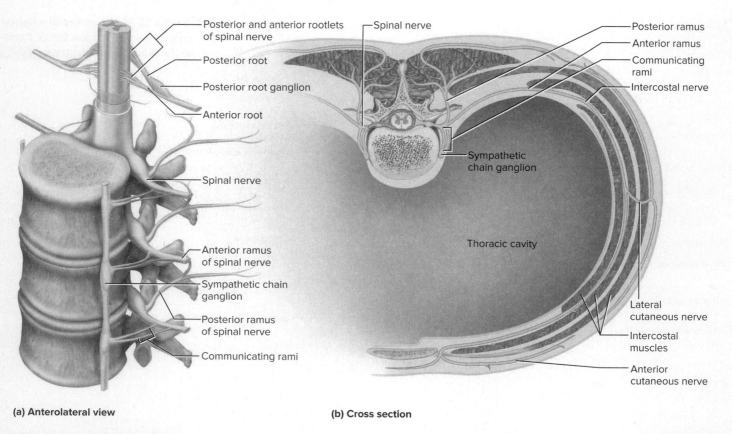

(a) Anterolateral view

(b) Cross section

FIGURE 13.14 Rami of the Spinal Nerves. (a) Anterolateral view of the spinal nerves and their subdivisions in relation to the spinal cord and vertebrae. (b) Cross section of the thorax showing innervation of muscles and skin of the chest and back. This section is cut through the intercostal muscles between two ribs.

As shown in figure 13.14, the anterior ramus also gives off a pair of *communicating rami,* which connect with a string of *sympathetic chain ganglia* alongside the vertebral column. These are seen only in spinal nerves T1 through L2. They are components of the sympathetic nervous system and are discussed more fully in section 15.2a.

13.2c Nerve Plexuses

Except in the thoracic region, the anterior rami branch and anastomose (merge) repeatedly to form five webs called nerve plexuses: the small **cervical plexus** in the neck, the **brachial plexus** near the shoulder, the **lumbar plexus** of the lower back, the **sacral plexus** immediately inferior to this, and finally, the tiny **coccygeal plexus** adjacent to the lower sacrum and coccyx. A general view of these plexuses is shown in figure 13.11; they are described in the next four tables, beginning with table 13.3. The spinal nerve roots that give rise to each plexus are indicated in violet in each illustration. Some of these roots give rise to smaller branches called *trunks, anterior divisions, posterior divisions,* and *cords,* which are color-coded and explained in the individual figures. Two of the nerves arising from these plexuses, the *radial*

and *sciatic,* are sites of unique nerve injuries described in Deeper Insight 13.4.

The nerves tabulated here have somatosensory and motor functions. *Somatosensory* means that they carry sensory signals from bones, joints, muscles, and the skin, in contrast to sensory input from the viscera or from special sense organs such as the eyes and ears. Somatosensory signals are for touch, heat, cold, stretch, pressure, pain, and other sensations. One of the most important somatosensory roles of these nerves is proprioception.

The motor function of these nerves is primarily to stimulate the contraction of skeletal muscles. They also innervate the bones of the corresponding regions, and carry autonomic fibers to some viscera and blood vessels, thus adjusting blood flow to local needs.

The following tables identify the areas of skin innervated by the sensory fibers and the muscle groups innervated by the motor fibers of the individual nerves. The muscle tables in chapter 10 provide a more detailed breakdown of the muscles supplied by each nerve and the actions they perform. You may assume that for each muscle, these nerves also carry sensory fibers from its proprioceptors. Throughout these tables, *nerve* is abbreviated *n.* and *nerves* as *nn.*

The Cervical Plexus

The cervical plexus on each side of the neck (**fig. 13.15**) receives fibers from the anterior rami of nerves C1 to C5 and gives rise to the nerves listed in **table 13.3,** in order from superior to inferior. The most important of these are the *phrenic*[22] *nerves* (FREN-ic), which travel down each side of the mediastinum, innervate the diaphragm, and play an indispensable role in breathing (see fig. 15.3). In addition to the major nerves listed in table 13.3, the cervical plexus gives off several motor branches that innervate the geniohyoid, thyrohyoid, scalene, levator scapulae, trapezius, and sternocleidomastoid muscles.

[22]*phren* = diaphragm

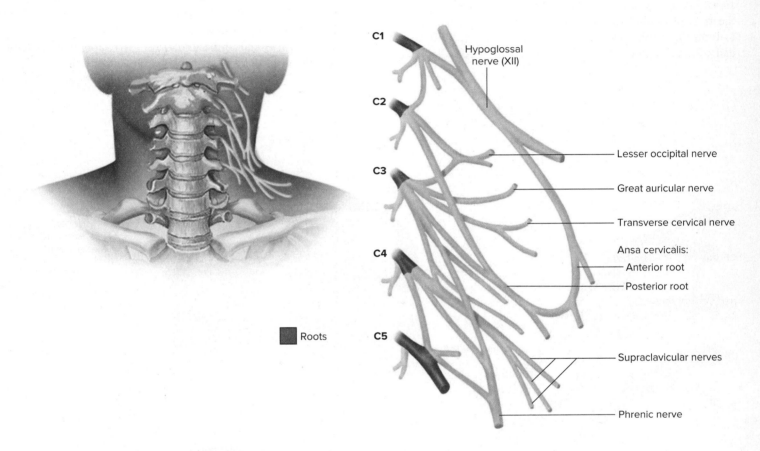

FIGURE 13.15 The Cervical Plexus. **APR**

❓ *Predict the consequences of a surgical accident in which a phrenic nerve is severed.*

TABLE 13.3	The Cervical Plexus		
Nerve	**Composition**	**Cutaneous and Other Sensory Innervation**	**Muscular Innervation (Motor and Proprioceptive)**
Lesser occipital n.	Somatosensory	Upper third of medial surface of external ear, skin posterior to ear, posterolateral neck	None
Great auricular n.	Somatosensory	Most of the external ear, mastoid region, region from parotid salivary gland (see fig. 10.7) to slightly inferior to angle of mandible	None
Transverse cervical n.	Somatosensory	Anterior and lateral neck, underside of chin	None
Ansa cervicalis	Motor	None	Omohyoid, sternohyoid, and sternothyroid muscles (see table 10.2)
Supraclavicular nn.	Somatosensory	Lower anterior and lateral neck, shoulder, anterior chest	None
Phrenic n.	Mixed	Diaphragm, pleura, and pericardium	Diaphragm (see table 10.4)

The Brachial Plexus

The **brachial plexus (figs. 13.16, 13.17)** is formed predominantly by the anterior rami of nerves C5 to T1 (C4 and T2 make small contributions). It passes over the first rib into the axilla and innervates the upper limb and some muscles of the neck and shoulder. This plexus is well known for its conspicuous M or W shape seen in cadaver dissections.

The subdivisions of this plexus are called *roots, trunks, divisions,* and *cords* (color-coded in fig. 13.16).The five **roots** are the anterior rami of C5 through T1. Roots C5 and C6 converge to form the **upper trunk;** C7 continues as the **middle trunk;** and C8 and T1 converge to form the **lower trunk.** Each trunk divides into an **anterior** and **posterior division.** As the body is dissected from the anterior side of the shoulder inward, the anterior divisions are found in front of the posterior ones. Finally, the six divisions merge to form three large fiber bundles—the **lateral, posterior,** and **medial cords.** From these cords arise the five major nerves listed in **table 13.4** in the order of figure 13.16 from superior to inferior.

Among other functions, these nerves carry sensory signals from the skin of the shoulder and upper limb to the spinal cord. **Figure 13.18** shows the regions of the wrist and hand innervated by the sensory fibers of these nerves (except for the axillary nerve, which doesn't extend to the forearm). The motor fibers of these nerves control muscles of the shoulder and upper limb.

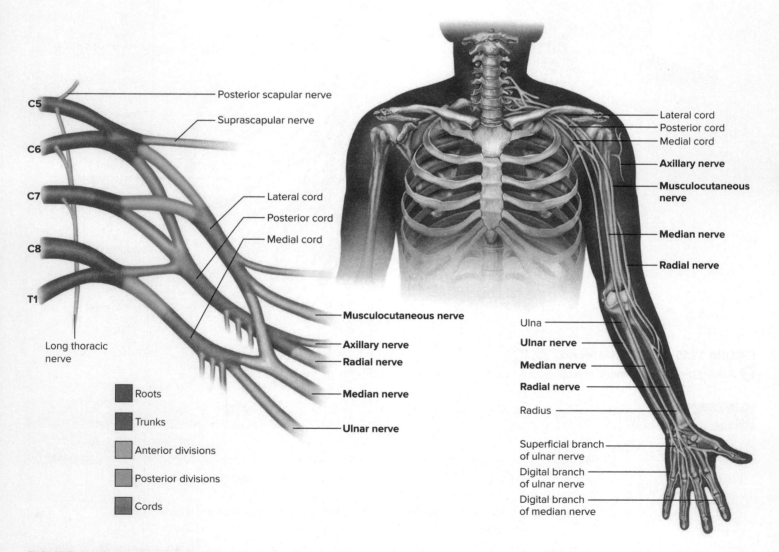

FIGURE 13.16 The Brachial Plexus. The labeled nerves innervate muscles tabulated in chapter 10, and those in boldface are further detailed in this table. **APR**

- Lateral cord
- Posterior cord
- Musculocutaneous nerve
- Axillary nerve
- Medial cord
- Radial nerve
- Median nerve
- Ulnar nerve
- Long thoracic nerve

FIGURE 13.17 Brachial Plexus of the Cadaver. Anterior view of the left shoulder. Most of the other structures resembling nerves in this photograph are blood vessels.

Christine Eckel/McGraw-Hill Education

TABLE 13.4	The Brachial Plexus			
Nerve	**Composition**	**Cord of Origin**	**Cutaneous and Joint Innervation (Sensory)**	**Muscular Innervation (Motor and Proprioceptive)**
Musculocutaneous n.	Mixed	Lateral	Skin of anterolateral forearm; elbow joint	Brachialis, biceps brachii, and coracobrachialis muscles (see tables 10.9, 10.10)
Axillary n.	Mixed	Posterior	Skin of lateral shoulder and arm; shoulder joint	Deltoid and teres minor muscles (see table 10.9)
Radial n.	Mixed	Posterior	Skin of posterior arm; posterior and lateral forearm and wrist; joints of elbow, wrist, and hand	Mainly extensor muscles of posterior arm and forearm (see tables 10.10, 10.11)
Median n.	Mixed	Lateral and medial	Skin of lateral two-thirds of hand; tips of digits I–IV; joints of hand	Mainly forearm flexors; thenar group and lumbricals I–II of hand (see tables 10.10–10.12)
Ulnar n.	Mixed	Medial	Skin of palmar and medial hand and digits III–V; joints of elbow and hand	Some forearm flexors; adductor pollicis; hypothenar group; interosseous muscles; lumbricals III–IV (see tables 10.11, 10.12)

- ■ Radial n.
- ■ Ulnar n.
- ■ Median n.
- ■ Musculocutaneous n.

(a) Anterior (palmar)   **(b) Posterior (dorsal)**

FIGURE 13.18 Cutaneous Innervation of the Hand by Nerves from the Brachial Plexus. (a) Anterior (palmar) view. (b) Posterior (dorsal) view.

DEEPER INSIGHT 13.3

CLINICAL APPLICATION

Shingles

Chickenpox *(varicella)*, a common disease of early childhood, is caused by the *varicella-zoster*[23] virus. It produces an itchy rash that usually clears up without complications. The virus, however, remains for life in the posterior root ganglia, kept in check by the immune system. If the immune system is compromised, however, the virus can travel along the sensory nerve fibers by *fast axonal transport* (see section 12.2d) and cause *shingles (herpes*[24] *zoster)*. The signs of shingles are a painful trail of skin discoloration and fluid-filled vesicles along the path of the nerve **(fig. 13.19).** These usually appear in the chest and waist, often on just one side of the body. In some cases, lesions appear on one side of the face, especially in and around the eye, and occasionally in the mouth.

There is no cure, and the vesicles usually heal spontaneously in 1 to 3 weeks. In the meantime, aspirin and steroidal ointments can help to relieve the pain and inflammation of the lesions. Antiviral drugs such as acyclovir can shorten the course of an episode of shingles, but only if taken within the first 2 to 3 days of outbreak. Even after the lesions disappear, however, some people suffer intense pain along the course of the nerve *(postherpetic neuralgia, PHN),* lasting for months or even years. PHN is difficult to treat, but pain relievers and antidepressants

FIGURE 13.19 Shingles Lesion Tracking the Course of a Sensory Nerve.

Franciscodiazpagador/Getty Images

are of some help. Shingles is particularly common after the age of 50. Childhood vaccination against varicella reduces the risk of shingles later in life. Adult vaccination is recommended in the United States for healthy persons over age 60.

---

[23]*varicella* = little spot; zoster = girdle
[24]*herpes* = creeping

The Lumbar Plexus

The **lumbar plexus (fig. 13.20)** is formed from the anterior rami of nerves L1 to L4 and some fibers from T12. With only five roots and two divisions, it is less complex than the brachial plexus. It gives rise to six major nerves, listed in **table 13.5** in anatomically descending order.

The sensory fibers of these nerves carry signals from the hip and knee joints and from skin of the genitalia, lower abdominal and gluteal regions, and the thigh, leg, and foot. The motor fibers control muscles of the hip, thigh, and scrotum.

TABLE 13.5	The Lumbar Plexus		
Nerve	**Composition**	**Cutaneous and Joint Innervation (Sensory)**	**Muscular Innervation (Motor and Proprioceptive)**
Iliohypogastric n.	Mixed	Skin of lower anterior abdominal and posterolateral gluteal regions	Internal and external oblique and transverse abdominal muscles (see table 10.5)
Ilioinguinal n.	Mixed	Skin of upper medial thigh; male scrotum and root of penis; female labia majora	Internal oblique
Genitofemoral n.	Mixed	Skin of middle anterior thigh; male scrotum; female labia majora	Male cremaster muscle (see fig. 27.7)
Lateral femoral cutaneous n.	Somatosensory	Skin of anterior and upper lateral thigh	None
Femoral n.	Mixed	Skin of anterior, medial, and lateral thigh and knee; skin of medial leg and foot; hip and knee joints	Iliacus, pectineus, quadriceps femoris, and sartorius muscles (see tables 10.13, 10.14)
Obturator n.	Mixed	Skin of medial thigh; hip and knee joints	Obturator externus; medial (adductor) thigh muscles (see table 10.13)

FIGURE 13.20 The Lumbar Plexus. APR

The Sacral and Coccygeal Plexuses

The **sacral plexus** is formed from the anterior rami of nerves L4, L5, and S1 through S4. It has six roots and anterior and posterior divisions. Since it is connected to the lumbar plexus by fibers that run through the *lumbosacral trunk,* the two plexuses are sometimes referred to collectively as the *lumbosacral plexus.* The **coccygeal plexus** is a tiny plexus formed from the anterior rami of S4, S5, and Co1 (**fig. 13.21; table 13.6**).

The *tibial* and *common fibular nerves* travel together through a connective tissue sheath; they are referred to collectively as the **sciatic nerve** (sy-AT-ic), a common focus of injury and pain (see Deeper Insight 13.4). The sciatic nerve passes

through the greater sciatic notch of the hip bone, extends for the length of the thigh, and ends at the popliteal fossa (back of the knee). Here, the tibial and common fibular nerves diverge and follow their separate paths into the leg. The tibial nerve descends through the leg and then gives rise to the *medial and plantar nerves* of the foot. The common fibular nerve divides into *deep* and *superficial fibular nerves.*

The sensory fibers of these nerves carry signals from joints of the hip, knee, and foot; from the gluteal and perineal regions and the genitals; and from skin from the gluteal region to the foot. The motor fibers control muscles of the gluteal and perineal regions, thigh, leg, and foot.

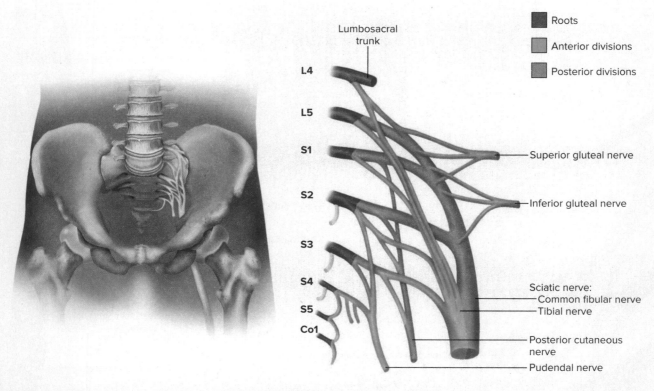

FIGURE 13.21 The Sacral and Coccygeal Plexuses. APR

TABLE 13.6	The Sacral and Coccygeal Plexuses		
Nerve	**Composition**	**Cutaneous and Joint Innervation (Sensory)**	**Muscular Innervation (Motor and Proprioceptive)**
Superior gluteal n.	Mixed	Hip joint	Gluteus minimus, gluteus medius, and tensor fasciae latae muscles (see table 10.13)
Inferior gluteal n.	Mixed	None	Gluteus maximus muscle (see table 10.13)
Posterior cutaneous n.	Somatosensory	Skin of gluteal region, perineum, posterior and medial thigh, popliteal fossa, and upper posterior leg	None
Tibial n.	Mixed	Skin of posterior leg; plantar skin; knee and foot joints	Hamstring muscles; posterior muscles of leg (see tables 10.14, 10.15); most intrinsic foot muscles (via plantar nerves) (see table 10.16)
Fibular (peroneal) nn. (common, deep, and superficial)	Mixed	Skin of anterior distal third of leg, dorsum of foot, and toes I–II; knee joint	Biceps femoris muscle; anterior and lateral muscles of leg; extensor digitorum brevis muscle of foot (see tables 10.14–10.16)
Pudendal n.	Mixed	Skin of penis and scrotum of male; clitoris, labia majora and minora, and lower vagina of female	Muscles of perineum (see table 10.7)

DEEPER INSIGHT 13.4

CLINICAL APPLICATION

Nerve Injuries

The radial and sciatic nerves are especially vulnerable to injury. The radial nerve, which passes through the axilla, may be compressed against the humerus by improperly adjusted crutches, causing *crutch paralysis.* A similar injury often resulted from the now-discredited practice of correcting a dislocated shoulder by putting a foot in a person's armpit and pulling on the arm. One consequence of radial nerve injury is *wrist drop*—the fingers, hand, and wrist are chronically flexed because the extensor muscles supplied by the radial nerve are paralyzed and cannot oppose the flexors.

Because of its position and length, the sciatic nerve of the hip and thigh is the most vulnerable nerve in the body. Trauma to this nerve produces *sciatica,* a sharp pain that travels from the gluteal region along the posterior side of the thigh and leg as far as the ankle. Ninety percent of cases result from a herniated intervertebral disc or osteoarthritis of the lower spine, but sciatica can also be caused by pressure from a pregnant uterus, dislocation of the hip, injections in the wrong area of the buttock, or sitting for a long time on the edge of a hard chair. Men sometimes suffer sciatica because of the habit of sitting on a wallet carried in the hip pocket.

13.2d Cutaneous Innervation and Dermatomes

Each spinal nerve except C1 receives sensory input from a specific area of skin called a **dermatome.**[25] A *dermatome map* (**fig. 13.22**) is a diagram of the cutaneous regions innervated by each spinal nerve. Such a map is oversimplified, however, because the dermatomes overlap at their edges by as much as 50%. Therefore, severance of one sensory nerve root does not entirely deaden sensation from a dermatome. It is necessary to sever or anesthetize three sequential spinal nerves to produce a total loss of sensation from one dermatome. Spinal nerve damage is assessed by testing the dermatomes with pinpricks and noting areas in which the patient has no sensation.

BEFORE YOU GO ON

Answer the following questions to test your understanding of the preceding section:

6. What is meant by the *anterior and posterior roots* of a spinal nerve? Which of these is sensory and which is motor?

7. Where are the neurosomas of the posterior root located? Where are the neurosomas of the anterior root?

8. List the five plexuses of spinal nerves and state where each one is located.

9. Identify which plexus gives rise to each of the following nerves: axillary, ilioinguinal, obturator, phrenic, pudendal, radial, and sciatic.

FIGURE 13.22 A Dermatome Map of the Anterior Aspect of the Body. Each zone of the skin is innervated by sensory branches of the spinal nerves indicated by the labels. Nerve C1 does not innervate the skin.

[25]*derma* = skin; *tome* = segment, part

13.3 Somatic Reflexes

Expected Learning Outcomes

When you have completed this section, you should be able to

a. define *reflex* and explain how reflexes differ from other motor actions;

b. describe the general components of a typical reflex arc; and

c. explain how the basic types of somatic reflexes function.

Most of us have had our reflexes tested with a little rubber hammer; a tap below the knee produces an uncontrollable jerk of the leg, for example. In this section, we discuss what reflexes are and how they are produced by an assembly of receptors, neurons, and effectors. We also survey the different types of neuromuscular reflexes and how they are important in motor coordination of our everyday tasks.

13.3a The Nature of Reflexes

Reflexes are quick, involuntary, stereotyped reactions of glands or muscles to stimulation. This definition sums up four important properties:

1. Reflexes *require stimulation*—they are not spontaneous actions like muscle tics but responses to sensory input.

2. Reflexes are *quick*—they generally involve only a few interneurons, or none, and minimum synaptic delay.

3. Reflexes are *involuntary*—they occur without intent, often without our awareness, and they are difficult to suppress. Given an adequate stimulus, the response is essentially automatic. You may become conscious of the stimulus that evoked a reflex, and this awareness may enable you to correct or avoid a potentially dangerous situation, but awareness is not a part of the reflex itself. It may come after the reflex action has been completed, and somatic reflexes can occur even if the spinal cord has been severed so that no stimuli reach the brain.

4. Reflexes are *stereotyped*—they occur in essentially the same way every time; the response is very predictable, unlike the variability of voluntary movement.

Reflexes include glandular secretion and contractions of all three types of muscle. The reflexes of skeletal muscle are called **somatic reflexes,** since they involve the somatic nervous system. Chapter 15 concerns the *visceral reflexes* of organs such as the heart and intestines. Somatic reflexes have traditionally been called *spinal reflexes,* but this is a misleading expression for two reasons: (1) Spinal reflexes are not exclusively somatic; visceral reflexes also involve the spinal cord. (2) Some somatic reflexes are mediated more by the brain than by the spinal cord.

A somatic reflex employs a **reflex arc,** in which signals travel along the following pathway **(fig. 13.23):**

1. *somatic receptors* in the skin, muscles, and tendons;

2. *afferent nerve fibers,* which carry information from these receptors to the posterior horn of the spinal cord or to the brainstem;

FIGURE 13.23 A Representative Reflex Arc.

3. an *integrating center,* a point of synaptic contact between neurons in the gray matter of the cord or brainstem;

4. *efferent nerve fibers,* which carry motor impulses to the muscles; and

5. *effectors,* the muscles that carry out the response.

In most reflex arcs, the integrating center includes one or more interneurons. Synaptic events in the integrating center determine whether the efferent neurons issue signals to the muscles. The more interneurons there are, the more complex the information processing can be, but with more synapses, there is a longer delay between input and output.

13.3b The Muscle Spindle

Many somatic reflexes involve stretch receptors called **muscle spindles** embedded in the muscles. These are among the body's **proprioceptors,** sense organs specialized to monitor the position and movement of body parts. The function of muscle spindles is to inform the brain of muscle length and body movements. This enables the brain to send motor commands back to the muscles that control muscle tone, posture, coordinated movement, and corrective reflexes (for example, to keep one's balance). Spindles are especially abundant in muscles that require fine control. Hand and foot muscles have 100 or more spindles per gram of muscle, whereas there are relatively few in large muscles with coarse movements, and none at all in the middle-ear muscles.

A muscle spindle is a bundle of usually seven or eight small, modified muscle fibers enclosed in an elongated fibrous capsule about 5 to 10 mm long **(fig. 13.24).** Spindles are especially concentrated at the ends of a muscle, near its tendons. The modified muscle fibers within the spindle are called

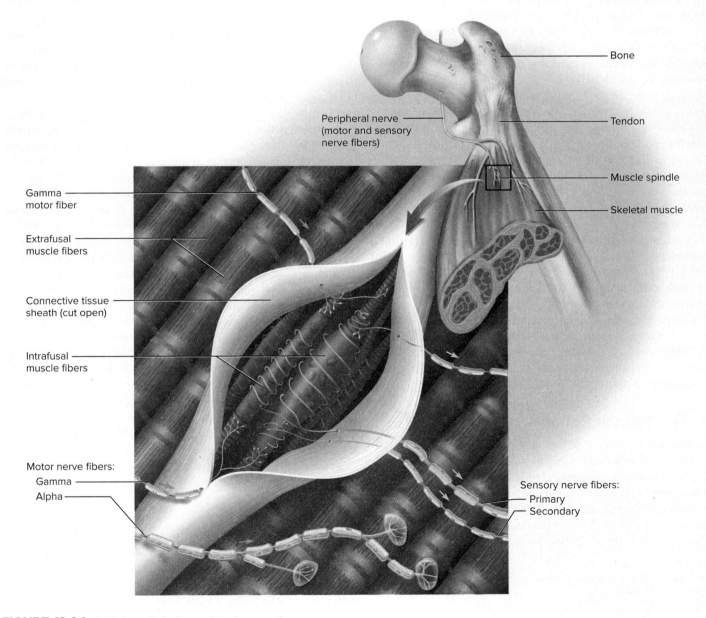

FIGURE 13.24 A Muscle Spindle and Its Innervation.

intrafusal[26] **fibers,** whereas those that make up the rest of the muscle and do its work are called **extrafusal fibers.**

Each end of an intrafusal fiber has a few sarcomeres. A **gamma motor neuron** of the spinal cord innervates each end and stimulates its contraction. This maintains tension and sensitivity of the intrafusal fiber, preventing it from going slack like an unstretched rubber band when a muscle shortens. Spinal motor neurons that supply the extrafusal muscle fibers are called **alpha motor neurons.** Up to now, we have studied only that type and the neuromuscular junctions they form with muscle, but nearly one-third of all spinal motor neurons are the gamma type—evidence of the great importance of muscle spindles.

The long midportion of an intrafusal fiber lacks sarcomeres and cannot contract, but is supplied by two types of sensory nerve fibers: *primary afferent fibers* that monitor muscle length and how rapidly it changes, which are therefore very responsive to sudden body movements; and *secondary afferent fibers* that monitor length only, not rate of change. Both of these sensory fiber types enter the posterior horn of the spinal cord, synapse on the alpha motor neurons and regulate their firing, and also send branches up the spinal cord to the brain. Through these fibers, the brain constantly but subconsciously monitors the length and tension of nearly every skeletal muscle throughout the body. This input is vital to the maintenance of posture, fine control of movements, and corrective reflexes.

Suppose, for example, you are standing on the deck of a boat that is gently rocking on the waves. At one moment, your body begins to tip forward. This stretches your calf muscles and their muscle spindles, setting off sensory signals to the spinal cord. The CNS responds to this by tensing your calf muscles to keep you from falling and to restore or maintain your upright posture. Then the boat rocks the other way and you begin to tip to the rear. The spindles in the calf muscles are now compressed and their signaling rate drops. Such input from the spindles inhibits the alpha motor neurons of the calf muscles, relaxing those muscles so they don't pull you farther backward. At the same time, your backward tilt stretches spindles in your anterior leg and thigh muscles, leading to their contraction and preventing you from falling over backward.

You can well imagine the importance of these reflexes to the coordination of such common movements as walking and dancing. In more subtle ways, all day long, your brain monitors input from the spindles of opposing muscles and makes fine adjustments in muscle tension to maintain your posture and coordination.

13.3c The Stretch Reflex

When a muscle is suddenly stretched, it "fights back"—it contracts, increases tone, and feels stiffer than an unstretched muscle. This response, called the **stretch (myotatic**[27]**) reflex,** helps to maintain equilibrium and posture, as we just saw in the rocking boat example. To take another case, if your head starts to tip forward, it stretches muscles at the back of your neck. This stimulates their muscle spindles, which send signals to the cerebellum by way of the brainstem. The

cerebellum integrates this information and relays it to the cerebral cortex, and the cortex sends signals back, via the brainstem, to the muscles. The muscles contract and raise your head.

Stretch reflexes often feed back not to a single muscle but to a set of synergists and antagonists. Since the contraction of a muscle on one side of a joint stretches the antagonist on the other side, the flexion of a joint creates a stretch reflex in the extensors, and extension creates a stretch reflex in the flexors. (Think of the way your biceps brachii is stretched when you extend your elbow, for example.) Consequently, stretch reflexes are valuable in stabilizing joints by balancing the tension of the extensors and flexors. They also dampen (smooth out) muscle action. Without stretch reflexes, a person's movements tend to be jerky. Stretch reflexes are especially important in coordinating vigorous and precise movements such as dance.

A stretch reflex is mediated primarily by the brain and is not, therefore, strictly a spinal reflex, but a weak component of it is spinal and occurs even if the spinal cord is severed from the brain. The spinal component can be more pronounced if a muscle is stretched very suddenly. This occurs in the reflexive contraction of a muscle when its tendon is tapped, as in the familiar *patellar* (knee-jerk) *reflex.* Tapping the patellar ligament with a reflex hammer abruptly stretches the quadriceps femoris muscle of the thigh **(fig. 13.25).** This stimulates numerous muscle spindles in the quadriceps and sends an intense volley of signals to the spinal cord, mainly by way of primary afferent fibers.

In the spinal cord, these fibers synapse directly with the alpha motor neurons that return to the muscle, thus forming **monosynaptic reflex arcs.** That is, there is only one synapse between the afferent and efferent neuron, so there is little synaptic delay and a very prompt response. The alpha motor neurons excite the quadriceps, making it contract and creating the knee jerk.

There are many other tendon reflexes. A tap on the calcaneal tendon causes plantar flexion of the foot, a tap on the triceps brachii tendon causes extension of the elbow, and a tap on the masseter causes clenching of the jaw. Testing somatic reflexes is valuable in diagnosing many diseases that cause exaggeration, inhibition, or absence of reflexes—for example, neurosyphilis and other infectious diseases, diabetes mellitus, multiple sclerosis, alcoholism, hormone and electrolyte imbalances, and lesions of the nervous system.

Stretch reflexes and other muscle contractions often depend on **reciprocal inhibition,** a reflex that prevents muscles from working against each other by inhibiting antagonists. In the knee jerk, for example, the quadriceps wouldn't produce much joint movement if its antagonists, the hamstring muscles, contracted at the same time. But reciprocal inhibition prevents that from happening. Some branches of the sensory fibers from the quadriceps muscle spindles stimulate spinal interneurons that, in turn, *inhibit* the alpha motor neurons of the hamstrings (fig. 13.25). The hamstrings remain relaxed and allow the quadriceps to extend the knee.

13.3d The Flexor (Withdrawal) Reflex

A **flexor reflex** is the quick contraction of flexor muscles resulting in the withdrawal of a limb from an injurious stimulus. For example, suppose you are wading in a lake and step on a broken bottle with your right foot **(fig. 13.26).** Even before you are consciously

---

[26]*intra* = within; *fus* = spindle
[27]*myo* = muscle; *tat* (from *tasis*) = stretch

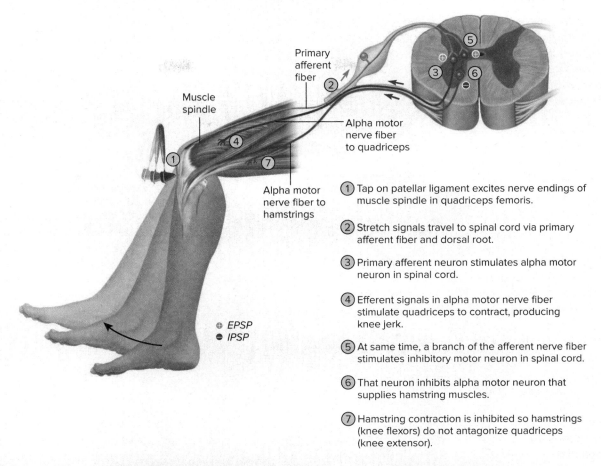

① Tap on patellar ligament excites nerve endings of muscle spindle in quadriceps femoris.

② Stretch signals travel to spinal cord via primary afferent fiber and dorsal root.

③ Primary afferent neuron stimulates alpha motor neuron in spinal cord.

④ Efferent signals in alpha motor nerve fiber stimulate quadriceps to contract, producing knee jerk.

⑤ At same time, a branch of the afferent nerve fiber stimulates inhibitory motor neuron in spinal cord.

⑥ That neuron inhibits alpha motor neuron that supplies hamstring muscles.

⑦ Hamstring contraction is inhibited so hamstrings (knee flexors) do not antagonize quadriceps (knee extensor).

⊕ EPSP
⊖ IPSP

FIGURE 13.25 The Patellar Tendon Reflex Arc and Reciprocal Inhibition of the Antagonistic Muscle. Plus signs indicate excitation of a postsynaptic cell (EPSP), and the minus sign indicates inhibition (IPSP). The tendon reflex occurs in the quadriceps femoris muscle, while the hamstring muscles exhibit reciprocal inhibition so they don't contract and oppose the quadriceps.

❓ *Why is no IPSP shown at point 7 if the contraction of this muscle is being inhibited?*

aware of the pain, you quickly pull your foot away before the glass penetrates any deeper. This action involves contraction of the flexors and relaxation of the extensors in that limb; the latter is another case of reciprocal inhibition.

The protective function of this reflex requires more than a quick jerk like a tendon reflex, so it involves more complex neural pathways. Sustained contraction of the flexors is produced by a parallel after-discharge circuit in the spinal cord (see fig. 12.32). This circuit is part of a **polysynaptic reflex arc**—a pathway in which signals travel over many synapses on their way back to the muscle. Some signals follow routes with only a few synapses and return to the flexor muscles quickly. Others follow routes with more synapses, and therefore more delay, so they reach the flexor muscles a little later. Consequently, the flexor muscles receive prolonged output from the spinal cord and not just one sudden stimulus as in a stretch reflex. By the time these efferent signals begin to die out, you will probably be consciously aware of the pain and begin taking voluntary action to prevent further harm.

13.3e The Crossed Extension Reflex

In the preceding situation, if *all* you did was to quickly lift the injured leg from the lake bottom, you would fall over. To prevent this and maintain your balance, other reflexes shift your center of gravity over the leg that is still planted on the ground. The **crossed extension reflex** is the contraction of extensor muscles in the limb opposite from the one that is withdrawn (fig. 13.26). It extends and stiffens that limb and enables you to keep your balance. To produce this reflex, branches of the afferent nerve fibers cross from the stimulated side of the body to the contralateral side of the spinal cord. There, they synapse with interneurons, which, in turn, excite or inhibit alpha motor neurons to the muscles of the contralateral limb.

In the ipsilateral leg (the side that was hurt), you would contract your flexors and relax your extensors to lift the leg from the ground. On the contralateral side, you would relax your flexors and contract the extensors to stiffen that leg, since it must suddenly support your entire body. At the same time, signals travel up the spinal cord and cause contraction of contralateral muscles of the hip and abdomen, such as your internal and external obliques, to shift your center of gravity over the extended leg. To a large extent, the coordination of all these muscles and maintenance of equilibrium are mediated by the cerebellum and cerebral cortex.

The flexor reflex employs an **ipsilateral reflex arc**—one in which the sensory input and motor output are on the same side of the spinal cord. The crossed extension reflex employs a **contralateral reflex arc,** in which the input and output are on opposite sides. An

② Sensory neuron activates multiple interneurons

③ Ipsilateral motor neurons to flexor excited

⑤ Contralateral motor neurons to extensor excited

④ Ipsilateral flexor contracts

⑥ Contralateral extensor contracts

① Stepping on glass stimulates pain receptors in right foot

Withdrawal of right leg (flexor reflex)

Extension of left leg (crossed extension reflex)

FIGURE 13.26 The Flexor and Crossed Extension Reflexes. A pain stimulus triggers a withdrawal reflex, which results in contraction of flexor muscles of the injured limb. At the same time, a crossed extension reflex results in contraction of extensor muscles of the opposite limb. The latter reflex aids in balance when the injured limb is raised. Note that for each limb, while the agonist contracts, the alpha motor neuron to its antagonist is inhibited, as indicated by the red minus signs in the spinal cord. **APR**

❓ *Would you expect this reflex arc to show more synaptic delay, or less, than the ones in figure 13.25? Why?*

intersegmental reflex arc is one in which the input and output occur at different levels (segments) of the spinal cord—for example, when pain to the foot causes contractions of abdominal and hip muscles higher up the body. Note that all of these reflex arcs can function simultaneously to produce a coordinated protective response to pain.

▶▶▶**APPLY WHAT YOU KNOW**

In section 12.6d, you read of serial and parallel processing. Which of these do you think best describes the stretch reflex, flexor reflex, and crossed extension reflex? Explain.

13.3f The Tendon Reflex

Tendon organs are proprioceptors located in a tendon near its junction with a muscle (**fig. 13.27**). A tendon organ is about 0.5 mm long. It consists of an encapsulated bundle of small, loose collagen fibers and one or more nerve fibers that penetrate the capsule and end in flattened leaflike processes between the collagen fibers. As long as the tendon is slack, its collagen fibers are slightly spread and put little pressure on the nerve endings. When muscle contraction pulls on the tendon, the collagen fibers come together like the two sides of a stretched rubber band and squeeze the nerve endings

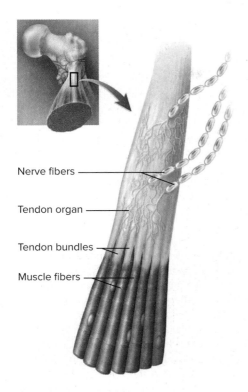

Nerve fibers

Tendon organ

Tendon bundles

Muscle fibers

FIGURE 13.27 A Tendon Organ.

between them. The nerve fiber sends signals to the spinal cord that provide the CNS with feedback on the degree of muscle tension at the joint.

The **tendon reflex** is a response to excessive tension on the tendon. It inhibits alpha motor neurons to the muscle so the muscle doesn't contract as strongly. This moderates muscle contraction before it tears a tendon or pulls it loose from the muscle or bone. Nevertheless, strong muscles and quick movements sometimes damage a tendon before the reflex can occur, causing such athletic injuries as a ruptured calcaneal tendon.

The tendon reflex also functions when some parts of a muscle contract more than others. It inhibits the muscle fibers connected with overstimulated tendon organs so their contraction is more comparable to the contraction of the rest of the muscle. This spreads the workload more evenly over the entire muscle, which is beneficial in such actions as maintaining a steady grip on a tool.

Table 13.7 and Deeper Insight 13.5 describe some injuries and other disorders of the spinal cord and spinal nerves.

BEFORE YOU GO ON

Answer the following questions to test your understanding of the preceding section:

10. Name five structural components of a typical somatic reflex arc. Which of these is absent from a monosynaptic arc?

11. State the function of each of the following in a muscle spindle: intrafusal fibers, gamma motor neurons, and primary afferent fibers.

12. Explain how nerve fibers in a tendon sense the degree of tension in a muscle.

13. Why must the withdrawal reflex, but not the stretch reflex, involve a polysynaptic reflex arc?

14. Explain why the crossed extension reflex must accompany a withdrawal reflex of the leg.

TABLE 13.7	Some Disorders of the Spinal Cord and Spinal Nerves
Guillain–Barré syndrome	An acute demyelinating nerve disorder often triggered by viral infection, resulting in muscle weakness, elevated heart rate, unstable blood pressure, shortness of breath, and sometimes death from respiratory paralysis
Neuralgia	General term for nerve pain, often caused by pressure on spinal nerves from herniated intervertebral discs
Paresthesia	Abnormal sensations of prickling, burning, numbness, or tingling in the absence of actual stimulation; a symptom of peripheral nerve disorders
Peripheral neuropathy	Any loss of sensory or motor function due to nerve injury; also called *nerve palsy*
Rabies (hydrophobia)	A disease usually contracted from animal bites, involving viral infection that spreads via somatic motor nerve fibers to the CNS and then autonomic nerve fibers; leads to seizures, coma, and death; invariably fatal if not treated before CNS symptoms appear
Spinal meningitis	Inflammation of the spinal meninges due to viral, bacterial, or other infection

You can find other spinal cord and peripheral nerve disorders described in the following places:

Carpal tunnel syndrome in Deeper Insight 10.5; *multiple sclerosis* and *Tay–Sachs disease* in Deeper Insight 12.2; *polio* and *amyotrophic lateral sclerosis* in Deeper Insight 13.2; *shingles* in Deeper Insight 13.3; *crutch paralysis* and *sciatica* in Deeper Insight 13.4; *spinal cord trauma and the forms of paralysis* in Deeper Insight 13.5; *leprosy* and *diabetic neuropathy* in section 16.2d, and the latter also in section 17.7e.

DEEPER INSIGHT 13.5

CLINICAL APPLICATION

Spinal Cord Trauma

In the United States, 10,000 to 12,000 people are newly paralyzed each year by spinal cord trauma, usually as a result of vertebral fractures. The greatest incidence is among males from 16 to 30 years old, because of their high-risk behaviors. Fifty-five percent of their injuries are from automobile and motorcycle accidents, 18% from sports, and 15% from gunshot and stab wounds. Elderly people are also at above-average risk because of falls, and in times of war, battlefield injuries account for many cases.

Effects of Injury

Complete *transection* (severance) of the spinal cord causes immediate loss of motor control at and below the level of the injury. Transection superior to segment C4 presents a threat of respiratory failure. Victims also lose sensation from the level of injury and below, although some patients temporarily feel burning pain within one or two dermatomes of the level of the lesion.

In the early stage, victims exhibit a syndrome called *spinal shock.* Muscles below the level of injury exhibit flaccid paralysis (inability to contract) and an absence of reflexes because of the lack of stimulation from higher levels of the CNS. For 8 days to 8 weeks after the accident, the patient typically lacks bladder and bowel reflexes and thus retains urine and feces. Lacking sympathetic stimulation to the blood vessels, a patient may exhibit *neurogenic shock* in which the vessels dilate and blood pressure drops dangerously low. Fever may occur because the hypothalamus cannot induce sweating to cool the body. Spinal shock can last from a few days to several weeks (usually 7 to 20 days).

As spinal shock subsides, somatic reflexes begin to reappear, at first in the toes and progressing to the feet and legs. Autonomic reflexes also reappear. Contrary to the earlier urinary and fecal retention, a patient now has the opposite problem, incontinence, as the rectum and bladder empty reflexively in response to stretch. Both the somatic and autonomic nervous systems typically exhibit exaggerated reflexes, a state called *hyperreflexia* or the *mass reflex reaction.* Stimuli such as a full bladder or cutaneous touch can trigger an extreme cardiovascular reaction. The systolic blood pressure, normally about 120 mm Hg, jumps to as high as 300 mm Hg. This causes intense headaches and sometimes a stroke. Pressure receptors in the major arteries sense this rise in blood pressure and activate a reflex that slows the heart, sometimes to a rate as low as 30 or 40 beats/minute *(bradycardia),* compared with a normal rate of 70 to 80. The patient may also experience profuse sweating and blurred vision.

Men at first lose the capacity for erection and ejaculation. They may recover these functions later and become capable of ejaculating and fathering children, but without sexual sensation. In females, menstruation may become irregular or cease.

The most serious permanent effect of spinal cord trauma is paralysis. The flaccid paralysis of spinal shock later changes to spastic paralysis as spinal reflexes are regained but lack inhibitory control from the brain. Spastic paralysis typically starts with chronic flexion of the hips and knees (flexor spasms) and progresses to a state in which the limbs become straight and rigid (extensor spasms). Three forms of muscle paralysis are *paraplegia,* a paralysis of both lower limbs resulting from spinal cord lesions at levels T1 to L1; *quadriplegia,* the paralysis of all four limbs resulting from lesions above level C5; and *hemiplegia,* paralysis of one side of the body, usually resulting not from spinal cord injuries but from a stroke or other brain lesion. Spinal cord lesions from C5 to C7 can produce a state of partial quadriplegia—total paralysis of the lower limbs and partial paralysis *(paresis,* or weakness) of the upper limbs.

Pathogenesis

Spinal cord trauma produces two stages of tissue destruction. The first is instantaneous—the destruction of cells by the traumatic event itself. The second stage, a wave of tissue death by necrosis and apoptosis, begins in minutes and lasts for days. It is far more destructive than the initial injury, typically converting a lesion in one spinal cord segment to a lesion that spans four or five segments, two above and two below the original site.

Microscopic hemorrhages appear in the gray matter and pia mater within minutes and grow larger over the next 2 hours. The white matter becomes edematous (swollen). Hemorrhaging and edema spread to adjacent segments of the cord and can fatally affect respiration or brainstem function when it occurs in the cervical region. *Ischemia* (iss-KEE-me-uh), the lack of blood, quickly leads to necrosis. The white matter regains circulation in about 24 hours, but the gray matter remains ischemic. Inflammatory cells (leukocytes and macrophages) infiltrate the lesion as the circulation recovers, and while they clean up necrotic tissue, they also contribute to the damage by releasing destructive free radicals and other toxic chemicals. The necrosis worsens, and is accompanied by another form of cell death, apoptosis. Apoptosis of the spinal oligodendrocytes, the myelinating glial cells of the CNS, results in demyelination of spinal nerve fibers, followed by death of the neurons.

In as little as 4 hours, this second wave of destruction, called *posttraumatic infarction,* consumes about 40% of the cross-sectional area of the spinal cord; within 24 hours, it destroys 70%. As many as five segments of the cord become transformed into a fluid-filled cavity, which is replaced with collagenous scar tissue over the next 3 to 4 weeks. This scar is one of the obstacles to the regeneration of lost nerve fibers.

Treatment

The first priority in treating a spinal injury patient is to immobilize the spine to prevent further trauma. Respiratory or other life support may also be required. Methylprednisolone, a steroid, dramatically improves recovery. Given within 3 hours of the trauma, it reduces injury to cell membranes and inhibits inflammation and apoptosis.

After these immediate requirements are met, reduction (repair) of the fracture is important. If a CT or MRI scan indicates spinal cord compression by the vertebral canal, a *decompression laminectomy* may be performed, in which vertebral laminae are removed from the affected region. CT and MRI have helped a great deal in recent decades for assessing vertebral and spinal cord damage, guiding surgical treatment, and improving recovery. Physical therapy is important for maintaining muscle and joint function as well as promoting the patient's psychological recovery.

Treatment strategies for spinal cord injuries are a vibrant field of contemporary medical research. Some current interests are the use of antioxidants to reduce free radical damage, and the implantation of pluripotent stem cells, which has produced significant (but not perfect) recovery from spinal cord lesions in rats. Public hopes have often been raised by promising studies reported in the scientific literature and news media, only to be dashed by the inability of other laboratories to repeat and confirm the results.

STUDY GUIDE

▶ Assess Your Learning Outcomes

To test your knowledge, discuss the following topics with a study partner or in writing, ideally from memory.

13.1 The Spinal Cord

1. Functions of the spinal cord
2. Skeletal landmarks that mark the extent of the adult spinal cord, and what occupies the vertebral canal inferior to the spinal cord
3. The four regions of the spinal cord and the basis for their names
4. What defines one segment of the cord
5. Two enlargements of the cord and why the cord is wider at these points
6. Names and structures of the three spinal meninges, in order from superficial to deep, and the relationships of the epidural and subarachnoid spaces to the meninges
7. The two types of ligaments that arise from the pia mater; where they are found and what purpose they serve
8. Organization of spinal gray and white matter as seen in cross sections of the cord; how gray and white matter differ in composition; and why they are called *gray* and *white matter*
9. The position of the posterior and anterior horns of the gray matter; where lateral horns are also found, and the functions of all three
10. The anatomical basis for dividing white matter into three funiculi on each side of the cord, and for dividing each funiculus into tracts
11. Names and functions of the ascending tracts of the spinal cord
12. The meanings of *first-* through *third-order* neurons in an ascending tract
13. Names and functions of the descending tracts
14. The locations and distinctions between upper and lower motor neurons in the descending tracts
15. Decussation and its implications for cerebral function in relation to sensation and motor control of the lower body, and for the effects of a stroke
16. What it means to say that the origin and destination of a tract, or any two body parts, are ipsilateral or contralateral

13.2 The Spinal Nerves

1. Structure of a nerve, especially the relationship of the endoneurium, perineurium, and epineurium to nerve fibers and fascicles
2. The basis for classifying nerve fibers as afferent or efferent, somatic or visceral, and special or general
3. The basis for classifying entire nerves as sensory, motor, or mixed
4. The definition and structure of a *ganglion*
5. The number of spinal nerves and their relationship to the spinal cord and intervertebral foramina
6. Anatomy of the posterior and anterior roots of a spinal nerve; the rootlets; and the posterior root ganglion
7. Anatomy of the anterior ramus, posterior ramus, and meningeal branch of a spinal nerve
8. What arises from the anterior ramus in the thoracic region as opposed to all other regions of the spinal cord
9. General structure of a spinal nerve plexus and the names and locations of the five plexuses
10. Distinctions between the roots, trunks, anterior and posterior divisions, and cords of a spinal nerve plexus; which of these five features occur in each of the five plexuses
11. Nerves that arise from each plexus and the body regions or structures to which each nerve provides sensory innervation, motor innervation, or both
12. Dermatomes and why they are relevant to the clinical diagnosis of nerve disorders

13.3 Somatic Reflexes

1. Four defining criteria of a reflex; how somatic reflexes differ from other types; and the flaw in calling somatic reflexes spinal reflexes
2. The pathway and constituents of a somatic reflex arc
3. The role of proprioceptors in somatic reflexes
4. Structure and function of muscle spindles
5. Stretch reflexes; one or more examples; the purpose they serve in everyday function; the mechanism of a stretch reflex; and an anatomical reason why stretch reflexes are often quicker than other types of somatic reflexes
6. Reciprocal inhibition and why it is important that it often accompany a stretch reflex
7. Flexor reflexes; a common purpose that they serve; and why it is beneficial for flexor reflexes to employ polysynaptic reflex arcs
8. Crossed extension reflexes and why it is important for this type of reflex to accompany a withdrawal reflex
9. The structure, location, and function of a tendon organ

STUDY GUIDE

▶ Testing Your Recall

Answers in Appendix A

1. Below L2, the vertebral canal is occupied by a bundle of spinal nerve roots called
   a. the terminal filum.
   b. the descending tracts.
   c. the gracile fasciculus.
   d. the medullary cone.
   e. the cauda equina.

2. The brachial plexus gives rise to all of the following nerves *except*
   a. the axillary nerve.
   b. the radial nerve.
   c. the obturator nerve.
   d. the median nerve.
   e. the ulnar nerve.

3. Nerve fibers that adjust the tension in a muscle spindle are called
   a. intrafusal fibers.
   b. extrafusal fibers.
   c. alpha motor neurons.
   d. gamma motor neurons.
   e. primary afferent fibers.

4. A stretch reflex requires the action of _____ to prevent an antagonistic muscle from interfering with the agonist.
   a. gamma motor neurons
   b. a withdrawal reflex
   c. a crossed extension reflex
   d. reciprocal inhibition
   e. a contralateral reflex

5. A patient has a gunshot wound that caused a bone fragment to nick the spinal cord. The patient now feels no pain or temperature sensations from that level of the body down. Most likely, the _____ was damaged.
   a. gracile fasciculus
   b. medial lemniscus

   c. tectospinal tract
   d. lateral corticospinal tract
   e. spinothalamic tract

6. Which of these is *not* a region of the spinal cord?
   a. cervical
   b. thoracic
   c. pelvic
   d. lumbar
   e. sacral

7. In the spinal cord, the somas of the lower motor neurons are found in
   a. the cauda equina.
   b. the posterior horns.
   c. the anterior horns.
   d. the posterior root ganglia.
   e. the fasciculi.

8. The outermost connective tissue wrapping of a nerve is called the
   a. epineurium.
   b. perineurium.
   c. endoneurium.
   d. arachnoid mater.
   e. dura mater.

9. The intercostal nerves between the ribs arise from which spinal nerve plexus?
   a. cervical
   b. brachial
   c. lumbar
   d. sacral
   e. none of them

10. All somatic reflexes share all of the following properties *except*
    a. they are quick.
    b. they are monosynaptic.

    c. they require stimulation.
    d. they are involuntary.
    e. they are stereotyped.

11. Outside the CNS, the somas of neurons are clustered in swellings called _____.

12. Distal to the intervertebral foramen, a spinal nerve branches into an anterior and posterior _____.

13. The cerebellum receives feedback from the muscles and joints by way of the _____ tracts of the spinal cord.

14. In the _____ reflex, contraction of flexor muscles in one limb is accompanied by the contraction of extensor muscles in the contralateral limb.

15. Modified muscle fibers serving primarily to detect stretch are called _____.

16. The _____ nerves arise from the cervical plexus and innervate the diaphragm.

17. The crossing of a nerve fiber or tract from the right side of the CNS to the left, or vice versa, is called _____.

18. The nonvisual awareness of the body's position and movements is called _____.

19. The _____ ganglion contains the somas of neurons that carry sensory signals to the spinal cord.

20. The sciatic nerve is a composite of two nerves, the _____ and _____.

▶ Building Your Medical Vocabulary

Answers in Appendix A

State a meaning of each word element, and give a medical term from this chapter that uses it or a slight variation of it.

1. arachno-

2. caudo-

3. contra-

4. cune-

5. ipsi-

6. phreno-

7. pia

8. proprio-

9. ram-

10. tecto-

STUDY GUIDE

▶ What's Wrong with These Statements?

Answers in Appendix A

Briefly explain why each of the following statements is false, or reword it to make it true.

1. The gracile fasciculus is a descending spinal tract.

2. The adult spinal cord terminates in the sacral canal.

3. Each spinal nerve rootlet is connected to its own segment of the spinal cord.

4. Some spinal nerves are sensory and others are motor.

5. The dura mater adheres tightly to the bone of the vertebral canal.

6. The anterior and posterior horns of the spinal cord are composed of white matter.

7. The corticospinal tracts carry sensory signals from the spinal cord to the cerebral cortex.

8. The dermatomes are nonoverlapping regions of skin innervated by different spinal nerves.

9. Somatic reflexes are those that do not involve the brain.

10. Ipsilateral reflex arcs are monosynaptic whereas contralateral arcs are polysynaptic.

▶ Testing Your Comprehension

1. Jillian is thrown from a horse. She strikes the ground with her chin, causing severe hyperextension of the neck. Emergency medical technicians properly immobilize her neck and transport her to a hospital, but she dies 5 minutes after arrival. An autopsy shows multiple fractures of vertebrae C1, C6, and C7 and extensive damage to the spinal cord. Explain why she died rather than being left quadriplegic.

2. Wallace is the victim of a hunting accident. A bullet grazed his vertebral column, and bone fragments severed the left half of his spinal cord at segments T8 through T10. Since the accident, Wallace has had a condition called *dissociated sensory loss,* in which he feels no sensations of deep touch or limb position on the *left* side of his body below the injury, and no sensations of pain or heat from the *right* side. Explain what spinal tract(s) the injury has affected and why these sensory losses are on opposite sides of the body.

3. Anthony gets into a fight between rival gangs. As an attacker comes at him with a knife, he turns to flee, but stumbles. The attacker stabs him on the medial side of the right gluteal fold and Anthony collapses. He loses all use of his right limb, being unable to extend his hip, flex his knee, or move his foot. He never fully recovers these lost functions. Explain what nerve injury Anthony has most likely suffered.

4. Stand with your right shoulder, hip, and foot firmly against a wall. Raise your left foot from the floor without losing contact with the wall at any point. What happens? Why? What principle of this chapter does this demonstrate?

5. When a patient needs a tendon graft, surgeons sometimes use the tendon of the palmaris longus, a relatively dispensable muscle of the forearm. The median nerve lies nearby and looks very similar to this tendon. There have been cases in which a surgeon mistakenly removed a section of this nerve instead of the tendon. What effects do you think such a mistake would have on the patient?

THE BRAIN AND CRANIAL NERVES

Diffusion tensor image of white matter tracts in a lateral view of the brain
Sherbrooke Connectivity Imaging Lab (SCIL)/Getty Images

CHAPTER OUTLINE

14.1 Overview of the Brain
14.1a Major Landmarks
14.1b Gray and White Matter
14.1c Embryonic Development

14.2 Meninges, Ventricles, Cerebrospinal Fluid, and Blood Supply
14.2a Meninges
14.2b Ventricles and Cerebrospinal Fluid
14.2c Blood Supply and the Brain Barrier System

14.3 The Hindbrain and Midbrain
14.3a The Medulla Oblongata
14.3b The Pons
14.3c The Midbrain
14.3d The Reticular Formation
14.3e The Cerebellum

14.4 The Forebrain
14.4a The Diencephalon
14.4b The Cerebrum

14.5 Integrative Functions of the Brain
14.5a The Electroencephalogram
14.5b Sleep
14.5c Cognition
14.5d Memory
14.5e Emotion
14.5f Sensation
14.5g Motor Control
14.5h Language
14.5i Cerebral Lateralization

14.6 The Cranial Nerves
14.6a Cranial Nerve Pathways
14.6b Cranial Nerve Classification
14.6c Cranial Nerve Survey

Study Guide

DEEPER INSIGHTS

14.1 Clinical Application: Meningitis

14.2 Clinical Application: Stroke

14.3 Clinical Application: Tracking the Tracts of White Matter

14.4 Clinical Application: Some Cranial Nerve Disorders

14.5 Clinical Application: Images of the Mind

Anatomy & Physiology
Revealed 4.0

Module 7: Nervous System

BRUSHING UP

- The anatomy of the brain and cranial nerves is, in many respects, described in relation to the skull. Therefore, it may be helpful to review the cranial bones and major foramina in section 8.2a.

- You must be familiar with neuron structure (see section 12.2c) and with glial cells and their functions (see section 12.3a).

- The brainstem contains extensions of the spinal cord tracts, so you will find it helpful to know those or refer back to table 13.1 as you study the brainstem.

- To best understand the cranial nerves, you should be familiar with the general structure of nerves and ganglia, afferent and efferent nerve fibers, and the distinction between sensory, motor, and mixed nerves (see section 13.2a).

The human brain has a high opinion of itself, often claiming to be the most complex object in the known universe. It would be hard to argue otherwise. It has a mystique that intrigues modern biologists and psychologists even as it did the philosophers of antiquity. Aristotle thought it was just a radiator for cooling the blood, but generations earlier, Hippocrates had expressed a more accurate view. "Men ought to know," he said, "that from the brain, and from the brain only, arise our pleasures, joy, laughter and jests, as well as our sorrows, pains, griefs and tears. Through it, in particular, we think, see, hear, and distinguish the ugly from the beautiful, the bad from the good, the pleasant from the unpleasant."

Brain function is so strongly associated with what it means to be alive and human that the cessation of brain activity is taken as a clinical criterion of death even when other organs of the body are still functioning. With its hundreds of neural pools and trillions of synapses, the brain performs sophisticated tasks beyond our present understanding. Still, all of our mental functions, no matter how complex, are ultimately based on the cellular activities described in chapter 12. The relationship of the mind and personality to the cellular function of the brain is a question that will provide fertile ground for scientific study and philosophical debate long into the future.

This chapter is a study of the brain and the cranial nerves directly connected to it. Here we will plumb some of the mysteries of motor control, sensation, emotion, thought, language, personality, memory, dreams, and plans. Brain circuitry and function easily fill many books the size of this one, and we can barely scratch the surface of this complex subject here. This coverage will, however, provide some intriguing insights and lay a foundation for further study in other courses.

14.1   Overview of the Brain

Expected Learning Outcomes

When you have completed this section, you should be able to

a. describe the major subdivisions and anatomical landmarks of the brain;

b. describe the locations of its gray and white matter; and

c. describe the embryonic development of the CNS and relate this to adult brain anatomy.

In the evolution of the central nervous system from the simplest vertebrates to humans, the spinal cord changed very little while the brain changed a great deal. In fishes and amphibians, the brain weighs about the same as the spinal cord, but in humans, it weighs 55 times as much. It averages about 1,600 g (3.5 lb) in men and 1,450 g in women. The difference between the sexes is proportional to body size, not intelligence. The Neanderthal people had larger brains than modern humans do.

Ours is the most sophisticated brain when compared to others in awareness of the environment, adaptability to environmental variation and change, quick execution of complex decisions, fine motor control and mobility of the body, and behavioral complexity. Over the course of human evolution, it has shown its greatest growth in areas concerned with vision, memory, abstract thought, and motor control of the prehensile hand.

14.1a Major Landmarks

We begin with a general overview of the major landmarks of the brain. These will provide important reference points as we progress through a more detailed study.

Two directional terms used in descriptions of CNS anatomy are *rostral* and *caudal*. **Rostral**[1] means "toward the nose" and **caudal**[2] means "toward the tail." These are apt descriptions for an animal such as a laboratory rat, on which so much brain research has been done. The terms are retained for human neuroanatomy as well, but in references to the brain, *rostral* means "toward the forehead" and *caudal* means "toward the spinal cord." In the spinal cord and brainstem, which are vertically oriented, *rostral* means "higher" and *caudal* means "lower."

We can conceptually divide the brain into three major portions—the *forebrain, cerebellum,* and *brainstem.* The most prominent part of the forebrain is the **cerebrum** (seh-REE-brum or SER-eh-brum). It constitutes about 83% of the brain's volume and consists of a pair of half globes called the **cerebral hemispheres (fig. 14.1a).** Each hemisphere is marked by thick folds called **gyri**[3] (JY-rye; singular, *gyrus*) separated by shallow grooves called **sulci**[4] (SUL-sye;

---

[1] *rostr* = nose
[2] *caud* = tail
[3] *gyr* = turn, twist
[4] *sulc* = furrow, groove

(a) Superior view

(b) Lateral view

(c) Lateral view

FIGURE 14.1 Surface Anatomy of the Brain. (a) Superior view of the cerebral hemispheres. (b) Left lateral view. (c) Lateral view of the cadaver brain. **APR**

singular, *sulcus*). A deep median groove, the **longitudinal cerebral fissure,** separates the right and left hemispheres from each other. At the bottom of this fissure, the hemispheres are connected by a thick bundle of nerve fibers called the **corpus callosum**[5]—a prominent landmark for anatomical description with a distinctive C shape in sagittal section **(fig. 14.2).**

The **cerebellum**[6] (SER-eh-BEL-um) occupies the posterior cranial fossa inferior to the cerebrum, separated from it by the **transverse cerebral fissure (fig. 14.1b, c;** also see fig. 8.9). It is also marked by fissures, sulci, and gyri (called *folia* in the cerebellum). The cerebellum is the second-largest region of the brain, constituting about 10% of its volume but containing over 50% of its neurons.

The **brainstem** is all of the brain except the forebrain and cerebellum. Its components, from rostral to caudal, are the *midbrain, pons,* and *medulla oblongata* (fig. 14.2).

In a living person, the brainstem is oriented like a vertical stalk with the forebrain perched on top like a mushroom cap. Postmortem changes give it a more oblique angle in the cadaver and, consequently, in many medical illustrations. Caudally, the brainstem ends at the foramen magnum of the skull, and the CNS continues below this as the spinal cord.

14.1b Gray and White Matter

The brain, like the spinal cord, is composed of gray and white matter (see figs. 14.5, 14.6c). Gray matter—the seat of the neurosomas, dendrites, and synapses—forms a surface layer called the **cortex** over the cerebrum and cerebellum, and deeper masses called **nuclei** surrounded by white matter. White matter lies deep to the cortical gray matter in most of the brain, opposite from the relationship of gray and white matter in the spinal cord. As in the spinal cord, white matter is composed of **tracts,** or bundles of axons, which here connect one part of the brain to another and to the spinal cord. These are described later in more detail.

14.1c Embryonic Development

Mature brain anatomy is often described in terms of *forebrain, midbrain,* and *hindbrain*—three terms that can be fully appreciated only with some awareness of the embryonic development of the CNS.

The nervous system develops from ectoderm, the outermost tissue layer of an embryo. Within the first 3 weeks, a *neural plate* forms along the dorsal midline of the embryo and sinks into the tissues to form a *neural groove,* with a raised *neural fold* along each side **(fig. 14.3).** The neural folds roll toward each other and fuse, somewhat like a closing zipper, beginning in the cervical

region and progressing both caudally and rostrally. By day 26, this process creates a hollow channel called the **neural tube.** Following closure, the neural tube separates from the overlying ectoderm, sinks a little deeper, and grows lateral processes that later form motor nerve fibers. The lumen of the neural tube becomes a fluid-filled space that later constitutes the *central canal* of the spinal cord and *ventricles* of the brain.

As the neural tube develops, some ectodermal cells that originally lay along the margin of the groove separate from the rest and form a longitudinal column on each side called the **neural crest.** Neural crest cells give rise to the two inner meninges (arachnoid mater and pia mater); most of the peripheral nervous system, including the sensory and autonomic nerves and ganglia and Schwann cells; and some other structures of the skeletal, integumentary, and endocrine systems.

By the fourth week, the neural tube exhibits three anterior dilations, or *primary vesicles,* called the **forebrain** (*prosencephalon*[7]) (PROSS-en-SEF-uh-lon), **midbrain** (*mesencephalon*[8]) (MES-en-SEF-uh-lon), and **hindbrain** (*rhombencephalon*[9]) (ROM-ben-SEF-uh-lon) **(fig. 14.4).** By the fifth week, it subdivides into five *secondary vesicles.* The forebrain divides into two of them, the **telencephalon**[10] (TEL-en-SEFF-uh-lon) and **diencephalon**[11] (DY-en-SEF-uh-lon); the midbrain remains undivided and retains the name **mesencephalon;** and the hindbrain divides into two vesicles, the **metencephalon**[12] (MET-en-SEF-uh-lon) and **myelencephalon**[13] (MY-el-en-SEF-uh-lon). The telencephalon has a pair of lateral outgrowths that later become the cerebral hemispheres. The diencephalon will become chiefly the *thalamus* and *hypothalamus,* and has a pair of cuplike lateral outgrowths, the *optic vesicles,* which will become the retinas of the eyes. The metencephalon will give rise to the pons and cerebellum, and the myelencephalon to the medulla oblongata. Figure 14.4 is color-coded to show which mature brain regions develop from each vesicle.

BEFORE YOU GO ON

Answer the following questions to test your understanding of the preceding section:

1. List the three major parts of the brain and describe their locations.

2. Define *gyrus* and *sulcus*.

3. Contrast the composition and locations of gray and white matter in the brain.

4. Explain how the five secondary brain vesicles arise from the neural tube.

---

[5]*corpus* = body; *call* = thick
[6]*cereb* = brain; *ellum* = little

[7]*pros* = before, in front; *encephal* = brain
[8]*mes* = middle
[9]*rhomb* = rhombus
[10]*tele* = end, remote
[11]*di* = through, between
[12]*met* = behind, beyond, distal to
[13]*myel* = spinal cord

FIGURE 14.2 Medial Aspect of the Brain. (a) Major anatomical landmarks of the medial surface. (b) Median section of the cadaver brain. **APR**

Photo: Christine Eckel/McGraw-Hill Education

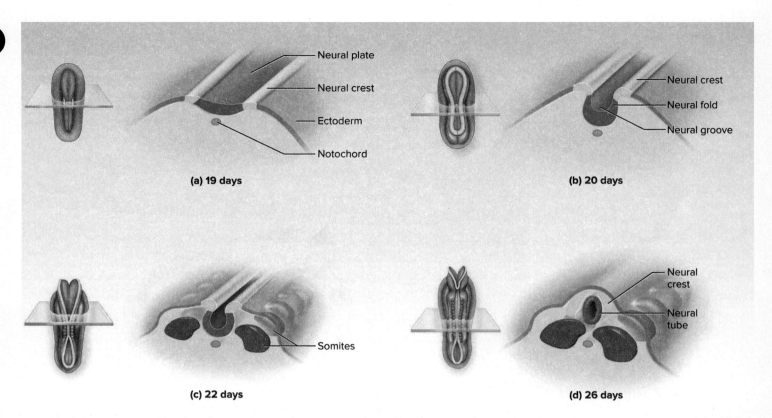

Neural plate
Neural crest
Ectoderm
Notochord

(a) 19 days

Neural crest
Neural fold
Neural groove

(b) 20 days

Somites

(c) 22 days

Neural crest
Neural tube

(d) 26 days

FIGURE 14.3 Formation of the Embryonic Neural Tube. The left-hand figure in each case is a dorsal view of the embryo, and the right-hand figure is a three-dimensional representation of the tissues at the indicated level of the respective embryo. (a) Neural plate at 19 days. (b) Neural groove at 20 days. (c) Neural groove closing at 22 days. (d) Neural tube at 26 days.

14.2 Meninges, Ventricles, Cerebrospinal Fluid, and Blood Supply

Expected Learning Outcomes

When you have completed this section, you should be able to

a. describe the meninges of the brain;
b. describe the fluid-filled chambers within the brain;
c. discuss the production, circulation, and function of the cerebrospinal fluid that fills these chambers; and
d. explain the significance of the brain barrier system.

14.2a Meninges

The brain is enveloped in three membranes, the *meninges,* which lie between the nervous tissue and bone. They protect the brain and provide a structural framework for its arteries and veins. As in the spinal cord, these are the *dura mater, arachnoid mater,* and *pia mater* **(fig. 14.5).** Their histological composition is described in section 13.1c and is similar in the cranium. However, the cranial dura mater consists of two layers—an outer *periosteal layer* equivalent to the periosteum of the cranial bones, and an inner *meningeal layer.* Only the meningeal layer continues into the vertebral canal, where it forms the dural sheath around the spinal cord. The cranial dura mater is pressed closely against the cranial bone, with no intervening epidural space like the one around the spinal cord. It isn't attached to the bone, however, except in limited places: around the foramen magnum, the sella turcica, the crista galli, and the sutures of the skull.

In some places, the two layers of dura are separated by **dural sinuses,** spaces that collect blood that has circulated through the brain. Two major, superficial ones are the **superior sagittal sinus,** found just under the calvaria along the median line, and the **transverse sinus,** which runs horizontally from the rear of the head toward each ear. These sinuses meet like an inverted T at the back of the brain and ultimately empty into the internal jugular veins of the neck. These and other sinuses of the brain are more fully described and pictured along with other cerebral blood vessels in section 20.7c.

In certain places, the meningeal layer of the dura folds inward to separate major parts of the brain from each other and limit brain movements within the cranium, as when one receives a jolt to the head. There are three of these: (1) the *falx*[14] *cerebri* (falks SER-eh-bry), which extends into the longitudinal cerebral fissure as a tough, crescent-shaped wall between the right and left cerebral hemispheres; (2) the *tentorium*[15] *cerebelli* (ten-TOE-ree-um), which stretches like a roof over the posterior cranial fossa and separates the cerebellum from the overlying cerebrum; and (3) the *falx cerebelli,* a vertical partition between the right and left halves of the cerebellum on the inferior side.

[14]*falx* = sickle
[15]*tentorium* = tent

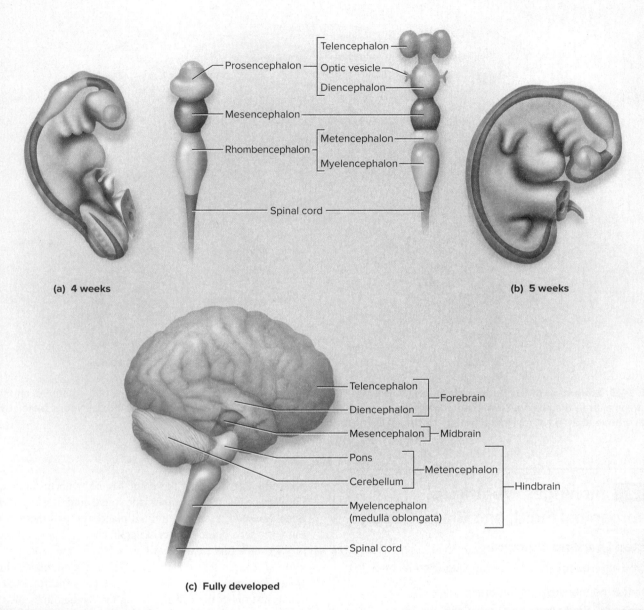

(a) 4 weeks

- Prosencephalon
  - Telencephalon
  - Optic vesicle
  - Diencephalon
- Mesencephalon
- Rhombencephalon
  - Metencephalon
  - Myelencephalon
- Spinal cord

(b) 5 weeks

- Telencephalon ⎤
- Diencephalon ⎦ Forebrain
- Mesencephalon — Midbrain
- Pons ⎤
- Cerebellum ⎦ Metencephalon ⎤
- Myelencephalon (medulla oblongata) ⎦ Hindbrain
- Spinal cord

(c) Fully developed

FIGURE 14.4 Primary and Secondary Vesicles of the Embryonic Brain. (a) The primary vesicles at 4 weeks. (b) The secondary vesicles at 5 weeks. (c) The fully developed brain, color-coded to relate its structures to the secondary embryonic vesicles.

The arachnoid mater and pia mater are similar to those of the spinal cord. The arachnoid mater is a transparent membrane over the brain surface, deep to the dura. A *subarachnoid space* separates it from the pia below; it contains the largest blood vessels of the cerebral surface. The pia mater is a very thin, delicate membrane, not usually visible without a microscope. Whereas the arachnoid meninx only overlies the sulci of the cerebral surface, the pia mater dips down into them, closely follows all the contours of the brain, and follows arteries for a short distance as they penetrate into the cerebrum.

14.2b Ventricles and Cerebrospinal Fluid

The brain has four internal chambers called **ventricles (fig. 14.6).** The largest and most rostral ones are the two **lateral ventricles,** which form an arc in each cerebral hemisphere. Through a tiny pore called the **interventricular foramen,** each lateral ventricle is connected to the **third ventricle,** a narrow median space inferior to the corpus callosum. From here, a canal called the **cerebral aqueduct** passes down the core of the midbrain and leads to the **fourth ventricle,** a small triangular chamber between the

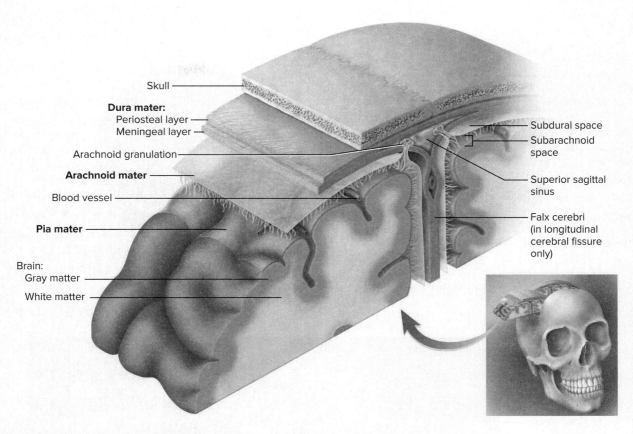

Skull

Dura mater:
Periosteal layer
Meningeal layer

Arachnoid granulation

Arachnoid mater

Blood vessel

Pia mater

Brain:
Gray matter
White matter

Subdural space

Subarachnoid
space

Superior sagittal
sinus

Falx cerebri
(in longitudinal
cerebral fissure
only)

FIGURE 14.5 The Meninges of the Brain. Frontal section of the head.

DEEPER INSIGHT 14.1

CLINICAL APPLICATION

Meningitis

Meningitis—inflammation of the meninges—is one of the most serious diseases of infancy and childhood. It occurs especially between 3 months and 2 years of age. Meningitis is caused by a variety of bacteria and viruses that invade the CNS by way of the nose and throat, often following respiratory, throat, or ear infections. The pia and arachnoid mater are most often affected, and from there the infection can spread to the adjacent nervous tissue. Meningitis can cause swelling of the brain, cerebral hemorrhaging, and sometimes death within mere hours of the onset of symptoms. Signs and symptoms include high fever, stiff neck, drowsiness, intense headache, and vomiting. Meningitis is diagnosed partly by examining the cerebrospinal fluid (CSF) for bacteria and white blood cells. The CSF is obtained by lumbar puncture (described in section 13.1c).

Death from meningitis can occur so suddenly that infants and children with a high fever should therefore receive immediate medical attention. Freshman college students show a slightly elevated incidence of meningitis, especially those living in crowded dormitories rather than off campus.

pons and cerebellum. Caudally, this space narrows and forms a **central canal** that extends through the medulla oblongata into the spinal cord.

On the floor or wall of each ventricle is a spongy mass of blood capillaries called a **choroid plexus** (CO-royd), named for its histological resemblance to a fetal membrane called the chorion. *Ependyma,* a type of neuroglia that resembles a cuboidal epithelium, lines the ventricles and canals and covers the choroid plexuses. It produces cerebrospinal fluid.

Cerebrospinal fluid (CSF) is a clear, colorless liquid that fills the ventricles and canals of the CNS and bathes its external surface. The brain produces about 500 mL of CSF per day, but the fluid is constantly reabsorbed at the same rate and only 100 to 160 mL is normally present at one time. About 40% of it is formed in the subarachnoid space external to the brain, 30% by the general ependymal lining of the brain ventricles, and 30% by the choroid plexuses. CSF production begins with the filtration of blood plasma through the capillaries of the brain. Ependymal cells modify the filtrate as it passes through them, so the CSF has more sodium chloride than blood plasma, but less potassium, calcium, and glucose and very little protein.

Caudal ← | → Rostral

(a) Lateral view

Lateral ventricles
Interventricular foramen
Third ventricle
Cerebral aqueduct
Fourth ventricle
Lateral aperture
Median aperture
Central canal

(b) Anterior view

Cerebrum
Lateral ventricle
Interventricular foramen
Third ventricle
Cerebral aqueduct
Fourth ventricle
Lateral aperture
Median aperture

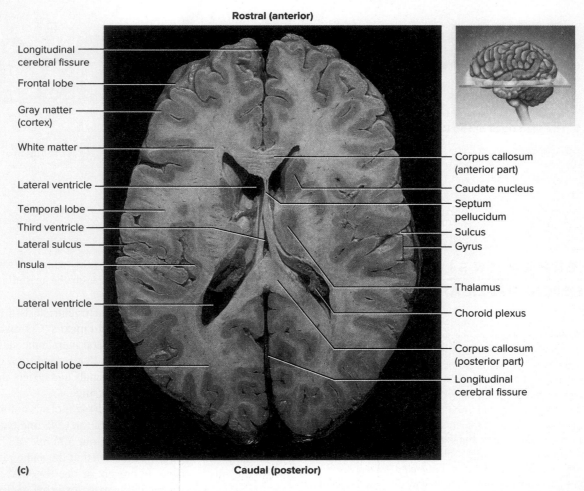

Rostral (anterior)

Longitudinal cerebral fissure
Frontal lobe
Gray matter (cortex)
White matter
Lateral ventricle
Temporal lobe
Third ventricle
Lateral sulcus
Insula
Lateral ventricle
Occipital lobe

Corpus callosum (anterior part)
Caudate nucleus
Septum pellucidum
Sulcus
Gyrus
Thalamus
Choroid plexus
Corpus callosum (posterior part)
Longitudinal cerebral fissure

(c)

Caudal (posterior)

FIGURE 14.6 Ventricles of the Brain. (a) Right lateral view. (b) Anterior view. (c) Superior view of a horizontal section of the cadaver brain, showing the lateral ventricles and some other features of the cerebrum. **APR**

CSF continually flows through and around the CNS, driven partly by its own pressure, partly by the beating of ependymal cilia, and partly by rhythmic pulsations of the brain produced by each heartbeat. The CSF of the lateral ventricles flows through the interventricular foramina into the third ventricle, then down the cerebral aqueduct to the fourth ventricle **(fig. 14.7).** The third and fourth ventricles and their choroid plexuses add more CSF along the way. A small amount of CSF fills the central canal of the spinal cord, but ultimately, all of it escapes through three pores in the fourth ventricle—a *median aperture* and two *lateral apertures.* These lead into the subarachnoid space on the brain and spinal cord surface. From there, CSF is reabsorbed by **arachnoid granulations,** extensions of the arachnoid meninx shaped like little sprigs of cauliflower, protruding through the dura mater into the superior sagittal sinus. CSF penetrates the walls of the granulations and mixes with blood in the sinus.

Cerebrospinal fluid serves three purposes:

1. **Buoyancy.** Because the brain and CSF are similar in density, the brain neither sinks nor floats in the CSF. It hangs from delicate strands of the arachnoid meninx. A human brain removed from the body has a dead weight of about 1,500 g, but when suspended in CSF, its effective weight is only about 50 g. By analogy, consider how much easier it is to lift another person when you're immersed in a lake than it is on land. This buoyancy allows the brain to attain considerable size without being impaired by its own weight. If the brain rested heavily on the floor of the cranium, the pressure would kill the nervous tissue.

2. **Protection.** CSF also protects the brain from striking the cranium when the head is jolted. If the jolt is severe, however, the brain still may strike the inside of the cranium or suffer shearing injury from contact with the angular surfaces of the cranial floor. This is one of the common findings in child abuse (shaken child syndrome) and in concussions and other *traumatic brain injury (TBI)* from auto accidents, boxing and football, and the like.

① CSF is secreted by choroid plexus in each lateral ventricle.

② CSF flows through interventricular foramina into third ventricle.

③ Choroid plexus in third ventricle adds more CSF.

④ CSF flows down cerebral aqueduct to fourth ventricle.

⑤ Choroid plexus in fourth ventricle adds more CSF.

⑥ CSF flows out two lateral apertures and one median aperture.

⑦ CSF fills subarachnoid space and bathes external surfaces of brain and spinal cord.

⑧ At arachnoid villi, CSF is reabsorbed into venous blood of dural venous sinuses.

Arachnoid villus
Superior sagittal sinus
Arachnoid mater
Subarachnoid space
Dura mater
Choroid plexus
Third ventricle
Cerebral aqueduct
Lateral aperture
Fourth ventricle
Median aperture
Central canal of spinal cord
Subarachnoid space of spinal cord

FIGURE 14.7 The Flow of Cerebrospinal Fluid.

3. **Chemical stability.** CSF rinses metabolic wastes from the nervous tissue and regulates its chemical environment. Slight changes in CSF composition can cause malfunctions of the nervous system. For example, a high glycine concentration disrupts the control of body temperature and blood pressure, and a high pH causes dizziness and fainting.

14.2c Blood Supply and the Brain Barrier System

The blood vessels that serve the brain are detailed in sections 20.7b and 20.7c. The brain constitutes only 2% of the adult body weight, yet it receives 15% of the blood flow (about 750 mL/min.) and consumes 20% of its oxygen and glucose. Because neurons have such a high demand for ATP, and therefore glucose and oxygen, the constancy of blood supply is especially critical to the nervous system. A mere 10-second interruption in blood flow can cause loss of consciousness; an interruption of 1 to 2 minutes can significantly impair neural function; and 4 minutes without blood usually causes irreversible brain damage (see Deeper Insight 14.2). For more on cerebral blood flow, see section 20.5a.

Despite its critical importance to the brain, blood is also a source of antibodies, macrophages, bacterial toxins, and other potentially harmful agents. Damaged brain tissue is essentially irreplaceable, and the brain therefore must be well protected. Consequently, there is a **brain barrier system (BBS)** that strictly regulates what can get from the bloodstream into the tissue fluid of the brain.

There are two potential points of entry that must be guarded: the blood capillaries throughout the brain tissue and the capillaries of the choroid plexuses. At the former site, the brain is well protected by the **blood–brain barrier,** which consists of tight junctions between the endothelial cells that form the capillary walls. In the developing brain, astrocytes reach out and contact the capillaries with their perivascular feet, stimulating the endothelial cells to form tight junctions that completely seal off the gaps between them. This ensures that anything leaving the blood must pass through the cells and not between them. The endothelial cells are more selective than gaps between them would be, and can exclude harmful substances from the brain tissue while allowing necessary ones to pass through. At the choroid plexuses, the brain is protected by a similar **blood–CSF barrier** formed by tight junctions between the ependymal cells. Tight junctions are absent from ependymal cells elsewhere, because it is important to allow exchanges between the brain tissue and CSF. That is, there is no brain–CSF barrier.

The BBS is highly permeable to water, glucose, and lipid-soluble substances such as oxygen, carbon dioxide, alcohol, caffeine, nicotine, and anesthetics. It is slightly permeable to sodium, potassium, chloride, and the waste products urea and creatinine. While the BBS is an important protective device, it is an obstacle to the delivery of medications such as antibiotics and cancer drugs, and thus complicates the treatment of brain diseases.

Trauma and inflammation sometimes damage the BBS and allow pathogens to enter the brain tissue. Furthermore, there are places called **circumventricular organs (CVOs)** in the third and fourth ventricles where the barrier is absent and the blood has direct access to brain neurons. These enable the brain to monitor and respond to fluctuations in blood glucose, pH, osmolarity, and other variables. Unfortunately, CVOs also afford a route of invasion by the human immunodeficiency virus (HIV).

DEEPER INSIGHT 14.2

CLINICAL APPLICATION

Stroke

A *stroke,* or *cerebral vascular accident (CVA),* is the sudden death of brain tissue resulting from an abrupt interruption of its blood supply. There are two kinds of stroke distinguished by cause: a *hemorrhagic stroke* resulting from the rupture of a cerebral or subarachnoid blood vessel, and more common, an *ischemic stroke* (iss-KEE-mic) resulting from the obstruction of a blood vessel, usually by a blood clot (thrombosis) or lipid deposit (atherosclerosis). Stroke is the second greatest cause of death after coronary artery disease; about one-half of those who have a stroke live less than one year longer. Stroke can occur at any age, but about two-thirds occur in people over 65 years old.

Aside from age, risk factors for stroke include hypertension; high blood cholesterol; obesity; diabetes mellitus; and such lifestyle choices as inadequate physical activity, overuse of alcohol, smoking, and using drugs such as amphetamines and cocaine. The risk of stroke can be reduced with statin drugs (to lower cholesterol level); antihypertensive drugs (to control blood pressure); and other lifestyle, medical, and surgical methods. Daily low-dose aspirin has not been found helpful for *primary prevention* of a first stroke, but may be beneficial for *secondary prevention* of recurrence in those who have had a previous stroke.

The signs and symptoms of a stroke depend on the function of the tissues downstream from the hemorrhage or obstruction, where neuron death occurs. They can include paralysis (usually on the side of the body opposite from a cerebral stroke), blindness or other loss of sensation, aphasia (loss of speech), or cognitive deficits (dysfunctions in awareness, memory, or reason). Similar symptoms lasting less than 24 hours are called *transient ischemic attack (TIA),* colloquially called a "mini-stroke." A TIA can foreshadow a true stroke later.

To be prepared to accurately recognize a stroke in another person and get quick help, think of the acronym **FAST: F**acial weakness (sagging or drooping on one side of the face), **A**rm drift (inability to prevent one arm from drifting down when asked to hold both arms above the head), **S**peech difficulty, and **T**ime to call emergency services. Some other signs include defects in the visual field, inability to stick the tongue out and move it left and right on command, and loss of memory (such as for one's name and the date). Irreversible brain damage (neuron death) occurs after about 3 hours and is therefore preventable with prompt emergency treatment—as stressed by the slogan, Time is brain!

Answer the following questions to test your understanding of the preceding section:

5. Name the three meninges from superficial to deep. How does the dura mater of the brain differ from that of the spinal cord?

6. Describe three functions of the cerebrospinal fluid.

7. Where does the CSF originate and what route does it take through and around the CNS?

8. Name the two components of the brain barrier system and explain the importance of this system.

14.3 The Hindbrain and Midbrain

Expected Learning Outcomes

When you have completed this section, you should be able to

a. list the components of the hindbrain and midbrain and their functions; and

b. describe the location and functions of the reticular formation.

The study of the brain in the following pages will be organized around the five secondary vesicles of the embryonic brain and their mature derivatives. We will proceed in a caudal to rostral direction, beginning with the hindbrain and its relatively simple functions and progressing to the forebrain, the seat of such complex functions as thought, memory, and emotion.

14.3a The Medulla Oblongata

As noted earlier, the embryonic hindbrain differentiates into two subdivisions: the myelencephalon and metencephalon (see fig. 14.4). The myelencephalon becomes just one adult structure, the **medulla oblongata** (meh-DULL-uh OB-long-GAH-ta).

The medulla (**fig. 14.8;** also see fig. 14.2) begins at the foramen magnum of the skull and extends for about 3 cm rostrally, ending at a transverse groove between the medulla and pons. It looks superficially like an extension of the spinal cord, but slightly wider. Significant differences are apparent, however, on closer inspection of its gross and microscopic anatomy. Externally, the anterior surface features a pair of ridges called the **pyramids.** Resembling side-by-side baseball bats, these are wider at the rostral end, taper caudally, and are separated by a longitudinal groove, the *anterior median fissure,* continuous with that of the spinal cord. Lateral to each pyramid is a prominent bulge called the **olive.** Posteriorly, the *gracile* and *cuneate fasciculi* of the spinal cord continue as two pairs of ridges on the medulla.

All nerve fibers connecting the brain to the spinal cord pass through the medulla. As we saw in the cord, some of these are ascending (sensory) and some are descending (motor) fibers. The ascending fibers include first-order sensory fibers of the gracile and cuneate fasciculi, which end in the **gracile** and **cuneate**

nuclei seen in **figure 14.9c.** Here, they synapse with second-order fibers that decussate and form the ribbonlike **medial lemniscus**[16] on each side. The second-order fibers rise to the thalamus, synapsing there with third-order fibers that complete the path to the cerebral cortex (compare fig. 13.6a).

The largest group of descending fibers is the pair of **corticospinal tracts** filling the pyramids on the anterior surface. These carry motor signals from the cerebral cortex on the way to the spinal cord, ultimately to stimulate the skeletal muscles. Any time you carry out a body movement below the neck, the signals en route to your muscles pass through here. About 90% of these fibers cross over at the *pyramidal decussation,* an externally visible point near the caudal end of the pyramids (fig. 14.8a). As a result, muscles below the neck are controlled by the contralateral side of the brain. A smaller *tectospinal tract* controls the neck muscles.

The medulla contains neural networks involved in a multitude of fundamental sensory and motor functions. The former include the senses of hearing, equilibrium, touch, pressure, temperature, taste, and pain; the latter include chewing, salivation, swallowing, gagging, vomiting, respiration, speech, coughing, sneezing, sweating, cardiovascular and gastrointestinal control, and head, neck, and shoulder movements. Signals for these functions enter and leave the medulla not only by way of the spinal cord, but also by four pairs of cranial nerves that begin or end here: cranial nerves VIII (in part), IX, X, and XII. At the level of the section in figure 14.9c, we see the origins of two of them, the vagus (X) and hypoglossal (XII). The names and functions of these nerves are detailed in section 14.6c.

Another feature seen in cross section is the wavy **inferior olivary nucleus,** a major relay center for signals going from many levels of the brain and spinal cord to the cerebellum. The *reticular formation,* detailed later, is a loose network of nuclei extending throughout the medulla, pons, and midbrain. In the medulla, it includes a **cardiac center,** which regulates the rate and force of the heartbeat; a **vasomotor center,** which regulates blood pressure and flow by dilating and constricting blood vessels; two **respiratory centers,** which regulate the rhythm and depth of breathing; and other nuclei involved in the aforementioned motor functions.

14.3b The Pons

The metencephalon develops into the pons and cerebellum. We will return to the cerebellum after finishing the brainstem. The **pons**[17] measures about 2.5 cm long. Most of it appears as a broad anterior bulge rostral to the medulla (see figs. 14.2, 14.8). Posteriorly, it consists mainly of two pairs of thick stalks called *cerebellar peduncles,* the cut edges in the upper half of **figure 14.9b.** They connect the cerebellum to the pons and midbrain (fig. 14.8b) and will be discussed with the cerebellum.

In cross section, the pons exhibits continuations of the previously mentioned reticular formation, medial lemniscus,

---

[16]*lemn* = ribbon; *iscus* = little
[17]*pons* = bridge

Diencephalon:
Thalamus
Infundibulum
Mammillary body

Optic tract

Cranial nerves:
Optic nerve (II)

Midbrain:
Cerebral peduncle

Oculomotor nerve (III)

Trochlear nerve (IV)

Trigeminal nerve (V)

Abducens nerve (VI)

Pons

Facial nerve (VII)

Vestibulocochlear nerve (VIII)

Glossopharyngeal nerve (IX)

Vagus nerve (X)

Medulla oblongata:
Pyramid
Anterior median fissure
Pyramidal decussation

Accessory nerve (XI)

Hypoglossal nerve (XII)

Spinal nerves

Spinal cord

Key
Diencephalon
Midbrain
Pons
Medulla oblongata

(a) Anterior view

Diencephalon:
Thalamus
Lateral geniculate body
Pineal gland
Medial geniculate body

Optic tract

Midbrain:
Superior colliculus
Inferior colliculus
Cerebral peduncle

Pons

Cerebellar peduncles
(cut edges):
Superior

Middle

Fourth ventricle

Inferior

**Medulla
oblongata**

Olive

Cuneate fasciculus

Gracile fasciculus

Spinal cord

(b) Posterolateral view

FIGURE 14.8 The Brainstem and Diencephalon. (a) Anterior view. (b) Posterolateral view. Color-coded to match the embryonic origins in figure 14.4. The boundary between the middle and inferior cerebellar peduncles is indistinct. **APR**

tectospinal tract, and other spinal tracts. The anterior half of the pons (lower half of fig. 14.9b) is dominated by tracts of white matter, including transverse fascicles that cross between left and right and connect the two hemispheres of the cerebellum, and longitudinal fascicles that carry sensory and motor signals up and down the brainstem.

Cranial nerves V to VIII begin or end in the pons, although we see only the trigeminal nerve (V) at the level of figure 14.9b.

The other three emerge from the groove between the pons and medulla. The functions of these four nerves, detailed in table 14.1, include sensory roles in hearing, equilibrium, and taste; facial sensations such as touch and pain; and motor roles in eye movement, facial expressions, chewing, swallowing, urination, and the secretion of saliva and tears. The reticular formation in the pons contains additional nuclei concerned with sleep, respiration, and posture.

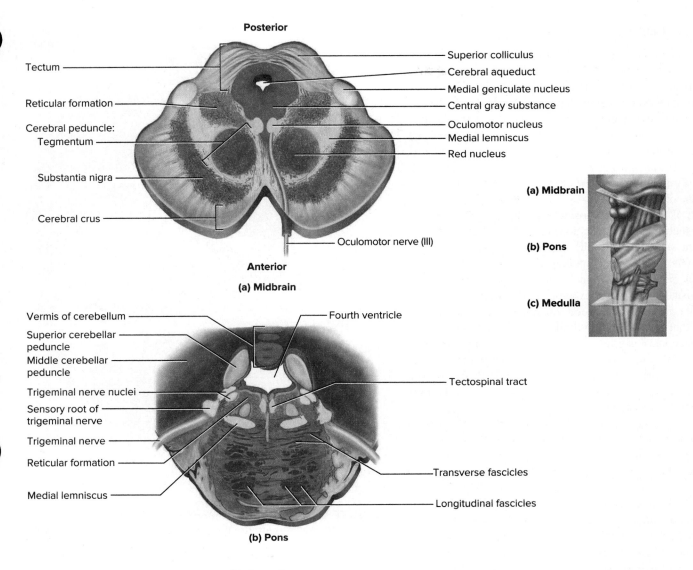

Posterior

Tectum

Reticular formation

Cerebral peduncle:
  Tegmentum

Substantia nigra

Cerebral crus

Superior colliculus

Cerebral aqueduct

Medial geniculate nucleus

Central gray substance

Oculomotor nucleus

Medial lemniscus

Red nucleus

Oculomotor nerve (III)

Anterior

(a) Midbrain

(a) Midbrain

(b) Pons

(c) Medulla

Vermis of cerebellum

Superior cerebellar peduncle

Middle cerebellar peduncle

Trigeminal nerve nuclei

Sensory root of trigeminal nerve

Trigeminal nerve

Reticular formation

Medial lemniscus

Fourth ventricle

Tectospinal tract

Transverse fascicles

Longitudinal fascicles

(b) Pons

Gracile nucleus

Cuneate nucleus

Reticular formation

Medial lemniscus

Hypoglossal nerve and nerve root

Corticospinal tract

Nucleus of hypoglossal nerve

Fourth ventricle

Nucleus of vagus nerve

Tectospinal tract

Inferior olivary nucleus

Olive

Pyramids of medulla

(c) Medulla oblongata

FIGURE 14.9   Cross Sections of the Brainstem. The level of each section is shown in the figure on the right. (a) The midbrain, cut obliquely to pass through the superior colliculi. (b) The pons. The straight edges indicate cut edges of the peduncles where the cerebellum was removed. (c) The medulla oblongata. **APR**

? *Trace the route taken through all three of these figures by fibers from the gracile and cuneate fasciculi described in section 13.1e.*

14.3c The Midbrain

The mesencephalon becomes just one mature brain structure, the **midbrain**—a short segment of brainstem that connects the hindbrain and forebrain (see figs. 14.2, 14.8). It contains the cerebral aqueduct, continuations of the medial lemniscus and reticular formation, and the motor nuclei of two cranial nerves that control eye movements: cranial nerves III (oculomotor) and IV (trochlear). Only the first of these is seen at the level of the cross section in **figure 14.9a.** (The *medial geniculate nucleus* seen in this figure is not part of the midbrain, but a part of the thalamus that happens to lie in the plane of this section.)

The part of the midbrain posterior to the cerebral aqueduct is a rooflike **tectum.**[18] It exhibits four bulges. The upper (rostral) pair, called the **superior colliculi**[19] (col-LIC-you-lye), control the *extrinsic muscles* of the eyes, enabling us to direct our gaze at a target; to track moving objects; and to turn the eyes and head in response to a visual stimulus (for example, to look at something that you catch sight of in your peripheral vision). Between the superior colliculi and thalamus are other visual control centers, the **pretectal nuclei,** that control pupillary diameter and focusing of the lens. The lower (caudal) pair, called the **inferior colliculi,** receives signals from the inner ear and relays them to other parts of the brain, especially the thalamus. These colliculi are sensitive to the time delays between sounds heard by the two ears and thus aid in locating the source of a sound in space. They also process fluctuations in pitch, which is important for such purposes as understanding another person's speech. They also mediate the reflexive turning of the head in response to a sound and one's tendency to jump when startled by a sudden noise.

Anterior to the cerebral aqueduct, the midbrain consists mainly of the **cerebral peduncles**—two stalks that anchor the cerebrum to the brainstem. Each peduncle has three main components: tegmentum, substantia nigra, and cerebral crus. The **tegmentum**[20] is dominated by the **red nucleus,** named for a pink color imparted by its high density of blood vessels. Fibers from the red nucleus form the *rubrospinal tract* in most mammals, but in humans its connections go mainly to and from the cerebellum, with which it collaborates in fine motor control. The **substantia nigra**[21] (sub-STAN-she-uh NY-gruh) is a dark gray to black nucleus pigmented with melanin. It is a motor center that relays inhibitory signals to the thalamus and basal nuclei (both of which are discussed later), suppressing unwanted body movement. Degeneration of the neurons in the substantia nigra leads to the muscle tremors of Parkinson disease (see Deeper Insight 12.4). The **cerebral crus** (plural, *crura*) is a bundle of nerve fibers that connect the cerebrum to the pons and carry the corticospinal nerve tracts.

The cerebral aqueduct is encircled by the **central (periaqueductal) gray substance.** This is involved with the reticular formation in controlling awareness of pain, as further described in section 16.2d.

FIGURE 14.10 The Reticular Formation. The formation consists of over 100 nuclei scattered throughout the brainstem. Red arrows indicate routes of input to the reticular formation; blue arrows indicate the radiating relay of signals from the thalamus to the cerebral cortex; and green arrows indicate output from the reticular formation to the spinal cord.

? *Locate components of the reticular formation in all three parts of figure 14.9.*

▶▶▶ APPLY WHAT YOU KNOW

Why are the inferior colliculi shown in figure 14.8b but not in figure 14.9a? How are these two figures related?

14.3d The Reticular Formation

The **reticular**[22] **formation** is a loose web of gray matter that runs vertically through all levels of the brainstem and the upper spinal cord, appearing at all three levels of figure 14.9. It occupies much of the space between the white fiber tracts and the more anatomically distinct brainstem nuclei, and has connections with many areas of the cerebrum (**fig. 14.10**). It consists of more than 100 small neural networks defined less by anatomical boundaries than

---

[18]*tectum* = roof, cover
[19]*colli* = hill; *cul* = little
[20]*tegmen* = cover
[21]*substantia* = substance; *nigra* = black

[22]*ret* = network; *icul* = little

by their use of different neurotransmitters. The functions of these networks include the following:

- **Somatic motor control.** Some motor neurons of the cerebral cortex send their axons to reticular formation nuclei, which then give rise to the *reticulospinal tracts* of the spinal cord. These tracts adjust muscle tension to maintain tone, balance, and posture, especially during body movements. The reticular formation also relays signals from the eyes and ears to the cerebellum so the cerebellum can integrate visual, auditory, and vestibular (balance and motion) stimuli into its role in motor coordination. Other motor nuclei include *gaze centers,* which enable the eyes to track and fixate objects, and *central pattern generators*—neural pools that produce rhythmic signals to the muscles of breathing and swallowing.

- **Cardiovascular control.** The reticular formation includes the previously mentioned cardiac and vasomotor centers of the medulla oblongata.

- **Pain modulation.** The reticular formation is one route by which pain signals from the lower body reach the cerebral cortex. It is also the origin of the *descending analgesic pathways* mentioned in the description of the reticulospinal tracts in section 13.1e. Under certain circumstances, the nerve fibers in these pathways act in the spinal cord to deaden one's awareness of pain.

- **Sleep and consciousness.** The reticular formation has projections to the thalamus and cerebral cortex that allow it some control over what sensory signals reach the cerebrum and come to our conscious attention. It plays a central role in states of consciousness such as alertness and sleep. Injury to the reticular formation can result in irreversible coma.

- **Habituation.** This is a process in which the brain learns to ignore repetitive, inconsequential stimuli while remaining sensitive to others. In a noisy city, for example, a person can sleep through traffic sounds but wake promptly to the sound of an alarm clock or a crying baby. Reticular formation nuclei that modulate activity of the cerebral cortex are called the *reticular activating system* or *extrathalamic cortical modulatory system.*

14.3e The Cerebellum

The cerebellum is the largest part of the hindbrain and second-largest part of the brain as a whole **(fig. 14.11).** It consists of right and left **cerebellar hemispheres** connected by a narrow worm-like bridge called the **vermis.**[23] Each hemisphere exhibits slender, transverse, parallel folds called **folia**[24] separated by shallow sulci. The cerebellum has a surface cortex of gray matter and a deeper layer of white matter. In a sagittal section, the white matter exhibits a branching, fernlike pattern called the **arbor vitae.**[25] Each hemisphere has four masses of gray matter called **deep nuclei** embedded in the white matter. All input to the cerebellum goes to the cortex and all of its output comes from the deep nuclei.

Although the cerebellum is only about 10% of the mass of the brain, it has about 60% as much surface area as the cerebral cortex and it contains more than half of all brain neurons. Its tiny, densely spaced **granule cells** are the most abundant type of neuron in the entire brain. Its most distinctive neurons, however, are the unusually large, globose **Purkinje**[26] **cells** (pur-KIN-jee). These have a tremendous profusion of dendrites compressed into a single plane like a flat tree (see fig. 12.5 and the photo on the opening page of chapter 12). The Purkinje cells are arranged in a single file, with these thick dendritic planes parallel to each other like books on a shelf. Their axons travel to the deep nuclei, where they synapse on output neurons that issue fibers to the brainstem.

The cerebellum is connected to the brainstem by three pairs of stalks called **cerebellar peduncles**[27] (peh-DUN-culs): a pair of *inferior peduncles* connected to the medulla oblongata, a pair of *middle peduncles* to the pons, and a pair of *superior peduncles* to the midbrain (see fig. 14.8b). These consist of thick bundles of nerve fibers that carry signals to and from the cerebellum. Connections between the cerebellum and brainstem are very complex, but overlooking some exceptions, we can draw a few generalizations. Most spinal input enters the cerebellum by way of the inferior peduncles; most input from the rest of the brain enters by way of the middle peduncles; and cerebellar output travels mainly by way of the superior peduncles.

The function of the cerebellum was unknown in the 1950s. By the 1970s, it had come to be regarded as a center for monitoring muscle contractions and aiding in motor coordination. People with cerebellar lesions exhibit serious deficits in coordination and locomotor ability; more will be said later in this chapter about the role of the cerebellum in movement. But cerebellar lesions also affect several sensory, linguistic, emotional, and other nonmotor functions. Recent studies by positron emission tomography (PET) and functional magnetic resonance imaging (fMRI) (both described in Deeper Insight 1.5), and behavioral studies of people with cerebellar lesions, have created a much more expansive view of cerebellar function. It appears that its general role is the evaluation of certain kinds of sensory input, and monitoring muscle movement is only part of its broader function.

The cerebellum is highly active when a person explores objects with the fingertips, for example, to compare the textures of two objects without looking at them. (Tactile nerve fibers from a rat's snout and a cat's forepaws also feed into the cerebellum.) Some spatial perception also resides here. The cerebellum is much more active when a person is required to solve a pegboard puzzle than when moving pegs randomly around the same puzzle board. People with cerebellar lesions also have difficulty identifying different views of a three-dimensional object as belonging to the same object.

The cerebellum is also a timekeeper. PET scans show increased cerebellar activity when a person is required to judge the elapsed time between two stimuli. People with cerebellar lesions have difficulty with rhythmic finger-tapping tasks and other tests of

---

[23]*verm* = worm
[24]*foli* = leaf
[25]*arbor* = tree; *vitae* = of life

[26]Johannes E. von Purkinje (1787–1869), Bohemian anatomist
[27]*ped* = foot; *uncle* = little

Superior colliculus
Inferior colliculus
Pineal gland
Posterior commissure
Cerebral aqueduct
Mammillary body
Midbrain
White matter (arbor vitae)
Oculomotor nerve
Gray matter
Fourth ventricle
Pons
Medulla oblongata

(a) Median section

Anterior
Vermis
Anterior lobe
Posterior lobe
Cerebellar hemisphere
Posterior
Folia

(b) Superior view

FIGURE 14.11 The Cerebellum.
(a) Median section, showing relationship to the brainstem.
(b) Superior view. **APR**

temporal judgment. An important aspect of cerebellar timekeeping is the ability to predict where a moving object will be in the next second or so. You can imagine the importance of this to a predator chasing its prey, to a tennis player, or in driving a car in heavy traffic. The cerebellum also helps to predict how much the eyes must move in order to compensate for head movements and remain fixed on an object.

Even hearing has some newly discovered and surprising cerebellar components. Cerebellar lesions impair a person's ability to judge differences in pitch between two tones and to distinguish between similar-sounding words such as *rabbit* and *rapid.* Language output also involves the cerebellum. If a person is given a noun such as *apple* and told to think of a related verb such as *eat,* the cerebellum shows higher PET activity than if the person is told merely to repeat the word *apple.*

People with cerebellar lesions also have difficulty planning and scheduling tasks. They tend to overreact emotionally and have difficulty with impulse control. Many children with

attention-deficit/hyperactivity disorder (ADHD) have abnormally small cerebellums.

BEFORE YOU GO ON

Answer the following questions to test your understanding of the preceding section:

9. List several visceral functions controlled by nuclei of the medulla. What general function would be lost if the pyramids of the medulla were severed?

10. List several sensory and motor functions of the pons.

11. What functions are served by the superior and inferior colliculi? To what portion of the brainstem do they belong?

12. Where is the reticular formation found? Define *reticular formation* in a single sentence.

13. List several functions of the cerebellum.

14.4 The Forebrain

Expected Learning Outcomes
When you have completed this section, you should be able to

a. name the three major components of the diencephalon and describe their locations and functions;

b. identify the five lobes of the cerebrum and their functions;

c. describe the three types of tracts in the cerebral white matter;

d. describe the distinctive cell types and histological arrangement of the cerebral cortex; and

e. describe the location and functions of the basal nuclei and limbic system.

The forebrain consists of the diencephalon and telencephalon. The diencephalon encloses the third ventricle and is immediately rostral to the brainstem. The telencephalon develops chiefly into the cerebrum.

14.4a The Diencephalon

Three structures arise from the embryonic diencephalon: the *thalamus, hypothalamus,* and *epithalamus.*

The Thalamus

Each side of the brain has a **thalamus,**[28] an ovoid mass perched at the superior end of the brainstem beneath the cerebral hemisphere (see figs. 14.6c, 14.8, 14.17). The two thalami form about four-fifths of the diencephalon. Laterally, they protrude into the lateral ventricles. Medially, they protrude into the third ventricle and are joined to each other by a narrow *interthalamic adhesion* in about 70% of people.

The thalamus consists of at least 23 nuclei, most of which fall into five groups: anterior, posterior, medial, lateral, and ventral. These groups and their functions are shown in **figure 14.12a.**

Broadly speaking, the thalamus is the "gateway to the cerebral cortex." Nearly all input to the cerebrum passes by way of synapses in the thalamic nuclei, including signals for taste, smell, hearing, equilibrium, vision, and such general senses as touch, pain, pressure, heat, and cold. (Some smell signals also get to the cerebrum by routes that bypass the thalamus.) The thalamic nuclei process this information, screen out much of it, and relay a small portion of it to the cerebral cortex.

The thalamus also serves in motor control by relaying signals from the cerebellum to the cerebrum and providing feedback loops between the cerebral cortex and the *basal nuclei* (deep cerebral motor centers). Finally, the thalamus is involved in the memory and emotional functions of the *limbic system,* a complex of structures that include some cerebral cortex of the temporal and frontal lobes and some of the anterior thalamic nuclei. The role of the thalamus in motor and sensory circuits is further discussed later in this chapter and chapter 16.

The Hypothalamus

The **hypothalamus** (see fig. 14.2) forms the floor and part of the walls of the third ventricle. It extends anteriorly to the *optic chiasm* (ky-AZ-um), where the optic nerves meet, and posteriorly to a pair of humps called the **mammillary**[29] **bodies.** Each mammillary body contains three or four *mammillary nuclei.* Their primary function is to relay signals from the limbic system to the thalamus. The pituitary gland is attached to the hypothalamus by a stalk *(infundibulum)* between the optic chiasm and mammillary bodies.

The hypothalamus is the major control center of the endocrine and autonomic nervous systems. It plays an essential role in the homeostatic regulation of nearly all organs of the body. Its nuclei include centers concerned with a wide variety of visceral functions **(fig. 14.12b):**

- **Hormone secretion.** The hypothalamus secretes hormones that control the anterior pituitary gland, thereby regulating growth, metabolism, reproduction, and stress responses. It also produces two hormones that are stored in the posterior pituitary gland, concerned with labor contractions, lactation, and water conservation. These relationships are explored especially in sections 17.2b and 17.2d.

- **Autonomic effects.** The hypothalamus is a major integrating center for the autonomic nervous system. It sends descending fibers to lower brainstem nuclei that influence heart rate, blood pressure, gastrointestinal secretion and motility, and pupillary diameter, among other functions.

- **Thermoregulation.** The *hypothalamic thermostat* consists of a collection of neurons, concentrated especially in the preoptic nucleus, that monitor body temperature. When the temperature deviates too much from its set point, this center activates mechanisms for lowering or raising the body temperature (see section 26.5c).

- **Food and water intake.** The hypothalamus regulates sensations of hunger and satiety. One nucleus in particular, the *arcuate nucleus,* contains receptors for hormones that increase hunger and energy expenditure, other hormones that reduce both, and hormones that exert long-term control over body mass (see section 26.1b). Hypothalamic neurons called *osmoreceptors* monitor blood osmolarity and stimulate water-seeking and drinking behavior when the body is dehydrated. Dehydration also stimulates the hypothalamus to produce *antidiuretic hormone,* which conserves water by reducing urine output.

- **Sleep and circadian rhythms.** The caudal part of the hypothalamus is part of the reticular formation. It contains nuclei that regulate the rhythm of sleep and waking.

---

[28]*thalamus* = chamber, inner room

[29]*mammill* = nipple

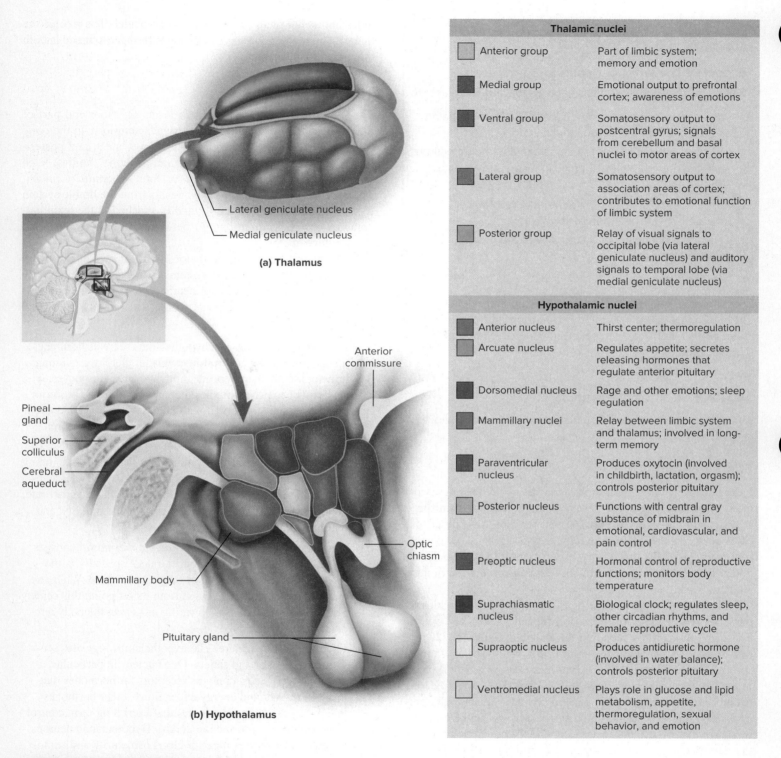

Thalamic nuclei	
Anterior group	Part of limbic system; memory and emotion
Medial group	Emotional output to prefrontal cortex; awareness of emotions
Ventral group	Somatosensory output to postcentral gyrus; signals from cerebellum and basal nuclei to motor areas of cortex
Lateral group	Somatosensory output to association areas of cortex; contributes to emotional function of limbic system
Posterior group	Relay of visual signals to occipital lobe (via lateral geniculate nucleus) and auditory signals to temporal lobe (via medial geniculate nucleus)

Hypothalamic nuclei	
Anterior nucleus	Thirst center; thermoregulation
Arcuate nucleus	Regulates appetite; secretes releasing hormones that regulate anterior pituitary
Dorsomedial nucleus	Rage and other emotions; sleep regulation
Mammillary nuclei	Relay between limbic system and thalamus; involved in long-term memory
Paraventricular nucleus	Produces oxytocin (involved in childbirth, lactation, orgasm); controls posterior pituitary
Posterior nucleus	Functions with central gray substance of midbrain in emotional, cardiovascular, and pain control
Preoptic nucleus	Hormonal control of reproductive functions; monitors body temperature
Suprachiasmatic nucleus	Biological clock; regulates sleep, other circadian rhythms, and female reproductive cycle
Supraoptic nucleus	Produces antidiuretic hormone (involved in water balance); controls posterior pituitary
Ventromedial nucleus	Plays role in glucose and lipid metabolism, appetite, thermoregulation, sexual behavior, and emotion

(a) Thalamus

(b) Hypothalamus

FIGURE 14.12 **The Diencephalon.** (a) Nuclei of the thalamus. (b) Nuclei of the hypothalamus. Only some of the nuclei of the thalamus and hypothalamus are shown, and some of their functions are listed. These lists are by no means complete. **APR**

Superior to the optic chiasm, the hypothalamus contains a *suprachiasmatic nucleus* that controls our 24-hour (circadian) rhythm of activity.

- **Memory.** The mammillary nuclei lie in the pathway of signals traveling from the hippocampus, an important memory center of the brain, to the thalamus. Thus, they are important in memory, and lesions to the mammillary nuclei cause memory deficits. (For a more complete discussion, see section 14.5d.)

- **Emotional behavior and sexual response.** Hypothalamic centers are involved in a variety of emotional responses including anger, aggression, fear, pleasure, and contentment; and in sexual drive, copulation, and orgasm.

The Epithalamus

The **epithalamus** is a very small mass of tissue composed mainly of the **pineal gland** (an endocrine gland discussed in section 17.3), the **habenula**[30] (a relay from the limbic system to the midbrain), and a thin roof over the third ventricle (see fig. 14.2a).

14.4b The Cerebrum

The embryonic telencephalon becomes the cerebrum, the largest and most conspicuous part of the human brain. Your cerebrum enables you to turn these pages, read and comprehend the words, remember ideas, talk about them with your peers, and take an examination. It is the seat of voluntary motor control and our most distinctly human mental processes. It is the most complex and challenging frontier of neurobiology.

Gross Anatomy

The cerebrum so dwarfs and conceals the other structures that people often think of "cerebrum" and "brain" as synonymous. Its major anatomical landmarks were described at the beginning of this chapter and should be reviewed if necessary (see figs. 14.1 and 14.2); most important are the two *cerebral hemispheres,* separated by the *longitudinal cerebral fissure* but connected by a prominent fiber tract, the *corpus callosum;* and the conspicuous wrinkles, or *gyri,* of each hemisphere, separated by grooves called *sulci.* The folding

---

[30]*haben* = strap, rein; *ula* = little

of the cerebral surface into gyri allows a greater amount of cortex to fit in the cranial cavity. The gyri give the cerebrum a surface area of about 2,500 cm^2. If the cerebrum were smooth-surfaced, it would have only one-third as much area and proportionally less information-processing capability. This extensive folding is one of the greatest differences between the human brain and the relatively smooth-surfaced brains of most other mammals.

Some gyri have consistent and predictable anatomy, while others vary from brain to brain and even from the right hemisphere to the left in the same person. Certain unusually prominent sulci divide each hemisphere into five anatomically and functionally distinct lobes, as follows. The first four are visible superficially and are named for the cranial bones overlying them; the fifth lobe *(insula)* is not visible from the surface.

It is difficult to summarize the functions of the cerebral lobes in any simple way; many functions such as vision, memory, speech, and emotion are distributed over multiple lobes, and different lobes contribute only part of the overall function. With that caveat, **figure 14.13** and the following list survey the anatomical distribution of many key cerebral functions.

1. The **frontal lobe** lies immediately behind the frontal bone, superior to the eyes. From the forehead, it extends caudally to a wavy vertical groove, the **central sulcus.** It is the seat of our conscious, abstract thought; explicit or declarative memory (see section 12.6e); cognitive and emotional processes such as mood, motivation, foresight, planning, decision making, emotional control, and judging socially

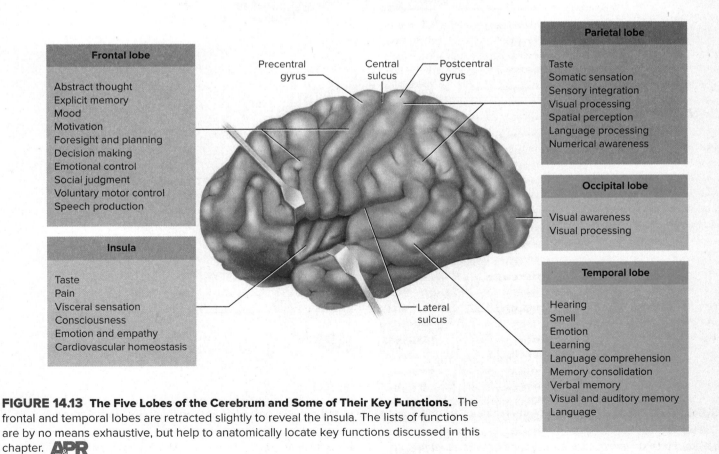

FIGURE 14.13 The Five Lobes of the Cerebrum and Some of Their Key Functions. The frontal and temporal lobes are retracted slightly to reveal the insula. The lists of functions are by no means exhaustive, but help to anatomically locate key functions discussed in this chapter. **A&PR**

appropriate behavior; and in speech production and other voluntary motor control.

2. The **parietal lobe** forms the uppermost part of the brain and underlies the parietal bone. Starting at the central sulcus, it extends caudally to the **parieto–occipital sulcus,** visible on the medial surface of each hemisphere (see fig. 14.2). It is concerned with taste, somatic sensation (such as touch, heat, and pain), and visual processing; multisensory integration such as correlating sights and sounds to holistically comprehend our sensory world; spatial perception and awareness of body orientation; language processing; and numerical awareness (a sense of the quantity of things we see before us).

3. The **occipital lobe** is at the rear of the head, caudal to the parieto–occipital sulcus and underlying the occipital bone. It is the principal visual center of the brain, where we first become aware of visual stimuli and process them to identify what we see.

4. The **temporal lobe** is a lateral, horizontal lobe deep to the temporal bone, separated from the frontal and parietal lobes above it by a deep **lateral sulcus.** It is concerned with hearing; smell; emotion; learning; language comprehension and memory of the grammar and vocabulary of the languages we speak; memory consolidation (formation of new long-term memories); and storage of verbal, visual, and auditory memories.

5. The **insula**[31] is a small mass of cortex deep to the lateral sulcus, made visible only by retracting or cutting away some of the overlying cerebrum (see fig. 14.6c). It plays roles in taste; pain; visceral sensation; consciousness; emotional responses and empathy (sympathetic awareness of the feelings of others); and cardiovascular homeostasis (such as heart rate and blood pressure responses to exercise).

[31]*insula* = island

(a) Sagittal section

(b) Frontal section

FIGURE 14.14 Tracts of Cerebral White Matter. (a) Sagittal section showing association and projection tracts. (b) Frontal section showing commissural and projection tracts.

? *What route can commissural tracts take between the right and left cerebral hemispheres other than the one shown here?*

The Cerebral White Matter

Most of the volume of the cerebrum is white matter. This is composed of glia and myelinated nerve fibers that transmit signals from one region of the cerebrum to another and between the cerebrum and lower brain centers. These fibers form bundles, or *tracts*, of three kinds (**fig. 14.14**):

1. **Projection tracts** extend vertically between higher and lower brain and spinal cord centers. They carry information between the cerebrum and the rest of the body. The corticospinal tracts, for example, carry motor signals from the cerebrum to the brainstem and spinal cord. Other projection tracts carry signals upward to the cerebral cortex. Such tracts form a broad, dense sheet called the *internal capsule* between the thalamus and basal nuclei (described shortly), then radiate in a diverging, fanlike array (the *corona radiata*[32]) to specific areas of the cortex.

2. **Commissural tracts** cross from one cerebral hemisphere to the other through bridges called **commissures** (COM-ih-shurs). The great majority of commissural tracts pass through the large corpus callosum (see fig. 14.2). A few tracts pass through the much smaller **anterior** and **posterior commissures**. Commissural tracts enable the two sides of the cerebrum to communicate with each other.

3. **Association tracts** connect different regions within the same cerebral hemisphere. *Long association fibers* connect different lobes of a hemisphere to each other, whereas *short association fibers* connect different gyri within a single lobe.

Among their roles, association tracts link perceptual and memory centers of the brain; for example, they enable you to see a rose, name it, and imagine its scent.

The Cerebral Cortex

Neural integration is carried out in the gray matter of the cerebrum, which is found in three places: the cerebral cortex, basal nuclei, and limbic system. We begin with the **cerebral cortex**,[33] a layer covering the surface of the hemispheres (see fig. 14.6c). Even though it is only 2 to 3 mm thick, the cortex constitutes about 40% of the mass of the brain and contains 14 to 16 billion neurons. It possesses two principal types of neurons (**fig. 14.15**) called *stellate cells* and *pyramidal cells*. **Stellate cells** have spheroidal neurosomas with short axons and dendrites projecting in all directions. They are concerned largely with receiving sensory input and processing information on a local level. **Pyramidal cells** are tall and conical. Their apex points toward the brain surface and has a thick dendrite with many branches and small, knobby *dendritic spines* (see fig. 12.5a). The base gives rise to horizontally oriented dendrites and an axon that passes into the white matter below. Pyramidal cells include the output neurons of the cerebrum—the only cerebral neurons whose fibers leave the cortex and connect with other parts of the CNS. Pyramidal cell axons have collaterals that synapse with other neurons in the cortex or in deeper regions of the brain.

About 90% of the human cerebral cortex is a six-layered tissue called **neocortex**[34] because of its relatively recent evolutionary origin. Although vertebrate animals have existed for about 600 million years, the neocortex didn't develop significantly until about 60 million years ago, when there was a sharp increase in the diversity

---

[32]*corona* = crown; *radiata* = radiating

[33]*cortex* = bark, rind
[34]*neo* = new

DEEPER INSIGHT 14.3

CLINICAL APPLICATION

Tracking the Tracts of White Matter

This chapter's opening photo was made with a relatively new application of magnetic resonance imaging (MRI) called *diffusion tensor imaging (DTI)*. DTI is a method of detecting the diffusion of water molecules to trace parallel bundles of myelinated nerve fibers in the white matter of the CNS. In most places, water diffusion is random and generates no useful signal. The parallel paths of nerve fibers and the insulating effect of their myelin sheaths, however, give a directionality to diffusion, since more water diffuses along the nerve fibers than across them. The vectors of diffusion can be color-coded so tracts passing between the left and right brain are colored red, anterior–posterior tracts are green, and superior–inferior tracts are blue. The most often mapped tract has been the corticospinal tract, the main output pathway for cerebral control of the muscles; the tract of second greatest interest is called the *arcuate fasciculus,* which connects the Wernicke and Broca speech areas (see section 14.5h). The mapping of fiber tracts by DTI is called *tractography.* The ultimate aim of cerebral tractography is to determine the brain's *connectome,* a complete map of the physical connections (circuitry) between brain regions.

DTI dates to the 1990s but has only more recently gained application to living patients, making it a highly sensitive imaging method for neuropathology. Its greatest use to date is the assessment of acute ischemic stroke, and the second greatest is diagnosis of brain tumors and their effects on movement, language, and vision. DTI is employed by neurosurgeons for locating tumors and lesions that don't show up on conventional MRI. This enables the surgeon to precisely target a problem area and avoid injury to other critical tracts nearby. It is helpful in monitoring the maturation of myelin in children, degenerative changes in old age, the progress of demyelinating diseases such as multiple sclerosis, the extent of traumatic brain injury (TBI), and brain responses to therapy. It has found clinical applications as well to Alzheimer disease, epilepsy, autism, and cocaine addiction. Clinicians hope that it will soon prove useful in the prognosis of diseases—predicting their developmental course to allow for earlier and more effective intervention. DTI has also been used in assessments of cardiac muscle, prostate disease, and athletic injuries to muscles and tendons.

Cortical surface

I

II

III

IV

V

VI

Small pyramidal cells

Stellate cells

Large pyramidal cells

White matter

FIGURE 14.15 Histology of the Neocortex. Neurons are arranged in six layers.

❓ *Note the long processes leading upward from each pyramidal cell body. Are those axons or dendrites? Explain your answer.*

of mammals. It attained its highest development by far in the primates. The six layers of neocortex, numbered in figure 14.15, vary from one part of the cerebrum to another in relative thickness, cellular composition, synaptic connections, size of the neurons, and destination of their axons. Layer IV is thickest in sensory regions and layer V in motor regions, for example. All axons that leave the cortex and enter the white matter arise from layers III, V, and VI.

Some regions of cerebral cortex have fewer layers. The earliest type of cortex to appear in vertebrate evolution was a one- to five-layered tissue called *paleocortex* (PALE-ee-oh-COR-tex), limited in humans to part of the insula and certain areas of the temporal lobe concerned with smell. The next to evolve was a three-layered *archicortex* (AR-kee-COR-tex), found in the human hippocampus, a memory-forming center in the temporal lobe. The neocortex was the last to evolve.

The Limbic System

The **limbic**[35] **system** is an important center of emotion and learning. It is a ring of cortex on the medial side of each hemisphere, encircling the corpus callosum and thalamus. Its most anatomically prominent components are the **cingulate**[36] **gyrus** (SING-you-let), which arches over the top of the corpus callosum in the frontal and parietal lobes; the **hippocampus**[37] in the medial temporal lobe **(fig. 14.16);** and the **amygdala**[38] (ah-MIG-da-luh) immediately rostral to the hippocampus, also in the temporal lobe. There are still differences of opinion on what structures to consider as parts of the limbic system, but these three are agreed upon. Other components include the mammillary bodies and other hypothalamic nuclei, some thalamic nuclei, parts of the basal nuclei, and parts of the frontal lobe called *prefrontal* and *orbitofrontal cortex.* Limbic system components are interconnected through a complex loop

[35]*limbus* = border
[36]*cingul* = girdle
[37]*hippocampus* = sea horse, named for its shape
[38]*amygdala* = almond

Cingulate gyrus

Corpus callosum

Medial prefrontal cortex

Orbitofrontal cortex

Basal nuclei

Amygdala

Temporal lobe

Fornix

Thalamic nuclei

Mammillary body

Hippocampus

FIGURE 14.16 The Limbic System. This ring of structures (shown in violet) includes important centers of learning and emotion. In the frontal lobe, there is no sharp rostral boundary to limbic system components.

of fiber tracts allowing for somewhat circular patterns of feedback among its nuclei and cortical neurons. All of these structures are bilaterally paired; there is a limbic system in each cerebral hemisphere.

The limbic system was long thought to be associated with smell because of its close association with olfactory pathways, but beginning in the early 1900s and continuing even now, experiments have abundantly demonstrated more significant roles in emotion and memory. Most limbic system structures have centers for both gratification and aversion. Stimulation of a gratification center produces a sense of pleasure or reward; stimulation of an aversion center produces unpleasant sensations such as fear or sorrow. Gratification centers dominate some limbic structures, such as the *nucleus accumbens* (not illustrated), while aversion centers dominate others such as the amygdala. The roles of the amygdala in emotion and the hippocampus in memory are described in section 14.5d.

The Basal Nuclei

The basal nuclei are masses of cerebral gray matter buried deep in the white matter, lateral to the thalamus (**fig. 14.17**). They are

FIGURE 14.17 The Basal Nuclei. (a) Frontal section of the brain. (b) Corresponding slice of fresh brain tissue. The basal nuclei are labeled in boldface. **APR**

Photo: Biophoto Associates/Science Source

often called *basal ganglia,* but the word *ganglion* is best restricted to clusters of neurons outside the CNS. Neuroanatomists disagree on how many brain centers to classify as basal nuclei, but agree on at least three: the **caudate**[39] **nucleus, putamen**[40] (pyu-TAY-men), and **globus pallidus.**[41] These three are collectively called the *corpus striatum*[42] because of their striped appearance. The putamen and globus pallidus together are also called the *lentiform*[43] *nucleus,* because they form a lens-shaped body. They are involved in motor control and are further discussed in a later section on that topic.

BEFORE YOU GO ON

Answer the following questions to test your understanding of the preceding section:

14. What are the three major components of the diencephalon? Which ventricle does the diencephalon enclose?

15. What is the role of the thalamus in sensory function?

16. List at least six functions of the hypothalamus.

17. Name the five lobes of the cerebrum and describe their locations, boundaries, and principal functions.

18. Distinguish between commissural, association, and projection tracts of the cerebrum.

19. Where is the limbic system located? What component of it is involved in emotion? What component is involved in memory?

20. Where are the basal nuclei located? What is their general function?

14.5 **Integrative Functions of the Brain**

Expected Learning Outcomes

When you have completed this section, you should be able to

a. list the types of brain waves and discuss their relationship to mental states;

b. describe the stages of sleep, their relationship to the brain waves, and the neural mechanisms of sleep;

c. identify the brain regions concerned with consciousness and thought, memory, emotion, sensation, motor control, and language; and

d. discuss the functional differences between the right and left cerebral hemispheres.

This section concerns such "higher" brain functions as sleep, memory, cognition, emotion, sensation, motor control, and language. These are associated especially with the cerebral cortex,

but not exclusively; they involve interactions between the cerebral cortex and such areas as the basal nuclei, brainstem, and cerebellum. It is impossible in many cases to assign these functions to one specific brain region; as we have already seen (see fig. 14.13), functions of the brain don't have such easily defined anatomical boundaries. Some functions overlap anatomically and some cross anatomical boundaries from one region to another, often remote region. Thus, we will consider these as *integrative* functions of the brain, focusing especially on the cerebrum but in many cases involving the combined action of multiple levels of the brain. Some of these present the most difficult challenges for neurobiology, but they are the most intriguing functions of the brain and involve its largest areas.

14.5a The Electroencephalogram

For research and clinical purposes, it is common to monitor electrical activity called **brain waves.** Recorded with electrodes on the scalp **(fig. 14.18a),** these are rhythmic voltage changes resulting predominantly from synchronized postsynaptic potentials (not action potentials) in the superficial layers of the cerebral cortex. The recording, called an **electroencephalogram**[44] **(EEG),** is useful in studying normal brain functions such as sleep and consciousness, and in diagnosing degenerative brain diseases, metabolic abnormalities, brain tumors, trauma, and so forth. States of consciousness ranging from high alertness to deep sleep are correlated with changes in the EEG. The complete and persistent absence of brain waves is a common clinical and legal criterion of brain death.

There are four types of brain waves **(fig. 14.18b),** distinguished by differences in amplitude (mV) and frequency. Frequency is expressed in hertz (Hz), or cycles per second:

1. **Alpha (α) waves** have a frequency of 8 to 13 Hz and are recorded especially in the parieto-occipital area. They dominate the EEG when a person is awake and relaxed, with the eyes closed and the mind wandering. They are suppressed when a person opens the eyes, receives specific sensory stimulation, or engages in a mental task such as performing mathematical calculations. They are absent during deep sleep.

2. **Beta (β) waves** have a frequency of 14 to 30 Hz and occur in the frontal to parietal region. They are accentuated during mental activity and sensory stimulation.

3. **Theta (θ) waves** have a frequency of 4 to 7 Hz. They are normal in children and in drowsy or sleeping adults, but a predominance of theta waves in awake adults suggests emotional stress or brain disorders.

4. **Delta (δ) waves** are high-amplitude "slow waves" with a frequency of less than 3.5 Hz. Infants exhibit delta waves when awake, and adults exhibit them in deep sleep. A predominance of delta waves in awake adults indicates serious brain damage.

---

[39]*caudate* = tailed, tail-like
[40]*putam* = pod, husk
[41]*glob* = globe, ball; *pall* = pale
[42]*corpus* = body; *striat* = stripe
[43]*lenti* = lens; *form* = shape

[44]*electro* = electricity; *encephalo* = brain; *gram* = record

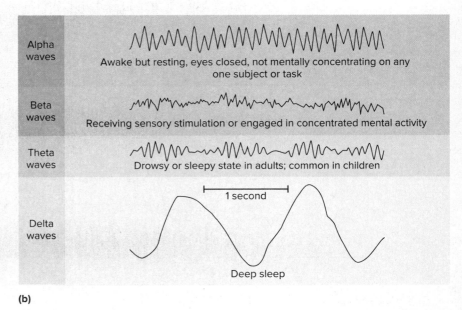

Alpha waves — Awake but resting, eyes closed, not mentally concentrating on any one subject or task

Beta waves — Receiving sensory stimulation or engaged in concentrated mental activity

Theta waves — Drowsy or sleepy state in adults; common in children

Delta waves — 1 second — Deep sleep

(a)   (b)

FIGURE 14.18 The Electroencephalogram (EEG). (a) An EEG is recorded from electrodes on the forehead and scalp. (b) Four classes of brain waves are seen in EEGs.

a: Larry Mulvehill/Science Source

14.5b Sleep

Sleep can be defined as a temporary, reversible loss of consciousness from which (in contrast to coma) one can awaken when stimulated. It is one of many bodily functions that occur in cycles called **circadian**[45] **rhythms** (sur-CAY-dee-an), so named because they are marked by events that recur at intervals of about 24 hours. Sleep superficially resembles other states of prolonged unconsciousness such as coma and animal hibernation, except that individuals cannot be aroused from those states by sensory stimulation.

Sleep occurs in distinct stages recognizable from changes in the EEG. In the first 30 to 45 minutes, the EEG waves drop in frequency but increase in amplitude as one passes through four sleep stages **(fig. 14.19a):**

- **Stage 1.** This stage usually lasts for 1 to 7 minutes. One feels drowsy, closes the eyes, and begins to relax. The EEG transitions from alpha waves to waves of mixed frequency. Thoughts come and go, often accompanied by a drifting sensation. One awakens easily if stimulated.

- **Stage 2.** This stage usually lasts for 10 to 25 minutes in the initial sleep cycle. One passes into light sleep. Brain waves have a lower frequency but higher voltage. They occasionally show 1 or 2 seconds of *sleep spindles,* high spikes resulting from interactions between neurons of the thalamus and cerebral cortex.

- **Stage 3.** This is moderate to deep sleep, typically beginning about 20 minutes after stage 1. Sleep spindles occur less

often, and theta and delta waves appear. The muscles relax, and the *vital signs* (body temperature, blood pressure, and heart and respiratory rates) fall.

- **Stage 4.** The muscles are now very relaxed, vital signs are at their lowest levels, and one becomes difficult to awaken. Stage 4 sleep usually occurs only in the first one-third of the night **(fig. 14.19b).**

About five times a night, a sleeper backtracks from stage 3 or 4 to stage 2 and exhibits bouts of **rapid eye movement (REM) sleep.** This is so named because the eyes oscillate back and forth as if watching a movie. It is also called *paradoxical sleep* because the EEG resembles the waking state, yet the sleeper is harder to arouse than in any other stage. Vital signs increase and the brain consumes even more oxygen than when awake. Sleep paralysis is especially strong during REM sleep and usually prevents the sleeper from acting out his or her dreams.

Bouts of REM sleep become longer and longer as the night progresses, and we usually wake in the morning from either stage 3 or REM sleep. Dreams occur in both REM and non-REM sleep, but REM dreams tend to be longer, more vivid, and more emotional than non-REM dreams. Most nightmares, however, occur during non-REM sleep. The parasympathetic nervous system is very active during REM sleep, causing constriction of the pupils and erection of the penis or clitoris. Erection, however, is seldom accompanied by sexual dream content.

Humans are a diurnal (daytime-active) species in whom falling light intensity induces sleepiness. As the light fades at the end of the day, certain retinal cells called *ganglion cells* transmit signals to part of the hypothalamus called the **suprachiasmatic nucleus (SCN)** (SOO-pra-KY-az-MAT-ic) (see fig. 14.12b).

[45]*circa* = approximately; *dia* = a day, 24 hours

(a) One sleep cycle

(b) Typical 8-hour sleep period

FIGURE 14.19 Sleep Stages and Brain Activity. (a) A single sleep cycle, from waking to deep sleep, followed by 10 minutes of REM sleep. (b) Stages of sleep over an 8-hour night in a typical young adult. Stage 4 is attained only in the first two cycles. Periods of REM sleep increase from about 10 minutes in the first cycle to as long as 50 minutes in the last hour of sleep. Most dreaming occurs during REM sleep.

Located just above the optic chiasm, the SCN is the body's master biological clock. It regulates not only sleep but also circadian rhythms of body temperature, urine production, hormone secretion, and other functions.

The SCN has two major output pathways leading to the nearby **dorsomedial nucleus (DMN)** and the pineal gland. The pineal gland responds to SCN input by secreting the hormone **melatonin.** Melatonin and the DMN together act on the reticular activating system (RAS) of the brainstem. The RAS now blocks signals from the thalamus to the cerebral cortex, so a drowsy person becomes less and less aware of environmental stimuli and lapses into sleep.

Waking is brought on by **orexins,** neuropeptides secreted by the hypothalamus. These stimulate the reticular activating

system, which then stimulates the thalamus. As a person begins to wake, electrical activity increases in the thalamus and then spreads through the cerebral cortex. Thus, the cerebral awareness of environmental stimuli increases and the sleeper wakes up. The excitatory pathway from RAS to thalamus and cortex involves multiple excitatory neurotransmitters including histamine; this is why antihistamines, antagonizing this pathway, can make one drowsy.

Orexins are absent or at low levels in a disorder called **narcolepsy,** in which a person experiences excessive daytime sleepiness and fatigue and may often fall asleep at work or school, with abnormally quick onset of REM sleep. Narcolepsy seems to be an autoimmune disease caused by antibody-mediated destruction of the orexin-producing neurons.

▶▶▶APPLY WHAT YOU KNOW

Some animals have been shown to exhibit narcolepsy as the result of a mutation in the gene for an orexin receptor. Looking ahead in the book to type 2 diabetes mellitus in section 17.7e, can you identify a common thread in these two disorders?

We spend about one-third of our lives asleep, and we might well regret such a terrible "waste of time." Yet sleep deprivation can cause widespread dysfunctions ranging from cognitive decline or hallucinations to weight loss, reduced immunity, and even death. Yet the reasons for sleep and dreaming remain elusive. Non-REM sleep has a restorative, refreshing effect, but it is unclear why quiet bed rest alone can't serve these purposes; why must we lose consciousness?

Sleep is associated with restorative anabolic processes in the immune, nervous, endocrine, and other systems. During waking hours, brain glycogen levels go down, while ATP consumption generates a sleep-inducing metabolite, adenosine (see Deeper Insight 15.2). Non-REM sleep reduces cerebral metabolic rate and ATP consumption, allowing glycogen and ATP levels to rebound while adenosine levels fall and other metabolites are cleared from the brain. Growth hormone secretion rises during slow-wave sleep.

REM sleep appears to be a time in which the brain either "consolidates" and strengthens memories by reinforcing synaptic connections, or purges superfluous information from memory by weakening or eliminating other synapses. When REM sleep is blocked in experimental animals without waking them, they forget tasks they had recently learned. Theta wave activity and the sleep spindles of stage 2 sleep are also associated with memory consolidation. Sleep research subjects who have just learned a new task show a higher density of sleep spindles than controls who have not.

14.5c Cognition

Cognition[46] is the range of mental processes by which we acquire and use knowledge—sensory perception, thought, reasoning, judgment, memory, imagination, and intuition. Such functions are widely distributed over regions of cerebral cortex called **association areas,** which constitute about 75% of all brain tissue. This is the most difficult area of brain research and the most incompletely understood aspect of cerebral function. Much of what we know about it has come from studies of patients with brain lesions—areas of tissue destruction resulting from cancer, stroke, and trauma. The many brain injuries incurred in World Wars I and II yielded an abundance of insights into regional brain functions. More recently, imaging methods such as PET and fMRI have yielded much more sophisticated insights. These methods allow a researcher to scan a person's brain during various cognitive or motor tasks and to see which brain regions are most active in different mental and task states (see Deeper Insight 14.5).

A few examples of the cognitive effects of cerebral lesions reveal some functions of the association areas:

- Parietal lobe lesions can cause people to become unaware of objects, or even their own limbs, on the other side of the body—a condition called *contralateral neglect syndrome.* In typical cases, men shave only half of the face, women apply makeup to only one side, patients dress only half of the body, and some people deny that one arm or leg belongs to them. Such patients are unable to find their way around—say, to describe the route from home to work or navigate within a familiar building.

- Temporal lobe lesions often result in *agnosia*[47] (ag-NO-zee-ah), the inability to recognize, identify, and name familiar objects. In *prosopagnosia,*[48] a person cannot remember familiar faces, even his or her own reflection in a mirror.

- Frontal lobe lesions are especially devastating to the qualities we think of as personality. The frontal lobe integrates information from sensory and motor regions of the cortex and from other association areas. It gives us a sense of our relationship to the rest of the world, enabling us to think about it and to plan and execute appropriate behavior. Lesions here may produce profound personality disorders and socially inappropriate behaviors.

14.5d Memory

Memory is one of the cognitive functions, but warrants special attention. We studied its forms and its neural and molecular mechanisms earlier (see section 12.6e). Now that you have been introduced to the gross anatomy of the brain, we can consider where those processes occur anatomically.

Our subject is really a little broader than memory per se. Information management by the brain entails learning (acquiring new information), memory proper (information storage and retrieval), and forgetting (eliminating trivial information). Brain-injured people are sometimes unable to recall things they once knew (**retrograde amnesia**) or unable to store new information (**anterograde amnesia**). *Amnesia* refers to defects in *explicit* memory (such as the ability to describe past events), not *implicit* memory (such as the ability to tie one's shoes).

The **hippocampus** of the limbic system is an important memory-forming center (see fig. 14.16). It doesn't store memories, but organizes sensory and cognitive experiences into a unified long-term memory. The hippocampus learns from sensory input while an experience is happening, but it has a short memory. Later, especially during sleep, it plays this memory repeatedly to the cerebral cortex, which is a "slow learner" but forms longer-lasting memories. This process of "teaching the cerebral cortex" until a long-term memory is established is called **memory consolidation.** Long-term memories are held in various areas of cortex. One's vocabulary and memory of faces and familiar

---

[46]*cognit* = to know

[47]*a* = without; *gnos* = knowledge
[48]*prosopo* = face, person

objects, for example, reside in the superior temporal lobe, and memories of one's plans and social roles are in the prefrontal cortex (see fig. 14.13).

Lesions of the hippocampus can cause profound anterograde amnesia. For example, in 1953, a famous patient known in the literature as H. M. (Henry Molaison, 1926–2008) underwent surgical removal of a large portion of both temporal lobes, including both hippocampi, to treat severe epilepsy. The operation had no adverse effect on his intelligence or explicit memory for things that had happened early in his life, but it left him with an inability to establish new explicit memories. He could hold a conversation with his psychologist, but a few minutes later deny that it had taken place. He worked with the same psychologist for more than 40 years after his operation, but couldn't remember who she was from day to day. He was nevertheless able to learn new motor skills, thus showing explicit and implicit memory to involve separate brain regions.

Other parts of the brain involved in memory include the cerebellum, with a role in learning motor skills, and the amygdala, with a role in emotional memory.

14.5e Emotion

Emotional feelings and memories are not exclusively cerebral functions, but result from an interaction between areas of the prefrontal cortex and diencephalon. Emotional control centers of the brain have been identified by studying people with brain lesions and by such techniques as surgical removal, ablation (destruction) of small regions with electrodes, and stimulation with electrodes and chemical implants, especially in experimental animals. Changes in behavior following such procedures give clues to the functions that a region performs. However, interpretation of the results is difficult and controversial because of the complex connections between the emotional brain and other regions.

The **prefrontal cortex (frontal association area)** is the most rostral part of the frontal lobe, just behind the forehead. It is well developed only in primates, especially humans. It is the seat of judgment, intent, and control over the expression of our emotions. However we may feel, it is here that we decide the appropriate way to show those feelings. But the feelings themselves, and emotional memories, arise from deeper regions of the brain, especially the hypothalamus and amygdala. Here lie the nuclei that stimulate us to recoil in fear from a rattlesnake or yearn for a lost love.

The amygdala is a major component of the limbic system described earlier (see fig. 14.16). It receives processed information from the general senses and from vision, hearing, taste, and smell. Such input enables it to mediate emotional responses to such stimuli as a disgusting odor, a foul taste, a beautiful sight, pleasant music, or a stomachache. It is especially important in the sense of fear, but also plays roles in decision making, food intake, sexual behavior, and drawing our attention to novel stimuli.

Output from the amygdala goes in two directions of special interest: (1) Some goes to the hypothalamus and lower brainstem and influences somatic and visceral motor systems. An emotional response to a stimulus may, through these connections, make one's heart race, raise the blood pressure, make the hair stand on end, or induce vomiting. (2) Other output goes to areas of the prefrontal cortex that mediate conscious control and expression of the emotions, such as the ability to express love, control anger, or overcome fear.

Many important aspects of personality depend on an intact, functional amygdala and hypothalamus. When specific regions are destroyed or artificially stimulated, humans and other animals exhibit either blunted or exaggerated expressions of anger, fear, aggression, self-defense, pleasure, pain, love, sexuality, and parental affection, as well as abnormalities in learning, memory, and motivation. Lesions of the amygdala, for example, can abolish the sense of fear.

Much of our behavior is shaped by learned associations between stimuli, our responses to them, and the rewards or punishments that result. Nuclei involved in feelings of reward and punishment have been identified in the hypothalamus of cats, rats, monkeys, and other animals. In a representative experiment, an electrode is implanted in an area of an animal's hypothalamus called the **median forebrain bundle (MFB).** The animal is placed in a chamber with a foot pedal wired to that electrode. When the animal steps on the pedal, it receives a mild electrical stimulus to the MFB. Apparently the sensation is strongly rewarding, because the animal soon starts to press the pedal over and over and may spend most of its time doing so—even to the point of neglecting food and water. Rats have been known to bar-press 5,000 to 12,000 times an hour, and monkeys up to 17,000 times an hour, to stimulate their MFBs.

These animals cannot tell us what they are feeling, but electrode implants have also been used to treat people who suffer otherwise incurable schizophrenia, pain, or epilepsy. These patients also repeatedly press a button to stimulate the MFB, but they do not report feelings of joy or ecstasy. Some are unable to explain why they enjoy the stimulus, and others report "relief from tension" or "a quiet, relaxed feeling." With electrodes misplaced in other areas of the hypothalamus, subjects report feelings of fear or terror when stimulated.

▶▶▶APPLY WHAT YOU KNOW

MRI scans show that John has a pea-size tumor in his hippocampus and Allan has a tumor of the same size in his amygdala. One of these patients is in prison for violent crimes in which he seemed to show no fear or sense of self-preservation. The other patient cannot remember the name of his new granddaughter, no matter how many times he is told. Which behavioral outcome do you associate with each patient? Why?

14.5f Sensation

A great deal of the cerebrum is concerned with the senses—most cortex of the insula and of the parietal, occipital, and temporal lobes. Regions called **primary sensory cortex** are the sites where sensory input is first received and one becomes conscious of a stimulus. Adjacent to these are association areas where this information is interpreted. For example, the primary visual cortex, which receives input from the eyes, is bordered by the visual association area, which interprets and makes cognitive sense of the visual stimuli so we know what we are looking at. Some association

Primary motor cortex

Motor association area

Broca area

Prefrontal cortex

Olfactory association area

Primary somatosensory cortex

Somatosensory association area

Primary gustatory cortex

Wernicke area

Visual association area

Primary visual cortex

Primary auditory cortex

Auditory association area

FIGURE 14.20 Some Functional Regions of the Left Cerebral Cortex. The Broca and Wernicke areas for language abilities are found in only one hemisphere, usually the left. The other regions shown here are mirrored in both hemispheres.

areas are *multimodal*—instead of processing information from a single sensory source, they receive input from multiple senses and integrate this into our overall perception of our surroundings. For example, in the frontal lobe just above the eyes, there is a patch of multimodal cortex called the *orbitofrontal cortex,* which receives taste, smell, and visual input to form our overall impression of the desirability of a particular food.

The sense organs and their signaling pathways in the CNS are the subject of chapter 16. Here we will examine only the areas of cerebral cortex involved in sensory perception.

The Special Senses

The **special senses** are limited to the head, and some employ relatively complex sense organs. They are vision, hearing, equilibrium, taste, and smell. Their primary cortices and association areas are located as follows **(fig. 14.20).**

- **Vision.** Visual signals are received by the **primary visual cortex** in the far posterior region of the occipital lobe. This is bordered anteriorly by the **visual association area,** which includes all the remainder of the occipital lobe, some of the posterior parietal lobe (concerned with spatial perception), and much of the inferior temporal lobe, where we recognize faces and other familiar objects.

- **Hearing.** Auditory signals are received by the **primary auditory cortex** in the superior region of the temporal lobe and in the nearby insula. The **auditory association area** occupies areas of temporal lobe inferior to the primary auditory cortex and deep within the lateral sulcus. This is where we become capable of recognizing spoken words, a familiar piece of music, or a voice on the telephone.

- **Equilibrium.** Signals from the inner ear for equilibrium project mainly to the cerebellum and several brainstem nuclei concerned with head and eye movements and visceral functions. Some fibers of this system, however, are routed through the thalamus to areas of association cortex in the roof of the lateral sulcus and near the lower end of the central sulcus. This is the seat of consciousness of our body movements and orientation in space.

- **Taste and smell.** Gustatory (taste) signals are received by the **primary gustatory cortex** in the inferior end of the postcentral gyrus of the parietal lobe (discussed shortly) and an anterior region of the insula. Olfactory (smell) signals are received by the **primary olfactory cortex** in the medial surface of the temporal lobe and inferior surface of the frontal lobe. The *orbitofrontal cortex* mentioned earlier serves as a multimodal association area for both taste and smell, and is thus highly important in our enjoyment or rejection of various foods and drinks.

The General Senses

The **general (somatosensory, somesthetic,**[49] **or somatic) senses** are distributed over the entire body and employ relatively simple receptors. They include such senses as touch, pressure, stretch, movement, heat, cold, and pain. Coming from the head, such signals reach the brain by way of certain cranial nerves, especially the trigeminal nerve; from the rest of the body, they ascend sensory tracts of the spinal cord such as the spinothalamic tract. In both routes, they decussate to the contralateral thalamus.

[49]*som* = body; *esthet* = feeling

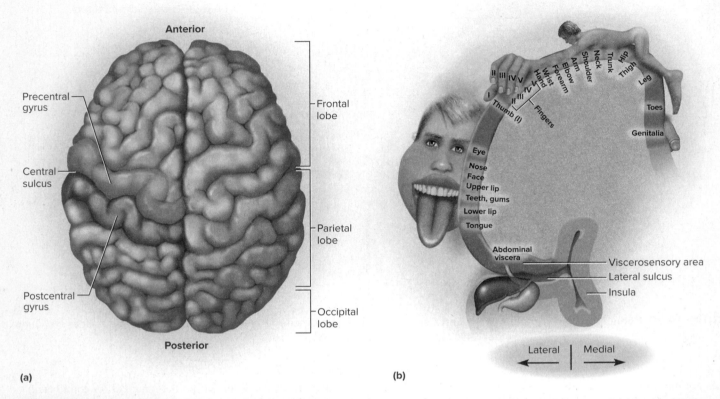

(a)   **(b)**

FIGURE 14.21 The Primary Somatosensory Cortex (Postcentral Gyrus). (a) Superior view of the brain showing the location of the postcentral gyrus (violet). (b) The sensory homunculus, drawn so that body parts are in proportion to the amount of cortex dedicated to their sensation. This gyrus also includes centers for visceral sensation from the intra-abdominal organs (the viscerosensory area). **APR**

The thalamus processes the input and selectively relays signals to the **postcentral gyrus.** This is a fold of the cerebrum that lies immediately caudal to the central sulcus and thus forms the rostral border of the parietal lobe **(fig. 14.21).** We can trace it from just above the lateral sulcus to the crown of the head and then downward into the longitudinal cerebral fissure. Its cortex is called the **primary somatosensory cortex.** Adjacent to it is a **somatosensory association area,** caudal to the gyrus and in the roof of the lateral sulcus (fig. 14.20). Awareness of stimulation occurs in the primary somatosensory cortex, but making cognitive sense of it is a function of the association area.

Because of the aforementioned decussation in sensory pathways, the right postcentral gyrus receives input from the left side of the body and vice versa. The primary somatosensory cortex is like an upside-down sensory map of the contralateral side of the body, traditionally diagrammed as a *sensory homunculus*[50] (fig. 14.21b). As the diagram shows, receptors in the lower limb project to superior and medial parts of the gyrus, and receptors in the face project to the inferior and lateral parts. Such point-for-point correspondence between an area of the body and an area of the CNS is called **somatotopy.**[51] The reason for the bizarre, distorted appearance of the homunculus is that the amount of cerebral tissue devoted to a given body region is proportional to

how richly innervated and sensitive that region is, not to its size. Thus, the hands and face are represented by a much larger region of somatosensory cortex than the trunk is.

14.5g Motor Control

The intention to contract a skeletal muscle begins in the **motor association (premotor) area** of the frontal lobe (fig. 14.20). This is where we plan our behavior—where neurons compile a program for the degree and sequence of muscle contractions required for an action such as dancing, typing, or speaking. The program is then transmitted to neurons of the **precentral gyrus (primary motor area),** which is the most posterior gyrus of the frontal lobe, immediately anterior to the central sulcus **(fig. 14.22).** Neurons here send signals to the brainstem and spinal cord, which ultimately results in muscle contractions.

The precentral gyrus, like the postcentral one, exhibits somatotopy. The neurons for toe movements, for example, are deep in the longitudinal cerebral fissure on the medial side of the gyrus. The summit of the gyrus controls the trunk, shoulder, and arm, and the inferolateral region controls the facial muscles. This map is diagrammed as a *motor homunculus* (fig. 14.22b). Like the sensory homunculus, it has a distorted look because the amount of cortex devoted to a given body region is proportional to the number of muscles and motor units in that region, not to the size of the region. The amount of cerebral tissue dedicated to the hands, face,

[50]*hom* = man; *unculus* = little
[51]*somato* = body; *topy* = place

FIGURE 14.22 The Primary Motor Cortex (Precentral Gyrus). (a) Superior view of the brain showing the location of the precentral gyrus (blue). (b) Motor homunculus, drawn so that body parts are in proportion to the amount of primary motor cortex dedicated to their control. **A&PR**

❓ *Which body regions are controlled by the largest areas of motor cortex—regions with a few large muscles or regions with numerous small muscles?*

and tongue reflects the importance of fine motor control in speech, facial expression, and use of the hands.

There is no exact point-for-point correspondence between an area of the precentral gyrus and a given muscle. A muscle is controlled by neurons at several points within a general area of the gyrus. Also, a given neuron in the gyrus may ultimately affect more than one muscle, such as muscles of shoulder and elbow movement that both contribute to the coordinated positioning of one's hand. Although the homunculus identifies cortical areas that are broadly responsible for motor control of a given region, the boundaries between these cortical areas overlap and are not sharply defined.

The pyramidal cells of the precentral gyrus are called **upper motor neurons.** Their fibers project caudally, with about 19 million fibers ending in nuclei of the brainstem and 1 million forming the corticospinal tracts. Most of these fibers decussate in the lower medulla oblongata (at the pyramidal decussation) and form the *lateral corticospinal tract* on each side of the spinal cord. A smaller number of fibers pass through the medulla without decussation and form the *anterior corticospinal tracts,* which cross over lower in the spinal cord. Each precentral gyrus thus controls muscles on the contralateral side of the body. In the brainstem or spinal cord, the fibers from the upper motor neurons synapse with **lower motor neurons** whose axons innervate the skeletal muscles (see fig. 13.7).

Other areas of the brain important in muscle control are the basal nuclei (see fig. 14.17) and cerebellum. Among the roles of the basal nuclei are determining the onset and cessation of intentional movements; the repetitive hip and shoulder movements that occur in walking; and highly practiced, learned behaviors that one carries out with little thought—for example, writing, driving a car, and tying one's shoes. The basal nuclei lie in a feedback circuit from the cerebrum to the basal nuclei to the thalamus and back to the cerebrum. They receive signals from the substantia nigra of the midbrain and from all areas of cerebral cortex, except for the primary visual and auditory cortices. The basal nuclei process these and issue their output to the thalamus, which relays these signals back to the midbrain and cerebral cortex—especially to the prefrontal cortex, motor association area, and precentral gyrus.

Lesions of the basal nuclei cause movement disorders called **dyskinesias.**[52] These are sometimes characterized by abnormal difficulty initiating movement, such as rising from a chair or beginning to walk, and by a slow shuffling walk. Such motor dysfunctions are seen in Parkinson disease (see Deeper Insight 12.4). Smooth, easy movements require the excitation of agonistic muscles and inhibition of their antagonists. In Parkinson disease, the antagonists

---

[52]*dys* = bad, abnormal, difficult; *kines* = movement

(a) Input to cerebellum

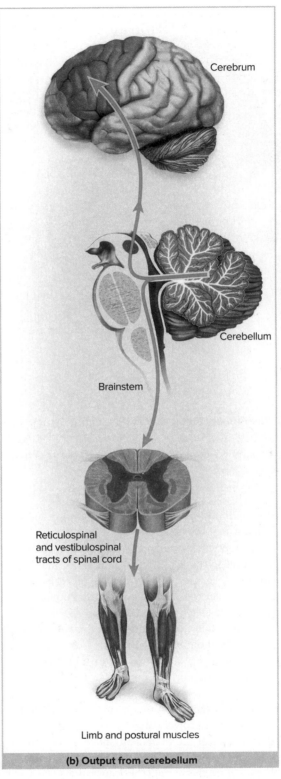

(b) Output from cerebellum

FIGURE 14.23 Motor Pathways Involving the Cerebellum. (a) Afferent (input) pathways from the cerebrum and sense organs to the cerebellum. (b) Efferent (output) pathways from the cerebellum to the skeletal muscles.

are not inhibited. Therefore, opposing muscles at a joint fight each other, making it a struggle to move as one wishes. Other dyskinesias are characterized by exaggerated or unwanted movements, such as flailing of the limbs *(ballismus)* in Huntington disease.

The cerebellum is highly important in motor coordination. In addition to its cognitive functions described earlier, it aids in learning motor skills, maintains muscle tone and posture, smooths muscle contractions, coordinates eye and body movements, and helps to coordinate the motions of different joints with each other (such as the shoulder and elbow in pitching a baseball). The cerebellum acts as a comparator in motor control. Through the middle peduncles, it receives information from the upper motor neurons of the cerebrum about the movements one intends to make, and information about body movement from the eyes and inner ears. Through the spinocerebellar tracts and inferior peduncles, it receives information from proprioceptors in the muscles and joints about the actual performance of the movement **(fig. 14.23a).** The Purkinje cells of the cerebellum compare the two. If there is a discrepancy between the intent and the performance, they signal the deep cerebellar nuclei. These, in turn, issue signals to the thalamus and lower brainstem, ultimately ascending to the motor association area of the cerebrum and the reticulospinal and vestibulospinal tracts of the spinal cord **(fig. 14.23b).** Output from these areas corrects the muscle performance to match the intent. Lesions of the cerebellum can result in a clumsy, awkward gait *(ataxia)* and make some tasks such as climbing stairs virtually impossible.

14.5h Language

Language includes several abilities—reading, writing, speaking, sign language, and understanding words—assigned to different regions of cerebral cortex **(fig. 14.24).** The **posterior speech area,** or **Wernicke[53] area** (WUR-ni-keh), is responsible for the recognition of spoken and written language. It lies just posterior to the lateral sulcus, usually in the left hemisphere, at the crossroad between visual, auditory, and somatosensory areas of cortex, receiving input from all these neighboring regions. The *angular gyrus,* part of the parietal lobe just caudal and superior to the Wernicke area, is important in the ability to read and write.

When we intend to speak, the Wernicke area formulates phrases according to learned rules of grammar and transmits a plan of speech to the **motor language area,** or **Broca[54] area,** located in the inferior

prefrontal cortex of the same hemisphere. PET scans show a rise in the metabolic activity of the Broca area as one prepares to speak or sign (see fig. 14.41). This area generates a motor program for the muscles of the larynx, tongue, cheeks, and lips to produce speech, as well as for the hand motions of signing. It transmits this program to the primary motor cortex, which executes it—that is, it issues commands to the lower motor neurons that supply the relevant muscles.

The emotional aspect of language is controlled by regions in the opposite hemisphere that mirror the Wernicke and Broca areas. Opposite the Broca area is the *affective language area.* Lesions to this area result in *aprosody*—flat, emotionless speech. The cortex opposite the Wernicke area is concerned with recognizing the emotional content of another person's speech. Lesions here can result in such problems as inability to understand a joke.

▶▶▶ APPLY WHAT YOU KNOW

Of all the language centers just described, which one best fits the concept of multimodal association cortex?

Aphasia[55] (ah-FAY-zee-uh) is any language deficit resulting from lesions in the hemisphere (usually the left) containing the Wernicke and Broca areas. The many forms of aphasia are difficult to classify. *Nonfluent (Broca) aphasia,* due to a lesion in the Broca area, results in slow speech, difficulty choosing words, or use of words that only approximate the correct word. For example, a person may say "tssair" when asked to identify a picture of a chair. In extreme cases, the person's entire vocabulary consists of two or three words, sometimes those that were being spoken when a stroke occurred. Such patients feel very frustrated with themselves and often maintain a tight-lipped reluctance to talk. A lesion to the Wernicke area may cause *fluent (Wernicke) aphasia,* in which a person speaks normally and sometimes excessively, but uses jargon and invented words that make little sense (for example, "choss" for chair). Such a person may be unable to comprehend

[53]Karl Wernicke (1848–1905), German neurologist
[54]Pierre Paul Broca (1824–80), French surgeon and anthropologist

[55]*a* = without; *phas* = speech

Anterior

Precentral gyrus

Speech center of primary motor cortex

Primary auditory cortex (in lateral sulcus)

Broca area

Posterior

Postcentral gyrus

Angular gyrus

Primary visual cortex

Wernicke area

FIGURE 14.24 Language Centers of the Left Hemisphere.

written and spoken words. In *anomic aphasia,* a person can speak normally and understand speech, but cannot identify written words or pictures. Shown a picture of a chair, the person may say, "I know what it is; I have a lot of them," but be unable to name the object.

This represents only a small sample of the complex and puzzling linguistic effects of brain lesions. Other lesions to small areas of cortex can cause impaired mathematical ability, a tendency to write only consonants, or difficulty understanding the second half of each word a person reads.

14.5i Cerebral Lateralization

The two cerebral hemispheres look identical at a glance, but close examination reveals a number of differences. For example, in many women the left temporal lobe is longer than the right. In left-handed people, the left frontal, parietal, and occipital lobes are usually wider than those on the right. The two hemispheres also differ in some of their functions (**fig. 14.25**). Neither hemisphere is "dominant," but each is specialized for certain tasks. This difference in function is called **cerebral lateralization.** The idea, however, that some people are "left-brained" (such as a mathematician or scientist) and others "right-brained" (such as a musician or artist) is only a discredited popular myth. Everyone uses the two hemispheres about equally.

One hemisphere, usually the left, is called the *categorical hemisphere.* It is specialized for spoken and written language and for the sequential and analytical reasoning employed in such fields as science and mathematics. This hemisphere seems to break information into fragments and analyze it in a linear way. The other hemisphere, usually the right, is called the *representational hemisphere.* It perceives information in a more integrated, holistic way. It is a seat of imagination and insight; musical and artistic skill;

perception of patterns and spatial relationships; and comparison of sights, sounds, smells, and tastes. It is better than the left hemisphere at recognizing the intonation, rhythm, and emotional content of other people's speech.

Cerebral lateralization is highly correlated with handedness. The left hemisphere is the categorical one in 96% of right-handed people, and the right hemisphere in 4%. Among left-handed people, the right hemisphere is categorical in 15% and the left in 70%, whereas in the remaining 15%, neither hemisphere is distinctly specialized.

Lateralization develops with age. In young children, if one cerebral hemisphere is damaged or removed (for example, because of brain cancer), the other hemisphere can often take over its functions. Adult males exhibit more lateralization than females and suffer more functional loss when one hemisphere is damaged. When the left hemisphere is damaged, men are three times as likely as women to become aphasic. The reason for this difference is not yet clear.

BEFORE YOU GO ON

Answer the following questions to test your understanding of the preceding section:

21. Suppose you are reading a novel and gradually fall asleep and begin to dream. How would your brain waves change during this sequence of events?

22. Describe the locations and functions of the somatosensory, visual, auditory, and frontal association areas.

23. Describe the somatotopy of the primary motor area and primary sensory area.

24. What are the roles of the Wernicke area, Broca area, and precentral gyrus in language?

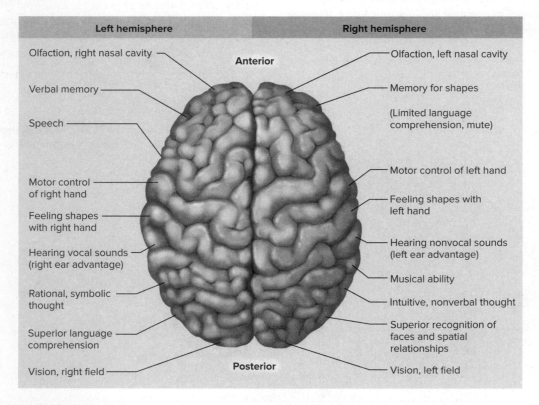

FIGURE 14.25 Lateralization of Cerebral Functions. The two cerebral hemispheres are not functionally identical.

14.6 The Cranial Nerves

Expected Learning Outcomes

When you have completed this section, you should be able to

a. list the 12 cranial nerves by name and number;

b. identify where each cranial nerve originates and terminates; and

c. state the functions of each cranial nerve.

To be functional, the brain must communicate with the rest of the body. Most of its input and output travels by way of the spinal cord, but it also communicates by way of 12 pairs of **cranial nerves.** These arise primarily from the base of the brain, exit the cranium through its foramina, and lead to muscles and sense organs located mainly in the head and neck. The cranial nerves are numbered I to XII starting with the most rostral pair **(fig. 14.26).** Each nerve also has a descriptive name such as *optic nerve* and *vagus nerve.*

14.6a Cranial Nerve Pathways

Most motor fibers of the cranial nerves begin in nuclei of the brainstem and lead to glands and muscles. The sensory fibers begin in receptors located mainly in the head and neck and lead mainly to the brainstem. These include the special senses such as vision and hearing, as well as general senses such as touch and proprioception. Pathways for the special senses are detailed in chapter 16. Sensory fibers for proprioception begin in the muscles innervated by motor fibers of the cranial nerves, but they often travel to the brain in a different nerve from the one that supplies the motor innervation.

Most cranial nerves carry fibers between the brainstem and ipsilateral receptors and effectors. Thus, a lesion in one side of the brainstem causes a sensory or motor deficit on the same side of the head. This contrasts with lesions of the motor and somatosensory cortex of the cerebrum, which, as we saw earlier, cause sensory and motor deficits on the *contralateral* side of the body. The exceptions are the optic nerve (II), where half the fibers decussate to the opposite side of the brain (see fig. 16.45), and the trochlear nerve (IV), in which all efferent fibers lead to a muscle of the contralateral eye.

Cranial nerves:

Frontal lobe

Olfactory tract

Optic chiasm

Temporal lobe

Infundibulum

Pons

Medulla

Cerebellum

Spinal cord

Olfactory bulb (from olfactory nerve, I)

Optic nerve (II)

Oculomotor nerve (III)

Trochlear nerve (IV)

Trigeminal nerve (V)

Abducens nerve (VI)

Facial nerve (VII)

Vestibulocochlear nerve (VIII)

Glossopharyngeal nerve (IX)

Vagus nerve (X)

Hypoglossal nerve (XII)

Accessory nerve (XI)

Frontal lobe

Longitudinal cerebral fissure

Olfactory tract

Temporal lobe

Optic chiasm

Pons

Medulla oblongata

Cerebellum

Spinal cord

(a)

(b)

FIGURE 14.26 The Cranial Nerves. (a) Base of the brain, showing the 12 cranial nerves. (b) Photograph of the cranial nerves. **APR**

b: Rebecca Gray/Don Kincaid/McGraw-Hill Education

14.6b Cranial Nerve Classification

Cranial nerves are traditionally classified as sensory (I, II, VIII), motor (III, IV, VI, XI, XII), or mixed (V, VII, IX, X). In reality, only cranial nerves I and II (for smell and vision) are purely sensory, whereas all of the rest contain both afferent and efferent fibers and are therefore mixed nerves. Those traditionally classified as motor not only stimulate muscle contractions but also contain sensory fibers of proprioception, which provide the brain with feedback for controlling muscle action and make one aware of such things as the position of the tongue and orientation of the head. Cranial nerve VIII, concerned with hearing and equilibrium, is traditionally classified as sensory, but it also has motor fibers that return signals to the inner ear and tune it to sharpen the sense of hearing. The nerves traditionally classified as mixed have sensory functions quite unrelated to their motor functions. For example, the facial nerve (VII) has a sensory role in taste and a motor role in controlling facial expressions.

The traditional classification of these nerves as sensory, motor, or mixed may remain relevant for such purposes as board examinations and comparisons to other books. Therefore, in order to teach the traditional classification while reminding you that all but the first two of these nerves are actually mixed, this section will characterize such nerves as *predominantly* sensory or motor rather than implying that they are exclusively so. Nerves described here as *mixed* or *sensory* are agreed upon by all authorities to be mixed or purely sensory. Nerves described as predominantly motor or sensory are traditionally classified that way but contain some fibers of the other type.

14.6c Cranial Nerve Survey

The following pages survey the key characteristics of each of the 12 pairs of cranial nerves. For each nerve, we begin with a general description of its functions and any unique characteristics it may have; an illustration of its anatomical course; and remarks on the effects of damage to that nerve and how neurologists assess its functionality or deficits. At the end of the descriptive survey, you will find a summary table and illustration, as well as a helpful mnemonic for remembering the names of these nerves in anatomical order.

The Olfactory Nerve (CN I)

The olfactory nerve **(fig. 14.27)** is a purely sensory nerve for smell. It consists of several separate fascicles that pass independently through the cribriform plate of the ethmoid bone in the roof of the

Eyeball

Optic nerve (II)

Optic chiasm
Optic tract

Pituitary gland

FIGURE 14.28 The Optic Nerve (II). **APR**

nasal cavity. It isn't visible on isolated brains because these fascicles are severed by removal of the brain from the skull, but on most such brains one can see the *olfactory bulbs* in which these nerves end.

The olfactory nerve can be damaged by sharp blows to the face that break the ethmoid bone, such as a person might suffer in striking an automobile dashboard in a collision. Such damage can cause **anosmia,** loss of the sense of smell. This not only diminishes some of life's pleasures, but can also be dangerous if a person fails to detect smoke, a gas leak, or spoiled food. A clinical sign of olfactory nerve damage is that a patient cannot smell even such strongly aromatic substances as coffee, vanilla, clove oil, or soap.

The Optic Nerve (CN II)

Four of the 12 cranial nerves (II, III, IV, and VI) are concerned exclusively with the eyes. The only purely sensory nerve among these is the **optic nerve (fig. 14.28),** which carries all visual signals from the retina to the brain. Damage to this nerve causes blindness in part or all of the visual field in the affected eye. Part of the optic nerve, the optic disc, can be seen and assessed in an eye examination with an ophthalmoscope (see fig. 16.30). Optic nerve function is also assessed by tests of visual acuity and range of peripheral vision.

The Oculomotor Nerve (CN III)

The **oculomotor**[56] **nerve (fig. 14.29)** is predominantly motor. It controls four of the six muscles of eye movement (see fig. 16.25), which direct the gaze up, down, and medially. It also controls the *levator palpebrae superioris muscle* of the upper eyelid and the internal eye muscles of focusing and pupillary diameter.

Clinical signs of damage to this nerve include inability to move the eye in the aforesaid directions or to track a moving object, a drooping eyelid, a fixed and dilated pupil, unequal size and shape of the

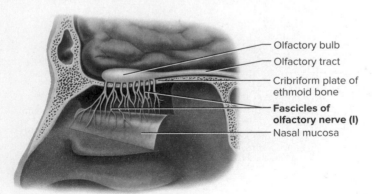

Olfactory bulb
Olfactory tract
Cribriform plate of ethmoid bone
Fascicles of olfactory nerve (I)
Nasal mucosa

FIGURE 14.27 The Olfactory Nerve (I). **A&PR**

[56]*oculo* = eye; *motor* = mover

FIGURE 14.29 The Oculomotor Nerve (III). A&PR

FIGURE 14.30 The Trochlear Nerve (IV). The tendon of the superior oblique muscle passes through a fibrous pulleylike sling, the trochlea (best seen in figure 16.25). A&PR

right and left pupils, lack of pupillary responses to light, double vision, difficulty focusing, and a tendency of the eye to rotate laterally at rest.

The Trochlear Nerve (CN IV)

The **trochlear**[57] **nerve** (TROCK-lee-ur) **(fig. 14.30),** like the oculomotor, is a predominantly motor nerve for eye movement, but its only action is to direct the gaze slightly downward and rotate the top of the eyeball toward the nose, especially in compensating for head movements (see fig. 16.25). This is the only cranial nerve that arises from the posterior side of the brainstem. It is also unique in being the only one to completely decussate; the trochlear nerve roots cross each other within the brainstem before emerging from the surface, so the left trochlear nerve controls the right eye and vice versa.

Signs of trochlear nerve damage include double vision and a weak ability to look downward. The affected eye often points superolaterally and a patient tends to tuck in the chin and tilt the

head downward to minimize the double vision. One assessment of trochlear nerve function is *Bielschowsky's head tilt test*—if the subject tilts the head toward one shoulder, the eye on that side shows an upward deviation.

The Trigeminal Nerve (CN V)

The **trigeminal**[58] **nerve** (tri-JEM-ih-nul) is the largest of all cranial nerves and the most important sensory nerve of the face. Its name refers to the fact that it divides into three prominent branches shortly after leaving the brainstem—the *ophthalmic (V₁), maxillary (V₂),* and *mandibular (V₃) divisions* **(fig. 14.31a).** V₁ and V₂ are sensory nerves, whereas V₃ is mixed. Damage to one of these divisions results in loss of sensation from the respective region of the face **(fig. 14.31b).** Division V₃ also has a motor function in mastication **(fig. 14.31c),** which is impaired by damage to this branch.

---

[57]trochlea = pulley (for a loop through which the muscle's tendon passes)

[58]*tri* = three; *gem* = born (*trigem* = triplets)

FIGURE 14.31 The Trigeminal Nerve (V).
Note the three prominent branches, V₁ to V₃, which give this nerve its name. (a) Anatomical course of the three branches of the trigeminal nerve. (b) Distribution of sensory fibers of each division. (c) Motor branches of the mandibular division (V₃). A&PR

(a) Anatomical course of the three branches of the trigeminal nerve

(b) Distribution of sensory fibers of each division

(c) Motor branches of the mandibular division (V₃)

One sign of damage to the V_1 branch is absence of the blink reflex, as when a neurologist touches the eye surface with a cotton wisp. V_2 damage is marked by loss of the senses of touch, pain (tested with pinpricks), and temperature (hot and cold). V_3 is assessed by palpating the masseter and temporalis muscles to feel for muscle tone when the subject clenches the teeth, and testing the ability to move the mandible from side to side and to open the mouth against resistance. The trigeminal nerve is the source of a common pain disorder called *trigeminal neuralgia* (see Deeper Insight 14.4).

The Abducens Nerve (CN VI)

The **abducens**[59] **nerve** (ab-DOO-senz) (**fig. 14.32**), like the oculomotor and trochlear, is also a predominantly motor nerve for

FIGURE 14.32 The Abducens Nerve (VI). By innervating the lateral rectus muscle, this nerve makes the eye turn away from the facial midline, as its name implies ("to turn away"). **APR**

eye movement. It controls a muscle, the *lateral rectus,* that directs the gaze laterally (see fig. 16.25). Damage to this nerve causes a medial deviation of the eye, as the medial rectus muscle is unopposed by the lateral rectus. Abducens nerve function is assessed by testing the subject's ability to direct the gaze laterally.

The Facial Nerve (CN VII)

The **facial nerve** (**fig. 14.33**) is the major motor nerve to the facial muscles (see table 10.1), but also contains sensory fibers from

---

[59]*ab* = away; *duc* = to lead or turn

Facial nerve (VII)
Geniculate ganglion
Pterygopalatine ganglion
Lacrimal (tear) gland
Chorda tympani branch (taste and salivation)
Submandibular ganglion
Sublingual gland
Parasympathetic fibers
Submandibular gland

Motor branch to muscles of facial expression

to (b)

(a)

Temporal
Zygomatic
Buccal
Mandibular
Cervical

(c)

Temporal
Zygomatic
Buccal
Mandibular
Cervical

(b)

FIGURE 14.33 The Facial Nerve (VII). (a) The facial nerve and associated organs. (b) The five major branches of the facial nerve. (c) A way to remember the distribution of the five major branches. **APR**

c: Joe DeGrandis/McGraw-Hill Education

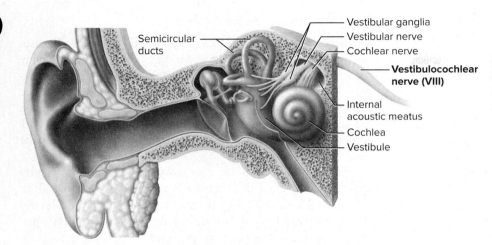

FIGURE 14.34 The Vestibulocochlear Nerve (VIII). Note how the nerve arises by convergence of the vestibular nerve and cochlear nerve, as its name implies. **APR**

taste buds of the anterior two-thirds of the tongue. It divides into five prominent branches: the *temporal, zygomatic, buccal, mandibular,* and *cervical branches.* Damage to the facial nerve can distort a person's sense of taste, especially for sweets, and cause sagging of facial muscles for lack of muscle tone (see Deeper Insight 14.4).

Motor functionality of the facial nerve is assessed by testing the subject's ability to smile, frown, whistle, raise the eyeballs, close the eyes, and perform other facial actions. Sensory function is assessed by testing sensitivity of the anterior tongue to substances such as sugar, salt, vinegar, and quinine, and testing the response of the tear glands to ammonia fumes.

The Vestibulocochlear Nerve (CN VIII)

The **vestibulocochlear**[60] **nerve** (vess-TIB-you-lo-COC-lee-ur) **(fig. 14.34)** is a predominantly sensory nerve of hearing and equilibrium, but it also has a motor effect on the inner ear to tune the sense of hearing (see section 16.4c). It forms by convergence of the *cochlear nerve* from the cochlea and *vestibular nerve* from the semicircular ducts and vestibule of the inner ear. Damage to this nerve can cause *sensorineural deafness* (see Deeper Insight 16.3), dizziness, nausea, and *nystagmus* (involuntary rhythmic oscillations of the eyes from side to side). CN VIII function is assessed by looking for nystagmus and testing a patient's hearing, balance, and ability to walk a straight line.

The Glossopharyngeal Nerve (CN IX)

The **glossopharyngeal**[61] **nerve** (GLOSS-oh-fah-RIN-jee-ul) **(fig. 14.35)** is a complex, mixed nerve with numerous sensory and motor functions in the head, neck, and thoracic region, including sensations from the tongue, throat, and outer ear; control of food intake; and even some aspects of cardiovascular and respiratory function. Signs of damage to this nerve include difficulty swallowing and loss of the bitter and sour taste sensations. It can be assessed by testing the rear of the tongue with bitter and sour substances, noting any speech impediments, and testing the gag reflex, swallowing, and coughing.

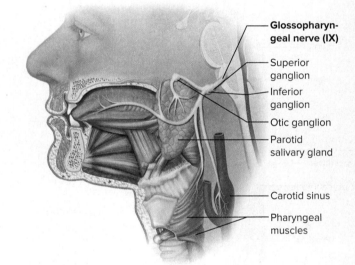

FIGURE 14.35 The Glossopharyngeal Nerve (IX). Note how the nerve branches go to the tongue and pharynx, as its name implies. **APR**

The Vagus Nerve (CN X)

The **vagus**[62] **nerve** (VAY-gus) **(fig. 14.36)** has the most extensive distribution of all cranial nerves, supplying organs not only in the head and neck but also most viscera of the thoracic and abdominopelvic cavities. It is a mixed nerve with major roles in the control of cardiac, pulmonary, digestive, and urinary functions. It carries 90% of all the fibers of the parasympathetic nervous system (see chapter 15).

Damage to both vagus nerves (right and left) is generally fatal. Signs of damage to just one vagus nerve include impaired swallowing and gastrointestinal motility, and hoarseness or loss of the voice. Vagal function is assessed by examining movements of the palate during speech; observing hoarseness or weakness of the voice or the inability to cough forcibly; and checking for swallowing difficulty or absence of the gag reflex.

---

[60]*vestibulo* = entryway (vestibule of inner ear); *cochlea* = conch, snail (cochlea of ear)
[61]*glosso* = tongue; *pharyng* = throat

[62]*vag* = wandering

The Accessory Nerve (CN XI)

The **accessory nerve (fig. 14.37)** takes an unusual path. It arises not from the brain but from the upper spinal cord; therefore, strictly speaking, it is not a true cranial nerve. It ascends alongside the spinal cord, enters the cranial cavity through the foramen magnum, then exits the cranium through the jugular foramen, bundled with the vagus and glossopharyngeal nerves. The accessory nerve is a predominantly motor nerve that controls mainly swallowing and the neck and shoulder muscles.

Signs of damage to the accessory nerve include difficulty moving the head, neck, and shoulders. Damage can paralyze the sternocleidomastoid muscle, causing the contralateral sternocleidomastoid to turn the head toward the injured side. Accessory nerve function can be assessed by testing the subject's ability to rotate the head and shrug the shoulders against resistance.

The Hypoglossal Nerve (CN XII)

The **hypoglossal**[63] **nerve** (HY-po-GLOSS-ul) **(fig. 14.38)** is a predominantly motor nerve controlling tongue movements of speech, food manipulation, and swallowing. Injury to the hypoglossal nerve impairs speech and swallowing. Damage to one nerve may cause the tongue to atrophy on that side, and the tongue may deviate laterally when the subject is asked to protrude it. Damage to both may render a person unable to protrude the tongue against resistance from a tongue depressor.

Cranial Nerve Learning Aid

Generations of biology and medical students have relied on mnemonic (memory-aiding) phrases and ditties, ranging from the sublimely silly to the unprintably ribald, to help them remember the cranial nerves. An old classic began, "On old Olympus' towering

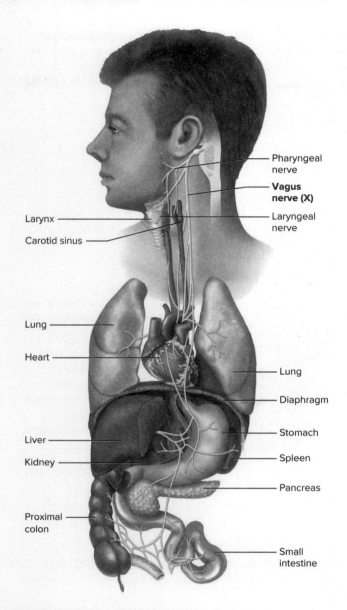

FIGURE 14.36 The Vagus Nerve (X). Note how this nerve "wanders" through much of the thoracic and abdominal cavities, as its name implies. **APR**

[63]*hypo* = below; *gloss* = tongue

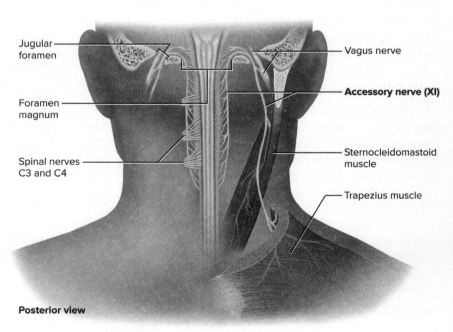

Posterior view

FIGURE 14.37 The Accessory Nerve (XI). This nerve used to be named the *spinal accessory,* referring to its origins in the spinal cord. **APR**

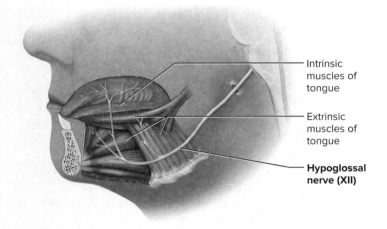

Intrinsic muscles of tongue

Extrinsic muscles of tongue

Hypoglossal nerve (XII)

FIGURE 14.38 The Hypoglossal Nerve (XII). Note how this nerve approaches the tongue from below, as its name implies.

APR

tops . . . ," with the first letter of each word matching the first letter of each cranial nerve (olfactory, optic, oculomotor, etc.). Some cranial nerves have changed names, however, since that mnemonic was invented. One of the author's former students, now a neurologist, devised the following "Old Opie" mnemonic that can remind you

of the first two to four letters (indicated in boldface) of most cranial nerve names:

Old	**ol**factory (I)
Opie	**op**tic (II)
occasionally	**oc**ulomotor (III)
tries	**tr**ochlear (IV)
trigonometry	**trig**eminal (V)
and	**a**bducens (VI)
feels	**f**acial (VII)
very	**ve**stibulocochlear (VIII)
gloomy,	**glo**ssopharyngeal (IX)
vague,	**vag**us (X)
and	**a**ccessory (XI)
hypoactive.	**hypo**glossal (XII)

Table 14.1 summarizes the cranial nerves with respect to their sensory, motor, or mixed composition, functions, origin and termination, and cranial passages (mostly foramina) through which the nerves pass on their way to or from the peripheral organs. **Figure 14.40** reviews their locations on the base of the brain and their sensory and motor connections with the peripheral organs.

DEEPER INSIGHT 14.4

CLINICAL APPLICATION

Some Cranial Nerve Disorders

Trigeminal neuralgia[64] *(tic douloureux*[65]*)* is a syndrome characterized by recurring episodes of intense stabbing pain on one side of the face, often compared to an electrical shock, with painless intervals between attacks. It is usually caused by a nearby blood vessel, but sometimes a tumor, putting pressure on the trigeminal nerve. It can be triggered by touch, toothbrushing, drinking, shaving, face washing, or even a breeze on the face. The pain lasts from a few seconds to a minute or two, but may strike up to 100 times a day; it can be so unbearable that it drove some sufferers to suicide before treatments became available. It affects women more often than men, and usually strikes after age 50. It can be treated with medications; vascular surgery to relieve pressure on the nerve; or gamma knife radiosurgery, which uses targeted gamma rays to deliberately injure a small spot on the nerve and stimulate a healing process.

Bell[66] *palsy* is a degenerative disorder of the facial nerve, probably due to the herpesvirus that causes cold sores. It is characterized by paralysis or weakness of the facial muscles on one side with resulting distortion of the facial features, such as sagging of the mouth or eyelid. **(fig. 14.39).** The paralysis may interfere with speech, prevent closure of the eye, and sometimes inhibit tear secretion, thus causing dry eyes. There may also be a partial loss of the sense of taste. Bell palsy may appear abruptly, sometimes overnight, and often disappears spontaneously within 3 to 5 weeks.

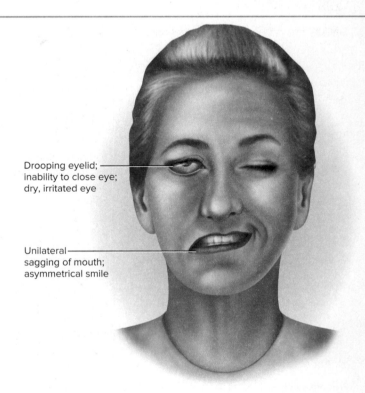

Drooping eyelid; inability to close eye; dry, irritated eye

Unilateral sagging of mouth; asymmetrical smile

FIGURE 14.39 Bell Palsy. Signs of damage to the right facial nerve.

[64]*neur* = nerve; *algia* = pain
[65]*tic* = twitch, spasm; *douloureux* = painful (French)
[66]Sir Charles Bell (1774–1842), Scottish physician

TABLE 14.1	The Cranial Nerves			
Composition	Function	Origin	Termination	Cranial Passage
I. Olfactory Nerve				
Sensory	Smell	Olfactory mucosa in nasal cavity	Olfactory bulbs	Cribriform foramina of ethmoid bone
II. Optic Nerve				
Sensory	Vision	Retina	Thalamus and midbrain	Optic foramen
III. Oculomotor Nerve				
Predominantly motor	Eye movements, opening of eyelid, pupillary constriction, focusing	Midbrain	Somatic fibers to levator palpebrae superioris; superior, medial, and inferior rectus muscles; and inferior oblique muscle of eye. Autonomic fibers enter eyeball and lead to constrictor of iris and ciliary muscle of lens.	Superior orbital fissure
IV. Trochlear Nerve				
Predominantly motor	Eye movements	Midbrain	Superior oblique muscle of eye	Superior orbital fissure
V. Trigeminal Nerve				
Ophthalmic division (V$_1$)				
Sensory	Touch, temperature, and pain sensations from upper face	Superior region of face as illustrated; surface of eyeball; lacrimal (tear) gland; superior nasal mucosa; frontal and ethmoidal sinuses	Pons	Superior orbital fissure
Maxillary division (V$_2$)				
Sensory	Same as V$_1$, lower on face	Middle region of face as illustrated; nasal mucosa; maxillary sinus; palate; upper teeth and gums	Pons	Foramen rotundum and infraorbital foramen
Mandibular division (V$_3$)				
Mixed	*Sensory:* Same as V$_1$ and V$_2$, lower on face *Motor:* Mastication	*Sensory:* Inferior region of face as illustrated; anterior two-thirds of tongue (but not taste buds); lower teeth and gums; floor of mouth; dura mater *Motor:* Pons	*Sensory:* Pons *Motor:* Anterior belly of digastric; masseter, temporalis, mylohyoid, and pterygoid muscles; tensor tympani muscle of middle ear	Foramen ovale
VI. Abducens Nerve				
Predominantly motor	Lateral eye movement	Inferior pons	Lateral rectus muscle of eye	Superior orbital fissure

TABLE 14.1	The Cranial Nerves *(continued)*			
VII. Facial Nerve				
Mixed	*Sensory:* Taste *Motor:* Facial expression; secretion of tears, saliva, nasal and oral mucus	*Sensory:* Taste buds of anterior two-thirds of tongue *Motor:* Pons	*Sensory:* Thalamus *Motor:* Somatic fibers to digastric muscle, stapedius muscle of middle ear, stylohyoid muscle, muscles of facial expression. Autonomic fibers to submandibular and sublingual salivary glands, tear glands, nasal and palatine glands.	Internal acoustic meatus and stylomastoid foramen
VIII. Vestibulocochlear Nerve				
Predominantly sensory	Hearing and equilibrium	*Sensory:* Cochlea, vestibule, and semicircular ducts of inner ear *Motor:* Pons	*Sensory:* Fibers for hearing end in medulla; fibers for equilibrium end at junction of medulla and pons *Motor:* Outer hair cells of cochlea of inner ear	Internal acoustic meatus
IX. Glossopharyngeal Nerve				
Mixed	*Sensory:* Taste; touch, pressure, pain, and temperature sensations from tongue and outer ear; regulation of blood pressure and respiration *Motor:* Salivation, swallowing, gagging	*Sensory:* Pharynx; middle and outer ear; posterior one-third of tongue (including taste buds); internal carotid artery *Motor:* Medulla oblongata	*Sensory:* Medulla oblongata *Motor:* Parotid salivary gland; glands of posterior tongue; stylo-pharyngeal muscle (which dilates pharynx during swallowing)	Jugular foramen
X. Vagus Nerve				
Mixed	*Sensory:* Taste; sensations of hunger, fullness, and gastrointestinal discomfort *Motor:* Swallowing, speech, deceleration of heart, broncho-constriction, gastrointestinal secretion and motility	*Sensory:* Thoracic and abdominopelvic viscera, root of tongue, pharynx, larynx, epiglottis, outer ear, dura mater *Motor:* Medulla oblongata	*Sensory:* Medulla oblongata *Motor:* Tongue, palate, pharynx, larynx, lungs, heart, liver, spleen, digestive tract, kidney, ureter	Jugular foramen
XI. Accessory Nerve				
Predominantly motor	Swallowing; head, neck, and shoulder movements	Spinal cord segments C1 to C6	Palate, pharynx, trapezius and sternocleidomastoid muscles	Jugular foramen
XII. Hypoglossal Nerve				
Predominantly motor	Tongue movements of speech, food manipulation, and swallowing	Medulla oblongata	Intrinsic and extrinsic muscles of tongue	Hypoglossal canal

FIGURE 14.40 Review of Cranial Nerve Pathways. Pathways concerned with sensory function are shown in green, and those concerned with motor function in red. All cranial nerves except I and II carry sensory fibers, but these functions are omitted in cases where the nerve is essentially motor and its sensory function is limited to proprioception from the muscles.

TABLE 14.2	Some Disorders Associated with the Brain and Cranial Nerves
Cerebral palsy	Muscular incoordination resulting from damage to the motor areas of the brain during fetal development, birth, or infancy; causes include prenatal rubella infection, drugs, or radiation exposure; oxygen deficiency during birth; and hydrocephalus
Concussion	Damage to the brain typically resulting from a blow, often with loss of consciousness, disturbances of vision or equilibrium, and short-term amnesia
Encephalitis	Inflammation of the brain, accompanied by fever, usually caused by mosquito-borne viruses or herpes simplex virus; causes neuronal degeneration and necrosis; can lead to delirium, seizures, and death
Epilepsy	Disorder causing sudden, massive discharge of neurons (seizures) resulting in motor convulsions, sensory and psychic disturbances, and often impaired consciousness; may result from congenital brain malformation birth trauma, later brain trauma, tumors, drug or alcohol abuse, or infections
Migraine headache	Recurring headaches often accompanied by nausea, vomiting, dizziness, and aversion to light, often triggered by such factors as weather changes, stress, hunger, red wine, or noise; more common in women and sometimes running in families
Schizophrenia	A thought disorder involving delusions, hallucinations, inappropriate emotional responses to situations, incoherent speech, and withdrawal from society, resulting from hereditary or developmental abnormalities in neural networks

You can find other brain and cranial nerve disorders described in the following places:

Hydrocephalus in Deeper Insight 8.2; brain tumors in Deeper Insight 12.1; *multiple sclerosis* and *Tay-Sachs disease* in Deeper Insight 12.2; *Parkinson* and *Alzheimer diseases* in Deeper Insight 12.4; *Parkinson disease, Huntington disease,* and *cerebellar ataxia* in section 14.5g; *meningitis* in Deeper Insight 14.1; *stroke* in Deeper Insight 14.2; *amnesia* in section 14.5d; *aprosody* and *aphasia* in section 14.5h; and *trigeminal neuralgia* and *Bell palsy* in Deeper Insight 14.4.

Like a machine with a great number of moving parts, the nervous system is highly subject to malfunctions. **Table 14.2** lists a few well-known brain and cranial nerve dysfunctions. The effects of aging on the CNS are described in section 29.4a.

BEFORE YOU GO ON

Answer the following questions to test your understanding of the preceding section:

25. List the purely sensory cranial nerves and state the function of each.

26. What is the only cranial nerve to extend beyond the head–neck region?

27. If the oculomotor, trochlear, or abducens nerve was damaged, the effect would be similar in all three cases. What would that effect be?

28. Which cranial nerve carries sensory signals from the greatest area of the face?

29. Name two cranial nerves involved in the sense of taste and describe where their sensory fibers originate.

DEEPER INSIGHT 14.5

CLINICAL APPLICATION

Images of the Mind

Enclosed as it is in the cranium, there is no easy way to observe a living brain directly. This has long frustrated neurobiologists, who once had to content themselves with glimpses of brain function afforded by electroencephalograms, patients with brain lesions, and patients who remained awake and conversant during brain surgery and consented to experimentation while the brain was exposed. Newer imaging methods, however, have yielded dramatic perspectives on brain function. Two of these—positron emission tomography (PET) and magnetic resonance imaging (MRI)—were explained in Deeper Insight 1.5. Both techniques rely on transient increases in blood flow to parts of the brain called into action to perform specific tasks. By monitoring these changes, neuroscientists can identify which parts of the brain are involved in specific tasks.

To produce a PET scan of the brain, the subject is given an injection of radioactively labeled glucose and a scan is made in a *control state* before any specific mental task is begun. Then the subject is given a task—for example, to read the word *car* and speak a verb related to it, such as *drive*. While the subject does so, new *task-state* scans are made. Neither control- nor task-state images are very revealing by themselves, but the computer subtracts the control-state data from the task-state data and presents a color-coded image of the difference.

In such images, the busiest areas of the brain "light up" on the computer screen from moment to moment as the task is performed **(fig. 14.41)**. This identifies the regions used for various stages of the task, such as reading the word, thinking of a verb to go with it, planning to say *drive,* and actually saying it. Among other things, such experiments demonstrate that the Broca and Wernicke areas aren't involved in simply repeating words; they are active, however, when a subject must evaluate a word and choose an appropriate response—that is, they function in formulating the new word the subject is going to say. PET scans also

show that different neural pools take over a task as we practice and become more proficient at it.

Functional magnetic resonance imaging (fMRI) depends on the role of astrocytes in brain metabolism. The main excitatory neurotransmitter secreted by cerebral neurons is glutamate. After a neuron releases glutamate and stimulates the next neuron, astrocytes quickly remove it from the synapse and convert it to glutamine. Astrocytes acquire the energy for this from the anaerobic fermentation of glucose. High activity in an area of cortex thus requires an increased blood flow to supply this glucose, but it doesn't elevate oxygen consumption from that blood. Thus, the oxygen supply exceeds demand in that part of the brain, and blood leaving the region contains more oxygen than the blood leaving less active regions. Since the magnetic properties of hemoglobin depend on how much oxygen is bound to it, fMRI can detect changes in brain circulation.

fMRI is more precise than PET and pinpoints regions of brain activity with a precision of 1 to 2 mm. It also has the advantages of requiring no injected substances and no exposure to radioisotopes. While it takes about 1 minute to produce a PET scan, fMRI produces images much more quickly, which makes it more useful for determining how the brain responds immediately to sensory input or mental tasks.

Cutting-edge advances in PET and fMRI technology have enhanced our knowledge of shifting patterns of brain activity associated with attention and consciousness, sensory perception, memory, emotion, motor control, reading, speaking, musical judgment, planning a chess strategy, and so forth. In addition to their contribution to basic neuroscience, these techniques are highly valuable to neurosurgery and psychopharmacology. They have enhanced our understanding of dysfunctions such as depression, schizophrenia, and attention-deficit/hyperactivity disorder (ADHD). Such methods ushered in an exciting era of safe visualization of brain function, producing pictures of the mind at work.

FIGURE 14.41 PET Scans of the Brain Made During the Performance of a Language Task.   These images show the cortical regions that are active when a person reads words and then speaks them. The most active areas are shown in red and less active areas in blue.

(all): Courtesy Dr. Marcus E. Raichle, MD, Washington University School of Medicine

▶ Assess Your Learning Outcomes

To test your knowledge, discuss the following topics with a study partner or in writing, ideally from memory.

14.1 Overview of the Brain

1. Typical mass of the adult brain
2. Meanings of the directional terms *rostral* and *caudal* in brain anatomy
3. Three principal divisions of the brain
4. Names of the tissue folds and shallow grooves in the cerebrum and cerebellum, and the names of the deep grooves between the two cerebral hemispheres and between the cerebrum and cerebellum
5. Locations of the gray and white matter in the cerebrum and cerebellum; the composition of gray and white matter and how this relates to their colors; and the meaning of *cortex, nucleus,* and *tract* in relation to brain gray and white matter
6. Embryonic development of the brain from neural plate to neural tube stage; differentiation into forebrain, midbrain, and hindbrain; development of the five embryonic brain vesicles; and the name and destiny of each vesicle

14.2 Meninges, Ventricles, Cerebrospinal Fluid, and Blood Supply

1. The three meninges of the brain; two subdivisions of the dura mater; and the relationship of the dural sinuses and subarachnoid space to the meninges
2. Ventricles of the brain, their names and locations, and the passages that connect them
3. The relationship of the ependymal cells and choroid plexuses to the brain ventricles and canals
4. The sources, flow, reabsorption, and functions of the cerebrospinal fluid (CSF)
5. The brain's demands for blood flow, glucose, and oxygen in relation to its percentage of the body weight
6. Structure and function of the blood–brain and blood–CSF barriers, and the clinical relevance of the brain barrier system
7. The locations, structural nature, function, and clinical relevance of circumventricular organs

14.3 The Hindbrain and Midbrain

1. The medulla oblongata: its location, gross anatomy, and general functions; which cranial nerves arise from it; and the specific functions of each of its tracts and nuclei
2. The pons: its location, gross anatomy, and functions; which cranial nerves arise from the pons and from the groove between pons and medulla; and its contributions to the anterolateral system and anterior spinocerebellar tract
3. The midbrain: its location, gross anatomy, and general functions; which cranial nerves arise from it; and the specific functions of its colliculi, substantia nigra, and central gray substance
4. How the reticular formation relates to the foregoing brainstem regions in location, structure, and functions
5. The cerebellum: its location, gross anatomy, unique neuron types, its three pairs of peduncles, and their relationship to cerebellar input and output
6. The classical view of cerebellar function and how this has lately expanded as a result of brain imaging studies and studies of people with cerebellar lesions

14.4 The Forebrain

1. The two major parts of the forebrain
2. Three principal parts of the diencephalon
3. Thalamic functions, location, and gross anatomy
4. Hypothalamic functions, location, gross anatomy, and anatomical and physiological relationships with the pituitary gland
5. The location, components, and functions of the epithalamus
6. Gross anatomy of the cerebral hemispheres, their five lobes, and the functions of each lobe
7. Three types of cerebral tracts and their defining anatomical and functional differences
8. Three locations of cerebral gray matter
9. The thickness, extent, and relative amount of tissue and number of neurons in the cerebral cortex
10. Two types of neurons in the cerebral cortex and their morphology
11. Evolutionary and structural distinctions between archicortex, paleocortex, and neocortex, and where each occurs in the human brain
12. The location, constituents, and functions of the limbic system
13. Names, locations, and functions of the basal nuclei

14.5 Integrative Functions of the Brain

1. Four kinds of brain waves in the EEG, the states in which each type normally dominates the EEG, and the clinical relevance of the EEG
2. Stages of sleep; physiological characteristics of each; roles of the hypothalamus, reticular formation, melatonin, and orexins in regulating the sleep cycle; and hypotheses on the functions of sleep
3. Association areas of the cerebral cortex; the involvement of some of these in cognitive function; and the contributions of brain injury patients and brain imaging methods to understanding the regional distribution of cognitive functions
4. Brain regions involved in memory; forms of amnesia
5. Brain regions involved in emotion, and insights into the neurobiology of emotion from brain trauma, ablation, and brain stimulation studies
6. Brain regions involved in the special and general senses
7. The functional relationship between primary sensory cortex and sensory association areas
8. Location of the postcentral gyrus; its somatosensory function; its somatotopy as charted in the sensory homunculus; and the effect of decussation on its function
9. Locations of the precentral gyrus and motor association area; their roles in motor control; the somatotopy of the gyrus as charted in the motor homunculus; and the effect of decussation on its function
10. Upper and lower motor neurons and the course of their axons
11. Roles of the basal nuclei and cerebellum in motor coordination and learned motor skills
12. Effects of Parkinson disease and basal nuclei lesions on motor control
13. Locations of the Wernicke and Broca areas; their roles in language; interaction of the Broca area and precentral gyrus in speech;

STUDY GUIDE

and forms of aphasia and other language deficits resulting from damage to the language centers
14. Cerebral lateralization; functional differences between the categorical and representational hemispheres; and why it cannot be said that a particular function is necessarily "right-brained" or "left-brained"

14.6 The Cranial Nerves

1. Names and numbers of the 12 pairs of cranial nerves, and their relationships to the brainstem and skull foramina
2. Which cranial nerves are purely sensory, which are mixed, which have traditionally been regarded as motor, and why it is not entirely accurate to simply call them motor nerves
3. For each cranial nerve, its location, functions, origin, termination, and passage through the skull
4. Signs of damage to each cranial nerve, and clinical methods of testing for damage

▶ Testing Your Recall

Answers in Appendix A

1. Which of these is caudal to the hypothalamus?
   a. the thalamus
   b. the optic chiasm
   c. the cerebral aqueduct
   d. the pituitary gland
   e. the corpus callosum

2. If the telencephalon was removed from a 5-week-old embryo, which of the following structures would fail to develop in the fetus?
   a. cerebral hemispheres
   b. the thalamus
   c. the midbrain
   d. the medulla oblongata
   e. the spinal cord

3. The blood–CSF barrier is formed by
   a. blood capillaries.
   b. endothelial cells.
   c. protoplasmic astrocytes.
   d. oligodendrocytes.
   e. ependymal cells.

4. The pyramids of the medulla oblongata contain
   a. descending corticospinal fibers.
   b. commissural fibers.
   c. ascending spinocerebellar fibers.
   d. fibers going to and from the cerebellum.
   e. ascending spinothalamic fibers.

5. Which of the following does *not* receive any input from the eyes?
   a. the hypothalamus
   b. the frontal lobe
   c. the thalamus
   d. the occipital lobe
   e. the midbrain

6. While studying in a noisy cafeteria, you get sleepy and doze off for a few minutes. You awaken with a start and realize that all the cafeteria sounds have just "come back." While you were dozing, this auditory input was blocked from reaching your auditory cortex by
   a. the temporal lobe.
   b. the hypothalamus.
   c. the reticular activating system.
   d. the medulla oblongata.
   e. the vestibulocochlear nerve.

7. Because of a brain lesion, a certain patient never feels full, but eats so excessively that she now weighs nearly 270 kg (600 lb). The lesion is most likely in her
   a. hypothalamus.
   b. amygdala.
   c. hippocampus.
   d. basal nuclei.
   e. pons.

8. The _____ is most closely associated with the cerebellum in embryonic development and remains its primary source of input fibers throughout life.
   a. telencephalon
   b. thalamus
   c. midbrain
   d. pons
   e. medulla

9. Damage to the _____ nerve could result in defects of eye movement.
   a. optic
   b. vagus
   c. trigeminal
   d. facial
   e. abducens

10. All of the following *except* the _____ nerve begin or end in the orbit.
   a. optic
   b. oculomotor
   c. trochlear
   d. abducens
   e. accessory

11. The right and left cerebral hemispheres are connected to each other by a thick C-shaped bundle of fibers called the _____.

12. The brain has four chambers called _____ filled with _____ fluid.

13. On a sagittal plane, the cerebellar white matter exhibits a branching pattern called the _____.

14. The only cranial nerve with direct control over the lower digestive tract is the _____.

15. Cerebrospinal fluid is secreted partly by a mass of blood capillaries called the _____ in each ventricle.

16. The primary motor area of the cerebrum is the _____ gyrus of the frontal lobe.

17. A lesion in which lobe of the cerebrum is most likely to cause a radical alteration of the personality?

18. Areas of cerebral cortex that identify or interpret sensory information are called _____.

19. Linear, analytical, and verbal thinking occurs in the _____ hemisphere of the cerebrum, which is on the left in most people.

20. The motor pattern for speech is generated in an area of cortex called the _____ and then transmitted to the primary motor cortex to be carried out.

STUDY GUIDE

▶ Building Your Medical Vocabulary

Answers in Appendix A

State a meaning of each word element, and give a medical term from this chapter that uses it or a slight variation of it.

1. -algia

2. cephalo-

3. cerebro-

4. corpo-

5. encephalo-

6. -gram

7. insulo-

8. oculo-

9. trochle-

10. -uncle

▶ What's Wrong with These Statements?

Answers in Appendix A

Briefly explain why each of the following statements is false, or reword it to make it true.

1. Eye movements are controlled mainly by the optic nerve and oculomotor nerve.

2. The cerebral hemispheres would fail to develop if the neural crests of the embryo were destroyed.

3. The midbrain is caudal to the pons.

4. The Broca area is contralateral to the Wernicke area.

5. Most of the cerebrospinal fluid is produced by the choroid plexuses.

6. Hearing is a function of the occipital lobe.

7. Respiration is controlled by nuclei in the hypothalamus.

8. The facial nerve carries sensory signals from a larger area of the face than any other cranial nerve does.

9. All cranial nerves innervate organs of the head–neck region.

10. Most of the brain's neurons are found in the cerebral cortex.

▶ Testing Your Comprehension

1. Which cranial nerve conveys pain signals to the brain in each of the following situations? (a) Sand blows into your eye; (b) you bite the back of your tongue; and (c) your stomach hurts from eating too much.

2. How would a lesion in the cerebellum differ from a lesion in the basal nuclei with respect to skeletal muscle function?

3. Suppose that a neuroanatomist performed two experiments on an animal with the same basic brainstem structure as a human's: In experiment 1, he selectively transected (cut across) the pyramids on the anterior side of the medulla oblongata; and in experiment 2, he selectively transected the gracile and cuneate fasciculi on the posterior side. How would the outcomes of the two experiments differ?

4. A person can survive destruction of an entire cerebral hemisphere but cannot survive destruction of the hypothalamus, which is a much smaller mass of brain tissue. Explain this difference and describe some ways that destruction of a cerebral hemisphere could affect one's quality of life.

5. What would be the most obvious effects of lesions that destroyed each of the following: (a) the hippocampus, (b) the amygdala, (c) the Broca area, (d) the occipital lobe, and (e) the hypoglossal nerve?

15

THE AUTONOMIC NERVOUS SYSTEM AND VISCERAL REFLEXES

Autonomic neurons in the enteric nervous system of the digestive tract

Biophoto Associates/Science Source

CHAPTER OUTLINE

15.1 General Properties of the Autonomic Nervous System

15.1a Visceral Reflexes

15.1b Divisions of the Autonomic Nervous System

15.1c Autonomic Output Pathways

15.2 Anatomy of the Autonomic Nervous System

15.2a The Sympathetic Division

15.2b The Adrenal Glands

15.2c The Parasympathetic Division

15.2d The Enteric Nervous System

15.3 Autonomic Effects on Target Organs

15.3a Neurotransmitters and Their Receptors

15.3b Dual Innervation

15.3c Control Without Dual Innervation

15.4 Central Control of Autonomic Function

Study Guide

DEEPER INSIGHTS

15.1 Clinical Application: Megacolon

15.2 Clinical Application: Drugs and the Nervous System

Anatomy & Physiology *Revealed* 4.0

Module 7: Nervous System

- The function of the autonomic nervous system is heavily grounded in the concepts of negative feedback and homeostasis (see section 1.6c).

- Figure 12.2 provides a helpful perspective on how the autonomic nervous system (ANS) fits into the scheme of the nervous system as a whole.

- It is important that you clearly understand neurotransmitters and receptors and why neurotransmitters can have either excitatory or inhibitory effects on different target cells (see sections 12.5c, 12.5d).

- Many effects of the ANS are carried out by smooth muscle. The innervation of smooth muscle will help in understanding the effects described in this chapter (see section 11.7b).

- You should be familiar with the anatomy and functions of the hypothalamus, a highly important autonomic integrating center (see section 14.4a).

- Autonomic signals travel to their target organs by way of the spinal nerves described in section 13.2b and cranial nerves III, VII, IX, and X, detailed in section 14.6c.

W e are consciously aware of many of the activities of our nervous system discussed in the preceding chapters—the general and special senses, our cognitive processes and emotions, and our voluntary movements. But there is another branch of the nervous system that operates in comparative secrecy, usually without our willing it, thinking about it, or even being able to consciously modify or suppress it.

This secretive agent is called the *autonomic nervous system (ANS)*. Its name means "self-governed," as it is almost fully independent of our will. Its job is to regulate such fundamental states and life processes as heart rate, blood pressure, body temperature, respiratory airflow, pupillary diameter, digestion, energy metabolism, defecation, urination, and sexual functions. In short, the ANS quietly manages a multitude of involuntary processes responsible for the body's homeostasis.

Walter Cannon (1871–1945), the American physiologist who coined such expressions as *homeostasis* and the *fight-or-flight reaction*, dedicated his career to the study of the autonomic nervous system. He found that an animal can live without a functional sympathetic nervous system (one of the two divisions of the ANS), but it must be kept warm and free of stress. It cannot regulate its body temperature, tolerate any strenuous exertion, or survive on its own. Indeed, the ANS is more necessary for survival than are many functions of the somatic nervous system; an absence of autonomic function is fatal because the body can't maintain homeostasis without it. Thus, for an understanding of bodily function, the mode of action of many drugs, and other aspects of health care, we must be especially aware of how the ANS works.

15.1 General Properties of the Autonomic Nervous System

Expected Learning Outcomes

When you have completed this section, you should be able to

a. explain how the autonomic and somatic nervous systems differ in form and function; and

b. explain how the two divisions of the autonomic nervous system differ in general function.

The **autonomic**[1] **nervous system (ANS)** can be narrowly defined as a motor nervous system that controls glands, cardiac muscle, and smooth muscle. It is also called the **visceral motor system** to distinguish it from the somatic motor system that controls the skeletal muscles. The primary target organs of the ANS are viscera of the thoracic and abdominopelvic cavities and some structures of the body wall, including cutaneous blood vessels, sweat glands, and arrector muscles (arrector pili) of the hairs.

The ANS usually carries out its actions involuntarily, without our intent or awareness, in contrast to the voluntary nature of the somatic motor system. This voluntary–involuntary distinction is not, however, as clear-cut as it may seem. Some skeletal muscle responses are quite involuntary, such as the somatic reflexes, and some muscles are difficult or impossible to control, such as the middle-ear muscles. On the other hand, therapeutic uses of biofeedback have shown that some people can learn to voluntarily control such visceral functions as blood pressure and heart rate.

Visceral effectors don't depend on the autonomic nervous system to function, but only to adjust their activity to the body's changing needs. The heart, for example, goes on beating even if all the nerves to it are severed, but the ANS normally modulates the heart rate in conditions of rest or exercise. If the somatic nerves to a skeletal muscle are severed, the muscle exhibits flaccid paralysis—it no longer functions. But if the autonomic nerves to cardiac or smooth muscle are severed, the muscle exhibits exaggerated responses (*denervation hypersensitivity*).

15.1a Visceral Reflexes

The ANS is responsible for the body's **visceral**[2] **reflexes**—unconscious, automatic, stereotyped responses to stimulation, much like the somatic reflexes discussed in section 13.3, but involving visceral receptors and effectors and slower responses. Some authorities regard the visceral afferent (sensory) pathways as part of the ANS, but most prefer to limit the term *ANS* to the efferent (motor) pathways. Regardless of this preference, however, autonomic activity involves a visceral reflex arc that includes (1) **receptors** (nerve endings that detect stretch, tissue damage, blood chemistry, body

---

[1]*auto* = self; *nom* = rule (self-governed)
[2]*viscero* = internal organs

Integrating center

② Glossopharyngeal nerve transmits signals to medulla oblongata

Receptor

① Baroreceptors sense increased blood pressure

Internal carotid artery

③ Vagus nerve transmits inhibitory signals to cardiac pacemaker

Effector

④ Heart rate decreases

FIGURE 15.1 An Autonomic Reflex Arc. This pathway produces the baroreflex, which compensates for blood pressure fluctuations in arteries above the heart.

temperature, and other internal stimuli), (2) **afferent neurons** leading to (3) an **integrating center** and interneurons in the CNS, (4) **efferent neurons** carrying motor signals away from the CNS, and finally (5) an **effector** that carries out the end response.

For example, high blood pressure activates a visceral *baro-reflex.*[3] It stimulates stretch receptors called *baroreceptors* in the internal carotid arteries and aorta, and they transmit signals via the glossopharyngeal nerves to the brainstem **(fig. 15.1).** The medulla integrates this with other information and transmits signals back to the heart by way of the vagus nerves. The vagus nerves slow down the heart and reduce blood pressure, thus completing a homeostatic negative feedback loop. A separate autonomic reflex accelerates the heart when blood pressure drops below normal—for example, when we move from a reclining to a standing position and gravity draws blood away from the upper body (see fig. 1.8).

[3]*baro* = pressure

15.1b Divisions of the Autonomic Nervous System

The ANS has two subsystems: sympathetic and parasympathetic. They differ in anatomy and function, but often innervate the same target organs and may have cooperative or contrasting effects on them. The **sympathetic division** adapts the body in many ways for physical activity—it increases alertness, heart rate, blood pressure, pulmonary airflow, blood glucose concentration, and blood flow to cardiac and skeletal muscle, but at the same time, it reduces blood flow to the skin and digestive tract. Cannon referred to extreme sympathetic responses as the *fight-or-flight reaction* because it comes into play when an animal must attack, defend itself, or flee from danger. In our own lives, this reaction occurs in situations involving arousal, exercise, competition, stress, danger, trauma, anger, or fear. Ordinarily, however, the sympathetic division has more subtle effects that we notice barely, if at all.

The **parasympathetic division,** by comparison, has a calming effect on many body functions. It is associated with reduced energy expenditure and normal bodily maintenance, including such functions as digestion and waste elimination. This is often called the *resting-and-digesting state.*

This does not mean that the body alternates between states where one system or the other is active. Normally, both systems are active simultaneously. They exhibit a background rate of activity called **autonomic tone,** and the balance between *sympathetic tone* and *parasympathetic tone* shifts in accordance with the body's changing needs. Parasympathetic tone, for example, maintains smooth muscle tone in the intestines and holds the resting heart rate down to about 70 to 80 beats/min. If the parasympathetic vagus nerves to the heart are cut, the heart beats at its own intrinsic rate of about 100 beats/min. Sympathetic tone keeps most blood vessels partially constricted and thus maintains blood pressure. A loss of sympathetic tone can cause such a rapid drop in blood pressure that a person goes into shock and may faint.

Neither division has universally excitatory nor inhibitory effects. The sympathetic division, for example, excites the heart but inhibits digestive and urinary functions, whereas the parasympathetic division has the opposite effects. We will later examine how differences in neurotransmitters and receptors account for these differences of effect.

15.1c Autonomic Output Pathways

Although usually defined only as part of the peripheral nervous system, the ANS has components in both the CNS and PNS. It includes control nuclei in the hypothalamus and other regions of the brainstem, motor neurons in the spinal cord and peripheral ganglia, and nerve fibers that travel through the cranial and spinal nerves you have already studied.

The autonomic motor pathway to a target organ differs significantly from somatic motor pathways. In somatic pathways, a motor neuron in the brainstem or spinal cord issues a myelinated axon that reaches all the way to a skeletal muscle. In autonomic pathways, the signal travels across *two* nerve fibers to get to the target organ, and it must cross a synapse where these two neurons meet in an autonomic ganglion **(fig. 15.2).** The first fiber is called the **preganglionic fiber.** It begins with a neurosoma in the brainstem or spinal cord. Its axon

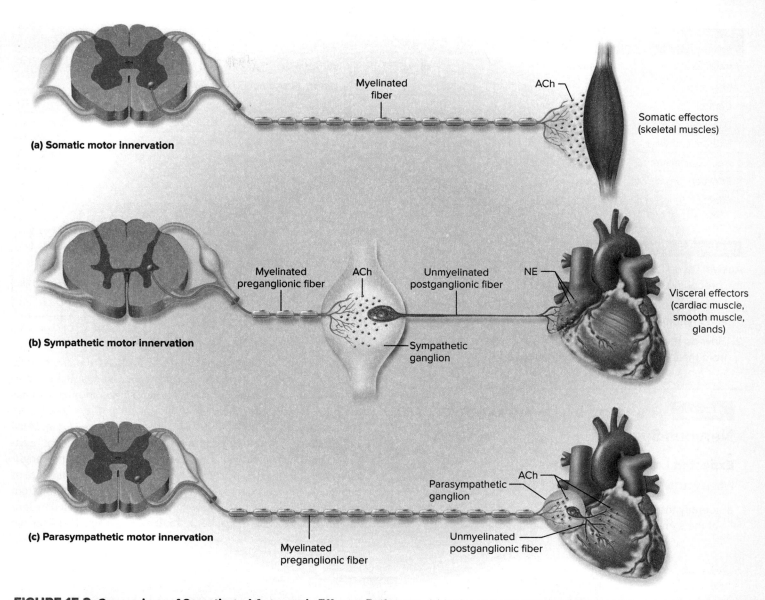

FIGURE 15.2 Comparison of Somatic and Autonomic Efferent Pathways. (a) In the somatic nervous system, each motor nerve fiber extends all the way from the CNS to a skeletal muscle. (b) In the sympathetic nervous system, a short preganglionic fiber extends from the spinal cord to a sympathetic ganglion usually near the vertebral column. A longer postganglionic fiber extends from there to the target organ. (c) In the parasympathetic nervous system, a long preganglionic fiber extends from the CNS to a ganglion in or near the target organ, and a short postganglionic fiber completes the path to specific effector cells in that organ (smooth or cardiac muscle or glands). (ACh = acetylcholine; NE = norepinephrine)

extends from there to an autonomic ganglion somewhere outside the CNS—often in a chain near the vertebral column or located in or near the target organ. Here, it meets the second neuron and secretes the neurotransmitter acetylcholine (ACh) to stimulate it. The axon from the second neuron, called the **postganglionic fiber,** leaves the ganglion and extends the rest of the way to the target organ or cells. Depending on the fiber type, it secretes either ACh or norepinephrine (NE). Autonomic preganglionic fibers are myelinated and the postganglionic fibers are unmyelinated. In contrast to somatic motor neurons, postganglionic fibers of the ANS don't usually synapse with a specific target cell, but end in a beadlike chain of varicosities that diffusely release neurotransmitter into the tissue and stimulate many cells simultaneously (see section 11.7b and fig. 11.21).

In summary, the autonomic nervous system is a division of the nervous system responsible for homeostasis, acting through mostly unconscious and involuntary control of glands, smooth muscle, and cardiac muscle. Its target organs are mainly thoracic and abdominopelvic viscera, but also include some cutaneous and other effectors. It acts through motor pathways that involve two nerve fibers, preganglionic and postganglionic, reaching from CNS to effector. The ANS has two divisions, sympathetic and parasympathetic, that often have cooperative or contrasting effects on the same target organ. Both divisions have excitatory effects on some target cells and inhibitory effects on others. These and other differences between the somatic and autonomic nervous systems are summarized in **table 15.1.**

TABLE 15.1	Comparison of the Somatic and Autonomic Nervous Systems	
Feature	Somatic	Autonomic
Effectors	Skeletal muscle	Glands, smooth muscle, cardiac muscle
Control	Usually voluntary	Usually involuntary
Distal nerve endings	Neuromuscular junctions	Varicosities
Efferent pathways	One nerve fiber from CNS to effector; no ganglia	Two nerve fibers from CNS to effector; synapse at a ganglion
Neurotransmitters	Acetylcholine (ACh)	ACh and norepinephrine (NE)
Effect on target cells	Always excitatory	Excitatory or inhibitory
Effect of denervation	Flaccid paralysis	Denervation hypersensitivity

---

BEFORE YOU GO ON

Answer the following questions to test your understanding of the preceding section:

1. How does the autonomic nervous system differ functionally and anatomically from the somatic motor system?

2. How do the general effects of the sympathetic division differ from those of the parasympathetic division?

15.2 Anatomy of the Autonomic Nervous System

Expected Learning Outcomes

When you have completed this section, you should be able to

a. identify the anatomical components and nerve pathways of the sympathetic and parasympathetic divisions;

b. discuss the relationship of the adrenal glands to the sympathetic nervous system; and

c. describe the enteric nervous system of the digestive tract and explain its significance.

15.2a The Sympathetic Division

All nerve fibers of the sympathetic division arise from the thoracic and lumbar regions of the spinal cord; hence, this system is also called the *thoracolumbar division*. Its preganglionic neurosomas are in the lateral horns and nearby gray matter of the spinal cord. Their axons exit the cord by way of spinal nerves T1 to L2 and lead to the nearby **sympathetic chain** of ganglia (**paravertebral**[4] **ganglia**). Thus, the sympathetic division has relatively short preganglionic nerve fibers. The sympathetic chain is a longitudinal series of ganglia that lie adjacent to both sides of the vertebral column from the cervical to the coccygeal level. They are interconnected by longitudinal nerve cords (**figs. 15.3, 15.4; see also fig. 13.14** for the

---

[4]*para* = next to; *vertebr* = vertebral column

FIGURE 15.3 The Sympathetic Chain Ganglia. Right lateral view of the thoracic cavity of a cadaver. (n. = nerve) **APR**

From *A Stereoscopic Atlas of Anatomy* by David L. Bassett. Courtesy of Dr. Robert A. Chase, M.D.

Cardiac n.
Thoracic ganglion
Communicating ramus
Sympathetic chain
Splanchnic n.
Phrenic n.
Vagus n.
Bronchi
Superior vena cava
Rib
Heart
Diaphragm

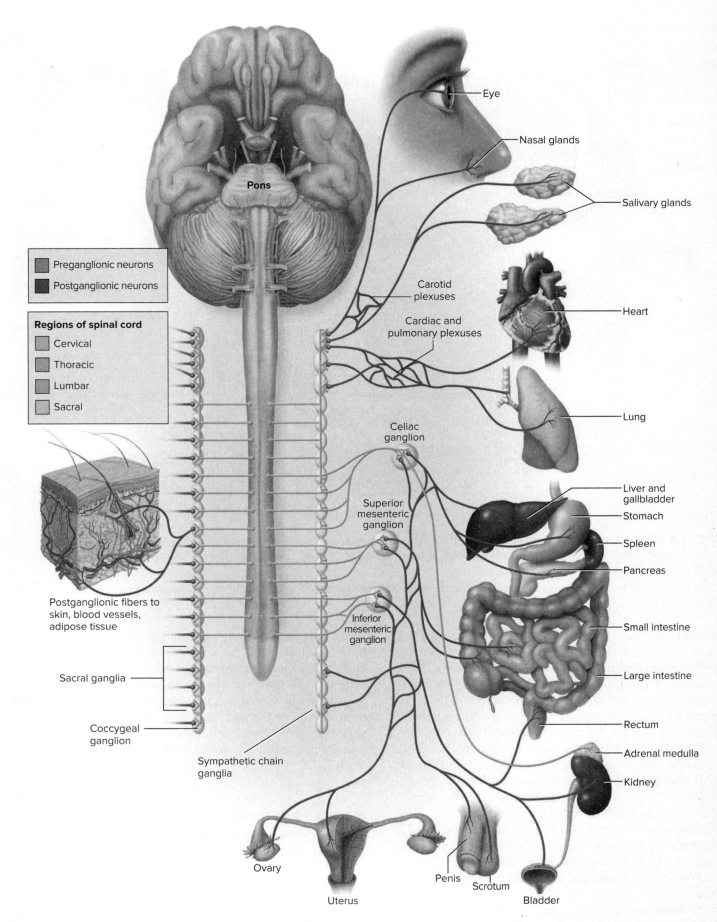

FIGURE 15.4 Schematic of the Sympathetic Nervous System. APR

❓ *Does the sympathetic innervation of the lungs make a person inhale and exhale? Explain.*

FIGURE 15.5 Neural Pathways Through the Sympathetic Chain Ganglia. Sympathetic fibers can follow any of the three numbered routes: (1) the spinal nerve route, (2) the sympathetic nerve route, or (3) the splanchnic nerve route. The somatic efferent pathway is shown on the left for comparison.

❓ *Name the parts of the spinal cord where the neurosomas of the sympathetic and somatic efferent neurons are located.*

relationship of this chain to the spinal nerves and vertebrae). The number of ganglia in this chain varies from person to person, but usually there are 3 cervical *(superior, middle,* and *inferior),* 11 thoracic, 4 lumbar, 4 sacral, and 1 coccygeal ganglion in each chain.

It may seem odd that sympathetic ganglia exist in the cervical, sacral, and coccygeal regions considering that sympathetic fibers arise only from the thoracic and lumbar regions of the spinal cord (levels T1 to L2). But as shown in figure 15.4, nerve cords from the thoracic region ascend to the ganglia in the neck, and cords from the lumbar region descend to the sacral and coccygeal ganglia. Consequently, sympathetic nerve fibers are distributed to every level of the body. As a general rule, the head receives sympathetic output arising from spinal cord segment T1, the neck from T2, the thorax and upper limbs from T3 to T6, the abdomen from T7 to T11, and the lower limbs from T12 to L2. There is considerable overlap and individual variation in this pattern, however.

In the thoracolumbar region, each paravertebral ganglion is connected to a spinal nerve by two branches called *communicating*

rami **(fig. 15.5).** The preganglionic fibers are small myelinated fibers that travel from the spinal nerve to the ganglion by way of the **white communicating ramus,**[5] which gets its color and name from the myelin. Unmyelinated postganglionic fibers leave the ganglion by way of the **gray communicating ramus,** named for its lack of myelin and duller color, and by other routes. This ramus returns to the spinal nerve. Postganglionic fibers, longer than the preganglionics, extend the rest of the way to the target organ.

▶▶▶**APPLY WHAT YOU KNOW**

Would autonomic postganglionic fibers have faster or slower conduction speeds than somatic motor fibers? Why? (See hints in section 12.3d.)

---

[5]*ramus* = branch

TABLE 15.2	Innervation to and from the Collateral Ganglia				
Sympathetic Ganglia	→	**Collateral Ganglia**	→	**Postganglionic Target Organs**	
Thoracic ganglion 5 to 9 or 10	→	Celiac ganglion	→	Stomach, spleen, liver, pancreas, small intestine, and kidneys	
Thoracic ganglia 9 to 12	→	Celiac and superior mesenteric ganglia	→	Small intestine, colon, and kidneys	
Lumbar ganglia	→	Inferior mesenteric ganglion	→	Rectum, urinary bladder, and reproductive organs	

After entering the sympathetic chain, preganglionic fibers may follow any of three courses:

- Some end in the ganglion they enter and synapse immediately with a postganglionic neuron.

- Some travel up or down the chain and synapse in ganglia at other levels. It is these fibers that link the paravertebral ganglia into a chain. They are the only route by which ganglia at the cervical, sacral, and coccygeal levels receive input.

- Some pass through the chain without synapsing and continue as *splanchnic nerves* (SPLANK-nic), to be considered shortly.

Nerve fibers leave the sympathetic chain by three routes. These are numbered in figure 15.5 to correspond to the following descriptions:

1. **The spinal nerve route.** Some postganglionic fibers exit a ganglion by way of the gray ramus, return to the spinal nerve or its subdivisions, and travel the rest of the way to the target organ. This is the route to most sweat glands, arrector muscles, and blood vessels of the skin and skeletal muscles.

2. **The sympathetic nerve route.** Other postganglionic fibers leave by way of sympathetic nerves that extend to the heart, lungs, esophagus, and thoracic blood vessels. These nerves form a **carotid plexus** around each carotid artery of the neck and issue fibers from there to effectors in the head—including sweat, salivary, and nasal glands; arrector muscles; blood vessels; and dilators of the iris. Some fibers from the superior and middle cervical ganglia form the *cardiac nerves* to the heart (which also contain parasympathetic fibers).

3. **The splanchnic[6] nerve route.** Some of the fibers that arise from spinal nerves T5 to T12 pass through the sympathetic ganglia without synapsing. Beyond the ganglia, they continue as **splanchnic nerves,** which lead to a second set of ganglia called **collateral ganglia.** Here the preganglionic fibers synapse with the postganglionics.

The collateral ganglia contribute to a network called the **abdominal aortic plexus** wrapped around the aorta **(fig. 15.6).**

There are three major collateral ganglia in this plexus—the **celiac, superior mesenteric,** and **inferior mesenteric ganglia**—located at points where arteries of the same names branch off the aorta. The postganglionic fibers accompany these arteries and their branches to the target organs. **Table 15.2** summarizes the innervation to and from the three major collateral ganglia.

The term *solar plexus* is used by some authorities as a collective name for the celiac and superior mesenteric ganglia, and by others as a synonym for the celiac ganglion only. The term comes from the nerves radiating from the ganglion like rays of the sun.

In summary, effectors in the muscles and body wall are innervated mainly by sympathetic fibers in the spinal nerves, effectors in the head and thoracic cavity by sympathetic nerves, and effectors in the abdominopelvic cavity by splanchnic nerves.

There is no simple one-to-one relationship between preganglionic and postganglionic fibers in the sympathetic division. For one thing, each postganglionic neuron may receive synapses from multiple preganglionic fibers, thus exhibiting the principle of *neural convergence.* Furthermore, each preganglionic fiber branches out to multiple postganglionic neurons, thus showing *neural divergence.* (See section 12.6d to review diverging and converging circuits.) Most sympathetic preganglionic fibers synapse with 10 to 20 postganglionic neurons. Therefore, when one preganglionic neuron fires, it can excite multiple postganglionic fibers leading to different target organs. The sympathetic division thus tends to have relatively widespread effects—as suggested by the name *sympathetic.*[7]

15.2b The Adrenal Glands

The paired **adrenal[8] (suprarenal) glands** rest like hats on the superior poles of the kidneys (fig. 15.6). Each adrenal is actually two glands with different functions and embryonic origins. The outer rind, the **adrenal cortex,** secretes steroid hormones discussed in section 17.3e. The inner core, the **adrenal medulla,** is essentially a sympathetic ganglion. It consists of modified postganglionic neurons without dendrites or axons. Sympathetic preganglionic fibers penetrate through the cortex and terminate on these cells. The sympathetic nervous system and adrenal medulla are so closely related in development and function that they are referred to collectively as the *sympathoadrenal system.*

---

[6]*splanchn* = viscera

[7]*sym* = together; *path* = feeling
[8]*ad* = near; *ren* = kidney

Diaphragm

Esophagus

Adrenal medulla

Adrenal cortex

(b)

Celiac ganglia

Adrenal gland

Celiac trunk

Renal plexus

First lumbar
sympathetic
ganglion

Aortic plexus

Aorta

Superior mesenteric
ganglion

Superior mesenteric
artery

Kidney

Inferior mesenteric artery

Inferior mesenteric
ganglion

Pelvic
sympathetic
chain

(a)

FIGURE 15.6 Abdominopelvic Components of the Sympathetic Nervous System. (a) Collateral ganglia, abdominal aortic plexus, and adrenal glands. (b) The adrenal gland, frontal section. Only the adrenal medulla plays a role in the sympathetic nervous system; the adrenal cortex has unrelated roles described in section 17.3e.

When stimulated, the adrenal medulla secretes a mixture of hormones into the bloodstream—about 85% epinephrine (adrenaline), 15% norepinephrine (noradrenaline), and a trace of dopamine. These hormones, the *catecholamines,* were briefly considered in section 12.5c because they also function as neurotransmitters.

15.2c The Parasympathetic Division

The parasympathetic division is also called the *craniosacral division* because its nerve fibers arise from the brain and sacral region of the spinal cord; they travel in certain cranial and sacral nerves. Neurosomas of the preganglionic neurons are located in the midbrain, pons, medulla oblongata, and segments S2 to S4 of the spinal cord (**fig. 15.7**). They issue long preganglionic fibers

that end in **terminal ganglia** in or near the target organ. (If a terminal ganglion is embedded in the wall of a target organ, it is also called an *intramural* [9] *ganglion.*) Thus, the parasympathetic division has long preganglionic fibers, reaching almost all the way to the target cells, and short postganglionic fibers that cover the rest of the distance.

There is some neural divergence in the parasympathetic division, but much less than in the sympathetic. The parasympathetic division has a ratio of fewer than five postganglionic fibers to every preganglionic fiber. Furthermore, the preganglionic fiber reaches the target organ before even this slight divergence occurs. The parasympathetic division is therefore relatively selective in its stimulation of target organs.

[9]*intra* = within; *mur* = wall

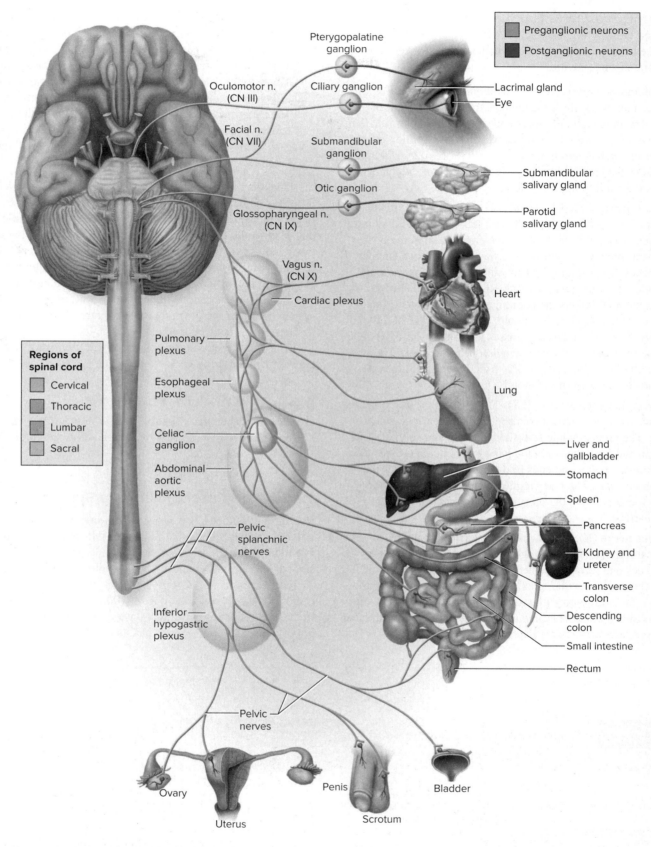

FIGURE 15.7 Schematic of the Parasympathetic Nervous System. A&PR

❓ *Which nerve carries the most parasympathetic nerve fibers?*

Parasympathetic fibers leave the brainstem in the following cranial nerves. The first three supply all parasympathetic innervation to the head, and the last one supplies viscera of the thoracic and abdominopelvic cavities.

1. **Oculomotor nerve (III).** The oculomotor nerve carries parasympathetic fibers that control the lens and pupil of the eye. The preganglionic fibers enter the orbit and terminate in the *ciliary ganglion* behind the eyeball. Postganglionic fibers enter the eyeball and innervate the *ciliary muscle,* which thickens the lens, and the *pupillary constrictor,* which narrows the pupil.

2. **Facial nerve (VII).** The facial nerve carries parasympathetic fibers that regulate the tear glands, salivary glands, and nasal glands. Soon after the facial nerve emerges from the pons, its parasympathetic fibers split away and form two smaller branches. The superior branch ends at the *pterygopalatine ganglion* near the junction of the maxilla and palatine bone. Postganglionic fibers then continue to the tear glands and glands of the nasal cavity, palate, and other areas of the oral cavity. The inferior branch crosses the middle-ear cavity and ends at the *submandibular ganglion* near the angle of the mandible. Postganglionic fibers from here supply salivary glands in the floor of the mouth.

3. **Glossopharyngeal nerve (IX).** The glossopharyngeal nerve also carries parasympathetic fibers concerned with salivation. The preganglionic fibers leave this nerve soon after its origin and form the *tympanic nerve.* This nerve crosses the middle-ear cavity and ends in the *otic*[10] *ganglion* near the foramen ovale. The postganglionic fibers then follow the trigeminal nerve to the *parotid salivary gland* just in front of the earlobe.

4. **Vagus nerve (X).** The vagus nerve carries about 90% of all parasympathetic preganglionic fibers. It travels down the neck and forms three networks in the mediastinum of the chest—the **cardiac plexus,** which supplies fibers to the heart; the **pulmonary plexus,** whose fibers accompany the bronchi and blood vessels into the lungs; and the **esophageal plexus,** whose fibers regulate swallowing.

At the lower end of the esophagus, these plexuses give off anterior and posterior **vagal trunks,** each of which contains fibers from both the right and left vagus nerves. These trunks penetrate the diaphragm, enter the abdominal cavity, and contribute to the extensive *abdominal aortic plexus* mentioned earlier. As we have seen, sympathetic fibers synapse here. The parasympathetic fibers, however, pass through the plexus without synapsing. They synapse farther along, in terminal ganglia in or near the liver, pancreas, stomach, small intestine, kidney, ureter, and proximal half of the colon.

The remaining parasympathetic fibers arise from levels S2 to S4 of the spinal cord. They travel a short distance in the anterior rami of the spinal nerves and then form **pelvic splanchnic nerves** that lead to the **inferior hypogastric plexus.** Some parasympathetic fibers synapse here, but most pass through this plexus and travel by way of **pelvic nerves** to the terminal ganglia in their target organs: the distal half of the colon, the rectum, urinary bladder, and reproductive organs. With few exceptions, the parasympathetic system doesn't innervate body wall structures (sweat glands, arrector muscles, or cutaneous blood vessels).

The sympathetic and parasympathetic divisions of the ANS are compared in **table 15.3.**

▶▶▶**APPLY WHAT YOU KNOW**

Would autonomic functions be affected if the anterior roots of the cervical spinal nerves were damaged? Why or why not?

15.2d The Enteric Nervous System

The digestive tract has a nervous network of its own called the **enteric**[11] **nervous system.** Unlike the ANS proper, it doesn't arise from the brainstem or spinal cord, but like the ANS, it innervates smooth muscle and glands. Thus, opinions differ on whether it should be considered part of the ANS. Pictured in this chapter's opening photo, it consists of about 100 million neurons embedded in the wall of the digestive tract—perhaps more neurons than there are in the spinal cord—and it has its own reflex arcs. The enteric

---

[10]*ot* = ear; *ic* = pertaining to

[11]*enter* = intestines; *ic* = pertaining to

TABLE 15.3	Comparison of the Sympathetic and Parasympathetic Divisions	
Feature	Sympathetic	Parasympathetic
Origin in CNS	Thoracolumbar	Craniosacral
Location of ganglia	Paravertebral ganglia adjacent to spinal column and prevertebral ganglia anterior to it	Terminal ganglia near or within target organs
Fiber lengths	Short preganglionic	Long preganglionic
	Long postganglionic	Short postganglionic
Neural divergence	Extensive	Minimal
Effects of system	Often widespread and general	More specific and local

nervous system regulates the motility of the esophagus, stomach, and intestines and the secretion of digestive enzymes and acid. To function normally, however, these digestive activities also require regulation by the sympathetic and parasympathetic systems. The enteric nervous system is discussed in more detail in section 25.1b. Its importance in intestinal motility becomes dramatically apparent when the system is absent (see Deeper Insight 15.1).

BEFORE YOU GO ON

Answer the following questions to test your understanding of the preceding section:

3. Explain why the sympathetic division is also called the thoracolumbar division even though its paravertebral ganglia extend all the way from the cervical to the sacral region.

4. Describe or diagram the structural relationships among the following: preganglionic fiber, postganglionic fiber, gray ramus, white ramus, and sympathetic ganglion.

5. Explain in anatomical terms why the parasympathetic division affects target organs more selectively than the sympathetic division does.

6. Trace the pathway of a parasympathetic fiber of the vagus nerve from the medulla oblongata to the small intestine.

DEEPER INSIGHT 15.1

CLINICAL APPLICATION

Megacolon

The importance of the enteric nervous system becomes vividly clear when it is absent. Such is the case in a hereditary defect called *Hirschsprung*[12] disease. During normal embryonic development, neural crest cells migrate to the large intestine and establish the enteric nervous system. In Hirschsprung disease, however, they fail to supply the distal parts of the large intestine, leaving the sigmoid colon and rectum (see fig. 25.33) without enteric ganglia. In the absence of these ganglia, the sigmoidorectal region lacks motility, constricts permanently, and obstructs the passage of feces. Feces accumulate and become impacted above the constriction, resulting in *megacolon*—a massive dilation of the bowel accompanied by abdominal distension and chronic constipation. The most life-threatening complications are colonic gangrene, perforation of the bowel, and bacterial infection of the peritoneum *(peritonitis)*. The treatment of choice is surgical removal of the affected segment and attachment of the healthy colon directly to the anal canal.

Hirschsprung disease is usually evident even in the newborn, which fails to have its first expected bowel movement. It affects four times as many infant boys as girls, and although its incidence in the general population is about 1 in 5,000 live births, it occurs in about 1 out of 10 infants with Down syndrome.

Hirschsprung disease is not the only cause of megacolon. In Central and South America, biting insects called *kissing bugs* transmit parasites called *trypanosomes* to humans. These parasites, similar to the ones that cause African sleeping sickness, cause *Chagas*[13] disease. Among other effects, they destroy the autonomic ganglia of the enteric nervous system, leading to a massively enlarged and often gangrenous colon.

[12]Harald Hirschsprung (1830–1916), Danish physician
[13]Carlos Chagas (1879–1934), Brazilian physician

15.3 Autonomic Effects on Target Organs

Expected Learning Outcomes

When you have completed this section, you should be able to

a. name the neurotransmitters employed at different synapses of the ANS;

b. name the receptors for these neurotransmitters and explain how they relate to autonomic effects;

c. explain how the ANS controls many target organs through dual innervation; and

d. explain how control is exerted in the absence of dual innervation.

15.3a Neurotransmitters and Their Receptors

As noted earlier, the divisions of the ANS often have contrasting effects on an organ. The sympathetic division accelerates the heartbeat and the parasympathetic division slows it down, for example. But each division of the ANS also can have contrasting effects on different organs. For example, the parasympathetic division contracts the wall of the urinary bladder but relaxes the male internal urethral sphincter; both actions are necessary for the expulsion of urine. It employs acetylcholine for both purposes. Similarly, the sympathetic division constricts most blood vessels but dilates the bronchioles of the lungs, and it achieves both effects with norepinephrine.

How can different autonomic neurons have such contrasting effects? There are two fundamental reasons: (1) Sympathetic and parasympathetic fibers secrete different neurotransmitters, and (2) target cells respond in different ways even to the same neurotransmitter depending on what type of receptors they have for it. All autonomic nerve fibers secrete either acetylcholine or norepinephrine, and each of these neurotransmitters has two major classes of receptors **(fig. 15.8)**.

- **Acetylcholine (ACh).** ACh is secreted by the preganglionic fibers in both divisions and the postganglionic fibers of the parasympathetic division **(table 15.4)**. A few sympathetic postganglionics also secrete ACh—those that innervate sweat glands and some blood vessels. Any nerve fiber that secretes ACh is called a **cholinergic fiber** (CO-lin-UR-jic), and any receptor that binds it is called a **cholinergic receptor**. There are two categories of cholinergic receptors:

  - **Muscarinic receptors** (MUSS-cuh-RIN-ic). These are named for muscarine, a mushroom toxin used in their discovery. All cardiac muscle, smooth muscle, and gland cells with cholinergic innervation have muscarinic receptors. There are different subclasses of muscarinic receptors with different effects; thus ACh excites intestinal

(a) Parasympathetic fiber

(b) Sympathetic adrenergic fiber

(c) Sympathetic cholinergic fiber

FIGURE 15.8 Neurotransmitters and Receptors of the Autonomic Nervous System. (a) All parasympathetic fibers are cholinergic. (b) Most sympathetic postganglionic fibers are adrenergic; they secrete norepinephrine (NE), and the target cell bears adrenergic receptors. (c) A few sympathetic postganglionic fibers are cholinergic; they secrete acetylcholine (ACh), and the target cell has cholinergic receptors of the muscarinic class.

smooth muscle by binding to one type of muscarinic receptor, and inhibits cardiac muscle by binding to a different type. Muscarinic receptors work through a variety of second-messenger systems.

- **Nicotinic receptors** (NIC-oh-TIN-ic). These are named for another botanical toxin helpful to their discovery—nicotine. They occur at all synapses in the autonomic ganglia, where the preganglionic fibers stimulate the

TABLE 15.4	Locations of Cholinergic and Adrenergic Fibers in the ANS	
Division	Preganglionic Fibers	Postganglionic Fibers
Sympathetic	Always cholinergic	Mostly adrenergic; a few cholinergic
Parasympathetic	Always cholinergic	Always cholinergic

postganglionic cells; on cells of the adrenal medulla; and at the neuromuscular junctions of skeletal muscle fibers. The binding of ACh to a nicotinic receptor is always excitatory. Nicotinic receptors work by opening ligand-gated ion channels and producing an excitatory postsynaptic potential in the target cell.

- **Norepinephrine (NE).** This neurotransmitter is secreted by nearly all sympathetic postganglionic fibers (table 15.4). Nerve fibers that secrete it are called **adrenergic fibers,** and the receptors for it are called **adrenergic receptors.** (NE is also called noradrenaline, the origin of the term *adrenergic.*) There are two principal categories of NE receptors:

  - α**-adrenergic receptors.** These usually have excitatory effects. For example, the binding of NE to α-adrenergic receptors promotes labor contractions, stimulates pilo-erection, and constricts dermal blood vessels, yet it inhibits intestinal motility. These contrasting effects result from the different actions of two subclasses of α-adrenergic receptors—α_1 and α_2. Receptors of the α_1 type act through calcium ions as a second messenger, whereas α_2 receptors inhibit the synthesis of cyclic AMP (cAMP).

  - β**-adrenergic receptors.** These are usually inhibitory. For example, NE relaxes and dilates the bronchioles (thus enhancing respiratory airflow) when it binds to β-adrenergic receptors of the smooth muscle. Yet when it binds to the β-adrenergic receptors of cardiac muscle, it has an excitatory effect. Such contrasting effects—increased pulmonary airflow and a stronger, faster heartbeat—are obviously appropriate to a state of exercise. Here again there are two receptor subclasses, β_1 and β_2, which mediate different effects. Both types, however, act through cAMP as a second messenger.

Table 15.5 summarizes the effects of sympathetic and parasympathetic stimulation on different target organs and shows how some of these effects hinge on the receptor type. Knowledge of these receptor types is also vital to the field of neuropharmacology (see Deeper Insight 15.2). Many naturally occurring drugs bind

TABLE 15.5	Effects of the Sympathetic and Parasympathetic Nervous Systems	
Target	**Sympathetic Effect and Receptor Type**	**Parasympathetic Effect (All Muscarinic)**
Adipose Tissue	Decreased fat breakdown (α)	No effect
	Increased fat breakdown (α, β)	
Eye		
Iris	Pupillary dilation (α)	Pupillary constriction
Ciliary muscle and lens	Relaxation for far vision (β)	Contraction for near vision
Lacrimal (tear) gland	None	Secretion
Integumentary System		
Merocrine sweat glands (cooling)	Secretion (muscarinic)	No effect
Apocrine sweat glands (scent)	Secretion (α)	No effect
Arrector pili muscles	Hair erection (α)	No effect
Endocrine System		
Adrenal Medulla	Hormone secretion (nicotinic)	No effect
Circulatory System		
Heart rate and force	Increased (β)	Decreased
Deep coronary arteries	Vasodilation (β)	Slight vasodilation
	Vasoconstriction (α)	
Blood vessels of most viscera	Vasoconstriction (α)	Vasodilation
Blood vessels of skin	Vasoconstriction (α)	Vasodilation, blushing
Platelets (blood clotting)	Increased clotting (α)	No effect
Respiratory System		
Bronchi and bronchioles	Bronchodilation (β)	Bronchoconstriction
Mucous glands	Decreased secretion (α)	No effect
	Increased secretion (β)	
Urinary System		
Kidneys	Reduced urine output (α)	No effect
Bladder wall	No effect	Contraction
Internal urinary sphincter	Contraction, urine retention (α)	Relaxation, urine release
Digestive System		
Salivary glands	Thick mucous secretion (α)	Thin serous secretion
Gastrointestinal motility	Decreased (α, β)	Increased
Gastrointestinal secretion	Decreased (α)	Increased
Liver	Glycogen breakdown (α, β)	Glycogen synthesis
Pancreatic enzyme secretion	Decreased (α)	Increased
Pancreatic insulin secretion	Decreased (α)	No effect
	Increased (β)	
Reproductive System		
Penile or clitoral erection	No effect	Stimulation
Glandular secretion	No effect	Stimulation
Orgasm, smooth muscle roles	Stimulation (α)	No effect
Uterus	Relaxation (β)	No effect
	Labor contractions (α)	

selectively to one or another class or subclass of receptor. Atropine binds only to muscarinic receptors and curare only to nicotinic receptors, for example. Many synthetic drugs are designed to be similarly selective.

The autonomic effects on glandular secretion are often the indirect results of action on blood vessels. Many glandular secretions begin as a filtrate of the blood, which is then modified by the gland cells. Increasing the blood flow through a gland (such as a salivary or sweat gland) tends to increase secretion, and reducing the blood flow reduces secretion.

Sympathetic effects tend to last longer than parasympathetic effects. After a parasympathetic fiber secretes ACh into a synapse, it is quickly broken down by acetylcholinesterase (AChE) and its effect lasts only a few seconds. The NE released by a sympathetic fiber, however, has various fates: (1) Some is reabsorbed by the nerve fiber, where it is either reused or broken down by an enzyme called *monoamine oxidase (MAO)*. (2) Some diffuses into the adjacent tissues, where it is degraded by another enzyme, *catechol-O-methyltransferase (COMT)*. (3) Much of it passes into the bloodstream, where MAO and COMT are absent. This NE and epinephrine from the adrenal gland circulate throughout the body and may exert their effects for several minutes before they are finally degraded by the liver.

ACh and NE are not the only neurotransmitters employed by the ANS. Although all autonomic fibers secrete one of these, many of them also secrete neuropeptides that modulate ACh or NE function. Sympathetic fibers may also secrete enkephalin, substance P, neuropeptide Y, somatostatin, neurotensin, or gonadotropin-releasing hormone. Some parasympathetic fibers relax blood vessels by stimulating the endothelial cells to release nitric oxide (NO), a gas. NO inhibits smooth muscle tone in the vessel wall, thus allowing the vessel to dilate. This increases the blood flow through the vessel. Among other functions, this mechanism is crucial to penile erection (see Deeper Insight 27.4).

▶▶▶APPLY WHAT YOU KNOW

Table 15.5 notes that the sympathetic nervous system has an α-adrenergic effect on blood platelets and promotes clotting. How can the sympathetic nervous system stimulate platelets, considering that platelets are drifting cell fragments in the bloodstream with no nerve fibers leading to them?

15.3b Dual Innervation

Most of the viscera receive nerve fibers from both the sympathetic and parasympathetic divisions and thus are said to have **dual innervation.** In such cases, the two divisions may have either *antagonistic* or *cooperative* effects on the same organ.

Antagonistic effects oppose each other. For example, the sympathetic division speeds up the heart and the parasympathetic division slows it down; the sympathetic division inhibits digestion and the parasympathetic division stimulates it; the sympathetic division dilates the pupil and the parasympathetic division constricts it. In some cases, these effects are exerted through dual

innervation of the same effector cells, as in the heart, where nerve fibers of both divisions stimulate some of the same cells. In other cases, antagonistic effects arise because each division innervates different effector cells with opposite effects on organ function. In the iris of the eye, for example, sympathetic fibers innervate pupillary dilator cells and parasympathetic fibers innervate constrictor cells **(fig. 15.9).**

Cooperative effects are seen when the two divisions act on different effectors to produce a unified overall effect. Salivation is a good example. The parasympathetic division stimulates serous cells of the salivary glands to secrete a watery, enzyme-rich secretion, while the sympathetic division stimulates mucous cells of the same glands to secrete mucus. The enzymes and mucus are both necessary components of the saliva.

FIGURE 15.9 Dual Innervation of the Iris. Shows antagonistic effects of the sympathetic (yellow) and parasympathetic (blue) divisions on the iris.

❓ *If a person is in a state of fear, would you expect the pupils to be dilated or constricted? Why?*

Even when both divisions innervate a single organ, they don't always innervate it equally or exert equal influence. For example, the parasympathetic division forms an extensive plexus in the wall of the digestive tract and exerts much more influence over it than the sympathetic division does. In the ventricles of the heart, by contrast, there is much less parasympathetic than sympathetic innervation.

15.3c Control Without Dual Innervation

Dual innervation isn't always necessary for the ANS to produce opposite effects on an organ. The adrenal medulla, arrector muscles, sweat glands, and many blood vessels receive only sympathetic fibers. The most significant example of control without dual innervation is regulation of blood pressure and routes of blood flow. The sympathetic fibers to a blood vessel have a baseline sympathetic tone, which keeps the vessels in a state of partial constriction called **vasomotor tone (fig. 15.10).** An increase in firing rate constricts a vessel by increasing smooth muscle contraction. A drop in firing frequency dilates a vessel by allowing the smooth muscle to relax. The blood pressure in the vessel, pushing outward on its wall, then dilates the vessel. Thus, the sympathetic division alone exerts opposite effects on the vessels.

Sympathetic control of vasomotor tone can shift blood flow from one organ to another according to the changing needs of the body. In times of emergency, stress, or exercise, the skeletal muscles and heart receive a high priority and the sympathetic division dilates the arteries that supply them. Processes such as digestion, nutrient absorption, and urine formation can wait; thus the sympathetic division constricts arteries to the gastrointestinal tract and kidneys. It also reduces blood flow through the skin, which may help to minimize bleeding in the event that the stress-producing situation leads to injury. Furthermore, since there isn't enough blood in the body to supply all the organ systems equally, it is necessary to temporarily divert blood away from some organs in order to supply an adequate amount to the muscular system.

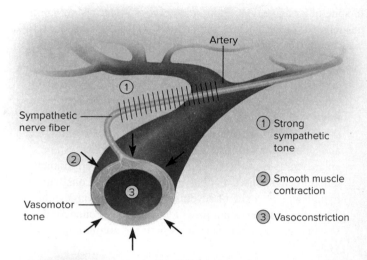

(a) Vasoconstriction

① Strong sympathetic tone

② Smooth muscle contraction

③ Vasoconstriction

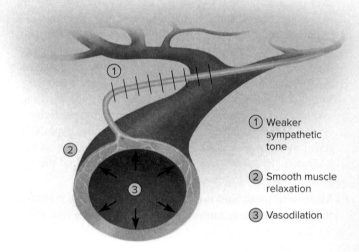

(b) Vasodilation

① Weaker sympathetic tone

② Smooth muscle relaxation

③ Vasodilation

FIGURE 15.10 Sympathetic and Vasomotor Tone.
(a) Vasoconstriction in response to a high rate of sympathetic nerve firing. (b) Vasodilation in response to a low rate of sympathetic nerve firing. Smooth muscle relaxation allows blood pressure within the vessel to push the vessel wall outward. Black lines crossing each nerve fiber represent action potentials, with a high firing frequency in part (a) and a lower frequency in part (b).

BEFORE YOU GO ON

Answer the following questions to test your understanding of the preceding section:

7. What neurotransmitters are secreted by adrenergic and cholinergic fibers?

8. Why do sympathetic effects last longer than parasympathetic effects?

9. How can the sympathetic division cause smooth muscle to relax in some organs but contract in others?

10. What are the two ways in which the sympathetic and parasympathetic systems can affect each other when they both innervate the same target organ? Give examples.

11. How can the sympathetic nervous system have contrasting effects in a target organ without dual innervation?

15.4 Central Control of Autonomic Function

Expected Learning Outcomes

When you have completed this section, you should be able to

a. identify what parts of the brain influence the autonomic nervous system;

b. give examples of emotional influences on autonomic function; and

c. give examples of spinal cord control over autonomic functions.

In spite of its name, the autonomic nervous system is not an independent nervous system. In this section we briefly consider how it is influenced by various levels of the central nervous system.

- **Cerebral cortex.** Even if we usually cannot consciously control the ANS, it is clear that the mind does influence it. Anger raises the blood pressure, fear makes the heart race, thoughts of good food make the stomach rumble, sexual thoughts or images increase blood flow to the genitals, and anxiety inhibits sexual function. The limbic system, an ancient part of the cerebral cortex, is involved in many emotional responses and has extensive connections with the hypothalamus, a site of several nuclei of autonomic control. Thus, the limbic system provides a pathway connecting sensory and mental experiences with the autonomic nervous system.

- **Hypothalamus.** Although the major site of CNS control over the somatic motor system is the primary motor cortex, the major control center of the visceral motor system is the hypothalamus. This small but vital region in the floor of the brain contains many nuclei for primitive functions, including hunger, thirst, thermoregulation, emotions, and sexuality. Artificial stimulation of different regions of the hypothalamus can activate the fight-or-flight response typical of the sympathetic nervous system or have the calming effects typical of the parasympathetic. Output from the hypothalamus travels largely to nuclei in more caudal regions of the brainstem, and from there to the cranial nerves and the sympathetic neurons in the spinal cord.

- **Midbrain, pons, and medulla oblongata.** These regions of the brainstem house numerous autonomic nuclei described in section 14.3: centers for cardiac and vasomotor control, salivation, swallowing, sweating, gastrointestinal secretion, bladder control, pupillary constriction and dilation, and other primitive functions. Many of these nuclei belong to the reticular formation, which extends from the medulla to the hypothalamus. Autonomic output from these nuclei travels by way of the spinal cord and the oculomotor, facial, glossopharyngeal, and vagus nerves.

- **Spinal cord.** Finally, the spinal cord integrates such autonomic reflexes as micturition (urination), defecation, erection, and ejaculation (details are in sections 23.7d, 25.7f, and 27.5). Fortunately, the brain is able to inhibit defecation and urination consciously, but when injuries sever the spinal cord from the brain, the autonomic spinal reflexes alone control the elimination of urine and feces.

Table 15.6 describes some dysfunctions of the autonomic nervous system.

BEFORE YOU GO ON

Answer the following questions to test your understanding of the preceding section:

12. What system in the brain connects our conscious thoughts and feelings with the autonomic control centers of the hypothalamus?

13. List some autonomic responses that are controlled by nuclei in the hypothalamus.

14. What are the roles of the midbrain, pons, and medulla in autonomic control?

15. Name some visceral reflexes controlled by the spinal cord.

TABLE 15.6	Some Disorders of the Autonomic Nervous System
Horner syndrome	Chronic unilateral pupillary constriction, sagging of the eyelid, withdrawal of the eye into the orbit, flushing of the skin, and lack of facial perspiration; results from lesions in the cervical ganglia, upper thoracic spinal cord, or brainstem that interrupt sympathetic innervation of the head.
Raynaud disease	Intermittent attacks of paleness, cyanosis, and pain in the fingers and toes, caused when cold or emotional stress triggers excessive vasoconstriction in the digits; most common in young women. In extreme cases, causes gangrene and may require amputation. Sometimes treated by severing sympathetic nerves to the affected regions.

You can find other autonomic nervous system disorders described in the following places:

The *mass reflex reaction* in Deeper Insight 13.5; *autonomic effects of cranial nerve injuries* in section 14.6c; *Hirschsprung disease* and *Chagas disease* in Deeper Insight 15.1; and *orthostatic hypotension* in table 20.2.

DEEPER INSIGHT 15.2

CLINICAL APPLICATION

Drugs and the Nervous System

Neuropharmacology is a branch of medicine that deals with the effects of drugs on the nervous system, especially drugs that mimic, enhance, or inhibit the action of neurotransmitters. A few examples will illustrate the clinical relevance of neurotransmitter and receptor functions.

A number of drugs work by stimulating adrenergic and cholinergic neurons or receptors. *Sympathomimetics*[14] are drugs that enhance sympathetic action. They stimulate adrenergic receptors or promote norepinephrine release. For example, phenylephrine, found in such cold medicines as Dimetapp and Sudafed PE, aids breathing by stimulating alpha receptors, dilating the bronchioles, and constricting nasal blood vessels, thus reducing swelling in the nasal mucosa. *Sympatholytics*[15] are drugs that suppress sympathetic action by inhibiting norepinephrine release or by binding to adrenergic receptors without stimulating them. Propranolol, for example, is a *beta-blocker.* It reduces hypertension partly by blocking β-adrenergic receptors. This interferes with the effects of epinephrine and norepinephrine on the heart and blood vessels. (It also reduces the production of *angiotensin II,* a hormone that stimulates vasoconstriction and raises blood pressure.)

Parasympathomimetics enhance parasympathetic effects. Pilocarpine, for example, relieves glaucoma (excessive pressure in the eyeball) by dilating a vessel that drains fluid from the eye. *Parasympatholytics* inhibit ACh release or block its receptors. Atropine, for example, blocks muscarinic receptors. It is sometimes used to dilate the pupils for eye examinations and to dry the mucous membranes of the respiratory tract before inhalation anesthesia. It is an extract of the deadly nightshade plant, *Atropa belladonna.*[16] Women of the Middle Ages used nightshade to dilate their pupils, which was regarded as a beauty enhancement.

The drugs we have mentioned so far act on the peripheral nervous system and its effectors. Many others act on the central nervous system. Strychnine, for example, blocks the inhibitory action of glycine on spinal motor neurons. The neurons then overstimulate the muscles, causing spastic paralysis and sometimes death by suffocation.

Sigmund Freud predicted that psychiatry would eventually draw upon biology and chemistry to deal with emotional problems once treated only by counseling and psychoanalysis. A branch of neuropharmacology called *psychopharmacology* has fulfilled his prediction. This field dates to the 1950s when chlorpromazine, an antihistamine, was incidentally found to relieve the symptoms of schizophrenia.

The management of clinical depression is one example of how contemporary psychopharmacology has supplemented counseling approaches. Some forms of depression result from deficiencies of the monoamine neurotransmitters. Thus, they yield to drugs that prolong the effects of the monoamines already present at the synapses. Prozac (fluoxetine), for example, blocks serotonin reuptake and prolongs its mood-elevating effect; thus Prozac is called a *selective serotonin reuptake inhibitor (SSRI).* It is also used to treat fear of rejection, excess sensitivity to criticism, lack of self-esteem, and inability to experience pleasure, all of which were long handled only through counseling, group therapy, or psychoanalysis. After monoamines are taken up from the synapse, they are degraded by monoamine oxidase (MAO). Drugs called *MAO inhibitors* slow down the breakdown of monoamine neurotransmitters and provide another pharmacological approach to depression.

Our growing understanding of neurochemistry also gives us more insight into the action of addictive drugs of abuse such as amphetamines

and cocaine. Amphetamines ("speed") chemically resemble norepinephrine and dopamine, two neurotransmitters associated with elevated mood. Dopamine is especially important in sensations of pleasure. Cocaine blocks dopamine reuptake and thus produces a brief rush of good feelings. But when dopamine is not reabsorbed by the neurons, it diffuses out of the synaptic cleft and is degraded elsewhere. Cocaine thus depletes the neurons of dopamine faster than they can synthesize it, so finally there is no longer an adequate supply to maintain normal mood. The postsynaptic neurons make new dopamine receptors as if "searching" for the neurotransmitter—all of which leads ultimately to anxiety, depression, and the inability to experience pleasure without the drug.

Caffeine exerts its stimulatory effect by competing with adenosine. Adenosine, which you know as a component of DNA, RNA, and ATP, also functions as a neuromodulator in the brain, inhibiting ACh release by cholinergic neurons. One theory of sleepiness is that it results when prolonged metabolic activity breaks down so much ATP that the accumulated adenosine has a noticeably inhibitory effect. Caffeine has enough structural similarity to adenosine **(fig. 15.11)** to bind to its receptors, but it doesn't produce the inhibitory effect. Thus, it prevents adenosine from exerting its effect, more ACh is secreted, and a person feels more alert.

FIGURE 15.11 Adenosine and Caffeine. Adenosine, a breakdown product of ATP and other chemicals, inhibits ACh release and produces a sense of sleepiness. Caffeine is similar enough to block the action of adenosine by binding to its receptors. This results in increased ACh release and heightened arousal.

left: Stockbyte/Getty Images; **right:** Duncan Walker/Getty Images

---

[14]*mimet* = imitate, mimic
[15]*lyt* = break down, destroy
[16]*bella* = beautiful, fine; *donna* = woman

STUDY GUIDE

▶ Assess Your Learning Outcomes

To test your knowledge, discuss the following topics with a study partner or in writing, ideally from memory.

15.1 General Properties of the Autonomic Nervous System

1. The fundamental function and effectors of the autonomic nervous system (ANS)
2. Why this system is called *autonomic;* how it differs from the somatic motor system
3. The fundamental, contrasting functions of the sympathetic and parasympathetic divisions of the ANS
4. Why it cannot be said that at any given moment, either the sympathetic or the parasympathetic division is active; the meaning of *autonomic tone*
5. Basic anatomical components of the ANS
6. How autonomic efferent pathways differ from somatic efferent pathways; the meanings of *preganglionic* and *postganglionic fibers*

15.2 Anatomy of the Autonomic Nervous System

1. Origin of the sympathetic preganglionic fibers and the routes they take to the sympathetic chain ganglia
2. Anatomy of the sympathetic chain; the number of ganglia at its various levels; and the body regions supplied by nerve fibers issuing from each group of ganglia
3. The gray and white communicating rami that connect the sympathetic ganglia to the spinal nerves; the reason they are named

gray and *white;* and the path that sympathetic nerve fibers take through these rami
4. Differences between the spinal nerve route, sympathetic nerve route, and splanchnic nerve route by which fibers leave the sympathetic chain
5. Various places in which a sympathetic preganglionic fiber may synapse with a postganglionic neuron
6. Locations of the celiac, superior mesenteric ganglia, and inferior mesenteric ganglia; the collective name for them; and the varied meanings of the expression *solar plexus*
7. The degree and significance of neural divergence in the sympathetic nervous system, and the effect of this on target organ stimulation
8. Why the adrenal medulla can be considered part of the sympathetic nervous system; what products it secretes when stimulated
9. Names and numbers of the cranial and spinal nerves that carry preganglionic fibers of the parasympathetic nervous system; which nerve carries the largest percentage of parasympathetic fibers
10. The path of the vagus nerve, and the names and locations of the plexuses and trunks to which it gives rise
11. Where, in general, terminal ganglia of the parasympathetic division are found, and therefore where the postganglionic fibers begin
12. The location and functions of the enteric nervous system

15.3 Autonomic Effects on Target Organs

1. Why the autonomic effect on a target cell depends on both the neurotransmitter released and the type of receptor on the target cell
2. The difference between cholinergic and adrenergic fibers and where each can be found in the ANS
3. What nicotinic and muscarinic receptors have in common, how they differ, and where they occur in the ANS
4. What α- and β-adrenergic receptors have in common, how they differ, and where they occur in the ANS
5. Neurotransmitter stability and how it relates to the duration of sympathetic versus parasympathetic effects
6. The variety of neurotransmitters and neuromodulators employed by them
7. Autonomic control of certain organs by dual innervation, and examples of antagonistic and cooperative effects on an organ
8. How the ANS can regulate organs that lack dual innervation

15.4 Central Control of Autonomic Function

1. Examples of the influence of the cerebral cortex, hypothalamus, midbrain, pons, medulla oblongata, and spinal cord on the autonomic nervous system, and their involvement in autonomic effects

▶ Testing Your Recall

Answers in Appendix A

1. The autonomic nervous system innervates all of these *except*
   a. cardiac muscle.
   b. skeletal muscle.
   c. smooth muscle.
   d. salivary glands.
   e. blood vessels.

2. Muscarinic receptors bind
   a. epinephrine.
   b. norepinephrine.
   c. acetylcholine.
   d. cholinesterase.
   e. neuropeptides.

3. All of the following cranial nerves *except* the _____ carry parasympathetic fibers.
   a. vagus
   b. facial
   c. oculomotor
   d. glossopharyngeal
   e. hypoglossal

4. Which of the following cranial nerves carries sympathetic fibers?
   a. oculomotor
   b. facial
   c. trigeminal
   d. vagus
   e. none of them

5. Which of these ganglia is *not* involved in the sympathetic division?
   a. intramural
   b. superior cervical
   c. paravertebral
   d. inferior mesenteric
   e. celiac

6. Epinephrine is secreted by
   a. sympathetic preganglionic fibers.
   b. sympathetic postganglionic fibers.
   c. parasympathetic preganglionic fibers.
   d. parasympathetic postganglionic fibers.
   e. the adrenal medulla.

7. The most significant autonomic control center within the CNS is
   a. the cerebral cortex.
   b. the limbic system.
   c. the midbrain.
   d. the hypothalamus.
   e. the sympathetic chain ganglia.

8. The gray communicating ramus contains
   a. visceral sensory fibers.
   b. parasympathetic motor fibers.
   c. sympathetic preganglionic fibers.
   d. sympathetic postganglionic fibers.
   e. somatic motor fibers.

9. Throughout the autonomic nervous system, the neurotransmitter released by the preganglionic fiber binds to _____ receptors on the postganglionic neuron.
   a. nicotinic
   b. muscarinic
   c. adrenergic
   d. alpha
   e. beta

10. Which of these does *not* result from sympathetic stimulation?
    a. dilation of the pupil
    b. acceleration of the heart
    c. digestive secretion
    d. enhanced blood clotting
    e. piloerection

11. Certain nerve fibers are called _____ fibers because they secrete norepinephrine.

12. _____ is a state in which a target organ receives both sympathetic and parasympathetic fibers.

13. _____ is a state of continual background activity of the sympathetic and parasympathetic divisions.

14. Most parasympathetic preganglionic fibers are found in the _____ nerve.

15. The digestive tract has a semi-independent nervous system called the _____ nervous system.

16. MAO and COMT are enzymes that break down _____ at certain ANS synapses.

17. The adrenal medulla consists of modified postganglionic neurons of the _____ nervous system.

18. The sympathetic nervous system has short _____ and long _____ nerve fibers.

19. Adrenergic receptors classified as α_1, β_1, and β_2 act by changing the level of _____ in the target cell.

20. Sympathetic fibers to blood vessels maintain a state of partial vasoconstriction called _____.

▶ Building Your Medical Vocabulary

Answers in Appendix A

State a meaning of each word element, and give a medical term from this chapter that uses it or a slight variation of it.

1. baro-

2. lyto-

3. muro-

4. nomo-

5. oto-

6. patho-

7. reno-

8. splanchno-

9. sym-

10. viscero-

▶ What's Wrong with These Statements?

Answers in Appendix A

Briefly explain why each of the following statements is false, or reword it to make it true.

1. The parasympathetic nervous system shuts down when the sympathetic nervous system is active, and vice versa.

2. The parasympathetic nervous system controls dilation and constriction of the blood vessels of the skin.

3. Voluntary control of the ANS is not possible.

4. The sympathetic nervous system stimulates digestion.

5. All sympathetic postganglionic fibers are adrenergic.

6. Urination and defecation cannot occur without signals from the brain to the bladder and rectum.

7. Some parasympathetic nerve fibers are adrenergic.

8. Sympathetic effects are more localized and organ-specific than parasympathetic effects.

9. The oculomotor, facial, hypoglossal, and vagus nerves carry the parasympathetic fibers that arise from the brain.

10. The two divisions of the ANS have cooperative effects on the iris.

STUDY GUIDE

▶ Testing Your Comprehension

1. You are dicing raw onions while preparing dinner, and the vapor makes your eyes water. Describe the afferent and efferent pathways involved in this response.

2. Suppose you are walking alone at night when you hear a dog growling close behind you. Describe the ways your sympathetic nervous system would prepare you to deal with this situation.

3. Suppose that the cardiac nerves were destroyed. How would this affect the heart and the body's ability to react to a stressful situation?

4. What would be the advantage to a wolf in having its sympathetic nervous system stimulate the arrector muscles? What happens in a human when the sympathetic system stimulates these muscles?

5. Pediatric literature has reported many cases of poisoning in children with Lomotil, an antidiarrheal medicine. Lomotil works primarily by means of the morphinelike effects of its chief ingredient, diphenoxylate, but it also contains atropine. Considering the mode of action described for atropine in Deeper Insight 15.2, why might it contribute to the antidiarrheal effect of Lomotil? In atropine poisoning, would you expect the pupils to be dilated or constricted? The skin to be moist or dry? The heart rate to be elevated or depressed? The bladder to retain urine or void uncontrollably? Explain each answer. Atropine poisoning is treated with physostigmine, a cholinesterase inhibitor. Explain the rationale of this treatment.

A vallate papilla of the tongue, where most taste buds are located (SEM)

Omikron/Science Source

CHAPTER OUTLINE

16.1 Properties and Types of Sensory Receptors
- 16.1a General Properties of Receptors
- 16.1b Classification of Receptors

16.2 The General Senses
- 16.2a Unencapsulated Nerve Endings
- 16.2b Encapsulated Nerve Endings
- 16.2c Somatosensory Projection Pathways
- 16.2d Pain

16.3 The Chemical Senses
- 16.3a Gustation—The Sense of Taste
- 16.3b Olfaction—The Sense of Smell

16.4 Hearing and Equilibrium
- 16.4a The Nature of Sound
- 16.4b Anatomy of the Ear
- 16.4c The Physiology of Hearing
- 16.4d The Physiology of Equilibrium

16.5 Vision
- 16.5a Light and Vision
- 16.5b Accessory Structures of the Orbit
- 16.5c Anatomy of the Eye
- 16.5d Formation of an Image
- 16.5e Sensory Transduction in the Retina
- 16.5f Light and Dark Adaptation
- 16.5g The Dual Visual System
- 16.5h Color Vision
- 16.5i Stereoscopic Vision
- 16.5j The Visual Projection Pathway

Study Guide

DEEPER INSIGHTS

16.1 Clinical Application: Middle-Ear Infection

16.2 Clinical Application: Deafness

16.3 Clinical Application: Causes of Blindness

16.4 Medical History: Anesthesia—From Ether Frolics to Modern Surgery

Anatomy & Physiology
Revealed 4.0

Module 7: Nervous System

BRUSHING UP

- Understanding how some senses work—including taste and smell—hinges on a knowledge of G proteins, cAMP, and second-messenger systems, explained in figure 3.8 and the associated text.

- This chapter's description of the hearing process assumes prior knowledge of mechanically gated membrane channels (see section 3.2a) and ion flow down an electrochemical gradient (see section 11.3c).

- Several concepts in section 12.6 are important for understanding sensory processes: postsynaptic potentials (EPSPs and IPSPs), temporal and spatial summation, neural coding, and converging neural circuits.

- You will need to remember the concepts of spinal cord tracts and decussation (see section 13.1e) to understand sensory pathways from lower-body receptors to the brain.

- Familiarity with certain cranial nerves is necessary for understanding other sensory pathways to the brain—especially the olfactory, optic, trigeminal, facial, vestibulocochlear, glossopharyngeal, and vagus nerves, all described in table 14.1.

- You must know the gross anatomy of the brain, especially lobes and gyri of the cerebral cortex, to understand where sensory signals are received and processed (see section 14.4b).

- To understand the anatomy of the ear, it will be helpful to recall the anatomy of the temporal bone (see "The Temporal Bones" in section 8.2a).

Anyone who enjoys music, art, fine food, or a good conversation appreciates the human senses. Yet their importance extends far beyond deriving pleasure from the environment. In the 1950s, behavioral scientists at Princeton University studied the methods used by Soviet Communists to extract confessions from political prisoners, including solitary confinement and sensory deprivation. Student volunteers were immobilized in dark soundproof rooms or suspended in dark chambers of water. In a short time, they experienced visual, auditory, and tactile hallucinations, incoherent thought patterns, deterioration of intellectual performance, and sometimes morbid fear or panic. Similar effects are seen in some burn patients who are immobilized and extensively bandaged (including the eyes) and thus suffer prolonged lack of sensory input. Patients connected to life-support equipment and confined under oxygen tents sometimes become delirious. Sensory input is vital to the integrity of personality and intellectual function.

Furthermore, much of the information communicated by the sense organs never comes to our conscious attention—blood pressure, body temperature, and muscle tension, for example. By monitoring such conditions, however, the sense organs initiate somatic and visceral reflexes that are indispensable to homeostasis and to our very survival in a ceaselessly changing and challenging environment.

16.1 Properties and Types of Sensory Receptors

Expected Learning Outcomes

When you have completed this section, you should be able to

a. define *receptor* and *sense organ*;

b. list the four kinds of information obtained from sensory receptors, and describe how the nervous system encodes each type; and

c. outline three ways of classifying receptors.

A sensory **receptor** is any structure specialized to detect a stimulus. Some receptors are simple, bare nerve endings, such as the receptors for heat and pain, whereas others are true sense organs. A **sense organ** is a structure that combines nervous tissue with other tissues that enhance its response to a certain type of stimulus. The accessory tissues may include epithelial, muscular, or connective tissue. Sense organs can be as complex as the eye and ear or as microscopic and simple as a dendrite wrapped in a little bit of connective tissue.

16.1a General Properties of Receptors

The fundamental purpose of any sensory receptor is **transduction,** the conversion of one form of energy to another—light, sound, heat, touch, vibration, or other forms of stimulus energy into nerve signals. (Any device that converts one energy form to another is a *transducer*—whether a sense organ, gasoline engine, or lightbulb.)

The initial effect of a stimulus on a sensory cell is a small local electrical change called a **receptor potential.** In many cases, such as the senses of touch and smell, the sensory cell is a neuron. If the receptor potential is strong enough, the neuron fires off a volley of action potentials, generating a nerve signal to the CNS. In other cases, such as taste and hearing, the sensory cell isn't a neuron but an epithelial cell. Nevertheless, it has synaptic vesicles at its base; it releases a neurotransmitter in response to a stimulus; and it stimulates an adjacent neuron. That neuron then generates signals to the CNS.

Not all sensory signals go to the brain, but when they do, we may experience a **sensation**—a subjective awareness of the stimulus. Yet most sensory signals delivered to the CNS produce no conscious sensation at all. Most of them are filtered out in the brainstem before reaching the cerebral cortex, a valuable function that keeps us from being driven mad by innumerable unimportant stimuli detected by the sense organs. Other nerve signals concern functions that don't require conscious awareness, such as monitoring blood pH and body temperature.

Sensory receptors transmit four kinds of information—*modality, location, intensity,* and *duration:*

1. **Modality** refers to the type of stimulus or the sensation it produces. Vision, hearing, and taste are examples of sensory modalities, as are subcategories of these such as red or blue light, bass or treble sounds. The action potentials for vision

are identical to the action potentials for taste or any other modality—so how can the brain tell a visual signal from a taste signal? Modality is determined by where the sensory signals end in the brain. Hypothetically, if we could rewire the brain so signals from the ear were routed to the visual cortex of the occipital lobe, we would perceive acoustic signals as light rather than sound.

2. **Location** is also encoded by which nerve fibers issue signals to the brain. Any sensory neuron detects stimuli within an area called its **receptive field.** In the sense of touch, for example, a single sensory neuron may cover an area of skin as large as 7 cm in diameter. No matter where the skin is touched within that field, it stimulates the same neuron. The brain may be unable to determine whether the skin was touched at "point A" or at some other point 1 or 2 cm away. If the skin is simultaneously touched at two places in the same field, it can feel like a single touch **(fig. 16.1a).** In some areas of the body such as the back, however, it isn't necessary to make finer distinctions. On the other hand, we must be able to localize touch sensations much more precisely with the fingertips. Here, each sensory neuron may cover a receptive field 1 mm or less in diameter—there is a much higher density of

tactile nerve fibers. Two points of contact just 2 mm apart will be felt as separate touches **(fig. 16.1b).** That is, the fingertips have finer *two-point touch discrimination* than the skin of the back. You can imagine the importance of this in reading braille, appreciating the texture of a fine fabric, or manipulating an object as small as a sesame seed. The receptive field concept pertains not only to touch, but also to other senses such as vision.

3. **Intensity** refers to whether a sound is loud or soft, a light is bright or dim, a pain is mild or excruciating, and so forth. It is encoded in three ways: (a) As stimulus intensity rises, the firing frequencies of sensory nerve fibers rise (see fig. 12.30); (b) intense stimuli recruit greater numbers of neurons to fire; and (c) weak stimuli activate only the most sensitive neurons, whereas strong stimuli can activate a less sensitive group of neurons with higher thresholds. Thus, the brain can distinguish intensities based on which neurons are firing, how many are doing so, and how fast they are firing.

4. **Duration,** or how long a stimulus lasts, is encoded by changes in firing frequency with the passage of time. All receptors exhibit the property of **sensory adaptation**—if the stimulus is prolonged, the firing of the neuron gets slower

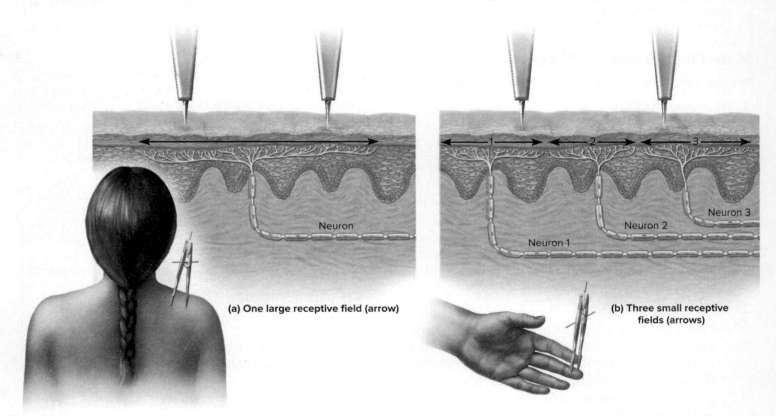

(a) One large receptive field (arrow)

(b) Three small receptive fields (arrows)

FIGURE 16.1 Receptive Fields. (a) A neuron with a large receptive field, as found in the skin of the back. If the skin is touched in two close-together places within this receptive field, the brain senses it as only one point of contact. (b) Neurons with small receptive fields, as found in the fingertips. Two close-together points of contact here are likely to stimulate two different neurons and be felt as separate touches.

? *If the receptive field in part (a) is 7 cm in diameter, is it ever possible for two touches 1 cm apart to be felt separately?*

over time, and with it, we become less aware of the stimulus. Adapting to hot bathwater is an example of this. Receptors are classified according to how quickly they adapt. **Phasic receptors** generate a burst of action potentials when first stimulated, then quickly adapt and sharply reduce or stop signaling even if the stimulus continues. Some of them fire again when the stimulus ceases. Phasic receptors are found in the senses of smell, hair movement, and cutaneous pressure and vibration. Their phasic nature explains why we may notice a suspicious odor such as a gas leak for several seconds, then the sensation fades in intensity even if the stimulus is still there. If we don't find the gas leak quickly, it becomes difficult to find it at all. **Tonic receptors** adapt more slowly and generate signals more steadily. Proprioceptors and pain receptors are among the most persistently responding tonic receptors because the brain must always be aware of body position, muscle tension, joint motion, or a pain.

▶▶▶ APPLY WHAT YOU KNOW

Although you may find it difficult to immerse yourself in a tub of hot water or a cold lake, you soon adapt and become more comfortable. In light of this, do you think cold and warm receptors are phasic or tonic? Explain.

16.1b Classification of Receptors

Receptors can be classified by various overlapping systems:

1. By stimulus modality:
   - **Thermoreceptors** respond to heat and cold.
   - **Photoreceptors,** the eyes, respond to light.
   - **Nociceptors,**[1] pain receptors, respond to tissue injury or situations that threaten to damage a tissue.
   - **Chemoreceptors** respond to chemicals, including odors, tastes, and body fluid composition.
   - **Mechanoreceptors** respond to physical deformation of a cell or tissue caused by vibration, touch, pressure, stretch, or tension. They include the organs of hearing and balance and many receptors of the skin, viscera, and joints.

2. By the origin of the stimulus:
   - **Exteroceptors** sense stimuli external to the body. They include the receptors for vision, hearing, taste, smell, and cutaneous sensations such as touch, heat, cold, and pain.
   - **Interoceptors** detect stimuli in the internal organs such as the stomach, intestines, and bladder, and produce feelings of stretch, pressure, visceral pain, and nausea.
   - **Proprioceptors** sense the position and movements of the body or its parts. They occur in muscles, tendons, and joint capsules.

3. By the distribution of receptors in the body. There are two broad classes of senses:
   - **General (somatosensory, somesthetic) senses** employ widely distributed receptors in the skin, muscles, tendons, joints, and viscera. These include touch, pressure, stretch, heat, cold, and pain, as well as many stimuli that we do not perceive consciously, such as blood pressure and composition. Their receptors are relatively simple—sometimes nothing more than a bare dendrite.
   - **Special senses** are limited to the head, are innervated by the cranial nerves, and employ relatively complex sense organs. The special senses are vision, hearing, equilibrium, taste, and smell.

BEFORE YOU GO ON

Answer the following questions to test your understanding of the preceding section:

1. Not every sensory *receptor* is a sense *organ*. Explain.

2. What does it mean to say sense organs are transducers? What form of energy do all receptors have as their output?

3. Not every sensory signal results in conscious awareness of a stimulus. Explain.

4. What is meant by the *modality* of a stimulus? Give some examples.

5. Three schemes of receptor classification were presented in this section. In each scheme, how would you classify the receptors for a full bladder? How would you classify taste receptors?

6. Nociceptors are tonic rather than phasic receptors. Speculate on why this is beneficial to homeostasis.

16.2 The General Senses

Expected Learning Outcomes

When you have completed this section, you should be able to

a. list several types of somatosensory receptors;

b. describe the projection pathways for the general senses; and

c. describe the types and mechanisms of pain, and explain how the brain modulates one's sensitivity to pain.

Receptors for the general senses are relatively simple in structure and physiology. They consist of one or a few sensory nerve fibers and, usually, a sparse amount of connective tissue. These receptors are shown in **figure 16.2** and described in **table 16.1.**

---

[1]*noci* = pain

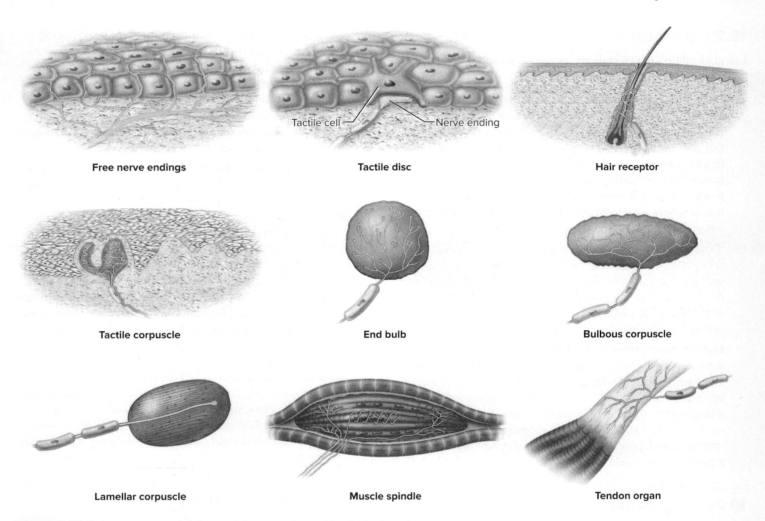

FIGURE 16.2 Receptors of the General Senses. See table 16.1 for functions.

TABLE 16.1	Receptors of the General Senses	
Receptor Type	**Locations**	**Modality**
Unencapsulated Endings		
Free nerve endings	Widespread, especially in epithelia and connective tissues	Pain, heat, cold
Tactile discs	Stratum basale of epidermis	Light touch, pressure
Hair receptors	Around hair follicle	Light touch, movement of hairs
Encapsulated Nerve Endings		
Tactile corpuscles	Dermal papillae of fingertips, palms, eyelids, lips, tongue, nipples, and genitals	Light touch, texture
End bulbs	Mucous membranes	Similar to tactile corpuscles
Bulbous corpuscles	Dermis, subcutaneous tissue, and joint capsules	Heavy continuous touch or pressure; joint movements
Lamellar corpuscles	Dermis, joint capsules, periosteum, breasts, genitals, and some viscera	Deep pressure, stretch, tickle, vibration
Muscle spindles	Skeletal muscles near tendon	Muscle stretch (proprioception)
Tendon organs	Tendons	Tension on tendons (proprioception)

16.2a Unencapsulated Nerve Endings

Unencapsulated nerve endings are dendrites with no connective tissue wrapping. They include the following:

- **Free nerve endings** include *warm receptors,* which respond to rising temperature; *cold receptors,* which respond to falling temperature; and *nociceptors* for pain. They are bare dendrites that have no special association with specific accessory cells or tissues. They are most abundant in the skin and mucous membranes. You can locate some of your cold receptors by gently touching your skin with the point of a graphite pencil, which conducts heat away from the skin. In spots where these receptors are located, this produces a sensation of cold.

- **Tactile discs** are tonic receptors for light touch, thought to sense textures, edges, and shapes. They are flattened nerve endings that terminate adjacent to specialized *tactile cells* in the basal layer of the epidermis. They respond to compression of the skin, which releases a chemical from the tactile cell that excites the associated nerve fiber.

- **Hair receptors (root hair plexuses)** are dendrites that coil around a hair follicle and respond to movements of the hair. They are stimulated when, for example, an ant walks across one's skin, bending one hair after another. However, they adapt quickly, so we aren't constantly irritated by the feel of clothing against the skin. Hair receptors are particularly important in the eyelashes, where the slightest touch evokes a protective blink reflex.

16.2b Encapsulated Nerve Endings

Encapsulated nerve endings are nerve fibers wrapped in glial cells or connective tissue. Most of them are mechanoreceptors for touch, pressure, and stretch. The connective tissue either enhances the sensitivity of the nerve fiber or makes it more selective with respect to which modality it responds to. The principal encapsulated nerve endings are as follows:

- **Tactile corpuscles** are phasic receptors for light touch and texture. They are tall, ovoid to pear-shaped, and consist of two or three nerve fibers meandering upward through a fluid-filled capsule of flattened Schwann cells. They are mechanically linked to the edges of the dermal papillae of the skin. Tactile corpuscles are especially concentrated in sensitive hairless areas such as the fingertips, palms, eyelids, lips, nipples, and genitals. Drag your fingernail lightly across the back of your hand, and then across your palm. The difference in sensation you feel is due to the high concentration of tactile corpuscles in the palmar skin. Tactile corpuscles enable you to tell the difference between silk and sandpaper, for example, by light strokes of your fingertips.

- **End bulbs** are functionally similar to tactile corpuscles, but instead of occurring in the skin, they are found in the mucous membranes of the lips and tongue, in the conjunctiva on the anterior surface of the eye, and in the epineurium of large nerves. They are ovoid bodies composed of a connective tissue sheath around a sensory nerve fiber.

- **Bulbous corpuscles** are tonic receptors for heavy touch, pressure, stretching of the skin, deformation of the fingernails, and joint movements. They aid in our perception of the shapes of objects held in the fingers. They are flattened, elongated capsules containing a few nerve fibers and are located in the dermis, subcutaneous tissue, and joint capsules. They connect the subcutaneous tissues to skin folds in the joints and palms and beneath the fingernails.

- **Lamellar (pacinian[2]) corpuscles** are phasic receptors chiefly for vibration. They are large ovoid receptors, up to 1 or 2 mm long, found in the periosteum of bones; in joint capsules; in the pancreas and some other viscera; and deep in the dermis, especially on the hands, feet, breasts, and genitals. They consist of a single dendrite running through the core of the organ, encapsulated by multiple concentric layers of cells. The innermost capsule layers are flattened Schwann cells, but the greater bulk of the capsule consists of concentric layers of fibroblasts with narrow fluid-filled spaces between them. In cross section, a lamellar corpuscle looks like an onion slice with a dot, the dendrite, in the middle. Lamellar corpuscles are phasic receptors that respond with one or two action potentials when pressure is applied to them and again when the pressure is taken away, but not in between. They are usually considered to function as vibration receptors, stimulated by such actions as stroking a surface with the fingertips. The friction ridges of the fingertips enhance the vibration and sensitivity of the lamellar corpuscles.

The other two encapsulated nerve endings in figure 16.2—muscle spindles and tendon organs—have already been discussed in relation to muscle reflexes in section 13.3.

16.2c Somatosensory Projection Pathways

Sensory projection is the transmission of information from a receptor, or a receptive field, to a specific locality in the cerebral cortex, enabling the brain to detect and identify the stimulus. The identifiable origin can be a small and specific area within a receptor, such as a small area of the retina or the skin. The pathways followed by sensory signals to their ultimate destinations in the CNS are called **projection pathways.** From the receptor to the final destination in the brain, most somatosensory signals travel by way of three neurons called the **first-, second-,** and **third-order neurons.** Their axons are called first- through third-order nerve fibers. The first-order fibers for touch, pressure, and proprioception are large, myelinated, and fast; those for heat and cold are small, unmyelinated or lightly myelinated, and slower.

Somatosensory signals from the head, such as facial sensations, travel by way of several cranial nerves (especially CN V, the trigeminal nerve) to the pons and medulla oblongata. In the brainstem, the first-order fibers of these neurons synapse with second-order neurons that decussate and lead to the contralateral thalamus. Third-order neurons then complete the projection pathway to the cerebrum. Proprioceptive signals from the head are an exception, as the second-order fibers carry these signals to the cerebellum.

Below the head, the first-order fibers enter the posterior horn of the spinal cord. Signals ascend the cord in the spinothalamic and

---

[2]Filippo Pacini (1812–83), Italian anatomist

other pathways detailed in table 13.1 and figure 13.6. These pathways decussate either at or near the point of entry into the cord, or in the brainstem, so the primary somatosensory cortex in each cerebral hemisphere receives signals from the contralateral side of the body.

Signals for proprioception below the head travel up the spinocerebellar tracts to the cerebellum. Signals from the thoracic and abdominal viscera travel to the medulla oblongata by way of sensory fibers in the vagus nerve (CN X). Visceral pain signals also ascend the spinal cord in the gracile fasciculus.

16.2d Pain

From a clinical standpoint, the sense of pain arguably deserves the most attention of all the senses we cover in this chapter, for it is the number one reason people seek medical attention. It is also one of the most difficult problems to treat.

Pain can be defined as an uncomfortable perception of tissue injury or noxious stimulation. It is not a disease in itself, but a symptom of an underlying condition. Its function is to motivate one to avoid, minimize, or escape danger—for example, by favoring a sprained ankle, withdrawing one's hand from a burn, or avoiding things that have caused pain in the past. Few of us enjoy pain and we may wish that no such thing existed, but it is an adaptive and necessary sensation—we'd be far worse off without it. We see evidence of this in such diseases as leprosy and diabetes mellitus, where nerve damage *(peripheral neuropathy)* can deaden the sense of pain and make a person unaware of minor injuries or not care enough about them. The resulting neglect can allow injuries to become gangrenous and cost people their fingers, toes, feet, or entire limbs.

Types of Pain

There are two clinical categories of pain—*nociceptive and neuropathic.* **Nociceptive pain** stems from tissue injury such as cuts, burns, and chemical irritation, detected by the nerve endings called nociceptors. These are especially abundant in the skin and mucous membranes and occur in nearly all organs, although not in the brain or liver. In some brain surgery, the patient must remain conscious and able to talk with the surgeon; such patients need only local anesthesia. Nociceptors occur in the meninges, however, and play an important role in headaches. Nociceptive pain is associated with tissue inflammation and responds to anti-inflammatory drugs such as aspirin, ibuprofen, and naprosyn.

Nociceptive pain is subdivided into three kinds according to origin:

1. **Visceral pain,** which arises from the internal organs of the body cavities. This type of pain is diffuse, dull, and hard to locate. Among its causes are stretch (such as an overfull bladder or stomach), chemical irritation (as in intestinal cramps and acid reflux), and ischemia (deficient blood flow, as in heart attacks and menstrual cramps). It is often associated with sensations of squeezing, cramping, and nausea.

2. **Deep somatic pain,** which arises from bones, joints, muscles, and related sources. Examples include the pain of arthritis, sprains, and bone fractures. It often results from excessive stretch (as in an ankle sprain or pulled muscle), but can have other causes.

3. **Superficial somatic pain,** which usually arises from the skin. Examples include the pain of cuts, burns, and insect stings.

Neuropathic pain stems from injuries to the nerves, spinal cord, meninges, or brain. It is often characterized by stabbing, burning, tingling, or "electrical" sensations. Examples include headache; phantom limb pain; fibromyalgia; and the pain of multiple sclerosis, postherpetic neuralgia (see Deeper Insight 13.3), and cancer (from tumor invasion of nerves or neurotoxic effects of chemotherapy).

Two types of nerve fibers mediate pain responses and contrasting qualities of pain called *fast pain* and *slow pain.* **Fast pain** is the immediate, sharp pain one feels from a cut, burn, or bee sting, for example. It is mediated by myelinated *A-delta (Aδ)* nerve fibers. These conduct signals at speeds of 12 to 30 m/s, so the pain is felt quickly upon injury. It is a sharp, localized, often stabbing feel. It is described as *discriminative pain* because we can usually tell exactly where it is coming from. **Slow pain** originates in small unmyelinated *type C* nerve fibers, which conduct at only 0.5 to 2.0 m/s. It has a delayed and more burning or aching feel, and it's harder to tell exactly where the pain is coming from. If you bang your shin on a piece of furniture, you may feel both of these: a sharp fast pain immediately, then after a few seconds, an even more intense and longer-lasting slow pain.

Projection Pathways for Pain

It is notoriously difficult for clinicians to locate the origin of a patient's pain, because pain travels by diverse and complex routes and the sensation can originate anywhere along the way. Pain signals reach the brain by two main pathways:

1. Signals from the head travel to the brainstem by way of four cranial nerves: mainly the trigeminal (CN V), but also the facial (CN VII), glossopharyngeal (CN IX), and vagus (CN X) nerves. Trigeminal fibers enter the pons and descend to synapses in the medulla. Pain fibers of the other three cranial nerves lead directly to the medulla. Second-order neurons arise in the medulla and ascend to the thalamus, which relays the message to the cerebral cortex, as detailed shortly.

2. Pain signals from the neck and below travel by way of three ascending spinal cord tracts: the spinothalamic tract, spinoreticular tract, and gracile fasciculus (**fig. 16.3;** see also section 13.1, fig. 13.6). The spinothalamic tract carries most of the somatic pain signals that ultimately reach the cortex and make us conscious of pain. The spinoreticular tract carries signals to the reticular formation of the brainstem, and these are ultimately relayed to the hypothalamus and limbic system. Here, they activate visceral, emotional, and behavioral reactions such as nausea, fear, and some reflexes. The gracile fasciculus carries signals to the thalamus for visceral pain, such as the pain of a stomachache or from passing a kidney stone.

When the thalamus receives pain signals from these sources, it relays most of them through third-order neurons to their final destination in the postcentral gyrus of the cerebrum. This is the site of discriminative pain perception—judgment of the location of injury and the quality and intensity of the sensation. Exactly what part of this gyrus receives the signals depends on where the pain originated, as described by the concept of somatotopy in section 14.5g (see fig. 14.21). Most of this gyrus is somatosensory. A region of the gyrus deep within the lateral sulcus of the brain, however, is a viscerosensory area, which receives the visceral signals conveyed by

FIGURE 16.3 Projection Pathways for Pain. A first-order neuron conducts a pain signal to the posterior horn of the spinal cord, a second-order neuron conducts it to the thalamus, and a third-order neuron conducts it to the cerebral cortex. Signals from the spinothalamic tract pass through the thalamus. Signals from the spinoreticular tract bypass the thalamus on the way to the sensory cortex.

the gracile fasciculus. The emotional aspects of pain register in the hypothalamus and the prefrontal cortex, limbic system, insula, and anterior cingulate gyrus of the cerebrum (see figs. 14.16, 14.17).

Pain in the viscera is often mistakenly thought to come from the skin or other superficial sites—for example, the pain of a heart attack is felt "radiating" along the left shoulder and medial side of the arm **(fig. 16.4)**. This phenomenon, called **referred pain,** results from the convergence of neural pathways in the CNS. In the case of cardiac pain, for example, spinal cord segments T1 to T5 receive input from the heart as well as from the chest and arm. Pain fibers from the heart and skin in this region converge and follow the same pathway from here to the thalamus and cerebral cortex. The brain can't distinguish which source the arriving signals are coming from. It acts as if it assumes they are most likely from the skin, since skin has more pain receptors than the heart and suffers injury more often. Knowledge of the origins of referred pain is important for the diagnosis of organ dysfunctions.

CNS Modulation of Pain

Regardless of expressions such as "pain in the joints" and "muscle pain," it must be realized that pain doesn't exist in these peripheral locations, but only in the brain; it is a conscious *perception* of signals from other organs. Because of this, it is subjective, highly variable, and influenced by a person's physical and mental state—for example, one's expectation of pain, fixation on it, or conversely, achieving pain relief by redirecting one's attention to distractors. This is why such techniques as acupuncture, meditation, hypnosis, placebos, exercise therapy, and Lamaze childbirth can effectively lessen pain.

Some examples show how pain originates in cerebral perception and not in the outlying organs. **Fibromyalgia** is a syndrome of sometimes excruciating, even disabling pain that seems to come from the muscles or bones; yet those distant organs are sending no abnormal nerve signals to the brain. The pain lies entirely in the brain's abnormal responses to normal incoming signals.

(a)   (b)

FIGURE 16.4 Referred Pain. (a) Areas of referred pain. Pain in these areas of the skin is often a symptom of problems in the indicated viscera. (b) The basis of referred pain in a heart attack.

Phantom limb pain is an eerie sensation of pain, sometimes intense, coming from a limb that has been lost by trauma or amputation. It is commonly accompanied by nonpainful *phantom limb sensations* such as the bulk and weight of the absent limb; or feelings of touch, itching, tingling, pressure, heat, or cold; or the illusion that one can even control the limb. Phantom pain can also occur when other body parts have been removed, such as after a tooth extraction or mastectomy. The neurological mechanism of phantom limb pain isn't known with certainty, and it doesn't respond to conventional pain medications. One might think that cutting nerves or even the spinal cord along pain pathways could eliminate the pain, but it doesn't. Battlefield trauma, in contrast to these other cases, shows strikingly how the brain can powerfully suppress the painful awareness of severe injury—many mortally wounded soldiers have reported little or no pain even in their dying hour. So as one neurologist has put it, the bane of pain lies mainly in the brain.

The central nervous system has **analgesic**[3] (pain-relieving) mechanisms related to the long-known analgesic effects of opium, morphine, and heroin. In 1974, neurophysiologists discovered receptor sites in the brain for these drugs. Since these opiates don't occur naturally in the body, physiologists wondered what normally binds to such receptors. They soon found two analgesic oligopeptides with 200 times the potency of morphine, and named them

enkephalins.[4] Larger analgesic neuropeptides, the **endorphins**[5] and **dynorphins**,[6] were discovered later. All three are known as **endogenous opioids** (which means "internally produced opium-like substances").

These opioids are secreted by the CNS, pituitary gland, digestive tract, and other organs in states of stress or exercise. In the CNS, they are found especially in the central gray substance of the midbrain (see fig. 14.9a) and posterior horn of the spinal cord. They are *neuromodulators* that can block the transmission of pain signals and produce feelings of pleasure and euphoria. Their secretion rises sharply in women giving birth, and they may even be responsible for the aforementioned battlefield reports. However, efforts to employ them in pain therapy have been disappointing.

How do these opioids block pain? For pain to be perceived, signals from the nociceptors must get beyond the posterior horn of the spinal cord and travel to the brain. Through mechanisms called **spinal gating,** pain signals can be stopped at the posterior horn. Two of these mechanisms are described here.

One of these involves **descending analgesic fibers**—nerve fibers that arise in the brainstem, travel down the spinal cord in the reticulospinal tract, and block pain signals from traveling to the brain. **Figure 16.5** depicts one relatively simple mechanism

---

[3]*an* = without; *alges* = pain

[4]*en* = within; *kephal* = head
[5]acronym from *endogenous morphine*
[6]*dyn* = pain

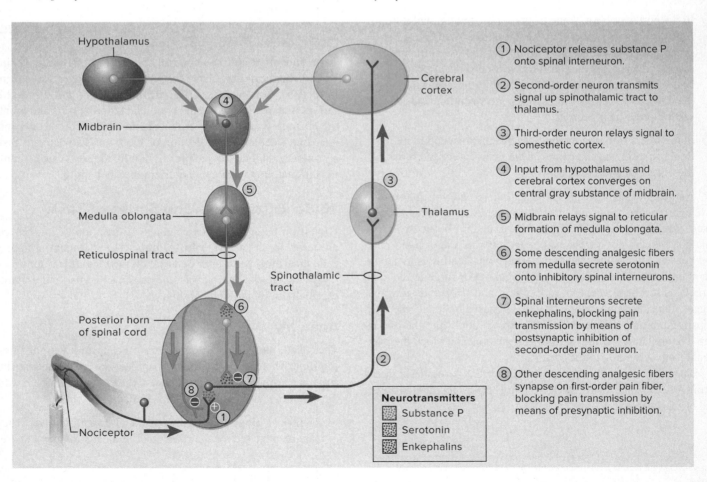

1. Nociceptor releases substance P onto spinal interneuron.

2. Second-order neuron transmits signal up spinothalamic tract to thalamus.

3. Third-order neuron relays signal to somesthetic cortex.

4. Input from hypothalamus and cerebral cortex converges on central gray substance of midbrain.

5. Midbrain relays signal to reticular formation of medulla oblongata.

6. Some descending analgesic fibers from medulla secrete serotonin onto inhibitory spinal interneurons.

7. Spinal interneurons secrete enkephalins, blocking pain transmission by means of postsynaptic inhibition of second-order pain neuron.

8. Other descending analgesic fibers synapse on first-order pain fiber, blocking pain transmission by means of presynaptic inhibition.

Neurotransmitters
- Substance P
- Serotonin
- Enkephalins

FIGURE 16.5 Spinal Gating of Pain Signals.

of pain modulation in the posterior horn. The normal route of pain transmission is indicated by the red arrows and steps 1 through 3:

① A nociceptor stimulates a second-order nerve fiber. The neurotransmitter at this synapse is called **substance P** (think *P* for "pain"[7]).

② The second-order nerve fiber transmits signals up the spinothalamic tract to the thalamus.

③ The thalamus relays the signals through a third-order neuron to the cerebral cortex, where one becomes conscious of pain.

The pathway for pain modulation is indicated by descending green arrows and steps 4 through 8:

④ Signals from the hypothalamus and cerebral cortex feed into the central gray substance of the midbrain, allowing both autonomic and conscious effects on pain perception.

⑤ The midbrain relays signals to certain nuclei in the reticular formation of the medulla oblongata.

⑥ The medulla issues descending, serotonin-secreting analgesic fibers to the spinal cord. These fibers travel the reticulospinal tract and end in the posterior horn at all levels of the cord.

⑦ In the posterior horn, some of these descending analgesic fibers synapse on short spinal interneurons, which in turn synapse on the second-order pain fiber. These interneurons secrete enkephalins to inhibit the second-order neuron. This is an example of postsynaptic inhibition, working on the downstream side of the synapse between the first- and second-order pain neurons.

⑧ Some fibers from the medulla also exert presynaptic inhibition, synapsing on the axons of the nociceptors and blocking the release of substance P.

Another mechanism of spinal gating is one you may often have consciously employed without knowing why it worked. Have you ever banged your elbow on the edge of a table, burned yourself on a hot cooking pot, or pinched your finger in a door and found that you could ease the pain by shaking your hand or massaging the injured area? This works because shaking or rubbing the area stimulates mechanoreceptors in the skin and deeper tissues and trigger signals in large type A myelinated nerve fibers. These lead to pain-inhibiting interneurons in the spinal cord, like the one at step 7, stimulating them to secrete enkephalins and block the pain signal from ascending to the brain.

The clinical management of pain has had a particularly interesting history, some of which is retold in Deeper Insight 16.4 at the end of this chapter.

BEFORE YOU GO ON

Answer the following questions to test your understanding of the preceding section:

7. What stimulus modalities are detected by free nerve endings?

8. Name any four encapsulated nerve endings and identify their stimulus modalities.

9. Where do most second-order somatosensory neurons synapse with third-order neurons?

10. Distinguish between nociceptive and neuropathic pain, and describe the three categories of nociceptive pain.

11. Outline the pathways that pain signals take to and within the brain.

12. Describe two ways that the CNS can modify a person's perception of incoming pain signals.

16.3   The Chemical Senses

Expected Learning Outcomes

When you have completed this section, you should be able to

a. explain how taste and smell receptors are stimulated; and

b. describe the receptors and projection pathways for these two senses.

Taste and smell are the chemical senses. In both cases, environmental chemicals stimulate sensory cells and trigger nerve signals in certain cranial nerves. Other chemoreceptors, not discussed in this section, are located in the brain and blood vessels and monitor the chemistry of the body fluids. Taste and smell not only influence our acceptance or rejection of food, but also activate gastric secretion and blood flow in preparation for digestion, and stimulate insulin secretion to promote nutrient metabolism.

16.3a Gustation—The Sense of Taste

Gustation begins with the chemical stimulation of sensory cells clustered in about 4,000 **taste buds.** The chemical stimuli are called **tastants.** Most taste buds are on the tongue, but some occur inside the cheeks and on the soft palate, pharynx, and epiglottis, especially in infants and children.

Anatomy

The visible bumps on the tongue are not taste buds but **lingual papillae.** There are four types of papillae—one without taste buds and three with taste buds just beneath the epithelial surface, where they are not visible to the eye (**fig. 16.6a**):

1. **Filiform**[8] **papillae** are tiny spikes without taste buds. They are responsible for the rough feel of a cat's tongue and are

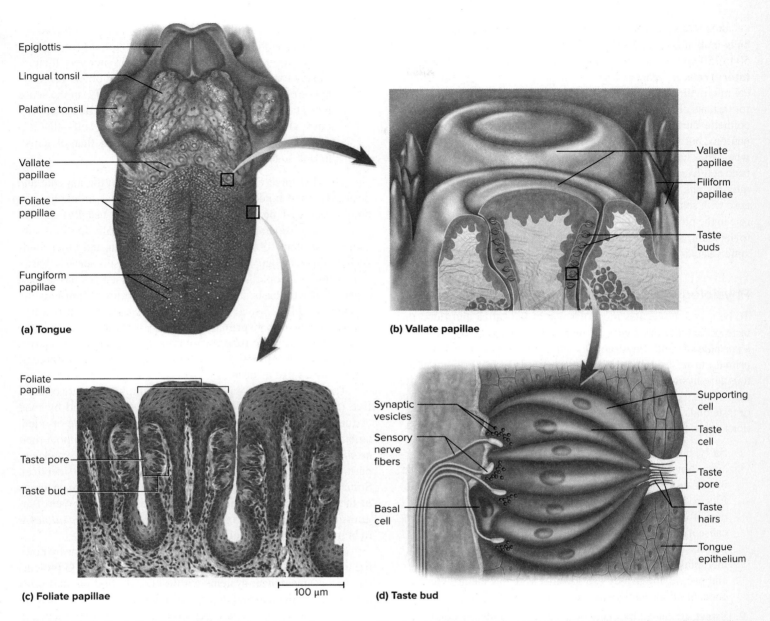

(a) Tongue

(b) Vallate papillae

(c) Foliate papillae

100 μm

(d) Taste bud

FIGURE 16.6 Gustatory (Taste) Receptors. (a) Superior surface of the tongue and locations of its papillae. (b) Detail of the vallate papillae (compare this chapter's opening photo). (c) Taste buds on the walls of two adjacent foliate papillae. (d) Structure of a taste bud. **APR**

c: Jose Luis Calvo/Shutterstock.com

important to many mammals for grooming the fur. They are the most abundant papillae on the human tongue, but they are small. They are innervated, however, and serve in one's sense of the texture of food, called *mouthfeel*.

2. **Foliate[9] papillae** are also weakly developed in humans. They form parallel ridges on the sides of the tongue about two-thirds of the way back from the tip, adjacent to the molar and premolar teeth, where most chewing occurs and most flavor chemicals are released from the food. Most of their taste buds degenerate by the age of 2 or 3 years. This may partially explain why children so often reject foods that are tolerated or enjoyed by adults.

3. **Fungiform[10] papillae** (FUN-jih-form) are shaped somewhat like mushrooms. Each has about three taste buds, located mainly on the apex. These papillae are widely distributed, but especially concentrated at the tip and sides of the tongue. These respond not only to taste but, like filiform papillae, also to food texture.

4. **Vallate[11] papillae** are large papillae arranged in a V at the rear of the tongue. Each is surrounded by a deep circular trench. There are only 7 to 12 vallate papillae, but they contain up to half of all taste buds—around 250 each, located on the walls of the papilla facing the trench **(fig. 16.6b).**

[9]*foli* = leaf; *ate* = like

[10]*fungi* = mushroom; *form* = shaped
[11]*vall* = wall; *ate* = like, possessing

Regardless of location and sensory specialization, all taste buds look alike **(fig. 16.6c, d).** They are lemon-shaped groups of 50 to 150 *taste cells, supporting cells,* and *basal cells.* **Taste (gustatory) cells** are more or less banana-shaped and have a tuft of apical microvilli called **taste hairs,** which serve as receptor surfaces for tastants. The hairs project into a pit called a **taste pore** on the epithelial surface of the tongue. Taste cells are epithelial cells, not neurons, but they synapse with sensory nerve fibers at their base, where they have synaptic vesicles and release neurotransmitters. A taste cell lives for only 7 to 10 days. **Basal cells** are stem cells that multiply and replace taste cells that have died, but they also synapse with sensory nerve fibers of the taste bud and may play some role in the processing of sensory information before the signal goes to the brain. **Supporting cells** resemble taste cells but have no synaptic vesicles or sensory role.

Physiology

To be tasted, molecules must dissolve in the saliva and flood the taste pore. On a dry tongue, sugar and salt have as little taste as a sprinkle of sand. Physiologists currently recognize at least five primary taste sensations. All of these are detected throughout the tongue; physiologists have abandoned the old concept of a taste map of the tongue that supposedly showed regional specialization of taste buds for different sensations. The five primary taste sensations are as follows:

1. **Salty,** produced by metal ions such as sodium and potassium. Since salts are vital electrolytes, there is obvious value in our ability to taste them and in having an appetite for salt. Electrolyte intake is also important in the osmotic regulation of the body's fluid balance. Electrolyte deficiencies can cause a craving for salt; many animals as diverse as insects, parrots, deer, and elephants seek salt deposits ranging from mud puddles and wet clay to the sweat on larger animals and people. Pregnancy can lower a woman's electrolyte concentrations and create a craving for salty food.

2. **Sweet,** produced by many organic compounds, especially sugars. Sweetness is associated with carbohydrates and foods of high caloric value. Many flowering plants have evolved sweet nectar and fruits that entice animals to eat them and disperse their pollen and seeds. Our fondness for fruit and sugar has coevolved with plant reproductive strategies.

3. **Umami** is a "meaty" taste produced by amino acids such as aspartic and glutamic acids—the savory taste of beef or chicken broth, cheese, and mushrooms. Pronounced "ooh-mommy," the word is Japanese slang for "delicious" or "yummy." The umami taste motivates protein intake.

4. **Sour,** usually associated with acids in such foods as citrus fruits. Sour is usually an aversive taste that helps one avoid spoiled foods and overloading the body's acid–base balance.

5. **Bitter,** associated with spoiled foods and alkaloids such as nicotine, caffeine, quinine, and morphine. Alkaloids are often poisonous, and the bitter taste sensation usually induces a human or animal to reject a food. While flowering

plants make their fruits temptingly sweet, they often load their leaves with bitter alkaloids to deter animals from eating them. However, many bitter compounds are not very toxic, and many toxic compounds in nature are not bitter. There is little correlation between bitterness and toxicity, so the sense of bitter taste is suspected of having other functions not yet discovered. The threshold for bitter is lowest of all—that is, we can taste lower concentrations of alkaloids than of acids, salts, and sugars.

The first three of these sensations reinforce the appetite and desire for three fundamental classes of nutrients—minerals, carbohydrates, and proteins. Sour and bitter are usually aversive stimuli that discourage the ingestion of potentially harmful substances, but many of us cultivate a taste for sour and bitter foods such as lemons, yogurt, turnip greens, and Brussels sprouts. Variations between people in the taste for such substances has at least a partial genetic basis. Two important categories of nutrients not included in this classical list are fats and water. Recent research suggests that two more primary tastes may warrant inclusion in the list, as they each have their own unique oropharyngeal receptors: **oleogustus,**[12] the proposed name for the taste of fats, and **water** (if one considers that a "taste").

The many flavors we perceive aren't simply a mixture of these five (or seven) primary tastes, but are also influenced by food texture, aroma, temperature, appearance, and one's state of mind, among other things. Many flavors depend on smell; without their aromas, cinnamon merely has a faintly sweet taste, coffee and peppermint are bitter, and apples and onions taste almost identical. Some flavors such as pepper are due to stimulation of free endings of the trigeminal nerve rather than taste buds. Filiform and fungiform papillae of the tongue are innervated by the *lingual nerve* (a branch of the trigeminal) and are sensitive to texture.

Sugars, alkaloids, and glutamate stimulate taste cells by binding to receptors on the cell surface, which then activate G proteins and second-messenger systems within the cell. Sodium and acids penetrate into the cell and depolarize it directly. By either mechanism, taste cells then release neurotransmitters that stimulate the sensory dendrites at their base.

Projection Pathways

The facial nerve (CN VII) collects sensory information from taste buds of the anterior two-thirds of the tongue, the glossopharyngeal nerve (CN IX) from the posterior one-third, and the vagus nerve (CN X) from taste buds of the palate, pharynx, and epiglottis. All taste fibers project to a site in the medulla oblongata called the *solitary nucleus.* Second-order neurons arise here and relay the signals to two destinations: (1) nuclei in the hypothalamus and amygdala that activate autonomic reflexes such as salivation, gagging, and vomiting, and that function in emotions and memory associated with taste; and (2) the thalamus, which relays signals to the primary gustatory cortex, located in the insula and roof of the lateral sulcus of the cerebrum, where we become conscious of the taste. Processed signals are further relayed to the orbitofrontal cortex (see fig. 14.16), where they

---

[12]*oleo* = oil; *gust* = taste

FIGURE 16.7 Olfactory (Smell) Receptors. (a) Location of the olfactory mucosa in the nasal cavity, and the olfactory bulb of the brain. (b) Neural pathways from the olfactory mucosa to the olfactory tract of the brain. (c) Detail of an olfactory cell.

(a)

Olfactory tract
Olfactory bulb
Olfactory nerve fascicle
Olfactory mucosa (reflected)

Olfactory bulb
Granule cell
Olfactory tract
Mitral cell
Tufted cell
Glomerulus
Olfactory nerve fascicle
Cribriform plate of ethmoid bone
Basal cell
Supporting cells
Olfactory cell
Olfactory gland
Olfactory hairs
Mucus
Odor molecules
Airflow

(b)

Axon
Neurosoma
Dendrite
Olfactory hairs
(c)

are integrated with signals from the nose and eyes and we form an overall impression of the flavor and palatability of food.

16.3b Olfaction—The Sense of Smell

Olfaction, the sense of smell, is a response to airborne chemicals called **odorants.** These are detected by receptor cells in a patch of epithelium, the **olfactory mucosa,** in the roof of the nasal cavity

(fig. 16.7). This location places the olfactory cells close to the brain, but it isn't as well ventilated as the lower nasal cavity; forcible sniffing is often needed to identify an odor or locate its source.

Anatomy

The olfactory mucosa covers about 5 cm² of the superior concha, cribriform plate, and nasal septum of each nasal fossa. It consists of 10 to 20 million **olfactory cells** as well as epithelial supporting cells and basal stem cells. The rest of the nasal cavity is lined by a nonsensory *respiratory mucosa.*

Unlike taste cells, which are epithelial, olfactory cells are neurons. They are shaped a little like bowling pins (fig. 16.7c). The widest part, the neurosoma, contains the nucleus. The neck and head of the cell are a modified dendrite with a swollen tip. The head bears 10 to 20 cilia called **olfactory hairs.** These cilia are immobile, but they have binding sites for odor molecules. They lie in a tangled mass embedded in a thin layer of mucus. The basal end of each cell tapers to become an axon. These axons collect into small fascicles that leave the nasal cavity through pores *(cribriform foramina)* in the ethmoid bone. Collectively, the fascicles are regarded as the olfactory nerve (CN I).

Olfactory cells are the only neurons directly exposed to the external environment. Apparently this is hard on them, because they have a life span of only 60 days. Unlike most neurons, however, they are replaceable. The basal cells continually divide and differentiate into new olfactory cells.

Physiology

Humans do not, as commonly supposed, have a poorer sense of smell than other mammals; that notion arose from certain unsubstantiated nineteenth-century assumptions of Paul Broca and Sigmund Freud and suffered from confirmation bias in twentieth-century and even more recent research. We have only about one-third as many functional olfactory genes as rodents do, but we have more complex olfactory bulbs and orbitofrontal cortex—the brain regions where odor signals are interpreted. Humans outperform rodents and dogs for some odors and underperform them for others. Our sense of smell is much more sensitive than our sense of taste; we can detect odor concentrations as low as a few parts per trillion.

Humans have about 400 types of odor receptors; each olfactory cell has only one receptor type and therefore binds only one odorant. With various combinations of input, however, most people can distinguish 2,000 to 4,000 different odors; some, such as world-class food and wine experts, can distinguish up to 10,000. Attempts to group these into odor classes have been inconclusive and controversial; the names for some suggested classes have included *pungent, floral, musky,* and *earthy.* Our olfactory sense is very subject to "top-down" influences of emotional and cognitive states, so the same odorant and concentration may smell different to us under different circumstances. On average, women are more sensitive to odors than men are, and they are measurably more sensitive to some odors near the time of ovulation as opposed to other phases of the menstrual cycle. Human olfactory communication is now known to play a role in family relationships, stress and anxiety levels, and responses to the reproductive status of others, even when we're not consciously aware of this information.

The first step in smell is that an odorant molecule must bind to a receptor on one of the olfactory hairs. Hydrophilic odorants diffuse freely through the mucus of the olfactory epithelium and bind directly to a receptor. Hydrophobic odorants are transported to the receptor by an *odorant-binding protein* in the mucus. When the receptor binds an odorant, it activates a G protein and, through it, the cyclic adenosine monophosphate (cAMP) second-messenger system.

The cAMP system ultimately opens ion channels in the plasma membrane, admitting cations (Na^+ or Ca^{2+}) into the cell and depolarizing it, creating a receptor potential. This triggers action potentials in the axon of the olfactory cell, and a signal is transmitted to the brain.

Some odorants, however, act on nociceptors of the trigeminal nerve rather than on olfactory cells. These include ammonia, menthol, chlorine, and the capsaicin of hot peppers. "Smelling salts" revive unconscious persons by stimulating the trigeminal nerve with ammonia fumes.

Projection Pathways

When olfactory fibers pass through the roof of the nose, they enter a pair of **olfactory bulbs** beneath the frontal lobes of the brain **(fig. 16.8).** Here they synapse with the dendrites of neurons called *mitral cells* and *tufted cells.* Olfactory cell axons reach up and mitral and tufted cell dendrites reach down to meet each other in spheroidal clusters called *glomeruli* (fig. 16.7b). All olfactory fibers leading to any one glomerulus come from cells with the same receptor type; thus each glomerulus is dedicated to a particular type of odor. Higher brain centers interpret complex odors such as chocolate, wine, perfume, or coffee by decoding signals from a combination of odor-specific glomeruli. This is similar to the way our visual system decodes all the colors of the spectrum using input from just three color-specific receptor cells of the eye.

The tufted and mitral cells carry output from the glomeruli. Their axons form bundles called **olfactory tracts,** which run caudally along the underside of the frontal lobes. Most fibers of the olfactory tracts end in various regions of the inferior surface of the temporal lobe regarded as the **primary olfactory cortex.** This is where one first becomes conscious of the odor. It is noteworthy that olfactory signals can reach the cerebral cortex directly, without first passing through the thalamus; thus, this is an extrathalamic[13] pathway. Except for the spinoreticular pain pathway (see fig. 16.3), no other sensory signals bypass the thalamus. Even in olfaction, however, some signals from the primary olfactory cortex continue to a relay in the thalamus on their way to olfactory association areas elsewhere. The primary olfactory cortex of each cerebral hemisphere relays signals to the contralateral hemisphere, so all further processing is mirrored on both sides of the brain even if an odor is initially detected in only one nasal fossa.

On both sides, signals are relayed to the amygdala, hippocampus, insula, and hypothalamus, which interact in complex ways to associate a present odor with olfactory memories and emotional responses. Thus, the odor of certain foods, a perfume, a hospital, or decaying flesh can evoke strong memories, emotional responses, and visceral reactions such as sneezing or coughing, the secretion of saliva and stomach acid, or vomiting. The orbitofrontal cortex, just above the eyes, also receives signals from the primary olfactory cortex. Here is where we identify and discriminate among odors. This region also integrates odor, taste, and vision into an overall impression of the desirability or acceptability of food.

Most areas of olfactory cortex also send fibers back to the olfactory bulbs by way of neurons called *granule cells.* Granule cells

---

[13]*extra* = outside; *thalam* = chamber, thalamus

1. Olfactory signals are received in olfactory bulbs.

2. Signals are relayed caudally in olfactory tracts.

3. Primary olfactory cortex in temporal lobe creates conscious perception of odor and relays signals to other brain destinations.

4. Signals from each temporal lobe are relayed to contralateral temporal lobe, so all processing is mirrored in both cerebral hemispheres.

5. Signals relayed to amygdala, hippocampus, insula, and hypothalamus relate odor to olfactory memory and produce emotional and visceral responses to odor.

6. Signals received in orbitofrontal cortex combine with other information to identify and discriminate among odors and integrate odor into sense of flavor.

7. Feedback to olfactory bulb modulates perception of odor according to circumstances such as hunger versus satiety.

Orbitofrontal cortex
Olfactory bulb
Olfactory tract
Insula
Amygdala
Hypothalamus
Primary olfactory cortex
Hippocampus

FIGURE 16.8 Olfactory Projection Pathways in the Brain. This shows only the extrathalamic pathways.

can inhibit the mitral and tufted cells. An effect of this feedback is that odors can change in quality and significance under different conditions. Food may smell more appetizing when you are hungry, for example, than when you have just eaten or when you are ill.

▶▶▶ APPLY WHAT YOU KNOW

Which taste sensations could be lost after damage to (1) the facial nerve or (2) the glossopharyngeal nerve? A fracture of which cranial bone would most likely eliminate the sense of smell?

BEFORE YOU GO ON

Answer the following questions to test your understanding of the preceding section:

13. What is the difference between a lingual papilla and a taste bud? Which is visible to the naked eye?

14. List the primary taste sensations and discuss their adaptive significance (survival value).

15. Which cranial nerves carry gustatory impulses to the brain?

16. What part of an olfactory cell binds odor molecules?

17. What brain regions serve the sense of smell and how do they differ in their olfactory functions?

16.4 Hearing and Equilibrium

Expected Learning Outcomes

When you have completed this section, you should be able to

a. identify the properties of sound waves that account for pitch and loudness;

b. describe the gross and microscopic anatomy of the ear;

c. explain how the ear converts vibrations to nerve signals and discriminates between sounds of different intensity and pitch;

d. explain how the vestibular apparatus enables the brain to interpret the body's position and movements; and

e. describe the pathways taken by auditory and vestibular signals to the brain.

Hearing is a response to vibrating air molecules, and **equilibrium** is the sense of body orientation, movement, and balance. Both senses reside in the inner ear, a maze of fluid-filled passages and sensory cells. We will study how the fluid is set in motion and how the sensory cells convert this motion into an informative nerve signal.

16.4a The Nature of Sound

To understand the physiology of hearing, it is necessary to know some basic properties of sound. **Sound** is any audible vibration of molecules. It can be transmitted through water, solids, or air, but not through a vacuum. Our discussion is limited to airborne sound.

Sound is produced by a vibrating object such as a tuning fork, a loudspeaker, or the vocal cords. Consider a loudspeaker. When the speaker cone moves forward, it pushes air molecules ahead of it. They collide with other molecules just ahead of them, and energy is transferred from molecule to molecule until it reaches the eardrum. No one molecule moves very far; they simply collide with each other like billiard balls until finally, some molecules collide with the eardrum and make it vibrate. The sensations we perceive as pitch and loudness are related to the physical properties of these vibrations.

Pitch

Pitch is our sense of whether a sound is "high" (treble) or "low" (bass). It is determined by the frequency at which the sound source, eardrum, and other parts of the ear vibrate. One movement of a vibrating object back and forth is called a *cycle,* and the number of cycles per second (or hertz, Hz) is called **frequency.** The lowest note on a piano, for example, is 27.5 Hz, middle C is 261 Hz, and the highest note is 4,176 Hz. The most sensitive human ears, particularly in children, can hear frequencies from 20 to 20,000 Hz, but most of us cannot hear over that broad a range. The *infrasonic* frequencies below 20 Hz aren't detected by the ear, but we sense them through vibrations of the skull and skin and they play a significant role in our appreciation of music. The inaudible vibrations above 20,000 Hz are *ultrasonic.* Human ears are most sensitive to frequencies ranging from 1,500 to 5,000 Hz. In this range, we can hear sounds of relatively low energy (volume), whereas sounds above or below this range must be louder to be audible, as reflected in the steeply rising curves at both ends of the violet range in **figure 16.9.**

Normal speech falls within the 1,500 to 5,000 Hz range. Most of the hearing loss suffered with age is in the range of 250 to 2,050 Hz.

Loudness

Loudness is the perception of sound energy, intensity, or the **amplitude** of vibration. In the speaker example, amplitude is a measure of how far forward and back the cone vibrates on each cycle and how much it compresses the air molecules in front of it. Loudness is expressed in decibels (dB), with 0 dB defined by a sound energy that corresponds to the threshold of human hearing. Every 10 dB step up the scale represents a sound with 10 times greater intensity. Thus, 10 dB is 10 times threshold, 20 dB is 100 times threshold, 30 dB is 1,000 times threshold, and so forth. Normal conversation has a loudness of about 60 dB. At most frequencies, the threshold of pain is 120 to 140 dB, approximately the intensity of a loud thunderclap. Prolonged exposure to sounds greater than 90 dB can cause permanent loss of hearing.

16.4b Anatomy of the Ear

The ear has three sections called the *outer, middle,* and *inner ear.* The first two are concerned only with transmitting sound to the inner ear, where vibration is converted to nerve signals.

Outer Ear

The **outer (external) ear** is essentially a funnel for conducting airborne vibrations to the *tympanic membrane* (eardrum). It begins with the fleshy **auricle (pinna)** on the side of the head, shaped and supported by elastic cartilage except for the earlobe, which is mostly adipose tissue. The auricle is an arrangement of named whorls and recesses that direct sound into the auditory canal **(fig. 16.10).**

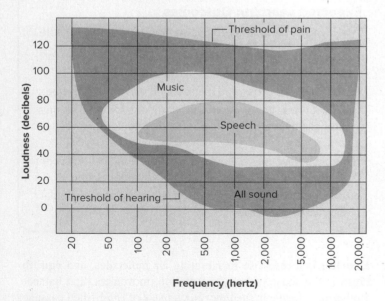

FIGURE 16.9 The Range of Human Hearing.

? *How would the shape of this graph change in a case of moderate hearing loss between 200 and 5,000 Hz?*

FIGURE 16.10 Anatomy of the Auricle (Pinna) of the Ear.

Joe DeGrandis/McGraw-Hill Education

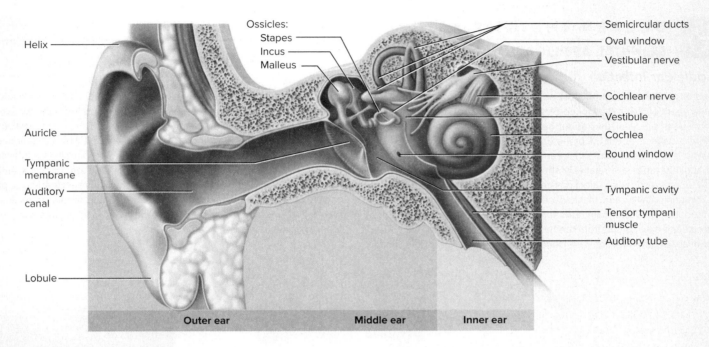

FIGURE 16.11 Internal Anatomy of the Ear.

The **auditory canal (external acoustic meatus)** is the passage leading through the temporal bone to the tympanic membrane. It follows a slightly S-shaped course for about 3 cm **(fig. 16.11).** It is lined with skin and supported by fibrocartilage at its opening and by the temporal bone for the rest of its length.

The outer end of the canal is protected by stiff **guard hairs.** The canal has ceruminous and sebaceous glands whose secretions mix with dead skin cells and form **cerumen** (earwax). Cerumen is sticky and coats the guard hairs, making them more effective in blocking foreign particles from the auditory canal. Its stickiness may also be a deterrent to insects, ticks, or other pests. In addition, it contains lysozyme and has a low pH, both of which inhibit bacterial growth; it waterproofs the canal and protects its skin and the tympanic membrane from absorbing water; and it keeps the tympanic membrane pliable. Cerumen normally dries and falls from the canal, but sometimes it becomes impacted and interferes with hearing.

Middle Ear

The **middle ear** is located in the **tympanic cavity** of the temporal bone. What we colloquially call the eardrum is anatomically known as the **tympanic[14] membrane.** It closes the inner end of the auditory canal and separates it from the middle ear. The membrane is shaped like a Chinese farmer (coolie) hat—about 1 cm in diameter and slightly concave on its outer surface. It is suspended in a ring-shaped groove in the temporal bone and vibrates freely in response to sound. It is innervated by sensory branches of the vagus and trigeminal nerves and is highly sensitive to pain.

Posteriorly (behind the ear), the tympanic cavity is continuous with the *mastoid cells*—air-filled spaces in the mastoid process

of the temporal bone. It is filled with air that enters by way of the **auditory (eustachian[15]** or **pharyngotympanic) tube,** a passageway to the nasopharynx. (Be careful not to confuse *auditory tube* with *auditory canal.*) The auditory tube serves to aerate and drain the middle ear. It is normally flattened and closed, but swallowing or yawning opens it and allows air to enter or leave the tympanic cavity. This equalizes air pressure on both sides of the tympanic membrane, allowing it to vibrate freely. Excessive pressure on one side or the other muffles the sense of hearing and may cause pain, as one commonly experiences in airline flight. Unfortunately, the auditory tube also allows throat infections to spread to the middle ear (see Deeper Insight 16.1).

The tympanic cavity, a space only 2 to 3 mm wide between the outer and inner ears, contains the three smallest bones and the two smallest skeletal muscles of the body. The bones, called the **auditory ossicles,[16]** connect the tympanic membrane to the inner ear. Progressing inward, the first is the **malleus,[17]** which has an elongated *handle* attached to the inner surface of the tympanic membrane, and a *head,* which articulates with the next ossicle. The second bone, the **incus,[18]** has a roughly triangular *body* that articulates with the malleus; a *long limb* that articulates with the stapes; and a *short limb* (not illustrated) suspended by a ligament from the wall of the cavity. The **stapes[19]** (STAY-peez), the body's smallest bone, is shaped like a stirrup. It has a *head* that articulates with the incus; two *limbs* that form an arch; and an elliptical *base* (*footplate*). The base is held by a ringlike ligament in an opening

---

[14]*tympan* = drum

[15]Bartholomeo Eustachio (1524–74), Italian anatomist
[16]*oss* = bone; *icle* = little
[17]*malleus* = hammer
[18]*incus* = anvil
[19]*stapes* = stirrup

DEEPER INSIGHT 16.1

CLINICAL APPLICATION

Middle-Ear Infection

About 80% of children get middle-ear infections, or *otitis*[20] *media,* in the first 10 years of life. In the United States, it accounts for about 30 million doctor visits per year and about half of all antibiotic prescriptions for children; it affects about 11% of the entire world population each year.

Why is it so common in children? Up to the age of about 8 years, the auditory tubes are relatively short and horizontal **(fig. 16.12).** They allow upper respiratory infections to spread easily from the throat to the tympanic cavity and mastoid cells. Fluid accumulation in the cavity produces pressure, pain, and impaired hearing. If otitis media goes untreated, it may spread from the mastoid cells and cause meningitis, a potentially deadly infection (see Deeper Insight 14.1). Chronic otitis media

can also cause fusion of the middle-ear bones and result in permanent hearing loss. It is sometimes necessary to drain fluid from the tympanic cavity by lancing the tympanic membrane and inserting a tiny drainage tube—a procedure called *tympanostomy.*[21] The tube, which is eventually sloughed out of the ear, relieves the pressure and permits the infection to heal. The great majority of children get a middle-ear infection in the first year of life. Holding an infant horizontally while breast feeding or putting it to bed with a bottle increases the risk of otitis media. To minimize the risk, infants should be held in an inclined, head-up position while feeding from the breast or bottle.

— Auditory tube —

(a) Pediatric   **(b) Adult**

FIGURE 16.12 The Auditory Tube of an Infant and Adult. (a) The shorter, nearly horizontal tube of infants and young children makes it easy for them to get middle-ear infections. (b) The longer and more oblique tube of the adult.

[20]*ot* = ear; *itis* = inflammation

[21]*tympano* = eardrum; *stomy* = making an opening

called the **oval window,** where the inner ear begins. On the other side of the base is an inner-ear liquid called *perilymph.*

The muscles of the middle ear are the stapedius and tensor tympani. The **stapedius** (sta-PEE-dee-us) arises from the posterior wall of the cavity and inserts on the stapes. The **tensor tympani** (TEN-sor TIM-pan-eye) arises from the wall of the auditory tube, travels alongside it, and inserts on the malleus. The function of these muscles is discussed under the physiology of hearing.

Inner Ear

The **inner (internal) ear** is housed in a maze of temporal bone passages called the **bony labyrinth,** which is lined by a system of fleshy tubes called the **membranous labyrinth (fig. 16.13).** Between the bony and membranous labyrinths is a cushion of fluid, the **perilymph** (PER-ih-limf), similar to cerebrospinal fluid. Within the membranous labyrinth is a fluid called **endolymph,** similar to intracellular fluid. The bony and membranous labyrinths form a tube-within-a-tube structure, somewhat like a bicycle inner tube within the tire.

The labyrinths begin with a chamber called the **vestibule,** which contains organs of equilibrium to be discussed later. The organ of hearing is the **cochlea**[22] (COC-lee-uh), a coiled tube that arises from the anterior side of the vestibule. In other vertebrates, the cochlea is straight or slightly curved. In most mammals, however, it assumes the form of a snail-like spiral, which allows a longer cochlea to fit in a compact space. In humans, the spiral is about 9 mm wide at the base and 5 mm high. Its apex points anterolaterally. The cochlea winds for about 2.5 coils around a screwlike bony axis called the **modiolus**[23] (mo-DY-oh-lus). The "threads of the screw" form a spiral platform that supports the fleshy tube of the cochlea.

A vertical section cuts through the cochlea about five times **(fig. 16.14a).** A single cross section looks like **figure 16.14b.** It is important to realize that the structures seen in cross section actually have the form of spiral strips winding around the modiolus from base to apex.

[22]*cochlea* = snail
[23]*modiolus* = hub

FIGURE 16.13 Anatomy of the Membranous Labyrinth. (a) Position and orientation within the petrous part of the temporal bone. (b) Structure of the membranous labyrinth and nerves. (c) Relationship of the perilymph (green) and endolymph (blue) to the labyrinth.

The cochlea has three fluid-filled chambers separated by membranes. The superior chamber is called the **scala**[24] **vestibuli** (SCAY-la vess-TIB-you-lye) and the inferior one is the **scala tympani.** These are filled with perilymph and communicate with each other through a narrow channel at the apex of the cochlea. The scala vestibuli begins near the oval window and spirals to the apex; from there, the scala tympani spirals back down to the base and ends at the **round window** (fig. 16.13). The round window is covered by a membrane called the *secondary tympanic membrane.*

The middle chamber is a triangular space, the **cochlear duct (scala media).** It is separated from the scala vestibuli above by a thin **vestibular membrane** and from the scala tympani below by a much thicker **basilar membrane.** Unlike those chambers, it is filled with endolymph rather than perilymph. The vestibular membrane separates the endolymph from the perilymph and helps to

maintain a chemical difference between them. Within the cochlear duct, supported on the basilar membrane, is the **spiral (acoustic) organ**—a thick epithelium of sensory and supporting cells and associated membranes **(fig. 16.14c).** This is the device that converts vibrations into nerve impulses, so we must pay particular attention to its structural details.

The spiral organ has an epithelium composed of **hair cells** and **supporting cells.** Hair cells are named for the long, stiff microvilli called **stereocilia**[25] on their apical surfaces. (Stereocilia aren't true cilia. They don't have an axoneme of microtubules as seen in cilia, and they don't move by themselves.) Resting on top of the stereocilia is a gelatinous **tectorial**[26] **membrane.**

The spiral organ has four rows of hair cells spiraling along its length **(fig. 16.15).** About 3,500 of these, called **inner hair cells (IHCs),** are arranged in a row on the medial side of the basilar

---

[24]*scala* = staircase

[25]*stereo* = solid
[26]*tect* = roof

FIGURE 16.14 Anatomy of the Cochlea. (a) Vertical section. In anatomical position, the apex of the cochlea faces downward and anterolaterally. (b) Detail of one section through the cochlea. (c) Detail of the spiral organ.

membrane (facing the modiolus). Each of these has a cluster of 50 to 60 stereocilia, graded from short to tall. Another 20,000 **outer hair cells (OHCs)** are neatly arranged in three rows across from the inner hair cells. Each OHC has about 100 stereocilia arranged in the form of a V, with their tips embedded in the tectorial membrane. All that we hear comes from the IHCs, which supply 90% to 95% of the sensory fibers of the cochlear nerve. The function of the OHCs is to adjust the response of the cochlea to different frequencies and enable the IHCs to work with greater precision. We will see later how this is done. Hair cells are not neurons, but synapse with nerve fibers at their base—the OHCs with both sensory and motor neurons and the IHCs with sensory neurons only.

16.4c The Physiology of Hearing

We can now examine the way sound affects the ear and produces action potentials. Sound waves enter the auditory canal on one side and nerve signals exit the inner ear on the other. Connecting these is the middle ear, so we begin with an analysis of its contribution.

The Middle Ear

One might wonder why we have a middle ear at all—why the tympanic membrane doesn't simply vibrate against the fluid-filled labyrinth of the inner ear. The reason is that the tympanic membrane, which moves in air, vibrates easily, whereas

Outer hair cells   Inner hair cells

10 μm

FIGURE 16.15 Apical Surfaces of the Cochlear Hair Cells. All signals that we hear come from the inner hair cells on the right.

SPL/Science Source

❓ *What is the function of the three rows of hair cells on the left?*

the stapes must push against the perilymph of the inner ear. This fluid resists motion much more than air does. If the tympanic membrane had air on one side and fluid on the other, the sound waves wouldn't have enough energy to move the fluid adequately. The tympanic membrane, however, has 18 times the area of the oval window. By concentrating the energy of the vibrating tympanic membrane on an area one-eighteenth that size, the ossicles create a greater force per unit area at the oval window and overcome the inertia of the fluid.

The auditory ossicles do not, however, provide any mechanical advantage, any amplification of sound. Vibrations of the stapes against the inner ear normally have the same amplitude as vibrations of the tympanic membrane against the malleus. Why, then, have a lever system composed of three ossicles? Why not simply have one ossicle concentrating the mechanical energy of the tympanic membrane directly on the inner ear?

The answer is that the ossicles serve at times to *lessen* the transfer of energy to the inner ear. They and their muscles have a protective function. In response to a loud noise, the tensor tympani pulls the tympanic membrane inward and tenses it, while the stapedius reduces the motion of the stapes. This **tympanic reflex** muffles the transfer of vibrations to the oval window. The reflex probably evolved in part for protection from loud but slowly building natural sounds such as thunder. The reflex has a latency of about 40 ms, which isn't quick enough to protect the inner ear from sudden artificial noises such as gunshots. The tympanic reflex also doesn't adequately protect the ears from sustained loud noises such as factory noise or loud music. Such noises can irreversibly damage the hair cells of the inner ear. It is therefore imperative to wear ear protection when using firearms or working in noisy environments (even symphony orchestras).

The middle-ear muscles also help to coordinate speech with hearing. Without them, the sound of your own speech would be so loud it could damage your inner ear, and it would drown out soft or high-pitched sounds from other sources. Just as you are about to speak, however, the brain signals these muscles to contract. This dampens the sense of hearing in phase with the inflections of your own voice and makes it easier to hear other people while you are speaking.

▶▶▶**APPLY WHAT YOU KNOW**

What type of muscle fibers—slow oxidative or fast glycolytic (see table 11.2)—do you think constitute the stapedius and tensor tympani? That is, which type would best suit the purpose of these muscles?

Stimulation of Cochlear Hair Cells

The next step in hearing is based on movement of the cochlear hair cells relative to stationary structures nearby. In this section, we will see how movements of the inner-ear fluids and basilar membrane move the hair cells and why it is important that the tectorial membrane near the hair cells remains relatively still.

A simple mechanical model of the ear can help in visualizing how this happens **(fig. 16.16).** (The vestibular membrane is omitted from the model for simplicity; it doesn't affect the mechanics of hearing.) As you listen to your favorite music, each inward movement of the tympanic membrane pushes the middle-ear ossicles inward. The stapes, in turn, pushes on the perilymph in the scala vestibuli. Perilymph, like other liquids, cannot be compressed, so it flows away from the stapes footplate. The resulting pressure in the scala vestibuli pushes the vestibular membrane downward; this pushes on the endolymph in the cochlear duct; the endolymph pushes down on the basilar membrane; the basilar membrane pushes on the perilymph in the scala tympani; and finally,

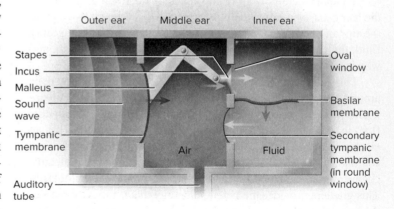

Outer ear   Middle ear   Inner ear

Stapes — Incus — Malleus — Sound wave — Tympanic membrane — Auditory tube

Oval window — Basilar membrane — Secondary tympanic membrane (in round window)

Air   Fluid

FIGURE 16.16 Mechanical Model of Hearing.

❓ *Why would high air pressure in the middle ear reduce the movements of the basilar membrane of the inner ear?*

Tip link

Mechanically gated K⁺ channel

Stereocilia

K⁺

Surface of hair cell

K⁺ gate closed

(a) Unstimulated

K⁺ gate open

K⁺

(b) Stimulated

FIGURE 16.17 Potassium Channels of the Cochlear Hair Cells. Each stereocilium has a K⁺ channel at its tip, shown at the small black box. (a) A hair cell at rest, unstimulated, with the potassium gates closed. (b) A stimulated hair cell, with the potassium gate of each stereocilium pulled open by the tip link leading to the next higher stereocilium.

the secondary tympanic membrane bulges outward to relieve the pressure. As the cycle of vibration continues, the stapes pulls back from the oval window and all of this happens in reverse.

In short, as the stapes goes in-out-in, the secondary tympanic membrane goes out-in-out, and the basilar membrane goes down-up-down. It is not difficult to see how this happens—the only thing hard to imagine is that it can happen as often as 20,000 times per second! The important thing about all this is that the hair cells, affixed to the basilar membrane, go along for the ride, bobbing up and down as the basilar membrane moves.

To understand how all of this leads to electrical excitation of the hair cells, we must seemingly digress for a moment to examine the tips of the hair cells. These are bathed in a high-potassium fluid, the endolymph. Why is this fluid so rich in potassium, and why is that important?

Potassium ions (K⁺) are secreted into the endolymph by cells around the circumference of the cochlear duct (on the wall opposite from the modiolus). The vestibular membrane retains this fluid in the duct. Relative to the perilymph, the endolymph has an electrical potential of about +80 mV, and the interior of the hair cell about –40 mV. Thus there is an exceptionally strong electrochemical gradient from the endolymph to the hair cell cytoplasm. This gradient provides the potential energy that ultimately enables the hair cell to work.

On the inner hair cells—the ones that generate all the signals we hear—each stereocilium has a single transmembrane protein at its tip that functions as a mechanically gated ion channel. A fine,

stretchy protein filament called a **tip link** extends like a spring from the ion channel of one stereocilium to the sidewall of the taller stereocilium next to it **(fig. 16.17).** The stereocilia increase in height progressively, so that every stereocilium but the tallest one has a tip link leading to a taller one beside it.

Now what of the tectorial membrane? This conspicuous structure of the spiral organ is anchored to the core of the cochlea and remains relatively still as the hair cells dance up and down to the beat of the music. Each time the basilar membrane rises upward toward the tectorial membrane, the hair cell stereocilia are pushed against that membrane and tilt toward the tallest one. As each taller stereocilium bends over, it pulls on the tip link of its neighbor. The tip link, connected to the ion gate of the next shorter stereocilium, pulls the gate open and allows ions to flood into the cell. The gate is nonselective, but since the predominant ion of the endolymph is K⁺, the primary effect of this gating is to allow a quick burst of K⁺ to flow into the hair cell. This depolarizes the hair cell while the gate is open, and when the basilar membrane drops and the stereocilium bends the other way, the gate closes and the cell becomes briefly hyperpolarized. During each moment of depolarization, a hair cell releases a burst of neurotransmitter from its base, exciting the sensory dendrite that synapses with the hair cell. Each depolarization thus generates a signal in the cochlear nerve.

To summarize: Each upward movement of the basilar membrane pushes the inner hair cells closer to the stationary tectorial membrane. This forces the stereocilia to bend in the direction of the tallest one. Each stereocilium has a tip link connecting it to an ion channel at the top of the next shorter stereocilium. When the taller one bends over, it pulls the channel open. Potassium ions flow into the hair cell and depolarize it. The hair cell releases a burst of neurotransmitter, exciting the sensory processes of the cochlear nerve cells below it. Thus a signal is generated in the cochlear nerve and transmitted to the brain.

Sensory Coding

For sounds to carry any meaning, we must distinguish differences in loudness and pitch. Our ability to do so stems from the fact that the cochlea responds differently to sounds of different amplitude and frequency. Variations in loudness (amplitude) cause variations in the intensity of cochlear vibration. A soft sound produces relatively slight up-and-down movements of the basilar membrane. Hair cells are stimulated only moderately, and a given sound frequency stimulates hair cells in a relatively limited, or focused, region of the cochlea. A louder sound makes the basilar membrane vibrate more vigorously. Hair cells respond more intensely, generating a higher firing frequency in the cochlear nerve. In addition, for a given frequency, a loud sound vibrates a longer segment of the basilar membrane and thus excites a greater number of hair cells. If the brain detects moderate firing rates associated with hair cells in relatively narrow bands of the cochlea, it interprets this as a soft sound. If it detects a high firing frequency in nerve fibers associated with broader bands, it interprets this as a louder sound.

Frequency discrimination is based on a structural gradient in the basilar membrane resembling a piano's gradient in sound frequency from the short strings to long strings. At its proximal end (the base of the cochlea), the membrane is attached, narrow, and stiff, like the piano's short strings. At its distal end (the apex of the cochlea), it is unattached, five times wider, and more flexible. Think of the basilar membrane as analogous to a wire stretched tightly between two posts. If you pluck the wire at one end, a wave of vibration travels down its length and back. This produces a standing wave, with some regions of the wire vertically displaced more than others. Similarly, a sound causes a standing wave in the basilar membrane. The peak amplitude of this wave is near the distal end in the case of low-frequency sounds and nearer the proximal end with sounds of higher frequencies. When the brain receives signals mainly from inner hair cells at the distal end, it interprets the sound as low-pitched; when signals come mainly from the proximal end, it interprets the sound as high-pitched **(fig. 16.18).** Speech, music, and other everyday sounds, of course, aren't pure tones—they create complex, ever-changing patterns of vibration in the basilar membrane that must be decoded by the brain.

Cochlear Tuning

Just as we tune a radio to receive a certain frequency, we also tune our cochlea to receive some frequencies better than others. The outer hair cells (OHCs) are supplied with a few sensory fibers (5% to 10% of those in the cochlear nerve), but more importantly, they receive motor fibers from the brain.

In response to sound, the OHCs send signals to the medulla oblongata by way of the sensory neurons, and the pons sends signals immediately back to the OHCs by way of the motor neurons. In response, the hair cells shorten as much as 15%. Remember that an OHC is anchored to the basilar membrane below and by its stereocilia to the tectorial membrane above. Therefore, tensing of an OHC reduces the basilar membrane's mobility. This results in some regions of the cochlea sending fewer signals to the brain than neighboring regions, so the brain can better distinguish between sound frequencies. When OHCs are experimentally incapacitated, the inner hair cells (IHCs) respond much less precisely to differences in pitch.

There is another mechanism of cochlear tuning involving the inner hair cells. The pons sends efferent fibers to the cochlea that synapse with the sensory nerve fibers near the base of the IHCs. The efferent fibers can inhibit the sensory fibers from firing in some areas of the cochlea, and thus enhance the contrast between signals from the more responsive and less responsive regions. Combined with the previously described role of the OHCs, this sharpens the tuning of the cochlea and our ability to discriminate pitch.

The Auditory Projection Pathway

The sensory nerve fibers beginning at the bases of the hair cells belong to bipolar sensory neurons. Their somas form a coil, the **spiral ganglion,** around the modiolus (see fig. 16.14), and their axons lead away from the cochlea as the **cochlear nerve.** This nerve joins the *vestibular nerve,* discussed later, and the two together become the *vestibulocochlear nerve* (CN VIII).

Each ear sends fibers to both sides of the medulla oblongata. There, they end in the *cochlear nuclei,* synapsing with second-order neurons that ascend to the *superior olivary nucleus* of the

DEEPER INSIGHT 16.2
CLINICAL APPLICATION

Deafness

Deafness means any hearing loss, from mild and temporary to complete and irreversible. *Conductive deafness* results from any condition that interferes with the transmission of vibrations to the inner ear. Such conditions include a damaged tympanic membrane, otitis media, blockage of the auditory canal, and otosclerosis. *Otosclerosis*[27] is fusion of the auditory ossicles to each other or fusion of the stapes to the oval window. Either way, it prevents the bones from vibrating freely. *Sensorineural (nerve) deafness* results from the death of hair cells or any of the nervous elements concerned with hearing. It is a common occupational disease of factory and construction workers, musicians, and other people exposed to frequent or sustained loud sounds. Deafness leads some people to develop delusions of being talked about, disparaged, or cheated. Beethoven said his deafness drove him nearly to suicide.

[27]*oto* = ear; *scler* = hardening; *osis* = process, condition

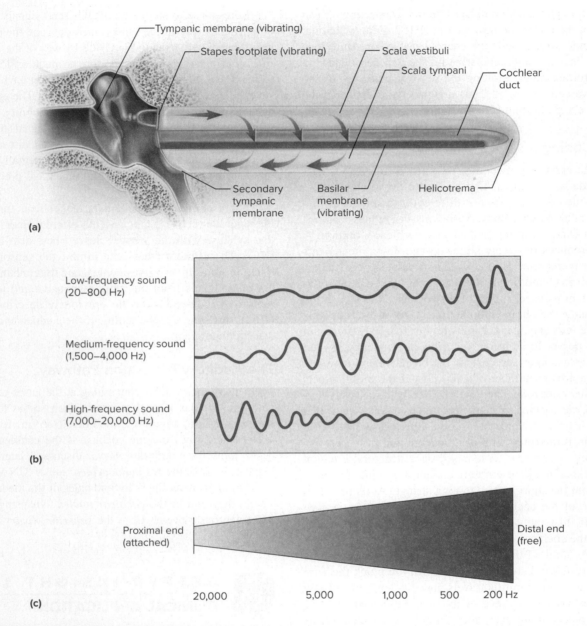

FIGURE 16.18 Frequency Response of the Basilar Membrane of the Cochlea. (a) The cochlea, uncoiled and laid out straight. (b) Sounds produce a standing wave of vibration along the basilar membrane. The peak amplitude of the wave varies with the frequency of the sound, as shown here. The amount of vibration is greatly exaggerated in this diagram to clarify the standing wave. (c) The taper of the basilar membrane and its correlation with sound frequencies. High frequencies (7,000–20,000 Hz) are best detected by hair cells near the narrow proximal end at the left, and low frequencies (20–800 Hz) by hair cells near the wider distal end at the right.

pons **(fig. 16.19).** By way of CN VIII, the superior olivary nucleus issues the efferent fibers back to the cochlea that are involved in cochlear tuning. By way of CNs V_3 and VII, it also issues motor fibers to the tensor tympani and stapedius muscles, respectively. The superior olivary nucleus also functions in **binaural**[28] **hearing**— comparing signals from the right and left ears to identify the direction from which a sound is coming.

Other fibers from the cochlear nuclei ascend to the inferior colliculi of the midbrain. The inferior colliculi help to locate the origin of a sound in space, process fluctuations in pitch that are important for such purposes as understanding another person's speech, and mediate the startle response and rapid head turning that occur in reaction to loud or sudden noises.

Third-order neurons begin in the inferior colliculi and lead to the thalamus. Fourth-order neurons complete the pathway from there to the primary auditory cortex; thus the auditory pathway, unlike most other sensory pathways, involves not three but four neurons from receptor to cerebral cortex. The primary auditory cortex lies in the superior margin of the temporal lobe and extends deeply into the lateral sulcus (see fig. 14.20). The temporal lobe

---

[28]*bin* = two; *aur* = ears

FIGURE 16.19 Auditory Pathways in the Brain. (a) Schematic. (b) Brainstem and frontal section of the cerebrum, showing the locations of auditory processing centers. (cranial nerve V_3 = trigeminal nerve, mandibular division; CN VII = facial nerve; CN VIII = vestibulocochlear nerve)

is the site of conscious perception of sound, and it completes the information processing essential to binaural hearing. Because of extensive decussation in the auditory pathway, damage to the right or left auditory cortex doesn't cause a unilateral loss of hearing.

16.4d The Physiology of Equilibrium

The original function of the ear in vertebrate evolution wasn't hearing, but equilibrium. Only later did vertebrates evolve the cochlea,

outer- and middle-ear structures, and auditory function of the ear. In humans, the receptors for equilibrium constitute the **vestibular apparatus,** which consists of three **semicircular ducts** and two chambers— an anterior **saccule**[29] and a posterior **utricle**[30] (see fig. 16.13).

The sense of equilibrium is divided into **static equilibrium,** the perception of the orientation of the head in space (whether it

---

[29]*saccule* = little sac
[30]*utricle* = little bag

FIGURE 16.20 **The Saccule and Utricle.** (a) Locations of the macula sacculi and macula utriculi. (b) Structure of a macula. (c) Action of the otolithic membrane on the hair cells when the head is tilted.

is erect or tilted in any direction), and **dynamic equilibrium,** the perception of motion or acceleration. There are two kinds of acceleration: (1) *linear acceleration,* a change in velocity in a straight line, as when riding in a car or elevator; and (2) *angular acceleration,* a change in the rate of rotation, as when your car turns a corner or you swivel in a rotating chair. The saccule and utricle are responsible for static equilibrium and the sense of linear acceleration; the semicircular ducts detect only angular acceleration.

The Saccule and Utricle

The saccule and utricle each contain a 2 × 3 mm patch of hair cells and supporting cells called a macula.[31] The **macula sacculi** lies almost vertically on the wall of the saccule, and the **macula utriculi** lies almost horizontally on the floor of the utricle **(fig. 16.20a).**

Each hair cell of a macula has 40 to 70 stereocilia and one true cilium called a **kinocilium.**[32] The tips of the stereocilia and kinocilium are embedded in a gelatinous **otolithic membrane.** This membrane is weighted with protein–calcium carbonate granules called **otoliths**[33] **(fig. 16.20b),** which add to the weight and inertia of the membrane and enhance the sense of gravity and motion.

Figure 16.20c shows how the macula utriculi detects tilt of the head. With the head erect, the otolithic membrane bears directly down on the hair cells, and stimulation is minimal. When you tilt your head down to read a book, however, the heavy otolithic membrane sags and bends the stereocilia, stimulating the hair cells. Any orientation of the head causes a combination of stimulation to the utricules and saccules of the two ears. The brain interprets head orientation by comparing these inputs to each other and to other input from the eyes

---

[31]*macula* = spot

[32]*kino* = moving
[33]*oto* = ear; *lith* = stone

and stretch receptors in the neck, thereby detecting whether only the head is tilted or the entire body is tipping.

The inertia of the otolithic membranes is especially important in detecting linear acceleration. Suppose you're sitting in a car at a stoplight and then begin to move. The membrane of the macula utriculi briefly lags behind the rest of the tissues, bends the stereocilia backward, and stimulates the cells. When you stop at the next light, the macula stops but the otolithic membrane keeps going for a moment, bending the stereocilia forward. The hair cells convert this stimulation to nerve signals, and the brain is thus advised of changes in your linear velocity.

The macula sacculi is nearly vertical and its hair cells therefore respond to vertical acceleration and deceleration. If you're standing in an elevator and it begins to move up, the membrane of the macula sacculi lags behind briefly and pulls down on the hairs. When the elevator stops, the membrane keeps going for a moment and bends the hairs upward. In both cases, the hair cells are stimulated and the brain is made aware of your vertical movements. These sensations are important in such ordinary actions as sitting down, and as your head bobs up and down during walking and running.

The Semicircular Ducts

The head also experiences rotary movements, such as when you spin in a rotating chair, walk down a hall and turn a corner, or bend forward to pick something up from the floor. Such movements are detected by the three semicircular ducts **(fig. 16.21),** housed in the bony *semicircular canals* of the temporal bone. The *anterior* and *posterior semicircular ducts* are oriented vertically at right angles to each other. The *lateral semicircular duct* is about 30° from the

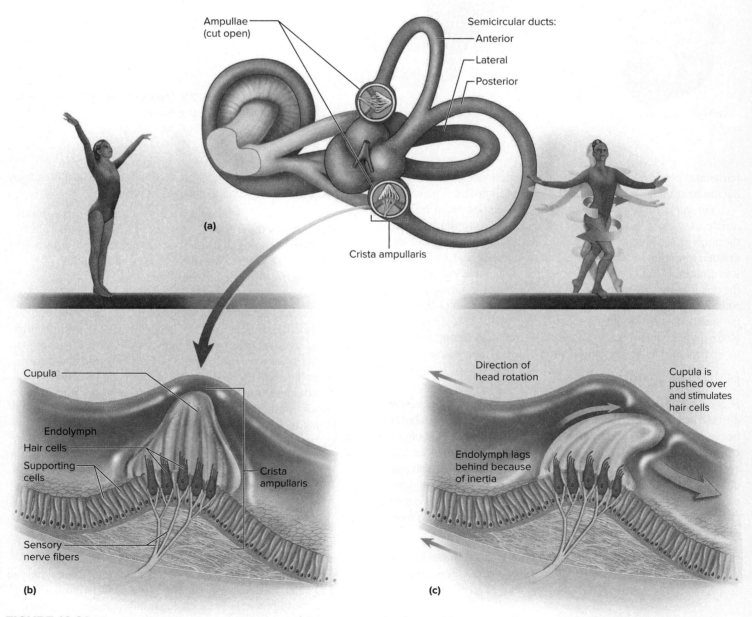

FIGURE 16.21 The Semicircular Ducts. (a) Structure of the semicircular ducts, with two ampullae opened to show the crista ampullaris and cupula. (b) Detail of the crista ampullaris. (c) Action of the endolymph on the cupula and hair cells when the head rotates.

FIGURE 16.22 Vestibular Projection Pathways in the Brain.

horizontal plane. The orientation of the ducts causes a different duct to be stimulated by rotation of the head in different planes.

The ducts are filled with endolymph. Each one opens into the utricle and has a dilated sac at one end called the **ampulla.**[34] Within the ampulla is a mound of hair cells and supporting cells called the **crista**[35] **ampullaris.** The hair cells have stereocilia and a kinocilium embedded in the **cupula,**[36] a gelatinous cap that extends from the crista to the roof of the ampulla. When the head turns, the duct rotates but the endolymph lags behind. It pushes the cupula, bends the stereocilia, and stimulates the hair cells. After 25 to 30 seconds of continual rotation, however, the endolymph catches up with the movement of the duct and stimulation of the hair cells ceases.

Projection Pathways

Hair cells of the macula sacculi, macula utriculi, and semicircular ducts synapse at their bases with sensory fibers of the **vestibular nerve.** This and the cochlear nerve merge to form the vestibulocochlear nerve (CN VIII). Fibers of the vestibular apparatus lead to a complex of four **vestibular nuclei** on each side of the pons and medulla. Nuclei on the right and left sides of the brainstem communicate extensively with each other, so each receives input from both the right and left ears. They process signals about the position and movement of the body and relay information to five targets (**fig. 16.22**):

1. The cerebellum, which integrates vestibular information into its control of head movements, eye movements, muscle tone, and posture.

2. The reticular formation, which is thought to adjust breathing and blood circulation to changes in posture.

3. The spinal cord, where fibers descend the two vestibulospinal tracts on each side (see fig. 13.5) and synapse on motor neurons that innervate the extensor (antigravity) muscles. This pathway allows you to make quick movements of the trunk and limbs to keep your balance.

4. The thalamus, which relays signals to two areas of the cerebral cortex. One is at the inferior end of the postcentral gyrus adjacent to sensory regions for the face. It is here that we become consciously aware of body position and movement. The other is slightly rostral to this, at the inferior end of the central sulcus in the transitional zone from primary sensory to motor cortex. This area is thought to be involved in motor control of the head and body.

5. Nuclei of the oculomotor, trochlear, and abducens nerves (CNs III, IV, and VI). These nerves produce eye movements that compensate for movements of the head (the *vestibulo–ocular reflex*). To observe this effect, hold a book in front of you at a comfortable reading distance and fix your gaze on the middle of the page. Move the book left and right about once per second, and you will be unable to read it. Now hold the book still and shake your head from side to side at the same rate. This time you will be able to read the page because the vestibulo–ocular reflex compensates for your head movements and keeps your eyes fixed on the target. This reflex enables you to keep your vision fixed on a distant object as you walk or run toward it.

---

[34]*ampulla* = little jar
[35]*crista* = crest, ridge
[36]*cupula* = little tub

BEFORE YOU GO ON

Answer the following questions to test your understanding of the preceding section:

18. What physical properties of sound waves correspond to the sensations of loudness and pitch?

19. What are the benefits of having auditory ossicles and muscles in the middle ear?

20. Explain how vibration of the tympanic membrane ultimately produces fluctuations of membrane voltage in a cochlear hair cell.

21. How does the brain recognize the difference between the musical notes high C and middle C? Between a loud sound and a soft one?

22. How does the function of the semicircular ducts differ from the function of the saccule and utricle?

23. How is sensory transduction in the semicircular ducts similar to that in the saccule and utricle?

16.5 Vision

Expected Learning Outcomes

When you have completed this section, you should be able to

a. describe the anatomy of the eye and its accessory structures;

b. discuss the structure of the retina and its receptor cells;

c. explain how the optical system of the eye creates an image on the retina;

d. discuss how the retina converts this image to nerve signals;

e. explain why different types of receptor cells and neural circuits are required for day and night vision;

f. describe the mechanism of color vision; and

g. trace the visual projection pathways in the brain.

16.5a Light and Vision

Vision (sight) is the perception of objects in the environment by means of the light they emit or reflect. **Light** is visible electromagnetic radiation. Human vision is limited to wavelengths ranging from about 400 to 700 nm. Most solar radiation of shorter and longer wavelengths is filtered out by ozone, carbon dioxide, and water vapor in the atmosphere. Therefore, the radiation that reaches the surface of the earth falls within that range, and vision is adapted to take advantage of the wavelenths available to us. The *ultraviolet (UV)* radiation just below 400 nm and the *infrared (IR)* radiation just above 700 nm are invisible to us. Furthermore, radiation in the ultraviolet range has such high energy that it destroys macromolecules rather than producing the controlled chemical reactions needed for vision, and radiation in the infrared range has such low energy that it merely warms the tissues, also failing to energize chemical reactions.

16.5b Accessory Structures of the Orbit

The eyeball occupies a bony socket called the **orbit.** This general area of the face, the *orbital region,* contains structures that protect and aid the eye **(figs. 16.23, 16.24):**

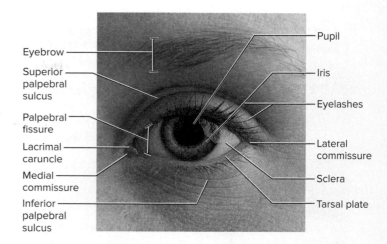

FIGURE 16.23 External Anatomy of the Orbital Region.
Joe DeGrandis/McGraw-Hill Education

- The **eyebrows** enhance facial expressions and nonverbal communication, but may also protect the eyes from glare and keep perspiration from running into the eye.

- The **eyelids,** or **palpebrae** (pal-PEE-bree), block foreign objects from the eye, prevent visual stimuli from disturbing one's sleep, and blink periodically to moisten the eye with tears and sweep debris from the surface. The eyelids are separated from each other by the **palpebral fissure** and meet each other at the corners called the **medial** and **lateral commissures.** The eyelid consists largely of the orbicularis oculi muscle covered with skin (fig. 16.24a). It also contains a supportive fibrous **tarsal plate** that is thickened along the margin of the eyelid. Within the plate are 20 to 25 **tarsal glands** that open along the edge of the eyelid. They secrete an oil that coats the eye and reduces tear evaporation. The **eyelashes** are guard hairs that help to keep debris from the eye. Touching the eyelashes stimulates hair receptors and triggers the blink reflex. They also disrupt airflow across the eye surface, thus protecting the eyes from drying. In windy or rainy conditions, we can squint so that the eyelashes protect the eyes even further, but without completely obstructing our vision.

- The **conjunctiva** (CON-junk-TY-vuh) is a transparent mucous membrane that covers the inner surface of the eyelid and anterior surface of the eyeball, except for the cornea. It secretes a thin mucous film that prevents the eyeball from drying. It is richly innervated and highly sensitive to pain. It is also very vascular, which is especially evident when the vessels are dilated and the eyes are "bloodshot." Because it is vascular and the cornea is not, the conjunctiva heals more readily than the cornea when injured.

- The **lacrimal**[37] **apparatus** (fig. 16.24b) consists of the lacrimal (tear) gland and a series of ducts that drain the tears into the nasal cavity. The **lacrimal gland,** about the size and shape of an almond, is nestled in a shallow fossa of the frontal bone in the superolateral corner of the orbit. About 12 short ducts lead from the gland to the surface of the conjunctiva. Tears cleanse and lubricate the eye surface, deliver oxygen and

---

[37]*lacrim* = tear

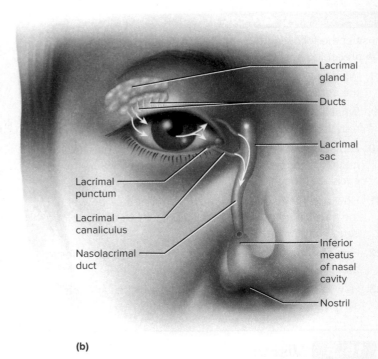

FIGURE 16.24 Accessory Structures of the Orbit. (a) Sagittal section of the eye and orbit. (b) The lacrimal apparatus. The arrows indicate the flow of tears from the lacrimal gland, across the front of the eye, into the lacrimal sac, and down the nasolacrimal duct.

? *What would be the effect of a blockage of the lacrimal puncta?*

nutrients to the conjunctiva, and contain an antibacterial enzyme, *lysozyme,* which protects the eye from infection. After washing across the eye, tears collect near the medial commissure and flow into a tiny pore, the **lacrimal punctum,**[38] on the margin of each eyelid. The punctum opens into a short **lacrimal canaliculus,** which leads to the **lacrimal sac** in the medial wall of the orbit. From this sac, a **nasolacrimal duct** carries the tears to the inferior meatus of the nasal cavity; thus an abundance of tears from crying or watery eyes can result in a runny nose. Once the tears enter the nasal cavity, they normally flow back to the throat and are swallowed. When you have a cold, the nasolacrimal ducts become swollen and obstructed, the tears cannot drain, and they may overflow from the brim of the eye.

- Six **extrinsic eye muscles** attach to the walls of the orbit and the external surface of the eyeball. *Extrinsic* means "arising externally"; it distinguishes these from the *intrinsic* muscles inside the eye that control the lens and pupil. The extrinsic muscles move the eye **(fig. 16.25).** They include four *rectus* ("straight") muscles and two *oblique* muscles. The **superior, inferior, medial,** and **lateral rectus** originate from a shared tendinous ring on the posterior wall of the orbit and insert on the anterior region of the eyeball, just beyond the visible "white of the eye." They move the eye up, down, medially, and laterally. The **superior oblique** travels along the medial wall of the orbit. Its tendon passes through a fibrocartilage ring, the **trochlea**[39] (TROCK-lee-uh), and inserts on the superolateral aspect of the eyeball. The **inferior oblique** extends from the

medial wall of the orbit to the inferolateral aspect of the eye. The oblique muscles slightly rotate the eyes when you tilt your head toward either shoulder, keeping the visual axes of the eyes aligned. People with weakness in the oblique muscles have to tilt their heads to fixate on objects; with the head erect, they have double vision. The oblique muscles also pull the eye forward, opposing tension in the rectus muscles that would otherwise pull the eye back more deeply into the orbit. The superior oblique muscle is innervated by the trochlear nerve (CN IV), the lateral rectus by the abducens (CN VI), and the other four muscles by the oculomotor nerve (CN III).

- **Orbital fat** surrounds the eye on the sides and back. It cushions the eye, allows it to move freely, and protects blood vessels and nerves in the rear of the orbit.

16.5c Anatomy of the Eye

The eyeball is a sphere about 24 mm in diameter **(fig. 16.26)** with three principal components: (1) three layers (tunics) that form its wall; (2) optical components that admit and focus light; and (3) neural components, the retina and optic nerve. The retina is not only a neural component but also part of the inner tunic. The cornea is part of the outer tunic as well as one of the optical components.

The Tunics

The three tunics of the eyeball are as follows:

- The outer **fibrous layer.** This is divided into two regions: sclera and cornea. The **sclera**[40] (white of the eye) covers most

---

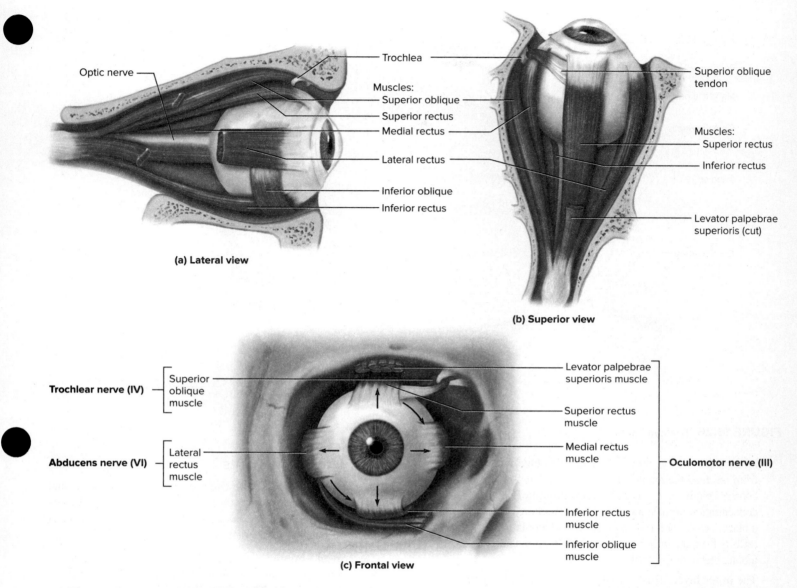

(a) Lateral view

(b) Superior view

(c) Frontal view

FIGURE 16.25 Extrinsic Muscles of the Eye. (a) Lateral view of the right eye. The lateral rectus muscle is cut to show a portion of the optic nerve. (b) Superior view of the right eye. (c) Innervation of the extrinsic muscles; arrows indicate the eye movement produced by each muscle.

What would cause the greatest loss of visual function—trauma to cranial nerve III, IV, or VI? Why?

of the eye surface and consists of dense collagenous connective tissue perforated by blood vessels and nerves. It serves as a tough fibrous protective cover for the eye and provides for attachment of the extrinsic muscles that move it. The **cornea** is the anterior transparent region of modified sclera that admits light into the eye. Most of it is composed of very compact layers of collagen fibrils and thin flat fibroblasts. It is covered by a thin stratified squamous epithelium anteriorly and a simple squamous epithelium posteriorly. Both epithelia pump sodium ions out of the corneal tissue. Water follows by osmosis, so this mechanism prevents the cornea from overhydrating, swelling, and losing transparency. When this ion pumping ceases at death, the cornea becomes cloudy. The anterior epithelium also is a source of stem cells that give the cornea a great capacity for regeneration if it is injured.

- The middle **vascular layer.** This is also called the **uvea**[41] (YOU-vee-uh) because it resembles a peeled grape in fresh dissection. It consists of three regions—the choroid, ciliary body, and iris. The **choroid** (CO-royd) is a highly vascular, deeply pigmented layer of tissue behind the retina. It gets its name from a histological resemblance to the chorion of the placenta. Its dense network of blood vessels provides all oxygenation, nourishment, and waste-removal services to the retina, which has no blood vessels of its own. The **ciliary body,** a thickened extension of the choroid, forms a muscular ring around the lens. It supports the iris and lens and secretes a fluid called aqueous humor. The **iris** is an adjustable diaphragm that controls the diameter of the **pupil,** its central opening. It has two pigmented

---

[41]*uvea* = grape

FIGURE 16.26 The Eye. Sagittal section.

layers: a posterior *pigment epithelium* that blocks stray light from reaching the retina, and the *anterior border layer,* which contains pigmented cells called **chromatophores.**[42] A high concentration of melanin in the chromatophores gives the iris a black, brown, or hazel color. If the melanin is scanty, light reflects from the posterior pigment epithelium and gives the iris a blue, green, or gray color.

- The **inner layer.** This consists of the retina and beginning of the optic nerve.

---

[42]*chromato* = color; *phore* = bearer

The Optical Components

The optical components of the eye are transparent elements that admit light rays, bend (refract) them, and focus images on the retina. They include the *cornea, aqueous humor, lens,* and *vitreous body.* The cornea has been described already.

- The **aqueous humor** is a serous fluid secreted by the ciliary body into a space called the **posterior chamber** between the iris and lens **(fig. 16.27).** It flows through the pupil into the **anterior chamber** between the iris and cornea. From here, it is reabsorbed by a circular vein called the **scleral venous sinus.** Normally the rate of reabsorption balances the rate of secretion (see Deeper Insight 16.3 for an important exception).

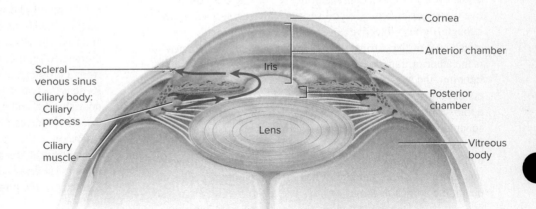

FIGURE 16.27 Production and Reabsorption of Aqueous Humor. Blue arrows indicate the flow of aqueous humor from the ciliary processes into the posterior chamber; through the pupil into the anterior chamber; and finally into the scleral venous sinus, the vein that reabsorbs the fluid.

- The **lens** is composed of flattened, tightly compressed, transparent cells called *lens fibers*. It is suspended behind the pupil by a ring of fibers called the **suspensory ligament (fig. 16.29;** also see fig. 16.26), which attaches it to the ciliary body. Tension on the ligament somewhat flattens the lens so it is about 9.0 mm in diameter and 3.6 mm thick at the middle.

- The **vitreous**[43] **body** is a transparent jelly that fills a space called the *vitreous chamber* behind the lens. The vitreous body maintains *intraocular pressure* within the eye, which supports its spherical shape and enables the extrinsic muscles to pull the eye in different directions without compressing and distorting it. It also keeps the retina smoothly pressed against the wall of the eye, which keeps it in close contact with its blood supply and is

essential for focusing images on the retina. An oblique channel through this body called the *hyaloid canal* is the remnant of an artery present in the embryo (see fig. 16.26).

The Neural Components

The neural components are the retina and optic nerve. The **retina** forms from a cup-shaped outgrowth of the diencephalon called the *optic vesicle* (see fig. 14.4b); it is actually a part of the brain—the only part that can be viewed without dissection. It is a thin transparent membrane attached to the rest of the eye at only two points: the **optic disc,** where the optic nerve leaves the rear (*fundus*) of the eye, and its scalloped anterior margin, the **ora serrata.**[44] The

[43]*vitre* = glassy

[44]*ora* = border, margin; *serrata* = notched, serrated

DEEPER INSIGHT 16.3

CLINICAL APPLICATION

Causes of Blindness

The most common causes of blindness are cataracts, glaucoma, macular degeneration, and diabetic retinopathy.

Cataracts are clouding of the lenses **(fig. 16.28a).** They occur as the lens fibers darken with age, fluid-filled clefts appear between them, and the clefts accumulate debris from degenerating fibers. Cataracts are a common complication of diabetes mellitus, but can also be induced by heavy smoking, ultraviolet radiation, radiation therapy, certain viruses and drugs, and other causes. They cause the vision to appear milky or as if one was looking from behind a waterfall.[45] Cataracts can be treated by replacing the natural lens with a plastic one. The implanted lens improves vision almost immediately, but glasses still may be needed for near vision.

Glaucoma is a state of elevated pressure within the eye that occurs when the scleral venous sinus is obstructed so aqueous humor isn't reabsorbed as fast as it is secreted. Pressure in the anterior and posterior chambers drives the lens back and puts pressure on the vitreous body. The vitreous body presses the retina against the choroid and compresses the blood vessels that nourish the retina. Without a

good blood supply, retinal cells die and the optic nerve may atrophy, producing blindness. Symptoms often go unnoticed until the damage is irreversible. Illusory flashes of light are an early symptom of glaucoma. Late-stage symptoms include dimness of vision,[46] a narrowed visual field **(fig. 16.28b),** and colored halos around artificial lights. Glaucoma can be detected early in the course of regular eye examinations by measurements of intraocular pressure and visual field. It can be halted with drugs or surgery, but lost vision can't be restored.

Macular degeneration is the death of receptor cells in the macula, the central part of the retina and location of the sharpest vision. It can develop so slowly that one fails to notice a change in vision, but eventually it results in loss of vision in the center of the visual field, often making it difficult or impossible to read, drive a car, or do fine daily tasks. It cannot yet be cured, but early detection can allow for treatments to slow its progression.

Diabetic retinopathy causes most adult blindness in the United States. It is a retinal degeneration caused by the effects of diabetes mellitus on the blood vessels that nourish the retina. With early detection and control of diabetes, blindness can be prevented in 90% of cases.

(a)

(b)

FIGURE 16.28 Cataracts and Glaucoma. (a) Lens clouded by cataracts. The visual effect of this is milky or blurry vision. (b) The visual field in advanced glaucoma, with tunnel vision.

a: sruilk/Shutterstock; b: Source: National Eye Institute, National Institutes of Health

[45]*cataract* = waterfall

[46]*glauco* = grayness

Suspensory ligament   Lens

2 mm

FIGURE 16.29 The Lens of the Eye (SEM). Posterior view of the lens and the suspensory ligament that anchors it to the ciliary body.

Ralph C Eagle Jr./Science Source/Getty Images Plus

retina can separate from the wall of the eyeball because of blows to the head or insufficient pressure from the vitreous body. Such a *detached retina* may cause blurry areas in the field of vision. It leads to blindness if the retina remains separated for too long from the blood supply in the choroid.

The retina is examined with an illuminating and magnifying instrument called an *ophthalmoscope* (**fig. 16.30**). Directly posterior to the center of the lens, on the visual axis of the eye, is a patch of cells called the **macula lutea**[47] about 3 mm in diameter. In the center of the macula is a tiny pit, the **fovea**[48] **centralis,** which produces the most finely detailed images for reasons explained later. About 3 mm medial to the macula lutea is the optic disc. Nerve fibers from all regions of the retina converge on this point and leave here in a bundle that constitutes the optic nerve. Blood vessels travel through the core of the optic nerve and enter and leave the eye at the optic disc. Eye examinations serve for more than evaluating the visual system; they allow for a direct, noninvasive examination of blood vessels for signs of hypertension, diabetes mellitus, atherosclerosis, and other vascular diseases.

The optic disc contains no receptor cells, so it produces a **blind spot** in the visual field of each eye. You can detect your blind spot and observe an interesting visual phenomenon with the help of **figure 16.31.** Close or cover your right eye and hold the image about 30 cm (1 ft) from your face. Fixate on the X with your left eye. Without taking your gaze off the X, move your head or the page slightly forward and back, or right and left, until the red dot disappears. This occurs because the image of the dot is falling on the blind spot of your left eye.

You should notice something else happen at the same time as the dot disappears—a phenomenon called **visual filling.** The green bar seems to fill in the space where the dot used to be. This occurs because the brain uses the image surrounding the blind spot to fill in the area

(a)

(b)

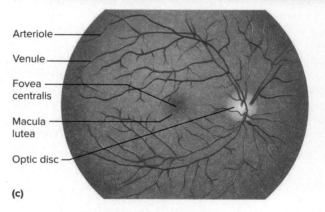

Arteriole

Venule

Fovea
centralis

Macula
lutea

Optic disc

(c)

FIGURE 16.30 The Fundus (Rear) of the Eye. (a) Use of the ophthalmoscope. (b) As seen with an ophthalmoscope. (c) Anatomical features of the fundus. Note the blood vessels diverging from the optic disc, where they enter the eye with the optic nerve. An eye examination also serves as a partial check on cardiovascular health.

a: Peter Dazeley/Getty Images; b: Lisa Klancher

? *Assuming these figures have not been flipped, do they show the subject's right or left eye? How can you tell?*

with similar but imaginary information. The brain acts as if it is better to assume that the unseen area probably looks like its surroundings than to allow a dark blotch to disturb your vision. This is one reason why we don't see a blind patch in the visual field. Another is that the eyes continually undergo minute flickering movements (**saccades**) that ensure that the same area of the visual field doesn't always project onto the same area of retina. What an eye doesn't see at one moment, it will see just milliseconds later when the saccades redirect its visual axis.

---

[47]*macula* = spot; *lutea* = yellow
[48]*fovea* = pit, depression

FIGURE 16.31 Demonstration of the Blind Spot and Visual Filling. See text for explanation of how to conduct this demonstration. Some adjustment of the distance from your face may be needed depending on the print or electronic medium in which you view the image.

Eye saccades are the fastest muscular movements in the human body. Without them, you'd be unable to read this page.

16.5d Formation of an Image

The visual process begins when light rays enter the eye, focus on the retina, and produce a tiny inverted image. When fully dilated, the pupil admits five times as much light as it does when fully constricted. Its diameter is controlled by two sets of contractile elements in the iris: (1) The **pupillary constrictor** consists of smooth muscle cells that encircle the pupil. When stimulated by the parasympathetic nervous system, it narrows the pupil and admits less light to the eye. (2) The **pupillary dilator** consists of a spokelike arrangement of contractile *myoepithelial cells.* When stimulated by the sympathetic nervous system, these cells contract, widen the pupil, and admit more light to the eye (see fig. 15.9). Pupillary constriction and dilation occur in three situations: in response to emotions, changes in light intensity, and a shift in one's gaze between distant and nearby objects. Constriction in response to a shift in gaze is part of the *near response* described shortly.

Pupillary constriction in response to light is called the **photopupillary reflex.** It is mediated by an autonomic reflex arc. When light intensity rises, signals are transmitted from the eye to the *pretectal region* of the upper midbrain. Preganglionic parasympathetic fibers travel by way of the oculomotor nerve from here to the *ciliary ganglion* in the orbit. From the ganglion, postganglionic fibers continue into the eye, where they stimulate the pupillary constrictor.

Sympathetic innervation to the pupil originates, like all other sympathetic efferent fibers, in the spinal cord. Preganglionic fibers ascend from the thoracic cord to the superior cervical ganglion. From there, postganglionic fibers follow the carotid arteries into the head and lead ultimately to the pupillary dilator.

Refraction

Image formation depends on **refraction,** the bending of light rays. Light travels at a speed of 300,000 km/s in a vacuum, but it slows down slightly in air, water, glass, and other media. The *refractive index* of a medium *(n)* is a measure of how much it retards light rays relative to air. The *refractive index* of air is arbitrarily set at $n = 1.00$. If light traveling through air strikes a medium of higher refractive index at a 90° *angle of incidence,* it slows down but doesn't change course—the light rays are not bent. If it strikes at any other angle, however, the light ray changes direction—it is refracted **(fig. 16.32a).** The greater the difference in refractive index between the two media, and the greater the angle of incidence, the stronger the refraction is.

As light enters the eye, it passes from air, with $n = 1.00$, to cornea, with $n = 1.38$. Light rays striking the very center of the cornea pass straight through, but because of the curvature of the cornea, rays striking off center are bent toward the center **(fig. 16.32b).** The aqueous humor has a refractive index of 1.33 and doesn't greatly alter

(a)

Air
$n = 1.00$

Lens
$n = 1.40$

Vitreous body
$n = 1.33$

Retina

Cornea
$n = 1.38$

Aqueous humor
$n = 1.33$

(b)

FIGURE 16.32 Principles of Refraction. (a) A refractive medium does not bend light rays that strike it at a 90° angle but does bend rays that enter or leave it at any other angle. (b) Refractive indices of the media from air to retina. The greater the difference between the refractive indices of two media, the more strongly light rays are refracted when passing from one to the next.

the path of the light. The lens has a refractive index of 1.40. As light passes from air to cornea, the refractive index changes by 0.38; but as it passes from aqueous humor to lens, the refractive index changes by only 0.07. Therefore, the cornea refracts light more than the lens does. The lens merely fine-tunes the image, especially as you shift your focus between near and distant objects.

The Near Response

Emmetropia[49] (EM-eh-TRO-pee-uh) is a state in which the eye is relaxed and focused on an object more than 6 m (20 ft) away, the light rays coming from that object are essentially parallel, and the rays are focused on the retina without effort. If the gaze shifts to something closer, light rays from the source are too divergent to be focused without effort. In other words, the eye is automatically focused on things in the distance unless you make an effort to focus elsewhere. For a wild animal or our prehistoric ancestors, this arrangement would be adaptive because it allows for alertness to predators or prey at a distance.

The **near response (fig. 16.33),** or adjustment to close-range vision, involves three processes to focus an image on the retina:

1. **Convergence of the eyes.** Move your finger gradually closer to a baby's nose and the baby will look cross-eyed at it. This **convergence** of the eyes orients the visual axis of each eye toward the object in order to focus its image on each fovea. If the eyes cannot converge accurately—for example, when the extrinsic muscles are weaker in one eye than in the other—double vision, or *diplopia,*[50] results. The images

[49]*em* = in; *metr* = measure; *opia* = vision
[50]*dipl* = double; *opia* = vision

Emmetropia

Convergence

(a)

Relatively thin lens

Relatively dilated pupil

Fovea

Emmetropia

Relatively thick lens

Relatively constricted pupil

Pupillary miosis and lens accommodation

(b)

FIGURE 16.33 Emmetropia and the Near Response. (a) Superior view of both eyes fixed on an object more than 6 m away (left), and both eyes fixed on an object closer than 6 m (right). (b) Lateral view of the eye fixed on a distant object (top) and nearby object (bottom).

fall on different parts of the two retinas and the brain sees two images. You can simulate this effect by pressing gently on the corner of one eyelid as you look at this text; the image of the text will fall on noncorresponding regions of the two eyes and cause you to see double.

2. **Constriction of the pupil.** Lenses cannot refract light rays at their edges as well as they can closer to the center. The image produced by any lens is therefore somewhat blurry around the edges; this *spherical aberration* is quite evident in an inexpensive microscope. It can be minimized by screening out the peripheral light rays and looking only at the better-focused center. In the eye, the pupil serves this

purpose by constricting as you focus on nearby objects. Like the diaphragm setting (f-stop) of a camera, the pupil thus has a dual purpose: to adjust the eye to variations in brightness and to reduce spherical aberration.

3. **Accommodation of the lens. Accommodation** is a change in the curvature of the lens that enables you to focus on a nearby object. When you look at something nearby, the ciliary muscle surrounding the lens contracts. This narrows the diameter of the ciliary body, relaxes the fibers of the suspensory ligament, and allows the lens to relax into a more convex shape **(fig. 16.34).** In emmetropia, the lens is about 3.6 mm thick at the center; in accommodation, it

Ciliary muscle relaxed

Suspensory ligament taut

Lens thins

(a) Distant vision (emmetropia)

Ciliary muscle contracted

Suspensory ligament relaxed

Lens thickens

(b) Near vision (accommodation)

FIGURE 16.34 Accommodation of the Lens. (a) In the emmetropic eye, the ciliary muscle is relaxed and dilated. It puts tension on the suspensory ligament and flattens the lens. (b) In accommodation, the ciliary muscle contracts and narrows in diameter. This reduces tension on the suspensory ligament and allows the lens to relax into a more convex shape.

thickens to as much as 4.5 mm. A more convex lens refracts light more strongly and focuses the divergent light rays onto the retina. The closest an object can be and still come into focus is called the **near point of vision.** It depends on the flexibility of the lens. The lens stiffens with age, so the near point averages about 9 cm at the age of 10 and 83 cm by the age of 60.

Some common defects in image formation are described in **table 16.2.**

▶▶▶**APPLY WHAT YOU KNOW**

Which extrinsic muscles of the eyes are the prime movers in convergence?

16.5e Sensory Transduction in the Retina

The conversion of light energy into action potentials occurs in the retina. We begin our exploration of this process with the cellular layout of the retina **(fig. 16.36).** From there we go to the pigments that absorb light and then to what happens when light is absorbed.

TABLE 16.2	Common Defects of Image Formation
Astigmatism[51]	Inability to simultaneously focus light rays that enter the eye on different planes. Focusing on vertical lines, such as the edge of a door, may cause horizontal lines, such as a tabletop, to go out of focus. Caused by a deviation in the shape of the cornea so that it is shaped like the back of a spoon rather than part of a sphere. Corrected with *cylindrical lenses,* which refract light more in one plane than another.
Hyperopia[52]	Farsightedness—a condition in which the eyeball is too short. The retina lies in front of the focal point of the lens, and the light rays have not yet come into focus when they reach the retina (see top of **fig. 16.35b**). Causes the greatest difficulty when viewing nearby objects. Corrected with *convex lenses,* which cause light rays to converge slightly before entering the eye, so they reach their focal point farther forward than usual, on the retina of the shortened eyeball.
Myopia[53]	Nearsightedness—a condition in which the eyeball is too long. Light rays come into focus before they reach the retina and begin to diverge again by the time they fall on it (see top of **fig. 16.35c**). Corrected with *concave lenses,* which cause light rays to diverge slightly before entering the eye, shifting the focal point posteriorly so that it falls on the retina of the elongated eyeball.
Presbyopia[54]	Reduced ability to accommodate for near vision with age. Caused by declining flexibility of the lens. Results in difficulty reading and doing close handwork. Corrected with *bifocal lenses* or reading glasses.

(a) Emmetropia (normal)   **(b) Hyperopia (farsightedness)**   **(c) Myopia (nearsightedness)**

FIGURE 16.35 Two Common Visual Defects and the Effects of Corrective Lenses. (a) The normal emmetropic eye, with light rays converging on the retina. (b) Hyperopia (farsightedness) and the corrective effect of a convex lens. (c) Myopia (nearsightedness) and the corrective effect of a concave lens.

---

[51]*a* = not; *stigma* = point; *ism* = condition
[52]*hyper* = excessive; *op* = eye; *ia* = condition

[53]*my* = closed; *op* = eye; *ia* = condition
[54]*presby* = old; *op* = eye; *ia* = condition

Back of eye

Sclera

Choroid

Pigment epithelium

Rod and cone outer segments

Rod and cone nuclei

Bipolar cells

Ganglion cells

Nerve fibers to optic nerve

Vitreous body

Front of eye

25 μm

(a)

Back of eye

Pigment epithelium

Photoreceptors:
Rod
Cone

Transmission of rod signals

Transmission of cone signals

Horizontal cell

Bipolar cell

Amacrine cell

Ganglion cell

To optic nerve ➡

Nerve fibers

Direction of light

(b)

FIGURE 16.36 Histology of the Retina. (a) Photomicrograph. (b) Schematic of the layers and synaptic relationships of the retinal cells.

a: Dennis Strete/McGraw-Hill Education

The most posterior part of the retina is the **pigmented layer,** which serves, like the black inside of a film camera, to absorb stray light so it doesn't reflect back through the retina and degrade the visual image.

The **neural layer** of the retina consists of three principal cell layers. Progressing from the rear of the eye forward, these are composed of *photoreceptor cells, bipolar cells,* and *ganglion cells:*

1. **Photoreceptor cells.** Photoreceptor cells absorb light and generate a chemical or electrical signal. There are three kinds: rods, cones, and certain ganglion cells. Only rods and cones produce visual images; the ganglion cells are discussed later. **Rods** and **cones** aren't neurons, but are related to the ependymal cells of the brain. Each rod or cone has an **outer segment** that points toward the wall of the eye and an **inner segment** facing the interior **(fig. 16.37).** The two segments are separated by a constriction containing nine pairs of microtubules; the outer segment is actually a

highly modified cilium specialized to absorb light. The inner segment contains mitochondria and other organelles. At its base, it gives rise to a cell body, which contains the nucleus, and to processes that synapse with retinal neurons in the next layer.

In a rod, the outer segment is cylindrical and resembles a stack of coins in a wrapper—there is a plasma membrane around the outside and a neatly arrayed stack of about 1,000 membranous discs inside. Each disc is densely studded with globular proteins—the visual pigment *rhodopsin.* The membranes hold these pigment molecules in a position that results in the most efficient light absorption. Rod cells are responsible for **night (scotopic[55]) vision** and produce images only in shades of gray **(monochromatic vision).**

[55] *scot* = dark; *op* = vision

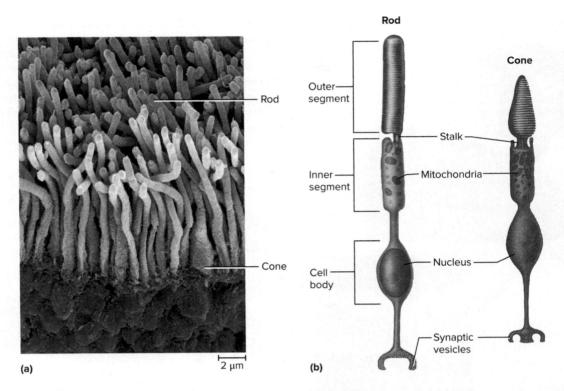

FIGURE 16.37 Rod and Cone Cells. (a) Rods and cones of a human retina (SEM). The outer segments are shown in tan (rods) and green (cones); cell bodies and inner segments are in violet. (b) Structure of human rods and cones. The rod outer segment is shown greatly shortened.

a: Steve Gschmeissner/Science Source

A cone cell is similar but shorter, its outer segment tapers to a point, and the discs are not detached from the plasma membrane but are parallel infoldings of it. Their visual pigment is called *photopsin*. Cones function in brighter light; they are responsible for **day (photopic[56]) vision** as well as **color (trichromatic) vision.**

Rods and cones continually renew their discs by addition of new ones at the proximal (basal) end of the outer segment, while old discs are shed from the distal tips of the cells and phagocytized by cells in the pigment epithelium.

2. **Bipolar cells.** Rods and cones synapse with the dendrites of **bipolar cells,** the first-order neurons of the visual pathway. They in turn synapse with the ganglion cells described next (fig. 16.36b).

3. **Ganglion cells. Ganglion cells** are the largest neurons of the retina, arranged in a single layer close to the vitreous body. They are the second-order neurons of the visual pathway. Most ganglion cells receive input from multiple bipolar cells. Their axons form the optic nerve. Some of the ganglion cells, however, are supplemental photoreceptors; they absorb light directly and transmit signals to brainstem nuclei that control pupillary diameter and the body's circadian rhythms. They don't contribute to visual images but detect only light intensity. Their sensory pigment is called **melanopsin.**

There are other retinal cells, but they don't form layers of their own. **Horizontal cells** and **amacrine**[57] **cells** form horizontal connections between rods, cones, and bipolar cells and intervene in the pathways from receptor cells to ganglion cells. They play diverse roles in enhancing the perception of contrast, the edges of objects, and changes in light intensity. In addition, much of the mass of the retina is composed of astrocytes and other types of glial cells.

There are approximately 130 million rods and 6.5 million cones in one retina, but only 1 million nerve fibers in the optic nerve. With a ratio of nearly 140 receptor cells to one optic nerve fiber, it is obvious that there must be substantial *neural convergence* and information processing in the retina itself before signals are conducted to the brain proper. Convergence occurs where multiple rods or cones synapse with one bipolar cell, and again where multiple bipolar cells feed into one ganglion cell. Later we will examine how convergence functions in visual resolution and night vision.

Visual Pigments

The visual pigment of the rods is called **rhodopsin** (ro-DOP-sin), or **visual purple.** It consists of two major parts (moieties): a protein called **opsin** and a vitamin A derivative called **retinene** or **retinal** (rhymes with "pal") **(fig. 16.38).** A dietary deficiency of vitamin A can lead to **night blindness,** a difficulty seeing in dim light. Opsin is embedded in the disc membranes of the rod's outer segment. All rods contain a single kind of rhodopsin with an absorption peak at

---

[56]*phot* = light; *op* = vision

[57]*a* = without; *macr* = long; *in* = fiber (lacking axons)

(b)

Disc
Cell membrane

(c)

Pigment molecule

(d)

Pigment molecule
Retinal
Opsin

(a)

FIGURE 16.38 Structure and Location of the Visual Pigments. (a) A rod cell. (b) Detail of the rod outer segment. (c) One disc of the outer segment showing the membrane studded with pigment molecules. (d) A pigment molecule, embedded in the unit membrane of the disc, showing the protein moiety, opsin, and the vitamin A derivative, retinal. (e) *Cis*-retinal, the isomer present in the absence of light. (f) *Trans*-retinal, the isomer produced when the pigment absorbs a photon of light.

(e)   *Cis*-retinal

(f)   *Trans*-retinal (bleached)

a wavelength of 500 nm. Rods are less sensitive to light of other wavelengths and cannot distinguish one color from another.

In cones, the pigment is called **photopsin.** Its retinal moiety is the same as that of rhodopsin, but the opsin moiety has a different amino acid sequence that determines which wavelengths of light the pigment absorbs. There are three kinds of cones, which are identical in appearance but optimally absorb different wavelengths of light. These differences enable us to perceive different colors.

Generating the Optic Nerve Signal

The events of sensory transduction are probably the same in rods and cones, but are better known in rods. In the dark, their retinal has a bent shape called *cis*-retinal. When it absorbs light, it changes to a straight form called *trans*-retinal and breaks away from the opsin **(fig. 16.39).** This is called **bleaching,** because purified rhodopsin changes from violet to colorless in the light. For a rod to continue functioning, *trans*-retinal has to be converted back to the *cis* form and reunited with opsin, producing functional rhodopsin at a rate that keeps pace with bleaching. Fifty percent of the bleached rhodopsin is regenerated in about 5 minutes. Cones are faster and their photopsin is 50% regenerated in about 90 seconds.

In the dark, rods don't sit quietly doing nothing. They steadily release the neurotransmitter glutamate from the basal end of the cell **(fig. 16.40a).** When a rod absorbs light, glutamate secretion ceases **(fig. 16.40b).** We won't delve into the mechanistic details of why this occurs. The important point is that the bipolar cells, next in line, are sensitive to these on and off pulses of glutamate secretion. Some bipolar cells are inhibited by glutamate and excited

when its secretion stops; these cells are therefore excited by rising light intensities. Other bipolar cells are excited by glutamate and therefore respond when light intensity drops. As your eye scans a scene, it passes areas of greater and lesser brightness. Their images on the retina cause a rapidly changing pattern of bipolar cell responses as the light intensity on a patch of retina rises and falls.

When bipolar cells detect fluctuations in light intensity, they stimulate ganglion cells either directly (by synapsing with them) or indirectly (via pathways that go through amacrine cells). Ganglion cells are the only retinal cells that produce action potentials; all other retinal cells produce only graded local potentials. Ganglion cells respond to the bipolar cells with rising and falling firing frequencies. Via the optic nerve, these changes provide visual signals to the brain.

16.5f Light and Dark Adaptation

The human eye can detect light intensities ranging from a single photon at threshold (near-total darkness) to 10 billion times as bright (the intensity of bright sunlight reflected from snow). **Light adaptation** is an adjustment in vision that occurs when you go from a dark or dimly lit area into brighter light. If you wake up in the night and turn on a lamp, at first you see a harsh glare; you may experience discomfort from the overstimulated retinas. Your pupils quickly constrict to reduce the intensity of stimulation, but color vision and visual acuity (the ability to see fine detail) remain below normal for 5 to 10 minutes—the time needed for pigment bleaching to adjust retinal sensitivity to this light intensity. The

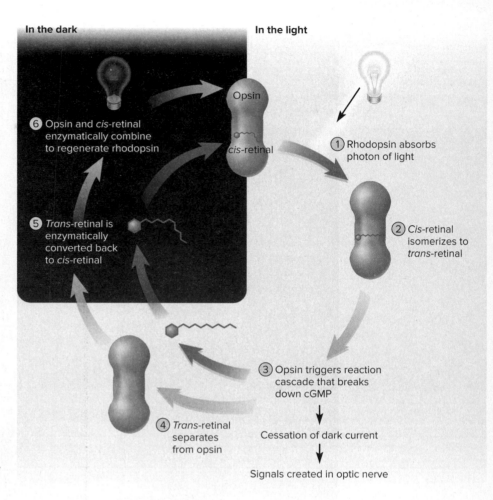

In the dark / **In the light**

(6) Opsin and *cis*-retinal enzymatically combine to regenerate rhodopsin

(5) *Trans*-retinal is enzymatically converted back to *cis*-retinal

Opsin / *cis*-retinal

(1) Rhodopsin absorbs photon of light

(2) *Cis*-retinal isomerizes to *trans*-retinal

(3) Opsin triggers reaction cascade that breaks down cGMP

(4) *Trans*-retinal separates from opsin

Cessation of dark current

Signals created in optic nerve

FIGURE 16.39 The Bleaching and Regeneration of Rhodopsin. Numbers 1 through 4 indicate the bleaching events that occur in the light; numbers 5 and 6 indicate the regenerative events that are independent of light. The regenerative events occur in light and dark, but in the light, they are outpaced by bleaching.

rods bleach quickly in bright light, and cones take over. Even in typical indoor light, rod vision is nonfunctional.

On the other hand, suppose you're sitting in a brightly lit room reading at night, and there is a power failure. Your eyes must undergo **dark adaptation** before you can see well enough to find your way in the dark. There isn't enough light for cone (photopic) vision, and it takes a little time for the rods (scotopic vision) to adjust to the dark. Your rod pigment was bleached by the lights in the room while the power was on, but now in the relative absence of light, rhodopsin regenerates faster than it bleaches. In a minute or two, scotopic vision begins to function, and after 20 to 30 minutes, the amount of regenerated rhodopsin is sufficient for your eyes to reach essentially maximum sensitivity. Dilation of the pupils also helps by admitting more light to the eye.

16.5g The Dual Visual System

You may wonder why we have both rods and cones. Why can't we simply have one type of receptor cell that produces detailed color vision, both day and night? The **duplicity theory** of vision holds that a single receptor system cannot produce both high sensitivity and high resolution. It takes one type of cell and neural circuit to provide sensitive night vision and a different type to provide high-resolution daytime vision.

The high sensitivity of rods in dim light stems partly from the extensive neural convergence that occurs between the rods and

ganglion cells. Up to 600 rods converge on each bipolar cell, and many bipolar cells converge on each ganglion cell. This allows for a high degree of *spatial summation* (**fig. 16.41a**). Weak stimulation of many rods can produce an additive effect on one bipolar cell, and several bipolar cells can collaborate to excite one ganglion cell. Thus, a ganglion cell can respond in dim light that only weakly stimulates any individual rod. Scotopic vision is functional even at a light intensity less than starlight reflected from a sheet of white paper. A shortcoming of this system is that it cannot resolve finely detailed images. One ganglion cell receives input from all the rods in about 1 mm^2 of retina—its receptive field. What the brain perceives is therefore a coarse, grainy image similar to an overenlarged photograph or low-resolution digital image.

Around the edges of the retina, receptor cells are especially large and widely spaced. If you fixate on the middle of this page, you will notice that you can't read the words near the margins. Visual acuity decreases rapidly as the image falls away from the fovea centralis. Our peripheral vision is a low-resolution system that serves mainly to alert us to motion in the periphery and to stimulate us to look that way to identify what is there.

When you look directly at something, its image falls on the fovea, which is occupied by about 4,000 tiny cones and no rods. The other neurons of the fovea are displaced to one side so they won't interfere with light falling on the cones. The smallness of these cones is like the smallness of the dots or pixels in a fine-grained (high-resolution) photograph; it is partially responsible for

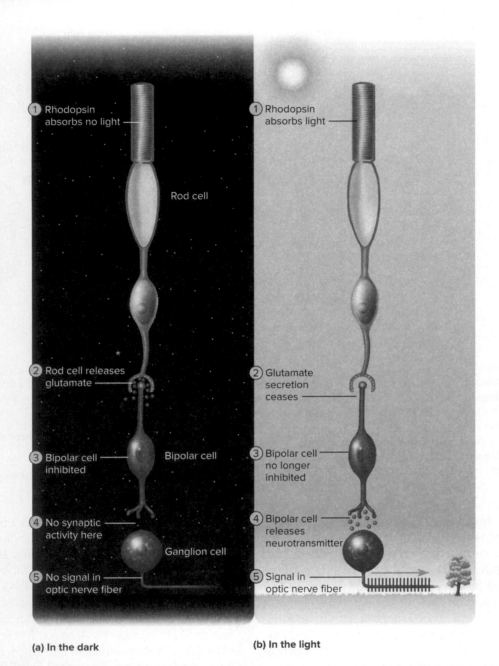

(a) In the dark

(b) In the light

FIGURE 16.40 Mechanism of Generating Visual Signals. (a) In complete darkness, rod cells are active whereas certain bipolar cells and the ganglion cells are not. (b) As rod cells absorb light, they are inhibited and these bipolar cells and ganglion cells are activated. Cones are believed to function similarly. Some bipolar cells behave differently from the type used in this example.

the high-resolution images formed at the fovea. In addition, the cones here show no neural convergence. Each cone synapses with only one bipolar cell and each bipolar cell with only one ganglion cell. This gives each foveal cone a "private line to the brain," and each ganglion cell of the fovea reports to the brain on a receptive field of just 2 μm^2 of retinal area **(fig. 16.41b).** Cones distant from the fovea exhibit some neural convergence but not nearly as much as rods do. The price of this lack of convergence at the fovea, however, is that cone cells are incapable of spatial summation, and the cone system is therefore less sensitive to light. The threshold of photopic (cone) vision lies between the intensity of starlight and moonlight reflected from white paper.

▶▶▶**APPLY WHAT YOU KNOW**

If you look directly at a dim star in the night sky, it disappears, and if you look slightly away, it reappears. Why?

16.5h Color Vision

Most nocturnal vertebrates have only rod cells, but many diurnal animals are endowed with cones and color vision. Color vision is especially well developed in primates for evolutionary reasons discussed in chapter 1 (see section 1.4b). It is based on three kinds of cones named for the absorption peaks of their photopsins: **short-wavelength (S) cones,** with peak sensitivity at a wavelength of

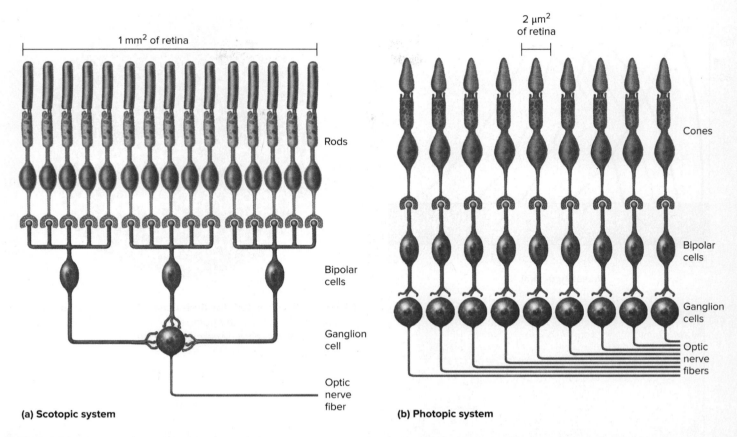

FIGURE 16.41 Retinal Circuitry and Visual Sensitivity. (a) In the scotopic (night vision) system, many rods converge on each bipolar cell and many bipolar cells converge on each ganglion cell. Note the large receptive field (1 mm^2), which makes this a low-resolution system. (b) In the photopic (day vision) system, there is little neural convergence. Note the small receptive field (2 μm^2), which makes this a high-resolution system.

420 nm; **medium-wavelength (M) cones,** which peak at 531 nm; and **long-wavelength (L) cones,** which peak at 558 nm. These were formerly called *blue, green,* and *red cones*—a less accurate terminology. "Red" cones, for example, don't peak in the red part of the spectrum (558 nm light is perceived as orange-yellow), but they are the only cones that respond at all to red light.

Our perception of colors is based on a mixture of nerve signals representing cones with different absorption peaks. In **figure 16.42,** note that light at 400 nm excites only S cones. At 500 nm, all three types of cones are stimulated; the L cones respond at 60% of their maximum capacity, M cones at 82% of their maximum, and S cones at 20%. The brain interprets this mixture as blue-green. The table in this figure shows how some other color sensations are generated by other response ratios.

Some individuals have a hereditary alteration or lack of one photopsin or another and thus exhibit **color blindness.** The most common form is *red–green color blindness,* which results from a lack of either L or M cones and causes difficulty distinguishing these and related shades from each other. For example, a person with normal *trichromatic* color vision sees **figure 16.43** as showing the number 74, whereas a person with red–green color blindness sees no number. Red–green color blindness is a sex-linked recessive trait. It occurs in about 8% of males and 0.5% of females.

16.5i Stereoscopic Vision

Stereoscopic vision (stereopsis) is depth perception—the ability to judge how far away objects are. It depends on having two eyes with overlapping visual fields, which allows each eye to look at the same object from a different angle. Stereoscopic vision contrasts with the *panoramic vision* of mammals such as rodents and horses, in which the eyes are on opposite sides of the head. Although stereoscopic vision covers a smaller visual field than panoramic vision and provides less alertness to sneaky predators, it has the advantage of depth perception. The evolutionary basis of our stereoscopic vision was also explained along with color vision in section 1.4b.

When you fixate on something within 30 m (100 ft), each eye views it from a slightly different angle and focuses its image on the fovea centralis. The point in space on which the eyes are focused is called the *fixation point.* Objects farther away than the fixation point cast an image somewhat medial to the foveas, and closer objects cast their images more laterally **(fig. 16.44).** The distance of an image from the two foveas provides the brain with information used to judge the position of other points relative to the fixation point.

16.5j The Visual Projection Pathway

The optic nerves arise from the retinal ganglion cells and leave each orbit through the optic canal. They converge to form an X, the

Wavelength (nm)	Percentage of maximum cone response (S : M : L)	Perceived hue
400	50 : 0 : 0	Violet
450	72 : 30 : 0	Blue
500	20 : 82 : 60	Blue-green
550	0 : 85 : 97	Green
625	0 : 3 : 35	Orange
675	0 : 0 : 5	Red

FIGURE 16.42 Absorption Spectra of the Retinal Cells. In the middle column of the table, each number indicates how strongly the respective cone cells respond as a percentage of their maximum capability for a given light intensity. At 550 nm, for example, L cones respond at 97% of their maximum, M cones at 85%, and S cones not at all. The result is a perception of green light.

❓ *If you were to add another row to this table for 600 nm, what would you enter in the middle and right-hand columns?*

optic chiasm[58] (ky-AZ-um), on the base of the brain anterior to the pituitary. Beyond this, the same fibers continue as a pair of **optic tracts.**

Within the chiasm, half of the fibers from each optic nerve cross over to the opposite side of the brain **(fig. 16.45).** This is called **hemidecussation,**[59] since only half of the fibers decussate. As a result, the right cerebral hemisphere sees objects in the left visual field, because their images fall on the right half of each retina (the medial half of the left eye and lateral half of the right eye). You can trace the nerve fibers from each half-retina in the figure to see that they lead to the right hemisphere. Conversely, the left hemisphere sees objects in the right visual field. Since the right brain controls motor responses on the left side of the body and vice versa, each side of the brain sees what is on the side of the body where it exerts motor control. In horses and other animals with panoramic vision, nearly 100% of the optic nerve fibers of the right eye decussate to the left brain and vice versa.

[58]*chiasm* = cross, X
[59]*hemi* = half; *decuss* = to cross, form an X

FIGURE 16.43 A Test for Red–Green Color Blindness. Persons with normal vision see the number 74. Persons with red–green color blindness see no discernible number.

Steve Allen/Getty Images

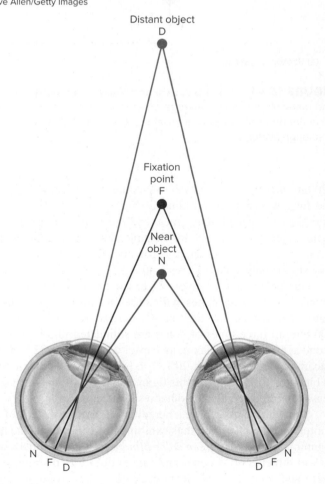

FIGURE 16.44 The Retinal Basis of Stereoscopic Vision (Depth Perception). When the eyes converge on the fixation point (F), more distant objects (D) are focused on the retinas medial to the fovea and the brain interprets them as being farther away than the fixation point. Nearby objects (N) are focused lateral to the fovea and interpreted as being closer.

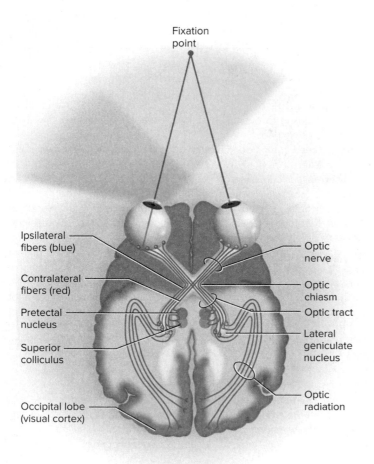

Fixation point

Ipsilateral fibers (blue)

Contralateral fibers (red)

Pretectal nucleus

Superior colliculus

Occipital lobe (visual cortex)

Optic nerve

Optic chiasm

Optic tract

Lateral geniculate nucleus

Optic radiation

FIGURE 16.45 The Visual Projection Pathway. Diagram of hemidecussation and projection to the primary visual cortex. Blue and yellow indicate the visual fields of the left and right eyes; green indicates the area of overlap and stereoscopic vision. Nerve fibers from the medial side of the right eye (red) descussate to the left side of the brain, while fibers from the lateral side remain on the right side of the brain. The converse is true of the left eye. The right occipital lobe thus monitors the left side of the visual field and the left occipital lobe monitors the right side.

? *If a stroke destroyed the optic radiation of the right cerebral hemisphere, how would it affect a person's vision? Would it affect the person's visual reflexes?*

Most axons of the optic tracts end in the **lateral geniculate**[60] **nucleus** (jeh-NIC-you-late) of the thalamus. Third-order neurons arise here and form the **optic radiation** of fibers in the white matter of the cerebrum. These project to the primary visual cortex of the occipital lobe, where conscious visual sensation

---

[60]*geniculate* = bent like a knee

occurs. Lesions along these tracts or in the occipital lobe can cause blindness even if the eyes are fully functional, with various effects on one of both eyes, and part or all of the visual field of either eye, depending on where the lesion occurs.

A few optic nerve fibers come from the photosensitive, melanopsin-containing ganglion cells and take a different route, ending in the superior colliculi and pretectal nuclei of the mid-brain. The superior colliculi control the visual reflexes of the ex-trinsic eye muscles, and the pretectal nuclei are involved in the photopupillary and accommodation reflexes.

Space doesn't allow us to consider much about the very complex processes of visual information processing in the brain. Some processing, such as contrast, brightness, motion, and stereopsis, begins in the retina. The primary visual cortex in the occipital lobe is connected by association tracts to association areas in the temporal and parietal lobes. The primary visual cortex relays information forward to association areas in the tempo-ral and parietal lobes for further processing. One pathway, called the **ventral stream,** runs forward through the lower temporal lobe. It carries signals mainly from cone cells and the central part of the retina. It is concerned with color vision, object rec-ognition, and visual memory, including our ability to read, rec-ognize faces, and identify other things we see. The other, called the **dorsal stream,** runs forward through the upper parietal lobe. It carries signals mainly from rods and the peripheral retina. It is concerned with recognizing the locations and spatial relation-ships of objects and with analysis of motion. What is yet to be learned about visual processing has important implications for biology, medicine, psychology, and even philosophy.

The effects of aging on the senses—especially taste, smell, hearing, and vision—are described at "Sense Organs" in sec-tion 29.4a.

BEFORE YOU GO ON

Answer the following questions to test your understanding of the preceding section:

24. Why can't we see wavelengths of 350 nm or 750 nm?

25. Why are light rays bent (refracted) more by the cornea than by the lens?

26. List as many structural and functional differences between rods and cones as you can.

27. Explain how the absorption of a photon of light leads to excitation of an optic nerve fiber.

28. Discuss the duplicity theory of vision, summarizing the advantage of having separate types of retinal photore-ceptor cells and neural circuits for photopic and scotopic vision.

DEEPER INSIGHT 16.4

MEDICAL HISTORY

Anesthesia—From Ether Frolics to Modern Surgery

Surgery is as old as civilization. People from the Stone Age to the pre-Columbian civilizations of the Americas practiced *trephination*—cutting a hole in the skull to let out "evil spirits" that were thought to cause headaches. The ancient Hindus were expert surgeons for their time, and the Greeks and Romans pioneered military surgery. But until the nineteenth century, surgery was a miserable and dangerous business, done only as a last resort and with little hope of the patient's survival. Surgeons rarely attempted anything more complex than amputations or kidney stone removal. A surgeon had to be somewhat indifferent to the struggles and screams of his patient. Most operations had to be completed in 3 minutes or less, and a strong arm and stomach were more important qualifications for a surgeon than extensive anatomical knowledge.

At least three things were needed for surgery to be more effective: better knowledge of anatomy, *asepsis*[61] for the control of infection, and *anesthesia*[62] for the control of pain. Early efforts to control surgical pain were crude and usually ineffective, such as choking a patient into unconsciousness and trying to complete the surgery before he or she awoke. Alcohol and opium were often used as anesthetics, but the dosage was poorly controlled; some patients were underanesthetized and suffered great pain anyway, and others died of overdoses. Often there was no alternative but for a few strong men to hold the struggling patient down as the surgeon worked. Charles Darwin originally intended to become a physician, but left medical school because he was sickened by observing "two very bad operations, one on a child," in the days before anesthesia.

In 1799, the English chemist Sir Humphry Davy suggested using nitrous oxide to relieve pain. His student, Michael Faraday, later a distinguished physicist, suggested ether. Neither of these ideas caught on for several decades, however. Nitrous oxide ("laughing gas") was a popular amusement in the 1800s, when traveling showmen went from town to town demonstrating its effects on volunteers from the audience. In 1841, at a medicine show in Georgia, some students were impressed with the volunteers' euphoric giggles and antics and asked a young local physician, Crawford W. Long, if he could make some nitrous oxide for them. Long lacked the equipment to synthesize it, but he recommended they try ether. Ether was commonly used in small oral doses for toothaches and "nervous ailments," but its main claim to popularity was its use as a party drug for so-called ether frolics. Long himself was a bit of a bon vivant who put on demonstrations for some of the young ladies, with the disclaimer that he could not be held responsible for whatever he might do under the influence of ether (such as stealing a kiss).

At these parties, Long noted that people sometimes fell and suffered considerable cuts and bruises without feeling pain. In 1842, he had a patient who was terrified of pain but needed a tumor removed from his neck. Long excised the tumor without difficulty as his patient sniffed ether from a towel. The operation created a sensation in town, but other physicians ridiculed Long and pronounced anesthesia dangerous. His medical practice declined as people grew afraid of him, but over the next 4 years he performed eight more minor surgeries on patients under ether. Struggling to overcome criticisms that the effects he saw were due merely to hypnotic suggestion or individual variation in pain sensitivity, Long even compared surgeries done on the same person with and without ether.

Long failed to publish his results quickly enough, and in 1844 he was scooped by a Connecticut dentist, Horace Wells, who had tried nitrous oxide as a dental anesthetic. Another dentist, William Morton of Boston, had tried everything from champagne to opium to kill pain in his patients. He too became interested in ether and gave a demonstration at Massachusetts General Hospital, where he etherized a patient and removed a tumor before a medical audience. Within a month of this successful and sensational demonstration, ether was being used in other cities of the United States and England. Morton patented a "secret formula" he called Morton's Letheon,[63] which smelled suspiciously of ether, but eventually he went broke in a greedy effort to monopolize ether anesthesia, and he died a pauper. His grave near Boston bears the epitaph:

WILLIAM T. G. MORTON
Inventor and Revealer of Anaesthetic Inhalation
Before Whom, in All Time, Surgery Was Agony.
By Whom Pain in Surgery Was Averted and Annulled.
Since Whom Science Has Control of Pain.

Wells, who had engaged in a bitter feud to establish himself as the inventor of ether anesthesia, committed suicide at the age of 33. Crawford Long ran a successful medical practice in Athens, Georgia, but to his death he remained disappointed that he had not received credit as the first to perform surgery on etherized patients. The Emory University Hospital Midtown in Atlanta was named the Crawford W. Long Memorial Hospital from 1931 to 2009 in his honor.

Ether and chloroform became obsolete when safer anesthetics were developed. *General anesthetics* such as isoflurane render a patient unconscious by crossing the blood–brain barrier and blocking nervous transmission through the brainstem. Certain general anesthetics deaden pain by activating GABA receptors and causing an inflow of Cl^-, which hyperpolarizes neurons and makes them less likely to fire. Diazepam (Valium) also employs this mechanism. *Local anesthetics* such as procaine (Novocaine) and tetracaine selectively deaden specific nerves. They decrease the permeability of membranes to Na^+, thereby reducing their ability to produce action potentials.

A sound knowledge of anatomy, control of infection and pain, and development of better tools converged to allow surgeons time to operate more carefully. As a result, surgery became more intellectually challenging and interesting. It attracted a more educated class of practitioner, which put it on the road to becoming the remarkable lifesaving approach that it is today.

---

[61]*a* = without; *sepsis* = infection
[62]*an* = without; *esthesia* = feeling, sensation

[63]*lethe* = oblivion, forgetfulness

STUDY GUIDE

▶ Assess Your Learning Outcomes

To test your knowledge, discuss the following topics with a study partner or in writing, ideally from memory.

16.1 Properties and Types of Sensory Receptors

1. The definition of *receptor* and the range of complexity in sensory receptors
2. The definition of *sensory transduction* and the relationship of neural action potentials to that concept
3. The production and role of the receptor potential in sensory transduction
4. Four kinds of stimulus information transmitted by sensory receptors
5. Five categories of receptors classified by stimulus modality
6. Three categories of receptors classified by origin of their stimuli
7. Differences between general (somatosensory) and special senses

16.2 The General Senses

1. Three types of unencapsulated sensory nerve endings, and what it means to say a nerve ending is *unencapsulated*
2. Six types of encapsulated nerve endings and how these differ from unencapsulated endings
3. The main routes for somatosensory signals from the head, and from the lower body, to the brainstem; the final destination of proprioceptive signals in the brain; and the final destination of most other sensory signals
4. Definitions of *pain* and *nociceptor*
5. Differences between fast pain and slow pain
6. Differences between nociceptive and neuropathic pain, and the three types of nociceptive pain
7. Differences between fast and slow pain, and the types of nerve fibers responsible for each
8. The general three-neuron pathway typically taken by pain signals to the cerebral cortex
9. Pain pathways from receptors in the head to the cerebral cortex, including the cranial nerves that carry pain signals
10. Pain pathways from the lower body to the cerebral cortex and reticular formation, including the spinal cord tracts that carry pain signals
11. Referred pain and its anatomical basis
12. Examples of the modulation of pain sensitivity by the brain

13. Spinal gating mechanisms that modify a person's sense of pain and awareness of body injury
14. Names of some analgesic neuropeptides and how they affect the sensation of pain

16.3 The Chemical Senses

1. Structure and locations of the taste buds
2. Types, locations, and functions of lingual papillae
3. Five primary taste sensations, and sensations other than taste that play a part in flavor
4. Mechanisms by which sugars, salts, alkaloids, acids, and glutamate excite taste cells
5. Which nerves carry taste signals, what routes they take to the brain, and what brain centers receive gustatory input
6. Structure and location of the olfactory mucosa and its receptor cells
7. How odor molecules excite olfactory cells
8. Which cranial nerve carries olfactory signals to the brain, and the route and point of termination of its nerve fibers
9. Sensory routes from the olfactory bulbs to the temporal lobes, insula, orbitofrontal cortex, hippocampus, amygdala, and hypothalamus; and the olfactory functions of these respective signal destinations
10. How the cerebral cortex influences olfactory bulb function and one's perception of smell

16.4 Hearing and Equilibrium

1. How sound is generated; what physical properties of a sound wave are measured in hertz and decibels; and what sensory qualities of sound correspond to those two physical properties
2. The total range of human hearing, in hertz, and the narrower range in which humans hear best
3. The decibel level of ordinary conversation and the thresholds of hearing and pain
4. Boundaries between the outer ear, middle ear, and inner ear
5. Anatomy of the outer ear and the function of its cerumen and guard hairs
6. Structure of the tympanic membrane and tympanic cavity; the names, structures, and anatomical arrangement of the auditory ossicles; the two middle-ear muscles; and anatomy of the auditory tube, its contribution to hearing, and its relevance to middle-ear infections

7. The bony and membranous labyrinths of the inner ear; the names and distribution of the two inner-ear fluids in relation to the labyrinths
8. Size and shape of the cochlea and its relationship to the modiolus
9. Cross-sectional anatomy of the cochlea
10. Structure of the spiral organ, especially the hair cells and tectorial membrane; differences between inner and outer hair cells
11. Function of the middle-ear ossicles and muscles; how the tympanic reflex works and how it protects one's hearing
12. How vibrations of the tympanic membrane lead to stimulation of the cochlear nerve
13. How the cochlea codes for differences in the pitch and loudness of sounds
14. How the outer hair cells tune the cochlea to improve its sensitivity to differences in pitch
15. The pathway from cochlear nerve to auditory centers of the brain; the feedback pathway from the pons back to the cochlea, and its purpose
16. Differences between static and dynamic equilibrium and between linear and angular acceleration
17. Structure of the saccule and utricle and the relevance of the spatial orientation of the macula in each one
18. How linear acceleration stimulates the hair cells of the saccule and utricle during linear acceleration; how the body senses the difference between vertical and horizontal acceleration
19. Structure of the semicircular ducts, especially the crista ampullaris and cupula
20. How acceleration stimulates hair cells of the crista ampullaris, and why the combined input of the six semicircular ducts enables the brain to sense tilting or rotation of the head in any direction
21. Why it cannot be said that the vestibular system senses motion of the head, but only changes in the rate of motion
22. The path taken by signals in the vestibular nerve to the cerebrum, cerebellum, reticular formation, spinal cord, and nuclei of the three cranial nerves for eye movements
23. Why it is important for eye movement to be coordinated with vestibular input

16.5 Vision

1. The definition of *vision* and the range of electromagnetic wavelengths over which human vision occurs
2. Six extrinsic eye muscles, their anatomy, the eye motions they produce, and the cranial nerves that control them
3. Components of the lacrimal apparatus and the route taken by tears as they wash over the eye and drain into the nasal cavity
4. Anatomy and functions of the eyebrow, eyelids, eyelashes, and conjunctiva
5. Three tunics of the eyeball and the structural components of each
6. Optical components of the eye
7. The secretion, flow, and reabsorption of aqueous humor
8. General structure of the retina; its two points of attachment to the wall of the eye; and the locations, structure, and functional significance of the optic disc, optic nerve, macula lutea, and fovea centralis
9. The cause of the blind spot and how the brain compensates for it

10. Structure and action of the pupillary constrictor and dilator; anatomy of their autonomic innervation; and the photopupillary reflex
11. Principles of refraction; points at which refraction occurs as light enters the eye; relative contributions of the cornea and lens to image formation, and the reason for the difference
12. The difference between emmetropia and the near response, and three mechanisms of the near response
13. Histological layers and cell types of the retina; three types of photoreceptor cells and their respective functions; functions of the retinal bipolar, ganglion, horizontal, and amacrine cells
14. The structures of rods and cones; where visual pigments are contained in these cells; the general structure of rhodopsin and photopsin; and how these two pigments differ
15. Differences in rod and cone function
16. How light absorption generates an optic nerve signal

17. Mechanisms of light and dark adaptation
18. Why a single retinal receptor system cannot achieve both low-threshold night vision and high-resolution day vision; why rod vision works in very low light but sacrifices resolution to do so; and why cone vision gives high resolution but sacrifices light sensitivity
19. Difference between S, M, and L cones; how neural coding and three cone types produce sensitivity to innumerable colors; and what causes color blindness
20. The retinal basis of stereoscopic vision
21. Projection pathways from the eyes to the occipital lobe and to the superior colliculi and pretectal nuclei of the midbrain
22. Hemidecussation, where it occurs, and how it determines what areas of the visual field are seen by the right and left occipital lobes
23. Brain centers of higher-order spatial and object recognition

▶ Testing Your Recall

Answers in Appendix A

1. Hot and cold stimuli are detected by
   a. free nerve endings.
   b. proprioceptors.
   c. end bulbs.
   d. lamellar corpuscles.
   e. tactile corpuscles.

2. _____ is a neurotransmitter that transmits pain sensations to second-order spinal neurons.
   a. Endorphin
   b. Enkephalin
   c. Substance P
   d. Acetylcholine
   e. Norepinephrine

3. _____ is a neuromodulator that blocks the conduction of pain signals by second-order spinal neurons.
   a. Endorphin
   b. Enkephalin
   c. Substance P
   d. Acetylcholine
   e. Norepinephrine

4. Most taste buds occur in
   a. the vallate papillae.
   b. the fungiform papillae.

   c. the filiform papillae.
   d. the palate.
   e. the lips.

5. The higher the frequency of a sound,
   a. the louder it sounds.
   b. the harder it is to hear.
   c. the more it stimulates the distal end of the spiral organ.
   d. the faster it travels through air.
   e. the higher its pitch.

6. Cochlear hair cells rest on
   a. the tympanic membrane.
   b. the secondary tympanic membrane.
   c. the tectorial membrane.
   d. the vestibular membrane.
   e. the basilar membrane.

7. The acceleration you feel when an elevator begins to rise is sensed by
   a. the anterior semicircular duct.
   b. the spiral organ.
   c. the crista ampullaris.
   d. the macula sacculi.
   e. the macula utriculi.

8. The color of light is determined by
   a. its velocity.
   b. its amplitude.
   c. its wavelength.
   d. refraction.
   e. how strongly it stimulates the rods.

9. The retina receives its oxygen supply from
   a. the hyaloid canal.
   b. the vitreous body.
   c. the choroid.
   d. the pigment epithelium.
   e. the scleral venous sinus.

10. Which of the following statements about photopic vision is false?
    a. It is mediated by the cones.
    b. It has a low threshold.
    c. It produces fine resolution.
    d. It does not function in starlight.
    e. It does not employ rhodopsin.

11. The most finely detailed vision occurs when an image falls on a pit in the retina called the _____.

12. The only cells of the retina that generate action potentials are the _____ cells.

STUDY GUIDE

13. The visual pigment of a cone cell is _____.

14. The gelatinous membranes of the macula sacculi and macula utriculi are weighted by protein–calcium carbonate granules called _____.

15. Three rows of _____ in the cochlea have V-shaped arrays of stereocilia and tune the frequency sensitivity of the cochlea.

16. The _____ is a tiny bone that vibrates in the oval window and thereby transfers sound vibrations to the inner ear.

17. The _____ of the midbrain receives auditory input and triggers the head-turning auditory reflex.

18. The function of the _____ in a taste bud is to replace dead taste cells.

19. Olfactory neurons synapse with mitral cells and tufted cells in the _____, which lies inferior to the frontal lobe.

20. In the phenomenon of _____, pain from the viscera is perceived as coming from an area of the skin.

▶ Building Your Medical Vocabulary

Answers in Appendix A

State a meaning of each word element, and give a medical term from this chapter that uses it or a slight variation of it.

1. bin-

2. decuss-

3. hemi-

4. lacrimo-

5. litho-

6. maculo-

7. noci-

8. scoto-

9. -sepsis

10. tricho-

▶ What's Wrong with These Statements?

Answers in Appendix A

Briefly explain why each of the following statements is false, or reword it to make it true.

1. The sensory (afferent) nerve fibers for touch end in the thalamus.

2. Things we touch with the left hand are perceived in the left cerebral hemisphere.

3. Things we see with the left eye are perceived only in the right cerebral hemisphere.

4. All chemoreceptors can be classified as exteroceptors.

5. The vitreous body occupies the posterior chamber of the eye.

6. Descending analgesic fibers prevent pain signals from reaching the spinal cord.

7. Signals for hearing are carried in cranial nerve VIII and signals for equilibrium in cranial nerve IX.

8. The tympanic cavity and membranous labyrinth are filled with endolymph.

9. Rods and cones release their neurotransmitter when stimulated by light, but not in total darkness.

10. All of the extrinsic muscles of the eye are controlled by the oculomotor nerve.

▶ Testing Your Comprehensionn

1. The principle of neural convergence is explained under "Neural Pools and Circuits" in section 12.6d. Discuss its relevance to referred pain and scotopic vision.

2. What type of cutaneous receptor enables you to feel an insect crawling through your hair? What type enables you to palpate a patient's pulse? What type enables a blind person to read braille?

3. Contraction of a muscle usually puts more tension on a structure, but contraction of the ciliary muscle puts less tension on the lens. Explain how.

4. In figure 16.45, suppose a stroke or trauma destroyed just the red fibers in the circled area labeled "Optic radiation." Predict how this would affect the subject's visual field for each eye. Predict the same for a lesion

that cut completely across the circled area labeled "Optic tract."

5. What would be the benefit of a drug that blocks the receptors for substance P?

Vascular corrosion cast of the blood vessels of the thyroid gland (SEM)
Susumu Nishinaga/Science Source

CHAPTER OUTLINE

17.1 Overview of the Endocrine System
- 17.1a Comparison of Endocrine and Exocrine Glands
- 17.1b Comparison of the Nervous and Endocrine Systems
- 17.1c Hormone Nomenclature

17.2 The Hypothalamus and Pituitary Gland
- 17.2a Anatomy
- 17.2b Hypothalamic Hormones
- 17.2c Anterior Pituitary Hormones
- 17.2d Posterior Pituitary Hormones
- 17.2e Control of Pituitary Secretion
- 17.2f A Further Look at Growth Hormone

17.3 Other Endocrine Glands
- 17.3a The Pineal Gland
- 17.3b The Thymus
- 17.3c The Thyroid Gland
- 17.3d The Parathyroid Glands
- 17.3e The Adrenal Glands
- 17.3f The Pancreatic Islets
- 17.3g The Gonads
- 17.3h Endocrine Functions of Other Tissues and Organs

17.4 Hormones and Their Actions
- 17.4a Hormone Chemistry
- 17.4b Hormone Synthesis
- 17.4c Hormone Secretion
- 17.4d Hormone Transport
- 17.4e Hormone Receptors and Modes of Action
- 17.4f Signal Amplification
- 17.4g Modulation of Target-Cell Sensitivity
- 17.4h Hormone Interactions
- 17.4i Hormone Clearance

17.5 Stress and Adaptation
- 17.5a The Alarm Reaction
- 17.5b The Stage of Resistance
- 17.5c The Stage of Exhaustion

17.6 Eicosanoids and Other Signaling Molecules

17.7 Endocrine Disorders
- 17.7a Hyposecretion and Hypersecretion
- 17.7b Pituitary Disorders
- 17.7c Thyroid and Parathyroid Disorders
- 17.7d Adrenal Disorders
- 17.7e Diabetes Mellitus

Connective Issues

Study Guide

DEEPER INSIGHTS

17.1 Clinical Application: Melatonin, SAD, and PMS

17.2 Clinical Application: Hormone Receptors and Therapy

17.3 Clinical Application: Anti-Inflammatory Drugs

17.4 Medical History: The Discovery of Insulin

**Anatomy &
Physiology**
Revealed 4.0

Module 8: Endocrine System

- The hypothalamus is an important control center for hormone action, so it is important to know its anatomy and functions (see section 14.4a).

- Many hormones are proteins and other peptides. In discussing the synthesis of this class of hormones, it will be assumed that you are familiar with protein structure (see section 2.4e) and synthesis (see section 4.2c).

- Many hormones exert their effects by binding to proteins of the plasma membranes of their target cells. It is important to be familiar with membrane proteins as receptors and with G proteins, cAMP, and second messengers (see fig. 3.8 and the associated text).

- A review of the transport maximum (T_m) of membrane proteins (see section 3.3e, as well as figs. 3.8 and 23.18) will enhance your understanding of diabetes mellitus in this chapter.

If the body is to function as an integrated whole, its organs must communicate with each other and coordinate their activities. Even simple organisms composed of only a few cells have mechanisms for intercellular communication, suggesting that such mechanisms evolved very early in the history of life. In humans, two such systems are especially prominent—the nervous and endocrine systems, which communicate with neurotransmitters and hormones, respectively.

Nearly everyone has heard of at least some hormones—growth hormone, thyroid hormone, estrogen, and insulin, for example. At least passingly familiar, too, are some of the glands that secrete them (such as the pituitary and thyroid glands) and some disorders that result from hormone deficiency, excess, or dysfunction (such as diabetes, goiter, and dwarfism).

This chapter is primarily about the *endocrine* (hormonal) *system* of communication. We will start with the relatively familiar and large-scale aspects of this system—a survey of the endocrine glands, their hormones, and the principal effects of those hormones. We will then work our way down to the finer and less familiar details—the chemical identity of hormones, how they are made and transported, and how they produce their effects on their target cells. Shorter sections at the end of the chapter discuss the role of the endocrine system in adapting to stress, some hormonelike *paracrine* secretions, and the pathologies that result from endocrine dysfunction.

17.1 Overview of the Endocrine System

Expected Learning Outcomes

When you have completed this section, you should be able to

a. define *hormone* and *endocrine system;*

b. name several organs of the endocrine system;

c. contrast endocrine with exocrine glands;

d. recognize the standard abbreviations for many hormones; and

e. describe similarities and differences between the nervous and endocrine systems.

The body has four principal avenues of communication from cell to cell:

1. **Gap junctions** join unitary smooth muscle, cardiac muscle, epithelial, and other cells to each other. They enable cells to pass nutrients, electrolytes, and signaling molecules directly from the cytoplasm of one cell to the cytoplasm of the next through pores in their plasma membranes (see fig. 5.28).

2. **Neurotransmitters** are released by neurons, diffuse across a narrow synaptic cleft, and bind to receptors on the surface of the next cell.

3. **Paracrines**[1] are secreted by one cell, diffuse to nearby cells in the same tissue, and stimulate their physiology. Some call them *local hormones.*

4. **Hormones,**[2] in the strict sense, are chemical messengers that are transported by the bloodstream and stimulate physiological responses in cells of another, often distant, tissue or organ. Certain hormones produced by the pituitary gland in the head, for example, act on organs as far away as the pelvic cavity.

This chapter is concerned mainly with hormones and, to some extent, paracrine secretions. The glands, tissues, and cells that secrete hormones constitute the **endocrine**[3] **system;** the study of this system and the diagnosis and treatment of its disorders is called **endocrinology.** The most familiar hormone sources are the organs traditionally recognized as **endocrine glands,** such as the pituitary, thyroid, and adrenal glands, among others **(fig. 17.1).** Growing knowledge of endocrinology has revealed, however, that hormones are also secreted by numerous organs and tissues not usually thought of as glands, such as the brain, heart, small intestine, bones, muscles, and adipose tissue.

17.1a Comparison of Endocrine and Exocrine Glands

In section 5.5b, we examined another category of glands, the *exocrine* glands. The classical distinction between exocrine and endocrine glands has been the presence or absence of ducts. Most exocrine glands secrete their products by way of a duct onto an epithelial surface such as the skin or the mucosa of the digestive tract. Endocrine glands, by contrast, are ductless and release their secretions into the bloodstream (see fig. 5.30). For this reason, hormones were originally called the body's "internal secretions"; the word *endocrine* still alludes to this fact. Exocrine secretions have extracellular effects such as the digestion of food, whereas endocrine secretions have intracellular effects—they alter cell metabolism.

---

[1]*para* = next to; *crin* = to secrete
[2]*hormone* = to excite, set in motion
[3]*endo* = into; *crin* = to secrete

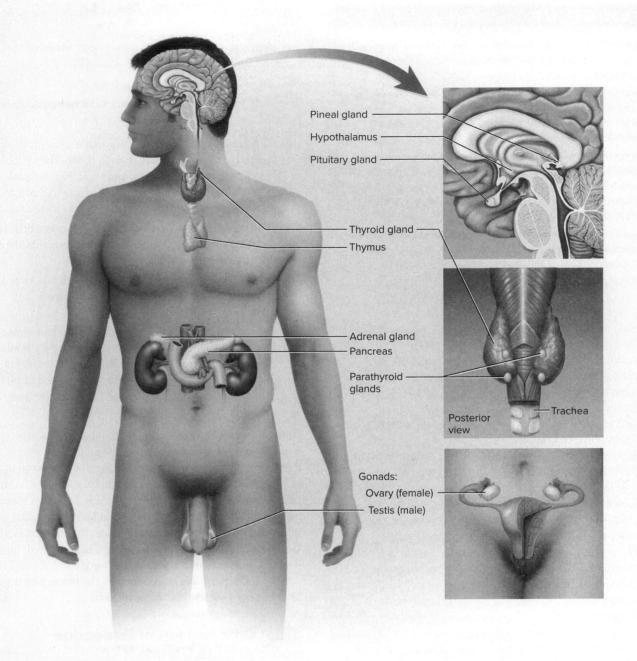

Pineal gland
Hypothalamus
Pituitary gland

Thyroid gland
Thymus

Adrenal gland
Pancreas

Parathyroid
glands

Trachea

Posterior
view

Gonads:
Ovary (female)
Testis (male)

FIGURE 17.1 Major Organs of the Endocrine System. This system also includes gland cells in many other organs not shown here.

After reading this chapter, name at least three hormone-secreting organs that are not shown in this illustration.

As we see in this chapter's opening photo, endocrine glands have an unusually high density of blood capillaries; these serve to pick up and carry away their hormones. Often, these vessels are an especially permeable type called *fenestrated capillaries,* which have patches of large pores in their walls allowing for the easy uptake of matter from the gland tissue (see fig. 20.5b).

Some glands and secretory cells defy simple classification as endocrine or exocrine. Liver cells, for example, behave as exocrine cells in the traditional sense by secreting bile into ducts that lead ultimately to the small intestine. However, they also secrete hormones into the blood, and in this respect they act as endocrine cells. They secrete albumin and blood-clotting factors directly into the blood as well. These do not fit the traditional concept of exocrine secretions, because they're not released by way of ducts or onto epithelial surfaces; nor do they fit the concept of endocrine secretions, because they're not hormones.

Liver cells are just one of nature's myriad ways of confounding our impulse to rigidly classify things into the fixed categories of our imagination.

17.1b Comparison of the Nervous and Endocrine Systems

The nervous and endocrine systems both serve for internal communication, but they're not redundant; they complement rather than duplicate each other's function (**table 17.1**). One important difference is that an efferent nerve fiber innervates only one organ and a limited number of cells within that organ, so its effects are usually precisely targeted and relatively specific (**fig. 17.2a**). Hormones, by contrast, are secreted into the bloodstream and circulate throughout the body (**fig. 17.2b**). Some of them, such as growth hormone and epinephrine, have more widespread effects than any one nerve fiber does.

Another difference is the speed with which the two systems start and stop responding to a stimulus. The nervous system typically responds within a few milliseconds, whereas it takes from several seconds to days for a hormone to act. When a stimulus ceases, the nervous system stops responding almost immediately, whereas hormonal effects can last for days or even longer. On the other hand, under long-term stimulation, most neurons quickly adapt and their response declines. The endocrine system shows more persistent responses.

But these differences shouldn't blind us to the similarities between the two systems. Both communicate chemically, and several chemicals function as both neurotransmitters and hormones—for example, norepinephrine, dopamine, and antidiuretic hormone (arginine vasopressin). Thus, a particular chemical such as dopamine can be considered a hormone when secreted by an endocrine cell but a neurotransmitter when secreted by a nerve cell. Another similarity is that some hormones

TABLE 17.1	Comparison of the Nervous and Endocrine Systems	
Nervous System	**Endocrine System**	
Communicates by means of electrical impulses and neurotransmitters	Communicates by means of hormones	
Releases neurotransmitters at synapses at specific target cells	Releases hormones into bloodstream for general distribution throughout body	
Usually has relatively local, specific effects	Sometimes has very general, widespread effects	
Reacts quickly to stimuli, usually within 1–10 ms	Reacts more slowly to stimuli, often taking seconds to days	
Stops quickly when stimulus stops	May continue responding long after stimulus stops	
Adapts relatively quickly to continual stimulation	Adapts relatively slowly; may respond for days to weeks	

and neurotransmitters produce identical effects on the same organ. For example, both norepinephrine and glucagon stimulate the liver to break down glycogen and release glucose. The nervous and endocrine systems continually regulate each other as they coordinate the activities of other organ systems. Some neurons trigger hormone secretion, and some hormones stimulate or inhibit neurons.

Some cells defy any attempt to rigidly classify them as neurons or gland cells. They act like neurons in many respects, but like endocrine cells, they release their secretions (such as oxytocin) into the bloodstream. Thus, we give them a hybrid name—**neuroendocrine cells.**

(a) Nervous system

FIGURE 17.2 Communication by the Nervous and Endocrine Systems. (a) A neuron has a long fiber that delivers its neurotransmitter to the immediate vicinity of its target cells. (b) Endocrine cells secrete a hormone into the bloodstream (left). At a point often remote from its origin, the hormone leaves the bloodstream and enters or binds to its target cells (right).

(b) Endocrine system

We have seen that neurotransmitters depend on receptors in the receiving cell; they can't exert any effect unless the receiving cell is equipped to bind and respond to them. This is true of hormones as well. When a hormone enters the bloodstream, it goes wherever the blood goes; there is no way to send it selectively to a particular organ. However, only certain **target organs** or **target cells** respond to it. Thyroid-stimulating hormone, for example, stimulates only the thyroid gland. In most cases, such selective responses are because only the target cells have receptors for a given hormone. They can also occur, however, because the circulating hormone is inactive and only the target cells have the enzyme needed to convert it to active form. Circulating testosterone, for example, is relatively inactive, but its target cells have an enzyme that converts it to dihydrotestosterone, which is much more potent.

17.1c Hormone Nomenclature

Many hormones are denoted by standard abbreviations that are used repeatedly in this chapter. These are listed alphabetically in **table 17.2** for use as a convenient reference while you work through the chapter. This list is by no means complete. It omits hormones that have no abbreviation, such as estrogen and insulin, and hormones that aren't discussed much in this chapter. Synonyms used by many authors are indicated in parentheses, but the first name listed is the one used in this book.

TABLE 17.2	Names and Abbreviations for Hormones	
Abbreviation	**Name**	**Source**
ACTH	Adrenocorticotropic hormone (corticotropin)	Anterior pituitary
ADH	Antidiuretic hormone (arginine vasopressin)	Posterior pituitary
CRH	Corticotropin-releasing hormone	Hypothalamus
DHEA	Dehydroepiandrosterone	Adrenal cortex
EPO	Erythropoietin	Kidneys, liver
FSH	Follicle-stimulating hormone	Anterior pituitary
GH	Growth hormone (somatotropin)	Anterior pituitary
GHRH	Growth hormone–releasing hormone	Hypothalamus
GnRH	Gonadotropin-releasing hormone	Hypothalamus
IGFs	Insulin-like growth factors (somatomedins)	Liver, other tissues
LH	Luteinizing hormone	Anterior pituitary
NE	Norepinephrine	Adrenal medulla
OT	Oxytocin	Posterior pituitary
PIH	Prolactin-inhibiting hormone (dopamine)	Hypothalamus
PRL	Prolactin	Anterior pituitary
PTH	Parathyroid hormone (parathormone)	Parathyroids
T_3	Triiodothyronine	Thyroid
T_4	Thyroxine (tetraiodothyronine)	Thyroid
TH	Thyroid hormone (T_3 and T_4 collectively)	Thyroid
TRH	Thyrotropin-releasing hormone	Hypothalamus
TSH	Thyroid-stimulating hormone (thyrotropin)	Anterior pituitary

BEFORE YOU GO ON

Answer the following questions to test your understanding of the preceding section:

1. Define the word *hormone* and distinguish a hormone from a neurotransmitter. Why is this an imperfect distinction?

2. Name some sources of hormones other than purely endocrine glands.

3. Describe some distinctions between endocrine and exocrine glands.

4. List some similarities and differences between the endocrine and nervous systems.

5. Discuss why the target-cell concept is essential for understanding hormone function.

17.2 The Hypothalamus and Pituitary Gland

Expected Learning Outcomes

When you have completed this section, you should be able to

a. describe the anatomical relationship between the hypothalamus and pituitary gland;

b. distinguish between the anterior and posterior lobes of the pituitary;

c. list the hormones produced by the hypothalamus and each lobe of the pituitary, and identify the functions of each hormone;

d. explain how the pituitary is controlled by the hypothalamus and its target organs; and

e. describe the effects of growth hormone.

There is no master control center that regulates the entire endocrine system, but the pituitary gland and hypothalamus have a more wide-ranging influence than any other part of the system, and several other endocrine glands cannot be adequately understood without first knowing how the pituitary influences them. This is therefore an appropriate place to begin a survey of the endocrine system.

17.2a Anatomy

The hypothalamus, shaped like a flattened funnel, forms the floor and walls of the third ventricle of the brain; we studied its structure and function in chapter 14 (see figs. 14.2 and 14.12b). It regulates primitive functions of the body ranging from water balance and thermoregulation to sex drive and childbirth. Many of its functions are carried out by way of the pituitary gland, which is closely associated with it both anatomically and physiologically.

The **pituitary gland (hypophysis[4])** is suspended from the floor of the hypothalamus by a **stalk (infundibulum[5])** and housed in a depression of the sphenoid bone, the sella turcica. The pituitary is usually about 1.3 cm wide and roughly the size and shape of a kidney bean; it grows about 50% larger in pregnancy. It is actually composed of two structures—the *anterior* and *posterior pituitary*—with independent origins and separate functions. The anterior pituitary arises from a pouch that grows upward from the embryonic pharynx, while the posterior pituitary arises as a bud growing downward from the brain **(fig. 17.3).** They come to lie side by side and are so closely joined that they look like a single gland.

The **anterior pituitary,** also called the *adenohypophysis[6]* (AD-eh-no-hy-POFF-ih-sis), constitutes about three-quarters of the pituitary as a whole; its main part is the large, bulging *anterior lobe* **(figs. 17.4a, 17.5a).** It has no nervous connection to the hypothalamus but is linked to it by a complex of blood vessels called the **hypophysial portal system** (hy-POFF-ih-SEE-ul) **(fig. 17.4b).** This system consists of a network of *primary capillaries* in the hypothalamus, a group of small veins called *portal venules* that travel down the stalk, and a complex of *secondary capillaries* in the anterior pituitary. The hypothalamus controls the anterior pituitary by secreting hormones that enter the primary capillaries, travel down the venules, and diffuse out of the secondary capillaries into the pituitary tissue. The hypothalamic hormones regulate secretion by various types of pituitary cells that we will study shortly.

The **posterior pituitary,** also called the *neurohypophysis,* constitutes the posterior one-quarter of the pituitary. The stalk that connects the pituitary to the hypothalamus is part of the posterior

---

[4]*hypo* = below; *physis* = growth
[5]*infundibulum* = funnel

---

[6]*adeno* = gland

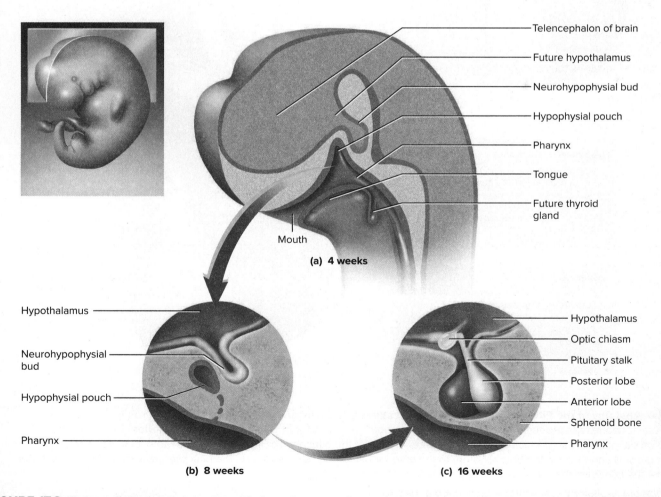

FIGURE 17.3 Embryonic Development of the Pituitary Gland. (a) Sagittal section of the head showing the neural and pharyngeal origins of the pituitary. (b) Pituitary development at 8 weeks. The hypophysial pouch, destined to become the anterior pituitary, has now separated from the pharynx. (c) Development at 16 weeks. The two lobes are now encased in bone and so closely associated they appear to be a single gland.

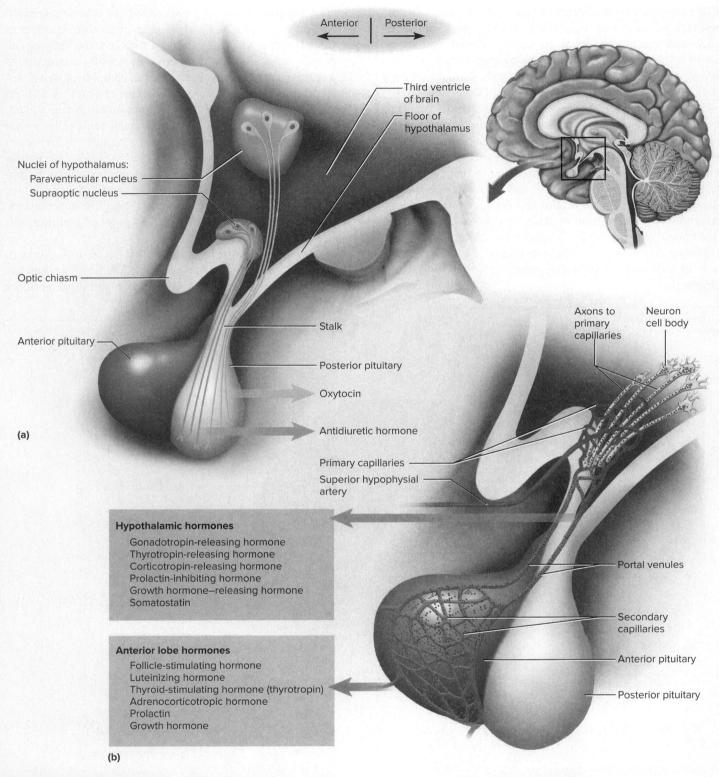

Anterior | Posterior

Third ventricle of brain

Floor of hypothalamus

Nuclei of hypothalamus:
Paraventricular nucleus
Supraoptic nucleus

Optic chiasm

Anterior pituitary

(a)

Stalk

Posterior pituitary

Oxytocin

Antidiuretic hormone

Primary capillaries

Superior hypophysial artery

Axons to primary capillaries

Neuron cell body

Hypothalamic hormones

Gonadotropin-releasing hormone
Thyrotropin-releasing hormone
Corticotropin-releasing hormone
Prolactin-inhibiting hormone
Growth hormone–releasing hormone
Somatostatin

Anterior lobe hormones

Follicle-stimulating hormone
Luteinizing hormone
Thyroid-stimulating hormone (thyrotropin)
Adrenocorticotropic hormone
Prolactin
Growth hormone

(b)

Portal venules

Secondary capillaries

Anterior pituitary

Posterior pituitary

FIGURE 17.4 Anatomy of the Pituitary Gland. (a) Major structures of the pituitary and hormones of the neurohypophysis. Note that these hormones are produced by two nuclei in the hypothalamus and later released from the posterior lobe of the pituitary. (b) The hypophysial portal system, which regulates the anterior lobe of the pituitary. The hormones in the violet box are secreted by the hypothalamus and travel in the portal system to the anterior pituitary. The hormones in the pink box are secreted by the anterior pituitary under the control of the hypothalamic releasers and inhibitors. **APR**

❓ *Which lobe of the pituitary is essentially composed of brain tissue?*

(a) Anterior pituitary

- Blood vessel
- Chromophobe
- Acidophil
- Basophil

(b) Posterior pituitary

- Hormone stored in nerve endings
- Glial cells (pituicytes)
- Nerve fibers

FIGURE 17.5 Histology of the Pituitary Gland. (a) The anterior lobe. Basophils include gonadotropic, thyrotropic, and corticotropic cells. Acidophils include somatotropic and prolactin cells. These subtypes are not distinguishable with this histological stain. The chromophobes resist staining; they may be either stem cells or cells that have been temporarily depleted of hormone. (b) The posterior lobe, composed of nervous tissue.

a: Victor P. Eroschenko; **b:** Biophoto Associates/Science Source

pituitary; the bulging part of the gland below the stalk is called the *posterior lobe*. The posterior pituitary is actually nervous tissue (nerve fibers and neuroglia, **fig. 17.5b**), not a true gland. Hormones of the posterior pituitary are made by certain neuroendocrine cells in the hypothalamus. Their axons pass down the stalk as a bundle called the **hypothalamo–hypophysial tract** and end in the posterior lobe (fig. 17.4a). Hormones are made in the neurosomas and move down the nerve fibers by axoplasmic flow to the posterior pituitary. Here they are stored in the nerve endings until a nerve signal coming down the same axons triggers their release.

17.2b Hypothalamic Hormones

To understand the physiology of the hypothalamic–pituitary system, we must begin with the hormones produced in the hypothalamus. There are eight of these (fig. 17.4)—six to regulate the anterior pituitary and two that are stored in the posterior pituitary and released on demand. Among the first six, those that stimulate the pituitary to release hormones of its own are called *releasing hormones;* those that suppress pituitary secretion are called *inhibiting hormones* **(table 17.3).** The releasing or inhibiting effect is identified in their names. *Somatostatin* is also called growth hormone–inhibiting hormone, though it also inhibits secretion of thyroid-stimulating hormone. Its name derives from *somatotropin,* a synonym for growth hormone, and *stat,* meaning to halt something (in this case, growth hormone secretion).

The other two hypothalamic hormones are *oxytocin (OT)* and *antidiuretic hormone (ADH).* These are stored and released by the posterior pituitary. OT comes mainly from neurons in the right and left **paraventricular**[7] **nuclei** of the hypothalamus, so called because they lie in the walls of the third ventricle. ADH comes mainly from the **supraoptic**[8] **nuclei,** named for their location just above the optic chiasm. Each nucleus also produces small quantities of the other hormone. ADH and OT are treated as posterior pituitary hormones for convenience even though the posterior lobe doesn't synthesize them.

17.2c Anterior Pituitary Hormones

The six main hormones of the anterior pituitary are as follows **(table 17.4).** The first two are collectively called **gonadotropins**[9] because they target the ovaries and testes (gonads). Their individual names (FSH and LH) refer to their female functions, although they are equally important in males.

1. **Follicle-stimulating hormone (FSH).** FSH is secreted by *gonadotropic cells.* In the ovaries, it stimulates the secretion of ovarian sex hormones and the development of the bubble-like *follicles* that contain the eggs. In the testes, it stimulates sperm production.

2. **Luteinizing hormone (LH).** LH is also secreted by the gonadotropic cells. In females, it stimulates *ovulation,* the release of an egg. It is named for the fact that after

[7]*para* = next to; *ventricular* = pertaining to the ventricle
[8]*supra* = above
[9]*gonado* = gonads; *trop* = to turn or change

TABLE 17.3	Hypothalamic Releasing and Inhibiting Hormones That Regulate the Anterior Pituitary
Hormone	**Principal Effects**
Thyrotropin-releasing hormone (TRH)	Promotes secretion of thyroid-stimulating hormone (TSH) and prolactin (PRL)
Corticotropin-releasing hormone (CRH)	Promotes secretion of adrenocorticotropic hormone (ACTH)
Gonadotropin-releasing hormone (GnRH)	Promotes secretion of follicle-stimulating hormone (FSH) and luteinizing hormone (LH)
Growth hormone–releasing hormone (GHRH)	Promotes secretion of growth hormone (GH)
Prolactin-inhibiting hormone (PIH)	Inhibits secretion of prolactin (PRL)
Somatostatin	Inhibits secretion of growth hormone (GH) and thyroid-stimulating hormone (TSH)

ovulation, the follicle becomes a yellowish body called the *corpus luteum.*[10] LH also stimulates the corpus luteum to secrete progesterone, a hormone important in pregnancy. In males, LH stimulates the testes to secrete testosterone.

3. **Thyroid-stimulating hormone (TSH),** or **thyrotropin.** TSH is secreted by pituitary cells called *thyrotropic cells.* It stimulates growth of the thyroid gland and the secretion of thyroid hormone, which has widespread effects on metabolic rate, body temperature, and other functions detailed later.

4. **Adrenocorticotropic hormone (ACTH),** or **corticotropin.** ACTH is secreted by cells called *corticotropic cells.* Its target organ and the basis for its name is the adrenal cortex. ACTH stimulates the cortex to secrete hormones called glucocorticoids (especially *cortisol*), which regulate glucose, protein, and fat metabolism and are important in the body's response to stress.

5. **Prolactin**[11] **(PRL).** PRL is secreted by pituitary cells called *prolactin cells.* The hormone and these cells are named for the role of PRL in lactation. During pregnancy, the lactotropes increase greatly in size and number, and PRL secretion rises proportionately, but it has no effect on the mammary glands until after a woman gives birth. Then, it stimulates them to synthesize milk.

6. **Growth hormone (GH),** or **somatotropin.** GH is secreted by *somatotropic cells,* the most numerous cells of the anterior pituitary. The pituitary produces at least a thousand times as much GH as any other hormone. The general effect of GH is to stimulate mitosis and cellular differentiation and thus to promote tissue growth throughout the body.

You can see that the anterior pituitary is involved in a chain of events linked by hormones: The hypothalamus secretes a releasing hormone; this induces a type of pituitary cell to secrete its hormone; that hormone is usually targeted to another endocrine gland elsewhere in the body; and finally that gland secretes a hormone with an effect of its own. For example, the hypothalamus secretes thyrotropin-releasing hormone (TRH); this induces the anterior pituitary to secrete thyroid-stimulating hormone (TSH); TSH, in turn, stimulates the thyroid gland to release thyroid hormone (TH); and finally, thyroid hormone exerts its metabolic effects throughout the body. Such a relationship between the hypothalamus, pituitary, and another downstream endocrine gland is called an **axis**—the *hypothalamo–pituitary–thyroid axis,* for example. **Figure 17.6** summarizes these relationships between the hypothalamus, anterior pituitary, and more downstream target organs.

17.2d Posterior Pituitary Hormones

The two posterior lobe hormones are ADH and OT (table 17.4). As we have already seen, they are synthesized in the hypothalamus, then transported to the posterior pituitary and stored until their release on command. Their functions are as follows:

1. **Antidiuretic hormone (ADH).** ADH increases water retention by the kidneys, reduces urine volume, and helps prevent dehydration. ADH also functions as a brain neurotransmitter and is usually called *arginine vasopressin (AVP)* in the neuroscience literature. This name refers to its ability to cause vasoconstriction, but this effect requires concentrations so unnaturally high for the human body that it is of doubtful significance except in pathological states. The ADH name refers to its more common function in everyday human physiology.

2. **Oxytocin (OT).** OT has a variety of reproductive functions in situations ranging from intercourse to breast-feeding. It surges in both sexes during sexual arousal and orgasm, possibly aiding in the propulsion of semen through the male reproductive tract and stimulating uterine contractions that help transport sperm up the female tract. OT also functions

---

[10]*corpus* = body; *lute* = yellow
[11]*pro* = favoring, promoting; *lact* = milk

TABLE 17.4	Pituitary Hormones	
Hormone	**Target Organ or Tissue**	**Principal Effects**
Anterior Pituitary		
Follicle-stimulating hormone (FSH)	Ovaries, testes	*Female:* Growth of ovarian follicles and secretion of estrogen *Male:* Sperm production
Luteinizing hormone (LH)	Ovaries, testes	*Female:* Ovulation, maintenance of corpus luteum *Male:* Testosterone secretion
Thyroid-stimulating hormone (TSH)	Thyroid gland	Growth of thyroid, secretion of thyroid hormone
Adrenocorticotropic hormone (ACTH)	Adrenal cortex	Growth of adrenal cortex, secretion of glucocorticoids
Prolactin (PRL)	Mammary glands	Milk synthesis
Growth hormone (GH)	Liver, bone, cartilage, muscle, fat	Widespread tissue growth, especially in the stated tissues
Posterior Pituitary		
Antidiuretic hormone (ADH)	Kidneys	Water retention
Oxytocin (OT)	Uterus, mammary glands	Labor contractions, milk release; possibly involved in ejaculation, sperm transport, sexual affection, and mother–infant bonding

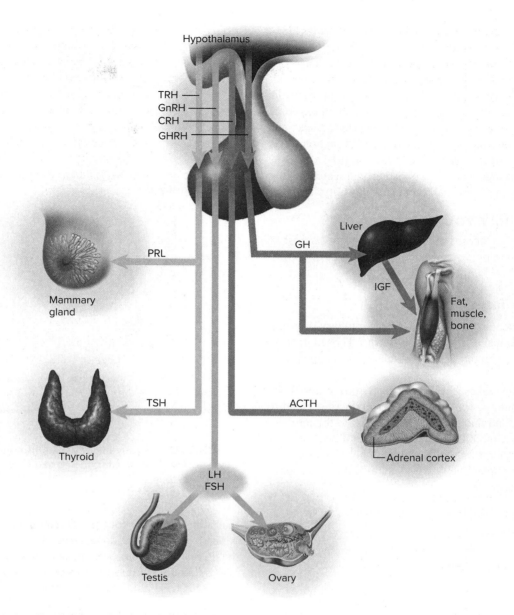

FIGURE 17.6 Hypothalamo–Pituitary–Target Organ Relationships. Hypothalamic releasing hormones, shown at the top, trigger secretion of all of the anterior pituitary hormones (bottom).

in feelings of sexual satisfaction and emotional bonding between partners. In childbirth, it stimulates labor contractions, and in lactating mothers, it stimulates the flow of milk from deep in the mammary gland to the nipple, where it is accessible to the infant. It may also promote emotional bonding between the mother and infant. In the absence of oxytocin, other female mammals tend to neglect their helpless infants, even allowing them to die.

17.2e Control of Pituitary Secretion

Pituitary hormones are not secreted at constant rates. GH is secreted mainly at night, LH peaks at the middle of the menstrual cycle, and OT surges during labor and breast-feeding, for example. The timing and amount of pituitary secretion are regulated by the hypothalamus, other brain centers, and feedback from the target organs.

Hypothalamic and Cerebral Control

Both lobes of the pituitary gland are strongly subject to control by the brain. Hypothalamic control enables the brain to monitor conditions within and outside the body and to stimulate or inhibit the release of anterior lobe hormones in response. For example, in times of stress, the hypothalamus triggers ACTH secretion, which leads to cortisol secretion and mobilization of materials needed for tissue repair. During pregnancy, the hypothalamus induces prolactin secretion so a woman will be prepared to lactate after the baby is born.

The posterior pituitary is controlled by **neuroendocrine reflexes**—the release of hormones in response to nerve signals. For example, dehydration raises the osmolarity of the blood, which is detected by hypothalamic neurons called *osmoreceptors*. They trigger ADH release, and ADH promotes water conservation. Excessive blood pressure, by contrast, stimulates stretch receptors in the heart and certain arteries. By another neuroendocrine reflex, this inhibits ADH release, increases urine output, and brings blood volume and pressure back to normal.

The suckling of an infant also triggers a neuroendocrine reflex mediated by oxytocin. Stimulation of the nipple sends nerve signals up the spinal cord and brainstem to the hypothalamus and from there to the posterior pituitary. This causes the release of oxytocin, which stimulates the release of milk. Neuroendocrine reflexes can also involve higher brain centers. For example, this lactation reflex can be triggered when the mother simply hears a baby cry—any baby—and emotional stress can affect gonadotropin secretion, thus disrupting ovulation, the menstrual rhythm, and fertility.

▶▶▶**APPLY WHAT YOU KNOW**

Which of the major themes at the end of chapter 1 is best exemplified by the neuroendocrine reflexes that govern ADH secretion?

Feedback from Target Organs

The regulation of other endocrine glands by the pituitary isn't simply a system of "command from the top down." Those target organs also regulate the pituitary and hypothalamus through various feedback loops.

Most often, this takes the form of **negative feedback inhibition**—the pituitary stimulates another endocrine gland to secrete its hormone, and that hormone feeds back to the pituitary or hypothalamus and inhibits further secretion of the pituitary hormone. **Figure 17.7** shows this in the hypothalamo–pituitary–thyroid axis as an example. The figure is numbered to correspond to the following description:

① The hypothalamus secretes thyrotropin-releasing hormone (TRH).

② TRH stimulates the anterior pituitary to secrete thyroid-stimulating hormone (TSH).

③ TSH stimulates the thyroid gland to secrete thyroid hormone (TH).

④ TH stimulates the metabolism of most cells throughout the body.

⑤ TH also *inhibits* the release of TSH by the pituitary.

⑥ To a lesser extent, TH also *inhibits* the release of TRH by the hypothalamus.

The negative feedback inhibition in this process consists of steps 5 and 6. It ensures that when the TH level is high, TSH secretion remains moderate. If thyroid hormone secretion drops, TSH secretion rises and stimulates the thyroid to secrete more hormone. This feedback keeps thyroid hormone levels oscillating around a set point in typical homeostatic fashion.

▶▶▶**APPLY WHAT YOU KNOW**

If the thyroid gland was removed from a cancer patient, would you expect the level of TSH to rise or fall? Why?

Feedback from a target organ isn't always inhibitory. As noted earlier, oxytocin triggers a positive feedback cycle during labor (see fig. 1.9). Uterine stretching sends a nerve signal to the brain

FIGURE 17.7 Negative Feedback Inhibition of the Pituitary Gland by the Thyroid Gland. See text for explanation of numbered steps.

that stimulates OT release. OT stimulates uterine contractions, which push the infant downward. This stretches the lower end of the uterus some more, which results in a nerve signal that stimulates still more OT release. This positive feedback cycle continues and intensifies until the infant is born.

17.2f A Further Look at Growth Hormone

The ultimate effects of most pituitary hormones are exerted through endocrine glands farther downstream, such as the thyroid, adrenal cortex, and gonads. Thus, we will study those effects in the following section as we survey those glands. Growth hormone is a different matter and warrants further exploration at this point.

Unlike the other pituitary hormones, GH is not targeted to just one or a few organs but has widespread effects on the body. In addition to directly stimulating diverse tissues such as muscle and bone, it induces the liver and other tissues to secrete growth stimulants called **insulin-like growth factors (IGF-I** and **IGF-II),** or **somatomedins,**[12] which then stimulate other target cells (fig. 17.6). Most of these effects are caused by IGF-I, but IGF-II is important in fetal growth.

One effect of IGF is to prolong the action of GH. All hormones have a **half-life,** the time required for 50% of the hormone to be cleared from the blood. GH is short-lived; its half-life is only 6 to

---

[12]acronym for *somatotropin mediating protein*

20 minutes. IGFs, by contrast, have half-lives of about 20 hours. The mechanisms of GH–IGF action include the following:

- **Protein synthesis.** Tissue growth requires protein synthesis, and protein synthesis needs two things: amino acids for building material, and messenger RNA (mRNA) for instructions. Within minutes of its secretion, GH boosts the translation of existing mRNA, and within a few hours, it also boosts the transcription of DNA and thus the production of more mRNA. GH also enhances amino acid transport into cells, and to ensure that protein synthesis outpaces breakdown, it suppresses protein catabolism.

- **Lipid metabolism.** To provide energy for growing tissues, GH stimulates adipocytes to catabolize fat and release fatty acids and glycerol into the blood. By providing these fuels, GH makes it unnecessary for cells to consume their own proteins. This is called the **protein-sparing effect.**

- **Carbohydrate metabolism.** GH also has a **glucose-sparing effect.** Its role in mobilizing fatty acids reduces the dependence of most cells on glucose so they won't compete with the brain, which is highly glucose-dependent. GH also stimulates glucose synthesis by the liver.

- **Electrolyte balance.** GH promotes Na^+, K^+, and Cl^- retention by the kidneys, enhances Ca^{2+} absorption by the small intestine, and makes these electrolytes available to the growing tissues.

The most conspicuous effects of GH are on bone, cartilage, and muscle growth, especially during childhood and adolescence. IGF-I accelerates bone growth at the epiphysial plates. It stimulates the multiplication of chondrocytes and osteogenic cells as well as protein deposition in the cartilage and bone matrix. In adulthood, it stimulates osteoblast activity and the appositional growth of bone; thus, it continues to influence bone thickening and remodeling.

GH secretion fluctuates greatly over the course of a day. It rises to 20 nanograms per milliliter of blood plasma (ng/mL) or higher during the first 2 hours of deep sleep, and may reach 30 ng/mL in response to vigorous exercise. GH secretion rises sharply in response to trauma, physical or emotional stress, hypoglycemia (low blood sugar), and other conditions. Small peaks occur after high-protein meals, but a high-carbohydrate meal tends to suppress GH secretion.

Another stimulus to GH secretion is **ghrelin**[13] (GRELL-in), a hormone secreted by the stomach, especially when it is empty. Ghrelin acts on the hypothalamus to produce the sensation of hunger as well as to stimulate release of growth hormone–releasing hormone (GHRH). This, in turn, stimulates a spike in GH secretion, priming diverse target organs to take advantage of the nutrients one is presumably about to consume.

GH level declines gradually with age. The serum level averages about 6 ng/mL in adolescence and one-quarter of that in very old age. The resulting decline in protein synthesis may contribute to aging of the tissues, including wrinkling of the skin and reduced muscular mass and strength. At age 30, the average adult body is 10% bone, 30% muscle, and 20% fat; at age 75, it averages 8% bone, 15% muscle, and 40% fat.

Answer the following questions to test your understanding of the preceding section:

6. What are two good reasons for considering the pituitary to be two separate glands?

7. Briefly contrast hypothalamic control of the anterior pituitary with its control of the posterior pituitary.

8. Name three anterior lobe hormones that have reproductive functions and three that have nonreproductive roles. What target organs are stimulated by each of these hormones?

9. In what sense does the pituitary "take orders" from the target organs under its command?

10. The liver does not secrete growth hormone, but does promote its action. How? How does GH affect the metabolism of proteins, fats, and carbohydrates?

17.3 Other Endocrine Glands

Expected Learning Outcomes

When you have completed this section, you should be able to

a. describe the structure and location of the remaining endocrine glands;

b. name the hormones these endocrine glands produce, what stimulates their secretion, and their functions; and

c. discuss the hormones produced by organs and tissues other than the classical endocrine glands.

17.3a The Pineal Gland

The **pineal gland** (PIN-ee-ul), named for its pine cone shape, is attached to the roof of the third ventricle of the brain, beneath the posterior end of the corpus callosum (see figs. 14.2 and 17.1). The philosopher René Descartes (1596–1650) thought it was the seat of the human soul. If so, children must have more soul than adults—a child's pineal gland is about 8 mm long and 5 mm wide, but after age 7 it regresses rapidly and is no more than a tiny shrunken mass of fibrous tissue in the adult. Such shrinkage of an organ is called **involution.**[14] Pineal secretion peaks between the ages of 1 and 5 years and declines 75% by the end of puberty.

We no longer look for the human soul in the pineal gland, but this little organ remains an intriguing mystery. It seems "wired" to respond to information from the eyes about the relative hours of light and darkness in the environment. Some optic nerve fibers, instead of going to the visual cortex of the brain, lead to the superior colliculi of the midbrain and synapse here with other neurons that, in sequence, carry the signal down the spinal cord, out through sympathetic nerve fibers to the sympathetic chain ganglia, and finally back into the head and specifically to the pineal gland. The pineal may play a role in establishing 24-hour *circadian rhythms* of

---

[13]named partly from *ghre* = growth, and partly as an acronym for *growth hormone–releasing hormone*

[14]*in* = inward; *volution* = rolling or turning

DEEPER INSIGHT 17.1

CLINICAL APPLICATION

Melatonin, SAD, and PMS

There seems to be a relationship between melatonin and mood disorders, including depression and sleep disturbances. Some people experience a mood dysfunction called *seasonal affective disorder (SAD)*, especially in winter when the days are short and people get less sunlight, and in extreme northern and southern latitudes where sunlight may be dim to nonexistent for months at a time. SAD thus affects about 20% of the population in Alaska but only 2.5% in Florida. The symptoms—which include depression, sleepiness, irritability, and carbohydrate craving—can be relieved by 2 or 3 hours of exposure to bright light each day *(phototherapy)*. Premenstrual syndrome (PMS) is similar to SAD and is also relieved by phototherapy. The melatonin level is elevated in both SAD and PMS and is reduced by phototherapy. However, there is also evidence that casts doubt on any causal link between melatonin and these mood disorders, so for now, "the jury is still out." Many people take melatonin for jet lag or as an aid to sleep, and it is quite effective, but it is also risky to use when we know so little, as yet, about its potential effect on reproductive function.

physiological function synchronized with the cycle of daylight and darkness. At night, it synthesizes **melatonin,** a monoamine, from serotonin. Melatonin secretion peaks about 3 to 5 hours after the onset of sleep each night, and fluctuates seasonally. In other animals with seasonal breeding, it regulates the gonads and sets the annual breeding season. Melatonin may suppress gonadotropin secretion in humans; removal of the pineal from animals causes premature sexual maturation. Some physiologists think that the human pineal gland may regulate the timing of puberty, but a clear demonstration of its role has remained elusive. Melatonin receptors are widespread in the body, including in the pancreatic islets. Melatonin receptor defects are correlated with a higher incidence of type 2 diabetes, and low melatonin levels reduce the body's insulin sensitivity. Melatonin has been implicated in jet lag and some mood and sleep disorders, although the evidence is inconclusive (see Deeper Insight 17.1).

17.3b The Thymus

The **thymus** plays a role in three systems: endocrine, lymphatic, and immune. It is a bilobed gland in the mediastinum superior to the heart, behind the sternal manubrium. In the fetus and infant, it is enormous in comparison to adjacent organs, sometimes extending between the lungs from near the diaphragm to the base of the neck **(fig. 17.8a).** It continues to grow until the age of 5 or 6 years. In adults, the gland weighs about 20 g up to age 60, but then it too, like the pineal, undergoes involution and becomes increasingly fatty and less glandular. In the elderly, it is a small fibrous and fatty remnant barely distinguishable from the surrounding mediastinal tissues **(fig. 17.8b).**

The thymus is a site of maturation for certain white blood cells called T cells that are critically important for immune defense. It secretes several hormones (**thymopoietin, thymosin,** and **thymulin**) that stimulate the development of other lymphatic organs and regulate the development and activity of T cells. Its histology and immune function are discussed more fully in chapter 21.

17.3c The Thyroid Gland

The **thyroid gland** is the largest adult gland to have a purely endocrine function, weighing about 25 g. It lies adjacent to the trachea immediately below the larynx, and is named for the nearby shield-like thyroid[15] cartilage of the larynx. It is shaped like a butterfly wrapped around the trachea, with two winglike lobes usually joined inferiorly by a narrow bridge of tissue, the **isthmus (fig. 17.9a).**

Histologically, the thyroid is composed mostly of sacs called **thyroid follicles (fig. 17.9b).** Each is filled with a protein-rich colloid and lined by a simple cuboidal epithelium of **follicular cells.** These cells secrete about 80 micrograms (μg) of **thyroid hormone (TH)** daily. About 90% of this is in a form called **thyroxine, T_4,** or **tetraiodothyronine** (TET-ra-EYE-oh-doe-THY-ro-neen), because its

---

[15]*thyr* = shield; *oid* = resembling

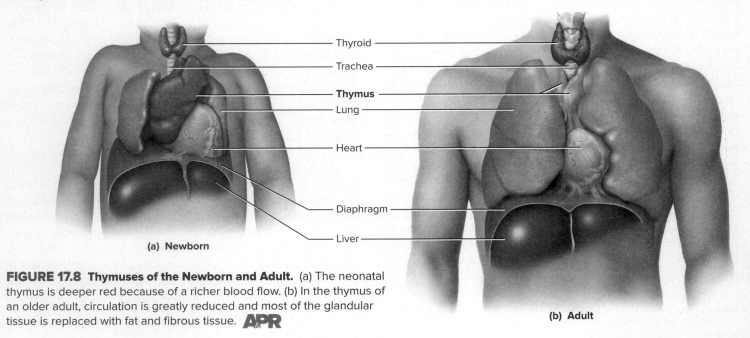

FIGURE 17.8 Thymuses of the Newborn and Adult. (a) The neonatal thymus is deeper red because of a richer blood flow. (b) In the thymus of an older adult, circulation is greatly reduced and most of the glandular tissue is replaced with fat and fibrous tissue. **APR**

Thyroid
Trachea
Thymus
Lung
Heart
Diaphragm
Liver

(a) Newborn

(b) Adult

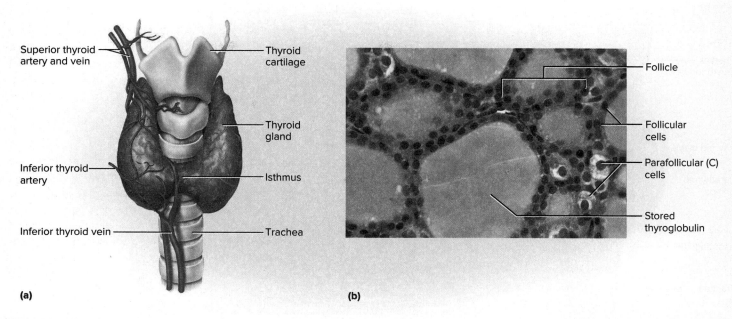

Superior thyroid artery and vein

Thyroid cartilage

Inferior thyroid artery

Thyroid gland

Isthmus

Inferior thyroid vein

Trachea

Follicle

Follicular cells

Parafollicular (C) cells

Stored thyroglobulin

(a)   **(b)**

FIGURE 17.9 Anatomy of the Thyroid Gland. (a) Gross anatomy, anterior view. (b) Histology, showing the saccular thyroid follicles (the source of thyroid hormone) and nests of C cells (the source of calcitonin). **APR**

b: Biophoto Associates/Science Source

structure includes four iodine atoms. The other 10% is a form with three iodine atoms, called **triiodothyronine** or T_3. That is, the expression *thyroid hormone* refers to T_3 and T_4 collectively.

Like other endocrine glands, the thyroid releases these hormones directly into the bloodstream. Each follicle is surrounded by a basketlike network of capillaries, the globular clusters of blood vessels seen in this chapter's opening photo. These are supplied by the superior and inferior thyroid arteries seen in figure 17.9a. The thyroid receives one of the body's highest rates of blood flow per gram of tissue and consequently has a dark reddish-brown color.

Thyroid hormone is secreted or inhibited in response to fluctuations in metabolic rate. The brain monitors the body's metabolic rate and stimulates TH secretion through the action of TRH and TSH as depicted in figure 17.7. The primary effect of TH is to increase one's metabolic rate. As a result, it raises oxygen consumption and has a **calorigenic**[16] **effect**—it increases heat production. To ensure an adequate blood and oxygen supply to meet this increased metabolic demand, thyroid hormone also raises the respiratory rate, heart rate, and strength of the heartbeat. It stimulates the appetite and accelerates the breakdown of carbohydrates, fats, and protein for fuel. Thyroid hormone also promotes alertness and quicker reflexes; growth hormone secretion; growth of the bones, skin, hair, nails, and teeth; and development of the fetal nervous system.

The thyroid gland also contains nests of **parafollicular cells,** also called **clear (C) cells,** at the periphery of the follicles. They respond to rising levels of blood calcium by secreting the hormone **calcitonin.** Calcitonin antagonizes parathyroid hormone (discussed shortly) and stimulates osteoblast activity, thus promoting calcium deposition and bone formation. Calcitonin is important

mainly in children, who have 10 times as many C cells as adults; it has little effect in adults (see section 7.4b).

17.3d The Parathyroid Glands

The **parathyroid glands** are ovoid glands, usually four in number, partially embedded in the posterior surface of the thyroid. Each is about 3 to 8 mm long and 2 to 5 mm wide, and is separated from the thyroid follicles by a thin fibrous capsule and adipose tissue **(fig. 17.10).** Often, they occur in other locations ranging from as high as the hyoid bone to as low as the aortic arch, and about 5% of people have more than four parathyroids. They secrete **parathyroid hormone (PTH),** which regulates blood calcium levels. Unlike the thyroid gland, the parathyroids are not regulated by the pituitary, but directly monitor blood composition and secrete PTH when the calcium level dips too low. PTH raises the calcium level by stimulating calcium reabsorption from the bones and reducing calcium losses in the urine. Calcium homeostasis is so crucial to neuromuscular and cardiovascular function that a person can die within just a few days if the parathyroids are removed without instituting hormone replacement therapy. This often happened to patients shortly following thyroid surgery before surgeons realized the existence and function of the tiny, nearly hidden parathyroids.

17.3e The Adrenal Glands

The **adrenal**[17] **(suprarenal) glands** sit like a cap on the superior pole of each kidney **(fig. 17.11).** Like the kidneys, they are retroperitoneal, lying outside the peritoneal cavity between the

---

[16]*calor* = heat; *genic* = producing

[17]*ad* = to, toward, near; *ren* = kidney; *al* = pertaining to

Pharynx (posterior view)

Thyroid gland

Parathyroid glands

Esophagus

Trachea

(a)

Parathyroid gland

Thyroid gland

(b)

FIGURE 17.10 The Parathyroid Glands. (a) Location of the usual four parathyroid glands partially embedded in the posterior surface of the thyroid gland. (b) Histology of a parathyroid gland in relation to its fibrous capsule and adjacent thyroid and adipose tissue. **APR**

b: Victor P. Eroschenko

Adrenal gland
Suprarenal vein
Kidney

Adrenal cortex
Adrenal medulla

FIGURE 17.11 The Adrenal Gland. (a) Location and gross anatomy. (b) Drawing and photomicrograph of adrenal histology. **APR**

b: Victor Eroschenko

Connective tissue capsule

Adrenal cortex

(a)

Zona glomerulosa

Zona fasciculata

Zona reticularis

Adrenal medulla

(b)

peritoneum and posterior body wall. The adult adrenal gland measures about 5 cm vertically, 3 cm wide, and 1 cm from anterior to posterior. It weighs about 8 to 10 g in the newborn, but loses half of this weight by the age of 2 years, mainly because of involution of its outer layer, the adrenal cortex. It remains at 4 to 5 g in adults.

Like the pituitary, the adrenal gland forms by the merger of two fetal glands with different origins and functions. Its inner core, the **adrenal medulla,** is 10% to 20% of the gland. Depending on blood flow, its color ranges from gray to dark red. Surrounding it is a much thicker **adrenal cortex,** constituting 80% to 90% of the gland and having a yellowish color due to its high concentration of cholesterol and other lipids.

The Adrenal Medulla

The **adrenal medulla** has a dual nature, acting as both an endocrine gland and a ganglion of the sympathetic nervous system (see section 15.2b). Sympathetic preganglionic nerve fibers penetrate through the cortex to reach **chromaffin cells** (cro-MAFF-in) in the medulla. Named for their tendency to stain brown with certain dyes, these cells are essentially sympathetic postganglionic neurons, but they have no dendrites or axon and they release their products into the bloodstream like any other endocrine gland. Thus they are considered to be neuroendocrine cells.

Upon stimulation by the nerve fibers—as in a situation of fear, pain, or other stress—the chromaffin cells release a mixture of catecholamines that we previously encountered as neurotransmitters (see fig. 12.22): about three-quarters **epinephrine,** one-quarter **norepinephrine,** and a trace of **dopamine.** Now we find them acting as hormones. They increase alertness and prepare the body in several ways for physical activity. They mobilize high-energy fuels such as lactate, fatty acids, and glucose. The liver boosts glucose levels by **glycogenolysis** (hydrolysis of glycogen to glucose) and **gluconeogenesis** (conversion of fats, amino acids, and other noncarbohydrates to glucose).

Epinephrine is said to have a **glucose-sparing effect.** It inhibits the secretion of insulin, so the muscles and other insulin-dependent organs absorb and consume less glucose. They fall back on alternative fuels such as fatty acids, while the blood glucose is left for use by the brain, which is more glucose-dependent but not insulin-dependent.

Adrenal catecholamines also raise the heart rate and blood pressure, stimulate circulation to the muscles, increase pulmonary airflow, and raise the metabolic rate. At the same time, they *inhibit* such temporarily inessential functions as digestion and urine production so that they do not compete for blood flow and energy.

The Adrenal Cortex

The **adrenal cortex** surrounds the medulla on all sides. It produces more than 25 steroid hormones, known collectively as the **corticosteroids** or **corticoids.** All of them are synthesized from cholesterol; this and other lipids impart a yellow color to the cortex. Only five corticosteroids are secreted in physiologically significant amounts; the others are either negligible in quantity or, if more abundant, are in chemically less active forms. The five most important corticosteroids fall into three categories: **mineralocorticoids,** which regulate the body's electrolyte balance; **glucocorticoids,** which regulate the

metabolism of glucose and other organic fuels; and **sex steroids,** with various developmental and reproductive functions.

The adrenal cortex has three layers of tissue (fig. 17.11b), which differ in their histology and hormone output.

1. The **zona glomerulosa**[18] (glo-MER-you-LO-suh) is a thin layer, less developed in humans than in many other mammals, located just beneath the capsule at the gland surface. The name *glomerulosa* ("full of little balls") refers to the arrangement of its cells in round clusters. The zona glomerulosa is the source of mineralocorticoids.

2. The **zona fasciculata**[19] (fah-SIC-you-LAH-ta) is a thick middle layer constituting about three-quarters of the cortex. Here the cells are arranged in parallel cords (fascicles), separated by blood capillaries, perpendicular to the gland surface. The cells are called **spongiocytes** because of a foamy appearance imparted by an abundance of cytoplasmic lipid droplets. The zona fasciculata secretes glucocorticoids and androgens.

3. The **zona reticularis**[20] (reh-TIC-you-LAR-iss) is the narrow, innermost layer, adjacent to the adrenal medulla. Its cells form a branching network for which the layer is named. Like the preceding layer, the zona reticularis also secretes glucocorticoids and androgens.

Aldosterone is the most significant mineralocorticoid, and is produced only by the zona glomerulosa. Falling blood pressure leads to increased aldosterone secretion by a mechanism called the *renin–angiotensin–aldosterone (RAA) system.* In brief, blood pressure sensors (baroreceptors) in major arteries near the heart detect falling blood pressure and activate a sympathetic reflex. This leads to the production of a hormone called angiotensin II. Among its other effects, angiotensin II stimulates the adrenal cortex to secrete aldosterone (see fig. 23.15 and the associated text for details). Aldosterone stimulates the kidneys to retain sodium. Water is retained with it by osmosis, so aldosterone helps to maintain blood volume and pressure.

Cortisol (also known clinically as *hydrocortisone*) is the most potent glucocorticoid, but the adrenals also secrete a weaker one called *corticosterone.* Glucocorticoids are secreted by the zona fasciculata and zona reticularis in response to ACTH from the pituitary. They stimulate fat and protein catabolism, gluconeogenesis, and the release of fatty acids and glucose into the blood. This helps the body adapt to stress and repair damaged tissues. Glucocorticoids also have an anti-inflammatory effect; hydrocortisone is widely used in ointments to relieve swelling and other signs of inflammation. Excessive glucocorticoid secretion or medical use, however, suppresses the immune system for reasons we will see in the discussion of stress physiology later in this chapter.

Androgens are the primary adrenal sex steroids, but the adrenals also produce small amounts of estrogen. ACTH regulates both adrenal androgen and cortisol secretion. The sex steroids, too, come from both the zona fasciculata and zona reticularis. The major androgen is **dehydroepiandrosterone (DHEA)** (de-HY-dro-EP-ee-an-DROSS-tur-own). It has little biological activity in

---

[18]*zona* = zone; *glomerul* = little balls; *osa* = full of
[19]*zona* = zone; *fascicul* = little cords; *ata* = possessing
[20]*zona* = zone; *reticul* = little network; *aris* = like

itself, but many tissues convert it to the more potent forms, *testosterone* and *dihydrotestosterone*. DHEA is produced in tremendous quantities by the large adrenal glands of the male fetus and plays an important role in the prenatal development of the male reproductive tract. At puberty in both sexes, androgens induce the growth of pubic and axillary hair and their associated apocrine sweat glands, and they stimulate the libido (sex drive) throughout adolescent and adult life. In men, the large amount of androgen secreted by the testes overshadows that produced by the adrenals. In women, however, the adrenal glands meet about 50% of the androgen requirement.

Estradiol is the main adrenal estrogen. It is normally of minor importance to women of reproductive age because its quantity is small compared with estrogen from the ovaries. After menopause, however, the ovaries no longer function and only the adrenals secrete estrogen. However, several other tissues, such as fat, convert androgens into additional estrogen. Both androgens and estrogens promote adolescent skeletal growth and help to sustain adult bone mass.

The medulla and cortex are not as functionally independent as once thought; each of them stimulates the other. Without stimulation by cortisol, the adrenal medulla atrophies significantly. Conversely, some chromaffin cells from the medulla extend into the cortex. When stress activates the sympathetic nervous system, these cells stimulate the cortex to secrete corticosterone and perhaps other corticosteroids.

▶▶▶ **APPLY WHAT YOU KNOW**

The zona fasciculata thickens significantly in pregnant women. What do you think would be the benefit of this phenomenon?

17.3f The Pancreatic Islets

The **pancreas** is an elongated, spongy gland located below and behind the stomach; most of it is retroperitoneal **(fig. 17.12).** It is primarily an exocrine digestive gland, and its gross anatomy is described in section 25.4c. Scattered throughout the exocrine tissue, however, are 1 to 2 million endocrine cell clusters called **pancreatic islets (islets of Langerhans**[21]**).** Although they are less than 2% of the pancreatic tissue, the islets secrete hormones of vital importance, especially in the regulation of **glycemia,** the blood glucose concentration. A typical islet measures about 75 × 175 μm and contains from a few to 3,000 cells. Its main cell types are alpha cells (20%), beta cells (70%), delta cells (5%), and a small number of PP and other cells. Islet cells respond directly to blood nutrient levels associated with the cycle of eating and fasting. Their functions are as follows:

- **Alpha (α) cells,** or **glucagon cells,** secrete **glucagon** between meals when the blood glucose concentration falls below 100 mg/dL. Glucagon exerts two primary actions on the

---

[21]Paul Langerhans (1847–88), German anatomist

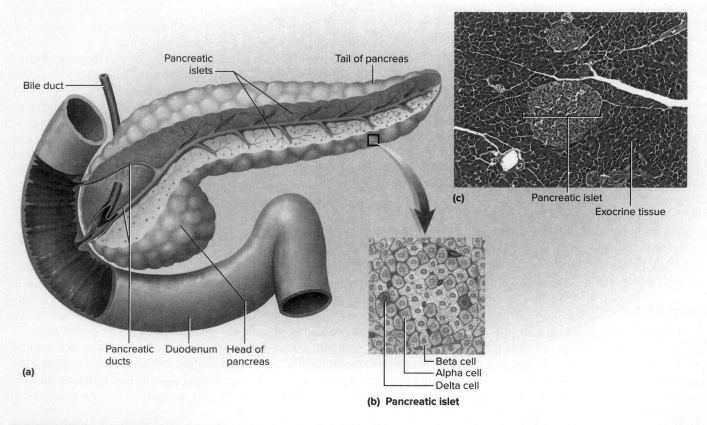

FIGURE 17.12 The Pancreas. (a) Gross anatomy and relationship to the duodenum. Pancreatic islets are most concentrated in the tail of the pancreas. (b) Cells of a pancreatic islet. (PP and G cells are not shown; they are few in number and cannot be distinguished with ordinary histological staining.) (c) Light micrograph of a pancreatic islet amid the darker exocrine acini, which produce digestive enzymes. **APR**

c: Al Telser/McGraw-Hill Education

liver: (1) glycogenolysis, the breakdown of glycogen into glucose; and (2) gluconeogenesis, the synthesis of glucose from fats and proteins (see fig. 17.23). These effects lead to the release of glucose into circulation, thus raising the blood glucose level. In adipose tissue, glucagon stimulates fat catabolism and the release of free fatty acids. Glucagon is also secreted in response to rising amino acid levels in the blood after a high-protein meal. It promotes amino acid absorption and thereby provides cells with the raw material for gluconeogenesis.

- **Beta (β) cells,** or **insulin cells,** secrete two hormones, *insulin* and *amylin.* **Insulin,** "the hormone of nutrient abundance," is secreted during and immediately following a meal when blood nutrient levels are rising; even the appetizing aroma of food stimulates insulin release in anticipation of eating. Osteocalcin and lipocalin 2, two hormones from the osteoblasts of bone, stimulate multiplication of beta cells, insulin secretion, and insulin sensitivity of other body tissues. The principal targets of insulin are the liver, skeletal muscles, and adipose tissue. In times of plenty, insulin stimulates cells to absorb glucose, fatty acids, and amino acids and to store or metabolize them; therefore, it lowers the level of blood glucose and other nutrients. It promotes the synthesis of glycogen, fat, and protein, thereby promoting the storage of excess nutrients for later use and enhancing cellular growth and differentiation. It also antagonizes glucagon, thus suppressing the use of already-stored fuels. The brain, liver, kidneys, and red blood cells absorb and use glucose without need of insulin, but insulin promotes glycogen synthesis in the liver. Insulin insufficiency or inaction is well known as the cause of diabetes mellitus, detailed later in this chapter. The beta cells also secrete another hormone, **amylin,** simultaneously with insulin. Amylin helps to reduce spikes in blood glucose by slowing the emptying of the stomach; modulating the secretion of gastric enzymes, acid, and bile; inhibiting glucagon secretion; and stimulating the sense of satiety (having had enough to eat).

- **Delta (δ) cells,** or **somatostatin cells,** secrete **somatostatin (growth hormone–inhibiting hormone)** concurrently with the release of insulin by the beta cells. Somatostatin helps to regulate the speed of digestion and nutrient absorption, and perhaps modulates the activity of other pancreatic islet cells.

- **PP cells** secrete **pancreatic polypeptide (PP)** for 4 or 5 hours after a meal. By acting on receptors in the brain, PP inhibits stimulation of the pancreas by the vagus nerve.

Any hormone that raises blood glucose concentration is called a *hyperglycemic hormone.* You may have noticed that glucagon is not the only hormone that does so; so do growth hormone, epinephrine, norepinephrine, cortisol, and corticosterone. Insulin is called a *hypoglycemic hormone* because it lowers blood glucose levels.

17.3g The Gonads

Like the pancreas, the **gonads** (ovaries and testes) function as both endocrine and exocrine glands. Their exocrine products are eggs and sperm, and their endocrine products are the *gonadal hormones,* most of which are steroids. Their gross anatomy is described in sections 27.2b and 28.1b. The gonads are hormonally inactive in childhood, but from puberty through adulthood they are regulated by the gonadotropic hormones of the anterior pituitary.

The ovaries secrete chiefly **estradiol, progesterone,** and **inhibin.** Each egg develops in its own follicle, which is lined by a wall of **granulosa cells** and surrounded by a capsule, the **theca (fig. 17.13a).** Theca cells synthesize the androgen *androstenedione,* and granulosa cells convert this to estradiol and lesser amounts of two other estrogens, *estriol* and *estrone.* In the middle of the monthly ovarian cycle, a mature follicle ovulates (ruptures and releases the egg). The remains of the follicle become the corpus luteum, which secretes progesterone for the next 12 days or so in a typical cycle (several weeks in the event of pregnancy).

The functions of estradiol and progesterone are detailed in section 28.3. In brief, they contribute to the development of the reproductive system and feminine physique, promote adolescent bone growth, regulate the menstrual cycle, sustain pregnancy, and prepare the mammary glands for lactation. Inhibin, which is also secreted by the follicle and corpus luteum, suppresses FSH secretion by means of negative feedback inhibition of the anterior pituitary.

The testis consists mainly of minute *seminiferous tubules* that produce sperm. Its endocrine secretions are **testosterone,** lesser amounts of weaker androgens and estrogens, and inhibin. Inhibin comes from **nurse cells** that form the walls of the seminiferous tubules. By limiting FSH secretion, it regulates the rate of sperm production. Nestled between the tubules are clusters of **interstitial endocrine cells,** the source of testosterone and the other sex steroids **(fig. 17.13b).** Testosterone stimulates development of the male reproductive system in the fetus and adolescent, the development of the masculine physique in adolescence, and the sex drive. It sustains sperm production and the sexual instinct throughout adult life.

17.3h Endocrine Functions of Other Tissues and Organs

Several other tissues and organs beyond the classical endocrine glands secrete hormones or hormone precursors:

- **The skin.** Keratinocytes of the epidermis convert a cholesterol-like steroid into **cholecalciferol** (COAL-eh-cal-SIF-er-ol), using energy from solar UV radiation. The liver and kidneys further convert cholecalciferol to a calcium-regulating hormone, **calcitriol** (see the following paragraphs).

- **The liver.** The liver is involved in the production of at least five hormones: (1) It converts the cholecalciferol from the skin into **calcidiol,** the next step in calcitriol synthesis. (2) It secretes a protein called **angiotensinogen,** which the kidneys, lungs, and other organs convert to a hormone called *angiotensin II,* a regulator of blood pressure (part of the renin–angiotensin–aldosterone system mentioned earlier). (3) It secretes about 15% of the body's **erythropoietin (EPO)** (er-RITH-ro-POY-eh-tin), a hormone that stimulates the red bone marrow to produce red blood cells (erythrocytes). The liver is therefore important in regulating the oxygen-carrying capacity of the blood. (4) It secretes **insulin-like growth factor I (IGF-I),** a hormone that mediates the action of growth hormone. (5) It secretes **hepcidin,** the principal hormonal mechanism of iron homeostasis. Hepcidin controls plasma iron level by inhibiting intestinal absorption of iron and

FIGURE 17.13 The Gonads. (a) Histology of an ovarian follicle. (b) Histology of the testis. The granulosa cells of the ovary and interstitial cells of the testis are endocrine cells. **APR**

a: Ed Reschke/Getty Images; b: Ed Reschke

mobilization of iron by the liver; thus, it prevents iron overload. In anemia, hepcidin levels fall and more iron is mobilized to support hemoglobin synthesis. In infections, hepcidin levels rise, reducing the level of free iron available to infectious microorganisms that need it for their reproduction.

- **The kidneys.** The kidneys play endocrine roles in the production of three hormones—calcitriol, angiotensin II, and erythropoietin. (1) They convert calcidiol into **calcitriol (vitamin D₃),** thus completing the three-step process begun by the skin and liver. Calcitriol raises the blood concentration of calcium by promoting its intestinal absorption and slightly inhibiting its loss in the urine. This makes more calcium available for bone deposition and other uses. (2) They secrete an enzyme called **renin** (REE-nin), which converts angiotensinogen to angiotensin I. As angiotensin I circulates through various organs, especially the lungs, *angiotensin-converting enzyme (ACE)* on the linings of the blood capillaries converts it to **angiotensin II,** a small peptide. This is a very potent hormone that constricts blood vessels throughout the body and thereby raises blood pressure. (3) The kidneys secrete about 85% of the body's erythropoietin. For this reason, renal failure tends to cause anemia as the EPO output of the failing kidneys declines.

- **The heart.** Rising blood pressure stretches the heart wall and stimulates cardiac muscle in the atria to secrete two similar

natriuretic[22] **peptides.** These hormones increase sodium excretion and urine output and oppose the action of angiotensin II, just described. Together, these effects lower the blood pressure.

- **The stomach and intestines.** The largest endocrine network of all is in the stomach and intestines. These have more than 15 kinds of *enteroendocrine cells,*[23] constituting about 1% of all the epithelial cells of the GI tract. Their hormones are collectively called **enteric hormones.** They coordinate different regions and glands of the digestive system with each other and affect feeding, digestion, gastrointestinal motility and secretion, and maintenance of the mucosa. Some of them are called **gut–brain peptides** because they originate in the digestive tract but stimulate the brain. **Peptide YY (PYY),** secreted by the small and large intestines, signals satiety (fullness) and tends to terminate eating. **Cholecystokinin**[24] **(CCK)** (CO-leh-SIS-toe-KY-nin), secreted by the small intestine in response to fats, stimulates the gallbladder to release bile and has an appetite-suppressing effect on the brain. **Ghrelin,**[25] as noted in the earlier remarks on growth hormone, is secreted

[22]*natri* = sodium; *uretic* = pertaining to urine
[23]*entero* = intestine
[24]*chole* = bile; *cysto* = sac (gallbladder); *kin* = action, motion
[25]named partly from *ghre* = growth, and partly as an acronym derived from growth hormone–releasing hormone

when the stomach is empty; it stimulates the appetite and the secretion of growth hormone–releasing hormone and thus, indirectly, growth hormone itself. **Gastrin**[26] is secreted by the stomach upon the arrival of food and stimulates other cells of the stomach to secrete hydrochloric acid. Gut–brain peptides and the endocrine regulation of hunger are further discussed in section 26.1b.

- **Adipose tissue.** Fat cells secrete at least three hormones that regulate carbohydrate and fat metabolism. The best-known one is **leptin,** which has long-term effects on appetite-regulating centers of the hypothalamus. A low level of leptin, signifying a deficiency of body fat, increases appetite and food intake, whereas a high level of leptin tends to blunt the appetite. Leptin also serves as a signal for the onset of puberty, which is delayed in persons with abnormally low

body fat. Leptin is treated with other enteric hormones in the aforementioned discussion of appetite.

- **Osseous tissue.** Osteoblasts secrete the hormones **osteocalcin** and **lipocalin 2.** Both of these hormones stimulate pancreatic beta cells and promote insulin secretion and action. Osseous tissue thus plays an important role in glucose metabolism and the regulation of blood glucose levels.

- **The placenta.** This organ performs many functions in pregnancy, including fetal nutrition, oxygenation, and waste removal. But it also secretes estrogen, progesterone, and other hormones that regulate pregnancy and stimulate development of the fetus and the mother's mammary glands.

You can see that the endocrine system is extensive. It includes numerous discrete glands as well as individual cells in the tissues of other organs. The endocrine organs and tissues other than the hypothalamus and pituitary are surveyed in **table 17.5.**

[26]*gastr* = stomach

TABLE 17.5	Hormones from Sources Other Than the Hypothalamus and Pituitary		
Source	**Hormone**	**Target Organs and Tissues**	**Principal Effects**
Pineal gland	Melatonin	Brain	Uncertain; may influence mood and sexual maturation
Thymus	Thymopoietin, thymosin, thymulin	Immune cells (T lymphocytes)	Stimulate T lymphocyte development and activity
Thyroid gland	Thyroxine (T_4) and triiodothyronine (T_3)	Most tissues	Elevate metabolic rate and heat production; increase respiratory rate, heart rate, and strength of heartbeat; stimulate appetite and accelerate breakdown of nutrients; promote alertness and quicken reflexes; stimulate growth hormone secretion and growth of skin, hair, nails, teeth, and fetal nervous system
	Calcitonin	Bone	Stimulates bone deposition, mainly in children
Parathyroid glands	Parathyroid hormone (PTH)	Bone, kidneys, small intestine	Raises blood Ca^{2+} level by stimulating bone resorption and inhibiting deposition, reducing urinary Ca^{2+} excretion, and enhancing calcitriol synthesis
Adrenal medulla	Epinephrine, norepinephrine, dopamine	Most tissues	Promote alertness; mobilize organic fuels; raise metabolic rate; stimulate circulation and respiration; increase blood glucose level; inhibit insulin secretion and glucose uptake by insulin-dependent organs (sparing glucose for brain)
Adrenal cortex	Aldosterone	Kidney	Promotes Na^+ and water retention and K^+ excretion; maintains blood pressure and volume
	Cortisol and corticosterone	Most tissues	Stimulate fat and protein catabolism, gluconeogenesis, stress resistance, and tissue repair
	Dehydroepiandrosterone	Bone, muscle, integument, brain, many other tissues	Precursor of testosterone; indirectly promotes growth of bones, pubic and axillary hair, apocrine glands, and fetal male reproductive tract; stimulates libido
Pancreatic islets	Glucagon	Primarily liver	Stimulates amino acid absorption, gluconeogenesis, glycogen and fat breakdown; raises blood glucose and fatty acid levels
	Insulin	Most tissues	Stimulates glucose and amino acid uptake; lowers blood glucose level; promotes glycogen, fat, and protein synthesis
	Somatostatin	Stomach, intestines, pancreatic islet cells	Modulates digestion, nutrient absorption, and glucagon and insulin secretion
	Gastrin	Stomach	Stimulates acid secretion and gastric motility

(continued)

TABLE 17.5	Hormones from Sources Other Than the Hypothalamus and Pituitary *(continued)*		
Source	**Hormone**	**Target Organs and Tissues**	**Principal Effects**
Ovaries	Estradiol	Many tissues	Stimulates female reproductive development and adolescent growth; regulates menstrual cycle and pregnancy; prepares mammary glands for lactation
	Progesterone	Uterus, mammary glands	Regulates menstrual cycle and pregnancy; prepares mammary glands for lactation
	Inhibin	Anterior pituitary	Inhibits FSH secretion
Testes	Testosterone	Many tissues	Stimulates fetal and adolescent reproductive development, musculoskeletal growth, sperm production, and libido
	Inhibin	Anterior pituitary	Inhibits FSH secretion
Skin	Cholecalciferol	—	Precursor of calcitriol (see kidneys)
Liver	Calcidiol	—	Precursor of calcitriol (see kidneys)
	Angiotensinogen	—	Precursor of angiotensin II (see kidneys)
	Erythropoietin	Red bone marrow	Promotes red blood cell production, increases oxygen-carrying capacity of blood
	Hepcidin	Small intestine, liver	Regulates plasma iron level
	Insulin-like growth factor I	Many tissues	Prolongs and mediates action of growth hormone
Kidneys	Angiotensin I	—	Precursor of angiotensin II, a vasoconstrictor
	Calcitriol	Small intestine	Increases blood calcium level mainly by promoting intestinal absorption of Ca^{2+}
	Erythropoietin	Red bone marrow	Promotes red blood cell production, increases oxygen-carrying capacity of blood
Heart	Natriuretic peptides	Kidney	Lower blood volume and pressure by promoting Na^+ and water loss
Stomach and small intestine	Cholecystokinin	Gallbladder, brain	Bile release; appetite suppression
	Gastrin	Stomach	Stimulates acid secretion
	Ghrelin	Brain	Stimulates hunger, initiates feeding
	Peptide YY	Brain	Produces sense of satiety, terminates feeding
	Other enteric hormones	Stomach, intestines	Coordinate secretion and motility in different regions of digestive tract
Adipose tissue	Leptin	Brain	Limits appetite over long term
Osseous tissue	Osteocalcin	Pancreas, adipose tissue	Stimulates pancreatic beta cells to multiply, increases insulin secretion, enhances insulin sensitivity of various tissues, and reduces fat deposition
	Lipocalin 2	Pancreatic beta cells	Promotes insulin secretion and action
Placenta	Estrogen, progesterone	Many tissues of mother and fetus	Stimulate fetal development and maternal bodily adaptations to pregnancy; prepare mammary glands for lactation

BEFORE YOU GO ON

Answer the following questions to test your understanding of the preceding section:

11. Identify three endocrine glands that are larger or more functional in infants or children than in adults. What is the term for the shrinkage of a gland with age?

12. Why does thyroid hormone elevate the body temperature?

13. Name a glucocorticoid, a mineralocorticoid, and a catecholamine secreted by the adrenal gland.

14. Does the action of glucocorticoids more closely resemble that of glucagon or insulin? Explain.

15. Define *hypoglycemic hormone* and *hyperglycemic hormone* and give an example of each.

16. What is the difference between a gonadal hormone and a gonadotropin?

17.4 Hormones and Their Actions

Expected Learning Outcomes

When you have completed this section, you should be able to

a. identify the chemical classes to which various hormones belong;

b. describe how hormones are synthesized and transported to their target organs;

c. describe how hormones stimulate their target cells;

d. explain how target cells regulate their sensitivity to circulating hormones;

e. describe how hormones affect each other when two or more of them stimulate the same target cells; and

f. discuss how hormones are removed from circulation after they have performed their roles.

Having surveyed the body's major hormones and their effects, we have established a cast of characters for the endocrine story, but we're left with some deeper questions: Exactly what is a hormone? How are hormones synthesized and transported to their destinations? How does a hormone produce its effects on a target organ? Thus, we now address endocrinology at the molecular and cellular levels.

17.4a Hormone Chemistry

Most hormones fall into three chemical classes: *steroids, monoamines,* and *peptides* (table 17.6, fig. 17.14).

1. **Steroid hormones** are derived from cholesterol. They include sex steroids produced by the testes and ovaries (such as estrogens, progesterone, and testosterone) and corticosteroids produced by the adrenal gland (such as cortisol, aldosterone, and DHEA). The gonads and adrenal cortex are

the only sites of steroid hormone synthesis. Calcitriol, the calcium-regulating hormone, isn't a steroid but is derived from one and has the same hydrophobic character and mode of action as the steroids, so it is commonly grouped with them.

2. **Monoamines (biogenic amines)** were introduced in section 12.5c, since this class also includes several neurotransmitters (see fig. 12.22). The monoamine hormones include dopamine, epinephrine, norepinephrine, melatonin, and thyroid hormone. The first three of these are also called *catecholamines*. Monoamines are made from amino acids

TABLE 17.6	Chemical Classification of Hormones
Steroids and Steroid Derivatives	**Monoamines**
Aldosterone	Dopamine
Androgens	Epinephrine
Calcitriol	Melatonin
Corticosterone	Norepinephrine
Cortisol	Thyroid hormone
Estrogens	
Progesterone	
Oligopeptides (3–10 Amino Acids)	
Angiotensin II	Gonadotropin-releasing hormone
Antidiuretic hormone	Oxytocin
Cholecystokinin	Thyrotropin-releasing hormone
Polypeptides (More Than 10 Amino Acids)	
Adrenocorticotropic hormone	Insulin
Calcitonin	Leptin
Corticotropin-releasing hormone	Natriuretic peptides
Gastrin	Pancreatic polypeptide
Ghrelin	Parathyroid hormone
Glucagon	Prolactin
Growth hormone	Somatostatin
Growth hormone–releasing hormone	Thymic hormones
Hepcidin	
Glycoproteins (Protein–Carbohydrate Complexes)	
Erythropoietin	Luteinizing hormone
Follicle-stimulating hormone	Thyroid-stimulating hormone
Inhibin	

FIGURE 17.14 The Chemical Classes of Hormones. (a) Two steroid hormones, defined by their four-membered rings derived from cholesterol. (b) Two monoamines, derived from amino acids and defined by their —NH— or —NH$_2$ (amino) groups. (c) A small peptide hormone, oxytocin, and a protein hormone, insulin, defined by their chains of amino acids (the yellow circles).

and retain an amino group, from which this hormone class gets its name.

3. **Peptide hormones** are chains of 3 to 200 or more amino acids. The two posterior pituitary hormones, oxytocin and antidiuretic hormone, are very similar oligopeptides of just nine amino acids. Probably the best-known large peptide (protein) hormone is insulin. Except for dopamine, the releasing and inhibiting hormones produced by the hypothalamus are polypeptides. Most hormones of the anterior pituitary are polypeptides or glycoproteins—polypeptides conjugated with short carbohydrate chains. Glycoprotein hormones usually have an identical alpha chain of 92 amino acids and a variable beta chain that distinguishes them from each other.

17.4b Hormone Synthesis

All hormones are made from either cholesterol or amino acids, with carbohydrate added in the case of glycoproteins.

Steroids

Steroid hormones are synthesized from cholesterol and differ mainly in the functional groups attached to the four-ringed steroid backbone. **Figure 17.15** shows the synthetic pathway for several steroid hormones. Notice that while estrogen and progesterone are typically thought of as "female" hormones and testosterone as a "male" hormone, these sex steroids are interrelated in their synthesis and have roles in both sexes.

Peptides

Peptide hormones are synthesized the same way as any other protein. The gene for the hormone is transcribed to form a molecule of mRNA, and ribosomes translate the mRNA and assemble amino acids in the right order to make the peptide. After the basic amino acid sequence is assembled, the rough endoplasmic reticulum and Golgi complex may further modify the peptide to form the mature hormone. Insulin, for example, begins as a single amino acid chain called *proinsulin*. A middle portion called the *connecting peptide* is removed to convert proinsulin to insulin, now composed of two polypeptide chains connected to each other by disulfide bridges **(fig. 17.16).**

▶▶▶ **APPLY WHAT YOU KNOW**

During the synthesis of glycoprotein hormones, where in the cell would the carbohydrate be added? (Hint: Look within section 4.2.)

FIGURE 17.15 The Synthesis of Steroid Hormones from Cholesterol. The ovaries secrete progesterone and estradiol, the adrenal cortex secretes cortisol and aldosterone, and the testes secrete testosterone.

Monoamines

Melatonin is synthesized from the amino acid tryptophan and all the other monoamines from the amino acid tyrosine. Thyroid hormone (TH) is unusual in that it is made of *two* tyrosines, and is the only process in the human body that uses iodine; a lack

FIGURE 17.16 The Synthesis of Insulin, a Representative Polypeptide Hormone. Proinsulin has a connecting (C) peptide, 31 amino acids long, that is removed to leave insulin. Insulin has two polypeptide chains, 30 and 21 amino acids long, joined by two disulfide bridges (—S—S—) represented by the yellow bars. A third disulfide bridge creates a loop in the short chain.

of dietary iodine causes a thyroid disorder called *goiter* (see fig. 17.26). **Figure 17.17** shows the steps in TH synthesis.

① Cells of the thyroid follicle begin the process by absorbing iodide (I⁻) ions from the blood of nearby capillaries. At the apical surface of the cells, facing the lumen of the follicle, they oxidize I⁻ to a reactive form of iodine represented by I* in the figure.

② In the meantime, the follicle cells also synthesize a large protein called **thyroglobulin (Tg).** Each Tg has 123 tyrosines among its amino acids, but only 4 to 8 of them are used to make TH. The cells release thyroglobulin by exocytosis from their apical surfaces into the lumen.

③ An enzyme at the cell surface adds iodine to a few of the tyrosines. Some tyrosines receive one iodine and become *monoiodotyrosine (MIT)* (MON-oh-eye-OH-do-TY-ro-seen); some receive two and become *diiodotyrosine (DIT)* (fig. 17.17b). Where Tg folds back on itself and two iodinated tyrosines meet, or where one tyrosine meets another on an adjacent Tg, the tyrosines link to each other through their side groups. If an MIT links up with a DIT, they form a complex with three iodines, destined to become the T_3 form of thyroid hormone; if two DITs unite, they form the forerunner of the T_4 form, with four iodines. One tyrosine then breaks away from

FIGURE 17.17 Thyroid Hormone Synthesis and Secretion. (a) TH synthesis in relation to the thyroid follicle. See text for explanation of numbered steps. (b) The meeting of an MIT and a DIT would produce the T_3 form of thyroid hormone, with three iodines. If both of these were DITs, it would produce the T_4 form with four iodines. (c) The two forms of mature thyroid hormone, triiodothyronine (T_3) and tetraiodothyronine (T_4).

its Tg, but for the time being, the hormone remains anchored to Tg through its other tyrosine. Tg is stored in the follicles until the thyroid gland receives a signal to release it. It is the pink-staining material in figure 17.9b.

④ When the follicle cells receive thyroid-stimulating hormone (TSH) from the anterior pituitary, they absorb droplets of Tg by pinocytosis. Within the cells, a lysosome contributes an enzyme that hydrolyzes the Tg chain, liberating thyroid hormone (TH).

⑤ TH is released from the basal side of the follicle cells into nearby blood capillaries. In the blood, it binds to various

transport proteins that carry it to its target cells. The released hormone is about 10% T_3 and 90% T_4 (fig. 17.17c); the respective roles of T_3 and T_4 are discussed shortly.

17.4c Hormone Secretion

Hormones aren't secreted at steady rates, nor do they have constant levels in the bloodstream throughout the day. Rather, they are secreted in some cases on a daily (circadian) rhythm, in other cases on a monthly rhythm (in a woman's ovarian cycle), or under the influence of stimuli that signify a need for them. These stimuli are of three kinds.

1. **Neural stimuli.** Nerve fibers supply some endocrine glands and elicit the release of their hormones. For example, the sympathetic nervous system stimulates the adrenal medulla to secrete epinephrine and norepinephrine in situations of stress. In childbirth, nerve signals originate from stretch receptors in the uterus, travel up the spinal cord and brainstem to the hypothalamus, and stimulate the release of oxytocin (see fig. 1.9).

2. **Hormonal stimuli.** Hormones from the hypothalamus regulate secretion by the anterior pituitary gland, and pituitary hormones stimulate other endocrine glands to release thyroid hormone, sex hormones, and cortisol.

3. **Humoral stimuli.** This refers to blood-borne stimuli. For example, rising blood glucose concentration stimulates the release of insulin, low blood osmolarity stimulates the secretion of aldosterone, and a low blood calcium level stimulates the secretion of parathyroid hormone.

Peptide hormones such as growth hormone and oxytocin, and catecholamines such as epinephrine and dopamine, are stored in secretory vesicles of the endocrine cell until needed and released by exocytosis when the cell receives a stimulus to do so. Steroid hormones such as estrogen and cortisol, however, are not stored in vesicles or released by exocytosis. They don't accumulate in the endocrine cell, but are released as fast as they're synthesized by diffusion through the cell surface. But this doesn't mean they're secreted at a constant rate, hour by hour and day by day. Stimuli such as FSH and ACTH can increase the synthesis and release of a steroid hormone by several-fold within hours. Thyroid hormone (TH) also diffuses freely through plasma membranes, but unlike steroids, it does accumulate in the gland awaiting a stimulus for secretion. It isn't stored in the endocrine cells, but in the extracellular spaces enclosed by the thyroid follicles. Until the stimulus (TSH) to secrete it arrives, it is bound to the protein thyroglobulin and cannot escape. The thyroid gland is thus able to store a large quantity of hormone, enough to meet the body's need for months even if TH synthesis ceases.

17.4d Hormone Transport

To get from an endocrine cell to a target cell, a hormone must travel in the blood, which is mostly water. Most of the monoamines and peptides are hydrophilic, so mixing with the blood plasma is no problem for them. Steroids and thyroid hormone, however, are hydrophobic. To travel in the watery bloodstream, they must bind to hydrophilic **transport proteins**—albumins and globulins synthesized by the liver. A hormone attached to a transport protein is called a **bound hormone,** and one that is not attached is an **unbound (free) hormone.** Only the unbound hormone can leave a blood capillary and get to a target cell.

Transport proteins not only carry the hydrophobic hormones, but also prolong their half-lives. They protect hormones from being broken down by enzymes in the blood plasma and liver and from being filtered out of the blood by the kidneys. Free hormone may be broken down or removed from the blood in a few minutes, whereas bound hormone may circulate for hours to weeks.

Thyroid hormone binds to three transport proteins in the blood plasma: *albumin;* an albumin-like protein called *thyretin;* and most of all, an alpha globulin named *thyroxine-binding globulin (TBG).* More than 99% of circulating TH is protein-bound. Bound TH serves as a long-lasting blood reservoir, so even if the thyroid is surgically removed (as for cancer), no signs of TH deficiency appear for about 2 weeks.

Steroid hormones bind to globulins such as *transcortin,* the transport protein for cortisol. Aldosterone is unusual. It has no specific transport protein but binds weakly to albumin and others. However, 85% of it remains unbound, and correspondingly, it has a half-life of only 20 minutes.

17.4e Hormone Receptors and Modes of Action

Hormones stimulate only those cells that have receptors for them—their *target cells.* The receptors are proteins or glycoproteins located on the plasma membrane, in the cytoplasm, or in the nucleus. They act like switches to turn certain metabolic pathways on or off when the hormones bind to them. A target cell usually has a few thousand receptors for a given hormone. Receptor defects lie at the heart of several endocrine diseases (see Deeper Insight 17.2).

Receptor–hormone interactions are similar to the enzyme–substrate interactions described in section 2.4f. Unlike enzymes, receptors do not chemically change their ligands, but they do exhibit enzymelike specificity and saturation. *Specificity* means that the receptor for one hormone will not bind other hormones. *Saturation* is the condition in which all the receptor molecules are occupied by hormone molecules. Adding more hormone cannot produce any greater effect.

Hormones fall into two broad classes according to where they act on their target cells—those that cannot enter the target cell but act through receptors at the cell surface, and those that can enter the target cell and act through receptors within it. The next two sections describe the mode of action of each of these classes.

DEEPER INSIGHT 17.2

CLINICAL APPLICATION

Hormone Receptors and Therapy

In treating endocrine disorders, it is essential to understand the role of hormone receptors. For example, a defect or deficiency of insulin receptors causes type 2 diabetes mellitus. No amount of insulin replacement can correct this. And while growth hormone is now abundantly available thanks to genetic engineering, it is useless to children with *Laron dwarfism,* who have a hereditary defect in their GH receptors. *Androgen insensitivity syndrome* is due to an androgen receptor defect or deficiency; it causes genetic males to develop feminine genitalia and other features (see Deeper Insight 27.1). Estrogen stimulates the growth of some malignant tumors with estrogen receptors. For this reason, estrogen replacement therapy should not be used for women with estrogen-dependent cancer.

Peptides and Catecholamines

Peptides and catecholamines are hydrophilic and cannot penetrate a target cell's plasma membrane, so they must stimulate its physiology indirectly. They bind to cell surface receptors, which are linked to second-messenger systems on the other side of the membrane **(fig. 17.18a).** The best-known second messenger is cyclic adenosine monophosphate (cAMP) **(fig. 17.19).** When glucagon binds to the surface of a liver cell, for example, its receptor activates a G protein, which in turn activates adenylate cyclase, the membrane enzyme that produces cAMP. cAMP leads ultimately to the activation of enzymes that hydrolyze glycogen stored in the cell, leading to the release of free glucose into the blood and fulfillment of the function of glucagon. Somatostatin, by contrast, *inhibits* cAMP synthesis. Second messengers don't linger in the cell for long. cAMP, for example, is quickly broken down by an enzyme called **phosphodiesterase,** and the hormonal effect is therefore short-lived.

Two other second-messenger systems begin with one of the phospholipids in the plasma membrane. When activated by certain hormones (blue box in **fig. 17.20**), the receptor activates a G protein linked to a nearby enzyme, *phospholipase,* in the plasma membrane. Phospholipase splits a membrane phospholipid into two fragments—a small phosphate-containing piece called **inositol triphosphate (IP$_3$)** (eye-NOSS-ih-tol), and a larger piece, the triglyceride backbone with two fatty acids still attached, called **diacylglycerol (DAG)** (di-ACE-ul-GLISS-ur-ol). IP$_3$ and DAG are the second messengers that go on to activate a wide variety of metabolic changes in the target cells, depending on what cells are involved and what internal signaling pathways they use.

DAG activates a protein kinase (PK), much like cAMP does in figure 17.19. PK phosphorylates various other enzymes, turning them on or off and thereby activating or suppressing equally various metabolic processes in the target cell. For example, we saw earlier that thyroid-stimulating hormone (TSH) binds to thyroid follicle cells and stimulates them to release thyroid hormone into the bloodstream. TSH works through the DAG second-messenger system. In other cells, DAG stimulates mitosis and cell proliferation. Some cancer-causing agents (carcinogens) act by mimicking this *mitogenic* effect of DAG.

IP$_3$ works by increasing the calcium (Ca^{2+}) concentration in the target cell. It can open Ca^{2+} channels in the plasma membrane, letting Ca^{2+} into the cell from the extracellular fluid, or it can open channels in the endoplasmic reticulum, causing it to release a flood of Ca^{2+} into the cytosol. Calcium then acts through several means to alter cell physiology:

1. It binds to certain calcium-dependent cytoplasmic enzymes that alter cell metabolism.

2. It can bind to a cytoplasmic calcium receptor, **calmodulin.** Not only is calcium-bound calmodulin the key to smooth muscle contraction, but it can also activate protein kinases with downstream effects just like cAMP or DAG.

3. It binds to membrane channels and changes their permeability to other solutes, in some cases altering the membrane potential (voltage) of the cell.

FIGURE 17.18 Hormone Actions on a Target Cell. Some process steps are omitted for simplicity. (a) Action of a peptide hormone through a surface receptor and second messenger. (b) Action of a steroid hormone by diffusing into the nucleus and binding to a nuclear receptor associated with a gene. Either process can lead to a great variety of effects on the target cell.

Childbirth affords one example of how an IP$_3$ second-messenger system works. Oxytocin (OT) from the pituitary gland binds to receptors on the smooth muscle cells of the uterus. It triggers the foregoing IP$_3$-releasing process, and IP$_3$, in turn, stimulates the sarcoplasmic reticulum to release Ca^{2+}. This initial burst of calcium then opens channels in the plasma membrane that admit still more calcium into the cell from the extracellular fluid. Calcium binds to calmodulin, and stimulates the labor contractions (see contraction of smooth muscle in section 11.7b).

The general point of all this is that hydrophilic hormones such as those listed in the blue box at the left side of figure 17.20 cannot enter the target cell. Yet by merely "knocking on the door" (binding to a surface receptor), they can initiate a flurry of metabolic activity within. The initial steps in this process are activation of a G protein and phospholipase. From there, divergent pathways are taken that involve DAG, IP$_3$, and Ca^{2+} as second messengers. Ultimately these pathways lead to metabolic pathways being switched on or off within the cell.

A given hormone doesn't always employ the same second messenger. ADH employs the IP$_3$–calcium system in smooth muscle but the cAMP system in kidney tubules. Insulin differs from all the foregoing mechanisms. Rather than using a second-messenger system, it binds to a plasma membrane enzyme, tyrosine kinase, which directly phosphorylates cytoplasmic proteins.

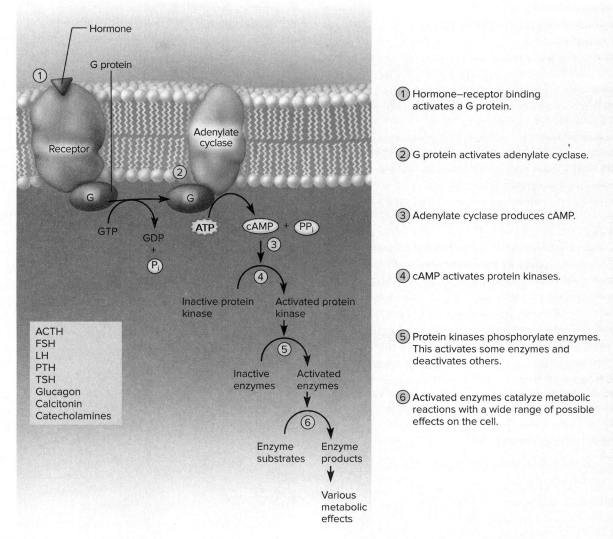

Hormone

G protein

①

Receptor

Adenylate cyclase

②

G

G

GTP

GDP
+
P$_i$

ATP

cAMP + PP$_i$

③

④

Inactive protein kinase

Activated protein kinase

ACTH
FSH
LH
PTH
TSH
Glucagon
Calcitonin
Catecholamines

Inactive enzymes

⑤

Activated enzymes

Enzyme substrates

⑥

Enzyme products

Various metabolic effects

① Hormone–receptor binding activates a G protein.

② G protein activates adenylate cyclase.

③ Adenylate cyclase produces cAMP.

④ cAMP activates protein kinases.

⑤ Protein kinases phosphorylate enzymes. This activates some enzymes and deactivates others.

⑥ Activated enzymes catalyze metabolic reactions with a wide range of possible effects on the cell.

FIGURE 17.19 Cyclic AMP (cAMP) as a Second Messenger. The green box lists some hormones that act in this manner.
❓ *Why are no steroid hormones listed in this box?*

Hormonal effects mediated through surface receptors tend to be relatively quick, because they don't depend on the cell synthesizing new proteins before anything else can happen. The oxytocin-induced labor contractions are a good example.

Steroids and Thyroid Hormone

Steroid hormones and thyroid hormone act more simply and directly. They are hydrophobic and pass easily through the plasma membrane of a target cell **(fig. 17.18b).** Most steroids go directly into the nucleus and bind to a receptor there; glucocorticoids, however, bind to a receptor in the cytosol, and the hormone–receptor complex is then transported into the nucleus. In either case, the receptor associates with the target gene in the nucleus and either activates or inhibits transcription of a gene for a metabolic enzyme or other protein. For a good example of this, see the *permissive effect* of estrogen on progesterone action on the uterus in forthcoming section 17.4h.

Thyroid hormone also acts on nuclear receptors but is handled a little differently. It enters the target cell by means of an ATP-dependent transport protein. Surprisingly, although 90% of the TH secreted by the thyroid gland is thyroxine (T_4), T_4 has little metabolic effect. Within the target-cell cytoplasm, an enzyme removes one iodine and converts it to the active form, T_3. This T_3, as well as a smaller amount of T_3 produced directly by the thyroid and absorbed from the blood, enters the nucleus and binds to receptors in the chromatin. One of the genes activated by T_3 is for the enzyme Na^+–K^+ ATPase (the Na^+–K^+ pump). One effect of this is to liberate heat from ATP, thus accounting for the calorigenic effect of thyroid hormone. T_3 also activates the transcription of genes for a norepinephrine receptor and part of the muscle protein myosin, thus enhancing the responsiveness of cells such as cardiac muscle to sympathetic stimulation and increasing the strength of the heartbeat.

Steroid and thyroid hormones typically require several hours to days to show an effect. This lag is due to the time required for genetic transcription, translation, and accumulation of enough protein product to have a significant effect on target-cell metabolism.

FIGURE 17.20 Diacylglycerol (DAG) and Inositol Triphosphate (IP$_3$) Second-Messenger System. These are employed by hormones listed in the blue box.

17.4f Signal Amplification

Hormones are extraordinarily potent chemicals. Through a mechanism called **signal amplification** (or the **cascade effect**), one hormone molecule triggers the synthesis of not just one enzyme molecule but an enormous number (**fig. 17.21**). To put it in a simplistic but illustrative way, suppose 1 glucagon molecule triggered the formation of 1,000 molecules of cAMP, cAMP activated a protein kinase, each protein kinase activated 1,000 other enzyme

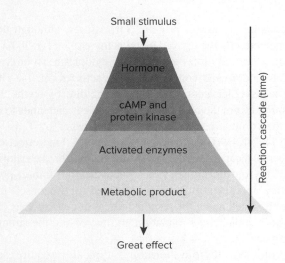

FIGURE 17.21 Signal Amplification. A single hormone molecule can trigger the production of many cAMP molecules and activation of many molecules of protein kinase. Amplification of the process at this and each succeeding step allows for a very small hormonal stimulus to cause a very large metabolic effect.

molecules, and each of those produced 1,000 molecules of a reaction product. These are modest numbers as chemical reactions go, yet even at this low estimate, each glucagon molecule would trigger the production of 1 billion molecules of reaction product. Whatever the actual numbers may be, you can see how signal amplification enables a very small stimulus to produce a very large effect. Hormones are therefore powerfully effective in tiny quantities. Their circulating concentrations are very low compared with other blood substances—on the order of nanograms per deciliter. Blood glucose, for example, is about 100 million times this concentrated. Because of signal amplification, target cells don't need a great number of hormone receptors.

17.4g Modulation of Target-Cell Sensitivity

Target cells can adjust their sensitivity to a hormone by changing the number of receptors for it. In **up-regulation,** a cell increases the number of receptors and becomes more sensitive to the hormone (**fig. 17.22a**). In late pregnancy, for example, the uterus produces oxytocin receptors, preparing itself for the surge of oxytocin that will occur during childbirth.

 Down-regulation is the process in which a cell reduces its receptor population and thus becomes less sensitive to a hormone

(a) Up-regulation

(b) Down-regulation

FIGURE 17.22 Modulation of Target-Cell Sensitivity. (a) Up-regulation, in which a cell produces more receptors and increases its own sensitivity to a hormone. (b) Down-regulation, in which a cell reduces the density of its receptors and lessens its sensitivity to a hormone.

(fig. 17.22b). This sometimes happens in response to long-term exposure to a high hormone concentration. For example, adipocytes down-regulate when exposed to high concentrations of insulin, and cells of the testis down-regulate in response to high concentrations of luteinizing hormone.

Hormone therapy often involves long-term use of abnormally high *pharmacological doses* of hormone, which may have undesirable side effects. Two ways in which adverse side effects can arise are: (1) Excess hormone may bind to receptor sites for other related hormones and mimic their effects; and (2) a target cell may convert one hormone into another, such as testosterone to estrogen. Thus, long-term high doses of testosterone can, paradoxically, have feminizing effects.

17.4h Hormone Interactions

No hormone travels in the bloodstream alone, and no cell is exposed to only one hormone. Rather, there are many hormones in the blood and tissue fluid at once. Cells ignore the majority of them because they have no receptors for them, but most cells are sensitive to more than one. In these cases, the hormones may have three kinds of interactive effects:

1. **Synergistic effects,** in which two or more hormones act together to produce an effect that is greater than the sum of their separate effects. For example, neither FSH nor testosterone alone stimulates much sperm production. When they act together, however, the testes produce some 300,000 sperm per minute.

2. **Permissive effects,** in which one hormone enhances the target organ's response to a second hormone to come later. For example, when estrogen from the ovaries enters cells of the uterine mucosa, it binds to its nuclear receptors and activates the gene for the progesterone receptor. The uterine cells then up-regulate their nuclear progesterone receptors. Progesterone comes later in the menstrual cycle, binds to these receptors, and stimulates transcription of the gene for a glycogen-synthesizing enzyme. The uterine cells then synthesize and accumulate glycogen for the nourishment of an embryo in the event of pregnancy. Progesterone has no effect on the uterine lining unless estrogen has been there earlier and prepared the way by inducing the synthesis of progesterone receptors. Estrogen thus has a permissive effect on progesterone action.

3. **Antagonistic effects,** in which one hormone opposes the action of another. For example, insulin lowers blood glucose level and glucagon raises it **(fig. 17.23).** During pregnancy, estrogen from the placenta inhibits the mammary glands from responding to prolactin; thus milk is not secreted until the placenta is shed at birth.

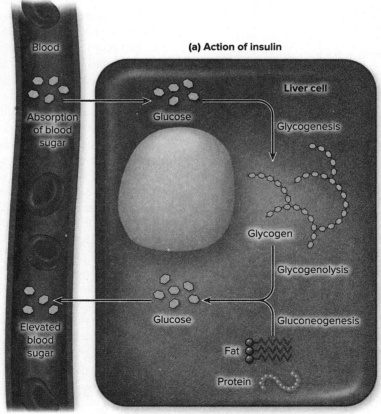

FIGURE 17.23 Antagonistic Effects of Insulin and Glucagon on the Liver. (a) Insulin action in promoting glucose uptake, glycogen storage, and lowering of blood glucose level. (b) Glucagon action in promoting glycogen breakdown, glucose synthesis, glucose release into the blood, and elevation of blood glucose level. Glycogenesis is the synthesis of glycogen; glycogenolysis is its hydrolysis (breakdown to glucose); and gluconeogenesis is the synthesis of glucose from noncarbohydrates, especially fats and proteins.

17.4i Hormone Clearance

Hormonal signals, like nervous signals, must be turned off when they have served their purpose. Most hormones are taken up and degraded by the liver and kidneys and then excreted in the bile or urine. Some are degraded by their target cells. As noted earlier, hormones that bind to transport proteins are removed from the blood much more slowly than hormones that don't employ transport proteins.

The rate of hormone removal is called the **metabolic clearance rate (MCR),** and the length of time required to clear 50% of the hormone from the blood is its **half-life.** The faster the MCR, the shorter is the half-life. Growth hormone, for example, uses no transport protein and has a half-life of only 6 to 20 minutes. Thyroxine, by contrast, is protected by transport proteins and maintains a physiologically effective level in the blood for up to 2 weeks after its secretion ceases.

BEFORE YOU GO ON

Answer the following questions to test your understanding of the preceding section:

17. What are the three chemical classes of hormones? Name at least one hormone in each class.

18. Why do corticosteroids and thyroid hormones require transport proteins to travel in the bloodstream?

19. Explain how MIT, DIT, T_3, and T_4 relate to each other structurally.

20. Where are hormone receptors located in target cells? Name one hormone that employs each receptor location.

21. Explain how one hormone molecule can activate millions of enzyme molecules.

17.5 Stress and Adaptation

Expected Learning Outcomes
When you have completed this section, you should be able to

a. give a physiological definition of *stress;* and

b. discuss how the body adapts to stress through its endocrine and sympathetic nervous systems.

Stress is defined as any situation that upsets homeostasis and threatens one's physical or emotional well-being. It affects us all from time to time, and we react to it in ways that are mediated mainly by the endocrine and sympathetic nervous systems. Physical causes of stress *(stressors)* include injury, surgery, hemorrhage, infection, intense exercise, temperature extremes, pain, and malnutrition. Emotional causes include anger, grief, depression, anxiety, fear, and guilt.

Whatever the cause, the body reacts to stress in a fairly consistent way called the **stress response** or **general adaptation syndrome (GAS).** The response typically involves elevated levels of epinephrine and cortisol; some physiologists define *stress* as any situation that raises the cortisol level. A pioneering researcher on stress physiology, Canadian biochemist Hans Selye, showed in 1936 that the GAS typically occurs in three stages, which he called the *alarm reaction,* the *stage of resistance,* and the *stage of exhaustion.*

17.5a The Alarm Reaction

The initial response to stress is the **alarm reaction,** mediated mainly by norepinephrine from the sympathetic nervous system and epinephrine from the adrenal medulla. In extreme cases, as in wild animals, these prepare the body to take action such as fighting or escaping danger. One of their effects, even in humans, is the consumption of stored glycogen, which is particularly important in the transition to the next stage of the stress response. Angiotensin and aldosterone levels also rise during the alarm reaction. Angiotensin helps to raise the blood pressure, and aldosterone promotes sodium and water conservation, which helps to offset possible losses by sweating and bleeding.

17.5b The Stage of Resistance

After a few hours, glycogen reserves are exhausted, yet the nervous system continues to demand glucose. If a stressful situation isn't resolved before the glycogen is gone, the body enters the **stage of resistance,** in which the first priority is to provide alternative fuels for metabolism. This stage is dominated by cortisol. The hypothalamus secretes corticotropin-releasing hormone (CRH); the pituitary responds by secreting adrenocorticotropic hormone (ACTH); and this, in turn, stimulates the adrenal cortex to secrete cortisol and other glucocorticoids. Cortisol promotes the breakdown of fat and protein into glycerol, fatty acids, and amino acids, providing the liver with raw material for gluconeogenesis. Like epinephrine, cortisol inhibits glucose uptake by most organs and thus has a glucose-sparing effect. It also inhibits protein synthesis, leaving the free amino acids available for gluconeogenesis.

Excessive cortisol secretion (or medical use of hydrocortisone) can, however, have some adverse effects. Cortisol suppresses the secretion of sex hormones such as estrogen, testosterone, and luteinizing hormone, causing disturbances of fertility and sexual function. Long-term elevation of cortisol secretion also compromises one's immunity (see Deeper Insight 21.4). It inhibits the synthesis of protective leukotrienes and prostaglandins (see section 17.6), suppresses antibody production, and kills immature T and B cells—two important families of immune cells. Wounds heal poorly, and a person under chronic stress is more susceptible to infections and some forms of cancer. Stress can aggravate peptic ulcers because of reduced resistance to the bacteria that cause them and because circulating epinephrine reduces secretion of the gastric mucus and pancreatic bicarbonate that normally protect the stomach lining.

17.5c The Stage of Exhaustion

The body's fat reserves can carry it through months of stress, but when fat is depleted, stress overwhelms homeostasis. The **stage of exhaustion** sets in, often marked by rapid decline and death. With its fat stores gone, the body now relies primarily on protein breakdown to meet its energy needs. Thus, there is a progressive wasting away of the muscles and weakening of the body. After prolonged stimulation, the adrenal cortex may stop producing glucocorticoids, making it all the more difficult to maintain glucose homeostasis. Aldosterone sometimes promotes so much water retention that it creates a state of hypertension, and while it conserves sodium, it hastens the elimination of potassium and hydrogen ions. This creates a state of hypokalemia (potassium deficiency in the blood) and alkalosis (excessively high blood pH), resulting in nervous and muscular system dysfunctions. Death frequently results from heart failure, kidney failure, or overwhelming infection.

BEFORE YOU GO ON

Answer the following questions to test your understanding of the preceding section:

22. Define *stress* from the standpoint of endocrinology.

23. Describe the stages of the general adaptation syndrome.

24. List six hormones that show increased secretion in the stress response. Describe how each one contributes to recovery from stress.

17.6 Eicosanoids and Other Signaling Molecules

Expected Learning Outcomes

When you have completed this section, you should be able to

a. explain what eicosanoids are and how they are produced;

b. identify some classes and functions of eicosanoids; and

c. describe several physiological roles of prostaglandins.

Neurotransmitters and hormones are not the only chemical messengers in the body. Others are *paracrine* and *autocrine* messengers—chemical signals that don't travel in the blood but stay and exert their effects locally. **Paracrine** signals diffuse only to nearby cells in the same tissue. Histamine, for example, is released by *mast cells* that lie alongside blood vessels in the connective tissues. It diffuses to the smooth muscle of the blood vessel, relaxing it and allowing vasodilation. Catecholamines such as epinephrine diffuse from the adrenal medulla to the cortex to stimulate corticosterone secretion. **Autocrine** signals stimulate the same cell that secretes them. For example, liver cells secrete a signaling molecule called *hepcidin* that controls our blood iron level (detailed in section 25.6f). Acting as a hormone, it travels in the blood to the small intestine and regulates its absorption of dietary iron. Acting as an autocrine signal, it also stimulates the very liver cells that secrete it and regulates their release of stored iron into the blood. Thus, a single chemical can be considered a hormone, paracrine, autocrine, or neurotransmitter depending on location and circumstances.

The **eicosanoids**[27] (eye-CO-sah-noyds) are an important family of paracrine secretions. They have 20-carbon backbones derived

---

[27]*eicosa* (variation of *icosa*) = 20

FIGURE 17.24 Eicosanoid Synthesis and Related Drug Actions. SAIDs are steroidal anti-inflammatory drugs such as hydrocortisone; NSAIDs are nonsteroidal anti-inflammatory drugs such as aspirin and ibuprofen. A representative leukotriene and a prostaglandin are shown at the bottom.

❓ *How would the body be affected by a drug that selectively inhibited lipoxygenase?*

from a fatty acid called **arachidonic acid** (ah-RACK-ih-DON-ic). Some peptide hormones and other stimuli liberate arachidonic acid from one of the phospholipids of the plasma membrane, and the following two enzymes then convert it to various eicosanoids **(fig. 17.24)**.

Lipoxygenase helps to convert arachidonic acid to **leukotrienes**—eicosanoids that mediate allergic and inflammatory reactions. **Cyclooxygenase** converts arachidonic acid to three other eicosanoids:

1. **Prostacyclin** is produced by the walls of the blood vessels, where it inhibits blood clotting and vasoconstriction.

2. **Thromboxanes** are produced by blood platelets. In the event of injury, they override prostacyclin and stimulate

TABLE 17.7	Some of the Roles of Prostaglandins
Prostaglandins	**Function**
Inflammatory	Promote fever and pain, two cardinal signs of inflammation
Endocrine	Mimic effects of TSH, ACTH, and other hormones; alter sensitivity of anterior pituitary to hypothalamic hormones; work with glucagon, catecholamines, and other hormones in regulation of fat mobilization
Nervous	Function as neuromodulators, altering the release or effects of neurotransmitters in the brain
Reproductive	Promote ovulation and formation of corpus luteum; induce labor contractions
Gastrointestinal	Inhibit gastric secretion
Vascular	Act as vasodilators and vasoconstrictors
Respiratory	Constrict or dilate bronchioles
Renal	Promote blood circulation through the kidney, increase water and electrolyte excretion

DEEPER INSIGHT 17.3
CLINICAL APPLICATION

Anti-Inflammatory Drugs

Cortisol and corticosterone are used as *steroidal anti-inflammatory drugs (SAIDs)*. They inhibit inflammation by blocking the release of arachidonic acid from the plasma membrane, thus inhibiting the synthesis of all eicosanoids. Their main disadvantage is that prolonged use causes side effects that mimic Cushing syndrome (described in the next section). Aspirin, ibuprofen (Motrin), and celecoxib (Celebrex) are *nonsteroidal anti-inflammatory drugs (NSAIDs)*, also called *COX inhibitors* because they block the action of cyclooxygenase (COX). Their advantage is that they do not affect lipoxygenase function or leukotriene production. One form of cyclooxygenase, COX-2, is specific to the cells of inflammation, so there has been a particular interest in developing safe COX-2 inhibitors that can treat inflammation without undesirable side effects such as gastrointestinal damage and cardiovascular events such as stroke and myocardial infarction.

COX inhibitors are also useful in the treatment of fever and thrombosis (undesirable blood clotting). Fever is thought to result from the action of prostaglandins on the hypothalamus. Aspirin exerts its antipyretic (fever-reducing) effect by inhibiting prostaglandin synthesis and its antithrombotic effect by inhibiting thromboxane synthesis.

vasoconstriction and clotting. Prostacyclin and thromboxanes are further discussed in chapter 18.

3. **Prostaglandins (PGs)** are the most diverse eicosanoids. They have a five-sided carbon ring in their backbone. They are named PG for *prostaglandin,* plus a third letter that indicates the type of ring structure (PGE, PGF, etc.) and a subscript that indicates the number of C=C double bonds in the side chain—such as the $PGF_{2\alpha}$ shown in figure 17.24. They were first found in bull semen and the prostate gland,

hence their name, but they are now thought to be produced in most organs of the body. The PGEs are usually antagonized by PGFs. For example, the PGE family relaxes smooth muscle in the bladder, intestines, bronchioles, and uterus and stimulates contraction of the smooth muscle of blood vessels. $PGF_{2\alpha}$ has precisely the opposite effects. Some other roles of prostaglandins are described in **table 17.7.**

Understanding the pathways of eicosanoid synthesis makes it possible to understand the action of some familiar drugs (see Deeper Insight 17.3). The roles of prostaglandins and other eicosanoids are further explored in later chapters on blood, immunity, and reproduction.

BEFORE YOU GO ON
Answer the following questions to test your understanding of the preceding section:

25. What are eicosanoids and how do they differ from neurotransmitters and hormones?

26. Distinguish between paracrine and endocrine effects.

27. State four functions of prostaglandins.

17.7 Endocrine Disorders

Expected Learning Outcomes
When you have completed this section, you should be able to

a. explain some general causes and examples of hormone hyposecretion and hypersecretion;

b. briefly describe some common disorders of pituitary, thyroid, parathyroid, and adrenal function; and

c. in more detail, describe the causes and pathology of diabetes mellitus.

As we saw in the discussion of signal amplification, a little hormone can have a great effect. Thus, it is necessary to tightly regulate hormone secretion and blood concentration. Variations in secretion and target-cell sensitivity often have very noticeable effects on the body. This section deals with some of the better-known dysfunctions of the endocrine system. The effects of aging on the endocrine system are described in section 29.4a.

17.7a Hyposecretion and Hypersecretion
Inadequate hormone release is called **hyposecretion.** It can result from tumors or lesions that destroy an endocrine gland or interfere with its ability to receive signals from another cell. For example, a fractured sphenoid bone can sever the hypothalamo–hypophysial tract and prevent the transport of oxytocin and antidiuretic hormone (ADH) to the posterior pituitary. The resulting ADH hyposecretion disables the water-conserving capability of the kidneys and leads to **diabetes insipidus,** an output of abundant but glucose-free urine. (*Insipidus* means "without taste" and

refers to the lack of sweetness of the urine, in contrast to the sugary urine of diabetes mellitus.) Autoimmune diseases can also lead to hormone hyposecretion when endocrine cells are attacked by one's own antibodies or immune cells. This is one of the causes of diabetes mellitus. A congenital absence or underdevelopment of the pituitary gland causes *panhypopituitarism,* wide-spectrum hyposecretion of multiple hormones affected by the pituitary, requiring lifelong multihormone replacement therapy.

Excessive hormone release, called **hypersecretion,** has multiple causes. Some tumors involve the overgrowth of functional endocrine tissue. A **pheochromocytoma** (FEE-o-CRO-mo-sy-TOE-muh), for example, is a tumor of the adrenal medulla that secretes excessive amounts of epinephrine and norepinephrine. Some tumors in nonendocrine organs produce hormones. For example, certain lung tumors secrete ACTH and overstimulate cortisol secretion by the adrenal gland. Whereas some autoimmune disorders can cause endocrine hyposecretion, others cause hypersecretion. An example of this is **toxic goiter** (Graves[28] disease), in which autoantibodies mimic the effect of TSH on the thyroid, activating the TSH receptor and causing thyroid hypersecretion. Endocrine hypersecretion disorders can also be mimicked by excess or long-term clinical administration of hormones such as cortisol.

Following are brief descriptions of some of the better-known disorders of the major endocrine glands. **Table 17.8** provides further details on some of these and lists some additional endocrine disorders.

17.7b Pituitary Disorders

The hypersecretion of growth hormone (GH) in childhood or adolescence, before the epiphysial plates (growth zones) of the long bones are depleted, causes **gigantism,** whereas childhood hyposecretion causes **pituitary dwarfism.** Now that growth

[28]Robert James Graves (1796–1853), Irish physician

hormone is plentiful, made by genetically engineered bacteria containing the human GH gene, pituitary dwarfism has become rare. In adulthood, after the epiphysial plates have closed, GH hyposecretion causes little if any problem, but hypersecretion causes **acromegaly**—thickening of the bones and soft tissues, with especially noticeable effects on the hands, feet, and face **(fig. 17.25).**

17.7c Thyroid and Parathyroid Disorders

Congenital hypothyroidism is thyroid hyposecretion present from birth. Severe or prolonged adult hypothyroidism can cause **myxedema** (MIX-eh-DEE-muh). Both syndromes are described in table 17.8, and both can be treated with oral thyroid hormone.

A more conspicuous, often striking abnormality of the thyroid is **endemic goiter (fig. 17.26),** caused by a deficiency of dietary iodine. Without iodine, the gland cannot synthesize TH. Without TH, the pituitary gland receives no feedback and acts as if the thyroid were understimulated. As if trying to stimulate it more strongly, the pituitary secretes extra TSH, which makes the thyroid produce more and more colloid (thyroglobulin). However, the colloid cannot be iodinated to make TH, so it simply accumulates in the gland, often producing visible swelling in the neck. The word *endemic* refers to the occurrence of a disease in a defined geographic locality. Endemic goiter was once common in the midwestern United States (the "goiter belt") because of iodine-poor soil and little access to iodine-rich seafood. It has now become almost nonexistent in developed countries, however, because of the addition of iodine to table salt, animal feeds, and fertilizers. It now occurs most often in localities that have neither these benefits nor access to seafood—notably central Africa and mountainous regions of South America, central Asia, and Indonesia.

The parathyroids, because of their location and small size, are sometimes accidentally removed in thyroid surgery or degenerate when neck surgeries cut off their blood supply. Without hormone replacement therapy, the resulting **hypoparathyroidism** causes a rapid decline in blood calcium level; in as little as 2 or 3 days, this can lead to a fatal, suffocating spasm of the muscles of the larynx *(hypocalcemic tetany).* **Hyperparathyroidism,** excess PTH secretion, is usually caused by a parathyroid tumor. It causes the bones to become soft, deformed, and fragile; it raises the blood levels of calcium and phosphate ions; and it promotes the formation of *renal calculi* (kidney stones) composed of calcium phosphate. Section 7.4b further describes the relationship among parathyroid function, blood calcium, and bone tissue.

FIGURE 17.25 Acromegaly, a Condition Caused by Growth Hormone Hypersecretion in Adulthood. The hand on the left exemplifies characteristic thickening compared to the normal hand on the right.

Barts Medical Library/Phototake

FIGURE 17.26 Endemic Goiter.
Karan Bunjean/Shutterstock

17.7d Adrenal Disorders

Cushing[29] syndrome is excess cortisol secretion owing to any of several causes: ACTH hypersecretion by the pituitary, ACTH-secreting tumors, or hyperactivity of the adrenal cortex independently of ACTH. Cushing syndrome disrupts carbohydrate and protein metabolism, leading to hyperglycemia, hypertension, muscular weakness, and edema. Muscle and bone mass are lost rapidly as protein is catabolized. Some patients exhibit abnormal fat deposition between the shoulders ("buffalo hump") or in the face ("moon face") **(fig. 17.27).** Long-term hydrocortisone therapy can have similar effects.

Adrenogenital syndrome (AGS), the hypersecretion of adrenal androgens, commonly accompanies Cushing syndrome. In children, AGS often causes enlargement of the penis or clitoris and the premature onset of puberty **(fig. 17.28).** Prenatal AGS can result in newborn girls exhibiting masculinized genitalia and being misidentified as boys. In women, AGS produces such masculinizing effects as increased body hair, deepening of the voice, and beard growth.

17.7e Diabetes Mellitus

The world's most prevalent metabolic disease is diabetes mellitus, affecting about 7% of the U.S. population and even more in areas such as Scandinavia and the Pacific Islands. It is the leading cause of adult blindness, renal failure, gangrene, and the necessity for limb amputations, and warrants a more extended discussion than the less common endocrine diseases. The pathology of DM is described here, and the chapter ends with an essay on the history of insulin (see Deeper Insight 17.4).

Diabetes mellitus[30] (DM) can be defined as a disruption of carbohydrate, fat, and protein metabolism resulting from the hyposecretion or inaction of insulin. The classic signs and

[29]Harvey Cushing (1869–1939), American physician
[30]*diabet* = to flow through; *melli* = honey

FIGURE 17.27 Cushing Syndrome. Note the "moon face" appearance characteristic of this syndrome.
Biophoto Associates/Science Source

FIGURE 17.28 Adrenogenital Syndrome (AGS). These are the genitals of a female with AGS, masculinized by prenatal hypersecretion of adrenal androgens. Note the resemblance of the labia majora to a scrotum and enlargement of the clitoris to resemble a penis. Infants with AGS are easily mistaken for boys and raised as such.
Medicimage RM/Medical Images

symptoms with which patients often first present to a physician are "the three polys": **polyuria**[31] (excessive urine output), **polydipsia**[32] (intense thirst), and **polyphagia**[33] (ravenous hunger). Blood and urine tests can confirm a diagnosis of DM by revealing three further signs: **hyperglycemia**[34] (elevated blood glucose), **glycosuria**[35] (glucose in the urine), and **ketonuria** (ketones in the urine). DM was originally named for the sweetness of the urine stemming from glycosuria. Before the advent of chemical tests for glucose, physicians tasted their patients' urine as part of their diagnostic process.

A little knowledge of kidney physiology is necessary to understand why glycosuria and polyuria occur. The kidneys filter blood plasma and convert the filtrate to urine. In a healthy person, the kidney tubules remove all glucose from the filtrate and return it to the blood, so there is little or no glucose in the urine. Water follows the glucose and other solutes by osmosis, so the tubules also reclaim most of the water in the filtrate.

But like any other carrier-mediated transport system, there is a limit to how fast the glucose transporters of the kidney can work. The maximum rate of reabsorption is called the *transport maximum*, T_m (see fig. 3.16). Because of the high blood glucose level in diabetes mellitus, glucose enters the kidney tubules so rapidly that it exceeds the T_m and the tubules can't reabsorb it fast enough. The excess passes through into the urine. Glucose and ketones in the tubules also raise the osmolarity of the tubular fluid and cause **osmotic diuresis**—water remains in the tubules with these solutes, so large amounts of water are passed in the urine. This accounts for the polyuria, dehydration, and thirst of diabetes. A person with untreated DM may pass 10 to 15 L of urine per day, compared with 1 or 2 L in a healthy person.

Types and Treatment

There are two forms of diabetes mellitus: type 1 and type 2. These were formerly called juvenile-onset or insulin-dependent DM, and adult-onset or non-insulin-dependent DM, respectively. These terms have lately been abandoned because they are too misleading. Although insulin is always used to treat type 1, it is frequently used for type 2 diabetes as well, and either type can occur at any age. Indeed, with the burgeoning problem of childhood obesity, nearly half of all new cases of childhood diabetes are now type 2.

Type 1 diabetes mellitus accounts for 5% to 10% of cases in the United States. What causes it? It begins with heredity. Several genes have been identified that predispose a person to type 1 DM. Then, when a genetically susceptible individual is infected by certain viruses (rubella, cytomegalovirus, or a few others), certain immune cells (CD4 and CD8 T cells discussed in section 21.4) attack and destroy pancreatic beta cells. To a great extent, this destruction is tolerated and produces no disease, but when 80% to 90% of the beta cells are gone, insulin falls to such a critically low level that it can no longer regulate glycemia, the blood glucose

level. Now comes the problematic hyperglycemia and all of its insidious complications. Type 1 diabetes is usually diagnosed before the age of 30, but may occur later. Its victims require insulin to survive—usually periodic injections or continual subcutaneous delivery by a small insulin pump worn on the body. A dry insulin inhaler is now available, but is not suitable for all patients. Meal planning, exercise, and self-monitoring of blood glucose levels are also important aspects of the treatment regimen.

Some 90% to 95% of diabetics, however, have **type 2 DM,** and a great many more are prediabetic. Here, the chief problem is not lack of insulin, but *insulin resistance*—unresponsiveness of the target cells to the hormone. The level of insulin may actually be very high in the early stage of the disease, although it later tends to fall to normal or subnormal levels. Again, heredity is one of the causes, although no one gene, or even a mere few, can be blamed for the disease; nearly 100 genes are known, so far, to elevate the risk of type 2 DM. There are great differences in prevalence from one ethnic group to another; it is relatively high, for example, among people of Native American, Hispanic, and Asian descent. It also has a tendency to run in families, and shows high concordance between genetically identical twins—if one twin develops type 2 DM, there is more than a 90% probability that the other will too. Other important risk factors are age, obesity, and a sedentary lifestyle. All of these are accompanied by the progressive replacement of muscular tissue with fat. Muscle plays a highly important role in absorbing blood glucose and buffering glycemia, so as muscle mass diminishes, a person becomes less and less able to regulate glycemia.

Type 2 DM develops slowly and is usually diagnosed after age 40, but it is becoming increasingly prevalent in young people because of early obesity. Aside from the loss of the glucose-buffering role of muscle, another apparent factor in type 2 DM is that adipose tissue secretes chemical signals that indirectly interfere with glucose transport into most cells—so the more body fat, the less efficient is glucose uptake. It is no surprise, then, that type 2 DM can often be successfully managed through a weight-loss program of diet and exercise, often supplemented with glycemia-lowering oral medications such as metformin and drugs to lower blood pressure and blood lipids. If these approaches prove inadequate, supplemental insulin therapy is also employed. About 12% of patients with type 2 DM are treated with insulin alone and another 14% with insulin combined with oral hypoglycemic agents (OHAs).

Pathogenesis

When cells cannot absorb glucose, they must get their energy someplace else; they metabolize fat and protein. In time, this leads to muscular atrophy, emaciation, and weakness. Before insulin therapy was introduced in 1922, the victims of type 1 DM wasted away to an astonishing extent (see Deeper Insight 17.4). Diabetes was described in the first century as "a melting down of the flesh and limbs into urine." Adult patients weighed as little as 27 to 34 kg (60–75 lb) and looked like victims of severe famine. Their breath had a disagreeable sweet ketone smell, like rotten apples. One typical patient was described by medical historian Michael Bliss as "barely able to lift his head from his pillow, crying most of the time from pain, hunger, and despair." In the terminal stage, patients became increasingly drowsy, gasped for air, became comatose,

---

[31]*poly* = much, excessive; *uri* = urine
[32]*poly* = much, excessive; *dipsia* = drinking
[33]*poly* = much, excessive; *phagia* = eating
[34]*hyper* = excess; *glyc* = sugar, glucose; *emia* = blood condition
[35]*glyco* = glucose, sugar; *uria* = urine condition

and died within a few hours. Most diabetic children lived less than 1 year after diagnosis—a year of utmost misery at that. Such was the natural course of the disease in the centuries before insulin therapy.

Rapid fat catabolism raises blood levels of free fatty acids and their breakdown products, the ketone bodies (acetoacetic acid, acetone, and β-hydroxybutyric acid). Ketonuria promotes polyuria, which flushes Na^+ and K^+ from the body. The resulting electrolyte deficiencies can cause abdominal pain, vomiting, irregular heartbeat, and neurological dysfunction. As acids, ketones lower the pH of the blood and produce a condition called **ketoacidosis.** This causes a deep, gasping breathing called *Kussmaul* [36] *respiration*, typical of terminal diabetes. It also depresses the nervous system and produces diabetic coma.

DM also leads to long-term degenerative cardiovascular and neurological diseases—signs that were seldom seen before insulin therapy, when patients died too quickly to show long-term effects. Through multiple, complex mechanisms, chronic hyperglycemia has devastating effects on small to medium blood vessels *(microvascular disease),* including *atherosclerosis,* the obstruction of vessels by plaques of lipid and overgrown smooth muscle. Both types of DM also thicken the basement membrane of the blood vessels, interfering with the delivery of nutrients and hormones to the tissues and with the removal of their wastes. This leads to irreversible tissue degeneration in many organs. Two of the common complications of long-term DM are blindness and renal failure, brought on by arterial degeneration in the retinas and kidneys. Death from kidney failure is much more common in type 1 DM than in type 2. In type 2, the most common cause of death is heart failure stemming from coronary artery disease.

Another complication is *diabetic neuropathy*—nerve damage resulting from impoverished blood flow. This can lead to erectile dysfunction, incontinence, and loss of sensation from affected areas of the body. Microvascular disease in the skin results in poor healing of skin wounds, so even a minor break easily becomes ulcerated, infected, and gangrenous. This is especially common in the feet, because people take less notice of foot injuries; circulation is poorer in the feet (farthest from the heart) than anywhere else; pressure on the feet makes them especially susceptible to tissue injury; and neuropathy may make a person unaware of skin lesions or reluctant to consent to amputation of toes, or more, from which they feel no pain. DM outweighs all other reasons for the amputation of gangrenous appendages.

Diabetes mellitus is not the only kind of diabetes. Diabetes insipidus, a disease with no relation to insulin, has already been mentioned in this chapter, and other forms are discussed with the urinary system in section 23.6b.

── BEFORE YOU GO ON ──

Answer the following questions to test your understanding of the preceding section:

28. Explain some causes of hormone hyposecretion, and give examples. Do the same for hypersecretion.

29. Why does a lack of dietary iodine lead to TSH hypersecretion? Why does the thyroid gland enlarge in endemic goiter?

30. In diabetes mellitus, explain the chain of events that lead to (a) osmotic diuresis, (b) ketoacidosis and coma, and (c) gangrene of the lower limbs.

[36]Adolph Kussmaul (1822–1902), German physician

TABLE 17.8	Some Disorders of the Endocrine System
Addison[37] disease	Hyposecretion of adrenal glucocorticoids and mineralocorticoids, causing hypoglycemia, hypotension, weight loss, weakness, loss of stress resistance, darkening of the skin, dehydration, hyponatremia (sodium deficiency), and other electrolyte imbalances; death can follow quickly from mineralocorticoid and electrolyte deficiencies.
Congenital hypothyroidism	Thyroid hormone hyposecretion present from birth, resulting in stunted physical development, thickened facial features, low body temperature, lethargy, and irreversible brain damage in infancy
Hyperinsulinism	Insulin excess caused by islet hypersecretion or injection of excess insulin, causing hypoglycemia, weakness, hunger, and sometimes *insulin shock*, which is characterized by disorientation, convulsions, or unconsciousness
Myxedema	A syndrome occurring in severe or prolonged adult hypothyroidism, characterized by low metabolic rate, sluggishness and sleepiness, weight gain, constipation, dry skin and hair, abnormal sensitivity to cold, and tissue swelling
Pheochromocytoma	A tumor of the adrenal medulla that secretes excess epinephrine and norepinephrine. Causes hypertension, elevated metabolic rate, nervousness, indigestion, hyperglycemia, and glycosuria.
Toxic goiter (Graves disease)	Thyroid hypertrophy and hypersecretion, occurring when autoantibodies mimic the effect of TSH and overstimulate the thyroid. Results in elevated metabolic rate and heart rate, nervousness, sleeplessness, weight loss, abnormal heat sensitivity and sweating, and bulging of the eyes (exophthalmos) resulting from eyelid retraction and edema of the orbital tissues.

You can find other endocrine system disorders described in the following places:

Acromegaly, adrenogenital syndrome, Cushing syndrome, diabetes inspidus, diabetes mellitus, endemic goiter, pituitary gigantism and *dwarfism,* and *hyper-* and *hypoparathyroidism* in section 17.7c; *gestational diabetes* in section 23.6b; and *androgen insensitivity syndrome* in Deeper Insight 27.1.

[37]Thomas Addison (1793–1860), English physician

DEEPER INSIGHT 17.4

MEDICAL HISTORY

The Discovery of Insulin

At the start of the twentieth century, physicians felt nearly helpless in the face of diabetes mellitus. They put patients on useless diets—the so-called oatmeal cure, potato cure, and others—or on starvation diets as low as 750 kcal per day so as not to "stress the system." They were resigned to the fact that their patients were doomed to die, and simple starvation seemed to produce the least suffering.

After the cause of diabetes was traced to the pancreatic islets in 1901, European researchers tried treating patients and experimental animals with extracts of pancreas, but became discouraged by the severe side effects of impurities in the extracts. They lacked the resources to pursue the problem to completion, and by 1913, the scientific community showed signs of giving up on diabetes.

But in 1920, Frederick Banting (1891–1941), a young Canadian physician with a faltering medical practice, became intrigued with a possible method for isolating the islets from the pancreas and testing extracts of the islets alone. He returned to his alma mater, the University of Toronto, to present his idea to Professor J. J. R. Macleod (1876–1935), a leading authority on carbohydrate metabolism. Macleod was unimpressed with Banting, finding his knowledge of the diabetes literature and scientific method superficial. Nevertheless, he felt Banting's idea was worth pursuing and thought that with his military surgical training, Banting might be able to make some progress where others had failed. He offered Banting laboratory space for the summer, giving him a marginal chance to test his idea. Banting was uncertain whether to accept, but when his fiancée broke off their engagement and an alternative job offer fell through, he closed his medical office, moved to Toronto, and began work. Little did either man realize that in 2 years' time, they would share a Nobel Prize and yet so thoroughly detest each other they would scarcely be on speaking terms.

A Modest Beginning

Macleod advised Banting on an experimental plan of attack and gave him an assistant, Charles Best (1899–1978). Best had just received his B.A. in physiology and looked forward to an interesting summer job with Banting before starting graduate school. Over the summer of 1921, they removed the pancreases from dogs to render them diabetic and tied off the pancreatic ducts in other dogs to make most of the pancreas degenerate while leaving the islets intact. Their plan was to treat the diabetic dogs with extracts made from the degenerated pancreases of the others.

It was a difficult undertaking. For a laboratory, they were assigned a tiny, filthy, unbearably hot, rooftop animal quarters that reeked of dog excrement. The tiny pancreatic ducts were very difficult to tie off, and it was hard to tell if all pancreatic tissue had been removed from the dogs intended to become diabetic. Several dogs died of overanesthesia, infection, and bleeding from Banting's clumsy surgical technique. Banting was also careless in reading his data and interpreting the results and had little interest in reading the literature to see what other researchers were doing. In Banting and Best's first publication, in early 1922, the data in their discussion disagreed with the data in the tables, and both disagreed with the data in their laboratory notebooks. These were not signs of promising researchers.

In spite of themselves, Banting and Best achieved modest positive results over the summer. Crude extracts brought one dog back from a diabetic coma and reduced the hyperglycemia and glycosuria of others. Buoyed by these results, Banting demanded a salary, a better laboratory,

and another assistant. Macleod grudgingly obtained salaries for the pair, but Banting began to loathe him for their disagreement over his demands, and he and Macleod grew in mutual contempt as the project progressed.

Success and Conflict

Macleod brought biochemist J. B. Collip (1892–1965) into the project in the fall of 1921 in hopes that he could produce purer extracts. More competent in experimental science, Collip was the first to show that pancreatic extracts could eliminate ketosis and restore the liver's ability to store glycogen. He obtained better and better results in diabetic rabbits until, by January 1922, the group felt ready for human trials. Banting was happy to have Collip on the team initially, but grew intensely jealous of him as Collip not only achieved better results than he had, but also developed a closer relationship with Macleod. Banting, who had no qualifications to perform human experiments, feared he would be pushed aside as the project moved to its clinical phase. At one point, the tension between Banting and Collip erupted into a near-fistfight in the laboratory.

Banting insisted that the first human trial be done with an extract he and Best prepared, not with Collip's. The patient was a 14-year-old boy who weighed only 29 kg (65 lb) and was on the verge of death. He was injected on January 11 with the Banting and Best extract, described by one observer as "a thick brown muck." The trial was an embarrassing failure, with only a slight lowering of the boy's blood glucose and a severe reaction to the impurities in the extract. On January 23, the same boy was treated again, but with Collip's cleaner extract. This time, his ketonuria and glycosuria were almost completely eliminated and his blood sugar dropped 77%. This was the first successful clinical trial of insulin. Six more patients were treated in February 1922 and quickly became stronger, more alert, and in better spirits. In April, the Toronto group began calling the product *insulin,* and at a medical conference in May, they gave the first significant public report of their success.

Banting felt increasingly sidelined from the project. He quit coming to the laboratory, drank heavily, and day-dreamed of leaving diabetes research to work on cancer. He remained only because Best pleaded with him to stay. Banting briefly operated a private diabetes clinic, but fearful of embarrassment over alienating the discoverer of insulin, the university soon lured him back with a salaried appointment and hospital privileges.

Banting had a number of high-profile, successful cases in 1922, such as 14-year-old Elizabeth Hughes, who weighed only 20 kg (45 lb) before treatment. She began treatment in August and showed immediate, dramatic improvement. She was a spirited, optimistic, and articulate girl who kept enthusiastic diaries of being allowed to eat bread, potatoes, and macaroni and cheese for the first time since the onset of her illness. "Oh it is simply too wonderful for words this stuff," she exuberantly wrote to her mother—even though the still-impure extracts caused her considerable pain and swelling. The world quickly beat a path to Toronto begging for insulin. The pharmaceutical firm of Eli Lilly and Company entered into an agreement with the University of Toronto for the mass production of insulin, and by the fall of 1923, over 25,000 patients were being treated at more than 60 Canadian and U.S. clinics.

The Bitter Fruits of Success

Banting's self-confidence was restored. He had become a public hero, and the Canadian Parliament awarded him an endowment generous enough to ensure a life of comfort. Several distinguished physiologists nominated Banting and Macleod for the 1923 Nobel Prize, and they won. When the award was announced, Banting was furious about having to share it with Macleod. At first, he threatened to refuse it, but when he cooled down, he announced that he would split his share of the prize money with Best. Macleod quickly announced that half of his share would go to Collip.

Interestingly, Romanian physiologist Nicolae Paulescu (1869–1931) succeeded in isolating insulin (which he called pancreine) and treating diabetic dogs with it in 1916, years before Banting even conceived or began his work. Paulescu published four papers on it in April 1921, 8 months before Banting and Best published their first, and he patented his method of isolating insulin in April 1922. Paulescu's work never advanced to clinical trials on humans, however, and was overlooked by the Nobel Committee.

As a recipient of Canada's first Nobel Prize, Banting basked in his stature as a national hero. He made life at the university so unbearable for Macleod, however, that Macleod left in 1928 to accept a university post in Scotland. Banting stayed on at Toronto. Although now wealthy and surrounded by admiring students, he achieved nothing significant in science for the rest of his career. He was killed in a plane crash in 1941. Best replaced Macleod on the Toronto faculty, led a distinguished career, and developed the anticoagulant heparin. Collip went on to play a lead role in the isolation of PTH, ACTH, and other hormones.

Insulin made an industry giant of Eli Lilly and Company. It became the first protein whose amino acid sequence was determined, for which Frederick Sanger received a Nobel Prize in 1958. Diabetics today no longer depend on a limited supply of insulin extracted from beef and pork pancreas. Human insulin is now in plentiful supply, made by genetically engineered bacteria. Paradoxically, while insulin has dramatically reduced the suffering caused by diabetes mellitus, it has increased the number of people who have the disease—because thanks to insulin, diabetics are now able to live long enough to raise families and pass on the genes for diabetes susceptibility.

CONNECTIVE ISSUES

Effects of the ENDOCRINE SYSTEM on Other Organ Systems

ALL SYSTEMS
The development and metabolism of most tissues are affected by growth hormone, insulin, insulin-like growth factors, thyroid hormone, and glucocorticoids.

INTEGUMENTARY SYSTEM
Sex hormones affect skin pigmentation, development of body hair and apocrine glands, and subcutaneous fat deposition.

SKELETAL SYSTEM
Skeletal growth and maintenance are regulated by numerous hormones—calcitonin, calcitriol, parathyroid hormone, growth hormone, estrogen, testosterone, and others.

MUSCULAR SYSTEM
Growth hormone and testosterone stimulate muscular growth; insulin regulates glucose uptake by muscle; other hormones regulate the electrolyte balances that are important in muscular contraction.

NERVOUS SYSTEM
Hormones exert negative feedback inhibition on the hypothalamus; several hormones affect nervous system development, mood, and behavior; hormones regulate the electrolyte balances that are important in neuron function.

CIRCULATORY SYSTEM
Angiotensin II, aldosterone, antidiuretic hormone, natriuretic peptides, and other hormones regulate blood volume and pressure; erythropoietin stimulates RBC production; thymic hormones stimulate WBC production; thrombopoietin stimulates platelet production; epinephrine, thyroid hormone, and other hormones affect the rate and force of the heartbeat.

LYMPHATIC AND IMMUNE SYSTEMS
Thymic hormones activate immune cells; glucocorticoids suppress immunity and inflammation.

RESPIRATORY SYSTEM
Epinephrine and norepinephrine dilate the bronchioles and increase pulmonary airflow.

URINARY SYSTEM
Antidiuretic hormone regulates urine volume; calcitriol, parathyroid hormone, aldosterone, and natriuretic peptides regulate electrolyte absorption by the kidneys.

DIGESTIVE SYSTEM
Insulin and glucagon regulate nutrient storage and metabolism; enteric hormones control gastrointestinal secretion and motility; gut–brain peptides affect appetite and regulate food intake and body weight.

REPRODUCTIVE SYSTEM
Gonadotropins and sex steroids regulate sexual development, spermatogenesis and oogenesis, the ovarian and uterine cycles, sex drive, pregnancy, fetal development, and lactation.

STUDY GUIDE

▶ Assess Your Learning Outcomes

To test your knowledge, discuss the following topics with a study partner or in writing, ideally from memory.

17.1 Overview of the Endocrine System

1. The importance of intercellular communication for survival, and the body's four mechanisms of intercellular communication
2. The general term for the cells and glands that secrete hormones, and the name of that branch of science and medicine that specializes in hormones
3. How endocrine glands differ from exocrine glands
4. Similarities, differences, and interactions between the nervous and endocrine systems
5. The term for organs or cells that are influenced by a given hormone, and why they are the only ones to respond to it even though the hormone travels throughout the body

17.2 The Hypothalamus and Pituitary Gland

1. Why the hypothalamus should be considered part of the endocrine system
2. The anatomical relationship of the hypothalamus to the pituitary gland; the two major parts of the pituitary; and how the hypothalamus communicates with each
3. Six hormones that are secreted by the hypothalamus to regulate the anterior pituitary, and their effects
4. Two hormones synthesized in the hypothalamus and stored in the posterior pituitary; how they get to the pituitary; and how their later release into the bloodstream is controlled
5. Six hormones secreted by the anterior pituitary, their abbreviations, and their target organs and functions
6. Two hormones secreted by the posterior pituitary, their abbreviations, their target organs and effects, and the role of neuroendocrine reflexes in their release
7. Examples and mechanisms of positive and negative feedback control of the hypothalamus and pituitary
8. The actions of growth hormone (GH) and the role of insulin-like growth factors in its effects

17.3 Other Endocrine Glands

1. Anatomy of the pineal gland; its involution; and its hormone and function
2. Anatomy of the thymus; its involution; and its hormones and functions
3. Anatomy of the thyroid gland; its hormones and functions; and the cells that produce each hormone
4. Anatomy of the parathyroid glands; their hormone and function
5. Anatomy of the adrenal glands, and structural differences between the cortex and medulla
6. Hormones and functions of the adrenal medulla
7. Three tissue zones of the adrenal cortex, the hormones of each zone, and their functions
8. Pancreatic islets and their cell types, hormones, and functions
9. Endocrine components of the ovaries and testes, and their hormones and functions
10. Hormones produced by the following tissues and organs, and their effects: the skin, liver, kidneys, heart, digestive tract, adipose tissue, osseous tissue, and placenta

17.4 Hormones and Their Actions

1. Three main chemical classes of hormones and examples of each
2. The synthesis of steroid hormones
3. The synthesis of peptide hormones such as insulin
4. Two amino acids that serve as hormone precursors and which hormones are produced from each
5. The types of stimuli that elicit hormone secretion, and examples of each
6. Thyroid hormone synthesis and secretion
7. The problem that must be overcome to transport thyroid hormone (TH) and steroid hormones in the blood, and how the transport mechanism affects their half-life
8. Where hormone receptors are located in the target cells, and differences between receptor systems for hydrophilic and hydrophobic hormones
9. Which hormones require second messengers to activate a target cell; how second messengers work, especially cAMP, DAG, and IP_3
10. How signal amplification enables small amounts of hormone to produce great physiological effects
11. How target cells modulate their hormone sensitivity
12. Three kinds of interactions that can occur when two or more of them act simultaneously on a target cell
13. How hormones are inactivated and cleared from the blood after completing their task

17.5 Stress and Adaptation

1. The physiological or medical definition of *stress*
2. Definition of the *stress response (general adaptation syndrome)*
3. The three stages of the stress response; the dominant hormones and physiological effects of each stage; and what marks the transition from one stage to the next

17.6 Eicosanoids and Other Signaling Molecules

1. Paracrine and autocrine secretions, examples, and how they compare and contrast with hormones
2. The general structure and metabolic precursor of eicosanoids
3. Synthesis and effects of leukotrienes
4. Synthesis and effects of the three cyclooxygenase (COX) products: prostacyclin, thromboxanes, and prostaglandins
5. How prostaglandins are named

17.7 Endocrine Disorders

1. Effects of growth hormone hyposecretion and hypersecretion, and how the effects differ between adult versus childhood onset
2. Myxedema, endemic goiter, and toxic goiter
3. Effects of hypo- and hyperparathyroidism
4. Cushing syndrome and adrenogenital syndrome
5. The "three polys" of diabetes mellitus (DM), and three clinical findings that typically confirm DM
6. The mechanism of glycosuria and osmotic diuresis typical of DM; how this relates to the transport maximum (T_m) of carrier-mediated transport
7. Differences between the cause, pathology, and treatment of types 1 and 2 DM
8. Consequences of inadequately treated DM and why each of its many pathological effects occurs

STUDY GUIDE

▶ Testing Your Recall

Answers in Appendix A

1. CRH secretion would *not* raise the blood concentration of
   a. ACTH.
   b. thyroxine.
   c. cortisol.
   d. corticosterone.
   e. glucose.

2. Which of the following hormones has the least in common with the others?
   a. adrenocorticotropic hormone
   b. follicle-stimulating hormone
   c. thyrotropin
   d. thyroxine
   e. prolactin

3. Which hormone would no longer be secreted if the hypothalamo–hypophysial tract were destroyed?
   a. oxytocin
   b. follicle-stimulating hormone
   c. growth hormone
   d. adrenocorticotropic hormone
   e. corticosterone

4. Which of the following is *not* a hormone?
   a. prolactin
   b. prolactin-inhibiting hormone
   c. thyroxine-binding globulin
   d. atrial natriuretic peptide
   e. cortisol

5. Where are the receptors for insulin located?
   a. in the pancreatic beta cells
   b. in the blood plasma
   c. on the target-cell membrane
   d. in the target-cell cytoplasm
   e. in the target-cell nucleus

6. What would be the consequence of defective ADH receptors?
   a. diabetes mellitus
   b. adrenogenital syndrome
   c. dehydration
   d. seasonal affective disorder
   e. none of these

7. Which of these has more exocrine than endocrine tissue?
   a. the pineal gland
   b. the adenohypophysis
   c. the thyroid gland
   d. the pancreas
   e. the adrenal gland

8. Which of these cells stimulate bone deposition?
   a. alpha cells
   b. beta cells
   c. C cells
   d. G cells
   e. T cells

9. Which of these hormones relies on cAMP as a second messenger?
   a. ACTH
   b. progesterone
   c. thyroxine
   d. testosterone
   e. estrogen

10. Prostaglandins are derived from
    a. phospholipase.
    b. cyclooxygenase.
    c. leukotriene.
    d. lipoxygenase.
    e. arachidonic acid.

11. The _____ develops from a pouch in the pharynx of the embryo.

12. Thyroxine (T_4) is synthesized by combining two iodinated molecules of the amino acid _____.

13. Growth hormone hypersecretion in adulthood causes a disease called _____.

14. The dominant hormone in the stage of resistance of the stress response is _____.

15. Adrenal steroids that regulate glucose metabolism are collectively called _____.

16. Testosterone is secreted by the _____ cells of the testis.

17. Target cells can reduce pituitary secretion by a process called _____.

18. Hypothalamic releasing factors are delivered to the anterior pituitary by way of a network of blood vessels called the _____.

19. A hormone is said to have a/an _____ effect when it stimulates the target cell to develop receptors for other hormones to follow.

20. _____ is a process in which a cell increases its number of receptors for a hormone, thus increasing its hormone sensitivity and response.

STUDY GUIDE

▶ Building Your Medical Vocabulary

Answers in Appendix A

State a meaning of each word element, and give a medical term from this chapter that uses it or a slight variation of it.

1. adeno-

2. chole-

3. diabet-

4. eicosa-

5. luteo-

6. -oid

7. -osa

8. pro-

9. tropo-

10. uri-

▶ What's Wrong with These Statements?

Answers in Appendix A

Briefly explain why each of the following statements is false, or reword it to make it true.

1. Castration would lower a man's blood gonadotropin concentration.

2. The receptors for glycoprotein hormones are usually in the target cell's cytoplasm or nucleus.

3. Thyroglobulin cannot be synthesized without iodine.

4. A tumor in an endocrine gland always results in the hypersecretion of its hormones.

5. All hormones are secreted by endocrine glands.

6. The great majority of cases of diabetes mellitus are caused by insulin deficiency.

7. The pineal gland and thymus are larger in adults than in children.

8. A deficiency of dietary iodine would lead to negative feedback inhibition of TSH synthesis.

9. The tissue at the center of the adrenal gland is called the zona reticularis.

10. Of the endocrine organs covered in this chapter, only the adrenal glands are paired; the rest are single.

▶ Testing Your Comprehension

1. Propose a model of signal amplification for the effect of a steroid hormone and construct a diagram similar to figure 17.21 for your model. It may help to review protein synthesis in chapter 4.

2. Suppose you were browsing in a health-food store and saw a product advertised: "Put an end to heart disease. This herbal medicine will rid your body of cholesterol!" Would you buy it? Why or why not? If the product were as effective as claimed, what are some other effects it would produce?

3. A person with toxic goiter tends to sweat profusely. Explain this in terms of homeostasis.

4. How is the action of a peptide hormone similar to the action of the neurotransmitter norepinephrine?

5. A young man is involved in a motorcycle accident that fractures his sphenoid bone. Shortly thereafter, he begins to excrete enormous amounts of urine, up to 30 L/day, and suffers intense thirst. His neurologist diagnoses the problem as diabetes insipidus. Explain how his head injury resulted in these effects on urinary function and thirst. Why would a sphenoid fracture be more likely than an occipital bone fracture to cause diabetes insipidus? What hormone imbalance resulted from this accident? Would you expect to find elevated glucose in the urine of this diabetic patient? Why or why not?

Red blood cells (red concave discs), white blood cells (blue), platelets (green), and filaments of the clotting protein fibrin (gray) (SEM)

Science Photo Library/Alamy Stock Photo

CHAPTER OUTLINE

18.1 Introduction

    18.1a Functions of the Circulatory System

    18.1b Components and General Properties of Blood

    18.1c Blood Plasma

    18.1d Blood Viscosity and Osmolarity

    18.1e How Blood Is Produced

18.2 Erythrocytes

    18.2a Erythrocyte Form and Function

    18.2b Hemoglobin

    18.2c Quantities of Erythrocytes and Hemoglobin

    18.2d The Erythrocyte Life History

    18.2e Erythrocyte Disorders

18.3 Blood Types

    18.3a The ABO Group

    18.3b The Rh Group

    18.3c Other Blood Groups

    18.3d Maternal–Fetal Mismatches

18.4 Leukocytes

    18.4a Leukocyte Form and Function

    18.4b Types of Leukocytes

    18.4c The Leukocyte Life History

    18.4d Leukocyte Disorders

18.5 Platelets and the Control of Bleeding

    18.5a Platelet Form and Function

    18.5b Platelet Production

    18.5c Hemostasis

    18.5d The Fate of Blood Clots

    18.5e Prevention of Inappropriate Clotting

    18.5f Clotting Disorders

Study Guide

DEEPER INSIGHTS

18.1 Clinical Application: Starvation and Plasma Protein Deficiency

18.2 Medical History: Charles Drew—Blood-Banking Pioneer

18.3 Clinical Application: Bone Marrow and Cord Blood Transplants

18.4 Clinical Application: The Complete Blood Count

18.5 Clinical Application: Liver Disease and Blood Clotting

18.6 Clinical Application: Clinical Management of Blood Clotting

Anatomy & Physiology
Revealed® 4.0

Module 9: Cardiovascular System

Blood has always had a special mystique. From time immemorial, people have seen blood flow from the body and, with it, the life of the individual. It is no wonder that blood was thought to carry a mysterious "vital force." Ancient Romans drank the blood of fallen gladiators in a belief that they could acquire a gladiator's vitality or that it could cure epilepsy. Even today, we become especially alarmed when we find ourselves bleeding, and the emotional impact of blood makes many people faint at the sight of it. From ancient Egypt to nineteenth-century America, physicians drained "bad blood" from their patients to treat everything from gout to headaches, from menstrual cramps to mental illness. It was long thought that hereditary traits were transmitted through the blood, and people still use such unfounded or metaphorical expressions as "I have one-quarter Cherokee blood."

Scarcely anything meaningful was known about blood until its cells were seen with the first microscopes. Even though blood is a uniquely accessible tissue, most of what we know about it dates only to the mid-twentieth century. Recent developments in this field have empowered us to save and improve the lives of countless people who would otherwise have suffered or died.

18.1 Introduction

Expected Learning Outcomes

When you have completed this section, you should be able to

a. describe the functions and major components of the circulatory system;

b. describe the components and physical properties of blood;

c. describe the composition of blood plasma;

d. explain the significance of blood viscosity and osmolarity; and

e. describe in general terms how blood is produced.

18.1a Functions of the Circulatory System

The **circulatory system** consists of the heart, blood vessels, and blood. The term **cardiovascular**[1] **system** refers only to the heart and vessels, which are the subject of the next two chapters. The study of blood, treated in this chapter, is called **hematology.**[2]

The fundamental purpose of the circulatory system is to transport substances from place to place in the body. Blood is the liquid medium in which these materials travel, blood vessels ensure the proper routing of blood to its destinations, and the heart is the pump that keeps the blood flowing.

More specifically, the functions of the circulatory system are as follows:

Transport

- Blood carries oxygen from the lungs to all of the body's tissues, while it picks up carbon dioxide from those tissues and carries it to the lungs to be removed from the body.

- It picks up nutrients from the digestive tract and delivers them to all of the body's tissues.

- It carries metabolic wastes to the kidneys for removal.

- It carries hormones from endocrine cells to their target organs.

- It transports a variety of stem cells from the bone marrow and other origins to the tissues where they lodge and mature.

Protection

- Blood plays several roles in inflammation, a mechanism for limiting the spread of infection.

- White blood cells destroy microorganisms and cancer cells and remove debris from the tissues.

- Antibodies and other blood proteins neutralize toxins and help to destroy pathogens.

- Platelets secrete factors that initiate blood clotting and other processes for minimizing blood loss, and contribute to tissue growth and blood vessel maintenance.

Regulation

- By absorbing or giving off fluid under different conditions, the blood capillaries stabilize fluid distribution in the body.

- By buffering acids and bases, blood proteins stabilize the pH of the extracellular fluids.

- Cutaneous blood flow is extremely important in dissipating metabolic heat from the body. Shifts in blood flow regulate body temperature by routing blood to the skin for heat loss or retaining it deeper in the body to conserve heat.

Considering the importance of efficiently transporting nutrients, wastes, hormones, and especially oxygen from place to place, it is easy to understand why an excessive loss of blood is quickly fatal, and why the circulatory system needs mechanisms for minimizing such losses.

---

[1]*cardio* = heart; *vas* = vessel
[2]*hem, hemato* = blood; *logy* = study of

18.1b Components and General Properties of Blood

Adults generally have about 4 to 6 liters of blood. It is a liquid connective tissue composed, like other connective tissues, of cells and an extracellular matrix. Its matrix is the blood **plasma,** a clear, light yellow fluid constituting a little over half of the blood volume. Suspended in the plasma are the **formed elements**—cells and cell fragments including the red blood cells, white blood cells, and platelets **(fig. 18.1).** The term *formed element* alludes to the fact that these are membrane-enclosed bodies with a definite structure visible with the microscope. Strictly speaking, they can't all be called *cells* because the platelets, as explained later, are merely fragments torn from certain bone marrow cells.

The formed elements are classified as follows:

> Erythrocytes[3] (red blood cells, RBCs)
>
> Platelets
>
> Leukocytes[4] (white blood cells, WBCs)
>> Granulocytes
>>> Neutrophils
>>>
>>> Eosinophils
>>>
>>> Basophils
>> Agranulocytes
>>> Lymphocytes
>>>
>>> Monocytes

Thus, there are seven kinds of formed elements: the erythrocytes, platelets, and five kinds of leukocytes. The five leukocyte types are divided into two categories, the *granulocytes* and *agranulocytes,* on grounds explained later.

Blood fractionation, the separation of blood into its basic components, is based on centrifugation and coagulation **(fig. 18.2).** First, a sample of blood in a tube is spun in a centrifuge for a few minutes. RBCs, the densest elements, settle to the bottom of the tube and typically constitute 37% to 52% of the total volume—a value called the **hematocrit** or **packed cell volume.** WBCs and platelets settle into a narrow cream- or buff-colored zone called the *buffy coat* just above the RBCs; they total 1% or less of the blood volume. At the top of the tube is the plasma, which is about 47% to 63% of the blood volume. If plasma is separated, allowed to coagulate (clot), and centrifuged again, the clotting proteins (mainly fibrin) settle to the bottom of the tube and the overlying fluid is then called **blood serum.** Serum is clinically valuable as a vehicle for vaccines, antivenins, and other therapies; for certain blood tests,

[3] *erythro* = red; *cyte* = cell
[4] *leuko* = white; *cyte* = cell

FIGURE 18.1 The Formed Elements of Blood.

❓ *Identify all the unlabeled formed elements by comparison to the labeled ones.*

Monocyte
Eosinophil
Platelet
Erythrocyte
Small lymphocyte
Young (band) neutrophil
Mature neutrophil
Large lymphocyte
Basophil

Withdraw blood

Centrifuge

Plasma (55% of whole blood)

Buffy coat: leukocytes and platelets (<1% of whole blood)

Erythrocytes (45% of whole blood)

Formed elements

(a)

Whole blood → Centrifugation → Formed elements / Plasma → Coagulation → Clotting proteins / Serum

(b)

FIGURE 18.2 The Components of Blood. (a) Centrifuging a sample of blood to separate formed elements from plasma and determine the hematocrit. (b) Fractionation of blood into formed elements, plasma, and serum.

TABLE 18.1	General Properties of Blood
Characteristic	**Typical Values for Healthy Adults***
Mean fraction of body weight	8%
Volume in adult body	Female: 4–5 L; male: 5–6 L
Volume/body weight	80–85 mL/kg
Mean temperature	38°C (100.4°F)
pH	7.35–7.45
Viscosity (relative to water)	Whole blood: 4.5–5.5; plasma: 2.0
Osmolarity	280–296 mOsm/L
Mean salinity (mainly NaCl)	0.9%
Hematocrit (packed cell volume)	Female: 37% to 48% Male: 45% to 52%
Hemoglobin	Female: 12–16 g/dL Male: 13–18 g/dL
Mean RBC count	Female: 4.2–5.4 million/µL Male: 4.6–6.2 million/µL
Platelet count	130,000–360,000/µL
Total WBC count	5,000–10,000/µL

*Values vary slightly depending on the testing methods used.

TABLE 18.2	Composition of Blood Plasma
Blood Component*	**Typical Values for Healthy Adults**
Water	92% by weight
Proteins	Total 6–9 g/dL
Albumin	60% of total protein, 3.2–5.5 g/dL
Globulins	36% of total protein, 2.3–3.5 g/dL
Fibrinogen	4% of total protein, 0.2–0.3 g/dL
Nutrients	
Glucose (dextrose)	70–110 mg/dL
Amino acids	33–51 mg/dL
Lactate	6–16 mg/dL
Total lipid	450–850 mg/dL
Cholesterol	120–220 mg/dL
Fatty acids	190–420 mg/dL
High-density lipoprotein (HDL)	30–80 mg/dL
Low-density lipoprotein (LDL)	62–185 mg/dL
Triglycerides (neutral fats)	40–150 mg/dL
Phospholipids	6–12 mg/dL
Iron	50–150 µg/dL
Trace elements	Traces
Vitamins	Traces
Electrolytes	
Sodium (Na^+)	135–145 mEq/L
Calcium (Ca^{2+})	9.2–10.4 mEq/L
Potassium (K^+)	3.5–5.0 mEq/L
Magnesium (Mg^{2+})	1.3–2.1 mEq/L
Chloride (Cl^-)	100–106 mEq/L
Bicarbonate (HCO_3^-)	23.1–26.7 mEq/L
Phosphate (HPO_4^{2-})	1.4–2.7 mEq/L
Sulfate (SO_4^{2-})	0.6–1.2 mEq/L
Nitrogenous Wastes	
Urea	10–20 mg/dL
Uric acid	1.5–8.0 mg/dL
Creatinine	0.6–1.5 mg/dL
Creatine	0.2–0.8 mg/dL
Ammonia	0.02–0.09 mg/dL
Bilirubin	0–1.0 mg/dL
Other Components	
Dissolved CO_2	2.62 mL/dL
Dissolved O_2	0.29 mL/dL
Dissolved N_2	0.98 mL/dL
Enzymes of diagnostic value	—
Hormones	—

*This table is limited to substances of greatest relevance to this and later chapters. Concentrations refer to plasma only, not to whole blood.

such as for hepatitis and prostate cancer; and for many clinical and research laboratory uses.

Table 18.1 lists several properties of blood. Some of the terms in that table are defined later in the chapter.

▶▶▶APPLY WHAT YOU KNOW

Based on your body weight, estimate the volume (in liters) and weight (in kilograms) of your own blood, using the data in table 18.1.

18.1c Blood Plasma

Even though blood plasma has no anatomy that we can study visually, we cannot ignore its importance as the matrix of this liquid connective tissue. Plasma is a complex mixture of water, proteins, nutrients, electrolytes, nitrogenous wastes, hormones, and gases **(table 18.2).** Protein is the most abundant plasma solute by weight, totaling 6 to 9 g/dL. Plasma proteins play a variety of roles including clotting, defense against pathogens, and transport of other solutes such as iron, copper, lipids, and hydrophobic hormones. There are three major categories of plasma proteins: albumin, globulins, and fibrinogen **(table 18.3).** Many other plasma proteins are indispensable to survival, but account for less than 1% of the total.

Albumin is the smallest and most abundant plasma protein. It serves to transport various solutes and buffer the pH of blood plasma. It also makes a major contribution to two physical

TABLE 18.3	Major Proteins of the Blood Plasma
Proteins	**Functions**
Albumin (60%)*	Responsible for colloid osmotic pressure; major contributor to blood viscosity; transports lipids, hormones, calcium, and other solutes; buffers blood pH
Globulins (36%)*	Transport and defense functions as itemized below
Alpha (α) Globulins	
Haptoglobulin	Transports hemoglobin released by dead erythrocytes
Ceruloplasmin	Transports copper
Prothrombin	Promotes blood clotting
Others	Transport lipids, fat-soluble vitamins, and hormones
Beta (β) Globulins	
Transferrin	Transports iron
Complement proteins	Aid in destruction of toxins and microorganisms
Others	Transport lipids
Gamma (γ) Globulins	Antibodies; combat pathogens
Fibrinogen (4%)*	Becomes fibrin, the major component of blood clots

*Mean percentage of the total plasma protein by weight

properties of blood: its *viscosity* and *osmolarity,* discussed shortly. Through its effects on these two variables, changes in albumin concentration can significantly affect blood volume, pressure, and flow. **Globulins** are divided into three subclasses; from smallest to largest in molecular weight, they are the alpha (α), beta (β), and gamma (γ) globulins. Globulins play various roles in solute transport, clotting, and immunity. **Fibrinogen** is a soluble precursor of *fibrin,* a sticky protein that forms the framework of a blood clot. Some of the other plasma proteins are enzymes involved in the clotting process.

The liver produces as much as 4 g of plasma protein per hour, contributing all of the major proteins except gamma globulins. The gamma globulins come from *plasma cells*—connective tissue cells that are descended from white blood cells called *B lymphocytes.*

▶▶▶**APPLY WHAT YOU KNOW**

How could a disease such as liver cancer or hepatitis result in impaired blood clotting?

In addition to protein, the blood plasma contains such nitrogen-containing compounds as free amino acids and nitrogenous wastes. **Nitrogenous wastes** are toxic end products of catabolism. The most abundant is *urea,* a product of amino acid

catabolism. These wastes are normally excreted by the kidneys at a rate that balances their production.

The plasma also transports nutrients absorbed by the digestive tract, including glucose, amino acids, fats, cholesterol, phospholipids, vitamins, and minerals. It transports dissolved oxygen, carbon dioxide, and nitrogen (see table 18.2). The dissolved nitrogen normally has no physiological role in the body (but see Deeper Insight 22.5).

Electrolytes are another important component of the blood plasma. Sodium ions constitute about 90% of the plasma cations. Sodium is more important than any other solute to the osmolarity of the blood. As such, it has a major influence on blood volume and pressure; people with high blood pressure are often advised to limit their sodium intake. Electrolyte concentrations are carefully regulated by the body and have rather stable concentrations in the plasma.

18.1d Blood Viscosity and Osmolarity

Two important properties of blood—viscosity and osmolarity—arise from the formed elements and plasma composition. **Viscosity** is the resistance of a fluid to flow, resulting from the cohesion of its particles. Loosely speaking, it is the thickness or stickiness of a fluid. At a given temperature, mineral oil is more viscous than water, for example, and honey is more viscous than mineral oil. Whole blood is 4.5 to 5.5 times as viscous as water, mainly because of the RBCs; plasma alone is 2.0 times as viscous as water, mainly because of its protein. Viscosity is important in circulatory function because it partially governs the flow of blood through the vessels. An RBC or protein deficiency reduces viscosity and causes blood to flow too easily, whereas an excess causes blood to flow too sluggishly. Either of these conditions puts a strain on the heart that may lead to serious cardiovascular problems if not corrected.

The **osmolarity** of blood, another important factor in cardiovascular function, is the total molarity of dissolved particles that cannot pass through the blood vessel wall. In order to nourish surrounding cells and remove their wastes, substances must pass between the bloodstream and tissue fluid through the capillary walls. This transfer of fluids depends on a balance between the filtration of fluid from the capillary and its reabsorption by osmosis (see section 20.3c). The rate of reabsorption is governed by the relative osmolarity of the blood versus the tissue fluid. If the blood osmolarity is too high, the bloodstream absorbs too much water. This raises the blood volume, resulting in high blood pressure and a potentially dangerous strain on the heart and arteries. If its osmolarity drops too low, too much water remains in the tissues. They become edematous (swollen) and the blood pressure may drop to dangerously low levels because of the water lost from the bloodstream.

It is therefore important that the blood maintain an optimal osmolarity. The osmolarity of the blood is a product mainly of its sodium ions, protein, and erythrocytes. The contribution of protein to blood osmotic pressure—called the **colloid osmotic pressure (COP)**—is especially important, as we see from the effects of extremely low-protein diets (see Deeper Insight 18.1).

DEEPER INSIGHT 18.1

CLINICAL APPLICATION

Starvation and Plasma Protein Deficiency

Several conditions can lead to *hypoproteinemia,* a deficiency of plasma protein: extreme starvation or dietary protein deficiency, liver diseases that interfere with protein synthesis, and protein loss through the urine or body surface in the cases of kidney disease and severe burns, respectively. As the protein content of the blood plasma drops, so does its osmolarity. The bloodstream loses more fluid to the tissues than it reabsorbs by osmosis. Thus, the tissues become edematous and a pool of fluid may accumulate in the abdominal cavity—a condition called *ascites* (ah-SY-teez) (see Deeper Insight 20.4).

Children who suffer severe dietary protein deficiencies often exhibit a condition called *kwashiorkor* (KWASH-ee-OR-cor) **(fig. 18.3).** The arms and legs are emaciated for lack of muscle, the skin is shiny and tight with edema, and the abdomen is swollen by ascites. *Kwashiorkor* is a Ghanian word for a "deposed" or "displaced" child who is no longer breast-fed. Symptoms appear when a child is weaned and placed on a diet consisting mainly of rice or other cereals. Children with kwashiorkor often die of diarrhea and dehydration.

FIGURE 18.3 Child with Kwashiorkor. Note the thin limbs and fluid-distended abdomen.

IMTSSA/CNRI/Science Source

18.1e How Blood Is Produced APR

We lose blood continually, not only from bleeding but also as blood cells grow old and die and plasma components are consumed or excreted from the body. Therefore, we must continually replace it. Every day, an adult typically produces 400 billion platelets, 200 billion RBCs, and 10 billion WBCs. The production of blood, especially its formed elements, is called **hematopoiesis**[5] (he-MAT-oh-poy-EE-sis). A knowledge of this process provides an indispensable foundation for understanding leukemia, anemia, and other blood disorders.

The tissues that produce blood cells are called **hematopoietic tissues.** The first hematopoietic tissues of the human embryo form in the *yolk sac,* a membrane associated with all vertebrate embryos. In most vertebrates (fish, amphibians, reptiles, and birds), this sac encloses the egg yolk, transfers its nutrients to the growing embryo, and produces the forerunners of the first blood cells. Even animals that don't lay eggs, however, have a yolk sac that retains its hematopoietic function. (It is also the source of cells that later produce eggs and sperm.) Cell clusters called *blood islands* form here by the third week of human development. They produce primitive stem cells that migrate into the embryo proper and colonize the bone marrow, liver, spleen, and thymus. Here, the stem cells multiply and give rise to blood cells throughout fetal development. The liver stops producing blood cells around the time of birth. The spleen stops producing RBCs soon after, but it continues to produce lymphocytes for life.

From infancy onward, the red bone marrow produces all seven kinds of formed elements, while lymphocytes are also produced in the lymphatic tissues and organs—especially the thymus, tonsils, lymph nodes, spleen, and mucous membranes. Blood formation in the bone marrow and lymphatic organs is called, respectively, **myeloid**[6] and **lymphoid hematopoiesis.**

All formed elements trace their origins to a common type of **hematopoietic stem cell (HSC)** in the bone marrow. In the stem-cell terminology of section 5.6c, HSCs would be classified as multipotent stem cells, destined to develop into multiple mature cell types. Hematologists, however, often call them *pluripotent stem cells (PPSCs).* (Stem-cell biology is a young science and specialists in different fields sometimes use different terminology.) HSCs multiply to maintain a small but persistent population in the bone marrow, but some of them go on to become a variety of more specialized cells called **colony-forming units (CFUs).** Each CFU is destined to produce one or another class of formed elements. The specific processes leading from an HSC to RBCs, WBCs, and platelets are described at later points in this chapter.

Blood plasma also requires continual replacement. It is composed mainly of water, which it obtains primarily by absorption from the digestive tract. Its electrolytes and organic nutrients are also acquired there, and its gamma globulins come from connective tissue plasma cells and its other proteins mainly from the liver.

BEFORE YOU GO ON

Answer the following questions to test your understanding of the preceding section:

1. Identify at least two each of the transport, protective, and regulatory functions of the circulatory system.

2. What are the two principal components of the blood?

3. List the three major classes of plasma proteins. Which one is absent from blood serum?

4. Define the *viscosity* and *osmolarity* of blood. Explain why each of these is important for human survival.

5. What does *hematopoiesis* mean? After birth, what one cell type is the starting point for all hematopoiesis?

---

[5]*hemato* = blood; *poiesis* = formation

[6]*myel* = bone marrow

18.2 Erythrocytes

Expected Learning Outcomes

When you have completed this section, you should be able to

a. discuss the structure and function of erythrocytes (RBCs);

b. describe the structure and function of hemoglobin;

c. state and define some clinical measurements of RBC and hemoglobin quantities;

d. describe the life history of erythrocytes; and

e. name and describe the types, causes, and effects of RBC excesses and deficiencies.

Erythrocytes, or **red blood cells (RBCs),** have two principal functions: (1) to pick up oxygen from the lungs and deliver it to tissues elsewhere, and (2) to pick up carbon dioxide from the tissues and unload it in the lungs. RBCs are the most abundant formed elements of the blood and therefore the most obvious things one sees upon its microscopic examination. They are also the most critical to survival; a severe deficiency of leukocytes or platelets can be fatal within a few days, but a severe deficiency of RBCs can be fatal within mere minutes. It is the lack of life-giving oxygen, carried by erythrocytes, that leads rapidly to death in cases of major trauma or hemorrhage.

18.2a Erythrocyte Form and Function

An erythrocyte is a discoidal cell with a biconcave shape—a thick rim and a thin sunken center. It is about 7.5 μm in diameter and 2.0 μm thick at the rim **(fig. 18.4).** Although most cells, including white blood cells, have an abundance of organelles, RBCs lose their nucleus and other organelles during maturation and are thus remarkably devoid of internal structure. When viewed with the transmission electron microscope, the interior of an RBC appears uniformly gray. Lacking mitochondria, RBCs rely exclusively on anaerobic fermentation to produce ATP. The lack of aerobic respiration prevents them from consuming the oxygen that they must transport to other tissues. If they were aerobic and consumed oxygen, they would be like a pizza delivery driver who ate a slice of your pizza on the way to your house. RBCs are made to deliver oxygen, not consume it. They are the only human cells that carry on anaerobic fermentation indefinitely.

The cytoplasm of an RBC consists mainly of a 33% solution of **hemoglobin** (about 280 million molecules per cell). This is the red pigment that gives an RBC its color and name. It is known especially for its role in oxygen transport, but it also aids in the transport of carbon dioxide and the buffering of blood pH. The cytoplasm also contains an enzyme, *carbonic anhydrase (CAH),* that catalyzes the reaction $CO_2 + H_2O \rightleftharpoons H_2CO_3$. The role of CAH in gas transport and pH balance is discussed in later chapters.

The plasma membrane of a mature RBC has glycolipids on the outer surface that determine a person's blood type. On its

Surface view

7.5 μm

2.0 μm

(a) Sectional view

(b)

Capillary wall

Erythrocytes

Lymphocyte

(c) 7 μm

FIGURE 18.4 The Structure of Erythrocytes. (a) Dimensions and shape of an erythrocyte. (b) SEM photo showing the biconcave shape of RBCs. (c) TEM photo of a lymphocyte (left) and two erythrocytes (right) in blood capillaries of the lung. Note how thin the RBCs are at the center, and their lack of organelles or other internal structure. **APR**

 Why are erythrocytes caved in at the center?

inner surface are two cytoskeletal proteins, *spectrin* and *actin,* that give the membrane resilience and durability. This is especially important when RBCs pass through small blood capillaries and sinusoids. Many of these passages are narrower than the diameter of an RBC, forcing the RBCs to stretch, bend, and fold as they squeeze through. When they enter larger vessels, RBCs spring back to their discoidal shape like an air-filled inner tube.

There has been appreciable, unresolved debate over whether the biconcave shape of the RBC has any functional advantage. Some suggest that it maximizes the ratio of cell surface area to volume and thereby promotes the quick diffusion of oxygen to all of the hemoglobin in the cell. This is hard to reconcile with the fact that the only place RBCs load oxygen is in the capillaries, and while squeezing through the tiny capillaries, they generally are not biconcave but compressed into ovoid or teardrop shapes. They spring back to the biconcave shape when reentering larger blood vessels, but no oxygen pickup occurs here. Another hypothesis is that the biconcave shape minimizes RBC spin (like a spinning ice-skater with her arms extended) and turbulence, enabling the dense slurry of RBCs to flow through the larger blood vessels with a smooth *laminar flow.* It has also been argued that it is simply the easiest, most stable shape for the cell and its cytoskeleton to relax into when the nucleus is removed, and it may have no physiological function at all.

18.2b Hemoglobin

Hemoglobin consists of four protein chains called **globins (fig. 18.5).** Two of these, the *alpha* (α) *chains,* are 141 amino acids long, and the other two, the *beta* (β) *chains,* are 146 amino acids long. Each chain is conjugated with a nonprotein moiety called the **heme** group, which binds oxygen to an iron atom (Fe) at its center. Each heme can carry one molecule of O_2; thus, the hemoglobin molecule as a whole can transport up to 4 O_2. About 5% of the CO_2 in the bloodstream is also transported by hemoglobin but is bound to the globin rather than to the heme. Gas transport by hemoglobin is discussed in detail in section 22.3c.

Hemoglobin exists in several forms with slight differences in the globin chains. The form just described is called *adult hemoglobin (HbA).* About 2.5% of an adult's hemoglobin, however, is of a form called HbA_2, which has two *delta* (δ) *chains* in place of the beta chains. The fetus produces a form called *fetal hemoglobin (HbF),* which has two *gamma* (γ) *chains* in place of the beta chains. The delta and gamma chains are the same length as the beta chains but differ in amino acid sequence. HbF binds oxygen more tightly than HbA does; this enables the fetus to extract oxygen from the mother's bloodstream.

18.2c Quantities of Erythrocytes and Hemoglobin

The RBC count and hemoglobin concentration are important clinical data because they determine the amount of oxygen the blood can carry. Three of the most common measurements are hematocrit, hemoglobin concentration, and RBC count. The

FIGURE 18.5 The Structure of Hemoglobin. (a) The hemoglobin molecule consists of two alpha proteins and two beta proteins, each conjugated to a nonprotein heme group. (b) Structure of the heme group. Oxygen binds to iron (Fe) at the center of the heme.

? *In what way does this exemplify a quaternary protein structure? What is the prosthetic group of hemoglobin? (Hint: See "Protein Structure" in section 2.4e.)*

hematocrit[7] **(packed cell volume, PCV)** is the percentage of whole blood volume composed of RBCs (see fig. 18.2). In men, it normally ranges between 42% and 52%; in women, between 37% and 48%. The **hemoglobin concentration** of whole blood is normally 13 to 18 g/dL in men and 12 to 16 g/dL in women. The RBC count is normally 4.6 to 6.2 million RBCs/μL in men and 4.2 to 5.4 million/μL in women. This is often expressed as cells per cubic millimeter (mm^3); 1 μL = 1 mm^3.

Notice that these values tend to be lower in women than in men. There are three physiological reasons for this: (1) Androgens stimulate RBC production, and men have higher androgen levels than women; (2) women of reproductive age have periodic menstrual losses; and (3) the hematocrit is inversely

[7]*hemato* = blood; *crit* = to separate

proportional to percentage body fat, which averages higher in women than in men. In men, the blood also clots faster and the skin has fewer blood vessels than in women. Such differences are not limited to humans. From the evolutionary standpoint, their adaptive value may lie in the fact that male animals fight more than females and suffer more injuries. These traits may serve to minimize or compensate for their blood loss.

▶▶▶ APPLY WHAT YOU KNOW

Explain why the hemoglobin concentration could appear deceptively high in a patient who is dehydrated, when in fact the patient does not have a hemoglobin or RBC excess.

18.2d The Erythrocyte Life History

An erythrocyte lives for an average of 120 days from the time it is produced in the red bone marrow until it dies and breaks up. In a state of balance and stable RBC count, the birth and death of RBCs amount to nearly 100 billion cells per day (1 million per second), or a packed cell volume of 20 mL/day.

Erythrocyte Production

Erythrocyte production is called **erythropoiesis** (eh-RITH-ro-poy-EE-sis). The process normally takes 3 to 5 days and involves four major developments: a reduction in cell size, an increase in cell number, the synthesis of hemoglobin, and the loss of the nucleus and other organelles.

Erythropoiesis begins when a hematopoietic stem cell (HSC) becomes an *erythrocyte colony-forming unit (ECFU)* **(fig. 18.6),** which has receptors for **erythropoietin (EPO),** a hormone secreted by the kidneys. EPO stimulates the ECFU to transform into an *erythroblast (normoblast).* Erythroblasts multiply, build up a large cell population, and synthesize hemoglobin. When this task is completed, the nucleus shrivels and is discharged from the cell. The cell is now called a *reticulocyte,* named for a temporary network (reticulum) of ribosome clusters transcribing the cell's remaining mRNA (polyribosomes; see the photo on the opening page of chapter 4).

Reticulocytes leave the bone marrow and enter the circulating blood. In a day or two, the last of the polyribosomes disintegrate

and disappear, and the cell is a mature erythrocyte. Normally, about 0.5% to 1.5% of the circulating RBCs are reticulocytes, but this percentage rises under certain circumstances. Blood loss, for example, stimulates accelerated erythropoiesis and leads to an increasing number of reticulocytes in circulation—as if the bone marrow is in such a hurry to replenish the lost RBCs that it rushes many developing RBCs into circulation a little early.

Iron Metabolism

Iron is a critical part of the hemoglobin molecule and therefore one of the key nutritional requirements for erythropoiesis. Men lose about 0.9 mg of iron per day through the urine, feces, and bleeding, and women of reproductive age lose an average of 1.7 mg/day through these routes and the added factor of menstruation. Since we absorb only a fraction of the iron in our food, we must consume 5 to 20 mg/day to replace our losses. A pregnant woman needs 20 to 48 mg/day, especially in the last 3 months, to meet not only her own need but also that of the fetus.

Dietary iron exists in two forms: ferric (Fe^{3+}) and ferrous (Fe^{2+}) ions. Stomach acid converts most Fe^{3+} to Fe^{2+}, the only form that the small intestine can absorb **(fig. 18.7).** A protein called **gastroferritin,**[8] produced by the stomach, then binds Fe^{2+} and transports it to the small intestine. Here, it is absorbed into the blood, binds to a plasma protein called **transferrin,** and travels to the bone marrow, liver, and other tissues. Bone marrow uses iron for hemoglobin synthesis; muscle uses it to make the oxygen-binding protein myoglobin; and nearly all cells use iron to make electron-transport molecules called cytochromes in their mitochondria. The liver binds surplus iron to a protein called *apoferritin,*[9] forming an iron-storage complex called **ferritin.** It releases Fe^{2+} into circulation when needed.

Some other nutritional requirements for erythropoiesis are vitamin B_{12} and folic acid, required for the rapid cell division and DNA synthesis that occurs in erythropoiesis, and vitamin C and copper, which are cofactors for some of the enzymes that synthesize hemoglobin.

---

[8]*gastro* = stomach; *ferrit* = iron; *in* = protein
[9]*apo* = separated from; *ferrit* = iron; *in* = protein

Hematopoietic stem cell (HSC)	Colony-forming unit (CFU)	Precursor cells		Mature blood cells
	Erythrocyte CFU	Erythroblasts	Reticulocytes	Erythrocytes

FIGURE 18.6 Erythropoiesis. Stages in the development of a red blood cell.

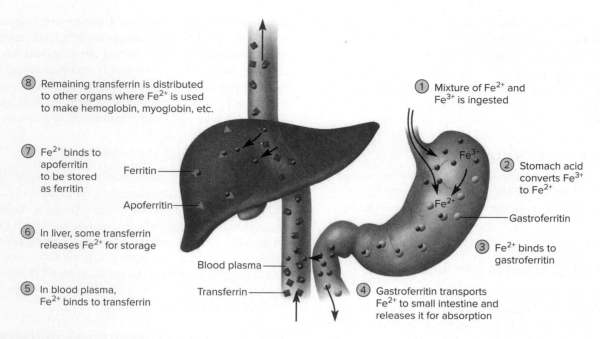

FIGURE 18.7 Iron Metabolism. Read clockwise from the upper right.

Erythrocyte Homeostasis

RBC count is maintained in a classic negative feedback manner
(fig. 18.8). If the count should drop (for example, because of hemorrhaging), it may result in a state of **hypoxemia**[10] (oxygen deficiency in the blood). The kidneys detect this and increase their
EPO output. Three or four days later, the RBC count begins to rise
and reverses the hypoxemia that started the process.

Hypoxemia has many causes other than blood loss. Another
is a low level of oxygen in the atmosphere. If you were to move
from Miami to Denver, for example, the lower O_2 level at the
high elevation of Denver would produce temporary hypoxemia
and stimulate EPO secretion and erythropoiesis. The blood of an
average adult has about 5 million RBCs/µL, but people who live
at high elevations may have counts of 7 to 8 million RBCs/µL.
Another cause of hypoxemia is an abrupt increase in the body's
oxygen consumption. If a formerly lethargic person takes up tennis or aerobics, for example, the muscles consume oxygen more
rapidly and create a state of hypoxemia that stimulates erythropoiesis. Endurance-trained athletes commonly have RBC counts
as high as 6.5 million RBCs/µL. "Doping" with erythropoietin
to build their RBC count and endurance has cost many athletes
their careers and honors.

Not all hypoxemia can be corrected by increasing erythropoiesis. In emphysema, for example, less lung tissue is available to
oxygenate the blood. Raising the RBC count cannot correct this,
but the kidneys and bone marrow have no way of knowing it. The
RBC count continues to rise in a futile attempt to restore homeostasis, resulting in a dangerous excess called *polycythemia,* discussed shortly.

**FIGURE 18.8 Correction of Hypoxemia by a Negative
Feedback Loop.**

Erythrocyte Death and Disposal

The life of an RBC is summarized in **figure 18.9.** As an RBC
ages and its membrane proteins (especially spectrin) deteriorate,
the membrane grows increasingly fragile. Without a nucleus or

[10]*hyp* = below normal; *ox* = oxygen; *emia* = blood condition

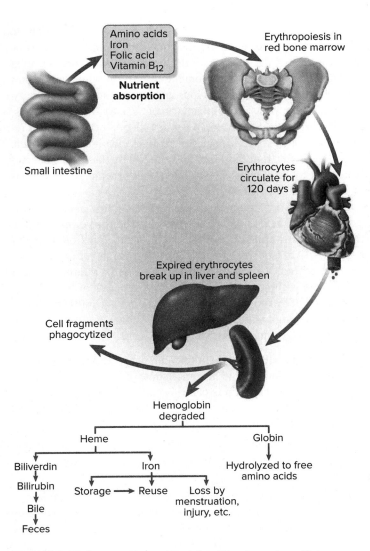

Amino acids
Iron
Folic acid
Vitamin B$_{12}$
Nutrient absorption

Erythropoiesis in red bone marrow

Small intestine

Erythrocytes circulate for 120 days

Expired erythrocytes break up in liver and spleen

Cell fragments phagocytized

Hemoglobin degraded

Heme   Globin

Biliverdin   Iron   Hydrolyzed to free amino acids

Bilirubin   Storage → Reuse   Loss by menstruation, injury, etc.

Bile

Feces

FIGURE 18.9 The Life and Death of Erythrocytes. Note especially the stages of hemoglobin breakdown and disposal.

ribosomes, an RBC cannot synthesize new spectrin. Many RBCs die in the spleen, which has been called the "erythrocyte graveyard." The spleen has channels as narrow as 3 μm that severely test the ability of old, fragile RBCs to squeeze through the organ. Old cells become trapped, broken up, and destroyed. An enlarged and tender spleen sometimes indicates diseases in which RBCs are rapidly breaking down.

Hemolysis[11] (he-MOLL-ih-sis), the rupture of RBCs, releases hemoglobin and leaves empty plasma membranes. The membrane fragments are easily digested by macrophages in the liver and spleen, but hemoglobin disposal is a bit more complicated. It must be disposed of efficiently, however, or it can block kidney tubules and cause renal failure. Macrophages begin the disposal process by separating the heme from the globin. They hydrolyze the globin into free amino acids, which can be used for energy-releasing catabolism or recycled for protein synthesis.

Disposing of the heme is another matter. First, the macrophage removes the iron and releases it into the blood, where it

combines with transferrin and is used or stored in the same way as dietary iron. The macrophage converts the rest of the heme into a greenish pigment called **biliverdin**[12] (BIL-ih-VUR-din), then further converts most of this to a yellow-green pigment called **bilirubin.**[13] Bilirubin is released by the macrophages and binds to albumin in the blood plasma. The liver removes it from the albumin and secretes it into the bile, to which it imparts a dark green color as the bile becomes concentrated in the gallbladder. Biliverdin and bilirubin are collectively known as **bile pigments.** The gallbladder discharges the bile into the small intestine, where bacteria convert bilirubin to *urobilinogen,* responsible for the brown color of the feces. Another hemoglobin breakdown pigment, *urochrome,* produces the yellow color of urine. A high level of bilirubin in the blood causes *jaundice,* a yellowish cast in the skin and whites of eyes. Jaundice may be a sign of rapid hemolysis or a liver disease or bile duct obstruction that interferes with bilirubin disposal.

18.2e Erythrocyte Disorders

Any imbalance between the rates of erythropoiesis and RBC destruction may produce an excess or deficiency of red cells. An RBC excess is called *polycythemia*[14] (POL-ee-sy-THEE-me-uh), and a deficiency of either RBCs or hemoglobin is called *anemia.*[15]

Polycythemia

Primary polycythemia (polycythemia vera) is due to cancer of the erythropoietic line of the red bone marrow. It can result in an RBC count as high as 11 million RBCs/μL and a hematocrit as high as 80%. Polycythemia from all other causes, called **secondary polycythemia,** is characterized by RBC counts as high as 6 to 8 million RBCs/μL. It can result from dehydration because water is lost from the bloodstream while erythrocytes remain and become abnormally concentrated. More often, it is caused by smoking, air pollution, emphysema, high altitude, excessive aerobic exercise, or other factors that create a state of hypoxemia and stimulate erythropoietin secretion.

The principal dangers of polycythemia are increased blood volume, pressure, and viscosity. Blood volume can double in primary polycythemia and cause the circulatory system to become tremendously engorged. Blood viscosity may rise to three times normal. Circulation is poor, the capillaries are congested with viscous blood, and the heart is dangerously strained. Chronic (long-term) polycythemia can lead to embolism, stroke, or heart failure. The deadly consequences of emphysema and some other lung diseases are due in part to polycythemia.

Anemia

The causes of **anemia** fall into three categories: (1) inadequate erythropoiesis or hemoglobin synthesis, (2) **hemorrhagic**

---

[11]*hemo* = blood; *lysis* = splitting, breakdown

[12]*bili* = bile; *verd* = green; *in* = substance
[13]*bili* = bile; *rub* = red; *in* = substance
[14]*poly* = many; *cyt* = cell; *hem* = blood; *ia* = condition
[15]*an* = without; *em* = blood; *ia* = condition

TABLE 18.4	Causes of Anemia
Categories of Anemia	**Causes or Examples**
Inadequate Erythropoiesis	
Iron-deficiency anemia	Dietary iron deficiency
Other nutritional anemias	Dietary folic acid, vitamin B_{12}, or vitamin C deficiency
Anemia due to renal insufficiency	Deficiency of EPO secretion
Pernicious anemia	Deficiency of intrinsic factor leading to inadequate vitamin B_{12} absorption
Hypoplastic and aplastic anemia	Destruction of myeloid tissue by radiation, viruses, some drugs and poisons (arsenic, benzene, mustard gas), or autoimmune disease
Anemia of old age	Declining erythropoiesis due to nutritional deficiencies, reduced physical activity, gastric atrophy (reduced intrinsic factor secretion), or renal atrophy (depressed EPO secretion)
Blood Loss (Hemorrhagic Anemia)	
From hereditary clotting deficiencies	Hemophilia
From nonhereditary causes	Trauma, aneurysm, menstruation, ulcer, etc.
RBC Destruction (Hemolytic Anemia)	
Drug reactions	Penicillin allergy
Poisoning	Mushroom toxins, snake and spider venoms
Parasitic infection	RBC destruction by malaria parasites
Hereditary hemoglobin defects	Sickle-cell disease, thalassemia
Blood type incompatabilities	Hemolytic disease of the newborn, transfusion reactions

anemia from bleeding, and (3) **hemolytic anemia** from RBC destruction. **Table 18.4** gives specific examples and causes for each category.

Anemia often results from kidney failure, because RBC production depends on erythropoietin, which is produced mainly by the kidneys. Erythropoiesis also declines with age, simply because the kidneys atrophy and produce less and less EPO as we get older. Compounding this problem, elderly people tend to get less exercise and eat less well, and both of these factors reduce erythropoiesis.

Nutritional anemia results from a dietary deficiency of any of the requirements for erythropoiesis discussed earlier. Its most common form is **iron-deficiency anemia,** characterized by small pale erythrocytes. Iron-deficiency anemia is usually caused by blood loss without getting enough dietary iron to compensate for it. A deficiency of vitamin B_{12} also causes anemia, but B_{12} is so abundant in meat that a deficiency is rare except in strict vegetarians. More

often, B_{12} deficiency occurs when glands of the stomach fail to produce a substance called **intrinsic factor** that the small intestine needs to absorb the vitamin. Elderly people sometimes develop **pernicious anemia,** an autoimmune disease in which antibodies destroy stomach tissue. Pernicious anemia can also be hereditary. Without proper postsurgical management, gastric-bypass and gastrectomy patients can develop a similar anemia because of the removal of stomach tissue or surgical rearrangement of the stomach, disconnecting it from the small intestine where the intrinsic factor is needed. Such anemias are treatable with vitamin B_{12} injections or oral B_{12} and intrinsic factor.

Hypoplastic[16] **anemia** is caused by a decline in erythropoiesis, whereas the complete failure or destruction of the myeloid tissue produces **aplastic anemia,** a complete cessation of erythropoiesis. Aplastic anemia leads to grotesque tissue necrosis and blackening of the skin. Barring successful treatment, most victims die within a year. About half of all cases of hypoplastic anemia are of unknown or hereditary cause, especially in adolescents and young adults.

Anemia has three potential consequences:

1. The tissues suffer **hypoxia** (oxygen deprivation). The individual is lethargic and becomes short of breath upon physical exertion. The skin is pallid because of the deficiency of hemoglobin. Severe anemic hypoxia can cause life-threatening necrosis of brain, heart, and kidney tissues.

2. Blood osmolarity is reduced. More fluid thus transfers from the bloodstream to the intercellular spaces, resulting in edema.

3. Blood viscosity is reduced. Because the blood puts up less resistance to flow, the heart beats faster than normal and cardiac failure may ensue. Blood pressure also drops because of the reduced volume and viscosity.

Sickle-Cell Disease

Sickle-cell disease and thalassemia (see table 18.8) are hereditary hemoglobin defects that occur mostly among people of African and Mediterranean descent, respectively. **Sickle-cell disease** afflicts about 1.3% of people of African-American heritage. It is caused by a recessive allele that modifies the hemoglobin. Sickle-cell hemoglobin (HbS) differs from normal HbA only in the sixth amino acid of the beta chain, where HbA has glutamic acid and HbS has valine. People who are homozygous for HbS exhibit sickle-cell disease. People who are heterozygous for it—about 7.7% of African-Americans—have *sickle-cell trait* but rarely have severe symptoms. However, if two carriers reproduce, each of their children has a 25% chance of being homozygous and having the disease.

HbS doesn't bind oxygen very well. At low oxygen concentrations, it becomes deoxygenated, polymerizes, and forms a gel that makes erythrocytes become elongated and pointed at the ends **(fig. 18.10),** hence the name of the disease. Sickled

---

[16]*hypo* = below normal; *plas* = formation; *tic* = pertaining to

7 μm

FIGURE 18.10 Sickle-Cell Disease. Shows one deformed, pointed erythrocyte and three normal erythrocytes.

Eye of Science/Science Source

erythrocytes are sticky; they **agglutinate**[17] (clump) and block small blood vessels, causing intense pain in oxygen-starved tissues. Blockage of the circulation can also lead to kidney or heart failure, stroke, severe joint pain, or paralysis. Hemolysis of the fragile cells causes anemia, which results in a hypoxemia that triggers further sickling in a deadly positive feedback loop. Chronic hypoxemia also causes fatigue, weakness, poor mental development, and deterioration of the heart and other organs. In a futile effort to counteract the hypoxemia, the hematopoietic tissues become so active that bones of the cranium and elsewhere become enlarged and misshapen. The spleen reverts to a hematopoietic role, while also disposing of dead RBCs, and becomes enlarged and fibrous. Sickle-cell disease is a prime example of *pleiotropy*—the occurrence of multiple phenotypic effects arising from a change in a single gene.

Without treatment, a child with sickle-cell disease has little chance of living to age 2. Advances in treatment, however, have steadily raised life expectancy to a little beyond age 50.

Why does sickle-cell disease exist? In Africa, where it originated, vast numbers of people die of malaria. Malaria is caused by a parasite that invades the RBCs and feeds on hemoglobin. Sickle-cell hemoglobin is detrimental to the parasites, and people heterozygous for sickle-cell disease are resistant to malaria. The lives saved by this gene far outnumber the deaths of homozygous individuals, so the gene persists in the population. The sickle-cell gene is less common in the United States and other essentially non-malarious regions than it is in Africa.

18.3 Blood Types

Expected Learning Outcomes

When you have completed this section, you should be able to

a. explain what determines a person's ABO and Rh blood types and how this relates to transfusion compatibility;

b. list some blood groups other than ABO and Rh and explain how they may be useful; and

c. describe the effects of a blood type incompatibility between mother and fetus.

Blood types and transfusion compatibility are a matter of interactions between the plasma and erythrocytes. Ancient Greek physicians attempted to transfuse blood from one person to another by squeezing it from a pig's bladder through a porcupine quill into the recipient's vein. Although some patients benefited from the procedure, it was fatal to others. The reason some people have compatible blood and some don't remained obscure until 1900, when Karl Landsteiner discovered blood types A, B, and O—a discovery that won him a Nobel Prize in 1930; type AB was discovered later. World War II stimulated great improvements in transfusions, blood banking, and blood substitutes (see Deeper Insight 18.2).

Blood types are based on large molecules called *antigens* and *antibodies.* Explained more fully in chapter 21, these will be introduced only briefly here. **Antigens** are complex molecules such as proteins, glycoproteins, and glycolipids that are genetically unique to each individual (except identical twins). They occur on the surfaces of all cells and enable the body to distinguish its own cells from foreign matter. When the body detects an antigen of foreign origin, it activates an immune response. This response consists partly of the *plasma cells,* mentioned earlier, secreting proteins called **antibodies.**

[17]*ag* = together; *glutin* = glue

DEEPER INSIGHT 18.2

MEDICAL HISTORY

Charles Drew—Blood-Banking Pioneer

Charles Drew **(fig. 18.11)** was a scientist remembered for his seminal contributions in hematology and civil rights, and for a sadly ironic end to his life. After receiving his M.D. from McGill University of Montreal in 1933, Drew became the first black person to pursue the advanced degree of Doctor of Science in Medicine, for which he studied transfusion and blood banking at Columbia University. He became the director of a new blood bank at Columbia Presbyterian Hospital in 1939 and organized numerous blood banks during World War II.

Drew saved countless lives by convincing physicians to use plasma rather than whole blood for battlefield and other emergency transfusions. Whole blood could be stored for only a week and given only to recipients with compatible blood types. Plasma could be stored longer and was less likely to cause transfusion reactions.

When the U.S. War Department issued a directive forbidding the storage of "Caucasian and Negro blood" in the same military blood banks, Drew denounced the order and resigned his position. He became a professor of surgery at Howard University in Washington, D.C., and later chief of staff at Freedmen's Hospital. He was a mentor for numerous young black physicians and campaigned to get them accepted into the medical community. The American Medical Association, however, refused to admit black members until the 1960s, excluding even Drew himself.

Late one night in 1950, Drew and three colleagues set out to volunteer their medical services to an annual free clinic in Tuskegee, Alabama. Drew fell asleep at the wheel and was critically injured in the resulting accident. Contrary to a myth that Drew was refused emergency treatment because of his race, doctors at the nearest hospital administered blood and attempted to revive him. Yet, for all the lives he saved through his pioneering work in transfusion, Drew himself bled to death at the age of 45.

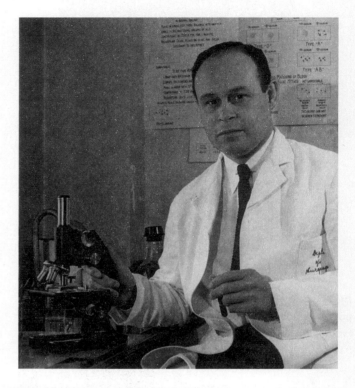

FIGURE 18.11 Charles Drew (1904–50).

Alfred Eisenstaedt/The LIFE Picture Collection/Getty Images

Antibodies bind to antigens and mark them, or the cells bearing them, for destruction. One method of antibody action is **agglutination,** in which each antibody molecule binds to two or more foreign cells and sticks them together. Repetition of this process produces large clumps of cells that can cause the complications of the transfusion reaction discussed shortly. The RBC surface antigens that trigger agglutination are called **agglutinogens** (ah-glue-TIN-oh-jens) and the plasma antibodies that bind to them are called **agglutinins** (ah-GLUE-tih-nins).

18.3a The ABO Group

Blood types A, B, AB, and O form the **ABO blood group** **(table 18.5).** One's ABO blood type is determined by the hereditary presence or absence of antigens A and B on the RBCs. The genetic determination of blood types is explained in section 4.4c. The antigens are glycolipids—membrane phospholipids with short carbohydrate chains bonded to them. **Figure 18.12** shows how the carbohydrate moieties of RBC surface antigens determine the ABO blood types.

Antibodies of the ABO group begin to appear in the plasma 2 to 8 months after birth. They reach their maximum concentrations between 8 and 10 years of age and then slowly decline for the rest of one's life. They are produced mainly in response to bacteria that inhabit the intestines, but they cross-react with RBC antigens and are therefore best known for their significance in transfusions.

Antibodies of the ABO group react against any A or B antigen except one's own. The antibody that reacts against antigen A is called *alpha agglutinin,* or *anti-A;* it is present in the plasma of people with type O or type B blood—that is, anyone who does *not* possess antigen A. The antibody that reacts against antigen B is *beta agglutinin,* or *anti-B,* and is present in type O and type A individuals—those who do not possess antigen B. Each antibody molecule has 10 binding sites where it can attach to either an A or B antigen. An antibody can therefore attach to several RBCs at once and agglutinate them **(fig. 18.13).**

A person's ABO blood type can be determined by placing one drop of blood in a pool of anti-A serum and another drop in a pool of anti-B. Blood type AB exhibits conspicuous agglutination in

TABLE 18.5	The ABO Blood Group			
	ABO Blood Type			
Characteristics	**Type O**	**Type A**	**Type B**	**Type AB**
Possible genotypes*	*ii*	$I^A I^A$ or $I^A i$	$I^B I^B$ or $I^B i$	$I^A I^B$
RBC antigen	Neither A nor B	A	B	Both A and B
Plasma antibody	Anti-A, anti-B	Anti-B	Anti-A	Neither
May safely receive RBCs of	Type O	Type O or A	Type O or B	Type O, A, B, or AB
May safely donate RBCs to	Type O, A, B, or AB	Type A or AB	Type B or AB	Type AB
Frequency in U.S. Population				
White	45%	40%	11%	4%
Black	49%	27%	20%	4%
Hispanic	63%	14%	20%	3%
Japanese	31%	38%	22%	9%
Native American	79%	16%	4%	<1%

*I^A is the dominant allele for agglutinogen A; I^B is the dominant allele for agglutinogen B; and allele *i* is recessive to both of these.

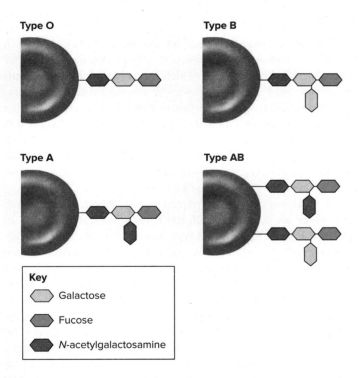

FIGURE 18.12 Chemical Basis of the ABO Blood Types. The terminal carbohydrates of the antigenic glycolipids are shown. All of them end with galactose and fucose (not to be confused with fructose). In type A, the galactose also has *N*-acetylgalactosamine added to it; in type B, it has another galactose; and in type AB, both of these chain types are present.

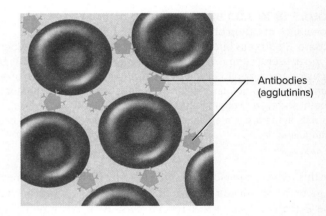

FIGURE 18.13 Agglutination of RBCs by an Antibody. Anti-A and anti-B have 10 binding sites, located at the 2 tips of each of the 5 Ys, and can therefore bind multiple RBCs to each other.

Type O blood is the most common and AB is the rarest in the United States. Percentages differ from one region of the world to another and among ethnic groups. People tend to marry within their locality and ethnic group, thus perpetuating statistical variations particular to that group (table 18.5).

In giving transfusions, it is imperative that the donor's RBCs not agglutinate as they enter the recipient's bloodstream. For example, if type B blood were transfused into a type A recipient, the recipient's anti-B would immediately agglutinate the donor's RBCs (**fig. 18.15**). A mismatch causes a **transfusion reaction**—the agglutinated RBCs block small blood vessels, hemolyze, and release their hemoglobin over the next few hours to days. Free hemoglobin can block the kidney tubules and cause death from acute renal failure within a week or so.

both antisera; type A or B agglutinates only in the corresponding antiserum; and type O doesn't agglutinate in either one (**fig. 18.14**).

FIGURE 18.14 ABO Blood Typing. Each row shows the appearance of a drop of blood mixed with anti-A and anti-B antisera. Blood cells become clumped if they possess the antigens for the antiserum (top row left, second row right, third row both) but otherwise remain uniformly mixed. Thus, type A agglutinates only in anti-A; type B agglutinates only in anti-B; type AB agglutinates in both; and type O agglutinates in neither of them. The antisera in the vials at the top are artificially colored to make them more easily distinguishable.

Claude Revey/Phototake

For this reason, a person with type A (anti-B) blood must never be given a transfusion of type B or AB blood. A person with type B (anti-A) must never receive type A or AB blood. Type O (anti-A and anti-B) individuals cannot safely receive type A, B, or AB blood.

Type AB is sometimes called the *universal recipient* because this blood type lacks both anti-A and anti-B antibodies; thus, it will not agglutinate donor RBCs of any ABO type. However, this overlooks the fact that the *donor's* plasma can agglutinate the *recipient's* RBCs if it contains anti-A, anti-B, or both. For similar reasons, type O is sometimes called the *universal donor.* The plasma of a type O donor, however, can agglutinate the RBCs of a type A, B, or AB recipient. There are procedures for reducing the risk of a transfusion reaction in certain mismatches, however, such as giving packed RBCs with a minimum of plasma.

Contrary to some people's belief, blood type is not changed by transfusion. It is fixed at conception and remains the same for life.

▶▶▶ APPLY WHAT YOU KNOW

Scientists have developed a method of enzymatically splitting N-acetylgalactosamine off the glycolipid of type A blood cells (see fig. 18.12). What clinical benefit do you think they saw as justifying their research effort?

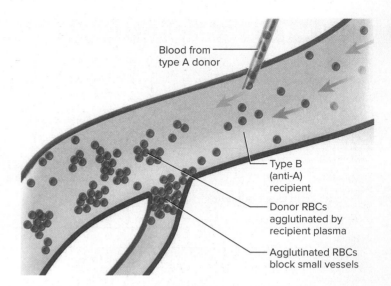

FIGURE 18.15 Effects of a Mismatched Transfusion. Donor RBCs become agglutinated in the recipient's blood plasma. The agglutinated RBCs lodge in smaller blood vessels downstream from this point and cut off the blood flow to vital tissues.

18.3b The Rh Group

The **Rh blood group** is named for the rhesus monkey, in which the Rh antigens were discovered in 1940. This group includes numerous RBC antigens, of which the principal types are antigens C, D, and E. Antigen D is by far the most reactive of these, so a person is considered **Rh-positive (Rh+)** if he or she has the D antigen (genotype *DD* or *Dd*) and **Rh-negative (Rh–)** if it is lacking (genotype *dd*). The Rh blood type is tested by using an anti-D reagent. The Rh type is often combined with the ABO type in a single expression such as O+ for type O, Rh-positive; or AB– for type AB, Rh-negative. Rh frequencies vary among ethnic groups just as ABO frequencies do. About 85% of white Americans are Rh+ and 15% are Rh–, whereas about 99% of Asians are Rh+. ABO blood type has no influence on Rh type, or vice versa. If the frequency of type O whites in the United States is 45%, and 85% of these are also Rh+, then the frequency of O+ individuals is the product of these separate frequencies: $0.45 \times 0.85 = 0.38$, or 38%.

In contrast to the ABO group, anti-D antibodies are not normally present in the blood. They form only in Rh– individuals who are exposed to Rh+ blood. If an Rh– person receives an Rh+ transfusion, the recipient produces anti-D. Since anti-D doesn't appear instantaneously, this presents little danger in the first mismatched transfusion. But if that person should later receive another Rh+ transfusion, his or her anti-D could agglutinate the donor's RBCs.

18.3c Other Blood Groups

ABO and Rh are not the only known blood groups. There are 33 groups in all, with more than 300 RBC antigens in addition to A, B, and D. Other groups include the Duffy, Kell, Kidd, Lewis, and MNS groups. These rarely cause transfusion reactions, but they are useful for such legal purposes as paternity and criminal cases and for research in anthropology and population genetics.

Now that DNA sequencing is more economical, however, it has replaced blood typing in many such applications.

18.3d Maternal–Fetal Mismatches

A condition called **hemolytic disease of the newborn (HDN),** or **erythroblastosis fetalis,** can occur when a woman has a baby with a mismatched blood type—most famously, but by no means all cases, when she is Rh– and carries an Rh+ fetus. The first pregnancy is likely to be uneventful because the placenta normally prevents maternal and fetal blood from mixing. However, if there is placental leakage during pregnancy, or at the time of birth, or if a miscarriage occurs, the mother is exposed to Rh+ fetal blood. She then begins to produce anti-D antibodies **(fig. 18.16).** If she becomes pregnant again with an Rh+ fetus, her anti-D antibodies may pass through the placenta and agglutinate the fetal erythrocytes. Agglutinated RBCs hemolyze, and the baby is born with hemolytic anemia, HDN.

If an Rh– woman has had one or more previous Rh+ pregnancies, her subsequent Rh+ children have about a 17% probability of being born with HDN. Such infants are often severely anemic. As the fetal hematopoietic tissues respond to the need for more RBCs, erythroblasts (immature RBCs) enter the circulation prematurely—hence the name *erythroblastosis fetalis.* Hemolyzed RBCs release hemoglobin, which is converted to bilirubin. High bilirubin levels can cause *kernicterus,* a syndrome of toxic brain damage that can be lethal or leave the child with motor, sensory, and mental deficiencies. HDN can be treated with *phototherapy*—exposing the infant to ultraviolet radiation, which degrades bilirubin as blood passes through capillaries of the skin. In more severe cases, an *exchange transfusion* may be given to completely replace the infant's Rh+ blood with Rh–. In time, the infant's hematopoietic tissues will replace the donor's RBCs with Rh+ cells, and by then the mother's antibody will have disappeared from the infant's blood.

Rh-based HDN, like so many other disorders, is easier to prevent than to treat. If an Rh– woman gives birth to (or miscarries) an Rh+ child, she can be given an *Rh immune globulin* (sold under such trade names as RhoGAM and Gamulin Rh). The immune globulin binds fetal RBC antigens so they cannot stimulate her immune system to produce anti-D. It is now common to give immune globulin at 28 to 32 weeks' gestation and at birth in any pregnancy in which the mother is Rh–.

Although an Rh mismatch produces the most severe HDN, it isn't the most common cause. Two out of three cases are due to ABO mismatches. In about 15% of U.S. pregnancies, the mother is type O and her fetus is type A, B, or AB. About 3% of these

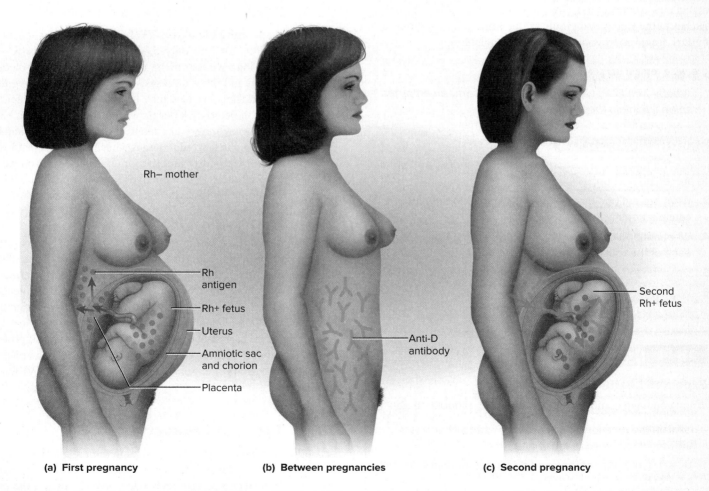

Rh– mother

Rh antigen

Rh+ fetus

Uterus

Amniotic sac and chorion

Placenta

Anti-D antibody

Second Rh+ fetus

(a) First pregnancy   **(b) Between pregnancies**   **(c) Second pregnancy**

FIGURE 18.16 Hemolytic Disease of the Newborn (HDN). (a) When an Rh– woman is pregnant with an Rh+ fetus, she is exposed to D (Rh) antigens, especially during childbirth. (b) Following that pregnancy, her immune system produces anti-D antibodies. (c) If she later becomes pregnant with another Rh+ fetus, her anti-D antibodies can cross the placenta and agglutinate the blood of that fetus, causing that child to be born with HDN.

DEEPER INSIGHT 18.3

CLINICAL APPLICATION

Bone Marrow and Cord Blood Transplants

A bone marrow transplant is one treatment option for leukemia, sickle-cell disease, and some other disorders. The principle is to eradicate the patient's original bone marrow with radiation or chemotherapy, including immune T cells that would attack transplanted marrow, and then replace with donor stem cells in hopes that they will rebuild a population of normal marrow and blood cells. Marrow is drawn from the donor's sternum or hip bone and injected into the recipient's circulation. Donor stem cells colonize the patient's marrow cavities and, ideally, build healthy marrow.

There are, however, drawbacks to this procedure. It is difficult to find compatible donors; surviving T cells in the patient may attack the donor marrow; and donor T cells may attack the patient's tissues (the graft-versus-host response). To inhibit graft rejection, the patient must take immunosuppressant drugs for life. These leave a person vulnerable to infection and have other adverse side effects. Infections are sometimes contracted from the donated marrow itself. In short, marrow transplant is

a high-risk procedure; up to one-third of patients die from complications of treatment (see the David Vetter case in fig 21.29).

An alternative is to use blood from placentas, which are normally discarded at every childbirth. Placental blood contains more stem cells than adult bone marrow, and is less likely to carry infectious microbes. With the parents' consent, it can be harvested from the umbilical cord with a syringe and stored almost indefinitely at cord blood banks. The immature immune cells in cord blood have less tendency to attack the recipient's tissues; thus, cord blood transplants have lower rejection rates and don't require as close a match between donor and recipient, meaning that more donors are available to a patient in need. Pioneered in the 1980s, cord blood transplants have successfully treated leukemia and a wide range of other blood diseases.

The use of cord blood may soon be overshadowed, however, by stem cell harvesting from peripheral blood (drawn from the blood vessels). Peripheral blood is more accessible than bone marrow, and with improvements in technique, it yields faster replacement of hematopoietic stem cells in the recipient.

---

pregnancies also result in HDN, but usually with mild effects. Mismatches in the Kell blood group are the third most common cause of HDN, followed by Kidd and Duffy mismatches.

▶▶▶ APPLY WHAT YOU KNOW

A baby with HDN typically has jaundice and an enlarged spleen. Explain these effects.

BEFORE YOU GO ON

Answer the following questions to test your understanding of the preceding section:

13. What are antibodies and antigens? How do they interact to cause a transfusion reaction?

14. What antibodies and antigens are present in people with each of the four ABO blood types?

15. Describe the causes, prevention, and treatment of HDN.

16. Why might someone be interested in determining a person's blood type other than ABO/Rh?

18.4 Leukocytes

Expected Learning Outcomes

When you have completed this section, you should be able to

a. explain the function of leukocytes in general and the individual role of each leukocyte type;

b. describe the appearance and relative abundance of each type of leukocyte;

c. describe the formation and life history of leukocytes; and

d. discuss the types, causes, and effects of leukocyte excesses and deficiencies.

18.4a Leukocyte Form and Function

Leukocytes, or **white blood cells (WBCs),** are the least abundant formed elements, totaling only 5,000 to 10,000 WBCs/μL. Yet we cannot live long without them, because they afford protection against infection and other diseases. WBCs are easily recognized in stained blood films because they have conspicuous nuclei that stain from light violet to dark purple with the most common blood stains. They are much more abundant in the body than their low number in blood films would suggest, because they spend only a few hours in the bloodstream, then migrate into the connective tissues and spend the rest of their lives there. It's as if the bloodstream were merely the subway that the WBCs take to work; in blood films, we see only the ones on their way to work, not the WBCs already at work in the tissues.

Leukocytes differ from erythrocytes in that they retain their organelles throughout life; thus, when viewed with the transmission electron microscope, they show a complex internal structure **(fig. 18.17).** Among their organelles are the usual instruments of protein synthesis—the nucleus, rough endoplasmic reticulum, ribosomes, and Golgi complex—for leukocytes must synthesize proteins in order to carry out their functions. Some of these proteins are packaged into lysosomes and other organelles, which appear as conspicuous cytoplasmic granules that distinguish one WBC type from another.

18.4b Types of Leukocytes

As outlined at the beginning of this chapter, there are five kinds of leukocytes. They are distinguished from each other by their relative size and abundance, the size and shape of their nuclei, the presence or absence of certain cytoplasmic granules, the coarseness and staining properties of those granules, and most importantly by their functions. Individual WBC types rise or fall in number in various disease conditions and physiological states.

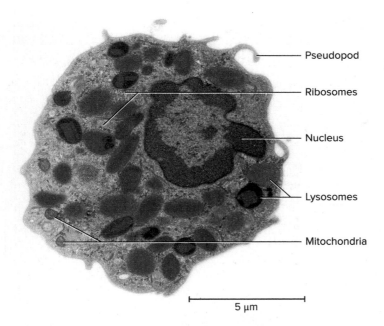

Pseudopod

Ribosomes

Nucleus

Lysosomes

Mitochondria

5 μm

FIGURE 18.17 The Structure of a Leukocyte (TEM). This example is an eosinophil. The lysosomes seen here in orange are the coarse pink granules seen in the eosinophil in table 18.6.

Scott Camazine/Alamy

All WBCs have lysosomes called **nonspecific granules** in the cytoplasm, also called *azurophilic*[18] *granules* because they absorb the blue or violet dyes of blood stains. Three of the five types—neutrophils, eosinophils, and basophils—are called **granulocytes** because they also have various kinds of **specific granules** that stain conspicuously and distinguish each cell type from the others. Basophils are named for the fact that their specific granules stain with methylene blue, a basic dye in a common blood-staining mixture called Wright's stain. Eosinophils are so named because they stain with eosin, an acidic dye in Wright's stain. The specific granules of neutrophils don't stain intensely with either basic or acidic stains. The colors in the following descriptions and table are those typically seen with Wright's stain, but may differ on slides you examine because of the use of other stains. Specific granules contain enzymes and other chemicals employed in defense against pathogens. The two remaining WBC types—monocytes and lymphocytes—are called **agranulocytes** because they lack specific granules. Nonspecific granules are inconspicuous to the light microscope, and these cells therefore have relatively clear-looking cytoplasm. **Table 18.6** shows the appearance and summarizes the characteristics and functions of each of the following WBC types.

Granulocytes

- **Neutrophils** (NEW-tro-fills) are the most abundant WBCs—generally about 4,150 cells/μL and constituting 60% to 70% of the circulating leukocytes. The nucleus is clearly visible and, in a mature neutrophil, typically consists of three to five lobes connected by slender nuclear strands. These strands are sometimes so delicate that they are scarcely visible, and

the neutrophil may seem as if it had multiple nuclei. Young neutrophils have an undivided band-shaped nucleus and are called *band cells*. Neutrophils are also called *polymorphonuclear leukocytes (PMNs)* because of their varied nuclear shapes. The cytoplasm contains fine reddish to violet specific granules, which contain lysozyme and other antimicrobial agents. The individual granules are barely visible with the light microscope, but their combined effect gives the cytoplasm a pale lilac color. Neutrophils are aggressively antibacterial cells. Their numbers rise—a condition called *neutrophilia*—in response to bacterial infections. They destroy bacteria in ways detailed in section 21.2b.

- **Eosinophils** (EE-oh-SIN-oh-fills) are harder to find in a blood film because they are only 2% to 4% of the WBC total, typically averaging about 170 cells/μL. Although relatively scanty in the blood, eosinophils are abundant in the mucous membranes of the respiratory, digestive, and lower urinary tracts. The eosinophil nucleus usually has two large lobes connected by a thin strand, and the cytoplasm has an abundance of coarse rosy to orange-colored specific granules. Eosinophils secrete chemicals that weaken or destroy relatively large parasites such as hookworms and tapeworms, too big for any one WBC to phagocytize. They also phagocytize and dispose of inflammatory chemicals, antigen–antibody complexes, and allergens (foreign antigens that trigger allergies). Allergies, parasitic infections, collagen diseases, and diseases of the spleen and central nervous system can cause an elevated eosinophil count called *eosinophilia*. The eosinophil count also fluctuates greatly from day to night, seasonally, and with the phase of the menstrual cycle.

- **Basophils** are the rarest of all formed elements. They average about 40 cells/μL and usually constitute less than 0.5% of the WBC count. They can be recognized mainly by an abundance of very coarse, dark violet specific granules. The nucleus is largely hidden from view by these granules, but is large, pale, and typically S- or U-shaped. Basophils secrete two chemicals that aid in the body's defense processes: (1) **histamine,** a vasodilator that widens the blood vessels, speeds the flow of blood to an injured tissue, and makes the blood vessels more permeable so that blood components such as neutrophils and clotting proteins can get into the connective tissues more quickly; and (2) **heparin,** an anticoagulant that inhibits blood clotting and thus promotes the mobility of other WBCs in the area. They also release chemical signals that attract eosinophils and neutrophils to a site of infection.

Agranulocytes

- **Lymphocytes** (LIM-fo-sites) are second to neutrophils in abundance and are thus quickly spotted when you examine a blood film. They average about 2,200 cells/μL and are 25% to 33% of the WBC count. They include the smallest WBCs; at 5 to 17 μm in diameter, they range from smaller than RBCs to two and a half times as large. They are sometimes classified into three size classes (table 18.6), but there are gradations between them. Medium and large lymphocytes are usually seen in fibrous connective tissues and only occasionally in the

[18]*azuro* = blue; *philic* = loving

TABLE 18.6	The White Blood Cells (Leukocytes)

Neutrophils

Differential count (% of WBCs)	60% to 70%
Mean absolute count	4,150 cells/µL
Diameter	9–12 µm

Appearance

Nucleus usually with 3–5 lobes in S- or C-shaped array

Fine reddish to violet specific granules in cytoplasm

Variations in Number

Increase in bacterial infections

Functions

Phagocytize bacteria

Release antimicrobial chemicals

Neutrophil   10 µm

Eosinophils

Differential count (% of WBCs)	2% to 4%
Mean absolute count	165 cells/µL
Diameter	10–14 µm

Appearance

Nucleus usually has two large lobes connected by thin strand

Large orange-pink specific granules in cytoplasm

Variations in Number

Fluctuate greatly from day to night, seasonally, and with phase of menstrual cycle

Increases in parasitic infections, allergies, collagen diseases, and diseases of spleen and central nervous system

Functions

Phagocytize antigen–antibody complexes, allergens, and inflammatory chemicals

Release enzymes that weaken or destroy parasites such as worms

Eosinophil   10 µm

Basophils

Differential count (% of WBCs)	< 0.5%
Mean absolute count	44 cells/µL
Diameter	8–10 µm

Appearance

Nucleus large and U- to S-shaped, but typically pale and obscured from view

Coarse, abundant, dark violet specific granules in cytoplasm

Variations in Number

Relatively stable

Increase in chickenpox, sinusitis, diabetes mellitus, myxedema, and polycythemia

Functions

Secrete histamine (a vasodilator), which increases blood flow to a tissue

Secrete heparin (an anticoagulant), which promotes mobility of other WBCs by preventing clotting

Basophils   10 µm

circulating blood (see fig. 18.1). The lymphocytes seen in blood films are mostly in the small size class. These are sometimes difficult to distinguish from basophils, but most basophils are conspicuously grainy, whereas the lymphocyte nucleus is uniform or merely mottled. Basophils also lack the rim of clear cytoplasm seen in most lymphocytes. Large lymphocytes are sometimes difficult to distinguish from monocytes. The lymphocyte

TABLE 18.6	The White Blood Cells (Leukocytes) *(continued)*

Lymphocytes

Differential count (% of WBCs)	25% to 33%
Mean absolute count	2,185 cells/µL
Diameter	
Small class	5–8 µm
Medium class	10–12 µm
Large class	14–17 µm

Appearance

Nucleus round, ovoid, or slightly dimpled on one side, of uniform or mottled dark violet color

In small lymphocytes, nucleus fills nearly all of the cell and leaves only a scanty rim of clear, light blue cytoplasm.

In larger lymphocytes, cytoplasm is more abundant; large lymphocytes may be hard to differentiate from monocytes.

Variations in Number

Increase in diverse infections and immune responses

Functions

Several functional classes usually indistinguishable by light microscopy

Destroy cancer cells, cells infected with viruses, and foreign cells

Present antigens to activate other cells of immune system

Coordinate actions of other immune cells

Secrete antibodies

Serve in immune memory

Lymphocyte · 10 µm

Monocytes

Differential count (% of WBCs)	3% to 8%
Mean absolute count	456 cells/µL
Diameter	12–15 µm

Appearance

Nucleus ovoid, kidney-shaped, or horseshoe-shaped; violet

Abundant cytoplasm with sparse, fine nonspecific granules

Sometimes very large with stellate or polygonal shapes

Variations in Number

Increase in viral infections and inflammation

Functions

Differentiate into macrophages (large phagocytic cells of the tissues)

Phagocytize pathogens, dead neutrophils, and debris of dead cells

Present antigens to activate other cells of immune system

Monocyte · 10 µm

nucleus is round, ovoid, or slightly dimpled on one side, and usually stains dark violet. In small lymphocytes, it fills nearly the entire cell and leaves only a narrow rim of light blue cytoplasm, often barely detectable, around the cell perimeter. The cytoplasm is more abundant in medium and large lymphocytes. There are several subclasses of lympho-cytes with different immune functions (see section 21.1b), but they cannot be distinguished by microscopic examina-tion of blood films. Collectively, they destroy cells that have been infected with viruses or turned malignant, and foreign cells (including parasites) that have been introduced into the body; they secrete chemicals that communicate with other

WBCs and coordinate their actions; they present antigens to activate immune responses; they give rise to the cells that secrete antibodies; and they provide long-term immunity to pathogens.

- **Monocytes** (MON-oh-sites) are usually the largest WBCs seen on a blood slide, often two or three times the diameter of an RBC. They average about 460 cells/μL and about 3% to 8% of the WBC count. The nucleus is large and clearly visible, often a relatively light violet, and typically ovoid, kidney-shaped, or horseshoe-shaped. The cytoplasm is abundant and contains sparse, fine granules. In prepared blood films, monocytes often assume sharply angular to spiky shapes (see fig. 18.1). The monocyte count rises in inflammation and viral infections. Monocytes go to work only after leaving the bloodstream and transforming into large tissue cells called **macrophages** (MAC-ro-fay-jez). Macrophages are highly phagocytic cells that consume dead or dying host and foreign cells, pathogenic chemicals and microorganisms, and other foreign matter equivalent to as much as 25% of their own volume per hour. They also chop up or process foreign antigens and display fragments of them on the cell surface to alert the immune system to the presence of a pathogen. Thus, they and a few other cells are called *antigen-presenting cells (APCs)*. The functions of macrophages are further detailed in chapter 21.

18.4c The Leukocyte Life History

Leukopoiesis (LOO-co-poy-EE-sis), the production of white blood cells, begins with the same hematopoietic stem cells (HSCs) as erythropoiesis. Some HSCs differentiate into distinct types of colony-forming units (CFUs) and then go on to produce the following cell lines (yellow zone in **fig. 18.18**), each of them now irreversibly committed to a certain outcome.

1. *Myeloblasts,* which ultimately differentiate into the three types of granulocytes (neutrophils, eosinophils, and basophils)

2. *Monoblasts,* which look identical to myeloblasts but lead ultimately to monocytes

3. *Lymphoblasts,* which produce all lymphocyte types

CFUs have receptors for *colony-stimulating factors (CSFs)*. Mature lymphocytes and macrophages secrete several types of CSFs in response to infections and other immune challenges. Each CSF stimulates a different WBC type to develop in response to specific needs. Thus, a bacterial infection may trigger the production of neutrophils, whereas an allergy stimulates eosinophil production, each process working through its own CSF.

The red bone marrow stores granulocytes and monocytes until they are needed and contains 10 to 20 times more of these cells than the circulating blood does. Lymphocytes begin developing in the bone marrow but don't stay there. Some types mature there and others migrate to the thymus to complete their development. Mature lymphocytes from both locations then colonize the spleen, lymph nodes, and other lymphatic organs and tissues.

Circulating leukocytes don't stay in the blood for very long. Granulocytes circulate for 4 to 8 hours and then migrate into the tissues, where they live another 4 or 5 days. Monocytes travel in the blood for 10 to 20 hours, then migrate into the tissues and transform into a variety of macrophages, which can live as long as a few years. Lymphocytes, responsible for long-term immunity, survive from a few weeks to decades; they leave the bloodstream for the tissues and eventually enter the lymphatic system, which empties them back into the bloodstream. Thus, they are continually recycled from blood to tissue fluid to lymph and back to the blood.

When leukocytes die, they are generally phagocytized and digested by macrophages. Dead neutrophils, however, are responsible for the creamy color of pus, and are sometimes disposed of by the rupture of a blister onto the skin surface.

▶▶▶ APPLY WHAT YOU KNOW

It is sometimes written that RBCs do not live as long as WBCs because RBCs do not have a nucleus and therefore cannot repair and maintain themselves. Explain the flaw in this argument.

18.4d Leukocyte Disorders

The total WBC count is normally 5,000 to 10,000 WBCs/μL. A count below this range, called **leukopenia**[19] (LOO-co-PEE-nee-uh), is seen in lead, arsenic, and mercury poisoning; radiation sickness; and such infectious diseases as measles, mumps, chickenpox, polio, influenza, typhoid fever, and AIDS. It can also be produced by glucocorticoids, anticancer drugs, and immunosuppressant drugs given to organ-transplant patients. Since WBCs are protective cells, leukopenia presents an elevated risk of infection and cancer. A count above 10,000 WBCs/μL, called **leukocytosis,**[20] usually indicates infection, allergy, or other diseases but can also occur in response to dehydration or emotional disturbances. More useful than a total WBC count is a *differential WBC count,* which identifies what percentage of the total WBC count consists of each type of leukocyte (see Deeper Insight 18.4).

Leukemia is a cancer of the hematopoietic tissues that usually produces an extraordinarily high number of circulating leukocytes and their precursors **(fig. 18.19).** Leukemia is classified as myeloid or lymphoid, acute or chronic. **Myeloid leukemia** is marked by uncontrolled granulocyte production, whereas **lymphoid leukemia** involves uncontrolled lymphocyte or monocyte production. **Acute leukemia** appears suddenly, progresses rapidly, and causes death within a few months if it is not treated. **Chronic leukemia** develops more slowly and may go undetected for many months; if untreated, the typical survival time is about 3 years. Both myeloid and lymphoid leukemia occur in acute and chronic forms. The greatest success in treatment and cure has been with acute lymphoblastic leukemia, the most common type of childhood cancer. Treatment employs chemotherapy and marrow transplants along with the control of side effects such as anemia, hemorrhaging, and infection.

As leukemic cells proliferate, they replace normal bone marrow and a person suffers from a deficiency of normal granulocytes,

---

[19]*leuko* = white; *penia* = deficiency
[20]*leuko* = white; *cyt* = cell; *osis* = condition

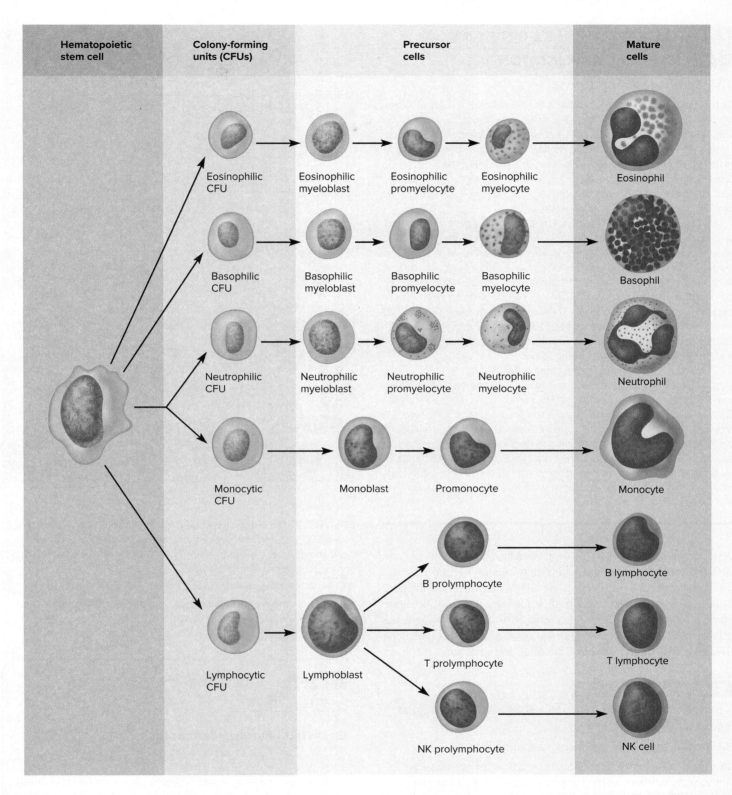

Hematopoietic stem cell	Colony-forming units (CFUs)	Precursor cells			Mature cells
	Eosinophilic CFU	Eosinophilic myeloblast	Eosinophilic promyelocyte	Eosinophilic myelocyte	Eosinophil
	Basophilic CFU	Basophilic myeloblast	Basophilic promyelocyte	Basophilic myelocyte	Basophil
	Neutrophilic CFU	Neutrophilic myeloblast	Neutrophilic promyelocyte	Neutrophilic myelocyte	Neutrophil
	Monocytic CFU	Monoblast	Promonocyte		Monocyte
	Lymphocytic CFU	Lymphoblast	B prolymphocyte		B lymphocyte
			T prolymphocyte		T lymphocyte
			NK prolymphocyte		NK cell

FIGURE 18.18 Leukopoiesis. Stages in the development of white blood cells. The hematopoietic stem cell at the left also is the ultimate source of red blood cells (see fig. 18.6) and platelet-producing cells.

❷ *Explain the meaning and relevance of the combining form* myelo- *seen in so many of these cell names.*

erythrocytes, and platelets. Although enormous numbers of leukocytes are produced and spill over into the bloodstream, they don't provide the usual protective functions of WBCs. They are like an army of children, present in vast numbers but too immature to perform a useful defensive role. The deficiency of competent WBCs leaves the patient vulnerable to **opportunistic infection**—the establishment of pathogenic organisms that usually cannot get a foothold in people with healthy immune systems. The RBC

DEEPER INSIGHT 18.4

CLINICAL APPLICATION

The Complete Blood Count

One of the most common laboratory tests in both routine medical examinations and the diagnosis of disease is a *complete blood count (CBC)*. The CBC yields a highly informative profile of data on multiple blood values: the number of RBCs, WBCs, and platelets per microliter of blood; the relative numbers (percentages) of each WBC type, called a *differential WBC count;* hematocrit; hemoglobin concentration; and various *RBC indices* such as RBC size *(mean corpuscular volume, MCV)* and hemoglobin concentration per RBC *(mean corpuscular hemoglobin, MCH)*.

RBC and WBC counts used to require the microscopic examination of films of diluted blood on a calibrated slide, and a differential WBC count required examination of stained blood films. Today, most laboratories use *electronic cell counters*. These devices draw a blood sample through a very narrow tube with sensors that identify cell types and measure cell sizes and hemoglobin content. These counters give faster and more accurate results based on much larger numbers of cells than the old visual methods. However, cell counters still misidentify some cells, and a medical technologist must review the results for suspicious abnormalities and identify cells that the instrument cannot.

The wealth of information gained from a CBC is too vast to give more than a few examples here. Various forms of anemia are indicated by low RBC counts or abnormalities of RBC size, shape, and hemoglobin content. A platelet deficiency can indicate an adverse drug reaction. A high neutrophil count suggests bacterial infection, and a high eosinophil count suggests an allergy or parasitic infection. Elevated numbers of specific WBC types or WBC stem cells can indicate various forms of leukemia. If a CBC does not provide enough information or if it suggests other disorders, additional tests may be done, such as coagulation time and bone marrow biopsy.

deficiency renders the patient anemic and fatigued, and the platelet deficiency results in hemorrhaging and impaired blood clotting. The immediate cause of death is usually hemorrhage or infection. Cancerous hematopoietic tissue often metastasizes from the bone marrow or lymph nodes to other organs of the body, where the cells displace or compete with normal cells. Metastasis to the bone tissue itself is common and leads to bone and joint pain.

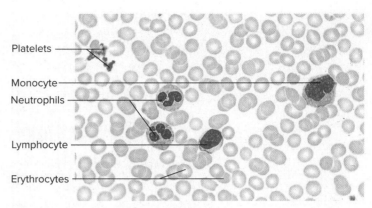

Platelets
Monocyte
Neutrophils
Lymphocyte
Erythrocytes

(a)

(b)   75 µm

FIGURE 18.19 Normal and Leukemic Blood. (a) A normal blood smear. (b) Blood from a patient with acute monocytic leukemia. Note the abnormally high number of white blood cells, especially monocytes, in part (b).

a: Ed Reschke; **b:** Leonard Lessin/Science Source

 With all these extra white cells, why isn't the body's infection-fighting capability increased in leukemia?

BEFORE YOU GO ON

Answer the following questions to test your understanding of the preceding section:

17. What is the overall function of leukocytes?

18. List the five kinds of leukocytes in order of abundance, identify whether each is a granulocyte or agranulocyte, and describe the functions of each one.

19. What does leukopoiesis have in common with erythropoiesis? How does it differ?

20. What can cause an abnormally high or low WBC count?

21. Suppose myeloblasts began multiplying out of control, but their subsequent development remained normal. What types of mature WBCs would be produced in excess? What types would not?

18.5 Platelets and the Control of Bleeding

Expected Learning Outcomes

When you have completed this section, you should be able to

a. describe the body's mechanisms for controlling bleeding;

b. list the functions of platelets;

c. describe two reaction pathways that produce blood clots;

d. explain what happens to blood clots when they are no longer needed;

e. explain what keeps blood from clotting in the absence of injury; and

f. describe some disorders of blood clotting.

Circulatory systems developed very early in animal evolution, and with them evolved mechanisms for stopping leaks, which are potentially fatal. **Hemostasis**[21] is the cessation of bleeding. Although hemostatic mechanisms may not stop a hemorrhage from a large blood vessel, they are quite effective at closing breaks in small ones. Platelets play multiple roles in hemostasis, so we begin with a consideration of their form and function.

18.5a Platelet Form and Function

Platelets are not cells but small fragments of marrow cells called *megakaryocytes.* They are the second most abundant formed elements, after erythrocytes; a normal platelet count in blood from a fingerstick ranges from 130,000 to 400,000 platelets/μL (averaging about 250,000). The platelet count can vary greatly, however, under different physiological conditions and in blood samples taken from various places in the body. In spite of their numbers, platelets are so small (2 to 4 μm in diameter) that they contribute even less than WBCs to the blood volume.

Platelets have a complex internal structure that includes lysosomes, mitochondria, microtubules, and microfilaments; **granules** filled with platelet secretions; and a system of channels called the **open canalicular system,** which opens onto the platelet surface **(fig. 18.20a).** They have no nucleus. When activated, they form pseudopods and are capable of ameboid movement.

Despite their small size, platelets have a greater variety of functions than any of the true blood cells:

- They secrete *vasoconstrictors,* chemicals that stimulate spasmodic constriction of broken vessels and thereby help to reduce blood loss.

- They stick together to form temporary *platelet plugs* that seal small breaks in injured blood vessels.

- They secrete *procoagulants,* or clotting factors, which promote blood clotting.

- They initiate the formation of a clot-dissolving enzyme that dissolves blood clots that have outlasted their usefulness.

- They secrete chemicals that attract neutrophils and monocytes to sites of inflammation.

- They internalize and destroy bacteria.

- They secrete *growth factors* that stimulate mitosis in fibroblasts and smooth muscle and thereby help to maintain and repair blood vessels.

18.5b Platelet Production

The production of platelets is a division of hematopoiesis called **thrombopoiesis** (THROM-bo-poy-EE-sis). (Platelets are occasionally called *thrombocytes.*[22]) Thrombopoiesis is stimulated by a hormone from the liver and kidneys called *thrombopoietin* (THROM-bo-POY-eh-tin). Some hematopoietic stem cells develop receptors for thrombopoietin and, under its influence, become *megakaryoblasts*—cells committed to the platelet-producing line.

(a)

(b)

FIGURE 18.20 Platelets. (a) Structure of blood platelets (TEM). (b) Platelets being produced by the shearing of proplatelets from a megakaryocyte. Note the sizes of the megakaryocyte and platelets relative to RBCs and WBCs. **APR**

a: NIBSC/Science Photo Library/Science Source

The megakaryoblast duplicates its DNA repeatedly without undergoing nuclear or cytoplasmic division. The result is a **megakaryocyte**[23] (MEG-ah-CAR-ee-oh-site), a gigantic cell up to 150 μm in diameter, visible to the naked eye, with a huge multilobed nucleus and multiple sets of chromosomes **(fig. 18.20b).** Most megakaryocytes live in the red bone marrow adjacent to blood-filled spaces called *sinusoids,* lined with a thin simple squamous epithelium called the *endothelium* (see fig. 21.8). Recent animal research has discovered, however, that megakaryocytes migrate freely between the lungs and bone marrow and produce most platelets while in the lungs. Whether this is also true in humans awaits confirmation.

A megakaryocyte sprouts long tendrils called *proplatelets* that protrude through the endothelium into the blood of the sinusoid. The blood flow shears off the proplatelets, which break up into platelets as they travel in the bloodstream. Much of this breakup is thought to occur when they pass through the small vessels of the lungs, because blood counts show more proplatelets entering the

---

[21]*hemo* = blood; *stasis* = stability
[22]*thrombo* = clotting; *cyte* = cell

[23]*mega* = giant; *karyo* = nucleus; *cyte* = cell

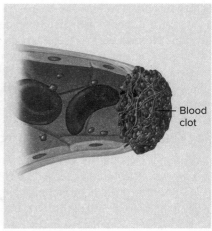

(a) Vascular spasm

(b) Platelet plug formation

(c) Coagulation

FIGURE 18.21 Hemostasis. (a) Vasoconstriction of a broken vessel reduces bleeding. (b) A platelet plug forms as platelets adhere to exposed collagen fibers of the vessel wall. The platelet plug temporarily seals the break. (c) A blood clot forms as platelets become enmeshed in fibrin threads. This forms a longer-lasting seal and gives the vessel a chance to repair itself.

❓ *How does a blood clot differ from a platelet plug?*

lungs than leaving and more platelets exiting. About 25% to 40% of the platelets are stored in the spleen and released as needed. The remainder circulate freely in the blood and live for about 5 to 6 days. Anything that interferes with platelet production can produce a dangerous platelet deficiency called **thrombocytopenia**[24] (see table 18.8).

18.5c Hemostasis

There are three hemostatic mechanisms—*vascular spasm, platelet plug formation,* and *blood clotting (coagulation)* (**fig. 18.21**). Platelets play an important role in all three.

Vascular Spasm

The most immediate protection against blood loss is **vascular spasm,** a prompt constriction of the broken vessel. Several things trigger this reaction. An injury stimulates pain receptors, some of which directly innervate nearby blood vessels and cause them to constrict. This effect lasts only a few minutes, but other mechanisms take over by the time it subsides. Injury to the smooth muscle of the blood vessel itself causes a longer-lasting vasoconstriction, and platelets release serotonin, a chemical vasoconstrictor. Thus, the vascular spasm is maintained long enough for the other two hemostatic mechanisms to come into play.

Platelet Plug Formation

Platelets don't adhere to the endothelium that lines healthy blood vessels and the heart. The endothelium is normally very smooth and coated with **prostacyclin,** a platelet repellent. When a vessel is broken, however, collagen fibers of its wall are exposed to

the blood. Upon contact with collagen or other rough surfaces, platelets grow long spiny pseudopods that adhere to the vessel and to other platelets; the pseudopods then contract and draw the walls of the vessel together. The mass of platelets thus formed, called a **platelet plug,** may reduce or stop minor bleeding. The platelet plug is looser and more delicate than the blood clot to follow; for this reason, a bleeding injury should be blotted with absorbent paper rather than wiped.

As platelets aggregate, they undergo **degranulation**—the exocytosis of their cytoplasmic granules and release of factors that promote hemostasis. Among these are serotonin, a vasoconstrictor; adenosine diphosphate (ADP), which attracts more platelets to the area and stimulates their degranulation; and **thromboxane A$_2$,** an eicosanoid that promotes platelet aggregation, degranulation, and vasoconstriction. Thus, a positive feedback cycle is activated that can quickly seal a small break in a blood vessel.

Coagulation

Coagulation (clotting) of the blood is the last but most effective defense against bleeding. It is important for the blood to clot quickly when a vessel has broken, but equally important for it not to clot in the absence of vessel damage. Because of this delicate balance, coagulation is one of the most complex processes in the body, involving over 30 chemical reactions. It is presented here in a very simplified form.

Perhaps clotting is best understood if we first consider its goal. The objective is to convert the plasma protein fibrinogen into **fibrin,** a sticky protein that adheres to the walls of a vessel. As blood cells and platelets arrive, they stick to the fibrin like insects in a spider web (**fig. 18.22,** inset). The resulting mass of fibrin, blood cells, and platelets ideally seals the break in the blood vessel. The complexity of clotting lies in how the fibrin is formed.

[24]*thrombo* = clotting; *cyto* = cell; *penia* = deficiency

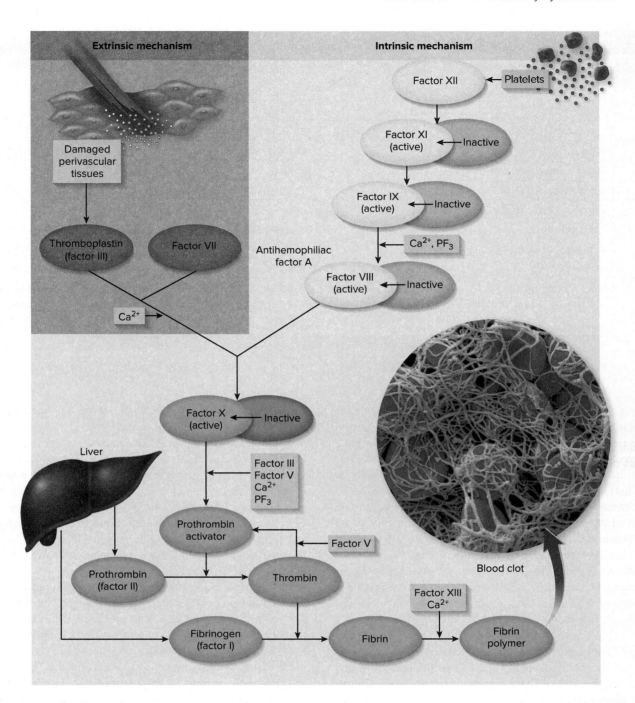

FIGURE 18.22 The Pathways of Coagulation. Most clotting factors act as enzymes that convert the next factor from an inactive form to an active form. One enzyme molecule at any given level activates many enzyme molecules at the next level down, so the overall effect becomes amplified at each step. Inset: Erythrocytes trapped in a mesh of sticky fibrin polymer.

(inset): Steve Gschmeissner/Science Photo Library/Getty Images

❓ *After you read about hemophilia C later in section 18.5f, explain whether it would affect the extrinsic mechanism, the intrinsic mechanism, or both.*

There are two reaction pathways to coagulation. One of them, the **extrinsic mechanism** (left side of fig. 18.22), is initiated by clotting factors released by the damaged blood vessel and peri-vascular[25] tissues. The word *extrinsic* refers to the fact that these factors come from sources external to the blood itself. Blood may also clot, however, without these tissue factors—for example, when platelets adhere to a fatty plaque of atherosclerosis or to a test tube. The reaction pathway in this case is called the **intrinsic mechanism** (right side of the figure) because it uses only clotting factors found in the blood itself. In most cases of bleeding, both the extrinsic and intrinsic mechanisms work simultaneously and interact with each other to achieve hemostasis.

—————————————————

[25]*peri* = around; *vas* = vessel; *cular* = pertaining to

TABLE 18.7	Clotting Factors (Procoagulants)		
Number	Name	Origin	Function
I	Fibrinogen	Liver	Precursor of fibrin
II	Prothrombin	Liver	Precursor of thrombin
III	Tissue thromboplastin	Perivascular tissue	Activates factor VII
V	Proaccelerin	Liver	Activates factor VII; combines with factor X to form prothrombin activator
VII	Proconvertin	Liver	Activates factor X in extrinsic pathway
VIII	Antihemophiliac factor A	Liver	Activates factor X in intrinsic pathway
IX	Antihemophiliac factor B	Liver	Activates factor VIII
X	Thrombokinase	Liver	Combines with factor V to form prothrombin activator
XI	Antihemophiliac factor C	Liver	Activates factor IX
XII	Hageman factor	Liver, platelets	Activates factor XI and plasmin; converts prekallikrein to kallikrein
XIII	Fibrin-stabilizing factor	Platelets, plasma	Cross-links fibrin filaments to make fibrin polymer and stabilize clot
PF_1	Platelet factor 1	Platelets	Same role as factor V; also accelerates platelet activation
PF_2	Platelet factor 2	Platelets	Accelerates thrombin formation
PF_3	Platelet factor 3	Platelets	Aids in activation of factor VIII and prothrombin activator
PF_4	Platelet factor 4	Platelets	Binds heparin during clotting to inhibit its anticoagulant effect

Clotting factors are called **procoagulants,** in contrast to the **anticoagulants** discussed later (see Deeper Insight 18.6). Most procoagulants are proteins produced by the liver **(table 18.7).** They are always present in the plasma in inactive form, but when one factor is activated, it functions as an enzyme that activates the next one in the pathway. That factor activates the next, and so on, in a sequence called a **reaction cascade**—a series of reactions, each of which depends on the product of the preceding one. Many of the clotting factors are identified by roman numerals, which indicate the order in which they were discovered, not the order of the reactions. Factors IV and VI are not included in this table. These terms were abandoned when it was found that factor IV was calcium and factor VI was activated factor V. The last four procoagulants in the table are called *platelet factors* (PF_1 through PF_4) because they are produced by the platelets.

Initiation of Coagulation The extrinsic mechanism is diagrammed on the top left side of figure 18.22. The damaged blood vessel and perivascular tissues release a lipoprotein mixture called **tissue thromboplastin**[26] **(factor III).** Factor III combines with factor VII to form a complex that, in the presence of Ca^{2+}, activates factor X. The extrinsic and intrinsic pathways differ only in how they arrive at active factor X. Therefore, before examining their common pathway from factor X to the end, let's consider how the intrinsic pathway reaches this step.

The intrinsic mechanism is diagrammed on the top right side of figure 18.22. Everything needed to initiate it is present in the plasma or platelets. When platelets degranulate, they release factor XII (Hageman factor, named for the patient in whom it was discovered). Through a cascade of reactions, this leads to activated factors XI, IX, and VIII, in that order—each serving as an enzyme that catalyzes the next step—and finally to factor X. This pathway also requires Ca^{2+} and PF_3.

Completion of Coagulation Once factor X is activated, the remaining events are identical in the intrinsic and extrinsic mechanisms (bottom half of fig. 18.22). Factor X combines with factors III and V in the presence of Ca^{2+} and PF_3 to produce *prothrombin activator.* This enzyme acts on a globulin called **prothrombin (factor II)** and converts it to the enzyme **thrombin.** Thrombin then converts fibrinogen into shorter strands of *fibrin monomer.* These monomers then covalently bond to each other end to end and form longer fibers of *fibrin polymer.* Factor XIII cross-links these strands to create a dense aggregation that forms the structural framework of the blood clot.

Once a clot begins to form, it launches a self-accelerating positive feedback process that seals off the damaged vessel more quickly. Thrombin works with factor V to accelerate the production of prothrombin activator, which in turn produces more thrombin.

The cascade of enzymatic reactions acts as an amplifying mechanism to ensure the rapid clotting of blood **(fig. 18.23).** Each activated enzyme in the pathway produces a larger number of enzyme molecules at the following step. One activated molecule of factor XII at the start of the intrinsic pathway, for example, very quickly produces thousands if not millions of fibrin molecules. Note the similarity of this process to the *signal amplification* that occurs in hormone action (see fig. 17.21).

Notice that the extrinsic mechanism requires fewer steps to activate factor X than the intrinsic mechanism does; it is a "short-cut" to coagulation. It takes 3 to 6 minutes for a clot to form by the intrinsic pathway but only 15 seconds or so by the extrinsic pathway. For this reason, when a small wound bleeds, you can stop the bleeding sooner by massaging the site. This releases thromboplastin from the perivascular tissues and activates or speeds up the extrinsic pathway.

After a clot has formed, spinous pseudopods of the platelets adhere to strands of fibrin and contract. This pulls on the fibrin

[26]*thrombo* = clot; *plast* = forming; *in* = substance

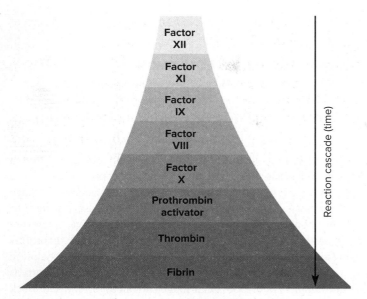

FIGURE 18.23 The Reaction Cascade in Blood Clotting. Each clotting factor produces many molecules of the next one, so the number of active clotting factors increases rapidly and a large amount of fibrin is quickly formed. The example shown here is for the intrinsic mechanism.

? *How does this compare with signal amplification in hormone action (compare fig. 17.21)?*

threads and draws the edges of the broken vessel together, like a drawstring closing a purse. Through this process of **clot retraction,** the clot becomes more compact within about 30 minutes.

A number of laboratory tests are used to evaluate the efficiency of coagulation. Normally, the bleeding of a fingerstick should stop within 2 to 3 minutes, and a sample of blood in a clean test tube should clot within 15 minutes. **Bleeding time** is most precisely measured by the *Ivy method*—inflating a blood pressure cuff on the arm to 40 mm Hg, making a 1 mm deep incision in the forearm, and measuring the time for it to stop bleeding. Normally it should stop in 1 to 9 minutes. Other techniques are available that can separately assess the effectiveness of the intrinsic and extrinsic mechanisms.

18.5d The Fate of Blood Clots

Once a clot has formed, a process of local tissue repair begins. Platelets and endothelial cells secrete a mitotic stimulant named **platelet-derived growth factor (PDGF).** PDGF stimulates fibroblasts and smooth muscle cells to multiply and repair the damaged blood vessel. Fibroblasts also invade the clot and produce fibrous connective tissue, which helps to strengthen and seal the vessel while the repairs take place.

Eventually, tissue repair is completed and the clot must be disposed of. **Fibrinolysis,** the dissolution of a clot, is achieved by a small cascade of reactions with a positive feedback component **(fig. 18.24).** In addition to promoting clotting, factor XII catalyzes the formation of a plasma enzyme called **kallikrein** (KAL-ih-KREE-in). Kallikrein, in turn, converts the inactive protein *plasminogen* into **plasmin,** a fibrin-dissolving enzyme that breaks up the clot. Thrombin also activates plasmin, and plasmin indirectly promotes the formation of more kallikrein, thus completing a positive feedback loop.

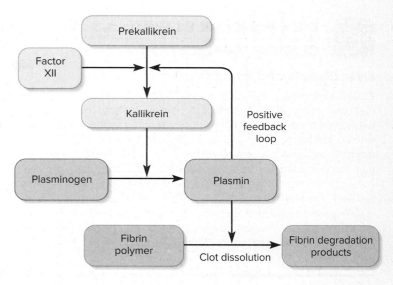

FIGURE 18.24 The Mechanism for Dissolving Blood Clots. Prekallikrein is converted to kallikrein. Kallikrein is an enzyme that catalyzes the formation of plasmin. Plasmin is an enzyme that dissolves the blood clot.

18.5e Prevention of Inappropriate Clotting

Precise controls are required to prevent coagulation when it isn't needed. These include the following:

- **Platelet repulsion.** As noted earlier, platelets don't adhere to the smooth prostacyclin-coated endothelium of healthy blood vessels.

- **Dilution.** Small amounts of thrombin form spontaneously in the plasma, but at normal rates of blood flow, the thrombin is diluted so quickly that a clot has little chance to form. If flow decreases, however, enough thrombin can accumulate to cause clotting. This can happen in circulatory shock, for example. When output from the heart is diminished and circulation slows down, widespread clotting throughout the circulatory system may occur.

- **Anticoagulants.** Thrombin formation is suppressed by anticoagulants in the plasma. **Antithrombin,** secreted by the liver, deactivates thrombin before it can act on fibrinogen. **Heparin,** secreted by basophils and mast cells, interferes with the formation of prothrombin activator, blocks the action of thrombin on fibrinogen, and promotes the action of antithrombin. Heparin is given by injection to patients with abnormal clotting tendencies.

18.5f Clotting Disorders

In a process as complex as coagulation, it's not surprising that things can go wrong. Clotting deficiencies can result from causes as diverse as malnutrition, leukemia, and gallstones (see Deeper Insight 18.5).

A deficiency of any clotting factor can shut down the coagulation cascade. This happens in **hemophilia,** a family of hereditary diseases characterized by deficiencies of one factor or another. Because of the sex-linked recessive mechanism of heredity, hemophilia occurs predominantly in males. They can inherit it only from their mothers, however, as happened with the descendants of Queen Victoria.

DEEPER INSIGHT 18.5

CLINICAL APPLICATION

Liver Disease and Blood Clotting

Proper blood clotting depends on normal liver function for two reasons. First, the liver synthesizes most of the clotting factors. Therefore, diseases such as hepatitis, cirrhosis, and cancer that degrade liver function result in a deficiency of clotting factors. Second, the synthesis of clotting factors II, VII, IX, and X require vitamin K. The absorption of vitamin K from the diet requires bile, a liver secretion. Gallstones can lead to a clotting deficiency by obstructing the bile duct and thus interfering with bile secretion and vitamin K absorption. Efficient blood clotting is especially important in childbirth, since both the mother and infant bleed from the trauma of birth. Therefore, pregnant women may be advised to take vitamin K supplements to ensure fast clotting, and newborn infants may be given vitamin K injections.

The lack of factor VIII causes *classical hemophilia (hemophilia A),* which accounts for about 83% of cases and afflicts 1 in 5,000 males worldwide. Lack of factor IX causes *hemophilia B,* which accounts for 15% of cases and occurs in about 1 out of 30,000 males. Factors VIII and IX are therefore known as *antihemophilic factors A* and *B.* A rarer form called *hemophilia C* (factor XI deficiency) is autosomal and not sex-linked, so it occurs equally in both sexes.

Before purified factor VIII became available in the 1960s, more than half of those with hemophilia died before age 5 and only 10% lived to age 21. Physical exertion causes bleeding into the muscles and joints. Excruciating pain and eventual joint immobility can result from intramuscular and joint **hematomas**[27] (masses of clotted blood in the tissues). Hemophilia varies in severity, however. Half of the normal level of clotting factor is enough to prevent the symptoms, and the symptoms are mild even in individuals with as little as 30% of the normal amount. Such cases may go undetected even into

adulthood. Bleeding can be relieved for a few days by transfusion of plasma or purified clotting factors.

▶▶▶APPLY WHAT YOU KNOW

Why is it important for people with hemophilia not to use aspirin? (Hint: See fig. 17.24, and do not say that aspirin "thins the blood," a figure of speech that is not literally true.)

Failure of the blood to clot takes far fewer lives, however, than unwanted clotting. **Thrombosis,** the abnormal clotting of blood in an unbroken blood vessel, becomes increasingly problematic in old age. About 25% of people over age 50 experience venous blockage by thrombosis, especially people who don't exercise regularly or who are confined to a bed or wheelchair. The blood clots especially easily in the veins, where blood flow is slowest. A **thrombus** (clot) may grow large enough to obstruct a small vessel, or a piece of it may break loose and begin to travel in the bloodstream as an **embolus.**[28] An embolus can lodge in a small artery and block blood flow from that point on. If that vessel supplies vital tissue of the heart, brain, lung, or kidney, *infarction* (tissue death) may result. About 650,000 Americans die annually of *thromboembolism* (traveling blood clots) in the cerebral, coronary, and pulmonary arteries. Most strokes and heart attacks are due to thrombosis, and pulmonary failure often results from thromboembolism.

Thrombosis is more likely to occur in veins than in arteries because blood flows more slowly in the veins and doesn't dilute thrombin and fibrin as rapidly. Most venous blood flows directly to the heart and then to the lungs. Therefore, blood clots arising in the limbs commonly lodge in the lungs and cause *pulmonary embolism.* When blood cannot circulate freely through the lungs, it cannot receive oxygen and a person may die of hypoxia.

Table 18.8 describes some additional disorders of the blood. The effects of aging on the blood are described in section 29.4a.

[27]*hemato* = blood; *oma* = mass

[28]*em* = in, within; *bolus* = ball, mass

TABLE 18.8	Some Disorders of the Blood
Disseminated intravascular coagulation (DIC)	Widespread clotting within unbroken vessels, limited to one organ or occurring throughout the body. Usually triggered by septicemia but also occurs when blood circulation slows markedly (as in cardiac arrest). Marked by widespread hemorrhaging, congestion of the vessels with clotted blood, and tissue necrosis in blood-deprived organs.
Infectious mononucleosis	Infection of B lymphocytes with Epstein–Barr virus, most commonly in adolescents and young adults. Usually transmitted by exchange of saliva, as in kissing. Causes fever, fatigue, sore throat, inflamed lymph nodes, and leukocytosis. Usually self-limiting and resolves within a few weeks.
Septicemia	Bacteremia (bacteria in the bloodstream) accompanying infection elsewhere in the body. Often causes fever, chills, and nausea, and may cause DIC or septic shock.
Thalassemia	A group of hereditary anemias most common in Greeks, Italians, and others of Mediterranean descent; shows a deficiency or absence of alpha or beta hemoglobin and RBC counts that may be less than 2 million/μL.
Thrombocytopenia	A platelet count below 100,000/mL. Causes include bone marrow destruction by radiation, drugs, poisons, or leukemia. Signs include small hemorrhagic spots in the skin or hematomas in response to minor trauma.

You can find other hematologic disorders described in the following places:

Hypoproteinemia in Deeper Insight 18.1; *anemia, polycythemia, hypoxemia,* and *sickle-cell disease* in section 18.2e; *transfusion reactions* and *hemolytic disease of the newborn* in sections 18.3b and 18.3d; *leukemia, leukocytosis,* and *leukopenia* in section 18.4d; and *hematoma, hemophilia, thrombosis,* and *thromboembolism* in section 18.5f.

BEFORE YOU GO ON

Answer the following questions to test your understanding of the preceding section:

22. What are the three basic mechanisms of hemostasis?

23. How do the extrinsic and intrinsic mechanisms of coagulation differ? What do they have in common?

24. In what respect does blood clotting represent a negative feedback loop? What part of it is a positive feedback loop?

25. Describe some of the mechanisms that prevent clotting in undamaged vessels.

26. Describe a common source and effect of pulmonary embolism.

DEEPER INSIGHT 18.6

CLINICAL APPLICATION

Clinical Management of Blood Clotting

For many cardiovascular patients, the goal of treatment is to prevent clotting or to dissolve clots that have already formed. Several strategies employ inorganic salts and products of bacteria, plants, and animals with anticoagulant and clot-dissolving effects.

Preventing Clots from Forming

Since calcium is an essential requirement for blood clotting, blood samples can be kept from clotting by adding a few crystals of sodium oxalate, sodium citrate, or EDTA[29]—salts that bind calcium ions and prevent them from participating in the coagulation reactions. Blood-collection equipment such as hematocrit tubes may also be coated with heparin, a natural anticoagulant whose action was explained earlier.

Since vitamin K is required for the synthesis of clotting factors, anything that antagonizes vitamin K usage makes the blood clot less readily. One vitamin K antagonist is *coumarin*[30] (COO-muh-rin), a sweet-smelling extract of tonka beans, sweet clover, and other plants, used in perfume. Taken orally by patients at risk for thrombosis, coumarin takes up to 2 days to act, but it has longer-lasting effects than heparin. A similar vitamin K antagonist is the pharmaceutical preparation *warfarin*[31] *(Coumadin),* which was originally developed as a pesticide—it makes rats bleed to death. Obviously, such anticoagulants must be used in humans with great care.

Many parasites feed on the blood of vertebrates and secrete anticoagulants to keep the blood flowing. Among these are aquatic worms known as leeches. Leeches secrete a local anesthetic that makes their bites painless; therefore, as early as 1567 BCE, physicians used them for bloodletting. This method was less painful and repugnant to their patients than *phlebotomy*[32]—cutting a vein—and indeed, leeching became very popular. In seventeenth-century France, it was quite the rage; tremendous numbers of leeches were used in ill-informed attempts to treat headaches, insomnia, whooping cough, obesity, tumors, menstrual cramps, mental illness, and almost anything else doctors or their patients imagined to be caused by "bad blood."

The first known anticoagulant was discovered in the saliva of the medicinal leech, *Hirudo medicinalis,* in 1884. Named *hirudin,* it is a polypeptide that prevents clotting by inhibiting thrombin. It causes the blood to flow freely while the leech feeds and for as long as an hour thereafter. While the doctrine of bad blood is now long discredited, leeches have lately reentered medical usage for other reasons **(fig. 18.25).** A major problem in reattaching a severed body part such as a finger or ear is that the tiny veins draining these organs are too small to reattach surgically. Since arterial blood flows into the reattached organ and can't flow out as easily, it pools and clots there. This inhibits the regrowth of veins and the flow of fresh blood through the organ, and often leads to necrosis. Some vascular

FIGURE 18.25 A Modern Use of Leeching. Two medicinal leeches are being used to remove clotted blood from a postsurgical hematoma. Despite their formidable size, the leeches secrete a natural anesthetic and produce a painless bite.
SPL/Science Source

❓ *How does the modern theory behind leeching differ from the theory of leeching that was popular a few centuries ago?*

surgeons now place leeches on the reattached part. Their anticoagulant keeps the blood flowing freely and allows new veins to grow. After 5 to 7 days, venous drainage is restored and leeching can be stopped.

Dissolving Clots That Have Already Formed

When a clot has already formed, it can be treated with clot-dissolving drugs such as *streptokinase,* an enzyme made by certain bacteria (streptococci). Intravenous streptokinase is used to dissolve blood clots in coronary vessels, for example. It is nonspecific, however, and digests almost any protein. *Tissue plasminogen activator (TPA)* works faster, is more specific, and is now made by transgenic bacteria. TPA converts plasminogen into the clot-dissolving enzyme plasmin. Some anticoagulants of animal origin also work by dissolving fibrin. A giant Amazon leech, *Haementeria,* produces one such anticoagulant named *hementin.* This, too, has been successfully produced by genetically engineered bacteria and used in cardiac patients to dissolve blood clots that do not yield to streptokinase or other drugs.

---

[29]ethylenediaminetetraacetic acid
[30]*coumaru* = tonka bean tree

[31]acronym from Wisconsin Alumni Research Foundation
[32]*phlebo* = vein; *tomy* = cutting

STUDY GUIDE

▶ Assess Your Learning Outcomes

To test your knowledge, discuss the following topics with a study partner or in writing, ideally from memory.

18.1 Introduction

1. Components of the circulatory system; the difference between the terms *circulatory system* and *cardiovascular system*
2. The diverse functions of blood; contributions of the blood to homeostasis
3. The two main components of whole blood; relative amounts of plasma and formed elements in the blood; and the three main categories of formed elements
4. The composition of blood plasma and serum
5. Importance of the viscosity and osmolarity of blood, what accounts for each, and the pathological effects of abnormal viscosity or osmolarity
6. The definition of *colloid osmotic pressure*
7. General aspects of hematopoiesis; where it occurs in the embryo, in the fetus, and after birth; and the stem cell with which all hematopoietic pathways begin

18.2 Erythrocytes

1. Erythrocyte (RBC) structure and function
2. The functions of hemoglobin and carbonic anhydrase
3. Hemoglobin structure and what parts of it bind O_2 and CO_2
4. Three ways of quantifying the RBCs and hemoglobin level of the blood; the definition and units of measurement of each; and reasons for the differences between male and female values
5. Stages of erythropoiesis and major transformations in each
6. Why iron is essential; how the stomach converts dietary iron to a usable form; and the roles of gastroferritin, transferrin, and ferritin in iron metabolism
7. Homeostatic regulation of erythropoiesis, including the origins and role of erythropoietin (EPO)
8. The life span of an RBC and how the body disposes of old RBCs
9. How the body disposes of the hemoglobin from expired RBCs and how this relates to the pigments of bile, feces, and urine

10. Excesses and deficiencies in RBC count and the forms, causes, and pathological consequences of each
11. Causes and effects of hemoglobin deficiencies and the pathology of sickle-cell disease and thalassemia

18.3 Blood Types

1. What determines a person's blood type; blood types of the ABO group and how they differ in genetics and RBC antigens
2. Why an individual does not have plasma antibodies against the ABO types at birth, but develops them during infancy; how these antibodies limit transfusion compatibility
3. The cause and mechanism of a transfusion reaction and why it can lead to renal failure and death; the meanings of *agglutination* and *hemolysis*
4. Blood types of the Rh group and how they differ in their genetics and RBC antigens
5. What can cause a person to develop antibodies against Rh-positive RBCs
6. Blood groups other than ABO and Rh, and their usefulness for certain purposes
7. Hemolytic disease of the newborn; why it seldom occurs in a woman's first susceptible child, but is more common in later pregnancies; and how it is treated and prevented

18.4 Leukocytes

1. The general function of all leukocytes (WBCs)
2. Three kinds of granulocytes, two kinds of agranulocytes, and what distinguishes granulocytes from agranulocytes as a class
3. The appearance, relative size and number, and functions of each WBC type, and the conditions under which each type increases in a differential WBC count
4. Three principal cell lines, the stages, and the anatomical sites of leukopoiesis
5. The relative length of time that WBCs travel in the bloodstream and spend in other tissues; which type recirculates into the blood and which types do not; and the relative life spans of WBCs
6. Causes and effects of leukopenia and leukocytosis

7. The naming and classification of various kinds of leukemia; why leukemia is typically accompanied by RBC and platelet deficiencies and elevated risk of opportunistic infection

18.5 Platelets and the Control of Bleeding

1. Platelet structure and functions, a typical platelet count, and why platelets are not considered to be cells
2. The sites and process of platelet production, and the hormone that stimulates it
3. Three mechanisms of hemostasis and their relative quickness and effectiveness
4. The general objective of coagulation; the end product of the coagulation reactions, and basic differences between the extrinsic and intrinsic mechanisms
5. Essentials of the extrinsic mechanism including the chemical that initiates it, other procoagulants involved, and the point at which it converges with the intrinsic mechanism at a common intermediate
6. Essentials of the intrinsic mechanism including the chemical that initiates it, other procoagulants involved, and the aforesaid point of convergence with the extrinsic mechanism
7. Steps in the continuation of coagulation from factor X to fibrin, including the procoagulants involved
8. The roles of positive feedback and enzyme amplification in coagulation
9. The processes of clot retraction, vessel repair, and fibrinolysis
10. Three mechanisms of preventing inappropriate coagulation in undamaged vessels
11. Causes of clotting deficiencies including the types, genetics, and pathology of hemophilia
12. Terms for unwanted or inappropriate clotting in a vessel, the clot itself, and a clot that breaks free and travels in the bloodstream
13. Why spontaneous clotting more often occurs in the veins than in the arteries; the danger presented by traveling blood clots; and why traveling clots so often lodge in the lungs even if they originate as far away as the lower limbs

STUDY GUIDE

▶ Testing Your Recall

Answers in Appendix A

1. Antibodies belong to a class of plasma proteins called
   a. albumins.
   b. gamma globulins.
   c. alpha globulins.
   d. procoagulants.
   e. agglutinins.

2. Serum is blood plasma minus its
   a. sodium ions.
   b. calcium ions.
   c. clotting proteins.
   d. globulins.
   e. albumin.

3. Which of the following conditions is most likely to cause hemolytic anemia?
   a. folic acid deficiency
   b. iron deficiency
   c. mushroom poisoning
   d. alcoholism
   e. hypoxemia

4. It is impossible for a type O+ baby to have a type _____ mother.
   a. AB–
   b. O–
   c. O+
   d. A+
   e. B+

5. Which of the following is *not* a component of hemostasis?
   a. platelet plug formation
   b. agglutination
   c. clot retraction
   d. vascular spasm
   e. degranulation

6. Which of the following contributes most to the viscosity of blood?
   a. albumin
   b. sodium
   c. globulins
   d. erythrocytes
   e. fibrin

7. Which of these is a granulocyte?
   a. a monocyte
   b. a lymphocyte
   c. a macrophage
   d. an eosinophil
   e. an erythrocyte

8. Excess iron is stored in the liver as a complex called
   a. gastroferritin.
   b. transferrin.
   c. ferritin.
   d. hepatoferritin.
   e. erythropoietin.

9. Pernicious anemia is a result of
   a. hypoxemia.
   b. iron deficiency.
   c. malaria.
   d. lack of intrinsic factor.
   e. Rh incompatibility.

10. The first clotting factor that the intrinsic and extrinsic pathways have in common is
    a. thromboplastin.
    b. Hageman factor.
    c. factor X.
    d. prothrombin activator.
    e. factor VIII.

11. Production of all the formed elements of blood is called _____.

12. The percentage of blood volume composed of RBCs is called the _____.

13. The extrinsic pathway of coagulation is activated by _____ from damaged perivascular tissues.

14. The RBC antigens that determine transfusion compatibility are called _____.

15. The hereditary lack of factor VIII causes a disease called _____.

16. The overall cessation of bleeding, involving several mechanisms, is called _____.

17. _____ results from a mutation that changes one amino acid in the hemoglobin molecule.

18. An excessively high RBC count is called _____.

19. Intrinsic factor enables the small intestine to absorb _____.

20. The kidney hormone _____ stimulates RBC production.

▶ Building Your Medical Vocabulary

Answers in Appendix A

State a meaning of each word element, and give a medical term from this chapter that uses it or a slight variation of it.

1. an-

2. -blast

3. erythro-

4. glutino-

5. hemo-

6. leuko-

7. -penia

8. phlebo-

9. -poiesis

10. thrombo-

STUDY GUIDE

▶ What's Wrong with These Statements?

Answers in Appendix A

Briefly explain why each of the following statements is false, or reword it to make it true.

1. Erythrocytes normally constitute most of the volume of the blood.

2. An abnormal increase in blood albumin concentration causes the blood to flow more sluggishly and therefore reduces blood pressure.

3. Anemia is caused by a low oxygen concentration in the blood.

4. *Hemostasis, coagulation,* and *clotting* are three terms for the same process.

5. If a type B+ woman has a baby with type O– blood, then her husband, who is type A+, cannot be the biological father.

6. Lymphocytes are the most abundant WBCs in the blood.

7. Potassium ions are a necessary cofactor at several steps in the blood-clotting process.

8. All formed elements of the blood except platelets come ultimately from the same pluripotent stem cells.

9. When RBCs die and break down, the globin moiety of hemoglobin is excreted and the heme is recycled to new RBCs.

10. Leukemia is a severe deficiency of white blood cells.

▶ Testing Your Comprehension

1. Why would erythropoiesis not correct the hypoxemia resulting from lung cancer?

2. People with chronic kidney disease often have hematocrits of less than half the normal value. Explain why.

3. An elderly white woman is hit by a bus and severely injured. Accident investigators are informed that she lives in an abandoned warehouse, where her few personal effects include several empty wine bottles and an expired driver's license indicating she is 72 years old. At the hospital, she is found to be severely anemic. List all the factors you can think of that may contribute to her anemia.

4. How is coagulation different from agglutination?

5. Although fibrinogen and prothrombin are equally necessary for blood clotting, fibrinogen is about 4% of the plasma protein whereas prothrombin is present only in small traces. In light of the roles of these clotting factors and your knowledge of enzymes, explain this difference in abundance.

THE CIRCULATORY SYSTEM: HEART

A three-dimensional CT scan of the heart; lateral view of a person facing left

Gondelon/Science Source

CHAPTER OUTLINE

19.1 Overview of the Cardiovascular System

   19.1a The Pulmonary and Systemic Circuits
   19.1b Position, Size, and Shape of the Heart
   19.1c The Pericardium

19.2 Gross Anatomy of the Heart

   19.2a The Heart Wall
   19.2b The Chambers
   19.2c The Valves
   19.2d Blood Flow Through the Chambers
   19.2e The Coronary Circulation
   19.2f Venous Drainage

19.3 Cardiac Muscle and the Cardiac Conduction System

   19.3a Structure of Cardiac Muscle
   19.3b Metabolism of Cardiac Muscle
   19.3c The Conduction System

19.4 Electrical and Contractile Activity of the Heart

   19.4a The Cardiac Rhythm
   19.4b Pacemaker Physiology
   19.4c Impulse Conduction to the Myocardium
   19.4d Electrical Behavior of the Myocardium
   19.4e The Electrocardiogram

19.5 Blood Flow, Heart Sounds, and the Cardiac Cycle

   19.5a Principles of Pressure and Flow
   19.5b Heart Sounds
   19.5c Phases of the Cardiac Cycle
   19.5d Overview of Volume Changes

19.6 Regulation of Cardiac Output

   19.6a Autonomic Innervation of the Heart
   19.6b Cardiac Output
   19.6c Heart Rate and Chronotropic Agents
   19.6d Stroke Volume and Inotropic Agents
   19.6e Exercise and Cardiac Output

Study Guide

DEEPER INSIGHTS

19.1 Clinical Application: Angina and Heart Attack

19.2 Clinical Application: Cardiac Arrhythmias

19.3 Clinical Application: Valvular Insufficiency

19.4 Clinical Application: Coronary Artery Disease

Anatomy & Physiology Revealed® 4.0

Module 9: Cardiovascular System

BRUSHING UP

- For the best understanding of cardiac muscle, be sure you are familiar with desmosomes and gap junctions (see section 5.5a) and with the structure of striated muscle cells (see section 11.2, which describes skeletal muscle fibers but also applies to cardiac muscle).

- You must be familiar with membrane resting potentials and action potentials (see sections 12.4b–12.4c) to understand cardiac pacemaker physiology and the excitation of cardiac muscle.

- Review excitation–contraction coupling in skeletal muscle (see fig. 11.9 and the associated text) for comparison with the process in cardiac muscle.

- The length–tension relationship of striated muscle (see fig. 11.12 and the associated text) helps to explain variation in the ejection of blood by the heart.

- Adjustment of cardiac output to states of rest and physical exertion hinges on understanding the anatomy and action of the sympathetic and parasympathetic nervous systems (see section 15.3a).

We are more conscious of our heart than we are of most organs, and more wary of its failure. Speculation about the heart is at least as old as written history. Some ancient Chinese, Egyptian, Greek, and Roman scholars correctly surmised that the heart is a pump for filling the vessels with blood. Aristotle's views, however, were a step backward. Perhaps because the heart quickens its pace when we're emotionally aroused, and because grief causes "heartache," he regarded it primarily as the seat of emotion, as well as a source of heat to aid digestion. During the Middle Ages, Western medical schools clung dogmatically to the ideas of Aristotle. Perhaps the only significant advance came from Arabic medicine, when thirteenth-century physician Ibn an-Nafis described the role of the coronary blood vessels in nourishing the heart. The sixteenth-century dissections and anatomical charts of Vesalius, however, greatly improved knowledge of cardiovascular anatomy and set the stage for a more scientific study of the heart and treatment of its disorders—the science we now call **cardiology.**[1]

In the early decades of the twentieth century, little could be recommended for heart disease other than bed rest. Then nitroglycerin was found to improve coronary circulation and relieve the pain resulting from physical exertion, digitalis proved helpful for treating abnormal heart rhythms, and diuretics were first used to reduce hypertension. Coronary bypass surgery; replacement of diseased valves; clot-dissolving enzymes; heart transplants; and artificial pacemakers, valves, and hearts have made cardiology one of today's most dramatic and attention-getting fields of medicine.

19.1 Overview of the Cardiovascular System

Expected Learning Outcomes

When you have completed this section, you should be able to

a. define and distinguish between the *pulmonary circuit* and *systemic circuit;*

b. describe the general location, size, and shape of the heart; and

c. describe the pericardium that encloses the heart.

The **cardiovascular**[2] **system** consists of the heart and blood vessels. The heart is a muscular pump that keeps blood flowing through the vessels. The vessels deliver the blood to all the body's organs and then return it to the heart. The broader term *circulatory system* also includes the blood, and some authorities use it to include the lymphatic system as well (described in section 21.1).

19.1a The Pulmonary and Systemic Circuits

The cardiovascular system has two major divisions: a **pulmonary circuit,** which carries blood to the lungs for gas exchange and returns it to the heart, and a **systemic circuit,** which supplies blood to every organ of the body, including other parts of the lungs and the wall of the heart itself **(fig. 19.1).**

The right half of the heart supplies the pulmonary circuit. It receives blood that has circulated through the body, unloaded its oxygen and nutrients, and picked up a load of carbon dioxide and other wastes. It pumps this oxygen-poor blood into a large artery, the *pulmonary trunk,* which immediately divides into right and left *pulmonary arteries.* These transport blood to the air sacs *(alveoli)* of the lungs, where carbon dioxide is unloaded and oxygen is picked up. The oxygen-rich blood then flows by way of the *pulmonary veins* to the left side of the heart.

The left side supplies the systemic circuit. Blood leaves it by way of another large artery, the *aorta.* The aorta turns like an inverted U, the *aortic arch,* and passes downward posterior to the heart. The arch gives off arteries that supply the head, neck, and upper limbs. The aorta then travels through the thoracic and abdominal cavities and issues smaller arteries to the other organs before branching into the lower limbs. After circulating through the body, the now deoxygenated systemic blood returns to the right side of the heart mainly by way of two large veins: the *superior vena cava* (draining the upper body) and *inferior vena cava* (draining everything below the diaphragm). The major arteries and veins entering and leaving the heart are called the *great vessels (great arteries* and *veins)* because of their relatively large diameters. These circulatory routes are detailed in the next chapter.

---

[1]*cardio* = heart; *logy* = study

[2]*cardio* = heart; *vas* = vessel

FIGURE 19.1 **General Schematic of the Cardiovascular System.**

❓ *Are the lungs supplied by the pulmonary circuit, the systemic circuit, or both? Explain.* **A&PR**

CO₂ O₂

Pulmonary circuit

O₂-poor, CO₂-rich blood

O₂-rich, CO₂-poor blood

Systemic circuit

CO₂ O₂

19.1b Position, Size, and Shape of the Heart

The heart lies within a thick partition called the **mediastinum** between the two lungs. It extends from a broad **base** at its uppermost end, where the great vessels are attached, to a bluntly pointed **apex** at the lower end, just above the diaphragm. It tilts toward the left from base to apex, so somewhat more than half the heart is to the left of the body's median plane. We can see this especially in a cross (horizontal) section through the thorax (**fig. 19.2;** see also figs. B.10, B.11 in atlas B).

The adult heart is about 9 cm (3.5 in.) wide at the base, 13 cm (5 in.) from base to apex, and 6 cm (2.5 in.) from anterior to posterior at its thickest point. Whatever one's body size, from child to adult, the heart is roughly the same size as the fist. It weighs about 300 g (10 ounces) in adults.

19.1c The Pericardium

The heart is enclosed in a double-walled sac called the **pericardium.**[3] The outer wall of the pericardium is a tough fibrous sac called the **fibrous pericardium.** It surrounds the heart but isn't attached to it. Deep to this is a thin membrane called the **serous pericardium.** This has two layers—a *parietal layer* that lines the inside of the fibrous pericardium, and a *visceral layer* that adheres to the heart surface and forms the outermost layer of the heart itself, the *epicardium* (**fig. 19.3**). The fibrous pericardium is anchored by ligaments to the diaphragm below and the sternum anterior to it, and more loosely anchored by fibrous connective tissue to mediastinal tissue posterior to the heart.

The space between the parietal and visceral layers of the serous pericardium is called the **pericardial cavity** (figs. 19.2b, 19.3). The heart isn't inside the pericardial cavity but enfolded by it. The relationship of the heart to the pericardium is often described by comparison to a fist pushed into an underinflated balloon (fig. 19.3c). The balloon surface in contact with the fist is like the epicardium; the outer balloon surface is like the parietal layer, and the air space between them is like the pericardial cavity.

The pericardial cavity contains 5 to 30 mL of **pericardial fluid,** exuded by the serous pericardium. The fluid lubricates the membranes and allows the heart to beat with minimal friction. In *pericarditis*—inflammation of the pericardium—the membranes may become roughened and produce a painful *friction rub* with each heartbeat. In addition to reducing friction, the pericardium isolates the heart from other thoracic organs and anchors it within the thorax. It allows the heart room to expand, yet resists excessive expansion (see *cardiac tamponade* in Deeper Insight A.1 in atlas A).

BEFORE YOU GO ON

Answer the following questions to test your understanding of the preceding section:

1. Distinguish between the pulmonary and systemic circuits and state which part of the heart supplies each one.

2. Predict the effect of a fibrous pericardium that fits too tightly around the heart. Predict the effect of a failure of the serous pericardium to secrete pericardial fluid.

---

[3]*peri* = around; *cardi* = heart

(a)

Sternum

Rib 3

Diaphragm

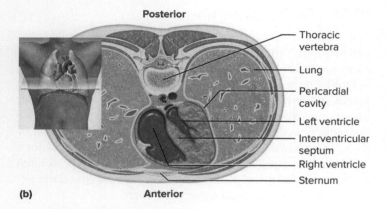

Posterior

Thoracic vertebra

Lung

Pericardial cavity

Left ventricle

Interventricular septum

Right ventricle

Sternum

(b)

Anterior

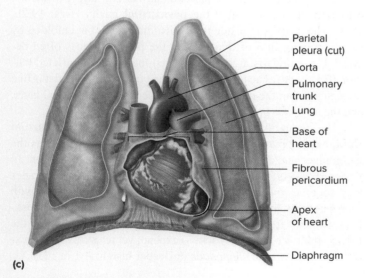

Parietal pleura (cut)

Aorta

Pulmonary trunk

Lung

Base of heart

Fibrous pericardium

Apex of heart

Diaphragm

(c)

FIGURE 19.2 Position of the Heart in the Thoracic Cavity.
(a) Relationship to the thoracic cage. (b) Cross section of the thorax at the level of the heart. (c) Frontal view with the lungs slightly retracted and the fibrous pericardium opened. **APR**

❓ *Does most of the heart lie to the right or left of the median plane?*

Pericardial cavity

Fibrous pericardium

Serous pericardium:

Parietal layer

Visceral layer (epicardium)

(b)

Myocardium

Endocardium

Epicardium

Fibrous pericardium

(a)   (c)

FIGURE 19.3 The Pericardium and Heart Wall. (a) Frontal section of the heart showing the three layers of the heart wall and relationship to the pericardium. (b) Detail of the pericardium and pericardial cavity. (c) A fist in a balloon shows, by analogy, how the double-walled pericardium wraps around the heart.

19.2 Gross Anatomy of the Heart

Expected Learning Outcomes

When you have completed this section, you should be able to

a. describe the three layers of the heart wall;

b. identify the four chambers of the heart;

c. identify the surface features of the heart and correlate them with its internal four-chambered anatomy;

d. identify the four valves of the heart;

e. trace the flow of blood through the four chambers and valves of the heart and adjacent blood vessels; and

f. describe the arteries that nourish the myocardium and the veins that drain it.

19.2a The Heart Wall

The heart wall consists of three layers: *epicardium, myocardium,* and *endocardium.*

The **epicardium**[4] (visceral layer of the serous pericardium) is a serous membrane of the external heart surface. It consists mainly of a simple squamous epithelium overlying a thin layer of areolar tissue. In some places, it also includes a thick layer of adipose tissue, whereas in other areas it is fat-free and translucent, so the muscle of the underlying myocardium shows through **(figs. 19.4a, 19.5).** The largest branches of the coronary blood vessels travel through the epicardium.

[4]*epi* = upon; *cardi* = heart

The **endocardium,**[5] a similar layer, lines the interior of the heart chambers (figs. 19.3, **19.4b**). Like the epicardium, this is a simple squamous epithelium overlying a thin areolar tissue layer; however, it has no adipose tissue. The endocardium covers the valve surfaces and is continuous with the endothelium of the blood vessels.

The **myocardium**[6] between these two is composed of cardiac muscle. This is by far the thickest layer and performs the work of the heart. Its thickness is proportional to the workload on the individual chambers. Its muscle is organized into bundles that spiral around the heart, forming the **vortex of the heart (fig. 19.6).** Consequently, when the ventricles contract, they exhibit a twisting or wringing motion that enhances the ejection of blood. Later we will examine the microscopic structure of the cardiac muscle cells, or *cardiomyocytes,* more closely.

The heart also has a framework of collagenous and elastic fibers that make up the **fibrous skeleton.** This tissue is especially concentrated in the walls between the heart chambers, in *fibrous rings (anuli fibrosi)* around the valves, and in sheets of tissue that interconnect these rings (fig. 19.8). The fibrous skeleton

[5]*endo* = internal, within; *cardi* = heart
[6]*myo* = muscle; *cardi* = heart

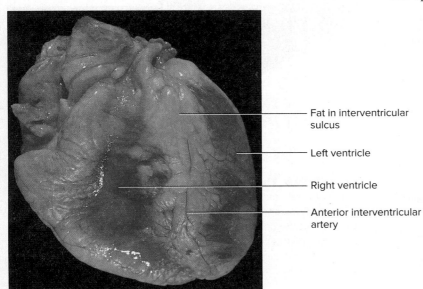

Fat in interventricular sulcus

Left ventricle

Right ventricle

Anterior interventricular artery

(a) Anterior view, external anatomy

Superior vena cava

Inferior vena cava

Interatrial septum

Left atrium

Left AV valve

Coronary blood vessels

Tendinous cords

Left ventricle

Endocardium

Myocardium

Epicardium

Base of heart

Right atrium

Opening of coronary sinus

Right AV valve

Trabeculae carneae

Right ventricle

Papillary muscles

Epicardial fat

Interventricular septum

Apex of heart

(b) Posterior view, internal anatomy

FIGURE 19.4 The Heart of a Human Cadaver. (a) Anterior view, external anatomy. (b) Posterior view, internal anatomy. **APR**

Aortic arch
Superior vena cava
Branches of the right pulmonary artery
Right pulmonary veins
Right auricle
Right atrium
Coronary sulcus
Right ventricle
Inferior vena cava

Ligamentum arteriosum
Ascending aorta
Left pulmonary artery
Pulmonary trunk
Left pulmonary veins
Left auricle
Anterior interventricular sulcus
Left ventricle
Apex of heart

(a) Anterior view

Aorta
Left pulmonary artery
Left pulmonary veins
Left atrium
Coronary sulcus
Coronary sinus
Fat
Left ventricle
Apex of heart

Superior vena cava
Right pulmonary artery
Right pulmonary veins
Right atrium
Inferior vena cava
Posterior interventricular sulcus
Right ventricle

(b) Posterior view

FIGURE 19.5 Surface Anatomy of the Heart. (a) Anterior view. (b) Posterior view. The coronary blood vessels on the heart surface are identified in figure 19.10. **APR**

has multiple functions: (1) It provides structural support for the heart, especially around the valves and the openings of the great vessels; it holds these orifices open and prevents them from excessively stretching when blood surges through them. (2) It anchors the cardiomyocytes and gives them something to pull against. (3) As a nonconductor of electricity, it serves as electrical insulation between the atria and the ventricles, so the atria cannot stimulate the ventricles directly. This insulation is important to the timing and coordination of electrical and contractile activity. (4) Some authorities think (though others disagree) that elastic recoil of the fibrous skeleton may aid in refilling the heart with blood after each beat in a manner described later.

(a)

(b)

FIGURE 19.6 The Vortex of the Heart. (a) Anterior view of the heart with the epicardium rendered transparent to expose the bundles of myocardial muscle. (b) View from the apex to show the way the muscle coils around the heart. This results in a twisting motion when the ventricles contract.

a, b: Photo and illustration by Roy Schneider, University of Toledo. Plastinated heart model for illustration courtesy of Dr. Carlos Baptista, University of Toledo

▶▶▶**APPLY WHAT YOU KNOW**

Parts of the fibrous skeleton sometimes become calcified in old age. How would you expect this to affect cardiac function?

19.2b The Chambers

The heart has four chambers, best seen in a frontal section (**fig. 19.7;** also see fig. 19.4b). The two superior chambers are the **right** and **left atria** (AY-tree-uh; singular, *atrium*[7]). They are receiving chambers for blood returning to the heart by way of the great veins. Most of the mass of each atrium is on the posterior side of the heart, so only a small portion is visible from an anterior view. Here, each atrium has an earlike flap called an **auricle**[8] that slightly increases its volume (fig. 19.5a). The atria exhibit thin flaccid walls corresponding to their light workload—all they do is pump blood into the ventricles immediately below. They are separated from each other by a wall called the **interatrial septum.** The right atrium and both auricles exhibit internal ridges of myocardium called **pectinate**[9] **muscles.**

[7]*atrium* = entryway
[8]*auricle* = little ear
[9]*pectin* = comb; *ate* = like

FIGURE 19.7 Internal Anatomy of the Heart. (a) Frontal section, anterior view. (b) Cross section, showing shapes and spatial relationship of ventricles. The right ventricle wraps around the left in a C shape, but both ventricles have the same volume. **A&PR**

? *Do the atrial pectinate muscles more nearly resemble the ventricular papillary muscles or the trabeculae carneae?*

The two inferior chambers, the **right** and **left ventricles,**[10] are the pumps that eject blood into the arteries and keep it flowing around the body. The right ventricle constitutes most of the anterior aspect of the heart, whereas the left ventricle forms the apex and inferoposterior aspect. Internally, the ventricles are separated by a thick muscular wall, the **interventricular septum.** The right ventricle pumps blood only to the lungs and back to the left atrium, so its wall is only moderately muscular. The wall of the left ventricle, including the septum, is two to four times as thick because it bears the greatest workload of all four chambers, pumping blood

through the entire body. The left ventricle is roughly circular in cross section, whereas the right ventricle wraps around the left and has a C shape (fig. 19.7b). To help visualize this, make a fist of your left hand and enfold it in your right hand; your left and right hands then approximate the shapes and spatial relationship of the left and right ventricles.

Both ventricles exhibit internal ridges called **trabeculae carneae**[11] (trah-BEC-you-lee CAR-nee-ee). It is thought that these ridges may serve to keep the ventricular walls from clinging to each other like suction cups when the heart contracts, and thus allow the

---

[10]*ventr* = belly, lower part; *icle* = little

[11]*trabecula* = little beam; *carne* = flesh, meat

chambers to expand more easily when they refill. If you wet your hands, press your palms firmly together, then pull them apart, you can appreciate how smooth wet surfaces cling to each other and how, without trabeculae, the heart walls might also do so.

On the surface, the boundaries of the four chambers are marked by three sulci (grooves), which are largely filled by fat and the coronary blood vessels (see fig. 19.5a). The **coronary**[12] **sulcus** encircles the heart near the base and separates the atria above from the ventricles below. It can be exposed by lifting the margins of the atria. The other two sulci extend obliquely down the heart from the coronary sulcus toward the apex—one on the front of the heart called the **anterior interventricular sulcus** and one on the back called the **posterior interventricular sulcus.** These sulci overlie the interventricular septum. The coronary sulcus and two interventricular sulci harbor the largest of the coronary blood vessels.

19.2c The Valves

To pump blood effectively, the heart needs valves that ensure a one-way flow. There is a valve between each atrium and its ventricle and another at the exit from each ventricle into its great artery (fig. 19.7), but the heart has no valves where the great veins empty into the atria. Each valve consists of two or three fibrous flaps of tissue called **cusps** or **leaflets,** covered with endocardium.

The **atrioventricular (AV) valves** regulate the openings between the atria and ventricles. The **right AV (tricuspid) valve** has three cusps and the **left AV valve** has two **(fig. 19.8).** The left AV valve is also known as the **mitral valve** (MY-trul) after its resemblance to a miter, the headdress of a church bishop; it has also formerly gone by the name of *bicuspid valve,* now considered inaccurate and obsolete. Stringy **tendinous cords (chordae tendineae)** (COR-dee ten-DIN-ee-ee), resembling the shroud lines of a parachute, connect the valve cusps to conical **papillary**[13] **muscles** on the floor of the ventricle. They prevent the AV valves from flipping inside out or bulging into the atria when the ventricles contract. Each papillary muscle has two or three basal attachments to the trabeculae carneae of the heart wall. Among other functions, these multiple attachments may govern the timing of electrical excitation of the papillary muscles, and they may distribute mechanical stress in a way similar to the weight of the Eiffel Tower supported on its four legs. The multiple attachments also provide some redundancy that protects an AV valve from complete mechanical failure should one attachment fail.

The **semilunar**[14] **valves** (pulmonary and aortic valves) regulate the flow of blood from the ventricles into the great arteries. The **pulmonary valve** controls the opening from the right ventricle into the pulmonary trunk, and the **aortic valve** controls the opening from the left ventricle into the aorta. Each has three cusps shaped like shirt pockets (fig. 19.8b). When blood is ejected from the ventricles, it pushes through these valves from below and

(a)

(b)

(c)

FIGURE 19.8 The Heart Valves. (a) Superior view of the heart with the atria removed. (b) The aortic valve, superior view, showing the three cusps meeting like a Y. One cusp is darkened by a blood clot. (c) Papillary muscle and tendinous cords seen from within the right ventricle. The upper ends of the cords are attached to the cusps of the right AV valve.

b: Biophoto Associates/Science Source; **c:** McGraw-Hill Education

---

[12]*coron* = crown; *ary* = pertaining to
[13]*papill* = nipple; *ary* = like, shaped
[14]*semi* = half; *lun* = moon

presses their cusps against the arterial walls. When the ventricles relax, arterial blood flows backward toward the ventricles, but quickly fills the cusps. The inflated pockets meet at the center and quickly seal the opening, so little blood flows back into the ventricles. Because of the way these valves are attached to the arterial wall, they cannot prolapse any more than a shirt pocket turns inside out if you jam your hand into it. Thus, they don't require or possess tendinous cords.

The valves don't open and close by any muscular effort of their own. The cusps are simply pushed open and closed by changes in blood pressure that occur as the heart chambers contract and relax. Later in this chapter, we will take a closer look at these pressure changes and their effect on the valves.

19.2d Blood Flow Through the Chambers

Until the sixteenth century, blood was thought to flow directly from the right ventricle into the left through invisible pores in the septum. This of course is not true. Blood is kept entirely separate on the right and left sides of the heart. **Figure 19.9** shows the pathway of the blood as it travels from the right atrium through the body and back to the starting point.

Blood that has been through the systemic circuit returns by way of the superior and inferior venae cavae to the right atrium. It flows directly from the right atrium, through the right AV (tricuspid) valve, into the right ventricle. When the right ventricle contracts, it ejects blood through the pulmonary valve into the pulmonary trunk, on its way to the lungs to exchange carbon dioxide for oxygen.

Blood returns from the lungs by way of two pulmonary veins on the left and two on the right; all four of these empty into the left atrium. Blood flows through the left AV (mitral) valve into the left ventricle. Contraction of the left ventricle ejects this blood through the aortic valve into the ascending aorta, on its way to another trip around the systemic circuit.

19.2e The Coronary Circulation

If your heart lasts for 80 years and beats an average of 75 times a minute, it will beat more than 3 billion times and each ventricle will pump more than 200 million liters of blood. It is, in short, a remarkably hardworking organ, and understandably, it needs an abundant supply of oxygen and nutrients. These needs aren't met to any appreciable extent by the blood in the heart chambers, because the diffusion of substances from there through the myocardium

1. Blood enters right atrium from superior and inferior venae cavae.

2. Blood in right atrium flows through right AV valve into right ventricle.

3. Contraction of right ventricle forces pulmonary valve open.

4. Blood flows through pulmonary valve into pulmonary trunk.

5. Blood is distributed by right and left pulmonary arteries to the lungs, where it unloads CO_2 and loads O_2.

6. Blood returns from lungs via pulmonary veins to left atrium.

7. Blood in left atrium flows through left AV valve into left ventricle.

8. Contraction of left ventricle (simultaneous with step 3) forces aortic valve open.

9. Blood flows through aortic valve into ascending aorta.

10. Blood in aorta is distributed to every organ in the body, where it unloads O_2 and loads CO_2.

11. Blood returns to right atrium via venae cavae.

Labels: Aorta, Left pulmonary artery, Superior vena cava, Pulmonary trunk, Right pulmonary veins, Left pulmonary veins, Left atrium, Aortic valve, Left AV valve, Right atrium, Left ventricle, Right AV valve, Right ventricle, Inferior vena cava

FIGURE 19.9 The Pathway of Blood Flow Through the Heart. The pathway from 4 through 6 is the pulmonary circuit, and the pathway from 9 through 11 is the systemic circuit. Violet arrows indicate oxygen-poor blood; orange arrows indicate oxygen-rich blood. **A&PR**

(a) Anterior view

- Right coronary artery (RCA)
- Small cardiac vein
- Right marginal branch of RCA
- Left coronary artery (LCA)
- Left auricle (reflected)
- Circumflex branch of LCA
- Great cardiac vein
- Anterior interventricular branch of LCA

(b) Posterior view

- Great cardiac vein
- Circumflex branch of LCA
- Coronary sinus
- Left marginal branch of LCA
- Left marginal vein
- Right coronary artery (RCA)
- Right marginal branch of RCA
- Posterior interventricular branch of RCA
- Posterior interventricular vein

(c)

FIGURE 19.10 The Principal Coronary Blood Vessels.
(a) Anterior view. (b) Posterior view. (c) A vascular corrosion cast of the coronary circulation. Note the great density of small vessels that service the hardworking myocardium.

c: Science Photo Library/Science Source

would be too slow. Instead, the myocardium has its own supply of arteries and capillaries that deliver blood to every muscle cell. The blood vessels of the heart wall constitute the **coronary circulation (fig. 19.10).**

At rest, the coronary blood vessels supply the myocardium with about 250 mL of blood per minute. This constitutes about 5% of the circulating blood going to meet the metabolic needs of the heart, even though the heart is only 0.5% of the body's weight. It receives 10 times its "fair share" to sustain its strenuous workload.

Arterial Supply

The coronary circulation is the most variable aspect of cardiac anatomy. The following description covers only the pattern seen in about 70% to 85% of persons, and only the few largest vessels. (Compare the great density of coronary blood vessels seen in fig. 19.10c.)

Immediately after the aorta leaves the left ventricle, it gives off a right and left coronary artery. The orifices of these two arteries lie deep in the pockets formed by two of the aortic valve cusps (see fig. 19.8a). The **left coronary artery (LCA)** travels through the coronary sulcus under the left auricle and divides into two branches:

1. The **anterior interventricular branch** travels down the anterior interventricular sulcus to the apex, rounds the bend, and travels a short distance up the posterior side of the heart. There it joins the posterior interventricular branch described shortly. Clinically, it is also called the *left anterior descending (LAD) branch.* This artery supplies blood to both ventricles and the anterior two-thirds of the interventricular septum.

2. The **circumflex branch** continues around the left side of the heart in the coronary sulcus. It gives off a **left marginal branch** that passes down the left margin of the heart and furnishes blood to the left ventricle. The circumflex branch then ends on the posterior side of the heart. It supplies blood to the left atrium and posterior wall of the left ventricle.

The **right coronary artery (RCA)** supplies the right atrium and sinuatrial node (pacemaker), continues along the coronary sulcus under the right auricle, and gives off two branches of its own:

1. The **right marginal branch** runs toward the apex of the heart and supplies the lateral aspect of the right atrium and ventricle.

2. The RCA continues around the right margin of the heart to the posterior side, sends a small branch to the atrioventricular node, then gives off a large **posterior interventricular branch.** This branch travels down the corresponding sulcus and supplies the posterior walls of both ventricles as well as the posterior portion of the interventricular septum. It ends by joining the anterior interventricular branch of the LCA.

The energy demand of the cardiac muscle is so critical that an interruption of the blood supply to any part of the myocardium can cause necrosis within minutes. A fatty deposit or blood clot in a coronary artery can cause a **myocardial infarction**[15] **(MI),** or heart attack (see Deeper Insight 19.1). To protect against this, some coronary arteries converge at various points and combine their blood flow to points farther downstream. Points where two arteries come together are called *arterial anastomoses* (ah-NASS-tih-MO-seez). They provide alternative routes of blood flow *(collateral circulation)* that can supply the heart tissue with blood if the primary route becomes obstructed.

In organs other than the heart, blood flow usually peaks when the heart contracts and ejects blood into the systemic arteries, and diminishes when the ventricles relax and refill. The opposite is true in the coronary arteries: Flow peaks when the heart relaxes. There are three reasons for this. (1) Contraction of the myocardium squeezes the coronary arteries and obstructs blood flow. (2) When the ventricles contract, the aortic valve is forced open and its cusps

cover the openings to the coronary arteries, blocking blood from flowing into them. (3) When they relax, blood in the aorta briefly surges back toward the heart. It fills the aortic valve cusps and some of it flows into the coronary arteries, like water pouring into a bucket and flowing out through a hole in the bottom. In the coronary blood vessels, therefore, blood flow increases during ventricular relaxation.

19.2f Venous Drainage

Venous drainage refers to the route by which blood leaves an organ. After flowing through capillaries of the heart wall, about 5% to 10% of the coronary blood empties from multiple tiny vessels called *small cardiac veins* directly into the heart chambers, especially the right ventricle. The rest returns to the right atrium by the following route:

- The **great cardiac vein** collects blood from the anterior aspect of the heart and travels alongside the anterior interventricular artery. It carries blood from the apex toward the coronary sulcus, then arcs around the left side of the heart and empties into the coronary sinus.

- The **posterior interventricular (middle cardiac) vein,** found in the posterior interventricular sulcus, collects blood from the posterior aspect of the heart. It, too, carries blood from the apex upward and drains into the same sinus.

- The **left marginal vein** travels from a point near the apex up the left margin, and also empties into the coronary sinus.

- The **coronary sinus,** a large transverse vein in the coronary sulcus on the posterior side of the heart, collects blood from all three of the aforementioned veins as well as some smaller ones. It empties blood into the right atrium.

 DEEPER INSIGHT 19.1

CLINICAL APPLICATION

Angina and Heart Attack

An obstruction of coronary blood flow can cause a chest pain known as *angina pectoris*[16] (an-JY-na PEC-toe-riss) or, more seriously, *myocardial infarction* (heart attack). Angina is a sense of heaviness or pain in the chest resulting from temporary and reversible *ischemia*[17] (iss-KEE-me-ah), or deficiency of blood flow to the cardiac muscle. It typically occurs when a partially blocked coronary artery constricts. The oxygen-deprived myocardium shifts to anaerobic fermentation, producing lactate, which stimulates pain receptors in the heart. The pain abates when the artery relaxes and normal blood flow resumes.

Myocardial infarction (MI), on the other hand, is the sudden death of a patch of myocardium resulting from long-term obstruction of the coronary circulation. Coronary arteries often become obstructed by a blood clot or fatty deposit called an *atheroma* (see Deeper Insight 19.4). As cardiac muscle downstream from the obstruction dies, the individual commonly feels a sense of heavy pressure or squeezing pain in the chest, often "radiating" to the shoulder and left arm. Some MIs are painless, "silent" heart attacks, especially in elderly or diabetic individuals. Infarctions weaken the heart wall and disrupt electrical conduction pathways, potentially leading to fibrillation and cardiac arrest. MI causes about 27% of deaths in the United States.

BEFORE YOU GO ON

Answer the following questions to test your understanding of the preceding section:

3. Name the three layers of the heart and describe their structural differences.

4. What are the functions of the fibrous skeleton?

5. Trace the flow of blood through the heart, naming each chamber and valve in order.

6. What are the three principal branches of the left coronary artery? Where are they located on the heart surface? What are the branches of the right coronary artery, and where are they located?

7. What is the medical significance of anastomoses in the coronary arterial system?

8. Why do the coronary arteries carry a greater blood flow during ventricular relaxation than they do during ventricular contraction?

9. What are the three major veins that empty into the coronary sinus?

[15]*infarct* = to stuff
[16]*angina* = to choke, strangle; *pectoris* = of the chest
[17]*isch* = holding back; *em* = blood; *ia* = condition

19.3 Cardiac Muscle and the Cardiac Conduction System

Expected Learning Outcomes

When you have completed this section, you should be able to

a. describe the unique structural and metabolic characteristics of cardiac muscle;

b. explain the nature and functional significance of the intercellular junctions between cardiac muscle cells; and

c. describe the heart's pacemaker and internal electrical conduction system.

The most obvious physiological fact about the heart is its rhythmicity. It contracts at regular intervals, typically about 75 beats per minute (bpm) in a resting adult. Among invertebrates such as clams, crabs, and insects, each heartbeat is triggered by a pacemaker in the nervous system. But in vertebrates from fish to humans, the heartbeat is said to be *myogenic*[18] because the signal originates within the heart itself. The heart is described as **autorhythmic**[19] because it doesn't depend on the nervous system for its rhythm. It has its own built-in pacemaker and electrical system. We now turn our attention to the cardiac muscle, pacemaker, and internal electrical system—the foundations for its electrical activity and rhythmic beat.

19.3a Structure of Cardiac Muscle

The heart is mostly muscle. Cardiac muscle is striated like skeletal muscle, but quite different from it in other structural and functional respects—and it has to be if we want it to pump infallibly, more than once every second, for at least eight or nine decades.

Cardiomyocytes, the muscle cells of the heart, are relatively short, thick, branched cells, typically 50 to 100 μm long and 10 to 20 μm wide **(fig. 19.11).** The ends of the cell are slightly branched, like a log with deep notches in the end. Through these branches, each cardiomyocyte contacts several others, so collectively they form a network throughout each pair of heart chambers—one network in the atria and one in the ventricles.

Most cardiomyocytes have a single, centrally placed nucleus, but up to one-third of them have two or more nuclei. The nucleus is often surrounded by a light-staining mass of glycogen. The sarcoplasmic reticulum is less developed than in skeletal muscle; it lacks terminal cisterns, although it does have footlike sacs associated with the T tubules. The T tubules are much larger than in skeletal muscle. During excitation of the cell, they admit calcium ions from the extracellular fluid to activate muscle contraction.

Cardiomyocytes are joined end to end by thick connections called **intercalated discs** (in-TUR-ka-LAY-ted). With the right histological stain, these appear as dark lines thicker than the

[18]*myo* = muscle; *genic* = arising from
[19]*auto* = self

(a)

(b)

(c)

FIGURE 19.11 Cardiac Muscle. (a) Light micrograph. (b) Structure of a cardiomyocyte and its relationship to adjacent cardiomyocytes. All of the colored area is a single cell. Note that it is notched at the ends and typically linked to two or more neighboring cardiomyocytes by the mechanical and electrical junctions of the intercalated discs. (c) Structure of an intercalated disc.

a: Ed Reschke

striations. An intercalated disc is a complex steplike structure with three distinctive features not found in skeletal muscle:

1. **Interdigitating folds.** The plasma membrane at the end of the cell is folded somewhat like the bottom of an egg carton. The folds of adjoining cells interlock with each other and increase the surface area of intercellular contact.

2. **Mechanical junctions.** The cells are tightly joined by two types of mechanical junctions: the fascia adherens and desmosomes. The *fascia adherens*[20] (FASH-ee-ah ad-HEER-enz) is the most extensive. It is a broad band, analogous to a strip of Velcro, in which the actin of the thin myofilaments is anchored to the plasma membrane and each cell is linked to the next via transmembrane proteins. The fascia adherens is interrupted here and there by *desmosomes*—patches of mechanical linkage that prevent the contracting cardiomyocytes from pulling apart. Both desmosomes and gap junctions (next) are described in greater detail in section 5.5a.

3. **Electrical junctions.** The intercalated discs also contain *gap junctions,* which form channels that allow ions to flow from the cytoplasm of one cardiomyocyte directly into the next. They enable each cardiomyocyte to electrically stimulate its neighbors. Thus, the entire myocardium of the two atria behaves almost like a single cell, as does the entire myocardium of the two ventricles. This unified action is essential for the effective pumping of a heart chamber.

Skeletal muscle contains satellite cells that can divide and replace dead muscle fibers to some extent. Cardiac muscle lacks these, however, so the repair of damaged cardiac muscle is almost

---

[20]*fascia* = band; *adherens* = adhering

entirely by fibrosis (scarring). Cardiac muscle has very limited capacity for mitosis and regeneration.

19.3b Metabolism of Cardiac Muscle

Cardiac muscle depends almost exclusively on aerobic respiration to make ATP. It is very rich in myoglobin (a short-term source of oxygen for aerobic respiration) and glycogen (for stored energy). Its huge mitochondria fill about 25% of the cell; skeletal muscle fibers, by comparison, have much smaller mitochondria that occupy only 2% of the fiber. Cardiac muscle is relatively adaptable with respect to the organic fuels used. At rest, the heart gets about 60% of its energy from fatty acids, 35% from glucose, and 5% from other fuels such as ketones, lactate, and amino acids. Cardiac muscle is more vulnerable to an oxygen deficiency than it is to the lack of any specific fuel. Because it makes little use of anaerobic fermentation or the oxygen debt mechanism, it is not prone to fatigue. You can easily appreciate this fact by squeezing a rubber ball in your fist once every second for a minute or two. You will soon feel weakness and fatigue in your hand muscles and perhaps feel all the more grateful that cardiac muscle can maintain a rhythm like this, without fatigue, for a lifetime.

19.3c The Conduction System

The heartbeat is coordinated by a **cardiac conduction system** composed of an internal pacemaker and nervelike conduction pathways through the myocardium. It generates and conducts rhythmic electrical signals in the following order **(fig. 19.12):**

1. The **sinuatrial (SA) node** is a patch of modified cardiomyocytes in the right atrium, just under the epicardium near the superior vena cava. This is the **pacemaker** that initiates each heartbeat and determines the heart rate.

Right atrium

Sinuatrial node (pacemaker)

Atrioventricular node

Atrioventricular bundle

Subendocardial conducting network

Left atrium

Subendocardial conducting network

Bundle branches

1. SA node fires.

2. Excitation spreads through atrial myocardium.

3. AV node fires.

4. Excitation spreads down AV bundle.

5. Subendocardial conducting network distributes excitation through ventricular myocardium.

FIGURE 19.12 The Cardiac Conduction System. Electrical signals travel along the pathways indicated by the arrows.

? *Which atrium is first to receive the signal that induces atrial contraction?*

② Signals from the SA node spread throughout the atria, as shown by the red arrows in the figure.

③ The **atrioventricular (AV) node** is located at the lower end of the interatrial septum near the right AV valve. This node acts as an electrical gateway to the ventricles; the fibrous skeleton acts as an insulator to prevent currents from getting to the ventricles by any other route.

④ The **atrioventricular (AV) bundle** is the pathway by which signals leave the AV node. The bundle soon forks into **right** and **left bundle branches,** which enter the interventricular septum and descend toward the apex.

⑤ The **subendocardial conducting network** (formerly called *Purkinje fibers*) consists of processes that arise from the lower end of the bundle branches. Although nervelike in their action, they are composed of modified cardiomyocytes specialized for electrical conduction rather than contraction. At the apex of the heart, they turn upward and ramify throughout the ventricular myocardium, distributing electrical excitation to the cardiomyocytes of the ventricles. They form a more elaborate network in the left ventricle than in the right. Once they have delivered the electrical signal to their limits, the cardiomyocytes themselves perpetuate it by passing ions from cell to cell through their gap junctions.

BEFORE YOU GO ON

Answer the following questions to test your understanding of the preceding section:

10. What organelle(s) is/are less developed in cardiac muscle than in skeletal muscle? What one(s) is/are more developed? What is the functional significance of these differences between muscle types?

11. What exactly is an intercalated disc, and what function is served by each of its components?

12. Cardiac muscle rarely uses anaerobic fermentation to generate ATP. What benefit do we gain from this fact?

13. Where is the pacemaker of the heart located? What is it called? Trace the path of electrical excitation from there to a cardiomyocyte of the left ventricle, naming each component of the conduction system along the way.

19.4 Electrical and Contractile Activity of the Heart

Expected Learning Outcomes

When you have completed this section, you should be able to

a. explain why the SA node fires spontaneously and rhythmically;

b. explain how the SA node excites the myocardium;

c. describe the unusual action potentials of cardiac muscle and relate them to the contractile behavior of the heart; and

d. interpret a normal electrocardiogram.

In this section, we examine how the electrical events in the heart produce its cycle of contraction and relaxation. Contraction is called **systole** (SIS-toe-lee) and relaxation is **diastole** (dy-ASS-toe-lee). These terms can refer to a specific part of the heart (for example, atrial systole), but if no particular chamber is specified, they usually refer to the more conspicuous and important ventricular action, which ejects blood from the heart.

19.4a The Cardiac Rhythm

The normal heartbeat triggered by the SA node is called the **sinus rhythm.** At rest, the adult heart typically beats about 70 to 80 times per minute, although heart rates from 60 to 100 bpm are not unusual.

Any region of spontaneous firing other than the SA node is called an **ectopic**[21] **focus.** If the SA node is damaged, an ectopic focus may take over the governance of the heart rhythm. The most common ectopic focus is the AV node, which produces a slower heartbeat of 40 to 50 bpm called a **nodal (junctional) rhythm.** If neither the SA nor AV node is functioning, other ectopic foci fire at rates of 20 to 40 bpm. The nodal rhythm is sufficient to sustain life, but a rate of 20 to 40 bpm provides too little flow to the brain to be survivable. This is one of the conditions that can call for an artificial pacemaker.

19.4b Pacemaker Physiology

Why does the SA node spontaneously fire at regular intervals? Unlike skeletal muscle or neurons, cells of the SA node don't have a stable resting membrane potential. Their membrane potential starts at about –60 mV and drifts upward, showing a gradual depolarization called the **pacemaker potential (prepotential) (fig. 19.13).** This results primarily from a slow inflow of Na^+ without a compensating outflow of K^+.

When the pacemaker potential reaches a threshold of –40 mV, voltage-gated calcium channels open and Ca^{2+} flows in from the extracellular fluid. This produces the rising (depolarizing) phase of the action potential, which peaks slightly above 0 mV. At that point, K^+ channels open and K^+ leaves the cell. This makes the cytosol increasingly negative and creates the falling (repolarizing) phase of the action potential. When repolarization is complete, the K^+ channels close and the pacemaker potential starts over, on its way to producing the next heartbeat. Each depolarization of the SA node sets off one heartbeat. When the SA node fires, it excites the other components in the conduction system; thus, the SA node serves as the system's pacemaker. At rest, it typically fires every 0.8 second or so, creating a heart rate of about 75 bpm.

---

[21]*ec* = out of; *top* = place

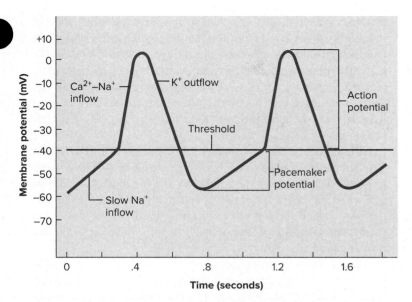

FIGURE 19.13 Pacemaker Potentials and Action Potentials of the SA Node.

19.4c Impulse Conduction to the Myocardium

Firing of the SA node excites atrial cardiomyocytes and stimulates the two atria to contract almost simultaneously. The signal travels at a speed of about 1 m/s through the atrial myocardium and reaches the AV node in about 50 ms. In the AV node, the signal slows down to about 0.05 m/s, partly because the cardiomyocytes here are thinner, but more importantly because they have fewer gap junctions over which the signal can be transmitted. This delays the signal at the AV node for about 100 ms—like

highway traffic slowing down at a small town. This delay is essential because it gives the ventricles time to fill with blood before they begin to contract.

The ventricular myocardium has a conduction speed of only 0.3 to 0.5 m/s. If this were the only route of travel for the excitatory signal, some cardiomyocytes would be stimulated much sooner than others. Ventricular contraction wouldn't be synchronized and the pumping effectiveness of the ventricles would be severely compromised. But signals travel through the AV bundle and subendocardial conducting network at a speed of 4 m/s, the fastest in the conduction system, owing to their very high density of gap junctions. Consequently, the entire ventricular myocardium depolarizes within 200 ms after the SA node fires, causing the ventricles to contract in near unison.

Ventricular systole begins at the apex of the heart, which is first to be stimulated, and progresses upward—pushing the blood upward toward the semilunar valves. Because of the spiral arrangement of the vortex of the heart, the ventricles twist slightly as they contract, like someone wringing out a towel.

▶▶▶ APPLY WHAT YOU KNOW

Some people have abnormal cords or bridges of myocardium that extend from atrium to ventricle, bypassing the AV node and other parts of the conduction system. How would you expect this to affect the cardiac rhythm?

19.4d Electrical Behavior of the Myocardium

The action potentials of cardiomyocytes are significantly different from those of neurons and skeletal muscle fibers (**fig. 19.14**).

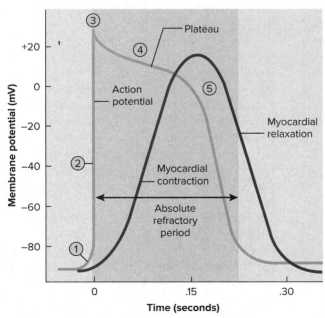

① Voltage-gated Na⁺ channels open.

② Na⁺ inflow depolarizes the membrane and triggers the opening of still more Na⁺ channels, creating a positive feedback cycle and a rapidly rising membrane voltage.

③ Na⁺ channels close when the cell depolarizes, and the voltage peaks at nearly +30 mV.

④ Ca²⁺ entering through slow Ca²⁺ channels prolongs depolarization of membrane, creating a plateau. Plateau falls slightly because of some K⁺ leakage, but most K⁺ channels remain closed until end of plateau.

⑤ Ca²⁺ channels close and Ca²⁺ is transported out of cell. K⁺ channels open, and rapid K⁺ outflow returns membrane to its resting potential.

FIGURE 19.14 Action Potential of a Ventricular Cardiomyocyte. The green curve is the action potential. The red curve represents rising and falling muscle tension as the myocardium contracts and relaxes.

❓ *What is the benefit of having such a long absolute refractory period in cardiac muscle?*

Cardiomyocytes have a stable resting potential of –90 mV and normally depolarize only when stimulated, unlike cells of the SA node. A stimulus opens voltage-gated sodium channels, causing a Na^+ inflow and depolarizing the cell to its threshold. The threshold voltage rapidly opens additional Na^+ channels and triggers a positive feedback cycle like the one seen in the firing of a neuron. The action potential peaks at nearly +30 mV. The Na^+ channels close quickly, and the rising phase of the action potential is very brief.

As action potentials spread over the plasma membrane, they open voltage-gated **slow calcium channels,** which admit a small amount of Ca^{2+} from the extracellular fluid into the cell. This Ca^{2+} binds to ligand-gated Ca^{2+} channels on the sarcoplasmic reticulum (SR), opening them and releasing a greater quantity of Ca^{2+} from the SR into the cytosol. This second wave of Ca^{2+} binds to troponin and triggers contraction in the same way as it does in skeletal muscle. The SR provides 90% to 98% of the Ca^{2+} needed for myocardial contraction.

In skeletal muscle and neurons, an action potential falls back to the resting potential within 2 ms. In cardiac muscle, however, the depolarization is prolonged for 200 to 250 ms (at a heart rate of 70–80 bpm), producing a long plateau in the action potential—perhaps because the Ca^{2+} channels of the SR are slow to close or because the SR is slow to remove Ca^{2+} from the cytosol. Cardiomyocytes remain contracted for as long as the action potential is in its plateau. Thus, in the figure, you can see the development of muscle tension (myocardial contraction) following closely behind the depolarization and plateau. Rather than showing a brief twitch like skeletal muscle, cardiac muscle has a more sustained contraction necessary to expel blood from the heart chambers. Both atrial and ventricular cardiomyocytes exhibit these plateaus, but they are more pronounced in the ventricles.

At the end of the plateau, Ca^{2+} channels close and K^+ channels open. Potassium diffuses rapidly out of the cell and Ca^{2+} is transported back into the extracellular fluid and SR. Membrane voltage drops rapidly, and muscle tension declines soon afterward.

Cardiac muscle has an *absolute refractory period* of 250 ms, compared with 1 to 2 ms in skeletal muscle. This long refractory period prevents wave summation and tetanus (in contrast to skeletal muscle, fig. 11.15), which would stop the pumping action of the heart if they occurred.

▶▶▶**APPLY WHAT YOU KNOW**

With regard to the ions involved, how does the falling (repolarization) phase of a myocardial action potential differ from that of a neuron's action potential? (See fig. 12.14.)

19.4e The Electrocardiogram

We can detect electrical currents in the heart by means of electrodes *(leads)* applied to the skin. An instrument called the *electrocardiograph* amplifies these signals and produces a record, usually on a moving paper chart, called an **electrocardiogram**[22] (**ECG** or **EKG**). To record an ECG,

electrodes are typically attached to the wrists, ankles, and six locations on the chest. Simultaneous recordings can be made from electrodes at different distances from the heart; collectively, they provide a comprehensive image of the heart's electrical activity. An ECG is a composite recording of all action potentials produced by the nodal and myocardial cells—it should not be misconstrued as a tracing of a single action potential.

Figure 19.15 shows a typical ECG, and **table 19.1** provides a functional interpretation of its components. The ECG has three principal deflections above and below the baseline: the *P wave, QRS complex,* and *T wave.* (These letters were arbitrarily chosen; they don't stand for any words.) **Figure 19.16** shows how these correspond to regions of the heart undergoing depolarization and repolarization.

The **P wave** is produced when a signal from the SA node spreads through the atria and depolarizes them. Atrial systole begins about 100 ms after the P wave begins, during the *PQ segment.* This segment is about 160 ms long and represents the time required for impulses to travel from the SA node to the AV node.

The **QRS complex** consists of a small downward deflection (Q), a tall sharp peak (R), and a final downward deflection (S). It is produced when the signal from the AV node spreads through the ventricular myocardium and depolarizes the muscle. This is the most conspicuous part of the ECG because it is produced mainly by depolarization of the ventricles, which constitute the largest muscle mass of the heart and generate the greatest electrical current. Its complex shape is due to the different sizes of the two ventricles and the different times required for them to depolarize. Ventricular systole begins shortly after the QRS complex, in the *ST segment.* The QRS interval is also a time of atrial repolarization and diastole, but atrial repolarization sends a relatively weak signal that is obscured by the electrical activity of the more muscular ventricles.

The PR interval represents the time it takes for a signal to pass through the AV node before activating the ventricles. It doesn't extend all the way to the R wave because the ventricular response begins at Q. Abnormalities in this interval can thus indicate defects that affect conduction time. The QT interval indicates how long the ventricles remain depolarized, and corresponds approximately to the duration of a cardiomyocyte action potential. It gets shorter during exercise. The ST segment corresponds to the plateau in the myocardial action potential and thus represents the time during which the ventricles contract and eject blood.

The **T wave** is generated by ventricular repolarization immediately before diastole. The ventricles take longer to repolarize than to depolarize; the T wave is therefore smaller and more spread out than the QRS complex, and it has a rounder peak. Even when the T wave is taller than the QRS complex, it can be recognized by its relatively rounded peak.

The ECG affords a wealth of information about the normal electrical activity of the heart. Deviations from normal—such as enlarged, inverted, or misshapen waves and abnormal time intervals between waves—are invaluable for diagnosing abnormalities

[22]*electro* = electricity; *cardio* = heart; *gram* = record of; the common abbreviation *EKG* comes from the German spelling, *elektrokardiogramm*

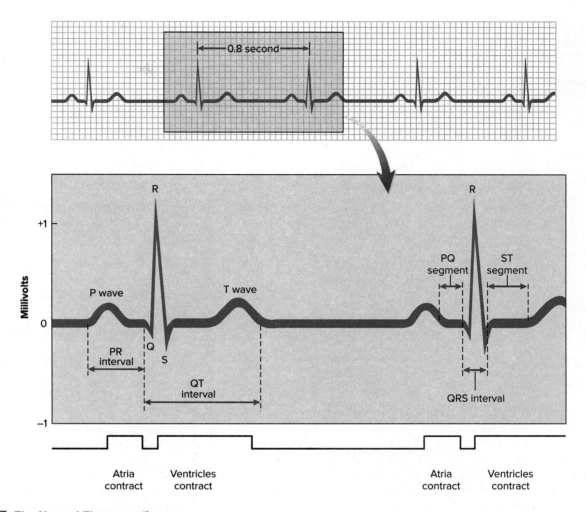

FIGURE 19.15 The Normal Electrocardiogram.

in the conduction pathways, myocardial infarction, enlargement of the heart, and electrolyte and hormone imbalances, among other disorders (see Deeper Insight 19.2).

TABLE 19.1	Interpretation of the Electrocardiogram
P wave	Atrial depolarization
QRS complex	Ventricular depolarization
T wave	Ventricular repolarization
PR interval	Signal conduction through AV node, before activating ventricles
QT interval	Duration of ventricular depolarization; shorter during exercise.
QRS interval	Atrial repolarization and diastole; repolarization concealed by QRS wave.
PQ segment	Signal conduction from SA node to AV node; atrial systole begins.
ST segment	Ventricular systole and ejection of blood; corresponds to plateau of cardiomyocyte action potential.

Any deviation from the regular, SA node–driven sinus rhythm of the heartbeat is called an **arrhythmia.** The most familiar and feared of these is **ventricular fibrillation (VF, VFib),**[23] the hallmark of a heart attack (myocardial infarction). Most cases occur in patients with a history of coronary artery disease. In striking contrast to the steady sinus rhythm, the ECG shows weak, chaotic ventricular depolarizations (compare **fig. 19.17a, b**) as electrical signals travel randomly about the myocardium and return to repeatedly restimulate the same area instead of dying out like a normal ventricular depolarization. To the surgeon's eye and hand, a fibrillating ventricle exhibits squirming, uncoordinated contractions often described as feeling "like a bag of worms." A fibrillating ventricle pumps no blood, so there is no coronary blood flow and myocardial tissue rapidly dies of ischemia, as does cerebral tissue. *Cardiac arrest* is the cessation of cardiac output, with the ventricles either motionless or in fibrillation.

Fibrillation kills quickly if it isn't stopped. *Defibrillation* is an emergency procedure in which the heart is given a strong electrical

[23]*fibril* = small fiber; *ation* = action, process

Key
- ■ Wave of depolarization
- ■ Wave of repolarization

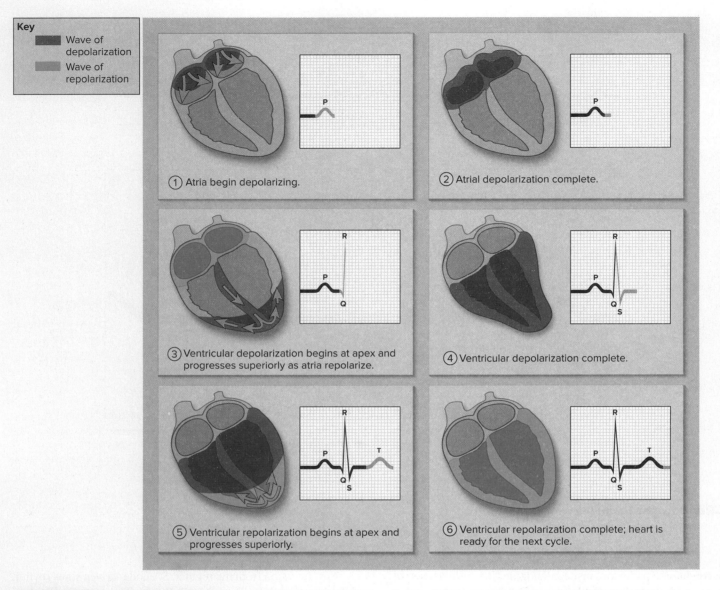

① Atria begin depolarizing.

② Atrial depolarization complete.

③ Ventricular depolarization begins at apex and progresses superiorly as atria repolarize.

④ Ventricular depolarization complete.

⑤ Ventricular repolarization begins at apex and progresses superiorly.

⑥ Ventricular repolarization complete; heart is ready for the next cycle.

FIGURE 19.16 Relationship of the Electrocardiogram (ECG) to Electrical Activity and Contraction of the Myocardium. Each heart diagram indicates the events occurring at the time of the colored segment of the ECG. Red indicates depolarizing or depolarized myocardium, and green indicates repolarizing or repolarized myocardium. Arrows indicate the direction in which a wave of depolarization or repolarization is traveling. **APR**

shock with a pair of paddle electrodes. The purpose is to depolarize the entire myocardium and stop the fibrillation, with the hope that the SA node will resume its sinus rhythm. This doesn't correct the underlying cause of the fibrillation, but it may sustain a patient's life long enough to allow for other corrective action.

BEFORE YOU GO ON

Answer the following questions to test your understanding of the preceding section:

14. Define *systole* and *diastole*.

15. How does the pacemaker potential of the SA node differ from the resting membrane potential of a neuron? Why is this important in creating the heart rhythm?

16. Why is it important that the AV node slow down signal conduction to the ventricles?

17. How does excitation–contraction coupling in cardiac muscle resemble that of skeletal muscle? How is it different?

18. What produces the plateau in the action potentials of cardiomyocytes? Why is this important to the pumping ability of the heart?

19. Identify the portion of the ECG that coincides with each of the following events: atrial depolarization, atrial systole, atrial repolarization, ventricular depolarization, ventricular systole, ventricular repolarization, ventricular diastole.

(a) Sinus rhythm (normal)

(b) Ventricular fibrillation

(c) Atrial fibrillation

(d) Heart block

(e) Premature ventricular contraction

FIGURE 19.17 Normal and Pathological Electrocardiograms. (a) Normal sinus rhythm. (b) Ventricular fibrillation, with grossly irregular waves of depolarization, as seen in a heart attack (myocardial infarction). (c) Atrial fibrillation; between heartbeats, the atria exhibit weak, chaotic, high-frequency depolarizations instead of normal P waves. (d) Heart block, in which some atrial depolarizations (P waves) are not conducted to the ventricles and not followed by ventricular QRS waves. (e) Premature ventricular contraction, or extrasystole (at arrow); note the absence of a P wave, the inverted QRS complex, and the misshapen QRS and elevated T.

DEEPER INSIGHT 19.2

CLINICAL APPLICATION

Cardiac Arrhythmias

Ventricular fibrillation **(fig. 19.17b)** is the most widely known arrhythmia, but others are not uncommon. *Atrial fibrillation (AF, AFib)* **(fig. 19.17c)** is a weak rippling contraction in the atria, manifested in the ECG by chaotic, high-frequency depolarizations (400–650/min.). Fibrillating atria fail to stimulate the ventricles, so we see a dissociation between the random atrial depolarizations and the ventricular QRS and T waves of the ECG. This is the most common atrial arrhythmia in the elderly. It can result from such causes as valvular disease, thyroid hormone excess, or myocardial inflammation, and is often seen in alcoholism. Blood continues to flow through the atria into the ventricles even during AF, so AF isn't immediately life-threatening. Blood flow is more sluggish, however, and there is a higher risk of blood clots causing pulmonary embolism or stroke.

Heart block is a failure of any part of the cardiac conduction system to conduct signals, usually as the result of disease and degeneration of conduction system fibers. In the ECG, one sees rhythmic atrial P waves, but the ventricles fail to receive the signal and no QRS wave follows the P (as in the second and third P waves of **fig. 19.17d**). A *bundle branch block* is a heart block resulting from damage to one or both branches of the AV bundle. Damage to the AV node causes *total heart block,* in which signals from the atria fail to reach the ventricles at all, and the ventricles beat at their own intrinsic rhythm of 20 to 40 bpm.

Premature ventricular contraction (PVC) is the result of a ventricular ectopic focus firing and setting off an extra beat *(extrasystole)* before the normal signal from the SA node arrives. The P wave is missing and the QRS wave is inverted and misshapen (see arrow in **fig. 19.17e**). PVCs can occur singly or in bursts. An occasional extra beat isn't serious, and may result from emotional stress, lack of sleep, or irritation of the heart by stimulants (nicotine, caffeine). Persistent PVCs, however, can indicate more serious pathology and sometimes lead to ventricular fibrillation and sudden death.

19.5 Blood Flow, Heart Sounds, and the Cardiac Cycle

Expected Learning Outcomes

When you have completed this section, you should be able to

a. explain why blood pressure is expressed in millimeters of mercury;

b. describe how changes in blood pressure operate the heart valves;

c. explain what causes the sounds of the heartbeat;

d. describe in detail one complete cycle of heart contraction and relaxation; and

e. relate the events of the cardiac cycle to the volume of blood entering and leaving the heart.

A **cardiac cycle** consists of one complete contraction and relaxation of all four heart chambers. We will examine these events in detail to see how they relate to the entry and expulsion of blood, but first we consider two related issues: (1) some general principles of pressure changes and how they affect the flow of blood, and (2) the heart sounds produced during the cardiac cycle, which we can then relate to the stages of the cycle.

19.5a Principles of Pressure and Flow

A fluid is a state of matter that can flow in bulk from place to place. In the body, this includes both liquids and gases—blood, lymph, air, and urine, among others. Certain basic principles of fluid movement *(fluid dynamics)* apply to all of these. In particular, flow is governed by two main variables: **pressure,** which impels a fluid to move, and **resistance,** which opposes flow. In this chapter, we will focus on how pressure changes govern the operation of the heart valves, the entry of blood into the heart chambers, and its expulsion into the arteries. In the next chapter, we will examine the roles of pressure and resistance in the flow of blood through the blood vessels, and later in the book, we apply the same principles to respiratory airflow. The flow of blood and of air down their pressure gradients are two applications of the general principle of gradients and flow described in section 1.6e.

Measurement of Pressure

Pressure is commonly measured by a device called a *manometer.* In simplest form, this is typically a J-shaped glass tube partially filled with mercury. The sealed upper end, above the mercury, contains a vacuum, whereas the lower end is open. Pressure applied at the lower end is measured in terms of how high it can push the mercury column up the evacuated end of the tube. In principle, any liquid would do, but mercury is used because it's so dense; it enables us to measure pressure with shorter columns than we would need with a less dense liquid such as water. Pressures are therefore commonly expressed in millimeters of mercury (mm Hg). Blood pressure, specifically, has been traditionally measured with a **sphygmomanometer**[24] (SFIG-mo-ma-NOM-eh-tur)—a calibrated mercury manometer with its open lower end attached to an inflatable pressure cuff wrapped around the arm (although mercury sphygmomanometers have been increasingly replaced by dial and digital devices). Blood pressure and the method of measuring it are discussed in greater detail in section 20.2a.

Pressure Gradients and Flow

A fluid flows only if it is subjected to more pressure at one point than at another. The difference creates a **pressure gradient,** and fluids always flow down their pressure gradients, from the high-pressure point to the low-pressure point. Before we relate this to blood flow, it may be easier to begin with an analogy—an air-filled syringe **(fig. 19.18).**

At rest, the air pressures within the syringe barrel and in the atmosphere surrounding it are equal. But for a given quantity (mass) of air, and assuming a constant temperature, pressure is inversely proportional to the volume of the container—the greater

FIGURE 19.18 Principles of Volume, Pressure, and Flow Illustrated with a Syringe. (a) As the plunger is pulled back, the volume of the enclosed space increases, its pressure falls, and pressure inside the syringe (P_1) is lower than the pressure outside (P_2). The pressure gradient causes air to flow inward until the pressures are equal. This is analogous to the filling of an expanding heart chamber. (b) As the plunger is depressed, the volume of the enclosed space decreases, P_1 rises above P_2, and air flows out until the pressures are equal. This is analogous to the ejection of blood from a contracting heart chamber. In both cases, fluids flow down their pressure gradients.

the volume, the lower the pressure, and vice versa. Suppose you pull back the plunger of the syringe (fig. 19.18a). This increases the volume and thus lowers the air pressure within the barrel. Now you have a pressure gradient, with pressure outside the syringe being greater than the pressure inside. Air will flow down its gradient into the syringe until the two pressures are equal. If you then push the plunger in (fig. 19.18b), pressure inside the barrel will rise above the pressure outside, and air will flow out—again going down its pressure gradient but in the reverse direction.

[24]*sphygmo* = pulse; *mano* = rare, sparse, roomy

The syringe barrel is analogous to a heart chamber such as the left ventricle. When the ventricle expands, its internal pressure falls. If the AV valve is open, blood flows into the ventricle from the atrium above. When the ventricle contracts, its internal pressure rises. When the aortic valve opens, blood is ejected from the ventricle into the aorta.

The opening and closing of the heart valves are governed by these pressure changes. Remember that the valves are just soft flaps of connective tissue with no muscle. They don't exert any effort of their own, but are passively pushed open and closed by the changes in blood pressure on the upstream and downstream sides of the valve.

When the ventricles are relaxed and their pressure is low, the AV valve cusps hang down limply and both valves are open

(fig. 19.19a). Blood flows freely from the atria into the ventricles even before the atria contract. As the ventricles fill with blood, the cusps float upward toward the closed position. When the ventricles contract, their internal pressure rises sharply and blood surges against the AV valves from below. This pushes the cusps together, seals the openings, and prevents blood from flowing back into the atria. The papillary muscles contract slightly before the rest of the ventricular myocardium and tug on the tendinous cords, preventing the valves from bulging excessively (prolapsing) into the atria or turning inside out like windblown umbrellas. (See *mitral valve prolapse* in Deeper Insight 19.3.)

The rising pressure in the ventricles also acts on the aortic and pulmonary valves. Up to a point, pressure in the aorta and pulmonary trunk opposes their opening, but when the ventricular

Atrioventricular valves open
(a)

Atrioventricular valves close
(first heart sound produced)

Atrium

Atrioventricular valve

Ventricle

Aorta

Pulmonary artery

Semilunar valve

Semilunar valves open
(b)

Semilunar valves close
(second heart sound produced)

FIGURE 19.19 Operation of the Heart Valves. These are correlated with the heart sounds described in section 19.5b. (a) The atrioventricular valves. When atrial pressure is greater than ventricular pressure, the valve opens and blood flows through (green arrows). When ventricular pressure rises above atrial pressure, the blood in the ventricle pushes the valve cusps closed. (b) The semilunar valves. When the pressure in the ventricles is greater than the pressure in the great arteries, the semilunar valves are forced open and blood is ejected. When ventricular pressure is lower than arterial pressure, arterial blood holds these valves closed.

? *What role do the tendinous cords play?*

DEEPER INSIGHT 19.3
CLINICAL APPLICATION

Valvular Insufficiency

Valvular insufficiency (incompetence) refers to any failure of a valve to prevent *reflux (regurgitation)*—the backward flow of blood. *Valvular stenosis*[25] is a form of insufficiency in which the cusps are stiffened and the opening is constricted by scar tissue. It frequently results from rheumatic fever, an autoimmune disease in which antibodies produced to fight a bacterial infection also attack the mitral and aortic valves. As the valves become scarred and constricted, the heart is overworked by the effort to force blood through the openings and may become enlarged. Regurgitation of blood through the incompetent valves creates turbulence that can be heard with a stethoscope as a *heart murmur.*

Mitral valve prolapse (MVP) is an insufficiency in which one or both mitral valve cusps bulge into the atrium during ventricular contraction. It is often hereditary and affects about 1 out of 40 people, especially young women. In many cases, it causes no serious dysfunction, but in some people it causes chest pain, fatigue, and shortness of breath.

In some cases, an incompetent valve can eventually lead to heart failure. A defective valve can be surgically repaired or replaced with an artificial valve or a valve transplanted from a pig heart.

---

pressure rises above the arterial pressure, it forces the valves open and blood is ejected from the heart **(fig. 19.19b).** Then as the ventricles relax again and their pressure falls below that in the arteries, arterial blood briefly flows backward and fills the pocketlike cusps of the semilunar valves. The three cusps meet in the middle of the orifice and seal it (see fig. 19.8b), thereby preventing arterial blood from reentering the heart.

▶▶▶ APPLY WHAT YOU KNOW

How would aortic valvular stenosis (see Deeper Insight 19.3) affect the amount of blood pumped into the aorta? How might this affect a person's physical stamina? Explain your reasoning.

19.5b Heart Sounds

As we follow events through the cardiac cycle, we will note the occurrence of two or three *heart sounds* audible with a stethoscope. Listening to sounds made by the body is called **auscultation** (AWS-cul-TAY-shun). The **first** and **second heart sounds,** symbolized S_1 and S_2, are often described as a "lubb-dupp"—S_1 is louder and longer and S_2 a little softer and sharper. In children and adolescents, it is normal to hear a **third heart sound** (S_3). This is rarely audible in people older than 30, but when it is, the heartbeat is said to show a *triple rhythm* or *gallop,* which may indicate an enlarged and failing heart. If the normal sounds are roughly simulated by drumming two fingers on a table, a triple rhythm sounds a little like drumming with three fingers. The heart

valves themselves operate silently, but S_1 and S_2 occur in conjunction with the closing of the valves as a result of turbulence in the bloodstream and movements of the heart wall.

19.5c Phases of the Cardiac Cycle

We now examine the phases of the cardiac cycle, the pressure changes that occur, and how the pressure changes and valves govern the flow of blood. Cardiovascular physiologist Carl J. Wiggers (1883–1963) devised an enormously informative chart, now known as the *Wiggers diagram* **(fig. 19.20),** for showing the major events that occur simultaneously at each moment throughout the cardiac cycle. Here, it is divided into colored and numbered bars to correspond to the following phases. Closely follow the figure as you study the text. Where to begin when describing a circular chain of events is somewhat arbitrary, but in this presentation, we begin with the filling of the ventricles. Remember that all these events are completed in less than 1 second.

1. **Ventricular filling.** During diastole, the ventricles expand and their pressure drops below that of the atria. As a result, the AV valves open and blood pours into the ventricles, raising the ventricular pressure and lowering atrial pressure. Ventricular filling occurs in three phases: **(1a)** The first one-third is *rapid ventricular filling,* when blood enters especially quickly. **(1b)** The second one-third, called *diastasis* (di-ASS-tuh-sis), is marked by slower filling. The P wave of the electrocardiogram occurs at the end of diastasis, marking the depolarization of the atria. **(1c)** In the last one-third, *atrial systole* completes the filling process. The right atrium contracts slightly before the left because it is the first to receive the signal from the SA node. At the end of ventricular filling, each ventricle contains an **end-diastolic volume (EDV)** of about 130 mL of blood. Only 40 mL (31%) of this is contributed by atrial systole.

2. **Isovolumetric contraction.** The atria repolarize, relax, and remain in diastole for the rest of the cardiac cycle. The ventricles depolarize, generate the QRS complex, and begin to contract. Wave Q marks the end of ventricular filling; R marks the transition from atrial systole to isovolumetric contraction of the ventricles; and S occurs during isovolumetric contraction. Pressure in the ventricles rises sharply and reverses the pressure gradient between atria and ventricles. The AV valves close as ventricular blood surges back against the cusps. Heart sound S_1 occurs at the beginning of this phase and is produced mainly by the left ventricle; the right ventricle is thought to make little contribution. Causes of the sound are thought to include tensing of the ventricular tissues and tendinous cords (like the twang of a suddenly stretched rubber band), turbulence in the blood as it surges against the closed AV valves, and impact of the heart against the chest wall.

This phase is called *isovolumetric*[26] because even though the ventricles contract, they don't eject blood yet and

---

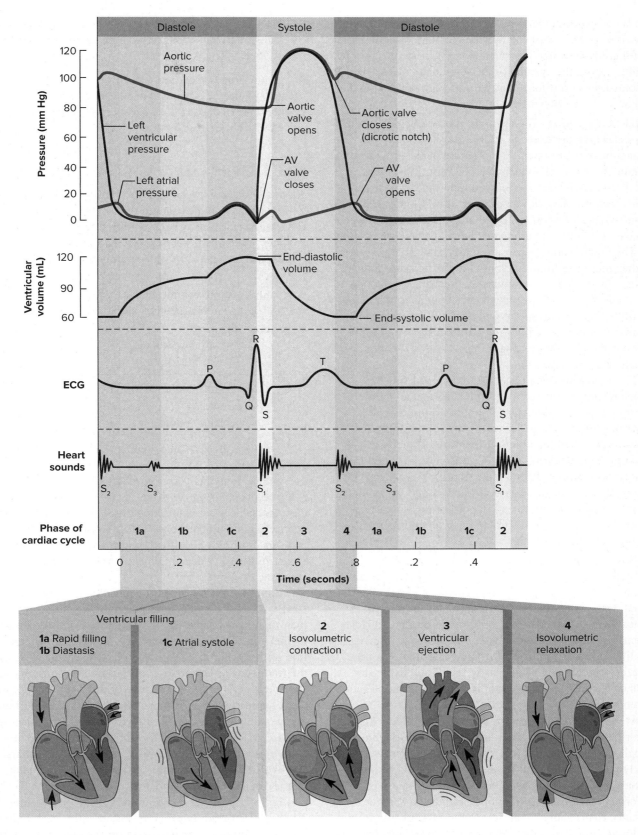

FIGURE 19.20   Modified Wiggers Diagram, Illustrating Events of the Cardiac Cycle. Two cycles are shown. The phases are numbered across the bottom to correspond to the text description.

❓ *Explain why the aortic pressure curve begins to rise abruptly at about 0.5 second.*

there is no change in their volume. This is because pressures in the aorta (80 mm Hg) and pulmonary trunk (10 mm Hg) are still greater than the pressures in the respective ventricles and thus oppose the opening of the semilunar valves. The cardiomyocytes exert force, but with all four valves closed, the blood can't go anywhere.

3. **Ventricular ejection.** The ejection of blood begins when ventricular pressure exceeds arterial pressure and forces the semilunar valves open. The pressure peaks at typically 120 mm Hg in the left ventricle and 25 mm Hg in the right. Blood spurts out of each ventricle rapidly at first *(rapid ejection),* then flows out more slowly under less pressure *(reduced ejection).* By analogy, suppose you were to shake up a bottle of carbonated beverage and remove the cap. The liquid would spurt out rapidly at high pressure and then more would dribble out at lower pressure, much like the blood leaving the ventricles. Ventricular ejection lasts about 200 to 250 ms, which corresponds to the plateau of the myocardial action potential but lags somewhat behind it (review the red tension curve in fig. 19.14). The T wave occurs late in this phase, beginning at the moment of peak ventricular pressure.

The ventricles don't expel all their blood. In an average resting heart, each ventricle contains an EDV of 130 mL. The amount ejected, about 70 mL, is called the **stroke volume (SV).** This is about 54% of the EDV, a percentage called the **ejection fraction.** The blood remaining behind, about 60 mL in this case, is called the **end-systolic volume (ESV).** Note that EDV − SV = ESV. In vigorous exercise, the ejection fraction may be as high as 90%. Ejection fraction is an important measure of cardiac health. A diseased heart may eject much less than 50% of the blood it contains.

4. **Isovolumetric relaxation.** This is early ventricular diastole, when the T wave ends and the ventricles begin to expand. There are competing hypotheses as to how they expand. One is that the blood flowing into the ventricles inflates them. Another is that contraction of the ventricles deforms the fibrous skeleton, which subsequently springs back like the rubber bulb of a turkey baster that has been squeezed and released. This elastic recoil and expansion would cause pressure to drop rapidly and suck blood into the ventricles.

At the beginning of ventricular diastole, blood from the aorta and pulmonary trunk briefly flows backward through the semilunar valves. The backflow, however, quickly fills the cusps and closes them, creating a slight pressure rebound that appears as the *dicrotic notch* of the aortic pressure curve (the top curve in the Wiggers diagram). Heart sound S_2 occurs as blood rebounds from the closed semilunar valves and the ventricles expand. This phase is called *isovolumetric* because the semilunar valves are closed, the AV valves haven't yet opened, and the ventricles are therefore taking in no blood. When the AV valves open, ventricular filling (phase 1) begins again.

Heart sound S_3, if it occurs, is thought to result from the transition from expansion of the empty ventricles to their sudden filling with blood.

In a resting person, atrial systole lasts about 0.1 second; ventricular systole, 0.3 second; and the *quiescent period* (when all four chambers are in diastole), 0.4 second. Total duration of the cardiac cycle is therefore 0.8 second (800 ms) in a heart beating at 75 bpm.

19.5d Overview of Volume Changes

An additional perspective on the cardiac cycle can be gained if we review the volume changes that occur. This "balance sheet" is from the standpoint of one ventricle; both ventricles have equal volumes. The volumes vary somewhat from one person to another and depend on a person's state of activity.

End-systolic volume (ESV) left from the previous heartbeat	60 mL
Passively added to the ventricle during atrial diastole	+ 30 mL
Added by atrial systole	+ 40 mL
Total: End-diastolic volume (EDV)	130 mL
Stroke volume (SV) ejected by ventricular systole	− 70 mL
Leaves: End-systolic volume (ESV)	60 mL

Notice that the ventricle pumps out as much blood as it received during diastole: 70 mL in this example. Both ventricles eject the same amount of blood even though pressure in the right ventricle is only about one-fifth the pressure in the left. Blood pressure in the pulmonary trunk is relatively low, so the right ventricle doesn't need to generate very much pressure to overcome it.

Equal output by the two ventricles is essential for homeostasis. If the right ventricle pumps more blood into the lungs than the left ventricle can handle on return, blood accumulates in the lungs, causing pulmonary hypertension, edema, and a risk of drowning in one's own body fluid **(fig. 19.21a).** One of the first signs of left ventricular failure is respiratory distress—shortness of breath and a sense of suffocation. Conversely, if the left ventricle pumps more blood than the right, blood accumulates in the systemic circuit, causing hypertension and widespread systemic edema **(fig. 19.21b).** Such systemic edema, once colloquially called *dropsy,* is marked by enlargement of the liver; distension of the jugular veins in the neck; swelling of the fingers, ankles, and feet; and **ascites** (ah-SITE-eez), a pooling of fluid in the abdominal cavity (see figures 18.3 and 20.38). It can lead to stroke or kidney failure. A failure of one ventricle increases the workload on the other, which stresses it and often leads to its eventual failure as well. In principle, if the output of the left ventricle were just 1% greater than output of the right, it would completely drain the lungs of blood in less than 10 minutes (although death would occur much sooner).

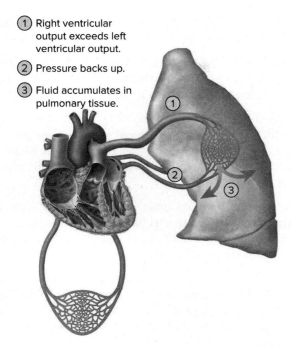

① Right ventricular output exceeds left ventricular output.

② Pressure backs up.

③ Fluid accumulates in pulmonary tissue.

(a) Pulmonary edema

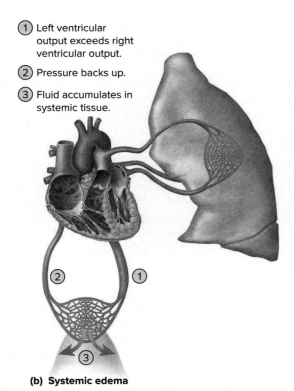

① Left ventricular output exceeds right ventricular output.

② Pressure backs up.

③ Fluid accumulates in systemic tissue.

(b) Systemic edema

FIGURE 19.21 The Necessity of Balanced Ventricular Output. (a) If the left ventricle pumps less blood than the right, blood pressure backs up into the lungs and causes pulmonary edema. (b) If the right ventricle pumps less blood than the left, pressure backs up in the systemic circulation and causes systemic edema. To maintain homeostasis, both ventricles must pump the same average amount of blood.

Fluid accumulation in either circuit due to insufficiency of ventricular pumping is called **congestive heart failure (CHF).** Common causes of CHF are myocardial infarction, chronic hypertension, valvular defects, and congenital (birth) defects in cardiac anatomy.

> **BEFORE YOU GO ON**
>
> Answer the following questions to test your understanding of the preceding section:
>
> **20.** Explain how a pressure gradient across a heart valve determines whether a ventricle ejects blood.
>
> **21.** What factors are thought to cause the first and second heart sounds? When do these sounds occur?
>
> **22.** What phases of the cardiac cycle are isovolumetric? Explain what this means.
>
> **23.** Define *end-diastolic volume* and *end-systolic volume;* explain verbally why stroke volume is the difference between these two.
>
> **24.** Why is it so important that neither the right nor left ventricle pump more blood than the other one for any prolonged time?

19.6 Regulation of Cardiac Output

Expected Learning Outcomes

When you have completed this section, you should be able to

a. trace the routes of sympathetic and parasympathetic nerves to their target cells in the heart;

b. define *cardiac output* and explain its importance;

c. identify the factors that govern cardiac output;

d. discuss some of the nervous and chemical factors that alter heart rate, stroke volume, and cardiac output;

e. explain how the right and left ventricles achieve balanced output; and

f. describe some effects of exercise on cardiac output.

The heart doesn't pump the same amount of blood every minute of every day, but varies its output according to states of rest, exercise, emotion, and other factors. We all know of circumstances that make our hearts beat faster or harder. In this section, we explore how the sympathetic and parasympathetic nervous systems, hormones and drugs, and other factors regulate the heart.

19.6a Autonomic Innervation of the Heart

Even though the heart has its own pacemaker, its rhythm and contraction strength are moderated by signals arising from two **cardiac centers** in the medulla oblongata of the brainstem. One of these, the **cardioacceleratory center,** communicates with the heart by way of right and left **cardiac nerves** carrying sympathetic postganglionic nerve fibers. The other is the nearby **cardioinhibitory center,** which communicates with the heart by way of the right and left **vagus nerves** carrying parasympathetic preganglionic nerve fibers.

The sympathetic pathway to the heart is shown on the left side of **figure 19.22,** numbered as follows:

1. Stimulatory signals from the cardioacceleratory center descend to the upper thoracic segments (T1–T4) of the spinal cord, ending on the sympathetic preganglionic neurons

FIGURE 19.22 Autonomic Innervation of the Heart.
For schematic purposes, the cardiac plexus is shown offset from the heart, not in its natural location. The sympathetic and parasympathetic pathways both occur on the right and left, even though the sympathetic chain and nerves are illustrated only on the left and both vagus nerves are illustrated on the right.

in the lateral horn. Preganglionic fibers arise here and travel to the adjacent sympathetic chain ganglia.

2. Some of these fibers synapse with postganglionic neurons at the level of entry into the chain, and others ascend to the cervical ganglia and meet the postganglionic neurons there.

3. Postganglionic fibers emerging from the sympathetic chain form the cardiac nerves. These lead to the **cardiac plexus,** a web of mixed sympathetic and parasympathetic fibers tucked in between the aortic arch, pulmonary trunk, and lower trachea.

4. Postganglionic fibers continue through the plexus without synapsing and end on several targets in and near the heart: the SA node, AV node, atrial and ventricular myocardium, aorta, pulmonary trunk, and coronary arteries.

Sympathetic stimulation increases the heart rate and contraction strength and dilates the coronary arteries to increase blood supply to the exercising myocardium.

The parasympathetic pathway is shown on the right side of the same figure. Parasympathetic signals from the cardioinhibitory center exit the medulla oblongata through the two vagus nerves. These nerves travel down the right and left sides of the mediastinum, but both are shown on the right side of the figure just for diagrammatic convenience. For a photo of the sympathetic chain and right vagus nerve in relation to the heart, see figure 15.3. The vagus nerves travel to their targets in the heart by the route numbered as follows in figure 19.22:

5. Preganglionic fibers of the vagus travel through the cardiac plexus, mingling with the sympathetic fibers. They synapse with short postganglionic fibers here in the plexus or in the epicardium of the atria, especially near the SA and AV nodes.

6. Short postganglionic fibers of the right vagus nerve lead mainly to the SA node and those of the left vagus lead mainly to the AV node, but each has some fibers that cross over to the other target cells. Furthermore, some parasympathetic fibers terminate on sympathetic fibers and inhibit them from stimulating the heart and opposing the parasympathetic effect. There is little or no parasympathetic innervation of the myocardium or ventricles.

In general, the sympathetic nerves dominate the control of contraction strength. Heart rate is strongly influenced by both divisions, but dominated by the vagus nerves. The cardiac nerves carry not only sympathetic efferent fibers, but also sensory (afferent) fibers from the heart to the CNS. These fibers are important in cardiovascular reflexes and the transmission of pain signals from the heart.

19.6b Cardiac Output

The entire point of all the cardiac physiology we have considered is to eject blood from the heart. The amount ejected by each ventricle in 1 minute is called the **cardiac output (CO).** If HR is heart rate (beats/min.) and SV is stroke volume (mL/beat),

$$CO = HR \times SV$$

At typical resting values,

$$CO = 75 \text{ beats/min.} \times 70 \text{ mL/beat} = 5{,}250 \text{ mL/min.}$$

Thus, the body's total volume of blood (4–6 L) passes through the heart every minute; or to look at it another way, an RBC leaving the left ventricle will, on average, arrive back at the left ventricle in about 1 minute.

Cardiac output varies with the body's state of activity. Vigorous exercise increases CO to as much as 21 L/min. in a person in good condition, and more than 40 L/min. in world-class athletes. The difference between the maximum and resting cardiac output is called **cardiac reserve.** People with severe heart disease may have little or no cardiac reserve and little tolerance of physical exertion.

Given that cardiac output equals HR × SV, you can see that there are only two ways to change it: Change the heart rate or change the stroke volume. We will consider factors that influence each of these variables, but bear in mind that heart rate and stroke volume are somewhat interdependent. They usually change together and in opposite directions. As heart rate goes up, stroke volume goes down, and vice versa.

19.6c Heart Rate and Chronotropic Agents

Heart rate is most easily measured by taking a person's **pulse** at some point where an artery runs close to the body surface, such as the *radial artery* in the wrist or *common carotid artery* in the neck. Each beat of the heart produces a surge of pressure that can be felt by palpating a superficial artery with the fingertips. Heart rate can be obtained by counting the number of pulses in 15 seconds and multiplying by 4 to get beats per minute. In newborn infants, the resting heart rate is commonly 120 bpm or greater. It declines steadily with age, averaging 72 to 80 bpm in young adult females and 64 to 72 bpm in young adult males. It rises again in the elderly.

Tachycardia[27] is a persistent, resting adult heart rate above 100 bpm. It can be caused by stress, anxiety, stimulants, heart disease, or fever. Heart rate also rises to compensate to some extent for a drop in stroke volume. Thus, the heart races when the body has lost a significant quantity of blood or when there is damage to the myocardium.

Bradycardia[28] is a persistent, resting adult heart rate below 60 bpm. It is common during sleep and in endurance-trained athletes. Endurance training enlarges the heart and increases its stroke volume, enabling it to maintain the same output with fewer beats. Hypothermia (low body temperature) also slows the heart and may be deliberately induced in preparation for cardiac surgery. Diving mammals such as whales and seals exhibit bradycardia during the dive, as do humans to some extent when the face is immersed in cool water.

Factors outside of the heart itself that raise the heart rate are called **positive chronotropic**[29] **agents,** and factors that lower it are **negative chronotropic agents.** We next consider some

chronotropic effects of the autonomic nervous system, hormones, electrolytes, and blood gases.

The Autonomic Nervous System

Although the nervous system doesn't initiate the heartbeat, it does modulate its rhythm and force through the aforementioned autonomic nerves. The sympathetic nervous system exerts its effects directly through the cardiac nerves and indirectly by stimulating the adrenal medulla. The nerve fibers secrete norepinephrine (NE) and the adrenal medulla secretes a mixture of about 85% epinephrine (Epi) and 15% NE—known collectively as *catecholamines.* Both of these have positive chronotropic *and* inotropic effects on the heart, but for the moment, we will focus on the chronotropic effect.

Epi and NE bind to β-adrenergic receptors in the heart and activate the cyclic adenosine monophosphate (cAMP) second-messenger system (see fig. 3.8). Cyclic AMP activates an enzyme that opens Ca^{2+} channels in the plasma membrane, admitting calcium from the extracellular fluid into the SA node and cardiomyocytes. This accelerates depolarization of the SA node and speeds up signal conduction through the AV node. Both of these quicken the contractions of the ventricular myocardium and speed up the heartbeat. In addition, cAMP accelerates the reuptake of Ca^{2+} by the sarcoplasmic reticulum of the cardiomyocytes. Quick Ca^{2+} reuptake shortens ventricular systole and the QT interval (see table 19.1), enabling the ventricles to relax and refill sooner than they do at rest. By accelerating *both* the contraction and relaxation of the heart, cAMP increases the heart rate.

Adrenergic stimulation can, in fact, raise the heart rate to as high as 230 bpm. This limit is set mainly by the refractory period of the SA node, which prevents it from firing any more frequently. Cardiac output peaks, however, at a rate of 160 to 180 bpm. At rates any higher than this, the ventricles have too little time to fill between beats. At a resting rate of 65 bpm, ventricular diastole lasts about 0.62 second, but at 200 bpm, it lasts only 0.14 second. Thus you can see that at excessively high heart rates, diastole is too brief to allow complete filling of the ventricles, and therefore stroke volume and cardiac output are reduced.

The parasympathetic vagus nerves, by contrast, have a negative chronotropic effect. Left to itself, the SA node and the heart as a whole have a resting rhythm of about 100 bpm; this is seen if all nerves to the heart are severed or if all sympathetic and parasympathetic action is pharmacologically blocked. But with intact, functional innervation, the vagus nerves have a steady background firing rate called **vagal tone** that holds the heart rate down to its usual 70 to 80 bpm at rest. The heart can be accelerated not only by sympathetic stimulation but also by reducing vagal tone and allowing the heart to "do its own thing." Extreme vagal stimulation can reduce the heart rate to as low as 20 bpm or even stop the heart briefly.

The postganglionic neurons of the vagus are cholinergic—they secrete acetylcholine (ACh) at the SA and AV nodes. ACh binds to muscarinic receptors and opens K^+ gates in the nodal cells. The resulting outflow of K^+ hyperpolarizes the cells, so the SA node fires less frequently and the heart slows down. ACh acts primarily on the SA node but also slows signal conduction through the AV

---

[27]*tachy* = speed, fast; *card* = heart; *ia* = condition
[28]*brady* = slow; *card* = heart; *ia* = condition
[29]*chrono* = time; *trop* = to change, to influence

node, thus delaying excitation of the ventricles. The vagus nerves have a faster-acting effect on the heart than the sympathetic nerves because ACh acts directly on membrane ion channels; sympathetic effects are slower because of the time taken for the cAMP system to open ion channels.

▶▶▶ APPLY WHAT YOU KNOW

Which of the intervals or segments of the electrocardiogram do you think is or are most affected by ACh? Explain your answer.

The Central Nervous System

There is a benefit to placing heart rate under the influence of cardiac centers in the medulla—these centers can receive input from many other sources and integrate it into a decision as to whether the heart should beat more quickly or slowly. Sensory and emotional stimuli can act on the cardiac centers by way of the cerebral cortex, limbic system, and hypothalamus; therefore, heart rate can climb even as you anticipate taking the first plunge on a roller coaster or competing in an athletic event, and it is influenced by emotions such as love and anger. The medulla also receives input from the following receptors in the muscles, joints, arteries, and brainstem:

- **Proprioceptors** in the muscles and joints provide information on changes in physical activity. Thus, the heart can increase its output even before the metabolic demands of the muscles rise.

- **Baroreceptors (pressoreceptors)** are pressure sensors in the aorta and internal carotid arteries (see figs. 15.1, 20.4). They send a continual stream of signals to the medulla. When the heart rate rises, cardiac output increases and raises the blood pressure at the baroreceptors. The baroreceptors increase their signaling to the medulla and, depending on circumstances, the medulla may issue vagal output to lower the heart rate. Conversely, the baroreceptors also inform the medulla of drops in blood pressure. The medulla can then issue sympathetic output to increase the heart rate, bringing cardiac output and blood pressure back up to normal (see fig. 1.8). Either way, a negative feedback loop usually prevents the blood pressure from deviating too far from normal.

- **Chemoreceptors** occur in the aortic arch, carotid arteries, and the medulla oblongata itself, and are sensitive to blood pH, CO_2, and O_2 levels. They are more important in respiratory control than in cardiovascular control, but they do influence the heart rate. If circulation to the tissues is too slow to remove CO_2 as fast as the tissues produce it, then CO_2 accumulates in the blood and cerebrospinal fluid (CSF) and produces a state of *hypercapnia* (CO_2 excess). Furthermore, CO_2 generates hydrogen ions by reacting with water: $CO_2 + H_2O \longrightarrow HCO_3^- + H^+$. The hydrogen ions lower the pH of the blood and CSF and may create a state of *acidosis* (pH < 7.35). Hypercapnia and acidosis stimulate the cardiac centers to increase the heart rate,

thus improving perfusion of the tissues and restoring homeostasis. The chemoreceptors also respond to extreme *hypoxemia* (oxygen deficiency), as in suffocation, but the effect is usually to slow down the heart, perhaps so the heart doesn't compete with the brain for the limited oxygen supply.

Such responses to fluctuations in blood chemistry and blood pressure, called **chemoreflexes** and **baroreflexes,** are good examples of negative feedback loops. They are discussed more fully in the next chapter.

Hormones, Drugs, and Other Chronotropic Chemicals

Heart rate is influenced by many other factors besides the autonomic nervous system. Thyroid hormone accelerates it by stimulating the up-regulation of β-adrenergic receptors, making the heart more sensitive to sympathetic stimulation; this is why tachycardia is one of the signs of hyperthyroidism. Glucagon, secreted by the alpha cells of the pancreatic islets, accelerates the heart by promoting cAMP production. Glucagon is sometimes given in cardiac emergencies to stimulate the heartbeat, and epinephrine is frequently given to support cardiac output and blood pressure in life-threatening allergic reactions.

Food-borne drugs and medications also have well-known chronotropic effects related to the catecholamine–cAMP mechanism. Nicotine accelerates the heart by stimulating catecholamine secretion. Caffeine and the related stimulants in tea and chocolate accelerate it by inhibiting cAMP breakdown, prolonging its adrenergic effect. Hypertension is often treated with drugs called *beta blockers,* which inhibit the binding of catecholamines to the β-adrenergic receptors and slow down the heart.

Electrolyte concentrations also strongly influence heart rate and contraction strength. The most powerful chronotropic effects are from potassium ions (K^+). In *hyperkalemia,*[30] a potassium excess, K^+ diffuses into the cardiomyocytes and keeps the membrane voltage elevated, inhibiting cardiomyocyte repolarization. The myocardium becomes less excitable, the heart rate becomes slow and irregular, and the heart may arrest in diastole. In *hypokalemia,* a potassium deficiency, K^+ diffuses out of the cardiomyocytes and they become hyperpolarized—the membrane potential is more negative than normal. This makes them harder to stimulate. These potassium imbalances are very dangerous and require emergency medical treatment.

Calcium also affects heart rate. A calcium excess (*hypercalcemia*) causes a slow heartbeat, whereas a calcium deficiency (*hypocalcemia*) elevates the heart rate. Such calcium imbalances are relatively rare, however, and when they do occur, their primary effect is on contraction strength, which is considered in the coming section on contractility. Section 24.2 further explores the causes and effects of imbalances in potassium, calcium, and other electrolytes.

---

[30]*hyper* = excess; *kal* = potassium (Latin, *kalium*); *emia* = blood condition

Increased heart rate (HR)

Positive chronotropic agents
  Sympathetic nervous system
  Epinephrine, norepinephrine
  Thyroid hormone
  Glucagon
  Nicotine, caffeine
  Hypocalcemia

Increased stroke volume (SV)

Increased preload (myocardial stretch)
Positive inotropic agents
  Sympathetic nervous system
  Epinephrine, norepinephrine
  Glucagon
  Digitalis
  Nicotine, caffeine
  Hypercalcemia

Reduced heart rate (HR)

Negative chronotropic agents
  Parasympathetic nervous system
  Acetylcholine
  Hypercalcemia
  Hypokalemia
  Beta blockers

Reduced stroke volume (SV)

Reduced preload
Reduced contractility
Increased afterload
Negative inotropic agents
  Hypocalcemia
  Hyperkalemia

Increased ⟶ **Reduced**

Cardiac Output
(CO = HR × SV)

FIGURE 19.23 Summary of Factors Affecting Cardiac Output. Factors on the left (green boxes) increase cardiac output, whereas factors on the right (red boxes) reduce it. These include the intrinsic state of the heart and adjacent arteries at a given time (preload, contractility, and afterload), and extrinsic agents that act on the heart (chronotropic and inotropic agents).

Several positive and negative chronotropic agents are shown in the top half of **figure 19.23,** summarizing how they affect cardiac output by altering heart rate. We move on now to the bottom half of the figure and inotropic agents, which act on cardiac output through their effects on contraction strength and stroke volume.

19.6d Stroke Volume and Inotropic Agents

The other factor in cardiac output is stroke volume. This is governed by three variables called *preload, contractility,* and *afterload.* Increased preload or contractility increases stroke volume, whereas increased afterload opposes the emptying of the ventricles and reduces stroke volume.

Preload

Preload is the amount of tension (stretch) in the ventricular myocardium immediately before it begins to contract. To understand how this influences stroke volume, imagine yourself engaged in heavy exercise. As active muscles massage your veins, they drive more blood back to the heart, increasing *venous return.* As more blood enters the heart, it stretches the myocardium. Because of the length–tension relationship of striated muscle (see fig. 11.12), moderate stretch enables the cardiomyocytes to generate more tension when they contract—that is, stretch increases preload. When the ventricles contract more forcefully, they expel more blood, thus adjusting cardiac output to the increase in venous return.

This principle is summarized by the **Frank–Starling law of the heart.**[31] In a concise, symbolic way, it states that SV ∝ EDV;

that is, stroke volume is proportional to the end-diastolic volume. In other words, the ventricles tend to eject as much blood as they receive. Within limits, the more they are stretched, the harder they contract on the next beat.

Although relaxed skeletal muscle is normally at an optimum length for the most forceful contraction, relaxed cardiac muscle is at less than optimum length. Additional stretch therefore produces a significant increase in contraction force on the next beat. This helps balance the output of the two ventricles. For example, if the right ventricle begins to pump an increased amount of blood, this soon arrives at the left ventricle, stretches it more than before, and causes it to increase its stroke volume and match that of the right.

Contractility

Contractility refers to how hard the myocardium contracts *for a given preload.* It doesn't describe the increase in tension produced by stretching the muscle, but rather an increase caused by factors that make the cardiomyocytes more responsive to stimulation. Factors that increase contractility are called **positive inotropic**[32] **agents,** and those that reduce it are **negative inotropic agents.**

Calcium has a strong, positive inotropic effect—it increases the strength of each contraction of the heart. This is not surprising, because Ca^{2+} not only is essential to the excitation–contraction coupling of muscle, but also prolongs the plateau of the myocardial action potential. Calcium imbalances therefore affect not only heart rate, as we have already seen, but also contraction strength. In hypercalcemia, extra Ca^{2+} diffuses into the cardiomyocytes and produces strong, prolonged contractions.

[31]Otto Frank (1865–1944), German physiologist; Ernest Henry Starling (1866–1927), English physiologist

[32]*ino* = fiber; *trop* = to change, to influence

In extreme cases, it can cause cardiac arrest in systole. In hypocalcemia, the cardiomyocytes lose Ca^{2+} to the extracellular fluid, leading to a weak, irregular heartbeat and potentially to cardiac arrest in diastole. However, severe hypocalcemia is likely to kill through skeletal muscle paralysis and suffocation before the cardiac effects are felt (see section 7.4b).

Agents that affect calcium availability have not only the chronotropic effects already examined, but also inotropic effects. We have already seen that norepinephrine increases calcium levels in the sarcoplasm; consequently, it increases not only heart rate but also contraction strength (as does epinephrine, for the same reason). The pancreatic hormone glucagon exerts an inotropic effect by stimulating cAMP production; a solution of glucagon and calcium chloride is sometimes used for the emergency treatment of heart attacks. Digitalis, a cardiac stimulant from the foxglove plant, also raises the intracellular calcium level and contraction strength; it is used to treat congestive heart failure.

Hyperkalemia has a negative inotropic effect because it reduces the strength of myocardial action potentials and thus reduces the release of Ca^{2+} into the sarcoplasm. The heart becomes dilated and flaccid. Hypokalemia, however, has little effect on contractility.

The vagus nerves have a negative inotropic effect on the atria, but they provide so little innervation to the ventricles that they have no significant effect on them.

▶▶▶ APPLY WHAT YOU KNOW

Suppose a person has a heart rate of 70 bpm and a stroke volume of 70 mL. A negative inotropic agent then reduces the stroke volume to 50 mL. What would the new heart rate have to be to maintain the same cardiac output?

Afterload

Afterload is the sum of all forces a ventricle must overcome before it can eject blood. The most significant contribution to afterload is the blood pressure in the aorta and pulmonary trunk immediately distal to the semilunar valves; it opposes the opening of these valves and thus limits stroke volume. For this reason, hypertension increases the afterload and opposes ventricular ejection. Anything that impedes arterial circulation, such as atherosclerotic plaque in the arteries, can also increase the afterload. In some lung diseases, scar tissue forms in the lungs and restricts pulmonary circulation. This increases the afterload in the pulmonary trunk. As the right ventricle works harder to overcome this resistance, it gets larger like any other muscle. Stress and hypertrophy of a ventricle can eventually cause it to weaken and fail. Right ventricular failure due to obstructed pulmonary circulation is called *cor pulmonale*[33] (CORE PUL-mo-NAY-lee). It is a common complication of emphysema, chronic bronchitis, and black lung disease (see section 22.4b).

19.6e Exercise and Cardiac Output

It is no secret that exercise makes the heart work harder, and it should come as no surprise that this increases cardiac output. The main reason the heart rate increases at the beginning of exercise is that proprioceptors in the muscles and joints transmit signals to the cardiac centers, signifying that the muscles are active and will quickly need an increased blood flow. Sympathetic output from the cardiac centers then increases cardiac output to meet the expected demand. As the exercise progresses, muscular activity increases venous return. This increases the preload on the right ventricle and is soon reflected in the left ventricle as more blood

---

[33]*cor* = heart; *pulmo* = lung

TABLE 19.2	Some Disorders of the Heart
Acute pericarditis	Inflammation of the pericardium, sometimes due to infection, radiation therapy, or connective tissue disease, causing pain and friction rub
Cardiomyopathy	Any disease of the myocardium not resulting from coronary artery disease, valvular dysfunction, or other cardiovascular disorders; can cause dilation and failure of the heart, thinning of the heart wall, or thickening of the interventricular septum
Infective endocarditis	Inflammation of the endocardium, usually due to bacterial infection, especially *Streptococcus* and *Staphylococcus*
Myocardial ischemia	Inadequate blood flow to the myocardium, usually because of coronary atherosclerosis; can lead to myocardial infarction
Pericardial effusion	Seepage of fluid from the pericardium into the pericardial cavity, often resulting from pericarditis and sometimes causing cardiac tamponade
Septal defects	Abnormal openings in the interatrial or interventricular septum, resulting in blood from the right atrium flowing directly into the left atrium, or blood from the left ventricle returning to the right ventricle; results in pulmonary hypertension, difficulty breathing, and fatigue; often fatal in childhood if uncorrected

You can find other cardiac disorders described in the following places:

Cardiac tamponade in Deeper Insight A.1; *friction rub* in section 19.1c; *angina pectoris* and *myocardial infarction* in Deeper Insight 19.1; *ventricular fibrillation, atrial fibrillation, premature ventricular contraction, bundle branch block, total heart block,* and *cardiac arrest* in Deeper Insight 19.2; *valvular stenosis, mitral valve prolapse,* and *heart murmur* in Deeper Insight 19.3; *congestive heart failure* in section 19.5d; *bradycardia and tachycardia* in section 19.6c; *cor pulmonale* in section 19.6d; and *atherosclerosis, arteriosclerosis,* and *coronary artery disease* in Deeper Insight 19.4.

flows through the pulmonary circuit and reaches the left heart. As the heart rate and stroke volume rise, cardiac output rises, which compensates for the increased venous return.

A sustained program of exercise causes hypertrophy of the ventricles, which increases their stroke volume. As explained earlier, this allows the heart to beat more slowly and still maintain a normal resting cardiac output. Some world-class, endurance-trained athletes have resting heart rates as low as 30 to 40 bpm, but because of the higher stroke volume, their resting cardiac output is about the same as that of an untrained person. Such athletes have greater cardiac reserve, so they can tolerate more exertion than a sedentary person can.

The effects of aging on the heart are discussed in section 29.4a, and some common heart diseases are listed in **table 19.2.** Disorders of the blood and blood vessels are described in chapters 18 and 20.

BEFORE YOU GO ON

Answer the following questions to test your understanding of the preceding section:

25. Explain why the vagus nerves have no significant inotropic effect on the ventricular myocardium.

26. Define *cardiac output* in words and with a simple mathematical formula.

27. Describe the cardiac center and innervation of the heart.

28. Explain what is meant by positive and negative chronotropic and inotropic agents. Give two examples of each.

29. How do preload, contractility, and afterload influence stroke volume and cardiac output?

30. Explain the principle behind the Frank–Starling law of the heart. How does this mechanism normally prevent pulmonary or systemic congestion?

DEEPER INSIGHT 19.4

CLINICAL APPLICATION

Coronary Artery Disease

Coronary artery disease (CAD) is a constriction of the coronary arteries usually resulting from *atherosclerosis*[34]—an accumulation of lipid deposits that degrade the arterial wall and obstruct the lumen. The most dangerous consequence of CAD is myocardial infarction (heart attack).

Pathogenesis

CAD begins when hypertension, diabetes, or other factors damage the arterial lining. Monocytes adhere to the lining, penetrate into the tissue, and become macrophages. Macrophages and smooth muscle cells absorb cholesterol and fat from the blood, which gives them a frothy appearance. They are then called *foam cells* and form visible *fatty streaks* on the arterial wall. Seen even in infants and children, these are harmless in themselves but have the potential to grow into atherosclerotic *plaques (atheromas*[35]*).*

Platelets adhere to these plaques and secrete a growth factor that stimulates local proliferation of smooth muscle and fibroblasts and deposition of collagen. The plaque grows into a bulging mass of lipid, fiber, and smooth muscle and other cells. When it obstructs 75% or more of the arterial lumen, it begins to cause symptoms such as angina pectoris. More seriously, inflammation of the plaque roughens its surface and creates a focal point for thrombosis. A blood clot can block what remains of the lumen, or break free and lodge in a smaller artery downstream. Sometimes a piece of plaque breaks free and travels as a *fatty embolus.* Furthermore, the plaque can contribute to spasms of the coronary artery, cutting off blood flow to the myocardium. If the lumen is already partially obstructed by a plaque and perhaps a blood clot, such a spasm can temporarily shut off the remaining flow and precipitate an attack of angina.

Over time, the resilient muscular and elastic tissue of an inflamed artery becomes increasingly replaced with scar tissue and calcium deposits, transforming an atheroma into a hard *complicated plaque* **(fig. 19.24).** Hardening of the arteries by calcified plaques is one cause of *arteriosclerosis.*[36] For reasons explained in the next chapter, this results in excessive surges of blood pressure that may weaken and rupture smaller arteries, leading to stroke and kidney failure.

Risk, Prevention, and Treatment

A paramount risk factor for CAD is excess *low-density lipoproteins (LDLs)* in the blood combined with defective LDL receptors in the arterial walls. LDLs are protein-coated droplets of cholesterol, fats, free fatty acids, and phospholipids (see "Cholesterol and Serum Lipoproteins" in section 26.1g). Most cells have LDL receptors that enable them to absorb these droplets from the blood so they can metabolize the cholesterol and other lipids. CAD can occur when the arterial cells have dysfunctional LDL receptors that "don't know when to quit," so the cells absorb and accumulate excess cholesterol.

Some risk factors for CAD are unavoidable—for example, heredity and aging. Most risk factors, however, are preventable—obesity, smoking, lack of exercise, and a personality fraught with anxiety, stress, and aggression, all conducive to the hypertension that initiates arterial damage. Diet, of course, is very significant. Eating animal fat raises one's LDL level and reduces the number of LDL receptors. Foods high in soluble fiber (such as beans, apples, and oat bran) lower blood cholesterol by an interesting mechanism: The liver normally converts cholesterol to bile acids and secretes them into the small intestine to aid fat digestion. The bile acids are reabsorbed farther down the intestine and recycled to the liver for reuse. Soluble fiber, however, binds bile acids and carries them out in the feces. To replace them, the liver synthesizes more, thus consuming more cholesterol.

CAD is often treated with a *coronary artery bypass graft (CABG).* Sections of the great saphenous vein of the leg or small thoracic arteries are used to construct a detour around the obstruction in the coronary artery. In *balloon angioplasty,*[37] a slender catheter is threaded into the coronary artery and then a balloon at its tip is inflated to press the atheroma against the arterial wall, widening the lumen. In *laser angioplasty,* the surgeon views the interior of the diseased artery with an illuminated catheter and vaporizes the atheroma with a laser. Angioplasty is less risky and expensive than bypass surgery, but is often followed by *restenosis*—atheromas grow back and reobstruct the artery months later. Insertion of a tube called a *stent* into the artery can prevent restenosis.

---

[34] *athero* = fat, fatty; *sclerosis* = hardening
[35] *athero* = fat, fatty; *oma* = mass, tumor
[36] *arterio* = artery; *sclerosis* = hardening

[37] *angio* = vessel; *plasty* = surgical repair

(a)

Lumen   Artery wall

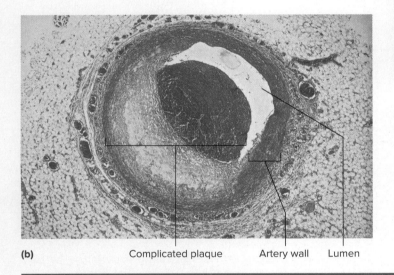

(b)

Complicated plaque   Artery wall   Lumen

(c)

FIGURE 19.24 Coronary Artery Disease. (a) Cross section of a healthy coronary artery. (b) Cross section of an artery with advanced coronary atherosclerosis. Most of the original lumen is obstructed by a *complicated plaque* composed of calcified scar tissue. The lumen is reduced to a small space that can easily be blocked by a stationary or traveling blood clot (thrombosis or thromboembolism, respectively) or by vasoconstriction.
(c) Coronary angiogram showing 60% obstruction, at arrow, of the anterior interventricular (left anterior descending, LAD) artery.

a: Ed Reschke/Getty Images; **b:** Image Source/Getty Images; **c:** kalewa/Shutterstock

STUDY GUIDE

▶ Assess Your Learning Outcomes

To test your knowledge, discuss the following topics with a study partner or in writing, ideally from memory.

19.1 Overview of the Cardiovascular System

1. Two subdivisions of the cardiovascular system and their respective functions
2. Names of the great vessels directly connected to the heart, and their relations to the heart chambers
3. The exact location of the heart, its size, and its base and apex
4. Anatomy and function of the pericardium and pericardial fluid

19.2 Gross Anatomy of the Heart

1. Three layers of the heart wall and their histological differences
2. Relative thickness of the myocardium in different chambers; the functional significance of those differences; and significance of the vortex of the heart
3. Structure and function of the fibrous skeleton of the heart
4. Anatomy and functions of the atria and ventricles; the internal septa that separate the four chambers and the external sulci that mark the chamber boundaries
5. Names and synonyms for all four valves of the heart
6. Structural differences between the valves; anatomy and function of the papillary muscles and tendinous cords
7. The path of blood flow through the heart chambers and valves
8. Anatomy of the coronary arteries and their main branches
9. Causes of myocardial infarction (MI) and how the collateral circulation in the coronary arteries reduces the risk of MI
10. Why coronary artery blood flow is greater when the heart relaxes than when it contracts, in contrast to the arterial system almost everywhere else in the body
11. Anatomy of the major veins that drain the myocardium, where this blood goes, and how the major veins are supplemented by the small cardiac veins

19.3 Cardiac Muscle and the Cardiac Conduction System

1. Structural properties of cardiomyocytes, how they differ from skeletal muscle, and how they relate to the unique function of cardiac muscle
2. Properties of cardiac muscle related to its nearly exclusive reliance on aerobic respiration
3. Components of the cardiac conduction system and the path traveled by electrical signals through the heart

19.4 Electrical and Contractile Activity of the Heart

1. The meanings of *systole* and *diastole*
2. Characteristics of the sinus rhythm of the heart; some causes of premature ventricular contraction; why an ectopic focus may take over control of the rhythm; how a nodal rhythm differs from the sinus rhythm; and the general term for any abnormal cardiac rhythm
3. The mechanism that causes cells of the SA node to depolarize rhythmically; a graph of the time course and voltages of the pacemaker potentials; how often this repeats itself in a normal resting heart; and the role of gated ion channels and specific ion inflows and outflows in creating the nodal rhythm
4. The spread of excitation through the atria, AV node, AV bundle and bundle branches, and subendocardial conducting network; changing conduction speeds at different points along this path, and why these changes are important; and the correlation of atrial and ventricular systole with the traveling wave of excitation
5. The twisting mode of ventricular contraction and the importance of the tendinous cords in preventing valvular prolapse
6. The cardiomyocyte resting potential; the actions of gated sodium, calcium, and potassium channels, and movements of these ions, in producing myocardial action potentials; how and why the shape of a myocardial action potential differs from that of a neuron; and how the plateau and unusually long refractory period of myocardial action potentials support the pumping effectiveness of the heart

7. Electrocardiograms and what happens in the heart during each ECG wave

19.5 Blood Flow, Heart Sounds, and the Cardiac Cycle

1. The principle on which a sphygmomanometer works, and why BP is expressed in millimeters of mercury (mm Hg)
2. The relationship of fluid volume, pressure, and flow, and how this relates to blood flow during the expansion and contraction of the heart chambers
3. Mechanisms that open and close the heart valves
4. The definition of *cardiac cycle,* and the names of its four phases
5. In each phase of the cardiac cycle, which chambers depolarize or repolarize, contract or relax; what each pair of valves does; what appears in the ECG; what accounts for the heart sounds; whether blood moves into or out of the atria or ventricles and where it is going when ejected; and blood volume and pressure changes in the left atrium, left ventricle, and aorta
6. The typical duration, in seconds, of atrial systole, ventricular systole, and the quiescent period, and how heart rate can be calculated from these values
7. The volume of blood typically found in each ventricle when it has finished filling; the volume ejected when a ventricle contracts; the percentage of ventricular blood that is ejected; the amount that remains behind when contraction is finished; and names of these four variables
8. Why it is necessary that each ventricle eject the same average amount of blood; what happens if either ventricle ejects more than the other over an extended period of time

19.6 Regulation of Cardiac Output

1. Pathways of sympathetic and parasympathetic innervation of the heart
2. The meaning of *cardiac output* and how to calculate it from heart rate and stroke volume
3. The typical resting heart rate and how it changes with age

STUDY GUIDE

4. Causes and terms for abnormally fast and slow resting heart rates
5. The definitions and cardiac effects of *positive* and *negative chronotropic agents*
6. Effects of the sympathetic and parasympathetic nervous systems on heart rate
7. The heart rate range associated with maximum cardiac output, and why cardiac output doesn't rise any farther at still faster heart rates
8. The intrinsic firing rate of the SA node and how vagal tone normally holds the resting heart rate below this

9. Control centers of the brainstem that regulate heart rate, and their mode of action
10. How proprioceptors, baroreceptors, and chemoreceptors influence heart rate
11. Effects of norepinephrine and acetylcholine on heart rate, and their mechanisms of action
12. Hormones and drugs that affect heart rate, and their mechanisms of action
13. Effects of potassium and calcium levels on heart rate, and their mechanisms of action
14. The meanings of *preload, contractility,* and *afterload,* and how each one affects stroke volume

15. How the Frank–Starling law of the heart matches stroke volume to venous return
16. The definitions and cardiac effects of *positive* and *negative inotropic agents*
17. How calcium and potassium affect myocardial contractility
18. How norepinephrine, glucagon, and digitalis affect contractility
19. How certain diseases reduce cardiac output by increasing afterload
20. The mechanisms by which exercise increases cardiac output
21. Why well-conditioned athletes may have unusually low resting heart rates

▶ Testing Your Recall

Answers in Appendix A

1. The cardiac conduction system includes all of the following *except*
   a. the SA node.
   b. the AV node.
   c. the bundle branches.
   d. the tendinous cords.
   e. the subendocardial conducting network.

2. To get from the right atrium to the right ventricle, blood flows through
   a. the pulmonary valve.
   b. the tricuspid valve.
   c. the bicuspid valve.
   d. the aortic valve.
   e. the mitral valve.

3. Assume that one ventricle of a child's heart has an EDV of 90 mL, an ESV of 60 mL, and a cardiac output of 2.55 L/min. What are the child's stroke volume (SV), ejection fraction (EF), and heart rate (HR)?
   a. SV = 60 mL; EF = 33%; HR = 85 bpm
   b. SV = 30 mL; EF = 60%; HR = 75 bpm
   c. SV = 150 mL; EF = 67%; HR = 42 bpm
   d. SV = 30 mL; EF = 33%; HR = 85 bpm
   e. Not enough information is given to calculate these.

4. A heart rate of 45 bpm and an absence of P waves suggest
   a. damage to the SA node.
   b. ventricular fibrillation.
   c. cor pulmonale.
   d. extrasystole.
   e. heart block.

5. There is/are _____ pulmonary vein(s) emptying into the right atrium of the heart.
   a. no
   b. one
   c. two
   d. four
   e. more than four

6. All of the following are positive chronotropic agents *except*
   a. caffeine.
   b. thyroid hormone.
   c. norepinephrine.
   d. acetylcholine.
   e. hypocalcemia.

7. The atria contract during
   a. the first heart sound.
   b. the second heart sound.
   c. the QRS complex.
   d. the PQ segment.
   e. the ST segment.

8. Cardiac muscle does not exhibit tetanus because it has
   a. fast Ca^{2+} channels.
   b. scanty sarcoplasmic reticulum.
   c. a long absolute refractory period.
   d. electrical synapses.
   e. exclusively aerobic respiration.

9. The blood contained in a ventricle during isovolumetric relaxation is
   a. the end-systolic volume.
   b. the end-diastolic volume.
   c. the stroke volume.
   d. the ejection fraction.
   e. none of these; the ventricle is empty then.

10. Drugs that increase myocardial contractility have a _____ effect.
    a. myogenic
    b. negative inotropic
    c. positive inotropic
    d. negative chronotropic
    e. positive chronotropic

11. The contraction of any heart chamber is called _____ and its relaxation is called _____.

12. The circulatory route from aorta to the venae cavae is the _____ circuit.

13. The circumflex artery travels in a groove called the _____.

14. The pacemaker potential of the SA node cells results from the slow inflow of _____.

15. Electrical signals pass quickly from one cardiomyocyte to another through the _____ of the intercalated discs.

16. Repolarization of the ventricles produces the _____ of the electrocardiogram.

17. The _____ nerves innervate the heart and tend to reduce the heart rate.

18. The death of cardiac tissue from lack of blood flow is commonly known as a heart attack, but clinically called _____.

19. Blood in the heart chambers is separated from the myocardium by a thin membrane called the _____.

20. The Frank–Starling law of the heart explains why the _____ of the left ventricle is the same as that of the right ventricle.

STUDY GUIDE

▶ Building Your Medical Vocabulary

Answers in Appendix A

State a meaning of each word element, and give a medical term from this chapter that uses it or a slight variation of it.

1. atrio-

2. brady-

3. cardio-

4. corono-

5. lun-

6. papillo-

7. semi-

8. tachy-

9. vaso-

10. ventro-

▶ What's Wrong with These Statements?

Answers in Appendix A

Briefly explain why each of the following statements is false, or reword it to make it true.

1. The blood supply to the myocardium is the coronary circulation; everything else is called the systemic circuit.

2. One-way valves prevent atrial systole from driving blood back into the venae cavae and pulmonary veins.

3. No blood can enter the ventricles until the atria contract.

4. The vagus nerves reduce the heart rate and the strength of ventricular contraction.

5. A high blood CO_2 level and low pH slow the heart rate.

6. The first heart sound occurs at the time of the P wave of the electrocardiogram.

7. If all nerves to the heart were severed, the heart would instantly stop beating.

8. If the two pulmonary arteries were clamped shut, pulmonary edema would soon follow.

9. Unlike skeletal muscle, cardiac muscle cells do not have a stable resting membrane potential.

10. An electrocardiogram is a tracing of the action potential of a cardiomyocyte.

▶ Testing Your Comprehension

1. Verapamil is a calcium channel blocker used to treat hypertension. It selectively blocks slow calcium channels. Would you expect it to have a positive or negative inotropic effect? Explain. (See Deeper Insight 3.1 to review calcium channel blockers.)

2. To temporarily treat tachycardia and restore the normal resting sinus rhythm, a physician may massage a patient's carotid artery near the angle of the mandible. Propose a mechanism by which this treatment would have the desired effect.

3. Becky, age 2, was born with a hole in her interventricular septum *(ventricular septal defect,* or *VSD).* Considering that the blood pressure in the left ventricle is significantly higher than blood pressure in the right ventricle, predict the effect of the VSD on Becky's pulmonary blood pressure, systemic blood pressure, and long-term changes in the ventricular walls.

4. In ventricular systole, the left ventricle is the first to begin contracting, but the right ventricle is the first to expel blood. Aside from the obvious fact that the pulmonary valve opens before the aortic valve, how can you explain this difference?

5. In dilated cardiomyopathy of the left ventricle, the ventricle can become enormously enlarged. Explain why this might lead to regurgitation of blood through the mitral valve (blood flowing from the ventricle back into the left atrium) during ventricular systole.

THE CIRCULATORY SYSTEM: BLOOD VESSELS AND CIRCULATION

Blood capillary beds
Biophoto Associates/Science Source

CHAPTER OUTLINE

20.1 General Anatomy of the Blood Vessels

    20.1a The Vessel Wall
    20.1b Arteries
    20.1c Capillaries
    20.1d Veins
    20.1e Circulatory Routes

20.2 Blood Pressure, Resistance, and Flow

    20.2a Blood Pressure
    20.2b Peripheral Resistance
    20.2c Regulation of Blood Pressure and Flow
    20.2d Two Purposes of Vasomotion

20.3 Capillaries and Fluid Exchange

    20.3a Diffusion
    20.3b Transcytosis
    20.3c Filtration and Reabsorption
    20.3d Edema

20.4 Venous Return and Circulatory Shock

    20.4a Mechanisms of Venous Return
    20.4b Venous Return and Physical Activity
    20.4c Circulatory Shock

20.5 Special Circulatory Routes

    20.5a Brain
    20.5b Skeletal Muscles
    20.5c Lungs

20.6 Anatomy of the Pulmonary Circuit

20.7 Systemic Vessels of the Axial Region

    20.7a The Aorta and Its Major Branches
    20.7b Arteries of the Head and Neck
    20.7c Veins of the Head and Neck
    20.7d Arteries of the Thorax
    20.7e Veins of the Thorax
    20.7f Arteries of the Abdominal and Pelvic Regions
    20.7g Veins of the Abdominal and Pelvic Regions

20.8 Systemic Vessels of the Appendicular Region

    20.8a Arteries of the Upper Limb
    20.8b Veins of the Upper Limb
    20.8c Arteries of the Lower Limb
    20.8d Veins of the Lower Limb

Connective Issues

Study Guide

DEEPER INSIGHTS

20.1 Clinical Application: Aneurysm

20.2 Clinical Application: Varicose Veins

20.3 Clinical Application: Air Embolism

20.4 Clinical Application: Portal Hypertension and Ascites

20.5 Clinical Application: Arterial Pressure Points

20.6 Clinical Application: Hypertension—"The Silent Killer"

Anatomy & Physiology
Revealed 4.0

Module 9: Cardiovascular System

BRUSHING UP

- The concepts of homeostatic set point and dynamic equilibrium should be reviewed (see section 1.6c) as background for understanding the control of blood pressure.

- Blood circulation is governed by the principle of flow down pressure gradients (see section 1.6e).

- The principles of blood volume, pressure, and flow discussed in this chapter hinge on the reasons explained in section 18.1d for the osmolarity and viscosity of blood.

- Familiarity with cardiac systole and diastole (see section 19.4) is necessary for understanding blood pressure in this chapter.

- The exchange of materials between the blood capillaries and surrounding tissues is based on the principles of filtration, osmosis and osmotic pressure, diffusion, and transcytosis introduced in section 3.3.

- Blood vessels of the limbs are described in this chapter with reference to muscle compartments (see section 10.1b, fig. 10.3) and interosseous membranes (see sections 8.4b, 8.5b; figs. 8.34, 8.40).

The route taken by the blood after it leaves the heart was a point of much confusion for many centuries. In Chinese medicine as early as 2650 BCE, blood was believed to flow in a complete circuit around the body and back to the heart, just as we know today. The Roman physician Claudius Galen (129–c. 199), however, argued that it flowed back and forth in the veins, like air in the bronchial tubes. He believed that the liver received food directly from the esophagus and converted it to blood, the heart pumped the blood through the veins to all other organs, and those organs consumed it. The arteries were thought to contain only a mysterious "vital spirit."

The Chinese view was right, but the first experimental demonstration of this didn't come for another 4,000 years. English physician William Harvey (1578–1657) studied the filling and emptying of the heart in snakes, tied off the vessels above and below the heart to observe the effects on cardiac filling and output, and measured cardiac output in a variety of living animals and estimated it in humans. He concluded that (1) the heart pumps more blood in half an hour than there is in the entire body, (2) not enough food is consumed to account for the continual production of so much blood, and therefore (3) the blood returns to the heart rather than being consumed by the peripheral organs. He couldn't explain how, since the microscope had yet to be developed to the point that enabled Antony van Leeuwenhoek (1632–1723) and Marcello Malpighi (1628–94) to discover the blood capillaries.

Harvey's work was the first experimental study of animal physiology and a milestone in the history of biology and medicine. But so entrenched were the ideas of Aristotle and Galen in the medical community, and so strange was the idea of doing experiments on living animals, that Harvey's contemporaries rejected his ideas. Indeed, some of them regarded him as a crackpot because his conclusion flew in the face of common sense—if the blood was continually recirculated and not consumed by the

tissues, they reasoned, then what purpose could it serve? We now know, of course, that he was right. Harvey's case is one of the most interesting in biomedical history, for it shows how empirical science overthrows old theories and spawns better ones, and how common sense and blind allegiance to authority can interfere with the acceptance of truth. But most importantly, Harvey's contributions represent the birth of experimental physiology.

20.1 General Anatomy of the Blood Vessels

Expected Learning Outcomes

When you have completed this section, you should be able to

a. describe the structure of a blood vessel;

b. describe the types of arteries, capillaries, and veins;

c. trace the general route usually taken by the blood from the heart and back again; and

d. describe some variations on this route.

There are three principal categories of blood vessels: arteries, veins, and capillaries (fig. 20.1). **Arteries** are defined as the efferent vessels of the cardiovascular system—that is, vessels that carry blood away from the heart. **Veins** are defined as the afferent vessels that carry blood back to the heart. (They are not defined by whether the blood in them is high or low in oxygen; see later discussion.) **Capillaries** are microscopic, thin-walled vessels that connect the smallest arteries to the smallest veins.

20.1a The Vessel Wall

Aside from their general location and direction of blood flow, the three categories of blood vessels also differ in the histological structure of their walls. The walls of arteries and veins are composed of three layers called *tunics* (fig. 20.2):

1. The **tunica interna (tunica intima)** lines the inside of the vessel and is exposed to the blood. It consists of a simple squamous epithelium called the **endothelium** overlying a basement membrane and a sparse layer of loose connective tissue; it is continuous with the endocardium of the heart. The endothelium acts as a selectively permeable barrier to materials entering or leaving the bloodstream; it secretes chemicals that stimulate dilation or constriction of the vessel; and it normally repels blood cells and platelets so that they flow freely without sticking to the vessel wall. When the endothelium is damaged, however, platelets may adhere to it and form a blood clot; and when the tissue around a vessel is inflamed, the endothelial cells produce *cell-adhesion molecules* that induce leukocytes to adhere to the surface. This causes leukocytes to congregate in tissues where their defensive actions are needed.

(a)

1 mm

(b)

100 μm

FIGURE 20.1 Micrographs of Blood Vessels. (a) A neurovascular bundle, composed of a small artery, vein, and nerve traveling together in a common sheath of connective tissue (LM). (b) A vascular cast of blood vessels of the thyroid gland (SEM). Each round mass of capillaries corresponds to one of the thyroid follicles seen in figure 17.9. These vessels pick up thyroid hormone for distribution throughout the body. a: Dennis Strete/McGraw-Hill Education; b: Susumu Nishinaga/Science Source

2. The **tunica media,** the middle layer, is usually the thickest. It consists of smooth muscle, collagen, and in some cases, elastic tissue. The relative amounts of smooth muscle and elastic tissue vary greatly from one vessel to another and form a basis for classifying vessels as described in the next section. The tunica media strengthens the vessels and prevents blood pressure from rupturing them, and it regulates the diameter of a blood vessel.

3. The **tunica externa (tunica adventitia**[1]**)** is the outermost layer. It consists of loose connective tissue that often merges with that of neighboring blood vessels, nerves, or other organs (fig. 20.1a). It anchors the vessel to adjacent tissues and provides passage for small nerves, lymphatic vessels, and smaller blood vessels that supply the tissues of the larger ones.

All blood vessels require nutrition, oxygenation, and waste-removal services for their own tissues, and the medium to large arteries and veins cannot have these needs fully met by the blood flowing through them. That blood flows too rapidly and the vessel wall is too thick for adequate exchange of chemicals between the blood and tissue fluid. Therefore, smaller vessels penetrate into the external surface of the large ones, gaining access through the tunica externa, and branch into capillaries that supply the deeper tissues of the larger vessel. The network of smaller vessels serving the larger one is called the **vasa vasorum**[2] (VAY-za vay-SO-rum). They are most conspicuous in the tunica externa because the loose organization of the tissue here doesn't hide them from view as much as tissue of the tunica media does. They supply blood to at least the outer half of the vessel wall. Tissues of the inner half are thought to be nourished by diffusion from blood in the lumen.

20.1b Arteries

Arteries are sometimes called the *resistance vessels* of the cardiovascular system because they have a relatively strong, resilient tissue structure. Each beat of the heart creates a surge of pressure in the arteries as blood is ejected into them, and arteries are built to withstand this. Being more muscular than veins, they retain their round shape even when empty, so they appear relatively circular or elliptical in tissue sections. They are divided into three classes by size, but of course there is a gradual transition from one class to the next.

1. **Conducting (elastic** or **large) arteries** are the biggest. Examples include the aorta, common carotid and subclavian arteries, pulmonary trunk, and common iliac arteries. They have a layer of elastic tissue called the *internal elastic lamina* at the border between the tunica interna and media, but microscopically, it is incomplete and difficult to distinguish from the elastic tissue of the media. The tunica media consists of 40 to 70 layers of elastic sheets, perforated like slices of Swiss cheese rolled into a tube, alternating with thin layers of smooth muscle, collagen, and elastic fibers. In histological sections, the view is dominated by this elastic tissue. The perforations allow for vasa vasorum and nerves to penetrate through all layers of the vessel and for smooth muscle cells to communicate with each other through gap junctions. There is an *external elastic lamina* at the border between the media and externa, but it, too, is difficult to distinguish from the elastic sheets of the tunica media. The tunica externa is quite

[1]*advent* = added to
[2]*vasa* = vessels; *vasorum* = of the vessels

FIGURE 20.2 Histology of the Blood Vessels.

? *Why do the arteries have so much more elastic tissue than the veins do?*

sparse in the largest arteries but is well supplied with vasa vasorum. Conducting arteries expand as they receive blood during ventricular systole, and recoil during diastole. As explained in the next section, this relieves smaller arteries downstream of excessive and harmful pressure surges. As arteries stiffen with age *(arteriosclerosis),* this protective effect declines, downstream vessels are subjected to greater stress, and the risks of aneurysm and hemorrhage rise (see Deeper Insight 20.1).

2. **Distributing (muscular** or **medium) arteries** are smaller branches that distribute blood to specific organs. You could compare a conducting artery to an interstate highway and distributing arteries to the exit ramps and state highways that serve individual towns. Most arteries that have specific anatomical names are in these first two size classes. Distributing arteries include the brachial, femoral, renal, and splenic arteries. Distributing arteries typically have up to 40 layers of smooth muscle constituting about three-quarters of the

DEEPER INSIGHT 20.1

CLINICAL APPLICATION

Aneurysm

An *aneurysm*[3] is a weak point in an artery or the heart wall. It forms a thin-walled, bulging sac that pulsates with each beat of the heart and may eventually rupture. In a *dissecting aneurysm,* blood accumulates between the tunics of an artery and separates them, usually because of degeneration of the tunica media. The most common sites of aneurysms are the abdominal aorta **(fig. 20.3)**, renal arteries, and the arterial circle at the base of the brain. Even without hemorrhaging, aneurysms can cause pain or death by putting pressure on brain tissue, nerves, adjacent veins, pulmonary air passages, or the esophagus. Other consequences include neurological disorders, difficulty in breathing or swallowing, chronic cough, or congestion of the tissues with blood. Aneurysms sometimes result from congenital weakness of the blood vessels and sometimes from trauma or bacterial infections such as syphilis. The most common cause, however, is the combination of arteriosclerosis and hypertension.

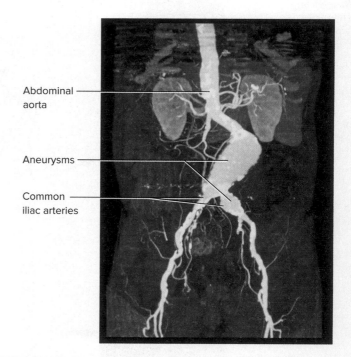

FIGURE 20.3 Aneurysms. A magnetic resonance angiogram (MRA) of the abdominopelvic region of a patient with hypertension, showing prominent bulges (aneurysms) of the inferior aorta and left common iliac artery.
Suttha Burawonk/Shutterstock

---

wall thickness. In histological sections, this smooth muscle is more conspicuous than the elastic tissue. Both the internal and external elastic laminae, however, are thick and often conspicuous.

3. **Resistance (small) arteries** are usually too variable in number and location to have individual names. They exhibit up to 25 layers of smooth muscle and relatively little elastic tissue. Compared to large arteries, they have a thicker tunica media in proportion to the lumen. The smallest of these arteries, up to 200 μm in diameter and with only one to three layers of smooth muscle, are called **arterioles.** Arterioles have very little tunica externa. They are the major point of control over how much blood an organ or tissue receives, as we shall see later.

In some places, such as the mesenteries, short vessels called **metarterioles**[4] **(thoroughfare channels)** link arterioles directly to venules and provide shortcuts through which blood can bypass the capillaries. They will be further discussed with capillary beds.

Arterial Sense Organs

Certain major arteries above the heart have sensory structures in their walls that monitor blood pressure and composition **(fig. 20.4).**

These receptors transmit information to the brainstem that serves to regulate the heartbeat, blood vessel diameters, and respiration. They are of three kinds:

1. **Carotid sinuses.** These are **baroreceptors**—sensors that monitor blood pressure. Ascending the neck on each side is a *common carotid artery,* which branches near the angle of the mandible, forming the *internal carotid artery* to the brain and *external carotid artery* to the face. The carotid sinuses are located in the wall of the internal carotid artery just above the branch point. The carotid sinus has a relatively thin tunica media and an abundance of glossopharyngeal nerve fibers in the tunica externa. The role of the baroreceptors in adjusting blood pressure, called the *baroreflex,* is described later in this chapter.

2. **Carotid bodies.** Also located near the branch of the common carotid arteries, these are oval receptors about 3 × 5 mm in size, innervated by sensory fibers of the glossopharyngeal nerves. They are **chemoreceptors**—sensors that monitor changes in blood composition. They primarily transmit signals to the brainstem respiratory centers, which adjust breathing to stabilize the blood pH and its CO_2 and O_2 levels.

3. **Aortic bodies.** These are one to three chemoreceptors located in the aortic arch near the arteries to the head and arms. They are structurally similar to the carotid bodies and have the same function, but transmit their signals to the brainstem via the vagus nerves.

---

[3]*aneurysm* = widening
[4]*meta* = beyond, next in a series

Right common carotid artery

Carotid body (containing chemoreceptors)

External carotid artery

Internal carotid artery

Carotid sinus (containing baroreceptors)

Left common carotid artery

Aortic bodies (containing chemoreceptors)

Left subclavian artery

Baroreceptors

Right subclavian artery

Aortic arch

FIGURE 20.4 Baroreceptors and Chemoreceptors in the Arteries Superior to the Heart. The structures shown here in the right carotid arteries are repeated in the left carotids. **APR**

20.1c Capillaries

For the blood to serve any purpose, materials such as nutrients, wastes, hormones, and leukocytes must pass between the blood and tissue fluids, through the walls of the vessels. There are only two places in the circulation where this occurs—the capillaries and some venules. We can think of these as the "business end" of the cardiovascular system, because all the rest of the system exists to serve the exchange processes that occur here. Since capillaries greatly outnumber venules, they are the more important of the two. Capillaries are sometimes called the *exchange vessels* of the cardiovascular system; the arterioles, capillaries, and venules are also called the **microvasculature (microcirculation).**

Capillaries (see figs. 20.1, 20.2) consist of only an endothelium and basal lamina. Their walls are as thin as 0.2 μm. They average about 5 μm in diameter at the proximal end (where they receive arterial blood), widen to about 9 μm at the distal end (where they empty into a small vein), and often branch along the way. Since erythrocytes are about 7.5 μm in diameter, they have

to stretch into elongated shapes to squeeze through the smallest capillaries.

Scarcely any cell in the body is more than 60 to 80 μm (about four to six cell widths) away from the nearest capillary. There are a few exceptions: Capillaries are scarce in tendons and ligaments and absent from epithelia, cartilage, and the cornea and lens of the eye.

Types of Capillaries

There are three types of capillaries, distinguished by the ease with which they allow substances to pass through their walls and by structural differences that account for their greater or lesser permeability.

1. **Continuous capillaries** (fig. 20.5a) occur in most tissues and organs, such as the skeletal muscles, lungs, and brain. Their endothelial cells, held together by tight junctions, form a continuous tube. A thin protein–carbohydrate layer, the **basal lamina,** surrounds the endothelium and separates it from the adjacent connective tissues. The endothelial cells are separated by narrow **intercellular clefts** about 4 nm wide. Small solutes such as glucose can pass through these clefts, but most plasma protein, other large molecules, and platelets and blood cells are held back. The continuous capillaries of the brain lack intercellular clefts and have more complete tight junctions that form the blood–brain barrier discussed in section 14.2c.

   Some continuous capillaries exhibit cells called **pericytes** that lie external to the endothelium. Pericytes have elongated tendrils that wrap around the capillary. They contain the same contractile proteins as muscle, and it is thought that they can contract and regulate blood flow through the capillaries. They also can differentiate into endothelial and smooth muscle cells and thus contribute to vessel growth and repair.

2. **Fenestrated capillaries** have endothelial cells riddled with patches of **filtration pores (fenestrations[5]) (fig. 20.5b, c).** These pores are about 20 to 100 nm in diameter, and are often spanned by a glycoprotein membrane that is much thinner than the cell's plasma membrane. They allow for the rapid passage of small molecules, but still retain most proteins and larger particles in the bloodstream. Fenestrated capillaries are important in organs that engage in rapid absorption or filtration—the kidneys, endocrine glands, small intestine, and choroid plexuses of the brain, for example.

3. **Sinusoids** are irregular blood-filled spaces in the liver, bone marrow, spleen, and some other organs (**fig. 20.6**). They are twisted, tortuous passageways, typically 30 to 40 μm wide, that conform to the shape of the surrounding tissue. The endothelial cells are separated by wide gaps

[5]*fenestra* = window

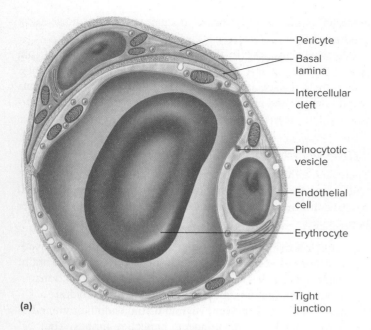

(a)

Pericyte

Basal lamina

Intercellular cleft

Pinocytotic vesicle

Endothelial cell

Erythrocyte

Tight junction

(c)

Nonfenestrated area

Filtration pores (fenestrations)

400 nm

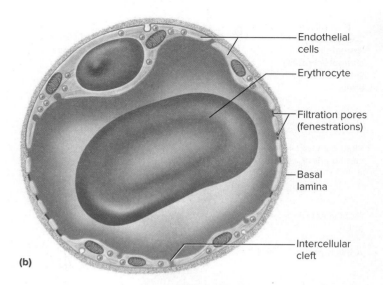

(b)

Endothelial cells

Erythrocyte

Filtration pores (fenestrations)

Basal lamina

Intercellular cleft

FIGURE 20.5 Types of Blood Capillaries. (a) Continuous capillary, cross section. (b) Fenestrated capillary, cross section. (c) Surface of a fenestrated endothelial cell showing patches of filtration pores (fenestrations) separated by nonfenestrated areas (SEM).

c: Courtesy of S. McNutt

? *Identify some organs that have fenestrated rather than continuous capillaries.*

Macrophage

Endothelial cells

Erythrocytes in sinusoid

Liver cell (hepatocyte)

Microvilli

Sinusoid

with no basal lamina, and the cells also frequently have especially large fenestrations through them. Even proteins and blood cells can pass through these pores; this is how albumin, clotting factors, and other proteins synthesized by the liver enter the blood, and how newly formed blood cells enter the circulation from the bone marrow and lymphatic organs. Some sinusoids contain macrophages or other specialized cells.

Capillary Beds

Capillaries are organized into webs called **capillary beds**—typically 10 to 100 capillaries supplied by a single arteriole or metarteriole (**fig. 20.7;** see also this chapter's opening photo). At their

FIGURE 20.6 A Sinusoid of the Liver. Large gaps between the endothelial cells allow blood plasma to directly contact the liver cells but retain blood cells in the lumen of the sinusoid.

(a) Sphincters open

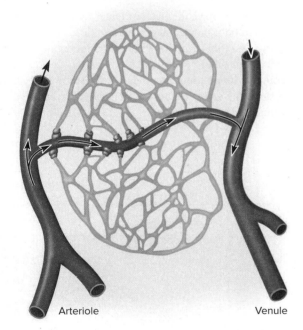

(b) Sphincters closed

FIGURE 20.7 Perfusion of a Capillary Bed. (a) Precapillary sphincters dilated and capillaries well perfused. (b) Precapillary sphincters closed, with most blood bypassing the capillaries.

distal end, capillaries transition to venules, gradually adding a thin tunica media. They may also drain into the distal end of a metarteriole, which then leads to a venule.

At any given time, about three-quarters of the body's capillaries are shut down because there isn't enough blood to supply all of them at once. In the skeletal muscles, for example, about 90% of the capillaries have little or no blood flow during periods of rest. During exercise, they receive an abundant flow while capillaries elsewhere—for example, in the skin and intestines—shut down to compensate. Capillary flow (perfusion) is usually regulated by the dilation or constriction of arterioles upstream from the capillary beds. In capillary beds supplied with metarterioles, there is often a single smooth muscle cell that wraps like a cuff around the opening to each capillary; it acts as a **precapillary sphincter** regulating blood flow. If the sphincters are relaxed, the capillaries are well perfused (fig. 20.7a). If many of the sphincters constrict, blood bypasses the capillaries, leaving them less perfused or even bloodless, and the blood takes a shortcut through the metarteriole directly to a nearby venule (fig. 20.7b).

20.1d Veins

Veins are regarded as the *capacitance vessels* of the cardiovascular system because they are relatively thin-walled and flaccid, and expand easily to accommodate an increased volume of blood; that is, they have a greater *capacity* for blood containment than arteries do. At rest, about 64% of the blood is found in the systemic veins as compared with only 13% in the systemic arteries **(fig. 20.8).** The reason veins are so thin-walled and accommodating is that, being distant from the ventricles of the heart, they are subjected to relatively low blood pressure. In large arteries, blood pressure averages 90 to 100 mm Hg and surges to 120 mm Hg during systole, whereas in veins it averages about 10 mm Hg. Furthermore, the blood flow in the veins is steady, rather than pulsating with the heartbeat like the flow in the arteries. Veins therefore don't require thick, pressure-resistant walls. They collapse when empty and thus have relatively flattened, irregular shapes in histological sections (see fig. 20.1a).

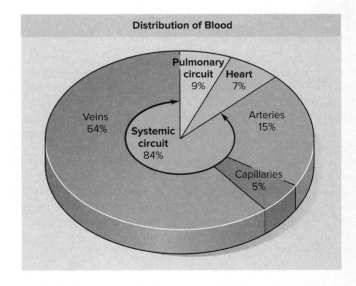

FIGURE 20.8 Typical Blood Distribution in a Resting Adult.

❓ *What anatomical fact allows the veins to contain so much more blood than the arteries do?*

As we trace blood flow in the arteries, we find it splitting off repeatedly into smaller and smaller *branches* of the arterial system. In the venous system, conversely, we find small veins merging to form larger and larger ones as they approach the heart. We refer to the smaller veins as *tributaries,* by analogy to the streams that converge and act as tributaries to rivers. In examining the types of veins, we will follow the direction of blood flow, working up from the smallest to the largest vessels.

1. **Postcapillary venules** are the smallest of the veins, beginning with diameters of about 10 to 20 µm. They receive blood from capillaries directly or by way of the distal ends of the metarterioles. They have a tunica interna with only a few fibroblasts around it and no muscle. Like capillaries, they are often surrounded by pericytes. Postcapillary venules are even more porous than capillaries; therefore, venules also exchange fluid with the surrounding tissues. Most leukocytes emigrate from the bloodstream through the venule walls.

2. **Muscular venules** receive blood from the postcapillary venules. They are up to 1 mm in diameter. They have a tunica media of one or two layers of smooth muscle, and a thin tunica externa.

3. **Medium veins** range up to 10 mm in diameter. Most veins with individual names are in this category, such as the radial and ulnar veins of the forearm and the small and great saphenous veins of the leg. Medium veins have a tunica interna with an endothelium, basement membrane, loose connective tissue, and sometimes a thin internal elastic lamina. The tunica media is much thinner than it is in medium arteries; it exhibits bundles of smooth muscle, but not a continuous muscular layer as seen in arteries. The muscle is interrupted by regions of collagenous, reticular, and elastic tissue. The tunica externa is relatively thick.

   Many medium veins, especially in the limbs, exhibit infoldings of the tunica interna that meet in the middle of the lumen, forming **venous valves** directed toward the heart (see fig. 20.21). The pressure in the veins isn't high enough to push all of the blood upward against the pull of gravity in a standing or sitting person. The upward flow of blood in these vessels depends partly on the massaging action of skeletal muscles and the ability of these valves to keep the blood from dropping down again when the muscles relax. When the muscles surrounding a vein contract, they force blood through these valves. The propulsion of venous blood by muscular massaging, aided by the venous valves, is a mechanism of blood flow called the *skeletal muscle pump.* Varicose veins result in part from the failure of the valves (see Deeper Insight 20.2).

4. **Venous sinuses** are veins with especially thin walls, large lumens, and no smooth muscle. Examples include the coronary sinus of the heart and the dural sinuses of the brain. Unlike other veins, they are not capable of vasoconstriction.

5. **Large veins** have diameters greater than 10 mm. They have some smooth muscle in all three tunics. They have a relatively thin tunica media with only a moderate amount of smooth muscle; the tunica externa is the thickest layer and contains longitudinal bundles of smooth muscle. Large veins include the venae cavae, pulmonary veins, internal jugular veins, and renal veins.

DEEPER INSIGHT 20.2

CLINICAL APPLICATION

Varicose Veins

In people who stand for long periods, such as barbers and cashiers, blood tends to pool in the lower limbs and stretch the veins. This is especially true of superficial veins, which are not surrounded by supportive tissue. Stretching pulls the cusps of the venous valves farther apart until the valves become incapable of sealing the vessel and preventing the backflow of blood. As the veins become further distended, their walls weaken and they develop into *varicose veins* with irregular dilations and twisted pathways **(fig. 20.9).** Obesity and pregnancy also promote development of varicose veins by putting pressure on large veins of the pelvic region and obstructing drainage from the lower limbs. Varicose veins sometimes develop because of hereditary weakness of the valves. With less drainage of blood, tissues of the leg and foot may become edematous and painful. *Hemorrhoids* are varicose veins of the anal canal.

FIGURE 20.9 Varicose Veins.
Jaroslav Moravcik/Shutterstock

20.1e Circulatory Routes

The simplest and most common route of blood flow is heart → arteries → capillaries → veins → heart. Blood usually passes through only one network of capillaries from the time it leaves the heart until the time it returns (**fig. 20.10a**), but there are exceptions, notably portal systems and anastomoses.

In a **portal system (fig. 20.10b)**, blood flows through two consecutive capillary networks before returning to the heart. Portal systems occur in the kidneys, connect the hypothalamus to the anterior pituitary gland, and connect the intestines to the liver (see section 20.7g).

An **anastomosis** is a point of convergence between two blood vessels other than capillaries. In an **arteriovenous anastomosis (shunt),** blood flows from an artery directly into a vein and bypasses the capillaries (**fig. 20.10c**). Shunts occur in the fingers, palms, toes, and ears, where they reduce heat loss in cold weather by allowing warm blood to bypass these exposed surfaces. Unfortunately, this makes these poorly perfused areas more susceptible to frostbite. The most common anastomoses are **venous anastomoses,** in which one vein empties directly into another (**fig. 20.10d**). These provide several alternative routes of drainage from an organ, so blockage of a vein is rarely as life-threatening as blockage of an artery. **Arterial anastomoses,** in which two arteries merge (**fig. 20.10e**), provide *collateral* (alternative) routes of blood supply to a tissue. Those of the coronary circulation were mentioned in section 19.2e. They are also common around joints where limb movement may temporarily compress an artery and obstruct one pathway. Several arterial and venous anastomoses are described later in this chapter.

(a) Simplest pathway
(1 capillary bed)

(b) Portal system
(2 capillary beds)

(c) Arteriovenous anastomosis
(shunt)

(d) Venous anastomoses

(e) Arterial anastomoses

FIGURE 20.10 Variations in Circulatory Pathways. (a) The simplest and most common pathway, with one capillary bed. (b) Portal systems, with two capillary beds in series. (c) Arteriovenous anastomoses, bypassing capillaries. (d) Venous anastomoses, with one vein feeding into another. (e) Arterial anastomoses, with one artery feeding into another.

❓ *After studying the blood vessels in sections 20.7 and 20.8, identify specific sites in the body where one can find arterial anastomoses, venous anastomoses, and portal systems.*

BEFORE YOU GO ON

Answer the following questions to test your understanding of the preceding section:

1. Name the three tunics of a typical blood vessel and explain how they differ from each other.

2. Contrast the tunica media of a conducting artery, arteriole, and venule and explain how the histological differences are related to the functional differences between these vessels.

3. Describe the differences between a continuous capillary, a fenestrated capillary, and a sinusoid.

4. Describe two routes by which substances can escape the bloodstream and pass through a capillary wall into the tissue fluid.

5. Describe the differences between a medium vein and a medium (muscular) artery. State the functional reasons for these differences.

6. Contrast an anastomosis and a portal system with the more typical pathway of blood flow.

20.2 Blood Pressure, Resistance, and Flow

Expected Learning Outcomes

When you have completed this section, you should be able to

a. explain the relationship between blood pressure, resistance, and flow;

b. describe how blood pressure is measured and expressed;

c. show how pulse pressure and mean arterial pressure are calculated;

d. describe three factors that determine resistance to blood flow;

e. explain how vessel diameter influences blood pressure and flow; and

f. describe some local, neural, and hormonal influences on vessel diameter.

To sustain life, the circulatory system must deliver oxygen and nutrients to the tissues, and remove their wastes, at a rate that keeps pace with tissue metabolism. Inadequate circulatory services to a tissue can lead within minutes to tissue necrosis and possibly death of the individual. Thus, it is crucial for the cardiovascular system to respond promptly to local needs and ensure that the tissues have an adequate blood supply at all times. This section of the chapter explores the mechanisms for achieving this.

The blood supply to a tissue can be expressed in terms of *flow* and *perfusion*. **Flow** is the amount of blood flowing through an organ, tissue, or blood vessel in a given time (such as mL/min.). **Perfusion** is the flow per given volume or mass of tissue, usually expressed in milliliters of blood per 100 grams of tissue per minute. Thus, a large organ such as the femur could have a *greater flow* but *less perfusion* than a small organ such as the ovary, because the ovary receives much more blood per gram of tissue.

In a resting individual, *total* flow is quite constant and is equal to cardiac output (typically 5.25 L/min.). Flow through individual organs, however, varies from minute to minute as blood is redirected from one organ to another. Digestion, for example, requires abundant flow to the intestines, and the cardiovascular system makes this available by reducing flow through other organs such as the kidneys. When digestion and nutrient absorption are over, blood flow to the intestines declines and a higher priority is given to the kidneys and other organs. Great variations in regional flow can occur with little or no change in total flow.

Hemodynamics, the physical principles of blood flow, are based mainly on pressure and resistance. These relationships can be concisely summarized by the formula

$$F \propto \Delta P/R.$$

In other words, the greater the pressure difference (ΔP) between two points, the greater the flow (F); the greater the resistance (R), the less the flow. Therefore, to understand the flow of blood, we must consider the factors that affect pressure and resistance.

20.2a Blood Pressure

Blood pressure (BP) is the force exerted by blood on a vessel wall. Contraction of the heart initiates a wave of pressure that sharply decreases as the blood flows farther and farther away from the ventricle of origin. Blood pressure measurement is a familiar part of a routine physical examination. It is typically measured at the **brachial artery** of the arm with a **sphygmomanometer** (SFIG-mo-meh-NOM-eh-tur). This device consists of an inflatable cuff connected to a rubber bulb for pumping air into it, and an aneroid dial gauge, digital sensor, or calibrated mercury column for measuring air pressure in the cuff. Mercury sphygmomanometers are

the "gold standard" for accuracy, but other devices have advantages for convenience and home use.

To take a patient's blood pressure with a mercury sphygmomanometer, the examiner wraps the cuff snugly around the patient's arm and inflates it with the bulb until it exceeds the systolic blood pressure. By squeezing the brachial muscles, this procedure collapses the brachial artery deep within. Even during systole, the heart can't force blood through the artery, and there is no blood flow distal to that point. The examiner now listens with a stethoscope at the bend of the elbow (cubital fossa) while slowly releasing air from the cuff.

At first there is no sound, but as soon as the systolic BP slightly exceeds the cuff pressure, each heartbeat forces the brachial artery open and allows a brief jet of blood to pass through. The vessel collapses again at diastole. This jet of blood and the subsequent surge of blood against the recollapsed artery cause turbulence that the examiner hears as a faint "bump" sound. The cuff pressure at the instant of the first bump is noted as the systolic BP. As the cuff pressure continues to decline, a bump can be heard every time the artery collapses—that is, once in each heartbeat. But soon the cuff pressure falls to a point that the brachial artery remains open even during diastole, and no further sounds occur. The point at which the last sound is detected is noted as the diastolic BP. Digital devices rely on detecting pulsations in the artery rather than detecting sounds from it.

Arterial blood pressure is expressed as a ratio of the **systolic pressure** generated by contraction (systole) of the left ventricle, to **diastolic pressure,** the minimum to which the BP falls when the ventricle is in diastole. Both are expressed in millimeters of mercury (mm Hg). A representative, healthy adult BP is 120/75 mm Hg (120 mm Hg systolic, 75 mm Hg diastolic).

The difference between systolic and diastolic pressure is called **pulse pressure** (not to be confused with pulse *rate*). For a blood pressure of 120/75, the pulse pressure (PP) would be

$$120 - 75 = 45 \text{ mm Hg.}$$

This is an important measure of the force that drives blood circulation and the maximum stress exerted on small arteries by the pressure surges generated by the heart.

Another measure of stress on the blood vessels is the **mean arterial pressure (MAP)**—the mean pressure you would obtain if you took measurements at several intervals (say every 0.1 second) throughout the cardiac cycle. MAP isn't simply an arithmetic mean of systolic and diastolic pressures, however, because the low-pressure diastole lasts longer than the high-pressure systole. A close estimate of MAP is obtained by adding diastolic pressure and one-third of the pulse pressure. For a blood pressure of 120/75,

$$\text{MAP} \approx 75 + 45/3 = 90 \text{ mm Hg.}$$

This is typical for vessels at the level of the heart, but *MAP* varies with the influence of gravity. In a standing adult, it is about 62 mm Hg in the major arteries of the head and 180 mm Hg in major arteries of the ankle.

It is the mean arterial pressure that most influences the risk of disorders such as **syncope** (SIN-co-pee) (fainting), atherosclerosis, kidney failure, edema, and aneurysm. The importance of preventing excessive blood pressure is therefore clear. One of the body's chief means of doing so is the ability of the arteries to stretch and recoil during the cardiac cycle. If the arteries were rigid tubes, pressure would rise much higher in systole and drop to nearly zero in diastole. Blood throughout the circulatory system would flow and stop, flow and stop, and put great stress on the small vessels. But healthy conducting (elastic) arteries expand with each systole, absorb some of the force of the ejected blood, and store potential energy. Then, when the heart is in diastole, their elastic recoil releases that as kinetic energy, exerts pressure on the blood, and maintains blood flow throughout the cardiac cycle. The elastic arteries thus smooth out pressure fluctuations and reduce stress on the smaller arteries.

▶▶▶ **APPLY WHAT YOU KNOW**

Explain how the histological structure of large arteries relates to their ability to stretch during systole and recoil during diastole.

Nevertheless, blood flow in the arteries is *pulsatile.* In the aorta, blood rushes forward at 120 cm/s during systole and has an average speed of 40 cm/s over the cardiac cycle. When measured farther away from the heart, systolic and diastolic pressures are lower and there is less difference between them **(fig. 20.11).** In capillaries and veins, the blood flows at a steady speed with little if any pulsation because the pressure surges have been damped out by the distance traveled and the elasticity of the arteries. This

is why an injured vein exhibits relatively slow, steady bleeding, whereas blood jets intermittently from a severed artery. In the inferior vena cava near the heart, however, venous flow fluctuates with the respiratory cycle for reasons explained later, and there is some fluctuation in the jugular veins of the neck.

As we get older, our arteries become less distensible and absorb less systolic force. This increasing stiffness of the arteries is called **arteriosclerosis**[6] ("hardening of the arteries"). The primary cause of it is cumulative damage by free radicals, which cause gradual deterioration of the elastic and other tissues of the arterial walls—much like old rubber bands that become less stretchy. Another contributing factor is **atherosclerosis,** the growth of lipid deposits in the arterial walls (see Deeper Insight 19.4). These deposits can become calcified *complicated plaques,* giving the arteries a hard, crunchy or bonelike consistency. As a result of these degenerative changes, blood pressure rises with age. Common blood pressures at the age of 20 are about 123/76 for males and 116/72 for females. For healthy persons at age 70, typical blood pressures are around 145/82 and 159/85 for the two sexes, respectively.

Hypertension (high BP) is commonly considered to be a chronic resting blood pressure higher than 140/90. (*Temporary* high BP resulting from emotion or exercise is not hypertension.) Among other effects, hypertension can weaken arteries and cause aneurysms, and it promotes the development of atherosclerosis (see Deeper Insight 20.6). **Hypotension** is chronic low resting BP. It may be a consequence of blood loss, dehydration, anemia, or other factors and is normal in people approaching death. There is no particular numerical criterion for hypotension.

Blood pressure is physiologically determined by three principal variables: cardiac output, blood volume, and resistance to flow. Cardiac output was discussed in section 19.6. Blood volume is regulated mainly by the kidneys, which have a greater influence than any other organ on blood pressure (assuming there is a beating heart). Their influence on blood pressure is discussed in chapters 23 and 24. Resistance to flow is our next topic of consideration.

20.2b Peripheral Resistance

Peripheral resistance is the opposition to flow that the blood encounters in vessels away from the heart. Moving blood would exert no pressure against a vessel wall unless it encountered at least some downstream resistance. Thus, pressure and resistance are not independent variables in blood flow—rather, pressure is affected by resistance, and flow is affected by both. Resistance, in turn, hinges on three variables that we consider now: blood viscosity, vessel length, and vessel radius.

Blood Viscosity

The viscosity of blood stems mainly from its plasma proteins (albumin) and erythrocytes (see section 18.1d). A deficiency of erythrocytes (anemia) or albumin (hypoproteinemia) reduces viscosity and speeds up blood flow. On the other hand, viscosity increases and flow declines in such conditions as polycythemia and dehydration.

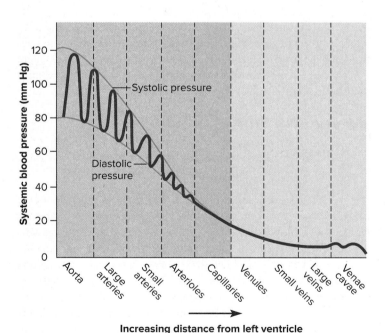

FIGURE 20.11 Changes in Blood Pressure Relative to Distance from the Heart. Because of arterial elasticity and the effect of friction against the vessel wall, all measures of blood pressure decline with distance—systolic pressure, diastolic pressure, pulse pressure, and mean arterial pressure. There is no pulse pressure beyond the arterioles, but there are slight pressure oscillations in the venae cavae caused by the respiratory pump described later in this chapter.

[Figure axis labels: Systemic blood pressure (mm Hg); values 0, 20, 40, 60, 80, 100, 120; Systolic pressure; Diastolic pressure; x-axis: Aorta, Large arteries, Small arteries, Arterioles, Capillaries, Venules, Small veins, Large veins, Venae cavae; Increasing distance from left ventricle]

---

[6]*arterio* = artery; *sclerosis* = hardening

Vessel Length

The farther a liquid travels through a tube, the more cumulative friction it encounters; pressure and flow therefore decline with distance. Partly for this reason, if you were to measure mean arterial pressure in a reclining person, you would obtain a higher value in the arm, for example, than in the ankle. In a reclining person, a strong pulse in the dorsal artery of the foot is a good sign of adequate cardiac output. If perfusion is good at that distance from the heart, it is likely to be good elsewhere in the systemic circulation.

Vessel Radius

Blood viscosity and vessel lengths don't change in the short term, of course. In a healthy individual, the only significant ways of controlling peripheral resistance from moment to moment are **vasoconstriction,** the narrowing of a vessel, and **vasodilation,** the widening of a vessel. Vasoconstriction occurs when the smooth muscle of the tunica media contracts. Vasodilation, however, is brought about not by any muscular effort to widen a vessel, but rather by muscular passivity—relaxation of the smooth muscle, allowing blood pressure to expand the vessel. Vasoconstriction and vasodilation are collectively called **vasomotion.** Vasomotion is controlled in part by a nucleus in the medulla oblongata of the brain called the **vasomotor center.**

The effect of vessel radius on blood flow stems from the friction of the moving blood against the vessel walls. Blood normally exhibits smooth, silent **laminar**[7] **flow.** That is, it flows in layers—faster near the center of a vessel, where it encounters less friction, and slower near the walls, where it drags against the vessel. You can observe a similar effect from the vantage point of a riverbank. The current may be very swift in the middle of a river but quite sluggish near shore, where the water encounters more friction against the riverbank and bottom. When a blood vessel dilates, a greater portion of the blood is in the middle of the stream and the average flow may be quite swift. When the vessel constricts, more of the blood is close to the wall and the average flow is slower **(fig. 20.12).**

Thus, the radius of a vessel markedly affects blood flow. Indeed, flow (F) is proportional not merely to vessel radius (r) but to the *fourth power* of radius—that is, $F \propto r^4$. This makes radius a very potent factor in the control of flow. For the sake of simplicity, consider a hypothetical blood vessel with a 1 mm radius when maximally constricted and a 3 mm radius when completely dilated. At a 1 mm radius, suppose the flow rate is 1 mL/min. By the formula $F \propto r^4$, consider how the flow would change as radius changed:

if r = 1 mm, then $r^4 = 1^4 = 1$, and F = 1 mL/min. (given);

if r = 2 mm, then $r^4 = 2^4 = 16$, and F = 16 mL/min.; and

if r = 3 mm, then $r^4 = 3^4 = 81$, and F = 81 mL/min.

These actual numbers don't matter; what matters is that a mere 3-fold increase in radius has produced an 81-fold increase in flow—a demonstration that vessel radius exerts a very powerful influence

(a)

(b)

FIGURE 20.12 Laminar Flow and the Effect of Vessel Radius. Blood flows more slowly near the vessel wall, as indicated by shorter arrows, than it does near the center of the vessel. Each arrow can be construed as the distance that a hypothetical blood cell would travel in a given amount of time, varying with its distance from the vessel wall. (a) When the vessel radius is large, the average velocity of flow is high. (b) When the radius is less, the average velocity is lower because a larger portion of the blood is slowed down by friction against the vessel wall.

over flow. Blood vessels are, indeed, capable of substantial changes in radius. **Figure 20.13,** for example, depicts an experiment in which a drop of epinephrine caused an arteriole to constrict to one-third of its relaxed diameter. Since blood viscosity and vessel length don't change from moment to moment, vessel radius is the most adjustable of all variables that govern peripheral resistance.

▶▶▶ APPLY WHAT YOU KNOW

Suppose a vessel with a radius of 1 mm had a flow of 3 mL/min., and then the vessel dilated to a radius of 5 mm. What would be the new flow rate?

To integrate this information, consider how the velocity of blood flow differs from one part of the systemic circuit to another **(table 20.1).** Flow is fastest in the aorta because it is a large vessel close to the pressure source, the left ventricle. From aorta to capillaries, velocity diminishes for three reasons: (1) The blood has traveled a greater distance, so friction has slowed it down. (2) The arterioles and capillaries have smaller radii and therefore put up more resistance. (3) Even though the radii of individual vessels become smaller as we progress farther from the heart, the number of vessels and their *total* cross-sectional area become greater and greater. The aorta has a cross-sectional area of 3 to 5 cm², whereas the total cross-sectional area of all the capillaries is about 4,500 to 6,000 cm². Thus, a given volume of aortic blood is distributed over a greater total area in the capillaries, which *collectively* form a wider path in the bloodstream. Just as water slows down when a narrow mountain stream flows into a lake, blood slows down as it enters pathways with a greater total area or volume.

---

[7]*lamina* = layer

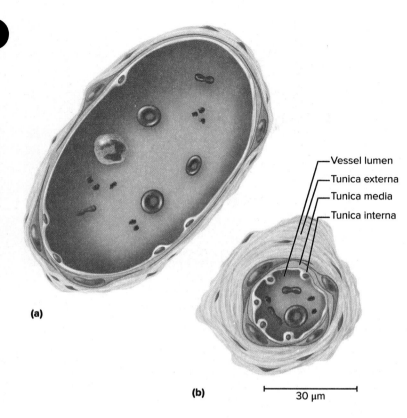

Vessel lumen
Tunica externa
Tunica media
Tunica interna

(a)

(b)

30 μm

FIGURE 20.13 The Capacity for Vasoconstriction in an Arteriole. (a) A dilated arteriole (cross section). (b) The same arteriole, at a point just 1 mm from the area shown in part (a). A single drop of epinephrine applied here caused the arteriole to constrict to about one-third of its dilated diameter. Drawn to scale from transmission electron micrographs.

TABLE 20.1	Blood Velocity in the Systemic Circuit	
Vessel	**Typical Lumen Diameter**	**Velocity***
Aorta	2.5 cm	1,200 mm/s
Arterioles	20–50 μm	15 mm/s
Capillaries	5–9 μm	0.4 mm/s
Venules	20 μm	5 mm/s
Inferior vena cava	3 cm	80 mm/s

*Peak systolic velocity in the aorta; mean or steady velocity in other vessels, assuming no upstream vasoconstriction adding to resistance

From capillaries to vena cava, velocity rises again. One reason for this is that the veins are larger than the capillaries, so they create less resistance. Furthermore, since many capillaries converge on one venule, and many venules on a larger vein, a large amount of blood is being forced into a progressively smaller channel—like water flowing from a lake into an outlet stream and thus flowing

faster again. Note, however, that blood in the veins never regains the velocity it had in the large arteries. This is because the veins are farther from the pressure head (the heart) and because they are more compliant than arteries—they stretch to accommodate more blood, and this reduces pressure and flow.

Arterioles are the most significant point of control over peripheral resistance and blood flow because (1) they are on the proximal sides of the capillary beds, so they are best positioned to regulate flow into the capillaries and thus regulate perfusion of the organs; (2) they greatly outnumber any other class of arteries and thus provide the most numerous control points; and (3) they are more muscular in proportion to their diameters than any other class of blood vessels and are highly capable of changing radius. Arterioles alone account for about half of the total peripheral resistance of the circulatory system. However, larger arteries and veins also influence peripheral resistance through their own constriction and dilation.

20.2c Regulation of Blood Pressure and Flow

Vasoreflexes, we have seen, are quick and powerful means of altering blood pressure and flow. There are three ways of controlling vasomotor activity: local, neural, and hormonal mechanisms. We now consider each of these influences in turn.

Local Control

Autoregulation is the ability of tissues to regulate their own blood supply. According to the *metabolic theory of autoregulation,* if a tissue is inadequately perfused, it becomes hypoxic and its metabolites (waste products) accumulate—CO_2, H^+, K^+, lactate, and adenosine, for example. These factors stimulate vasodilation, which increases blood flow. As the bloodstream delivers oxygen and carries away the metabolites, the vessels reconstrict. Thus, a homeostatic dynamic equilibrium is established that adjusts perfusion to the tissue's metabolic needs.

In addition, platelets, endothelial cells, and the perivascular tissues secrete a variety of **vasoactive chemicals** that stimulate vasodilation under such conditions as trauma, inflammation, and exercise. These include histamine, bradykinin, and prostaglandins. The drag of blood flowing against the endothelial cells creates a *shear stress* (like rubbing your palms together) that stimulates them to secrete prostacyclin and nitric oxide, which are vasodilators.

If a tissue's blood supply is cut off for a time and then restored, it often exhibits **reactive hyperemia**—an increase above the normal level of flow. This may be due to the accumulation of metabolites during the period of ischemia. Reactive hyperemia is seen when the skin flushes (reddens) after a person comes in from the cold. It also occurs in the forearm if a blood pressure cuff is inflated for too long and then loosened.

Over a longer time, a hypoxic tissue can increase its own perfusion by **angiogenesis**[8]—the growth of new blood vessels. (This

---
[8]*angio* = vessels; *genesis* = production of

term also refers to embryonic development of blood vessels.) Three situations in which this is important are the regrowth of the uterine lining after each menstrual period, the development of a higher density of blood capillaries in the muscles of well-conditioned athletes, and the growth of arterial bypasses around obstructions in the coronary circulation. There is great clinical importance in determining how growth factors and inhibitors control angiogenesis. Malignant tumors secrete growth factors that stimulate a dense network of vessels to grow into them and provide nourishment to the cancer cells. Oncologists are interested in finding a way to block *tumor angiogenesis,* thus to choke off a tumor's blood supply and perhaps shrink or kill it.

Neural Control

In addition to local control, the blood vessels are under remote control by the central and autonomic nervous systems. The vasomotor center of the medulla oblongata exerts sympathetic control over blood vessels throughout the body. (Precapillary sphincters have no innervation, however, and respond only to local and hormonal stimuli.) Sympathetic nerve fibers stimulate most blood vessels to constrict, and allow for vasodilation by reducing the nerve firing rate *(sympathetic tone).* The role of sympathetic and vasomotor tone in controlling vessel diameter is explained in section 15.3c.

The vasomotor center is an integrating center for three autonomic reflexes—*baroreflexes, chemoreflexes,* and the *medullary ischemic reflex.* A **baroreflex**[9] is a negative feedback response to changes in blood pressure (see fig. 15.1). The changes are detected by baroreceptors of the carotid sinuses (see fig. 20.4). Glossopharyngeal nerve fibers from these sinuses transmit signals continually to the brainstem. When blood pressure rises, their signaling rate rises. This *inhibits* the sympathetic cardiac and vasomotor neurons and reduces sympathetic tone, and it *excites* the vagal fibers to the heart. Thus, it reduces the heart rate and cardiac output, dilates the arteries and veins, and reduces blood pressure **(fig. 20.14).** When blood pressure drops below normal, on the other hand, the opposite reactions occur and BP rises back to normal.

Baroreflexes are important chiefly in short-term regulation of BP, for example in adapting to changes in posture. Perhaps you have jumped quickly out of bed and felt a little dizzy for a moment. This occurs because gravity draws the blood into the large veins of the abdomen and lower limbs when you stand, which reduces venous return to the heart and cardiac output to the brain. Normally, the baroreceptors respond quickly to this drop in pressure and restore cerebral perfusion (see fig. 1.8). Baroreflexes are not effective in correcting chronic hypertension, however. Within 2 days or less, they adjust their set point to the higher BP and maintain dynamic equilibrium at this new level.

A **chemoreflex** is an autonomic response to changes in blood chemistry, especially its pH and concentrations of O_2 and CO_2. It is initiated by the chemoreceptors called *aortic bodies* and *carotid bodies* described earlier. The primary role of chemoreflexes is to adjust respiration to changes in blood chemistry, but they have a secondary role in vasoreflexes. Hypoxemia (blood O_2 deficiency), hypercapnia

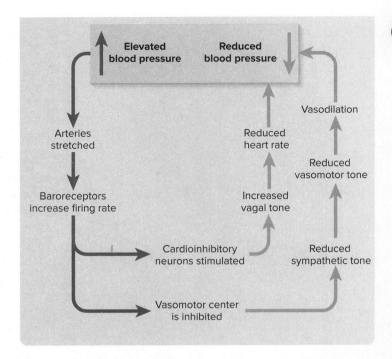

FIGURE 20.14 Negative Feedback Control of Blood Pressure. High blood pressure activates this cycle of reactions that ideally return blood pressure to normal.

(CO_2 excess), and acidosis (low blood pH) stimulate the chemoreceptors and act through the vasomotor center to induce widespread vasoconstriction. This increases overall BP, thus increasing perfusion of the lungs and the rate of gas exchange. Chemoreceptors also stimulate breathing, so increased ventilation of the lungs matches their increased perfusion. Increasing one without the other (airflow without blood flow, or vice versa) would be of little use.

The **medullary ischemic reflex** (iss-KEE-mic) is an autonomic response to reduced perfusion of the brain. The medulla oblongata monitors its own blood supply and activates corrective reflexes when it senses a state of ischemia (insufficient perfusion). Within seconds of a drop in perfusion, the cardiac and vasomotor centers of the medulla send sympathetic signals to the heart and blood vessels that accelerate the heart and constrict the vessels. These actions raise the blood pressure and ideally restore normal cerebral perfusion. The cardiac and vasomotor centers also receive input from other brain centers, so stress, anger, and arousal can raise the blood pressure. The hypothalamus acts through the vasomotor center to redirect blood flow in response to exercise or changes in body temperature.

Hormonal Control

Hormones are another means of remote control over perfusion. All of the following hormones influence blood pressure, some through their vasoactive effects and some through means such as regulating water balance:

- **Angiotensin II.** This is a potent vasoconstrictor that raises the blood pressure. Its synthesis and action are depicted in figure 23.15. Its synthesis requires *angiotensin-converting*

---

[9]*baro* = pressure

enzyme (ACE). Hypertension is often treated with drugs called *ACE inhibitors,* which block the action of this enzyme, thus lowering angiotensin II levels and blood pressure.

- **Aldosterone.** This "salt-retaining hormone" primarily promotes Na^+ retention by the kidneys. Since water follows sodium osmotically, Na^+ retention promotes water retention, thereby supporting blood pressure.

- **Natriuretic peptides.** These hormones, secreted by the heart, antagonize aldosterone. They increase Na^+ excretion by the kidneys, thus reducing blood volume and pressure. They also have a generalized vasodilator effect that helps to lower blood pressure.

- **Antidiuretic hormone.** ADH primarily promotes water retention, but at pathologically high concentrations it is also a vasoconstrictor—hence its alternate name, *arginine vasopressin.* Both of these effects raise blood pressure.

- **Epinephrine and norepinephrine.** These adrenal and sympathetic catecholamines bind to α-adrenergic receptors on the smooth muscle of most blood vessels. This stimulates vasoconstriction and raises the blood pressure.

▶▶▶ APPLY WHAT YOU KNOW

Renin inhibitors are drugs used to treat hypertension. Explain how you think they would produce the desired effect.

20.2d Two Purposes of Vasomotion

Vasomotion (vasoconstriction and vasodilation) serves two physiological purposes: a generalized raising or lowering of blood pressure throughout the body, and selectively modifying the perfusion of a particular organ and rerouting blood from one region of the body to another.

A generalized increase in blood pressure requires centralized control—an action on the part of the medullary vasomotor center or by hormones that circulate throughout the system, such as angiotensin II or epinephrine. Widespread vasoconstriction raises the overall blood pressure because the whole "container" (the blood vessels) squeezes on a fixed amount of blood, like water pressure rising if you squeeze a plastic water bottle. This can be important in supporting cerebral perfusion in situations such as hemorrhaging or dehydration, in which blood volume has significantly fallen. Conversely, generalized vasodilation lowers BP throughout the system.

The rerouting of blood and changes in the perfusion of individual organs can be achieved by either central or local control. For example, during periods of exercise, the sympathetic nervous system can selectively reduce flow to the kidneys and digestive tract. Yet as we saw earlier, metabolite accumulation in a tissue can stimulate local vasodilation and increase perfusion of that tissue without affecting circulation elsewhere in the body.

If a specific artery constricts, pressure downstream from the constriction drops and pressure upstream from it rises. If blood can travel by either of two routes and one route puts up more resistance than the other, most blood follows the path of least resistance. This mechanism enables the body to redirect blood from one organ to another.

For example, if you are dozing in an armchair after a big meal **(fig. 20.15a),** vasoconstriction shuts down blood flow to 90% or more of the capillaries in the muscles of your lower limbs (and muscles elsewhere). This raises the BP above the limbs, where the aorta gives off a branch, the superior mesenteric artery, supplying the small intestine. High resistance in the circulation of the limbs and low resistance in the superior mesenteric artery route blood to the small intestine, where it is needed to absorb the nutrients you are digesting.

(a) Resting **(b) Exercise**

FIGURE 20.15 Redirection of Blood Flow in Response to Changing Metabolic Needs. (a) After a meal, the intestines receive priority and the skeletal muscles receive relatively little flow. (b) During exercise, the muscles receive higher priority. The vasomotion that redirects such blood flow occurs mainly at the level of the arterioles in the respective organs.

On the other hand, during vigorous exercise, the arteries in your lungs, coronary circulation, and muscles dilate. To increase the circulation in these routes, vasoconstriction must occur elsewhere, such as the kidneys and digestive tract (**figs. 20.15b, 20.16**). That reduces their perfusion for the time being, making more blood available to the organs important in sustaining exercise. Thus, local changes in peripheral resistance can shift blood flow from one organ system to another to meet the changing metabolic priorities of the body.

BEFORE YOU GO ON

Answer the following questions to test your understanding of the preceding section:

7. Explain why a drop in diastolic pressure would raise one's pulse pressure even if systolic pressure remained unchanged. How could this rise in pulse pressure adversely affect the blood vessels?

8. Explain why arterial blood flow is pulsatile and venous flow is not.

9. What three variables affect peripheral resistance to blood flow? Which of these is most able to change from one minute to the next?

10. What are the three primary mechanisms for controlling vessel radius? Briefly explain each.

11. Explain how the baroreflex serves as an example of homeostasis and negative feedback.

12. Explain how the body can shift the flow of blood from one organ system to another.

20.3 Capillaries and Fluid Exchange

Expected Learning Outcomes

When you have completed this section, you should be able to

a. describe how materials get from the blood into the surrounding tissues;

b. describe and calculate the forces that enable capillaries to give off and reabsorb fluid; and

c. describe the causes and effects of edema.

Only 250 to 300 mL (5%) of the blood is in the capillaries at any given time. This, however, is the most important blood in the body, for it is mainly across capillary walls that exchanges occur between the blood and surrounding tissues. **Capillary exchange** refers to this two-way movement of fluid.

Chemicals given off by the capillary blood to the perivascular tissues include oxygen, glucose and other nutrients, antibodies, and hormones. Chemicals taken up by the capillaries include carbon dioxide and other wastes, and many of the same substances as they give off: glucose and fatty acids released from storage in the liver and adipose tissue; calcium and other minerals released from bone; antibodies secreted by immune cells; and hormones secreted by the endocrine glands. Thus, many chemicals have a two-way traffic between the blood and connective tissue, leaving the capillaries at one point and entering at another. Along with all these solutes, there is substantial movement of water into and out of the bloodstream across the capillary walls. Significant exchange also occurs across

(a)

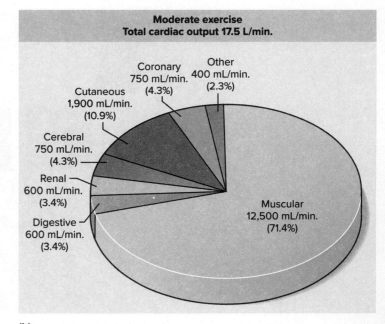

(b)

FIGURE 20.16 Differences in Systemic Blood Flow According to States of Physical Activity. (a) Relative blood flow at rest. (b) Relative blood flow during exercise.

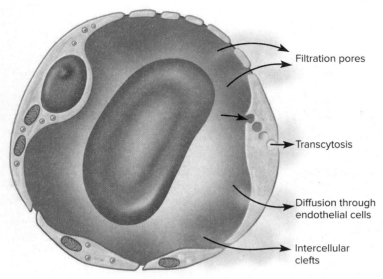

FIGURE 20.17 Routes of Fluid Exchange Across the Capillary Wall. Materials move through the capillary wall via filtration pores (in fenestrated capillaries only), by transcytosis, by diffusion through the endothelial cells, and via intercellular clefts.

the walls of the venules, but capillaries are the more important exchange site because they so greatly outnumber the venules.

The mechanisms of capillary exchange are difficult to study quantitatively because it is hard to measure pressure and flow in such small vessels. For some representative cutaneous capillaries, however, BP has been measured at 32 mm Hg at the arterial end and 15 mm Hg at the venous end, 1 mm away. Capillary BP drops rapidly because of the substantial friction the blood encounters in such narrow vessels. It takes 1 to 2 seconds for an RBC to pass through a typical capillary, traveling about 0.7 mm/s.

Chemicals pass through the capillary wall by three routes **(fig. 20.17):**

1. the endothelial cell cytoplasm;
2. intercellular clefts between the endothelial cells; and
3. filtration pores of the fenestrated capillaries.

The mechanisms of movement through the capillary wall are *diffusion, transcytosis, filtration,* and *reabsorption.*

20.3a Diffusion

The most important mechanism of exchange is diffusion. Glucose and oxygen, being more concentrated in the systemic blood than in the tissue fluid, diffuse out of the blood. Carbon dioxide and other wastes, being more concentrated in the tissue fluid, diffuse into the blood. (O_2 and CO_2 diffuse in the opposite directions in the lungs.) Such diffusion is possible only if the solute can either permeate the plasma membranes of the endothelial cells or find passages large enough to pass through—namely, the filtration pores and intercellular clefts. Such lipid-soluble substances as steroid hormones, O_2, and CO_2 diffuse easily through the plasma membranes. Substances insoluble in lipids, such as glucose and electrolytes, must

pass through membrane channels, filtration pores, or intercellular clefts. Large molecules such as proteins are usually held back.

20.3b Transcytosis

Transcytosis is a process in which endothelial cells pick up material on one side of the plasma membrane by pinocytosis or receptor-mediated endocytosis, transport the vesicles across the cell, and discharge the material on the other side by exocytosis (see fig. 3.22). This probably accounts for only a small fraction of solute exchange across the capillary wall, but fatty acids, albumin, and some hormones such as insulin move across the endothelium by this mechanism.

20.3c Filtration and Reabsorption

Capillary fluid exchange is driven in large part by the equilibrium between filtration and osmosis discussed in section 3.3. Typically, fluid filters out of the arterial end of a capillary and osmotically reenters it at the venous end. This fluid delivers materials to the cells and rinses away their metabolic wastes. It may seem odd that a capillary could give off fluid at one point and reabsorb it at another. This comes about as the result of a shifting balance between osmosis and hydrostatic pressure **(fig. 20.18). Hydrostatic pressure** is the physical force exerted by a liquid against a surface such as a capillary wall. Blood pressure is one example.

A typical capillary has a blood pressure of about 30 mm Hg at the arterial end. The hydrostatic pressure of the interstitial space has been difficult to measure, but a typical value accepted by many authorities is –3 mm Hg. The negative value indicates that this is a slight suction, which helps draw fluid out of the capillary. (This force will be represented hereafter as 3_{out}.) In this case, the positive hydrostatic pressure within the capillary and the negative interstitial pressure work in the same direction, creating a total outward force of about 33 mm Hg.

These forces are opposed by **colloid osmotic pressure (COP),** the portion of the osmotic pressure due to protein. The blood has a COP of about 28 mm Hg, due mainly to albumin. Tissue fluid has less than one-third the protein concentration of blood plasma and has a COP of about 8 mm Hg. The difference between the COP of blood and tissue fluid is called **oncotic pressure:** $28_{in} - 8_{out} = 20_{in}$. Oncotic pressure tends to draw water into the capillary by osmosis, opposing hydrostatic pressure.

These opposing forces produce a **net filtration pressure (NFP)** of 13 mm Hg out, as follows:

Hydrostatic pressure

Blood pressure		30_{out}
Interstitial pressure	+	3_{out}
Net hydrostatic pressure		33_{out}

Colloid osmotic pressure (COP)

Blood COP		28_{in}
Tissue fluid COP	–	8_{out}
Net COP (oncotic pressure)		20_{in}

FIGURE 20.18 The Forces of Capillary Filtration and Reabsorption. Note the shift from net filtration at the arterial end (left) to net reabsorption at the venous end (right). **A&PR**

Net filtration pressure (NFP)

Net hydrostatic pressure	33_{out}
Oncotic pressure	$-$ 20_{in}
Net filtration pressure	13_{out}

The NFP of 13 mm Hg causes about 0.5% of the blood plasma to leave the capillaries at the arterial end.

At the venous end, however, capillary blood pressure is lower—about 10 mm Hg. All the other pressures are essentially unchanged. Thus, we get the following:

Hydrostatic pressure

Blood pressure		10_{out}
Interstitial pressure	$+$	3_{out}
Net hydrostatic pressure		13_{out}

Net reabsorption pressure

Oncotic pressure	20_{in}
Net hydrostatic pressure	$-$ 13_{out}
Net reabsorption pressure	7_{in}

The prevailing force is inward at the venous end because osmotic pressure overrides filtration pressure. The **net reabsorption pressure** of 7 mm Hg inward causes the capillary to reabsorb fluid at this end.

Now you can see why a capillary gives off fluid at one end and reabsorbs it at the other. The only pressure that changes significantly from the arterial end to the venous end is the capillary blood pressure, and this change is responsible for the shift from filtration to reabsorption. With a reabsorption pressure of 7 mm Hg and a net filtration pressure of 13 mm Hg, it might appear that far more fluid would leave the capillaries than reenter them. However, since capillaries branch along their length, there are more of them at the venous end than at the arterial end, which partially compensates for the difference between filtration and reabsorption pressures. They also typically have nearly twice the diameter at the venous end than they have at the arterial end, so there is more capillary surface area available to reabsorb fluid than to give it off. Consequently, capillaries in most places reabsorb about 85% of the fluid they filter. The rest is absorbed and returned to the blood by way of the lymphatic system, as described in the next chapter.

Of course, water is not the only substance that crosses the capillary wall by filtration and reabsorption. Chemicals dissolved in the water are "dragged" along with it and pass through the capillary wall if they are not too large. This process, called **solvent drag,** will be important in our discussions of kidney function in chapter 23.

Variations in Capillary Filtration and Reabsorption

The figures in the preceding discussion are only examples; circumstances differ from place to place in the body and from time to time in the same capillaries. Capillaries usually reabsorb most of the fluid they filter, but not always. The kidneys have capillary networks called *glomeruli* in which there is little or no reabsorption; they are entirely devoted to filtration. Alveolar capillaries of the lungs, by contrast, are almost entirely dedicated to absorption so fluid doesn't fill the air spaces.

Capillary activity also varies from moment to moment. In a resting tissue, most precapillary sphincters are constricted and the capillaries are collapsed. Capillary BP is very low (if there is any flow at all), and reabsorption predominates. When a tissue becomes more metabolically active, its capillary flow increases. In active muscles, capillary pressure rises to the point that filtration overrides reabsorption along the entire length of the capillary. Fluid accumulates in the muscle and increases muscular bulk by as much as 25%. Capillary permeability is also subject to chemical influences. Traumatized tissue releases such chemicals as substance P, bradykinin, and histamine, which increase permeability and filtration.

20.3d Edema

Edema is the accumulation of excess fluid in a tissue. It often shows as swelling of the face, fingers, abdomen, or ankles **(fig. 20.19),** but also occurs in internal organs where its effects are hidden from view. Edema occurs when fluid filters into a tissue faster than it is reabsorbed. It has three fundamental causes:

1. **Increased capillary filtration.** Numerous conditions can increase the rate of capillary filtration and accumulation of fluid in the tissues. Kidney failure, for example, leads to water retention and hypertension, raising capillary blood pressure and filtration rate. Histamine dilates arterioles and raises

capillary pressure and makes the capillary wall more permeable. Capillaries generally become more permeable in old age as well, putting elderly people at increased risk of edema. Capillary blood pressure also rises in cases of poor venous return—the flow of blood from the capillaries back to the heart. As we will see in the next section, good venous return depends on muscular activity. Therefore, edema is a common problem among people confined to bed or a wheelchair.

 Failure of the right ventricle of the heart tends to cause pressure to back up in the systemic veins and capillaries, thus resulting in systemic edema. Failure of the left ventricle causes pressure to back up in the lungs, causing pulmonary edema.

2. **Reduced capillary reabsorption.** Capillary reabsorption depends on oncotic pressure, which is proportional to the concentration of blood albumin. Therefore, a deficiency of albumin (hypoproteinemia) produces edema by reducing the reabsorption of tissue fluid. Since albumin is produced by the liver, liver diseases such as cirrhosis tend to lead to hypoproteinemia and edema. Edema is commonly seen in regions of famine due to dietary protein deficiency (see Deeper Insight 18.1). Hypoproteinemia and edema also commonly result from severe burns, owing to the loss of protein from body surfaces no longer covered with skin, and from kidney diseases that allow protein to escape in the urine.

3. **Obstructed lymphatic drainage.** The lymphatic system, described in detail in the next chapter, is a network of one-way vessels that collect fluid from the tissues and return it to the bloodstream. Obstruction of these vessels or the surgical removal of lymph nodes can interfere with fluid drainage and lead to the accumulation of tissue fluid distal to the obstruction.

Edema has multiple pathological consequences. As the tissues become congested with fluid, oxygen delivery and waste removal are impaired and the tissues may begin to die. Pulmonary edema presents a threat of suffocation as fluid replaces air in the lungs,

FIGURE 20.19 Lymphedema. On the right is a 52-year-old woman with severe lymphedema of the legs and feet; on the left, for comparison, is a 21-year-old woman without edema. Blockage of lymphatic vessels is one of several causes of edema.
MedicImage/Alamy

and cerebral edema can produce headaches, nausea, and sometimes delirium, seizures, and coma. In severe edema, so much fluid can transfer from the blood vessels to the tissue spaces that blood volume and pressure drop low enough to cause circulatory shock (described in the next section).

Answer the following questions to test your understanding of the preceding section:

13. List the three mechanisms of capillary exchange and relate each one to the structure of capillary walls.

14. What forces favor capillary filtration? What forces favor reabsorption?

15. How can a capillary shift from a predominantly filtering role at one time to a predominantly reabsorbing role at another?

16. State the three fundamental causes of edema and explain why edema can be dangerous.

20.4 Venous Return and Circulatory Shock

Expected Learning Outcomes

When you have completed this section, you should be able to

a. explain how blood in the veins is returned to the heart;

b. discuss the importance of physical activity in venous return;

c. discuss several causes of circulatory shock; and

d. name and describe the stages of shock.

Hieronymus Fabricius (1537–1619) discovered the valves of the veins but did not understand their function. That was left to his student, William Harvey, who performed simple experiments on the valves that you can easily reproduce. In **figure 20.20,** by Harvey, the experimenter has pressed on a vein at point H to block flow from the wrist toward the elbow. With another finger, he has milked the blood out of it up to point O, the first valve proximal to H. When he tries to force blood downward, it stops at that valve. It can go no farther, and it causes the vein to swell at that point. Blood can flow from right to left through that valve but not from left to right. So as Harvey correctly surmised, the valves serve to ensure a one-way flow of blood toward the heart.

You can easily demonstrate the action of these valves in your own hand. Hold your hand still, below waist level, until veins stand up on the back of it. (Do not apply a tourniquet!) Press on a vein close to your knuckles, and while holding it down, use another finger to milk that vein toward the wrist. It collapses as you force the blood out of it, and if you remove the second finger, it will not refill. The valves prevent blood from flowing back into it from

FIGURE 20.20 An Illustration from William Harvey's *De Motu Cordis* (1628). Such experiments demonstrated the existence of one-way valves in veins of the arms. See text for explanation.
UniversalImagesGroup/Getty Images

❓ *In the space between O and H, what (if anything) would happen if the experimenter lifted his finger from point O? What if he lifted his finger from point H? Why?*

above. When you remove the first finger, however, the vein fills from below.

20.4a Mechanisms of Venous Return

The flow of blood back to the heart, called **venous return,** is achieved by five mechanisms:

1. **The pressure gradient.** Pressure generated by the heart is the most important force in venous flow, even though it is substantially weaker in the veins than in the arteries. Pressure in the venules ranges from 12 to 18 mm Hg, and pressure at the point where the venae cavae enter the heart, called **central venous pressure,** averages 4.6 mm Hg. Thus, there is a venous pressure gradient (ΔP) of about 7 to 13 mm Hg favoring the flow of blood toward the heart. The pressure gradient and venous return increase when blood volume increases. Venous return also increases in the event of generalized, widespread vasoconstriction because this reduces the volume of the circulatory system and raises blood pressure and flow.

2. **Gravity.** When you are sitting or standing, blood from your head and neck returns to the heart simply by flowing downward through the large veins above the heart. Thus, the large veins of the neck are normally collapsed or nearly so, and their venous pressure is close to zero. The dural sinuses of the brain, however, have more rigid walls and cannot collapse. Their pressure is as low as –10 mm Hg, creating a risk of *air embolism* if they are punctured (see Deeper Insight 20.3).

3. **The skeletal muscle pump.** In the limbs, the veins are surrounded and massaged by the muscles. Contracting muscles squeeze the blood out of the compressed part of a vein, and the valves ensure that this blood can go only toward the heart **(fig. 20.21).**

4. **The thoracic (respiratory) pump.** This mechanism aids the flow of venous blood from the abdominal to the thoracic cavity. When you inhale, your thoracic cavity

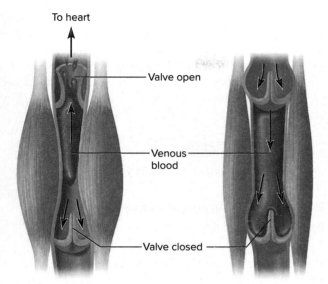

To heart

Valve open

Venous blood

Valve closed

(a) Contracted skeletal muscles   **(b) Relaxed skeletal muscles**

FIGURE 20.21 The Skeletal Muscle Pump. (a) Muscle contraction squeezes the deep veins and forces blood through the next valve in the direction of the heart. Valves below the point of compression prevent backflow. (b) When the muscles relax, blood flows back downward under the pull of gravity but can flow only as far as the nearest valve.

expands and its internal pressure drops, while downward movement of the diaphragm raises the pressure in your abdominal cavity. The *inferior vena cava (IVC),* your largest vein, is a flexible tube passing through both of these cavities. If abdominal pressure on the IVC rises while thoracic pressure on it drops, then blood is squeezed upward toward the heart. It isn't forced back into the lower limbs because the valves there prevent this. Because of the thoracic pump, central venous pressure fluctuates from 2 mm Hg when you inhale to 6 mm Hg when you exhale, and blood flows faster when you inhale. This is what produces the slight fluctuations in blood pressure at the right end of the graph in figure 20.11.

5. **Cardiac suction.** During ventricular systole, the tendinous cords pull the AV valve cusps downward, slightly expanding the atrial space. This creates a slight suction that draws blood into the atria from the venae cavae and pulmonary veins.

20.4b Venous Return and Physical Activity

Exercise increases venous return for multiple reasons. The heart beats faster and harder, increasing cardiac output and blood pressure. Blood vessels of the skeletal muscles, lungs, and coronary circulation dilate, increasing flow. The increase in respiratory rate and depth enhances the action of the thoracic pump. Muscle contractions increase venous return by means of the skeletal muscle pump. Increased venous return then increases cardiac output, which is important in perfusion of the muscles just when they need it most.

Conversely, when a person is still, blood accumulates in the limbs because venous pressure isn't high enough to override the weight of the blood and drive it upward. Such accumulation of blood is called **venous pooling.** To demonstrate this effect, hold one hand below your waist and hold the other hand over your head. After about 1 minute, quickly bring your two hands together and compare the palms. The hand held overhead will likely appear pale because its blood has drained out of it; the hand held below the waist will likely be redder than normal because of venous pooling in its veins and capillaries. Venous pooling can be troublesome to people who must stand for prolonged periods—such as cashiers, barbers and hairdressers, members of a choir, and people in military service—and when sitting still for too long, as in a cramped seat on a long airline flight. If enough blood accumulates in the limbs, cardiac output may fall so low that the brain is inadequately perfused and a person may experience dizziness or syncope. This can usually be prevented by periodically tensing the calf and other muscles to keep the skeletal muscle pump active. Military jet pilots often perform maneuvers that could cause the blood to pool in the abdomen and lower limbs, causing loss of vision or consciousness. To prevent this, they wear pressurized *G suits* that inflate and tighten on the lower limbs during these maneuvers; in addition, they can tense their abdominal muscles to prevent venous pooling and blackout.

▶▶▶ **APPLY WHAT YOU KNOW**

Why is venous pooling not a problem when you are sleeping and the skeletal muscle pump is inactive?

20.4c Circulatory Shock

Circulatory shock (not to be confused with electrical or spinal shock) is any state in which cardiac output is insufficient to meet the body's metabolic needs. All forms of circulatory shock fall into two categories: (1) **cardiogenic shock,** caused by inadequate pumping by the heart, usually as a result of myocardial infarction; and (2) **low venous return (LVR) shock,** in which cardiac output is low because too little blood is returning to the heart.

There are three principal forms of LVR shock:

1. **Hypovolemic shock,** the most common form, is produced by a loss of blood volume as a result of hemorrhage, trauma, bleeding ulcers, burns, or dehydration. Dehydration is a major cause of death from heat exposure. In hot weather, the body excretes as much as 1.5 L of sweat per hour. Water transfers from the bloodstream to replace tissue fluid lost in the sweat, and blood volume may drop too low to maintain adequate circulation.

2. **Obstructed venous return shock** occurs when any object, such as a growing tumor or aneurysm, compresses a vein and impedes its blood flow.

3. **Venous pooling shock** occurs when the body has a normal total blood volume, but too much of it accumulates in the lower body. This can result from long periods of standing

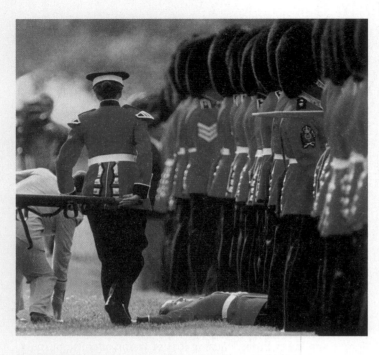

FIGURE 20.22 Venous Pooling Shock. This is a common consequence of prolonged, rigid standing in ceremonial attention.
Carlo Allegri/AFP/Getty Images

or sitting (**fig. 20.22**) or from widespread vasodilation. **Neurogenic shock** is a form of venous pooling shock that results from a sudden loss of vasomotor tone, allowing the vessels to dilate. This can result from causes as severe as brainstem trauma or as slight as an emotional shock.

Elements of both venous pooling and hypovolemic shock are present in certain cases, such as septic shock and anaphylactic shock, which involve both vasodilation and a loss of fluid through abnormally permeable capillaries. **Septic shock** occurs when bacterial toxins trigger vasodilation and increased capillary permeability. **Anaphylactic shock,** discussed more fully in section 21.6, results from exposure to an antigen to which a person is allergic, such as bee venom. Antigen–antibody complexes trigger the release of histamine, which causes generalized vasodilation and increased capillary permeability.

Responses to Circulatory Shock

Shock is clinically described according to severity as compensated or decompensated. In **compensated shock,** several homeostatic mechanisms bring about spontaneous recovery. The hypotension resulting from low cardiac output triggers the sympathetic baroreflex and the production of angiotensin II, both of which counteract shock by stimulating vasoconstriction. Furthermore, if a person faints and falls to a horizontal position, gravity restores blood flow to the brain. Even quicker recovery is achieved if the person's feet are elevated to promote drainage of blood from the legs.

If these mechanisms prove inadequate, **decompensated shock** ensues and several life-threatening positive feedback loops occur. Poor cardiac output results in myocardial ischemia and infarction, which further weaken the heart and reduce output. Slow circulation of the blood can lead to disseminated intravascular coagulation

(DIC) (see table 18.8). As the vessels become congested with clotted blood, venous return grows even worse. Ischemia and acidosis of the brainstem depress the vasomotor and cardiac centers, causing loss of vasomotor tone, further vasodilation, and further drop in BP and cardiac output. Before long, damage to the cardiac and brain tissues may be unsurvivable. About half of those who go into decompensated shock die from it.

BEFORE YOU GO ON

Answer the following questions to test your understanding of the preceding section:

17. Explain how respiration aids venous return.
18. Explain how muscular activity and venous valves aid venous return.
19. Define *circulatory shock.* What are some of the causes of low venous return shock?

20.5 Special Circulatory Routes

Expected Learning Outcomes
When you have completed this section, you should be able to

a. explain how the brain maintains stable perfusion;
b. discuss the causes and effects of strokes and transient ischemic attacks;
c. explain the mechanisms that increase muscular perfusion during exercise; and
d. contrast the blood pressure of the pulmonary circuit with that of the systemic circuit, and explain why the difference is important in pulmonary function.

Certain circulatory pathways have special physiological properties adapted to the functions of their organs. Two of these are described in other chapters: the coronary circulation (see fig. 19.10) and fetal and placental circulation (see fig. 29.10). Here we take a closer look at the circulation to the brain, skeletal muscles, and lungs.

20.5a Brain

Total blood flow to the brain fluctuates less than that of any other organ (about 700 mL/min. at rest). Such constancy is important because even a few seconds of oxygen deprivation causes loss of consciousness, and 4 or 5 minutes of anoxia is time enough to cause irreversible damage. Although total cerebral perfusion is fairly stable, blood flow can be shifted from one part of the brain to another in a matter of seconds as different parts engage in motor, sensory, or cognitive functions (see fig. 14.41).

The brain regulates its own blood flow in response to changes in BP and chemistry. The cerebral arteries dilate when the systemic BP drops and constrict when it rises, thus minimizing fluctuations in cerebral BP. Cerebral blood flow therefore remains quite stable even when mean arterial pressure (MAP) fluctuates from 60 to 140 mm Hg. However, an MAP below 60 mm Hg produces syncope and an MAP above 160 mm Hg causes cerebral edema.

The main chemical stimulus for cerebral autoregulation is pH. Poor perfusion allows CO_2 to accumulate in the brain. This lowers the pH of the tissue fluid and triggers local vasodilation, which improves perfusion. Extreme hypercapnia, however, depresses neural activity. The opposite condition, hypocapnia, raises the pH and stimulates vasoconstriction, thus reducing perfusion and giving CO_2 a chance to rise to a normal level. Hyperventilation (exhaling CO_2 faster than the body produces it) induces hypocapnia, which leads to cerebral vasoconstriction, ischemia, dizziness, and sometimes syncope.

Brief episodes of cerebral ischemia produce **transient ischemic attacks (TIAs),** characterized by temporary dizziness, loss of vision or other senses, weakness, paralysis, headache, or aphasia. A TIA may result from spasms of diseased cerebral arteries. It lasts from just a moment to a few hours and is often an early warning of an impending stroke. People with TIAs should receive prompt medical attention to identify the cause using brain imaging and other diagnostic means. Immediate treatment should be initiated to prevent a stroke. **Stroke,** or **cerebrovascular accident (CVA),** is far more serious, entailing death (infarction) of brain tissue and often irreversible loss of bodily functions. Stroke is discussed more extensively in Deeper Insight 14.2.

20.5b Skeletal Muscles

In contrast to the brain, the skeletal muscles receive a highly variable blood flow depending on their state of exertion. At rest, the arterioles are constricted, most of the capillary beds are shut down, and total flow through the muscular system is about 1 L/min. During exercise, the arterioles dilate in response to muscle metabolites such as lactate, nitric oxide (NO), adenosine, CO_2, and H^+. Blood flow through the muscles can increase more than 20-fold during strenuous exercise, which requires that blood be diverted from other organs such as the digestive tract and kidneys to meet the needs of the working muscles.

Muscular contraction compresses the blood vessels and impedes flow. For this reason, isometric contraction causes fatigue more quickly than intermittent isotonic contraction. If you squeeze a rubber ball as hard as you can without relaxing your grip, you feel the muscles fatigue more quickly than if you intermittently squeeze and relax.

20.5c Lungs

After birth, the pulmonary circuit is the only route in which the arteries carry oxygen-poor blood and the veins carry oxygen-rich blood; the opposite situation prevails in the systemic circuit. The pulmonary arteries have thin distensible walls with less elastic tissue than the systemic arteries. Thus, they have a BP of only 25/10 mm Hg. Capillary hydrostatic pressure is about 10 mm Hg in the pulmonary circuit as compared with an average of 17 mm Hg in systemic capillaries. This lower pressure has two implications for pulmonary circulation: (1) Blood flows more slowly through the pulmonary capillaries, and therefore it has more time for gas exchange; and (2) oncotic pressure overrides hydrostatic pressure, so these capillaries are engaged almost entirely in absorption. This prevents fluid accumulation in the alveolar walls and lumens, which would compromise gas exchange. In a condition such as

mitral valve stenosis, however, blood may back up in the pulmonary circuit, raising the capillary hydrostatic pressure and causing pulmonary edema, congestion, and hypoxemia.

Another unique characteristic of the pulmonary arteries is their response to hypoxia. Systemic arteries dilate in response to local hypoxia and improve tissue perfusion. By contrast, pulmonary arteries constrict. Pulmonary hypoxia indicates that part of the lung is not being ventilated well, perhaps because of mucous congestion of the airway or a degenerative lung disease. Vasoconstriction in poorly ventilated regions of the lung redirects blood flow to better ventilated regions.

▶▶▶ **APPLY WHAT YOU KNOW**

What abnormal skin coloration would result from pulmonary edema?

BEFORE YOU GO ON
Answer the following questions to test your understanding of the preceding section:

20. In what conspicuous way does perfusion of the brain differ from perfusion of the skeletal muscles?

21. How does a stroke differ from a transient ischemic attack? Which of these bears closer resemblance to a myocardial infarction?

22. How does the low hydrostatic blood pressure in the pulmonary circuit affect the fluid dynamics of the capillaries there?

23. Contrast the vasomotor response of the lungs with that of skeletal muscles to hypoxia.

20.6 Anatomy of the Pulmonary Circuit

Expected Learning Outcome
When you have completed this section, you should be able to

a. trace the route of blood through the pulmonary circuit.

The next three sections of this chapter center on the names and pathways of the principal arteries and veins. The pulmonary circuit is described here, and the systemic arteries and veins are described in the two sections that follow.

The pulmonary circuit (**fig. 20.23**) begins with the **pulmonary trunk,** a large vessel that ascends diagonally from the right ventricle and branches into the right and left **pulmonary arteries.** As it approaches the lung, the right pulmonary artery branches in two, and both branches enter the lung at a medial indentation called the *hilum* (see fig. 22.9). The upper branch is the **superior lobar artery,** serving the superior lobe of the lung. The lower branch divides again within the lung to form the **middle lobar** and **inferior lobar arteries,** supplying the lower two lobes of that lung. The left pulmonary artery is much more variable. It gives off several superior lobar arteries to the superior lobe before entering

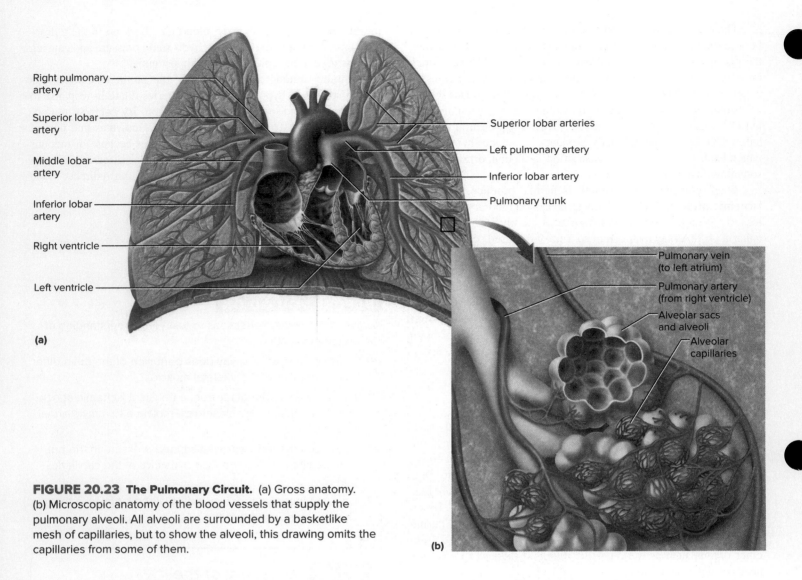

Right pulmonary artery

Superior lobar artery

Middle lobar artery

Inferior lobar artery

Right ventricle

Left ventricle

Superior lobar arteries

Left pulmonary artery

Inferior lobar artery

Pulmonary trunk

(a)

Pulmonary vein (to left atrium)

Pulmonary artery (from right ventricle)

Alveolar sacs and alveoli

Alveolar capillaries

(b)

FIGURE 20.23 The Pulmonary Circuit. (a) Gross anatomy. (b) Microscopic anatomy of the blood vessels that supply the pulmonary alveoli. All alveoli are surrounded by a basketlike mesh of capillaries, but to show the alveoli, this drawing omits the capillaries from some of them.

the hilum, then enters the lung and gives off a variable number of inferior lobar arteries to the inferior lobe.

In both lungs, these arteries lead ultimately to small basketlike capillary beds that surround the pulmonary alveoli (air sacs). This is where the blood unloads CO_2 and picks up O_2. After leaving the alveolar capillaries, the pulmonary blood flows into venules and veins, ultimately leading to the main **pulmonary veins** that exit the lung at the hilum. The left atrium of the heart receives two pulmonary veins on each side (see fig. 19.5b).

The purpose of the pulmonary circuit is primarily to exchange CO_2 for O_2. The lungs also receive a separate systemic blood supply by way of the *bronchial arteries* (see section 20.7d).

BEFORE YOU GO ON

Answer the following questions to test your understanding of the preceding section:

24. Trace the flow of an RBC from right ventricle to left atrium and name the vessels along the way.

25. The lungs have two separate arterial supplies. Explain their functions.

20.7 Systemic Vessels of the Axial Region

Expected Learning Outcomes

When you have completed this section, you should be able to

a. identify the principal systemic arteries and veins of the axial region; and

b. trace the flow of blood from the heart to any major organ of the axial region and back to the heart.

The systemic circuit (**figs. 20.24, 20.25**) supplies oxygen and nutrients to all organs and removes their metabolic wastes. The coronary circulation, already described, is part of this. This section surveys the remaining arteries and veins of the axial region—the head, neck, and trunk. The following sections, 20.7a through 20.7g, trace arterial outflow and venous return, region by region. They outline only the most common circulatory pathways; there is

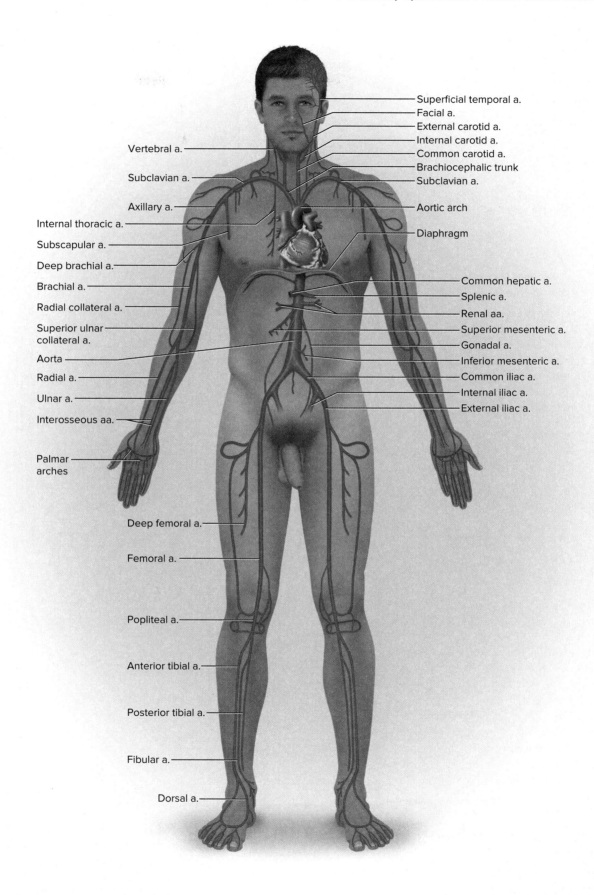

FIGURE 20.24 The Major Systemic Arteries. Different arteries are illustrated on the left than on the right for clarity, but nearly all of those shown occur on both sides.

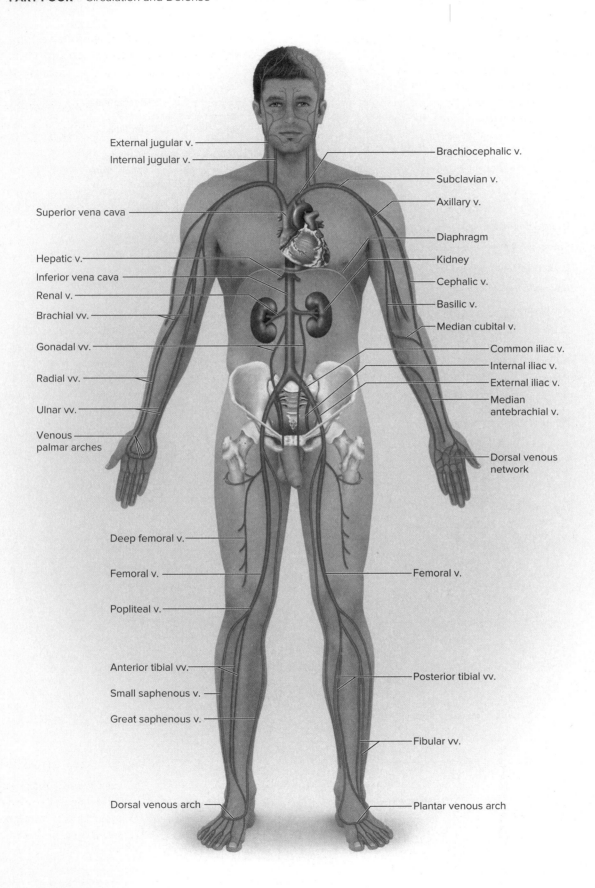

External jugular v.

Internal jugular v.

Superior vena cava

Hepatic v.

Inferior vena cava

Renal v.

Brachial vv.

Gonadal vv.

Radial vv.

Ulnar vv.

Venous palmar arches

Deep femoral v.

Femoral v.

Popliteal v.

Anterior tibial vv.

Small saphenous v.

Great saphenous v.

Dorsal venous arch

Brachiocephalic v.

Subclavian v.

Axillary v.

Diaphragm

Kidney

Cephalic v.

Basilic v.

Median cubital v.

Common iliac v.

Internal iliac v.

External iliac v.

Median antebrachial v.

Dorsal venous network

Femoral v.

Posterior tibial vv.

Fibular vv.

Plantar venous arch

FIGURE 20.25 **The Major Systemic Veins.** Different veins are illustrated on the left than on the right for clarity, but nearly all of those shown occur on both sides.

a great deal of anatomical variation in the circulatory system from one person to another.

The names of the blood vessels often describe their location by indicating the body region traversed (as in the *axillary artery* and *brachial veins*), an adjacent bone (as in *temporal artery* and *ulnar vein*), or the organ supplied or drained by the vessel (as in *hepatic artery* and *renal vein*). In many cases, an artery and adjacent vein have similar names (*femoral artery* and *femoral vein,* for example).

As you trace blood flow in these sections, it is important to refer frequently to the illustrations. Verbal descriptions alone are likely to seem obscure if you don't make full use of the explanatory illustrations. Throughout the remaining figures, the abbreviations *a.* and *aa.* mean *artery* and *arteries,* and *v.* and *vv.* mean *vein* and *veins.*

20.7a The Aorta and Its Major Branches

All systemic arteries arise from the aorta, which has three principal regions, often compared to the shape of an umbrella handle **(fig. 20.26).**

1. The **ascending aorta** rises for about 5 cm above the left ventricle. Its only branches are the coronary arteries, which

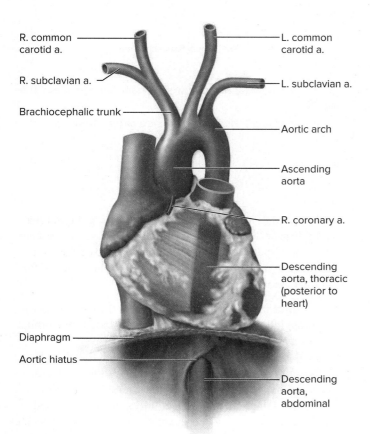

R. common carotid a.
R. subclavian a.
Brachiocephalic trunk
L. common carotid a.
L. subclavian a.
Aortic arch
Ascending aorta
R. coronary a.
Descending aorta, thoracic (posterior to heart)
Diaphragm
Aortic hiatus
Descending aorta, abdominal

FIGURE 20.26 The Thoracic Aorta. (L. = left; R. = right)

arise behind two cusps of the aortic valve. They are the origins of the coronary circulation (see section 19.2e).

2. The **aortic arch** curves to the left like an inverted U superior to the heart. It gives off three major arteries in this order: the **brachiocephalic**[10] **trunk** (BRAY-kee-oh-seh-FAL-ic), **left common carotid artery** (cah-ROT-id), and **left subclavian**[11] **artery** (sub-CLAY-vee-un). These are further traced in sections 20.7b and 20.8a.

3. The **descending aorta** passes downward posterior to the heart, at first to the left of the vertebral column and then anterior to it, through the thoracic and abdominal cavities. It is called the **thoracic aorta** above the diaphragm and the **abdominal aorta** below it. It ends in the lower abdominal cavity by forking into the right and left *common iliac arteries.*

20.7b Arteries of the Head and Neck

All blood supply to the head, neck, and upper limbs comes from the aortic arch by way of the following arteries.

Origins of the Head–Neck Arteries

The head and neck (including the brain) receive blood from four pairs of arteries **(fig. 20.27):**

1. The **common carotid arteries,** probably the best-known avenues to the head, ascend the anterolateral region of the neck alongside the trachea. Shortly after leaving the aortic arch, the brachiocephalic trunk divides into the **right common carotid artery** to the head and the *right subclavian artery* leading to the right arm (further traced in section 20.7d). The **left common carotid artery** arises independently from the aortic arch a little distal to the brachiocephalic trunk.

2. The **vertebral arteries** arise from the right and left subclavian arteries, travel up the neck through the transverse foramina of vertebrae C1 through C6, and enter the cranial cavity through the foramen magnum.

3. The **thyrocervical**[12] **trunks** are tiny arteries arising from the subclavians lateral to the vertebral arteries. They supply the thyroid gland and some scapular muscles.

4. The **costocervical**[13] **trunks** arise from the subclavian arteries a little farther laterally. They supply the deep neck muscles and some of the intercostal muscles of the superior rib cage.

Continuation of the Common Carotid Arteries

The common carotid arteries have the most extensive distribution of all head–neck arteries. Near the laryngeal prominence ("Adam's apple"), each common carotid branches into an *external* and *internal carotid artery,* with further branches as follows:

---

[10]*brachio* = arm; *cephal* = head
[11]*sub* = below; *clavi* = clavicle, collarbone
[12]*thyro* = thyroid gland; *cerv* = neck
[13]*costo* = rib; *cerv* = neck

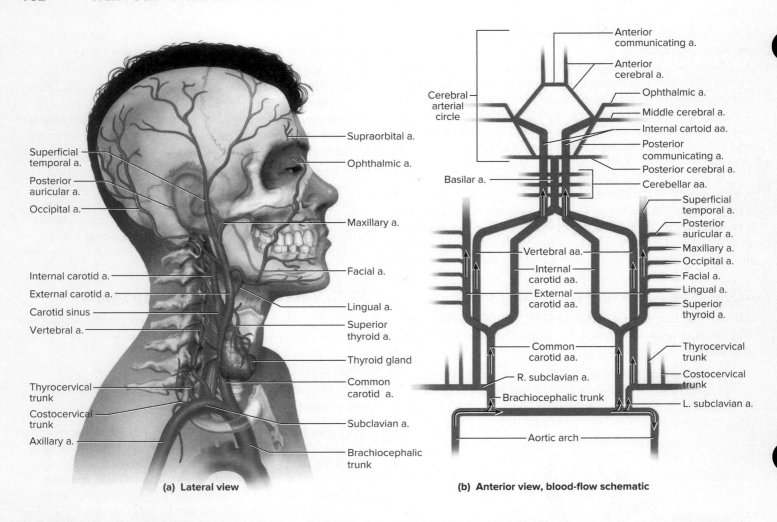

(a) Lateral view

(b) Anterior view, blood-flow schematic

FIGURE 20.27 Superficial (Extracranial) Arteries of the Head and Neck. (a) Lateral view. (b) Blood-flow schematic, anterior view. The upper part of the schematic depicts the cerebral circulation in figure 20.28. **APR**

❓ *List the arteries, in order, that an erythrocyte must travel to get from the left ventricle to the skin of the left side of the forehead.*

1. The **external carotid artery** ascends the side of the head external to the cranium and supplies most extracranial structures except the orbits. It gives rise to the following arteries in ascending order:

   a. the **superior thyroid artery** to the thyroid gland and larynx;

   b. the **lingual artery** to the tongue;

   c. the **facial artery** to the skin and muscles of the face;

   d. the **occipital artery** to the posterior scalp;

   e. the **maxillary artery** to the teeth, maxilla, oral cavity, and external ear; and

   f. the **superficial temporal artery** to the chewing muscles, nasal cavity, lateral aspect of the face, most of the scalp, and the dura mater.

2. The **internal carotid artery** passes medial to the angle of the mandible and passes through the carotid canal of the temporal bone into the cranial cavity. It supplies the orbits

and about 80% of the cerebrum. Compressing the internal carotids near the mandible can therefore cause loss of consciousness (but is dangerous and must never be done for amusement, as some people do). Principal branches of the internal carotid artery (**fig. 20.28**) are

   a. the **ophthalmic artery,** which leaves the cranial cavity through the optic canal and supplies blood to the orbit, nose, and forehead;

   b. the large **middle cerebral artery,** which travels in the lateral sulcus of the cerebrum, supplies the insula, then issues numerous branches to the lateral region of the frontal, temporal, and parietal lobes of the brain; and

   c. the smaller **anterior cerebral artery,** which travels anteriorly, then turns back and arches posteriorly over the corpus callosum as far as the posterior limit of the parietal lobe; it gives off extensive branches to the frontal and parietal lobes.

Cerebral arterial circle:
  Anterior communicating a.
  Anterior cerebral a.
  Posterior communicating a.
  Posterior cerebral a.
Basilar a.
Vertebral a.
Spinal aa.

Internal carotid a.
Middle cerebral a.

Caudal | Rostral

Anterior cerebral a.

Cerebellar aa.:
  Superior
  Anterior inferior
  Posterior inferior

Posterior cerebral a.

(a) Inferior view

(b) Median section

FIGURE 20.28 The Cerebral Blood Supply. (a) Inferior view of the brain showing the blood supply to the brainstem, cerebellum, and cerebral arterial circle. (b) Median section of the brain showing the more distal branches of the anterior and posterior cerebral arteries. Branches of the middle cerebral artery are distributed over the lateral surface of the cerebrum (not illustrated). **APR**

Continuation of the Vertebral Arteries

The vertebral arteries give rise to small branches that supply the spinal cord and its meninges, the cervical vertebrae, and deep muscles of the neck. They next enter the foramen magnum, supply the cranial bones and meninges, and continue as follows:

1. They converge to form a single median **basilar artery** at the junction of the medulla oblongata and pons. The basilar artery runs along the anterior aspect of the pons and gives off branches to the cerebellum, pons, and inner ear. At or near the pons–midbrain border, it divides into the following arteries.

2. Right and left **posterior cerebral arteries** arise from the basilar artery, sweep posteriorly to the rear of the brain, and serve the inferior and medial regions of the temporal and occipital lobes, the midbrain, and the thalamus.

The Cerebral Arterial Circle

In figure 20.28a and the upper part of 20.27b, you can see that a number of these arteries form a loop that encircles the pituitary gland and optic chiasm. This is called the **cerebral arterial circle** (*circle of Willis*[14]). Its components include the basilar artery;

internal carotid arteries; anterior, middle, and posterior cerebral arteries; and the following anastomoses that complete the loop:

1. one **anterior communicating artery** medially connecting the right and left anterior cerebral arteries; and

2. a slender, right and left **posterior communicating artery,** connecting the internal carotid and posterior cerebral arteries.

Only about 20% of people, however, have a complete cerebral arterial circle. Usually, one or more components are absent or so constricted as to provide no significant blood flow. Knowledge of the distribution of the arteries arising from the cerebral arterial circle is crucial for understanding the effects of blood clots, aneurysms, and strokes on brain function. Cerebral aneurysms occur most often in this complex of arteries or in the basilar artery proximal to the circle.

20.7c Veins of the Head and Neck

The head and neck are drained mainly by three pairs of veins—the *internal* and *external jugular veins* and *vertebral veins*. We will trace these from their origins to the *subclavian veins.*

Dural Venous Sinuses

After blood circulates through the brain, it collects in large, thin-walled, modified veins called **dural venous sinuses**—blood-filled spaces between the layers of the dura mater (**fig. 20.29**).

[14]Thomas Willis (1621–75), English anatomist

Corpus callosum

Great cerebral vein

Straight sinus

Confluence of sinuses

Transverse sinus

Sigmoid sinus

Internal jugular v.

Superior sagittal sinus

Inferior sagittal sinus

(a) Dural venous sinuses, medial view

Superior ophthalmic vein

Cavernous sinus

Sigmoid sinus

Transverse sinus

Confluence of sinuses

Superficial middle cerebral vein

To internal jugular v.

Straight sinus

(b) Dural venous sinuses, inferior view

Superficial temporal v.

Occipital v.

Vertebral v.

External jugular v.

Internal jugular v.

Axillary v.

Superior ophthalmic v.

Facial v.

Superior thyroid v.

Thyroid gland

Subclavian v.

Brachiocephalic v.

FIGURE 20.29 Veins of the Head and Neck. (a) Dural venous sinuses seen in a median section of the cerebrum. (b) Dural venous sinuses seen in an inferior view of the cerebrum. (c) Superficial (extracranial) veins of the head and neck. **AP R**

(c) Superficial veins of the head and neck

A reminder of the structure of the dura mater may be helpful in understanding these sinuses. This tough membrane between the brain and cranial bones has a *periosteal layer* against the bone and a *meningeal layer* against the brain. In a few places, a space exists between these layers to accommodate a blood-collecting sinus. Between the two cerebral hemispheres is a vertical, sickle-shaped wall of dura called the *falx cerebri,* which contains two of the sinuses. There are about 13 dural venous sinuses in all; we survey only the few most prominent ones here.

1. The **superior sagittal sinus** is contained in the superior margin of the falx cerebri and overlies the longitudinal cerebral fissure (fig. 20.29a; see also figs. 14.5, 14.7). It begins anteriorly near the crista galli of the skull and extends posteriorly to the level of the posterior occipital protuberance. Here it bends, usually to the right, and drains into a *transverse sinus.*

2. The **inferior sagittal sinus** is contained in the inferior margin of the falx cerebri and arches over the corpus callosum, deep in the longitudinal cerebral fissure. Posteriorly, it joins the *great cerebral vein* and their union forms the **straight sinus,** which continues to the rear of the head. There, the superior sagittal and straight sinuses meet in a space called the *confluence of the sinuses.*

3. Right and left **transverse sinuses** lead away from the confluence and encircle the inside of the occipital bone, leading toward the ears (fig. 20.29b); their path is marked by grooves on the inner surface of the occipital bone (see fig. 8.5b). The right transverse sinus receives blood mainly from the superior sagittal sinus, and the left one drains mainly the straight sinus. Laterally, each transverse sinus makes an S-shaped bend, the **sigmoid sinus,** then exits the cranium through the jugular foramen. From here, blood flows down the *internal jugular vein* of the neck.

4. The **cavernous sinuses** are honeycombs of blood-filled spaces on each side of the body of the sphenoid bone (fig. 20.29b). They receive blood from the *superior ophthalmic vein* of the orbit and the *superficial middle cerebral vein* of the brain, among other sources. They drain through several outlets including the transverse sinus, internal jugular vein, and facial vein. They are clinically important because infections can pass from the face and other superficial sites into the cranial cavity by this route. Also, inflammation of a cavernous sinus can injure important structures that pass through it, including the internal carotid artery and cranial nerves III to VI.

Major Veins of the Neck

Blood flows down the neck mainly through three veins on each side, all of which empty into the subclavian vein (fig. 20.29c).

1. The **internal jugular**[15] **vein** (JUG-you-lur) courses down the neck deep to the sternocleidomastoid muscle. It receives

---

[15]*jugul* = neck, throat

DEEPER INSIGHT 20.3

CLINICAL APPLICATION

Air Embolism

Injury to the dural sinuses or jugular veins presents less danger from loss of blood than from air sucked into the circulatory system. The presence of air in the bloodstream is called *air embolism* **(fig. 20.30).** This is an important concern to neurosurgeons, who sometimes operate with the patient in a sitting position. If a dural sinus is punctured, air can be sucked into the sinus and accumulate in the heart chambers, which blocks cardiac output and causes sudden death. Smaller air bubbles in the systemic circulation can cut off blood flow to the brain, lungs, myocardium, and other vital tissues. As little as 0.5 mL of air in a coronary artery can cause cardiac arrest.

Embolism blocking common carotid artery

Embolism blocking pulmonary valve

FIGURE 20.30 Air Embolism.

---

most of the blood from the brain; picks up blood from the **facial vein, superficial temporal vein,** and **superior thyroid vein** along the way; passes behind the clavicle; and joins the subclavian vein (which is further traced in section 20.7e).

2. The **external jugular vein** courses down the side of the neck superficial to the sternocleidomastoid muscle and empties into the subclavian vein. It drains tributaries from

the parotid salivary gland, facial muscles, scalp, and other superficial structures. Some of this blood also follows venous anastomoses to the internal jugular vein.

3. The **vertebral vein** travels with the vertebral artery in the transverse foramina of the cervical vertebrae. Although the companion artery leads to the brain, the vertebral vein doesn't come from there. It drains the cervical vertebrae, spinal cord, and some of the small deep muscles of the neck, and empties into the subclavian vein.

Section 20.7e traces this blood flow the rest of the way to the superior vena cava and heart.

20.7d Arteries of the Thorax

The thorax is supplied by several arteries arising directly from the descending aorta and from the subclavian and axillary arteries. The thoracic aorta begins distal to the aortic arch and ends at the **aortic hiatus** (hy-AY-tus), a passage through the diaphragm. Along the way, it sends off numerous small branches to the thoracic viscera and body wall **(fig. 20.31)**.

Visceral Branches of the Thoracic Aorta

Visceral branches of the thoracic aorta include the following:

1. The **bronchial arteries** vary in number and arrangement; there are usually two on the left and one on the right. The right bronchial artery usually arises from one of the left bronchial arteries or from a *posterior intercostal artery.* The bronchial arteries supply the visceral pleura, pericardium, and esophagus, and enter the lungs to supply the bronchi, bronchioles, and larger pulmonary blood vessels.

2. The **esophageal arteries** are four or five unpaired arteries that come off the anterior surface of the aorta and supply the esophagus.

3. The **mediastinal arteries** (not illustrated) are numerous small vessels that supply structures of the posterior mediastinum.

▶▶▶ APPLY WHAT YOU KNOW

Both the pulmonary arteries and bronchial arteries supply blood to the lungs, but when this blood emerges from the lungs, the pulmonary blood is richer in oxygen and the bronchial blood is poorer in oxygen. Explain why.

Parietal Branches of the Thoracic Aorta

The following branches supply chiefly the muscles, bones, and skin of the chest; only the first are illustrated.

1. The **posterior intercostal arteries** are nine pairs of vessels arising from the posterior side of the aorta. They course around the posterior side of the rib cage between ribs 3 through 12, then anastomose with the anterior intercostal arteries described in the next section. They supply the inter-

costal, pectoralis, serratus anterior, and some abdominal muscles, as well as the vertebrae, spinal cord, meninges, breasts, skin, and subcutaneous tissue.

2. One pair of **subcostal arteries** arises from the aorta inferior to rib 12. They supply the posterior intercostal tissues, vertebrae, spinal cord, and deep muscles of the back.

3. A variable number of **superior phrenic**[16] **arteries** (FREN-ic) arise at the aortic hiatus and supply the superior and posterior regions of the diaphragm.

▶▶▶ APPLY WHAT YOU KNOW

The posterior intercostal arteries are larger in lactating women than in men and nonlactating females. Explain why.

Branches of the Subclavian and Axillary Arteries

The thoracic wall is also supplied by branches of the subclavian arteries and their continuations, the *axillary arteries,* in the shoulder and axillary regions. Recall that the right subclavian artery arises from the brachiocephalic trunk and the left subclavian artery arises directly from the aortic arch (see fig. 20.26). On both the right and left, the subclavian artery gives off the internal thoracic artery, then continues as the axillary artery. Further divisions of these arteries are as follows:

1. The **internal thoracic (mammary) artery** supplies the breast and anterior thoracic wall. It issues the following branches:

    a. The **pericardiophrenic artery** supplies the pericardium and diaphragm.

    b. The **anterior intercostal arteries** arise from the internal thoracic artery as it descends alongside the sternum. They travel between the ribs; supply the ribs and the intercostal and pectoral muscles, breast, and skin; and, finally, anastomose with the posterior intercostal arteries. Each sends one branch along the lower margin of the rib above it and another, parallel branch along the upper margin of the rib below it.

2. The **thoracoacromial**[17] **trunk** (THOR-uh-co-uh-CRO-me-ul) provides branches to the superior shoulder and pectoral regions.

3. The **lateral thoracic artery** supplies the pectoral, serratus anterior, and subscapularis muscles. It also issues branches to the breast and is larger in females than in males.

4. The **subscapular artery** is the largest branch of the axillary artery. It supplies the scapula and the latissimus dorsi, serratus anterior, teres major, deltoid, triceps brachii, and intercostal muscles.

---

[16]*phren* = diaphragm
[17]*thoraco* = chest; *acr* = tip, apex; *om* = shoulder

(a) Major arteries

(b) Blood-flow schematic

FIGURE 20.31 Arteries of the Thorax. (a) Major thoracic arteries. (b) Blood-flow schematic of the thoracic arteries.

20.7e Veins of the Thorax

The superior vena cava receives all venous drainage not only from the head and neck as already described, but also from all thoracic organs and tissues.

Tributaries of the Superior Vena Cava

The most prominent veins of the upper thorax carry blood from the shoulder region to the heart **(fig. 20.32);** they are as follows:

1. The **subclavian vein** drains the upper limb (see "Deep Veins" in section 20.8b). It begins at the lateral margin of the first rib and travels posterior to the clavicle. It receives the external jugular and vertebral veins, then ends (changes names) where it receives the internal jugular vein.

2. The **brachiocephalic vein** forms by the union of the subclavian and internal jugular veins. The right brachiocephalic is only about 2.5 cm long and the left is about 6 cm long. They

(a)

FIGURE 20.32 Venous Drainage of the Posterior Wall of the Thorax and Abdomen. (a) The azygos system of the thoracic wall. This system provides venous drainage from the wall and viscera of the thorax, but the visceral tributaries are not illustrated. (b) Blood-flow schematic of the thoracic and abdominal drainage. The components above the diaphragm constitute the azygos system. There is a great deal of individual variation in this anatomy. **APR**

(b)

receive tributaries from the vertebrae, thyroid gland, and upper thoracic wall and breast, then converge to form the next vein.

3. The **superior vena cava (SVC)** forms by the union of the right and left brachiocephalic veins. It travels inferiorly for about 7 cm and empties into the right atrium of the heart. Its main tributary is the *azygos vein*. It drains all structures superior to the diaphragm except the pulmonary circuit and coronary circulation. It also receives drainage from the abdominal cavity by way of the azygos system, described next.

The Azygos System

The principal venous drainage of the thoracic organs is by way of the *azygos system* (AZ-ih-goss) (fig. 20.32). The most prominent vein of this system is the **azygos**[18] **vein,** which ascends the right side of the posterior thoracic wall. It is named for the lack of a mate on the left. It receives the following tributaries, then empties into the superior vena cava at the level of vertebra T4.

1. The right **ascending lumbar vein** drains the right abdominal wall, then penetrates the diaphragm and enters the thoracic cavity. The azygos vein begins where the right ascending lumbar vein meets the right **subcostal vein** beneath rib 12.

2. The right **posterior intercostal veins** drain the intercostal spaces. The first (superior) one empties into the right brachiocephalic vein; intercostal veins 2 and 3 join to form the *right superior intercostal vein* before emptying into the azygos vein; and intercostal veins 4 through 11 each enter the azygos vein separately.

3. The right **esophageal, mediastinal, pericardial,** and **bronchial veins** (not illustrated) drain their respective organs into the azygos vein.

4. The **hemiazygos**[19] **vein** ascends the posterior thoracic wall on the left. It begins where the left ascending lumbar vein, having just penetrated the diaphragm, joins the subcostal vein below rib 12. The hemiazygos then receives the lower three posterior intercostal veins, esophageal veins, and mediastinal veins. At the left of vertebra T9, it crosses to the right and empties into the azygos vein.

5. The **accessory hemiazygos vein** descends the posterior thoracic wall on the left. It receives drainage from posterior intercostal veins 4 through 8 and sometimes the left bronchial veins. It crosses to the right at the level of vertebra T8 and empties into the azygos vein.

The left posterior intercostal veins 1 to 3 are the only ones on this side that do not ultimately drain into the azygos vein. The first one usually drains directly into the left brachiocephalic vein. The second and third unite to form the *left superior intercostal vein,* which empties into the left brachiocephalic vein.

20.7f Arteries of the Abdominal and Pelvic Regions

After passing through the aortic hiatus, the aorta descends through the abdominal cavity and ends at the level of vertebra L4, where it branches into the right and left common iliac arteries. The abdominal aorta is retroperitoneal.

Major Branches of the Abdominal Aorta

The abdominal aorta gives off arteries in the order listed here, from superior to inferior **(fig. 20.33).** Those named in the plural are paired left and right, whereas those named in the singular are solitary median arteries.

1. The **inferior phrenic arteries** supply the inferior surface of the diaphragm. They may arise from the aorta, celiac trunk, or renal artery. Each issues two or three small **superior suprarenal arteries** to the ipsilateral adrenal (suprarenal) gland.

2. The **celiac**[20] **trunk** (SEE-lee-ac) supplies the upper abdominal viscera (see next section).

3. The **superior mesenteric artery** supplies the intestines.

Inferior phrenic a.
Aortic hiatus
Celiac trunk
Suprarenal aa. — Superior / Middle / Inferior
Superior mesenteric a.
Renal a.
Lumbar aa.
Gonadal a.
Inferior mesenteric a.
Common iliac a.
Internal iliac a.
Median sacral a.

FIGURE 20.33 The Abdominal Aorta and Its Major Branches. APR

---

4. The **middle suprarenal arteries** arise laterally from the aorta, usually at the same level as the superior mesenteric artery. They supply the adrenal glands.

5. The **renal arteries** supply the kidneys and issue a small **inferior suprarenal artery** to each adrenal gland.

6. The **gonadal arteries—ovarian arteries** of the female and **testicular arteries** of the male—are long, slender arteries that arise from the midabdominal aorta and descend along the posterior body wall to the female pelvic cavity or male scrotum. They supply the gonads. The gonads begin their embryonic development near the kidneys, and the gonadal arteries are then quite short. As the gonads descend to the pelvic cavity, these arteries grow and acquire their peculiar length and course.

7. The **inferior mesenteric artery** supplies the distal end of the large intestine.

8. The **lumbar arteries** arise from the lower aorta in four pairs. They supply the posterior abdominal wall (muscles, joints, and skin) and the spinal cord and other tissues in the vertebral canal.

9. The **median sacral artery,** a tiny median artery at the inferior end of the aorta, supplies the sacrum and coccyx.

10. The **common iliac arteries** arise as the aorta forks at its lower end. We will further trace these in later discussions of arteries of the pelvic region and lower limb.

Branches of the Celiac Trunk

The celiac circulation to the upper abdominal viscera is the most complex route off the abdominal aorta. Because it has numerous anastomoses, the bloodstream doesn't follow a simple linear path but divides and rejoins itself at several points. As you study the following description, locate these branches in **figure 20.34** and identify the points of anastomosis. The small size and complex pathways of these arteries are major obstacles to surgical treatment of pancreatic cancer.

The short, stubby celiac trunk, barely more than 1 cm long, is a median branch of the aorta just below the diaphragm. It immediately gives rise to three branches—the *common hepatic, left gastric,* and *splenic arteries.*

1. The **common hepatic artery** passes to the right and issues two main branches:

   a. The **gastroduodenal artery** gives off the **right gastro-omental artery** to the stomach. It then continues as the **pancreaticoduodenal artery** (PAN-cree-AT-ih-co-dew-ODD-eh-nul), which splits into two branches that pass around the anterior and posterior sides of the head of the pancreas. These anastomose with the two branches of the *inferior pancreaticoduodenal artery,* discussed under "Mesenteric Blood Supply," following shortly.

   b. The **hepatic artery proper** ascends toward the liver. It gives off the **right gastric artery,** then branches into the

right and **left hepatic arteries.** The right hepatic artery issues a **cystic artery** to the gallbladder, then the two hepatic arteries enter the liver from below.

2. The **left gastric artery** supplies the stomach and lower esophagus, arcs around the *lesser curvature* (superomedial margin) of the stomach, and anastomoses with the right gastric artery (fig. 20.34b). Thus, the right and left gastric arteries approach from opposite directions and supply this margin of the stomach. The left gastric also has branches to the lower esophagus, and the right gastric also supplies the duodenum.

3. The **splenic artery** supplies blood to the spleen, but gives off the following branches on the way there.

   a. Several small **pancreatic arteries** supply the pancreas.
   b. The **left gastro-omental artery** arcs around the *greater curvature* (inferolateral margin) of the stomach and anastomoses with the right gastro-omental artery. These two arteries stand off about 1 cm from the stomach itself and travel through the superior margin of the greater omentum, a fatty membrane suspended from the greater curvature (see figs. B.4 in atlas B, 25.3). They furnish blood to both the stomach and omentum.
   c. The **short gastric arteries** supply the upper portion *(fundus)* of the stomach.

Mesenteric Blood Supply

The mesentery is a translucent sheet that suspends the intestines and other abdominal viscera from the posterior body wall (see figs. A.8 in atlas A, 25.3). It contains numerous arteries, veins, and lymphatic vessels that supply and drain the intestines. The arterial supply arises from the *superior* and *inferior mesenteric arteries;* numerous anastomoses between these ensure adequate collateral circulation to the intestines even if one route is temporarily obstructed.

The **superior mesenteric artery (fig. 20.35a)** is the most significant intestinal blood supply, serving nearly all of the small intestine and the proximal half of the large intestine. It arises medially from the upper abdominal aorta and gives off the following branches:

1. The **inferior pancreaticoduodenal artery** branches to pass around the anterior and posterior sides of the pancreas and anastomose with the two branches of the superior pancreaticoduodenal artery.

2. Twelve to 15 **jejunal** and **ileal arteries** fan out through the mesentery to supply nearly all of the small intestine (portions called the *jejunum* and *ileum*).

3. The **ileocolic artery** (ILL-ee-oh-CO-lic) supplies the ileum, appendix, and parts of the large intestine *(cecum* and *ascending colon).*

4. The **right colic artery** also supplies the ascending colon.

5. The **middle colic artery** supplies most of the *transverse colon.*

(a) Branches of the celiac trunk

(b) Celiac circulation to the stomach

(c) Blood-flow schematic

FIGURE 20.34 **Branches of the Celiac Trunk.**
(a) Anatomy of the celiac system with the stomach removed to expose the more posterior arteries.
(b) Arterial supply to the stomach. (c) Blood-flow schematic of the celiac system. **APR**

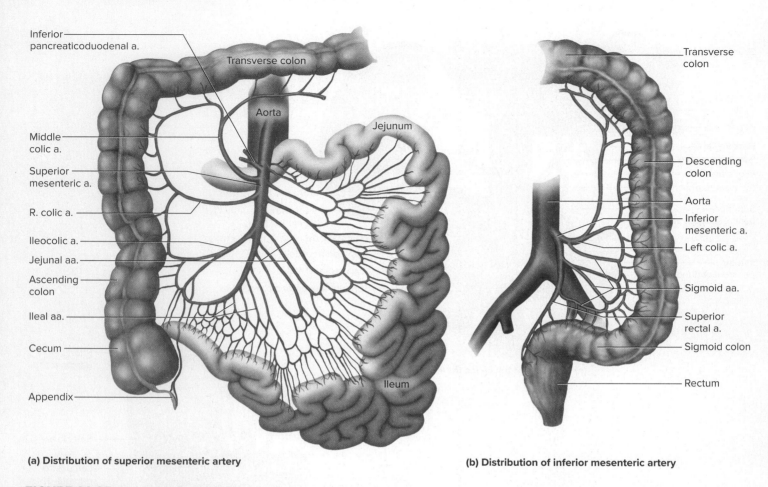

Inferior pancreaticoduodenal a.

Transverse colon

Aorta

Middle colic a.

Superior mesenteric a.

R. colic a.

Ileocolic a.

Jejunal aa.

Ascending colon

Ileal aa.

Cecum

Appendix

Jejunum

Ileum

Transverse colon

Descending colon

Aorta

Inferior mesenteric a.

Left colic a.

Sigmoid aa.

Superior rectal a.

Sigmoid colon

Rectum

(a) Distribution of superior mesenteric artery

(b) Distribution of inferior mesenteric artery

FIGURE 20.35 The Mesenteric Arteries. (a) Distribution of the superior mesenteric artery. (b) Distribution of the inferior mesenteric artery. **APR**

The **inferior mesenteric artery** arises from the lower abdominal aorta and serves the distal part of the large intestine **(fig. 20.35b).** It gives off three main branches:

1. The **left colic artery** supplies the transverse and *descending colon.*

2. The **sigmoid arteries** supply the descending and *sigmoid colon.*

3. The **superior rectal artery** supplies the rectum.

Arteries of the Pelvic Region

The two common iliac arteries arise by branching of the aorta, descend for another 5 cm, then at the level of the sacroiliac joint, each divides into an external and internal iliac artery. The *external iliac artery* supplies mainly the lower limb (see section 20.8c). The **internal iliac artery** supplies mainly the pelvic wall and viscera. Its branches are shown only in schematic form in figure 20.42. Shortly after its origin, it divides into anterior and posterior trunks. The **anterior trunk** produces the following branches:

1. The **superior vesical**[21] **artery** supplies the urinary bladder and distal end of the ureter. It arises indirectly from the anterior trunk by way of a short *umbilical artery,* a remnant of the artery that travels through the fetal umbilical cord. The rest of the umbilical artery becomes a closed fibrous cord after birth.

2. In men, the **inferior vesical artery** supplies the bladder, ureter, prostate, and seminal vesicle. In women, the corresponding vessel is the **vaginal artery,** which supplies the vagina and part of the bladder and rectum.

3. The **middle rectal artery** supplies the rectum.

4. The **obturator artery** exits the pelvic cavity through the obturator foramen and supplies the adductor muscles of the medial thigh.

5. The **internal pudendal**[22] **artery** (pyu-DEN-dul) serves the perineum and erectile tissues of the penis and clitoris; it

[21]*vesic* = bladder
[22]*pudend* = literally, "shameful parts"; the external genitals

supplies blood for the vascular engorgement and erection of these organs during sexual arousal.

6. In women, the **uterine artery** is the main blood supply to the uterus and supplies some blood to the vagina. It enlarges substantially in pregnancy and is the chief source of blood to the placenta, and thus critically important to fetal development. It passes up the uterine margin, then turns laterally at the uterine tube and anastomoses with the ovarian artery, thus supplying blood to the ovary as well (see fig. 28.7).

7. The **inferior gluteal artery** supplies the gluteal muscles and hip joint.

The **posterior trunk** produces the following branches:

1. The **iliolumbar artery** supplies the lumbar body wall and pelvic bones.

2. The **lateral sacral arteries** lead to tissues of the sacral canal, skin, and muscles posterior to the sacrum. There are usually two of these, superior and inferior.

3. The **superior gluteal artery** supplies the skin and muscles of the gluteal region and the muscle and bone tissues of the pelvic wall.

20.7g Veins of the Abdominal and Pelvic Regions

The most significant route of venous drainage from all of the body below the diaphragm is the **inferior vena cava (IVC).** This is the body's largest blood vessel, having a diameter of about 3.5 cm. It forms by the union of the right and left common iliac veins at the level of vertebra L5 and drains many of the abdominal viscera as it ascends the posterior body wall. It is retroperitoneal and lies immediately to the right of the aorta.

Tributaries of the Inferior Vena Cava

The IVC picks up blood from numerous tributaries in the following ascending order **(fig. 20.36):**

1. The **internal iliac veins** drain the gluteal muscles, the medial aspect of the thigh, the urinary bladder, and rectum; the prostate and ductus deferens of the male; and the uterus and vagina of the female. They unite with the *external iliac veins,* which drain the lower limb and are described in section 20.8d. Their union forms the **common iliac veins,** which then converge to form the IVC.

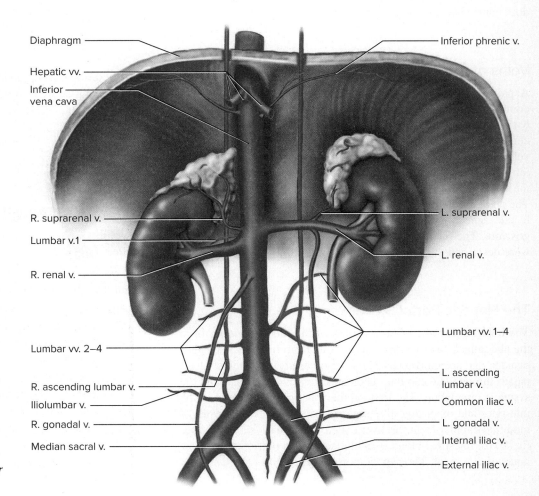

FIGURE 20.36 The Inferior Vena Cava and Its Tributaries. Compare the blood-flow schematic in figure 20.32b.

❓ *Why do the veins that drain the ovaries and testes terminate so far away from the gonads?*

Labels in figure:
Diaphragm
Hepatic vv.
Inferior vena cava
Inferior phrenic v.
R. suprarenal v.
Lumbar v.1
R. renal v.
L. suprarenal v.
L. renal v.
Lumbar vv. 1–4
Lumbar vv. 2–4
R. ascending lumbar v.
Iliolumbar v.
R. gonadal v.
Median sacral v.
L. ascending lumbar v.
Common iliac v.
L. gonadal v.
Internal iliac v.
External iliac v.

2. Four pairs of **lumbar veins** empty into the IVC as well as into the ascending lumbar veins described in the next section.

3. The **gonadal veins—ovarian veins** of the female and **testicular veins** of the male—drain the gonads. Like the gonadal arteries, and for the same reason, these are long slender vessels that end far from their origins. The left gonadal vein empties into the left renal vein, whereas the right gonadal vein empties directly into the IVC.

4. The **renal veins** drain the kidneys into the IVC. The left renal vein also receives blood from the left gonadal vein and left suprarenal veins. It is up to three times as long as the right renal vein, since the IVC lies to the right of the midline of the body.

5. The **suprarenal veins** drain the adrenal (suprarenal) glands. The right suprarenal empties directly into the IVC and the left suprarenal empties into the left renal vein.

6. The **inferior phrenic veins** drain the inferior aspect of the diaphragm.

7. Three **hepatic veins** drain the liver, extending a short distance from its superior surface to the IVC.

After receiving these inputs, the IVC penetrates the diaphragm and enters the right atrium of the heart from below (see fig. 19.5). It does not receive any thoracic drainage.

Veins of the Abdominal Wall

A pair of **ascending lumbar veins** receives blood from the common iliac veins below and from the aforementioned lumbar veins of the posterior body wall (see fig. 20.32b). The ascending lumbar veins give off anastomoses with the inferior vena cava beside them as they ascend to the diaphragm. The left ascending lumbar vein passes through the diaphragm via the aortic hiatus and continues as the hemiazygos vein above. The right ascending lumbar vein passes through the diaphragm to the right of the vertebral column and continues as the azygos vein. The further paths of the azygos and hemiazygos veins were described earlier.

The Hepatic Portal System

The **hepatic portal system** receives all the blood draining from the abdominal digestive tract, as well as from the pancreas, gallbladder, and spleen **(fig. 20.37).** Like other portal systems, blood passes through two capillary beds in series on one trip around the systemic circulation. The first capillary bed in this case is in the intestines and other digestive organs; the second is a network of modified capillaries in the liver called the *hepatic sinusoids* (see fig. 20.6). Intestinal blood is richly laden with nutrients for a few hours after a meal. The hepatic portal system gives the liver first

claim to these nutrients before they are distributed to the rest of the body. It also allows the liver to cleanse the blood of bacteria and toxins picked up from the intestines—an important function of the liver. The principal veins of the hepatic portal system are as follows:

1. The **inferior mesenteric vein** receives blood from the rectum and distal colon. It converges in a fanlike array in the mesentery and empties into the splenic vein.

2. The **superior mesenteric vein** receives blood from the entire small intestine, ascending colon, transverse colon, and stomach. It, too, exhibits a fanlike arrangement in the mesentery, then joins the splenic vein to form the hepatic portal vein.

3. The **splenic vein** drains the spleen and travels across the upper abdominal cavity toward the liver. Along the way, it picks up **pancreatic veins** from the pancreas, then picks up the inferior mesenteric vein, and ends where it meets the superior mesenteric vein.

4. The **hepatic portal vein** is the continuation beyond the convergence of the splenic and superior mesenteric veins. It travels about 8 cm upward and to the right, receives the **cystic vein** from the gallbladder, then enters the inferior surface of the liver. In the liver, it ultimately leads to the innumerable microscopic hepatic sinusoids. Blood from the sinusoids collects in the hepatic veins, and they empty into the IVC. Circulation within the liver is described in more detail in section 25.4.

5. The left and right **gastric veins** form an arc along the lesser curvature of the stomach and empty into the hepatic portal vein.

BEFORE YOU GO ON

Answer the following questions to test your understanding of the preceding section:

26. Concisely contrast the destinations of the external and internal carotid arteries.

27. Briefly state the organs or parts of organs that are supplied with blood by (a) the cerebral arterial circle, (b) the celiac trunk, (c) the superior mesenteric artery, and (d) the internal iliac artery.

28. If you were dissecting a cadaver, where would you look for the internal and external jugular veins? What muscle would help you distinguish one from the other?

29. Trace the path of a blood cell from the left lumbar body wall to the superior vena cava, naming the vessels through which it would travel.

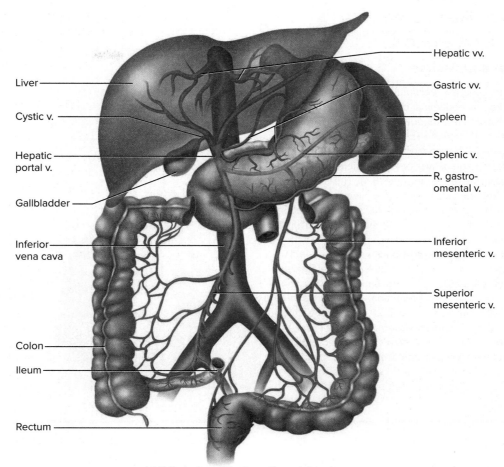

(a) Tributaries of the hepatic portal system

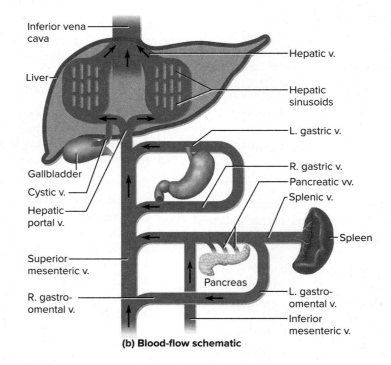

(b) Blood-flow schematic

FIGURE 20.37 The Hepatic Portal System. (a) Tributaries of the hepatic portal system. (b) Blood-flow schematic of the system. **A&PR**

DEEPER INSIGHT 20.4

CLINICAL APPLICATION

Portal Hypertension and Ascites

Ascites (ah-SY-teez) is the abnormal accumulation of serous fluid in the peritoneal cavity, marked by abdominal distension **(fig. 20.38).** It is usually associated with alcoholism, although it can have other causes such as malnutrition (see fig. 18.3), heart failure, infection, cancer, and chronic hepatitis. Excessive alcohol consumption causes *cirrhosis* of the liver—inflammation, destruction of liver cells, and their replacement by fatty and fibrous tissue (see Deeper Insight 26.4). The degenerating liver puts up excess resistance to blood flow, leading to *portal hypertension*—high blood pressure in the hepatic portal circulation. Since the spleen drains into the hepatic portal system, blood pressure backs up into the spleen as well. Both liver and spleen become enlarged (*hepatomegaly* and *splenomegaly*) and "weep" serous fluid into the peritoneal cavity. The failing liver also fails to adequately clear hormones from the blood, so aldosterone and antidiuretic hormone accumulate and stimulate increased fluid retention, further worsening the ascites. A person can retain as much as 10 to 20 liters of abdominal fluid, pressing up on the other abdominal viscera and diaphragm and producing shortness of breath. An infection called *spontaneous bacterial peritonitis (SBP)* is a frequent complication of ascites. The prognosis for recovery from ascites, especially with SBP, is poor; about 25% of patients die within one year of diagnosis, often because of continued heavy drinking.

FIGURE 20.38 Ascites. The abdomen is distended with accumulated serous fluid that has filtered from the liver, spleen, and intestinal blood vessels.
Mediscan/Alamy

20.8 Systemic Vessels of the Appendicular Region

Expected Learning Outcomes

When you have completed this section, you should be able to

a. identify the principal systemic arteries and veins of the limbs; and

b. trace the flow of blood from the heart to any region of the upper or lower limb and back to the heart.

The principal vessels of the appendicular region are detailed in the following sections, 20.8a through 20.8d. Although the appendicular arteries are usually deep and well protected, the veins occur in both deep and superficial groups.

Venous pathways have more anastomoses than arterial pathways, so the route of flow is often not as clear. If all the anastomoses were illustrated, many of these venous pathways would look more like confusing networks than a clear route back to the heart. Therefore, most anastomoses—especially the highly variable and unnamed ones—are omitted from the figures to allow you to focus on the more general course of blood flow. The blood-flow schematics in several figures will also help, like road maps, to clarify these routes.

20.8a Arteries of the Upper Limb

The upper limb is supplied by a prominent artery that changes name along its course from *subclavian* to *axillary* to *brachial artery,* then issues branches to the arm, forearm, and hand **(fig. 20.39).**

The Shoulder and Arm

The shoulder and arm, like the head and most thoracic viscera, also receive their blood supply from arteries arising from the aortic arch. These pathways begin with the subclavian arteries, named for their proximity to the clavicles.

1. The brachiocephalic trunk arises from the aortic arch and branches into the right common carotid artery and **right subclavian artery.** The **left subclavian artery** arises directly from the aortic arch slightly distal to the brachiocephalic trunk. Each subclavian arches over the respective lung, rising as high as the base of the neck slightly superior to the clavicle. It then passes posterior to the clavicle, downward over the first rib, and ends in name only at the rib's lateral margin. In the shoulder, it gives off several small branches to the thoracic wall and viscera, which were described in section 20.7d.

2. As the artery continues past the first rib, it is named the **axillary artery.** It continues through the axillary region, gives off small thoracic branches, and ends, again in name

Common carotid a.

Subclavian a.

Brachiocephalic trunk

Axillary a.

Circumflex humeral aa.

Brachial a.

Deep brachial a.

Superior ulnar collateral a.

Radial collateral a.

Interosseous aa.:
Common
Posterior
Anterior

Radial a.

Ulnar a.

Deep palmar arch

Superficial palmar arch

(a) Major arteries

Lateral | Medial

Subclavian a.

Axillary a.

Circumflex humeral aa.

Brachial a.

Deep brachial a.

Superior ulnar collateral a.

Radial collateral a.

Ulnar a.

Radial a.

Interosseous aa.:
Common
Posterior
Anterior

Dorsal carpal arch

Deep palmar arch

Superficial palmar arch

(b) Blood-flow schematic

FIGURE 20.39 Arteries of the Upper Limb. (a) The major arteries, anterior view. (b) Blood-flow schematic of the upper limb.

? *Why are arterial anastomoses especially common at joints such as the shoulder and elbow?*

only, at the neck of the humerus. Here, it gives off a pair of **circumflex**[23] **humeral arteries,** which encircle the humerus, anastomose with each other laterally, and supply blood to the shoulder joint and deltoid muscle. Beyond this loop, it continues as the next artery.

3. The **brachial artery** (BRAY-kee-ul) continues down the median and anterior sides of the humerus and ends just distal to the elbow, supplying the anterior flexor muscles of the brachium along the way. This artery is the usual site of blood pressure measurement with the sphygmomanometer.

4. The **deep brachial artery** arises from the proximal end of the brachial artery and supplies the humerus and triceps brachii muscle. About midway down the arm, it continues as the next artery.

5. The **radial collateral artery** arises from the deep brachial artery, descends the lateral side of the arm, and empties into the radial artery slightly distal to the elbow.

6. The **superior ulnar collateral artery** arises about midway along the brachial artery and descends in the medial side of the arm. It empties into the ulnar artery slightly distal to the elbow.

The Forearm and Hand

Just distal to the elbow, the brachial artery forks into the *radial* and *ulnar arteries.*

1. The **radial artery** descends the forearm laterally, alongside the radius, nourishing the lateral forearm muscles. The most common place to take a pulse is at the radial artery just proximal to the thumb.

2. The **ulnar artery** descends the forearm medially, alongside the ulna, nourishing the medial forearm muscles.

3. The **interosseous**[24] **arteries** of the forearm lie between the radius and ulna. They begin with a short **common interosseous artery** branching from the upper end of the ulnar artery. The common interosseous quickly divides into anterior and posterior branches. The **anterior interosseous artery** travels down the anterior side of the interosseous membrane, nourishing the radius, ulna, and deep flexor muscles. It ends distally by passing through the interosseous membrane to join the posterior interosseous artery. The **posterior interosseous artery** descends along the posterior side of the interosseous membrane and nourishes mainly the superficial extensor muscles.

4. Two U-shaped **palmar arches** arise by anastomosis of the radial and ulnar arteries at the wrist. The **deep palmar arch** is fed mainly by the radial artery and the **superficial palmar arch** mainly by the ulnar artery. The arches issue smaller arteries to the palmar region and fingers.

20.8b Veins of the Upper Limb

Both superficial and deep veins drain the upper limb **(fig. 20.40).** We will trace these veins in the order of blood flow, from the hand to the axillary and subclavian veins, which, in turn, lead off toward the heart.

Superficial Veins

The superficial veins travel through the subcutaneous tissue of the limb; you may be able to see several of them through the skin of your hand, forearm, and arm **(fig. 20.41).** They are larger in diameter and carry more blood than the deep veins.

1. The **dorsal venous network** is a plexus of veins often visible through the skin on the back of the hand. It empties into the major superficial veins of the forearm, the cephalic and basilic veins.

2. The **cephalic**[25] **vein** (sef-AL-ic) arises from the lateral side of the network, travels up the lateral side of the forearm and arm to the shoulder, and joins the axillary vein there. Intravenous fluids are often administered through the distal end of this vein.

3. The **basilic**[26] **vein** (bah-SIL-ic) arises from the medial side of the network, travels up the posterior side of the forearm, and continues into the arm. It turns deeper about midway up the arm and joins the *brachial vein* at the axilla. As an aid to remembering which vein is cephalic and which is basilic, visualize your arm held straight away from the torso (abducted) with the thumb up. The cephalic vein runs along the upper side of the arm closer to the head (as suggested by *cephal,* "head"), and the name *basilic* is suggestive of the lower (basal) side of the arm (although not named for that reason; *basilic* means "prominent," as in *basilica*).

4. The **median cubital vein** is a short anastomosis between the cephalic and basilic veins that obliquely crosses the cubital fossa (anterior bend of the elbow). It is often clearly visible through the skin and is a common site for drawing blood.

5. The **median antebrachial vein** drains a network of blood vessels in the hand called the **superficial palmar venous network.** It travels up the medial forearm and terminates at the elbow, emptying variously into the basilic, median cubital, or cephalic vein.

Deep Veins

Deep veins run parallel to the arteries and often have similar names (*brachial veins* and *brachial artery,* for example). In some cases, the deep veins occur in pairs flanking the corresponding artery (such as the two *radial veins* traveling alongside the *radial artery*).

1. The **deep** and **superficial venous palmar arches** receive blood from the fingers and palmar region. They are anastomoses that join the radial and ulnar veins.

2. A pair of **radial veins** arise from the lateral side of the palmar arches and course up the forearm alongside the radius. Slightly distal to the elbow, they converge and give rise to one of the brachial veins.

---

[23]*circum* = around; *flex* = to bend
[24]*inter* = between; *osse* = bones

[25]*cephal* = head
[26]*basilic* = prominent, important

(a) Major veins

(b) Blood-flow schematic

FIGURE 20.40 Veins of the Upper Limb. (a) The major veins, anterior view. (b) Blood-flow schematic of the upper limb. Variations on this pattern are highly common. Many venous anastomoses are omitted for clarity. **APR**

❓ *Name three veins that are often visible through the skin of the upper limb.*

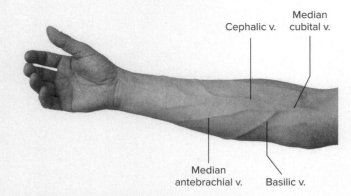

Cephalic v.

Median cubital v.

Median antebrachial v.

Basilic v.

FIGURE 20.41 Superficial Veins of the Lower Arm and Upper Forearm. Anterior view. These prominent and accessible veins, especially the basilic and median cubital, are commonly used for drawing blood specimens and administering I.V. fluids and transfusions, and for cardiac catheterization.

Taborsk/Getty Images

3. A pair of **ulnar veins** arise from the medial side of the palmar arches and course up the forearm alongside the ulna. They unite near the elbow to form the other brachial vein.

4. The two **brachial veins** continue up the brachium, flanking the brachial artery, and converge into a single vein just before the axillary region.

5. The **axillary vein** forms by the union of the brachial and basilic veins. It begins at the lower margin of the teres major muscle and passes through the axillary region, picking up the cephalic vein along the way. At the lateral margin of the first rib, it changes name to the subclavian vein.

6. The **subclavian vein** continues into the shoulder posterior to the clavicle and ends where it meets the internal jugular vein of the neck. There it becomes the brachiocephalic vein described earlier. The right and left brachiocephalics converge and form the superior vena cava, which empties into the right atrium of the heart.

20.8c Arteries of the Lower Limb

As we have already seen, the aorta forks at its lower end into the right and left common iliac arteries, and each of these soon divides again into an internal and external iliac artery. We traced the internal iliac artery in section 20.7f, and we now trace the external iliac as it supplies the lower limb **(figs. 20.42, 20.43).**

Arteries from the Pelvic Region to the Knee

1. The **external iliac artery** issues small branches to the skin and muscles of the abdominal wall and pelvis, then passes deep to the inguinal ligament and becomes the femoral artery.

2. The **femoral artery** passes through the *femoral triangle* of the upper medial thigh, where its pulse can be palpated (see Deeper Insight 20.5). In the triangle, it gives off sev-

eral small arteries to the skin, then produces the following branches before descending the rest of the way to the knee.

   a. The **deep femoral artery** arises from the lateral side of the femoral, within the triangle. It is the largest branch and is the major arterial supply to the thigh muscles.

   b. Two **circumflex femoral arteries** arise from the deep femoral, encircle the head of the femur, and anastomose laterally. They supply mainly the femur, hip joint, and hamstring muscles.

3. The **popliteal artery** is a continuation of the femoral artery in the popliteal fossa at the back of the knee. It begins where the femoral artery emerges from an opening *(adductor hiatus)* in the tendon of the adductor magnus muscle and ends where it splits into the *anterior* and *posterior tibial arteries*. As it passes through the popliteal fossa, it gives off anastomoses called **genicular**[27] **arteries** that supply the knee joint.

Arteries of the Leg and Foot

In the leg proper, the three most significant arteries are the anterior tibial, posterior tibial, and fibular arteries.

1. The **anterior tibial artery** arises from the popliteal artery and immediately penetrates through the interosseous membrane of the leg to the anterior compartment. There, it travels lateral to the tibia and supplies the extensor muscles. Upon reaching the ankle, it gives off the following:

   a. The **dorsal artery of the foot** traverses the ankle and upper medial surface of the foot and gives rise to the arcuate artery.

   b. The **arcuate artery** sweeps across the foot from medial to lateral and gives off small arteries that supply the toes.

2. The **posterior tibial artery** is a continuation of the popliteal artery that passes down the leg, deep in the posterior compartment, supplying flexor muscles along the way. Inferiorly, it passes behind the medial malleolus of the ankle and into the plantar region of the foot. It gives rise to the following:

   a. The **median** and **lateral plantar arteries** originate by branching of the posterior tibial artery at the ankle. The medial plantar artery supplies mainly the great toe. The lateral plantar artery sweeps across the sole of the foot and becomes the deep plantar arch.

   b. The **deep plantar arch** gives off another set of small arteries to the toes.

3. The **fibular (peroneal) artery** arises from the proximal end of the posterior tibial artery near the knee. It descends through the lateral side of the posterior compartment, supplying lateral muscles of the leg along the way, and ends in a network of arteries in the heel.

---

[27]*genic* = of the knee

FIGURE 20.42 Arteries of the Lower Limb. (a) The major arteries in anterior view, with the foot strongly plantar flexed so its upper (dorsal) surface faces the viewer. (b) The major arteries in posterior view, with the sole of the foot facing the viewer. **APR**

20.8d Veins of the Lower Limb

We will follow venous drainage of the lower limb from the toes to the inferior vena cava **(figs. 20.44, 20.45).** As in the upper limb, there are deep and superficial veins with anastomoses between them. Most of the anastomoses are omitted from the illustrations.

Superficial Veins

The superficial veins of the lower thigh and leg lie in the subcutaneous tissue; thus, they are often visible through the skin and accessible with relative ease for such purposes as I.V. fluid therapy. They are particularly prone to becoming varicose veins (see fig. 20.9).

1. The **dorsal venous arch** (fig. 20.44a) is often visible through the skin on the dorsum of the foot. It collects blood from the toes and more proximal part of the foot, and has numerous anastomoses similar to the dorsal venous network of the hand. It gives rise to the following:

2. The **small (short) saphenous**[28] **vein** (sah-FEE-nus) arises from the lateral side of the arch and passes up that side of the leg as far as the knee. There, it drains into the popliteal vein.

---

[28]*saphen* = standing

Aorta

Common iliac aa.

External iliac a.

Femoral a.

Deep femoral a.

Circumflex femoral aa.

Popliteal a.

Anterior tibial a.

Posterior tibial a.

Fibular a.

Deep plantar arch

Internal iliac a.:

Anterior trunk

Posterior trunk

Iliolumbar a.

Lateral sacral aa.

Superior gluteal a.

Umbilical a.

Superior vesical a.

Inferior vesical a. (male)

Vaginal a. (female)

Middle rectal a.

Obturator a.

Internal pudendal a.

Uterine a. (female)

Inferior gluteal a.

Dorsal a.

Arcuate a.

FIGURE 20.43 Arterial Schematic of the Pelvic Region and Lower Limb (Anterior View). The pelvic schematic on the right is stretched for clarity. These arteries are not located as far inferiorly as the arteries depicted adjacent to them on the left.

3. The **great (long) saphenous vein,** the longest vein in the body, arises from the medial side of the arch and travels all the way up the leg and thigh to the inguinal region. It empties into the femoral vein slightly inferior to the inguinal ligament. It is commonly used as an access site for the long-term administration of intravenous fluids and for cardiac catheterization. It is a relatively accessible vein in infants and in patients in shock whose veins have collapsed. Portions of this vein are commonly used as grafts in coronary artery bypass surgery. The great and small saphenous veins are among the most common sites of varicose veins.

Deep Veins

Deep veins of the lower limb accompany the deep arteries. Some of them mirror the arrangement seen in the forelimb, with paired veins flanking the corresponding artery (such as the two *posterior tibial veins* running alongside the *posterior tibial artery*).

1. The **deep plantar venous arch** receives blood from the toes and gives rise to **lateral** and **medial plantar veins** on the respective sides of the foot. The lateral plantar vein gives

off the *fibular veins,* then crosses over to the medial side and approaches the medial plantar vein (but does not join it). The two plantar veins pass behind the medial malleolus of the ankle and continue as a pair of parallel *posterior tibial veins.*

2. The two **posterior tibial veins** pass up the leg embedded deep in the calf muscles. They converge like an inverted Y into a single vein about two-thirds of the way up the tibia.

3. The two **fibular (peroneal) veins** ascend the back of the leg and similarly converge like an inverted Y.

4. The **popliteal** vein begins near the knee by convergence of these two Ys. It passes through the popliteal fossa at the back of the knee.

5. The two **anterior tibial veins** travel up the anterior compartment of the leg between the tibia and fibula (fig. 20.44a). They arise from the medial side of the dorsal venous arch, converge just distal to the knee, then flow into the popliteal vein.

6. The **femoral vein** is a continuation of the popliteal vein into the thigh. It drains blood from the deep thigh muscles and femur.

Lateral | Medial

Medial | Lateral

- Inferior vena cava
- Common iliac v.
- Internal iliac v.
- External iliac v.

Circumflex femoral vv.

Circumflex femoral vv.

- Deep femoral v.
- Femoral v.
- Great saphenous v.
- Popliteal v.

- Anterior tibial v.

Small saphenous v.

Small saphenous v.

Superficial veins
Deep veins

- Fibular vv.
- Posterior tibial vv.

Anterior tibial vv.

Dorsal venous arch

- Medial plantar v.
- Lateral plantar v.
- Deep plantar venous arch

(a) Anterior view

(b) Posterior view

FIGURE 20.44 Veins of the Lower Limb.
(a) The major veins in anterior view, with the foot strongly plantar flexed so its upper (dorsal) surface faces the viewer. (b) The major veins in posterior view, with the sole of the foot facing the viewer. **APR**

7. The **deep femoral vein** drains the femur and muscles of the thigh supplied by the deep femoral artery. It receives tributaries along the shaft of the femur, then a pair of *circumflex femoral veins* that encircle the upper femur. It finally drains into the upper femoral vein.

8. The **external iliac vein** forms by the union of the femoral and great saphenous veins near the inguinal ligament.

9. The **internal iliac vein** follows the course of the internal iliac artery and its distribution. Its tributaries drain the gluteal muscles; the medial aspect of the thigh; the urinary bladder, rectum, prostate, and ductus deferens of the male; and the uterus and vagina of the female.

10. The **common iliac vein** forms by the union of the external and internal iliac veins. The right and left

common iliacs then unite to form the inferior vena cava.

▶▶▶ **APPLY WHAT YOU KNOW**

There are certain similarities between the arteries of the hand and foot. What arteries of the wrist and hand are most comparable in arrangement and function to the arcuate artery and deep plantar arch of the foot?

Deeper Insight 20.5 exemplifies the relevance of vascular anatomy to emergency first aid. The most common cardiovascular diseases are atherosclerosis and hypertension. **Table 20.2** provides links to further information on these and several other vascular diseases. See "Circulatory System" in section 29.4a for the effects of aging on the blood vessels.

Lateral | Medial

Common iliac v.

Inferior vena cava

External iliac v.

Internal iliac v.

Femoral v.

Deep femoral v.

Circumflex
femoral vv.

Great saphenous v.

Popliteal v.

Small saphenous v.

Anterior tibial vv.

Superficial veins

Deep veins

Fibular vv.

Posterior tibial vv.

Dorsal venous arch

Medial plantar v.

Lateral plantar v.

Deep plantar venous arch

**FIGURE 20.45 Venous Schematic of
the Lower Limb (Anterior View).**

TABLE 20.2	Some Vascular Disorders
Dissecting aneurysm	Splitting of the layers of an arterial wall from each other because of the accumulation of blood between layers. Results from either a tear in the tunica interna or rupture of the vasa vasorum.
Fat embolism	The presence of fat globules traveling in the bloodstream. Globules originate from bone fractures, fatty degeneration of the liver, and other causes and may block cerebral or pulmonary blood vessels.
Orthostatic hypotension	A decrease in blood pressure that occurs when one stands, often resulting in blurring of vision, dizziness, and syncope (fainting). Results from sluggish or inactive baroreflexes.

You can find other vascular disorders described in the following places:

Thrombosis and *thromboembolism* in section 18.5f; *air embolism* in Deeper Insight 20.3; *atherosclerosis* and *arteriosclerosis* in Deeper Insight 19.4 and section 20.2a; *aneurysm* in Deeper Insight 20.1; *hypertension* and *hypotension* in section 20.2a and Deeper Insight 20.6; *varicose veins* in Deeper Insight 20.2; *circulatory shock* in section 20.4c; *transient ischemic attack (TIA)* and *stroke* in section 20.5a; and *edema* in section 20.3d.

DEEPER INSIGHT 20.5

CLINICAL APPLICATION

Arterial Pressure Points

In some places, major arteries come close enough to the body surface to be palpated. These places can be used to take a pulse, and they can serve as emergency *pressure points,* where firm pressure can be applied to temporarily reduce arterial bleeding **(fig. 20.46a).** One of these points is the *femoral triangle* of the upper medial thigh

(fig. 20.46b, c). This is an important landmark for arterial supply, venous drainage, and innervation of the lower limb. Its boundaries are the sartorius muscle laterally, the inguinal ligament superiorly, and the adductor longus muscle medially. The femoral artery, vein, and nerve run close to the surface at this point.

FIGURE 20.46 Arterial Pressure Points. (a) Areas where arteries lie close enough to the surface that a pulse can be palpated or pressure can be applied to reduce arterial bleeding. (b) Structures in the femoral triangle. (c) The three boundaries that define the femoral triangle.

▶▶▶ **APPLY WHAT YOU KNOW**

From the preceding discussion of arteries and veins of the limbs, identify an artery or vein with special relevance to each of the following clinical applications: (1) the artery from which blood pressure is usually measured; (2) the artery where a patient's pulse is most often taken; (3) a vessel where pressure should be applied to stop arterial bleeding from a laceration of the thigh; (4) upper and lower limb veins where intravenous fluid is often administered; (5) a vein where blood samples are commonly drawn; and (6) a vein from which a portion may be removed and used for a coronary artery bypass graft.

BEFORE YOU GO ON

Answer the following questions to test your understanding of the preceding section:

30. Trace one possible path of a red blood cell from the left ventricle to the toes.

31. Trace one possible path of a red blood cell from the fingers to the right atrium.

32. The subclavian, axillary, and brachial arteries are really one continuous artery. What is the reason for giving it three different names along its course?

33. State two ways in which the great saphenous vein has special clinical significance. Where is this vein located?

 DEEPER INSIGHT 20.6

CLINICAL APPLICATION

Hypertension—"The Silent Killer"

Hypertension, the most common cardiovascular disease, affects about 30% of Americans over age 50, and 50% by age 74. It is a "silent killer" that can wreak its destructive effects for 10 to 20 years before the symptoms are first noticed. Hypertension is the major cause of heart failure, stroke, and kidney failure. It damages the heart because it increases the afterload, which makes the ventricles work harder to expel blood. The myocardium enlarges up to a point (the *hypertrophic response*), but eventually it becomes excessively stretched and less efficient. Hypertension strains the blood vessels and tears the endothelium, thereby creating lesions that become focal points of atherosclerosis. Atherosclerosis then worsens the hypertension and establishes an insidious positive feedback cycle.

Another positive feedback cycle involves the kidneys. Their arterioles thicken in response to the stress, their lumens become narrower, and renal blood flow declines. In response to the resulting drop in blood pressure, the kidneys release renin, which leads to the formation of the vasoconstrictor angiotensin II and the release of aldosterone, a hormone that promotes salt retention (see "The Renin–Angiotensin–Aldosterone Mechanism" in section 23.3d). These effects worsen the hypertension that already existed. If diastolic pressure exceeds 120 mm Hg, the kidneys and heart may deteriorate rapidly, blood vessels of the eye hemorrhage, blindness may ensue, and death usually follows within 2 years.

Primary hypertension, which accounts for 90% of cases, results from such a complex web of behavioral, hereditary, and other factors that it is difficult to sort out any specific underlying cause. It was once considered such a normal part of the "essence" of aging that it continues to be called by another name, *essential hypertension*. That term suggests a fatalistic resignation to hypertension as a fact of life, but this need not be. Many risk factors have been identified, and most of them are controllable.

One of the chief culprits is obesity. Each pound of extra fat requires miles of additional blood vessels to serve it, and all of this added vessel length increases peripheral resistance and blood pressure. Just carrying around extra weight, of course, also increases the workload on the heart. Even a small weight loss can significantly reduce blood pressure. Sedentary behavior is another risk factor. Aerobic exercise helps to reduce hypertension by controlling weight, reducing emotional tension, and stimulating vasodilation.

Dietary factors are also significant contributors to hypertension. Diets high in saturated fat contribute to atherosclerosis. Potassium and magnesium reduce blood pressure; thus, diets deficient in these minerals promote hypertension. The relationship of salt intake to hypertension has been a controversial subject. The kidneys compensate so effectively for excess salt intake that dietary salt has little effect on the blood pressure of most people. Reduced salt intake may, however, help to control hypertension in older people and in people with reduced renal function.

Nicotine makes a particularly devastating contribution to hypertension because it stimulates the myocardium to beat faster and harder, while it stimulates vasoconstriction and increases the afterload against which the myocardium must work. Just when the heart needs extra oxygen, nicotine causes coronary vasoconstriction and promotes myocardial ischemia.

Some risk factors cannot be changed at will—ancestry, heredity, and sex. Hypertension runs in some families. A person whose parents or siblings have hypertension is more likely than average to develop it. The incidence of hypertension is about 30% higher, and the incidence of strokes about twice as high, among blacks as among whites. From ages 18 to 54, hypertension is more common in men, but above age 65, it is more common in women. Even people at risk from these factors, however, can minimize their chances of hypertension by changing risky behaviors.

Treatments for primary hypertension include weight loss, diet, and certain drugs. Diuretics lower blood volume and pressure by promoting urination. ACE inhibitors block the formation of the vasoconstrictor angiotensin II. Beta-blockers such as propranolol also lower angiotensin II level, but do it by inhibiting the secretion of renin. Calcium channel blockers such as verapamil and nifedipine inhibit the inflow of calcium into cardiac and smooth muscle, thus inhibiting their contraction, promoting vasodilation, and reducing cardiac workload.

Secondary hypertension, which accounts for about 10% of cases, is high blood pressure that results from other identifiable disorders. These include kidney disease (which may cause renin hypersecretion), atherosclerosis, hyperthyroidism, Cushing syndrome, and polycythemia. Secondary hypertension is corrected by treating the underlying disease.

Effects of the CIRCULATORY SYSTEM on Other Organ Systems

ALL SYSTEMS
Blood delivers O_2 to all tissues and organs and removes CO_2 and other wastes from them, distributes nutrients and hormones throughout the body, and carries heat from deeper organs to the body surface for elimination.

INTEGUMENTARY SYSTEM
Dermal blood flow strongly affects overall body temperature.

LYMPHATIC AND IMMUNE SYSTEMS
Blood vessels produce tissue fluid, which becomes the lymph; blood contains the WBCs and plasma proteins employed in immunity.

SKELETAL SYSTEM
Blood delivers the minerals needed for bone deposition; delivers hormones that regulate skeletal growth; and delivers hormones to the bone marrow that stimulate RBC, WBC, and platelet production.

RESPIRATORY SYSTEM
Blood picks up O_2 from the lungs and releases CO_2 to be exhaled; low capillary blood pressure and blood oncotic pressure keep alveoli from filling with fluid.

MUSCULAR SYSTEM
Blood removes the heat generated by exercise.

URINARY SYSTEM
Urine production begins with blood filtration; blood carries away the water and solutes reabsorbed by the kidneys; blood pressure maintains renal function.

NERVOUS SYSTEM
Endothelial cells of the blood vessels maintain the blood–brain barrier and play a role in production of cerebrospinal fluid.

DIGESTIVE SYSTEM
Blood picks up absorbed nutrients and helps in reabsorption and recycling of bile salts and minerals from the intestines.

ENDOCRINE SYSTEM
Blood is the medium in which all hormones travel to their target organs.

REPRODUCTIVE SYSTEM
Blood delivers the hormones that affect reproductive function; vasodilation produces erection in sexual response; blood provides nutrients, oxygen, and other metabolic needs to the fetus and removes its wastes.

▶ Assess Your Learning Outcomes

To test your knowledge, discuss the following topics with a study partner or in writing, ideally from memory.

20.1 General Anatomy of the Blood Vessels

1. Definitions of *arteries, veins,* and *capillaries* with respect to the path of blood flow
2. Tunics of an artery or vein, and their general histological differences
3. Structure and functions of the endothelium
4. Location and function of the vasa vasorum
5. Three size classes of arteries; how and why they differ not just in diameter, but also histologically
6. The relationship of metarterioles to a capillary bed, and the function of the precapillary sphincters of a metarteriole
7. Location, structure, and function of the carotid sinuses, carotid bodies, and aortic bodies
8. Histology of the three types of capillaries and how it relates to their functions
9. Organization of a capillary bed and how its perfusion is regulated
10. Why veins are called *capacitance vessels* and how this relates to the structural difference between veins and arteries
11. What capillaries and postcapillary venules have in common with respect to fluid exchange
12. Structural differences between muscular venules, medium veins, and large veins
13. Structure and purpose of the venous valves, where they occur, and the reason certain veins have valves but arteries of corresponding size do not
14. How venous sinuses differ from other veins, and where they occur
15. How portal systems and anastomoses differ from simpler routes of blood flow; types of anastomoses and their purposes

20.2 Blood Pressure, Resistance, and Flow

1. The difference between blood flow and perfusion
2. How blood flow is related to resistance and pressure differences; the mathematical expression of these relationships
3. How to determine systolic pressure, diastolic pressure, and pulse pressure; how to estimate mean arterial pressure (MAP), and why MAP differs from head to foot

4. The meanings of *hypertension* and *hypotension*
5. Why arterial expansion and recoil during the cardiac cycle reduce pulse pressure and ease the strain on small arteries
6. Why arterial flow is pulsatile but capillary and venous flow are not
7. Why blood pressure rises with age
8. Variables that determine blood pressure
9. Variables that determine peripheral resistance; whether each one is directly or inversely proportional to resistance; and which of them is most changeable from moment to moment
10. Terms for widening and narrowing of a blood vessel by muscular contraction and relaxation
11. The mathematical relationship between peripheral resistance and vessel radius; why this is related to the laminar flow of blood; and why it makes vasoreflexes such a powerful influence on blood flow
12. Why blood velocity declines from aorta to capillaries and rises again from capillaries to veins, but never rises as high in veins as it was in the aorta
13. Why arterioles exert a greater influence than any other category of blood vessels on tissue perfusion
14. Three levels of control over blood pressure and flow
15. Short- and long-term mechanisms of local control of blood flow; examples of vasoactive chemicals and how they can cause reactive hyperemia
16. Angiogenesis and its importance for cancer therapy
17. The role of the vasomotor center of the medulla oblongata in controlling blood flow; baroreflexes, chemoreflexes, and the medullary ischemic reflex
18. Mechanisms of action by angiotensin II, aldosterone, natriuretic peptides, antidiuretic hormone, epinephrine, and norepinephrine on blood pressure
19. How vasomotion can change systemwide blood pressure or redirect blood flow from one region to another; circumstances that call for redirection of blood flow

20.3 Capillaries and Fluid Exchange

1. The meaning of *capillary exchange,* and substances involved in the process

2. Three routes and four mechanisms by which materials pass through capillary walls
3. Substances exchanged by simple diffusion; factors that determine whether a substance can diffuse through a capillary wall
4. Capillary transcytosis and some substances exchanged this way
5. In capillary filtration, three forces that draw fluid out of the capillaries and one force that draws fluid into them
6. The values and net effects of capillary exchange forces at the arterial and venous ends of a capillary, and how they enable a capillary to give off fluid at one end and reabsorb it at the other
7. Relative amounts of fluid given off and reabsorbed by a model capillary, and what compensates for the difference between filtration and reabsorption
8. The role of solvent drag in capillary exchange
9. Why the dynamics of capillary absorption can change from moment to moment or differ in various places in the body; examples of places where the capillaries are engaged entirely in net filtration or reabsorption
10. Chemicals that affect capillary permeability and filtration
11. Three causes of edema, and its pathological consequences

20.4 Venous Return and Circulatory Shock

1. The definition of *venous return,* and five mechanisms that drive it
2. How the skeletal muscle pump works and why it depends on venous valves
3. Why exercise increases venous return
4. Why physical inactivity can lead to venous pooling; consequences of venous pooling
5. Definition of *circulatory shock*
6. Two basic categories of circulatory shock, three forms of low venous return (LVR) shock, and situations in which each form of shock may occur
7. Why septic and anaphylactic shock cannot be strictly classified into any single category of LVR shock
8. Differences between compensated and decompensated shock

20.5 Special Circulatory Routes

1. A typical value for cerebral blood flow and why its constancy is important

STUDY
GUIDE

2. How the brain regulates its blood flow and what chemical stimulus is the most potent in activating its regulatory mechanisms

3. The causes, effects, and difference between a transient ischemic attack (TIA) and cerebral vascular accident (stroke)

4. Variability of skeletal muscle perfusion; what stimuli increase perfusion to meet the demands of exercise; and why isometric contraction causes fatigue more quickly than isotonic contraction does

5. How pulmonary circulation differs from systemic circulation with respect to blood pressure, capillary exchange, relative oxygenation of arterial and venous blood, and the vasomotor response to hypoxia

20.6 Anatomy of the Pulmonary Circuit

1. The route of blood flow in the pulmonary circuit

2. Where the capillaries of the pulmonary circuit are found and the function they serve

3. How the function of the pulmonary circuit differs from that of the bronchial arteries, which also supply the lungs

20.7 Systemic Vessels of the Axial Region

1. For all named blood vessels in this outline, their anatomical location; the vessel from which they arise; the course they follow; and the organs, body regions, or other blood vessels they supply

2. The ascending aorta, aortic arch, and descending aorta, and the thoracic and abdominal segments of the descending aorta (section 20.7a)

3. Branches that arise from the ascending aorta and aortic arch: the coronary arteries, brachiocephalic trunk, left common carotid artery, and left subclavian artery (section 20.7a)

4. Four principal arteries of the neck: the common carotid, vertebral artery, thyrocervical trunk, and costocervical trunk (section 20.7b)

5. The external and internal carotid arteries; branches of the external carotid (superior thyroid, lingual, facial, occipital, maxillary, and superficial temporal arteries); and branches of the internal carotid (ophthalmic, anterior cerebral, and middle cerebral arteries) (section 20.7b)

6. Convergence of the vertebral arteries to form the basilar artery; the posterior cerebral arteries and arteries to the cerebellum, pons, and inner ear arising from the basilar artery (section 20.7b)

7. The location and constituents of the cerebral arterial circle (section 20.7b)

8. Dural venous sinuses; the superior sagittal, inferior sagittal, transverse, and cavernous sinuses; outflow from the sinus system into the internal jugular veins (section 20.7c)

9. The internal jugular, external jugular, and vertebral veins of the neck (section 20.7c)

10. Visceral branches (bronchial, esophageal, and mediastinal arteries) and parietal branches (posterior intercostal, subcostal, and superior phrenic arteries) of the thoracic aorta (section 20.7d)

11. Arteries of the thorax and shoulder that arise from the subclavian artery and its continuation, the axillary artery: the internal thoracic artery, thoracoacromial trunk, lateral thoracic artery, and subscapular artery (section 20.7d)

12. The subclavian vein, brachiocephalic vein, and superior vena cava; landmarks that define the transition from one to another (section 20.7e)

13. The azygos system of thoracic veins, especially the azygos, hemiazygos, and accessory hemiazygos veins; their tributaries, including the posterior intercostal, subcostal, esophageal, mediastinal, pericardial, bronchial, and ascending lumbar veins (section 20.7e)

14. Branches of the abdominal aorta: inferior phrenic arteries; celiac trunk; and superior mesenteric, middle suprarenal, renal, gonadal (ovarian or testicular), inferior mesenteric, lumbar, median sacral, and common iliac arteries (section 20.7f)

15. The general group of organs supplied by the celiac trunk; its three primary branches—the common hepatic, left gastric, and splenic arteries—and smaller branches given off by each of these (section 20.7f)

16. Branches of the superior mesenteric artery: inferior pancreaticoduodenal, jejunal, ileal, and right and middle colic arteries (section 20.7f)

17. Branches of the inferior mesenteric artery: left colic, sigmoid, and superior rectal arteries (section 20.7f)

18. Two main branches of the common iliac artery, the posterior and anterior trunks of the internal iliac artery, and the organs supplied by those trunks (section 20.7f)

19. Convergence of the internal and external iliac veins to form the common iliac vein; convergence of the right and left common

iliac veins to form the inferior vena cava (IVC) (section 20.7g)

20. Abdominal tributaries of the IVC: lumbar, gonadal (ovarian or testicular), renal, suprarenal, hepatic, and inferior phrenic veins (section 20.7g)

21. The ascending lumbar veins, their drainage in the abdomen, and their continuation into the thorax (section 20.7g)

22. The hepatic portal system and its tributaries: the splenic vein; the pancreatic, inferior mesenteric, and superior mesenteric veins draining into it; continuation of the splenic vein as the hepatic portal vein; the cystic vein and gastric veins draining into the hepatic portal vein; hepatic sinusoids in the liver; and hepatic veins (section 20.7g)

20.8 Systemic Vessels of the Appendicular Region

1. The main artery to the upper limb, which changes name along its course from subclavian to axillary to brachial artery; branches of the brachial artery in the arm (deep brachial and superior ulnar collateral arteries); and the radial collateral artery (section 20.8a)

2. Brachial artery branches that supply the forearm: radial and ulnar arteries; anterior and posterior interosseous arteries; and deep and superficial palmar arches (section 20.8a)

3. The dorsal venous network of the hand; median antebrachial vein; and median cubital vein (section 20.8b)

4. The venous palmar arches, and brachial, basilic, axillary, and subclavian veins (section 20.8b)

5. Continuation of the external iliac artery as the femoral artery; deep femoral and circumflex femoral branches of the femoral artery; popliteal artery; and anterior and posterior tibial arteries (section 20.8c)

6. The dorsal pedal and arcuate arteries that arise from the anterior tibial artery; fibular, medial plantar, and lateral plantar arteries; and deep plantar arch (section 20.8c)

7. The superficial dorsal venous arch, small and great saphenous veins, and popliteal vein (section 20.8d)

8. The deep plantar venous arch; the lateral and medial plantar veins; fibular and posterior tibial veins; and anterior tibial veins (section 20.8d)

9. The femoral, deep femoral, and external iliac veins (section 20.8d)

STUDY GUIDE

► Testing Your Recall

Answers in Appendix A

1. Blood often flows into a capillary bed from
   a. the distributing arteries.
   b. the conducting arteries.
   c. a metarteriole.
   d. an arteriovenous anastomosis
   e. the venules.

2. Plasma solutes enter the tissue fluid most easily from
   a. continuous capillaries.
   b. fenestrated capillaries.
   c. arteriovenous anastomoses.
   d. collateral vessels.
   e. venous anastomoses.

3. A blood vessel adapted to withstand a high pulse pressure would be expected to have
   a. an elastic tunica media.
   b. a thick tunica interna.
   c. one-way valves.
   d. a flexible endothelium.
   e. a rigid tunica media.

4. The substance most likely to cause a rapid drop in blood pressure is
   a. epinephrine.
   b. norepinephrine.
   c. angiotensin II.
   d. serotonin.
   e. histamine.

5. A person with a systolic blood pressure of 130 mm Hg and a diastolic pressure of 85 mm Hg would have a mean arterial pressure of about
   a. 85 mm Hg.
   b. 100 mm Hg.
   c. 108 mm Hg.
   d. 115 mm Hg.
   e. 130 mm Hg.

6. The velocity of blood flow decreases if
   a. vessel radius increases.
   b. blood pressure increases.
   c. viscosity increases.
   d. viscosity decreases.
   e. afterload increases.

7. Blood flows faster in a venule than in a capillary because venules
   a. have one-way valves.
   b. are more muscular.
   c. are closer to the heart.
   d. have higher blood pressures.
   e. have larger diameters.

8. In a case where interstitial hydrostatic pressure is negative, the only force causing capillaries to reabsorb fluid is
   a. colloid osmotic pressure of the blood.
   b. colloid osmotic pressure of the tissue fluid.
   c. capillary hydrostatic pressure.
   d. interstitial hydrostatic pressure.
   e. net filtration pressure.

9. Intestinal blood flows to the liver by way of
   a. the superior mesenteric artery.
   b. the celiac trunk.
   c. the inferior vena cava.
   d. the azygos system.
   e. the hepatic portal system.

10. The brain receives blood from all of the following vessels *except* the _____ artery or vein.
    a. basilar
    b. vertebral
    c. internal carotid
    d. internal jugular
    e. anterior communicating

11. The highest arterial blood pressure attained during ventricular contraction is called _____ pressure. The lowest attained during ventricular relaxation is called _____ pressure.

12. The capillaries of skeletal muscles are of the structural type called _____.

13. _____ shock occurs as a result of exposure to an antigen to which one is hypersensitive.

14. The role of breathing in venous return is called the _____.

15. The difference between the colloid osmotic pressure of blood and that of the tissue fluid is called _____.

16. Movement across the capillary endothelium by the uptake and release of fluid droplets is called _____.

17. All efferent fibers of the vasomotor center belong to the _____ division of the autonomic nervous system.

18. The pressure sensors in the major arteries near the head are called _____.

19. Most of the blood supply to the brain comes from a ring of arterial anastomoses called the _____.

20. The major superficial veins of the arm are the _____ on the medial side and _____ on the lateral side.

► Building Your Medical Vocabulary

Answers in Appendix A

State a meaning of each word element, and give a medical term from this chapter that uses it or a slight variation of it.

1. angio-
2. brachio-
3. celi-
4. fenestra-
5. jugulo-
6. -orum
7. sapheno-
8. sub-
9. thoraco-
10. vesico-

STUDY GUIDE

▶ What's Wrong with These Statements?

Answers in Appendix A

Briefly explain why each of the following statements is false, or reword it to make it true.

1. Blood cannot get from an artery to a vein without first passing through some capillaries.

2. Blood returns from the brain to the heart by way of the external jugular veins.

3. The body's longest blood vessel is the inferior vena cava.

4. Arteries have a series of valves that ensure a one-way flow of blood.

5. If the radius of a blood vessel doubles and all other factors remain the same, blood flow through that vessel also doubles.

6. The femoral triangle is bordered by the inguinal ligament, sartorius muscle, and rectus femoris muscle.

7. The lungs receive blood exclusively from the pulmonary circuit of the circulatory system.

8. If blood capillaries fail to reabsorb all the fluid they emit, edema will occur.

9. An aneurysm is a ruptured blood vessel.

10. In the baroreflex, a drop in arterial blood pressure triggers a corrective vasodilation of the systemic blood vessels.

▶ Testing Your Comprehension

1. It is a common lay perception that systolic blood pressure should be 100 plus a person's age. Evaluate the validity of this statement.

2. Calculate the net filtration or reabsorption pressure at a point in a hypothetical capillary assuming a hydrostatic blood pressure of 28 mm Hg, an interstitial hydrostatic pressure of −2 mm Hg, a blood COP of 25 mm Hg, and an interstitial COP of 4 mm Hg. Give the magnitude (in mm Hg) and direction (in or out) of the net pressure.

3. Aldosterone secreted by the adrenal gland must be delivered to the kidney immediately below. Trace the route that an aldosterone molecule must take from the adrenal gland to the kidney, naming all major blood vessels in the order traveled.

4. People in shock commonly exhibit paleness, cool skin, tachycardia, and a weak pulse. Explain the physiological basis for each of these signs.

5. Discuss why it is advantageous to have baroreceptors in the aortic arch and carotid sinus rather than in some other location such as the common iliac arteries.

THE LYMPHATIC AND IMMUNE SYSTEMS

Natural killer (NK) cells (orange) attacking a human cancer cell (red)
Eye of Science/Science Source

CHAPTER OUTLINE

21.1 The Lymphatic System

  21.1a Lymph and the Lymphatic Vessels
  21.1b Lymphatic Cells
  21.1c Lymphatic Tissues
  21.1d Lymphatic Organs

21.2 Innate Immunity

  21.2a External Barriers
  21.2b Leukocytes and Macrophages
  21.2c Antimicrobial Proteins
  21.2d Natural Killer Cells
  21.2e Fever
  21.2f Inflammation

21.3 Adaptive Immunity—General Aspects

  21.3a Forms of Adaptive Immunity
  21.3b Antigens
  21.3c Antibodies
  21.3d Lymphocytes
  21.3e Antigen-Presenting Cells

21.4 Cellular Immunity

  21.4a Recognition
  21.4b Attack
  21.4c Memory

21.5 Humoral Immunity

  21.5a Recognition
  21.5b Attack
  21.5c Memory

21.6 Immune System Disorders

  21.6a Hypersensitivity
  21.6b Autoimmune Diseases
  21.6c Immunodeficiency Diseases

Connective Issues
Study Guide

DEEPER INSIGHTS

21.1 Clinical Application: Lymph Nodes and Metastatic Cancer

21.2 Clinical Application: Reye Syndrome

21.3 Clinical Application: Asthma

21.4 Clinical Application: Neuroimmunology and the Mind–Body Connection

Anatomy & Physiology
Revealed® 4.0

Module 10: Lymphatic System

BRUSHING UP

- Lymphatic organs have several features in common with exocrine glands—capsule, septa, stroma, and parenchyma. Brushing up on the meaning of those terms and their structural relationships may help you better understand anatomy of the lymphatic organs (see section 5.5b).

- The mechanisms of lymph flow are similar to those for the venous return of blood (see section 20.4a).

- Leukocytes are deeply involved in immunity and defense in various ways. You can brush up on leukocyte types, appearances, and functions most easily in table 18.6.

- The actions of immune cells against disease agents involve the processes of phagocytosis, receptor-mediated endocytosis, and exocytosis described in section 3.3f.

- The eicosanoids introduced in section 17.6 play multiple important roles in the immune processes described in this chapter.

It may come as a surprise to know that the human body harbors at least as many bacterial cells as human cells. But it shouldn't be surprising. After all, human homeostasis works wonderfully not only to sustain our lives, but also to provide a predictable, warm, wet, nutritious habitat for our internal guests. It's a wonder that the body isn't overrun and consumed by microorganisms—which indeed quickly happens when one dies and homeostasis ceases.

Many of these guest microorganisms are beneficial or even necessary to human health, but some have the potential to cause disease if they get out of hand. Furthermore, we're constantly exposed to new invaders through the food and water we consume, the air we breathe, and even the surfaces we touch. We must have a means of keeping such would-be colonists in check.

One of these defenses was discovered in 1882 by a moody, intense, Russian zoologist, Elie Metchnikoff (1845–1916). When studying the tiny transparent larvae of starfish, he observed mobile cells wandering throughout their bodies. He thought at first that they must be digestive cells, but when he saw similar cells in sea anemones ingest nonnutritive dye particles that he injected, he thought they must play a defensive role. Metchnikoff knew that mobile cells also exist in human blood and pus and quickly surround a splinter introduced through the skin, so he decided to experiment to see if the starfish cells would do the same. He impaled a starfish larva on a rose thorn, and the next morning he found the thorn crawling with cells that seemed to be trying to devour it. He later saw similar cells devouring and digesting infectious yeast in tiny transparent crustaceans called water fleas. He coined the word *phagocytosis* for this reaction and named the wandering cells *phagocytes*—terms we still use today.

Metchnikoff showed that animals from simple sea anemones and starfish to humans actively defend themselves against disease agents. His observations marked the founding of cellular and comparative immunology, and won him the scientific respect he had so long coveted. Indeed, he shared the 1908 Nobel Prize for Physiology or Medicine with Paul Ehrlich (1854–1915), who had developed the theory of humoral immunity, a process also discussed in this chapter.

This chapter focuses largely on the *immune system,* which is not an organ system but rather a population of cells that inhabit all of our organs and defend the body from agents of disease. But immune cells are especially concentrated in a true organ system, the *lymphatic system.* This is a network of organs and veinlike vessels that recover tissue fluid, inspect it for disease agents, activate immune responses, and return the fluid to the bloodstream. It is with the lymphatic system that we begin this chapter's exploration.

21.1 The Lymphatic System APR

Expected Learning Outcomes

When you have completed this section, you should be able to

a. list the functions of the lymphatic system;

b. explain how lymph forms and returns to the bloodstream;

c. name the major cells of the lymphatic system and state their functions;

d. name and describe the types of lymphatic tissue; and

e. describe the structure and function of the red bone marrow, thymus, lymph nodes, tonsils, and spleen.

The **lymphatic system (fig. 21.1)** consists of a network of vessels that penetrate nearly every tissue of the body, and a collection of tissues and organs that produce immune cells. These include the lymph nodes, spleen, thymus, tonsils, and red bone marrow.

The lymphatic system has three functions:

1. **Fluid recovery.** Fluid continually filters from blood capillaries into the tissue (interstitial) spaces. The capillaries reabsorb about 85% of it on average, but the 15% they don't absorb would, over the course of a day, amount to 2 to 4 L of water and one-quarter to one-half of the protein in the blood plasma. One would die of circulatory failure within hours if this water and protein were not returned to the bloodstream. One task of the lymphatic system is to reabsorb this excess and return it to the blood. Even partial interference with lymphatic drainage can lead to severe **lymphedema** (see fig. 20.19).

2. **Immunity.** As the lymphatic system recovers tissue fluid, it also picks up foreign cells and chemicals from the tissues. On its way back to the bloodstream, the fluid passes through lymph nodes, where immune cells stand guard against foreign matter. When they detect anything potentially harmful, they activate a protective immune response.

3. **Lipid absorption.** In the small intestine, special lymphatic vessels called *lacteals* absorb dietary lipids that are not absorbed by the blood capillaries.

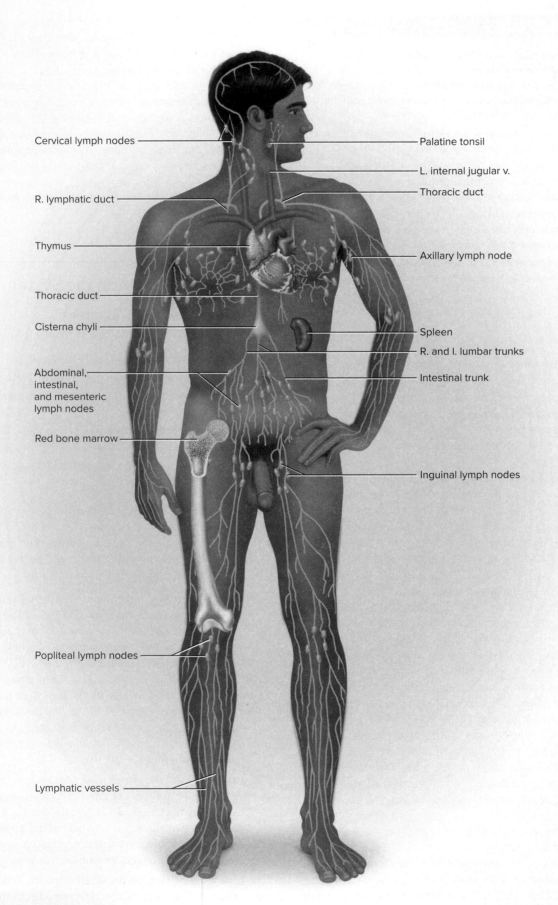

Cervical lymph nodes

Palatine tonsil

L. internal jugular v.

R. lymphatic duct

Thoracic duct

Thymus

Axillary lymph node

Thoracic duct

Cisterna chyli

Spleen

R. and l. lumbar trunks

Abdominal, intestinal, and mesenteric lymph nodes

Intestinal trunk

Red bone marrow

Inguinal lymph nodes

Popliteal lymph nodes

Lymphatic vessels

FIGURE 21.1 The Lymphatic System.

The components of the lymphatic system are (1) *lymph,* the recovered fluid; (2) *lymphatic vessels,* which transport the lymph; (3) *lymphatic tissue,* composed of aggregates of lymphocytes and macrophages that populate many organs of the body; and (4) *lymphatic organs,* in which these cells are especially concentrated and which are set off from surrounding organs by connective tissue capsules.

21.1a Lymph and the Lymphatic Vessels

Lymph is usually a clear, colorless fluid, similar to blood plasma but low in protein. It originates as tissue fluid taken up by the lymphatic vessels. Its composition varies substantially from one place and time to another. After a meal, for example, lymph draining from the small intestine has a milky appearance because of its lipid content. Lymph leaving the lymph nodes contains a large number of lymphocytes—indeed, this is the main supply of lymphocytes to the bloodstream. Lymph may also contain macrophages, hormones, bacteria, viruses, cellular debris, or even traveling cancer cells.

Lymphatic Vessels

Lymph flows through a system of **lymphatic vessels (lymphatics)** similar to blood vessels. These begin with microscopic **lymphatic capillaries (terminal lymphatics),** which penetrate nearly every tissue of the body but are absent from cartilage, bone, bone marrow, and the cornea. They are closely associated with blood capillaries, but unlike them, they are closed at one end **(fig. 21.2)** and there is no through-flow of fluid like there is for capillary blood; lymph flow begins here. A lymphatic capillary consists of a sac of thin endothelial cells that loosely overlap each other like the shingles of a roof. The cells are tethered to surrounding tissue by *anchoring filaments* that prevent the sac from collapsing.

Unlike the endothelial cells of blood capillaries, lymphatic endothelial cells aren't joined by tight junctions, nor do they have a continuous basal lamina; indeed, the gaps between them are so large that bacteria, lymphocytes, and other cells and particles can enter along with the tissue fluid. Thus, the composition of lymph arriving at a lymph node is like a report on the state of the upstream tissues.

The overlapping edges of the endothelial cells act as valves that can open and close. When tissue fluid pressure is high, it pushes the flaps inward (open) and fluid flows into the capillary. When pressure is higher in the lymphatic capillary than in the tissue fluid, the flaps are pressed outward (closed).

▶▶▶APPLY WHAT YOU KNOW

Contrast the structure of a lymphatic capillary with that of a continuous blood capillary. Explain why their structural difference is related to their functional difference.

Lymphatic vessels form in the embryo by budding from the veins, so it's not surprising that the larger ones have a similar histology. They have a *tunica interna* with an endothelium and valves **(fig. 21.3),** a *tunica media* with elastic fibers and smooth muscle, and a thin outer *tunica externa.* Their walls are thinner and their valves are closer together than those of the veins.

(a)

(b)

FIGURE 21.2 Lymphatic Capillaries. (a) Relationship of the lymphatic capillaries to a bed of blood capillaries. (b) Uptake of tissue fluid by a lymphatic capillary.

? *Why can metastasizing cancer cells get into the lymphatic system more easily than they can enter the bloodstream?*

As the lymphatic vessels converge along their path, they become larger and larger vessels with changing names. The route from the tissue fluid back to the bloodstream is: lymphatic capillaries → collecting vessels → 11 lymphatic trunks → two collecting ducts → subclavian veins. Thus, there is a continual recycling of fluid from blood to tissue fluid to lymph and back to the blood **(fig. 21.4).**

The lymphatic capillaries converge to form **collecting vessels.** These often travel alongside veins and arteries and share a common connective tissue sheath with them. At irregular intervals, they empty into lymph nodes. The lymph trickles slowly through each node, where bacteria are phagocytized and immune cells monitor the fluid for foreign antigens. It leaves the other side of

(a)

Valve

(b)

Lymph

Lymph flows forward through open valves

Closed valves prevent backflow

FIGURE 21.3 Valves in the Lymphatic Vessels. (a) Photograph of a lymphatic valve. (b) Operation of the valves to ensure a one-way flow of lymph.

a: Dennis Strete/McGraw-Hill Education

 What would be the consequence if these valves did not exist?

Lymphatic system

Cardiovascular system

Lymphatic capillaries

Lymph nodes

Lymphatic trunks

Collecting duct

Collecting vessels

Lymph flow

Lymphatic capillaries

Pulmonary circuit

Subclavian vein

Superior vena cava

Blood flow

Systemic circuit

FIGURE 21.4 Fluid Exchange Between the Circulatory and Lymphatic Systems. Blood capillaries lose fluid to the tissue spaces. The lymphatic system picks up excess tissue fluid and returns it to the bloodstream.

 Identify two benefits in having lymphatic capillaries pick up tissue fluid that is not reclaimed by the blood capillaries.

the node through another collecting vessel, traveling on and often encountering additional lymph nodes before it finally returns to the bloodstream.

Eventually, the collecting vessels converge to form larger **lymphatic trunks,** each of which drains a major portion of the body. There are 11 of these, whose names indicate their locations and parts of the body they drain: a solitary *intestinal trunk* and paired *jugular, subclavian, bronchomediastinal, intercostal,* and *lumbar trunks.* The lumbar trunks drain not only the lumbar region but also the lower limbs.

The lymphatic trunks converge to form two **collecting ducts,** the largest of the lymphatic vessels (**fig. 21.5**):

1. The **right lymphatic duct** is formed by the convergence of the right jugular, subclavian, and bronchomediastinal trunks in the right thoracic cavity. It receives lymphatic drainage from the right arm and right side of the thorax and head and empties into the right subclavian vein.

2. The **thoracic duct,** on the left, is larger and longer. It begins just below the diaphragm anterior to the vertebral column at the level of the second lumbar vertebra. Here, the two lumbar trunks and the intestinal trunk join and form a sac called the **cisterna chyli** (sis-TUR-nuh KY-lye), named for the large amount of *chyle* (fatty intestinal lymph) that it collects after a meal. The thoracic duct then passes through the diaphragm with the aorta and ascends the mediastinum adjacent to the vertebral column. As it passes through the thorax, it receives additional lymph from the left bronchomediastinal, left subclavian, and left jugular trunks, then empties into the left subclavian vein. Collectively, this duct therefore drains all of the body below the diaphragm, and the left upper limb and left side of the head, neck, and thorax.

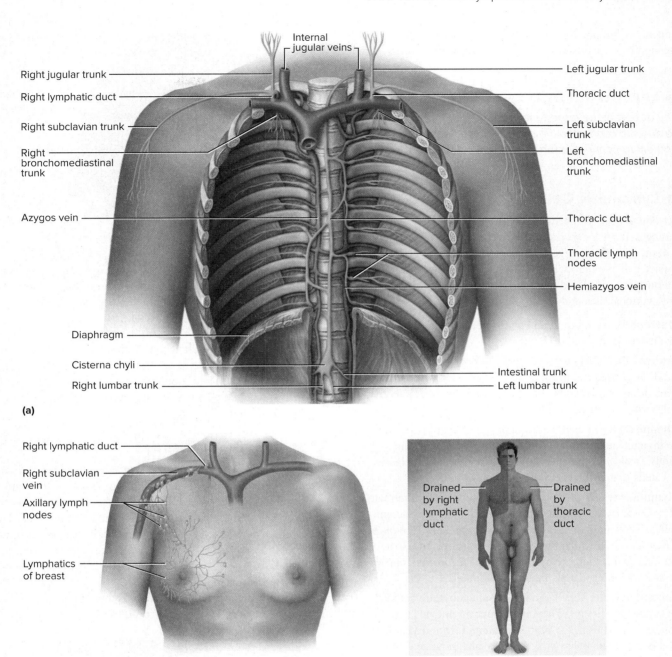

(a)

(b)

(c)

FIGURE 21.5 Lymphatics of the Thoracic Region. (a) Lymphatics of the thorax and upper abdomen and their relationship to the subclavian veins, where the lymph returns to the bloodstream. (b) Lymphatic drainage of the right mammary and axillary regions. (c) Regions of the body drained by the right lymphatic duct and thoracic duct. **APR**

❓ *Why are the axillary lymph nodes often biopsied in cases of suspected breast cancer?*

Flow of Lymph

Lymph flows under forces similar to those that govern venous return (see section 20.4a), except that the lymphatic system has no pump like the heart and lymph flows at even lower pressure and speed than venous blood. The primary mechanism of flow is rhythmic contractions of the lymphatic vessels themselves, which contract when the fluid stretches them. The valves of lymphatic vessels, like those of veins, prevent the fluid from flowing backward. Lymph flow is also produced by skeletal muscles squeezing the lymphatic vessels, like the skeletal muscle pump that moves venous blood. Since lymphatic vessels are often wrapped with an artery in a common connective tissue sheath, arterial pulsation also rhythmically squeezes the adjacent lymphatic vessels and contributes to lymph flow. A thoracic (respiratory) pump promotes the flow of lymph from the abdominal to the thoracic cavity as one inhales, just as it does in venous return. Finally, at the point where the collecting ducts empty into the subclavian veins, the rapidly flowing bloodstream draws the lymph into it. Considering these mechanisms of lymph flow, it should be apparent why physical exercise significantly increases the rate of lymphatic return.

▶▶▶**APPLY WHAT YOU KNOW**

Why does it make more functional sense for the collecting ducts to connect to the subclavian veins than it would for them to connect to the subclavian arteries?

21.1b Lymphatic Cells **A&PR**

Another component of the lymphatic system is *lymphatic tissue,* which ranges from loosely scattered cells in the mucous membranes of the respiratory, digestive, urinary, and reproductive tracts to compact cell populations encapsulated in lymphatic organs. These tissues are composed of a variety of lymphocytes and other cells with various roles in defense and immunity:

1. **Neutrophils** are aggressively antibacterial leukocytes and are described in section 18.4b.

2. **Natural killer (NK) cells** are large lymphocytes that attack and destroy bacteria, transplanted tissues, and *host cells* (cells of one's own body) that have either become infected with viruses or turned cancerous.

3. **T lymphocytes (T cells)** are lymphocytes that mature in the thymus and later depend on thymic hormones; the *T* stands for *thymus-dependent.* There are several subclasses of T cells that will be introduced later.

4. **B lymphocytes (B cells)** are lymphocytes that differentiate into *plasma cells*—connective tissue cells that secrete antibodies. They are named for an organ in chickens (the *bursa of Fabricius*[1]) in which they were first discovered. However, you may find it more helpful to think of *B* for *bone marrow,* the site where these cells mature in humans.

5. **Macrophages** are very large, avidly phagocytotic cells of the connective tissues. They arise from monocytes that have emigrated from the bloodstream and from division of preexisting tissue macrophages. They phagocytize tissue debris, dead neutrophils, bacteria, and other foreign matter **(fig. 21.6).** They also process foreign matter and display antigenic fragments of it to certain T cells, thus alerting the immune system to the presence of an enemy. Macrophages and other cells that do this are collectively called **antigen-presenting cells (APCs).**

6. **Dendritic cells** are branched, mobile APCs found in the epidermis, mucous membranes, and lymphatic organs. They play an important role in alerting the immune system to pathogens that have breached the body surfaces. They engulf foreign matter by receptor-mediated endocytosis rather than phagocytosis, but otherwise function like macrophages. After internalizing an antigen, they migrate to a nearby lymph node and activate an immune reaction to it.

[1]Hieronymus Fabricius (Girolamo Fabrizzi) (1537–1619), Italian anatomist

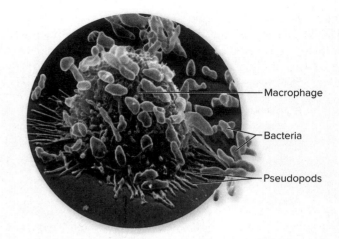

FIGURE 21.6 Macrophages Phagocytizing Bacteria. Filamentous pseudopods of the macrophages snare the rod-shaped bacteria and draw them to the cell surface, where they are phagocytized.

David M. Phillips/Science Source

7. **Reticular cells** are branched, stationary APCs that contribute to the connective tissue framework (stroma) of the lymphatic organs. (They should not be confused with reticular *fibers,* which are fine, branched collagen fibers common in lymphatic organs.)

21.1c Lymphatic Tissues

Lymphatic (lymphoid) tissues are aggregations of lymphocytes in the connective tissues of mucous membranes and various organs. The simplest form is **diffuse lymphatic tissue,** in which the lymphocytes are scattered rather than densely clustered. It is particularly prevalent in body passages that are open to the exterior—the respiratory, digestive, urinary, and reproductive tracts—where it is called **mucosa-associated lymphatic tissue (MALT).**

In some places, lymphocytes and macrophages congregate in dense masses called **lymphatic nodules (follicles) (fig. 21.7),**

FIGURE 21.7 Lymphatic Nodule in the Mucous Membrane of the Small Intestine. This nodule is part of the mucosa-associated lymphatic tissue (MALT).

Garry DeLong/Science Source

which come and go as pathogens invade the tissues and the immune system answers the challenge. Abundant lymphatic nodules are, however, a relatively constant feature of the lymph nodes (see fig. 21.10), tonsils, and appendix. In the ileum, the distal portion of the small intestine, they form clusters called **aggregated lymphoid nodules** (formerly named *Peyer patches*).

21.1d Lymphatic Organs

In contrast to the diffuse lymphatic tissue, **lymphatic (lymphoid) organs** have well-defined anatomical sites and at least a partial connective tissue capsule that separates the lymphatic tissue from neighboring tissues. These organs include the red bone marrow, thymus, lymph nodes, tonsils, and spleen. The red bone marrow and thymus are regarded as *primary lymphatic organs* because they are the sites where B and T lymphocytes, respectively, become *immunocompetent*—that is, able to recognize and respond to antigens. The lymph nodes, tonsils, and spleen are called *secondary lymphatic organs* because immunocompetent lymphocytes migrate to these organs only after they mature in the primary lymphatic organs.

Red Bone Marrow

Red bone marrow may not seem to be an organ; when aspirated from the bones for the purpose of biopsy or transfusion, it simply looks like extra-thick blood. Yet a careful microscopic examination of less disturbed marrow shows that it has a surprising degree of structure and is composed of multiple tissues, meeting the criteria of an organ, even if a very soft one.

There are two kinds of bone marrow: yellow and red. Yellow bone marrow is mainly adipose tissue and can be disregarded for present purposes, but red bone marrow is involved in hematopoiesis (blood formation) and immunity. In children, it occupies the medullary spaces of nearly all bones. In adults, it is limited to the axial skeleton and the proximal heads of the humerus and femur. Red bone marrow is an important supplier of lymphocytes to the immune system.

Red bone marrow is soft, loosely organized, highly vascular material, separated from osseous tissue by the endosteum of the bone. It produces all classes of formed elements of the blood; its red color comes from the abundance of erythrocytes. Numerous small arteries enter *nutrient foramina* on the bone surface, penetrate the bone, and empty into large *sinusoids* (45 to 80 μm wide) in the marrow (**fig. 21.8**). The sinusoids drain into a *central longitudinal vein* that exits the bone via the same route that the arteries entered. The sinusoids are lined by endothelial cells, like other blood vessels, and are surrounded by reticular cells and reticular fibers. The blood-forming cells are attached in various ways to the reticular cells and other elements of the framework (stroma) of the marrow. The reticular cells secrete colony-stimulating factors and other signals that induce stem cells to develop into leukocytes, erythrocytes, and platelet-forming megakaryocytes. In the long bones of the limbs, aging reticular cells accumulate fat and transform into adipose cells, eventually replacing red bone marrow with yellow bone marrow.

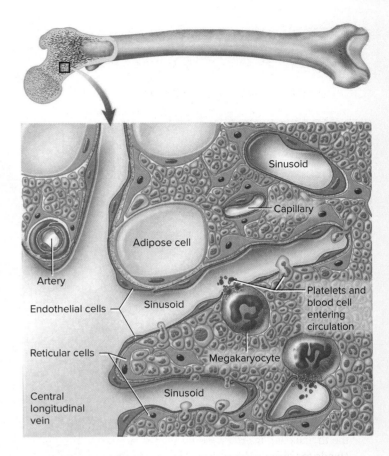

FIGURE 21.8 Histology of the Red Bone Marrow. The formed elements of blood squeeze through the endothelial cells into the sinuses, which converge on a central longitudinal vein at the lower left.

The spaces between the sinusoids are occupied by *islands (cords)* of hematopoietic tissue, composed of macrophages and blood cells in all stages of development. The macrophages destroy malformed blood cells and the nuclei discarded by developing erythrocytes. As blood cells mature, they push their way through the reticular and endothelial cells to enter the sinusoid and flow away in the bloodstream.

Thymus

The **thymus** is a member of the endocrine, lymphatic, and immune systems. It houses developing lymphocytes and secretes hormones that regulate their later activity. It is a bilobed organ located between the sternum and aortic arch in the superior mediastinum. The thymus shows a remarkable degree of degeneration (involution) with age, as described and illustrated earlier (see fig. 17.8).

The fibrous capsule of the thymus gives off trabeculae (septa) that divide the gland into several angular lobules. Each lobule has a light central *medulla* populated by T lymphocytes, surrounded by a dense, darker *cortex* (**fig. 21.9**). Both cortex and medulla have a branching network of interconnected **epithelial cells** that play important roles in lymphocyte development discussed later; their roles differ in the two regions, so they are separately described as *cortical epithelial cells* and *medullary epithelial cells*.

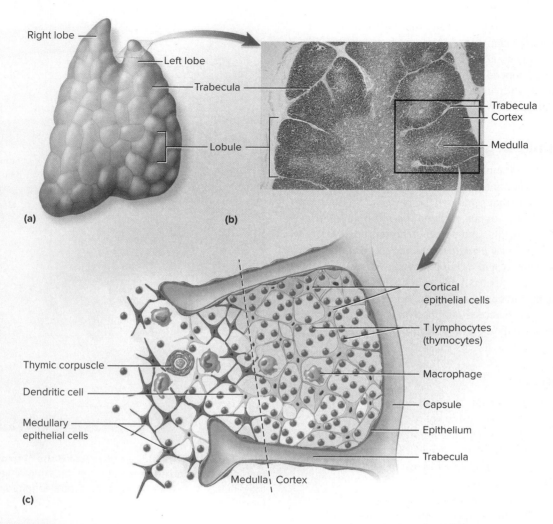

Right lobe

Left lobe

Trabecula

Lobule

(a)

Trabecula
Cortex

Medulla

(b)

Cortical
epithelial cells

T lymphocytes
(thymocytes)

Thymic corpuscle

Dendritic cell

Macrophage

Medullary
epithelial cells

Capsule

Epithelium

Trabecula

Medulla Cortex

(c)

FIGURE 21.9 The Thymus. (a) Gross anatomy. (b) Histology. (c) Cellular architecture of a lobule. **APR**

b: Dennis Strete/McGraw-Hill Education

In the cortex, cortical epithelial cells and capillary *pericytes* (see fig. 20.5a) surround the blood capillaries and form a *blood–thymus barrier,* which isolates developing lymphocytes from premature exposure to blood-borne antigens. After developing in the cortex, T cells migrate to the medulla, where they spend another 3 weeks. There is no blood–thymus barrier in the medulla; mature T cells enter blood or lymphatic vessels here and leave the thymus. Some of the medullary epithelial cells are arranged in keratinized whorls called *thymic corpuscles,* which may play a role in the immune system's *self-tolerance* (restraint from attacking the body's own tissues) and are useful for identifying the thymus histologically.

Epithelial cells of the thymus secrete several signaling molecules that promote the development and action of T cells both locally (as paracrines) and systemically (as hormones); these include *thymosin, thymopoietin, thymulin, interleukins,* and *interferon.* If the thymus is removed from newborn mammals, they waste away and never develop immunity. Other lymphatic organs also seem to depend on thymosins or T cells and develop poorly in thymectomized animals. The role of the thymus in T cell development is discussed later.

Lymph Nodes

Lymph nodes (fig. 21.10) are the most numerous lymphatic organs, numbering about 450 in a typical adult. They serve two functions: to cleanse the lymph and to act as a site of T and B cell activation. A lymph node is an elongated or bean-shaped structure, usually less than 3 cm long, often with an indentation called the *hilum* on one side. It is enclosed in a fibrous capsule with trabeculae that partially divide the interior of the node into compartments. Between the capsule and parenchyma is a narrow, relatively clear space called the *subcapsular sinus,* which contains reticular fibers, macrophages, and dendritic cells. Deep to this, the gland consists mainly of a stroma of reticular connective tissue and a parenchyma of lymphocytes and antigen-presenting cells.

The parenchyma is divided into an outer C-shaped **cortex** that encircles about four-fifths of the organ, and an inner **medulla** that extends to the surface at the hilum. The cortex consists mainly of ovoid to conical lymphatic nodules. When the lymph node is fighting a pathogen, these nodules acquire light-staining **germinal centers** where B cells multiply and differentiate into plasma cells. The medulla consists largely of a branching network of *medullary*

FIGURE 21.10 Anatomy of a Lymph Node. (a) Partially bisected lymph node showing pathway of lymph flow. (b) Detail of the boxed region in part (a). (c) Reticular fiber stroma and immune cells in a medullary sinus (SEM). **A&PR**

c: Francis Leroy, Biocosmos/Science Source

cords composed of lymphocytes, plasma cells, macrophages, reticular cells, and reticular fibers. The cortex and medulla also contain lymph-filled sinuses continuous with the subcapsular sinus.

Several **afferent lymphatic vessels** lead into the node along its convex surface. Lymph flows from these vessels into the subcapsular sinus, percolates slowly through the sinuses of the cortex and medulla, and leaves the node through one to three **efferent lymphatic vessels** that emerge from the hilum. No other lymphatic organs have afferent lymphatic vessels; lymph nodes are the only organs that filter lymph as it flows along its course. The lymph node is a bottleneck that slows down lymph flow and allows time for cleansing it of foreign matter. Macrophages and reticular cells of the sinuses remove about 99% of the impurities before the lymph leaves the node. On its way to the bloodstream, lymph flows through one lymph node after another and thus becomes quite thoroughly cleansed of impurities.

Blood vessels also penetrate the hilum. Arteries follow the medullary cords and give rise to capillary beds in the medulla

and cortex. In the *deep cortex* near the junction with the medulla, lymphocytes emigrate from the bloodstream into the parenchyma of the node. Most lymphocytes in the deep cortex are T cells.

Lymph nodes are widespread but especially concentrated in the following locations (see fig. 21.1 for an overview):

- *Cervical lymph nodes* occur in deep and superficial groups in the neck, and monitor lymph coming from the head and neck.

- *Axillary lymph nodes* are concentrated in the armpit (axilla) and receive lymph from the upper limb and breast (see fig. 21.5b).

- *Thoracic lymph nodes* occur in the thoracic cavity, especially in the mediastinum, and receive lymph from the mediastinum, lungs, and airway.

- *Abdominal lymph nodes* occur in the posterior abdominopelvic wall and receive lymph from the urinary and reproductive systems.

FIGURE 21.11 Some Areas of Lymph Node Concentration. (a) Mesenteric lymph nodes associated with the large intestine.
(b) Inguinal lymph nodes in the groin of a female cadaver. **APR**

b: Rebecca Gray/Don Kincaid/McGraw-Hill Education

- *Intestinal* and *mesenteric lymph nodes* are found in the mesenteries **(fig. 21.11a)** and adjacent to the appendix and intestines; they receive lymph from the digestive tract.

- *Inguinal lymph nodes* occur in the groin **(fig. 21.11b)** and receive lymph from the entire lower limb.

- *Popliteal lymph nodes* occur at the back of the knee and receive lymph from the foot and leg proper.

When a lymph node is under challenge by an antigen, it may become swollen and painful to the touch—a condition called **lymphadenitis**[2] (lim-FAD-en-EYE-tis). Physicians routinely palpate the accessible lymph nodes of the cervical, axillary, and inguinal regions for swelling. The collective term for all lymph node diseases is **lymphadenopathy**[3] (lim-FAD-eh-NOP-a-thee). Lymph nodes are common sites of metastatic cancer (see Deeper Insight 21.1).

DEEPER INSIGHT 21.1

CLINICAL APPLICATION

Lymph Nodes and Metastatic Cancer

Metastasis is a phenomenon in which cancerous cells break free of the original *primary tumor,* travel to other sites in the body, and establish new tumors (see fig. 4.22). Because of the high permeability of lymphatic capillaries, metastasizing cancer cells easily enter them and travel in the lymph. They tend to lodge in the first lymph node they encounter and multiply there, eventually destroying the node. Cancerous lymph nodes are swollen but relatively firm and usually painless. Cancer of a lymph node is called *lymphoma.*

Once a tumor is well established in one node, cells may emigrate from there and travel to the next. However, if the metastasis is detected early enough, cancer can sometimes be eradicated by removing not only the primary tumor, but also the nearest lymph nodes downstream from that point. For example, breast cancer is often treated with a combination of lumpectomy or mastectomy along with removal of the nearby axillary lymph nodes.

---

[2]*lymph* = water; *adeno* = gland; *itis* = inflammation
[3]*lymph* = water; *adeno* = gland; *pathy* = disease

Tonsils

The **tonsils** are patches of lymphatic tissue located at the entrance to the pharynx, where they guard against ingested and inhaled pathogens. Each is covered by an epithelium and has deep pits called **tonsillar crypts** lined by lymphatic nodules **(fig. 21.12).** The crypts often contain food debris, dead leukocytes, bacteria, and antigenic chemicals. Below the crypts, the tonsils are partially separated from underlying connective tissue by an incomplete fibrous capsule.

There are three main sets of tonsils: (1) a single median **pharyngeal tonsil (adenoids)** on the wall of the pharynx just behind the nasal cavity; (2) a pair of **palatine tonsils** at the posterior margin of the oral cavity; and (3) numerous **lingual tonsils,** each with a single crypt, concentrated in patches embedded in each side of the root of the tongue (see fig. 25.5a).

The palatine tonsils are the largest and most often infected. *Tonsillitis* is an acute inflammation of the palatine tonsils, usually caused by a viral or bacterial infection. Their surgical removal, called *tonsillectomy,* used to be one of the most common surgical procedures performed on children, but is less common today. Tonsillitis is now usually treated with antibiotics.

Spleen

The **spleen,** the body's largest lymphatic organ, measures up to 12 cm long and usually weighs about 150 g. It is located in the left hypochondriac region, just inferior to the diaphragm and posterolateral to the stomach **fig. 21.13;** see also atlas B, fig. B.6). It is

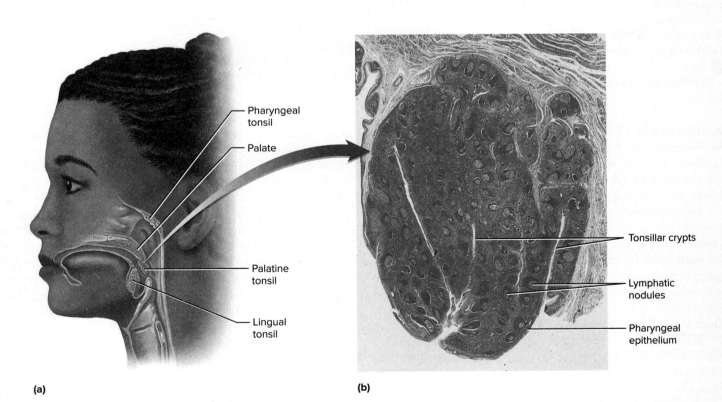

(a)

(b)

FIGURE 21.12 The Tonsils. (a) Locations of the tonsils. (b) Histology of the palatine tonsil. **APR**

b: Biophoto Associates/Science Source

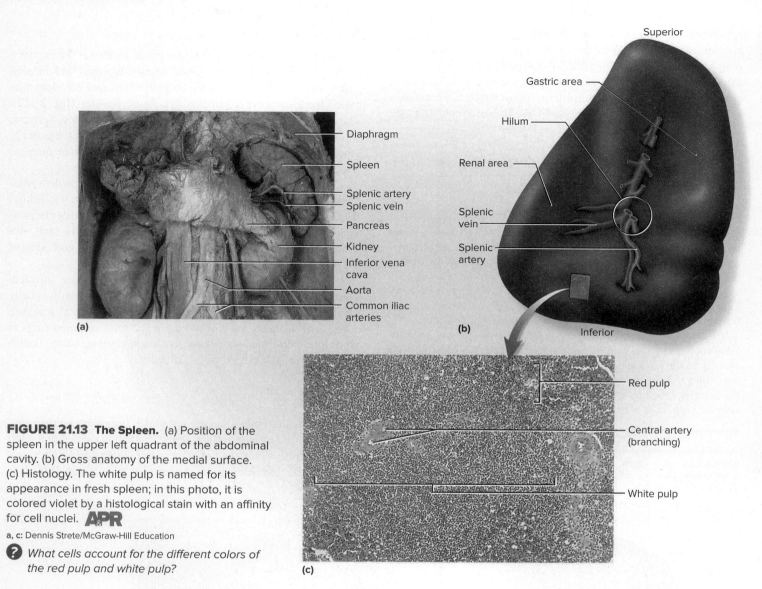

FIGURE 21.13 The Spleen. (a) Position of the spleen in the upper left quadrant of the abdominal cavity. (b) Gross anatomy of the medial surface. (c) Histology. The white pulp is named for its appearance in fresh spleen; in this photo, it is colored violet by a histological stain with an affinity for cell nuclei. **A&PR**

❓ *What cells account for the different colors of the red pulp and white pulp?*

---

protected by ribs 10 through 12. The spleen fits snugly between the diaphragm, stomach, and kidney and has indentations called the *gastric area* and *renal area* where it presses against these adjacent viscera. It has a medial hilum penetrated by the splenic artery, splenic vein, and lymphatic vessels.

The parenchyma exhibits two types of tissue named for their appearance in fresh specimens (not in stained sections): **red pulp,** which consists of sinuses gorged with concentrated erythrocytes; and **white pulp,** which consists of lymphocytes and macrophages aggregated like sleeves along small branches of the splenic artery. In tissue sections, white pulp appears as an ovoid mass of lymphocytes with an arteriole passing through it. However, the three-dimensional shape is not egglike but cylindrical.

These two tissue types reflect the multiple functions of the spleen. Its blood capillaries are very permeable; they allow red blood cells (RBCs) to leave the bloodstream, accumulate in the sinuses of the red pulp, and reenter the bloodstream later. The spleen is an "erythrocyte graveyard"—old, fragile RBCs rupture as they squeeze through the capillary walls into the sinuses. Macrophages phagocytize their remains, just as they dispose of

blood-borne bacteria and other cellular debris. The spleen produces blood cells in the fetus and can resume this role in adults in the event of extreme anemia. Lymphocytes and macrophages of the white pulp monitor the blood for foreign antigens, much like the lymph nodes do the lymph. The spleen is a reservoir for a large "standing army" of monocytes, waiting in a state of emergency preparedness. In such events as microbial infection, myocardial infarction, or gaping wounds, angiotensin II stimulates the spleen to release great numbers of monocytes into the bloodstream. The monocytes help to combat pathogens and repair damaged tissues. The spleen also helps to stabilize blood volume by transferring excess plasma from the bloodstream into the lymphatic system.

The spleen is highly vascular and vulnerable to trauma and infection. A ruptured spleen can hemorrhage fatally, but is difficult to repair surgically. Therefore, a once-common procedure in such cases was its removal, *splenectomy*. A person can live without a spleen, but people with splenectomies are more susceptible to septicemia, other infections, and premature death. Like tonsillectomy, splenectomy is now performed less commonly than it used to be.

BEFORE YOU GO ON

Answer the following questions to test your understanding of the preceding section:

1. List the primary functions of the lymphatic system. What do you think would be the most noticeable effect of clamping the right lymphatic duct closed?

2. How does fluid get into the lymphatic system? What prevents it from draining back out?

3. What do NK, T, and B cells have in common? How do their functions differ?

4. List five major cell types of lymphatic tissues and state the function of each.

5. Predict the relative seriousness of removing the following organs from a 2-year-old child: (a) a lymph node, (b) the spleen, (c) the thymus, (d) the palatine tonsils.

21.2 Innate Immunity

Expected Learning Outcomes

When you have completed this section, you should be able to

a. define *pathogen* and *immune system;*

b. identify the body's three lines of defense against pathogens;

c. contrast innate and adaptive immunity;

d. describe the defensive functions of various types of leukocytes and macrophages;

e. describe the immune roles of interferons and the complement system;

f. explain how fever serves a protective role against pathogens; and

g. describe the process of inflammation and explain what accounts for its cardinal signs.

For all living organisms, one of the greatest survival challenges is coping with **pathogens**[4]—viruses, bacteria, fungi, and other microorganisms that cause disease. However, the body must also defend itself from nonliving disease agents such as poison ivy toxin and allergy-producing chemicals in food, cosmetics, and latex gloves.

We can think of the body as having three lines of defense against these environmental threats:

1. The **first line of defense** consists of epithelial barriers, notably the skin and mucous membranes, which are impenetrable to most of the pathogens that daily assault us.

2. The **second line of defense** consists of protections against pathogens that break through those external barriers. These defenses include leukocytes and macrophages, antimicrobial proteins, natural killer cells, fever, and inflammation.

3. The **third line of defense** is adaptive immunity, a group of mechanisms that not only defeat a pathogen but leave the body with a memory of it, enabling one to defeat it so quickly in future encounters that the pathogen causes no illness.

Such defenses collectively compose the **immune**[5] **system,** which consists of a widely distributed population of cells that inhabit nearly every organ of the body; diverse chemicals they produce to neutralize and destroy pathogens; physical barriers to invasion such as the skin and mucous membranes; and physiological processes such as fever and inflammation. Note that unlike other body systems, this is not an organ system but a collective term for numerous defensive cells and mechanisms.

Our mechanisms of defense are divided into two broad forms called innate and adaptive immunity, although there is such extensive interaction between the two that they're not entirely separable or independent. **Innate immunity** consists of defenses we're born with (hence *innate*) and that protect us from a broad spectrum of disease agents; it encompasses the first and second lines of defense just described. **Adaptive immunity,** by contrast, constitutes the third line of defense and is distinguished by its capacity for immune memory. It *adapts* the body to the presence of a pathogen so we become less vulnerable to the illness it could otherwise cause.

Innate immunity has three characteristics that differentiate it from adaptive immunity:

1. It is a **local** effect, in most cases, warding off a pathogen at the point of invasion (such as the itchy inflammation of a rash or mosquito bite) with little effect anywhere else. Fever is an exception, having a *systemic* (body-wide) effect.

2. It is **nonspecific.** Each mechanism of innate immunity, such as the physical barrier of the skin and the antiviral effect of fever, acts against a broad spectrum of disease agents, not against one particular pathogen. Innate immunity used to be called *nonspecific defense* for this reason.

3. It **lacks memory** of any prior exposure to a pathogen, so it is no easier to defeat that pathogen on later exposures than it was the first time.

Much of our innate immunity employs three basic kinds of defense: (1) *protective proteins* such as keratin, interferons, and complement; (2) *protective cells* such as neutrophils and macrophages; and (3) *protective processes* such as fever and inflammation. We will examine these in the coming sections.

21.2a External Barriers

Our first line of defense is the skin and mucous membranes—the physical barriers to microbial invasion. When the skin is broken by a scrape or animal bite or destroyed by a burn, one of the most urgent treatment concerns is the prevention of infection. This attests to the importance of intact skin as a barrier. Its surface is composed mainly of keratin, a tough protein that few pathogens can penetrate. Furthermore, with exceptions such as the axillary and pubic areas, it is too dry and poor in nutrients to support much microbial growth. Even

[4]*patho* = disease, suffering; *gen* = producing

[5]*immuno* = free

those microorganisms that do adhere to the epidermis are continually cast off as we shed dead surface keratinocytes (see fig. 5.12).

The skin also is coated with diverse antimicrobial chemicals. Sweat and sebum coat it with a protective **acid mantle**—a thin film of lactic and fatty acids that inhibit bacterial growth. Sweat also contains an antibacterial peptide called **dermicidin.** Keratinocytes, neutrophils, macrophages, and other cells also produce peptides called **defensins** and **cathelicidins**[6] (ca-THEL-ih-SY-dins) that destroy bacteria, viruses, and fungi. The effects of these defenses are enhanced by vitamin D (calcitriol), pointing to the benefit of a moderate amount of sunlight exposure for one's resistance to infection. (See section 7.4b for a reminder of the role of sunlight in vitamin D synthesis.)

The digestive, respiratory, urinary, and reproductive tracts are open to the exterior, making them vulnerable to invasion, but they're protected by mucous membranes. Sticky mucus physically ensnares microorganisms. Those trapped in the respiratory mucus are moved by cilia to the pharynx, swallowed, and destroyed by stomach acid. Microorganisms are flushed from the upper digestive tract by saliva and from the lower urinary tract by urine. Mucus, tears, and saliva also contain **lysozyme,** an enzyme that destroys bacteria by dissolving their cell walls.

Beneath the epithelia of the skin and mucous membranes is a layer of areolar tissue. Its ground substance contains a giant glycosaminoglycan called **hyaluronic acid** (HI-ul-yur-ON-ic), which gives it a viscous consistency. It is normally difficult for microorganisms to migrate through this sticky tissue gel. Some pathogens overcome this obstacle, however, by producing an enzyme called *hyaluronidase,* which breaks it down to a thinner consistency that is more easily penetrated. Hyaluronidase occurs in some snake venoms and bacterial toxins and is produced by some parasitic protozoans to facilitate their invasion of the connective tissues.

21.2b Leukocytes and Macrophages

The skin and mucous membranes obviously aren't invincible barriers. They suffer frequent cuts, nicks, and scrapes and are daily breached even by such mundane activities as shaving and brushing our teeth. But when microorganisms get past these epithelial barriers, they're attacked by **phagocytes** (phagocytic cells) that have a voracious appetite for foreign matter. Leukocytes and macrophages play especially important roles in both innate and adaptive immunity and, therefore, in both the second and third lines of defense.

Leukocytes

The five types of leukocytes are illustrated and described in table 18.6. We will now examine their contributions to resistance and immunity in more detail.

1. **Neutrophils** spend most of their lives wandering in the connective tissues killing bacteria. One of their methods is simple phagocytosis and digestion—engulfing microorganisms with their pseudopods and destroying them with lysosomal enzymes (see figs. 3.12, 3.20). The other is a more complex process that produces a cloud of bactericidal chemicals.

When a neutrophil detects bacteria in the immediate area, its lysosomes migrate to the cell surface and *degranulate,* or discharge their enzymes into the tissue fluid. Here, the enzymes catalyze a reaction called the **respiratory burst:** The neutrophil rapidly absorbs oxygen and reduces it to *superoxide anions* ($O_2 \cdot^-$), which react with H^+ to form hydrogen peroxide (H_2O_2). Another lysosomal enzyme produces hypochlorite (HClO), the active ingredient in chlorine bleach, using chloride ions in the tissue fluid. Superoxide, hydrogen peroxide, and hypochlorite are highly toxic; they form a chemical **killing zone** around the neutrophil that destroys far more bacteria than the neutrophil can destroy by phagocytosis alone. Unfortunately for the neutrophil, it too is killed by these chemicals. These potent oxidizing agents can also damage connective tissues and sometimes contribute to rheumatoid arthritis.

2. **Eosinophils** are found especially in the mucous membranes, standing guard against parasites, allergens (allergy-causing antigens), and other foes. They congregate especially at sites of allergy, inflammation, or parasitic infection. They help to kill parasites such as tapeworms and roundworms, which are too big to phagocytize, by producing superoxide, hydrogen peroxide, and various toxic proteins including a neurotoxin. These agents promote the action of basophils and mast cells (see next paragraph). Eosinophils phagocytize and degrade antigen–antibody complexes. Finally, they secrete enzymes that degrade and limit the action of histamine and other inflammatory chemicals that, unchecked, could cause tissue damage.

3. **Basophils** secrete chemicals that aid the mobility and action of other leukocytes: *leukotrienes* that activate and attract neutrophils and eosinophils; the vasodilator *histamine,* which increases blood flow and speeds the delivery of leukocytes to the area; and the anticoagulant *heparin,* which inhibits the formation of blood clots that would impede leukocyte mobility. These substances are also produced by **mast cells,** a type of connective tissue cell similar to basophils. Eosinophils promote basophil and mast cell action by stimulating them to release these secretions.

4. **Lymphocytes** all look more or less alike in blood films, but there are several functional types. Three basic categories have already been mentioned: natural killer (NK) cells, T cells, and B cells. In the circulating blood, about 80% of the lymphocytes are T cells, 15% B cells, and 5% NK and stem cells. The roles of these lymphocyte types are too diverse for easy generalizations here, but are described in later sections on NK cells and adaptive immunity. NK cells are part of our innate immunity and the others function mainly in adaptive immunity. Certain lymphocytes called *helper T cells* function in both innate and adaptive immunity.

5. **Monocytes** are leukocytes that emigrate from the blood into the connective tissues and transform into macrophages. All of the body's avidly phagocytic cells except leukocytes are called the **macrophage system.** Dendritic cells are included even though they come from different stem cells

---

[6]*catheli* = universal; *cid* = kill; *in* = substance, protein

than macrophages and employ receptor-mediated endocytosis instead of phagocytosis to internalize foreign matter. Some phagocytes are wandering cells that actively seek pathogens, whereas reticular cells and others are fixed in place and phagocytize only those pathogens that come to them—although they are strategically positioned for this to occur. Macrophages are widely distributed in the loose connective tissues, but there are also specialized forms with more specific localities: *microglia* in the central nervous system, *alveolar macrophages* in the lungs, and *stellate macrophages* in the liver, for example.

21.2c Antimicrobial Proteins

Multiple types of proteins inhibit microbial reproduction and provide short-term, nonspecific resistance to pathogenic bacteria and viruses. We have already considered several at the skin surface, and now turn our attention to two families of blood-borne antimicrobial proteins.

Interferons

When certain cells (especially leukocytes) are infected with viruses, they secrete proteins called **interferons.** These are of little benefit to the cell that secretes them, but are like its "dying words" that alert neighboring cells and protect them from becoming infected. They bind to surface receptors on those cells and activate second-messenger systems within. This induces the synthesis of dozens of antiviral proteins that defend a cell by such means as breaking down viral genes and preventing viral replication. Interferons also activate NK cells and macrophages, which destroy infected cells before they can liberate a swarm of newly replicated viruses. Interferons also confer resistance to cancer, since the activated NK cells destroy malignant cells.

Complement System

The **complement system** is a group of 30 or more globulins that make powerful contributions to both innate and adaptive immunity. Immunology pioneer Paul Ehrlich (1854–1915) named it *complement* because it "completes the action of antibody," and indeed this is the principal means of pathogen destruction in antibody-mediated immunity. Since Ehrlich's time, however, the complement system has also been found important in innate immunity.

Complement proteins are synthesized mainly by the liver. They circulate in the blood in inactive form and are activated in the presence of pathogens. The inactive proteins are named with the letter *C* and a number, such as C3. Activation splits them into fragments, which are further identified by lowercase letters (C3a and C3b, for example).

Activated complement contributes to pathogen destruction by four methods: inflammation, immune clearance, phagocytosis, and cytolysis. We will examine the pathways of activation with a view to understanding how each of these goals is achieved. There are three such routes **(fig. 21.14):** the classical, alternative, and lectin pathways.

The **classical pathway** requires an antibody to get it started; thus it is part of adaptive immunity. The antibody binds to an antigen on the surface of a microorganism and changes shape, exposing a pair of *complement-binding sites* (see fig. 21.19). Complement C1 binds to these sites and sets off a reaction cascade. Like the cascade of blood-clotting reactions, each step generates an enzyme that catalyzes the production of many more molecules at the next step; each step is an amplifying process, so many molecules of product result from a small beginning. In the classical pathway, the cascade is called **complement fixation,** since it results in the attachment of a chain of complement proteins to the antibody.

The alternative and lectin pathways require no antibodies and thus belong to our innate immunity. Complement C3 slowly and spontaneously breaks down in the blood into C3a and C3b. In the **alternative pathway,** C3b binds directly to targets such as human tumor cells, viruses, bacteria, and yeasts. This, too, triggers a reaction cascade—this time with an *autocatalytic effect* in which C3b leads to the accelerated splitting of more C3 and production of even more C3b.

Lectins are plasma proteins that bind to carbohydrates. In the **lectin pathway,** a lectin binds to certain sugars of a microbial cell surface and sets off yet another reaction cascade leading to C3b production.

As we can see, the splitting of C3 into C3a and C3b is an intersection where all three pathways converge. These two C3 fragments then produce, directly or indirectly, the end results of the complement system:

1. **Inflammation.** C3a stimulates mast cells and basophils to secrete histamine and other inflammatory chemicals. It also activates and attracts neutrophils and macrophages, the two key cellular agents of pathogen destruction in inflammation. The exact roles of these chemicals and cells are explained in the section on inflammation to follow.

2. **Immune clearance.** C3b binds antigen–antibody (Ag–Ab) complexes to red blood cells. As these RBCs circulate through the liver and spleen, the macrophages of those organs strip off and destroy the Ag–Ab complexes, leaving the RBCs unharmed. This is the principal means of clearing foreign antigens from the bloodstream.

3. **Phagocytosis.** Bacteria, viruses, and other pathogens are phagocytized and digested by neutrophils and macrophages. However, those phagocytes can't easily internalize "naked" microorganisms. C3b assists them by means of **opsonization**[7]—it coats microbial cells and serves as binding sites for phagocyte attachment. The way Elie Metchnikoff described this, opsonization "butters up" the foreign cells to make them more appetizing to phagocytes.

4. **Cytolysis.**[8] C3b splits another complement protein, C5, into C5a and C5b. C5a joins C3a in its proinflammatory actions, but C5b plays a more important role in pathogen destruction. It binds to the enemy cell and then attracts complements C6, C7, and C8. This conglomeration of proteins (now called C5b678) goes on to bind up to 17 molecules of complement C9, which form a ring called the *membrane*

---

[7]*opson* = to prepare food
[8]*cyto* = cell; *lysis* = split apart, break down

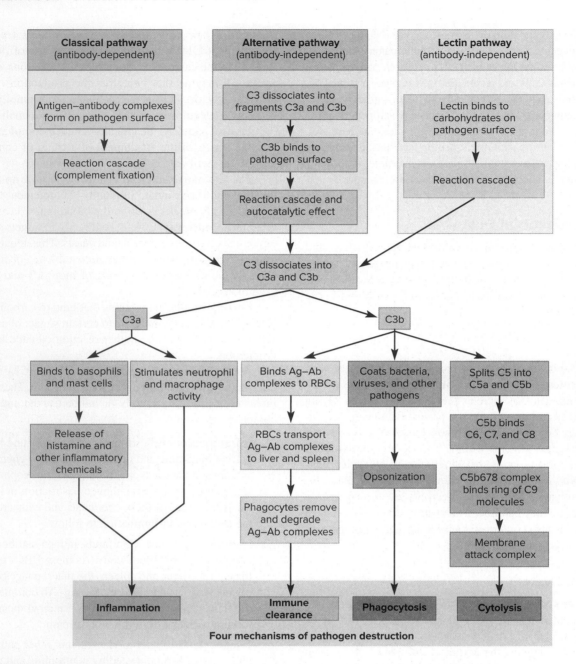

FIGURE 21.14 Complement Activation. The classical, alternative, and lectin pathways all lead to the cleavage of complement C3 into C3a and C3b. Those two fragments activate processes that lead to enhanced inflammation, immune clearance, phagocytosis, and cytolysis.

attack complex (**fig. 21.15**). The complex forms a hole in the target cell up to 10 nm wide, about the diameter of a single protein molecule. The cell can no longer maintain homeostasis; electrolytes leak out, water flows rapidly in, and the cell ruptures.

21.2d Natural Killer Cells

Natural killer (NK) cells continually patrol the body "on the lookout" for pathogens or diseased host cells. They attack and destroy bacteria, cells of transplanted organs and tissues, cells infected with viruses, and cancer cells. Upon recognition of an enemy cell, the

NK cell binds to it and releases proteins called **perforins,** which polymerize in a ring and create a hole in its plasma membrane (**fig. 21.16**). This is a kiss of death, for the hole allows a rapid inflow of water and salts. That alone may kill the target cell, but the NK cell also secretes a group of protein-degrading enzymes called **granzymes.** These enter the pore made by the perforins, destroy the target cell's enzymes, and induce apoptosis (programmed cell death).

21.2e Fever

Fever is an abnormal elevation of body temperature. It is also known as **pyrexia,** and the term *febrile* means pertaining to fever

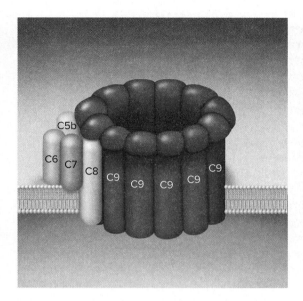

FIGURE 21.15 The Membrane Attack Complex. Complement proteins C5b, C6, C7, and C8 form an organizing center around which many C9 molecules create a ring that opens a lethal hole in the enemy cell membrane.

❓ *In what way does the action of the membrane attack complex resemble the action of perforin?*

(as in "febrile attack"). Fever results from trauma, infections, drug reactions, brain tumors, and several other causes. Because of variations in human body temperature, there is no exact criterion for what constitutes a fever—a temperature that is febrile for one person may be normal for another.

Fever is commonly regarded as an undesirable side effect of illness, and we make efforts to reduce it for the sake of comfort. However, it is usually a beneficial defense mechanism that does more good than harm. People with colds, for example, recover more

quickly and are less infective to others when they allow a fever to run its course rather than using **antipyretic** (fever-reducing) medications such as aspirin and ibuprofen. Fever is beneficial in that it (1) promotes interferon activity, (2) inhibits reproduction of bacteria and viruses, and (3) elevates metabolic rate and accelerates tissue repair.

Fever is typically initiated by **exogenous pyrogens**[9]—fever-producing agents originating outside the body, such as the surface glycolipids of bacteria and viruses. As neutrophils and macrophages attack such pathogens, they secrete a variety of polypeptides that act as **endogenous pyrogens.**[10] These in turn stimulate neurons of the anterior hypothalamus to raise the set point for body temperature—say, to 39°C (102°F) instead of the usual 37°C **(fig. 21.17).** Prostaglandin E_2 (PGE_2), secreted in the hypothalamus, enhances this effect. Aspirin and ibuprofen reduce fever by inhibiting prostaglandin synthesis, but in some circumstances, using aspirin to control fever can have deadly consequences (see Deeper Insight 21.2).

When the set point rises, a person shivers to generate heat and the cutaneous arteries constrict to reduce heat loss. In the stage of fever called *onset,* one has a rising temperature, yet experiences chills and feels cold and clammy to another's touch. In the next stage, *stadium,* the temperature oscillates around the higher set point for as long as the pathogen is present. The elevated temperature enhances the action of interferons and other antimicrobial proteins, and it inhibits bacterial reproduction.

When the infection is defeated, pyrogen secretion ceases and the hypothalamic thermostat is set back to normal. This activates heat-losing mechanisms, especially cutaneous vasodilation and

---

[9]*exo* = from outside; *genous* = arising; *pyro* = fire, heat; *gen* = producing
[10]*endo* = from within; *genous* = arising; *pyro* = fire, heat; *gen* = producing

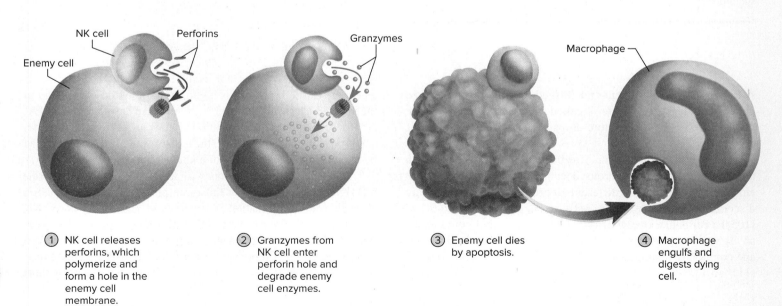

① NK cell releases perforins, which polymerize and form a hole in the enemy cell membrane.

② Granzymes from NK cell enter perforin hole and degrade enemy cell enzymes.

③ Enemy cell dies by apoptosis.

④ Macrophage engulfs and digests dying cell.

FIGURE 21.16 The Action of a Natural Killer (NK) Cell.

FIGURE 21.17   The Course of a Fever.

DEEPER INSIGHT 21.2

CLINICAL APPLICATION

Reye Syndrome

In children younger than 15, an acute viral infection such as chickenpox or influenza is sometimes followed by a serious disorder called *Reye*[11] *syndrome* (pronounced "rye"). First recognized in 1963, this disease is characterized by swelling of brain neurons and fatty infiltration of the liver and other viscera. Neurons die from hypoxia and the pressure of the swelling brain, which results in nausea, vomiting, disorientation, seizures, and coma. About 30% of victims die, and the survivors sometimes suffer mental retardation. Reye syndrome can be triggered by the use of aspirin to control fever; parents are strictly advised never to give aspirin to children with chickenpox or flulike symptoms.

sweating. The skin is warm and flushed during this phase. The phase of falling temperature is called *defervescence* in general, *crisis (flush)* if the temperature drops abruptly, or *lysis* if it falls slowly.

Even though most fevers are beneficial, excessively high temperature can be dangerous because it speeds up different enzymatic pathways to different degrees, causing metabolic discoordination and cellular dysfunction. Fevers above 40.5°C (105°F) can make one delirious. Convulsions and coma ensue at higher temperatures, and death or irreversible brain damage commonly results from fevers that range from 44° to 46°C (111° to 115°F).

21.2f Inflammation

Inflammation is a local defensive response to tissue injury of any kind, including trauma and infection. Its general purposes are (1) to limit the spread of pathogens and ultimately destroy them, (2) to remove the debris of damaged tissue, and (3) to initiate tissue repair. Inflammation is characterized by four **cardinal signs:** redness, swelling, heat, and pain. Some authorities list impaired use as a fifth sign, but this may or may not occur and when it does, it is mostly because of the pain.

▶▶▶APPLY WHAT YOU KNOW

In spite of the expression, the four "cardinal signs" of inflammation are actually a mixture of signs and symptoms. With the aid of the glossary definitions, identify which are signs and which are symptoms.

Words ending in the suffix *-itis* denote inflammation of specific organs and tissues: *arthritis, encephalitis, peritonitis, gingivitis,* and *dermatitis,* for example. Inflammation can occur anywhere in the body, but it is most common and observable in the skin, which is subject to more trauma than any other organ. Examples of cutaneous inflammation include an itchy mosquito bite, sunburn, a poison ivy rash, and the redness and blistering produced by manual labor, tight shoes, or a kitchen burn.

The following discussion will account for the four cardinal signs and explain how the three purposes of inflammation are achieved. Inflammation is mediated by several types of cells and chemicals summarized in **table 21.1.** Many of the chemicals that regulate inflammation and immunity are in a class called **cytokines**[12]—small proteins that serve as a chemical communication network among immune cells. Cytokines usually act at short

TABLE 21.1	Agents of Inflammation
Cellular Agents	
Basophils	Secrete histamine, heparin, leukotrienes, and kinins
Endothelial cells	Produce cell-adhesion molecules to recruit leukocytes; secrete platelet-derived growth factor
Eosinophils*	Produce antiparasitic oxidizing agents and toxic proteins; stimulate basophils and mast cells; limit action of histamine and other inflammatory chemicals
Fibroblasts	Rebuild damaged tissue by secreting collagen, ground substance, and other tissue components
Helper T cells*	Secrete chemotactic factors and colony-stimulating factors
Macrophages*	Clean up tissue damage; phagocytize bacteria, tissue debris, dead and dying leukocytes, and pathogens
Mast cells	Same actions as basophils
Monocytes	Emigrate into inflamed tissue and become macrophages
Neutrophils	Phagocytize bacteria; secrete bactericidal oxidizing agents; secrete cytokines that activate more leukocytes
Platelets	Secrete clotting factors and platelet-derived growth factor
Chemical Agents	
Bradykinin	A kinin that stimulates vasodilation and capillary permeability, stimulates pain receptors, and is a neutrophil chemotactic factor
Chemotactic factors	Chemicals that provide a trail that neutrophils and other leukocytes can follow to specific sites of infection and tissue injury; include bradykinin, leukotrienes, and some complement proteins
Colony-stimulating factors	Hormones that raise the WBC count by stimulating leukopoiesis
Complement*	Proteins that promote phagocytosis and cytolysis of pathogens, activate and attract neutrophils and macrophages, and stimulate basophils and mast cells to secrete inflammatory chemicals
Cytokines	Small proteins produced mainly by WBCs; have autocrine, paracrine, and endocrine roles in cellular communication; act especially in inflammation and immune responses; include interleukins, interferons, colony-stimulating factors, chemotactic factors, and others
Fibrinogen	A plasma clotting protein that filters into inflamed tissue and forms fibrin, thus coagulating tissue fluid, sequestering pathogens, and forming a temporary scaffold for tissue rebuilding
Heparin	A polysaccharide secreted by basophils and mast cells that inhibits clotting of tissue fluid within the area walled off by fibrin, thus promoting free mobility of leukocytes that attack infectious microorganisms
Histamine	An amino acid derivative produced by basophils and mast cells that stimulates vasodilation and capillary permeability
Kinins	Plasma proteins that are activated by tissue injury and stimulate vasodilation, capillary permeability, and pain; include bradykinin
Leukotrienes	Eicosanoids that stimulate vasodilation, capillary permeability, and neutrophil chemotaxis, especially in inflammation and allergy
Prostaglandins	Eicosanoids that stimulate pain, fever, and vasodilation; promote neutrophil diapedesis; and enhance histamine and bradykinin action
Selectins	Cell-adhesion molecules of endothelial cells that adhere to circulating leukocytes, promoting margination and diapedesis

* These agents of inflammation have additional roles in adaptive immunity described in table 21.4.

range, either on neighboring cells (a *paracrine*[13] effect) or on the same cell that secretes them (an *autocrine*[14] effect); these terms are distinguished from the long-distance *endocrine* effects of hormones. Cytokines include interferons, interleukins, tumor necrosis factor, chemotactic factors, and other chemicals you will soon encounter in this discussion.

Inflammation involves three major processes: mobilization of the body's defenses, containment and destruction of pathogens, and tissue cleanup and repair.

Mobilization of Defenses

The most immediate requirement for dealing with tissue injury is to get defensive leukocytes to the site quickly. Damaged tissues release cytokines that stimulate the red bone marrow to release neutrophils into circulation, raising the blood neutrophil count within hours. In addition, certain cells secrete *vasoactive* chemicals

---

[13]*para* = next to; *crin* = to secrete
[14]*auto* = self; *crin* = to secrete

that dilate the blood vessels in the area of injury. Among these are histamine, leukotrienes, and other cytokines secreted by basophils, mast cells, and cells damaged by the pathogens that triggered the inflammation. The resulting increase in local blood flow is called **hyperemia.** Hyperemia not only results in the more rapid delivery of leukocytes, but also washes toxins and metabolic wastes from the tissue more rapidly.

In addition to dilating local blood vessels, the vasoactive chemicals stimulate endothelial cells of the blood capillaries and venules to contract slightly, widening the gaps between them and increasing capillary permeability. This allows for the easier movement of fluid, leukocytes, and plasma proteins from the bloodstream into the surrounding tissue. Among the helpful proteins filtering from the blood are complement, antibodies, and clotting factors, all of which aid in combating pathogens.

Endothelial cells actively recruit leukocytes. In the area of injury, they produce cell-adhesion molecules called **selectins,** which make their membranes sticky, and snag leukocytes arriving in the bloodstream. Leukocytes adhere loosely to the selectins and slowly tumble along the endothelium, sometimes coating it so thickly they obstruct blood flow. This adhesion to the vessel wall is called **margination.** The leukocytes then crawl through the gaps between the endothelial cells—an action called **diapedesis**[15] or **emigration**—and enter the tissue fluid of the damaged tissue **(fig. 21.18).** Most diapedesis occurs across the walls of the postcapillary venules. Cells and chemicals that have left the bloodstream are said to be *extravasated.*[16]

In the events that have already transpired, we can see the basis for the four cardinal signs of inflammation: (1) The heat results from the hyperemia; (2) redness is also due to hyperemia and in some cases, such as sunburn, to extravasated erythrocytes in the tissue; (3) swelling (edema) is due to the increased fluid filtration from the capillaries; and (4) pain results from direct injury to the nerves, pressure on the nerves from the edema, and stimulation of pain receptors by prostaglandins, some bacterial toxins, and a kinin called **bradykinin.**

▶▶▶ APPLY WHAT YOU KNOW

Review eicosanoid synthesis (fig. 17.24) and explain why aspirin eases the pain of inflammation.

Containment and Destruction of Pathogens

One priority in inflammation is to prevent pathogens from spreading through the body. The fibrinogen that filters into the tissue fluid clots in areas adjacent to the injury, forming a sticky mesh that sequesters (walls off and isolates) bacteria and other microorganisms. Heparin, the anticoagulant, prevents clotting in the immediate area of the injury, so bacteria or other pathogens are essentially trapped in a fluid pocket surrounded by a gelatinous

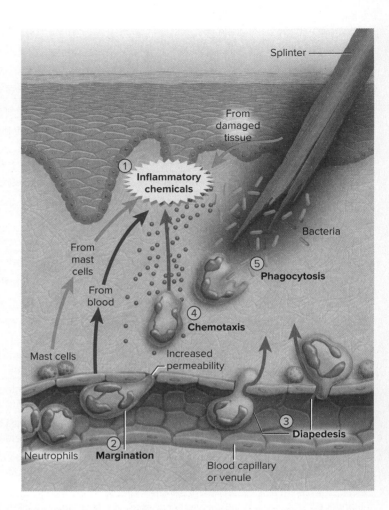

FIGURE 21.18 Neutrophil Behavior in Inflammation. Chemical messengers are released by basophils, mast cells, blood plasma, and damaged tissue. These inflammatory chemicals stimulate leukocyte margination (adhesion to the blood vessel wall), diapedesis (crawling through gaps in the wall), chemotaxis (movement toward the source of the inflammatory chemicals), and phagocytosis (engulfing bacteria or other pathogens).

capsule of clotted fluid. They are attacked by antibodies, phagocytes, and other defenses, while the surrounding areas of clotted tissue fluid prevent them from easily escaping this onslaught.

The chief enemies of bacteria are neutrophils, which accumulate in the inflamed tissue within an hour. After emigrating from the bloodstream, they exhibit **chemotaxis**—attraction to chemicals *(chemotactic factors)* that guide them to the site of injury or infection. These include bradykinin, leukotrienes, ATP released by dying cells, and ADP released by platelets. As they encounter bacteria, neutrophils avidly phagocytize and digest them, and destroy many more by the respiratory burst described earlier. The major stages of neutrophil action are summarized in figure 21.18.

Neutrophils also recruit macrophages and additional neutrophils by secreting cytokines, like shouting "Over here!" to bring in reinforcements. Activated macrophages and T cells in the inflamed tissue secrete cytokines called *colony-stimulating factors,* which promote the production of more leukocytes (leukopoiesis) by the red bone marrow. Within a few hours of onset, the neutrophil count

---

[15]*dia* = through; *pedesis* = stepping
[16]*extra* = outside; *vas* = vessel

in the blood can rise from the normal 4,000 or 5,000 cells/μL to as high as 25,000 cells/μL, a condition called **neutrophilia.** In the case of allergy or parasitic infection, an elevated eosinophil count, or **eosinophilia,** may also occur. The task of eosinophils was described earlier.

Tissue Cleanup and Repair

Monocytes are major agents of tissue cleanup and repair. They arrive within 8 to 12 hours, emigrate from the bloodstream, and turn into macrophages. Macrophages engulf and destroy bacteria, damaged host cells, and dead and dying neutrophils. They also act as antigen-presenting cells, activating the processes of adaptive immunity.

Edema also contributes to tissue cleanup. The swelling compresses veins and reduces venous drainage, while it forces open the valves of lymphatic capillaries and promotes lymphatic drainage. The lymphatics can collect and remove bacteria, dead cells, proteins, and tissue debris better than blood capillaries or venules can.

As the battle progresses, all of the neutrophils and most of the macrophages die. These dead cells, other tissue debris, and tissue fluid form a pool of yellowish fluid called **pus,** which accumulates in a tissue cavity called an **abscess.**[17] Pus is usually absorbed, but sometimes it forms a blister between the epidermis and dermis and may be released by its rupture.

Blood platelets and endothelial cells in an area of injury secrete **platelet-derived growth factor,** an agent that stimulates fibroblasts to multiply and synthesize collagen. At the same time, hyperemia delivers oxygen, amino acids, and other necessities of protein synthesis, while the heat of inflamed tissue increases metabolic rate and the speed of mitosis and tissue repair. The fibrin clot in the tissue may provide a scaffold for reconstruction. Pain also contributes importantly to recovery. It is an important alarm signal that calls our attention to the injury and makes us limit the use of a body part so it has a chance to rest and heal.

BEFORE YOU GO ON

Answer the following questions to test your understanding of the preceding section:

6. Earlier it was stated that innate immunity employs protective proteins, protective cells, and protective processes. Give three examples of each.

7. What are macrophages? Give four examples and state where they are found.

8. How do interferons and the complement system protect against disease?

9. Summarize the benefits of fever and the limits of these benefits.

10. List the cardinal signs of inflammation and state the cause of each.

---

[17]*ab* = away; *scess* (from *cedere*) = to go

21.3 Adaptive Immunity— General Aspects

Expected Learning Outcomes

When you have completed this section, you should be able to

a. identify the three distinguishing characteristics of adaptive immunity;

b. contrast cellular and humoral immunity, active and passive immunity, and natural and artificial immunity;

c. describe the chemical properties of antigens;

d. describe the structure and five types of antibodies;

e. describe and contrast the development of T and B lymphocytes; and

f. describe the general roles played by lymphocytes, antigen-presenting cells, and interleukins in the immune response.

The remainder of this chapter is concerned with *adaptive immunity* (the third line of defense). Adaptive immunity was noticed even in the fifth century BCE, when Greek historian Thucydides remarked that people who recovered from a disease often became immune to that one but remained susceptible to others. Adaptive immunity is now defined by three characteristics that distinguish it from the three that were itemized earlier (local, nonspecific, and lacking memory) for innate immunity:

1. It has a **systemic effect.** When an adaptive response is mounted against a particular threat such as a bacterial infection, it acts throughout the body to defeat that pathogen wherever it may be found.

2. It exhibits **specificity.** Adaptive immunity is directed against a specific pathogen. Immunity to one disease such as chickenpox does not confer immunity to others such as tetanus.

3. It has a **memory.** When reexposed to the same pathogen, the body reacts so quickly that there is no noticeable illness. The reaction time for inflammation and other innate defenses, by contrast, is just as long for later exposures as for the initial one.

21.3a Forms of Adaptive Immunity

In the late 1800s, it was discovered that immunity can be transferred from one animal to another by way of the blood serum. In the mid-1900s, however, it was found that serum doesn't always confer immunity; sometimes only donor lymphocytes do so. Thus, biologists came to recognize two types of adaptive immunity, called cellular and humoral immunity, although the two interact extensively and often respond to the same pathogen.

Cellular (cell-mediated) immunity employs lymphocytes that directly attack and destroy foreign cells or diseased host cells. It is a means of ridding the body of pathogens that reside inside human cells, where they are inaccessible to antibodies: intracellular viruses, bacteria, yeasts, and protozoans, for example. Cellular immunity also acts against parasitic worms, cancer cells, and cells of transplanted tissues and organs.

Humoral (antibody-mediated) immunity employs antibodies, which don't directly destroy pathogens but tag them for destruction by mechanisms described later. The expression *humoral* refers to antibodies dissolved in the body fluids ("humors"). Humoral immunity is effective against extracellular viruses, bacteria, yeasts, protozoans, and molecular (noncellular) disease agents such as toxins, venoms, and allergens. In the unnatural event of a mismatched blood transfusion, it also destroys foreign erythrocytes.

Note that humoral immunity works mainly against the *extracellular* stages of infectious microorganisms. When such microorganisms invade host cells, they are usually sheltered from attack by serum antibodies. However, the *intracellular* stages are still vulnerable to cellular immunity, which destroys them by killing the cells that harbor them. Furthermore, certain antibodies in the IgE class (explained later) bind to parasitic worms and aid in their destruction. Thus, humoral and cellular immunity sometimes attack the same microorganism in different ways or at different points in its life cycle. After our discussion of the details of the two processes, you will find cellular and humoral immunity summarized and compared in table 21.5.

Other ways of classifying immunity are active versus passive and natural versus artificial. In *active immunity,* the body makes its own antibodies or T cells against a pathogen, whereas in *passive immunity,* the body acquires them from another person or an animal that is immune to the pathogen. Either type of immunity can occur naturally or, for treatment and prevention purposes, it can be induced artificially. Thus we can recognize four classes of immunity under this scheme:

1. **Natural active immunity.** This is the production of one's own antibodies or T cells as a result of natural exposure to an antigen.

2. **Artificial active immunity.** This is the production of one's own antibodies or T cells as a result of **vaccination** against diseases such as smallpox, tetanus, or influenza. A **vaccine** consists of either dead or *attenuated* (weakened) pathogens that can stimulate an immune response but cause little or no discomfort or disease. In some cases, periodic *booster shots* are given to restimulate immune memory and maintain a high level of protection (tetanus boosters, for example). Vaccination has eliminated smallpox worldwide and greatly reduced the incidence of life-threatening childhood diseases, but many people continue to die from influenza and other diseases that could be prevented by vaccination. Even polio, once nearly eradicated worldwide, has lately made a comeback in some countries because of antivaccination superstition and politics.

3. **Natural passive immunity.** This is a temporary immunity that results from acquiring antibodies produced by another person. The only natural ways for this to happen are for a fetus to acquire antibodies from the mother through the placenta before birth, or for a baby to acquire them during breast-feeding.

4. **Artificial passive immunity.** This is a temporary immunity that results from the injection of an *immune serum* obtained from another person or from animals (such as horses) that have antibodies against a certain pathogen. Immune serum is used for emergency treatment of snakebites, botulism, tetanus, rabies, and other diseases.

Only the two forms of active immunity involve memory and thus provide future protection. Passive immunity typically lasts for only 2 or 3 weeks, until the acquired antibody is degraded. The remaining discussion is based on natural active immunity.

21.3b Antigens

An **antigen**[18] (**Ag**) is any molecule that triggers an immune response. Some antigens are free molecules such as venoms, toxins, and food-borne substances; others are components of plasma membranes and bacterial cell walls. Small universal molecules such as glucose and amino acids aren't antigenic; if they were, our immune systems would attack the nutrients and other molecules essential to our very survival. Most antigens have molecular weights over 10,000 amu and have enough structural complexity and variability to be unique to each individual: proteins, polysaccharides, glycoproteins, and glycolipids. Their uniqueness enables the body to distinguish its own ("self") molecules from those of any other individual or organism ("nonself"). The immune system learns to distinguish self-antigens from nonself-antigens so that it normally attacks only nonself-antigens.

Only certain regions of an antigen molecule, called **epitopes (antigenic determinants),** stimulate immune responses. One antigen molecule typically has several different epitopes, however, that can stimulate the simultaneous production of different antibodies.

Some molecules, called **haptens**[19] **(incomplete antigens),** are too small to be antigenic in themselves, but they can stimulate an immune response by binding to a host macromolecule and creating a unique complex that the body recognizes as foreign. After the first exposure, the hapten alone may stimulate an immune response without needing to bind to a host molecule. Many people are allergic to haptens in cosmetics, detergents, industrial chemicals, poison ivy, and animal dander. The most common drug allergy is to penicillin—a hapten that binds to host proteins in allergic individuals, creating a complex that binds to mast cells and triggers massive release of histamine and other inflammatory chemicals. This can cause death from *anaphylactic shock.*

21.3c Antibodies

Antibodies (Abs), also called **immunoglobulins (Igs),** are proteins in the gamma globulin class that play a variety of roles in defense. Some of them are integral proteins in the plasma membranes of basophils and mast cells and thus function in innate immunity. Others, with roles in adaptive immunity, are membrane proteins of B lymphocytes or *soluble antibodies* dissolved in body fluids such as blood plasma, lymph, mucus, saliva, intestinal secretions, tears, and breast milk. We have said much about antibodies already in chapter 18 (blood types) and this chapter; it is now time to take a closer look at what they are and how they work.

The basic structural unit of an antibody, called an **antibody monomer,** is composed of four polypeptides linked by disulfide (—S—S—) bonds **(fig. 21.19).** These include two **heavy chains** about 400 amino acids long and two **light chains** about half that long. Each heavy chain has a hinge region where the antibody is bent, giving the monomer a T or Y shape.

---

[18]acronym from *anti*body *gen*erating
[19]from *haptein* = to fasten

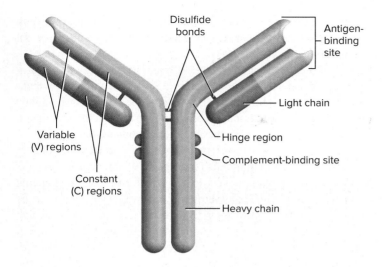

FIGURE 21.19 Antibody Structure. A molecule of IgG, a monomer.

All four chains have a **variable (V) region** that gives an antibody its uniqueness. The V regions of each heavy and light chain pair combine to form an **antigen-binding site** on each arm. These sites are where the antibody attaches to the epitopes of antigen molecules. The rest of each chain is a **constant (C) region,** which has the same amino acid sequence, or nearly so, in all antibodies of a given class (within one person). The C region determines the mechanism of an antibody's action—for example, whether it can bind complement proteins.

There are five classes of antibodies named **IgA, IgD, IgE, IgG,** and **IgM** after the structures of their C regions (*alpha, delta, epsilon, gamma,* and *mu*). As shown in **table 21.2,** IgD, IgE, and IgG are monomers; IgA has a monomer form as well as a *dimer* composed of two cojoined monomers; and IgM has a monomer form as well as a *pentamer* composed of five monomers. The surface antigen receptors of B cells are IgD and IgM molecules. IgG is particularly important in the immunity of the newborn because it crosses the placenta with relative ease. Thus, it transfers immunity from the mother to her fetus. In addition, an infant acquires some maternal IgA through breast milk and colostrum (the fluid secreted for the first 2 or 3 days of breast-feeding).

The human immune system is believed capable of producing at least 10 billion and perhaps up to 1 trillion different antibodies. Any one person has a much smaller subset of these, but such an enormous potential helps to explain why we can deal with the tremendous diversity of antigens that exist in our environment. Yet such huge numbers may seem puzzling, because we're accustomed to thinking of each protein in the body being encoded by one gene, and we have only slightly more than 20,000 genes, most of which have nothing to do with immunity. How can so few genes generate so many antibodies? Obviously there cannot be a different gene for each one. One means of generating diversity is that the genome contains several hundred DNA segments that are shuffled and combined in various ways to produce antibody genes unique to each clone of B cells. This process is called **somatic recombination,** because it forms new combinations of DNA base sequences in somatic (nonreproductive) cells. Another mechanism

TABLE 21.2	**The Five Classes of Antibodies**		
Class	**Structure**		**Location and Function**
IgA	Monomer	Dimer	Found as a monomer in blood plasma and mainly as a dimer in mucus, tears, milk, saliva, and intestinal secretions. Sometimes also forms trimers and tetramers. Prevents pathogens from adhering to epithelia and penetrating underlying tissues. Provides passive immunity to the newborn.
IgD	Monomer		A transmembrane protein of B cells; functions in activation of B cells by antigens.
IgE	Monomer		A transmembrane protein of basophils and mast cells. Stimulates them to release histamine and other mediators of inflammation and allergy; important in immediate hypersensitivity reactions and in attracting eosinophils to sites of parasitic infection.
IgG	Monomer		Constitutes about 80% of circulating antibodies in blood plasma. The predominant antibody secreted in the secondary immune response. IgG and IgM are the only antibodies with significant complement-fixation activity. Crosses placenta and confers temporary immunity on the fetus. Includes the anti-D antibodies of the Rh blood group.
IgM	Monomer	Pentamer	Constitutes about 10% of circulating antibodies in plasma. Monomer is a transmembrane protein of B cells, where it functions as part of the antigen receptor. Pentamer occurs in blood plasma and lymph. The predominant antibody secreted in the primary immune response; very strong agglutinating and complement-fixation abilities; includes the anti-A and anti-B agglutinins of the ABO blood group.

of generating diversity is that B cells in the germinal centers of lymphatic nodules undergo exceptionally high rates of mutation, a process called **somatic hypermutation**—not just recombining preexisting DNA but creating wholly new DNA sequences. These and other mechanisms explain how we can produce such a tremendous variety of antibodies with a limited number of genes.

21.3d Lymphocytes

The major cells of the adaptive immune system are lymphocytes, macrophages, and dendritic cells, which are especially concentrated at strategic places such as the lymphatic organs, skin, and mucous membranes. Lymphocytes fall into three classes: natural killer (NK) cells, T lymphocytes, and B lymphocytes. We have already discussed the role of NK cells in innate immunity. Here, we must take a closer look at T and B lymphocytes, the principal agents of adaptive immunity.

T Lymphocytes (T Cells)

The life history of a T cell basically involves three stages and three anatomical stations in the body **(fig. 21.20)**. We can loosely think of these stages as their "birth," their "training" or maturation, and finally their "deployment" to locations where they will carry out their immune function.

T cells are produced in the red bone marrow by the hematopoietic stem cells described in section 18.1e. Newborn T cells enter the bloodstream and travel to the thymus—the "school" where they mature into fully functional T cells and face two harsh "graduation exams" that test their usefulness to the immune system.

On arrival, T cells go first to the thymic cortex and cluster on the cortical epithelial cells (see fig. 21.9). The epithelial cells test these young lymphocytes to see which ones will be capable of later recognizing foreign antigens and antigen-presenting cells (APCs). If a T cell possesses the proper receptors to recognize them, it receives a protective signal from the epithelial cell that spares its life. T cells that can't recognize them receive no life-sparing signal. They do, however, get a chance to "retake the test"; they can reshuffle the DNA for their antigen receptors and try again. But if they fail once more, they're doomed. Within 3 or 4 days, these useless T cells die by apoptosis and the cortical macrophages phagocytize them. T cells that pass the test, demonstrating their ability to respond to antigens, are called **immunocompetent.**

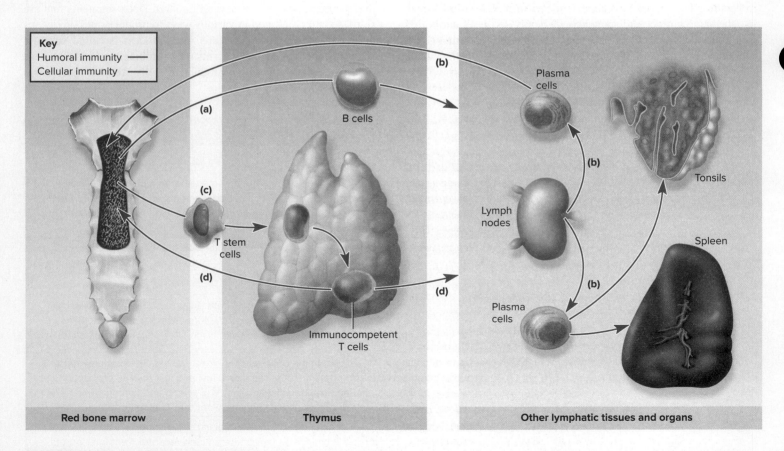

FIGURE 21.20 The Life History and Migrations of B and T Cells. Humoral immunity is represented by the violet pathways and cellular immunity by the red. (a) B cells achieve immunocompetence in the red bone marrow (left), and many emigrate to lymphatic tissues and organs, including the lymph nodes, tonsils, and spleen (right). (b) Plasma cells develop in the lymph nodes (among other sites) and emigrate to the bone marrow and other lymphatic organs, where they spend a few days secreting antibodies. (c) T stem cells emigrate from the bone marrow and attain immunocompetence in the thymus. (d) Immunocompetent T cells emigrate from the thymus and recolonize the bone marrow or colonize various lymphatic organs (right).

The cortical process of selecting the few T cells that prove their immunocompetence is called **positive selection.** Only a small percentage of them pass this cortical "exit exam" and get the opportunity to move on to "graduate school," the medulla. This explains why the medulla of the thymus looks so much lighter than the cortex in histological sections (see fig. 21.9b), somewhat like graduate school enrollments compared to undergraduate.

But like graduate students, the T cells in the medulla face yet another difficult test. Even more T cells are doomed to fail and possibly die. Here, their examiners are macrophages and reticular cells. Unlike the cortical epithelial cells, these are derived from bone marrow and test the T cells in a different way. Their role is to weed out T cells that react to the antigens of one's own body or to one's own APCs even in the absence of antigen. Such overreactive T cells could be a future threat if they attacked one's own tissues. Owing to the imperfections of the immune system, this sometimes happens anyway, causing notorious *autoimmune diseases* described later in section 21.6b. But to minimize that risk, macrophages and reticular cells present the T cells with self-antigens and antigen-binding proteins called MHCs (explained shortly). In one theory, T cells that react too strongly to these are killed by apoptosis. There is also evidence, however, that the thymic corpuscles secrete a signal (cytokine) that renders these self-reactive T cells permanently inactive (a state called **anergy**[20]) or converts them to *regulatory T cells* that moderate the activity of the *cytotoxic (killer) T cells* that actually attack foreign cells. In any event, elimination or conversion of self-reactive T cells is called **negative selection,** and it leaves the immune system in a state of **self-tolerance**—restraint from attacking one's own tissues.

Only about 2% of the T cells survive both positive and negative selection; 98% are eliminated, especially by positive selection. (And you think *your* final exams are tough!) The few survivors multiply and form *clones* of identical T cells programmed to respond to a particular antigen. These cells, which are immunocompetent but haven't yet encountered an enemy (foreign antigen), constitute the **naive lymphocyte pool.** Naive T cells leave the thymus by way of blood and lymphatic vessels (there is no blood–thymus barrier in the medulla), disperse throughout the body, and colonize lymphatic tissues and organs everywhere (the lymph nodes, tonsils, bone marrow, and so forth) (fig. 21.20d). They become especially concentrated in the deep cortex of the lymph nodes. They are now ready to do battle.

B Lymphocytes (B Cells)

B cell maturation occurs entirely within the red bone marrow. Adults produce about 50 million B cells per day, but only 10% enter the general circulation. The other 90% are apparently destroyed in the course of positive and negative selection in the bone marrow. Self-tolerant B cells that survive selection go on to multiply and generate immunocompetent B cell clones. These cells disperse throughout the body and colonize the same organs

as T cells (fig. 21.20a, b). They are abundant in the lymphatic nodules of the lymph nodes and in the spleen, bone marrow, and mucous membranes.

21.3e Antigen-Presenting Cells

Although the function of T cells is to recognize and attack foreign antigens, they usually can't do this on their own. They require the help of **antigen-presenting cells (APCs).** In addition to their other roles, dendritic cells, macrophages, reticular cells, and B cells function as APCs.

APC function hinges on a family of genes called the **major histocompatibility complex (MHC)** on chromosome 6. These genes code for **MHC proteins**—proteins on the APC surface that are shaped a little like hotdog buns, with an elongated groove for holding the "hotdog" of the foreign antigen. MHC proteins are structurally unique to every person except for identical twins. They act as "identification tags" that label every cell of your body as belonging to you.

When an APC encounters an antigen, it internalizes it by endocytosis, digests it into molecular fragments, and displays the relevant fragments (its epitopes) in the grooves of the MHC proteins **(fig. 21.21a).** These steps are called **antigen processing.** Wandering T cells regularly inspect APCs for displayed antigens **(fig. 21.21b).** If an APC displays a self-antigen, the T cells disregard it. If it displays a nonself-antigen, however, they initiate an attack. APCs thus alert the immune system to the presence of a foreign antigen. The key to a successful defense is then to quickly mobilize immune cells against it.

With so many cell types involved in immunity, it's not surprising that they require chemical messengers to coordinate their activities. Lymphocytes and APCs talk to each other with cytokines called **interleukins**[21]—chemical signals from one leukocyte (or leukocyte derivative) to another.

With this introduction to the main actors in immunity, we can now look at the more specific features of cellular and humoral immunity. Since the terminology of immune cells and chemicals is quite complex, you may find it helpful to refer often to table 21.4 as you read the following discussions.

BEFORE YOU GO ON

Answer the following questions to test your understanding of the preceding section:

11. How does adaptive immunity differ from innate immunity?
12. How does humoral immunity differ from cellular immunity?
13. Contrast active and passive immunity. Give natural and artificial examples of each.
14. What structural properties distinguish antigenic molecules from those that are not antigenic?
15. Describe the structure of an antibody monomer and state what part of it binds to antigens.

[20]*an* = without; *erg* = action, work

[21]*inter* = between; *leuk* = leukocytes

16. What is an immunocompetent lymphocyte? What does a lymphocyte have to produce in order to become immunocompetent?

17. What role does the thymus play in the life history of a T cell?

18. What role does an antigen-presenting cell play in the activation of a T cell?

(a)

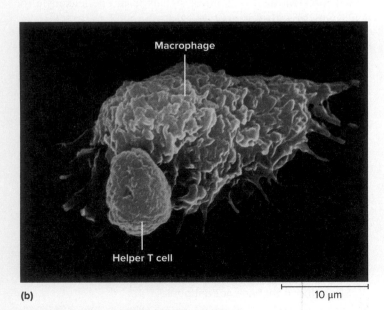

(b)

10 μm

FIGURE 21.21 The Action of Antigen-Presenting Cells (APCs).
(a) Stages in the processing and presentation of an antigen by an APC such as a macrophage. (b) Macrophage presenting processed antigen to helper T cell.
b: CNRI/Science Source

21.4 Cellular Immunity APR

Expected Learning Outcomes

When you have completed this section, you should be able to

a. list the types of lymphocytes involved in cellular immunity and describe the roles they play;

b. describe the process of antigen presentation and T cell activation;

c. describe how T cells destroy enemy cells; and

d. explain the role of memory cells in cellular immunity.

Cellular (cell-mediated) immunity is a form of adaptive immunity in which T lymphocytes directly attack and destroy diseased or foreign cells, and the immune system remembers the antigens of those invaders and prevents them from causing disease in the future. There are at least 15 kinds of T cells with new ones being discovered continually, but the 4 principal types important to our discussion are these:

1. **Cytotoxic T (T_C) cells** are the "effectors" of cellular immunity that carry out the attack on foreign cells.

2. **Helper T (T_H) cells** promote the action of T_C cells as well as play key roles in humoral and innate immunity. All other T cells are involved in cellular immunity only.

3. **Regulatory T (T_R) cells,** or **T-regs,** limit the immune response by inhibiting multiplication and cytokine secretion by other T cells. Once a pathogen is defeated, it is important to down-regulate T_C activity, because otherwise T_C cells would continue to produce inflammatory cytokines that could cause extensive tissue damage in the long term. T_R cells prevent this and reduce the risk of developing autoimmune diseases.

4. **Memory T (T_M) cells** are descended from T_C cells and are responsible for memory in cellular immunity. Thus, they sustain the body's vigilance against a given pathogen without causing the damage that a persistent population of T_C cells would do.

T_C cells are also known as T8, CD8, or CD8+ cells because they have a surface glycoprotein called CD8. T_H and T_R cells are also known as T4, CD4, or CD4+ cells, after their glycoprotein, CD4. (*CD* stands for *cluster of differentiation,* a classification system for many cell surface molecules.) These glycoproteins are cell-adhesion molecules that enable T cells to bind to other cells in the events to be described shortly.

Both cellular and humoral immunity occur in three stages that we can think of as recognition, attack, and memory (or "the three *R*s of immunity"—recognize, react, and remember). In cellular immunity, the events of each stage are as follows.

21.4a Recognition

The recognition phase has two aspects: antigen presentation and T cell activation.

Antigen Presentation

When an antigen-presenting cell (APC) encounters and processes an antigen, it typically migrates to the nearest lymph node and displays it to the T cells. Cytotoxic and helper T cells patrol the lymph nodes and other tissues as if looking for trouble. When they encounter a cell displaying an antigen on an MHC protein, they initiate an immune response. T cells respond to two classes of MHC proteins:

1. *MHC-I proteins* occur on every nucleated cell of the body (not erythrocytes). These proteins are constantly produced by the cell and transported to the plasma membrane. Along the way, they pick up small peptides in the cytoplasm and display them once they are installed in the membrane. If the peptides are normal self-antigens, they don't elicit a T cell response. If they are viral proteins or abnormal antigens made by cancer cells, however, they do. In this case, the Ag–MHC protein complex is like a tag on the host cell that says, "I'm diseased; kill me." Infected or malignant cells are then destroyed before they can do further harm to the body.

2. *MHC-II proteins* (also called *human leukocyte antigens, HLAs*) occur only on APCs and display only foreign antigens.

As T cells mature in the thymus and undergo the positive and negative selection described earlier, they become limited to recognizing a specific class of MHC protein. This process, called **MHC restriction,** results in T_C cells responding only to MHC-I proteins and T_H cells only to MHC-II **(table 21.3).**

T Cell Activation

T cell activation is shown in **figure 21.22.** It begins when a T_C or T_H cell binds to an MHC protein displaying an epitope that the T cell is programmed to recognize. Before the response can go any further, the T cell must bind to another protein, related to interleukins, found on the surface of APCs in damaged or infected tissues. In a sense, the T cell has to check twice to see if it really has bound to an APC displaying a suspicious antigen. This signaling process, called **costimulation,** helps to ensure that the immune system doesn't launch an attack in the absence of an enemy, which could turn against one's own body with injurious consequences.

Lack of costimulation drives the T cell into *anergy,* a state of inactivity and ineffectiveness. Successful costimulation, in

TABLE 21.3	Comparison of the Responses of Cytotoxic and Helper T Cells	
Characteristic	T_C Cells	T_H Cells
Cells capable of stimulating a response	Any nucleated cell	Antigen-presenting cells
MHC protein	MHC-I	MHC-II

① **Antigen recognition**
T cell binds to an APC displaying an antigen fragment (epitope).

② **Costimulation**
T cell binds to a second protein on the APC.

③ **Clonal selection**
T cell undergoes repeated mitosis and produces a large number of effector cells and memory T cells.

④ **Attack**
Effector T_C cells attack and destroy abnormal cells with a lethal hit; T_H cells secrete interleukins that stimulate multiple forms of attack.

FIGURE 21.22 T Cell Activation.

contrast, activates the process of **clonal selection:** The T cell undergoes repeated mitosis, giving rise to a clone of identical T cells programmed against the same epitope. Some cells in the clone become effector cells that carry out an immune attack, and some become memory T (T_M) cells.

21.4b Attack

Helper and cytotoxic T cells play different roles in the attack phase. Helper T (T_H) cells are necessary for most immune responses. They play a central coordinating role in both humoral and cellular immunity **(fig. 21.23).** When a T_H cell recognizes an Ag–MHC protein complex, it secretes interleukins that exert three effects: (1) to attract neutrophils and natural killer cells; (2) to attract macrophages, stimulate their phagocytic activity, and inhibit them from leaving the area; and (3) to stimulate T and B cell mitosis and maturation.

Cytotoxic T (T_C) cells are the only T lymphocytes that directly attack and kill other cells **(fig. 21.24).** When a T_C cell recognizes a complex of antigen and MHC-I protein on a diseased or foreign cell, it "docks" on that cell, delivers a **lethal hit** of chemicals that will destroy it, and goes off in search of other enemy cells while the chemicals do their work. Among these chemicals are

- *perforin* and *granzymes,* which kill the target cell in the same manner as we saw earlier for NK cells (see fig. 21.16);

- *interferons,* which inhibit viral replication and recruit and activate macrophages, among other effects; and

- *tumor necrosis factor (TNF),* which aids in macrophage activation and kills cancer cells.

FIGURE 21.24 Destruction of a Cancer Cell by Cytotoxic T Cells.
Steve Gschmeissner/Science Source

As more and more cells are recruited by helper T cells, the immune response exerts an overwhelming force against the pathogen. The **primary response,** seen on first exposure to a particular pathogen, peaks in about a week and then gradually declines.

▶▶▶ APPLY WHAT YOU KNOW

How is a cytotoxic T cell like a natural killer (NK) cell? How are they different?

21.4c Memory

The primary response is followed by immune memory. Following clonal selection, some T cells become memory cells. These cells are long-lived and much more numerous than naive T cells. Aside from their sheer numbers, they also require fewer steps to be activated, and therefore respond to antigens more rapidly. Upon reexposure to the same pathogen later in life, memory cells mount a quick attack called the **T cell recall response.** This time-saving response destroys a pathogen so quickly that no noticeable illness occurs—that is, the person is immune to the disease.

BEFORE YOU GO ON

Answer the following questions to test your understanding of the preceding section:

19. Name four types of lymphocytes involved in cellular immunity. Which of these is also essential to humoral immunity?

20. What are the three phases of an immune response?

21. Explain why cytotoxic T cells are activated by a broader range of host cells than are helper T cells.

22. Describe some ways in which cytotoxic T cells destroy target cells.

FIGURE 21.23 The Central Role of Helper T Cells in Innate, Humoral, and Cellular Immunity.

❓ *Why does AIDS reduce the effectiveness of all three defenses listed across the bottom of the figure?* (Hint: *See section 21.6c.*)

21.5 Humoral Immunity

Expected Learning Outcomes

When you have completed this section, you should be able to

a. explain how B cells recognize and respond to an antigen;

b. describe the actions of antibodies against disease agents;

c. explain the mechanism of memory in humoral immunity; and

d. compare and contrast cellular and humoral immunity.

Humoral immunity is a more indirect method of defense than cellular immunity. Instead of directly attacking enemy cells, the B lymphocytes of humoral immunity produce antibodies that bind to antigens and tag them for destruction by other means. But like cellular immunity, humoral immunity works in three stages: recognition, attack, and memory.

21.5a Recognition

An immunocompetent B cell has thousands of surface receptors for one antigen. B cell activation begins when an antigen binds to several of these receptors, links them together, and is taken into the cell by receptor-mediated endocytosis (compare fig. 3.21). One reason small molecules are not antigenic is that they're too small to link multiple receptors together. After endocytosis, the B cell processes (digests) the antigen, links some of the epitopes to its MHC-II proteins, and displays these on the cell surface. Thus, the B cell itself acts as an antigen-presenting cell.

Usually, the B cell response goes no further unless a helper T cell binds to this Ag–MHC protein complex. (Some B cells are directly activated by antigens without the help of a T_H cell.) When a T_H cell binds to the complex, it secretes interleukins that activate the B cell. This triggers clonal selection—B cell mitosis giving rise to a battalion of identical B cells programmed against that antigen **(fig. 21.25).**

Most cells of the clone differentiate into **plasma cells.** These are larger than B cells and contain an abundance of rough endoplasmic reticulum **(fig. 21.26).** Plasma cells develop mainly in the germinal centers of the lymphatic nodules of the lymph nodes. About 10% of them remain in the lymph nodes, but the rest leave the nodes, take up residence in the bone marrow and elsewhere, and there produce antibodies until they die.

The job of a plasma cell is to synthesize and secrete antibodies, which it churns out at the remarkable rate of 2,000 molecules per second over a life span of 4 to 5 days. These antibodies travel throughout the body in the blood and other body fluids. The first time you're exposed to a particular antigen, your plasma cells produce mainly an antibody class called IgM. In later exposures to the same antigen, they produce mainly IgG.

21.5b Attack

Once released by a plasma cell, antibodies use four mechanisms to render antigens harmless:

1. **Neutralization** means the masking of critical regions of an antigen molecule by antibodies. Only certain regions of an antigen are pathogenic—for example, the parts of a toxin molecule or virus that enable these agents to bind to human cells. Antibodies can neutralize an antigen by masking these active regions.

2. **Complement fixation** is an action in which antibodies bind complement proteins to an enemy cell, leading to its destruction. IgM and IgG bind to enemy cells and change shape, exposing their complement-binding sites (see fig. 21.19). This initiates the binding of complement to the enemy cell surface and leads to inflammation, phagocytosis, immune clearance, and cytolysis, as described earlier (see fig. 21.14). Complement fixation is the primary mechanism of defense against foreign cells such as bacteria and mismatched erythrocytes. It opsonizes bacteria and makes it easier for phagocytes to ingest and destroy them.

3. **Agglutination** is the clumping of enemy cells by antibodies, described earlier in the discussion of ABO and Rh blood types (see section 18.3a). It is effective not only in mismatched blood transfusions, but more importantly as a defense against bacteria. An antibody molecule has 2 to 10 binding sites; thus, it can bind to antigen molecules on two or more enemy cells at once and stick them together **(fig. 21.27a).** This immobilizes microbes and other alien cells and prevents them from spreading through the tissues. Further, neutrophils and macrophages can phagocytize agglutinated clusters of bacteria more efficiently than phagocytizing bacteria one at a time.

4. **Precipitation** is a similar process in which antigen molecules (not whole cells) are clumped by adhesion to antibodies **(fig. 21.27b).** This creates large Ag–Ab complexes that can be removed by immune clearance or phagocytized by eosinophils.

You will note that antibodies don't directly destroy an antigen in any of these mechanisms. They render it harmless by the mechanisms just stated and mark it for destruction by other agents such as complement, macrophages, or eosinophils.

▶▶▶ APPLY WHAT YOU KNOW

Explain why IgM has a stronger power of agglutination than antibodies of any other class.

21.5c Memory

When a person is exposed to a particular antigen for the first time, the humoral immune reaction is called the **primary response.** The appearance of protective antibodies is delayed

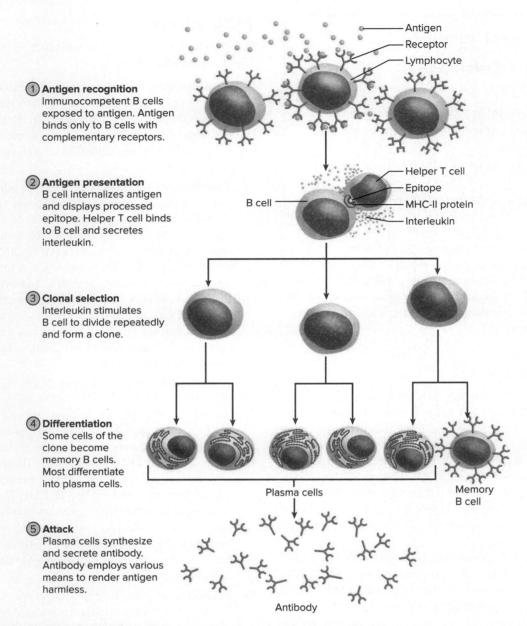

FIGURE 21.25 Clonal Selection and Ensuing Events of the Humoral Immune Response.

for 3 to 6 days while naive B cells multiply and differentiate into plasma cells. As the plasma cells begin secreting antibody, the **antibody titer** (level in the blood plasma) begins to rise **(fig. 21.28).** IgM appears first, peaks in about 10 days, and soon declines. IgG levels rise as IgM declines, but even the IgG titer drops to a low level within a month.

The primary response, however, leaves one with an immune memory of the antigen. During clonal selection, some members of the clone become **memory B cells** rather than plasma cells (see fig. 21.25). Memory B cells, found mainly in the germinal centers of the lymph nodes, mount a very quick **secondary,** or **anamnestic,**[22] **response** (an-am-NESS-tic) if reexposed to the

same antigen. Plasma cells form within hours, so the IgG titer rises sharply and peaks within a few days. The response is so rapid that the antigen has little chance to exert a noticeable effect on the body, and no illness results. A low level of IgM is also secreted and quickly declines, but IgG remains elevated for weeks to years, conferring lasting protection. Memory doesn't last as long in humoral immunity, however, as it does in cellular immunity.

Table 21.4 summarizes many of the cellular and chemical agents involved in humoral and cellular immunity. **Table 21.5** compares the main features of humoral and cellular immunity. Remember that these two processes often occur simultaneously, and in conjunction with inflammation as a three-pronged attack on the same pathogen.

---

[22]*ana* = back; *mnes* = remember

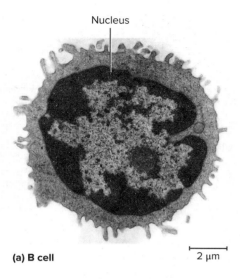

Nucleus

(a) B cell   2 µm

Mitochondria   Rough endoplasmic reticulum

(b) Plasma cell   2 µm

FIGURE 21.26 B Cell and Plasma Cell. (a) B cells have little cytoplasm and scanty organelles. (b) A plasma cell, which differentiates from a B cell, has an abundance of rough endoplasmic reticulum.

a, b: Don W. Fawcett/Science Source

? *What does this endoplasmic reticulum do in the plasma cell?*

BEFORE YOU GO ON

Answer the following questions to test your understanding of the preceding section:

23. What is the difference between a B cell and a plasma cell?

24. Describe four ways in which an antibody acts against an antigen.

25. Why does the secondary immune response prevent a pathogen from causing disease, while the primary immune response does not?

Antibodies (IgM)

(a)

Antigens

Antibody monomers

(b)

FIGURE 21.27 Agglutination and Precipitation by Antibodies. (a) Agglutination of foreign erythrocytes by IgM, a pentamer. (b) A precipitated antigen–antibody complex involving a free molecular antigen and an antibody monomer such as IgG.

FIGURE 21.28 The Primary and Secondary (Anamnestic) Responses in Humoral Immunity. (a) The primary response, seen on first exposure to a given antigen. (b) The secondary response, seen in an immune person on reexposure. The individual is exposed to antigen on day 0 in both cases. Note the differences in the speed of response, the height of the antibody titer, and the rate of decline in antibody titer.

TABLE 21.4	Agents of Adaptive Immunity
Cellular Agents	
B cells	Serve as antigen-presenting cells in humoral immunity; differentiate into antibody-secreting plasma cells
CD4 (T4) cells*	T lymphocytes with CD4 surface glycoproteins; helper and regulatory T cells
CD8 (T8) cells	T lymphocytes with CD8 surface glycoproteins; cytotoxic T cells
Cytotoxic T (killer T, T_C, or CD8) cells	Effectors of cellular immunity; directly attack and destroy enemy cells; produce perforin, granzymes, interferon, tumor necrosis factor, and other cytokines
Dendritic cells	Branched, mobile APCs of the skin, mucous membranes, and lymphatic tissues; internalize antigen, migrate to lymph nodes, and present it to T_H and T_C cells
Eosinophils*	Phagocytize and degrade Ag–Ab complexes
Helper T (T_H) cells*	Play a central regulatory role in innate, humoral, and cellular immunity; recognize antigen fragments displayed by APCs with MHC-II proteins; secrete interleukins that activate B, T_C, and NK cells, neutrophils, and macrophages
Macrophages*	Phagocytize pathogens and expended or damaged host cells; act as antigen-presenting cells (APCs)
Memory B cells	Activated B cells that do not immediately differentiate into plasma cells; act as a pool of B cells that can execute a quick secondary response upon reexposure to the same antigen that initially activated them
Memory T cells	Activated T cells that do not immediately differentiate into effector T cells; act as a pool of T cells that can execute a quick T cell recall response upon reexposure to the same antigen that initially activated them
Naive lymphocytes	Immunocompetent lymphocytes that are capable of responding to an antigen but have not yet encountered one
Plasma cells	Develop from B cells that have been activated by helper T cells; synthesize and secrete antibodies
Regulatory T (T_R) cells (T-regs)	T cells that inhibit other T cells from multiplying or secreting cytokines; important especially in preventing autoimmune diseases but may also play defensive roles against infection and cancer and protect the fetus from maternal immune responses
Reticular cells	Fixed cells in the stroma of the lymph nodes and other lymphatic organs; act as APCs in the thymus, with a role in maturation of T cells
Chemical Agents	
Antibody (Ab)	A gamma globulin produced by plasma cells in response to an antigen; counteracts antigen by means of complement fixation, neutralization of toxins, agglutination, or precipitation
Antigen (Ag)	Molecule capable of triggering an immune response; usually a protein, polysaccharide, glycolipid, or glycoprotein
Complement*	Group of plasma proteins that help to destroy pathogens by cytolysis, phagocytosis, immune clearance, or innate immunity (inflammation)
Granzyme	Proteolytic enzyme produced by NK and T_C cells; enters the pore made by perforins, degrades enzymes of the enemy cell, and induces apoptosis
Hapten (incomplete antigen)	Small molecule initially unable to trigger an immune response by itself but able to bind to host molecules and produce a complex that is antigenic; may subsequently activate an immune response without binding to a host molecule
Interleukin	Cytokine produced by leukocytes and macrophages to stimulate other leukocytes
Perforin	A protein produced by NK and T_C cells that binds to target cells, produces a hole, and admits granzymes into the cell
Tumor necrosis factor (TNF)	Cytokine secreted by T_C cells that activates macrophages and kills cancer cells

*These agents have additional roles in inflammation described in table 21.1.

TABLE 21.5	Some Comparisons Between Cellular and Humoral Immunity	
Characteristic	**Cellular Immunity**	**Humoral Immunity**
Disease agents	Intracellular viruses, bacteria, yeasts, and protozoans; parasitic worms; cancer cells; transplanted tissues and organs	Extracellular viruses, bacteria, yeasts, and protozoans; toxins, venoms, and allergens; mismatched RBCs
Effector cells	Cytotoxic T cells	Plasma cells (develop from B cells)
Other cells involved in attack	Helper T cells	Helper T cells
Antigen-presenting cells	B cells, macrophages, dendritic cells, nearly all cells	B cells
MHC proteins	MHC-I and MHC-II	MHC-II only
Chemical agents of attack	Perforins, granzymes, interferons, tumor necrosis factor	Antibodies, complement
Mechanisms of counteracting or destroying pathogens	Cytolysis, phagocytosis, apoptosis	Cytolysis, phagocytosis, immune clearance, inflammation, neutralization, agglutination, precipitation
Memory	T cell recall response	Secondary (anamnestic) response

21.6 Immune System Disorders

Expected Learning Outcomes

When you have completed this section, you should be able to

a. distinguish between the four classes of immune hypersensitivity and give an example of each;

b. explain the cause of anaphylaxis and distinguish local anaphylaxis from anaphylactic shock;

c. state some reasons immune self-tolerance may fail, and give examples of the resulting diseases; and

d. describe the pathology of immunodeficiency diseases, especially AIDS.

Because the immune system involves complex cellular interactions controlled by numerous chemical messengers, there are many points at which things can go wrong. The immune response may be too vigorous, too weak, or misdirected against the wrong targets. A few disorders are summarized here to illustrate the consequences.

21.6a Hypersensitivity

Hypersensitivity is an excessive, harmful immune reaction to antigens. It includes reactions to tissues transplanted from another person *(alloimmunity),* abnormal reactions to one's own tissues *(autoimmunity),* and **allergies,**[23] which are reactions to environmental antigens. Such antigens, called **allergens,** occur in mold, dust, pollen, vaccines, bee and wasp venoms, animal dander, toxins from poison ivy and other plants, and foods such as nuts, milk, eggs, and shellfish. Drugs such as penicillin, tetracycline, and insulin are allergenic to some people.

One classification system recognizes four kinds of hypersensitivity, distinguished by the types of immune agents (antibodies or T cells) involved and their methods of attack on the antigen. In this system, type I is also characterized as *acute (immediate)*

hypersensitivity because the response is very rapid, whereas types II and III are characterized as *subacute* because they exhibit a slower onset (1–3 hours after exposure) and last longer (10–15 hours). Type IV is a delayed cell-mediated response, whereas the other three are quicker antibody-mediated responses.

- **Type I (acute) hypersensitivity** includes the most common allergies. Some authorities use the word *allergy* for type I reactions only, and others use it for all four types. Type I is an IgE-mediated reaction that begins within seconds of exposure and usually subsides within 30 minutes, although it can be severe and even fatal. Allergens bind to IgE on the membranes of basophils and mast cells and stimulate them to secrete histamine and other inflammatory and vasoactive chemicals. These chemicals trigger glandular secretion, vasodilation, increased capillary permeability, smooth muscle spasms, and other effects. The clinical signs include local edema, mucus hypersecretion and congestion, watery eyes, a runny nose, hives (red itchy skin), and sometimes cramps, diarrhea, and vomiting. Some examples of type I hypersensitivity are food allergies and **asthma,**[24] a local inflammatory reaction to inhaled allergens (see Deeper Insight 21.3).

  Anaphylaxis[25] (AN-uh-fih-LAC-sis) is an immediate and intense type I reaction. Local anaphylaxis can be relieved with antihistamines. **Anaphylactic shock** is a severe, widespread acute hypersensitivity that occurs when an allergen such as bee venom or penicillin is introduced to the bloodstream of an allergic individual, or when a person ingests certain foods (such as peanuts) to which he or she is allergic. It is characterized by bronchoconstriction, dyspnea (labored breathing), widespread vasodilation, circulatory shock, and sometimes sudden death. Antihistamines are inadequate by themselves to counter anaphylactic shock, but epinephrine relieves the symptoms by dilating the bronchioles, increasing cardiac output, and restoring blood pressure. Fluid therapy and respiratory support are sometimes required.

---

[23]*allo* = altered; *erg* = action, reaction

[24]*asthma* = panting
[25]*ana* = against; *phylax* = prevention

DEEPER INSIGHT 21.3

CLINICAL APPLICATION

Asthma

Asthma, a type I hypersensitivity, is the most common chronic illness of children, especially boys. It is the leading cause of school absenteeism and childhood hospitalization in the United States. About half of all cases develop before age 10 and only 15% after age 40. In the United States, it affects about 5% of adults and up to 10% of children, and takes about 3,400 lives per year. Moreover, asthma is on the rise; there are many more cases and deaths now than there were decades ago.

In *allergic (extrinsic) asthma,* the most common form, a respiratory crisis is triggered by allergens in pollen, mold, animal dander, food, dust mites, or cockroaches. The allergens stimulate plasma cells to secrete IgE, which binds to mast cells of the respiratory mucosa. Reexposure to the allergen causes the mast cells to release a complex mixture of inflammatory chemicals, which trigger intense airway inflammation. *Nonallergic (intrinsic) asthma* isn't caused by allergens but can be triggered by infections, drugs, air pollutants, cold dry air, exercise, or emotions. This form is more common in adults than in children, but the effects are much the same.

Within minutes, the bronchioles constrict spasmodically *(bronchospasm),* causing severe coughing, wheezing, and sometimes suffocation. A second respiratory crisis often occurs 6 to 8 hours later. Interleukins attract eosinophils to the bronchial tissue, where they secrete proteins that paralyze the respiratory cilia, severely damage the epithelium, and

lead to scarring and extensive long-term damage to the lungs. The bronchioles also become edematous and plugged with thick, sticky mucus. People who die of asthmatic suffocation typically show airways so plugged with gelatinous mucus that they couldn't exhale. The lungs remain hyperinflated even at autopsy.

Asthma is treated with epinephrine and other β-adrenergic stimulants to dilate the airway and restore breathing, and with inhaled corticosteroids or nonsteroidal anti-inflammatory drugs to minimize airway inflammation and long-term damage. The treatment regimen can be very complicated, often requiring more than eight different medications daily, and compliance is therefore difficult for children and patients with low income or educational attainment.

Asthma runs in families and seems to result from a combination of heredity and environmental irritants. In the United States, asthma is most common, paradoxically, in two groups: (1) inner-city children who are exposed to crowding, poor sanitation, and poor ventilation, and who don't often go outside or get enough exercise; and (2) children from extremely clean homes, perhaps because they have had too little opportunity to develop normal immunities. Asthma is also more common in countries where vaccines and antibiotics are widely used. It is less common in developing countries and in farm children of the United States.

- **Type II (antibody-dependent cytotoxic) hypersensitivity** occurs when IgG or IgM attacks antigens bound to cell surfaces. The reaction leads to complement activation and either lysis or opsonization of the target cell. Macrophages phagocytize and destroy opsonized platelets, erythrocytes, or other cells. Examples of cell destruction by type II reactions are blood transfusion reactions and some drug reactions. In some other type II responses, an antibody binds to cell surface receptors and either interferes with their function as in myasthenia gravis, or overstimulates the cell as in toxic goiter. (See table 21.6 for further information on these diseases.)

- **Type III (immune complex) hypersensitivity** occurs when IgG or IgM forms antigen–antibody complexes that precipitate beneath the endothelium of the blood vessels or in other tissues. At the sites of deposition, these complexes activate complement and trigger intense inflammation, causing tissue destruction. Two examples of type III hypersensitivity are the autoimmune diseases acute glomerulonephritis and systemic lupus erythematosus, a widespread inflammation of the connective tissues (see table 21.6).

- **Type IV (delayed) hypersensitivity** is a cell-mediated reaction in which the signs appear about 12 to 72 hours after exposure. It begins when APCs in the lymph nodes display antigens to helper T cells, and these T cells secrete interferon and other cytokines that activate cytotoxic T cells and macrophages. The result is a mixture of nonspecific and immune responses. Type IV reactions include allergies to haptens in cosmetics and poison ivy; graft rejection; the tuberculosis skin test; and the beta cell destruction that causes type 1 diabetes mellitus.

21.6b Autoimmune Diseases

Autoimmune diseases are failures of self-tolerance—the immune system fails to distinguish self-antigens from foreign ones and produces **autoantibodies** that attack the body's own tissues. Autoimmunity is usually prevented by negative selection of developing T and B cells, but there are at least three reasons why self-tolerance may fail:

1. **Cross-reactivity.** Some antibodies against foreign antigens react to similar self-antigens. In rheumatic fever, for example, a streptococcus infection stimulates production of antibodies that react not only against the bacteria but also against antigens of the heart tissue. It often results in scarring and stenosis (narrowing) of the mitral and aortic valves.

2. **Abnormal exposure of self-antigens to the blood.** Some of our native antigens are normally not exposed to the blood. For example, a blood–testis barrier (BTB) normally isolates sperm cells from the blood. Breakdown of the BTB can cause sterility when sperm first form in adolescence and activate the production of autoantibodies.

3. **Change in the structure of self-antigens.** Viruses and drugs may change the structure of self-antigens and cause the immune system to perceive them as foreign. One theory of type 1 diabetes mellitus is that a viral infection alters the antigens of the insulin-producing beta cells of the pancreatic islets, which leads to an autoimmune attack on the cells.

Evidence is emerging that not all self-reactive T cells are eliminated by clonal deletion in the thymus. We all apparently have at least some T cells poised to attack our own tissues, but regulatory T (T_R) cells keep them in check and normally prevent autoimmune disease.

21.6c Immunodeficiency Diseases

In the foregoing diseases, the immune system reacts too vigorously or directs its attack against the wrong targets. In immunodeficiency diseases, by contrast, the immune system fails to respond vigorously enough.

Severe Combined Immunodeficiency Disease

Severe combined immunodeficiency disease (SCID) is a group of disorders caused by recessive alleles that result in a scarcity or absence of both T and B cells. Children with SCID are highly vulnerable to opportunistic infections and must live in protective enclosures. The most publicized case was David Vetter, who spent his life in sterile plastic chambers and suits **(fig. 21.29),** finally succumbing at age 12 to cancer triggered by a viral infection. Children with SCID are sometimes helped by transplants of bone marrow or fetal thymus, but in some cases the transplanted cells fail to survive and multiply, or transplanted T cells attack the patient's tissues (the *graft-versus-host response*). David contracted the fatal virus from his sister through a bone marrow transplant.

Acquired Immunodeficiency Syndrome

Other immunodeficiency diseases are nonhereditary and contracted after birth. The best-known example is **acquired immunodeficiency syndrome (AIDS),** a group of conditions in which infection with the **human immunodeficiency virus (HIV)** severely depresses the immune response.

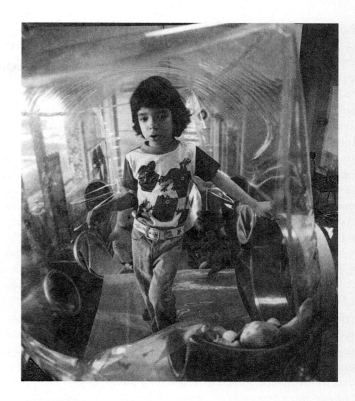

FIGURE 21.29 Boy with Severe Combined Immunodeficiency Disease. David Vetter lived with SCID from birth (1971) to age 12 and had to spend his entire life in sterile enclosures.

Source: Texas Children's Hospital

HIV **(fig. 21.30)** has an inner core consisting of a protein *capsid* that encloses two molecules of RNA, two molecules of an enzyme called *reverse transcriptase,* and a few other enzyme molecules. The capsid is enclosed in another layer of viral protein, the *matrix.* External to this is a *viral envelope* composed of phospholipids and glycoproteins derived from the host cell.

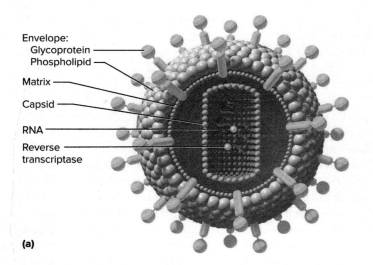

Envelope:
Glycoprotein
Phospholipid

Matrix

Capsid

RNA

Reverse
transcriptase

(a)

Protoplasmic
blebs of dying
T cell

Emerging
viruses

(b)

1 μm

FIGURE 21.30 The Human Immunodeficiency Virus (HIV). (a) Structure of the virus. (b) Viruses emerging from a dying helper T cell. Each virus can now invade a new helper T cell and produce a similar number of descendants.

b: NIBSC/SPL/Science Source

Like other viruses, HIV can be replicated only by a living host cell. It invades helper T (CD4) cells, dendritic cells, and macrophages. HIV adheres to a target cell by means of one of its envelope glycoproteins and "tricks" the target cell into internalizing it by receptor-mediated endocytosis. Within the host cell, reverse transcriptase uses the viral RNA as a template to synthesize DNA—the opposite of the usual process of genetic transcription. Viruses that carry out this RNA → DNA reverse transcription are called *retroviruses*.[26] The new DNA is inserted into the host cell's DNA, where it may lie dormant for months to years. When activated, however, it induces the host cell to produce new viral RNA, capsid proteins, and matrix proteins. As the new viruses emerge from the host cell (fig. 21.30b), they are coated with bits of the cell's plasma membrane, forming the new viral envelope. The new viruses then adhere to more host cells and repeat the process.

By destroying T_H cells, HIV strikes at a central coordinating agent of innate, humoral, and cellular immunity (see fig. 21.23). After an incubation period ranging from a few months to 12 years, the patient begins to experience flulike episodes of chills and fever as HIV attacks T_H cells. At first, antibodies against HIV are produced and the T_H count returns nearly to normal. As the virus destroys more and more cells, however, the signs and symptoms become more pronounced: night sweats, fatigue, headache, extreme weight loss, and lymphadenitis.

A normal T_H count is 600 to 1,200 cells/μL, but a criterion of AIDS is a count less than 200/μL. One can be HIV-positive for 10 years or longer without having AIDS. As T_H cells become severely depleted, however, a person falls vulnerable to opportunistic infections with such pathogens as *Toxoplasma* (a protozoan previously known mainly for causing birth defects), *Pneumocystis* (a group of respiratory fungi), herpes simplex virus, cytomegalovirus (which can cause blindness), and tuberculosis bacteria. White patches may appear in the mouth, caused by *Candida* (thrush) or Epstein–Barr[27] virus (leukoplakia). A cancer called Kaposi[28] sarcoma, rare in the general population but common in AIDS patients, originates in the endothelial cells of the blood vessels and causes bruiselike purple lesions visible in the skin, mouth, respiratory tract, and elsewhere (**fig. 21.31**).

Patients with AIDS show no response to standard skin tests for delayed hypersensitivity. Slurred speech, loss of motor and cognitive functions, and dementia may occur as HIV invades the brain by way of infected phagocytes (microglia) and induces them to release toxins that destroy neurons and astrocytes.

HIV is transmitted through blood, semen, vaginal secretions, and breast milk. It can be transmitted from mother to fetus through the placenta or from mother to infant during childbirth or nursing. HIV occurs in saliva and tears, but is apparently not transmitted by those fluids. The most common means of transmission are sexual intercourse (vaginal, anal, or oral), contaminated blood products, and drug injections with contaminated needles. Worldwide, about 75% of HIV infections are acquired

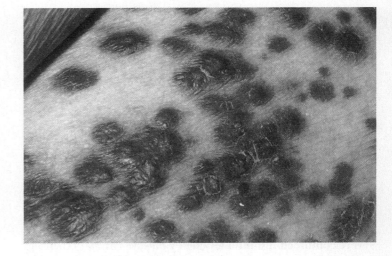

FIGURE 21.31 Kaposi Sarcoma. Typical skin lesions.
Source: National Cancer Institute (NCI)

through heterosexual, predominantly vaginal intercourse. In the United States, most cases occur in men who have sex with other men, but adolescents are the fastest-rising group of AIDS patients because of the exchange of unprotected sexual intercourse for drugs. The sharing of needles for drug use remains the chief means of transmission in urban ghettos. Since 1984, all blood donated for transfusion has been tested for HIV, and the risk of infection from transfusions is now less than 1%. HIV cannot be contracted by donating blood.

AIDS is not known to be transmitted through casual contact—for example, to family members, friends, coworkers, classmates, or medical personnel in charge of AIDS patients. It is not transmitted by kissing or by mosquitoes or other blood-sucking arthropods.

HIV survives poorly outside the human body. It is destroyed by laundering; dishwashing; exposure to heat (50°C [135°F] for at least 10 minutes); chlorination of swimming pools and hot tubs; and disinfectants such as bleach, Lysol, hydrogen peroxide, rubbing alcohol, and germicidal skin cleansers such as Betadine and Hibiclens. A properly used, undamaged latex condom is an effective barrier to HIV, but animal membrane condoms have gaps too large to block HIV transmission.

The AIDS epidemic triggered an effort of unprecedented intensity to find a vaccine or cure. HIV is a difficult pathogen to attack. Since it hides within host cells, it usually escapes recognition by the immune system. In the brain, it is protected by the blood–brain barrier. About 1% of HIV's genes mutate every year. Such rapid mutation is a barrier to both natural immunity and development of a vaccine. Even when immune cells do become sensitized to HIV, the virus soon evolves new surface antigens that escape recognition. The high mutation rate also would quickly make today's vaccine ineffective against tomorrow's strain of the virus.

The first anti-HIV drug approved by the Food and Drug Administration (FDA) was azidothymidine (AZT, or Retrovir), which inhibited reverse transcriptase and prolonged the lives of some HIV-positive individuals. In 1996, a family of drugs called protease inhibitors became available, typically used in a "triple cocktail" combining these with two reverse transcriptase inhibitors.

[26]*retr* = an acronym from reverse transcription
[27]M. A. Epstein (1921–), British physician; Y. M. Barr (1932–2016), British virologist
[28]Moritz Kaposi (1837–1902), Austrian physician

But by 1997, HIV had evolved resistance even to those drugs and they were failing in more than half of all patients. Two newer families of drugs for AIDS are *fusion inhibitors,* which interfere with the ability of HIV to invade host cells, and *integrase inhibitors,* which block the incorporation of HIV genes into human DNA. Today, at least 24 anti-HIV drugs are on the market, typically used in combinations of three or more. Such drug combinations have substantially reduced AIDS morbidity and mortality (disease and death) and reduced the hospital and hospice census. There is no cure for AIDS yet, but for those with access to antiretroviral therapy (ART), life expectancy at diagnosis is now close to that of uninfected persons. ART has saved the lives of millions of patients, but it requires lifelong drug therapy and a complicated regimen. Compliance is difficult for some, especially the homeless, drug abusers, the poorly nourished, and persons unable to pay for treatment. The major unresolved questions in AIDS therapy today are when to start drug treatment, which drugs to use, when to switch drugs, how to improve patient compliance, and how to make therapy available in impoverished nations where AIDS is especially rampant and people can't adhere well to complicated treatment regimens.

One obstacle to treatment and prevention is the lack of animal models for vaccine and drug research and development. Most animals aren't susceptible to HIV. The chimpanzee is an exception, but chimpanzees are difficult to maintain, and there are economic barriers and ethical controversies surrounding their use.

There also remain a number of unanswered questions about the basic biology of HIV. It is still unknown, for example, why there are such strikingly different patterns of heterosexual versus homosexual transmission in different countries, and why some people succumb so rapidly to infection but others can be HIV-positive for years without developing AIDS. AIDS remains a stubborn problem sure to challenge virologists and epidemiologists for many years to come.

We have surveyed the major classes of immune system disorders and a few particularly notorious immune diseases. A few additional lymphatic and immune system disorders are described in **table 21.6**. The effects of aging on the immune system are described in section 29.4a.

BEFORE YOU GO ON

Answer the following questions to test your understanding of the preceding section:

26. How does subacute hypersensitivity differ from acute hypersensitivity? Give an example of each.

27. Aside from the time required for a reaction to appear, how does delayed hypersensitivity differ from the acute and subacute types?

28. State some reasons why antibodies may begin attacking self-antigens to which they did not previously respond. What are these self-reactive antibodies called?

29. What is the distinction between a person who has an HIV infection and a person who has AIDS?

30. How does a reverse transcriptase inhibitor such as AZT slow the progress of AIDS?

TABLE 21.6	Some Disorders of the Lymphatic and Immune Systems
Contact dermatitis	A form of delayed hypersensitivity that produces skin lesions limited to the site of contact with an allergen or hapten; includes responses to poison ivy, cosmetics, latex, detergents, industrial chemicals, and some topical medicines.
Hives (urticaria[29])	An allergic skin reaction characterized by a "wheal-and-flare" reaction—white blisters (wheals) surrounded by reddened areas (flares), usually with itching. Caused by local histamine release in response to allergens. Can be triggered by food or drugs, but sometimes by nonimmunological factors such as cold, friction, or emotional stress.
Hodgkin[30] disease	A lymph node malignancy, with early symptoms including enlarged painful nodes, especially in the neck, and fever; often progresses to neighboring lymph nodes. Radiation and chemotherapy cure about 75% of patients.
Splenomegaly[31]	Enlargement of the spleen, sometimes without underlying disease but often indicating infections, autoimmune diseases, heart failure, cirrhosis, Hodgkin disease, and other cancers. The enlarged spleen may "hoard" erythrocytes, causing anemia, and may become fragile and subject to rupture.
Systemic lupus erythematosus[32]	Formation of autoantibodies against DNA and other nuclear antigens, resulting in accumulation of antigen–antibody complexes in blood vessels and other organs, where they trigger widespread connective tissue inflammation. Named for skin lesions once likened to a wolf bite. Causes fever, fatigue, joint pain, weight loss, intolerance of bright light, and a "butterfly rash" across the nose and cheeks. Death may result from renal failure.

You can find other lymphatic and immune system disorders described in the following places:

Rheumatoid arthritis in Deeper Insight 9.5; *myasthenia gravis* in Deeper Insight 11.4; *toxic goiter* in section 17.7a; *type 1 diabetes mellitus* in section 17.7e; *rheumatic fever* in Deeper Insight 19.3; *edema* in section 20.3d; *asthma* in Deeper Insight 21.3; *lymphadenitis* in section 21.1d; *allergy, anaphylaxis, severe combined immunodeficiency disease (SCID),* and *acquired immunodeficiency syndrome (AIDS)* in section 21.6; and *acute glomerulonephritis* in table 23.3.

[29]*urtica* = nettle
[30]Thomas Hodgkin (1798–1866), British physician
[31]*megaly* = enlargement
[32]*lupus* = wolf; *erythema* = redness

DEEPER INSIGHT 21.4

CLINICAL APPLICATION

Neuroimmunology and the Mind–Body Connection

Neuroimmunology is a relatively new branch of medicine concerned with the relationship between mind and body in health and disease. It is attempting especially to understand how a person's state of mind influences health and illness through a three-way communication between the nervous, endocrine, and immune systems.

The sympathetic nervous system issues nerve fibers to the spleen, thymus, lymph nodes, and aggregated lymphoid nodules of the ileum, where nerve fibers contact T cells, B cells, and macrophages. These immune cells have adrenergic receptors for norepinephrine and many other neurotransmitters such as neuropeptide Y, substance P, and vasoactive intestinal peptide (VIP). These neurotransmitters have been shown to influence immune cell activity in various ways. Epinephrine, for example, reduces the lymphocyte count and inhibits NK cell activity, thus suppressing both innate and adaptive immunity. Cortisol, another stress hormone, inhibits T cell and macrophage activity, antibody production, and the secretion of inflammatory chemicals. It also induces atrophy of the thymus, spleen, and lymph nodes and reduces the number of circulating lymphocytes, macrophages, and eosinophils. Thus, it is not surprising that prolonged stress increases susceptibility to illnesses such as infections and cancer. The existence of lymphatic vessels in the central nervous system, discovered only recently, has intriguing implications for direct effects of the brain on the immune system.

The immune system also sends messages to the nervous and endocrine systems. Immune cells synthesize numerous hormones and neurotransmitters that we normally associate with endocrine and nerve cells. B lymphocytes produce adrenocorticotropic hormone (ACTH) and enkephalins; T lymphocytes produce growth hormone, thyroid-stimulating hormone, luteinizing hormone, and follicle-stimulating hormone; and monocytes secrete prolactin, VIP, and somatostatin. The interleukins and tumor necrosis factor (TNF) secreted by immune cells produce feelings of fatigue and lethargy when we are sick, and stimulate the hypothalamus to secrete corticotropin-releasing hormone, leading to ACTH and cortisol secretion. It remains uncertain and controversial whether the quantities of some of these substances produced by immune cells are enough to have far-reaching effects on the body, but it seems increasingly possible that immune cells influence nervous and endocrine functions in ways that affect recovery from illness.

Although neuroimmunology has met with some skepticism among physicians, there is less and less room for doubt about the importance of a person's state of mind to immune function. People under stress, such as medical students during examination periods and people caring for relatives with Alzheimer disease, show more respiratory infections than other people and respond less effectively to hepatitis and flu vaccines. The attitudes, coping abilities, and social support systems of patients significantly influence survival time even in such serious diseases as AIDS and breast cancer. Women with breast cancer die at significantly higher rates if their husbands or partners cope poorly with stress and respond to the patient's illness with indifference, denial, or worse. Attitudes such as optimism, cheer, depression, resignation, or despair in the face of disease significantly affect immune function. Religious beliefs can also influence the prospect of recovery. Indeed, ardent believers in voodoo sometimes die just from the belief that someone has cast a spell on them. The stress of hospitalization can counteract the treatment one gives to a patient, and neuroimmunology has obvious implications for treating patients in ways that minimize their stress and thereby promote recovery.

Effects of the **LYMPHATIC AND IMMUNE SYSTEMS** on Other Organ Systems

ALL SYSTEMS

Lymphatic vessels drain excess tissue fluid, prevent edema, and remove cellular debris and pathogens. The immune system monitors all organs and provides defense against pathogens. Natural killer cells patrol the body and protect against cancer. All systems are subject to a variety of hypersensitivity and autoimmune disorders.

INTEGUMENTARY SYSTEM

The immune system includes dendritic cells of the epidermis, which guard against pathogens. The skin is a common site of inflammation. Autoimmunity causes pemphigus vulgaris and some other skin diseases. Hypersensitivity causes skin eruptions such as hives and dermatitis.

SKELETAL SYSTEM

Autoimmunity causes rheumatoid arthritis.

MUSCULAR SYSTEM

Autoimmunity causes myasthenia gravis, leading to muscle weakness and paralysis.

NERVOUS SYSTEM

Microglia cells scour the central nervous system for pathogens and tissue debris.

ENDOCRINE SYSTEM

Lymph transports some hormones. Autoimmunity is a factor in type 1 diabetes mellitus. Type II hypersensitivity causes toxic goiter of the thyroid. Immune cells secrete numerous hormones.

CIRCULATORY SYSTEM

Lymphatic vessels return fluid and lymphocytes to the bloodstream; the spleen disposes of expired RBCs; lymphatic organs filter pathogens and debris from the blood. Autoimmunity damages heart valves in rheumatic fever, and immune hypersensitivity causes circulatory failure in anaphylactic shock.

RESPIRATORY SYSTEM

Alveolar macrophages remove debris from the lungs; pulmonary lymphatic vessels are especially abundant and are needed to prevent fluid accumulation in the lungs. Immune hypersensitivity has effects ranging from respiratory congestion to asthma.

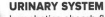

URINARY SYSTEM

Lymphatics absorb fluid and proteins in the kidneys, which is essential to enabling kidneys to concentrate the urine and conserve water. Autoimmunity causes acute glomerulonephritis.

DIGESTIVE SYSTEM

Lymphatic vessels called lacteals in the small intestine absorb nearly all dietary lipids and fat-soluble vitamins.

REPRODUCTIVE SYSTEM

Immunity to cells that are genetically different from other body cells requires the testes and ovaries to have barriers that protect sperm and eggs from immune destruction. A mismatch in Rh type can cause antibodies to attack fetal RBCs, causing hemolytic disease of the newborn.

STUDY GUIDE

▶ Assess Your Learning Outcomes

To test your knowledge, discuss the following topics with a study partner or in writing, ideally from memory.

21.1 The Lymphatic System

1. Functions and basic constituents of the lymphatic system
2. The definition, appearance, and composition of lymph
3. How lymph is produced; characteristics of lymphatic capillaries that allow cells and other large particles to enter the lymph
4. Lymphatic collecting vessels, trunks, and collecting ducts; the similarity of lymphatic vessels to some blood vessels; and their relationship to the lymph nodes
5. Names of the 11 lymphatic trunks and two collecting ducts; the body regions drained by them; and the two points at which lymph empties into the bloodstream
6. Mechanisms that propel the flow of lymph
7. Six types of cells found in lymphatic tissue, and their functions
8. The nature of diffuse lymphatic tissue and where it is found; what *MALT* stands for and where it is found
9. How lymphatic nodules differ from diffuse lymphatic tissue; the name of the clusters of lymphatic nodules found in the distal small intestine
10. How lymphatic organs differ from diffuse lymphatic tissue and lymphatic nodules; the two primary lymphatic organs and three secondary lymphatic organs, and why they are called this
11. Structure and function of red bone marrow
12. The location, gross anatomy, and histology of the thymus; the functional difference between its cortex and medulla; the functions of its epithelial cells; and the necessity of the thymus to immunity
13. Structure and function of lymph nodes; the significance of lymph nodes having both afferent and efferent lymphatic vessels, unlike any other lymphatic organs; the approximate number of lymph nodes and seven regions in which they are especially concentrated; and the meaning of *lymphadenitis* and *lymphadenopathy*
14. Types of tonsils, where they are located, and their structure and function; the most common cause of tonsillitis

15. Location, gross anatomy, and histology of the spleen; the difference between the red and white pulp; and functions associated with each type of pulp

21.2 Innate Immunity

1. Three lines of defense against pathogens
2. The definition of *immune system;* components of the system
3. Differences between innate and adaptive immunity; defining characteristics of the former
4. Three properties of the skin that make it an effective barrier to pathogens; the roles of organic acids and antimicrobial peptides in its barrier function
5. How the mucous membranes resist microbial invasion
6. Mechanisms by which each WBC type combats pathogens and illness, including the chemicals that some of them secrete in the performance of these functions
7. The one type of lymphocyte that is involved in innate immunity
8. Types of macrophages; their origin and functions
9. Interferons, their source, and how they oppose the spread of viruses
10. Complement proteins, their source, and how they are named
11. Three pathways of complement activation; how each is initiated; which pathways function in innate and adaptive immunity; and four mechanisms of pathogen destruction aided by complement
12. The actions of natural killer cells and the roles of perforins and granzymes in defense
13. Benefits of fever (pyrexia) and why the body's defenses may be compromised by antipyretic drugs; sources of pyrogens and how they trigger the onset of fever; the stages and course of a fever and how it combats pathogens; and the danger of excessive fever
14. Four cardinal signs of inflammation; chemicals that mobilize the body's defenses and initiate inflammation; and specific actions of these chemicals
15. The neutrophil actions of margination, diapedesis, chemotaxis, phagocytosis, the respiratory burst, and cytokine secretion
16. Other mechanisms of pathogen containment and destruction in inflammation

17. Examples of inflammatory cytokines and their roles
18. How hyperemia, bradykinin, and other factors account for the four cardinal signs of inflammation
19. The recruitment and action of macrophages
20. The formation, composition, and fate of pus
21. Mechanisms of tissue repair carried out after a pathogen is defeated

21.3 Adaptive Immunity—General Aspects

1. The three defining characteristics of adaptive immunity
2. Two basic forms of adaptive immunity, the differences between them, and the types of pathogens each one attacks
3. How adaptive immunity is classified as active or passive and as natural or artificial; which types result in immune memory and lasting protection, and which do not
4. The definition of *antigen* and the chemical characteristics of antigens
5. The role of the epitope in the antigenicity of a molecule
6. Haptens and how they become antigenic
7. The definition of *antibody* and where antibodies occur
8. Structure of the antibody monomer
9. The five antibody classes and their structural and functional differences
10. The life history of T cells including their origin; migration to the thymus; the sites, processes, and purposes of positive and negative selection; and dispersal of the naive lymphocyte pool
11. The meanings of *immunocompetence* and *self-tolerance,* how these are attained, and their importance to immunity
12. The life history of B cells including their origin, positive and negative selection, and dispersal
13. The necessity of antigen-presenting cells (APCs) to immunity; cell types that serve as APCs; the mechanism of antigen processing; and the role of MHC proteins in antigen presentation
14. Interleukins and their role in immunity

21.4 Cellular Immunity

1. Four classes of T lymphocytes involved in cellular immunity, and the function of each
2. Three fundamental stages of cellular immunity

STUDY GUIDE

3. What an APC does when it detects a foreign antigen; functional differences between MHC-I and MHC-II proteins; and the meaning of *MHC restriction*

4. Antigen recognition, costimulation, and clonal selection of a T cell; differentiation of selected T cells into effector cells and memory cells

5. How activated T_H cells stimulate neutrophils, NK cells, and macrophages

6. How activated T_C cells destroy target cells; the roles of interferons, perforin, granzymes, and tumor necrosis factor

7. Characteristics of immune memory and the T cell recall response in cellular immunity

21.5 Humoral Immunity

1. Similarities and differences between humoral and cellular immunity

2. How an immunocompetent B cell responds when it encounters a foreign antigen; the roles of MHC-II proteins and a T_H cell in its response

3. Clonal selection of activated B cell and differentiation of memory cells and plasma cells; the difference between a plasma cell and a B cell

4. Four mechanisms by which antibodies combat antigens

5. The secondary (anamnestic) response in humoral immune memory

21.6 Immune System Disorders

1. Three principal things that can go wrong with immune function

2. Hypersensitivity; names and characteristics of its four types, and examples of disorders of each type

3. The basic cause of autoimmune diseases; what normally prevents them; and three reasons why an autoimmune disease may appear, with an example of each

4. The basic cause of immunodeficiency diseases and the specific cause of severe combined immunodeficiency disease (SCID)

5. The pathology of acquired immunodeficiency syndrome (AIDS), including the structure of the human immunodeficiency virus (HIV); its mode of action; its effect on helper T cell count; the diseases to which AIDS makes a person more susceptible; how HIV is and is not transmitted; and treatment approaches to AIDS

▶ Testing Your Recall

Answers in Appendix A

1. The only lymphatic organ with both afferent and efferent lymphatic vessels is
   a. the spleen.
   b. a lymph node.
   c. a tonsil.
   d. a lymphatic follicle.
   e. the thymus.

2. Which of the following cells are involved in innate immunity but not in adaptive immunity?
   a. helper T cells
   b. cytotoxic T cells
   c. natural killer cells
   d. B cells
   e. plasma cells

3. The respiratory burst is used by _____ to kill bacteria.
   a. neutrophils
   b. basophils
   c. mast cells
   d. NK cells
   e. cytotoxic T cells

4. Which of these is a macrophage?
   a. a microglial cell
   b. a plasma cell
   c. a reticular cell
   d. a helper T cell
   e. a mast cell

5. The cytolytic action of the complement system is most similar to the action of
   a. interleukin-1.
   b. platelet-derived growth factor.
   c. granzymes.
   d. perforin.
   e. IgE.

6. _____ become antigenic by binding to larger host molecules.
   a. Epitopes
   b. Haptens
   c. Interleukins
   d. Pyrogens
   e. Cell-adhesion molecules

7. Which of the following correctly states the order of events in humoral immunity? Let
   1 = antigen display,
   2 = antibody secretion,
   3 = secretion of interleukin,
   4 = clonal selection, and
   5 = endocytosis of an antigen.
   a. 3–4–1–5–2
   b. 5–3–1–2–4
   c. 3–5–1–4–2
   d. 5–3–1–4–2
   e. 5–1–3–4–2

8. The cardinal signs of inflammation include all of the following *except*
   a. redness.
   b. swelling.
   c. heat.
   d. fever.
   e. pain.

9. A helper T cell can bind only to another cell that has
   a. MHC-II proteins.
   b. an epitope.
   c. an antigen-binding site.
   d. a complement-binding site.
   e. a CD4 protein.

10. Which of the following results from a lack of self-tolerance?
    a. SCID
    b. AIDS
    c. systemic lupus erythematosus
    d. anaphylaxis
    e. asthma

11. Any microorganism capable of causing disease is called a/an _____.

12. Mucous membranes contain an antibacterial enzyme called _____.

13. _____ is a condition in which one or more lymph nodes are swollen and painful to the touch.

STUDY GUIDE

14. The movement of leukocytes through a capillary or venule wall is called _____.

15. In the process of _____, complement proteins coat bacteria and serve as binding sites for phagocytes.

16. Any substance that triggers a fever is called a/an _____.

17. The chemical signals produced by leukocytes to stimulate other leukocytes are called _____.

18. Part of an antibody called the _____ binds to part of an antigen called the _____.

19. Self-tolerance results from a process called _____, in which lymphocytes programmed to react against self-antigens die.

20. Any disease in which antibodies attack one's own tissues is called a/an _____ disease.

▶ Building Your Medical Vocabulary

Answers in Appendix A

State a meaning of each word element, and give a medical term from this chapter that uses it or a slight variation of it.

1. ana-

2. crino-

3. extra-

4. -genous

5. immuno-

6. kino-

7. lympho-

8. -megaly

9. -pathy

10. pyro-

▶ What's Wrong with These Statements?

Answers in Appendix A

Briefly explain why each of the following statements is false, or reword it to make it true.

1. Some bacteria employ lysozyme to liquefy the tissue gel and make it easier for them to get around.

2. The thymus is the principal birthplace of all lymphocytes.

3. Interferons are named that because they interfere with the mechanisms of inflammation.

4. T lymphocytes are involved only in cell-mediated immunity.

5. The fundamental purpose of negative selection is to eliminate T cells that cannot recognize or respond to foreign antigens.

6. One function of the thymus and spleen is to filter incoming lymph and remove pathogens and other impurities from it.

7. All mechanisms of antibody action ultimately depend on complement fixation as an intermediate step.

8. Anyone who is HIV-positive is considered to have AIDS.

9. Anergy is a frequent cause of autoimmune diseases.

10. B cells that circulate freely in the bloodstream are called plasma cells.

▶ Testing Your Comprehension

1. David Vetter, the boy with SCID discussed in section 21.6c, died of infection with the Epstein–Barr virus, which had lurked unsuspected in his sister's bone marrow. (a) Why was David's sister chosen as a bone marrow donor rather than someone else? (b) Considering that the virus was fatal to David, why had it not caused any illness in his sister and, indeed, existed in her with no one being aware of it? (c) Do you think a cord blood transfusion (see Deeper Insight 18.3) may have been a better option for David than the marrow transplant? Discuss your reasoning.

2. In treating a woman for malignancy in the right breast, the surgeon removes some of her axillary lymph nodes. Following surgery, the patient experiences edema of her right arm. Explain why.

3. A girl with a defective heart receives a new heart transplanted from another child who was killed in an accident. The patient is given an antilymphocyte serum containing antibodies against her lymphocytes. The transplanted heart is not rejected, but the patient dies of an overwhelming bacterial infection. Explain why the antilymphocyte serum was given and why the patient was so vulnerable to infection.

4. A burn research center uses mice for studies of skin grafting. To prevent graft rejection, the mice are thymectomized at birth. Even though B cells don't develop in the thymus, these mice show no humoral immune response and are very susceptible to infection. Explain why the removal of the thymus would improve the success of skin grafts but adversely affect humoral immunity.

5. Contrast the structure of a B cell with that of a plasma cell, and explain how their structural difference relates to their functional difference.

CHAPTER

22

THE RESPIRATORY SYSTEM

The bronchial trees with colors indicating bronchopulmonary segments

Mediscan/Alamy

CHAPTER OUTLINE

22.1 Anatomy of the Respiratory System

    22.1a The Nose

    22.1b The Pharynx

    22.1c The Larynx

    22.1d The Trachea

    22.1e The Lungs and Bronchial Tree

    22.1f The Pleurae

22.2 Pulmonary Ventilation

    22.2a The Respiratory Muscles

    22.2b Neural Control of Breathing

    22.2c Pressure, Resistance, and Airflow

    22.2d Alveolar Ventilation

    22.2e Measurement of Pulmonary Ventilation

    22.2f Variations in the Respiratory Rhythm

22.3 Gas Exchange and Transport

    22.3a Composition of Air

    22.3b Alveolar Gas Exchange

    22.3c Gas Transport

    22.3d Systemic Gas Exchange

    22.3e Alveolar Gas Exchange Revisited

    22.3f Adjustment to the Metabolic Needs of Tissues

    22.3g Blood Gases and the Respiratory Rhythm

22.4 Respiratory Disorders

    22.4a Oxygen Imbalances

    22.4b Chronic Obstructive Pulmonary Diseases

    22.4c Smoking and Lung Cancer

Connective Issues

Study Guide

DEEPER INSIGHTS

22.1 Clinical Application: Tracheotomy

22.2 Clinical Application: Ondine's Curse

22.3 Medical History: The Flight of the Zenith

22.4 Clinical Application: Carbon Monoxide Poisoning

22.5 Clinical Application: Diving Physiology and Decompression Sickness

Anatomy & Physiology
Revealed® 4.0

Module 11: Respiratory System

BRUSHING UP

- The study of respiratory anatomy will draw on your knowledge of mucous and serous membranes (see section 5.5c); parietal and visceral layers of thoracic membranes (see atlas A, section A.3b); and simple squamous and ciliated pseudostratified columnar epithelia (see table 5.2).

- In reading on pulmonary ventilation, you may wish to refer back to the primary muscles of respiration (see table 10.4) and the accessory muscles that aid in breathing (see tables 10.5, 10.6, 10.7).

- Brushing up on the medulla oblongata and pons of the brainstem will help in understanding the respiratory control centers located there (see sections 14.3a, 14.3b).

- Respiratory airflow and gas exchange hinge on principles of flow down gradients (see section 1.6e) and effects of pressure and resistance on the flow of fluids (see section 19.5a).

- Diffusion and the factors that affect its rate are central to the exchange of gases in the lungs and peripheral tissues (see section 3.3b).

- Knowledge of hemoglobin structure is necessary for understanding oxygen transport in the blood (see section 18.2b).

Breath represents life. The first breath of a baby and the last gasp of a dying person are two of the most dramatic moments of human experience. But why do we breathe? It comes down to the fact that most of our metabolism directly or indirectly requires ATP. Most ATP synthesis requires oxygen and generates carbon dioxide—thus driving the need to breathe in order to supply the former and eliminate the latter. The respiratory system consists essentially of tubes that deliver air to the lungs, where oxygen diffuses into the blood and carbon dioxide diffuses out.

The respiratory and cardiovascular systems collaborate to deliver oxygen to tissues throughout the body and transport carbon dioxide to the lungs for elimination. Not only do these two systems have a close spatial relationship in the thoracic cavity, but they also have such a close functional relationship that they're often jointly called the *cardiopulmonary system.* A disorder that affects the lungs has direct and pronounced effects on the heart, and vice versa. As discussed in the next two chapters, the respiratory system also collaborates closely with the urinary system to regulate the body's acid–base balance, which is why we consider these systems consecutively in this group of chapters.

22.1   Anatomy of the Respiratory System

Expected Learning Outcomes

When you have completed this section, you should be able to

a. state the functions of the respiratory system;

b. name and describe the organs of this system;

c. trace the flow of air from the nose to the pulmonary alveoli; and

d. relate the function of any portion of the respiratory tract to its gross and microscopic anatomy.

The term *respiration* can mean ventilation of the lungs (breathing) or the use of oxygen in cellular metabolism. In this chapter, we are concerned with the first process. Cellular respiration was introduced in chapter 2 and is considered more fully in chapter 26.

The **respiratory system** is an organ system that rhythmically takes in air and expels it from the body, thereby supplying the body with oxygen and expelling the carbon dioxide that it generates. However, it has a broader range of functions than is commonly supposed:

1. **Gas exchange.** It provides for oxygen and carbon dioxide exchange between the blood and air.

2. **Communication.** It serves for speech and other vocalization (laughing, crying).

3. **Olfaction.** It provides the sense of smell, which is important in social interactions, food selection, and avoiding danger (such as a gas leak or spoiled food).

4. **Acid–base balance.** By eliminating CO_2, it helps to control the pH of the body fluids. Excess CO_2 reacts with water and generates carbonic acid; therefore, if respiration doesn't keep pace with CO_2 production, acid accumulates and the body fluids have an abnormally low pH *(acidosis).*

5. **Blood pressure regulation.** The lungs carry out a step in synthesizing *angiotensin II,* which helps to regulate blood pressure.

6. **Platelet production.** More than half of one's blood platelets are made by megakaryocytes in the lungs (not in the bone marrow).

7. **Blood and lymph flow.** Breathing creates pressure gradients between the thorax and abdomen that promote the flow of lymph and venous blood.

8. **Blood filtration.** The lungs filter small blood clots from the bloodstream and dissolve them, preventing clots from obstructing more vital pathways such as the coronary, cerebral, and renal circulation.

9. **Expulsion of abdominal contents.** Breath-holding and abdominal contraction help to expel abdominal contents during urination, defecation, and childbirth.

The principal organs of the respiratory system are the nose, pharynx, larynx, trachea, bronchi, and lungs (**fig. 22.1**). Within the lungs, air flows along a dead-end pathway consisting essentially of bronchi → bronchioles → alveoli (with some refinements to be introduced later). Incoming air stops in the *alveoli* (millions of tiny, thin-walled air sacs), exchanges gases with the bloodstream through the alveolar wall, and then flows back out.

The **conducting zone** of the respiratory system consists of those passages that serve only for airflow, essentially from the nostrils through the major bronchioles. The walls of these passages

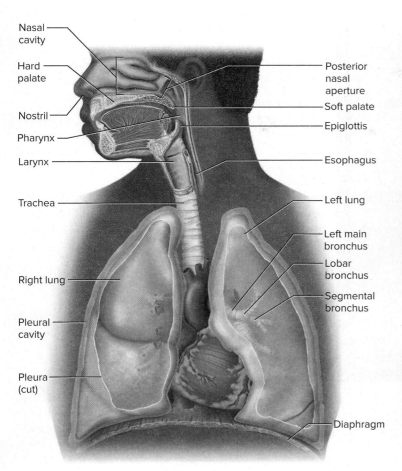

FIGURE 22.1 The Respiratory System. APR

(a)

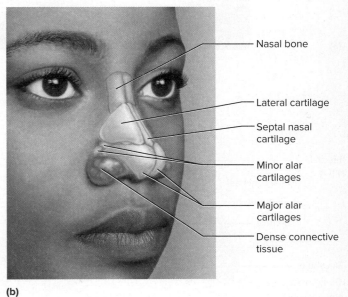

(b)

FIGURE 22.2 Anatomy of the Nasal Region. (a) Surface anatomy. (b) Connective tissues that shape the nose.

Joe DeGrandis/McGraw-Hill Education

are too thick for adequate diffusion of oxygen from the air into the blood. The **respiratory zone** consists of the alveoli and other gas-exchange regions of the distal airway. The airway from the nose through the larynx is often called the **upper respiratory tract** (that is, the respiratory organs in the head and neck), and the regions from the trachea through the lungs compose the **lower respiratory tract** (the respiratory organs of the thorax). However, these are inexact terms and various authorities place the dividing line between the upper and lower tracts at different points.

22.1a The Nose

The **nose** has several functions: It warms, cleanses, and humidifies inhaled air; it detects odors; and it serves as a resonating chamber that amplifies the voice. It extends from a pair of anterior openings called the **nostrils,** or **nares** (NAIR-eze) (singular, *naris*), to a pair of posterior openings called the **posterior nasal apertures,** or **choanae**[1] (co-AH-nee).

The facial part of the nose is shaped by bone and hyaline cartilage. Its superior half is supported by a pair of small nasal bones medially and the maxillae laterally. The inferior half is supported by the **lateral** and **alar cartilages (fig. 22.2).** By palpating your own nose, you can easily find the boundary between the bone above and

the more flexible cartilage below. The flared portion on each side of the lower end of the nose, called the **ala nasi**[2] (AIL-ah NAZE-eye), is shaped by the alar cartilages and dense connective tissue.

The internal chamber of the nose, called the **nasal cavity,** is divided into right and left halves called **nasal fossae** (FAW-see) **(fig. 22.3).** The dividing wall is a vertical plate, the **nasal septum,** composed of bone and hyaline cartilage. The vomer forms the inferior part of the septum, the perpendicular plate of the ethmoid bone forms its superior part, and the *septal cartilage* forms its anterior part. The ethmoid and sphenoid bones compose the roof of

[1]*choana* = funnel

[2]*ala* = wing; *nasi* = of the nose

(a)

(b)

(c)

FIGURE 22.3 Anatomy of the Upper Respiratory Tract. (a) Median section of the head. (b) Internal anatomy. (c) The nasal septum and regions of the pharynx.

a: Rebecca Gray/McGraw-Hill Education

❓ *Draw a line across part (a) of this figure to indicate the boundary between the upper and lower respiratory tract.*

the nasal cavity, and the hard palate forms its floor. The palate separates the nasal cavity from the oral cavity and allows you to breathe while chewing food. The paranasal sinuses and the naso-lacrimal ducts of the orbits drain into the nasal cavity (see figs. 8.8, 16.24, respectively).

The nasal cavity begins with a small dilated chamber called the **vestibule** just inside the nostril, bordered by the ala nasi. This space is lined with stratified squamous epithelium like the facial skin, and has stiff **guard hairs,** or **vibrissae** (vy-BRISS-ee), that block debris from entering the nose. Posterior to the vestibule, the nasal cavity expands into a much larger chamber, but it doesn't have much open space. Most of it is occupied by three folds of tissue—the **superior, middle,** and **inferior nasal conchae**[3] (CON-kee), or **turbinates**—that project from the lateral walls toward the septum (fig. 22.3; see also fig. 8.7 for a frontal view). Beneath each concha is a narrow air passage called a **meatus** (me-AY-tus). The narrowness of these passages and the turbulence caused by the conchae ensure that most air contacts the mucous membrane on its way through. As it does, most dust in the air sticks to the mucus and the air picks up moisture and heat from the mucosa. The conchae thus enable the nose to cleanse, warm, and humidify the air more effectively than if the air had an unobstructed flow through a cavernous space.

Past the vestibule, the mucosa (mucous membrane) of the nasal cavity consists of a ciliated pseudostratified columnar epithelium overlying a loose connective tissue lamina propria (see figs. 3.6, 3.31). Over most of the mucosa, the epithelium is called the **respiratory epithelium.** Its **ciliated cells** are capped with a fringe of about 200 motile cilia per cell and coated with a layer of mucus. The second most abundant cells of the epithelium are the wine-glass-shaped **goblet cells,** which secrete most of the mucus. In lesser numbers, the respiratory epithelium also contains endocrine cells, chemosensory *brush cells*, and basal stem cells.

Inhaled dust, pollen, bacteria, and other foreign matter are trapped in the sticky blanket of mucus covering the epithelium. The cilia of the epithelium beat in waves that drive this debris-laden mucus posteriorly to the pharynx, where it is swallowed. The particulate debris is either digested or passes through the digestive tract rather than contaminating the lungs.

A small area of nasal mucosa has an **olfactory epithelium,** concerned with the sense of smell. It covers about 5 cm[2] in the roof of the nasal fossa and adjacent parts of the septum and superior concha. Its structure and function are detailed in section 17.2b (see fig. 17.7). A notable contrast with the respiratory epithelium is that the cilia of the olfactory epithelium are immobile. They lie flattened against the mucosal surface like a plate of spaghetti noodles and serve to bind odor molecules, not to propel mucus.

The lamina propria of the nasal cavity is loose (areolar) connective tissue. In the respiratory mucosa, it contains glands that supplement the mucus produced by the goblet cells. In the olfactory mucosa, the lamina propria has large serous **olfactory glands.** These secrete a watery serous fluid that bathes the olfactory cilia and facilitates the diffusion of odor molecules from inhaled air to their receptors on the cilia.

The lamina propria contains large blood vessels that help to warm the air. The inferior concha has an especially extensive venous plexus called the **erectile tissue (swell body).** Every 30 to 60 minutes, the erectile tissue on one side swells with blood and restricts airflow through that fossa. Most air is then directed through the other nostril, allowing the engorged side time to recover from drying. Thus, the preponderant flow of air shifts between the right and left nostrils once or twice each hour. If one nostril is blocked and the other nasal fossa is over-ventilated for several days, its pseudostratified columnar epithelium changes to stratified squamous, which better resists drying. This is an example of the *metaplasia* explained in section 5.6b.

22.1b The Pharynx

The **pharynx** (FAIR-inks, FAR-inks) is a muscular funnel extending about 13 cm (5 in.) from the posterior nasal apertures to the larynx. It has three regions: the *nasopharynx, oropharynx,* and *laryngopharynx* (fig. 22.3c).

The **nasopharynx** is distal to the posterior nasal apertures and superior to the soft palate. It receives the auditory (pharyngotympanic or eustachian) tubes from the middle ears and houses the pharyngeal tonsil. Inhaled air turns 90° downward as it passes through the nasopharynx. Relatively large particles (>10 μm) generally can't make the turn because of their inertia. They collide with the wall of the nasopharynx and stick to the mucosa near the tonsil, which is well positioned to respond to airborne pathogens.

The **oropharynx** is a space between the posterior margin of the soft palate and the epiglottis.

The **laryngopharynx** (la-RIN-go-FAIR-inks) lies mostly posterior to the larynx, extending from the superior margin of the epiglottis to the inferior margin of the cricoid cartilage. The esophagus begins at that point.

The nasopharynx passes only air and is lined by pseudostratified columnar epithelium, whereas the oropharynx and laryngopharynx pass air, food, and drink and are lined by more abrasion-resistant stratified squamous epithelium. Muscles of the pharynx play necessary roles in swallowing and speech.

22.1c The Larynx

The **larynx** (LAIR-inks) is a cartilaginous chamber about 4 cm (1.5 in.) long **(fig. 22.4).** Its primary function is to keep food and drink out of the airway, but it evolved the additional role of sound production *(phonation)* in many animals; hence, we colloquially think of it as the "voice box."

The superior opening of the larynx is guarded by a flap of tissue called the **epiglottis**[4] just posterior to the root of the tongue. At rest, the epiglottis stands almost vertically. During swallowing, however, *extrinsic muscles* of the larynx pull the larynx upward toward the epiglottis, the tongue pushes the epiglottis downward to meet it, and the epiglottis closes the airway and directs food and drink into the esophagus behind it. The *vestibular folds* of the larynx, discussed shortly, play a greater role in keeping food and drink out of the airway, however.

---

[3]*concha* = seashell

[4]*epi* = above, upon; *glottis* = back of the tongue

FIGURE 22.4 Anatomy of the Larynx. (a) Anterior aspect. (b) Posterior aspect. (c) Median section with most muscles removed in order to show the cartilages. **APR**

❓ *Which three cartilages in this figure are more mobile than any of the others?*

In infants, the larynx is relatively high in the throat and the epiglottis touches the soft palate. This creates a more or less continuous airway from the nasal cavity to the larynx and allows an infant to breathe continually while swallowing. The epiglottis deflects milk away from the airstream, like rain running off a tent while it remains dry inside. By age 2, the root of the tongue becomes more muscular and forces the larynx to descend to a lower position. It then becomes impossible to breathe and swallow at the same time without choking.

The framework of the larynx consists of nine cartilages. The first three are solitary and relatively large. The most superior one, the **epiglottic cartilage,** is a spoon-shaped supportive plate of elastic cartilage in the epiglottis. The largest, the **thyroid[5] cartilage,** is named for its shieldlike shape. It broadly covers the anterior and lateral aspects of the larynx. The "Adam's apple" is an anterior peak of the thyroid cartilage called the *laryngeal prominence.* Testosterone stimulates the growth of this prominence, which is therefore larger in males than in females. Inferior to the thyroid cartilage is a ringlike **cricoid[6] cartilage** (CRY-coyd). The thyroid and cricoid cartilages essentially constitute the "box" of the voice box.

The remaining cartilages are smaller and occur in three pairs. Posterior to the thyroid cartilage are the two **arytenoid[7] cartilages** (AR-ih-TEE-noyd), and attached to their upper ends is a pair of little horns, the **corniculate[8] cartilages** (cor-NICK-you-late). The arytenoid and corniculate cartilages function in speech, as explained shortly. A pair of **cuneiform[9] cartilages** (cue-NEE-ih-form) supports the soft tissues between the arytenoids and the epiglottis.

A group of fibrous ligaments binds the cartilages of the larynx together and forms a suspension system for the upper airway. A broad sheet called the **thyrohyoid membrane** suspends the larynx from the hyoid bone above it. Below, the **cricothyroid ligament** suspends the cricoid cartilage from the thyroid cartilage. This is the most clinically important of all these ligaments, as this is where emergency incisions are made in a *tracheotomy* to restore breathing when the airway above it is obstructed (see Deeper Insight 22.1). The **cricotracheal ligament** suspends the trachea from the cricoid cartilage. All of these are collectively called the *extrinsic ligaments* because they link the larynx to other organs. The *intrinsic ligaments* are contained entirely within the larynx and link its nine cartilages to each other; they include ligaments of the vocal cords and vestibular folds.

The interior wall of the larynx has two folds on each side that stretch from the thyroid cartilage in front to the arytenoid cartilages in back. The superior **vestibular folds** (fig. 22.4c) play no role in speech but close the larynx during swallowing. They are supported by the **vestibular ligaments.** The inferior **vocal cords (vocal folds)** produce sound when air passes between them. They

---

[5]*thyr* = shield; *oid* = resembling
[6]*crico* = ring; *oid* = resembling
[7]*aryten* = ladle; *oid* = resembling

[8]*corni* = horn; *cul* = little; *ate* = possessing
[9]*cune* = wedge; *form* = shape

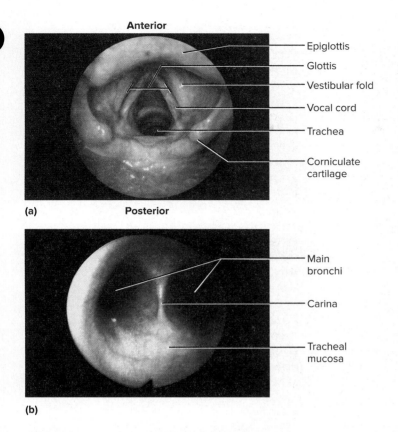

Anterior

Epiglottis
Glottis
Vestibular fold
Vocal cord
Trachea
Corniculate cartilage

(a)

Posterior

Main bronchi
Carina
Tracheal mucosa

(b)

FIGURE 22.5 Endoscopic Views of the Respiratory Tract.
(a) Superior view of the larynx, seen with a laryngoscope. (b) Lower end of the trachea, where it forks into the two primary bronchi, seen with a bronchoscope.

a: CNRI/Science Photo Library; b: BSIP/Newscom

contain the **vocal ligaments** and are covered with stratified squamous epithelium, best suited to endure vibration and contact between the cords. The vocal cords and the opening between them are collectively called the **glottis (fig. 22.5a).**

The walls of the larynx are quite muscular. The superficial *extrinsic muscles* connect the larynx to the hyoid bone and elevate the larynx during swallowing. Also called the *infrahyoid group,* they are named and described in table 10.2.

The deeper *intrinsic muscles* control the vocal cords by pulling on the corniculate and arytenoid cartilages, causing the cartilages to pivot. Depending on their direction of rotation, the arytenoid cartilages abduct or adduct the vocal cords **(fig. 22.6).** Air forced between the adducted vocal cords vibrates them, producing a high-pitched sound when the cords are relatively taut and a lower-pitched sound when they are more slack. In adult males, the vocal cords are usually longer and thicker, vibrate more slowly, and produce lower-pitched sounds than in females. Loudness is determined by the force of the air passing between the vocal cords. Although the vocal cords alone produce sound, they don't produce intelligible speech; some anatomists have likened their sound to a hunter's duck call. The crude sounds from the larynx are formed into words by actions of the pharynx, oral cavity, tongue, and lips.

22.1d The Trachea

The **trachea** (TRAY-kee-uh), or "windpipe," is a tube about 12 cm (4.5 in.) long and 2.5 cm (1 in.) in diameter, anterior to the esophagus **(fig. 22.7a).** It is supported by 16 to 20 C-shaped rings of hyaline cartilage. The trachea is named for the corrugated texture imparted

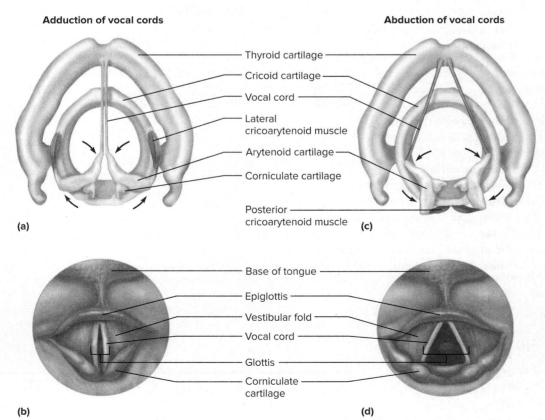

Adduction of vocal cords

Abduction of vocal cords

Thyroid cartilage
Cricoid cartilage
Vocal cord
Lateral cricoarytenoid muscle
Arytenoid cartilage
Corniculate cartilage
Posterior cricoarytenoid muscle

Anterior
Posterior

(a)
(c)

Base of tongue
Epiglottis
Vestibular fold
Vocal cord
Glottis
Corniculate cartilage

(b)
(d)

FIGURE 22.6 Action of Some of the Intrinsic Laryngeal Muscles on the Vocal Cords.
(a) Adduction of the vocal cords by the *lateral cricoarytenoid muscles.* (b) Adducted vocal cords seen with the laryngoscope. (c) Abduction of the vocal cords by the *posterior cricoarytenoid muscles.* (d) Abducted vocal cords seen with the laryngoscope. The intrinsic muscles are much more numerous and complex than these two pairs isolated for illustration.

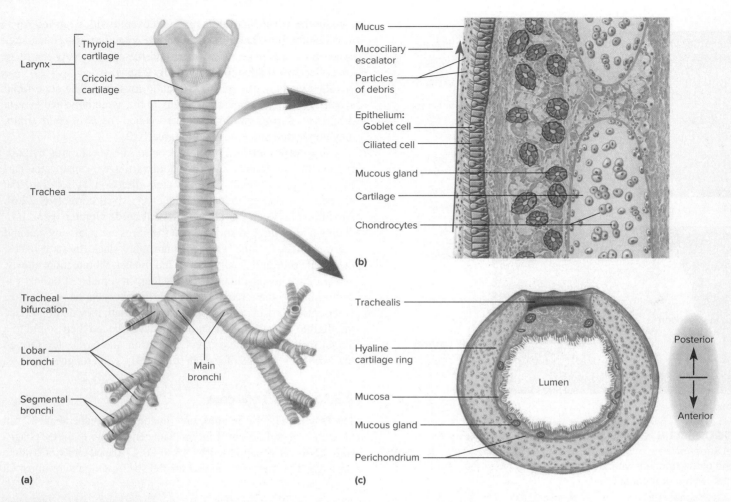

FIGURE 22.7 Anatomy of the Lower Respiratory Tract. (a) Anterior view. (b) Longitudinal section of the trachea showing the action of the mucociliary escalator. (c) Cross section of the trachea showing the C-shaped tracheal cartilage. **APR**

 Why do inhaled objects more often go into the right main bronchus than into the left?

by these rings;[10] you should be able to palpate a few of these between your larynx and sternum. Like the wire spiral in a vacuum cleaner hose, the cartilage rings reinforce the trachea and keep it from collapsing when you inhale. The open part of the C faces posteriorly, where it is spanned by a smooth muscle, the **trachealis (fig. 22.7c).** The gap in the C allows room for the esophagus to expand as swallowed food passes by. The trachealis contracts or relaxes to adjust airflow.

The inner lining of the trachea is a pseudostratified columnar epithelium composed mainly of mucus-secreting goblet cells, ciliated cells, and short basal stem cells **(figs. 22.7b, 22.8).** The mucus traps inhaled particles and the upward beating of the cilia drives the debris-laden mucus toward the pharynx, where it is swallowed. This mechanism of debris removal is called the **mucociliary escalator.**

The connective tissue beneath the tracheal epithelium contains lymphatic nodules, mucous and serous glands, and the tracheal cartilages. The outermost layer of the trachea, called the **adventitia,** is fibrous connective tissue that blends into the adventitia of other organs of the mediastinum, especially the esophagus.

[10]*trache* = rough

DEEPER INSIGHT 22.1
CLINICAL APPLICATION

Tracheotomy

The functional importance of the nasal cavity becomes especially obvious when it is bypassed. If the upper airway is obstructed, it may be necessary to make a temporary opening in the trachea inferior to the larynx and insert a tube to allow airflow—a procedure called *tracheotomy.* (A permanent opening is called a *tracheostomy.*) Life-saving tracheotomies are performed hundreds of times per day in the United States alone. They prevent asphyxiation, but the inhaled air bypasses the nasal cavity and thus is not humidified. If the opening is left for long, the mucous membranes of the respiratory tract dry out and become encrusted, interfering with the clearance of mucus from the tract and promoting infection. When a patient is on a ventilator and air is introduced directly into the trachea *(intubation)*, the air must be filtered and humidified by the apparatus to prevent respiratory tract damage.

Cilia

Goblet cell

4 µm

FIGURE 22.8 The Tracheal Epithelium Showing Ciliated Cells and Nonciliated Goblet Cells (SEM). The small bumps on the goblet cells are microvilli.

Prof. P.M. Motta/Univ. "La Sapienza," Rome/Science Source

? *What is the function of the goblet cells?*

At the level of the sternal angle, the trachea ends at a fork called the *tracheal bifurcation,* where it gives off the right and left *main bronchi.* The lowermost tracheal cartilage has an internal median ridge called the **carina**[11] (ca-RY-na) that directs the airflow to the right and left **(fig. 22.5b).**

22.1e The Lungs and Bronchial Tree

Each **lung** is a somewhat conical organ with a broad, concave **base** resting on the diaphragm and a blunt peak called the **apex** projecting slightly above the clavicle **(fig. 22.9).** The broad **costal surface** is pressed against the rib cage, and the smaller concave **mediastinal surface** faces medially. The mediastinal surface exhibits a slit called the **hilum** through which the lung receives the main bronchus, blood vessels, lymphatics, and nerves. These structures constitute the **root** of the lung.

The lungs are crowded by adjacent organs and neither fill the entire rib cage, nor are they symmetrical **(fig. 22.10).** Inferior to the lungs and diaphragm, much of the space within the rib cage is occupied by the liver, spleen, and stomach (see fig. B.5 in atlas B). The right lung is shorter than the left because the liver rises higher on the right. The left lung, although taller, is narrower than the right because the heart tilts toward the left and occupies more space on this side of the mediastinum. On the medial surface, the left lung has an indentation called the **cardiac impression** where

---
[11]*carina* = keel

the heart presses against it; part of this is visible anteriorly as a crescent-shaped **cardiac notch** in the margin of the lung. The right lung has three lobes—**superior, middle,** and **inferior.** A deep groove called the **horizontal fissure** separates the superior and middle lobes, and a similar **oblique fissure** separates the middle and inferior lobes. The left lung has only a **superior** and **inferior lobe** and a single oblique fissure.

The Bronchial Tree

Each lung has a branching system of air tubes called the **bronchial tree,** extending from the main bronchus to about 65,000 *terminal bronchioles.* Arising from the fork in the trachea, the **right main (primary) bronchus** (BRON-cus) is 2 to 3 cm long. It is slightly wider and more vertical than the left one; consequently, *aspirated* (inhaled) foreign objects lodge in the right bronchus more often than in the left. The right main bronchus gives off three branches—the **superior, middle,** and **inferior lobar (secondary) bronchi**—one to each lobe of the lung. The **left main bronchus** is about 5 cm long and slightly narrower and more horizontal than the right. It gives off superior and inferior lobar bronchi to the two lobes of the left lung.

In both lungs, the lobar bronchi branch into **segmental (tertiary) bronchi.** There are 10 of these in the right lung and 8 in the left. Each one ventilates a functionally independent unit of lung tissue called a **bronchopulmonary segment.** This chapter's opening photo shows a cast of the bronchial trees made by the corrosion cast technique described in chapter 3, with each bronchopulmonary segment injected with a different color of resin.

The main bronchi are supported, like the trachea, by rings of hyaline cartilage, whereas the cartilages transition to overlapping crescent-shaped plates by the time we reach the lobar and segmental bronchi. All of the bronchi are lined with ciliated pseudostratified columnar epithelium, but the cells grow shorter and the epithelium thinner as we progress distally. The lamina propria has an abundance of mucous glands and lymphatic nodules *(mucosa-associated lymphatic tissue, MALT),* favorably positioned to intercept inhaled pathogens. All divisions of the bronchial tree have a substantial amount of elastic connective tissue, which contributes to the recoil that expels air from the lungs in each respiratory cycle. The mucosa also has a well-developed layer of smooth muscle, the *muscularis mucosae,* which contracts or relaxes to constrict or dilate the airway, thus regulating airflow.

Bronchioles (BRON-kee-olz) are continuations of the airway that lack supportive cartilage and are 1 mm or less in diameter. The portion of the lung ventilated by one bronchiole is called a **pulmonary lobule.** Bronchioles have a ciliated cuboidal epithelium and a well-developed layer of smooth muscle in their walls. Spasmodic contractions of this muscle at death cause the bronchioles to exhibit a wavy lumen in most histological sections.

Each bronchiole divides into 50 to 80 **terminal bronchioles,** the final branches of the conducting zone **(fig. 22.11).** These measure 0.5 mm or less in diameter and have no mucous glands or goblet cells. They do have cilia, however, so that mucus draining into them from the more proximal air passages can be driven back by the mucociliary escalator, preventing congestion of the terminal bronchioles and alveoli.

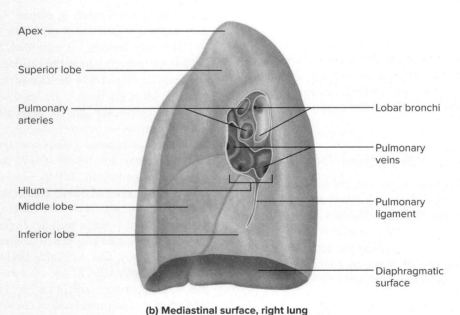

FIGURE 22.9 Gross Anatomy of the Lungs. (a) Anterior view, with frontal section through right lung. (b) Mediastinal surface of right lung, showing hilum. **APR**

Each terminal bronchiole gives off two or more smaller **respiratory bronchioles,** which have alveoli budding from their walls. They are considered the beginning of the respiratory zone because their alveoli participate in gas exchange. Their walls have scanty smooth muscle, and the smallest of them are nonciliated. Each respiratory bronchiole divides into 2 to 10 elongated, thin-walled passages called **alveolar ducts,** which also have alveoli along their

walls. The alveolar ducts and smaller divisions have nonciliated simple squamous epithelia. The ducts end in **alveolar sacs,** which are clusters of alveoli arrayed around a central space called the **atrium.** The distinction between an alveolar duct and atrium is their shape—an elongated duct, or an atrium with about equal length and width. It is often a subjective judgment whether to regard a space as an alveolar duct or atrium.

FIGURE 22.10 Cross Section Through the Thoracic Cavity. This photograph is oriented the same way as the reader's body. The pleural cavity is especially evident where the left lung has shrunken away from the thoracic wall, but in a living person the lung fully fills this space, the parietal and visceral pleurae are pressed together, and the pleural cavity is only a potential space between the membranes, as on the right side of this photograph.

Rebecca Gray/Don Kincaid/McGraw-Hill Education

FIGURE 22.11 Histology of the Lung. (a) Light micrograph. (b) Scanning electron micrograph. Note the spongy texture of the lung.

a: MICROSCAPE/Science Source; b: Biophoto Associates/Science Source

Branches of the *pulmonary artery* closely follow the bronchial tree on their way to the alveoli. The *bronchial arteries* service the bronchi, bronchioles, and some other pulmonary and thoracic tissues (see section 20.7d); they don't extend to the alveoli.

In summary, the path of airflow is as follows. The first several passages belong to the conducting zone, where there are no alveoli and the tissue walls are too thick for any significant exchange of oxygen or carbon dioxide with the blood: nasal cavity → pharynx → trachea → main bronchus → lobar bronchus → segmental bronchus → bronchiole → terminal bronchiole. Then begins the respiratory zone, where all of the passages have alveoli along their walls (or are themselves alveoli) and thus engage in gas exchange: respiratory bronchiole → alveolar duct → atrium → alveolus.

Alveoli

The functional importance of human lung structure is best appreciated by comparison to the lungs of a few other animals. In frogs and other amphibians, the lung is a simple hollow sac, like an inflated balloon, lined with blood vessels. This is sufficient to meet the oxygen needs of animals with low metabolic rates. Mammals, with their high metabolic rates, could never have evolved with such a simple lung. Rather than consisting of one large sac, each human lung is a spongy mass composed of 150 million little sacs, the alveoli. These provide about 70 m² of gas-exchange surface per lung—about equal to the floor area of a handball court or a room 8.4 m (25 ft) square.

An **alveolus** (AL-vee-OH-lus) is a pouch about 0.2 to 0.5 mm in diameter (**fig. 22.12**). Thin, broad cells called **squamous (type I) alveolar cells** cover about 95% of the alveolar surface area. Their thinness allows for rapid gas diffusion between the air and blood. The other 5% is covered by round to cuboidal **great (type II) alveolar cells.** Squamous alveolar cells cover so much more surface area because they're so thin and spread out, even though they're

far outnumbered by the great alveolar cells. If you visualize a ball of dough as representing a great alveolar cell, the same amount of dough rolled out into a thin sheet would be analogous to a squamous alveolar cell. Great alveolar cells have two functions: (1) They repair the alveolar epithelium when the squamous cells are damaged; and (2) they secrete *pulmonary surfactant,* a mixture of phospholipids and protein that coats the alveoli and smallest bronchioles and prevents the bronchioles from collapsing when one exhales. This surfactant function is later explained in greater detail.

The most numerous of all lung cells are **alveolar macrophages (dust cells),** which wander the lumens of the alveoli and connective tissue between them. These cells keep the alveoli free of debris by phagocytizing dust particles that escape entrapment by mucus in the more proximal parts of the respiratory tract. In lungs that are infected or bleeding, the macrophages also phagocytize bacteria and loose blood cells. As many as 100 million alveolar macrophages perish each day as they ride up the mucociliary escalator to be swallowed and digested, thus ridding the lungs of their load of debris.

Each alveolus is surrounded by a web of blood capillaries supplied by small branches of the pulmonary artery. The barrier between the alveolar air and blood, called the **respiratory membrane,** consists only of the squamous alveolar cell, the squamous endothelial cell of the capillary, and their shared basement

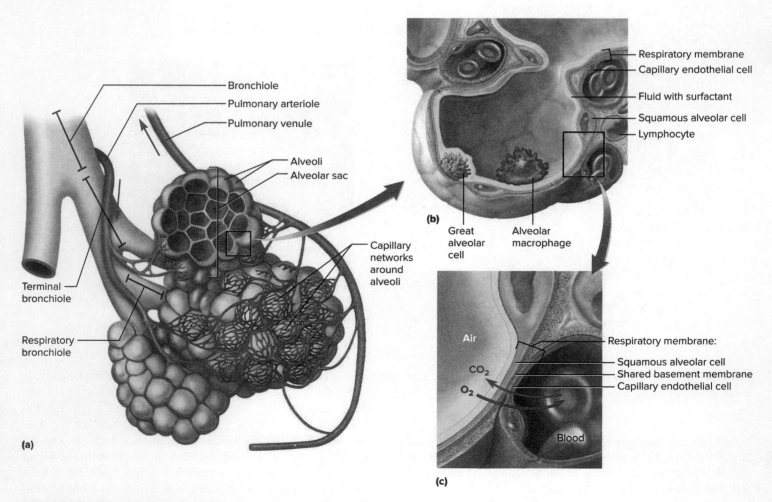

FIGURE 22.12 Pulmonary Alveoli. (a) Clusters of alveoli and their blood supply. (b) Structure of an alveolus. (c) Structure of the respiratory membrane.

membrane. These have a total thickness of only 0.5 μm, just 1/15 the diameter of a single erythrocyte.

It is crucial to prevent fluid from accumulating in the alveoli, because gases diffuse too slowly through liquid to sufficiently aerate the blood. Except for a thin film of moisture on the alveolar wall, the alveoli are kept dry by the absorption of excess water by the blood capillaries. The mean blood pressure in these capillaries is only 10 mm Hg compared to 30 mm Hg at the arterial end of the average capillary elsewhere. This low blood pressure is greatly overridden by the oncotic pressure that retains fluid in the capillaries (see the principles of capillary fluid exchange in section 20.3), so the osmotic uptake of water overrides filtration and keeps the alveoli free of excess fluid. The low capillary blood pressure also prevents rupture of the delicate respiratory membrane. The lungs also have a more extensive lymphatic drainage than any other organ in the body.

22.1f The Pleurae

A serous membrane, the **pleura** (PLOOR-uh), lines the thoracic wall and forms the surface of the lung. It has two layers, visceral and parietal. The **visceral pleura** forms the surface of the lung and extends even into the fissures between the lobes. At the hilum, it turns back on itself and forms the **parietal pleura,** which adheres to the mediastinum, inner surface of the rib cage, and superior surface of the diaphragm (see fig. 22.10). An extension of the parietal pleura, the *pulmonary ligament,* connects it to the diaphragm.

The space between the parietal and visceral pleurae is called the **pleural cavity.** The pleural cavity doesn't *contain* the lung; rather, it wraps around the lung, much like the pericardium wraps around the heart. The pleural cavity contains nothing but a thin film of lubricating **pleural fluid;** the cavity is only a *potential space,* meaning there is normally no room between the membranes. However, under pathological conditions such as chest wounds or seepage of fluid into the space *(pleural effusion),* the pleural cavity can fill with air or liquid, as discussed later in this chapter.

The pleurae and pleural fluid have three functions:

1. **Reduction of friction.** Pleural fluid acts as a lubricant that enables the lungs to expand and contract with minimal friction. Infection of the pleurae can produce a condition called *pleurisy,* in which the pleurae roughen and rub together, making each breath a painful experience.

2. **Creation of a pressure gradient.** The pleurae play a role, explained later, in the creation of a pressure gradient that expands the lungs when one inhales.

3. **Compartmentalization.** The pleurae, mediastinum, and pericardium compartmentalize the thoracic organs and prevent infections of one organ from spreading easily to neighboring organs.

BEFORE YOU GO ON

Answer the following questions to test your understanding of the preceding section:

1. A dust particle is inhaled and gets into an alveolus without being trapped along the way. Describe the path it takes, naming all air passages from external naris to alveolus. What would happen to it after arrival in the alveolus?

2. Describe the histology of the epithelium and lamina propria of the nasal cavity and the functions of the cell types present.

3. Palpate two of your laryngeal cartilages and name them. Name the ones that cannot be palpated on a living person.

4. Describe the roles of the intrinsic muscles, corniculate cartilages, and arytenoid cartilages in speech.

5. Contrast the epithelium of the bronchioles with that of the alveoli and explain how the structural difference is related to their functional difference.

6. Explain why it is fallacious to say, as some sources do, that the lungs are contained in the pleural cavities.

22.2 Pulmonary Ventilation

Expected Learning Outcomes

When you have completed this section, you should be able to

a. name the muscles of respiration and describe their roles in breathing;

b. describe the brainstem centers that control breathing and the inputs they receive from other levels of the nervous system;

c. explain how pressure gradients account for the flow of air into and out of the lungs, and how those gradients are produced;

d. identify the sources of resistance to airflow and discuss their relevance to respiration;

e. explain the significance of anatomical dead space to alveolar ventilation;

f. define the clinical measurements of pulmonary volume and capacity; and

g. define terms for various deviations from the normal pattern of breathing.

With the foregoing anatomical background, our next objective is to understand how the lungs are ventilated. Breathing, or pulmonary ventilation, consists of a repetitive cycle of **inspiration** (inhaling) and **expiration** (exhaling). One complete breath, in and out, is called a **respiratory cycle.**

We must distinguish at times between quiet and forced respiration. **Quiet respiration** refers to relaxed, unconscious, automatic breathing, the way one would breathe when reading a book or listening to a class lecture and not thinking about breathing. **Forced respiration** means unusually deep or rapid breathing, as in a state of exercise or when singing, playing a wind instrument, blowing up a balloon, coughing, or sneezing.

The lungs don't ventilate themselves. The only muscle they contain is smooth muscle in the walls of the bronchi and bronchioles. This muscle adjusts the diameter of the airway and affects the speed of airflow, but it doesn't create the airflow. That job belongs to the skeletal muscles of the trunk.

Air, like other fluids, flows down a pressure gradient from a point of higher pressure to one of lower pressure. Recall the syringe analogy (see fig. 19.18), in which we saw how an increase in the volume of a space reduces its pressure and results in an inflow of fluid. The action of the respiratory muscles is much like that of the syringe plunger—at one moment, to increase the volume and lower the pressure in the thoracic cavity, so air flows in; at the next moment, to reduce thoracic volume and raise pressure, so air flows out. We will next examine these muscular actions, how the nervous system controls them, and how the variables of pressure and resistance affect airflow and pulmonary ventilation.

22.2a The Respiratory Muscles

The principal muscles of respiration are the diaphragm and intercostal muscles (**fig. 22.13**). The diaphragm is the prime mover (see table 10.4); it alone produces about two-thirds of the pulmonary

airflow. When relaxed, it bulges upward to its farthest extent, pressing against the base of the lungs. The lungs are at their minimum volume. When the diaphragm contracts, it tenses and flattens somewhat, dropping about 1.5 cm in relaxed inspiration and as much as 12 cm in deep breathing. Not only does its descent enlarge the superior-to-inferior dimension of the thoracic cage, but its flattening also pushes outward on the sternum and ribs and enlarges the anterior-to-posterior dimension. Enlargement of the thoracic cavity lowers its internal pressure and produces an inflow of air. When the diaphragm relaxes, it bulges upward again, compresses the lungs, and expels air.

Several other muscles aid the diaphragm as synergists. Chief among these are the internal and external intercostal muscles between the ribs. Their primary function is to stiffen the thoracic cage during respiration and prevent it from collapsing when the diaphragm descends. However, they also contribute to enlargement and contraction of the thoracic cage and add about one-third of the air that ventilates the lungs. During quiet breathing, the scalene muscles of the neck fix ribs 1 and 2 (hold them stationary),

Inspiration

Sternocleidomastoid
(elevates sternum)

Scalenes
(fix or elevate ribs 1–2)

External intercostals
(elevate ribs 2–12,
widen thoracic cavity)

Pectoralis minor (cut)
(elevates ribs 3–5)

Internal intercostals,
intercartilaginous part
(aid in elevating ribs)

Diaphragm
(descends and
increases depth
of thoracic cavity)

Forced expiration

**Internal intercostals,
costal part**
(depress ribs 1–11,
narrow thoracic cavity)

Diaphragm
(ascends and
reduces depth
of thoracic cavity)

Rectus abdominis
(depresses lower ribs,
pushes diaphragm upward
by compressing
abdominal organs)

External oblique
(same effects as
rectus abdominis)

FIGURE 22.13 The Respiratory Muscles. Boldface indicates the principal respiratory muscles; the others are accessory. Arrows indicate the direction of muscle action. Muscles listed on the left are active during inspiration and those on the right are active during forced expiration. Note that the diaphragm is active in both phases, and different parts of the internal intercostal muscles serve for inspiration and expiration. Some other accessory muscles not shown here are discussed in the text.

while the external intercostal muscles pull the other ribs upward. Since most ribs are anchored at both ends—by their attachment to the vertebral column at the proximal (posterior) end and their attachment through the costal cartilage to the sternum at the distal (anterior) end—they swing upward like the handles on a bucket and thrust the sternum forward. These actions increase both the transverse (left to right) and anteroposterior diameters of the chest. In deep breathing, the anteroposterior dimension can increase as much as 20% as the chest swells.

Other muscles of the chest and abdomen also aid in breathing, especially during forced respiration; thus they are considered *accessory muscles* of respiration. Deep inspiration is aided by the erector spinae, which arches the back and increases chest diameter, and by several muscles that elevate the upper ribs: the sternocleidomastoids and scalenes of the neck; the pectoralis minor, pectoralis major, serratus anterior, and serratus posterior superior of the chest; and the *intercartilaginous part* of the internal intercostals (the anterior part between the costal cartilages). Although the scalenes merely fix the upper ribs during quiet respiration, they elevate them during forced inspiration.

Normal expiration is an energy-saving passive process achieved by the elasticity of the lungs and thoracic cage. The bronchial tree, the attachments of the ribs to the spine and sternum, and the tendons of the diaphragm and other respiratory muscles spring back when the muscles relax. As these structures recoil, the thoracic cage diminishes in size, the air pressure in the lungs rises above the atmospheric pressure outside, and the air flows out. The only muscular effort involved in normal expiration is a braking action—that is, the muscles relax gradually rather than abruptly, thus preventing the lungs from recoiling too suddenly. This makes the transition from inspiration to expiration smoother.

In forced expiration, the rectus abdominis pulls down on the sternum and lower ribs, while the *interosseous part* of the internal intercostals (the lateral part between the ribs proper) and the serratus posterior inferior pull the other ribs downward. These actions reduce the chest dimensions and expel air more rapidly and thoroughly than usual. Other lumbar, abdominal, and even pelvic muscles contribute to forced expiration by raising the pressure in the abdominal cavity and pushing some of the viscera, such as the stomach and liver, up against the diaphragm. This increases the pressure in the thoracic cavity and thus helps to expel air. Such "abdominal breathing" is particularly important in singing and public speaking.

Not only does abdominal pressure affect thoracic pressure, but the opposite is also true. Depression of the diaphragm raises abdominal pressure and helps to expel the contents of certain abdominal organs, thus aiding in childbirth, urination, defecation, and vomiting. During such actions, we often consciously or unconsciously employ the **Valsalva**[12] **maneuver.** This consists of taking a deep breath, holding it by closing the glottis, and then contracting the abdominal muscles to raise abdominal pressure and push the organ contents out.

[12]Antonio Maria Valsalva (1666–1723), Italian anatomist

22.2b Neural Control of Breathing

The heartbeat and breathing are the two most conspicuously rhythmic processes in the body. The heart, we have seen, has an internal pacemaker, but the lungs do not. No autorhythmic pacemaker cells for respiration have been found that are analogous to those of the heart, and the exact mechanism for setting the rhythm of respiration remains obscure. But we do know that breathing depends on repetitive stimuli from the brain. It ceases if the nerve connections to the thoracic muscles are severed or if the spinal cord is severed high on the neck. There are two reasons for this dependence on the brain: (1) Skeletal muscles, unlike cardiac muscle, can't contract without nervous stimulation. (2) Breathing involves the well-orchestrated action of multiple muscles and thus requires a central coordinating mechanism.

Breathing is controlled at two levels of the brain. One is cerebral and conscious, enabling us to inhale or exhale at will. The other is unconscious and automatic. Most of the time, we breathe without thinking about it—fortunately, for we otherwise couldn't go to sleep without fear of respiratory arrest (see Deeper Insight 22.2).

Brainstem Respiratory Centers

The automatic, unconscious cycle of breathing is controlled by three pairs of respiratory centers in the reticular formation of the medulla oblongata and pons **(fig. 22.14).** There is one of each on the right and left sides of the brainstem; the two sides communicate with each other so that the respiratory muscles contract symmetrically.

1. The **ventral respiratory group (VRG)** is the primary generator of the respiratory rhythm. It is an elongated neural network in the medulla with two commingled webs of neurons—**inspiratory (I) neurons** and **expiratory (E) neurons**—each forming a reverberating neural circuit (see fig. 12.32). In quiet breathing (called *eupnea*), the I neuron circuit fires for about 2 seconds at a time, issuing nerve signals to integrating centers in the spinal cord. The spinal centers relay signals by way of the phrenic nerves to the diaphragm and by way of intercostal nerves to the external intercostal muscles. Contraction of these muscles enlarges the thoracic cage and causes inspiration. As long as the I neurons are firing, they also inhibit the E neurons. Eventually, however, the I neurons stop firing, either because of fatigue or because they're inhibited by signals from an outside source. As their activity wanes, the E neurons begin firing. They further inhibit the I neurons, allowing the inspiratory muscles to relax. Elastic recoil of the thoracic cage expels air from the lungs. Relaxed expiration normally lasts about 3 seconds. Then the E neuron activity wanes, the I neurons resume firing, and the cycle repeats itself. In eupnea, this oscillating pattern of neural activity, alternating between the I neuron and E neuron circuits, produces a respiratory rhythm of about 12 breaths per minute.

2. The **dorsal respiratory group (DRG)** is one of the mechanisms that modifies this basic respiratory rhythm. It is a web of neurons that extends for much of the length of the medulla between the VRG and the central canal of the brainstem. Obviously, we don't always breathe at the same rate. Breathing can be faster or slower, shallower or deeper, as the DRG and other mechanisms modulate the VRG's activity. The DRG is an integrating center that receives input from several sources detailed in the coming discussion: a respiratory center in the pons (the PNG discussed next); a chemosensitive center of the anterior medulla oblongata; chemoreceptors in certain major arteries; stretch and irritant receptors in the airway; and higher brainstem centers that allow for emotional influences on breathing. The DRG issues output to the VRG that modifies the respiratory rhythm to adapt to varying conditions.

3. The **pontine respiratory group (PRG)** on each side of the pons receives input from higher brain centers including the hypothalamus, limbic system, and cerebral cortex, and issues output to both the DRG and VRG. By acting on those centers in the medulla, it hastens or delays the transition from inspiration to expiration, making each breath shorter and shallower, or longer and deeper. The PRG adapts breathing to special circumstances such as sleep, exercise, vocalization, and emotional responses (for example, in crying, gasping, or laughing).

Key
— Inputs to respiratory centers of medulla
— Outputs to spinal centers and respiratory muscles

Output from hypothalamus, limbic system, and higher brain centers

Pons

Pontine respiratory group (PRG)

Dorsal respiratory group (DRG)

Ventral respiratory group (VRG)

Spinal integrating centers

Central chemoreceptors

Glossopharyngeal n.

Vagus n.

Medulla oblongata

Intercostal nn.

Phrenic n.

Diaphragm and intercostal muscles

Accessory muscles of respiration

FIGURE 22.14 Respiratory Control Centers in the Central Nervous System. See text for explanation of these centers and control pathways.

DEEPER INSIGHT 22.2

CLINICAL APPLICATION

Ondine's Curse

In German legend, there was a water nymph named Ondine who took a mortal lover. When her lover proved unfaithful, the king of the nymphs put a curse on him that took away his automatic physiological functions. Consequently, he had to remember to take each breath, and he couldn't go to sleep or he would die of suffocation—which, as exhaustion overtook him, was indeed his fate.

Some people suffer a disorder called *Ondine's curse,* in which the automatic respiratory functions are disabled. This can result from accidents of neurosurgery, brainstem damage from polio, or mutation of a gene that codes for chemoreceptors in the medulla oblongata and blunts the response to CO_2 accumulation. Victims of Ondine's curse must remember to take each breath and can't go to sleep without the aid of a mechanical ventilator.

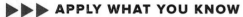

▶▶▶ APPLY WHAT YOU KNOW

Some authorities refer to the respiratory rhythm as an autonomic function. Discuss whether you think this is an appropriate word for it. What are the effectors of the autonomic nervous system? (See chapter 15.) What are the effectors that ventilate the lungs? What bearing might this have on the question?

Central and Peripheral Input to the Respiratory Centers

Variations in the respiratory rhythm are possible because the respiratory centers of the medulla and pons receive input from several other levels of the nervous system and therefore respond to the body's varying physiological needs. For example, anxiety can trigger a bout of uncontrollable *hyperventilation* in some people, a state in which breathing is so rapid that it expels CO_2 from the body faster than it's produced. As blood CO_2 levels drop, the pH rises and causes the cerebral arteries to constrict. This reduces cerebral perfusion and may cause dizziness or fainting. Hyperventilation can be brought under control by having a person rebreathe the expired CO_2 from a paper bag held over the nose and mouth.

Multiple sensory receptors also provide information to the respiratory centers:

- **Central chemoreceptors** are brainstem neurons that respond especially to changes in the pH of the cerebrospinal fluid. They are concentrated on each side of the medulla oblongata at a point only 0.2 mm beneath its anterior surface. The pH of the CSF reflects the CO_2 level in the blood, so by regulating respiration to maintain a stable CSF pH, the respiratory centers also ensure a stable blood CO_2 level.

- **Peripheral chemoreceptors** are located in the carotid and aortic bodies of the large arteries above the heart **(fig. 22.15).** They respond to the O_2 and CO_2 content of the blood, but most of all to pH. The carotid bodies communicate with the

FIGURE 22.15 The Peripheral Chemoreceptors of Respiration. Peripheral chemoreceptors in the aortic arch and carotid bodies monitor blood gas concentrations and blood pH. They send signals about blood chemistry to the dorsal respiratory group of the medulla oblongata through sensory fibers in the vagus and glossopharyngeal nerves.

brainstem by way of the glossopharyngeal nerves, and the aortic bodies by way of the vagus nerves. Sensory fibers in these nerves enter the medulla and synapse with neurons of the DRG.

- **Stretch receptors** are found in the smooth muscle of the bronchi and bronchioles and in the visceral pleura. They respond to inflation of the lungs and signal the DRG by way of the vagus nerves. Excessive inflation triggers the **inflation (Hering–Breuer)[13] reflex,** a protective somatic reflex that strongly inhibits the I neurons and stops inspiration. In infants, this may be a normal mechanism of transition from inspiration to expiration, but after infancy it is activated only by extreme stretching of the lungs.

- **Irritant receptors** are nerve endings amid the epithelial cells of the airway. They respond to smoke, dust, pollen,

---

[13]Heinrich Ewald Hering (1866–1948), German physiologist; Josef Breuer (1842–1925), Austrian physician

chemical fumes, cold air, and excess mucus. They transmit signals by way of the vagus nerves to the DRG, and the DRG returns signals to the respiratory and bronchial muscles, resulting in such protective reflexes as broncho-constriction, shallower breathing, breath-holding *(apnea),* or coughing.

Voluntary Control of Breathing

Voluntary control of breathing is important in singing, speaking, breath-holding, and other circumstances. Such control originates in the motor cortex of the cerebrum. The output neurons send impulses down the corticospinal tracts to the integrating centers in the spinal cord, bypassing the brainstem centers. There are limits to voluntary control. Temperamental children may threaten to hold their breath until they die, but it's impossible to do so. Holding one's breath raises the CO_2 level of the blood until a *breaking point* is reached when automatic controls override one's will. This forces a person to resume breathing even if he or she has lost consciousness.

22.2c Pressure, Resistance, and Airflow

Now that we know the neuromuscular aspects of breathing, we'll explore how the expansion and contraction of the thoracic cage produce airflow into and out of the lungs.

Understanding pulmonary ventilation, the transport of gases in the blood, and the exchange of gases with the tissues draws on certain *gas laws* of physics. These are named after their discoverers and are not intuitively easy to remember by name. **Table 22.1** lists them for your convenience and may be a helpful reference as you progress through respiratory physiology.

TABLE 22.1	The Gas Laws of Respiratory Physiology
Boyle's law[14]	The pressure of a given quantity of gas is inversely proportional to its volume (assuming a constant temperature).
Charles's law[15]	The volume of a given quantity of gas is directly proportional to its absolute temperature (assuming a constant pressure).
Dalton's law[16]	The total pressure of a gas mixture is equal to the sum of the partial pressures of its individual gases.
Henry's law[17]	At the air–water interface, the amount of gas that dissolves in water is determined by its solubility in water and its partial pressure in the air (assuming a constant temperature).

---

[14]Robert Boyle (1627–91), Anglo–Irish physicist and chemist
[15]Jacques A. C. Charles (1746–1823), French physicist
[16]John Dalton (1766–1844), English physicist and chemist
[17]William Henry (1774–1836), English chemist

Pressure and Airflow

Respiratory airflow is governed by the same principles of flow, pressure, and resistance as blood flow. As we saw in section 20.2, the flow (F) of a fluid is directly proportional to the pressure difference between two points (ΔP) and inversely proportional to resistance (R):

$$F \propto \Delta P/R.$$

For the moment, we'll focus especially on ΔP, the pressure gradient that produces airflow. We'll deal with resistance later.

The pressure that drives inspiration is **atmospheric (barometric) pressure**—the weight of the air above us. At sea level, this averages 760 mm Hg, or by definition, *1 atmosphere (1 atm).* It fluctuates with the weather from day to day and is lower at higher elevations, but we'll use the average sea-level value as a reference point for discussion.

One way to change the pressure of an enclosed gas is to change the volume of its container. This fact is summarized by **Boyle's law:** Assuming a constant temperature, *the pressure of a given quantity of gas is inversely proportional to its volume.* If the lungs contain a quantity of gas and lung volume increases, their internal pressure (**intrapulmonary pressure**) falls. Conversely, if lung volume decreases, intrapulmonary pressure rises. (Compare this to the syringe analogy.)

If the intrapulmonary pressure falls below the atmospheric pressure, then air tends to flow down its pressure gradient into the lungs. Conversely, if intrapulmonary pressure rises above atmospheric pressure, air flows out. Therefore, all we have to do to breathe is to cyclically raise and lower the intrapulmonary pressure, employing the neuromuscular mechanisms recently described.

When dealing with respiratory airflow and thoracic pressures generated by the foregoing muscle actions, we must use a new unit of measurement, different from what you're accustomed to. In recent chapters, we used *millimeters of mercury (mm Hg)* as a measure of blood pressure, and we'll use it again later in this chapter when we speak of atmospheric pressure and blood gases. Millimeters of mercury is a measure of how high up a vacuum tube a force such as blood pressure or the weight of the atmosphere can push a column of mercury. Mercury is a very heavy liquid, so we use it because pressures can be measured with a relatively short column of mercury, as in the sphygmomanometer of a doctor's office. But the pressures in respiratory airflow are so small that they couldn't move a mercury column much at all; mercury-based instruments aren't sensitive enough. Respiratory physiologists therefore traditionally used water columns, which are more sensitive, and we measure these pressures in *centimeters of water (cm H_2O).* (1 mm Hg $\approx$ 13.6 mm H_2O $\approx$ 1.4 cm H_2O.) Small pressure changes will move a column of water more than a column of mercury; one can see them and measure them more accurately (although now they're measured with electronic instruments rather than water-filled tubes).

Since respiratory airflow is driven by a *difference* between surrounding (ambient) atmospheric pressure and pressures in the chest, the following discussion is based on *relative* pressures. If we speak of pressure in the pulmonary alveoli reaching –2 cm H_2O during inspiration, we mean it falls 2 cm H_2O below the ambient atmospheric pressure; if it rises to +3 cm H_2O during expiration, it is 3 cm H_2O above ambient atmospheric pressure.

Inspiration

Now consider the flow of air into the lungs—inspiration. **Figure 22.16** traces the events and pressure changes that occur throughout a respiratory cycle.

At the beginning (step ① in the figure), there is no movement of the thoracic cage, no difference between the air pressure within the lungs and the ambient atmospheric pressure, and no airflow. What happens when the thoracic cage expands? Why don't the lungs remain the same size and simply occupy less space in the chest? Consider the two layers of the pleura: the parietal pleura lining the rib cage and the visceral pleura on the lung surface. They aren't anatomically attached to each other along their surfaces, but they're wet and cling together like sheets of wet paper. The space between them (*pleural cavity*) is only about 10 to 30 μm wide (about the width of one typical cell). At the end of a normal expiration, the chest wall (including the parietal pleura) tends to expand outward because of its elasticity while the lungs, because of their elasticity, tend to recoil inward. Thus, the lungs and chest wall are pulling in opposite directions. This creates a negative **intrapleural pressure,** averaging about –5 cm H_2O, between the parietal and visceral pleurae.

When the ribs swing up and out during inspiration (step ② in the figure), the parietal pleura follows. If not for the cohesion of water, this could pull the chest wall away from the lungs; but because the two wet membranes cling to each other, it only reduces the intrapleural pressure a little more, to about –8 cm H_2O. As the visceral pleura (lung surface) is pulled outward, it stretches the alveoli just below the surface of the lung. Those alveoli are mechanically linked to deeper ones by their walls, and stretch the deeper ones as well. Thus the entire lung expands along with the thoracic cage. As in the syringe analogy, the alveoli increase in volume and decrease in pressure. Pressure within the alveoli, the **intrapulmonary (alveolar) pressure,** drops to an average of –1 cm H_2O. So now there is a pressure gradient from the ambient pressure at the nostrils to the negative pressure in the alveoli. Air flows down its gradient and ventilates the lungs. In short, we inhale.

Yet this isn't the only force that expands the lungs. Another is warming of the inhaled air. As we see from **Charles's law** (see table 22.1), the volume of a given quantity of gas is directly proportional to its absolute temperature. Inhaled air is warmed to 37°C (99°F) by the time it reaches the alveoli. This means that on a cool North American day when the outdoor temperature is, say, 16°C (60°F), the air temperature would increase by 21°C (39°F) during inspiration, becoming comparable to air temperature in the tropics. An inhaled volume of 500 mL will expand to 536 mL and this thermal expansion will contribute to inflation of the lungs.

When the respiratory muscles stop contracting, the inflowing air quickly achieves an intrapulmonary pressure equal to atmospheric pressure, and flow stops. In quiet breathing, the dimensions of the thoracic cage increase by only a few millimeters in each direction, but this is enough to increase its total volume by

about 500 mL. Typically, about 500 mL of air therefore flows into the respiratory tract.

▶▶▶ **APPLY WHAT YOU KNOW**

When you inhale, does your chest expand because your lungs inflate, or do your lungs inflate because your chest expands? Explain.

Expiration

Relaxed expiration is a passive process achieved, as we have seen, mainly by the elastic recoil of the thoracic cage. This recoil compresses the lungs and raises the intrapulmonary pressure to about +1 cm H_2O (step ③ in the figure). Air thus flows down its pressure gradient, out of the lungs. In forced breathing, the accessory muscles raise intrapulmonary pressure as high as +40 cm H_2O.

The effect of pulmonary elasticity is evident in a pathological state of pneumothorax and atelectasis. **Pneumothorax** is the presence of air in the pleural cavity. If the thoracic wall is punctured between the ribs, for example, inspiration sucks air through the wound into the pleural cavity and the visceral and parietal pleurae separate; what was a *potential space* between them becomes an air-filled cavity. Without the negative intrapleural pressure to keep the lungs inflated, the lungs recoil and collapse. The collapse of part or all of a lung is called **atelectasis**[18] (AT-eh-LEC-ta-sis). Atelectasis can also result from airway obstruction—for example, by a lung tumor, aneurysm, swollen lymph node, or aspirated object. Blood absorbs gases from the alveoli distal to the obstruction, and that part of the lung collapses because it can't be reinflated.

Resistance to Airflow

Pressure is one determinant of airflow; the other is resistance. The greater the resistance, the slower the flow. But what governs resistance? Two factors are of particular importance: diameter of the bronchioles and pulmonary compliance.

Like arterioles, the large number of bronchioles, their small diameter, and their ability to change diameter make them the primary means of controlling resistance. The trachea and bronchi can also change diameter to a degree, but are more constrained by the supporting cartilages in their walls. An increase in the diameter of a bronchus or bronchiole is called **bronchodilation** (BRON-co-dy-LAY-shun) and a reduction in diameter is called **bronchoconstriction.** Epinephrine and the sympathetic nerves (norepinephrine) stimulate bronchodilation and increase airflow. Histamine, parasympathetic nerves (acetylcholine), cold air, and chemical irritants are among the factors that stimulate bronchoconstriction. Many people have suffocated from the extreme bronchoconstriction brought on by anaphylactic shock or asthma (see Deeper Insight 21.3).

---

[18]*atel* = imperfect, incomplete; *ectasis* = expansion

No airflow

Pleural cavity

Intrapulmonary pressure 0 cm H$_2$O

Diaphragm

Intrapleural pressure –5 cm H$_2$O

Ribs swing upward like bucket handles during inspiration.

① **At rest, atmospheric and intrapulmonary pressures are equal, and there is no airflow.**

Ribs swing downward like bucket handles during expiration.

④ **Pause**

Airflow

Airflow

Intrapleural pressure –8 cm H$_2$O

Intrapleural pressure –5 cm H$_2$O

Intrapulmonary pressure –1 cm H$_2$O

Intrapulmonary pressure +1 cm H$_2$O

Diaphragm rises

Diaphragm flattens

② **Inspiration**

③ **Expiration**

Rib

Rib

Sternum

Sternum

Rib

Rib

Sternum

Sternum

Ribs elevated, thoracic cavity expands laterally

Sternum swings up, thoracic cavity expands anteriorly

Ribs depressed, thoracic cavity narrows

Sternum swings down, thoracic cavity contracts posteriorly

② In inspiration, the thoracic cavity expands laterally, vertically, and anteriorly; intrapulmonary pressure drops 1 cm H$_2$O below atmospheric pressure, and air flows into the lungs.

③ In expiration, the thoracic cavity contracts in all three directions; intrapulmonary pressure rises 1 cm H$_2$O above atmospheric pressure, and air flows out of the lungs.

FIGURE 22.16 The Respiratory Cycle. All pressures given here are relative to atmospheric pressure external to the body, which is considered to be zero as a point of reference. Note that pressures governing respiratory airflow are measured in cm H$_2$O (centimeters of water), not mm Hg. Like bucket handles, each rib is attached at both ends (to the spine and sternum) and swings up and down during inspiration and expiration. **APR**

Pulmonary compliance means the ease with which the lungs expand, or more exactly, the change in lung volume relative to a given pressure change. Inspiratory effort may produce the same intrapleural pressure in two people, but the lungs will expand less in a person with poorer compliance (stiffer lungs), or at least that person must expend more effort to inflate the lungs to the same degree as the other. Think of the difference in effort it takes to blow up a brand-new party balloon (with low compliance) versus blowing up one that has been inflated before. Pulmonary compliance is reduced by degenerative lung diseases such as tuberculosis and black lung disease, in which the lungs are stiffened by scar tissue. In such conditions, the thoracic cage expands normally but the lungs expand relatively little.

A major limitation on pulmonary compliance is the thin film of water on the respiratory epithelium, especially from the respiratory bronchioles to the alveoli. This film is necessary for gas exchange, but creates a potential problem for pulmonary ventilation. Water molecules are attracted to each other by hydrogen bonds, creating surface tension, as we saw in section 2.2a. You can appreciate the strength of this attraction if you reflect on the difficulty of separating two sheets of wet paper compared with two sheets of dry paper. Surface tension draws the walls of the airway inward toward the lumen. If it went unchecked, parts of the airway would collapse with each expiration and would strongly resist reinflation. This is especially so in small airways such as the respiratory bronchioles and alveolar ducts leading to the alveoli.

The solution to this problem takes us back to the great alveolar cells and their surfactant. A surfactant is an agent that disrupts the hydrogen bonds of water and reduces surface tension; soaps and detergents are everyday examples. The pulmonary surfactant is composed of amphipathic proteins and phospholipids. These molecules are partially hydrophobic, so they spread out over the surface of the water film, partially embedded in it like ice cubes floating in a bowl of water. As the small airways deflate, the surfactants are squeezed closer together, like the ice cubes being pushed together into a smaller area. If the air spaces were covered with a film of water only, they could continue collapsing, because water molecules can pile up into a thicker film of moisture. The physical structure of the surfactants resists compression, however. They can't pile up into a thicker layer because their hydrophilic regions resist separation from the water below. As they become crowded into a small area and resist layering, they retard and then halt the collapse of the airway.

Deep breathing spreads pulmonary surfactant throughout the small airways. Patients recently out of surgery are encouraged to breathe deeply, even though it may hurt, in order to promote this spread of surfactant up the alveolar ducts and small bronchioles. Those who don't adhere to their breathing exercises can experience collapse of portions of the lung that are not adequately coated with surfactant.

The importance of this surfactant is especially apparent when it's lacking. Premature infants often have a surfactant deficiency and great difficulty breathing (see section 29.3b). The resulting *infant respiratory distress syndrome (IRDS)* can be treated by administering artificial surfactant.

22.2d Alveolar Ventilation

Air that actually enters the alveoli becomes available for gas exchange, but not all inhaled air gets that far. About 150 mL of it fills the conducting zone of the airway. Since this air cannot exchange gases with the blood, the conducting zone is called the **anatomical dead space.** The dead space is about 1 mL per pound of body weight in a healthy person. In some pulmonary diseases, however, it can be substantially greater. Some alveoli may be unable to exchange gases because they lack blood flow or because the pulmonary membrane is thickened by edema or fibrosis. **Physiological (total) dead space** is the sum of anatomical dead space and any pathological dead space that may exist.

The anatomical dead space varies with circumstances. In a state of relaxation, parasympathetic stimulation keeps the airway somewhat constricted. This minimizes the dead space so more of the inhaled air ventilates the alveoli. In a state of arousal or exercise, by contrast, the sympathetic nervous system dilates the airway, which increases airflow. The increased airflow outweighs the air that is wasted by filling the increased dead space.

If a person inhales 500 mL of air and 150 mL of it stays in the dead space, then 350 mL ventilates the alveoli. Multiplying this by the respiratory rate gives the **alveolar ventilation rate (AVR)**—for example,

$$350 \text{ mL/breath} \times 12 \text{ breaths/min.} = 4{,}200 \text{ mL/min.}$$

Of all measures of pulmonary ventilation, this one is most directly relevant to the body's ability to get oxygen to the tissues and dispose of carbon dioxide.

The air current taken in by inspiration never actually enters the alveoli. This *bulk flow* of air gets only as far as the terminal bronchioles. Beyond this point, the air passages are so narrow and their resistance to flow is so great that bulk flow ceases. Oxygen completes its journey to the alveoli, and carbon dioxide leaves them, by simple diffusion. Pulsation of the pulmonary arteries generated by the heartbeat probably compresses adjacent airways and accelerates this process, however.

The lungs never completely empty during expiration. There is always some leftover air called the *residual volume,* typically about 1,300 mL that one cannot exhale even with maximum effort. Residual air mixes with fresh air arriving on the next inspiration, so the same oxygen-depleted air doesn't remain in the lungs cycle after cycle. It takes about 90 seconds, or approximately 18 breaths at an average rate and depth of breathing, to completely replace all pulmonary air.

22.2e Measurement of Pulmonary Ventilation

Clinicians often measure a patient's pulmonary ventilation in order to assess the severity of a respiratory disease or monitor the patient's improvement or deterioration. The process of making such measurements is called **spirometry.**[19] It entails

---

[19] *spiro* = breath; *metry* = process of measuring

having the subject breathe into a device called a **spirometer (fig. 22.17a),** which recaptures the expired breath and records such variables as the rate and depth of breathing, speed of expiration, and rate of oxygen consumption. Representative measurements for a healthy adult male are given in **table 22.2** and explained in **figure 22.17b.** Female values are somewhat lower because of smaller average body size.

Four of these values are called *respiratory volumes:* tidal volume, inspiratory reserve volume, expiratory reserve volume, and residual volume. **Tidal volume (TV)** is the amount of air inhaled and exhaled in one cycle; in quiet breathing, it averages about 500 mL. Beyond the amount normally inhaled, it's typically possible to inhale another 3,000 mL with maximum effort; this is the **inspiratory reserve volume (IRV).** Similarly, with maximum

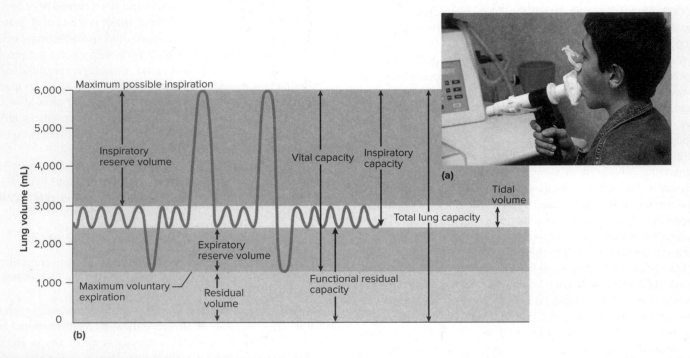

FIGURE 22.17 Respiratory Volumes and Capacities. (a) Subject breathing into a spirometer. (b) An idealized spirogram. The wavy line indicates inspiration when it rises and expiration when it falls. Compare with table 22.2.

a: BSIP/Science Source

? *Why is a nose clip necessary for obtaining a valid spirogram?*

TABLE 22.2	Respiratory Volumes and Capacities for an Average Young Adult Male	
Measurement	**Typical Value**	**Definition**
Respiratory Volumes		
Tidal volume (TV)	500 mL	Amount of air inhaled and exhaled in one cycle during quiet breathing
Inspiratory reserve volume (IRV)	3,000 mL	Amount of air in excess of tidal volume that can be inhaled with maximum effort
Expiratory reserve volume (ERV)	1,200 mL	Amount of air in excess of tidal volume that can be exhaled with maximum effort
Residual volume (RV)	1,300 mL	Amount of air remaining in the lungs after maximum expiration; the amount that can never be voluntarily exhaled
Respiratory Capacities		
Vital capacity (VC)	4,700 mL	The amount of air that can be inhaled and then exhaled with maximum effort; the deepest possible breath (VC = ERV + TV + IRV)
Inspiratory capacity (IC)	3,500 mL	Maximum amount of air that can be inhaled after a normal tidal expiration (IC = TV + IRV)
Functional residual capacity (FRC)	2,500 mL	Amount of air remaining in the lungs after a normal tidal expiration (FRC = RV + ERV)
Total lung capacity (TLC)	6,000 mL	Maximum amount of air the lungs can contain (TLC = RV + VC)

effort, one can normally exhale another 1,200 mL beyond the normal amount; this difference is the **expiratory reserve volume (ERV).** Even after a maximum voluntary expiration, there remains a **residual volume (RV)** of about 1,300 mL. This air allows gas exchange with the blood to continue even between the times one inhales fresh air.

Four other measurements, called *respiratory capacities,* are obtained by adding two or more of the respiratory volumes: **vital capacity (VC)** (ERV + TV + IRV), **inspiratory capacity** (TV + IRV), **functional residual capacity** (RV + ERV), and **total lung capacity** (RV + VC). Vital capacity, the maximum ability to ventilate the lungs in one breath, is an especially important measure of pulmonary health.

Spirometry helps to assess and distinguish between *restrictive* and *obstructive* lung disorders. **Restrictive disorders** are those that reduce pulmonary compliance, thus limiting the amount to which the lungs can be inflated. They show in spirometry as a reduced vital capacity. Any disease that produces pulmonary fibrosis has a restrictive effect: black lung disease and tuberculosis, for example. **Obstructive disorders** are those that interfere with airflow by narrowing or blocking the airway. They make it harder to inhale or exhale a given amount of air. Asthma and chronic bronchitis are the most common examples. Obstructive disorders can be measured by having the subject exhale as rapidly as possible into a spirometer and measuring **forced expiratory volume (FEV)**—the volume of air or the percentage of the vital capacity that can be exhaled in a given time interval. A healthy adult should be able to expel 75% to 85% of the vital capacity in 1.0 second (a value called the $FEV_{1.0}$). At home, asthma patients and others can monitor their respiratory function by blowing into a handheld meter that roughly measures **peak flow,** the maximum speed of expiration.

The amount of air inhaled per minute is the **minute respiratory volume (MRV).** MRV largely determines the alveolar ventilation rate. It can be measured directly with a spirometer or obtained by multiplying tidal volume by respiratory rate. For example, if a person has a tidal volume of 500 mL per breath and a rate of 12 breaths per minute, the MRV would be $500 \times 12 = 6,000$ mL/min. During heavy exercise, MRV may be as high as 125 to 170 L/min. This is called **maximum voluntary ventilation (MVV),** formerly called *maximum breathing capacity.*

22.2f Variations in the Respiratory Rhythm

Relaxed, quiet breathing is called **eupnea**[20] (YOOP-nee-uh). It is typically characterized by a tidal volume of about 500 mL and a respiratory rate of 12 to 15 breaths per minute. Conditions ranging from exercise or anxiety to various disease states can cause deviations such as abnormally fast, slow, or labored breathing. **Table 22.3** defines the clinical terms for several such variations. You should familiarize yourself with these before proceeding, because later discussions in this chapter assume a working knowledge of some of these terms.

TABLE 22.3	Variations in the Respiratory Rhythm
Apnea[21] (AP-nee-uh)	Temporary cessation of breathing (one or more skipped breaths)
Dyspnea[22] (DISP-nee-uh)	Labored, gasping breathing; shortness of breath
Hyperpnea[23] (HY-purp-NEE-uh)	Increased rate and depth of breathing in response to exercise, pain, or other conditions
Hyperventilation	Increased pulmonary ventilation in excess of metabolic demand, frequently associated with anxiety; expels CO_2 faster than it is produced, thus lowering the blood CO_2 concentration and raising the blood pH
Hypoventilation[24]	Reduced pulmonary ventilation; leads to an increase in blood CO_2 concentration if ventilation is insufficient to expel CO_2 as fast as it is produced
Kussmaul[25] respiration	Deep, rapid breathing often induced by acidosis; seen in diabetes mellitus
Orthopnea[26] (or-thop-NEE-uh)	Dyspnea that occurs when a person is lying down or in any position other than standing or sitting erect; seen in heart failure, asthma, emphysema, and other conditions
Respiratory arrest	Permanent cessation of breathing (unless there is medical intervention)
Tachypnea[27] (tack-ip-NEE-uh)	Accelerated respiration

Other variations in pulmonary ventilation serve the purposes of speaking, expressing emotion (laughing, crying), yawning, hiccuping, expelling noxious fumes, coughing, sneezing, and expelling abdominal contents. Coughing is induced by irritants in the lower respiratory tract. To cough, we close the glottis and contract the respiratory and abdominal muscles, producing high pressure in the lower respiratory tract. We then suddenly open the glottis and release an explosive burst of air at speeds over 900 km/h (600 mi./h). This drives mucus and foreign matter toward the pharynx and mouth. Sneezing is triggered by irritants in the nasal cavity. Its mechanism is similar to coughing except that the glottis is continually open, the soft palate and tongue block the flow of air while thoracic pressure builds, and then the soft palate is depressed to direct part of the airstream through the nose. These actions are coordinated by coughing and sneezing centers in the medulla oblongata.

[21]*a* = without; *pnea* = breathing
[22]*dys* = difficult, abnormal, painful; *pnea* = breathing
[23]*hyper* = above normal
[24]*hypo* = below normal
[25]Adolph Kussmaul (1822–1902), German physician
[26]*ortho* = straight, erect; *pnea* = breathing
[27]*tachy* = fast; *pnea* = breathing

[20]*eu* = easy, normal; *pnea* = breathing

Answer the following questions to test your understanding of the preceding section:

7. Explain why contraction of the diaphragm causes inspiration but contraction of the transverse abdominal muscle causes expiration.

8. Which brainstem respiratory nucleus is indispensable to respiration? What do the other nuclei do?

9. Explain why Boyle's law is relevant to the action of the respiratory muscles.

10. Explain why eupnea requires little or no action by the muscles of expiration.

11. Identify a benefit and a disadvantage of normal (nonpathological) bronchoconstriction.

12. Suppose a healthy person has a tidal volume of 650 mL, an anatomical dead space of 160 mL, and a respiratory rate of 14 breaths per minute. Calculate her alveolar ventilation rate.

13. Suppose a person has a total lung capacity of 5,800 mL, a residual volume of 1,200 mL, an inspiratory reserve volume of 2,400 mL, and an expiratory reserve volume of 1,400 mL. Calculate his tidal volume.

22.3 Gas Exchange and Transport

Expected Learning Outcomes

When you have completed this section, you should be able to

a. define *partial pressure* and discuss its relationship to a gas mixture such as air;

b. contrast the composition of inspired and alveolar air;

c. discuss how partial pressure affects gas transport by the blood;

d. describe the mechanisms of transporting O_2 and CO_2;

e. describe the factors that govern gas exchange in the lungs and systemic capillaries;

f. explain how gas exchange is adjusted to the metabolic needs of different tissues; and

g. discuss the effect of blood gases and pH on the respiratory rhythm.

Ultimately, respiration is about gases, especially oxygen and carbon dioxide. We will turn our attention now to the behavior of these gases in the human body: how oxygen is obtained from inspired air and delivered to the tissues, and how carbon dioxide is removed from the tissues and released into the expired air. First, however, it is necessary to understand the composition of the air we inhale and how gases behave in contact with the water film that lines the alveoli.

22.3a Composition of Air

Air consists of about 78.6% nitrogen; 20.9% oxygen; 0.04% carbon dioxide; several quantitatively minor gases such as argon, neon, helium, methane, and ozone; and a variable amount of water vapor. Water vapor constitutes from 0% to 4%, depending on temperature and humidity; we will use a value of 0.5%, typical of a cool clear day.

Total atmospheric pressure is a sum of the contributions of these individual gases—a principle known as **Dalton's law** (see table 22.1). The separate contribution of each gas in a mixture is called its **partial pressure** and is symbolized with a P followed by the formula of the gas, such as P_{O_2}. As we are now concerned with atmospheric pressures and how they influence the partial pressures of blood gases, we return to mm Hg as our unit of measurement (not cm H_2O as when we were considering pulmonary ventilation). If we assume the average sea-level atmospheric pressure of 760 mm Hg and oxygen is 20.9% of the atmosphere, then P_{O_2} is simply

$$0.209 \times 760 \text{ mm Hg} = 159 \text{ mm Hg}.$$

Applying Dalton's law to the aforementioned mixture of gases in the air gives us the following:

$$P_{N_2} + P_{O_2} + P_{H_2O} + P_{CO_2}$$

$$\approx 597 + 159 + 3.7 + 0.3$$

$$= 760.0 \text{ mm Hg}.$$

These values change dramatically at higher altitude (see Deeper Insight 22.3).

This is the composition of the air we inhale, but it's not the composition of air in the alveoli. Alveolar air can be sampled with an apparatus that collects the last 10 mL of expired air. As we see in **table 22.4,** its composition differs from that of the atmosphere because of three influences: (1) It is humidified by contact with the mucous membranes, so its P_{H_2O} is more than 10 times higher than that of the inhaled air. (2) Freshly inspired air mixes with residual air left from the previous respiratory cycle, so its oxygen is diluted and it is enriched with CO_2 from the residual air. (3) Alveolar air

TABLE 22.4	Composition of Inspired (Atmospheric) and Alveolar Air			
Gas	Inspired Air*		Alveolar Air	
N_2	78.6%	597 mm Hg	74.9%	569 mm Hg
O_2	20.9%	159 mm Hg	13.7%	104 mm Hg
H_2O	0.5%	3.7 mm Hg	6.2%	47 mm Hg
CO_2	0.04%	0.3 mm Hg	5.3%	40 mm Hg
Total	100%	760 mm Hg	100%	760 mm Hg

* Typical values for a cool clear day; values vary with temperature and humidity. Other gases present in small amounts are disregarded.

DEEPER INSIGHT 22.3

MEDICAL HISTORY

The Flight of the Zenith

As any aviator or mountain hiker knows, the composition of air changes dramatically from sea level to high altitude. This was all too sadly realized by French physician Paul Bert (1833–86), who is commonly recognized as the founder of aerospace medicine. He invented the first pressure chamber capable of simulating the effects of high altitude, and undertook a variety of experiments on human and animal subjects to test the effects of variation in oxygen partial pressure. In 1875, balloonist Gaston Tissandier set out from Paris in a hot-air balloon named the *Zenith* with two of Bert's protégés in physiology. Their aim was to investigate the effects of the low oxygen pressures attainable only by ascending to a very high altitude. They ignored Bert's advice to breathe supplemental oxygen continually, rather than only when they felt the need for it. As they ascended, they observed each other and took notes. Seeing no ill effects, they continued to throw out ballast and go higher, eventually to 28,000 feet. But after passing 24,000 feet, they experienced stupefaction, muscular paralysis, euphoria, and finally unconsciousness. The balloon eventually descended on its own with two of the three men dead; only Tissandier lived to write about it.

---

exchanges O_2 and CO_2 with the blood. Thus, the Po_2 of alveolar air is about 65% that of inhaled air, and its Pco_2 is more than 130 times higher.

▶▶▶APPLY WHAT YOU KNOW

Expired air considered as a whole (not just the last 10 mL) is about 15.3% O_2 and 4.2% CO_2. Why would these values differ from the ones for alveolar air?

22.3b Alveolar Gas Exchange APR

Air in the alveolus is in contact with the film of water covering the alveolar epithelium. For oxygen to get into the blood, it must dissolve in this water and pass through the respiratory membrane separating the air from the bloodstream. For carbon dioxide to leave the blood, it must pass the other way and diffuse out of the water film into the alveolar air. This back-and-forth traffic of O_2 and CO_2 across the respiratory membrane is called **alveolar gas exchange.**

The reason O_2 diffuses in one direction and CO_2 in the other is that each gas diffuses down its own partial pressure gradient. Whenever air and water are in contact with each other, gases diffuse down their gradients until the partial pressure of each gas in the air is equal to its partial pressure in the water. If a gas has a greater partial pressure in the water than in the air, it diffuses into the air; the smell of chlorine near a swimming pool is evidence of

this. If the partial pressure of the gas is greater in the air, it diffuses into the water.

Henry's law states that *at the air–water interface, for a given temperature, the amount of gas that dissolves in the water is determined by its solubility in water and its partial pressure in the air* **(fig. 22.18).** Thus, the greater the Po_2 in the alveolar air, the more O_2 the blood picks up. And since blood arriving at an alveolus has a higher Pco_2 than air, it releases CO_2 into the alveolar air. At the alveolus, the blood is said to *unload* CO_2 and *load* O_2. Each gas in a mixture behaves independently; the diffusion of one gas doesn't influence the diffusion of another.

Both O_2 loading and CO_2 unloading involve erythrocytes (RBCs). The efficiency of these processes therefore depends on

(a) Oxygen

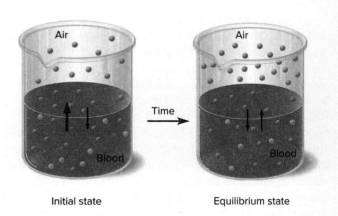

(b) Carbon dioxide

FIGURE 22.18 Henry's Law and Its Relationship to Alveolar Gas Exchange. (a) The Po_2 of alveolar air is initially higher than the Po_2 of the blood arriving at an alveolus. Oxygen diffuses into the blood until the two are in equilibrium. (b) The Pco_2 of the arriving blood is initially higher than the Pco_2 of alveolar air. Carbon dioxide diffuses into the alveolus until the two are in equilibrium.

how long an RBC spends in an alveolar capillary compared with how long it takes for each gas to be fully loaded or unloaded—that is, for them to reach equilibrium concentrations in the capillary blood. It takes about 0.25 second to reach equilibrium. At rest, when blood circulates at its slowest speed, an RBC takes about 0.75 second to pass through an alveolar capillary—plenty of time to pick up a maximum load of oxygen. Even in vigorous exercise, when the blood flows faster, an erythrocyte is in the alveolar capillary for about 0.3 second, which is still adequate.

Several variables affect the efficiency of alveolar gas exchange, and under abnormal conditions, some of these can prevent the complete loading and unloading of gases:

- **Pressure gradients of the gases.** P_{O_2} is about 104 mm Hg in alveolar air and 40 mm Hg in blood arriving at an alveolus. Oxygen therefore diffuses from the air into the blood, where it reaches a P_{O_2} of 104 mm Hg. Before the blood leaves the lung, however, this drops to about 95 mm Hg. This oxygen dilution occurs because the pulmonary veins anastomose with bronchial veins in the lungs, so there is some mixing of the oxygen-rich pulmonary blood with oxygen-poor systemic blood. Even though the blood is 100% saturated with oxygen as it leaves the alveolar capillaries, it is nearly impossible for it to remain 100% saturated by the time it leaves the lungs.

  P_{CO_2} is about 46 mm Hg in blood arriving at the alveolus and 40 mm Hg in alveolar air. CO_2 therefore diffuses from the blood into the alveoli. These changes are summarized here and in the yellow zone of **figure 22.19.**

Blood entering lungs		*Blood leaving lungs*	
P_{O_2}	40 mm Hg	P_{O_2}	95 mm Hg
P_{CO_2}	46 mm Hg	P_{CO_2}	40 mm Hg

  These gradients differ under special circumstances such as high elevation and *hyperbaric oxygen therapy* (treatment with oxygen at >1 atm of pressure) **(fig. 22.20).** At high elevations, the partial pressures of all atmospheric gases are lower. Atmospheric P_{O_2}, for example, is 159 mm Hg at sea level but only 110 mm Hg at 3,000 m (10,000 feet). The O_2 gradient from air to blood is proportionately less, so as we can predict from Henry's law, less O_2 diffuses into the blood. In a hyperbaric oxygen chamber, by contrast, a patient is exposed to 3 to 4 atm of oxygen to treat such conditions as gangrene (to kill anaerobic bacteria) and carbon monoxide poisoning (to displace the carbon monoxide from hemoglobin). The P_{O_2} ranges from 2,300 to 3,000 mm Hg. Thus, there is a very steep gradient of P_{O_2} from alveolus to blood and diffusion into the blood is accelerated.

- **Solubility of the gases.** Gases differ in their ability to dissolve in water. Carbon dioxide is about 20 times as soluble as oxygen, and oxygen is about twice as soluble as nitrogen. Even though the pressure gradient of O_2 is much greater than that of CO_2 across the respiratory membrane, equal amounts of the two gases are exchanged because CO_2 is so much more soluble and diffuses more rapidly.

- **Membrane thickness.** The respiratory membrane between the blood and alveolar air is only 0.5 μm thick in most places—much less than the 7 to 8 μm diameter of a single RBC. Thus, it presents little obstacle to diffusion **(fig. 22.21a).** In such heart conditions as left ventricular failure, however, blood pressure builds up in the lungs and promotes capillary filtration into the connective tissues, causing the respiratory membranes to become edematous and thickened (similar to their condition in pneumonia; **fig. 22.21b).** The gases have farther to travel between blood and air, so oxygen can't get to the RBCs quickly enough to fully load their hemoglobin. Under these circumstances, blood leaving the lungs has an unusually low P_{O_2} and high P_{CO_2}.

- **Membrane area.** In good health, each lung has about 70 m^2 of respiratory membrane available for gas exchange. Since the alveolar capillaries contain a total of only 100 mL of blood at any one time, this blood is spread very thinly. Several pulmonary diseases, however, decrease the alveolar surface area and thus lead to low blood P_{O_2}—for example, emphysema **(fig. 22.21c)** and lung cancer.

- **Ventilation–perfusion coupling.** Gas exchange requires not only good ventilation of the alveoli but also good perfusion of their capillaries. *Ventilation–perfusion coupling* refers to physiological responses that match airflow to blood flow and vice versa. For example, if part of a lung were poorly ventilated because of tissue destruction or an airway obstruction, it would be pointless to direct much blood to that tissue. Poor ventilation leads to a low P_{O_2} in that region of the lung. This stimulates local vasoconstriction, rerouting the blood to better-ventilated areas of the lung where it can pick up more oxygen (left side of **fig. 22.22a).** In contrast, increased ventilation raises the local blood P_{O_2} and this stimulates vasodilation, increasing blood flow to that region to take advantage of the oxygen availability (right side of same figure). These reactions of the pulmonary arteries are opposite from the reactions of systemic arteries, which dilate in response to hypoxia. Furthermore, changes in the blood flow to a region of a lung stimulate bronchoconstriction or dilation, adjusting ventilation so that air is directed to the best-perfused parts of the lung **(fig. 22.22b).**

22.3c Gas Transport

Gas transport is the process of carrying gases from the alveoli to the systemic tissues and vice versa. This section explains how the blood loads and transports O_2 and CO_2.

Oxygen

Arterial blood carries about 20 mL of oxygen per deciliter. About 98.5% of it is bound to hemoglobin in the RBCs and 1.5% is dissolved in the blood plasma. Hemoglobin is specialized for oxygen transport. It consists of four protein (globin) chains, each with one heme group (see fig. 18.5). Each heme can bind 1 O_2 to the iron atom at its center; thus, one

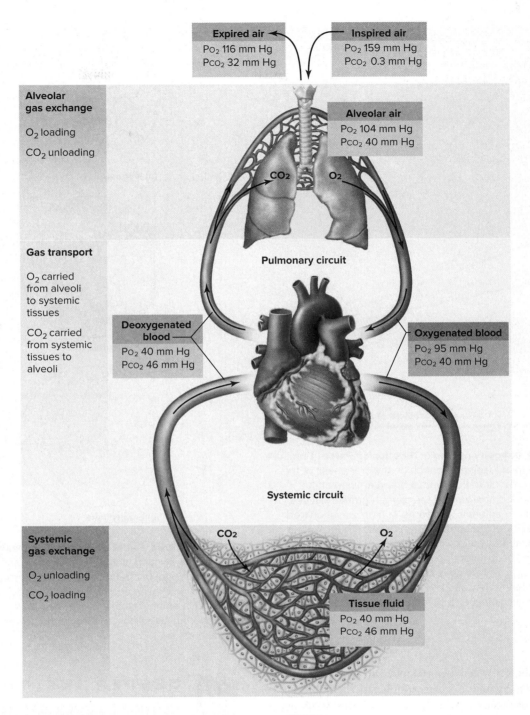

FIGURE 22.19 Changes in Po$_2$ and Pco$_2$ Along the Circulatory Route. **APR**

❓ *Trace the partial pressure of oxygen from inspired air to expired air and explain each change in Po$_2$ along the way. Do the same for Pco$_2$.*

hemoglobin molecule can carry up to 4 O$_2$. If one or more molecules of O$_2$ are bound to hemoglobin, the compound is called **oxyhemoglobin (HbO$_2$),** whereas hemoglobin with no oxygen bound to it is **deoxyhemoglobin (HHb).** When hemoglobin is 100% saturated, every molecule of it carries 4 O$_2$; if it is 75%

saturated, there is an average of 3 O$_2$ per hemoglobin molecule; if it is 50% saturated, there is an average of 2 O$_2$ per hemoglobin; and so forth. The poisonous effect of carbon monoxide stems from its competition for the O$_2$ binding site (see Deeper Insight 22.4).

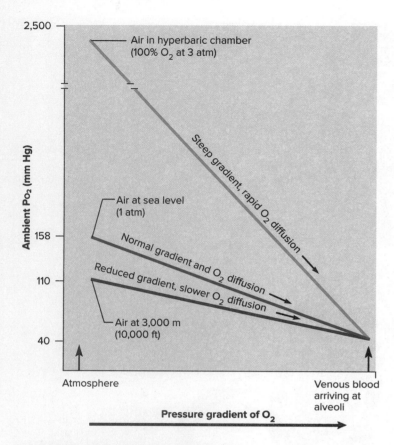

FIGURE 22.20 Oxygen Loading in Relation to Partial Pressure Gradient. The rate of loading depends on the steepness of the gradient from alveolar air to the venous blood arriving at the alveolar capillaries. Compared with the oxygen gradient at sea level (blue line), the gradient is less steep at high elevation (red line) because the P_{O_2} of the atmosphere is lower. Thus, oxygen loading of the pulmonary blood is slower. In a hyperbaric chamber with 100% oxygen, the gradient from air to blood is very steep (green line), and oxygen loading is correspondingly rapid. This is an illustration of Henry's law and has important effects in diving, aviation, mountain climbing, and oxygen therapy.

The relationship between hemoglobin saturation and ambient P_{O_2} is shown by the **oxyhemoglobin dissociation curve (fig. 22.23).** As you can see, it isn't a simple linear relationship. At low P_{O_2}, the curve rises slowly; then there is a rapid increase in oxygen loading as P_{O_2} rises farther. This reflects the way hemoglobin loads oxygen. When the first heme group binds O_2, hemoglobin changes shape in a way that facilitates uptake of the second O_2 by another heme group. This, in turn, promotes the uptake of the third and then the fourth O_2—hence the rapidly rising midportion of the curve. At high P_{O_2} levels, the curve levels off because the hemoglobin approaches 100% saturation and cannot load much more oxygen.

▶▶▶**APPLY WHAT YOU KNOW**

Is oxygen loading a positive or negative feedback process? Explain.

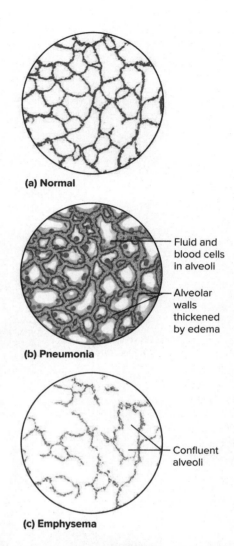

(a) Normal

(b) Pneumonia

Fluid and blood cells in alveoli

Alveolar walls thickened by edema

(c) Emphysema

Confluent alveoli

FIGURE 22.21 Pulmonary Alveoli in Health and Disease. (a) In a healthy lung, the alveoli are small and have thin respiratory membranes. (b) In pneumonia, the respiratory membranes (alveolar walls) are thick with edema, and the alveoli contain fluid and blood cells. (c) In emphysema, alveolar membranes break down and neighboring alveoli join to form larger, fewer alveoli with less total surface area.

 DEEPER INSIGHT 22.4

CLINICAL APPLICATION

Carbon Monoxide Poisoning

The lethal effect of carbon monoxide (CO) is well known. This colorless, odorless gas occurs in cigarette smoke, engine exhaust, and fumes from gas furnaces and space heaters. It binds to the iron of hemoglobin to form *carboxyhemoglobin (HbCO).* Thus, it competes with oxygen for the same binding site. Not only that, but it binds 210 times as tightly as oxygen. Thus, CO tends to tie up hemoglobin for a long time. Less than 1.5% of the hemoglobin is occupied by carbon monoxide in most nonsmokers, but this figure rises to as much as 3% in residents of heavily polluted cities and 10% in heavy smokers. An ambient concentration of 0.1% CO is enough to bind 50% of a person's hemoglobin, and a concentration of 0.2% is quickly lethal.

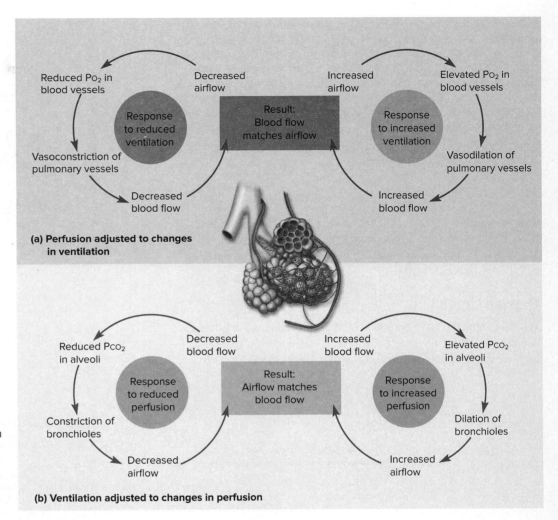

FIGURE 22.22 Ventilation–Perfusion Coupling. Negative feedback loops adjust airflow and blood flow to each other. (a) Blood circulation can be adjusted to match above- or below-normal ventilation of a part of the lung. (b) Ventilation can be adjusted to match above- or below-normal blood circulation to a part of the lung.

(a) Perfusion adjusted to changes in ventilation

(b) Ventilation adjusted to changes in perfusion

Carbon Dioxide

Carbon dioxide is transported in three forms: carbonic acid, carbamino compounds, and dissolved gas.

1. About 90% of the CO_2 is hydrated (reacts with water) to form **carbonic acid,** which then dissociates into bicarbonate and hydrogen ions:

$$CO_2 + H_2O \longrightarrow H_2CO_3 \longrightarrow HCO_3^- + H^+.$$

   More will be said about this reaction shortly.

2. About 5% binds to the amino groups of plasma proteins and hemoglobin to form **carbamino compounds—**chiefly **carbaminohemoglobin (HbCO₂).** The reaction with hemoglobin can be symbolized $Hb + CO_2 \longrightarrow HbCO_2$. Carbon dioxide doesn't compete with oxygen because CO_2 and O_2 bind to different sites on the hemoglobin molecule—oxygen to the heme moiety and CO_2 to the polypeptide chains. Hemoglobin can therefore transport both O_2 and CO_2 simultaneously. As we will see, however, each gas somewhat inhibits transport of the other.

3. The remaining 5% of the CO_2 is carried in the blood as dissolved gas, like the CO_2 in sparkling wines and carbonated beverages.

The relative amounts of CO_2 exchanged between the blood and alveolar air differ from the percentages just given. About 70% of the *exchanged* CO_2 comes from carbonic acid, 23% from carbamino compounds, and 7% from the dissolved gas. That is, blood gives up the dissolved CO_2 gas and CO_2 from the carbamino compounds more easily than it gives up the CO_2 in bicarbonate.

22.3d Systemic Gas Exchange

Systemic gas exchange is the unloading of O_2 and loading of CO_2 at the systemic capillaries (see fig. 22.19, bottom; **fig. 22.24**).

Carbon Dioxide Loading

Aerobic respiration produces a molecule of CO_2 for each O_2 it consumes. The tissue fluid therefore contains a relatively high P_{CO_2} and there is typically a CO_2 gradient of $46 \rightarrow 40$ mm Hg from tissue fluid to blood. Consequently, CO_2 diffuses into the bloodstream, where it is carried in the three forms already noted. Most of it reacts with water to produce bicarbonate ions (HCO_3^-) and hydrogen ions (H^+). This reaction occurs slowly in the blood plasma but much faster in the RBCs, where it is catalyzed by the enzyme *carbonic anhydrase.* An antiport called the *chloride–bicarbonate exchanger*

FIGURE 22.23 The Oxyhemoglobin Dissociation Curve. This curve shows the relative amount of hemoglobin that is saturated with oxygen (*y*-axis) as a function of ambient (surrounding) oxygen partial pressure (*x*-axis). As it passes through the alveolar capillaries where the P_{O_2} is high, hemoglobin becomes saturated with oxygen. As it passes through the systemic capillaries where the P_{O_2} is low, it typically gives up about 22% of its oxygen (color bar at top of graph). This percentage is called the *utilization coefficient.*

> **?** *What would be the approximate utilization coefficient if the systemic tissues had a P_{O_2} of 30 mm Hg?*

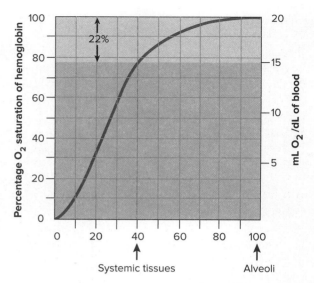

then pumps most of the HCO_3^- out of the RBC in exchange for Cl^- from the blood plasma. This exchange is called the **chloride shift.** Most of the H^+ binds to hemoglobin or oxyhemoglobin, which thus buffers the intracellular pH.

Oxygen Unloading

When H^+ binds to oxyhemoglobin (HbO_2), it reduces the affinity of hemoglobin for O_2 and tends to make hemoglobin release it. Oxygen consumption by respiring tissues keeps the P_{O_2} of tissue

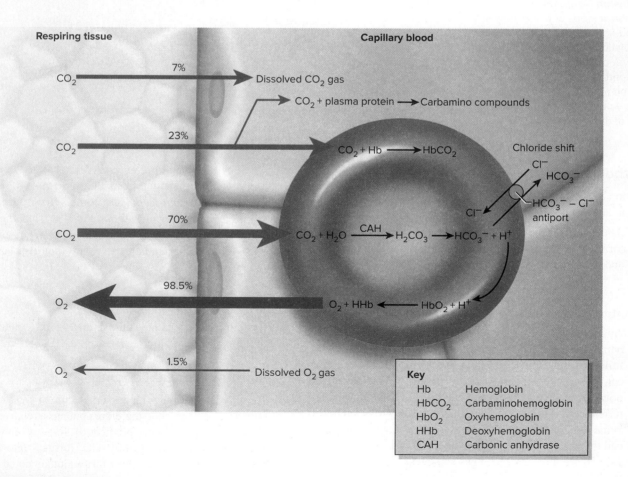

FIGURE 22.24 Systemic Gas Exchange. Blue arrows show the three mechanisms of CO_2 loading and transport; their thickness represents the relative amounts of CO_2 loaded in each of the three forms. Red arrows show the two mechanisms of O_2 unloading; their thickness indicates the relative amounts unloaded by each mechanism. Note that CO_2 loading releases hydrogen ions in the erythrocyte, and hydrogen ions promote O_2 unloading.

fluid relatively low, so there is typically a pressure gradient of about 95 → 40 mm Hg of oxygen from the arterial blood to the tissue fluid. Thus, the liberated oxygen—along with some that was carried as dissolved gas in the plasma—diffuses from the blood into the tissue fluid.

As blood arrives at the systemic capillaries, its oxygen concentration is about 20 mL/dL and the hemoglobin is about 97% saturated. As it leaves the capillaries of a typical resting tissue, its oxygen concentration is about 15.6 mL/dL and the hemoglobin is about 75% saturated. Thus, it has given up 4.4 mL/dL—about 22% of its oxygen load. This fraction is called the **utilization coefficient** (see fig. 22.23). The oxygen remaining in the blood after it passes through the capillary bed provides a **venous reserve** of oxygen, which can sustain life for 4 to 5 minutes even in the event of respiratory arrest. At rest, the circulatory system releases oxygen to the tissues at an overall rate of about 250 mL/min.

22.3e Alveolar Gas Exchange Revisited

The processes illustrated in figure 22.24 make it easier to understand alveolar gas exchange more fully. As shown in **figure 22.25,** the reactions that occur in the lungs are essentially the reverse of

systemic gas exchange. As hemoglobin loads oxygen, its affinity for H^+ declines. Hydrogen ions dissociate from the hemoglobin and bind with bicarbonate ions (HCO_3^-) transported from the plasma into the RBCs. Chloride ions are transported back out of the RBC (a reverse chloride shift). The reaction of H^+ and HCO_3^- reverses the hydration reaction and generates free CO_2. This diffuses into the alveolus to be exhaled—as does the CO_2 released from carbaminohemoglobin and CO_2 gas that was dissolved in the plasma.

22.3f Adjustment to the Metabolic Needs of Tissues

Hemoglobin doesn't unload the same amount of oxygen to all tissues. Some tissues need more and some less, depending on their state of activity. Hemoglobin responds to such variations and unloads more oxygen to the tissues that need it most. In exercising skeletal muscles, for example, the utilization coefficient may be as high as 80%. Four factors adjust the rate of oxygen unloading to the metabolic rates of different tissues:

1. **Ambient P_{O_2}.** Since an active tissue consumes oxygen rapidly, the P_{O_2} of its tissue fluid remains low. From the

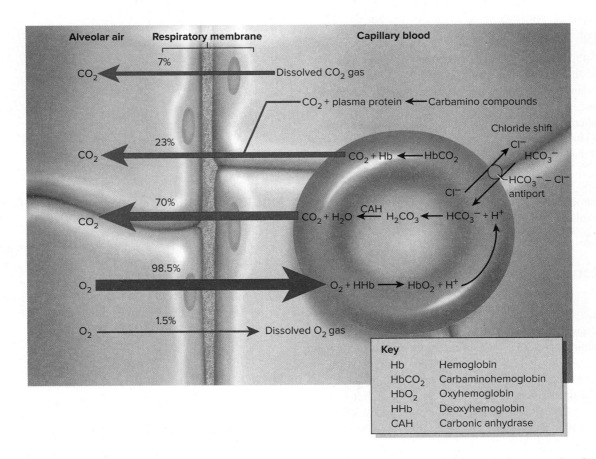

FIGURE 22.25 Alveolar Gas Exchange. Arrow colors and thicknesses represent the same variables as in the preceding figure. Note that O_2 loading promotes the decomposition of carbonic acid into H_2O and CO_2, and most exhaled CO_2 comes from the erythrocytes.

❓ *In what fundamental way does this differ from the preceding figure? Following alveolar gas exchange, will the blood contain a higher or lower concentration of bicarbonate ions than it did before?*

oxyhemoglobin dissociation curve (see fig. 22.23), you can see that at a low Po_2, HbO_2 releases more oxygen.

2. **Temperature.** When temperature rises, the oxyhemoglobin dissociation curve shifts to the right **(fig. 22.26a)**; in other words, elevated temperature promotes oxygen unloading. Active tissues are warmer than less active ones and thus extract more oxygen from the blood passing through them.

3. **Ambient pH.** Active tissues also generate extra CO_2, which raises the H^+ concentration and lowers the pH of the blood. Hydrogen ions weaken the bond between hemoglobin and oxygen and thereby promote oxygen unloading—a phenomenon called the **Bohr**[28] **effect.** This can also be seen in the oxyhemoglobin dissociation curve, where a drop in pH shifts the curve to the right **(fig. 22.26b)**. The effect is less pronounced at the high Po_2 present in the lungs, so pH has relatively little effect on pulmonary oxygen loading. In the systemic capillaries, however, Po_2 is lower and the Bohr effect is more pronounced.

4. **BPG.** Erythrocytes have no mitochondria and meet their energy needs solely by anaerobic fermentation. One of their metabolic intermediates is **bisphosphoglycerate (BPG),** which binds to hemoglobin and promotes oxygen unloading. An elevated body temperature (as in fever) stimulates BPG synthesis, as do thyroxine, growth hormone, testosterone, and epinephrine. All of these hormones thus promote oxygen unloading to the tissues.

The rate of CO_2 loading is also adjusted to varying needs of the tissues. A low level of oxyhemoglobin (HbO_2) enables the blood to transport more CO_2, a phenomenon known as the **Haldane**[29] **effect.** This occurs for two reasons: (1) HbO_2 doesn't bind CO_2 as well as deoxyhemoglobin (HHb) does. (2) HHb binds more hydrogen ions than HbO_2 does, and by removing H^+ from solution, HHb shifts the carbonic acid reaction to the right:

$$H_2O + CO_2 \longrightarrow HCO_3^- + H^+.$$

A high metabolic rate keeps oxyhemoglobin levels relatively low and thus allows more CO_2 to be transported by these two mechanisms.

22.3g Blood Gases and the Respiratory Rhythm

Normally the systemic arterial blood has a Po_2 of 95 mm Hg, a Pco_2 of 40 mm Hg, and a pH of 7.40 ± 0.05. The rate and depth of breathing are adjusted to maintain these values. This is possible because the brainstem respiratory centers receive input from central and peripheral chemoreceptors that monitor the composition of the blood and CSF, as described earlier in this chapter. Of these three chemical stimuli, the most potent stimulus for breathing is pH, followed by CO_2; perhaps surprisingly, the least significant is O_2.

(a) Effect of temperature

(b) Effect of pH

FIGURE 22.26 Effects of Temperature and pH on Oxyhemoglobin Dissociation. (a) For a given Po_2, hemoglobin unloads more oxygen at higher temperatures. (b) For a given Po_2, hemoglobin unloads more oxygen at lower pH (the Bohr effect). Both mechanisms cause hemoglobin to release more oxygen to tissues with higher metabolic rates.

❓ *Why is it physiologically beneficial to the body that the curves in part (a) shift to the right as temperature increases?*

Hydrogen Ions

Ultimately, pulmonary ventilation is adjusted to maintain the pH of the brain. The central chemoreceptors in the medulla oblongata mediate about 75% of the change in respiration induced by pH shifts, and yet H^+ doesn't cross the blood–brain barrier very easily. However, CO_2 does, and once it's in the CSF, it reacts with water to produce carbonic acid, and the carbonic acid dissociates into bicarbonate and hydrogen ions. The CSF contains relatively little protein to buffer the hydrogen ions, so most H^+ remains free, and it strongly stimulates the central chemoreceptors. Hydrogen ions are also a potent stimulus to the peripheral chemoreceptors, which mediate about 25% of the respiratory response to pH changes.

How do we know it is the H^+ that stimulates the central chemoreceptors and not primarily the CO_2 that diffuses into the CSF? Experimentally, it is possible to vary the pH or the P_{CO_2} of the CSF while holding the other variable steady. When pH alone changes, there is a strong effect on respiration; when P_{CO_2} alone changes, the effect is weaker. Therefore, even though these two variables usually change together, we can see that the chemoreceptors react primarily to the H^+.

A blood pH lower than 7.35 is called **acidosis** and a pH greater than 7.45 is called **alkalosis.** The P_{CO_2} of arterial blood normally ranges from 37 to 43 mm Hg. A P_{CO_2} less than 37 mm Hg is called **hypocapnia,**[30] and is the most common cause of alkalosis. The most common cause of acidosis is **hypercapnia,** a P_{CO_2} greater than 43 mm Hg. When these pH imbalances occur because the rate of pulmonary ventilation doesn't match the rate of CO_2 production, they are called *respiratory acidosis* and *respiratory alkalosis* (further discussed under "Disorders of Acid–Base Balance" in section 24.3).

The corrective homeostatic response to acidosis is hyperventilation, "blowing off" CO_2 faster than the body produces it. As CO_2 is eliminated from the body, the carbonic acid reaction shifts to the left:

$$CO_2 + H_2O \longleftarrow H_2CO_3 \longleftarrow HCO_3^- + H^+.$$

Thus, the H^+ on the right is consumed, and as H^+ concentration declines, the pH rises and ideally returns the blood from the acidotic range to normal.

The corrective response to alkalosis is hypoventilation, which allows CO_2 to accumulate in the body fluids faster than we exhale it. Hypoventilation shifts the reaction to the right, raising the H^+ concentration and lowering the pH to normal:

$$CO_2 + H_2O \longrightarrow H_2CO_3 \longrightarrow HCO_3^- + H^+.$$

Although pH changes usually result from P_{CO_2} changes, they can have other causes. In diabetes mellitus, for example, rapid fat oxidation releases acidic ketone bodies, causing an abnormally low pH called *ketoacidosis.* Ketoacidosis tends to induce a form of dyspnea called *Kussmaul respiration* (see table 22.3). Hyperventilation cannot reduce the level of ketone bodies in the blood, but by blowing off CO_2, it reduces the concentration of CO_2-generated H^+ and compensates to some degree for the H^+ released by the ketone bodies.

Carbon Dioxide

Although the arterial P_{CO_2} has a strong influence on respiration, we have seen that it's mostly an indirect one, mediated through its effects on the pH of the CSF. Yet the experimental evidence described earlier shows that CO_2 has some effect even when pH remains stable. At the beginning of exercise, the rising blood CO_2 level may directly stimulate the peripheral chemoreceptors and trigger an increase in ventilation more quickly than the central chemoreceptors do.

Oxygen

The partial pressure of oxygen usually has little effect on respiration. Even in eupnea, the hemoglobin is at least 97% saturated with O_2, so little can be added by increasing pulmonary ventilation. Arterial P_{O_2} significantly affects respiration only if it drops below 60 mm Hg. At low elevations, such a low P_{O_2} seldom occurs even in prolonged holding of the breath. A moderate drop in P_{O_2} does stimulate the peripheral chemoreceptors, but another effect overrides this: As the level of HbO_2 falls, hemoglobin binds more H^+ (see fig. 22.24). This raises the blood pH, which inhibits respiration and counteracts the effect of low P_{O_2}.

At about 10,800 feet (3,300 m), arterial P_{O_2} falls to 60 mm Hg and the stimulatory effect of hypoxemia on the carotid bodies overrides the inhibitory effect of the pH increase. This produces heavy breathing in people who aren't acclimated to high elevation. Long-term hypoxemia can lead to a condition called **hypoxic drive,** in which respiration is driven more by the low P_{O_2} than by CO_2 or pH. This occurs in situations such as emphysema and pneumonia, which interfere with alveolar gas exchange, and in mountain climbing of at least 2 or 3 days' duration.

Respiration and Exercise

It is common knowledge that we breathe more heavily during exercise, and it's tempting to think this occurs because exercise raises CO_2 levels, lowers the blood pH, and lowers blood O_2 levels. However, this is not true; all these values remain essentially the same in exercise as they do at rest. It appears that the increased respiration has other causes: (1) When the brain sends motor commands to the muscles (via the lower motor neurons of the spinal cord), it also sends this information to the respiratory centers, so they increase pulmonary ventilation in anticipation of the needs of the exercising muscles. In contrast to homeostasis by negative feedback, this is considered a *feed-forward* mechanism, in which signals are transmitted to the effectors (brainstem respiratory centers) to produce a change *in anticipation* of need. (2) Exercise stimulates proprioceptors of the muscles and joints, and they transmit excitatory signals to the brainstem respiratory centers. Thus, the respiratory centers increase breathing because they're informed that the muscles have been told to move or are actually moving. The increase in pulmonary ventilation keeps blood gas values at their normal levels in spite of the elevated O_2 consumption and CO_2 generation by the muscles.

In summary, the main chemical stimulus to pulmonary ventilation is the H^+ in the CSF and tissue fluid of the brain. These hydrogen ions arise mainly from CO_2 diffusing into the CSF and brain and generating H^+ through the carbonic acid reaction. Therefore, the P_{CO_2} of the arterial blood is an important driving force in respiration, even though its action on the chemoreceptors is indirect. Ventilation is adjusted to maintain arterial pH at about 7.40 and arterial P_{CO_2} at about 40 mm Hg. This automatically ensures that the blood is at least 97% saturated with O_2 as well.

[30]*capn* = smoke

Under ordinary circumstances, arterial P_{O_2} has relatively little effect on respiration. When it drops below 60 mm Hg, however, it excites the peripheral chemoreceptors and stimulates an increase in ventilation. This can be significant at high elevations and in certain lung diseases. The increase in respiration during exercise results from the expected or actual activity of the muscles, not from any change in blood gas pressures or pH.

BEFORE YOU GO ON

Answer the following questions to test your understanding of the preceding section:

14. Why is the composition of alveolar air different from that of the atmosphere?

15. What four factors affect the efficiency of alveolar gas exchange?

16. Explain how perfusion of a pulmonary lobule changes if it is poorly ventilated.

17. How is most oxygen transported in the blood, and why does carbon monoxide interfere with this?

18. What are the three ways in which blood transports CO_2?

19. Give two reasons why highly active tissues can extract more oxygen from the blood than less active tissues do.

20. Define *hypocapnia* and *hypercapnia*. Name the pH imbalances that result from these conditions and explain the relationship between P_{CO_2} and pH.

21. What is the most potent chemical stimulus to respiration, and where are the most effective chemoreceptors for it located?

22. Explain how changes in pulmonary ventilation can correct pH imbalances.

22.4 Respiratory Disorders

Expected Learning Outcomes
When you have completed this section, you should be able to

a. describe the forms and effects of oxygen deficiency and oxygen excess;

b. describe the chronic obstructive pulmonary diseases and their consequences; and

c. explain how lung cancer begins, progresses, and exerts its lethal effects.

The delicate lungs are exposed to a wide variety of inhaled pathogens and debris, so it's not surprising that they're prone to a host of diseases. Several already have been mentioned in this chapter and some others are briefly described in **table 22.5.** The effects of aging on the respiratory system are discussed in section 29.4a.

22.4a Oxygen Imbalances

Hypoxia is a deficiency of oxygen in a tissue or the inability to use oxygen. It's not a respiratory disease in itself but is often a consequence of respiratory diseases. Hypoxia is classified according to cause:

- **Hypoxemic hypoxia,** a state of low arterial P_{O_2}, is usually due to inadequate pulmonary gas exchange. Some of its root causes include atmospheric deficiency of oxygen at high elevations; impaired ventilation, as in drowning or aspiration of foreign matter; respiratory arrest; and degenerative lung diseases. It also occurs in carbon monoxide poisoning, which prevents hemoglobin from transporting oxygen.

- **Ischemic hypoxia** results from inadequate blood circulation, as in congestive heart failure.

- **Anemic hypoxia** is due to anemia and the resulting inability of the blood to carry adequate oxygen.

- **Histotoxic hypoxia** occurs when a metabolic poison such as cyanide prevents the tissues from using the oxygen delivered to them.

Hypoxia is often marked by **cyanosis,** blueness of the skin. Whatever the cause, its primary danger is the necrosis of oxygen-starved tissues. This is especially critical in organs with the highest metabolic demands, such as the brain, heart, and kidneys.

An oxygen excess is also dangerous. It is safe to breathe 100% oxygen at 1 atm for a few hours, but **oxygen toxicity** rapidly develops when pure oxygen is breathed at 2.5 atm or greater. Excess oxygen generates hydrogen peroxide and free radicals that destroy enzymes and damage nervous tissue; thus, it can lead to seizures, coma, and death. This is why scuba divers breathe a mixture of oxygen and nitrogen rather than pure compressed oxygen (see Deeper Insight 22.5). Hyperbaric oxygen was formerly used to treat premature infants for respiratory distress syndrome, but it caused retinal deterioration and blinded many infants before the practice was discontinued.

22.4b Chronic Obstructive Pulmonary Diseases

Chronic obstructive pulmonary diseases (COPDs) are defined by a long-term obstruction of airflow and substantial reduction of pulmonary ventilation. The major COPDs are *chronic bronchitis* and *emphysema.* COPDs are leading causes of adult mortality in the United States. They are almost always caused by cigarette smoking, but occasionally result from air pollution, occupational exposure to airborne irritants, or a hereditary defect. Most COPD patients exhibit mixed chronic bronchitis and emphysema, but one form or the other often predominates.

Chronic bronchitis is severe, persistent inflammation of the lower respiratory tract. Goblet cells of the bronchial mucosa enlarge and secrete excess mucus, while at the same time, the cilia are immobilized and unable to discharge it. Thick, stagnant mucus accumulates in the lungs and furnishes a growth medium for bacteria. Furthermore, tobacco smoke incapacitates the alveolar

TABLE 22.5	Some Disorders of the Respiratory System
Acute rhinitis	The common cold. Caused by many types of viruses that infect the upper respiratory tract. Symptoms include congestion, increased nasal secretion, sneezing, and dry cough. Transmitted especially by contact of contaminated hands with mucous membranes; not transmitted orally.
Adult respiratory distress syndrome	Acute lung inflammation and alveolar injury stemming from trauma, infection, burns, aspiration of vomit, inhalation of noxious gases, drug overdoses, and other causes. Alveolar injury is accompanied by severe pulmonary edema and hemorrhaging, followed by fibrosis that progressively destroys lung tissue. Fatal in about 40% of cases under age 60 and in 60% of cases over age 65.
Pneumonia	A lower respiratory infection caused by any of several viruses, fungi, or protozoans, but most often the bacterium *Streptococcus pneumoniae*. Causes filling of alveoli with fluid and dead leukocytes and thickening of the respiratory membrane, which interferes with gas exchange and causes hypoxemia. Especially dangerous to infants, the elderly, and people with compromised immune systems, such as AIDS and leukemia patients.
Sleep apnea	Cessation of breathing for 10 seconds or longer during sleep; sometimes occurs hundreds of times per night, often accompanied by restlessness and alternating with snoring. Can result from altered function of CNS respiratory centers, airway obstruction, or both. Over time, may lead to daytime drowsiness, hypoxemia, polycythemia, pulmonary hypertension, congestive heart failure, and cardiac arrhythmia. Most common in obese people and men.
Tuberculosis (TB)	Pulmonary infection with the bacterium *Mycobacterium tuberculosis*, which invades the lungs by way of air, blood, or lymph. Stimulates the lung to form fibrous nodules called tubercles around the bacteria. Progressive fibrosis compromises the elastic recoil and ventilation of the lungs and causes pulmonary hemorrhaging as it invades blood vessels. Especially common among impoverished and homeless people and becoming increasingly common among people with AIDS.

You can find other respiratory system disorders described in the following places:

Cystic fibrosis in Deeper Insight 3.2; *pulmonary hypertension* and *pulmonary edema* in section 19.5d; *cor pulmonale* in sections 19.6d and 22.4b; *asthma* in Deeper Insight 21.3; *Ondine's curse* in Deeper Insight 22.2; *pneumothorax* and *atelectasis* in section 22.2c; *apnea, dyspnea, orthopnea, hyperpnea, tachypnea, hyper-* and *hypoventilation, Kussmaul respiration,* and *respiratory arrest* in table 22.3; *carbon monoxide poisoning* in Deeper Insight 22.4; *hypoxia, COPDs (emphysema* and *chronic bronchitis),* and *lung cancer* in section 22.4; *decompression sickness* in Deeper Insight 22.5; *respiratory acidosis* and *alkalosis* in section 24.3d; and *infant respiratory distress syndrome* in section 29.3b.

macrophages and reduces one's defense against respiratory infection. Smokers with chronic bronchitis develop a chronic cough and bring up a thick mixture of mucus and cellular debris called **sputum** (SPEW-tum). Since blood flowing through congested areas of the lung cannot load a normal amount of oxygen, the ventilation–perfusion ratio is reduced and such patients commonly exhibit hypoxemia and cyanosis.

In **emphysema**[31] (EM-fih-SEE-muh), alveolar walls break down and alveoli converge into fewer and larger spaces (see fig. 22.21c). Thus, there is much less respiratory membrane available for gas exchange. In severe cases, the lungs are flabby and cavitated with spaces as big as grapes or even ping-pong balls. The severity of the disease may not be fully appreciated by looking only at histological specimens, since such large spaces are not seen on microscope slides. The lungs also become fibrotic and less elastic. The air passages open adequately during inspiration, but they tend to collapse and obstruct the outflow of air. Air becomes trapped in the lungs, and over a period of time a person becomes barrel-chested. The overly stretched thoracic muscles contract weakly, which further contributes to the difficulty of expiration. Since proportionate amounts of alveolar wall and capillaries are both destroyed, the ventilation–perfusion ratio of the lung is relatively normal, and

persons with emphysema don't necessarily show the cyanosis that typifies chronic bronchitis. People with emphysema can become exhausted and emaciated because they expend three to four times the normal amount of energy just to breathe. Even slight physical exertion, such as walking across a room, can cause severe shortness of breath.

▶▶▶**APPLY WHAT YOU KNOW**

Explain how the length–tension relationship of skeletal muscle (see fig. 11.12 and associated text) accounts for the weakness of the respiratory muscles in emphysema.

COPD tends to reduce vital capacity and causes hypoxemia, hypercapnia, and respiratory acidosis. Hypoxemia stimulates the kidneys to secrete erythropoietin, which leads to accelerated erythrocyte production and polycythemia (see fig. 18.8). As the blood thickens and lung tissue deteriorates, the right ventricle has to work harder to force blood through the lungs. Like any heavily used muscle, it enlarges and may eventually fail. Such ventricular hypertrophy and failure resulting from obstructed pulmonary circulation is called **cor pulmonale.**[32]

---

[31]*emphys* = inflamed

[32]*cor* = heart; *pulmo* = lung

22.4c Smoking and Lung Cancer

Lung cancer (**fig. 22.27**) accounts for more deaths than any other form of cancer. The most important cause of lung cancer is cigarette smoking, distantly followed by air pollution. Cigarette smoke contains at least 60 carcinogenic compounds. Lung cancer commonly follows or accompanies COPD.

There are three forms of lung cancer, the most common of which is **squamous-cell carcinoma.** In its early stage, basal cells of the bronchial epithelium multiply and the ciliated pseudostratified epithelium transforms into the stratified squamous type. As the dividing epithelial cells invade the underlying tissues of the bronchial wall, the bronchus develops bleeding lesions. Dense masses of keratin and malignant squamous cells appear in the lung parenchyma and replace functional respiratory tissue.

A second form of lung cancer, nearly as common, is **adenocarcinoma,**[33] which originates in the mucous glands of the lamina propria. The least common (10% to 20% of lung malignancies) but most dangerous form is **small-cell (oat-cell) carcinoma,** named for clusters of cells that resemble oat grains. This originates in the main bronchi but invades the mediastinum and metastasizes quickly to other organs.

[33]*adeno* = gland; *carcino* = cancer; *oma* = tumor

Over 90% of lung tumors originate in the mucous membranes of the large bronchi. As a tumor invades the bronchial wall and grows around it, it compresses the airway and may cause atelectasis (collapse) of more distal parts of the lung. Growth of the tumor produces a cough, but coughing is such an everyday occurrence among smokers it seldom causes much alarm. Often, the first sign of serious trouble is coughing up blood. Lung cancer metastasizes so rapidly that it has usually spread to other organs by the time it's diagnosed. Common sites of metastasis are the pericardium, heart, bones, liver, lymph nodes, and brain. The chance of recovery is poor relative to other cancers, but is improving with advances in treatment. About 18% of patients survive for 5 years after diagnosis.

The popular use of recreational marijuana and its legalization in some states has sparked attention to it as a risk factor for lung cancer. Marijuana smokers inhale more deeply and hold it longer than cigarette smokers do, and they smoke the marijuana "joint" down to its last nub, loaded with tar and other carcinogens. Studies to date have been too limited to establish a significant correlation between marijuana use and lung cancer. Ongoing investigations aim to improve study methods and survey older people with a long history of marijuana use. The growing popularity of electronic cigarettes ("vaping") is also cause for concern. It has been associated with a serious lung disease called *popcorn lung,* but the composition of vaping liquids has undergone changes and the jury is still out on the seriousness of this public health risk.

(a) Healthy lung, mediastinal surface

Tumors

(b) Smoker's lung with carcinoma

FIGURE 22.27 Effects of Smoking. (a) Photograph of a healthy lung, mediastinal surface. (b) Photograph of a smoker's lung with carcinoma.

a: Dennis Strete/McGraw-Hill Education; **b:** Biophoto Associates/Science Source

BEFORE YOU GO ON

Answer the following questions to test your understanding of the preceding section.

23. Describe the four classes of hypoxia.

24. Name and compare the two COPDs and describe some pathological effects that they have in common.

25. In what lung tissue does lung cancer originate? How does it kill?

DEEPER INSIGHT 22.5

CLINICAL APPLICATION

Diving Physiology and Decompression Sickness

Because of the popularity of scuba diving, many people know something about the scientific aspects of breathing under high pressure. But diving is by no means a new fascination. As early as the fifth century BCE, Aristotle described divers using snorkels and taking containers of air underwater in order to stay down longer. Some Renaissance artists depicted divers many meters deep breathing from tubes to the water surface. In reality, this would be physically impossible. For one thing, such tubes would have so much dead space that fresh air from the surface wouldn't reach the diver. The short snorkels used today are about the maximum length that will work for surface breathing. Another reason snorkels can't be used at greater depths is that water pressure increases by 1 atm for every 10 m of depth, and even at 1 m the pressure is so great that a diver can't expand the chest muscles without help. This is one reason why scuba divers use pressurized air tanks. The tanks create a positive intrapulmonary pressure and enable the diver to inhale with only slight assistance from the thoracic muscles. Scuba tanks also have regulators that adjust the outflow pressure to the diver's depth and the opposing pressure of the surrounding water.

But breathing pressurized (hyperbaric) gas presents its own problems. Divers can't use pure oxygen because of the problem of oxygen toxicity. Instead, they use compressed air—a mixture of 21% oxygen and 79% nitrogen. On land, nitrogen presents no physiological problems; it dissolves poorly in blood and it is physiologically inert. But under hyperbaric conditions, larger amounts of nitrogen dissolve in the blood. (Which of the gas laws in table 22.1 applies here?) Even more dissolves in adipose tissue and the myelin of the brain, since nitrogen is more soluble in lipids. In the brain, it causes *nitrogen narcosis,* or what scuba inventor Jacques Cousteau (1910–1997) termed "rapture of the deep." A diver can become dizzy, euphoric, and dangerously disoriented; for every 15 to 20 m of depth, the effect is said to be equivalent to that of one martini on an empty stomach.

Strong currents, equipment failure, and other hazards sometimes make scuba divers panic, hold their breath, and quickly swim to the surface (a *breath-hold ascent*). Ambient (surrounding) pressure falls rapidly as a diver ascends, and the air in the lungs expands just as rapidly. (Which gas law is demonstrated here?) It is imperative that an ascending diver keep his or her airway open to exhale the expanding gas; otherwise it's likely to cause *pulmonary barotrauma*—ruptured alveoli. Then, when the diver takes a breath of air at the surface, alveolar air goes directly into the bloodstream and causes air embolism (see Deeper Insight 20.3).

After passing through the heart, the emboli enter the cerebral circulation because the diver is head-up and air bubbles rise in liquid. The resulting cerebral embolism can cause motor and sensory dysfunction, seizures, unconsciousness, and drowning.

Barotrauma can be fatal even at the depths of a backyard swimming pool. In one case, children trapped air in a bucket 1 m underwater and then swam under the bucket to breathe from the air space. Because the bucket was under water, the air in it was compressed. One child filled his lungs under the bucket, did a "mere" 1 m breath-hold ascent, and his alveoli ruptured. He died in the hospital, partly because the case was mistaken for drowning and not treated for what it really was. This would not have happened to a person who inhaled at the surface, did a breath-hold dive, and then resurfaced—nor is barotrauma a problem for those who do breath-hold dives to several meters. (Why? What is the difference?)

Even when not holding the breath, but letting the expanding air escape from the mouth, a diver must ascend slowly and carefully to allow for decompression of the nitrogen that has dissolved in the tissues. *Decompression tables* prescribe safe rates of ascent based on the depth and the length of time a diver has been down. When pressure drops, nitrogen dissolved in the tissues can go either of two places—it can diffuse into the alveoli and be exhaled, or it can form bubbles like the CO_2 in a bottle of soda when the cap is removed. The diver's objective is to ascend slowly, allowing for the former and preventing the latter. If a diver ascends too rapidly, nitrogen "boils" from the tissues—especially in the 3 m just below the surface, where the relative pressure change is greatest. A diver may double over in pain from bubbles in the joints, bones, and muscles—a disease called the *bends* or *decompression sickness (DCS)*. Nitrogen bubbles in the pulmonary capillaries cause *chokes*—substernal pain, coughing, and dyspnea. DCS is sometimes accompanied by mood changes, seizures, numbness, and itching. These symptoms usually occur within an hour of surfacing, but they are sometimes delayed for up to 36 hours. DCS is treated by putting the individual in a hyperbaric chamber to be recompressed and then *slowly* decompressed.

DCS is also called *caisson disease*. A caisson is a watertight underwater chamber filled with pressurized air. Caissons are used in underwater construction work on bridges, tunnels, ships' hulls, and so forth. Caisson disease was first reported in the late 1800s among workmen building the foundations of the Brooklyn Bridge.

Effects of the **RESPIRATORY SYSTEM**
on Other Organ Systems

ALL SYSTEMS
Delivers oxygen to the tissues and removes their carbon dioxide; maintains proper acid–base balance in the tissues

INTEGUMENTARY SYSTEM
Respiratory disorders can cause such skin discolorations as the cyanosis of hypoxemia or the cherry-red color of carbon monoxide poisoning.

SKELETAL SYSTEM
Any respiratory disorder that causes hypoxemia stimulates accelerated erythropoiesis in the red bone marrow.

MUSCULAR SYSTEM
Acid–base imbalances of respiratory origin can affect neuromuscular function.

NERVOUS SYSTEM
Respiration affects the pH of the cerebrospinal fluid, which in turn affects neural function with effects ranging from hyperexcitability to depressed excitability and coma.

ENDOCRINE SYSTEM
Lungs produce angiotensin-converting enzyme (ACE), which converts angiotensin I to the hormone angiotensin II; hypoxemia stimulates secretion of erythropoietin.

CIRCULATORY SYSTEM
Thoracic pump aids venous return of blood; proplatelets break up into platelets in the lungs; angiotensin II, produced in the lungs, stimulates vasoconstriction and helps regulate blood volume and pressure; respiration strongly influences blood pH; obstruction of pulmonary circulation can lead to right-sided heart failure; lungs filter blood clots and emboli and prevent them from obstructing vital arteries elsewhere.

LYMPHATIC AND IMMUNE SYSTEMS
Thoracic pump promotes lymph flow and its return to the bloodstream.

URINARY SYSTEM
Valsalva maneuver aids in urination; urinary and respiratory systems collaborate in acid–base balance and compensate for each other's deficiencies in maintaining normal pH; hypoxemia stimulates kidneys to secrete erythropoietin.

DIGESTIVE SYSTEM
Valsalva maneuver aids in defecation.

REPRODUCTIVE SYSTEM
Valsalva maneuver aids in childbirth.

STUDY GUIDE

▶ Assess Your Learning Outcomes

To test your knowledge, discuss the following topics with a study partner or in writing, ideally from memory.

22.1 Anatomy of the Respiratory System

1. Two meanings of the word *respiration*
2. Functions of the respiratory system
3. The distinction between the respiratory and conducting divisions of this system, and constituents of each division
4. The distinction between the upper and lower respiratory tract and the dividing line between them
5. The extent of the nasal cavity, names of its anterior and posterior openings, and names of the two chambers separated by the nasal septum, and histology of its mucosa
6. Names and functions of the scroll-like folds that arise from each lateral wall of the nasal cavity
7. Anatomy and functions of the pharynx, larynx, and trachea
8. Gross anatomy of the lungs; how the right and left lungs differ; and the structures that enter or leave through the hilum
9. Divisions of the bronchial tree from main bronchus to segmental bronchi, and histological changes along the way
10. How bronchioles differ from bronchi; two types of bronchioles; and how the two differ in histology and function
11. Alveolar ducts and alveoli; cell types of the alveoli and their functions; the relationship of pulmonary blood vessels to the alveoli; and the structure of the respiratory membrane in relation to its function
12. The parietal and visceral pleurae, pleural cavity, and pleural fluid

22.2 Pulmonary Ventilation

1. The two phases of the respiratory cycle
2. Actions of the respiratory muscles; the prime mover and synergists of respiration
3. Locations and functions of the brainstem respiratory centers; their connections with each other and with other levels of the CNS; and routes of CNS output to the respiratory muscles
4. Locations and roles of the central and peripheral chemoreceptors, stretch receptors, and irritant receptors in modulating the respiratory rhythm

5. The neural pathway for voluntary control of respiration
6. The mathematical relationship of airflow, pressure, and resistance
7. Actions of the sternum and rib cage during the respiratory cycle
8. Why the pressures driving respiratory airflow are measured in cm H_2O rather than mm Hg like other pressures considered in this and previous chapters
9. How and why intrapulmonary pressure changes relative to atmospheric pressure during inspiration; how Boyle's and Charles's laws relate to pulmonary ventilation
10. The role of elastic recoil of the thorax in expiration; how and why intrapulmonary pressure changes relative to atmospheric pressure in expiration
11. How pulmonary ventilation is affected by bronchodilation, bronchoconstriction, pulmonary compliance, and alveolar surfactant
12. A typical adult tidal volume; how much of this ventilates the alveoli and how much remains in the anatomical dead space; and how to calculate alveolar ventilation rate
13. Use of the spirometer to measure pulmonary ventilation; the meanings and typical values of the four respiratory volumes and four capacities
14. How to determine forced expiratory volume, minute respiratory volume, and maximum voluntary ventilation
15. The difference between restrictive and obstructive disorders of respiration, their respective effects on certain respiratory volumes and capacities, and examples of each
16. Definitions of *eupnea, dyspnea, hyperpnea, hyperventilation, hypoventilation, Kussmaul respiration, orthopnea, respiratory arrest,* and *tachypnea*

22.3 Gas Exchange and Transport

1. Composition of the atmosphere and average partial pressures of its constituent gases at sea level; the application of Dalton's law to partial pressures and total atmospheric pressure
2. Differences between the composition of atmospheric air and alveolar air, and reasons for the differences

3. Why gas exchange depends on the ability of the gases to dissolve in water; the application of Henry's law to the air–water interface in the alveoli
4. Four variables that determine the rate of O_2 loading and CO_2 unloading by blood passing through the alveolar capillaries
5. How ventilation–perfusion coupling matches pulmonary airflow to blood flow for optimal gas exchange
6. The two modes of O_2 transport in the blood and the relative amounts of O_2 transported by each; where O_2 binds to the hemoglobin molecule; how much O_2 a hemoglobin molecule can carry; and what hemoglobin is called when O_2 is bound to it
7. Interpretation of the oxyhemoglobin dissociation curve, including the reason for its shape and how it can be used to show the amount of O_2 unloading as hemoglobin passes through a typical systemic tissue
8. Three modes of CO_2 transport in the blood and the relative amounts of CO_2 transported by each; where CO_2 binds to hemoglobin; and what hemoglobin is called when CO_2 is bound to it
9. How carbonic anhydrase (CAH) and the chloride shift aid in the loading of CO_2 from the tissue fluids; the reaction catalyzed by CAH
10. How the loading of CO_2 in systemic tissues influences the unloading of O_2; the meaning of the *utilization coefficient* and a typical resting value
11. How the loading of O_2 in the lungs influences the unloading of CO_2
12. Four mechanisms that adjust the amount of O_2 unloaded by hemoglobin to the needs of individual tissues
13. How the Haldane effect modifies CO_2 loading in relation to the metabolic rates of individual tissues
14. Three factors that stimulate the central and peripheral chemoreceptors, and their relative influences on breathing

STUDY GUIDE

15. The normal pH range of the blood; terms for deviations above and below this range; terms for the CO_2 imbalances that cause these pH deviations; and how the body homeostatically regulates blood pH

16. The mechanism by which exercise increases respiration

22.4 Respiratory Disorders

1. The definition of *hypoxia;* its four varieties and the cause of each; and the consequences of uncorrected hypoxia

2. The mechanism and effects of oxygen toxicity

3. The names, most common cause, and pathology of the two chronic obstructive pulmonary diseases (COPDs)

4. The most common cause of lung cancer, and the names and pathological differences between the three forms of lung cancer

▶ Testing Your Recall

Answers in Appendix A

1. The nasal cavity is divided by the nasal septum into right and left
   a. nares.
   b. vestibules.
   c. fossae.
   d. choanae.
   e. conchae.

2. The intrinsic laryngeal muscles regulate speech by rotating
   a. the extrinsic laryngeal muscles.
   b. the corniculate cartilages.
   c. the arytenoid cartilages.
   d. the hyoid bone.
   e. the vocal cords.

3. The largest air passages that engage in gas exchange with the blood are
   a. the respiratory bronchioles.
   b. the terminal bronchioles.
   c. the primary bronchi.
   d. the alveolar ducts.
   e. the alveoli.

4. Respiratory arrest would most likely result from a tumor of the
   a. pons.
   b. midbrain.
   c. thalamus.
   d. cerebellum.
   e. medulla oblongata.

5. Which of these values is normally highest?
   a. tidal volume
   b. inspiratory reserve volume
   c. expiratory reserve volume
   d. residual volume
   e. vital capacity

6. The _____ protects the lungs from injury by excessive inspiration.
   a. pleura
   b. rib cage
   c. inflation reflex
   d. Haldane effect
   e. Bohr effect

7. According to _____, the warming of air as it is inhaled helps to inflate the lungs.
   a. Boyle's law
   b. Charles's law
   c. Dalton's law
   d. the Bohr effect
   e. the Haldane effect

8. Poor blood circulation causes _____ hypoxia.
   a. ischemic
   b. histotoxic
   c. hemolytic
   d. anemic
   e. hypoxemic

9. Most of the CO_2 that diffuses from the blood into an alveolus comes from
   a. dissolved gas.
   b. carbaminohemoglobin.
   c. carboxyhemoglobin.
   d. carbonic acid.
   e. expired air.

10. The duration of an inspiration is set by
    a. the pneumotaxic center.
    b. the phrenic nerves.
    c. the vagus nerves.
    d. the I neurons.
    e. the E neurons.

11. The superior opening into the larynx is guarded by a tissue flap called the _____.

12. Within each lung, the airway forms a branching complex called the _____.

13. The great alveolar cells secrete a phospholipid–protein mixture called _____.

14. Intrapulmonary pressure must be lower than _____ pressure for inspiration to occur.

15. _____ disorders reduce the speed of airflow through the airway.

16. Some inhaled air does not participate in gas exchange because it fills the _____ of the respiratory tract.

17. Inspiration depends on the ease of pulmonary inflation, called _____, whereas expiration depends on _____, which causes pulmonary recoil.

18. Inspiration is caused by the firing of I neurons in the _____ of the medulla oblongata.

19. The matching of airflow to blood flow in any region of the lung is called _____.

20. A blood pH > 7.45 is called _____ and can be caused by a CO_2 deficiency called _____.

STUDY GUIDE

▶ Building Your Medical Vocabulary

Answers in Appendix A

State a meaning of each word element, and give a medical term from this chapter that uses it or a slight variation of it.

1. atel-

2. capni-

3. carcino-

4. corni-

5. eu-

6. -meter

7. naso-

8. -pnea

9. spiro-

10. thyro-

▶ What's Wrong with These Statements?

Answers in Appendix A

Briefly explain why each of the following statements is false, or reword it to make it true.

1. The phrenic nerves fire during inspiration only.

2. There is one segmental bronchus for each lobe of the lung.

3. The most abundant cells in the lung are the squamous alveolar cells.

4. If you increase the volume of a given quantity of gas, its pressure increases.

5. Atelectasis is always preceded by pneumothorax.

6. The greatest effect of accelerated breathing is to increase the oxygen content of the blood.

7. Oxygen is the only gas we inhale from the atmosphere, since we can't use the nitrogen or other gases in it.

8. Most of the air one inhales never makes it to the alveoli.

9. The lower the P_{CO_2} of the blood is, the lower its pH is.

10. Most of the CO_2 transported by the blood is in the form of dissolved gas.

▶ Testing Your Comprehension

1. Discuss how the different functions of the conducting division and respiratory division relate to differences in their histology.

2. State whether hyperventilation would raise or lower each of the following—the blood P_{O_2}, P_{CO_2}, and pH—and explain why. Do the same for emphysema.

3. Some competitive swimmers hyperventilate before a race, thinking they can "load up extra oxygen" and hold their breaths longer underwater. While they can indeed hold their breaths longer, it is not for the reason they think. Furthermore, some have lost consciousness and drowned because of this practice. What is wrong with this thinking, and what accounts for the loss of consciousness?

4. Consider a man in good health with a 650 mL tidal volume and a respiratory rate of 11 breaths per minute. Report his minute respiratory volume in liters per minute. Assuming his anatomical dead space is 185 mL, calculate his alveolar ventilation rate in liters per minute.

5. An 83-year-old woman is admitted to the hospital, where a critical care nurse attempts to insert a nasoenteric tube ("stomach tube") for feeding. The patient begins to exhibit dyspnea, and a chest X-ray reveals air in the right pleural cavity and a collapsed right lung. The patient dies 5 days later from respiratory complications. Name the conditions revealed by the X-ray and explain how they could have resulted from the nurse's procedure.

CHAPTER 23

THE URINARY SYSTEM

The renal glomerulus, a mass of capillaries where the kidney filters the blood (SEM of a vascular cast)

SPL/Science Source

CHAPTER OUTLINE

23.1 Functions of the Urinary System
- 23.1a Functions of the Kidneys
- 23.1b Nitrogenous Wastes
- 23.1c Excretion

23.2 Anatomy of the Kidney
- 23.2a Position and Associated Structures
- 23.2b Gross Anatomy
- 23.2c Renal Circulation
- 23.2d The Nephron
- 23.2e Renal Innervation

23.3 Urine Formation I: Glomerular Filtration
- 23.3a The Filtration Membrane
- 23.3b Filtration Pressure
- 23.3c Glomerular Filtration Rate
- 23.3d Regulation of Glomerular Filtration

23.4 Urine Formation II: Tubular Reabsorption and Secretion
- 23.4a The Proximal Convoluted Tubule
- 23.4b The Nephron Loop
- 23.4c The Distal Convoluted Tubule and Collecting Duct

23.5 Urine Formation III: Water Conservation
- 23.5a The Collecting Duct
- 23.5b Control of Water Loss
- 23.5c The Countercurrent Multiplier
- 23.5d The Countercurrent Exchange System

23.6 Urine and Renal Function Tests
- 23.6a Composition and Properties of Urine
- 23.6b Urine Volume
- 23.6c Renal Function Tests

23.7 Urine Storage and Elimination
- 23.7a The Ureters
- 23.7b The Urinary Bladder
- 23.7c The Urethra
- 23.7d Voiding Urine

Connective Issues

Study Guide

DEEPER INSIGHTS

23.1 Evolutionary Medicine: The Kidney and Life on Dry Land

23.2 Clinical Application: Kidney Stones

23.3 Clinical Application: Urinary Tract Infection (UTI)

23.4 Clinical Application: Urination and Spinal Cord Injuries

23.5 Clinical Application: Renal Insufficiency and Hemodialysis

Anatomy & Physiology Revealed 4.0

Module 13: Urinary System

BRUSHING UP

- Many aspects of kidney physiology are governed by the principle of flow down pressure, osmotic, and electrochemical gradients (see section 1.6e).

- You must be familiar with the forces of capillary fluid exchange to understand glomerular filtration, the first step in urine production (see section 20.3c).

- The concepts of osmosis, tonicity, and osmolarity are central to understanding the stages of urine production and the role of the kidneys in regulating fluid balance and urine concentration (see sections 3.3c, 3.3d; "Osmolarity and Osmolality" in appendix B).

- Refresh your memory of carrier-mediated transport in section 3.3e, especially symports and antiports, in order to best understand how the kidney processes a blood plasma filtrate to produce urine.

To live is to metabolize, and metabolism unavoidably produces a variety of waste products that are not merely unneeded by the body, but indeed toxic if allowed to accumulate. We rid the body of some of these wastes through the respiratory and digestive tracts and the sweat glands, but the urinary system is the principal means of waste excretion. The kidneys are glands that separate metabolic wastes from the blood. The rest of the urinary system serves only for the transport, storage, and elimination of urine. Most of our focus in this chapter is therefore on the kidneys.

Their task goes far beyond waste excretion. As we will see, the kidneys also play indispensable roles in regulating blood volume, pressure, and composition. In performing these tasks, they have a very close physiological relationship with the endocrine, circulatory, and respiratory systems, covered in recent chapters.

Anatomically, the urinary system is closely associated with the reproductive system. In many animals, the eggs and sperm are emitted through the urinary tract, and the two systems have a shared embryonic development and adult anatomical relationship. This is reflected in humans, where the systems develop together in the embryo and, in the male, the urethra continues to serve as a passage for both urine and sperm. Thus, the urinary and reproductive systems are often collectively called the *urogenital (U–G) system,* and *urologists* treat both urinary and male reproductive disorders. We examine the anatomical relationship between the urinary and reproductive systems in chapter 27, but the physiological link to the circulatory and respiratory systems is more important to consider at this time.

23.1 Functions of the Urinary System

Expected Learning Outcomes

When you have completed this section, you should be able to

a. name and locate the organs of the urinary system;

b. list several functions of the kidneys in addition to urine formation;

c. name the major nitrogenous wastes and identify their sources; and

d. define *excretion* and identify the systems that excrete wastes.

The **urinary system** consists of six principal organs: two **kidneys,** two **ureters,** the **urinary bladder,** and the **urethra (fig. 23.1).** The urinary tract has important spatial relationships with the vagina and uterus in females and the prostate in males. These relationships are best appreciated from the sagittal views of figures 27.10 and 28.1 and can be seen in the cadaver in figure B.14 of atlas B.

23.1a Functions of the Kidneys APR

Although the primary function of the kidneys is excretion, they play more roles than are commonly realized:

- They filter the blood and excrete the toxic metabolic wastes.

- They regulate blood volume, pressure, and osmolarity by regulating water output.

- They regulate the electrolyte and acid–base balance of the body fluids.

- They secrete the hormone *erythropoietin,* which stimulates the production of red blood cells and thus supports the oxygen-carrying capacity of the blood.

- They help to regulate calcium homeostasis and bone metabolism by participating in the synthesis of calcitriol.

- They clear hormones and drugs from the blood and thereby limit their action.

- They detoxify free radicals.

- In conditions of extreme starvation, they help to support the blood glucose level by synthesizing glucose from amino acids.

In view of such diverse roles, it is easy to see why renal failure can lead to the collapse of many other physiological functions as well.

23.1b Nitrogenous Wastes

A **waste** is any substance that is useless to the body or present in excess of the body's needs. A **metabolic waste,** more specifically, is a waste substance produced by the body. The food residue in feces, for example, is a waste but not a metabolic waste, since it wasn't produced by the body and, indeed, never entered the body's tissues.

Among the most toxic of our metabolic wastes are small nitrogen-containing compounds called **nitrogenous wastes (fig. 23.2).** About 50% of the nitrogenous waste is urea, a by-product of protein catabolism. Proteins are hydrolyzed to amino acids, then the $-NH_2$ group is removed from each amino acid. The $-NH_2$ forms ammonia, which is exceedingly toxic but

Diaphragm
11th and 12th ribs
Adrenal gland
Renal artery
Renal vein
Kidney
Vertebra L2
Aorta
Inferior vena cava
Ureter
Urinary bladder
Urethra

(a) Anterior view

(b) Posterior view

FIGURE 23.1 The Urinary System. (a) Anterior view. (b) Posterior view. Organs of the urinary system are indicated in boldface.

H
H—N—H

Ammonia

O
||
H₂N—C—NH₂

Urea

Uric acid

Creatinine

FIGURE 23.2 The Major Nitrogenous Wastes.

❓ *How is each of these wastes produced in the body?*

Other nitrogenous wastes in the urine include **uric acid** and **creatinine** (cree-AT-ih-neen), produced by the catabolism of nucleic acids and creatine phosphate, respectively. Although less toxic than ammonia and less abundant than urea, these too are potentially harmful and must be eliminated.

The level of nitrogenous waste in the blood is typically expressed as **blood urea nitrogen (BUN).** The normal concentration of blood urea is 10 to 20 mg/dL. An elevated BUN is called **azotemia**[1] (AZ-oh-TEE-me-uh) and may indicate renal insufficiency. It can progress to **uremia** (you-REE-me-uh), a syndrome of diarrhea, vomiting, dyspnea, and cardiac arrhythmia stemming from the toxicity of the nitrogenous wastes. Convulsions, coma, and death can follow within a few days. Deeper Insight 23.5 discusses treatments for renal insufficiency.

23.1c Excretion

Excretion is the process of separating wastes from the body fluids and eliminating them from the body. It is carried out by four organ systems:

1. The respiratory system excretes carbon dioxide, small amounts of other gases, and water.

which the liver quickly converts to urea, $CO(NH_2)_2$, a somewhat less toxic waste:

$$2\ NH_3 + CO_2 \longrightarrow H_2N\underset{\overset{\|}{O}}{-}C-NH_2 + H_2O$$

---

[1]*azot* = nitrogen; *emia* = blood condition

2. The integumentary system excretes water, inorganic salts, lactate, and urea in the sweat.

3. The digestive system not only *eliminates* food residue (which is not a process of excretion) but also actively *excretes* water, salts, carbon dioxide, lipids, bile pigments, cholesterol, and other metabolic wastes.

4. The urinary system excretes a broad variety of metabolic wastes, toxins, drugs, hormones, salts, hydrogen ions, and water.

BEFORE YOU GO ON

Answer the following questions to test your understanding of the preceding section:

1. State at least four functions of the kidneys other than forming urine.

2. List four nitrogenous wastes and their metabolic sources.

3. Name some wastes eliminated by three systems other than the urinary system.

23.2 Anatomy of the Kidney APR

Expected Learning Outcomes

When you have completed this section, you should be able to

a. describe the location and general appearance of the kidneys;

b. identify the external and internal features of the kidney;

c. trace the flow of blood through the kidney;

d. trace the flow of fluid through the renal tubules; and

e. describe the nerve supply to the kidney.

23.2a Position and Associated Structures

The kidneys lie against the posterior abdominal wall at the level of vertebrae T12 to L3. The right kidney is slightly lower than the left because of the space occupied by the large right lobe of the liver above it. Rib 12 crosses the approximate middle of the left kidney. The kidneys are retroperitoneal, along with the ureters, urinary bladder, renal artery and vein, and the adrenal glands **(fig. 23.3).**

23.2b Gross Anatomy

Each kidney weighs about 150 g and measures about 11 cm long, 6 cm wide, and 3 cm thick—about the size of a bar of bath soap. The lateral surface is convex, and the medial surface is concave and has a slit, the **hilum,** that admits the renal nerves, blood vessels, lymphatics, and ureter.

The kidney is protected by three layers of connective tissue (fig. 23.3b): (1) A fibrous **renal fascia,** immediately deep to the parietal peritoneum, binds the kidney and associated organs to the abdominal wall; (2) the **perirenal fat capsule,** a layer of adipose tissue, cushions the kidney and holds it in place; and (3) the **fibrous capsule** encloses the kidney like a cellophane wrapper

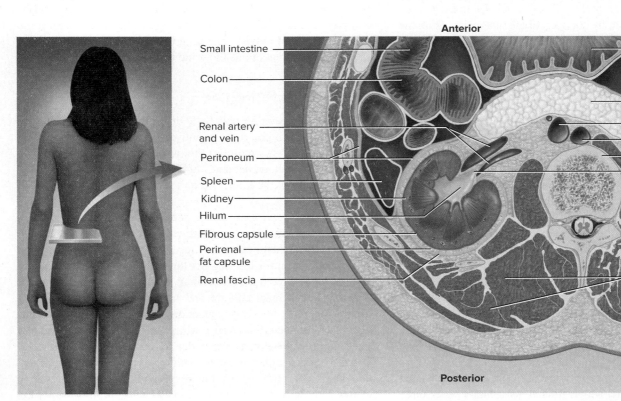

FIGURE 23.3 Retroperitoneal Position of the Kidney. Cross section of the abdomen at the level of vertebra L1.

❓ *If the kidney were not retroperitoneal, where on this figure would you have to relocate it?*

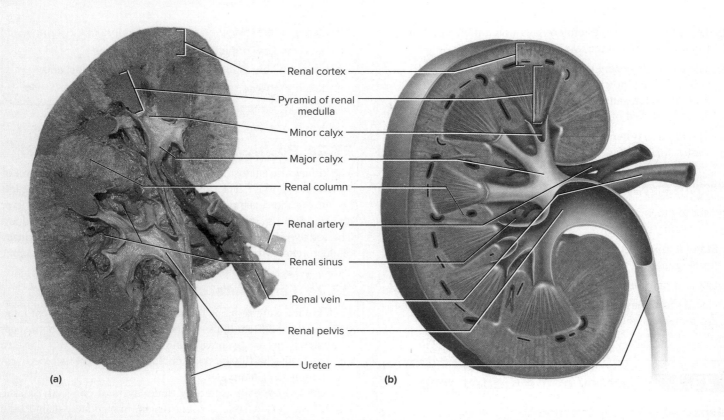

FIGURE 23.4 Gross Anatomy of the Kidney. Posterior views. (a) Photograph of frontal section. (b) Major anatomical features. The adipose tissue that normally fills in the renal sinus is omitted to better reveal the other structures that occupy the sinus. **APR**

a: Rebecca Gray/McGraw-Hill Education

anchored at the hilum, and protects it from trauma and infection. The kidneys are suspended by collagen fibers that extend from the fibrous capsule, through the fat, to the renal fascia. The renal fascia is fused with the peritoneum anteriorly and with the fascia of the lumbar muscles posteriorly. In spite of all this, the kidneys drop about 3 cm when one goes from lying down to standing, as when getting out of bed in the morning. Under some circumstances, they become detached and drift even lower, with pathological results (see nephroptosis, or "floating kidney," in table 23.3).

The renal parenchyma—the glandular tissue that forms the urine—appears C-shaped in frontal section **(fig. 23.4).** It encircles a medial cavity, the **renal sinus,** occupied by blood and lymphatic vessels, nerves, and urine-collecting structures. Adipose tissue fills the remaining space in the sinus and holds these structures in place.

The parenchyma is divided into two zones: an outer **renal cortex** about 1 cm thick and an inner **renal medulla** facing the sinus. The boundary between the cortex and medulla is called the **corticomedullary junction.** Extensions of the cortex called **renal columns** project toward the sinus and divide the medulla into 6 to 10 **renal pyramids.** Each pyramid is conical, with a broad base facing the cortex and a blunt point called the **renal papilla** facing the sinus. One pyramid and the overlying cortex constitute one *lobe* of the kidney.

The papilla of each renal pyramid is nestled in a cup called a **minor calyx**[2] (CAY-lix), which collects its urine. Two or three

minor calyces (CAY-lih-seez) converge to form a **major calyx,** and two or three major calyces converge in the sinus to form the funnel-like **renal pelvis.**[3] The ureter is a tubular continuation of the renal pelvis that drains the urine down to the urinary bladder.

23.2c Renal Circulation

The kidneys account for only 0.4% of the body weight, but receive about 1.1 liters of blood per minute, or 21% of the cardiac output (the *renal fraction*)—more for the purpose of waste removal than to meet the metabolic demands of the kidney tissue. This is a hint of how important the kidneys are in regulating blood volume and composition.

The larger divisions of the renal circulation are shown in **figure 23.5.** Each kidney is supplied by a **renal artery** arising from the aorta. Just before or after entering the hilum, the renal artery divides into a few **segmental arteries,** and each of these further divides into a few **interlobar arteries.** An interlobar artery penetrates each renal column and travels between the pyramids toward the corticomedullary junction. Along the way, it branches again to form **arcuate arteries,** which make a sharp 90° bend and travel along the base of the pyramid. Each arcuate artery gives rise to several **cortical radiate arteries,** which pass upward into the cortex.

The finer branches of the renal circulation are shown in **figure 23.6.** As a cortical radiate artery ascends through the cortex,

---

[2]*calyx* = cup

[3]*pelvis* = basin

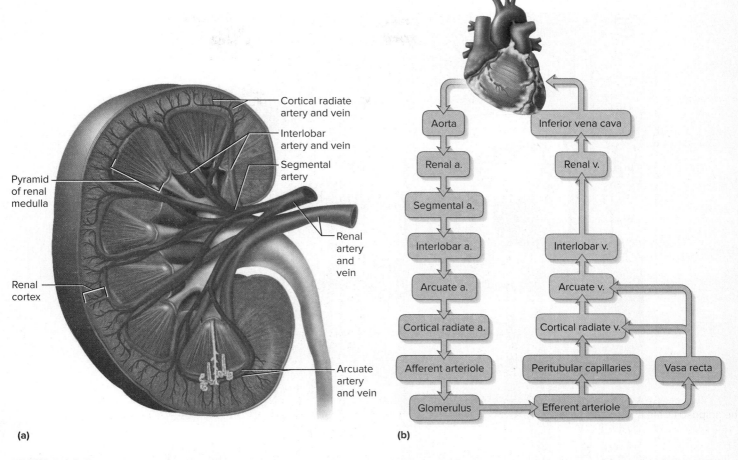

(a)

(b)

FIGURE 23.5 Renal Circulation. (a) The larger blood vessels of the kidney. (b) Flowchart of renal circulation. The pathway through the vasa recta (instead of peritubular capillaries) applies only to the juxtamedullary nephrons. Pink indicates the arterial supply to the nephrons; violet indicates the portal system of connected capillary beds; and blue indicates venous drainage of the kidney.

a series of **afferent arterioles** arise from it at nearly right angles like limbs arising from the trunk of a pine tree. Each afferent arteriole supplies one functional unit of the kidney called a **nephron**[4] (NEF-ron). The afferent arteriole leads to a ball of capillaries called a **glomerulus**[5] (glo-MERR-you-lus), enclosed in a sphere called the *glomerular capsule.* Blood leaves the glomerulus by way of an **efferent arteriole.**

The efferent arteriole usually leads to a plexus of **peritubular capillaries,** named for the fact that they form a network around another part of the nephron, the *renal tubule.* The renal tubule reabsorbs most of the water and solutes that filter out of the blood at the glomerulus and returns these to the bloodstream by way of these peritubular capillaries. The peritubular capillaries carry it away to the **cortical radiate veins, arcuate veins, interlobar veins,** and the **renal vein,** in that order. These veins travel parallel to the arteries of the same names. (There are, however, no segmental veins corresponding to the segmental arteries.) The renal vein leaves the hilum and drains into the inferior vena cava.

The renal medulla receives only 1% to 2% of the total renal blood flow, supplied by a network of vessels called the **vasa recta**[6] (VAH-za REC-ta). These arise from the efferent arterioles of the nephrons closest to the medulla. Here, the efferent arterioles descend immediately into the medulla and give rise to the vasa recta instead of peritubular capillaries. Capillaries of the vasa recta lead into venules that ascend and empty into the arcuate and cortical radiate veins. Capillaries of the vasa recta are wedged into the tight spaces between the medullary parts of the renal tubule, and carry away water and solutes reabsorbed by those sections of the tubule. Figure 23.5b summarizes the route of renal blood flow.

23.2d The Nephron APR

Each kidney has about 1.2 million nephrons. To understand how just one of these works is to understand nearly everything about how the whole kidney works. Each nephron is composed of two principal parts: a *renal corpuscle,* which filters the blood

---

[4]*nephro* = kidney
[5]*glomer* = ball; *ulus* = little

[6]*vasa* = vessels; *recta* = straight

FIGURE 23.6 Microcirculation of the Kidney. For clarity, vasa recta are shown only on the left and peritubular capillaries only on the right. In the juxtamedullary nephron (left), the efferent arteriole gives rise to the vasa recta of the medulla. In the cortical nephron (right), the nephron loop barely dips into the renal medulla and the efferent arteriole gives rise to peritubular capillaries. (DCT = distal convoluted tubule; PCT = proximal convoluted tubule)

plasma, and a long coiled *renal tubule,* which converts the filtrate to urine.

The Renal Corpuscle

The **renal corpuscle (fig. 23.7)** consists of the glomerulus described earlier and a two-layered **glomerular capsule** (formerly called the *Bowman capsule*) that encloses it. The parietal (outer) layer is a simple squamous epithelium, and the visceral (inner) layer consists of elaborate cells called **podocytes**[7] wrapped around the capillaries of the glomerulus. The two layers are separated by a filtrate-collecting **capsular space.** In tissue sections, this space appears as an empty circular or C-shaped space around the glomerulus.

Opposite sides of the renal corpuscle are called the vascular and urinary poles. At the **vascular pole,** the afferent arteriole enters the capsule, bringing blood to the glomerulus, and the efferent arteriole leaves the capsule and carries blood away. The afferent arteriole is significantly larger than the efferent arteriole. Thus, the glomerulus has a large inlet and a small outlet—a point whose functional significance will become apparent later. At the **urinary pole,** the parietal wall of the capsule turns away from the corpuscle and gives rise to the renal tubule. The simple squamous epithelium of the capsule becomes simple cuboidal in the tubule.

---

[7]*podo* = foot; *cyte* = cell

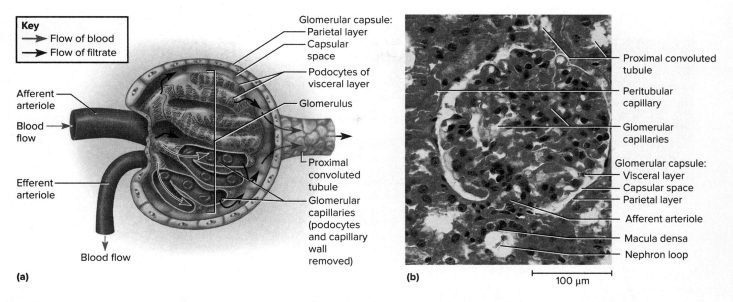

Key
→ Flow of blood
→ Flow of filtrate

Afferent arteriole
Blood flow
Efferent arteriole
Blood flow

Glomerular capsule:
Parietal layer
Capsular space
Podocytes of visceral layer
Glomerulus
Proximal convoluted tubule
Glomerular capillaries (podocytes and capillary wall removed)

(a)

Proximal convoluted tubule
Peritubular capillary
Glomerular capillaries
Glomerular capsule:
Visceral layer
Capsular space
Parietal layer
Afferent arteriole
Macula densa
Nephron loop

(b)   100 μm

FIGURE 23.7 The Renal Corpuscle. (a) Anatomy of the corpuscle. (b) Light micrograph of the renal corpuscle and sections of the surrounding renal tubule. **APR**

b: Alvin Telser/McGraw-Hill Education

The Renal Tubule

The **renal tubule** is a duct that leads away from the glomerular capsule and ends at the tip of a medullary pyramid. It is about 3 cm long and divided into four regions: the *proximal convoluted tubule, nephron loop, distal convoluted tubule,* and *collecting duct* **(fig. 23.8).** The first three of these are parts of one nephron; the collecting duct receives fluid from many nephrons. Each region has unique physiological properties and roles in the production of urine.

1. The **proximal convoluted tubule (PCT)** arises from the glomerular capsule. It is the longest and most coiled of the four regions and therefore dominates histological sections of renal cortex. It has a simple cuboidal epithelium with prominent microvilli (a brush border), which attests to the great deal of absorption that occurs here. The microvilli give the epithelium a distinctive shaggy look.

2. The **nephron loop** (formerly called the *loop of Henle*) is a long U-shaped portion of the renal tubule found mostly in the medulla. It begins where the PCT straightens out and dips toward or into the medulla, forming the **descending limb** of the loop. At its deep end, the loop turns 180° and forms the **ascending limb,** which returns to the cortex, traveling parallel and close to the descending limb. The loop is divided into thick and thin segments. The **thick segments** have a simple cuboidal epithelium. They form the initial part of the descending limb and part or all of the ascending limb. The cells here are heavily engaged in active transport of salts, so they have very high metabolic

activity and are loaded with mitochondria, accounting for their thickness. The **thin segment** has a simple squamous epithelium. It forms most of the descending limb, and in some nephrons, it rounds the bend and continues partway up the ascending limb. The cells here have low metabolic activity, but the thin segment of the descending limb is very permeable to water.

3. The **distal convoluted tubule (DCT)** begins shortly after the ascending limb reenters the cortex. It is shorter and less coiled than the proximal convoluted tubule, so fewer sections of it are seen in histological sections. It has a cuboidal epithelium with smooth-surfaced cells nearly devoid of microvilli. The DCT is the end of the nephron.

4. The **collecting duct** receives fluid from the DCTs of several nephrons as it passes back into the medulla. Numerous collecting ducts converge toward the tip of a medullary pyramid, and near the papilla, they merge to form a larger **papillary duct.** About 30 papillary ducts end in pores at the conical tip of each papilla. Urine drains from these pores into the minor calyx that encloses the papilla. The collecting and papillary ducts are lined with simple cuboidal epithelium.

The flow of fluid from the point where the glomerular filtrate is formed to the point where urine leaves the body is: glomerular capsule → proximal convoluted tubule → nephron loop → distal convoluted tubule → collecting duct → papillary duct → minor calyx → major calyx → renal pelvis → ureter → urinary bladder → urethra.

FIGURE 23.8 Microscopic Anatomy of the Nephron. (a) Location of the nephrons in one wedge-shaped lobe of the kidney. (b) Structure of a nephron. For clarity, the nephron is stretched out to separate the convoluted tubules. The nephron loop is greatly shortened for the purpose of illustration. (c) The true proportions of the nephron loops relative to the convoluted tubules. Three nephrons are shown. Their proximal and distal convoluted tubules are commingled in a single tangled mass in each nephron. Note the extreme lengths of the nephron loops. **AP&R**

Juxtamedullary and Cortical Nephrons

Not all nephrons are identical. Those close to the medulla are called **juxtamedullary**[8] **nephrons.** They have very long nephron loops that extend as far as the apex of the renal pyramid. The remaining nephrons, farther from the medulla, are called **cortical nephrons.** They have relatively short nephron loops that dip only slightly into the outer medulla before turning back (see fig. 23.6) or turn back even before leaving the cortex. Some of them have no nephron loops at all. As you will see later, nephron loops are responsible for maintaining an osmotic gradient in the medulla that helps the body conserve water. Although only 15% of the nephrons are juxtamedullary, they are almost solely responsible for maintaining this gradient.

23.2e Renal Innervation

Wrapped around each renal artery is a **renal plexus** of nerves and ganglia (see fig. 15.6). The plexus follows branches of the renal artery into the parenchyma of the kidney, issuing nerve fibers to the blood vessels and convoluted tubules of the nephrons. The renal plexus carries sympathetic innervation from the abdominal aortic plexus (especially its superior mesenteric and celiac ganglia) as well as afferent pain fibers from the kidneys en route to the spinal cord. Stimulation by the sympathetic fibers of the renal plexus tends to reduce glomerular blood flow and therefore the rate of urine production, although these rates are influenced by other factors as well. Another role of the sympathetic fibers is to respond to falling blood pressure by stimulating the kidneys to secrete *renin,* an enzyme that activates hormonal mechanisms for restoring blood pressure. The kidneys also receive parasympathetic innervation from branches of the vagus nerve, but its function is unknown.

BEFORE YOU GO ON

Answer the following questions to test your understanding of the preceding section:

4. Arrange the following in order from the most numerous to the least numerous structures in a kidney: glomeruli, major calyces, minor calyces, cortical radiate arteries, interlobar arteries.

5. Trace the path taken by one red blood cell from the renal artery to the renal vein.

6. Consider one molecule of urea in the urine. Trace the route that it took from the point where it left the bloodstream to the point where it left the body.

---

[8]*juxta* = next to

23.3 Urine Formation I: Glomerular Filtration APR

Expected Learning Outcomes

When you have completed this section, you should be able to

a. describe the process by which the kidney filters the blood plasma, including the relevant cellular structure of the glomerulus;

b. Explain the forces that promote and oppose filtration, and calculate the filtration pressure if given the magnitude of these forces; and

c. describe how the nervous system, hormones, and the nephron itself regulate filtration.

The kidney converts blood plasma to urine in four stages: glomerular filtration, tubular reabsorption, tubular secretion, and water conservation **(fig. 23.9).** These are the themes of the next three sections of the chapter. As we trace fluid through the nephron, we will refer to it by different names that reflect its changing composition: (1) The fluid in the capsular space, called **glomerular filtrate,** is similar to blood plasma except that it has almost no protein. (2) The fluid from the proximal convoluted tubule through the distal convoluted tubule is called **tubular fluid.** It differs from the glomerular filtrate because of substances removed and added by the tubule cells. (3) The fluid is called **urine** once it enters the collecting duct, since it undergoes little alteration beyond that point except for a change in water content.

23.3a The Filtration Membrane

Glomerular filtration, discussed in this section, is a process in which water and some solutes in the blood plasma pass from capillaries of the glomerulus into the capsular space of the nephron. To do so, fluid passes through three barriers that constitute a **filtration membrane (fig. 23.10):**

1. **The fenestrated endothelium of the capillary.** Endothelial cells of the glomerular capillaries are honeycombed with large filtration pores about 70 to 90 nm in diameter (see fig. 20.5b). Like fenestrated capillaries elsewhere, these are highly permeable, although their pores are small enough to exclude blood cells from the filtrate.

2. **The basement membrane.** This consists of a proteoglycan gel. Passing large molecules through it would be like trying to grind sand through a kitchen sponge: A few grains may penetrate its small spaces and reach the other side, but most would be held back. On the basis of size alone, the basement membrane excludes molecules larger than 8 nm. Even some smaller molecules, however, are held back by a negative charge on the proteoglycans. Blood albumin is slightly

① **Glomerular filtration**
Creates a plasmalike
filtrate of the blood

② **Tubular reabsorption**
Removes useful solutes
from the filtrate, returns
them to the blood

③ **Tubular secretion**
Removes additional
wastes from the blood,
adds them to the filtrate

④ **Water conservation**
Removes water from the
urine and returns it to
blood; concentrates
wastes

Blood flow

Renal corpuscle

Flow of filtrate

Peritubular
capillaries

Renal tubule

H_2O

H_2O

H_2O

Urine

FIGURE 23.9 Basic Stages of Urine Formation.

smaller than 7 nm, but it is also negatively charged and thus repelled by the basement membrane. Although the blood plasma is 7% protein, the glomerular filtrate is only 0.03% protein. It has traces of albumin and smaller polypeptides, including some hormones.

3. **Filtration slits.** A podocyte of the glomerular capsule is shaped somewhat like an octopus, with a bulbous cell body and several thick arms. Each arm has numerous extensions called **foot processes (pedicels[9])** that wrap around the capillaries and interdigitate with each other, like wrapping your hands around a pipe and lacing your fingers together. The foot processes have negatively charged **filtration slits** about 30 nm wide between them, which are an additional obstacle to large anions.

Almost any molecule smaller than 3 nm can pass freely through the filtration membrane into the capsular space **(fig. 23.11)**. This includes water, electrolytes, glucose, fatty acids, amino acids, nitrogenous wastes, and vitamins. Such solutes have about the same concentration in the glomerular filtrate as in the blood plasma. Some solutes of low molecular weight are retained in the bloodstream, however, because they're bound to plasma proteins that can't get through the membrane. For example, most calcium, iron, and thyroid hormone in the blood are bound to proteins that retard their filtration by the kidneys. The small fraction that is unbound, however, passes freely through the membrane and appears in the urine.

Kidney infections and trauma can damage the filtration membrane and allow albumin or blood cells to filter through. Kidney disease is sometimes marked by the presence of protein (especially albumin) or blood in the urine—conditions called **proteinuria (albuminuria)** and **hematuria,**[10] respectively. Distance runners and competitive swimmers often experience temporary proteinuria and hematuria. Strenuous exercise greatly reduces perfusion of the kidneys, and the glomerulus deteriorates under the prolonged hypoxia, thus leaking protein and sometimes blood into the filtrate.

23.3b Filtration Pressure

Glomerular filtration follows the same principles that govern filtration in other blood capillaries (see section 20.3c), but there are significant differences in the magnitude of the forces involved:

- The blood hydrostatic pressure (BHP) is much higher here than elsewhere—about 60 mm Hg compared with 10 to 15 mm Hg in most other capillaries. This results from the fact that the afferent arteriole is substantially larger than the efferent arteriole, giving the glomerulus a large inlet and small outlet (fig. 23.10a).

- The hydrostatic pressure in the capsular space is about 18 mm Hg, compared with the slightly negative interstitial pressures elsewhere. This results from the high rate of filtration and continual accumulation of fluid in the capsule.

- The colloid osmotic pressure (COP) of the blood is about the same here as anywhere else, 32 mm Hg.

- The glomerular filtrate is almost protein-free and has no significant COP. (This can change markedly in kidney diseases that allow protein to filter into the capsular space.)

On balance, then, we have a high outward pressure of 60 mm Hg, opposed by two inward pressures of 18 and 32 mm Hg **(fig. 23.12),** giving a **net filtration pressure (NFP)** of

$$60_{out} - 18_{in} - 32_{in} = 10 \text{ mm Hg}_{out}.$$

Efferent arteriole

Podocyte cell body

Glomerulus

Pedicels (separated by narrow filtration slits)

Afferent arteriole

(a) — 100 µm

(b) — 5 µm

Capsular space

Blood plasma

Erythrocyte

Podocytes

Foot processes

Basement membrane

Endothelial cell

Filtration pores

Filtration slits

(c) — 0.5 µm

FIGURE 23.10 Structure of the Glomerulus. (a) A vascular cast of the glomerulus and associated arterioles (SEM). Note the difference in diameters between the afferent and efferent arterioles. (b) Blood capillaries of the glomerulus closely wrapped in the spidery podocytes that form the visceral layer of the glomerular capsule (SEM). (c) A blood capillary and podocytes showing filtration pores of the fenestrated capillary and filtration slits between foot processes of the podocytes (TEM).

a: Steve Gschmeissner/Getty Images; **b:** Don W. Fawcett/Science Source; **c:** Biophoto Associates/Science Source

? *Which is larger, the efferent arteriole or the afferent arteriole? How does the difference affect the function of the glomerulus?*

In most blood capillaries, the BHP drops low enough at the venous end that osmosis overrides filtration and the capillaries reabsorb fluid. Although BHP also drops along the course of the glomerular capillaries, it remains high enough that these capillaries are engaged solely in filtration. They reabsorb little or no fluid.

The high blood pressure in the glomeruli makes the kidneys especially vulnerable to hypertension, which can have devastating effects on renal function. Hypertension ruptures glomerular capillaries and leads to scarring of the kidneys *(nephrosclerosis).* It promotes atherosclerosis of the renal blood vessels just as it does elsewhere in the body and thus diminishes renal blood supply. Over time, hypertension often leads to renal failure and renal failure leads to worsening hypertension in an insidious positive feedback loop.

23.3c Glomerular Filtration Rate

Glomerular filtration rate (GFR) is the amount of filtrate formed per minute by the two kidneys combined. For every 1 mm Hg of net filtration pressure, the kidneys of a young adult male produce about 12.5 mL of filtrate per minute. This value, called the *filtration coefficient (K_f),* depends on the permeability and surface area of the filtration barrier. K_f is about 10% lower in women than in men. For males,

$$\text{GFR} = \text{NFP} \times K_f = 10 \times 12.5 = 125 \text{ mL/min.}$$

In young adult females, the GFR is about 105 mL/min.

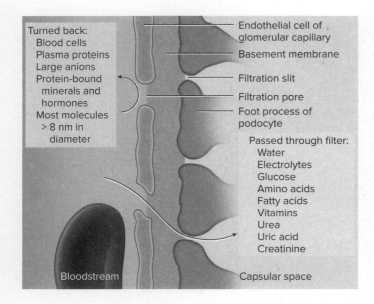

Turned back:
  Blood cells
  Plasma proteins
  Large anions
  Protein-bound
    minerals and
    hormones
  Most molecules
    > 8 nm in
    diameter

Endothelial cell of glomerular capillary

Basement membrane

Filtration slit

Filtration pore

Foot process of podocyte

Passed through filter:
  Water
  Electrolytes
  Glucose
  Amino acids
  Fatty acids
  Vitamins
  Urea
  Uric acid
  Creatinine

Bloodstream

Capsular space

FIGURE 23.11 The Glomerular Filtration Membrane.
Small solutes pass through the membrane into the filtrate (right) and larger particles are turned back and retained in the bloodstream (left).

BHP 60 out

COP 32 in

CP 18 in

NFP 10 out

Blood hydrostatic pressure *(BHP)*	60 mm Hg_{out}
Colloid osmotic pressure *(COP)*	−32 mm Hg_{in}
Capsular pressure *(CP)*	−18 mm Hg_{in}
Net filtration pressure *(NFP)*	10 mm Hg_{out}

FIGURE 23.12 The Forces Involved in Glomerular Filtration.

These rates are equivalent to 180 L/day in males and 150 L/day in females—impressive numbers considering that this is about 60 times the amount of blood in the body and 50 to 60 times the amount of filtrate produced by all other capillaries combined. Obviously, only a small portion of this is eliminated as urine. An average adult reabsorbs 99% of the filtrate and excretes 1 to 2 L of urine per day.

23.3d Regulation of Glomerular Filtration

GFR must be precisely controlled. If it's too high, fluid flows through the renal tubules too rapidly for them to reabsorb the usual amount of water and solutes. Urine output rises and creates a threat of dehydration and electrolyte depletion. If GFR is too low, fluid flows sluggishly through the tubules, they reabsorb wastes that should be eliminated in the urine, and azotemia may occur.

The only way to adjust GFR from moment to moment is to change glomerular blood pressure. This is achieved by three homeostatic mechanisms: renal autoregulation, sympathetic control, and hormonal control.

Renal Autoregulation

Renal autoregulation is the ability of the nephrons to adjust their own blood flow and GFR without external (nervous or hormonal) control. It enables them to maintain a relatively stable GFR in spite of changes in arterial blood pressure. If the mean arterial pressure (MAP) rose from 100 to 125 mm Hg and there were no renal autoregulation, urine output would increase from the normal 1 to 2 L/day to more than 45 L/day. Because of renal autoregulation, however, urine output increases only a few percent even if MAP rises as high as 160 mm Hg. Renal autoregulation thus helps to ensure stable fluid and electrolyte balance in spite of the many circumstances that substantially alter one's blood pressure. There are two mechanisms of autoregulation: the myogenic mechanism and tubuloglomerular feedback.

The Myogenic[11] Mechanism   This mechanism of stabilizing the GFR is based on the tendency of smooth muscle to contract when stretched. When arterial blood pressure rises, it stretches the afferent arteriole. The arteriole constricts and prevents blood flow into the glomerulus from changing very much. Conversely, when blood pressure falls, the afferent arteriole relaxes and allows blood to flow more easily into the glomerulus. Either way, glomerular blood flow and filtration remain fairly stable.

Tubuloglomerular Feedback   This is a mechanism by which the glomerulus receives feedback on the status of the downstream tubular fluid and adjusts filtration to regulate its composition, stabilize nephron performance, and compensate for fluctuations in blood pressure. It involves a structure, the **juxtaglomerular apparatus** (JUX-tuh-glo-MER-you-lur), found at the very end of the nephron loop where it has just reentered the renal cortex. Here, the loop contacts the afferent and efferent arterioles at the vascular pole of the renal corpuscle **(fig. 23.13).**

---

[11]*myo* = muscle; *genic* = produced by

FIGURE 23.13 The Juxtaglomerular Apparatus.

Labels: Sympathetic nerve fiber; Podocytes; Mesangial cells; Efferent arteriole; Granular cells; Afferent arteriole; Smooth muscle cells; Macula densa; Nephron loop

Tubuloglomerular feedback begins with the **macula densa,**[12] a patch of slender, closely spaced sensory cells on one side of the loop. When the GFR is very high, the sodium chloride level in the nephron loop is also sharply elevated. Cells of the macula densa absorb Na^+, K^+, and Cl^-, and water follows osmotically; the cells swell and secrete ATP from their basal surfaces. The ATP is metabolized to adenosine by nearby **mesangial**[13] **cells** (mez-AN-jee-ul), which pack the spaces between the arterioles and within the glomerulus.

Adenosine acts as a paracrine messenger that stimulates nearby **granular (juxtaglomerular) cells.** These are modified smooth muscle cells wrapped around the afferent arteriole and to a lesser extent the efferent arteriole. They respond to a rising adenosine level by constricting the afferent arteriole. That reduces blood flow into the glomerulus, thereby reducing GFR and completing the negative feedback loop (**fig. 23.14**). The mesangial cells of the glomerulus may also contract, constricting the glomerular capillaries and reducing filtration. Mesangial cells also form a supportive matrix for the glomerulus and phagocytize tissue debris, preventing the glomerular filtration membrane from clogging up.

Just beyond the juxtaglomerular apparatus, the distal convoluted tubule has a limited capacity for NaCl reabsorption. The tubuloglomerular feedback process may help to prevent overloading the distal convoluted tubule with NaCl, and thus prevent excessive NaCl and water loss in the urine.

Granular cells are named for the fact that they contain granules of *renin,* which they secrete in response to a drop in blood pressure. This initiates a renin–angiotensin–aldosterone mechanism, explained shortly, which restores blood pressure and supports blood volume.

Two important points must be noted about renal autoregulation. First, it doesn't completely prevent changes in GFR. Like any other homeostatic mechanism, it maintains a *dynamic equilibrium;* the GFR fluctuates within narrow limits. Changes in blood pressure *do* affect GFR and urine output. Second, renal autoregulation cannot compensate for extreme blood pressure variations. Over an MAP range of 90 to 180 mm Hg, the GFR remains quite stable. Below 70 mm Hg, however, glomerular filtration and urine output cease. This can happen in hypovolemic shock.

Sympathetic Control

Sympathetic nerve fibers richly innervate the renal blood vessels. In strenuous exercise or acute conditions such as circulatory shock, sympathetic stimulation and adrenal epinephrine constrict the afferent arterioles. This reduces GFR and urine output while redirecting blood from the kidneys to the heart, brain, and skeletal muscles, where it is more urgently needed. Under such conditions, GFR may be as low as a few milliliters per minute.

The Renin–Angiotensin–Aldosterone Mechanism

Any substantial drop in blood pressure due to bleeding, dehydration, or other cause is detected by the baroreceptors in the aorta

Diagram labels: High *GFR*; Reduced *GFR*; Increased NaCl load in nephron loop; Afferent arteriole constricts; Macula densa secretes ATP; Adenosine stimulates granular cells; Mesangial cells metabolize ATP to adenosine

FIGURE 23.14 Tubuloglomerular Feedback.

---

[12]*macula* = spot, patch; *densa* = dense
[13]*mes* = in the middle; *angi* = vessel

and carotid arteries. They transmit a signal to the brainstem, leading to various corrective sympathetic reflexes (see fig. 20.14). One such response is that sympathetic fibers stimulate the granular cells to secrete the enzyme **renin** (REE-nin). Renin acts on *angiotensinogen,* a protein in the blood plasma, to split off a 10-amino-acid peptide called angiotensin I. In the lungs and kidneys, **angiotensin-converting enzyme (ACE)** removes two more amino acids, converting it to **angiotensin II,** a hormone that acts in several ways to restore fluid volume and blood pressure **(fig. 23.15):**

- It is a potent vasoconstrictor. Widespread vasoconstriction raises the mean arterial blood pressure throughout the body.

- In the kidneys, it constricts the efferent arterioles and to a somewhat lesser degree, the afferent arterioles. By constricting the glomerular outlet more than the inlet, it raises glomerular blood pressure and GFR, or at least prevents a drastic reduction in GFR, thus ensuring continued filtration of wastes from the blood even when blood pressure has fallen.

- Constriction of the efferent arteriole lowers the blood pressure (BP) in the peritubular capillaries downstream from them. Since capillary BP normally opposes fluid reabsorption, this reduction in BP strongly enhances the reabsorption of NaCl and water from the nephron. More water is returned to the bloodstream instead of being lost in the urine.

- Angiotensin II stimulates the adrenal cortex to secrete aldosterone, which promotes sodium and water reabsorption by the distal convoluted tubule and collecting duct. Angiotensin II also directly stimulates sodium and water reabsorption in the proximal convoluted tubule.

- It stimulates the posterior pituitary gland to secrete antidiuretic hormone, which promotes water reabsorption by the collecting duct.

- It stimulates the sense of thirst and encourages water intake.

Some of these effects are explained more fully later in this chapter and in section 24.2a. Collectively, they raise blood pressure by reducing water loss, encouraging water intake, and constricting blood vessels.

▶▶▶APPLY WHAT YOU KNOW

Would you expect ACE inhibitors to increase or reduce urine output? Why? (ACE inhibitors are explained under "Hormonal Control" in section 20.2c.)

To summarize the events thus far: Glomerular filtration occurs because the high blood pressure of the glomerular capillaries overrides the colloid osmotic pressure of the blood. The filtration membrane allows most plasma solutes into the capsular space while retaining blood cells and protein in the bloodstream. Glomerular filtration is maintained at a fairly steady rate of about 105 to 125 mL/min. (female and male, respectively) in spite of variations in systemic blood pressure. Stability is

FIGURE 23.15 The Renin–Angiotensin–Aldosterone Mechanism. This chain of events is activated by a drop in blood pressure and acts to raise it again.

achieved by renal autoregulation, sympathetic control, and hormonal control.

BEFORE YOU GO ON

Answer the following questions to test your understanding of the preceding section:

7. Name the four major processes in urine production.

8. Trace the movement of a urea molecule from the blood to the capsular space, and name the barriers it passes through.

9. Calculate the net filtration pressure in a patient whose blood COP is only 10 mm Hg because of hypoproteinemia. Assume other relevant variables to be normal.

10. Assume a person is moderately dehydrated and has low blood pressure. Describe the homeostatic mechanisms that would help the kidneys maintain a normal GFR.

23.4 Urine Formation II: Tubular Reabsorption and Secretion

Expected Learning Outcomes

When you have completed this section, you should be able to

a. describe how the renal tubules reabsorb useful solutes from the glomerular filtrate and return them to the blood;

b. describe how the tubules secrete solutes from the blood into the tubular fluid; and

c. describe how the nephron regulates water excretion.

Conversion of the glomerular filtrate to urine involves the removal and addition of chemicals by tubular reabsorption and secretion, to be described in this section. Here we trace the course of the tubular fluid through the nephron, from proximal convoluted tubule through distal convoluted tubule, and see how the filtrate is modified at each point along the way. Refer to figure 23.9 to put these processes into perspective.

23.4a The Proximal Convoluted Tubule

The proximal convoluted tubule (PCT) reabsorbs about 65% of the glomerular filtrate, while it also removes some substances from the blood and secretes them into the tubule for disposal in the urine. The importance of the PCT is reflected in its relatively great length and prominent microvilli, which increase its absorptive surface area. Its cells also contain abundant large mitochondria that provide ATP for active transport. Of all the calories you consume and ATP you use each day, about 6% goes just for driving these processes in the PCTs of your kidneys.

Tubular Reabsorption

Tubular reabsorption is the process of reclaiming water and solutes from the tubular fluid and returning them to the blood. The PCT reabsorbs a greater variety of chemicals than any other part of the nephron.

There are two routes of reabsorption: (1) the **transcellular**[14] **route,** in which substances pass through the cytoplasm and out the base of the epithelial cells; and (2) the **paracellular**[15] **route,**

in which substances pass through gaps between the cells. The "tight" junctions between the epithelial cells are quite leaky and allow significant amounts of water to pass through. As it travels through the epithelium, water carries with it a variety of dissolved solutes—a process called **solvent drag.** Whether by the transcellular or paracellular route, water and solutes enter the tissue fluid at the base of the epithelium, and from there they are taken up by the peritubular capillaries. In the following discussion and **figure 23.16,** we examine mechanisms of reabsorption.

Sodium Chloride Sodium reabsorption is the key to almost everything else, because it creates an osmotic and electrical gradient that drives the reabsorption of water and the other solutes. It is the most abundant cation in the glomerular filtrate, with a concentration of 140 mEq/L in the fluid entering the proximal convoluted tubule and only 12 mEq/L in the cytoplasm of the epithelial cells. This is a very steep concentration gradient favoring its transport into the epithelial cells.

Two types of transport proteins in the apical cell surface are responsible for sodium uptake: (1) various *symports* that simultaneously bind Na^+ and another solute such as glucose, amino acids, or lactate; and (2) an Na^+–H^+ *antiport* that pulls Na^+ into the cell while pumping H^+ out of the cell into the tubular fluid. This antiport is a means not only of reabsorbing sodium, but also of eliminating acid from the body fluids. Angiotensin II activates the Na^+–H^+ antiport and thereby exerts a strong influence on sodium reabsorption.

Sodium is prevented from accumulating in the epithelial cells by Na^+–K^+ pumps in the basal domain of the plasma membrane, which pump Na^+ out into the extracellular fluid. From there, it is picked up by the peritubular capillaries and returned to the bloodstream. These Na^+–K^+ pumps, like those anywhere, are ATP-consuming active transport pumps. Although the sodium-transporting symports in the apical membrane don't consume ATP, they are considered an example of **secondary active transport** because of their dependence on the Na^+–K^+ pumps at the base of the cell.

Chloride ions, being negatively charged, follow Na^+ because they're electrically attracted to it. There also are various antiports in the apical cell membrane that absorb Cl^- in exchange for other anions that they eject into the tubular fluid. Chloride and potassium ions are driven out through the basal cell surface by a K^+–Cl^- symport. Both Na^+ and Cl^- also pass through the tubule epithelium by the paracellular route.

Other Electrolytes Potassium, magnesium, and phosphate ions pass through the paracellular route with water. Phosphate is also cotransported into the epithelial cells with Na^+. Roughly 52% of the filtered calcium is reabsorbed by the paracellular route and 14% by the transcellular route in the PCT. Calcium absorption here is independent of hormonal influence, but another 33% of the calcium is reabsorbed later in the nephron under the influence of parathyroid hormone, to be discussed shortly. The remaining 1%, normally, is excreted in the urine.

---

[14]*trans* = across

[15]*para* = next to

FIGURE 23.16 Reabsorption in the Proximal Convoluted Tubule. Water and solutes in the tubular fluid (right) are carried through the tubule epithelium by various means including symports, antiports, aquaporins, and the paracellular route between cells. They enter the tissue fluid at the base of the epithelium and are picked up by the peritubular capillaries (left). Many other solutes not shown here are reabsorbed by similar means.

? *How would increased Na^+ reabsorption affect the pH of the urine? Where is the clue to this in the figure?*

Glucose   Glucose is cotransported with Na^+ by symports called **sodium–glucose transporters (SGLTs).** It is then removed from the basolateral surface of the cell by facilitated diffusion. Normally, all glucose in the tubular fluid is reabsorbed and there is none in the urine.

Nitrogenous Wastes   Urea passes through the epithelium with water. The nephron as a whole reabsorbs 40% to 60% of the urea in the tubular fluid, but since it reabsorbs 99% of the water, urine has a substantially higher urea concentration than blood or glomerular filtrate. When blood enters the kidney, its urea concentration is about 20 mg/dL; when it leaves the kidney, it is typically down to 10.4 mg/dL. Thus, the kidney removes about half of the urea, keeping its concentration down to a safe level but not completely clearing the blood of it.

The PCT reabsorbs nearly all the uric acid entering it, but later parts of the nephron secrete it back into the tubular fluid. Creatinine isn't reabsorbed at all, but stays in the tubule and is all passed in the urine.

Water   The kidneys reduce about 180 L of glomerular filtrate to 1 or 2 L of urine each day, so water reabsorption is a significant function. About two-thirds of the water is reabsorbed by the PCT. The reabsorption of all the salt and organic solutes as just described makes the tubule cells and tissue fluid hypertonic to the tubular fluid. Water follows the solutes by osmosis through both the paracellular and transcellular routes. Transcellular absorption occurs by way of water channels called **aquaporins** in the apical and basolateral domains of the plasma membrane, enabling water to enter the tubule cells at the apical surface and leave them (to return to the blood) via the basolateral surface.

Because the PCT reabsorbs proportionate amounts of solutes and water, the osmolarity of the tubular fluid remains unchanged here. Elsewhere in the nephron, water reabsorption varies continually because of the action of hormones responding to the body's state of hydration. In the PCT, however, water is reabsorbed at a constant rate called **obligatory water reabsorption.**

Uptake by the Peritubular Capillaries

After water and solutes leave the basal surface of the tubule epithelium, they are reabsorbed by the peritubular capillaries. The mechanisms of capillary absorption are osmosis and solvent drag. Three factors promote osmosis into these capillaries: (1) The accumulation of reabsorbed fluid on the basal side of the epithelium cells creates a high tissue fluid pressure that physically drives water into the capillaries. (2) The narrowness of the efferent arteriole lowers the blood hydrostatic pressure (BHP) from 60 mm Hg in the glomerulus to only 8 mm Hg in the peritubular capillaries, so there is less resistance to reabsorption here than in most systemic capillaries. (3) As blood passes through the glomerulus, a lot of water is filtered out but nearly all of the protein remains in the blood. Therefore, the blood has an elevated colloid osmotic pressure (COP) by the time it leaves the glomerulus. With a high COP and low BHP in the capillaries and a high hydrostatic pressure in the tissue fluid, the balance of forces in the peritubular capillaries strongly favors reabsorption. This tendency is even further accentuated by angiotensin II. By constricting the afferent and efferent arterioles, this hormone reduces blood pressure in the

(1) Angiotensin II secreted

(2) Constricts afferent and especially efferent arterioles

(3) Maintains or increases glomerular blood pressure and glomerular filtration

and

(4) Reduces blood pressure in peritubular capillary

H₂O

(5) Reduces resistance to tubular reabsorption

(6) Tubular reabsorption increases

(7) Urine volume is less but concentration is high

FIGURE 23.17 The Effect of Angiotensin II on Tubular Reabsorption.

Normoglycemia

Hyperglycemia

Glomerular filtration

Glucose-transport protein

Glucose reabsorption

(a) Normal urine volume, glucose-free

(b) Increased urine volume, with glycosuria

FIGURE 23.18 The Transport Maximum. (a) At a normal blood glucose concentration (normoglycemia), all glucose filtered by the glomerulus is reabsorbed by glucose-transport proteins in the proximal convoluted tubule, and the urine is glucose-free. (b) At a high blood glucose concentration (hyperglycemia), more glucose is filtered than the transport proteins, now saturated, are able to absorb. The glomerular filtration of glucose now exceeds the transport maximum (T_m) of the renal tubule. Excess glucose escapes reabsorption and appears in the urine (glycosuria).

peritubular capillaries and thereby reduces their resistance to fluid reabsorption **(fig. 23.17).**

The Transport Maximum

There is a limit to the amount of solute that the renal tubule can reabsorb because there are limited numbers of transport proteins in the plasma membranes. If all the transporters are occupied as solute molecules pass through, some solute will escape reabsorption and appear in the urine. The maximum rate of reabsorption is the **transport maximum (T_m)**, which is reached when the transporters are saturated (see fig. 3.16).

Each solute reabsorbed by the renal tubule has its own T_m. Glucose, for example, has a T_m of 320 mg/min. It normally enters the renal tubule at a rate of 125 mg/min., well within the T_m; thus all of it is reabsorbed. But at blood glucose levels above 220 mg/dL, glucose is filtered faster than the renal tubule can reabsorb it, and the excess passes in the urine—a condition called **glycosuria**[16] (GLY-co-SOOR-ee-uh) **(fig. 23.18).** In untreated diabetes mellitus, the plasma glucose concentration may exceed 400 mg/dL, so glycosuria is one of the classic signs of this disease.

Tubular Secretion

Tubular secretion is a process in which the renal tubule extracts chemicals from the capillary blood and secretes them into the tubular fluid (see fig. 23.9). In the proximal convoluted tubule and nephron loop, it serves three purposes: (1) It contributes to acid–base balance by secreting varying proportions of hydrogen to bicarbonate ions, as detailed in section 24.3. (2) It extracts wastes from the blood, including urea, uric acid, bile acids, ammonia, and a little creatinine. Uric acid secretion compensates for its reabsorption earlier in the PCT and accounts for all of the uric acid in the urine. (3) It clears drugs and contaminants from the blood, such as morphine, penicillin, and aspirin. One reason why so many drugs must be taken three or four times a day is to keep pace with this

---

[16]*glycos* = sugar; *uria* = urine condition

clearance and maintain a therapeutically effective drug concentration in the blood.

23.4b The Nephron Loop

The primary function of the nephron loop is to generate an osmotic gradient that enables the collecting duct to concentrate the urine and conserve water, as discussed later. But in addition, the loop reabsorbs about 25% of the Na^+, K^+, and Cl^- and 15% of the water in the glomerular filtrate. Cells in the thick segment of the ascending limb of the loop have proteins in the apical membranes that simultaneously bind 1 Na^+, 1 K^+, and 2 Cl^- from the tubular fluid and cotransport them into the cytoplasm. These ions leave the basolateral cell surfaces by active transport of Na^+ and diffusion of K^+ and Cl^-. Potassium reenters the cell by means of the Na^+–K^+ pump and then reenters the tubular fluid, but NaCl remains in the tissue fluid of the renal medulla. The thick segment is impermeable to water; thus water cannot follow the reabsorbed electrolytes, and tubular fluid becomes very dilute (200 mOsm/L) by the time it passes from the nephron loop into the distal convoluted tubule.

23.4c The Distal Convoluted Tubule and Collecting Duct

Fluid arriving in the DCT still contains about 20% of the water and 7% of the salts from the glomerular filtrate. If this were all passed as urine, it would amount to 36 L/day, so a great deal of fluid reabsorption is still to come. The DCT and collecting duct reabsorb variable amounts of water and salts and are regulated by several hormones—particularly aldosterone, natriuretic peptides, antidiuretic hormone, and parathyroid hormone.

There are two kinds of cells in the DCT and collecting duct. The **principal cells** are the more abundant; they have receptors for the foregoing hormones and are involved chiefly in salt and water balance. **Intercalated cells** are fewer in number. They reabsorb K^+ and secrete H^+ into the tubule and are involved mainly in acid–base balance, as discussed in section 24.3c. The major hormonal influences on the DCT and collecting duct are as follows.

Aldosterone

Aldosterone, the "salt-retaining hormone," is a steroid secreted by the adrenal cortex when the blood Na^+ concentration falls or its K^+ concentration rises. A drop in blood pressure also induces aldosterone secretion, but indirectly—it stimulates the kidney to secrete renin; this leads to the production of angiotensin II; and angiotensin II stimulates aldosterone secretion (see fig. 23.15).

Aldosterone acts on the thick segment of the ascending limb of the nephron loop, on the DCT, and on the cortical portion of the collecting duct. It stimulates these segments of the nephron to reabsorb Na^+ and secrete K^+. Water and Cl^- follow the Na^+, so the net effect is that the body retains NaCl and water, urine volume is reduced, and the urine has an elevated K^+ concentration. Water retention helps to maintain blood volume and pressure. Section 24.2a deals further with the action of aldosterone.

Natriuretic Peptides

The heart secretes natriuretic peptides in response to high blood pressure. These hormones exert four actions that result in the excretion of more salt and water in the urine, thereby reducing blood volume and pressure:

1. They dilate the afferent arteriole and constrict the efferent arteriole, which increases the GFR.
2. They antagonize the renin–angiotensin–aldosterone mechanism by inhibiting renin and aldosterone secretion.
3. They inhibit the secretion of antidiuretic hormone and its action on the kidney.
4. They inhibit NaCl reabsorption by the collecting duct.

Antidiuretic Hormone (ADH)

Dehydration, loss of blood volume, and rising blood osmolarity stimulate arterial baroreceptors and hypothalamic osmoreceptors. In response, the posterior pituitary gland secretes ADH. ADH makes the collecting duct more permeable to water, so water in the tubular fluid reenters the tissue fluid and blood rather than being lost in the urine. Its mechanisms of doing so are described in the next section (23.5).

Parathyroid Hormone (PTH)

A calcium deficiency (hypocalcemia) stimulates the parathyroid glands to secrete PTH, which acts in several ways to restore calcium homeostasis. Its effect on bone metabolism was described in chapter 7 (section 7.4b). In the kidney, it acts on the PCT to inhibit phosphate reabsorption and acts on the DCT and thick segment of the nephron loop to increase calcium reabsorption. On average, about 25% of the filtered calcium is reabsorbed by the thick segment and 8% by the DCT. PTH therefore increases the phosphate content of the urine and lowers its calcium content. This helps minimize further calcium loss. Because phosphate isn't retained along with the calcium, the calcium ions stay in circulation rather than precipitating into the bone tissue as calcium phosphate. Calcitriol and calcitonin have similar but weaker effects on the DCT. PTH also stimulates calcitriol synthesis by the epithelial cells of the PCT.

In summary, the PCT reabsorbs about 65% of the glomerular filtrate and returns it to the blood of the peritubular capillaries. Much of this reabsorption occurs by osmotic and cotransport mechanisms linked to the active transport of sodium ions. The nephron loop reabsorbs another 25% of the filtrate, although its primary role is to aid the function of the collecting duct. The DCT reabsorbs more sodium chloride and water, but its rates of reabsorption are subject to control by hormones, especially aldosterone and natriuretic peptides. These tubules also extract drugs, wastes, and some other solutes from the blood and secrete them into the tubular fluid. The DCT essentially completes the process of determining the chemical composition of the urine. The principal function left to the collecting duct is to conserve water.

BEFORE YOU GO ON

Answer the following questions to test your understanding of the preceding section:

11. The reabsorption of water, Cl^-, and glucose by the PCT is linked to the reabsorption of Na^+, but in three very different ways. Contrast these three mechanisms.

12. Explain why a substance appears in the urine if its rate of glomerular filtration exceeds the T_m of the renal tubule.

13. Contrast the effects of aldosterone and natriuretic peptides on the renal tubule.

23.5 Urine Formation III: Water Conservation

Expected Learning Outcomes

When you have completed this section, you should be able to

a. explain how the collecting duct and antidiuretic hormone regulate the volume and concentration of urine; and

b. explain how the kidney maintains an osmotic gradient in the renal medulla that enables the collecting duct to function.

The kidney serves not just to eliminate metabolic waste from the body but also to prevent excessive water loss in doing so, and thus to support the body's fluid balance. As the kidney returns water to the tissue fluid and bloodstream, the fluid remaining in the renal tubule, and ultimately passed as urine, becomes more and more concentrated. In this section, we examine the kidney's mechanisms for conserving water and concentrating the urine.

23.5a The Collecting Duct

The collecting duct (CD) begins in the cortex, where it receives tubular fluid from numerous nephrons. As it passes through the medulla, it usually reabsorbs water and concentrates the urine. When urine enters the upper end of the CD, it is isotonic with blood plasma (300 mOsm/L), but by the time it leaves the lower end, it can be up to four times as concentrated—that is, highly hypertonic to the plasma. This ability to concentrate wastes and control water loss was crucial to the evolution of terrestrial animals such as ourselves (see Deeper Insight 23.1).

DEEPER INSIGHT 23.1

EVOLUTIONARY MEDICINE

The Kidney and Life on Dry Land

Physiologists first suspected that the nephron loop played a role in water conservation because of their studies of a variety of animal species. Animals that must conserve water have longer, more numerous nephron loops than animals with little need to conserve it. Fish and amphibians lack nephron loops and produce urine that is isotonic to their blood plasma. Aquatic mammals such as beavers have short nephron loops and only slightly hypertonic urine.

But the kangaroo rat, a desert rodent, provides an instructive contrast. It lives on seeds and other dry foods and can go its whole life without drinking water. Its kidneys are so efficient that it can live entirely on the water produced by aerobic respiration. They have extremely long nephron loops and produce urine that is 10 to 14 times as concentrated as the blood plasma (compared with 4 times, at most, in humans).

Comparative studies thus suggested a hypothesis for the function of the nephron loop and prompted many years of difficult research that led to the discovery of the countercurrent multiplier mechanism for water conservation. This shows how comparative anatomy provides suggestions and insights into function and why physiologists don't study human function in isolation from other species.

Two facts enable the collecting duct to produce such hypertonic urine: (1) The osmolarity of the extracellular fluid is four times as high in the lower medulla as it is in the cortex, and (2) the medullary portion of the CD is more permeable to water than to solutes. Therefore, as urine passes down the CD through the increasingly hypertonic medulla, water leaves the tubule by osmosis, most NaCl and other wastes remain behind, and the urine becomes more and more concentrated (**fig. 23.19**).

23.5b Control of Water Loss

Just how concentrated the urine becomes depends on the body's state of hydration. For example, if you drink a large volume of water, you soon produce a large volume of hypotonic urine—a response called *water diuresis*[17] (DY-you-REE-sis). Under such conditions, the cortical portion of the CD reabsorbs NaCl but is impermeable to water. Salt is removed from the urine, water stays in it, and urine osmolarity may be as low as 50 mOsm/L.

Dehydration, on the other hand, causes the urine to be scanty and more concentrated. The high blood osmolarity of a dehydrated person stimulates the pituitary to release ADH. ADH increases water reabsorption (reduces urine output) by two mechanisms. (1) Within seconds, cells of the collecting duct transfer aquaporins from storage vesicles in the cytoplasm to the apical cell surface and the cells begin taking up more water from the tubular fluid. (2) If the ADH level remains elevated for ≥24 hours, it induces the cell to transcribe the aquaporin gene and manufacture more aquaporins, further raising the water permeability of the collecting duct.

---

[17]*diuresis* = passing urine

Tubular fluid
(300 mOsm/L)

300   Cortex

Medulla

600   H₂O

900   H₂O

1,200 H₂O

Osmolarity of tissue fluid (mOsm/L)

Collecting
duct

Nephron
loop

Urine
(up to 1,200 mOsm/L)

FIGURE 23.19 Water Reabsorption by the Collecting Duct. Note that the osmolarity of the tissue fluid increases fourfold from 300 mOsm/L in the cortex to 1,200 mOsm/L deep in the medulla. Urine concentration increases proportionately as water leaves the duct through its aquaporins.

By contrast, when you're well hydrated, ADH secretion falls; the tubule cells remove aquaporins from the plasma membrane and store them in cytoplasmic vesicles. The duct is then less permeable to water, so more water remains in the duct and you produce abundant, dilute urine.

In extreme cases, the blood pressure of a dehydrated person is low enough to significantly reduce the glomerular filtration rate. When the GFR is low, fluid flows more slowly through the renal tubules and there is more time for tubular reabsorption. Less salt remains in the urine as it enters the CD, so there is less opposition to the osmosis of water out of the duct and into the ECF. More water is reabsorbed and less urine is produced.

23.5c The Countercurrent Multiplier

The ability of the CD to concentrate urine depends on the osmotic gradient of the renal medulla. It may seem surprising that the ECF osmolarity is four

times as great deep in the medulla as in the cortex. We would expect salt to diffuse toward the cortex until it was evenly distributed through the kidney. However, there is a mechanism that overrides this: The nephron loop acts as a **countercurrent multiplier,** which continually recaptures salt and returns it to the deep medullary tissue. It is called a *multiplier* because it multiplies the osmolarity deep in the medulla, and a *countercurrent* mechanism because it is based on fluid flowing in opposite directions in two adjacent tubules—downward in the descending limb and upward in the ascending limb.

Figure 23.20 shows how this works. Steps ② through ⑤ form a positive feedback loop. As fluid flows down the descending limb of the nephron loop, it passes through an environment of increasing osmolarity. Most of the descending limb is very permeable to water but not to NaCl; therefore, water passes by osmosis from the tubule into the ECF, leaving NaCl behind. The tubule contents increase in osmolarity, reaching about 1,200 mOsm/L by the time the fluid rounds the bend at the lower end of the loop.

Most or all of the ascending limb (its thick segment), by contrast, is impermeable to water, but has pumps that cotransport Na⁺, K⁺, and Cl⁻ into the ECF. This keeps the osmolarity of the renal medulla high. Since water remains in the tubule, the tubular fluid becomes more and more dilute as it approaches the cortex and is only about 100 mOsm/L at the top of the loop.

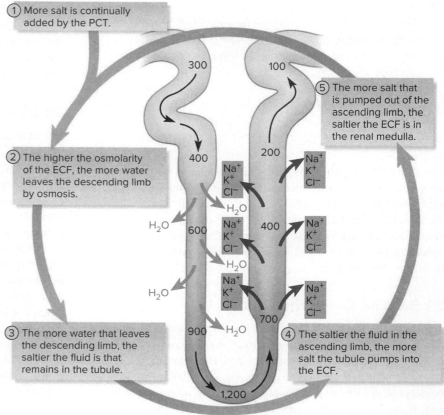

① More salt is continually added by the PCT.

② The higher the osmolarity of the ECF, the more water leaves the descending limb by osmosis.

③ The more water that leaves the descending limb, the saltier the fluid is that remains in the tubule.

④ The saltier the fluid in the ascending limb, the more salt the tubule pumps into the ECF.

⑤ The more salt that is pumped out of the ascending limb, the saltier the ECF is in the renal medulla.

FIGURE 23.20 The Countercurrent Multiplier of the Nephron Loop. The numbers in the tubule are in mOsm/L.

FIGURE 23.21 Functional Relationship of the Nephron Loop, Vasa Recta, and Collecting Duct. These three structures work together to maintain a gradient of osmolarity in the renal medulla. The numbers in the tubule and vasa recta are in mOsm/L.

About 40% of the high osmolarity in the deep medullary tissue, however, is due not to NaCl and KCl, but to urea—which continually cycles from the collecting duct to the nephron loop and back (**fig. 23.21**). The lower end of the collecting duct is somewhat permeable to urea, permitting it to diffuse into the ECF. Some of it enters the descending limb of the loop and travels through the loop and DCT back to the collecting duct. Neither the thick segment nor the DCT is permeable to urea. Combined with new urea constantly added by glomerular filtration, urea remains concentrated in the collecting duct and some of it always diffuses out into the medulla, adding to its osmolarity.

23.5d The Countercurrent Exchange System

The large volume of water reabsorbed by the collecting duct must be returned to the bloodstream. It is picked up and carried away by the vasa recta, but this poses a problem: Why don't the vasa recta also carry away the urea and salt needed to maintain the high osmolarity of the medulla? The answer is that the vasa recta form a **countercurrent exchange** system that prevents this from happening. Blood flows in opposite directions in adjacent parallel capillaries. As it flows downward into the medulla, the vessels exchange water for salt—water diffuses out of the

capillaries and salt diffuses in. But as the blood flows back up to the cortex, the opposite occurs; the vasa recta give up salt and absorb water. Indeed, they absorb more water on the way out than they give up on the way in. Thus, they carry away water absorbed by the kidney but don't subtract from the osmolarity of the medulla.

To summarize what we have studied in this section, the collecting duct can adjust water reabsorption to produce urine as hypotonic (dilute) as 50 mOsm/L or as hypertonic (concentrated) as 1,200 mOsm/L, depending on the body's need for water conservation or removal. In a state of hydration, ADH is not secreted and the cortical part of the CD reabsorbs salt without reabsorbing water; the water remains to be excreted in the dilute urine. In a state of dehydration, ADH is secreted, the medullary part of the CD reabsorbs water, and the urine is more concentrated. The CD is able to do this because it passes through an osmotic gradient in the medulla from 300 mOsm/L near the cortex to 1,200 mOsm/L near the papilla. This gradient is maintained by the countercurrent multiplier of the nephron loop, which concentrates NaCl in the lower medulla, and by the diffusion of urea from the collecting duct into the medulla. The vasa recta are arranged as a countercurrent exchange system that enables them to remove water from the medulla without subtracting from its osmotic gradient.

TABLE 23.1	Hormones Affecting Renal Function	
Hormone	**Renal Targets**	**Effects**
Aldosterone	Nephron loop, DCT, CD	Promotes Na^+ reabsorption and K^+ secretion; indirectly promotes Cl^- and H_2O reabsorption; maintains blood volume and reduces urine volume
Angiotensin II	Afferent and efferent arterioles, PCT	Reduces water loss, stimulates thirst and encourages water intake, and constricts blood vessels, thus raising blood pressure. Reduces GFR; stimulates PCT to reabsorb NaCl and H_2O; stimulates aldosterone and ADH secretion
Antidiuretic hormone	Collecting duct	Promotes H_2O reabsorption; reduces urine volume, increases concentration
Natriuretic peptides	Afferent and efferent arterioles, collecting duct	Dilate afferent arteriole, constrict efferent arteriole, increase GFR; inhibit secretion of renin, ADH, and aldosterone; inhibit NaCl reabsorption by collecting duct; increase urine volume and lower blood pressure
Calcitonin	DCT	Weak effects similar to those of parathyroid hormone
Calcitriol	DCT	Weak effects similar to those of parathyroid hormone
Epinephrine and norepinephrine	Juxtaglomerular apparatus, afferent arteriole	Induce renin secretion; constrict afferent arteriole; reduce GFR and urine volume
Parathyroid hormone	PCT, DCT, nephron loop	Promotes Ca^{2+} reabsorption by loop and DCT; increases phosphate excretion by PCT; promotes calcitriol synthesis

FIGURE 23.22 Solutes Reabsorbed and Secreted in Each Portion of the Renal Tubule.

Figure 23.22 summarizes the major solutes reabsorbed and secreted in each part of the renal tubule. **Table 23.1** summarizes the hormones that affect renal function.

BEFORE YOU GO ON

Answer the following questions to test your understanding of the preceding section:

14. Predict how ADH hypersecretion would affect the sodium concentration of the urine, and explain why.

15. Concisely contrast the role of the countercurrent multiplier with that of the countercurrent exchanger.

16. How would the function of the collecting duct change if the nephron loop didn't exist?

23.6 Urine and Renal Function Tests

Expected Learning Outcomes

When you have completed this section, you should be able to

a. describe the composition and properties of urine; and

b. carry out some calculations to evaluate renal function.

Medical diagnosis often rests on determining the current and recent physiological state of the tissues. No two fluids are as valuable for this purpose as blood and urine. **Urinalysis,** the examination of the physical and chemical properties of urine, is therefore one of the most routine procedures in medical examinations. The principal characteristics of urine and certain tests used to evaluate renal function are described here.

23.6a Composition and Properties of Urine

The basic composition and properties of urine are as follows:

- **Appearance.** Urine varies from almost colorless to deep amber, depending on the body's state of hydration. The yellow color of urine is due to **urochrome,**[18] a pigment produced by the breakdown of hemoglobin from expired erythrocytes. Pink, green, brown, black, and other colors can result from certain foods, vitamins, drugs, and metabolic diseases. Urine is normally clear but turns cloudy upon standing because of bacterial growth. Pus in the urine (**pyuria**[19]) makes it cloudy and suggests kidney infection. Blood in the urine (**hematuria**) may be due to a urinary tract infection, trauma, or kidney stones. Cloudiness or blood in a urine specimen sometimes, however, simply indicates contamination with semen or menstrual fluid.

- **Odor.** Fresh urine has a distinctive but not necessarily repellent odor. As it stands, however, bacteria multiply, degrade urea to ammonia, and produce the pungent odor typical of stale wet diapers. Asparagus and other foods can impart distinctive aromas to the urine. Diabetes mellitus gives it a sweet, fruity odor of acetone. A mousy odor suggests phenylketonuria (PKU), and a rotten odor may indicate urinary tract infection.

- **Specific gravity.** This is a ratio of the density (g/mL) of a substance to the density of distilled water. Distilled water has a specific gravity of 1.000, and urine ranges from 1.001 when it is very dilute to 1.028 when it is very concentrated. Multiplying the last two digits of the specific gravity by a proportionality constant of 2.6 gives an estimate of the grams of solid matter per liter of urine. For example, a specific gravity of 1.025 indicates a solute concentration of $25 \times 2.6 = 65$ g/L.

- **Osmolarity.** Urine can have an osmolarity as low as 50 mOsm/L in a very hydrated person or as high as 1,200 mOsm/L in a dehydrated person. Compared with the osmolarity of blood (300 mOsm/L), then, urine can be either hypotonic or hypertonic.

- **pH.** The body constantly generates metabolic acids and gets rid of them by excreting mildly acidic urine, usually with a pH of about 6.0 (but ranging from 4.5 to 8.2). The next chapter (section 24.3) explains how the kidneys regulate urine pH.

- **Chemical composition.** Urine averages 95% water and 5% solutes by volume (**table 23.2**). Normally, the most abundant solute is urea, followed by sodium chloride, potassium chloride, and lesser amounts of creatinine, uric acid, phosphates, sulfates, and traces of calcium, magnesium, and sometimes bicarbonate. Urine contains urochrome

TABLE 23.2	Properties and Composition of Urine	
Physical Properties		
Specific gravity	1.001–1.028	
Osmolarity	50–1,200 mOsm/L	
pH	6.0 (range 4.5–8.2)	
Solute	**Concentration***	**Output****
Inorganic Ions		
Chloride	533 mg/dL	6.4 g/day
Sodium	333 mg/dL	4.0 g/day
Potassium	166 mg/dL	2.0 g/day
Phosphate	83 mg/dL	1 g/day
Ammonia	60 mg/dL	0.68 g/day
Calcium	17 mg/dL	0.2 g/day
Magnesium	13 mg/dL	0.16 g/day
Nitrogenous Wastes		
Urea	1.8 g/dL	21 g/day
Creatinine	150 mg/dL	1.8 g/day
Uric acid	40 mg/dL	0.5 g/day
Urobilin	125 µg/dL	1.52 mg/day
Bilirubin	20 µg/dL	0.24 mg/day
Other Organics		
Amino acids	288 µg/dL	3.5 mg/day
Ketones	17 µg/dL	0.21 mg/day
Carbohydrates	9 µg/dL	0.11 mg/day
Lipids	1.6 µg/dL	0.02 mg/day

*Typical values for a young adult male
**Assuming a urine output of 1.2 L/day

and a trace of bilirubin from the breakdown of hemoglobin and related products, and urobilin, a brown oxidized derivative of bilirubin. It is abnormal to find glucose, free hemoglobin, albumin, ketones, or bile pigments in the urine; their presence is an important indicator of disease.

23.6b Urine Volume

An average adult produces 1 to 2 L of urine per day. An output of >2 L/day is called **diuresis** or **polyuria**[20] (POL-ee-YOU-ree-uh). Fluid intake and some drugs can temporarily increase output to as much as 20 L/day. Chronic diseases such as diabetes (see next) can do so over a long term. **Oliguria**[21] (oll-ih-GURE-ee-uh)

---

[18]*uro* = urine; *chrom* = color
[19]*py* = pus; *ur* = urine; *ia* = condition

[20]*poly* = many, much; *ur* = urine; *ia* = condition
[21]*oligo* = few, a little; *ur* = urine; *ia* = condition

is an output of <500 mL/day, and **anuria**[22] is an output of 0 to 100 mL/day. Low output can result from kidney disease, dehydration, circulatory shock, prostate enlargement, and other causes. If urine output drops to <400 mL/day, the body cannot maintain a safe, low concentration of wastes in the blood plasma. The result is azotemia.

Diabetes

Diabetes[23] is any metabolic disorder resulting in chronic polyuria. There are at least four forms of diabetes: *diabetes mellitus*[24] *type 1* and *type 2, gestational diabetes,* and *diabetes insipidus.*[25] In most cases, the polyuria results from a high concentration of glucose in the renal tubule. Glucose osmotically retains water in the tubule, so more water passes in the urine *(osmotic diuresis)* and a person may become severely dehydrated. In diabetes mellitus and gestational diabetes, the high glucose level in the tubular fluid is a result of hyperglycemia, a high level in the blood. About 1% to 3% of pregnant women experience gestational diabetes, in which pregnancy reduces the mother's insulin sensitivity, resulting in hyperglycemia and glycosuria. Diabetes insipidus results from ADH hyposecretion. Without ADH, the collecting duct doesn't reabsorb much water, so more water passes in the urine.

Diabetes mellitus and gestational diabetes are characterized by glycosuria. Before chemical tests for urine glucose were developed, physicians diagnosed diabetes mellitus by tasting the patient's urine for sweetness. Tests for glycosuria are now as simple as dipping a chemical test strip into the urine specimen—an advance in medical technology for which urologists are no doubt grateful. In diabetes insipidus, the urine contains no glucose and, by the old diagnostic method, would not taste sweet.

Diuretics

A **diuretic** is any chemical that increases urine volume. Some diuretics, such as caffeine, act by dilating the afferent arteriole and thus increasing glomerular filtration rate. Others act by reducing tubular reabsorption of water. Alcohol, for example, inhibits ADH secretion and thereby reduces reabsorption in the collecting duct. **Loop diuretics** such as furosemide (Lasix) act on the nephron loop to inhibit the Na^+–K^+–Cl^- symport. This impairs the countercurrent multiplier, thus reducing the osmotic gradient in the renal medulla and making the collecting duct unable to reabsorb as much water as usual. Diuretics are commonly administered to treat hypertension and congestive heart failure by reducing the body's fluid volume and blood pressure.

---

[22]*an* = without; *ur* = urine; *ia* = condition
[23]*diabetes* = passing through
[24] *melli* = honey, sweet
[25] *insipid* = tasteless

23.6c Renal Function Tests

There are several tests for diagnosing kidney diseases, evaluating their severity, and monitoring their progress. Here we examine two methods used to determine renal clearance and glomerular filtration rate.

Renal Clearance

Renal clearance is the volume of blood plasma from which a particular waste is completely removed in 1 minute. It represents the net effect of three processes:

> Glomerular filtration of the waste
> + Amount added by tubular secretion
> − Amount removed by tubular reabsorption
> _____
> Renal clearance

In principle, we could determine renal clearance by sampling blood entering and leaving the kidney and comparing their waste concentrations. In practice, it isn't practical to draw blood samples from the renal artery and vein, but clearance can be assessed indirectly by collecting samples of blood and urine, measuring the waste concentration in each, and measuring the rate of urine output.

Suppose the following values were obtained for urea:

U (urea concentration in urine)	= 6.0 mg/mL
V (rate of urine output)	= 2 mL/min.
P (urea concentration in plasma)	= 0.2 mg/mL

The renal clearance (C) of urea would be:

$$C = UV/P$$
$$= (6.0 \text{ mg/mL})(2 \text{ mL/min.})/0.2 \text{ mg/mL}$$
$$= 60 \text{ mL/min.}$$

This means the equivalent of 60 mL of blood plasma is completely cleared of urea per minute. If this person has a normal GFR of 125 mL/min., then the kidneys will clear urea from 60/125 = 48% of the glomerular filtrate. This is a normal rate of urea clearance and is sufficient to maintain safe levels of urea in the blood.

▶▶▶**APPLY WHAT YOU KNOW**

What would you expect the value of renal clearance of glucose to be in a healthy individual? Why?

Glomerular Filtration Rate

Assessment of kidney disease often calls for a measurement of GFR. We cannot determine GFR from urea excretion for two reasons: (1) Some of the urea in the urine is secreted by the renal tubule, not filtered by the glomerulus; and (2) much of the urea filtered by the glomerulus is reabsorbed by the tubule. To measure GFR ideally requires a substance that isn't secreted or reabsorbed at all, so all of it in the urine gets there by glomerular filtration.

There doesn't appear to be a single urine solute produced by the body that isn't secreted or reabsorbed to some degree. However, several plants, including garlic and artichoke, produce a polysaccharide called inulin (IN-you-lin; not to be confused with insulin) that is useful for GFR measurement. All inulin filtered by the glomerulus remains in the renal tubule and appears in the urine; none is reabsorbed, nor does the tubule secrete it. GFR can be measured by injecting inulin and subsequently measuring the rate of urine output and the concentrations of inulin in the blood and urine.

For inulin, GFR is equal to the renal clearance. Suppose, for example, that a patient's plasma concentration of inulin is $P = 0.5$ mg/mL, the urine concentration is $U = 30$ mg/mL, and urine output is $V = 2$ mL/min. This person has a normal GFR:

$$\text{GFR} = \text{UV/P}$$

$$= (30 \text{ mg/mL})(2 \text{ mL/min.})/0.5 \text{ mg/mL}$$

$$= 120 \text{ mL/min.}$$

In clinical practice, GFR is more often estimated from creatinine excretion. This has a small but acceptable error of measurement, and is an easier procedure than injecting and measuring inulin.

A solute that is reabsorbed by the renal tubules will have a renal clearance *less* than the GFR (provided its tubular secretion is less than its rate of reabsorption). This is why the renal clearance of urea is about 60 mL/min. A solute that is secreted by the renal tubules will have a renal clearance *greater* than the GFR (provided its reabsorption does not exceed its secretion). Creatinine, for example, has a renal clearance of 140 mL/min.

BEFORE YOU GO ON

Answer the following questions to test your understanding of the preceding section:

17. Define *oliguria* and *polyuria*. Which of these is characteristic of diabetes?

18. Identify a cause of glycosuria other than diabetes mellitus.

19. How is the diuresis produced by furosemide like the diuresis produced by diabetes mellitus? How are they different?

20. Explain why GFR cannot be determined by measuring the amount of NaCl in the urine.

23.7 Urine Storage and Elimination

Expected Learning Outcomes
When you have completed this section, you should be able to

a. describe the functional anatomy of the ureters, urinary bladder, and male and female urethra; and

b. explain how the nervous system and urethral sphincters control the voiding of urine.

Urine is produced continually, but fortunately it doesn't drain continually from the body. Urination is episodic—occurring when we allow it. This is made possible by an apparatus for storing urine and neural controls for its timely release.

23.7a The Ureters

The renal pelvis funnels urine into the ureter, a retroperitoneal, muscular tube that extends to the urinary bladder. The ureter is about 25 cm long and reaches a maximum diameter of about 1.7 cm near the bladder. The ureters pass posterior to the bladder and enter it from below, passing obliquely through its muscular wall and opening onto its floor. A small flap of mucosa acts as a valve at the opening of each ureter into the bladder, preventing urine from backing up into the ureter when the bladder contracts.

The ureter has three layers: an adventitia, muscularis, and mucosa. The adventitia is a connective tissue layer that binds it to the surrounding tissues. The muscularis consists of two layers of smooth muscle over most of its length, but a third layer appears in the lower ureter. The mucosa has a urothelium (transitional epithelium) that begins in the minor calyces of the kidney and extends from there through the bladder.

When urine enters the ureter and stretches it, the muscularis contracts and initiates a peristaltic wave that milks the urine from the renal pelvis down to the bladder. These contractions occur every few seconds to few minutes, proportional to the rate at which urine enters the ureter. The lumen of the ureter is very narrow and is easily obstructed or injured by kidney stones (see Deeper Insight 23.2).

23.7b The Urinary Bladder

The urinary bladder (**fig. 23.23**) is a muscular sac on the floor of the pelvic cavity, inferior to the peritoneum and posterior to the pubic symphysis. It is covered by parietal peritoneum on its flattened superior surface and by a fibrous adventitia elsewhere.

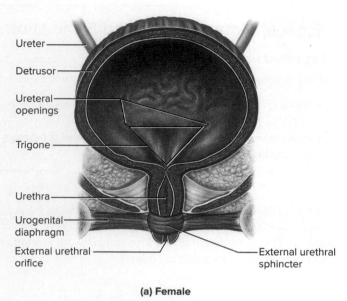

Ureter
Detrusor
Ureteral openings
Trigone
Urethra
Urogenital diaphragm
External urethral orifice
External urethral sphincter

(a) Female

FIGURE 23.23 **The Urinary Bladder and Urethra (Frontal Sections).** (a) Female. (b) Male. **APR**

 Why are women more susceptible than men to bladder infections?

Ureter
Rugae
Ureteral openings
Trigone
Prostate
Prostatic urethra
Membranous urethra
Bulbourethral gland
Spongy (penile) urethra
Penis
Detrusor
Internal urethral sphincter
Urogenital diaphragm
External urethral sphincter

External urethral orifice

(b) Male

DEEPER INSIGHT 23.2

CLINICAL APPLICATION

Kidney Stones

A *renal calculus*[26] (kidney stone) is a hard granule of calcium or phosphate salts and sometimes other components such as uric acid **(fig 23.24).** The presence of kidney stones is called *nephrolithiasis*. Renal calculi form in the renal pelvis and are usually small enough to pass unnoticed in the urine flow. Some, however, grow as large as a few centimeters and block the renal pelvis or ureter, which can lead to the destruction of nephrons as pressure builds in the kidney. A large, jagged calculus passing down the ureter stimulates strong contractions that can be excruciatingly painful. It can also tear the ureter and cause hematuria. Causes of renal calculi include hypercalcemia (excess calcium in the blood), dehydration, pH imbalances, frequent urinary tract infections, or an enlarged prostate gland causing urine retention. The most common treatment for renal calculi in the United States is called *shock wave lithotripsy*[27] *(SWL)*. The patient lies on an operating table under general anesthesia with a water-filled cushion on the abdomen or behind the kidney. One to two thousand underwater electrical sparks produce focused shock waves that pulverize the calculi into "stone dust" that the patient can subsequently pass in the urine. Very large stones (>2 cm) may require surgical removal, however, and SWL is not used for pregnant women, morbidly obese patients, and some others.

FIGURE 23.24 **Kidney Stones.**
remik44992/Shutterstock

Its muscularis, called the **detrusor**[28] (deh-TROO-zur), consists of three layers of smooth muscle. The openings of the two ureters and the urethra mark a smooth-surfaced triangular area called the **trigone**[29] (TRY-goan) on the bladder floor. This is a common site of bladder infection (see Deeper Insight 23.3).

The mucosa is lined with urothelium (transitional epithelium), whose unique surface *umbrella cells* protect it from the hypertonic and acidic urine as described in section 5.2b. The epithelium is thicker here than anywhere else in the urinary tract, since it is subject to the most prolonged exposure to stored urine. When the bladder is empty, the mucosa is typically five or six cells thick and exhibits wrinkles called **rugae**[30] (ROO-jee). As the bladder fills, the rugae flatten and the epithelium thins to two or three cells thick. A moderately full bladder contains about 500 mL of urine and extends about 12.5 cm from top to bottom. The maximum capacity is 700 to 800 mL.

23.7c The Urethra

The urethra conveys urine out of the body. In the female, it is a tube 3 to 4 cm long bound to the anterior wall of the vagina by fibrous connective tissue. Its opening, the **external urethral orifice,** lies between the vaginal orifice and clitoris. The male urethra is about 18 cm long and has three regions: (1) The **prostatic urethra** begins at the urinary bladder and passes for about 2.5 cm through the prostate. During orgasm, it receives semen from the reproductive glands. (2) The **membranous urethra** is a short (0.5 cm), thin-walled portion where the urethra passes through the muscular floor of the pelvic cavity. (3) The **spongy (penile) urethra** is about 15 cm long and passes through the penis to the external urethral orifice. It is named for the *corpus spongiosum* of the penis, through which it passes. The male urethra assumes an S shape: It passes downward from the bladder, turns anteriorly as it enters the root of

the penis, and then turns about 90° downward again as it enters the external, pendant part of the penis. The mucosa has a urothelium near the bladder, pseudostratified epithelium for most of its length, and finally stratified squamous epithelium near the external urethral orifice. There are mucous **urethral glands** in its wall.

In males, smooth muscle around the neck of the bladder and nearby urethra is of a different physiological type than the detrusor, richly innervated by the sympathetic nervous system and thickened into a collar of muscle called the **internal urethral sphincter.** It compresses the urethra when it contracts. Its function seems to be to prevent the reflux of semen into the bladder during ejaculation. There is no evidence of an internal urethral sphincter in females. In both sexes, however, there is an **external urethral sphincter** of skeletal muscle encircling the urethra where it passes through the pelvic floor. This provides voluntary control over the voiding of urine.

23.7d Voiding Urine

Between acts of urination, when the bladder is filling, it is important for the detrusor to relax and the urethral sphincters to remain closed. This is ensured by sympathetic pathways that originate in the upper lumbar spinal cord. Postganglionic fibers travel through the hypogastric nerve to the detrusor and the male internal urethral sphincter. The detrusor relaxes to allow the bladder to fill, but the sphincters contract to ensure that urine is retained in the bladder until voluntarily voided. The external urethral sphincter is held closed by somatic motor fibers that travel from the sacral spinal cord via the pudendal nerve to the sphincter.

To understand the act of urination, also called **micturition**[33] (MIC-too-RISH-un), let's begin with infants and very young children who haven't yet developed voluntary bladder control. Here, it is controlled by a relatively simple spinal **micturition reflex** depicted in steps ① through ④ of **figure 23.25.** Filling of the bladder to about 200 mL or more excites stretch receptors in the bladder wall. These issue signals by way of sensory fibers in the pelvic nerve to the sacral spinal cord (usually segments S2–S3). Efferent signals from the spinal cord travel back to the bladder muscles by way of motor fibers in the pelvic nerve and a parasympathetic ganglion in the bladder wall. Output from the ganglion stimulates the detrusor to contract. When there is no voluntary control over urination, this reflex alone releases urine from the bladder. This is the case, too, in people with spinal cord injuries that disconnect the brain from the lower spinal cord (see Deeper Insight 23.4). The inability to consciously control urination is called **urinary incontinence,** and can result not only from spinal cord injuries but from a number of disease conditions as well as simple aging (see table 23.3).

Normally, however, one also has voluntary control over urination. These mechanisms are shown in steps ⑤ through ⑧ of figure 23.25. There are still uncertainties in the scientific community, however, about some of the details. The figure depicts the male, with both internal and external urethral sphincters.

In both sexes, signals from the stretch receptors of the bladder ascend the spinal cord to a nucleus in the pons called the **micturition center.** This nucleus integrates information about bladder tension

[28]*de* = down; *trus* = push
[29]*tri* = three; *gon* = angle
[30]*rugae* = folds, wrinkles
[31]*cyst* = bladder; *itis* = inflammation
[32]*pyel* = pelvis; *itis* = inflammation

[33]*mictur* = to urinate

Involuntary micturition reflex

1. Stretch receptors detect filling of bladder, transmit afferent signals to spinal cord.

2. Signals return to bladder from spinal cord segments S2 and S3 via parasympathetic fibers in pelvic nerve.

3. Efferent signals excite detrusor muscle.

4. Efferent signals relax internal urethral sphincter. Urine is involuntarily voided if not inhibited by brain.

Voluntary control

5. For voluntary control, micturition center in pons receives signals from stretch receptors.

6. If it is timely to urinate, pons returns signals to spinal interneurons that excite detrusor and relax internal urethral sphincter. Urine is voided.

7. If it is untimely to urinate, signals from cerebrum excite spinal interneurons that keep external urethral sphincter contracted. Urine is retained in bladder.

8. If it is timely to urinate, signals from cerebrum inhibit sacral neurons that keep external sphincter closed. External urethral sphincter relaxes and urine is voided.

FIGURE 23.25 Neural Control of Micturition in the Male. Control in the female differs slightly because of the lack of an internal urethral sphincter.

with information from other brain centers such as the amygdala and cerebrum. Thus, urination can be prompted by fear (via the amygdala) or inhibited by cerebral awareness that the circumstances are inappropriate for urination. Fibers from the micturition center descend the spinal cord through the reticulospinal tracts. Some of these fibers inhibit sympathetic neurons that innervate and constrict the male's internal urethral sphincter. Others descend farther to the sacral spinal cord and excite the parasympathetic neurons that relax the internal urethral sphincter and stimulate the detrusor. The initial contraction of the detrusor raises pressure in the bladder, further exciting the stretch receptors that started the process; thus, a positive feedback loop is established that intensifies bladder contraction as urination proceeds.

For the bladder to begin emptying, however, one final obstacle must be overcome—the external urethral sphincter. Nerve fibers from the cerebral cortex descend by way of the corticospinal tracts and inhibit sacral somatic motor neurons that normally keep that sphincter constricted. It is this voluntary component of micturition that gives a person conscious control of when to urinate and the ability to stop urination in midstream. Voluntary control develops as the nervous system

DEEPER INSIGHT 23.4

CLINICAL APPLICATION

Urination and Spinal Cord Injuries

Knowledge of the neural control of micturition is particularly important for understanding and treating persons with spinal cord injuries. Transection of the spinal cord, as in many cervical fractures, disconnects the *supraspinal* control centers (cerebrum and pons) from the spinal cord circuits that control urination. During the period of spinal shock (see Deeper Insight 13.5), a person is generally incontinent—lacking any control over urination. Bladder control returns as the spinal cord recovers, but is limited to the involuntary micturition reflex. The bladder often cannot empty completely, and there is consequently an increased incidence of cystitis.

matures in early childhood. Males expel the last few milliliters of urine by voluntarily contracting the bulbocavernosus muscle that ensheathes the root of the penis. This helps to reduce the retention of urine in the longer male urethra.

TABLE 23.3	Some Disorders of the Urinary System
Acute glomerulone-phritis	An autoimmune inflammation of the glomeruli, often following a streptococcus infection. Results in destruction of glomeruli leading to hematuria, proteinuria, edema, reduced glomerular filtration, and hypertension. Can progress to chronic glomerulonephritis and renal failure, but most individuals recover from acute glomerulonephritis without lasting effect.
Acute renal failure	An abrupt decline in renal function, often due to traumatic damage to the nephrons or a loss of blood flow stemming from hemorrhage or thrombosis.
Chronic renal failure	Long-term, progressive, irreversible loss of nephrons; see Deeper Insight 23.5 for a variety of causes. Requires a kidney transplant or hemodialysis.
Hydronephrosis[34]	Increase in fluid pressure in the renal pelvis and calyces owing to obstruction of the ureter by kidney stones, nephroptosis, or other causes. Can progress to complete cessation of glomerular filtration and atrophy of nephrons.
Nephroptosis[35] (NEFF-rop-TOE-sis) ("floating kidney")	Slippage of the kidney to an abnormally low position. Occurs in people with too little body fat to hold the kidney in place and in people who subject the kidneys to prolonged vibration, such as truck drivers, equestrians, and motorcyclists. Can twist or kink the ureter, which causes pain, obstructs urine flow, and potentially leads to hydronephrosis.
Nephrotic syndrome	Excretion of large amounts of protein in the urine ($\geq$3.5 g/day) due to glomerular injury. Can result from trauma, drugs, infections, cancer, diabetes mellitus, lupus erythematosus, and other diseases. Loss of plasma protein leads to edema, ascites, hypotension, and susceptibility to infection (because of immunoglobulin loss).
Urinary incontinence	Inability to hold the urine; involuntary leakage from the bladder. Can result from aging; incompetence of the urinary sphincters; bladder irritation; pressure on the bladder in pregnancy; an obstructed urinary outlet so that the bladder is constantly full and dribbles urine (overflow incontinence); uncontrollable urination due to brief surges in bladder pressure, as in laughing or coughing (stress incontinence); and neurological disorders such as spinal cord injuries.

You can find other urinary system disorders described in the following places:

Uremia and *azotemia* in section 23.1b; *hematuria, proteinuria,* and *nephrosclerosis* in sections 23.3a, 23.3b; *pyuria, polyuria, oliguria,* and *anuria* in section 23.6b; *kidney stones* in Deeper Insight 23.2; *urinary tract infection* in Deeper Insight 23.3; and *renal insufficiency* in Deeper Insight 23.5.

In females, the external urethral sphincter may be the only mechanism for urine retention. Retention therefore depends on the strength of the voluntary pelvic floor muscles, which can be compromised by a history of pregnancy or obesity. Urinary incontinence is therefore more common in women than in men.

If the urge to urinate arises at an inconvenient time and one must suppress it, the stretch receptors fatigue and stop firing. As bladder tension increases, however, the signals return with increasing frequency and persistence. Conversely, there are times when the bladder isn't full enough to trigger the micturition reflex, but one wishes to "go" anyway because of a long drive or lecture coming up. In this case, the Valsalva maneuver (see section 22.2a) is used to compress the bladder and excite the stretch receptors early, thereby getting the reflex started. The Valsalva maneuver also aids in emptying the bladder.

The effects of aging on the urinary system are discussed in section 29.4a. Some disorders of this system are briefly described in **table 23.3**.

BEFORE YOU GO ON

Answer the following questions to test your understanding of the preceding section:

21. Describe the location and function of the detrusor.

22. Compare and contrast the functions of the internal and external urethral sphincters.

23. In males, the sympathetic nervous system triggers ejaculation and, at the same time, stimulates constriction of the internal urethral sphincter. What purpose is served by the latter action?

---

[34]*hydro* = water; *nephr* = kidney; *osis* = medical condition
[35]*nephro* = kidney; *ptosis* = sagging, falling

DEEPER INSIGHT 23.5

CLINICAL APPLICATION

Renal Insufficiency and Hemodialysis

Renal insufficiency is a state in which the kidneys cannot maintain homeostasis due to extensive destruction of their nephrons. Some causes of nephron destruction include

- hypertension;
- chronic or repetitive kidney infections;
- trauma from such causes as blows to the lower back or continual vibration from machinery;
- prolonged ischemia and hypoxia, as in long-distance runners and swimmers;
- poisoning by heavy metals such as mercury and lead and solvents such as carbon tetrachloride, acetone, and paint thinners. These are absorbed into the blood from inhaled fumes or by skin contact and then filtered by the glomeruli. They kill renal tubule cells.
- blockage of renal tubules with proteins small enough to be filtered by the glomerulus—for example, myoglobin released by skeletal muscle damage and hemoglobin released by a transfusion reaction;
- atherosclerosis, which reduces blood flow to the kidney; and
- glomerulonephritis, an autoimmune disease of the glomerular capillaries.

Nephrons can regenerate and restore kidney function after short-term injuries. Even when some of the nephrons are irreversibly destroyed, others hypertrophy and compensate for their lost function. Indeed, a person can survive on as little as one-third of one kidney. When 75% of the nephrons are lost, however, urine output may be as low as 30 mL/h

compared with the normal rate of 50 to 60 mL/h. This is insufficient to maintain homeostasis and is accompanied by azotemia and acidosis. Uremia develops when there is 90% loss of renal function. Renal insufficiency also tends to cause anemia because the diseased kidneys produce too little erythropoietin (EPO), the hormone that stimulates red blood cell formation.

Hemodialysis is a procedure for artificially clearing wastes from the blood when the kidneys aren't adequately doing so **(fig. 23.26).** Blood is pumped from the radial artery to a *dialysis machine* (artificial kidney) and returned to the patient by way of a vein. In the dialysis machine, the blood flows through a semipermeable cellophane tube surrounded by dialysis fluid. Urea, potassium, and other solutes that are more concentrated in the blood than in the dialysis fluid diffuse through the membrane into the fluid, which is discarded. Glucose, electrolytes, and drugs can be administered by adding them to the dialysis fluid so they will diffuse through the membrane into the blood. People with renal insufficiency accumulate substantial amounts of excess body water between treatments, and dialysis serves also to remove it. Patients are typically given erythropoietin (EPO) to compensate for the lack of EPO from the failing kidneys.

Hemodialysis patients typically have three sessions per week for 4 to 8 hours per session. In addition to inconvenience, hemodialysis carries risks of infection and thrombosis. Blood tends to clot when exposed to foreign surfaces, so an anticoagulant such as heparin is added during dialysis. Unfortunately, this inhibits clotting in the patient's body as well, and dialysis patients sometimes suffer internal bleeding.

A procedure called *continuous ambulatory peritoneal dialysis (CAPD)* is more convenient. It can be carried out at home by the patient, who is provided with plastic bags of dialysis fluid. Fluid is introduced into the abdominal cavity through an indwelling catheter. Here, the peritoneum provides over 2 m² of blood-rich semipermeable membrane. The fluid is left in the body cavity for 15 to 60 minutes to allow the blood to equilibrate with it; then it is drained, discarded, and replaced with fresh dialysis fluid. The patient is not limited by a stationary dialysis machine and can go about most normal activities. CAPD is less expensive and promotes better morale than conventional hemodialysis, but it is less efficient in removing wastes and it is more often complicated by infection.

FIGURE 23.26 Hemodialysis. Blood is pumped into a dialysis chamber, where it flows through a selectively permeable membrane surrounded by dialysis fluid. Blood leaving the chamber passes through a bubble trap to remove air before it is returned to the patient's body. The fluid picks up excess water and metabolic wastes from the patient's blood and may contain medications that diffuse into the blood.

Photo: Hank Morgan/Science Source

Effects of the **URINARY SYSTEM** on Other Organ Systems

ALL SYSTEMS
Excretes metabolic wastes to prevent poisoning of the tissues; maintains fluid, electrolyte, and acid–base balance necessary for homeostasis

INTEGUMENTARY SYSTEM
Fluid balance maintained by the kidneys is essential for normal secretion of sweat.

SKELETAL SYSTEM
Calcitriol synthesis and other roles of the kidneys in calcium and phosphate homeostasis are necessary for normal bone deposition and maintenance.

MUSCULAR SYSTEM
Renal control of Na$^+$, K$^+$, and Ca^{2+} balance is important for muscle excitability and contractility.

NERVOUS SYSTEM
Renal control of Na$^+$, K$^+$, and Ca^{2+} balance is important for neuron signal generation, conduction, and synaptic transmission.

ENDOCRINE SYSTEM
Kidneys secrete erythropoietin, initiate the synthesis of angiotensin II, indirectly stimulate aldosterone secretion, and clear hormones and their metabolites from the body.

CIRCULATORY SYSTEM
Kidneys affect blood pressure more than any other organ but the heart and regulate blood composition; renal dysfunction can cause electrolyte imbalances that affect the cardiac rhythm.

LYMPHATIC AND IMMUNE SYSTEMS
Acidity of urine provides innate immunity against urinary tract infections; renal failure burdens lymphatic system by creating fluid retention and edema.

RESPIRATORY SYSTEM
Respiratory rhythm is sensitive to acid–base imbalances that may result from renal dysfunction.

DIGESTIVE SYSTEM
Kidneys excrete toxins absorbed by intestines; kidneys excrete metabolites generated by the liver; calcitriol secreted by the kidneys stimulates calcium absorption by the small intestine.

REPRODUCTIVE SYSTEM
Male urethra serves as common passage for urine and semen; maternal urinary system excretes fetal wastes.

STUDY GUIDE

▶ Assess Your Learning Outcomes

To test your knowledge, discuss the following topics with a study partner or in writing, ideally from memory.

23.1 Functions of the Urinary System

1. The six main organs of the urinary system
2. Six or more functions of the kidneys
3. Four main nitrogenous wastes and their metabolic sources; how metabolic wastes differ from other wastes
4. Blood urea nitrogen, azotemia, and uremia, and the pathological effects of uremia
5. The meaning of *excretion,* and four human organ systems that perform it

23.2 Anatomy of the Kidney

1. Location of the kidneys in relation to adjacent tissues and organs
2. Structures that enter and leave the kidney through the hilum
3. Three layers of tissue that surround and encapsulate the kidney
4. Arrangement of the renal parenchyma around the renal sinus; organization of the cortex, medulla, and renal pyramids
5. The relationship of a renal pyramid to a minor calyx, and of the minor calyces to the major calyces, renal pelvis, and ureter
6. The term for the microscopic functional units of the kidney, and their approximate number per kidney
7. Blood flow from the renal artery through the kidney to the renal vein, including circulation through both the cortex and the medulla
8. Structure of the renal corpuscle
9. Fluid flow from the point where it filters from the blood to the point where it leaves the kidney; those parts of the renal tubule that belong to a single nephron and the part that is shared by multiple nephrons
10. The distinction between glomerular filtrate, tubular fluid, and urine in relation to the progress of the fluid through the renal tubule
11. Differences in the structure and function of cortical nephrons and juxtamedullary nephrons
12. Innervation of the kidney and two effects of sympathetic stimulation on renal function

23.3 Urine Formation I: Glomerular Filtration

1. Four basic stages of urine formation
2. Structure of the glomerular filtration membrane; roles of the capillary fenestrations, basement membrane, and filtration slits in producing glomerular filtrate; the difference between the filtrate and blood plasma
3. Definitions of *proteinuria* and *hematuria,* and their causes
4. Forces of capillary filtration that account for the net filtration pressure (NFP) in the glomerulus; the magnitude of each and of the NFP
5. How glomerular filtration rate can be calculated from the NFP and filtration coefficient; typical GFR values
6. The meaning of *renal autoregulation*
7. How the myogenic mechanism of renal autoregulation works
8. Structure and function of the juxtaglomerular apparatus and how tubuloglomerular feedback works
9. How the sympathetic nervous system regulates GFR
10. The renin–angiotensin–aldosterone mechanism of regulating GFR; multiple effects of angiotensin II on the kidney and on the rest of the body
11. Effects of antidiuretic hormone and aldosterone on the kidney and how they reduce water loss from the body

23.4 Urine Formation II: Tubular Reabsorption and Secretion

1. The percentage of the glomerular filtrate that is eliminated as urine, and percentages of the water in the filtrate that are reabsorbed by PCT and nephron loop
2. Transcellular and paracellular routes of reabsorption, and the role of solvent drag
3. How the PCT reabsorbs NaCl, other electrolytes, glucose, urea, and water; why the reabsorption of water and solutes depends directly or indirectly on the sodium–glucose transporter (SGLT) and sodium reabsorption
4. How tubular reabsorption is limited by the transport maximum (T_m); how this relates to glycosuria in diabetes mellitus

5. Substances added to the tubular fluid by tubular secretion in the PCT
6. The primary functions of the nephron loop and DCT, and what they reabsorb
7. Effects of aldosterone, natriuretic peptides, parathyroid hormone, calcitonin, and calcitriol on nephron function

23.5 Urine Formation III: Water Conservation

1. Function of the collecting duct (CD) and the range of urine osmolarities it can produce
2. How the osmotic gradient of the renal medulla and selective permeability of the CD act to concentrate the urine
3. The effect of antidiuretic hormone (ADH) on the CD and the role of aquaporins in this effect
4. Function of the countercurrent multiplier and how it performs that role
5. Function of the countercurrent exchanger and how it performs that role

23.6 Urine and Renal Function Tests

1. Why urine is yellow; why the shade of yellow varies; and some causes of other, unusual colors
2. Normal ranges of urine specific gravity, osmolarity, and pH
3. Normal and abnormal odors of urine and some reasons for the latter
4. The three most abundant solutes in urine, and some causes of other, unusual solutes
5. Typical daily output of urine, and terms for abnormally low and high outputs
6. The defining sign of diabetes in general; four forms of diabetes and their causes
7. The general effect of diuretics; modes of diuretic action of caffeine, alcohol, and loop diuretics
8. Methods of measuring a person's GFR and renal clearance; ability to calculate these if given the necessary data

23.7 Urine Storage and Elimination

1. The route and mechanism of urine transport from the kidney to the urinary bladder; anatomy and histology of the ureters and their relationship to the bladder

STUDY GUIDE

2. Anatomy and histology of the urinary bladder; the detrusor, mucosal epithelium, rugae, and trigone
3. Anatomy of the female urethra and external urethral sphincter
4. Anatomy of the male urethra, its three segments, and the internal and external urethral sphincters
5. The mechanism of the spinal micturition reflex and the neural anatomy involved
6. Mechanisms of brainstem and cerebral control of micturition

▶ Testing Your Recall

Answers in Appendix A

1. Micturition occurs when the _____ contracts.
   a. detrusor
   b. internal urethral sphincter
   c. external urethral sphincter
   d. muscularis of the ureter
   e. all of the above

2. The compact ball of capillaries in a nephron is called
   a. the nephron loop.
   b. the peritubular plexus.
   c. the renal corpuscle.
   d. the glomerulus.
   e. the vasa recta.

3. Which of these is the most abundant nitrogenous waste in the blood?
   a. uric acid
   b. urea
   c. ammonia
   d. creatinine
   e. albumin

4. Which of these lies closest to the renal cortex?
   a. the parietal peritoneum
   b. the renal fascia
   c. the fibrous capsule
   d. the perirenal fat capsule
   e. the renal pelvis

5. Most sodium is reabsorbed from the glomerular filtrate by
   a. the vasa recta.
   b. the proximal convoluted tubule.
   c. the distal convoluted tubule.
   d. the nephron loop.
   e. the collecting duct.

6. A glomerulus and glomerular capsule make up one
   a. renal capsule.
   b. renal corpuscle.
   c. kidney lobule.
   d. kidney lobe.
   e. nephron.

7. The kidney has more _____ than any of the other structures listed.
   a. arcuate arteries
   b. minor calyces
   c. medullary pyramids
   d. afferent arterioles
   e. collecting ducts

8. The renal clearance of _____ is normally zero.
   a. sodium
   b. potassium
   c. uric acid
   d. urea
   e. amino acids

9. Beavers have relatively little need to conserve water and could therefore be expected to have _____ than humans do.
   a. fewer nephrons
   b. longer nephron loops
   c. shorter nephron loops
   d. longer collecting ducts
   e. longer convoluted tubules

10. Increased ADH secretion should cause the urine to have
    a. a higher specific gravity.
    b. a lighter color.
    c. a higher pH.
    d. a lower urea concentration.
    e. a lower potassium concentration.

11. The _____ reflex is an autonomic reflex activated by pressure in the urinary bladder.

12. _____ is the ability of a nephron to adjust its GFR independently of nervous or hormonal influences.

13. The two ureters and the urethra form the boundaries of a smooth area called the _____ on the floor of the urinary bladder.

14. The _____ is a group of epithelial cells of the nephron loop that monitors the composition of the tubular fluid.

15. To enter the capsular space, filtrate must pass between foot processes of the _____, cells that form the visceral layer of the glomerular capsule.

16. Glycosuria occurs if the rate of glomerular filtration of glucose exceeds the _____ of the proximal convoluted tubule.

17. _____ is a hormone that regulates the amount of water reabsorbed by the collecting duct.

18. The _____ sphincter of the male is under involuntary control and relaxes during the micturition reflex.

19. Very little _____ is found in the glomerular filtrate because it is negatively charged and is repelled by the basement membrane of the glomerulus.

20. Blood flows through the _____ arteries just before entering the cortical radiate arteries.

STUDY GUIDE

▶ Building Your Medical Vocabulary

Answers in Appendix A

State a meaning of each word element, and give a medical term from this chapter that uses it or a slight variation of it.

1. azoto-

2. cysto-

3. glomer-

4. juxta-

5. meso-

6. nephro-

7. podo-

8. -ptosis

9. pyelo-

10. recto-

▶ What's Wrong with These Statements?

Answers in Appendix A

Briefly explain why each of the following statements is false, or reword it to make it true.

1. The proximal convoluted tubule is not subject to hormonal influence.

2. Sodium is the most abundant solute in the urine.

3. The collecting ducts outnumber the renal corpuscles of the kidney.

4. Tight junctions prevent material from leaking between the epithelial cells of the renal tubule.

5. All forms of diabetes are characterized by glucose in the urine.

6. If all other conditions remain the same, dilation of the efferent arteriole raises the glomerular filtration rate.

7. Angiotensin II promotes urine output.

8. The minimum osmolarity of urine is 300 mOsm/L, equal to the osmolarity of the blood.

9. The amount of glucose in the urine is directly proportional to the amount of sodium.

10. Micturition depends on relaxation of the detrusor.

▶ Testing Your Comprehension

1. How would the glomerular filtration rate be affected by kwashiorkor (see Deeper Insight 18.1)?

2. A patient produces 55 mL of urine per hour. Urea concentration is 0.25 mg/mL in her blood plasma and 8.6 mg/mL in her urine. (a) What is her rate of renal clearance for urea? (b) About 95% of adults excrete urea at a rate of 12.6 to 28.6 g/day. Is this patient above, within, or below this range? Show how you calculated your answers.

3. A patient with poor renal perfusion is treated with an ACE inhibitor and goes into renal failure. Explain the reason for the renal failure.

4. Drugs called *renin inhibitors* are used to treat hypertension. Explain how they would have this effect.

5. Discuss how the unity of form and function is exemplified by each of the following comparisons: (a) the thin and thick segments of the nephron loop; (b) the proximal and distal convoluted tubules; and (c) the afferent and efferent arterioles.

Water balance begins at the renal glomeruli like the one shown here

Michael Abbey/Science Source

FLUID, ELECTROLYTE, AND ACID–BASE BALANCE

CHAPTER OUTLINE

24.1 Fluid Balance

   24.1a Fluid Compartments

   24.1b Water Gain and Loss

   24.1c Regulation of Intake

   24.1d Regulation of Output

   24.1e Disorders of Fluid Balance

24.2 Electrolyte Balance

   24.2a Sodium

   24.2b Potassium

   24.2c Calcium

   24.2d Magnesium

   24.2e Chloride

   24.2f Phosphates

   24.2g Summary of Imbalances

24.3 Acid–Base Balance

   24.3a Acids, Bases, and Buffers

   24.3b Respiratory Control of pH

   24.3c Renal Control of pH

   24.3d Disorders of Acid–Base Balance

   24.3e Compensation for Acid–Base Imbalances

   24.3f Acid–Base Imbalances in Relation to Electrolyte and Fluid Imbalances

Study Guide

DEEPER INSIGHTS

24.1 Clinical Application: Fluid Balance in Cold Weather

24.2 Clinical Application: Fluid Replacement Therapy

Anatomy & Physiology
Revealed 4.0

Module 13: Urinary System

Cellular function requires a fluid medium with a carefully controlled composition. If the quantity, osmolarity, electrolyte concentration, or pH of this medium is altered, life-threatening disorders of cellular function may result. Consequently, the body has several mechanisms for keeping these variables within narrow limits and maintaining three types of homeostatic stability: fluid, electrolyte, and acid–base balance.

These balances are maintained by the collective action of the urinary, respiratory, digestive, integumentary, endocrine, nervous, cardiovascular, and lymphatic systems. This chapter describes the homeostatic regulation of fluid, electrolyte, and acid–base balance and shows the close relationship of these variables to each other. These balances are so crucial that fluid therapy, aimed at restoring one or more of these, is often a critical aspect of patient care. Problems of fluid and electrolyte balance also arise frequently in sports and recreational medicine, in situations ranging from summer football practice to backcountry hiking.

24.1 Fluid Balance

Expected Learning Outcomes

When you have completed this section, you should be able to

a. name the major fluid compartments and explain how water moves from one to another;

b. list the body's sources of water and routes of water loss;

c. describe the mechanisms of regulating water intake and output; and

d. describe some conditions in which the body has a deficiency or excess of water or an improper distribution of water among the fluid compartments.

We enter the world in rather soggy condition, having swallowed, excreted, and floated in amniotic fluid for months. At birth, a baby's weight is as much as 75% water; infants normally lose a little weight in the first day or two as they excrete the excess. Young adult men average 55% to 60% water; women average slightly less because they have more adipose tissue, which is nearly free of water. Obese and elderly people are as little as 45% water by weight. The **total body water (TBW)** content of a 70 kg (150 lb) young male is about 40 L.

24.1a Fluid Compartments

Body water is distributed among certain **fluid compartments,** areas separated by selectively permeable membranes and differing from each other in chemical composition. The major fluid compartments are

65% *intracellular fluid (ICF)* and

35% *extracellular fluid (ECF),* subdivided into

    25% *tissue (interstitial) fluid,*

    8% *blood plasma* and *lymph,* and

    2% *transcellular fluid,* a catch-all category for cerebrospinal, synovial, peritoneal, pleural, and pericardial fluids; vitreous and aqueous humors of the eye; bile; and fluid in the digestive, urinary, and respiratory tracts.

Fluid is continually exchanged between compartments by way of capillary walls and plasma membranes **(fig. 24.1).** Water moves by osmosis from the digestive tract to the bloodstream and by capillary filtration from the blood to the tissue fluid. From the tissue fluid, it may be reabsorbed by the capillaries, osmotically absorbed into cells, or taken up by the lymphatic system, which returns it to the bloodstream.

Because water moves so easily through plasma membranes, osmotic gradients between the ICF and ECF never last for very long. If a local imbalance arises, osmosis usually restores the balance within seconds so that intracellular and extracellular osmolarity are equal. If the osmolarity of the tissue fluid rises, water moves out of the cells; if it falls, water moves into the cells.

Osmosis from one fluid compartment to another is determined by the relative concentration of solutes in each compartment. The most abundant solute particles by far are the electrolytes—especially sodium salts in the ECF and potassium salts in the ICF. Electrolytes play the principal role in governing the body's water distribution and total water content; the subjects of fluid and electrolyte balance are therefore inseparable.

24.1b Water Gain and Loss

A person is in a state of **fluid balance** when daily gains and losses are equal and fluids are properly distributed in the body. We typically gain and lose about 2,500 mL/day **(fig. 24.2).** The gains come from two sources. One of these is **metabolic water** (about 200 mL/day), which is produced as a by-product of dehydration synthesis reactions and aerobic respiration:

$$C_6H_{12}O_6 + 6\,O_2 \longrightarrow 6\,CO_2 + 6\,H_2O.$$

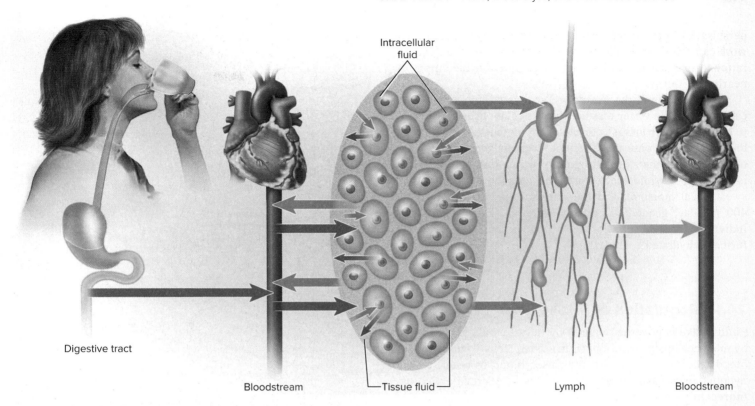

Intracellular
fluid

Digestive tract

Bloodstream   Tissue fluid   Lymph   Bloodstream

FIGURE 24.1 The Movement of Water Between the Major Fluid Compartments. Ingested water is absorbed by the bloodstream. There is a two-way exchange of water between the blood and tissue fluid and between the tissue fluid and intracellular fluids. Excess tissue fluid is picked up by the lymphatic system, which returns it to the bloodstream.

❓ *In which of these compartments would fluid accumulate in edema?*

The other source is **preformed water,** which is ingested in food (700 mL/day) and drink (1,600 mL/day).

The routes of water loss are more varied:

- 1,500 mL/day is excreted as urine.

- 200 mL/day is eliminated in the feces.

- 300 mL/day is lost in the expired breath. You can easily visualize this by breathing onto a cool surface such as a mirror.

- 100 mL/day of sweat is secreted by a resting adult at an ambient air temperature of 20°C (68°F).

- 400 mL/day is lost as **cutaneous transpiration,**[1] water that diffuses through the epidermis and evaporates. This is not the same as sweat; it's not a glandular secretion. A simple way to observe it is to cup the palm of your hand for a minute against a cool nonporous surface such as a laboratory benchtop or mirror. When you take your hand away, you will notice the water that transpired through the skin and condensed on that surface, even in places that were not in contact with your skin.

Water loss varies greatly with physical activity and environmental conditions. Respiratory loss increases in cold weather, for example, because cold air is drier and absorbs more body water from the respiratory tract. Hot, humid weather slightly reduces the respiratory loss but increases

Gains
2,500 mL/day

Losses
2,500 mL/day

Metabolic water
200 mL

Food
700 mL

Drink
1,600 mL

Feces
200 mL

Expired air
300 mL

Cutaneous
transpiration
400 mL

Sweat 100 mL

Urine
1,500 mL

FIGURE 24.2 Typical Daily Water Gains and Losses in a State of Fluid Balance.

---

[1]*trans* = through; *spiration* = breathing

perspiration to as much as 1,200 mL/day. Prolonged, heavy work can raise the respiratory loss to 650 mL/day and perspiration to as much as 5 L/day, though it reduces urine output by nearly two-thirds.

Output through the breath and cutaneous transpiration is called **insensible water loss** because we're not usually aware of it. **Sensible water loss** is noticeable output, particularly through the urine and in case of sufficient sweating to produce obvious wetness of the skin. **Obligatory water loss** is output that is relatively unavoidable: expired air, cutaneous transpiration, sweat, fecal moisture, and the minimum urine output, about 400 mL/day, needed to prevent azotemia. Even dehydrated individuals cannot prevent such losses; thus, they become further dehydrated.

24.1c Regulation of Intake

Fluid intake is governed mainly by thirst, which is controlled by the mechanisms shown in **figure 24.3.** Dehydration reduces blood volume and pressure and raises blood osmolarity. The hypothalamus has at least three groups of neurons called **osmoreceptors** that respond to angiotensin II and rising osmolarity of the ECF—both of which are signs that the body has a water deficit. The osmoreceptors communicate with other hypothalamic neurons that produce antidiuretic hormone (ADH), thus promoting water conservation; they apparently communicate also with the cerebral cortex to produce a conscious sense of thirst. A mere 2% to 3% increase in plasma osmolarity makes a person intensely thirsty, as does a 10% to 15% blood loss.

When we're thirsty, we salivate less. There are two reasons for this: (1) The osmoreceptor response leads to sympathetic output from the hypothalamus that inhibits the salivary glands. (2) Saliva is produced primarily by capillary filtration, but in a dehydrated person, this is opposed by the lower capillary blood pressure and higher osmolarity of the blood. Reduced salivation produces a dry, sticky-feeling mouth and a desire to drink, but it is by no means certain that this is the primary motivation to drink. Some people don't secrete saliva, yet they don't drink any more than normal individuals except when eating, when they need water to moisten the food. The same is true of experimental animals that have the salivary ducts tied off.

Long-term satiation of thirst depends on absorbing water from the small intestine and lowering the osmolarity of the blood. Reduced osmolarity stops the osmoreceptor response, promotes capillary filtration, and makes the saliva more abundant and watery. However, these changes require 30 minutes or longer to take effect, and it would be rather impractical if we had to drink that long while waiting to feel satisfied. Water intake would be grossly excessive, indeed potentially fatal. Fortunately, there are mechanisms that act more quickly to temporarily quench the thirst and allow time for the change in blood osmolarity to occur.

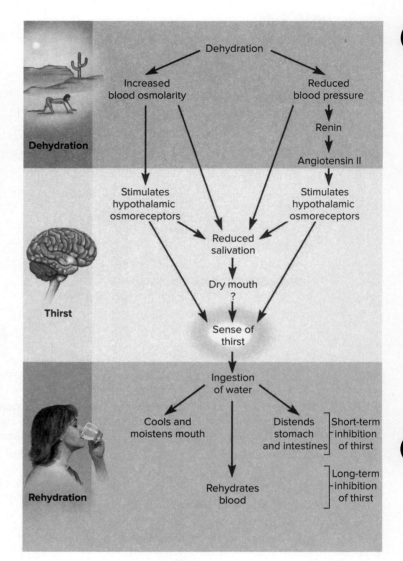

FIGURE 24.3 Dehydration, Thirst, and Rehydration. See the text concerning the uncertainty about a dry mouth contributing to the sense of thirst.

Experiments with rats and dogs have isolated the stimuli that quench the thirst. One of these is cooling and moistening the mouth; rats drink less if their water is cool than if it is warm, and simply moistening the mouth temporarily satisfies an animal even if the water is drained from its esophagus before it reaches the stomach. Distension of the stomach and small intestine is another inhibitor of thirst. If a dog is allowed to drink while the water is drained from its esophagus but its stomach is inflated with a balloon, its thirst is satisfied for a time. If the water is drained away but the stomach isn't inflated, satiation doesn't last as long. Such fast-acting stimuli as coolness, moisture, and filling of the stomach stop an animal (and presumably a human) from drinking an excessive amount of liquid, but they're effective for only 30 to 45 minutes. If they're not soon followed by absorption of water into the bloodstream, the thirst soon returns. Only a drop in blood osmolarity produces a lasting effect.

24.1d Regulation of Output

The only way to control water output significantly is through variations in urine volume. It must be realized, however, that the kidneys can't completely prevent water loss, nor can they replace lost water or electrolytes. Therefore, they never restore fluid volume or osmolarity, but in dehydration they can support existing fluid levels and slow down the rate of loss until water and electrolytes are ingested.

To understand the effect of the kidneys on fluid and electrolyte balance, it is also important to bear in mind that if a substance is reabsorbed by the kidneys, it is kept in the body and returned to the ECF, where it will affect fluid volume and composition. If a substance is filtered by the glomerulus or secreted by the renal tubules and not reabsorbed, then it is excreted in the urine and lost from the body fluids.

Changes in urine volume are usually linked to adjustments in sodium reabsorption. As sodium is reabsorbed or excreted, proportionate amounts of water accompany it. The total volume of fluid remaining in the body may change, but its osmolarity remains stable. Controlling fluid balance by controlling sodium excretion is best understood in the context of electrolyte balance, discussed later in the chapter.

Antidiuretic hormone (ADH), however, provides a means of controlling water output independently of sodium (**fig. 24.4**). In true dehydration (defined shortly), blood volume declines and sodium concentration rises. The increased osmolarity of the blood stimulates the hypothalamic osmoreceptors, which stimulate the posterior pituitary to release ADH. In response to ADH, cells of the collecting ducts of the kidneys synthesize the proteins called aquaporins. When installed in the plasma membrane, these serve as channels that allow water to diffuse out of the duct into the hypertonic tissue fluid of the renal medulla. The kidneys then reabsorb more water and produce less urine. Sodium continues to be excreted, so the *ratio* of sodium to water in the urine increases (the urine becomes more concentrated). By helping the kidneys retain water, ADH slows down the decline in blood volume and the rise in its osmolarity. Thus, the ADH mechanism forms a negative feedback loop.

Conversely, if blood volume and pressure are too high or blood osmolarity is too low, ADH release is inhibited. The renal tubules reabsorb less water, urine output increases, and total body water declines. This is an effective way of compensating for hypertension. Since the lack of ADH increases the ratio of water to sodium in the urine, it raises the sodium concentration and osmolarity of the blood.

24.1e Disorders of Fluid Balance

The body is in a state of fluid imbalance if there is an abnormality of total *volume, concentration,* or *distribution* of water among the compartments.

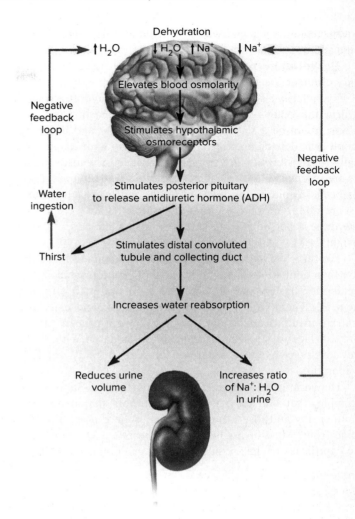

FIGURE 24.4 The Action of Antidiuretic Hormone. Pathways shown in red represent negative feedback.

Fluid Deficiency

Fluid deficiency arises when output exceeds intake over a long enough period of time. The two kinds of deficiency—volume depletion and dehydration—differ in the relative loss of water and electrolytes and the resulting osmolarity of the ECF. This important distinction calls for different strategies of fluid replacement therapy (see Deeper Insight 24.2 at the end of the chapter).

Volume depletion (hypovolemia[2]) occurs when proportionate amounts of water *and* sodium are lost without replacement. Total body water declines but osmolarity remains normal. Volume depletion occurs in cases of hemorrhage, severe burns, and chronic vomiting or diarrhea. A less common cause is aldosterone

---

[2]*hypo* = below normal; *vol* = volume; *emia* = blood condition

hyposecretion (Addison disease), which results in inadequate sodium and water reabsorption by the kidneys.

Dehydration (negative water balance) occurs when the body eliminates significantly more water than sodium, so the ECF osmolarity rises. The simplest cause of dehydration is lack of drinking water—for example, when stranded in a desert or at sea. It can be a serious problem for elderly and bedridden people who depend on others to provide them with water, especially for those who can't express their need or whose caretakers are insensitive to it. Diabetes mellitus, ADH hyposecretion (diabetes insipidus), profuse sweating, and overuse of diuretics are additional causes of dehydration. Cold weather can dehydrate a person just as much as hot weather (see Deeper Insight 24.1).

For three reasons, infants are more vulnerable to dehydration than adults: (1) Their high metabolic rate produces toxic metabolites faster, and they excrete more water to eliminate them. (2) Their kidneys are not fully mature and can't concentrate urine as effectively. (3) They have a greater ratio of body surface to volume; consequently, compared with adults, they lose twice as much water per kilogram of body weight by evaporation.

Dehydration affects all fluid compartments. Suppose, for example, that you play a strenuous tennis match on a hot summer day and lose a liter of sweat. Where does this fluid come from? Most of it filters out of the bloodstream through the capillaries of the sweat glands. In principle, 1 L of sweat

would amount to about one-third of the blood plasma. However, as the blood loses water, its osmolarity rises and water from the tissue fluid enters the bloodstream to balance the loss. This raises the osmolarity of the tissue fluid, so water moves out of the cells to balance that **(fig. 24.5).** Ultimately, all three fluid compartments (the ICF, blood, and tissue fluid) lose water. To excrete 1 L of sweat, about 300 mL of water would come from the tissue fluid and 700 mL from the ICF. Immoderate exercise without fluid replacement can lead to losses greater than 1 L per hour.

The most serious effects of fluid deficiency are circulatory shock due to loss of blood volume and neurological dysfunction due to dehydration of brain cells. Volume depletion by diarrhea is a major cause of infant mortality, especially under unsanitary conditions that lead to intestinal infections such as cholera.

FIGURE 24.5 Effects of Profuse Sweating on the Fluid Compartments. In extreme dehydration, the loss of intracellular fluid can cause cellular shrinkage and dysfunction.

DEEPER INSIGHT 24.1

CLINICAL APPLICATION

Fluid Balance in Cold Weather

Hot weather and profuse sweating are obvious threats to fluid balance, but so is cold weather. The body conserves heat by constricting the blood vessels of the skin and subcutaneous tissue, thus forcing blood into the deeper circulation. This raises the blood pressure, which inhibits the secretion of antidiuretic hormone and increases the secretion of natriuretic peptides. These hormonal changes increase urine output and reduce blood volume. In addition, cold air is relatively dry and increases respiratory water loss. This is why exercise causes the respiratory tract to "burn" more in cold weather than in warm.

These cold-weather respiratory and urinary losses can cause significant hypovolemia. Furthermore, the onset of exercise stimulates vasodilation in the skeletal muscles. In a hypovolemic state, there may not be enough blood to supply them, and a person may experience weakness, fatigue, or fainting (hypovolemic shock). In winter sports and other activities such as snow shoveling, it is important to maintain fluid balance. Even if you don't feel thirsty, it is beneficial to take ample amounts of warm liquids such as soup or cider. Coffee, tea, and alcohol, however, have diuretic effects that defeat the purpose of fluid intake.

Fluid Excess

Fluid excess is less common than fluid deficiency because the kidneys are highly effective at compensating for excessive intake by excreting more urine **(fig. 24.6).** Renal failure and other causes, however, can lead to excess fluid retention.

Fluid excesses are of two types called volume excess and hypotonic hydration. In **volume excess,** both sodium and water are retained and the ECF remains isotonic. This can result from aldosterone hypersecretion or renal failure. In **hypotonic hydration** (also called **water intoxication** or **positive water balance**), more water than sodium is retained or ingested and the ECF becomes hypotonic. This can occur if you lose a large amount of water *and* salt through urine and sweat and you replace it by drinking plain water. Without a proportionate intake of electrolytes, water dilutes the ECF, makes it hypotonic, and causes cellular swelling. ADH hypersecretion can cause hypotonic hydration by stimulating excessive water retention as sodium continues to be excreted. Among the most serious effects of either type of fluid excess are pulmonary and cerebral edema and death.

The four preceding forms of fluid imbalance are summarized and compared in **table 24.1.**

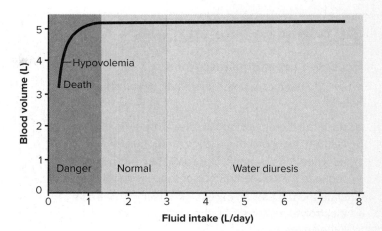

FIGURE 24.6 The Relationship of Blood Volume to Fluid Intake. The kidneys cannot compensate very well for inadequate fluid intake. Below an intake of about 1 L/day, blood volume drops significantly (left) and there may be a threat of death from hypovolemic shock. The kidneys compensate very well, on the other hand, for abnormally high fluid intake (right); they eliminate the excess by water diuresis and maintain a stable blood volume.

▶▶▶ **APPLY WHAT YOU KNOW**

Some tumors of the brain, pancreas, and small intestine secrete ADH. What type of fluid imbalance would this produce? Explain why.

Fluid Sequestration

Fluid sequestration[3] (seh-ques-TRAY-shun) is a condition in which excess fluid accumulates in a particular location. Total body water and osmolarity may be normal, but the volume of circulating blood may drop to the point of causing circulatory shock. The most common form of sequestration is *edema,* the abnormal accumulation of fluid in the interstitial spaces, causing swelling of a tissue (see detailed discussion in section 20.3d). Hemorrhage can be another cause of fluid sequestration; blood that pools and clots in the tissues is lost to circulation. Yet another example is *pleural effusion,* caused by some lung infections, in which as much as several liters of fluid accumulate in the pleural cavity.

TABLE 24.1	Forms of Fluid Imbalance	
Form	Total Body Water	Osmolarity
Fluid Deficiency		
Volume depletion (hypovolemia)	Reduced	Isotonic (normal)
Dehydration (negative water balance)	Reduced	Hypertonic (elevated)
Fluid Excess		
Volume excess	Elevated	Isotonic (normal)
Hypotonic hydration (positive water balance, water intoxication)	Elevated	Hypotonic (reduced)

BEFORE YOU GO ON

Answer the following questions to test your understanding of the preceding section:

1. List five routes of water loss. Which one accounts for the greatest loss? Which one is most controllable?

2. Explain why even a severely dehydrated person inevitably experiences further fluid loss.

3. Suppose there were no mechanisms to stop the sense of thirst until the blood became sufficiently hydrated. Explain why we would routinely suffer hypotonic hydration.

4. Summarize the effect of ADH on total body water and blood osmolarity.

5. Name and define the four types of fluid imbalance, and give an example of a situation that could produce each type.

[3]*sequestr* = to isolate

24.2 Electrolyte Balance

Expected Learning Outcomes

When you have completed this section, you should be able to

a. describe the physiological roles of sodium, potassium, calcium, magnesium, chloride, and phosphates;

b. describe the hormonal and renal mechanisms that regulate the concentrations of these electrolytes; and

c. state the term for an excess or deficiency of each electrolyte and describe the consequences of these imbalances.

Electrolyte balance is a state in which the amount of electrolytes absorbed by the small intestine balances the amount lost from the body, chiefly through the urine, and in which electrolyte concentrations in the body fluids are regulated within homeostatic limits. Electrolytes are physiologically important for multiple reasons: They are chemically reactive and participate in metabolism, they determine the electrical potential (charge difference) across cell membranes, and they strongly affect the osmolarity of the body fluids and the body's water content and distribution. Strictly speaking, electrolytes are salts such as sodium chloride, not just sodium or chloride ions. In common usage, however, the individual ions are often loosely referred to as electrolytes. The major cations of the electrolytes are sodium (Na^+), potassium (K^+), calcium (Ca^{2+}), magnesium (Mg^{2+}), and hydrogen (H^+); the major anions are chloride (Cl^-), bicarbonate (HCO_3^-), and phosphates (P_i). Hydrogen and bicarbonate regulation are discussed later under acid–base balance. Here we focus on the other six.

The typical concentrations of these ions in the blood plasma versus intracellular fluid are compared in **figure 24.7.** Notice that in spite of great differences in electrolyte concentrations, the two fluid compartments have the same osmolarity (300 mOsm/L). Blood plasma is the most accessible fluid for measurements of electrolyte concentration, so excesses and deficiencies are defined with reference to normal plasma concentrations. Concentrations in the tissue fluid differ only slightly from those in the plasma. The prefix *normo-* denotes a normal electrolyte concentration (for example, *normokalemia*), and *hyper-* and *hypo-* denote concentrations that are, respectively, sufficiently above or below normal to cause physiological disorders.

In the following sections, we'll examine the functions, homeostatic regulation, and imbalances of the principal electrolytes and then review the causes and effects of the cation imbalances in a summary table.

24.2a Sodium

Functions

Sodium is one of the principal ions responsible for the resting membrane potentials of cells, and the inflow of sodium through membrane channels is an essential event in the depolarization

(a) Blood plasma

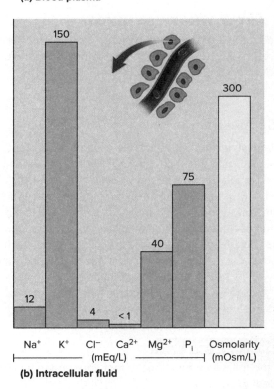

(b) Intracellular fluid

FIGURE 24.7 Electrolyte Concentrations in the Blood Plasma and Intracellular Fluid. Concentrations in the blood plasma (a) are shown directly above concentrations of the same ions in the intracellular fluid (b) for contrast. P_i collectively represents inorganic phosphates in the forms of PO_4^{3-}, HPO_4^{2-}, and $H_2PO_4^-$. Electrolyte concentrations (green bars) are given in mEq/L. Concentrations in mmol/L are the same for Na^+, K^+, and Cl^-; one-half of the illustrated values for Ca^{2+} and Mg^{2+}; and one-third for PO_4^{3-}. Osmolarity (yellow bars) is given in mOsm/L.

that underlies nerve and muscle function. Sodium is the principal cation of the ECF; sodium salts account for 90% to 95% of its osmolarity. Sodium is therefore the most significant solute in determining total body water and the distribution of water among fluid compartments. Sodium ions bound to the proteoglycans of cartilage retain water, ensuring that cartilages are well hydrated and able to act as effective cushions and shock absorbers. Sodium gradients across the plasma membrane provide the potential energy that is tapped to cotransport other solutes such as glucose, potassium, and calcium. The Na^+–K^+ pump is an important mechanism for generating body heat. Sodium bicarbonate ($NaHCO_3$) plays a major role in buffering the pH of the ECF.

Homeostasis

An adult needs about 0.5 g of sodium per day, whereas the typical American diet contains 3 to 7 g/day. Thus a dietary sodium deficiency is rare, and the primary concern is adequate excretion of the

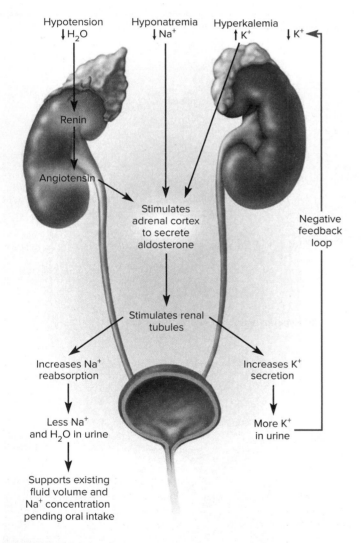

FIGURE 24.8 The Action of Aldosterone. The pathway shown in red represents negative feedback.

❓ *What is required, in addition to aldosterone, to increase blood volume?*

excess. This is one of the most important roles of the kidneys. There are multiple mechanisms for controlling sodium concentration, tied to its effects on blood pressure and osmolarity and coordinated by aldosterone, antidiuretic hormone, and the natriuretic peptides.

Aldosterone, the "salt-retaining hormone," plays the primary role in adjustment of sodium excretion. Hyponatremia and hyperkalemia directly stimulate the adrenal cortex to secrete aldosterone, and hypotension stimulates its secretion by way of the renin–angiotensin–aldosterone mechanism (**fig. 24.8**).

Only cells in the ascending limb of the nephron loop, the distal convoluted tubule, and the cortical part of the collecting duct have aldosterone receptors. Aldosterone, a steroid, binds to nuclear receptors and activates transcription of a gene for the Na^+–K^+ pump. In 10 to 30 minutes, enough Na^+–K^+ pumps are synthesized and installed in the plasma membrane to produce a noticeable effect—sodium concentration in the urine begins to fall and potassium concentration rises as the tubules reabsorb more sodium and secrete more hydrogen and potassium ions. Water and chloride passively follow sodium. Thus, the primary effects of aldosterone are that the urine contains less NaCl and more potassium and has a lower pH. An average adult male excretes 5 g of sodium per day, but the urine can be virtually sodium-free when aldosterone level is high. Although aldosterone strongly influences sodium reabsorption, it has little effect on plasma sodium *concentration* because reabsorbed sodium is accompanied by a proportionate amount of water.

Elevated blood pressure inhibits the renin–angiotensin–aldosterone mechanism. The kidneys then reabsorb almost no sodium beyond the proximal convoluted tubule (PCT), and the urine contains up to 30 g of sodium per day.

Aldosterone has only slight effects on urine volume, blood volume, and blood pressure in spite of the tendency of water to follow sodium osmotically. Even in aldosterone hypersecretion, blood volume is rarely more than 5% to 10% above normal. An increase in blood volume increases blood pressure and glomerular filtration rate (GFR). Even though aldosterone increases the tubular reabsorption of sodium and water, this is offset by the rise in GFR and there is only a small drop in urine output.

Antidiuretic hormone modifies water excretion independently of sodium excretion. Thus, unlike aldosterone, it can change sodium *concentration*. A high concentration of sodium in the blood stimulates the posterior lobe of the pituitary gland to release ADH. The kidneys then reabsorb more water, which slows down any further increase in blood sodium concentration. ADH alone cannot lower the concentration; this requires water ingestion to dilute the existing sodium. A drop in sodium concentration, by contrast, inhibits ADH release. More water is excreted, thereby raising the relative amount of sodium that remains in the blood.

The natriuretic peptides inhibit sodium and water reabsorption and the secretion of renin and ADH. The kidneys then eliminate more sodium and water and lower the blood pressure. Angiotensin II, by contrast, activates the Na^+–H^+ antiport in the PCT and thereby increases sodium reabsorption and reduces urinary sodium output.

Several other hormones also affect sodium homeostasis. Estrogen mimics the effect of aldosterone and causes women to retain water during pregnancy and part of the menstrual cycle. Progesterone reduces sodium reabsorption and has a diuretic effect. High levels of glucocorticoids promote sodium reabsorption and edema.

In some cases, sodium homeostasis is achieved by regulation of salt intake. A craving for salt occurs in people who are depleted of sodium—for example, by blood loss or Addison disease. Pregnant women sometimes develop a craving for salty foods. Salt craving is not limited to humans; many animals ranging from elephants to butterflies seek out wet salty soil where they can obtain this vital mineral.

Imbalances

True imbalances in sodium concentration are relatively rare because sodium excess or depletion is almost always accompanied by proportionate changes in water volume. **Hypernatremia**[4] is a plasma sodium concentration in excess of 145 mEq/L. It can result from the administration of intravenous saline (see Deeper Insight 24.2 at the end of this chapter). Its major consequences

are water retention, hypertension, and edema. Extreme hypertonicity shrinks cells, which damages their cytoskeleton, breaks DNA, and induces cell death by apoptosis. **Hyponatremia**[5] (less than 130 mEq/L) is usually the result of excess body water rather than excess sodium excretion, as in the case mentioned earlier of a person who loses large volumes of sweat or urine and replaces it by drinking plain water. Usually, hyponatremia is quickly corrected by excretion of the excess water, but if uncorrected it produces the symptoms of hypotonic hydration described earlier.

24.2b Potassium

Functions

Potassium is the most abundant cation of the ICF and is the greatest determinant of intracellular osmolarity and cell volume. Along with sodium, it produces the resting membrane potentials and action potentials of nerve and muscle cells **(fig. 24.9a).**

---

[4]*hyper* = excess, above normal; *natr* = sodium; *emia* = blood condition

[5]*hypo* = deficiency, below normal; *natr* = sodium; *emia* = blood condition

Elevated extracellular K+ concentration

Less diffusion of K+ out of cell

Elevated RMP (cells partially depolarized)

Cells more excitable

(b) Hyperkalemia

K+ concentrations in equilibrium

Equal diffusion into and out of cell

Normal resting membrane potential (RMP)

(a) Normokalemia

Reduced extracellular K+ concentration

Greater diffusion of K+ out of cell

Reduced RMP (cells hyperpolarized)

Cells less excitable

(c) Hypokalemia

FIGURE 24.9 Effects of Potassium Imbalances on Membrane Potentials. The circular diagram above each cell represents the voltage measured across the plasma membrane. (a) Normokalemia, with a normal resting membrane potential (RMP). (b) Hyperkalemia, with an elevated RMP. (c) Hypokalemia, with a depressed RMP.

Potassium is as important as sodium to the Na^+–K^+ pump and its functions of cotransport and thermogenesis (heat production). It is an essential cofactor for protein synthesis and some other metabolic processes.

Homeostasis

Potassium homeostasis is closely linked to that of sodium. Regardless of the body's state of potassium balance, about 90% of the potassium filtered by the glomerulus is reabsorbed by the PCT and the rest is excreted in the urine. Variations in potassium excretion are controlled later in the nephron by changing the amount of potassium returned to the tubular fluid by the distal convoluted tubule and cortical portion of the collecting duct (CD). When potassium concentration is high, these tubules secrete more potassium into the filtrate and the urine may contain more potassium than the glomerulus filters from the blood. When blood potassium level is low, the tubules secrete less. The distal convoluted tubule and collecting duct reabsorb potassium through their intercalated cells.

Aldosterone regulates potassium balance along with sodium (fig. 24.8). A rise in potassium concentration stimulates the adrenal cortex to secrete aldosterone. Aldosterone stimulates renal secretion of potassium at the same time that it stimulates reabsorption of sodium. The more sodium there is in the urine, the less potassium, and vice versa.

Imbalances

Potassium imbalances are the most dangerous of all electrolyte imbalances. **Hyperkalemia**[6] (>5.5 mEq/L) can have completely opposite effects depending on whether potassium concentration rises quickly or slowly. It can rise quickly when, for example, a crush injury or hemolytic anemia releases large amounts of potassium from ruptured cells. This can also result from a transfusion with outdated, stored blood because potassium leaks from erythrocytes into the plasma during storage. A sudden increase in extracellular potassium tends to make nerve and muscle cells abnormally excitable. Normally, potassium continually passes into and out of cells at equal rates—leaving by diffusion and reentering by the Na^+–K^+ pump. But in hyperkalemia, there is less concentration difference between the ICF and ECF, so the outward diffusion of potassium is reduced. More potassium remains in the cell than normal, and the plasma membrane therefore has a less negative resting potential and is closer to the threshold at which it will set off action potentials **(fig. 24.9b).** This is a very dangerous condition that can quickly produce cardiac arrest. High-potassium solutions are sometimes used by veterinarians to euthanize animals and are used in some states as a lethal injection for capital punishment.

Hyperkalemia can also have a slower onset stemming from such causes as aldosterone hyposecretion, renal failure, or acidosis. (The relationship of acid–base imbalances to potassium

imbalances is explained later.) Paradoxically, if the extracellular potassium concentration rises slowly, nerve and muscle become *less* excitable. *Slow* depolarization of a cell inactivates voltage-gated sodium channels, and the channels don't become excitable again until the membrane repolarizes. Inactivated sodium channels cannot produce action potentials.

Hypokalemia (<3.5 mEq/L) rarely results from a dietary deficiency, because most diets contain ample amounts of potassium; it can occur, however, in people with depressed appetites. Hypokalemia more often results from heavy sweating, chronic vomiting or diarrhea, excessive use of laxatives, aldosterone hypersecretion, or alkalosis. As ECF potassium concentration falls, more potassium moves from the ICF to the ECF. With the loss of these cations from the cytoplasm, cells become hyperpolarized and nerve and muscle cells are less excitable **(fig. 24.9c).** This is reflected in muscle weakness, loss of muscle tone, depressed reflexes, and irregular electrical activity of the heart.

▶▶▶APPLY WHAT YOU KNOW

Some tumors of the adrenal cortex secrete excess aldosterone and may cause paralysis. Explain this effect and identify the electrolyte and fluid imbalances you would expect to observe in such a case.

24.2c Calcium

Functions

Calcium lends strength to the skeleton, activates the sliding filament mechanism of muscle contraction, serves as a second messenger for some hormones and neurotransmitters, activates exocytosis of neurotransmitters and other cellular secretions, and is an essential factor in blood clotting. Cells maintain a very low intracellular calcium concentration because they require a high concentration of phosphate ions (for reasons discussed shortly). If calcium and phosphate were both very concentrated in a cell, calcium phosphate crystals would precipitate in the cytoplasm. To maintain a high phosphate concentration but avoid crystallization of calcium phosphate, cells pump out Ca^{2+} and keep it at a low intracellular concentration, or else sequester Ca^{2+} in the smooth ER and release it only when needed. Cells that store Ca^{2+} often have a protein called *calsequestrin,* which binds the stored Ca^{2+} and keeps it chemically unreactive.

Homeostasis

Calcium concentration is regulated chiefly by parathyroid hormone, calcitriol, and in children, calcitonin (see section 7.4b). These hormones work through their effects on bone deposition and resorption, intestinal absorption of calcium, and urinary excretion. Figure 7.17 presents two flowcharts that summarize calcium homeostasis.

---

[6]*hyper* = above normal; *kal* = potassium; *emia* = blood condition

Imbalances

Hypercalcemia (>5.8 mEq/L) can result from alkalosis, hyperparathyroidism, or hypothyroidism. It reduces the sodium permeability of plasma membranes and inhibits the depolarization of nerve and muscle cells. At concentrations ≥12 mEq/dL, hypercalcemia causes muscular weakness, depressed reflexes, and cardiac arrhythmia.

Hypocalcemia (<4.5 mEq/L) can result from vitamin D deficiency, diarrhea, pregnancy, lactation, acidosis, hypoparathyroidism, or hyperthyroidism. It increases the sodium permeability of plasma membranes, causing the nervous and muscular systems to be overly excitable. Tetany occurs when calcium concentration drops to 6 mg/dL and may be lethal at 4 mg/dL (2 mEq/L) due to laryngospasm and suffocation.

24.2d Magnesium

Functions

About 54% of the body's magnesium (Mg^{2+}) is in the bone tissue and 45% in the intracellular fluid, especially in the skeletal muscles. Magnesium is the second most abundant intracellular cation after potassium. Most ICF Mg^{2+} is complexed with ATP, but Mg^{2+} is also a necessary cofactor for many enzymes, membrane transport proteins, and nucleic acids. Magnesium imbalances can therefore have wide-ranging effects on membrane transport, membrane electrical potentials, cell metabolism, and DNA replication.

Homeostasis

Magnesium levels in the blood plasma normally range from 1.5 to 2.0 mEq/L, whereas intracellular concentrations are quite variable from one tissue to another, but range up to 40 mEq/L in skeletal muscle. Dietary intake of magnesium is typically 140 to 360 mg/day, but only 30% to 40% of it is absorbed by the small intestine and the rest passes through unused. Its intestinal absorption is regulated mainly by vitamin D.

About two-thirds of the body's Mg^{2+} loss is via the feces and one-third via the urine. Retention or loss of plasma Mg^{2+} is regulated mainly by the thick segment of the ascending limb of the nephron loop, where about 70% of the filtered Mg^{2+} is reabsorbed; smaller amounts are reabsorbed in other segments of the nephron. Reabsorption is mainly by the paracellular route (between tubule epithelial cells), driven by the positive electrical potential of the tubular fluid repelling the positive magnesium ions. Parathyroid hormone governs the rate of reabsorption and is the primary regulator of plasma Mg^{2+} level.

Imbalances

Magnesium imbalances are usually due to excessive loss from the body rather than dietary deficiency. **Hypermagnesemia,** an excess (>2.0 mEq/L), is rare except in renal insufficiency. It tends to have a sedative effect, with lethargy, muscle weakness, and weak reflexes; and it can cause respiratory depression or failure, hypotension due to lack of vasomotor tone, and flaccid, diastolic cardiac arrest.

Hypomagnesemia, a plasma Mg^{2+} deficiency (<1.5 mEq/L), can result from intestinal malabsorption, vomiting, diarrhea, or renal disease. It results in hyperirritability of the nervous and muscular systems; muscle tremors, spasms, or tetanus; hypertension resulting from excessive vasoconstriction; and tachycardia and ventricular arrhythmia.

24.2e Chloride

Functions

Chloride ions are the most abundant anions of the ECF and thus make a major contribution to its osmolarity. Chloride ions are required for the formation of stomach acid (HCl), and they are involved in the chloride shift that accompanies carbon dioxide loading and unloading by the erythrocytes (see figs. 22.24, 22.25). By a similar mechanism explained later, chloride ions play a major role in the regulation of body pH.

Homeostasis

Chloride ions are strongly attracted to Na^+, K^+, and Ca^{2+}. It would require great expenditure of energy to keep chloride ions separate from these cations, so chloride homeostasis is achieved primarily as a side effect of sodium homeostasis—as sodium is retained or excreted, chloride ions passively follow.

Imbalances

Hyperchloremia (>105 mEq/L) is usually the result of dietary excess or administration of intravenous saline. **Hypochloremia** (<95 mEq/L) is usually a side effect of hyponatremia but sometimes results from hyperkalemia or acidosis. In the latter case, the kidneys retain potassium by excreting more sodium, and sodium takes chloride with it. The primary effects of chloride imbalances are disturbances in acid–base balance, but this works both ways—a pH imbalance arising from some other cause can also produce a chloride imbalance. Chloride balance is therefore discussed further in connection with acid–base balance in section 24.3.

24.2f Phosphates

Functions

The inorganic phosphates (P_i) of the body fluids are an equilibrium mixture of phosphate (PO_4^{3-}), monohydrogen phosphate (HPO_4^{2-}), and dihydrogen phosphate ($H_2PO_4^-$) ions. Phosphates are relatively concentrated in the ICF, where they are generated by the hydrolysis of ATP and other phosphate compounds. They are a component of phospholipids, DNA, RNA, ATP, GTP, cAMP, creatine phosphate, and related compounds. Every process that depends on ATP depends on phosphate ions. Phosphates activate many metabolic pathways by phosphorylating enzymes and substrates such as glucose. They are also important as buffers that help stabilize the pH of the body fluids.

Homeostasis

The average diet provides ample amounts of phosphate, which is readily absorbed by the small intestine. Plasma phosphate concentration is usually maintained at about 4 mEq/L, with continual loss of excess phosphate by glomerular filtration. If plasma phosphate concentration drops much below this level, however, the renal tubules reabsorb all filtered phosphate.

Parathyroid hormone increases the excretion of phosphate as part of the mechanism for increasing the concentration of free calcium ions in the ECF. Lowering the ECF phosphate concentration minimizes the formation of calcium phosphate and thus helps support plasma calcium concentration. Rates of phosphate excretion are also strongly affected by the pH of the urine, as discussed in section 24.3.

Imbalances

Phosphate homeostasis is not as critical as that of other electrolytes. The body can tolerate broad variations several times above or below the normal concentration with little immediate effect on physiology.

24.2g Summary of Imbalances

Table 24.2 summarizes some causes and effects of sodium, potassium, calcium, and magnesium imbalances. Imbalances in the two anions previously discussed—chloride and phosphate—are important mainly in relation to acid–base balance, discussed in the next section.

The causes and effects of electrolyte balances are so complex and interconnected as to be almost bewildering. They sometimes appear even contradictory, such as vomiting and diarrhea being among the causes or effects of *both* excess and deficiency of the same electrolyte. To fully explain these would require a chapter in itself. Table 24.2 therefore presents only a few of the especially common causes and

TABLE 24.2	Causes and Effects of Some Key Electrolyte Imbalances	
Imbalance	**Causes**	**Clinical Manifestations**
Sodium		
Hypernatremia (Na+ > 145 mEq/L)	Water loss from chronic vomiting or diarrhea; burns; diuretics; excessive sweating; diabetes insipidus. Dehydration from lack of access to water or inadequate thirst and water consumption. I.V. hypertonic saline infusion. Excess dietary sodium. Corticosterone hypersecretion.	Thirst and polyuria. Hypertension. Muscle spasms. CNS dysfunction due to brain cell shrinkage: confusion, lethargy or excitability, cerebral hemorrhage, seizures, coma.
Hyponatremia (Na+ < 130 mEq/L)	Excess consumption of plain water. Vomiting, diarrhea, burns, diuretics, polydipsia. Hormone imbalances including cortisol, aldosterone, or thyroid hormone deficiency; diabetes mellitus (with hyperglycemia); ADH excess. Kidney disease.	Usually CNS dysfunction due to cerebral edema: headache, confusion, disorientation, personality changes, lethargy, stupor, seizures, coma. Pulmonary and cerebral edema. Tachycardia.
Potassium		
Hyperkalemia (K+ > 5.5 mEq/L)	Kidney failure, burns, GI bleeding, crush injuries, rhabdomyolysis, hemolysis, transfusion with outdated blood, rapid I.V. KCl infusion, acidosis, hyperglycemia, aldosterone hyposecretion.	Usually begin with cardiotoxic effects: ventricular fibrillation, bradycardia, cardiac arrest. Irritability. Muscle weakness, flaccid paralysis. Nausea, vomiting, diarrhea.
Hypokalemia (K+ < 3.5 mEq/L)	Laxative or diuretic overuse; chronic vomiting or diarrhea; heavy sweating. Glucocorticoid or aldosterone hypersecretion. Alkalosis.	Polyuria, polydipsia. Fatigue. Muscle pain, cramps, weakness, loss of muscle tone, depressed reflexes, tetanus, rhabdomyolysis. Cardiac hyperexcitability, arrhythmia. Nausea, vomiting, alkalosis, confusion. Respiratory arrest.
Calcium		
Hypercalcemia (Ca2+ > 5.8 mEq/L)	Usually excess bone resorption; metastatic bone cancer and some other cancers; Paget disease; parathyroid hormone or vitamin D excess; thyroid hormone deficiency; immobility. Excessive use of calcium carbonate antacids. Renal failure. Alkalosis.	Anorexia, nausea, vomiting, constipation. Polydipsia, polyuria, nocturia. Depression, personality changes. Muscle weakness, depressed reflexes. Bone pain. Kidney stones. Cardiac arrhythmia, arrest. Confusion, delirium, stupor, coma.
Hypocalcemia (Ca2+ < 4.5 mEq/L)	Dietary deficiency of calcium, magnesium, or vitamin D; vitamin D malabsorption; inadequate sunlight; parathyroid hormone deficiency or thyroid hormone excess. Chronic diarrhea. Pregnancy, lactation. Kidney disease. Acidosis.	Bone weakness, osteomalacia, fractures. Muscle cramps, tremors, tetanus, seizures; laryngospasm and asphyxiation. Paresthesia. Heart failure.
Magnesium		
Hypermagnesemia (Mg2+ > 2.0 mEq/L)	Excessive intake, as in magnesium-based antacids. Deficiency of aldosterone or thyroid hormone. Renal failure.	Muscle weakness, depressed reflexes, lethargy. Respiratory depression or failure. Hypotension, cardiac arrest.
Hypomagnesemia (Mg2+ < 1.5 mEq/L)	Alcoholism; intestinal malabsorption; malnutrition. Chronic vomiting or diarrhea. Some diuretics. Lactation. Renal failure.	Anorexia, lethargy, nausea, vomiting. Muscle weakness, tremor, spasms, fasciculations, tetanus. Hypertension, tachycardia, cardiac arrhythmia.

effects. It omits many drugs and clinical procedures (such as gastric suction and fluid therapy) that can trigger imbalances. (See Deeper Insight 24.2 for further discussion of fluid therapy.) With few exceptions, we focus on spontaneous illnesses and patient behaviors—for example, the common problem of surreptitious overuse of laxatives and diuretics by people obsessed with weight loss, often throwing themselves into dangerous fluid, electrolyte, and acid–base disturbances.

The clinical manifestations tabulated here include both **signs** (effects on a patient that can be observed by others, such as seizures and coma) and **symptoms** (effects that can be felt only by the patient, such as headache, nausea, and malaise). The tabulated manifestations don't occur in every case of imbalance, or even most. They vary greatly with the magnitude of imbalance and from one population to another—such as infants, healthy adults, and people who are elderly or disabled—and imbalances are often asymptomatic or produce signs discovered only in blood work or electrocardiograms.

For some of the terminology here, you may need to refer back to hormone imbalances (hyper- and hyposecretion, section 17.7), renal functions (section 23.6), or other earlier discussions in this book. *Acidosis* and *alkalosis,* associated with electrolyte imbalances as both cause and effect, are detailed in the next section (24.3).

Anorexia is lack of appetite. *Malaise* is a general sense of feeling uncomfortable or unwell, as if "coming down with something." *Polydipsia* means excessive thirst and drinking; *polyuria,* often associated with it, is excessive urine output. *Nocturia* is being awakened and having to urinate during the night and, sometimes, bed-wetting in children and incontinent adults.

Because electrolytes are so important in cell membrane potentials and nerve and muscle action, imbalances often have neuromuscular manifestations: *paresthesia,* abnormal sensation in the absence of external stimulation, such as numbness, burning, or tingling of the fingers; *muscle fasciculations,* involuntary twitching of individual fascicles of a muscle; or more extreme muscular reactions such as convulsions and *tetanus*—not the bacterial disease but the sustained contraction of a muscle, without relaxation. If it affects the respiratory muscles, tetanus can cause respiratory arrest; if it causes tightening of the laryngeal muscles, it can cause death by asphyxiation. Some electrolyte imbalances cause *rhabdomyolysis,* the death of muscle fibers, releasing their contents into the bloodstream, potentially clogging the kidneys with myoglobin and causing renal failure.

BEFORE YOU GO ON

Answer the following questions to test your understanding of the preceding section:

6. Which do you think would have the most serious effect, and why—a 5 mEq/L increase in the plasma concentration of sodium, potassium, chloride, or calcium?

7. Answer the same question for a 5 mEq/L decrease.

8. Explain why ADH is more likely than aldosterone to change the osmolarity of the blood plasma.

9. Explain why aldosterone hyposecretion could cause hypochloremia.

10. Magnesium sulfate, commonly sold as Epsom salt, can be used as a bath salt to relax tight, aching muscles. Explain this effect in view of what you have learned in this chapter section.

11. Why are more phosphate ions required in the ICF than in the ECF? How does this affect the distribution of calcium ions between these fluid compartments?

24.3   Acid–Base Balance

Expected Learning Outcomes

When you have completed this section, you should be able to

a. define *buffer* and write chemical equations for the bicarbonate, phosphate, and protein buffer systems;

b. discuss the relationship between pulmonary ventilation, pH of the extracellular fluids, and the bicarbonate buffer system;

c. explain how the kidneys secrete hydrogen ions and how these ions are buffered in the tubular fluid;

d. identify some types and causes of acidosis and alkalosis, and describe the effects of these pH imbalances; and

e. explain how the respiratory and urinary systems correct acidosis and alkalosis, and compare their effectiveness and limitations.

As we saw in section 2.4f, metabolism depends on the functioning of enzymes, and enzymes are very sensitive to pH. Slight deviations from the normal pH can shut down metabolic pathways as well as alter the structure and function of other macromolecules. Consequently, acid–base balance is one of the most important aspects of homeostasis.

The blood and tissue fluid normally have a pH of 7.35 to 7.45. **Acid–base balance** is a state in which the pH of the body fluids is homeostatically regulated within this range. Such a narrow range of variation is remarkable considering that our metabolism constantly produces acid: lactic acid from anaerobic fermentation, phosphoric acids from nucleic acid catabolism, fatty acids and ketones from fat catabolism, and carbonic acid from carbon dioxide. These acids are a constant challenge to our enzyme function, homeostasis, and survival. Here we examine *buffering* mechanisms for stabilizing internal pH and maintaining acid–base balance.

24.3a Acids, Bases, and Buffers

The pH of a solution is determined solely by its hydrogen ions (H^+). An acid is any chemical that releases H^+ in solution. A **strong acid** such as hydrochloric acid (HCl) ionizes freely, gives up most of its hydrogen ions, and can markedly lower the pH of a solution. A **weak acid** such as carbonic acid (H_2CO_3) ionizes only slightly and keeps most hydrogen in a chemically bound form that doesn't affect pH. A base is any chemical that accepts H^+. A **strong base** such as the hydroxide ion (OH^-) has a strong tendency to bind H^+ and raise the pH, whereas a **weak base** such as the bicarbonate ion (HCO_3^-) binds less of the available H^+ and has less effect on pH.

A **buffer,** broadly speaking, is any mechanism that resists pH changes by converting a strong acid or base to a weak one. The body has both physiological and chemical buffers. A **physiological buffer** is a system—namely, the respiratory or urinary system—that stabilizes pH by controlling the body's output of acids, bases, or CO_2. Of all buffer systems, the urinary system buffers the greatest quantity of acid or base, but it requires several hours to days to exert an effect. The respiratory system exerts an effect within a few minutes but cannot alter the pH as much as the urinary system can.

A **chemical buffer** is a substance that binds H^+ and removes it from solution as its concentration begins to rise, or releases H^+ into solution as its concentration falls. Chemical buffers can restore normal pH within a fraction of a second. They function as mixtures called **buffer systems** composed of a weak acid and a weak base. We have three major chemical buffer systems—the bicarbonate, phosphate, and protein systems.

The amount of acid or base that can be neutralized by a chemical buffer system depends on two factors: the concentration of the buffers and the pH of their working environment. Each system has an optimum pH at which it functions best; its effectiveness is greatly reduced if the pH of its environment deviates too far from this. The relevance of these factors will become apparent as you study the following buffer systems.

The Bicarbonate Buffer System

The **bicarbonate buffer system** is a solution of carbonic acid and bicarbonate ions. As we can see in the carbonic acid reaction, carbonic acid (H_2CO_3) forms by the reaction of carbon dioxide with water, then dissociates into bicarbonate (HCO_3^-) and H^+:

$$CO_2 + H_2O \rightleftharpoons H_2CO_3 \rightleftharpoons HCO_3^- + H^+.$$

This is a reversible reaction. When it proceeds to the right, carbonic acid acts as a weak acid by releasing H^+ and lowering pH. When the reaction proceeds to the left, bicarbonate acts as a weak base by binding H^+, removing the ions from solution, and raising pH.

At a pH of 7.4, the bicarbonate system wouldn't ordinarily have a particularly strong buffering capacity outside of the body. This is too far from its optimum pH of 6.1. If a strong acid was added to a beaker of carbonic acid–bicarbonate solution at pH 7.4, the preceding reaction would shift only slightly to the left. Much surplus H^+ would remain and the pH would be substantially lower. In the body, by contrast, the bicarbonate system works quite well because the lungs and kidneys constantly remove CO_2 and prevent an equilibrium from being reached. This keeps the reaction moving to the left, and more H^+ is neutralized. Conversely, if there is a need to lower the pH, the kidneys excrete HCO_3^- and keep this reaction moving to the right, which elevates the H^+ concentration of the ECF. Thus, you can see that the physiological and chemical buffers of the body function together in maintaining acid–base balance.

▶▶▶**APPLY WHAT YOU KNOW**

In the systemic circulation, arterial blood has a mean pH of 7.40 and venous blood has a mean of 7.35. What do you think causes this difference?

The Phosphate Buffer System

The **phosphate buffer system** is a solution of HPO_4^{2-} and $H_2PO_4^-$. It works in much the same way as the bicarbonate system. The following reaction can proceed to the right to liberate H^+ and lower pH, or it can proceed to the left to bind H^+ and raise pH:

$$H_2PO_4^- \rightleftharpoons HPO_4^{2-} + H^+.$$

The optimal pH for this system is 6.8, closer to the actual pH of the ECF (7.4). Thus, the phosphate buffer system has a stronger buffering effect than an equal amount of bicarbonate buffer. However, phosphates are much less concentrated in the ECF than bicarbonate, so they are less important in buffering the ECF. They are more important in the renal tubules and ICF, where not only are they more concentrated, but the pH is lower and closer to their functional optimum. In the ICF, the constant production of metabolic acids creates pH values ranging from 4.5 to 7.4, probably averaging 7.0. The reason for the low pH in the renal tubules is discussed later.

The Protein Buffer System

Proteins are more concentrated than either bicarbonate or phosphate buffers, especially in the ICF. The **protein buffer system** accounts for about three-quarters of all chemical buffering in the body fluids. The buffering ability of proteins is due to certain side groups of their amino acid residues. Some have carboxyl side groups (—COOH), which release H^+ when pH begins to rise and thus lower pH:

$$-COOH \longrightarrow -COO^- + H^+.$$

Others have amino side groups ($-NH_2$), which bind H^+ when pH falls too low, thus raising pH toward normal:

$$-NH_2 + H^+ \longrightarrow -NH_3{}^+.$$

▶▶▶ APPLY WHAT YOU KNOW

What protein do you think is the most important buffer in blood plasma? In erythrocytes?

24.3b Respiratory Control of pH

The respiratory buffer system adjusts the pH of the body fluids by raising or lowering the rate and depth of breathing. The equation for the bicarbonate buffer system shows that the addition of CO_2 to the body fluids raises H^+ concentration and lowers pH, while the removal of CO_2 has the opposite effects. This is the basis for the strong buffering capacity of the respiratory system. Indeed, this system can neutralize two or three times as much acid as the chemical buffers can.

Carbon dioxide is constantly produced by aerobic metabolism and is normally eliminated by the lungs at an equivalent rate. As explained in section 22.3g, rising CO_2 concentration and falling pH stimulate peripheral and central chemoreceptors, which stimulate an increase in pulmonary ventilation. This expels excess CO_2 and thus reduces H^+ concentration. The free H^+ becomes part of the water molecules produced by this reaction:

$$HCO_3{}^- + H^+ \longrightarrow H_2CO_3 \longrightarrow CO_2 \text{ (expired)} + H_2O.$$

Conversely, a drop in H^+ concentration raises pH and reduces pulmonary ventilation. This allows metabolic CO_2 to accumulate in the ECF faster than it is expelled, thus lowering pH to normal.

These are classic negative feedback mechanisms that result in acid–base homeostasis. Respiratory control of pH has some limitations, however, that are discussed later under acid–base imbalances.

24.3c Renal Control of pH

The kidneys can neutralize more acid or base than either the respiratory system or the chemical buffers. They do so by varying the amount of acid eliminated in the urine. The essence of this mechanism is that the renal tubules secrete H^+ into the tubular fluid, where most of it binds to bicarbonate, ammonia, and phosphate buffers. Bound and free H^+ are then excreted in the urine. Thus the kidneys, in contrast to the lungs, actually expel H^+ itself from the body. The other buffer systems only reduce its concentration by binding it to another chemical.

Figure 24.10 shows how the renal tubules secrete and neutralize H^+. The hydrogen ions are colored so you can trace them from the blood (step ①) to the tubular fluid (step ⑥). Notice that it is not a simple matter of transporting free H^+ across the tubule cells; rather, the H^+ travels in the form of carbonic acid and water molecules.

The tubular secretion of H^+ takes place at step ⑥, where the ion is pumped out of the tubule cell into the tubular fluid. This can happen only if there is a steep enough concentration gradient between a high H^+ concentration within the cell and a lower concentration in the tubular fluid. If the pH of the tubular fluid drops any lower than 4.5, H^+ concentration in the fluid is so high that tubular secretion ceases. Thus, pH 4.5 is the **limiting pH** for tubular secretion. This has added significance later in our discussion.

In a person with normal acid–base balance, all bicarbonate ions ($HCO_3{}^-$) in the tubular fluid are consumed by neutralizing H^+; thus there is no $HCO_3{}^-$ in the urine. Bicarbonate ions are filtered by the glomerulus, gradually disappear from the tubular fluid, and appear in the peritubular capillary blood. It *appears* as if $HCO_3{}^-$ is reabsorbed by the renal tubules, but this is not the case; indeed, the renal tubules are incapable of reabsorbing $HCO_3{}^-$ directly. The cells of the proximal convoluted tubule, however, have carbonic anhydrase (CAH) on their brush borders facing the lumen. This breaks down the H_2CO_3 in the tubular fluid to $CO_2 + H_2O$ (step ⑩). It is the CO_2 that is reabsorbed, not the bicarbonate. For every CO_2 reabsorbed, however, a *new* bicarbonate ion is formed in the tubule cell and released into the blood (step ⑤). The effect is the same as if the tubule cells had reabsorbed bicarbonate itself.

Note that for every bicarbonate ion that enters the peritubular capillaries, a sodium ion does too. Thus, Na^+ reabsorption by the renal tubules is part of the process of neutralizing acid. The more acid the kidneys excrete, the less sodium the urine contains.

The tubules secrete somewhat more H^+ than the available bicarbonate can neutralize. The urine therefore contains a slight excess of free H^+, which gives it a pH of about 5 to 6. Yet if all of the excess H^+ secreted by the tubules remained in this free ionic form, the pH of the tubular fluid would drop far below the limiting pH of 4.5, and H^+ secretion would stop. This must be prevented, and there are additional buffers in the tubular fluid to do so.

The glomerular filtrate contains Na_2HPO_4 (dibasic sodium phosphate), which reacts with some of the H^+ **(fig. 24.11)**. A hydrogen ion replaces one of the sodium ions in the buffer, forming NaH_2PO_4 (monobasic sodium phosphate). This is passed in the urine, and the displaced Na^+ is transported into the tubule cell and from there to the bloodstream.

In addition, tubule cells catabolize certain amino acids and release ammonia (NH_3) as a product. Ammonia diffuses into the tubular fluid, where it acts as a base to neutralize acid. It reacts with H^+ and Cl^- (the most abundant anion in the glomerular filtrate) to form ammonium chloride (NH_4Cl), which is passed in the urine.

Since there is so much chloride in the tubular fluid, you may ask why H^+ isn't simply excreted as hydrochloric acid (HCl). Why involve ammonia? The reason is that HCl is a strong acid—it dissociates almost completely, so most of its hydrogen would

① H^+ in blood reacts with HCO_3^- to form H_2CO_3.

② H_2CO_3 decomposes into H_2O and CO_2, which enter the tubule cell.

③ Tubule cells acquire CO_2 from blood, tubular fluid, and their own aerobic respiration.

④ Carbonic anhydrase (CAH) combines H_2O and CO_2 to re-form H_2CO_3.

⑤ H_2CO_3 ionizes to form HCO_3^- (which returns to the blood) and H^+.

⑥ Na^+–H^+ antiport exchanges H^+ for Na^+.

⑦ $NaHCO_3$ from glomerular filtrate decomposes into Na^+ and HCO_3^-. Na^+ is pumped into tubule cell.

⑧ Na^+ is removed by Na^+–K^+ pump at the base of the cell.

⑨ HCO_3^- reacts with H^+ from tubule cell to form H_2CO_3.

⑩ CAH on brush border decomposes H_2CO_3 to H_2O and CO_2 again.

⑪ CO_2 enters the tubular cell and H_2O passes in the urine (carrying the H^+ that was originally in the blood).

FIGURE 24.10 Secretion and Neutralization of Hydrogen Ions in the Kidneys. The colored hydrogen symbols allow you to trace hydrogen from H^+ in the blood to H_2O in the urine.

? *If the pH of the tubular fluid went down, how would its Na^+ concentration change?*

be in the form of free H^+. The pH of the tubular fluid would drop below the limiting pH and prevent excretion of more acid. Ammonium chloride, by contrast, is a weak acid—most of its hydrogen remains bound to it and doesn't lower the pH of the tubular fluid.

24.3d Disorders of Acid–Base Balance

Figure 24.12 represents acid–base balance with an instructive metaphor to show its dependence on the bicarbonate buffer system. At a normal pH of 7.4, the ECF has a 20:1 ratio of HCO_3^- to H_2CO_3. Excess hydrogen ions convert HCO_3^- to H_2CO_3 and tip the balance to a lower pH. A pH below 7.35 is considered to be a state of **acidosis.** On the other hand, a H^+ deficiency causes H_2CO_3 to dissociate into H^+ and HCO_3^-, thus tipping the balance to a higher pH. A pH above 7.45 is a state of **alkalosis.** Either of these imbalances has potentially fatal effects. A person can't live more than a few hours if the blood pH is below 7.0 or above 7.7, and a pH below 6.8 or above 8.0 is quickly fatal.

In acidosis, H^+ diffuses down its concentration gradient into cells, and to maintain electrical balance, K^+ diffuses out

(**fig. 24.13a**). The H^+ is buffered by intracellular proteins, so this exchange results in a net loss of cations from the cell. This makes the resting membrane potential more negative than usual (hyperpolarized) and makes nerve and muscle cells more difficult to stimulate. This is why acidosis depresses the central nervous system and causes such symptoms as confusion, disorientation, and coma.

In alkalosis, the extracellular H^+ concentration is low. Hydrogen ions diffuse out of the cells and K^+ diffuses in to replace them (**fig. 24.13b**). The net gain in positive intracellular charges shifts the membrane potential closer to firing level and makes the nervous system hyperexcitable. Neurons fire spontaneously and overstimulate skeletal muscles, causing muscle spasms, tetanus, convulsions, or respiratory paralysis.

Acid–base imbalances fall into two categories, respiratory and metabolic (**table 24.3**). **Respiratory acidosis** occurs when the rate of alveolar ventilation fails to keep pace with the body's rate of CO_2 production. Carbon dioxide accumulates in the ECF and lowers its pH. This occurs in such conditions as emphysema, in which there is a severe reduction in

Blood of peritubular capillary

Renal tubule cells (proximal convoluted tubule)

Tubular fluid

Key
- Antiport
- Diffusion through a membrane channel
- Diffusion through the membrane lipid

FIGURE 24.11 Acid Buffering in the Urine. Reactions in the tubule cells are the same as in figure 24.10 but are simplified in this diagram. The essential differences are the buffering mechanisms shown in the tubular fluid. Hydrogen symbols are colored to allow tracing them from carbonic acid to the urine.

FIGURE 24.12 The Relationship of Bicarbonate–Carbonic Acid Ratio to pH. At a normal pH of 7.40, there is a 20:1 ratio of bicarbonate ions (HCO_3^-) to carbonic acid (H_2CO_3) in the blood plasma. An excess of HCO_3^- tips the balance toward alkalosis, whereas an excess of H_2CO_3 tips it toward acidosis.

the number of functional alveoli. **Respiratory alkalosis** results from hyperventilation, in which CO_2 is eliminated faster than it is produced.

Metabolic acidosis can result from increased production of organic acids, such as lactic acid in anaerobic fermentation and ketone bodies in alcoholism and diabetes mellitus. It can also result from the excessive ingestion of acidic drugs such as aspirin or from the loss of base due to chronic diarrhea or overuse of laxatives. Dying persons also typically exhibit acidosis. **Metabolic alkalosis** is rare but can result from overuse of bicarbonates (such as oral antacids and intravenous bicarbonate solutions) or from the loss of stomach acid by chronic vomiting.

24.3e Compensation for Acid–Base Imbalances

In **compensated** acidosis or alkalosis, either the kidneys compensate for pH imbalances of respiratory origin, or the respiratory system compensates for pH imbalances of metabolic origin. **Uncompensated** acidosis or alkalosis is a pH imbalance that the body cannot correct without clinical intervention.

In **respiratory compensation,** changes in pulmonary ventilation correct the pH of the body fluids by expelling or retaining

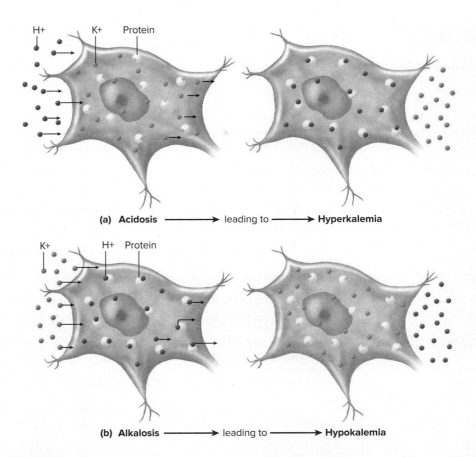

(a) Acidosis ⟶ leading to ⟶ **Hyperkalemia**

(b) Alkalosis ⟶ leading to ⟶ **Hypokalemia**

FIGURE 24.13 The Relationship Between Acid–Base Imbalances and Potassium Imbalances. (a) In acidosis, H^+ diffuses into the cells and drives out K^+, elevating the K^+ concentration of the ECF. (b) In alkalosis, H^+ diffuses out of the cells and K^+ diffuses in to replace it, lowering the K^+ concentration of the ECF.

? *How would you change part (a) to show the effect of hyperkalemia on the pH of the ECF?*

TABLE 24.3	Some Causes of Acidosis and Alkalosis	
	Acidosis	**Alkalosis**
Respiratory	Hypoventilation, apnea, or respiratory arrest; asthma; emphysema; cystic fibrosis; chronic bronchitis; narcotic overdose	Hyperventilation due to pain or emotions such as anxiety; oxygen deficiency (as at high elevation)
Metabolic	Excess production of organic acids as in diabetes mellitus and starvation; long-term anaerobic fermentation; hyperkalemia; chronic diarrhea; excessive alcohol consumption; drugs such as aspirin and laxatives	Rare but can result from chronic vomiting; overuse of bicarbonates (antacids); aldosterone hypersecretion

CO_2. If there is a CO_2 excess (hypercapnia), pulmonary ventilation increases to expel CO_2 and bring the blood pH back up to normal. If there is a CO_2 deficiency (hypocapnia), ventilation is reduced to allow CO_2 to accumulate in the blood and lower the pH to normal.

This is very effective in correcting pH imbalances due to abnormal P_{CO_2} but not very effective in correcting other causes of acidosis and alkalosis. In diabetic acidosis, for example, the lungs can't reduce the concentration of ketone bodies in the blood, although one can somewhat compensate for the H^+ that ketones release by increasing pulmonary ventilation and exhausting extra CO_2. The respiratory system can adjust a blood pH of 7.0 back to 7.2 or 7.3 but not all the way back to the normal 7.4. Although the respiratory system has a very powerful buffering effect, its ability to stabilize pH is therefore limited.

Renal compensation is an adjustment of pH by changing the rate of H^+ secretion by the renal tubules. The kidneys are slower to respond to pH imbalances but better at restoring a fully normal pH. Urine usually has a pH of 5 to 6, but in acidosis it may fall as low as 4.5 because of excess H^+, whereas in alkalosis it may rise as high as 8.2 because of excess HCO_3^-. The kidneys can't act quickly enough to compensate for short-term pH imbalances, such as the acidosis that could result from an asthmatic attack lasting an hour or two, or the alkalosis resulting from a brief episode of emotional hyperventilation. They are effective, however, at compensating for pH imbalances that last for a few days or longer.

TABLE 24.4		Some Relationships Among Fluid, Electrolyte, and Acid–Base Imbalances	
Cause		**Potential Effect**	**Reason**
Acidosis	→	Hyperkalemia	H^+ diffuses into cells and displaces K^+ (see fig. 24.13a). As K^+ leaves the ICF, K^+ concentration in the ECF rises.
Hyperkalemia	→	Acidosis	Opposite from the above; high K^+ concentration in the ECF causes less K^+ to diffuse out of the cells than normally. H^+ diffuses out to compensate, and this lowers the extracellular pH.
Alkalosis	→	Hypokalemia	H^+ diffuses from ICF to ECF. More K^+ remains in the ICF to compensate for the H^+ loss, causing a drop in ECF K^+ concentration (see fig. 24.13b).
Hypokalemia	→	Alkalosis	Opposite from the preceding line; low K^+ concentration in the ECF causes K^+ to diffuse out of cells. H^+ diffuses in to replace K^+, lowering the H^+ concentration of the ECF and raising its pH.
Acidosis	→	Hypochloremia	More Cl^- is excreted as NH_4Cl to buffer the excess acid in the renal tubules, leaving less Cl^- in the ECF.
Alkalosis	→	Hyperchloremia	More Cl^- is reabsorbed from the renal tubules, so ingested Cl^- accumulates in the ECF rather than being excreted.
Hyperchloremia	→	Acidosis	More H^+ is retained in the blood to balance the excess Cl^-, causing hyperchloremic acidosis.
Hypovolemia	→	Alkalosis	More Na^+ is reabsorbed by the kidney. Na^+ reabsorption is coupled to H^+ secretion (see fig. 24.10), so more H^+ is secreted and pH of the ECF rises.
Hypervolemia	→	Acidosis	Less Na^+ is reabsorbed, so less H^+ is secreted into the renal tubules. H^+ retained in the ECF causes acidosis.
Acidosis	→	Hypocalcemia	Acidosis causes more Ca^{2+} to bind to plasma protein and citrate ions, lowering the concentration of free, ionized Ca^{2+} and causing symptoms of hypocalcemia.
Alkalosis	→	Hypercalcemia	Alkalosis causes more Ca^{2+} to dissociate from plasma protein and citrate ions, raising the concentration of free Ca^{2+}.

In acidosis, the renal tubules increase the rate of H^+ secretion. The extra H^+ in the tubular fluid must be buffered; otherwise, the fluid pH could exceed the limiting pH and H^+ secretion would stop. Therefore, in acidosis, the renal tubules secrete more ammonia to buffer the added H^+, and the amount of ammonium chloride in the urine may rise to 7 to 10 times normal.

▶▶▶ APPLY WHAT YOU KNOW

Suppose you measured the pH and ammonium chloride concentration of urine from a person with emphysema and urine from a healthy individual. How would you expect the two to differ, and why?

In alkalosis, the bicarbonate concentration and pH of the urine are elevated. This is partly because there is more HCO_3^- in the blood and glomerular filtrate and partly because there isn't enough H^+ in the tubular fluid to neutralize all the HCO_3^- in the filtrate.

24.3f Acid–Base Imbalances in Relation to Electrolyte and Fluid Imbalances

The foregoing discussion once again underscores a point made early in this chapter—we can't understand or treat imbalances of fluid, electrolyte, or acid–base balance in isolation from each other,

because each of these frequently affects the other two. **Table 24.4** itemizes and explains some of these interactions. This is by no means a complete list of how fluid, electrolytes, and pH affect each other, but it does demonstrate their interdependence. Note that many of these relationships are reciprocal—for example, acidosis can cause hyperkalemia, and conversely, hyperkalemia can cause acidosis.

BEFORE YOU GO ON

Answer the following questions to test your understanding of the preceding section:

12. Write two chemical equations that show how the bicarbonate buffer system compensates for acidosis and alkalosis and two equations that show how the phosphate buffer system compensates for these imbalances.

13. Why are phosphate buffers more effective in the cytoplasm than in the blood plasma?

14. Renal tubules cannot reabsorb HCO_3^-; yet HCO_3^- concentration in the tubular fluid falls while in the blood plasma it rises. Explain this apparent contradiction.

15. In acidosis, the renal tubules secrete more ammonia. Why?

DEEPER INSIGHT 24.2

CLINICAL APPLICATION

Fluid Replacement Therapy

One of the most significant problems in the treatment of seriously ill patients is the restoration and maintenance of proper fluid volume, composition, and distribution among the fluid compartments. Fluids may be administered to replenish total body water, restore blood volume and pressure, shift water from one fluid compartment to another, or restore and maintain electrolyte and acid–base balance.

Drinking water is the simplest method of fluid replacement, but it doesn't replace electrolytes. Heat exhaustion can occur when you lose water and salt in the sweat and replace the fluid by drinking plain water. Broths, juices, and sports drinks replace water, carbohydrates, and electrolytes.

Patients who can't take fluids by mouth must be treated by alternative routes. Some fluids can be given by enema and absorbed through the colon. All routes of fluid administration other than the digestive tract are called *parenteral*[7] routes. The most common of these is the intravenous (I.V.) route, but for various reasons, including inability to find a suitable vein, fluids are sometimes given by subcutaneous (sub-Q), intramuscular (I.M.), or other parenteral routes. Many kinds of sterile solutions are available to meet the fluid replacement needs of different patients.

In cases of extensive blood loss, there may not be time to type and cross-match blood for a transfusion. The more urgent need is to replenish blood volume and pressure. *Normal saline* (isotonic, 0.9% NaCl) is a relatively quick and simple means of raising blood volume while maintaining normal osmolarity, but it has significant shortcomings. It takes three to five times as much saline as whole blood to rebuild normal volume because much of the saline escapes the circulation into the interstitial fluid compartment or is excreted by the kidneys. In addition, normal saline can induce hypernatremia and hyperchloremia, because the body excretes the water but retains much of the NaCl. Hyperchloremia can, in turn, produce acidosis. Normal saline also lacks potassium, magnesium, and calcium. Indeed, it dilutes those electrolytes that are already present and creates a risk of cardiac arrest from hypocalcemia. Saline also dilutes plasma albumin and RBCs, creating still greater risks for patients who have suffered extensive blood loss. Nevertheless, the emergency maintenance of blood volume sometimes takes temporary precedence over these other considerations.

Fluid therapy is also used to correct pH imbalances. Acidosis may be treated with *Ringer's lactate solution,* which includes sodium to rebuild ECF volume, potassium to rebuild ICF volume, lactate to balance the cations, and enough glucose to make the solution isotonic. Alkalosis can be treated with potassium chloride. This must be administered very carefully, because potassium ions can cause painful venous spasms, and

even a small potassium excess can cause cardiac arrest. High-potassium solutions should never be given to patients in renal failure or whose renal status is unknown, because in the absence of renal excretion of potassium, they can bring on lethal hyperkalemia. Ringer's lactate or potassium chloride also must be administered very cautiously, with close monitoring of blood pH, to avoid causing a pH imbalance opposite the one that was meant to be corrected. Too much Ringer's lactate causes alkalosis and too much KCl causes acidosis.

Plasma volume expanders are hypertonic solutions or colloids that are retained in the bloodstream and draw interstitial water into it by osmosis. They include albumin, sucrose, mannitol, and dextran. Plasma expanders are also used to combat hypotonic hydration by drawing water out of swollen cells, averting such problems as seizures and coma. A plasma expander can draw several liters of water out of the intracellular compartment within a few minutes.

Patients who can't eat are often given isotonic 5% dextrose (glucose). A fasting patient loses as much as 70 to 85 g of protein per day from the tissues as protein is broken down to fuel the metabolism. Giving 100 to 150 g of I.V. glucose per day reduces this by half and is said to have a *protein-sparing effect.* More than glucose is needed in some cases—for example, if a patient hasn't eaten for several days and can't be fed by nasogastric tube (due to lesions of the digestive tract, for example) or if large amounts of nutrients are needed for tissue repair following severe trauma, burns, or infections. In *total parenteral nutrition (TPN),* or *hyperalimentation,*[8] a patient is provided with complete I.V. nutritional support, including a protein hydrolysate (amino acid mixture), vitamins, electrolytes, 20% to 25% glucose, and on alternate days, a fat emulsion.

The water from parenteral solutions is normally excreted by the kidneys. If the patient has renal insufficiency, however, excretion may not keep pace with intake, and there is a risk of hypotonic hydration. Intravenous fluids are usually given slowly, by *I.V. drip,* to avoid abrupt changes or overcompensation for the patient's condition. In addition to pH, the patient's heart rate, blood pressure, hematocrit, and plasma electrolyte concentrations are monitored, and the patient is examined periodically for respiratory sounds indicating pulmonary edema.

The delicacy of fluid replacement therapy underscores the close relationships among fluids, electrolytes, and pH. It is dangerous to manipulate any one of these variables without close attention to the others. Parenteral fluid therapy is usually used for persons who are seriously ill. Their homeostatic mechanisms are already compromised and leave less room for error than in a healthy person.

---

[7]*para* = beside; *enter* = intestine

[8]*hyper* = above normal; *aliment* = nourishment

STUDY GUIDE

▶ Assess Your Learning Outcomes

To test your knowledge, discuss the following topics with a study partner or in writing, ideally from memory.

24.1 Fluid Balance

1. Fluid compartments; total body water content of a typical young adult; and what percentages of this are in the intracellular and extracellular fluid compartments
2. How water moves from one fluid compartment to another; which differs more from one compartment to another—osmolarity or chemical composition—and why
3. What it means to be in a state of fluid balance
4. The meanings of *sensible, insensible,* and *obligatory water loss*
5. Typical daily water gain and loss; sources of water gain and avenues of water loss, and the typical amounts of each
6. How the hypothalamus senses the body's state of hydration and how it promotes fluid intake and water conservation when necessary
7. Short- and long-term mechanisms by which thirst is satiated; why it is important to have the short-term mechanisms
8. The role of antidiuretic hormone in regulating the body's rate of fluid loss
9. The meaning of *fluid deficiency;* two forms of fluid deficiency and how they differ; and potential consequences of fluid deficiency
10. The meaning of *fluid excess;* two forms of fluid excess and how they differ; and potential consequences of fluid excess

11. The meaning of *fluid sequestration;* examples and potential consequences of fluid sequestration

24.2 Electrolyte Balance

1. Functions of electrolytes in general; the body's common electrolytes; and the relative ECF and ICF concentrations of Na^+, K^+, Cl^-, Ca^{2+}, Mg^{2+}, and phosphate ions
2. Physiological functions of sodium; how it is regulated by aldosterone, antidiuretic hormone, and natriuretic peptides; and causes and effects of hyper- and hyponatremia
3. Physiological functions of potassium; how it is regulated by the kidneys and aldosterone; and causes and effects of hyper- and hypokalemia
4. Physiological functions of calcium; why cells usually must maintain a low intracellular calcium level; how calcium homeostasis is regulated by parathyroid hormone, calcitriol, and calcitonin; and causes and effects of hyper- and hypocalcemia
5. Physiological functions of magnesium; how vitamin D and parathyroid hormone regulate magnesium level; and causes and effects of magnesium deficiency and excess
6. Physiological functions of chloride; why chloride levels are determined mainly by the sodium-regulating mechanisms; and causes and effects of hyper- and hypochloremia
7. Three forms of inorganic phosphate ions (P_i); the physiological functions of phosphate; and how phosphate homeostasis is regulated

24.3 Acid–Base Balance

1. The normal pH range of the blood and most other ECF, and why acid–base balance is so crucial to homeostasis
2. Strong and weak acids and bases, and examples
3. How a buffer resists pH changes; the body's principal physiological and chemical buffer systems
4. How the bicarbonate, phosphate, and protein buffer systems neutralize excess acid or base; their effectiveness relative to each other and to the physiological buffer systems
5. How the respiratory system buffers pH
6. How the renal tubule secretes acid; why urine is normally bicarbonate-free; why acid secretion is linked to sodium reabsorption; and the roles of Na_2HPO_4 and NH_3 in buffering urinary acid
7. The ratio of H_2CO_3 to HCO_3^- at normal blood pH, and how this ratio changes in acidosis and alkalosis; the pH levels at which acidosis and alkalosis are soon fatal
8. Mechanisms behind the neuromuscular effects of acidosis and alkalosis
9. Common causes of respiratory and metabolic acidosis and alkalosis
10. The difference between compensated and uncompensated acidosis and alkalosis, and how the respiratory and urinary systems compensate for pH imbalances
11. Examples of how fluid, electrolyte, and acid–base imbalances can each cause, or be caused by, imbalances in the other two categories

▶ Testing Your Recall

Answers in Appendix A

1. The greatest percentage of the body's water is in
   a. the blood plasma.
   b. the lymph.
   c. the intracellular fluid.
   d. the interstitial fluid.
   e. the extracellular fluid.

2. Hypertension is likely to increase the secretion of
   a. natriuretic peptide.
   b. antidiuretic hormone.
   c. bicarbonate ions.
   d. aldosterone.
   e. ammonia.

3. _____ increases water reabsorption without increasing sodium reabsorption.
   a. Antidiuretic hormone
   b. Aldosterone
   c. Natriuretic peptide
   d. Parathyroid hormone
   e. Calcitonin

4. Hypotonic hydration can result from
   a. ADH hypersecretion.
   b. ADH hyposecretion.
   c. aldosterone hypersecretion.
   d. aldosterone hyposecretion.
   e. *a* and *d* only.

5. Tetanus is most likely to result from
   a. hypernatremia.
   b. hypokalemia.
   c. hyperkalemia.
   d. hypocalcemia.
   e. *c* and *d* only.

STUDY GUIDE

6. The principal determinant of intracellular osmolarity and cellular volume is
   a. protein.
   b. phosphate.
   c. potassium.
   d. sodium.
   e. chloride.

7. Increased excretion of ammonium chloride in the urine most likely indicates
   a. hypercalcemia.
   b. hyponatremia.
   c. hypochloremia.
   d. alkalosis.
   e. acidosis.

8. The most effective buffer in the intracellular fluid is
   a. phosphate.
   b. protein.
   c. bicarbonate.
   d. carbonic acid.
   e. ammonia.

9. Tubular secretion of hydrogen is directly linked to
   a. tubular secretion of potassium.
   b. tubular secretion of sodium.
   c. tubular reabsorption of potassium.
   d. tubular reabsorption of sodium.
   e. tubular secretion of chloride.

10. Hyperchloremia is most likely to result in
    a. alkalosis.
    b. acidosis.
    c. hypernatremia.
    d. hyperkalemia.
    e. hypovolemia.

11. The most abundant cation in the ECF is _____.

12. The two most abundant cations in the ICF are _____ and _____.

13. Water produced by the body's chemical reactions is called _____.

14. The skin loses water by two processes, sweating and _____.

15. Any abnormal accumulation of fluid in a particular place in the body is called _____.

16. An excessive concentration of potassium ions in the blood is called _____.

17. A deficiency of sodium ions in the blood is called _____.

18. A blood pH of 7.2 caused by inadequate pulmonary ventilation would be classified as _____.

19. Tubular secretion of hydrogen ions ceases if the acidity of the tubular fluid falls below a value called the _____.

20. Long-term satiation of thirst depends on a reduction of the _____ of the blood.

▶ Building Your Medical Vocabulary

Answers in Appendix A

State a meaning of each word element, and give a medical term from this chapter that uses it or a slight variation of it.

1. aliment-

2. -emia

3. entero-

4. kali-

5. natri-

6. para-

7. sequestr-

8. spiro-

9. trans-

10. vol-

▶ What's Wrong with These Statements?

Answers in Appendix A

Briefly explain why each of the following statements is false, or reword it to make it true.

1. Hyperkalemia lowers the resting membrane potentials of nerve and muscle cells and makes them less excitable.

2. Aldosterone promotes sodium and water retention and can therefore greatly increase blood pressure.

3. Injuries that rupture a lot of cells, such as the mangling of an arm by farm or factory machinery, tend to elevate the Na^+ concentration of the ECF and can thereby cause cardiac arrest.

4. The phosphate level in the ECF is very crucial to survival and must be hormonally regulated within a narrow safe range.

5. Parathyroid hormone promotes calcium and phosphate reabsorption by the kidneys.

6. The bicarbonate system buffers more acid than any other chemical buffer.

7. The more sodium the renal tubules reabsorb, the less hydrogen ion appears in the urine and the higher the urine pH tends to be.

8. The body compensates for respiratory alkalosis by increasing the respiratory rate.

9. In true dehydration, the body fluids remain isotonic although total body water is reduced.

10. Long-term quenching of thirst results primarily from wetting and cooling of the mouth when one drinks.

STUDY GUIDE

▶ Testing Your Comprehension

1. A duck hunter is admitted to the hospital with a shotgun injury to the abdomen. He has suffered extensive blood loss but is conscious. He complains of being intensely thirsty. Explain the physiological mechanism connecting his injury to his thirst.

2. A woman living at poverty level finds bottled water at the grocery store next to the infant formula. The label on the water states that it is made especially for infants, and she construes this to mean that it can be used as a nutritional supplement. The water is much cheaper than formula, so she gives her baby several ounces of bottled water a day as a substitute for formula. After several days the baby has seizures and is taken to the hospital, where it is found to have edema, acidosis, and a plasma sodium concentration of 116 mEq/L. The baby is treated with anticonvulsants followed by normal saline and recovers. Explain each of the signs.

3. Explain why the respiratory and urinary systems are both necessary for the bicarbonate buffer system to work effectively in the blood plasma.

4. A 4-year-old child is caught up in tribal warfare in Africa. In a refugee camp, the only drinking water is from a sewage-contaminated pond. The child soon develops severe diarrhea from cholera and dies 10 days later of cardiac arrest. Explain the possible physiological cause(s) of his death.

5. The left column indicates some increases or decreases in blood plasma values. In the right column, replace the question mark with an up or down arrow to indicate the expected effect. Explain each effect.

Cause	Effect
a. ↑ H_2O	? Na^+
b. ↑ Na^+	? Cl^-
c. ↓ K^+	? H^+
d. ↑ H^+	? K^+
e. ↓ Ca^{2+}	? PO_4^{3-}

Mucosa of the small intestine

Steve Gschmeissner/Science Source

25

THE DIGESTIVE SYSTEM

CHAPTER OUTLINE

25.1 General Anatomy and Digestive Processes

    25.1a Digestive Function
    25.1b General Anatomy
    25.1c Relationship to the Peritoneum
    25.1d Regulation of the Digestive Tract

25.2 The Mouth Through Esophagus

    25.2a The Mouth
    25.2b Mastication
    25.2c Saliva and the Salivary Glands
    25.2d The Pharynx
    25.2e The Esophagus
    25.2f Swallowing

25.3 The Stomach

    25.3a Gross Anatomy
    25.3b Innervation and Circulation
    25.3c Microscopic Anatomy
    25.3d Gastric Secretions
    25.3e Gastric Motility
    25.3f Vomiting
    25.3g Digestion and Absorption
    25.3h Protection of the Stomach
    25.3i Regulation of Gastric Function

25.4 The Liver, Gallbladder, and Pancreas

    25.4a The Liver
    25.4b The Gallbladder and Bile

    25.4c The Pancreas
    25.4d Regulation of Secretion

25.5 The Small Intestine

    25.5a Gross Anatomy
    25.5b Circulation
    25.5c Microscopic Anatomy
    25.5d Intestinal Secretion
    25.5e Intestinal Motility

25.6 Chemical Digestion and Absorption

    25.6a Carbohydrates
    25.6b Proteins
    25.6c Lipids
    25.6d Nucleic Acids
    25.6e Vitamins
    25.6f Minerals
    25.6g Water

25.7 The Large Intestine

    25.7a Gross Anatomy
    25.7b Circulation
    25.7c Microscopic Anatomy
    25.7d Intestinal Microbes and Gas
    25.7e Absorption and Motility
    25.7f Defecation

Connective Issues

Study Guide

DEEPER INSIGHTS

25.1 Clinical Application: Tooth and Gum Disease

25.2 Clinical Application: Peptic Ulcer

25.3 Clinical Application: Gallstones

25.4 Clinical Application: Lactose Intolerance

25.5 Clinical Application: Diverticulosis and Diverticulitis

25.6 Clinical Application: Colorectal Cancer

25.7 Medical History: The Man with a Hole in His Stomach

Anatomy & Physiology *Revealed* 4.0

Module 12: Digestive System

- All chemical digestion consists of hydrolysis reactions, so you should understand this process and its bearing on the relationship between monomers and polymers (see section 2.4b).

- You must also know the basic structure of polysaccharides, disaccharides, triglycerides, and proteins to understand how these are digested (see section 2.4c).

- To best understand the activation and deactivation of various digestive enzymes in different regions of the digestive tract, refresh your memory on the effect of pH on enzyme action (see section 2.4f).

- The intestinal absorption of nutrients employs multiple modes of membrane transport discussed in section 3.3.

Most of the nutrients we eat cannot be used in the form found in our food. They must be broken down into smaller components, such as amino acids and monosaccharides, that are universal to all species. Consider what happens if you eat a piece of beef, for example. The myosin of beef differs very little from that of your own muscles, but the two are not identical, and even if they were, beef myosin could not be absorbed, transported in the blood, and properly installed in your muscle cells. Like any other dietary protein, it must be broken down into amino acids before it can be used. Since beef and human proteins are made of the same 20 amino acids, those of beef proteins could indeed become part of your own myosin but could equally well end up in your insulin, fibrinogen, collagen, or any other protein.

The digestive system is essentially a disassembly line—its primary purpose is to break nutrients down into forms that can be used by the body and to absorb them so they can be distributed to the tissues. The study of the digestive tract and the diagnosis and treatment of its disorders is called **gastroenterology.**[1]

25.1   General Anatomy and Digestive Processes

Expected Learning Outcomes

When you have completed this section, you should be able to

a. list the functions and major physiological processes of the digestive system;

b. distinguish between mechanical and chemical digestion;

c. identify the basic chemical process that underlies all chemical digestion, and name the major substrates and products of this process;

d. list the regions of the digestive tract and the accessory organs of the digestive system;

e. identify the layers of the digestive tract and describe its relationship to the peritoneum; and

f. describe the general nervous and chemical controls over digestive function.

25.1a Digestive Function **APR**

The **digestive system** is the organ system that processes food, extracts nutrients from it, and eliminates the residue. It does this in five stages:

1. **ingestion,** the selective intake of food;

2. **digestion,** the mechanical and chemical breakdown of food into a form usable by the body;

3. **absorption,** the uptake of nutrient molecules into the epithelial cells of the digestive tract and then into the blood or lymph;

4. **compaction,** absorbing water and consolidating the indigestible residue into feces; and finally,

5. **defecation,** the elimination of feces.

The digestion stage itself has two facets, mechanical and chemical. **Mechanical digestion** is the physical breakdown of food into smaller particles. It is achieved by the cutting and grinding action of the teeth and the churning contractions of the stomach. Mechanical digestion exposes more food surface to the action of digestive enzymes. **Chemical digestion** is a series of *hydrolysis* reactions that break dietary macromolecules into their monomers: polysaccharides into monosaccharides, proteins into amino acids, fats into monoglycerides and fatty acids, and nucleic acids into nucleotides. It is carried out by digestive enzymes produced by the salivary glands, stomach, pancreas, and small intestine. Some nutrients are already present in usable form in the ingested food and are absorbed without being digested: vitamins, free amino acids, minerals, cholesterol, and water.

25.1b General Anatomy

The digestive system has two anatomical subdivisions, the digestive tract and the accessory organs **(fig. 25.1).** The **digestive tract (alimentary[2] canal)** is a muscular tube extending from mouth to anus. It measures about 5 m (16 ft) long in a living person, but about 9 m (30 ft) in the cadaver due to the loss of muscle tone at death. It includes the mouth, pharynx, esophagus, stomach, small intestine, and large intestine. The stomach and intestines constitute the *gastrointestinal (GI) tract.* The teeth, tongue, salivary glands, liver, gallbladder, and pancreas are considered **accessory organs** of the digestive system.

The digestive tract is open to the environment at both ends. Most of the material in it hasn't entered any body tissues and is considered to be external to the body until it's absorbed by epithelial cells of the alimentary canal. In the strict sense, defecated food residue was never in the body.

[1] *gastro* = stomach; *entero* = intestines; *logy* = study of

[2] *aliment* = food

Most of the digestive tract follows the basic structural plan shown in **figure 25.2,** with a wall composed of the following tissue layers in order from the inner to the outer surface:

Mucosa
>   Epithelium
>
>   Lamina propria
>
>   Muscularis mucosae

Submucosa

Muscularis externa
>   Inner circular layer
>
>   Outer longitudinal layer

Serosa
>   Areolar tissue
>
>   Mesothelium

Slight variations on this theme are found in different regions of the tract.

The inner lining of the tract, called the **mucosa** or **mucous membrane,** consists of an epithelium, a loose connective tissue layer called the **lamina propria,** and a thin layer of smooth muscle called the **muscularis mucosae** (MUSS-cue-LERR-is mew-CO-see). The epithelium is simple columnar in most of the digestive tract, but stratified squamous from the oral cavity through the esophagus and in the lower anal canal, where the tract is subject to more abrasion. The muscularis mucosae tenses the mucosa, creating grooves and ridges that enhance its surface area and contact with food. This improves the efficiency of digestion and nutrient absorption. The mucosa exhibits an abundance of lymphocytes and lymphatic nodules—the *mucosa-associated lymphatic tissue (MALT)* (see section 21.1c).

The **submucosa** is a thicker layer of loose connective tissue containing blood vessels and lymphatics, a nerve plexus, and in some places, glands that secrete lubricating mucus into the lumen. The MALT extends into the submucosa in some parts of the GI tract.

The **muscularis externa** consists of usually two layers of muscle near the outer surface. Cells of the inner layer encircle the tract while those of the outer layer run longitudinally. In some places, the circular layer is thickened to form valves (sphincters) that regulate the passage of material through the tract. The muscularis externa is responsible for the motility that propels food and residue through the digestive tract.

The **serosa** is composed of a thin layer of areolar tissue topped by a simple squamous mesothelium. The serosa begins in the lower 3 to 4 cm of the esophagus and ends just before the rectum. The pharynx, most of the esophagus, and the rectum have no serosa but are surrounded by a fibrous connective tissue layer called the **adventitia,** which blends into the adjacent connective tissue of other organs.

The esophagus, stomach, and intestines have a nervous network called the **enteric**[3] **nervous system,** which regulates

---

[3]*enter* = intestine

FIGURE 25.1 The Digestive System.

digestive tract motility, secretion, and blood flow. This system is thought to have over 100 million neurons—more than the spinal cord! It can function independently of the central nervous system, although the CNS usually exerts a significant influence on its action. It is usually regarded as part of the autonomic nervous system, but opinions on this vary.

The enteric nervous system is composed of two networks of neurons: the **submucosal plexus** in the submucosa and the **myenteric plexus** of parasympathetic ganglia and nerve fibers between layers of the muscularis externa. Parasympathetic preganglionic fibers of the vagus nerves terminate in the ganglia of the myenteric plexus. Postganglionic fibers arising in this plexus not only innervate the muscularis externa, but also pass through its inner circular layer and contribute to the submucosal plexus.

Diaphragm

Esophageal hiatus

Mucosa:
    Stratified squamous
    epithelium
    Lamina propria
    Muscularis mucosae

Enteric nervous system:
    Myenteric plexus

    Submucosal plexus

    Parasympathetic ganglion of
    myenteric plexus

Submucosa:
    Esophageal gland

Lumen

Muscularis externa:
    Inner circular layer
    Outer longitudinal layer

Serosa

Blood vessels

FIGURE 25.2 Tissue Layers of the Digestive Tract. Cross section of the esophagus just below the diaphragm where it meets the stomach.

The myenteric plexus controls peristalsis and other contractions of the muscularis externa, and the submucosal plexus controls movements of the muscularis mucosae and glandular secretion of the mucosa. The enteric nervous system also includes sensory neurons that monitor tension in the gut wall and conditions in the lumen.

25.1c Relationship to the Peritoneum

In processing food, the stomach and intestines undergo such strenuous contractions that they need freedom to move in the abdominal cavity. They're not tightly bound to the abdominal wall, but over most of their length, loosely suspended from it by connective tissue sheets called **mesenteries.** Mesenteries hold the abdominal viscera in their proper relationship to each other, provide passage for the blood vessels and nerves that supply the digestive tract, and contain many lymph nodes and lymphatic vessels. The small intestine is especially mobile, squirming and coiling during digestion. The mesentery prevents it from becoming twisted and tangled by its own contractions or changes in body position. Intestinal

movements are eased by the slippery peritoneal fluid that lubricates the mesenteries.

The parietal peritoneum is a serous membrane that lines the wall of the abdominal cavity. Along the posterior (dorsal) midline of the body, it turns inward and forms the **posterior mesentery,** a translucent two-layered membrane extending to the digestive tract. Upon reaching an organ such as the stomach or small intestine, the two layers of the mesentery separate and pass around opposite sides of the organ, forming the serosa. In some places, the two layers come together again on the far side of that organ and continue as another sheet of tissue, the **anterior mesentery.** This mesentery may hang freely in the abdominal cavity or attach to the anterior abdominal wall or other organs. The relationship between the mesenteries and the serosa is shown in atlas A, figures A.7 and A.8.

Two anterior mesenteries called *omenta* are associated with the stomach **(fig. 25.3).** The **lesser omentum** extends the short distance from the liver to the right superior margin *(lesser curvature)* of the stomach. The bile duct is embedded in the lesser omentum along its course between the liver and duodenum, as are

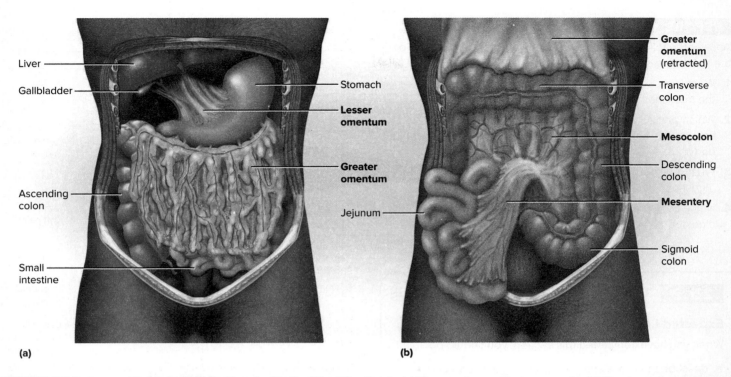

Liver

Gallbladder

Stomach

Lesser omentum

Greater omentum

Ascending colon

Jejunum

Small intestine

(a)

Greater omentum (retracted)

Transverse colon

Mesocolon

Descending colon

Mesentery

Sigmoid colon

(b)

FIGURE 25.3 Serous Membranes Associated with the Digestive Tract. (a) The greater and lesser omenta. (b) Greater omentum and small intestine retracted to show the mesocolon and mesentery. These membranes contain the mesenteric arteries and veins. **APR**

some major blood vessels leading to the liver. A much larger and fatty **greater omentum** hangs like an apron from the left inferior margin (*greater curvature*) of the stomach, loosely covering the small intestine. At its inferior margin, the greater omentum turns back on itself and passes upward, behind the superficial layer; the two layers adhere firmly to each other along most of their course. At the superior margin, the upturned layer continues as a serosa that encloses the spleen and transverse colon. From the transverse colon, it continues to the posterior abdominal wall and anchors the colon. This mesentery of the colon is called the **mesocolon.** The omenta have a loosely organized, lacy appearance due partly to their irregular distribution of adipose tissue. In obesity, much of the abdominal fat is in these mesenteries (*mesenteric fat;* see fig. 26.1).

The omenta adhere to perforations or inflamed areas of the stomach or intestines, contribute immune cells to the site, and isolate infections that might otherwise give rise to peritonitis. The greater omentum is part of the body's first line of defense against toxins and infections. Clusters of lymphatic tissue called *milky spots* in the omentum perform a "policing" function—collecting bacteria, other cells, and antigens from the peritoneal fluid and, if they detect a threat, initiating an immune response. Unfortunately, however, they fail to recognize metastatic cancer cells, which can aggregate in the omentum and turn it into a breeding ground for gastrointestinal, ovarian, and other cancers.

When an organ is enclosed by mesentery (serosa) on all sides, it is considered to be within the peritoneal cavity, or **intraperitoneal.** When an organ lies against the posterior body wall and is covered by peritoneum on the anterior side only, it is said to be outside the peritoneal cavity, or **retroperitoneal.** The duodenum, most of the pancreas, and parts of the large intestine are retroperitoneal. The stomach, liver, and other parts of the small and large intestines are intraperitoneal.

25.1d Regulation of the Digestive Tract

The motility and secretion of the digestive tract are controlled by nervous, hormonal, and paracrine mechanisms. The neural controls include short and long autonomic reflexes. In **short (myenteric) reflexes,** stretching or chemical stimulation of the digestive tract acts through the myenteric plexus to stimulate contractions in nearby regions of the muscularis externa, such as the peristaltic contractions of swallowing. **Long (vagovagal) reflexes** act through autonomic nerve fibers that carry sensory signals from the digestive tract to the brainstem and motor commands back to the digestive tract. Parasympathetic fibers of the vagus nerves are especially important in stimulating digestive motility and secretion by way of these long reflexes.

The digestive tract also produces numerous hormones (such as *gastrin* and *secretin*) and paracrine secretions (such as *histamine* and *prostaglandins*) that stimulate digestive function. The hormones are secreted into the blood and stimulate relatively distant parts of the digestive tract. The paracrine secretions diffuse through the tissue fluids and stimulate nearby target cells.

Answer the following questions to test your understanding of the preceding section:

1. What is the term for the serous membrane that suspends the intestines from the abdominal wall?

2. Which physiological process of the digestive system truly moves a nutrient from the outside to the inside of the body?

3. What one type of reaction is the basis of all chemical digestion?

4. Name some nutrients that are absorbed without being digested.

25.2 The Mouth Through Esophagus

Expected Learning Outcomes

When you have completed this section, you should be able to

a. describe the gross anatomy of the digestive tract from the mouth through the esophagus;

b. describe the composition and functions of saliva; and

c. describe the nervous control of salivation and swallowing.

25.2a The Mouth

The mouth is also known as the **oral,** or **buccal, cavity** (BUCK-ul). Its functions include ingestion (food intake), taste and other sensory responses to food, mastication (chewing), chemical digestion (starch is partially digested in the mouth), swallowing, speech, and respiration. The mouth is enclosed by the cheeks, lips, palate, and tongue **(fig. 25.4).** Its anterior opening between the lips is the **oral fissure** and its posterior opening into the throat is the **fauces**[4] (FAW-seez). The mouth is lined with stratified squamous epithelium. This epithelium is keratinized in areas subject to the greatest food abrasion, such as the gums and hard palate, and nonkeratinized in other areas such as the floor of the mouth, the soft palate, and the inside of the cheeks and lips.

The Cheeks and Lips

The cheeks and lips retain food and push it between the teeth for chewing. They are essential for articulate speech and for sucking and blowing actions, including suckling by infants. Their fleshiness is due mainly to subcutaneous fat, the buccinator muscles of the cheeks, and the orbicularis oris muscle of the lips. A median fold called the **labial frenulum**[5] attaches each lip

FIGURE 25.4 The Oral Cavity. For a photographic medial view, see figure B.2 in atlas B. **APR**

to the gum, anteriorly between the incisors. The **vestibule** is the space between the cheeks or lips and the teeth—the space where you insert your toothbrush when brushing the outer surfaces of your teeth.

The lips are divided into three areas: (1) The *cutaneous area* is colored like the rest of the face and has hair follicles and sebaceous glands; on the upper lip, this is where a mustache grows. (2) The *red area (vermilion)* is the hairless region where the lips meet (where one might apply lipstick). It has unusually tall dermal papillae, which allow blood capillaries and nerve endings to come closer to the epidermal surface. Thus, this area is redder and more sensitive than the cutaneous area. (3) The *labial mucosa* is the inner surface of the lip, facing the gums and teeth. It contains mucous *labial glands* that contribute to the saliva.

The Tongue

The tongue, although muscular and bulky, is a remarkably agile, sensitive, and versatile organ. It aids in food intake; it is sensitive enough to feel a stray hair or grain of sand in a bite of food and has sensory receptors for taste, texture, and temperature that are important in the acceptance or rejection of food; it compresses and breaks up food; it maneuvers food between the teeth for mastication while it avoids being bitten; it dislodges food particles from the teeth after a meal; it secretes mucus and enzymes; it compresses the chewed food into a soft mass, or *bolus,* that is easier to swallow; it initiates swallowing; and it is necessary for articulate speech. Its surface is covered with nonkeratinized

---

[4]*fauces* = throat
[5]*labi* = lip; *frenulum* = little bridle

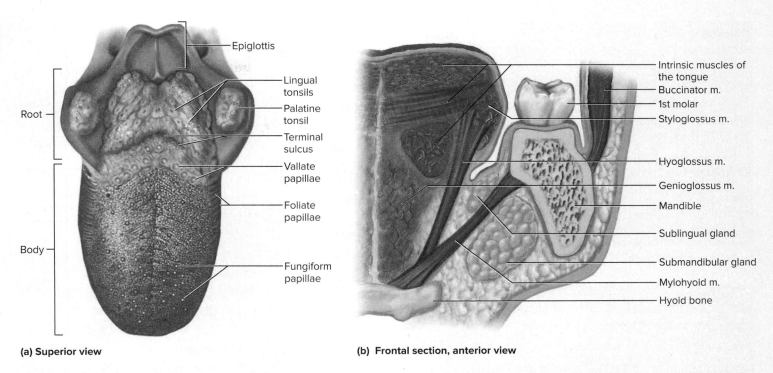

(a) **Superior view**

(b) **Frontal section, anterior view**

FIGURE 25.5 **The Tongue.** (a) Superior (dorsal) view. (b) Frontal section, anterior view. For a photographic sagittal section, see figure B.2 in atlas B.

stratified squamous epithelium and exhibits bumps and projections called **lingual papillae,** the site of most taste buds. The types of papillae and sense of taste are discussed in section 16.3a, and the general anatomy of the tongue is shown in **figure 25.5.**

▶▶▶**APPLY WHAT YOU KNOW**

How does proprioception protect the tongue from being bitten?

The anterior two-thirds of the tongue, called the **body,** occupies the oral cavity; and the posterior one-third, the **root,** occupies the oropharynx. The boundary between them is marked by a V-shaped row of **vallate papillae** and, behind these, a groove called the **terminal sulcus.** The body is attached to the floor of the mouth by a median fold called the **lingual frenulum.**

The muscles of the tongue, which compose most of its mass, are described in table 10.2. The **intrinsic muscles,** contained entirely within the tongue, produce the relatively subtle tongue movements of speech. The **extrinsic muscles** arise from attachments outside the tongue and produce most of the stronger tongue movements of food manipulation. The extrinsic muscles include the *genioglossus, hyoglossus, palatoglossus,* and *styloglossus* (fig. 25.5b; see also fig. 10.10). Within the tongue, amid the muscles, are serous and mucous **lingual glands,** which secrete a portion of the saliva. The **lingual tonsils** are contained in the root.

The Palate

The palate, separating the oral cavity from the nasal cavity, makes it possible to breathe while chewing food. Its anterior portion, the **hard (bony) palate,** is supported by the palatine processes of the maxillae and by the smaller palatine bones. It has transverse ridges called *palatine rugae* that aid the tongue in holding and manipulating food. Posterior to this is the **soft palate,** which has a more spongy texture and is composed mainly of skeletal muscle and glandular tissue, but no bone. It has a conical median projection, the **uvula,**[6] visible at the rear of the mouth. The uvula helps to retain food in the mouth until one is ready to swallow.

At the rear of the mouth, two muscular arches on each side begin at the roof near the uvula and descend to the floor. The anterior one is the **palatoglossal arch** and the posterior one is the **palatopharyngeal arch.** The latter arch marks the beginning of the pharynx. The palatine tonsils are located on the wall between the arches.

The Teeth

The teeth are collectively called the **dentition.** They serve to *masticate* the food, breaking it into smaller pieces. This not only makes the food easier to swallow, but also exposes more surface area to the action of digestive enzymes and thus speeds up chemical digestion. Adults normally have 16 teeth in the mandible (lower jaw) and 16 in the maxilla (upper jaw). From the midline to the rear

---

[6]*uvula* = little grape

(a) Deciduous (baby) teeth

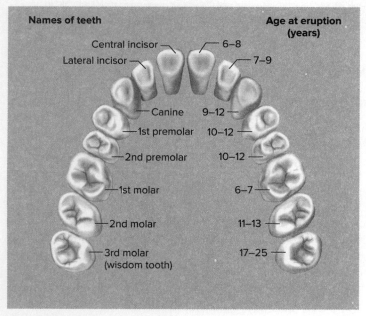

(b) Permanent teeth

FIGURE 25.6 The Dentition. (a) Deciduous (baby) teeth. (b) Permanent (adult) teeth. Each figure shows only the upper teeth. The ages at eruption are composite ages for the corresponding upper and lower teeth. Generally, the lower (mandibular) teeth erupt somewhat earlier than their upper (maxillary) counterparts. **APR**

❓ *Which teeth are absent from a 3-year-old child?*

of each jaw, there are two incisors, a canine, two premolars, and up to three molars on each side **(fig. 25.6).** The **incisors** are anterior, chisel-like cutting teeth used to bite off a piece of food. The **canines** are more pointed and act to puncture and shred it. They serve as weapons in many mammals but became reduced in the course of human evolution until they now project barely beyond the other teeth. The **premolars** and **molars** have relatively broad, lumpy surfaces adapted for crushing, shredding, and grinding; they are often informally called the *grinders.*

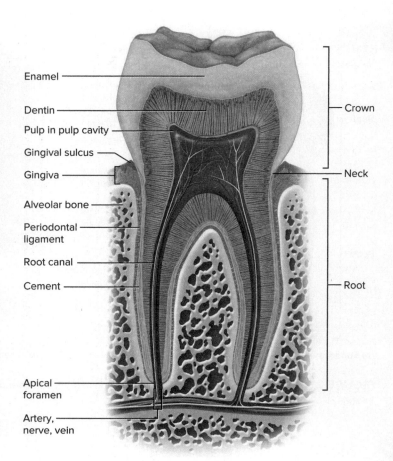

FIGURE 25.7 Structure of a Tooth and Its Alveolus. Shows typical anatomy of the tooth and periodontal tissues. This particular example is a molar.

❓ *Of all the components shown here, which is or are not living tissue(s)?*

Each tooth is embedded in a socket called an **alveolus,** forming a joint called a *gomphosis* between the tooth and bone **(fig. 25.7).** The alveolus is lined by a **periodontal ligament** (PERR-ee-oh-DON-tul), a modified periosteum whose collagen fibers penetrate into the bone on one side and into the tooth on the other. This anchors the tooth firmly in the alveolus but allows for slight movement under the stress of chewing. The ligament is innervated by proprioceptive nerve fibers that enable one to sense tooth movements and bite force. This sense is lost when a person has a tooth replaced by a dental implant. The gum, or **gingiva** (JIN-jih-vuh), covers the alveolar bone. Regions of a tooth are defined by their relationship to the gingiva: The **crown** is the portion above the gum attachment; the **root** is the portion below the gum, embedded in alveolar bone; and the **neck** is the point where the crown, root, and gum meet. The space between the tooth and gum is the **gingival sulcus.** The hygiene of this sulcus is especially important to dental health (see Deeper Insight 25.1). With some variation, the incisors and canines have a single root; the premolars have one or two roots; the first and second molars have two to three roots; and in the third molar, the roots are usually fused into one.

DEEPER INSIGHT 25.1

CLINICAL APPLICATION

Tooth and Gum Disease

The human mouth is home to more than 700 species of microorganisms, especially bacteria. Bacteria and sugars form a sticky residue on the teeth called *plaque*. If plaque isn't thoroughly removed by brushing and flossing, bacteria multiply, metabolize the sugars, and release lactic and other acids. These acids dissolve the minerals of enamel and dentin, and the bacteria enzymatically digest the collagen and other organic components. The eroded "cavities" of the tooth are known as *dental caries*.[7] If not repaired, caries may fully penetrate the dentin and spread to the pulp cavity. This requires either extraction of the tooth or *root canal therapy,* in which the pulp is removed and replaced with inert material.

When plaque calcifies on the tooth surface, it is called *calculus (tartar).* Calculus in the gingival sulcus wedges the tooth and gum apart and allows bacterial invasion of the sulcus. This leads to *gingivitis,* or gum inflammation. Nearly everyone has gingivitis at some time. In some cases, bacteria spread from the sulcus into the alveolar bone and begin to dissolve it, producing *periodontal disease.* About 86% of people over age 70 have some degree of periodontal disease and many suffer tooth loss as a result. This accounts for 80% to 90% of adult tooth loss.

Most of a tooth consists of hard yellowish tissue called **dentin,** covered with **enamel** in the crown and **cement** in the root. Dentin and cement are living connective tissues with cells or cell processes embedded in a calcified matrix. Cells of the cement *(cementocytes)* are scattered more or less randomly and occupy tiny cavities similar to the lacunae of bone. Cells of the dentin *(odontoblasts)* line the pulp cavity and have slender processes that travel through tiny parallel tunnels in the dentin. Enamel is not a tissue but a cell-free secretion produced before the tooth erupts above the gum. Damaged dentin and cement can regenerate, but damaged enamel cannot—it must be artificially repaired.

Internally, a tooth has a dilated **pulp cavity** in the crown and a narrow **root canal** in the lower root. These spaces are occupied by **pulp**—a mass of loose connective tissue, blood and lymphatic vessels, and nerves. These nerves and vessels enter the tooth through a pore, the **apical foramen,** at the basal end of each root canal.

The meeting of the teeth when the mouth closes is called **occlusion** and the surfaces where they meet are called the **occlusal surfaces** (ah-CLUE-zul). The occlusal surfaces of the premolars and molars have rounded bumps called **cusps**—two on each of the premolars and four to five on the molars. Cusps of the upper and lower premolars and molars mesh when the jaws are closed and slide over each other as the jaw makes lateral chewing motions.

This grinds and tears food more effectively than if the occlusal surfaces were flat.

Teeth develop beneath the gums and **erupt** (emerge) in predictable order. Twenty **deciduous teeth** *(milk teeth* or *baby teeth)* erupt from the ages of 6 to 30 months, beginning with the incisors (fig. 25.6a). Between 6 and 25 years of age, these are replaced by the 32 **permanent teeth.** As a permanent tooth grows below a deciduous tooth **(fig. 25.8),** the root of the deciduous tooth dissolves and leaves little more than the crown by the time it falls out. The third molars *(wisdom teeth)* erupt around ages 17 to 25, if at all. Over the course of human evolution, the face became flatter and the jaws shorter, leaving little room for the third molars. Thus, they often remain below the gum and become *impacted*—so crowded against neighboring teeth and bone that they cannot erupt.

25.2b Mastication

Mastication (chewing) breaks food into pieces small enough to be swallowed and exposes more surface to the action of digestive enzymes. It is the first step in mechanical digestion. Mastication requires little thought because food stimulates oral receptors that trigger an automatic chewing reflex. The tongue, buccinator, and orbicularis oris muscles manipulate food and push it between the teeth. The masseter and temporalis muscles produce the up-and-down crushing action of the teeth, and the lateral and medial pterygoid muscles and masseters produce side-to-side grinding action (see fig. 10.11).

FIGURE 25.8 Permanent and Deciduous Teeth in a Child's Skull. This dissection shows erupted deciduous teeth and, deep to them and marked with asterisks, the permanent teeth waiting to erupt.

Rebecca Gray/Don Kincaid/McGraw-Hill Education

---

[7]*caries* = rottenness

25.2c Saliva and the Salivary Glands

Saliva moistens and cleanses the mouth, inhibits bacterial growth, dissolves molecules so they can stimulate the taste buds, digests a little starch and fat, and makes swallowing easier by binding the food particles into a soft mass (bolus) and lubricating it with mucus. It is a hypotonic solution of 97.0% to 99.5% water, a pH of 6.8 to 7.0, and the following solutes:

- **mucus,** which binds and lubricates the food bolus;
- **electrolytes,** salts of Na^+, K^+, Cl^-, phosphate, and bicarbonate;
- **lysozyme,** an enzyme that kills bacteria;
- **immunoglobulin A (IgA),** an antibacterial antibody;
- **salivary amylase,** an enzyme that begins starch digestion in the mouth; and
- **lingual lipase,** an enzyme that begins fat digestion in the mouth (but mainly after the food is swallowed).

The Salivary Glands

There are two kinds of salivary glands, intrinsic and extrinsic. The **intrinsic (minor) salivary glands** are an indefinite number of small glands dispersed amid the other oral tissues—*lingual glands* in the tongue, *labial glands* on the inside of the lips, *palatine glands* of the palate, and *buccal glands* on the inside of the cheeks. They secrete saliva at a fairly constant rate whether we're eating or not, but in relatively small amounts. This saliva contains lingual lipase and lysozyme and moistens the mouth when we're not eating.

The **extrinsic (major) salivary glands** are three pairs of larger, more discrete organs located outside of the oral mucosa. They are compound tubuloacinar glands (see fig. 5.31) with a tree-like duct system leading to the oral cavity **(fig. 25.9).** The secretory acini at the twig ends of the tree are in some cases purely mucous, in others purely serous, and in mixed acini, composed of both mucous and serous cells **(fig. 25.10).** Serous cells secrete a watery fluid rich in enzymes and electrolytes. The three pairs of extrinsic salivary glands are as follows:

1. The **parotid**[8] **glands** lie just beneath the skin anterior to the earlobes. The parotid duct passes superficially over the masseter muscle, pierces the buccinator muscle, and opens into the mouth opposite the second upper molar tooth. *Mumps* is a viral inflammation and swelling of the parotid glands.

2. The **submandibular glands** are located halfway along the body of the mandible, medial to its margin, just deep to the mylohyoid muscle. The submandibular duct empties into the mouth at a papilla on the side of the lingual frenulum, near the lower central incisors.

FIGURE 25.9 The Extrinsic Salivary Glands. Part of the mandible is omitted to expose the sublingual gland medial to it. **A&PR**

3. The **sublingual glands** are in the floor of the mouth. They have multiple ducts that empty into the mouth posterior to the papillae of the submandibular ducts.

Salivation

The extrinsic salivary glands secrete 1.0 to 1.5 L of saliva per day, mainly in response to food in the mouth. Salivation is controlled mainly by parasympathetic nerve fibers that originate in a group of **salivatory nuclei,** located in the brainstem near the junction of the pons and medulla oblongata. Preganglionic fibers travel from here by way of the facial and glossopharyngeal nerves to ganglia in and near the salivary glands; short postganglionic fibers then finish the path from there to the gland cells themselves (see fig. 15.7). Sympathetic nerve fibers also innervate the salivary glands, originating in the superior cervical ganglion of the sympathetic chain and following blood vessels to the glands (see fig. 15.4). Sympathetic stimulation, however, produces only slight increases in salivation, and may even inhibit it, as when fear or nervousness produces a dry mouth feeling.

---

[8]*par* = next to; *ot* = ear

FIGURE 25.10 Microscopic Anatomy of the Salivary Glands.
(a) Duct and acini of a generalized salivary gland with a mixture of mucous and serous cells. Serous cells often form crescent-shaped caps called serous demilunes over the ends of mucous acini.
(b) Histology of the sublingual salivary gland.

b: Dennis Strete/McGraw-Hill Education

The salivatory nuclei respond to a wide variety of stimuli: the feel of food in the mouth; the flavor of food sensed by the taste buds, olfactory receptors, and free nerve endings in the oral mucosa; and the aroma, sight, or thought of food even when food isn't actually present in the mouth. Irritants in the pharynx, esophagus, stomach, and upper duodenum—spicy foods, stomach acid, and toxins, for example—also stimulate salivation. This helps to dilute and wash away irritating substances.

Parasympathetic stimulation dilates the blood vessels to the salivary glands and increases blood flow. Cells of the gland acini filter water and electrolytes from the blood and add their own secretory products to the filtrate—amylase, lingual lipase, lysozyme, and mucus. Cells of the ducts modify the electrolyte composition of the saliva. Salivary amylase begins to digest starch as the food

is chewed; lingual lipase begins fat digestion to a slight extent; and the mucus binds the masticated food particles into an easily swallowed bolus. Without mucus, one must drink much more liquid to swallow food.

25.2d The Pharynx

The pharynx, described in section 22.1b, is a muscular funnel that connects the oral cavity to the esophagus and the nasal cavity to the larynx; thus, it is a point where the digestive and respiratory tracts intersect. It has a deep layer of longitudinally oriented skeletal muscle and a superficial layer of circular skeletal muscle. The circular muscle is divided into superior, middle, and inferior **pharyngeal constrictors,** which force food downward during swallowing. When food is not being swallowed, the inferior constrictor remains contracted to exclude air from the esophagus. This constriction is regarded as the **upper esophageal sphincter,** although it isn't an anatomical feature of the esophagus. It disappears at the time of death when the muscle relaxes. Thus, it is regarded as a *physiological sphincter* rather than a constant anatomical structure.

25.2e The Esophagus

The **esophagus** is a straight muscular tube 25 to 30 cm long (see figs. 25.1, 25.2). It begins at a level between vertebra C6 and the cricoid cartilage, inferior to the larynx and posterior to the trachea. After passing downward through the mediastinum, it penetrates the diaphragm at an opening called the *esophageal hiatus,* continues another 3 to 4 cm, and meets the stomach at the level of vertebra T7. Its opening into the stomach is called the **cardial orifice** (named for its proximity to the heart). Food pauses briefly at this point before entering the stomach because of a constriction called the **lower esophageal sphincter (LES).** The LES prevents stomach contents from regurgitating into the esophagus, thus protecting the esophageal mucosa from the erosive effect of stomach acid. "Heartburn" has nothing to do with the heart, but is the burning sensation produced by acid reflux into the esophagus.

The wall of the esophagus is organized into the tissue layers described earlier, with some regional specializations. The mucosa has a nonkeratinized stratified squamous epithelium. The submucosa contains **esophageal glands** that secrete lubricating mucus into the lumen. When the esophagus is empty, the mucosa and submucosa are deeply folded into longitudinal ridges, giving the lumen a starlike shape in cross section.

The muscularis externa is composed of skeletal muscle in the upper one-third of the esophagus, a mixture of skeletal and smooth muscle in the middle one-third, and only smooth muscle in the lower one-third.

Most of the esophagus is in the mediastinum. Here, it is covered with a connective tissue adventitia that merges into the adventitias of the trachea and thoracic aorta. The short segment below the diaphragm is partially covered by a serosa.

25.2f Swallowing

Swallowing, or **deglutition** (DEE-glu-TISH-un), is a complex action involving over 22 muscles in the mouth, pharynx, and esophagus, coordinated by the **swallowing center,** a pair of nuclei in the medulla oblongata. This center communicates with muscles of the pharynx and esophagus by way of the trigeminal, facial, glossopharyngeal, and hypoglossal nerves (cranial nerves V, VII, IX, and XII). Swallowing occurs in three phases **(fig. 25.11).**

1. The **oral phase** is under voluntary control. During chewing, the tongue collects food, presses it against the palate to form a bolus, and pushes it posteriorly. Food thus accumulates in the oropharynx in front of the blade of the epiglottis. When the bolus reaches a critical size, the epiglottis tips posteriorly and the bolus slides around it, through a space on each side, into the laryngopharynx.

2. The **pharyngeal phase** is involuntary. The soft palate and root of the tongue block food and drink from entering the nasal cavity or reentering the mouth. To prevent choking, breathing is automatically suspended, the infrahyoid muscles pull the larynx up to meet the epiglottis and cover the laryngeal opening, and the vocal cords adduct to close the airway. These actions also widen the upper esophagus to receive the food. The pharyngeal constrictors contract in order from superior to middle to inferior, driving the bolus downward into the esophagus.

3. The **esophageal phase** is a wave of involuntary contractions called **peristalsis,** controlled jointly by the brainstem swallowing center and the myenteric plexus in the esophageal wall. The bolus stimulates stretch receptors that feed into the plexus, which transmits signals to the muscularis externa above and below the bolus. The circular muscle layer above the bolus constricts and pushes the food downward. Below the bolus, the circular muscle relaxes while the longitudinal muscle contracts. The latter action pulls the wall of the esophagus slightly upward, making it a little shorter and dilating it to receive the descending food.

   When one is standing or sitting upright, most food and liquid drop through the esophagus by gravity faster than the peristaltic wave can keep up with them. Peristalsis, however, propels more solid food pieces and ensures that you can swallow regardless of the body's position—even standing on your head! Liquid normally reaches the stomach in 1 to 2 seconds and a food bolus in 4 to 8 seconds. As a bolus reaches the lower end of the esophagus, the lower esophageal sphincter relaxes to let it pass into the stomach.

1. **Oral phase.** The tongue forms a food bolus and pushes it into the laryngopharynx.

2. **Pharyngeal phase.** The palate, tongue, vocal cords, and epiglottis block the oral and nasal cavities and airway while pharyngeal constrictors push the bolus into the esophagus.

Esophagus

Stomach

3. **Esophageal phase.** Peristalsis drives the bolus downward, and relaxation of the lower esophageal sphincter admits it into the stomach.

FIGURE 25.11 Swallowing. See numbered steps in the text for further explanation. **APR**

❓ *What actions prevent the pharynx from forcing food back into the mouth or nose?*

Answer the following questions to test your understanding of the preceding section:

5. List as many functions of the tongue as you can.

6. Imagine a line from the root canal of a tooth to the adjacent bone of the mandible. Name the tissues, in order, through which this line would pass.

7. What is the difference in function and location between intrinsic and extrinsic salivary glands? Name the extrinsic salivary glands and describe their locations.

8. Identify at least two histological features of the esophagus that are especially tied to its role in swallowing.

9. Describe the mechanisms that prevent food from entering the nasal cavity and larynx during swallowing.

25.3 The Stomach APR

Expected Learning Outcomes

When you have completed this section, you should be able to

a. describe the gross and microscopic anatomy of the stomach;

b. state the function of each type of epithelial cell in the gastric mucosa;

c. identify the secretions of the stomach and state their functions;

d. explain how the stomach produces hydrochloric acid and pepsin;

e. describe the contractile responses of the stomach to food; and

f. describe the three phases of gastric function and how gastric activity is activated and inhibited.

The stomach is a muscular sac in the upper left abdominal cavity immediately inferior to the diaphragm. It functions primarily as a food storage organ, with an internal volume of about 50 mL when empty and 1.0 to 1.5 L after a typical meal. When extremely full, it may hold up to 4 L and extend nearly as far as the pelvis.

Well into the nineteenth century, authorities regarded the stomach as essentially a grinding chamber, fermentation vat, or cooking pot. Some even attributed digestion to a supernatural spirit in the stomach. We now know that it mechanically breaks up food particles, liquefies the food, and begins the chemical digestion of proteins and fat. This produces an acidic, soupy or pasty mixture of semidigested food called **chyme**[9] (pronounced "kime"). Most digestion occurs after the chyme passes on to the small intestine.

25.3a Gross Anatomy

The stomach is J-shaped **(fig. 25.12),** relatively vertical in tall people and more nearly horizontal in short people. It is divided into four regions: (1) The **cardial part (cardia)** is a small area within about 3 cm of the cardial orifice. (2) The **fundic region (fundus)** is the dome superior to the esophageal attachment. (3) The **body** is the greatest part distal to the cardial orifice. (4) The **pyloric part** is a slightly narrower pouch at the inferior end; it is subdivided into a funnel-like **antrum**[10] and a narrower **pyloric canal.** The latter terminates at the **pylorus,**[11] a narrow passage into the duodenum. The pylorus is surrounded by a thick ring of smooth muscle, the **pyloric sphincter,** which regulates the passage of chyme into the duodenum.

Between the esophagus and duodenum, the stomach has two margins called the **greater** and **lesser curvatures.** The greater curvature is the long way around, about 40 cm, along the inferolateral surface. The greater omentum, overhanging the small intestine, is suspended from the greater curvature. The lesser curvature is the shorter distance from esophagus to duodenum, about 10 cm, along the superomedial margin facing the liver. The lesser omentum spans the space between the liver and lesser curvature (see fig. 25.3).

25.3b Innervation and Circulation

The stomach receives parasympathetic nerve fibers from the vagus nerves and sympathetic fibers from the celiac ganglia (see table 15.2). It is supplied with blood by branches of the celiac trunk (see section 20.7f, fig. 20.34c). All blood drained from the stomach and intestines enters the hepatic portal circulation and filters through the liver before returning to the heart.

25.3c Microscopic Anatomy

The stomach wall has tissue layers similar to those of the esophagus, with some variations. The mucosa is covered with a simple columnar glandular epithelium **(fig. 25.13).** The apical regions of its cells are filled with mucin; after it is secreted, mucin swells with water and becomes mucus. The mucosa and submucosa are flat and smooth when the stomach is full, but as it empties, these layers form conspicuous longitudinal wrinkles called **gastric rugae**[12] (ROO-jee). The lamina propria is almost entirely occupied by tubular glands, to be described shortly. The muscularis externa has three layers, rather than two: outer longitudinal, middle circular, and inner oblique layers (fig. 25.12).

---

[9]*chyme* = juice

[10]*antrum* = cavity
[11]*pylorus* = gatekeeper
[12]*rugae* = folds, creases

(a)

(b)

FIGURE 25.12 The Stomach. (a) Gross anatomy. (b) Photograph of the internal surface. **APR**

b: Rebecca Gray/McGraw-Hill Education

? *How does the muscularis externa of the stomach differ from that of the esophagus?*

Lumen of stomach

Epithelium

Gastric pit

Pyloric gland

Gastric gland

Mucosa

Mucous
neck cell

Mucous cell

Parietal cell

G cell

Lamina
propria

Submucosa

Muscularis
mucosae

Chief cell

Lymphatic
vessel

Lymphatic
nodule

Artery

Vein

(a) Stomach wall

Mucous neck cell

Parietal cell

Mucous cell

Chief cell

G cell

(b) Pyloric gland

(c) Gastric gland

(d) Gastric pits

FIGURE 25.13 Mucosa of the Stomach Wall. (a) Layers of the mucosa. (b) A pyloric gland from the inferior end of the stomach. Note the absence of chief cells and relatively few parietal cells. (c) A gastric gland, the most widespread type in the stomach. (d) Openings of the gastric pits into the stomach, surrounded by the rounded apical surfaces of the columnar epithelial cells of the mucosa (SEM). **APR**

d: Steve Gschmeissner/Science Source

▶▶▶**APPLY WHAT YOU KNOW**

Contrast the epithelium of the esophagus with that of the stomach. Why is each epithelial type best suited to the function of its respective organ?

The gastric mucosa is pocked with depressions called **gastric pits** lined with the same columnar epithelium as the surface (fig. 25.13). Two or three tubular glands open into the bottom of each gastric pit and span the rest of the lamina propria. In the cardial and pyloric regions, they are called **cardial glands** and **pyloric glands,** respectively. In the rest of the stomach, they are called **gastric glands.** These three glands differ in cellular composition, but collectively have the following cell types:

- **Mucous cells,** which secrete mucus, predominate in the cardial and pyloric glands. In gastric glands, they are called *mucous neck cells* and are concentrated in the narrow *neck* of the gland, where it opens into the gastric pit.

- **Regenerative (stem) cells,** found in the base of the pit and neck of the gland, divide rapidly and produce a continual supply of new cells. Newly generated cells migrate upward to the gastric surface as well as downward into the glands to replace cells that die.

- **Parietal cells,** found mostly in the upper half of the gland, secrete *hydrochloric acid*, *intrinsic factor,* and an appetite-stimulating hormone called *ghrelin*. They are found mostly in the gastric glands, but a few occur in the pyloric glands.

- **Chief cells,** so named because they are the most numerous, secrete the enzymes *gastric lipase* and *pepsinogen*. They dominate the lower half of the gastric glands but are absent from cardial and pyloric glands.

- **Enteroendocrine cells,** concentrated especially in the lower end of a gland, secrete hormones and paracrine messengers that regulate digestion. They occur in all regions of the stomach, but are most abundant in the gastric and pyloric glands. There are at least eight kinds of enteroendocrine cells in the stomach, each of which produces a different chemical messenger.

In general, the cardial and pyloric glands secrete mainly mucus; acid and enzyme secretion occur predominantly in the gastric glands; and hormones are secreted throughout the stomach.

25.3d Gastric Secretions

The gastric glands produce 2 to 3 L of **gastric juice** per day, composed mainly of water, hydrochloric acid, and pepsin.

Hydrochloric Acid

Gastric juice has a high concentration of hydrochloric acid (HCl) and a pH as low as 0.8. This is more acidic than the battery acid of a car and low enough to dissolve tooth enamel or cause a chemical burn on the skin. How, then, does the stomach produce and tolerate such acidity?

FIGURE 25.14 The Mechanism of Hydrochloric Acid Secretion. The parietal cell combines water with CO_2 from the blood to form carbonic acid (bottom line of figure). Carbonic acid breaks down into bicarbonate ion (HCO_3^-) and hydrogen ion (H^+). HCO_3^- returns to the blood. In exchange, Cl^- enters the lumen with H^+ and the two form hydrochloric acid.

❓ *What role does active transport play in this process?*

The reactions that produce HCl (**fig. 25.14**) may seem familiar by now because they have been discussed in previous chapters—most recently in connection with renal excretion of H^+ in section 24.3c. Parietal cells contain carbonic anhydrase (CAH), which catalyzes the first step in the following reaction:

$$CO_2 + H_2O \xrightarrow{CAH} H_2CO_3 \longrightarrow HCO_3^- + H^+.$$

Parietal cells pump the H^+ from this reaction into the lumen of a gastric gland by an active-transport protein similar to the Na^+–K^+ pump, called **H^+–K^+ ATPase.** This is an antiport that uses the energy of ATP to pump H^+ out of the cell and K^+ into it. HCl secretion doesn't affect the pH within the parietal cell because H^+ is pumped out as fast as it is generated. The bicarbonate ions (HCO_3^-) are exchanged for chloride ions (Cl^-) from the blood plasma—the same *chloride-shift* process that occurs in the renal tubules and red blood cells—and the Cl^- is pumped into the lumen of the gastric gland to join the H^+.

Thus, HCl accumulates in the stomach while bicarbonate ions accumulate in the blood. Because of the bicarbonate, blood leaving the stomach has a higher pH when digestion is occurring than when the stomach is empty. This high-pH blood is called the *alkaline tide*.

Stomach acid has several functions: (1) It activates the enzymes pepsin and lingual lipase, as discussed shortly. (2) It breaks up connective tissues and plant cell walls, helping to liquefy food and form chyme. (3) It converts ingested ferric ions (Fe^{3+}) to ferrous ions (Fe^{2+}), a form of iron that can be absorbed and used for hemoglobin synthesis. (4) It contributes to innate immunity by destroying most ingested pathogens.

Pepsin

Several digestive enzymes are secreted as inactive proteins called **zymogens** and then converted to active enzymes by the removal of some of their amino acids. In the stomach, chief cells secrete a zymogen called **pepsinogen.** Hydrochloric acid removes some of its amino acids and converts it to **pepsin.** Since pepsin digests protein, and pepsinogen itself is a protein, pepsin has an *autocatalytic* effect—as some pepsin is formed, it converts pepsinogen into more pepsin **(fig. 25.15).** The ultimate function of pepsin, however, is to digest dietary proteins to shorter peptide chains, which then pass to the small intestine, where their digestion is completed.

Gastric Lipase

The chief cells also secrete **gastric lipase.** This enzyme and lingual lipase, which plays a minor role, digest 10% to 15% of the dietary fat in the stomach. The remainder is digested in the small intestine.

Intrinsic Factor

Parietal cells secrete a glycoprotein called **intrinsic factor** that is essential for the absorption of vitamin B_{12} by the small intestine. Intrinsic factor binds vitamin B_{12} and the intestinal cells then absorb this complex by receptor-mediated endocytosis. Without vitamin B_{12}, hemoglobin cannot be synthesized and anemia develops. The secretion of intrinsic factor is the only indispensable function of the stomach. Digestion can continue following removal of the stomach (*gastrectomy*), but a person usually must then take

vitamin B_{12} by injection, or vitamin B_{12} and intrinsic factor orally. As we age, the gastric mucosa atrophies and less intrinsic factor is secreted, increasing the risk of anemia. Some people, especially in old age, develop *pernicious anemia,* the result of an autoimmune disease that destroys gastric mucosa and reduces intrinsic factor secretion.

Chemical Messengers

The gastric and pyloric glands have various kinds of enteroendocrine cells that collectively produce as many as 20 hormones and paracrine signals. Several of them are peptides produced in both the digestive tract and the central nervous system; thus, they are called **gut–brain peptides.** These include substance P, vasoactive intestinal peptide (VIP), secretin, gastric inhibitory peptide (GIP), cholecystokinin, and neuropeptide Y (NPY). The functions of some of these peptides in digestion will be explained in the following sections, and their roles in appetite regulation are discussed in the next chapter (section 26.1b).

Several of the gastric secretions are summarized in **table 25.1.** Some of the functions listed there are explained later in the chapter.

25.3e Gastric Motility

As you begin to swallow, food stimulates mechanoreceptors in the pharynx and they transmit signals to the medulla oblongata. The medulla relays signals to the stomach by way of the vagus nerves. The stomach reacts with a **receptive-relaxation response,** in

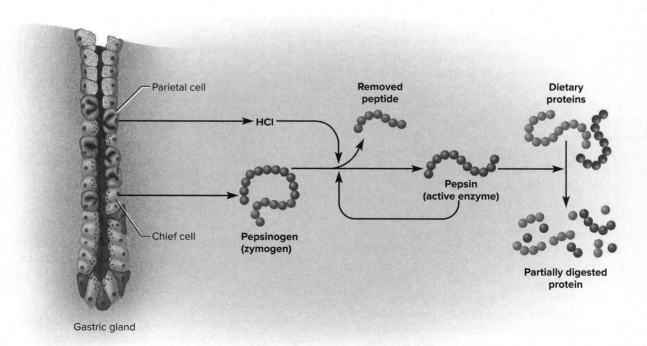

FIGURE 25.15 The Production and Action of Pepsin. The chief cells secrete pepsinogen and the parietal cells secrete HCl. HCl removes some of the amino acids from pepsinogen and converts it to pepsin. Pepsin catalyzes the production of more pepsin (an autocatalytic effect), as well as partially digesting dietary protein.

TABLE 25.1	Major Secretions of the Gastric Glands	
Secretory Cells	**Secretion**	**Function**
Mucous neck cells	Mucus	Protects mucosa from HCl and enzymes
Parietal cells	Hydrochloric acid	Activates pepsin and lingual lipase; helps liquefy food; reduces dietary iron to usable form (Fe^{2+}); destroys ingested pathogens
	Intrinsic factor	Enables small intestine to absorb vitamin B_{12}
Chief cells	Pepsinogen	Converted to pepsin, which digests protein
	Gastric lipase	Digests fat
Enteroendocrine cells	Gastrin	Stimulates gastric glands to secrete HCl and enzymes; stimulates intestinal motility; relaxes ileal papilla
	Serotonin	Stimulates gastric motility
	Histamine	Stimulates HCl secretion
	Somatostatin	Inhibits gastric secretion and motility; delays emptying of stomach; inhibits secretion by pancreas; inhibits gallbladder contraction and bile secretion; reduces blood circulation and nutrient absorption in small intestine
	Gut–brain peptides	Various roles in short- and long-term appetite regulation and energy balance

which it briefly resists stretching but then relaxes to accommodate the arriving food.

Soon, the stomach shows a rhythm of peristaltic contractions governed by a *basic electrical rhythm* set off by **enteric pacemaker cells** in the muscularis externa. The upper stomach (fundus) doesn't participate in these, but below the fundus, around midbody, a tight ring of constriction appears about every 20 seconds and progresses downward toward the antrum, becoming stronger as it goes. After food has been in the stomach for 30 minutes or so, these contractions become especially intense. They churn the food, mix it with gastric juice, and promote its physical breakup and chemical digestion.

The antrum holds about 30 mL of chyme at a time. The muscularis is thickest here, and acts as a strong *antral pump* that breaks up semidigested food into smaller particles and prepares it for the small intestine. A leading wave of antral contraction proceeds to the pyloric valve and closes it tightly. A trailing wave then comes along, churning and breaking up the chyme. Chyme jets backward through the trailing constriction into the gastric body, where it awaits the next wave of contraction to drive it down again. The repetitive downward propulsion and reverse jetting of chyme break the food into smaller and smaller particles. Food particles are not allowed to pass into the duodenum until they are reduced to 1 to 7 mm in size, and only about 3 mL of chyme is squirted into the duodenum at a time. Allowing only small amounts into the duodenum enables the duodenum to neutralize the stomach acid and digest nutrients little by little. If the duodenum becomes overfilled, it inhibits gastric motility and postpones receiving more chyme; the mechanism for this is discussed later. A typical meal is emptied from the stomach in about 4 hours, but it takes less time if the meal is more liquid, longer if the stomach contents are more acidic, and as long as 6 hours if the meal is high in fat.

25.3f Vomiting

Vomiting is the forceful ejection of stomach and intestinal contents (chyme) from the mouth. It involves multiple muscular actions integrated by the **emetic**[13] **center** of the medulla oblongata. Vomiting is commonly induced by overstretching of the stomach or duodenum; chemical irritants such as alcohol and bacterial toxins; visceral trauma (especially to the pelvic organs); intense pain; or psychological and sensory stimuli that activate the emetic center (thus, vomiting can be induced by repugnant sights, smells, and thoughts).

Vomiting is usually preceded by nausea and retching. In **retching,** thoracic expansion and abdominal contraction create a pressure difference that dilates the esophagus. The lower esophageal sphincter relaxes while the stomach and duodenum contract spasmodically. Chyme enters the esophagus but then drops back into the stomach as the muscles relax; it doesn't get past the upper esophageal sphincter. Retching is often accompanied by tachycardia, profuse salivation, and sweating. Vomiting occurs when abdominal contraction and rising thoracic pressure force the upper esophageal sphincter open, the esophagus and body of the stomach relax, and chyme is driven out of the stomach and mouth by strong abdominal contraction combined with reverse peristalsis of the gastric antrum and duodenum. **Projectile vomiting** is sudden vomiting with no prior nausea or retching. It may be caused by neurological lesions but is also common in infants after feeding.

Chronic vomiting can cause dangerous fluid, electrolyte, and acid–base imbalances. In cases of frequent vomiting, as in the eating disorder *bulimia,* the tooth enamel becomes severely eroded

---
[13]*emet* = vomiting

by the hydrochloric acid in the chyme. Aspiration (inhalation) of this acid is very destructive to the respiratory tract. Many have died from aspiration of vomit when they were unconscious or semiconscious. This is the reason that surgical anesthesia, which may induce nausea, must be preceded by fasting until the stomach and small intestine are empty.

25.3g Digestion and Absorption

Salivary and gastric enzymes partially digest protein and lesser amounts of starch and fat in the stomach, but most digestion and nearly all nutrient absorption occur after the chyme passes into the small intestine. The stomach doesn't absorb any significant amount of nutrients but does absorb aspirin and some lipid-soluble drugs. Alcohol is absorbed mainly by the small intestine, so its intoxicating effect depends partly on how rapidly the stomach is emptied.

25.3h Protection of the Stomach

One might think that the stomach would be its own worst enemy; it is, after all, made of meat. Some people enjoy haggis and tripe, dishes made from animal stomachs, and have no difficulty digesting those. Why, then, doesn't the human stomach digest itself? The answer is that the living stomach is protected

in three ways from the harsh acidic and enzymatic environment it creates:

1. **Mucous coat.** The thick, highly alkaline mucus resists the action of acid and enzymes.
2. **Tight junctions.** The epithelial cells are joined by tight junctions that prevent gastric juice from seeping between them and digesting the connective tissue below.
3. **Epithelial cell replacement.** In spite of these other protections, the stomach's epithelial cells live only 3 to 6 days and are then sloughed off into the chyme and digested with the food. They are replaced just as rapidly, however, by cell division in the gastric pits.

The breakdown of these protective mechanisms can result in inflammation and peptic ulcer (see Deeper Insight 25.2).

25.3i Regulation of Gastric Function

The nervous and endocrine systems collaborate to increase gastric secretion and motility when food is eaten and to suppress them as the stomach empties. Gastric activity is divided into three stages called the *cephalic, gastric,* and *intestinal phases,* based on whether the stomach is being controlled by the brain, by itself,

DEEPER INSIGHT 25.2

CLINICAL APPLICATION

Peptic Ulcer

Inflammation of the stomach, called *gastritis,* can lead to a *peptic ulcer* as pepsin and hydrochloric acid erode the stomach wall **(fig. 25.16)**. Peptic ulcers occur even more commonly in the duodenum and occasionally in the esophagus. If untreated, they can perforate the organ and cause fatal hemorrhaging, or perforate the organ and cause peritonitis—both potentially fatal. Most such fatalities occur in people over age 65.

There is no evidence to support the popular belief that peptic ulcers result from psychological stress. Hypersecretion of acid and pepsin is sometimes involved, but even normal secretion can cause ulceration if the mucosal defense is compromised by other causes. Most ulcers involve an acid-resistant bacterium, *Helicobacter pylori,* that invades the mucosa of the stomach and duodenum and opens the way to chemical damage to the tissue. Other risk factors include smoking and the use of aspirin and other nonsteroidal anti-inflammatory drugs (NSAIDs). NSAIDs suppress the synthesis of prostaglandins, which normally stimulate the secretion of protective mucus and acid-neutralizing bicarbonate.

At one time, the most widely prescribed drug in the United States was cimetidine (Tagamet), which was designed to treat peptic ulcers by reducing acid secretion. Histamine stimulates acid secretion by binding to sites on the parietal cells called H_2 *receptors;* cimetidine, an H_2 *blocker,* prevents this binding. Lately, however, ulcers have been treated more successfully with antibiotics against *Helicobacter* combined with bismuth suspensions such as Pepto-Bismol. This is a much shorter and less expensive course of treatment and permanently cures about 90% of peptic ulcers, as compared with a cure rate of only 20% to 30% for H_2 blockers.

(a) Normal   **(b) Peptic ulcer**

FIGURE 25.16 Endoscopic Views of the Gastroesophageal Junction. The esophagus can be seen opening into the cardial stomach. (a) A view of the cardial orifice from above, showing a healthy esophageal mucosa. The small white spots are reflections of light from the endoscope. (b) A bleeding peptic ulcer. A peptic ulcer typically has an oval shape and yellow-white color. Here the yellowish floor of the ulcer is partially obscured by black blood clots, and fresh blood is visible around the margin of the ulcer.

a, b: CNRI/Science Source

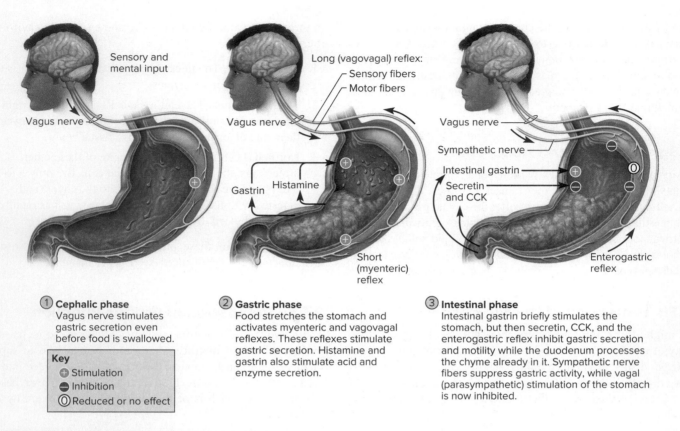

① **Cephalic phase**
Vagus nerve stimulates gastric secretion even before food is swallowed.

Key
⊕ Stimulation
⊖ Inhibition
Ⓞ Reduced or no effect

② **Gastric phase**
Food stretches the stomach and activates myenteric and vagovagal reflexes. These reflexes stimulate gastric secretion. Histamine and gastrin also stimulate acid and enzyme secretion.

③ **Intestinal phase**
Intestinal gastrin briefly stimulates the stomach, but then secretin, CCK, and the enterogastric reflex inhibit gastric secretion and motility while the duodenum processes the chyme already in it. Sympathetic nerve fibers suppress gastric activity, while vagal (parasympathetic) stimulation of the stomach is now inhibited.

FIGURE 25.17 Neural and Hormonal Control of the Stomach.

or by the small intestine, respectively **(fig. 25.17)**. These phases overlap and all three can occur simultaneously.

The Cephalic Phase

The **cephalic phase** is the stage in which the stomach responds to the mere sight, smell, taste, or thought of food, even before food enters the stomach. These sensory and mental inputs converge on the hypothalamus, which relays signals to the medulla oblongata. Vagus nerve fibers from the medulla stimulate the enteric nervous system of the stomach, which, in turn, stimulates secretion by the parietal cells (acid) and G cells (gastrin). About 40% of the stomach's acid secretion occurs in the cephalic phase. This prepares the stomach to receive and process food.

The Gastric Phase

The **gastric phase** is a period in which swallowed food and semidigested protein (peptides and amino acids) activate gastric activity. About one-half of acid secretion and two-thirds of total gastric secretion occur during this phase. Ingested food stimulates gastric activity in two ways: by stretching the stomach and by raising the pH of its contents. Stretch activates two reflexes: a short reflex mediated through the myenteric plexus and a long reflex mediated through the vagus nerves and brainstem.

Gastric secretion is stimulated chiefly by three chemicals: acetylcholine (ACh), histamine, and gastrin. ACh is secreted by

parasympathetic nerve fibers of both the short and long reflex pathways. Histamine is a paracrine secretion from enteroendocrine cells in the gastric glands. **Gastrin** is a hormone produced by enteroendocrine **G cells** in the pyloric glands. All three of these signals stimulate parietal cells to secrete hydrochloric acid and intrinsic factor. The chief cells secrete pepsinogen in response to gastrin and especially ACh, and ACh also stimulates mucus secretion.

As dietary protein is digested, it breaks down into smaller peptides and amino acids, which directly stimulate the G cells to secrete even more gastrin—a positive feedback loop that accelerates protein digestion **(fig. 25.18)**. Small peptides also buffer stomach acid so the pH doesn't fall excessively low. But as digestion continues and these peptides are emptied from the stomach, the pH drops lower and lower. Below pH of 2, stomach acid inhibits the parietal cells and G cells—a negative feedback loop that winds down the gastric phase as the need for pepsin and HCl declines.

The Intestinal Phase

The **intestinal phase** begins when chyme starts arriving in the duodenum; it enables the duodenum to control the rate of gastric emptying so the duodenum has time to process the chyme and is not overwhelmed by a sudden overload. This phase is mediated by both hormonal and nervous reflexes.

FIGURE 25.18 Feedback Control of Gastric Secretion. Note the positive feedback loop: Pepsin digests dietary protein; partially digested protein and ingested food raise the pH in the stomach; the elevated pH stimulates G cells to secrete gastrin; gastrin stimulates the gastric glands to secrete more pepsinogen and hydrochloric acid (HCl); pepsin production increases; and food digestion is accelerated. In addition to the feedback mechanism shown here, acetylcholine from parasympathetic fibers of the vagus nerves and histamine from the G cells of the gastric glands also powerfully stimulate pepsinogen and HCl secretion.

Initially, the duodenum enhances gastric secretion. Stretching of the duodenum accentuates vagovagal reflexes that stimulate the stomach, and peptides and amino acids in the chyme stimulate G cells of the duodenum to secrete **intestinal gastrin,** which further stimulates the stomach.

Soon, however, the acid and semidigested fats in the duodenum trigger the **enterogastric reflex**—the duodenum sends inhibitory signals to the stomach by way of the enteric nervous system, and sends signals to the medulla that (1) inhibit the vagal nuclei, thus reducing vagal stimulation of the stomach; and (2) stimulate sympathetic neurons, which send inhibitory signals to the stomach. Chyme also stimulates duodenal enteroendocrine cells to release **secretin** and **cholecystokinin (CCK)**

(CO-leh-SIS-toe-KY-nin). They primarily stimulate the pancreas and gallbladder, but also suppress gastric secretion and motility. The effect of this is that gastrin secretion declines and the pyloric sphincter contracts tightly to limit the admission of more chyme into the duodenum. This gives the duodenum time to work on the chyme it has already received before being loaded with more.

The enteroendocrine cells also secrete **glucose-dependent insulinotropic peptide (GIP).** Originally called *gastric-inhibitory peptide* (the original source of the GIP abbreviation), it is no longer thought to have a significant effect on the stomach, but seems more concerned with stimulating insulin secretion in preparation for processing the nutrients about to be absorbed by the small intestine.

Answer the following questions to test your understanding of the preceding section:

10. Name four types of epithelial cells of the gastric and pyloric glands and state what each one secretes.

11. Explain how the gastric glands produce hydrochloric acid and why this produces an alkaline tide.

12. What positive feedback cycle can you identify in the formation and action of pepsin?

13. How does food in the duodenum inhibit motility and secretion in the stomach?

25.4 The Liver, Gallbladder, and Pancreas

Expected Learning Outcomes

When you have completed this section, you should be able to

a. describe the gross and microscopic anatomy of the liver, gallbladder, bile duct system, and pancreas;

b. describe the digestive secretions and functions of the liver, gallbladder, and pancreas; and

c. explain how hormones regulate secretion by the liver and pancreas.

The small intestine receives not only chyme from the stomach but also secretions from the liver and pancreas, which enter the digestive tract near the junction of the stomach and small intestine. These secretions are so important to the digestive processes of the small intestine that it is necessary to understand them before we move on to intestinal physiology.

25.4a The Liver A&PR

The liver **(fig. 25.19)** is a reddish brown gland located immediately inferior to the diaphragm, filling most of the right hypochondriac and epigastric regions. It is the body's largest gland, weighing about 1.4 kg (3 lb). It has a tremendous variety of functions, but only one of them, the secretion of bile, contributes to digestion. Others are discussed in the following chapter (see table 26.6), which provides a more thorough physiological basis for understanding nondigestive liver functions.

Gross Anatomy

The liver has four lobes called the right, left, quadrate, and caudate lobes. From an anterior view, we see only a large **right lobe** and smaller **left lobe.** They are separated from each other by the **falciform**[14] **ligament,** a sheet of mesentery that suspends the liver from the diaphragm and anterior abdominal wall. The **round ligament (ligamentum teres),** also visible anteriorly, is a fibrous remnant of the umbilical vein, which carries blood from the umbilical cord to the liver of a fetus.

From the inferior view, we also see a squarish **quadrate lobe** next to the gallbladder and a tail-like **caudate**[15] **lobe** posterior to that. An irregular opening between these lobes, the **porta hepatis,**[16] is a point of entry for the hepatic portal vein and proper hepatic artery and a point of exit for the bile passages, all of which travel in the lesser omentum. The gallbladder adheres to a depression on the inferior surface of the liver between the right and quadrate lobes. The posterior aspect of the liver has a deep sulcus that accommodates the inferior vena cava. The superior surface has a *bare area* where it is attached to the diaphragm. The rest of the liver is covered by a serosa.

Microscopic Anatomy

The interior of the liver is filled with innumerable cylinders called **hepatic lobules,** about 2 mm long by 1 mm in diameter. A lobule consists of a **central vein** passing down its core, surrounded by radiating plates of cuboidal cells called **hepatocytes (fig. 25.20).** Imagine spreading a book wide open until its front and back covers touch. The pages of the book would fan out around the spine somewhat like the plates of hepatocytes fan out from the central vein of a liver lobule.

Each plate of hepatocytes is an epithelium one or two cells thick. The spaces between the plates are blood-filled channels called **hepatic sinusoids.** These are lined by a fenestrated endothelium that separates the hepatocytes from the blood cells, but allows blood plasma into the space between the hepatocytes and endothelium. The hepatocytes have a brush border of microvilli that project into this space. Blood filtering through the sinusoids comes directly from the stomach and intestines. After a meal, the hepatocytes absorb glucose, amino acids, iron, vitamins, and other nutrients from it for metabolism or storage. They also remove and degrade hormones, toxins, bile pigments, and drugs. At the same time, they secrete albumin, lipoproteins, clotting factors, angiotensinogen, and other products into the blood. Between meals, they break down stored glycogen and release glucose into the circulation. The sinusoids also contain phagocytic cells called **stellate macrophages,** which remove bacteria and debris from the blood.

The liver secretes bile into narrow channels, the **bile canaliculi,** between the back-to-back layers of hepatocytes within each plate. Bile passes from there into small **bile ductules** between the lobules, and these converge to ultimately form **right** and **left hepatic ducts.** They converge on the inferior side of the liver to form the **common hepatic duct.** A short distance farther on, this is joined by the **cystic duct** coming from the gallbladder **(fig. 25.21).** Their union forms the **bile duct,** which descends through the lesser omentum toward the duodenum. Near the duodenum, the bile duct joins the duct of the pancreas

---

[14]*falci* = sickle; *form* = shape

[15]*caud* = tail

[16]*porta* = gateway, entrance; *hepatis* = of the liver

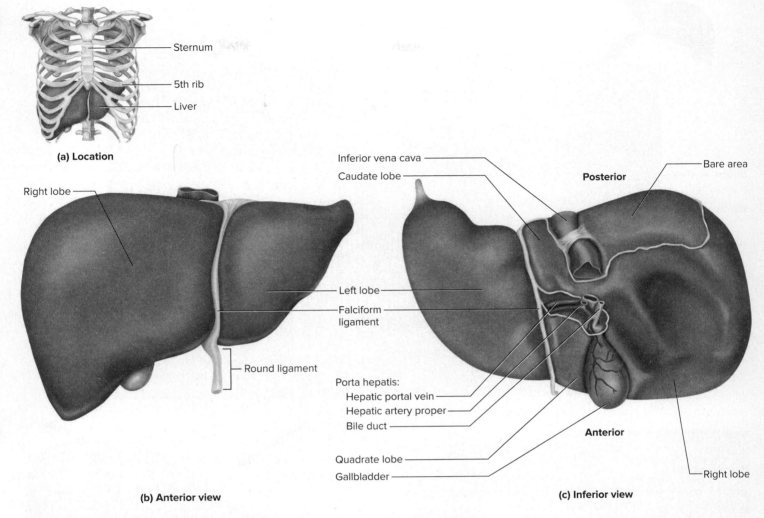

(a) Location

- Sternum
- 5th rib
- Liver

- Right lobe

- Inferior vena cava
- Caudate lobe
- **Posterior**
- Bare area

- Left lobe
- Falciform ligament

- Round ligament

Porta hepatis:
- Hepatic portal vein
- Hepatic artery proper
- Bile duct

Anterior

- Quadrate lobe
- Gallbladder

- Right lobe

(b) Anterior view

(c) Inferior view

FIGURE 25.19 Gross Anatomy of the Liver. (a) Location relative to thoracic cage. (b) Anterior view, showing right and left lobes. (c) Inferior view, showing all four lobes. **APR**

and forms an expanded chamber called the **hepatopancreatic ampulla.** The ampulla terminates at a fold of tissue, the **major duodenal papilla,** on the duodenal wall. This papilla contains a muscular **hepatopancreatic sphincter,** which regulates the passage of bile and pancreatic juice into the duodenum. Between meals, this sphincter is closed and prevents the release of bile into the intestine.

The hepatic lobules are separated by a sparse connective tissue stroma. In cross sections, the stroma is especially visible in the triangular areas where three or more lobules meet. Here there is often a **hepatic triad** consisting of a bile ductule and two blood vessels—branches of the hepatic artery proper and the hepatic portal vein.

Circulation

The liver receives blood from two sources: about 70% from the **hepatic portal vein** and 30% from the **hepatic arteries.** The hepatic portal vein receives blood from the stomach, intestines,

pancreas, and spleen, and carries it into the liver at the porta hepatis; see the *hepatic portal system* in section 20.7g. All nutrients absorbed by the small intestine reach the liver by this route except for lipids (transported in the lymphatic system). Arterial blood bound for the liver exits the aorta at the celiac trunk and follows the route shown in figure 20.34: celiac trunk → common hepatic artery → hepatic artery proper → right and left hepatic arteries, which enter the liver at the porta hepatis. These arteries deliver oxygen and other materials to the liver.

Both the hepatic portal vein and hepatic arteries travel to the liver through the lesser omentum. Within the liver, their finer branches meet each other in the spaces between the liver lobules, and both drain into the liver sinusoids. Hence, there is an unusual mixing of venous and arterial blood in the sinusoids. After processing by the hepatocytes, the blood collects in the central vein at the core of the lobule. Blood from the central veins ultimately converges on a few hepatic veins that exit the superior surface of the liver and empty into the nearby inferior vena cava.

FIGURE 25.20 Microscopic Anatomy of the Liver. (a) The hepatic lobules and their relationship to the blood vessels and bile tributaries. (b) Histological section of the liver. (c) A hepatic sinusoid.

b: Dennis Strete/McGraw-Hill Education

❓ *Identify two blood vessels in tables 20.7 and 20.8 that supply blood to the hepatic sinusoids.* **APR**

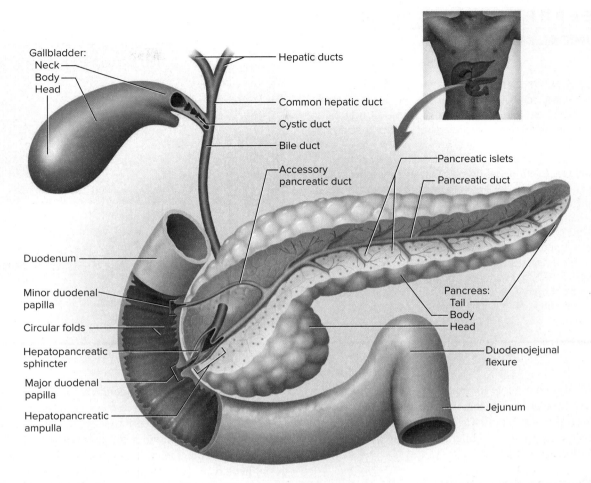

FIGURE 25.21 Gross Anatomy of the Gallbladder, Pancreas, and Bile Passages. The liver is omitted to show more clearly the gallbladder, which adheres to its inferior surface, and the hepatic ducts, which emerge from the liver tissue. **APR**

25.4b The Gallbladder and Bile

The **gallbladder** is a pear-shaped sac on the underside of the liver that serves to store and concentrate bile. It is about 10 cm long and internally lined by a highly folded mucosa with a simple columnar epithelium. Its head *(fundus)* usually projects slightly beyond the inferior margin of the liver. Its neck *(cervix)* leads into the cystic duct, which leads in turn to the bile duct.

The gallbladder doesn't secrete anything of its own except for a little mucus. It only receives, stores, and concentrates the bile produced by the liver. **Bile** is a light yellow-green color when secreted by the liver, but becomes a deep, intense green when concentrated in the gallbladder. It is a watery solution of minerals, cholesterol, neutral fats, phospholipids, bile pigments, bile acids, and lipid-transport vesicles called *micelles* (explained in section 25.6c). The principal bile pigment is **bilirubin,** derived from the decomposition of hemoglobin. Bacteria of the intestine metabolize bilirubin to a colorless product called **urobilinogen.** About half of this is reabsorbed in the small intestine and excreted by the kidneys. It is

converted to **urobilin,** which gives urine its yellow color. Urobilinogen remaining in the intestine is converted to **stercobilin,** from which feces get their brown color. In the absence of bile secretion, the feces are grayish white and marked with streaks of undigested fat *(acholic feces).* **Bile acids (bile salts)** are steroids synthesized from cholesterol. Bile acids, micelles, and **lecithin,** a phospholipid, aid in fat digestion and absorption, as discussed later. All other components of the bile are wastes destined for excretion in the feces. When these waste products become excessively concentrated, they may form gallstones (see Deeper Insight 25.3).

Bile gets into the gallbladder by first filling the bile duct, then overflowing into the gallbladder. Between meals, the gallbladder absorbs water and electrolytes from the bile and concentrates it by a factor of 5 to 20 times. The liver secretes about 500 to 1,000 mL of bile per day.

About 80% of the bile acids are reabsorbed in the ileum, the last portion of the small intestine, and returned to the liver, where the hepatocytes absorb and resecrete them. This route of secretion, reabsorption, and resecretion, called the *enterohepatic circulation,*

DEEPER INSIGHT 25.3

CLINICAL APPLICATION

Gallstones

Gallstones (biliary calculi) **(fig. 25.22)** are hard masses in the gallbladder or bile ducts, usually composed of cholesterol, calcium carbonate, and bilirubin. *Cholelithiasis,*[17] the formation or presence of gallstones, is most common in obese women over the age of 40, but certainly not limited to them; it occurs also in men and sometimes even children. It can occur when cholesterol becomes too concentrated to stay in solution and begins to precipitate out as crystals that steadily grow in size. The gallbladder may contain dozens of gallstones or even more, some over 1 cm in diameter. Gallstones can cause excruciating pain when they obstruct the bile ducts or when the gallbladder or bile ducts contract. When they block the flow of bile into the duodenum, they cause jaundice (yellowing of the skin due to bile pigment accumulation), poor fat digestion, inflammation of the pancreas *(pancreatitis)*, and impaired absorption of fat-soluble vitamins. Gallstones are now usually removed by minimally invasive laparoscopic surgery. Only the gallbladder can concentrate bile enough to cause gallstones, so it is often removed along with the gallstones to prevent recurrence.

FIGURE 25.22 Gallstones.
Biophoto Associates/Science Source

---

reuses the bile acids two or more times during the digestion of an average meal. The 20% of the bile that isn't reabsorbed is excreted in the feces. This is the body's only way of eliminating excess cholesterol. The liver synthesizes new bile acids from cholesterol to replace the quantity lost in the feces.

▶▶▶ APPLY WHAT YOU KNOW

Certain drugs designed to reduce blood cholesterol work by blocking the reabsorption of bile acids in the ileum. Explain why they would have this cholesterol-lowering effect.

25.4c The Pancreas

The pancreas (fig. 25.21) is a flattened, spongy, retroperitoneal gland pressed between the body wall and the greater curvature of the stomach. It measures 12 to 15 cm long and about 2.5 cm thick. It has a globose *head* encircled by the duodenum, a midportion called the *body,* and a blunt, tapered *tail* on the left. The pancreas is both an endocrine and exocrine gland. Its endocrine part is the pancreatic islets, which secrete insulin and glucagon (see section 17.3f). About 99% of the pancreas is exocrine tissue, which secretes 1,200 to 1,500 mL of **pancreatic juice** per day. Pancreatic islets are relatively concentrated in the tail of the pancreas, whereas the head is more exocrine. Over 90% of pancreatic cancers arise from the ducts of the exocrine portion *(ductal carcinomas)*, so cancer is most common in the head of the gland.

The cells of the secretory acini exhibit a high density of rough endoplasmic reticulum and secretory vesicles *(zymogen granules)* **(fig. 25.23).** The acini open into a system of branched ducts that eventually converge on the main **pancreatic duct.** This duct runs lengthwise through the middle of the gland and joins the bile duct at the hepatopancreatic ampulla. The hepatopancreatic sphincter thus controls the release of both bile and pancreatic juice into the duodenum. Most people, however, also have a smaller **accessory pancreatic duct** that branches from the main pancreatic duct and opens independently into the duodenum at the **minor duodenal papilla,** proximal to the major papilla. The accessory duct bypasses the sphincter and allows pancreatic juice to be released into the duodenum even when bile is held back.

Pancreatic juice is an alkaline mixture of water, enzymes, zymogens, sodium bicarbonate, and other electrolytes. The acini secrete the enzymes and zymogens, whereas the ducts secrete the sodium bicarbonate. The bicarbonate buffers HCl arriving from the stomach.

The pancreatic zymogens are **trypsinogen** (trip-SIN-oh-jen), **chymotrypsinogen** (KY-mo-trip-SIN-o-jen), and **procarboxypeptidase** (PRO-car-BOC-see-PEP-tih-dase). When trypsinogen is secreted into the intestinal lumen, it is converted to **trypsin** by **enteropeptidase,** an enzyme on the brush border of the duodenum **(fig. 25.24).** Trypsin is autocatalytic—it converts trypsinogen into still more trypsin. It also converts the other two zymogens into **chymotrypsin** and **carboxypeptidase,** in addition to its primary role of digesting dietary protein.

Other pancreatic enzymes include **pancreatic amylase,** which digests starch; **pancreatic lipase,** which digests fat; and **ribonuclease** and **deoxyribonuclease,** which digest RNA and DNA, respectively. Unlike the zymogens, these enzymes are not altered after secretion. They become fully active, however, only upon exposure to bile or ions in the intestinal lumen.

---

[17]*chole* = bile, gall; *lith* = stone; *iasis* = medical condition

(a)

Acinar cells

Zymogen granules

Stroma

Ducts

Exocrine acinar cells

Vein

(b)

50 μm

FIGURE 25.23 Microscopic Anatomy of the Pancreas. (a) An acinus. (b) Histological section of the exocrine tissue and some of the connective tissue stroma.

b: Dennis Strete/McGraw-Hill Education

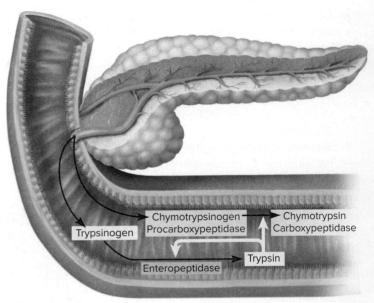

FIGURE 25.24 The Activation of Pancreatic Enzymes in the Small Intestine. The pancreas secretes trypsinogen, and enteropeptidase on the duodenal brush border converts it to trypsin. Trypsin not only digests dietary protein but also catalyzes the production of more trypsin and activates two other pancreatic zymogens—chymotrypsinogen and procarboxypeptidase.

TABLE 25.2	Exocrine Secretions of the Pancreas
Secretion	**Function**
Sodium Bicarbonate	Neutralizes HCl
Zymogens	Converted to active digestive enzymes after secretion
Trypsinogen	Becomes trypsin, which digests protein
Chymotrypsinogen	Becomes chymotrypsin, which digests protein
Procarboxypeptidase	Becomes carboxypeptidase, which hydrolyzes the terminal amino acid from the carboxyl (—COOH) end of small peptides
Enzymes	
Pancreatic amylase	Digests starch
Pancreatic lipase	Digests fat
Ribonuclease	Digests RNA
Deoxyribonuclease	Digests DNA

The exocrine secretions of the pancreas are summarized in **table 25.2.** Their specific digestive functions are explained later in more detail.

25.4d Regulation of Secretion

Three stimuli are chiefly responsible for the release of pancreatic juice and bile.

1. **Acetylcholine (ACh)** comes from the vagus nerves and enteric neurons. ACh stimulates the pancreatic acini to

secrete their enzymes even during the cephalic phase of gastric control, before food is swallowed. The enzymes remain stored in the pancreatic acini and ducts, however, in preparation for release later when chyme enters the duodenum.

2. **Cholecystokinin**[18] **(CCK)** is secreted by the mucosa of the duodenum and proximal jejunum (the next segment of the small intestine), primarily in response to fats in the small intestine. CCK also stimulates the pancreatic acini to secrete enzymes, but it is named for its strongly stimulatory effect on the gallbladder. It induces contractions of the gallbladder and relaxation of the hepatopancreatic sphincter, discharging bile into the duodenum.

3. **Secretin** is produced by the same regions of the small intestine, mainly in response to the acidity of chyme from the stomach. Secretin stimulates the ducts of both the liver and pancreas to secrete an abundant sodium bicarbonate solution. In the pancreas, this flushes the enzymes into the duodenum. Sodium bicarbonate buffers the hydrochloric acid arriving from the stomach, with the reaction

$$HCl + NaHCO_3 \longrightarrow NaCl + H_2CO_3 \text{ (carbonic acid)}.$$

The carbonic acid then breaks down to carbon dioxide and water. CO_2 is absorbed into the blood and ultimately exhaled. What is left in the small intestine, therefore, is salt water—NaCl and H_2O. Sodium bicarbonate is therefore important in protecting the intestinal mucosa from HCl as well as raising the intestinal pH to the level needed for activity of the pancreatic and intestinal digestive enzymes.

▶▶▶**APPLY WHAT YOU KNOW**

Draw a negative feedback loop showing how secretin influences duodenal pH.

BEFORE YOU GO ON

Answer the following questions to test your understanding of the preceding section:

14. What does the liver contribute to digestion?

15. Trace the pathway taken by bile acids from the liver and back. What is this pathway called?

16. Name two hormones, four enzymes, and one buffer secreted by the pancreas, and state the function of each.

17. What stimulates cholecystokinin (CCK) secretion, and how does CCK affect other parts of the digestive system?

25.5 The Small Intestine

Expected Learning Outcomes

When you have completed this section, you should be able to

a. describe the gross and microscopic anatomy of the small intestine;

b. state how the mucosa of the small intestine differs from that of the stomach, and explain the functional significance of the differences;

c. define *contact digestion* and describe where it occurs; and

d. describe the types of movement that occur in the small intestine.

Nearly all chemical digestion and nutrient absorption occur in the small intestine. To perform these roles efficiently and thoroughly, the small intestine is the longest part of the digestive tract—about 5 m long (range 3–7 m) in a living person; in the cadaver, where there is no muscle tone, it is up to 8 m long. The term *small* intestine refers not to its length but to its diameter—about 2.5 cm (1 in.).

25.5a Gross Anatomy

The small intestine is a coiled tube filling most of the abdominal cavity inferior to the stomach and liver. It is divided into three regions **(fig. 25.25):** the *duodenum, jejunum,* and *ileum.*

The **duodenum** (dew-ODD-eh-num, DEW-oh-DEE-num) constitutes the first 25 cm (10 in.). Its name refers to its length, about equal to the width of 12 fingers.[19] It begins at the pyloric valve, arcs around the head of the pancreas and passes to the left, and ends at a sharp bend called the **duodenojejunal flexure.**

---

[19]*duoden* = 12

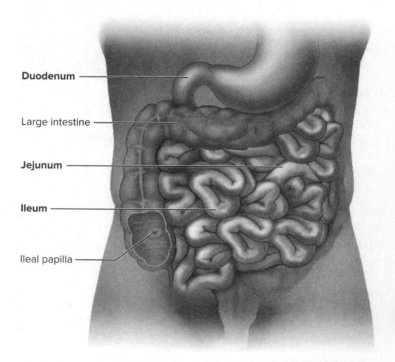

FIGURE 25.25 Gross Anatomy of the Small Intestine. APR

Labels: Duodenum; Large intestine; Jejunum; Ileum; Ileal papilla

---

[18]*chole* = bile; *cysto* = bladder (gallbladder); *kin* = action

Slightly distal to the pyloric valve, it exhibits the previously described wrinkles called the major and minor duodenal papillae, where it receives the pancreatic duct and accessory pancreatic duct, respectively. Along with the pancreas, most of the duodenum is retroperitoneal. It receives the stomach contents, pancreatic juice, and bile. Stomach acid is neutralized here, fats are physically broken up (emulsified) by the bile acids, pepsin is inactivated by the elevated pH, and pancreatic enzymes take over the job of chemical digestion.

The **jejunum** (jeh-JOO-num), by definition, is the first 40% of the small intestine beyond the duodenum—about 1.0 to 1.7 m in a living person. Its name refers to the fact that early anatomists typically found it to be empty.[20] The jejunum begins in the upper left quadrant of the abdomen but lies mostly within the umbilical region (defined in atlas A, fig. A.4). Its wall is thick and muscular, and it has an especially rich blood supply, which gives it a relatively red color. Most digestion and nutrient absorption occur here.

The **ileum**[21] forms the last 60% of the postduodenal small intestine (about 1.6 to 2.7 m). It occupies mainly the hypogastric region and part of the pelvic cavity. Compared with the jejunum, its wall is thinner, less muscular, less vascular, and has a paler pink color. On the side opposite from its mesenteric attachment, the ileum has prominent lymphatic nodules in clusters called **aggregated lymphoid nodules** (*Peyer*[22] *patches*), which are readily visible to the naked eye and become progressively larger approaching the large intestine.

The end of the small intestine is the **ileocecal junction** (ILL-ee-oh-SEE-cul), where the ileum joins the *cecum* of the large intestine. The muscularis of the ileum is thickened at this point to form a valve called the **ileal papilla,** which protrudes into the cecum. The papilla consists of a pair of lips with an opening, the **ileal orifice,** between them. The ileal papilla regulates the passage of food residue into the large intestine and prevents feces from backing up into the ileum.

25.5b Circulation

The small intestine receives nearly all of its blood supply from the *superior mesenteric artery,* which fans out through the mesentery to give rise to 12 to 15 *jejunal* and *ileal arteries* leading to the intestinal wall (see fig. 20.35a). Branches of these arteries travel through the submucosa and give rise to capillary beds in the villi (described shortly), where the blood picks up all absorbed nutrients except lipids. Blood from here converges on another fanlike array of mesenteric veins, which leave by way of the *superior mesenteric vein* (see fig. 20.37a). This joins the splenic vein and then flows into the hepatic portal system, headed for the liver with its load of nutrients (see section 20.7g).

25.5c Microscopic Anatomy

The tissue layers of the small intestine are reminiscent of those in the esophagus and stomach with modifications appropriate for nutrient digestion and absorption. The lumen is lined with simple columnar epithelium. The muscularis externa is notable for a thick inner circular layer and a thinner outer longitudinal layer. The jejunum and ileum are intraperitoneal and thus covered on all sides with a serosa, which is continuous with the complex, folded mesentery that suspends the small intestine from the posterior abdominal wall. Most of the duodenum is retroperitoneal and has a serosa only on its anterior surface; its other surfaces are covered by adventitia.

Effective digestion and absorption require the small intestine to have a large internal surface area. This is provided by its relatively great length and by three kinds of internal folds or projections: the *circular folds, villi,* and *microvilli.* If the mucosa were smooth, like the inside of a hose, it would have a surface area of about 0.3 to 0.5 m^2, but with these surface elaborations, its actual surface area is about 200 m^2—clearly a great advantage for nutrient absorption. The circular folds increase the surface area by a factor of 2 to 3, the villi by a factor of 10, and the microvilli by a factor of 20.

▶▶▶ APPLY WHAT YOU KNOW

The small intestine exhibits some of the same structural adaptations as the proximal convoluted tubule of the kidney (section 23.2d), and for the same reasons. Discuss what they have in common, the reasons for it, and how this relates to this book's theme of the unity of form and function.

Circular folds, the largest of these elaborations, are transverse to spiral ridges up to 1 cm high (see fig. 25.21). These involve only the mucosa and submucosa; they are not visible on the external surface, which is smooth. They slow the progress of the chyme and make it flow on a somewhat spiral path, which increases its contact with the mucosa and promotes more thorough mixing and nutrient absorption. Circular folds begin in the duodenum. In the jejunum, they are especially large, tall, and closely spaced. They become smaller and more sparse in the ileum. These changes are correlated with the relative amount of nutrient absorption occurring in each region. Circular folds are absent from the distal half of the ileum, but most nutrient absorption is completed by that point.

Villi (VIL-eye; singular, *villus*) are projections that give the inner lining of the intestine a fuzzy texture, like a terry cloth towel. They are about 0.5 to 1.0 mm high, with tongue- to fingerlike shapes **(fig. 25.26).** Villi are largest in the duodenum and become progressively smaller in more distal regions of the intestine. Villi are covered with two kinds of epithelial cells: columnar **enterocytes (absorptive cells)** and mucus-secreting **goblet cells.** Like epithelial cells of the stomach, those of the small intestine are joined by tight junctions that prevent digestive enzymes from seeping between them and eroding the underlying tissue.

The core of a villus is filled with areolar tissue of the lamina propria and contains an arteriole, blood capillaries, a venule, and a lymphatic capillary called a **lacteal** (LAC-tee-ul). The blood capillaries absorb most nutrients, but the lacteal absorbs most lipids.

---

[20]*jejun* = empty, dry
[21]from *eilos* = twisted
[22]Johann K. Peyer (1653–1712), Swiss anatomist

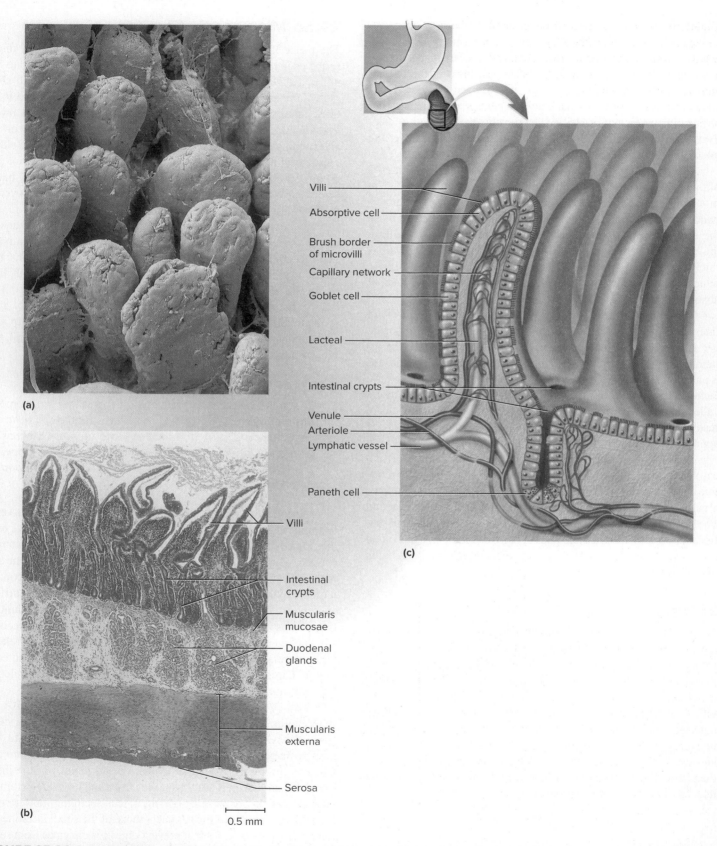

(a)

(b)

0.5 mm

Villi

Absorptive cell

Brush border
of microvilli

Capillary network

Goblet cell

Lacteal

Intestinal crypts

Venule

Arteriole

Lymphatic vessel

Paneth cell

(c)

Villi

Intestinal
crypts

Muscularis
mucosae

Duodenal
glands

Muscularis
externa

Serosa

FIGURE 25.26 Intestinal Villi. (a) Villi (SEM). Each villus is about 1 mm high. (b) Histological section of the duodenum showing villi, intestinal crypts, and duodenal glands. (c) Structure of a villus.

a: Meckes/Ottawa/Science Source; b: Dennis Strete/McGraw-Hill Education

Lipids give its contents a milky appearance for which the lacteal is named.[23] The core of the villus also has a few smooth muscle cells that contract periodically. This enhances mixing of the chyme in the intestinal lumen and milks lymph down the lacteal to larger lymphatics in the submucosa.

Microvilli are much smaller plasma membrane extensions, about 1 μm high, that form a fuzzy **brush border** on the surface of each enterocyte. In addition to increasing surface area, they contain **brush border enzymes** in the plasma membrane. These enzymes carry out some of the final stages of chemical digestion. They're not secreted into the lumen; instead, the chyme must contact the brush border for digestion to occur. This process, called **contact digestion,** is one reason why it is so important that intestinal contractions churn the chyme and ensure that it all contacts the mucosa.

On the floor of the small intestine, between the bases of the villi, there are numerous pores that open into tubular glands called **intestinal crypts.** These crypts, similar to the gastric glands, extend as far as the muscularis mucosae. In the upper half, they consist of enterocytes and goblet cells like those of the villi. The lower half is dominated by dividing stem cells. In its life span of 3 to 6 days, an epithelial cell migrates up the crypt to the tip of the villus, where it is sloughed off and digested. A few **Paneth**[24] **cells** are clustered at the base of each crypt. They secrete lysozyme, phospholipase, and defensins—defensive proteins that resist bacterial invasion of the mucosa.

The duodenum has prominent **duodenal glands** in the submucosa. They secrete an abundance of bicarbonate-rich mucus, which neutralizes stomach acid and shields the mucosa from its erosive effects. The mucus also contains signaling molecules that influence immune cells of the GI tract (dendritic cells) to tolerate food antigens and beneficial bacteria. Throughout the small intestine, the lamina propria and submucosa have a large population of lymphocytes that intercept pathogens before they can invade the bloodstream. In some places, such as the ileum, these are aggregated into conspicuous lymphatic nodules.

25.5d Intestinal Secretion

The intestinal crypts secrete 1 to 2 L of **intestinal juice** per day, especially in response to acid, hypertonic chyme, and distension of the intestine. This fluid has a pH of 7.4 to 7.8. It contains water and mucus but relatively little enzyme. Most enzymes that function in the small intestine are found in the brush border and pancreatic juice.

25.5e Intestinal Motility

Contractions of the small intestine serve three functions: (1) to mix chyme with intestinal juice, bile, and pancreatic juice, allowing these fluids to neutralize acid and digest nutrients more effectively; (2) to churn chyme and bring it into contact with the mucosa for contact digestion and nutrient absorption; and (3) to move residue toward the large intestine.

Segmentation is a movement in which stationary ringlike constrictions appear at several places along the intestine and then relax as new constrictions form elsewhere **(fig. 25.27a).** This is the most common type of intestinal contraction. Its effect is to knead or churn the contents. Pacemaker cells of the muscularis externa set the rhythm of segmentation, with contractions about 12 times per minute in the duodenum and 8 to 9 times per minute in the ileum. Since the contractions are less frequent distally, segmentation causes slow progression of the chyme toward the colon. The intensity (but not frequency) of contractions is modified by nervous and hormonal influences.

---

[23]*lact* = milk
[24]Josef Paneth (1857–90), Austrian physician

(a) Segmentation

(b) Peristalsis

FIGURE 25.27 Types of Intestinal Motility. (a) Segmentation, showing progressive stages from top to bottom. Circular constrictions of the intestine cut into the contents, churning and mixing them as represented by the progressive blending of the two colors. (b) The migrating motor complex of peristalsis, in which successive waves of peristalsis overlap each other. Each wave travels partway down the intestine and milks the contents toward the colon.

When most nutrients have been absorbed and little remains but undigested residue, segmentation declines and peristalsis begins. The duodenum secretes a hormone called **motilin** that triggers a peristaltic wave beginning in the duodenum. The wave travels 10 to 70 cm and dies out, only to be followed by another wave that begins a little farther down the tract than the first one **(fig. 25.27b).** These successive, overlapping waves of contraction are called a **migrating motor complex.** They milk the chyme toward the colon over a period of about 2 hours. A second complex then expels residue and bacteria from the small intestine, thereby helping to limit bacterial colonization. Refilling of the stomach at the next meal suppresses peristalsis and reactivates segmentation as new chyme enters the small intestine.

The ileal papilla is usually closed. Food in the stomach, however, triggers both the release of gastrin and the **gastroileal reflex,** both of which enhance segmentation in the ileum and relax the papilla. As the cecum fills with residue, the pressure pinches the papilla shut and prevents the reflux of cecal contents into the ileum.

Answer the following questions to test your understanding of the preceding section:

18. What three structures increase the absorptive surface area of the small intestine?

19. Sketch a villus and label its epithelium, brush border, lamina propria, blood capillaries, and lacteal.

20. Distinguish between segmentation and the migrating motor complex of the small intestine. How do these differ in function?

25.6 Chemical Digestion and Absorption

Expected Learning Outcomes

When you have completed this section, you should be able to

a. describe how each major class of nutrients is chemically digested, name the enzymes involved, and discuss the functional differences among these enzymes; and

b. describe how each type of nutrient is absorbed by the small intestine.

Chemical digestion and nutrient absorption are essentially finished by the time food residue leaves the small intestine and enters the cecum. But before going on to the functions of the large intestine, we trace each major class of nutrients—especially carbohydrates, proteins, and fats—from the mouth through the small intestine to see how it is chemically degraded and absorbed. The illustration series in this section depicts the details of the digestion and absorption of each major class of organic nutrients.

25.6a Carbohydrates

Most digestible dietary carbohydrate is starch. Cellulose is indigestible and is not considered here, although its importance as dietary fiber is discussed in section 26.1f. The amount of glycogen in the diet is negligible, but it is digested in the same manner as starch.

Starch is digested first to oligosaccharides up to eight glucose residues long, then into the disaccharide maltose, and finally to glucose, which is absorbed by the small intestine. The process begins in the mouth, where salivary amylase breaks starch down into shorter segments (oligosaccharides). Salivary amylase functions best at pH 6.8 to 7.0, typical of the oral cavity. It is quickly denatured upon contact with stomach acid, but it can digest starch for as long as 1 to 2 hours in the stomach as long as it is in the middle of a food mass and escapes contact with the acid. Amylase therefore works longer when the meal is larger, especially in the fundus, where gastric motility is weakest and a food bolus takes longer to break up. As acid, pepsin, and the churning contractions of the stomach break up the bolus, amylase is denatured; it doesn't function at a pH any lower than 4.5. Being a protein, amylase is then digested by pepsin along with the dietary proteins.

About 50% of the dietary starch is digested before it reaches the small intestine. Its digestion resumes in the small intestine when the chyme mixes with pancreatic amylase **(fig. 25.28).** Starch is entirely converted to oligosaccharides and maltose within 10 minutes. Its digestion is completed as the chyme contacts the brush border of the enterocytes. Two brush border enzymes, **dextrinase** and **glucoamylase,** hydrolyze oligosaccharides that are

FIGURE 25.28 Starch Digestion in the Small Intestine. Pancreatic amylase digests starch into maltose and small oligosaccharides. Brush border enzymes (maltase, dextrinase, and glucoamylase) digest these to glucose, which is absorbed by the epithelial cells.

three or more residues long, while **maltase** hydrolyzes maltose. The end product of all of these is glucose, which is then absorbed.

Maltose is also present in some foods, but the major dietary disaccharides are sucrose (cane sugar) and lactose (milk sugar). They are digested by the brush border enzymes **sucrase** and **lactase,** respectively, and the resulting monosaccharides are immediately absorbed (glucose and fructose from the former; glucose and galactose from the latter). In most of the world population, however, lactase production ceases or declines to a low level after age 4 and lactose becomes indigestible (see Deeper Insight 25.4).

DEEPER INSIGHT 25.4

CLINICAL APPLICATION

Lactose Intolerance

Homo sapiens is a strange species indeed—the only mammal that continues drinking milk after infancy, and even odder, drinks the milk of other species! To digest milk sugar (lactose) requires the enzyme lactase. Understandably, lactase production normally ceases after weaning in mammals, including most humans worldwide. Its cessation results in *lactose intolerance,* the inability to digest lactose. This occurs in about 90% of American blacks, who are predominantly descended from African tribes that historically didn't raise livestock for milk; 15% of American whites; 70% or more of Mediterraneans; and nearly all people of Asian descent, including those of us descended from the native migrants into the Americas.

If a person with lactose intolerance consumes milk, lactose passes undigested into the large intestine, increases the osmolarity of the intestinal contents, and causes colonic water retention and diarrhea. In addition, lactose fermentation by colonic bacteria produces gas, resulting in painful cramps and flatulence. People with lactose intolerance can, however, consume products such as yogurt and cheese, in which bacteria have broken down the lactose, and they can digest milk and ice cream with the aid of lactase tablets.

Lactase is encoded by a gene named *LCT* on chromosome 2. In most people, after the age of weaning, another gene on that chromosome, *MCM6,* shuts down *LCT.* Lactase production wanes and finally ceases, rendering the individual no longer able to digest lactose. Many people who *can* freely consume dairy products have a mutation in the *MCM6* gene that makes it unable to turn off *LCT.* This confers on them a state of *lactase persistence,* the continued production of lactase in adulthood, with a lasting ability to digest lactose. A DNA analysis of ancient European skeletons suggests that the *MCM6* mutation appeared about 4,300 years ago. Other mutations resulting in lactase persistence occurred independently in at least three populations of Africa. Today, lactase persistence is found only in people of western and northern Europe, Mongolia, a few pastoral tribes of Africa, and their descendants in the Americas and elsewhere. They have an ancestral history of milking domestic animals that goes back as much as 10,000 years.

The causes of lactase persistence are more complex than simple genetics, however. There are various degrees of lactose tolerance and intolerance, and environmental factors seem to play a role. Some people, such as immigrant populations, acquire tolerance over time through exposure to dairy products.

FIGURE 25.29 Monosaccharide Absorption by the Small Intestine. Glucose and galactose are absorbed by SGLT symports (violet) in the apical membranes of the absorptive cells (right). Glucose is also absorbed along with water through the paracellular route (between cells) by solvent drag. Fructose is absorbed separately by facilitated diffusion. Most fructose is converted to glucose within the epithelial cell. The monosaccharides pass through the basal membrane of the cell by facilitated diffusion (left) and are then absorbed by the blood capillaries of the villus.

The plasma membrane of the enterocytes has transport proteins that absorb monosaccharides as fast as they are produced by the foregoing enzymes **(fig. 25.29).** About 80% of the absorbed sugar is glucose, which is taken up by a sodium–glucose transporter (SGLT) like that of the kidney tubules (see section 23.4a). The glucose is subsequently transported out the base of the cell into the extracellular fluid (ECF). Sugar entering the ECF increases its osmolarity, and this draws water osmotically from the lumen of the intestine, through the now-leaky tight junctions between the epithelial cells. Water carries more glucose and other nutrients with it by *solvent drag,* much as it does in the kidney. After a high-carbohydrate meal, solvent drag absorbs two to three times as much glucose as the SGLT.

The SGLT also absorbs galactose, whereas fructose is absorbed by facilitated diffusion using a separate carrier that doesn't depend on Na^+. Inside the enterocyte, most fructose is converted to glucose. Glucose, galactose, and the small amount of remaining fructose are then transported out the base of the cell by facilitated diffusion and absorbed by the blood capillaries of the villus. The hepatic portal system delivers them to the liver; the next chapter (section 26.2) follows the fate of these sugars from there.

25.6b Proteins

The amino acids absorbed by the small intestine come from three sources: (1) dietary proteins, (2) digestive enzymes digested by each other, and (3) sloughed epithelial cells digested by these enzymes. Amino acids from the last two sources total about 30 g/day, compared with about 44 to 60 g/day from the diet.

Enzymes that digest proteins are called **proteases (peptidases).** They are absent from the saliva but first encountered in the stomach. Here, pepsin hydrolyzes any peptide bond between tyrosine and phenylalanine, thereby digesting 10% to 15% of the dietary protein into shorter polypeptides and a small amount of free amino acids **(fig. 25.30).** Pepsin has an optimal pH of 1.5 to 3.5, so it is inactivated when it passes into the duodenum and mixes with the alkaline pancreatic juice (pH 8).

In the small intestine, the pancreatic enzymes trypsin and chymotrypsin take over protein digestion by hydrolyzing polypeptides into even shorter oligopeptides. Finally, these are taken apart one amino acid at a time by three more enzymes: (1) **Carboxypeptidase** removes amino acids from the —COOH end of the chain; (2) **aminopeptidase** removes them from the —NH₂ end; and (3) **dipeptidase** splits dipeptides in the middle and releases the last two free amino acids. The last two of these are brush border enzymes, whereas carboxypeptidase is a pancreatic secretion.

Amino acid absorption is similar to that of monosaccharides. Enterocytes have several sodium-dependent amino acid cotransporters for different classes of amino acids. Dipeptides and tripeptides can also be absorbed, but they are hydrolyzed within the enterocytes before their amino acids are released to the bloodstream. At the basal surfaces of the cells, amino acids behave like the monosaccharides discussed previously—they leave the cell by facilitated diffusion, enter the capillaries of the villus, and are carried away in the hepatic portal circulation.

The absorptive cells of infants can take up intact proteins by pinocytosis and release them to the blood by exocytosis. This allows IgA from breast milk to pass into an infant's bloodstream and confer passive immunity from mother to infant. It has the disadvantage, however, that intact proteins entering the infant's blood are detected as foreign antigens and sometimes trigger food allergies. As the intestine matures, its ability to pinocytose protein declines but never completely ceases.

25.6c Lipids

The hydrophobic quality of lipids makes their digestion and absorption more complicated than that of carbohydrates and proteins **(fig. 25.31).** Fats are digested by enzymes called **lipases.** *Lingual lipase,* secreted by the intrinsic salivary glands of the tongue, digests a small amount of fat while food is still in the mouth, but becomes more active at the acidic pH of the stomach. Here it is joined by *gastric lipase,* which makes a much larger contribution to preduodenal fat digestion. About 10% to 15% of dietary fat is digested before the chyme passes on to the duodenum.

Being hydrophobic, ingested fat takes the form of large globules that, without further physical processing, could be attacked by these lipases only at their surface. This would result in rather slow, inefficient digestion. The stomach's vigorous *antral pumping*

described earlier, however, breaks the fat up into small droplets dispersed through the watery chyme—that is, it *emulsifies* the fat, exposing much more of its surface to enzymatic action. The resulting **emulsification droplets** are promptly passed on to the duodenum and coated by certain components of the bile—lecithin and bile acids. These agents have hydrophobic regions attracted to the surface of a fat droplet and hydrophilic regions attracted to the surrounding water. The agitation produced by intestinal segmentation breaks the fat up further into droplets as small as 1 μm, and the coating of lecithin and bile acids keeps it broken up, preventing the droplets from coalescing into larger globules.

There is enough *pancreatic lipase* in the small intestine after a meal to digest the average daily fat intake in as little as 1 or 2 minutes. When lipase acts on a triglyceride, it removes the first and third fatty acids from the glycerol backbone and usually leaves the middle one. The products of lipase action are therefore two free fatty acids (FFAs) and a monoglyceride. Being smaller than triglycerides, these are more soluble in the enterocyte plasma membrane and thus easier to absorb.

The absorption of lipids depends on minute droplets in the bile called **micelles**[25] (my-SELLS). Micelles, made in the liver, consist of 20 to 40 bile acid molecules aggregated with their hydrophilic side groups facing outward and their hydrophobic steroid rings facing inward. Bile phospholipids and cholesterol diffuse into the center of the micelle to form its core. The micelles pass down the bile duct into the duodenum, where they absorb fat-soluble vitamins, more cholesterol, and the fatty acids and monoglycerides produced by fat digestion. Because of their charged, hydrophilic surfaces, micelles remain suspended in water more easily than free lipids do. When they contact the enterocytes, they release their lipid cargo. Some of the lipids simply diffuse through the plasma membrane into the enterocytes, but these cells also have specific carrier proteins that facilitate their uptake. The micelles are reused, picking up another cargo of lipids and ferrying them to the enterocytes. Without micelles, the small intestine absorbs only about 40% to 50% of the dietary fat and almost no cholesterol.

Within the enterocytes, fatty acids and monoglycerides are transported into the smooth endoplasmic reticulum and resynthesized into triglycerides. The Golgi complex combines these with a small amount of cholesterol and coats the complex with a film of phospholipid and protein, forming droplets 75 to 1,200 nm in diameter called **chylomicrons**[26] (KY-lo-MY-crons). It packages chylomicrons into secretory vesicles that migrate to the basal surface of the cell and release their contents into the core of the villus. Although some free fatty acids enter the blood capillaries, chylomicrons are too large to penetrate the endothelium. They are taken up instead by the more porous lacteals into the lymph. This fatty, milk-white intestinal lymph, called *chyle* (pronounced "kile"), flows through larger and larger lymphatic vessels of the mesenteries, eventually passing through the cisterna chyli (see fig. 21.1) to the thoracic duct, then entering the bloodstream at the left subclavian vein. The further fate of dietary fat is described in the next chapter (section 26.3a).

---

[25]*mic* = grain, crumb; *elle* = little
[26]*chyl* = juice; *micr* = small

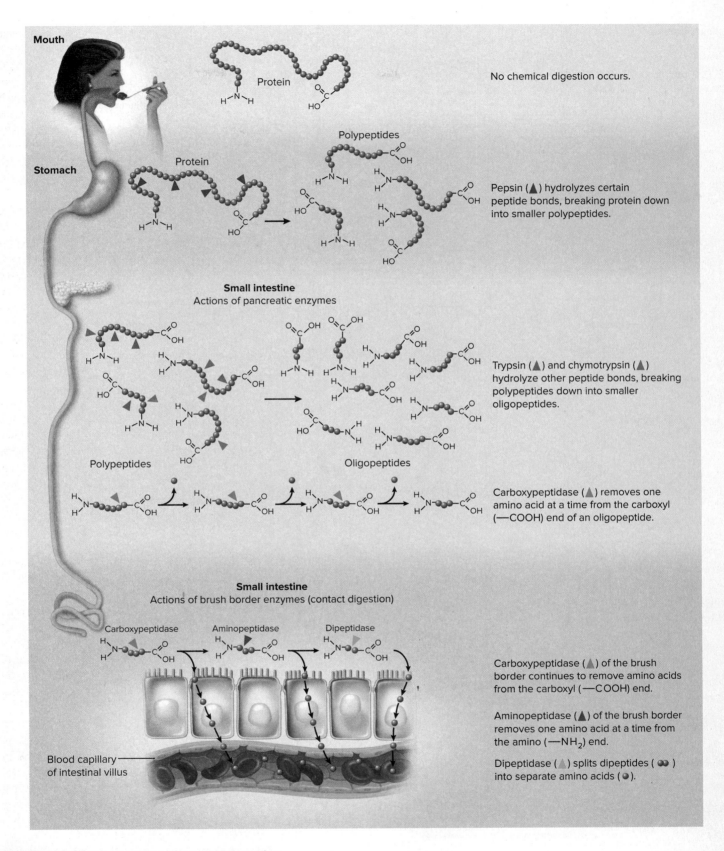

Mouth

Protein

No chemical digestion occurs.

Stomach

Protein

Polypeptides

Pepsin (▲) hydrolyzes certain peptide bonds, breaking protein down into smaller polypeptides.

Small intestine
Actions of pancreatic enzymes

Polypeptides

Oligopeptides

Trypsin (▲) and chymotrypsin (▲) hydrolyze other peptide bonds, breaking polypeptides down into smaller oligopeptides.

Carboxypeptidase (▲) removes one amino acid at a time from the carboxyl (—COOH) end of an oligopeptide.

Small intestine
Actions of brush border enzymes (contact digestion)

Carboxypeptidase   Aminopeptidase   Dipeptidase

Blood capillary of intestinal villus

Carboxypeptidase (▲) of the brush border continues to remove amino acids from the carboxyl (—COOH) end.

Aminopeptidase (▲) of the brush border removes one amino acid at a time from the amino (—NH$_2$) end.

Dipeptidase (▲) splits dipeptides (●●) into separate amino acids (●).

FIGURE 25.30 Protein Digestion and Absorption.

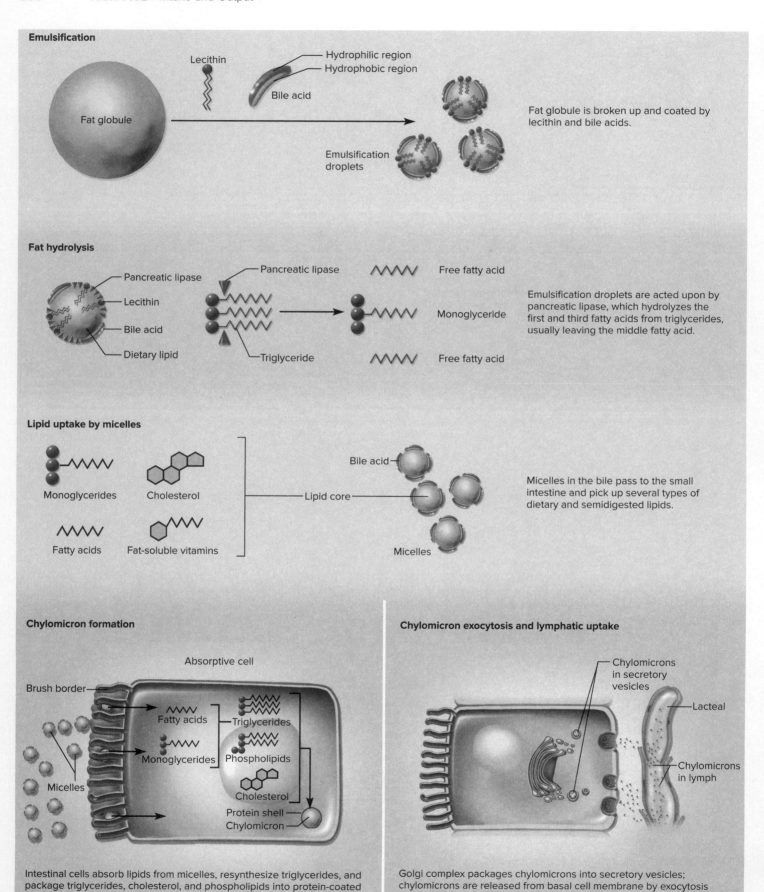

Emulsification

Lecithin

Hydrophilic region
Hydrophobic region

Bile acid

Fat globule

Emulsification droplets

Fat globule is broken up and coated by lecithin and bile acids.

Fat hydrolysis

Pancreatic lipase

Lecithin

Bile acid

Dietary lipid

Pancreatic lipase

Triglyceride

Free fatty acid

Monoglyceride

Free fatty acid

Emulsification droplets are acted upon by pancreatic lipase, which hydrolyzes the first and third fatty acids from triglycerides, usually leaving the middle fatty acid.

Lipid uptake by micelles

Monoglycerides   Cholesterol

Fatty acids   Fat-soluble vitamins

Lipid core

Bile acid

Micelles

Micelles in the bile pass to the small intestine and pick up several types of dietary and semidigested lipids.

Chylomicron formation

Absorptive cell

Brush border

Fatty acids   Triglycerides

Monoglycerides   Phospholipids

Cholesterol

Micelles

Protein shell
Chylomicron

Intestinal cells absorb lipids from micelles, resynthesize triglycerides, and package triglycerides, cholesterol, and phospholipids into protein-coated chylomicrons.

Chylomicron exocytosis and lymphatic uptake

Chylomicrons in secretory vesicles

Lacteal

Chylomicrons in lymph

Golgi complex packages chylomicrons into secretory vesicles; chylomicrons are released from basal cell membrane by exocytosis and enter the lacteal (lymphatic capillary) of the villus.

FIGURE 25.31 Fat Digestion and Absorption.

Explain why the right lymphatic duct does not contribute any dietary fat to the bloodstream.

Figure 25.32 summarizes the major aspects of carbohydrate, protein, and fat digestion and absorption.

25.6d Nucleic Acids

The nucleic acids, DNA and RNA, are present in much smaller quantities than the polymers discussed previously. The **nucleases** (ribonuclease and deoxyribonuclease) of pancreatic juice hydrolyze these to their constituent nucleotides. **Nucleosidases** and **phosphatases** of the brush border then decompose the nucleotides into phosphate ions, nitrogenous bases, and simple sugars (ribose from RNA and deoxyribose from DNA). These products are transported across the intestinal epithelium by membrane carriers and enter the capillary blood of the villus.

25.6e Vitamins

Vitamins are not digested, but absorbed unchanged. The fat-soluble vitamins A, D, E, and K are absorbed with other lipids as just described. Therefore, if they are ingested without fat-containing food, such as by simply popping vitamin tablets with a glass of water, they're not absorbed at all but passed in the feces and wasted. Water-soluble vitamins (the B complex and vitamin C) are absorbed by simple diffusion. An exception is vitamin B_{12}, an unusually large molecule that is absorbed poorly unless bound to the intrinsic factor (IF) secreted by the stomach's parietal cells. Then as it passes down the small intestine, the B_{12}–IF complex binds to receptors on absorptive cells of the distal ileum, where it is taken up by receptor-mediated endocytosis.

25.6f Minerals

Minerals (electrolytes) are absorbed along the entire length of the small intestine. Sodium ions are cotransported with sugars and amino acids. Chloride ions are actively transported in the distal ileum by a pump that exchanges them for bicarbonate ions, reversing the chloride–bicarbonate exchange that occurs in the stomach. Potassium ions are absorbed by simple diffusion. The K^+ concentration of chyme rises as water is absorbed, creating a gradient favorable to K^+ absorption. In diarrhea, when water absorption is hindered, potassium ions remain in the intestine and pass with the feces; therefore, chronic diarrhea can lead to hypokalemia.

Most minerals are absorbed at fairly constant rates regardless of need, leaving it to the kidneys to excrete any excess. Iron is one exception; its absorption is hormonally regulated. Intestinal enterocytes bind ferrous ions (Fe^{2+}) and take them in by active transport; they cannot absorb ferric ions (Fe^{3+}), but stomach acid (HCl) reduces most Fe^{3+} to absorbable Fe^{2+}. Fe^{2+} is transported to the basal surface of the cell and there taken up by the extracellular protein *transferrin.* The transferrin–iron complex diffuses into the blood and is carried to such places as the bone marrow for hemoglobin synthesis, muscular tissue for myoglobin synthesis, and the liver for storage (see fig. 18.7). Excess dietary iron, if absorbed, binds irreversibly to ferritin in the enterocyte and is

held there until that cell sloughs off and passes in the feces. Iron absorption and mobilization are regulated by the liver hormone **hepcidin.** An iron overload is dangerously toxic (indeed, a leading cause of death in young children who get into a parent's iron supplement pills), but hepcidin normally prevents overload. It *inhibits* intestinal iron absorption and the mobilization of iron from the liver, thus preventing the blood iron level from rising too high. Anemia and hypoxia reduce hepcidin synthesis, removing its inhibitory effect and thus allowing increased absorption of dietary iron and mobilization of stored iron so it becomes available for hemoglobin synthesis.

Young adult women have four times as many iron transport proteins in the intestinal mucosa as men have. Can you explain this in terms of functional significance?

The small intestine absorbs nearly all dietary phosphate, predominantly by active transport. By contrast, it absorbs only about 40% of the dietary calcium, leaving the rest to pass in the feces. In the duodenum, calcium is absorbed by the transcellular route. It enters the enterocytes through calcium channels in the apical plasma membrane and binds to a cytoplasmic protein called *calbindin.* This keeps the intracellular concentration of free calcium low, maintaining a gradient that favors uptake. What free calcium exists in the cytoplasm is then pumped out the basal side of the cell by active transport, using a protein called *calcium–ATPase* as well as a *sodium–calcium antiport.* From there, it enters the blood capillaries of the villus.

Transcellular calcium uptake is under hormonal influence. Parathyroid hormone is secreted in response to a drop in blood calcium level. It stimulates the kidneys to synthesize vitamin D from the precursors made by the epidermis and liver (see fig. 7.16). Vitamin D then affects the absorptive cells of the duodenum in three ways: It increases the number of calcium channels in the apical membrane, the amount of calbindin in the cytoplasm, and the number of calcium–ATPase pumps in the basal membrane. Thus, it increases absorption of dietary calcium and raises the level of calcium in the blood.

Because of their much greater length, the jejunum and ileum absorb much more calcium than the duodenum does, but here it is by the paracellular route (passing between cells) and is independent of hormones. Most absorbed calcium is from meat and dairy products. Although green leafy vegetables are high in calcium, little of this is absorbed because they also contain an agent, *oxalate,* that binds calcium and makes it unavailable for absorption. Dietary fat retards calcium absorption by reacting with it to form poorly absorbed calcium soaps.

25.6g Water

The digestive system is one of several systems involved in fluid balance. The digestive tract receives about 9 L of water per day—0.7 L in food, 1.6 L in drink, and 6.7 L in the gastrointestinal secretions: saliva, gastric juice, bile, pancreatic juice, and intestinal juice. About 8 L of this is absorbed by the small intestine and 0.8 L by the large intestine, leaving 0.2 L voided in the daily fecal

Site	(a) Carbohydrates	(b) Proteins	(c) Fats

FIGURE 25.32 Summary of the Digestion and Absorption of the Major Organic Nutrients. Enzyme names are indicated in red.

(pasta): John A. Rizzo/Getty Images; (eggs, cheeses): Pixtal/agefotostock

❓ *As they leave the small intestine, sugars and amino acids arrive at the liver before any other organ, but lipids do not. Why not? Trace the route that a dietary fat would have to take to reach the liver, using information from section 21.1a and table 20.7.*

output. Water is absorbed by osmosis, following the absorption of salts and organic nutrients that create an osmotic gradient from the intestinal lumen to the ECF.

Diarrhea occurs when the large intestine absorbs too little water. This occurs when the intestine is irritated by bacteria and feces pass through too quickly for adequate reabsorption, or when the feces contain abnormally high concentrations of a solute such as lactose that opposes osmotic absorption of water. *Constipation* occurs when fecal movement is slow, too much water is reabsorbed, and the feces become hardened. This can result from lack of dietary fiber, lack of exercise, emotional upset, or long-term laxative abuse.

▶▶▶**APPLY WHAT YOU KNOW**

Magnesium sulfate (Epsom salt) is poorly absorbed by the intestines. In light of this, explain why it has a laxative effect.

BEFORE YOU GO ON

Answer the following questions to test your understanding of the preceding section:

21. What three classes of nutrients are most abundant? What are the end products of enzymatic digestion of each?

22. What two digestive enzymes occur in the saliva? Why is one of these more active in the stomach than in the mouth?

23. Name as many enzymes of the intestinal brush border as you can, and identify the substrate or function of each.

24. Explain the distinctions between an emulsification droplet, a micelle, and a chylomicron.

25. What happens to digestive enzymes after they have done their job? What happens to dead epithelial cells that slough off the gastrointestinal mucosa? Explain.

25.7 The Large Intestine

Expected Learning Outcomes

When you have completed this section, you should be able to

a. describe the gross anatomy of the large intestine;

b. summarize the functions of the large intestine;

c. contrast the mucosa of the colon with that of the small intestine;

d. state the physiological significance of intestinal bacteria;

e. discuss the types of contractions that occur in the colon; and

f. explain the neurological control of defecation.

The large intestine receives about 500 mL of indigestible food residue per day, reduces it to about 150 mL of feces by absorbing water and salts, and eliminates the feces by defecation.

25.7a Gross Anatomy

The large intestine measures about 1.5 m (5 ft) long and 6.5 cm (2.5 in.) in diameter in the cadaver. It is named for its relatively large diameter, not its length. It consists of four regions: the cecum, colon, rectum, and anal canal (**fig. 25.33**).

The **cecum**[27] is a blind pouch in the lower right abdominal quadrant inferior to the ileal papilla. Attached to its lower end is the **appendix,** a blind tube 2 to 7 cm long. The appendix is densely populated with lymphocytes and is a significant source of immune cells.

The **colon** is that part of the large intestine between the ileocecal junction and rectum (not including the cecum, rectum, or anal canal). It is divided into the ascending, transverse, descending, and sigmoid regions. The **ascending colon** begins at the ileal papilla and passes up the right side of the abdominal cavity. It makes a 90° turn at the **right colic (hepatic) flexure,** near the right lobe of the liver, and becomes the **transverse colon.** This passes horizontally across the upper abdominal cavity and turns 90° downward at the **left colic (splenic) flexure** near the spleen. Here it becomes the **descending colon,** which passes down the left side of the abdominal cavity. Ascending, transverse, and descending colons thus form a squarish, three-sided frame around the small intestine; this, however, is highly variable from person to person and between ethnic groups.

The pelvic cavity is narrower than the abdominal cavity, so at the hip bone, the colon turns medially and travels along the iliac fossa before turning downward at the pelvic inlet into the pelvic cavity. (See fig. 8.35 for review of the skeletal landmarks.) The resulting S-shaped portion of the tract is called the **sigmoid**[28] **colon.** (Visual examination of this region is performed with an instrument called a *sigmoidoscope.*) In the pelvic cavity, the large intestine continues as the **rectum,**[29] about 15 cm long. Despite its name, the rectum is not quite straight but has three lateral curves as well as an anteroposterior curve. It has three infoldings called **transverse rectal folds (rectal valves),** which enable it to retain feces while passing gas.

The final 3 cm of the large intestine is the **anal canal** (fig. 25.33b), which passes through the levator ani muscle of the pelvic floor and terminates at the anus. Here, the mucosa forms longitudinal ridges called **anal columns** with depressions between them called **anal sinuses.** As feces pass through the canal, they press the sinuses and cause them to exude extra mucus and lubricate the canal during defecation. Prominent **hemorrhoidal veins** form superficial plexuses in the anal columns and around the orifice. Unlike veins in the limbs, they lack valves and are particularly subject to distension and venous pooling. *Hemorrhoids* are permanently distended veins that protrude into the anal canal or form bulges external to the anus. They can result from the impaired venous return that occurs in obesity and pregnancy.

The muscularis externa of the colon is unusual. Although it completely encircles the colon just as it does the small intestine, its

---

[27]*cec* = blind
[28]*sigm* = sigma or S; *oid* = resembling
[29]*rect* = straight

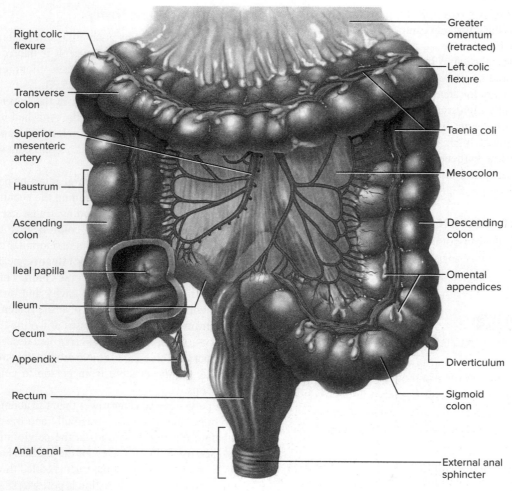

Right colic flexure

Transverse colon

Superior mesenteric artery

Haustrum

Ascending colon

Ileal papilla

Ileum

Cecum

Appendix

Rectum

Anal canal

Greater omentum (retracted)

Left colic flexure

Taenia coli

Mesocolon

Descending colon

Omental appendices

Diverticulum

Sigmoid colon

External anal sphincter

(a) Gross anatomy

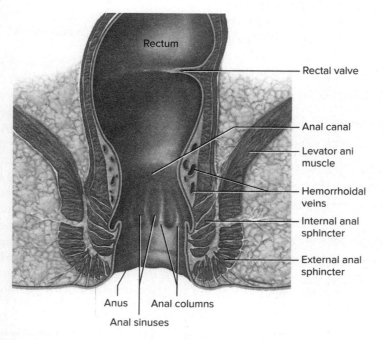

Rectum

Rectal valve

Anal canal

Levator ani muscle

Hemorrhoidal veins

Internal anal sphincter

External anal sphincter

Anus   Anal columns

Anal sinuses

(b) Anal canal

FIGURE 25.33
The Large Intestine.
(a) Gross anatomy.
(b) Anatomy of the anal canal. **APR**

? *Which anal sphincter is controlled by the autonomic nervous system? Which is controlled by the somatic nervous system? Explain the basis for your answers.*

longitudinal fibers are especially concentrated in three thickened, ribbonlike strips. Each strip is called a **taenia coli**[30] (TEE-nee-ah CO-lye) (plural, *taeniae coli*). The muscle tone of the taeniae coli contracts the colon lengthwise and causes its wall to bulge, forming pouches called **haustra**[31] (HAW-stra; singular, *haustrum*). Haustra are conspicuous in colonic X-rays of living patients; they are not evident in a cadaver, however, because they disappear when muscle tone is lost at death. In the rectum and anal canal, the longitudinal muscle forms a continuous sheet and haustra are absent. The anus is regulated by two sphincters: an **internal anal sphincter** composed of smooth muscle of the muscularis externa and an **external anal sphincter** composed of skeletal muscle of the pelvic diaphragm.

The ascending and descending colon are retroperitoneal and have a serosa only on the anterior surface, whereas the transverse and sigmoid colon are entirely enclosed in serosa and anchored to the posterior abdominal wall by the mesocolon. The serosa of the transverse through sigmoid colon often has **omental (fatty) appendices,** clublike fatty pouches of peritoneum of unknown function. In older people, the serosal surface of the colon often shows outpocketings called *diverticula* (see Deeper Insight 25.5).

DEEPER INSIGHT 25.5

CLINICAL APPLICATION

Diverticulosis and Diverticulitis

Older people often exhibit a condition called *diverticulosis,* the presence of pouches *(diverticula)* on the wall of the colon (see fig. 25.33a). This occurs in about one-half of Americans and Canadians over 60 years old. It is often attributed to a low-fiber diet; when the stool is lacking in bulk, contractions of the colon are thought to put more pressure on the wall and cause it to balloon out, especially in the sigmoid region. This is an unproven hypothesis, however, and other factors are known or suspected. The wall of the colon weakens with age, for one thing, and there is also a hereditary component. In a pair of identical twins, if one gets diverticulosis, there is a 40% chance that the other will too, regardless of diet.

Diverticulosis can cause dull pain, mild cramps, constipation, and blood in the stool, but it is often asymptomatic and discovered only when a person has a colonoscopy or CT scan. Symptoms can be relieved by increasing dietary fiber intake. The most serious consequences of diverticulosis are that it increases the risk of colon cancer and that about one in five cases progresses to diverticulitis.

Diverticulitis is inflammation of diverticula, possibly caused by impacted feces in the pouches. This causes more intense abdominal pain, tenderness in the lower left quadrant (but on the right in Asians), and sometimes nausea, vomiting, and fever. Diverticula can perforate and leak feces into the pelvic cavity, causing life-threatening peritonitis. Diverticulitis is treated with antibiotics and pain relievers, but sometimes requires surgical resection of the colon (removal of the affected region).

25.7b Circulation

The large intestine is served by mesenteric arteries and veins much like the small intestine. Branches of the superior mesenteric artery fan out to supply the ascending colon and most of the transverse colon; the inferior mesenteric artery supplies the rest of the transverse colon as well as the descending and sigmoid colon and the rectum (see section 20.7f). The superior and inferior mesenteric veins drain the same parts of the large intestine as the correspondingly named arteries and drain into the hepatic portal system (see section 20.7g).

25.7c Microscopic Anatomy

The mucosa of the large intestine (**fig. 25.34**) has a simple columnar epithelium in all regions except the lower half of the anal canal, where it has a nonkeratinized stratified squamous epithelium. The latter resists abrasion by the passage of feces. There are no circular folds or villi in the large intestine, but there are intestinal crypts. They are deeper than in the small intestine and have a greater density of goblet cells; mucus is their only significant secretion. The lamina propria and submucosa have an abundance of lymphatic tissue, providing protection from the bacteria that densely populate the large intestine.

25.7d Intestinal Microbes and Gas

The large intestine harbors about 800 species of bacteria collectively called the **gut microbiome.** We have a mutually beneficial relationship with many of these. We provide them with room and board while they provide us with nutrients from our food that we're not equipped to extract on our own. For example, they digest cellulose, pectin, and other plant polysaccharides for which we have no digestive enzymes, and we absorb the resulting sugars. Thus, we get more nutrition from our food because of these bacteria than we would get without them. Indeed, one person may get more calories than another from the same amount of food because of differences in their bacterial populations. Some bacteria also synthesize B vitamins and vitamin K, which are absorbed by the colon. This vitamin K is especially important because the diet alone usually doesn't provide enough to ensure adequate blood clotting.

One of the less desirable and sometimes embarrassing products of these bacteria is intestinal gas. The large intestine contains about 7 to 10 L of gas, expelling about 500 mL/day as **flatus** and reabsorbing the rest. Much of this is swallowed air that has worked its way through the digestive tract, but the gut microbes add to it. Painful cramping can result when undigested nutrients pass into the colon and furnish an abnormal substrate for bacterial action, so the bacteria produce excess gas—for example, in lactose intolerance. Flatus is composed mostly of nitrogen (N_2), carbon dioxide (CO_2), hydrogen (H_2), methane (CH_4), hydrogen sulfide (H_2S), and two amines: indole and skatole. Indole, skatole, and H_2S produce most of the odor of flatus and feces, whereas the others are odorless. The hydrogen gas is combustible and has been known to explode during the use of electrical cauterization in surgery.

---

[30]*taenia* = ribbon; *coli* = of the colon
[31]*haustr* = to draw, like a purse string

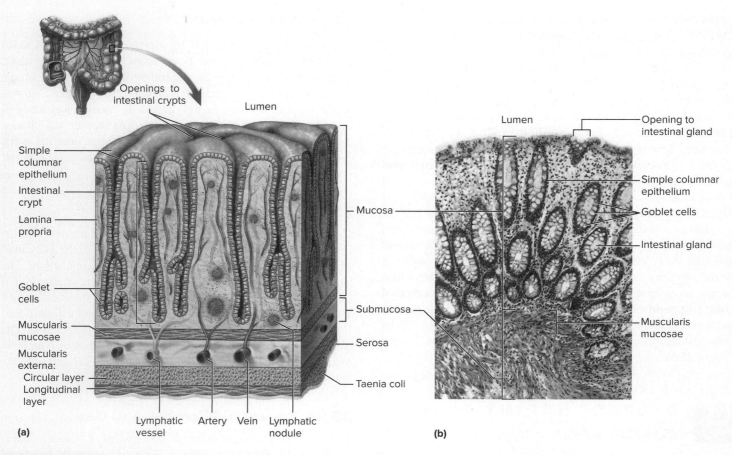

FIGURE 25.34 Microscopic Anatomy of the Colon. (a) A block of tissue showing all layers from mucosa (top) to serosa (bottom). (b) Light micrograph of the mucosa.

b: Jose Luis Calvo/Shutterstock

25.7e Absorption and Motility

The large intestine takes 36 to 48 hours to reduce the residue of a meal to feces, with the residue spending the greatest time (about 24 h) in the transverse colon. The colon doesn't chemically change the residue, but reabsorbs water and electrolytes (especially NaCl) from it. Feces usually consist of about 75% water and 25% solids. The solids are about 30% bacteria, 30% undigested dietary fiber, 10% to 20% fat, and smaller amounts of protein, sloughed epithelial cells, salts, mucus, and other digestive secretions. The fat is not from the diet but from bacteria and broken-down epithelial cells.

The most common type of colonic motility is a type of segmentation called **haustral contractions,** which occur about every 30 minutes. Distension of a haustrum with feces stimulates it to contract. This churns and mixes the residue, promotes water and salt absorption, and passes the residue distally to another haustrum. Stronger contractions called **mass movements** occur one to three times a day. They last about 15 minutes and move residue several centimeters at a time. They are often triggered by the **gastrocolic and duodenocolic reflexes,** in which filling of the stomach and duodenum stimulates motility of the colon. Mass movements occur especially in the transverse to sigmoid colon, often within an hour after breakfast, moving the feces that accumulated and stretched the colon overnight.

25.7f Defecation

Stretching of the rectum stimulates the defecation reflexes, which account for the urge to defecate that is often felt soon after a meal. You may find it enlightening to compare the following discussion of the defecation reflexes with the similar mechanisms controlling urination (see fig. 23.25). The predictability of the defecation reflexes is useful in house training pets and toilet training children. The overall process involves two reflexes:

1. The **intrinsic defecation reflex.** This reflex is mediated entirely by the myenteric plexus. Stretch signals travel through the plexus to the muscularis of the descending and sigmoid colon and the rectum. This activates a peristaltic wave that drives feces downward, and it relaxes the internal anal sphincter. This reflex is relatively weak, however, and usually requires the cooperative action of the following reflex.

2. The **parasympathetic defecation reflex.** This is a spinal reflex. Its principal events **(fig. 25.35)** are that stretch signals are transmitted to the spinal cord; motor signals return by way of the pelvic nerves; and these signals intensify peristalsis in the descending and sigmoid colon and rectum while they relax the internal anal sphincter.

Impulses from
cerebral cortex

Sensory
fibers

Parasympathetic
motor fibers

Stretch
receptors

Voluntary
motor
fibers

Sigmoid
colon

Stretch
receptors

Rectum

Anal canal

Internal anal
sphincter

External anal
sphincter

① Feces stretch the rectum and stimulate stretch
receptors, which transmit signals to the spinal cord.

② A spinal reflex stimulates contraction of the rectum.

③ The spinal reflex also relaxes the internal anal sphincter.

④ Impulses from the brain prevent untimely defecation
by keeping the external anal sphincter contracted.
Defecation occurs only if this sphincter also relaxes.

FIGURE 25.35 Neural Control of Defecation.

These reflexes are involuntary and are the sole means of controlling defecation in infants and some people with transecting spinal cord injuries. However, the external anal sphincter, like the external urethral sphincter controlling urination, is under voluntary control, enabling one to limit defecation to appropriate circumstances. Voluntary retention of feces is also aided by the **puborectalis muscle,** which loops around the rectum like a sling and creates a sharp anorectal angle that blocks the passage of feces. Defecation normally occurs only when the external anal sphincter and puborectalis muscle are voluntarily relaxed. The kink in the rectum then straightens out and the sphincter opens to allow the feces to fall away. Defecation is also aided by the voluntary Valsalva maneuver, in which a breath hold and contraction of the abdominal muscles increase abdominal pressure, compress the rectum, and squeeze the feces from it. This maneuver can also initiate the defecation reflex by forcing feces from the descending colon into the rectum. The external anal sphincter and external urethral sphincter are controlled together by inhibitory signals from the brainstem, so as this inhibition is released, defecation is usually accompanied by urination. Some spinal cord injuries and diseases abolish the voluntary control of the external anal sphincter. The resulting inability to voluntarily retain the feces is called *fecal incontinence.*

If the defecation urge is suppressed, contractions cease in a few minutes and the rectum relaxes. The defecation reflexes recur a few hours later or when another mass movement propels more feces into the rectum.

The effects of aging on the digestive system are discussed in section 29.4a. The most life-threatening digestive disorder in older age is colorectal cancer (see Deeper Insight 25.6). Other disorders of this system are described in **table 25.3.**

BEFORE YOU GO ON

Answer the following questions to test your understanding of the preceding section:

26. How does the mucosa of the large intestine differ from that of the small intestine? How does the muscularis externa differ?

27. Name and briefly describe two types of contractions that occur in the colon and nowhere else in the alimentary canal.

28. Describe the reflexes that cause defecation in an infant. Describe the additional neural controls that function following toilet training.

DEEPER INSIGHT 25.6

CLINICAL APPLICATION

Colorectal Cancer

Cancer of the colon and rectum *(colorectal cancer)* is the second most common cause of cancer deaths in the United States. It usually arises from mucus-secreting goblet cells in a colon polyp and, if neglected, readily metastasizes to distant organs including the liver, lungs, and brain. Early signs of colorectal cancer include persistent diarrhea, constipation, and abdominal pain; blood in the stool; and weakness, fatigue, and weight loss.

Some risk factors for colorectal cancer are avoidable: smoking, heavy alcohol use, physical inactivity, and obesity. The connection with diet is uncertain, but diets low in fruit and vegetable fiber and high in red meats, char-grilled meat, and processed meats such as hot dogs and sandwich meats, have been implicated as possible risk factors. Other factors are unavoidable. Persons of African and Ashkenazi Jewish ancestry and persons with a family history of colorectal cancer are at greater than average risk. It is associated also with a history of colorectal polyps, inflammatory bowel disease (IBS), and diverticulosis (see Deeper Insight 25.5). It is more common among prosperous than poorer people, perhaps because of dietary and other lifestyle choices.

All colorectal cancer is genetic, but only a small percentage of cases are hereditary—in other words, it is seldom passed from parent to offspring, but almost always arises by new mutations during one's lifetime. About 85% of cases involve mutation of tumor suppressor gene *p53* on chromosome 17 (see Deeper Insight 4.3), but additional mutations on other chromosomes are also needed for progression to malignancy. For this reason, colorectal cancer is slow to develop and usually appears after age 50, but it can occur at any age. Colorectal cancer mortality is declining worldwide because of improved awareness, better screening, and better treatment; yet it is rising among younger people for unknown reasons.

The best safeguards against colorectal cancer are regular screening and early detection. The most common screening methods are fecal exams for traces of nonvisible (occult) blood; *sigmoidoscopy* (endoscopic examination of the rectum and sigmoid colon); *and colonoscopy* (endoscopy of the entire colon). Polyps observed in endoscopy are removed before they can turn malignant. The frequency and starting age of such screenings vary with medical opinion, but a common recommendation is to have fecal exams annually (mail-in kits for at-home sampling are available for this), sigmoidoscopy every 5 years, and colonoscopy every 10 years, starting at age 50.

TABLE 25.3	Some Digestive System Diseases
Acute pancreatitis	Severe pancreatic inflammation perhaps caused by trauma leading to leakage of pancreatic enzymes into parenchyma, where they digest tissue and cause inflammation and hemorrhage.
Appendicitis	Inflammation of the appendix, with swelling, pain, and sometimes gangrene, perforation, and peritonitis.
Cancers	Malignant tumors especially of the esophagus, stomach, colon, liver, and pancreas, with colorectal and pancreatic cancer being among the leading causes of cancer death in the United States.
Crohn disease	Inflammation of small and large intestines, similar to ulcerative colitis. Produces granular lesions and fibrosis of intestine, diarrhea, and lower abdominal pain. Often hereditary.
Dysphagia	Difficulty swallowing. Can result from esophageal obstructions (tumors, constrictions) or impaired peristalsis (due to neuromuscular disorders).
Gluten-sensitive enteropathy	Formerly called *sprue* or *celiac disease.* Atrophy of intestinal villi triggered in genetically susceptible individuals by *gluten,* the protein component of cereal grains. Onset is usually in infancy or early childhood. Results in severe malabsorption of most nutrients, causing watery or fatty diarrhea, abdominal pain, diminished growth, and multiple problems tied to nutritional deficiencies. Treatable with intensive dietary management.
Hiatal hernia	Protrusion of part of the stomach into the thoracic cavity, where the negative thoracic pressure may cause it to balloon. Often causes gastroesophageal reflux (especially when a person is supine) and esophagitis (inflammation of the esophagus).
Ulcerative colitis	Chronic inflammation resulting in ulceration of the large intestine, especially the sigmoid colon and rectum. Tends to be hereditary but exact causes are not well known.
You can find other digestive system disorders described in the following places:	

Portal hypertension and *ascites* in Deeper Insight 20.4; *dental caries, gingivitis,* and other *periodontal disease* in Deeper Insight 25.1; *impacted molars* in section 25.2a; *heartburn* in section 25.2e; *peptic ulcer* in Deeper Insight 25.2; *gallstones* in Deeper Insight 25.3; *lactose intolerance* in Deeper Insight 25.4; *diverticulosis* and *diverticulitis* in Deeper Insight 25.5; *colorectal cancer* in Deeper Insight 25.6; *constipation* and *diarrhea* in section 25.6g; *hemorrhoids* in section 25.7a; and *hepatitis* and *hepatic cirrhosis* in Deeper Insight 26.3.

DEEPER INSIGHT 25.7

MEDICAL HISTORY

The Man with a Hole in His Stomach

Perhaps the most famous episode in the history of digestive physiology began with a grave accident in 1822 on Mackinac Island in northern Michigan. Alexis St. Martin, a 28-year-old Canadian voyageur **(fig. 25.36),** was standing outside a trading post when he was accidentally hit by a shotgun blast from 3 feet away. A frontier Army doctor stationed at Fort Mackinac, William Beaumont, was summoned to examine St. Martin. As Beaumont later wrote, "a portion of the lung as large as a turkey's egg" protruded through St. Martin's lacerated and burnt flesh. Below that was a portion of the stomach with a puncture in it "large enough to receive my forefinger." Beaumont did his best to pick out bone fragments and dress the wound, though he did not expect St. Martin to survive.

Surprisingly, he lived. Over a period of months, the wound extruded pieces of bone, cartilage, gunshot, and gun wadding. As the wound healed, a fistula (hole) remained in the stomach, so large that Beaumont had to cover it with a compress to prevent food from coming out. The opening remained, covered only by a loose flap of skin, for the rest of St. Martin's life. A fold of tissue later grew over the fistula, but it was easily opened. A year later, St. Martin was still feeble. Town authorities decided they could no longer support him on public funds and wanted to ship him 1,500 miles to his home. Beaumont, however, was imbued with a passionate sense of destiny. Very little was known about digestion, and he saw the accident as a unique opportunity to learn. He took St. Martin in at his personal expense and performed 238 experiments on him over several years. Beaumont had never attended medical school and had little idea how scientists work, yet he proved to be an astute experimenter. Under crude frontier conditions and with almost no equipment, he discovered many of the basic facts of gastric physiology discussed in this chapter.

"I can look directly into the cavity of the stomach, observe its motion, and almost see the process of digestion," Beaumont wrote. "I can pour in water with a funnel and put in food with a spoon, and draw them out again with a siphon." He put pieces of meat on a string into the stomach and removed them hourly for examination. He sent vials of gastric juice to the leading chemists of America and Europe, who could do little but report that it contained hydrochloric acid. He proved that digestion required HCl and could even occur outside the stomach, but he found that HCl alone didn't digest meat; gastric juice must contain some other digestive ingredient. Theodor Schwann, one of the founders of the cell theory, identified that ingredient as pepsin. Beaumont also demonstrated that gastric juice is secreted only in response to food; it didn't accumulate between meals as previously thought. He disproved the idea that hunger is caused by the walls of the empty stomach rubbing against each other.

Now disabled from wilderness travel, St. Martin agreed to participate in Beaumont's experiments in exchange for room and board—though he felt helpless and humiliated by it all. The fur trappers taunted him as "the man with a hole in his stomach," and he longed to return to his work in the wilderness. He had a wife and daughter in Canada whom he rarely got to see, and he ran away repeatedly to join them. He was once gone for 4 years before poverty made him yield to Beaumont's financial enticement to come back. Beaumont despised St. Martin's drunkenness and profanity and was quite insensitive to his embarrassment and discomfort over the experiments. Yet St. Martin's temper enabled Beaumont to make the first direct observations of the relationship between emotion and digestion. During St. Martin's emotional outbursts, Beaumont noted that little digestion occurred—as we now know, the sympathetic nervous system inhibits digestive activity.

Beaumont published a book in 1833 that laid the foundation for modern gastric physiology and dietetics. It was enthusiastically received by the medical community and had no equal until Russian physiologist Ivan Pavlov (1849–1936) performed his celebrated experiments on digestion in animals. Building on the methods pioneered by Beaumont, Pavlov received the 1904 Nobel Prize for Physiology or Medicine.

In 1853, Beaumont slipped on some ice, suffered a blow to the base of his skull, and died a few weeks later. St. Martin continued to tour medical schools and submit to experiments by other physiologists, whose conclusions were often less correct than Beaumont's. Some, for example, attributed chemical digestion to lactic acid instead of hydrochloric acid. St. Martin lived in wretched poverty in a tiny shack with his wife and several children, and died 28 years after Beaumont. By then he was senile and believed he had been to Paris, where Beaumont had often promised to take him.

William Beaumont (1785–1853)   **Alexis St. Martin (1794–1880)**

FIGURE 25.36 Doctor and Patient in a Pioneering Study of Digestion. Today's standards of scientific ethics would prohibit Beaumont's experiments on St. Martin.

(both): Source: National Library of Medicine

CONNECTIVE ISSUES

Effects of the DIGESTIVE SYSTEM on Other Organ Systems

ALL SYSTEMS
Digestion provides all systems with nutrients in a form usable for cellular metabolism and tissue growth and maintenance.

INTEGUMENTARY SYSTEM
Liver disease can cause the skin discoloration of jaundice; excess dietary fat is deposited in dermal and subcutaneous adipose tissue.

SKELETAL SYSTEM
Bone deposition and maintenance depend on calcium and phosphate absorption by the small intestine.

MUSCULAR SYSTEM
The liver promotes recovery from muscle fatigue by metabolizing lactic acid generated by the muscles; the liver and intestinal epithelium store iron and release it as needed for myoglobin synthesis.

NERVOUS SYSTEM
Gut–brain peptides produced in the stomach and small intestine stimulate appetite-controlling centers in the brain; chronic vomiting or diarrhea can cause electrolyte and pH imbalances that lead to nervous system dysfunction.

ENDOCRINE SYSTEM
The liver degrades hormones and limits their action; many hormones are produced by endocrine cells of the stomach, intestines, pancreas, and liver.

CIRCULATORY SYSTEM
Intestinal fluid absorption supports blood volume; the liver degrades the heme from dead erythrocytes; the liver synthesizes the albumin, most blood-clotting factors, and other plasma proteins; the liver stabilizes blood glucose level; the liver and intestinal epithelium store iron and release it as needed for hemoglobin synthesis; the liver secretes erythropoietin, which stimulates RBC production.

LYMPHATIC AND IMMUNE SYSTEMS
The small intestine is the source of lipids transported by lymphatic vessels; the intestinal mucosa is a major source of lymphocytes; acid, enzymes, and lysozyme provide innate immunity against ingested pathogens; infants acquire passive immunity by intestinal absorption of IgA from breast milk.

RESPIRATORY SYSTEM
Contraction of the abdominal muscles pushes the stomach against the diaphragm and aids in forced expiration.

URINARY SYSTEM
The liver synthesizes the urea excreted by the kidneys; this urea also contributes to the osmotic gradient that enables the kidneys to concentrate urine; intestines complement the kidneys in water and electrolyte reabsorption.

REPRODUCTIVE SYSTEM
Digestion provides nutrients for fetal growth; certain aspects of egg fertilization depend on calcium absorbed by the small intestine.

STUDY GUIDE

▶ Assess Your Learning Outcomes

To test your knowledge, discuss the following topics with a study partner or in writing, ideally from memory.

25.1 General Anatomy and Digestive Processes

1. Three functions of the digestive system and the five stages in which it carries these out
2. The difference between the digestive tract and the accessory organs of the digestive system; the organs that belong in each category
3. Tissue layers typical of most regions of the digestive tract
4. Functions of the enteric nervous system; its two subdivisions, their locations, and their respective functions
5. Functions of the mesenteries and their relationship to the abdominal digestive organs
6. Hormones, paracrines, and visceral reflexes that regulate motility and secretion in the digestive tract

25.2 The Mouth Through Esophagus

1. Seven functions of the oral cavity
2. Anatomical boundaries of the mouth; variations in its mucosal epithelium; and all digestive system organs contained in it
3. Anatomy of the cheeks and lips; the three regions of the lip
4. Anatomy and functions of the tongue; what forms the border between the body and root of the tongue; its intrinsic and extrinsic muscles; the lingual glands and lingual tonsils
5. Anatomy of the hard and soft palates; the two arches that mark the border between the oral cavity and pharynx
6. The structure of a typical tooth and periodontal tissues; the four kinds of teeth and number and position of each; the mode of replacement of deciduous teeth by permanent teeth; and the functions of mastication
7. Six functions of saliva; its composition and pH; general histology of salivary glands; names and function of the intrinsic salivary glands; the three pairs of extrinsic salivary glands; and how the nervous system regulates salivation

8. Anatomy of the pharynx; the pharyngeal constrictor muscles and upper esophageal sphincter, and the functions of these muscles
9. Gross anatomy and histology of the esophagus; the distribution of skeletal and smooth muscle in the esophageal wall; the location and function of the esophageal glands; the cardial orifice and lower esophageal sphincter
10. The physiology of swallowing; the swallowing center and the cranial nerves involved in the process; what occurs in the buccal and pharyngoesophageal phases of swallowing; how peristalsis is controlled; and to what extent swallowing depends on peristalsis or occurs independently of it

25.3 The Stomach

1. Anatomy and functions of the stomach; features that mark its beginning and end; and the volume of the empty stomach and its full capacity
2. Innervation of the stomach; its blood supply and relation to the hepatic portal system
3. Structure of the gastric mucosa including the gastric pits; the glands that open into them; and differences in the spatial distribution and functions of gastric, cardial, and pyloric glands; and five cell types of these glands and their respective functions
4. The composition of gastric juice
5. The cells that secrete hydrochloric acid, how they do so, and the functions of the acid
6. The difference between a zymogen and an active enzyme
7. The source of pepsinogen; how it is converted to pepsin; and the function of pepsin
8. The source and function of gastric lipase
9. The source and function of intrinsic factor; the effect of hyposecretion of intrinsic factor; and how a deficiency of intrinsic factor is treated
10. Hormones and paracrine secretions that regulate gastrointestinal function; why some of these are called gut–brain peptides
11. The nature and functions of the receptive-relaxation response and peristalsis in the stomach; the reason very little chyme is passed into the duodenum at a time

12. The degree of digestion that occurs in the stomach; what is absorbed by the stomach and what is not
13. How the stomach is protected from its own acid and enzymes
14. How gastric activity is controlled; the regulatory mechanisms of the cephalic, gastric, and intestinal phases

25.4 The Liver, Gallbladder, and Pancreas

1. The location, gross anatomy, and digestive function of the liver
2. Structure and spatial arrangement of the hepatic lobules
3. The route of blood flow through the liver
4. Composition and functions of the bile, the route of bile flow from the hepatocytes to the duodenum; and the recycling of bile acids and its relationship to the elimination of cholesterol from the body
5. Structure and function of the gallbladder, and its connection to the bile duct
6. Location, gross anatomy, and digestive functions of the pancreas
7. Structure of the pancreatic acini, the duct system, and its connection to the duodenum
8. Composition and digestive functions of pancreatic juice; the names and functions of its digestive zymogens and enzymes
9. Hormones that regulate the secretion of bile and pancreatic juice

25.5 The Small Intestine

1. Structures that mark the beginning and end of the small intestine; the three regions of the small intestine and their respective lengths and histological differences
2. How the small intestine is protected from the erosive effect of stomach acid
3. The importance of surface area for the function of the small intestine, and four features that give it a large surface area
4. Histology of the intestinal villi and crypts; the cell types found in each, and their respective functions; the relationship of the lacteal to the lymphatic system, and its function
5. Brush border enzymes of the small intestine and their functions
6. Two types of intestinal motility and their functional difference

STUDY GUIDE

25.6 Chemical Digestion and Absorption

1. Steps in carbohydrate digestion from the mouth to small intestine; the enzymes involved at each step and their respective contributions to carbohydrate hydrolysis

2. Mechanisms of monosaccharide absorption by the intestinal mucosa

3. Steps in protein digestion from the stomach to small intestine; the enzymes involved at each step and their respective contributions to peptide hydrolysis

4. Mechanisms of amino acid absorption by the intestinal mucosa

5. Steps in fat digestion from the stomach to small intestine; the enzymes involved at each step and their respective contributions to peptide hydrolysis

6. The necessity of emulsification for efficient fat digestion; the role of bile acids and lecithin in this process

7. Mechanisms of lipid absorption by the intestinal mucosa; the role of the lacteals; and how and why lipid absorption and transport differ from the absorption and transport of sugars and amino acids

8. Differences between emulsification droplets, micelles, and chylomicrons in lipid processing

9. Digestion of DNA and RNA in the small intestine, and the mode of absorption of their decomposition products

10. Nutrients that are not digested but simply absorbed; special aspects of the absorption of fat-soluble vitamins and vitamin B_{12}

11. Modes of absorption of water and minerals, particularly Na^+, K^+, Cl^-, Fe^{2+}, and Ca^{2+}; the roles of hepcidin, calcitriol, and parathyroid hormone in iron and calcium absorption

25.7 The Large Intestine

1. Gross anatomy, histology, and functions of the large intestine, including its six segments and total length

2. The gut microbiome of the colon; the composition and sources of flatus

3. Control of colonic motility; differences between its haustral contractions and mass movements; and the role of gastrocolic and duodenocolic reflexes

4. Mechanisms of the intrinsic and parasympathetic defecation reflexes, and of voluntary control over defecation; the neuroanatomy involved in these control mechanisms; and how the Valsalva maneuver can trigger and aid defecation

▶ Testing Your Recall

Answers in Appendix A

1. Which of the following enzymes acts in the stomach?
   a. chymotrypsin
   b. lingual lipase
   c. carboxypeptidase
   d. enteropeptidase
   e. dextrinase

2. Which of the following enzymes does *not* digest any nutrients?
   a. chymotrypsin
   b. lingual lipase
   c. carboxypeptidase
   d. enteropeptidase
   e. dextrinase

3. Which of the following is *not* an enzyme?
   a. chymotrypsin
   b. enteropeptidase
   c. secretin
   d. pepsin
   e. nucleosidase

4. The substance in question 3 that is *not* an enzyme is
   a. a zymogen.
   b. a nutrient.
   c. an emulsifier.
   d. a neurotransmitter.
   e. a hormone.

5. The lacteals absorb
   a. chylomicrons.
   b. micelles.
   c. emulsification droplets.
   d. amino acids.
   e. monosaccharides.

6. All of the following contribute to the absorptive surface area of the small intestine *except*
   a. its length.
   b. the brush border.
   c. haustra.
   d. circular folds.
   e. villi.

7. Which of the following is a periodontal tissue?
   a. the gingiva
   b. the enamel
   c. the cement
   d. the pulp
   e. the dentin

8. Anatomically, the _____ of the stomach most closely resemble the _____ of the small intestine.
   a. gastric pits, intestinal crypts
   b. pyloric glands, intestinal crypts
   c. rugae, aggregated lymphoid nodules
   d. parietal cells, goblet cells
   e. gastric glands, duodenal glands

9. Which of the following cells secrete digestive enzymes?
   a. chief cells
   b. mucous neck cells
   c. parietal cells
   d. goblet cells
   e. enteroendocrine cells

10. What phase of gastric regulation includes inhibition by the enterogastric reflex?
    a. the intestinal phase
    b. the gastric phase
    c. the buccal phase
    d. the cephalic phase
    e. the pharyngoesophageal phase

11. Cusps are a feature of the _____ surfaces of the molars and premolars.

12. The acidity of the stomach halts the action of _____ but promotes the action of _____, both of which are salivary enzymes.

13. The _____ salivary gland is named for its proximity to the ear.

14. The submucosal and myenteric plexuses collectively constitute the _____ nervous system.

15. Nervous stimulation of gastrointestinal activity is mediated mainly through the parasympathetic fibers of the _____ nerves.

STUDY GUIDE

16. Food in the stomach causes G cells to secrete _____, which in turn stimulates the secretion of HCl and pepsinogen.

17. Stellate macrophages occur in blood-filled spaces of the liver called _____.

18. The brush border enzyme that finishes the job of protein digestion, splitting the last two amino acids apart, is called _____.

19. Fats entering the duodenum are coated with lecithin and _____ to keep them emulsified and enhance the efficiency of their digestion.

20. Within the absorptive cells of the small intestine, ferritin binds the nutrient _____.

▶ Building Your Medical Vocabulary

Answers in Appendix A

State a meaning of each word element, and give a medical term from this chapter that uses it or a slight variation of it.

1. antro-

2. chylo-

3. -elle

4. emet-

5. freno-

6. hepato-

7. jejuno-

8. porto-

9. pyloro-

10. sigmo-

▶ What's Wrong with These Statements?

Answers in Appendix A

Briefly explain why each of the following statements is false, or reword it to make it true.

1. Fat is not digested until it reaches the duodenum.

2. A tooth is composed mostly of enamel.

3. Hepatocytes secrete bile into the hepatic sinusoids.

4. The ileal papilla regulates the passage of chyme from the stomach into the duodenum.

5. Lipids enter the circulation when the intestinal lacteals take up the micelles formed by the small intestine.

6. Hepcidin increases iron availability by promoting its intestinal absorption.

7. The only monosaccharide absorbed by the small intestine is glucose.

8. Most of the water in the digestive tract is reabsorbed in the large intestine.

9. Secretin is a hormone that stimulates the stomach to secrete digestive enzymes.

10. Tight junctions of the small intestine prevent anything from leaking between the epithelial cells.

▶ Testing Your Comprehension

1. On Monday, David has a waffle with syrup for breakfast at 7:00, and by 10:30 he feels hungry again. On Tuesday, he has bacon and eggs and feels satisfied until his noon lunch hour approaches. Assuming an equal quantity of food on both days and all other factors to be equal, explain this difference.

2. Which of these do you think would have the most severe effect on digestion: surgical removal of the stomach, gallbladder, or pancreas? Explain.

3. What do carboxypeptidase and aminopeptidase have in common? Identify as many differences between them as you can.

4. What do micelles and chylomicrons have in common? Identify as many differences between them as you can.

5. Explain why most dietary lipids must be absorbed by the lacteals rather than by the blood capillaries of a villus.

Mitochondria (green) and smooth endoplasmic reticulum in a cell of the ovary (SEM)

P.M. Motta/S. Makabe & T. Nagur/SPL/Science Source

CHAPTER OUTLINE

26.1 Nutrition

26.1a Body Weight and Energy Balance

26.1b Appetite

26.1c Calories

26.1d Nutrients

26.1e Carbohydrates

26.1f Fiber

26.1g Lipids

26.1h Proteins

26.1i Minerals and Vitamins

26.2 Carbohydrate Metabolism

26.2a Glucose Catabolism

26.2b Glycolysis

26.2c Anaerobic Fermentation

26.2d Aerobic Respiration

26.2e Overview of ATP Production

26.2f Glycogen Metabolism

26.3 Lipid and Protein Metabolism

26.3a Lipids

26.3b Proteins

26.3c Liver Functions in Metabolism

26.4 Metabolic States and Metabolic Rate

26.4a The Absorptive State

26.4b The Postabsorptive State

26.4c Metabolic Rate

26.5 Body Heat and Thermoregulation

26.5a Body Temperature

26.5b Heat Production and Loss

26.5c Thermoregulation

26.5d Disturbances of Thermoregulation

Study Guide

DEEPER INSIGHTS

26.1 Clinical Application: Obesity

26.2 Evolutionary Medicine: Evolution of the Sweet Tooth

26.3 Clinical Application: Hepatitis and Cirrhosis

26.4 Clinical Application: Alcohol and Alcoholism

Anatomy & Physiology Revealed 4.0

Module 12: Digestive System

- Understanding metabolism requires prior familiarity with the basic concepts of metabolism in section 2.3d and metabolic pathways in section 2.4f.

- You should be familiar with the chemistry of cholesterol and saturated and unsaturated fats before reading on LDLs, HDLs, and other blood lipoproteins in this chapter (see section 2.4d).

- Lipoprotein processing involves receptor-mediated endocytosis, introduced in section 3.3f.

- The catabolic reactions discussed in this chapter are predominantly ways of making ATP. You must understand the structure and functions of ATP (see fig. 2.29 and the associated text).

- Insulin and glucagon are particularly important in regulating metabolism; brush up on these hormones in section 17.3f if necessary.

- In the regulation of body temperature, heat transfer follows the principle of flow down thermal gradients introduced in section 1.6e.

Nutrition is the starting point and basis for all human form and function. From the time a single-celled, fertilized egg divides in two, nutrition provides the matter needed for cell division, growth, and development. It is the source of fuel that provides the energy for all biological work, and of the raw materials for replacement of worn-out biomolecules and cells. The fact that nutrients are only the raw materials means, further, that chemical change—metabolism—lies at the foundation of form and function. In chapter 25, we saw how the digestive system breaks nutrients down into usable form and absorbs them into the blood and lymph. We now consider these nutrients in more depth, follow their fate after absorption, and explore related issues of metabolism and body heat.

26.1 Nutrition

Expected Learning Outcomes

When you have completed this section, you should be able to

a. describe some factors that regulate hunger and satiety;

b. define *nutrient* and list the six major categories of nutrients;

c. state the function of each class of macronutrients, the approximate amounts required in the diet, and some major dietary sources of each;

d. name the blood lipoproteins, state their functions, and describe how they differ from each other; and

e. name the major vitamins and minerals required by the body and the general functions they serve.

26.1a Body Weight and Energy Balance

The subject of nutrition quickly brings to mind the subject of body weight and the popular desire to control it. Weight is determined by one's energy balance—if energy intake and output are equal, body weight is stable. We gain weight if intake exceeds output and lose weight if output exceeds intake. Body weight usually remains quite stable over many years' time and seems to have a homeostatic set point. This has been experimentally demonstrated in animals. If an animal is force-fed until it becomes obese and then allowed to feed at will, it voluntarily reduces intake and soon stabilizes at its former weight. Similarly, if an animal is undernourished until it loses much of its weight and then allowed to feed at will, it increases its intake and again quickly stabilizes at its former weight.

In humans, the set point varies greatly from person to person, and body weight results from a combination of hereditary and environmental influences. From studies of identical twins and other people, it appears that about 30% to 50% of the variation in human weight is due to heredity, and the rest to factors such as eating and exercise habits.

26.1b Appetite

The struggle for weight control often seems to be a struggle against the appetite. Since the early 1990s, physiologists have discovered a still-growing list of peptide hormones and regulatory pathways that control short- and long-term appetite and body weight. Some of the hormones have been called *gut–brain peptides* because they act as two-way chemical signals between the gastrointestinal tract and brain. A few will be described here to give some idea of regulatory mechanisms known to date and where a great deal of research is currently focused.

Short-Term Regulators of Appetite

The following peptides work over periods of minutes to hours, making one feel hungry and begin eating, then feel satiated and end a meal:

- **Ghrelin.**[1] This is secreted by parietal cells in the gastric fundus, especially when the stomach is empty. It produces the sensation of hunger and stimulates the hypothalamus to secrete growth hormone–releasing hormone, priming the body to take advantage of the nutrients about to be absorbed. Within an hour after eating, ghrelin secretion ceases.

- **Peptide YY (PYY).** This hormone is secreted by enteroendocrine cells in the ileum and colon, which sense that food has arrived even as it enters the stomach. They secrete PYY long before the chyme reaches the ileum, and in quantities proportional to the calories consumed. The primary effect of PYY is to signal satiety and terminate eating. Thus, ghrelin is one of the signals that begins a meal, and PYY is one of the signals that ends it. PYY remains elevated well after a meal. It acts as an *ileal brake* that prevents the stomach from emptying too quickly, and therefore prolongs the sense of satiety.

[1] named partly from *ghre* = growth, and partly as an acronym derived from growth hormone–releasing hormone

- **Cholecystokinin (CCK).** CCK is secreted by enteroendocrine cells in the duodenum and jejunum. It stimulates the secretion of bile and pancreatic enzymes, but also stimulates the brain and sensory fibers of the vagus nerves, producing an appetite-suppressing effect. Thus, it joins PYY as a signal to stop eating.

- **Amylin.** This hormone from the beta cells of the pancreatic islets also produces a feeling of satiety and winds down the digestive activities of the stomach.

Long-Term Regulators of Appetite

Other peptides regulate appetite, metabolic rate, and body weight over the longer term, thus governing one's average rate of caloric

intake and energy expenditure over periods of weeks to years. The following two members of this group work as "adiposity signals," informing the brain of how much adipose tissue the body has and activating mechanisms for adding or reducing fat.

- **Leptin.**[2] Leptin is secreted by adipocytes throughout the body. Its level is proportional to one's fat stores, so this is the brain's primary way of knowing how much body fat we have. Adipose tissue is innervated by thick bundles of

---

[2]*lept* = thin

DEEPER INSIGHT 26.1

CLINICAL APPLICATION

Obesity

Obesity is clinically defined as a weight more than 20% above the recommended norm for one's age, sex, and height. In the United States, about 30% of the population is obese and another 35% is overweight; there has lately been an alarming increase in the number of children who are morbidly obese by the age of 10. You can judge whether you are overweight or obese by calculating your *body mass index (BMI)*. If W is your weight in kilograms and H is your height in meters, BMI = W/H^2. (Or if using weight in pounds and height in inches, BMI = $703W/H^2$.) A BMI of 20 to 25 kg/m^2 is considered to be optimal for most people. A BMI over 27 kg/m^2 is considered overweight, and above 30 kg/m^2 is considered obese **(fig. 26.1).**

Excess weight shortens life expectancy and increases a person's risk of atherosclerosis, hypertension, diabetes mellitus, joint pain and degeneration, kidney stones, and gallstones; cancer of the breast, uterus, and liver in women; and cancer of the colon, rectum, and prostate in men. Excess thoracic fat impairs breathing and results in increased blood Pco_2, sleepiness, and reduced vitality. Obesity is also a significant impediment to some surgeries.

Heredity plays as much a role in obesity as in height, and even more than in many other disorders generally acknowledged to be hereditary. However, a predisposition to obesity is often greatly worsened by overfeeding in infancy and childhood. Consumption of excess calories in childhood causes adipocytes to increase in size and number. In adulthood, adipocytes don't multiply except in some extreme weight gains; their number remains constant while weight gains and losses result from changes in cell size (cellular hypertrophy).

As so many dieters learn, it is very difficult to substantially reduce one's adult weight. Most diets are unsuccessful over the long run as dieters lose and regain the same weight over and over. From an evolutionary standpoint, this is not surprising. The body's appetite- and weight-regulating mechanisms have evolved more to limit weight loss than weight gain, for a scarcity of food was surely a more common problem than a food surplus for our prehistoric ancestors. Were it not for the mechanisms that thwart weight loss, our ancestors might not have made it through the lean eons and we might not be here; but now that we're surrounded with a glut of tempting food, these survival mechanisms have become mechanisms of pathology.

Understandably, pharmaceutical companies are keenly interested in developing effective weight-control drugs. There could be an enormous profit, for example, in a drug that inhibits ghrelin signaling

or enhances or mimics leptin or melanocortin signaling. Such efforts have so far met with little success, but clearly a prerequisite to drug development is a better understanding of appetite-regulating peptides and their receptors.

FIGURE 26.1 MRI Images of Persons with Obesity and Normal Weight. The individual on the left is a 40-year-old female 168 cm (5 ft 6 in.) tall, weighing 113 kg (250 lb), with a BMI of 40.3 kg/m^2. The adipose tissue is yellowish-white. Note the large amount of *subcutaneous fat* between the skin and muscle and *mesenteric fat* in the abdominal cavity.

Marty Chobot/National Geographic Creative

sympathetic nerve fibers. Leptin stimulates these fibers to secrete norepinephrine, which, in turn, stimulates fat breakdown *(lipolysis)*. Animals with a leptin deficiency or a defect in leptin receptors exhibit *hyperphagia* (overeating) and extreme obesity. With few exceptions, however, obese humans are not leptin-deficient or aided by leptin injections. *Leptin insensitivity* is a more common factor in obesity—a receptor defect rather than a hormone deficiency. Adipose tissue is increasingly seen as an important source of multiple hormones that influence the body's energy balance.

• **Insulin.** Insulin is secreted by the pancreatic beta cells. It stimulates glucose and amino acid uptake and promotes glycogen and fat synthesis. But it also has receptors in the brain and functions, like leptin, as an index of the body's fat stores. It has a weaker effect on appetite than leptin does, however.

An important brain center for appetite regulation is the **arcuate nucleus** of the hypothalamus. The aforementioned peptides have receptors in the arcuate nucleus, although they act on other target cells in the body as well. The arcuate nucleus has two neural networks involved in hunger. One group secretes **neuropeptide Y (NPY),** itself a potent appetite stimulant. The other secretes **melanocortin,** which inhibits eating. Ghrelin stimulates NPY secretion, whereas insulin, PYY, and leptin inhibit it. Leptin also stimulates melanocortin secretion **(fig. 26.2)** and inhibits the secretion of appetite stimulants called *endocannabinoids,* named for their resemblance to the tetrahydrocannabinol (THC) of marijuana.

▶▶▶ APPLY WHAT YOU KNOW

A friend encourages you to invest in a company that proposes to produce tablets of leptin and CCK to be taken as oral diet pills. Would you consider this a wise investment? Why or why not?

Gut–brain peptides certainly aren't the whole story behind appetite regulation. Hunger is also stimulated partly by gastric peristalsis. Mild **hunger contractions** begin soon after the stomach is emptied and increase in intensity over a period of hours. They can become quite a painful and powerful incentive to eat, yet they don't affect the amount of food consumed—this remains much the same even when nervous connections to the stomach and intestines are severed to cut off all conscious perception of hunger contractions. Food intake is terminated not only by PYY and CCK, but also in ways similar to the way water intake slakes the thirst (see section 24.1c). Merely chewing and swallowing food briefly satisfy the appetite, even if the food is removed through an esophageal fistula (opening) before reaching the stomach. Inflating the stomach with a balloon inhibits hunger even in an animal that hasn't actually swallowed any food. Satiety produced by these mechanisms, however, is very short-lived. Lasting satiety depends on the absorption of nutrients into the blood.

Appetite isn't merely a question of *how much* but also *what kind* of food is consumed. Even animals shift their diets from one kind of food to another, apparently because some foods provide nutrients that others do not. In humans, different neurotransmitters also seem to govern the appetite for different classes of nutrients.

For example, *norepinephrine* stimulates the appetite for carbohydrates, *galanin* for fatty foods, and *endorphins* for protein.

26.1c Calories

One calorie is the amount of heat that will raise the temperature of 1 g of water 1°C. One thousand calories is called a Calorie (capital *C*) in dietetics and a **kilocalorie (kcal)** in biochemistry and physiology. The relevance of calories to physiology is that they are a measure of the capacity to do biological work.

Nearly all dietary calories come from carbohydrates, proteins, and fats. Carbohydrates and proteins yield about 4 kcal/g when completely oxidized, and fats yield about 9 kcal/g. Alcohol (7.1 kcal/g) and sugary foods promote malnutrition by providing "empty calories"—they suppress the appetite but fail to provide other nutrients the body requires (see Deeper Insight 26.4). In sound nutrition, the body's energy needs are met by more complex foods that simultaneously meet the need for proteins, lipids, vitamins, and other nutrients.

When a chemical is described as **fuel** in this chapter, we mean it is oxidized solely or primarily to extract energy from it. The extracted energy is usually used to make ATP, which then transfers the energy to other physiological processes (see fig. 2.30).

26.1d Nutrients

A **nutrient** is any ingested chemical that is absorbed into the tissues and used for growth, repair, or maintenance of the body. Nutrients fall into six major classes: water, carbohydrates, lipids, proteins, minerals, and vitamins. Water, carbohydrates, lipids, and proteins are considered **macronutrients** because they must be consumed in relatively large quantities. Minerals and vitamins are called **micronutrients** because only small quantities are required. We measure dietary requirements in terms of grams per day for macronutrients but only milligrams to micrograms per day for micronutrients.

Many nutrients can be synthesized by the body when they are unavailable from the diet. The body is incapable, however, of synthesizing minerals, most vitamins, eight of the amino acids, and one to three of the fatty acids. These are called **essential nutrients** because it is essential that they be included in the diet.

Table 26.1 summarizes the functions and recommended daily intake of the macronutrient classes. (Micronutrients are discussed and tabulated later in this section.) The tabulated data are daily intakes recommended by the U.S. National Academy of Medicine for persons 19 to 50 years old. The Academy uses the term **recommended dietary allowance (RDA)** when there are enough sound data to say what intake should meet the needs of 97% or more of men and women in a given age group. In cases of less certainty, owing to less data, the Academy gives values called **adequate intake (AI),** the daily consumption thought to meet the needs of persons in a given age range. RDA and AI are components of a broader system of nutritional recommendations called *dietary reference intakes (DRIs),* which also include recommendations for special classes of individuals such as infants, children, and pregnant or lactating women, and tolerable upper

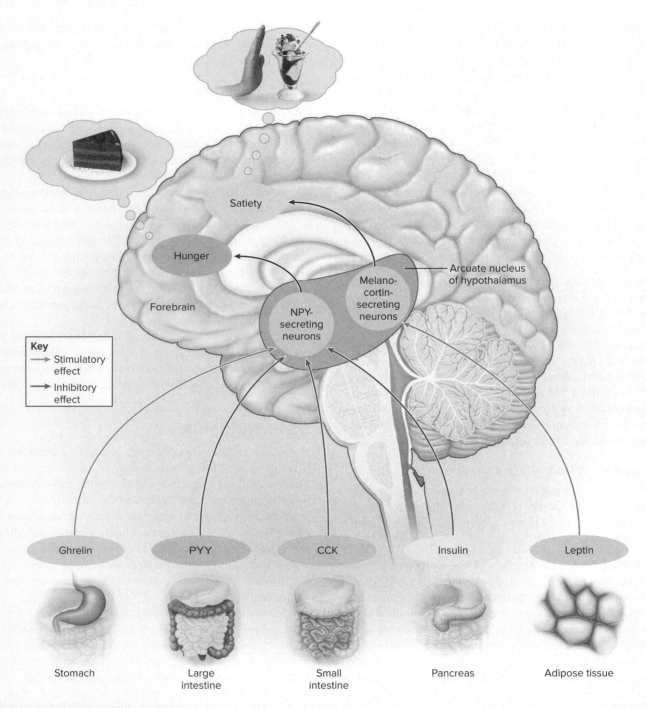

FIGURE 26.2 Principal Pathways of Appetite Regulation by Gut–Brain Peptides. Tissues and organs at the bottom of the figure are sources of peptides that stimulate or inhibit appetite-regulating neurons in the arcuate nucleus of the hypothalamus. Depending on the balance of stimulation and inhibition, those neurons secrete NPY or melanocortin to create a conscious sensation of hunger or satiety, respectively. (The arcuate nucleus is shown far larger than its real size.) (PYY = peptide YY; CCK = cholecystokinin; NPY = neuropeptide Y)

limits on intake to avoid toxic excesses such as iron overload and hypervitaminosis (explained shortly).

26.1e Carbohydrates

A well-nourished average adult has about 440 g of carbohydrate in the body, most of it in three places: 325 g of muscle glycogen, 90 to 100 g of liver glycogen, and 15 to 20 g of blood glucose.

Sugars function as a structural component of other molecules including glycoproteins, glycolipids, ATP and related nucleotides (GTP, cAMP, etc.), and nucleic acids; they can also be converted to amino acids and fats. Most of the body's carbohydrate, however, serves as fuel—an easily oxidized source of chemical energy. Most cells meet their energy needs from a combination of carbohydrates and fats, but some cells, such as neurons and erythrocytes, depend almost exclusively on carbohydrates. Even a brief period

TABLE 26.1	Macronutrient Requirements and Functions		
Nutrient	**RDA/AI, Males**	**RDA/AI, Females**	**Representative Functions**
Water	3.7 L	2.7 L	Solvent; coolant; reactant or product in many metabolic reactions (especially hydrolysis and condensation); dilutes and eliminates metabolic wastes; supports blood volume and pressure
Carbohydrate	130 g	130 g	Fuel; a component of nucleic acids, ATP and other nucleotides, glycoproteins, and glycolipids
Protein	56 g	46 g	Muscle contraction; ciliary and flagellar motility; structure of cellular membranes and extracellular material; enzymes; major component of connective tissues; transport of plasma lipids; some hormones; oxygen-binding and -transport pigments; blood-clotting factors; blood viscosity and osmolarity; antibodies; immune recognition; neuromodulators; buffers; emergency fuel
Fat	20% to 35% of total calories		Fuel; plasma membrane structure; myelin sheaths of nerve fibers; hormones; eicosanoids; bile acids; insulation; protective padding around organs; absorption of fat-soluble vitamins; vitamin D synthesis; some blood-clotting factors

of **hypoglycemia**[3] (deficiency of blood glucose) causes nervous system disturbances felt as weakness or dizziness.

Blood glucose concentration is therefore carefully regulated, mainly through the interplay of insulin and glucagon. Among other effects, these hormones regulate the balance between glycogen and free blood glucose. If blood glucose concentration drops too low, the body draws on its stores of glycogen to meet its energy needs. If glycogen stores are depleted, physical endurance is greatly reduced. Thus, it is important to consume enough carbohydrate to ensure that the body maintains adequate stores of glycogen for periods of exercise and fasting (including sleep).

Carbohydrate intake also influences the metabolism of other nutrients. When glucose and glycogen levels are too low to meet our energy needs, we oxidize fat as fuel; conversely, excess carbohydrate is converted to fat. This is why the consumption of starchy and sugary foods has a pronounced effect on body weight. It is unwise, however, to try to "burn off fat" by excessively reducing carbohydrate intake. The complete and efficient oxidation of fats depends on adequate carbohydrate intake and the presence of certain intermediates of carbohydrate metabolism. If these are lacking, fats are incompletely oxidized to ketone bodies, which may cause metabolic acidosis.

Requirements

Because carbohydrates are rapidly oxidized, they are required in greater amounts than any other nutrient except water. The RDA for both sexes is about 130 g. The brain alone consumes about 120 g of glucose per day. Most Americans get about 40% to 50% of their calories from carbohydrates, but highly active people should get up to 60%.

The National Academy of Medicine recommends that no more than 25% of one's daily calories come from added sugars beyond those that occur naturally in food. Carbohydrate consumption in the United States has become excessive over the past century, however, because of a combination of fondness for sweets,

DEEPER INSIGHT 26.2

EVOLUTIONARY MEDICINE

Evolution of the Sweet Tooth

Our craving for sugar doubtlessly originated in our prehistoric ancestors. Not only did they have to work much harder to survive than we do, but high-calorie foods were scarce and people were at constant risk of starvation. Those who were highly motivated to seek and consume sugary, high-calorie foods passed their "sweet tooth" on to us, their descendants—along with a similarly adaptive appetite for other rare but vital nutrients, namely fat and salt. The tastes that were essential to our ancestors' survival can now be a disadvantage in a culture in which salty, fatty, and sugary foods are all too easy to obtain; the food industry eagerly capitalizes on these tastes.

increased use of sugar in processed foods, and reduced physical activity (see Deeper Insight 26.2). A century ago, Americans consumed an average of 1.8 kg (4 lb) of sugar per year. Now, with sucrose and high-fructose corn syrup so widely used in processed foods and beverages, the average American ingests 200 to 300 g of carbohydrate per day and the equivalent of 27 kg (60 lb) of table sugar and 21 kg (46 lb) of corn syrup per year. A single 355 mL (12 oz) serving of nondiet soft drink contains 38 to 43 g of sugar (about 8 teaspoons).

Dietary carbohydrates come in three principal forms: monosaccharides, disaccharides, and polysaccharides (complex carbohydrates). The only nutritionally significant polysaccharide is starch. Although glycogen is a polysaccharide, only trivial amounts of it are present in meat. Cellulose, another polysaccharide, isn't considered a nutrient because it is not digested and never enters the human tissues. Its importance as dietary fiber, however, is discussed shortly.

The three major dietary disaccharides are sucrose, lactose, and maltose. The major monosaccharides are glucose, galactose, and fructose, which arise mainly from the digestion of starch and disaccharides. The small intestine and liver convert fructose and

[3]*hypo* = below normal; *glyc* = sugar; *emia* = blood condition

galactose to glucose, so ultimately all carbohydrate digestion generates glucose. Outside of the hepatic portal system, glucose is the only monosaccharide present in the blood in significant quantity; thus, it is known as *blood sugar*. Its concentration is normally maintained at 70 to 110 mg/dL in peripheral venous blood.

▶▶▶ APPLY WHAT YOU KNOW

Glucose concentration is about 15 to 30 mg/dL higher in arterial blood than in most venous blood. Explain why.

The effect of a dietary carbohydrate on one's blood glucose level can be expressed as the **glycemic index (GI).** The effect of ingesting 50 g of glucose on blood glucose level over the next 2 hours is set at 100, and the effects of other carbohydrates are expressed in relation to this. A carbohydrate with a GI of 50, for example, would produce half the effect of pure glucose. Carbohydrates with a high GI (≥70) are quickly digested and absorbed and rapidly raise blood glucose. Such carbohydrates are found, for example, in white bread, white rice, baked white potatoes, and many processed breakfast cereals. High-GI carbohydrates stimulate a high insulin demand and raise the risk of obesity and type 2 diabetes mellitus. Carbohydrates with a low GI (≤55) are digested more slowly and raise blood glucose more gradually. These include the carbohydrates found in most fruits and vegetables, legumes, milk, and grainy bread and pasta. The glycemic index of a given food varies, however, from person to person and even in the same person from day to day, and depends as well on how the food is cooked.

Ideally, most carbohydrate intake should be in the form of starch. This is partly because foods that provide starch also usually provide other nutrients. Simple sugars not only provide empty calories but also promote tooth decay. A typical American, however, now obtains only 50% of his or her carbohydrates from starch and the other 50% from sucrose and corn syrup.

Dietary Sources

Nearly all dietary carbohydrates come from plants—particularly grains, legumes, fruits, and root vegetables. Sucrose is refined from sugarcane and sugar beets. Fructose is present in fruits, honey, and corn syrup. Maltose is present in some foods such as germinating cereal grains. Lactose is the most abundant solute in cow's milk (about 4.6% lactose by weight).

26.1f Fiber

Dietary fiber refers to all fibrous materials of plant and animal origin that resist digestion. Most is plant matter—the carbohydrates cellulose and pectin and such noncarbohydrates as gums and lignin. Although it's not a nutrient, fiber is an essential component of the diet. The RDA of fiber is about 25 g for females and 38 g for males, but average intake varies greatly from country to country—from 40 to 150 g/day in India and Africa to only 12 g/day in the United States.

Water-soluble fiber includes pectin and certain other carbohydrates found in oats, beans, peas, carrots, brown rice, and fruits. It reduces blood cholesterol and low-density lipoprotein (LDL) levels (see Deeper Insight 19.4). **Water-insoluble fiber**

includes cellulose, hemicellulose, and lignin. It apparently has no effect on cholesterol or LDL levels, but it absorbs water and swells, thereby softening the stool and increasing its bulk by 40% to 100%. This effect stretches the colon and stimulates peristalsis, thereby quickening the passage of feces. In doing so, water-insoluble fiber reduces the risk of constipation and diverticulitis (see table 25.3).

Studies are contradictory as to whether dietary fiber has a clear effect on the incidence of colorectal cancer. Excess fiber can actually have a deleterious effect on health by interfering with the absorption of iron, calcium, magnesium, phosphorus, and some trace elements.

26.1g Lipids

Healthy young men and women average, respectively, about 15% and 25% fat by weight. Fat accounts for most of the body's stored energy. Lesser amounts of phospholipid, cholesterol, and other lipids also play vital structural and physiological roles.

A well-nourished adult meets 80% to 90% of his or her resting energy needs from fat. Fat is superior to carbohydrates for energy storage for two reasons: (1) Carbohydrates are hydrophilic, absorb water, and thus expand and occupy more space in the tissues. Fat, however, is hydrophobic, contains almost no water, and is a more compact energy storage substance. (2) Fat is less oxidized than carbohydrate and contains over twice as much energy (9 kcal/g of fat compared with 4 kcal/g of carbohydrate). A man's typical fat reserves contain enough energy for 119 hours of running, whereas his carbohydrate stores would last for only 1.6 hours.

Fat has **glucose-sparing** and **protein-sparing effects**—as long as enough fat is available to meet the energy needs of the tissues, protein isn't catabolized for fuel and glucose is spared for consumption by cells that can't use fat, such as neurons.

Vitamins A, D, E, and K are fat-soluble vitamins, which depend on dietary fat for their absorption by the intestine. People who ingest less than 20 g of fat per day are at risk of vitamin deficiency because there isn't enough fat in the intestine to transport these vitamins into the tissues.

Phospholipids and cholesterol are major structural components of plasma membranes and myelin. Cholesterol is also important as a precursor of steroid hormones, bile acids, and vitamin D. Thromboplastin, an essential blood-clotting factor, is a lipoprotein. Two fatty acids—arachidonic acid and linoleic acid—are precursors of prostaglandins and other eicosanoids.

In addition to its metabolic and structural roles, fat has important protective and insulating functions described in section 5.3c.

Requirements

Fat should account for no more than 35% of one's daily caloric intake, and no more than 10% of fat intake should be saturated fat. A typical American consumes 30 to 150 g of fat per day and obtains 40% to 50% of his or her calories from fat. It is important that the diet include certain **essential fatty acids**—so named because the body cannot synthesize these from other precursors and therefore must obtain them from food. These include linoleic acid

FIGURE 26.3 Serum Lipoproteins. (a) Structure of a lipoprotein. (b) Lipid and protein composition of the lipoprotein classes.

and possibly linolenic and arachidonic acids; there are differences of opinion about the body's ability to synthesize the last two. As long as 1% to 2% of the total energy intake comes from linoleic acid, people show no signs of essential fatty acid deficiency. In the typical Western diet, linoleic acid provides about 6% of the calories. Other fatty acids can be synthesized in the body if lacking from the diet.

Sources

Saturated fats are predominantly of animal origin. They occur in meat, egg yolks, and dairy products but also in some plant products such as coconut and palm oils (common in nondairy coffee creamers and other products). Processed foods such as hydrogenated oils and vegetable shortening are also high in saturated fat, which is therefore abundant in many baked goods. Unsaturated fats predominate in nuts, seeds, and most vegetable oils. The essential fatty acids are amply provided by the vegetable oils in mayonnaise, salad dressings, and margarine and by whole grains and vegetables. Excessive consumption of saturated and unsaturated fats is a risk factor for diabetes mellitus, cardiovascular disease, and breast and colon cancer.

The richest source of cholesterol is egg yolks, but it is also prevalent in milk products; shellfish (especially shrimp); organ meats such as kidneys, liver, and brains; and other mammalian meat. Foods of plant origin contain only insignificant traces of cholesterol. The serum cholesterol level is strongly influenced by the types and quantity of fatty acids in the diet.

Cholesterol and Serum Lipoproteins

Lipids are an important part of the diet and must be transported to all cells of the body, yet they are hydrophobic and don't dissolve in the aqueous blood plasma. This problem is overcome by complexes called **lipoproteins**—tiny droplets with a core of

cholesterol and triglycerides and a coating of proteins and phospholipids **(fig. 26.3a).** The coating not only enables the lipids to remain suspended in the blood, but also serves as a recognition marker for cells that absorb them. The complexes are often referred to as *serum lipoproteins* because their concentrations are expressed in terms of a volume of blood serum, not whole blood.

Lipoproteins are classified into four major categories (and some lesser ones) by their density: **chylomicrons, high-density lipoproteins (HDLs), low-density lipoproteins (LDLs),** and **very-low-density lipoproteins (VLDLs) (fig. 26.3b).** The higher the proportion of lipid to protein, the lower the density; this is easy to remember if you bear in mind that oil and fat float in water because of their relatively low density. These particles also differ considerably in size: Chylomicrons range widely from 75 to 1,200 nm in diameter, but the others diminish in size from VLDLs (30–80 nm), to LDLs (18–25 nm), to HDLs (5–12 nm). Their most important differences, however, are in composition and function.

Figure 26.4 shows the three primary pathways by which blood lipoproteins are made and processed. Chylomicrons form in the absorptive cells of the small intestine and then pass into the lymphatic system and ultimately the bloodstream. Endothelial cells of the blood capillaries have a surface enzyme called **lipoprotein lipase** that hydrolyzes chylomicron triglycerides into monoglycerides and free fatty acids (FFAs). These products can then pass through the capillary walls into adipocytes, where they are resynthesized into storage triglycerides. Some FFAs, however, remain in the blood plasma bound to albumin. The remainder of a chylomicron after the triglycerides have been extracted, called a *chylomicron remnant,* is removed and degraded by the liver.

VLDLs, produced by the liver, transport lipids to the adipose tissue for storage. When their triglycerides are removed in the adipose tissue, the VLDLs become LDLs and contain mostly cholesterol. Cells that need cholesterol (usually for membrane structure or steroid hormone synthesis) absorb LDLs by receptor-mediated

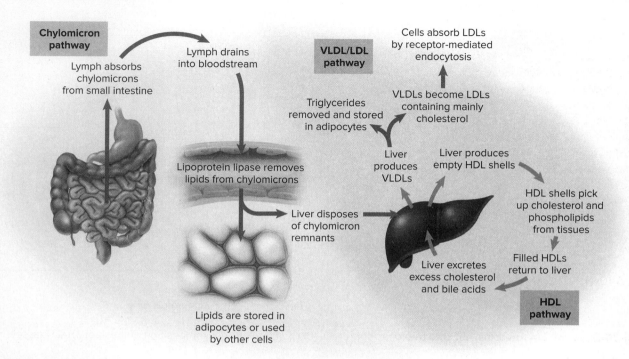

FIGURE 26.4 Lipoprotein Processing. Pathways of processing chylomicrons, VLDL/LDLs, and HDLs.

❓ *Why is a high HDL:LDL ratio healthier than a high LDL:HDL ratio?*

endocytosis, digest them with lysosomal enzymes, and release the cholesterol for cellular use.

HDL production begins in the liver, which produces an empty, collapsed protein shell. This shell travels in the blood and picks up cholesterol and phospholipids from other organs. The next time it circulates through the liver, the liver removes the cholesterol and eliminates it in the bile as either cholesterol or bile acids. HDLs are therefore a vehicle for removing excess cholesterol from the body.

It is desirable to maintain a total cholesterol concentration of 200 mg/dL or less in the blood plasma. From 200 to 239 mg/dL is considered borderline high, and levels over 240 mg/dL are pathological. Most of the body's cholesterol is endogenous (internally synthesized) rather than dietary, and the body compensates for variations in dietary intake. High intake steps down cholesterol synthesis by the liver, whereas a low dietary intake steps it up. Thus, lowering dietary cholesterol has little effect on blood cholesterol levels. The Academy of Medicine recommends consuming no more than 300 mg of cholesterol per day, which is only a little more than the amount in one egg yolk (240 mg). However, recent studies have found no correlation between dietary cholesterol intake and levels of serum cholesterol, and dietary cholesterol has lately been dismissed as no longer a concern for public health.

More important is the fact that certain saturated fatty acids (SFAs) raise serum cholesterol level. For example, palmitic acid, a 16-carbon SFA, raises serum cholesterol by blocking its uptake by the tissues (yet stearic acid, an 18-carbon SFA, doesn't raise it). Some food advertising is deceptive on this point. It may truthfully advertise a food as being cholesterol-free, but neglect to mention

that it contains SFAs that may raise the consumer's cholesterol level anyway. A moderate reduction of saturated fatty acid intake can lower blood cholesterol by 15% to 20%—considerably more effective than reducing dietary cholesterol itself. An unsaturated fatty acid, linoleic acid, has a cholesterol-lowering effect.

Vigorous exercise also lowers blood cholesterol levels. The mechanism is somewhat roundabout: Exercise reduces the sensitivity of the right atrium of the heart to blood pressure, so the heart secretes less natriuretic peptide. Consequently, the kidneys excrete less sodium and water, and the blood volume rises. This dilutes the lipoproteins in the blood, and the adipocytes compensate by producing more lipoprotein lipase. The adipocytes therefore consume more blood triglycerides. This shrinks VLDL particles, which shed some of their cholesterol in the process, and HDLs pick up this free cholesterol for removal by the liver.

Blood cholesterol isn't the only important measure of healthy lipid concentrations. A high LDL concentration is a warning sign because, as you can see from the function of LDLs described previously, it signifies a high rate of cholesterol deposition in the arteries. In the popular media, LDLs are often called "bad cholesterol"—a misleading expression, since these lipoproteins are made of more than just cholesterol, and there is only one kind of cholesterol, which does more overall good than harm. LDLs are elevated not only by saturated fats but also by cigarette smoking, coffee, and stress. A high proportion of HDL—so-called "good cholesterol"—is beneficial because it indicates that cholesterol is being removed from the arteries and transported to the liver for disposal. Thus it is desirable to increase your ratio of HDL to LDL. This is best done with a diet low in calories and saturated fats and is promoted by regular aerobic exercise.

26.1h Proteins

Protein constitutes about 12% to 15% of the body mass; 65% of it is in the skeletal muscles. Proteins are responsible for muscle contraction and the motility of cilia and flagella. They are a major structural component of all cellular membranes, with multiple roles such as membrane receptors, pumps, ion channels, and cell-identity markers. Fibrous proteins such as collagen, elastin, and keratin make up much of the structure of bone, cartilage, tendons, ligaments, skin, hair, and nails. Globular proteins include antibodies, hormones, neuromodulators, hemoglobin, myoglobin, and about 2,000 enzymes that control nearly every aspect of cellular metabolism. They also include the albumin and other plasma proteins that maintain blood viscosity and osmolarity and transport lipids and some other plasma solutes. Proteins buffer the pH of body fluids and contribute to the resting membrane potentials of all cells. No other class of biomolecules has such a broad variety of functions.

Requirements

For persons of average weight, the RDA of protein is 46 to 56 g, depending on age and sex. Multiplying your weight in pounds by 0.37 or your weight in kilograms by 0.8 gives an estimate of your protein RDA in grams. A higher intake is recommended, however, under conditions of stress, infection, injury, and pregnancy. Infants and children require more protein than adults relative to body weight. Excessive protein intake, however, overloads the kidneys with nitrogenous waste and can cause renal damage. This is a risk in certain high-protein fad diets.

Total protein intake isn't the only significant measure of dietary adequacy. The nutritional value of a protein depends on whether it supplies the right amino acids in the proportions needed for building human proteins. Adults can synthesize 12 of the 20 amino acids from other organic compounds when they're not available from the diet, but there are 8 **essential amino acids** that we can't synthesize: isoleucine, leucine, lysine, methionine, phenylalanine, threonine, tryptophan, and valine. (Infants also require histidine.) In addition, 2 amino acids can be synthesized only from essential amino acids: cysteine from methionine and tyrosine from phenylalanine. The other 10 (9 in infants) are called **inessential amino acids**—not because the body doesn't need them but because it can synthesize its own when the diet doesn't supply them.

When a protein is to be synthesized, all of the necessary amino acids must be present at once; if even one is missing, the protein cannot be made. High-quality **complete proteins** are those that provide all essential amino acids in the necessary proportions for human tissue growth, maintenance, and nitrogen balance. Lower-quality **incomplete proteins** lack one or more essential amino acids. For example, cereals are low in lysine and legumes are low in methionine.

Protein quality is also determined by **net protein utilization**—the percentage of the amino acids in a protein that the human body uses. We typically use 70% to 90% of animal protein but only 40% to 70% of plant protein. It therefore takes a larger serving of plant protein than animal protein to meet our needs—for example, we need 400 g (about 14 oz) of rice and beans to provide as much usable protein as 115 g (about 4 oz) of lean hamburger. However, reducing meat intake and increasing plant intake have advantages. Among other considerations, plant foods provide more vitamins, minerals, and fiber; less saturated fat; no cholesterol; and less pesticide. In an increasingly crowded world, it must also be borne in mind that it requires far more land (and often habitat destruction) to produce meat than to produce food crops.

Dietary Sources

The animal proteins of meat, eggs, and dairy products closely match human proteins in amino acid composition. Thus, animal products provide high-quality complete protein, whereas plant proteins are incomplete. Nevertheless, this doesn't mean that your dietary protein *must* come from meat; indeed, about two-thirds of the world's population receives adequate protein nutrition from diets containing very little meat. We can combine plant foods so that one provides what another lacks—beans and rice, for example, are a complementary combination of legume and cereal. Beans provide the isoleucine and lysine lacking in grains, while rice provides the tryptophan and cysteine lacking in beans.

Nitrogen Balance

Proteins are our chief dietary source of nitrogen. **Nitrogen balance** is a state in which the rate of nitrogen ingestion equals the rate of excretion (chiefly as nitrogenous wastes in the urine). Growing children exhibit a state of **positive nitrogen balance** because they ingest more than they excrete, thus retaining protein for tissue growth. Pregnant women and athletes in resistance training also show positive nitrogen balance. When excretion exceeds ingestion, a person is in a state of **negative nitrogen balance.** This indicates that body proteins are being broken down and used as fuel. Proteins of the muscles and liver are more easily broken down than others; therefore, negative nitrogen balance tends to be associated with muscle atrophy. Negative nitrogen balance may occur if carbohydrate and fat intake are insufficient to meet the need for energy. Carbohydrates and fats are said to have a protein-sparing effect because they prevent protein catabolism when present in sufficient amounts to meet one's energy needs.

Nitrogen balance is affected by some hormones. Growth hormone and sex steroids promote protein synthesis and positive nitrogen balance during childhood, adolescence, and pregnancy. Glucocorticoids, on the other hand, promote protein catabolism and negative nitrogen balance in states of stress.

▶▶▶ **APPLY WHAT YOU KNOW**

Would you expect a person recovering from a long infectious disease to be in a state of positive or negative nitrogen balance? Why?

26.1i Minerals and Vitamins

Minerals are inorganic elements that plants extract from soil or water and introduce into the food web. Vitamins are small dietary organic compounds that are necessary to metabolism. Neither is used as fuel, but both are essential to our ability to use other nutrients. With the exception of a few vitamins, these nutrients can't be synthesized by the body and must be included in the diet. They are required in relatively small quantities, and thus classified as micronutrients. Nevertheless, they have very potent effects on physiology. Indeed, excessive amounts are toxic and potentially lethal.

RDAs have not been firmly established for all vitamins and minerals. Values in the following tables are RDAs where known, but where an RDA isn't firmly established, they are measures of adequate intake (AI) as explained earlier. The tabulated values are for average healthy adults from 19 to 50 years old.

Minerals

Minerals constitute about 4% of the body mass, with three-quarters of this being the calcium and phosphorus in the bones and teeth. Phosphorus is also a key structural component of phospholipids, DNA, RNA, ATP, cAMP, GTP, and creatine phosphate, and is the basis of the phosphate buffer system (see section 24.3a). Calcium, iron, magnesium, and manganese function as cofactors for enzymes. Iron is essential to the oxygen-carrying capacity of hemoglobin and myoglobin. Chlorine is a component of stomach acid (HCl). Many mineral salts function as electrolytes and govern the function of nerve and muscle cells, osmotically regulate the content and distribution of water in the body, and maintain blood volume (see chapter 24).

Table 26.2 summarizes adult mineral requirements and dietary sources. Broadly speaking, the best sources of minerals are vegetables, legumes, milk, eggs, fish, shellfish, and some other meats. Cereal grains are a relatively poor source, but processed cereals are often mineral-fortified.

Throughout history, sodium chloride has been both a prized commodity and a curse. Animal tissues contain relatively large amounts of salt, and carnivores rarely lack ample salt in their diets. Plants, however, are relatively poor in salt, so herbivores often must supplement their diet by ingesting salt from the soil. As humans developed agriculture and became more dependent on plants, they also became increasingly dependent on supplemental salt. Salt was once used as a common form of payment for goods and services—the word *salary* comes from *sal* (salt). Our fondness for salt and high sensitivity to it undoubtedly stem from its physiological importance and its scarcity in a largely vegetarian diet.

Whether salt intake is significantly correlated with hypertension and heart disease has been a subject of great controversy for decades. The kidneys are remarkably capable of excreting excess salt, and multiple studies have found no elevated risk of hypertension in healthy people on high-salt diets. Other investigators, however, dispute this. The greatest concern about salt intake arises for patients with renal insufficiency, diabetes, hypertension, and people with a hereditary elevated sensitivity to salt. Hypertension is a leading cause of death among black Americans, for example, who have twice the risk of hypertension and 10 times the risk of dying from it that white Americans have. An evolutionary theory for this is that people of West African ancestry, which includes the majority of black Americans, have kidneys with an especially strong tendency to retain salt. For adults in good health with no hypersensitivity to salt, the Academy of Medicine recommends a daily sodium intake of 1.5 g/day and an upper limit of 2.3 g/day

TABLE 26.2	Mineral Requirements and Some Dietary Sources		
Mineral	RDA/AI, Males	RDA/AI, Females	Some Dietary Sources*
Major Minerals			
Potassium	4,700 mg	4,700 mg	Red meat, poultry, fish, cereals, spinach, squash, bananas, apricots, tomatoes, orange juice, beans, peas, nuts
Chloride	2,300 mg	2,300 mg	Table salt, some vegetables; usually present in excess
Sodium	1,500 mg	1,500 mg	Table salt, processed foods, egg whites, miso; usually present in excess
Calcium	1,000 mg	1,000 mg	Milk, cheese, fish, shellfish, greens, tofu, orange juice, fortified cereals
Phosphorus	700 mg	700 mg	Red meat, poultry, fish, eggs, milk, legumes, whole grains, nuts
Magnesium	400–420 mg	310–320 mg	Milk, greens, whole grains, nuts, legumes, dark chocolate
Trace Minerals			
Zinc	11 mg	8 mg	Red meat, seafood, nuts, wheat germ, legumes, cereals, yeast
Iron	8 mg	15 mg	Red meat, liver, shellfish, eggs, dried fruits, nuts, legumes, molasses
Fluoride	4 mg	3 mg	Fluoridated water and toothpaste, tea, seafood, seaweed
Manganese	2.3 mg	1.8 mg	Greens, fruits, legumes, whole grains, nuts
Copper	900 µg	900 µg	Red meat, liver, shellfish, legumes, whole grains, nuts, cocoa
Iodine	150 µg	150 µg	Marine fish, fish oils, shellfish, iodized salt
Selenium	55 µg	55 µg	Red meats, organ meats, fish, shellfish, eggs, cereals
Molybdenum	45 µg	45 µg	Beans, whole grains, nuts
Chromium	35 µg	25 µg	Meats, liver, cheese, eggs, whole grains, yeast, wine

*"Red meat" refers to mammalian muscle such as beef and pork. "Organ meat" refers to brain, pancreas, heart, kidney, etc. Liver is specified separately and refers to beef, pork, and chicken livers, which are similar for most nutrients.

(sodium only; this is 5.8 g of NaCl); the average American intake is about 3.4 g/day (8.5 g NaCl), about 77% of which comes from presalted packaged and restaurant foods.

Vitamins

Vitamins were originally named with letters in the order of their discovery, but they also have chemically descriptive names such as ascorbic acid (vitamin C) and riboflavin (vitamin B_2). Most vitamins must be obtained from the diet **(table 26.3)**, but the body synthesizes some of them from precursors called *provitamins*—niacin from the amino acid tryptophan; vitamin D from cholesterol; and vitamin A from carotene, which is abundantly present in carrots, squash, and other yellow vegetables and fruits. Vitamin K, pantothenic acid, biotin, and folic acid are produced by the bacteria of the large intestine. Indeed, the feces contain more biotin than food does.

Vitamins are classified as water-soluble or fat-soluble. **Water-soluble vitamins** are absorbed with water from the small intestine, dissolve freely in the body fluids, and are quickly excreted by the kidneys. They can't be stored in the body and therefore seldom accumulate to excess. The water-soluble vitamins are ascorbic acid and the B vitamins. Ascorbic acid promotes hemoglobin synthesis, collagen synthesis, and sound connective tissue structure; it also is an antioxidant that scavenges free radicals and possibly reduces the risk of cancer. The B vitamins function as coenzymes or parts of coenzyme molecules; they assist enzymes by transferring electrons from one metabolic reaction to another, making it possible for enzymes to catalyze these reactions. Some of their functions arise later in this chapter as we consider carbohydrate metabolism.

Fat-soluble vitamins are incorporated into lipid micelles in the small intestine and absorbed with dietary lipids. They vary in function more than water-soluble vitamins. Vitamin A is a component of the visual pigments and promotes proteoglycan synthesis and epithelial maintenance. Vitamin D promotes calcium absorption and bone mineralization. Vitamin K is essential to prothrombin synthesis and blood clotting. Vitamins A and E are antioxidants, like ascorbic acid.

It is common knowledge that various diseases result from vitamin deficiencies, but it is less commonly known that **hypervitaminosis** (vitamin excess) also causes disease. A *deficiency* of vitamin A, for example, can result in night blindness, dry skin and hair, a dry conjunctiva and cloudy cornea, and increased incidence of urinary, digestive, and respiratory infections. This is the world's most common vitamin deficiency. An *excess* of vitamin A, however, may cause anorexia, nausea and vomiting, headache, pain and fragility of the bones, hair loss, an enlarged liver and spleen, and birth defects. Vitamins B_6, C, D, and E have also been implicated in toxic hypervitaminosis.

Some people take *megavitamins*—doses 10 to 1,000 times the RDA—thinking that they will improve athletic performance. Since vitamins are not burned as fuel, and small amounts fully meet the body's metabolic needs, there is no evidence that vitamin supplements improve performance except when used to correct a dietary deficiency. Megadoses of fat-soluble vitamins can be especially harmful because they are the most likely to accumulate in the body to toxic levels.

TABLE 26.3	Vitamin Requirements and Some Dietary Sources		
Vitamin	**RDA/AI, Males**	**RDA/AI, Females**	**Some Dietary Sources***
Water-Soluble Vitamins			
Ascorbic acid (C)	90 mg	75 mg	Citrus fruits, strawberries, tomatoes, greens, cabbage, cauliflower, broccoli, brussels sprouts
Niacin (nicotinic acid, B_3)	16 mg	14 mg	Readily synthesized from tryptophan, which is present in any diet with adequate protein; red meat, liver, other organ meats, poultry, fish, apricots, legumes, whole grains, mushrooms
Panthothenic acid (B_5)	5 mg	5 mg	Widely distributed, and deficiencies are rare; red meat, liver, other organ meats, eggs, green and yellow vegetables, legumes, whole grains, mushrooms, yeast
Riboflavin (B_2)	1.3 mg	1.1 mg	Widely distributed, and deficiencies are rare; all types of meat, milk, eggs, greens, whole grains, apricots, legumes, mushrooms, yeast
Pyridoxine (B_6)	1.3 mg	1.3 mg	Red meat, fish, liver, other organ meats, greens, apricots, legumes, whole grains, seeds
Thiamin (B_1)	1.2 mg	1.1 mg	Red meat, liver, other organ meats, eggs, greens, asparagus, legumes, whole grains, seeds, yeast
Folic acid (folacin, B_9)	400 μg	400 μg	Eggs, liver, greens, citrus fruits, legumes, whole grains, seeds
Biotin (B_7)	30 μg	30 μg	Red meat, liver, other organ meats, eggs, cheese, cabbage, cauliflower, bananas, legumes, nuts
Cobalamin (B_{12})	2.4 μg	2.4 μg	Red meat, liver, other organ meats, shellfish, eggs, milk; absent from food plants
Fat-Soluble Vitamins			
α-tocopherol (E)	15 mg	15 mg	Fish oils, greens, seeds, wheat germ, vegetable oils, margarine, nuts
Retinol (A)	900 μg	700 μg	Fish oils, eggs, cheese, milk, greens, other green and yellow vegetables and fruits, margarine
Phylloquinone (K)	120 μg	90 μg	Liver, greens, cabbage, cauliflower; most of requirement is supplied by intestinal bacteria
Calcitriol (D)	5 μg	5 μg	Fish, fish oils, fortified milk; formed by exposure of skin to sunlight

*See footnote in table 26.2.

Answer the following questions to test your understanding of the preceding section:

1. Name two hormones that regulate short-term hunger and satiety. How does leptin differ from these in its effects?

2. Explain the following statement: Cellulose is an important part of a healthy diet but it is not a nutrient.

3. What class of nutrients provides most of the calories in the diet? What class of nutrients provides the body's major reserves of stored energy?

4. Contrast the functions of VLDLs, LDLs, and HDLs. Explain how this is related to the fact that a high blood HDL level is desirable, but a high VLDL–LDL level is undesirable.

5. Why do some proteins have more nutritional value than others?

26.2 Carbohydrate Metabolism

Expected Learning Outcomes

When you have completed this section, you should be able to

a. describe the principal reactants and products of each major step of glucose oxidation;

b. contrast the functions and products of anaerobic fermentation and aerobic respiration;

c. explain where and how cells produce ATP; and

d. describe the production, function, and use of glycogen.

Most dietary carbohydrate is burned as fuel within a few hours of absorption. Although three monosaccharides are absorbed from digested food—glucose, galactose, and fructose—the last two are quickly converted to glucose, and all oxidative carbohydrate metabolism is essentially a matter of glucose catabolism. The overall reaction for this is

$$C_6H_{12}O_6 + 6 O_2 \longrightarrow 6 CO_2 + 6 H_2O.$$

The function of this reaction is not to produce carbon dioxide and water but to transfer energy from glucose to ATP.

Along the pathway of glucose oxidation are several links through which other nutrients—especially fats and amino acids—can also be oxidized as fuel. Carbohydrate catabolism therefore provides a central vantage point from which we can view the catabolism of all fuels and the generation of ATP.

26.2a Glucose Catabolism

If the preceding reaction were carried out in a single step, it would generate a short, intense burst of heat—like the burning of paper, which has the same chemical equation. Not only would this be useless to the body's metabolism, it would kill the cells. In the body, however, the process is carried out in a series of small steps, each controlled by a separate enzyme. Energy is released in small

manageable amounts, and as much as possible is transferred to ATP. The rest is released as heat.

There are three major pathways of glucose catabolism:

1. **glycolysis,**[4] which splits a glucose molecule into two molecules of pyruvate;

2. **anaerobic fermentation,** which reduces pyruvate to lactate without using oxygen; and

3. **aerobic respiration,** which requires oxygen and oxidizes pyruvate to carbon dioxide and water.

You may find it helpful to review figure 2.31 for a broad overview of these processes and their relationship to ATP production. The next four figures examine these processes in closer detail.

Coenzymes are vitally important to these reactions. Enzymes remove electrons (as hydrogen atoms) from the intermediate compounds of these pathways, but they don't bind them. Instead, they transfer the hydrogen atoms to coenzymes, and the coenzymes donate them to other compounds later in one of the reaction pathways. Thus, the enzymes of glucose catabolism can't function without their coenzymes.

The two coenzymes of special importance to glucose catabolism are **NAD$^+$** (nicotinamide adenine dinucleotide) and **FAD** (flavin adenine dinucleotide). Both are derived from B vitamins—NAD$^+$ from niacin and FAD from riboflavin. Hydrogen atoms are removed from metabolic intermediates in pairs—that is, two protons and two electrons (2 H$^+$ and 2 e$^-$) at a time—and transferred to a coenzyme. This produces a reduced coenzyme with a higher free energy content than it had before the reaction. Coenzymes thus become the temporary carriers of the energy extracted from glucose metabolites. The reactions for this are

$$FAD + 2 H \longrightarrow FADH_2$$

and

$$NAD^+ + 2 H \longrightarrow NADH + H^+.$$

FAD binds two protons and two electrons to become FADH$_2$. NAD$^+$, however, binds the two electrons but only one of the protons to become NADH. The other proton remains a free hydrogen ion, H$^+$ (or H$_3$O$^+$ when it combines with water, but it is represented in this chapter as H$^+$).

26.2b Glycolysis

Upon entering a cell, glucose begins a series of conversions called glycolysis (**fig. 26.5**). The numbered steps in this figure correspond to the numbered explanations in the following text.

① **Phosphorylation.** The enzyme *hexokinase* transfers an inorganic phosphate (P$_i$) group from ATP to glucose, producing glucose 6-phosphate (G6P). This has two effects:

- It keeps the intracellular concentration of glucose low, maintaining a concentration gradient that favors the continued diffusion of more glucose into the cell.

---

[4]*glyco* = sugar; *lysis* = splitting

- Phosphorylated compounds can't pass through the plasma membrane, so this prevents the sugar from leaving the cell. In most cells, step 1 is irreversible because the cells lack the enzyme to convert G6P back to glucose. The few exceptions are cells that must be able to release free glucose to the blood: absorptive cells of the small intestine, proximal convoluted tubule cells in the kidney, and liver cells.

G6P is a versatile molecule that can be converted to fat or amino acids, polymerized to form glycogen for storage, or further oxidized to extract its energy. For now, we are mainly concerned with its further oxidation (glycolysis), the general effect of which is to split G6P (a six-carbon sugar, C_6) into two three-carbon (C_3) molecules of **pyruvate.** Continue tracing these steps in figure 26.5 as you read.

② **Priming.** G6P is rearranged (isomerized) to form fructose 6-phosphate, which is phosphorylated again to form fructose 1,6-diphosphate. This "primes" the process by providing activation energy, somewhat like the heat of a match used to light a fireplace. Two molecules of ATP have already been consumed, but just as a fire gives back more heat than it takes to start it, aerobic respiration eventually gives back far more ATP than it takes to prime glycolysis.

③ **Cleavage.** The "lysis" part of glycolysis occurs when fructose 1,6-diphosphate splits into two three-carbon (C_3) molecules. Through a slight rearrangement of one of them (not shown in the figure), this generates two molecules of **PGAL (phosphoglyceraldehyde,** also called **glyceraldehyde 3-phosphate).**

④ **Oxidation.** Each PGAL is then oxidized by removing a pair of hydrogen atoms. The electrons and one proton are picked up by NAD^+ and the other proton is released into the cytosol, yielding NADH + H^+. At this step, a phosphate (P_i) group is also added to each of the C_3 fragments. Unlike the earlier steps, this P_i isn't supplied by ATP but comes from the cell's pool of free phosphate ions.

⑤ **Dephosphorylation.** In the next two steps, phosphate groups are taken from the glycolysis intermediates and transferred to ADP, converting it to ATP. The C_3 compound becomes pyruvate.

The net end products of glycolysis are therefore

$$2 \text{ pyruvate} + 2 \text{ NADH} + 2 \text{ H}^+ + 2 \text{ ATP.}$$

Note that 4 ATP are actually produced (step 5), but 2 ATP were consumed to initiate glycolysis (steps 1 and 2), so the net gain is 2 ATP per glucose. Some of the energy originally in the glucose is contained in this ATP, some is in the NADH, and some is lost as heat. Most of the energy, however, remains in the pyruvate.

26.2c Anaerobic Fermentation

The fate of pyruvate depends on whether or not oxygen is available. In an exercising muscle, the demand for ATP may exceed the supply of oxygen. The only ATP the cells can make under these circumstances is the 2 ATP produced by glycolysis. Cells without

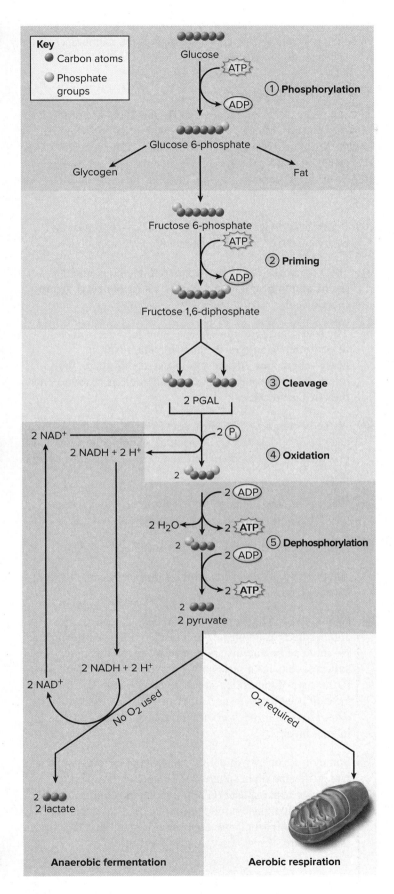

FIGURE 26.5 Glycolysis and Anaerobic Fermentation.
Numbered reaction steps are explained in the text.

mitochondria, such as erythrocytes, are restricted to making ATP by this method.

But glycolysis would quickly come to a halt if the reaction stopped at pyruvate. Why? Because it would use up the supply of NAD^+, which is needed to accept electrons at step 4 and keep glycolysis going. NAD^+ must be replenished. To achieve that in the absence of oxygen, a cell resorts to a one-step reaction called anaerobic fermentation. (This is often inaccurately called *anaerobic respiration,* but strictly speaking, human cells don't carry out anaerobic respiration; that is a process found only in certain bacteria.) In this pathway, NADH donates a pair of electrons to pyruvate, thus reducing it to **lactate** and regenerating NAD^+.

▶▶▶ **APPLY WHAT YOU KNOW**

Does lactate have more free energy than pyruvate or less? Explain.

Lactate leaves the cells that generate it and travels by way of the bloodstream to the liver. When oxygen becomes available again, the liver oxidizes lactate back to pyruvate, which can then enter the aerobic pathway described shortly. The oxygen required to do this is part of the excess postexercise oxygen consumption (EPOC) created by exercising skeletal muscles (see section 11.6c). The liver can also convert lactate back to G6P and can do either of two things with that: (1) Polymerize it to form glycogen for storage, or (2) remove the phosphate group and release free glucose into the blood.

Although anaerobic fermentation keeps glycolysis running a little longer, it has some drawbacks. One is that it is wasteful, because most of the energy of glucose is still in the lactate and has contributed no useful work. The other is that excess lactate is toxic.

Skeletal muscle is relatively tolerant of anaerobic fermentation, and cardiac muscle is less so. The brain employs almost no anaerobic fermentation. During birth, when the infant's blood supply is cut off, almost all of its organs switch to anaerobic fermentation so they don't compete with the brain for the limited supply of oxygen.

▶▶▶ **APPLY WHAT YOU KNOW**

At what point in figure 26.5 would the reaction stop, and what reaction intermediate would accumulate, if NAD^+ was unavailable to the cell? What process replenishes the NAD^+ supply?

26.2d Aerobic Respiration

Most ATP is generated in the mitochondria, which require oxygen as the final electron acceptor. In the presence of oxygen, pyruvate enters the mitochondria and is oxidized by aerobic respiration. This occurs in two principal steps:

- a group of reactions we will call the **matrix reactions,** because their controlling enzymes are in the fluid of the mitochondrial matrix; and

- reactions we will call the **membrane reactions,** because their controlling enzymes are bound to the membranes of the mitochondrial cristae.

The Matrix Reactions

The matrix reactions are shown in **figure 26.6,** where the reaction steps are numbered to resume where the previous figure ended. Most of the matrix reactions constitute a series called the **citric acid (Krebs[5]) cycle.** Preceding this, however, are three steps that prepare pyruvate to enter the cycle and thus link glycolysis to it.

⑥ Pyruvate is *decarboxylated;* that is, CO_2 is removed and pyruvate, a C_3 compound, becomes a C_2 compound.

⑦ NAD^+ removes hydrogen atoms from the C_2 compound (an oxidation reaction) and converts it to an **acetyl group (acetic acid).**

⑧ The acetyl group binds to coenzyme A, a derivative of pantothenic acid (a B vitamin). The result is **acetyl-coenzyme A (acetyl-CoA).** At this stage, the C_2 remnant of the original glucose molecule is ready to enter the citric acid cycle.

⑨ At the beginning of the citric acid cycle, CoA hands off the acetyl (C_2) group to a C_4 compound, **oxaloacetic acid.** This produces the C_6 compound **citric acid,** for which the cycle is named.

⑩ Water is removed and the citric acid molecule is reorganized, but still retains its six carbon atoms.

⑪ Hydrogen atoms are removed and accepted by NAD^+.

⑫ Another CO_2 is removed and the substrate becomes a five-carbon chain.

⑬–⑭ Steps 11 and 12 are essentially repeated, generating another free CO_2 molecule and leaving a four-carbon chain. No more carbon atoms are removed beyond this point; the substrate remains a series of C_4 compounds from here back to the start of the cycle. The three carbon atoms of pyruvate have all been removed as CO_2 at steps 6, 12, and 14. These *decarboxylation reactions* are the source of most of the CO_2 in your breath.

⑮ Some of the energy in the C_4 substrate goes to phosphorylate guanosine diphosphate (GDP) and to convert it to guanosine triphosphate (GTP), a molecule similar to ATP. GTP quickly transfers the P_i group to ADP to make ATP. Coenzyme A participates again in this step but is not shown in the figure.

---

[5]Sir Hans Krebs (1900–1981), German biochemist

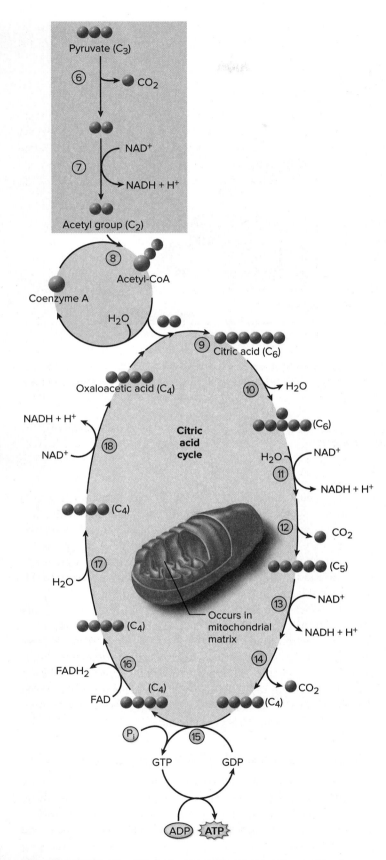

FIGURE 26.6 The Mitochondrial Matrix Reactions. Numbered reaction steps begin where figure 26.5 left off and are explained in the text.

(16) Two hydrogen atoms are removed and accepted by the coenzyme FAD.

(17) Water is added.

(18) Two final hydrogen atoms are removed and transferred to NAD⁺. This reaction generates oxaloacetic acid, which is available to start the cycle all over again.

It is important to remember that for every glucose molecule that entered glycolysis, all of these matrix reactions occur twice (once for each pyruvate). The matrix reactions can be summrized in four equations.

$$2 \text{ pyruvate} + 6 \text{ H}_2\text{O} \longrightarrow 6 \text{ CO}_2$$

$$+ \ 2 \text{ ADP} + 2 \text{ P}_i \longrightarrow 2 \text{ ATP}$$

$$+ \ 8 \text{ NAD}^+ + 8 \text{ H}_2 \longrightarrow 8 \text{ NADH} + 8 \text{ H}^+$$

$$+ \ 2 \text{ FAD} + 2 \text{ H}_2 \longrightarrow 2 \text{ FADH}_2$$

There is nothing left of the organic matter of the glucose; its carbon atoms have all been carried away as CO_2. Although still more of its energy is lost as heat along the way, some is stored in the additional 2 ATP, and most of it, by far, is in the reduced coenzymes: 8 NADH and 2 $FADH_2$ molecules generated by the matrix reactions and 2 NADH generated by glycolysis. These must be oxidized to extract the energy from them.

The citric acid cycle not only oxidizes glucose metabolites but is also a pathway and source of intermediates for the synthesis of fats and nonessential amino acids. The connections between the citric acid cycle and the metabolism of other nutrients are discussed later.

The Membrane Reactions

The membrane reactions have two purposes: (1) to further oxidize NADH and $FADH_2$ and transfer their energy to ATP and (2) to regenerate NAD⁺ and FAD and make them available again to earlier reaction steps. The membrane reactions are carried out by a series of compounds called the **mitochondrial electron-transport chain (fig. 26.7).** Most members of the chain are bound to the inner mitochondrial membrane. They are arranged in a precise order that enables each one to receive a pair of electrons from the member on one side of it (or, in two cases, from NADH and $FADH_2$) and pass these electrons along to the member on the other side—like a row of people passing along a hot potato. By the time the "potato" reaches the last member in the chain, it is relatively "cool"—its energy has been used to make ATP.

The members of this transport chain are as follows:

- **Flavin mononucleotide (FMN),** a derivative of riboflavin similar to FAD, bound to a membrane protein. FMN accepts electrons from NADH.

- **Iron–sulfur (Fe–S) centers,** complexes of iron and sulfur atoms bound to membrane proteins.

- **Coenzyme Q (CoQ),** which accepts electrons from $FADH_2$. Unlike the other members, this is a relatively small, mobile molecule that moves about in the membrane.
- **Copper (Cu) ions** bound to two membrane proteins.
- **Cytochromes,**[6] five enzymes with iron cofactors, so named because they are brightly colored in pure form. In order of participation in the chain, they are cytochromes b, c_1, c, a, and a_3.

Electron Transport   Figure 26.7 shows the order in which electrons are passed along the chain. Hydrogen atoms are split apart as they transfer from coenzymes to the chain. The protons are pumped into the intermembrane space **(fig. 26.8),** and the electrons travel in pairs (2 e^-) along the transport chain. Each electron carrier in the chain becomes reduced when it receives an electron pair and oxidized again when it passes the electrons along to the next carrier. Energy is liberated at each transfer.

The final electron acceptor in the chain is oxygen. Each oxygen atom (half of an O_2 molecule) accepts two electrons (2 e^-) from cytochrome a_3 and two protons (2 H^+) from the mitochondrial matrix. The result is a molecule of water:

$$1/2\ O_2 + 2\ e^- + 2\ H^+ \longrightarrow H_2O$$

This is the body's primary source of the *metabolic water* mentioned in section 24.1b—water synthesized in the body rather than ingested in food and drink. This reaction also explains why the body requires oxygen. Without it, this reaction stops and, like a traffic jam, stops all the other processes leading to it. As a result, a cell produces too little ATP to sustain life, and death ensues within a few minutes.

The Chemiosmotic Mechanism   Of primary importance is what happens to the energy liberated by the electrons as they pass along the chain. Some of it is unavoidably lost as heat, but some of it drives the **respiratory enzyme complexes.** The first complex includes FMN and five or more Fe–S centers; the second complex includes cytochromes b and c_1 and an Fe–S center; and the third complex includes two copper centers and cytochromes a and a_3. Each complex collectively acts as a **proton pump** that removes H^+ from the mitochondrial matrix and pumps it into the space between the inner and outer mitochondrial membranes. Coenzyme Q is a shuttle that transfers electrons from the first pump to the second, and cytochrome c shuttles electrons from the second pump to the third.

FIGURE 26.7 The Mitochondrial Electron-Transport Chain. Transport molecules are grouped into three enzyme complexes, each of which acts as a proton pump. Molecules at the upper left of the figure have a relatively high free energy content, and molecules at the lower right are relatively low in energy.

❓ *What two molecules import energy into this reaction chain, supplying the energy that becomes stored in ATP?*

These pumps create a very high H^+ concentration (low pH) and positive charge between the membranes compared with a low H^+ concentration and negative charge in the mitochondrial matrix. That is, they create a steep electrochemical gradient across the inner mitochondrial membrane. If the inner membrane were freely permeable to H^+, these ions would have a strong tendency to diffuse down this gradient and back into the matrix.

The inner membrane, however, is permeable to H^+ only through specific channel proteins called **ATP synthase** (separate from the electron-transport system). As H^+ flows through these channels, it creates an electrical current (which, you may recall, is simply moving charged particles). ATP synthase harnesses the energy of this current to drive ATP synthesis. This process is called the **chemiosmotic**[7] **mechanism,** which suggests the "push" created by the electrochemical H^+ gradient.

26.2e Overview of ATP Production

NADH releases its electron pairs (as hydrogen atoms) to FMN in the first proton pump of the electron-transport system. From there to the end of the chain, this generates enough energy to synthesize 2.5 ATP molecules per electron pair. $FADH_2$ releases its electron pairs to coenzyme Q, the shuttle between the first and second proton pumps. Therefore, it enters the chain at a point beyond the first pump and doesn't contribute energy to

---

[6]*cyto* = cell; *chrom* = color

[7]*chemi* = chemical; *osmo* = push

FIGURE 26.8 The Chemiosmotic Mechanism of ATP Synthesis. Each enzyme complex pumps hydrogen ions into the space between the mitochondrial membranes. These hydrogen ions diffuse back into the matrix by way of ATP synthase, which taps their energy to synthesize ATP.

that pump. Each $FADH_2$ contributes enough energy to synthesize 1.5 ATP.

With that in mind, we can draw up an energy balance sheet to see how much ATP is produced by the complete aerobic oxidation of glucose to CO_2 and H_2O and where the ATP comes from; see also **figure 26.9.** This summary refers back to the reaction steps 1 to 18 in figures 26.5 and 26.6. For each glucose molecule, there are

	10	NADH produced at steps 4, 7, 11, 13, and 18
	× 2.5	ATP per NADH produced by the electron-transport chain
	= 25	**ATP** generated by NADH
Plus:	2	$FADH_2$ produced at step 16
	× 1.5	ATP per $FADH_2$ produced by the electron-transport chain
	= 3	**ATP** generated by $FADH_2$
Plus:	**2**	**ATP** net amount generated by glycolysis (step 5 offset by step 2)
	2	**ATP** generated by the matrix reactions (step 15)
Total:	**32**	**ATP** per glucose.

This should be viewed as a theoretical maximum. There is some uncertainty about how much H^+ must be pumped between the mitochondrial membranes to generate each ATP, and some of the energy from the H^+ current is consumed by pumping ATP from the mitochondrial matrix into the cytosol and exchanging it for more raw materials (ADP and P_i) pumped from the cytosol into the mitochondria.

Furthermore, the NADH generated by glycolysis can't enter the mitochondria and donate its electrons directly to the electron-transport chain. In liver, kidney, and heart cells, NADH passes its electrons to *malate,* a shuttle molecule that delivers the electrons to the beginning of the electron-transport chain and generates the maximum amount of ATP. In skeletal muscle and brain cells, however, the glycolytic NADH transfers its electrons to *glycerol phosphate,* a different shuttle that donates the electrons farther down the electron-transport chain and generates less ATP. Therefore, the amount of ATP produced per NADH differs from one cell type to another and is still unknown for others.

But if we assume the maximum ATP yield, every mole (180 g) of glucose releases enough energy to synthesize up to 32 moles of ATP. Glucose has an energy content of 686 kcal/mole and ATP has 7.3 kcal/mole (233.6 kcal in 32 moles). This means

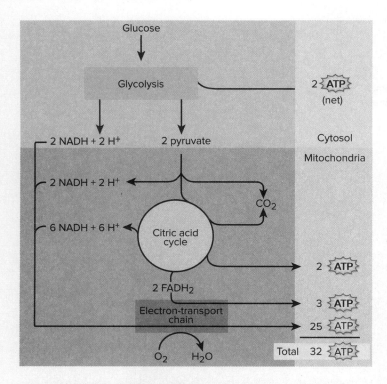

FIGURE 26.9 Summary of the Sources of ATP Generated by the Complete Oxidation of Glucose.

that aerobic respiration has an **efficiency** (a ratio of energy output to input) of up to 233.6/686 kcal = 34%. The other 66% is lost as body heat.

The pathways of glucose catabolism are summarized in **table 26.4.** The aerobic respiration of glucose can be represented in the summary equation:

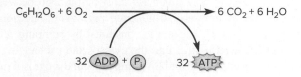

$$C_6H_{12}O_6 + 6\,O_2 \longrightarrow 6\,CO_2 + 6\,H_2O$$

$$32\,\text{ADP} + P_i \longrightarrow 32\,\text{ATP}$$

26.2f Glycogen Metabolism

ATP is quickly used after it is synthesized—it is an *energy-transfer* molecule, not an *energy-storage* molecule. Therefore, if the body has an ample amount of ATP and there is still more glucose in the blood, it doesn't produce and store excess ATP but converts the glucose to other compounds better suited for energy storage—namely, glycogen and fat. Fat synthesis is considered in section 26.3. Here we consider the synthesis and use of glycogen **(fig. 26.10),** as well as the generation of glucose from other sources. The average adult body contains about 400 to 450 g of glycogen: nearly one-quarter of it in the liver, three-quarters of it in the skeletal muscles, and small amounts in cardiac muscle and other tissues.

Glycogenesis, the synthesis of glycogen, is stimulated by insulin. Glucose 6-phosphate (G6P) is isomerized to glucose 1-phosphate (G1P). The enzyme *glycogen synthase* then cleaves off the phosphate group and attaches the glucose to a growing polysaccharide chain, thus assembling glycogen one glucose at a time.

Glycogenolysis, the hydrolysis of glycogen, releases glucose between meals when new glucose is not being ingested. The process is stimulated by glucagon and epinephrine. The enzyme *glycogen phosphorylase* begins by phosphorylating a glucose residue and splitting it off the glycogen molecule as G1P. This is isomerized to G6P, which can then enter the pathway of glycolysis.

G6P usually cannot leave the cells that produce it. Liver cells, however, have an enzyme called *glucose 6-phosphatase,* which removes the phosphate group and produces free glucose. This can diffuse out of the cell into the blood, where it is available to any cells in the body. Although muscle cells can't directly release glucose into the blood, they contribute indirectly to blood glucose concentration because they release pyruvate and lactate, which are converted to glucose by the liver.

Gluconeogenesis[8] is the synthesis of glucose from noncarbohydrates such as glycerol and amino acids. It occurs chiefly in the liver, but after several weeks of fasting, the kidneys also undertake

---

[8]*gluco* = sugar, glucose; *neo* = new; *genesis* = production of

TABLE 26.4	Pathways of Glucose Catabolism		
Stage	**Principal Reactants**	**Principal Products**	**Purpose**
Glycolysis	Glucose, 2 ADP, 2 P_i, 2 NAD$^+$	2 pyruvate, 2 ATP, 2 NADH, 2 H_2O	Reorganizes glucose and splits it in two in preparation for further oxidation by the mitochondria; sole source of ATP in anaerobic conditions
Anaerobic fermentation	2 pyruvate, 2 NADH	2 lactate, 2 NAD$^+$	Regenerates NAD$^+$ so glycolysis can continue to function (and generate ATP) in the absence of oxygen
Aerobic respiration			
Matrix reactions	2 pyruvate, 8 NAD$^+$, 2 FAD, 2 ADP, 2 P_i, 8 H_2O	6 CO_2, 8 NADH, 2 FADH$_2$, 2 ATP, 2 H_2O	Remove electrons from pyruvate and transfer them to coenzymes NAD$^+$ and FAD; produce some ATP
Membrane reactions	10 NADH, 2 FADH$_2$, 6 O_2	Up to 28 ATP, 12 H_2O	Finish oxidation and produce most of the ATP of cellular respiration

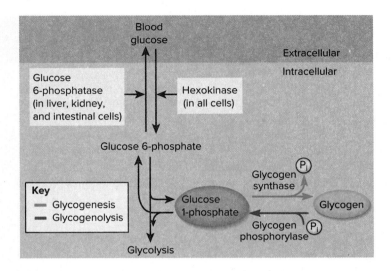

FIGURE 26.10 Major Pathways of Glucose Storage and Use. In most cells, the glucose 1-phosphate generated by glycogenolysis can undergo only glycolysis. In liver, kidney, and intestinal cells, it can be converted back to free glucose and released into circulation.

TABLE 26.5	Some Terminology Related to Glucose and Glycogen Metabolism
Anabolic (Synthesis) Reactions	
Glycogenesis	The synthesis of glycogen by polymerizing glucose
Gluconeogenesis	The synthesis of glucose from noncarbohydrates such as glycerol and amino acids
Catabolic (Breakdown) Reactions	
Glycolysis	The splitting of glucose into two molecules of pyruvate in preparation for anaerobic fermentation or aerobic respiration
Glycogenolysis	The hydrolysis of glycogen to release free glucose or glucose 1-phosphate

this process and eventually produce just as much glucose as the liver does.

Table 26.5 summarizes the distinctions among those similar terms.

BEFORE YOU GO ON

Answer the following questions to test your understanding of the preceding section:

6. Identify the reaction steps in figures 26.5 and 26.6 at which vitamins are needed for glucose catabolism.

7. In the laboratory, glucose can be oxidized in a single step to CO_2 and H_2O. Why do living cells do it in so many little steps?

8. Explain the origin of the word *glycolysis* and why this is an appropriate name for the function of that reaction pathway.

9. What are two advantages of aerobic respiration over anaerobic fermentation?

10. What important enzyme is found in the inner mitochondrial membrane other than those of the electron-transport chain? Explain how its function depends on the electron-transport chain.

11. Describe how the liver responds to (a) an excess and (b) a deficiency of blood glucose.

26.3 Lipid and Protein Metabolism

Expected Learning Outcomes

When you have completed this section, you should be able to

a. describe the processes of lipid catabolism and anabolism;

b. describe the processes of protein catabolism and anabolism;

c. explain the metabolic source of ammonia and how the body disposes of it; and

d. describe the many functions of the liver and explain why liver failure is so serious.

In the foregoing discussion, glycolysis and the mitochondrial reactions were treated from the standpoint of carbohydrate oxidation. These pathways also serve for the oxidation of proteins and lipids as fuel and as a source of metabolic intermediates that can be used for protein and lipid synthesis. Here we examine these related metabolic pathways.

26.3a Lipids

Triglycerides are stored primarily in the body's adipocytes, where the average fat molecule remains for about 2 to 3 weeks. Although the total amount of stored fat remains quite constant, there is a continual turnover as lipids are released, transported in the blood, and either oxidized for energy or redeposited in other adipocytes. Synthesizing fats from other types of molecules is called **lipogenesis,** and breaking down fat for fuel is called **lipolysis** (lih-POL-ih-sis).

Lipogenesis

It's common knowledge that a diet high in sugars causes us to put on fat. Lipogenesis employs compounds such as sugars and amino acids to synthesize glycerol and fatty acids, the triglyceride precursors. PGAL, one of the intermediates of glycolysis, can be converted to glycerol. As glucose and amino acids enter the citric acid cycle by way of acetyl-CoA, the acetyl-CoA can also be diverted to make fatty acids. The glycerol and fatty acids can then be condensed to form a triglyceride, which can be stored in the adipose tissue or converted to other lipids. These pathways are summarized in **figure 26.11.**

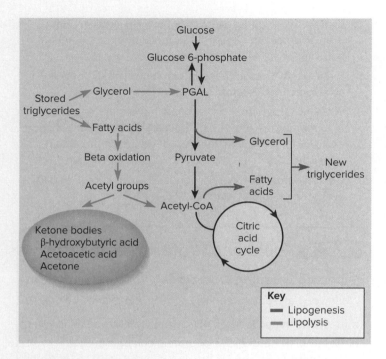

FIGURE 26.11 Pathways of Lipolysis and Lipogenesis in Relation to Glycolysis and the Citric Acid Cycle.

❓ *Name the acid–base imbalance that results from the accumulation of the ketone bodies shown in the oval.*

Lipolysis

Lipolysis, also shown in figure 26.11, begins with the hydrolysis of a triglyceride into glycerol and fatty acids—a process stimulated by epinephrine, norepinephrine, glucocorticoids, thyroid hormone, and growth hormone. The glycerol and fatty acids are further oxidized by separate pathways. Glycerol is easily converted to PGAL and thus enters the pathway of glycolysis. It generates only half as much ATP as glucose, however, because it is a C_3 compound compared with glucose (C_6); thus, it leads to the production of only half as much pyruvate.

The fatty acid component is catabolized in the mitochondrial matrix by a process called **beta oxidation,** which removes 2 carbon atoms at a time. The resulting acetyl (C_2) groups are bonded to coenzyme A to make acetyl-CoA—the entry point into the citric acid cycle. A fatty acid of 16 carbon atoms can yield 129 molecules of ATP—obviously a much richer source of energy than a glucose molecule; remember, too, that each fat molecule generates 3 fatty acids.

Excess acetyl groups can be metabolized by the liver in a process called **ketogenesis.** Two acetyl groups are condensed to form acetoacetic acid, and some of this is further converted to β-hydroxybutyric acid and acetone. These three products are the **ketone bodies.** Some cells convert acetoacetic acid back to acetyl-CoA and thus feed the C_2 fragments into the citric acid cycle to extract their energy. When the body is rapidly oxidizing fats, however, excess ketone bodies accumulate. This causes the ketoacidosis typical of type 1 diabetes mellitus, in which cells must oxidize fats because they can't absorb glucose.

Acetyl-CoA cannot go backward up the glycolytic pathway and produce glucose, because this pathway is irreversible past the point of pyruvate. Although glycerol can be used for gluconeogenesis, fatty acids cannot.

It was mentioned earlier that fats can't be completely oxidized when there isn't enough carbohydrate in the diet. This is because the mitochondrial reactions can't proceed without oxaloacetic acid as a "pickup molecule" in the citric acid cycle. When carbohydrate is unavailable, oxaloacetic acid is converted to glucose and becomes unavailable to the citric acid cycle. Fat oxidation then produces excess ketones, leading to elevated blood ketones *(ketosis)* and potentially to a resulting pH imbalance *(ketoacidosis).* Ketosis can be a serious risk in extreme low-carbohydrate diets.

26.3b Proteins

About 100 g of tissue protein breaks down each day into free amino acids. These combine with the amino acids from the diet to form an **amino acid pool** that cells can draw upon to make new proteins. The fastest rate of tissue protein turnover is in the intestinal mucosa, where epithelial cells are replaced at a very high rate. Dead cells are digested along with the food and thus contribute to the amino acid pool. Of all the amino acid absorbed by the small intestine, about 50% is from the diet, 25% from dead epithelial cells, and 25% from enzymes that have digested each other.

Some amino acids in the pool can be converted to others. Free amino acids also can be converted to glucose and fat or directly used as fuel. Such conversions involve three processes: (1) **deamination,** the removal of an amino group ($—NH_2$); (2) **amination,** the addition of $—NH_2$; or (3) **transamination,** the transfer of $—NH_2$ from one molecule to another. The following discussion shows how these processes are involved in amino acid metabolism.

Use as Fuel

The first step in using amino acids as fuel is to deaminate them. After the $—NH_2$ group is removed, the remainder of the molecule is called a *keto acid.* Depending on which amino acid is involved, the resulting keto acid may be converted to pyruvate, acetyl-CoA, or one of the acids of the citric acid cycle **(fig. 26.12).** It is important to note that some of these reactions are reversible. When there is a deficiency of amino acids in the body, citric acid cycle intermediates can be aminated and converted to amino acids, which are then available for protein synthesis. In gluconeogenesis, keto acids are used to synthesize glucose, essentially through a reversal of the glycolysis reactions.

Transamination, Ammonia, and Urea

When an amino acid is deaminated, its amino group is transferred to a citric acid cycle intermediate, α-ketoglutaric acid, converting it to glutamic acid. Such transamination reactions are the route by which several amino acids enter the citric acid cycle.

Glutamic acid can travel from any of the body's cells to the liver. Here, its —NH_2 group is removed, converting it back to α-ketoglutaric acid. The —NH_2 becomes ammonia (NH_3), which is extremely toxic to cells and cannot be allowed to accumulate. In a pathway called the **urea cycle,** the liver quickly combines ammonia with carbon dioxide to produce a less toxic waste, urea. Urea is then excreted in the urine as one of the body's nitrogenous wastes. Other nitrogenous wastes and their sources are described in section 23.1b. When a diseased liver can't carry out the urea cycle, NH_3 accumulates in the blood and death from *hepatic coma* may ensue within a few days.

Protein Synthesis

Protein synthesis, described in detail in section 4.2c, is a complex process involving DNA, mRNA, tRNA, ribosomes, and often the rough ER. It is stimulated by growth hormone, thyroid hormones, and insulin. It requires a supply of all the amino acids necessary for a particular protein. The liver can make many of these amino acids from other amino acids or from citric acid cycle intermediates by transamination reactions. The essential amino acids, however, must be obtained from the diet.

26.3c Liver Functions in Metabolism

You may notice that the liver plays a wide variety of roles in the processes discussed in this chapter—especially carbohydrate, lipid, and protein metabolism. Although it is connected to the digestive tract and regarded as a digestive gland, most of its functions are nondigestive **(table 26.6).** Except for phagocytosis, all of them are performed by the hepatocytes described in section 25.4a. Such functional diversity is remarkable in light of the uniform structure of these cells. Because of the numerous critical functions performed by the liver, degenerative liver diseases such as hepatitis, cirrhosis, and liver cancer are especially life-threatening (see Deeper Insight 26.3).

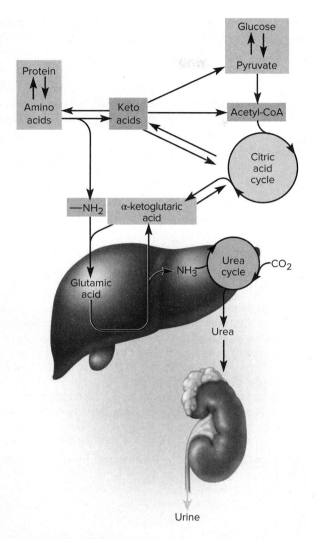

FIGURE 26.12 Pathways of Amino Acid Metabolism in Relation to Glycolysis and the Citric Acid Cycle.

❓ *Find a pathway for gluconeogenesis in this diagram.*

TABLE 26.6	Functions of the Liver

Carbohydrate Metabolism
Converts dietary fructose and galactose to glucose. Stabilizes blood glucose concentration by storing excess glucose as glycogen (glycogenesis), releasing glucose from glycogen when needed (glycogenolysis), and synthesizing glucose from fats and amino acids (gluconeogenesis) when glucose demand exceeds glycogen reserves. Receives lactate generated by anaerobic fermentation in skeletal muscle and other tissues and converts it back to pyruvate or glucose 6-phosphate.

Lipid Metabolism
Degrades chylomicron remnants. Carries out most of the body's lipogenesis (fat synthesis) and synthesizes cholesterol and phospholipids; produces VLDLs to transport lipids to adipose tissue and other tissues for storage or use; and stores fat in its own cells.

Carries out most beta oxidation of fatty acids; produces ketone bodies from excess acetyl-CoA. Produces HDL shells, which pick up excess cholesterol from other tissues and return it to the liver; excretes the excess cholesterol in bile.

Protein and Amino Acid Metabolism
Carries out most deamination and transamination of amino acids. Removes —NH_2 from glutamic acid and converts the resulting ammonia to urea by means of the urea cycle. Synthesizes nonessential amino acids by transamination reactions.

Synthesis of Plasma Proteins
Synthesizes nearly all the proteins of blood plasma, including albumin, alpha and beta globulins, fibrinogen, prothrombin, and several other clotting factors. (Does not synthesize plasma enzymes or gamma globulins.)

(continued)

TABLE 26.6	Functions of the Liver (continued)

Vitamin and Mineral Metabolism
Converts vitamin D_3 to calcidiol, a step in the synthesis of calcitriol; stores a 3- to 4-month supply of vitamin D. Stores a 10-month supply of vitamin A and enough vitamin B_{12} to last from one to several years. Secretes hepcidin to regulate iron absorption; stores iron in ferritin and releases it as needed. Excretes excess calcium by way of the bile.

Digestion
Synthesizes bile acids and lecithin, which emulsify fat and promote its digestion; produces micelles, which aid in absorption of dietary lipids.

Disposal of Drugs, Toxins, and Hormones
Detoxifies alcohol, antibiotics, and many other drugs. Metabolizes bilirubin from RBC breakdown and excretes it as bile pigments. Deactivates thyroxine and steroid hormones and excretes them or converts them to a form more easily excreted by the kidneys.

Phagocytosis
Macrophages cleanse blood of bacteria and other foreign matter.

DEEPER INSIGHT 26.3

CLINICAL APPLICATION

Hepatitis and Cirrhosis

Hepatitis, inflammation of the liver, is usually caused by one of the six strains of hepatitis viruses (HVA to HVF). They differ in mode of transmission, severity of the resulting illness, affected age groups, and the best strategies for prevention. Hepatitis A *(infectious hepatitis)* is common and mild. Over 45% of people in urban areas of the United States have had it. It spreads rapidly in such settings as day-care centers and residential institutions for psychiatric patients, and it can be acquired by eating uncooked seafood such as oysters, as well as by sewage contamination of food and water and by hand-to-mouth fecal transmission. Hepatitis A can cause up to 6 months of illness, but most people recover and then have permanent immunity to it. Hepatitis E, with a similar mode of transmission, is uncommon in the United States but a significant cause of water-borne epidemics and mortality in less economically developed countries.

Hepatitis B and C are far more serious in the United States. Both are transmitted sexually and through blood and other body fluids; the incidence of hepatitis C has surpassed AIDS as a sexually transmitted disease. Initial signs and symptoms of hepatitis include fatigue, malaise, nausea, vomiting, and weight loss. The liver becomes enlarged and tender. Jaundice, or yellowing of the skin (see fig. 6.6b), tends to follow as hepatocytes are destroyed, bile passages are blocked, and bile pigments accumulate in the blood. Hepatitis B and C often lead to chronic hepatitis, which can progress to cirrhosis or liver cancer. More liver transplants are necessitated by hepatitis C than by any other cause.

Cirrhosis is an irreversible inflammatory liver disease. Most cases result from alcohol abuse, but hepatitis, gallstones, pancreatic inflammation, and other conditions can also bring it about. It develops slowly over a period of years, but has a high mortality rate and is one of the leading causes of death in the United States. It is characterized by a disorganized liver histology in which regions of scar tissue alternate with nodules of regenerating cells, giving the liver a lumpy or knobby appearance and hardened texture **(fig. 26.13).** As in hepatitis, blockage of the bile passages results in jaundice. Protein synthesis declines as the liver deteriorates, leading to ascites, impaired blood clotting, and other cardiovascular effects (see Deeper Insight 26.4). Obstruction of the hepatic circulation by scar tissue leads to *angiogenesis,* the growth of new blood vessels to bypass the liver. Deprived of blood, the condition of the liver worsens, with increasing necrosis and, often, liver failure. The prognosis for recovery is often poor.

FIGURE 26.13 Liver with Cirrhosis.
Biophoto Associates/Getty Images

BEFORE YOU GO ON

Answer the following questions to test your understanding of the preceding section:

12. Which of the processes in table 26.5 is most comparable to lipogenesis? Which is most comparable to lipolysis? Explain.

13. When fats are converted to glucose, only the glycerol component is used in this way, not the fatty acid. Explain why, and state what happens to the fatty acids.

14. What metabolic process produces ammonia? How does the body dispose of ammonia?

26.4 Metabolic States and Metabolic Rate

Expected Learning Outcomes

When you have completed this section, you should be able to

a. define the *absorptive* and *postabsorptive states;*

b. explain what happens to carbohydrates, fats, and proteins in each of these states;

c. describe the hormonal and nervous regulation of each state;

d. define *metabolic rate* and *basal metabolic rate;* and

e. describe some factors that alter the metabolic rate.

One's metabolism changes from hour to hour depending on how long it has been since the last meal. The **absorptive (fed) state** prevails for about 4 hours during and after a meal. This is a time in which nutrients are being absorbed and may be stored or used immediately to meet energy and other needs. The **postabsorptive (fasting) state** prevails in the late morning, late afternoon, and overnight. During this time, the stomach and small intestine are empty and the body's energy needs are met from stored fuels. The two states are compared in **table 26.7** and explained in the following discussion.

26.4a The Absorptive State

In the absorptive state, blood glucose is readily available for ATP synthesis. It serves as the primary fuel and spares the body from having to draw on stored fuels. The status of major nutrient classes during this phase is as follows:

- **Carbohydrates.** Absorbed sugars are transported by the hepatic portal system to the liver. Most glucose passes through the liver and becomes available to cells everywhere in the body. Glucose in excess of immediate need, however, is absorbed by the liver and may be converted to glycogen

or fat. Most fat synthesized in the liver is released into the circulation; its further fate is comparable to that of dietary fats, discussed next.

- **Fats.** Fats enter the lymphatic system as chylomicrons and initially bypass the liver on their way to the subclavian veins, where they are dumped into the bloodstream. As described earlier, lipoprotein lipase removes fats from the blood-borne chylomicrons for uptake by the tissues, especially adipose and muscular tissue. The liver disposes of the chylomicron remnants. Fats are the primary energy substrate for hepatocytes, adipocytes, the heart, and the muscles.

- **Amino acids.** Amino acids, like sugars, circulate first to the liver. Most pass through and become available to other cells for protein synthesis. Some, however, are removed by the liver and have one of the following fates: (1) to be used for protein synthesis; (2) to be deaminated and used as fuel for ATP synthesis; or (3) to be deaminated and used for fatty acid synthesis.

Regulation of the Absorptive State

The absorptive state is regulated largely by insulin, which is secreted in response to elevated blood glucose and amino acid levels and to the intestinal hormones gastrin, secretin, cholecystokinin, and glucose-dependent insulinotropic peptide (GIP). Insulin regulates the rate of glucose uptake by nearly all cells except neurons, kidney cells, and erythrocytes, which have independent rates of uptake. With those exceptions, insulin has the following effects on its target cells:

- Within minutes, it increases the cellular uptake of glucose by as much as 20-fold. As cells absorb glucose, the blood glucose concentration falls.

- It stimulates glucose oxidation, glycogenesis, and lipogenesis.

- It inhibits gluconeogenesis, which makes sense because blood glucose concentration is already high and there is no immediate need for more.

TABLE 26.7	Major Aspects of the Absorptive and Postabsorptive States	
	Absorptive	**Postabsorptive**
Regulatory hormones	Principally insulin Also gastrin, secretin, CCK, and GIP	Principally glucagon Also epinephrine, growth hormone
Carbohydrate metabolism	Blood glucose rising Glucose uptake Glucose stored by glycogenesis Gluconeogenesis suppressed	Blood glucose falling Glucose released by glycogenolysis Gluconeogenesis stimulated
Lipid metabolism	Lipogenesis occurring Lipid uptake from chylomicrons Lipid storage in fat and muscle	Lipolysis occurring Fatty acids oxidized as fuel Glycerol used for gluconeogenesis
Protein metabolism	Amino acid uptake, protein synthesis Excess amino acids oxidized as fuel	Proteins oxidized as fuel if glycogen and fat stores are inadequate for energy needs

- It stimulates the active transport of amino acids into cells and promotes protein synthesis.
- It acts on the brain as an adiposity signal, an index of the body's fat stores.

Following a high-protein, low-carbohydrate meal, it may seem that the amino acids would stimulate insulin secretion; insulin would accelerate both amino acid and glucose uptake; and since there was relatively little glucose in the ingested food, this would create a risk of hypoglycemia. In actuality, this is prevented by the fact that a high amino acid level stimulates the secretion of *both* insulin and glucagon. Glucagon, you may recall, is an insulin antagonist. It supports an adequate level of blood glucose to meet the needs of the brain.

26.4b The Postabsorptive State

The essence of the postabsorptive state is to homeostatically regulate plasma glucose concentration within about 90 to 100 mg/dL. This is especially critical to the brain, which cannot use alternative energy substrates except in cases of prolonged fasting. The postabsorptive status of major nutrients is as follows:

- **Carbohydrates.** Glucose is drawn from the body's glycogen reserves (glycogenolysis) or synthesized from other compounds (gluconeogenesis). The liver usually stores enough glycogen after a meal to support 4 hours of postabsorptive metabolism before significant gluconeogenesis occurs.
- **Fats.** Adipocytes and hepatocytes hydrolyze fats and convert the glycerol to glucose. Free fatty acids (FFAs) can't be converted to glucose, but they can favorably affect blood glucose concentration. As the liver oxidizes them to ketone bodies, other cells absorb and use these, or use FFAs directly, as their source of energy. By switching from glucose to fatty acid catabolism, they leave glucose for use by the brain (the glucose-sparing effect). After 4 to 5 days of fasting, the brain begins to use ketone bodies as supplemental fuel.
- **Proteins.** If glycogen and fat reserves are depleted, the body begins to use proteins as fuel. Some proteins are more resistant to catabolism than others. Collagen is almost never broken down for fuel, but muscle protein goes quickly. People with cancer and certain other chronic diseases sometimes exhibit **cachexia**[9] (ka-KEX-ee-ah), an extreme wasting away due to altered metabolism and loss of appetite (anorexia).

Regulation of the Postabsorptive State

Postabsorptive metabolism is more complex than the absorptive state. It is regulated mainly by the sympathetic nervous system

and glucagon, but several other hormones are involved. As blood glucose level drops, insulin secretion declines and the pancreatic alpha cells secrete glucagon. Glucagon promotes glycogenolysis and gluconeogenesis, raising the blood glucose level, and it promotes lipolysis and a rise in FFA levels. Thus, it makes both glucose and lipids available for fuel.

The sympathoadrenal system also promotes glycogenolysis and lipolysis, especially under conditions of injury, fear, anger, and other forms of stress. Adipose tissue is richly innervated by the sympathetic nervous system, while adipocytes, hepatocytes, and muscle cells also respond to epinephrine from the adrenal medulla. In circumstances in which there is likely to be tissue injury and a need for repair, the sympathoadrenal system therefore mobilizes stored energy reserves and makes them available to meet the demands of tissue repair. Stress also stimulates the release of cortisol, which promotes fat and protein catabolism and gluconeogenesis.

Growth hormone is secreted in response to a rapid drop in blood glucose level and in states of prolonged fasting. It opposes insulin and raises the blood glucose concentration.

26.4c Metabolic Rate

Metabolic rate means the amount of energy liberated in the body per unit of time, expressed in such terms as kcal/h or kcal/day. Metabolic rate can be measured directly by putting a person in a **calorimeter,** a closed chamber with water-filled walls that absorb the heat given off by the body. The rate of energy release is measured from the temperature change of the water. Metabolic rate can also be measured indirectly with a spirometer (see fig. 22.17), which measures the amount of oxygen a person consumes. For every liter of oxygen, approximately 4.82 kcal of energy is released from organic nutrients. This is only an estimate, because the number of kilocalories per liter of oxygen varies slightly with the type of nutrients the person is oxidizing at the time of measurement.

Metabolic rate depends on physical activity, mental state, absorptive or postabsorptive status, thyroid hormone and other hormones, and other factors. The **basal metabolic rate (BMR)** is a baseline or standard of comparison that minimizes the effects of such variables. It is the metabolic rate when one is awake but relaxed, in a room at comfortable temperature, in a postabsorptive state 12 to 14 hours after the last meal. It isn't the minimum metabolic rate needed to sustain life. When one is asleep, the metabolic rate is slightly lower than the BMR. **Total metabolic rate (TMR)** is the sum of BMR and energy expenditure for voluntary activities, especially muscular contractions.

The BMR of an average adult is about 2,000 kcal/day for a male and slightly less for a female. Roughly speaking, one must therefore consume at least 2,000 kcal/day to fuel essential metabolic tasks—active transport, muscle tone, brain activity, cardiac and respiratory rhythms, renal function, and other essential processes. Even a relatively sedentary lifestyle requires another 500 kcal/day to support a low level of physical activity, and someone who does hard physical labor may require as much as 5,000 kcal/day.

---

[9]*cac* = bad; *exia* = body condition

Aside from physical activity, some factors that raise the TMR and caloric requirements include pregnancy, anxiety (which stimulates epinephrine release and muscle tension), fever (TMR rises about 14% for each 1°C of body temperature), eating (TMR rises after a meal), and the catecholamine and thyroid hormones. TMR is relatively high in children and declines with age. Therefore, as we reach middle age we often find ourselves gaining weight with no apparent change in food intake.

Some factors that lower TMR include apathy, depression, and prolonged starvation. In weight-loss diets, loss is often rapid at first and then goes more slowly. This is partly because the initial loss is largely water and partly because the TMR drops over time, fewer dietary calories are "burned off," and there is more lipogenesis even with the same caloric intake. As one reduces food intake, the body reduces its metabolic rate to conserve body mass—thus making weight loss all the more difficult.

BEFORE YOU GO ON

Answer the following questions to test your understanding of the preceding section:

15. Define *absorptive* and *postabsorptive states.* In which state is the body storing excess fuel? In which state is it drawing from these stored fuel reserves?

16. What hormone primarily regulates the absorptive state, and what are the major effects of this hormone?

17. Explain why triglycerides have a glucose-sparing effect.

18. List a variety of factors and conditions that raise a person's total metabolic rate above basal metabolic rate.

26.5 Body Heat and Thermoregulation

Expected Learning Outcomes

When you have completed this section, you should be able to

a. identify the principal sources of body heat;

b. describe some factors that cause variations in body temperature;

c. define and contrast the different forms of heat loss;

d. describe how the hypothalamus monitors and controls body temperature; and

e. describe conditions in which the body temperature is excessively high or low.

The enzymes that control our metabolism depend on an optimal, stable working temperature. In order to maintain this, the rates of metabolic heat generation and heat loss from the body must match. An excessively low body temperature, called **hypothermia,** can slow down the metabolism to the point that it can't sustain life. Conversely, an excessively high body temperature, called **hyperthermia,** can make some metabolic pathways race ahead of others and disrupt their coordination to the point that this, too, can lead to death. **Thermoregulation,** the balance between heat production and loss, is therefore a critically important aspect of homeostasis.

26.5a Body Temperature

"Normal" body temperature depends on when, where, and in whom it is measured. Body temperature fluctuates about 1°C (1.8°F) in a 24-hour cycle. It tends to be lowest in the early morning and highest in the late afternoon. Temperature also varies from place to place in one body. The body is warmest in its deep *core* and cooler near the surface, the *shell.*

The most important body temperature is the **core temperature**—the temperature of organs in the cranial, thoracic, and abdominal cavities. The best estimate of core temperature obtainable with ease is rectal temperature: usually 37.2° to 37.6°C (99.0° to 99.7°F), but as high as 38.5°C (101°F) in active children and some adults. **Shell temperature** is the temperature closer to the surface, especially skin and oral temperature. Here, heat is lost from the body and temperatures are slightly lower than rectal temperature. Adult oral temperature is typically 36.6° to 37.0°C (97.9° to 98.6°F) but may be as high as 40°C (104°F) during hard exercise. Shell temperature fluctuates as a result of processes that serve to maintain a stable core temperature.

Blood circulation is crucial to thermoregulation. We depend on blood flow to carry metabolic heat from the body core to the shell, where it can be dissipated into the environment. Without this "radiator effect," we would soon die of hyperthermia as metabolic heat raised the core temperature beyond survivable range.

26.5b Heat Production and Loss

Most body heat comes from exergonic (energy-releasing) chemical reactions such as nutrient oxidation and ATP use. A little heat is generated by joint friction, blood flow, and other movements. At rest, most heat is generated by the brain, heart, liver, and endocrine glands; the skeletal muscles contribute about 20% to 30% of the total resting heat. Increased muscle tone or exercise greatly increases heat generation in the muscles, however; in vigorous exercise, they produce 30 to 40 times as much heat as all the rest of the body.

The body loses heat in four ways: radiation, conduction, convection, and evaporation:

1. **Radiation** is the emission of infrared (IR) rays by moving molecules. In essence, *heat* means molecular motion, and all molecular motion produces IR radiation. When an object absorbs IR rays, its molecular motion and temperature increase. Therefore, IR radiation removes heat from its source and adds heat to anything that absorbs it. The heat lamps in bathrooms and restaurants work on this principle. Our bodies continually receive IR from the objects around us and give off IR to our surroundings. Since we're usually warmer than the objects around us, we usually lose more heat this way than we gain.

2. **Conduction** is the transfer of kinetic energy from molecule to molecule as they collide with one another. Heat generated in the body core is conducted to the surface through the tissues, then lost from the body by conduction from the skin to any cooler objects or medium in contact with it. The warmth of your body adds to the molecular motion of your clothes, the chair you sit in, the air around you, or water if you go swimming or sit in a cool bath. With this transfer of kinetic energy to your surroundings, you lose body heat. You can also gain heat by conduction, as on a very hot day when the air temperature is greater than your shell temperature, or when you use a heating pad for sore muscles, bask in a hot tub, or lie on hot sand at the beach.

3. **Convection** is the transfer of heat to a moving fluid—blood, air, or water. Most of the heat generated by metabolism in the body core is carried by convection in the bloodstream to the body surface. At the skin surface, body heat warms the adjacent air. Warm air is less dense than cool air, so it rises from the body and is replaced by cooler air from below. This can be visualized by a technique called *schlieren photography* (**fig. 26.14**). Water does this as well if one stands still enough in a cool lake. Such movement of a fluid caused entirely by its change in temperature and density is called **natural convection.** When air movement is forced by a fan or the wind, even if the air itself is no cooler, it carries heat away from the body more rapidly. This effect, called **forced convection,** is the reason why, even at the same temperature, we feel cooler on a windy day than on a day when the air is still. It is the basis for the *windchill factor* of a cold, windy day. Forced convection increases heat loss by both conduction and evaporation (discussed next), but has no effect on radiation.

4. **Evaporation** is the change from a liquid to a gaseous state. The cohesion of water molecules hampers their vibrations in response to heat input. If the temperature of water is raised sufficiently, however, its molecular motion becomes great enough for molecules to break free and evaporate. Evaporation of water thus carries a substantial amount of heat with it (0.58 kcal per gram of water). This is the significance of perspiration. Sweat wets the skin surface and its evaporation carries heat away. In extreme conditions, the body can lose 2 L or more of sweat and dissipate up to 600 kcal of heat per hour by evaporative loss. Evaporative heat loss is increased by forced convection, as you can readily feel when you are sweaty and stand in front of a fan or a refreshing breeze begins to blow.

The relative amounts of heat lost by different methods depend on prevailing conditions. A nude body in still air at 21°C (70°F) loses about 60% of its heat by radiation, 18% by conduction and convection, and 22% by evaporation. If air temperature is higher than skin temperature, evaporation becomes the only means of heat loss because radiation and conduction add more heat to the body than they remove from it. Hot, humid weather hinders even

Key
→ Conduction
〜➔ Convection

(a)   (b)

FIGURE 26.14 Heat Loss by Conduction and Convection.
(a) Heat transfers from the body to surrounding air molecules by conduction. Warm air then rises from the body by convection, carrying heat away. Cool air replaces this warm air from below.
(b) Schlieren photograph of the column of warm air rising from the body.
b: Dr. Ray Clark FRPS/Science Source

evaporative cooling because there is less of a humidity gradient from skin to air. Such conditions increase the risk of heatstroke (discussed shortly).

26.5c Thermoregulation

Thermoregulation is achieved through several negative feedback loops that govern cutaneous blood flow and other mechanisms for raising or lowering body temperature. The preoptic area of the hypothalamus (anterior to the optic chiasm) functions as a **hypothalamic thermostat.** It monitors blood temperature and receives signals also from **peripheral thermoreceptors** located mainly in the skin. In turn, it sends signals either to the **heat-loss center,** a nucleus still farther anterior in the hypothalamus, or to the **heat-promoting center,** a more posterior nucleus near the mammillary bodies.

When the heat-loss center senses that the blood temperature is too high, it activates heat-losing mechanisms. The first and simplest of these is cutaneous vasodilation, which increases blood flow to the body shell (skin and subcutaneous tissue) and thus promotes heat loss. In warm conditions, the shell, or zone of relatively cool tissue, may be less than 1 cm thick **(fig. 26.15a).** If cutaneous vasodilation fails to restore normal temperature, the heat-loss center triggers sweating. It also inhibits the heat-promoting center.

When the blood temperature is too low, the heat-promoting center activates mechanisms to conserve body heat or generate more. By way of the sympathetic nervous system, it causes cutaneous vasoconstriction. This diverts warm blood from the shell and retains it deeper in the body so less heat is lost through the skin. The cooler shell may then be several centimeters thick **(fig. 26.15b).** In other mammals, the sympathetic nervous system also stimulates the arrector muscles, which make the hair stand on end. This traps an insulating blanket of still air near the skin. The human sympathetic nervous system attempts to do this as well, but since our body hair is so slight, usually the only noticeable effect of this is goose bumps.

If dermal vasoconstriction can't restore or maintain normal core temperature, the body resorts to **shivering thermogenesis.** If you leave a warm house on a very cold day, you may notice that your muscles become tense, sometimes even painfully taut, and you begin to shiver. Shivering involves a spinal reflex that causes tiny alternating contractions in antagonistic muscle pairs. Every muscle contraction releases heat from ATP, and shivering can increase the body's heat production as much as fourfold.

Nonshivering thermogenesis is a longer-term mechanism for generating heat, occurring especially in the colder seasons of the year. The sympathetic nervous system raises the metabolic rate as much as 30% after several weeks of cold weather. More nutrients are burned as fuel; we consume more calories to "stoke the furnace"; and consequently, we have more appetite in the winter than in the summer. Infants generate heat by breaking down *brown fat,* a tissue in

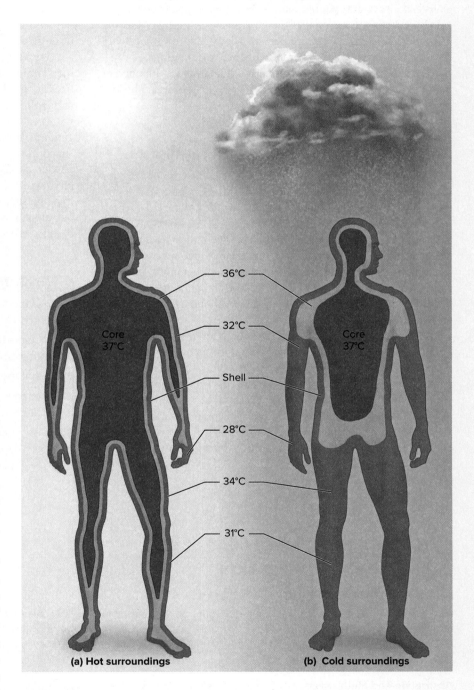

FIGURE 26.15 Distribution of Body Heat in Hot and Cold Surroundings. (a) In a hot environment, cutaneous vasodilation carries heat close to the body surface and the shell is relatively thin. (b) In a cold environment, cutaneous vasoconstriction retains heat deeper in the body and the insulating shell is significantly thicker. It is easy to see from this why frostbite so often affects extremities such as fingers and toes.

which lipolysis is not linked to ATP synthesis, so all the energy released from the fat is in the form of heat.

In addition to these physiological mechanisms, and of even greater importance, humans and other animals practice **behavioral thermoregulation**—behaviors that raise or lower the body's heat gains and losses. Just getting out of the sun greatly cuts down heat gain by radiation, for example; shedding heavy clothing or kicking off a blanket at night helps to cool the body; and dressing more warmly maintains body temperature in cold surroundings.

In summary, you can see that thermoregulation is a function of multiple organs: the brain, autonomic nerves, thyroid gland, skin, blood vessels, and skeletal muscles.

26.5d Disturbances of Thermoregulation

Fever is a normal protective mechanism against infection (see section 21.2e); it should be allowed to run its course if it is not too high. A body temperature above 42° to 43°C (108° to 110°F), however, is dangerous. The high temperature elevates the metabolic rate and the body generates heat faster than its heat-losing mechanisms can disperse it. Therefore, the elevated metabolic rate increases the fever and the fever increases the metabolic rate in a dangerous positive feedback loop. A core temperature of 44° to 45°C (111° to 113°F) can produce fatal metabolic dysfunction and neurological damage.

Exposure to excessive heat causes heat cramps, heat exhaustion, and heatstroke. **Heat cramps** are painful muscle spasms that result from electrolyte loss in the sweat. They occur especially when one begins to relax after strenuous exertion and heavy sweating. **Heat exhaustion** results from more severe water and electrolyte loss and is characterized by hypotension, dizziness, vomiting, and sometimes fainting. Prolonged heat waves, especially if accompanied by high humidity, bring on many deaths from **heatstroke (sunstroke).** The body gains heat by radiation and conduction, but the humidity retards evaporative cooling. Heatstroke is clinically defined as a state in which the core body temperature is over 40°C (104°F); the skin is hot and dry; and the subject exhibits nervous system dysfunctions such as delirium, convulsions, or coma. It is also accompanied by tachycardia, hyperventilation, inflammation, and multiorgan dysfunction; it is often fatal.

Hypothermia can result from exposure to cold weather or immersion in icy water. It, too, entails life-threatening positive feedback loops. If the core temperature falls below 33°C (91°F), the metabolic rate drops so low that heat production can't keep pace with heat loss, and the temperature falls even more. Death from cardiac fibrillation may occur below 32°C (90°F), but some people survive body temperatures as low as 29°C (84°F) in a state of suspended animation. A body temperature below 24°C (75°F) is usually fatal. It is dangerous to give alcohol to someone in a state of hypothermia; the chemical "burn" produces an illusion of warmth, but alcohol actually accelerates heat loss by dilating cutaneous blood vessels.

BEFORE YOU GO ON

Answer the following questions to test your understanding of the preceding section:

19. What is the primary source of body heat? What are some lesser sources?

20. What mechanisms of heat loss are aided by convection?

21. Describe the major heat-promoting and heat-losing mechanisms of the body.

22. Describe the positive feedback loops that can cause death from hyper- and hypothermia.

DEEPER INSIGHT 26.4

CLINICAL APPLICATION

Alcohol and Alcoholism

Alcohol not only is a popular mind-altering drug but also is regarded in many cultures as a food staple. As a source of empty calories, an addictive drug, and a toxin, it can have a broad spectrum of adverse effects on the body when it is abused.

Absorption and Metabolism

Alcohol is rapidly absorbed from the digestive tract—about 10% of it from the stomach and 90% from the proximal small intestine. Carbonation, as in beer and sparkling wines, increases its rate of absorption by moving it more quickly to the small intestine, whereas food reduces its absorption by delaying gastric emptying. Alcohol is soluble in both water and fat, so it is rapidly distributed to all body tissues and easily crosses the blood–brain barrier to exert its intoxicating effects on the brain.

Alcohol is detoxified by the hepatic enzyme *alcohol dehydrogenase,* which oxidizes it to acetaldehyde. This enters the citric acid cycle and is oxidized to CO_2 and H_2O. The average adult male can clear the blood of about 10 mL of 100% (200 proof) alcohol per hour—the amount in about 30 mL (1 oz) of whiskey or 355 mL (12 oz) of beer. Women have less alcohol dehydrogenase and clear alcohol from the bloodstream more slowly. They are also more vulnerable to alcohol-related illnesses such as cirrhosis of the liver.

Tolerance to alcohol, the ability to "hold your liquor," results from two factors: behavioral modification, such as giving in less readily to lowered inhibitions, and increased levels of alcohol dehydrogenase in response to routine alcohol consumption. Alcohol dehydrogenase also deactivates other drugs, and drug dosages must be adjusted to compensate for this when treating alcoholics for other diseases.

Physiological Effects

Nervous System Alcohol is a depressant that inhibits the release of norepinephrine and disrupts the function of GABA receptors. In low doses, it depresses inhibitory synapses and creates sensations of confidence, euphoria, and giddiness. As the dosage rises, however, the breakdown products of ethanol enhance the diffusion of K^+ out of neurons, hyperpolarizing them and making them less responsive to neurotransmitters. Thus, the timing and coordination of communication between neurons are impaired, resulting in such symptoms of intoxication as slurred speech, poor coordination, and slower reaction time. These symptoms begin to become significant at a blood alcohol level of 80 to 100 mg/dL—the legal criterion of intoxication in many states. Above 400 mg/dL, alcohol can so disrupt the electrophysiology of neurons as to induce coma and death.

Liver The liver's role in metabolizing alcohol makes it especially susceptible to long-term toxic effects. Heavy drinking stresses the liver with a high load of acetaldehyde and acetate; this depletes its oxidizing agents and reduces its ability to catabolize these intermediates as well as fatty acids. Alcoholism often produces a greatly enlarged and fatty liver for multiple reasons: The calories provided by alcohol make it unnecessary to burn fat as fuel, fatty acids are poorly oxidized, and acetaldehyde is converted to new fatty acids. Acetaldehyde also causes inflammation of the liver and pancreas *(hepatitis* and *pancreatitis),* leading to disruption of digestive function. Acetaldehyde and other toxic intermediates destroy hepatocytes faster than they can be regenerated, leading to cirrhosis (see Deeper Insight 26.3). Many symptoms of alcoholism stem from deterioration of liver functions. Hepatic coma may occur as the liver becomes unable to produce urea, thus allowing ammonia to accumulate in the blood. Jaundice is a sign of the liver's inability to excrete bilirubin.

Circulatory System Deteriorating liver functions exert several effects on the blood and cardiovascular system. Blood clotting is impaired because the liver cannot synthesize clotting factors adequately. Edema results from inadequate synthesis of blood albumin. Cirrhosis obstructs the hepatic portal blood circulation. Portal hypertension results, and combined with hypoproteinemia, this causes the liver and other organs to "weep" serous fluid into the peritoneal cavity. This leads to *ascites—* swelling of the abdomen with as much as several liters of serous fluid (see Deeper Insight 20.4). The combination of hypertension and impaired clotting often leads to hemorrhaging. *Hematemesis,*[10] the vomiting of blood, may occur as enlarged veins of the esophagus hemorrhage. Alcohol abuse also destroys myocardial tissue, reduces contractility of the heart, and causes cardiac arrhythmia.

Digestive System and Nutrition Alcohol breaks down the protective mucous barrier of the stomach and the tight junctions between its epithelial cells. Thus, it may cause gastritis and bleeding. Heavy drinking, especially in combination with smoking, increases the incidence of esophageal cancer. Malnutrition is a typical complication of alcoholism, partly because the empty calories of alcohol suppress the appetite for more nutritious foods. The average American gets about 4.5% of his or her calories from alcohol (more when nondrinkers are excluded), but heavy drinkers may obtain half or more of their calories from alcohol and have less appetite for foods that would meet their other nutritional requirements. In addition, acetaldehyde interferes with vitamin absorption and use. Thiamine deficiency is common in alcoholism, and thiamine is routinely given to alcoholics in treatment.

Addiction

Alcohol is the most widely available addictive drug in America. In many respects it is almost identical to barbiturates in its toxic effects, its potential for tolerance and dependence, and the risk of overdose. The difference is that obtaining barbiturates usually requires a prescription, while obtaining alcohol requires, at most, proof of age.

Alcoholism is defined by a combination of criteria, including the pathological changes just described; physiological tolerance of high concentrations; impaired physiological, psychological, and social functionality; and withdrawal symptoms occurring when intake is reduced or stopped. Heavy drinking followed by a period of abstinence—for example, when a patient is admitted to the hospital and cannot get access to alcohol—may trigger *delirium tremens (DT),* characterized by restlessness, insomnia, confusion, irritability, tremors, incoherent speech, hallucinations, convulsions, and coma. DT has a 5% to 15% mortality rate.

Most alcoholism (type I) sets in after age 25 and is usually associated with stress or peer pressure. These influences lead to increased drinking, which can start a vicious cycle of illness, reduced job performance, family and social problems, arrest, and other stresses leading to still more drinking. A smaller number of alcoholics have type II alcoholism, which is at least partially hereditary. Most people with type II alcoholism are men who become addicted before age 25, especially the sons of other type II alcoholics. Type II alcoholics show abnormally rapid increases in blood acetaldehyde levels when they drink, and they have unusual brain waves (EEGs) even when not drinking. Children of alcoholics have a higher than average incidence of becoming alcoholic even when raised by nonalcoholic foster parents. It is by no means inevitable that such people will become alcoholic, but stress or peer pressure can trigger alcoholism more easily in those who are genetically predisposed to it.

Alcoholism is treated primarily through behavior modification—abstinence, peer support, avoidance or correction of the stresses that encourage drinking, and sometimes psychotherapy. Drugs such as disulfiram (Antabuse) have been used to support behavior modification programs by producing unpleasant effects from alcohol consumption, but drug treatment has been fraught with potentially dangerous side effects and little evidence of effectiveness.

---

[10]*hemat* = blood; *emesis* = vomiting

STUDY GUIDE

▶ Assess Your Learning Outcomes

To test your knowledge, discuss the following topics with a study partner or in writing, ideally from memory.

26.1 Nutrition

1. Evidence of a homeostatic set point for body weight; relative contributions of heredity and behavior to differences in body weight
2. The meaning of *gut–brain peptides*
3. Sources and actions of ghrelin, peptide YY, cholecystokinin, and amylin as short-term regulators of appetite
4. Sources and actions of leptin and insulin as adiposity signals and long-term regulators of appetite
5. Roles of the arcuate nucleus, neuropeptide YY, and melanocortin in appetite regulation
6. The role of hunger contractions in the onset of feeding; factors that satiate hunger over the short and long terms, and how these resemble the satiation of thirst
7. Hormones that stimulate an appetite for specific classes of nutrients such as carbohydrates, fats, and proteins
8. The definition of *calorie* and how this relates to dietary Calories (kilocalories)
9. Principal dietary sources of calories; the relative yield from fats as compared to carbohydrates and proteins; and the meaning of *empty calories*
10. The definition of *nutrient;* why some nutrients are not digested and yield no calories; and why some things are not considered nutrients even though they are important components of a healthy diet
11. The difference between macronutrients and micronutrients; the nutrients in each category
12. Why some substances are considered to be essential nutrients
13. Forms and amounts of stored and mobile carbohydrates in the body; how the body uses carbohydrates; how dietary carbohydrates influence the metabolism of fats
14. Hormones that regulate the balance between blood glucose and stored glycogen; the normal range of blood glucose concentration
15. The recommended daily intake of carbohydrates; the percentage of calories that come from carbohydrates in a typical U.S. diet; the forms in which carbohydrates exist in the diet, and their relative amounts; dietary sources of carbohydrates

16. The meaning of *dietary fiber;* the recommended daily intake of fiber and how actual consumption varies around the world; the forms of dietary fiber; which forms are classified as soluble and insoluble fiber, and differences in the health benefits of these two classes; the detrimental effects of too much dietary fiber
17. Two reasons why the body stores more energy as fat than as carbohydrate; the caloric yield from fat compared to carbohydrate; the typical percentages of body fat in normal males and females
18. Why fat is said to have glucose-sparing and protein-sparing effects
19. Metabolically important lipids other than fat, and their uses in the body
20. The recommended daily intake of fat, and how the typical U.S. diet compares to this.
21. Why linoleic acid is called an essential fatty acid; two other fatty acids that might be essential
22. Dietary sources of saturated and unsaturated fats, essential fatty acids, and cholesterol; the health risks of excessive saturated and unsaturated fats
23. Types of lipoproteins found in the bloodstream—chylomicrons, VLDLs, LDLs, and HDLs—and the differences in their source, composition, and functions; the relevance of LDLs and HDLs to cardiovascular health and how these relate to the colloquial expressions "good cholesterol" and "bad cholesterol"
24. Functions of proteins in the body, and a typical percentage of the body mass composed of protein
25. The recommended daily intake of protein; how that can be estimated from body weight; good dietary sources of protein; conditions that call for a protein intake greater than normal; and the risks from excessive dietary protein
26. Why nutritional value of a protein depends on its amino acid composition; the eight essential amino acids; the difference between complete and incomplete proteins; the meaning of net protein utilization and why it is different for dietary proteins of plant and animal origin; and some dietary and ecological advantages of plant protein
27. The meanings of *positive* and *negative nitrogen balance* and the conditions under which each of them occurs

28. The definition and ultimate source of *dietary minerals;* the functions of minerals in the body; the most abundant minerals in the body; and, in general, good and poor dietary sources of minerals
29. The general history of human salt consumption; the recommended daily intake of sodium and some reasons why most U.S. diets greatly exceed this; and the consequences of excessive sodium intake
30. The definition of *vitamins;* dietary and nondietary sources; the functions of vitamins in the body; the difference between water-soluble and fat-soluble vitamins

26.2 Carbohydrate Metabolism

1. The summary equation for the complete aerobic oxidation of glucose
2. Function of the coenzymes NAD^+ and FAD in glucose oxidation
3. The general process and outcome of glycolysis, and its net ATP yield
4. Anaerobic fermentation and its primary purpose
5. The cellular site of aerobic respiration, its end products, and its advantages
6. The citric acid cycle, where it occurs, the fate of the carbon atoms that originated in the glucose, and the cycle's yield of ATP, NADH, and $FADH_2$
7. Mitochondrial membrane reactions; where they occur in the organelle; the components of the mitochondrial electron-transport chain; how mitochondria transport electrons from NADH and $FADH_2$ to oxygen; and how the reactions produce metabolic water
8. Mitochondrial proton pumps, the chemiosmotic mechanism, and ATP synthase in producing ATP
9. The net ATP yield of glycolysis and aerobic respiration; the amount of ATP produced at each step from glucose to H_2O; and why the yield varies slightly between different cell types
10. The efficiency of aerobic respiration and how to calculate this
11. How excess glucose is converted to glycogen; the body's typical glycogen store and where these reserves are located; and the processes and purposes of glycogenesis, glycogenolysis, and gluconeogenesis

STUDY GUIDE

26.3 Lipid and Protein Metabolism

1. What cells are primarily responsible for storing and releasing triglycerides; the essence of the lipogenesis and lipolysis carried out by these and other cells
2. The process of lipolysis including the hydrolysis of triglycerides and the beta oxidation of fatty acids; the ATP yield from complete oxidation of a typical fatty acid
3. The meaning of *ketogenesis;* the metabolic use of ketone bodies and the pathological effects of excessive ketone levels; common circumstances in which excessive ketogenesis occurs
4. A typical daily rate of protein turnover in the body; where and why the fastest rate of protein turnover occurs; and the dietary and nondietary sources of the amino acids absorbed by the small intestine
5. The uses of free amino acids in the amino acid pool
6. What occurs in the deamination, amination, and transamination processes in amino acid metabolism; the body's uses of deaminated amino acids
7. How amino acids are shuttled into the citric acid cycle for oxidation as fuel
8. How the liver produces urea
9. Other nondigestive functions of the liver

26.4 Metabolic States and Metabolic Rate

1. When the body is in its absorptive state; what things occur in this state with respect to carbohydrate, fat, and protein metabolism
2. The main hormone that regulates the absorptive state; its primary metabolic effects in this state; and what antagonist modulates its effects
3. When the body is in its postabsorptive state; what things occur in this state with respect to carbohydrate, fat, and sometimes protein metabolism
4. Hormones that regulate the postabsorptive state, their effects, and the role of the sympathetic nervous system in its regulation
5. The meaning of *metabolic rate;* how it is measured and in what units of measurement it is expressed; what factors cause it to vary; and how the basal metabolic rate (BMR) differs from total metabolic rate (TMR)
6. Typical values for the BMR and TMR under different conditions of physical exertion

26.5 Body Heat and Thermoregulation

1. The meaning of *thermoregulation;* terms for abnormally low and high body temperatures; and reasons why those two conditions can be fatal

2. Typical core and shell body temperatures and how the two are measured
3. How most body heat is produced, and which organs are the most important sources of body heat at rest and in exercise
4. Four mechanisms by which body heat is lost, and the percentages of total heat loss attributable to each of them at rest and at a comfortable ambient temperature (21°C)
5. How the hypothalamic thermostat monitors core and shell body temperature
6. Two mechanisms for lowering body temperature and two mechanisms of raising it; regions of the hypothalamus involved in each
7. The meanings of *nonshivering thermogenesis* and *behavioral thermoregulation,* and examples of the latter
8. Differences between heat cramps, heat exhaustion, and heatstroke; the role of positive feedback loops in hyperthermia; and how and at what temperature range heatstroke can lead to death
9. The role of positive feedback loops in hypothermia, and how and at what temperature range hypothermia can lead to death

▶ Testing Your Recall

Answers in Appendix A

1. _____ are not used as fuel and are required in relatively small quantities.
   a. Micronutrients
   b. Macronutrients
   c. Essential nutrients
   d. Proteins
   e. Lipids

2. The only significant digestible polysaccharide in the diet is
   a. glycogen.
   b. cellulose.
   c. starch.
   d. maltose.
   e. fiber.

3. Which of the following store(s) the greatest amount of energy for the smallest amount of space in the body?
   a. glucose
   b. triglycerides
   c. glycogen
   d. proteins
   e. vitamins

4. The lipoproteins that remove cholesterol from the tissues are
   a. chylomicrons.
   b. lipoprotein lipases.
   c. VLDLs.
   d. LDLs.
   e. HDLs.

5. Which of the following is most likely to make you hungry?
   a. leptin
   b. ghrelin
   c. cholecystokinin
   d. peptide YY
   e. melanocortin

6. The primary function of B-complex vitamins is to act as
   a. structural components of cells.
   b. sources of energy.
   c. components of pigments.
   d. antioxidants.
   e. coenzymes.

7. FAD is reduced to $FADH_2$ in
   a. glycolysis.
   b. anaerobic fermentation.
   c. the citric acid cycle.
   d. the electron-transport chain.
   e. beta oxidation of lipids.

8. The primary, direct benefit of anaerobic fermentation is to
   a. regenerate NAD^+.
   b. produce $FADH_2$.
   c. produce lactate.
   d. dispose of pyruvate.
   e. produce more ATP than glycolysis does.

9. Which of these occurs in the mitochondrial matrix?
   a. glycolysis
   b. chemiosmosis
   c. the cytochrome reactions
   d. the citric acid cycle
   e. anaerobic fermentation

10. When the body emits more infrared energy than it absorbs, it is losing heat by
    a. convection.
    b. forced convection.
    c. conduction.
    d. radiation.
    e. evaporation.

11. A/an _____ protein lacks one or more essential amino acids.

12. In the postabsorptive state, glycogen is hydrolyzed to liberate glucose. This process is called _____ .

13. Synthesis of glucose from amino acids or triglycerides is called _____ .

14. The major nitrogenous waste resulting from protein catabolism is _____ .

15. The organ that synthesizes the nitrogenous waste in question 14 is the _____ .

16. The absorptive state is regulated mainly by the hormone _____ .

17. The temperature of organs in the body cavities is called _____ .

18. The appetite hormones ghrelin, leptin, CCK, and others act on part of the hypothalamus called the _____ nucleus.

19. The brightly colored, iron-containing, electron-transfer molecules of the inner mitochondrial membrane are called _____ .

20. The flow of H^+ from the intermembrane space to the mitochondrial matrix creates an electrical current used by the enzyme _____ to make _____ .

► Building Your Medical Vocabulary

Answers in Appendix A

State a meaning of each word element, and give a medical term from this chapter that uses it or a slight variation of it.

1. asco-

2. cac-

3. chromo-

4. -genesis

5. glyco-

6. -ites

7. lepto-

8. -lysis

9. neo-

10. osmo-

► What's Wrong with These Statements?

Answers in Appendix A

Briefly explain why each of the following statements is false, or reword it to make it true.

1. Ghrelin and leptin are two hormones that stimulate the appetite.

2. Water is not considered a nutrient because we don't digest it and it contains no calories.

3. A gram of carbohydrates contains twice as much energy as a gram of fat.

4. Most of the body's cholesterol comes from the diet.

5. There is no harm in maximizing one's daily protein intake.

6. Extremely low-fat fad diets tend to produce potentially dangerous high levels of blood ketones.

7. Reactions occurring in the mitochondrial matrix produce more ATP than glycolysis and the mitochondrial membrane reactions combined.

8. Gluconeogenesis occurs especially in the absorptive state during and shortly after a meal.

9. Brown fat generates more ATP than white fat and is therefore especially important for thermoregulation.

10. At an air temperature of 21°C (70°F), the body loses more heat as conduction to the surrounding air than by any other means.

► Testing Your Comprehension

1. Cyanide blocks the transfer of electrons from cytochrome a_3 to oxygen. In light of this, explain why it is so lethal. Also explain whether cyanide poisoning could be treated by giving a patient supplemental oxygen, and justify your answer.

2. Chapter 17 defines and describes some hormone actions that are synergistic and antagonistic. Identify some synergistic and antagonistic hormone interactions in the postabsorptive state of metabolism.

3. Ms. Jones, a 42-year-old woman, complains, "Everything I eat goes to fat. But my husband and my son eat twice as much as I do, and they're both as skinny as can be." How would you explain this to her?

4. A television advertisement proclaims, "Feeling tired? Need more energy? Order your supply of Zippy Megavitamins and feel better fast!" Your friend Cathy is about to send in her order, and you try to talk her out of wasting her money. Summarize the argument you would use.

5. Explain why a patient whose liver has been extensively damaged by hepatitis could show elevated concentrations of thyroid hormone and bilirubin in the blood.

CHAPTER

27

THE MALE REPRODUCTIVE SYSTEM

Seminiferous tubules, where sperm are produced. Sperm tails are seen as hairlike masses in the center of the tubules (SEM).

CNRI/SPL/Science Source

CHAPTER OUTLINE

27.1 Sexual Reproduction and Development

27.1a The Two Sexes

27.1b Overview of the Reproductive System

27.1c Chromosomal Sex Determination

27.1d Prenatal Hormones and Sexual Differentiation

27.1e Development of the External Genitalia

27.1f Descent of the Gonads

27.2 Male Reproductive Anatomy

27.2a The Scrotum

27.2b The Testes

27.2c The Spermatic Ducts

27.2d The Accessory Glands

27.2e The Penis

27.3 Puberty, Hormonal Control, and Climacteric

27.3a Puberty and Adolescence

27.3b Hormonal Control of Male Reproductive Function

27.3c Aging and Sexual Function

27.4 Sperm and Semen

27.4a Meiosis

27.4b Sperm Production

27.4c The Spermatozoon

27.4d Semen

27.5 Male Sexual Response

27.5a Anatomical Foundations

27.5b Excitement and Plateau

27.5c Orgasm and Ejaculation

27.5d Resolution

Study Guide

DEEPER INSIGHTS

27.1 Clinical Application: Androgen Insensitivity Syndrome

27.2 Clinical Application: Prostate Diseases

27.3 Clinical Application: Reproductive Effects of Pollution

27.4 Clinical Application: Treating Erectile Dysfunction

27.5 Clinical Application: Sexually Transmitted Diseases

Anatomy & Physiology
Revealed® 4.0

Module 14: Reproductive System

From all we have learned of the structural and functional complexity of the human body, it seems a wonder that it works at all! The fact is, however, that even with modern medicine we can't keep it working forever. The body suffers various degenerative changes as we age, and eventually our time is up and we must say good-bye. Yet our genes live on in new containers—our offspring. The production of offspring is the subject of these last three chapters. In this chapter, we examine some general aspects of human reproductive biology and then focus on the role of the male in reproduction. The next two chapters deal, respectively, with female reproductive function and on the embryonic development of humans and changes at the other end of the life span—the changes of old age.

27.1 Sexual Reproduction and Development

Expected Learning Outcomes

When you have completed this section, you should be able to

a. identify the most fundamental biological distinction between male and female;

b. define *primary sex organs, secondary sex organs,* and *secondary sex characteristics;*

c. explain the role of the sex chromosomes in determining sex;

d. explain how the Y chromosome determines the response of the fetal gonad to prenatal hormones;

e. identify which of the male and female external genitalia are homologous to each other; and

f. describe the descent of the gonads and explain why it is important.

27.1a The Two Sexes

The essence of sexual reproduction is that it is biparental—the offspring receive genes from two parents, so they are not genetically identical to either one. To achieve this, the parents must produce **gametes**[1] (sex cells) that meet and combine their genes in a **zygote**[2] (fertilized egg). Two things are necessary for reproduction to be successful: (1) gamete motility so they can achieve contact, and (2) enough cytoplasm to divide up into the first cells of a developing embryo, with intracellular nutrients to sustain it until it begins to receive external nutrition. A single cell cannot perform both of these roles optimally, because to contain ample cytoplasm means to be relatively large and heavy, and this is inconsistent with the need for motility. Therefore, these tasks are usually apportioned to two kinds of gametes. The small motile one—little more than DNA with a propeller—is the **sperm (spermatozoon),** and the large nutrient-laden one is the **egg (ovum).**

In any sexually reproducing species, by definition, an individual that produces eggs is female and one that produces sperm is male. These criteria are not always that simple, as we see in certain abnormalities in sexual development. Genetically, however, any human with a Y sex chromosome is classified as male and anyone lacking a Y is classified as female. In contrast to these chromosomally determined sexes, *gender* is a person's psychological sense of feminine, masculine, or other sexual identity. One's gender can be different from one's biological sex.

In mammals, the female is also the parent that provides a sheltered internal environment for the development and nutrition of the embryo. For fertilization and development to occur in the female, the male must have a copulatory organ, the penis, for introducing his gametes into the female reproductive tract, and the female must have a copulatory organ, the vagina, for receiving the sperm. This is the most obvious difference between the sexes, but appearances can be deceiving (see fig. 17.28 and Deeper Insight 27.1).

27.1b Overview of the Reproductive System

The **reproductive system** in the male serves to produce sperm and introduce them into the female body. The female reproductive system produces eggs, receives the sperm, provides a place for the union of these gametes, harbors the fetus, gives birth, and nourishes the offspring.

The reproductive system consists of primary and secondary sex organs. The **primary sex organs,** or **gonads,**[3] are organs that produce the gametes—**testes** of the male and **ovaries** of the female. The **secondary sex organs** are organs other than gonads that are necessary for reproduction. In the male, they constitute a system of ducts, glands, and the penis, concerned with sperm storage, survival, and delivery. In the female, they include the uterine tubes, uterus, and vagina, concerned with uniting the sperm and egg and harboring the fetus.

According to location, the reproductive organs are classified as **external** and **internal genitalia (table 27.1).** The external genitalia are located in the perineum (see fig. 27.6). Most of them are externally visible, except for the subcutaneous accessory glands of the female perineum. The internal genitalia are located mainly in the pelvic cavity, except for the male testes and some associated ducts contained in the scrotum.

---

[1]*gam* = marriage, union
[2]*zygo* = yoke, union
[3]*gon* = seed

DEEPER INSIGHT 27.1

CLINICAL APPLICATION

Androgen Insensitivity Syndrome

Occasionally, a child who was always assumed to be a girl shows the usual changes of puberty except that she fails to menstruate. Medical examination reveals testes in the abdomen and the XY karyotype of a male. The testes produce a normal male level of testosterone, but the target cells lack receptors for it and the testosterone therefore has no effect. This condition is called *androgen insensitivity syndrome (AIS)* or *testicular feminization.*

Figure 27.1 shows three siblings with XY chromosomes, but feminized by AIS. In such individuals, the external genitals exhibit female anatomy from birth, as if no testosterone was present. At puberty, feminine breasts and other secondary sex characteristics develop because testosterone can't override the effect of the small amounts of estrogen that testes normally secrete. Despite external appearances, there are no ovaries, vagina, or uterus. If the abdominal testes are not removed, an AIS person has an elevated risk of testicular cancer.

FIGURE 27.1 Androgen Insensitivity Syndrome.

Courtesy Mihaly Bartalos, from M. Bartalos and T.A. Baramki

 In what way is androgen insensitivity syndrome similar to type 2 diabetes mellitus?

TABLE 27.1	The External and Internal Genitalia
External Genitalia	**Internal Genitalia**
Male	
Penis	Testes (singular, *testis*)
Scrotum	Epididymides (singular, *epididymis*)
	Ductus deferentes (singular, *ductus deferens*)
	Seminal vesicles
	Prostate
	Bulbourethral glands
Female	
Mons pubis	Ovaries
Labia majora (singular, *labium majus*)	Uterine tubes
	Uterus
Labia minora (singular, *labium minus*)	Vagina
Clitoris	
Vaginal orifice	
Vestibular bulbs	
Vestibular glands	
Paraurethral glands	

Secondary sex characteristics are features that further distinguish the sexes and play a role in mate attraction. They typically appear only as an animal approaches sexual maturity (during adolescence in humans). From the call of a bullfrog to the tail of a peacock, these are well known in the animal kingdom. In humans, the physical attributes that contribute to sexual attraction are so culturally conditioned that it's harder to identify what secondary sex characteristics are biologically fundamental. Generally accepted as such in both sexes are the pubic and axillary hair and their associated scent glands, and the pitch of the voice. Other traits commonly regarded as male secondary sex characteristics are the facial hair, relatively coarse and visible hair on the torso and limbs, and the relatively muscular physique. In females, they include the distribution of body fat, enlargement of the breasts (independently of lactation), and relatively hairless appearance of the skin.

27.1c Chromosomal Sex Determination

What determines whether a zygote develops into a male or female? The distinction begins with the combination of sex chromosomes bequeathed to the zygote. Most of our cells have 23 pairs of chromosomes: 22 pairs of *autosomes* and 1 pair of *sex chromosomes* (see fig. 4.16). A sex chromosome can be either a large X chromosome or a small Y chromosome. Every egg contains an X chromosome, but half of the sperm carry an X and the other half carry a Y. If an egg is fertilized with an X-bearing sperm, it produces

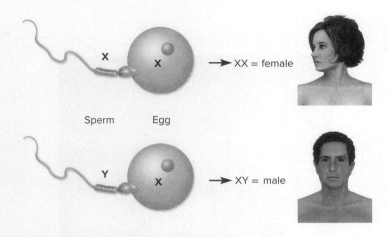

FIGURE 27.2 Chromosomal Sex Determination. All eggs carry the X chromosome. The sex of a child is determined by whether the egg is fertilized by an X-bearing sperm or a Y-bearing sperm.

an XX zygote that is destined to become a female. If it is fertilized with a Y-bearing sperm, it produces an XY zygote destined to become a male. Thus, the sex of a child is determined at conception (fertilization), and not by the mother's egg but by the sperm that fertilizes it **(fig. 27.2).**

27.1d Prenatal Hormones and Sexual Differentiation

Sex determination doesn't end with fertilization, however. It requires an interaction between genes and hormones produced by the mother and fetus. Just as we have seen with other hormones, those involved here require specific receptors on their target cells to exert an effect.

Up to a point, a fetus is sexually undifferentiated, or "noncommittal" as to which sex it will become. Its gonads begin to develop at 5 to 6 weeks as *gonadal ridges,* each lying alongside a primitive kidney, the *mesonephros,* which later degenerates. Adjacent to each gonadal ridge are two ducts: the **mesonephric**[4] **duct** (MEZ-oh-NEF-ric), which originally serves the mesonephros, and the **paramesonephric**[5] **duct.** In males, the mesonephric ducts develop into the reproductive tract and the paramesonephric ducts degenerate. In females, the opposite occurs **(fig. 27.3).**

But why? The Y chromosome has a gene called **SRY** (sex-determining region of the Y) that codes for a protein called **testis-determining factor (TDF).** TDF then interacts with genes on some of the other chromosomes, including a gene on the X chromosome for androgen receptors and genes that initiate the development of male anatomy. By 8 to 9 weeks, the male gonadal ridge has become a rudimentary testis that begins to secrete testosterone. Testosterone stimulates the mesonephric duct on its own side to develop into the system of male reproductive

ducts. By this time, the testis also secretes a hormone called **müllerian-inhibiting factor (MIF)** that causes atrophy of the paramesonephric duct (formerly called the *müllerian duct*) on that side. Even an adult male, however, retains a tiny Y-shaped vestige of the paramesonephric ducts, like a vestigial uterus and uterine tubes, in the area of the prostatic urethra.

It may seem as if androgens should induce the formation of a male reproductive tract and estrogens induce a female reproductive tract. However, the estrogen level is always high during pregnancy, so if this mechanism were the case, it would feminize all fetuses. Thus, the development of a female results from the absence of androgens, not the presence of estrogens.

27.1e Development of the External Genitalia

You perhaps regard the external genitals as the most definitive characteristics of a male or female, yet there is more similarity between the sexes than most people realize. In the embryo, the genitals begin developing from identical structures in both sexes. By 6 weeks, the embryo has the following:

- the **genital tubercle,** an anterior median bud;
- **urogenital folds,** a pair of medial tissue folds slightly posterior to the genital tubercle; and
- **labioscrotal folds,** a larger pair of tissue folds lateral to the urogenital folds.

By the end of week 9, the fetus begins to show sexual differentiation, and either male or female genitalia are distinctly formed by the end of week 12 **(fig. 27.4).** In the female, the three structures just listed become the clitoral glans, labia minora, and labia majora, respectively; all of these are more fully described in the next chapter. In the male, the genital tubercle elongates to form the *phallus;* the urogenital folds fuse to enclose the urethra, joining the phallus to form the penis; and the labioscrotal folds fuse to form the scrotum, a sac that will later contain the testes.

Male and female organs that develop from the same embryonic structure are said to be **homologous.** Thus, the penis is homologous to the clitoris and the scrotum is homologous to the labia majora. This becomes strikingly evident in some abnormalities of sexual development. In the presence of excess androgen, the clitoris may become greatly enlarged and resemble a small penis. In other cases, the ovaries descend into the labia majora as if they were testes descending into a scrotum. Such abnormalities sometimes result in mistaken identification of the sex of an infant at birth.

27.1f Descent of the Gonads

Both male and female gonads initially develop high in the abdominal cavity, near the kidneys, and migrate into the pelvic cavity (ovaries) or scrotum (testes). In the embryo, a connective tissue cord called the *gubernaculum*[6] (GOO-bur-NACK-you-lum)

---

[4]*meso* = middle; *nephr* = kidney
[5]*para* = next to; *meso* = middle; *nephr* = kidney

[6]*gubern* = rudder, to guide

FIGURE 27.3 Embryonic Development of the Male and Female Reproductive Tracts. Note that the male tract develops from the mesonephric duct and the female tract from the paramesonephric duct; the other duct in each sex degenerates.

anchors the gonad to the floor of the abdominopelvic cavity. As the male gubernaculum continues to grow, it passes between the internal and external oblique muscles and into the scrotal swelling. Independently of migration of the testis, the peritoneum also develops a fold that extends into the scrotum as the *vaginal*[7] *process.* The gubernaculum and vaginal process create a path of low resistance through the groin, anterior to the pubic symphysis, called the **inguinal canal**—the most common site

[7]*vagin = sheath*

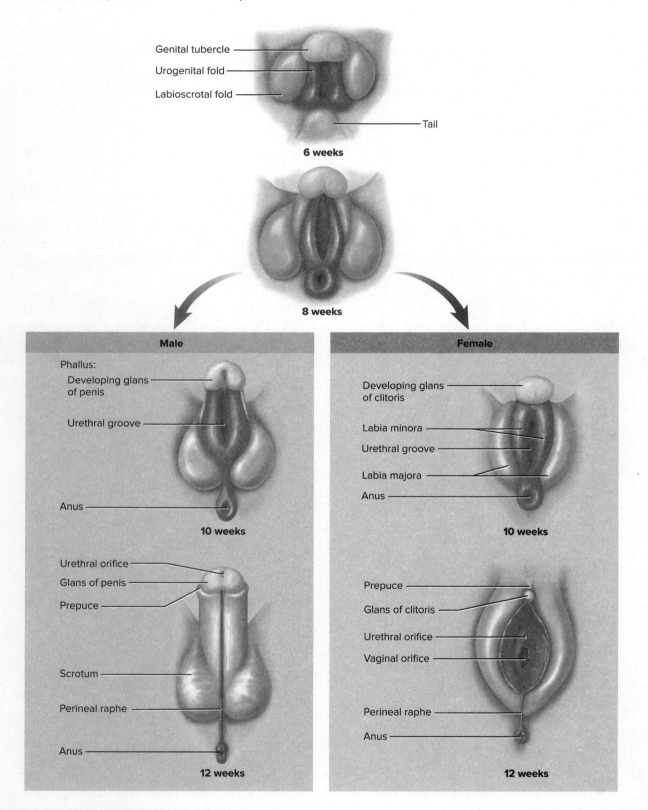

FIGURE 27.4 **Development of the External Genitalia.** By 6 weeks, the embryo has three primordial structures—the genital tubercle, urogenital folds, and labioscrotal folds—which will become the male or female genitalia. At 8 weeks, these structures have grown but the sexes are still indistinguishable. Slight sexual differentiation is noticeable at 10 weeks, and the sexes are fully distinguishable by 12 weeks. Matching colors identify homologous structures of the male and female.

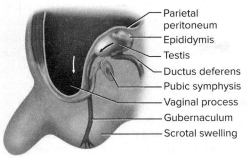

- Parietal peritoneum
- Epididymis
- Testis
- Ductus deferens
- Pubic symphysis
- Vaginal process
- Gubernaculum
- Scrotal swelling

(a) 3-month fetus

- Muscular wall of abdomen
- Inguinal canal
- Vaginal process
- Penis

(b) 8-month fetus

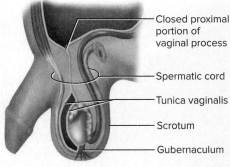

- Closed proximal portion of vaginal process
- Spermatic cord
- Tunica vaginalis
- Scrotum
- Gubernaculum

(c) 1-month-old infant

FIGURE 27.5 Descent of the Testis. (a) Abdominal position of the testes in a 3-month fetus. (b) Descending testes in an 8-month fetus. (c) Scrotal position of the testes in a 1-month-old infant. Note that the testis and spermatic ducts are retroperitoneal. An extension of the peritoneum called the *vaginal process* passes through the inguinal canal and becomes the tunica vaginalis.

❓ *Why is this structure of male anatomy called the* tunica vaginalis?

of herniation in boys and men (*inguinal hernia;* see Deeper Insight 10.3).

The **descent of the testes (fig. 27.5)** begins as early as week 6. The superior part of the embryonic gonad degenerates and its inferior part migrates downward, guided by the gubernaculum. In the seventh month, the testes abruptly pass through the inguinal canals into the scrotum. As they descend, they are accompanied by ever-elongating testicular arteries and veins and by lymphatic vessels, nerves, spermatic ducts, and extensions of the internal oblique muscle. The vaginal process becomes separated from the peritoneal cavity and persists as a sac, the *tunica vaginalis,* enfolding the anterior and lateral sides of the testis and protecting the testis from abrasion against the scrotum with movements of the body.

The actual mechanism of descent remains obscure, although multiple hypotheses have been offered. Testosterone stimulates descent, but it is unknown how. The reason this descent is necessary, however, will be explained in the next section.

About 3% of boys are born with undescended testes, or **cryptorchidism.**[8] In most such cases, the testes descend within the first year of infancy, but if they don't, the condition can usually be corrected with a testosterone injection or a fairly simple surgery to dilate the inguinal canals and draw the testes into the scrotum. Uncorrected cryptorchidism, however, leads inevitably to sterility and sometimes to testicular cancer.

The ovaries also descend, but to a much lesser extent. The female gubernaculum extends from the inferior pole of the ovary to the labioscrotal fold. The ovaries eventually lodge just inferior to the brim of the lesser pelvis. Each gubernaculum becomes a pair of ligaments that support the ovary and uterus.

BEFORE YOU GO ON

Answer the following questions to test your understanding of the preceding section:

1. Define *gonad* and *gamete.* Explain the relationship between the terms.

2. Define *male, female, sperm,* and *egg.*

3. What are mesonephric and paramesonephric ducts? What factors determine which one develops and which one regresses in the fetus?

4. What male structure develops from the genital tubercle and urogenital folds? What develops from the labioscrotal folds?

5. Describe the pathway taken during descent of the male gonad.

27.2 **Male Reproductive Anatomy** **APR**

Expected Learning Outcomes

When you have completed this section, you should be able to

a. describe the anatomy of the scrotum, testes, and penis;

b. describe the pathway taken by a sperm cell from its formation to its ejaculation, naming all the passages it travels; and

c. state the names, locations, and functions of the male accessory reproductive glands.

We will survey the male reproductive system in order according to the sites of sperm formation, transport, and emission—therefore beginning with the scrotum and testes, continuing through the spermatic ducts and accessory glands associated with them, and ending with the penis.

27.2a The Scrotum

The scrotum and penis constitute the external genitalia of the male and occupy the **perineum**[9] (PERR-ih-NEE-um). This is a

---

[8]*crypto* = hidden; *orchid* = testis; *ism* = condition

[9]*peri* = around; *neum,* from *inan* = to empty out

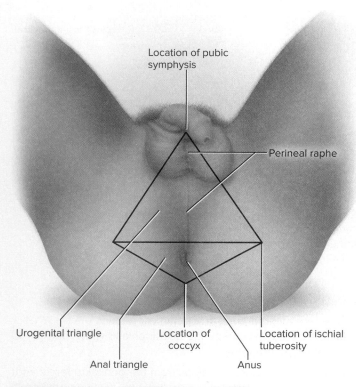

Location of pubic symphysis

Perineal raphe

Urogenital triangle

Location of coccyx

Anal triangle

Location of ischial tuberosity

Anus

FIGURE 27.6 The Male Perineum. APR

diamond-shaped area between the thighs bordered by the pubic symphysis, ischial tuberosities, and coccyx **(fig. 27.6).**

The **scrotum**[10] is a pendulous pouch of skin, muscle, and fibrous connective tissue, containing the testes **(fig. 27.7).** The skin has sebaceous glands, sparse hair, rich sensory innervation, and somewhat darker pigmentation than skin elsewhere. Just deep to the skin is a layer of smooth muscle and connective tissue called the **dartos**[11] **fascia;** its muscle is called the **dartos muscle.** The muscle is normally somewhat contracted, giving the skin a wrinkled texture. The scrotum is divided into right and left compartments by an internal **median septum,** which prevents infections of one testis from spreading to the other one. The location of the septum is externally marked by a seam called the **perineal raphe** (RAY-fee), which also extends anteriorly along the ventral side of the penis and posteriorly as far as the margin of the anus. The left testis is usually suspended lower than the right so the two aren't compressed against each other between the thighs.

Posteriorly, the scrotum contains the **spermatic cord,** a bundle of fibrous connective tissue containing the *ductus deferens* (a sperm duct), lymphatic vessels, testicular nerves, the testicular artery, and a complex of veins called the **pampiniform**[12] **plexus** closely surrounding the artery. It passes upward behind and superior to the testis, where it is easily palpated through the skin of the

[10]*scrotum* = bag
[11]*dartos* = skinned
[12]*pampin* = tendril; *form* = shape

Ureter

Urinary bladder

Testicular artery

Testicular vein

Ductus deferens

Spermatic cord (opened)

Pampiniform plexus

Ductus deferens

Tunica vaginalis

Epididymis

Testis

Scrotum

Penis

Inguinal ligament

Superficial inguinal ring

Spermatic cord

Fascia of spermatic cord

Cremaster

Dartos fascia

Perineal raphe

FIGURE 27.7 The Scrotum and Spermatic Cord. APR

scrotum. It continues across the anterior side of the pubis and into a 4 cm **inguinal canal,** which leads through the muscles of the groin and emerges into the pelvic cavity. The inferior entrance into the inguinal canal is called the *superficial inguinal ring,* and its superior exit into the pelvic cavity is the *deep inguinal ring.* Within the scrotum, the spermatic cord is enmeshed in strips of internal oblique muscle called the **cremaster.**[13]

The original reason that a scrotum evolved is a subject of debate among reproductive biologists; it is absent from some mammals, such as elephants, sea cows, and whales, so it isn't universally necessary for mammalian sperm production. For whatever reason human testes reside in the scrotum, however, they are now adapted to this cooler environment and can't produce sperm at the core body temperature of 37°C; they must be held at about 35°C. The scrotum has three mechanisms for regulating the temperature of the testes:

1. **The cremaster.** When it is cold, the cremaster contracts and draws the testes closer to the body to keep them warm. When it is warm, the cremaster relaxes and the testes are suspended farther from the body.

2. **The dartos muscle.** This, too, contracts when it is cold, and the scrotum becomes taut and even more wrinkled. The tautness of the scrotum helps to hold the testes snugly against the warm body and it reduces the surface area of the scrotum, thus reducing heat loss.

3. **The pampiniform plexus.** Without this venous plexus, warm arterial blood would heat the testis and inhibit sperm production. The pampiniform plexus, however, prevents this by acting as a *countercurrent heat exchanger* **(fig. 27.8).** The relatively cool venous blood ascending the spermatic cord toward the pelvic cavity draws heat out of the descending arterial blood headed for the testis, so by the time it reaches the testis the blood is 1.5° to 2.5°C cooler than the core body temperature.

27.2b The Testes

The testes (testicles) **(fig. 27.9)** are combined endocrine and exocrine glands that produce sex hormones and sperm. The testis is oval and slightly flattened, about 4 cm long, 3 cm from anterior to posterior, and 2.5 cm wide. Its anterior and lateral surfaces are covered by the tunica vaginalis. The testis itself has a white fibrous capsule called the **tunica albuginea**[14] (TOO-nih-ca AL-byu-JIN-ee-uh). Connective tissue septa extend from the capsule into the parenchyma, dividing it into 250 to 300 wedge-shaped lobules. Each lobule contains one to three **seminiferous**[15] **tubules** (SEM-ih-NIF-er-us)—slender ducts up to 70 cm long in which the sperm are produced. Between the seminiferous tubules are clusters of **interstitial**[16] **endocrine cells,** the source of testosterone.

FIGURE 27.8 The Countercurrent Heat Exchanger. Although illustrated as one vessel, the pampiniform plexus is a network of veins surrounding the testicular artery (see fig. 27.7). Note the cooling of the arterial blood as it descends the artery and its heat transfers to the adjacent, ascending venous blood.

A seminiferous tubule has a narrow lumen lined by a thick **germinal epithelium.** The epithelium consists of several layers of germ cells in the process of becoming sperm, and a much smaller number of tall **nurse cells (supporting cells, sustentocytes**[17]**),** which protect the germ cells and promote their development. The germ cells depend on the nurse cells for nutrients, waste removal, growth factors, and other needs. The nurse cells also secrete two proteins called *androgen-binding protein* and *inhibin,* both of which regulate sperm production in ways described later.

A nurse cell is shaped a little like a tree trunk whose roots spread out over the basement membrane, forming the boundary of the tubule, and whose thick trunk reaches to the tubule lumen. Tight junctions between adjacent nurse cells form a **blood–testis barrier (BTB),** which prevents antibodies and other large molecules in the blood and intercellular fluid from getting to the germ cells. This is important because the germ cells, being genetically different from other cells of the body, would otherwise be attacked by the immune system. Some cases of sterility occur when the BTB fails to form adequately in adolescence and the immune system produces autoantibodies against the germ cells.

---

[13]*cremast* = to suspend
[14]*tunica* = coat; *alb* = white
[15]*semin* = seed, sperm; *fer* = to carry
[16]*inter* = between; *stit* = placed

[17]*sustent* = support

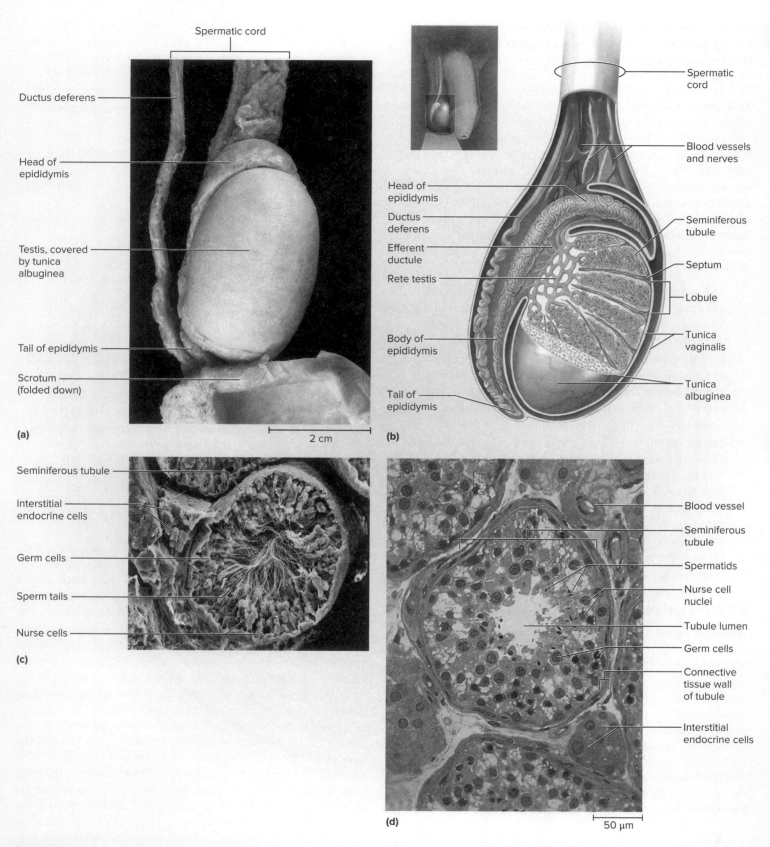

FIGURE 27.9 The Testis and Associated Structures. (a) The scrotum is opened and folded downward to reveal the testis and associated organs. (b) Anatomy of the testis, epididymis, and spermatic cord. (c) Scanning electron micrograph of the seminiferous tubules. (d) Light micrograph. This region of the tubule did not have mature sperm in it at the time. **APR**

a: Dennis Strete/McGraw-Hill Education; c: Steve Gschmeissner/Science Source; d: Ed Reschke/Getty Images

▶▶▶ **APPLY WHAT YOU KNOW**

Would you expect to find blood capillaries in the walls of the seminiferous tubules? Why or why not?

The seminiferous tubules lead into a network called the **rete**[18] **testis** (REE-tee), embedded in the capsule on the posterior side of the testis. Sperm partially mature in the rete. They are moved along by the flow of fluid secreted by the nurse cells and by the cilia on some rete cells. Sperm don't swim while they're in the male reproductive tract.

Each testis is supplied by a **testicular artery** that arises from the abdominal aorta just below the renal artery. This is a very long, slender artery that winds its way down the posterior abdominal wall before passing through the inguinal canal into the scrotum (see fig. 27.7). Its blood pressure is very low, and indeed this is one of the few arteries to have no pulse. Consequently, blood flow to the testes is quite meager and the testes receive a poor oxygen supply. In response to this, the sperm develop unusually large mitochondria, which may precondition them for survival in the hypoxic environment of the female reproductive tract.

Blood leaves the testis by way of the pampiniform plexus of veins. As these veins pass through the inguinal canal, they converge and form the **testicular vein.** The right testicular vein drains into the inferior vena cava and the left one drains into the left renal vein. Lymphatic vessels also drain each testis. They travel through the inguinal canal with the veins and lead to lymph nodes adjacent to the lower aorta. Lymph from the penis and scrotum, however, travels to lymph nodes adjacent to the iliac arteries and veins and in the inguinal region.

Testicular nerves lead to the gonads from spinal cord segments T10 and T11. They are mixed sensory and motor nerves containing predominantly sympathetic but also some parasympathetic fibers. The sensory fibers are concerned primarily with pain and the autonomic fibers are predominantly vasomotor, for regulation of blood flow.

27.2c The Spermatic Ducts

After leaving the testis, the sperm travel through a series of *spermatic ducts* to reach the urethra **(fig. 27.10).** These include the following:

- **Efferent ductules.** About 12 small efferent ductules arise from the posterior side of the testis and carry sperm to the epididymis. They have clusters of ciliated cells that help drive the sperm along.

- **Duct of the epididymis.** The **epididymis**[19] (EP-ih-DID-ih-miss; plural, *epididymides*) is a site of sperm maturation and storage. It adheres to the posterior side of the testis (fig. 27.9). The superior clublike portion where it receives the efferent ductules is called the *head;* the long middle portion below the last efferent ductule is the *body;* and the slender inferior end is the *tail.* It contains a single coiled

duct embedded in connective tissue. The duct is about 6 m (20 ft) long, but it is so slender and highly coiled that it is packed into an epididymis only 7.5 cm long. The duct reabsorbs about 90% of the fluid secreted by the testis. Sperm are physiologically immature when they leave the testis but mature as they travel through the head and body of the epididymis. In 20 days or so, they reach the tail. They are stored here and in the adjacent portion of the ductus deferens pending ejaculation, and remain viable for 40 to 60 days. If they become too old without being ejaculated, they disintegrate and the epididymis reabsorbs them.

- **Ductus deferens.**[20] The duct of the epididymis straightens out at the tail, turns 180°, and becomes the ductus deferens, also called the *vas deferens* (from which comes the term *vasectomy*). This is a muscular tube about 45 cm long and 2.5 mm in diameter. It passes upward through the spermatic cord and inguinal canal and enters the pelvic cavity. There, it turns medially and approaches the urinary bladder. After passing between the bladder and ureter, the duct turns downward behind the bladder and widens into a terminal **ampulla.** The ductus deferens ends by uniting with the duct of an accessory gland, the seminal vesicle. The duct has a very narrow lumen and a thick wall of smooth muscle well innervated by sympathetic nerve fibers. Vasectomy, the surgical method of male contraception, consists of cutting out a short portion of the ductus (vas) deferens to interrupt the passage of sperm.

- **Ejaculatory duct.** Where the ductus deferens and duct of the seminal vesicle meet, they form a short (2 cm) ejaculatory duct, which passes through the prostate and empties into the urethra. The ejaculatory duct is the last of the spermatic ducts.

The male urethra is shared by the reproductive and urinary systems. It is about 18 cm long and consists of three regions: the *prostatic, membranous,* and *spongy (penile) urethra.* Although it serves both urinary and reproductive roles, it cannot pass urine and semen simultaneously for reasons explained in section 27.5c.

27.2d The Accessory Glands

There are three sets of *accessory glands* in the male reproductive system—the seminal vesicles, prostate, and bulbourethral glands:

1. The **seminal vesicles (seminal glands)** are a pair of glands posterior to the urinary bladder; one is associated with each ductus deferens. A seminal vesicle is about 5 cm long, with approximately the dimensions of one's little finger. It has a connective tissue capsule and underlying layer of smooth muscle. The secretory portion is a very convoluted duct with numerous branches that form a complex labyrinth. It empties into the ejaculatory duct. The yellowish secretion of the seminal vesicles constitutes about 65% to 75% of the semen.

---

[18]*rete* = network
[19]*epi* = upon; *didym* = twins, testes

[20]*de* = away from; *fer* = to carry

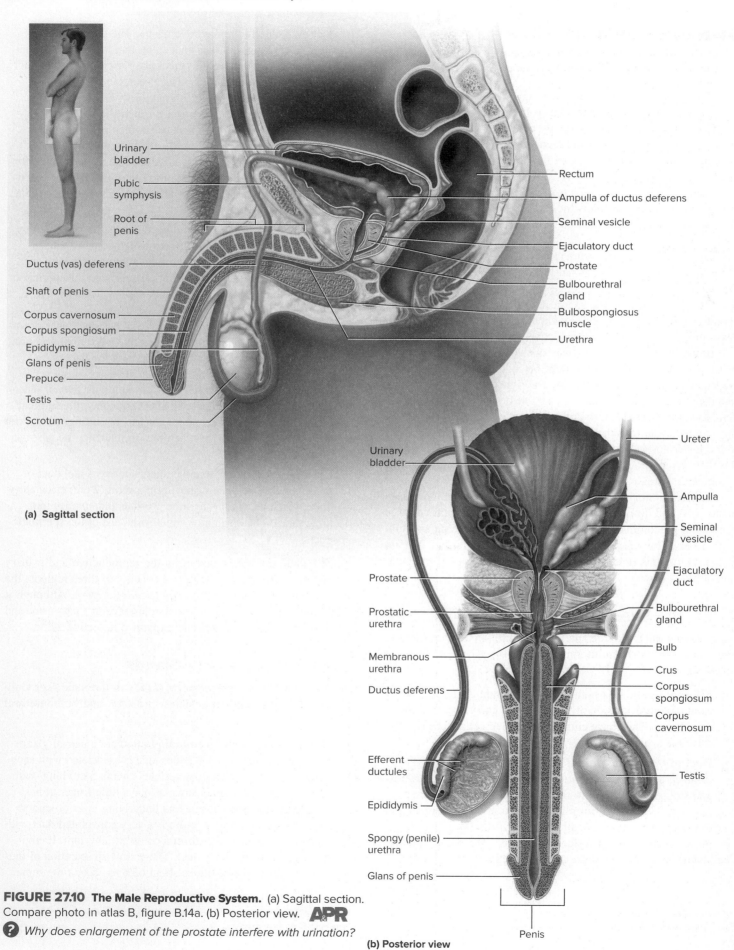

Urinary bladder

Pubic symphysis

Root of penis

Ductus (vas) deferens

Shaft of penis

Corpus cavernosum

Corpus spongiosum

Epididymis

Glans of penis

Prepuce

Testis

Scrotum

Rectum

Ampulla of ductus deferens

Seminal vesicle

Ejaculatory duct

Prostate

Bulbourethral gland

Bulbospongiosus muscle

Urethra

(a) Sagittal section

Urinary bladder

Prostate

Prostatic urethra

Membranous urethra

Ductus deferens

Efferent ductules

Epididymis

Spongy (penile) urethra

Glans of penis

Ureter

Ampulla

Seminal vesicle

Ejaculatory duct

Bulbourethral gland

Bulb

Crus

Corpus spongiosum

Corpus cavernosum

Testis

Penis

(b) Posterior view

FIGURE 27.10 The Male Reproductive System. (a) Sagittal section. Compare photo in atlas B, figure B.14a. (b) Posterior view. **APR** A&P

❓ *Why does enlargement of the prostate interfere with urination?*

2. The **prostate**[21] (PROSS-tate) surrounds the urethra and ejaculatory ducts immediately inferior to the urinary bladder (see Deeper Insight 27.2). It measures about 2 × 4 × 3 cm and is an aggregate of 30 to 50 compound tubuloacinar glands enclosed in a single fibrous capsule. These glands empty through about 20 pores in the urethral wall. The stroma of the prostate consists of connective tissue and smooth muscle, like that of the seminal vesicles. The milky secretion of the prostate constitutes about 25% to 30% of the semen.

3. The **bulbourethral glands** are named for their position near a dilated bulb at the inner end of the penis and their association with the urethra. They are brownish, spherical glands about 1 cm in diameter, with a 2.5 cm duct to the urethra. During sexual arousal, they produce *preejaculate,* a clear slippery fluid that lubricates the head of the penis in preparation for intercourse. More importantly, though, it protects the sperm by neutralizing the acidity of the male's urethra and female's vagina, and it creates a relatively low-viscosity pathway for sperm to migrate from the vagina, through the cervical canal, into the uterus.

DEEPER INSIGHT 27.2

CLINICAL APPLICATION

Prostate Diseases

The prostate weighs about 20 g by age 20, remains at that weight until age 45 or so, and then begins to grow slowly again. By age 70, over 80% of men show some degree of *benign prostatic hyperplasia (BPH)—* noncancerous enlargement of the gland. The major complication of this is that it compresses the urethra, obstructs the flow of urine, makes it harder to completely empty the bladder, and sometimes promotes bladder and kidney infections. BPH is more common in men who are obese, physically inactive, and consume alcohol to excess.

Prostate cancer is the second most common cancer in men (after lung cancer); it affects about 9% of men over the age of 50. Prostate tumors tend to form near the periphery of the gland, where they don't obstruct urine flow; therefore, they often go unnoticed until they cause pain. Prostate cancer often metastasizes to nearby lymph nodes and then to the lungs and other organs. It is more common among black Americans than among white Americans and uncommon among people of Japanese descent.

The position of the prostate immediately anterior to the rectum allows it to be palpated through the rectal wall to check for tumors. This procedure is called *digital rectal examination (DRE).* Prostate cancer can also be diagnosed from elevated levels of *serine protease* (also known as *prostate-specific antigen, PSA*) and *acid phosphatase* (another prostatic enzyme) in the blood. Up to 80% of men with prostate cancer survive when it is detected and treated early, but only 10% to 50% survive if it spreads beyond the prostatic capsule. It is such a slow-growing cancer, however, that if discovered late in life, the risks of surgery may outweigh the benefits and the doctor and patient may reasonably elect not to treat it. It can often be managed until the elderly patient dies normally of some other cause.

27.2e The Penis

The **penis**[22] serves to deposit semen in the vagina. Half of it is an internal **root** and half is the externally visible **shaft** and **glans**[23] (**fig. 27.11,** fig. 27.10). The glans is the expanded head at the distal end of the penis with the external urethral orifice at its tip. The external portion of the penis is about 8 to 10 cm (3–4 in.) long and 3 cm in diameter when flaccid (nonerect); the typical dimensions of an erect penis are 13 to 18 cm (5–7 in.) long and 4 cm in diameter.

Directional terminology may be a little confusing in the penis, because the *dorsal* side is the one that faces anteriorly, at least when the penis is flaccid, whereas the *ventral* side of the penis faces posteriorly. This is because in most mammals, the penis is horizontal, held against the abdomen by skin, and it points anteriorly. The urethra passes through its lower, more obviously ventral, half. Directional terminology in the human penis follows the same convention as for other mammals, even though our bipedal posture and more pendulous penis change these anatomical relationships.

The skin is loosely attached to the penile shaft, allowing for movement and expansion during erection. It continues over the glans as the **prepuce (foreskin).** The inner surface of the prepuce and facing surface of the glans are covered with a thin mucous membrane similar to the inner surface of the eyelid. At birth and for at least a few years after, these two membranes are firmly fused to each other; attempts to retract the foreskin of an infant or child can therefore be injurious. The membranes separate gradually over a period of several years; this is usually complete by age 17. The adult prepuce remains anchored to the proximal margin of the glans by a ventral fold of tissue called the **frenulum.**[24] Contrary to some sources, there are no glands in the prepuce; in some males, however, exfoliated epithelial cells and fluid accumulate beneath the prepuce as a creamy secretion called *smegma.*[25] The mucosa of the adult prepuce contains dendritic cells, a component of the immune system (see section 21.1b), thus providing protection from infection at this body orifice. However, these dendritic cells and those of the vagina afford a potential route of entry for HIV, the AIDS virus. The prepuce is the most densely innervated and sensitive region of the penis, with an abundance of tactile corpuscles and nerves concentrated in ridges on its proximal inner surface. Circumcision thus removes the most sensitive part of the penis. The exposed mucous membrane of the glans then transforms to a thin, drier, and less sensitive epidermis.

The shaft of the penis consists mainly of three cylindrical bodies called **erectile tissues,** which fill with blood during sexual arousal and account for its enlargement and erection. A single erectile body, the **corpus spongiosum,** passes along the ventral side of the penis and encloses the penile urethra. It expands at the distal end to fill the entire glans. Proximal to the glans, the dorsal side of the penis has a **corpus cavernosum** (plural, *corpora*

---

[21]*pro* = before; *stat* = to stand; commonly misspelled and mispronounced "prostrate"

[22]*penis* = tail
[23]*glans* = acorn
[24]*fren* = bridle; *ulum* = little
[25]*smegma* = unguent, ointment, soap

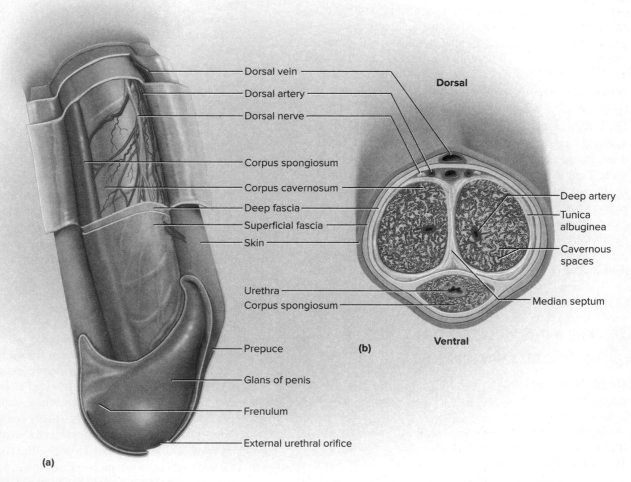

FIGURE 27.11 Anatomy of the Penis. (a) Superficial dissection of shaft, lateral view. (b) Cross section at midshaft. **APR**

❓ *What is the functional benefit of the corpus spongiosum not having a tunica albuginea?*

cavernosa) on each side. Each is ensheathed in a fibrous **tunica albuginea,** and they are separated from each other by a **median septum.** (Be careful not to confuse these with the tunica albuginea of the testis and median septum of the scrotum.)

All three cylinders of erectile tissue are spongy in appearance and contain numerous tiny blood sinuses called **cavernous spaces (lacunae).** The partitions between these spaces, called **trabeculae,** are composed of connective tissue and smooth **trabecular muscle.** In the flaccid penis, muscle tone in the trabeculae collapses the cavernous spaces, which appear as tiny slits in the tissue.

At the body surface, the penis turns 90° and continues inward as the root. The corpus spongiosum terminates internally as a dilated **bulb,** which is ensheathed in the bulbospongiosus muscle and attached to the lower surface of the muscular pelvic floor within the urogenital triangle (see fig. 10.21). The corpora cavernosa diverge like the arms of a Y. Each arm, called a *crus* (pronounced "cruss"; plural, *crura*), attaches the penis to the pubic arch (ischiopubic ramus) and perineal membrane on its respective

side. Each crus is enveloped by an ischiocavernosus muscle. The innervation and blood supply to the penis are discussed later in connection with the mechanism of erection.

BEFORE YOU GO ON

Answer the following questions to test your understanding of the preceding section:

6. State the names and locations of two muscles that help regulate the temperature of the testes.

7. Name three types of cells in the testes, and describe their locations and functions.

8. Name all the ducts that the sperm follow, in order, from the time they form in the testis to the time of ejaculation.

9. Describe the locations and functions of the seminal vesicles, prostate, and bulbourethral glands.

10. Name the erectile tissues of the penis, and describe their locations relative to each other.

27.3 Puberty, Hormonal Control, and Climacteric

Expected Learning Outcomes

When you have completed this section, you should be able to

a. describe the hormonal control of puberty;

b. describe the resulting changes in the male body; and

c. define and describe *male climacteric* and the effect of aging on male reproductive function.

Unlike any other organ system, the reproductive system remains dormant for several years after birth. At puberty, however, a hormonal surge awakens the reproductive system and begins preparing it for adult reproductive function.

27.3a Puberty and Adolescence

Definitions of *puberty* and *adolescence* vary. This book regards **puberty**[26] in boys to be the period from the onset of pituitary gonadotropin secretion (usually around age 10 to 12) until the first ejaculation of viable sperm (typically around age 14). The following chapter describes the criteria and age span for girls. **Adolescence**[27] is the broader period of time including puberty and extending until a person attains full adult height. Hormonal mechanisms explained in the next section—especially the rising levels of *testosterone* and *dihydrotestosterone (DHT),* the two **androgens**—bring about the many bodily changes familiar to all who have endured adolescence:

- They stimulate growth of the sex organs. The first visible sign of puberty is usually enlargement of the testes and scrotum around age 13. The penis continues to grow for about 2 more years after the testes attain their mature size. Internally, the ducts and accessory glands also enlarge.

- Testosterone stimulates sperm production. If testosterone secretion ceases, the sperm count and semen volume decline rapidly and a male becomes sterile.

- Testosterone stimulates a burst of generalized body growth—the limb bones elongate rapidly, a boy grows taller, and he develops more muscle mass. The larynx enlarges, deepening the voice and making the thyroid cartilage more prominent on the front of the neck. Even erythropoiesis is accelerated, resulting in a higher RBC count and hematocrit in males than in females. Along with all this growth, the basal metabolic rate increases, accompanied by an increase in appetite.

- Testosterone also stimulates the brain and awakens the **libido** (sex drive)—although, perhaps surprisingly, the neurons convert it to estrogen, which is what directly

affects the behavior. With increasing libido and sensitivity to stimulation, erections occur frequently and ejaculation often occurs during sleep (nocturnal emissions, or "wet dreams").

- DHT stimulates development of the pubic hair, axillary hair, and later the facial hair. The skin becomes darker and thicker and secretes more sebum, which often leads to acne; acne patients have 2 to 20 times the normal level of DHT in their skin. The apocrine scent glands of the perineal, axillary, and beard areas develop in conjunction with the hair in those regions.

27.3b Hormonal Control of Male Reproductive Function

The testes secrete substantial amounts of testosterone in the first trimester (3 months) of fetal development. Even in the first few months of infancy, testosterone levels are about as high as they are in midpuberty, but then the testes become dormant for the rest of infancy and childhood.

It is still unknown why puberty begins at age 10 to 12 and not earlier. There seems to be a childhood brake on the secretion of **gonadotropin-releasing hormone (GnRH).** Release from this restraint in late childhood brings on GnRH secretion. GnRH travels to the anterior pituitary and stimulates cells called *gonadotropic cells.* These cells secrete the two pituitary *gonadotropins:* **follicle-stimulating hormone (FSH)** and **luteinizing hormone (LH)** (see section 17.2c). From now through adulthood, reproductive function is regulated by hormonal interactions between the hypothalamus, pituitary gland, and gonads—the *hypothalamo–pituitary–gonadal axis.*

Luteinizing hormone (LH) is also sometimes called *interstitial cell–stimulating hormone (ICSH)* because it stimulates the interstitial endocrine cells of the testes to secrete androgens, mainly testosterone. Among the many effects of testosterone previously itemized, it promotes sperm production. That effect, however, requires that the testes concentrate testosterone in the sites where the sperm develop. This is where the other gonadotropin comes into play: follicle-stimulating hormone (FSH, named for its female function). FSH stimulates the nurse cells of the testes to secrete **androgen-binding protein (ABP),** which binds and accumulates testosterone in the seminiferous tubules and epididymis. Testosterone has no effect on the testes if FSH and ABP are absent. Developing sperm have no androgen receptors and don't respond directly to it.

Thus, we can see that testicular development and sperm production depend on signals from the pituitary and ultimately the hypothalamus, but this isn't the whole story. The testes also send feedback signals to those sites and regulate gonadotropin output **(fig. 27.12).** Nurse cells secrete not only ABP but also a hormone called **inhibin,** which selectively suppresses the secretion of FSH by the pituitary. A lower FSH level results in a reduced rate of sperm production. If, however, sperm production is too slow (reflected by a sperm count below 20 million sperm per milliliter of semen), inhibin levels drop, FSH level rises, and sperm production increases. Inhibin thus provides a means for the pituitary

[26]*puber* = grown up
[27]*adolesc* = to grow up

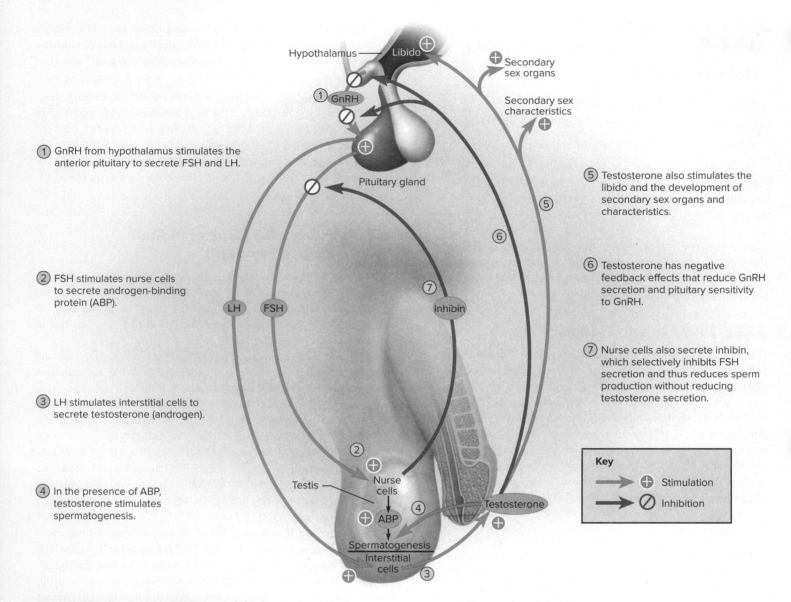

① GnRH from hypothalamus stimulates the anterior pituitary to secrete FSH and LH.

② FSH stimulates nurse cells to secrete androgen-binding protein (ABP).

③ LH stimulates interstitial cells to secrete testosterone (androgen).

④ In the presence of ABP, testosterone stimulates spermatogenesis.

⑤ Testosterone also stimulates the libido and the development of secondary sex organs and characteristics.

⑥ Testosterone has negative feedback effects that reduce GnRH secretion and pituitary sensitivity to GnRH.

⑦ Nurse cells also secrete inhibin, which selectively inhibits FSH secretion and thus reduces sperm production without reducing testosterone secretion.

Key

→ ⊕ Stimulation

→ ⊘ Inhibition

FIGURE 27.12 Hormonal Relationships Between the Hypothalamus, Pituitary Gland, and Testis.

to monitor and adjust the rate of sperm production, but without reducing the production of testosterone (which depends on LH output, not FSH). Testosterone, however, also has negative feedback effects on the hypothalamus, reducing the output of GnRH and the pituitary's sensitivity to it. This enables the hypothalamo–pituitary system to monitor and regulate testosterone secretion.

▶▶▶ **APPLY WHAT YOU KNOW**

If a male animal is castrated, would you expect FSH and LH levels to rise, fall, or be unaffected. Why?

27.3c Aging and Sexual Function

Menopause is a well-known period of declining reproductive function in women, but whether men experience any such decline—sometimes called **andropause** or **male climacteric**—is a controversial concept not accepted by all authorities. Testosterone

secretion peaks around age 20 at about 7 mg/day, then steadily declines to as little as one-fifth of this level by age 80. There is a corresponding decline in the number and secretory activity of the interstitial endocrine cells and nurse cells. Along with the declining testosterone level, the sperm count and libido diminish, as do the volume and force of ejaculation. By age 65, sperm count is typically about one-third of what it was in a man's 20s. Older men may become less interested in sex, but nevertheless remain capable of fathering a child throughout old age.

As testosterone and inhibin levels decline, so does feedback inhibition of the pituitary. The less-inhibited pituitary secretes elevated levels of FSH and LH. These gonadotropins may cause mood changes, hot flashes, or even illusions of suffocation—symptoms similar to those in perimenopausal women. Most men, however, notice few or no such effects as they pass through this age. Despite comedic or sardonic references to "male menopause," the term *menopause* refers to the cessation of menstruation and therefore makes no sense in the context of male physiology.

About 20% of men in their 60s and 50% of men in their 80s experience *erectile dysfunction (impotence),* the frequent inability to produce or maintain an erection sufficient for intercourse (see table 27.3). Erectile dysfunction (ED) and declining sexual activity can have a major impact on older people's perception of the quality of life. Over 90% of men with ED, however, remain able to ejaculate.

BEFORE YOU GO ON

Answer the following questions to test your understanding of the preceding section:

11. State the source, target organ, and effect of GnRH.
12. Identify the target cells and effects of FSH and LH.
13. Explain how testicular hormones affect the secretion of FSH and LH.
14. Describe the major effects of androgens on the body.
15. Define *male climacteric.*

27.4 Sperm and Semen

Expected Learning Outcomes

When you have completed this section, you should be able to

a. describe the stages of meiosis and contrast meiosis with mitosis;
b. describe the sequence of cell types in spermatogenesis, and relate these to the stages of meiosis;
c. describe the role of nurse cells in spermatogenesis;
d. describe or draw and label a sperm cell; and
e. describe the composition of semen and functions of its components.

Spermatogenesis is the process of sperm production. It occurs in the seminiferous tubules and involves three principal events: (1) division and remodeling of a relatively large germ cell into four small, mobile cells with flagella; (2) reduction of the chromosome number by one-half; and (3) a shuffling of the genes so that each chromosome of the sperm carries new gene combinations that didn't exist in the chromosomes of the parents. This ensures genetic variety in the offspring, further accentuated by the independent assortment of chromosomes into different sperm, and the randomness of fertilization. The genetic recombination and reduction in chromosome number are achieved through a form of cell division called **meiosis,** which produces four daughter cells that subsequently differentiate into sperm.

27.4a Meiosis

In nearly all living organisms except bacteria, there are two forms of cell division: mitosis and meiosis. Mitosis, described in section 4.3d, is the basis for division of the single-celled fertilized egg,

growth of an embryo, and all postnatal growth and tissue repair. It is essentially the splitting of a cell with a distribution of chromosomes that results in two genetically identical daughter cells. It consists of four stages: prophase, metaphase, anaphase, and telophase.

You may find it beneficial to review figure 4.15 because of the important similarities and differences between mitosis and meiosis. There are three important differences:

1. In mitosis, each double-stranded chromosome divides into two single-stranded ones, but each daughter cell still has 46 chromosomes (23 pairs). Meiosis, by contrast, reduces the chromosome number by half. The parent cell is **diploid (2*n*),** meaning it has 46 chromosomes in 23 homologous pairs, whereas the daughter cells are **haploid (*n*),** with 23 unpaired chromosomes.

2. In mitosis, the chromosomes don't change their genetic makeup. In an early stage of meiosis, however, the chromosomes of each homologous pair join and exchange portions of their DNA. This creates new combinations of genes, so the chromosomes we pass to our offspring aren't the same ones that we inherited from our parents.

3. In mitosis, each parent cell produces only two daughter cells. In meiosis, it produces four. In the male, four sperm therefore develop from each original germ cell. The situation is somewhat different in the female (see section 28.3a).

Why use such a relatively complicated process for gametogenesis? Why not use mitosis, as we do for all other cell replication in the body? The answer is that sexual reproduction is, by definition, biparental. If we're going to combine gametes from two parents to make a child, there must be a mechanism for keeping the chromosome number constant from generation to generation. Mitosis would produce eggs and sperm with 46 chromosomes each. If these gametes combined, the zygote and the next generation would have 92 chromosomes per cell, the generation after that would have 184, and so forth. To prevent the chromosome number from doubling in every generation, the number is reduced by half during gametogenesis. Meiosis[28] is sometimes called *reduction division* for this reason.

The stages of meiosis are fundamentally the same in both sexes. Briefly, it consists of two cell divisions in succession and occurs in the following phases: prophase I, metaphase I, anaphase I, telophase I, interkinesis, prophase II, metaphase II, anaphase II, and telophase II. These events are detailed in **figure 27.13,** but let us note some of the unique and important aspects of meiosis.

In prophase I, pairs of homologous chromosomes line up side by side and form **tetrads** (*tetra* denoting the four chromatids). One chromosome of each tetrad is from the individual's father (the paternal chromosome) and the other is from the mother (the maternal chromosome). The paternal and maternal chromosomes

---

[28]*meio* = less, fewer

FIGURE 27.13 Meiosis. For simplicity, the cell is shown with only two pairs of homologous chromosomes. Human cells begin meiosis with 23 pairs.

❓ *Although we pass the same genes to our offspring as we inherit from our parents, we do not pass on the same chromosomes. What process in this figure accounts for the latter fact?*

exchange segments of DNA in a process called **crossing-over.** This creates new combinations of genes and thus contributes to genetic variety in the offspring.

After crossing-over, the chromosomes line up at the midline of the cell in metaphase I, they separate at anaphase I, and the cell divides in two at telophase I. This looks superficially like mitosis, but there is an important difference: The centromeres don't divide and the chromatids don't separate from each other at anaphase I; rather, each homologous chromosome parts company with its twin. Therefore, at the conclusion of meiosis I, each chromosome is still double-stranded, but each daughter cell has only 23 chromosomes— it has become haploid.

Meiosis II is more like mitosis—the chromosomes line up on the cell equator again at metaphase II, the centromeres divide, and each chromosome separates into two chromatids. These chromatids are drawn to opposite poles of the cell at anaphase II. At the end of meiosis II, there are four haploid cells, each containing 23 single-stranded chromosomes. Fertilization combines 23 chromosomes from the father with 23 chromosomes from the mother and reestablishes the diploid number of 46 in the zygote.

27.4b Sperm Production

Now we will relate meiosis to spermatogenesis (**fig. 27.14**). The first stem cells specifically destined to become sperm are **primordial germ cells.** Like the first blood cells, these form in the yolk sac, a membrane associated with the developing embryo. In the fifth to sixth week of development, they crawl into the embryo itself and colonize the gonadal ridges. Here they become stem cells called **spermatogonia.** They remain dormant through childhood, lying along the periphery of the seminiferous tubule near the basement membrane, outside the blood–testis barrier (BTB).

At puberty, testosterone secretion rises, reactivates the spermatogonia, and brings on spermatogenesis. The essential steps of spermatogenesis are as follows, numbered to match figure 27.14.

(1) Spermatogonia divide by mitosis. One daughter cell from each division remains near the tubule wall as a stem cell called a *type A spermatogonium.* Type A spermatogonia serve as a lifetime supply of stem cells, so men normally remain fertile even in old age. The other daughter cell, called a *type B spermatogonium,* migrates slightly away from the wall on its way to producing sperm.

FIGURE 27.14 Spermatogenesis. 2*n* indicates diploid cells and *n* indicates haploid cells. The process proceeds from the bottom of the figure to the top. The daughter cells from secondary spermatocytes through spermatids remain connected by slender cytoplasmic processes until spermiogenesis is complete and individual spermatozoa are released. See text for explanation of the five process steps.

? *Why must the primary spermatocyte move through the blood–testis barrier before undergoing meiosis?*

Cross section of seminiferous tubules

Lumen of seminiferous tubule

Sperm

(5) Spermiogenesis

Spermatid

(4) Meiosis II

Secondary spermatocyte

(3) Meiosis I

Blood–testis barrier

Primary spermatocyte

Nurse cell

(2)

Type B spermatogonium

Tight junction

(1)

Type A spermatogonium

Basement membrane of seminiferous tubule

(2) The type B spermatogonium enlarges and becomes a **primary spermatocyte.** Since this cell is about to undergo meiosis and become genetically different from other cells of the body, it must be protected from the immune system. Ahead of the primary spermatocyte, the tight junction between two nurse cells is dismantled, opening a door for the movement of the spermatocyte toward the lumen. Behind it, a new tight junction forms like a door closing between the spermatocyte and the blood supply in the periphery of the tubule. The spermatocyte moves forward, like an astronaut passing through the double-door airlock of a spaceship, and is now protected by the BTB closing behind it.

(3) Now safely isolated from blood-borne antibodies, the primary spermatocyte undergoes meiosis I, which gives rise to two equal-size, haploid, and genetically unique **secondary spermatocytes.**

(4) Each secondary spermatocyte undergoes meiosis II, dividing into two **spermatids**—a total of four for each spermatogonium.

(5) A spermatid divides no further, but undergoes a transformation called **spermiogenesis,** in which it differentiates into a single spermatozoon **(fig. 27.15).** The spermatid sprouts a tail (flagellum) and discards most of its cytoplasm, making the sperm a lightweight, self-propelled cell. It won't move under its own power, however, until ejaculation.

Each stage of this process is a little closer to the lumen than the earlier stages. All stages on the lumenal side of the BTB are bound to the nurse cells by tight junctions and gap junctions, and are closely enveloped in tendrils of the nurse cells. Throughout their meiotic divisions, the daughter cells never completely separate, but remain connected to each other by cytoplasmic bridges. Eventually, however, the mature sperm separate from each other, depart from their supportive nurse cells, and are washed down the tubule by a slow flow of fluid. Spermatogenesis occurs in cycles, progressing down the tubule in *spermatogenic waves.* At any particular time, therefore, cells in a given portion of the tubule are at about the same stage of spermatogenesis, not evenly distributed among all stages.

FIGURE 27.15 Spermiogenesis and Sperm Structure. (a) Stages of spermiogenesis, with mature sperm structure at stage 4. (b) Light micrograph of human sperm (1,000×).

b: Alicia Gonzalez/Alamy

It takes about 70 days for a type B spermatogonium to become mature spermatozoa. A young man produces about 300,000 sperm per minute, or 400 million per day.

27.4c The Spermatozoon

The spermatozoon is an example, par excellence, of the unity of form and function—shaped by evolution for lightness, streamlining, motility, and the effective delivery of its cargo of DNA. It has two parts: a pear-shaped head and a long tail (fig. 27.15, stage 4). The **head,** about 4 to 5 μm long and 3 μm wide at its broadest part, contains three structures: a nucleus, acrosome, and flagellar basal body. The most important of these is the nucleus, which fills most of the head and contains a haploid set of condensed, genetically inactive chromosomes. The **acrosome**[29] is a lysosome in the form of a thin cap covering the apical half of the nucleus. It contains enzymes that are later used to penetrate the egg if the sperm is successful. The basal body of the tail flagellum is nestled in an indentation at the basal end of the nucleus. This is essentially a solitary centriole surrounded by a centrosome, like the centrioles of other cells. It plays an important role in fertilization (see section 29.1b).

The **tail** is divided into three regions called the middle piece, principal piece, and endpiece. The **middle piece,** a cylinder about 5 to 9 μm long and half as wide as the head, is the thickest part. It contains numerous large mitochondria that coil tightly around the axoneme of the flagellum. They produce the ATP needed for the beating of the tail when the sperm migrates up the female reproductive tract. The **principal piece,** 40 to 45 μm long, constitutes most of the tail and consists of the axoneme surrounded by a sheath of supportive fibers. The **endpiece,** 4 to 5 μm long, consists of the axoneme only and is the narrowest part of the sperm.

27.4d Semen

The fluid expelled during orgasm is called **semen**[30] (seminal fluid). A typical ejaculation discharges 2 to 5 mL of semen, composed of about 2% to 5% sperm and spermatic duct secretions, 25% to 30% prostatic fluid, 65% to 70% seminal vesicle fluid, and a trace of bulbourethral fluid. Most of the sperm emerge in the first one or two jets of semen. The semen usually has a **sperm count** of 50 to 120 million sperm/mL. A sperm count any lower than 20 to 25 million sperm/mL is usually associated with **infertility** (sterility), the inability to fertilize an egg (see table 27.3).

Table 27.2 lists the major constituents of semen with their glandular sources and functions. This isn't an exhaustive list; other components with still-unknown functions are not tabulated.

A well-known property of semen is its stickiness, an adaptation that promotes fertilization. It arises when the clotting enzyme from the prostate activates *prosemenogelin,* converting it to a sticky fibrinlike protein, **semenogelin.** This process is very similar to blood clotting. Semenogelin entangles the sperm, sticks to the walls of the inner vagina and cervix, and ensures that the semen doesn't simply drain back out of the vagina. It may also promote the uptake of sperm-laden clots of semen into the uterus. Twenty to 30 minutes after ejaculation, the **serine protease** of the prostatic fluid breaks down semenogelin and liquifies the semen. Serine protease is also known clinically as **prostate-specific antigen (PSA);** an elevated blood PSA level is an indicator of possible prostate cancer (see Deeper Insight 27.2).

Fructose and **citrate** are the two most significant nutrients in semen, metabolized by the sperm mitochondria to generate ATP for flagellar motility. **Calcium** is also needed for flagellar beating, and plays an additional role in releasing enzymes from the acrosome when the sperm penetrates an egg. **Zinc** in the semen

---

[29]*acro* = tip, peak; *some* = body

[30]*semen* = seed

TABLE 27.2	**Composition of Semen**		
Source	**Fraction of Semen Volume**	**Constituents**	**Functions**
Testes	2% to 5%	50–120 million sperm per mL	Fertilization of egg
Seminal vesicles	65% to 75%	Fructose, citrate, ascorbic acid	Nutrients for sperm
		Prostaglandins	Stimulate peristalsis of ductus deferens and uterus; inhibit sperm rejection by female immune system
		Prosemenogelin	Fibrinogen-like clotting protein; coagulates semen
		Calcium	Activates sperm motility; triggers release of enzymes from acrosome for penetration of egg
Prostate	25% to 30%	Clotting enzyme	Acts on prosemenogelin, coagulates semen
		Serine protease	Liquifies coagulated semen after delay
		Citrate	Nutrient for sperm
		Bicarbonate and phosphate buffers	Neutralize acidity of urethra and vagina
		Zinc	Stabilizes sperm chromatin; deficiency causes sperm fragility
Bulbourethral glands	<1%	Mucus	Lubricates urethra; forms low-viscosity channel for sperm migration through vagina and cervix
		Buffers	Neutralize acidity of urethra and vagina

stabilizes the chromatin in the sperm head; a dietary deficiency of zinc can result in fragile sperm and infertility.

Sperm encounter hostile conditions as soon as they leave the safety of the spermatic ducts. Residual urine makes the male urethra acidic, and the vagina has a film of lactic acid to protect it from microbial overgrowth. The prostate and bulbourethral glands secrete bicarbonate and phosphate buffers to neutralize this acidity and protect the sperm. Yet another jeopardy to the sperm is attack by the female's immune system, which recognizes the sperm as foreign. **Prostaglandins** from the seminal vesicles inhibit this immune response and minimizes sperm loss. (Prostaglandins are named for the fact that they were first discovered in the fluid from the bovine prostate.) Prostaglandins also stimulate peristalsis in both sexes—in the ductus deferens of the male, driving sperm toward the penis; and in the uterus of the female, possibly serving to draw up semen from the vagina. Peristaltic waves in the uterus and uterine tubes may also help spread sperm through the female reproductive tract.

Two requirements must be met for sperm motility: an elevated pH and an energy source. The pH of the vagina is about 3.5 to 4.0, and the male spermatic ducts are also quite acidic. The sperm remain motionless at such low pHs. But as the prostatic fluid buffers the vaginal and seminal acidity, the pH rises to about 7.5 and the sperm become active at the time of ejaculation. The activated sperm now thrash with their tails and crawl up the mucosa of the vagina and uterus.

DEEPER INSIGHT 27.3

CLINICAL APPLICATION

Reproductive Effects of Pollution

In recent decades, there has been a great deal of interest within the scientific community and popular media concerning *endocrine disrupting chemicals (EDCs)*—environmental agents that interfere with our natural hormones and may disrupt reproduction, development, and homeostasis.

Proven or suspected EDCs occur in industrial solvents and lubricants, pesticides, plastics (including the plastic lining of food cans), prescription drugs and medical devices, and even infant soy milk formulas. We're exposed to EDCs through food, drinking water, contaminated air and soil, and household chemicals, and in certain occupations such as farming and manufacturing that use such chemicals. Some EDCs may affect not only the exposed person, but also that person's descendants through the DNA methylation and epigenetic effects (see section 4.4h). Your grandchildren could be genetically affected by your chemical exposures today.

EDCs act by several mechanisms. They or their metabolites can mimic the effects of estrogen by activating its receptors; antagonize the effects of androgens; alter gene expression; or disrupt positive and negative feedback loops that regulate the body's secretion of its own estrogens and androgens.

EDCs are suspected or implicated, at least tentatively, in a broad spectrum of reproductive abnormalities—in males, cryptorchidism, hypospadias (a urethra opening ventrally on the penis instead of at the tip), low sperm count and reduced motility, and testicular cancer; in females, premature breast development, premature menopause, breast cancer, uterine fibroids, endometriosis, and disrupted ovulation and lactation.

But in many such cases, the data are contradictory, weak, or in dispute. The evidence, however disturbing, is often vague and indirect. It is enormously difficult to prove a link between a particular reproductive disorder and a suspected EDC. For one thing, we can't experiment on this with humans, and the results of animal experiments often don't translate to humans. For another, the effects may be delayed by years, decades, or perhaps even generations; and there are so many variables in human life that they cover up the trail of causation—prenatal or infant exposure with no effects visible until adulthood; changes in occupation, residence, and environmental exposure; migration and international adoption; and exposure to complex mixtures of environmental chemicals—making it impossible to single out any one cause or to know how they interact to produce effects that no one of them would produce alone. Despite such difficulties, EDCs pose a compelling problem under continuing investigation.

BEFORE YOU GO ON

Answer the following questions to test your understanding of the preceding section:

16. Name the stages of spermatogenesis from spermatogonium to spermatozoa. How do they differ in the number of chromosomes per cell and chromatids per chromosome?

17. Describe the two major parts of a spermatozoon and state what organelles or cytoskeletal components are contained in each.

18. List the major contributions of the seminal vesicles and prostate to the semen, and state the functions of these components.

27.5   Male Sexual Response

Expected Learning Outcomes

When you have completed this section, you should be able to

a. describe the blood and nerve supply to the penis; and

b. explain how these govern erection and ejaculation.

Sexual intercourse is also known as **coitus, coition,**[31] or **copulation.**[32] The physiology of sexual intercourse was unexplored territory before the 1950s because of repressive attitudes toward the subject. British psychologist Havelock Ellis published the groundbreaking six-volume *Studies in the Psychology of Sex* (1897–1910), only to have it banned for several years as too controversial for victorian tastes. In the 1950s, William Masters and Virginia Johnson daringly launched the first physiological studies of sexual response in the laboratory. In 1966, they published *Human Sexual Response,* which detailed measurements and observations on more than 10,000 sexual acts by nearly 700 volunteer men and women. Masters and Johnson then turned their attention to disorders of sexual function and pioneered modern therapy for sexual dysfunctions. They divided intercourse into

---

[31]*coit* = to come together
[32]*copul* = to link, bond

four recognizable phases, which they called *excitement, plateau, orgasm,* and *resolution.* The following discussion is organized around this model, although other authorities have modified it or proposed alternatives.

27.5a Anatomical Foundations

To understand male sexual function, we must give closer attention to the blood circulation and nerve supply to the penis.

Each internal iliac artery gives rise to an **internal pudendal (penile) artery,** which enters the root of the penis and divides in two. One branch, the **dorsal artery,** travels dorsally along the penis not far beneath the skin (see fig. 27.11), supplying blood to the skin, fascia, and corpus spongiosum. The other branch, the **deep artery,** travels through the core of the corpus cavernosum and gives off smaller **helicine**[33] **arteries,** which penetrate the trabeculae and empty into the cavernous spaces. When the penis is flaccid, most of its blood supply comes from the dorsal arteries. When the deep artery dilates, the cavernous spaces fill with blood and the penis becomes erect. There are numerous anastomoses between the dorsal and deep arteries, so neither of them is the exclusive source of blood to any one erectile tissue. A median **deep dorsal vein** drains blood from the penis. It runs between the two dorsal arteries beneath the deep fascia and empties into a plexus of prostatic veins.

The penis is richly innervated by sensory and motor nerve fibers. The glans and foreskin have an abundance of tactile, pressure, and temperature receptors, especially on the frenulum and proximal margin of the glans. The receptors lead by way of a pair of prominent **dorsal nerves** to the **internal pudendal nerves,** then via the sacral plexus to segments S2 to S4 of the spinal cord. Sensory fibers of the shaft, scrotum, perineum, and elsewhere are also highly important to sexual stimulation.

Both autonomic and somatic motor fibers carry impulses from integrating centers in the spinal cord to the penis and other pelvic organs. Sympathetic fibers arise from levels T12 to L2, pass through the hypogastric and pelvic plexuses, and innervate the penile arteries, trabecular muscle, spermatic ducts, and accessory glands. They dilate the penile arteries and can induce erection even when the sacral region of the spinal cord is damaged. They also initiate erection in response to input to the special senses and to sexual thoughts.

Parasympathetic fibers extend from segments S2 to S4 of the spinal cord through the pudendal nerves to the arteries of the penis. They are involved in an autonomic reflex arc that causes erection in response to direct stimulation of the penis and perineal region.

27.5b Excitement and Plateau

The **excitement phase** is characterized by **vasocongestion** (swelling of the genitals with blood); **myotonia** (muscle tension); and increases in heart rate, blood pressure, and pulmonary ventilation **(fig. 27.16).** The bulbourethral glands secrete their fluid during this phase. The excitement phase can be initiated by a broad spectrum of erotic stimuli—sights, sounds, aromas, touch—and even

by dreams or thoughts. Conversely, emotions can inhibit sexual response and make it difficult to function when a person is anxious, stressed, or preoccupied with other thoughts.

The most obvious manifestation of male sexual arousal is **erection** of the penis, which makes entry of the vagina possible. Erection is an autonomic reflex mediated predominantly by parasympathetic nerve fibers that travel alongside the deep and helicine arteries of the penis. These fibers trigger the secretion of nitric oxide (NO), which leads to the relaxation of the deep arteries and cavernous spaces (see Deeper Insight 27.4). Whether this is enough to cause erection, or whether it is also necessary to block the outflow of blood from the penis, is still debated. According to one hypothesis, as cavernous spaces near the deep arteries fill with blood, they compress the spaces closer to the periphery of the erectile tissue. This is where blood leaves the erectile tissues, so the compression of the peripheral spaces helps retain blood in the penis. Their compression is aided by the fact that each corpus cavernosum is wrapped in a tunica albuginea, which fits over the erectile tissue like a tight fibrous sleeve and contributes to its tension and firmness. In addition, the bulbospongiosus and ischiocavernosus muscles aid in erection by compressing the root of the penis and forcing blood forward into the shaft.

As the corpora cavernosa expand, the penis becomes enlarged, rigid, and elevated to an angle conducive to entry of the vagina. Longitudinal expansion of the penis retracts the foreskin, exposing the glans. Once **intromission** (entry) is achieved, the tactile and pressure sensations produced by vaginal massaging of the penis further accentuate the erection reflex.

The corpus spongiosum has neither a central artery nor a tunica albuginea. It swells and becomes more visible as a cordlike ridge along the ventral surface of the penis, but it doesn't become nearly as engorged and hardened as the corpora cavernosa. Vasocongestion isn't limited to the penis; the testes also become as much as 50% larger during excitement.

In the **plateau phase,** variables such as respiratory rate, heart rate, and blood pressure are sustained at a high level, or rise slightly, for a few seconds to a few minutes before orgasm. This phase may be marked by increased vasocongestion and myotonia.

▶▶▶APPLY WHAT YOU KNOW

Why is it important that the corpus spongiosum not become as engorged and rigid as the corpora cavernosa?

27.5c Orgasm and Ejaculation

The **orgasm,**[34] or **climax,** is a short but intense reaction that lasts 3 to 15 seconds and usually is marked by the discharge of semen. The heart rate increases to as high as 180 beats/min., blood pressure rises proportionally, and the respiratory rate becomes as high as 40 breaths/min. From the standpoint of producing offspring, the most significant aspect of male orgasm is the **ejaculation**[35] of semen into the vagina.

---

[33]*helic* = coil, helix

[34]*orgasm* = swelling
[35]*e* = *ex* = out; *jacul* = to throw

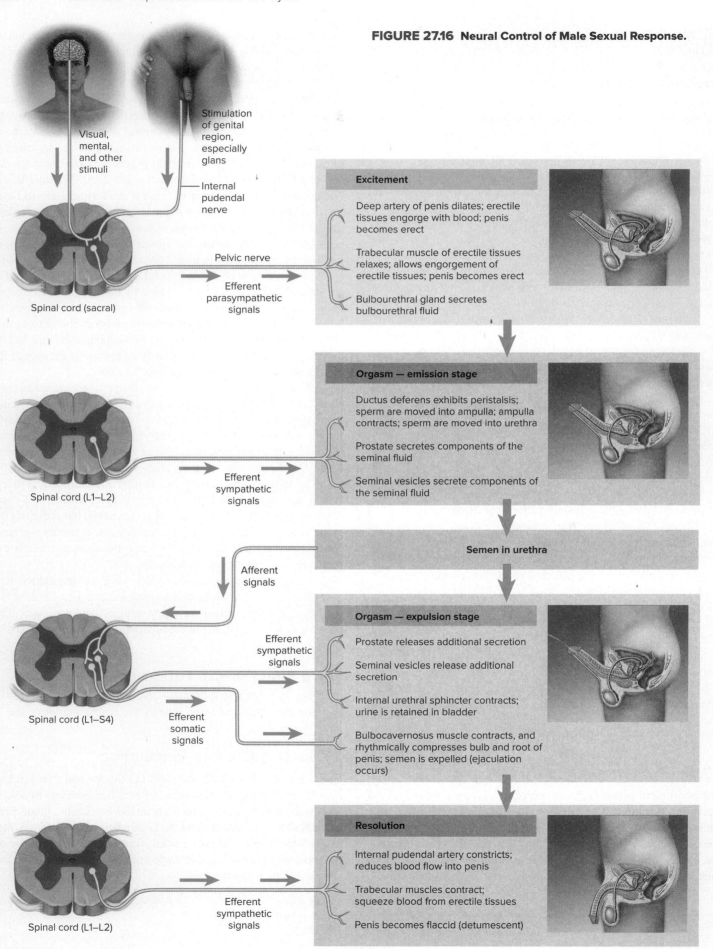

FIGURE 27.16 **Neural Control of Male Sexual Response.**

Visual, mental, and other stimuli

Stimulation of genital region, especially glans

Internal pudendal nerve

Pelvic nerve

Efferent parasympathetic signals

Spinal cord (sacral)

Excitement

Deep artery of penis dilates; erectile tissues engorge with blood; penis becomes erect

Trabecular muscle of erectile tissues relaxes; allows engorgement of erectile tissues; penis becomes erect

Bulbourethral gland secretes bulbourethral fluid

Orgasm — emission stage

Ductus deferens exhibits peristalsis; sperm are moved into ampulla; ampulla contracts; sperm are moved into urethra

Prostate secretes components of the seminal fluid

Seminal vesicles secrete components of the seminal fluid

Spinal cord (L1–L2)

Efferent sympathetic signals

Semen in urethra

Afferent signals

Efferent sympathetic signals

Spinal cord (L1–S4)

Efferent somatic signals

Orgasm — expulsion stage

Prostate releases additional secretion

Seminal vesicles release additional secretion

Internal urethral sphincter contracts; urine is retained in bladder

Bulbocavernosus muscle contracts, and rhythmically compresses bulb and root of penis; semen is expelled (ejaculation occurs)

Resolution

Internal pudendal artery constricts; reduces blood flow into penis

Trabecular muscles contract; squeeze blood from erectile tissues

Penis becomes flaccid (detumescent)

Spinal cord (L1–L2)

Efferent sympathetic signals

DEEPER INSIGHT 27.4

CLINICAL APPLICATION

Treating Erectile Dysfunction

Among the most lucrative drugs developed in the 1990s were the popular treatments for erectile dysfunction: sildenafil (Viagra), vardenafil (Levitra), and tadalafil (Cialis). The basis for developing them was a seemingly unrelated discovery: the role of nitric oxide in cell signaling. When sexual stimulation triggers NO secretion, NO activates the enzyme guanylate cyclase. Guanylate cyclase produces cyclic guanosine monophosphate (cGMP). cGMP then relaxes the smooth muscle of the deep arteries and cavernous spaces of the corpora cavernosa, increasing blood flow into these erectile tissues and bringing about an erection **(fig. 27.17)**.

The erection subsides when cGMP is broken down by the enzyme *phosphodiesterase type 5 (PDE5)*. The problem for many men and their partners is that it subsides too soon, and the solution is to prevent cGMP from breaking down so fast. The aforementioned drugs are in a family called *phosphodiesterase inhibitors*. By slowing down the action of PDE5, they prolong the life of cGMP and the duration of the erection.

FIGURE 27.17 The Action of Viagra and Other Phosphodiesterase Inhibitors.

Ejaculation occurs in two stages called emission and expulsion. In **emission,** the sympathetic nervous system stimulates peristalsis in the smooth muscle of the ductus deferens, which propels sperm from the tail of the epididymis, along the ductus, and into the ampulla. Contractions of the ampulla propel the sperm into the prostatic urethra, and contractions of smooth muscle in the prostate force prostatic fluid into the urethra. Secretions of the seminal vesicles join the semen soon after the prostatic secretion. The contractions and seminal flow of this phase create an urgent sensation that ejaculation is inevitable.

Semen in the urethra activates somatic and sympathetic reflexes that result in its **expulsion.** Sensory signals travel to the spinal cord via the internal pudendal nerve and reach an integrating center in the upper lumbar region. Sympathetic nerve fibers carry motor signals from here out to the prostate and seminal vesicles, causing the smooth muscle in their walls to express more fluid into the urethra. The sympathetic reflex also constricts the internal urethral sphincter so urine can't enter the urethra and semen can't enter the bladder.

Somatic motor signals leave the third and fourth sacral segments of the cord and travel to the bulbospongiosus, ischiocavernosus, and levator ani muscles. The bulbospongiosus, which envelops the root of the penis (see fig. 10.20a), undergoes five or six strong, spasmodic contractions that compress the urethra and forcibly expel the semen. Most sperm are ejected in the first milliliter of semen, mixed primarily with prostatic fluid. The seminal vesicle secretion follows and flushes most remaining sperm from the ejaculatory ducts and urethra. Some sperm may seep from the penis prior to ejaculation, and pregnancy can therefore result from genital contact even without orgasm.

Orgasm is accompanied by an intense feeling of release from tension. Ejaculation and orgasm are not the same. Although they usually occur together, it is possible to have all of the sensations of orgasm without ejaculating, and ejaculation occasionally occurs with little or no sensation of orgasm.

27.5d Resolution

Immediately following orgasm comes the **resolution** phase. Discharge of the sympathetic nervous system constricts the internal pudendal artery and reduces the flow of blood into the penis. It also causes contraction of the trabecular muscles, which squeeze blood from the erectile tissues. The penis may remain semierect long enough to continue intercourse, which may be important to the female's attainment of climax, but gradually the penis undergoes **detumescence**—it becomes soft and flaccid again. The resolution phase is also a time in which cardiovascular and respiratory functions return to normal. Many people break out in sweat during the resolution phase. In men, resolution is followed by a **refractory period,** lasting anywhere from 10 minutes to a few hours, in which it is usually impossible to attain another erection and orgasm.

Men and women have many similarities and a few significant differences in sexual response. The response cycle of women is described in section 28.4.

Two of the most common concerns related to sex are sexually transmitted diseases (STDs) and contraception. Understanding most contraceptive methods requires a prior understanding of female anatomy and physiology, so contraceptive techniques for both sexes are discussed at the end of the next chapter, while STDs are discussed in this chapter (see Deeper Insight 27.5). This is not to imply, of course, that STDs are only a male concern and contraception only a female concern.

Reproductive disorders specific to males and females are briefly summarized in **table 27.3** and table 28.5, respectively. The effects of aging on the reproductive system are described in section 29.4a.

TABLE 27.3	Some Male Reproductive Disorders
Breast cancer	Accounts for 0.2% of male cancers in the United States; usually seen after age 60 but sometimes in children and adolescents. (For every male who gets breast cancer, about 175 females do so.) Usually felt as a lump near the nipple, often with crusting and discharge from nipple. Often quite advanced by the time of diagnosis, with poor prospects for recovery, because of denial and delay in seeking treatment.
Erectile dysfunction (impotence)	Inability to maintain an erection adequate for vaginal entry in half or more of one's attempts. Can stem from aging and declining testosterone level as well as cardiovascular and neurological diseases, diabetes mellitus, medications, fear of failure, depression, and other causes.
Hypospadias[36] (HY-po-SPAY-dee-us)	A congenital defect in which the urethra opens on the ventral side or base of the penis rather than at the tip; usually corrected surgically at about 1 year of age.
Infertility	Inability to fertilize an egg because of a low sperm count (lower than 20 to 25 million/mL), poor sperm motility, or a high percentage of deformed sperm (two heads, defective tails, etc.). May result from malnutrition, gonorrhea and other infections, toxins, or testosterone or zinc deficiency.
Penile cancer	Accounts for 1% of male cancers in the United States; most common in black males ages 50 to 70 and of low income. Most often seen in men with nonretractable foreskins (phimosis) combined with poor penile hygiene; least common in men circumcised at birth.
Testicular cancer	The most common solid tumor in men 15 to 34 years old, especially white males of middle to upper economic classes. Typically begins as a painless lump or enlargement of the testis. Highly curable if detected early. Men should routinely palpate the testes for normal size and smooth texture.
Varicocele (VAIR-ih-co-seal)	Abnormal dilation of veins of the spermatic cord, so that they resemble a "bag of worms." Occurs in 10% of males in the United States. Caused by absence or incompetence of venous valves. Reduces testicular blood flow and often causes infertility.
You can find other male reproductive disorders described in the following places:	

Androgen insensitivity syndrome in Deeper Insight 27.1; *benign prostatic hyperplasia* and *prostate cancer* in Deeper Insight 27.2; and *sexually transmitted diseases (gonorrhea, chlamydia, syphilis, genital herpes,* and *genital warts)* in Deeper Insight 27.5.

BEFORE YOU GO ON

Answer the following questions to test your understanding of the preceding section:

19. Explain how penile blood circulation changes during sexual arousal and why the penis becomes enlarged and stiffened.

20. State the roles of the sympathetic, parasympathetic, and somatic nervous systems in male sexual response.

[36]*hypo* = below; *spad* = to draw off (the urine)

DEEPER INSIGHT 27.5

CLINICAL APPLICATION

Sexually Transmitted Diseases

Sexually transmitted diseases (STDs) have been well known since the writings of Hippocrates. Here we discuss three bacterial STDs—gonorrhea, chlamydia, and syphilis—and three viral STDs—genital herpes, genital warts, and hepatitis. AIDS is discussed in section 21.6.

All of these STDs have an *incubation period* in which the pathogen multiplies without symptoms, and a *communicable period* in which one can transmit the disease to others, even in the absence of symptoms. STDs often cause fetal deformity, stillbirth, and neonatal death.

Gonorrhea (GON-oh-REE-uh) is caused by the bacterium *Neisseria gonorrhoeae.* Galen, thinking the pus discharged from the penis was semen, named the disease *gonorrhea* ("flow of seed"). Gonorrhea causes abdominal discomfort, genital pain and discharge, painful urination, and abnormal uterine bleeding, but most infected women are asymptomatic. It can cause scarring of the uterine tubes, resulting in infertility. Gonorrhea is treated with antibiotics.

Nongonococcal urethritis (NGU) is any urethral inflammation caused by agents other than the gonorrhea bacterium. NGU often produces pain or discomfort on urination. The most common bacterial NGU is *chlamydia,* caused by *Chlamydia trachomatis.* Most chlamydia infections are asymptomatic, but they may cause urethral discharge and pain in the testes or pelvic region. Gonorrhea and chlamydia frequently occur together.

Pelvic inflammatory disease (PID) is acute, painful inflammation of the female pelvic organs, usually due to infection with *Chlamydia* or *Neisseria.* It often causes irregular menstrual bleeding or pain on urination. It can result in sterility and may require surgical removal of infected uterine tubes or other organs. PID is responsible for many cases of ectopic pregnancy (see Deeper Insight 29.2).

Syphilis (SIFF-ih-liss) is caused by a spiral bacterium named *Treponema pallidum.* After an incubation period of 2 to 6 weeks, a small, hard lesion called a *chancre* (SHAN-kur) appears at the site of infection—in males, usually on the penis, but in females, sometimes out of sight in the vagina. It disappears in 4 to 6 weeks, ending the first stage of syphilis and often creating an illusion of recovery. A second stage ensues, however, with a widespread pink rash, other skin eruptions, fever, joint pain, and hair loss. This subsides in 3 to 12 weeks, but symptoms can come and go for up to 5 years. A person is contagious even when symptoms are not present. The disease may progress to a third stage, *tertiary syphilis (neurosyphilis),* with cardiovascular damage and brain lesions that can cause paralysis and dementia. Syphilis is treatable with antibiotics.

Genital herpes is the most common STD in the United States, with 20 to 40 million infected people. It is usually caused by the *herpes simplex virus type 2 (HSV-2).* After an incubation period of 4 to 10 days, the virus causes blisters on the penis of the male; on the labia, vagina, or cervix of the female; and sometimes on the thighs and buttocks of either sex. Over 2 to 10 days, these blisters rupture, seep fluid, and begin to form scabs. The initial infection may be painless or may cause intense pain, urethritis, and watery discharge from the penis or vagina. The lesions heal in 2 to 3 weeks and leave no scars.

During this time, however, HSV colonizes sensory nerves and ganglia. Here it can lie dormant for years, later migrating along the nerves and causing epithelial lesions anywhere on the body. The movement from place to place is the basis of the name *herpes.*[37] Most patients have five to seven recurrences, ranging from several years apart to several times a year. An infected person is contagious to a sexual partner when the lesions are present and sometimes even when they are not. HSV may increase the risk of cervical cancer and AIDS.

Genital warts (condylomas) are caused by various strains of *human papillomavirus (HPV).* In most cases, HPV infection goes away on its own and does not cause any health problems. If it persists, however, it can cause genital warts or worse. In the male, warts usually appear on the penis, perineum, or anus; and in the female, they are usually on the cervix, vaginal wall, perineum, or anus. Lesions are sometimes small and almost invisible. Genital warts can be treated with cryosurgery (freezing and excision), laser surgery, or interferon. An HPV vaccine became available in 2006 and infection rates in young people seem to be declining as a result of vaccination. Certain strains of HPV different from those that usually cause genital warts cause cancer of the penis, vagina, cervix, anus, and throat; nearly all cervical cancer and about 70% of throat cancer is caused by HPV. There are about 14 million new cases of HPV infection in the United States per year. Given the alarming incidence of sexual activity and HPV among young adolescents, the U.S. Food and Drug Administration recommends vaccination of girls and boys at the age of puberty in hopes of immunizing most before the onset of sexual activity.

Hepatitis B and *C* are inflammatory liver diseases caused by the hepatitis B and C viruses (HBV, HCV), introduced in Deeper Insight 26.3. Although they can be transmitted by means other than sex, they are becoming increasingly common as STDs. Hepatitis C threatens to become a major epidemic of the twenty-first century. It already far surpasses the prevalence of AIDS and is the leading reason for liver transplants in the United States.

---

[37]*herp* = to creep

STUDY GUIDE

▶ Assess Your Learning Outcomes

To test your knowledge, discuss the following topics with a study partner or in writing, ideally from memory.

27.1 Sexual Reproduction and Development

1. Essential characteristics of sexual reproduction; what defines *male* and *female* in any sexual species; and the names and defining characteristics of their respective gametes and gonads
2. Multiple functions of the male and female reproductive systems
3. Which organs of each sex are considered to be primary and secondary sex organs; which are considered to be internal and external genitalia; and how secondary sex characteristics differ from secondary sex organs
4. The difference between sex chromosomes and autosomes; the number of each; names of the sex chromosomes; and how males and females differ chromosomally
5. The male-determining gene on the Y chromosome; the name of the protein encoded by it; and the effect of that protein on embryonic development
6. Fates of the mesonephric and paramesonephric ducts in the male and female embryos, and why this is indirectly determined by the *SRY* gene
7. Structures of the embryonic genital tubercle, urogenital folds, and labioscrotal folds; and mature structures of the male and female that arise from each
8. Descent of the gonads; similarities and differences in this process in the male and female fetus; and consequences for the male if it fails to occur to completion

27.2 Male Reproductive Anatomy

1. Anatomy and functions of the scrotum and spermatic cord
2. Why the testes must be kept cooler than the core body temperature, and three mechanisms for achieving this
3. Anatomy of the testis and functions of its seminiferous tubules and interstitial endocrine cells
4. The germinal epithelium of the seminiferous tubule; its cell types and their functions; the necessity and structure of the blood–testis barrier

5. The arterial supply and venous drainage of the testis; testicular nerves and lymphatic vessels
6. Gross anatomy of the epididymis; the series of spermatic ducts from the efferent ductules to the ejaculatory duct; and differences in their anatomy and relationships to adjacent organs
7. Three sets of male accessory glands; their anatomy, functions, and relationships to adjacent organs
8. The root, shaft, and glans of the penis; the prepuce and frenulum distally; the bulb and the crura proximally
9. Erectile tissues of the penis; their histological structure; the cavernous spaces, trabeculae, and trabecular muscle, differences between the corpus spongiosum and the corpora cavernosa

27.3 Puberty, Hormonal Control, and Climacteric

1. Definitions of *puberty* and *adolescence;* the typical age range of each
2. The hormonal trigger for the onset of puberty; the roles of GnRH, FSH, LH, androgens, and androgen-binding protein in male adolescence
3. Bodily changes of male adolescence, onset of the libido, and their respective hormonal causes
4. The source of inhibin and its role in male sexual physiology
5. Changes in the levels of testosterone, inhibin, FSH, and LH over the male life span; how these relate to negative feedback inhibition of the pituitary; and the effects experienced by some men in male climacteric
6. Effects of old age on male sexual function

27.4 Sperm and Semen

1. The meaning of *spermatogenesis* and distinction between spermatogenesis and spermiogenesis
2. Why meiosis is necessary in sexually reproducing species, how it affects the chromosome number of gametes, and why chromosome number remains constant from one generation to the next
3. The nine stages of meiosis—four in meiosis I, interkinesis, and four in meiosis II; events

of each stage; the final number of functional daughter cells, their chromosome number, and final chromosome structure in the male
4. Multiple reasons that meiosis results in genetic diversity of the gametes
5. The origin of primordial germ cells, their migration to the embryonic gonads, and the stage they have reached by the time of birth
6. The mode and location of production of type A and type B spermatogonia and the difference between them
7. Stages in the transformation of spermatogonia to sperm; how the germ cells migrate through the germinal epithelium as this is occurring; when they must pass through the blood–testis barrier, why, and how they do so
8. The process and effects of spermiogenesis
9. Morphology of a mature sperm and the functions of its parts
10. Composition of semen and percentages of it that come from its three main sources; functions of the fructose, semenogelin, clotting enzyme, prostaglandins, and serine protease; and the identity and clinical relevance of prostate-specific antigen (PSA)
11. The typical sperm count and the minimum necessary for fertility

27.5 Male Sexual Response

1. Blood vessels of the penis and how they function in the flaccid penis and in erection
2. The nerve supply of the penis and its relationship to the sacral plexus and spinal cord; three types of nerve fibers that the penis receives
3. Physiological changes that the male undergoes during the excitement phase of the sexual response
4. The mechanism of erection including the roles of parasympathetic stimulation, nitric oxide, and vasodilation; why the corpora cavernosa become more engorged than the corpus spongiosum
5. The physiological state of the male in the plateau phase of sexual response
6. The mechanism of male orgasm, including the stages of ejaculation
7. Characteristics of the male resolution phase; the mechanism of penile detumescence; and the refractory period

STUDY GUIDE

▶ Testing Your Recall

Answers in Appendix A

1. The ductus deferens develops from the
   _____ of the embryo.
   a. mesonephric duct
   b. paramesonephric duct
   c. phallus
   d. labioscrotal folds
   e. urogenital folds

2. The protein that clots and causes the stickiness of the semen is
   a. semenogelin.
   b. prostaglandin.
   c. fibrin.
   d. phosphodiesterase.
   e. serine protease.

3. The expulsion of semen occurs when the
   bulbospongiosus muscle is stimulated by
   a. somatic efferent neurons.
   b. somatic afferent neurons.
   c. sympathetic efferent neurons.
   d. parasympathetic efferent neurons.
   e. prostaglandins.

4. Prior to ejaculation, sperm are stored
   primarily in
   a. the seminiferous tubules.
   b. the cavernous spaces.
   c. the epididymis.
   d. the seminal vesicles.
   e. the ejaculatory ducts.

5. The penis is attached to the pubic arch by
   crura of
   a. the corpora cavernosa.
   b. the corpus spongiosum.
   c. the perineal membrane.
   d. the bulbospongiosus.
   e. the ischiocavernosus.

6. The first hormone secreted at the onset of
   puberty is
   a. follicle-stimulating hormone.
   b. interstitial cell–stimulating hormone.
   c. human chorionic gonadotropin.
   d. gonadotropin-releasing hormone.
   e. testosterone.

7. When it is necessary to reduce sperm
   production without reducing testosterone
   secretion, the nurse cells secrete
   a. dihydrotestosterone.
   b. androgen-binding protein.
   c. LH.
   d. FSH.
   e. inhibin.

8. Four spermatozoa arise from each
   a. primordial germ cell.
   b. type A spermatogonium.
   c. type B spermatogonium.
   d. secondary spermatocyte.
   e. spermatid.

9. The point in meiosis at which sister chromatids separate from each other is
   a. prophase I.
   b. metaphase I.
   c. anaphase I.
   d. anaphase II.
   e. telophase II.

10. Blood is forced out of the cavernous spaces
    of the penis by contraction of the _____
    a. bulbospongiosus muscle.
    b. ischiocavernosus muscle.
    c. cremaster.
    d. trabecular muscles.
    e. dartos muscle.

11. Under the influence of androgens, the
    embryonic _____ duct develops into the
    male reproductive tract.

12. A deficiency of _____ can cause fragile
    sperm and male infertility.

13. The _____, a network of veins in the spermatic cord, helps keep the testes cooler than
    the core body temperature.

14. All germ cells beginning with the _____
    are genetically different from the rest of the
    body cells and therefore must be protected
    by the blood–testis barrier.

15. The corpora cavernosa, as well as the testes,
    have a fibrous capsule called the _____.

16. Over half of the semen consists of secretions from a pair of glands called the
    _____.

17. The blood–testis barrier is formed by tight
    junctions between the _____ cells.

18. The earliest haploid stage of spermatogenesis is the _____.

19. Erection of the penis occurs when nitric
    oxide causes the _____ arteries to dilate.

20. A sperm penetrates the egg by means of
    enzymes in its _____.

▶ Building Your Medical Vocabulary

Answers in Appendix A

*State a meaning of each word element, and give
a medical term from this chapter that uses it or a
slight variation of it.*

1. crypto-

2. didymo-

3. e-

4. gamo-

5. -ism

6. meio-

7. orchido-

8. rete

9. -some

10. vagino-

STUDY GUIDE

▶ What's Wrong with These Statements?

Answers in Appendix A

Briefly explain why each of the following statements is false, or reword it to make it true.

1. Testosterone is secreted by the nurse cells of the testis.

2. Sperm carry most of their DNA in the thick middle piece just behind the head of the cell.

3. Spermiogenesis is a stage in which the male sex cells complete meiosis II and become mature sperm cells.

4. The testes and penis are the primary sex organs of the male.

5. A high testosterone level makes a fetus develop a male reproductive system, and a high estrogen level makes it develop a female reproductive system.

6. Most of the volume of the semen is produced by the testes.

7. The pampiniform plexus serves to keep the testes warm enough for sperm to develop.

8. Developing sperm must bind testosterone in order to complete their development.

9. Male menopause is the cessation of sperm production around the age of 55.

10. Erection is caused by sympathetic stimulation of the erectile tissues.

▶ Testing Your Comprehension

1. Explain why testosterone may be considered both an endocrine and a paracrine secretion of the testes. (Review paracrines in section 17.6 if necessary.)

2. A young man is in a motorcycle accident that severs his spinal cord at the neck and leaves him paralyzed from the neck down. When informed of the situation, his wife asks the physician if her husband will be able to have erections and father any children. What should the doctor tell her? Explain your answer.

3. Considering the temperature in the scrotum, would you expect hemoglobin to unload more or less oxygen to the testes than it unloads in the warmer internal organs? Why? (*Hint:* See fig. 22.26.) How would you expect this fact to influence sperm development?

4. Why is it possible for spermatogonia to be outside the blood–testis barrier, yet necessary for primary spermatocytes and later stages to be within the barrier, isolated from the blood?

5. A 68-year-old man taking medication for hypertension complains to his physician that it has made him impotent. Explain why this could be an effect of antihypertension drugs.

MRI scan of a 36-week-old fetus in the uterus

Simon Fraser/Science Source

THE FEMALE REPRODUCTIVE SYSTEM

CHAPTER OUTLINE

28.1 Female Reproductive Anatomy

    28.1a Sexual Differentiation

    28.1b The Genitalia

    28.1c The Breasts and Mammary Glands

28.2 Puberty and Menopause

    28.2a Puberty

    28.2b Climacteric and Menopause

28.3 Oogenesis and the Sexual Cycle

    28.3a Oogenesis

    28.3b Folliculogenesis

    28.3c The Sexual Cycle

28.4 Female Sexual Response

    28.4a Excitement and Plateau

    28.4b Orgasm and Resolution

28.5 Pregnancy and Childbirth

    28.5a Prenatal Development

    28.5b Hormones of Pregnancy

    28.5c Adjustments to Pregnancy

    28.5d Childbirth

    28.5e The Puerperium

28.6 Lactation

    28.6a Development of the Mammary Glands

    28.6b Colostrum and Milk Synthesis

    28.6c Milk Ejection

    28.6d Breast Milk

Connective Issues

Study Guide

DEEPER INSIGHTS

28.1 Clinical Application: Pap Smears and Cervical Cancer

28.2 Evolutionary Medicine: The Evolution of Menopause

28.3 Clinical Application: Signs of Ovulation

28.4 Clinical Application: Endometriosis

28.5 Clinical Application: Methods of Contraception

Anatomy & Physiology
Revealed 4.0

Module 14: Reproductive System

BRUSHING UP

- Female reproductive anatomy is best understood from the perspective of the embryonic development of the reproductive system described in sections 27.1d to 27.1e and figures 27.3 and 27.4.

- You will need a good understanding of gonadotropin-releasing hormone, follicle-stimulating hormone, and luteinizing hormone; the hypothalamo–pituitary–ovarian axis; and negative feedback inhibition of the hypothalamus and pituitary (see section 17.2) to understand several aspects of ovarian development, pregnancy, and lactation.

- Hormonal interactions in the reproductive cycle and pregnancy involve the synergistic, permissive, and antagonistic effects described in section 17.4h.

- Oogenesis (egg production) involves the same stages of meiosis as spermatogenesis in the male (see section 27.4b); only the differences between oogenesis and spermatogenesis are presented in this chapter.

- Female sexual response is much like that of the male (see section 27.5); only the differences between female and male are presented in this chapter.

The female reproductive system is more complex than the male's because it serves more purposes. Whereas the male needs only to produce and deliver gametes, the female must do this as well as provide nutrition and safe harbor for fetal development, then give birth and nourish the infant. Furthermore, female reproductive physiology is more conspicuously cyclic, and female hormones are secreted in a more complex sequence compared with the relatively steady, simultaneous secretion of regulatory hormones in the male.

This chapter discusses the anatomy of the female reproductive system; the production of gametes and how it relates to the ovarian and menstrual cycles; the female sexual response; and the physiology of pregnancy, birth, and lactation. Embryonic and fetal development are treated in the next chapter.

28.1 Female Reproductive Anatomy A&PR

Expected Learning Outcomes

When you have completed this section, you should be able to

a. describe the structure of the ovary;

b. trace the female reproductive tract and describe the gross anatomy and histology of each organ;

c. identify the ligaments that support the female reproductive organs;

d. describe the blood supply to the female reproductive tract;

e. identify the external genitalia of the female; and

f. describe the structure of the breast.

28.1a Sexual Differentiation

The female reproductive system (**fig. 28.1**) is obviously different from that of the male, but as we saw earlier, the two sexes are indistinguishable for the first 8 to 10 weeks of development (see fig. 27.4). The female reproductive tract develops from the paramesonephric duct not because of the positive action of any hormone, but because of the absence of testosterone and müllerian-inhibiting factor (MIF). Without testosterone, the mesonephric duct degenerates while the genital tubercle becomes the glans of the clitoris, the urogenital folds develop into labia minora, and the labioscrotal folds develop into labia majora. Without MIF, the paramesonephric ducts develop into the uterine tubes, uterus, and vagina (see fig. 27.3). This developmental pattern can be disrupted, however, by abnormal hormonal exposure before birth, as happens in adrenogenital syndrome (see fig. 17.28).

28.1b The Genitalia

The **internal genitalia** include the ovaries and a duct system that runs from the vicinity of each ovary to the outside of the body—the uterine tubes, uterus, and vagina. The **external genitalia** include principally the clitoris, labia minora, labia majora, and associated subcutaneous glands and erectile tissues. These occupy the perineum, which is defined by the same skeletal landmarks as in the male (see fig. 27.6). The ovaries are the primary sex organs, and the other internal and external genitalia are the secondary sex organs.

The Ovaries

The female gonads are the **ovaries,**[1] which produce egg cells (ova) and sex hormones. The ovary is an almond-shaped organ nestled in the *ovarian fossa,* a depression in the posterior pelvic wall. It measures about 3 cm long, 1.5 cm wide, and 1 cm thick. Its capsule, like that of the testis, is called the **tunica albuginea.** The interior of the ovary is divided into a central **medulla** and an outer **cortex (fig. 28.2),** with no sharp boundary between them. The medulla is a core of fibrous connective tissue occupied by the largest arteries and veins within the ovary. The cortex is the site of the many thousands of ovarian **follicles.** A follicle consists of a spheroidal mass of small *follicular cells* surrounding one ovum. As a follicle matures, a capsule develops around it, the follicular cells become multilayered, and a fluid cavity forms at the center. The ovary has no system of tubules like the testis; eggs are released one at a time by the bursting of the follicles *(ovulation).* In childhood, the ovaries are smooth-surfaced. During the reproductive years, they become more corrugated because growing follicles of various ages produce bulges in the surface. After menopause, the ovaries are shrunken and composed mostly of scar tissue.

---

[1]*ov* = egg; *ary* = place for

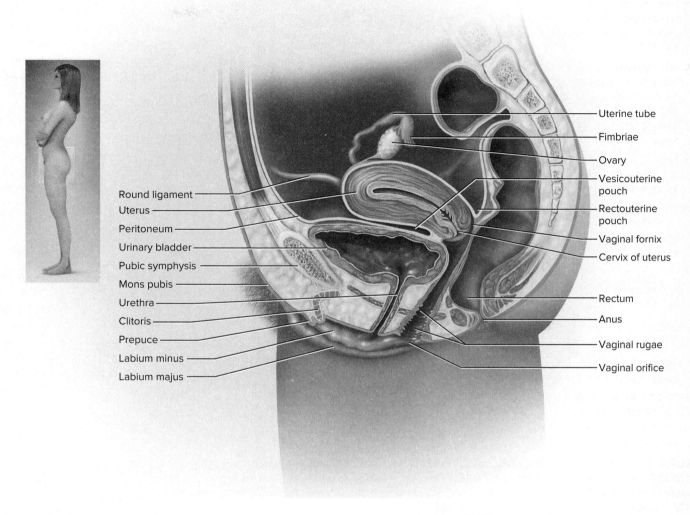

FIGURE 28.1 The Female Reproductive System. Compare photo in atlas B, figure B.14b. **A&PR**

Several connective tissue ligaments hold the ovaries and other internal genitalia in place **(fig. 28.3).** The medial pole of the ovary is attached to the uterus by the **ovarian ligament** and its lateral pole is attached to the pelvic wall by the **suspensory ligament.** The anterior margin of the ovary is anchored by a peritoneal fold called the **mesovarium.**[2] This ligament extends to a sheet of peritoneum called the *broad ligament,* which flanks the uterus and encloses the uterine tube in its superior margin. If you picture these ligaments as a sideways T, the vertical bar would represent the broad ligament, flanking the uterus and enfolding the uterine tube at its superior end, and the horizontal bar would represent the mesovarium enfolding the ovary at its free end.

The ovary receives blood from two arteries: the **ovarian artery,** which passes through the suspensory ligament and approaches the lateral pole of the ovary, and the **ovarian branch**

of the uterine artery, which passes through the mesovarium and approaches the medial pole. The ovarian artery is the female equivalent of the male testicular artery, arising high on the aorta and traveling down the posterior body wall to the gonad. The ovarian and uterine arteries anastomose along the margin of the ovary and give off multiple small arteries that enter the ovary on that side. Ovarian veins, lymphatics, and nerves also travel through the suspensory ligament. The veins and lymphatics follow courses similar to those described for the testes (see section 27.2b).

The Uterine Tubes

The **uterine tubes,** also called **oviducts** or **fallopian**[3] **tubes,** are canals about 10 cm long leading from each ovary to the uterus. At the distal (ovarian) end, the tube flares into a trumpet-shaped

---

[2]*mes* = middle; *ovari* = ovary

[3]Gabriele Fallopio (1523–62), Italian anatomist and physician

FIGURE 28.2 Structure of the Ovary. Arrows indicate the developmental sequence of an ovarian follicle; they are not meant to imply that follicles migrate around the ovary. The follicles are not drawn to scale; indeed the mature follicle at the top and ovulating follicle at the lower right have diameters about equal to the entire 3 cm length of the ovary, from ovarian ligament to suspensory ligament, and bulge from the ovarian surface like enormous blisters. **APR**

infundibulum[4] with feathery projections called **fimbriae**[5] (FIM-bree-ee); the middle and longest part of the tube is the **ampulla;** and near the uterus it forms a narrower **isthmus.** The uterine tube is enfolded in the **mesosalpinx**[6] (MEZ-oh-SAL-pinks), which is the superior margin of the broad ligament.

The wall of the uterine tube is well endowed with smooth muscle. Its mucosa is highly folded into longitudinal ridges and has an epithelium of ciliated cells and a smaller number of secretory cells **(fig. 28.4).** The cilia beat toward the uterus and, with the help of muscular contractions of the tube, convey the egg in that direction.

The Uterus

The **uterus**[7] is a thick muscular chamber that opens into the roof of the vagina and usually tilts forward over the urinary bladder

(see fig. 28.1). Its function is to harbor the fetus, nourish it, and expel the fetus at the end of its development. It is somewhat pear-shaped, with a broad superior curvature called the **fundus,** a midportion called the **body,** and a cylindrical inferior end called the **cervix** (fig. 28.3). The uterus measures about 7 cm from cervix to fundus, 4 cm wide at its broadest point on the fundus, and 2.5 cm thick, but it is somewhat larger in women who have been pregnant.

The lumen of the uterus is roughly triangular, with its two upper corners opening into the uterine tubes. In the nonpregnant uterus, the lumen isn't a hollow cavity but rather a *potential space;* the mucous membranes of the opposite walls are pressed against each other with little room between them. The lumen communicates with the vagina by way of a narrow passage through the cervix called the **cervical canal.** The superior opening of this canal into the body of the uterus is the *internal os*[8]

---

[4]*infundibulum* = funnel
[5]*fimbria* = fringe
[6]*meso* = mesentery; *salpin* = trumpet
[7]*uterus* = womb

[8]*os* = mouth

FIGURE 28.3 **The Female Reproductive Tract and Supportive Ligaments.** (a) Posterior view of the reproductive tract. (b) Relationship of the uterine tube and ovary to the supporting ligaments. (c) Anterior view of the major female reproductive organs from a cadaver. **APR**

c: Rebecca Gray/McGraw-Hill Education

|← 4 μm →|

FIGURE 28.4 Epithelial Lining of the Uterine Tube. Secretory cells are shown in red and green, and cilia of the ciliated cells in yellow (SEM). **APR**

SPL/Science Source

❓ *What purpose do these cilia serve?*

(pronounced "oz" or "ose") and its opening into the vagina is the *external os.* The canal contains **cervical glands** that secrete mucus, thought to prevent the spread of microorganisms from the vagina into the uterus. Near the time of ovulation, the mucus becomes thinner than usual and allows easier passage for sperm.

Uterine Wall   The uterine wall consists of three layers. The outermost layer, the **perimetrium,** is a thin serosa of simple squamous epithelium and loose connective tissue.

The middle and thickest layer is the **myometrium,**[9] about 1.25 cm thick in the nonpregnant uterus. It is composed mainly of bundles of smooth muscle that sweep downward from the fundus and spiral around the body of the uterus. The myometrium is

less muscular and more fibrous near the cervix; the cervix itself is almost entirely collagenous. The muscle cells of the myometrium are about 40 μm long immediately after menstruation, but they are twice this long at the middle of the menstrual cycle (near the time of ovulation) and 10 times as long in pregnancy. The function of the myometrium is to produce the labor contractions that help to expel the fetus.

The innermost layer is a mucosa called the **endometrium.**[10] It has a simple columnar epithelium, compound (branching) tubular glands, and a lamina propria populated by leukocytes, macrophages, and other cells **(fig. 28.6).** The superficial half to two-thirds of it, called the **functional layer (stratum functionalis),** is shed in each menstrual period. The deeper layer, called the **basal layer (stratum basalis),** stays behind and regenerates a new functional layer in the next cycle (see fig. 28.16). When pregnancy occurs, the endometrium is the site of attachment of the embryo and forms the maternal part of the *placenta* from which the fetus is nourished.

Uterine Ligaments and Peritoneum   The uterus is supported by the muscular floor of the pelvis and folds of peritoneum that form supportive ligaments around the organ, as they do for the ovary and uterine tube (see fig. 28.3a). The **broad ligament** has two parts: the *mesosalpinx* mentioned earlier and the *mesometrium* on each side of the uterus. The cervix and superior part of the vagina are supported by **cardinal (lateral cervical) ligaments** extending to the pelvic wall. A pair of **uterosacral ligaments** attaches the posterior side of the uterus to the sacrum, and a pair of **round ligaments** arises from the anterior surface of the uterus, passes through the inguinal canals, and terminates in the labia majora, much like the gubernaculum of the testis terminating in the scrotum.

As the peritoneum folds around the various pelvic organs, it creates several dead-end recesses and pouches (extensions of the peritoneal cavity). Two major ones are the *vesicouterine*[11] *pouch,* which forms the space between the uterus and urinary bladder, and *rectouterine pouch* between the uterus and rectum (see fig. 28.1).

Blood Supply   The uterine blood supply is particularly important to the menstrual cycle and pregnancy. A **uterine artery** arises from each internal iliac artery and travels through the broad ligament to the uterus **(fig. 28.7).** It gives off several branches that penetrate into the myometrium and lead to **arcuate arteries.** Each arcuate artery travels in a circle around the uterus and anastomoses with the arcuate artery on the other side. Along its course, it gives rise to smaller arteries that penetrate the rest of the way through the myometrium, into the endometrium, and give off **spiral arteries.** The spiral arteries wind tortuously between the endometrial glands toward the surface of the mucosa (see fig. 28.16). They rhythmically constrict and dilate, making the mucosa alternately blanch and flush with blood.

---

[9]*myo* = muscle; *metr* = uterus

[10]*endo* = inside; *metr* = uterus
[11]*vesico* = bladder

DEEPER INSIGHT 28.1

CLINICAL APPLICATION

Pap Smears and Cervical Cancer

Cervical cancer usually occurs between the ages of 30 and 50, especially in women who smoke, who began sexual activity at an early age, and who have histories of frequent sexually transmitted diseases or cervical inflammation. It is almost always caused by the human papillomavirus (HPV), a sexually transmitted pathogen (see Deeper Insight 27.5). Cervical cancer typically begins in the epithelial cells of the lower cervix, develops slowly, and remains a local, easily removed lesion for several years. If the cancerous cells spread to the subepithelial connective tissue, however, the cancer is said to be *invasive* and potentially life-threatening.

Next to condoms, the best protection against cervical cancer is early detection by means of a *Pap*[12] *smear*—a procedure in which loose cells are removed from the cervix and vagina with a small flat stick and cervical brush, then microscopically examined. The pathologist looks for cells with signs of *dysplasia* (abnormal development) or carcinoma **(fig. 28.5).** One system of staging Pap smears classifies abnormal results into three grades of *cervical intraepithelial neoplasia (CIN)*. Findings are rated on the following scale, and further vigilance or treatment is planned accordingly:

CIN I—mild dysplasia with cellular changes typically associated with HPV
CIN II—moderate dysplasia with precancerous lesions
CIN III—severe dysplasia, *carcinoma in situ* (preinvasive carcinoma of surface cells)

A rating of ASCUS or CIN I calls for a repeat Pap smear and visual examination of the cervix *(colposcopy)* in 3 to 6 months. CIN II calls for a biopsy, often done with an "electric scalpel" in a procedure called LEEP (loop electrosurgical excision procedure). A cone of tissue is removed to evaluate the depth of invasion by the malignant or premalignant cells. This in itself may be curative if all margins of the specimen are normal, indicating all abnormal cells were removed. CIN III may be cause for radiation therapy or *hysterectomy*[13] (removal of the uterus).

From age 21 to 65, women are advised to have a Pap smear every 3 years. Pap smears are not recommended before age 21 or after 65.

(a) Normal cells

20 μm

(b) Malignant (CIN III) cells

FIGURE 28.5 Pap Smears. These are smears of squamous epithelial cells scraped from the cervix. (a) Normal cells. Note the small nuclei and large volume of cytoplasm. (b) Malignant (CIN III) cells. Note the greatly enlarged nuclei and proportionately less cytoplasm.

a: Komsan Loonprom/Shutterstrock; b: Parviz M. Pour/Science Source

The Vagina

The **vagina**[14] is a tube about 8 to 10 cm long that allows for the discharge of menstrual fluid, receipt of the penis and semen, and birth of a baby. The vaginal wall is thin but stretches easily. It consists of an outer adventitia, a middle muscularis, and an inner mucosa. The vagina tilts posteriorly between the urethra and rectum; the urethra is bound to its anterior wall. The vagina has no glands, but it is lubricated by the seepage of serous fluid through its walls *(transudation)* and by mucus from the cervical glands above it. The vagina extends slightly beyond the cervix as a blind pouch called the *fornix*[15] (see figs. 28.1, 28.3a).

---

[12]George N. Papanicolaou (1883–1962), Greek–American physician and cytologist
[13]*hyster* = uterus; *ectomy* = cutting out
[14]*vagina* = sheath

[15]*fornix* = arch, vault

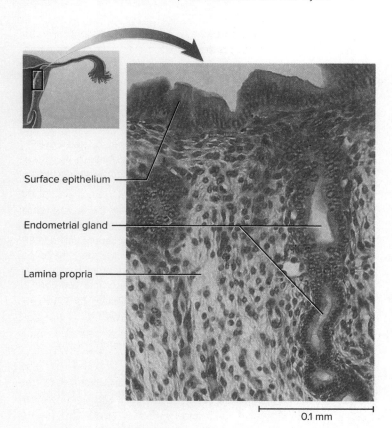

FIGURE 28.6 Functional Layer of the Endometrium. APR

Ed Reschke

The lower end of the vagina has transverse friction ridges, or **vaginal rugae,** which contribute to both male and female stimulation during intercourse. At the vaginal orifice, the mucosa folds inward and forms a membrane, the **hymen,** which stretches across the opening. The hymen has one or more openings to allow for discharge of menstrual fluid. It is often torn during childhood or adolescence by ordinary physical activity, the use of tampons, or medical examination with a speculum, or it may tear at the first act of sexual intercourse.

The vaginal epithelium is simple cuboidal in childhood, but the estrogens of puberty transform it into a stratified squamous epithelium. This is an example of *metaplasia,* the transformation of one tissue type to another. The epithelial cells are rich in glycogen. Bacteria ferment this to lactic acid, which produces a low vaginal pH (about 3.5–4.0) that inhibits the growth of pathogens. This acidity is neutralized by the semen so it doesn't harm the sperm. The mucosa also has antigen-presenting **dendritic cells,** which aid in immunity at this body orifice, but unfortunately are also a route by which HIV can invade the female body.

▶▶▶ APPLY WHAT YOU KNOW

For what functional reason do you think the vaginal epithelium changes type at puberty? Of all types of epithelium it might become, why stratified squamous?

---

[16]*vulva* = covering
[17]*pudend* = shameful

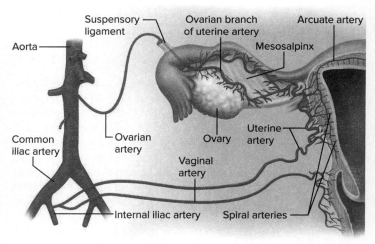

FIGURE 28.7 Blood Supply to the Female Reproductive Tract. The vaginal, uterine, and ovarian arteries are exaggerated in length by the perspective of the drawing, moving the aorta away from the uterus for clarity.

The External Genitalia

The external genitalia occupy most of the perineum and are collectively called the **vulva**[16] or **pudendum;**[17] they include the mons pubis, labia majora and minora, clitoris, vaginal orifice, and accessory glands and erectile tissues (**fig. 28.8**). These structures lie within the same anatomical boundaries as described for the male perineum (see fig. 27.6): the pubic symphysis anteriorly, coccyx posteriorly, and ischial tuberosities laterally. Muscles of the pelvic floor in this region are especially important in pregnancy and obstetrics and in supporting the internal reproductive organs (see section 10.3d and fig. 10.21).

The **mons**[18] **pubis** consists mainly of an anterior mound of adipose tissue overlying the pubic symphysis, bearing most of the pubic hair (see fig. A.3a in atlas A). The **labia majora**[19] (singular, *labium majus*) are a pair of thick folds of skin and adipose tissue inferior to the mons; the fissure between them is the *pudendal cleft.* Pubic hair grows on the lateral surfaces of the labia majora at puberty, but the medial surfaces remain hairless. Medial to the labia majora are the much thinner, entirely hairless **labia minora**[20] (singular, *labium minus*). The area enclosed by them, called the **vestibule,**[21] contains the urinary and vaginal orifices.

The **clitoris**[22] (CLIT-er-is, cli-TOR-is) is structured like the penis in many respects, but has no urinary role. Its function is entirely sensory, serving as the primary center of sexual stimulation. Unlike the penis, it is almost entirely internal, it has no corpus spongiosum, and it doesn't enclose the urethra. Essentially, it is a pair of corpora cavernosa enclosed in connective tissue. Its head, the **glans,** is the only externally visible portion.

---

[18]*mons* = mountain
[19]*labi* = lip; *major* = larger, greater
[20]*minor* = smaller, lesser
[21]*vestibule* = entryway
[22]origin uncertain; possibly from *kleis* = door key, or *klei* + *ein* = to close

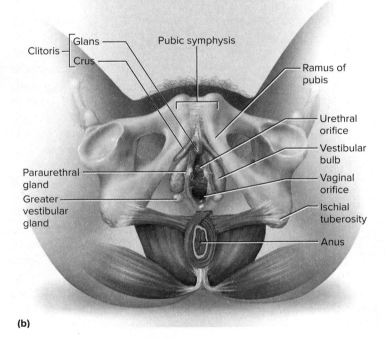

FIGURE 28.8 The Female Perineum. (a) Surface anatomy.
(b) Subcutaneous structures. **A&PR**

❓ *The male prostate is homologous with what gland(s) of the female?*

It is partially covered by a hoodlike **prepuce** where the labia minora meet. The **body** passes internally, inferior to the pubic symphysis (see fig. 28.1). At its internal end, the corpora cavernosa diverge like a Y as a pair of **crura,** which, like those of the penis, attach the clitoris to each side of the pubic arch. The circulation and innervation of the clitoris are much like those of the penis (see section 27.5a).

Just deep to the labia majora, a pair of subcutaneous erectile tissues called the **vestibular bulbs** bracket the vagina like parentheses. On each side of the vagina is a pea-size **greater vestibular (Bartholin[23]) gland** with a short duct opening into the vestibule or lower vagina (fig. 28.8b). These glands are homologous to the bulbourethral glands of the male. They keep the vulva moist, and during sexual excitement they provide most of the lubrication for intercourse. The vestibule is also lubricated by a number of **lesser vestibular glands.** A pair of mucous **paraurethral glands** opens into the vestibule near the external urethral orifice. These are homologous to the male prostate and are also called the **female prostate** or, formerly, the Skene[24] glands. A discussion of the physiology of intercourse later in this chapter details the functions of these organs.

28.1c The Breasts and Mammary Glands

The **breast (fig. 28.9)** is a mound of tissue overlying the pectoralis major muscle. It enlarges at puberty and remains so for life, but most of this time it contains very little mammary gland. The **mammary gland** develops within the breast during pregnancy, remains active in the lactating breast, and atrophies when a woman ceases to nurse.

The breast has two principal regions: the conical to pendulous **body,** with the nipple at its apex, and an extension toward the armpit called the **axillary tail.** Lymphatics of the axillary tail are especially important as a route of breast cancer metastasis.

The nipple is surrounded by a circular zone, the **areola,** usually darker than the rest of the breast. Dermal blood capillaries and nerves come closer to the surface here than in the surrounding skin, accentuating the color and sensitivity of the areola. Pregnancy increases melanin deposition in the areola and nipple, making them a more visible "target" to the indistinct vision of a nursing infant. Sensory nerve fibers of the areola are important in triggering a *milk ejection reflex* when an infant nurses. The areola has sparse hairs and **areolar glands,** visible as small bumps on the surface. These glands are intermediate between sweat glands and mammary glands in their degree of development. When a woman is nursing, secretions of the areolar glands and sebaceous glands protect the areola and nipple from chapping and cracking. The dermis of the areola has smooth muscle fibers that contract in response to cold, touch, and sexual arousal, wrinkling the skin and erecting the nipple.

Internally, the nonlactating breast consists mostly of adipose and collagenous tissue (fig. 28.9b). Breast size is determined by the amount of adipose tissue and has no relationship to the amount of milk the mammary gland can produce. **Suspensory ligaments** attach the breast to the dermis of the overlying skin and to the fascia of the underlying muscle (pectoralis major). Although the nonlactating breast contains little glandular tissue, it does have a system of ducts branching through its fibrous stroma and converging on the nipple.

When the mammary gland develops during pregnancy, it exhibits 15 to 20 lobes arranged radially around the nipple, separated

---

[23]Caspar Bartholin (1655–1738), Danish anatomist
[24]Alexander Skene (1837–1900), American gynecologist

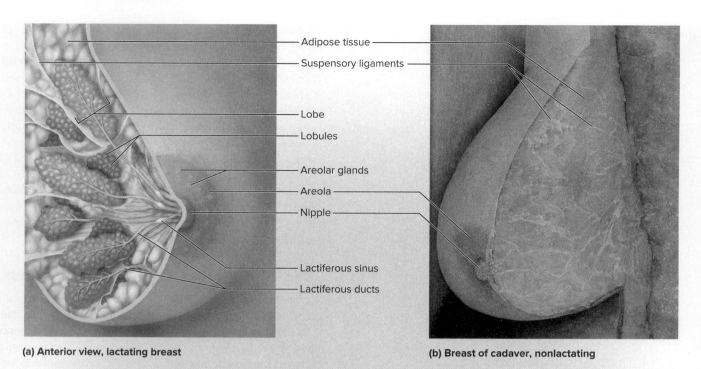

(a) Anterior view, lactating breast

Adipose tissue
Suspensory ligaments
Lobe
Lobules
Areolar glands
Areola
Nipple
Lactiferous sinus
Lactiferous ducts

(b) Breast of cadaver, nonlactating

Rib
Intercostal muscles
Pectoralis minor
Pectoralis major
Fascia
Suspensory ligament
Lobules
Lobe
Adipose tissue
Nipple
Lactiferous sinus
Lactiferous duct
Myoepithelial cells

Secretory cells

(c) Sagittal section, lactating breast

(d) Secretory acini, lactating breast, SEM

FIGURE 28.9 The Breast. Parts (a), (c), and (d) depict the breast in a lactating state. Some of the features in parts (a) and (c) are absent from the nonlactating breast in part (b). The cluster of lobules boxed in part (c) would contain numerous microscopic acini like the ones in part (d). **APR**

❓ *What is the function of the myoepithelial cells in part (d)?*

FIGURE 28.10 Breast Cancer Screening and Treatment. (a) Radiologic technologist assisting a patient in mammography. (b) Mammogram of a breast with a tumor visible at the arrow (left), compared with the appearance of normal fibrous connective tissue of the breast (right). (c) Patient following mastectomy of the right breast. (d) The same patient following surgical breast reconstruction. **APR**

a: Tony Stone Images/Getty Images; b: The Image Bank/Getty Images; c, d: Biophoto Associates/Science Source

❓ *What is the diagnostic benefit of compressing the breast for a mammogram?*

from each other by stroma (fig. 28.9a). Each lobe is drained by a **lactiferous**[25] **duct,** which dilates to form a **lactiferous sinus** opening onto the nipple. Distally, each duct branches repeatedly with the finest branches ending in sacs called acini. The acini are organized into grapelike clusters (lobules) within each lobe of the breast. Each acinus consists of a sac of conical secretory cells arranged around a central lumen (see fig. 5.30). Like an orange in a mesh bag, the acinus is surrounded by a network of contractile **myoepithelial cells** (fig. 28.9d). Their role in milk release and other aspects of the lactating breast are described later in this chapter.

Breast Cancer

Breast cancer **(fig. 28.10)** occurs in one out of every eight or nine American women and is the most common and deadly cancer of women worldwide. Breast tumors begin with cells of the mammary

ducts and may metastasize to other organs by way of the mammary and axillary lymphatics. Signs of breast cancer include a palpable lump (the tumor), puckering of the skin, changes in skin texture, and discharge from the nipple.

Two genes called *BRCA1* and *BRCA2* have been implicated in some breast cancers, as have some rarer genes that code for mechanisms of DNA repair, but most breast cancer is nonhereditary. Some breast tumors are stimulated by estrogen. Consequently, breast cancer is more common among women who begin menstruating early in life and who reach menopause relatively late—that is, women who have a long period of fertility and estrogen exposure. Its incidence is declining in the United States and many other western countries, but rising in Latin America, Russia, and East Asia owing to changes in diet, excessive alcohol and fat intake, obesity, smoking, exposure to carcinogenic chemicals, delayed childbearing, longer life expectancy, and denial—delaying medical consultation and treatment. Over 70% of cases, however, lack any identifiable risk factors.

[25]*lact* = milk; *fer* = to carry

The majority of tumors are discovered during breast self-examination (BSE), which should be a monthly routine for all women. Instructions for this are widely available on the Internet. *Mammograms* (breast X-rays), however, can detect tumors too small to be noticed by BSE. Although opinions vary, a schedule commonly recommended is to have a baseline mammogram at age 40 or 50 and then biennially through the mid-70s. Annual screening is deemed to produce too many false positive results and unnecessary biopsies.

Treatment of breast cancer is usually by *lumpectomy* (removal of the tumor only) or *simple mastectomy* (removal of the breast tissue only or breast tissue and some axillary lymph nodes). *Radical mastectomy,* invented in the late 1800s and rarely done since 1985, involved the removal of not only the breast but also the underlying muscle, fascia, and lymph nodes. The concept behind this was to "root out" all vestiges of the cancer (the word *radical* means "root") by amputating not just the breast but all underlying tissues to which the cancer might have spread. But this was before oncologists realized that cancer doesn't simply radiate out from the primary tumor, but spreads (metastasizes) through the lymphatic system or blood. By the time a radical mastectomy was performed, the breast cancer usually had already spread to far-distant parts of the body. Radical mastectomy terribly disfigured and disabled countless women, yet proved no more effective than simple mastectomy or lumpectomy.

Surgery is generally followed by radiation or chemotherapy, and estrogen-sensitive tumors may also be treated with an estrogen blocker such as tamoxifen. A natural-looking breast can often be reconstructed from skin, fat, and muscle from other parts of the body.

BEFORE YOU GO ON

Answer the following questions to test your understanding of the preceding section:

1. How do the site of female gamete production and mode of release from the gonad differ from those in the male?

2. How is the structure of the uterine tube mucosa related to its function?

3. Contrast the function of the endometrium with that of the myometrium.

4. List the subcutaneous erectile tissues and glands of the female perineum and state the function of each.

28.2 Puberty and Menopause

Expected Learning Outcomes
When you have completed this section, you should be able to

a. name the hormones that regulate female reproductive function, and state their roles;

b. describe the principal signs of puberty;

c. describe the hormonal changes of female climacteric and their effects; and

d. define and describe *menopause,* and distinguish menopause from climacteric.

Puberty and menopause are physiological transitions at the beginning and end of a female's reproductive years.

28.2a Puberty

Puberty begins at ages 8 to 10 for most well-nourished girls in affluent countries, but significantly later in many countries. Yet, a 1997 study of more than 17,000 girls in the United States showed 3% of black girls and 1% of white girls beginning puberty by age 3, and 27% and 7%, respectively, by age 7. The youngest mother known to history was Lina Medina of Peru, who began menstruating at 8 months of age, was impregnated at age 4, and gave birth at age 5, in 1939. You can find her story online.

Puberty is triggered by the same hypothalamic and pituitary hormones in girls as in boys (see section 27.3b), but with one important difference. Body fat is crucial to a female's ability to sustain the caloric demands of pregnancy and lactation, so girls and women don't ovulate if they have too little of it. A girl must attain about 17% body fat before she can begin to ovulate and menstruate, and adult women commonly cease if they drop below 22% fat. This is an evolutionary adaptation stemming from times when adequate nutrition was by no means a sure thing, serving to spare females from pregnancies that likely couldn't lead to successful birth and lactation. How does the brain know if fat reserves are adequate for reproduction? **Leptin,** the hormone secreted by fat cells, acts as an adiposity signal to the brain. If the leptin level is too low, the hypothalamus doesn't secrete **gonadotropin-releasing hormone (GnRH).** Puberty is therefore delayed in undernourished girls, unusually thin girls, and in other girls with low body fat due to such vigorous and habitual physical activity as gymnastics and ballet. Ovulation and menstruation may cease even in adult women who become underweight.

But if fat stores are adequate, then the GnRH level rises in late childhood and stimulates the anterior pituitary to secrete follicle stimulating hormone (FSH) and luteinizing hormone (LH). FSH, especially, stimulates development of the ovarian follicles, which, in turn, secrete estrogens, progesterone, inhibin, and a small amount of androgens. These hormone levels rise gradually from ages 8 to 12 and then more sharply in the early teens. The **estrogens**[26] are feminizing hormones with widespread effects on the body. They include *estradiol* (the most abundant), *estriol,* and *estrone.* Most of the visible changes at puberty result from estradiol and androgens.

The earliest noticeable sign of puberty in girls is **thelarche**[27] (theel-AR-kee), the onset of breast development. Estrogen, progesterone, and prolactin initially induce the formation of lobules and ducts in the breast. Duct development is completed under the influence of glucocorticoids and growth hormone, while adipose and fibrous tissue enlarge the breast. Breast development is complete around age 20, but they undergo minor changes in each menstrual cycle and major changes in pregnancy.

Thelarche is soon followed by **pubarche** (pyu-BAR-kee), the appearance of pubic and axillary hair, sebaceous glands, and axillary glands. Androgens from the ovaries and adrenal cortex stimulate

---

[26]*estro* = desire, frenzy; *gen* = to produce
[27]*thel* = breast, nipple; *arche* = beginning

pubarche as well as the libido (sex drive). Women secrete about 0.5 mg of androgens per day, compared with 6 to 8 mg/day in men.

Next comes **menarche**[28] (men-AR-kee), the first menstrual period. In Europe and America, the average age at menarche declined from age 16.5 in 1860 to age 12.5 today, probably due mostly to improved nutrition, although many other genetic, geographic, and ethnic variables enter into this.

Menarche doesn't necessarily signify fertility. A girl's first few menstrual cycles are typically *anovulatory* (no egg is ovulated). Most girls don't ovulate regularly until a year or more after they begin menstruating. This probably reflects the fact, detailed later in this chapter, that it takes about 290 days for ovarian follicles to mature and ovulate.

Estradiol stimulates many other changes of puberty. It causes the vaginal metaplasia described earlier. It stimulates growth of the ovaries and secondary sex organs. It stimulates growth hormone secretion and causes a rapid increase in height and widening of the pelvis. Estradiol is largely responsible for the feminine physique because it stimulates fat deposition in the mons pubis, labia majora, hips, thighs, buttocks, and breasts. It makes a girl's skin thicken, but the skin remains thinner, softer, and warmer than in males of corresponding age.

Progesterone[29] acts primarily on the uterus, preparing it for possible pregnancy in the second half of each menstrual cycle and playing roles in pregnancy discussed later. Estrogens and progesterone also regulate FSH and LH secretion through negative feedback inhibition of the anterior pituitary. **Inhibin** selectively regulates FSH secretion.

Thus, we see many hormonal similarities in males and females from puberty onward. The sexes differ less in the identity of the hormones present than in their relative amounts—high androgen levels and low estrogen levels in males and the opposite in females. Another difference is that these hormones are secreted more or less continually and simultaneously in males, whereas in females, secretion is distinctly cyclic and the hormones are secreted in sequence. This will be very apparent as you read about the ovarian and menstrual cycles later in this chapter.

28.2b Climacteric and Menopause

Women, like men, go through a midlife change in hormone secretion called the **climacteric.** In women, it is accompanied by **menopause,** the cessation of menstruation (see Deeper Insight 28.2).

A female is born with up to 2 million eggs in her ovaries, each in its own follicle. The older she gets, the fewer follicles remain. Climacteric begins not at any specific age, but when she has only about 1,000 follicles left. Even the remaining ones are less responsive to gonadotropins, so they secrete less estrogen and progesterone. Without these steroids, the uterus, vagina, and breasts atrophy. Intercourse may become uncomfortable, and vaginal infections more common, as the vagina becomes thinner, less distensible, and drier. The skin becomes thinner, cholesterol levels rise (increasing the risk of cardiovascular disease), and bone mass declines (increasing the risk of osteoporosis). Blood vessels constrict and dilate in response to shifting hormone balances, and the sudden dilation of cutaneous

arteries often causes **hot flashes**—a spreading sense of heat from the abdomen to the thorax, neck, and face, accompanied by sweating. Hot flashes may occur several times a day, sometimes accompanied by headaches resulting from the sudden vasodilation of arteries in the head. In some people, the changing hormonal profile also causes mood changes. Many physicians prescribe hormone replacement therapy (HRT)—low doses of estrogen and progesterone usually taken orally or by a skin patch—to relieve some of these symptoms. The risks and benefits of HRT are under debate.

▶▶▶**APPLY WHAT YOU KNOW**

FSH and LH secretions rise at climacteric and these hormones attain high concentrations in the blood. Explain this using the preceding information and what you know about the pituitary–gonadal relationship.

Menopause is the cessation of menstrual cycles, usually occurring between the ages of 45 and 55. The average age has increased steadily in the last century and is now about 52. It is difficult to precisely establish the time of menopause because the menstrual periods can stop for several months and then begin again. Menopause is generally considered to have occurred when there has been no menstruation for a year or more.

BEFORE YOU GO ON

Answer the following questions to test your understanding of the preceding section:

5. Describe the similarities and differences between male and female puberty.

6. Describe the major changes that occur in female climacteric and the principal cause of these changes.

7. What is the difference between climacteric and menopause?

DEEPER INSIGHT 28.2
EVOLUTIONARY MEDICINE

The Evolution of Menopause

There has been considerable speculation about why women don't remain fertile to the end of their lives, as men do. Some theorists argue that menopause served a biological purpose for our prehistoric foremothers. Human offspring take a long time to rear. Beyond a certain point, the frailties of age make it unlikely that a woman could rear another infant to maturity or even survive the stress of pregnancy. She might do better in the long run to become infertile and finish rearing her last child, or help to rear her grandchildren, instead of having more. In this view, menopause was biologically advantageous for our ancestors—in other words, an evolutionary adaptation.

Others argue against this "grandmother hypothesis" on the grounds that Pleistocene (Ice Age) skeletons indicate that early hominids rarely lived past age 40. If this is true, menopause setting in at 45 to 55 years of age could have served little purpose. In this view, Pleistocene women may indeed have been fertile to the end of their lives; menopause now may be just an artifact of modern nutrition and medicine, which have made it possible for us to live much longer than our ancestors did.

[28]*men* = monthly; *arche* = beginning
[29]*pro* = favoring; *gest* = pregnancy; *sterone* = steroid hormone

28.3 Oogenesis and the Sexual Cycle APR

Expected Learning Outcomes

When you have completed this section, you should be able to

a. describe the process of egg production (oogenesis);

b. describe changes in the ovarian follicles (folliculogenesis) in relation to oogenesis;

c. describe the hormonal events that regulate the ovarian cycle;

d. describe how the uterus changes during the menstrual cycle; and

e. construct a chart of the phases of the monthly sexual cycle showing the hormonal, ovarian, and uterine events of each phase.

The reproductive lives of women are conspicuously cyclic. They include the **reproductive cycle,** which encompasses the sequence of events from fertilization to giving birth and returning to a state of fertility, and the **sexual cycle,** which encompasses the events that recur every month when pregnancy doesn't intervene. The sexual cycle, in turn, consists of two interrelated cycles controlled by shifting patterns of hormone secretion: the **ovarian cycle,** consisting of events in the ovaries, and the **menstrual cycle,** consisting of parallel changes in the uterus.

Before we delve into the familiar 28-day sexual cycle, let us look at the developmental stages that the eggs and their follicles go through. Then we can integrate that with the controlling hormones and the monthly rhythm of ovulation and menstruation.

28.3a Oogenesis

Egg production is called **oogenesis**[30] (OH-oh-JEN-eh-sis) **(fig. 28.11).** Like spermatogenesis, it produces a haploid gamete by means of meiosis. There are, however, numerous differences between the two. The most obvious, perhaps, is that males produce sperm continually at a rate of about 400 million/day, whereas oogenesis is a distinctly cyclic event that usually releases only one egg per month. Oogenesis is accompanied by cyclic changes in hormone secretion and in the histological structure of the ovaries and uterus; the uterine changes result in the monthly menstrual flow.

Oogenesis begins before a girl is born. The first germ cells arise, like those of the male, from the embryonic yolk sac. They migrate into the embryo and colonize the gonadal ridges in the first 5 to 6 weeks of development and then differentiate into **oogonia** (OH-oh-GO-nee-uh). Oogonia multiply until the fifth month, reach 6 to 7 million in number, then go into a state of arrested development until shortly before birth. At that time, they differentiate into **primary oocytes** and go as far as early meiosis I. All of them reach this stage by birth; no oogonia remain after that.

[30]*oo* = egg; *genesis* = production

Any stage from the primary oocyte to the time of fertilization can be called an **egg,** or **ovum.**

Most primary oocytes undergo a process of degeneration called **atresia** (ah-TREE-zhee-uh) before a girl is born. As many as 2 million oocytes remain at the time of birth; 90% of those undergo atresia during childhood, and by puberty, only about 200,000 remain. This is the female's lifetime supply of gametes, but it's more than ample; even if she ovulated one egg every 28 days from the age of 14 to 50, she would ovulate only 480 eggs. All of the other oocytes undergo atresia between puberty and menopause.

Egg development resumes in adolescence. Each month for about 30 years, except during pregnancy, a cohort (developmental group) of about 2 dozen arrested follicles is recruited to resume development. (After 30 years of ovulation, the process becomes more irregular, and at menopause, it ceases.) The 28-day menstrual cycle creates a common misconception that eggs and ovarian follicles take 28 days to develop from one ovulation to the next. In reality, though, from the time a cohort of follicles is activated, it takes about 290 days for one of them to fully mature and ovulate. The reason for the 28-day cycle is that one cohort follows behind another, in overlapping waves, with one cohort reaching maturity each month.

Meiosis I is completed on the day of ovulation and produces two daughter cells—a large one called the **secondary oocyte** and a much smaller one called the **first polar body.** The polar body may pointlessly complete meiosis II, but ultimately it or its daughter cells degenerate. It is merely a means of discarding the extra haploid set of chromosomes. The secondary oocyte proceeds as far as metaphase II, then stops. If it isn't fertilized, it dies and never finishes meiosis. If fertilized, it completes meiosis II and casts off a second polar body, which disposes of one chromatid from each chromosome. The chromosomes of the large remaining egg then unite with those of the sperm. Further development of the fertilized egg is discussed in the next chapter.

Note the contrast between spermatogenesis and oogenesis. In spermatogenesis, a primary spermatocyte gives rise to four equal-size sperm. But in oogenesis, a primary oocyte gives rise to only one mature egg; the other three daughter cells (at most) are tiny polar bodies that die. In oogenesis it is important to produce an egg with as much cytoplasm as possible, because if fertilized it must divide repeatedly and produce numerous daughter cells. Splitting each oocyte into four equal but small parts would run counter to this purpose of providing enough "raw material" for early embryonic development.

28.3b Folliculogenesis

As an egg undergoes oogenesis, the follicle around it undergoes **folliculogenesis,** passing through the following stages.

1. **Primordial follicles.** A primordial follicle is about 40 μm in diameter. It consists of a primary oocyte in early meiosis, surrounded by a single layer of squamous follicular cells and a basement membrane separating the follicle from the surrounding stroma of the ovary **(fig. 28.12a).** Primordial follicles are concentrated in the ovarian cortex, close to the capsule. All of them form in the sixth through ninth month of gestation, but they persist into adulthood. In the adult

Development of egg (oogenesis)

Development of follicle (folliculogenesis)

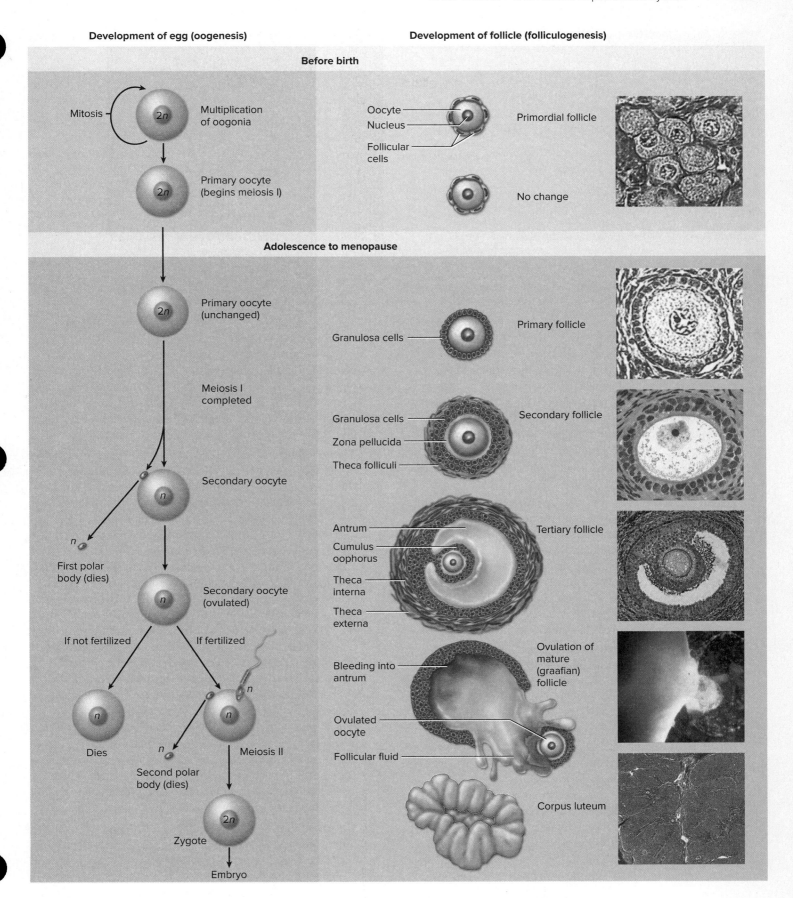

FIGURE 28.11 Oogenesis (Left) and Corresponding Development of the Follicle (Right). A&PR

FIGURE 28.12 Ovarian Follicles. (a) Note the very thin layer of squamous cells around the oocyte in a primordial follicle, and the single layer of cuboidal cells in a primary follicle. (b) A mature (graafian) follicle. Just before ovulation, this follicle will grow to as much as 25 to 30 mm in diameter. **A&PR**

a: Ed Reschke; b: Ed Reschke/Getty Images

ovary, 90% to 95% of the follicles are primordial. Those that survive childhood atresia must wait at least 13 years—and some as long as 50 years—before they are reactivated, resume their development, and have a chance to ovulate. **Recruitment (primordial follicle activation)** awakens about 2 dozen primordial follicles each month to begin a 290-day march to maturity; usually, only one of these eventually ovulates and the rest die.

2. **Primary follicles.** About 140 days into the cycle, the recruited primordial follicles have become primary follicles about 100 μm in diameter. These have a larger oocyte and their follicular cells are now cuboidal, but still form only a single layer. Development from primordial to primary follicle is controlled by local **paracrine** growth factors, not by gonadotropins (FSH or LH), for which they still have no receptors. The follicular cells, however, soon develop FSH receptors and the follicle depends on FSH to progress to the next stage.

3. **Secondary follicles.** These appear about 170 days into the cycle, still about six menstrual cycles before ovulation. They are distinguished by follicular cells that have multiplied and piled atop each other to form two or more layers; they are now called **granulosa cells.** Secondary follicles are about 200 μm in diameter and the oocyte alone is about 120 μm—five times its original size in preparation for development of the early embryo. The oocyte is surrounded by a layer of glycoprotein gel, the **zona pellucida.**[31]

    The follicle is now enclosed in a tough husk called the **theca**[32] **folliculi** (THEE-ca fol-IC-you-lye). The theca is richly supplied with blood vessels, which deliver nutrients, hormones, and cholesterol. It divides into an outer *theca externa* of innervated smooth muscle and an inner *theca interna* of steroid-synthesizing cells. LH and insulin stimulate the theca interna to absorb cholesterol from the blood and convert it to androgens—androstenedione and a lesser amount of testosterone. These androgens diffuse inward to the granulosa cells, where FSH stimulates them to convert it to estrogens, especially estradiol.

4. **Tertiary follicles.** About 60 days (2 menstrual cycles) before ovulation, the granulosa cells begin secreting **follicular fluid,** which pools in the follicle wall. The presence of these pools defines the tertiary follicle. As they enlarge, the pools merge and become a single fluid-filled cavity, the **antrum (fig. 28.12b).** Because of this, tertiary and mature follicles are called *antral follicles,* whereas the earlier stages are called *preantral.* On one side of the antrum, a mound of granulosa cells called the **cumulus oophorus**[33] covers the oocyte and secures it to the follicle wall. Antral follicles are big enough to be seen even in ultrasound scans of the ovaries. This is one approach to evaluating female fertility.

    The innermost layer of cells of the cumulus, surrounding the zona pellucida and oocyte, is called the **corona radiata.**[34] These cells and the oocyte sprout microvilli that reach out and interdigitate like the fingers of two hands laced together. Gap junctions form between them, enabling the cells to pass nutrients, wastes, and chemical signals to each other. Some granulosa microvilli penetrate deeply into the oocyte and almost reach its nucleus. Nothing can get to the oocyte except by going through the corona radiata, which forms a protective barrier around the egg functionally similar to the blood–testis barrier described in section 27.2b.

5. **Mature follicles.** Normally only one follicle in each month's cohort becomes a mature follicle (fig. 28.12b), destined to ovulate while the rest degenerate. About 20 days before ovulation (that is, late in the previous menstrual cycle), one

---

[31]*zona* = zone; *pellucid* = clear, transparent
[32]*theca* = box, case

[33]*cumulus* = little mound; *oo* = egg; *phor* = to carry
[34]*corona* = crown; *radiata* = radiating

follicle in the cohort is selected to be the dominant follicle, the one destined to ovulate. It captures and sequesters FSH, while other follicles fail to accumulate FSH and their development slows down. By the time a woman's menstrual period begins—day 1 of the cycle in which this follicle will ovulate—the dominant follicle is about 5 mm in diameter and 2 weeks away from ovulation. By 5 days before ovulation, it grows to 20 mm in diameter and is considered a **preovulatory (graafian[35]) follicle;** it attains an astonishing size up to 30 mm by the time of ovulation, about as wide as the whole ovary is long. The preovulatory follicle bulges from the ovary like a huge fluid-filled blister. As ovulation approaches, the cumulus oophorus swells and its attachment to the follicle wall constricts until the oocyte is attached to the wall by only a narrow stalk. In the last day or so, the oocyte and cumulus break away and float freely in the antrum, ready for that momentous event in its life—ovulation.

28.3c The Sexual Cycle

We can now correlate these changes in the egg cell and follicle with the rhythms of the ovaries and uterus—the sexual cycle. The cycle averages 28 days in length, so we will use that as the basis for the timetable described in the following discussion. It commonly varies from 20 to 45 days, however, so be aware that the timetable given in this discussion may differ from person to person and month to month. As you study this cycle, bear in mind that it is regulated by the **hypothalamo–pituitary–ovarian axis:** Hormones of the hypothalamus regulate the pituitary gland; pituitary hormones regulate the ovaries; and the ovaries, in turn, secrete hormones that regulate the uterus. That is, the basic hierarchy of control can be represented: hypothalamus → pituitary → ovaries → uterus. However, the ovaries also exert positive and negative feedback controls over the hypothalamus and pituitary, so these hormonal relationships are more like a communication loop than a straight-line chain of command.

Let us start with a brief preview of the sexual cycle as a whole **(fig. 28.13).** It begins with a 2-week *follicular phase.* The first 3 to 5 days of this are marked by menstruation, the vaginal discharge of blood and endometrial tissue. The uterus then replaces the lost tissue by mitosis. During this endometrial reconstruction, a cohort of tertiary follicles grows until the dominant follicle ovulates around day 14. After ovulation, the remainder of that follicle becomes a body called the *corpus luteum.* Over the next 2 weeks, called the *luteal phase,* the corpus luteum stimulates endometrial secretion, making the endometrium thicken still more, up to about day 26. If pregnancy doesn't occur, the endometrium breaks down again in the last 2 days. As loose tissue and blood accumulate, menstruation begins and the cycle starts over.

The Ovarian Cycle

We'll now see, in three principal steps, what happens in the ovaries and in their relationship to the hypothalamus and pituitary gland. This is the ovarian cycle, depicted in the top half of figure 28.13.

1. **Follicular phase.** The follicular phase of the cycle extends from the beginning of menstruation until ovulation—that is, from day 1 to day 14 in an average cycle. The portion from the *end* of menstruation until ovulation is also called the *preovulatory phase.* The follicular phase is the most variable part of the cycle, so unfortunately for family planning or pregnancy avoidance, it is seldom possible to reliably predict the date of ovulation.

   During the follicular phase, FSH stimulates continued growth of all follicles in the cohort, but of the dominant follicle above all. FSH stimulates the granulosa cells of the antral follicles to secrete estradiol. In response to estradiol, the dominant follicle up-regulates its receptors for FSH, LH, and estradiol itself, thereby becoming increasingly sensitive to these hormones. At the same time, estradiol inhibits the secretion of gonadotropin-releasing hormone (GnRH) by the hypothalamus. The anterior pituitary gland secretes less and less FSH, but an increasing amount of LH. Most antral follicles suffer from the reduced FSH level and degenerate (undergo atresia). The dominant follicle, however, has the richest blood supply and the greatest density of FSH receptors, so it becomes the mature, preovulatory follicle. The ovary, at this stage, also exhibits follicles in many other stages, belonging to other cohorts trailing behind the lead cohort like freshmen to juniors trailing behind the senior class.

   Dramatic changes occur in the last day or so preceding ovulation. Estradiol from the ovary stimulates the hypothalamus and anterior pituitary **(fig. 28.14).** Responding to GnRH from the hypothalamus and to the estradiol directly, the anterior pituitary produces a surge of LH and a lesser spike in FSH (see the midpoint of fig. 28.13a). LH induces several momentous events. The primary oocyte completes meiosis I, producing a haploid secondary oocyte and the first polar body. Follicular fluid builds rapidly; the follicle swells to as much as 25 or 30 mm in diameter, contains up to 7 mL of fluid, and bulges from the ovary like a blister. Macrophages and leukocytes are attracted to the area and secrete enzymes that weaken the follicular wall and adjacent ovarian tissue. A nipplelike **stigma** appears on the ovarian surface over the follicle. With mounting internal pressure and a weakening wall, the mature follicle approaches rupture.

   Meanwhile, the uterine tube prepares to catch the oocyte when it emerges. It swells with edema; its fimbriae envelop and caress the ovary in synchrony with the woman's heartbeat; and its cilia create a gentle current in the nearby peritoneal fluid.

2. **Ovulation.** Ovulation itself takes only 2 or 3 min. The stigma seeps follicular fluid for 1 or 2 min., and then the follicle bursts. The remaining fluid oozes out, carrying the oocyte and cumulus oophorus **(fig. 28.15).** These are normally swept up by the ciliary current and taken into the uterine tube, although many oocytes fall into the pelvic cavity and die.

3. **Luteal phase.** Days 15 to 28, from just after ovulation to the onset of menstruation, are called the **luteal (postovulatory) phase.** Assuming pregnancy doesn't occur, the major events of this phase are as follows.

---

[35]Reijnier de Graaf (1641–73), Dutch physiologist and histologist

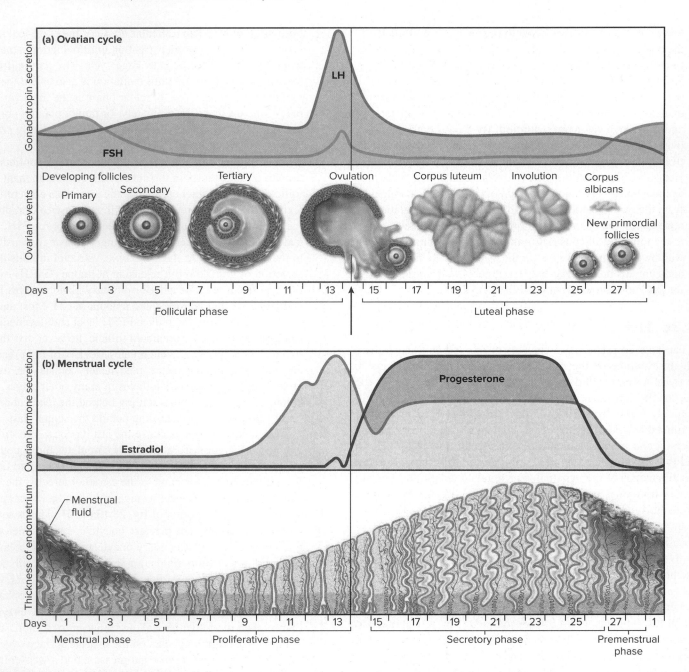

FIGURE 28.13 The Female Sexual Cycle. (a) The ovarian cycle (events in the ovary). (b) The menstrual cycle (events in the uterus). The two hormone levels in part (a) are drawn to the same scale, but those in part (b) are not. The peak progesterone concentration is about 17 times as high as the peak estradiol concentration. **APR**

When the follicle ruptures, it collapses and bleeds into the antrum. As the clotted blood is slowly absorbed, granulosa and theca interna cells multiply and fill the antrum, and a dense bed of blood capillaries grows amid them. The ovulated follicle has now become a structure called the **corpus luteum,**[36] named for a yellow lipid that accumulates in the theca interna cells. These cells are now called **lutein cells.** The transformation from ruptured follicle to corpus luteum is regulated by LH; hence, LH is also called *luteotropic hormone.*

LH stimulates the corpus luteum to continue growing and to secrete rising levels of estradiol and progesterone. There is a 10-fold increase in progesterone level (fig. 28.14), which has a crucial role in preparing the uterus for the possibility of pregnancy. Notwithstanding its luteinizing role, however, LH secretion declines steadily over the rest of the cycle, as does FSH. This is because the high levels of estradiol and progesterone, along with inhibin from the corpus luteum, have a negative feedback effect on the pituitary. (This is the basis for hormonal birth control; see Deeper Insight 28.5.)

If pregnancy doesn't occur, the corpus luteum begins a process of **involution,** or shrinkage, beginning around

---

[36]*corpus* = body; *lute* = yellow

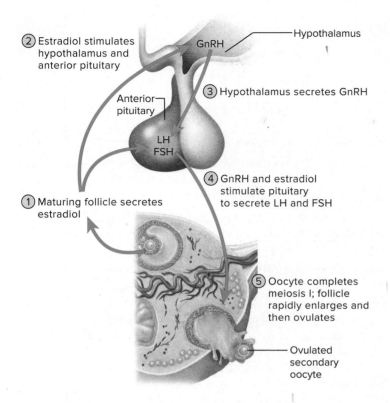

① Maturing follicle secretes estradiol

② Estradiol stimulates hypothalamus and anterior pituitary

③ Hypothalamus secretes GnRH

④ GnRH and estradiol stimulate pituitary to secrete LH and FSH

⑤ Oocyte completes meiosis I; follicle rapidly enlarges and then ovulates

Hypothalamus

GnRH

Anterior pituitary

LH FSH

Ovulated secondary oocyte

FIGURE 28.14 Hormonal Control of Ovulation.

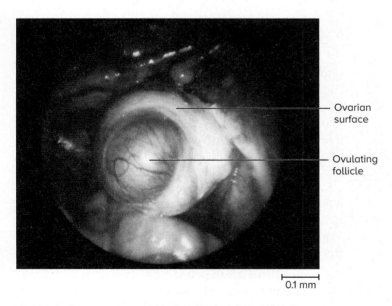

Ovarian surface

Ovulating follicle

0.1 mm

FIGURE 28.15 Endoscopic Photo of Human Ovulation.

Dr Patrice Langlois/CNRI/Science Source

day 22 (8 days after ovulation). By day 26, involution is complete and the corpus luteum has become an inactive bit of scar tissue called the **corpus albicans.**[37] With the waning of ovarian steroid secretion, the pituitary is no longer inhibited and FSH levels begin to rise again, stimulating a new cohort of follicles.

Signs of Ovulation

If a couple is attempting to conceive a child or avoid pregnancy, it is important to be able to tell when ovulation occurs. The signs are subtle but detectable. For one, the cervical mucus becomes thinner and more stretchy. Also, the resting body temperature *(basal temperature)* rises 0.2° to 0.3°C (0.4° to 0.6°F). This is best measured first thing in the morning, before rising from bed; the change can be detected if basal temperatures are recorded for several days before ovulation in order to see the difference. The LH surge that occurs about 24 hours before ovulation can be detected with a home testing kit. Finally, some women experience twinges of ovarian pain known by the German name, *mittelschmerz,*[38] which lasts from a few hours to a day or so at the time of ovulation. The most likely time to become pregnant is within 24 hours after the cervical mucus changes consistency and the basal temperature rises.

All of these events repeat themselves every month, but bear in mind that the follicles engaged in each monthly cycle began their development prenatally, as much as 50 years earlier, then lay dormant through childhood; the oocyte that ovulates each month began ripening 290 days earlier, not in the cycle when it ovulates. Ovulation normally occurs in only one ovary per cycle, and the ovaries usually alternate from month to month. **Table 28.1** summarizes the main events of the ovarian cycle and correlates them with events of the menstrual cycle, which we examine next.

The Menstrual Cycle

The menstrual cycle runs concurrently with the ovarian cycle. It consists of a buildup of the endometrium through most of the sexual cycle, followed by its breakdown and vaginal discharge. It is divided into a *proliferative phase, secretory phase, premenstrual phase,* and *menstrual phase* (see fig. 28.13b). The menstrual phase averages 5 days long, and the first day of noticeable vaginal discharge is defined as day 1 of the sexual cycle. But even though it begins our artificial timetable for the cycle, menstruation is best understood after you become acquainted with the buildup of endometrial tissue that precedes it. Thus, we begin our survey of the cycle with the proliferative phase.

1. **Proliferative phase.** The functional layer of endometrial tissue lost in the last menstruation is rebuilt during the proliferative phase. At the end of menstruation, around day 5, the endometrium is about 0.5 mm thick and consists only of the basal layer. But as a new cohort of follicles develops, they secrete more and more estrogen. Estrogen stimulates mitosis in the basal layer and the prolific regrowth of blood vessels, thus regenerating the functional layer **(fig. 28.16a).** By day 14, the endometrium is 2 to 3 mm thick. Estrogen also stimulates endometrial cells to produce progesterone receptors, priming them for the progesterone-dominated secretory phase to follow.

TABLE 28.1	Phases of the Ovarian Cycle	

Days	Phase	Major Features
1–14	**Follicular Phase**	Development of ovarian follicles and secretion primarily of estradiol. Coincides with menstrual and proliferative phases of the menstrual cycle.
	Primordial follicle	Formed prenatally and many persist into adulthood. Consists of an oocyte surrounded by a single layer of squamous follicular cells.
	Primary follicle	Consists of an oocyte surrounded by one layer of cuboidal follicular cells.
	Secondary follicle	Follicular cells stratify, become granulosa cells, and secrete a zona pellucida. Theca folliculi forms around follicle.
	Tertiary follicle	Develops from a secondary follicle in each cycle. Forms an antrum filled with follicular fluid and exhibits a cumulus oophorus, corona radiata, zona pellucida, and bilayered theca.
	Dominant follicle	The tertiary follicle that is destined to ovulate. Present by the end of the menstrual phase. Hormonally dominates the rest of the cycle, while other follicles in the cohort undergo atresia. Secretes mainly estradiol. Coincides with the proliferative phase of the menstrual cycle, in which the uterine endometrium thickens by mitosis.
	Mature follicle	The dominant follicle just prior to ovulation. Attains a diameter of 20 to 30 mm and builds to high internal fluid pressure as adjacent ovarian wall weakens.
14	**Ovulation**	Rupture of mature follicle and release of oocyte.
15–28	**Luteal (Postovulatory) Phase**	Dominated by corpus luteum. Coincides with secretory and premenstrual phases of the menstrual cycle.
	Corpus luteum	Develops from ovulated follicle by proliferation of granulosa and theca interna cells. Progesterone stimulates thickening of endometrium by secretion (secretory phase of the menstrual cycle). Begins to involute by day 22 in the absence of pregnancy; involution complete by day 26.
	Corpus albicans	Scar tissue left by involution of corpus luteum; not hormonally active. In the absence of progesterone, endometrium exhibits ischemia, necrosis, and sloughing of tissue. Necrotic endometrial tissue mixes with blood and forms menstrual fluid.

2. **Secretory phase.** The endometrium thickens still more during the **secretory phase,** but as a result of secretion and fluid accumulation rather than mitosis. This phase extends from day 15 (after ovulation) to day 26 of a typical cycle. After ovulation, the corpus luteum secretes mainly progesterone. This hormone stimulates the endometrial glands to secrete glycogen. The glands grow wider, longer, and more coiled, and the lamina propria swells with tissue fluid **(fig. 28.16b).** By the end of this phase, the endometrium is 5 to 6 mm thick—a soft, wet, nutritious bed available for embryonic development in the event of pregnancy.

3. **Premenstrual phase.** The last 2 days or so of the cycle are the **premenstrual phase,** a period of endometrial degeneration. As we have already seen, when there is no pregnancy, the corpus luteum atrophies and the progesterone level falls sharply. The drop in progesterone triggers spasmodic contractions of the spiral arteries of the endometrium, causing endometrial ischemia (interrupted blood flow). The premenstrual phase is therefore also called the **ischemic phase** (iss-KEE-mic). The lack of blood flow brings on death of endometrial tissue in the functional layer. Arterial spasms in the deeper myometrium are responsible for menstrual cramps, as any muscle deprived of blood flow hurts, and the uterus is mostly muscle. As the endometrial glands, stroma, and blood vessels degenerate, pools of blood accumulate in the functional layer. Necrotic endometrium falls away from the uterine wall, mixes with blood and

serous fluid in the lumen, and forms the **menstrual fluid (fig. 28.16c).**

4. **Menstrual phase.** When enough menstrual fluid accumulates in the uterus, it begins to be discharged from the vagina for a period called the **menstrual phase (menses).** The first day of discharge marks day 1 of a new cycle. The average woman expels about 40 mL of blood and 35 mL of serous fluid over a 5-day period. Menstrual fluid contains fibrinolysin, so it doesn't clot. The vaginal discharge of clotted blood may indicate uterine pathology rather than normal menstruation. Menstruation sheds the functional layer of endometrium, leaving once again only the basal layer.

To summarize the female sexual cycle, the ovaries go through a follicular phase characterized by growing follicles, then ovulation, and then a postovulatory (mostly luteal) phase dominated by the corpus luteum. The uterus, in the meantime, goes through a menstrual phase in which it discharges its functional layer; then a proliferative phase in which it replaces that tissue by mitosis; then a secretory phase in which the endometrium thickens by the accumulation of secretions; and finally, a premenstrual (ischemic) phase in which the functional layer breaks down again. The first half of the cycle is governed largely by follicle-stimulating hormone (FSH) from the pituitary gland and estrogen from the ovaries. Ovulation is triggered by luteinizing hormone (LH) from the pituitary, and the second half of the cycle is governed mainly by LH and progesterone, the latter secreted by the corpus luteum of the ovary.

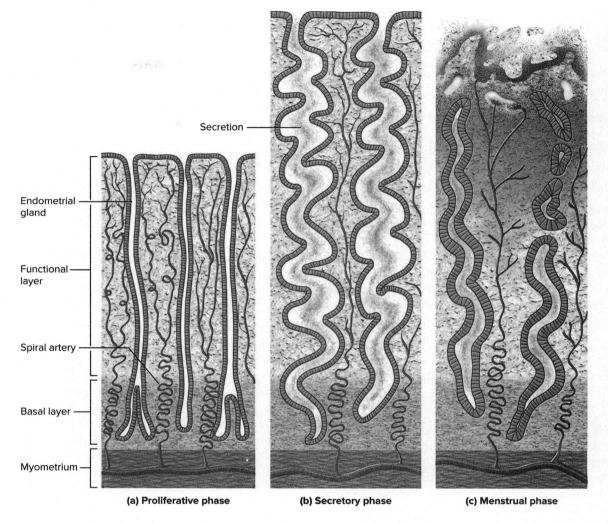

Secretion

Endometrial gland

Functional layer

Spiral artery

Basal layer

Myometrium

(a) Proliferative phase   **(b) Secretory phase**   **(c) Menstrual phase**

FIGURE 28.16 Endometrial Changes Through the Menstrual Cycle. (a) Late proliferative phase. The endometrium is 2 to 3 mm thick and has relatively straight, narrow endometrial glands. Spiral arteries penetrate upward between the endometrial glands. (b) Secretory phase. The endometrium has thickened to 5 to 6 mm by accumulating glycogen and mucus. The endometrial glands are much wider and more distinctly coiled, showing a zigzag or "sawtooth" appearance in histological sections. (c) Menstrual phase. Ischemic tissue has begun to die and fall away from the uterine wall, with bleeding from broken blood vessels and pooling of blood in the tissue and uterine lumen.

BEFORE YOU GO ON

Answer the following questions to test your understanding of the preceding section:

8. Name the sequence of cell types in oogenesis and identify the ways in which oogenesis differs from spermatogenesis.

9. Distinguish between primordial, primary, secondary, and tertiary follicles. Describe the major structures of a mature follicle.

10. Describe what happens in the ovary during the follicular and postovulatory phases.

11. Describe what happens in the uterus during the menstrual, proliferative, secretory, and premenstrual phases.

12. Describe the effects of FSH and LH on the ovary.

13. Describe the effects of estrogen and progesterone on the uterus, hypothalamus, and anterior pituitary.

 DEEPER INSIGHT 28.4

CLINICAL APPLICATION

Endometriosis

Any growth of endometrium in a site other than the uterus is called *endometriosis*. Endometrial tissue may lodge and grow in the uterine tubes, surface of the ovaries, urinary bladder, vagina, pelvic cavity, peritoneum, small or large intestine, or even as far away as the lungs and pleural cavity. Endometrium in these ectopic locations acts the same as in the uterus, thickening with each premenstrual phase and being shed in synchrony with menstruation. The shed tissue has nowhere to go, and causes irritation and formation of scar tissue and adhesions in the body cavities.

Endometriosis affects 6% to 10% of women, often causing pelvic pain; dysmenorrhea (painful menstruation); pain associated with intercourse, urination, or defecation; abnormal vaginal bleeding; and sometimes infertility. The most widely accepted theory of endometriosis is *retrograde menstruation*—menstrual fluid flowing backward and exiting through the uterine tube instead of the vagina. Management options include hormone therapy and surgery; in severe cases, hysterectomy may be required as a last resort. No cure is known at present.

28.4 Female Sexual Response

Expected Learning Outcomes

When you have completed this section, you should be able to

a. describe the female sexual response at each phase of intercourse; and

b. compare and contrast the female and male sexual responses.

Female sexual response, the physiological changes that occur during intercourse, may be viewed in terms of the four phases identified by Masters and Johnson and discussed in section 27.5: excitement, plateau, orgasm, and resolution (**fig. 28.17**). The neurological and vascular controls of the female response are essentially the same as in the male and need not be repeated here. The emphasis here is on ways the female response differs from that of the male.

28.4a Excitement and Plateau

Excitement is marked by myotonia; vasocongestion; and increased heart rate, blood pressure, and respiratory rate. Although vasocongestion works by the same mechanism in both sexes, its effects are quite different in females. The labia minora become congested and often protrude beyond the labia majora. The labia majora redden and enlarge, then flatten and spread away from the vaginal orifice. The lower one-third of the vagina constricts to form a narrow passage called the **orgasmic platform,** owing partly to vasocongestion of the vestibular bulbs that bracket the vaginal orifice (see fig. 28.8b). The narrower canal and the vaginal rugae (friction ridges) enhance stimulation and help induce orgasm in both partners. The upper end of the vagina, in contrast, dilates and becomes cavernous.

Increased blood flow in the vaginal wall turns it purple and produces a serous fluid, the **vaginal transudate,** that seeps through the wall into the canal. Along with secretions of the greater vestibular glands, this moistens the vestibule and provides lubrication.

The uterus, which normally tilts forward over the urinary bladder, stands more erect during excitement and the cervix withdraws from the vagina. In plateau, the uterus is nearly vertical and extends into the greater pelvis. This is called the **tenting effect.**

Although the vagina is the female copulatory organ, the clitoris is more comparable to the penis in structure, physiology, and importance as the primary focus of sexual stimulation. It has a high concentration of sensory nerve endings, which, by contrast, are relatively scanty in the vagina. Recall that the penis and clitoris are homologous structures. Both have a pair of corpora cavernosa with *deep arteries* and become engorged by the same mechanism. The glans and shaft of the clitoris swell to two or three times their unstimulated size, but since the clitoris cannot swing upward away from the body like the penis, it tends to withdraw beneath the prepuce. Thrusting of the penis in the vagina tugs on the labia minora and, by extension, pulls on the prepuce and stimulates the clitoris. The clitoris may also be stimulated by pressure between the pubic symphyses of the partners.

The breasts also become congested and swollen during the excitement phase, and the nipples become erect. Stimulation of the breasts also enhances sexual arousal.

28.4b Orgasm and Resolution

Late in plateau, many women experience involuntary pelvic thrusting, followed by 1 to 2 seconds of "suspension" or "stillness" preceding orgasm. Orgasm is commonly described as an intense sensation spreading from the clitoris through the pelvis, sometimes with pelvic throbbing and a spreading sense of warmth. The orgasmic platform gives three to five strong contractions about 0.8 second apart, while the cervix plunges spasmodically into the vagina and pool of semen, should this be present. The uterus exhibits peristaltic waves of contraction, which may help to draw semen from the vagina. The anal and urethral sphincters constrict, and the paraurethral glands, homologous to the prostate, sometimes expel fluid similar to prostatic fluid. Tachycardia and hyperventilation occur; the breasts enlarge still more and the areolae often become engorged; and in many women, a reddish, rashlike flush appears on the lower abdomen, chest, neck, and face.

During resolution, the uterus drops forward to its resting position. The orgasmic platform quickly relaxes, while the inner end of the vagina returns more slowly to its normal dimensions. The flush disappears quickly and the areolae and nipples undergo rapid detumescence (loss of vascular congestion), but it may take 5 to 10 minutes for the breasts to return to their normal size. In many women (and men), there is a postorgasmic outbreak of perspiration. Unlike men, women do not have a refractory period and may quickly experience additional orgasms.

BEFORE YOU GO ON

Answer the following questions to test your understanding of the preceding section:

14. What are the female sources of lubrication in coitus?

15. What female tissues and organs become vasocongested?

16. Describe the actions of the uterus throughout the sexual response cycle.

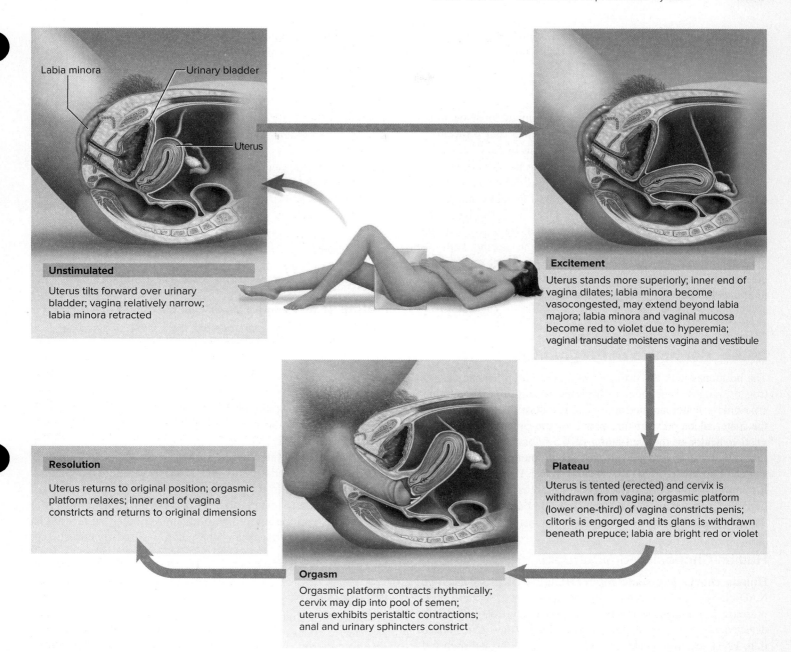

Unstimulated

Uterus tilts forward over urinary bladder; vagina relatively narrow; labia minora retracted

Labia minora — Urinary bladder

Uterus

Excitement

Uterus stands more superiorly; inner end of vagina dilates; labia minora become vasocongested, may extend beyond labia majora; labia minora and vaginal mucosa become red to violet due to hyperemia; vaginal transudate moistens vagina and vestibule

Plateau

Uterus is tented (erected) and cervix is withdrawn from vagina; orgasmic platform (lower one-third) of vagina constricts penis; clitoris is engorged and its glans is withdrawn beneath prepuce; labia are bright red or violet

Orgasm

Orgasmic platform contracts rhythmically; cervix may dip into pool of semen; uterus exhibits peristaltic contractions; anal and urinary sphincters constrict

Resolution

Uterus returns to original position; orgasmic platform relaxes; inner end of vagina constricts and returns to original dimensions

FIGURE 28.17 Stages of the Female Sexual Response. Anatomy is shown in the supine position.

28.5 Pregnancy and Childbirth

Expected Learning Outcomes

When you have completed this section, you should be able to

a. list the major hormones that regulate pregnancy and explain their roles;

b. describe a woman's bodily adaptations to pregnancy;

c. identify the physical and chemical stimuli that increase uterine contractility in late pregnancy;

d. describe the mechanism of labor contractions;

e. name and describe the three stages of labor; and

f. describe the physiological changes that occur in a woman during the weeks following childbirth.

This section treats pregnancy from the maternal standpoint—that is, adjustments of the woman's body to pregnancy and the mechanism of childbirth. Development of the fetus is described in the next chapter.

Gestation (pregnancy) lasts an average of 266 days from conception to childbirth, but the gestational calendar is usually measured from the first day of the woman's last menstrual period (LMP). Thus, the birth is predicted to occur 280 days (40 weeks) from the LMP. The duration of pregnancy, called its *term,* is commonly described in 3-month intervals called **trimesters.**

28.5a Prenatal Development

A few fundamental facts of fetal development must be introduced as a foundation for understanding maternal physiology. All the products of conception—the embryo or fetus as well as the placenta and membranes associated with it—are collectively called the **conceptus.** The developing individual is a hollow ball called a *blastocyst* for much of the first 2 weeks, an *embryo* from day 16 through week 8, and a *fetus* from the beginning of week 9 until birth. The fetus is attached by way of an *umbilical cord* to a disc-shaped *placenta* on the uterine wall. The placenta provides fetal nutrition and waste disposal, and secretes hormones that regulate pregnancy, mammary development, and fetal development. For the first 4 weeks after birth, the infant is called a *neonate.*[39]

28.5b Hormones of Pregnancy

The hormones with the strongest influences on pregnancy are estrogens, progesterone, human chorionic gonadotropin, and human chorionic somatomammotropin. The levels of these hormones in the maternal blood over the course of the pregnancy provide a good indicator of the well-being of the fetus. They are secreted primarily by the placenta, but the corpus luteum is an important source of hormones in the first several weeks. If the corpus luteum is removed before week 7, the pregnancy almost always miscarries (aborts). From weeks 7 to 17, the corpus luteum degenerates and the placenta takes over its endocrine functions.

Human Chorionic Gonadotropin

Human chorionic gonadotropin (HCG) (CORE-ee-ON-ic go-NAD-oh-TRO-pin) is secreted by the blastocyst and placenta. Its presence in the urine is the basis of pregnancy tests and can be detected with home testing kits as early as 8 or 9 days after conception. HCG secretion peaks around 10 to 12 weeks and then falls to a relatively low level for the rest of gestation (**fig. 28.18**). Like LH, it stimulates growth of the corpus luteum, which doubles in size and secretes increasing amounts of progesterone and estrogen. Without HCG, the corpus luteum would atrophy and the uterus would expel the conceptus.

Estrogens

Estrogen secretion increases to about 30 times the usual amount by the end of gestation. The corpus luteum is an important source of estrogen for the first 12 weeks; after that, it comes mainly from the placenta. The adrenal glands of the mother and fetus secrete androgens, which the placenta converts to

FIGURE 28.18 Hormone Levels over the Course of Pregnancy.

? *How does the changing ratio of estradiol to progesterone relate to the onset of labor contractions?*

estrogens. The most abundant estrogen of pregnancy is estriol, but its effects are relatively weak; estradiol is less abundant but 100 times as potent.

Estrogen stimulates tissue growth in the fetus and mother. It causes the mother's uterus and external genitalia to enlarge, the mammary ducts to grow, and the breasts to increase to nearly twice their former size. It makes the pubic symphysis more elastic and the sacroiliac joints more limber, so the pelvis widens during pregnancy and the pelvic outlet expands during childbirth.

Progesterone

The placenta secretes a great deal of progesterone, and early in the pregnancy, so does the corpus luteum. Progesterone and estrogen suppress pituitary secretion of FSH and LH, thereby preventing more follicles from developing during pregnancy. Progesterone also suppresses uterine contractions so the conceptus isn't prematurely expelled. It prevents menstruation and promotes the proliferation of *decidual cells* of the endometrium, on which the blastocyst feeds. Once estrogen has stimulated growth of the mammary ducts, progesterone stimulates development of the secretory acini—another step toward lactation.

Human Chorionic Somatomammotropin

Human chorionic somatomammotropin (HCS) (so-MAT-oh-MAM-oh-TRO-pin) is secreted in amounts several times that of all the other hormones combined, yet its function is the least understood. The placenta begins secreting HCS around the fifth week and output increases steadily from then until term, in proportion to the size of the placenta.

---

[39]*neo* = new; *nate* = born, birth

HCS is sometimes called *human placental lactogen* because, in other mammals, it causes mammary development and lactation; however, it doesn't induce lactation in humans. Its effects seem similar to those of growth hormone, but weaker. It also seems to reduce the mother's insulin sensitivity and glucose usage such that the mother consumes less glucose and leaves more of it for use by the fetus. HCS promotes the release of free fatty acids from the mother's adipose tissue, providing an alternative energy substrate for her cells to use in lieu of glucose.

Other Hormones

Many other hormones induce additional bodily changes in pregnancy (**table 28.2**). A pregnant woman's pituitary gland grows about 50% larger and produces markedly elevated levels of thyrotropin, prolactin, and ACTH. The thyroid gland also enlarges about 50% under the influence of HCG, pituitary thyrotropin, and *human chorionic thyrotropin* from the placenta. Elevated thyroid hormone secretion increases the metabolic rate of the mother and fetus. The parathyroid glands enlarge and stimulate osteoclast activity, liberating calcium from the mother's bones for fetal use. ACTH stimulates glucocorticoid secretion, which may serve primarily to mobilize amino acids for fetal protein synthesis. Aldosterone secretion rises and promotes fluid retention, contributing to the mother's increased blood volume. The corpus luteum and placenta secrete *relaxin,* which relaxes the pubic symphysis in other animals but doesn't seem to have this effect in humans. In humans, it synergizes progesterone in stimulating the multiplication of decidual cells in early pregnancy and promotes the growth of blood vessels in the pregnant uterus.

28.5c Adjustments to Pregnancy

Pregnancy places considerable stress on a woman's body and requires adjustments in nearly all the organ systems. A few of the major adjustments and effects of pregnancy are described here.

Digestive System, Nutrition, and Metabolism

For many women, one of the first signs of pregnancy is **morning sickness**—nausea, especially after rising from bed—in the first few months of gestation. The cause is unknown. One hypothesis is that it stems from the reduced intestinal motility caused by the steroids of pregnancy. Another is that it is an evolutionary adaptation to protect the fetus. The fetus is most vulnerable to toxins at the same time that morning sickness peaks. Women with morning sickness tend to prefer bland foods and avoid spicy and pungent foods, which are highest in compounds that could be toxic to the fetus. Women who experience morning sickness have fewer infants with birth defects than women who don't have it. In some women, the nausea progresses to vomiting. Occasionally, this is severe enough to require hospitalization (see *hyperemesis gravidarum* in table 28.5).

Constipation and heartburn are common in pregnancy. The former is another result of reduced intestinal motility. The latter is due to the enlarging uterus pressing upward on the stomach, causing the reflux of gastric contents into the esophagus.

The basal metabolic rate rises about 15% in the second half of gestation. Pregnant women often feel overheated because of this and the effort of carrying the extra weight. The appetite may be

TABLE 28.2	The Hormones of Pregnancy
Hormone	**Effects**
Human chorionic gonadotropin (HCG)	Prevents involution of corpus luteum and stimulates its growth and secretory activity; basis of pregnancy tests
Estrogens	Stimulate maternal and fetal tissue growth, including enlargement of uterus and maternal genitalia; stimulate development of mammary ducts; soften pubic symphysis and sacroiliac joints, facilitating pelvic expansion in pregnancy and childbirth; suppress FSH and LH secretion
Progesterone	Suppresses premature uterine contractions; prevents menstruation; stimulates proliferation of decidual cells, which nourish embryo; stimulates development of mammary acini; suppresses FSH and LH secretion
Human chorionic somatomammotropin (HCS)	Has weak growth-stimulating effects similar to growth hormone, and a glucose-sparing effect on mother, making glucose more available to fetus; mobilizes fatty acids for use as maternal fuel
Pituitary thyrotropin	Stimulates thyroid activity and metabolic rate
Human chorionic thyrotropin	Same effect as pituitary thyrotropin
Parathyroid hormone	Stimulates osteoclasts and mobilizes maternal calcium for fetal use
Adrenocorticotropic hormone	Stimulates glucocorticoid secretion; thought to mobilize amino acids for fetal protein synthesis
Aldosterone	Causes fluid retention, contributing to increased maternal blood volume
Relaxin	Promotes development of decidual cells and blood vessels in the pregnant uterus

strongly stimulated, but a pregnant woman needs only 300 extra kcal/day even in the last trimester. Some women overeat, however, and gain as much as 34 kg (75 lb) of weight compared with a healthy average of 11 kg (24 lb). Maternal nutrition should emphasize the quality of food eaten, not quantity.

During the last trimester, the fetus needs more nutrients than the mother's digestive tract can absorb. In preparation for this, the placenta stores nutrients early in gestation and releases them in the final trimester. The demand is especially high for protein, iron, calcium, and phosphates. A pregnant woman needs an extra 600 mg/day of iron for her own added hematopoiesis and 375 mg/day for the fetus. She is likely to become anemic if she doesn't ingest enough iron during late pregnancy. Supplemental vitamin K is often given late in pregnancy to promote prothrombin synthesis in the fetus. In the United States, newborns are routinely given an injection of vitamin K to minimize the risk of neonatal hemorrhage, especially in the brain, caused by the stresses of birth. A vitamin D supplement helps to ensure adequate calcium absorption to meet fetal demands. Supplemental folic acid reduces the risk of neurological disorders in the fetus, such as spina bifida and anencephaly (failure of the cerebrum, cerebellum, and calvaria to develop), but it is effective only if taken habitually prior to conception.

Circulatory System

By full term, the placenta receives about 625 mL/min. of blood from the mother. The mother's blood volume rises about 30% during pregnancy because of fluid retention and hematopoiesis; she eventually has about 1 to 2 L of extra blood. Cardiac output rises about 30% to 40% above normal by 27 weeks, but for unknown reasons, it falls almost to the nonpregnant level in the last 8 weeks. As the pregnant uterus puts pressure on the large pelvic blood vessels, it interferes with venous return from the legs and pelvic region. This can cause blood pressure to back up into the pelvic region and lower limbs, resulting in hemorrhoids, varicose veins, and edema of the ankles and feet.

Respiratory System

Over the course of pregnancy, respiratory rate remains constant but the tidal volume and minute ventilation increase about 40%. There are two reasons for this: (1) The oxygen demand rises in proportion to the woman's increased metabolic rate and the increasing needs of the fetus. (2) Progesterone increases the sensitivity of the woman's respiratory chemoreceptors to carbon dioxide, and ventilation is adjusted to keep her arterial P_{CO_2} lower than normal. Low maternal P_{CO_2} promotes the diffusion of CO_2 from the fetal bloodstream through the placenta into the maternal blood. As pregnancy progresses, many women feel an increasing "air hunger" (dyspnea) and make more conscious efforts to breathe. This sensation apparently stems from increased CO_2 sensitivity and, late in pregnancy, pressure on the diaphragm from the growing uterus. In the last month, however, the pelvis usually expands enough for the fetus to drop lower in the abdominopelvic cavity, taking some pressure off the diaphragm and allowing one to breathe more easily.

Urinary System

Aldosterone and the steroids of pregnancy promote water and salt retention by the kidneys. Nevertheless, the glomerular filtration rate increases by 50% and urine output is slightly elevated. This enables a woman to dispose of both her own and the fetus's metabolic wastes. As the pregnant uterus compresses the bladder and reduces its capacity, urination becomes more frequent and some women experience uncontrollable leakage of urine (incontinence).

Integumentary System

The skin grows to accommodate expansion of the abdomen and breasts and the added fat deposition in the hips and thighs. Stretching of the dermis often tears the connective tissue and causes **striae** (stretch marks) (STRY-ee). These appear reddish at first but fade after pregnancy. Melanocyte activity increases in some areas and darkens the areolae and linea alba. The latter often becomes a dark line, the **linea nigra**[40] (LIN-ee-uh NY-gruh), from the umbilical to the pubic region. Some women also acquire a temporary blotchy darkening of the skin over the nose and cheeks called the "mask of pregnancy," or **chloasma**[41] (clo-AZ-muh), which usually disappears when the pregnancy is over.

Uterine Growth and Weight Gain

The uterus weighs about 50 g when a woman is not pregnant and about 900 g by the end of pregnancy. Its growth is monitored by palpating the fundus, which eventually reaches almost to the xiphoid process **(fig. 28.19). Table 28.3** shows the distribution of weight gain in a typical healthy pregnancy.

28.5d Childbirth

In the seventh month of gestation, the fetus normally turns into a head-down *vertex position*. Consequently, most babies are born head first, the head acting as a wedge that widens the mother's cervix, vagina, and vulva during birth. The ancients thought that the fetus kicked against the uterus and pushed itself out head first. The fetus, however, is a rather passive player in its own birth; its expulsion is achieved only by the contractions of the mother's uterine and abdominal muscles. Yet there is evidence that the fetus may play some role in its birth by chemically stimulating labor contractions and perhaps even sending chemical signals that signify when it is developed enough to be born.

[40]*linea* = line; *nigra* = black
[41]*chlo* = green; *asma* = to be

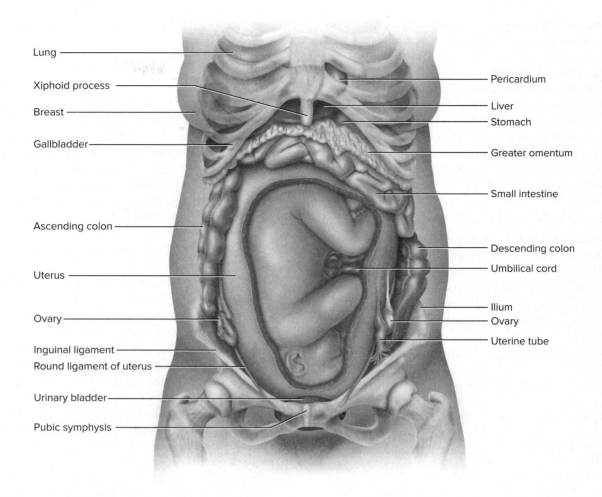

Lung

Xiphoid process

Breast

Gallbladder

Ascending colon

Uterus

Ovary

Inguinal ligament

Round ligament of uterus

Urinary bladder

Pubic symphysis

Pericardium

Liver

Stomach

Greater omentum

Small intestine

Descending colon

Umbilical cord

Ilium

Ovary

Uterine tube

FIGURE 28.19 The Full-Term Fetus in Vertex Position. Note the displacement and compression of the abdominal viscera.

TABLE 28.3	Typical Distribution of Weight Gain in Pregnancy
Fetus	3 kg (7 lb)
Placenta, fetal membranes, and amniotic fluid	1.8 kg (4 lb)
Blood and tissue fluid	2.7 kg (6 lb)
Fat	1.4 kg (3 lb)
Uterus	0.9 kg (2 lb)
Breasts	0.9 kg (2 lb)
Total	**11 kg (24 lb)**

Uterine Contractility

Over the course of gestation, the uterus exhibits relatively weak **Braxton Hicks**[42] **contractions.** These become stronger in late pregnancy and often send women rushing to the hospital with "false labor." At term, however, these contractions transform suddenly into the more powerful **labor contractions.** True labor contractions mark the onset of **parturition** (PAR-too-RISH-un), the process of giving birth.

Progesterone and estrogen (estradiol) balance may be one factor in this pattern of increasing contractility. Both hormone levels increase over the course of gestation. Progesterone inhibits uterine contractions, but its secretion levels off or declines slightly after 6 months, while estradiol secretion continues to rise (fig. 28.18). Estradiol stimulates uterine contractions and may be a factor in the irritability of the uterus in late pregnancy.

Also, as the pregnancy nears full term, the posterior pituitary releases more oxytocin (OT) and the uterine muscle equips itself with more OT receptors by up-regulation. Oxytocin promotes labor in two ways: (1) It directly stimulates muscle of the myometrium; and (2) it stimulates the fetal membranes to secrete prostaglandins, which are synergists of OT in producing labor contractions. Labor is prolonged if OT or prostaglandins are lacking, and it may be induced or accelerated by giving intravenous OT or a vaginal prostaglandin suppository.

The conceptus itself may produce chemical stimuli promoting its own birth. Fetal cortisol secretion rises in late pregnancy and may enhance estrogen secretion by the placenta. The fetal

[42]John Braxton Hicks (1823–97), British gynecologist

pituitary gland also produces oxytocin, which doesn't enter the maternal circulation but may promote prostaglandin secretion as noted earlier.

Uterine stretching is also thought to play a role in initiating labor. Stretching any smooth muscle increases its contractility, and movements of the fetus produce the sort of intermittent stretch that is especially stimulatory to the myometrium. Twins are born an average of 19 days earlier than solitary infants, probably because of the greater stretching of the uterus. When a fetus is in the vertex position, its head pushes against the cervix, which is especially sensitive to stretch.

Labor Contractions

Labor contractions begin about 30 minutes apart. As labor progresses, they become more intense and eventually occur every 1 to 3 minutes. It is important that they be intermittent rather than one long, continual contraction. Each contraction compresses the arteries in the uterine wall and sharply reduces maternal blood flow to the placenta, so the uterus must periodically relax to restore flow and oxygen delivery to the fetus. Contractions are strongest in the fundus and body of the uterus and weaker near the cervix, thus pushing the fetus downward.

According to the **positive feedback theory of labor,** labor contractions are induced by stretching of the cervix. This triggers a reflex contraction of the uterine body that pushes the fetus downward and stretches the cervix still more. Thus, there is a self-amplifying cycle of stretch and contraction. In addition, cervical stretching induces a neuroendocrine reflex through the spinal cord, hypothalamus, and posterior pituitary. The posterior pituitary releases oxytocin, which is carried in the blood and stimulates the uterine muscle both directly and through the action of prostaglandins. This, too, is a positive feedback cycle: cervical stretching → oxytocin secretion → uterine contraction → cervical stretching (see fig. 1.9).

As labor progresses, a woman feels a growing urge to "bear down." A reflex arc extends from the uterus to the spinal cord and back to the skeletal muscles of the abdomen. Contraction of these muscles—partly reflexive and partly voluntary—aids in expelling the fetus, especially when combined with the Valsalva maneuver for increasing intra-abdominal pressure.

The pain of labor is due at first mainly to ischemia of the myometrium—muscle hurts when deprived of blood, and each labor contraction temporarily restricts uterine circulation. As the fetus enters the vaginal canal, the pain becomes stronger because of stretching of the cervix, vagina, and perineum and sometimes the tearing of tissue. At this stage, the obstetrician may perform an *episiotomy*—an incision in the vulva to widen the vaginal orifice and prevent random tearing. The pain of human childbirth, compared with the relative ease with which other mammals give birth, is an evolutionary product of two factors: the unusually large brain and head of the human infant, and the narrowing of the pelvic outlet, which helped to adapt hominids to bipedal locomotion.

Stages of Labor

Labor occurs in three stages **(fig. 28.20).** The duration of each tends to be longer in a **primipara**[43] (a woman giving birth for the first time) than in a **multipara**[44] (a woman who has previously given birth).

1. **Dilation (first) stage.** This is the longest stage, lasting 8 to 24 hours in a primipara but as little as a few minutes in a multipara. It is marked by the **dilation** (widening) of the cervical canal and **effacement** (thinning) of the cervix. The cervix reaches a maximum diameter of about 10 cm (the diameter of the baby's head). During dilation, the fetal membranes usually rupture and the *amniotic fluid* is discharged (the "breaking of the waters"). Dilation in labor is often called *dilatation* (DIL-a-TAY-shun), meaning dilation of an organ beyond the normal limits seen in other circumstances.

2. **Expulsion (second) stage.** This stage typically lasts about 30 to 60 minutes in a primipara and as little as 1 minute in a multipara. It begins when the baby's head enters the vagina and lasts until the baby is entirely expelled. The baby is said to be **crowning** when the top of its head is visible, stretching the vulva. Delivery of the head is the most difficult part, with the rest of the body following more easily. An episiotomy may be performed during this stage. An attendant often uses a suction bulb to remove mucus from the baby's mouth and nose even before it is fully delivered. When the baby is fully expelled, an attendant drains the blood of the umbilical vein into the baby, clamps the umbilical cord in two places, and cuts the cord between the clamps.

3. **Placental (third) stage.** The uterus continues to contract after expulsion of the baby. The placenta, however, is a nonmuscular organ that can't contract, so it buckles away from the uterine wall. About 350 mL of blood is typically lost at this stage, but contractions of the myometrium compress the blood vessels and prevent more extensive bleeding. The placenta, amnion, and other fetal membranes are expelled by uterine contractions, which may be aided by a gentle pull on the umbilical cord. The membranes *(afterbirth)* must be carefully inspected to be sure everything has been expelled. If any of these structures remain in the uterus, they can cause postpartum hemorrhaging. The umbilical blood vessels are counted because an abnormal number in the cord may indicate cardiovascular abnormalities in the infant.

---

[43]*primi* = first; *para* = birth
[44]*multi* = many; *para* = births

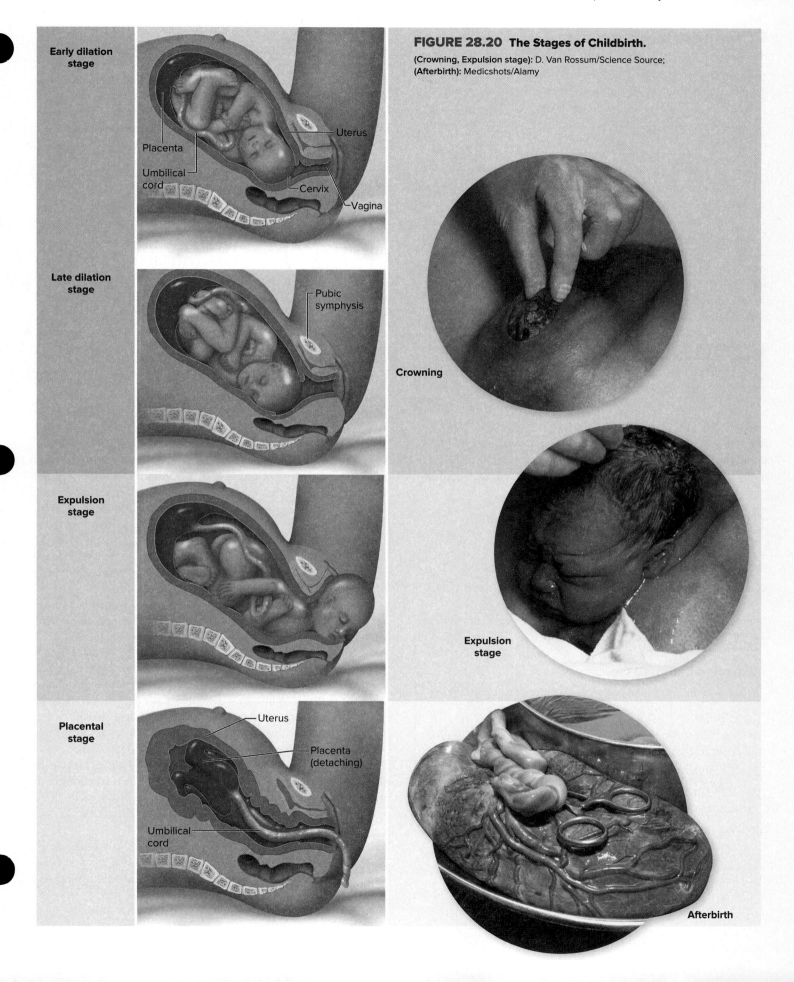

Early dilation stage

Placenta

Umbilical cord

Uterus

Cervix

Vagina

Late dilation stage

Pubic symphysis

Expulsion stage

Placental stage

Uterus

Placenta (detaching)

Umbilical cord

FIGURE 28.20 The Stages of Childbirth.
(Crowning, Expulsion stage): D. Van Rossum/Science Source;
(Afterbirth): Medicshots/Alamy

Crowning

Expulsion stage

Afterbirth

28.5e The Puerperium

The first 6 weeks **postpartum** (after birth) are called the **puerperium**[45] (PYU-er-PEER-ee-um), a period in which the mother's anatomy and physiology stabilize and the reproductive organs return nearly to their condition prior to pregnancy. Shrinkage of the uterus during this period is called **involution.** In a lactating woman, it loses about 50% of its weight in the first week and is nearly at its pregravid weight in 4 weeks. Involution is achieved by **autolysis** (self-digestion) of uterine cells by their own lysosomal enzymes. For about 10 days, this produces a vaginal discharge called **lochia,** which is bloody at first and then turns clear and serous. Breast-feeding promotes involution because (1) it suppresses estrogen secretion, which would otherwise cause the uterus to remain more flaccid; and (2) it stimulates oxytocin secretion, which causes the myometrium to contract and firm up the uterus sooner. It is important for the puerperium to be undisturbed, as emotional upset can inhibit lactation in some women.

BEFORE YOU GO ON

Answer the following questions to test your understanding of the preceding section:

17. List the roles of HCG, estrogen, progesterone, and HCS in pregnancy.

18. What is the role of the corpus luteum in pregnancy? What eventually takes over this role?

19. List and briefly explain the special nutritional requirements of pregnancy.

20. How much weight does the average woman gain in pregnancy? What contributes to this weight gain other than the fetus?

21. Describe the positive feedback theory of labor.

22. What major events define the three stages of labor?

28.6   Lactation

Expected Learning Outcomes

When you have completed this section, you should be able to

a. describe development of the breasts in pregnancy;

b. describe the shifting hormonal balance that regulates the onset and continuation of lactation;

c. describe the mechanism of milk ejection;

d. contrast colostrum with breast milk; and

e. discuss the benefits of breast-feeding.

---

[45]*puer* = child; *per* (from *par*) = birth

Lactation is the synthesis and ejection of milk from the mammary glands. It lasts for as little as 1 week in women who do not breast-feed, but it can continue for many years as long as the breast is stimulated by a nursing infant or mechanical device (breast pump). Numerous studies conducted before the widespread marketing of artificial infant formulas suggest that worldwide, women traditionally nursed their infants until a median age of about 2.8 years.

28.6a Development of the Mammary Glands

The high estrogen level in pregnancy stimulates the ducts of the mammary glands to grow and branch extensively. Growth hormone, insulin, glucocorticoids, and prolactin also contribute to this development. Once the ducts are complete, progesterone stimulates the budding and development of acini at the ends of the ducts (see fig. 28.9).

28.6b Colostrum and Milk Synthesis

In late pregnancy, the mammary acini and ducts are distended with a secretion called **colostrum.** This is similar to breast milk in protein and lactose content but contains about one-third less fat. It is the infant's only natural source of nutrition for the first 1 to 3 days postpartum. Colostrum has a thin watery consistency and a cloudy yellowish color. The amount of colostrum secreted per day is at most 1% of the amount of milk secreted later, but since infants are born with excess body water and ample fat, they don't require high calorie or fluid intake at first. A major benefit of colostrum is that it contains immunoglobulins, especially IgA. IgA resists digestion and may protect the infant from gastroenteritis. It is also thought to be pinocytosed by the small intestine and to confer wider, systemic immunity to the neonate.

Milk synthesis is promoted by **prolactin,** a hormone of the anterior pituitary gland. In the nonpregnant state, **dopamine (prolactin-inhibiting hormone)** from the hypothalamus inhibits prolactin secretion. Prolactin secretion begins 5 weeks into the pregnancy, and by full term it is 10 to 20 times its normal level. Even so, prolactin has little effect on the mammary glands until after birth. While the steroids of pregnancy prepare the mammary glands for lactation, they antagonize prolactin and suppress milk synthesis. When the placenta is discharged at birth, the steroid levels abruptly drop and allow prolactin to have a stronger effect. Milk is synthesized in increasing quantity over the following week. Milk synthesis also requires the action of growth hormone, cortisol, insulin, and parathyroid hormone to mobilize the necessary amino acids, fatty acids, glucose, and calcium.

At the time of birth, baseline prolactin secretion drops to the nonpregnant level. Every time the infant nurses, however, it jumps to 10 to 20 times this level for the next hour and

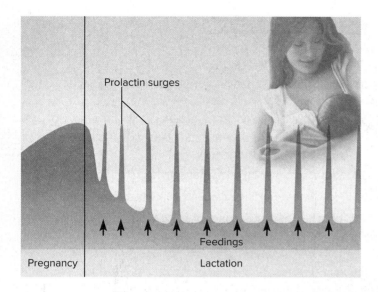

FIGURE 28.21 Prolactin Secretion in the Lactating Female.
Each time the infant nurses, maternal prolactin secretion surges.
This stimulates synthesis of the milk that will be available at the
next feeding.

stimulates the synthesis of milk for the next feeding **(fig. 28.21).**
These prolactin surges are accompanied by smaller increases in
estrogen and progesterone secretion. If the mother doesn't nurse
or these hormone surges are absent (due to pituitary damage, for
example), the mammary glands stop producing milk in about
a week. Even if she does nurse, milk production declines after
7 to 9 months.

Only 5% to 10% of women become pregnant again while
breast-feeding. Apparently, either prolactin or nerve signals from
the breast inhibit GnRH secretion, which, in turn, results in reduced
gonadotropin secretion and ovarian cycling. This mechanism may
have evolved as a natural means of spacing births, but breast-
feeding isn't a reliable means of contraception. Even in women
who breast-feed, the ovarian cycle sometimes resumes several
months postpartum. In those who don't breast-feed, the cycles
resume in a few weeks, but for the first 6 months they are usually
anovulatory.

28.6c Milk Ejection

Milk is continually secreted into the mammary acini, but it
doesn't flow easily into the ducts. Its flow, called **milk ejection
(letdown),** is controlled by a neuroendocrine reflex. The infant's
suckling stimulates nerve endings of the nipple and areola,
which in turn signal the hypothalamus and posterior pituitary
to release oxytocin. Oxytocin stimulates the myoepithelial
cells that enmesh each gland acinus (see fig. 28.9d). These cells
are of epithelial origin, but are packed with actin and contract
like smooth muscle to squeeze milk from the acinus into the
duct. The infant doesn't get any milk for the first 30 to 60 seconds

of suckling, but milk soon fills the ducts and lactiferous sinuses
and is then easily sucked out.

▶▶▶ **APPLY WHAT YOU KNOW**

*When a woman is nursing her baby at one breast, would
you expect only that breast, or both breasts, to eject
milk? Explain why.*

28.6d Breast Milk

Table 28.4 compares the composition of colostrum, human milk,
and cow's milk. Breast milk changes composition over the first
2 weeks, varies from one time of day to another, and changes even
during the course of a single feeding. For example, at the end of a
feeding there is less lactose and protein in the milk, but six times
as much fat, as there is at the beginning.

Cow's milk is not a good substitute for human milk. It has
one-third less lactose but three to five times as much protein and
minerals. The excess protein forms a harder curd in the infant's
stomach, so cow's milk isn't digested and absorbed as efficiently as
mother's milk. It also increases the infant's nitrogenous waste excretion, which increases the incidence and severity of diaper rash,
particularly as bacteria in the diaper break urea down to ammonia,
a skin irritant.

Colostrum and milk have a laxative effect that helps to clear
the neonatal intestine of *meconium,* a greenish black, sticky
fecal matter composed of bile, epithelial cells, and other wastes
that accumulated during fetal development. By clearing bile and
bilirubin from the body, breast-feeding also reduces the incidence
and degree of jaundice in neonates. Breast milk promotes
colonization of the neonatal intestine with beneficial bacteria
and supplies antibodies that lend protection against infection by
pathogenic bacteria. Breast-feeding also tends to promote a closer
bond between mother and infant.

TABLE 28.4	A Comparison of Colostrum, Human Milk, and Cow's Milk*		
Nutrient	**Human Colostrum**	**Human Milk**	**Cow's Milk**
Total protein (g/L)	22.9	10.6	30.9
Lactalbumin (g/L)	—	3.7	25.0
Casein (g/L)	—	3.6	2.3
Immunoglobulins (g/L)	19.4	0.09	0.8
Fat (g/L)	29.5	45.4	38.0
Lactose (g/L)	57	71	47
Calcium (mg/L)	481	344	1,370
Phosphorus (mg/L)	157	141	910

*Colostrum data are for the first day postpartum, and human milk data are for "mature
milk" at about 15 days postpartum.

A woman nursing one baby eventually produces about 1.5 L of milk per day; women with twins produce more. Lactation places a great metabolic demand on the mother. It is equivalent to losing 50 g of fat, 100 g of lactose (made from her blood glucose), and 2 to 3 g of calcium phosphate per day. A woman is at greater risk of bone loss when breast-feeding than when she is pregnant, because much of the infant's skeleton is still cartilage at birth and becomes mineralized at her expense in the first year postpartum. If a nursing mother doesn't have enough calcium and vitamin D in her own diet, lactation stimulates parathyroid hormone secretion and osteoclast activity, taking calcium from her bones to supply her baby.

To conclude this chapter, **table 28.5** briefly describes some common female reproductive disorders.

BEFORE YOU GO ON

Answer the following questions to test your understanding of the preceding section:

23. Why is little or no milk secreted while a woman is pregnant?

24. How does a lactating breast differ from a nonlactating breast in structure? What stimulates these differences to develop during pregnancy?

25. What is colostrum and what is its significance?

26. How does suckling stimulate milk ejection?

27. Why is breast milk superior to cow's milk for an infant?

TABLE 28.5	Some Female Reproductive Disorders
Amenorrhea	Absence of menstruation. Normal in pregnancy, lactation, early adolescence, and perimenopausal years, but can also result from gonadotropin hyposecretion, genetic disorders, CNS disorders, or insufficient body fat.
Dysmenorrhea	Painful menstruation in the absence of pelvic disease, caused by excessive endometrial prostaglandin secretion. Prostaglandins stimulate painful contractions of myometrium and uterine blood vessels. Usually begins around age 15 or 16 and affects up to 75% of women from 15 to 25 years old at times. May accompany leiomyoma.
Leiomyoma (uterine fibroids)	Benign tumor of the uterine smooth muscle. Usually small and asymptomatic, but can become very large and may cause abnormal uterine bleeding, heavy menstruation, and dysmenorrhea.
Disorders of Pregnancy	
Abruptio placentae[46]	Premature separation of the placenta from the uterine wall, often associated with preeclampsia or cocaine use. May require birth by cesarean section.
Ectopic[47] pregnancy	Implantation of the conceptus anywhere other than the uterus; usually in the uterine tube (tubal pregnancy) but occasionally in the abdominopelvic cavity. See Deeper Insight 29.2 for further details.
Gestational diabetes	A form of diabetes mellitus that develops in about 1% to 3% of pregnant women, characterized by insulin insensitivity, hyperglycemia, glycosuria, and a risk of excessive fetal size and birth trauma. Glucose metabolism often returns to normal after delivery of the infant, but 40% to 60% of women with gestational diabetes develop diabetes mellitus within 15 years after the pregnancy.
Hyperemesis gravidarum[48]	Severe vomiting, dehydration, alkalosis, and weight loss in early pregnancy, often requiring hospitalization to stabilize fluid, electrolyte, and acid–base balance; sometimes associated with liver damage.
Placenta previa[49]	Blockage of the cervical canal by the placenta, preventing birth of the infant before the placenta separates from the uterus. Requires birth by cesarean section.
Preeclampsia[50]	Gestational hypertension and proteinuria, often with edema of the face and hands, occurring especially in the third trimester in primiparas. Correlated with abnormal development of placental arteries, potentially leading to widespread thrombosis and organ dysfunction in the mother. Occurs in 5% to 8% of pregnancies. Sometimes progresses to *eclampsia* (seizures), which can be fatal to the mother, fetus, or both. Eclampsia may occur postpartum.
Spontaneous abortion	Occurs in 10% to 15% of pregnancies, usually because of fetal deformities or chromosomal abnormalities incompatible with survival, but may also result from maternal abnormalities, infectious disease, and drug abuse.
You can find other female reproductive disorders described in the following places:	
Sexually transmitted diseases in Deeper Insight 27.5; *breast cancer* in section 28.1c; *cervical cancer* in Deeper Insight 28.1; and *endometriosis* in Deeper Insight 28.4.	

[46]*ab* = away; *rupt* = to tear; *placentae* = of the placenta

[47]*ec* = out of; *top* = place

[48]*hyper* = excessive; *emesis* = vomiting; *gravida* = pregnant woman

[49]*pre* = before; *via* = the way (obstructing the way)

[50]*ec* = forth; *lampsia* = shining

DEEPER INSIGHT 28.5

CLINICAL APPLICATION

Methods of Contraception

The term *contraception* is used here to mean any procedure or device intended to prevent pregnancy (the presence of an implanted conceptus in the uterus). This essay describes the most common methods of contraception, some issues involved in choosing among them, and their relative reliability. Several options are shown in **figure 28.22.**

Behavioral Methods

Abstinence (refraining from intercourse) is obviously a completely reliable method if used consistently. The *fertility awareness-based method* (formerly called the *rhythm method*) is based on avoiding intercourse near the time of expected ovulation. Among typical users, it has a 25% failure rate, partly due to lack of restraint and partly because it is difficult to predict the exact date of ovulation. Intercourse must be avoided for at least 7 days before ovulation so there will be no surviving sperm in the reproductive tract when the egg is ovulated, and for at least 2 days after ovulation so there will be no fertile egg present when sperm are introduced. This method is valuable, however, for couples who are trying to conceive a child by having intercourse at the time of apparent ovulation.

Withdrawal (coitus interruptus) requires the male to withdraw the penis before ejaculation. This often fails because of lack of willpower, because sperm are occasionally present in the preejaculatory fluid, and because sperm ejaculated anywhere in the vulva can potentially get into the reproductive tract.

Barrier and Spermicidal Methods

Barrier methods are designed to prevent sperm from getting into or beyond the vagina. They are most effective when used with chemical *spermicides,* available as nonprescription foams, creams, and jellies.

The *male condom* is a sheath of latex, rubber, or animal membrane (lamb intestine) that is unrolled over the erect penis and collects the semen. It is inexpensive, convenient, and very reliable when used carefully. About 25% of American couples who use contraceptives use only condoms, which rank second to birth-control pills in popularity.

The *female condom* is less used. It is a polyurethane sheath with a flexible ring at each end. The inner ring fits over the cervix and the outer ring covers the external genitalia. Male and female condoms are the only contraceptives that also protect against disease transmission. Animal membrane condoms, however, are porous to HIV and hepatitis B viruses and do not afford dependable protection from disease.

The *diaphragm* is a latex or rubber dome that is placed over the cervix to block sperm migration. It requires a physical examination and prescription to ensure proper fit, but is otherwise comparable to the condom in convenience and reliability, provided it is used with a spermicide. Without a spermicide, it isn't very effective.

The *sponge* is a foam disc inserted before intercourse to cover the cervix. It is impregnated with a spermicide and acts by trapping and killing the sperm. It requires no prescription or fitting. The sponge provides

Male condom

Female condom

Diaphragm with contraceptive jelly

Contraceptive foam with vaginal applicator

Birth-control pills

Vaginal ring

Copper IUD

FIGURE 28.22 Contraceptive Devices.

(Male, Female Condom, Diaphragm, Contraceptive foam, Birth Control Pills, & IUD): Bob Coyle/McGraw-Hill Education; **(Vaginal ring):** vario images GmbH & Co.KG/Alamy

(continued)

protection for up to 24 hours, and must be left in place for at least 6 hours after the last act of intercourse.

Contraceptive sponges and other barrier methods date to antiquity. Ancient Egyptians and Greeks used vaginal sponges soaked in lemon juice, which had a mild spermicidal effect. Some Egyptian women used vaginal pessaries made of crocodile dung and honey, but crocodile dung is difficult to find in pharmacies these days, limiting the modern usefulness of this idea.

Hormonal Methods

Most hormonal methods of contraception are aimed at preventing ovulation. Efforts to develop a hormonal contraceptive for men have so far been unsuccessful, but are continuing. In women, hormonal contraceptives mimic the negative feedback effect of ovarian hormones, inhibiting FSH and LH secretion so follicles do not mature. For most women, they are highly effective. Their differences lie in convenience, method of application, and to some extent, reliability and risk of complications.

The oldest and most widely used hormonal method is the *combined oral contraceptive,* or *birth-control pill.* "The pill" is composed of estrogen and progestin (a synthetic progesterone-like hormone). It must be taken daily, at the same time of day, for 21 days each cycle. It comes in 28-day packets, marked day by day, with the last 7 pills being plain sugar just to keep the user in the habit of taking one every day. The 7-day withdrawal from hormones allows for menstruation. Side effects include an elevated risk of heart attack or stroke in smokers and in women with a history of diabetes, hypertension, or clotting disorders.

Other hormonal methods avoid the need to remember a daily pill. One option is a skin patch that releases estrogen and progestin transdermally. It is changed at 7-day intervals (3 patches per month and 1 week without). Also available is a soft, flexible vaginal ring that releases estrogen and progestin for absorption through the vaginal mucosa. It must be worn continually for 3 weeks and removed for the fourth week of each cycle. *Medroxyprogesterone* is a progestin administered by injection every 3 months. It provides highly reliable, long-term contraception, although in some women it causes headaches, nausea, or weight gain, and fertility may not return immediately when its use is discontinued.

Some drugs can be taken orally after intercourse to prevent implantation of a conceptus. These are called emergency contraceptive pills (ECPs), or "morning-after pills." An ECP is a high dose of estrogen and progestin or a progestin alone. It can be taken up to 120 hours after intercourse and induces menstruation within 2 weeks. ECPs work on several fronts: inhibiting ovulation; inhibiting sperm or egg transport in the uterine tube; and preventing implantation. They don't work if a blastocyst is already implanted. ECPs are available without a prescription in most places.

Mifepristone (RU-486) is a progesterone antagonist. It is used less as a contraceptive than as an *abortifacient;* in high doses, it induces abortion up to 2 months into pregnancy. But at a dose of 2 mg/day, it prevents ovulation like other steroidal contraceptives, and a single 10 mg dose can also be used as an emergency "morning-after" contraceptive if taken after intercourse but before ovulation.

The Intrauterine Device

Intrauterine devices (IUDs) are springy, often T-shaped devices inserted through the cervical canal into the uterus. Some IUDs act by releasing a synthetic progesterone, but most have a copper wire wrapping or copper sleeve. IUDs act by irritating the uterine lining and interfering with blastocyst implantation, and copper IUDs also inhibit sperm motility. A copper IUD is the most effective after-sex contraceptive even when installed up to 10 days after intercourse. IUDs, however, have relatively high up-front cost. An IUD can be left in place for 5 to 12 years.

Surgical Sterilization

People who are confident that they don't want more children (or any) often elect to be surgically sterilized. This entails the cutting and tying or clamping of the genital ducts, thus blocking the passage of sperm or eggs. Surgical sterilization has the advantage of convenience, since it requires no further attention. Its initial cost is higher, however, and for people who later change their minds, surgical reversal is much more expensive than the original procedure and is often unsuccessful. *Vasectomy* is the severing of the ductus (vas) deferens, done through a small incision in the scrotum. In *tubal ligation,*[51] the uterine tubes are cut. This can be done through small abdominal incisions to admit a cutting instrument and laparoscope (viewing device).

Issues in Choosing a Contraceptive

Many issues enter into the appropriate choice of a contraceptive, including personal preference, pattern of sexual activity, medical history, religious views, convenience, initial and ongoing costs, and disease prevention. For most people, however, the two primary issues are safety and reliability.

Table 28.6 shows the expected rates of failure for several types of contraception as reported by the World Health Organization (WHO). Each column shows the number of sexually active women who typically become pregnant within 1 year while they or their partners are using the indicated contraceptives. The lowest rate (perfect use) is for those who use the method correctly and consistently, whereas the higher rate (typical use) is based on random surveys of users and takes human error (lapses and incorrect usage) into account.

We have not considered all the currently available methods of contraception or all the issues important to the choice of a contraceptive. No one contraceptive method can be recommended as best for all people. Further information necessary to a sound choice and proper use of contraceptives should be sought from a health department, college health service, physician, Planned Parenthood, or other such sources.

TABLE 28.6	**Failure Rates of Contraceptive Methods**	
Method	Rate of Failure (Pregnancies per 100 Users)	
	Perfect Use	Typical Use
No protection	85	85
Rhythm method	3–5	25
Withdrawal	4	27
Spermicide alone	18	26
Condom alone (male or female)	2–5	15–21
Diaphragm with spermicide	6	16
Vaginal sponge	9–20	16–32
Birth-control pill, patch, or vaginal ring	0.3–0.5	8
Medroxyprogesterone	0.3	3
Intrauterine device	0.2–0.6	0.2–0.8
Vasectomy	0.10	0.15
Tubal ligation	0.5	0.5

[51]*ligat* = to tie

Effects of the REPRODUCTIVE SYSTEM on Other Organ Systems

INTEGUMENTARY SYSTEM
At puberty, androgens stimulate development of body hair and apocrine glands and increased sebaceous secretion; estrogens stimulate fat deposition and breast development; pregnancy necessitates growth of skin, especially in abdominal and mammary regions, and may cause pigmentation changes and stretch marks.

SKELETAL SYSTEM
Androgens and estrogens stimulate bone deposition and adolescent skeletal growth and maintain adult bone mass.

MUSCULAR SYSTEM
Androgens stimulate muscle growth; sexual climax and childbirth involve contractions of specific skeletal muscles.

NERVOUS SYSTEM
Sex steroids stimulate the brain and libido; gonadal and placental hormones exert negative feedback control on the hypothalamus.

ENDOCRINE SYSTEM
The gonads and placenta secrete androgens, estrogens, progesterone, and hormones of pregnancy.

CIRCULATORY SYSTEM
Androgens stimulate erythropoiesis; estrogens inhibit atherosclerosis in females; pregnancy increases blood volume and cardiac output and may cause varicose veins.

LYMPHATIC AND IMMUNE SYSTEMS
Barriers in the testis and ovary protect germ cells from antibodies; androgens somewhat inhibit immunity and increase susceptibility to infectious diseases.

RESPIRATORY SYSTEM
Sexual arousal increases pulmonary ventilation; pregnancy increases CO_2 sensitivity of respiratory chemoreceptors and increases tidal volume and minute ventilation.

URINARY SYSTEM
Sexual arousal constricts the male internal urinary sphincter, which prevents reflux of semen into the male urinary bladder; prostatic hyperplasia can impede urine flow; pregnancy crowds the bladder, reduces its capacity, and may cause incontinence; pregnancy promotes salt and water reabsorption by the kidneys and increases glomerular filtration rate and urine output.

DIGESTIVE SYSTEM
A growing fetus crowds the stomach and intestines and may cause heartburn; pregnancy is often associated with constipation and nausea.

STUDY GUIDE

▶ Assess Your Learning Outcomes

To test your knowledge, discuss the following topics with a study partner or in writing, ideally from memory.

28.1 Female Reproductive Anatomy

1. Why the female paramesonephric duct develops into a reproductive tract whereas the male's paramesonephric duct does not; what mature female structures arise from the duct

2. What mature female structures arise from the embryonic genital tubercle, urogenital folds, and labioscrotal folds

3. Internal structure of the ovary; its supportive ligaments; and its blood and nerve supplies

4. General structure of ovarian follicles; their function and how they compare and contrast with the male's seminiferous tubules; and their location in the ovary

5. Gross anatomy of the uterine (fallopian) tube; its three segments; its supportive ligament; and the structure of its mucosa and relationship of that structure to its function

6. Gross anatomy of the uterus; its supportive ligaments; its relationship to the vesicouterine and rectouterine pouches; its blood supply; and function of the cervical glands

7. Tissue layers of the uterine wall; histology of the endometrium; functions of the endometrial sublayers

8. The tilt of the vagina and its relationship to adjacent organs; histology of its mucosa in childhood and adulthood; significance of its dendritic cells; sources of its lubrication; and anatomy of the hymen

9. Anatomy of the vulva including the mons pubis, labia majora and minora, clitoris and prepuce, vaginal and urethral orifices, accessory glands, and erectile tissues

10. Anatomy of the mature breast in the resting and lactating states

11. The prevalence of breast cancer; its diagnostic signs; genetic and other risk factors for breast cancer; preventive breast care; and treatment options

28.2 Puberty and Menopause

1. The typical age of onset of female puberty in the United States and Europe

2. The hormonal trigger for the onset of female puberty; roles of GnRH, FSH, LH, inhibin, androgens, and estrogens; and three types of estrogen

3. Thelarche, pubarche, and menarche as signs of female puberty; their hormonal causes; and the hormonal basis of the libido

4. The typical age of menarche and its relationship to the onset of ovulation

5. Bodily effects of estradiol, progesterone, and inhibin in female puberty

6. Ovarian and hormonal changes that bring on female climacteric and menopause; effects commonly experienced by perimenopausal women; and differences between female and male climacteric

7. The criterion for determining that a woman has passed through menopause, and the reason one cannot identify the exact time of menopause

28.3 Oogenesis and the Sexual Cycle

1. Meanings of and distinctions between the female *sexual cycle, ovarian cycle,* and *menstrual cycle*

2. The meaning of *oogenesis* and ways in which it differs from spermatogenesis

3. Prenatal development of oogonia and primary oocytes; the peak number of oogonia typically attained in the fetus; why this number is so much less at the time of birth and again by the onset of puberty; and the name for the prenatal and childhood degeneration of female germ cells

4. Ways in which meiosis in the female differs from that in the male; why male gametogenesis produces four functional gametes per stem cell and female gametogenesis produces only one; what happens to the other three meiotic daughter cells in the female

5. How far meiosis has progressed by the time the egg is ovulated, and what must happen thereafter for meiosis to be completed

6. Development of a follicle from primary to mature (graafian) types; the structural differences between the stages; and how folliculogenesis is correlated with oogenesis

7. Structural details of a mature follicle

8. The timetable of oogenesis and folliculogenesis; why the ovarian cycle is considered to average 28 days long whereas any given egg and follicle take much longer to mature; and what event marks day 1 of a cycle

9. How many follicles begin to develop in each cycle of folliculogenesis; how many of them normally ovulate; what happens to the rest, and what that fate is called

10. Roles of FSH and LH in regulating the ovarian cycle

11. The process of ovulation, the day on which it occurs in a typical cycle, and how the egg gets into the uterine tube

12. Production, structure, function, and eventual involution of the corpus luteum; why days 15 through 28 of a typical cycle are called the luteal phase; what remains after a corpus luteum has fully involuted

13. Four phases of the menstrual cycle; what happens histologically to the endometrium in each phase; what days of the cycle are spanned by each phase; and what hormones regulate these changes

28.4 Female Sexual Response

1. Ways in which the female sexual response differs from the male's in the excitement phase; vasocongestion of the labia, clitoris, and breasts; sources of lubrication of the vagina and vulva; anatomical changes in the vagina and uterus during sexual arousal

2. Physiological responses of orgasm and resolution in the female and how they differ from those of the male; absence of a refractory period and potential for multiple orgasms

28.5 Pregnancy and Childbirth

1. The timetable of gestation and how the date of birth is predicted

2. What is included in the conceptus

3. Human chorionic gonadotropin (HCG), its source and effects, the time course of its rising and falling secretion during pregnancy, and its usefulness in pregnancy tests

4. Sources and effects of estrogen, progesterone, and human chorionic gonadotropin in pregnancy

5. Effects of thyroid hormone, parathyroid hormone, glucocorticoids, aldosterone, and relaxin in pregnancy

6. Causes of morning sickness, constipation, and heartburn in pregnancy; the change in basal metabolic rate and the related nutritional needs of pregnancy

7. Effects of pregnancy on blood volume and cardiac output; how pregnancy can cause edema, hemorrhoids, and varicose veins

8. Effects of pregnancy on respiratory function; the mechanism for enhancing diffusion of carbon dioxide from the fetal blood into the maternal blood of the placenta

STUDY GUIDE

9. Effects of pregnancy on glomerular filtration, urine output, and the capacity of the bladder

10. Effects of pregnancy on the skin; causes of striae (stretch marks), the linea nigra, and chloasma

11. The vertex position and the developmental age at which the fetus typically assumes it

12. The nature and possible cause of Braxton Hicks contractions, when they occur, and how they differ from true labor contractions

13. Factors that stimulate the onset of labor contractions; the roles of oxytocin, positive feedback, and the voluntary abdominal muscles in labor

14. Events that mark each stage of labor, and the names of the stages

15. The puerperium, its time course, the postpartum changes in a woman's body during this time

28.6 Lactation

1. Influences of estrogen, growth hormone, insulin, glucocorticoids, and prolactin on mammary gland development during pregnancy

2. The fluid secreted by the mammary glands for the first few days postpartum, how it differs from breast milk, and its benefits to the neonate

3. Why prolactin stimulates milk synthesis after birth but not during pregnancy

4. The neuroendocrine reflex stimulated by the suckling of an infant, and the roles of oxytocin and prolactin in breast-feeding

5. Composition of breast milk in comparison to colostrum and cow's milk; reasons why cow's milk is less healthy than breast milk for an infant

6. How breast milk varies in composition from one time to another; which components of the milk are released early, and which are released nearer the end, of a single feeding

7. The daily quantity of breast milk typically produced (eventually) and its nutritional demands on the mother

▶ Testing Your Recall

Answers in Appendix A

1. Of the following organs, the one(s) most comparable to the penis in structure is/are
   a. the clitoris.
   b. the vagina.
   c. the vestibular bulbs.
   d. the labia minora.
   e. the prepuce.

2. The ovaries secrete all of the following *except*
   a. estrogens.
   b. progesterone.
   c. androgens.
   d. follicle-stimulating hormone.
   e. inhibin.

3. The first haploid stage in oogenesis is
   a. the oogonium.
   b. the primary oocyte.
   c. the secondary oocyte.
   d. the second polar body.
   e. the zygote.

4. The secondary oocyte is shielded from immune attack by
   a. the zona pellucida.
   b. the theca folliculi.
   c. the cumulus oophorus.
   d. follicular fluid.
   e. inhibin.

5. The hormone that most directly influences the secretory phase of the menstrual cycle is
   a. HCG.
   b. FSH.
   c. LH.
   d. estrogen.
   e. progesterone.

6. The ischemic phase of the uterus results from
   a. rising progesterone levels.
   b. falling progesterone levels.
   c. stimulation by oxytocin.
   d. stimulation by prostaglandins.
   e. stimulation by estrogens.

7. Before secreting milk, the mammary glands secrete
   a. prolactin.
   b. colostrum.
   c. lochia.
   d. meconium.
   e. chloasma.

8. One is less likely to become pregnant while nursing than at other times because _____ inhibits GnRH secretion.
   a. FSH
   b. prolactin
   c. prostaglandin
   d. oxytocin
   e. HCG

9. Smooth muscle cells of the myometrium and myoepithelial cells of the mammary glands are the target cells for
   a. prostaglandins.
   b. LH.
   c. oxytocin.
   d. progesterone.
   e. FSH.

10. Which of these is *not* true of the luteal phase of the sexual cycle?
    a. Progesterone level is high.
    b. The endometrium stores glycogen.

   c. Ovulation occurs.
   d. Fertilization may occur.
   e. The endometrial glands enlarge.

11. Each egg cell develops in its own fluid-filled space called a/an _____.

12. The mucosa of the uterus is called the _____.

13. A girl's first menstrual period is called _____.

14. A yellowish structure called the _____ secretes progesterone during the secretory phase of the menstrual cycle.

15. The layer of cells closest to a mature secondary oocyte is the _____.

16. A tertiary follicle differs from a primary follicle in having a cavity called the _____.

17. Menopause occurs during a midlife period of changing hormone secretion called _____.

18. All the products of fertilization, including the embryo or fetus, the placenta, and the embryonic membranes, are collectively called the _____.

19. The funnel-like distal end of the uterine tube is called the _____ and has feathery processes called _____.

20. Postpartum uterine involution produces a vaginal discharge called _____.

STUDY GUIDE

▶ Building Your Medical Vocabulary

Answers in Appendix A

State a meaning of each word element, and give a medical term from this chapter that uses it or a slight variation of it.

1. -arche

2. -arum

3. cumulo-

4. gesto-

5. hystero-

6. lacto-

7. ligat-

8. metri-

9. oo-

10. primi-

▶ What's Wrong with These Statements?

Answers in Appendix A

Briefly explain why each of the following statements is false, or reword it to make it true.

1. After ovulation, a follicle begins to move down the uterine tube to the uterus.

2. Human chorionic gonadotropin is secreted by the granulosa cells of the follicle.

3. A follicle ovulates shortly after its oocyte completes meiosis II.

4. A slim girl who is active in dance and gymnastics is likely to begin menstruating at a younger age than an overweight inactive girl.

5. There are more future egg cells in the ovary at puberty than there are at birth.

6. Women do not lactate while they are pregnant because prolactin is not secreted until after birth.

7. Colostrum contains more fat than mature milk, ensuring the neonate of adequate caloric intake in its first month of separation from placental nutrition.

8. Normally, only one primary follicle develops to the secondary follicle stage in each month's ovarian cycle.

9. Progesterone promotes the uterine contractions of childbirth.

10. The entire endometrium is shed in each menstrual period.

▶ Testing Your Comprehension

1. Would you expect puberty to create a state of positive or negative nitrogen balance? Explain. (See "Nitrogen Balance" in section 26.1h for background.)

2. Aspirin and ibuprofen can inhibit the onset of labor and are sometimes used to prevent premature birth. Review your knowledge of these drugs and the mechanism of labor, and explain this effect.

3. At 6 months postpartum, a nursing mother is in an automobile accident that fractures her skull and severs the hypophysial portal vessels. How would you expect this to affect her milk production? How would you expect it to affect her future ovarian cycles? Explain the difference.

4. If the ovaries are removed in the first 6 weeks of pregnancy, the embryo will be aborted. If they are removed later in pregnancy, the pregnancy can go to a normal full term. Explain the difference.

5. A breast-feeding woman leaves her baby at home and goes shopping. There, she hears another woman's baby crying and notices her blouse becoming wet with a little exuded milk. Explain the physiological link between hearing that sound and the ejection of milk.

Boy meets girl: the union of sperm and egg (SEM)

Francis Leroy, Biocosmos/Science Source

CHAPTER

29

HUMAN DEVELOPMENT AND AGING

CHAPTER OUTLINE

29.1 Fertilization and the Preembryonic Stage

   29.1a Sperm Migration and Capacitation

   29.1b Fertilization

   29.1c Major Stages of Prenatal Development

   29.1d The Preembryonic Stage

29.2 The Embryonic and Fetal Stages

   29.2a Embryonic Folding and Organogenesis

   29.2b Embryonic Membranes

   29.2c Prenatal Nutrition

   29.2d Fetal Development

29.3 The Neonate

   29.3a Adapting to Life Outside the Uterus

   29.3b Premature Infants

   29.3c Birth Defects

29.4 Aging and Senescence

   29.4a Senescence of the Organ Systems

   29.4b Exercise, Senescence, and Life Expectancy

   29.4c Theories of Senescence

   29.4d Death

Study Guide

DEEPER INSIGHTS

29.1 Clinical Application: Twins

29.2 Clinical Application: Ectopic Pregnancy

29.3 Clinical Application: Neonatal Assessment

29.4 Clinical Application: Cleft Lip and Palate

29.5 Clinical Application: Assisted Reproductive Technology

Anatomy & Physiology
Revealed 4.0

Module 14: Reproductive System

BRUSHING UP

- Understanding the fertilization process requires familiarity with sperm structure (see section 27.4c).

- Understanding fertilization and the events immediately following it also depends on a knowledge of oogenesis (see section 28.3a) and the structure of the mature ovarian follicle (see fig. 28.12).

- Understanding embryonic implantation in the uterus and the development of the placenta requires a knowledge of the histology of the uterine endometrium (see "Uterine Wall" in section 28.1b, and fig. 28.16).

Perhaps the most dramatic, miraculous-seeming aspect of human life is the transformation of a one-celled fertilized egg into an independent, fully developed individual. From the beginning of recorded thought, people have pondered how a baby forms in the mother's body and how two parents can produce another human being who, although unique, possesses characteristics of each. Aristotle, in his quest to understand prenatal development, dissected bird embryos and established the sequence in which their organs appeared and took shape. He also speculated that the hereditary traits of a child resulted from the mixing of the male's semen with the female's menstrual blood. Such misconceptions about human development persisted for many centuries. In the seventeenth century, scientists thought that the features of the infant existed in a preformed state in the egg or sperm, and simply unfolded and expanded as the embryo developed. Some thought that the head of the sperm had a miniature human curled up in it, while others thought that the miniature person existed in the egg and the sperm were parasites in the semen.

The modern science of **embryology**—the study of prenatal development—wasn't born until the nineteenth century, largely because darwinism at last gave biologists a systematic framework for asking the right questions and discovering unifying themes in the development of diverse species of animals, including humans. It was in that era, too, that the human egg was first observed. Embryology is now a part of **developmental biology,** a broader science that embraces changes in form and function from fertilized egg through old age. A rapidly expanding area of developmental biology today is the genetic regulation of development.

In this book's closing chapter, it is fitting that we reflect on the closing chapter of life as well. Why do our bodies wear out? Isn't there something we can do about that? Is there any cure for old age on the horizon? Our scope of discussion here embraces the human life span from conception to death.

29.1 Fertilization and the Preembryonic Stage

Expected Learning Outcomes

When you have completed this section, you should be able to

a. describe the processes of sperm migration and fertilization;

b. explain how an egg prevents fertilization by more than one sperm;

c. describe the major events that transform a fertilized egg into an embryo; and

d. describe the implantation of the preembryo in the uterine wall.

Authorities attach different meanings to the word **embryo.** Some use it to denote stages beginning with the fertilized egg or at least with the two-celled stage produced by its first division. Others first apply the word *embryo* to an individual 16 days old, when it consists of three **primary germ layers** called the *ectoderm, mesoderm,* and *endoderm.* The events leading up to that stage are called *embryogenesis,* so the first 16 days after fertilization are called the *preembryonic stage.* This is the sense in which we will use such terms in this book. We begin with the process in which a sperm locates and fertilizes the egg.

29.1a Sperm Migration and Capacitation

If it is to survive, an egg must be fertilized within 12 to 24 hours of ovulation; yet it takes about 72 hours for an egg to reach the uterus. Therefore, in order to fertilize an egg before it dies, sperm must encounter it somewhere in the distal one-third of the uterine tube. The vast majority of sperm never make it that far. Many are destroyed by vaginal acid or drain out of the vagina. Others fail to penetrate the mucus of the cervical canal, and those that do are often destroyed by leukocytes in the uterus. Of those that get past the uterus, probably half go up the wrong uterine tube. Finally, about 200 spermatozoa reach the vicinity of the egg—not many of the 300 million that were ejaculated.

Sperm migrate mainly by means of the snakelike lashing of their tails as they crawl along the female mucosa, but they are assisted by certain aspects of female physiology. Strands of mucus guide them through the cervical canal. Although female orgasm isn't required for fertilization, orgasm does involve uterine contractions that may suck semen from the vagina and spread it throughout the uterus, like hand lotion pressed between your palms. The egg itself may release a chemical that attracts sperm from a short distance; this has been demonstrated for some animals but remains unproven for humans.

Sperm can reach the distal uterine tube in half an hour or less after ejaculation, but they can't fertilize an egg for about 10 hours. While migrating, they must undergo a process called **capacitation** that makes them capable of penetrating an egg. In fresh sperm, the plasma membrane is toughened by cholesterol. This prevents the premature release of acrosomal enzymes while sperm are still in the male, and thus avoids both wastage of sperm and enzymatic damage to the spermatic ducts. After ejaculation, however, fluids of the female reproductive tract leach cholesterol from the plasma membrane and dilute other inhibitory factors in the semen. The membrane of the sperm head becomes more fragile and more permeable to calcium ions, which diffuse into the sperm and stimulate more powerful lashing of the tail.

Sperm remain viable for up to 6 days after ejaculation, so there is little chance of pregnancy from intercourse occurring more than a week before ovulation. Fertilization also is unlikely if intercourse takes place more than 14 hours after ovulation, because the egg would no longer be viable by the time the sperm became capacitated. For those wishing to conceive a child, the optimal window of opportunity is therefore from a few days before ovulation to 14 hours after. Those wishing to avoid pregnancy, however, should allow a wider margin of safety for variations in sperm and egg longevity, capacitation time, and time of ovulation—variations that make the fertility-awareness based (rhythm) method of contraception so unreliable.

29.1b Fertilization

When the sperm encounters an egg, it undergoes an **acrosomal reaction**—exocytosis of the acrosome, releasing the penetration enzymes. But the first sperm to reach an egg isn't the one to fertilize it. Sperm must first penetrate the granulosa cells and zona pellucida that surround it **(fig. 29.1).** It may require numerous sperm to clear a path for the one that penetrates the egg proper.

Two of the acrosomal enzymes are **hyaluronidase,** which digests the hyaluronic acid that binds granulosa cells together, and **acrosin,** a protease similar to the trypsin of pancreatic juice. When a path has been cleared through the granulosa cells, a sperm

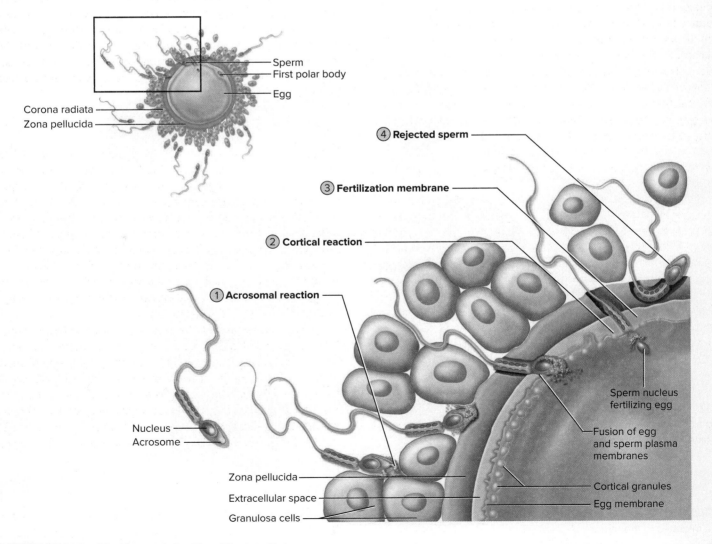

FIGURE 29.1 Fertilization and the Slow Block to Polyspermy.

binds to the zona pellucida and releases its enzymes, digesting a pathway through the zona until it contacts the egg itself. The sperm head and middle piece enter the egg, but the egg destroys the sperm mitochondria and passes only maternal mitochondria on to the offspring.

Fertilization combines the haploid *(n)* set of sperm chromosomes with the haploid set of egg chromosomes, producing a diploid *(2n)* set. Fertilization by two or more sperm, called **polyspermy,** would produce a triploid *(3n)* or larger set of chromosomes and the egg would die of a "gene overdose." Thus it is important for the egg to prevent this, and it has two mechanisms for doing so: a fast block and slow block to polyspermy. In the **fast block,** binding of the sperm to the egg opens Na^+ channels in the egg membrane. The rapid inflow of Na^+ depolarizes the membrane and inhibits the attachment of any more sperm. The **slow block** involves secretory vesicles called **cortical granules** just beneath the membrane. Sperm penetration releases an inflow of Ca^{2+}; this, in turn, stimulates a **cortical reaction** in which the cortical granules release their secretion beneath the zona pellucida. The secretion swells with water, pushes any remaining sperm away from the egg, and creates an impenetrable **fertilization membrane** between the egg and zona pellucida.

▶▶▶ **APPLY WHAT YOU KNOW**

What similarity can you see between the slow block to polyspermy and the release of acetylcholine from the synaptic vesicles of a neuron? (Compare "Synaptic Transmission," section 12.5d.)

A secondary oocyte begins meiosis II before ovulation and completes it only if fertilized. Through the formation of a second polar body, the fertilized egg discards one chromatid from each chromosome. The sperm and egg nuclei then swell and become **pronuclei.** The sperm centrosome sprouts microtubules and pushes the sperm pronucleus toward the center of the egg, aiding in the union of the male and female pronuclei. A mitotic spindle forms between the pronuclei, each of them ruptures, and the chromosomes of the two gametes mix into a single diploid set **(fig. 29.2).** The fertilized egg, now called a **zygote,** is ready for its first mitotic division.

As in section 28.5a, we will use the term *conceptus* for everything that arises from this zygote—not only the developing individual but also the placenta, umbilical cord, and membranes associated with the embryo and fetus.

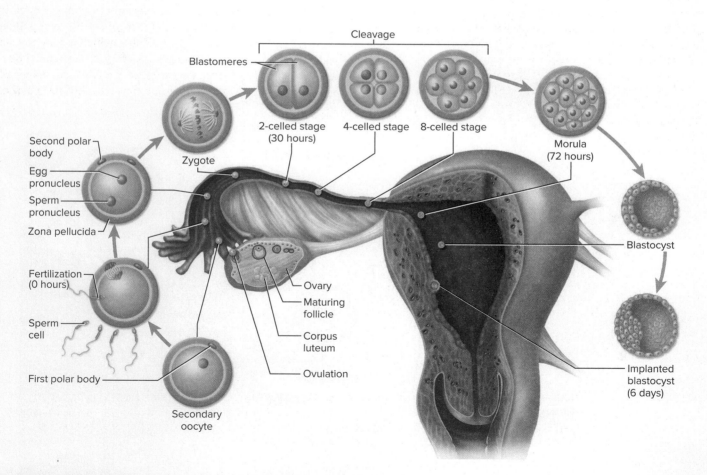

FIGURE 29.2 Migration of the Conceptus. The egg is fertilized in the distal end of the uterine tube, and the preembryo begins cleavage as it migrates to the uterus.

❓ *Why can't the egg be fertilized in the uterus?*

29.1c Major Stages of Prenatal Development

Clinically, the course of a pregnancy is divided into 3-month intervals called **trimesters:**

1. The **first trimester** extends from fertilization through the first 12 weeks. This is the most precarious stage, for more than half of all embryos die then. The conceptus is most vulnerable to stress, drugs, and nutritional deficiencies during this time, but may also die early because of chromosomal defects or other factors.

2. The **second trimester** (weeks 13 through 24) is a period in which the organs complete most of their development. It becomes possible with sonography to see good anatomical detail in the fetus. By the end of this trimester, the fetus looks distinctly human, and with intensive care, infants born at the end of the second trimester have a chance of survival.

3. In the **third trimester** (week 25 to birth), the fetus grows rapidly and the organs achieve enough cellular differentiation to support life outside the womb. Some organs, such as the brain, liver, and kidneys, however, require further differentiation after birth to become fully functional. At 35 weeks from fertilization, the fetus typically weighs about 2.5 kg (5.5 lb). It is considered mature at this weight, and usually survives if born early. Most twins are born at about 35 weeks and solitary infants at 40 weeks.

From a more biological than clinical standpoint, human development is divided into three stages called the *preembryonic, embryonic,* and *fetal stages.* The timetable and landmark events that distinguish them are outlined in **table 29.1** and described in the following pages.

29.1d The Preembryonic Stage

The **preembryonic stage** comprises the first 16 days of development, culminating in the existence of an embryo. This is a period in which the zygote divides into hundreds of cells, the cells organize themselves into the primary germ layers, and the conceptus becomes firmly attached to the uterine wall. It can be summarized in three words: *cleavage, implantation,* and *embryogenesis.*

Cleavage

Cleavage refers to mitotic divisions that occur in the first 3 days, while the conceptus migrates down the uterine tube (fig. 29.2). The first cleavage occurs about 30 hours after fertilization and produces the first two daughter cells, or **blastomeres.**[1] These divide simultaneously at shorter and shorter time intervals, doubling the number of blastomeres each time. By the time the conceptus arrives in the uterus, about 72 hours after ovulation, it consists of 16 or more cells and somewhat resembles a mulberry—hence, it is called a **morula.**[2] The morula is no larger than the zygote; cleavage merely produces smaller and smaller blastomeres and a larger number of cells from which to form different embryonic tissues. At times, the morula splits into two cell masses and produces identical twins (see Deeper Insight 29.1).

The morula lies free in the uterine cavity for 4 to 5 days and divides into 100 cells or so. Meanwhile, the zona pellucida disintegrates and releases the conceptus, which is now at a stage called the **blastocyst**—a hollow sphere with an outer layer of squamous cells called the **trophoblast,**[3] an inner cell mass called the **embryoblast,** and an internal cavity called the **blastocoel**

---

[1]*blast* = bud, precursor; *mer* = segment, part
[2]*mor* = mulberry; *ula* = little
[3]*troph* = food, nourishment

TABLE 29.1	The Stages of Prenatal Development	
Stage	**Age from Time of Fertilization**	**Major Developments and Defining Characteristics**
Preembryonic Stage		
Zygote	0–30 hours	A single diploid cell formed by the union of egg and sperm
Cleavage	30–72 hours	Mitotic division of the zygote into smaller, identical blastomeres
Morula	3–4 days	A spheroidal stage consisting of 16 or more blastomeres
Blastocyst	4–16 days	A fluid-filled, spheroidal stage with an outer mass of trophoblast cells and inner mass of embryoblast cells; becomes implanted in the endometrium; the inner cell mass forms an embryonic disc and differentiates into the three primary germ layers
Embryonic Stage	16 days–8 weeks	A stage in which the primary germ layers differentiate into organs and organ systems; ends when all organ systems are present
Fetal Stage	8–38 weeks	A stage in which organs grow and mature at a cellular level to the point of being capable of supporting life independently of the mother

DEEPER INSIGHT 29.1

CLINICAL APPLICATION

Twins

There are two ways in which twins are produced (and, by extension, other multiple births). About two-thirds of twins are *dizygotic (DZ)*—produced when two eggs are ovulated and fertilized by separate sperm. They are no more or less genetically similar than any other siblings and may be of different sexes. Multiple ovulation can also result in triplets, quadruplets, or even greater numbers of offspring. DZ twins implant separately on the uterine wall and each forms its own placenta **(fig. 29.3)**, although their placentas may fuse if they implant close together.

Monozygotic (MZ) twins are produced when a single egg is fertilized and, up to 14 days later, the cell mass divides into two at the morula or embryoblast stage. When the split is early in gestation, each twin usually forms its own amnion and placenta. If it comes late, they may share a single amnion and placenta. Such twins have a high mortality rate because they become entangled in each other's umbilical cords. MZ twins that split still later are at risk of developing as conjoined twins (fused at the head, chest, abdomen, or pelvis). MZ twins are genetically identical, or nearly so, and GE are therefore of the same sex and nearly identical appearance. Identical triplets and quadruplets occasionally result from the splitting of a single embryoblast.

Reproductive biologists are beginning to question whether MZ twins are truly genetically identical. They have suggested that blastomeres may undergo mutation in the course of DNA replication, and the splitting of the embryoblast may represent an attempt of each cell mass to reject the other one as genetically different and seemingly foreign.

FIGURE 29.3 Dizygotic Twins with Separate Placentas.
A three-dimensional sonogram.

Bernard Benoit/GE Healthcare/Science Source

(BLAST-oh-seal) **(fig. 29.4a).** The trophoblast is destined to form part of the placenta and play an important role in nourishment of the embryo, whereas the embryoblast is destined to become the embryo itself.

Implantation

About 6 days after ovulation, the blastocyst attaches to the endometrium, usually on the fundus or posterior wall of the uterus. The process of attachment, called **implantation,** begins when the blastocyst adheres to the endometrium. The trophoblast cells on this side separate into two layers. In the superficial layer, in contact with the endometrium, the plasma membranes break down and the trophoblast cells fuse into a multinucleate mass called the **syncytiotrophoblast**[4] (sin-SISH-ee-oh-TRO-fo-blast). (A *syncytium* is any body of cytoplasm containing multiple nuclei.) The deep layer, close to the embryoblast, is called the **cytotrophoblast** because it retains individual cells divided by membranes **(fig. 29.4b).**

The syncytiotrophoblast grows into the uterus like little roots, digesting endometrial cells along the way. The endometrium reacts to this injury by growing over the blastocyst and eventually covering it, so the conceptus becomes completely buried in endometrial tissue **(fig. 29.4c).** Implantation takes about a week and is completed about the time the next menstrual period would have occurred if the woman had not become pregnant.

Another role of the trophoblast is to secrete human chorionic gonadotropin (HCG). HCG stimulates the corpus luteum to secrete estrogen and progesterone, and progesterone suppresses menstruation. The level of HCG in the mother's blood rises until the end of the second month. During this time, the trophoblast develops into a membrane called the *chorion,* which takes over the role of the corpus luteum and makes HCG unnecessary. The ovaries then become inactive for the rest of the pregnancy, but estrogen and progesterone levels rise dramatically as they are secreted by the ever-growing chorion (see fig. 28.18).

Embryogenesis

During implantation, the embryoblast undergoes **embryogenesis**—arrangement of the blastomeres into the three primary germ layers: *ectoderm, mesoderm,* and *endoderm.* At the beginning of this phase, the embryoblast separates slightly from the trophoblast, creating a narrow space between them called the **amniotic cavity.** The embryoblast flattens into an **embryonic disc** composed initially of two layers: the *epiblast* facing the amniotic cavity and the *hypoblast* facing away. Some hypoblast cells multiply and form a membrane called the *yolk sac* enclosing the blastocoel. Now the embryonic disc is flanked by two spaces: the amniotic cavity on one side and the yolk sac on the other (fig. 29.4c).

Meanwhile, the embryonic disc elongates and, around day 15, a thickened cell layer called the **primitive streak** forms along the midline of the epiblast, with a **primitive groove** running down its middle **(fig. 29.5).** These events make the embryo bilaterally symmetric and define its future right and left sides, dorsal and ventral surfaces, and cephalic and caudal (head and tail) ends.

---

[4]*syn* = together; *cyt* = cell

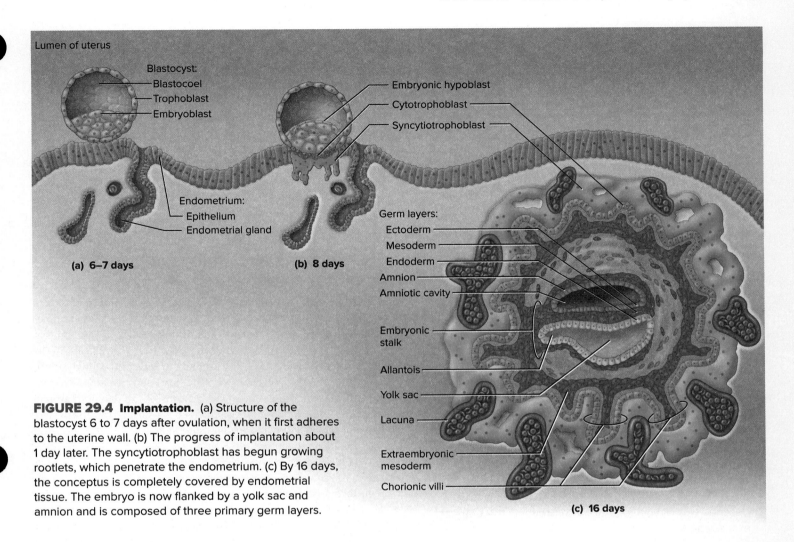

FIGURE 29.4 Implantation. (a) Structure of the blastocyst 6 to 7 days after ovulation, when it first adheres to the uterine wall. (b) The progress of implantation about 1 day later. The syncytiotrophoblast has begun growing rootlets, which penetrate the endometrium. (c) By 16 days, the conceptus is completely covered by endometrial tissue. The embryo is now flanked by a yolk sac and amnion and is composed of three primary germ layers.

 DEEPER INSIGHT 29.2

CLINICAL APPLICATION

Ectopic Pregnancy

In about 1 out of 300 pregnancies, the blastocyst implants somewhere other than the uterus, producing an *ectopic*[5] *pregnancy*. Most cases are *tubal pregnancies,* implantation in the uterine tube. This usually occurs because the conceptus encounters a constriction resulting from such causes as earlier pelvic inflammatory disease, tubal surgery, previous ectopic pregnancies, or repeated miscarriages. The uterine tube can't expand enough to accommodate the growing conceptus for long; if the situation isn't detected and treated early, the tube usually ruptures within 12 weeks, potentially with fatal hemorrhaging. Occasionally, a conceptus implants in the abdominopelvic cavity, producing an *abdominal pregnancy.* It can grow anywhere it finds an adequate blood supply—for example, on the broad ligament or the outside of the uterus, colon, or bladder. About 1 pregnancy in 7,000 is abdominal. This is a serious threat to the mother's life and usually requires therapeutic abortion, but about 9% of abdominal pregnancies result in live birth by cesarean section.

---

[5]*ec* = outside; *top* = place

The next step is **gastrulation**—multiplying epiblast cells migrate medially toward the primitive groove and down into it. They replace the original hypoblast with a layer now called **endoderm,** which will become the inner lining of the digestive tract among other things. A day later, migrating epiblast cells form a third layer between the first two, called **mesoderm.** Once this is formed, the remaining epiblast is called **ectoderm.** Thus, all three primary germ layers arise from the original epiblast. Some mesoderm overflows the embryonic disc and becomes an extensive *extraembryonic mesoderm,* which contributes to formation of the placenta (fig. 29.4c).

The ectoderm and endoderm are epithelia composed of tightly joined cells, but the mesoderm is a more loosely organized tissue. It later differentiates into skeletal muscle and a fetal connective tissue called **mesenchyme**—a loose network of wispy *mesenchymal cells* embedded in a gelatinous ground substance. Mesenchyme, in turn, differentiates into such tissues as smooth and cardiac muscle, cartilage, bone, and blood.

Once the three primary germ layers are formed, embryogenesis is complete and the individual is considered an embryo. It is about 2 mm long and 16 days old at this point.

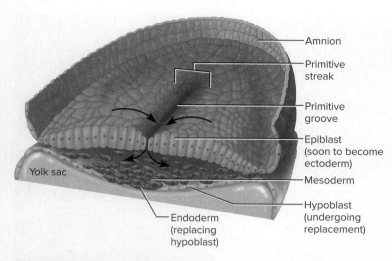

Amnion

Primitive streak

Primitive groove

Epiblast (soon to become ectoderm)

Mesoderm

Hypoblast (undergoing replacement)

Yolk sac

Endoderm (replacing hypoblast)

FIGURE 29.5 Formation of the Primary Germ Layers (Gastrulation). Composite view of the embryosnic disc at 15 to 16 days. Epiblast cells migrate over the surface and down into the primitive groove, first replacing the hypoblast cells with endoderm, then filling the space with mesoderm. Upon completion of this process, the uppermost layer is considered ectoderm.

BEFORE YOU GO ON

Answer the following questions to test your understanding of the preceding section:

1. How soon can a sperm reach an egg after ejaculation? How soon can it fertilize an egg? What accounts for the difference?

2. Describe two ways a fertilized egg prevents the entry of excess sperm.

3. In the blastocyst, what are the cells called that eventually give rise to the embryo? What are the cells that carry out implantation?

4. What major characteristic distinguishes an embryo from a preembryo?

29.2 The Embryonic and Fetal Stages

Expected Learning Outcomes

When you have completed this section, you should be able to

a. describe the formation and functions of the placenta;

b. explain how the conceptus is nourished before the placenta takes over this function;

c. describe the embryonic membranes and their functions;

d. identify the major tissues derived from the primary germ layers;

e. describe the major events of fetal development; and

f. describe the fetal circulatory system.

Sixteen days after conception, the germ layers are present and the **embryonic stage** of development begins. Over the next 6 weeks, a placenta forms on the uterine wall and becomes the embryo's primary means of nutrition, while the germ layers differentiate into organs and organ systems—a process called **organogenesis (table 29.2).** Although these organs are still far from functional, it is their presence at 8 weeks that marks the transition from the embryonic stage to the fetal stage. In the following pages, we will examine the transformation from embryo to fetus, how the membranes collectively known as the "afterbirth" develop around the fetus, and how the conceptus is nourished throughout its gestation.

29.2a Embryonic Folding and Organogenesis

One of the major transformations to occur in the embryonic stage is conversion of the flat embryonic disc of figure 29.5 into a somewhat cylindrical form. This occurs during week 4 as the embryo rapidly grows and folds around the yolk sac **(fig. 29.6).** As the cephalic and caudal ends curve around the ends of the yolk sac, the embryo becomes C-shaped, with the head and tail almost touching. At the same time, the lateral margins of the disc fold around the sides of the yolk sac to form the ventral surface of the embryo. This lateral folding encloses a longitudinal channel, the *primitive gut,* which later becomes the digestive tract.

As a result of embryonic folding, the entire surface is covered with ectoderm, which later produces the epidermis of the skin. In the meantime, the mesoderm splits into two layers. One of them adheres to the ectoderm and the other to the endoderm, opening a body cavity between them called the **coelom** (SEE-loam) (fig. 29.6c). The coelom divides into the thoracic cavity and peritoneal cavity separated by a wall, the diaphragm. By the end of week 5, the thoracic cavity further subdivides into pleural and pericardial cavities.

Two more especially significant events in organogenesis are the appearance of a **neural tube** (fig. 29.6b), which will later become the brain and spinal cord, and segmentation of the mesoderm into blocks of tissue called **somites** (fig. 29.7a, b). These will give rise to the vertebral column, trunk muscles, and dermis of the skin.

We cannot delve at greater length into development of all the organ systems, but this description is at least enough to see how some of them begin to form. Some have also been described in earlier chapters. Some highlights of prenatal development through the end of gestation are summarized in **table 29.3.** Prenatal growth is charted by weight and body length. Body length is customarily measured from the crown of the head to the curve of the buttocks (*crown-to-rump length, CRL),* thus excluding the lower limbs. **Figure 29.7** illustrates some key features of embryonic and fetal development. In the following discussion, we will see how these features unfold.

▶▶▶ **APPLY WHAT YOU KNOW**

List the four primary tissue types of the adult body (see section 5.1) and identify which of the three primary germ layers of the embryo predominantly gives rise to each.

TABLE 29.2	Derivatives of the Three Primary Germ Layers
Germ Layer	**Major Derivatives**
Ectoderm	Epidermis; hair follicles and arrector muscles; cutaneous glands; nervous system; adrenal medulla; pineal and pituitary glands; lens, cornea, and intrinsic muscles of the eye; internal and external ear; salivary glands; epithelia of nasal cavity, oral cavity, and anal canal
Mesoderm	Bones; bone marrow; cartilage; skeletal, cardiac, and most smooth muscle; adrenal cortex; middle ear; dermis; blood; blood and lymphatic vessels; lymphatic tissue; epithelium of kidneys, ureters, gonads, and genital ducts; mesothelium of abdominal and thoracic cavities
Endoderm	Most mucosal epithelium of digestive and respiratory tracts; mucosal epithelium of urinary bladder and parts of urethra; epithelial components of accessory reproductive and digestive glands (except salivary glands); thyroid and parathyroid glands; thymus

FIGURE 29.6 Embryonic Folding. The right-hand figures are cross sections cut about midway along the figures on the left. Part (a) corresponds to figure 29.4c at a slightly later stage of development. Note the general trend for the cephalic and caudal (head and tail) ends of the embryo to curl toward each other (left-hand figures) until the embryo assumes a C shape, and for the flanks of the embryo to fold laterally (right-hand figures), converting the flat embryonic disc into a more cylindrical body and eventually enclosing a body cavity (c).

FIGURE 29.7 The Developing Human. Parts (a) through (d) show development through the end of the embryonic stage. Parts (e) and (f) represent the fetal stage of development.

a: Source: Human Developmental Anatomy Center, National Museum of Health and Medicine, Silver Spring, MD, USA; b: Anatomical Travelogue/Science Source; c: ISM/Medical Images; d: Dr G. Moscoso/Science Source; e: John Watney/Science Source; f: Biophoto Associates/Science Source

(a) 3 weeks

Neural plate

Neural groove

Somites

Amnion (cut edge)

Primitive streak

0.1 cm

(b) 4 weeks

Future lens

Pharyngeal arches

Heart bulge

Arm bud

Tail

Leg bud

Somites

0.3 cm

(c) 7 weeks

Ear

Eye

Digital rays

Liver bulge

Umbilical cord

Foot plate

Tail

1.0 cm

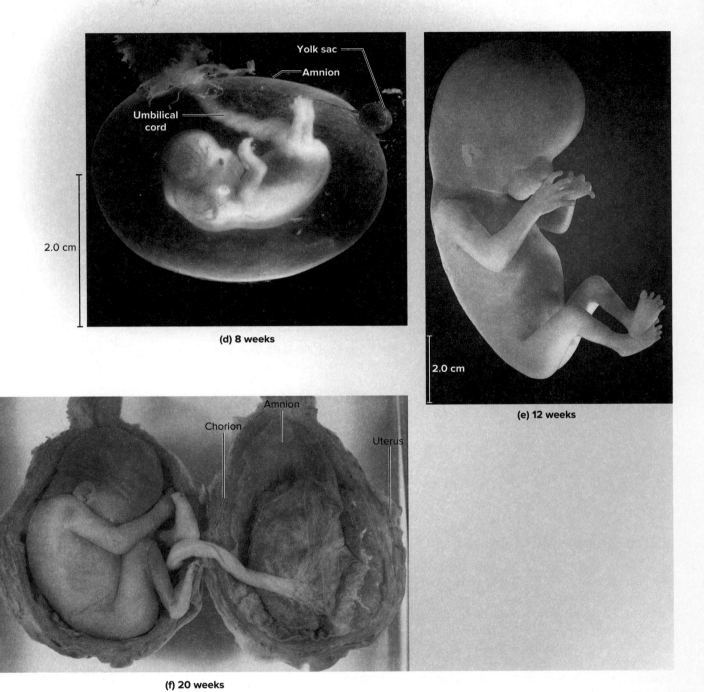

(d) 8 weeks

(e) 12 weeks

(f) 20 weeks

TABLE 29.3		Major Events of Prenatal Development, with Emphasis on the Fetal Stage
End of Week	**Crown-to-Rump Length; Weight**	**Developmental Events**
4	0.6 cm; <1 g	Vertebral column and central nervous system begin to form; limbs represented by small limb buds; heart begins beating around day 22; no visible eyes, nose, or ears
8	3 cm; 1 g	Eyes form, eyelids fused shut; nose flat, nostrils evident but plugged with mucus; head nearly as large as the rest of the body; brain waves detectable; bone calcification begins; limb buds form paddlelike hands and feet with ridges called **digital rays,** which then separate into distinct fingers and toes; blood cells and major blood vessels form; genitals present but sexes not yet distinguishable
12	9 cm; 45 g	Eyes well developed, facing laterally; eyelids still fused; nose develops bridge; external ears present; limbs well formed, digits exhibit nails; fetus swallows amniotic fluid and produces urine; fetus moves, but too weakly for mother to feel it; liver is prominent and produces bile; palate is fusing; sexes can be distinguished
16	14 cm; 200 g	Eyes face anteriorly, external ears stand out from head, face looks more distinctly human; body larger in proportion to head; skin is bright pink, scalp has hair; joints forming; lips exhibit sucking movements; kidneys well formed; digestive glands forming and **meconium**[6] (fetal feces) accumulating in intestine; heartbeat can be heard with a stethoscope
20	19 cm; 460 g	Body covered with fine **downy hair (lanugo**[7]**)** and cheeselike sebaceous secretion called **vernix caseosa,**[8] which protects it from amniotic fluid; skin bright pink; brown fat forms and will be used for postpartum heat production; fetus is now bent forward into "fetal position" because of crowding; **quickening** occurs—mother can feel fetal movements
24	23 cm; 820 g	Eyes partially open; skin wrinkled, pink, and translucent; lungs begin producing surfactant; rapid weight gain
28	27 cm; 1,300 g	Eyes fully open; skin wrinkled and red; full head of hair present; eyelashes formed; fetus turns into upside-down **vertex position;** testes begin to descend into scrotum; marginally viable if born at 28 weeks
32	30 cm; 2,100 g	Subcutaneous fat deposition gives fetus a more plump, babyish appearance, with lighter, less wrinkled skin; testes descending; twins usually born at this stage
36	34 cm; 2,900 g	More subcutaneous fat deposited, body plump; downy hair is shed; nails extend to fingertips; limbs flexed; firm hand grip
38	36 cm; 3,400 g	Prominent chest, protruding breasts; testes in inguinal canal or scrotum; fingernails extend beyond fingertips

29.2b Embryonic Membranes

Several accessory organs develop alongside the embryo: a *placenta; umbilical cord;* and four embryonic membranes called the *amnion, yolk sac, allantois,* and *chorion* (figs. 29.6, 29.7). To understand these membranes, it helps to realize that all mammals evolved from egg-laying reptiles. Within the shelled, self-contained egg of a reptile, the embryo rests atop a yolk, which is enclosed in the yolk sac; it floats in a little sea of liquid contained in the amnion; it stores its toxic wastes in the allantois; and to breathe, it has a chorion permeable to gases. All of these membranes persist in mammals, including humans, but are modified in their functions.

The **amnion** is a transparent sac that develops from cells of the epiblast (see figs. 29.4c, 29.5). It grows to completely enclose the embryo and is penetrated only by the umbilical cord. The amnion fills with **amniotic fluid** (fig. 29.7d), which protects the embryo

from trauma, infection, and temperature fluctuations; allows the freedom of movement important to muscle development; enables the embryo to develop symmetrically; prevents body parts from adhering to each other, such as an arm to the trunk; and stimulates lung development as the fetus "breathes" the fluid. At first, the amniotic fluid forms by filtration of the mother's blood plasma, but beginning at 8 to 9 weeks, the fetus urinates into the amniotic cavity about once an hour and contributes substantially to the fluid volume. The volume grows slowly, however, because the fetus swallows amniotic fluid at a comparable rate. At term, the amnion contains 700 to 1,000 mL of fluid.

The **yolk sac** arises from hypoblast cells opposite the amnion. It is a small sac suspended from the ventral side of the embryo. It contributes to the formation of the digestive tract and produces the first blood cells and forerunners of the future egg or sperm cells.

The **allantois** (ah-LON-toe-iss) begins as an outpocketing of the yolk sac (see fig. 29.4c), but eventually becomes an outgrowth of the caudal end of the gut. It forms the foundation for the umbilical cord and becomes part of the urinary bladder. It can be seen in cross sections of umbilical cord cut near the fetal end.

---

[6]*mecon* = poppy juice, opium; refers to an appearance similar to black tar opium
[7]*lan* = down, wool
[8]*vernix* = varnish; *caseo* = cheese

The **chorion** is the outermost membrane, enclosing all the rest of the membranes and the embryo (fig. 29.7f). Initially, it has shaggy outgrowths called **chorionic villi** around its entire surface, but as the pregnancy advances, the villi of the placental region grow and branch while the rest of them degenerate. At the placental attachment, the chorion is then called the *villous chorion,* and the rest is called the *smooth chorion.* The villous chorion forms the fetal portion of the placenta, discussed shortly.

29.2c Prenatal Nutrition

Over the course of gestation, the conceptus is nourished in three different, overlapping ways: by *uterine milk, trophoblastic nutrition,* and *placental nutrition.*

Uterine milk is a glycogen-rich secretion of the uterine tubes and endometrial glands. The conceptus absorbs this fluid as it travels down the tube and lies free in the uterine cavity before implantation. The accumulating fluid forms the blastocoel in figure 29.4a.

As it implants, the conceptus makes a transition to **trophoblastic nutrition,** in which it consumes so-called **decidual[9] cells** of the endometrium. Progesterone from the corpus luteum stimulates these cells to proliferate and accumulate a store of glycogen, proteins, and lipids. As the conceptus burrows into the endometrium, the syncytiotrophoblast digests the decidual cells and supplies the nutrients to the embryoblast. Trophoblastic nutrition is the only mode of nutrition for the first week after implantation. It remains the dominant source of nutrients through the end of week 8; the period from implantation through week 8 is therefore called the **trophoblastic phase** of the pregnancy. Trophoblastic nutrition wanes as placental nutrition takes over, and ceases entirely by the end of week 12 **(fig. 29.8).**

The **placenta[10]** is the fetus's life-support system—a disc-shaped organ attached to the uterine wall on one side, and on the other, attached to the fetus by way of the **umbilical cord (fig. 29.9).** It is the fetus's source of oxygen and nutrients, and its means of waste disposal. The diffusion of nutrients from the mother's blood through the placenta into the fetal blood is called **placental nutrition.**

The placenta begins to develop about 11 days after conception, becomes the dominant mode of nutrition around the beginning of week 9, and is the sole mode of nutrition from the end of week 12 until birth. The period from week 9 until birth is called the **placental phase** of the pregnancy.

Figure 29.4 depicts the early development of the placenta, or *placentation.* The process begins during implantation, as extensions of the syncytiotrophoblast penetrate more and more deeply into the endometrium, like the roots of a tree penetrating into the nourishing "soil" of the uterus. These roots are the early chorionic villi. As they penetrate uterine blood vessels, they become surrounded by *lacunae,* or endometrial spaces filled with maternal blood (see fig. 29.4c). The lacunae eventually merge to form a single blood-filled cavity, the **placental sinus.** Exposure to maternal blood stimulates increasingly rapid growth of the villi, which

FIGURE 29.8 The Timetable of Trophoblastic and Placental Nutrition. Trophoblastic nutrition peaks at 2 weeks and ends by 12 weeks. Placental nutrition begins at 2 weeks and becomes increasingly important until birth, 37 weeks after implantation. The two modes of nutrition overlap up to the twelfth week, but the *trophoblast phase* is the period in which most nutrients are supplied by trophoblastic nutrition, and the *placental phase* is the period in which most (eventually all) nutrition comes from the placenta.

? *At what point do the two modes contribute equally to prenatal nutrition?*

become branched and treelike. Extraembryonic mesoderm grows into the villi and gives rise to the blood vessels that connect to the embryo by way of the umbilical cord.

When fully developed, the placenta is about 20 cm in diameter, 3 cm thick, and weighs about one-sixth as much as the newborn infant. The surface attached to the uterine wall is rough and consists of chorionic villi embedded in the endometrium. The surface facing the fetus is smooth and gives rise to the umbilical cord (fig. 29.9).

The umbilical cord contains two **umbilical arteries** and one **umbilical vein.** Pumped by the fetal heart, blood flows into the placenta by way of the umbilical arteries and then returns to the fetus by way of the umbilical vein. The chorionic villi are *filled with* fetal blood and *surrounded by* maternal blood (fig. 29.9a); the two bloodstreams don't mix unless there is damage to the placental barrier. The barrier, however, is only 3.5 μm thick—half the diameter of a red blood cell. Early in development, the villi have thick membranes that are not very permeable to nutrients and wastes, and their total surface area is relatively small. As the villi grow and branch, their surface area increases and the membranes become thinner and more permeable. This brings about a dramatic increase in *placental conductivity,* the rate at which substances diffuse through the membrane. Materials diffuse from the side of the membrane where they are more concentrated to the side where they are less so. Therefore, oxygen and nutrients pass from the maternal blood to the fetal blood, while fetal wastes pass the other way to be eliminated by the mother. Unfortunately, the placenta is also permeable to nicotine, alcohol, and other drugs that may be present in the maternal bloodstream.

---

[9]*decid* = falling off
[10]*placenta* = flat cake

FIGURE 29.9 **The Placenta and Embryonic Membranes.**
(a) The placenta at 12 weeks of gestation, showing the relationship between fetal and maternal circulation and relationship of the placenta to the fetus and embryonic membranes attached to the uterine wall. (b) The fetal side of the placenta after birth, showing blood vessels, the umbilical cord, and some of the amniotic sac attached to the lower left margin. (c) The maternal (uterine) side, where chorionic villi give the placenta a rougher texture.

b, c: Dr. Kurt Benirschke

Table 29.4 summarizes the nutritional, excretory, and other functions of the placenta.

29.2d Fetal Development

By the end of 8 weeks, all of the organ systems are present, the individual is about 3 cm long, and it is now considered a **fetus** (see fig. 29.7d). The bones have just begun to calcify and the skeletal muscles exhibit spontaneous contractions, although these are too weak to be felt by the mother. The heart, beating since the fourth week, now circulates blood. The heart and liver are very large and form the prominent ventral bulge seen in figure 29.7c. The head is nearly half the total body length. The fetus is the final stage of prenatal development, from the start of the ninth week until birth. The organs that formed during the embryonic stage now undergo growth and cellular differentiation, acquiring the functional capability to support life outside the mother.

The circulatory system shows the most conspicuous anatomical changes from a prenatal state, dependent on the placenta, to the independent neonatal (newborn) state **(fig. 29.10).** The unique aspects of fetal circulation are the umbilical–placental circuit and the presence of three circulatory shortcuts called *shunts*. The internal iliac arteries give rise to the umbilical arteries, which pass on either side of the bladder into the umbilical cord. The blood in these arteries is low in oxygen and high in carbon dioxide and other fetal wastes; thus, they are depicted in blue in figure 29.10a. The arterial blood discharges its wastes in the placenta, loads oxygen and nutrients, and returns to the fetus by way of a single umbilical vein, which leads toward the liver. The umbilical vein is depicted in red because of its well-oxygenated blood. Some of this venous blood filters through the liver to nourish it. However, the immature liver is not capable of performing many of its postpartum functions, so it doesn't require a great deal of perfusion before birth. Most of the venous blood therefore bypasses it by way of a shunt called the **ductus venosus,** which leads directly to the inferior vena cava.

In the inferior vena cava, placental blood mixes with venous blood from the fetus's body and flows to the right atrium of the heart. After birth, the right ventricle pumps all of its blood into the lungs, but there is little need for this in the fetus because the lungs aren't yet functional. Therefore, most fetal blood bypasses the pulmonary circuit. Some goes directly from the right atrium to the left through the **foramen ovale,** a hole in the interatrial septum. Some also goes into the right ventricle and is pumped into the pulmonary trunk, but most of this is routed to the aorta by way of a short shunt called the **ductus arteriosus.** This occurs because the collapsed state of the fetal lungs creates high resistance and blood pressure in the pulmonary circuit, so blood in the pulmonary trunk takes a path of less resistance—through the ductus into the aorta, where the blood pressure is lower. The lungs receive only a trickle of blood, sufficient to meet their metabolic needs during development. Blood leaving the left ventricle enters the general systemic circulation, and some of this returns

TABLE 29.4	Functions of the Placenta
Nutritional roles	Permits nutrients such as glucose, amino acids, fatty acids, minerals, and vitamins to diffuse from the maternal blood to the fetal blood; stores nutrients such as carbohydrates, protein, iron, and calcium in early pregnancy and releases them to the fetus later, when daily fetal demand is greater than the mother can absorb from the diet
Excretory roles	Permits nitrogenous wastes such as ammonia, urea, uric acid, and creatinine to diffuse from the fetal blood to the maternal blood so the mother can excrete them
Respiratory roles	Permits O_2 to diffuse from mother to fetus and CO_2 from fetus to mother
Endocrine roles	Secretes estrogens, progesterone, relaxin, human chorionic gonadotropin, and human chorionic somatomammotropin; allows other hormones synthesized by the conceptus to pass into the mother's blood and maternal hormones to pass into the fetal blood
Immune roles	Transfers maternal antibodies (especially IgG) into the fetal blood to confer passive immunity on the fetus

to the placenta. This circulatory pattern changes dramatically at birth, when the neonate is cut off from the placenta and the lungs expand with air. Those changes will be described later.

Full-term fetuses have an average crown-to-rump length of about 36 cm (14 in.) and average weight of about 3.0 to 3.4 kg (6.6–7.5 lb). The fetus gains about 50% of its birth weight in the last 10 weeks.

Review table 29.3 and figure 29.7 for additional aspects of fetal development.

BEFORE YOU GO ON

Answer the following questions to test your understanding of the preceding section:

5. Distinguish between trophoblastic and placental nutrition.

6. Identify the two sources of blood to the placenta. Where do these two bloodstreams come closest to each other? What keeps them separated?

7. State the functions of the placenta, amnion, chorion, yolk sac, and allantois.

8. What developmental characteristic distinguishes a fetus from an embryo? At what gestational age is this attained?

9. Identify the three circulatory shunts of the fetus. Why does the blood take these "shortcuts" before birth?

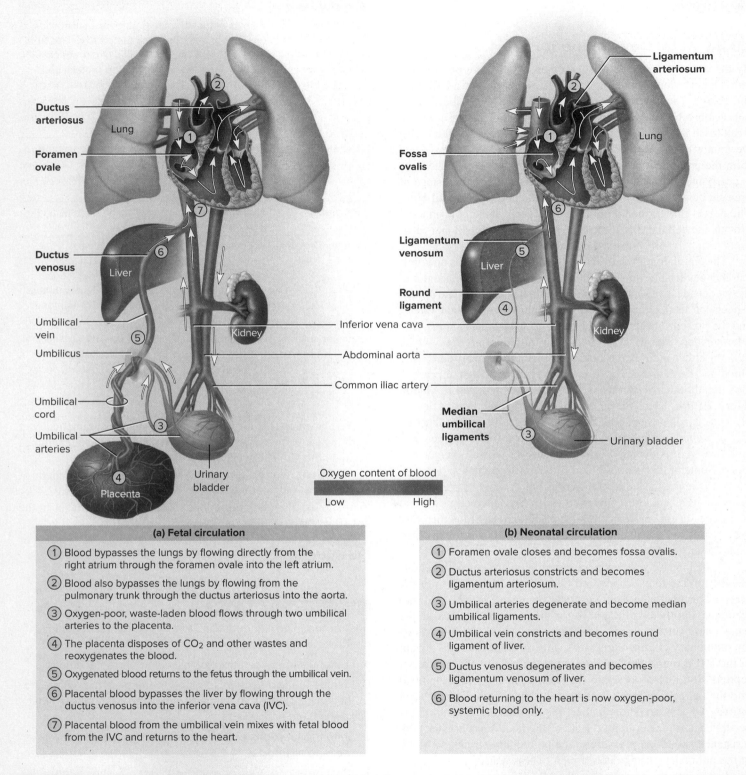

Ductus arteriosus

Lung

Foramen ovale

Ductus venosus

Liver

Umbilical vein

Umbilicus

Kidney

Umbilical cord

Umbilical arteries

Placenta

Urinary bladder

Ligamentum arteriosum

Lung

Fossa ovalis

Ligamentum venosum

Liver

Round ligament

Kidney

Inferior vena cava

Abdominal aorta

Common iliac artery

Median umbilical ligaments

Urinary bladder

Oxygen content of blood

Low   High

(a) Fetal circulation

1. Blood bypasses the lungs by flowing directly from the right atrium through the foramen ovale into the left atrium.

2. Blood also bypasses the lungs by flowing from the pulmonary trunk through the ductus arteriosus into the aorta.

3. Oxygen-poor, waste-laden blood flows through two umbilical arteries to the placenta.

4. The placenta disposes of CO_2 and other wastes and reoxygenates the blood.

5. Oxygenated blood returns to the fetus through the umbilical vein.

6. Placental blood bypasses the liver by flowing through the ductus venosus into the inferior vena cava (IVC).

7. Placental blood from the umbilical vein mixes with fetal blood from the IVC and returns to the heart.

(b) Neonatal circulation

1. Foramen ovale closes and becomes fossa ovalis.

2. Ductus arteriosus constricts and becomes ligamentum arteriosum.

3. Umbilical arteries degenerate and become median umbilical ligaments.

4. Umbilical vein constricts and becomes round ligament of liver.

5. Ductus venosus degenerates and becomes ligamentum venosum of liver.

6. Blood returning to the heart is now oxygen-poor, systemic blood only.

FIGURE 29.10 Blood Circulation in the Fetus and Newborn. Boldface terms in part (a) indicate the three shunts in the fetal circulation, which allow most blood to bypass the liver and lungs. Boldface terms in part (b) indicate the postpartum vestiges of fetal structures.

29.3 The Neonate

Expected Learning Outcomes

When you have completed this section, you should be able to

a. describe how and why the circulatory system changes at birth;

b. explain why the first breaths of air are relatively difficult for a neonate;

c. describe the major physiological problems of a premature infant; and

d. discuss some common causes of birth defects.

The period immediately following birth is a crisis in which the neonate suddenly must adapt to life outside the mother's body. The first 6 to 8 hours are a **transitional period** in which the heart and respiratory rates increase and the body temperature falls. Physical activity then declines and the baby sleeps for about 3 hours. In its second period of activity, the baby often gags on mucus and debris in the pharynx. The baby then sleeps again, becomes more stable, and begins a cycle of waking every 3 to 4 hours to feed. The first 4 weeks of life constitute the **neonatal period** and the infant at this stage is called a **neonate.**

29.3a Adapting to Life Outside the Uterus

The most dramatic and suspenseful event at birth is for the neonate to begin breathing on its own. Breathing, of course, is useless unless most or all of the blood circulates through the lungs to exchange gases there. Until birth, most blood has bypassed the lungs, so the onset of breathing also necessitates a radical change in blood circulation. The respiratory and circulatory systems therefore figure prominently in the neonate's adaptations to life outside the uterus, but other systems also must adapt to this new and challenging lifestyle.

Respiratory Adaptations

It is an old misconception that a neonate must be spanked to stimulate it to breathe. During birth, CO_2 accumulates in the baby's blood and strongly stimulates the respiratory chemoreceptors. Unless the infant is depressed by oversedation of the mother, it normally begins breathing spontaneously. It requires a great effort, however, to take the first few breaths and inflate the collapsed alveoli. For the first 2 weeks, a baby takes about 45 breaths per minute, but subsequently stabilizes at about 12 breaths per minute.

Circulatory Adaptations

When the lungs expand with air, resistance and blood pressure in the pulmonary circuit drop rapidly and pressure in the right heart falls below that in the left. Blood flows briefly from the left atrium to the right through the foramen ovale (opposite from its prenatal flow) and pushes two flaps of tissue into place to close this shunt. In most people, these flaps fuse and permanently seal the foramen during the first year, leaving a depression, the *fossa ovalis,* in the interatrial septum (see fig. 19.7). In about 25% of people, however, the foramen ovale remains unsealed and the flaps are held in place only by the relatively high blood pressure in the left atrium. Pressure changes in the pulmonary trunk and aorta also cause the ductus arteriosus to collapse. It closes permanently around 3 months of age and leaves a permanent cord, the *ligamentum arteriosum,* between the two vessels.

After the umbilical cord is clamped and cut, the umbilical arteries and vein collapse and become fibrotic. The proximal part of each umbilical artery becomes the *superior vesical artery,* which remains to supply the bladder. Other obliterated vessels become fibrous cords or ligaments: The distal parts of the umbilical arteries become the *median umbilical ligaments* of the abdominal wall; the umbilical vein becomes the *round ligament (ligamentum teres)* of the liver; and the ductus venosus (a former shunt around the liver) becomes the *ligamentum venosum* on the inferior surface of the liver (fig. 29.10b).

Immunologic Adaptations

Cellular immunity begins to appear early in fetal development, but the immune responses of the neonate are still weak. Fortunately, an infant is born with a near-adult level of IgG acquired from the mother through the placenta. This maternal IgG breaks down rapidly after birth, declining to about half the initial level in the first month and to essentially none by 10 months. Nevertheless, maternal IgG levels remain high enough for 6 months to protect the infant from measles, diphtheria, polio, and most other infectious diseases (but not whooping cough). By 6 months, the infant's own IgG reaches about half the typical adult level. The lowest total (maternal + infant) level of IgG exists around 5 to 6 months of age, and respiratory infections are especially common at that age. The IgA in the colostrum also gives a breast-fed neonate immune protection against gastroenteritis.

Other Adaptations

Thermoregulation and fluid balance are also critical aspects of neonatal physiology. An infant has a larger ratio of surface area to volume than an adult does, so it loses heat more easily. One of its defenses against hypothermia is brown fat, a special adipose tissue deposited from weeks 17 to 20 of fetal development. The mitochondria of brown fat release all the energy of fatty acids as heat rather than using it to make ATP; thus, this is a heat-generating tissue. As a baby grows, its metabolic rate increases and it accumulates even more subcutaneous fat, thus producing and retaining more heat. Nevertheless, body temperature is more variable in infants and children than in adults.

The kidneys aren't fully developed at birth and can't concentrate the urine as much as a mature kidney can. Consequently, infants have a relatively high rate of water loss and require more fluid intake, relative to body weight, than adults do.

In addition, the liver is still not fully functional at birth, most joints aren't yet ossified, and myelination of the nervous system isn't completed until adolescence. Indeed, humans are born in a very immature state compared with other mammals—a fact necessitated by the narrow outlet of the female pelvis, which in turn was a product of the evolution of bipedal locomotion (see Deeper Insight 8.5).

CLINICAL APPLICATION

Neonatal Assessment

A newborn infant is immediately evaluated for general appearance, vital signs (temperature, pulse, and respiratory rate), weight, length, and head circumference and other dimensions; and it is screened for congenital disorders such as phenylketonuria (PKU). At 1 minute and 5 minutes after birth, the heart rate, respiratory effort, muscle tone, reflexes, and skin color are noted and given a score of 0 (poor), 1, or 2 (excellent). The total (0–10), called the Apgar[11] score, is a good predictor of infant survival. Infants with low Apgar scores may have neurological damage and need immediate attention if they are to survive. A low score at 1 minute suggests asphyxiation and may demand assisted ventilation. A low score at 5 minutes indicates a high probability of death.

29.3b Premature Infants

Infants born before the start of week 37 are generally considered **premature.** They have multiple difficulties in respiration, thermoregulation, excretion, digestion, and liver function. Neonates born at that age average about 2.6 kg (5.8 lb). Most neonates weighing 1.5 to 2.5 kg are viable, but with difficulty. Those weighing under 500 g rarely survive.

The respiratory system is adequately developed by 7 months of gestation to support independent life. Infants born before this have a deficiency of pulmonary surfactant, causing **infant respiratory distress syndrome (IRDS),** also called *hyaline membrane disease.* The alveoli collapse each time the infant exhales, and a great effort is needed to reinflate them. The infant becomes very fatigued by the high energy demand of breathing. IRDS may be treated by ventilating the lungs with oxygen-enriched air at a positive pressure to keep the lungs inflated between breaths, and by administering surfactant as an inhalant. Nevertheless, IRDS remains the most common cause of neonatal death.

A premature infant has an incompletely developed hypothalamus and therefore cannot thermoregulate effectively. Body temperature must be controlled by placing the infant in a warmer.

It is difficult for premature infants to ingest milk because of their small stomach volume and underdeveloped sucking and swallowing reflexes. Some must be fed by nasogastric or nasoduodenal tubes. Most of them, however, can tolerate human milk or formulas. Infants under 1.5 kg (3.3 lb) require nutritional supplements of calcium, phosphorus, and protein.

The liver is also poorly developed, and bearing in mind its very diverse functions (see table 26.6), you can probably understand why this would have several serious consequences. The liver synthesizes inadequate amounts of albumin, so the baby suffers hypoproteinemia. This upsets the balance between capillary filtration and reabsorption and leads to edema. The infant bleeds easily because of a deficiency of the clotting factors synthesized by the liver. This is true to some degree even in full-term infants, however,

because the baby's intestines are not yet colonized by the bacteria that synthesize vitamin K, which is essential for the synthesis of clotting factors. Vitamin K injections are now routine (but somewhat controversial) for newborns in the United States. Jaundice is common in neonates, especially premature babies, because the liver can't dispose of bile pigments such as bilirubin efficiently. A moderately elevated bilirubin level *(neonatal hyperbilirubinemia)* is normal and desirable in all infants, however. Bilirubin plays a valuable role as an antioxidant until the infant can develop its other antioxidant systems.

29.3c Birth Defects

A birth defect, or **congenital anomaly,**[12] is the abnormal structure or position of an organ at birth, resulting from a defect in prenatal development. The study of birth defects is called **teratology.**[13] Birth defects are the most common cause of infant mortality in North America. Not all of them are noticeable at birth; some are detected months to years later. Thus, by the age of 2 years, 6% of children are diagnosed with congenital anomalies, and by age 5 the incidence is 8% because of those that have made their presence known in the interim. The following sections discuss some known causes of congenital anomalies, but in 50% to 60% of cases, the cause is unknown.

Teratogens

Teratogens[14] are agents that cause anatomical deformities in the fetus. They fall into three major classes: drugs and other chemicals, infectious diseases, and radiation such as X-rays. The effect of a teratogen depends on the genetic susceptibility of the embryo, the dosage of the teratogen, and the time of exposure. Teratogen exposure during the first 2 weeks usually doesn't cause birth defects, mainly because the affected embryo doesn't survive to birth; most abort. Teratogens can exert destructive effects at any stage of development, but the period of greatest vulnerability is weeks 3 through 8. Different organs have different critical periods. For example, limb abnormalities are most likely to result from teratogen exposure at 24 to 36 days, and brain abnormalities from exposure at 3 to 16 weeks.

Perhaps the most notorious teratogenic drug is thalidomide, a sedative first marketed in West Germany in 1957. Thalidomide was taken by many women in early pregnancy to relieve morning sickness, and by others as a sleeping aid even before they knew they were pregnant. By the time it was removed from the market in 1961, it had affected an estimated 10,000 to 20,000 babies worldwide, many of them born with unformed arms or legs **(fig. 29.11)** and often with defects of the ears, heart, and intestines. The U.S. Food and Drug Administration never approved thalidomide for market, but many American women obtained it from foreign sources or by participation in clinical drug trials. Thalidomide has recently been reintroduced and used under more tightly controlled conditions for more limited purposes such as treating leprosy.

---

[11]Virginia Apgar (1909–74), American anesthesiologist

[12]*con* = with; *gen* = born; *a* = without; *nomaly* = evenness, regularity
[13]*terato* = monster; *logy* = study of
[14]*terato* = monster; *gen* = producing

FIGURE 29.11 Schoolboy Showing the Effect of Thalidomide on Upper Limb Development.

Time & Life Pictures/Getty Images

People in some Third World countries still take it in a misguided attempt to treat AIDS and other diseases, resulting in an upswing in severe birth defects. A general lesson to be learned from the thalidomide tragedy and other cases is that pregnant women should avoid all sedatives, barbiturates, and opiates. Even the acne medicine isotretinoin (Accutane) has caused severe birth defects. Many teratogens produce less obvious or delayed effects, including physical or mental impairment, inattention, hyperirritability, strokes, seizures, respiratory arrest, crib death, and cancer.

Alcohol causes more birth defects than any other drug. A pregnant woman taking even one drink a day can noticeably affect fetal and childhood development, with some effects not noticed until a child begins school. Alcohol abuse during pregnancy can cause **fetal alcohol syndrome (FAS),** characterized by a small head, malformed facial features, cardiac and central nervous system defects, stunted growth, and behavioral signs such as hyperactivity, nervousness, and a poor attention span. Cigarette smoking also contributes to fetal and infant mortality, ectopic pregnancy, anencephaly (failure of the cerebrum to develop), cleft lip and palate, and cardiac abnormalities. Diagnostic X-rays should be avoided during pregnancy because radiation can have teratogenic effects.

Infectious diseases are largely beyond the scope of this book, but it must be noted at least briefly that several microorganisms can cross the placenta and cause serious congenital anomalies, stillbirth, or neonatal death. Common viral infections of the fetus and newborn include herpes simplex, rubella, cytomegalovirus, and human immunodeficiency virus (HIV). Many infants were born in 2016, especially in Brazil, with *microcephaly* (abnormally small

heads) attributed to the mosquito-borne Zika virus. Congenital bacterial infections include gonorrhea and syphilis. *Toxoplasma,* a protozoan contracted from meat, unpasteurized milk, and house cats, is another common cause of fetal deformity. Some of these pathogens have relatively mild effects on adults, but because of its immature immune system, the fetus is vulnerable to devastating effects such as blindness, hydrocephalus, cerebral palsy, seizures, and profound physical and mental retardation. These diseases are treated in greater detail in microbiology textbooks.

Mutagens and Genetic Anomalies

Genetic anomalies are the most common known cause of birth defects, accounting for an estimated one-third of all cases and 85% of those with an identifiable cause. One cause of genetic defects is **mutations,** or changes in DNA structure. Among other disorders, mutations cause achondroplastic dwarfism (see Deeper Insight 7.2), cleft lip and palate (see Deeper Insight 29.4),

---

 DEEPER INSIGHT 29.4

CLINICAL APPLICATION

Cleft Lip and Palate

Cleft lip and palate (CLP) is one of the most common congenital malformations, occurring in 1 or 2 out of 1,000 live births. Cleft palate (*palatoschizis,* PAL-ah-to-SKIZ-is) results from failure of the fetal maxillae to join on the midline, leaving a median fissure in the roof of the mouth between the oral and nasal cavities **(fig. 29.12).** It is often accompanied by an upper lip cleft (*cheiloschizis,* KY-lo-SKIZ-is) on one or both sides. A cleft palate makes it difficult for an infant to generate the suction needed for nursing, and may be accompanied by frequent ear infections, abnormal tooth development, and difficulties in hearing and speech. In recent years, at least 10 genes on 8 chromosomes have been identified that cause CLP, but it can also have environmental causes such as tobacco, alcohol, and corticosteroid use during pregnancy. Cleft lip and palate can be corrected by maxillofacial surgery with good cosmetic results, but may require follow-up speech therapy and dental surgery, often placing a heavy long-term burden on patients and families. Correction by 18 months of age can improve language acquisition and avoid the psychosocial problems often faced by school-age children with CLP.

FIGURE 29.12 An Infant with Cleft Lip and Palate. (a) Cleft lip. (b) Cleft palate. Note the fissure between the oral and nasal cavities, which creates problems for nursing infants in suckling and swallowing.

microcephaly, stillbirth, and childhood cancer. Mutations can occur through errors in DNA replication during the cell cycle or under the influence of environmental agents called **mutagens,** including some chemicals, viruses, and radiation.

Some of the most common genetic disorders result not from mutagens, but from the failure of homologous chromosomes to separate during meiosis. Recall that homologous chromosomes pair up during prophase I and normally separate from each other at anaphase I (see section 27.4a). This separation, called *disjunction,* produces daughter cells with 23 chromosomes each.

In **nondisjunction,** a pair of chromosomes fails to separate. Both chromosomes go to the same daughter cell, which receives 24 chromosomes while the other cell receives 22. **Aneuploidy**[15] (AN-you-PLOY-dee), the presence of an extra chromosome or lack of one, is the most common cause of miscarriages, accounting for about 50% of cases. The lack of a chromosome, leaving one chromosome without a match, is called **monosomy,** whereas the presence of one extra chromosome, producing a triple set, is called **trisomy.** Aneuploidy can be detected prior to birth by **amniocentesis,** the examination of cells in a sample of amniotic fluid, or by **chorionic villus sampling (CVS),** the biopsy of cells from the chorion.

Figure 29.13 compares normal disjunction of the X chromosomes with some effects of nondisjunction. In nondisjunction, an egg may receive both X chromosomes. If it is fertilized by an X-bearing sperm, the result is an XXX zygote and a set of anomalies called the **triplo-X syndrome.** Some triplo-X females are infertile, have mild intellectual impairments, or both. If an XX egg is fertilized by a Y-bearing sperm, the result is an XXY combination and **Klinefelter**[16] **syndrome.** People with Klinefelter syndrome are sterile males, usually of average intelligence, but with undeveloped testes, sparse body hair, unusually long arms and legs, and enlarged breasts (*gynecomastia*[17]). This syndrome often goes undetected until puberty, when failure to develop the secondary sex characteristics may prompt genetic testing.

The other possible outcome of X chromosome nondisjunction is that an egg may receive no X chromosome (both X chromosomes are discarded in the first polar body). If fertilized by a Y-bearing sperm, it dies for lack of the indispensable genes on the X chromosome. If it is fertilized by an X-bearing sperm, however, the result

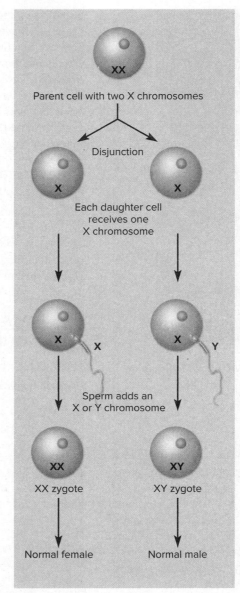

(a) Normal disjunction of X chromosomes

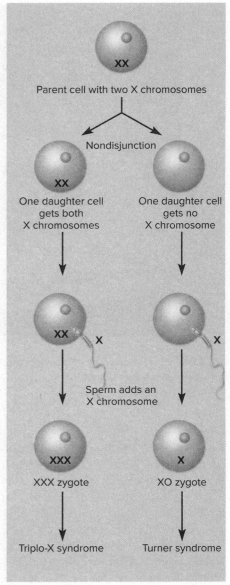

(b) Nondisjunction of X chromosomes

FIGURE 29.13 Disjunction and Nondisjunction. (a) The outcome of normal disjunction and fertilization by X- or Y-bearing sperm. (b) Two of the possible outcomes of nondisjunction followed by fertilization with an X-bearing sperm.

❓ *In the right half of the figure, what would the two outcomes be if the sperm carried a Y chromosome instead of an X?*

is a female with **Turner**[18] **syndrome,** with an XO combination (O represents the absence of one sex chromosome). About 97% of fetuses with Turner syndrome die before birth. Survivors show no serious impairments as children, but tend to have a webbed neck and widely spaced nipples. At puberty, secondary sex characteristics fail to develop (**fig. 29.14**). The ovaries are nearly absent, the girl remains sterile, and she usually has a short stature.

---

[15]*an* = not, without; *eu* = true, normal; *ploid,* from *diplo* = double, paired

[16]Harry F. Klinefelter Jr. (1912–90), American physician

[17]*gyneco* = female; *mast* = breast; *ia* = condition

[18]Henry H. Turner (1892–1970), American endocrinologist

FIGURE 29.14 Turner Syndrome. A 22-year-old woman with Turner syndrome, with an XO karyotype (see figure 29.13, far right). Note her short stature (about 145 cm, or 4 ft 9 in.), lack of sexual development, webbed neck, and widely spaced nipples.

Courtesy Mihaly Bartalos

intellectual development; short stature; a relatively flat face with a flat nasal bridge; low-set ears; *epicanthal folds* at the medial corners of the eyes; an enlarged, protruding tongue; stubby fingers; and a short broad hand with only one palmar crease (**fig. 29.15**). People with Down syndrome tend to have outgoing, affectionate personalities. Cognitive impairment is common and sometimes severe, but doesn't always occur.

About 75% of fetuses with trisomy-21 die before birth, and 20% of infants born with it die before the age of 10 from such causes as immune deficiency and abnormalities of the heart or kidneys. For those who survive beyond that age, modern medical care has extended life expectancy to the 60s. After the age of 40, however, about 75% of these people develop early-onset Alzheimer disease, linked to a gene on chromosome 21.

Aneuploidy is far more common in humans than in any other species, and 90% of cases are of maternal rather than paternal origin. These facts seem to result from the extraordinarily long time it takes for human oocytes to complete meiosis—as long as 50 years. For various reasons, including defects in the mitotic spindle and in chromosomal crossing-over, aging eggs become less and less able to separate their chromosomes into two identical sets. This is evident in the statistics of Down syndrome: The chance of having a child with Down syndrome is about 1 in 3,000 for a woman under 30, 1 in 365 by age 35, and 1 in 9 by age 48.

BEFORE YOU GO ON

Answer the following questions to test your understanding of the preceding section.

10. How does inflation of the lungs at birth affect the route of blood flow through the heart?

11. Why is respiratory distress syndrome common in premature infants?

12. Define *nondisjunction* and explain how it causes aneuploidy. Name two syndromes resulting from aneuploidy.

The other 22 pairs of chromosomes (the *autosomes*) are also subject to nondisjunction. Only three autosomal trisomies are survivable: those involving chromosomes 13, 18, and 21. The reason is that these three chromosomes are relatively gene-poor. In all other autosomal cases, trisomy gives the embryo a lethal "overdose" of genes. Even the nonlethal trisomies are the leading genetic cause of mental and developmental abnormalities.

Nondisjunction of chromosomes 13 and 18 results in *Patau syndrome (trisomy-13)* and *Edward syndrome (trisomy-18),* respectively. Nearly all fetuses with these trisomies die before birth. Infants born with these syndromes are severely deformed, and fewer than 5% survive for 1 year.

The most survivable trisomy, and therefore the most common among children and adults, is **Down**[19] **syndrome (trisomy-21).** Down syndrome occurs in about 1 out of 700 to 800 live births in the United States. Its signs include impaired physical and

---

29.4 Aging and Senescence

Expected Learning Outcomes

When you have completed this section, you should be able to

a. define *senescence* and distinguish it from aging;

b. describe some major changes that occur with aging in each organ system;

c. summarize some current theories of senescence; and

d. be able to explain how exercise and other factors can slow the rate of senescence.

Like Ponce de León searching for the legendary fountain of youth in Florida, people yearn for a way to preserve their youthful appearance and function. Our real concern, however, is not aging but senescence. The term **aging** is used in various ways but is taken here to mean all changes that occur in the body with the passage of time—including the growth, development, and increasing

---

[19]John Langdon H. Down (1828–96), British physician

FIGURE 29.15 Down Syndrome. (a) A child with Down syndrome (center) and her sisters, who are unaffected by the trisomy. (b) The karyotype (chromosomal chart) of a male with Down syndrome, showing the trisomy of chromosome 21 (oval with arrow). (c) Characteristics of the hand in Down syndrome. (d) The epicanthal fold over the medial commissure (canthus) of the left eye.

a: Denys Kuvaiev/Alamy; b: Zuzana Egertova/Alamy Stock Photo

❓ *What was the sex of the person from whom the karyotype in part (b) was obtained?*

functional efficiency that occur from childhood to adulthood, as well as the degenerative changes that occur later in life. **Senescence** is the degeneration that occurs in an organ system after the age of peak functional efficiency. It includes a gradual loss of reserve capacities, reduced ability to repair damage and compensate for stress, and increased susceptibility to disease.

Senescence is not just a personal concern but an important issue for health-care providers. In the early twentieth century, most Americans died before age 65, often from infectious diseases. Now, about three-quarters of deaths in the United States are at ages 65 and older, and one in seven living Americans is in that age bracket. As the average age of the population rises, health-care professionals find themselves increasingly occupied by the prevention and treatment of the diseases of age. The leading causes of death change markedly with age. Accidents, homicide, suicide,

and AIDS figure prominently in the deaths of people 18 to 34 years old, whereas the major causes of death after age 55 are clearly related to senescence of the organ systems: in descending order, heart disease, cancer, stroke diabetes, chronic obstructive pulmonary disease (COPD), and infectious lung diseases (pneumonia and influenza). The causes of senescence, however, remain as much a scientific mystery today as cancer was 50 years ago and heredity was 100 years ago.

As we survey the senescence of the organ systems, you should notice many points relevant not only to caring for an aging population but also to personal health and fitness practices that can lessen the effects of senescence and improve the quality of life in your own later years. In addition, the study of senescence calls for renewed attention to the multiple interactions among organ systems. As you will see, the senescence of one organ system typically contributes to the senescence of others. Your study of this topic will bring together many concepts introduced in earlier chapters of the book. You may find the glossary helpful in refreshing your memory of concepts revisited in the following discussion.

29.4a Senescence of the Organ Systems

Organ systems don't all degenerate at the same rate. For example, from ages 30 to 80, the speed of nerve conduction declines only 10% to 15%, but the number of functional glomeruli in the kidneys declines about 60%. Some physiological functions show only moderate changes at rest but more pronounced differences when tested under exercise conditions. The organ systems also vary widely in the age at which senescence becomes noticeable. There are forerunners of atherosclerosis, for example, even in infants; visual and auditory sensitivities begin to decline soon after puberty. By contrast, the female reproductive system doesn't show significant senescence until menopause and then its decline is relatively abrupt. Aside from these unusual examples, most physiological measures of performance peak between the late teens and age 30, then decline at a rate influenced by the level of use of the organs.

Integumentary System

Two-thirds of people age 50 and over, and nearly all people over age 70, have medical concerns or complaints about their skin. Senescence of the integumentary system often becomes noticeable by the late 40s. The hair turns grayer and thinner as melanocytes die out, mitosis slows down, and dead hairs are not replaced. The atrophy of sebaceous glands leaves the skin and hair drier. As epidermal mitosis declines and collagen is lost from the dermis, the skin becomes almost paper-thin and translucent. It becomes looser because of a loss of elastic fibers and flattening of the dermal papillae, which normally form a stress-resistant corrugated boundary between the dermis and epidermis. If you pinch a fold of skin on the back of a child's hand, it quickly springs back when you let go; do the same on an older person and the skinfold remains longer. Because of its loss of elasticity, aged skin sags to various degrees and may hang loosely from the arm and other places.

Aged skin has fewer blood vessels than younger skin, and those that remain are more fragile. The skin can become reddened as broken vessels leak into the connective tissue. Many older people exhibit **rosacea**—patchy networks of tiny, dilated blood vessels visible especially on the nose and cheeks. Because of the fragility of the dermal blood vessels, aged skin bruises more easily. Injuries to the skin are more common and severe in old age, partly because the cutaneous nerve endings decline by two-thirds from age 20 to age 80, leaving one less aware of touch, pressure, and injurious stimuli. Injured skin heals slowly in old age because of poorer circulation and a relative scarcity of immune cells and fibroblasts. Antigen-presenting dendritic cells decline by as much as 40% in the aged epidermis, leaving the skin more susceptible to recurring infections.

Thermoregulation can be a serious problem in old age because of the atrophy of cutaneous blood vessels, sweat glands, and subcutaneous fat. Older people are more vulnerable to hypothermia in cold weather and heatstroke in hot weather. Heat waves and cold spells take a disproportionate toll among elderly poor individuals who suffer from a combination of reduced homeostasis and inadequate housing.

These are all "normal" changes in the skin, or **intrinsic aging**—changes that occur more or less inevitably with the passage of time. In addition, there is **photoaging**—degenerative changes in proportion to a person's lifetime exposure to ultraviolet radiation. UV radiation accounts for more than 90% of the integumentary changes that people find medically troubling or cosmetically disagreeable: skin cancer; yellowing and mottling of the skin; age spots *(solar lentigines)* (len-TIJ-ih-neez), which resemble enlarged freckles on the back of the hand and other sun-exposed areas; and wrinkling, which affects the face, hands, and arms more than areas of the body that receive less exposure. A lifetime of outdoor activity can give the skin a leathery, deeply wrinkled, "outdoorsy" appearance **(fig. 29.16),** but beneath this rugged exterior is a less happy histological appearance. The sun-damaged skin shows many malignant and premalignant cells, extensive damage to the dermal blood vessels, and dense masses of coarse, frayed elastic fibers underlying the surface wrinkles and creases.

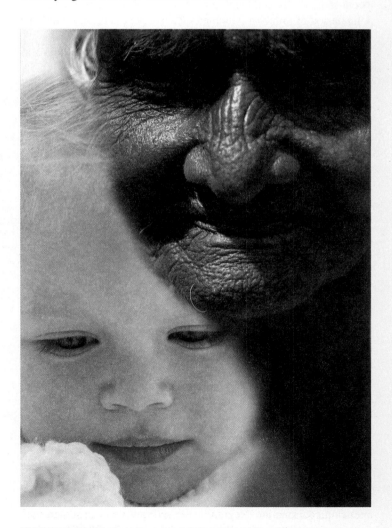

FIGURE 29.16 Senescence of the Skin. The skin exhibits both intrinsic aging and photoaging. The deep creases seen in the old woman's face result mainly from photoaging.

(young child): Stephen Flint/Alamy; **(old person):** Mark Girvan Portraits/Alamy

Senescence of the skin has far-reaching effects on other organ systems. Cutaneous vitamin D production declines as much as 75% in old age. This is all the more significant because the elderly spend less time outdoors; and because of increasing lactose intolerance, they often avoid dairy products, the only dietary source of vitamin D. Consequently, the elderly are at high risk of calcium deficiency, which, in turn, contributes to bone loss, muscle weakness, and impaired glandular secretion and synaptic transmission.

Skeletal System

After age 30, osteoblasts become less active than osteoclasts. This imbalance results in **osteopenia,** the loss of bone; when the loss is severe enough to compromise a person's physical activity and health, it is called *osteoporosis* (see Deeper Insight 7.4). After age 40, women lose about 8% of their bone mass per decade and men about 3%. Bone loss from the jaws is a contributing factor in tooth loss.

Not only does bone density decline with age, but the bones become more brittle as the cells synthesize less protein. Fractures occur more easily and heal more slowly. A fracture may impose a long period of immobility, which makes a person more vulnerable to pneumonia and other life-threatening infectious diseases.

People notice more stiffness and pain in the synovial joints as they age, and degenerative joint diseases affect the lifestyle of 85% of people over age 75. Synovial fluid is less abundant and the articular cartilage is thinner or absent. Exposed bone surfaces abrade each other and cause friction, pain, and reduced mobility. *Osteoarthritis* is the most common joint disease of older people and one of the most common causes of physical disability (see Deeper Insight 9.5). Even breathing becomes more difficult and tiring in old age because expansion of the thorax is restricted by calcification of the sternocostal joints. Degeneration of the intervertebral discs causes back pain and stiffness, but herniated discs are less common in old age than in youth because the discs become more fibrous and stronger, with less nucleus pulposus.

Muscular System

One of the most noticeable changes we experience with age is the replacement of lean body mass (muscle) with fat. The change is dramatically exemplified by CT scans of the thigh. In a young well-conditioned male, muscle accounts for 90% of the cross-sectional area of the midthigh, whereas in a frail 90-year-old woman, it is only 30%. Muscular strength and mass peak in the 20s; by the age of 80, most people have only half as much strength and endurance. Many people over age 75 can't lift a 4.5 kg (10 lb) weight with their arms; such simple tasks as carrying a sack of groceries into the house may become impossible. The loss of strength is a major contributor to falls, fractures, and dependence on others for the routine activities of daily living. Fast-twitch fibers exhibit the earliest and most severe atrophy, thus increasing reaction time and reducing coordination. Regular exercise, however, can remarkably preserve and restore muscle size and strength **(fig. 29.17).**

There are multiple reasons for the loss of strength. Aged muscle fibers have fewer myofibrils, so they're smaller and weaker. The sarcomeres are increasingly disorganized, and muscle

mitochondria are smaller and have reduced quantities of oxidative enzymes. Aged muscle has less ATP, creatine phosphate, glycogen, and myoglobin; consequently, it fatigues quickly. Muscles also exhibit more fat and fibrosis with age, which limits their movement and blood circulation. With reduced circulation, muscle injuries heal more slowly and with more scar tissue.

But the weakness and easy fatigue of aged muscle also stem from the senescence of other organ systems. There are fewer motor neurons in the spinal cord, and some muscle shrinkage may represent denervation atrophy. The remaining neurons produce less acetylcholine and show less efficient synaptic transmission, which makes the muscles slower to respond to stimulation. As muscle atrophies, motor units have fewer muscle fibers per motor neuron, and more motor units must be recruited to perform a given task. Tasks that used to be easy, such as buttoning the clothes or eating a meal, take more time and effort. The sympathetic nervous system is also less efficient in old age; consequently, blood flow to the muscles doesn't respond efficiently to exercise and this contributes to their rapid fatigue.

Nervous System

The nervous system reaches its peak development around age 30. After age 40, the brain loses about 5% of its mass every 10 years,

Quadriceps femoris · Subcutaneous fat · Adductor muscles · Gracilis · Sciatic nerve · Femur · Hamstring muscles

(a)

(b)

FIGURE 29.17 Muscle Senescence and Its Prevention.
Drawings based on CT scans of cross sections of the thighs.
(a) Thighs of a 74-year-old triathlete, showing how well muscle mass can be maintained in old age with exercise. (b) Thighs of a sedentary 74-year-old, showing extensive muscle atrophy and replacement by subcutaneous fat.

and possibly faster after age 70. The cerebral gyri are narrower, the sulci are wider, the cortex is thinner, and there is more space between the brain and meninges. The remaining cortical neurons have fewer synapses, and for multiple reasons, synaptic transmission is less efficient: The neurons produce less neurotransmitter, they have fewer receptors, and the neuroglia around the synapses is more leaky and allows neurotransmitter to diffuse away. The degeneration of myelin sheaths with age also slows down nerve conduction. The losses are not uniform throughout the brain, but fastest in the prefrontal cortex—the area concerned with abstract thought, judgment, planning, personality, and explicit memory— and in the hippocampus, the seat of memory formation.

Neurons exhibit less rough ER and Golgi complex with age, which indicates that their metabolism is slowing down. Old neurons accumulate lipofuscin pigment and show more neurofibrillary tangles—dense mats of cytoskeletal elements in their cytoplasm. In the extracellular material, plaques of fibrillar protein (amyloid) appear, especially in people with Down syndrome and Alzheimer disease (AD). AD is the most common nervous disability of old age (see Deeper Insight 12.4).

Not all functions of the central nervous system are equally affected by senescence. Motor coordination, intellectual function, and short-term memory decline more than language skills and long-term memory. Elderly people are often better at remembering things in the distant past than remembering recent events.

The sympathetic nervous system loses adrenergic receptors with age and becomes less sensitive to norepinephrine. This contributes to a decline in homeostatic control of such variables as body temperature and blood pressure. Many elderly people experience *orthostatic hypotension*—a drop in blood pressure when they stand, which sometimes results in dizziness, loss of balance, or fainting. Impaired balance and neuromuscular reflexes, slow baroreflexes, and fragile bones often combine tragically in a fall that breaks a hip and leads to death from pneumonia and other complications of immobility.

Sense Organs

Some sensory functions decline shortly after adolescence. Presbyopia (loss of flexibility in the lenses) makes it more difficult for the eyes to focus on nearby objects. Visual acuity declines and often requires corrective lenses at a young age. Cataracts (cloudiness of the lenses) are more common in old age. Night vision is impaired as more and more light is needed to stimulate the retina. This has several causes: There are fewer receptor cells in the retina, the lens and vitreous body becomes less transparent, and the pupil becomes narrower as the pupillary dilators atrophy. Even without cataracts, the retina receives only one-third as much light at age 60 as it does at age 20. Dark adaptation takes longer as the enzymatic reactions of the photoreceptor cells become slower. Changes in the structure of the iris, ciliary body, or lens can block the reabsorption of aqueous humor, thereby increasing the risk of glaucoma. Having to give up reading and driving can be among the most difficult changes of lifestyle in old age.

Auditory sensitivity peaks in adolescence and declines afterward. The tympanic membrane and the joints between the auditory ossicles become stiffer, so vibrations are transferred less effectively

to the inner ear, creating a degree of conductive deafness. Nerve deafness occurs as the number of cochlear hair cells and auditory nerve fibers declines. The greatest hearing loss occurs at high frequencies and in the frequency range of most conversation. The death of receptor cells in the semicircular ducts, utricle, and saccule, and of nerve fibers in the vestibular nerve and neurons in the cerebellum, results in dizziness and impaired balance—another factor in falls and bone fractures.

The senses of taste and smell are blunted as the taste buds, olfactory cells, and second-order neurons in the olfactory bulbs decline in number. Food may lose its appeal, and declining sensory function can therefore be a factor in malnutrition.

Endocrine System

The endocrine system degenerates less than any other organ system. The reproductive hormones drop sharply and growth hormone and thyroid hormone levels decline steadily after adolescence. The pineal gland becomes increasingly calcified in old age and its melatonin output falls. Since melatonin is a sleep-inducing hormone, this may be a reason for the insomnia often experienced by the elderly. Other hormones continue to be secreted at fairly stable levels even into old age. Target-cell sensitivity declines, however, so some hormones have less effect. For example, the pituitary gland is less sensitive to negative feedback inhibition by adrenal glucocorticoids; consequently, the response to stress is more prolonged than usual. Diabetes mellitus is more common in old age, largely because target cells have fewer insulin receptors. In part, this is an effect of the greater percentage of body fat in the elderly. The more fat at any age, the less sensitive other cells are to insulin. Body fat increases as the muscles atrophy, and muscle is one of the body's most significant glucose-buffering tissues. Because of the blunted insulin response, glucose levels remain elevated longer than normal after a meal.

Circulatory System

Cardiovascular disease is a leading cause of death in old age. Senescence has multiple effects on the blood, heart, arteries, and veins. Anemia may result from nutritional deficiencies, inadequate exercise, disease, and other causes. The factors that cause anemia in older people are so complicated it's almost impossible to control them enough to determine whether aging alone causes it. Evidence suggests that there is no change in the baseline rate of erythropoiesis in old age. Hemoglobin concentration, cell counts, and other variables are about the same among healthy people in their 70s as in the 30s. However, older people don't adapt well to stress on the hematopoietic system, perhaps because of the senescence of other organ systems. As the gastric mucosa atrophies, for example, it produces less of the intrinsic factor needed for vitamin B_{12} absorption. This increases the risk of anemia. As the kidneys age and the number of nephrons declines, less erythropoietin is secreted. There may also be a limit to how many times the hematopoietic stem cells can divide and continue giving rise to new blood cells. Whatever its cause, anemia limits the amount of oxygen that can be transported and thus contributes to the atrophy of tissues everywhere in the body.

▶▶▶ **APPLY WHAT YOU KNOW**

Draw a positive feedback loop showing how anemia and senescence of the kidneys could affect each other.

Nearly everyone is affected to some degree in old age by arteriosclerosis and atherosclerosis (see the distinction in section 20.2a). Coronary atherosclerosis leads to the degeneration of myocardial tissue; angina pectoris and myocardial infarction become more common; the heart wall becomes thinner and weaker; and stroke volume, cardiac output, and cardiac reserve decline. Like other connective tissues, the fibrous skeleton of the heart becomes less elastic. This limits cardiac distension and reduces the force of systole. Degenerative changes in the SA and AV nodes and conduction pathways of the heart lead to a higher incidence of cardiac arrhythmia and heart block. Physical endurance is compromised by the drop in cardiac output.

Arteries stiffened by arteriosclerosis cannot expand as effectively to accommodate the pressure surges of cardiac systole. Blood pressure therefore rises steadily with age, although it is uncertain to what extent this is an inherent biological effect of aging versus an effect of culture and lifestyle. It doesn't occur in all societies. Atherosclerosis also narrows the arteries and reduces the perfusion of most organs. The effects of reduced circulation on the skin, skeletal muscles, and brain have already been noted. The combination of atherosclerosis and hypertension weakens the arteries and increases the risk of aneurysm and stroke.

Atherosclerotic plaques trigger thrombosis, especially in the lower limbs, where flow is relatively slow and the blood clots more easily. About 25% of people over age 50 experience venous blockage by thrombosis—especially people who don't exercise regularly.

Degenerative changes in the veins are most evident in the limbs. The valves become weaker and less able to stop the backflow of blood. Blood pools in the legs and feet, raises capillary blood pressure, and causes edema. Chronic stretching of the vessels often produces varicose veins and hemorrhoids. Support hose can reduce edema by compressing the tissues and forcing tissue fluid to return to the bloodstream, but physical activity is even more important in promoting venous return.

Immune System

The amounts of lymphatic tissue and red bone marrow decline with age; consequently, there are fewer hematopoietic stem cells, disease-fighting leukocytes, and antigen-presenting cells (APCs). Also, the lymphocytes produced in these tissues often fail to mature and become immunocompetent. Both humoral and cellular immunity depend on APCs and helper T cells, and therefore both types of immune response are blunted. As a result, an older person is less protected against cancer and infectious diseases. It becomes especially important in old age to be vaccinated against influenza and other acute seasonal infections.

Respiratory System

Pulmonary ventilation declines steadily after the 20s and is one of several factors in the gradual loss of stamina. The costal cartilages and joints of the thoracic cage become less flexible, the lungs have less elastic tissue, and the lungs have fewer alveoli. Vital capacity, minute respiratory volume, and forced expiratory volume fall. The elderly are also less capable of clearing the lungs of irritants and pathogens and are therefore increasingly vulnerable to respiratory infections. Pneumonia causes more deaths than any other infectious disease and is often contracted in hospitals and nursing homes.

The chronic obstructive pulmonary diseases (COPDs)—emphysema and chronic bronchitis—are more common in old age since they represent the cumulative effects of a lifetime of degenerative change. They are among the leading causes of death in old age, but are avoidable; they are much less common among nonsmokers than among smokers. Pulmonary obstruction also contributes to cardiovascular disease, hypoxemia, and hypoxic degeneration in all the organ systems. Respiratory health is therefore a major concern in aging.

Urinary System

The kidneys exhibit a striking degree of atrophy with age. From ages 25 to 85, the number of nephrons declines 30% to 40% and up to a third of the remaining glomeruli become atherosclerotic, bloodless, and nonfunctional. The kidneys of a 90-year-old are 20% to 40% smaller than those of a 30-year-old and receive only half as much blood. The glomerular filtration rate is proportionately lower and the kidneys are less efficient at clearing wastes from the blood. Although baseline renal function is adequate even in old age, there is little reserve capacity; thus, other diseases can lead to surprisingly rapid renal failure. Drug doses often need to be reduced in old age because the kidneys can't clear drugs from the blood as rapidly; this is a contributing factor in overmedication among the aged.

Water balance is more precarious in old age because the kidneys are less responsive to antidiuretic hormone and because the sense of thirst is sharply reduced. Even when given free access to water, many elderly people don't drink enough to maintain normal blood osmolarity. Dehydration is therefore common.

Voiding and bladder control become problematic for both men and women. About 80% of men over the age of 80 are affected by benign prostatic hyperplasia. The enlarged prostate compresses the urethra and interferes with emptying of the bladder. Urine retention may cause pressure to back up in the kidneys, aggravating the failure of the nephrons. Older women are subject to incontinence (leakage of urine), especially if their history of pregnancy and childbearing has weakened the pelvic muscles and urethral sphincter. Senescence of the sympathetic nervous system and nervous disorders such as stroke and Alzheimer disease can also cause incontinence.

Digestive System and Nutrition

Restaurants commonly offer a "seniors' menu" and reduced prices to customers over 60 or 65 years old, knowing that they tend to eat less. There are multiple reasons for this reduced appetite. Older people have lower metabolic rates and tend to be less active than younger people and, hence, need fewer calories. The stomach atrophies with age, and it takes less to fill it up. For many, food has less esthetic appeal in old age because of losses in the senses of smell, taste, and even vision. In addition, older people secrete less saliva, making food less flavorful and swallowing more difficult. Reduced salivation also makes the teeth more prone to caries. Dentures are an unpleasant fact of life for many people over 65 who have lost their teeth to caries and periodontitis. Atrophy of the epithelium of the oral cavity and esophagus makes these surfaces more subject to abrasion and may further detract from the ease of chewing and swallowing. Even the ability to chew declines; the muscles of mastication lose about 40% of their mass and the mandibular bone loses about 20% over the course of a lifetime.

The reduced mobility of old age makes shopping and meal preparation more troublesome, and with food losing its sensory appeal, some decide that it simply isn't worth the trouble. However, one's protein, vitamin, and mineral requirements remain essentially unchanged, so vitamin and mineral supplements may be needed to compensate for reduced food intake and poorer intestinal absorption. Malnutrition is common among older people and is an important factor in anemia and reduced immunity.

As the gastric mucosa atrophies, it secretes less acid and intrinsic factor. Acid deficiency reduces the absorption of calcium, iron, zinc, and folic acid. Heartburn becomes more common as the weakening lower esophageal sphincter fails to prevent the reflux of stomach contents into the esophagus. The most common digestive complaint of older people is constipation, which results from reduced muscle tone and weaker peristalsis of the colon. This seems to stem from a combination of factors: atrophy of the muscularis externa, reduced sensitivity to neurotransmitters that promote motility, less fiber and water in the diet, and less exercise. The liver, gallbladder, and pancreas show only slightly reduced function, but the drop in liver function reduces the rate of drug deactivation and can contribute to overmedication.

Reproductive System

In men, the senescent changes in the reproductive system are relatively gradual; they include declining testosterone secretion, sperm count, semen volume, and libido. By age 65, sperm count is about one-third of what it was in a man's 20s. Men remain fertile (capable of fathering a child) well into old age, but impotence (inability to maintain an erection) can occur because of atherosclerosis, hypertension, medication, and psychological reasons.

In women, the changes are more pronounced and develop more rapidly. Over the course of menopause, the ovarian follicles are used up, gametogenesis ceases, and the ovaries stop producing sex steroids. This may result in vaginal dryness, genital atrophy, and reduced libido and may make sex less enjoyable. With the loss of ovarian steroids, a postmenopausal woman has an elevated risk of osteoporosis and atherosclerosis.

29.4b Exercise, Senescence, and Life Expectancy

Other than the mere passage of time, senescence results from obesity and insufficient exercise more than from any other causes.

Conversely, good nutrition and exercise are the best ways to slow its progress.

People who exercise moderately for as little as 10 to 15 minutes a day have been found to live, on average, 3 years longer than sedentary people, although it is difficult to separate causation from correlation in such studies. That is, it remains unclear how much of this is an *effect* of the exercise and how much results from *correlation* with other healthy lifestyle choices. There is no question, however, that regular exercise maintains endurance, strength, and joint mobility, and reduces the incidence and severity of hypertension, osteoporosis, obesity, and diabetes mellitus. This is especially true if you begin a program of regular physical exercise early in life and make a lasting habit of it. If you stop exercising regularly after middle age, the body rapidly becomes deconditioned, although appreciable reconditioning can be achieved even when an exercise program is begun late in life. A person in his or her 90s can increase muscle strength two- or threefold in 6 months with as little as 40 minutes of isometric exercise a week. The improvement results from a combination of muscle hypertrophy and neural efficiency.

Resistance exercises may be the most effective way of reducing accidental injuries such as bone fractures, whereas endurance exercises reduce body fat and increase cardiac output and maximum oxygen uptake. A general guideline for ideal endurance training is to have three to five periods of aerobic exercise per week, each 20 to 60 minutes long and vigorous enough to reach 60% to 90% of your maximum heart rate. The maximum is best determined by a stress test but averages about 220 beats per minute minus one's age in years.

An exercise program should ideally be preceded by a complete physical examination and stress test. Warm-up and cool-down periods are especially important in avoiding soft tissue injuries and undue cardiovascular stress. Because of their lower capacity for thermoregulation, older people must be careful not to overdo exercise, especially in hot weather. At the outset of a new exercise program, it is best to "start low and go slow."

29.4c Theories of Senescence

Why do our organs wear out? Why must we die? There still is no general theory on this. The question actually comes down to two issues: (1) What are the mechanisms that cause the organs to deteriorate with age? (2) Why has natural selection not eliminated these and produced bodies capable of longer life?

Mechanisms of Senescence

Numerous hypotheses have been proposed and discarded to explain why organ function degenerates with age. Some authorities maintain that senescence is an intrinsic process governed by inevitable or even preprogrammed changes in cellular function. Others attribute senescence to extrinsic (environmental) factors that progressively damage our cells over the course of a lifetime.

There is good evidence of a hereditary component to longevity. Unusually long and short lives tend to run in families. Monozygotic (identical) twins are more likely than dizygotic

FIGURE 29.18 Progeria. This is a genetic disorder in which senescence appears to be greatly accelerated. The individuals here, from left to right, are 15, 12, and 26 years old. Few people with progeria live as long as the woman on the right.

Bettmann/Getty Images

twins to die at a similar age. One striking genetic defect called *progeria*[20] is characterized by greatly accelerated senescence **(fig. 29.18).** Symptoms begin to appear by age 2. The child's growth rate declines, the muscles and skin become flaccid, most victims lose their hair, and most die in early adolescence from advanced atherosclerosis. In Werner syndrome, caused by a defective gene on chromosome 8, people show marked senescence beginning in their 20s and usually die by age 50. There is some controversy over the relevance or similarity of these syndromes to normal senescence, but they do demonstrate that many of the changes associated with old age can be brought on by a genetic anomaly.

Knowing that senescence is partially genetic, however, doesn't answer the question about why tissues degenerate. Quite likely, no one theory explains all forms of senescence, but let's briefly examine some of them.

Replicative Senescence Normal organ function usually depends on a rate of cell renewal that keeps pace with cell death. There is a limit, however, to how many times cells can divide. Human cells cultured in the laboratory divide 80 to 90 times if taken from a fetus, but only 20 to 30 times if taken from older people. After reaching their maximum number of divisions, cultured cells degenerate and die. This decline in mitotic potential with age is called **replicative senescence.**

Why this occurs is a subject of lively research. Much of the evidence points to the **telomere,**[21] a "cap" on each end of a chromosome analogous to the plastic tip of a shoelace. In humans, it consists of a noncoding nucleotide sequence CCCTAA repeated 1,000 times or more. One of its functions may be to stabilize the chromosome and prevent it from unraveling or sticking to other chromosomes. Telomeres also prevent a cell from falsely recognizing the ends of a chromosome as broken and activating DNA repair mechanisms that would do more harm than good. Also, during DNA replication, DNA polymerase can't reproduce the very ends of the DNA molecule. If there were functional genes at the end, they wouldn't get duplicated. The telomere may therefore provide a bit of "disposable" DNA at the end, so that DNA polymerase doesn't fail to replicate genes that would otherwise be there. Every time DNA is replicated, 50 to 100 bases are lost from the telomere. In old age, the telomere may be exhausted and the polymerase may then indeed fail to replicate some of the terminal genes. Old chromosomes may therefore be more vulnerable to damage, replication errors, or both, causing old cells to be increasingly dysfunctional. The long life of stem cells and "immortality" of cancer cells results from an enzyme called *telomerase,* lacking from healthy differentiated cells. Telomerase enables cancerous and stem cells to repair telomere damage and escape the limit on number of cell divisions.

DNA Damage Theory Telomere damage and replicative senescence are clearly not the entire answer to why organs degenerate. Skeletal muscle fibers and brain neurons exhibit pronounced senescence, yet these cells are nonmitotic. Another leading theory of senescence is unrepaired DNA damage. DNA suffers an astonishing 10,000 to 100,000 damaging events per day in the average mammalian cell. Most of it is fixed by DNA repair enzymes, but this is not 100% efficient. Some damage persists and accumulates as cells age, especially nondividing cells like neurons, cardiomyocytes, and skeletal muscle fibers. Such cumulative damage has been shown to cause age-related decline in liver and kidney function; cardiac and muscular strength; and brain functions related to neuronal survival, synaptic plasticity, learning, and memory.

Cross-Linking Theory About one-fourth of the body's protein is collagen. With age, collagen molecules become cross-linked by more and more disulfide bridges, thus making the fibers less soluble and more stiff. This is thought to be a factor in several of the most noticeable changes of the aging body, including stiffening of the joints, lenses, and arteries. Similar cross-linking of DNA and enzyme molecules could progressively impair their functions as well.

Other Protein Abnormalities Not only collagen but also many other proteins exhibit increasingly abnormal structure in older

---

[20]*pro* = before; *ger* = old age

[21]*telo* = end; *mer* = piece, part, segment

tissues and cells. The changes are not in amino acid sequence—therefore not attributable to DNA mutations—but lie in the way the proteins are folded and other moieties such as carbohydrates are attached to them. This is another reason that cells accumulate more dysfunctional proteins as they age.

Autoimmune Theory Some of the altered macromolecules described previously may be recognized as foreign antigens and stimulate lymphocytes to mount an immune response against the body's own tissues. Autoimmune diseases do, in fact, become more common in old age.

Evolution and Senescence

If certain genes contribute to senescence, it raises an evolutionary question: Why doesn't natural selection eliminate them? In an attempt to answer this, biologists once postulated that senescence and death were for the good of the species—a way for older, worn-out individuals to make way for younger, healthier ones. We can see the importance of death for the human population by imagining that science had put an end to senescence and people died at an annual rate of only 1 per 1,000 regardless of age (the rate at which American 18-year-olds now die). If so, the median age of the population would be 163, and 13% of us would live to be 2,000 years old. The implications for world population, social order, and competition for resources would be staggering. Thus, it is easy to understand why death was once interpreted as a self-sacrificing phenomenon for the good of the species.

But this hypothesis has several weaknesses. One of them is the fact that natural selection works exclusively through the effects of genes on the reproductive rates of individuals. A species evolves only because some members reproduce more than others. A gene that doesn't affect reproductive rate can be neither eliminated nor favored by natural selection. Genes for disorders such as Alzheimer disease have little or no effect until a person is past reproductive age. Our prehistoric and even fairly recent ancestors usually died of accidents, predation, starvation, weather, and infectious diseases at an early age. Few people lived long enough to be affected by atherosclerosis, colon cancer, or Alzheimer disease. Natural selection would have been "blind" to such death-dealing genes, which would escape the selection process and remain with us today.

29.4d Death

Life expectancy, the average length of life in a given population, has steadily increased in industrialized countries. People born in the United States at the beginning of the twentieth century had a life expectancy of only 45 to 50 years; nearly half of them died of infectious disease. As of 2017, by contrast, the average boy born in the United States that year could expect to live 76.1 years and the average girl 81.1 years. This trend is due mostly to victories over infant and child mortality, not to advances at the other end of the life span. **Life span,** the maximum age attainable by humans, increased significantly from 1950 to 2010. As yet, however, there is no verifiable record of anyone living past the age of 122 years—Jeanne Calment (1875–1997) of Arles, France, whose fascinating story you can read on the Internet.

There is no definable instant of biological death. Some organs function for an hour or more after the heart stops beating. During this time, even if a person is declared legally dead, living organs may be removed for transplantation. For legal purposes, death was once defined as the loss of a spontaneous heartbeat and respiration. Now that cardiopulmonary functions can be artificially maintained for years, this criterion is less distinct. Clinical death is now widely defined in terms of **brain death**—a lack of cerebral activity indicated by a flat electroencephalogram for 30 minutes to 24 hours (depending on state laws), accompanied by a lack of reflexes or lack of spontaneous respiration and heartbeat.

Death usually results from the failure of a particular organ, which then has a cascading effect on other organs. Kidney failure, for example, leads to the accumulation of toxic wastes in the blood, which in turn leads to loss of consciousness, brain function, respiration, and heartbeat.

Ninety-nine percent of us will die before age 100, and there is little chance that this outlook will change within our lifetimes. We cannot presently foresee any "cure for old age" or significant extension of the human life span. The real issue is to maintain the best possible quality of life, and when the time comes to die, to be able to do so in comfort and dignity.

BEFORE YOU GO ON

Answer the following questions to test your understanding of the preceding section.

13. Define *aging* and *senescence*.

14. List some tissues and organs in which changes in collagenous and elastic connective tissues contribute to senescence.

15. Many older people have difficulty with mobility and self-maintenance tasks such as dressing and cooking. Name some organ systems whose senescence is most relevant to these limitations.

16. Explain why both endurance and resistance exercise are important in old age.

17. Summarize five cellular or biochemical mechanisms that may contribute to senescence.

DEEPER INSIGHT 29.5

CLINICAL APPLICATION

Assisted Reproductive Technology

Fertile heterosexual couples who have frequent intercourse and use no contraception have a 90% chance of conceiving within 1 year. About one in six American couples, however, is *infertile*—unable to conceive. For this and other reasons, individuals and couples often resort to clinical assistance to achieve parenthood. Reasons include avoidance of genetic defects when a parent is known to be a carrier of a harmful allele; male or female infertility; a history of repeated miscarriages; desire to store eggs or sperm for delayed parenthood or to preserve healthy gametes before undergoing chemotherapy, ovarian or testicular cancer surgery, or other procedures; and the desire for children among single persons and same-sex couples. In the United States alone, there are more than 140,000 trials and 55,000 births per year by the methods to be described here. *Assisted reproductive technology (ART)* is defined by the U.S. Public Health Service and World Health Organization as techniques in which both eggs and sperm are manipulated.

Intrauterine Insemination

Intrauterine insemination (IUI) was formerly called artificial insemination (AI). It technically lies outside the definition of ART, since it involves only sperm manipulation, but it is the oldest and least expensive clinical means of overcoming male infertility to achieve parenthood. If a man's infertility is only a matter of low sperm output (oligospermia[22]), he can achieve biological fatherhood by a process of pooling and storing sperm from multiple ejaculates, washing them to remove seminal fluid (which can inflame the female uterus), then introducing them into the uterus with a cervical catheter. If he produces no sperm at all (aspermia) or is at risk of transmitting a genetic defect, sperm from other donors is used. Men planning vasectomies sometimes donate sperm for storage as insurance against the death of a child, divorce and remarriage, or a change in family planning. AI was first performed in 1790, and for more than a century, the sperm donors were often the physicians themselves or medical students who donated sperm for pay. Cryogenic (frozen) storage of semen in liquid nitrogen was developed in 1953, techniques of insemination refined, and IUI became much more common. The first commercial sperm banks opened in 1970. Most women undergoing IUI use sperm from anonymous donors, but are able to select from a catalog that specifies the donors' physical and intellectual traits. The success rate for IUI ranges from 1 in 6 to 20 trials.

Oocyte Donation

The counterpart to sperm donation is *oocyte donation,* in which fresh oocytes are obtained from a donor, fertilized, and transplanted to the uterus of the client. A woman may choose this procedure for a variety of reasons: being past menopause, having had her ovaries removed, or having a hereditary disorder she doesn't want to pass to her children, for example. The donated oocytes are sometimes provided by a relative or may be left over from another woman's in vitro fertilization (see next section). The first baby conceived by oocyte donation was born in 1984. This procedure has a success rate of 20% to 50%.

In Vitro Fertilization

In vitro[23] *fertilization (IVF)* is an option for couples with such issues as male infertility and women whose uterine tubes cannot pass an egg or preembryo because of scars from pelvic inflammatory disease or other defects, but who have a healthy uterus. It is called *in vitro* for the fact that sperm and eggs are introduced to each other in a clinical vessel, with fertilization occurring there before the preembryo is introduced to the body. In preparation for IVF, a woman is treated with gonadotropins to induce the superovulation of multiple eggs. A physician views the ovary with a laparoscope and removes eggs by suction. These are placed in a solution that mimics the chemical environment of her reproductive tract, and sperm are added to the dish. In some cases, fertilization is assisted by piercing the zona pellucida before the sperm are added *(zona drilling)* or by injecting sperm directly into the egg through a micropipette **(fig. 29.19).** The latter method, called *intracytoplasmic sperm injection (ICSI),* is standard procedure in some clinics.

Preembryos reach the 8- to 16-cell stage about a day after fertilization. If there is a concern about parental chromosome abnormalities such as nondisjunction, one or two cells can be safely removed at this stage and inspected to ensure absence of defects. The clinician then introduces a preembryo into the uterus and monitors the woman's HCG level over the following days for signs of implantation. In cases where a woman has lost her ovaries to disease, the oocytes may be provided by another donor, sometimes a relative. Excess IVF preembryos may be donated to other infertile couples, frozen and used for later implantation attempts, or donated for such purposes as fertility and stem-cell research. One record pregnancy was achieved with an embryo that had been frozen 20 years earlier.

The chance of success in each IVF attempt is 20% to 35%—highest in younger women. IVF has been used in animal breeding since the 1950s, but the first child conceived this way was born (amid much controversy) in 1978. The practice grew rapidly after that, and at least 8 million babies conceived by IVF had been born by 2018. The inventor of IVF,

Oocyte and first polar body

Zona pellucida

Sperm in micropipette

Holding pipette

FIGURE 29.19 Intracytoplasmic Sperm Injection (ICSI). The egg is held with gentle suction on the pipette at left while a sperm is injected into its cytoplasm through the micropipette on the right.
Phanie/Alamy

---

[22]*oligo* = few

[23]*in vitro* = in glass

Robert Edwards, received the 2010 Nobel Prize for Physiology or Medicine for this achievement.

Surrogate Mothers

IVF is an option only for women who have a functional uterus. A woman who has had a hysterectomy or is otherwise unable to become pregnant or maintain a pregnancy may contract with a *surrogate mother* who provides a "uterus for hire." Some surrogates are both genetic and gestational mothers, and others gestational only. In the former case, the surrogate is artificially inseminated by a man's sperm and agrees to give the baby to the man and his partner at birth. In the latter case, oocytes are collected from one woman's ovaries, fertilized in vitro, and the preembryos are placed in the surrogate's uterus. This is typical of cases in which a woman has functional ovaries but no functional uterus. Several hundred babies have been produced this way in the United States. In at least one case, a woman carried the child of her infertile daughter, thus giving birth to her own granddaughter.

Gamete Intrafallopian Transfer

The limited success rate of IVF led to a search for more reliable and cost-effective techniques. *Gamete intrafallopian transfer (GIFT)* was developed in the mid-1980s on the conjecture that pregnancy would be more successful if the oocyte was fertilized and began cleavage in a more natural environment. Eggs are obtained from a woman after a weeklong course of ovulation-inducing drug treatment. The most active sperm cells are isolated from the semen, and the eggs and sperm are introduced into her uterine tube proximal to any existing obstruction. In a modification called *zygote intrafallopian transfer (ZIFT),* fertilization occurs in vitro and the preembryo is introduced into the uterine tube. Traveling down the uterine tube seems to improve the chance of implantation when the conceptus reaches the uterus. The chance of success in both GIFT and ZIFT is about 22%.

Embryo Adoption

Embryo adoption is used when a woman has malfunctioning ovaries but a normal uterus. A man's sperm are used to artificially inseminate another woman. A few days later, the preembryo is flushed from the donor's uterus before it implants and is transferred to the uterus of the woman who wishes to have a child.

Issues in Choice of Method

Many issues, sometimes complicated, can enter into one's choice of ART options. The methods described here vary widely in success rates and cost. Expenses range from less than $1,000 for IUI to more than $20,000 for GIFT and ZIFT—and this is cost *per trial,* with as little as 15% to 35% chances of success. There is much variation in whether insurance policies or national health services will cover the costs.

Like many other advances in medicine, reproductive technology also has its own ethical and legal dilemmas, some of which are especially confounding. Perhaps the most common problem is the surrogate mother who changes her mind. Surrogates enter into a contract to surrender the baby to a couple at birth, but after carrying a baby for 9 months and giving birth, they sometimes feel differently. This raises questions about the definition of motherhood, especially if she is the gestational but not the genetic mother.

The converse problem is illustrated by a case in which the child had hydrocephalus and neither the contracting couple nor the surrogate mother wanted it. In this case, genetic testing showed that the child actually had been fathered by the surrogate's husband, not the man who had contracted for her service. The surrogate then accepted the baby as her own. Nevertheless, the case raised the question of whether the birth of a genetically defective child constituted fulfillment of the contract and obligated the contracting couple to accept the child, or whether such a contract implied that the surrogate mother must produce a healthy child.

In another case, a wealthy couple was killed in an accident and left frozen preembryos in an IVF clinic. A lawsuit was filed on behalf of the preembryos on the grounds that they were heirs to the couple's estate and should be carried to birth by a surrogate mother so they could inherit it. The court ruled against the suit and the preembryos were allowed to perish. In still another widely publicized case, a man sued his wife for custody of their frozen preembryos as part of a divorce settlement.

IVF also creates a question of what to do with the excess preembryos. Some people view their disposal as a form of abortion, even if the preembryo is only a mass of 8 to 16 undifferentiated cells. On the other hand, there are those who see such excess preembryos as a research opportunity to obtain information that can't be obtained in any other way. In 1996, an IVF clinic in England was allowed to destroy 3,300 unclaimed preembryos, but only after heated public controversy.

It is common for scientific advances to require new advances in law and ethics. The parallel development of these disciplines is necessary if we are to benefit from the developments of science and ensure that knowledge is applied in an ethical and humane manner.

STUDY GUIDE

▶ Assess Your Learning Outcomes

To test your knowledge, discuss the following topics with a study partner or in writing, ideally from memory.

29.1 Fertilization and the Preembryonic Stage

1. Why sperm must meet an egg near the distal end of the uterine tube; factors that may aid in sperm migration

2. Why freshly ejaculated sperm cannot fertilize an egg; the process in which they acquire that ability

3. The acrosomal reaction, barriers to fertilization, and why multiple sperm must collaborate to fertilize an egg

4. The term for fertilization of an egg by two or more sperm, and how the egg normally prevents this

5. Events that occur between penetration by a sperm and the mingling of sperm and egg chromosomes; the term for a fertilized egg

6. The division of pregnancy into three trimesters and into preembryonic, embryonic, and fetal stages

7. Duration of the preembryonic stage; the three major events that occur in this stage; and the end product of this stage

8. The meaning of *cleavage;* the term for the resulting cells; development and characteristics of the morula and blastocyst; and structural and functional distinctions between the trophoblast and embryoblast

9. The process of implantation; the cytotrophoblast and syncytiotrophoblast; and the role of the latter in implantation and nourishment of the conceptus

10. The source and function of human chorionic gonadotropin

11. The process of gastrulation; the formation and names of the three primary germ layers; and why the process is called embryogenesis

29.2 The Embryonic and Fetal Stages

1. Major events that occur in the embryonic stage and the developmental ages at which it begins and ends

2. Longitudinal and lateral folding of the embryo; how this gives rise to the body cavity and gut; and the term for the differentiation of the embryonic organs

3. Structure and function of the four extraembryonic membranes associated with the embryo and fetus

4. The source and functions of amniotic fluid

5. How the conceptus is nourished prior to implantation

6. The mode of trophoblastic nutrition; what the trophoblast digests for its nourishment at this stage; the developmental age at which this yields to the placenta as the dominant mode of nutrition; and the age at which it ends entirely

7. When and how the placenta begins to form; the structure of the placenta; the method of transfer of nutrients and wastes between the maternal and fetal blood

8. Structure of the umbilical cord and the origin and termination of the umbilical blood vessels in the fetal body

9. When the individual is considered to be a fetus, and by what criterion

10. Developments that occur in the fetal stage; the meaning of *digital rays, meconium, lanugo, vernix caseosa, quickening,* and *vertex position*

11. Why the fetal circulatory system differs from the neonatal system, and names of the circulatory shunts in the fetus

29.3 The Neonate

1. What occurs in the neonate in the transitional period of 6 to 8 hours after birth; the duration of the neonatal period

2. What happens to the umbilical arteries and fetal circulatory shunts after birth; names of the adult remnants of the shunts

3. Why breathing is so difficult for the neonate

4. Sources of neonatal immunity; the age at which an infant produces ample amounts of antibody on its own

5. Why neonatal thermoregulation is so critical; the functional importance of brown fat in the neonate

6. Why the neonate needs more fluid intake relative to body weight than adults do

7. Why premature infants are at risk of infant respiratory distress syndrome, hypothermia, hypoproteinemia, edema, clotting deficiency, and jaundice

8. Four principal causes of congenital anomalies (birth defects)

9. Three classes of teratogens, with examples of each; why organ systems vary in age of maximum susceptibility to teratogenesis

10. Some of the more common and serious infectious diseases of the newborn

11. Nondisjunction and how it gives rise to triplo-X, Klinefelter, Turner, and Down syndromes; the signs of these syndromes; the term for any condition in which there are more or fewer than two sex chromosomes or two copies of each autosome

29.4 Aging and Senescence

1. The difference in meaning between *aging* and *senescence*

2. Differences in the leading causes of death in old age compared to those of young adulthood

3. Senescent changes in the integumentary system; differences between intrinsic aging and photoaging of the skin; and why integumentary senescence raises the risk of lactose intolerance, bone loss, muscle weakness, and poorer glandular secretion and synaptic transmission

4. Senescent changes in the skeletal system and how males and females differ in this respect

5. Senescent changes in the muscular system, and how they arise partially from nervous system senescence

6. Senescent changes in the nervous system and why this has especially widespread effects on homeostasis

7. Forms of declining sensory function in old age

8. The relatively slight senescence of most endocrine functions except for reproductive hormones; reasons for increased risk of diabetes mellitus as one ages

9. Senescent changes in the circulatory system including the blood, heart, and blood vessels; how circulatory system senescence contributes to atrophy in many other organs

10. Why senescent losses of immune function raise the risk and severity of infectious diseases and cancer

11. Senescent changes in the respiratory system, including the thoracic cage; why pneumonia and chronic obstructive pulmonary diseases become such prevalent causes of death

STUDY GUIDE

12. Senescent changes in the urinary system; how this affects fluid needs and drug therapy in old age; and reasons for increasing urine retention in elderly men and incontinence in elderly women
13. Senescent changes in the digestive system and their effects on nutrition
14. Senescent changes in the male and female reproductive systems and the hormonal effects of this senescence on other organ systems

15. The relative contributions of regular exercise to longevity and quality of life in old age, and disorders for which the risk is reduced by regular exercise
16. Theories of senescence including replicative senescence and the telomere theory; protein and DNA cross-linking; protein misfolding and other structural defects; and the autoimmune theory

17. The hereditary contribution to life expectancy; the reason natural selection cannot eliminate genes that cause the degenerative diseases of old age
18. The difference between life span and life expectancy, and why medical science has been able to extend one but not the other
19. Clinical and legal criteria of death and the issue of how clinical death differs from complete biological death

▶ Testing Your Recall

Answers in Appendix A

1. When a conceptus arrives in the uterus, it is at what stage of development?
   a. zygote
   b. morula
   c. blastomere
   d. blastocyst
   e. embryo

2. The entry of a sperm nucleus into an egg must be preceded by
   a. the cortical reaction.
   b. the acrosomal reaction.
   c. the fast block.
   d. implantation.
   e. cleavage.

3. The stage of the conceptus that implants in the uterine wall is
   a. a blastomere.
   b. a morula.
   c. a blastocyst.
   d. an embryo.
   e. a zygote.

4. Chorionic villi develop from
   a. the zona pellucida.
   b. the endometrium.
   c. the syncytiotrophoblast.
   d. the embryoblast.
   e. the corona radiata.

5. Which of these results from aneuploidy?
   a. Turner syndrome
   b. fetal alcohol syndrome
   c. nondisjunction
   d. progeria
   e. rubella

6. Fetal urine accumulates in the _____ and contributes to the fluid there.
   a. placental sinus
   b. yolk sac
   c. allantois
   d. chorion
   e. amnion

7. Cells may be limited in how many times they can divide because of shortening of the _____ of their chromosomes at each division.
   a. blastomeres
   b. centromeres
   c. telomeres
   d. life expectancy
   e. autosomes

8. Photoaging is a major factor in the senescence of
   a. the integumentary system.
   b. the eyes.
   c. the nervous system.
   d. the skeletal system.
   e. the cardiovascular system.

9. Which of these is *not* a common effect of senescence?
   a. reduced synthesis of vitamin D
   b. atrophy of the kidneys
   c. atrophy of the cerebral gyri
   d. increased risk of intervertebral disc herniation
   e. reduced pulmonary vital capacity

10. For the first 8 weeks of gestation, a conceptus is nourished mainly by
    a. the placenta.
    b. amniotic fluid.

    c. colostrum.
    d. decidual cells.
    e. yolk cytoplasm.

11. Viruses and chemicals that cause congenital anatomical deformities are called _____.

12. Aneuploidy is caused by _____, the failure of two homologous chromosomes to separate in meiosis.

13. The maximum age attainable by a member of the human species is called the _____.

14. The average age attained by humans in a given population is called the _____.

15. Fetal blood flows through growths called _____, which project into the placental sinus.

16. The enzymes with which a sperm penetrates an egg are contained in an organelle called the _____.

17. Stiffening of the arteries, joints, and lenses in old age may be a result of cross-linking between _____ molecules.

18. An enlarged tongue, epicanthal folds of the eyes, and mental retardation are characteristic of a genetic anomaly called _____.

19. The fossa ovalis is a remnant of a fetal shunt called the _____.

20. A developing individual is first classified as a/an _____ when the three primary germ layers have formed.

STUDY GUIDE

▶ Building Your Medical Vocabulary

Answers in Appendix A

State a meaning of each word element, and give a medical term from this chapter that uses it or a slight variation of it.

1. con-
2. decidu-
3. gero-
4. mor-
5. placento-
6. -ploid
7. senesc-
8. syn-
9. telo-
10. terato-

▶ What's Wrong with These Statements?

Answers in Appendix A

Briefly explain why each of the following statements is false, or reword it to make it true.

1. The ability of sperm to fertilize an egg declines sharply in the first 5 hours after ejaculation.

2. Fertilization normally occurs in the lumen of the uterus.

3. An egg is usually fertilized by the first sperm that contacts it.

4. The zygote begins dividing into multiple embryonic cells as soon as it reaches the uterus.

5. The conceptus is first considered a fetus when all three primary germ layers have formed.

6. As the placenta develops, the membranes of its chorionic villi become thicker and more selective with respect to what blood-borne substances it allows to pass through to the fetus.

7. The kidneys show significantly less senescence in old age than most other organs.

8. Fetal blood bypasses the nonfunctional liver by passing through the foramen ovale.

9. Blood in the umbilical vein has a lower P_{O_2} than blood in the umbilical arteries.

10. The gradual destruction of telomeres by telomerase is thought to be a major contributor to senescence and cell death.

▶ Testing Your Comprehension

1. Suppose a woman had a mutation resulting in a tough zona pellucida that did not disintegrate after the egg was fertilized. How would this affect her fertility? Why?

2. Suppose a drug were developed that could slow down the rate of collagen cross-linking with age. What diseases of old age could be made less severe with such a drug?

3. Some health-food stores market the enzyme superoxide dismutase (SOD) as an oral antioxidant to retard senescence. Explain why it would be a waste of your money to buy it.

4. In some children, the ductus arteriosus fails to close after birth—a condition that eventually requires surgery. Predict how this condition would affect (a) pulmonary blood pressure, (b) systemic diastolic pressure, and (c) the right ventricle of the heart.

5. Only one sperm is needed to fertilize an egg, yet a man who ejaculates fewer than 10 million sperm is usually infertile. Explain this apparent contradiction. Supposing 10 million sperm were ejaculated, predict how many would come within close range of the egg. How likely is it that any one of these sperm would fertilize it?

APPENDIX A

ANSWER KEYS

This appendix provides answers to questions in the figure legends and the end-of-chapter questions in the "Testing Your Recall," "Building Your Medical Vocabulary," and "What's Wrong with These Statements?" sections. Answers to "Apply What You Know" and "Testing Your Comprehension" questions are available to instructors on the text Connect site at www.mcgrawhillconnect.com.

CHAPTER 1

Figure Legend Questions

1.7   Vasodilation allows more blood to flow close to the body surface and to lose heat through the skin; thus, it cools the body.

1.9   Yes; one could say that pregnancy activates a series of events leading to childbirth, the termination of the pregnancy. Thus, it has the qualities of a negative feedback loop.

1.11 MRI is better than X-rays for visualizing nervous tissue, since X-rays do not penetrate bone very well. It also shows better contrast than X-rays in visualization of other soft tissues. X-rays are better than PET scans for visualizing bones, teeth, and other hard or dense tissues, as PET scans have relatively low resolution and do not serve well to visualize tissues with little regional variation in metabolic rate.

Testing Your Recall

1. a	8. c	15. homeostasis
2. e	9. d	16. set point
3. d	10. b	17. negative feedback
4. a	11. dissection	18. organ
5. c	12. gradient	19. stereoscopic
6. c	13. deduction	20. prehensile, opposable
7. a	14. psychosomatic	

Building Your Medical Vocabulary (Answers may vary; these are acceptable examples.)

1. listen—auscultation	6. nature—physiology
2. apart—dissection	7. cut—dissection
3. the same—homeostasis	8. to stay—homeostasis
4. change—metabolism	9. solid—stereoscopic
5. touch—palpation	10. to cut—tomography

What's Wrong with These Statements?

1. Auscultation means listening to body sounds, not inspecting the body by touch.
2. MRI does not involve ionizing radiation and has no known risk to a fetus.
3. Positive feedback is beneficial in limited cases, but more often it causes rapid departure from the homeostatic set point and may cause illness or death.
4. Each cell has many organelles, so organelles far outnumber cells.
5. Matter will move spontaneously down a gradient without the need for application of external energy.
6. Leeuwenhoek was a textile merchant who built microscopes to examine fabric.

7. A scientific theory is founded on a large body of evidence and summarizes what is already known.
8. Both the treatment and control groups consist of volunteer patients.
9. Evolutionary biologists do not believe humans evolved from monkeys, but that humans and apes evolved from the same ancestor.
10. Negative feedback is a self-corrective process with a beneficial effect on the body.

ATLAS A

Figure Legend Questions

A.4   Right lower quadrant (RLQ)

A.8   No, it is inferior to the peritoneal cavity, since the peritoneum passes over its superior surface.

Testing Your Recall

1. d	8. d	15. hand, foot
2. c	9. b	16. meninges
3. e	10. d	17. retroperitoneal
4. d	11. mesenteries	18. medial
5. d	12. parietal	19. inferior
6. a	13. mediastinum	20. cubital, popliteal
7. a	14. nuchal	

Building Your Medical Vocabulary (Answers may vary; these are acceptable examples.)

1. before—antebrachium	6. within—intraperitoneal
2. neck—cervical	7. wall—parietal
3. above—epigastric	8. around—peritoneum
4. below—hypochondriac	9. behind—retroperitoneal
5. groin—inguinal	10. arrow—sagittal

What's Wrong with These Statements?

1. A sagittal section could only pass between the lungs or through one lung, not through both of them.
2. A frontal section passes from left to right and could include both eyes.
3. The knee is proximal to the ankle (tarsal region).
4. The diaphragm is inferior to the lungs.
5. The esophagus is superior to the stomach.
6. The liver extends from the hypochondriac to the epigastric region, superior to the lumbar region.
7. The heart is enfolded by the pericardial cavity but not contained within it.
8. The kidneys are retroperitoneal.
9. The peritoneum lines the outside of the stomach and intestines.
10. The sigmoid colon is in the lower left quadrant.

CHAPTER 2

Figure Legend Questions

2.8   Because water molecules are attracted to each other, it requires more thermal energy for any one of them to break free and evaporate.

2.12 Decomposition

2.26 No, the amount of energy released is the same with or without an enzyme.

Testing Your Recall

1. a	9. b	16. -ose, -ase
2. c	10. d	17. phospholipids
3. a	11. cation	18. cyclic adenosine
4. c	12. free radicals	monophosphate
5. a	13. catalyst, enzymes	19. anaerobic
6. e	14. anabolism	fermentation
7. b	15. dehydration	20. substrate
8. c	synthesis	

Building Your Medical Vocabulary (Answers may vary; these are acceptable examples.)

1. not—atom	6. water—hydrolysis
2. oxygen—aerobic	7. part—polymer
3. both—amphipathic	8. one—monomer
4. heat—calorie	9. few—oligosaccharide
5. glue—colloid	10. loving—hydrophilic

What's Wrong with These Statements?

1. The monomers of a polysaccharide are monosaccharides (simple sugars).
2. ATP is not an energy-storage molecule. Most of our reserve energy is stored in fat.
3. Such molecules are called isomers, not isotopes.
4. Catabolism produces products with less energy than the reactants.
5. Peptide bonds join amino acids together, not sugars.
6. A saturated fat is one to which no more hydrogen can be added.
7. Enzymes, like all catalysts, are not consumed by the reactions they catalyze.
8. Above a certain temperature, enzymes denature and cease working.
9. These solutes have different molecular weights, so 2% solutions would not contain the same number of molecules per unit volume.
10. A solution with pH 8 has one-tenth the hydrogen ion concentration of one with pH 7.

CHAPTER 3

Figure Legend Questions

3.8   Adenylate cyclase is a transmembrane protein. The G protein is peripheral.
3.19 The Na^+–K^+ pump requires ATP, whereas osmosis does not. ATP is quickly depleted after a cell dies.
3.22 Transcytosis is a combination of endocytosis and exocytosis.
3.27 Large molecules such as enzymes and RNA must pass through the nuclear pores, but pores in the plasma membrane must be small enough to prevent such large molecules from escaping the cell.
3.33 A centriole is composed of a cylinder of nine groups of microtubules, but in a centriole, there are three microtubules in each group and in an axoneme there are only two. Also, an axoneme usually has a central pair of microtubules, whereas a centriole does not.

Testing Your Recall

1. e	9. d	16. exocytosis
2. b	10. b	17. Ribosomes,
3. d	11. micrometers	proteasomes
4. b	12. second messenger	18. smooth ER,
5. e	13. Voltage-gated	peroxisomes
6. e	14. hydrostatic	19. ligand-gated
7. a	pressure	channel
8. c	15. hypertonic	20. cistern

Building Your Medical Vocabulary (Answers may vary; these are acceptable examples.)

1. opposite—antiport	6. easy—facilitated
2. color—chromatin	7. spindle—fusiform
3. together—cotransport	8. study of—cytology
4. cell—cytoplasm	9. process—pinocytosis
5. into, within—endocytosis	10. eat—phagocytosis

What's Wrong with These Statements?

1. Osmosis does not require ATP, so it can continue even after cell death.
2. Some cells have two or more nuclei and some cells have none.
3. Second messengers activate enzymes in the cell; they are not transport proteins.
4. Peroxisomes are produced by the endoplasmic reticulum, not by the Golgi complex.
5. A channel could not move material from the outside of a cell to the inside unless it extended all the way across the membrane; it must be a transmembrane protein.
6. The plasma membrane consists primarily of phospholipid molecules.
7. The brush border is composed of microvilli.
8. Cells in hypertonic saline will lose water and shrivel.
9. Osmosis is not a carrier-mediated process and therefore not subject to a transport maximum.
10. Many ribosomes lie free in the cytosol, not attached to the ER or nucleus.

CHAPTER 4

Figure Legend Questions

4.1 The helix would bulge where two purines were paired and would be constricted where two pyrimidines were paired.
4.8 The ribosome would have no way of holding the partially completed peptide in place while adding the next amino acid.

Testing Your Recall

1. a	8. d	14. polyribosome
2. e	9. d	15. RNA polymerase
3. c	10. a	16. genome
4. c	11. cytokinesis	17. 46, 92, 92
5. e	12. cyclin-dependent	18. ribosome
6. b	kinases	19. growth factors
7. a	13. genetic code	20. autosomes

Building Your Medical Vocabulary (Answers may vary; these are acceptable examples.)

1. different—allele	6. nucleus—karyotype
2. finger—polydactyly	7. next in a series—metaphase
3. double—diploid	8. shape—polymorphism
4. half—haploid	9. change—mutation
5. different—heterozygous	10. many—polydactyly

What's Wrong with These Statements?

1. There are no ribosomes on the Golgi complex; they are on the rough ER.
2. There are no genes for steroids, carbohydrates, or phospholipids, but only for proteins.
3. RNA is half the weight of a DNA of the same length because it has only one nucleotide strand, whereas DNA has two.
4. Each amino acid is represented by a triplet, a sequence of three bases, not one base pair.
5. A single gene can code for multiple proteins.

6. This law describes the pairing of bases between the two strands of DNA, not between mRNA and tRNA.
7. Only about 2% of the human DNA codes for proteins.
8. Mutations can be harmful, beneficial, or neutral.
9. Males have only one X chromosome, but have two sex chromosomes (the X and Y).
10. Several RNA polymerase molecules at once can transcribe a gene.

CHAPTER 5

Figure Legend Questions

5.2 They are longitudinal sections. In a cross section, both the egg white and yolk would look circular. In an oblique section, the white would look elliptical but the yolk would still look circular.
5.12 The epithelia of the tongue, oral cavity, esophagus, and anal canal would look similar to this.
5.28 Gap junctions
5.30 Exocytosis
5.31 The sketch would look like one of the purple sacs in the middle figure, budding directly from the epithelial surface with no duct.
5.32 Holocrine glands, because entire cells break down to become the secretion, and these must be continually replaced.

Testing Your Recall

1.	a	9.	b	17.	basement
2.	b	10.	b		membrane
3.	c	11.	necrosis	18.	matrix (extracellular material)
4.	e	12.	mesothelium		
5.	c	13.	lacunae	19.	multipotent
6.	a	14.	fibers	20.	simple
7.	b	15.	collagen		
8.	e	16.	skeletal muscle		

Building Your Medical Vocabulary (Answers may vary; these are acceptable examples.)

1. away—apoptosis
2. cartilage—chondrocyte
3. outer—ectoderm
4. producing—collagen
5. tissue—histology
6. whole—holocrine
7. glassy—hyaline
8. dead—necrosis
9. formed—neoplasia
10. scale—squamous

What's Wrong with These Statements?

1. The esophageal epithelium is nonkeratinized.
2. All of them contact the basement membrane.
3. There are a few cases of skeletal muscles not attached to bones.
4. Glandular secretions are produced by cells of the parenchyma; the stroma is nonsecretory supportive connective tissue.
5. Adipose tissue is an exception; cells constitute most of its volume.
6. Adipocytes are also found in areolar tissue, either singly or in small clusters.
7. Tight junctions serve mainly to restrict the passage of material between cells.
8. Neoplasia is abnormal tissue growth, such as tumors; development of mature tissue types from nonspecialized tissues is called differentiation.
9. Excitability is characteristic of all living cells, but most highly developed in nerve and muscle cells.
10. Perichondrium is lacking from fibrocartilage and from hyaline articular cartilage.

CHAPTER 6

Figure Legend Questions

6.5 Keratinocytes
6.8 Cuticle
6.11 Asymmetry (A), irregular border (B), and color (C). The photo does not provide enough information to judge the diameter of the lesion (D).

Testing Your Recall

1.	d	8.	a	14.	cyanosis
2.	c	9.	a	15.	dermal papillae
3.	d	10.	d	16.	earwax
4.	b	11.	Insensible	17.	sebaceous glands
5.	a		perspiration	18.	anagen
6.	e	12.	arrector muscle	19.	dermal papilla
7.	c	13.	debridement	20.	third-degree

Building Your Medical Vocabulary (Answers may vary; these are acceptable examples.)

1. substance—melanin
2. white—albinism
3. skin—dermatology
4. through—diaphoresis
5. same—homograft
6. injure—lesion
7. black—melanoma
8. tumor—carcinoma
9. nail—eponychium
10. hair—piloerector

What's Wrong with These Statements?

1. Basal cell carcinoma is the most common form of skin cancer.
2. The number of melanocytes is about the same in all skin colors; dark skin results from the accumulation of melanin in keratinocytes.
3. Keratin is the protein of the epidermis; the dermis is composed mainly of collagen.
4. Vitamin D synthesis begins in the keratinocytes.
5. Epidermal cell multiplication occurs in the stratum basale.
6. Cells of the cortex are also dead; the only living hair cells are in and near the hair bulb.
7. The hypodermis is not considered to be a layer of the skin.
8. Different races have about the same density of melanocytes but different amounts of melanin.
9. A genetic lack of melanin causes albinism, not pallor. Pallor is a temporary, nonhereditary paleness of the skin.
10. Apocrine sweat glands develop at puberty.

CHAPTER 7

Figure Legend Questions

7.1 The wider epiphyses provide surface area for muscle attachment and bone articulation, while the narrowness of the diaphysis minimizes weight.
7.6 Places where bone comes close to the skin, such as the sternum and hips
7.7 Temporal bone, parietal bone, and several others
7.9 Humerus, radius, ulna, femur, tibia, fibula
7.10 An infant's joints are still cartilaginous.
7.12 The zones of cell proliferation and cell hypertrophy

Testing Your Recall

1.	e	8.	e	15.	hypocalcemia
2.	a	9.	b	16.	Osteoblasts
3.	d	10.	d	17.	calcitriol
4.	c	11.	hydroxyapatite	18.	osteoporosis
5.	d	12.	canaliculi	19.	metaphysis
6.	a	13.	appositional	20.	osteomalacia
7.	d	14.	solubility product		

Building Your Medical Vocabulary (Answers may vary; these are acceptable examples.)

1. calcium—hypocalcemia
2. destroy—osteoclast
3. softening—osteomalacia
4. marrow—osteomyelitis
5. straight—orthopedics
6. bone—osseous
7. bone—osteocyte
8. growth—diaphysis
9. dart—spicule
10. place—ectopic

What's Wrong with These Statements?

1. Flat cranial bones have a middle layer of spongy bone called the diploe.
2. Cartilage is removed and replaced by bone, not calcified and transformed into bone.
3. The most common bone disease is osteoporosis, not fractures.
4. Bones elongate at the epiphysial plate, not the articular cartilage.
5. Osteoclasts develop from stem cells in the bone marrow, not from osteoblasts.
6. Osteoblasts give rise only to osteocytes and are therefore unipotent.
7. Hydroxyapatite is the major mineral of bone; the major protein is collagen.
8. Osteons have blood vessels in their central canals, not in the canaliculi.
9. The major effect of vitamin D is bone resorption, though it also promotes deposition.
10. Parathyroid hormone indirectly promotes bone resorption, not deposition.

CHAPTER 8

Figure Legend Questions

8.10 Any five of these: the occipital, parietal, sphenoid, zygomatic, and palatine bones, and the mandible and maxilla

8.12 Any five of these: the frontal, lacrimal, and sphenoid bones, and the vomer, maxilla, and inferior concha

8.25 Rupture of this ligament allows the atlas to slip anteriorly and the dens of the axis to tear into the spinal cord.

8.35 The adult hand lacks epiphysial plates, the growth zones of a child's long bones.

8.41 The three cuneiforms and the cuboid bone of the tarsus are arranged in a row similar to the distal carpal bones (trapezium, trapezoid, capitate, and hamate), with the trapezium corresponding to the median cuneiform (proximal to digit I in each case). In the proximal row, the navicular bone of the tarsus is somewhat similar to the scaphoid of the carpus, being proximal to digit I and articulating with three bones of the distal row, but the calcaneus and talus are very different, being adapted to their load-bearing role.

Testing Your Recall

1. b
2. e
3. a
4. d
5. a
6. e
7. c
8. b
9. e
10. b
11. fontanelles
12. temporal
13. sutures
14. sphenoid
15. anulus fibrosus
16. dens
17. auricular
18. styloid
19. foramina
20. medial longitudinal

Building Your Medical Vocabulary (Answers may vary; these are acceptable examples.)

1. rib—intercostal
2. helmet—cranium
3. tough—dura mater
4. tongue—hypoglossa
5. little—ossicle
6. breast—mastoid
7. foot—bipedal
8. wing—pterygoid
9. above—supraorbital
10. ankle—metatarsal

What's Wrong with These Statements?

1. It passes out the jugular foramen.
2. Each hand and foot has 14 phalanges.
3. The female pelvis is wider and shallower than the male's.
4. The carpal bones are in the base of the hand, not the narrow wrist region.
5. Muscles of the infraspinous fossa are easily palpated on the back, inferior to the scapular spine.
6. You would be resting your elbow on the olecranon.
7. The lumbar vertebrae have transverse processes but no transverse costal facets.
8. The most frequently broken bone is the clavicle.
9. *Arm* refers to the region containing only the humerus; *leg* refers to the region containing the tibia and fibula.
10. These extra bones in the cranium are called sutural bones, not sesamoid bones.

CHAPTER 9

Figure Legend Questions

9.2 The gomphosis, because a tooth is not a bone

9.4 The pubic symphysis consists of the cartilaginous interpubic disc and the adjacent parts of the two pubic bones.

9.5 Interphalangeal joints are not subjected to a great deal of compression.

9.7 $MA = 1.0$. Shifting the fulcrum to the left would increase the MA of this lever, while the lever would remain first class.

9.19 The atlas (C1)

Testing Your Recall

1. c
2. b
3. a
4. e
5. c
6. c
7. a
8. d
9. b
10. d
11. synovial fluid
12. bursa
13. pivot
14. Kinesiology
15. gomphosis
16. serrate
17. extension
18. range of motion
19. articular disc
20. talus

Building Your Medical Vocabulary (Answers may vary; these are acceptable examples.)

1. away—abduction
2. joint—arthritis
3. characterized by—cruciate
4. letter X—cruciate
5. leg—talocrural
6. to lead—adduction
7. movement—kinesiology
8. moon—meniscus
9. to lay back—supination
10. to pull—protraction

What's Wrong with These Statements?

1. Osteoarthritis occurs in almost everyone after a certain age; rheumatoid arthritis is less common.
2. A kinesiologist studies joint movements; a rheumatologist treats arthritis.
3. Synovial joints are diarthroses and amphiarthroses, but never synarthroses.
4. There is no meniscus in the elbow.
5. This action involves hyperextension of the shoulder.
6. The cruciate ligaments are in the knee.
7. The round ligament is somewhat slack and probably does not secure the femoral head.
8. The knuckles are diarthroses.

9. Synovial fluid is secreted by the synovial membrane of the joint capsule and fills the bursae.
10. A tooth is not a bone.

CHAPTER 10

Figure Legend Questions

10.4 The brachialis and lateral head of the triceps brachii have direct attachments; the biceps brachii and long head of the triceps brachii have indirect attachments.

10.8 The zygomaticus major, levator palpebrae superioris, and orbicularis oris

10.16 Pectoralis minor, serratus anterior, and all three layers of the upper intercostal muscles

10.27 *Teres* refers to the round or cordlike shape of the first muscle, and *quadratus* refers to the four-sided shape of the second.

10.28 Part (c) represents a cross section cut too high on the forearm to include these muscles.

10.34 Climbing stairs, walking, running, or riding a bicycle

10.38 The soleus

Testing Your Recall

1. b	9. d	16.	urogenital
2. b	10. c		triangle
3. a	11. buccinator	17.	linea alba
4. c	12. fascicle	18.	synergist
5. e	13. prime mover	19.	bipennate
6. e	(agonist)	20.	sphincter
7. b	14. hamstring		
8. a	15. flexor retinacula		

Building Your Medical Vocabulary (Answers may vary; these are acceptable examples.)

1. head—splenius capitis
2. work—synergist
3. bundle—fascicle
4. lip—levator labii superioris
5. lower back—quadratus lumborum
6. mouse—muscle
7. muscle—perimysium
8. shoulder—omohyoid
9. feather—bipennate
10. third—peroneus tertius

What's Wrong with These Statements?

1. The connective tissue that encloses individual muscle fibers is endomysium.
2. The orbicularis was once thought to be a sphincter but is no longer interpreted as such.
3. The biceps brachii is fusiform.
4. A synergist aids an agonist in its function; it does not oppose the agonist.
5. Many skeletal muscles are innervated by cranial nerves (those in the head and neck).
6. Normally, no muscular effort is needed to exhale.
7. The hamstrings flex the knee and therefore would not aid in lifting the body to the next step.
8. The facial nerve innervates 16 of the facial muscles tabulated in this chapter. The trigeminal innervates 6 tabulated muscles, but they are muscles of chewing and swallowing, not facial expression.
9. The adductor pollicis is by far the strongest of the three listed muscles (important in the strength of the hand grip), but is not listed first.
10. They are on opposite sides of the tibia and act as antagonists.

ATLAS B

Figure Legend Questions

B.1 Orbicularis oris; trapezius

B.5 The lungs, heart, liver, stomach, gallbladder, and spleen, among others

B.8 Sternocleidomastoids

B.11 Posterior

B.13 Fat (adipose tissue)

B.18 Five: one tendon proximally (at the humerus) and four distally (in the hand)

B.19 The mark would belong close to where the leader for the styloid process of the radius ends in figure B.19b.

B.20 The mark would belong close to where the leader for the rectus femoris ends; the vastus intermedius is deep to this.

B.21 The fibula

B.24 There is no such bone; digit I (the great toe) has only a proximal and distal phalanx.

B.25 Answers to the muscle test are as follows:

1. f	11. x	21. k
2. b	12. m	22. d
3. k	13. n	23. f
4. p	14. e	24. b
5. h	15. g	25. a
6. y	16. v	26. u
7. z	17. f	27. j
8. w	18. c	28. i
9. c	19. x	29. g
10. a	20. w	30. q

CHAPTER 11

Figure Legend Questions

11.1 The striations distinguish it from smooth muscle; the multiple nuclei adjacent to the plasma membrane and the parallel fibers distinguish it from both cardiac and smooth muscle.

11.2 The electrical excitation spreading down the T tubule must excite the opening of calcium gates in the terminal cisterns.

11.13 ATP is needed to pump Ca^{2+} back into the sarcoplasmic reticulum by active transport and to induce each myosin head to release actin so the sarcomere can relax.

11.16 The gluteus maximus and quadriceps femoris

11.17 The muscle tension curve would drop gradually while the muscle length curve would rise.

Testing Your Recall

1. a	8. c	15.	acetylcholine
2. c	9. e	16.	myoglobin
3. b	10. b	17.	Z discs
4. d	11. threshold	18.	varicosities
5. c	12. oxidative	19.	muscle tone
6. c	13. terminal cisterns	20.	Isometric
7. e	14. myosin		contraction

Building Your Medical Vocabulary (Answers may vary; these are acceptable examples.)

1. weak—myasthenia
2. self—autorhythmic
3. abnormal—dystrophy
4. same—isometric
5. length—isometric
6. muscle—myocyte
7. flesh—sarcolemma
8. time—temporal
9. tension—isotonic
10. growth—dystrophy

What's Wrong with These Statements?

1. A motor neuron may supply 1,000 or more muscle fibers; a motor unit consists of one motor neuron and all the muscle fibers it innervates.
2. Somatic motor neurons do not innervate cardiac muscle.
3. Fast glycolytic fibers fatigue relatively quickly.
4. Thin myofilaments extend well into the A bands, where they overlap with thick myofilaments.
5. Thin and thick myofilaments do not shorten, but glide over each other.
6. Thick and thin myofilaments are present but not arranged in a way that produces striations.
7. Under natural conditions, a muscle seldom or never attains complete tetanus.
8. Even excitation would be impossible without ATP, because ATP drives the active transport (Na^+–K^+) pumps that maintain membrane potential and excitability.
9. A muscle produces most of its ATP during this time by anaerobic fermentation, which generates lactate; it does not consume lactate.
10. Autorhythmicity is limited to unitary smooth muscle.

CHAPTER 12

Figure Legend Questions

12.9 Its conduction speed is relatively slow, but it has a small diameter and contributes relatively little bulk to the nervous tissue.

12.25 One EPSP is a voltage change of only 0.5 mV or so. A change of about 15 mV is required to reach threshold and make a neuron fire.

12.30 The CNS interprets a stimulus as more intense if it receives signals from high-threshold sensory neurons than if it receives signals only from low-threshold neurons.

12.32 A reverberating circuit, because a neuron early in the circuit is continually restimulated

Testing Your Recall

1. e	9. d	16. myelin sheath gaps
2. c	10. b	
3. d	11. afferent	17. axon hillock, initial segment
4. a	12. conductivity	
5. c	13. absolute refractory period	18. norepinephrine
6. e		19. facilitated zone
7. d	14. dendrites	20. Neuromodulators
8. a	15. oligodendrocytes	

Building Your Medical Vocabulary (Answers may vary; these are acceptable examples.)

1. forward—anterograde	6. knot—ganglion
2. to touch—synapse	7. to walk—retrograde
3. star—astrocyte	8. nerve—neuroglia
4. tree—dendrite	9. hard—sclerosis
5. carry—afferent	10. body—somatic

What's Wrong with These Statements?

1. A neuron never has more than one axon.
2. Oligodendrocytes of the brain perform the same function as Schwann cells of the PNS.
3. The extracellular concentration of Na^+ is greater than its intracellular concentration.
4. Only a small fraction of the neuron's Na^+ and K^+ exchange places across the plasma membrane.
5. The threshold stays the same but an EPSP brings the membrane potential closer to the threshold.
6. The absolute refractory period sets an upper limit on firing frequency.
7. The effect of a neurotransmitter varies from place to place depending on the type of receptor present.
8. The signals travel rapidly through the internodal segments and slow down at each myelin sheath gap.
9. Learning involves modification of the synapses of existing neurons, not an increase in the neuron population.
10. Neurons cannot undergo mitosis to replace those that are lost, although limited replacement occurs through multiplication and differentiation of stem cells.

CHAPTER 13

Figure Legend Questions

13.5 If it were T10, there would be no cuneate fasciculus; that exists only from T6 up.

13.10 They are in the anterior horn of the spinal cord.

13.13 They are afferent, because they arise from the posterior root of the spinal nerve.

13.15 Severing one phrenic nerve paralyzes the diaphragm on the ipsilateral side; severing both of them paralyzes the entire diaphragm and causes respiratory arrest.

13.25 Motor neurons are capable only of exciting skeletal muscle (endplate potentials are always excitatory). To inhibit muscle contraction, it is necessary to inhibit the motor neuron at the CNS level (point 6).

13.26 They would show more synaptic delay, because there are more synapses in the pathway.

Testing Your Recall

1. e	8. a	15. intrafusal fibers
2. c	9. e	16. phrenic
3. d	10. b	17. decussation
4. d	11. ganglia	18. proprioception
5. e	12. ramus	19. posterior root
6. c	13. spinocerebellar	20. tibial, common fibular
7. c	14. crossed extension	

Building Your Medical Vocabulary (Answers may vary; these are acceptable examples.)

1. spider—arachnoid	6. diaphragm—phrenic
2. tail—cauda equina	7. tender—pia mater
3. opposite—contralateral	8. oneself—proprioception
4. wedge—cuneate	9. branch—ramus
5. same—ipsilateral	10. roof—tectospinal

What's Wrong with These Statements?

1. The gracile fasciculus is an ascending (sensory) tract.
2. It terminates in the upper lumbar region of the vertebral column.
3. Each segment of the spinal cord has several rootlets.
4. All spinal nerves are mixed nerves; none are purely sensory or motor.
5. The dura is separated from the bone by a fat-filled epidural space.
6. The horns of the spinal cord are gray matter.
7. Corticospinal tracts are descending (motor) tracts, not sensory.
8. Dermatomes overlap each other by as much as 50%.
9. Some somatic reflexes are mediated primarily through the brainstem and cerebellum.
10. Many ipsilateral reflex arcs are also polysynaptic.

CHAPTER 14

Figure Legend Questions

14.9   Signals in the cuneate fasciculus ascend to the cuneate nucleus in part (c), and signals in the gracile fasciculus ascend to the nearby gracile nucleus. Both of them decussate together to the contralateral medial lemniscus in parts (b) and (a) and travel this route to the thalamus.

14.10   The reticular formation is labeled on all three parts of the figure.

14.14   Commissural tracts also cross through the anterior and posterior commissures shown in figure 14.2.

14.15   Dendrites; the axons project downward into the white matter.

14.22   Regions with numerous small muscles

Testing Your Recall

1. c	8. d	14. vagus nerve
2. a	9. e	15. choroid plexus
3. e	10. e	16. precentral
4. a	11. corpus callosum	17. frontal
5. b	12. ventricles,	18. association areas
6. c	cerebrospinal	19. categorical
7. a	13. arbor vitae	20. Broca area

Building Your Medical Vocabulary (Answers may vary; these are acceptable examples.)

1. pain—neuralgia
2. head—hydrocephalus
3. brain—cerebrospinal
4. body—corpus callosum
5. brain—encephalitis
6. record of—electroencephalogram
7. island—insula
8. eye—oculomotor
9. pulley—trochlea
10. little—peduncle

What's Wrong with These Statements?

1. The optic nerve is purely sensory. Eye movements are controlled by the oculomotor, trochlear, and abducens nerves.
2. The cerebral hemispheres do not develop from neural crest tissue.
3. The midbrain is rostral to the pons.
4. The Broca and Wernicke areas are ipsilateral, both in the left hemisphere in most people.
5. The choroid plexuses produce only 30% of the CSF.
6. Hearing is a temporal lobe function; vision resides in the occipital lobe.
7. Respiration is controlled by nuclei in the pons and medulla oblongata.
8. The trigeminal nerve carries sensory signals from the largest area of the face.
9. The vagus nerve (cranial nerve X) innervates organs of the thoracic and abdominopelvic cavities.
10. The cerebellum contains more than half of all brain neurons.

CHAPTER 15

Figure Legend Questions

15.4   No; inhaling and exhaling are controlled by the somatic motor system and skeletal muscles.

15.5   The neurosoma of the somatic efferent neuron is in the anterior horn and the neurosoma of the sympathetic preganglionic neuron is in the lateral horn.

15.7   The vagus nerve

15.9   The pupils dilate because fear increases sympathetic output, which induces dilation.

Testing Your Recall

1. b	3. e	5. a
2. c	4. e	6. e

7. d	12. Dual innervation	17. sympathetic
8. d	13. Autonomic tone	18. preganglionic,
9. a	14. vagus	postganglionic
10. c	15. enteric	19. cAMP
11. adrenergic	16. norepinephrine	20. vasomotor tone

Building Your Medical Vocabulary (Answers may vary; these are acceptable examples.)

1. pressure—baroreflex
2. dissolve—sympatholytic
3. wall—intramural
4. rule—autonomic
5. ear—otic
6. feeling—parasympathetic
7. kidney—adrenal
8. internal organs—splanchnic
9. together—sympathetic
10. internal organs—visceral

What's Wrong with These Statements?

1. Both systems are always simultaneously active.
2. Cutaneous blood vessels receive only sympathetic fibers.
3. In biofeedback and other circumstances, limited voluntary control of the ANS is possible.
4. The sympathetic division inhibits digestion.
5. The sympathetic division has a few cholinergic postganglionic fibers, although most are adrenergic.
6. Waste elimination can occur by autonomic spinal reflexes without necessarily involving the brain.
7. All parasympathetic fibers are cholinergic.
8. The sympathetic division has more neural divergence and therefore more widespread, less organ-specific, effects than the parasympathetic division.
9. The hypoglossal nerve carries no parasympathetic fibers.
10. They have antagonistic effects on the iris.

CHAPTER 16

Figure Legend Questions

16.1   Yes; two touches are felt separately if they straddle the boundary between two separate receptive fields.

16.9   The lower margin of the violet zone ("all sound") would be higher in that frequency range.

16.15   They are the outer hair cells, which function to "tune the cochlea" and improve discrimination between sounds of different pitches.

16.16   It would oppose the inward movement of the tympanic membrane and, thus, reduce the amount of vibration transferred to the inner ear.

16.24   This would prevent tears from draining into the lacrimal canals, resulting in more watery eyes.

16.25   Cranial nerve III, because it controls more eye movements than IV or VI.

16.30   It is the right eye. The optic disc is always medial to the fovea, so this has to be a view of the observer's left and the subject's right.

16.42   Approximately 68:20:0, and yellow

16.45   It would cause blindness in the left half of the visual field. It would not affect the visual reflexes.

Testing Your Recall

1. a	8. c	15. outer hair cells
2. c	9. c	16. stapes
3. b	10. b	17. inferior colliculi
4. a	11. fovea centralis	18. basal cells
5. e	12. ganglion	19. olfactory bulb
6. e	13. photopsin	20. referred pain
7. d	14. otoliths	

Building Your Medical Vocabulary (Answers may vary; these are acceptable examples.)

1. two—binaural
2. cross over—hemidecussation
3. half—hemidecussation
4. tears—lacrimal
5. stone—otolithic
6. spot—macula sacculi
7. pain—nociceptor
8. dark—scotopic
9. infection—asepsis
10. hair—peritrichial

What's Wrong with These Statements?

1. These fibers end in the medulla oblongata.
2. They are perceived in the right hemisphere because of decussation.
3. Because of hemidecussation, each hemisphere receives signals from both eyes.
4. Chemoreceptors that monitor blood chemistry, for example, are interoceptors.
5. The posterior chamber, the space between iris and lens, is filled with aqueous humor.
6. Descending analgesic fibers block signals that have reached the dorsal horn of the spinal cord.
7. Cranial nerve VIII carries both auditory and equilibrium signals.
8. The tympanic cavity is filled with air.
9. Rods release glutamate in the dark and stop when they are illuminated.
10. The trochlear and abducens nerves control the superior oblique and lateral rectus, respectively.

CHAPTER 17

Figure Legend Questions

17.1 Heart, liver, stomach, small intestine, placenta (any three)
17.4 The posterior pituitary (neurohypophysis)
17.19 Steroids enter the target cell; they do not bind to membrane receptors or activate second messengers.
17.24 Such a drug would block leukotriene synthesis and thus inhibit allergic and inflammatory responses.

Testing Your Recall

1. b
2. d
3. a
4. c
5. c
6. c
7. d
8. c
9. a
10. e
11. adenohypophysis
12. tyrosine
13. acromegaly
14. cortisol
15. glucocorticoids
16. interstitial endocrine
17. negative feedback inhibition
18. hypophysial portal system
19. permissive
20. Up-regulation

Building Your Medical Vocabulary (Answers may vary; these are acceptable examples.)

1. gland—adenohypophysis
2. bile—cholecystokinin
3. flow through—diabetes
4. twenty—eicosanoid
5. yellow—luteinizing
6. resembling—thyroid
7. full of—glomerulosa
8. favoring—progesterone
9. turn—gonadotropin
10. urine—glycosuria

What's Wrong with These Statements?

1. Gonadotropin secretion would rise because of a lack of negative feedback inhibition from the testes.
2. Glycoproteins cannot enter the target cell; they bind to surface receptors of the plasma membrane.
3. Thyroglobulin can be synthesized, but it cannot be iodinated, so no thyroid hormone is produced.
4. Tumors can cause hypersecretion but can also destroy endocrine cells and lead to hyposecretion.
5. Hormones are also secreted by the heart, liver, kidneys, and other organs not generally regarded as glands.
6. Most cases of diabetes mellitus are caused by insensitivity to insulin, not a lack of insulin.
7. The pineal gland and thymus undergo involution with age; they are larger in children than they are later in life.
8. Without iodine, there is no thyroid hormone (TH); without TH, there can be no negative feedback inhibition.
9. The tissue at the center is the adrenal medulla.
10. There are also two testes, two ovaries, and four parathyroid glands.

CHAPTER 18

Figure Legend Questions

18.1 (Answering this requires labeling the illustration.)
18.4 The sunken center represents the former location of the nucleus.
18.5 Hemoglobin consists of a noncovalent association of four protein chains. The prosthetic group is the heme moiety of each of the four chains.
18.18 *Myelo-* refers to the bone marrow, where these cells develop.
18.19 Although numerous, these WBCs are immature and incapable of performing their defensive roles.
18.21 A platelet plug lacks the fibrin mesh that a blood clot has.
18.22 It would affect only the intrinsic mechanism.
18.23 In both blood clotting and hormonal signal amplification, the product of one reaction step is an enzyme that catalyzes the production of many more molecules of the next product. Thus, there is a geometric increase in the number of product molecules at each step and ultimately, a large final result from a small beginning.
18.25 The older theory of leeching was that many disorders are caused by "bad blood," which could be removed painlessly by medicinal leeches. The modern practice is to take advantage of the anticoagulants in the leech saliva to promote blood flow to a tissue or to dissolve and remove clots that have already formed.

Testing Your Recall

1. b
2. c
3. c
4. a
5. b
6. d
7. d
8. c
9. d
10. c
11. hematopoiesis
12. hematocrit (packed cell volume)
13. thromboplastin
14. agglutinogens
15. hemophilia
16. hemostasis
17. Sickle-cell disease
18. polycythemia
19. vitamin B_{12}
20. erythropoietin

Building Your Medical Vocabulary (Answers may vary; these are acceptable examples.)

1. without—anemia
2. producing—erythroblast
3. red—erythrocyte
4. aggregate—agglutination
5. blood—hemostasis
6. white—leukocyte
7. deficiency—leukopenia
8. vein—phlebotomy
9. formation—hemopoiesis
10. clot—thrombosis

What's Wrong with These Statements?

1. Most of the volume is usually plasma.
2. Hyperproteinemia causes retention of more water in the circulatory system and raises blood pressure.
3. Oxygen deficiency is the result of anemia, not its cause.
4. Clotting (coagulation) is one mechanism of hemostasis, but hemostasis includes other mechanisms (vascular spasm and platelet plug), so it is not synonymous with the other two terms.
5. He can be the father if he is heterozygous for type A *($I^A i$)*, the mother is heterzygous for type B *($I^B i$)*, and both are heterzygous for

Rh type *(Dd).* The baby could then inherit *ii* and *dd,* and have phenotype O–.

6. The most abundant WBCs are neutrophils.
7. Blood clotting requires Ca^{2+} at several steps.
8. Even platelets arise ultimately from hematopoietic stem cells.
9. The heme is excreted; the globin is broken down into amino acids that can be reused.
10. In leukemia, there is an excess of WBCs. A WBC deficiency is leukopenia.

CHAPTER 19

Figure Legend Questions

19.1 Both; they receive pulmonary arteries from the pulmonary circuit and bronchial arteries from the systemic circuit.
19.2 To the left
19.7 The trabeculae carneae
19.12 The right atrium
19.14 It ensures that wave summation and tetanus will not occur, thus ensuring relaxation and refilling of the heart chambers.
19.19 They prevent prolapse of the AV valves during ventricular systole.
19.20 This is the point at which the aortic valve opens and blood is ejected into the aorta, raising its blood pressure.

Testing Your Recall

1. d	9. a	16. T wave
2. b	10. c	17. vagus
3. d	11. systole, diastole	18. myocardial
4. a	12. systemic	infarction
5. a	13. atrioventricular	19. endocardium
6. d	(coronary) sulcus	20. cardiac output
7. d	14. Na^+	
8. c	15. gap junctions	

Building Your Medical Vocabulary (Answers may vary; these are acceptable examples.)

1. entryway—atrium
2. slow—bradycardia
3. heart—cardiology
4. crown—coronary
5. moon—semilunar
6. nipple—papillary
7. semi—semilunar
8. fast—tachycardia
9. vessel—vasomotor
10. belly—ventricle

What's Wrong with These Statements?

1. The coronary circulation is part of the systemic circuit; the other division is the pulmonary circuit.
2. There are no valves at the openings of the venae cavae.
3. The first two-thirds of ventricular filling occurs before the atria contract. The atria add only about 31% of the blood that fills the ventricles.
4. The vagus nerves affect heart rate but not contraction strength.
5. High CO_2 and low pH accelerate the heart rate.
6. The first heart sound occurs at the time of the QRS complex.
7. The heart has its own internal pacemaker and would continue beating; the nerves only alter the heart rate.
8. That would result from clamping the pulmonary veins, not the pulmonary arteries.
9. Cardiomyocytes do have a stable resting potential when they are at rest.
10. The ECG is a composite record of the electrical activity of the entire myocardium, not a record from a single myocyte. It looks much different from an action potential.

CHAPTER 20

Figure Legend Questions

20.2 Veins are subjected to less pressure than arteries and have less need of elasticity.
20.5 Endocrine glands, kidneys, the small intestine, and choroid plexuses of the brain
20.8 Veins have less muscular and elastic tissue, so they expand more easily than arteries.
20.10 Arterial anastomoses: the arterial circle of the brain, the celiac circulation, encircling the heads of the humerus and femur, and the arterial arches of the hand and foot. Venous anastomoses: the jugular veins, the azygos system, the mesenteric veins, and venous networks of the hand and foot. Portal systems: the hepatic portal system and (outside of this chapter) the hypophysial portal system. Answers may vary.
20.20 Nothing would happen if he lifted his finger from point O because the valve at that point would prevent blood from flowing downward and filling the vein. If he lifted his finger from point H, blood would flow upward, fill the vein, and the vein between O and H would stand out.
20.27 Aorta → left common carotid a. → external carotid a. → superficial temporal a.
20.36 The ovaries and testes begin their embryonic development near the kidneys. The gonadal veins elongate as the gonads descend to the pelvic cavity and scrotum.
20.39 Joint movements may temporarily compress an artery. Anastomoses allow for continued blood flow through alternative routes to more distal regions.
20.40 The cephalic, basilic, and median cubital vv.

Testing Your Recall

1. c	9. e	16. transcytosis
2. b	10. d	17. sympathetic
3. a	11. systolic, diastolic	18. baroreceptors
4. e	12. continuous	19. cerebral arterial
5. b	capillaries	circle
6. c	13. Anaphylactic	20. basilic, cephalic
7. e	14. thoracic pump	
8. a	15. oncotic pressure	

Building Your Medical Vocabulary (Answers may vary; these are acceptable examples.)

1. vessel—angiogenesis
2. arm—brachiocephalic
3. abdomen—celiac
4. window—fenestrations
5. neck—jugular
6. belonging to—vasa vasorum
7. standing—saphenous
8. below—subclavian
9. chest—thoracoacromial
10. bladder—vesical

What's Wrong with These Statements?

1. Blood can bypass capillaries by flowing through an arteriovenous anastomosis.
2. Blood drains from the brain by way of the internal jugular veins.
3. The longest blood vessel is the great saphenous vein.
4. Some veins have valves, but arteries do not.
5. By the formula $F \propto r^4$, the flow increases in proportion to the fourth power of radius, or 16-fold.
6. The femoral triangle is bordered by the inguinal ligament, sartorius muscle, and adductor longus muscle.
7. The lungs also receive blood from the bronchial arteries of the systemic circuit.
8. Most capillaries reabsorb only a portion (typically about 85%) of the fluid they filter; the rest is absorbed by the lymphatic system.

9. An aneurysm is a weak, bulging vessel that *may* rupture.
10. The response to falling blood pressure is a corrective vasoconstriction.

CHAPTER 21

Figure Legend Questions

21.2 There are much larger gaps between the endothelial cells of lymphatic capillaries than between those of blood capillaries.

21.3 There would be no consistent one-way flow of lymph. Lymph and tissue fluid would accumulate, especially in the lower regions of the body.

21.4 (1) Prevention of excess tissue fluid accumulation and (2) monitoring the tissue fluids for pathogens

21.5 Lymph flows from the breast to the axillary lymph nodes. Therefore, metastatic cancer cells tend to lodge first in those nodes.

21.13 Erythrocytes in the red pulp; lymphocytes and macrophages in the white pulp

21.15 Both of these produce a ring of proteins in the target-cell plasma membrane, opening a hole in the membrane through which the cell contents escape.

21.23 All three defenses depend on the action of helper T cells, which are destroyed by HIV.

21.26 The ER is the site of antibody synthesis.

Testing Your Recall

1. b	9. a	16. pyrogen
2. c	10. c	17. interleukins
3. a	11. pathogen	18. antigen-binding
4. a	12. lysozyme	site, epitope
5. d	13. Lymphadenitis	19. clonal deletion
6. b	14. diapedesis (emi-	20. autoimmune
7. e	gration)	
8. d	15. opsonization	

Building Your Medical Vocabulary (Answers may vary; these are acceptable examples.)

1. apart—anaphylactic	6. set in motion—cytokine
2. secrete—paracrine	7. water—lymphatic
3. outside—extravasated	8. enlargement—splenomegaly
4. arising—endogenous	9. disease—lymphadenopathy
5. freedom—immunology	10. fire—pyrogen

What's Wrong with These Statements?

1. Lysozyme is a bacteria-killing enzyme.
2. The principal birthplace of lymphocytes is the red bone marrow; T lymphocytes migrate from there to the thymus.
3. Interferons promote inflammation.
4. Helper T cells are also necessary to humoral and innate immunity.
5. Negative selection serves to eliminate or deactivate self-reactive T cells (not unresponsive ones), thus preventing immune attack on one's own tissues.
6. The thymus and spleen do not receive or filter any incoming lymph; only lymph nodes do this.
7. Only antibodies of the IgG and IgM classes employ complement fixation.
8. One can be HIV-positive without having AIDS. AIDS is defined by a low helper T cell count ($< 200/\mu L$), not by the presence of the virus.
9. Anergy is a loss of lymphocyte activity, whereas autoimmune diseases result from misdirected activity.
10. Plasma cells are antibody-synthesizing cells of the connective tissues that develop from B cells; they are not found in the blood plasma.

CHAPTER 22

Figure Legend Questions

22.3 The line would cross the figure just slightly above the trachea label.

22.4 Epiglottic, corniculate, and arytenoid

22.7 The right main bronchus is slightly wider and more vertical than the left, making it easier for aspirated objects to fall into the right.

22.8 To secrete mucus

22.17 Any airflow through the nose would not be registered by the spirometer, and the spirometer could not give a correct reading of pulmonary ventilation.

22.19 P_{O_2} drops from 104 to 95 mm Hg on its way out of the lungs because of some mixing with systemic blood. It drops farther to 40 mm Hg when the blood gives up O_2 to respiring tissues, and remains at this level until the blood is reoxygenated back in the lungs. P_{CO_2} is 40 mm Hg leaving the lungs and rises to 46 mm Hg when CO_2 is picked up from respiring tissues. It remains at that level until the blood returns to the lungs and unloads CO_2.

22.23 About 60%

22.25 In the alveoli, CO_2 leaves the blood, O_2 enters, and all the chemical reactions are the reverse of those in figure 22.23. The blood bicarbonate concentration will be reduced following alveolar gas exchange because bicarbonate is taken up by the chloride shift antiport and converted to CO_2 and water.

22.26 A higher temperature suggests a relatively high metabolic rate and, thus, an elevated need for oxygen. Comparison of these curves shows that for a given P_{O_2}, hemoglobin gives up more oxygen at warmer temperatures.

Testing Your Recall

1. c	10. a	17. compliance,
2. c	11. epiglottis	elasticity
3. a	12. bronchial tree	18. ventral respiratory
4. e	13. pulmonary	group
5. e	surfactant	19. ventilation–
6. c	14. atmospheric	perfusion coupling
7. b	15. Obstructive	20. alkalosis,
8. a	16. anatomical dead	hypocapnia
9. d	space	

Building Your Medical Vocabulary (Answers may vary; these are acceptable examples.)

1. imperfect—atelectasis	6. measuring device—spirometer
2. smoke—hypercapnia	7. nose—nasofacial
3. cancer—carcinoma	8. breathing—dyspnea
4. horn—corniculate	9. breath—spirometry
5. true—eupnea	10. shield—thyroid

What's Wrong with These Statements?

1. They also fire during expiration (although at a lower rate) to exert a braking action on the diaphragm.
2. The two lungs have a total of 18 segmental bronchi but only 5 lobes.
3. The most abundant cells are alveolar macrophages.
4. When volume increases, pressure decreases.
5. Atelectasis can have other causes such as airway obstruction.
6. The greatest effect is to lower the CO_2 level; accelerated breathing has little effect on blood O_2.
7. We inhale all the gases of the atmosphere, whether the body uses them or not; there is no way to separate one gas from another as we inhale.

8. In an average 500 mL tidal volume, 350 mL reaches the alveoli.
9. The lower the P_{CO_2}, the higher the pH.
10. Most blood CO_2 is transported as bicarbonate ions.

CHAPTER 23

Figure Legend Questions

23.2 Ammonia is produced by the deamination of amino acids; urea is synthesized from ammonia and carbon dioxide; uric acid is produced from nucleic acids; and creatinine is produced from creatine phosphate.

23.3 It would be in the dark space at the top of the figure, where the spleen, colon, and small intestine are shown.

23.10 The afferent arteriole is larger. The relatively large inlet to the glomerulus and its small outlet result in high blood pressure in the glomerulus. This is the force that drives glomerular filtration.

23.16 It lowers the urine pH; the more Na^+ that is reabsorbed, the more H^+ is secreted into the tubular fluid. This is seen at the Na^+–H^+ antiport along the right margin of the figure.

23.23 The relatively short female urethra is less of an obstacle for bacteria traveling from the perineum to the urinary bladder.

Testing Your Recall

1. a	9. c	16. transport maximum
2. d	10. a	
3. b	11. micturition	17. Antidiuretic hormone
4. c	12. Renal autoregulation	
5. b		18. internal urethral
6. b	13. trigone	19. protein
7. d	14. macula densa	20. arcuate
8. e	15. podocytes	

Building Your Medical Vocabulary (Answers may vary; these are acceptable examples.)

1. nitrogen—azotemia
2. bladder—cystitis
3. ball—glomerulus
4. next to—juxtaglomerular
5. middle—mesangial
6. kidney—nephron
7. foot—podocyte
8. sagging—nephroptosis
9. pus—pyelonephritis
10. straight—vasa recta

What's Wrong with These Statements?

1. Calcium and sodium reabsorption by the PCT are influenced by parathyroid hormone and angiotensin II.
2. Urine contains more urea and chloride than sodium.
3. There is one renal corpuscle per nephron and many nephrons drain into each collecting duct, so renal corpuscles substantially outnumber collecting ducts.
4. "Tight" junctions of the renal tubule are quite leaky, and a substantial amount of tubular fluid passes through them to be reabsorbed by the paracellular route.
5. Diabetes insipidus shows no glucose in the urine.
6. Dilation of the efferent arteriole reduces resistance and thus lowers the glomerular blood pressure and filtration rate.
7. Angiotensin II stimulates aldosterone secretion and Na^+ reabsorption, thus reducing urine output.
8. Urine can be as dilute as 50 mOsm/L.
9. Normally there is abundant sodium but no glucose in the urine.
10. Micturition is caused by contraction of the detrusor.

CHAPTER 24

Figure Legend Questions

24.1 The tissue fluid compartment
24.8 Ingestion of water
24.10 It would decrease along with the pH.
24.13 Reverse both arrows to point to the left.

Testing Your Recall

1. c	9. d	16. hyperkalemia
2. a	10. b	17. hyponatremia
3. a	11. Na^+	18. respiratory acidosis
4. a	12. K^+ and Mg^{2+}	
5. d	13. metabolic water	19. limiting pH
6. c	14. cutaneous transpiration	20. osmolarity
7. e		
8. b	15. fluid sequestration	

Building Your Medical Vocabulary (Answers may vary; these are acceptable examples.)

1. food—hyperalimentation
2. blood—hypoxemia
3. intestine—parenteral
4. potassium—hyperkalemia
5. sodium—hyponatremia
6. next to—parenteral
7. isolate—sequestration
8. breathing—transpiration
9. across—transpiration
10. volume—hypovolemia

What's Wrong with These Statements?

1. This is an effect of hypokalemia, not hyperkalemia.
2. Aldosterone has only a small influence on blood pressure.
3. Such injuries elevate the ECF potassium level.
4. Phosphate concentration is less critical than that of other electrolytes and can safely vary over a relatively broad range.
5. PTH promotes calcium absorption but phosphate excretion.
6. Protein buffers more acid than bicarbonate or phosphates do.
7. Increased sodium reabsorption increases urinary H^+ excretion and lowers the urine pH.
8. Alkalosis tends to cause a reduction of respiratory rate and pulmonary ventilation so the ECF pH will rise again.
9. More water than salt is lost in true dehydration, so the body fluids become hypertonic.
10. Oral wetting and cooling have only a short-term effect on thirst.

CHAPTER 25

Figure Legend Questions

25.6 The first and second premolars and third molar

25.7 The enamel is a secretion, not a tissue; all the rest are living tissues.

25.11 Blockage of the mouth by the root of the tongue and blockage of the nose by the soft palate

25.12 The muscularis externa of the esophagus has two layers of muscle, with skeletal muscle in the upper to middle regions and smooth muscle in the middle to lower regions. In the stomach, it has three layers of muscle, all of which are smooth muscle.

25.14 It exchanges H^+ for K^+ (H^+–K^+ ATPase is an active transport pump).

25.20 The hepatic artery and the hepatic portal vein

25.32 Lipids do not enter the hepatic portal system that leads directly to the liver. Dietary fats are absorbed into the lacteals of the small intestine, then would have to travel the following route, at a minimum, to reach the liver: intestinal trunk → thoracic duct → left subclavian vein → heart → aorta → celiac trunk → common hepatic artery → hepatic artery proper → hepatic arteries → liver.

25.33 The internal anal sphincter is composed of smooth muscle and therefore controlled by the autonomic nervous system. The external anal sphincter is composed of skeletal muscle and therefore controlled by the somatic nervous system.

Testing Your Recall

1. b	8. a	15. vagus
2. d	9. a	16. gastrin
3. c	10. a	17. sinusoids
4. e	11. occlusal	18. dipeptidase
5. a	12. amylase, lipase	19. bile acids
6. c	13. parotid	20. iron
7. a	14. enteric	

Building Your Medical Vocabulary (Answers may vary; these are acceptable examples.)

1. cavity—antrum	6. liver—hepatocyte
2. juice—chylomicron	7. dry—jejunum
3. little—micelle	8. gateway—portal
4. vomiting—emetic	9. gateway—pyloric
5. bridle—frenulum	10. S-shaped—sigmoid

What's Wrong with These Statements?

1. Fat digestion begins in the stomach.
2. Most of the tooth is dentin.
3. Hepatocytes secrete bile into the bile canaliculi.
4. The ileal papilla regulates the passage of residue from the ileum of the small intestine into the cecum of the large intestine.
5. The lacteals take up chylomicrons, not micelles.
6. Hepcidin inhibits iron absorption and prevents iron overload.
7. The small intestine absorbs not only glucose but also fructose and galactose.
8. Most of the water is absorbed by the small intestine.
9. Secretin stimulates the liver and pancreas to secrete bicarbonate and inhibits gastric secretion.
10. Water, glucose, and other nutrients pass between cells, through the tight junctions.

CHAPTER 26

Figure Legend Questions

26.4 A high HDL:LDL ratio indicates that excess cholesterol is being transported to the liver for removal from the body. A high LDL:HDL ratio indicates a high rate of cholesterol deposition in the walls of the arteries.
26.7 NADH and FADH$_2$
26.11 Acidosis, ketoacidosis, or metabolic acidosis
26.12 Amino acids → keto acids → pyruvate → glucose

Testing Your Recall

1. a	8. a	15. liver
2. c	9. d	16. insulin
3. b	10. d	17. core temperature
4. e	11. incomplete	18. arcuate
5. b	12. glycogenolysis	19. cytochromes
6. e	13. gluconeogenesis	20. ATP synthase,
7. c	14. urea	ATP

Building Your Medical Vocabulary (Answers may vary; these are acceptable examples.)

1. bag—ascites	6. like—ascites
2. bad—cachexia	7. thin—leptin
3. color—cytochrome	8. splitting—glycolysis
4. producing—lipogenesis	9. new—gluconeogenesis
5. sugar—hypoglycemia	10. push—chemiosmotic

What's Wrong with These Statements?

1. Leptin suppresses the appetite.
2. A nutrient is a dietary substance that is absorbed into the tissues and becomes part of the body; water meets this criterion and is considered a nutrient.
3. Fat has more than twice as many calories per gram as carbohydrate does.
4. Most of the cholesterol is endogenous, not dietary.
5. Excessive protein intake generates excess nitrogenous waste and can cause renal damage.
6. Excessive ketone production is an effect of high-fat diets.
7. The membrane reactions produce up to 28 ATP per glucose, whereas glycolysis and the matrix reactions produce only 4 ATP.
8. Gluconeogenesis is a postabsorptive phenomenon.
9. Brown fat does not generate ATP.
10. At 21°C, the body loses about 60% of its heat by radiation.

CHAPTER 27

Figure Legend Questions

27.1 Both disorders result from defects in hormone receptors rather than from a lack of the respective hormone.
27.5 The word *vagina* means "sheath." The tunica vaginalis partially ensheathes the testis.
27.10 An enlarged prostate compresses the urethra and interferes with emptying of the bladder.
27.11 A tunica albuginea would allow excessive pressure to build in the corpus spongiosum, compressing the urethra and interfering with ejaculation.
27.13 The crossing-over in prophase I results in a mixture of maternal and paternal genes in each chromosome.
27.14 The next cell stage in meiosis, the secondary spermatocyte, is genetically different from the other cells of the body and would be subject to immune attack if not isolated from the antibodies in the blood.

Testing Your Recall

1. a	9. d	15. tunica albuginea
2. a	10. d	16. seminal vesicles
3. a	11. mesonephric	17. nurse
4. c	12. zinc	18. secondary
5. a	13. pampiniform	spermatocyte
6. d	plexus	19. deep
7. e	14. secondary	20. acrosome
8. c	spermatocytes	

Building Your Medical Vocabulary (Answers may vary; these are acceptable examples.)

1. hidden—cryptorchidism	6. reduction—meiosis
2. twins—epididymis	7. testis—cryptorchidism
3. out—ejaculation	8. network—rete testis
4. union—gamete	9. body—acrosome
5. condition—cryptorchidism	10. sheath—tunica vaginalis

What's Wrong with These Statements?

1. Testosterone is secreted by the interstitial endocrine cells.
2. The sperm DNA is contained in the nucleus in the sperm head.
3. Meiosis II is completed before spermiogenesis, which is the conversion of the four spermatids to mature sperm.
4. Only the testes are primary sex organs; the penis is a secondary sex organ.

5. Female fetal development results from a low testosterone level, not from estrogen.
6. The seminal vesicles contribute about 60% of the semen.
7. The pampiniform plexus prevents the testes from overheating.
8. Sperm have no testosterone receptors; testosterone binds to androgen-binding protein in the seminiferous tubules.
9. There is no such phenomenon as male menopause, and sperm production normally continues throughout old age.
10. Erection is the result of parasympathetic stimulation of the blood vessels of the erectile tissues.

CHAPTER 28

Figure Legend Questions

28.4   To move the egg or conceptus toward the uterus
28.8   Paraurethral glands
28.9   They cause milk to flow from the acinus into the ducts of the mammary gland.
28.10 This results in a clearer image since the X-rays do not have to penetrate such a thick mass of tissue.
28.18 The rising ratio of estradiol to progesterone makes the uterus more irritable.

Testing Your Recall

1. a	8. b	15. corona radiata
2. d	9. c	16. antrum
3. c	10. c	17. climacteric
4. a	11. follicle	18. conceptus
5. e	12. endometrium	19. infundibulum,
6. b	13. menarche	fimbriae
7. b	14. corpus luteum	20. lochia

Building Your Medical Vocabulary (Answers may vary; these are acceptable examples.)

1. beginning—menarche	6. milk—lactation
2. of—gravidarum	7. to tie—tubal ligation
3. mound—cumulus	8. uterus—endometrium
4. pregnancy—progesterone	9. egg—oogenesis
5. uterus—hysterectomy	10. first—primipara

What's Wrong with These Statements?

1. Only the ovum and cumulus oophorus cells enter the uterine tube, not the whole follicle.
2. HCG is secreted by the placenta.
3. Meiosis II is not completed until after ovulation, and only if the egg is fertilized.
4. Such girls often have lower than average body fat and delayed menarche because of this.
5. Many eggs and follicles undergo atresia during childhood, so their number is greatly reduced by the age of puberty.
6. Prolactin is secreted during pregnancy but does not induce lactation then.
7. Colostrum is lower in fat than milk is.
8. Up to two dozen follicles mature to the secondary follicle stage in each cycle, but usually only one of them ovulates.

9. Progesterone inhibits uterine contractions.
10. Only the superficial (functional) layer is shed.

CHAPTER 29

Figure Legend Questions

29.2   An unfertilized egg dies long before it reaches the uterus.
29.8   About 8 weeks
29.13 XXY (Klinefelter syndrome) and YO (a zygote that would not survive)
29.15 Female, as seen from the two X chromosomes at the lower right

Testing Your Recall

1. b	8. a	15. chorionic villi
2. b	9. d	16. acrosome
3. c	10. d	17. collagen
4. c	11. teratogens	18. Down syndrome
5. a	12. nondisjunction	(trisomy-21)
6. e	13. life span	19. foramen ovale
7. c	14. life expectancy	20. embryo

Building Your Medical Vocabulary (Answers may vary; these are acceptable examples.)

1. with—congenital	6. double—aneuploidy
2. falling off—decidual	7. aging—senescence
3. old age—progeria	8. together—syncytiotrophoblast
4. mulberry—morula	9. end—telomere
5. flat cake—placenta	10. monster—teratogen

What's Wrong with These Statements?

1. The ability to fertilize an egg increases in the first 10 hours as sperm become capacitated.
2. Fertilization occurs in the uterine tube.
3. Several early-arriving sperm clear a path for the one that fertilizes the egg.
4. Early cell divisions occur while the conceptus is still in the uterine tube, about 2 days before arrival in the uterus.
5. The individual is considered an embryo then; it is not a fetus until all of the major organs have formed.
6. As they develop, the chorionic villi become thinner and more permeable.
7. The kidneys exhibit one of the greatest degrees of senescence, shrinking 60% to 80% by age 90.
8. The foramen ovale is a shunt in the heart by which blood bypasses the lungs; the shunt that bypasses the liver is the ductus venosus.
9. Blood in the umbilical vein is returning from the placenta, where it has picked up oxygen and thus has a high P_{O_2}.
10. Telomerase is a telomere-repairing enzyme that prolongs cell life, as in stem cells and cancer cells.

APPENDIX D

Table D.2 Question
Asn-Ile-Tyr-Val-Arg-Asp

APPENDIX B

SYMBOLS, WEIGHTS, AND MEASURES

Units of Length

m	meter
km	kilometer (10^3 m)
cm	centimeter (10^{-2} m)
mm	millimeter (10^{-3} m)
μm	micrometer (10^{-6} m)
nm	nanometer (10^{-9} m)

Units of Mass and Weight

amu	atomic mass unit
MW	molecular weight
mole	MW in grams
g	gram
kg	kilograms (10^3 g)
mg	milligrams (10^{-3} g)
μg	micrograms (10^{-6} g)

Units of Pressure

atm	atmospheres (1 atm = 760 mm Hg)
mm Hg	millimeters of mercury
P_X	partial pressure of gas X (as in Po_2)

Conversion Factors

1 in. = 2.54 cm	1 cm = 0.394 in.
1 fl oz = 29.6 mL	1 mL = 0.034 fl oz
1 qt = 0.946 L	1 L = 1.057 qt
1 g = 0.0035 oz	1 oz = 28.38 g
1 lb = 0.45 kg	1 kg = 2.2 lb
°C = (5/9)(°F − 32)	°F = (9/5)(°C) + 32

Units of Volume

L	liter
dL	deciliter (= 100 mL) (10^{-1} L)
mL	milliliter (10^{-3} L)
μL	microliter (= 1 mm³) (10^{-6} L)

Units of Heat

cal	"small" calorie
kcal	kilocalorie (Calorie; 1 kcal = 1,000 cal)
Cal	"large" (dietary) calorie (1 Cal = 1,000 cal)

Greek Letters

α	alpha
β	beta
γ	gamma
Δ	delta (uppercase)
δ	delta (lowercase)
η	eta
θ	theta
λ	lambda
μ	mu
π	pi

Units of Concentration

Chemical concentrations—the amounts of solute in a given volume of solution—are expressed in different ways for different scientific or clinical purposes. Some of these are explained here, particularly those used in this book.

WEIGHT PER VOLUME

A simple way to express concentration is the weight of solute in a given volume of solution. For example, intravenous (I.V.) saline typically contains 8.5 grams of NaCl per liter of solution (8.5 g/L). For many biological purposes, however, we deal with smaller quantities such as *milligrams per deciliter (mg/dL)*. For example, a typical serum cholesterol concentration may be 200 mg/dL, also expressed as 200 mg/100 mL or 200 milligram-percent (mg-%).

PERCENTAGES

Percentage concentration is also simple to compute, but it is necessary to specify whether the percentage refers to the weight or to the volume of solute in a given volume of solution. For example, if we begin with 5 g of dextrose (an isomer of glucose) and add enough water to make 100 mL of solution, the resulting concentration will be 5% weight per volume (w/v). A commonly used intravenous fluid is D5W, which stands for 5% w/v dextrose in distilled water.

If the solute is a liquid, such as ethanol, percentages refer to volume of solute per volume of solution. Thus, 70 mL of ethanol diluted with water to 100 mL of solution produces 70% volume per volume (70% v/v) ethanol.

MOLARITY

Percentage concentrations are easy to prepare, but that unit of measurement is inadequate for many purposes. The physiological effect of a chemical depends on how many molecules of it are present in a given volume, not the weight of the chemical. Five percent glucose, for example, contains almost twice as many sugar molecules as the same volume of 5% sucrose. Each solution contains 50 g of sugar per liter, but glucose has a molecular weight (MW) of 180 and sucrose has a MW of 342. Since each molecule of glucose is lighter, 50 g of glucose contains more molecules than 50 g of sucrose.

To produce solutions with a known number of molecules per volume, we must factor in the molecular weight. If we know the MW and weigh out that many grams of the substance, we have a quantity known as 1 *mole*. One mole of glucose is 180 g and 1 mole of sucrose is 342 g. Each quantity contains the same number of molecules of the respective sugar—a number known as *Avogadro's number,* 6.023×10^{23} molecules per mole.

Molarity (M) is the number of moles of solute per liter of solution. A one-molar (1.0 M) solution of glucose contains 180 g/L, and 1.0 M solution of sucrose contains 342 g/L. Both have the same number of solute molecules in a given volume. Body fluids and laboratory solutions usually are less concentrated than 1 M, so biologists and clinicians more often work with *millimolar* (mM, 10^{-3} M) and *micromolar* (μM, 10^{-6} M) concentrations.

OSMOLARITY AND OSMOLALITY

The osmotic concentration of body fluids has a great effect on cellular function, body water distribution, nutrient absorption, urinary water loss, the sense of thirst, and blood pressure. Thus, it is important to quantify osmotic concentrations in physiology and in clinical practice (as when giving I.V. fluid therapy). Physiologists and clinicians usually express osmotic concentration in *milliosmoles per liter* (mOsm/L).

One *osmole* is 1 mole of dissolved particles. If a solute does not ionize in water, then 1 mole of the solute yields 1 osmole (osm) of dissolved particles. A solution of 1 M glucose, for example, is also 1 osm/L. If a solute does ionize, it yields two or more dissolved particles in solution. A 1 M solution of NaCl, for example, contains 1 mole/L of sodium ions and 1 mole/L of chloride ions; 1 M calcium chloride ($CaCl_2$) would (if it completely dissociated) yield 1 mole of calcium ions and 2 moles of chloride ions. All ions equally affect osmosis and must be separately counted in a measure of osmotic concentration. Thus,

$$1 \text{ M NaCl} = 2 \text{ osm/L, and}$$

$$1 \text{ M CaCl}_2 = 3 \text{ osm/L.}$$

Osmolality is the number of osmoles of solute per kilogram of water, and *osmolarity* is the number of osmoles per liter of solution. Most clinical calculations are based on osmolarity, since it is easier to measure the volume of a solution than the weight of water it contains. The difference between osmolality and osmolarity can be important in experimental work, but at the concentrations of human body fluids, there is less than 1% difference between the two, and the two are essentially interchangeable for clinical purposes.

All body fluids and many clinical solutions are mixtures of many chemicals. The osmolarity of such a solution is the total osmotic concentration of all of its dissolved particles.

A concentration of 1 osm/L is substantially higher than we find in most body fluids, so physiological concentrations are usually expressed in terms of milliosmoles per liter (mOsm/L) (1 mOsm/L = 10^{-3} osm/L). Blood plasma, tissue fluid, and intracellular fluid measure about 300 mOsm/L.

MILLIEQUIVALENTS PER LITER

Electrolytes are important for their chemical, physical (osmotic), and electrical effects on the body. Their electrical effects, which determine such things as nerve, heart, and muscle actions, depend not only on their concentration but also on their electrical charge. A calcium ion (Ca^{2+}) has twice the electrical effect of a sodium ion (Na^+), for example, because it carries twice the charge. In measuring electrolyte concentrations, one must take the charges into account.

One *equivalent* (Eq) of an electrolyte is the amount that would electrically neutralize 1 mole of hydrogen ions (H^+) or hydroxide ions (OH^-). For example, 1 mole (58.4 g) of NaCl yields 1 mole, or 1 Eq, of Na^+ in solution. Thus, an NaCl solution of 58.4 g/L contains 1 equivalent of Na^+ per liter (1 Eq/L). One mole (98 g) of sulfuric acid (H_2SO_4) yields 2 moles of positive charges (H^+). Thus, 98 g of sulfuric acid per liter would be a solution of 2 Eq/L.

The electrolytes in human body fluids have concentrations less than 1 Eq/L, so we more often express their concentrations in *milliequivalents per liter* (mEq/L). If you know the millimolar concentration of an electrolyte, you can easily convert this to mEq/L by multiplying it by the valence (charge) of the ion:

$$1 \text{ mM Na}^+ = 1 \text{ mEq/L}$$

$$1 \text{ mM Ca}^{2+} = 2 \text{ mEq/L}$$

$$1 \text{ mM Fe}^{3+} = 3 \text{ mEq/L}$$

ACIDITY AND ALKALINITY (PH)

Acidity is expressed in terms of pH, a measure derived from the molarity of H^+. Molarity is represented by square brackets, so the molarity of H^+ is symbolized $[H^+]$. pH is the negative logarithm of hydrogen ion molarity; that is,

$$pH = -\log [H^+].$$

In pure water, 1 in 10 million molecules of H_2O ionizes into hydrogen and hydroxide ions: $H_2O \rightleftharpoons H^+ + OH^-$. Pure water has a neutral pH because it contains equal amounts of H^+ and OH^-. Since 1 in 10 million molecules ionize, the molarity of H^+ and the pH of water are as follows:

$$[H^+] = 0.0000001 \text{ M} = 10^{-7} \text{ M}$$

$$\log [H^+] = -7$$

$$pH = -\log [H^+] = 7$$

The pH scale ranges from 0 to 14 and is logarithmic, so each integer up or down the scale represents a 10-fold difference in $[H^+]$. This is exemplified by the following three strongly acidic pH values:

If $[H^+] = 0.1$ M	$pH = -\log 10^{-1} = 1.0$
If $[H^+] = 0.01$ M	$pH = -\log 10^{-2} = 2.0$
If $[H^+] = 0.001$ M	$pH = -\log 10^{-3} = 3.0$

The less concentrated the H^+, the higher the pH. pH values below 7.0 are considered acidic. Above 7.0, they are basic or alkaline. The following three values lie at the basic end of the pH scale:

If $[H^+] = 0.000000000001$ M	$pH = -\log 10^{-12} = 12.0$
If $[H^+] = 0.0000000000001$ M	$pH = -\log 10^{-13} = 13.0$
If $[H^+] = 0.00000000000001$ M	$pH = -\log 10^{-14} = 14.0$

PERIODIC TABLE OF THE ELEMENTS

Nineteenth-century chemists discovered that when they arranged the known elements by atomic weight, certain properties reappeared periodically. In 1869, Russian chemist Dmitri Mendeleev published the first modern periodic table of the elements, leaving gaps for those that had not yet been discovered. He accurately predicted properties of the missing elements, which helped other chemists discover and isolate them.

Each row in the table is a *period* and each column is a *group (family)*. Each period has one electron shell more than the period above it, and as we progress from left to right within a period, each element has one more proton and electron than the one before. The dark steplike line from boron (5) to astatine (85) separates the metals to the left of it (except hydrogen) from the nonmetals to the right. Each period begins with a soft, light, highly reactive *alkali metal,* with one valence electron, in family

IA. Progressing from left to right, the metallic properties of the elements become less and less pronounced. Elements in family VIIA are highly reactive gases called *halogens,* with seven valence electrons. Elements in family VIIIA, called *noble (inert) gases,* have a full valence shell of eight electrons, which makes them chemically unreactive.

Ninety-one of the elements occur naturally on earth. Physicists have created elements up to atomic number 118 in the laboratory, but the International Union of Pure and Applied Chemistry has established formal names only through element 112 to date.

The 24 elements with normal roles in human physiology are color-coded according to their relative abundance in the body (see table 2.1). Others, however, may be present as contaminants with very destructive effects (such as arsenic, lead, and radiation poisoning).

THE GENETIC CODE AND AMINO ACIDS

Table D.1 lists the 20 amino acids involved in protein structure, with the three-letter and one-letter symbols used in writing out amino acid sequences (as in fig. 2.25, for example). **Table D.2** is the genetic code, relating mRNA codons to the amino acids. To determine what amino acid is encoded by a given mRNA codon, choose the first letter of the codon from the column on the left, the second letter from the row across the top, and the third letter from the column on the right. The amino acid symbol is at the intersection of these three. For example, codon GCU codes for alanine (Ala) and CAC codes for histidine (His). Codons UAA, UAG, and UGA are *stop codons,* which do not code for any amino acid but indicate the end of the message to be translated, like the period at the end of a sentence.

TABLE D.1	The 20 Amino Acids and Their Symbols				
Alanine	Ala	A	Leucine	Leu	L
Arginine	Arg	R	Lysine	Lys	K
Asparagine	Asn	N	Methionine	Met	M
Aspartate	Asp	D	Phenylalanine	Phe	F
Cysteine	Cys	C	Proline	Pro	P
Glutamate	Glu	E	Serine	Ser	S
Glutamine	Gln	Q	Threonine	Thr	T
Glycine	Gly	G	Tryptophan	Trp	W
Histidine	His	H	Tyrosine	Tyr	Y
Isoleucine	Ile	I	Valine	Val	V

? The following genetic code represents a well-known hormone, angiotensin II: UUU-CCC-CAC-AUA-UAU-GUA-AGG-GAC. The amino acid sequence represented by this code begins Phe-Pro. . . . Write the rest of the sequence. (Answer is found in appendix A.)

TABLE D.2	The Genetic Code				
First Position	**Second Position**				**Third Position**
	A	G	C	U	
A	Lys	Arg	Thr	Ile	A
	Lys	Arg	Thr	Met	G
	Asn	Ser	Thr	Ile	C
	Asn	Ser	Thr	Ile	U
G	Glu	Gly	Ala	Val	A
	Glu	Gly	Ala	Val	G
	Asp	Gly	Ala	Val	C
	Asp	Gly	Ala	Val	U
C	Gln	Arg	Pro	Leu	A
	Gln	Arg	Pro	Leu	G
	His	Arg	Pro	Leu	C
	His	Arg	Pro	Leu	U
U	STOP	STOP	Ser	Leu	A
	STOP	Trp	Ser	Leu	G
	Tyr	Cys	Ser	Phe	C
	Tyr	Cys	Ser	Phe	U

$H_2N-\overset{\displaystyle H}{\underset{\displaystyle CH_3}{C}}-COOH$	$H_2N-\overset{\displaystyle H}{\underset{\displaystyle (CH_2)_3 \\ NH \\ C=N^+H_2 \\ NH_2}{C}}-COOH$	$H_2N-\overset{\displaystyle H}{\underset{\displaystyle CH_2 \\ H_2N-C=O}{C}}-COOH$	$H_2N-\overset{\displaystyle H}{\underset{\displaystyle CH_2 \\ COOH}{C}}-COOH$
Alanine	**Arginine**	**Asparagine**	**Aspartic acid**
$H_2N-\overset{\displaystyle H}{\underset{\displaystyle CH_2 \\ SH}{C}}-COOH$	$H_2N-\overset{\displaystyle H}{\underset{\displaystyle CH_2 \\ CH_2 \\ COOH}{C}}-COOH$	$H_2N-\overset{\displaystyle H}{\underset{\displaystyle CH_2 \\ CH_2 \\ H_2N-C=O}{C}}-COOH$	$H_2N-\overset{\displaystyle H}{\underset{\displaystyle H}{C}}-COOH$
Cysteine	**Glutamic acid**	**Glutamine**	**Glycine**
Histidine	**Isoleucine**	**Leucine**	**Lysine**
Methionine	**Phenylalanine**	**Proline**	**Serine**
Threonine	**Tryptophan**	**Tyrosine**	**Valine**

MEDICAL WORD ROOTS AND AFFIXES

a- no, not, without (atom, agranulocyte)
ab- away (abducens, abduction)
acetabulo- small cup (acetabulum)
acro- tip, extremity, peak (acromion, acromegaly)
ad- to, toward, near (adsorption, adrenal)
adeno- gland (lymphadenitis, adenohypophysis)
aero- air, oxygen (aerobic, anaerobic)
af- toward (afferent)
ag- together (agglutination)
-al pertaining to (parietal, pharyngeal, temporal)
ala- wing (ala nasi)
albi- white (albicans, linea alba, albino)
algi- pain (analgesic, myalgia)
aliment- nourishment (alimentary)
allo- other, different (allele, allograft)
amphi- both, either (amphiphilic, amphiarthrosis)
an- without (anaerobic, anemic)
ana- 1. up, build up (anabolic, anaphylaxis)
   2. apart (anaphase, anatomy)
   3. back (anastomosis)
andro- male (androgen)
angi- vessel (angiogram, angioplasty, hemangioma)
ante- before, in front (antebrachium)
antero- forward (anterior, anterograde)
anti- against (antidiuretic, antibody, antagonist)
apo- from, off, away, above (apocrine, aponeurosis)
arbor- tree (arboreal, arborization)
artic- 1. joint (articulation)
   2. speech (articulate)
-ary pertaining to (axillary, coronary)
-ase enzyme (polymerase, kinase, amylase)
ast-, astro- star (aster, astrocyte)
-ata, -ate 1. possessing (hamate, corniculate)
   2. plural of -a (stomata, carcinomata)
athero- fat (atheroma, atherosclerosis)
atrio- entryway (atrium, atrioventricular)
auri- ear (auricle, binaural)
auto- self (autolysis, autoimmune)
axi- axis, straight line (axial, axoneme, axon)
baro- pressure (baroreceptor, hyperbaric)
bene- good, well (benign, beneficial)
bi- two (bipedal, biceps, bifid)
bili- bile (biliary, bilirubin)
bio- life, living (biology, biopsy, microbial)
blasto- precursor, bud, producer (fibroblast, osteoblast, blastomere)
brachi- arm (brachium, brachialis, antebrachium)
brady- slow (bradycardia, bradypnea)
bucco- cheek (buccal, buccinator)
burso- purse (bursa, bursitis)
calc- calcium, stone (calcaneus, hypocalcemia)
callo- thick (callus, callosum)
calori- heat (calorie, calorimetry, calorigenic)
calv-, calvari- bald, skull (calvaria)

calyx cup, vessel, chalice (glycocalyx, renal calyx)
capito- head (capitis, capitate, capitulum)
capni- smoke, carbon dioxide (hypocapnia)
carcino- cancer (carcinogen, carcinoma)
cardi- heart (cardiac, cardiology, pericardium)
carot- 1. carrot (carotene)
   2. stupor (carotid)
carpo- wrist (carpus, metacarpal)
case- cheese (caseosa, casein)
cata- down, break down (catabolism)
cauda- tail (cauda equina, caudate nucleus)
-cel little (pedicel)
celi- belly, abdomen (celiac)
centri- center, middle (centromere, centriole)
cephalo- head (cephalic, encephalitis)
cervi- neck, narrow part (cervix, cervical)
chiasm- cross, X (optic chiasm)
choano- funnel (choana)
chole- bile (cholecystokinin, cholelithotripsy)
chondro- 1. grain (mitochondria)
   2. cartilage, gristle (chondrocyte, perichondrium)
chromo- color (dichromat, chromatin, cytochrome)
chrono- time (chronotropic, chronic)
cili- eyelash (cilium, superciliary)
circ- about, around (circadian, circumduction)
cis- cut (incision, incisor)
cistern, cisterna reservoir (Golgi cistern, cisterna chyli)
clast- break down, destroy (osteoclast)
clavi- hammer, club, key (clavicle, supraclavicular)
-cle little (tubercle, corpuscle)
cleido- clavicle (sternocleidomastoid)
cnemo- lower leg (gastrocnemius)
co- together (coenzyme, cotransport)
collo- 1. hill (colliculus)
   2. glue (colloid, collagen)
contra- opposite (contralateral)
corni- horn (cornified, corniculate, cornu)
corono- crown (coronary, corona, coronal)
corpo- body (corpus luteum, corpora quadrigemina)
corti- bark, rind (cortex, cortical)
costa- rib (intercostal, subcostal)
coxa- hip (os coxae, coxal)
crani- helmet (cranium, epicranius)
cribri- sieve, strainer (cribriform, area cribrosa)
crino- separate, secrete (holocrine, endocrinology)
crista- crest (crista galli, mitochondrial crista)
crito- to separate (hematocrit)
cruci- cross (cruciate ligament)
-cule, -culus small (canaliculus, trabecula, auricular)
cune- wedge (cuneiform, cuneatus)
cutane-, cuti- skin (subcutaneous, cuticle)
cysto- bladder (cystitis, cholecystectomy)
cyto- cell (cytology, cytokinesis, monocyte)

de- down (defecate, deglutition, dehydration)
demi- half (demifacet, demilune)
den-, denti- tooth (dentition, dens, dental)
dendro- tree, branch (dendrite, oligodendrocyte)
derma-, dermato- skin (dermatology, hypodermic)
desmo- band, bond, ligament (desmosome, syndesmosis)
dia- 1. across, through, separate (diaphragm, dialysis)
    2. day (circadian)
dis- 1. apart (dissect, dissociate)
    2. opposite, absence (disinfect, disability)
diure- pass through, urinate (diuretic, diuresis)
dorsi- back (dorsal, dorsum, latissimus dorsi)
duc- to carry (duct, adduction, abducens)
dys- bad, abnormal, painful (dyspnea, dystrophy)
e- out (ejaculate, eversion)
-eal pertaining to (arboreal, pineal)
ec-, ecto- outside, out of, external (ectopic, ectoderm, splenectomy)
ef- out of (efferent, effusion)
-elle small (fontanelle, organelle, micelle)
electro- electricity (electrocardiogram, electrolyte)
em- in, within (embolism, embedded)
emesi-, emeti- vomiting (emetic, hyperemesis)
-emia blood condition (anemia, hypoxemia)
en- in, into (enzyme, parenchyma)
encephalo- brain (encephalitis, telencephalon)
enchymo- poured in (mesenchyme, parenchyma)
endo- within, into, internal (endocrine, endocytosis)
entero- gut, intestine (mesentery, myenteric)
epi- upon, above (epidermis, epiphysis, epididymis)
ergo- work, energy, action (allergy, adrenergic, synergist)
eryth-, erythro- red (erythema, erythrocyte)
esthesio- sensation, feeling (anesthesia, somesthetic)
eu- good, true, normal, easy (eupnea, aneuploidy)
exo- out (exopeptidase, exocytosis, exocrine)
facili- easy (facilitated)
fasci- band, bundle (fascia, fascicle)
fenestr- window (fenestrated)
fer- to carry (efferent, uriniferous)
ferri- iron (ferritin, transferrin)
fibro- fiber (fibroblast, fibrosis)
fili- thread (myofilament, filiform)
flagello- whip (flagellum)
foli- leaf (folic acid, folia)
-form shape (cuneiform, fusiform)
fove- pit, depression (fovea)
funiculo- little rope, cord (funiculus)
fusi- 1. spindle (fusiform)
    2. pour out (perfusion)
gamo- marriage, union (monogamy, gamete)
gastro- belly, stomach (gastrointestinal, digastric)
-gen, -genic, -genesis producing, giving rise to (pathogen, carcinogenic, glycogenesis)
genio- chin (geniohyoid, genioglossus)
germi- 1. sprout, bud (germinal, germinativum)
    2. microbe (germicide)
gero- old age (progeria, geriatrics, gerontology)
gesto- 1. to bear, carry (ingest).
    2. pregnancy (gestation, progesterone)
glia- glue (neuroglia, microglia)
globu- ball, sphere (globulin, hemoglobin)
glom- ball (glomerulus)

glosso- tongue (glossopharyngeal, hypoglossal)
glyco- sugar (glycogen, glycolysis, hypoglycemia)
gono- 1. angle, corner (trigone)
    2. seed, sex cell, generation (gonad, oogonium, gonorrhea)
gradi- walk, step (retrograde, gradient)
-gram recording of (electrocardiogram, sonogram)
-graph recording instrument (sonograph, electrocardiograph)
-graphy recording process (sonography, radiography)
gravi- severe, heavy (gravid, myasthenia gravis)
gyro- turn, twist (gyrus)
hallu- great toe (hallucis)
hemato- blood (hematology, hematopoiesis)
hemi- half (hemidesmosome, hemisphere)
-hemia blood condition (polycythemia)
hemo- blood (hemophilia, hemoglobin)
hetero- different, other, various (heterozygous)
histo- tissue, web (histology, histone)
holo- whole, entire (holistic, holocrine)
homeo- constant, unchanging, uniform (homeostasis, homeothermic)
homo- same, alike (homologous, homozygous)
hyalo- clear, glassy (hyaline, hyaluronic acid)
hydro- water (dehydration, hydrolysis, hydrophobic)
hyper- above, above normal, excessive (hyperkalemia, hypertonic)
hypo- below, below normal, deficient (hypogastric, hyponatremia, hypophysis)
-ia condition (anemia, hypocalcemia, osteomalacia)
-ial pertaining to (hypophysial, epiphysial)
-ic pertaining to (isotonic, hemolytic, antigenic)
-icle, -icul small (ossicle, canaliculus, reticular)
ilia- flank, loin (ilium, iliac)
-illa, -illus little (papilla)
-in protein, substance (trypsin, fibrin, melanin)
infra- below (infraspinous, infrared)
ino- fiber (inotropic, inositol)
insulo- island (insula, insulin)
inter- between (intercellular, intervertebral)
intra- within (intracellular, intraocular)
iono- ion (ionotropic, cationic)
ischi- to hold back (ischium, ischemia)
-ism 1. process, state, condition (metabolism, rheumatism)
    2. doctrine, belief, theory (holism, reductionism, naturalism)
iso- same, equal (isometric, isotonic, isomer)
-issimus most, greatest (latissimus, longissimus)
-ite little (dendrite, somite)
-itis inflammation (dermatitis, gingivitis)
jug- to join (conjugated, jugular)
juxta- next to (juxtamedullary, juxtaglomerular)
kali- potassium (hypokalemia)
karyo- seed, nucleus (megakaryocyte, karyotype)
kerato- horn (keratin, keratinocyte)
kine- motion, action (kinetic, kinase, cytokinesis)
labi- lip (labium, levator labii)
lacera- torn, cut (foramen lacerum, laceration)
lacrimo- tear, cry (lacrimal gland, nasolacrimal)
lacto- milk (lactose, lactation, prolactin)
lamina- layer (lamina propria, laminar flow)
latero- side (bilateral, ipsilateral)
lati- broad (fascia lata, latissimus dorsi)
-lemma husk (sarcolemma, neurilemma)
lenti- lens (lentiform)
-let small (platelet)

leuko- white (leukocyte, leukemia)
levato- to raise (levator labii, elevation)
ligo- to bind (ligand, ligament)
line- line (linea alba, linea nigra)
litho- stone (otolith, lithotripsy)
-logy study of (histology, physiology, hematology)
lucid- light, clear (stratum lucidum, zona pellucida)
lun- moon, crescent (lunate, lunule, semilunar)
lute- yellow (macula lutea, corpus luteum)
lyso-, lyto- split apart, break down (lysosome, hydrolysis, electrolyte, hemolytic)
macro- large (macromolecule, macrophage)
macula- spot (macula lutea, macula densa)
mali- bad (malignant, malocclusion, malformed)
malle- hammer (malleus, malleolus)
mammo- breast (mammary, mammillary)
mano- hand (manus, manipulate)
manubri- handle (manubrium)
masto- breast (mastoid, gynecomastia)
medi- middle (medial, mediastinum, intermediate)
medullo- marrow, pith (medulla)
mega- large (megakaryocyte, hepatomegaly)
melano- black (melanin, melanocyte, melancholy)
meno- month (menstruation, menopause)
mento- chin (mental, mentalis)
mero- part, segment (isomer, centromere, merocrine)
meso- in the middle (mesoderm, mesentery)
meta- beyond, next in a series (metaphase, metacarpal)
metabol- change (metabolism, metabolite)
-meter measuring device (calorimeter, spirometer)
metri- 1. length, measure (isometric, emmetropic).
    2. uterus (endometrium)
micro- small (microscopic, microcytic, microglia)
mito- thread, filament (mitochondria, mitosis)
mono- one (monocyte, monogamy, mononucleosis)
morpho- form, shape, structure (morphology, amorphous)
muta- change (mutagen, mutation)
myelo- 1. spinal cord (poliomyelitis, myelin)
    2. bone marrow (myeloid, myelocytic)
myo-, mysi- muscle (myoglobin, myosin, epimysium)
natri- sodium (hyponatremia, natriuretic)
neo- new (neonatal, gluconeogenesis)
nephro- kidney (nephron, hydronephrosis)
neuro- nerve (aponeurosis, neurosoma, neurology)
nucleo- nucleus, kernel (nucleolus, nucleic acid)
oo- egg (oogenesis, oocyte)
ob- 1. life (aerobic, microbe)
    2. against, toward, before (obstetrics, obturator, obstruction)
oculo- eye (oculi, oculomotor)
odonto- tooth (odontoblast, periodontal)
-oid like, resembling (colloid, sigmoid, ameboid)
-ole small (arteriole, bronchiole, nucleolus)
oligo- few, a little, scanty (oligopeptide, oliguria)
-oma tumor, mass (carcinoma, hematoma)
omo- shoulder (omohyoid, acromion)
onycho- nail, claw (hyponychium, onychomycosis)
op- vision (optics, myopia, photopic)
-opsy viewing, to see (biopsy, rhodopsin)
or- mouth (oral, orbicularis oris)
orbi- circle (orbicularis, orbit)
organo- tool, instrument (organ, organelle)

ortho- straight (orthopnea, orthodontics, orthopedics)
-ose 1. full of (adipose)
    2. sugar (sucrose, glucose)
-osis 1. process (osmosis, exocytosis)
    2. condition, disease (cyanosis, thrombosis).
    3. increase (leukocytosis)
osmo- push (osmosis, chemiosmotic)
osse-, oste- bone (osseous, osteoporosis)
oto- ear (otolith, otitis, parotid)
-ous 1. full of (nitrogenous, edematous)
    2. pertaining to (mucous, nervous)
    3. like, characterized by (squamous, filamentous)
ovo- egg (ovum, ovary, ovulation)
oxy- 1. oxygen (hypoxia, oxyhemoglobin)
    2. sharp, quick (oxytocin)
palli- pale (pallor, globus pallidus)
palpebro- eyelid (palpebrae)
pan- all (panhypopituitarism, pancreas)
panni- cloth, rag (pannus, panniculus)
papillo- nipple (papilla, papillary)
par- birth (postpartum, parturition, multiparous)
para- next to (parathyroid, parotid)
parieto- wall (parietal)
patho- 1. disease (pathology, pathogen)
    2. feeling (sympathetic)
pecto- 1. chest (pectoralis)
    2. comblike (pectineus)
pedi- 1. foot (bipedal, pedicle)
    2. child (pediatrics)
pelvi- basin (pelvis, pelvic)
-penia deficiency (leukopenia, thrombocytopenia)
penna- feather (unipennate, bipennate)
peri- around (periosteum, peritoneum, periodontal)
perone- fibula (peroneus tertius, peroneal nerve)
phago- eat (phagocytosis, macrophage)
philo- loving, attracted to (hydrophilic, amphiphilic)
phobo- fearing, repelled by (hydrophobic)
phor- to carry, bear (diaphoresis, electrophoresis)
phragm- partition (diaphragm)
phreno- diaphragm (phrenic nerve)
physio- nature, natural cause (physiology, physician)
-physis growth (diaphysis, hypophysis)
pilo- hair (piloerection)
pino- drink, imbibe (pinocytosis)
planto- sole of foot (plantaris, plantar wart)
plasi- growth (hyperplasia, neoplasia)
plasm- shaped, molded (cytoplasm, endoplasmic)
plasti- form (thromboplastin)
platy- flat (platysma)
pnea- breath, breathing (eupnea, dyspnea)
pneumo- air, breath, lung (pneumonia, pneumothorax)
podo- foot (pseudopod, podocyte)
poies- forming (hematopoiesis, erythropoietin)
poly- many, much, excessive (polypeptide, polyuria)
primi- first (primary, primipara, primitive)
pro- 1. before, in front, first (prokaryote, prophase, prostate)
    2. promote, favor (progesterone, prolactin)
pseudo- false (pseudopod)
psycho- mind (psychosis, psychosomatic)
ptero-, pterygo- wing (pterygoid)
-ptosis dropping, falling, sagging (apoptosis, nephroptosis)

puncto- point (puncta)
pyro- fire (pyrogen, antipyretic)
quadri- four (quadriceps, quadratus)
quater- fourth (quaternary)
radiat- radiating (corona radiata)
rami- branch (ramus)
recto- straight (rectus abdominis, rectum)
reno- kidney (renal, renin)
reti- network (reticular, rete testis)
retinac- retainer, bracelet (retinaculum)
retro- behind, backward (retroperitoneal, retrovirus)
rhombo- rhombus (rhomboideus, rhombencephalon)
rubo-, rubro- red (bilirubin, rubrospinal)
rugo- fold, wrinkle (ruga, corrugator)
sacculo- little sac (saccule)
sarco- flesh, muscle (sarcoplasm, sarcomere)
scala- staircase (scala tympani)
sclero- hard, tough (sclera, sclerosis)
scopo- see (microscope, endoscopy)
secto- cut (section, dissection)
semi- half (semilunar, semimembranosus)
sepsi- infection (asepsis, septicemia)
-sis process (diapedesis, amniocentesis)
sole- sandal, sole of foot, flatfish (sole, soleus)
soma-, somato- body (somatic, somatotropin)
spheno- wedge (sphenoid)
spiro- breathing (inspiration, spirometry)
splanchno- viscera (splanchnic)
spleno- 1. bandage (splenius capitis)
    2. spleen (splenic artery)
squamo- scale, flat (squamous, desquamation)
stasi-, stati- put, remain, stay the same (hemostasis, homeostatic)
steno- narrow (stenosis)
ster-, stereo- solid, three-dimensional (steroid, stereoscopic)
sterno- breast, chest (sternum, sternocleidomastoid)
stria- stripe (striated, corpus striatum)
sub- below (subcutaneous, subclavicular)
sulc- furrow, groove (sulcus)
supra- above (supraspinous, supraclavicular)
sura- calf of leg (triceps surae)
sym- together (sympathetic, symphysis)
syn- together (synostosis, syncytium)
tachy- fast (tachycardia, tachypnea)

tarsi- ankle (tarsus, metatarsal)
tecto- roof, cover (tectorial membrane, tectum)
telo- last, end (telophase, telencephalon)
tempo- time (temporal)
terti- third (tertiary)
theli- nipple, female, tender (epithelium, polythelia)
thermo- heat (thermogenesis, thermoregulation)
thrombo- blood clot (thrombosis, thrombin)
thyro- shield (thyroid, thyrohyoid)
-tion process (circulation, pronation)
toci- birth (oxytocin)
tomo- 1. cut (tomography, atom, anatomy)
    2. segment (dermatome, myotome, sclerotome)
tono- force, tension (isotonic, tonus, myotonia)
topo- place, position (isotope, ectopic)
trabo- plate (trabecula)
trans- across (transpiration, transdermal)
trapezi- 1. table, grinding surface (trapezium)
    2. trapezoidal (trapezius)
tri- three (triceps, triglyceride)
tricho- hair (trichosiderin, peritrichial)
trocho- wheel, pulley (trochlea)
troph- 1. food, nourishment (trophic, trophoblast)
    2. growth (dystrophy, hypertrophy)
tropo- to turn, change (metabotropic, gonadotropin)
tunica- coat (tunica intima, tunica vaginalis)
tympano- drum, eardrum (tympanic, tensor tympani)
-ul small (trabecula, tubule, capitulum, glomerulus)
-uncle, -unculus small (homunculus, caruncle)
uni- one (unipennate, unipolar)
uri- urine (glycosuria, urinalysis, diuretic)
utriculo- little bag (utriculus)
vagino- sheath (invaginate, tunica vaginalis)
vago- wander (vagus)
vaso- vessel (vascular, vas deferens, vasa recta)
ventro- belly, lower part (ventral, ventricle)
vermi- worm (vermis, vermiform appendix)
vertebro- spine (vertebrae, intervertebral)
vesico- bladder, blister (vesical, vesicular)
villo- hair, hairy (microvillus)
vitre- glass (in vitro, vitreous humor)
vivi- life, alive (in vivo, revive)
zygo- union, join, mate (zygomatic, zygote, azygos)

GLOSSARY

A

abdominal cavity The body cavity between the diaphragm and pelvic brim. fig. A.5

abduction (ab-DUC-shun) Movement of a body part away from the median plane, as in raising an arm away from the side of the body. fig. 9.13

absorption 1. Process in which a chemical passes through a membrane or tissue surface and becomes incorporated into a body fluid or tissue. **2.** Any process in which one substance passes into another and becomes a part of it. *Compare* adsorption.

acetylcholine (ACh) (ASS-eh-till-CO-leen) A neurotransmitter released by somatic motor fibers, parasympathetic fibers, and some other neurons, composed of choline and an acetyl group. fig. 12.22

acetylcholinesterase (AChE) (ASS-eh-till-CO-lin-ESS-ter-ase) An enzyme that hydrolyzes acetylcholine, thus halting signal transmission at a cholinergic synapse.

acid A proton (H+) donor; a chemical that releases protons into solution.

acidosis An acid–base imbalance in which the blood pH is lower than 7.35.

acinus (ASS-ih-nus) A sac of secretory cells at the inner end of a gland duct. fig. 5.30

actin A filamentous intracellular protein that provides cytoskeletal support and interacts with other proteins, especially myosin, to cause cellular movement; important in muscle contraction and membrane actions such as phagocytosis, ameboid movement, and cytokinesis. fig. 11.3

action The movement produced by contraction of a particular muscle, or its role in preventing an unwanted movement; the function of a muscle.

action potential A rapid voltage change in which a plasma membrane briefly reverses electrical polarity; has a self-propagating effect that produces a traveling wave of excitation in nerve and muscle cells.

active site The region of a protein that binds to a ligand, such as the substrate-binding site of an enzyme or the hormone-binding site of a receptor.

active transport Transport of particles through a selectively permeable membrane, up their concentration gradient, with the aid of a carrier that consumes ATP.

acute Pertaining to a disease with abrupt onset, intense symptoms, and short duration. *Compare* chronic.

adaptation 1. An evolutionary process leading to the establishment of species characteristics that favor survival and reproduction. **2.** Any characteristic of anatomy, physiology, or behavior that promotes survival and reproduction. **3.** A sensory process in which a receptor adjusts its sensitivity or response to the prevailing level of stimulation, such as dark adaptation of the eye.

adaptive immunity A system of pathogen-specific defenses, developed upon initial exposure to an antigen and entailing cellular memory and accelerated responses to the pathogen on later reexposure, thereby preventing or minimizing disease. Includes various T and B lymphocytes, plasma cells, and antibodies. *Compare* innate immunity.

adduction (ad-DUC-shun) Movement of a body part toward the median plane, such as bringing the feet together from a spread-legged position. fig. 9.13

adenosine triphosphate (ATP) (ah-DEN-oh-seen tri-FOSS-fate) A molecule composed of adenine, ribose, and three phosphate groups that functions as a universal energy-transfer molecule; yields adenosine diphosphate (ADP) and an inorganic phosphate group (Pi) upon hydrolysis. fig. 2.29a

adenylate cyclase (ah-DEN-ih-late SY-clase) An enzyme of the plasma membrane that makes cyclic adenosine monophosphate (cAMP) by removing two phosphate groups from ATP; important in the activation of the cAMP second-messenger system.

adipocyte (AD-ih-po-site) A fat cell.

adipose tissue A connective tissue composed predominantly of adipocytes; fat.

adrenergic (AD-ren-UR-jic) Pertaining to norepinephrine (NE) and epinephrine, as in adrenergic nerve fibers that secrete NE, adrenergic receptors that bind it, and adrenergic effects on a target organ.

adrenocorticotropic hormone (ACTH) A hormone secreted by the anterior pituitary gland that stimulates the adrenal cortex to secrete cortisol and other glucocorticoids; one of the stress hormones.

adsorption The binding of one substance to the surface of another without becoming a part of the latter. *Compare* absorption.

aerobic exercise (air-OH-bic) Exercise in which oxygen is used to produce ATP; endurance exercise.

aerobic respiration Oxidation of organic compounds in a reaction series that requires oxygen and produces ATP.

afferent (AFF-ur-ent) Carrying toward, as in a blood vessel that carries blood toward a tissue or a nerve fiber that conducts signals toward the central nervous system.

agglutination (ah-GLUE-tih-NAY-shun) Clumping of cells by antibodies.

albumin (al-BYU-min) A class of small proteins constituting about 60% of the protein fraction of the blood plasma; plays roles in blood viscosity, colloid osmotic pressure, and solute transport.

aldosterone (AL-doe-steh-RONE, al-DOSS-teh-rone) A steroid hormone secreted by the adrenal cortex that acts on the kidneys to promote sodium retention and potassium excretion.

alkalosis An acid–base imbalance in which the blood pH is higher than 7.45.

allele (ah-LEEL) Any of the alternative forms that one gene can take, such as dominant and recessive alleles.

alveolus (AL-vee-OH-lus) **1.** A microscopic air sac of the lung. **2.** A gland acinus. **3.** A tooth socket. **4.** Any small anatomical space.

amino acid A small organic molecule with an amino group and a carboxyl group; amino acids are the monomers of which proteins are composed, and function also as neurotransmitters and in other roles.

amnion A transparent sac that encloses the fetus and amniotic fluid.

amphipathic (AM-fih-PATH-ic) Pertaining to a molecule that has both hydrophilic and hydrophobic regions, such as phospholipids, bile acids, and some proteins.

ampulla (am-PULL-uh) A wide or saclike portion of a tubular organ such as a semicircular duct or uterine tube.

anabolism (ah-NAB-oh-lizm) Any metabolic reactions that consume energy and construct more complex molecules with higher free energy from less complex molecules with lower free energy; for example, the synthesis of proteins from amino acids. *Compare* catabolism.

anaerobic fermentation (AN-err-OH-bic) A reduction reaction independent of oxygen that converts pyruvate to lactate and enables glycolysis to continue under anaerobic conditions.

anastomosis (ah-NASS-tih-MO-sis) An anatomical convergence, the opposite of a branch; a point where two blood vessels merge and combine their bloodstreams or where two nerves or ducts converge. fig. 20.10

anatomical position A reference posture that allows for standardized anatomical terminology. A subject in anatomical position is standing with the feet flat on the floor, arms down to the sides, and the palms and eyes directed forward. fig. A.1

androgen (AN-dro-jen) Testosterone or a related steroid hormone. Stimulates bodily changes at puberty in both sexes, adult libido in both sexes, development of male anatomy in the fetus and adolescent, and spermatogenesis.

aneurysm (AN-you-rizm) A weak, bulging point in the wall of a heart chamber or blood vessel that presents a threat of hemorrhage.

angiogenesis (AN-jee-oh-GEN-eh-sis) The growth of new blood vessels.

angiotensin II (AN-jee-oh-TEN-sin) A hormone produced from angiotensinogen (a plasma protein) by the kidneys and lungs; raises blood pressure by stimulating vasoconstriction and stimulating the adrenal cortex to secrete aldosterone.

anion (AN-eye-on) An ion with more electrons than protons and consequently a net negative charge.

antagonist 1. A muscle that opposes the agonist at a joint. **2.** Any agent, such as a hormone or drug, that opposes another.

antagonistic effect 1. An effect in which two hormones, neurotransmitters, or divisions of the nervous system oppose each other and produce opposite effects on a target cell or organ. **2.** An effect in which a drug blocks or otherwise opposes the action of one of the body's own signaling mechanisms.

antebrachium (AN-teh-BRAY-kee-um) The region from elbow to wrist; the forearm.

anterior Pertaining to the front (facial-abdominal aspect) of the body; ventral.

anterior root The branch of a spinal nerve that emerges from the anterior side of the spinal cord and carries efferent (motor) nerve fibers; often called *ventral root*. fig. 13.2b

antibody A protein of the gamma globulin class that reacts with an antigen and aids in protecting the body from its harmful effects; found in the blood plasma, in other body fluids, and on the surfaces of certain leukocytes and their derivatives.

antidiuretic hormone (ADH) (AN-tee-DYE-you-RET-ic) A hormone released by the posterior lobe of the pituitary gland in response to low blood pressure; promotes water retention by the kidneys. Also known as *arginine vasopressin.*

antigen (AN-tih-jen) Any large molecule capable of binding to an antibody or immune cells and triggering an immune response; usually a protein, glycoprotein, or glycolipid.

antigen-presenting cell (APC) A cell that phagocytizes an antigen and displays fragments of it on its surface for recognition by other cells of the immune system; chiefly macrophages and B lymphocytes.

antiport A cotransport protein that moves two or more solutes in opposite directions through a cellular membrane, such as the Na^+–K^+ pump.

apical surface The uppermost surface of an epithelial cell, usually exposed to the lumen of an organ. fig. 3.4

apocrine Pertaining to certain sweat glands with large lumens and relatively thick, aromatic secretions and to similar glands such as the mammary gland; named for a mistaken belief that they form secretions by pinching off bits of apical cytoplasm.

apoptosis (AP-op-TOE-sis) Programmed cell death; the normal death of cells that have completed their function. *Compare* necrosis.

appendicular (AP-en-DIC-you-lur) Pertaining to the limbs and their supporting skeletal girdles. fig. 8.1

areolar tissue (AIR-ee-OH-lur) A fibrous connective tissue with loosely organized, widely spaced fibers and cells and an abundance of fluid-filled space; found under nearly every epithelium, among other places. fig. 5.14

arrector A smooth muscle associated with each hair follicle that serves to erect the hair. fig. 6.7

arteriole (ar-TEER-ee-ole) A small artery that empties into a metarteriole or capillary.

arteriosclerosis (ar-TEER-ee-o-sclair-O-sis) Stiffening of the arteries correlated with age or disease processes, caused primarily by cumulative free radical damage and tissue deterioration. *Compare* atherosclerosis.

articular cartilage A thin layer of hyaline cartilage covering the articular surface of a bone at a synovial joint, serving to reduce friction and ease joint movement. fig. 9.5

articulation A skeletal joint; any point at which two bones meet; may or may not be movable.

aspect A particular view of the body or one of its structures, or a part that faces in a particular direction, such as the anterior aspect.

atherosclerosis (ATH-ur-oh-skleh-ROE-sis) A degenerative disease of the blood vessels characterized by the presence of lipid deposits; often leading to calcification of the vessel wall and obstruction of coronary, cerebral, or other vital arteries. *Compare* arteriosclerosis.

atrioventricular (AV) node (AY-tree-oh-ven-TRIC-you-lur) A group of autorhythmic cells in the interatrial septum of the heart that relays excitation from the atria to the ventricles.

atrioventricular (AV) valves The mitral (left) and tricuspid (right) valves between the atria and ventricles of the heart; the left AV valve was formerly known as the *bicuspid valve.*

atrophy (AT-ro-fee) Shrinkage of a tissue due to age, disuse, or disease.

autoantibody An antibody that fails to distinguish the body's own molecules from foreign molecules and thus attacks host tissues, causing autoimmune diseases.

autoimmune disease Any disease in which antibodies fail to distinguish between foreign and self-antigens and attack the body's own tissues; for example, systemic lupus erythematosus and rheumatic fever.

autolysis (aw-TOLL-ih-sis) Digestion of cells by their own internal enzymes.

autonomic nervous system (ANS) (AW-toe-NOM-ic) A motor division of the nervous system that innervates glands, smooth muscle, and cardiac muscle; consists of sympathetic and parasympathetic divisions and functions largely without voluntary control. *Compare* somatic nervous system.

autoregulation The ability of a tissue to adjust its own blood supply through vasomotion or angiogenesis.

autosome (AW-toe-some) Any chromosome except the sex chromosomes. Genes on the autosomes are inherited without regard to the sex of the individual.

avascular Devoid of blood vessels, as in epithelia and cartilage.

axial (AC-see-ul) Pertaining to the head, neck, and trunk; the part of the body excluding the appendicular portion. fig. 8.1

axillary (ACK-sih-LERR-ee) Pertaining to the armpit.

axon terminal The swollen tip at the distal end of an axon; the site of synaptic vesicles and neurotransmitter release. fig. 12.21

axoneme The central core of a cilium or flagellum, composed of microtubules; these are arranged in a circular array of nine microtubule pairs, and in flagella and mobile cilia, another microtubule pair at the center of the circle.

B

baroreceptor (BER-oh-re-SEP-tur) A cardiovascular pressure sensor that triggers autonomic reflexes in response to fluctuations in blood pressure; baroreceptors are located in the heart, aortic arch, and carotid sinuses.

basal surface The lowermost surface of an epithelial cell, attached to either a basement membrane or an underlying epithelial cell. fig. 3.4

base 1. A chemical that binds protons from solution; a proton acceptor. **2.** Any of the purines or pyrimidines of a nucleic acid (adenine, thymine, guanine, cytosine, or uracil) serving in part to code for protein structure. **3.** The broadest part of a tapered organ such as the uterus or heart or the inferior aspect of an organ such as the brain.

basement membrane A thin layer of glycoproteins, collagen, and glycosaminoglycans beneath the deepest cells of an epithelium, serving to bind the epithelium to the underlying tissue. fig. 5.33

basophil (BAY-so-fill) A granulocyte with coarse cytoplasmic granules that produces heparin, histamine, and other chemicals involved in inflammation. table 18.6

belly The thick part of a skeletal muscle between its origin and insertion. fig. 10.4

bicarbonate buffer system An equilibrium mixture of carbonic acid, bicarbonate ions, and hydrogen ions ($H_2CO_3 \rightleftharpoons HCO_3^- + H^+$) that stabilizes the pH of the body fluids.

bicarbonate ion An anion, HCO_3^-, that functions as a base in the buffering of body fluids.

biogenic amines A class of chemical messengers with neurotransmitter and hormonal functions, synthesized from amino acids and retaining an amino group; also called *monoamines*. Examples include epinephrine and thyroxine.

bipedalism The habit of walking on two legs; a defining characteristic of the family Hominidae that underlies many skeletal and other characteristics of humans.

blood–brain barrier (BBB) A barrier between the bloodstream and nervous tissue of the CNS that is impermeable to many blood solutes and thus prevents them from affecting the brain tissue; formed by the tight junctions between capillary endothelial cells, the basement membrane of the endothelium, and the perivascular feet of astrocytes.

B lymphocyte A lymphocyte that functions as an antigen-presenting cell and, in humoral immunity, differentiates into an antibody-producing plasma cell; also called a *B cell*.

body 1. The entire organism. **2.** Part of a cell, such as a neuron, containing the nucleus and most other organelles. **3.** The largest or principal part of an organ such as the stomach or uterus; also called the *corpus*.

brachial (BRAY-kee-ul) Pertaining to the arm proper, the region from shoulder to elbow.

bradykinin (BRAD-ee-KY-nin) An oligopeptide produced in inflammation that stimulates vasodilation, increases capillary permeability, and stimulates pain receptors.

brainstem The stalklike lower portion of the brain, composed of the medulla oblongata, pons, and midbrain. fig. 14.8

bronchiole (BRON-kee-ole) A pulmonary air passage that is usually 1 mm or less in diameter and lacks cartilage but has relatively abundant smooth muscle, elastic tissue, and a simple cuboidal, usually ciliated epithelium.

bronchus (BRONK-us) A relatively large pulmonary air passage with supportive cartilage in the wall; any passage beginning with the main bronchus at the fork in the trachea and ending with segmental bronchi, from which air continues into the bronchioles.

brush border A fringe of microvilli on the apical surface of an epithelial cell, serving to enhance surface area and promote absorption. fig. 5.6

bursa A sac filled with synovial fluid at a synovial joint, serving to facilitate muscle or joint action. fig. 9.6

C

calcification The hardening of a tissue due to the deposition of calcium salts.

calorie The amount of thermal energy that will raise the temperature of 1 g of water by 1°C. Also called a *small calorie* to distinguish it from a dietary Calorie (capital *C*), or kilocalorie.

Calorie *See* kilocalorie.

canaliculus (CAN-uh-LIC-you-lus) A microscopic canal, as in osseous tissue. fig. 7.4

capillary (CAP-ih-LERR-ee) The narrowest type of vessel in the cardiovascular and lymphatic systems; engages in fluid exchanges with surrounding tissues.

capillary exchange The process of fluid transfer between the bloodstream and tissue fluid through the walls of the blood capillaries.

capsule The fibrous covering of a structure such as the spleen or a synovial joint.

carbohydrate A hydrophilic organic compound composed of carbon and a 2:1 ratio of hydrogen to oxygen; includes sugars, starches, glycogen, and cellulose.

carbonic anhydrase An enzyme found in erythrocytes and kidney tubule cells that catalyzes the decomposition of carbonic acid into carbon dioxide and water or the reverse reaction ($H_2CO_3 \rightleftharpoons CO_2 + H_2O$).

carcinogen (car-SIN-oh-jen) An agent capable of causing cancer, including certain chemicals, viruses, and ionizing radiation.

cardiac output *(CO)* The amount of blood pumped by each ventricle of the heart in 1 minute.

cardiomyocyte A cardiac muscle cell.

cardiovascular system An organ system consisting of the heart and blood vessels, serving for the transport of blood. *Compare* circulatory system.

carpal Pertaining to the wrist (carpus).

carrier 1. A protein in a cellular membrane that performs carrier-mediated transport. **2.** A person who is heterozygous for a recessive allele and does not exhibit the associated phenotype, but may transmit this allele to his or her children; for example, a carrier for sickle-cell disease.

carrier-mediated transport Any process of transporting materials through a cellular membrane that involves reversible binding to a transport protein.

catabolism (ca-TAB-oh-lizm) Any metabolic reactions that release energy and break relatively complex molecules with high free energy into less complex molecules with lower free energy; for example, digestion and glycolysis. *Compare* anabolism.

catecholamine (CAT-eh-COAL-uh-meen) A subclass of biogenic amines that includes epinephrine, norepinephrine, and dopamine. fig. 12.22

cation (CAT-eye-on) An ion with more protons than electrons and consequently a net positive charge.

caudal (CAW-dul) **1.** Pertaining to a tail or narrow tail-like part of an organ. **2.** Pertaining to the inferior part of the trunk of the body, where the tail of other animals arises. *Compare* cranial. **3.** Relatively distant from the forehead, especially in reference to structures of the brain and spinal cord; for example, the medulla oblongata is caudal to the pons. *Compare* rostral.

celiac (SEEL-ee-ac) Pertaining to the abdomen.

central nervous system (CNS) The brain and spinal cord.

centriole (SEN-tree-ole) An organelle composed of a short cylinder of nine triplets of microtubules, usually paired with another centriole perpendicular to it; origin of the mitotic spindle; identical to the basal body of a cilium or flagellum. fig. 3.33

cephalic (seh-FAL-ic) Pertaining to the head.

cerebellum (SER-eh-BELL-um) A large portion of the brain posterior to the brainstem and inferior to the cerebrum, responsible for equilibrium, motor coordination, and memory of learned motor skills. fig. 14.11

cerebrospinal fluid (CSF) (SERR-eh-bro-SPY-nul, seh-REE-bro-SPY-nul) A liquid that fills the ventricles of the brain, the central canal of the spinal cord, and the space between the CNS and dura mater.

cerebrum (seh-REE-brum, SERR-eh-brum) The largest and most superior part of the brain, divided into two convoluted cerebral hemispheres separated by a deep longitudinal fissure.

cervical (SUR-vih-cul) Pertaining to the neck or a narrow part (cervix) of certain organs.

cervix (SUR-vix) **1.** The neck. **2.** A narrow or necklike part of an organ such as the uterus and gallbladder. fig. 28.3

channel protein A transmembrane protein that has a pore through it for the passage of materials between the cytoplasm and extracellular fluid. fig. 3.7

chemical bond A force that attracts one atom to another, such as their opposite charges or the sharing of electrons.

chemical synapse A meeting of a nerve fiber and another cell with which the neuron communicates by releasing neurotransmitters. fig. 12.21

chemoreceptor An organ or cell specialized to detect chemicals, as in the carotid bodies and taste buds.

chief cells The majority type of cell in an organ or tissue such as the parathyroid glands or gastric glands.

cholecystokinin (CCK) (CO-leh-SIS-toe-KY-nin) A polypeptide employed as a hormone and neurotransmitter, secreted by some brain neurons and cells of the digestive tract. fig. 12.22

cholesterol (co-LESS-tur-ol) A steroid that functions as part of the plasma membrane and as a precursor for all other steroids in the body.

cholinergic (CO-lin-UR-jic) Pertaining to acetylcholine (ACh), as in cholinergic nerve fibers that secrete ACh, cholinergic receptors that bind it, or cholinergic effects on a target organ.

chondrocyte (CON-dro-site) A cartilage cell; a former chondroblast that has become enclosed in a lacuna in the cartilage matrix. figs. 5.19 to 5.21

chorion (CO-ree-on) An embryonic membrane external to the amnion; forms part of the placenta and has diverse functions including fetal nutrition, waste removal, and hormone secretion. fig. 29.7f

chromatin (CRO-muh-tin) Filamentous material in the interphase nucleus, composed of DNA and associated proteins.

chromosome A complex of DNA and protein carrying the genetic material of a cell's nucleus. Normally there are 46 chromosomes in the nucleus of each cell except germ cells. fig. 4.5

chronic 1. Long-lasting. 2. Pertaining to a disease that progresses slowly and has a long duration. *Compare* acute.

chronic bronchitis A chronic obstructive pulmonary disease characterized by damaged and immobilized respiratory cilia, excessive mucus secretion, infection of the lower respiratory tract, and bronchial inflammation; caused especially by cigarette smoking. *See also* chronic obstructive pulmonary disease.

chronic obstructive pulmonary disease (COPD) Certain lung diseases (chronic bronchitis and emphysema) that result in long-term obstruction of airflow and substantially reduced pulmonary ventilation; one of the leading causes of death in old age.

cilium (SILL-ee-um) A hairlike process, with an axoneme, projecting from the apical surface of an epithelial cell; often motile and serving to propel matter across the surface of an epithelium, but sometimes nonmotile and serving sensory roles. fig. 3.10

circulatory shock A state of cardiac output inadequate to meet the metabolic needs of the body.

circulatory system An organ system consisting of the heart, blood vessels, and blood. *Compare* cardiovascular system.

circumduction A joint movement in which one end of an appendage remains relatively stationary and the other end is moved in a circle. fig. 9.16

cirrhosis (sih-RO-sis) A degenerative liver disease characterized by replacement of functional parenchyma with fibrous and adipose tissue; causes include alcohol, other poisons, and viral and bacterial inflammation.

cistern (SIS-turn) A fluid-filled space or sac, such as the cisterna chyli of the lymphatic system and cisterns of the endoplasmic reticulum and Golgi complex. fig. 3.28

climacteric A period in the lives of men and women, usually in the early 50s, marked by changes in the level of reproductive hormones; a variety of somatic and psychological effects; and, in women, cessation of ovulation and menstruation (menopause).

clone A population of cells that are mitotically descended from the same parent cell and are identical to each other genetically or in other respects, such as a B- or T-cell clone.

coagulation (co-AG-you-LAY-shun) The clotting of blood, lymph, tissue fluid, or semen.

coenzyme (co-EN-zime) A small organic molecule, usually derived from a vitamin, that is needed to make an enzyme catalytically active; acts by accepting electrons from an enzymatic reaction and transferring them to a different reaction chain.

cofactor A nonprotein such as a metal ion or coenzyme needed for an enzyme to function.

cohesion The clinging of identical molecules such as water to each other.

collagen (COLL-uh-jen) The most abundant protein in the body, forming the fibers of many connective tissues in places such as the dermis, tendons, and bones.

colloid An aqueous mixture of particles that are too large to pass through most selectively permeable membranes but small enough to remain evenly dispersed through the solvent by the thermal motion of solvent particles; for example, the proteins in blood plasma.

colloid osmotic pressure *(COP)* A portion of the osmotic pressure of a body fluid that is due to its protein. *Compare* oncotic pressure.

column *See* funiculus.

columnar A cellular shape that is significantly taller than it is wide. fig. 5.6

commissure (COM-ih-shur) 1. A bundle of nerve fibers that crosses from one side of the brain or spinal cord to the other. fig. 14.2 2. A corner or angle at which the eyelids, lips, or genital labia meet; in the eye, also called the *canthus*. fig. 16.23

complement 1. To complete or enhance the structure or function of something else, as in the coordinated action of two hormones. 2. A system of plasma proteins involved in defense against pathogens.

computerized tomography (CT) A method of medical imaging that uses X-rays and a computer to create an image of a thin section of the body; also called a *CT scan*.

concentration gradient A difference in chemical concentration from one point to another, as on two sides of a plasma membrane.

conception The fertilization of an egg, producing a zygote.

conceptus All products of conception, ranging from a fertilized egg to the full-term fetus with its embryonic membranes, placenta, and umbilical cord. *Compare* embryo, fetus, preembryo.

condyle (CON-dile) A rounded knob on a bone serving to smooth the movement of a joint. fig. 8.2

conformation The three-dimensional structure of a protein that results from interaction among its amino acid side groups, its interactions with water, and the formation of disulfide bonds.

congenital Present at birth; for example, an anatomical defect, a syphilis infection, or a hereditary disease.

conjugated A state in which one organic compound is bound to another compound of a different class, such as a protein conjugated with a carbohydrate to form a glycoprotein.

connective tissue A tissue usually composed of more extracellular than cellular volume and usually with a substantial amount of extracellular fiber; forms supportive frameworks and capsules for organs, binds structures together, holds them in place, stores energy (as in adipose tissue), or transports materials (as in blood).

contractility 1. The ability to shorten. 2. The amount of force that a contracting muscle fiber generates for a given stimulus; may be increased by epinephrine, for example, while stimulus strength remains constant. 3. The amount of force that a contracting heart chamber generates for a given preload.

contralateral On opposite sides of the body, as in reflex arcs in which the stimulus comes from one side of the body and a response is given by muscles on the other side. *Compare* ipsilateral.

cooperative effect Effect in which two hormones, or both divisions of the autonomic nervous system, work together to produce a single overall result.

corona A halo- or crownlike structure, as in the corona radiata of the brain and ovaries, coronary circulation of the heart, and coronal suture of the skull.

corona radiata 1. An array of nerve tracts in the brain that arise mainly from the thalamus and fan out to different regions of the cerebral cortex. **2.** The first layer of cuboidal cells immediately external to the zona pellucida around an egg cell.

coronal plane *See* frontal plane.

coronary circulation A system of blood vessels that serve the wall of the heart. fig. 19.10

corpus 1. Body or mass, such as the corpus callosum and corpus luteum. **2.** The main part of an organ such as the stomach or uterus, as opposed to such regions as the head, tail, or cervix.

corrosion cast *See* vascular corrosion cast.

cortex The outer layer of some organs such as the adrenal glands, kidneys, cerebrum, lymph nodes, ovaries, and hairs; usually covers or encloses tissue called the *medulla*.

corticosteroid (COR-tih-co-STERR-oyd) Any steroid hormone secreted by the adrenal cortex, such as aldosterone, cortisol, and sex steroids.

costal (COSS-tul) Pertaining to the ribs.

cotransport A form of carrier-mediated transport in which a membrane protein transports two solutes simultaneously or within the same cycle of action by either facilitated diffusion or active transport; for example, the sodium–glucose transporter and the Na⁺–K⁺ pump.

countercurrent A situation in which two fluids flow side by side in opposite directions, as in the countercurrent multiplier of the kidney and the countercurrent heat exchanger of the scrotum.

cranial (CRAY-nee-ul) **1.** Pertaining to the cranium of the skull. **2.** In a position relatively close to the head or a direction toward the head. *Compare* caudal.

cranial nerve Any of 12 pairs of nerves connected to the base of the brain and passing through foramina of the cranium.

crista (plural, *cristae*) An anatomical crest, such as the crista galli of the ethmoid bone or the crista of a mitochondrion.

cross section A cut perpendicular to the long axis of the body or an organ.

crural (CROO-rul) Pertaining to the leg proper or to the crus (leg) of a organ.

crus (pronounced cruss; plural, *crura*) **1.** A leglike extension of an organ such as the penis or clitoris. figs. 27.10b, 28.8 **2.** The leg proper (crural region, from knee to ankle) of the lower limb.

cuboidal (cue-BOY-dul) A cellular shape that is roughly like a cube or in which the height and width are about equal; typically looks squarish in tissue sections. fig. 5.5

current A moving stream of charged particles such as ions or electrons.

cusp 1. One of the flaps of a valve of the heart, veins, and lymphatic vessels. **2.** A conical projection on the occlusal surface of a premolar or molar tooth.

cutaneous (cue-TAY-nee-us) Pertaining to the skin.

cyanosis (SY-uh-NO-sis) A bluish color of the skin and mucous membranes due to ischemia or hypoxemia.

cyclic adenosine monophosphate (cAMP) A cyclic molecule produced from ATP by the enzymatic removal of two phosphate groups; serves as a second messenger in many hormone and neurotransmitter actions. fig. 2.29b

cytolysis (sy-TOL-ih-sis) The rupture and destruction of a cell by such agents as complement proteins and hypotonic solutions.

cytoplasm The contents of a cell between its plasma membrane and its nuclear envelope, consisting of cytosol, organelles, inclusions, and the cytoskeleton.

cytoskeleton A system of protein microfilaments, intermediate filaments, and microtubules in a cell, serving in physical support, cellular movement, and the routing of molecules and organelles to their destinations within the cell. fig. 3.24

cytosol A clear, featureless, gelatinous colloid in which the organelles and other internal structures of a cell are embedded.

cytotoxic T cell A T lymphocyte that directly attacks and destroys infected body cells, cancerous cells, and the cells of transplanted tissues.

D

daughter cells Cells that arise from a parent cell by mitosis or meiosis.

deamination (dee-AM-ih-NAY-shun) Removal of an amino group from an organic molecule; a step in the catabolism of amino acids.

decomposition reaction A chemical reaction in which a larger molecule is broken down into smaller ones. *Compare* synthesis reaction.

decussation (DEE-cuh-SAY-shun) The crossing of nerve fibers from the right side of the central nervous system to the left or vice versa, especially in the spinal cord, medulla oblongata, and optic chiasma.

deep Relatively far from the body surface; opposite of *superficial*. For example, most bones are deep to the skeletal muscles.

degranulation Exocytosis and disappearance of cytoplasmic granules, especially in platelets and granulocytes.

denaturation A change in the three-dimensional conformation of a protein that destroys its enzymatic or other functional properties, usually caused by extremes of temperature or pH.

dendrite Extension of a neuron that receives information from other cells or from environmental stimuli and conducts signals to the soma. Dendrites are usually shorter, more branched, and more numerous than the axon and are incapable of producing action potentials. fig. 12.4

dendritic cell An antigen-presenting cell of the epidermis and mucous membranes. fig. 6.3

denervation atrophy The shrinkage of skeletal muscle that occurs when the motor neuron dies or is severed from the muscle.

dense connective tissue A connective tissue with a high density of fiber, relatively little ground substance, and scanty cells; seen in tendons and the dermis, for example.

depolarization A shift in the electrical potential across a plasma membrane to a value less negative than the resting membrane potential; associated with excitation of a nerve or muscle cell. *Compare* hyperpolarization.

dermal papilla A bump or ridge of dermis that extends upward to interdigitate with the epidermis and create a wavy boundary that resists stress and slippage of the epidermis.

dermis The deeper of the two layers of the skin, underlying the epidermis and composed of fibrous connective tissue.

desmosome (DEZ-mo-some) A patchlike intercellular junction that mechanically links two cells together. fig. 5.28

diabetes (DY-uh-BEE-teez) Any disease characterized by chronic polyuria of metabolic origin; diabetes mellitus unless otherwise specified.

diabetes insipidus (in-SIP-ih-dus) A form of diabetes that results from hyposecretion of antidiuretic hormone; unlike other forms, it is not characterized by hyperglycemia or glycosuria.

diabetes mellitus (DM) (mel-EYE-tus) A form of diabetes that results from hyposecretion of insulin or from a deficient target-cell response to it; signs include hyperglycemia and glycosuria.

diaphysis (dy-AF-ih-sis) The shaft of a long bone. fig. 7.1

diarthrosis (DY-ar-THRO-sis) *See* synovial joint.

diastole (dy-ASS-toe-lee) A period in which a heart chamber relaxes and fills with blood; especially ventricular relaxation.

diastolic pressure (DY-ah-STAHL-ic) The minimum arterial blood pressure measured during the interval between heartbeats

diencephalon (DY-en-SEFF-uh-lon) A portion of the brain between the midbrain and corpus callosum; composed of the thalamus, epithalamus, and hypothalamus. fig. 14.12

differentiation Development of a relatively unspecialized cell or tissue into one with a more specific structure and function.

diffusion Spontaneous net movement of particles from a place of high concentration to a place of low concentration (down a concentration gradient).

diploid (2n) In humans, having 46 chromosomes in 23 homologous pairs, with one member of each chromosome pair coming from each parent.

disaccharide (dy-SAC-uh-ride) A carbohydrate composed of two simple sugars (monosaccharides) joined by a glycosidic bond; for example, lactose, sucrose, and maltose. fig. 2.16

disseminated intravascular coagulation (DIC) Widespread clotting of the blood within unbroken vessels, leading to hemorrhaging, congestion of the vessels with clotted blood, and ischemia and necrosis of organs.

distal Relatively distant from a point of origin or attachment; for example, the wrist is distal to the elbow. *Compare* proximal.

disulfide bond A covalent bond that links two cysteine residues through their sulfur atoms, serving to join one peptide chain to another or to hold a single chain in its three-dimensional conformation.

diuretic (DY-you-RET-ic) A chemical that increases urine output.

dominant 1. Pertaining to a genetic allele that is phenotypically expressed in the presence of any other allele. **2.** Pertaining to a trait that results from a dominant allele.

dopamine (DOE-puh-meen) An inhibitory catecholamine neurotransmitter of the central nervous system, especially of the basal nuclei, where it acts to suppress unwanted motor activity. fig. 12.22

dorsal Toward the back (spinal) side of the body; in humans, usually synonymous with *posterior.*

dorsal root *See* posterior root.

dorsiflexion (DOR-sih-FLEC-shun) A movement of the ankle that reduces the joint angle and raises the toes. fig. 9.22

duodenum (DEW-oh-DEE-num, dew-ODD-eh-num) The first portion of the small intestine extending for about 25 cm from the pyloric valve of the stomach to a sharp bend called the *duodenojejunal flexure;* receives chyme from the stomach and secretions from the liver and pancreas. fig. 25.25

dynein (DINE-een) A motor protein involved in the beating of cilia and flagella and in the movement of molecules and organelles within cells, as in retrograde transport in a nerve fiber.

E

eccrine (ECK-rin) Pertaining to gland cells that release their product by exocytosis; also called *merocrine.*

ectoderm The outermost of the three primary germ layers of an embryo; gives rise to the epidermis and nervous system.

ectopic (ec-TOP-ic) In an abnormal location; for example, ectopic pregnancy and ectopic pacemakers of the heart.

edema (eh-DEE-muh) Abnormal accumulation of tissue fluid resulting in swelling of the tissue.

effector A molecule, cell, or organ that carries out a response to a stimulus.

efferent (EFF-ur-ent) Carrying away or out, as in a blood vessel that carries blood away from a tissue or a nerve fiber that conducts signals away from the central nervous system.

elastic fiber A connective tissue fiber, composed of the protein elastin, that stretches under tension and returns to its original length when released; responsible for the resilience of organs such as the skin and lungs.

elasticity The tendency of a stretched structure to return to its original dimensions when tension is released.

electrical synapse A gap junction that enables one cell to stimulate another directly, without the intermediary action of a neurotransmitter; such synapses connect the cells of cardiac muscle and single-unit smooth muscle.

electrolyte A salt that ionizes in water and produces a solution that conducts electricity; loosely speaking, any ion that results from the dissociation of such salts, such as sodium, potassium, calcium, chloride, and bicarbonate ions.

elevation A joint movement that raises a body part, as in hunching the shoulders or closing the mouth.

embolism (EM-bo-lizm) The obstruction of a blood vessel by an embolus.

embolus (EM-bo-lus) Any abnormal traveling object in the bloodstream, such as agglutinated bacteria or blood cells, a blood clot, or an air bubble.

embryo A developing individual from the sixteenth day of gestation when the three primary germ layers have formed, through the end of the eighth week when all of the organ systems are present. *Compare* conceptus, fetus, preembryo.

emphysema (EM-fih-SEE-muh) A degenerative lung disease characterized by a breakdown of alveoli and diminishing surface area available for gas exchange; occurs with aging of the lungs but is greatly accelerated by smoking or air pollution.

endocrine gland (EN-doe-crin) A ductless gland that secretes hormones into the bloodstream; for example, the thyroid and adrenal glands. *Compare* exocrine gland.

endocytosis (EN-doe-sy-TOE-sis) Any process in which a cell forms vesicles from its plasma membrane and takes in large particles, molecules, or droplets of extracellular fluid; for example, phagocytosis and pinocytosis.

endoderm The innermost of the three primary germ layers of an embryo; gives rise to the mucosae of the digestive and respiratory tracts and to their associated glands.

endogenous (en-DODJ-eh-nus) Originating internally, such as the endogenous cholesterol synthesized in the body in contrast to the exogenous cholesterol coming from the diet. *Compare* exogenous.

endometrium (EN-doe-MEE-tree-um) The mucosa of the uterus; the site of implantation and source of menstrual discharge.

endoplasmic reticulum (ER) (EN-doe-PLAZ-mic reh-TIC-you-lum) An extensive system of interconnected cytoplasmic tubules or channels; classified as rough ER or smooth ER depending on the presence or absence of ribosomes on its membrane. fig. 3.28

endothelium (EN-doe-THEEL-ee-um) A simple squamous epithelium that lines the lumens of the blood vessels, heart, and lymphatic vessels.

enteric (en-TERR-ic) Pertaining to the small intestine, as in enteric hormones and enteric nervous system.

eosinophil (EE-oh-SIN-oh-fill) A granulocyte with a large, often bilobed nucleus and coarse cytoplasmic granules that stain with eosin; phagocytizes antigen–antibody complexes, allergens, and inflammatory chemicals and secretes enzymes that combat parasitic infections. table 18.6

epidermis A stratified squamous epithelium that constitutes the superficial layer of the skin, overlying the dermis. fig. 6.3

epinephrine (EP-ih-NEFF-rin) A catecholamine that functions as a neurotransmitter in the sympathetic nervous system and as a hormone secreted by the adrenal medulla; also called *adrenaline.* fig. 12.22

epiphysial plate (EP-ih-FIZ-ee-ul) A plate of hyaline cartilage between the epiphysis and diaphysis of a long bone in a child or adolescent, serving as a growth zone for bone elongation. figs. 7.9, 7.11

epiphysis (eh-PIF-ih-sis) **1.** The head of a long bone. fig. 7.1 **2.** The pineal gland (epiphysis cerebri).

epithelium A type of tissue consisting of one or more layers of closely adhering cells with little intercellular material and no blood vessels; forms the coverings and linings of many organs and the parenchyma of the glands. Also known as *epithelial tissue*.

erectile tissue A tissue that functions by swelling with blood, as in the penis and clitoris and inferior concha of the nasal cavity.

erythrocyte (eh-RITH-ro-site) A red blood cell.

erythropoiesis (eh-RITH-ro-poy-EE-sis) The production of erythrocytes.

erythropoietin (eh-RITH-ro-POY-eh-tin) A hormone that is secreted by the kidneys and liver in response to hypoxemia and stimulates erythropoiesis.

estrogens (ESS-tro-jenz) A family of steroid hormones known especially for producing female secondary sex characteristics and regulating various aspects of the menstrual cycle and pregnancy; major forms are estradiol, estriol, and estrone.

evolution A change in the relative frequencies of alleles in a population over a period of time; the mechanism that produces adaptations in human form and function. *See also* adaptation.

excitability The ability of a cell to respond to a stimulus, especially the ability of nerve and muscle cells to produce membrane voltage changes in response to stimuli.

excitation–contraction coupling Events that link the synaptic stimulation of a muscle cell to the onset of contraction.

excitatory postsynaptic potential (EPSP) A partial depolarization of a postsynaptic neuron or muscle cell in response to a neurotransmitter, making it more likely to reach threshold and produce an action potential.

excretion The process of eliminating metabolic waste products from a cell or from the body. *Compare* secretion.

exocrine gland (EC-so-crin) A gland that secretes its products into another organ or onto the body surface, usually by way of a duct; for example, salivary and gastric glands. *Compare* endocrine gland.

exocytosis (EC-so-sy-TOE-sis) A process in which a vesicle in the cytoplasm of a cell fuses with the plasma membrane and releases its contents from the cell; used in the elimination of cellular wastes, in the release of gland products and neurotransmitters, and for the replacement of membrane removed by endocytosis.

exogenous (ec-SODJ-eh-nus) Originating externally, such as exogenous (dietary) cholesterol; extrinsic. *Compare* endogenous.

expiration 1. Exhaling. **2.** Dying.

extension Movement of a joint that increases the angle between articulating bones (straightens the joint). fig. 9.12 *Compare* flexion.

extracellular fluid (ECF) Any body fluid that is not contained in the cells; for example, blood, lymph, and tissue fluid.

extrinsic (ec-STRIN-sic) **1.** Originating externally, such as extrinsic blood-clotting factors; exogenous. **2.** Not fully contained within a specified organ or region but acting on it, such as the extrinsic muscles of the hand and eye. *Compare* intrinsic.

F

facilitated diffusion The process of transporting a chemical through a cellular membrane, down its concentration gradient, with the aid of a carrier that does not consume ATP; enables substances to diffuse through the membrane that would do so poorly, or not at all, without a carrier.

facilitation Making a process more likely to occur, such as the firing of a neuron, or making it occur more easily or rapidly, as in facilitated diffusion.

fascia (FASH-ee-uh) A layer of connective tissue between the muscles or separating the muscles from the skin. fig. 10.1

fascicle (FASS-ih-cul) A bundle of muscle or nerve fibers ensheathed in connective tissue; multiple fascicles bound together constitute a muscle or nerve as a whole. figs. 10.1, 13.9

fat 1. A triglyceride molecule. **2.** Adipose tissue.

fatty acid An organic molecule composed of a chain of an even number of carbon atoms with a carboxyl group at one end and a methyl group at the other; one of the structural subunits of triglycerides and phospholipids.

fenestrated (FEN-eh-stray-ted) Perforated with holes or slits, as in fenestrated blood capillaries and the elastic sheets of large arteries. fig. 20.5

fetus In human development, an individual from the beginning of the ninth week when all of the organ systems are present, through the time of birth. *Compare* conceptus, embryo, preembryo.

fibrin (FY-brin) A sticky fibrous protein formed from fibrinogen in blood, tissue fluid, and lymph; forms the matrix of a blood clot.

fibroblast A connective tissue cell that produces collagen fibers and ground substance; the only type of cell in tendons and ligaments.

fibrosis Replacement of damaged tissue with fibrous scar tissue rather than by the original tissue type; scarring. *Compare* regeneration.

fibrous connective tissue Any connective tissue with a preponderance of fiber, such as areolar, reticular, dense regular, and dense irregular connective tissues.

filtration A process in which hydrostatic pressure forces a fluid through a selectively permeable membrane (especially a capillary wall).

fire To produce an action potential, as in nerve and muscle cells.

fix 1. To hold a structure in place; for example, by fixator muscles that prevent unwanted joint movements. **2.** To preserve a tissue by means of a fixative such as formalin.

flexion A joint movement that, in most cases, decreases the angle between two bones. fig. 9.12 *Compare* extension.

fluid balance A state in which average daily water gains (by intake and synthesis) equal water losses, and water is properly distributed among the body's fluid compartments. *Also called* water balance.

fluid compartment Any of the major categories of fluid in the body, separated by selectively permeable membranes and differing from each other in chemical composition. Primary examples are the intracellular fluid, tissue fluid, blood, and lymph.

follicle (FOLL-ih-cul) **1.** A small space, such as a hair follicle, thyroid follicle, or ovarian follicle. **2.** An aggregation of lymphocytes in a lymphatic organ or mucous membrane.

follicle-stimulating hormone (FSH) A hormone secreted by the anterior pituitary gland that stimulates development of the ovarian follicles and egg cells in females and sperm production in males.

foramen (fo-RAY-men) (plural, *foramina*) A hole through a bone or other organ, in many cases providing passage for blood vessels and nerves.

formed element An erythrocyte, leukocyte, or platelet; any cellular component of blood or lymph as opposed to the extracellular fluid component.

fossa (FOSS-uh) A depression in an organ or tissue, such as the fossa ovalis of the heart, cranial fossa of the skull, or olecranon fossa of the elbow.

fovea (FO-vee-uh) A small pit, such as the fovea capitis of the femur or fovea centralis of the retina.

free energy The potential energy in a chemical that is available to do work.

free radical A particle derived from an atom or molecule, having an unpaired electron that makes it highly reactive and destructive to cells; produced by intrinsic processes such as aerobic respiration and by extrinsic agents such as chemicals and ionizing radiation.

frenulum (FREN-you-lum) A fold of tissue that attaches a movable structure to a relatively immovable one, such as the lip to the gum or the tongue to the floor of the mouth. fig. 25.4

frontal plane An anatomical plane that passes through the body or an organ from right to left and superior to inferior, dividing the body or organ into anterior and posterior portions; also called a *coronal plane*. fig. A.1

fundus The base, the broadest part, or the part farthest from the opening of certain viscera such as the stomach and uterus.

funiculus (few-NICK-you-lus) A bundle of nerve fibers in the spinal cord, arranged in three pairs constituting the spinal white matter; subdivided into tracts (fasciculi). fig. 13.3

fusiform (FEW-zih-form) Shaped like the spindle of a spinning wheel; elongated, thick in the middle, and tapered at both ends, such as the shape of a smooth muscle cell or a muscle spindle.

G

gamete (GAM-eet) An egg or sperm cell.

gametogenesis (geh-ME-to-JEN-eh-sis) The production of eggs or sperm.

ganglion (GANG-glee-un) A cluster of nerve cell bodies in the peripheral nervous system, often resembling a knot in a string.

gangrene Tissue necrosis resulting from ischemia.

gap junction A junction between two cells consisting of a pore surrounded by a ring of proteins in the plasma membrane of each cell; allows solutes to diffuse from the cytoplasm of one cell to the next; functions include cell-to-cell nutrient transfer in the developing embryo and electrical communication between cells of cardiac and smooth muscle. *See also* electrical synapse. fig. 5.28

gastric Pertaining to the stomach.

gate A protein channel in a cellular membrane that can open or close in response to chemical, electrical, or mechanical stimuli, thus controlling when substances are allowed to pass through the membrane. May occur in the plasma membrane as well as the membranes of cytoplasmic organelles.

gene An information-containing segment of DNA that codes for the production of a molecule of RNA, which in most cases goes on to play a role in the synthesis of one or more proteins.

gene locus The site on a chromosome where a given gene is located.

genome (JEE-nome) All the genes of one individual, estimated at about 20,000 genes in humans.

genotype (JEE-no-type) The pair of alleles possessed by an individual at one gene locus on a pair of homologous chromosomes; strongly influences the individual's phenotype for a given trait.

germ cell A gamete (sperm or egg) or any precursor cell destined to become a gamete, such as a primary oocyte or spermatogonium.

germ layer Any of the first three tissue layers of an embryo: ectoderm, mesoderm, or endoderm.

gestation (jess-TAY-shun) Pregnancy.

globulin (GLOB-you-lin) A globular protein such as an enzyme, antibody, or albumin; especially a family of proteins in the blood plasma that includes albumin, antibodies, fibrinogen, and prothrombin.

glomerular capsule (glo-MERR-you-lur) A double-walled capsule around each glomerulus of the kidney; receives glomerular filtrate and empties into the proximal convoluted tubule. fig. 23.7

glomerulus (glo-MERR-you-lus) **1.** A spheroidal mass of blood capillaries in the kidney that filters plasma and produces glomerular filtrate, which is further processed to form the urine. fig. 23.7 **2.** A spheroidal mass of nerve endings in the olfactory bulb where olfactory neurons from the nose synapse with mitral and dendritic cells of the bulb. fig. 16.7

glucagon (GLUE-ca-gon) A hormone secreted by alpha cells of the pancreatic islets in response to hypoglycemia; promotes glycogenolysis and other effects that raise blood glucose concentration.

glucocorticoid (GLUE-co-COR-tih-coyd) Any hormone of the adrenal cortex that affects carbohydrate, fat, and protein metabolism; chiefly cortisol and corticosterone.

gluconeogenesis (GLUE-co-NEE-oh-JEN-eh-sis) The synthesis of glucose from noncarbohydrates such as fats and amino acids.

glucose A monosaccharide ($C_6H_{12}O_6$) also known as blood sugar; glycogen, starch, cellulose, and maltose are made entirely of glucose, and glucose constitutes half of a sucrose or lactose molecule. The isomer involved in human physiology is also called *dextrose*.

glucose-sparing effect An effect of fats or other energy substrates in which they are used as fuel by most cells, so that those cells do not consume glucose; this makes more glucose available to cells such as neurons that cannot use alternative energy substrates.

glycocalyx (GLY-co-CAY-licks) A layer of carbohydrate molecules covalently bonded to the phospholipids and proteins of a plasma membrane; forms a surface coat on all human cells.

glycogen (GLY-co-jen) A glucose polymer synthesized by liver, muscle, uterine, and vaginal cells that serves as an energy-storage polysaccharide.

glycogenesis (GLY-co-JEN-eh-sis) The synthesis of glycogen.

glycogenolysis (GLY-co-jeh-NOLL-ih-sis) The hydrolysis of glycogen, releasing glucose.

glycolipid (GLY-co-LIP-id) A phospholipid molecule with a carbohydrate covalently

bonded to it, found in the plasma membranes of cells.

glycolysis (gly-COLL-ih-sis) A series of anaerobic oxidation reactions that break a glucose molecule into two molecules of pyruvate and produce a small amount of ATP.

glycoprotein (GLY-co-PRO-teen) A protein molecule with a smaller carbohydrate covalently bonded to it; found in mucus and the glycocalyx of cells, for example.

glycosaminoglycan (GAG) (GLY-co-seh-ME-no-GLY-can) A polysaccharide composed of modified sugars with amino groups; the major component of a proteoglycan. GAGs are largely responsible for the viscous consistency of tissue gel and the stiffness of cartilage.

glycosuria (GLY-co-SOOR-ee-uh) The presence of glucose in the urine, typically indicative of a kidney disease, diabetes mellitus, or other endocrine disorder.

goblet cell A mucus-secreting gland cell, shaped somewhat like a wineglass, found in the epithelia of many mucous membranes. fig. 5.33a

Golgi complex (GOAL-jee) An organelle composed of several parallel cisterns, somewhat like a stack of saucers, that modifies and packages newly synthesized proteins and synthesizes carbohydrates. fig. 3.29

Golgi vesicle A membrane-bounded vesicle pinched from the Golgi complex, containing its chemical product; may be retained in the cell as a lysosome or become a secretory vesicle that releases the product by exocytosis.

gonad The ovary or testis.

gonadotropin (go-NAD-oh-TRO-pin) A pituitary hormone that stimulates the gonads; specifically FSH and LH.

G protein A protein of the plasma membrane that is activated by a membrane receptor and, in turn, opens an ion channel or activates an intracellular physiological response; important in linking ligand–receptor binding to second-messenger systems.

gradient A difference or change in any variable, such as pressure or chemical concentration, from one point in space to another; provides a basis for molecular movements such as gas exchange, osmosis, and facilitated diffusion, and for bulk movements such as the flow of blood, air, and heat.

gray matter A zone or layer of tissue in the central nervous system where the neuron cell bodies, dendrites, and synapses are found; forms the cerebral cortex and basal nuclei; cerebellar cortex and deep nuclei; nuclei of the brainstem; and core of the spinal cord. fig. 14.6c

gross anatomy Bodily structure that can be observed without magnification.

growth factor A chemical messenger that stimulates mitosis and differentiation of target cells that have receptors for it; important in such processes as fetal development, tissue maintenance and repair, and hematopoiesis; sometimes a contributing factor in cancer.

growth hormone (GH) A hormone of the anterior pituitary gland with multiple effects on many tissues, generally promoting tissue growth.

gustation (gus-TAY-shun) The sense of taste.

gyrus (JY-rus) A wrinkle or fold in the cortex of the cerebrum or cerebellum.

H

hair cell Sensory cell of the cochlea, semicircular ducts, utricle, and saccule, with a fringe of surface microvilli that respond to the relative motion of a gelatinous membrane at their tips; responsible for the senses of hearing, body position, and motion.

hair follicle An epidermal pit that contains a hair and extends into the dermis or hypodermis.

half-life ($T_{1/2}$) 1. The time required for one-half of a quantity of a radioactive element to decay to a stable isotope *(physical half-life)* or to be cleared from the body through a combination of radioactive decay and physiological excretion *(biological half-life)*. **2.** The time required for one-half of a quantity of hormone to be cleared from the bloodstream.

haploid (*n*) In humans, having 23 unpaired chromosomes instead of the usual 46 chromosomes in homologous pairs; in any organism or cell, having half the normal diploid number of chromosomes for that species.

helper T cell A type of lymphocyte that performs a central coordinating role in humoral and cellular immunity; target of the human immunodeficiency virus (HIV).

hematocrit (he-MAT-oh-crit) The percentage of blood volume that is composed of erythrocytes; also called *packed cell volume.*

hematoma (HE-muh-TOE-muh) A mass of clotted blood in the tissues; forms a bruise when visible through the skin.

hematopoiesis (he-MAT-o-poy-EE-sis) Production of any of the formed elements of blood.

hematopoietic stem cell (HSC) A cell of the red bone marrow that can give rise, through a series of intermediate cells, to erythrocytes, platelets, various kinds of macrophages, and any type of leukocyte.

heme (pronounced "heem") The nonprotein, iron-containing prosthetic group of hemoglobin or myoglobin; oxygen binds to its iron atom. fig. 18.5

hemoglobin (HE-mo-GLO-bin) The red gas transport pigment of an erythrocyte.

heparin (HEP-uh-rin) A polysaccharide secreted by basophils and mast cells that inhibits blood clotting.

hepatic (heh-PAT-ic) Pertaining to the liver.

hepatic portal system A network of blood vessels that connect capillaries of the intestines to capillaries (sinusoids) of the liver, thus delivering newly absorbed nutrients directly to the liver.

heterozygous (HET-er-oh-ZY-gus) Having nonidentical alleles at the same gene locus of two homologous chromosomes.

high-density lipoprotein (HDL) A lipoprotein of the blood plasma that is about 50% lipid and 50% protein; functions to transport phospholipids and cholesterol from other organs to the liver for disposal. A high proportion of HDL to low-density lipoprotein (LDL) is desirable for cardiovascular health.

hilum (HY-lum) A point on the surface of an organ where blood vessels, lymphatic vessels, or nerves enter and leave, usually marked by a depression and slit; the midpoint of the concave surface of any organ that is roughly bean-shaped, such as the lymph nodes, kidneys, and lungs. Also called the *hilus.* fig. 22.9b

histamine An amino acid derivative secreted by basophils, mast cells, and some neurons; functions as a paracrine secretion and neurotransmitter to stimulate effects such as gastric secretion, bronchoconstriction, and vasodilation. fig. 12.22

histology 1. The microscopic structure of tissues and organs. **2.** The study of such structure.

homeostasis (HO-me-oh-STAY-sis) The tendency of a living body to maintain relatively stable internal conditions in spite of greater changes in its external environment.

homologous (ho-MOLL-uh-gus) **1.** Having the same embryonic or evolutionary origin but not necessarily the same function, such as the scrotum and labia majora. **2.** Pertaining to two chromosomes with identical structures and gene loci but not necessarily identical alleles; each member of the pair is inherited from a different parent.

homozygous (HO-mo-ZY-gus) Having identical alleles at the same gene locus of two homologous chromosomes.

hormone A chemical messenger that is secreted by an endocrine gland or isolated gland cell, travels in the bloodstream, and triggers a physiological response in distant cells with receptors for it.

host cell Any cell belonging to the human body, as opposed to foreign cells introduced to it by such causes as infections and tissue transplants.

human chorionic gonadotropin (HCG) A hormone of pregnancy secreted by the chorion that stimulates continued growth of the corpus luteum and secretion of its hormones. HCG in urine is the basis for pregnancy testing.

human immunodeficiency virus (HIV) A virus that infects human helper T cells and other cells, suppresses immunity, and causes AIDS.

hyaline cartilage (HY-uh-lin) A form of cartilage with a relatively clear matrix and fine collagen fibers but no conspicuous elastic fibers or coarse collagen bundles as in other types of cartilage.

hyaluronic acid (HI-ul-yur-ON-ic) A glycosaminoglycan that is particularly abundant in connective tissues, where it becomes hydrated and forms the tissue gel.

hydrogen bond A weak attraction between a slightly positive hydrogen atom on one molecule and a slightly negative oxygen or nitrogen atom on another molecule, or between such atoms on different parts of the same molecule; responsible for the cohesion of water and the coiling of protein and DNA molecules, for example.

hydrolysis (hy-DROL-ih-sis) A chemical reaction that breaks a covalent bond in a molecule by adding an —OH group to one side of the bond and —H to the other side, thus consuming a water molecule.

hydrophilic (HY-dro-FILL-ic) Pertaining to molecules that attract water or dissolve in it because of their polar nature.

hydrophobic (HY-dro-FOE-bic) Pertaining to molecules that do not attract water or dissolve in it because of their nonpolar nature; such molecules tend to dissolve in lipids and other nonpolar solvents.

hydrostatic pressure The physical force exerted against a surface by a liquid such as blood or tissue fluid, as opposed to osmotic and atmospheric pressures.

hypercalcemia (HY-per-cal-SEE-me-uh) An excess of calcium ions in the blood.

hypercapnia (HY-pur-CAP-nee-uh) An excess of carbon dioxide in the blood.

hyperextension A joint movement that increases the angle between two bones beyond 180°. fig. 9.12

hyperglycemia (HY-pur-gly-SEE-me-uh) An excess of glucose in the blood.

hyperkalemia (HY-pur-ka-LEE-me-uh) An excess of potassium ions in the blood.

hypernatremia (HY-pur-na-TREE-me-uh) An excess of sodium ions in the blood.

hyperplasia (HY-pur-PLAY-zhuh) The growth of a tissue through cellular multiplication, not cellular enlargement. *Compare* hypertrophy.

hyperpolarization A shift in the electrical potential across a plasma membrane to a value more negative than the resting membrane potential, tending to inhibit a nerve or muscle cell. *Compare* depolarization.

hypersecretion Excessive secretion of a hormone or other gland product; can lead to endocrine disorders such as Addison disease or gigantism, for example.

hypertension Excessively high blood pressure; criteria vary but it is often considered to be a condition in which systolic pressure exceeds 140 mm Hg or diastolic pressure exceeds 90 mm Hg at rest.

hypertonic Having a higher osmotic pressure than human cells or some other reference solution and tending to cause osmotic shrinkage of cells.

hypertrophy (hy-PUR-truh-fee) The growth of a tissue through cellular enlargement, not cellular multiplication; for example, the growth of muscle under the influence of exercise. *Compare* hyperplasia.

hypocalcemia (HY-po-cal-SEE-me-uh) A deficiency of calcium ions in the blood.

hypocapnia (HY-po-CAP-nee-uh) A deficiency of carbon dioxide in the blood.

hypodermis (HY-po-DUR-miss) A layer of connective tissue deep to the skin; also called *superficial fascia, subcutaneous tissue,* or when it is predominantly adipose, *subcutaneous fat.*

hypoglycemia (HY-po-gly-SEE-me-uh) A deficiency of glucose in the blood.

hypokalemia (HY-po-ka-LEE-me-uh) A deficiency of potassium ions in the blood.

hyponatremia (HY-po-na-TREE-me-uh) A deficiency of sodium ions in the blood.

hyposecretion Inadequate secretion of a hormone or other gland product; can lead to endocrine disorders such as diabetes mellitus or pituitary dwarfism, for example.

hypothalamic thermostat A nucleus in the hypothalamus that monitors body temperature and sends afferent signals to hypothalamic heat-promoting or heat-losing centers to maintain thermal homeostasis.

hypothalamus (HY-po-THAL-uh-muss) The inferior portion of the diencephalon of the brain, forming the walls and floor of the third ventricle and giving rise to the posterior pituitary gland; controls many fundamental physiological functions such as appetite, thirst, and body temperature and exerts many of its effects through the endocrine and autonomic nervous systems. fig. 14.12b

hypothesis An informed conjecture that is capable of being tested and potentially falsified by experimentation or data collection. *See also* theory.

hypotonic Having a lower osmotic pressure than human cells or some other reference solution and tending to cause osmotic swelling and lysis of cells.

hypoxemia (HY-pock-SEE-me-uh) A deficiency of oxygen in the bloodstream.

hypoxia (hy-POCK-see-uh) A deficiency of oxygen in any tissue.

I

immune system A system of diverse defenses against disease, including leukocytes and other immune cells, defensive chemicals such as antibodies, physiological processes such as fever and inflammation, and physical barriers to infection such as the skin and mucous membranes; not an organ system in itself but an inclusive term for defensive components of multiple organ systems. *See also* adaptive immunity; innate immunity.

immunity The ability to ward off a specific infection or disease by means of any of the body's innate or adaptive immune mechanisms.

immunoglobulin (IM-you-no-GLOB-you-lin) *See* antibody.

implantation The attachment of a conceptus to the endometrium of the uterus.

inclusion Any visible object in the cytoplasm of a cell other than an organelle or cytoskeletal element; usually a foreign body or a stored cell product, such as a virus, dust particle, lipid droplet, glycogen granule, or pigment.

infarction (in-FARK-shun) **1.** The sudden death of tissue from a lack of blood perfusion. **2.** An area of necrotic tissue produced by this process; also called an *infarct.*

inferior Lower than another structure or point of reference from the perspective of anatomical position; for example, the stomach is inferior to the diaphragm.

inflammation A complex of tissue responses to trauma or infection serving to ward off a pathogen and promote tissue repair; recognized by the cardinal signs of redness, heat, swelling, and pain.

inguinal (IN-gwih-nul) Pertaining to the groin.

inhibitory postsynaptic potential (IPSP) Hyperpolarization of a postsynaptic neuron in response to a neurotransmitter, making it less likely to reach threshold and fire.

innate immunity Nonspecific defenses against infection or disease that are present and functional from birth, work equally against multiple disease agents, do not require prior exposure, and do not possess immune memory. Includes epithelial barriers, natural killer cells, other leukocytes, macrophages, antimicrobial proteins, and the processes of inflammation and fever. *See also* adaptive immunity.

innervation (IN-ur-VAY-shun) The nerve supply to an organ.

insertion Traditionally, the attachment of a skeletal muscle to a bone or other structure that moves when the muscle contracts. This term is now being abandoned by authorities in human anatomy. *Compare* origin.

inspiration **1.** Inhaling. **2.** The stimulus that resulted in this book.

integral protein A protein of the plasma membrane that penetrates into or all the way through the phospholipid bilayer. fig. 3.6

integration A process in which a neuron receives input from multiple sources and their combined effects determine its output; the cellular basis of information processing by the nervous system.

intercalated disc (in-TUR-ka-LAY-ted) A complex of fascia adherens, gap junctions, and desmosomes that join two cardiac muscle cells end to end, microscopically visible as a dark line that helps to histologically distinguish this muscle type; functions as a mechanical and electrical link between cells. fig. 19.11

intercellular Between cells.

intercostal (IN-tur-COSS-tul) Between the ribs, as in the intercostal muscles, arteries, veins, and nerves.

interdigitate (IN-tur-DIDJ-ih-tate) To fit together like the fingers of two folded hands; for example, at the dermal–epidermal boundary, intercalated discs of the heart, and foot processes of the podocytes in the kidney. fig. 23.10b

interleukin (IN-tur-LOO-kin) A hormonelike chemical messenger from one leukocyte to another, serving as a means of communication and coordination during immune responses.

interneuron (IN-tur-NEW-ron) A neuron that is contained entirely in the central nervous system and, in the path of signal conduction, lies anywhere between an afferent pathway and an efferent pathway.

interosseous membrane (IN-tur-OSS-ee-us) A fibrous membrane that connects the radius to the ulna and the tibia to the fibula along most of the shaft of each bone. fig. 8.34

interphase That part of the cell cycle between one mitotic phase and the next, from the end of cytokinesis to the beginning of the next prophase.

interstitial (IN-tur-STISH-ul) **1.** Pertaining to the extracellular spaces in a tissue. **2.** Located between other structures, as in the interstitial cells of the testis and interstitial (extracellular) fluid of the tissues.

intervertebral disc A cartilaginous pad between the bodies of two adjacent vertebrae.

intracellular Within a cell.

intracellular fluid (ICF) The fluid contained in the cells; one of the body's major fluid compartments.

intravenous (I.V.) **1.** Present or occurring within a vein, such as an intravenous blood clot. **2.** Introduced directly into a vein, such as an intravenous injection or I.V. drip.

intrinsic (in-TRIN-sic) **1.** Arising from within, such as intrinsic blood-clotting factors; endogenous. **2.** Fully contained within a specified organ or region, such as the intrinsic muscles of the hand and eye. *Compare* extrinsic.

involuntary Not under conscious control, including tissues such as smooth and cardiac muscle and events such as reflexes.

involution (IN-vo-LOO-shun) Shrinkage of a tissue or organ by autolysis, such as involution of the thymus after childhood and of the uterus after pregnancy.

ion A chemical particle with unequal numbers of electrons or protons and consequently a net negative or positive charge; it may have a single atomic nucleus as in a sodium ion or a few atoms as in a bicarbonate ion, or it may be a large molecule such as a protein.

ionic bond The force that binds a cation to an anion.

ionizing radiation High-energy electromagnetic rays that eject electrons from atoms or molecules and convert them to ions, frequently causing cellular damage; for example, X-rays and gamma rays.

ipsilateral (IP-sih-LAT-ur-ul) On the same side of the body, as in reflex arcs in which a muscular response occurs on the same side of the body as the stimulus. *Compare* contralateral.

ischemia (iss-KEE-me-uh) Insufficient blood flow to a tissue, typically resulting in metabolite accumulation and sometimes tissue death.

isometric Pertaining to a form of muscle contraction in which internal tension increases but the muscle does not change length or move a resisting object.

isotonic **1.** Having the same osmotic pressure as human cells or some other reference solution. **2.** Pertaining to a form of muscle contraction in which a muscle changes length but maintains a constant amount of tension.

J

jaundice (JAWN-diss) A yellowish color of the skin, corneas, mucous membranes, and body fluids due to an excessive concentration of bilirubin; usually indicative of a liver disease, obstructed bile secretion, or hemolytic disease.

K

ketone (KEE-tone) Any organic compound with a carbonyl ($C=O$) group covalently bonded to a two-carbon backbone.

ketone bodies Certain ketones (acetone, acetoacetic acid, and β-hydroxybutyric acid) produced by the incomplete oxidation of fats, especially when fats are being rapidly catabolized. *See also* ketosis.

ketonuria (KEE-toe-NEW-ree-uh) The abnormal presence of ketones in the urine; a sign of diabetes mellitus but also occurring in other conditions that entail rapid fat oxidation.

ketosis (kee-TOE-sis) An abnormally high concentration of ketone bodies in the blood, occurring in pregnancy, starvation, diabetes mellitus, and other conditions; tends to cause acidosis and to depress the nervous system.

kilocalorie The amount of heat energy needed to raise the temperature of 1 kg of water by 1°C; 1,000 calories. Also called a *Calorie* or *large calorie. See also* calorie.

kinase Any enzyme that adds an inorganic phosphate (P_i) group to another organic molecule. Also called a *phosphokinase*.

L

labium (LAY-bee-um) A lip, such as those of the mouth and the labia majora and minora of the vulva.

lactate A small organic acid produced as an end product of the anaerobic fermentation of pyruvate; called *lactic acid* in its nonionized form.

lacuna (la-CUE-nuh) (plural, *lacunae*) A small cavity or depression in a tissue such as bone or cartilage; called a *cavernous space* in erectile tissues of the penis and clitoris.

lamella (la-MELL-uh) A little plate, such as a lamella of bone. fig. 7.4

lamina (LAM-ih-nuh) A thin layer, such as the lamina of a vertebra or the lamina propria of a mucous membrane. fig. 8.22

lamina propria (LAM-ih-nuh PRO-pree-uh) A thin layer of areolar tissue immediately deep to the epithelium of a mucous membrane. fig. 5.33a

larynx (LAIR-inks) A cartilaginous chamber in the neck containing the vocal cords; colloquially called the voicebox.

latent period The interval between a stimulus and response, especially in the action of nerve and muscle cells.

lateral Away from the midline of an organ or median plane of the body; toward the side. *Compare* medial.

law A verbal or mathematical description of a predictable natural phenomenon or of the relationship between variables; for example, Boyle's law of gases and the law of complementary base pairing in DNA.

lesion A circumscribed zone of tissue injury, such as a skin abrasion or myocardial infarction.

leukocyte (LOO-co-site) A white blood cell.

leukotriene (LOO-co-TRY-een) An eicosanoid that promotes allergic and inflammatory responses such as vasodilation and neutrophil chemotaxis; secreted by basophils, mast cells, and damaged tissues.

libido (lih-BEE-do) Sex drive; psychological motivation to engage in sex.

ligament A collagenous band or cord that binds one organ to another, especially one bone to another, and serves to hold organs in place; for example, the cruciate ligaments of the knee, broad ligament of the uterus, and falciform ligament of the liver.

ligand (LIG-and, LY-gand) A chemical that binds reversibly to a receptor site on a protein, such as a neurotransmitter that binds to a membrane receptor or a substrate that binds to an enzyme.

ligand-gated channel A channel protein in a plasma membrane that opens or closes when another chemical (ligand) binds to it, enabling the ligand to determine when substances can enter or leave the cell.

light microscope (LM) A microscope that produces images with visible light.

linea (LIN-ee-uh) An anatomical line, such as the linea alba of the abdomen or linea aspera of the femur.

lingual (LING-gwul) Pertaining to the tongue, as in lingual papillae.

lipase (LY-pace) An enzyme that hydrolyzes a triglyceride into fatty acids and glycerol.

lipid A hydrophobic organic compound composed mainly of carbon and a high ratio of hydrogen to oxygen; includes fatty acids, fats, phospholipids, steroids, and prostaglandins.

lipoprotein (LIP-oh-PRO-teen) A protein-coated lipid droplet in the blood plasma or lymph, serving as a means of lipid transport; for example, chylomicrons and high- and low-density lipoproteins.

load **1.** To pick up a gas for transport in the bloodstream. **2.** The resistance acted upon by a muscle.

lobule (LOB-yool) A small subdivision of an organ or of a lobe of an organ, especially of a gland.

locus The site on a chromosome where a given gene is located.

long bone A bone such as the femur or humerus that is markedly longer than wide and that generally serves as a lever.

longitudinal Oriented along the longest dimension of the body or of an organ.

loose connective tissue *See* areolar tissue.

low-density lipoprotein (LDL) A lipoprotein of the blood plasma that is about 80% lipid (mainly cholesterol) and 20% protein; functions to transport cholesterol to target cells. A high proportion of LDL to high-density lipoprotein (HDL) is a risk factor for cardiovascular disease.

lumbar Pertaining to the lower back and sides, between the thoracic cage and pelvis.

lumen (LOO-men) The internal space of a hollow organ such as a blood vessel or the esophagus, or a space surrounded by secretory cells as in a gland acinus.

luteinizing hormone (LH) (LOO-tee-in-ize-ing) A hormone secreted by the anterior pituitary gland that stimulates ovulation in females and testosterone secretion production in males.

lymph The fluid contained in lymphatic vessels and lymph nodes, produced by the absorption of tissue fluid.

lymphatic system (lim-FAT-ic) An organ system consisting of lymphatic vessels, lymph nodes, the tonsils, spleen, and thymus; functions include tissue fluid recovery and immunity.

lymph node A small organ found along the course of a lymphatic vessel that filters the lymph and contains lymphocytes and macrophages, which respond to antigens in the lymph. fig. 21.11

lymphocyte (LIM-fo-site) A relatively small agranulocyte with numerous types and roles in innate, humoral, and cellular immunity. table 18.6

lysosome (LY-so-some) A membrane-bounded organelle containing a mixture of enzymes with a variety of intracellular and extracellular roles in digesting foreign matter, pathogens, and expired organelles.

lysozyme (LY-so-zime) An enzyme found in tears, milk, saliva, mucus, and other body fluids that destroys bacteria by digesting their cell walls. Also called *muramidase*.

M

macromolecule Any molecule of large size and high molecular weight, such as a protein, nucleic acid, polysaccharide, or triglyceride.

macrophage (MAC-ro-faje) Any cell of the body, other than a leukocyte, that is specialized for phagocytosis; usually derived from a blood monocyte and often functioning as an antigen-presenting cell.

macula (MAC-you-luh) A patch or spot, such as the macula lutea of the retina and macula sacculi of the inner ear.

malignant (muh-LIG-nent) Pertaining to a cell or tumor that is cancerous; capable of metastasis.

mast cell A connective tissue cell, similar to a basophil, that secretes histamine, heparin, and other chemicals involved in inflammation; often concentrated along the course of blood capillaries.

matrix 1. The extracellular material of a tissue. **2.** The fluid within a mitochondrion containing enzymes of the citric acid cycle. **3.** The substance or framework within which other structures are embedded, such as the fibrous matrix of a blood clot. **4.** A mass of epidermal cells from which a hair root or nail root develops.

mechanoreceptor A sensory nerve ending or organ specialized to detect mechanical stimuli such as touch, pressure, stretch, or vibration.

medial Toward the midline of an organ or median plane of the body. *Compare* lateral.

median plane The sagittal plane that divides the body or an organ into equal right and left halves; also called *midsagittal plane*. fig. A.1

mediastinum (MEE-dee-ah-STY-num) The thick median partition of the thoracic cavity that separates one pleural cavity from the other and contains the heart, great blood vessels, esophagus, trachea, and thymus. fig. A.5

medulla (meh-DULE-uh, meh-DULL-uh) Tissue deep to the cortex of certain two-layered organs such as the lymph nodes, adrenal glands, hairs, and kidneys.

medulla oblongata (meh-DULL-uh OB-long-GAH-ta) The most caudal part of the brainstem, immediately superior to the foramen magnum of the skull, connecting the spinal cord to the rest of the brain. figs. 14.2, 14.8

meiosis (my-OH-sis) A form of cell division in which a diploid cell divides twice and produces four haploid daughter cells; occurs only in gametogenesis.

melanocyte A cell of the stratum basale of the epidermis that synthesizes melanin and transfers it to the keratinocytes.

meninges (meh-NIN-jeez) (singular, *meninx*) Three fibrous membranes between the central nervous system and surrounding bone: the dura mater, arachnoid mater, and pia mater. fig. 14.5

merocrine (MERR-oh-crin) *See* eccrine.

mesenchyme (MES-en-kime) A gelatinous embryonic connective tissue derived from the mesoderm; differentiates into all permanent connective tissues and cardiac and smooth muscle.

mesentery (MESS-en-tare-ee) A serous membrane that binds the intestines together and suspends them from the abdominal wall; the visceral continuation of the peritoneum. fig. 25.3

mesoderm (MES-oh-durm) The middle layer of the three primary germ layers of an embryo; gives rise to muscle and connective tissue.

metabolism (meh-TAB-oh-lizm) Chemical reactions within a living organism.

metabolite (meh-TAB-oh-lite) Any chemical produced by metabolism.

metaplasia Transformation of one mature tissue type into another; for example, a change from pseudostratified to stratified squamous epithelium in an overventilated nasal cavity.

metastasis (meh-TASS-tuh-sis) The spread of cancer cells from the original tumor to a new location, where they seed the development of a new tumor.

microtubule An intracellular cylinder composed of the protein tubulin, forming centrioles, the axonemes of cilia and flagella, and part of the cytoskeleton.

microvillus An outgrowth of the plasma membrane that increases the surface area of a cell and functions in absorption and some sensory processes; distinguished from cilia and flagella by its smaller size and lack of an axoneme.

milliequivalent One-thousandth of an equivalent, which is the amount of an electrolyte that would neutralize 1 mole of H^+ or OH^-. Electrolyte concentrations are commonly expressed in milliequivalents per liter (mEq/L).

mitochondrion (MY-toe-CON-dree-un) An organelle specialized to synthesize ATP, enclosed in a double unit membrane with infoldings of the inner membrane called cristae.

mitosis (my-TOE-sis) A form of cell division in which a cell divides once and produces two genetically identical daughter cells; sometimes used to refer only to the division of the genetic material or nucleus and not to include cytokinesis, the subsequent division of the cytoplasm.

moiety (MOY-eh-tee) A chemically distinct subunit of a macromolecule, such as the heme and globin moieties of hemoglobin or the lipid and carbohydrate moieties of a glycolipid.

molarity A measure of chemical concentration expressed as moles of solute per liter of solution.

mole The mass of a chemical equal to its molecular weight in grams, containing 6.023×10^{23} molecules.

monocyte (MON-oh-site) An agranulocyte specialized to migrate into the tissues and transform into a macrophage. table 18.6

monomer (MON-oh-mur) **1.** One of the identical or similar subunits of a larger molecule in the dimer to polymer range; for example, the glucose monomers of starch, the amino acids of a protein, or the nucleotides of DNA. **2.** One subunit of an antibody molecule, composed of four polypeptides.

monosaccharide (MON-oh-SAC-uh-ride) A simple sugar, or sugar monomer; chiefly glucose, fructose, and galactose.

motor neuron A neuron that transmits signals from the central nervous system to any effector (muscle or gland cell); its axon is an efferent nerve fiber.

motor protein Any protein that produces movements of a cell or its components owing to its ability to undergo quick repetitive changes in conformation and to bind reversibly to other molecules; for example, myosin, dynein, and kinesin.

motor unit One motor neuron and all the skeletal muscle fibers innervated by it.

mucosa (mew-CO-suh) A tissue layer that forms the inner lining of an anatomical tract that is open to the exterior (the respiratory, digestive, urinary, and reproductive tracts). Composed of epithelium, connective tissue (lamina propria), and often smooth muscle (muscularis mucosae). fig. 5.33a

mucous membrane See *mucosa*.

multipotent Pertaining to a stem cell that has the potential to develop into two or more types of fully differentiated, functional cells, but not into an unlimited variety of cell types.

muscle fiber One skeletal muscle cell; a myofiber.

muscle tone A state of continual, partial contraction of resting skeletal or smooth muscle.

muscular system An organ system composed of the skeletal muscles, specialized mainly for maintaining postural support and producing movements of the bones.

muscularis externa The external muscular wall of certain viscera such as the esophagus and small intestine. fig. 25.2

muscularis mucosae (MUSK-you-LERR-iss mew-CO-see) A layer of smooth muscle immediately deep to the lamina propria of a mucosa. fig. 5.33a

mutagen (MEW-tuh-jen) Any agent that causes a mutation, including viruses, chemicals, and ionizing radiation.

mutation Any change in the structure of a chromosome or a DNA molecule, often resulting in a change of organismal structure or function.

myelin (MY-eh-lin) A lipid sheath around a nerve fiber, formed from closely spaced spiral layers of the plasma membrane of a Schwann cell or oligodendrocyte. fig. 12.8

myelin sheath gap A short unmyelinated segment between Schwann cells of a myelinated nerve fiber; site of action potential generation during saltatory conduction; also called *node of Ranvier*. fig. 12.4

myocardium (MY-oh-CAR-dee-um) The middle, muscular layer of the heart.

myoepithelial cell An epithelial cell that has become specialized to contract like a muscle cell; important in dilation of the pupil and ejection of secretions from gland acini.

myofilament A protein microfilament responsible for the contraction of a muscle cell, composed mainly of myosin or actin. fig. 11.2

myoglobin (MY-oh-GLO-bin) A red oxygen-storage pigment of muscle; supplements hemoglobin in providing oxygen for aerobic muscle metabolism.

myosin A motor protein that constitutes the thick myofilaments of muscle and has globular, mobile heads of ATPase that bind to actin molecules; also serves contractile functions in other cell types. fig. 11.3

N

necrosis (neh-CRO-sis) Pathological tissue death due to such causes as infection, trauma, or hypoxia. *Compare* apoptosis.

negative feedback A self-corrective mechanism that underlies most homeostasis, in which a bodily change is detected and responses are activated that reverse the change and restore stability and preserve normal body function.

negative feedback inhibition A mechanism for limiting the secretion of a pituitary tropic hormone. The tropic hormone stimulates another endocrine gland to secrete its own hormone, and that hormone inhibits further release of the tropic hormone.

neonate (NEE-oh-nate) A newborn infant up to 4 weeks old.

neoplasia (NEE-oh-PLAY-zhuh) Abnormal growth of new tissue, such as a tumor, with no useful function.

nephron (NEF-ron) One of approximately 1 million blood-filtering, urine-producing units in each kidney; consists of a glomerulus, glomerular capsule, proximal convoluted tubule, nephron loop, and distal convoluted tubule. fig. 23.8

nerve A cordlike organ of the peripheral nervous system composed of multiple nerve fibers ensheathed in connective tissue.

nerve fiber The axon of a single neuron.

nerve impulse A wave of self-propagating action potentials traveling along a nerve fiber.

nervous tissue A tissue composed of neurons and neuroglia.

net filtration pressure A net force favoring filtration of fluid from a capillary or venule when all the hydrostatic and osmotic pressures of the blood and tissue fluids are taken into account.

neural pool A group of interconnected neurons of the central nervous system that perform a single collective function; for example, the vasomotor center of the brainstem and speech centers of the cerebral cortex.

neural tube A dorsal hollow tube in the embryo that develops into the central nervous system. fig. 14.3

neuroglia (noo-ROG-lee-uh) All cells of nervous tissue except neurons; cells that perform various supportive and protective roles for the neurons.

neuromuscular junction A synapse between a nerve fiber and a muscle fiber; also called a *motor end plate*. fig. 11.7

neuron (NOOR-on) A nerve cell; an electrically excitable cell specialized for producing and conducting action potentials and secreting chemicals that stimulate adjacent cells.

neuropeptide A peptide secreted by a neuron, often serving to modify the action of a neurotransmitter; for example, endorphins, enkephalin, and cholecystokinin. fig. 12.22

neurotransmitter A chemical released at the distal end of an axon that stimulates an adjacent cell; for example, acetylcholine, norepinephrine, or serotonin.

neutrophil (NEW-tro-fill) A granulocyte, usually with a multilobed nucleus, that serves especially to destroy bacteria by means of phagocytosis, intracellular digestion, and secretion of bactericidal chemicals. table 18.6

nitrogenous base (ny-TRODJ-eh-nus) An organic molecule with a single or double carbon–nitrogen ring that forms one of the building blocks of ATP, other nucleotides, and nucleic acids; the basis of the genetic code. fig. 4.1

nitrogenous waste Any nitrogen-containing substance produced as a metabolic waste and excreted in the urine; chiefly ammonia, urea, uric acid, and creatinine.

nociceptor (NO-sih-SEP-tur) A nerve ending specialized to detect tissue damage and produce a sensation of pain; pain receptor.

norepinephrine (NE) (nor-EP-ih-NEF-rin) A catecholamine that functions as a neurotransmitter and adrenal hormone, especially in the sympathetic nervous system. fig. 12.22

nuclear envelope A pair of membranes enclosing the nucleus of a cell, with prominent pores allowing traffic of molecules between the nucleoplasm and cytoplasm. fig. 3.27

nucleic acid (new-CLAY-ic) An acidic polymer of nucleotides found or produced in the nucleus, functioning in heredity and protein synthesis; of two types, DNA and RNA.

nucleotide (NEW-clee-oh-tide) An organic molecule composed of a nitrogenous base, a monosaccharide, and a phosphate group; the monomer of a nucleic acid.

nucleus (NEW-clee-us) **1.** A cell organelle containing DNA and surrounded by a double membrane. **2.** A mass of neurons (gray matter) surrounded by white matter of the brain, including the basal nuclei and brainstem nuclei. **3.** The positively charged core of an atom, consisting of protons and neutrons. **4.** A central structure, such as the nucleus pulposus of an intervertebral disc.

nurse cell A supporting cell in the seminiferous tubules of the testes, acting to enfold and protect developing germ cells and promote the production of sperm; also called *Sertoli cell; sustentacular cell.*

O

olfaction (ole-FAC-shun) The sense of smell.

oncotic pressure (on-COT-ic) The difference between the colloid osmotic pressure of the blood and that of the tissue fluid, usually favoring fluid absorption by the blood capillaries. *Compare* colloid osmotic pressure.

oocyte (OH-oh-site) In the development of an egg cell, any haploid stage between meiosis I and fertilization.

oogenesis (OH-oh-JEN-eh-sis) The production of a fertilizable egg cell through a series of mitotic and meiotic cell divisions; female gametogenesis.

opposition A movement of the thumb in which it approaches or touches any fingertip of the same hand.

orbit The eye socket of the skull.

organ Any anatomical structure that is composed of at least two different tissue types, has recognizable structural boundaries, and has a discrete function different from the structures around it. Many organs are microscopic and many organs contain smaller organs, such as the skin containing numerous microscopic sense organs.

organelle Any structure within a cell that carries out one of its metabolic roles, such as mitochondria, centrioles, endoplasmic reticulum, and the nucleus; an intracellular structure other than the cytoskeleton and inclusions.

origin Traditionally, the relatively stationary attachment of a skeletal muscle to a bone or other structure. This term is now being abandoned by authorities in human anatomy. *Compare* insertion.

osmolality (OZ-mo-LAL-ih-tee) The molar concentration of dissolved particles in 1 kg of water.

osmolarity (OZ-mo-LERR-ih-tee) The molar concentration of dissolved particles in 1 L of solution.

osmoreceptor (OZ-mo-re-SEP-tur) A neuron of the hypothalamus that responds to changes in the osmolarity of the extracellular fluid.

osmosis (oz-MO-sis) The net flow of water through a selectively permeable membrane,

resulting from either a chemical concentration difference or a mechanical force across the membrane.

osmotic pressure The amount of pressure that would have to be applied to one side of a selectively permeable membrane to stop osmosis; proportional to the concentration of nonpermeating solutes on that side and therefore serving as an indicator of solute concentration.

osseous (OSS-ee-us) Pertaining to bone.

ossification (OSS-ih-fih-CAY-shun) Bone formation.

osteoblast Bone-forming cell that arises from an osteogenic cell, deposits bone matrix, and eventually becomes an osteocyte.

osteoclast Macrophage of the bone surface that dissolves the matrix and returns minerals to the extracellular fluid.

osteocyte A mature bone cell formed when an osteoblast becomes surrounded by its own matrix and entrapped in a lacuna.

osteon A structural unit of compact bone consisting of a central canal surrounded by concentric cylindrical lamellae of matrix. fig. 7.4

osteoporosis (OSS-tee-oh-pore-OH-sis) A degenerative bone disease characterized by a loss of bone mass, increasing susceptibility to spontaneous fractures, and sometimes deformity of the vertebral column; causes include aging, estrogen hyposecretion, and insufficient resistance exercise.

ovulation (OV-you-LAY-shun) The release of a mature oocyte by the bursting of an ovarian follicle.

ovum Any stage of the female gamete from primary oocyte until fertilization; a primary or secondary oocyte; an egg.

oxidation A chemical reaction in which one or more electrons are removed from a molecule, lowering its free energy content; opposite of reduction and always linked to a reduction reaction.

P

pancreatic islet (PAN-cree-AT-ic EYE-let) A small cluster of endocrine cells in the pancreas that secretes insulin, glucagon, somatostatin, and other intercellular messengers; also called *islet of Langerhans.* fig. 17.12

papilla (pa-PILL-uh) A conical or nipplelike structure, such as a lingual papilla of the tongue or the papilla of a hair bulb.

papillary (PAP-ih-lerr-ee) **1.** Pertaining to or shaped like a nipple, such as the papillary muscles of the heart. **2.** Having papillae, such as the papillary layer of the dermis.

paracrine (PAIR-uh-crin) **1.** A chemical messenger similar to a hormone whose effects

are restricted to the immediate vicinity of the cells that secrete it; sometimes called a *local hormone.* **2.** Pertaining to such a secretion, as opposed to *endocrine.*

parasympathetic nervous system (PERR-uh-SIM-pa-THET-ic) A division of the autonomic nervous system that issues efferent fibers through the cranial and sacral nerves and exerts cholinergic effects on its target organs.

parathyroid hormone (PTH) A hormone secreted by the parathyroid glands that raises blood calcium concentration by stimulating bone resorption by osteoclasts, promoting intestinal absorption of calcium, and inhibiting urinary excretion of calcium.

parenchyma (pa-REN-kih-muh) The tissue that performs the main physiological functions of an organ, especially a gland, as opposed to the tissues (stroma) that mainly provide structural support.

parietal (pa-RY-eh-tul) **1.** Pertaining to a wall, as in the parietal cells of the stomach and parietal bone of the skull. **2.** Pertaining to the outer or more superficial layer of a two-layered membrane such as the pleura, pericardium, or glomerular capsule. *Compare* visceral. fig. A.6

pathogen Any disease-causing microorganism.

pedicle (PED-ih-cul) A small footlike process, as in the vertebrae and renal podocytes; also called a *pedicel.*

pelvic cavity The space enclosed by the true (lesser) pelvis, containing the urinary bladder, rectum, and internal reproductive organs.

pelvis A basinlike structure such as the pelvic girdle of the skeleton or the urine-collecting space near the hilum of the kidney. figs. 8.36, 23.4

peptide Any chain of two or more amino acids. *See also* polypeptide, protein.

peptide bond A group of four covalently bonded atoms (a —C=O group bonded to an —NH group) that links two amino acids in a protein or other peptide. fig. 2.23b

perfusion The amount of blood supplied to a given mass of tissue in a given period of time (such as mL/g/min.).

perichondrium (PERR-ih-CON-dree-um) A layer of fibrous connective tissue covering the surface of hyaline or elastic cartilage.

perineum (PERR-ih-NEE-um) The region between the thighs bordered by the coccyx, pubic symphysis, and ischial tuberosities; contains the orifices of the urinary, reproductive, and digestive systems. figs. 27.6, 28.8

periosteum (PERR-ee-OSS-tee-um) A layer of fibrous connective tissue covering the surface of a bone. fig. 7.1

peripheral nervous system (PNS) A subdivision of the nervous system composed of all nerves and ganglia; all of the nervous system except the central nervous system.

peripheral protein A protein of the plasma membrane that clings to its intracellular or extracellular surface but does not penetrate into the phospholipid bilayer.

peristalsis (PERR-ih-STAL-sis) A wave of constriction traveling along a tubular organ such as the esophagus or ureter, serving to propel its contents.

peritoneum (PERR-ih-toe-NEE-um) A serous membrane that lines the peritoneal cavity of the abdomen and covers the mesenteries and viscera.

perivascular (PERR-ih-VASS-cue-lur) Pertaining to the region surrounding a blood vessel.

phagocytosis (FAG-oh-sy-TOE-sis) A form of endocytosis in which a cell surrounds a foreign particle with pseudopods and engulfs it, enclosing it in a cytoplasmic vesicle called a *phagosome.*

phalanx (FAY-lanks) (plural, *phalanges*) Any of the bones in the fingers or toes; there are two in the thumb and great toe and three in each of the other digits.

pharynx (FAIR-inks, FAR-inks) A muscular passage in the throat at which the respiratory and digestive tracts cross.

phospholipid An amphipathic molecule composed of two fatty acids and a phosphate-containing group bonded to the three carbons of a glycerol molecule; composes most of the molecules of the plasma membrane and other cellular membranes.

phosphorylation Addition of an inorganic phosphate (P_i) group to an organic molecule.

piloerector *See* arrector.

pinocytosis (PIN-oh-sy-TOE-sis) A form of endocytosis in which the plasma membrane sinks inward and imbibes droplets of extracellular fluid.

plantar (PLAN-tur) Pertaining to the sole of the foot.

plaque A small scale or plate of matter, such as dental plaque, the fatty plaques of atherosclerosis, and the amyloid plaques of Alzheimer disease.

plasma The noncellular portion of the blood; its liquid matrix, usually constituting slightly over one-half of its volume.

plasma membrane The membrane that encloses a cell and controls the traffic of molecules in and out of the cell. fig. 3.5

platelet A formed element of the blood derived from a megakaryocyte, known especially for its role in stopping bleeding, but with additional roles in dissolving blood clots, stimulating inflammation, promoting tissue growth and blood vessel maintenance, and destroying bacteria.

pleura (PLOOR-uh) A double-walled serous membrane that encloses each lung.

plexus A network of blood vessels, lymphatic vessels, or nerves, such as a choroid plexus of the brain or brachial plexus of nerves.

pluripotent Pertaining to a stem cell of the inner cell mass of a blastocyst that is capable of developing into any type of embryonic cell, but not into cells of the accessory organs of pregnancy.

polymer A molecule that consists of a long chain of identical or similar subunits, such as protein, DNA, and starch.

polypeptide Any chain of more than 10 or 15 amino acids. *See also* protein.

polysaccharide (POL-ee-SAC-uh-ride) A polymer of simple sugars; for example, glycogen, starch, and cellulose. fig. 2.17

polyuria (POL-ee-YOU-ree-uh) Excessive output of urine.

popliteal (pop-LIT-ee-ul) Pertaining to the posterior aspect of the knee.

positron emission tomography (PET) A method of producing a computerized image of the physiological state of a tissue using injected radioisotopes that emit positrons.

posterior Near or pertaining to the back or spinal side of the body; dorsal.

posterior root The branch of a spinal nerve that enters the posterior side of the spinal cord and carries afferent (sensory) nerve fibers; often called *dorsal root.* fig. 13.2b

postganglionic (POST-gang-glee-ON-ic) Pertaining to a neuron that conducts signals from a ganglion to a more distal target organ.

postsynaptic (POST-sih-NAP-tic) Pertaining to a neuron or other cell that receives signals from the presynaptic neuron at a synapse. fig. 12.19

potential A difference in electrical charge from one point to another, especially on opposite sides of a plasma membrane; usually measured in millivolts.

preembryo A developing individual from the time of fertilization to the time, at 16 days, when the three primary germ layers have formed. *Compare* conceptus, embryo, fetus.

preganglionic (PRE-gang-glee-ON-ic) Pertaining to a neuron that conducts signals from the central nervous system to a ganglion.

presynaptic (PRE-sih-NAP-tic) Pertaining to a neuron that conducts signals to a synapse. fig. 12.19

primary germ layers The ectoderm, mesoderm, and endoderm; the three tissue layers of an early embryo from which all later tissues and organs arise.

prime mover The muscle that produces the most force in a given joint action; agonist.

programmed cell death *See* apoptosis.

prolactin (PRL) A hormone secreted by the anterior pituitary gland that stimulates the mammary glands to secrete milk.

pronation (pro-NAY-shun) A rotational movement of the forearm that turns the palm downward or posteriorly. fig. 9.18

proprioception (PRO-pree-oh-SEP-shun) The nonvisual perception, usually subconscious, of the position and movements of the body, resulting from input from proprioceptors and the vestibular apparatus of the inner ear.

proprioceptor (PRO-pree-oh-SEP-tur) A sensory receptor of the muscles, tendons, and joint capsules that detects muscle contractions and joint movements.

prostaglandin (PROSS-ta-GLAN-din) An eicosanoid with a five-sided carbon ring in the middle of a hydrocarbon chain, playing a variety of roles in inflammation, neurotransmission, vasomotion, reproduction, and metabolism. fig. 2.21

prostate (PROSS-tate) A male reproductive gland that encircles the urethra immediately inferior to the bladder and contributes to the semen. (Avoid the mispronunciation "prostrate.") fig. 27.10

protein A large polypeptide; while criteria for a protein are somewhat subjective and variable, polypeptides over 50 amino acids long are generally classified as proteins.

proteoglycan (PRO-tee-oh-GLY-can) A large molecule composed of a bristlelike arrangement of glycosaminoglycans surrounding a protein core in a shape resembling a bottle brush. Binds cells to extracellular materials and gives the tissue fluid a gelatinous consistency.

proximal Relatively near a point of origin or attachment; for example, the shoulder is proximal to the elbow. *Compare* distal.

pseudopod (SOO-doe-pod) A temporary cytoplasmic extension of a cell used for locomotion (ameboid movement) and phagocytosis.

pseudostratified columnar epithelium A type of epithelium with tall columnar cells reaching the free surface and shorter basal cells that do not reach the surface, but with all cells resting on the basement membrane; creates a false appearance of stratification. fig. 5.7

pulmonary circuit A route of blood flow that supplies blood to the pulmonary alveoli for gas exchange and then returns it to the heart; all blood vessels between the right ventricle and the left atrium of the heart.

pyrogen (PY-ro-jen) A fever-producing agent.

R

ramus (RAY-mus) An anatomical branch, as in a nerve or in the pubis.

receptor **1.** A cell or organ specialized to detect a stimulus, such as a taste cell or the eye. **2.** A protein molecule that binds and responds to a chemical such as a hormone, neurotransmitter, or odor molecule.

receptor-mediated endocytosis A process in which certain molecules in the extracellular fluid bind to receptors in the plasma membrane, these receptors gather together, the membrane sinks inward at that point, and the molecules are incorporated into vesicles in the cytoplasm. fig. 3.21

receptor potential A variable, local change in membrane voltage produced by a stimulus acting on a receptor cell; generates an action potential if it reaches threshold.

recessive 1. Pertaining to a genetic allele that is not phenotypically expressed in the presence of a dominant allele. **2.** Pertaining to a trait that results from a recessive allele.

reduction 1. A chemical reaction in which one or more electrons are added to a molecule, raising its free energy content; opposite of *oxidation* and always linked to an oxidation reaction. **2.** Treatment of a fracture by restoring the broken parts of a bone to their proper alignment.

reflex A stereotyped, automatic, involuntary response to a stimulus; includes somatic reflexes, in which the effectors are skeletal muscles, and visceral (autonomic) reflexes, in which the effectors are usually visceral muscle, cardiac muscle, or glands.

reflex arc A simple neural pathway that mediates a reflex; involves a receptor, an afferent nerve fiber, often one or more interneurons, an efferent nerve fiber, and an effector.

refractory period 1. A period of time after a nerve or muscle cell has responded to a stimulus in which it cannot be reexcited by a threshold stimulus. **2.** A period of time after male orgasm when it is not possible to reattain erection or ejaculation.

regeneration Replacement of damaged tissue with new tissue of the original type. *Compare* fibrosis.

renin (REE-nin) An enzyme secreted by the kidneys in response to hypotension; converts the plasma protein angiotensinogen to angiotensin I, leading indirectly to a rise in blood pressure.

repolarization Reattainment of the resting membrane potential after a nerve or muscle cell has depolarized.

residue Any one of the amino acids in a protein or other peptide.

resistance 1. A force that opposes the flow of a fluid such as air or blood. **2.** A force, or load, that opposes the action of a muscle or lever.

resting membrane potential (RMP) A stable voltage across the plasma membrane of an unstimulated nerve or muscle cell.

reticular cell (reh-TIC-you-lur) A delicate, branching phagocytic cell found in the reticular connective tissue of the lymphatic organs.

reticular fiber A fine, branching collagen fiber coated with glycoprotein, found in the stroma of lymphatic organs and some other tissues and organs.

reticular tissue A connective tissue composed of reticular cells and reticular fibers, found in bone marrow, lymphatic organs, and in lesser amounts elsewhere.

ribosome A granule found free in the cytoplasm or attached to the rough endoplasmic reticulum or nuclear envelope composed of ribosomal RNA and enzymes; specialized to read the nucleotide sequence of messenger RNA and assemble a corresponding sequence of amino acids to make a protein.

risk factor Any environmental factor or characteristic of an individual that increases one's chance of developing a particular disease; includes such intrinsic factors as age, sex, and race and such extrinsic factors as diet, smoking, and occupation.

rostral Relatively close to the forehead, especially in reference to structures of the brain and spinal cord; for example, the frontal lobe is rostral to the parietal lobe. *Compare* caudal.

ruga (ROO-ga) (plural *rugae,* ROO-jee) **1.** An internal fold or wrinkle in the mucosa of a hollow organ such as the stomach and urinary bladder; typically present when the organ is empty and relaxed but not when the organ is full and stretched. **2.** Tissue ridges in such locations as the hard palate and vagina. fig. 25.12

S

saccule (SAC-yule) A saclike receptor in the inner ear with a vertical patch of hair cells, the macula sacculi; senses the orientation of the head and responds to vertical acceleration, as when riding in an elevator or standing up. fig. 16.20

sagittal plane (SADJ-ih-tul) Any plane that extends from anterior to posterior and cephalic to caudal and that divides the body into right and left portions. *Compare* median plane.

sarcomere (SAR-co-meer) In skeletal and cardiac muscle, the portion of a myofibril from one Z disc to the next, constituting one contractile unit. fig. 11.5

sarcoplasmic reticulum (SR) The smooth endoplasmic reticulum of a muscle cell, serving as a calcium reservoir. fig. 11.2

scanning electron microscope (SEM) A microscope that uses an electron beam in

place of light to form high-resolution, three-dimensional images of the surfaces of objects; capable of much higher magnifications than a light microscope.

sclerosis (scleh-RO-sis) Hardening or stiffening of a tissue, as in multiple sclerosis of the central nervous system or atherosclerosis of the blood vessels.

sebum (SEE-bum) An oily secretion of the sebaceous glands that keeps the skin and hair pliable.

secondary active transport A mechanism in which solutes are moved through a plasma membrane by a carrier that does not itself use ATP but depends on a concentration gradient established by an active transport pump elsewhere in the cell.

second messenger A chemical that is produced within a cell (such as cAMP) or that enters a cell (such as calcium ions) in response to the binding of a messenger to a membrane receptor, and that triggers a metabolic reaction in the cell.

secretion 1. A chemical released by a cell to serve a physiological function, such as a hormone or digestive enzyme. **2.** The process of releasing such a chemical, often by exocytosis. *Compare* excretion.

selectively permeable membrane A membrane that allows some substances to pass through while excluding others; for example, the plasma membrane and dialysis membranes.

semicircular ducts Three ring-shaped, fluid-filled tubes of the inner ear that detect angular accelerations of the head; each is enclosed in a bony passage called the semicircular canal. fig. 16.21

semilunar valve A valve that consists of crescent-shaped cusps, including the aortic and pulmonary valves of the heart and valves of the veins and lymphatic vessels. fig. 19.8

semipermeable membrane *See* selectively permeable membrane.

senescence (seh-NESS-ense) Degenerative changes that occur with age.

sensation Conscious perception of a stimulus; pain, taste, and color, for example, are not stimuli but sensations resulting from stimuli.

sensory nerve fiber An axon that conducts information from a receptor to the central nervous system; an afferent nerve fiber.

serosa (seer-OH-sa) A thin epithelial membrane composed of a simple squamous epithelium overlying a thin layer of areolar tissue; covers the external surfaces of viscera such as the lungs, stomach, and intestines, and forms membranes such as the peritoneum, pleura, and pericardium, or a portion of such membranes. Also called *serous membrane*.

serous fluid (SEER-us) A watery, low-protein fluid similar to blood serum, formed as a filtrate of the blood or tissue fluid or as a secretion of serous gland cells; moistens the serous membranes.

serous membrane A membrane such as the peritoneum, pleura, or pericardium that lines a body cavity or covers the external surfaces of the viscera; composed of a simple squamous mesothelium and a thin layer of areolar connective tissue. Also called *serosa*. fig. 5.33b

sex chromosomes The X and Y chromosomes, which determine the sex of an individual.

shock 1. Circulatory shock, a state of cardiac output that is insufficient to meet the body's physiological needs, with consequences ranging from fainting to death. **2.** Insulin shock, a state of severe hypoglycemia caused by administration of insulin. **3.** Spinal shock, a state of depressed or lost reflex activity inferior to a point of spinal cord injury. **4.** Electrical shock, the effect of a current of electricity passing through the body, often causing muscular spasm and cardiac arrhythmia or arrest.

sign An objective manifestation of illness that any observer can see, such as cyanosis or edema. *Compare* symptom.

simple epithelium An epithelium in which all cells rest directly on the basement membrane; includes simple squamous, cuboidal, and columnar types, and pseudostratified columnar. fig. 5.3

sinuatrial node The pacemaker of the heart; a patch of autorhythmic cells in the right atrium that initiates each heartbeat.

sinus 1. An air-filled space in the cranium. **2.** A modified, relatively dilated vein that lacks smooth muscle and is incapable of vasomotion, such as the dural sinuses of the cerebral circulation and coronary sinus of the heart. **3.** A small fluid-filled space in an organ such as the spleen and lymph nodes. **4.** Pertaining to the sinuatrial node of the heart, as in *sinus rhythm*.

sodium–glucose transporter (SGLT) A symport that simultaneously transports Na⁺ and glucose into a cell.

sodium–potassium (Na⁺–K⁺) pump An active transport mechanism in the plasma membrane that uses energy from ATP to expel three sodium ions from the cell and import two potassium ions into the cell for each cycle of the pump; used to drive secondary active transport processes, regulate cell volume, maintain an electrical charge gradient across the plasma membrane, and generate body heat.

somatic 1. Pertaining to the body as a whole. **2.** Pertaining to the skin, bones, and skeletal muscles as opposed to the viscera. **3.** Pertaining to cells other than germ cells.

somatic nervous system A division of the nervous system that includes efferent fibers mainly from the skin, muscles, and skeleton and afferent fibers to the skeletal muscles. *Compare* autonomic nervous system.

somatosensory 1. Pertaining to widely distributed *general senses* in the skin, muscles, tendons, joint capsules, and viscera, as opposed to the *special senses* found in the head only; also called *somesthetic*. **2.** Pertaining to the cerebral cortex of the postcentral gyrus, which receives input from such receptors.

somite One segment in a linear series of mesodermal masses that form on each side of the neural tube and give rise to trunk muscles, vertebrae, and dermis. fig. 29.7

spermatogenesis (SPUR-ma-toe-JEN-eh-sis) The production of sperm cells through a series of mitotic and meiotic cell divisions; male gametogenesis.

spermatozoon (spur-MAT-oh-ZO-on) A sperm cell.

sphincter (SFINK-tur) A ring of muscle that opens or closes an opening or passageway; found, for example, in the eyelids, around the urinary orifice, and at the beginning of a blood capillary.

spinal nerve Any of the 31 pairs of nerves that arise from the spinal cord and pass through the intervertebral foramina or through the gap between the spine and cranium.

spindle 1. An elongated structure that is thick in the middle and tapered at the ends (fusiform). **2.** A football-shaped complex of microtubules that guide the movement of chromosomes in mitosis and meiosis. fig. 4.15 **3.** A stretch receptor in the skeletal muscles. fig. 13.24

spine 1. The vertebral column. **2.** A pointed process or sharp ridge on a bone, such as the styloid process of the cranium and spine of the scapula.

splanchnic (SPLANK-nic) Pertaining to the digestive tract.

squamous (SKWAY-mus) **1.** Thin and flat. **2.** A cellular shape that is flat or scaly; pertains especially to a class of epithelial cells. figs. 5.4, 5.12

stem cell Any undifferentiated cell that can divide and differentiate into more functionally specific cell types such as blood cells and germ cells.

stenosis (steh-NO-sis) The narrowing of a passageway such as a heart valve or uterine tube; a permanent, pathological constriction as opposed to physiological constriction of a passageway.

steroid (STERR-oyd, STEER-oyd) A lipid molecule that consists of four interconnected carbon rings; cholesterol and several of its derivatives.

stimulus A chemical or physical agent in a cell's surroundings that is capable of creating a physiological response in the cell; especially agents detected by sensory cells, such as chemicals, light, and pressure.

strain The extent to which a body, such as a bone, is deformed when subjected to stress. *Compare* stress.

stratified epithelium A type of epithelium in which some cells rest on top of others instead of on the basement membrane; includes stratified squamous, cuboidal, and columnar types, and urothelium. fig. 5.3

stress 1. A mechanical force applied to any part of the body; important in stimulating bone growth, for example. *Compare* strain. **2.** A condition in which any environmental influence disturbs the homeostatic equilibrium of the body and stimulates a physiological response, especially involving the increased secretion of certain adrenal hormones.

stroke volume The volume of blood ejected by one ventricle of the heart in one contraction.

stroma The connective tissue framework of a gland, lymphatic organ, or certain other viscera, as opposed to the tissue (parenchyma) that performs the physiological functions of the organ.

subcutaneous (SUB-cue-TAY-nee-us) Beneath the skin.

substrate 1. A chemical that is acted upon and changed by an enzyme. **2.** A chemical used as a source of energy, such as glucose and fatty acids.

substrate specificity The ability of an enzyme to bind only one substrate or a limited range of related substrates.

sulcus (SUL-cuss) A groove in the surface of an organ, as in the cerebrum or heart.

summation 1. A phenomenon in which multiple stimuli combine their effects on a cell to produce a response; seen especially in nerve and muscle cells. **2.** A phenomenon in which multiple muscle twitches occur so closely together that a muscle fiber cannot fully relax between twitches but develops more tension than a single twitch produces. fig. 11.15

superficial Relatively close to the surface; opposite of deep. For example, the ribs are superficial to the lungs.

superior Higher than another structure or point of reference from the perspective of anatomical position; for example, the lungs are superior to the diaphragm.

supination (SOO-pih-NAY-shun) A rotational movement of the forearm that turns the palm so that it faces upward or forward. fig. 9.18

surfactant (sur-FAC-tent) A chemical that reduces the surface tension of water and enables it to penetrate other substances more

effectively. Examples include pulmonary surfactant and bile acids.

sympathetic nervous system A division of the autonomic nervous system that issues efferent fibers through the thoracic and lumbar nerves and usually exerts adrenergic effects on its target organs; includes a chain of paravertebral ganglia adjacent to the vertebral column, and the adrenal medulla.

symphysis (SIM-fih-sis) A joint in which two bones are held together by fibrocartilage; for example, between bodies of the vertebrae and between the right and left pubic bones.

symport A cotransport protein that moves two solutes simultaneously through a plasma membrane in the same direction, such as the sodium–glucose transporter.

symptom A subjective manifestation of illness that only the ill person can sense, such as dizziness or nausea. *Compare* sign.

synapse (SIN-aps) **1.** A junction at the end of an axon where it stimulates another cell. **2.** A gap junction between two cardiac or smooth muscle cells at which one cell electrically stimulates the other; called an *electrical synapse.*

synaptic cleft (sih-NAP-tic) A narrow space between an axon terminal and the membrane of the postsynaptic cell, across which a neurotransmitter diffuses. fig. 12.21

synaptic vesicle A spheroidal organelle in an axon terminal containing neurotransmitter. fig. 12.21

syndrome A suite of related signs and symptoms stemming from a specific pathological cause.

synergist (SIN-ur-jist) A muscle that works with the prime mover (agonist) to contribute to the same overall action at a joint.

synergistic effect An effect in which two agents working together (such as two hormones) exert an effect that is greater than the sum of their separate effects. For example, neither follicle-stimulating hormone nor testosterone alone stimulates significant sperm production, but the two of them together stimulate production of vast numbers of sperm.

synovial fluid A lubricating fluid similar to egg white in consistency, found in the synovial joint cavities and bursae.

synovial joint A point where two bones are separated by a narrow, encapsulated space filled with lubricating synovial fluid; most such joints are relatively mobile. Also called *diarthrosis.*

synthesis reaction A chemical reaction in which relatively small molecules are combined to form a larger one. *Compare* decomposition reaction.

systemic (sis-TEM-ic) Widespread or pertaining to the body as a whole, as in the systemic circulation.

systemic circuit All blood vessels that convey blood from the left ventricle to all organs of the body and back to the right atrium of the heart; all of the cardiovascular system except the heart and pulmonary circuit.

systole (SIS-toe-lee) The contraction of any heart chamber; ventricular contraction unless otherwise specified.

systolic pressure (sis-TOLL-ic) The peak arterial blood pressure measured during ventricular systole.

T

target cell A cell acted upon by a nerve fiber, hormone, or other chemical messenger.

tarsal Pertaining to the ankle (tarsus).

T cell A type of lymphocyte involved in innate immunity, humoral immunity, and cellular immunity; occurs in several forms including helper, cytotoxic, and suppressor T cells and natural killer cells.

tendon A collagenous band or cord associated with a muscle, usually attaching it to a bone and transferring muscular tension to it.

tetanus 1. A state of sustained muscle contraction produced by temporal summation as a normal part of contraction; also called *tetany.* **2.** Spastic muscle paralysis produced by the toxin of the bacterium *Clostridium tetani* or other causes.

thalamus (THAL-uh-muss) The largest part of the diencephalon, located immediately inferior to the corpus callosum and bulging into each lateral ventricle; a point of synaptic relay of nearly all signals passing from lower levels of the CNS to the cerebrum. figs. 14.8, 14.12a

theory An explanatory statement, or set of statements, that concisely summarizes the state of knowledge on a phenomenon and provides direction for further study; for example, the fluid-mosaic theory of the plasma membrane and the sliding filament theory of muscle contraction. *See also* hypothesis.

thermogenesis The production of heat, for example, by shivering or by the action of thyroid hormones.

thermoreceptor A neuron specialized to respond to heat or cold, found in the skin and mucous membranes, for example.

thermoregulation Homeostatic regulation of the body temperature within a narrow range by adjustments of heat-promoting and heat-losing mechanisms.

thorax A region of the trunk between the neck and the diaphragm; the chest.

threshold 1. The minimum voltage to which the plasma membrane of a nerve or muscle cell must be depolarized before it produces an action potential. **2.** The minimum combination

of stimulus intensity and duration needed to generate an afferent signal from a sensory receptor.

thrombosis (throm-BO-sis) The formation or presence of a thrombus.

thrombus A clot that forms in a blood vessel or heart chamber; may break free and travel in the bloodstream as a thromboembolus. *Compare* embolus.

thyroid hormone Either of two similar hormones, thyroxine and triiodothyronine, synthesized from iodine and tyrosine.

thyroid-stimulating hormone (TSH) A hormone of the anterior pituitary gland that stimulates the thyroid gland; also called *thyrotropin.*

thyroxine (thy-ROCK-seen) The thyroid hormone secreted in greatest quantity, with four iodine atoms; also called *tetraiodothyronine.* fig. 17.17

tight junction A region in which adjacent cells are bound together by fusion of the outer phospholipid layer of their plasma membranes; forms a zone that encircles each cell near its apical pole and reduces or prevents flow of material between cells. fig. 5.28

tissue An aggregation of cells and extracellular materials, usually forming part of an organ and serving some discrete aspect of that organ's function; all tissues belong to one of the four primary classes—epithelial, connective, muscular, and nervous tissue.

totipotent Pertaining to a stem cell of the early preembryo, prior to development of a blastocyst, that has the potential to develop into any type of embryonic, extraembryonic, or adult cell.

trabecula (tra-BEC-you-la) A thin plate or layer of tissue, such as the calcified trabeculae of spongy bone or the fibrous trabeculae that subdivide a gland. fig. 7.4

trachea (TRAY-kee-uh) A cartilage-supported tube from the inferior end of the larynx to the origin of the main bronchi; conveys air to and from the lungs; colloquially called the *windpipe.*

translation The process in which a ribosome reads an mRNA molecule and synthesizes the protein specified by its genetic code.

transmembrane protein An integral protein that extends through a plasma membrane and contacts both the extracellular and intracellular fluid. fig. 3.6

transmission electron microscope (TEM) A microscope that uses an electron beam in place of light to form high-resolution, two-dimensional images of ultrathin slices of cells or tissues; capable of extremely high magnification.

triglyceride (try-GLISS-ur-ide) A lipid composed of three fatty acids joined to a glycerol; also called a *triacylglycerol* or *neutral fat*. fig. 2.18

trunk 1. That part of the body excluding the head, neck, and appendages. **2.** A major blood vessel, lymphatic vessel, or nerve that gives rise to smaller branches; for example, the pulmonary trunk and spinal nerve trunks.

T tubule Transverse tubule; a tubular extension of the plasma membrane of a muscle cell that conducts action potentials into the sarcoplasm and excites the sarcoplasmic reticulum. fig. 11.2

tunic (TOO-nic) A layer that encircles or encloses an organ, such as the tunics of a blood vessel or eyeball.

tympanic membrane The eardrum.

U

ultrastructure Fine details of tissue and cell structure, as far down as the molecular level, revealed by the electron microscope.

ultraviolet radiation Invisible, ionizing, electromagnetic radiation with shorter wavelength and higher energy than violet light; causes skin cancer and photoaging of the skin but is required in moderate amounts for the synthesis of vitamin D.

unipotent Pertaining to a stem cell that has the potential to develop into only one type of fully differentiated, functional cell, such as an epidermal stem cell that can become only a keratinocyte.

unmyelinated (un-MY-eh-lih-NAY-ted) Lacking a myelin sheath. fig. 12.9

urea (you-REE-uh) A nitrogenous waste produced from two ammonia molecules and carbon dioxide; the most abundant nitrogenous waste in the blood and urine. fig. 23.2

urothelium A type of stratified epithelium that lines much of the urinary tract, characterized by domed *umbrella cells* at the surface, protecting underlying cells from the acidity and hypertonicity of urine. Also called *transitional epithelium*. fig. 5.11

uterine tube A duct that extends from the ovary to the uterus and conveys an egg or conceptus to the uterus; also called *fallopian tube* or *oviduct*.

utricle (YOU-tri-cul) A saclike receptor in the inner ear with a nearly horizontal patch of hair cells, the macula utriculi; senses the orientation of the head and responds to horizontal acceleration, as when accelerating or decelerating in a car. fig. 16.20

V

van der Waals force A weak attraction between two atoms occurring when a brief fluctuation in the electron cloud density of one atom induces polarization of an adjacent atom; important in association of lipids with each other, protein folding, and protein–ligand binding.

varicose vein A vein that has become permanently distended and convoluted due to a loss of competence of the venous valves; especially common in the lower limb, esophagus, and anal canal (where they are called hemorrhoids).

vas (vass) (plural, *vasa*) A vessel or duct.

vascular Pertaining to blood vessels.

vascular corrosion cast A technique for visualizing the blood vessels of organs and tissues, especially the microvasculature, by flushing blood from the vessels, injecting a resin and letting it solidify, then digesting the actual tissue away with a corrosive agent, leaving only the resin cast; this is then viewed with a scanning electron microscope. figs. 20.1b, 23.10a

vasoconstriction (VAY-zo-con-STRIC-shun) The narrowing of a blood vessel due to muscular contraction of its tunica media.

vasodilation (VAY-zo-dy-LAY-shun) The widening of a blood vessel due to relaxation of the muscle of its tunica media and the outward pressure of the blood exerted against the wall.

vasomotion Collective term for vasoconstriction and vasodilation; any change in the diameter of a blood vessel.

vasomotor center A nucleus in the medulla oblongata that transmits signals to the blood vessels and regulates vessel diameter.

ventral Pertaining to the front of the body, the regions of the chest and abdomen; anterior.

ventral root *See* anterior root.

ventricle (VEN-trih-cul) A fluid-filled chamber of the brain or heart.

venule (VEN-yule) The smallest type of vein, receiving drainage from capillaries.

vertebra (VUR-teh-bra) One of the bones of the vertebral column.

vertebral column (VUR-teh-brul) A posterior series of usually 33 vertebrae; encloses the spinal cord, supports the skull and thoracic cage, and provides attachment for the limbs and postural muscles. Also called *spine* or *spinal column*.

vesicle A fluid-filled tissue sac or an organelle such as a synaptic or secretory vesicle.

vesicular transport The movement of particles or fluid droplets through the plasma membrane by the process of endocytosis or exocytosis.

viscera (VISS-er-uh) (singular, *viscus*) The organs contained in the body cavities, such as the brain, heart, lungs, stomach, intestines, and kidneys.

visceral (VISS-er-ul) **1.** Pertaining to the viscera. **2.** The inner or deeper layer of a two-layered membrane such as the pleura, pericardium, or glomerular capsule. *Compare* parietal. fig. A.6

visceral muscle Unitary smooth muscle found in the walls of blood vessels and the digestive, respiratory, urinary, and reproductive tracts.

viscosity The resistance of a fluid to flow; the thickness or stickiness of a fluid.

voluntary muscle Muscle that is usually under conscious control; skeletal muscle.

vulva The female external genitalia; the mons, labia majora, and all superficial structures between the labia majora.

W

water balance An equilibrium between fluid intake and output or between the amounts of fluid contained in the body's different fluid compartments.

white matter White myelinated nervous tissue deep to the cortex of the cerebrum and cerebellum and superficial to the gray matter of the spinal cord. fig. 14.6

X

X chromosome The larger of the two sex chromosomes; males have one X chromosome and females have two in each somatic cell.

X-ray 1. A high-energy, penetrating electromagnetic ray with wavelengths in the range of 0.1 to 10 nm; used in diagnosis and therapy. **2.** A photograph made with X-rays; radiograph.

Y

Y chromosome Smaller of the two sex chromosomes, found only in males and having little if any genetic function except development of the testis.

yolk sac An embryonic membrane that encloses the yolk in vertebrates that lay eggs and serves in humans as the origin of the first blood and germ cells.

Z

zygomatic arch An arch of bone anterior to the ear, formed by the zygomatic processes of the temporal, frontal, and zygomatic bones; origin of the masseter muscle.

zygote A single-celled, fertilized egg.

Note: Page numbers in *italics* indicate illustrations and their captions; page numbers followed by *t* indicate material in tables.

A

A (alpha) cells, of pancreatic islets, *628,* 628–629
A band, *387,* 389, *389*
ABCD rule, of skin cancer, 191
"Abdominal breathing," 839
Abdominal cavity, 32, 32*t,* 33, 34
  frontal view of, *374*
  muscles of abdominal wall, *321–323,* 322*t*
  pressure in, and hernia, 328, *328*
  veins of abdominal wall, *765*
Abdominal lymph nodes, *784,* 791
Abdominal pressure, 328, 839
Abdominal region, *30,* 31
  arteries of, *759*
  expulsion of contents, 826
  frontal view of, *374*
  quadrants of, 31, *31*
  regions of, 31, *31*
  surface anatomy of, *371*
  transverse section through, *34, 374*
  veins of, *763, 765*
Abdominopelvic cavity, 32, *32,* 32*t,* 33–34, *34*
Abduction, *277,* 280, *280*
  hand and digit, 283–284, *284*
ABO blood group, 130, 131, 667–668, 668–670, *669, 670*
Abortifacient, 1070
Abortion, spontaneous, 1068*t*
Abruptio placentae, 1068*t*
Abscess, 803
Absolute refractory period, 436, 704
Absorption
  alcohol, 943, 1002
  epithelial, 141
  nutrient, 926, 956–963, *962*
    gastric, 943
    large intestine, 966
    mineral (electrolyte), 961
    small intestine, 952–956
    vitamin, 961
  transdermal, 176
Absorptive cells, of small intestine, 953, *954*
Absorptive state, 997*t,* 997–998
Abstinence, 1069

Acceleration, and dynamic equilibrium, 588
Accessory glands, of male reproductive system, 1017–1019
Accessory organs
  of digestive system, 926
  of skin, 183
Accessory pancreatic duct, 950
Accommodation, visual, *598,* 598–599
Accutane, 1093
ACE. *See* Angiotensin-converting enzyme
Acetabular labrum, *289,* 290
Acetabulum, *254, 255, 255*
Acetic acid, 988
Acetyl group, 988
Acetylcholine (ACh), 393
  in Alzheimer disease, 454
  in autonomic nervous system, 553–554, *554,* 554*t,* 556
  caffeine and, 559, *559*
  in cardiac physiology, 715
  in digestion, 944
  discovery of, 441
  drugs affecting, 556
  in muscle excitation, 394, *395*
  in muscle relaxation, 396, *399*
  in myasthenia gravis, 415
  as neurotransmitter, 435, 441–445, *443,* 443*t, 444,* 553–554
  in pancreatic regulation, 951
  in Parkinson disease, 454
  in smooth muscle, 412
Acetylcholine (ACh) receptors, 393, 553–554, *554,* 554*t*
Acetylcholinesterase, 393, 396, *399,* 445, 556
Acetyl-coenzyme A (acetyl-CoA), 988, 993–994
ACh. *See* Acetylcholine
Achilles tendon
  anatomy of, 259, 293, *293,* 303, *307, 350, 352, 353, 380, 381, 382*
  rupture of, 352
Achondroplastic dwarfism, 211, *211,* 1093
Acid(s), 50, 914–920. *See also specific acids*
  pH scale of, 50
  strong, 915
  weak, 915
Acid mantle, 175, 189, 796

Acid reflux, 935
Acid–base balance, 902, 914–920
  bicarbonate–carbonic acid ratio and, 917
  bone and, 199
  buffers and, 915–916
  compensatory mechanisms in, 918–920
  disorders of, 917–920
  fluid replacement therapy and, 921
  relationship with electrolyte and fluid imbalances, 920, 920*t*
  renal regulation of, 883, 916–917, *917,* 919
  respiratory regulation of, 826, 857, 916, 917, 919*t*
Acidic solutions, 50
Acidosis, 716, 738, 857, 917–920, *918*
  compensated and uncompensated, 918
  fluid replacement therapy for, 921
  metabolic, 918, 919*t*
  relationship with fluid and electrolyte imbalances, 920, 920*t*
  respiratory, 857, 917, 919*t*
Acinar cells, of pancreas, 950, *951*
Acinar glands, 162, *164*
Acinus (acini)
  of exocrine gland, 162, *163*
  of mammary glands, *164, 1046, 1047, 1060, 1067*
  of pancreas, 950
Acne
  description of, 193*t*
  medication for, 1093
Acholic feces, 949
Acoustic (spiral) organ, 581, *582*
Acquired immunodeficiency syndrome (AIDS), 817–819
Acromegaly, 645, *645*
Acromial end, of clavicle, 249, *249*
Acromial region, *30*
Acromion, 250, *250, 371, 372, 377*
Acronyms, 20
Acrosin, 1077
Acrosomal reaction, 1077, *1077*
Acrosome, *1026,* 1027, 1077, *1077*
Actin, 84, *84,* 388, *388*
  distribution and functions of, 389
  in erythrocytes, 662

  in skeletal muscle contraction, 396, *397, 398*
  in skeletal muscle relaxation, 396
  in skeletal muscle striation, 389
  in smooth muscle contraction, 413, *413*
Actin–myosin cross-bridge, 396, *398*
Action, of skeletal muscle, 304–305, 401–404
Action potentials, 394, *435,* 435–436, *437*
  all-or-none law of, 436
  in cardiac muscle, *703,* 703–704
  local potentials vs., 435*t*
  nerve signal vs., 438
  neural coding of, 448–450, *449*
  refractory period of, 436, *438,* 449, 704
  in retina (vision), 599
  in skeletal muscle contraction, 396, *397, 398*
  in skeletal muscle excitation, 394, *395*
  in smooth muscle, 412
  threshold for, 401, 435
Activation energy, 65, *66*
Active immunity, 804
Active site, 65, *66,* 388
Active transport, 87
  primary, 91, 96*t*
  secondary, 91–92, *92,* 96*t,* 881
Acute glomerulonephritis, 895*t*
Acute hypersensitivity, 815
Acute leukemia, 676
Acute pancreatitis, 968*t*
Acute pericarditis, 718*t*
Acute renal failure, 895*t*
Acute rhinitis, 859*t*
Adam's apple, 154, 315, 830
Adaptation, 9–12
  basic primate, *10,* 10–11
  to bipedalism, 11, 262–263, *262–263*
  bone tissue, to mechanical stress, 205, *206,* 210
  definition of, 9
  light and dark, in vision, 602–603, 1099
  neonatal, 1091
  sensory, 565–566
Adaptive immunity, 795, 803–814
  agents of, 814*t*
  cellular (cell-mediated), 803, 808–810, 815*t*
  forms of, 803–804

humoral (antibody-mediated), 804, 811–814, 815*t*
lymphocytes in, 806–807
specificity of, 803
systemic effect of, 803
Addiction, 1003
Addison disease, 648*t*
Adduction, 280, *280*
hand and digit, 283, *284*
A-delta (Aδ) nerve fibers, 569
Adenine, 67, 110–111, *111,* 116
Adenocarcinoma, *74,* 860
Adenohypophysis. *See* Anterior pituitary gland
Adenoids, 793, *793*
Adenosine, 87, 114, 559, *559*
Adenosine diphosphate (ADP), 67
in ATP synthesis, 405–406, 986–987
in hemostasis, 680
in muscle fatigue, 406
Adenosine triphosphatases (ATPases), 67, 396, 940, *941,* 961
Adenosine triphosphate (ATP), 67–69
brain supply of, 502
caffeine and, 559, *559*
in cardiac muscle, 701
depletion of, in muscle fatigue, 406
as energy currency, 67
as energy-transfer molecule, 992
in fast-twitch muscle fibers, 407*t,* 407–408
hydrolysis of, 67
immediate energy from, 405
long-term energy from, 406
in membrane transport, 87, 91–92
production of, 67, *69,* 404–406, *405,* 986–987
in aerobic respiration, 986, 988–990, *989, 990*
in anaerobic fermentation, 68, *69,* 404–406, *405,* 986, *987,* 987–988
chemiosmotic mechanism of, 990, *991*
in citric acid (Krebs) cycle, 988, *989*
coenzyme (NAD⁺) in, 67, *67*
in erythrocytes, 661
in glycolysis, 68, *69,* 406, 986–987, *987*
in mitochondria, 69, 103, 987, 989–990
in mitochondrial electron-transport chain, 988–990, *990*
overview of, 990–992, *992*
respiratory (oxygen) requirements of, 826
short-term energy from, 406

in skeletal muscle contraction, 396, *398,* 404–406
in skeletal muscle metabolism, 300, 305, 385, 404–406
sleep and, 519
in slow-twitch muscle fibers, 407, 407*t*
in smooth muscle contraction, 413
structure of, 67, *68*
uses of, *68*
Adenylate cyclase, 82, *83*
Adequate intake (AI), 977, 979*t*
ADHD, 508, 538
Adhesion, 48, 65, 82
Adhesive glycoproteins, 149
Adipocytes, 148, 153, 641
Adipose tissue, 60, 148, 153, *153,* 153*t*
aging and, 1097
breast, 1045, *1046*
endocrine function of, 631, 632*t*
and female puberty, 1048
hypertrophy of, 166
mesenteric, 929
mons pubis, 1044
obesity, 776, 976, 1101
omental, 929
orbital, *230,* 592
perirenal, *374,* 869
subcutaneous, 179–180
Adjectival forms, of medical terms, 20–21
Adolescence, male, 1021
Adrenal cortex, 549, *550, 626,* 627–628
adrenal medulla and, 627–628
hormones of, 627–628, 631*t*
layers of, *626,* 627
Adrenal glands, *369,* 549–550, *550, 614,* 625–632, *626*
disorders of, 646
in stress response, 642
Adrenal medulla, 549–550, *550, 626,* 627
adrenal cortex and, 627–628
hormones of, 550, 627, 631*t*
sympathetic innervation of, *547*
Adrenaline. *See* Epinephrine
Adrenergic receptors, 554, *554,* 554*t,* 715, 1099
Adrenergic synapse, 444, *445*
Adrenocorticotropic hormone (ACTH), 63, 616*t,* 620, 620*t,* 633*t*
in glucocorticoid stimulation, 627
hypersecretion of, 646
immune system and, 820
in pregnancy, 1061, 1061*t*
secretion of, 621
in steroid stimulation, 637
in stress response, 642
tumors secreting, 645

Adrenogenital syndrome (AGS), 646, *646*
Adult hemoglobin (HbA), 662
Adult respiratory distress syndrome, 859*t*
Adult stem cells, 166
Adventitia
of blood vessel, 726, *726, 727*
of digestive system, 927
of esophagus, 935
of small intestine, 953
of trachea, 832
Aerobic respiration, 67, *69,* 404–406, *405,* 986, 988–990
in cardiac muscle, 701
coenzyme (NAD⁺) in, 67, *67*
matrix reactions in, 988–990, *989,* 992*t*
membrane reactions in, 988, 992*t*
Aerospace medicine, 849
Affective language area, 525
Afferent arterioles, 871, *871, 872, 873, 874, 877, 879*
Afferent division, of peripheral nervous system, 422
Afferent lymphatic vessels, 791, *791*
Afferent nerve fibers, 470, 471*t,* 482, *482,* 484
Afferent (sensory) neurons, 422–423, *423,* 544
Affixes, 20
Afterbirth, 1064, *1065*
After-discharge, 451
Afterload, 718
Age, bone, 210
Age spots, 1097
Agglutination, 667, 668–669, *669,* 811
Agglutinins, 668, *669*
Agglutinogens, 668
Aggregated lymphoid nodules, 789, 953
Aging, 1095–1103
accelerated (progeria), 1102, *1102*
blood pressure and, 735, 1099, 1100
circulatory system and, 1099–1100
definition of, 1095
digestive system/nutrition and, 1100–1101
endocrine system and, 1099
evolution and, 1103
exercise and, 1101
female reproductive system and, 1038, 1049, 1101
health-care impact of, 1096
immune system and, 1100
integumentary system and, 1097–1098
intrinsic, 1097
life expectancy and, 1101, 1103

male reproductive system and, 1022–1023, 1101
nervous system and, 1098–1099
organ systems and, 1097–1101
respiratory system and, 1100
sense organs and, 1099
skeletal system and, 219, 1098
theories of, 1101–1103
thermoregulation and, 1097, 1101
urinary system and, 1100
Agnosia, 519
Agonist muscles, 305
Agranulocytes, 657, *657,* 673–674
life history of, 676
production of, 676, *677*
AIDS, 817–819
Air, composition of, 848*t,* 848–849
Air embolism, 755
"Air hunger," in pregnancy, 1062
Air pressure, in syringe, 708
Airflow, pulmonary, 837–848
expiration, 837, 843, *844*
inspiration, 837, 843, *844*
path of, 835–836
pressure and, 842–843
resistance to, 843–845
Ala(e)
nasi, *364,* 827, *827*
sacral, 246, *246*
Alarm reaction, 642
Albinism, 182
Albumin, 272, 637, 658, 659*t,* 876
Albuminuria, 876
Alcohol, 943, 1002–1003
absorption of, 943, 1002
fetal effects of, 1093
Alcoholism, 1002–1003
Aldosterone, 627, 631*t*
in blood pressure control, 739
chemical classification of, 633*t*
in potassium homeostasis, 911
in pregnancy, 1061, 1061*t,* 1062
in renal regulation, 879–880, *880,* 884, 888*t*
in sodium homeostasis, 909, *909*
in stress response, 642
synthesis of, *635*
transport of, 637
Alimentary canal. *See* Digestive tract
Alkaline solutions, 50
Alkaline tide, 940
Alkaloids, 574
Alkalosis, 857, 917–920, *918*
compensated and uncompensated, 918
fluid replacement therapy for, 921
metabolic, 918, 919*t*
relationship with fluid and electrolyte imbalances, 920, 920*t*
respiratory, 857, 918, 919*t*
stress and, 643
Alkaptonuria, *130,* 130–131

Allantois, *1081, 1083,* 1086, *1088*
Allele(s), 128–130, *129*
Allergens, 815
Allergies, 815
Allograft, 194
Alloimmunity, 815
All-or-none law, of action
        potentials, 436
Alopecia, 186
Alpha (α) adrenergic receptors, 554
Alpha (α) cells, of pancreatic islets,
        628, 628–629
Alpha (α) chains, of hemoglobin,
        662, *662*
Alpha (α) globulins, 659*t*
Alpha (α) helix, 63, *64*
Alpha (α) particles, 43
Alpha (α) waves, 516, *517*
Alpha agglutinin, 668
Alternative pathway, of complement
        system, 797, *798*
Alternative splicing, 117, *117*
Altitude, and air composition, 849
Alveolar cells
        great (type II), 836, *836*
        squamous (type I), 836, *836*
Alveolar ducts, 834
Alveolar gas exchange, 849–850,
        852, 855, 855
Alveolar (intrapulmonary) pressure,
        843, *844*
Alveolar macrophages, 797, 836, *836*
Alveolar sacs, 834, *836*
Alveolar ventilation, 845
Alveolar ventilation rate, 845
Alveolus (alveoli)
        of bone, *226,* 226*t*
        of exocrine gland, 162
        of lung, 690, 836–837
        of tooth, 236, 932, *932*
Alzheimer, Alois, 453
Alzheimer disease, *453,* 453–454,
        1099
Amacrine cells, *600,* 601
Amenorrhea, 1068*t*
Amination, 994
Amine(s), biogenic, 441, *443,* 443*t,*
        633*t,* 633–634
        synthesis of, 635–636
        transport of, 637
Amino acid(s), 56, *62,* 62–63
        absorptive state, 997
        digestion of, 958, *959, 962*
        essential, 983
        inessential, 983
        metabolism of, 994, *995,* 995*t*
        as neurotransmitters, 441,
                443, 443*t*
        sequences of, 650
        translation to, 117–120, *119*
Amino acid pool, 994
Amino group, 55, *55,* 62
Amino sugars, 149
Aminoacyl site, 118

Aminopeptidase, 958, *959, 962*
Ammonia, 50, 867–868, *868,* 994
Amnesia, 519
Amniocentesis, 1094
Amnion, 20, *1081–1085,* 1086, *1088*
Amniotic cavity, 1080, *1081, 1083,*
        1088
Amniotic fluid, 1064, 1086, *1088*
Amoeba, 87, *87*
Amphetamines, 559
Amphiarthrosis. *See* Cartilaginous
        joints
Amphipathic substances, 61
Amplitude, 578, 585
Ampulla
        crista ampullaris, *589,* 590
        of ductus deferens, 1017, *1018*
        of ear, *581, 589,* 590
        hepatopancreatic, 947, *948,* 950
        of uterine tubes, 1040, *1041*
Amygdala, 514, *514*
        in emotion, 520
        in memory, 520
        in olfaction, 576, *577*
        in taste, 574
Amylase, 65
        pancreatic, 950, 951*t,* 956, *962*
        salivary, 67, 934–935, 956
Amylin, 629, 976
Amyloid protein, 453, 1099
Amyotrophic lateral sclerosis
        (ALS), 469
Anabolic reactions. *See* Anabolism
Anabolic steroids, 216
Anabolic–androgenic steroids, 69–70
Anabolism, 54, 54*t*
Anaerobic fermentation, 68, *69,* 404
        in carbohydrate metabolism, 986,
                987, 987–988, 992*t*
        in cardiac physiology, 699, 701
        in erythrocytes, 661
        in skeletal muscle metabolism,
                404–406, *405*
Anaerobic threshold, 406
Anagen, 186
Anal canal, 963, *964, 967*
Anal columns, 963, *964*
Anal sinuses, 963, *964*
Anal triangle, *326,* 327*t, 1014*
Analgesic fibers, descending, 467,
        507, 571, *571*
Analgesic mechanisms, of CNS, 571
Anamnestic response, in humoral
        immunity, 812, *813*
Anaphase
        I, in meiosis, 1023, *1024*
        II, in meiosis, 1023, *1024*
        in mitosis, *126,* 127
Anaphylactic shock, 746, 804, 815
Anaphylaxis, 815
Anastomoses, 733
        arterial, 699, 733, *733*
        arteriovenous, 733, *733*
        venous, 733, *733,* 766

Anatomical dead space, 845
Anatomical neck, of humerus, 251,
        251
Anatomical snuffbox, *378*
Anatomical terminology, 19–21,
        28–29
        analyzing, 19–20
        directional, 29, 29*t*
        history of, 19
        obscure origins in, 20
        planes, *28,* 28–29
        plural, adjectival, and possessive
                forms in, 20*t,* 20–21
        positional, 28, *28*
        pronunciation of, 21
Anatomy
        comparative, 2
        definition of, 2
        gross, 2
        hierarchy of complexity in,
                11–12, *12,* 21
        microscopic, 2, 138
        origins and history of, 3–6
        reductionism vs. holism, 12
        regional, 363
        scope of, 2
        surface anatomy, 363
        techniques for studying, 2
        variations in, 13, *14*
Anaxonic neurons, 425, *425*
Androgen(s), 627–628, 633*t*
        erythrocyte production and, 662
        in female puberty, 1048–1049
        hypersecretion of, 646, *646*
        in male puberty, 1021
Androgen insensitivity syndrome,
        637, 1009
Androgen-binding protein, 1015,
        1021, *1022*
Androgenic effect, 70
Androgenital syndrome, 1038
Andropause, 1022
Anemia, 665–666, 666*t*
        aging and, 1099
        aplastic, 666
        bone marrow in, 205
        hemolytic, 666, 666*t*
        hemorrhagic, 665–666, 666*t*
        hepcidin in, 630
        hypoplastic, 666
        iron-deficiency, 666
        pernicious, 666, 941
        sickle-cell. *See* Sickle-cell
                disease
Anemic hypoxia, 666, 858
Anencephaly, 1062
Anergy, 807, 809
Anesthesia, 608
        epidural, 462
        general, 608
        local, 608
Aneuploidy, 1094–1095
Aneurysms, 1100
        description of, 728

dissecting, 728, 774*t*
Angina pectoris, 699
Angiogenesis, 737
Angiography, *22,* 23
Angioplasty, 719
Angiotensin, 642, 879–880, *880*
Angiotensin I, 632*t, 880*
Angiotensin II, 559, 627, 629,
        738–739, 826, 880,
        880, 882, 883, 888*t*
Angiotensin-converting enzyme
        (ACE), 630, 738–739, 880
Angiotensin-converting enzyme
        (ACE) inhibitors, 739, 776
Angiotensinogen, 629, 632*t,* 880,
        880
Angle (bone)
        mandible, 238, *238*
        rib, 247, *248*
        scapular, 250, *250*
        sternal, 246, *247, 371*
Angle of incidence, 597
Angular acceleration, 588
Angular gyrus, 525, *525*
"Animal electricity," 401
Animal model, 10
Anions, 44, 45*t,* 908
Anisotropic (A) bands, *387,* 389, *389*
Ankle, 31
        bones of, 259–261, *260*
        joint (talocrural), 275, 284–285,
                292–293, *293*
        ligaments supporting, 293, *293*
        movements of, 284–285
        range of motion, 275
Ankle sprain, 293
Ankylosis, 294
Anomic aphasia, 525
Anorexia, 914
Anovulatory cycles, 1049
ANS. *See* Autonomic nervous
        system
Ansa cervicalis, 315, *475,* 475*t*
Antabuse, 1003
Antagonist muscles, *304,* 305
Antagonistic effects
        of ANS, 556
        of hormones, 641, *641*
Antagonistic pair of muscles, 305
Antebrachial region (antebrachium),
        30, 31, 250. *See also*
        Forearm
Anterior, 28, 29, 29*t*
Anterior border layer, of iris, 594
Anterior chamber, of eye, 594, *594*
Anterior mesentery, 34, 928
Anterior nasal spine, *236*
Anterior pituitary gland
        anatomy of, 617, *618*
        histology of, 617, *619*
        hormones of, *618,* 619–620, 620*t*
        hypothalamic control of, 509,
                617, *620*
Anterograde amnesia, 519

Anterograde transport, 426
Anterolateral system, 465, *465, 467*
Anti-A, 668, *669*
Anti-B, 668, *669*
Antibodies, 667–668, 804–806, 814*t*
    agglutination by, 811, *813*
    attack by, 811
    in blood types, 667–672
    classes of, 805, 805*t*
    complement system and, 797, *798*
    in hypersensitivity, 816
    neutralization by, 811
    precipitation by, 811, *813*
    production of, 805–806
    soluble, 804
    structure of, 63, 804–806, *805*
Antibody monomer, 804
Antibody-dependent cytotoxic hypersensitivity, 816
Antibody-mediated immunity, 804, 811–814
    attack phase of, 811, *812*
    cellular immunity vs., 815*t*
    lymphocytes in, *810,* 811–814, *812*
    memory in, 811–812, *812*
    primary response in, 811, *813*
    recognition phase of, 811, *812*
    secondary (anamnestic) response in, 812, *813*
Anticoagulants, 682, 683, 685
Anticodons, 118
Anti-D, 670
Antidepressants, 559
Antidiuretic hormone (ADH), 509, 615, 616*t, 618,* 619–620, 620*t*
    in blood pressure control, 739
    chemical classification of, 634
    in fluid balance, 905, *905*
    hyposecretion of, 644
    in renal function, 884–887, 888*t,* 890
    second messenger for, 638, *640*
    secretion of, 621–622
    in sodium homeostasis, 909
Antigen(s), 148, 667, 804, 814*t*
    in autoimmune disease, 816
    in blood types, 667–672
    in cellular immunity, 808–810
    definition of, 804
    in humoral immunity, 811–814
    hypersensitivity to, 816
    incomplete, 804, 814*t*
    neutralization of, 811
    primary response to, 812, *813*
Antigen D, 670
Antigen presentation, 807, *809,* 811
Antigen processing, 807
Antigen-binding site, 805, *805*
Antigenic determinants, 804
Antigen-presenting cells (APCs), 676, 788, 807, *809,* 811

Antigravity muscles, 300
Antihelix, *578*
Antihemophilic factor A (factor VIII), *681,* 682, 682*t,* 684
Antihemophilic factor B (factor IX), *681,* 682, 682*t,* 684
Antihemophilic factor C (factor XI), *681,* 682, 682*t,* 683
Anti-inflammatory drugs, *643,* 644
Antimicrobial proteins, 797–798
Antioxidants, 45, 453, 985
Antiport, 91, 881, 940, 961
Antithrombin, 683
Antitragus, *578*
Antral follicles, 1052
Antral pump, 942
Antrum
    of ovarian follicles, *1051,* 1052, *1052*
    of stomach, 937, *938,* 942
Anuclear cells, 97
Anulus fibrosus, 242, *242,* 693
Anuria, 890
Anus, 375, *927,* 964, *1014*
Aorta, 373, 374, 690, *694, 695, 749, 751, 771, 772, 1044*
    abdominal, 369, 751, 759, *1090*
    anatomical variations in, *14*
    aortic arch, 368, 690, *694,* 716, *749, 751, 752, 757*
    ascending, *694, 751*
    baroreceptors of, 716
    blood flow in, 735, 736, 737*t*
    descending, *751, 757*
    thoracic, *369*
Aortic bodies, 728, *729,* 738, 841, *841*
Aortic hiatus, 759
Aortic valve, *695, 696, 696, 697, 709, 711*
Apex
    of fibula, 258, *259*
    of heart, 691, *692, 693, 694*
    of lung, 833, *834*
    of nose, *364*
    of patella, 256, *258*
Apgar score, 1092
Aphasia, 525
Aplastic anemia, 666
Apnea, 847*t,* 859*t*
Apocrine glands, 163, *164, 188, 188–189,* 190*t*
Apoferritin, 663, *664*
Aponeurosis, 304
    abdominal, 320–322, 328
    biceps brachii, *338*
    galea aponeurotica, 309, *310*
    palmar, 304, *338*
    plantar, *356*
Apoptosis (programmed cell death), 162, 169, 806
Appendages. *See* Lower limb; Upper limb; *specific anatomy*
Appendicitis, 968*t*

Appendicular lymph nodes, 792
Appendicular region, 31
    blood vessels of, 766–774
Appendicular skeleton, 224, 224*t, 225*
Appendix, *368, 927,* 963, *964*
Appetite, 975–977
    aging and, 1100
    gut–brain peptide pathways in, *978*
    long-term regulators of, 976–977
    short-term regulators of, 975–976
Appositional growth, 210
Aprosody, 525
Aquaporins, 88, 882
Aqueous humor, 593–594, *594*
Arachidonic acid, 61, 643, 981
Arachnoid granulations, *499,* 501
Arachnoid mater
    brain, 497, 498, *499*
    spinal cord, *461, 462,* 462–463
Arachnoid membrane, 462
Arbor vitae, 507, *508*
Arboreal habitat, 10
Arches
    foot, 261, *261, 380, 381, 382*
    palatoglossal, *930,* 931
    palatopharyngeal, *930,* 931
    pharyngeal, *1084*
    vertebral, 242, *242, 243,* 244
    zygomatic, *229, 230,* 232, *237, 310*
Archicortex, 514
Arcuate line, *255*
Areola, *371,* 1045, *1046*
Areolar glands, 1045, *1046*
Areolar tissue, 150, *151*
Arginine vasopressin. *See* Antidiuretic hormone
Aristotle, 3, 12, 690, 725, 861
Arm, *30,* 31
    arteries of, *767*
    bone of, 250, *251*
    movements of, 277
    muscles acting on, 330–331, *333,* 334*t*–336*t*
Arm bud, *1084*
Armpit, *376*
Aromatase, 70
Arrhythmias, 705, *707*
Arterial anastomosis, 699, 733, *733*
Arteries, generally, 725–738
    abdominal region, *759, 762*
    aging and, 1100
    arterial pressure points, 775, *775*
    arterial sense organs, 728, *729*
    blood flow in, 735, 736, 737*t*
    blood pressure in, 735
    conducting (elastic or large), 726–727, *727*
    coronary circulation, 697–699, *698*
    disorders of, 774*t*
        aneurysms, 728, 774*t*

    arteriosclerosis, 212, 719, 727, 735, 1100
    atherosclerosis, 69, 648, 719, *720,* 735, 747, 773, 776, 877, 896, 1100
    coronary artery disease, *720,* 737
    fat embolism, 719, 774*t*
    orthostatic hypertension, 774*t,* 1099
    distributing (muscular or medium), 727, 727–728
    fetal and neonatal, 1089, *1090*
    great, 690
    head and neck, *752*
    lower limb, *771, 772*
    major systemic, *749*
    micrograph of, *726*
    nomenclature for, 751
    pelvic region, *772*
    pulmonary circulation, 690, *691, 747–748, 748*
    resistance (small), 728
    as resistance vessels, 726
    systemic circuit, 690, *691,* 697, *748–774, 749*
    thorax, *757*
    tunics of, 725–726, *726, 727*
    upper limb, *767*
Arteries, specific
    aorta and branches. *See* Aorta
    arcuate, *771, 772,* 870, *871, 1042, 1044*
    auricular, posterior, *752*
    axillary, 368, *749, 752, 757, 767*
    basilar, *752, 753*
    brachial, 368, 734, *749, 767,* 775
        deep, *749, 767*
    brachiocephalic, 751
    bronchial, 748, *757,* 835
    carotid
        common, *367, 370,* 715, 728, *729, 749, 751, 752, 757, 767,* 775
        external, 728, *729, 749, 752*
        internal, 716, 728, *729, 749, 752, 753*
    central, of retina, *594*
    cerebellar, *752, 753*
        anterior inferior, *753*
        posterior inferior, *753*
        superior, *753*
    cerebral
        anterior, *752, 753*
        middle, *752, 753*
        posterior, *753*
    circumflex, 698, *698*
    colic, *762*
        left, *762*
        middle, *762*
        right, *762*
    communicating
        anterior, *752, 753*
        posterior, *753*

Arteries, specific *(continued)*
coronary, 719
left, 698, *698*
right, *698*, 698–699
cortical radiate, 871, *871, 872*
cystic, *761*
deep, of clitoris, 1058
deep, of penis, 1029
dorsal, of penis, *1020*, 1029
dorsal carpal arch, 767
dorsal pedal, *749, 771, 772, 775*
esophageal, *757*
facial, *749, 752, 775*
femoral, *367, 749, 771, 772, 775*
circumflex, *771, 772*
deep, *749, 771, 772*
fibular (peroneal), *749, 771, 772*
gastric, *761*
left, *761*
right, *761*
short, *761*
gastroduodenal, *761*
gastroepiploic, *761*
gastro-omental, *761*
left, *761*
right, *761*
genicular, *771*
gluteal
inferior, *772*
superior, *772*
gonadal, *749, 759*
helicine, 1029
hepatic, 947
common, *749, 761, 947*
left, 947
proper, *761, 947, 948*
right, 947
humeral, circumflex, *767*
ileal, *762*, 953
ileocolic, *762*
iliac
common, *369, 749, 759, 771,
772, 1044, 1090*
external, *749, 771, 772*
internal, *749, 759, 771, 772,
1044*, 1089
iliolumbar, *772*
intercostal
anterior, *757*
posterior, *757*
interlobar, 870, *871*
interosseous, *749, 767*
anterior, *767*
common, *767*
posterior, *767*
interventricular
anterior, *693*, 698, *698*
posterior, *698*, 699
jejunal, *762*, 953
lingual, *752*
lobar, *747, 748*
middle, *747, 748*
superior, *747, 748*
lumbar, *759*

marginal
left, *698, 698*
right, *698, 698*
maxillary, *752*
mesenteric, *369, 374*, 762, 965
inferior, *369, 374, 749, 759,
761, 762*, 965
superior, *369, 374, 749, 759,
761, 762*, 953, *964*, 965
obturator, *771, 772*
occipital, *752*
ophthalmic, *752*
ovarian, 1039, *1041, 1044*
palmar arches, *749, 767*
pancreatic, *761*
pancreaticoduodenal, *761, 762*
pericardiophrenic, *757*
phrenic, inferior, *759*
plantar
lateral, *771*
medial, *771*
plantar arches, *771, 772*
popliteal, *749, 771, 772, 775*
pudendal, internal, *772*, 1029
pulmonary, 690, *694, 695, 709,
747, 748, 834*, 835
left, *694, 695*
right, *694, 695*
radial, 715, *749, 775*
radial collateral, *749, 767*
rectal
middle, *772*
superior, *762*
renal, 13, *749, 759*, 870, *870*
sacral
lateral, *772*
medial, *759*
median, *759*
segmental, 870, *871*
sigmoid, *762*
spinal, *753*
spiral, 1042, *1044, 1057*
splenic, *369, 749, 761*
subclavian, *368, 370, 749, 751,
752, 757, 767*
branches of, *757*
left, *751, 752, 757*
right, *757*
subcostal, *757*
subscapular, *749, 757*
supraorbital, *752*
suprarenal
inferior, *759*
middle, *759*
superior, *759*
tarsal
lateral, *771*
medial, *771*
temporal, superficial, *749, 752,
775*
testicular, *1014, 1015*, 1017
thoracic
internal, *749, 757*
lateral, *757*

thyroid, superior, *752*
tibial
anterior, *749, 771, 772*
posterior, *749, 772, 775*
ulnar, *749, 767*
ulnar collateral, superior, *749,
767*
umbilical, *772*, 1087, *1088*, 1089,
1090
uterine, *772*, 1042, *1044*
ovarian branch of, 1039, *1044*
vaginal, *772, 1044*
vertebral, 245, *749, 752, 753, 757*
vesical
inferior, *772*
superior, *772*, 1091
Arterioles, 727, 728
afferent, 871, *871, 872, 873, 874,
877, 879*
blood flow in, 737, 737*t*
efferent, 871, *871, 872, 873, 874,
877, 879*
Arteriosclerosis, 212, 719, 727, 735,
1100
Arteriovenous anastomosis, 733, *733*
Arthritis, 294*t*, 294–295, *295*, 1098
Arthrology, 268
Arthroplasty, 294–295, *295*
Arthroscope, 292
Arthroscopic surgery, 292
Arthroscopy, 292
Articular capsule. *See* Capsule,
articular (joint)
Articular cartilage, 200, *200*, 272,
273
Articular disc, 273
Articular facets
inferior, 247
of cervical vertebra, 244
of lumbar vertebra, *244*
of thoracic vertebra, 244, *244*
patellar, 256, *258*
superior, *242*, 247
of cervical vertebra, *243*, 244
of lumbar vertebra, *244*
of thoracic vertebra, 244, *244*
Articulated skeleton, 227
Articulation, 226*t*. *See also* Joint(s)
Artifact, 178
Artificial immunity, 804
Artificial insemination (AI), 1104
Artificial joints, *267*, 294–295, *295*
Artificial skin, 194
Artificial tissues, 170
Ascending aorta, *694*, 751
Ascending colon, *927, 929, 963, 964*
Ascending limb, of nephron loop,
873, *874*, 884
Ascending tracts, of spinal cord,
463–466, *465, 466t, 467*
Ascites, 712, 766, *766*
Ascorbic acid. *See* Vitamin C
Asepsis, 608
Aspartate, 441, *443*, 443*t*

Aspirin, 52, 644, 799, 943
Assisted reproductive technology
(ART), 167, 1104–1105
Association areas, 520, *521*
Association tracts, *512*, 513
Aster, *126*, 127
Asthma, 815, 816, 845
Astigmatism, 599*t*
Astrocytes, *427*, 427*t*, 428, 502
Astrocytosis, 428
Ataxia, 525
Atelectasis, 843
Atheromas, 719
Atherosclerosis, 719, *720*, 773, 776
aging and, 1100
anabolic steroids and, 69
blood pressure and, 735, 776
diabetes and, 648
renal vessel, 877, 896
stroke caused by, 747
Atherosclerotic plaques, 719, *720*,
735, 1100
Athletes
angiogenesis in, 738
bone density and mass in, 210
cardiac physiology of, 715, 719
doping by, 664
energy use in, 406
flexibility of, *268*
Marfan syndrome in, 150
muscles of, 300, 408
nitrogen balance in, 983
oxygen uptake in, 406
red blood cell counts of, 664
steroid use by, 69–70, 216
Athlete's foot, 193*t*
Athletic injuries, 358
Atlas (C1 vertebra), 243–244
Atmospheric pressure, 842–843, 848
Atom(s), 12, *12*, 42, *42*
Atomic mass, 42
Atomic mass, relative, 43
Atomic mass unit, 42
Atomic number, 41
Atomic structure, 42, *42*
Atomic weight, 43
ATP. *See* Adenosine triphosphate
ATP synthase, 990, *991*
Atresia, 1050
Atria, of heart
anatomy of, *373*, 693–695, 694
blood flow through, 697, *697*
Atrial fibrillation, 707, *707*
Atrial systole, 710–712, *711*
Atrioventricular (AV) bundle,
701, 702
Atrioventricular (AV) node, *701*, 702
Atrioventricular (AV) valves, *693,
695, 696*, 696–697, 709,
709, 711
Atrium of bronchial tree, 834
Atropa belladonna, 559
Atrophy, 167
denervation, 391, 432, 1098

disuse, 414*t*
Atropine, 556, 559
Attachment
  of muscles, 303–304, *304*
  of ribs, 245, 247–248, *248, 249t*
Attack phase, of immunity, 808, 811, *812*
Attention-deficit/hyperactivity disorder (ADHD), 508, 538
Attenuated vaccine, 804
Auditory association area, 521, *521*
Auditory cortex, primary, 521, *521, 586, 587*
Auditory projection pathway, 585–587, *587*
Auditory tube, 579, *580*
  hearing function of, *583*
  infant vs. adult, *581*
Auricle
  of ear, *364, 578, 579*
  of heart, 694, *694, 698*
Auricular region, *364*
Auricular surface
  of hip bone, 254, *255*
  of sacrum, 245, *246,* 254
Auscultation, 2, 710
Australopithecus, 11
Autoantibodies, 645, 816
Autocatalytic effect, 797, 941, 950
Autocrine messengers, 643, 801
Autograft, 194
Autoimmune diseases, 645, 807, 816–817
Autoimmune theory, of senescence, 1103
Autoimmunity, 815
Autolysis, 101, 1066
Autonomic nervous system (ANS), 422, 542–572
  aging and, 1099
  anatomy of, 546–553
  antagonistic effects in, 556
  blood vessel control by, 738
  cardiac innervation/regulation by, *547, 549, 551,* 715–716
  central control of, 557–558
  contrasting effects of, 553
  control without dual innervation, 557, *557*
  cooperative effects in, 556
  disorders of, 558*t*
  divisions of, 422, 544
  drugs affecting, 554–556, 559
  dual innervation by, *556, 556–557*
  enteric innervation by, *542, 552–553, 927, 928*
  gastric innervation by, 937
  general properties of, 543–546
  mind–body connection in, 820
  neurotransmitters and receptors of, 553–556, *554*
  output pathways of, 544–545

penile innervation by, 1029
reflexes of, 482–483, 543–544, *544*
renal innervation/regulation by, 875
somatic nervous system vs., 544–545, *545, 546t*
target organs and effects of, 543, 553–557, 555*t*
Autonomic tone, 544
Autophagy, 101
Autoregulation, 737
  metabolic theory of, 737
  renal, 878–879
Autorhythmicity
  of cardiac muscle, 410, 700
  of smooth muscle, 412
Autosomes, 128, 1009, 1095
Avicenna, 4, *4*
Axes of rotation, joint, 275–277, *277*
Axial region
  blood vessels of, 748–766
  definition of, 31
Axial skeleton, 224, 224*t, 225*
Axilla, *376*
Axillary border, of scapula, 250, *250*
Axillary folds, *376*
Axillary lymph nodes, *784, 787,* 791
Axillary region, *30, 376*
Axillary tail, of breast, *371,* 1045
Axis, 620
  hypothalamo–pituitary–ovarian, 1053
  hypothalamo–pituitary–thyroid, 620, *621, 622, 622,* 1021
Axis (C2 vertebra), *243,* 244
Axoaxonic synapses, *440,* 441
Axodendritic synapses, *440,* 441
Axolemma, 423, *424*
Axon(s), 157, *158,* 158*t,* 423, *424, 425*
  bundles, in nerves, 469–470
  conduction speed of, 429–431
  first- through third-order, 465
  initial segment of, *424,* 429
  myelinated, *427, 430,* 440
  signal conduction in, 437–438, *438, 439*
  structure of, 423, *424,* 425
  trigger zone of, *424,* 429
  unmyelinated, *430, 431, 437–438, 438*
Axon collateral, 423, *424*
Axon hillock, 423, *424*
Axon terminal, 393, *424,* 425, 441, *441*
Axonal transport, 425–426, 478
Axoneme, *85,* 86
Axoplasm, 423, *424*
Axosomatic synapses, *440,* 441
Azidothymidine (AZT), 818
Azotemia, 868
Azurophilic granules, 673
Azygos system, *758*

B

B (beta) cells, of pancreatic islets, *628,* 629
B cells. *See* B lymphocytes
B lymphocytes, 788
  activation of, 790, 811
  as antigen-presenting cells, 807, 811
  hormonal production by, 820
  in humoral immunity, 811–814, *812,* 814*t*
  immunocompetent, 789, *806, 807,* 811
  in innate immunity, 796
  life history of, *806, 807*
  maturation of, 788, *806,* 807
  memory, *812,* 814*t*
  plasma cells from, 659, 788, *806,* 811, *813*
  production of, 676, *677,* 807
  self-tolerant, 807
Baby (deciduous) teeth, *932,* 933, *933*
Back
  injuries of, 326
  muscles of, *323, 324, 325*
  surface anatomy of, *372*
Background radiation, 44
Bacon, Francis, 6–7
Bacteria
  immune/inflammatory response to, 783, 795, 802
  mitochondrial evolution from, 103
  in normal flora, 783, 965
Balance. *See* Equilibrium (sense)
Baldness, pattern, 186
Ballismus, 524
Balloon angioplasty, 719
Band cells, 673
Banting, Frederick, 649–650
Barber-surgeon, 5
Bare area, of liver, *947*
Bare zone, of myosin molecule, 388
Barium sulfate, 22
Barometric pressure, 842–843, 848
Baroreceptor(s), 16, *16,* 544, *544,* 627, 716, 728, *729*
Baroreflex, 16, 544, *544,* 716, 728, 738
Barotrauma, 861
Barrier(s)
  blood–brain, 428, 502, 729
  blood–CSF, 502
  blood–testis, 816, 1015, *1025*
  blood–thymus, 790
  contraception method, *1069,* 1069–1070
  immune, 795–796
Bartholin gland, 1045
Basal body, *85,* 86, 104, 105*t*
Basal cell(s), 574
Basal cell carcinoma, 190, *191*

Basal laminae, 393
  of capillaries, 729, *730*
  of muscle (neuromuscular junction), *389, 392,* 393
  of nerve fiber, 469
Basal layer, of endometrium, 1042, *1057*
Basal metabolic rate (BMR), 998
Basal temperature, 1055
Base
  anatomical
    of cranium, 231
    of heart, 691, *692, 693*
    of lung, *834*
    of metacarpal bone, 252, *253*
    of patella, 256, *258*
    of stapes, 579
  chemical, 50, *51,* 914–920, 933. (*See also* Acid–base balance)
    strong, 915
  nitrogenous, 67, 110, *111,* 114, 123
Base pairs, 8, 111, 123
Base triplet, 115, 116*t*
Baseball finger, 358
Basement membrane, *79, 98,* 141
  alveolar, shared, *836,* 836–837
  blood vessel, 725
  diabetes and, 648
  glomerular, 875–876, *877*
  ovarian follicle, 1050
Basic (alkaline) solutions, 50
Basilar membrane
  anatomy of, 581, *582*
  frequency response of, 584, *586*
  hearing function of, *583,* 584–585, *586*
Basilar part, of occipital bone, 233
Basophils, 154, 657, *657,* 673–674, 674*t*
  in inflammation, 801*t,* 802
  in innate immunity, 796
  production of, 676, *677*
Battery, polarized, 433
Beaumont, William, 969, *969*
Becquerel, Henri, 203
Bed sore, 169
Beethoven, Ludwig, 585
Behavioral contraception, 1069
Behavioral thermoregulation, 1002
Bell palsy, 533
Belladonna, 559
"Bends," 861
Benign prostatic hyperplasia (BPH), 1019, 1100
Bernard, Claude, 15
Bert, Paul, 849
Best, Charles, 649–650
Beta (ß) adrenergic blocking agents, 559
Beta (ß) adrenergic receptors, 554, 715

Beta (ß) cells, of pancreatic islets, 628, 629
Beta (ß) chains, of hemoglobin, 662, 662
Beta (ß) globulins, 659t
Beta (ß) particles, 43
Beta (ß) sheet (ß-pleated sheet), 63, 64
Beta (ß) waves, 516, 517
Beta agglutinin, 668, 669
Beta (ß)-amyloid protein, 453
Beta blockers, 559, 716, 776
Beta oxidation, 994
Bias, experimenter, 8
Bicarbonate, 915, 940, 941, 950, 951t
Bicarbonate buffer system, 915, 918
Bicuspid valve. See Mitral valve
Bielschowsky's head tilt test, 529
Bile, 949–950
Bile acids (bile salts), 58t, 949, 958, 960
Bile duct, 927, 928, 946, 948, 949
Bile ductules, 946, 948
Bile pigments, 665, 665
Bilirubin, 182, 665, 665, 949, 1092
Biliverdin, 665, 665
Binaural hearing, 586
Biochemistry, 41
Biogenic amines, 441, 443, 443t, 633t, 633–634
    synthesis of, 635–636
    transport of, 637
Biological half-life, 44
Biomedical science
    origins of, 3–6
    revolution in, 6
Biopsy
    bone marrow, 678
    core, 147
    CT guided needle, 147
    definition of, 147
    excisional, 147
    incisional, 147
    needle aspiration, 147
Bioreactor, 170
Biotin, 985, 985t
Bipedalism, 11, 262–263, 262–263, 291
Bipolar cells, of retina, 600, 601–604, 604, 605
Bipolar neurons, 425, 425
Birth. See Childbirth
Birth control. See Contraception
Birth defects, 1092–1095. See also specific birth defect
    mutagens and genetics in, 1093–1095
    teratogens and, 1092–1093
Birth-control pills, 1069, 1070, 1070t
Birthmarks, 183
Bisphosphoglycerate (BPG), 856
Bisphosphonates, 219
Bitter taste, 574

Black hair, 185, 185
Black lung disease, 845
Bladder. See Gallbladder; Urinary bladder
Blagden, Charles, 16
Blastocoel, 1079, 1081
Blastocyst, 166, 1060, 1078, 1079, 1079t, 1081
Blastomeres, 1078, 1079
Bleaching, of visual pigment, 602, 603
Blebbing, 169
Bleeding
    cessation of. See Hemostasis
    coagulation of. See Coagulation
Bleeding time, 683
Blind spot, 594, 596
Blindness, 595
    color, 131, 131, 605, 606
    night, 601, 985
    occipital lobe lesions and, 607
Bliss, Michael, 647
Blister, 179
Blocker's arm, 358
Blonde hair, 185, 185
Blood, 154, 156, 156t, 157, 655–688
    calcium concentration in, 212
    cardiac output of, 713–719
    centrifuging, 657, 657
    circulation of. See Circulation
    coagulation. See Coagulation
    composition of, 154, 657, 657
    deoxygenated, 690, 691
    distribution, in adult, 731, 731
    filtration of, 826
    flow. See Blood flow
    formed elements of, 154, 157, 657, 657
    fractionation of, 657
    functions of, 148, 154, 656
    general properties of, 657–658, 658t
    historical and cultural view of, 656, 725
    osmolarity of, 659
    oxygenated, 690, 691
    pH of, 914
    phosphate concentration in, 215
    plasma. See Plasma
    production of, 660
    regulatory function of, 656
    study of (hematology), 656
    transport in
        carbon dioxide, 661, 690, 691, 853–854
        hormone, 637
        iron, 663, 664
        oxygen, 661–662, 690, 691, 850–855
    viscosity of, 659, 735
    volume of. See Blood volume
Blood banking, 668
Blood cells, 154, 157, 655, 657. See also specific types

Blood clot. See also Coagulation
    clinical management of, 685
    fate of, 683, 683
    formation of, 679–680, 680, 681
    in fracture healing, 216
    in tissue repair, 167, 168
Blood disorders, 684t
    anemia, 205, 630, 665–666, 666t, 941, 1099
    coagulation, 683–684
    disseminated intravascular coagulation, 684t
    eosinophilia, 673
    hemophilia, 683–684
    infectious mononucleosis, 684t
    leukemia, 676, 678
    leukocyte, 676–678
    leukocytosis, 676
    leukopenia, 676
    neutrophilia, 673
    polycythemia, 664–665, 665
    septicemia, 684t
    sickle-cell disease, 666–667, 667
    thalassemia, 684t
    thrombocytopenia, 684t
    thrombosis, 684
Blood flow, 733–740. See also Circulation
    blood viscosity and, 735
    to brain, 746–747
    circulatory routes of, 733, 733
    exercise and, 739, 739, 740
    hemodynamics of, 734
    in inflammation, 802
    laminar, 662, 736, 736
    to lungs, 747
    metabolic needs and, 739, 739
    per volume or mass (perfusion), 734, 739–740
    in pregnancy, 1062
    pulsatile, 735
    regulation of, 737–739
        local, 737–738
        neural, 738
    resistance to, 735–737
    respiratory function and, 826
    to skeletal muscle, 747
    special routes of, 746–747
    in thermoregulation, 999, 1001, 1001
    total, in resting person, 734
    variations in systemic circuit, 736, 737t
    vasomotion and, 739–740
    venous return in, 744–746
    vessel length and, 736
    vessel radius and, 736–737
Blood gases, 856–858. See also Gas exchange
Blood hydrostatic pressure (BHP), 876–877, 878, 882
Blood islands, 660, 789
Blood pressure, 733–740
    aging and, 735, 1099, 1100

    capillary, 741, 743
    definition of, 734
    diastolic, 734, 735
    distance from heart and, 735, 735
    elevated. See Hypertension
    heart rate and, 716
    low, 735
    measurement of, 708, 734–735
    in pregnancy (preeclampsia), 1068t
    regulation of, 15–17, 16, 737–739
        baroreceptors in, 627
        baroreflexes/baroreceptors in, 16, 16, 544, 544, 716, 728, 729, 738
        chemoreflexes in, 738
        hormonal, 738–739
        local, 737–738
        negative feedback in, 738, 738
        neural, 738
        renal, 739, 776, 875, 878–880, 882–883
        renin–angiotensin–aldosterone system and, 627, 879–880, 880
        respiratory, 826
        sympathetic, 557, 557
    resistance and, 735–737
    systolic, 734, 735
    vasomotion and, 739–740
Blood smear, 156, 156t
Blood sugar, 980. See also Glucose, blood levels of
Blood transfusions
    adverse reactions in, 669, 670
    blood type in, 668–670
    Drew's pioneering work in, 668
Blood types, 667–672
    ABO, 667–670, 669, 670
    antibodies in, 667–668
    antigens in, 667
    genetics of, 130, 131
    glycolipids in, 83
    maternal–fetal mismatches in, 671, 671–672
    mismatch, in transfusion, 669, 670
    Rh, 670–672
    universal recipients and donors, 670
Blood urea nitrogen (BUN), 868
Blood vessels, 724–781. See also Arteries; Capillaries; Veins
    aging and, 1100
    appendicular region, 766–774
    axial region, 748–766
    disorders of, 774t
    endothelium of, 163, 725, 727
    general anatomy of, 725–733
    length of, 736
    micrographs of, 726

nomenclature for, 751
pregnancy and, 1062
radius of, *736,* 736–737
tunics of, 165, 725–726,
    726, 727
vascular corrosion cast of, 77,
    612
wall of, 725–726, *726, 727*
Blood volume
blood pressure and, 735
in cardiac cycle, 710–712,
    711, 713
in hypovolemia, 905, 907*t,* 920*t*
in hypovolemic shock, 745
in volume excess, 907, 907*t*
Blood–brain barrier, 428, 502, 729
Blood–CSF barrier, 502
Bloodletting, 685
Blood–testis barrier, 816, 1015, *1025*
Blood–thymus barrier, 790
Body cavities, 31–34, *32,* 32*t*
abdominopelvic, *32,* 32*t,*
    32–34, *34*
cranial, 32, *32,* 32*t,* 227
pericardial, 32, *32,* 32*t, 33, 373,*
    691, 692, 835
peritoneal, 33, *34, 374*
potential spaces, 35
thoracic, 32, *32,* 32*t*
    compartmentalization of, 837
    cross section through, *835*
vertebral canal, 32, *32,* 32*t*
Body (corpus)
of breast, *371,* 1045
of clitoris, 1045
of epididymis, *1016*
of pancreas, 950
of stomach, 937, *938*
of uterus, 1040, *1041*
Body heat, 999–1002
distribution in hot and cold,
    1001, *1001*
hair and retention of, 186
loss of, 999–1000, *1000*
production of, 999
Body mass, aging and, 1098
Body mass index (BMI), 976
Body odor, 188
Body (of bone)
hyoid, 238, *238*
ischial, *254,* 255, *255*
mandible, 238, *238*
metacarpal, 252, *253*
pubis, *254, 255,* 256
sphenoid, 234, *234*
sternum, 246, *247*
vertebra, 241, *242, 243, 244*
Body of nail, 187, *187*
Body of tongue, 931, *931*
Body regions, 29–31, *30*
Body temperature, 999–1002. *See
    also* Fever
core, 999
normal, 999

ovulation and, 1055
regulation of, 999–1002
    aging and, 1097, 1101
    behavioral, 1002
    disturbances in, 1002, 1097
    by hypothalamus, 509, 799,
        1001
    negative feedback in, 15, *16,*
        1001
    neonatal, 1091
    by skin, 176
shell, 999
Body weight, 975
appetite and, 975–977
diabetes and, 647
gain, in pregnancy, 1062, 1063*t*
obesity, 776, 976, 1101
set point in, 975
weight-control drugs and, 976
Bohr, Niels, 42
Bohr effect, 856
Bond(s). *See* Chemical bond(s)
Bone(s), generally, 154
agents affecting metabolism of,
    215*t*
aging and, 224, 1098
anatomical features of, *226,* 226*t,*
    226–227
bipedalism adaptations, 262–263,
    262–263
blood supply to, 203
calcification of, 199, 206, 209,
    209, 211
composition of, 154, 199
development of, 206–211, 1089
elongation of, 208–210
fetal, *207,* 208
functions of, 148, 199
general features of, 199–201,
    200, 201
growth of, 208–211, 215–216
histology of. *See* Bone tissue
loss of (osteoporosis), 211, 219,
    1098
lower limb, 256–263
mechanical stress adaptation by,
    205, *205,* 210
metabolism of, 215*t*
mineralization of, 199, 206, 211
ossification of
    ectopic, 212
    endochondral, *207,* 207–208
    intramembranous, 206–207,
        207, 210
physiology of, 211–216
    calcium homeostasis in,
        212–215
    hormonal control of, *214,*
        215*t,* 215–216
    phosphate homeostasis of,
        215
radioactive contamination of, 203
remodeling of, 202, 210–211,
    217, *217*

skull, 224*t, 225*
soft drinks and, 216
upper limb, 250–252
widening and thickening of, 210
Bone(s), specific
alveolar, *932*
auditory ossicles, 224*t,* 238, *579,*
    579–580, *581,* 583, *583*
    hearing function of, 583, *583*
    incus, 238, 583, *583*
    malleus, 238, 579, *579,*
        583, 583
    stapes, 238, *579,* 579–580,
        581, 583, 583
calcaneus, 259, *260,* 380,
    381, 382
capitate, 252, *253*
carpal, *225,* 252, *253*
clavicle, 206, *225, 247, 249,*
    249–250, *364, 366,*
    367, 371
    intramembranous ossification
        of, 206
    muscles acting on, *329,* 330*t*
coccyx, *225,* 240, 246, *246,*
    372, 1014
cranial, 199, *200,* 206–207, *207,*
    231–235
    in infancy and childhood,
        238–239, *239*
    intramembranous ossification
        of, 206–207, *207*
cuboid, 259, *260*
cuneiforms, 259, *260*
ethmoid, *227*–229, 235, *235, 236*
    cribriform plate of, *228, 229,*
        235, *235, 236*
    labyrinth of, 235
    perpendicular plate of, *228,*
        235, *235*
facial, 224*t,* 235–238
femur, *225,* 256, *258, 370, 380*
    adaptations to bipedalism,
        262, *262*
    bone tissue of, *200*
    head of, 256, *258*
    muscles acting on, *344, 345,*
        346*t*–347*t*
    neck of, 256, *258*
    patellar surface of, 256, *258*
    popliteal surface of, 256, *258*
    rotation of, *281*
    surface anatomy of, *380*
    trochanters of, 256, *258, 372*
fibula, *225,* 258, *259, 380,*
    381, 382
    head of, 258, *259, 380*
    surface anatomy of, *380,*
        381, 382
foot, *225,* 259–261, *260*
frontal, *225, 227*–229, 231–232,
    236
    fetal, *239*
    fusion in childhood, 239

glabella of, *226,* 232
markings of, 231–232
hamate, 252, *253*
hip (ossa coxae), *225, 254,*
    254–256, *255*
humerus, *225,* 251, *251,* 368
    anatomical neck of, 251, *251*
    head of, 251, *251*
    rotation of, *281,* 281–282
    surgical neck of, 251, *251*
hyoid, 238, *238, 931*
ilium, *254, 255, 255*
    anterior inferior spine of, *254,*
        255, *376*
    anterior superior spine of, 31,
        254, 255, 366, 371
    crest, *254, 255, 255, 371, 372*
    posterior inferior spine of,
        255
    posterior superior spine of,
        255, *376*
incus, 238, 583, *583*
ischium, *254, 255,* 255–256
lacrimal, *227*–229, 235, *236,* 237
lunate, 252
malleus, 238, 579, *579,* 583, *583*
mandible, *225, 227*–230, 235,
    238, *238, 931, 934*
    angle of, 238, *238*
    body of, 238, *238*
    depression of, 281
    fetal, *239*
    fusion in childhood, 238
    movements of, 278, 283, *284*
    protraction of, 283, *284*
    ramus of, 238, *238*
    teeth of, *932*
maxilla, *225, 227*–230, *236,*
    236–237
    alveolar processes of, 236
    fetal, *239*
    frontal process of, 236
    orbital surface of, *236*
    palatine process of, *228, 229,*
        237
    teeth of, *932*
metacarpal, 207, *207, 225,* 250,
    252, *253*
metatarsal, *225,* 259–261, *260,*
    380, 381, 382
modiolus, *310,* 311, 580
mylohyoid, *931*
nasal, *227*–229, 235, *236,* 238,
    239
navicular, 259, *260*
occipital, *225, 228*–230,
    233–234, *236*
    basilar part of, 233
    fetal, *239*
orbit, 227, *230, 236, 591, 591*
palatine, *229,* 235, *236,* 237
    horizontal plate of, 237
    perpendicular plate of, 237
parietal, *225, 227*–230, 232

Bone(s), specific (*continued*)
fetal, *239*
markings of, 232
patella, *225,* 256, *258, 379, 380*
pectoral girdle, *247,* 249–250
pelvic girdle, 254–256
phalanges
of foot, *225,* 259, *260*
of hand, *225,* 252, *253*
pisiform, 252, *253*
pubis, *254,* 256, *257*
radius, *225,* 250, *251,* 251–252,
377, 378
head of, 251, *251, 377*
neck of, *251,* 252
surface anatomy of, *377, 378*
ribs, *225, 247,* 247–248, *248,*
370, 373
angle of, 247, *248*
attachments of, 245, 247–248,
248, 249*t*
false, *247,* 248
floating, *247,* 248
head of, 247, *248*
neck of, 247, *248*
shaft of, 247, *248*
true, *247,* 248
tubercle of, 247, *248,* 249*t*
sacrum, *225,* 240, 245–246,
246, 375
alae of, 246, *246*
anterior surface of, 245
auricular surface of, 245,
246, 254
pelvic surface of, 254, *254*
sexual dimorphism in, *257*
surface anatomy of, *372*
scaphoid, 252, *253*
scapula, *225, 247,* 249–250, *250*
depression of, 281, *329,* 330*t*
elevation of, 281, *329,* 330*t*
medial border of, *372*
muscles acting on, *329,* 330*t*
protraction of, 281, *329,* 330*t*
skull, 224*t, 225,* 227–229
sphenoid, *227–229, 234,*
234–235, *236*
body of, 234, *234*
fetal, *239*
greater and lesser wings of,
234, *234, 236*
stapes, 238, *579,* 579–580, *581,*
583, *583*
sternum, *225,* 246–247, *247, 364,*
367, 371, 373
articulation with ribs,
247–248, 249*t*
body of, 246, *247, 371*
talus, 259, *260*
tarsal, *225,* 259, *260*
temporal, *227–229,* 232–233,
233, 236, 581
fetal, *239*
mastoid part of, 233, *233*

petrous part of, *229,* 233, *233,*
581
squamous part of, 232, *233*
styloid process of, *228, 229,*
233, *233,* 238, *238*
tympanic part of, 232–233,
233
zygomatic process of, 232,
233
thoracic cage, 246–248
tibia, *225,* 256, *258, 259, 380,*
381, 382
trapezium, 252, *253*
trapezoid, 252, *253*
triquetrum, 252, *253*
ulna, *225,* 250, *251,* 252, *377,*
378
head of, *251,* 252
surface anatomy of, *377, 378*
vertebrae. *See* Vertebrae
vomer, *227–229,* 235, 238
zygomatic, *227–229,* 235, *236,*
237, *239*
Bone age, 210
Bone cancer, 203, 218*t*
Bone cells, 201–202, *202. See also*
specific types
Bone deposition, zone of, 209, *209*
Bone disorders, 216–219, 218*t*
fractures, 216–217, *217*
osteitis deformans, 211, 218*t*
osteogenesis imperfecta, 203,
218*t*
osteomalacia, 213
osteomyelitis, 218*t*
osteoporosis, 211, 219, *219,* 1098
osteosarcoma, 203, 218*t*
Bone markings, 226, *226,* 226*t*
Bone marrow, 154, 199, *200, 204,*
205, *375*
adult distribution of, 205, *205,*
789
aging and, 1100
anemia and, 205
B lymphocyte maturation in, 788,
806, 807
biopsy of, 678
blood cell production in, 660
cancer of, 132, 665
dendritic cells from, 177
hematopoiesis in, 154, 199, 660,
663, 676
histology of, 789, *789*
leukocyte storage in, 676
long bone, 166
as lymphatic organ, 783,
784, 789
plasma cells in, 811
red. *See* Red bone marrow
reticular tissue of, 150
sinusoids of, 679, *679,* 730, 789
spongy bone, 205
stem cells of, 166, 167, 202, *202,*
656, 660, 672, 676

transplant of, 672, 817
yellow, *200,* 205, *205,* 789
Bone tissue, 198–222
adaptation to mechanical stress,
205, *205,* 210
cells of, 201–202, *202*
compact, 154, *156,* 156*t,* 200,
200, 201, 203, *204, 205*
endocrine function of, 631, 632*t*
histology of, 154, 201–205, *204*
matrix of, 202–203
mineralization of, 199, 206,
211–212
physiology of, 211–216
calcium homeostasis in,
212–215
hormonal control of, *214,*
215*t,* 215–216
phosphate homeostasis of,
215
spongy, 154, 200, *200, 201,*
205, *205*
Bony (hard) palate, *230,* 237, *827,*
828, 930, 931
Bony labyrinth, 580
Bony palate. *See* Hard palate
Booster shots, 804
Botox Cosmetic, 393
Botulinum toxin, 393
Botulism, 393
Bound hormones, 637
Boyle's law, 8, 842, 842*t*
Brachial region, *30,* 31, 250
Brachiocephalic trunk, *370, 749,*
751, 757, 767
Bradycardia, 488, 715
Bradykinin, 801*t,* 802
BRAF oncogene, 190
Brain, 493–526. *See also* specific
anatomy
aging and, 1099
auditory pathways in, 585–587,
587
barrier around, 428, 502, 729
blood supply of, 502, 746–747
buoyancy of, 501
cognitive function of, 519
communication via cranial
nerves, 527. (*See also*
Cranial nerves)
disorders of, 537*t*
agnosia, 519
Alzheimer's disease, *453,*
453–454, 1099
anencephaly, 1062
cancer (tumors), 428
cerebral palsy, 537*t*
concussion, 537*t*
contralateral deficits, 519, 527
contralateral neglect
syndrome, 519
encephalitis, 537*t*
epilepsy, 7, 451, 520, 537*t,*
656

meningitis, 499, 580
migraine headaches, 537*t*
Parkinson disease, 454,
523–524
schizophrenia, 537*t*
trauma (injuries), 519
electrophysiology of, 516, *517*
embryonic development of, 495,
497, 498, 1082
emotional function of, 520
evolution of, 493, 513
gray matter of, 495, *499, 500,*
505, 506, 513, 515
imaging of, 538, *538*
integrative functions of, 516–526
language function of, *525,*
525–526, 538, *538*
lateralization of functions, 526,
526
lobes of, *494, 496, 500, 511,*
511–512
major landmarks of, 493–495
medial aspect of, *496*
memory function of, 519
meninges of, 497–498, *499*
motor control function of,
522–525
nuclei of, 495
overview of, 493–495
respiratory control by, 839–841,
840
sensation functions of, 520–522,
564
sexual dimorphism in, 493
surface anatomy, *494*
tracts of, 495
ventricles of, 495, *496,* 498–502
visual pathways in, 605–607, *607*
white matter of, *492,* 495, *499,*
500, 512, 513
Brain barrier system (BBS), 502
Brain death, 1103
Brain tumors, 428
Brain waves, 516, *517*
Brainstem, *365,* 464, *494,* 495, *504.*
See also Medulla
oblongata
in autonomic control, 558
cardiac centers of, *16,* 16–17,
503, 507, 713
cross sections of, *505*
respiratory centers of, 503,
839–841, *840,* 857
Braxton Hicks contractions, 1063
BRCA1, 1047
BRCA2, 1047
Breast(s), *366, 371,* 1045–1048,
1046
development of, 1048
enlarged male, 70, 1094
fat of, *373*
lactating, 1066–1068
lymphatics of, *787,* 1045
mammary glands vs., 189

pregnancy and, 1045, 1047
  sexual response of, 1058
Breast cancer, 1047–1048
  genetics of, 1047
  male, 1032*t*
  neuroimmunology of, 820
  screening for, *1047*, 1048
  treatment of, *1047*, 1048
Breast milk, 1067*t*, 1067–1068
Breast self-examination (BSE), 1048
Breastfeeding, 1045, 1066–1068
  colostrum in, 1066, 1067, 1067*t*,
    1091
  hormones of, 621–622, 622,
    1066–1067, *1067*
  maternal metabolism and
    nutrition in, 1068
  milk composition in, 1067*t*,
    1067–1068
  milk ejection in, 1045, 1067
  milk synthesis in, 1066–1067
  uterine involution promoted by,
    1066
Breath-hold ascent, 861
Breathing. *See* Pulmonary ventilation
Bridge of nose, *364*
Brim of pelvis, *254, 255, 370*
Brittle bone disease, 203
Broca aphasia, 525
Broca area, *521, 525, 525, 538, 538*
Bromhidrosis, 188
Bronchial tree, *825,* 833–836
Bronchioles, 833, *836*
  respiratory, 833–834, *836*
  terminal, 833–834, *836*
Bronchitis, chronic, 858–859, 1100
Bronchoconstriction, 843
Bronchodilation, 843
Bronchomediastinal trunk, *786, 787*
Bronchopulmonary segments, 833
Bronchus (bronchi), *369,* 833–836
  hyaline cartilage of, *155,* 833
  lobar, *827, 832, 834*
  main, *827, 831, 832,* 833
  segmental, *827, 832,* 833
Brown fat, 153
Brown hair, 185
Brush border, 84, *954, 955*
Brush border enzymes, 955, 956,
  956
Buccal cavity. *See* Oral cavity
Buccal glands, 934
Buccal region, *364*
"Buffalo hump," 646
Buffer, 50, 915–916
Buffer systems, 915–916
Bulb(s)
  end, *567,* 567*t,* 568
  hair, 184, *184*
  olfactory, 235, *575, 576, 577*
  penis, *375, 1018,* 1020
  vestibular, of vagina, 1045, *1045*
Bulbourethral glands, *1018,* 1019
Bulbous corpuscle, *567,* 567*t,* 568

Bulge, of hair follicle, *184,* 185
Bulimia, 942
Bulk flow, of air, 845
Bundle branches, *701,* 702
Bungarotoxin, 415
Buoyancy, of brain, 501
Burns, 191–192, *192,* 194
Bursae, 273, *273*
  deep infrapatellar, 292, *299*
  of Fabricius, 788
  olecranon, *288,* 289
  popliteal, 292
  prepatellar, 292, *299*
  semimembranosus, 292
  shoulder, *287,* 289
  subacromial, *287,* 288
  subcoracoid, *287,* 288
  subdeltoid, *287,* 288
  subscapular, *287,* 288
  superficial infrapatellar,
    292, *299*
  suprapatellar, 292, *299*
  tendon sheaths as, 273, *273*
Bursitis, 294*t*
Byers, Eben, 203

C

Cachexia, 133, 998
CAD. *See* Coronary artery disease
Cadaver, 2
Caffeine, 559, *559,* 716
Caisson disease, 861
Calbindin, 961
Calcaneal region, *30*
Calcaneal tendon
  anatomy of, 259, 293, *293,* 303,
    307, 350, 352, 353, 380,
    381, 382
  rupture of, 352
Calcidiol, 213, 629, 632*t*
Calcification
  of bone, 199, 206, 209, *209,* 211
  zone of, 209, *209*
Calcitonin, 625, 631*t,* 633*t*
  in calcium homeostasis, 213,
    214, 215*t,* 625, 911
  in renal function, 884, 888*t*
Calcitriol, 630
  in calcium homeostasis, 213,
    214, 215*t,* 911
  in renal function, 884, 888*t*
  as steroid hormone, 632*t,* 633
Calcium, 911–912
  absorption of, 961
  acid–base balance and, 920*t*
  agents affecting metabolism of,
    215*t*
  as anticoagulant target, 685
  blood concentration of, 212
  body content of, 41, 41*t,* 212
  breastfeeding and, 1068
  calcitonin and, 213, *214,* 215*t,*
    625, 911

calcitriol and, 213, *214,* 215*t,*
    629, 911
  in cardiac physiology, *703,*
    703–704, 716
  deficiency of, 212, *214,* 716, 884,
    912, 913*t,* 920*t*
  dietary, 984, 984*t*
  excess of, 212, *214,* 716, 912,
    913*t,* 920*t*
  functions of, 212, 911
  homeostasis of, 212–215, *214,*
    215*t,* 867, 911
  hormonal control of, *214,* 215
  in long-term memory, 452
  parathyroid hormone and, *214,*
    215, 215*t,* 625, 884, 911
  in pregnancy, 1062
  renal reabsorption of, 881, 884
  in rigor mortis, 400
  as second messenger, 452,
    554, 638
  in skeletal muscle contraction,
    387, 396, *397*
  in skeletal muscle excitation, 394
  in skeletal muscle relaxation,
    396, *399*
  in smooth muscle contraction,
    413
  soft drinks and, 216
  storage in bones, 199, 212
Calcium channel blockers, 82, 776
Calcium channels, *703,* 704
Calcium chloride, 45*t*
Calcium–ATPase, 961
Calculus, 212
  dental (tartar), 933
  gallstones, 950
  renal, 892
Callus, 179, 216–217, *217*
Calmodulin, 413, 638
Calorie (cal), 48, 977
Calorigenic effect, 625
Calorimeter, 998
Calsequestrin, 387, 911
Calvaria, *230,* 231–232
Calyces
  major, 870, *870*
  minor, 870, *870*
cAMP. *See* Cyclic adenosine
    monophosphate
Canal(s), 226, 226*t*
  alimentary. *See* Digestive tract
  anal, *375, 927, 963, 964,* 967
  carotid, *229,* 231*t,* 233
  central
    of bone, 154, *156,* 203, *204*
    of spinal cord, 495, 499, *500,*
      501
  cerebral aqueduct, *496, 498, 500,*
    501, 505, 508, 510
  cervical, 1040, *1041,* 1042, 1064
  condylar, *229,* 233
  hyaloid, *594,* 595
  hypoglossal, *228, 229,* 231*t,* 233

inguinal, 328, *328,* 1011, 1013,
    1013, 1015
  optic, *229,* 231*t,* 234, *234, 236*
  osteonic, 154, *156*
  perforating (Volkmann), 203, *204*
  pyloric, 937, *938*
  root, *932,* 933
  sacral, 245, *246*
  semicircular, *581,* 589–590
  uterine. *See* Uterine tubes
  vertebral, 32, *32,* 32*t,* 242
Canaliculi
  bile, 946, *948*
  bone, 154, *156,* 201
  lacrimal apparatus, 592, *592*
Cancellous bone. *See* Spongy bone
Cancer, 132–133
  adenocarcinoma, *74,* 860
  bone, 203, 218*t*
  bone marrow, 132, 665
  brain, 428
  breast, 820, 1032*t, 1047,*
    1047–1048
  cervical, 1043, *1043*
  colorectal, 968, 980
  diacylglycerol and, 638
  digestive system, 968*t*
  estrogen-dependent, 637
  genetics of, 132–133
  HIV-related, 818, *818*
  leukemia, 132, 676, *678*
  lung, 860, *860*
  lymphoma, 132
  metastasis of, 132, *133,* 793
  natural killer cell attack on, *782*
  pancreatic, 950
  penile, 1032*t*
  prostate, 1019
  sarcoma, 132, 203, 218*t*
  skin, 132, 190–191, *191*
  testicular, 1032*t*
  tumor angiogenesis in, 738
Candida, 818
Canine teeth, 932
Cannon, Walter, 15, 543, 544
Canon of Medicine, The
    (Avicenna), 4
Capacitance, 439
Capacitance vessels, veins as, 731
Capacitation of sperm, 1076–1077
Capillaries, blood, 725, 729–731
  blood flow in, 735, 737, 737*t*
  blood pressure in, 741, 743
  choroid plexus, *496, 499, 500,*
    501, 501
  continuous, 729, *730*
  as exchange vessels, 729. (*See
    also* Capillary exchange)
  fenestrated, 614, 729, *730,*
    741, 875
  glomerular, *873,* 875
  in hypophyseal portal system,
    617
  micrograph of, *726*

Capillaries, blood (continued)
   peritubular, 871, 871–873,
     882–883
   primary, 617
   secondary, 617, 618
   types of, 729–730
   vasa recta, 871, 871, 872, 887,
     887
Capillaries, lymphatic, 785, 785
Capillary beds, 724, 730–731, 731
Capillary exchange, 729, 740–744
   blood pressure in, 741
   chemicals given off and taken up
     in, 740
   diffusion in, 741, 741
   edema and, 743–744
   filtration in, 87, 88, 89, 741,
     741–743, 742, 743
   reabsorption in, 741–743, 742,
     743
   transcytosis in, 741, 741
Capillary hemangiomas, 183
Capitulum, 251, 251
Capsid, 817, 817
Capsular pressure (CP), 878
Capsular space, renal, 872,
    873, 877
Capsule
   articular (joint), 272, 272, 275
     ankle, 292–293
     elbow, 288, 288–289
     fibrous, 272, 272
     knee, 290–292, 299
     shoulder, 286–288
     strength and tautness of, 275
     temporomandibular joint, 286
   gland, 161, 163
   internal, 513, 515
   lymph node, 790, 791
   renal, 869–870, 874
     fibrous, 869–870
     glomerular, 871, 872,
      873, 874
     perirenal fat, 869
   thymus, 789, 790
Carbamino compounds, 853
Carbaminohemoglobin, 853
Carbohydrates, 55, 56–58
   absorptive state, 997
   chemical formula of, 56
   conjugated, 57–58, 58t
   dietary, 978–980, 979t
     calories from, 977
     glycemic index of, 980
     requirements for, 979t,
      979–980
     sources of, 980
   digestion of, 956–957, 962
   energy from, 57
   functions of, 58t
   growth hormone and, 623
   metabolism of, 986–993, 993t.
     (See also specific
     processes)

   in aerobic respiration, 986,
     988–990, 989, 990
   in anaerobic fermentation,
     986, 987, 987–988
   in glycolysis, 986–987, 987
   liver's role in, 995t
   overview of, 990–992, 992
   naming of, 56
   postabsorptive state, 998
   types of, 56–58, 57
Carbon, 55
   atomic number of, 41
   atomic structure of, 42
   body percentage of, 41, 41t
   isotopes of, 43
Carbon-12, 43
Carbon backbones, 55
Carbon dioxide
   air content of, 848, 848t
   blood gas levels of, 856–857
   covalent bonds of, 46, 46
   excess of (hypercapnia), 716,
     738, 857
   hyperventilation and, 841, 857
   loading of, 853–854, 856
   low (hypocapnia), 857
   partial pressure of, 848, 851
   respiratory exchange of,
     848–858
   transport in blood, 661, 690, 691,
     853–854
Carbon monoxide, as
     neurotransmitter, 442
Carbon monoxide poisoning, 851,
    852
Carbon (organic) compounds, 55–70.
     See also specific types
Carbonic acid, 853, 856, 915
Carbonic acid/bicarbonate buffer
     system, 915, 918
Carbonic anhydrase, 65, 661, 853,
    940
Carboxyl group, 55, 55, 62
Carboxypeptidase, 950, 951, 958,
    959, 962
Carcinogens, 43, 132, 638
Carcinoma, 132. See also Cancer
Cardia, of stomach, 937
Cardiac arrest, 707
Cardiac arrhythmias, 705, 707
Cardiac centers, 16, 16–17, 503,
    507, 713
Cardiac conduction system, 701,
    701–702
Cardiac cycle, 707–708, 710–712
   volume changes in, 712–713, 713
   Wiggers diagram of, 710, 711
Cardiac impression, 833, 834
Cardiac muscle, 157–158, 159, 159t,
    409–410, 700–702
   absolute refractory period of, 704
   action potentials in, 703,
     703–704
   autorhythmicity of, 410, 700

   cells of, 409, 693, 700, 700–701
   contraction and relaxation of
     cardiac cycle of, 707–708,
     710–712, 711
     electrophysiology of, 702–706
   electrical junctions in, 700, 701
   mechanical junctions in, 700, 701
   metabolism of, 410, 701
   properties of, 409
   repair (fibrosis) of, 701
   skeletal and smooth muscle
     vs., 410t
   striations of, 158, 159, 159t, 409,
     700, 700
   structure of, 700, 700–701
Cardiac notch, 833
Cardiac output (CO), 713–719
   ANS and, 715
   blood pressure and, 735
   chemicals and, 716–717
   circulatory shock and, 746
   exercise and, 715, 718–719
   heart rate in, 714–715
   in pregnancy, 1062
   renal fraction of, 870
   stroke volume in, 715
Cardiac reserve, 715
Cardiac suction, 745
Cardiac tamponade, 718t
Cardial glands, 940
Cardial orifice, 935
Cardinal signs, of inflammation,
    800, 802
Cardiogenic shock, 745
Cardioinhibitory center, 714
Cardiology, 690
Cardiomyocytes
   action potentials of, 703,
     703–704
   description of, 158, 409, 693,
     700, 700–701
Cardiomyopathy, 718t
Cardiopulmonary system, 35, 826
Cardiovascular health
   aging and, 1099
   anabolic steroids and, 69
   cholesterol and, 62, 719, 776
   coronary artery disease and, 719,
     720
   diabetes mellitus and, 648
   hypertension and, 776
   trans fats and, 60, 60
Cardiovascular system, 656. See also
    Blood; Blood vessels;
    Heart
   aging and, 1099
   components of, 690
   overview of, 690–691, 691
   pulmonary circuit in, 690, 691
   systemic circuit in, 690, 691
Caries, dental, 933, 1101
Carina, 831, 832, 833
Carotene, 181, 985
Carotenoids, 45

Carotid bodies, 728, 729, 738,
    841, 841
Carotid sinus, 728, 729, 738, 752
Carpal region, 30, 31, 250
Carpal tunnel, 336, 341, 341
Carpal tunnel syndrome, 336, 341
Carrier proteins, 82, 90–93
   enzyme vs., 90–91
   saturation of, 90, 90
Carrier-mediated transport, 87,
    90–93, 96t
Carriers, genetic, 129
Cartilage, generally, 154, 199
   articular, 200, 200, 272, 273
   cells of, 154
   elastic, 154, 155, 155t
   endochondral ossification in, 207,
     207–208
   fibrocartilage, 154, 155, 155t,
     272–273
     of hip, 290
     of interpubic disc, 271, 271
     of intervertebral disc, 271
     of shoulder, 286
     of symphysis, 271
     of synovial joint, 272–273
   functions of, 148, 154
   growth of, 209, 209
   hyaline, 154, 155, 155t
     articular, 200, 200, 272, 273
     of bronchus, 155, 833
     endochondral ossification in,
      207, 207–208
     growth of, 209, 210
     of symphyses, 271
     of synchondroses, 271
     of synovial joints, 272
     of trachea, 831, 832
     zone of reserve, 209, 209
   interstitial growth of, 210
Cartilage, specific
   acetabular labrum, 289, 290
   arytenoid, 830, 830
   corniculate, 830, 830, 831
   costal, 225, 247, 247, 249t, 367
   cricoid, 830, 830, 832
   cuneiform, 830, 830
   epiglottic, 830, 830
   glenoid labrum, 286, 287
   laryngeal, 830, 830–831, 832
   nasal, 827, 827
   of nose, 236, 827
     alar, 827, 827
     lateral, 827, 827
     septal, 827
   septal, 237
   thyroid, 154, 315, 368, 371, 830,
     830, 832
   tracheal, 830, 831–832, 832
Cartilaginous joints, 271, 271–272
   symphyses, 271, 271–272
     intervertebral discs,
      271, 271
     mental, 238

pubic, *254,* 255, *255, 257,*
271, *271, 375*
synchondrosis, 271, *271*
Cascade effect, 640, *640*
Casein, 121
Catabolism, 54, 54*t*
glucose, 986–987, *992*
Catagen, 186
Catalase, 101
Catalysts, 54
enzyme, 66–67
protein, 65
Cataracts, 595, 1099
Catecholamines, 441
adrenal secretion of, 550, 627
in blood pressure control, 739
in cardiac physiology, 715
as hormones, 627, 633*t,* 633–634,
638–639
mode of action, 638–639
in stress response, 642
Catechol-O-methyltransferase
(COMT), 556
Categorical hemisphere, 526
Cathedra, 5
Cathelicidins, 796
Cations, 44, 45*t,* 908
Cauda equina, 461, *461*
Caudal, 29, 29*t,* 493, *494*
Caudate lobe, of liver, 946, *947*
Caveolae, 413
Cavernous hemangiomas, 183
Cavernous sinuses, *754*
Cavities. *See also specific cavities*
body. *See* Body cavities
bone, 226*t*
skull, 227–231
Cavities, dental (caries), 933, 1101
CD4 cells, 808, 814*t*
CD8 cells, 808, 814*t*
Cecum, *368, 374, 927,* 963, *964*
Celebrex (celecoxib), 644
Celiac trunk, *759, 761*
Cell(s), 12, *12*
basic components of, 77–78, *79*
as characteristic of life, 13
crenation of, 89, *90*
cytoskeleton of, 78, 81, 97, *98*
electrically excitable, 393–394
interior of, 96–104, *98,* 105*t*
microscopy of, *77,* 77–78,
78*t,* 105*t*
organelles of, 78, 97–104, 105*t.*
(See also specific
organelles)
shapes of, 75–77, *76*
size of, 76, 78*t*
structure of, 75–78, *79*
study of (cytology), 2, 75
surface area of, 76, *77,* 88
surface extensions of, 83–87
surface of, *79,* 79–87
ultrastructure of, *77,* 77–78
volume, regulation of, 92

Cell body, 157, *158, 158t,* 423,
424, 425
Cell counters, electronic, 678
Cell cycle, *125,* 125–127
Cell cycle checkpoints, 127
Cell death, programmed, *162,*
169, 806
Cell division, 113–114, *114*
contact inhibition of, 127
cytokinesis, 127
meiosis, 128
mitosis, 125, *126,* 127
regulation of, 127–128
Cell hypertrophy, zone of, 209, *209*
Cell junctions, 160, *161*
Cell membrane. *See* Plasma (cell)
membrane
Cell proliferation, zone of, 209, *209*
Cell suicide, 101. *See also* Apoptosis
Cell theory, 6, 9, 21, 75
Cell-adhesion molecules, 82, *82,* 725
Cell-identity markers, 82, *82*
Cellulae, 5, 75
Cellular (cell-mediated) immunity,
803, 808–810
attack phase of, 808
humoral immunity vs., 815*t*
memory phase of, 808
recognition phase of, 808–810
Cellular hypertrophy, 976
Cellulose, 57, 58*t*
Cementum, *932,* 933
Central chemoreceptors, and
ventilation, *840,* 841,
856–857
Central fatigue, 406
Central fiber, 425
Central gray matter, *505,* 506
Central nervous system (CNS), 421,
421. See also Brain;
Spinal cord
in autonomic control, 557–558
cells of, 422–432. (*See also*
Neuroglia; Neurons)
directional terms for, 493, *494*
drugs affecting, 559
embryonic development of, 460,
495, *497*
myelin/myelination in, 428–429
in pain modulation, 570–572
Central pattern generators,
460, 507
Central sulcus, *494, 496,* 511, *511*
Central venous pressure, 744
Centrifugal myelination, 429
Centriolar organization center, 104
Centrioles, *79,* 104, *104,* 105*t,*
126, 127
Centripetal myelination, 429
Centromere, 114, *114*
Centrosomes, *79,* 97, *98,* 104, 105*t*
Centrum, of vertebra, 241, *242, 243,*
244
Cephalic, 29, 29*t*

Cephalic phase, of gastric function,
943, *944*
Cephalic region, *30. See also* Head
(cephalic region)
Cerebellar hemispheres, 507, *508*
Cerebellar peduncles, 503, 507
Cerebellum, *365,* 464, *494, 495, 496,*
507–508, *508*
equilibrium function of, 590
functions of, 507–508
lesions of, 508
motor control by, 523–525, *524*
vermis of, 507, *508*
Cerebral angiogram, *22*
Cerebral aqueduct, *496, 498, 500,*
501, 505, 508, 510
Cerebral arterial circle, *752, 753*
Cerebral cortex, 495, 513
in autonomic control, 558
evolution of, 513
histology of, 513, *514*
language function of, *525,*
525–526
neocortex of, 513, *514*
in pain modulation, *571,* 572
sensation functions of,
520–522
synapses of, 446
thalamus as gateway to, 509
Cerebral crus, *505,* 506
Cerebral hemispheres, 464, 493, 511,
526, *526,* 607
Cerebral infarction, 169
Cerebral palsy, 537*t*
Cerebral peduncles, *504, 505,* 506
Cerebrospinal fluid (CSF), 427, *427,*
499–502
flow of, 501, *501*
lumbar puncture of, 464, *464*
production of, 499
purposes of, 501–502
reabsorption of, 499
sampling of, 462–463, 499
Cerebrovascular accident (CVA).
See Stroke
Cerebrum, *365,* 493, *494, 500,*
511–516
cortex of. *See* Cerebral cortex
gray matter of, 513, 515
gross anatomy of, *511,* 511–512
lateralization of functions,
526, *526*
sensation functions of, 520–522
surface area of, 511
white matter of, *512,* 513
Ceruloplasmin, 659*t*
Cerumen, 189, 579
Ceruminous glands, 189, 190*t*
Cervical curvature, *240,* 241
Cervical enlargement, 460, *461*
Cervical glands, 1042
Cervical intraepithelial neoplasia
(CIN), 1043
Cervical lymph nodes, *784,* 791

Cervical region, *30,* 31, 460.
See also Neck
(cervical region)
Cervix of gallbladder, 949
Cervix of uterus, *375, 1039,* 1040,
1041
cancer of, 1043, *1043*
effacement, in childbirth, 1064,
1065
sexual response of, 1058, *1059*
Cesium-137, 44
Chagas disease, 553
Chambers
of eye, *594,* 594–595
of heart, 694–696. (*See also*
Atria; Ventricles, of heart)
Chancre, 1033
Channel(s), 81. *See also specific*
channel
Channel proteins, 81, *82*
Channelopathies, 81
Chaperone, 120
Characteristics of life, 13–15
Charles's law, 842*t,* 843
Charley horse, 358
Charnley, Sir John, 294
Checkpoints in cell cycle, 127
Cheek, 930
functions of, 930
muscles acting on, 312*t*
Cheiloschisis, 1093
Chemical bond(s), 45–47, 46*t*
covalent, *46,* 46*t,* 46–47
hydrogen, 46, 46*t,* 47, *47*
ionic, 46, 46*t*
peptide, 62–63
Chemical buffer, 933
Chemical digestion, 926, 930, 952,
956–963
Chemical energy, 52
Chemical equation, 52
Chemical reactions, 52–55
activation energy for, 65
in body (metabolism), 54*t,* 54–55
catalysts for, 54
classification of, 52–53, *53*
definition of, 52
energy and, 52–55
equilibrium in, 53
rates of, 54
Chemical senses, 572–577
Chemical synapses, 441, *442*
Chemiosmotic mechanism, 990, *991*
Chemoreceptor(s), 566, 716, 728,
729
central, *840,* 841, 856–857
peripheral, 841, *841,* 856–858
Chemoreflexes, 716, 738
Chemotactic factors, 801*t,* 802
Chemotaxis, 802
Chewing, 930, 933, 936
aging and, 1101
appetite and, 977
cheeks and lips in, 930

Chewing *(continued)*
  muscles of, *310, 313,* 315*t*–316*t,*
    933, 1101
  teeth in, 931–933
  tongue in, 930–931, 933, 936,
    936
Chickenpox, 478, 800
Chief cells, of stomach, *939,* 940,
  941, 942*t*
Childbirth, 1062–1065
  crowning in, 1064, *1065*
  hormones in, 638, 1063–1064
  labor contractions in, 1064
  neural stimuli in, 637
  positive feedback in, 17, *17,* 1064
  stages of, 1064, *1065*
Chimpanzees, evolution of, 10–11
Chin, cleft, 129, *129*
Chin (mentum), *364*
Chinese medicine, 4
Chlamydia, 1033
Chlamydia trachomatis, 1033
Chloasma, 1062
Chloride, 912
  dietary, 984*t*
  functions of, 912
  homeostasis of, 912
  imbalances in, 912, 920*t*
  renal reabsorption of, 881, 884
Chloride ions, 961. *See also*
    Chloride
Chloride pumps, 86
Chloride shift, 854, 940
Chloride–bicarbonate exchanger,
  853–854
Chlorine
  body percentage of, 41*t,* 42
  dietary, 984
Chlorine ions, 44, *44*
Chloroform, 608
Chlorpromazine, 559
Choanae, *229,* 235, *828*
Cholecalciferol, 213, 629, 632*t.*
    See also Vitamin D
Cholecystokinin (CCK), 442, 630,
  632*t, 941*
  appetite regulation by, 976, *978*
  gastric inhibition by, *944,* 945
  as neurotransmitter, 443*t*
  pancreatic regulation by, 952
Cholelithiasis, 950
Cholesterol, 61–62
  bile acid synthesis from, 949
  blood levels of, *981,* 982
  cardiovascular effects of, 62,
    719, 776
  dietary, 62, 981–982
  good vs. bad, 62
  in plasma membrane, 80, *80*
  steroid synthesis from, 633,
    634, *635*
  structure of, *61,* 61–62
  in testosterone synthesis, 122,
    123

Choline, 61
Cholinergic crisis, 393
Cholinergic fiber, 553
Cholinergic receptors, 393, 553–554,
  554, 554*t*
Cholinergic synapse, 442, 444, *444*
Chondroblasts, 154
Chondrocytes, 154
  in bone growth, 209, *209*
  in endochondral ossification, *207,*
    208
Chondroitin sulfate, 149
Chordae tendineae, *693, 695,* 696,
  696
Chorion, 1080, *1085, 1087, 1088*
Chorionic villi, *1081, 1087, 1088*
Chorionic villus sampling (CVS),
  1094
Choroid, 593, *594, 600*
Chromaffin cells, 627
Chromatic aberration, 6
Chromatids, sister, 113
Chromatin, 97, *99,* 112, *113*
Chromatophilic substance, 423, *424*
Chromatophores, 594
Chromium, 41*t,* 984*t*
Chromophobe, *619*
Chromosomal disorders, *1094,*
  1094–1095
Chromosome(s), 110, 112–114,
    128–132
  diploid number of, 128, 1023,
    1078
  disjunction of, 1094, *1094*
  genes and alleles on, 128–129,
    129
  haploid number of, 128, 1023,
    1050, 1078
  homologous, 128, 1023–1025,
    1024
  karyotype of, 128, *128*
  meiosis and, 1023–1025, *1024*
  mitosis and, *126,* 127
  nondisjunction of, 1094, *1094*
  paternal and maternal, 128
  sex, 128, 131, *131,* 1009–1010,
    1010
  structure at metaphase, 114, *114*
Chromosome territory, 113
Chronic bronchitis, 858–859, 1100
Chronic leukemia, 676
Chronic obstructive pulmonary
    disease (COPD), 858–859,
    1100
Chronic renal failure, 895*t*
Chronotropic agents, 715, 718
Chronotropic effects
  of ANS, 715
  of chemicals, 716–717
Chyle, 786, 958
Chylomicron remnant, 981
Chylomicrons, 958, *960, 962,* 981,
  981
Chyme, 937

gastric, 937, 942–943, *944*
  intestinal, 955
  vomiting of, 942–943
Chymotrypsin, 950, *951,* 958,
  959, 962
Chymotrypsinogen, 950, *951,* 951*t*
Cialis, 1031
Ciba Pharmaceutical Company, 69
Cilia, 84–86, 105*t*
  action of, 85, *86*
  defects in, 84–85
  distribution in body, 85
  primary, 84
  structure of, *85,* 86
  uterine tube, 85, *85,* 1040
Ciliary body, 593, *594*
Ciliopathies, 85
Cimetidine, 943
Cingulate gyrus, *496,* 514, *514*
Circadian rhythms, 509–510, 518,
    623–624, 636
Circle of Willis, *752, 753*
Circular folds, 953
Circular muscle(s). *See* Sphincters
Circulation
  collateral, 699, 733
  enterohepatic, 949
  fetal, 1089, *1090*
  historical and cultural view of,
    725
  maternal–fetal, *1088*
  microcirculation, 729
  neonatal, 1089, *1090,* 1091
  pulmonary, 690, *691,* 747–748,
    748
  renal, 870–871, *871, 872*
  routes of, 733, *733*
  special routes of, 746–747
  systemic, 690, *691,* 697,
    748–774, *749–750*
Circulatory shock, 683, 745, 906
Circulatory system, 35, *37,* 690.
    See also Blood; Blood
    vessels; Heart
  aging and, 1099–1100
  components of, 656
  effects on other systems, 777
  fluid exchange with lymphatic
    system, 785, *786*
  functions of, 656
  parasympathetic effects on, 555*t*
  pregnancy and, 1062
  sympathetic effects on, 555*t*
Circumcision, 1019
Circumduction, 281, *281*
Circumferential lamellae, 203,
  204, 210
Circumventricular organs (CVOs),
  502
Cirrhosis, 996
cis-retinal, 602, *602*
Cisterna chyli, *784,* 786, *787*
Cisterns (cisternae), 100, *100*
Citric acid, 988

Citric acid (Krebs) cycle, 988–989,
  989
Classical hemophilia, 684
Classical pathway, of complement
    system, 797, *798*
Clathrin, 93, *94,* 120
Clathrin-coated vesicle, 93, *94*
Clavicular notches, 246, *247*
Clear (C) cells, 213, 625
Clear margins, 147
Clearance, hormone, 642
Cleavage, *1078,* 1079*t,* 1079–1080
Cleavage furrow, *126,* 127
Cleft chin, genetics of, 129, *129*
Cleft lip and palate (CLP), 237,
  1093, *1093*
Cleft palate, 237
Climacteric
  female, 1049
  male, 1022
Climax. *See* Orgasm
Clitoris, *1039,* 1044–1045, *1045*
  development of, 1010, *1012,*
    1038
  masculinization of, 646, *646*
  sexual response of, 1058
Cloaca, *1011*
Clonal selection, *809,* 810, *812*
Closed reduction of fracture, 217
Clostridium, 169
Clostridium botulinum, 393
Clostridium tetani, 393
Clot. *See* Blood clot
Clot retraction, 683
Clotting. *See* Coagulation
Club hair, 186
Clubbed fingers, 187
Cluster of differentiation, 808
CNS. *See* Central nervous system
Coagulation, *680,* 680–683
  anticoagulants and, 683, 685
  clinical management of, 685
  completion of, *681,* 682–683
  complexity of, 680
  disorders of, 683–684
  extrinsic mechanism of, 681, *681*
  fate of blood clot in, 683, *683*
  inappropriate, prevention of, 683
  initiation of, *681,* 682
  intrinsic mechanism of, *681,*
    681–682
  laboratory tests of, 683
  liver disease and, 684
  pathways of, *681*
  reaction cascade in, 682, *683*
  in tissue repair, 167, *168*
Coagulation factor(s)
  deficiency of, 683–684
  description of, 679, *681,* 682,
    682*t*
Cobalamin. *See* Vitamin B$_{12}$
Cobalt, 41*t*
Cocaine, 559
Cochlea

anatomy of, *579,* 580–582, *581, 582*
    hearing function of, *583, 584,* 585
    sensory coding in, 585
Cochlear duct, 581, *581, 582*
Cochlear tuning, 585
Coconut oil, 60
Codominant alleles, 130
Codons, 116
Coelom, 1082, *1083*
Coenzyme(s), 67, *67,* 986
Coenzyme Q (CoQ), 990
Cofactors, 67, *67*
Cognition, 519
    definition of, 519
    lesion effects on, 519
Cohen, Stanley, 432
Cohesion, 48
Coition. *See* Sexual intercourse
Coitus. *See* Sexual intercourse
Coitus interruptus, 1069
Cola, and bones, 216
Cold receptors, 568
Cold weather, and fluid balance, 903, 906
Colic flexure
    left, 963, *964*
    right, 963, *964*
Collagen, 63
    aging and, 1097
    in areolar tissue, 150
    in basal lamina, 393
    in basement membrane, 141
    in blood vessels, 726
    in bone tissue, 200, 202–203, *204,* 212, 216
    in breast, 1045
    in dermis, *177, 179, 180,* 1097
    in eyeball, 593
    in fibrocartilage, 154
    in fracture healing, 216, *217*
    in hair follicle, 185
    in heart, 693
    joints and, 270
    in kidney, 870
    in meninges, 462
    in mesenchyme, 139
    in mucous membrane, *165*
    muscle and, 303, 386
    parathyroid hormone and, 215
    platelets and, 680
    sacrificial bonds of, 203
    in tendon organ, 486
    in tendons, 386
    in uterus, 1042
Collagenous fibers, 149, *149*
Collateral circulation, 699, 733
Collecting ducts, 786, *787,* 873
    functional relationship with nephron loop, 887, *887*
    reabsorption by, 881, 884–887
    water conservation by, 885–887, *886*

Collecting vessels, 785
Colliculi
    inferior, *504, 506, 508,* 586, *587*
    superior, *504, 505, 506, 508, 510,* 607, *607*
Collip, J. B., 649
Colloid osmotic pressure (COP), 659, 741, *742,* 876, *878*
Colloids, 49, *50,* 51*t*
Colon, 963–967
    absorption in, 966
    ascending, *927, 929,* 963, *964*
    descending, *374, 551, 927, 929,* 963, *964*
    massive dilation of (megacolon), 553
    motility of, 966
    parasympathetic innervation of, *551*
    sigmoid, *374, 375, 927, 929,* 963, *964, 967*
    transverse, *374, 551, 927, 929,* 963–965, *964*
Colonoscopy, 968
Colony-forming units (CFUs), 660, 663, *663,* 676, *677*
Colony-stimulating factors (CSFs), 676, 801*t,* 802
Color blindness, 131, *131,* 605, *606*
Color vision, 601, 602, 604–605, *606*
Colorectal cancer, 968, 980
Colostrum, 1066, 1067, 1067*t,* 1091
Column(s)
    anal, 963, *964*
    renal, 870, *870*
    spinal cord, *462, 473*
    vertebral. *See* Vertebral column
Columnar cells, 75, *76,* 142
Coma, hepatic, 995
Combined oral contraceptive, *1069,* 1070
Combining vowels, 19
Comminuted fracture, 216, *217*
Commissural tracts, *512,* 513
Commissures, 513
    anterior, *496, 510,* 513
    gray, *462,* 463
    lateral, *364, 591, 591*
    medial, *364, 591, 591*
    posterior, *496, 508,* 513
Communicable period, 1033
Communicating (gap) junctions, 160, *161, 162,* 410, 613, *700,* 701
Communication
    cellular, 613
    cranial nerve, 527
    endocrine, 613
    endocrine system vs. nervous system, *615,* 615*t,* 615–616
    neuronal, 437–438, *438*
    nonverbal, 176, *177,* 186, 591
    respiratory functions in, 826

Compact bone, 154, *156,* 156*t,* 200, *200, 201,* 203, *204, 205*
Compaction, in digestive system, 926
Comparative anatomy, 2
Comparative physiology, 3
Compartment syndrome, 303
Compartments
    fluid, 902, *903*
    forearm, 336
    leg, 354*t, 355*
    muscle, 302–303, *303*
    thigh, 349*t*
    thoracic cavity, 837
Compensated acidosis or alkalosis, 918
Compensated shock, 746
Complement fixation, 797, *798,* 811
Complement proteins, 659*t*
Complement system, 797, *798,* 801*t,* 814*t*
Complementary base pairs, 8, 111, 123
Complement-binding sites, 797
Complete blood count (CBC), 678
Complete (fused) tetanus, 402, *403*
Complete proteins, 983
Compliance, pulmonary, 845
Complicated plaque, 719, *720,* 735
Composite, 203
Compound(s), 45
    mixtures vs., 48
    organic, 55–70
Compound exocrine glands, 162, *164*
Compound microscope, 5, 5–6
Compressor urethrae, *326,* 327*t*
Computed tomography (CT)
    biopsy guided using, 147
    description of, *22, 23, 689*
Computer-assisted design and manufacture (CAD/CAM), 295
Concentration, reactant, 54
Concentration, solution, 51
Concentration gradients, 18, *18,* 87–88, 91, *91*
Concentric contractions, 402, *403,* 404
Concentric lamellae, 154, *156,* 203, *204*
Conceptus, 1060, 1078, *1078, 1081*
Concussion, 537*t*
Condensation, 56
Condom
    female, 1069, *1069,* 1070*t*
    male, 1069, *1069,* 1070*t*
Conducting arteries, 726–727, *727*
Conducting zone, of respiratory system, 826
Conduction
    cardiac, *701,* 701–702
    heat, 1000, *1000*
    signal, in neurons, 437–438, *438, 439*
Conductive deafness, 585

Conductivity
    muscle, 385
    neuron, 422
    placental, 1087
    spinal cord, 460
Condyles, *226,* 226*t*
    femoral, 256, *258*
    humeral, 251, *251*
    mandibular, *228,* 238, *238*
    occipital, *229,* 233
    tibial, 258, *259,* 380
Condylomas, 1033
Cone(s), *600,* 600–602, *601,* 604–605
    in color vision, 601, 602, 604–605, *606*
    in day (photopic) vision, 601, 603, 604, *605*
    in dual visual system, 603–604
    evolution of, 604
    in light and dark adaptation, 603
    long-wavelength (L), 605, *606*
    medium-wavelength (M), 605, *606*
    short-wavelength (S), 604–605, *606*
Confluence of sinuses, *754*
Conformation, 63
Congenital anomaly, 1092–1095
Congenital hypothyroidism, 645, 648*t*
Congestive heart failure (CHF), 713
Conjugated carbohydrates, 57–58, 58*t*
Conjugated proteins, 63
Conjunctiva, 591, *592*
Connecting peptide, 634, *635*
Connective tissue, 138, 138*t,* 148–157
    areolar, 150, *151*
    blood as, 154, 157
    bone as, 154
    cartilage as, 154
    dense, 149–150
        irregular, 150, *152,* 152*t*
        regular, 150, *152,* 152*t*
    fibrous, 148–150
        cells of, 149
        components of, 148–149
        types of, 149–153
    functions of, 148
    loose, 149, *151*
    muscle, *301,* 301–313, 386
    reticular, 150, *151*
Connective tissue disease, 150
Connective tissue root sheath, *184,* 185
Connectome, 513
Connexons, 160
Consciousness, 507
Consolidation, of memory, 519
Constant (C) region, of antibody, 805, *805*
Constipation, 963, 1101

Contact dermatitis, 193*t*, 819*t*
Contact digestion, 955
Contact inhibition, 127
Continuous ambulatory peritoneal dialysis (CAPD), 896
Continuous capillaries, 729, *730*
Continuous conduction, 437–438, *438*
Contraception, 1031, 1069–1070
  barrier methods of, *1069*, 1069–1070
  behavioral methods of, 1069
  breastfeeding and, 1067
  definition of, 1069
  failure rates of, 1070, 1070*t*
  hormonal methods of, *1069*, 1070
  issues in choosing, 1070
Contractile proteins, 388
Contractility
  of muscle, 385
  of myocardium, 717–718
  of uterus, 1063–1064
Contraction(s), 396
  labor, 1064
  muscle. *See* Muscle contraction
Contraction phase, of muscle twitch, 401, *401*
Contracture, 414*t*
Contralateral, 29, 29*t*, 318*t*, 465, 527
Contralateral deficits, lesion, 519, 527
Contralateral neglect syndrome, 519
Contralateral reflex arc, 485
Control center, in negative feedback, 16–17
Control group, 8
Controls, experimental, 8
Convection, 1000, *1000*
Convergence
  eyes, 597–598
  neural, 549, 601, 604, *605*
Convergent muscles, 302
Converging circuit, *450*, 451
Cooperative effects, of ANS, 556
COPD (chronic obstructive pulmonary disease), 858–859
Copper, 41*t*, 663, 984*t*, 990
Copulation. *See* Sexual intercourse
Cor pulmonale, 718, 859
Cord(s)
  of brachial plexus, *476*, 477, 477*t*–478*t*
  of spinal nerves, 476
Cord blood transplant, 672
Core biopsy, 147
Core particle, of nucleosome, 113, *113*
Core temperature, 999
Cornea, *592*, 593, *594*
Corns, 179
Cornua (horns)
  of coccyx, 246, *246*
  of hyoid bone, 238

of spinal cord, *462*, 463
Corona radiata, 513, 1052, *1077*
Coronal plane, *28*, 28–29
Coronary arteries, 719
  left, 698, *698*
  right, *698*, 698–699
Coronary artery bypass graft, 736
Coronary artery bypass surgery, 690
Coronary artery disease (CAD), 719, *720*
Coronary circulation, 697–699, *698*
Coronary sinus, *693, 694, 698,* 699, 732
Coronary sulcus, *694*, 696
Coronavirus, *40*
Corpus albicans, *1040, 1054,* 1055, 1056*t*
Corpus callosum, 495, *496, 500,* 511
Corpus cavernosum, *375, 1018,* 1019–1020, *1020*, 1029, *1044*, 1058
Corpus luteum, 620, 629, *1040, 1051*, 1054, *1054*, 1056*t*, 1060, *1078*
Corpus spongiosum, *375, 1018,* 1019, *1020*, 1029
Corpus striatum, *515*, 516
Corrective lenses, *599*, 599*t*
Cortex
  of adrenal glands. *See* Adrenal cortex
  of cerebrum. *See* Cerebral cortex
  of hair, 184, *184*
  of kidney, 870, *871, 872, 874*
  of lymph node, 790, *791, 791*
  of ovary, 1038, *1040*
  of thymus, 789, *790*
Cortical granules, *1077*, 1078
Cortical nephrons, *872, 874*, 875
Cortical reaction, *1077*, 1078
Corticoids, 627
Corticomedullary junction, 870, *872*
Corticospinal tract, 466, 503, *505*
  anterior, *465*, 466, 466*t*, *468*, 523
  lateral, *465*, 466, 466*t*, *468*, 523
Corticosteroids, 627
Corticosterone, 627, 631*t*, 633*t*
Corticotropin. *See* Adrenocorticotropic hormone
Corticotropin-releasing hormone (CRH), 616*t*, 620, 633*t*, 642
Cortisol, 620, 627, 631*t*, 633*t*
  excess, 642
  fetal, and childbirth, 1063
  hypersecretion of, 645, 646, *646*
  in stress response, 642
  synthesis of, *635*
Costal facets, *244*, 245, *247, 248,* 249*t*
Costal groove, 247, *248*
Costal margin, 246, *247, 371*
Costal surface, of lung, 833, *834*

Costimulation, in immunity, 809, *809*
Costocervical trunks, *752, 757*
Cotransport, 91, 96*t*
Coughing, 847
Coumadin, 685
Coumarin, 685
Countercurrent exchange system, 887–888
Countercurrent heat exchanger, 1015, *1015*
Countercurrent multiplier, *886*, 886–887
Countertransport, 91, 96*t*
Cousteau, Jacques, 861
Covalent bonds, 46, 46*t*, 46–47
  double, 46, 46*t*
  nonpolar, 46, 46*t*, *47*
  polar, 46*t*, 46–47, *47*
  single, 46, 46*t*
Cow's milk, human milk vs., 1067, 1067*t*
COX inhibitors, 644
Coxal region, *30*
Cramps, 414*t*
Cranial cavity, 32, *32*, 32*t*, 227
Cranial fossae, 231, *232*
Cranial nerves, 305, *504, 527,* 527–537
  abducens (VI), *504*, 527, 530, 534*t*, *536*, 590, *593*
  accessory (XI), *504*, 527, 532, 535*t*, *536*
  classification of, 528
  disorders of, 533, 537*t*
  facial (VII), *504*, 527, 528, 530, 530–531, *536*
    buccal branch of, *530*, 531
    cervical branch of, *530*, 531
    mandibular branch of, *530*, 531
    in pain, 569
    parasympathetic fibers in, *551*, 552
    in taste, 574
    temporal branch of, *530*, 531
    zygomatic branch of, *530*, 531
  glossopharyngeal (IX), *504*, 527, 531, 535*t*, *536*, 738, 934
    in pain, 569
    parasympathetic fibers in, *551*, 552
    in taste, 574
  hypoglossal (XII), 233, *504, 505,* 527, 533, 535*t*, *536*
  mnemonic for, 532–533
  motor, 528
  numbering of, 527, 528
  oculomotor (III), *504, 505, 508,* 527, 529, 534*t*, *536*, 551, 552, 590, *593*
    equilibrium function of, 590

    parasympathetic fibers in, *551*, 552
  olfactory (I), *527*, 528, *528*, 534*t*, *536*
  optic (II), 230, *504*, 527, *527*, 528, *528*, 534*t*, *536*, 593, *594*, 595, 602, *604*, 605–607, *607*
    generating signal of, 602, *604*
    visual pathway of, 605–607, *607*
  parasympathetic fibers in, *551*, 552
  pathways of, 527, *536*
  sensory, 528
  survey of, 528–533
  trigeminal (V), *504, 505*, 527, 529, 534*t*, *536*
    maxillary division of, *529*, 534*t*
    ophthalmic division of, *529*, 534*t*
    in pain, 569
  trochlear (IV), *504*, 527, *527*, *529*, 534*t*, *536*, 593
    equilibrium function of, 590
  vagus (X), 441, *504, 505*, 527, 535*t*, *536*
    in blood pressure regulation, 738
    in cardiac physiology, 715
    in digestion control, 944, *944*
    enteric innervation by, 929
    in pain, 569
    parasympathetic fibers in, *551*, 552, 715
  vestibulocochlear nerve (CN VIII), *504*, 527, 531, 535*t*, *536*, 585–586
Cranial region, *30*
Craniosacral division, of ANS. *See* Parasympathetic division, of ANS
Cranium, 231–232, *365*
Creatine kinase (CK), 65, 405, *405*
Creatine phosphate, 405
Creatinine, 868, *868*, 882
Crenation, 89, *90*
Crest
  of bone, 226, 226*t*
    iliac, *254*, 255, *255, 371, 372*
    intertrochanteric, 256, *258*
    lateral sacral, 245, *247*
    median sacral, 245, *247*
    neural, 495, *497*
Cribriform plate, *228, 229,* 235, *235, 236*
Crick, Francis, 112, *112*
Crisis, febrile, 800
CRISPR-Cas9, 115
Crista ampullaris, *589*, 590
Crista galli, *228, 229, 230,* 235, *235, 236*
Cristae, of mitochondria, 103, *103*

Crohn disease, 968t
Cross section, 140, *140*
Cross-bridge, actin–myosin, 396, *398*
Crossed extension reflex, 485–486, *486*
Crossing-over, 1025
Cross-linking theory, 1102
Cross-reactivity, 816
Cross-training, 409
Crown, of tooth, 932, *932*
Crowning, 1064, *1065*
Crown-to-rump length (CRL), 1082, *1086t, 1089*
Crural region, *30*, 31
Crus (crura)
    cerebral, *505, 506*
    of clitoris, 1045, *1045*
    of penis, *1018,* 1020
Crush syndrome, 414t
Crutch paralysis, 481
Cryptorchidism, 1013
Crypts
    intestinal, *954, 955, 965*
    tonsillar, 793, *793*
CSF. *See* Cerebrospinal fluid
CT scans, *22, 23, 689*
Cubital region, *30*
Cuboidal cells, 75, 76, *76, 142*
Cultured epithelial autograft, 194
Cumulus oophorus, *1051,* 1052, *1052, 1055*
Cuneate fasciculus, 465, *465, 466t, 467, 473,* 503, *504*
Cupula, of ear, *589,* 590
Curare, 393, 556
Curie, Marie, 43, *43,* 203
Curie, Pierre, 203
Curly hair, 185, *185*
Current, electrical, 433
Curvatures, of vertebral column, *240,* 241
Cushing syndrome, 646, *646*
Cusps
    of heart valves, 696
    of teeth, 933
Cutaneous area, of lips, 930
Cutaneous glands, 183, 187–189, *188, 190t*
Cutaneous innervation, 481, *481*
Cutaneous membrane, 163
Cutaneous transpiration, 189, 903–904
Cutaneous vasoconstriction, 176
Cutaneous vasodilation, 176
Cuticle
    of hair, 184
    of nail, 187, *187*
Cyanosis, 181, 858
Cyclic adenosine monophosphate (cAMP), *68,* 69, 82, *83*
    in cardiac physiology, 715, 716
    in hormone action, 638, *639,* 640
    in olfaction, 576
    signal amplification of, 640, *640*

in synaptic transmission, 444, *445*
Cyclin(s), 127
Cyclin-dependent kinases (Cdks), 127
Cyclooxygenase 2 (COX-2), 644
Cyclooxygenase (COX), 643, *643*
Cysteine, 63
Cystic duct, 946, *948*
Cystic fibrosis, 86
Cystitis, 893
Cytochromes, 990
Cytogenetics, 110
Cytokine(s), 800–802, 801t
Cytokinesis, 127
Cytology, 2, 75, 142
Cytolysis, complement system and, 797–798
Cytomegalovirus, 818
Cytoplasm, 77–78, *78*
Cytosine, 110, *111,* 114, 132
Cytoskeleton, 78, 81, 97, *98*
Cytosol, 78, 120
Cytotoxic T cells, 808–810, *809, 809t, 810,* 814t
Cytotrophoblast, 1080, *1081*

D

D (delta) cells, of pancreatic islets, *628,* 629
DAG-IP$_3$, second messenger system, 638, *640*
Dalton, John, 42
Dalton's law, 842t, 848
Dander, 179
Dandruff, 179
Dark adaptation, in eyes, 603, 1099
Dartos fascia, 1014, *1014*
Darwin, Charles, 9, 608
Daughter chromosome, *126,* 127
Daughter DNA, 123, *124*
Davy, Sir Humphry, 608
Day vision, 601, 603, 604, *605*
De Humani Corporis Fabrica (On the Structure of the Human Body) (Vesalius), 5
De Motu Cordis (On the Motion of the Heart) (Harvey), 5
Dead space
    anatomical, 845
    physiological (total), 845
Deafness, 585, 1099
Deamination, 994
Death, 1103
Debridement, 192
Decay, of radioisotopes, 43
Decidual cells, of endometrium, 1060, 1087
Deciduous teeth, *932,* 933, *933*
Declarative memory, 452, 519
Decompensated shock, 746
Decomposition reactions, 53, *53,* 54t
Decompression laminectomy, 488

Decompression sickness, 861
Decompression tables, 861
Deconditioned muscle, 409
Decremental potentials, 434, 435t
Decubitus ulcers, 169
Decussation, 465, 503, *504*
Deep, 29, 29t
Defecation, 926, 966–967, *967*
Defecation reflexes, 558, 966–967
Defensins, 175, 796
Defervescence, 800, *819*
Defibrillation, 705
Degeneration, tissue, 169
Degenerative joint disease, 294t, 294–295, *295,* 1098
Deglutition (swallowing), 936
    muscles of, *310, 313,* 315t–316t
    phases of, 936, *936*
Degranulation
    of neutrophils, 796
    of platelets, 680
Degrees of freedom, 277
Dehydration, 509, 885–886, 890, 904, *904,* 906, 907t, 1100
Dehydration synthesis, 48, 56, *56*
Dehydroepiandrosterone (DHEA), 616t, 628, 631t
Delayed hypersensitivity, 816
Delayed-onset muscle soreness, 414t
Delta (δ) cells, of pancreatic islets, *628,* 629
Delta (δ) chains, in hemoglobin, 662
Delta (δ) waves, 516, *517*
Democritus, 42
Denaturation, 63
Dendrites, 157, *158,* 158t, 425, *425,* 441
Dendritic cells, 177, *177,* 788, *790,* 796, 1044
Dendritic spines, 452, 513
Denervation atrophy, 391, 432, 1098
Denervation hypersensitivity, 543
Dens, *243,* 244
Dense bodies, of smooth muscle, 412, *413*
Dense bone. *See* Compact bone
Dense connective tissue, 149–150, *152,* 152t
    irregular, 150, *152,* 152t
    regular, 150, *152,* 152t
Dense-core vesicles, 442
Dental anesthesia, 608
Dental caries, 933
Dental disease, 933
Dentin, *932,* 933
Dentition, 931, *932. See also* Teeth
Dentures, 1101
Deoxygenated blood, 690, *691*
Deoxyhemoglobin, 851
Deoxyribonuclease, 950, 951t, 961
Deoxyribonucleic acid. *See* DNA
Deoxyribose, 110
Dephosphorylation, 987, *987*

Depolarization, of membrane, 394, 434, *434*
Deprenyl, 454
Depression (bone), 226t
Depression (mood), 559
Depression (movement), *280,* 280–281, 283
    mandibular, 281
    scapular, 280–281, *329,* 330t
Depth perception, 605, *606*
Dermal papilla, *177, 178,* 179, 184, *184*
Dermatitis, 193t
    contact, 193t, 819t
    seborrheic, 193t
Dermatology, 175
Dermatome map, 481, *481*
Dermatomes, 481
Dermcidin, 175, 796
Dermis, 163, 175, *176,* 179, *180,* 180t
    aging and, 1097
    pregnancy and, 1062
Descartes, René, 6–7, 433, 623
Descending analgesic fibers, 467, 507, 571
Descending aorta, *751, 757*
Descending colon, *374, 551, 927, 929,* 963, *964*
Descending limb, of nephron loop, 873
Descending tracts, of spinal cord, 464, *465, 466t,* 466–468, *468*
Descent of Man, The (Darwin), 9
Desmosomes, *79, 98,* 160, *161, 700, 701*
Desquamation, 142, *147*
Detached retina, 596
Detumescence, 1031
Deuterium, 43
Development, 14–15, 1076–1095
    bone, 206–211, 1089
    brain, 495, *497, 498,* 1082
    embryonic stage, 1079, 1079t, 1082–1089, *1084–1085*
    fertilization, *1075,* 1076–1078, *1077, 1078*
    fetal stage, 1079, 1079t, *1084–1085,* 1086t, 1089
    germ layers and, 139, 1076, 1080–1081, *1081, 1082, 1083,* 1083t
    gland, 160, *162*
    lymphatic vessel, 785
    neonatal, 1091–1095
    pituitary gland, 617, *617*
    preembryonic stage, 1076, 1079t, 1079–1081
    prenatal, 1060, 1079–1090, 1086t
    reproductive system, 1010–1013, *1011, 1012,* 1038
    spinal cord, 460, 1082
    tissue, 139, 166

Developmental biology, 1076
Developmental plasticity, 166
Dextrinase, 956, *956*
Dextrocardia, 13
Diabetes insipidus, 644–645, 648
Diabetes mellitus, 300, 645–650, 890
    aging and, 1099
    definition of, 646
    gestational, 890, 1068*t*
    glycosuria in, 883
    insulin therapy for, 647–650
    pathogenesis of, 647–648
    signs and symptoms of, 646–647
    type 1, 647, 890
    type 2, 637, 647, 890
    urine odor in, 889
Diabetic nephropathy, 648
Diabetic neuropathy, 648
Diacylglycerol (DAG), 638, *640*
Dialysis, 896, *896*
Dialysis machine, 896
Dialysis membranes, 88
Diapedesis, 802
Diaphoresis, 189
Diaphragm (contraceptive), 1069,
    1069, 1070*t*
Diaphragm (pelvic), *326*, 327*t*
Diaphragmatic surface, of lung, *834*
Diaphysis, 200, *200*, 269
Diarrhea, 906, 961, 963
Diarthrosis, 272. *See also* Synovial
    joints
Diastasis, 710, *711*
Diastole, 702, 710–712, *711*
Diastolic pressure, 734, *735*
Diazepam, 608
Dicrotic notch, *711*, 712
Diencephalon, 495, *498, 504,*
    509–511
Dietary fiber, 980
Dietary reference intakes (DRIs),
    977
Differential WBC count, 676, 678
Differentiation, 14, 166
Diffuse lymphatic tissue, 788
Diffusion
    capillary, 741, *741*
    facilitated, 91, *91, 96t*
    simple, 87–88, *96t*
Diffusion tensor imaging (DTI), 513
Digestion, 926, *962*
    Beaumont's studies of, 969
    carbohydrate, 956–957, *962*
    cephalic phase of, 943, *944*
    chemical, 926, 930, 952,
        956–963
    chyme in, 937, 942–943
    contact, 955
    fat (lipid), 943, 950, 958, *960,*
        962
    gastric function in, 937, 943–945,
        944
    gastric phase of, 943, *944*
    intestinal phase of, *944,* 944–945

mechanical, 926, 933
nucleic acid, 926, 950, 961
protein, 926, 940, *941,* 944, *945,*
    958, *959, 962*
salivation and, 935
small intestinal function in, 952
starch, 67, 930, 935, 943, 950,
    956, 956–957
water, 961, 963
Digestive system, 35, *37,* 925–973
    accessory organs of, 926
    aging and, 1100–1101
    anatomy of, 926–928, *927*
    disorders of, 968*t*
        acute pancreatitis, 968*t*
        appendicitis, 968*t*
        cancer, 950, 968*t*
        constipation, 963, 1101
        Crohn disease, 968*t*
        diarrhea, 906, 961, 963
        diverticulitis, 965
        dysphagia, 968*t*
        gallstones, 950
        gastritis, 943
        gluten-sensitive enteropathy,
            968*t*
        heartburn, 935, 1101
        hiatal hernia, 328, 968*t*
        lactose intolerance, 957
        peptic ulcer, 943, *943*
        ulcerative colitis, 968*t*
    effect on other systems, 970
    enzymes of, 926
    epithelial tissue of, 927, *928*
    excretion by, 869
    function of, 926
    hormones of, 929
    liver, gallbladder, and pancreas,
        946–952
    mesenteries of, 928–929, *929*
    mouth through esophagus,
        930–937
    nervous system of, 552–553,
        927–928, *928*
    parasympathetic effects on, 555*t*
    peristalsis in, 928, 936, *936,* 942
    pregnancy and, 1061–1062
    regulation of, 929
    stomach, 937–946
    sympathetic effects on, 555*t*
    tissue layers of, 927, *928*
Digestive tract, 926–927
    epithelial tissue of, 927, *928*
    measurement of, 926
    openings of, 926–927
    regulation of, 929
    tight junctions of, 160
    tissue layers of, 927, *927*
Digit(s), 31
    athletic injuries to, 358
    bones of, 250, 252, *253*
    movements of, 283–284, *284*
    surface anatomy of, *377*
Digital rays, *1084,* 1086*t*

Digital rectal examination
    (DRE), 1019
Digitalis, 718
Dihydrotestosterone, 628, 1021
Diiodotyrosine (DIT), 635, *636*
Dilation stage, of labor, 1064, *1065*
Dipeptidase, 958, *959, 962*
Dipeptides, 63
Diploe, 200, *201, 229, 232*
Diploid cell, 128, 1023, 1078
Diplopia, 597
Direct (fleshy) attachment, of
    muscle, 303, *304*
Directional terms, 29, 29*t,* 493, *494*
Disaccharides, 56, *57, 58t*
Disarticulated skeleton, 227
Discharge zone, of neural pool, 450,
    450
Discoid cells, *76*
Discoidal cells, 76
Discriminative pain, 569
Disjunction, 1094, *1094*
Dislocation, 294*t*
    shoulder, 287
Disodium phosphate, 45*t*
Displaced fracture, 216
Dissecting aneurysm, 728, 774*t*
Dissection, 2
Disseminated intravascular
    coagulation (DIC), 684*t*
Distal, 29, 29*t*
Distal convoluted tubule (DCT), *872,*
    873, *873, 874,* 884–885
Distal phalanx, 252, *253*
Distributing arteries, *727,* 727–728
Disulfide bridges, 63, *65*
Disulfiram, 1003
Disuse atrophy, 167, 414*t*
Diuresis, 647, 885, 889
Diuretics, 776, 890
Diverging circuit, 450, *450*
Diverticulitis, 965
Diverticulosis, 965
Diving, physiology in, 861
Dizygotic twins, 1080, *1080*
DNA, 69, 110–114
    aging and, 1102
    base sequence of, *120*
    complementary base pairs of, 8
    damage to, 113
    digestion of, 950, 961
    double helix of, 110–112, *111*
    errors in, 124, 132
    evolution of, 9
    functions of, 69, 112
    gene segment of. *See* Gene(s)
    indirect control of testosterone
        synthesis, 122, *123*
    length of molecule, 110
    mitochondrial, 103
    mutations of, 15, 124, 132, 1094
    nuclear packaging of, 113–114
    nucleotides of, 56, 110, *111,* 116
    peptide structure and, *120*

replication of, 123–124, *124*
    RNA vs., 114*t*
    transcription of, 116–117
DNA damage theory, 1102
DNA helicase, 123, *124*
DNA hybridization, 9
DNA ligase, 123, *124*
DNA methylation, 132
DNA polymerase, 123, *124*
Dominant alleles, 128–129, *129,* 131
Dominant ovarian follicles, 1053,
    1056*t*
Dopamine
    adrenal secretion of, 550
    as hormone, 615, 627, 631*t,* 633*t,*
        633–634, 637
    as neurotransmitter, 441, *443,*
        443*t,* 446, 615
    in Parkinson disease, 454
    as prolactin-inhibiting hormone,
        616*t,* 1066
    in substance abuse, 559
Doping, blood, 664
Dorsal, 29, 29*t*
Dorsal (posterior) mesentery, 34, *34,*
    928, *1083*
Dorsal respiratory group (DRG),
    840, *840*
Dorsal venous network, 750, *769*
Dorsiflexion, 284, *285*
Dorsum nasi, 364
Dorsum sellae, 234, *234*
Double covalent bonds, 46, 46*t*
Double helix, 110–112, *111*
Double vision, 597
Double-blind method, 8
Double-jointed, 275
Dowager's hump. *See* Kyphosis
Down syndrome, 454, 1095, *1096,*
    1099
Down the gradient, 17, *18,* 87, 91
Down-regulation, *640,* 640–641
Dreams, 517
Drew, Charles, 668, *668*
Dropsy, 712
Drugs. *See also specific drug*
    abuse/misuse of, 559
    aging and, 1100
    anticoagulant, 683, 685
    anti-inflammatory, *643,* 644
    hepatic metabolism of, 996*t*
    neuropharmacology of, 554–556,
        559
    pH and, 52
    psychopharmacology of, 559
    renal clearance of, 883–884,
        1100
    transdermal, 176
    weight-control, 976
Dry cells, 401
Dry gangrene, 169
Dual innervation, autonomic, *556,*
    556–557
Dual visual system, 603–604

Duchenne muscular dystrophy (DMD), 415
Duct(s), 160–161, *163, 164,* 613. *See also specific ducts*
Ductal carcinomas, 950
Ductus arteriosus, 1089, *1090*
Ductus deferens, *368,* 1014, *1014, 1016,* 1017, *1018*
  development of, *1011*
  vasectomy, 1070, 1070*t*
Ductus venosus, 1089, *1090,* 1091
Duffy blood group, 670, 672
Duodenal glands, *954,* 955
Duodenal papilla
  major, 947, *948*
  minor, 950
Duodenocolic reflex, 966
Duodenojejunal flexure, 952
Duodenum, *369, 374, 938, 952,* 952–956
  in gastric regulation, 942, 945
  secretions of, 955
Duplicity theory of vision, 603
Dura mater, 165, 231
  brain, 497, *499*
  spinal cord, 461, *461, 462*
Dural sheath, spinal cord, 461, *462*
Dural sinuses, 497, *499*
Dural venous sinuses, 732, *754*
Duration, of stimulus, 565–566
Dust cells (alveolar macrophages), 797, 836, *836*
Dwarfism
  achondroplastic, 211, *211,* 1093
  Laron, 637
  pituitary, 211, 645
Dynamic equilibrium, 15, 588, 879
Dynein, *85, 86, 99,* 426
Dynorphins, 571
Dyskinesias, 523
Dysmenorrhea, 1068*t*
Dysphagia, 968*t*
Dyspnea, 847*t*
  in pregnancy, 1062
Dystrophin, 388, *389,* 415

E

Ear
  aging and, 1099
  anatomy of, 578–582
  development of, *1084*
  equilibrium function of, 577, 587–591
  evolution of, 587
  hair cells of, *582,* 583–585, *584*
  hearing function of, 521, 578, 582–587
  sensory coding in, 585
Ear infections, 580
Eardrum. *See* Tympanic membrane
Earlobe, *578*
Earwax, 189, 579
Eccentric contractions, 402, *403,* 404

Eccrine (merocrine) glands, 162, *164, 188,* 189, 190*t*
ECG. *See* Electrocardiogram
Echocardiography, 23
Eclampsia, 1068*t*
Ectoderm, 139, 495, *497,* 1076, 1080–1081, *1082, 1083,* 1083*t*
Ectopic focus, 702
Ectopic ossification, 212
Ectopic pregnancy, 1068*t*
Eczema, 193*t*
ED. *See* Erectile dysfunction
Edema, 712, 743–744, 907, 909
  in inflammation, 800, 802, 803
Edward syndrome, 1095
Edwards, Robert, 1105
EEG. *See* Electroencephalogram
Effacement, 1064
Effector, 17, 422, *482, 483,* 543, 544
Efferent arterioles, 871, *871, 872, 873, 874,* 877, 879
Efferent division, of peripheral nervous system, 422
Efferent ductules, 85, *1016,* 1017, *1018*
Efferent lymphatic vessels, 791, *791*
Efferent (motor) neurons, 423, *423,* 544
  alpha, *483,* 484
  gamma, *483,* 484
  lower, 466, *468,* 523
  somatic, 391
  upper, 466, *468,* 523
Efferent nerve fibers, 470, 471*t,* *482,* 483
Efficiency, 992
Effort arm, 274, *274,* 276
Eggs (ova), 629, *630,* 1008
  atresia of, 1050
  donation of, 1104
  fertilization of, *1075,* 1076–1078, *1077*
  haploid chromosomes of, 128
  number of, 1049, 1050
  production of, 1050, *1051*
  release of, 1038, *1051,* 1053, *1054, 1055,* 1056
Ehrlich, Paul, 783, 797
Eicosanoids, 58, 58*t,* 61, *643,* 643–644
Ejaculation, 1027, 1029–1031, *1030*
Ejaculatory duct, 1017, *1018*
Ejection, ventricular, *711,* 712
Ejection fraction, 712
EKG. *See* Electrocardiogram
Elastic arteries, 726–727
Elastic cartilage, 154, *155,* 155*t*
Elastic fibers, 149, 150
  aging and, 1097
  in areolar tissue, 150
  in arteries, 726
  in dermis, 179, 1097
  in elastic cartilage, 154

  in heart, 693
  in lymphatic vessels, 785
  in Marfan syndrome, 150
  in meninges, 462
  in mucous membrane, *165*
Elastic filaments, 388, *389*
Elastic lamina
  external, 726
  internal, 726
Elastic tissue, 150
Elasticity, 385
Elastin, 149
Elbow, 277, *288,* 288–289, 358
Electrical current, 433
Electrical energy, 52
Electrical gradients, 18, *18*
Electrical potentials, 394, 433. *See also* Action potentials; Resting membrane potential
  local, 434–435, 435*t*
  postsynaptic, *446,* 446–447, *447*
Electrical synapses, 441
Electrically excitable cells, 393–394
Electrocardiogram (ECG, EKG), 704–706, *705, 706, 707*
Electrochemical gradients, 18, 394
Electroencephalogram (EEG), 516, *517, 518*
Electrolyte(s), 45, 45*t*
  absorption of, 961
  channels for, 81
  growth hormone and, 623
  intracellular fluid, 908, *908*
  pancreatic juice, 950
  physiological importance of, 908
  plasma, 658*t,* 658–659, 908, *908*
  renal reabsorption of, 881, 884
  saliva, 934
  storage in bones, 199
Electrolyte balance, 902, 908–914, 913*t*
  acid–base imbalances and, 920, 920*t*
  fluid balance/imbalances and, 902, 920, 920*t*
  fluid replacement therapy and, 921
  relationship with fluid and acid–base imbalances, 920, 920*t*
Electromagnetic energy, 52
Electromagnetic radiation, 591
Electron(s), 42, *42*
  sharing, in covalent bonds, 46, *47*
  valence, 42
Electron acceptor (oxidizing agent), 54
Electron donor (reducing agent), 54
Electron shells, 42
Electronic cell counters, 678
Electron-transport chain, 989–990, *990*
Electrophysiology, 393

  of brain, 516, *517*
  of heart, 703–706
  of neurons, 433–440
Eleidin, 178
Elements, 41–42
  in human body, 41*t,* 41–42
  periodic table of, 41
  symbols for, 41
Elevation, *280,* 281
  mandibular, 281
  scapular, 281, *329,* 330*t*
Eli Lilly and Company, 649–650
Ellis, Havelock, 1028
Elongation of bone, 208–210
Embolus/embolism, 684
  air, 755
  fat, 719, 774*t*
  pulmonary, 684
Embryo, 1060, 1076, 1080–1082
Embryo adoption, 1105
Embryoblast, 166, 1079, *1081*
Embryogenesis, 1076, 1080–1081
Embryology, 1076
Embryonic disc, 1080, *1081,* 1082
Embryonic folding, 1082
Embryonic limb rotation, 260, *260*
Embryonic membranes, 1086–1087, *1088*
Embryonic stage, of development, 1079, 1079*t,* 1082–1089, *1084–1085*
Embryonic stalk, *1081, 1083*
Embryonic stem cells, 166, 167
Embryonic tissue, 139, 166
Emergency contraceptive pills (ECPs), 1070
Emetic center, 942
Emigration (diapedesis), 802
Emission, in ejaculation, *1030,* 1031
Emmetropia, 597, *598, 599*
Emotion(s), 520
  hypothalamus and, 510, 520
  limbic system and, 514–515, 520
  olfaction and, 576
Emotional memories, 452, 520
Emphysema, 859, 1100
Empty magnification, 77
Emulsification, 958, *960*
Emulsification droplets, 958, *960*
Emulsion, 49–50
Enamel, *932,* 933
Encapsulated nerve endings, 567, 567*t,* 568
Encephalitis, 537*t*
End bulb, *567,* 567*t,* 568
End product, 67
End-diastolic volume, 710, *711,* 712
Endemic goiter, 645, *646*
Endergonic reactions, 54, 54*t*
Endocannabinoids, 977
Endocarditis, 718*t*
Endocardium, 165, *692,* 693, *693, 695*

Endochondral ossification, *207,* 207–208

Endocrine cells, interstitial, 629, *630,* 1015, *1016*

Endocrine disorders, 644–648, 648*t.* *See also specific disorders*
adrenal, 646
diabetes mellitus, 646–650
hypersecretion, 644–645
hyposecretion, 644–645
parathyroid, 645
pituitary, 645
thyroid, 645

Endocrine disrupting chemicals (EDCs), 1028

Endocrine glands, 160–161, 613–615. *See also specific endocrine glands*
development of, 160–161, *162*
structure of, 160–161, *163*

Endocrine system, 35, *37,* 421, 612–654. *See also specific glands*
aging and, 1099
effects on other systems, 650
fenestrated capillaries of, 614
immune system and, 820, 821
negative feedback inhibition in, 622, *622*
nervous system vs., *615,* 615*t,* 615–616
overview of, 613–615

Endocrinology, 2–3, 613

Endocytosis, 93–95, *94,* 96*t*

Endoderm, 139, 1076, 1080–1081, *1082, 1083,* 1083*t*

Endogenous opioids, 571

Endogenous pyrogens, 799

Endolymph, 580, 581, *581, 582, 583–584, 589,* 590

Endometriosis, 1057

Endometrium, *1041,* 1042
basal layer of, 1042
decidual cells of, 1060, 1087
functional layer of, 1042, *1044, 1057*
menstrual cycle and, *1054,* 1055–1056, *1057*
shedding of layer, 1042

Endomysium, 301, *301,* 386

Endoneurium, 429, *430, 431,* 432, 469, *470*

Endoplasmic reticulum, 99–101, *100*
aging and, 1099
protein processing and secretion in, 120, *121*
rough, *79, 99,* 100, *100,* 105*t*
smooth, *79, 100,* 100–101, 105*t, 974*

Endorphins, 442, 443*t,* 445, 571, 977

Endosteum, 200, *200, 204*

Endothelial cells, 785, 801*t,* 802

Endothelium, 163, 679, *679, 725, 727*

Endpiece, of sperm, *1026,* 1027

End-plate potential (EPP), 394

End-systolic volume, *711,* 712

Endurance, muscle, 406

Endurance exercise, 409

Energy, 52–55
activation, 65, *66*
ATP as currency of, 67
carbohydrates as source of, 57
chemical, 52
in chemical reactions, 52–55
definition of, 52
electrical, 52
electromagnetic, 52
free, 52
kinetic, 52
in muscle metabolism, 404–406
potential, 52
storage of, in adipose tissue, 153
transfer in body (metabolism), 54*t,* 54–55

Energy balance, 975

Energy levels (electron shells), 42

Engram, 451

Enkephalins, 443*t,* 445, 556, 571, *571,* 820

ENS. *See* Enteric nervous system

Enteric hormones, 630, 632*t*

Enteric nervous system (ENS), *542,* 552–553, 927–928, *928*

Enteric pacemaker cells, 942, 955

Enterocytes, 953

Enteroendocrine cells, 630, 940, 941, 942*t,* 944

Enterogastric reflex, *944,* 945

Enterohepatic circulation, 949

Enteropeptidase, 950, *951*

Envelope proteins, 179

Enzyme(s), 65–67. *See also specific enzymes*
active site of, 65
bactericidal, 592
blood, as disease markers, 65
brush border, 955, 956
carriers vs., 90–91
cofactors for, 67, *67*
digestive, 926
lysosomal, 101
metabolic pathways of, 67
naming of, 65
pancreatic, 950, *951,* 951*t,* 957, 958, *959*
in plasma membrane, 81, *82*
reaction products of, 66, *67*
structure and action of, 65–67, *66*
substrate of, 65–66, *66*
synthesis reactions of, 122

Enzyme–substrate complex, 66, *66*

Enzyme–substrate specificity, 66

Eosin, 673

Eosinophil(s), 154, 657, *657,* 673, 674*t*
in inflammation, 801*t*
in innate immunity, 796

production of, 676, *677*

Eosinophilia, 673, 803

Ependymal cells, 427, *427,* 427*t,* 499, 502

Epiblast, 1080, *1082*

Epicanthal folds, 1095, *1096*

Epicardial fat, *693*

Epicardium, *692,* 693, *693, 695*

Epicondyles, *226,* 226*t*
femoral, 256, *258, 380*
humeral, 251, *251, 377*

Epidermal keratinocytes, 213

Epidermal ridges, 179

Epidermal water barrier, 179

Epidermis, 163, 175, *176, 177–179,* 180*t*
aging and, 1097
cells of, 177, *177*
keratinocyte life cycle in, 178–179
layers of, *177,* 177–178, 180*t*

Epididymis, *368, 375, 1014,* 1017, *1018*
development of, *1011*
duct of, 1017
head of, *1016,* 1017
tail of, *1016,* 1017

Epidural anesthesia, 462

Epidural space, 461, *461*

Epigastric region, 31, *31*

Epigenetics, 131–132

Epiglottis, *365, 827, 828,* 829, *830, 831,* 936

Epilepsy, 7, 451, 520, 537*t,* 656

Epimysium, *301,* 302, 386, 390*t*

Epinephrine
adrenal secretion of, 550, 627
in blood pressure control, 739
in cardiac physiology, 715, 716
drugs affecting, 559
as first messenger, 82
as hormone, 627, 631*t,* 633*t,* 633–634, *634,* 637
hypersecretion of, 645
immune effects of, 820
as neurotransmitter, 441, *443,* 443*t*
in pulmonary ventilation, 843
in renal function, 888*t*
in stress response, 642

Epineurium, 469, *470*

Epiphyseal closure, 210

Epiphyseal line, 200, *200, 207,* 210

Epiphyseal plate, 200, *207,* 207–209, *208*

Epiphysis, 200, *200, 207,* 207–209, 269

Episiotomy, 1064

Epithalamus, *496,* 511

Epithelial cells
cortical, 789, *790*
digestive tract, 926–927
gastric, replacement of, 943
intestinal, 955

medullary, 789, *790*
renal tubule, 881
shapes of, *142*
thymic, 789, *790,* 806

Epithelial root sheath, *184,* 185

Epithelial tissue, 138, 138*t,* 141–147
apical surface of, 141
avascularity of, 141
basal surface of, 141
cornified, 142
of digestive tract, 927, *928*
of esophagus, 142, 935
functions of, 141
germinal, 1015
histology of, 141–147
keratinized, 142, *145,* 145*t*
location of, 138*t,* 141
mesothelium, 163
metaplasia of, 166
mitosis or repair, 141
of mouth (oral cavity), 930
nonkeratinized, 142, *145,* 145*t*
olfactory, 829
pigment, of retina, 594, 599, *600*
pseudostratified columnar, 141–142, *142, 144,* 144*t*
of renal corpuscle, 872
of renal tubules, 873, 881
respiratory, 829
simple, 141–142, *142*
columnar, 141–142, *144,* 144*t*
cuboidal, 141, *143,* 143*t*
squamous, 141, *143,* 143*t*
of small intestine, 953
sole of foot, *145*
squamous, *142*
simple, 141, *143,* 143*t*
stratified, 142, *145,* 145*t*
of stomach, 937, *939*
stratified, 141, *142,* 142–147
columnar, 142
cuboidal, 142, *146*
squamous, 142, *145,* 145*t*
of tongue, 931
of trachea, 832, *832, 853*
transitional, 142–147, 891, 893
of ureters, 891
of urethra, 893
of urinary bladder, 893
of uterine tubes, 1040, *1042*
of uterus, 1042
of vagina, 142, *145, 147,* 166, 1044

Epithelium. *See* Epithelial tissue

Epitopes, 804

Eponychium, 187, *187*

Eponyms, 19

Epstein-Barr virus, 818

Equilibrium
in chemical reactions, 53
in internal state of body, 15
in resting membrane potential, 434

Equilibrium (sense), 521, 577, 587–591
 projection pathways of, 590, *590*
Erectile dysfunction, 1023, 1031, 1032*t*
Erectile tissue
 female, 1045
 male, 1019–1020. (*See also* Penis)
 nasal, 829
Erection of penis, 556, 558, 1019, 1029
Errors, in DNA, 124, 132
Eruption, of teeth, *932,* 933, *933*
Erythema, 181–182
Erythroblast, 663, *663*
Erythroblastosis fetalis, 671, *671*
Erythrocyte(s), 154, *156, 655,* 657, *657,* 661–667
 agglutination of, 667, 668–669, *669*
 anuclear, 97
 biconcave shape of, *661,* 661–662
 death and disposal of, 664–665, *665,* 794
 deficiency of. *See* Anemia
 disorders of, 665–667
 excess of, 665
 functions of, 661
 gender differences in, 662–663
 hemolysis of, 665
 homeostasis of, *664,* 664–665
 life history of, 663–665, *665*
 metabolism of, 661
 numbers of, 663, 676
 osmosis in, 88
 as percentage of whole blood, 657, *657,* 662
 production of, 660, 663, *663*
 quantities of, 663
 structure of, *661,* 661–662
 surface antigens of, 667–668, *669*
 tonicity and, 89, *90*
 volume of (hematocrit), 657, 662
Erythrocyte colony-forming unit (ECFU), 663, *663*
Erythropoiesis, 660, 663, *663*
Erythropoietin (EPO), 616*t,* 629–630, 632*t,* 633*t,* 663, 666, 867, 896, 1099
Eschar, 191
Esophageal glands, *928,* 935
Esophageal hiatus, *928,* 935
Esophageal phase, of swallowing, 936
Esophagus, *365, 369, 827, 828,* 926, *927,* 935, *938*
 epithelial tissue of, 142, 935
 innervation of, 927
 smooth muscle of, 158, 411, *411,* 935
 swallowing function of, 936, *936*
Essential amino acids, 983

Essential fatty acids, 59, 980
Essential hypertension, 776
Essential nutrients, 977
Estradiol, 628, 629, 632*t,* 1048–1049
 chemical classification of, *634*
 in childbirth, 1063
 in menstrual cycle, 1053–1054, *1054*
 in ovarian cycle, 1053–1054, *1055*
 synthesis of, *635*
Estriol, 1048
Estrogen, 633*t*
 adrenal secretion of, 628
 in bone physiology, 215*t,* 216
 breastfeeding and suppression of, 1066
 in childbirth, 1063
 in hormonal contraception, 1070
 interaction with progesterone, 641
 in menstrual cycle, *1054,* 1055
 in ovarian cycle, 1053–1054, *1055, 1056*
 ovarian secretion of, 629
 placental secretion of, 631, 632*t*
 in pregnancy, 1060, *1060,* 1061*t,* 1066
 in puberty, 1048–1049
 in sodium homeostasis, 909
 synthesis of, 634, *635*
Estrogen blockers, 1048
Estrogen receptors, 637
Estrogen-dependent cancer, 637
Estrone, 1048
Ether, 608
Ethmoidal cells, 235, *235*
Ethmoidal sinus, 227, *230,* 235
Eumelanin, 180, 185, *185*
Eupnea, 847
Eustachian tube. *See* Auditory tube
Evaporation, 1000
Eversion, of foot, 285, *285*
Evolution, 9–12, 21. *See also* Evolutionary medicine
 of bipedalism, 11, 262–263, *262–263*
 of brain, 493, 513
 as characteristic of life, 15
 definition of, 9
 of olfaction, 576
 of photoreceptor cells, 604
 of primates, *10,* 10–11
 of scrotum, 1014
 and senescence, 1103
Evolutionary medicine, 11
 kidney/water conservation, 885
 menopause, 1049
 mitochondria, 103
 skeletal adaptations to bipedalism, 262–263, *262–263*
 skin color, 182
 sweet tooth, 979

vestigial organs, 10
Excess postexercise oxygen consumption (EPOC), 406–407, 988
Exchange reactions, 53
Exchange transfusion, for hemolytic disease of newborn, 671
Exchange vessels, capillaries as, 729. *See also* Capillary exchange
Excisional biopsy, 147
Excitability, 14, 385, 393–394, 422
Excitable tissues, 157–159
Excitation
 of neurons, *434,* 434–435
 of skeletal muscle fibers, 394, *395*
 of smooth muscle, 412–413
Excitation–contraction coupling, 396, *397*
Excitatory adrenergic synapse, 444–445, *445*
Excitatory cholinergic synapse, 442, 444, *444*
Excitatory neurotransmitters, 434–435, 442, 444, *444,* 445, 447
Excitatory postsynaptic potentials (EPSP), *446,* 446–447, *447*
Excitatory potentials, 434–435
Excitement phase, of sexual response, 1029
 female, 1058, *1059*
 male, 1029, *1030*
Excretion, 868–869
 definition of, 868
 epithelial, 141
 glandular, 160
 urinary (renal), 867, 869
Excursion, of mandible, 283, *284*
Exercise
 aging and, 1101
 articular cartilage in, 273
 blood flow in, 739, *739, 740*
 cardiac output in, 715, 718–719
 endurance, 409
 energy/muscle metabolism in, 404–406, *405*
 muscle fatigue/endurance in, 406
 resistance, 409, 1101
 respiration in, 857–858
 venous return in, 745
Exergonic reactions, 54
Exfoliate cytology, 142
Exfoliation, 142, *147,* 179
Exhaustion stage, in stress response, 642–643
Exocrine glands, 160–162, 613–615
 compound, 162, *164*
 development of, 160–161, *162*
 ducts of, 160–161, *163, 164,* 613
 secretions of, 613
 simple, 162, *164*

structure of, 162, *163*
 types of, 162, *164*
Exocytosis, 93, *94,* 95, 96*t*
Exogenous pyrogens, 799
Exons, 116, *117*
Experimental design, 7–8
Experimenter bias, 8
Expiration, 837, 843, *844*
Expiratory (E) neurons, 839, *840*
Expiratory reserve volume (ERV), *846,* 846*t,* 847
Explicit (declarative) memory, 452, 519
Exploratory surgery, 2
Expressed allele, 129
Expulsion, in ejaculation, *1030,* 1031
Expulsion stage, of childbirth, 1064, *1065*
Extensibility, 385
Extension, *279,* 279–280
 hand and digit, 283–284, *284*
 head and trunk, 282, *283*
Extensions, of bone, 226*t*
External os, *1041,* 1042
External (outer) ear, 578, *578*
External rotation, *281,* 282
External tension, muscle, 401
Exteroceptors, 566
Extracellular face, of plasma membrane, 79, *80*
Extracellular fluid (ECF), 78, 139, 902
 membrane polarization/potential and, 393–394, 433–434
 tonicity and, 89–90
Extracellular material (matrix), 138, 141
Extraembryonic mesoderm, 1081, *1081,* 1087
Extrathalamic cortical modulatory system, 507
Extravasated cells, 802
Extremities. *See* Lower limb; Upper limb; *specific anatomy*
Extrinsic mechanism, of coagulation, 681, *681*
Extrinsic muscles, 304
 of eye, 592, *592, 593*
 of larynx, 829
 of tongue, 931
Extrinsic proteins. *See* Peripheral proteins
Extrinsic salivary glands, 934, *934*
Eye(s)
 aging and, 599*t,* 1099
 anatomy of, 592–597, *594*
 chambers of, *594,* 594–595
 convergence of, 597–598
 corrective lenses for, *599,* 599*t*
 development of, *1084*
 evolution of, 604
 extrinsic muscles of, 592, *592, 593*
 innervation of, *593*

Eye(s) *(continued)*
 dual, 556, *556*
  parasympathetic, *551, 552*
  sympathetic, *547,* 597
 movement, muscles of, *230, 309*
 neural components of, 592, 595–597
 optical components of, 592, 594–595
 parasympathetic effects on, 555*t*
 primate, 10, *10*
 saccades of, 596
 sympathetic effects on, 555*t*
 tunics of, 592–594
 vision function of, 591–607
Eye color, 130, 131, *131,* 594
Eyebrows, 186, 591, *591*
Eyelashes, 591, *591*
Eyelids, 591, *591*
Eyepiece (ocular lens), 5

F

F actin, 388
Fabricius, Hieronymus, 744
Facet, of bone, 226*t*
Facial (Bell) palsy, 533
Facial bones, 224*t,* 235–238
Facial expression
 eyebrow function in, 186, 591
 muscles of, *308, 309, 310,* 311*t*–312*t*
 skin function in, 176, *177*
Facial muscles, *308*
Facial region, *30, 364*
Facilitated diffusion, 91, *91,* 96*t*
Facilitated zone, of neural pool, 450, *450*
Facilitation
 presynaptic, 447, *448*
 synaptic, 452
Fact, 8
Factor I (fibrinogen), 658*t*–659*t,* 659, *681,* 682, 682*t*
Factor II (prothrombin), 659*t, 681,* 682, 682*t*
Factor III, *681,* 682, 682*t*
Factor V, *681,* 682, 682*t*
Factor VII, *681,* 682, 682*t*
Factor VIII, *681,* 682, 682*t,* 684
Factor IX, *681,* 682, 682*t,* 684
Factor X, *681,* 682, 682*t*
Factor XI, *681,* 682, 682*t,* 683
Factor XII, *681,* 682, 682*t,* 683, *683*
Factor XIII, *681,* 682*t*
FAD, 986
FADH, 986
Fallen arches, 261
Fallopian tubes. *See* Uterine tubes
False pelvis, 255
False ribs, *247,* 248
Falsifiability, 7
Falx cerebelli, 497
Falx cerebri, 497, *499*

Faraday, Michael, 608
Farsightedness, *599,* 599*t*
Fascia, *301, 302, 303*
 adherens, *700,* 701
 breast, *1046*
 dartos, 1014, *1014*
 lata, 346*t*
 renal, 869
 superficial (hypodermis), 175, *176,* 179–180, 180*t*
 thoracolumbar, *323*
Fascicles
 brainstem, *505*
 muscle, *301, 302, 302,* 390*t*
 nerve, 469, *470*
Fasciculus (fasciculi), 463
 cuneate, 465, *465,* 466*t, 467, 473,* 503, *504*
 gracile, 465, *465,* 466*t, 467, 473,* 503, *504*
Fasciitis, plantar, 357
Fasciotomy, 303
Fast anterograde transport, 426
Fast axonal transport, 426, 478
Fast block, of polyspermy, 1078
Fast (first) pain, 569
Fast retrograde transport, 426
Fast-twitch (fast-glycolytic) fibers, 407*t,* 407–408, 408*t*
Fat(s)
 absorptive state, 997
 body. *See* Adipose tissue
 dietary, 977, 979*t,* 980–981
  requirements for, 979*t,* 980–981
  sources of, 981
 digestion of, 943, 950, 958, *960, 962*
 emulsification of, 958, *960*
 function of, 60
 metabolism of, 993–994, *994,* 995*t*
 neutral, 60
 orbital, 592
 polyunsaturated, 60
 postabsorptive state, 998
 saturated, 60, 981
 stress response and, 642
 taste of, 574
 trans, 60, *60*
Fat cells (adipocytes), 148, 153
Fat embolus/embolism, 719, 774*t*
Fatigue
 central, 406
 muscle, 406, 409
Fat-soluble vitamins, 58*t,* 961, 985, 985*t*
Fatty acids, 58*t,* 58–59
 essential, 59, 980
 polyunsaturated, 59
 saturated, 59, 981
 unsaturated, 59

Fatty streaks, in atherosclerosis, 719
Fauces, 930
Febrile, 798–799
Fecal incontinence, 967
Feces, 963, 966–967
 acholic, 949
 composition of, 966
 defecation of, 966–967, *967*
Fed (absorptive) state, 997*t,* 997–998
Feedback loops, 15. *See also* Negative feedback; Positive feedback
Female(s)
 adult body, *30*
 blood values in, 658*t,* 662
 iron metabolism in, 663
 muscles of pelvic floor, *326,* 327*t*
 nutritional requirements of, 977, 979*t*
 pelvic structure of, 256, *257,* 257*t, 375*
 spinal cord injury in, 488
 superficial anatomy of, *366, 369–371*
 urethra of, *370, 375,* 893, *1039*
Female condom, 1069, *1069,* 1070*t*
Female reproductive system, 35, *37,* 1008, 1037–1074
 aging and, 1038, 1049
 anatomy of, 1038–1048, *1039*
 blood supply of, 1042, *1044*
 childbirth in, 1062–1065
 complexity of, 1038
 disorders of, 1068*t*
  abruptio placentae, 1068*t*
  amenorrhea, 1068*t*
  breast cancer, 820, *1047,* 1047–1048
  cervical cancer, 1043, *1043*
  dysmenorrhea, 1068*t*
  ectopic pregnancy, 1068*t*
  endometriosis, 1057
  gestational diabetes, 890, 1068*t*
  hyperemesis gravidarum, 1061, 1068*t*
  leiomyoma, 1068*t*
  morning sickness, 1061
  pelvic inflammatory disease, 1033
  placenta previa, 1068*t*
  preeclampsia, 1068*t*
  sexually transmitted diseases, 1031, 1033
  spontaneous abortion, 1068*t*
 effects on other systems, 1071
 embryonic development of, 1010, *1011, 1012,* 1038
 external genitalia of, 1038, 1044–1045
 menopause in, 1038, 1049
 menstrual cycle in, 1042, *1054,* 1055–1056, *1057*
 oogenesis in, 1050

 ovarian cycle in, 1050, 1053–1055, *1054,* 1056, 1056*t*
 physiology of, 1038, 1048–1068
 pregnancy in, 1059–1062
 puberty in, 1048–1049
 sexual cycle in, 1050, 1053–1057
Female sexual response, 1058, *1059*
Femoral region, *30, 31, 299*
Femoral triangle, 775
Fenestrated capillaries, 614, 729, *730,* 741, 875
Fenestrations, 729, *730*
Fermentation
 anaerobic, 68, *69,* 404
  in carbohydrate metabolism, 986, *987,* 987–988, 992*t*
  in cardiac physiology, 699, 701
  in erythrocytes, 661
  in skeletal muscle metabolism, 404–406, *405*
 lactic acid, 68
Ferric ions, 44, 663, *664*
Ferritin, 663, *664*
Ferrous ions, 44, 663, *664*
Fertilization
 description of, *1075,* 1076–1078, *1077, 1078*
 in vitro, 167, 170, 1104–1105
Fertilization membrane, *1077, 1078*
Fetal alcohol syndrome (FAS), 1093
Fetal hemoglobin, 662
Fetal stage, of development, 1079, 1079*t, 1084–1085,* 1086*t,* 1089
Fetus, *1037,* 1060, 1089
 bones of, *207,* 208
 circulation of, 1087, *1088,* 1089, *1090*
 crown-to-rump length of, 1089
 development of, 1079, 1079*t, 1084–1085,* 1086*t,* 1089
 expulsion of. *See* Childbirth
 Rh mismatch with mother, *671,* 671–672
 sexual differentiation of, 1009–1010
 skull of, 239, *239*
 sonography of, 23, *23*
 spinal cord of, 460
 vertex position of, 1062, *1063,* 1086*t*
Fever, 644, 798–800, 1002
 course of, 799–800, *819*
 excessively high, 800
 positive feedback in, 17
Fiber(s)
 association, 513
 central, 425
 collagenous, 149, *149*
 elastic, 149
 fibrous connective tissue, 149
 lens, 595

muscle. *See* Muscle fibers
nerve, 157, 469, 471*t*. (*See also* Axon(s))
perforating, 200, 203, *204*
peripheral, 425
postganglionic. *See* Postganglionic fibers
preganglionic. *See* Preganglionic fibers
Purkinje, 702
reticular, 149
somatic motor, 391
Fiber, dietary, 980
Fibrillin, 149
Fibrin, *655*, 659, *681*, 682
Fibrin polymer, *681*, 682
Fibrinogen, 658*t*–659*t*, 659, *681*, 682, 682*t*, 801*t*, 802
Fibrinolysis, 683, *683*
Fibrin-stabilizing factor (factor XIII), *681*, 682*t*
Fibroblast(s), 148, 167, *168*, 801*t*
Fibroblastic phase, of tissue repair, 167
Fibrocartilage, 154, *155*, 155*t*
    of hip, 290
    of interpubic disc, 271, *271*
    of intervertebral disc, *271*
    of shoulder, 286
    of symphysis, 271
    of synovial joint, 272–273
Fibroids, uterine, 1068*t*
Fibromyalgia, 414*t*, 570
Fibrosis, 167, *168*
Fibrous cells, 76
Fibrous connective tissue, 148–150
    cells of, 148
    components of, 148–149
    types of, 149–153
Fibrous (F) actin, 388
Fibrous layer
    of eye, 592–593
    of parietal pericardium, *692*
    of periosteum, 200
Fibrous pericardium, 691
Fibrous proteins, 63
Fibrous rings, around heart valves, 693
Fibrous skeleton, of heart, 693, *696*
Fight-or-flight response, 543, 544
Filaggrin, 179
Filament(s)
    elastic, 388, *389*
    myofilaments, *387*, 387–388, *388*, 390*t*
    thick, 387–388, *388, 389*, 390*t*, 413
    thin, 97, 388, *388, 389*, 390*t*, 413
Filiform papillae, 572–573, *573*
Filtration, 87, 96*t*
    blood, 826
    capillary, 87, *88*, 89, *741, 741–743, 742*, 743
    epithelial, 141

glomerular, 875–881, 1100
Filtration membrane, 875–876, *878*
Filtration pores, 87, 729, *730*, 877
Filtration pressure, 741, *742*, 876–877, *878*
Filtration slits, 876
Fimbriae, of uterine tube, *1039, 1040, 1041*
Finger(s), 31
    athletic injuries to, 358
    bones of, 250, 252, *253*
    movements of, 283–284, *284*
    surface anatomy of, *377*
First heart sound, 710
First line of defense, 795
First polar body, 1050, *1051, 1077, 1078*
First trimester, of pregnancy, 1079
First-class lever, 275, *276*
First-degree burns, 191, *192*
First-order neuron, 465, *467*, 568
Fissures, 226*t*
    anterior median, 460, *462*, 503, *504*
    horizontal, of lung, 833, *834*
    longitudinal, *494, 495, 500*, 511
    oblique, of lung, 833, *834*
    oral, 930
    orbital
        inferior, 231*t, 236*, 237
        superior, 231*t, 234, 234, 236*, 237
    palpebral, 591, *591*
    transverse cerebral, *494*, 495
Fistula, 969
Fixation point, 605, *606*
Fixative, 140
Fixator muscles, 305
Fixed ends, of skeletal muscles. *See* Origin
Flaccid paralysis, 393, 488
Flagella, 86, 105*t*
Flat bones, 199, *201*
Flat feet, 261
Flatus, 965
Flavin mononucleotide (FMN), 989
Fleshy attachment, of muscle, 303, *304*
Flexibility, of joints, 268, *268*
Flexion, 277, 279, 279–280
    foot, 284, *285*
    hand and digit, 283–284, *284*
    head and trunk, 282, *283*
    lateral, 282, *283*
Flexion lines (creases), 182, *377, 378*
Flexor reflex, 484–485, *486*
Flexor retinaculum, 252, 336, *338*
Flexor (withdrawal) reflex. *See* Withdrawal (flexor) reflex
Floating kidney, 895*t*
Floating ribs, *247*, 248
Flow, 17–19, 21, 708–710. *See also* Airflow; Blood flow
    pressure gradient and, 708

resistance to, 708
    syringe illustration of, *708*, 708–709
Fluent (Wernicke) aphasia, 525
Fluid(s). *See also specific fluids*
    lymphatic–circulatory exchange of, 785, *786*
    replacement of, 921
Fluid balance, 902–907
    aging and, 1100
    antidiuretic hormone and, 904, *905*
    cold weather and, 903, 906
    definition of, 902
    digestive function in, 961
    disorders of, 905–907, 907*t*
    intake regulation in, 904–905
    kidneys and, 885–888
    lymphatic system and, 783, 902, *903*
    neonatal, 1091
    output regulation in, 905
    relationship with electrolyte and acid–base imbalances, 920, 920*t*
    water gain and loss in, 902–904, *903*
Fluid compartments, 902, *903*
Fluid deficiency, 905–906, 907*t*. *See also* Dehydration
Fluid dynamics, 708–709
Fluid excess, 907, *907*, 907*t*
Fluid replacement therapy, 921
Fluid sequestration, 907
Fluid-mosaic theory of cell membranes, 9
Fluoride, 984*t*
Fluorine, 41*t*
Fluoxetine, 559
Foam cells, 719
Folia, 507, *508*
Foliate papillae, 573, *573*, 931
Folic acid, 663, 985, 985*t*
Follicle(s)
    in folliculogenesis, 1052
    hair, *174, 184*, 184–185
    ovarian. *See* Ovarian follicles
    thyroid, 624, *625*
Follicle-stimulating hormone (FSH), 616*t*, 619, 620*t*, 633*t*, 637
    in folliculogenesis, 1052–1053
    inhibin and, 629, 1021–1022, 1049
    in males, 1021–1022, *1022*
    in ovarian cycle, 1053–1055, *1055*, 1056
    in puberty (female), 1048–1049
Follicular cells, 624, *1051, 1052*
Follicular fluid, *1051, 1052, 1055*
Follicular phase, of ovarian cycle, 1053, *1054*, 1056*t*
Folliculogenesis, 1050–1053, *1051, 1052*
Fontanelles, 238, *239*

anterior, 238, *239*
anterolateral, 238, *239*
mastoid, 239, *239*
posterior, 238, *239*
posterolateral, 239, *239*
sphenoid, 238, *239*
Foot, 31, 344
    adaptations to bipedalism, *262*, 262–263
    arches of, 261, *261, 380, 381, 382*
    bones of, 225, 259–261, *260*
    intrinsic muscles of, *356, 357*
    movements of, 284–285, *285*
    muscles acting on, *350, 352, 353, 355, 356*, 357*t*
    sole, epithelium of, *145*
    surface anatomy of, *380–382*
Foot plate, 260, *1084*
Foot processes, of glomerulus, 876, 877
Foramina, *226*, 226*t*, 231, 231*t*
    apical, *932*, 933
    cribriform (olfactory), 229, 235, *235, 236*, 576
    of frontal bone, 231*t*, 232
    greater palatine, 229, 237
    incisive, 229, *236*, 237
    infraorbital, *226*, 228, 231*t, 236*, 237
    interventricular, 498, *500*
    intervertebral, 242, *243*
    jugular, 228, 229, 231*t*, 233
    lacerum, 229, 235
    magnum, 229, 231, 231*t*, 233, *365*
    mandibular, 231*t*, 238, *238*
    mastoid, 229, 233
    maxillary, 231*t*
    mental, *226*, 228, 231*t*, 238, *238*
    nutrient, 200, *200*, 203, 789
    obturator, 254, 255, *255*
    of occipital bone, 231*t*, 233
    ovale, 229, 231*t, 234, 234*, 1089, *1090*, 1091
    parietal, *230*, 232
    rotundum, 229, 231*t, 234, 234*
    sacral (pelvic)
        anterior, 245, *246*
        posterior, 245, *246*
    of skull, 231, 231*t*
    of sphenoid bone, 231*t*, 233, *234, 234*–235
    spinosum, 229, *234*, 234–235
    stylomastoid, 229, 233
    supraorbital, *226*, 231*t*, 232, *236*
    of temporal bone, 231*t*, 233
    transverse, *243, 244*, 245
    vertebral, 242, *242, 243, 244*, 245
    zygomaticofacial foramen, *228*, 237
Forced convection, 1000
Forced expiratory volume (FEV), 847

Forced respiration, 837
Forearm, *30, 31*
   arteries of, *767*
   bones of, 250, *251, 251*–252
   compartments of, 336, 339
   muscles acting on, 304, *335,*
      335t–336t
   supination and pronation of, 282,
      282, 335, 335t–336t
   surface anatomy of, *377*
Forebrain, 495, 509–516
Foregut, *1083*
Forehead, *364*
Foreskin (prepuce), *1012, 1014,*
   1018, 1019, *1020*
Forgetting, 453
Formed elements, of blood, 154,
   157, 657, *657. See also*
   specific elements
   as percentage of whole blood,
      657, *657*
   production of, 660
Fornices, vaginal, 1043
   anterior, *1039*
   lateral, *1041*
   posterior, *1039*
Fossae, *226, 226t*
   coronoid, 251, *251*
   cranial, 231, *232*
   cubital, *377*
   hypophyseal, *234*
   hypophysial, 234
   iliac, *254, 255, 255*
   incisive, 237
   infraspinous, 250, *250*
   intercondylar, 256, *258*
   lacrimal, 237
   mandibular, *229, 232, 233*
   nasal, 235, 827
   olecranon, 251, *251*
   ovalis, *695, 1090,* 1091
   ovarian, 1038
   popliteal, *379, 381*
   radial, 251, *251*
   subscapular, 250, *250*
   supraclavicular, *364, 371*
   supraspinous, 250, *250*
   triangular, *578*
Fourth ventricle, *496, 498, 500,* 501,
   501, 504, 505, 508
Fourth-degree burns, 191
Fovea, *226, 226t*
   capitis, *256, 258, 289,* 290
   centralis, *594, 596, 596,* 604
Fracture(s), 216–217
   aging and, 1098
   comminuted, 216, *217*
   displaced, 216
   greenstick, 216, *217*
   healing of, 216–217, *217*
   nondisplaced, *217*
   pathological, 216
   stress, 216
   treatment of, 217, *218*

   X-rays of, *217*
Fracture hematoma, 216, *217*
Franklin, Rosalind, 112, *112*
Frank–Starling law of the heart, 717
Freckles, 181, 182
Free edge, of nail, 187, *187*
Free energy, 52
Free nerve endings, *567, 567t,* 568
Free radicals, 45
Free ribosomes, *79*
Free (unbound) hormones, 637
Frenulum
   labial, 930, *930*
   lingual, *930,* 931, *934*
   penis, 1019, *1020*
Frequency of sound, *578, 578,*
   585, *586*
Freud, Sigmund, 559
Friction, pleural fluid and, 837
Friction ridges, 179, 182
Friction rub, 691
Frons (forehead), *364*
Frontal association area. *See*
   Prefrontal cortex
Frontal lobe, *494, 496, 500,* 511, *511*
   functions of, *511,* 511–512
   lesions of, 519
   motor control by, 522
Frontal plane, *28,* 28–29
Frontal region, *364*
Frontal section, *28*
Frontal sinus, 227, *228, 230, 236,*
   365, 828
Fructose, 56, *57, 58t,* 957, *962*
Fuel, 977
Fulcrum, *274, 274,* 276
Full-thickness burns, 191, *192*
Functional groups, 55, *55,* 304–305
Functional layer, of endometrium,
   1042, *1044, 1057*
Functional MRI (fMRI), 23, 507,
   519, 538
Functional residual capacity, *846,*
   846t, 847
Fundic region, of stomach, 937, *938*
Fundus
   of eye, 595, *596*
   of gallbladder, 949
   of stomach, 937, *938*
   of uterus, 1040, *1041*
Fungiform papillae, *573, 573,* 931
Funiculi (columns), of spinal cord,
   462, 463, 473
"Funny bone," 251
Fused (complete) tetanus, *402, 403*
Fusiform cells, *76, 76*
Fusion inhibitors, 819

G

G actin, *388, 388*
G cells, *939,* 944, *945*
G$_1$ checkpoint, 127
G$_0$ (G-zero) phase, 125

G$_1$ phase, 125, *125*
G$_2$ phase, 125, *125*
G protein(s), 82, *83,* 441, *445*
   in olfaction, 576
GABA (τ-aminobutyric acid), 441,
   443, 443t, 444, 447
GABA-ergic synapse, 444
Galactose, 56, *57, 58t,* 957, *962,* 979
Galanin, 977
Galea aponeurotica, 309, *310*
Galen, Claudius, 3, 5, 433, 725
Gallbladder, *367, 374,* 926, *927, 929,*
   949–950
   aging and, 1101
   parasympathetic innervation of,
      551
   sympathetic innervation of, *547*
Gallop, in heart sounds, 710
Gallstones, 950
Galvani, Luigi, 401, 433
Gamete intrafallopian transfer
   (GIFT), 1105
Gametes (sex cells), 1008. *See also*
   Eggs; Sperm
Gamma (τ) chains, in fetal
   hemoglobin, 662
Gamma (τ) globulins, 659, 659t, 804
Gamma (τ) rays, 43
Gamma motor neurons, *483,* 484
Gamma-aminobutyric acid. *See*
   GABA
Gamulin Rh, 671
Ganglia, 422, 470, *470*
   autonomic, 545, *545*
   basal (basal nuclei), 454, 509,
      514, 515, 515–516, 523
   celiac, *547,* 549, 549t, *550, 551,*
      937
   cervical, 548
   ciliary, *551,* 552, 597
   coccygeal, *547,* 548
   collateral, *548,* 549, 549t
   dorsal root, *470, 471, 473*
   geniculate, *530*
   inferior, *531*
   intramural, 550
   lumbar, 548
   mesenteric, *547,* 549, 549t, *550*
   otic, *531, 551,* 552
   parasympathetic, *545, 893, 894,*
      927, 928
   posterior root, *470, 471, 473, 474*
   pterygopalatine, *551, 552*
   sacral, *547,* 548
   spiral (cochlear), *581, 582,* 585
   submandibular, *551, 552*
   superior, *531*
   sympathetic chain, *473, 474, 546,*
      546–549, *547, 548*
   terminal, 550
   thoracic, 548
   trigeminal, *529*
Ganglion cells, *600,* 601–604, *604,*
   605, 607

Gangrene, 70, 169
Gap junctions, 160, *161, 162,* 410,
   613, *700,* 701
Gap phases of cell cycle, 125, *125*
Gas(es). *See also specific gases*
   blood, 856–858
   as neuromodulators, 445
   as neurotransmitters, 442
   partial pressure of, 848–850, *851*
   pressure gradients of, 850, *851*
   solubility of, 850
Gas, intestinal (flatus), 965
Gas exchange, 826, 848–858
   alveolar, 849–850, *852,* 855, *855*
   carbon dioxide loading in,
      853–854, 856
   exercise and, 857–858
   Henry's law and, 849, *849*
   metabolic adjustments in,
      855–856
   oxygen unloading in, 854–855
   pH and, *856,* 857–858
   pressure gradients and, 850, *851,*
      852
   systemic, 853–855, *854*
   temperature and, 856, *856*
Gas gangrene, 169
Gas laws, 841, 842t
Gas transport, 850–853
Gastrectomy, 941
Gastric area, of spleen, *794, 794*
Gastric glands, *939, 940,* 942t, *945*
Gastric inhibitory peptide (GIP), 941
Gastric lipase, 941, 942t, *962*
Gastric motility, 941–942
Gastric phase, of gastric function,
   944, *944*
Gastric pits, *939, 940*
Gastric rugae, 937, *938*
Gastric secretions, 940–941, 942t,
   945
Gastrin, 631, 631t, 632t, 633t, 929,
   942t, *945, 945*
Gastritis, 943
Gastrocolic reflex, 966
Gastroenterology, 20, 926
Gastroferritin, 663, *664*
Gastroileal reflex, 956
Gastrointestinal (GI) tract, 926. *See*
   also Digestive system
Gastrulation, 1081
Gates (gated channels), 81, *82*
Gaze centers, 507
Gehrig, Lou, 469
Gene(s), 112, 114–122, 128–132
   activation of, *122*
   alleles of, 128–130, *129*
   cancer, 133
   definition of, 115
   environment and, 131, *132*
   expression of, 129, 131
   locus of, 128
   multiple, in inheritance, 130, *130*
   pleiotropy of, *130,* 130–131

"Gene overdose," 1078
Gene pool, 129
Gene regulation, 120–122, *122*
General adaptation syndrome (GAS), 642
General anesthetics, 608
General nerve fibers, 470, 471*t*
General senses, 521–522, 566–572
Gene-substitution therapy, 115
Genetic anomalies, 1093–1095
Genetic code, 116, 116*t*
Genetic counselors, 129
Genetic engineering
    anticoagulants, 685
    growth hormone, 637, 645
Genetic Revolution, 6
Genetics
    alleles in, 128–130, *129*
    cancer, 132–133
    environmental influences in, 131, *132*
    genotype in, 129, 131
    Mendelian, 110
    molecular, 110
    penetrance in, 131
    phenotype in, 129, 131, *132*
    pleiotropy in, *130*, 130–131
    polygenic inheritance in, 130, *130*
    Punnett square in, 129
    sex-linked traits in, 131, *131*
Genital herpes, 1033
Genital tubercle, 1010, *1012*, 1038
Genital warts, 1033
Genitalia, external, 1008, 1009*t*
    development of, 1010, *1012*
    female, 1038, 1044–1045
    male. *See* Penis
    masculinization of, 646, *646*
Genitalia, internal, 1008, 1009*t*
    female, 1038
Genitourinary system, 35
Genome, 110, 115
Genomic medicine, 110, 115
Genomics, 115
Genotype, 129, 131
Germ cells
    description of, 128, *1016*
    primordial, 1025
Germ layers, 139, 1076, 1080–1081, *1081, 1082, 1083*, 1083*t*
Germinal center, of lymph node, *791*
Germinal epithelium, 1015
Gestation, 1059. *See also* Pregnancy
Gestational diabetes, 890, 1068*t*
GFR. *See* Glomerular filtration rate
Ghrelin, 623, 630–631, 632*t*, 633*t*, 940, 975, 977, *978*
Gigantism, 645
Gingiva, 932, *932*
Gingival sulcus, 932, *932*
Gingivitis, 933
Glabella, *226*, 232
Gladiolus, 246

Glands, 160–163. *See also* Endocrine glands; Exocrine glands; *specific glands*
    cutaneous, 183, 187–189, *188, 190t*
    development of, 160–161, *162*
    mixed, 162
    secretions of, 160, 162–163
    structure of, 162, *163*
    unicellular, 161
Glans of clitoris, 1044, *1045*, 1058
Glans of penis, *375, 1012, 1014, 1018,* 1019, *1020*
Glaucoma, 559, 595
Glenoid cavity, 250, *250*
Glenoid labrum, 286, *287*
Glial cell. *See* Neuroglia
Gliomas, 428
Globular (G) actin, 388, *388*
Globular proteins, 63, 983
Globulins, 637, 658*t*–659*t*, 659, 797, *798*
Globus pallidus, 454, *515*, 516
Glomerular capillaries, *873*, 875
Glomerular capsule, 871, 872, *873, 874*
Glomerular filtrate, 875
Glomerular filtration, 875–881, *876*
    aging and, 1100
    disorders of, 876
    pressure in, 876–877, *878*
    regulation of, 878–880
Glomerular filtration membrane, 875–876, *878*
Glomerular filtration rate (GFR), 877–878, 891, 1100
Glomerulonephritis, acute, 895*t*
Glomerulus
    olfactory, *575*, 576
    renal, *866*, 871, *872, 873, 874, 877, 901*
Glottis, 831, *831*
Glucagon, 615, 628–629, 631*t*, 633*t*, 638, 950
    in absorptive state, 997
    antagonism with insulin, 641, *641*
    inotropic effect of, 718
    in postabsorptive state, 998
    signal amplification by, 640, *640*
Glucoamylase, 956, *956*
Glucocorticoids, 620, 627, 639, 643, 910
Gluconeogenesis, 627, 992–993, 993*t*
Glucose, 56, *57, 58t*
    aging and, 1099
    blood levels of, 300, 628–629, 640, *641,* 647–648, 890, 979, 998
    brain supply of, 502
    catabolism of, 986, *992,* 992*t*
        in aerobic respiration, 986, *989, 990*

    in anaerobic fermentation, 986, *987*
    in glycolysis, 986–987, *987*
        pathways of, 986, 992
    dietary, 979
    digestion and absorption, 956, *962*
    insulin–glucagon interaction and, 641, *641*
    metabolic states and, 998
    renal reabsorption of, 882, 883
    renal regulation of, 867
    storage and use, pathways of, 992, *993*
    in stress response, 642–643
    transport of, 882
Glucose 6-phosphatase, 992, *993*
Glucose 6-phosphate (G6P), 986–987, *987*
Glucose transporters, 647
Glucose-dependent insulinotropic peptide (GIP), 945
Glucose-sparing effect, 623, 627, 642, 980
Glutamate, 441, *443,* 443*t,* 452
    in vision, 602, *604*
Gluteal fold, *372*
Gluteal region, *30, 372*
Gluten-sensitive enteropathy, 968*t*
Glycemia, 628
    hyperglycemia, 647, 890
    hypoglycemia, 979, 998
Glycemic index (GI), 980
Glyceraldehyde 3-phosphate, 987
Glycerol phosphate, 991
Glycine, 435, 441, *443,* 443*t,* 447
    strychnine and, 559
Glycocalyx, 80, 83, 83*t, 84*
Glycogen, 57, *58,* 58*t,* 386
    in cardiac muscle, 410, 701
    metabolism of, 992–993, 993*t*
    in neurons, 423
    in skeletal muscle, *405,* 406
    sleep and, 519
    stress response and, 642
    in uterine milk, 1087
Glycogen phosphorylase, 992, *993*
Glycogen synthase, 992, *993*
Glycogenesis, 992, 993*t*
Glycogen–lactate system, *405,* 406
Glycogenolysis, 627, 992, *993,* 993*t*
Glycolipids, 57, 58*t*
    in glycocalyx, 83
    in plasma membrane, *80*
Glycolysis, 986–987, *987, 993,* 993*t*
    ATP production in, 68, *69,* 406, 986–987, *987*
    coenzyme (NAD^+) in, 67, *67,* 986
Glycoprotein(s), 57, 58*t*
    adhesive, 149
    assembly of, 120
    in glycocalyx, 83
    in plasma membrane, *80,* 81

Glycoprotein hormones, 633*t,* 634
Glycosaminoglycans, 149, 154
Glycosuria, 647, 883, 890
G2/M checkpoint, 127
GM_2 ganglioside, 429
Goblet cells, 141, 161, 162, 829, *832, 853, 953, 954*
Goiter
    endemic, 645, *646*
    toxic, 644, 648*t*
"Goiter belt," of U.S., 645
Golgi, Camillo, 433
Golgi complex, *79,* 101, *101,* 105*t*
    aging and, 1099
    protein processing and secretion in, 120, *121*
Golgi vesicles, *79,* 101, *101,* 105*t,* 120
Gomphoses, *269,* 269–270
Gomphosis, 932
Gonad(s), *614,* 629, *630,* 1008. *See also* Ovaries; Testes
    descent of, 1010–1013, *1013*
    development of, 1010, *1011*
Gonadal hormones, 629, 632*t*
Gonadal ridges, 1010, *1011*
Gonadotropes, 1021
Gonadotropin(s), 619–620
Gonadotropin-releasing hormone (GnRH), 556, 616*t,* 619*t*
    in male reproductive function, 1021–1022, *1022*
    in ovarian cycle, 1053, *1055*
    in puberty (female), 1048
Gonadotropins, 619
Gonorrhea, 1033, 1093
Gout, 294*t*
Graafian (mature) follicle, *1040, 1051,* 1052–1053, 1056*t*
Gracile fasciculus, 465, *465,* 466*t, 467, 473,* 503, *504*
Graded potentials, 434, 435*t*
Gradients, 17–19, 21
    concentration, 18, *18,* 87–88, 91, *91*
    definition of, 17
    down the gradient, 17, *18,* 87, 91
    electrical, 18, *18*
    electrochemical, 18
    pressure, 17, *18*
    thermal, 18, *18*
    up the gradient, 17, 91
Graft-versus-host response, 672, 817
Grandmother hypothesis, 1049
Granular cells, of juxtaglomerular apparatus, 879, *879*
Granulation tissue, 167, *168,* 216
Granule cells, 507, *575,* 576
Granules, 679
Granulocytes, 657, *657,* 673
    life history of, 676
    production of, 676, *677*
Granulosa cells, 629, *630, 1051, 1052, 1052,* 1077, *1077*

Granzymes, 798, 810, 814*t*
Gratification centers, 515
Graves disease, 645, 648*t*
Gravity, and venous return, 744
Gray hair, 185, *185*
Gray matter
  brain, 495, *499, 500, 505,* 506,
    513, 515
  central, *505, 506*
  spinal cord, *462,* 463
Gray's Anatomy, 5
Great (type II) alveolar cells, 836,
  836
Great vessels, 690
Greater curvature, of stomach, 929,
  937, *938*
Greater omentum, 34, *34, 367,* 929,
  929, 937, 938, 964
Greater pelvis, 255
Greater sciatic notch, 255, *255*
Greater trochanter, 256, *258, 372*
Greater vestibular gland, 1045, *1045*
Greater wing, of sphenoid bone, 234,
  234, 236
Greek medicine, 3, 19
Greenstick fracture, 216, *217*
Grinders (teeth), 932
Groin pull, 358
Gross anatomy, 2
Ground substance, 139, 149
Growth, 14–15
  appositional, 210
  bone, 208–211, 215–216
  cartilage, *209,* 210
  hair, 185–186
  interstitial, 210
  tissue, 166
Growth factors, 127, 215*t,* 432, 679
  insulin-like growth, 616*t,*
    622–623, 629, 632*t*
  nerve, 169, 428, 432, 454
  platelet-derived, 683, 803
Growth hormone (GH), 215*t,* 616*t,*
  620, 620*t,* 622–623
  age and, 623
  chemical classification of, 633*t*
  growth stimulants induced by,
    622–623
  half-life of, 622–623, 642
  hypersecretion of, 645, *645*
  hyposecretion of, 645
  receptors for, 637
  secretion of, 621, 623, 637
  synthetic, 637, 645
  widespread effects of, 622–623
Growth hormone-inhibiting hormone
    (GHIH), 629
Growth hormone-releasing hormone
    (GHRH), 616*t,* 619*t,* 623,
    633*t*
Growth plate. *See* Epiphyseal plate
Guanine, 110, *111,* 114, 116
Guanosine triphosphate (GTP), 69
Guard hairs, 186, 579, *828,* 829

Gubernaculum, 1010–1013, *1013*
Guillain–Barré syndrome, 487*t*
Gum disease, 933
Gum (gingiva), 932, *932*
Gustation (taste), 521, 572–575
  aging and, 1099
  anatomy of, 572–574
  physiology of, 574
  projection pathways of, 574–575
Gustatory cells, *573,* 574
Gustatory cortex, primary, 521, *521,*
  574
Gut microbiome, 965
Gut–brain peptides, 442, 630–631,
    941, 942*t,* 975–977, *978*
Gynecomastia, 70, 1094
Gyrus (gyri), *494, 496, 500,* 511,
  511
  angular, 525, *525*
  cingulate, *496,* 514, *514*
  postcentral, *511,* 522, *522, 525*
  precentral, *494, 511,* 522–523,
    523

H

H. M. (brain study subject), 520
H band, 389, *389*
Habenula, *496,* 511
Habituation, 507
Hageman factor (factor XII), *681,*
  682, 682*t*
Hair, 175, 183–187
  aging and, 1097
  distribution of, 183–184
  functions of, 186
  growth and loss of, 185–186
  layers of, 184, *184*
  structure of, *184,* 184–185
  texture of, 185, *185*
  types of, 183–184
Hair bulb, *184,* 184–185
Hair cells of ear, 581–582, *582*
  equilibrium function of, *588,*
    588–589, *589*
  hearing function of, 583–585,
    584
  tip link of, 584, *584*
Hair color, 185, *185,* 1097
Hair cycle, 185–186
Hair follicle, *174, 184,* 184–185
Hair matrix, 184, *184*
Hair receptors, *184,* 185, 186, *567,*
  567*t,* 568
Hair root, *184,* 184–185
Hair shaft, 184, *184*
Haldane effect, 856
Half-life
  of hormones, 622–623, 637, 642
  of radioisotopes, 44
Hallux, *381*
Hamburger, Viktor, 432
Hamulus, 252, *253*
Hand, 31

  arteries of, *767*
  bones of, *225,* 250, 252, *253*
  intrinsic muscles of, 304, 343*t*
  movements of, 283–284, *284*
  muscles acting on, *337, 338,*
    339t–340t, 341, *342,* 343*t*
  surface anatomy of, *377, 378*
  tendon sheaths and bursae of,
    273
Hand plate, 260
Handedness, 526
Handle, of malleus, 579
Haploid cells, 128, 1023, 1050, 1078
Haptens, 804, 814*t*
Haptoglobulin, 659*t*
Hard callus, 217, *217*
Hard keratin, 183
Hard palate, 230, 237, *827, 828,*
  930, 931
Harvey, William, 4–5, 725, 744, *744*
Hassall's corpuscles. *See* Thymic
    corpuscles
Haustra, *964,* 965
Haustral contractions, 966
Haversian system (osteon), *156,* 203,
  204
Hawking, Stephen, 469, *469*
HDL. *See* High-density lipoproteins
Head
  of bone, 226, 226*t*
    of femur, 256, *258*
    of fibula, 258, *259, 380*
    of humerus, 251, *251*
    of malleus, 579
    of metacarpal bone, 252, *253*
    of metatarsal bone, *380, 381,*
      382
    of radius, *251, 252, 377*
    of rib, 247, *248*
    of stapes, 579
    of ulna, *251,* 252
  of epididymis, *1016,* 1017
  of gallbladder, 949
  of lipids, 79
  of muscle, 303
  of myosin molecule, 388, *388*
  of pancreas, *628,* 950
  of phospholipids, 61, 79
  of sperm, *1026,* 1027
Head (cephalic region), *30,* 31
  adaptations to bipedalism, 263,
    263
  arteries of, *752*
  frontal section of, *230*
  median section of, *365*
  movements of, 282, *283,*
    588–590
  muscles of, 308–318
  surface anatomy of, *364*
  veins of, *754*
Head tilt test, 529
Headaches, migraine, 537*t*
Healing
  fracture, 216–217, *217*

  stages of, 167, *168*
  tissue, 167
Hearing, 521, 578
  aging and, 1099
  binaural, 586
  cochlear tuning in, 585
  human, range of, 578, *578*
  mechanical model of, 583, *583*
  nature of sound and, 578, 585
  physiology of, 582–587
  projection pathway of, 585–587,
    587
  sensory coding in, 585
Hearing loss, 585, 1099
Heart, *368, 373,* 689–723
  anatomical variations in, 13, *14*
  atria of, *373,* 693–695, *694,* 697,
    697
  autonomic innervation of,
    713–714, *714*
  autonomic nervous system and,
    547, 549, *551,* 715–716
  blood flow through, 697, *697*
  blood output of, 713–719
  cadaver, *693*
  conduction system of, *701,*
    701–702
  coronary circulation of, 697–699,
    698
  CT scan of, *689*
  development of, *1083, 1084,*
    1089
  electrical and contractile activity
    of, 702–706
  electrocardiogram of, 704–706,
    705, 706, 707
  endocrine function of, 630, 632*t*
  fibrous skeleton of, 693, *696*
  Frank–Starling law of, 717
  gross anatomy of, 692–699
  historical and cultural view of,
    690
  internal anatomy of, *695*
  nerve supply to, *547,* 549, *551*
  position of, 691, *692*
  rhythm of, 410, 700, 702, 710
  shape of, 691
  size of, 691
  spinal cord trauma and, 488
  study of (cardiology), 690
  surface anatomy of, *694*
  venous drainage of, 699
  ventricles of, *373,* 693–695,
    695–696, 697, *697,*
    709, *711*
Heart attack (myocardial infarction),
    169, 684, 699, 705
Heart block, 707, *707*
Heart bulge, *1084,* 1089
Heart disorders, 718*t*
  angina pectoris, 699
  arrhythmias, 705, *707*
  cardiac arrest, 705
  cardiac tamponade, 718*t*

cardiomyopathy, 718*t*
congestive heart failure, 713
cor pulmonale, 718
coronary artery disease, 719, *720*
infective endocarditis, 718*t*
left ventricular failure, 712
myocardial infarction, 169, 684,
    699, 705
myocardial ischemia, 718*t*
pericardial effusion, 718*t*
pericarditis, 691, 718*t*
septal defects, 718*t*
valvular insufficiency, 710
Heart failure, 712
Heart murmurs, 710
Heart rate (HR), 715
    ANS on, 715
    chemicals and, 716–717
    chronotropic agents and, 715
Heart sounds, 710
Heart surgery, 690, 719
Heart tube, *1083*
Heart valves, 696–697. *See also*
        specific valves
    blood flow through, 697
    cusps or leaflets of, 696
    fibrous rings around, 693
    insufficiency of, 710
    operation of, *709,* 710–712, *711*
    repair or replacement of, 690,
        710
    stenosis of, 710
Heart wall, *692,* 692–694
Heartburn, 935, 1101
Heat
    body. *See* Body heat
    in inflammation, 800, 802
    as kinetic energy, 52
    unit of measurement, 48
Heat capacity of water, 48
Heat cramps, 1002
Heat exhaustion, 1002
Heat loss, 999–1000, *1000*
Heat production, 999
Heat shock proteins, 120
Heat-loss center, 1001
Heat-promoting center, 1001
Heatstroke, 1002, 1097
Heavy chains, of antibody, 804, *805*
Helicobacter pylori, 943
Helix, of ear, *578*
Helper T cells, 796, 801*t,* 808–810,
        809, 809*t, 810,* 814*t*
Hemangiomas, 183
Hematocrit, 657, 662
Hematology, 656
Hematoma, 182, 216, *217,* 684
Hematopoiesis, 660
    aging and, 1099
    erythrocyte, 663, *663*
    leukocyte, 676, *677*
    lymphocyte, 676, *677, 806,*
        806–807
    lymphoid, 660

myeloid, 660
platelet, 679, *679*
Hematopoietic stem cell (HSC), 660,
    663, 676
Hematopoietic tissue, 205
Hematuria, 876, 889
Heme, 662, *662,* 665, 850
Hementin, 685
Hemidecussation, 606
Hemidesmosomes, *79, 98,* 160, *161*
Hemiplegia, 488
Hemodialysis, 896, *896*
Hemodynamics, 734
Hemoglobin, 181–182, 661–663,
    850–853
    adult (HbA), 662
    aging and, 1099
    disposal of, 665, *665*
    fetal (HbF), 662
    HbA₂, 662
    mean corpuscular, 678
    metabolic adjustments in,
        855–856
    pH and, *856*
    sickle-cell (HbS), 666
    structure of, 63, *64, 662, 662,*
        850–851
    temperature and, 856, *856*
Hemoglobin concentration, 662
Hemolysis, 665
Hemolytic anemia, 666, 666*t*
Hemolytic disease of the newborn
        (HDN), *671,* 671–672
Hemophilia, 683–684
Hemophilia A, 684
Hemophilia C, 684
Hemopoietic tissue, 660. *See also*
        Red bone marrow
Hemorrhage. *See also* Hemostasis
    erythrocytes in, 661
    fluid sequestration in, 907
    neonatal, 1062
Hemorrhagic anemia, 665–666, 666*t*
Hemorrhoids, 732, 963, *964,* 1062
Hemostasis, 678–685, *680*
    clinical management of, 685
    coagulation in, *680,* 680–683,
        681
    definition of, 679
    disorders of, 683–684
    platelet plugs in, 679, 680, *680*
    reaction cascade in, 682, *683*
    vascular spasm in, 680, *680*
Henry's law, 842*t,* 849, *849*
Heparin, 148–149, 673, 683, 685,
    802
    development of, 650
    in inflammation, 801*t,* 802
    in innate immunity, 796
Hepatic coma, 995
Hepatic ducts
    common, 946, *948*
    left, 946, *948*
    right, 946, *948*

Hepatic flexure, 963, *964*
Hepatic lobules, 946, *948*
Hepatic macrophages, 797, 946, *948*
Hepatic portal system, *765,* 947, 953
Hepatic sinusoids, 729, *730, 765,*
    947, *948*
Hepatic triad, 947, *948*
Hepatitis, 996
Hepatitis A, 996
Hepatitis B, 996, 1033
Hepatitis C, 996, 1033
Hepatitis E, 996
Hepatocytes, 946, *948*
Hepatopancreatic ampulla, 947, *948,*
    950
Hepcidin, 629–630, 632*t,* 633*t,* 643,
    961
Heredity, 110, 128. *See also*
        Chromosome(s); DNA;
        Gene(s)
Hering–Breuer reflex, 841
Hernia, 328
    hiatal, 328, 968*t*
    inguinal, 328, *328*
    umbilical, 328
Herniated disc, 242, *242*
Heroin, 571
Herpes simplex, 818
Herpes simplex type 2 (HSV-2),
    1033
Herpes zoster, 478
Heterozygous, 129
Hexokinase, 986, *993*
Hiatal hernia, 328, 968*t*
Hiatus
    aortic, *759*
    esophageal, *928,* 935
    sacral, 245
Hierarchy of complexity, 11–12, *12,*
    21
High-density lipoproteins (HDLs),
    62, 981, *981–982*
Hilum
    of kidney, 869
    of lung, 747, 833, *834*
    of lymph node, 790
Hindbrain, 495, 503–508
Hindgut, *1083*
Hip
    bones of (ossa coxae), *225, 254,*
        254–256, *255*
    joint, 275, *289,* 289–290
    ligaments supporting, *289,* 290
    muscles acting on, 344, *344, 345,*
        346*t*–347*t*
Hip replacement, 294–295, *295*
Hippocampus, 514, *514*
    lesions of, 520
    in memory, 519–520
    in olfaction, 576, *577*
Hippocrates, 3, 132, 287, 493
Hippocratic Oath, 3
Hirschsprung disease, 553
Hirsutism, 186

Hirudin, 685
Histamine, 148, 167, 441, *443,* 443*t,*
    446, 643, 673
    in digestive system, 929, 942*t,*
        943, 944, *944*
    in inflammation, 801*t,* 802
    in innate immunity, 796
Histamine receptors, 943
Histological sections, 139–140
Histology, 2, 138
    anterior pituitary gland,
        617, *619*
    blood vessel, 725–726, *727*
    bone tissue. *See* Bone tissue
    cell junction, 160, *161*
    cerebral cortex, 513, *514*
    connective tissue, 148–157
    epithelial tissue, 141–147
    gland, 160–163
    gonad, 629, *630*
    large intestine, 963
    liver, 946, *948*
    lung, 833–836, *835*
    lymphatic system, 788
    membrane, 163, 165
    muscular tissue. *See* Muscular
        tissue
    nervous tissue, 157
    ovary, 629, *630, 974*
    pancreas, 950, *951*
    posterior pituitary gland, 619,
        619
    red bone marrow, 789, *789*
    retina, *600,* 600–601, *601*
    skeletal muscle. *See* Skeletal
        muscle
    small intestine, *954,* 972
    stem cell, 166
    stomach, *939,* 940
    testis, 629, *630,* 1015–1017, *1016*
    thymus, 789–790, *790*
    thyroid gland, 624–625, *625*
    tissue development, 139, 166
    tissue growth, 166
    tissue repair, 141, 167
Histones, 113, 124
Histopathology, 2
Histotoxic hypoxia, 858
"Hitting the wall," 406
HIV, 817–819
Hives, 819*t*
H⁺–K⁺ ATPase, 940, *941*
Hodgkin disease, 819*t*
Holism, 12
Holocrine glands, 163, *164*
Homeostasis, 14, 15–17, 21, 543.
        See also specific
        substances and processes
    acid–base balance in, 902,
        914–920
    blood pressure, 15–17, *16*
    calcium, 212–215, *214,* 215*t,* 867
    definition of, 15
    disturbance of (stress), 642–643

Homeostasis (continued)
electrolyte balance in, 902, 908–914
erythrocyte, 663–664, 664
fluid balance in, 902–907
negative feedback in, 15–17
phosphate, 215, 912–913
potassium, 911
sodium, 909–910
temperature, 15, 16
Homo, 11
Homo erectus, 11
Homo sapiens, 9, 11
Homologous chromosomes, 128, 1023–1025, 1024
Homologous organs, 1010
Homozygous, 129
Hooke, Robert, 5–6, 75
Horizontal cells, of retina, 600, 601
Horizontal plane, 28, 29
Horizontal plate, of palatine bone, 237
Hormonal methods, of contraception, 1069, 1070
Hormone(s), 161, 613–642, 631t–632t. See also specific hormones
aging and, 1099
antagonistic effects of, 641, 641
blood concentration of, 640
in blood pressure control, 738–739
in bone physiology, 214, 215, 215t, 215–216
bound and unbound, 637
chemistry of, 633t, 633–634, 634
clearance of, 642
definition of, 161
digestive system, 929
enteric, 630, 632t
glycoprotein, 633t, 634
half-life of, 622–623, 637, 642
hormonal stimuli for, 637
humoral stimuli for, 637
hypersecretion of, 645
hyposecretion of, 644–645
immune system and, 820
inhibiting, 619
interactions of, 641
mode of action, 637–639, 638
monoamine, 633t, 633–634, 634
neural stimuli for, 637
neurotransmitters vs., 615–616
nomenclature for, 616, 616t
peptide, 633t, 634, 634
permissive effects of, 641
pharmacological doses of, 641
polypeptide, 633t, 634
pregnancy, 1060, 1060–1061, 1061t, 1066
prenatal, 1010
releasing, 619
second messengers and, 638, 639, 640

secretion of, 636–637
signal amplification of, 640, 640
steroid, 633, 633t, 634
synergistic effects of, 641
synthesis of, 534–636
target organs or cells of, 615, 616, 637, 638
target-cell sensitivity of, modulation of, 640, 640–641
transport of, 637
Hormone receptors, 637–639
down-regulation of, 640, 640–641
saturation of, 637
specificity of, 637
therapeutic implications of, 637
up-regulation of, 640, 640
Hormone replacement therapy (HRT), 1049
Hormone therapy, 641
Horner syndrome, 558t
Horns. See Cornua
Hot flashes, 1049
HPV (human papillomavirus), 1033, 1043
Hughes, Elizabeth, 649
Human body. See also specific anatomy
anatomic variations, 13, 14
characteristics of life, 13–15
colorized MRI scan of, 1
functions of, 13–18
gradients and flow, 17–19
homeostasis and negative feedback in, 15–17
physiological variations in, 15
positive feedback in, 17
structural hierarchy of, 11–12, 12
Human chorionic gonadotropin (HCG), 1060, 1060, 1061t, 1080
Human chorionic somatomammotropin (HCS), 1060–1061, 1061t
Human chorionic thyrotropin, 1061, 1061t
Human immunodeficiency virus (HIV), 817, 817–819
Human leukocyte antigens (HLAs), 809
Human origins, 9–12. See also Evolution
Human papillomavirus (HPV), 1033, 1043
Human placental lactogen, 1061
Human Sexual Response (Masters and Johnson), 1028
Humoral immunity, 804, 811–814
attack phase of, 811, 812
cellular immunity vs., 815t
lymphocytes in, 810, 811–814, 812
memory in, 811–812, 812

primary response in, 811, 813
recognition phase of, 811, 812
secondary (anamnestic) response in, 812, 813
Hunger, 975–977
growth hormone and, 623
hypothalamus and, 509
Hunger contractions, 977
Huntington disease, 524
Hyaline cartilage, 154, 155, 155t
articular, 200, 200, 272, 273
of bronchus, 155, 833
endochondral ossification in, 207, 207–208
growth of, 209, 210
of symphyses, 271
of synchondroses, 271
of synovial joints, 272
of trachea, 831, 832
zone of reserve, 209, 209
Hyaluronic acid, 149, 272, 796
Hyaluronidase, 796, 1077
Hydration spheres, 48, 49, 88
Hydrochloric acid, 940, 941, 942t, 945, 969
Hydrocortisone. See Cortisol
Hydrogen
body percentage of, 41, 41t
covalent bonds of, 46
isotopes of, 43, 43
Hydrogen bonds, 46t, 47, 47
Hydrogen ions, 50, 856–857
Hydrogen peroxide, 101, 796
Hydrolysis, 48, 56
of ATP, 67
in digestive system, 926
Hydronephrosis, 895t
Hydrophilic heads, of phospholipids, 61, 79
Hydrophilic substances, 48
Hydrophobia (rabies), 487t
Hydrophobic substances, 48
Hydrophobic tails, of phospholipids, 61, 80
Hydrostatic pressure, 89, 89, 741, 742, 876–877, 882
Hydroxide ions, 50
Hydroxyapatite, 202, 211
Hydroxyl group, 55, 55
Hyman, Flo, 150
Hymen, 1011, 1044, 1045
Hyper-, 908
Hyperabduction, 280
Hyperadduction, 280
Hyperalimentation, 921
Hyperbaric oxygen therapy, 850
Hyperbilirubinemia, 1092
Hypercalcemia, 212, 214, 716, 912, 913t, 920t
Hypercapnia, 716, 738, 857
Hyperchloremia, 912, 920t
Hyperemesis gravidarum, 1061, 1068t
Hyperemia, 802

Hyperemia, reactive, 737
Hyperextension, 279, 280, 282
Hyperglycemia, 647, 890
Hyperglycemic hormone, 629
Hyperinsulinism, 648t
Hyperkalemia, 716, 910, 911, 913t, 919, 920t
Hyperkyphosis, 219, 241, 241
Hyperlordosis, 241
Hypermagnesemia, 912, 913t
Hypernatremia, 910, 913t
Hyperopia, 599, 599t
Hyperparathyroidism, 645
Hyperphagia, 977
Hyperplasia, 166, 411
Hyperpnea, 847t
Hyperpolarization, 435–436
Hyperreflexia, 488
Hypersensitivity
denervation, 543
description of, 815–816
Hypertension, 735, 776, 877, 984
aging and, 1100
beta-blockers for, 559
essential, 776
portal, 766
in pregnancy, 1068t
primary, 776
secondary, 776
Hyperthermia, 999
Hyperthyroidism, 645, 648t
Hypertonic solutions, 89
Hypertrophic response, myocardial, 776
Hypertrophy, 166
adipose tissue, 166
cell, zone of, 209, 209
skeletal muscle, 166
smooth muscle, 411
Hyperventilation, 841, 847t, 857
Hypervitaminosis, 985
Hypervolemia (volume excess), 907, 907t, 920t
Hypo-, 908
Hypoblast, 1080, 1081, 1082
Hypocalcemia, 212, 214, 716, 884, 912, 913t, 920t
Hypocapnia, 857
Hypochloremia, 912, 920t
Hypochlorite, 796
Hypochondriac region, 31, 31
Hypodermis, 175, 176, 179–180, 180t
Hypogastric region, 31, 31
Hypoglycemia, 979, 998
Hypoglycemic hormone, 629
Hypokalemia, 643, 716, 910, 911, 913t, 919, 920t, 961
Hypomagnesemia, 912, 913t
Hyponatremia, 19, 910, 913t
Hyponychium, 187
Hypoparathyroidism, 645
Hypophysial portal system, 617
Hypophysis. See Pituitary gland

Hypoplastic anemia, 666
Hypopolarization. *See*
        Depolarization
Hypoproteinemia, 660
Hypospadias, 1032*t*
Hypotension
    definition of, 735
    orthostatic, 774*t*, 1099
Hypothalamic thermostat, 509, 1001
Hypothalamo–hypophysial tract, 619
Hypothalamo–pituitary–ovarian axis,
        1053
Hypothalamo–pituitary–thyroid axis,
        620, *621,* 622, *622,* 1021
Hypothalamus, *496,* 509–510,
        610–623, *614, 618*
    in autonomic control, 509, 558
    in emotion, 510, 520
    hormones of, 509, 617, *618,* 619,
        619*t*
    in hunger and satiety, 509, 623,
        977, *978*
    nuclei of, *618,* 619
    in olfaction, 576, *577*
    in ovarian cycle, 1053–1055,
        1055
    in pain modulation, *571,* 572
    in pituitary control, 509
    pituitary regulation by, 509, 619,
        619*t,* 621–622
    in sexual response, 510
    in sleep, 509, 517
    in stress response, 642
    in taste, 574
    in thermoregulation, 509, 799,
        1001
    in thirst and fluid balance, 904,
        904
    visceral functions of, 509–510
Hypothenar eminence, 343*t, 377,*
        378
Hypothermia, 999, 1002, 1097
Hypothesis, 7
Hypothetico-deductive method, 7
Hypothyroidism
    congenital, 645, 648*t*
    endemic goiter, 645, *646*
Hypotonic hydration, 907, 907*t*
Hypotonic solutions, 89
Hypoventilation, 847*t*
Hypovolemia, 905–906, 907*t,* 920*t*
Hypovolemic shock, 745
Hypoxemia, 187, 664, *664,* 667, 716,
        738
Hypoxemic hypoxia, 858
Hypoxia, 666, 747, 858
Hypoxic drive, 857
Hysterectomy, 1043

I

I band(s), *387,* 389, *389*
Ibn Sina (Avicenna), 4
Ibuprofen, 644, 799

Ileocecal junction, 953
Ileocecal valve, 953, *964*
Ileocolic lymph nodes, *792*
Ileum, *952,* 953, *964*
Iliotibial tract, *307, 345, 347, 348,*
        379, 380
Image formation
    defects of, 599*t*
    description of, 597–599
Imaging, 2, *22,* 22–23, 538
Immediate hypersensitivity, 815
Immediate memory, 452
Immune complex hypersensitivity,
        816
Immune serum, 804
Immune system, 783, 795–821.
        See also Immunity
    aging and, 1100
    effects on other systems, 821
    first line of defense in, 795
    mind–body connection in, 820
    second line of defense in, 795
    self-tolerance in, 790, 816–817
    third line of defense in, 795
Immune system disorders,
        815–819
    anaphylactic shock, 746, 804,
        815
    asthma, 815, 816, 845
    autoimmune diseases, 645, 807,
        816–817
    contact dermatitis, 819*t*
    hives (urticaria), 819*t*
    Hodgkin disease, 819*t*
    hypersensitivity, 815–816
    immunodeficiency diseases,
        817–819
    splenomegaly, 766, 819*t*
    systemic lupus erythematosus,
        819*t*
Immunity
    active, 804
    adaptive, 795, 803–814, 814*t*
    artificial, 804
    attack phase of, 808, 811
    cellular (cell-mediated), 803,
        808–810, 815*t*
    connective tissue and, 148
    costimulation in, 809
    humoral (antibody-mediated),
        804, *810,* 811–814, 815*t*
    inflammatory response in,
        800–803
    innate, 795–803
    lymphatic system and, 783
    memory in, 803, 810, 811–812,
        812
    natural, 804
    neonatal, 1091
    passive, 804
    recognition phase of, 808–810,
        811, *812*
    skin and, 175, 795–796
    three *R*s of, 808

Immunocompetent lymphocytes,
        789, 806, *806,* 811
Immunodeficiency diseases,
        817–819
Immunoglobulin(s) (Ig), 130,
        804–806. *See also*
        Antibodies
Immunoglobulin A (IgA), 805, 805*t,*
        934
Immunoglobulin D (IgD), 805, 805*t*
Immunoglobulin E (IgE), 805, 805*t,*
        815
Immunoglobulin G (IgG), 805, 805*t,*
        816, 1091
Immunoglobulin M (IgM), 805,
        805*t,* 816
Immunosuppressants, 194
Impacted teeth, 933
Implantation, 1080, *1081*
Implicit memory, 452, 519
Impotence, 1023, 1031, 1032*t*
In vitro fertilization (IVF), 167,
        1104–1105
Incisional biopsy, 147
Incisors, 932, *932*
Inclusions, cellular, 78, 104, 105*t*
Incomplete antigens, 804
Incomplete dominance, 130
Incomplete proteins, 983
Incomplete tetanus, 402, *403*
Incontinence, urinary, 893,
        895*t,* 1100
Incubation period, 1033
Incus, 238, 579, *579*
Indirect attachment, of muscle,
        303–304, *304*
Inductive method, 7
Inessential amino acids, 983
Infant. *See* Neonate
Infant respiratory distress syndrome,
        845, 1092
Infarction
    cerebral, 169
    myocardial, 169, 684, 699, 705
    posttraumatic, 488
Infection(s). *See also specific*
        infection
    congenital, 1093
    febrile response to, 798–800
    opportunistic
        with cancer, 133
        with HIV/AIDS, 818, *818*
        with leukemia, 677
    urinary tract, 893
Infectious hepatitis, 996
Infectious mononucleosis, 684*t*
Infective endocarditis, 718*t*
Inferior, 29, 29*t*
Infertility
    male, 1027, 1032*t*
    reproductive technology and,
        167, 1104–1105
Inflammation, 800–803
    agents of, 801*t*

cardinal signs of, 800, 802
complement system in, 797
definition of, 800
mobilization of defenses in,
        801–802
neutrophil behavior in, 801–803,
        802
nomenclature for, 800
tissue cleanup and repair in, 803
Inflation reflex, 841
Infrared radiation, 591, 999
Infrasonic frequencies, 578
Infundibulum
    of pituitary, *504,* 617
    of uterine tubes, 1040, *1041*
Ingestion, 926
Inguinal hernia, 328, *328*
Inguinal lymph nodes, *784,* 792, *792*
Inguinal region, 30, 31, *31*
Inguinal ring
    external, *1014,* 1015
    internal, 1015
Inhibin, 629, 632*t,* 633*t,* 1015,
        1021–1022, 1049, 1054
Inhibiting hormones, 619
Inhibition
    negative feedback, 622, *622*
    presynaptic, 448, *449*
    reciprocal, 484
Inhibitory GABA-ergic synapse, 444
Inhibitory neurotransmitters,
        434–435, 444, *445,* 447
Inhibitory postsynaptic potentials
        (IPSP), *446,* 447
Inhibitory potentials, 434–435
Initial segment, of axons, *424,* 429
Injuries
    athletic, 358
    back, 326
    brain, 519
    dural sinus or jugular vein, 755
    knee, 292, *292*
    nerve, 481
    spinal cord, 461, 488, 894
Innate immunity, 795–803
    antimicrobial proteins in,
        797–798
    barrier defenses in, 795–796
    characteristics of, 795
    complement system in, 797, *798*
    fever in, 798–800
    inflammation in, 800–803
    leukocytes in, 796–797
    local effect of, 795
    lymphocytes in, 796, *810*
    nonspecific mechanisms of, 795
Inner cell mass, 166
Inner ear
    anatomy of, *579,* 580–582
    equilibrium function of, 587–591
    hearing function of, *583,*
        583–585, *584*
Inner hair cells of ear, 581, *582,* 585
Inner layer, of eye, 594

Inner segment, of photoreceptors, 600, *600, 601*
Inner-ear cavity, 227
Innervation, 305. *See also specific nerves and structures*
Inositol triphosphate (IP₃), 638, *640*
Inotropic agents, 717–718
Insensible perspiration, 189
Insensible water loss, 904
Insertion, of muscle, 303
Inspection, 2
Inspiration, 837, 843, *844*
Inspiratory capacity, *846,* 846*t,* 847
Inspiratory (I) neurons, 839, *840*
Inspiratory reserve volume (IRV), *846,* 846*t,* 847
Insula, *500,* 511, *511,* 512, *515,* 521
  in olfaction, 576, *577*
  in taste, 574
Insulin, 629, 631*t,* 647–650, 950
  in absorptive state, 997
  aging and, 1099
  antagonism with glucagon, 641, *641*
  appetite regulation by, 977, *978*
  in bone physiology, 215*t*
  chemical classification of, 633*t,* 634, *634*
  discovery of, 649–650
  genetically engineered, 650
  hypersecretion of, 648*t*
  mode of action, 641
  modulation of target-cell sensitivity, 641
  in postabsorptive state, 998
  posttranslational modification of, 120
  receptor-mediated endocytosis of, 93
  structure of, *65*
  synthesis of, 120, 634, *635*
Insulin inhaler, 647
Insulin receptors, 637
Insulin resistance, 647
Insulin therapy, 647–650
Insulin-like growth factors (IGFs), 616*t,* 622–623, 629, 632*t*
Integrase inhibitors, 818
Integrating (control) center, 16–17, *482,* 483, 544
Integration, neural, 446–453, 460
Integrative functions of brain, 516–526
  cerebral lateralization and, 526, *526*
  cognition, 519
  electroencephalogram and, 516, *517, 518*
  emotion, 520
  language, 525–526
  memory, 519–520
  motor control, 522–525
  sensation, 520–522
  sleep, 517–519

Integument. *See* Skin
Integumentary system, 35, *36,* 174–197. *See also* Hair; Nail; Skin
  aging and, *1097,* 1097–1098
  components of, 175
  disorders of, 190–192, 193*t*
  effects on other organ systems, 195
  excretion by, 869
  parasympathetic effects on, 555*t*
  pregnancy and, 1062
  sympathetic effects on, 555*t*
Intercalated cells, of distal convoluted tubule, 884
Intercalated discs, 158, *159,* 159*t,* 409, *700,* 700–701
Intercellular clefts, 729, *730, 741*
Intercellular face, 78*t*
Intercondylar eminence, 258, *259*
Intercostal trunk, 786
Intercourse. *See* Sexual intercourse
Interdigitating folds, 701
Interferon(s), 790, 797, 801, 810
Interleukin(s), 790, 807, *809,* 814*t,* 820
Intermaxillary suture, *226, 229,* 236
Intermediate cuneiform, 259, *260*
Intermediate fibers, 408
Intermediate filaments, *79,* 97, *98,* 105*t*
Intermediates, metabolic, 67
Internal os, 1040, *1041,* 1042
Internal rotation, *277,* 281, *281*
Internal tension, muscle, 401
Interneurons, 423, *423*
Interoceptors, 566
Interosseous margin, *251,* 252
Interosseous membrane (IM), *251, 252,* 271
Interpubic disc, *254,* 254–255, *271*
Interscapular region, *30*
Intersegmental reflex arc, 486
Interstitial cell-stimulating hormone (ICSH), 1021. *See also* Luteinizing hormone
Interstitial endocrine cells, 629, *630,* 1015, *1016*
Interstitial fluid, 78, 139, 902, *903*
Interstitial growth of cartilage, 210
Interstitial lamellae, 203
Intertrochanteric line, 256, *258*
Intertubercular line, 31, *31*
Intervertebral discs, 240, 242, *242, 243,* 271, *271,* 365
  aging and, 1098
  herniated, 242, *242*
  structure of, 242, *242*
Intestinal crypts, *954,* 955, 965
Intestinal gas, 965
Intestinal gastrin, 945
Intestinal juice, 955
Intestinal lymph nodes, *784,* 792, *792*

Intestinal phase, of gastric function, *944,* 944–945
Intestinal trunks, *784, 786, 787*
Intestines. *See* Large intestine; Small intestine
Intracellular face, of plasma membrane, 79, *80*
Intracellular fluid (ICF), 78, 902
  electrolyte concentration in, 908, *908*
  membrane polarization/potential and, 393–394, 433–434
  tonicity and, 89
Intracytoplasmic sperm injection (ICSI), 1104, *1104*
Intrafusal muscle fibers, *483,* 484
Intramembranous ossification, 206–207, *207,* 210
Intramuscular (I.M.) fluids, 921
Intraperitoneal organs, 33, 929
Intrapleural pressure, 843, *844*
Intrapulmonary pressure, 843, *844*
Intrauterine device (IUD), *1069,* 1070, 1070*t*
Intravenous (I.V.) fluids, 921
Intrinsic aging, 1097
Intrinsic defecation reflex, 966
Intrinsic factor, 666, 941, 942*t,* 961
Intrinsic mechanism, of coagulation, *681,* 681–682
Intrinsic muscles, 304
  of foot, *356, 357*
  of hand, 304, 343*t*
  of larynx, 831, *831*
  of tongue, 931, *931*
Intromission, 1029
Introns, 117, *117*
Inulin, 891
Inversion, of foot, 285, *285*
Involuntary muscle, 158, 386, 409
Involution, 623
  of adrenal cortex, 627
  of corpus luteum, 1054, *1054*
  of pineal gland, 623
  of thymus, 623, 789
  of uterus, 1066
Iodine, 41*t,* 984*t*
Ion(s), *44,* 44–45, 908. *See also specific ions*
  formation of, 44, 45
  in membrane polarization/ potentials, 393–394, 433–434
  in muscle excitation, 394
  valence of, 44
Ion channels, 81. *See also specific types*
Ion trapping, 52
Ionic bonds, 46, 46*t*
Ionization, 44, *44*
Ionizing radiation, 43–44
IP₃. *See* Inositol triphosphate
Ipsilateral, 29, 29*t,* 318*t,* 465
Ipsilateral reflex arc, 485

Iris, *591,* 593–594, *594*
  dual innervation of, 556
Iron
  absorption of, 961
  body percentage of, 41*t,* 42
  deficiency of, 187
  dietary, 984, 984*t*
  metabolism of, 663, *664*
  in pregnancy, 1062
Iron ions, 44
Iron-deficiency anemia, 666
Iron–sulfur (Fe-S) centers, 989
Irregular bones, 200
Irreversible action potentials, 436
Irritability, 14
Irritant receptors, 841–842
Ischemia
  cardiac, 699
  in compartment syndrome, 303
  in spinal cord trauma, 488
Ischemic hypoxia, 858
Ischemic phase, of menstrual cycle, 1056
Ischiadic notch. *See* Sciatic notch
Islets of Langerhans. *See* Pancreatic islets
Isomers, 45
Isometric contractions, 402, *403, 404, 404*
Isotonic contractions, 402, *403,* 404, *404*
Isotonic solutions, 89
Isotopes, 42–44, *43*
Isotretinoin, 1093
Isotropic (I) bands, *387,* 389, *389*
Isovolumetric contraction, 710, *711*
Isovolumetric relaxation, *711,* 712
Isthmus
  of thyroid gland, 624, *625*
  of uterine tubes, *1041*
IUD (intrauterine device), *1069,* 1070, 1070*t*
IVF. *See In vitro* fertilization
Ivy method, 683

J

James, Jesse, 103
Jaundice, 182, 665, 996, 1067, 1092
Jejunum, *929, 952,* 953
Jewish medicine, 4
Johnson, William, 1028, 1058
Joint(s), generally, 267–298
  aging and, 1098
  artificial, 294, *295*
  ball-and-socket, 256, 277, *277, 278*
  biaxial, 277
  bony, 269
  capsule of, 272, *272,* 275
  cartilaginous, *271,* 271–272
  classification of, 268–269
  condylar, 277, *278*
  definition of, 268

double-jointed, 275
ellipsoid, 277, *278*
fibrous, *269*, 269–271
    gomphoses, *269*, 269–270
    sutures. *See* Sutures
    syndesmosis, 269, *269*,
        270–271
flexibility of, 268, *268*
gliding, 277, *278*
hinge, 277, *278*
as lever systems, *274*, 274–276,
    276
monaxial, 277
multiaxial, 277
naming of, 268
pivot, 277, *278*
plane, 277, *278*
replacement of, *267*, 294–295,
    295
saddle, 277, *278*
study of, 268
synovial, 272–285
    accessory structures of, 273
    aging and, 1098
    axes of rotation, 275–277,
        277
    classes of, 277–278, *278*
    exercise and, 273
    general anatomy of, *272*,
        272–273
    lever systems and, *274*,
        274–276, *276*
    movements of, 279–285
    range of motion, 275
Joint(s), specific
    acromioclavicular, *247*, 249, 272
    ankle (talocrural), 275, 284–285,
        292–293, *293*
    atlantoaxial, *243*, 244, 277
    atlanto–occipital, 244, 268
    elbow, 277, *288*, 288–289, 358
    glenohumeral (humeroscapular),
        249, 268, 277, *278*,
        286–288, *287*
    hip, 275, *289*, 289–290
        ligaments supporting, *289*,
            290
        muscles acting on, 344, *344*,
            345, 346*t*–347*t*
        total replacement, 294–295,
            295
    humeroradial, *288*, 288
    humeroulnar, *278*, *288*, 288
    intercarpal, *278*
    interphalangeal, 275, 277, 339*t*,
        377, *378*
        muscles acting on, 339*t*
        range of motion, 275
        surface anatomy of, *377*, *378*
    knee (tibiofemoral), *278*,
        290–292, *291*, *299*
        adaptation to bipedalism, 291
        artificial, *295*
        bursae of, 292, *299*

fibrocartilage of, 272–273
injury to, 292, *292*
menisci of, 273, 290, *291*,
    292, *292*, *299*
muscles acting on, *347*, *348*,
    349*t*
range of motion, 275
surface anatomy of, *379*
metacarpophalangeal, 275, 277,
    278, 339*t*, *377*, *378*
    muscles acting on, 339*t*
    surface anatomy of, *377*, *378*
patellofemoral, 290
proximal radioulnar, *288*, 289
radiocarpal, 277
radioulnar, 268, 277, *278*, *288*,
    289
sacroiliac, 245–246, 254, *254*
sternoclavicular, *246*, 249, 272,
    277, *364*
temporomandibular (TMJ), 238,
    272, 285–286
trapeziometacarpal, 277, *278*
Joint cavity, 272, *272*
    glenoid, 250, *250*
    synovial, 272, *272*
Joint disorders, 294*t*
    aging and, 1098
    arthritis, 294*t*, 294–295, *295*,
        1098
    bursitis, 294*t*
    dislocation, 287, 294*t*
    gout, 294*t*
    osteoarthritis, 294, 1098
    rheumatism, 294*t*
    rheumatoid arthritis, 294, *295*
    sprain, 293, 294*t*
    strain, 294*t*
    synovitis, 294*t*
    tendinitis, 294*t*
Jugular notch, 246, *364*, *371*
Jugular trunks, 786, *787*
Junctional folds, 393
Junctional (nodal) rhythm, 702
Juxtaglomerular apparatus, 878–879,
    879
Juxtaglomerular cells, 879, *879*
Juxtamedullary nephrons, *872*, *874*,
    875

K

Kallikrein, 683, *683*
Kaposi sarcoma, 818, *818*
Karyotype, 128, *128*
Kearns–Sayre syndrome (KSS), 103
Kell blood group, 670, 672
Keratin, 63, 97, 142, 175, 183, 795
Keratinocytes, 177, *177*, 178–179,
    213, 629, 796
Keratohyalin granules, 178
Kernicterus, 671
Keto acid, 994, *995*
Ketoacidosis, 648, 857

Ketogenesis, 994
Ketone(s), diabetes mellitus and, 647
Ketone bodies, 857, 994
Ketonuria, 647, 648
Kidd blood group, 670, 672
Kidney(s), *369*, *374*, *868*, 869–875
    aging and, 1100
    anatomical variations in, 13, *14*
    autoregulation of, 878–879
    blood pressure and, 739, 776,
        875, 878–880, 882–883
    blood supply of, 870–871, *871*,
        872
    diabetes mellitus and, 647, 648
    in erythrocyte homeostasis, 664
    evolution of, 885
    filtration in, 875–881
    fluid excess and, 907, *907*
    fluid flow in, 873, *874*
    function tests of, 890
    functional unit of, 871–875
    functions of, 867
        endocrine, 161, 630, 867
        exocrine, 161
    gross anatomy of, 869–870, *870*
    hormones affecting, 888*t*
    innervation of, *547*, *551*, 875,
        879, 879–880
    neonatal, 1091
    perineal fat of, *374*
    pH (acid–base) control by,
        916–917, *917*, 919
    position of, 869, *869*
    pregnancy and, 1062
    reabsorption in, *876*, 881–885,
        882, 888*t*, 905
    renin–angiotensin–aldosterone
        mechanism and, 879–880,
        880, 884, 909
    renin–angiotensin–aldosterone
        system and, 627
    water conservation in, *876*,
        885–888, 905
    weight and dimensions of, 869
Kidney disease
    acute glomerulonephritis, 895*t*
    anemia in, 665–666
    hydronephrosis, 895*t*
    incontinence, 893, 895*t*
    kidney stones, 892
    nephroptosis, 895*t*
    nephrotic syndrome, 895*t*
    renal failure, 896, 1103
        acute, 895*t*
        chronic, 895*t*
        heart failure and, 712
    renal insufficiency, 896
Kidney stones, 892
Killer T cells, 807, 808–810, *809*,
    809*t*, *810*, 814*t*
Killing zone, 796
Kilocalorie (kcal), 977
Kinases, 67, 82, *83*
    creatine, 65, 405, *405*

cyclin-dependent, 127
hexokinase, 986, *993*
myokinase, 405, *405*
myosin light-chain, 413
protein, 452, 638
streptokinase, 685
thrombokinase (factor X), *681*,
    682, 682*t*
Kinesin, *98*, 426
Kinesiology, 268, 279
Kinetic energy, 52
Kinetochore, 114, *114*, 125, *126*
Kinins, 801*t*
Kinocilium, 588
Kissing bugs, 553
Klinefelter syndrome, 1094
Knee, 278, 290–292, *291*, *299*
    adaptation to bipedalism, 291
    artificial, *267*, 295
    bursae of, 292, *299*
    fibrocartilage of, 273
    injury to, 292, *292*
    menisci of, 273, 290, *291*, 292,
        292, *299*
    muscles acting on, *347*, *348*, 349*t*
    range of motion, 275
    surface anatomy of, *379*
Knee-jerk reflex, 484, *485*
Krebs cycle, 988–989, *989*
Kussmaul respiration, 648, 847*t*, 857
Kwashiorkor, 660, *660*
Kyphosis, 240

L

L cones, 605, *606*
Labeled line code, 448
Labia majora
    anatomy of, *375*, *1039*, 1044,
        1045
    development of, 1010, *1012*,
        1038
Labia minora
    anatomy of, *375*, *1039*, 1044,
        1045
    development of, 1010, *1012*,
        1038
Labial frenulum, 930, *930*
Labial glands, 930
Labial mucosa, 930
Labioscrotal folds, 1010, *1012*
Labor
    contractions of, 1064
    pain of, 1064
    positive feedback theory of, 1064
    stages of, 1064, *1065*
Labyrinth
    of ear, 580, *581*
    of ethmoid bone, 235
Lacrimal apparatus, 591–592
Lacrimal caruncle, *591*
Lacrimal glands, *551*, 552, 591–592,
    592
Lacrimal punctum, 592, *592*

Lacrimal sac, 237, 592, *592*
Lactase, 957
Lactate dehydrogenase (LDH), 65
Lactate threshold, 406
Lactation (breastfeeding), 1045, 1066–1068
  colostrum in, 1066–1067, 1067*t*, 1091
  hormones of, 622, 1066–1067, *1067*
  maternal metabolism and nutrition in, 1068
  milk composition in, 1067*t*, 1067–1068
  milk ejection in, 1045, 1067
  milk synthesis in, 1066–1067
  uterine involution promoted by, 1066
Lacteals, 783, 953, *954, 962*
Lactic acid fermentation, 68
Lactiferous ducts, *1046*, 1047
Lactiferous sinus, *1046*, 1047
Lactose, 56, *57*, 58*t*, 957, *962*, 979
Lactose intolerance, 957
Lactotropes, 620
Lacunae
  of bone, *156*, 201
  of cartilage, 154
  of conceptus, *1081*, 1087
  of penis, 1020, *1020*
Lamella(e)
  circumferential, 203, *204*, 210
  concentric, 154, *156*, 203, *204*
  interstitial, 203
Lamellar corpuscles, 182, 272, *567*, 567*t*, 568
Lamina, 242, *242, 243*
  of cervical vertebra, *243, 244*
  of thoracic vertebra, *244*
Lamina propria, 163, *165, 411*, 927, *928*
  of large intestine, 965
  of small intestine, 953
  of stomach, 937, *939*
  of uterus, 1042, *1044*
Laminar flow, of blood, 662, 736, *736*
Laminectomy, decompression, 488
Landsteiner, Karl, 667
Langerhans, islets of. *See* Pancreatic islets
Language, 508, *525*, 525–526, 538, *538*
Lanugo, 184, 1086*t*
Large arteries, 726–727, *727*
Large intestine, *367, 368, 374*, 926, 963–967
  absorption in, 966
  circulation of, 965
  endocrine function of, 630–631
  gross anatomy of, 963–965, *964*
  innervation of, 927
  microbes and gas in, 965
  microscopic anatomy of, 965

  motility of, 966
  sympathetic innervation of, *547*
Large motor units, 391, *391*, 408
Large subunit of ribosomes, 118
Large veins, *727*, 732
Laron dwarfism, 637
Laryngeal prominence, 830, *830*
Laryngopharynx, *828*, 829
Larynx, *365*, 827, *828*, 829–831, *830, 832*
  cartilages of, *830*, 830–831, *832*
  ligaments of, 830, *830*
  muscles of, 316*t*, 829, 831, *831*
Laser angioplasty, 719
Last menstrual period (LMP), 1059
Latch-bridge mechanism, 413
Latent period, of muscle twitch, 401, *401*
Lateral, 29, 29*t*
Lateral apertures, *500, 501, 501*
Lateral excursion, of mandible, 283, *284*
Lateral (external) rotation, *281*, 281–282
Lateral flexion, 282, *283*
Lateral longitudinal arch of foot, 261, *261, 381, 382*
Lateral mass, of cervical vertebra, *243*, 244
Lateralization, cerebral, 526, *526*
Laughing gas, 608
Law(s), 8–9
Law of complementary base pairing, 8, 111, 123
Law of mass action, 53
Law of nature, 8
LDL. *See* Low-density lipoproteins
Leader sequence, in translation, 118
Leaflets, of heart valves, 696
Leak channels, 81
Learning strategy
  for cranial nerves, 532–533
  for muscles, 305, 308, 363
  for regional and surface anatomy, 363
Leber hereditary optic neuropathy, 103
Lecithin, 61, *61*, 949, *960*
Lectin(s), 797
Lectin pathway, of complement system, 797, *798*
Leeches, 685, *685*
LEEP procedure, 1043
Leeuwenhoek, Antony van, 5–6, 725
Left arteries, 726–727, *727*
Left atrioventricular valve, *693*, 696, *696*, 697, *697*
Left atrium, *693, 694, 694, 695*, 697, *697*
Left bundle branch, 702
Left lower quadrant (LLQ), 31, *31*
Left rotation, 282, *283*
Left upper quadrant (LUQ), 31, *31*
Left ventricle, *693, 694*, 695, *695*, 697, *697*

Left ventricular failure, 712
Left-brained, 526
Leg, 31, 344. *See also* Lower limb
  compartments of, 354*t*, *355*
  muscles acting on, *347, 348*, 349*t*
  surface anatomy of, *380–381*
Leg bud, *1084*
Leiomyoma, 1068*t*
Lemniscus, medial, *467*, 503, *505*
Length–tension relationship, 326, 385, 400, *400*, 408
Lens, *594, 595, 596*
  accommodation of, *598*, 598–599
Lens fibers, 595
Leptin, 631, 632*t*, 633*t*, 976–977, *978*, 1048
Lesions, 190
Lesser curvature, of stomach, 937, *938*
Lesser omentum, 34, *34*, 928, *929*, 937, *938*
Lesser pelvis, 255
Lesser sciatic notch, 255, *255*
Lesser trochanter, 256, *258*
Lesser vestibular glands, 1045
Lesser wing, of sphenoid bone, 234, *234, 236*
Lethal hit, in immunity, 810
Leukemia, 132, 676, *678*
Leukocyte(s), 148, 154, *156, 157*, 655, 657, 672–678. *See also specific types*
  aging and, 1100
  diapedesis of, 802
  disorders of, 676–678
  functions of, 672
  in inflammation, 801–803
  in innate immune, 796–797
  life history of, 676, *677*
  numbers of, 672, 676
  production of, 660, 676, *677*
  structure of, 672, *673*
  types of, 657, *657*, 672–676, 674*t*–675*t*
Leukocytosis, 676
Leukopenia, 676
Leukoplakia, 818
Leukopoiesis, 676, *677*
Leukotrienes, 643, *643*, 796, 801*t*, 802
Lever systems, *274*, 274–276, *276*
Levi-Montalcini, Paola, 432
Levi-Montalcini, Rita, 432
Levodopa, for Parkinson disease, 454
Lewis blood group, 670
Libido, 1021
Life
  characteristics of, 13–15
  clinical and legal criteria of, 15
Life expectancy, 1101, 1103
Life span, 1103
Lifting, and back injury, 326
Ligaments, generally, 199, *272*, 273
  collagenous fibers of, 149, *149*

  function of, 148
  retinaculum, 303
  tautness of, 275
Ligaments, specific
  anular, *288*, 289
  broad, 1039, *1041*, 1042
  calcaneofibular, *293*
  cardinal (lateral cervical), *1041*, 1042
  coccygeal, 463
  coracohumeral, *287*, 288
  cricotracheal, 830, *830*
  cruciate
    anterior, 290–292, *291, 292, 292, 299*
    posterior, 290, *299*
  denticulate, *462*, 463
  extensor retinaculum, *350*
  extracapsular, of knee, 290
  falciform, 946, *947*
  flexor retinaculum, *341, 342*
  glenohumeral, *287*, 288
  iliofemoral, *289*, 290
  inguinal, *321, 322, 366, 371*
  intracapsular, of knee, 290
  ischiofemoral, *289*, 290
  laryngeal
    extrinsic, 830, *830*
    intrinsic, 830
  lateral, TMJ, 286, *286*
  lateral collateral, 293, *293*
    of ankle, 293, *293*
    fibular, 290, *299*
    radial, *288*, 289
  ligamentum arteriosum, *694*, *1090*, 1091
  ligamentum teres, *289*, 290, 946, 1091
  ligamentum venosum, *1090*
  medial collateral
    tibial, 290, *292, 299*
    ulnar, *288*, 289
  medial (deltoid), 293, *293*
  median umbilical, *1090*, 1091
  nuchal, 234, 244–245
  ovarian, 1039, *1040*, 1042
  patellar, 256, 290, *291, 292, 292, 299, 348, 379, 380*
  periodontal, 270, 932, *932*
  plantar, long, *293*
  posterior cruciate, 290, *299*
  posterior talofibular, *293*
  posterior tibiofibular, 293, *293*
  pubofemoral, *289*, 290
  pulmonary, *834*, 837
  round
    of hip, *289*, 290
    of liver, 946, *1090*, 1091
    of ovary, *1039*
    of uterus, *1041*, 1042
  sphenomandibular, 286, *286*
  stylohyoid, 238
  stylomandibular, 286, *286*
  suspensory

of breast, 1045, *1046*
of eye, *594, 595, 596*
of ovary, 1039, *1041, 1044*
talofibular, *293*
talonavicular, *293*
dorsal, *293*
thyrohyoid, 830, *830*
tibiofibular, 293, *293*
transverse
acetabular, *289,* 290
of cervical vertebra, *243,* 244
humeral, *287,* 288
of knee, 290
uterosacral, *1041,* 1042
vestibular, 830
vocal, 831
Ligand(s), 63, 90
Ligand-gated channels, 81
Ligation, tubal, 1070, 1070*t*
Light
definition of, 591
refraction of, 597, *597*
visible, 591
and vision, 591
Light adaptation, in eyes, 602–603
Light chains, of antibody, 804, *805*
Light microscopes, 77, *77–78,* 78*t*
Light therapy, 624
for hemolytic disease of newborn, 671
for melatonin/mood disorders, 624
Limb rotation, embryonic, 260, *260*
Limbic system, 509, *514,* 514–515
aversion center of, 515
in emotion, 514–515, 520
functions of, 514–515
gratification centers of, 515
and memory, 519–520
Limiting pH, 916
Line(s) of bone, *226,* 226*t*
epiphyseal, 200, *200, 207,* 210
intertrochanteric, 256, *258*
intertubercular, 31, *31*
linea aspera, 256, *258*
midclavicular, 31, *31*
nuchal, *229,* 234
spiral (pectineal), *255,* 256, *258*
subcostal, 31, *31*
supracondylar, 256, *258*
temporal, *228,* 232
Linea alba, *321,* 322*t, 371*
Linea nigra, 1062
Linea semilunaris, *321, 371*
Linear acceleration, 588
Lingual frenulum, *930,* 931, *934*
Lingual glands, 931, 934
Lingual lipase, 934–935, 940–941, 958, *962*
Lingual papillae, 572, *573,* 931
Lingual tonsils, 793, *793, 828,* 931, *931*
Linoleic acid, *59,* 62, 980
Linolenic acid, 981

Lip(s), *364,* 930, *930*
cleft, 237
cutaneous area of, 930
functions of, 930
mucosa of, 930
muscles acting on, 311
red area (vermilion) of, 930
Lipase
gastric, 941, 942*t,* 962
lingual, 934–935, 940–941, 958, *962*
lipoprotein, 981, *982*
pancreatic, 950, 951*t,* 958, *960, 962*
Lipid(s), 55, 58–62
absorption of, 783
absorptive state, 997
composition of, 58
dietary, 977, 979*t,* 980–982
requirements for, 979*t,* 980–981
sources of, 981
digestion of, 943, 950, 958, *960, 962*
emulsification of, 958, *960*
functions of, 58*t*
growth hormone and, 623
metabolism of, 993–994, *994,* 995*t*
in plasma membrane, *80,* 80–81
postabsorptive state, 998
types of, 58
Lipid bilayer, 79
Lipid rafts, 147
Lipocalin-2, 629
Lipofuscin, 423, 1099
Lipogenesis, 993, *994*
Lipolysis, 977, 993, 994, *994*
Lipoprotein(s), 62, 981, *981–982*
high-density, 62, 981, *981–982*
low-density, 62, 93, 719, 981, *981–982*
very-low-density, 981–982, *981–982*
Lipoprotein lipase, 981, *982*
Lipoxygenase, 643, *643*
Lithotripsy, 892
Liver, *367, 374,* 926, *927, 929,* 946–947
aging and, 1101
bile secretion by, 949–950
circulation in, 947
development of, *1083, 1084,* 1089
endocrine function of, 614–615, 629–630, 632*t,* 643, 961
exocrine function of, 614–615
as exocrine gland, 161
functions of, 995, 995*t*–996*t*
gross anatomy of, 946, *947*
lobes of, 946, *947*
metabolic role of, 995, 995*t*–996*t*
microscopic anatomy of, 946–947, *948*

neonatal, 1091
parasympathetic innervation of, *551*
plasma production in, 659
sinusoids of, 729, *730, 765,* 947, *948*
sympathetic innervation of, *547*
Liver bud, *1083*
Liver bulge, *1084,* 1089
Liver disease
cirrhosis, 996
clotting deficiencies in, 684
hepatitis, 996, 1033
portal hypertension, 766
Lobar bronchi, *827, 832, 834*
Lobes
of brain, *494, 496, 500, 511,* 511–512. (*See also specific lobe*)
of breast, *1046*
of glands, 162
of kidney, 870
of liver, 946, *947*
of lung, 368, 833, *834*
of pituitary gland
anterior, 617, *618*
posterior, *618,* 619
Lobules
of breast, *1046*
of ear, *578–579*
of glands, 162
of liver, 946, *948*
of lung, 833
of testis, *1016*
of thymus, 789, *790*
Local anesthetics, 608
Local potentials, 434–435, 435*t*
Location, sensory information on, 565, *565*
Lochia, 1066
Lockjaw (tetanus), 393
Locomotion
bipedalism and, 11, 262–263, 291
spinal cord in, 460
Locus of gene, 128
Loewi, Carl, 441
Long, Crawford W., 608
Long association fibers, 513
Long bones, 199, *200*
Long limb, of incus, 579
Longitudinal arches, of foot, 261, *261, 380, 381, 382*
Longitudinal fascicles, *505*
Longitudinal section, 140, *140*
Long-term depression (LTD), 453
Long-term memory, 452
Long-term potentiation (LTP), 452
Long-wavelength cones, 605, *606*
Loop diuretics, 890
Loose connective tissue, 149, *151*
Lou Gehrig's disease. *See* Amyotrophic lateral sclerosis

Loudness, 578, 585
Low venous return (LVR) shock, 745
Low-density lipoproteins (LDLs), 62, 93, 719, 981, *981–982*
Lower limb, *30,* 31. *See also* Leg
adaptations to bipedalism, 262–263, *262–263*
arteries of, *771, 772*
bones of, 224*t, 225,* 256–263
muscles acting on, 344–358
surface anatomy of, *379–382*
veins of, *773, 774*
Lower motor neurons, 466, *468,* 523
Lower respiratory tract, 827, *832*
Lubb-dupp, 710
Lumbar cistern, 462
Lumbar curvature, *240,* 241
Lumbar enlargement, *461*
Lumbar puncture, 463, 464, *464,* 499
Lumbar region, *30,* 31, *31, 299,* 460
Lumbar trunks, 786, *787*
Lumbosacral enlargement, 460
Lumen
of digestive tract, 927, *928*
of esophagus, 935
of intestines, 950, 953, 957, *957, 962,* 963
of mammary gland, 1047
of stomach, *939, 940, 941*
of uterus, 35
Lumpectomy, 1048
Lung(s), *367, 369, 373, 374,* 827, 833–837, *836*
aging and, 1100
airflow path in, 835–836
alveoli of, 690, 836–837
anatomy of, 833–837, *834*
blood flow to and from, 690, *691,* 747–748, *748*
bronchial tree of, 833–836
compliance of, 845
disorders of, 858–861
histology of, 833–836, *835*
lobes of, *368,* 833, *834*
lobules of, 833
obstructive disorders of, 847
parasympathetic innervation of, *551*
restrictive disorders of, 847
space occupied by, 833
species comparisons of, 836
sympathetic innervation of, *547*
ventilation of, 826, 837–848
volumes and capacities of, *846,* 846*t,* 846–847
Lung cancer, 860, *860*
Lunule, 187, *187*
Luteal phase, of ovarian cycle, 1053–1055, *1054,* 1056*t*
Lutein cells, 1054
Luteinizing hormone (LH), 616*t,* 619–620, 620*t,* 633*t*
in folliculogenesis, 1053–1054
in males, 1021, *1022*

Luteinizing hormone (continued)
　modulation of target-cell
　　sensitivity, 641
　in ovarian cycle, 1053–1054,
　　1055, 1056
　in puberty (female), 1048–1049
　secretion of, 621
Luteotropic hormone. See
　　Luteinizing hormone
Lymph, 785–788, 902, 903
　composition of, 785
　flow of, 787–788, 791
　respiratory function and, 826
Lymph node(s), 789, 790–793
　abdominal, 784, 791
　appendicular, 792
　axillary, 784, 787, 791
　cervical, 784, 791
　functions of, 790
　hematopoiesis in, 660
　ileocolic, 792
　inguinal, 784, 792, 792
　intestinal, 784, 792, 792
　mesenteric, 784, 792, 792
　mesocolic, 792
　metastatic cancer and, 793
　popliteal, 784, 792
　structure of, 790–791, 791
　swollen, 793
　thoracic, 787, 791
Lymphadenitis, 793
Lymphadenopathy, 793
Lymphatic capillaries, 785, 785
Lymphatic cells, 788
Lymphatic ducts, 784, 787
Lymphatic nodules, 788, 788, 791
Lymphatic organs, 785, 789–795.
　　See also specific organs
　lymph nodes, 789, 790–793
　red bone marrow, 783, 784, 789
　spleen, 784, 789, 793–794, 794
　thymus, 784, 789–790, 790
　tonsils, 789, 793, 793
Lymphatic system, 35, 36, 783–795,
　　784
　components of, 785
　effects on other systems, 821
　fluid exchange with circulatory
　　system, 785, 786
　fluid regulation by, 783, 902, 903
　functions of, 783–785
　histology of, 788
　immune function of, 783
　obstruction in, 743
Lymphatic tissue, 660, 785,
　　788–789
Lymphatic trunks, 786
Lymphatic vessels, 784, 785–786
　afferent, 791, 791
　efferent, 791, 791
　embryonic development of, 785
　structure of, 785
　valves of, 785, 785
Lymphedema, 743, 783

Lymphoblasts, 676, 677
Lymphocytes, 148, 156, 157, 657,
　　657, 673–676, 675t, 788,
　　806–807. See also B
　　lymphocytes; T
　　lymphocytes
　in adaptive immunity, 806–807
　in cellular immunity, 808–810
　colony-stimulating factor
　　secretion by, 676
　in humoral immunity, 810,
　　811–814
　immunocompetent, 789, 806,
　　806, 811
　in innate immunity, 796, 810
　life history of, 806, 806–807
　marrow storage of, 676
　naïve pool of, 807, 814t
　production of, 676, 677
Lymphoid hematopoiesis, 660
Lymphoid leukemia, 676
Lymphoid nodules, aggregated, 789,
　　953
Lymphoma, 132
Lysis, febrile, 800
Lysosomal enzymes, 101
Lysosomes, 77, 79, 94, 98, 101, 102,
　　105t, 120, 673
Lysozymes, 592, 796, 935

M

M cones, 605, 606
M line, 388, 389, 389
M phase of cell cycle, 125, 125
Macleod, J. J. R., 649–650
Macromolecules, 12, 12, 55–56
Macronutrients, 977, 979t
Macrophage system, 796–797
Macrophages, 148, 167, 168, 676,
　　788
　in adaptive immunity, 814t
　alveolar, 797, 836, 836
　as antigen-presenting cells, 676,
　　788, 807
　colony-stimulating factor
　　secretion by, 676
　in erythrocyte disposal, 665
　hepatic, 797, 946, 948
　in HIV infection, 818
　in inflammation, 801t, 802
　phagocytosis by, 676, 788, 788,
　　796–797
　pseudopods of, 87
　stellate, 797
　in T-cell selection, 807
Macula, 588
Macula densa, 873, 879, 879
Macula lutea, 594, 596, 596
Macula sacculi, 588, 588–589
Macula utriculi, 588, 588
Magnesium, 912
　body percentage of, 41t, 42
　deficiency of, 913t

dietary, 984, 984t
　excess of, 913t
　functions of, 912
　homeostasis of, 912
　imbalances in, 912
　renal reabsorption of, 881, 912
Magnesium chloride, 45t
Magnetic resonance imaging (MRI),
　　22, 23
　colorized, of human body, 1
　fetal, 1037
　lumbar, pelvic, and femoral
　　region, 299
Maimonides, 4
Main bronchus, 827, 831, 832, 833
Major calyx, 870, 870
Major duodenal papilla, 947, 948
Major histocompatibility complex
　　(MHC), 807, 809, 809
Malaise, 914
Malate, 991
Male(s)
　adult body, 30
　blood values in, 658t, 662
　hemophilia in, 683
　iron metabolism in, 663
　muscles of pelvic floor, 326, 327t
　nutritional requirements of, 977,
　　979t
　pelvic structure of, 256, 257,
　　257t, 375
　spinal cord injury in, 488
　superficial anatomy of, 367–368,
　　371–372
　urethra of, 893, 1017, 1018, 1020
　urinary bladder of, 1018
Male climacteric, 1022
Male condom, 1069, 1069, 1070t
Male reproductive system, 35, 37,
　　1008, 1013–1032
　accessory glands of, 1017–1019
　aging and, 1022–1023
　anatomy of, 1013–1020, 1018
　disorders of, 1032t
　　androgen-insensitivity
　　　syndrome, 1009, 1009
　　breast cancer, male, 1032t
　　cryptorchidism, 1013
　　erectile dysfunction, 1023,
　　　1031, 1032t
　　hypospadias, 1032t
　　infertility, 1032t
　　penile cancer, 1032t
　　prostate diseases, 1019
　　sexually transmitted diseases,
　　　1031, 1033
　　testicular cancer, 1032t
　　varicocele, 1032t
　embryonic development of,
　　1010–1013, 1011, 1012
　hormonal control of, 1021–1022,
　　1022
　physiology of, 1021–1032
　puberty and adolescence, 1021

sperm and semen in, 1023–1028
　spermatogenesis in, 1023, 1025,
　　1025–1027
Male sexual response, 1028–1031
　anatomical foundations of, 1029
　excitement phase of, 1029, 1030
　neural control of, 1030
　orgasm in, 1030, 1031
　plateau phase of, 1029
　refractory period of, 1031
　resolution phase of, 1030, 1031
Male urethra, 893
Malleoli, 382
　lateral, 258, 259, 380, 381, 382
　medial, 258, 259, 381
Malnutrition, in elderly, 1101
Malpighi, Marcello, 725
MALT. See Mucosa-associated
　　lymphoid tissue
Maltase, 956, 957
Maltose, 56, 57, 58t, 956, 957, 962,
　　979
Mammary glands, 162–163, 164,
　　189, 190t, 1045–1048,
　　1066–1068
Mammary ridges, 189
Mammillary body, 496, 504, 508,
　　510, 514, 514
Mammograms, 1047, 1048
Mandibular branch
　of facial nerve, 530, 531
　of trigeminal nerve, 529, 534t
Mandibular notch, 238, 238
Manganese, 41t, 984, 984t
Manometer, 708, 734
Manubrium, 246, 247, 371
Marfan syndrome, 150
Margination, 802
Marrow. See Bone marrow
Marrow cavity, 200, 200
　primary, 207, 208
　secondary, 208
"Mask of pregnancy," 1062
Mass, atomic, 42, 43
Mass action, law of, 53
Mass movements, 966
Mass reflex reaction, 488
Mast cells, 148, 167, 796, 801t
Mastectomy, 1047, 1048
Masters, William, 1028, 1058
Mastication, 930, 933, 936
　aging and, 1101
　appetite and, 977
　cheeks and lips in, 930
　muscles of, 310, 313, 315t–316t,
　　933, 1101
　teeth in, 931–933
　tongue in, 930–931, 933, 936,
　　936
Mastoid notch, 229, 233, 233
Mastoid part, of temporal bone, 233,
　　233
Maternal age, and Down syndrome,
　　1095

Matrix
  of bone tissue, 202–203
  extracellular, 138, 141
  hair, 184, *184*
  of HIV, 817, *817*
  of mitochondria, 103, *103*
  of nail, 187, *187*
Matrix reactions, in aerobic
      respiration, 988–989, *989,*
      992t
Maturation phase, of tissue repair,
      167
Mature ovarian follicle, *1040, 1051,*
      1052–1053, 1056*t*
Maxillary sinus, 227, *230*
Maximum oxygen uptake (VO₂max),
      406
Maximum voluntary ventilation
      (MVV), 847
Mean arterial pressure (MAP), 734,
      878
Mean corpuscular hemoglobin
      (MCH), 678
Mean corpuscular volume (MCV),
      678
Meatus
  acoustic
    external, *228, 229,* 231*t,* 232,
        233, 578, 579
    internal, *228, 229,* 233, *233*
  of bone, *226,* 226*t*
  nasal cavity, *828,* 829
Mechanical advantage (MA), *274,*
      274–275
Mechanical digestion, 926, 933
Mechanical junctions, in cardiac
      muscle, *700,* 701
Mechanical stress, bone adaptation
      to, 205, *205,* 210
Mechanically gated channels, 81
Mechanoreceptors, 566, 568
Meconium, 1067, 1086*t*
Medial, 29, 29*t*
Medial excursion, of mandible, 283,
      284
Medial (internal) rotation, *277,* 281,
      281
Medial lemniscus, *467,* 503, *505*
Medial longitudinal arch of foot,
      261, *261, 380, 381, 382*
Median aperture, *500, 501, 501*
Median forebrain bundle (MFB), 520
Median plane, 28
Mediastinal surface, of lung, 833,
      834
Mediastinum, 32, *32,* 691
Medical art
  evolution of, 4, *4*
  Gray's Anatomy, 5
  Vesalius atlas, 5
Medical history
  anesthesia, 608
  "animal electricity," 401
  blood banking, 668

double helix, 110–112
flight of the Zenith, 849
homeostasis demonstration, 16
insulin, 649–650
man with hole in stomach, 969
nerve growth factor, 432
radiation, 43
radioactive bones, 203
word origins, 20
Medical imaging, 2, *22,* 22–23
Medical terminology, 19–20
Medications. *See* Drugs
Medicine
  evolutionary, 11
  genomic, 110, 115
  modern, birth of, 3–6
  origins of, 3–6
  revolution in, 6
  terminology of, 19–20
Medieval medicine, 3–4
Medina, Lina, 1048
Medium arteries, *727,* 727–728
Medium veins, *727, 732*
Medium-wavelength (M) cones,
      605, *606*
Medroxyprogesterone, 1070, 1070*t*
Medulla
  of adrenal glands. *See* Adrenal
      medulla
  of hair, 184, *184*
  of kidney, 870, *871, 872, 874*
  of lymph node, 790–791, *791*
  of ovary, 1038, *1040*
  of thymus, 789, *790*
Medulla oblongata, *494, 495, 496,*
      503, *504, 508*
  in autonomic control, 558
  in blood pressure control, 738
  cardiac centers of, 503, 507, 713
  in digestion control, 941, 944
  emetic center of, 942
  in hearing, 585
  in pain modulation, 572
  respiratory centers of, 503,
      839–841, *840*
  in taste, 574
  vasomotor center of, 503, 507,
      736, 737–739
Medullary cavity, 200, *200*
Medullary cone, 461, *461*
Medullary cords, of lymph node,
      790–791, *791*
Medullary ischemic reflex, 738
Medullary sinus, of lymph
      node, *791*
Megacolon, 553
Megakaryoblasts, 679
Megakaryocytes, 679, *679*
Megavitamins, 985
Meiosis, 128
  in oogenesis, 1050, *1051*
  in spermatogenesis, 1023–1025,
      1024
Meiosis II, 1078

Melanin, 131, 180–182, *181,* 185,
      185, 423, 594
Melanocortin, 977, *978*
Melanocyte(s), 177, *177,* 180, 190,
      1097
Melanocyte-stimulating hormone
      (MSH), 190
Melanoma, 132, 190–191, *191*
Melanopsin, 601, 607
Melanosomes, 177
Melatonin, 518, 624, 631*t,* 633*t,*
      633–634
  synthesis of, 635
Membrane(s), 31–34, 32*t,* 163, 165,
      165. See also specific
      types
Membrane attack complex (MAC),
      797–798, *799*
Membrane potentials, 93, 157. *See*
      also Action potentials
  local, 434–435, 435*t*
  resting, 93, 394, *433,* 433–434
Membrane proteins, *80, 81,*
      81–82, *82*
Membrane reactions, in aerobic
      respiration, 988, 992*t*
Membrane skeleton (terminal web),
      79, 84, 97, *98*
Membrane transport, 87–96, 96*t*
  active, 87, 91–92, 96*t*
  carrier-mediated, 87, 90–93, 96*t*
  by filtration, 87, 96*t*
  osmolarity and tonicity in, 89–90
  by osmosis, 88–89, *89,* 96*t*
  passive, 87
  by simple diffusion, 87–88, 96*t*
  vesicular, 93–95, 96*t*
Membranous labyrinth, 580, *581*
Membranous organelles, 97
Membranous urethra, 893, 1017,
      1018
Memory, 451–453, 519–520
  aging and, 1099
  Alzheimer disease and, 453, 1099
  brain lesions and, 520
  emotional, 452, 520
  explicit (declarative), 452, 519
  forgetting, 453
  frontal lobe and, 511
  hippocampus and, 520
  hypothalamus and, 510
  immediate, 452
  immune, 803, 810, 811–812, *812*
  implicit, 452, 519
  long-term, 452
  olfaction and, 576
  short-term, 452
  sleep and, 519
  working, 452
Memory B cells, *812,* 814*t*
Memory consolidation, 519
Memory T cells, 808, *809,* 810, 814*t*
Memory trace, 452
Menarche, 1049

Melanin, 131, 180–182, *181,* 185,
Mendel, Gregor, 110
Mendelian genetics, 110
Meningeal layer, of dura mater, 497,
      499
Meninges, 32, 231
  brain, 497–498, *499*
  spinal cord, *461,* 461–463, *462*
Meningitis, 487*t,* 499, 580
Meniscal injuries, 292, *292*
Meniscus (menisci), 273, 290, *291,*
      292, *292,* 299
Menopause, 1038, 1049, 1101
  evolution of, 1049
  hirsutism in, 186
  hot flashes in, 1049
Menses, 1056
Menstrual cycle, 1042, 1050, *1054,*
      1055–1056, *1057*
  anovulatory, 1049
  disorders of, 1068*t*
  ischemic phase of, 1056
  menstrual phase of, *1054,* 1056
  pheromones and, 576
  premenstrual phase of, *1054,*
      1056
  proliferative phase of, *1054,*
      1055, *1057*
  secretory phase of, *1054,* 1056,
      1057
Menstrual fluid, 1056
Menstrual phase, *1054,* 1056, *1057*
Menstruation. *See also* Menstrual
      cycle
  cessation of. *See* Menopause
  first, 1049
Mental cleft, 312*t*
Mental protuberance, *226, 238, 238*
Mental region, *364*
Mental spines, *228*
Mental symphysis, 238
Mentolabial sulcus, *364*
Mentum (chin), *364*
Mercury, millimeters of, 842
Merocrine (eccrine) glands, 162,
      164, 188, 189, 190*t*
Mesangial cells, 879, *879*
Mesencephalon, 495, *498,* 503
Mesenchymal cells, 139, 1081
Mesenchyme, 139, 148, 206–207,
      1081
Mesenteric lymph nodes, *784,* 792,
      792
Mesenteries, 34, *34, 374, 375,*
      928–929, *929*
  anterior (ventral), 34, 928
  posterior (dorsal), 34, *34,* 928,
      1083
Mesocolic lymph nodes, 792
Mesocolon, 34, 929, *929, 964,* 965
Mesoderm, 139, 1076, 1080–1081,
      1081, 1082, 1083, 1083*t,*
      1087
Mesometrium, *1041,* 1042
Mesonephric duct, 1010, *1011*

Mesonephros, 1010, *1011*
Mesosalpinx, 1040, *1041*, 1042, *1044*
Mesothelium, 163
Mesovarium, 1039, *1041*
Messenger RNA (mRNA), *109*, 114
    DNA transcription to, 116–117
    translation of, 117–120, *119*
Metabolic acidosis, 918, 919*t*
Metabolic alkalosis, 918, 919*t*
Metabolic clearance rate (MCR), 642
Metabolic pathways, 67
Metabolic rate, 998–999
Metabolic states, 997*t*, 997–999
Metabolic theory of autoregulation, 737
Metabolic water, 902
Metabolism, 14, 54*t*, 54–55, 975.
        See also specific processes
    aging and, 1100
    alcohol, 1002
    breastfeeding and, 1068
    carbohydrate, 986–993, 993*t*
    enzymes and, 65–67
    lipid, 993–994
    liver's role in, 995, 995*t*–996*t*
    pregnancy and, 1061–1062
    protein, 994–995
Metaphase, *114*
    I, in meiosis, 1023, *1024*, 1025
    II, in meiosis, 1023, *1024*
    in mitosis, *126*, 127
Metaphysis, 208, 209, *209*
Metaplasia, 166, 829, 1044
Metarterioles, 728
Metastasis, 132, *133*, 793
Metatarsal bones, *225*, 259–261, *260*
    surface anatomy of, *380*, *381*, *382*
Metchnikoff, Elie, 783
Metencephalon, 495, *498*, 503
Methyl group, 55, *55*, 132
Methylation, DNA, 132
Methylene blue, 673
Methylprednisolone, for spinal cord injury, 488
Miacalcin. *See* Calcitonin
Micelles, 949, 958, *960*, *962*, 985
Microcephaly, 1093
Microcirculation, 729
Microfilaments (thin filaments), *79*, *97*, *98*, 105*t*, 388, *388*, *389*, 390*t*, 413
Microglia, *427*, 427*t*, 427–428, 797
Micrographia (Hooke), 5
Micrometer (μm), 76
Micronutrients, 977
Microscopes
    development of, 5–6
    Hooke's compound, *5*, 5–6
    Leeuwenhoek's simple, 5–6
    light, *77*, 77–78, 78*t*
    scanning electron, 77, *85*
    size of structures under, 78*t*

transmission electron, *77*, 77–78, 78*t*, 105*t*
Microscopic anatomy, 2, 138
Microtubules, *79*, *85*, 86, 97, *98*, *99*, *104*, 105*t*
Microvascular disease, 648
Microvasculature, 729
Microvilli, *79*, *84*, *84*, *85*, *98*, 105*t*, 953, 955
Micturition, 893–894, *894*
Micturition center, 893
Micturition reflex, 558, 893, *894*
Midbrain, 495, *496*, 503–506, *504*, *508*
    in autonomic control, 558
    in pain modulation, *571*, 572
Midclavicular line, 31, *31*
Middle cerebellar peduncle, *505*
Middle ear
    anatomy of, *579*, 579–580
    cavity of, 227
    hearing function of, 582–583
    infection of, 580
Middle peduncles, 507
Middle phalanx, 252, *253*
Middle pharyngeal constrictor, *313*, 316*t*
Midpiece, of sperm, *1026*, 1027
Midsagittal plane, 28
Miescher, Johann Friedrich, 110
Mifepristone (Mifeprex), 1070
Migraine headaches, 537*t*
Migrating motor complex, 956
Milk, breast (human), 1067*t*, 1067–1068
Milk, uterine, 1087
Milk ejection (letdown), 1067
Milk ejection reflex, 1045, 1067
Milk lines, 189
Milk production. *See* Lactation
Milky spots, 929
Milliequivalents per liter, 51
Millimeters of mercury (mm Hg), 842
Milliosmoles per liter (mOsm/L), 89
Millisievert (mSV), 44
Mind–body connection, 820
Mineral(s), 42
    absorption of, 961
    dietary, 977, 983–984, 984*t*
    hepatic metabolism of, 996*t*
Mineral resorption, 211–212
Mineralization, 199, 206, 211
Mineralocorticoids, 627
Minor calyx, 870, *870*, 874
Minor duodenal papilla, 950
Minute respiratory volume (MRV), 847
Miscarriage (spontaneous abortion), 1068*t*
Mitochondria, *79*, *98*, 103, *103*, 105*t*, *974*
    ATP production in, 69, 103, 987, 989–990, *990*

clinical significance of, 103
    evolution of, 103
Mitochondrial DNA, 103
Mitochondrial electron-transport chain, 989–990, *990*
Mitochondrial myopathy, 103
Mitosis, 125, *126*, 141
    aging and, 1102
    in sperm, 1025
Mitotic phase of cell cycle, 125, *125*
Mitotic spindle, *126*, 127
Mitral cells, *575*, 576
Mitral valve, *693*, 696, *696*, 697, *697*
Mitral valve prolapse, 710
Mittelschmerz, 1055
Mixed glands, 162
Mixed nerves, 470, 528
Mixtures, 48–52
    compounds vs., 48
    types of, 49–50, 51*t*
MNS blood group, 670
Modality, sensory, 564–565
Model, animal, 10
Moiety, 58
Molaison, Henry, 520
Molarity, 51
Molars, *931*, 932, *932*
Mole (skin), 181, 182, 190–191
Molecular clock, 103
Molecular formula, 45, *45*
Molecular genetics, 110
Molecular motors, 65
Molecular weight (MW), 45, 88
Molecules, 12, *12*, 45–48
    representations of, 45, *45*
Molybdenum, 984*t*
Monoamine(s), 441, *443*, 443*t*, 559, 633*t*, 633–634, *634*
    synthesis of, 635–636
    transport of, 637
Monoamine oxidase (MAO), 445, 556, 559
Monoamine oxidase inhibitors, 445, 454, 559
Monoblasts, 676, *677*
Monochromatic vision, 600
Monocytes, 148, *156*, 157, 657, *657*, 673, 675*t*, 676
    hormonal production by, 820
    in inflammation, 801*t*, 803
    in innate immunity, 796–797
    life history of, 676
    production of, 676, *677*
    in tissue cleanup and repair, 803
Monoiodotyrosine (MIT), 635, *636*
Monomers, 56
Mononucleosis, infectious, 684*t*
Monosaccharides, 56, *57*, 58*t*
Monosomy, 1094
Monosynaptic reflex arcs, 484
Monozygotic twins, 1080, 1101
Mons pubis, *366*, *1039*, 1044, *1045*
"Moon face," 646, *646*
Morning sickness, 1061

Morning-after contraception, 1070
Morphine, 571
Morton, William, 608
Morton's Letheon, 608
Morula, *1078*, 1079, 1079*t*
Motilin, 956
Motor association area, *521*, 522
Motor control. *See* Movement; *specific anatomy*
Motor cortex, primary, *521*, 522–523, *523*
Motor division, of peripheral nervous system, 422
Motor (efferent) nerve fibers, 470, 471*t*, *482*, 483
Motor (efferent) neurons, 423, *423*, 544
    gamma, *483*, 484
    lower, 466, *468*, 523
    somatic, 391
    upper, 466, *468*, 523
Motor end plate, 392–393
Motor homunculus, 522
Motor nerves, 470
Motor neurons, *483*, 484
Motor proteins, 65
Motor unit summation, multiple, 402
Motor units, *391*, 391–392
    active, size of, 408
    aging and, 1098
    definition of, 391
    recruitment of, 402
    size principle and, 402
    slow- and fast-twitch fibers of, 408
    summation of, 402, *403*, 408
Motrin (ibuprofen), 644
Mouth (oral cavity), *930*, 930–933
    anatomy of, 930, *930*
    epithelial tissue of, 930
    functions of, 930
    muscles acting on, 311
Mouthfeel, 573
Movement
    brain function in, 522–525
    connective tissue and, 148
    contralateral, 318*t*
    ipsilateral, 318*t*
    muscles and, 300
    reticular formation and, 507
    skeleton (bones) and, 199
    study of, 268
Moving attachment, of muscle, 303–304
MRI, *22*, 23
    colorized, of human body, *1*
    fetal, *1037*
    functional, 23, 507, 519, 538
    lumbar, pelvic, and femoral region, *299*
Mucin, 162
Mucociliary escalator, 832
Mucosa. *See* Mucous membranes

Mucosa-associated lymphoid tissue (MALT), 788, *788*, 833, 927
Mucous cells
of salivary glands, *935*
of stomach, *939, 940, 945*
Mucous coat, of stomach, 943
Mucous glands, 162
Mucous membranes, 163, *165, 411*
of digestive tract, 927, *928*
of esophagus, 935
hematopoiesis in, 660
of immune system, 796
of labia, 930
of large intestine, 963
olfactory, 575, *575*
respiratory, 576
of small intestine, *925*, 953
of stomach, *939, 940*
of trachea, *831*, 832
of uterine tubes, 1040
of uterus, 1042
of vagina, 1044
Mucous neck cells, *939, 940, 942t*
Mucus, 163
in saliva, 935
Müllerian duct, 1010, *1011*
Müllerian-inhibiting factor (MIF), 1010
Multimodal association areas, 521
Multinuclear cells, 97
Multipara, 1064
Multipennate muscles, 302, *302*
Multiple alleles, 129
Multiple births, 1080
Multiple motor unit (MMU) summation, 402, 408
Multiple sclerosis (MS), 429
Multiple-gene inheritance, *130,* 130–131
Multipolar neurons, 425, *425*
Multipotent stem cells, 166
Multiunit smooth muscle, 412, *412*
Mumps, 934
Murmurs, heart, 710
Muscarinic receptors, 553–554, *554,* 715
Muscle(s), generally, 157–158
adaptations to bipedalism, *262, 263*
antagonist, *304,* 305
antigravity, 300
attachments of, 303–304, *304*
cardiac. *See* Cardiac muscle
circular. *See* Sphincters
conductivity of, 385
connective tissue of, *301,* 301–313, 386
contractility of, 385
contraction. *See* Muscle contraction
deconditioned, 409
definition of term, 300
elasticity of, 385

excitability of, 385
extensibility of, 385
extrinsic, 304
of eye, 592, *592, 593*
of larynx, 829, 830
of tongue, 931
fascicle orientation of, 302, *302*
fixator, 305
functional groups of, 305
functions of, 300–301
head and neck, 308–318
innervation of, 305, *392,* 392–393
insertion of, 304
intrinsic, 304
of foot, *356, 357*
of hand, 304, *343t*
of larynx, 831, *831*
of tongue, 931, *931*
involuntary, 158, 386, 409
learning strategy for, 305, 308, 363
length–tension relationship of, 326, 400, *400,* 408
lower limb, 344–358
major, *306–307*
metabolism of, 300, 385, 404–409
naming of, 301, 305, 308
nerve relationship with, 391–394
origin of, 304
parallel, 302, *302*
pennate, 302, *302*
prime mover (agonist), 305
quadrilateral, 302
range of motion and, 275
respiratory, *319,* 320t, *838,* 838–839
rigor mortis of, 400
size of, 408
skeletal. *See* Skeletal muscle
smooth. *See* Smooth muscle
strength and conditioning of, 408–409
striations of, 158, *159,* 159t, 386, 409
structural hierarchy of, 390t
synergist, *304,* 305
test of recognition, *383*
triangular (convergent), 302, *302*
trunk, 318–328
universal characteristics of, 385
upper limb, 329–343
voluntary, 158, 386
Muscle(s), specific
abdominal oblique
external, *306, 307, 320, 321, 322, 322t, 323, 324, 333, 366, 367, 371*
internal, *306, 307, 320, 321, 322, 322t, 323, 324, 367*
abductor digiti minimi, 342, 343t, *356, 357t, 381*

abductor hallucis, *356,* 357t, *380, 381*
abductor pollicis brevis, *338, 342,* 343t
abductor pollicis longus, 340t, *342*
adductor, *306, 344, 346t, 355, 368, 369, 370*
adductor brevis, *344,* 346t, *348, 355, 369, 370*
adductor hallucis, *356,* 357t
adductor longus, *306, 344, 346t, 347, 348, 355, 366, 368, 369, 370*
adductor magnus, *307, 344, 345,* 346t, *348, 355, 368, 370*
adductor pollicis, *342,* 343t, *378*
anconeus, 336t, *337*
arrector pili, *176, 184,* 185
arytenoid, *830*
axial, of arm, 334t
biceps brachii, 275, *287, 288, 302, 302,* 305, *306, 331, 333, 335,* 335t, *337, 366, 372, 376, 377*
biceps femoris, *307, 345, 349t, 355, 379, 380*
brachialis, 303, *304,* 305, *306, 331,* 335t, *337*
brachioradialis, *306, 331,* 336t, *337, 372, 377*
buccinator, *310,* 311, 312t, *930, 931, 933*
bulbospongiosus, *326,* 327t, *1018,* 1031
ciliary, 552, *594*
compressor urethrae, *326,* 327t
coracobrachialis, *306, 331,* 334t, *337, 367, 368*
corrugator supercilii, *309, 310,* 311t
cremaster, *1014,* 1015
cricoarytenoid, *831*
deltoid, 251, *287, 288,* 302, *302, 306, 307, 323, 331, 333,* 334t, *337, 366, 371, 372, 376, 377*
depressor anguli oris, *308, 309, 310,* 312t
depressor labii inferioris, *308, 309, 310,* 312t
detrusor, 893
diaphragm, 32, *32, 33, 34,* 319, 320t, *367, 373, 374, 827, 838, 838, 927, 928*
digastric, *314,* 316t
erector spinae, *307, 323, 324, 325, 372, 374,* 839
iliocostalis group, *324, 325*
longissimus group, *324, 325*
spinalis group, *324, 325*
surface anatomy of, *372*

extensor carpi radialis brevis, *337, 338,* 340t
extensor carpi radialis longus, *307, 337, 338,* 340t, *377*
extensor carpi ulnaris, *307, 337, 338,* 340t, *377*
extensor digiti minimi, *337, 338,* 340t
extensor digitorum, *307, 337, 338,* 340t, *377*
extensor digitorum brevis, *350,* 357t, *381, 382*
extensor digitorum longus, *306, 350, 351,* 354t, *355*
extensor hallucis longus, *351,* 354t, *355*
extensor indicis, *338,* 340t
extensor pollicis brevis, *338,* 340t
extensor pollicis longus, *338,* 340t
fibularis (peroneus) brevis, *350, 351, 352, 353,* 354t, *355*
fibularis (peroneus) longus, *306, 307, 350, 351, 352, 353,* 354t, *355, 380, 381*
fibularis (peroneus) tertius, *350,* 354t
flexor carpi radialis, *306, 337, 338,* 339t, *377*
flexor carpi ulnaris, *307, 337, 338,* 339t, *372, 377*
flexor digiti minimi brevis, *342,* 343t, *356,* 357t
flexor digitorum brevis, *356,* 357t
flexor digitorum longus, *307, 352, 353,* 354t, *355*
flexor digitorum profundus, *306, 337, 338,* 339t
flexor digitorum superficialis, *337, 338,* 339t, *342*
flexor hallucis brevis, *356,* 357t
flexor hallucis longus, *307, 352, 353,* 354t, *355*
flexor pollicis brevis, *342,* 343t
flexor pollicis longus, *306, 337, 338,* 339t
frontalis, *186, 306, 308, 309, 310*
fusiform, 302, *302*
gastrocnemius, 302, 305, *307, 345, 350, 351, 352, 353,* 354t, *355, 379, 380, 381,* 408
gemellus inferior, *345,* 346t
gemellus superior, *345,* 346t
genioglossus, *313,* 315t, 931, *931*
geniohyoid, *313,* 316t
gluteals
adaptations to bipedalism, *262,* 263
gluteus maximus, 256, *262, 263, 307, 323, 326, 345,* 346t, *372*
gluteus medius, *262, 263, 307, 323, 345,* 346t, *370, 372*

Muscle(s), specific (*continued*)
  gluteals (*continued*)
    gluteus minimus, 262, 263,
      307, 323, 345, 346t
    surface anatomy of, *372*
  gracilis, *306, 307, 344, 345, 346t,*
    347, 348, 355, 366, 368,
    369, 370, 379
  hamstrings, 275, *345, 348, 349t,*
    358, *372*
  hyoglossus, 931, *931*
  hyoid, 316t
    infrahyoid group, *314, 316t,*
      831
    suprahyoid, *314,* 316t
  hypoglossus, *313, 314, 315t*
  hypothenar group, 343t
  iliacus, 255, *344, 346t, 348, 370*
  iliococcygeus, *326*
  iliocostalis group, *324, 325*
  iliocostalis lumborum, *325*
  iliocostalis thoracis, *325*
  iliopsoas, *344, 346t, 347, 348*
  infrahyoid, *314, 316t,* 831
  infraspinatus, 287, 288, *307, 323,*
    330, 331, 333, 334t, 372
  intercostal, 319, 320t, *325,* 367,
    370, 838, 839, *1046*
    external, *319,* 320t, *325,*
      367, 370
    innermost, 319
    internal, *319,* 320t, *367, 370*
  interosseous
    dorsal, *342,* 343t, *356,* 357t
    palmar, 302, *302, 342,* 343t
    plantar, *356,* 357t
  ischiocavernosus, *326,* 327t, 1031
  ischiococcygeus, *326*
  lateral rectus, 530, *592, 592, 593*
  lateral rotator, *307, 323, 345,*
    346t–347t
  latissimus dorsi, *307, 321, 323,*
    325, 331, 333, 334t, 337,
    372, 376
  levator anguli oris, *309,* 312t
  levator ani, *326,* 327t, *964,* 1031
  levator labii superioris, *308, 309,*
    310, 312t
  levator palpebrae superioris, *309,*
    311t, *528, 592, 593*
  levator scapulae, *307, 310, 314,*
    323, 329, 330t
  longissimus capitis, *317, 324*
  longissimus cervicis, *317*
  longissimus group, *324, 325*
  longissimus thoracis, *324, 325*
  lumbrical, *342,* 343t, *356,* 357t
  masseter, *230,* 238, 302, *306,*
    308, 310, 313, 933, 934
  medial rectus, *592, 593*
  mentalis, *309, 310,* 311, 312t
  midpalmar group, 343t
  multifidus, *324,* 325t
  mylohyoid, *313, 314, 316t*

  nasalis, *308, 309, 310,* 311t
  oblique, of eye, 592, *593*
  obturator externus, *344, 345, 346t*
  obturator internus, *345,* 347t
  occipitalis, *307, 310*
  occipitofrontalis, 311t
  omohyoid, *310, 314,* 316t, *367*
  opponens digiti minimi, *342,* 343t
  opponens pollicis, *342,* 343t
  orbicularis oculi, 302, *302, 306,*
    308, 309, 310, 311t, *592*
  orbicularis oris, *306, 308, 310,*
    311t, *933*
  palatoglossus, *313,* 315t, *931*
  palmar interosseous, 302, *302,*
    342, 343t
  palmaris longus, 13, *337, 338,*
    339t, *377*
  papillary, *693, 695, 696,*
    696, 709
  pectinate, 694, *695*
  pectineus, *344, 346t, 347, 348,*
    368, 369
  pectoralis major, 302, *302, 306,*
    321, 331, 333, 334t, 337,
    366, 371, 373, 376, 377,
    839, 1046
  pectoralis minor, *321, 329,* 330t,
    838, 839, 1046
  perineal, *326,* 327t
  pharyngeal constrictor, *310, 313,*
    314, 316t, *935*
  piloerector, *176, 184,* 185
  piriformis, *326, 344, 345,* 347t
  plantar interosseous, *356,* 357t
  plantaris, 13, *352, 353*
  platysma, *306, 308, 309, 310,*
    311, 312t, *366*
  popliteus, 291, 349t, *353*
  pronator quadratus, *306, 335,*
    336t
  pronator teres, *335,* 336t, *337*
  psoas major, *344, 346t, 348, 370*
  pterygoid, *313, 933*
  pubococcygeus, *326*
  puborectalis, 967
  pupillary constrictor, 552, 597
  pupillary dilator, 597
  quadratus lumborum, *324,* 325t,
    370
  quadratus plantae, *356,* 357t
  quadriceps femoris, 256, 305,
    347, 347t, *348,* 349t, 408,
    408t
  rectus
    of eye, 592, *592, 593*
    lateral, 592, *592, 593*
  rectus abdominis, 302, *302, 306,*
    320, 321, 322, 322t, *371,*
    374, 838, 839
  rectus femoris, 302, *302, 306,*
    347, 348, 349t, 355, 366,
    368, 369, 379
  rhomboids, 305

  rhomboid major, *307, 323,*
    329, 330t, *333*
  rhomboid minor, *307, 323,*
    329, 330t, *333*
  risorius, *309, 310,* 312t
  rotator cuff, 287, 288, *330,* 334t
  sartorius ("tailor's muscle"), 302,
    306, 347, 348, 349t, 355,
    366, 369
  scalenes, *314,* 317, 318t, *838,*
    839
  semimembranosus, 302, *307,*
    345, 349t, *355, 379, 380*
  semispinalis capitis, *307,* 317,
    317, 318t, *323, 324*
  semispinalis cervicis, 317, 318t,
    324
  semispinalis thoracis, *324,* 325t
  semitendinosus, *307, 345,* 349t,
    355, 379
  serratus anterior, *306, 307, 321,*
    323, 329, 330t, *333, 371,*
    839
  serratus posterior, *307, 323,* 324,
    839
  soleus, *306, 307, 350, 351, 352,*
    353, 355, 380, 381, 408
  spinalis, *324, 325*
  spinalis group (erector spinae),
    324, 325
  spinalis thoracis, *325*
  splenius capitis, *307, 314,* 317,
    318t, *323, 324*
  splenius cervicis, 317, 318t, *324*
  stapedius, 580
  sternocleidomastoid, *306, 307,*
    310, 314, 317, *317,* 318t,
    323, 371, 838, 839
  sternohyoid, *310, 314,* 316t
  sternothyroid, *310, 314,* 316t
  styloglossus, *313,* 315t, *931, 931*
  stylohyoid, 238, *314,* 316t
  subclavius, *321*
  subscapularis, 287, 288, *330,*
    331, 334t, *367*
  superior oblique, 592, *593*
  superior rectus, 592, *592, 593*
  supinator, *306, 335,* 336t, *337*
  suprahyoid, *314,* 316t
  supraspinatus, 287, 288, *307,*
    323, 330, 331, 334t
  temporalis, *230,* 232, *310, 313,*
    933
  tensor fasciae latae, *306,* 346t,
    348, 366, 368, 369, 379
  tensor tympani, 580
  teres major, *307, 323, 331, 333,*
    334t, *337, 372*
  teres minor, 287, 288, *307, 323,*
    331, 333, 334t
  thenar group, 343t
  thyrohyoid, *310, 314,* 316t
  tibialis anterior, *306, 350, 351,*
    354t, *355, 380, 381*

  tibialis posterior, *307, 353,* 354t,
    355
  trabecular, 1020
  trachealis, 832, *832*
  transverse abdominal, *306, 320,*
    321, 322, 322t, *367*
  transverse perineal, *326,* 327t
  trapezius, *306, 307, 314,* 317,
    317, 318t, *323, 325, 329,*
    330t, *366, 371, 372, 377*
  triceps brachii, 303, *304, 307,*
    331, 333, 335t, *337, 372,*
    376, 377
  vastus intermedius, *306, 348,*
    349t, *355, 369*
  vastus lateralis, *306, 345, 347,*
    348, 349t, *355, 366, 369,*
    379, 380
  vastus medialis, *306, 347, 348,*
    349t, *355, 369, 379, 380*
  zygomaticus major, 302, *306,*
    308, 309, 310, 312t
  zygomaticus minor, *309, 310,*
    312t
Muscle compartments, 302–303, *303*
Muscle contraction
  cardiac muscle. *See* Cardiac
    muscle
  skeletal muscle, 396, *398*
    calcium in, 387, 396, *397*
    concentric, 402, *403,* 404
    eccentric, 402, *403,* 404
    energy for, 404–406
    excitation for, 394, 400
    excitation–contraction
      coupling for, 396, *397*
    force or strength of, 400, *400,*
      401–402, *402*
    isometric, 402, *403,* 404, *404*
    isotonic, 402, *403,* 404, *404*
    myogram of, 401, *401*
    nerve stimulation in, 391–394
    neurological control of,
      522–524
    sliding filament theory of, 9,
      396, *398*
    stimulus intensity and, 402,
      402, 403
    whole-muscle behavior in,
      401–404
  smooth muscle, *413,* 413–414
Muscle endurance, 406
Muscle fascicle, *301,* 302, *302,*
  390t, 408
Muscle fasciculations, 914
Muscle fatigue, 406, 409
Muscle fibers, 157, *159,* 159t, *301,*
  386, *386,* 386–391, *387,*
  390t, 409
  aging and, 1098
  behavior of, 394–400
  of cardiac muscle, 409, *693, 700,*
    700–701, 703, 703
  contraction of, 396, *398*

excitation of, 394, *395*
excitation–contraction coupling of, 396, *397*
extrafusal, *483,* 484
fast-twitch (type II), 407*t,* 407–408, 408*t*
intermediate, 408
intrafusal, *483,* 484
physiological classes of, 407*t,* 407–408
relaxation of, 396–400, *399*
slow-twitch (type I), 407*t,* 407–408, 408*t*
of smooth muscle, 409, 412
Muscle soreness, delayed-onset, 414*t*
Muscle spindles, *483,* 483–484, *567,* 567*t,* 568
Muscle tone
skeletal muscle, 275, 400
smooth muscle, 413
Muscle twitch, 401–404
contraction phase of, 401, *401*
contraction strength of, 401–402, *402*
latent period of, 401, *401*
relaxation phase of, 401, *401*
stimulus intensity and, 402, *402, 403*
summation of, 402, *403*
Muscular arteries, 727–728
Muscular disorders, 414*t*
athletic injuries, 358
carpal tunnel syndrome, 339*t,* 341
compartment syndrome, 303
contracture, 414*t*
cramps, 414*t*
crush syndrome, 414*t*
delayed-onset soreness, 414*t*
disuse atrophy, 167, 414*t*
fibromyalgia, 414*t,* 570
hernia, 328, *328,* 968*t*
mitochondrial myopathy, 103
muscular dystrophy, 388, 415
myasthenia gravis, 393, 415, *415*
myositis, 414*t*
paralysis. *See* Paralysis
Muscular dystrophy, 388, 415
Muscular system, 35, *36,* 299–361, *306–307. See also* Muscle(s)
aging and, 1098
effects on other systems, *416*
hand and wrist, *337, 338,* 339*t*–340*t,* 341, *342,* 343*t*
head and neck, 308–318
lower limb, 344–358
trunk, 318–328
upper limb, 329–343
Muscular tissue, 385–419. *See also specific types*
cells of, 385, 386–387, 409
characteristics of, 385

histology of, 138, 138*t,* 157–159, *159,* 159*t,* 386–391
nerve relationship with, 391–394
striations of, 158, *159,* 159*t,* 386, *389,* 389–391, 409
structural hierarchy of, 390*t*
Muscular venules, 732
Muscularis externa, *411,* 927, *928*
of colon, 963
of esophagus, 935
of small intestine, 953, *954*
of stomach, 937, 942
Muscularis mucosae, 163, *165, 411,* 833
of digestive tract, 927, *928*
of small intestine, *954,* 955
of stomach, *939*
Musculoskeletal system, 35
Muslim culture and medicine, 4
Mussolini, Benito, 432
Mutagen(s), 1094
Mutagenic radiation, 43
Mutations, 15, 124, 132, 1093–1094
Myasthenia gravis, 393, 415, *415*
Myelencephalon, 495, *498,* 503
Myelin, *427,* 428–429, *430,* 438–440, 1099
Myelin sheath, *424,* 429, *430*
Myelin sheath disorders, 429
Myelin sheath gap, 429
Myelinated nerve fibers, *427,* 429, *430,* 438–440
Myelination, 428–429, *430*
Myeloblasts, 676, *677*
Myeloid hematopoiesis, 660
Myeloid leukemia, 676
Myeloid tissue, 205. *See also* Red bone marrow
Myenteric reflexes, 929, 944, *944*
Myoblasts, 386
Myocardial infarction, 169, 684, 699, 705
Myocardial ischemia, 718*t*
Myocardial vortex, 693, *694*
Myocardium, *692, 693,* 693–694, *695*
contractility of, 717–718
electrical behavior of, *703,* 703–704
hypertrophic response of, 776
impulse conduction to, 703
Myocytes, 158, 409–410, 411–412
of cardiac muscle, 409, 693, *700,* 700–701, *703,* 703–704
of smooth muscle, 409, 411–412
Myoepithelial cells, 189, 597, *1046,* 1047
Myofibrils, 386–387, *387,* 390*t*
Myofilaments, *387,* 387–388, *388,* 390*t*
Myogenic mechanism, in renal autoregulation, 878
Myoglobin, 386, 410, 701
Myoglobinuria, 303

Myogram, 401, *401*
Myokinase, 405, *405*
Myology, 300
Myometrium, *1041, 1042, 1057*
Myopathy, mitochondrial, 103
Myopia, *599,* 599*t*
Myosin, 387–388, *388*
digestion of, 926
distribution and functions of, 389
in skeletal muscle contraction, 396, *397, 398*
in skeletal muscle relaxation, 396
in skeletal muscle striation, 389
in smooth muscle contraction, 413
Myosin ATPase, 396
Myosin light-chain kinase, 413
Myositis, 414*t*
Myotatic (stretch) reflex, 484
Myotonia, in male sexual response, 1029
Myxedema, 645

N

NAD+, 67, *67,* 986
Nail(s), 175, 187
diagnostic value of, 187
structure of, 187, *187*
Nail bed, 187, *187*
Nail body, 187, *187*
Nail cuticle, 187, *187*
Nail fold, 187, *187*
Nail groove, 187, *187*
Nail matrix, 187, *187*
Nail plate, 187, *187*
Nail root, 187, *187*
Naive lymphocyte pool, 807, 814*t*
Narcolepsy, 518
Nares, 827, *827,* 828
Nasal aperture, posterior, *229,* 235, 827, *827,* 828
Nasal cavity, 227, *230, 236, 365,* 827, *827*
Nasal concha, *230,* 235, *236*
inferior, *226,* 235, *236,* 237, *828,* 829
middle, *226,* 235, *235, 236, 828,* 829
superior, 235, *235, 236, 828,* 829
Nasal glands, parasympathetic innervation of, *551,* 552
Nasal region, *364,* 827
Nasal spine, anterior, *236*
Nasolacrimal duct, 592, *592*
Nasopharynx, *828,* 829
Nasus. *See* Nose
Natal cleft, *372*
Natriuretic peptides, 630, 632*t,* 633*t,* 739, 884, 888*t,* 909
Natural convection, 1000
Natural immunity
active, 804

passive, 804
Natural killer (NK) cells, *782,* 788, 796, 798, *799,* 806
Natural selection, 9
Nature, law of, 8
Nausea, 942
Navicular, 259, *260*
Near point of vision, 599
Near response, 597–599, *598*
Nearsightedness, *599,* 599*t*
Neck
of femur, 256, *258*
of gallbladder, 949
of humerus, 251, *251*
of radius, *251,* 252
of rib, 247, *248*
of tooth, 932, *932*
Neck (cervical region), 30, 31, *364*
arteries of, *752*
muscles moving, 308–318
surface anatomy of, *364*
veins of, *754*
Necrosis, 169
Needle aspiration biopsy, 147
Negative chronotropic agents, 715
Negative feedback, 15–17
in blood pressure control, 738, *738*
in endocrine system, 622, *622*
in erythrocyte homeostasis, 664, *664*
in thermoregulation, 15, *16,* 1001
Negative feedback inhibition, 622, *622*
Negative inotropic agents, 717
Negative nitrogen balance, 983
Negative selection, of T cells, 807
Negative water balance, 906, 907*t. See also* Dehydration
Neglect syndrome, contralateral, 518
Neisseria gonorrhoeae, 1033
Neocortex, 513
Neonatal period, 1091
Neonate, 1060, 1091–1095
adaptations of, 1091
assessment of, 1092
birth defects in, 1092–1095
circulation of, 1089, *1090*
cranial assessment of, 239
immunity of, 1091
premature, 1092
respiration of, 1091, 1092
skull of, 239, *239*
transitional period in, 1091
Neoplasia, 166
Neoplasm, 166. *See also* Cancer
Nephron(s), 871–875
aging and, 1100
cortical, *872, 874,* 875
juxtamedullary, *872, 874,* 875
microscopic anatomy of, *874*

Nephron loop, *872*, 873, *873*, *874*, *879*
  countercurrent multiplier of, *886*, 886–887
  functional relationship with collecting duct, 887, *887*
  reabsorption in, 884
Nephroptosis, 895*t*
Nephrosclerosis, 877
Nephrotic syndrome, 895*t*
Nerve(s), generally, 469–470, *470*
  cutaneous innervation by, 481, *481*
  disorders of
    chickenpox (varicella), 478, 800
    diabetic neuropathy, 648
    multiple sclerosis, 429
    neuralgia, 478, 487*t*, 533
    shingles, 478
    Tay–Sachs disease, 429
  mixed, 470, 528
  motor, 470
  motor functions of, 474
  muscle innervation by, 305
  peripheral, 469
  regeneration of, *431*, 431–432
  sensory, 470
  somatosensory functions of, 474
Nerve(s), specific
  alveolar, *529*
  ansa cervicalis, 315, 475, 475*t*
  anterior cutaneous, *474*
  axillary, 476, 477, 488*t*
  buccal, *530*, 531
  cardiac, 549
  cervical, *461*, 471, *472*, 481, *530*, 531
  coccygeal, 471, *472*
  cochlear, *579*, 581, 582, 585, 588
  cranial. *See* Cranial nerves
  cutaneous
    anterior, *474*
    lateral, *474*
    lateral femoral, 477, 478*t*
    musculocutaneous, 476, 477, 488*t*
    posterior, *480*, 480*t*
  dorsal, of penis, *1020*, 1029
  femoral, *367*, 477, 478*t*
  fibular (peroneal), *480*, 480*t*
  genitofemoral, 477, 478*t*
  great auricular, 475, 475*t*
  iliohypogastric, 477, 478*t*
  ilioinguinal, 477, 478*t*
  inferior gluteal, *480*, 480*t*
  intercostal, *472*, 472, *474*
  laryngeal, *532*
  lateral femoral cutaneous, 477, 478*t*
  lesser occipital, 475, 475*t*
  lingual, *529*, 574
  long thoracic, *477*
  lumbar, *461*, 471, *472*, 481

  mandibular, *530*, 531
  median, 476, 477, 488*t*
  musculocutaneous, 476, 477, 488*t*
  obturator, 477, 478*t*
  occipital, lesser, 475, 475*t*
  pelvic, *551*, 552
  pharyngeal, *532*
  phrenic, 475, 475*t*
  pudendal, 480, 480*t*, 1029
  radial, 474, 476, 477, 481, 488*t*
  sacral, *461*, 471, *472*, 481
  sciatic, *459*, 472, 474, 480, 480*t*, 481
  spinal. *See* Spinal nerve(s)
  splanchnic, 548, 549, *551*, 552
  superior gluteal, *480*, 480*t*
  supraclavicular, 475, 475*t*
  temporal, *530*, 531
  testicular, 1017
  thoracic, *461*, 471, *472*, 481
  tibial, *480*, 480*t*
  transverse cervical, 475, 475*t*
  tympanic, 552
  ulnar, 476, 477, 488*t*
  vestibular, *579*, 581, 585, 588, 590
  zygomatic, *530*, 531
Nerve cells. *See* Neurons
Nerve deafness, 585, 1099
Nerve endings
  encapsulated, 567, 567*t*, 568
  unencapsulated, 567, 567*t*, 568
Nerve fibers. *See* Axon(s)
Nerve growth factors, 167, 428, 432, 454
Nerve impulse, 394. *See also* Action potentials
Nerve injuries, 481
Nerve plexuses, 474
  abdominal aortic, 549, *550*, 551, 552
  brachial, *368*, 472, 474, 476, 477, 477*t*
  cardiac, *547*, 551
  carotid, *547*, 549
  cervical, 472, 474–475, *475*, 475*t*
  coccygeal, 472, 474, *480*, 480*t*
  esophageal, *551*, 552
  hypogastric, inferior, *551*, 552
  lumbar, *472*, 474, 477, 478*t*
  myenteric, 927, *928*
  pulmonary, *547*, 551, 552
  renal, 875
  sacral, *472*, 474, *480*, 480*t*
  "solar plexus," 549
  submucosal, 927, *928*
Nerve signal, 394. *See also* Action potentials
Nerve–muscle relationship, 391–394
Nervous system, 35, *37*
  aging and, 1098–1099
  autonomic. *See* Autonomic nervous system

  cells of, 422–432. (*See also* Neuroglia; Neurons)
  central. *See* Central nervous system
  coordinating task of, steps in, 421
  drug effects on, 554–556, 559
  effects on other systems, 455
  enteric, *542*, 552–553, 927–928, *928*
  immune system and, 820, 821
  overview of, *421*, 421–422
  parasympathetic. *See* Parasympathetic division, of ANS
  peak of development, 1098
  peripheral. *See* Peripheral nervous system
  study of, 421
  subdivisions of, *421*, 421–422
  sympathetic. *See* Sympathetic division, of ANS
Nervous tissue, 138, 138*t*, 157, *158*, 158*t*, 420–458
  action potentials in. *See* Action potentials
  histology of, 157
  neural integration in, 446–453, 460
  supporting cells. *See* Neuroglia
  synapses. *See* synapses
Net filtration pressure (NFP), 741, *742*, 876, 878
Net hydrostatic pressure, 741–742
Net protein utilization, 983
Net reabsorption pressure, 742, *742*
Neural circuits, *450*, 450–451
Neural coding, 448–450, *450*
Neural control. *See also specific anatomy and processes*
  of blood pressure, 737–739
  of male sexual response, *1030*
  of micturition, 893–894, *894*
  of pulmonary ventilation, 839–841, *840*
Neural convergence, 549, 601, 604, *605*
Neural crest, 495, *497*
Neural divergence, 549, *550*
Neural folds, 495, *497*
Neural groove, 495, *497*, *1083*, *1084*
Neural integration, 446–453, 460
Neural plate, 495, *497*, *1084*
Neural pools, *450*, 450–451
Neural tube, 495, *497*, 1082
Neuralgia, 487*t*
  postherpetic, 478
  trigeminal, 533
Neurilemma, *424*, 429, *430*
Neurilemmocytes. *See* Schwann cells
Neurobiology, 421
Neuroendocrine cells, 615
Neuroendocrine reflexes, 621–622
Neurofibril(s), 423, *424*

  Neurofibrillary tangles, 453, *453*, 1099
Neurogenic shock, 488, 746
Neuroglia, 157, *158*, 158*t*, *427*, 427–428
  aging and, 1099
  and brain tumors, 428
  of CNS, *427*, 427*t*, 427–428
  of PNS, *427*, 427*t*, 428
  types of, 427*t*, 427–428
Neurohypophysis. *See* Posterior pituitary gland
Neuroimmunology, 820
Neuromodulators, 442, 445, 571
Neuromuscular junction, 392–393, 440
Neuromuscular toxins, 393
Neurons, 157, *158*, 158*t*, 422–426. *See also* Motor neurons; Sensory neurons
  action potentials of, 435–436
  aging and, 1099
  anaxonic, 425, *425*
  autonomic system, 544–545
  bipolar, 425, *425*
  circuits of, *450*, 450–451
  conductivity of, 422
  electrophysiology of, 433–440
  energy needs of, 502
  enteric system, 927
  excitability of, 422
  excitation of, *434*, 434–435
  expiratory, 839, *840*
  first-order, 465, *467*, 568
  functional classes of, 422–423, *423*
  gland cells vs., 615
  inclusions of, 423
  information-processing ability of, 446–453
  inspiratory, 839, *840*
  multipolar, 425, *425*
  neuromodulators of, 442, 445
  number of, 426
  parallel processing by, 451
  pools of, *450*, 450–451
  postganglionic
    parasympathetic, *551*, 552
    pupillary, 597
    sympathetic, *547*, 548, 549
  postsynaptic, 440, *440*, *441–442*
  preganglionic, 544–545
    parasympathetic, 550–552, *551*
    pupillary, 597
    sympathetic, 546–549, *547*, *548*
  presynaptic, 440, *440*, *441–442*
  presynaptic facilitation of, 447–448, *448*
  presynaptic inhibition of, 448, *449*
  pseudounipolar, 425

refractory period of, 436, *438,* 449

second-order, 465, *467,* 568

secretions of, 422

serial processing by, 451

signal conduction in, 437–438, *438, 439*

structural variations in, 425, *425*

structure of, 423, *424, 425*

synapses of, *440,* 440–446, *442*

third-order, 465, *467,* 568

unipolar, 425, *425*

Neuropathy

diabetic, 648

Leber hereditary optic, 103

peripheral, 487*t,* 569

Neuropeptide Y, 556, 941, 977, *978*

Neuropeptides, 442, *443,* 443*t,* 445, 571

Neuropharmacology, 554–556, 559

Neurophysiology, 2

Neurosoma, 157, *158,* 158*t,* 423, *424,* 425

Neurosyphilis, 1033

Neurotensin, 556

Neurotransmitters, 393, 422, 440–445, 443*t,* 613

aging and, 1099

in Alzheimer disease, 454

in autonomic nervous system, 553–556, *554*

definition of, 441

degradation of, 445

diffusion of, 445

discovery of, 441

excitatory, 434–435, 442, 444, *444, 445,* 447

hormone vs., 615–616

immune effects of, 820

inhibitory, 434–435, 444, *445,* 447

in neuropharmacology, 554–556, 559

in Parkinson disease, 454

postsynaptic potentials produced by, *446,* 446–447

reuptake of, 445

Neutral fats, 60

Neutral solution, 50

Neutralization of antigen, 811

Neutrons, 42, *42*

Neutrophil(s), 148, 154, *156,* 657, *657,* 673, 674*t*

chemotaxis of, 802

in inflammation, 801*t,* 801–803, *802*

in innate immunity, 796

life history of, 676

production of, 676, *677*

Neutrophilia, 673, 803

Nevus (mole), 181, 182, 190–191

Niacin, 985, 985*t*

Nicotine, cardiovascular effects of, 716, 776

Nicotinic receptors, 554, *554*

Night blindness, 601, 985

Night vision, 600, 603, *605,* 1099

Nightmares, 517

Nightshade, 559

9 + 2 structure, 86

Nipple, *371,* 1045, *1046*

Nitric oxide (NO), 442, 445, 452, 556

Nitrogen

air content of, 848, 848*t*

atomic structure of, *42*

body percentage of, 41*t*

Nitrogen balance, 983

Nitrogen narcosis, 861

Nitrogenous base, 67, 110, *111,* 114, 123

Nitrogenous wastes, 658*t,* 659, 867–868, *868,* 882, 889*t*

Nitroglycerin, for heart problems, 690

Nitrous oxide, 608

NMDA receptors, 452

Nociceptors, 566, 568, 569, *571,* 572

Nocturia, 914

Nocturnal emissions, 1021

Nodal (junctional) rhythm, 702

Node of Ranvier, *424,* 429, 438

Nondecremental action potentials, 436

Nondisjunction, 1094, *1094*

Nondisplaced fractures, 216, *217*

Nonfluent (Broca) aphasia, 525

Nongonococcal urethritis (NGU), 1033

Nonpolar covalent bonds, 46, 46*t,* 47

Non-REM sleep, 517, *518,* 519

Nonshivering thermogenesis, 1001

Nonspecific granules, 673

Nonspecific immunity. *See* Innate immunity

Nonsteroidal anti-inflammatory drugs (NSAIDs), *643,* 644, 943

Nonverbal communication, 176, *177,* 186, 591

Noradrenaline. *See* Norepinephrine

Norepinephrine

adrenal secretion of, 550

aging and, 1099

in appetite, 977

in autonomic nervous system, 554, *554,* 554*t,* 554–556, 715

in blood pressure control, 739

in cardiac physiology, 715

drugs affecting, 559

as hormone, 615, 616*t,* 627, 631*t*

hypersecretion of, 645

as neurotransmitter, 441, *443,* 443*t,* 444, *445,* 447, 554, 615

in pulmonary ventilation, 843

in stress response, 642

Norepinephrine receptors, 554, *554,* 554*t*

Normal saline, 90, 921

Normo-, 908

Normoblast (erythroblast), 663, *663*

Normokalemia, 908, *910*

Nose, *827,* 827–829

apex of, *364*

bridge of, *364*

epithelium of, 829

functions of, 827

olfactory role of, 575–577, 827

root of, *364*

Nostrils, 827, *827, 828*

Notochord, *497*

Novocaine (procaine), 608

NSAIDs. *See* Nonsteroidal anti-inflammatory drugs

Nuchal lines

inferior, *229,* 234

superior, *229,* 234

Nuchal region, *30, 364*

Nuclear division, 110

Nuclear envelope, *79, 97, 99*

Nuclear lamina, 97, *99*

Nuclear pores, 97, *99*

Nucleases, 950, 951*t,* 961

Nuclei

of atoms, 42

of brain, 495, 507

accumbens, 515

anterior group, *510*

arcuate, *510,* 977, *978*

basal, 454, 509, *514, 515,* 515–516, 523

caudate, *500, 515,* 516

cochlear, 585, *587*

corpus striatum, *515,* 516

cuneate, *467, 505*

deep, of cerebellum, 507

dorsomedial, *510,* 518

gracile, *467, 505*

hypothalamic group, *618,* 619

inferior olivary, 503, *505*

lateral geniculate, *504, 510, 607, 607*

lentiform, *515,* 516

mammillary, *510, 510*

medial geniculate, 503, *504, 505, 510*

paraventricular, *510, 618,* 619

posterior, *510*

preoptic, *510*

pretectal, 607, *607*

red, *505, 506*

salivatory, 934–935

solitary, 574

subthalamic, 454, *515*

superior olivary, 585–586, *587*

suprachiasmatic, 510, *510,* 517–518

supraoptic, *510, 618,* 619

thalamic group, 454, 503, *504,* 509, *510,* 514, *514, 607*

ventral intermediate, 454

ventromedial, *510*

vestibular, 590, *590*

of cells, *77, 79,* 97–99, *99,* 105*t*

DNA packaging in, 113–114

of neuron, 423, *424*

of Schwann cell, *424*

of sperm and egg (pronuclei), 1078, *1078*

Nucleic acids, 55, 69, 110–114. *See also* DNA; RNA

digestion of, 926, 950, 961

Nuclein, 110

Nucleolus, *79,* 99, *99*

Nucleoplasm, 97, *99*

Nucleosidases, 961

Nucleotides, 67, 69

conversion, in translation, 117–120, *119*

DNA, 56, 110, *111,* 116

genetic code, 116, 116*t*

RNA, 114

Nucleus pulposus, *242*

Nurse cells, 629, *630,* 1015, *1016,* 1021, *1022, 1025*

Nutrient(s), 977–986

definition of, 977

digestion and absorption of, 926, 956–963, *962*

essential, 977

metabolic states and, 997–999

plasma transport of, 658*t,* 659

prenatal, 1061–1062, 1087

Nutrition, 974–986

aging and, 1100–1101

breastfeeding and, 1068

calories in, 977

guidelines and requirements for, 977, 979*t*

lack of (starvation or malnutrition), 660, 867, 1101

metabolic states and, 997–999

NuvaRing, 1070*t*

prenatal, 1061–1062, 1087

O

Oat-cell carcinoma, 860

Obesity, 776, 976, 1101

Objective lens, 5

Obligatory water loss, 904

Obligatory water reabsorption, 882

Oblique section, 140, *140*

Obstructed venous return shock, 745

Obstructive lung disorders, 847

Obturator membrane, 255

Occipital lobe, *494, 496, 500, 511,* 512

functions of, *511,* 512, 521

lesions of, 607

in vision, 607, *607*

Occipital protuberance, external, 229, 234
Occipital region, 364
Occlusal surfaces, 933
Occlusion, 933
Ocular lens, 5
Odontoblasts, 933
Odontoid process. *See* Dens
Odor
  body, 188
  urine, 889
Odorant-binding protein, 576
Odorants, 575–577
Oil(s), 60
Olecranon, 251, 252, 372, 376, 377
Oleogustus, 574
Olfaction, 515, 521, 572, 575–577
  aging and, 1099
  anatomy of, 575, 575–576
  evolution of, 576
  physiology of, 576
  projection pathways of, 576–577, 577
  respiratory function and, 826
Olfactory association area, 521
Olfactory bulb, 235, 528, 575, 576, 577
Olfactory cells, 575, 576
Olfactory cortex, primary, 521, 576, 577
Olfactory hairs, 575, 576
Olfactory mucosa, 575, 575
Olfactory receptors, 576
Olfactory tract, 575, 576, 577
Oligodendrocytes, 427, 427, 427t
  disorders of, 429
  myelin sheath formed by, 428, 430
Oligopeptides, 63
Oligosaccharides, 56–57
Oliguria, 889–890
Olives, of medulla oblongata, 503, 504, 505
Omental appendages, 964, 965
Omentum
  greater, 34, 34, 367, 929, 929, 937, 938, 964
  lesser, 34, 34, 928, 929, 937, 938
On the Motion of the Heart (Harvey), 5
On the Origin of Species by Means of Natural Selection (Darwin), 9
On the Parts of Animals (Aristotle), 3
Oncogenes, 133
  BRAF, 190
  BRCA1 and *BRCA2*, 1047
Oncology, 132
Oncotic pressure, 741, 742
Ondine's curse, 841
Onset of fever, 799, 819
Oocyte(s), 630, 1040, 1051, 1052, 1055

donation of, 1104
  primary, 1050, 1051
  secondary, 1050, 1051, 1078
Oogenesis, 1050, 1051
Oogonia, 1050, 1051
Open canalicular system, of platelets, 679, 679
Open reduction of fracture, 217, 218
Ophthalmoscope, 596, 596
Opiates, 571
Opioids, endogenous, 571
Opium, 571
Opportunistic infections
  with cancer, 133
  with HIV/AIDS, 818, 818
  with leukemia, 677
Opposable thumbs, 10, 10
Opposition, of thumb, 284, 284
Opsin, 601, 602
Opsonization, 797
Optic chiasm, 496, 606, 607
Optic disc, 594, 595–596, 596
Optic pressure, 595
Optic radiation, 607, 607
Optic tracts, 504, 515, 606, 607
Optic vesicles, 495, 498, 595
Ora serrata, 594, 595
Oral cavity, 227, 230, 365, 927, 930, 930–933
  anatomy of, 930, 930
  epithelial tissue of, 930
  functions of, 930
Oral phase, of swallowing, 936, 936
Oral region, 364
Orbit, 227, 230, 236, 591
  accessory structures of, 591–592, 592
  external anatomy of, 591
Orbital plate, of ethmoid bone, 235, 235, 236
Orbital region, 364
Orbitofrontal cortex, 514, 514, 521
  in olfaction, 576, 577
  in taste, 574
Orexins, 518
Organ(s), 12, 12
  development of, 1082
  intraperitoneal, 33, 929
  retroperitoneal, 33, 869, 869, 929
Organ systems. *See also specific systems*
  aging and, 1097–1101
  description of, 12, 12, 35, 36–37
Organelles, 12, 12, 78, 97–104, 105t.
    See also specific organelles
Organic compounds, 55–70
Organism, 12, 12
Organization, 13
Organogenesis, 1082
Orgasm
  ejaculation in, 1027
  female, 1058, 1059
  male, 1030, 1031

Orgasmic platform, 1058, 1059
Origin, of muscle, 304
Oropharynx, 828, 829
Orthopedics, 216
Orthopnea, 847t
Orthostatic hypotension, 774t, 1099
Os
  external, 1041, 1042
  internal, 1040, 1041, 1042
Osmolarity, 89–90
  blood, 659
  urine, 889
Osmoreceptors, 509, 621, 904
Osmosis, 88–89, 96t
  capillary, 88, 89, 89, 741–743
  conceptual model of, 89, 89
  fluid balance and, 902
  imbalances in, 89
Osmotic diuresis, 890
Osmotic diuretics, 647
Osmotic pressure
  colloid (COP), 659, 741, 742, 876, 878
  definition of, 89, 89
Osseous (bony) labyrinth, 580
Osseous tissue. *See* Bone tissue
Ossicles, auditory, 579, 579–580, 581
  hearing function of, 583
Ossification, 206–208
  bony joint, 269
  ectopic, 211
  endochondral, 207, 207–208
  fontanelle, 239
  intramembranous, 206–207, 207, 210
Ossification center
  primary, 207, 207–208
  secondary, 207, 208
Osteitis deformans, 211, 218t
Osteoarthritis, 294, 1098
Osteoblasts, 201, 202
  in bone growth, 209, 209
  in bone remodeling, 210
  in endochondral ossification, 208
  endocrine function of, 631
  in intramembranous ossification, 206, 207
  in mineralization, 211
  stimulation, in calcium homeostasis, 213, 214
Osteocalcin, 201, 629, 631, 632t
Osteoclasts, 202, 202
  in endochondral ossification, 208
  inhibition, calcium homeostasis, 213, 214
  in remodeling, 210
  in resorption, 202, 211
Osteocytes, 198, 201, 202, 205–206, 207
Osteogenesis, 206. *See also* Ossification
Osteogenesis imperfecta, 203, 218t
Osteogenic cells, 201, 202

Osteogenic layer, of periosteum, 200
Osteoid tissue, 206, 207
Osteology, 199
Osteomalacia, 213
Osteomyelitis, 218t
Osteon(s), 156, 203, 204
Osteopenia, 1098
Osteoporosis, 211, 219, 219, 1098, 1101
Osteosarcoma, 203, 218t
Otitis media, 580
Otolith(s), 588, 588
Otolithic membrane, 588, 588
Otosclerosis, 585
Outer cell mass, 166
Outer ear, 578, 578–579
Outer hair cells of ear, 582, 582, 585
Outer segment, of photoreceptors, 600, 600, 601
Oval window, 579, 580, 582, 583
Ovarian cycle, 1050, 1053–1055, 1054, 1056
  follicular phase of, 1053, 1054, 1056t
  luteal phase of, 1053–1055, 1054, 1056t
  ovulation in, 1053, 1054, 1055, 1056t
Ovarian follicles, 629, 630, 1038, 1050–1053, 1052
  development and stages, 1040, 1050–1053, 1051, 1054
  dominant, 1053, 1056t
  mature, 1040, 1051, 1052–1053, 1056t
  ovarian cycle and, 1053–1055, 1054, 1056t
  primary, 1040, 1051, 1052, 1052, 1054, 1056t
  primordial, 1040, 1050–1052, 1051, 1052, 1054, 1056t
  recruitment of, 1052
  secondary, 1040, 1051, 1052, 1054, 1056t
  tertiary, 1040, 1051, 1052, 1054, 1056t
Ovaries, 369, 614, 1008, 1038–1039, 1039, 1040
  aging and, 1038
  blood supply of, 1039, 1044
  descent of, 1010
  development of, 1010, 1011
  endocrine function of, 629, 632t
  exocrine function of, 629
  histology of, 629, 630, 974
  hormonal stimulation of, 619–620
  parasympathetic innervation of, 551
  puberty and, 1038
  stigma of, 1053, 1055
  structure of, 1040
  sympathetic innervation of, 547
Ovoid cells, 76

Ovulation, 629, 1038, *1051,* 1053, *1054,* 1056, 1056*t, 1078*
  endoscopic view of, *1055*
  signs of, 1055
  stimulation of, 619–620
Ovum (ova), 629, *630, 1008*
  atresia of, 1050
  donation of, 1104
  fertilization of, *1075,* 1076–1078, *1077*
  haploid chromosomes of, 128
  number of, 1049, 1050
  production of, 1050, *1051*
  release of, 1038, *1051,* 1053, *1054, 1055,* 1056
Oxaloacetic acid, 988
Oxidation, 54, 54*t*
Oxidation–reduction (redox) reactions, 54
Oxidative stress, in Alzheimer disease, 453
Oxidizing agent, 54
Oxygen
  air content of, 848, 848*t*
  blood gas levels of, 856–858
  body percentage of, 41, 41*t*
  brain supply of, 502
  imbalances in, 858
  isotopes of, 43
  maximum uptake of, 406
  metabolic adjustments in, 855–856
  partial pressure of, 848, 850, *851*
  postexercise consumption of, 406–407
  respiratory exchange of, 848–858
  transport in blood, 661–662, 690, *691,* 850–855
  unloading of, 854–855
  utilization coefficient of, *854,* 855
  venous reserve of, 855
Oxygen debt, 406
Oxygen therapy, hyperbaric, 850
Oxygen toxicity, 858
Oxygenated blood, 690, *691*
Oxyhemoglobin, 851–854
Oxyhemoglobin dissociation curve, 852, *854,* 856, *856*
Oxytocin, 17, *17,* 63, 412, 616*t, 618,* 619, 620*t,* 620–621
  in breastfeeding, 622, 1067
  chemical classification of, 634, *634*
  in childbirth, 638, 640, 1063–1064
  second messenger for, 638, *640*
  secretion of, 622, 637
  up-regulation of, 640

P

P wave, 704, *705*
Pacemaker, 410, 701, *701,* 702, *703*

Pacemaker cells, enteric, 942, 955
Pacemaker potential, 702, *703*
Pacinian (lamellar) corpuscles, 182, 272, *567,* 567*t,* 568
Packed cell volume (hematocrit), 657, 662
Paget disease, 211, 218*t*
Pain, 569–572
  central nervous system modulation of, 570–572
  descending analgesic pathways and, 467, 507, 571, *571*
  epidural anesthesia for, 462
  fast (first), 569
  in inflammation, 800, 802, 803
  labor, 1064
  phantom limb, 571
  projection pathways for, 569–570, *570*
  referred, 570, *571*
  reticular formation and, 507, 572
  slow (second), 569
  somatic, 569
  spinal gating of, *571,* 571–572
  surgical anesthesia for, 608
  visceral, 569
Pain receptors, 566, 568, 569, *571, 572*
Palate, 237, *365,* 931
  cleft, 237
  hard, *230,* 237, *827, 828, 930,* 931
  soft, 237, *827, 828, 930,* 931
Palatine glands, 934
Palatine rugae, 931
Palatine tonsils, *784,* 793, *793, 828, 930,* 931
Paleocortex, 514
Pallidotomy, 454
Pallor, 182
Palmar abduction, 284, *284*
Palmar aponeurosis, 304, *338*
Palmar region, *30*
Palmar venous network, superficial, *769*
Palmitic acid, *59,* 62
Palpation, 2
Palpebrae, 591, *591*
Pancreas, *369, 374, 614, 628,* 628–629, 926, *927,* 950–951
  aging and, 1101
  dysfunction, in diabetes mellitus, 649
  endocrine function of, 161, 628–629, 631*t,* 950
  exocrine function of, 161, 950, 951*t*
  microscopic anatomy of, 950, *951*
  parasympathetic innervation of, *551*
  regulation of secretions, 951–952
  sympathetic innervation of, *547*

Pancreatic amylase, 950, 951*t,* 956, *962*
Pancreatic cancer, 950
Pancreatic duct, 950
Pancreatic enzymes, 950, *951,* 951*t,* 957, 958, *959*
Pancreatic islets, *628,* 628–629, 631*t,* 649, 950
Pancreatic juice, 950
Pancreatic lipase, 950, 951*t,* 958, *960, 962*
Pancreatic polypeptide, 633*t*
Pancreatitis, acute, 968*t*
Pancreine, 650
Paneth cells, *954,* 955
Panhypopituitarism, 645
Panoramic vision, 605–606
Pantothenic acid, 985, 985*t*
Pap smear, 142, 1043, *1043*
Papillae
  dermal, *177, 178,* 179, 184, *184*
  duodenal, 947, *948*
  filiform, 572–573, *573*
  foliate, 573, *573, 931*
  fungiform, 573, *573, 931*
  lingual, 572, *573,* 931
  renal, 870, *874*
  tongue, *563,* 572–574, *573,* 931, *931*
  vallate, *563, 573, 573,* 931, *931*
Papillary duct, 873, *874*
Papillary layer, of dermis, 179, *180,* 180*t*
Paracellular route, of tubular reabsorption, 881
Paracrines, 613, 643, 801, 929, 1052
Paradoxical sleep, 517
Parafollicular cells, of thyroid gland, 213, 625, *625*
Parallel after-discharge circuit, *450,* 451
Parallel muscles, 302, *302*
Parallel processing, 451
Paralysis
  crutch, 481
  flaccid, 393, 488
  infantile. *See* Poliomyelitis
  neuromuscular toxins and, 393
  partial, 488
  sleep, 517
  spastic, 488
  spinal cord trauma and, 488
Paralysis agitans. *See* Parkinson disease
Paramesonephric duct, 1010, *1011,* 1038
Paranasal sinuses, 227, *230, 365*
Paraplegia, 488
Parasympathetic defecation reflex, 966
Parasympathetic division, of ANS, 422, 544–545, 550–552, *551*
  blood vessel control by, 738

cardiac innervation/regulation by, *551,* 715
  drugs affecting, 554–556, 559
  in dual innervation, *556,* 556–557
  effects on target organs, 553–557, 555*t*
  enteric innervation by, 927, 929
  gastric innervation by, 937
  neurotransmitters and receptors of, 553–556, *554*
  penile innervation by, 1029
  salivary gland innervation by, 935
  sympathetic division vs., 552*t*
Parasympathetic tone, 544
Parasympatholytics, 559
Parasympathomimetics, 559
Parathyroid glands, *614,* 625, *626,* 631*t*
  pregnancy and, 1061
Parathyroid hormone (PTH), 616*t,* 625, 631*t,* 633*t*
  in calcium homeostasis, *214,* 215, 215*t,* 625, 884–885
  deficiency of, 645
  excess of, 645
  osteoblast stimulation using, 219
  in pregnancy, 1061*t*
  in renal function, 884–885, 888*t*
Paraurethral glands, 1045, *1045*
Parenchyma, 162
  of kidney, 870
  of lymph nodes, 790
  of spleen, 794
Parental DNA, 123, *124*
Parenteral routes, for fluids, 921
Paresis, 488
Paresthesia, 487*t,* 914
Parietal cells, of stomach, *939,* 940, *941,* 942*t,* 944, *945*
Parietal layer, 32
Parietal lobe, *494, 496, 511,* 512
  functions of, *511,* 512, 521
  lesions of, 519
Parietal pericardium, 32, *33,* 691, *692*
Parietal peritoneum, 33, *34,* 928
Parietal pleura, 33, *33,* 35, *835,* 837, 843
Parieto–occipital sulcus, *496,* 512
Parkinson disease, 454, *454,* 523–524
Parotid duct, 934
Parotid gland, *551,* 552, *927,* 934, *934*
Partial paralysis, 488
Partial pressure of gases, 848–850, *851*
Partially hydrogenated oil, 60
Partial-thickness burns, 191, *192*
Parturition, 1063. *See also* Childbirth
Passive immunity, 804
  artificial, 804
  natural, 804

Passive transport, 87
Pasteur, Louis, 75
Patau syndrome, 1095
Patch, contraceptive, 1070, 1070t
Patellar reflex, 484, 485
Patellar region, 30
Patellar retinacula, 290
Patellar surface, of femur, 256, 258
Pathogens
    containment and destruction of, 802–803
    definition of, 795
Pathological fracture, 216
Pathophysiology, 3
Pattern baldness, 186
Paulescu, Nicolae, 650
Pavlov, Ivan, 969
Peak flow, 847
Pectineal line, 255, 256, 258
Pectoral girdle
    bones of, 224t, 225, 247, 249–250
    muscles acting on, 330, 330t
Pectoral region, 30
Pedal region, 30
Pedicels, of glomerulus, 876, 877
Pedicle, 242, 242
    of cervical vertebra, 243
    of lumbar vertebra, 244
Peer review, 8
Pelvic brim, 254, 255, 370
Pelvic cavity, 32, 32t, 33, 375
Pelvic curvature, 240, 241
Pelvic diaphragm, 326, 327t
Pelvic floor, muscles of, 326, 327t
Pelvic girdle, 224t, 254–256
Pelvic inflammatory disease (PID), 1033
Pelvic inlet, 254, 255
Pelvic kyphosis, 240
Pelvic outlet, 254, 255
Pelvic region, 299
    arteries of, 772
    veins of, 763, 765
Pelvis, 225, 254, 255–256
    adaptations to bipedalism, 262–263, 263
    landmarks of, 376
    sexual dimorphism of, 256, 257, 257t
    surface anatomy of, 376
Penetrance, 131
Penicillin, 52
Penile urethra, 893, 1017, 1018
Penis, 367, 368, 375, 1019–1020, 1020
    blood supply of, 1029
    bulb of, 375, 1018, 1020
    cancer of, 1032t
    crus of, 1018, 1020
    detumescence of, 1031
    development of, 1010, 1011, 1012
    erection of, 556, 558, 1019, 1029

fascia of, 1014, 1014
glans of, 375, 1012, 1014, 1018, 1019, 1020
    innervation of, 1029
    parasympathetic innervation of, 551
    puberty and, 1021
    root of, 375, 1018, 1019
    sexual response of, 1028–1031
    shaft of, 375, 1018, 1019–1020
    sympathetic innervation of, 547
Pennate muscle, 302, 302
Pepsin, 65, 67, 941, 941, 945, 958, 959, 962, 969
Pepsinogen, 941, 941, 942t, 945
Peptic ulcer, 943, 943
Peptidases. See Protease(s)
Peptide(s), 62, 62–63
    gut–brain, 442, 630–631, 941, 942t, 975–977, 978
    natriuretic, 630, 632t, 633t, 739, 884, 888t, 909
    neuropeptides, 442, 443, 443t, 445, 571
Peptide bonds, 62
Peptide hormones, 633t, 634, 634
    mode of action, 638, 638–639
    secretion of, 637
    synthesis of, 634
Peptide YY (PYY), 630, 632t, 975, 977, 978
Peptidyl site, 118
Pepto-Bismol, 943
Percentage, concentration, 51
Percussion, 2
Perforating fibers, 200, 203, 204
Perforins, 798, 810, 814t
Perfusion, 734, 739–740
Periaqueductal (central) gray matter, 505, 506
Pericardial cavity, 32, 32, 32t, 33, 373, 691, 692, 835
Pericardial effusion, 718t
Pericardial fluid, 32, 691
Pericardial sac, 33, 691, 692
Pericarditis, 691, 718t
Pericardium, 32, 367, 691
    parietal, 32, 33, 691, 692
    visceral, 32, 33, 691, 693
Perichondrium, 154
Pericytes, 729, 730, 790
Perilymph, 580, 581, 581, 582, 583–584
Perimetrium, 1041, 1042
Perimysium, 301, 302, 386, 390t
Perineal body, 326, 327t
Perineal raphe, 1012, 1014, 1045
Perineal region, 30
Perineal space, 326, 327t
Perineum, 326, 327, 469
    female, 326, 1038, 1044, 1045
    male, 326, 1013–1014, 1014
Perineurium, 470
Periodic table of elements, 41

Periodontal disease, 933
Periodontitis, 1101
Periosteal layer, of dura mater, 497, 499
Periosteum, 154, 165, 200, 200, 204
Peripheral chemoreceptors, and ventilation, 841, 841, 856–858
Peripheral fiber, 425
Peripheral nerves, 469
Peripheral nervous system (PNS), 421, 422. See also Autonomic nervous system
    cells of, 422–432. (See also Neuroglia; Neurons)
    divisions of, 422
    drugs affecting, 559
    myelin/myelination in, 428–429
    regeneration in, 431, 431
Peripheral neuropathy, 487t, 569
Peripheral proteins, in plasma membrane, 80, 81, 81
Peripheral resistance (PR), 735–737
Peripheral thermoreceptors, 1001
Peripheral vision, 603
Peristalsis, 414, 928
    in stomach, 942
    in swallowing, 936, 936
Peritoneal cavity, 33, 34, 374
Peritoneal dialysis, 896
Peritoneal fluid, 33
Peritoneum, 33, 374
    parietal, 33, 34, 928
    visceral, 33, 34
Peritonitis, 35, 553
Peritrichial endings (hair receptors), 567, 567t
Peritubular capillaries, 871, 871–873, 882–883
Perivascular feet, 427, 428
Permanent teeth, 932, 933, 933
Permeability, membrane, 88
Permissive effects, of hormones, 641
Pernicious anemia, 666, 941
Peroxisomes, 101, 102, 105t
Perpendicular plate
    of ethmoid bone, 228, 235, 235
    of palatine bone, 237
Personality, 520
Pes planus, 261
PET. See Positron emission tomography
Petrous part, of temporal bone, 229, 233, 233, 581
Peyer patches (aggregated lymphoid nodules), 789, 953
pH, 50, 914–920
    bicarbonate–carbonic acid ratio and, 918
    of blood, 914
    compensatory mechanisms for, 918–919
    and drug action, 52

and enzyme action, 66
    fluid replacement therapy and, 921
    and gas exchange, 856, 856–858
    limiting, 916
    optimum, 914–915
    osseous tissue and, 212
    relationship with fluid and electrolyte imbalances, 920, 920t
    renal control of, 916–917, 917, 919
    respiratory control of, 826, 857, 916, 917, 919t
    of stomach, 940, 944, 945
    of tissue fluid, 914
    of urine, 889
    of vagina, 1028, 1044
pH partitioning, 52
pH scale, 50, 51
Phagocytes, 783, 796
Phagocytosis, 93, 94, 96t, 167, 783, 796–797
    complement system and, 797
    by macrophages, 676, 788, 788, 796–797
Phagosome, 93, 94
Phalanx (phalanges)
    foot, 225, 259, 260
    hand, 225, 252, 253
Phallus, 1010. See also Penis
Phantom limb pain, 571
Pharyngeal phase, of swallowing, 936, 936
Pharyngeal tonsils, 793, 793, 828
Pharyngotympanic tube. See Auditory tube
Pharynx, 365, 827, 829, 926, 927, 935
    muscles of, 313, 315t, 316t, 935
    swallowing function of, 936, 936
Phasic receptors, 566
Phenobarbital, 52
Phenotype, 129, 131, 132
Phenylephrine, 559
Phenylketonuria (PKU), 889
Pheochromocytoma, 645, 648t
Pheomelanin, 180, 185, 185
Pheromones, 188
Philtrum, 364
Phlebotomy, 685
Phonation, 829
Phosphagen system, 405, 405
Phosphatases, 961
Phosphate(s), 912–913
    absorption of, 961
    functions of, 912
    homeostasis of, 215, 912–913
    imbalances in, 913
    in muscle fatigue, 406
    plasma concentration of, 215
    in pregnancy, 1062
    renal reabsorption of, 881
    storage in bones, 199, 215

Phosphate buffer system, 915–916
Phosphate group, 55, *55*
Phosphodiesterase, 638
Phosphodiesterase type 5 (PDE5)
    inhibitors, 1031
Phosphoglyceraldehyde (PGAL), 987
Phospholipase, 638
Phospholipids, 58, 58*t*, 60–61, *61*
    in plasma membrane, *80,* 80–81
Phosphorus
    body percentage of, 41, 41*t*
    dietary, 984, 984*t*
Phosphorylation, 67, 986–987, *987*
Photoaging, 1097, *1097*
Photons, 52
Photopic vision, 601, 603–604, *605*
Photopsin, 602
Photopupillary reflex, 597
Photoreceptor cells, 566, *600,*
    600–601, *601*
    in color vision, 601, 602,
        604–605, *606*
    in day (photopic) vision, 601,
        603–604, *605*
    in dual visual system, 603–604
    evolution of, 604
    in light and dark adaptation, 603
    in night (scotopic) vision, 600,
        603
Phototherapy
    for hemolytic disease of newborn,
        671
    for melatonin/mood disorders,
        624
Phylloquinone. *See* Vitamin K
Physical half-life, 44
Physiological buffer, 915
Physiological dead space, 845
Physiology. *See also specific*
        processes
    characteristics of life in, 13–15
    comparative, 3
    definition of, 2
    gradients and flow in, 17–19,
        18, 21
    homeostasis in, 14, 21
    negative feedback in, 15–17, *16*
    origins and history of, 3–6
    positive feedback in, 17, *17*
    scope of, 2–3
    subdisciplines of, 2–3
    variations in, 15
Pia mater
    brain, 497, 498, *499*
    spinal cord, *461, 462,* 463
Pig, skin graft from, 194
Pigment(s)
    hair, 185, *185,* 1097
    skin, 180–182, *181*
    visual, 601–602, *602, 603*
Pilocarpine, 559
Pilus, 183
Pineal gland, *496, 508, 510,* 511,
    614, 623–624, 631*t*

Pinna, *364,* 578, *579*
Pinocytic vesicles, 93
Pinocytosis, 93, 96*t*
Pitch, 578, 585
Pitcher's arm, 358
Pituitary dwarfism, 211, 645
Pituitary gigantism, 645
Pituitary gland, *496, 510,* 610–623,
    614
    absence or underdevelopment of,
        645
    aging and, 1099
    anatomy of, 616–619, *618*
    anterior
        anatomy of, 617, *618*
        histology of, 617, *619*
        hormones of, 619–620, 620*t*
        hypothalamic control of, 509,
            619, 620*t*
    disorders of, 645
    embryonic development of, 617,
        617
    negative feedback inhibition of,
        622, *622*
    in ovarian cycle, 1053–1054,
        1055
    posterior
        anatomy of, 617, *618,* 619
        histology of, 619, *619*
        hormones of, 620*t,* 620–621
        neuroendocrine reflexes and,
            621–622
    pregnancy and, 1061
Placebo, 8
Placenta, 1042, 1060, 1086, 1087,
    1088
    blood flow of, 1087
    development of, *1081,* 1087
    endocrine function of, 631, 632*t,*
        1060, 1089*t*
    expulsion of, 1064, *1065*
    functions of, 1089*t*
Placenta previa, 1068*t*
Placental abruption, 1068*t*
Placental blood, stem cells in, 672
Placental conductivity, 1087
Placental nutrition, 1087, *1087*
Placental phase, of pregnancy, 1087
Placental sinus, 1087, *1088*
Placental stage, of childbirth, 1064,
    1065
Placentation, *1081,* 1087
Planes, anatomical, *28,* 28–29
Planes of section, 140, *140*
Planetary model of atom, 42, *42*
Plantar fasciitis, 357
Plantar flexion, 284, *285*
Plaques
    atherosclerotic, 719, *720,* 735,
        1100
    complicated, 719, *720,* 735
    dental, 933
    senile, 453, *453*
Plasma, 154, 658–659, 902

    calcium in, 212
    composition of, 658*t,* 658–659
    electrolytes in, 658*t,* 658–659,
        908, *908*
    nitrogenous wastes in, 658*t,* 659
    nutrients in, 658*t,* 659
    as percentage of whole blood,
        657, *657*
    phosphate in, 215
    production of, 659
    protein deficiency in, 660
    proteins in, 658*t,* 658–659
Plasma (cell) membrane, 78, *79,*
    79–82
    composition of, 78
    electron microscopy of, 79, *80*
    extracellular face of, *79, 80*
    fluid-mosaic theory of, 9
    functions of, 78, 79, 105*t*
    intracellular face of, *79, 80*
    lipids of, *80,* 80–81
    permeability of, 88
    potential/polarization of, 393–
        394, 433–434
    proteins of, *80, 81,* 81–82
    TEM appearance of, 105*t*
    transport across, 87–96
    umbrella cell, 147
Plasma cells, 148, 659, 667, 788,
    806, 811–814, *812, 813,*
    814*t*
Plasma proteins, 658*t,* 658–659, 995*t*
    deficiency of, 660
Plasma transfusion, 668
Plasma volume expanders, 921
Plasmapheresis, 415
Plasmin, 683, *683*
Plasminogen, 683, *683*
Plasticity, 166
    smooth muscle, 414
    synaptic, 451–452
Plate(s)
    cribriform, *228, 229,* 235, *235,*
        236
    epiphyseal, 200, *207, 208,*
        208–209
    horizontal, of palatine bone, 237
    orbital, of ethmoid bone, 235,
        235, 236
    perpendicular
        of ethmoid bone, *228,* 235,
            235
        of palatine bone, 237
    pterygoid, *229, 234,* 235
    tarsal, 591, *591, 592*
Plateau phase, in sexual response,
    1029
    female, 1058, *1059*
    male, 1029
Platelet(s), *156,* 157, *655,* 657, *657,*
    678–680
    degranulation of, 680
    functions of, 679
    in hemostasis, 678–680, 683

    in inflammation, 801*t,* 803
    life history of, 679–680
    numbers of, 679
    open canalicular system of, *679,*
        679
    production of, *679,* 679–680
    pseudopods of, *679, 679*
    repulsion of, 683
    structure of, *679, 679*
Platelet count, 679
Platelet factors, *681,* 682, 682*t*
Platelet plugs, 679, 680, *680*
Platelet-derived growth factor
    (PDGF), 683, 803
Pleiotropy, *130,* 130–131, 667
Pleura, 33, *367, 827, 837,* 843
    parietal, 33, *33,* 35, *835, 837,* 843
    potential space between, 35
    visceral, 33, *33,* 35, *835, 837,*
        843
Pleural cavity, *32,* 32*t,* 33, *33, 369,*
    373, 827, 835, 837, 843,
    844
Pleural effusion, 907
Pleural fluid, 33, 837
Pleurisy, 837
Plexuses. *See* Nerve plexuses;
        Vascular plexuses; *specific*
        plexus
Plicae circulares (circular folds), 953
Plural forms, of medical terms, 20,
    20*t*
Pluripotent stem cells, 166, 660
Pneumocystis, 818
Pneumonia, 859*t,* 1100
Pneumothorax, 843
Podocytes, 872, *873,* 876, 877, *879*
Polar body
    first, 1050, *1051, 1077, 1078*
    second, *1051, 1078, 1078*
Polar covalent bonds, 46, 46*t,* 47
Polarized battery, 433
Polarized plasma membrane,
    393–394, 433
Poliomyelitis, 469
Pollex. *See* Thumb
Polycythemia, 664–665
Polycythemia vera, 665
Polydactyly, 131
Polydipsia, 647, 914
Polygenic inheritance, 130, *130*
Polygonal cells, 75, *76*
Polymerization, 56
Polymers, 56
Polypeptide(s), 63
Polypeptide hormones, 633*t,* 634
Polyphagia, 647
Polyribosome, 118, *119*
Polysaccharides, 56–57, 58*t*
Polyspermy, *1077,* 1078
Polysynaptic reflex arc, 485
Polythelia, 189
Polyunsaturated fats, 60
Polyunsaturated fatty acids, 59

Polyuria, 647, 889, 914
Pons, *494, 495, 496,* 503–504, *504, 508*
  in auditory pathway, 586
  in autonomic control, 558
  micturition center of, 893
  in respiratory control, 839–841
Pontine respiratory group (PRG), 840, *840*
Popcorn lung, 860
Popliteal lymph nodes, *784, 792*
Popliteal region, *30, 290*
Popliteal surface, of femur, 256, *258*
Porous-coated prostheses, 295, *295*
Porta hepatis, 946, 947, *947*
Portal hypertension, 766
Portal system, 733, *733*
  hepatic, *765, 947,* 953
  hypophysial, 617
Portal venules, 617
Port-wine stain, 183
Position, anatomical, 28, *28*
Positive chronotropic agents, 715
Positive feedback, 17
  in childbirth, 17, *17,* 1064
  in digestion, 944, *945*
  in harmful processes, 17
Positive inotropic agents, 717
Positive nitrogen balance, 983
Positive selection, of T cells, 807
Positive water balance, 907, *907t*
Positron emission tomography (PET), 20, *22, 23,* 538, *538*
  of Broca area, 525
  of cerebellum, 507–508
  of cognitive function, 519
Possessive forms, of medical terms, 20
Postabsorptive state, 997, *997t,* 998
Postcapillary venules, 732
Postcentral gyrus, *511, 522, 522,* 525
Posterior, 28, 29, *29t*
Posterior chamber, of eye, 594, *594*
Posterior mesentery, 34, *34,* 928, *1083*
Posterior nasal apertures, *229,* 235, 827, *827, 828*
Posterior pituitary gland
  anatomy of, 617, *618,* 619
  histology of, 619
  hormones of, *620t,* 620–621
  neuroendocrine reflexes and, 621–622
Posterior spinocerebellar tract, *465,* 466, *466t*
Posterior superior iliac spine, 255, *255, 376*
Posterior triangle. *See* Anal triangle
Postganglionic fibers, 545
  enteric, 927
  parasympathetic, *551,* 552
  pupillary, 597
  sympathetic, *547, 548,* 549

Postherpetic neuralgia, 478
Postovulatory (luteal) phase, 1053–1055, *1054,* 1056t
Postpartum period, 1066
Postsynaptic inhibition, 572
Postsynaptic neuron, 440, *440, 441–442*
Postsynaptic potentials, 444, *446, 446–447, 447*
Posttetanic potentiation, 452
Posttranslational modification, 120
Posttraumatic infarction, 488
Potassium, 910–911
  absorption of, 961
  in acid–base imbalance, 917, *919,* 920t
  in action potential, 436, *437*
  body percentage of, 41t, 42
  in cardiac physiology, *703,* 704, 716
  deficiency of, 643, 716, *910,* 911, *913t,* 917, *919,* 920t, 961
  dietary, 984t
  in ear (hearing), 584, *584*
  excess of, 716, *910,* 911, 913t, 917, *919,* 920t
  functions of, 910–911
  homeostasis of, 911
  in membrane polarization/ potentials, *433,* 433–434, *910,* 911
  in muscle fatigue, 406
  in neuronal signal conduction, 437
  renal reabsorption of, 881, 884
  in stress response, 643
Potassium channels, *433,* 433–434, *437, 703,* 704, 715
Potassium chloride, 45t
Potential(s). *See specific potential*
Potential difference, 393–394, 433. *See also* Resting membrane potential
Potential energy, 52
Potential spaces, 35, 1040
Potentiation
  posttetanic, 452
  synaptic, 451
Power stroke
  of cilia, 85, *86*
  of myosin, 396
PP cells, of pancreas, 629
PQ segment, 704, *705*
Prader–Willi syndrome, 132
Preantral follicles, 1052
Prebone (osteoid) tissue, 206, *207*
Precentral gyrus, *494, 511,* 522–523, *523*
Precipitation, by antibodies, 811, *813*
Preeclampsia, 1068t
Preembryonic stage, 1076, 1079t, 1079–1081
Prefixes, 19
Preformed water, 903

Prefrontal cortex, 514, *514,* 520, *521*
  aging and, 1099
  in emotion, 520
Preganglionic fibers, 544
  enteric, 927
  parasympathetic, 550–552, *551*
  pupillary, 597
  sympathetic, 546–549, *547, 548*
Pregnancy, 1059–1062
  adjustments to, 1061–1062
  assisted reproductive technology and, 167, 1104–1105
  breast changes in, 1045, 1047
  circulatory system in, 1062
  diabetes in, 890, 1068t
  digestive system, nutrition, and metabolism in, 1061–1062, 1087
  disorders of, 1068t
  ectopic, 1068t
  hormones of, *1060,* 1060–1061, 1061t, 1066
  maternal–fetal Rh mismatch in, *671,* 671–672
  multiple gestation, 1080
  nitrogen balance in, 983
  prevention of. *See* Contraception
  respiratory system in, 1062
  trimesters of, 1060, 1079
  urinary system in, 1062
  uterine growth and weight gain in, 1062, 1063t
Prehensile hands, 10, *10*
Preload, 717
Premature infants, 1092
Premature ventricular contractions (PVCs), 707, *707*
Premenstrual phase, *1054,* 1056
Premenstrual syndrome (PMS), 624
Premolars, 932, *932*
Premotor area, *521,* 522
Pre-mRNA, 117, *117*
Prenatal development, 1060, 1079–1090
  embryonic stage of, 1079, 1079t, 1082–1089, *1084–1085*
  fetal stage of, 1079, 1079t, *1084–1085,* 1086t, 1089
  major events of, 1086t
  nutrition and, 1087
  preembryonic stage of, 1076, 1079t, 1079–1081
Prenatal nutrition, 1087
Preovulatory follicle, 1053
Preovulatory phase, 1053
Prepuce
  female, *375, 1012, 1039,* 1045, *1045*
  male, *1012, 1014, 1018,* 1019, *1020*
Presbyopia, 599t, 1099
Pressure
  abdominal, 328, 839

airflow (ventilation) and, 842–843
atmospheric (barometric), 842–843, 848
blood. *See* Blood pressure
capsular, *878*
central venous, 744
filtration, 741, *742,* 876–877, *878*
hydrostatic, 89, *89,* 741, *742,* 876–877, 882
intrapleural, 843, *844*
intrapulmonary, 843, *844*
measurement of, 708
oncotic, 741, *742*
optic, 595
osmotic, 89, *89*
  colloid, 659, 741, *742,* 876, *878*
partial, of gases, 848–850, *851*
principles of, 708–710
pulse, 734
reabsorption, 742, *742*
syringe illustration of, *708,* 708–709
thoracic, 839, 842, *844*
Pressure gradient, 17, *18,* 708
  and gas exchange, 850, *851, 852*
Pressure points, arterial, 775, *775*
Pressure sores, 169
Presynaptic facilitation, 447–448, *448*
Presynaptic inhibition, 448, *449*
Presynaptic neuron, 440, *440, 441–442*
Pretectal region, 597
Primary active transport, 91, 96t
Primary afferent fibers, 484
Primary auditory cortex, 521, *521,* 586, *587*
Primary capillaries, 617
Primary cilium, 84
Primary curvatures, 241
Primary germ layers, 139, 1076, 1080–1081, *1081, 1082, 1083,* 1083t
Primary gustatory cortex, 521, *521,* 574
Primary hypertension, 776
Primary lymphatic organs, 789
Primary (main) bronchus, *827, 831, 832,* 833
Primary marrow cavity, *207,* 208
Primary motor cortex, *521,* 522–523, 525
Primary olfactory cortex, 521, 576, *577*
Primary oocytes, 1050, *1051*
Primary ossification center, *207,* 207–208
Primary ovarian follicles, *1040, 1051,* 1052, *1052*
Primary polycythemia, 665
Primary response, in humoral immunity, 811, *813*

Primary sensory cortex, 520
Primary sex organs. *See* Gonad(s)
Primary somatosensory cortex, *521, 522, 522*
Primary spermatocyte, *1025,* 1026
Primary structure of protein, 63, *64,* 120
Primary vesicles, 495, *498*
Primary visual cortex, 521, *521, 607, 607*
Primates
    basic adaptations of, *10,* 10–11
    bipedalism of, 11
Prime mover muscles, 305
Primipara, 1064
Primitive groove, 1080, *1082*
Primitive gut, 1082, *1083*
Primitive streak, 1080, *1082, 1084*
Primordial follicles, *1040, 1050–1052, 1051, 1052, 1054,* 1056*t*
Primordial germ cells, *1025*
Principal cells, of distal convoluted tubule, 884
Principal piece, of sperm, 1027
Proaccelerin (factor V), *681,* 682, 682*t*
Procaine, 608
Procarboxypeptidase, 950, 951*t*
Procedural memory, 452
Processes of bone, *226,* 226*t*
    alveolar, 238
    clinoid, anterior, 234, *234*
    condylar, 238, *238*
    coracoid, 250, *250*
    coronoid, 238, *238,* 251, *251*
    frontal, of maxilla, *236*
    inferior articular, *243*
        of cervical vertebra, *243, 244*
    mastoid, *228, 229,* 233, *233*
    odontoid, 244
    palatine, *228, 229,* 237
    pterygoid, *234,* 235
    spinous, 242, *242, 243*
        of cervical vertebra, *243, 244,* 245
        of lumbar vertebra, *244,* 245
        of thoracic vertebra, *244,* 245
    styloid
        of radius, *251,* 252, *377, 378*
        of temporal bone, *228, 229, 233, 233,* 238, *238*
        of ulna, *251,* 252, *377, 378*
    superior articular, 242, *243*
        of cervical vertebra, *243*
        of lumbar vertebra, *244*
        of sacral vertebra, 246, *246*
    temporal process, of zygomatic bone, *228*
    transverse, 242, *242, 243*
        of cervical vertebra, *243, 244*
        of coccygeal vertebra, *246*
        of lumbar vertebra, *244*
        of thoracic vertebra, *244*

xiphoid, 246, *247, 371*
zygomatic process, *228,* 232, *233*
    of frontal bone, *236*
    of temporal bone, 232, *233*
Procoagulants, 679, 682
Proconvertin (factor VII), *681,* 682, 682*t, 683*
Products
    of chemical reactions, 52
    of enzyme reactions, 66, *66*
    of metabolic pathways, 67
Progeria, 1102, *1102*
Progesterone, 629, 632*t,* 633*t,* 1049
    in childbirth, 1063
    interaction with estrogen, 641
    in menstrual cycle, *1054,* 1056
    in ovarian cycle, 1055–1056, 1056
    placental secretion of, 631, 632*t*
    in pregnancy, 1060, *1060,* 1061*t,* 1062, 1066
    in sodium homeostasis, 930
    synthesis of, 634, *635*
Progestin, 1070
Programmed cell death, *162,* 169, 806
Proinsulin, 634, *635*
Projectile vomiting, 942
Projection(s), of bone, 226*t*
Projection pathways, 568–569
    for equilibrium, 590, *590*
    for hearing, 585–587, *587*
    for olfaction, 576–577, *577*
    for pain, 569–570, *570*
    for taste, 574–575
    for vision, 605–607, *607*
Projection tracts, *512,* 513
Prolactin, 121, 616*t,* 620, 620*t,* 633*t*
    in lactation, 1066–1067, *1067*
    in pregnancy, 1061
Prolactin-inhibiting hormone (PIH), 616*t,* 619*t,* 1066
Proliferative phase, of menstrual cycle, *1054,* 1055, *1057*
Pronation, 282, *282,* 285, *335,* 336*t*
Prone, 28
Pronuclei, 1078
Pronunciation, 21
Prophase, 113
    I, in meiosis, 1023–1025, *1024*
    II, in meiosis, 1023, *1024*
    in mitosis, 125, *126*
Proplatelets, 679, *679*
Propranolol, 559, 776
Proprioception, 465, 474, 528
Proprioceptors, 483, 486, 566, 716
Prosemenogelin, 1027
Prosencephalon, 495, *498*
Prosopagnosia, 519
Prostacyclin, *643,* 644, 680
Prostaglandin(s), 61, *61, 643,* 644, 801*t*
    in childbirth, 1063–1064
    in digestive system, 929

functions of, 644, 644*t*
Prostaglandin E₂ (PGE₂), 799
Prostate, *375, 1018, 1019*
Prostate cancer, 1019
Prostate diseases, 1019, 1100
Prostate-specific antigen (PSA), 1019, 1027
Prostatic fluid, 1027
Prostatic urethra, 893, 1017, *1018*
Prostheses, joint, *267,* 295, *295*
Prosthetic group, 63
Protease(s), 212, 958
Protease inhibitors, 818
Proteasomes, 102, *102,* 105*t,* 127, 453
Protection
    by cerebrospinal fluid, 501
    by circulatory system, 656
    by connective tissue, 148
    by epithelial tissue, 141
    by innate immunity, 795. (*See also* Innate immunity)
    by skeleton, 199
    by skin, 175, 795–796
Protein(s), 55, 62–65
    absorption of, 958, *959, 962*
    absorptive state, 997
    aging and, 1102–1103
    antimicrobial, 797–798
    axonal transport of, 425–426
    carrier, 82
    channel, 81–82, *82*
    chaperone, 120
    complement, 659*t*
    complete, 983
    composition of, 56, 62
    conformation of, 63
    conjugated, 63
    contractile, 388
    degradation of, 102, *102*
    denaturation of, 63
    diabetes mellitus and, 646
    dietary, 977, 979*t,* 983
        calories from, 977
        nitrogen balance and, 983
        requirements for, 979*t,* 983
        sources of, 983
    digestion of, 926, 940, *941,* 944, *945, 958, 959, 962*
    envelope, 179
    enzymes as, 65–67. (*See also* Enzyme(s))
    fibrous, 63
    folding of, 120
    functions of, 63–65
    globular, 63, 983
    incomplete, 983
    membrane, *80, 81,* 81–82, *82*
    metabolism of, 994–995, 995*t,* 1012
    MHC, 807, 810
    net utilization of, 983
    newly synthesized, destinations and functions of, 121*t*

origin of word, 62
packaging or transport of, 118
peptide bonds of, 62
peripheral, *80,* 81, *81*
plasma, 658–659, 659*t,* 995*t*
    deficiency of, 660
postabsorptive state, 998
posttranslational modification of, 120
    in pregnancy, 1062
processing and secretion of, 120
protective, 795
receptor, 81, *82*
regulatory, 122, 388
structure of, 63, *64, 65,* 120
synthesis of, 116–120, 623, 995
transport, 65, 637
Protein buffer system, 915–916
Protein kinases, 452, 638
Protein phosphatases, 453
Protein-sparing effect, 623, 921, 980
Proteinuria, 876
Proteoglycans, 58, 58*t,* 149
Prothrombin, 659*t, 681,* 682, 682*t*
Prothrombin activator, *681,* 682
Protofilaments, 97, *99*
Proton(s), 42, *42*
Proton acceptor, base as, 50
Proton donor, acid as, 50
Proton pump, 990, *990*
Protraction, 281, *281*
    mandible, 283, *284*
    scapular, 281, *329,* 330*t*
Protuberance, of bone, 226*t*
Provitamins, 985
Proximal, 29, 29*t*
Proximal convoluted tubule (PCT), *872,* 873, *873, 874,* 881–882, *882*
Proximal phalanx, 252, *253*
Prozac (fluoxetine), 559
"Prune fingers," 179
Pseudopods, 86–87, *87,* 679, *679*
Pseudounipolar neurons, 425
Psoriasis, 193*t*
Psychic pneuma, 433
Psychopharmacology, 559
Psychosomatic effects, 8
Pterygoid plates
    lateral, *229, 234,* 235
    medial, *229, 234,* 235
Pubarche, 1048–1049
Puberty
    female, 1048–1049
    male, 1021
Pubic region, *30,* 31, *31*
Pubic symphysis, *254,* 255, *255, 257, 271, 271, 375*
    female, *1039, 1045*
    male, *1014, 1018*
Pudendal cleft, 1044
Pudendum, 1044
Puerperium, 1066
Pulled groin, 358

Pulled hamstring, 358
Pulmonary barotrauma, 861
Pulmonary circuit, 690, *691*, 747–748
Pulmonary compliance, 845
Pulmonary disorders. *See* Respiratory disorders
Pulmonary embolism, 684
Pulmonary function tests, *846*, 846*t*
Pulmonary lobule, 833
Pulmonary surfactant, 836, *836*, 845
Pulmonary trunk, 690, *694*, 697, *697*, 709, 747, *748*
Pulmonary valve, *695*, 696, *696*, 697
Pulmonary ventilation, 826, 837–848
   aging and, 1100
   assessment of, 845–847
   blood gases and, 856–858
   chemoreceptors in, *840*, 841, *841*
   exercise and, 857–858
   gas exchange in, 826, 848–858
   gas laws and, 842, 842*t*
   metabolic adjustments in, 855–856
   muscles in, *838*, 838–839
   neural control of, 839–841, *840*
   pH and, *856*, 856–858
   pressure and airflow in, 842–843
   resistance to airflow in, 843–845
   rhythm variations in, 847, 847*t*
   temperature and, 856
   voluntary control of, 842
Pulmonary venule, *836*
Pulp
   of spleen
      red, 794, *794*
      white, 794, *794*
   of tooth, *932*, 933
Pulp cavity, of tooth, 933
Pulsatile blood flow, 735
Pulse, 715
Pulse pressure, 734
Pumps, 82
   antral, 942
   chloride, 86
   proton, 990, *990*
   skeletal muscle, 732, 744, *745*
   sodium–potassium, 91–93, *92*, 434, 881, 884, 909, 911, 940
   thoracic (respiratory), 744–745
Punnett square, 129
Pupil, *591*, 593, *594*
   constriction of, 597, 598
   dilation of, 597
   dual innervation of, 556, *556*
   sympathetic innervation of, 597
Pupillary constrictor, 552, 597
Pupillary dilator, 597
Purines, 110, 441
Purkinje cells, *420*, 507, 525
Purkinje fibers, 702
Pus, 803
Putamen, *515*, 516

Pyloric glands, *939*, 940, 941, *945*
Pyloric part, of stomach, 937, *938*
Pyloric valve, 302
Pylorus, 937, *938*
Pyramid(s)
   of kidney, 870, *870*
   of medulla oblongata, 503, *504*, *505*
Pyramidal cells, 446, 513, *514*, 523
Pyramidal decussation, 503, *504*
Pyrexia, 798. *See also* Fever
Pyridoxine, 985*t*
Pyridoxine (vitamin B₆), 985, 985*t*
Pyrimidines, 110, *111*
Pyrogens, 799
Pyruvate, 68, *69*, 987, *987*
Pyuria, 889

Q

QRS complex, 704, *705*
Quadrants, of abdomen, 31, *31*
Quadrate lobe, of liver, 946, *947*
Quadrilateral muscles, 302
Quadriplegia, 488
Qualitative information, neural coding of, 449
Quantitative information, neural coding of, 449, *449*
Quaternary structure of protein, 63, *64*, 120
Quickening, 1086*t*
Quiescent period, cardiac, 712
Quiet respiration, 837

R

Rabies, 487*t*
Radial abduction, 284
Radial artery pulse, 715
Radial flexion, 283, *284*
Radial notch, *251*, 252
Radiation
   background, 44
   electromagnetic, 591
   heat loss via, 999
   infrared, 591, 999
   ionizing, 43–44
   measurement of, 44
   ultraviolet. *See* Ultraviolet radiation
Radiation therapy, 44
Radical group, 62
Radical mastectomy, 1048
Radioactivity, 42–44
   bone contamination from, 203
   decay of, 43
   discovery of, 43
   half-life of, 44
Radiography, *22*, 22–23
Radioisotopes, 43–44
Radiology, 2
Radium, 203
Radon, 44

Ramón y Cajal, Santiago, 433, 441, 446
Ramus (rami)
   communicating, *473*, 474, *474*, *546*, 548, *548*
      gray, 548, *548*
      white, 548, *548*
   of ischium, *254*, *255*, 256
   of mandible, 238, *238*
   of pubis, *254*, *255*, 256, *1045*
   of spinal nerves, 305, 472, *473*, 474, *474*, *546*, 548, *548*
Range of motion (ROM), 275
Ranvier, node of, *424*, 429, 438
Rapid ejection, ventricular, 712
Rapid eye movement (REM) sleep, 517, *518*, 519
Rapid ventricular filling, 710, *711*
Rate of reactions, 54
Raynaud disease, 558*t*
Reabsorption
   capillary, 741–743, *742*, 743, 882–883
   tubular (renal), 876, 881–885, *882*, *888*, 905
Reabsorption pressure, 742, *742*
Reactant(s), 53, 67
Reactant concentration, 54
Reaction cascade, 682, *683*
Reactions, chemical. *See* Chemical reactions
Reactive hyperemia, 737
Receptive field, 565, *565*
Receptive–relaxation response, 414, 941
Receptor(s), 16. *See also specific receptor*
   ACh, 393, 553–554, *554*, 554*t*
   adrenergic, 554, *554*, 554*t*
   autonomic nervous system, 553–556, *554*
   hair, *184*, 185, 186
   hormone, 637–639
   irritant, 841–842
   in neuropharmacology, 554–556, 559
   phasic, 566
   protein, 81, *82*
   sensory, 176, 422, 564–572
   somatic, 482, *482*
   stretch, 841. (*See also* Stretch receptors)
   tonic, 566
Receptor potential, 564
Receptor-mediated endocytosis, 93–95, *94*, 96*t*
Recessive alleles, 128–129, *129*, 131
Reciprocal inhibition, 484
Recognition, in immunity, 808–810, 811, *812*
Recommended dietary allowance (RDA), 977, 979*t*
Reconstructive phase, of tissue repair, 167

Recovery stroke
   of cilia, 85, *86*
   of myosin, 396
Recruitment
   definition of, 449
   of motor units, 402
   of ovarian follicles, 1052
Rectal valves, 963, *964*
Rectouterine pouch, *1039*, 1042
Rectum, *370*, *375*, *927*, 963, *964*
   defecation from, 966–967, *967*
   parasympathetic innervation of, *551*
   sympathetic innervation of, *547*
Rectus sheath, *320*, *321*, 366
Red, ability to see, *606*
Red area, of lips, 930
Red blood cell(s). *See* Erythrocyte(s)
Red blood cell count, 662–663, 678
Red bone marrow, 199, *200*, 205, *375*
   adult distribution of, 205, *205*, 789
   aging and, 1100
   B-cell maturation in, *806*, 807
   cancer of, 665
   hematopoiesis in, 199, 660, 663, 676
   histology of, 789, *789*
   in inflammation, 801, 802
   leukocyte storage in, 676
   as lymphatic organ, 783, *784*, 789
   sinusoids of, 679, *679*, 789
   stimulation of, 629, 801
   T-cell production in, 807
   vertebral, 241
Red hair, 185, *185*
Red muscles, 408
Red pulp, of spleen, 794, *794*
Red (slow-twitch) fibers, 407*t*, 407–408, 408*t*
Red–green color blindness, 605, *606*
Redness, in inflammation, 800, 802
Redox reactions, 54
Reduced ejection, ventricular, 712
Reducing agent, 54
Reduction, 54, 54*t*
Reduction division. *See* Meiosis
Reductionism, 12
Referred pain, 570
Reflex arc, *482*, 482–483
   autonomic (visceral), 543–544, *544*
   contralateral, 485
   intersegmental, 486
   ipsilateral, 485
   monosynaptic, 484
   polysynaptic, 485
Reflex(es)
   crossed extension, 485–486, *486*
   defecation, 558, 966–967
   definition of, 482
   duodenocolic, 966

enterogastric, *944,* 945
flexor (withdrawal), 484–485, *486*
gastrocolic, 966
gastroileal, 956
inflation (Hering-Breuer), 841
long (vasovagal), 929, *944,* 945
medullary ischemic, 738
micturition, 558, 893, *894*
milk ejection, 1045, 1067
neuroendocrine, 621–622
patellar, 484, *485*
photopupillary, 597
properties of, 482
short (myenteric), 929, 944, *944*
somatic, 460, 482–487
spinal cord trauma and, 488
stretch, 484
tendon, 486–487
tympanic, 583, *587*
vestibulo–ocular, 590
visceral, 482, 543–544, *544*
Reflexive memory. *See* Implicit memory
Reflux, valvular, 710
Refraction, 597, *597*
Refractive index, 597
Refractory period
    in action potentials, 436, *438,* 449, 704
    in cardiac muscle, 704
    in male sexual response, 1031
Regeneration, 167, *168*
    nerve, *431,* 431–432
    visual pigment, 602, *603*
Regeneration tube, 432
Regenerative medicine, 170
Region(s), of body, 29–31, *30*
Regional anatomy, 363
Regulatory proteins, 122, 388
Regulatory T cells, 807, 808, 814*t,* 817
Regurgitation, valvular, 710
Reissner membrane. *See* Vestibular membrane
Relative atomic mass, 43
Relative refractory period, 436
Relaxation, muscle, 396–400, *399*
Relaxation phase, of muscle twitch, 401, *401*
Relaxin, 1061, 1061*t*
Release factor, 118
Releasing hormones, 619
Remodeling, bone, 202, 210–211, 217, *217*
Remodeling phase, of tissue repair, 167
Renal area, of spleen, 794, *794*
Renal autoregulation, 878–879
Renal calculus, 892
Renal circulation, 870–871, *871, 872*
Renal clearance, 890
Renal compensation, 919
Renal corpuscle, 871–872, *873, 874*

Renal cortex, 870, *870, 871, 872, 874*
Renal failure, 896, 1103
    acute, 895*t*
    chronic, 895*t*
Renal fascia, 869
Renal fraction, of cardiac output, 870
Renal function tests, 890
Renal insufficiency, 896
Renal medulla, 870, *871, 872, 874*
Renal papillae, 870, *874*
Renal pelvis, 870, *870*
Renal pyramids, 870, *870*
Renal sinus, 870, *870*
Renal tubule, 872, *872, 873, 873, 874*
Renin, 630, 776, 875, 879–880
Renin–angiotensin–aldosterone (RAA) mechanism, 879–880, *880,* 884, 909
Renin–angiotensin–aldosterone (RAA) system, 627
Replication, DNA, 123–124, *124*
Replication fork, 123, *124*
Replicative senescence, 1102
Repolarization, of membrane, 394
Reposition, of thumb, 284, *284*
Representational hemisphere, 526
Reproduction, 15, 1008
Reproductive cycle, 1050
Reproductive hormones, 1021–1022. *See also specific hormone*
Reproductive system, 35, *37,* 1008–1013
    aging and, 1022–1023, 1038, 1049, 1101
    descent of gonads in, 1010–1013, *1013*
    development of external genitalia in, 1010, *1012*
    disorders of
        androgen-insensitivity syndrome, 1009, *1009*
        breast cancer, male, 1032*t*
        cryptorchidism, 1013
        erectile dysfunction, 1023, 1031, 1032*t*
        hypospadias, 1032*t*
        infertility, male, 1032*t*
        male, 1032*t*
        pelvic inflammatory disease, 1033
        penile cancer, 1032*t*
        prostate diseases, 1019
        sexually transmitted diseases, 1031, 1033
        testicular cancer, 1032*t*
        varicocele, 1032*t*
    embryonic development of, 1010–1013, *1011, 1012,* 1038
    environmental agents and, 1028
    female. *See* Female reproductive system

homologous organs of, 1010
male. *See* Male reproductive system
overview of, 1008–1009
parasympathetic effects on, 555*t*
sexual differentiation in, 1010, 1038
sympathetic effects on, 555*t*
urinary system association with, 867
Reproductive technology, 167, 1104–1105
Reserve cartilage, zone of, 209, *209*
Residual volume (RV), 845, *846,* 846*t,* 847
Resistance, 708
    to airflow, 863–865
    to blood flow, 735–737
Resistance arm, 274, *274, 276*
Resistance arteries, 728
Resistance exercise, 409, 1101
Resistance stage, in stress response, 642
Resistance vessels, arteries as, 726
Resolution, in microscopy, 77
Resolution phase, in sexual response, 1029
    female, 1058, *1059*
    male, *1030,* 1031
Resorption bays, 202
Resorption of bone, 202, 211–212
Respiration
    aerobic. *See* Aerobic respiration
    aging and, 1100
    cerebral control of, 503
    definition of, 826
    neonatal, 1091, 1092
    pulmonary. *See* Pulmonary ventilation
Respiratory acidosis, 857, 917, 919*t*
Respiratory alkalosis, 857, 918, 919*t*
Respiratory arrest, 847*t*
Respiratory bronchioles, 834, *836*
Respiratory burst, 796
Respiratory centers, 503, 839–841, *840,* 857
Respiratory compensation, 918–919
Respiratory cycle, 837, *844*
Respiratory disorders, 858–861
    acute rhinitis, 859*t*
    adult respiratory distress syndrome, 859*t*
    atelectasis, 843
    black lung disease, 845
    chronic bronchitis, 858–859, 1100
    chronic obstructive pulmonary diseases, 858–859, 1100
    cyanosis, 858
    decompression sickness, 861
    emphysema, 859, 1100
    hypoxia, 666, 747, 858
    infant respiratory distress syndrome, 845

lung cancer, 860, *860*
    obstructive, 847
    oxygen imbalances, 858
    oxygen toxicity, 858
    pneumonia, 859*t,* 1100
    pneumothorax, 843
    restrictive, 847
    sleep apnea, 859*t*
    tuberculosis, 818, 845, 859*t*
Respiratory distress, in left ventricular failure, 712
Respiratory distress syndrome
    adult, 859*t*
    infant, 845
Respiratory enzyme complexes, 990, *990*
Respiratory membrane, 836, *836,* 850
Respiratory mucosa, 576
Respiratory muscles, *319,* 320*t, 838,* 838–839
Respiratory rhythm, 841
    variations in, 847, 847*t*
Respiratory system, 35, *36,* 825–865
    aging and, 1100
    anatomy of, 826–837, *827*
    conducting zone of, 826–827
    effects on other systems, 862
    endoscopic view of, *831*
    excretion by, 868–869
    functions of, 826
    gas exchange in, 826, 848–858
    parasympathetic effects on, 555*t*
    pH (acid–base) control by, 826, 857, 916, 917, 919*t*
    pregnancy and, 1062
    pulmonary ventilation in, 826, 837–848
    respiratory zone of, 827
    sympathetic effects on, 555*t*
    water loss via, 903–904
Respiratory (thoracic) pump, 744–745
Respiratory volumes and capacities, *846,* 846*t,* 846–847
Respiratory zone, 827
Responsiveness, 14
Restenosis, after angioplasty, 719
Resting membrane potential (RMP), 93, 394, *433,* 433–434
Resting-and-digesting state, 544
Restrictive lung disorders, 847
Retching, 942
Rete testis, *1016*
Reticular activating system (RAS), 507, 518
Reticular cells, 788, 789, *789,* 807, 814*t*
Reticular fibers, 149
    in areolar tissue, 150
    in dermis, 179
    of lymph node, 790, *791*
    in lymph nodes, 790–791, *791*
    in red bone marrow, 790
    in reticular tissue, 150

Reticular formation, 466, 503, *505, 506,* 506–507
  in equilibrium, 590
  in pain, 507, 572
Reticular layer, of dermis, 179, *180, 180t*
Reticular tissue, 150, *151*
Reticulocytes, 663, *663*
Reticulospinal tract, 507
  lateral, *465,* 466*t,* 467
  medial, *465,* 466*t,* 467
Retina, *594,* 595–596
  detached, 596
  histology of, *600,* 600–601, *601*
  image formation in, 597–599
  ophthalmoscopic examination of, 596, *596*
  pigment epithelium of, 594, 599, *600*
  sensory transduction in, 599–602
Retinaculum, 304
Retinal, 601, *602*
Retinene, 601
Retraction (movement), 281, *281*
  mandible, 283, *284*
  scapular, 281, *329,* 330*t*
Retraction of blood clot, 683
Retrograde amnesia, 519
Retrograde transport, 426
Retroperitoneal organs, 33, 869, 929
Retrovir, 818
Retroviruses, 818
Reuptake, 445
Reverberating circuits, *450,* 451
Reverse osmosis, 89
Reverse transcriptase, 817, *817*
Reverse transcriptase inhibitors, 818
Reversible potentials, 434, 435*t*
Reversible reactions, 53
Reye syndrome, 800
Rh blood group, 670–672
  maternal–fetal mismatch in, *671,* 671–672
Rh immune globulin, 671
Rhabdomyolysis, 914
Rhenium-188, 190
Rheumatic fever, 710
Rheumatism, 294*t*
Rheumatoid arthritis, 294, *295*
Rheumatoid factor, 294
Rheumatologists, 294
Rhinitis, acute, 859*t*
Rh-negative, 670
Rhodopsin, 600, 601–602, *603, 604*
RhoGAM, 671
Rhombencephalon, 495, *498*
Rhomboids, 305
Rh-positive, 670
Rhythm method, of contraception, 1069, 1070*t*
Riboflavin, 985, 985*t*
Ribonuclease, 950, 951*t,* 961
Ribonucleic acid. *See* RNA
Ribose, 67

Ribosomal RNA (rRNA), 114
Ribosomes, 99, 100, *100,* 105*t, 109*
  free, *79*
  large subunit of, 118
  small subunit of, 118
  in translation, 117–120, *119*
RICE treatment, 358
Rickets, 203, 213
Right atrioventricular valve, *693, 695,* 696, *696,* 697, *697*
Right atrium, *693,* 694, *694, 695,* 697, *697*
Right bundle branch, 702
Right lower quadrant (RLQ), 31
Right lymphatic duct, *784,* 786, *787*
Right main bronchus, 833
Right rotation, 282, *283*
Right upper quadrant (RUQ), 31, *31*
Right ventricle, *693,* 694, 695, *695,* 697, *697*
Right-brained, 526
Rigor mortis, 400
Ringer's lactate solution, 921
Ringworm, 193*t*
RNA, 69, 114
  digestion of, 950, 961
  DNA transcription to, 116–117
  DNA vs., 114*t*
  function of, 114
  messenger. *See* Messenger RNA
  nucleotides of, 114
  ribosomal, 114
  transfer. *See* Transfer RNA
  translation of, 117–120, *119*
RNA polymerase, 116, 122
Rod(s), *600,* 600–603, *601*
  in dual visual system, 603–604
  in light and dark adaptation, 603
  in night (scotopic) vision, 600, 603, *605*
  in signal generation, 602, *604*
Roman medicine, 3, 19
Root(s)
  of brachial plexus, 488*t*
  of hair, *184,* 184–185
  of lumbar plexus, *477,* 478*t*
  of lung, 833
  of nail, 187, *187*
  of nose, *364*
  of penis, *375, 1018,* 1019
  of spinal nerve, *470,* 471–472, *473, 474*
  of tongue, 931, *931*
  of tooth, 932, *932*
Root canal of tooth, *932,* 933
Root (word), 19
Rootlets, of spinal nerve, *470,* 471, *473, 474*
Rosacea, 193*t,* 1097
Rostral, 29, 29*t,* 493, *494*
Rotation, 281–282
  axes of, joint, 275–277, *277*
  lateral (external), *281,* 282
  left, 282, *283*

medial (internal), *277, 281, 281*
  right, 282, *283*
  scapular, *329,* 330*t*
Rotator cuff
  anatomy of, *287,* 288, *330,* 334*t*
  injury to, 335
Rough endoplasmic reticulum, *79, 99,* 100, *100,* 105*t*
Round window, *579,* 581, *581*
RU-486, 1070
Rubrospinal tracts, 506
Ruffled border, of osteoclast, 202, *202*
Rugae
  gastric, 937, *938*
  palatine, 931
  urinary bladder, *892,* 893
  vaginal, *1039,* 1044

S

S cones, 604–605
S phase of cell cycle, 125, *125*
Saccades, 596
Saccule, *581,* 587–589, *588*
Sacral hiatus, 245, *246*
Sacral promontory, 246, *246*
Sacral region, *30,* 460
Sacral vertebrae, 240, *240,* 245, *246*
Sacrificial bonds, 203
Sagittal plane, 28, *28*
Sagittal section, *28*
Sagittal sinuses
  inferior, *754*
  superior, 497, *499, 754*
Saline, normal, 921
Saliva, 904, 934–935
Salivary amylase, 67, 934–935, 956
Salivary duct, *935*
Salivary glands, 904, 926, 934–935
  extrinsic (major), 934
  innervation of, *547, 551,* 552, 934–935
  intrinsic (minor), 934
  microscopic anatomy of, 934, *935*
  parotid, *551,* 552, *927,* 934, *934*
  salivation by, 934–935
  sublingual, *927, 931,* 934, *934*
  submandibular, *551, 927, 931,* 934, *934*
Salivation, 934–935, 1101
Salt. *See also* Sodium chloride
  dietary, 776, 910, 984*t*
Saltatory conduction, 438–440, *439*
Salty taste, 574
Sample size, 8
Sanger, Frederick, 650
Sarcolemma, 386, *387,* 390*t*
Sarcoma, 132, 203, 218*t*
Sarcomeres, *389,* 389–391, 390*t*
  contraction of, 396, *398*
  relaxation of, 396, *399*
Sarcoplasm, 386, *387*

Sarcoplasmic reticulum, 387, 390*t*
Satellite cells, 386, 427*t,* 428
Satiety, hypothalamic regulation of, 509
Saturated fats, 60, 981
Saturated fatty acids, 59, 981
Saturation
  carrier, 90, *90*
  hormone receptor, 637
Scala media (cochlear duct), 581, *581, 582*
Scala tympani, 581, *581, 582*
Scala vestibuli, 581, *581, 582,* 583
Scalp, *365*
Scalp muscles, *310*
Scanning electron microscope (SEM), 77, *85*
Scapular angles, 250, *250, 372*
Scapular muscles, 334*t*
Scapular region, *30*
Schizophrenia, 537*t,* 559
Schleiden, Matthias, 6
Schlieren photography, 1000, *1000*
Schwann, Theodor, 6, 75, 969
Schwann cells, *424,* 427*t,* 428, 469
  in myelin sheath formation, 429, *430*
  in nerve regeneration, *431,* 432
Sciatic notch
  greater, 255, *255*
  lesser, 255, *255*
Sciatica, 481
Scientific method, 6–9
Sclera, *591,* 592–593, *594, 600*
Scleral venous sinus, 594, *594*
Sclerosis, 428
Scoliosis, *241*
Scotopic vision, 600, 603, *605,* 1099
Scrotum, *367, 368, 375,* 1013–1015, *1014, 1016, 1018*
  countercurrent heat exchanger in, 1015, *1015*
  development of, 1010–1011, *1012*
  evolution of, 1014
  parasympathetic innervation of, *551*
  puberty and, 1021
  sympathetic innervation of, *547*
Scuba diving, 861
Seasonal affective disorder (SAD), 624
Sebaceous glands, *164, 188,* 189, 190*t,* 1097
Seborrheic dermatitis, 193*t*
Sebum, 179, 189
Second heart sound, 710
Second line of defense, 795
Second messenger(s), 81, 82, *83*
  calcium as, 452, 554, 638
  cAMP, 69, 82, *83.* (*See also* Cyclic adenosine monophosphate)
  in cardiac physiology, 715

DAG-IP$_3$, 638, *640*
in hormone action, 638, *638, 639, 640*
in immune system, 797
in long-term memory, 452
in olfaction, 576
in smooth muscle, 413
in synaptic transmission, 442, 444, *445*
in taste, 576
Second polar body, *1051, 1078, 1078*
Second trimester, of pregnancy, 1079
Secondary active transport, 91–92, *92, 96t*, 881
Secondary afferent fibers, 484
Secondary capillaries, 617, *618*
Secondary curvatures, 241
Secondary hypertension, 776
Secondary lymphatic organs, 789
Secondary marrow cavity, 208
Secondary oocyte, 1050, *1051, 1078*
Secondary ossification center, 208
Secondary ovarian follicles, *1040, 1051,* 1052
Secondary polycythemia, 665
Secondary response, in humoral immunity, 812, *813*
Secondary sex characteristics, 1009
Secondary sex organs, 1008
Secondary spermatocyte, *1025,* 1026
Secondary structure of protein, 63, *64,* 120
Secondary tympanic membrane, 581, *581, 583,* 584
Secondary vesicles, 495, *498*
Second-class lever, 275, *276*
Second-degree burns, 191, *192*
Second-order neuron, 465, *467,* 568
Secretin, 929, *944,* 945, 952
Secretion(s), 160, 162–163. *See also specific secretions and organs*
endocrine, 613–615
epithelial, 141
exocrine, 613
modes of, 162–163, *164*
neuronal, 422
paracrine, 613, 643
types of, 162
Secretory granules, 442
Secretory phase, of menstrual cycle, *1054,* 1056, *1057*
Secretory vesicles, *79, 98,* 101, 120
Segment(s)
bronchopulmonary, 833
of limb, 31
of neurons, 429
of spinal nerves, 460
Segmental bronchi, *827, 832,* 833
Segmentation, of small intestine, 955, *955*
Selectins, 801t, 802
Selection, natural, 9
Selection pressures, 9

Selective estrogen-receptor modulators (SERM), 219
Selective permeable membrane, 87
Selective serotonin reuptake inhibitors (SSRIs), 559
Selenium, 45, 984t
Self-antigens, 816
Self-tolerance, 790, 807, 816–817
Self-tolerant B cells, 807
Sella turcica, *228, 229,* 234, *234, 236*
Selye, Hans, 642
Semen, 1027t, 1027–1028
Semenogelin, 1027
Semicircular ducts, *579, 581,* 587, *588, 589,* 589–590
Semiconservative replication, 124, *124*
Semilunar valves, 696, 697, *697, 709,* 710, *711*
Seminal vesicles, *375,* 1017–1019, *1018*
development of, *1011*
semen production in, 1027
Seminiferous tubules, 629, *630, 1007, 1016,* 1017
spermatogenesis in, 1023, *1025,* 1025–1027
Senescence, 1097. *See also* Aging
accelerated (progeria), 1102, *1102*
definition of, 1096
evolution and, 1103
and exercise, 1101
health-care impact of, 1096
and life expectancy, 1101, 1103
of organ systems, 1097–1101
replicative, 1102
theories of, 1101–1103
Senile atrophy, 167
Senile plaques, 453, *453*
Sensation, 520–522, 564
epithelial, 141
skin, 175–176
Sense organs, 563–611. *See also specific organs*
aging and, 1099
arterial, 728
definition of, 564
Senses
chemical, 572–577
general, 521–522, 566–572
special, 521–522, 566
Sensible water loss, 904
Sensorineural deafness, 585
Sensory adaptation, 565–566
Sensory (afferent) nerve fibers, 470, 471t, 482, *482,* 484
Sensory (afferent) neurons, 422–423, *423,* 544
Sensory (ascending) tracts, 463–466, *465,* 466t, *467*
Sensory coding, 448
Sensory cortex, primary, 520

Sensory deprivation, 564
Sensory division, of PNS, 422
Sensory homunculus, 522
Sensory nerves, 470
Sensory neurons, 422–423, *423,* 544
Sensory projection, 568
Sensory receptor(s), 176, 422, 564–572. *See also specific types*
classification of, 566
duration information for, 565–566
general properties of, 564–566
for general senses, 566–572, *567,* 567t
intensity information for, 565
location information for, 565, *565*
modality information for, 564–565
projection pathways of, 568–569
Septal defects, 718t
Septic shock, 746
Septicemia, 684t
Septum (septa)
gland, 161, *163*
interatrial, *693,* 694, *695*
intermuscular, 303, *303*
interventricular, *693,* 695, *695*
median, of penis, 1020, *1020*
median, of scrotum, 1014, *1014*
nasal, *230,* 827, *828*
pellucidum, *500*
of testis, 1015, *1016*
Serial processing, 451
Serine protease, 1027
Serotonin, 441, *443,* 443t, 446
in depression/depression therapy, 559
gastric secretion of, 942t
in hemostasis, 680
in pain, *571,* 572
Serous cells, of salivary glands, 935
Serous fluid, 163, *165*
Serous glands, 162
Serous membranes (serosa), 32
digestive tract, 927, *928, 929*
of esophagus, 935
histology of, 163, *165*
of small intestine, 953, *954*
visceral peritoneum, 34, *34*
Serous pericardium, 691, 693
Serum, 657
Serum lipoproteins, 981
Servetus, Michael, 5
Sesamoid bones, 224
Set point, 15, 799, 975
Severe combined immune deficiency (SCID), 817, *817*
Sex chromosomes, 128, 131, *131,* 1009–1010, *1010*
Sex determination, 1009–1010, *1010*
Sex drive (libido), 1021
Sex hormones, 1021–1022. *See also specific hormones*

Sex organs
primary. *See* Gonad(s)
secondary, 1008
Sex pheromones, 188
Sex steroids, 627–628, 633
Sex-influenced traits, 186
Sex-linked traits, 131, *131*
Sexual cycle, 1050, 1053–1057, *1054*
Sexual differentiation, 1010, 1038
Sexual dimorphism
of brain, 493
in erythrocyte production, 662
of pelvis (skeleton), 256, *257,* 257t
Sexual function. *See also* Sexual response
hypothalamus and, 510
spinal cord trauma and, 488
Sexual intercourse, 1028
female sexual response, 1058, *1059*
male sexual response in, 1028–1031
Sexual response
aging and, 1022–1023
anatomical foundations of, 1029
excitement phase of, 1029, *1030*
female, 1058, *1059*
male, 1028–1031
neural control of, *1030*
orgasm in, *1030,* 1031
plateau phase of, 1029
refractory period of, 1031
resolution phase of, *1030,* 1031
Sexually transmitted diseases (STDs), 1031, 1033
Shaft
of clitoris, 1058
of femur, *258*
of hair, 184, *184*
of penis, *375, 1018,* 1019–1020
of rib, 247, *248*
of tibia, 258, *259*
Sharpey's fibers. *See* Perforating fibers
Shear stress, in blood flow, 737
Shell temperature, 999
Shin splints, 351
Shingles, 478
Shivering thermogenesis, 1001
Shock
anaphylactic, 746, 804, 815
cardiogenic, 745
circulatory, 683, 745, 906
compensated, 746
decompensated, 746
hypovolemic, 745
low venous return, 745
neurogenic, 488, 746
obstructed venous return, 745
septic, 746
spinal, 488
venous pooling, 745–746

Short association fibers, 513
Short bones, 200
Short limb, of incus, 579
Short (myenteric) reflexes, 929
Short-term memory, 452
Short-wavelength (S) cones,
        604–605, 606
Shoulder
    arteries of, 767
    bursae of, 287, 288
    dislocation of, 287
    joint of, 249, 268, 276, 278,
        286–288, 287
    ligaments supporting, 287, 288
    movements of, 277, 329
    muscles acting on, 286–288, 287,
        328, 330t
Shoulder girdle. See Pectoral girdle
Shunt, 733, 733, 1089
Sickle-cell disease, 124, 131,
        666–667, 667
Sickle-cell trait, 666
Sievert (Sv), 44
Sight, sense of. See Vision
Sigmoid colon, 374, 375, 927, 929,
        963, 964, 967
Sigmoid sinus, 755
Sigmoidoscopy, 968
Signal amplification, 444, 640, 640,
        682
Signal conduction, in neurons,
        437–438, 438, 439
Signs, 914
Sildenafil, 1031
Silent killer, hypertension as, 776
Simple diffusion, 87–88, 96t
Simple exocrine glands, 162, 164
Simple mastectomy, 1048
Simple microscopes, 5–6
Single covalent bond, 46, 46t
Single-nucleotide polymorphisms,
        115
Single-unit smooth muscle, 412, 412
Sinoatrial (SA) node, 701, 701–702,
        702, 703
Sinus rhythm, 702, 707
Sinuses, 226, 226t
    anal, 964, 965
    carotid, 728, 729, 738, 752
    cavernous, 754
    confluence of, 754
    coronary, 693, 694, 698, 699, 732
    dural, 497, 499
    dural venous, 732, 754
    ethmoidal, 227, 230, 235
    frontal, 227, 228, 230, 236, 365,
        828
    lactiferous, 1046, 1047
    maxillary, 227, 230
    medullary (lymph node), 791,
        791
    paranasal, 227, 230, 365
    placental, 1087, 1088
    renal, 870, 870

sagittal
    inferior, 754
    superior, 497, 499, 754
scleral venous, 594, 594
sigmoid, 754
sphenoid, 227, 228, 230, 236,
        365
straight, 754
subscapular (lymph node), 790,
        791
transverse, 497, 754
urogenital, 1011
Sinusoids
    hepatic, 729, 730, 765, 947, 948
    red bone marrow, 679, 679, 789,
        789
Sister chromatids, 113
Situs inversus, 13
Situs perversus, 13
Situs solitus, 13
Size principle, 402
Skeletal muscle(s), 157–158, 159,
        159t, 299–361, 386
    action of, 304–305, 401–404
    aging and, 218, 1098
    antagonist, 304, 305
    atrophy of, 167
    attachments of, 303–304, 304
    blood flow to, 747
    cardiac and smooth muscle vs.,
        410t
    circular, 302, 302
    compartments of, 301–302,
        302–303, 303
    conductivity of, 385
    connective tissue of, 301,
        301–313, 386
    contractility of, 385
    contractions. See Skeletal muscle
        contraction
    deconditioned, 409
    elasticity of, 385
    excitability of, 385
    extensibility of, 385
    extrinsic, 304
    fascicle orientation of, 302, 302
    fatigue of, 406, 409
    fibers. See Skeletal muscle fibers
    fixator, 305
    functional groups of, 305
    functions of, 300–301
    fusiform, 302, 302
    head and neck, 308–318
    hypertrophy of, 166
    innervation of, 305, 392,
        392–393
    insertion of, 304
    intrinsic, 304
    learning strategy for, 305, 308,
        363
    length–tension relationship of,
        326, 400, 400, 408
    lower limb, 344–358
    major, 306–307

metabolism of, 300, 385,
        404–409
microscopic anatomy of,
        386–391
motor units of, 391, 391–392
naming of, 301
nerve relationship with, 391–394
nonverbal communication by, 176
origin of, 304
parallel, 302, 302
pennate, 302, 302
prime mover (agonist), 305
rigor mortis of, 400
size of, 408
somatic reflexes of, 482–487
strength and conditioning of,
        408–409
striations of, 158, 159, 159t, 386,
        389, 389–391
structural hierarchy of, 390t
study of, 300
synergist, 304, 305
test of recognition, 383
triangular (convergent), 302, 302
trunk, 318–328
upper limb, 329–343
Skeletal muscle contraction, 396,
        398
    calcium in, 387, 396, 397
    concentric, 402, 403, 404
    eccentric, 402, 403, 404
    energy for, 404–406
    excitation for, 394, 395
    excitation–contraction coupling
        for, 396, 397
    force or strength of, 400, 400,
        401–402, 402
    isometric, 402, 403, 404, 404
    isotonic, 402, 403, 404, 404
    myogram of, 401, 401
    nerve stimulation in, 391–394
    neurological control of, 522–524
    sliding filament theory of, 9,
        396, 398
    stimulus intensity and, 402,
        402, 403
    whole-muscle behavior in,
        401–404
Skeletal muscle fibers, 157, 159,
        159t, 386, 386–391, 387,
        390t
    behavior of, 394–400
    contraction of, 396
    excitation in, 394
    excitation–contraction coupling
        of, 396
    extrafusal, 483, 484
    fast-twitch (type II), 407t,
        407–408, 408t
    intermediate, 408
    intrafusal, 483, 484
    physiological classes of, 407t,
        407–408
    relaxation of, 396–400, 399

slow-twitch (type I), 407t,
        407–408, 408t
Skeletal muscle pump, 732, 744, 745
Skeletal system, 35, 36, 199,
        223–266. See also Bone(s)
    aging and, 1098
    bones of, 224t, 224–227, 225
    effects on other systems, 220
Skeleton, 199, 224, 225
    adaptations to bipedalism,
        262–263, 262–263
    appendicular, 224, 224t, 225
    articulated, 227
    axial, 224, 224t, 225
    disarticulated, 227
    fetal, 207, 208
    sexual dimorphism in, 256, 257,
        257t
Skin, 175–183
    absorption by, 176
    accessory organs of, 183
    aging and, 1097, 1097–1098
    body area and weight of, 175
    collagenous fibers of, 149
    as cutaneous membrane, 163
    endocrine function of, 629, 632t
    functions of, 175–176
    glands of, 183, 187–189, 188,
        190t
    healing (repair) of, 167, 168
    immune function of, 175,
        795–796
    innervation of, 481, 481
    nonverbal communication by,
        176, 177
    pregnancy and, 1062
    sensation in, 175–176
    structure of, 163, 175, 176, 180t
    thermoregulation by, 176
    thickness of, 175
    transpiration via, 189, 903–904
    ultraviolet radiation and, 175,
        181, 182, 190, 193, 1097
Skin color, 180–182
    diagnostic value of, 181–182
    evolution of, 182
    polygenic inheritance of, 130
    variation in, 181, 181
Skin disorders, 190–192, 193t
    acne, 193t
    basal cell carcinoma, 190, 191
    burns, 191–192, 192, 194
    cancer, 132, 190–191, 191
    contact dermatitis, 193t, 819t
    dermatitis, 193t, 819t
    eczema, 193t
    lesions, 190
    melanoma, 132, 190–191, 191
    psoriasis, 193t
    ringworm, 193t
    rosacea, 193t
    seborrheic dermatitis, 193t
    squamous cell carcinoma,
        190, 191

tinea, 193*t*
Skin grafts, 194
Skin markings, 182–183
Skull, 224*t*, 225, 227–229
 adaptations to bipedalism, 263, *263*
 anterior view of, *226*
 bones associated with, 238
 cavities of, 227–231
 frontal section of, *230*
 in infancy and childhood, 238–239, *239*
 intramembranous ossification of, 206–207, *207*
 lateral view of, *228*
 median section of, *228*
 superior view of, *230*
 sutures of, 227, *227–230*, 232, 270, *270*
 trephination of, 608
Skullcap (calvaria), *230*, 231–232
Sleep, 517–519
 definition of, 517
 hypothalamus and, 509, 517
 and memory, 519
 non-REM, 517, *518*, 519
 purposes of, 519
 rapid eye movement (REM), 517, *518*, 519
 reticular formation and, 507
 stages of, 517, *518*
Sleep apnea, 859*t*
Sleep paralysis, 517
Sleep spindles, 517
Sliding filament theory, of muscle contraction, 9, 396, *398*
Slow axonal transport, 426
Slow block, of polyspermy, 1078
Slow calcium channels, 704
Slow (second) pain, 569
Slow-twitch (slow-oxidative) fibers, 407*t*, 407–408, 408*t*
Slow-wave sleep, 519
Small arteries, 728
Small intestine, *368, 374, 375, 926, 927, 929*, 952–956
 absorption in, 941, 952–956
 circular folds of, 953
 circulation of, 952
 endocrine function of, 161, 630–631, 632*t*, 953
 in gastric regulation, 942, 944
 gross anatomy of, *952*, 952–953
 innervation of, 927
 length and diameter of, 952
 lining of, *137*
 mesenteries and movement of, 928
 microscopic anatomy of, 953, *954*
 motility of, *955*, 955–956
 mucosa of, *925*, 953
 parasympathetic innervation of, *551*

secretions affecting, 946
secretions of, 955
sympathetic innervation of, *547*
tight junctions of, 160
villi of, 953, *954*
Small motor units, 391, *391*, 408
Small subunit of ribosomes, 118
Small veins, 732
Small-cell carcinoma, 860
Smallpox vaccine, 804
Smears, 140
 blood, *156*, 156*t*
 spinal cord, *158*, 158*t*
Smell (olfaction), 515, 521, 572, 575–577
 aging and, 1099
 anatomy of, *575, 576*
 evolution of, 576
 physiology of, 576
 projection pathways of, 576–577, *577*
 respiratory function and, 826
Smoking
 and COPD, 858–859
 and lung cancer, 860, *860*
Smooth chorion, 1087
Smooth endoplasmic reticulum, *79, 100*, 100–101, 105*t*, *974*
Smooth muscle(s), 157–158, *159*, 159*t*, 409, 410–414
 autorhythmicity of, 412
 contraction of, *413*, 413–414
 dense bodies of, 412, *413*
 excitation of, 412–413
 hyperplasia of, 411
 hypertrophy of, 411
 innervation of, 411, *411*
 multiunit, 412, *412*
 myocytes of, 409, 411–412
 plasticity of, 414
 response to stretch, 414
 single-unit (unitary), 412, *412*
 skeletal and cardiac muscle vs., 410*t*
 types of, 412
 visceral, 158, *411*, 412
Smooth muscle tone, 413
Sneezing, 847
Snuffbox, anatomical, *378*
Sodium, 908–910
 absorption of, 961
 in action potential, 436, *437*
 aldosterone and, 909, *909*
 body percentage of, 41*t*, 42
 deficiency of, 910, 913*t*
 dietary, 984, 984*t*
 excess of, 910, 913*t*
 functions of, 908–909
 homeostasis of, 909–910
 in membrane polarization/ potentials, *433*, 433–434
 in neuronal signal conduction, 437
 plasma, 658*t*, 659

renal reabsorption of, 881, 884, 905
 in stress response, 643
Sodium bicarbonate, 45*t*, 909
Sodium channels, *433*, 433–434, *437*
Sodium chloride (NaCl), 45*t*
 dietary, 984*t*
 ionic bonds of, 46
 renal reabsorption of, 881, 884
Sodium ion(s), 44. *See also* Sodium
Sodium–calcium antiport, 961
Sodium–glucose transporters (SGLTs), 91–92, *92*, 882, 957
Sodium–potassium pump, 91–93, *92*, 434, 881, 884, 909, 911, 940
Soft callus, 216, *217*
Soft drinks, and bone, 216
Soft keratin, 183
Soft palate, 237, *827, 828, 930, 931*
Soft spots. *See* Fontanelles
Solar lentigines, 1097
Solubility, of gases, 850
Soluble antibodies, 804
Solute, 49
Solution(s), 49–50, 51*t*
 acidic, 50
 basic (alkaline), 50
 concentration of, 51
 neutral, 50
 properties of, 49
 tonicity of, 89–90, *90*
Solvent, 49
 water as, 48
Solvent drag, 742, 881, 957
Soma (cell body), 157, *158*, 158*t*, 423, *424*, 425
Somatic cells, 128
Somatic hypermutation, 806
Somatic motor division, of peripheral nervous system, 422
Somatic motor fibers, 391
Somatic motor neurons, 391
Somatic nerve fibers, 470, 471*t*
Somatic pain, 569
Somatic receptors, 482, *482*
Somatic recombination, 805
Somatic reflexes, 460, 482–487
 crossed extension, 485–486, *486*
 flexor (withdrawal), 484–485, *486*
 muscle spindles of, *483*, 483–484
 patellar, 484, *485*
 reflex arc of, *482*, 482–483
 stretch, 484
 tendon, 486–487
Somatic senses, 521
Somatic sensory division, of peripheral nervous system, 422
Somatomedins, 616*t*, 622–623

Somatosensory association area, *521*, 522
Somatosensory cortex, primary, *521, 522, 522*
Somatosensory projection pathways, 568–569
Somatosensory senses, 521, 566
Somatostatin, 556, 619, 619*t*, 629, 631*t*, 633*t*, 638, 942*t*
Somatotopy, 522
Somatotropic cells, 620
Somatotropin. *See* Growth hormone
Somesthetic senses, 521, 566
Somites, *497*, 1082, *1084*
Sonogram, 23, *23*
Sonography, 23, *23*
Sound
 definition of, 578
 frequency of, 578, *578*, 585, *586*
 nature of, 578, 585
Sound waves, 582
Sour taste, 574
Space, muscular atrophy in, 167
Spastic paralysis, 488
Spatial summation, 447, *447*, 603
Special nerve fibers, 470, 471*t*
Special senses, 521–522, 566. *See also specific senses*
Specific granules, 673
Specific gravity of urine, 889
Specific immunity. *See* Adaptive immunity
Specificity
 adaptive immunity, 803
 carrier, 90
 enzyme–substrate, 66
 receptor–hormone, 637
Spectrin, 662, 664
Spelling, 21
Sperm count, 1027
Sperm (spermatozoon), 1008, 1023–1028
 acrosomal reaction of, 1077, *1077*
 acrosome of, *1026*, 1027, 1077, *1077*
 aging and, 1022, 1101
 artificial insemination by, 1104
 capacitation of, 1076–1077
 cortical reaction of, *1077*, 1078
 fertilization by, *1075*, 1076–1078, *1077*
 haploid chromosomes of, 128, 1023
 meiosis and, 1023–1025, *1024*
 migration of, 1076
 motility of, 1027–1028
 production of, 1015, 1021–1022, 1023, *1025*, 1025–1027
 puberty and, 1021
 reabsorption of, 1017
 storage of, 1017
 structure of, *1026*, 1027
 tail of, *1007, 1016, 1026*, 1027

Sperm washing, 1104
Spermatic cord, *1013, 1014,*
1014–1015, 1016
Spermatic ducts, 1017
Spermatids, *1025,* 1026
Spermatocytes
primary, *1025,* 1026
secondary, *1025,* 1026
Spermatogenesis, 1023, *1025,*
1025–1027
Spermatogonia, *1025,* 1025–1027
Spermicides, 1069–1070, 1070*t*
Spermiogenesis, *1025,* 1026
SPF (sun protection factor), 193
Sphenoid sinus, 227, *228, 230,*
236, 365
Spherical aberration, 6, 598
Spheroidal cells, 76
Sphincters, 157, 300, 302, *302,* 927
anal, 302
external, 302, *326,* 327*t, 964,*
965, 967, *967*
internal, *964,* 965, 967, *967*
esophageal
lower, 935
upper, 935
hepatopancreatic, 947, *948*
orbicularis oculi, 302, *302, 306,*
308, 309, 310, 592
physiological, 935
precapillary, 731, *731,* 738
pyloric, 302, 937, *938,* 945
urethral
external, 302, 893–894
internal, 893
Sphygmomanometer, 708, 734
Spicules, 205
Spina bifida, 1062
Spinal cord, *365, 373, 374,* 460–468,
461
aging and, 1098
in autonomic control, 558
cervical enlargement of, 460, *461*
conduction in, 460
cross-sectional anatomy of,
462–463, 463
development of, 460, 1082
disorders of, 487*t*
functions of, 460
lumbar enlargement of, *461*
lumbosacral enlargement of, 460
meninges of, *461,* 461–463, *462*
neural integration in, 460
regions of, 460
surface anatomy of, 460–461
white matter of, *462,* 463
Spinal cord injuries, 461, 488, 894
Spinal cord smear, *158,* 158*t*
Spinal curvatures, *240,* 241
Spinal gating, *571,* 571–572
Spinal meningitis, 487*t*
Spinal nerve(s), 305, 460, *461, 462,*
471–481, *504*
dermatomes of, 481, *481*

distal branches of, 472, 474
ganglia of, 471, *471, 473, 474*
meningeal branch of, 472, *473*
motor functions of, 474
nerve plexuses. *See* Nerve
plexuses
proximal branches of, 471–472
rami of, 305, *473,* 474, *474, 546,*
548, *548*
rootlets of, *470,* 471, *473, 474*
roots of, *470,* 471–472, *473, 474*
somatosensory functions of, 474
sympathetic fibers to, 549
Spinal nerve disorders, 487*t*
amyotrophic lateral sclerosis, 469
crutch paralysis, 481
Guillain–Barré syndrome, 487*t*
Hirschsprung disease, 553
neuralgia, 478, 487*t*
paresthesia, 487*t*
peripheral neuropathy, 487*t,* 569
poliomyelitis, 469
rabies (hydrophobia), 487*t*
sciatica, 481
spinal meningitis, 487*t*
Spinal osteoporosis, 219, *219*
Spinal reflexes, 460
Spinal shock, 488
Spinal tap, 463, 464, *464*
Spinal tracts, 463–468
ascending, 463–466, 466*t, 467*
decussation of, 465
descending, 464, *465,* 466*t,*
466–468, *468*
Spindle fibers, 125, *126*
Spine. *See* Vertebrae; Vertebral
column
Spine of bone, 226, 226*t*
iliac, *254,* 255, *255, 366, 371,*
376
anterior inferior, *254,* 255,
376
anterior superior, 31, *254,*
255, *366, 371*
posterior inferior, 255
posterior superior, 255, *376*
ischial, *254,* 255, *255*
mental, of mandible, *228,* 238
scapular, 250, *250*
Spinocerebellar tract
anterior, *465,* 466, 466*t*
posterior, *465,* 466, 466*t*
Spinoreticular tract, *465,* 466*t*
Spinothalamic tract, *465,* 465–466,
466*t, 467*
Spiral ganglion of cochlea, *581*
Spiral organ, 581, *582*
Spiral (pectineal) line, *255,* 256, *258*
Spirometer, 846
Spirometry, 845–847, *846,* 846*t*
Spleen, *368, 369, 784,* 789, 793–794,
794
erythrocyte death in, 665, *665,*
794

gastric area of, 794, *794*
hematopoiesis in, 660, 794
parasympathetic innervation of,
551
red pulp of, 794, *794*
renal area of, 794, *794*
ruptured, 794
sympathetic innervation of, *547*
white pulp of, 794, *794*
Splenectomy, 794
Splenic flexure, 963, *964*
Splenomegaly, 766, 819*t*
Split-skin graft, 194
Sponge, contraceptive, 1069–1070,
1070*t*
Spongiocytes, 627
Spongy bone, 154, 200, *200, 201,*
205, *205*
Spongy urethra, 893, 1017, *1018*
Spontaneous abortion, 1068*t*
Spontaneous bacterial peritonitis
(SBP), 766
Spontaneous generation, 75
Sports injuries, 358
Sprain, 293, 294*t*
Spreads, 140
Sputum, 859
Squamous cell(s), 75, *76,* 142
Squamous cell carcinoma, 190, *191,*
860
Squamous epithelial tissue, *142*
simple, 141, *143,* 143*t*
stratified, 142, *145,* 145*t*
Squamous part, of temporal bone,
232, *233*
Squamous (type I) alveolar cells,
836, *836*
SRY gene, 1010
St. Martin, Alexis, 969, *969*
ST segment, 704, *705*
Stability, muscles for, 300
Stadium, fever, 799, *819*
Stage of exhaustion, in stress
response, 642–643
Stage of resistance, in stress
response, 642
Stains, 140
Starch, 57, 58*t*
dietary intake of, 980
digestion of, 67, 930, 935, 943,
950, *956,* 956–957
Start checkpoint in cell cycle, 127
Start codon, 116
Starvation, 660, 867
Static equilibrium, 587
Stationary attachment, of muscle,
304
Statistical tests, 8
Stearic acid, *59*
Stellate cells, 75, *76,* 513, *514*
Stellate macrophages, 797
Stem cell(s), 166
adult, 166
aging and, 1100

embryonic, 166, 167
epidermal, 177, *177*
gastric, 940
hematopoietic, 660, 663, 676
multipotent, 166
muscle, 386
pluripotent, 166, 660
totipotent, 166
unipotent, 166
Stem cell transplant, 672
Stem (root), 19
Stenosis, valvular, 710
Stents, for arteries, 719
Stercobilin, 949
Stereocilia, 581
Stereopsis, 605, *606*
Stereoscopic vision, 10, *10,* 605, *606*
Stereotyped reflexes, 482
Sterilization, surgical, 1070
Sternal angle, 246, *247, 371*
Sternal end, of clavicle, 249, *249*
Sternal region, *30*
Steroidal anti-inflammatory drugs
(SAIDs), *643,* 644
Steroids, 58, 58*t, 61,* 61–62,
627–628, 633, 633*t, 634.*
See also specific types
anabolic, 216
anabolic–androgenic, 69–70
in bone physiology, 216
mode of action, 637, *638*
secretion of, 636–637
for spinal cord injury, 488
synthesis of, 122, *123,* 634, *635*
transport of, 637
Stigma of ovary, 1053, *1055*
Stimuli/stimulus, 14
duration of, 565–566
for sensory receptors, 564–566
Stimulus intensity
and muscle contraction, 402, *402,*
403
and neural coding, *449,* 450
and sensory receptors, 565
Stomach, *367, 368, 374,* 926, *927,*
929, 937–946
absorption of, 943
aging and, 1100–1101
blood supply of, 937
body of, 937, *938*
cardial part (cardia) of, 937, *938*
digestive function of, 937,
944–945, *945*
endocrine function of, 161,
630–631, 632*t,* 940
fundic region (fundus) of, 937,
938
greater curvature of, 929, 937,
938
gross anatomy of, 937, *938*
historical perspective on, 937
innervation of, 927, 937
lesser curvature of, 937, *938*
mesenteries associated with, 928

microscopic anatomy of, 937, *939, 940*
motility of, 941–942
parasympathetic innervation of, *551*
pH of, 940, 944, *945*
protection of, 943
pyloric part of, 937, *938*
regulation of, *944,* 944–945
secretions of, 940–941, 942*t, 945*
sympathetic innervation of, *547*
tight junctions of, 160, 943, 1003
volume of, 937
Stop codon, 116
Strabismus, 415
Straight hair, 185, *185*
Straight sinus, *754*
Strain, 294*t*
Stratum basale, 177, *177,* 180*t*
Stratum basalis, 1042, *1057*
Stratum corneum, 175, 176, *177,* 180*t*
Stratum functionalis, 1042, *1044, 1057*
Stratum granulosum, 177, *177,* 180*t*
Stratum lucidum, *177,* 178, 180*t*
Stratum spinosum, *177,* 177–178, 180*t*
Strawberry birthmarks, 183
Streptokinase, 685
Stress, 642–643
    definition of, 642
    mind–body connection in, 820
Stress fracture, 216
Stress proteins, 120
Stress response, 642–643
Stressors, 642
Stress–relaxation response, 414
Stretch, smooth muscle response to, 414
Stretch marks, 179, 1062
Stretch receptors
    in blood pressure, 16, *16,* 544, *544*
    in childbirth, 637
    in defecation, *967*
    in equilibrium, 589
    in micturition, 893–894, *894*
    in neural circuit, 451
    in pulmonary ventilation, 841
    in skeletal muscles, 302
    in somatic reflexes, 483
    in swallowing, 936
Stretch reflex, 484
Striae, 179, 1062
Striations
    cardiac muscle, 158, *159,* 159*t,* 409, 700, *700*
    skeletal muscle, 157–158, *159,* 159*t,* 386, *389,* 389–391
Stroke, 502, 684, 712, 747, 1100
Stroke volume (SV), 712, 715, 717
Stroma, 162
Strong acids and bases, 915

Structural formula, 45, *45*
Structural hierarchy, 11–12, *12,* 21
Strychnine, 559
Studies in the Psychology of Sex (Ellis), 1028
Subarachnoid space, *461,* 462, *462,* 498, *499*
Subclavian trunks, 786, *787*
Subcostal line, 31, *31*
Subcutaneous fat, 179–180
Subcutaneous (sub-Q) fluids, 921
Subcutaneous tissue, 175, *176,* 179–180, 180*t*
Subdural space, *499*
Subendocardial conducting network, *701,* 702
Sublingual ducts, *934*
Sublingual gland, *927, 931, 934, 934*
Submandibular duct, *934*
Submandibular gland, *551, 552, 927, 931, 934, 934*
Submucosa, *927, 928*
    of esophagus, 935
    of large intestine, 965
    of stomach, 937
Subscapular sinus, 790, *791*
Substance abuse, 559
Substance P, 443*t,* 556, *571, 572,* 941
Substantia nigra, *505,* 506
Substrate, 65, 66, *66*
Sucrase, 957
Sucrose, 56, *57,* 58*t, 962,* 979
Sudoriferous glands. *See* Sweat glands
Suffixes, 19–20
Sugar(s), 56–57, 58*t*
    amino, 149
    dietary, 978–980
    digestion of, 926, 957
Sulcus (sulci)
    of bone, 226*t*
        intertubercular, 251, *251*
    of brain, 493, *494,* 495, *500, 511,* 511–512
        central, *494, 496,* 511, *511*
        lateral, *494, 500, 511,* 512
        parieto–occipital, *496,* 512
    gingival, 932, *932*
    of heart, 696
        coronary, *694,* 696
        interventricular, *693, 694,* 696
    mentolabial, *364*
    palpebral
        inferior, *364*
        superior, *364*
    of spinal cord, posterior median, 460, *461, 462, 473*
    terminal, 931, *931*
Sulfur, body percentage of, 41*t,* 42
Summation, 402, *403*
    multiple motor unit, 402, 408

postsynaptic potential, 446–447, *447*
    spatial, 447, *447,* 603
    temporal, 402, 408, 447, *447*
    wave, 402
Sun protection factor (SPF), 193
Sunscreens, 193
Sunstroke, 1002
Superciliary ridge, *364*
Superficial, 29, 29*t*
Superior, 29, 29*t*
Superoxide anion, 45, 796
Superoxide dismutase (SOD), 45
Supination, 282, *282,* 285, *335,* 335*t–336t*
Supine, 28
Supporting cells
    in ear, 581, *582, 588, 589*
    in nervous system. *See* Neuroglia
    in olfaction, *575,* 576
    in taste, *573,* 574
Supracondylar lines, 256, *258*
Supracondylar ridge
    lateral, 251, *251*
    medial, 251, *251*
Supraorbital margin, *226,* 231
Supraorbital notch, 231
Suprarenal glands. *See* Adrenal glands
Suprascapular notch, 250, *250*
Supraspinal control centers, 894
Suprasternal (jugular) notch, 246, *247, 364, 371*
Surface anatomy, *30,* 363. *See also specific anatomy*
    abdominal region, *371*
    back, *372*
    brain, *494*
    head and neck, *364*
    lower limb, *379–382*
    spinal cord, 460–461
    upper limb, *377–378*
Surface area, of cells, 76, *77,* 88
Surface film, 48
Surface tension, 48
Surfactant, 836, *836,* 845
Surgical anesthesia, 608
Surgical neck, of humerus, 251, *251*
Surrogate mothers, 1105
Suspension, 49, *50, 51t*
Sustentocytes (nurse cells), 629, *630,* 1015, *1016,* 1021, *1022, 1025*
Sutural bones, 225, *230,* 232
Sutures, 227, 270, *270*
    butt, 270, *270*
    coronal, *226, 228, 230,* 231, *239*
    intermaxillary, *226, 229,* 236
    lambdoid, *228, 229, 230,* 231, *239*
    lap, 270, *270*
    metopic, 239
    plane, 270, *270*
    sagittal, *230,* 231, *239*

serrate, 270, *270*
    squamous, *226, 228,* 231, *239,* 270, *270*
Swallowing, 829, 936
    appetite and, 977
    difficulty (dysphagia), 968*t*
    muscles of, *313, 314,* 315*t–316t*
    phases of, 936, *936*
Swallowing center, 936
Sweat glands, 163, *188,* 188–189, 190*t,* 1097
Sweat (sweating), 903–904, 906, *906,* 1002
Sweet taste, 574, 979
"Sweet tooth," 979
Swell body, nasal, 829
Swelling. *See* Edema
Sympathetic division, of ANS, 422, 544–545
    abdominopelvic components of, 549, *550*
    aging and, 1099
    anatomy of, 546–549, *547*
    blood vessel control by, 738
    cardiac innervation/regulation by, *547,* 549, 714
    drugs affecting, 559
    in dual innervation, *556,* 556–557
    effects on target organs, 553–557, 555*t*
    exit route of nerve fibers in, 549
    gastric innervation by, 937
    mind–body connection in, 820
    neurotransmitters and receptors of, 553–556, *554*
    parasympathetic division vs., 552*t*
    penile innervation by, 1028
    renal innervation/regulation by, 875, 879, *879*
    salivary gland innervation by, 934
    in vasomotor tone, 557, *557*
Sympathetic tone, 544, 738
Sympathetic trunk, *548*
Sympathoadrenal system, 549
Sympatholytics, 559
Sympathomimetics, 559
Symphyses, *271,* 271–272
    mental, 238
    pubic, *254, 255, 255, 257,* 271, *271, 375*
Symport, 91, 881
Symptoms, 914
Synapses, 392, 425, *440,* 440–446, *442*
    aging and, 1099
    autonomic system, 544–545, 549
    axoaxonic, *440,* 441
    axodendritic, *440,* 441
    axosomatic, *440,* 441
    chemical, 441, *442*
    electrical, 441

Synapses (continued)
excitatory adrenergic, 444–445, 445
excitatory cholinergic, 442, 444, 444
inhibitory GABA-ergic, 444
memory and, 451–453
numbers of, 441, 446
structure of, 441, 442
sympathetic, 549
Synaptic cleft, 441, 442
Synaptic delay, 444
Synaptic facilitation, 452
Synaptic plasticity, 451–453
Synaptic potentiation, 451
Synaptic transmission, 442–445
Synaptic vesicles, 393, 425, 441, 442
Synarthrosis, 269–271
Synchondrosis, 271, 271
Syncope, 734
Syncytiotrophoblast, 1080, 1081
Syncytium, 1080
Syndesmosis, 269, 270, 270–271
Synergist muscles, 304, 305
Synergistic effects, of hormones, 641
Synostosis, 269
Synovial fluid, 272, 272, 1098
Synovial joints, 272–285
accessory structures of, 273
aging and, 1098
axes of rotation, 275–277, 277
classes of, 277–278, 278
exercise and, 273
general anatomy of, 272, 272–273
as lever systems, 274–276, 276, 280
lever systems and, 274–276, 276, 280
movements of, 279–285
abduction, 277, 280, 280, 283–284, 284
adduction, 280, 280, 283, 284
ankle, 284–285
circumduction, 281, 281
depression, 280, 280–281, 283
elevation, 280, 280–281, 283
excursion, 283, 284
extension. See Extension
flexion. See Flexion
foot, 284–285, 285
forearm, 282, 282, 335, 335t–336t
hand and digit, 283–284, 284
head and trunk, 282, 283
lateral flexion, 282, 283
pronation, 282, 282, 285, 335, 336t
protraction, 281, 281, 283, 284
retraction, 281, 281, 283, 284
rotation. See Rotation

supination, 282, 282, 285, 335, 335t–336t
zero position, 279
range of motion, 275
Synovial membrane, 165, 272, 272
Synovial (tendon) sheaths, 273, 273
Synovitis, 294t
Synthesis phase of cell cycle, 125, 125
Synthesis reactions, 53, 53, 54t
Syphilis, 1033, 1093
Syringe, volume, pressure, and flow in, 708, 708–709
Systemic circuit, 690, 691, 697, 748–774, 749–750
Systemic gas exchange, 853–855, 854
Systemic lupus erythematosus (SLE), 819t
Systole, 702, 710–712, 711
Systolic pressure, 734, 735

T

T₃. See Triiodothyronine
T₄. See Tetraiodothyronine
T cell recall response, 810
T cells. See T lymphocytes
T lymphocytes, 788, 806–807
activation of, 790, 809, 810
in adaptive immunity, 806–810, 814t
aging and, 1100
antigen presentation to, 807, 809, 809
apoptosis of, 806
in autoimmune disease, 816–817
in cellular immunity, 808–810, 810
clonal selection of, 809, 810
cytotoxic (killer), 807, 808–810, 809, 809t, 810, 814t
helper, 796, 801t, 808–810, 809, 809t, 810, 814t
hormonal production by, 820
in humoral immunity, 810, 811
in hypersensitivity, 816
immunocompetent, 789, 806, 806
in immunodeficiency diseases, 817–819
in innate immunity, 796, 810
life history of, 806, 806–807
maturation of, 624, 788, 789, 790, 790, 806
memory, 808, 809, 810, 814t
naive pool of, 807
negative selection of, 807
number of (T-cell count), 818
positive selection of, 807
production of, 676, 677, 806
recall response of, 810
regulatory, 807, 808, 814t
T (transverse) tubules, 387, 387, 390t, 700

T wave, 704, 705
Table salt. See Sodium chloride
Tachycardia, 715
Tachypnea, 847t
Tactile cells, 177, 177, 568
Tactile corpuscles, 567, 567t, 568
Tactile discs, 177, 567, 567t, 568
Tadalafil, 1031
Taenia coli, 964, 965
Tail
of breast, 371, 1045
of embryo, 1084
of epididymis, 1017
of myosin molecule, 388, 388
of pancreas, 628, 950
of phospholipids, 61, 79
of sperm, 1007, 1016, 1026, 1027
Tamoxifen, 1048
Target cells, of hormones, 615, 616, 637, 638, 640, 640–641
Target organs
of ANS, 543, 553–557, 555t
of pituitary gland, feedback from, 622, 622
Tarsal glands, 591, 592
Tarsal plate, 591, 591, 592
Tarsal region, 30, 31
Tartar, 933
Tastants, 572–574
Taste, 521, 572–575
aging and, 1099
anatomy of, 572, 573
physiology of, 574
projection pathways of, 574–575
Taste buds, 574, 931
Taste cells, 573, 574
Taste hairs, 573, 574
Taste pore, 573, 574
Tay–Sachs disease, 429
Tears, 591–592
Tectorial membrane
anatomy of, 581, 582
hearing function of, 584
Tectospinal tract, 465, 466t, 466–467, 503, 505
Tectum, 466, 505, 506
Teeth, 926, 927, 931–933, 932
aging and, 1101
alveoli of, 236, 932, 932
deciduous, 932, 933, 933
disease of, 933
eruption of, 932, 933, 933
function of, 931–933
impacted, 933
meeting of (occlusion), 933
permanent, 932, 933, 933
structure of, 932, 932–933
wisdom, 932, 933
Tegmentum, 505, 506
Telencephalon, 495, 498, 511
Telogen, 185
Telomerase, 1102
Telomere, 1102
Telophase

I, in meiosis, 1023, 1024
II, in meiosis, 1023, 1024
in mitosis, 126, 127
Temperature. See also Body temperature
and diffusion, 88
and enzyme action, 66
and gas exchange, 856, 856
and negative feedback, 15, 16
and reaction rate, 54
and smooth muscle, 412
Temporal lines, 228, 232
Temporal lobe, 494, 496, 500, 511, 512
functions of, 511, 512, 521
in hearing, 586
lesions of, 519
Temporal region, 364
Temporal summation, 402, 408, 447, 447
Temporomandibular joint (TMJ) dysfunction, 286
Tendinitis, 294t
Tendinous cords, 693, 695, 696, 696
Tendinous intersections, 321, 322, 371
Tendon(s), generally, 199, 273, 304, 304
actions of, 275
athletic injuries to, 358
collagenous fibers of, 149, 149
function of, 148
Tendon(s), specific
abdominal oblique, 322t
abductor pollicis longus, 342
biceps brachii, 275, 287, 288, 288, 303, 335t
biceps femoris, 381
calcaneal (Achilles), 259, 293, 293, 303, 307, 350, 352, 352, 353, 380, 381, 382
central, of diaphragm, 319, 320t
common flexor, 338
extensor carpi radialis, 338
extensor carpi ulnaris, 342
extensor digiti minimi, 342
extensor digitorum, 338, 340t, 342, 377, 378
extensor digitorum longus, 350, 381, 382
extensor hallucis longus, 350, 381
extensor indicis, 338, 342
extensor pollicis brevis, 342, 378
extensor pollicis longus, 342, 378
fibularis (peroneus) brevis, 293, 380
fibularis (peroneus) longus, 293, 380
fibularis (peroneus) tertius, 350
flexor carpi radialis, 273, 303, 339t, 341, 342, 378
flexor carpi ulnaris, 341, 342
flexor digitorum longus, 356

flexor digitorum profundus, *273, 338, 341, 342*
flexor digitorum superficialis, *273, 338, 341, 342*
flexor hallucis longus, *356*
flexor pollicis longus, *273, 341, 342*
gastrocnemius, *352*
iliotibial band (tract), *307, 348, 379, 380*
infraspinatus, *287, 288*, 334*t*
latissimus dorsi, *337*
palmaris longus, 303, 339*t, 341, 342, 378*
patellar, *347*
patellar retinacula, 290
plantaris, *352*
quadriceps femoris, 256, 290, *299, 348, 379*
semimembranosus, 290, *380*
semitendinosus, *380, 381*
subscapularis, *287, 288*, 334*t*
supraspinatus, *287, 288*, 334*t*
teres minor, *287, 288*, 334*t*
tibialis anterior, *293, 380*
tibialis posterior, *293*
transverse abdominal, 322*t*
triceps brachii, *288*
Tendon organs, 486–487, *487, 567, 567t*, 568
Tendon reflex, 486–487
Tendon sheaths, 273, *273*
Tennis elbow, 358
Tenting effect, 1058
Tentorium cerebelli, 497
Teratogens, 1092–1093
Teratology, 1092
Terminal arborization, *424*, 425
Terminal bronchioles, 833, *836*
Terminal button, *424*, 425
Terminal cisterns, 387, *387*
Terminal filum, *461*, 463
Terminal hairs, 184
Terminal lymphatics, 785
Terminal sulcus, 931, *931*
Terminal web, *79, 84, 97, 98*
Terminologia Anatomica (TA), 19
Terminology, anatomical, 19–21, 28–29
   analyzing, 19–20
   directional, 29, 29*t*
   history of, 19
   obscure origins in, 20
   planes, *28*, 28–29
   plural, adjectival, and possessive forms in, 20*t*, 20–21
   positional, 28
   pronunciation of, 21
Tertiary ovarian follicles, *1040, 1051*, 1052
Tertiary (segmental) bronchi, *827, 832*, 833
Tertiary structure of protein, 63, *64*, 120

Tertiary syphilis, 1033
Testes, *368, 375, 614*, 1008, *1014,* 1015–1017, *1016, 1018*
   associated structures, *1016*
   barrier for, 816, 1015, *1025*
   blood supply of, 1017
   cilia of, 85
   descent of, 1010–1013, *1013*
   development of, 1010, *1011*
   endocrine function of, 629, 632*t*, 1015, 1021–1022
   epithelial tissue of, 1015
   exocrine function of, 629
   histology of, 629, *630*, 1015–1017, *1016*
   innervation of, 1017
   origin of term, 20
   puberty and, 1021
   testosterone synthesis of, 122, *123*
   undescended, 1013
Testicles. *See* Testes
Testicular cancer, 1032*t*
Testis-determining factor (TDF), 1010
Testosterone, 629, 632*t*, 1021–1022, *1022*
   aging and, 1022, 1101
   and baldness, 186
   in bone physiology, 215*t*, 216
   chemical classification of, *634*
   DHEA conversion to, 628
   FSH interaction with, 641
   in puberty, 1022
   synthesis of, 122, *123*, 633, *635*
Tetanic stimulation, 452
Tetanus, muscle, 402, 413, 914
   completed (fused), 402, *403*
   incomplete, 402, *403*
Tetanus toxin (lockjaw), 393
Tetany, 212
Tetrads, 1023
Tetraiodothyronine (T$_4$, thyroxine), 616*t*, 624–625, 631*t*, 642
Texture
   hair, 185, *185*
   mouthfeel, 573
   sensation of, 507, 565, *567, 567t,* 568
Thalamus, *496, 500, 504*, 509, *510*
   in equilibrium, 590, *590*
   functions of, 509
   as gateway to cerebral cortex, 509
   nuclei of, 454, *504*, 509, *510,* 514, *514, 607*
   in pain modulation, *571*, 572
   in taste, 574
   in vision, *607*
Thalassemia, 684*t*
Thalidomide, 1092–1093, *1093*
Theca, 629, *630*
Theca externa, *1051*, 1052

Theca folliculi, *1051*, 1052, *1052*
Theca interna, *1051*, 1052
Thelarche, 1048
Thenar eminence, 343*t, 377, 378*
Theory, 8. *See also specific theories*
Thermal gradient, 18, *18*
Thermal stability of water, 48–49
Thermogenesis
   nonshivering, 1001
   shivering, 1001
Thermoreceptors, 566, 568, 1001
Thermoregulation, 999–1002
   aging and, 1097, 1101
   behavioral, 1002
   definition of, 999
   disturbances in, 1002, 1097
   by hypothalamus, 509, 799, 1001
   negative feedback in, 15, *16,* 1001
   neonatal, 1091
   by skin, 176
Theta (Θ) waves, 516, *517*
Thiamin, 985*t*
Thiamine. *See* Vitamin B$_1$
Thick filaments, 387–388, *388, 389, 390t*, 413
Thick segment, of nephron loop, 873, *874*
Thick skin, 175
Thigh, 31
   compartments of, 346*t*, 349*t*
   muscles acting on, *344, 345, 346t–347t*
   surface anatomy of, *379*
Thin filaments, *79, 97, 98*, 105*t*, 388, *388, 389, 390t*, 413
Thin segment, of nephron loop, 873, *874*
Thin skin, 175
Third heart sound, 710
Third line of defense, 795
Third trimester, of pregnancy, 1079
Third ventricle, 498, *500*, 501, *501*
Third-class lever, 275, *276*
Third-degree burns, 191, *192*
Third-order neuron, 465, *467*, 568
Thirst, 904, *904*
Thoracic aorta, *369, 751*
Thoracic cage, 225, 246–248, *247*
   respiratory role of, 839, 842
   ribs. *See* Bone(s), specific
   sternum. *See* Bone(s), specific
Thoracic cavity, 32, 32*t*, 32–33
   compartmentalization of, 837
   cross section through, *835*
Thoracic curvature, 240, *241*
Thoracic duct, *784*, 786, *787*
Thoracic kyphosis, 240
Thoracic lymph nodes, *787*, 791
Thoracic pressure, 839, 842, *844*
Thoracic pump, 744–745
Thoracic region (chest), 30, 460, *787*
   arteries of, *757*

surface anatomy of, *371*
transverse section of, *373*
veins of, *758*
Thoracoacromial trunk, 757
Thoracolumbar division, of ANS. *See* Sympathetic division, of ANS
Thoracolumbar fascia, 323
Thoroughfare channels, 728
Three *R*s of immunity, 808
Three-dimensional interpretation, *139*, 140
Threshold
   action potential, 401, 435
   anaerobic, 406
   lactate, 406
Throat. *See* Pharynx
Thrombin, *681*, 682, 683
Thrombocytes. *See* Platelet(s)
Thrombocytopenia, 684*t*
Thromboembolism, 684
Thrombokinase (factor X), *681*, 682, 682*t*
Thromboplastin, *681*, 682, 682*t*, 980
Thrombopoiesis, *679*, 679–680
Thrombopoietin, 679
Thrombosis, 684
Thromboxane A$_2$, 680, 685
Thromboxanes, *643*, 643–644
Thrombus, 684
Thrush, 818
Thucydides, 803
Thumb
   movements of, 283–284, *284*
   opposable, 10, *10*
   surface anatomy of, *377, 378*
Thymectomy, 415
Thymic corpuscles, 790, *790*
Thymine, 110, *111*, 115
Thymopoietin, 624, 631*t*, 790
Thymosin, 624, 631*t*, 790
Thymulin, 624, 631*t*, 790
Thymus, *614*, 624, *624, 784,* 789–790, *790*
   histology of, 789–790, *790*
   hormones of, 624, 631*t*, 633*t*
   lymphocyte maturation in, 624, 660, 788, 789, *790*, 806, *806*
   secretions of, 790
   structure of, 789–790, *790*
Thyretin, 637
Thyrocervical trunks, *752, 757*
Thyroglobulin (Tg), 635–636
Thyroid disorders, 645, 648*t*. *See also specific disorders*
Thyroid follicles, 624, *625*
Thyroid gland, *368, 614*, 624–625, *625*
   histology of, 624–625, *625*
   negative feedback to pituitary gland, 622, *622*
   pregnancy and, 1061

Thyroid hormones, 616t, 620, 622, 622, 624–625
action of, 637
in bone physiology, 215t
in cardiac physiology, 716
deficiency of, 645, 646
excess of, 645, 648t
secretion of, 637
synthesis of, 635–636, 636
transport of, 637
Thyroid-stimulating hormone (TSH), 616, 616t, 620, 620t, 622, 622, 625, 633t, 637, 638
in pregnancy, 1061, 1061t
Thyrotropin. See Thyroid-stimulating hormone
Thyrotropin-releasing hormone (TRH), 616t, 619t, 622, 622, 625
Thyroxine, 616t, 624–625, 631t, 642
Thyroxine-binding globulin (TBG), 637
Tic douloureux, 533
Tidal volume (TV), 846, 846, 846t
Tight junctions, 160, 161, 502
in blood–brain barrier, 502, 730
in blood–testis barrier, 1015
in carbohydrate digestion, 957, 957
in epidermis, 178, 179
in renal tubules, 881, 882
in seminiferous tubules, 1025
in small intestine, 953
in spermatogenesis, 1025, 1026
in stomach, 160, 943, 1003
Tinea, 193t
Tip link, 584, 584
Tissandier, Gaston, 849
Tissue(s), 12. See also specific types
definition of, 138
degeneration and death of, 167, 169
development of, 140, 166
excitable, 157–159
growth of, 166
hematopoietic, 660
lymphatic. See Lymphatic tissue
primary classes of, 138t, 138–139
repair of, 141, 167
study of, 138–140
Tissue engineering, 170
Tissue (interstitial) fluid, 78, 139, 902, 903
Tissue plasminogen activator (TPA), 685
Tissue sections, 139–141
planes of, 140, 140
three-dimensional interpretation of, 139, 140
Tissue specimens, 2, 139–141
fixative for, 140
sections, 139–141, 140
smears, 140

blood, 156, 156t
spinal cord, 158, 158t
spreads, 140
stains for, 140
three-dimensional interpretation of, 139, 140
Tissue thromboplastin, 681, 682, 682t
Titin, 388
Toe(s), 31
Toe-off, 262, 284, 354t
Tone
autonomic, 544
muscle (skeletal muscle), 275, 400
smooth muscle, 413
sympathetic, 738
vagal, 715
vasomotor, 557, 557
Tongue, 230, 365, 828, 926, 927, 930, 930–931, 934
anatomy of, 931, 931
epithelial tissue of, 142, 931
function of, 572–574, 930–931, 933, 936, 936
muscles of, 313, 315t, 931, 931
papillae of, 563, 572–574, 573, 931, 931
taste buds of, 573, 574, 931
Tonic receptors, 566
Tonicity, 89–90, 90
Tonsil(s), 789, 793, 793
hematopoiesis in, 660
lingual, 793, 793, 828, 931, 931
palatine, 784, 793, 793, 828, 930, 931
pharyngeal, 793, 793, 828
Tonsillar crypts, 793, 793
Tonsillectomy, 793
Tonsillitis, 793
Tooth. See Teeth
Total body water (TBW), 902
Total dead space, 845
Total heart block, 707
Total hip replacement (THR), 294–295, 295
Total lung capacity, 846, 846t
Total metabolic rate (TMR), 998–999
Total parenteral nutrition (TPN), 921
Totipotent stem cells, 166
Touch, sense of
receptive field for, 565
receptors for, 568
Toxic goiter, 645, 648t
Toxoplasma, 818
Toxoplasma, 1093
Trabeculae
of bone, 201, 204, 205, 205–206, 207
of glands, 162, 163
of lymph node, 790, 791
of penis, 1020
of thymus, 789, 790

Trabeculae carneae, 693, 695, 695
Trace elements, 41t, 42
Trachea, 365, 369, 827, 828, 830, 831–833, 832
endoscopic view of, 831
epithelial tissue of, 832, 832, 853
mucociliary escalator of, 832, 832
mucosa of, 831
Tracheostomy, 832
Tracheotomy, 830, 832
Tractography, 513
Tracts. See also specific tracts
brain, 495, 513
digestive. See Digestive tract
gastrointestinal, 926
spinal cord, 463–468
Tragus, 578
Trans fats, 60, 60
Transamination, 994–995
Transcellular fluid, 902
Transcellular route, of tubular reabsorption, 881
Transcortin, 637
Transcription, 116–117, 122
Transcription activator, 122
Transcytosis, 95, 95, 741, 741
Transdermal absorption, 176
Transduction, 564
Transection of spinal cord, 488
Transfer RNA (tRNA), 113, 117, 117–118, 119
Transferrin, 659t, 663, 664, 961
Transfusion, blood
blood types and, 668–670
Drew's pioneering work in, 668
Transfusion reaction, 669, 670
Transient ischemic attacks (TIAs), 747
Transitional period, in neonate, 1091
Translation, 117–120, 119
elongation in, 118
initiation of, 118, 119
termination of, 118, 119
Transmembrane proteins, 80, 81, 81–82, 82
Transmission electron microscope (TEM), 77, 77–78, 78t, 105t
Transpiration, cutaneous, 189, 903–904
Transplantation
bone marrow, 672, 817
umbilical cord blood, 672
Transport, 96t
active, 87, 91–92, 96t, 881
anterograde, 426
axonal, 426, 478
by blood, 148
carrier-mediated, 87, 90–93, 96t
circulatory, 656
by filtration, 87, 88, 96t
gas, 850–853
hormone, 637

membrane, 87–96
osmolarity and tonicity in, 89–90
by osmosis, 88–89, 89, 96t
passive, 87
retrograde, 426
by simple diffusion, 87–88, 96t
vesicular, 93–95, 96t
Transport maximum (T$_m$), 90, 90, 647, 883, 883
Transport proteins, 65, 637
Transport vesicle, 118, 120
Transporters. See Carrier proteins
trans-retinal, 602, 602
Transudation, 1043
Transverse arch of foot, 261, 261
Transverse colon, 374, 551, 927, 929, 963–965, 964
Transverse lines, 246
Transverse plane, 28, 29
Transverse rectal folds (rectal valves), 963, 964
Transverse section, 28
Transverse sinus, 497, 754
Transverse (T) tubules, 387, 387, 390t, 700
Trauma. See Injuries
Traumatic brain injury (TBI), 501
Treatment group, 8
T-regs, 808, 814t, 817
Trephination, 608
Treponema pallidum, 1033
Triad, 387
Triangular body, of ear, 579
Triangular muscle, 302, 302
Tributaries, of veins, 732
Trichromatic vision, 605. See also Color vision
Tricuspid valve, 693, 695, 696, 696, 697, 697
Trigeminal neuralgia, 533
Trigger zone, 424, 429, 435–436
Triglycerides, 58, 58t, 60
metabolism of, 993
serum levels of, 981, 981
structure of, 60
synthesis of, 59, 60
Trigone, 893
Triiodothyronine (T$_3$), 616t, 625, 631t
Trimesters, of pregnancy, 1060, 1079
Tripeptides, 63
Triple rhythm, of heart, 710
Triple-X syndrome, 1094
Trisomy, 1095
Trisomy-13, 1095
Trisomy-18, 1095
Trisomy-21, 454, 1095, 1096
Tristearin, 58
Tritium, 42
Trochanter
description of, 226, 226t
of femur, 256, 258, 372
Trochlea

of humerus, 251, *251*
of orbit, 592, *593*
Trochlear notch, *251,* 252
Trochlear surface, of talus, 259, *260*
Trophoblast, 1079, *1081*
Trophoblastic nutrition, 1087, *1087*
Trophoblastic phase, of pregnancy, 1087
Tropomyosin, 388, *388,* 396, *399*
Troponin, *388,* 396, 399
True pelvis, 255
True ribs, *247,* 248
Trunk, 31
    of brachial plexus, *476,* 488*t*
    movements of, 282, *283*
    muscles of, 318–328
    of spinal nerves, 476
    surface anatomy of, *366*
Trypanosomes, 553
Trypsin, 65, 67, 950, *951,* 958, *959, 962*
Trypsinogen, 950, *951,* 951*t*
Tryptophan, 635
Tubal ligation, 1070, 1070*t*
Tubercle, of bone, *226,* 226*t*
    conoid, 249, *249*
    humeral, 251, *251*
    pelvic, 31
    rib, 247, *248,* 249*t*
    vertebral, *243,* 244
Tuberculosis, 818, 845, 859*t*
Tuberculum sellae, 234
Tuberosities, *226,* 226*t*
    deltoid, 251, *251*
    gluteal, 256, *258*
    ischial, 255, *255, 1045*
    radial, *251,* 252
    tibial, 258, *259, 379*
Tubular fluid, 875
Tubular glands, 162, *164*
Tubular reabsorption, *876,* 881–885, *882, 888,* 905
Tubular secretion, *876,* 881, 883–884
Tubules
    renal, 872, *872, 873, 873, 874*
    seminiferous, 629, *630, 1007, 1016,* 1017, *1025*
    transverse (T), 387, *387,* 390*t,* 700
Tubulin, 97, *99*
Tubuloacinar gland, 162, *164*
Tubuloglomerular feedback, 878–879, *879*
Tufted cells, *575,* 576
Tumor(s), 166
    brain, 428
    hormonal hypersecretion and, 645
    malignant. *See* Cancer
Tumor angiogenesis, 132, 737
Tumor necrosis factor (TNF), 801, 810, 814*t,* 820
Tumor-suppressor genes, 133
Tunics (tunica)

albuginea, 1015, *1016,* 1020, *1020,* 1038, *1040, 1052*
blood vessel, 165, 725–726, *726, 727*
externa (adventitia)
    of blood vessels, 726, *726, 727*
    of digestive system, 927
    of esophagus, 935
    of small intestine, 953
    of trachea, 832
eyeball, 592–594
fibrous layer, 592–593
interna (intima)
    of blood vessels, 165, 725, *726, 727*
    of lymphatic vessels, 785
lymphatic vessel, 785
media
    of blood vessels, 726, *726, 727*
    of lymphatic vessels, 785
vaginalis, 1013, *1013, 1014, 1016*
Turbinates, 829
Turner syndrome, 1094, *1095*
Twins, 1080, *1080,* 1101–1102
Twitch, 401. *See also* Muscle twitch
Two-dimensional sections, *139,* 140
Two-point touch discrimination, 565
Tympanic cavity, 579, *579*
Tympanic membrane
    aging and, 1099
    anatomy of, 578, *579, 581*
    hearing function of, 582, *583*
    secondary, 581, *581, 583,* 584
Tympanic part, of temporal bone, 232–233, *233*
Tympanic reflex, 583, *587*
Tympanostomy, 580
Type 1 diabetes mellitus, 647, 890
Type 2 diabetes mellitus, 637, 647, 890
Type A spermatogonia, 1025, *1025*
Type B spermatogonia, 1025, *1025*
Type I (acute) hypersensitivity, 815
Type I muscle fibers, 407*t,* 407–408, 408*t*
Type II hypersensitivity, 816
Type II muscle fibers, 407*t,* 407–408, 408*t*
Type III hypersensitivity, 816
Type IV hypersensitivity, 816
Tyrosine, 635–636

U

Ulcer(s)
    decubitus, 169
    peptic, 943, *943*
Ulcerative colitis, 968*t*
Ulnar flexion, 283, *284*
Ulnar notch, of radius, *251,* 252
Ultrasonic frequencies, 578

Ultrasound, 23, *23*
Ultrastructure, of cell, 2, *77,* 77–78
Ultraviolet radiation, 43, 591
    protection against, 193
    and skin, 175, 181, 182, 190, 193, 1097
    UVA, 193
    UVB, 193
Umami, 574
Umbilical cord, 1060, *1084, 1085,* 1087, *1088, 1090*
Umbilical cord blood transplant, 672
Umbilical hernia, 328
Umbilical region, *30, 31, 32*
Umbilicus, *366, 371, 1090*
Umbrella cells, 147, 893
Unbound hormones, 637
Uncompensated acidosis or alkalosis, 918
Undescended testes, 1013
Unencapsulated nerve endings, *567,* 567*t,* 568
Unhappy triad, of knee injuries, 292, *292*
Unicellular glands, 161
Unipennate muscle, 302, *302*
Unipolar neurons, 425, *425*
Uniport, 91
Unipotent stem cells, 166
Unitary smooth muscle, 412, *412*
Unity of form and function, 2, 21
Universal donor, of blood, 670
Universal recipient, of blood, 670
Universal solvent, water as, 48
Unmyelinated nerve fibers, 429, *430, 431,* 437–438, *438*
Unsaturated fatty acids, 59
Up the gradient, 17, 91
Upper limb, 31
    adaptations to bipedalism, 263
    arteries of, 766, 768
    bones of, 224*t, 225,* 250–252
    muscles acting on, 329–343
    surface anatomy of, *377–378*
    veins of, *769*
Upper motor neurons, 466, *468,* 523
Upper respiratory tract, 827, *828*
Up-regulation, 640, *640*
Upright posture, 11
Uracil, 113, 116
Urea, 658*t,* 659, 867–868, *868,* 882, 995
Urea cycle, 995
Uremia, 868
Ureter(s), *369,* 867, *868, 870,* 891
    parasympathetic innervation of, *551*
    variations in, 13
Urethra, 867, *868,* 893
    development of, 1010
    female, *370, 375,* 893, *1039*
    male, 893, 1017, *1018, 1020*
Urethral glands, 893
Urethral orifice, 893, *1045*

Urethritis, nongonococcal, 1033
Uric acid, 868, *868,* 883
Urinalysis, 888–889
Urinary bladder, *367, 369, 375,* 867, *868,* 891–895
    aging and, 1100
    female, *1039*
    infection of, 893
    male, *1018*
    parasympathetic innervation of, *551*
    pregnancy and, 1062
    sympathetic innervation of, *547*
    transitional epithelium of, 142–147, 893
Urinary incontinence, 893, 895*t,* 1100
Urinary pole, of renal corpuscle, 872
Urinary system, 35, *36,* 866–900
    aging and, 1100
    components of, 867, *868*
    effects on other systems, 897
    parasympathetic effects on, 555*t*
    pregnancy and, 1062
    reproductive system association with, 867
    sympathetic effects on, 555*t*
Urinary system disorders
    acute glomerulonephritis, 895*t*
    hydronephrosis, 895*t*
    incontinence, 893, 895*t*
    kidney stones, 892
    nephroptosis, 895*t*
    nephrotic syndrome, 895*t*
    renal failure, 895*t,* 896, 1003
    renal insufficiency, 896
    urinary tract infection, 893
Urinary tract, transitional epithelium of, 142–147
Urinary tract infection (UTI), 892, 893
Urination, 891–895
Urine, 875
    appearance of, 889
    composition and properties of, 889, 889*t*
    formation of, 875–888, *876*
        glomerular filtration in, 875–881
        tubular reabsorption and secretion in, *876,* 881–884, 905
        tubular secretion in, *876*
        water conservation in, *876,* 885–888, 905
    odor of, 889
    osmolarity of, 889
    pH of, 889
    specific gravity of, 889
    storage and elimination of, 891–895
    voiding of, 893–895, *894*
    volume of, 889–890, 905
Urine tests, 888–891

Urobilin, 949
Urobilinogen, 665, 949
Urochrome, 665, 889
Urogenital folds, 1010, *1012*, 1038
Urogenital sinus, *1011*
Urogenital system, 35, 867
Urogenital triangle, *326*, 327, *1014*
Uronic acid, 149
Uroplakins, 147
Urothelium, *146*
Urticaria, 819*t*
U.S. National Academy of Medicine, 977
Uterine fibroids, 1068*t*
Uterine milk, 1087
Uterine tubes, 369, *1039*, 1039–1040, *1041*
   cilia of, 85, *85*, 1040
   development of, *1011*, 1039–1040
   epithelial lining of, 1040, *1042*
   fimbriae of, *1039*, 1040, *1040*, *1041*
   ligation of, 1070, 1070*t*
Uterus, 369, 375, *1039*, 1040–1042
   blood supply of, 1042, *1044*
   cervix of. *See* Cervix
   contractility of, 1063–1064
   development of, *1011*
   fetal expulsion from. *See* Childbirth
   function of, 1040
   growth, in pregnancy, 1062
   involution of, 1066
   ligaments supporting, *1041*, 1042
   lumen of, 35
   menstrual cycle and, *1054*, 1055–1056, *1056*
   parasympathetic innervation of, *551*
   potential space of, 1040
   sexual response of, 1058, *1059*
   stretching of, 1064
   surrogate mother, 1105
   sympathetic innervation of, *547*
   wall, layers of, *1041*, 1042
Utilization coefficient, *854*, 855
Utricle, *581*, 587–589, *588*
Uvea, 593
Uvula, *828*, *930*, 931

V

Vaccination, 804, 1100
Vaccine, 804
Vagal tone, 715
Vagal trunks, 552
Vagina, *370*, *375*, *1041*, 1043–1044
   development of, *1011*, *1012*, 1038
   epithelial tissue of, 142, *145*, *147*, 166, 1044
   fetal passage through, 1064, *1065*
   fornices of, *1039*, *1041*, 1043

   function of, 1043
   pH of, 1028, 1044
   postpartum discharge from, 1066
   sexual response of, 1058, *1059*
   transudation of, 1043
Vaginal orifice, *1039*, *1045*
Vaginal process, 1011, *1013*
Vaginal rugae, *1039*, 1044
Vaginal transudate, 1058
Vagovagal reflexes, 929, *944*, 945
Valence, of ions, 45
Valence electrons, 42
Valium (diazepam), 608
Vallate papillae, *563*, 573, *573*, 931, *931*
Valsalva maneuver, 839, 895, 967, 1064
Valves. *See also specific valves*
   heart. *See* Heart valves
   lymphatic vessels, 785
   venous, *727*, 732, 744, *744*
Valvular insufficiency, 710
Valvular stenosis, 710
van der Waals forces, 46, 46*t*, 47, 63
Variable (V) region, of antibody, 805, *805*
Variations
   anatomical, 13, *14*
   physiological, 15
Varicella-zoster virus (chickenpox), 478, 800
Varicocele, 1032*t*
Varicose veins, 732, 1062
Varicosities, 411, *411*, 425, 545
Vas deferens. *See* Ductus deferens
Vasa recta, 871, *871*, *872*, 887, *887*
Vasa vasorum, 726, *727*
Vascular corrosion cast, 77, *612*
Vascular layer, of eye, 593
Vascular plexuses
   choroid, *496*, 499, *500*, 501, *501*
   pampiniform, *1014*, 1014–1015, *1015*
Vascular pole, of renal corpuscle, 872
Vascular spasm, 680, *680*
Vasectomy, 1070, 1070*t*
Vasoactive chemicals, 737, 801–802
Vasoactive intestinal peptide (VIP), 941
Vasocongestion
   in female sexual response, 1058
   in male sexual response, 1029
Vasoconstriction, 15, 176, 557, *557*, 680, *680*, 736, *736*, 739–740
Vasoconstrictors, 679
Vasodilation, 15, 176, 557, *557*, 736, 739–740
Vasomotion, 736, 739–740
Vasomotor center, 503, 507, 736, 737–739
Vasomotor tone, 413, 557, *557*

Vasopressin. *See* Antidiuretic hormone
Veins, generally, 725, 731–732
   abdominal and pelvic region, *763*, 765
   aging and, 1100
   blood flow in, 735
   as capacitance vessels, 731
   disorders of, 774*t*
   fetal and neonatal, 1089, *1090*
   great, 690
   head and neck, *754*
   hemorrhoidal, 732, 963, *964*
   large, 727, 732
   lower limb, *773*, 774
   major systemic, *750*
   medium, 727, 732
   micrograph of, *726*
   nomenclature for, 751
   renal, *763*
   small, 732
   thoracic, *758*
   tributaries of, 732
   tunics of, 725–726, *726*, *727*
   upper limb, *769*
   varicose, 732, 1062
Veins, specific
   arcuate, 871, *871*
   axillary, *368*, *750*, *769*
   azygos, *758*, *787*
   basilic, *750*, *769*
   brachial, *368*, *750*, *769*
   brachiocephalic, *368*, *750*, *754*, *758*
   cardiac
     great, *698*, 699
     middle, 699
     small, *698*
     smallest, 699
   central, of liver, 946–947, *948*
   central, of retina, 594
   central longitudinal, 789, *789*
   cephalic, *366*, *368*, *377*, *750*, *769*, *770*
   cerebral, great, 755
   coronary sinus, *693*, 694, *698*, 699, 732
   cortical radiate, 871, *871*, *872*
   cystic, *765*
   dorsal, of penis, *1020*
     deep, 1029
   facial, *754*
   femoral, *366*, *367*, *750*, *773*, *774*, 775
     circumflex, *773*, *774*
     deep, *750*, *773*, *774*
   fibular (peroneal), *750*, *773*, *774*
   gastric, *765*
   gastro-omental, *765*
   gonadal, *763*
   hemiazygos, *758*, *787*
     accessory, *758*
   hepatic, *369*, *750*, *763*
   hepatic portal, *765*, 947, *947*, *948*

   iliac
     common, *750*, *758*, *763*, *773*, 774
     external, *750*, *763*, *773*, *774*
     internal, *750*, *763*, *773*, *774*
   iliolumbar, *758*
   intercostal
     posterior, *758*
     superior, *758*
   interlobar, 871, *871*
   interventricular, posterior, 699
   jugular
     external, *367*, *750*, *754*, *769*
     internal, *367*, *373*, *750*, *754*, *769*, *784*
   lumbar, *758*
     ascending, *758*
   marginal, left, *698*, 699
   median antebrachial, *750*, *769*
   median cubital, *377*, *750*, *769*
   mediastinal, *758*
   mesenteric, *369*, *374*, 965
     inferior, *765*, 965
     superior, *369*, *374*, *765*, *952*, 965
   occipital, *754*
   ophthalmic, superior, *754*
   ovarian, *1041*
   palmar, *750*, *769*
   pampiniform plexus, *1014*, 1014–1015, *1015*
   pancreatic, *765*
   pericardial, *758*
   phrenic, inferior, *763*
   plantar
     lateral, *773*, *774*
     medial, *773*, *774*
   popliteal, *750*, *773*, *774*
   pulmonary, 690, 694, 695, 697, *697*, 748, *748*, *834*
     left, 694, *695*, 697
     right, 694, *695*, 697
   radial, *769*
   renal, *750*, *758*, *763*, 870, 871, *871*
   saphenous
     great, *366*, *750*, *773*, *774*, 775
     small, *750*, *773*, *774*
   splenic, *765*, *952*
   subclavian, *368*, *373*, *750*, *758*, *769*, *785*, *786*, *787*
   subcostal, *758*
   suprarenal, *763*
   temporal, superficial, *754*
   testicular, 1017
   thyroid, superior, *754*
   tibial
     anterior, *750*, *773*, *774*
     posterior, *750*, *773*, *774*
   ulnar, *750*, *770*
   umbilical, 1087, *1088*, 1089, *1090*
   vena cava

inferior, *374,* 690, *693, 694, 695,* 697, *697,* 737, *737t,* 744, *750, 758, 763, 765, 773,* 1089, *1090*
    superior, *368, 369,* 690, *694, 695,* 697, *697, 750, 758, 769*
venous arches
    dorsal, *773, 774*
    palmar, *750, 769*
    plantar, *750, 773, 774*
vertebral, 245, *754*
Vellus hairs, 184
Venous anastomosis, 733, *733,* 766
Venous drainage, 699
Venous pooling, 745
Venous pooling shock, 745–746
Venous pressure, central, 744
Venous reserve, 855
Venous return, 744–746. *See also* Veins
    mechanisms of, 744–745
    physical activity and, 745
Venous sinuses, *229,* 732
    confluence of, *754*
    dural, 732, *754*
    scleral, 594, *594*
Venous valves, *727,* 732, 744, *744*
Ventilation, 826, 837–848
    aging and, 1100
    alveolar, 845
    assessment of, 845–847
    blood gases and, 856–858
    exercise and, 857–858
    gas exchange in, 826, 848–858
    metabolic adjustments in, 855–856
    neural control of, 839–841, *840*
    pH and, *856,* 856–858
    rhythm variations in, 847, *847t*
    temperature and, 856, *856*
Ventilation–perfusion coupling, 850, *853*
Ventral, 29, *29t*
Ventral (anterior) mesentery, 34, 928
Ventral respiratory group (VRG), 839, *840*
Ventricles
    of brain, 495, *496,* 498–502, *500*
    of heart
        anatomy of, *373,* 693–695, 695–696
        blood flow through, 697, *697*
        ejection from, 710, *711*
        filling of, 710, *711*
Ventricular ejection, *711,* 712
Ventricular fibrillation, 705–706, *707*
Ventricular filling, 710, *711*
Ventricular systole, 710–712
Venules, *727*
    blood flow in, 737, *737t*
    muscular, 732
    portal, 617

postcapillary, 732
pulmonary, *836*
Vermilion, 930
Vermis, 507, *508*
Vermis of cerebellum, *505*
Vernix caseosa, *1086t*
Vertebra prominens (C7), 245
Vertebrae, *224t, 225,* 239–246, *373, 374, 375. See also* Spine
    anatomical variations in, 240
    articulation with ribs, 245, 247–248, *248, 249t*
    body (centrum) of, 241, *242, 243, 244,* 245
    cervical, 240, *240,* 243, 243–245, *244*
        posterior arch of, *243,* 244
        X-ray of, *223*
    coccygeal, 240, *240,* 246, *246*
    compression of, 240
    general structure of, 241–242
    lumbar, 240, *240,* 245
    regional characteristics of, *243,* 243–246, *244*
    sacral, 240, *240,* 245, *246*
    thoracic, 240, *240, 244,* 245
Vertebral border, of scapula, 250, *250*
Vertebral column, 239–246, *240, 365. See also* Vertebrae
    adaptations to bipedalism, 263, *263*
    bones of, *224t, 225,* 239–246
    curvatures of, *240,* 241
    flexibility of, 268
    flexion of, 282, *283*
    function of, 240
    general features of, 240–241
    muscles acting on, *323, 324, 325*
Vertebral furrow, *372*
Vertebral notch
    inferior, 242
    superior, 242
Vertebral region, *30*
Vertebral ribs. *See* Floating ribs
Vertebrosternal ribs. *See* True ribs
Vertex position, of fetus, 1062, *1063, 1086t*
Very-low-density lipoproteins (VLDLs), 981, *981–982*
Vesalius, Andreas, 4–5, 690
Vesicles, 93
    clathrin-coated, 93, *94*
    dense-core, 442
    Golgi, *79,* 101, *101, 105t,* 120
    optic, 495, *498,* 595
    pinocytic, 93
    primary, 495, *498*
    secondary, 495, *498*
    secretory, *79, 98,* 101, 120
    seminal, *375,* 1017–1019, *1018*
        development of, *1011*
        semen production in, 1027
    synaptic, 393, 425, 441, *442*

transport, 118, 120
Vesicouterine pouch, *1039,* 1042
Vesicular transport, 93–95, *96t*
Vestibular apparatus, 587, 590
Vestibular bulbs, 1045, *1045*
Vestibular folds, *828,* 829, 830, *830, 831*
Vestibular glands
    greater, 1045, *1045*
    lesser, 1045
Vestibular membrane, 581, *582*
Vestibular sense. *See* Equilibrium (sense)
Vestibule
    of ear, *579, 580, 581*
    of nasal cavity, *828,* 829
    of oral cavity, 930, *930*
    of vagina, 1044, *1045*
Vestibulo–ocular reflex, 590
Vestibulospinal tract, 590, *590*
    lateral, *465, 466t,* 467–468
    medial, *465, 466t,* 467–468
Vestigial organs, 10
Vetter, David, 817, *817*
Viagra, 1031
Vibrissae, 186, 829
Victoria (queen), 683
Villi
    arachnoid, *501*
    chorionic, *1081,* 1087, *1088*
    intestinal, 953, 955
    microvilli, *79,* 84, *84, 85, 98, 105t,* 953
Villous chorion, 1087
Viral envelope, 817, *817*
Viscera, 31
Visceral effectors, 543, 544
Visceral layer, 32
Visceral motor division, of peripheral nervous system, 422
Visceral motor system. *See* Autonomic nervous system
Visceral muscle, 158, *411,* 412
Visceral nerve fibers, 470, *471t*
Visceral pain, 569
Visceral pericardium, 32, *33,* 691, 693
Visceral peritoneum, 33, *34*
Visceral pleura, 33, *33, 35, 835,* 837, 843
Visceral reflexes, 482, 543–544, *544*
Visceral sensory division, of peripheral nervous system, 422
Viscosity
    blood, 659, 735
    definition of, 659
Visible light, 591
Vision, 521, 591–607
    accommodation in, *598,* 598–599
    aging and, *599t,* 1099
    color, 601, 602, 604–605, *606*

corrective lenses for, *599, 599t*
day (photopic), 601, 603
double, 597
dual system in, 603–604
duplicity theory of, 603
generating optic nerve signal in, 602, *604*
image formation in, 597–599, *599t*
light and, 591
light and dark adaptation in, 603, 1099
monochromatic, 600
near point of, 599
near response in, 597–599, *598*
neural convergence in, 601, 604, *605*
night (scotopic), 600, 603, *605,* 1099
panoramic, 605–606
peripheral, 603
primate, 10, *10*
projection pathway of, 605–607, *607*
refraction in, 597, *597*
sensory transduction in, 599–602
spatial summation in, 603
stereoscopic, 605, *606*
vitamin A and, 601, *602,* 985
Visual acuity, 603
Visual association area, 521, *521,* 607
Visual cortex, primary, 521, *521,* 607, *607*
Visual filling, 596
Visual pigments, 601–602, *602, 603*
Visual purple (rhodopsin), 601
Vital capacity (VC), *846, 846t,* 847
Vital force, 656
Vital signs, during sleep, 517
Vitamin(s), 977, 983–985
    absorption of, 961
    deficiency of, 985
    excess of, 985
    fat-soluble, *58t,* 961, 985, *985t*
    hepatic metabolism of, *996t*
    nomenclature for, 985
    precursors of, 985
    in pregnancy, 1062
    requirements for, *985t*
    sources of, *985t*
    supplements, 985
    water-soluble, 961, 985, *985t*
Vitamin A, 985
    absorption of, 961
    in bone physiology, *215t*
    deficiency of, 985
    excess of, 985
    requirements and sources, *985t*
    in vision, 601, *602,* 985
Vitamin B, 965, 985
Vitamin B$_1$ (thiamin), *985t*
Vitamin B$_2$ (riboflavin), 985, *985t*
Vitamin B$_3$ (niacin), 985, *985t*

Vitamin B₅ (pantothenic acid), 985, 985*t*

Vitamin B₆ (pyridoxine), 985, 985*t*

Vitamin B₁₂, 663, 666, 941, 961, 985*t*, 1099

Vitamin C, 45, 985
  absorption of, 961
  in bone physiology, 215*t*
  excess of, 985
  requirements and sources, 985*t*

Vitamin D, 985. *See also* Calcitriol
  absorption of, 961
  aging and, 1098
  breastfeeding and, 1068
  in calcium homeostasis, 213, *214*, 215*t*
  in immunity, 796
  in phosphate homeostasis, 213
  in pregnancy, 1062
  requirements and sources, 985*t*
  synthesis of, 175, 182, 213, *214*

Vitamin E, 45, 985
  absorption of, 961
  excess of, 985
  requirements and sources, 985*t*

Vitamin K, 961, 965, 985, 985*t*, 1062, 1092

Vitamin K antagonists, 685

Vitelline duct, *1083*

Vitreous body. *See* Vitreous humor

Vitreous chamber, 595

Vitreous humor, *594, 600*

VLDL. *See* Very low-density lipoproteins

Vocal cords (folds), *365, 828, 830, 830–831, 831*

Voice, 829
  change, in males, 1021

Voice box, 829

Voiding urine, 893–895, *894*

Volta, Alessandro, 401

Voltage, 394, 433

Voltage-gated channels, 81

Voltaic cell, 401

Volume, syringe illustration of, 708–709

Volume depletion (hypovolemia), 905–906, 907*t*, 920*t*

Volume excess, 907, 907*t*

Voluntary muscles, 158, 386

VO₂max (maximum oxygen uptake), 406

Vomiting, 942–943

Vowels, combining, 19

Vulva, 1044

W

Walking upright, 11, 262–263, *262–263*

Warfarin, 685

Warm receptors, 568

Warts, genital, 1033

Waste
  definition of, 867
  excretion of, 141, 160, 868–869
  metabolic, 867
  nitrogenous, 658*t*, 659, 867–868, *868*, 882, 889*t*

Wasting, in diabetes mellitus, 647

Water, 48–49
  adhesion of, 48
  air content of, 848, 848*t*
  chemical reactivity of, 48
  cohesion of, 48
  digestion of, 961, 963
  gain and loss, in fluid balance, 902–904, *903*
  heat capacity of, 48–49
  hydration spheres of, 48, *49,* 88
  hydrogen bonds of, 47, *47,* 48
  metabolic, 903
  as nutrient, 977, 979*t*
  osmosis of, 88–89, *89,* 902
  preformed, 903
  renal reabsorption of, 882, 885–888
  solvency of, 48
  as taste, 574
  thermal stability of, 48–49
  total body, 902

Water balance
  aging and, 1100
  negative, 906, 907*t*. (*See also* Dehydration)
  positive, 907, 907*t*

Water barrier, epidermal, 179

Water conservation, renal, *876,* 885–888, *886,* 905
  control of water loss, 885–886
  countercurrent exchange in, 887

countercurrent multiplier in, 886, *886*
  evolutionary role of, 885

Water diuresis, 885

Water intoxication, 907

Water molecule, 48

Water-insoluble fiber, 980

Water-soluble fiber, 980

Water-soluble vitamins, 961, 985, 985*t*

Watson, James, 112, *112*

Wave summation, 402

Wavy hair, 185, *185*

Weak acids and bases, 915

Wear-and-tear arthritis, 294

Weight. *See* Body weight

Weight per volume, 51

Weight-control drugs, 976

Wells, Horace, 608

Wernicke aphasia, 525

Wernicke area, *521, 525, 525,* 538, *538*

"Wet dreams," 1021

Wet gangrene, 169

White blood cell count, 678

White blood cells. *See* Leukocyte(s)

White (collagenous) fibers, 149, *149*

White (fast-twitch) fibers, 407*t*, 408, 408*t*

White fat, 153

White hair, 185

White matter
  brain, *492, 495, 499, 500, 512,* 513
  spinal cord, *462, 463*

White muscles, 408

White pulp, of spleen, 794, *794*

Wiggers, Carl J., 710

Wiggers diagram, 710, *711*

Wilkins, Maurice, 112

Wilms tumor, *133*

Windchill factor, 1000

Windpipe. *See* Trachea

Wisdom teeth, *932,* 933

Withdrawal (flexor) reflex, 484–485, *486*

Withdrawal method, of contraception, 1069, 1070*t*

Wolff's law of bone, 210

Work, 52

Working memory, 452

Wound healing, 167, *168*

Wright's stain, 673

Wrist, *30,* 31
  arteries of, *767*
  muscles acting on, *337, 338,* 339*t*–340*t*
  surface anatomy of, *378*

Wrist drop, 481

X

X chromosomes, 128, 131, *131, 1009–1010, 1010,* 1094

Xenograft, 194

X-linked traits, 131, *131*

X-ray(s), 22, 22–23, 43

X-ray diffraction, 112, *112*

Y

Y chromosomes, 128, 131, *131, 1009–1010, 1010,* 1094

Yellow bone marrow, *200,* 205, *205,* 789

Yolk sac, 660, 1080, *1081,* 1082, *1083, 1085,* 1086, *1088*

Z

Z disc, *387,* 388, *389,* 389–391, 390*t*

Zenith (hot-air balloon), 849

Zero position, 279

Ziegler, John, 69–70

Zika virus, 1093

Zinc, dietary, 984*t*

Zona drilling, 1104

Zona fasciculata, *626,* 627

Zona glomerulosa, *626,* 627

Zona pellucida, *1051,* 1052, *1052, 1077,* 1077–1078, *1078*

Zona reticularis, *626,* 627

Zone of bone deposition, 209, *209*

Zone of calcification, 209, *209*

Zone of cell hypertrophy, 209, *209*

Zone of cell proliferation, 209, *209*

Zone of reserve cartilage, 209, *209*

Zygote, 1008, 1078, *1078,* 1079*t*

Zygote intrafallopian transfer (ZIFT), 1105

Zymogen(s), 941, *941,* 950, 951*t*

Zymogen granules, 950, *951*